| Name | Symbol | Atomic number | Atomic Mass* | Notes | Name | Symbol | Atomic number | Atomic Mass* | Notes |
|---|---|---|---|---|---|---|---|---|---|
| Actinium | Ac | 89 | [227] | 5 | Molybdenum | Mo | 42 | 95.94(2) | 1 |
| Aluminum | Al | 13 | 26.9815386(8) | | Neodymium | Nd | 60 | 144.242(3) | 1 |
| Americium | Am | 95 | [243] | 5 | Neon | Ne | 10 | 20.1797(6) | 1, 3 |
| Antimony | Sb | 51 | 121.760(1) | 1 | Neptunium | Np | 93 | [237] | 5 |
| Argon | Ar | 18 | 39.948(1) | 1, 2 | Nickel | Ni | 28 | 58.6934(2) | |
| Arsenic | As | 33 | 74.92160(2) | | Niobium | Nb | 41 | 92.90638(2) | |
| Astatine | At | 85 | [210] | 5 | Nitrogen | N | 7 | 14.0067(2) | 1, 2 |
| Barium | Ba | 56 | 137.327(7) | | Nobelium | No | 102 | [259] | 5 |
| Berkelium | Bk | 97 | [247] | 5 | Osmium | Os | 76 | 190.23(3) | 1 |
| Beryllium | Be | 4 | 9.012182(3) | | Oxygen | O | 8 | 15.9994(3) | 1, 2 |
| Bismuth | Bi | 83 | 208.98040(1) | | Palladium | Pd | 46 | 106.42(1) | 1 |
| Bohrium | Bh | 107 | [272] | 5 | Phosphorus | P | 15 | 30.973762(2) | |
| Boron | B | 5 | 10.811(7) | 1, 2, 3 | Platinum | Pt | 78 | 195.084(9) | |
| Bromine | Br | 35 | 79.904(1) | | Plutonium | Pu | 94 | [244] | 5 |
| Cadmium | Cd | 48 | 112.411(8) | 1 | Polonium | Po | 84 | [209] | 5 |
| Calcium | Ca | 20 | 40.078(4) | 1 | Potassium | K | 19 | 39.0983(1) | 1 |
| Californium | Cf | 98 | [251] | 5 | Praseodymium | Pr | 59 | 140.90765(2) | |
| Carbon | C | 6 | 12.0107(8) | 1, 2 | Promethium | Pm | 61 | [145] | 5 |
| Cerium | Ce | 58 | 140.116(1) | 1 | Protactinium | Pa | 91 | 231.03588(2) | 5 |
| Cesium | Cs | 55 | 132.9054519(2) | | Radium | Ra | 88 | [226] | 5 |
| Chlorine | Cl | 17 | 35.453(2) | 3 | Radon | Rn | 86 | [222] | 5 |
| Chromium | Cr | 24 | 51.9961(6) | | Rhenium | Re | 75 | 186.207(1) | |
| Cobalt | Co | 27 | 58.933195(5) | | Rhodium | Rh | 45 | 102.90550(2) | |
| Copernicium | Cn | 112 | 277 | | Roentgenium | Rg | 111 | [280] | 5 |
| Copper | Cu | 29 | 63.546(3) | 2 | Rubidium | Rb | 37 | 85.4678(3) | 1 |
| Curium | Cm | 96 | [247] | 5 | Ruthenium | Ru | 44 | 101.07(2) | 1 |
| Darmstadtium | Ds | 110 | [281] | 5 | Rutherfordium | Rf | 104 | [267] | 5 |
| Dubnium | Db | 105 | [268] | 5 | Samarium | Sm | 62 | 150.36(2) | 1 |
| Dysprosium | Dy | 66 | 162.500(1) | 1 | Scandium | Sc | 21 | 44.955912(6) | |
| Einsteinium | Es | 99 | [252] | 5 | Seaborgium | Sg | 106 | [271] | 5 |
| Erbium | Er | 68 | 167.259(3) | 1 | Selenium | Se | 34 | 78.96(3) | |
| Europium | Eu | 63 | 151.964(1) | 1 | Silicon | Si | 14 | 28.0855(3) | 2 |
| Fermium | Fm | 100 | [257] | 5 | Silver | Ag | 47 | 107.8682(2) | 1 |
| Fluorine | F | 9 | 18.9984032(5) | | Sodium | Na | 11 | 22.98976928(2) | |
| Francium | Fr | 87 | [223] | 5 | Strontium | Sr | 38 | 87.62(1) | 1, 2 |
| Gadolinium | Gd | 64 | 157.25(3) | 1 | Sulfur | S | 16 | 32.065(5) | 1, 2 |
| Gallium | Ga | 31 | 69.723(1) | | Tantalum | Ta | 73 | 180.94788(2) | |
| Germanium | Ge | 32 | 72.64(1) | | Technetium | Tc | 43 | [98] | 5 |
| Gold | Au | 79 | 196.966569(4) | | Tellurium | Te | 52 | 127.60(3) | 1 |
| Hafnium | Hf | 72 | 178.49(2) | | Terbium | Tb | 65 | 158.92535(2) | |
| Hassium | Hs | 108 | [270] | 5 | Thallium | Tl | 81 | 204.3833(2) | |
| Helium | He | 2 | 4.002602(2) | 1, 2 | Thorium | Th | 90 | 232.03806(2) | 1, 5 |
| Holmium | Ho | 67 | 164.93032(2) | | Thulium | Tm | 69 | 168.93421(2) | |
| Hydrogen | H | 1 | 1.00794(7) | 1, 2, 3 | Tin | Sn | 50 | 118.710(7) | 1 |
| Indium | In | 49 | 114.818(3) | | Titanium | Ti | 22 | 47.867(1) | |
| Iodine | I | 53 | 126.90447(3) | | Tungsten | W | 74 | 183.84(1) | |
| Iridium | Ir | 77 | 192.217(3) | | Ununbium | Uub | 112 | [285] | 5, 6 |
| Iron | Fe | 26 | 55.845(2) | | Ununhexium | Uuh | 116 | [293] | 5, 6 |
| Krypton | Kr | 36 | 83.798(2) | 1, 3 | Ununoctium | Uuo | 118 | [294] | 5, 6 |
| Lanthanum | La | 57 | 138.90547(7) | 1 | Ununpentium | Uup | 115 | [288] | 5, 6 |
| Lawrencium | Lr | 103 | [262] | 5 | Ununquadium | Uuq | 114 | [289] | 5, 6 |
| Lead | Pb | 82 | 207.2(1) | 1, 2 | Ununtrium | Uut | 113 | [284] | 5, 6 |
| Lithium | Li | 3 | 6.941(2) | 1, 2, 3, 4 | Uranium | U | 92 | 238.02891(3) | 1, 3, 5 |
| Lutetium | Lu | 71 | 174.967(1) | 1 | Vanadium | V | 23 | 50.9415(1) | |
| Magnesium | Mg | 12 | 24.3050(6) | | Xenon | Xe | 54 | 131.293(6) | 1, 3 |
| Manganese | Mn | 25 | 54.938045(5) | | Ytterbium | Yb | 70 | 173.04(3) | 1 |
| Meitnerium | Mt | 109 | [276] | 5 | Yttrium | Y | 39 | 88.90585(2) | |
| Mendelevium | Md | 101 | [258] | 5 | Zinc | Zn | 30 | 65.409(4) | |
| Mercury | Hg | 80 | 200.59(2) | | Zirconium | Zr | 40 | 91.224(2) | 1 |

* A number in parentheses at the end of an atomic mass is the uncertainty in the preceding digit.

NOTES

**1** Geological specimens are known in which the element has an isotopic composition outside the limits for normal material. The difference between the atomic mass of the element in such specimens and that given in the Table may exceed the stated uncertainty.

**2** Range in isotopic composition of normal terrestrial material prevents a more precise value being given; the tabulated value should be applicable to any normal material.

**3** Modified isotopic compositions may be found in commercially available material because it has been subject to an undisclosed or inadvertant isotopic fractionation. Substantial deviations in atomic mass of the element from that given in the Table can occur.

**4** Commercially available Li materials have atomic masses that range between 6.939 and 6.996; if a more accurate value is required, it must be determined for the specific material.

**5** Element has no stable nuclides. The value enclosed in brackets, e.g. [209], indicates the mass number of the longest-lived isotope of the element. However three such elements (Th, Pa, and U) do have a characteristic terrestrial isotopic composition, and for these an atomic mass is tabulated.

**6** The names and symbols for elements 113-118 are under review. The temporary system recommended by J Chatt, *Pure Appl. Chem.*, 51, 381-384 (1979) is used above.

## ABOUT THE COVER

Shown on the cover of this book is a canola field, one of the most colourful agricultural crops in Canada. Canola is a Canadian invention and comprises of plants of the genus *Brassica* that have been specifically bred to have low levels of eurcic acid and certain glucosinolates in their oil. These molecules would otherwise make the oil derived from the plants unsuitable for human consumption. This selective breeding program was carried out by researchers at the University of Manitoba in Winnipeg and at the Agriculture Canada Research Station in Saskatoon. The result of this program was a crop that produces an oil high in levels of unsaturated fatty acids that can be used in a variety of food products. Plant-derived unsaturated fatty acids have been shown to be a healthier alternative to the fats derived from animal sources.

The images on the cover show an increasingly microscopic view of the canola crop, starting with the familiar yellow field. An individual canola plant is shown in the top right, giving us a view of the individual flower petals. Below that is the harvested canola seed before it is crushed for the oil that it contains. The jars contain the result of the crushing process, the now-ubiquitous canola oil used in many cooking processes. The final panel shows how a chemist views canola oil, by considering the chemical structure of the individual oil molecules.

Canola oil consists primarily of fatty acids that contain 18 carbon atoms with one, two or three carbon—carbon double bonds (unsaturations) in their carbon chains. The structures shown on the front cover are stylized versions of the canola oil molecules showing the three types of fatty acids that make up canola oil, oleic, linolenic and α-linolenic. These fatty acids are shown as being combined with a molecule of glycerol to form an ester derivative (a triacyl glyceride), which is how they are found in the plant. One depiction of the oil molecule is the line drawing commonly used by chemists to represent the chemical structure of organic molecules. Another version is the "ball and stick" model that better represents the overall three-dimensional structure of the molecule. In Chapter 10 you will learn about organic molecules and the different ways of depicting them. As well you will learn the various functional groups (such as esters or carboxylic acids) that are commonly encountered in the molecules studied by organic chemists.

Throughout the book, we continually relate the way that chemists view the world, as atoms bonded together in molecules, to the macroscopic world around us. You will discover that the physical and chemical properties of molecules are determined by the way that their atoms are bonded together. These molecular properties in turn are reflected in the macroscopic world with which you interact on a daily basis. By developing an understanding of the molecular world, you will begin to view the macroscopic world the way a chemist does!

# ALL THE HELP, RESOURCES, AND PERSONAL SUPPORT YOU AND YOUR STUDENTS NEED!

2-Minute Tutorials and all of the resources you & your students need to get started
www.wileyplus.com/firstday

Student support from an experienced student user
Ask your local representative for details!

Collaborate with your colleagues, find a mentor, attend virtual and live events, and view resources
www.WhereFacultyConnect.com

Pre-loaded, ready-to-use assignments and presentations
www.wiley.com/college/quickstart

Technical Support 24/7 FAQs, online chat, and phone support
www.wileyplus.com/support

Your WileyPLUS Account Manager
Training and implementation support
www.wileyplus.com/accountmanager

www.wileyplus.com

# MAKE IT YOURS!

## Robert C. (Bob) Burk

Robert C. (Bob) Burk is an Associate Professor of Chemistry at Carleton University in Ottawa. He received his B.Sc. and M.Sc. degrees from Carleton. Both involved work in the area of radioanalytical chemistry. Bob worked for six years in the nuclear industry, doing research on the production of nuclear fuels, then returned to Carleton University to do a Ph.D. in the area of supercritical fluid extraction. He has been a member of the Carleton Chemistry faculty since 1993. His research involves the use of supercritical fluids for separation purposes, as well as development of analytical methods for organics in water using novel solid phase materials such as carbon nanotubes.

Bob is especially interested in the use of technology for teaching chemistry. His lectures are recorded and available for viewing on i-Tunes, and parts of them appear on Youtube. After-hours, Bob uses instant messaging systems, and more recently social networking tools, to communicate with students while they are studying. He was a winner of the 2004 OCUFA teaching award, and a 3M teaching fellowship in 2006.

When not at work, Bob loves to sail in the summer, cross-country ski in the winter with his wife Ewa and two children, Ashley and Adam, as well as renovate houses and build period furniture all year round.

(Photo courtesy of Ashley M. Burk.)

## John Olmsted III

John Olmsted III is Professor Emeritus of Chemistry at California State University, Fullerton, from which he retired in 2003 after nearly 40 years of teaching and research in general and experimental physical chemistry. John was honored as the CSUF Outstanding Professor in 1997–98 and served as department chair from 1998 to 2001. In addition to 25 years at CSUF, he taught for 12 years at the American University of Beirut. He had visiting teaching/research appointments at UCLA and the University of North Carolina at Chapel Hill and did research at the Max-Planck-Institut für Biophysikalische Chemie (Göttingen, Germany), the University of California at San Diego, and Sandia National Laboratory in Albuquerque, NM.

John received his BS degree in chemistry from Carnegie Institute of Technology (now Carnegie-Mellon University) and his PhD in physical chemistry from UC Berkeley, where he also did postdoctoral work at the Lawrence Berkeley Laboratory. He has more than 30 refereed research publications and has also published regularly on chemical education topics in the *Journal of Chemical Education*.

In his retirement, besides continuing to write chemistry textbooks, John keeps busy with his interests in gardening, photography, and the philosophy of chemistry. He and his wife Eileen enjoy traveling, dancing, and visiting with their three married children and two grandchildren.

(Photo courtesy of the Friends of the Fullerton Arboretum.)

## Greg Williams

Greg Williams is an Adjunct Professor of Chemistry at the University of Oregon. He earned an undergraduate degree in chemistry at UCLA and a Ph.D. in inorganic chemistry at Princeton University. He has taught and conducted research at the University of Oregon, California State University, Fullerton, UCLA, and the University of California, Irvine. Outside the classroom, Greg's professional work is concentrated on developing graphics, digital animation, and interactive multimedia for teaching chemistry.

When he is not teaching or writing about chemistry, Greg can be found somewhere in the western United States, backpacking, climbing, skiing, fly fishing, or kayaking. He also sings low bass with the Eugene Vocal Arts Ensemble. Greg lives in Eugene, Oregon, with his wife Trudy Cameron, a Professor of Economics at the University of Oregon, and their daughters, Casey and Perry. He absolutely insists on enjoying life.

(Photo courtesy of Fred E. Skinner.)

# CHEMISTRY

## CANADIAN EDITION

**John Olmsted III**
*California State University, Fullerton*

**Greg Williams**
*University of Oregon*

**Robert C. Burk**
*Carleton University, Ottawa*

**WILEY**

John Wiley & Sons Canada, Ltd.

**Library and Archives Canada Cataloguing in Publication**

Olmsted, John.

Chemistry / John A. Olmsted, Gregory M. Williams, Robert C. Burk.— 1st Canadian ed.

ISBN 978-0-470-15579-0

1. Chemistry—Textbooks. I. Williams, Gregory M II. Burk, Robert Charles, 1958-  III. Title.

QD33.2.O46 2010               540               C2009-906948-2

**Production Credits**

ACQUISITIONS EDITOR: Darren Lalonde
VICE PRESIDENT & PUBLISHER: Veronica Visentin
VICE PRESIDENT, PUBLISHING SERVICES: Karen Bryan
CREATIVE DIRECTOR, PUBLISHING SERVICES: Ian Koo
MARKETING MANAGER: Patty Maher
EDITORIAL MANAGER: Karen Staudinger
DEVELOPMENTAL EDITOR: Theresa Fitzgerald
PROJECT AND MANUFACTURING COORDINATOR: Tegan Wallace
EDITORIAL ASSISTANT: Laura Hwee
PERMISSIONS COORDINATOR: Jessie Coffey
TYPESETTING: Preparé Inc.
COVER DESIGN: Laserwords
COVER IMAGES: Canola field: © Malcom Fife/Corbis; Canola flower: © Todd Gipstein/CORBIS;
        Canola seed: © Visuals Unlimited/Corbis; Canola oil: © Image Studios/Getty Images;
        Molecular model/chemical formula: Laserwords
PRINTING AND BINDING: World Color Press Inc.

Printed and bound in the United States.
1 2 3 4 5  WC  14 13 12 11 10

John Wiley & Sons Canada, Ltd.
6045 Freemont Blvd.
Mississauga, Ontario L5R 4J3
Visit our website at www.wiley.ca

# PREFACE

The subject of this book is obvious from its title, and the general plan is clear from the table of contents. The goal of this preface is to explain our purpose and clarify the plan. We want to help our potential readers discover how the first Canadian edition of *Chemistry* can guide students to a confident mastery of the fundamentals of chemistry.

## THE MOLECULAR PERSPECTIVE: INFORMING QUANTITATIVE REASONING

The purpose of our book is to describe the fundamentals of chemistry as a chemist thinks about the subject. Matter is composed of atoms and molecules, and chemists explain the behaviour of matter by describing how atoms and molecules behave and interact. We believe that students will master chemistry more quickly and thoroughly when they grasp this molecular perspective; hence our general theme: "Think molecules."

When the first US edition of *Chemistry, the Molecular Science* was published, no major general chemistry textbook emphasized the molecular approach. Now almost every textbook shows some molecular pictures. Our text, which is based on the fourth US edition of this book, maintains the uniqueness in reinforcing each chemical topic with molecular-level explanations. Our molecular approach to concepts appears throughout the book, but it can be seen most clearly in the following chapters:

- Chapter 1, where we emphasize the molecular underpinnings of stoichiometry and other fundamental aspects of chemistry
- Chapter 2, where we introduce the ideal gas equation by describing how molecular kinetic behaviour varies with macroscopic variables
- Chapters 6 and 7, where the theories of chemical bonding in the molecules themselves are described using copious diagrams
- Chapter 12, where we develop chemical kinetics starting with reaction mechanisms
- Chapters 13 and 14, where we emphasize the importance of identifying the species in solution as the first step in understanding aqueous chemical equilibria
- Chapter 16, where we present molecular descriptions of electrode surfaces to support the chemical basis of electrochemical cells

Because chemists work quantitatively just as frequently as they think molecularly, we use the molecular approach to facilitate successful quantitative reasoning, not to replace it. To see one example of this approach, explore our presentation (Section 13.6) of the species in aqueous solution as an essential prologue to stoichiometric and equilibrium calculations involving solution phase reactions. Look, too, at how we introduce the ideal gas equation (Section 2.2): we use a qualitative description of how gaseous molecules behave to develop the quantitative statement.

## EMPHASIS ON PROBLEM-SOLVING

Students usually attempt to solve problems using easy algorithms. Instructors want their students to reason through problems rather than relying mindlessly on algorithms. We believe there is validity to both views, because whereas reasoning is at the heart of any science, every experienced scientist regularly employs algorithms as "shortcuts." Reflecting this view, we have developed stepwise approaches to problem-solving that blend sound reasoning with the efficient application of algorithms. We use a seven-step problem-solving template consistently for quantitative problems, including a version designed specifically for attacking equilibrium problems. We also provide step-by-step procedures for the construction of Lewis structures and for the balancing of redox reactions. Beyond this, for each Example problem, we provide a "strategy" of attack, designed to encourage students to reason their way to solutions rather than search aimlessly for algorithms. Each Example problem is paired with a Practice Exercise of the same type.

## HIGHLIGHTS OF THE FIRST CANADIAN EDITION

Although chemistry is, in many ways, a mature science, our understanding of how best to teach chemistry continues to evolve. The challenge to authors of general chemistry textbooks is to incorporate contemporary teaching and learning strategies while maintaining the established core of subject matter. As was true in the fourth US edition, our electronic materials—web-based—are integrated with the textbook. We have participated closely in the development of the media components, to ensure that these supplements present and develop concepts in the same manner that we use in the textbook.

We have made several organizational changes in this edition:

- The first four chapters were condensed into Chapter 1. This chapter is review material for graduates of most Canadian final-year high school chemistry courses.
- What was Chapter 5, *The Behaviour of Gases*, is now Chapter 2. This allows students to immediately begin thinking using the molecular approach and to do quantitative problem-solving on material related to the simplest form of matter.
- What was Chapter 6, *Energy and Its Conservation*, is now Chapter 3, providing early, concentrated coverage of this essential topic.
- Chapter 10, *Organic Chemistry* has been added to this edition in response to the needs of the Canadian market.
- Chapter 13 on macromolecules was moved to Chapter 17, in recognition that not all instructors will choose to cover it.

### Canadian Content

The first Canadian edition has been produced specifically for Canadian students. There is a long history of chemical research in Canada, and the Canadian chemical industry is vibrant. We therefore have included many examples of Canadian successes in the laboratory that have resulted in significant advancement of the science. Nobel Prize winners such as Dr. John Polayni at the University of Toronto are featured, as is the work of current researchers such as Dr. R.W. Gillham at the University of Waterloo. Canadian production figures are included wherever possible to give the student a sense of the magnitude of the chemical industry in this country.

Another significant feature of this Canadian edition is the exclusive use of SI units and constants. This simplifies calculations and makes many unit conversions unnecessary. Furthermore, all symbols, values of constants, and even organic nomenclature are done in accordance with the guidelines of the International Union of Pure and Applied Chemistry (IUPAC). This standardizes the presentation throughout the book, and teaches students to communicate in the same fashion as most chemical researchers.

## Coverage of Energy

In this Canadian edition, the chapter entitled *Energy and Its Conservation* was moved up to Chapter 3 in order to emphasize the importance of energy to chemistry. This chapter has sections on Properties of Energy, Energy Transfers, Energy Changes in Chemical Processes, Measuring Energy Changes: Calorimetry, Enthalpy, and Energy Sources.

Moving this material forward has two advantages. First, an early chapter that focuses explicitly on energy emphasizes to students the centrality of energy to chemistry. Second, an early description of calorimetry makes it possible to include calorimetry experiments early in the laboratory portion of the course. To accommodate those who prefer a more integrated approach, we have organized Chapter 3 so that it is possible to defer calorimetry and enthalpy until later in the course.

## Molecular Structure

We have reorganized our coverage of molecular structure into two separate parts— facts and theories. Chapter 6, *Fundamentals of Molecular Structure*, concentrates on more concrete, readily grasped features of molecules. We use Lewis structures to organize the facts of molecular properties: molecular shapes (VSEPR) and bond lengths, strengths, and polarities. Chapter 7, *Theories of Chemical Bonding,* introduces the more abstract concepts of quantum chemistry: orbital overlap, hybridization, molecular orbitals, delocalization, and band theory. The presentation of facts before theory allows students to become comfortable with tangible features of Lewis structures and molecular shapes before confronting the abstractions inherent in bonding theory.

## Organic Chemistry

Previous editions of this book did not contain a chapter dedicated to organic chemistry. Although there are numerous organic examples throughout the book, Chapter 10 now provides a concentrated look at nomenclature, functional groups, stereochemistry, and three general classes of reactions: substitution, elimination, and addition. The chapter is organized such that instructors can cover only the general material on hydrocarbons, nomenclature, and functional groups, if they wish. Alternatively, they may want to add the section on stereochemistry to their course. The three sections on organic reactions can then be added alone or together. The coverage is thus more than extensive enough for courses where organic chemistry is one of many topics covered. $S_N1$ and $S_N2$ reactions are covered in some detail and can be used as an introduction to reaction mechanisms, which are covered in more detail in Chapter 12, *Kinetics: Mechanisms and Rates of Reactions.*

## Problems

Problem-solving is a strong feature of this book, as it was in the fourth US edition. Specifically, the book incorporates the following:
- A Practice Exercise accompanies and reinforces every Example done in the text.

- Examples illustrate how to calculate yields in limiting reactant problems (Chapter 1), how to determine reaction orders using the initial rate method (Chapter 12), how to treat polyprotic acid titrations (Chapter 14), and how to analyze the common-ion effect on solubility (Chapter 15).
- Our supporting materials include a greatly expanded electronic homework feature and additional Interactive Learning Ware problem-solving tutorials.
- More elaborate Group Study Problems provide platforms for group learning.
- Qualitative and visualization problems encourage the development of conceptual and visual skills.

## OTHER UNIQUE FEATURES

The first US edition of *Chemistry, the Molecular Science* introduced features that were not present in other general chemistry textbooks but that the original authors felt reflected how contemporary chemistry should be taught. These resonated well with our users, so we have retained them: molecular approaches to traditional topics (described above), use of organic and biochemical examples equally with inorganic ones, a "rational" approach to Lewis structures, and a separate chapter on macromolecules.

### Organic and Biochemical Examples

Whereas most other general chemistry textbooks restrict their use of organic and biochemical molecules to a separate chapter, we feel that general chemistry is just that—a general introduction to principles that apply equally well to all categories of chemical compounds. Chapter 10, *Organic Chemistry*, presents an in-depth treatment of organic chemistry on its own, but students encounter many organic and biochemical substances in their daily lives, so the use of organic examples in other chapters can help excite student interest. We introduce organic line structures in Chapter 1, allowing us to use interesting molecules as we introduce basic chemical concepts.

### Macromolecules

Macromolecules, both natural and synthetic, are everywhere in the contemporary world. In recognition of their importance, we devote Chapter 17 to describing macromolecule formation and structure. The chapter builds on the structure and bonding information of earlier chapters, but it can be omitted without loss of continuity. We place this chapter with three other chapters that can be covered at the will of the instructor: *The Transition Metals, The Main Group Elements, and Nuclear Chemistry and Radiochemistry.*

## PROVIDING WHAT STUDENTS NEED

Our discussions with students and colleagues, over the many years that we have taught general chemistry, have convinced us that relevance, timeliness, readability, and problem-solving strategies all are important for student success.

Most students who take general chemistry do not intend to major in chemistry. They are pursuing careers in the health sciences, biology, agriculture, engineering, geology, physics, or even non-science-related disciplines. Many of these students need to be convinced that a sound understanding of chemical principles is essential to their field of interest. Our chapter introductions offer short descriptions of topics that have "real-world" importance and are related to material that appears in the chapter. Focussed material provides further examples under five general themes: *Chemistry and Life, Chemistry and the Environment, Chemistry and Technology, Chemical Milestones,* and *Tools for Discovery*. In addition, some chapters end with sections that may interest a particular group of students but can be omitted without loss of continuity. For examples, see Sections 3.6, Energy Sources; 11.6, Bioenergetics; 15.5, Complexation Equilibria; and 18.4, Transition Metals in Biology. Parts of Chapter 10, *Organic Chemistry*, can be incorporated into a course or not, as described above.

## LEARNING RESOURCES

Learners have a variety of styles, which are best served by supplements to textual and oral explanations. The supplementary resources that accompany this first Canadian edition can be divided into three categories: *features in the textbook, tools designed for the student, and ancillaries addressed to the instructor*.

### Text Features

**Learning Objectives:** Every section of every chapter has a clear learning objective associated with it. These are listed on the Chapter opener as well as in the Visual Summary at the end of each chapter. Other elements of each section, such as the Skills to Master and Key Concepts (see below for more information on these), are tied to these learning objectives. It is common for university-level programs and courses to have learning objectives associated with them, so it makes sense to extend to the textbook this logical way of organizing material.

**Copious Examples:** To provide students with early reinforcement, each worked example in the text ends with a Practice Exercise. In addition, each section of the text ends with three Section Exercises designed to test comprehension of concepts introduced in that section. Answers to the practice exercises and the section exercises appear at the end of each chapter.

**Does the Result Make Sense?** Every instructor has anecdotes about student answers that are completely nonsensical. To help students learn how to recognize whether or not a result is reasonable, we make a check for sensibility the last step in our quantitative problem-solving strategy. To reinforce this habit, every worked example ends with a brief analysis showing that the result makes sense.

**Common Pitfalls:** Students often make common mistakes on a given type of problem. The "Common Pitfalls" caution students, and in many cases explain exactly how to avoid these mistakes. Often, this is simply a matter of reminding students to follow the steps detailed in the chapter (e.g., steps for drawing a Lewis structure, or steps for balancing a redox reaction). In other cases, it is a reminder of a fundamental concept that a student must bear in mind, such as deciding whether the heat is flowing into or out of the system.

**Visual Summaries:** Each chapter features a two-page Visual Summary of its most important elements, presented in pictorial fashion. These concept maps provide students with what we believe will be more powerful learning tools than standard Chapter Summaries can provide. The Visual Summaries include the following visual reminders of important ideas developed in the chapter:

- Sequential **organization by section** that shows the logical flow of the chapter
- A reminder of the **Learning Objectives** for each section
- **Skills to Master:** In addition to learning to think critically and assimilating key concepts, students of chemistry must master many skills. Each of these skills and the page number where each skill is first introduced are listed in the visual summary of each section.
- **Key Concepts:** Certain ideas in chemistry, such as the laws of conservation, are both essential and powerful. We identify these as Key Concepts and list them for each section in the Visual Summaries.
- **All the important equations**, colour-highlighted in yellow as they are in the text
- **Molecular views** of chemical processes, emphasizing our molecular approach to chemistry
- **Flowcharts** that summarize the logic of analytical reasoning
- **Tabular summaries**, providing visual representations of how data can be organized
- Representative **chemical reactions**, because reactions are ultimately what chemistry is about

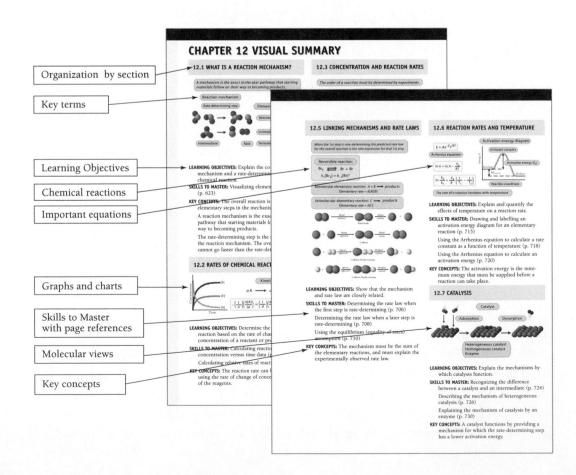

**Chapter Problems:** Our chapter Problems have been praised for their variety and quality. After a set of paired problems specific to each section of the chapter, we present Additional Paired Problems that are not section-specific. These are followed by Additional Problems that are unpaired. These include Group Study Problems that encourage exploration and collaboration in finding solutions to more complex problems.

**Index of Equations:** All of the key equations introduced in the text are provided inside the back cover for quick reference.

## Student Tools

Technology-based resources have become essential parts of every general chemistry toolkit, but all too often these have not been well integrated with the textbook. Our technology components, found at *www.wiley.com/canada/olmsted*, fall into two distinct categories: Study & Practice and Assessment.

## Study & Practice

To enhance visualization, molecular understanding, and problem solving effectively, we believe that computer-based materials for student use should closely mirror the molecular and problem-solving approach of the textbook. We have worked actively with the software developers to ensure products that truly complement our text.

The **website** that accompanies the text (*www.wiley.com/canada/olmsted*) features animations, video clips, and online problem tutorials that we designed and edited. These address *key concepts* including identification of molecular/ionic species in solution, construction of Lewis structures, properties of atomic orbitals, and other molecular visualizations. Also included are a number of simple *molecular visualization* clips to aid students in learning how to "think molecules."

## WileyPlus with CATALYST

 **catalyst** *WileyPlus* with CATALYST partners with the instructor to teach students how to think their way through a problem, rather than rely on a list of memorized equations, by placing a strong emphasis on developing problem-solving skills and conceptual understanding. *WileyPlus* with CATALYST incorporates an online learning system designed to facilitate dynamic learning and retention of learned concepts. CATALYST was developed by Dr. Patrick Wegner (California State University, Fullerton) to promote conceptual understanding and visualization of chemical phenomena in undergraduate chemistry courses.

CATALYST assignments have multiple levels of parameterization and test on key concepts from multiple points of view (visual, symbolic, graphical, quantitative). Hundreds of end-of-chapter problems are available for assignment, and all are available with multiple forms of problem-solving support.

## WileyPLUS with CATALYST resources

### For Students

**Office Hours Videos** are video worked examples based on end-of-chapter problems, prepared by Canadian author, Bob Burk, to ensure consistency and to better tie in the problem-solving strategy presented in the text examples to a selection of chapter problems.

**Chem FAQs** give alternate presentations for topics in the book using interactive examples, virtual experiments, animations, and learning games. They guide students to construct knowledge themselves through a series of questions.

**Skill Building Tutorials** are animated presentations of key concepts, problem-solving strategies, and fundamental tools for approaching general chemistry

**Interactive LearningWare** are interactive exercises that allow students to work through the solutions to key end of chapter problems

**Study Guide.** The Study Guide further enhances understanding of text concepts and contains additional worked examples giving detailed steps involved in solving them, additional problems and questions, chapter overviews, alternate problem-solving approaches, and extensive review exercises.

**Video Demonstrations** are short video clips illustrating chemical principles and laboratory procedures.

### For Instructors

A description of each of the following resources can be found on the following page.

- Instructor's Solutions Manual
- Test Bank
- Digital Image Archive
- PowerPoint Lecture Slides

### Assessment

- The text's website also includes *Interactive Learning Ware (ILW)*, a step-by-step problem solving tutorial program that guides students through selected problems from the book. The program includes text excerpts relevant to each ILW problem, which students can access while on-line if they have difficulty understanding the underlying concepts. When appropriate, students can also link to molecular representations of the problems. The ILW problems are representative of those that students frequently find most difficult and they reinforce students' critical thinking and problem solving skills. An ILW icon identifies each ILW problem in the textbook.

- *WileyPLUS* includes all of the companion website's assets and more. In classes using *WileyPLUS*, students are able to complete homework and receive immediate feedback on their progress, as well as have their worked tracked by their instructor.
- The *GO Tutorials* lead students through a homework problem in multiple steps, and give hints when wrong answers are submitted. Some online tutorial systems trap a student when wrong answers are given which can generate a lot of frustration. GO Tutorials are not traps because, at any step along the way, a student can return to the main problem.
- The *Student Study Guide* includes chapter overviews, learning objectives, sample exercises with worked out solutions, and self-test questions for each chapter.
- The *Student Solutions Manual* provides worked-out solutions for the odd-numbered problems in the text.

### Instructor Ancillaries

These ancillaries provide resources both for lecture and course management.

- The *Instructor's Solutions Manual* is available on the Instructors' website and permits instructors to select, collate, and print solutions to the specific set of chapter problems that have been assigned as homework.
- The *PowerPoint Presentations* including selected illustrations from the text, are editable so instructors can modify the content to suit their own preferences. Also, the art has been optimized for large-scale projection in the lecture hall.
- The *Instructor's Manual*, available on the web, provides chapter overviews and lecture outlines.
- The *Test Bank*, revised and expanded to meet the needs of the Canadian market, includes a significant number of molecular reasoning problems.
- The *Personal Response Systems/"Clicker" Questions* constitute a bank of questions is available for anyone using personal response systems technology in their classroom.

## ACKNOWLEDGMENTS

Darren Lalonde, Acquisitions Editor at Wiley, conceived the idea of bringing this book to the Canadian market, and set the project in motion. It was Darren's idea to create an Editorial Advisory Board composed of Canadian academics, and to allow the Board to make key decisions regarding the content of the book before allowing any changes to the (already excellent) fourth edition. The result is a book with a familiar, yet distinctive, feel that Canadian teachers and students will appreciate. I am indebted to Carolyn Wells, Director of Market Development at Wiley, for bringing me into the project, and for her early work on gathering issues in the Canadian market.

I greatly appreciate the efforts of the Developmental Editor, Theresa Fitzgerald. Her thoroughness, attention to detail, and organizational skills were essential for this project. Theresa tied together the team of people responsible for all aspects of the project from reviewers to copy editors, and from solutions authors to artists. She facilitated effective communication, adherence to schedules, and accuracy of the finished product. The book would not exist but for Theresa's hard work.

John Sorensen from the University of Manitoba was essential for development of the new chapter on organic chemistry. John, in cooperation with the Editorial Advisory Board, mapped out the content of the chapter and wrote the end of chapter exercises. The cover design concept was also John's.

The sales and marketing team, headed by Patty Maher, continues to play a vital role by furnishing valuable information about the needs of the Canadian market for this book, and by explaining the strengths of the finished product to this market.

Adrian So designed the interior of this edition, making it functional and giving it a very readable style. May Look did a wonderful job as Copy Editor, catching chemistry as well as style and grammatical errors. Jessie Coffey, Permissions Researcher, secured the beautiful new images in this edition. Alan Yoshioka revised and expanded the index.

I want to acknowledge and thank my wife Ewa. She has put up with many evenings and weekends without me because of this project.

Finally, I would like to acknowledge the efforts of our reviewers and Editorial Advisory Board members. Their many thoughts and suggestions helped to improve the final version of the book.

Robert C. Burk

# CONTENTS IN BRIEF

# CONTENTS

# Chapter 10
**Organic Chemistry   530**

# Chapter 11
**Spontaneity of Chemical Processes   608**

# Contents

# FUNDAMENTAL CONCEPTS OF CHEMISTRY

## OUR WORLD IS MOLECULAR

How would you describe the differences between a cup of coffee and a cup of hot water? What probably come to mind are the aroma, the dark colour, and the taste of a good cup of coffee. Coffee's action as a stimulant is another obvious difference. These properties come from the chemical compounds that hot water dissolves from ground coffee beans. These compounds are molecules constructed from different atoms bound together in very specific arrangements. The molecule that makes coffee a stimulant is caffeine. Our background photo is a magnification of crystals of pure caffeine, and the inset is a ball-and-stick model of this molecule.

The properties of a particular molecule are due to the types and number of atoms it contains and how those atoms are arranged in space. Caffeine is a stimulant because it has the same shape as one part of cyclic adenosine monophosphate (cyclic AMP), a molecule that helps to regulate the supply of energy in the brain. Once caffeine has been absorbed into the blood and carried to the brain, it binds to an enzyme

that normally binds with and controls the supply of cyclic AMP. As a result, the enzyme can no longer bind cyclic AMP, the brain's supply of this energy-regulating molecule is increased, and we feel stimulated.

Molecular sizes and shapes play key roles in determining chemical and physical properties. The immense variety of chemical and physical properties displayed by substances in the natural world mirrors an equally immense variety of different types of molecules. However, variety need not come from a large number of different elements. The molecules that make up a cup of coffee are made up almost entirely of atoms of just five elements: hydrogen, carbon, oxygen, nitrogen, and sulphur. Carbon, in particular, is capable of combining in many different ways, generating molecules with elaborate structures.

Even simple molecules often have strikingly different properties. For example, carbon and oxygen are the base elements of two different simple compounds. Whereas a molecule of carbon monoxide contains one oxygen atom and one carbon atom, carbon dioxide

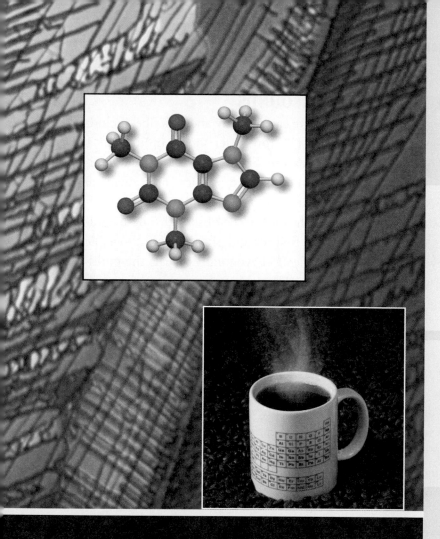

# 1

contains two atoms of oxygen and one atom of carbon. Although these molecules have some common properties (both are colourless, odourless gases), the difference in chemistry caused by a change of one atom is profound. We produce and exhale carbon dioxide as a natural by-product of metabolism. This compound is relatively harmless to humans. In contrast, carbon monoxide is a deadly poison, even at very low concentrations.

This chapter introduces the ideas of atoms and molecules, how to quantify them, and the basics of chemical reactions.

## CHAPTER CONTENTS

## LEARNING OBJECTIVES

*Upon completion of this chapter you should be able to:*

- recognize elemental symbols and names of the elements
- name compounds from molecular pictures
- recognize the SI units commonly used in chemistry, and perform some common unit conversions
- determine the correct number of significant figures in the result of a calculation
- solve mass–number- and molecular–weight-type problems
- calculate concentrations in solutions and of diluted solutions
- balance chemical reactions and solve stoichiometry-type problems
- calculate reaction yields
- identify limiting–reagent-type problems

## 1.1 ATOMS, MOLECULES, AND COMPOUNDS

Every substance has physical properties that we can measure and describe, including shape, colour, and texture. For example, the iron girder shown in Figure 1-1 has a lustrous silvery colour and a smooth texture. Substances also have chemical properties, such as their flammability. **Macroscopic** physical and chemical properties can be observed with the eye. The underlying or **microscopic** structure of a chemical substance requires magnification to be observed. The magnified view of iron shown in Figure 1-1 reveals fissures and pits that are not visible in the macroscopic view. Still further magnification eventually reveals the building blocks of matter. This, the **molecular** or **atomic view,** is an essential part of every chemist's thinking.

### Atoms

The fundamental unit of a chemical substance is called an **atom.** The word is derived from the Greek word *atomos,* meaning "uncuttable." Atoms are extremely small, with diameters of the order of one ten billionth of a metre. To give you some idea of just how small that is, a sample of carbon the size of the period at the end of this sentence contains more atoms than the number of stars in the Milky Way. Any sample of matter large enough for us to see or feel contains an unfathomable number of atoms.

### Molecules

Atoms combine to make all the substances in the world around us, but they do so in very orderly ways. Most substances that we encounter in day-to-day life are made up of small units called **molecules.**

The simplest molecules contain just two atoms and are termed **diatomic.** For example, a molecule of hydrogen is made up of two hydrogen atoms. Figure 1-2 represents a diatomic hydrogen molecule as two spheres connected together.

Because chemistry deals mostly with the behaviour of molecules, this book emphasizes chemistry's molecular foundation. Throughout this book, you will see many figures that represent molecules, with each atom represented by a sphere. Although there are more than 100 different types of atoms, only about 20 are encountered frequently in our world. Many of the molecules described in this book are made up of just 10 different types of atoms: hydrogen, carbon, nitrogen, oxygen, phosphorus, sulphur, fluorine, chlorine, bromine, and iodine. Figure 1-3 shows the colour scheme that we use to represent these atoms.

With practice, you can recognize simple molecules, such as carbon dioxide and water, just by looking at their models. Figure 1-4 shows scale models of a few molecules whose names may be familiar to you. The structures of larger molecules, however, are too complex to be recognized at a glance. Consequently, chemists have created a shorthand language of symbols, formulas, and equations that convey information about atoms and molecules in a simple manner. A specific letter or pair of letters designates each type of atom. These symbols, in turn, are combined into formulas that describe the compositions of more complex

**Figure 1-1**

Iron appears different at the macroscopic, microscopic, and atomic levels.

**Figure 1-2**

Two hydrogen atoms

One hydrogen molecule

A hydrogen molecule can be represented by connecting two spheres together, with each sphere representing one hydrogen atom.

The name *hydrogen* refers to both atoms and molecules. To minimize confusion, we refer to *atomic hydrogen* when we mean hydrogen atoms and *molecular hydrogen* when we mean hydrogen molecules.

**Figure 1-3**

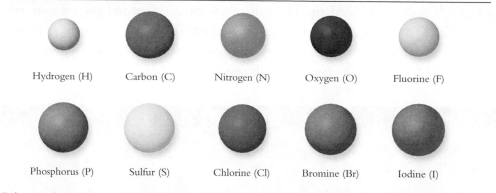

Colour-coded scale models of 10 types of atoms that appear frequently in this book.

**Figure 1-4**

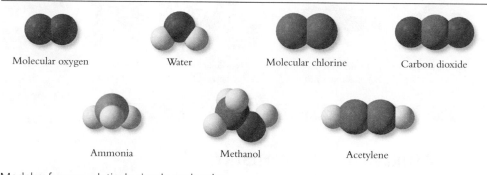

Models of seven relatively simple molecules.

chemical substances. Formulas can then be used to write chemical equations that describe how molecules change during a chemical reaction.

## Chemical Formulas

A chemical **compound** is a substance that contains more than one element. The relative amounts of the elements in a particular compound do not change: every molecule of a particular chemical substance contains a characteristic number of atoms of its constituent elements. For example, every water molecule contains two hydrogen atoms and one oxygen atom. To describe this atomic composition, chemists write the **chemical formula** for water as $H_2O$.

The chemical formula for water shows how formulas are constructed. The formula lists the symbols of all elements found in the compound, in this case H (hydrogen) and O (oxygen). A subscript number after an element's symbol denotes how many atoms of that element are present in the molecule. The subscript 2 in the formula for water indicates that each molecule contains two hydrogen atoms. No subscript is used when only one atom is present, as is the case for the oxygen atom in a water molecule. Molecules always contain whole numbers of atoms, and so the subscripts in chemical formulas of molecular substances are always integers.

Molecules vary considerably in complexity. Molecular oxygen is made up of two oxygen atoms, so its chemical formula is $O_2$. A carbon monoxide molecule

Molecular oxygen
$O_2$

Water
$H_2O$

Carbon monoxide
CO

Methane
$CH_4$

contains one atom of carbon and one atom of oxygen, so its chemical formula is CO. Each molecule of methane, the major constituent of natural gas, contains one carbon atom and four hydrogen atoms, so its formula is $CH_4$. You will encounter still more complex structures, such as methanol (made up of three different elements, $CH_4O$) as you progress through this book.

## SECTION EXERCISES

● **1.1.1** What are the elemental symbols for cerium, cesium, copper, calcium, and carbon?

● **1.1.2** What are the names of the elements represented by the symbols Zr, Ni, Sn, W, Se, Be, and Au?

● **1.1.3** Molecular pictures of some common molecules are shown here. What are their chemical formulas?

Hydrogen peroxide (a common disinfectant and bleaching agent)     Sulfur dioxide (a common air pollutant)     Dinitrogen oxide (laughing gas)     Acetic acid (vinegar)     Ethylene (used to make polyethylene plastic)

## 1.2 MEASUREMENTS IN CHEMISTRY

The knowledge that allows chemists to describe, interpret, and predict the behaviour of chemical substances is gained by making careful experimental measurements. The properties of a sample can be divided into **physical properties**, which can be measured without observing a chemical reaction, and **chemical properties**, which are displayed only during a chemical transformation. Physical properties include familiar attributes such as size, colour, and mass. Some chemical properties also are familiar to us. As examples, bleach reacts chemically with many coloured substances to destroy their colours, and molecular oxygen reacts chemically with many fuels to generate heat.

### Physical Properties

Length, area, and volume measure the size of an object. In addition to its volume, every object possesses a certain quantity of matter, called its **mass** ($m$). Mass measurements are particularly important in chemistry. Consequently, highly accurate balances are essential instruments in chemistry laboratories. A modern balance, whether in a delicatessen or in a chemistry laboratory (Figure 1-5), provides a quick and accurate digital readout of an object's mass.

Chemists measure **time** ($t$) because they want to know how long it takes for chemical transformations to occur. Some chemical reactions, such as the conversion of green plants into petroleum, may take millions of years. Other chemical processes, such as an explosion of dynamite, are incredibly fast. Whereas wristwatches typically measure time only to the nearest second, chemists have

## FOCUS 1-1 TOOLS FOR DISCOVERY: ATOMIC-LEVEL MICROSCOPY

Atoms and molecules are much too tiny to see, even with the most powerful light microscopes. In recent years, however, scientists have developed a set of immensely powerful magnifying techniques that make it possible to visualize how individual atoms are arranged in solids. These techniques are scanning tunnelling microscopy (STM) and atomic force microscopy (AFM).

Scanning tunnelling microscopy and AFM generate images of atoms and molecules using highly miniaturized lever arms with atomically sharp tips. These probes respond to the contours of individual atoms. In AFM, the tip can be moved with extreme precision across a surface. The tip responds to individual atoms on that surface, moving up and down by tiny amounts as it passes over each atom. A laser beam focused on the edge of the lever arm detects this tiny motion, and an optical detector creates an image of the surface.

At their highest sensitivities, STM and AFM generate images that show how atoms are arranged on the surfaces they probe. At first, scientists used these tools to explore how atoms are arranged on surfaces. The example (a) shows individual atoms on the surface of nickel metal.

Increasingly, scientists are using STM and AFM to explore chemical reactions. For example, molecules often undergo chemical reactions when they adhere to a metal surface. Scanning tunnelling microscopy can be used to monitor the atomic changes that take place during such chemical reactions. It is even possible to use STM to manipulate individual atoms. Figures (b) and (c) show how iron atoms can be arranged on the surface of copper metal to make an "atomic corral."

Whereas STM is best suited for imaging atoms, AFM is more appropriate for larger structures. Deoxyribonucleic acid (DNA) is visible as a strand (figure d), surrounded by coloured regions that are protein molecules bound to the DNA.

The next generation of molecular probes combine magnetic resonance imaging (MRI) with the atomic force microscope. Physicians use MRI to do brain scans to pinpoint tumours or blood clots. By coupling MRI with an atomic force scanner and cooling the sample to extremely low temperatures, it is now possible to detect an individual spinning electron in an atom.

(a) Nickel metal surface

(b) Iron atoms forming a ring

(c) A complete "corral" of iron atoms

(d) A strand of DNA

**Figure 1-5**

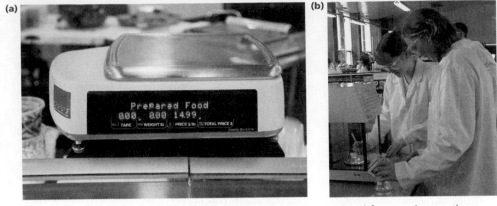

Masses are determined using balances that compare the gravitational force acting on the object being weighed to that on a standard mass. (*a*) Delicatessen balance. (*b*) Students wearing protective glasses while using an analytical balance.

**Figure 1-6**

Dimensions of objects known to humans span more than 30 orders of magnitude, from interstellar distances ($>10^{16}$ m) to the diameter of a hydrogen nucleus ($\sim 10^{-15}$ m).

developed instruments that make it possible to study processes that occur in less than $10^{-14}$ (0.000 000 000 000 01) second.

Most of us associate **temperature** (*T*) with the concepts of hot and cold. More accurately, however, temperature is the property of matter that determines the direction of heat flow. Heat flows naturally from a warm object to a cool object, from higher temperature to lower temperature. Heat is a form of energy, and because energy changes in chemical systems have important consequences, chemists are interested in temperature changes that occur during chemical transformations.

All experimental sciences rely on quantitative measurements of properties. Every measurement gives a numerical result that has three aspects: a numerical **magnitude**; an indicator of scale, called a **unit**; and a **precision**. Each aspect is essential, and all three must be reported to make a measurement scientifically valuable.

## Magnitude

The magnitudes of experimental values in science range from vanishingly small to astronomically large, as summarized in Figure 1-6. To simplify manipulating very large and very small numbers, we use **scientific notation** (see Appendix A for a review). For example, the diameter of a carbon atom is 0.000 000 0003 metres. This number is more conveniently expressed as $3 \times 10^{-10}$ m.

Scientists routinely study objects whose sizes extend far beyond the narrow range encountered in daily life. Physicists, for example, study atomic nuclei measuring $10^{-15}$ m across, and astronomers study our universe, which spans about $10^{23}$ m. Chemists are most often interested in matter on the smaller side of this range. Length measurements in the laboratory vary from metres to subatomic sizes, $10^{-12}$ m. To further simplify the use of very large and very small numbers, scientists use prefixes that change the unit sizes by multiples of 10. For example,

**TABLE 1-1 FREQUENTLY USED PREFIXES FOR MAGNITUDES**

| Prefix | Symbol | Number | Exponential Notation |
|---|---|---|---|
| atto | a | 0.000 000 000 000 000 001 | $10^{-18}$ |
| femto | f | 0.000 000 000 000 001 | $10^{-15}$ |
| pico | p | 0.000 000 000 001 | $10^{-12}$ |
| nano | n | 0.000 000 001 | $10^{-9}$ |
| micro | $\mu$ | 0.000 001 | $10^{-6}$ |
| milli | m | 0.001 | $10^{-3}$ |
| centi | c | 0.01 | $10^{-2}$ |
| kilo | k | 1000 | $10^3$ |
| mega | M | 1 000 000 | $10^6$ |
| giga | G | 1 000 000 000 | $10^9$ |
| tera | T | 1 000 000 000 000 | $10^{12}$ |

the prefix pico means "$10^{-12}$." The symbol for picometre is pm. The diameter of a carbon atom is $3 \times 10^{-10}$ m, which is 300 pm:

$$\text{Carbon diameter} = 3 \times 10^{-10} \text{ m} = 300 \times 10^{-12} \text{ m}$$
$$= 300 \text{ pm} \qquad \uparrow \text{ pico}$$

The most common magnitude prefixes are listed in Table 1-1.

## Units

The units associated with a numerical value are just as important as the value itself. It is *essential* to include a unit with every experimental value. A unit of measurement is an agreed-upon standard with which other values are compared. Scientists use the metre as the standard unit of length. The metre was originally chosen to be $10^{-7}$ times the length of a line from the North Pole to the equator. It is now defined as the distance travelled by light through a vacuum in $^1/_{299,792,458}$ of a second. Volume is measured in litres or cubic metres. Temperature is measured in degrees Celsius (°C), or kelvins (K).

The international scientific community prefers to work exclusively with a single set of units, the Système International (SI), which expresses each fundamental physical quantity in decimally (power of 10) related units. The seven base units of the SI are listed in Table 1-2. Other units are based on some combination of these base units. The unit for volume, for example, is obtained from the base unit for length: a cube that measures 1 metre on a side has a volume of 1 cubic metre.

## TABLE 1-2 BASE SI UNITS

| Quantity | Unit | Symbol |
|---|---|---|
| Mass | kilogram | kg |
| Length | metre | m |
| Time | second | s |
| Temperature | kelvin | K |
| Amount | mole | mol |
| Electric current | ampere | A |
| Luminous intensity | candela | cd |

## Unit Conversions

It is often necessary to convert measurements from one set of units to another. Chemists frequently need to convert volumes from one unit to another. One enormous benefit of using the SI system is that conversions from one unit to another are simple, often requiring only multiplying by some power of 10. For example, the SI unit of volume is the cubic metre, but chemists usually work with much smaller volumes. Hence, chemists often express volume using the litre (L), which is defined to be exactly $10^{-3}$ m$^3$. Another volume unit in common use is the millilitre (mL), or $10^{-3}$ L. The millilitre is the same as the cubic centimetre (cm$^3$).

$$1\ L = 10^{-3}\ m^3,\ 1\ mL = 10^{-3}\ L = 1\ cm^3$$

Some commonly used units are not in the SI system, however, and so it is useful to be able to perform unit conversions. A unit conversion starts with an equality between different units. For example, the equivalence between two commonly used pressure units, the bar and the Torr, is

$$1\ bar = 750.06\ Torr$$

Each equality gives two conversion ratios. The equality between bar and Torr can be rearranged to give two ratios:

$$1 = \frac{1\ bar}{750.06\ Torr} \quad and \quad 1 = \frac{750.06\ Torr}{1\ bar}$$

**Figure 1-7**

100°C — Water boils — 373.15 K
100°C — 100 K
0°C — Water freezes — 273.15 K

−273.15°C — Absolute zero — 0 K

Celsius       Kelvin

Celsius and Kelvin temperature scales. Kelvins and degrees Celsius have the same unit size but different zero points. "Room temperature" is typically about 295 K (22 °C).

To convert from one unit to another, we multiply by the ratio that leads to an appropriate cancellation of units. For example, a pressure of exactly two bar is expressed in Torr as follows:

$$2\ bar\left(\frac{750.06\ Torr}{1\ bar}\right) = 1\ 500.12\ Torr$$

A unit equality may link SI units and non-SI units (1 bar = 750.06 Torr), decimally related units ($10^6$ cm$^3$ = 1 m$^3$), or base units and derived units (1 L = $10^{-3}$ m$^3$). Some of the more common unit equalities are given on the inside back cover of this text. Scientists use two units for temperature, the Celsius (°C) scale and the Kelvin (K) scale. These scales are shown schematically in Figure 1-7. Unlike other scientific units, the unit size of the Celsius and Kelvin scales is the same, but their zero points differ. For both scales, the difference in temperature between the freezing and boiling points of water is defined to be 100 units. However, the temperature at which ice melts to liquid water is 0 °C and 273.15 K.

The conversion between kelvins and degrees Celsius is therefore straightforward. A temperature change of 1 °C is the same as a temperature change of 1 K. To convert from one scale to the other, add or subtract 273.15:

$$T(K) = T(°C) + 273.15 \quad or \quad T(°C) = T(K) - 273.15$$

## Precision and Accuracy

The exactness of a measurement is expressed by its **precision**. This concept can be explained with an example. Suppose three swimmers are discussing the temperature of a swimming pool. The first dips a finger in the water and says that the temperature is "about 24 °C." The second examines an immersed pool thermometer and reports the temperature to be 26 °C. The third swimmer, who has been monitoring daily variations in the pool's temperature, uses a portable precision digital thermometer and reports, "According to my precision thermometer, the pool temperature is 25.8 °C."

The swimmers have measured the water temperature using different measuring devices with different levels of precision. The first swimmer's "finger test" is precise to about 3 °C: $T = 24 \pm 3$ °C (read "twenty-four plus or minus three degrees"). The pool thermometer gives a reading that is precise to the nearest degree: $T = 26 \pm 1$ °C. The precision thermometer measures to the nearest tenth of a degree: $T = 25.8 \pm 0.1$ °C. An actual temperature between 25.7 and 25.9 °C falls within the precision ranges of all three measurements, so all three are correct to within their stated limits of precision.

Whereas precision describes the exactness of a measurement, **accuracy** describes how close a measurement is to the true value. Figure 1-8 illustrates the difference between precision and accuracy. An expert archer shoots arrows with high precision, with all the arrows hitting the target near the same spot (*a* and *b*). Under calm conditions, the expert is also quite accurate, shooting all the arrows into the bull's-eye (*a*). In a strong wind, however, the archer may be precise but not accurate, regularly missing the bull's-eye (*b*). In contrast, a novice shooter is not very precise (*c* and *d*): the arrows scatter over a large area. The novice usually is not accurate (*c*), but an occasional arrow may accurately find the bull's-eye (*d*).

The goal of any measurement in science is to be as accurate as possible. However, determining the accuracy of a measurement is much harder than determining precision. An archer can examine the target to see if the arrows found their mark, but a scientific researcher studying a new phenomenon does not know the correct value. A scientist can assess precision by making additional measurements to find out how closely repeated measurements agree with one another. In contrast, assessing accuracy requires careful attention to the design of an experiment and the instruments used in that experiment. The instruments must be used to measure some known value in order to assess their ability to accurately measure unknown quantities.

**Figure 1-8**

Patterns of arrows striking a target illustrate the notions of precision and accuracy. (*a*) is precise and accurate. (*b*) is precise but inaccurate. (*c*) is neither precise nor accurate. (*d*) is an imprecise pattern with one accurate shot.

## Significant Figures

Scientific measurements should always include both magnitude and precision. Indicating the precision limits with a plus/minus ($\pm$) statement is cumbersome, particularly when many numerical values are reported at one time. Scientists have agreed to simplify the reporting of precision. Experimental measurements are written so that there is an uncertainty of up to one unit in the last reported digit. For example, a temperature reported as 26 °C is greater than 25 °C but less than 27 °C, or 26 $\pm$ 1 °C. A temperature reported as 25.8 °C is 25.8 $\pm$ 0.1 °C. The number of digits expressed in a numerical value is called the number of **significant figures**. The value 26 has two significant figures, whereas 25.8 has three significant figures.

Zeros can present a problem when determining the precision of a numerical value. This is because zeros are needed both to locate the decimal point and to express precision. From the statement that the Sun is 150 000 000 km from the Earth, we cannot tell whether the measurement is precise to nine significant figures or whether the zeros are there only to put the decimal point in the right place. Scientific notation eliminates this ambiguity because a power of 10 locates the decimal point, leaving us free to indicate the precision by the number of digits. For example, a distance of $1.5 \times 10^8$ km means that the number is precise to $\pm 0.1 \times 10^8$ km; the number has two significant figures. If this distance is known to a precision of $\pm 0.01 \times 10^8$ km, then it is written as $1.50 \times 10^8$ km, with three significant figures. Zeros at the end of a number with a decimal point always indicate increased precision. Writing "0.010" in scientific notation clarifies that this number has two significant figures rather than three or four: $0.010 = 1.0 \times 10^{-2}$.

To determine how many significant figures there are in a particular numerical value, read the number from left to right and count all the digits, starting with the first digit that is not zero. To avoid ambiguities about trailing zeros, we place a decimal point after a value when its trailing zeros are significant. For example, "110" has only two significant figures (110 $\pm$ 10), whereas "110." has three significant figures (110 $\pm$ 1). Here are some examples of significant figures:

| Number | Significant Figures | Number | Significant Figures |
|---|---|---|---|
| 500 | 1 | 505 | 3 |
| 0.05 | 1 | $5.00 \times 10^3$ | 3 |
| 55 | 2 | 5.000 | 4 |
| 50. | 2 | 505.0 | 4 |

## SECTION EXERCISES

● **1.2.1** Convert the following measurements to scientific notation and express in base SI units: 0.000 463 L, 17 935 km, and 260 000 h (precise to three significant figures).

● **1.2.2** One light-year is the distance light travels in exactly one year. The speed of light is $1.08 \times 10^9$ km/h. Express the speed of light and the length of one light-year in SI units.

● **1.2.3** Convert each of these measurements to SI units: 155 pounds (mass of a typical person, 1 kg = 2.2 pounds), 110.0 yards (full length of a Canadian football field, 1 yard = 91.44 cm), 39 °C (body temperature of someone with a slight fever), and 365.2422 days (length of one year).

## 1.3. CHEMICAL PROBLEM SOLVING

Many students have difficulty solving chemistry problems, but problem solving is a skill that you can learn and master. To help you, we devote extra attention to setting up and solving problems. Throughout this book, the Examples include information about methods for finding solutions.

Each Example includes a strategy that describes how to approach the problem. After describing the solution, we offer a Practice Exercise designed to help you master the particular type of problem while the example is fresh in your mind. The answers to these Practice Exercises appear at the end of each chapter, along with the answers to the Section Exercises.

Chemical problems that call for calculations often involve several tasks. It can be hard to recognize where to start and how to negotiate these tasks. A stepwise procedure can help you master these types of calculations. We recommend the sequence of steps shown in the box below. When you see a problem that appears complex, apply the stepwise procedure.

The first step in any problem is to identify what you are asked to find. Next, try to visualize what is going on. Think about the chemistry and draw pictures to help you see what is taking place. Once you have a good idea about the problem, determine which equations apply to the calculation and organize the data that are provided. We identify the most important equations by giving them numbers and screening them in yellow, and we recommend that you memorize these equations. Once you have organized the equations and data, you can perform appropriate manipulations and substitutions to carry out the calculation correctly. After you have finished the problem, check to see that your answer makes sense. Checking whether your answer is reasonable will keep you from making careless errors.

Example 1-1 provides a first illustration of this stepwise procedure.

**Solving Quantitative Problems**

Step 1 **Determine what is asked for.**
What type of problem is it? What are you asked to find?

Step 2 **Visualize the problem.**
Draw pictures that illustrate what takes place.
If chemical changes occur, draw molecular pictures.

Step 3 **Organize the data.**
What data are available?
How are the data related to what is asked for?

Step 4 **Identify a process to solve the problem.**
What concepts are required? What equations apply?

Step 5 **Manipulate the equations.**
If necessary, do calculations in steps. Solve for what is asked for.

Step 6 **Substitute and calculate.**
Keep track of units.
Use the correct number of significant figures.

Step 7 **Does the result make sense?**
Are the units consistent? Is the result sensible?

## EXAMPLE 1-1 PROBLEM-SOLVING STRATEGY

Spoons are electroplated with silver to give them an attractive finish.

Electroplating is a process by which a metal such as copper is coated with another metal such as silver or chromium. The transfer of metal atoms is driven by an electrical current. In an electroplating process, a spoon is coated with silver from a silver rod. In the process, $1.0 \times 10^{21}$ atoms are transferred from the rod to the spoon, and the rod loses 0.179 g of its mass. Use this information and the density of silver ($10.50 \text{ g/cm}^3$) to estimate the volume occupied by one silver atom.

**Strategy:** Apply the seven-step approach. The first two steps are part of the strategy.

1. Determine what is asked for. The problem asks for the volume of a silver atom.

2. Visualize the problem. Draw a picture that shows the setup and the information provided in the problem. A picture often helps to interpret and summarize a problem. In this case, the figure should show the spoon, the silver wire, and some indication that atoms are transferred from the wire to the spoon.

**Solution:**

3. Organize the data. First, what data are given in the problem?

$$\text{Mass of silver} = 0.179 \text{ g}$$

$$\text{Number of atoms transferred} = 1.0 \times 10^{21} \text{ atoms}$$

$$\text{Density of silver metal} = 10.50 \text{ g/cm}^3$$

Silver atoms
0.179 g silver
$1.0 \times 10^{21}$ atoms transferred

Rod          Spoon

4. Identify a process to solve the problem. The question asks about the volume of one silver atom. Mass and volume are related through density: $\rho = m/V$. From this equation, we can calculate the total volume of the silver atoms. The problem also gives the total number of silver atoms transferred from the wire to the spoon. The volume of a single atom is the total volume divided by the number of atoms. Oftentimes, a flowchart helps to summarize the process:

5. Manipulate the equations as needed. The flowchart shows two division steps.
   To begin the calculations, we use the mass and density to calculate the total volume:

$$\rho = m/V \quad \text{so} \quad V = m/\rho$$

This is $V_{\text{total}}$, which we divide by the number of atoms to find the volume of a single atom:

$$V_{\text{atom}} = \frac{V_{\text{total}}}{\text{Number of atoms}}$$

6. Substitute and calculate.

$$V_{\text{total}} = \frac{m}{\rho} = \frac{0.179 \text{ g}}{10.50 \text{ g/cm}^3} = 1.70 \times 10^{-2} \text{ cm}^3$$

$$V_{\text{atom}} = \frac{V_{\text{total}}}{\text{\# of atoms}} = \frac{1.70 \times 10^{-2} \text{ cm}^3}{1.0 \times 10^{21} \text{ atoms}} = 1.7 \times 10^{-23} \text{ cm}^3/\text{atom}$$

**7. Does the Result Make Sense?** Is this a "reasonable" value for the volume of an atom? At this stage of your study of chemistry, perhaps you cannot easily answer this question. Atoms are unimaginably small, however, and a volume of $10^{-23}$ cm$^3$ certainly is unimaginably small. Also, the units are right: cm$^3$ is a measure of volume, and the calculation yields units of volume per atom, as the question asked.

**Common Pitfalls:** For this, and for any problem, the most common pitfall is not using a structured approach to solving it. We strongly suggest you follow the seven-step procedure as often as possible when solving problems.

**Practice Exercise 1.1**
A cylindrical jar has an inside diameter of 3.00 cm and a height of 8.00 cm. The empty jar weighs 185.65 g. Filled with gasoline, the same jar weighs 225.40 g. Find the density of gasoline.

## SECTION EXERCISES

**●1.3.1** The standard unit of length in the land of Ferdovia is the frud (2.000 frud = $10^{-2}$ m). Fast-food restaurants in Ferdovia sell fruit drinks in rectangular cartons measuring 50.00 by 40.00 by 80.0 fruds. The Frod family left a freshly opened carton on their kitchen table, and their pet ferret (Fred) drank some of the fruit drink. Mrs. Frod found the height of the remaining liquid to be 61 fruds. How many cubic metres of fruit drink did Fred drink?

**●1.3.2** A technician used a section of glass tubing to measure the density of a liquid that the laboratory needed to identify. The inside diameter of the tubing was known to be 0.87 mm. An empty piece of the tubing weighed 0.785 g. When liquid was drawn into the tubing to a height of 4.0 cm, the tubing and liquid weighed 0.816 g. Find the density of the liquid.

## 1.4 COUNTING ATOMS: THE MOLE

Chemistry is a quantitative science, and chemists frequently measure amounts of matter, which means measuring the numbers of atoms. Counting atoms is difficult, but we can easily measure the mass of a sample of matter. To convert a mass measurement into a statement about the number of atoms in a sample, we must know the mass of an individual atom.

There is an enormous difference between the mass of one atom and the masses of samples measured in the laboratory. For example, a good laboratory balance measures mass values from about $10^{-6}$ to $10^3$ g. An atom, on the other hand, has a mass between $10^{-24}$ and $10^{-21}$ g. Even a tiny mass of $10^{-6}$ g contains approximately $10^{16}$ atoms.

### The Mole and Avogadro's Number

To avoid having to work with unimaginably large numbers, chemists use a convenient unit called the **mole (mol):** *one mole is the number of atoms in exactly 12 g of the pure isotope carbon-12.*

Using mass spectrometers, scientists have determined that the mass of a $^{12}$C atom is $1.992\,646 \times 10^{-23}$ g. This experimental mass combined with the definition of the mole gives the number of atoms in 1 mole:

$$\left(\frac{12 \text{ g}}{1 \text{ mol}}\right)\left(\frac{1 \text{ atom}}{1.992646 \times 10^{-23} \text{ g}}\right) = 6.022142 \times 10^{23} \text{ atoms/mol}$$

The number of items in 1 mole is important enough to be given its own name and symbol. It is known as **Avogadro's number**, commonly symbolized $N_A$.

**Figure 1-9**

Different units are useful for measuring different-sized items. Eggs are sold by the dozen, paper is sold by the ream, and atoms are measured by the mole. Twenty-four pre-1982 Canadian pennies contain about 1 mole of copper atoms.

Although Avogadro's number is known with seven-figure accuracy, four figures are enough for most calculations: $N_A = 6.022 \times 10^{23}$ items/mol.

The mole is a convenient unit for chemical amounts because the mass of most typical laboratory samples contains between 0.01 and 10 moles of atoms or molecules. For example, 1.20 g of $^{12}$C contains $6.02 \times 10^{22}$ atoms of carbon, which is 0.100 mole. The convenience of using a larger unit to describe many small items is not limited to atoms and molecules. Eggs are sold by the dozen, and paper is sold by the ream; chemists "count" atoms and molecules by the mole. There are 12 eggs in a dozen eggs, 500 sheets of paper in a ream of paper, and $6.022 \times 10^{23}$ atoms in a mole of copper (Figure 1-9).

How large is a mole? If a computer were to count items at one million per second, it would take about 19 billion years to reach Avogadro's number. If every person on the planet (population = 6 billion) were to spend 1 million dollars per second, it would take about 300 years to spend 1 mole of dollars. One mole of Canadian one-dollar coins would cover the surface of the Earth to a depth of over 1,000 km!

## Molar Mass

The **molar mass (M)** of any substance is the mass of 1 mole of that substance. Each isotope of a particular element has a different mass. Therefore, the mass of 1 mole of any isotope has a unique value, its *isotopic molar mass*. This characteristic molar mass can be found by multiplying the mass of one atom of that isotope by Avogadro's number. For example, mass spectrometry experiments reveal that one atom of carbon-13 has a mass of $2.15928 \times 10^{-23}$, from which we can calculate the isotopic molar mass of $^{13}$C atoms

$$(2.15928 \times 10^{-23} \text{ g})\left(\frac{6.022142 \times 10^{23}}{1 \text{ mol}}\right) = 13.0035 \text{ g/mol}$$

A sample of a chemical element is usually a mixture of all of its stable isotopes. The carbon in a sample of graphite is 99.892% $^{12}$C and 1.108% $^{13}$C; elemental titanium is a mixture of five stable isotopes, and tin has 10 isotopes, each with its own particular mass. Every element has a characteristic distribution of isotopes, and each isotope of the element contributes proportionally to the molar mass of the element. The mass contributed by each isotope is proportional to its fractional abundance, as given by Equation 1-1:

Elemental molar mass

$$= \Sigma(\text{Fractional abundance})(\text{Isotopic molar mass}) \qquad (1\text{-}1)$$

The symbol $\Sigma$ is the Greek letter sigma. It means "take the sum of the following quantities."

Equation 1-1 combines multiplication and addition steps. The table below shows the fractional abundance and molar mass of each of the titanium isotopes.

| Isotope | Isotopic Molar Mass (g/mol) | Fractional Abundance |
|---|---|---|
| $^{46}$Ti | 45.95263 | 0.0825 |
| $^{47}$Ti | 46.95176 | 0.0744 |
| $^{48}$Ti | 47.94795 | 0.7322 |
| $^{49}$Ti | 48.94787 | 0.0541 |
| $^{50}$Ti | 49.94479 | 0.0518 |

Using Equation 1-1, the elemental molar mass of titanium is:

$$M_{\text{Ti}} = (0.0825)(45.95263) + (0.0744)(46.95176) + (0.732)(47.94795)$$
$$+ (0.0541)(48.94787) + (0.0518)(49.94479) = 47.62 \text{ g/mol}$$

For purposes of chemical bookkeeping, it is unnecessary to know the isotopic molar masses and isotopic distributions of the elements. All we need to know is the mass of 1 mole of an element containing its natural composition of isotopes. These molar masses usually are included in the periodic table, and they appear on the inside front and back covers of this textbook.

## Mass–Mole–Atom Conversions

Molar mass can be thought of as a conversion factor between the mass in grams and the number of moles. These conversions are essential in chemistry, because chemists count amounts of substances in moles but routinely measure masses in grams. To determine the amount (number of moles) in a sample, we measure the mass of the sample and then divide by the molar mass of the substance:

$$n = \frac{m}{M} \qquad (1\text{-}2)$$

If we know the moles and wish to convert to mass, we rearrange Equation 1-2 and multiply that number of moles by the molar mass: $m = nM$.

Suppose, for example, that a bracelet contains 168 g of silver (Ag). To determine the number of moles of silver in the bracelet, divide the mass by the molar mass of Ag, which is 107.87 g/mol:

$$n = \frac{m}{M} \quad so \quad \frac{168 \text{ g Ag}}{107.87 \text{ g/mol}} = 1.56 \text{ mol Ag}$$

If, for example, a chemical reaction calls for 0.250 mole of silver, we multiply moles by molar mass to determine the mass of Ag that should be used:

$$m = nM \quad so \quad m = (0.250 \text{ mol Ag})(107.87 \text{ g/mol}) = 27.0 \text{ g Ag}$$

Avogadro's number is the conversion factor that links the number of moles with the number of individual particles. To determine the number of atoms in a sample of an element, we multiply the number of moles by Avogadro's number: $\# = n\, N_A$. Likewise, if a sample contains a certain number of atoms, the number of moles in the sample can be calculated by dividing by Avogadro's number:

$$n = \frac{\#}{N_A} \tag{1-3}$$

The number of silver atoms in the bracelet that contains 1.56 moles of Ag is found by multiplying the amount in moles by Avogadro's number:

$\# = nN_A$   so

$\# = (1.56 \text{ mol Ag})(6.022 \times 10^{23} \text{ atoms/mol}) = 9.39 \times 10^{23} \text{ atoms Ag}$

Radioactivity detectors, which "count" the number of nuclei that decay, provide an example of a conversion from number of particles to moles. If a radioactivity measurement on a blood sample from a patient receiving radiation treatment indicates that $1.75 \times 10^4$ nuclei of $^{99M}$Tc decay in one hour, dividing by Avogadro's number converts to moles:

$$n = \frac{\#}{N_A} \quad \text{so} \quad n = \frac{1.75 \times 10^4 \text{ nuclei } {}^{99m}\text{Tc}}{6.022 \times 10^{23} \text{ nuclei/mol}} = 2.91 \times 10^{-20} \text{ mol } {}^{99m}\text{Tc}$$

Avogadro's number and molar mass make it possible to convert readily among the mass of a pure element, the number of moles, and the number of atoms in the sample. These conversions are represented schematically in the flowchart shown in Figure 1-10.

**Figure 1-10**

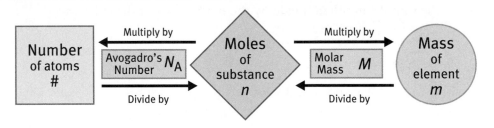

This flowchart shows the conversion processes among the number of moles, the mass, and the number of atoms for a sample of an element.

Moles occupy the central position of this flowchart because the mole is the unit that chemists use in almost all chemical calculations. When you set out to solve a chemical problem, first interpret the question on the atomic/molecular level. The second part of chemical problem solving often involves quantitative calculations, which usually require working with moles.

**SECTION EXERCISES**

● **1.4.1** Use the data in the following table to calculate the molar mass of naturally occurring sulphur.

| Isotope | Isotopic Molar Mass (g/mol) | Abundance (%) |
|---|---|---|
| $^{32}S$ | 31.97207 | 95.02 |
| $^{33}S$ | 32.97146 | 0.75 |
| $^{34}S$ | 33.96786 | 4.21 |
| $^{36}S$ | 35.96709 | 0.02 |

● **1.4.2** Calculate the number of atoms present in 2.55 g of each of these elements: (a) lithium; (b) titanium; and (c) gold.

● **1.4.3** Recently, there has been concern about pollution in the home from radon, a radioactive gas whose elemental molar mass is 222 g/mol. The U.S. Environmental Protection Agency believes that a level of radon of $3.6 \times 10^{-17}$ g/L of air is unhealthy. At this level, how many moles of radon would there be in a living room of volume 2455 L? How many atoms is this?

## 1.5 AMOUNTS OF COMPOUNDS

As described in the section above, chemists use the mole to describe amounts of substances. The procedures introduced in that section work equally well for compounds as for elements. To apply these procedures, we need to be able to determine the molar masses of chemical compounds. We begin by detailing a procedure for quickly drawing organic molecules. Knowing how to do this saves time when discussing larger molecules.

## Line Structures

The chemistry of living processes is complex, and many carbon-based molecules found in living organisms have complex structures. Because of this complexity, chemists have developed **line structures**, which are compact representations of the structural formulas of carbon compounds. Line structures are constructed according to the following guidelines:

### Guidelines for Line Structures

1. All bonds except C—H bonds are shown as lines.
2. C—H bonds are not shown in the line structure.
3. Single bonds are shown as single lines; double bonds are shown as two lines; triple bonds are shown as three lines.
4. Carbon atoms are not labelled.
5. All atoms except carbon and hydrogen are labelled with their elemental symbols.
6. Hydrogen atoms are labelled when they are attached to any atom other than carbon.

Example 1-2 illustrates these guidelines for three relatively simple molecules.

**EXAMPLE 1-2 DRAWING LINE STRUCTURES**

Construct line structures for the compounds with the following structural formulas:

$C_5H_{12}O$
Methyl tert-butyl ether
(a key antiknock ingredient
in gasoline)

$C_3H_8$
2-Propanol
(rubbing alcohol)

$C_5H_8$
Isoprene
(the building block of
natural rubber)

**Strategy:** A line structure is built by using information about how atoms are connected in the molecule. The structural formula is the basis for the line structure. Apply the guidelines to convert the structural formula representation into a line-structure representation.

**Solution:** The first two guidelines state that bonds are represented by lines and that all C—H bonds are ignored. Thus, we remove the C—H bonds from the structural formula, leaving a bond framework. According to the third guideline, the double bonds remain as two lines:

According to the fourth guideline, carbon atoms are not labelled, so we remove the C's:
These are the line structures for the three molecules:

**Does the Result Make Sense?** To check that the line structures represent the correct substances, count the number of intersections and line ends, which should equal the number of C atoms in the compound. The first line structure has five, the second has three, and the third has five, matching the chemical formulas.

**Common Pitfalls:** When learning line structures, it is common to miss the fact that there is a carbon atom at the end of each line (that does not have an atom explicitly shown) and at the intersection of two or more lines.

**Practice Exercise 1.2**

Draw the line structures for the following compounds:

(a)
Propane
$C_3H_8$

(b)
Dimethyl ether
$(CH_3)_2O$

(c)
Ethanol
$CH_3CH_2OH$

Ethane,
$C_2H_6$

Ethylene,
$C_2H_4$

Acetylene,
$C_2H_2$

Although line structures are convenient, a chemist must be able to convert them back into chemical formulas. The reconstruction of a complete formula from a line structure relies on the most important general feature of carbon chemistry: in all neutral molecules containing carbon, each carbon atom has *four* chemical bonds. Keep in mind that carbon atoms are not shown in a line structure, so the first step in constructing a structural formula from a line structure is to place a C at every line intersection and at the end of every line. Then add singly bonded hydrogen atoms (—H) until every carbon atom has four bonds. Example 1-3 illustrates this procedure.

## EXAMPLE 1-3 CONVERTING LINE STRUCTURES

Construct the structural formulas and determine the chemical formulas from the following line drawings:

**Strategy:** Line drawings show all structural features except carbon atoms and C—H bonds. Convert a line drawing into a structural formula in two steps. First, place a C at any unlabelled line end and at each line intersection. Second, add C—H bonds until each carbon atom has four bonds. The chemical formula is then obtained by counting the atoms of each element.

**Solution:** Place a C at each intersection and line end:

Now add C—H bonds until each carbon atom has a total of four bonds. Each carbon atom in the first structure has two bonds, so each carbon needs two C—H bonds to complete the structural formula:

The second structure has four carbon atoms. The two end carbons have just one bond, so each needs three C—H bonds. The carbon with the double bond to oxygen already has its complete set of four bonds. The other inner carbon atom has a bond to C and a bond to O, so it needs two C—H bonds:

The third structure contains a triple bond. The end carbon needs one C—H bond, but the other carbon atom of the triple bond already has four bonds. The next carbon has two bonds, one to its carbon neighbour and one to the oxygen. Two C—H bonds are needed to give this atom its usual set of four bonds:

These are the correct structural formulas. Now it is a simple matter to count the number of atoms of each element and write the chemical formulas of the compounds: $C_2H_4Cl_2$, $C_4H_8O_2$, and $C_3H_4O$.

**Does the Result Make Sense?** Check the consistencies of the structural formulas by counting the number of bonds associated with each carbon atom. If you have converted the line structure correctly, each carbon atom will have four bonds.

**Common Pitfalls:** It is easy at first to forget that each carbon makes a total of four bonds.

**Practice Exercise 1.3**
Convert the following line structure into structural and chemical formulas:

The structural formulas of the compounds in these examples are not very complicated, yet their line structures are clearer than other representations. In ever-larger carbon-containing compounds, the simplification provided by line drawings is a great help to chemists. Line structures are not cluttered by the C—H bonds in the molecule, making them easier both to draw and to interpret.

## Molar Masses of Chemical Compounds

Just as each element has a characteristic molar mass, so does every chemical compound. Chemical compounds are composed of atoms bound together into molecules, or ions clustered together in electrically neutral aggregates. In either case, a chemical formula describes the atomic composition of a compound.

One mole of any chemical compound is 1 mole of its chemical formula unit. Here are some examples:

- One mole of diatomic oxygen ($O_2$) is 1 mole of $O_2$ molecules. Each molecule contains two oxygen atoms, so 1 mole of $O_2$ molecules contains 2 moles of oxygen atoms.

- One mole of methane ($CH_4$) is 1 mole of $CH_4$ molecules, so it contains 1 mole of carbon atoms and 4 moles of hydrogen atoms.
- One mole of sodium chloride (NaCl) contains 1 mole each of $Na^+$ cations and $Cl^-$ anions.
- One mole of magnesium chloride ($MgCl_2$) contains 1 mole of $Mg^{2+}$ and 2 moles of $Cl^-$.

When atoms combine to form molecules, the atoms retain their atomic identities and characteristic molar masses. Thus, we can add elemental molar masses to obtain the molar mass of any compound.

One mole of $O_2$ molecules contains 2 moles of O atoms, so the molar mass of molecular $O_2$ is calculated as follows:

$$\left(\frac{2 \text{ mol O}}{1 \text{ mol } O_2}\right)\left(\frac{15.999 \text{ g}}{1 \text{ mol O}}\right) = 31.998 \text{ g/mol } O_2$$

$$M_{O_2} = 31.998 \text{ g/mol}$$

You can verify that the molar masses of $CH_4$, $NaNO_3$, and $MgCl_2$ are 16.043 g/mol, 84.994 g/mol, and 95.211 g/mol, respectively.

As chemical formulas become more complicated, the calculation of molar mass requires more steps, as shown in Examples 1-4 and 1-5.

## EXAMPLE 1-4 MOLAR MASS OF A MOLECULE

The line drawing and chemical formula of Sevin, a common insecticide, appear in the margin. Determine the molar mass of Sevin.

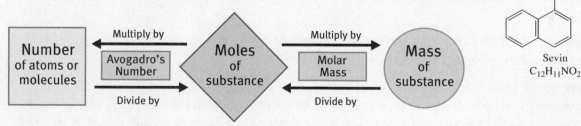

Sevin
$C_{12}H_{11}NO_2$

**Strategy:** To determine the molar mass of a substance, we need its chemical formula and elemental molar masses. From the chemical formula, determine the number of moles of each element contained in 1 mole of the substance. Multiply each elemental molar mass by the number of moles of that element and add the results for each element.

**Solution:** The line structure demonstrates how the atoms are connected in the molecule, but we do not need this information because the chemical formula is provided. We use the chemical formula to calculate the mass of each element in 1 mole of the compound:

$$\left(\frac{12 \text{ mol C}}{1 \text{ mol } C_{12}H_{11}NO_2}\right)\left(\frac{12.011 \text{ g C}}{\text{mol C}}\right) = \left(\frac{144.13 \text{ g C}}{1 \text{ mol } C_{12}H_{11}NO_2}\right)$$

$$\left(\frac{11 \text{ mol H}}{1 \text{ mol } C_{12}H_{11}NO_2}\right)\left(\frac{1.0079 \text{ g H}}{\text{mol H}}\right) = \left(\frac{11.087 \text{ g H}}{1 \text{ mol } C_{12}H_{11}NO_2}\right)$$

$$\left(\frac{1\ mol\ N}{1\ mol\ C_{12}H_{11}NO_2}\right)\left(\frac{14.007\ g\ C}{mol\ N}\right) = \left(\frac{14.007\ g\ N}{1\ mol\ C_{12}H_{11}NO_2}\right)$$

$$\left(\frac{2\ mol\ O}{1\ mol\ C_{12}H_{11}NO_2}\right)\left(\frac{15.999\ g\ O}{mol\ O}\right) = \left(\frac{31.998\ g\ O}{1\ mol\ C_{12}H_{11}NO_2}\right)$$

Now add the individual components to find the molar mass of the compound:

$$M = \left(\frac{144.13\ g\ C}{1\ mol\ C_{12}H_{11}NO_2}\right) + \left(\frac{11.087\ g\ H}{1\ mol\ C_{12}H_{11}NO_2}\right) + \left(\frac{14.007\ g\ N}{1\ mol\ C_{12}H_{11}NO_2}\right) + \left(\frac{31.998\ g\ O}{1\ mol\ C_{12}H_{11}NO_2}\right)$$

$$= 201.22\ g/mol$$

**Does the Result Make Sense?** We can do a quick check to see that the magnitude of the molar mass is about right by limiting the use of significant figures. Sevin contains 12 C and 11 H atoms, which contribute $(12 \times 12) + (11 \times 1) = 155$ g/mol to its molar mass. Add the contributions of two O atoms and one N atom: $(2 \times 16) + (1 \times 14) = 46$, giving 201 g/mol.

### Practice Exercise 1.4

Determine the molar mass of Malathion, an insecticide whose line structure and chemical formula appear below.

Malathion
$C_{10}H_{19}O_6PS_2$

## EXAMPLE 1-5 MOLAR MASS OF AN IONIC COMPOUND

Metal salts often exist as hydrates. One example is iron(II) nitrate hexahydrate. Determine the molar mass of this compound.

**Strategy:** Before we can calculate a molar mass, we need a chemical formula. Then we can calculate the masses of each of the elements in 1 mole of the compound.

**Solution:** Iron(II) has a charge of $+2$, and the nitrate anion carries a $-1$ charge, so there must be two nitrates for every one iron ion, and *hexa-* stands for 6, so the chemical formula is $Fe(NO_3)_2 \cdot 6H_2O$. It is convenient to decompose the formula into a list of elements, removing the parentheses and including the atoms in the waters of hydration.

One mole of $Fe(NO_3)_2 \cdot 6H_2O$ contains 1 mol Fe, 12 mol H, 2 mol N, and 12 mol O:

$$(1\ mol\ Fe)(55.845\ g/mol) = 55.845\ g\ Fe$$
$$(12\ mol\ H)(1.0079\ g/mol) = 12.095\ g\ H$$
$$(2\ mol\ N)(14.007\ g/mol) = 28.014\ g\ N$$
$$\underline{(12\ mol\ O)(15.999\ g/mol) = 191.99\ g\ O}$$
$$M\ Fe(NO_3)_2 \cdot 6H_2O = 287.94\ g/mol$$

**Does the Result Make Sense?** Another way to calculate this molar mass is by working with the individual components of the formula. Each formula unit of the compound contains one iron(II) cation, two nitrate anions, and six waters of hydration: $M\ Fe(NO_3)_2 \cdot 6H_2O = M\ Fe^{2+} + 2\ M\ NO_3^- + 6\ M\ H_2O$. Verify that this procedure gives the same result.

**Common Pitfalls:** It is easy to miss the parentheses around a group of atoms, but it is important not to do so.

**Practice Exercise 1.5**
Determine the molar mass of $Al(NO_3)_3 \cdot 9H_2O$.

## Mass–Mole–Number Conversions for Compounds

As emphasized in Section 1.4, many of the calculations in chemistry involve converting back and forth among the mass of a substance, the number of moles, and the number of atoms and/or molecules. These calculations are all centred on the mole. The connections shown in Figure 1-10 apply to chemical compounds as well as to atoms of pure elements. Molar mass and Avogadro's number provide links between the mass of a sample, the number of moles, and the number of molecules, according to Equations 1-2 and 1-3.

For example, the following calculation shows how many moles and molecules of Sevin are in a 5.0-g sample (the molar mass of the insecticide Sevin was calculated in Example 1-4):

$$n = \frac{m}{M} = \frac{5.0 \text{ g}}{201.22 \text{ g/mol}} = 0.0248 \text{ mol Sevin}$$

$$\# = nN_A = (0.0248 \text{ mol Sevin})(6.022 \times 10^{23} \text{ molecules/mol}) =$$
$$1.5 \times 10^{22} \text{ molecules of Sevin}$$

Notice that, once again, moles are at the centre of the scheme. To repeat, *calculations of chemical amounts centre on the mole.*

Mass–mole–number calculations often involve atoms *within* a compound as well as the compound itself. The chemical formula provides the link between the moles of a compound and the number of moles of the compound's individual elements:

Example 1-6 illustrates mass–mole–number conversions involving both elements and compounds.

## EXAMPLE 1-6  ELEMENTAL CONTENT

Ammonium nitrate ($NH_4NO_3$) is used as a fertilizer because it is a good source of nitrogen atoms. It is such a good source, in fact, that it ranks among the top industrial chemicals produced yearly in the world. How many moles of nitrogen atoms are present in a 454-g bag of $NH_4NO_3$ fertilizer? How many atoms is this?

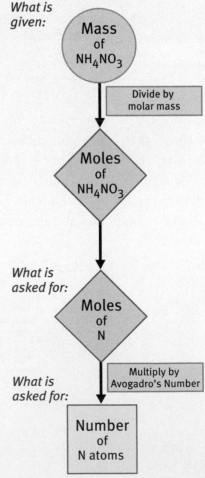

$NH_4^+$    $NO_3^-$
Ammonium    Nitrate

**Strategy:** First think about the chemistry of the problem, and then construct a mathematical solution. Use the seven-step procedure.

1. The problem asks for the number of moles and atoms of nitrogen in a 454-g sample of $NH_4NO_3$.

2. Visualize $NH_4NO_3$. (Always think atoms and molecules.) Ammonium and nitrate are common polyatomic ions whose chemical formulas you should remember.

**Solution:**

3. The problem gives the mass of the sample:

$$\text{Mass of sample} = 454 \text{ g } NH_4NO_3$$

4. Equations 1-2 and 1-3 link mass, moles, and atoms:

$$n = \frac{m}{M} \quad \text{and} \quad n = \frac{\#}{N_A}$$

5. Often, it helps to draw a flowchart that organizes the steps necessary to analyze and solve a problem. A flowchart for this problem appears in the margin.

6. Now we are ready to work through the calculations. Determine the molar mass of $NH_4NO_3$:

$$M = (2 \text{ mol N})(14.007 \text{ g/mol}) + (4 \text{ mol H})(1.0079 \text{ g/mol})$$
$$+ (3 \text{ mol O})(15.999 \text{ g/mol})$$
$$= 80.043 \text{ g/mol}$$

Divide the sample mass by the molar mass to determine moles of $NH_4NO_3$:

$$n = \frac{m}{M} = \left( \frac{454 \text{ g } NH_4NO_3}{80.043 \text{ g/mol}} \right) = 5.672 \text{ mol } NH_4NO_3$$

The chemical formula reveals that every mole of $NH_4NO_3$ contains 2 moles of nitrogen atoms:

$$(5.672 \text{ mol } NH_4NO_3)\left( \frac{2 \text{ mol N atoms}}{1 \text{ mol } NH_4NO_3} \right) = 11.34 \text{ mol N atoms}$$

As a final answer, this value should be rounded to three significant figures to match the data (454 g): 11.3 mol N atoms. For the next step of the overall calculation, however, the value should retain a fourth digit to avoid possible rounding errors. Finish the problem by rearranging Equation 1-3, and use Avogadro's number to convert from the number of moles to the number of atoms:

$$\# = nN_A = (11.34 \text{ mol N})(6.022 \times 10^{23} \text{ atoms/mol})$$
$$= 6.83 \times 10^{24} \text{ N atoms}$$

*What is given:*

**Mass** of $NH_4NO_3$

Divide by molar mass

**Moles** of $NH_4NO_3$

**Moles** of N

*What is asked for:*

Multiply by Avogadro's Number

*What is asked for:*

**Number** of N atoms

The final result is rounded to three significant figures to match the three significant figures given for the mass of the bag of fertilizer.

**7. Does the Result Make Sense?** Although $10^{24}$ is an immense number of atoms, we know that 1 mole contains $6 \times 10^{23}$ atoms. Numbers of the order of 10 are reasonable when calculating the number of atoms in a sample of everyday size.

**Common Pitfalls:** It is tempting to simply use the number of significant digits shown on your calculator, but make sure you round off the answer appropriately.

**Practice Exercise 1.6**
How many moles and atoms of hydrogen are present in 125 g of water ($H_2O$)?

## SECTION EXERCISES

**1.5.1** Calculate the molar masses of the following compounds:

| Compound | Name | Uses |
|---|---|---|
| (a) $NaAsO_2$ | Sodium arsenite | Insecticide, herbicide, fungicide |
| (b) $Na_4SiO_4$ | Sodium silicate | Soaps, adhesives, flame retardant |
| (c) $Ca(ClO)_2$ | Calcium hypochlorite | Bactericide, bleach, pool cleaner |
| (d) $Ba(NO_3)_2$ | Barium nitrate | Green fireworks |
| (e) $Tl_2SO_4$ | Thallium sulphate | Rat and ant poisons |

**1.5.2** Twenty different amino acids are the essential building blocks of proteins. Calculate the molar masses of these three:

(a) Histidine $C_6H_9N_3O_2$

(b) Cysteine $C_3H_7NO_2S$

(c) Asparagine $C_4H_8N_2O_3$

**1.5.3** Adenosine triphosphate (ATP, shown below) is the body's principal energy storage molecule. The formula of ATP is $C_{10}H_{16}N_5O_{13}P_3$. (a) How many moles of ATP are in a 50.0-mg sample? (b) How many molecules of ATP is this? (c) How many atoms of phosphorus are in the sample? (d) How many atoms of nitrogen are there? (e) What is the mass of oxygen in 50.0 mg of ATP? (f) How many total atoms are there in 50.0 mg of ATP?

## 1.6 AQUEOUS SOLUTIONS

We developed the concept of the mole in terms of pure chemical substances, but many chemical reactions take place in solution. To treat solution reactions quantitatively, we need ways to apply the mole concept to solutions. A substance used to dissolve solutes is a **solvent**, and a pure substance dissolved in solution is a **solute**. Most of the time, the solvent is a liquid and is present in much larger quantities than any solutes.

Chemists use many different liquid solvents, but we focus most of our attention on water. When water is the solvent, the solution is said to be **aqueous**. A rich array of chemistry occurs in aqueous solution, including many geological and biochemical processes. Aqueous solutions dominate the chemistry of the Earth and the biosphere. The oceans, for example, are aqueous solutions of various cations and anions, with sodium and chloride being the most abundant. The oceans can be thought of as huge aqueous solvent vessels for the remarkably complex chemistry of our world. Blood is an aqueous system that contains many ionic species, most notably carbonate, sodium, and potassium. In fact, the human body is mostly water, and much of the biochemistry of life takes place in aqueous solution.

## Molarity

Any solution contains at least two chemical species: the solvent and one or more solutes. The mass of a solution is the sum of the masses of the solvent and all dissolved solutes. To answer questions such as "How much is there?" about solutions, we need to know the amount of each solute present in a specified volume of solution. The amount of a solute in a solution is given by the **concentration** ($c$), which is the ratio of the amount of solute to the amount of solution. In chemistry the most common measure of concentration is **molarity**, which is the number of *moles of solute ($n$)* divided by the total *volume of the solution ($V$)* in litres. In this case,

$$c = \frac{\text{Moles of solute}}{\text{Total volume of solution}} \quad \text{or} \quad c = \frac{n}{V} \tag{1-4}$$

Equation 1-4 defines concentration in terms of molarity, but note that many other units of concentration are used in chemistry. (Several of these are detailed in Chapter 11.) If a solution has a concentration of, for example, 2.00 mol/L, we would write $c = 2.00$ M. **Be careful not to confuse this unit of concentration ($M = $ mol/L) with the symbol for molar mass ($M = $ g/mol).** Confusion also arises because many chemists will refer to the "molarity of the solution" instead of the "concentration of the solution, expressed as molarity."

An aqueous solution is prepared by dissolving a measured quantity of solid or liquid in enough water to give a desired final volume. For example, we might dissolve 2.05 g of table sugar (sucrose) in just enough water to make the final volume 25.0 mL. To determine the concentration of the sugar solution, we must convert the mass of sucrose into moles and convert the solution volume into litres.

To begin, use the formula of sucrose, which is $C_{12}H_{22}O_{11}$, to calculate the molar mass, 342 g/mol. Next, convert mass to moles in the usual manner:

$$n = \frac{m}{M} = \frac{2.05 \text{ g sugar}}{342 \text{ g/mol}} = 5.994 \times 10^{-3} \text{ mol sugar}$$

Then, convert the volume of the solution from millilitres to litres:

$$(25.0 \text{ mL})(10^{-13} \text{ L/mL}) = 2.50 \times 10^{-2} \text{ L}$$

Finally, divide moles of sugar by volume in litres to obtain concentration in mol/L:

$$c = \frac{n}{V} = \frac{5.994 \times 10^{-3} \text{ mol}}{2.50 \times 10^{-2} \text{ L}} = 0.240 \text{ M}$$

Concentrations are used so frequently in chemistry that shorthand notations are essential. For example, the concentration of a species is expressed by enclosing its formula in square brackets: $[C_{12}H_{22}O_{11}] = 0.240 \text{ M}$ for the sugar solution example above.

## EXAMPLE 1-7   PREPARING A SOLUTION

What mass of nickel(II) chloride hexahydrate is required to prepare 250 mL of aqueous solution with a concentration of 0.255 M?

**Strategy:** As with all calculations of chemical amounts, we must work with moles. Because grams are asked for, we must do a mole–mass conversion; this requires the molar mass of the substance, which in turn requires that we know the chemical formula.

**Solution:** Begin by finding the chemical formula from the name, nickel(II) chloride hexahydrate. Chloride ion carries a −1 charge, and the (II) indicates that nickel is a +2 cation. Electrical neutrality requires two chloride ions for every nickel ion. The name also tells us that the salt contains six water molecules for each unit of nickel(II) chloride. Thus, each formula unit of the salt contains one $Ni^{2+}$ cation, two $Cl^-$ anions, and six water molecules: $NiCl_2 \cdot 6H_2O$.

To determine the mass of $NiCl_2 \cdot 6H_2O$ required to prepare the solution, first calculate the number of moles of the salt required, and then use the molar mass to determine the mass in grams:

$$n = cV = \left(\frac{0.255 \text{ mol}}{L}\right)\left(\frac{10^{-3} \text{ L}}{1 \text{ L}}\right)(250. \text{ mL}) = 6.375 \times 10^{-2} \text{ mol } NiCl_2 \cdot 6H_2O$$

We need the molar mass of $NiCl_2 \cdot 6H_2O$ to convert moles to mass. The molar mass of $NiCl_2 \cdot 6H_2O$ is 237.69 g/mol:

$$m = nM = (6.375 \times 10^{-2} \text{ mol})(237.69 \text{ g/mol}) = 15.15 \text{ g}$$

To prepare the desired solution, we would first weigh 15.2 g of solid using a balance. (Rounding to three significant figures is consistent with the target molarity stated in the problem.) We would transfer this solid into a 250-mL volumetric flask, add enough water to dissolve the solid, and then continue adding water and mixing until the solution level matched the 250 mL mark on the flask. Note that molarity is moles per litre of *solution*, and making this solution will require slightly less than 250 mL of water.

**Does the Result Make Sense?** The mass is substantially less than the molar mass of the compound. Given that the volume is considerably less than 1 L and the target concentration is considerably less than 1 M, this is a reasonable result.

**Practice Exercise 1.7**

Calculate the mass of ammonium nitrate needed to prepare 100 mL of aqueous solution with a concentration of 0.400 M.

## Species in Solution

To understand the chemical behaviour of solutions, we must "think molecules." Before working any problem involving aqueous solutions, begin with the question, "What chemical species are present in the solution?" There will always be an abundance of water molecules in an aqueous solution. In addition, there will be solute species, which may be molecules or ions.

An aqueous solution of a molecular substance such as sugar ($C_{12}H_{22}O_{11}$) or ethanol ($C_2H_5OH$) contains individual molecules in a sea of water molecules (Figure 1-11). We know from measurements of electrical conductivity that these solutes dissolve as neutral molecules. Figure 1-12 shows that pure water does not conduct electricity, and neither does a solution of sugar in water. This result shows that these solutions contain no mobile charged particles. Sugar and ethanol dissolve as neutral molecules.

In contrast to sugar, solid NaCl dissolves in water to give a liquid that conducts electricity. Figure 1-12 shows that a solution of NaCl is a good conductor. When an ionic compound dissolves in water, its component cations and anions are free to move about in the solution. Mixing leads to a uniform distribution of $Na^+$ and $Cl^-$ ions throughout the entire solution, with each ion surrounded by a sheath of water molecules, as shown in Figure 1-13.

The presence of ions in solution is what gives a NaCl solution the ability to conduct electricity. If positively and negatively charged wires are dipped into the solution, the ions in the solution respond to the charges on the wires. Chloride anions move toward the positive wire, and sodium cations move toward the neg-

**Figure 1-11**

Molecular picture of an aqueous solution of ethanol, which contains ethanol molecules and water molecules.

**Figure 1-12**

Pure water *(left)* and a solution of sugar *(right)* do not conduct electricity because they contain virtually no ions. A solution of salt *(centre)* conducts electricity well because it contains mobile cations and anions.

**Figure 1-13**

Molecular picture of a solution of sodium chloride in water. All the molecules and ions move freely about, but overall electrical neutrality is maintained because the total amount of anionic charge equals the total amount of cationic charge.

**Figure 1-14**

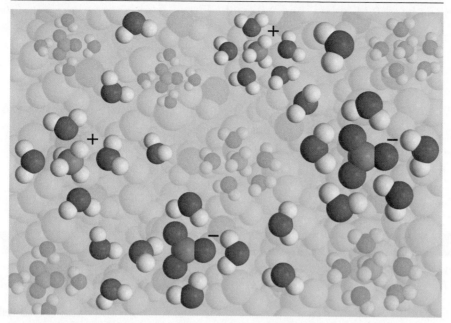

Molecular view of an aqueous solution of ammonium nitrate. Ammonium ions separate from nitrate ions, but each of these species remains intact as polyatomic clusters of atoms.

ative wire. This directed movement of ions in solution is a flow of electrical current. Pure water, which has virtually no dissolved ions, does not conduct electricity. Any solution formed by dissolving an ionic solid in water conducts electricity. Ordinary tap water, for example, contains ionic impurities that make it an electrical conductor.

Notice that the molecular view in Figure 1-13 includes ions as well as neutral molecules. "Molecular view" means a view of how matter looks at the atomic/molecular level. Thus, such illustrations may contain ions and individual atoms as well as molecules.

When a salt containing polyatomic ions dissolves in water, the cations separate from the anions, but each polyatomic ion remains intact. An example is ammonium nitrate, composed of $NH_4^+$ polyatomic cations and $NO_3^-$ polyatomic anions. Ammonium nitrate dissolves in water to give a solution containing $NH_4^+$ cations and $NO_3^-$ anions, as Figure 1-14 illustrates.

Because aqueous solutions containing ions play a central role in the world around us, the chemistry of such solutions is discussed in depth in several chapters of this book. To understand the chemistry of aqueous solutions containing ions, *it is essential that you learn to recognize the common ions at a glance.* This is especially true of the polyatomic ions.

## Concentrations of Ionic Solutions

The ratio of cations and anions in a solution of a salt is determined by its chemical formula. For example, a solution of NaCl contains $Na^+$ cations and $Cl^-$ anions in 1 : 1 ratio: a 1.0 M solution of NaCl has 1.0 M each of $Na^+$ and $Cl^-$. In contrast, $NiCl_2$ contains 2 moles of $Cl^-$ anions for every 1 mole of $Ni^{2+}$ cations. This

The *(s)* and *(aq)* in the equation indicate phases: *(s)* designates solid and *(aq)* designates aqueous solution.

ratio is maintained when nickel(II) chloride dissolves in water; a 1.0 M solution of $NiCl_2$ has 1.0 M of $Ni^{2+}$ and 2.0 M of $Cl^-$:

$$NiCl_2 \ (s) \xrightarrow{\ H_2O\ } Ni^{2+} \ (aq) + Cl^- \ (aq) + Cl^- \ (aq)$$

$$1 \text{ mol} \Uparrow \qquad\qquad 1 \text{ mol} \Uparrow \qquad\qquad 2 \text{ mol} \Uparrow$$

Notice that preserving the cation–anion ratio ensures that the solution is electrically neutral overall. The total positive charge carried by 1 mole of $Ni^{2+}$ equals the total negative charge carried by 2 moles of $Cl^-$.

Example 1-8 treats concentrations of polyatomic ions.

## EXAMPLE 1-8 MOLARITY OF IONS IN SOLUTION

Find the concentrations (in mol/L) of the ionic species present in 250 mL of an aqueous solution containing 1.75 g of ammonium sulphate, $(NH_4)_2SO_4$.

**Strategy:** The chemical formula identifies the ions that are present in the final solution. The formula also tells us how many moles of each ion are present in 1 mole of the salt. Use mass, molar mass, and volume to calculate molarity.

**Solution:** Looking at the formula, we recognize the ammonium cation, $NH_4^+$, and the sulphate anion, $SO_4^{2-}$. When ammonium sulphate dissolves in water, it dissociates into its component polyatomic ions. Each mole of salt will produce 2 mol $NH_4^+$ and 1 mol $SO_4^{2-}$.

After identifying what species are present in solution, we can calculate molarities. Begin with the molar mass of the salt:

$$(2 \text{ mol N})(14.01 \text{ g/mol}) = 28.02 \text{ g N}$$
$$(8 \text{ mol H})(1.008 \text{ g/mol}) = 8.064 \text{ g H}$$
$$(1 \text{ mol S})(32.06 \text{ g/mol}) = 32.06 \text{ g S}$$
$$\underline{(4 \text{ mol O})(16.00 \text{ g/mol}) = 64.00 \text{ g O}}$$
$$1 \text{ mol } (NH_4)_2SO_4 = 132.14 \text{ g}$$

Use the molar mass to determine the number of moles of salt and, in turn, the number of moles of each ion. Notice that even though we are aiming for molarities of ionic species, we determine the moles of the salt. This is because the mass that is provided is the *mass of the salt*. To find the moles of salt (*n*), this mass must be divided by the *molar mass of the salt*:

$$n = \frac{m}{M} = \frac{1.75 \text{ g}}{132.14 \text{ g/mol}} = 1.324 \times 10^{-2} \text{ mol } (NH_4)_2SO_4$$

From the number of moles of salt, calculate the number of moles of each ion:

$$(1.324 \times 10^{-2} \text{ mol } (NH_4)_2SO_4)\left[\frac{2 \text{ mol } NH_4^+}{1 \text{ mol } (NH_4)_2SO_4}\right] = 2.648 \times 10^{-2} \text{ mol } NH_4^+$$

$$(1.324 \times 10^{-2} \text{ mol } (NH_4)_2SO_4)\left[\frac{1 \text{ mol } SO_4^{2-}}{1 \text{ mol } (NH_4)_2SO_4}\right] = 1.324 \times 10^{-2} \text{ mol } SO_4^{2-}$$

Now find the molarities of the individual ions:

$$(250. \text{ mL})(10^{-3} \text{ L/mL}) = 0.250 \text{ L}$$

$$[NH_4^+] = \frac{2.648 \times 10^{-2} \text{ mol}}{0.250 \text{ L}} = 1.06 \times 10^{-1} \text{ M}$$

$$[SO_4^{2-}] = \frac{1.324 \times 10^{-2} \text{ mol}}{0.250 \text{ L}} = 5.30 \times 10^{-2} \text{ M}$$

**Does the Result Make Sense?** The concentration of ammonium ions is twice that of sulphate ions. This is consistent with the 2 : 1 ratio of ammonium to sulphate in the chemical formula. You should understand that when we describe this solution as "0.0530 M $(NH_4)_2SO_4$" stating moles of salt per litre of solution, we are describing a solution that actually contains a *mixture of ions each, with its own molarity.*

**Common Pitfalls:** Do not forget that salts dissociate in solution. When we say that a 1.0 M solution of a salt is prepared, it means we dissolve 1.0 mole of salt in 1 L of water, but there will be at least 2.0 moles or more of ions in this solution.

**Practice Exercise 1.8**
Calculate the molarities of the ions present in the solution of nickel(II) chloride hexahydrate prepared in Example 1-7.

> Remember to carry an extra significant figure until the final calculation.
>
> Use dimensional analysis to obtain correct conversion factors. Rearrange 1 mL = $10^{-3}$ L to give 1 = ($10^{-3}$ L/mL) with mL in the denominator so that it cancels mL in the numerator.

## Dilutions

One of the common methods for preparing a solution of known concentration is to dilute an existing solution of higher concentration by adding more solvent. In a **dilution**, the *amount of solute* remains the same, but the *volume of the solution* increases. Thus, a dilution results in a solution of lower molarity.

The most important feature of a dilution is that *the number of moles of solute does not change during the dilution.* Consider the dilution of an aqueous solution of nickel(II) chloride, as shown in Figure 1-15. The photo shows that the green

**Figure 1-15**

When a concentrated solution of $NiCl_2$ (*left*) is diluted by adding more solvent, the resulting solution (*right*) contains the same numbers of $Ni^{2+}$ and $Cl^-$ ions. The solution is more dilute, however, because the ions are spread around in a larger volume of solvent.

colour, which is due to the $Ni^{2+}$ ions, fades when the solution is diluted. This fading is visible evidence that the concentration has decreased. At the molecular level, there are the same *numbers* of $Ni^{2+}$ and $Cl^-$ ions in the solution before the dilution and after the dilution. However, because the volume of the solution increases, the **concentration** of the new solution is lower than the concentration of the original solution.

The quantitative aspects of dilutions can be found from the fact that the number of moles of solute does not change during a dilution from the initial (i) solution to the final (f) solution:

$$n_i = n_f$$

Because we can arrange Equation 1-4 to give $n = cV$, this leads to a simple equation:

$$c_i V_i = c_f V_f \qquad (1\text{-}5)$$

Equation 1-5 is convenient for dilution calculations. If any three of the quantities are known, we can calculate the fourth, as Example 1-9 shows.

## EXAMPLE 1-9 DILUTION OF A SOLUTION

Aqueous hydrochloric acid, HCl, is usually sold as a 12.0 M solution, commonly referred to as *concentrated HCl*. A chemist needs to prepare 2.00 L of 1.50 M HCl for a number of different applications. What volume of concentrated HCl solution should the chemist use in the dilution?

**Strategy:** The seven-step problem-solving approach is appropriate.

1. The question asks for the volume of concentrated HCl required to prepare a dilute solution.

2. Visualization: A concentrated solution is diluted to a larger volume. The chemist will remove a sample from the bottle of concentrated HCl solution and mix it with water.

**Solution:**

3. The information available is as follows:

   Concentrated HCl = 12.0 M    Diluted HCl = 1.50 M

   Final volume = 2.00 L

4. This is a dilution process, so we use Equation 1-5.

5. Rearrange Equation 1-5 to solve for $V_i$, the volume required: $V_i = \dfrac{c_f V_f}{c_i}$.

6. Do the calculation. In Step 3, we identified the appropriate values:

   $c_f = 1.50$ M,    $V_f = 2.00$ L,    $c_i = 12.0$ M

$$V_i = \frac{(1.50 \text{ M})(2.00 \text{ L})}{12.0 \text{ M}} = 0.250 \text{ L}$$

**7. Does the Result Make Sense?** The volume has the correct units, and the amount required is smaller than the total volume. This is reasonable.

**Common Pitfalls:** Make sure that if the solution is diluted, its concentration decreases. If your calculation says otherwise, you have incorrectly rearranged Equation 1-5.

**Practice Exercise 1.9**
What volume (in mL) of concentrated ammonia (14.8 M) should be used to prepare 0.500 L of a 0.350 M solution?

When preparing acid solutions, water should never be added directly to concentrated acid. Addition of water often causes rapid local heating that may lead to a dangerous acid splash. Instead, concentrated acid is always added to the water. Additional water can then be added safely to the less concentrated solution that has resulted. The solution in Example 1-9 is prepared by using a volume-measuring device, such as a graduated cylinder, to measure 0.250 L (250 mL) of concentrated HCl. About 0.5 L of water is added to a 2.00 L volumetric flask, and then the concentrated acid from the graduated cylinder is poured slowly into the water. Filling the volumetric flask to the mark with additional water completes the dilution.

Chemists often need to work with solutions with a variety of concentrations. Often, a relatively concentrated "stock" solution is prepared by weighing. Then, a portion of this solution is diluted volumetrically. In procedures of this kind, it is important to keep track of the volumes, as illustrated in Example 1-10 and shown in the flowchart in Figure 1-16.

## EXAMPLE 1-10  PREPARING SOLUTIONS

An agricultural chemist wished to study the effect of varying fertilizer applications on the growth of tomato plants. The chemist prepared a stock aqueous solution of urea, $(NH_2)_2CO$, by dissolving 1.75 g of this compound in water to make 1.00 L of solution. Then, she prepared a series of more dilute solutions to apply to her tomato plants. One of these solutions contained 5.00 mL of stock solution diluted to give a final volume of 25.00 mL. What was the concentration of urea in this diluted solution?

**Strategy:** Use the seven-step method.

1. We are asked to determine the concentration of the dilute solution. This requires us first to determine the concentration of the stock solution, so we need to do two separate calculations.

2 and 3.  A flowchart, shown below, helps us to visualize the problem and organizes the data. The flowchart shows that the calculation has two parts. First, we treat the stock solution. Then, we use the result to treat the diluted solution.

**Solution:**

4. We have mass and volume information for the stock solution, so we use Equations 1-2 and 1-4:

$$n = \frac{m}{M} \quad \text{and} \quad c = \frac{n}{V}$$

For the diluted solution, we use Equation 1-5:

$$c_i V_i = c_f V_f$$

5. Manipulate the equations to obtain equations for the molarity of the stock and diluted solutions:

$$c_{stock} = \frac{m}{(M)(V_{stock})} \quad \text{and} \quad c_{final} = \frac{(c_{stock})(V_{removed})}{V_{final}}$$

We need the molar mass of urea:

$$M = (4 \text{ mol H})(1.0079 \text{ g/mol}) + (2 \text{ mol N})(14.007 \text{ g/mol})$$
$$+ (1 \text{ mol C})(12.011 \text{ g/mol}) + (1 \text{ mol O})(15.999 \text{ g/mol})$$
$$= 60.056 \text{ g/mol}$$

6. Substitute and calculate:

$$c_{stock} = \frac{m}{(M)(V_{stock})} = \frac{1.75 \text{ g}}{(60.056 \text{ g/mol})(1.00 \text{ L})} = 2.914 \times 10^{-2} \text{ M}$$

$$c_{final} = \frac{(c_{stock})(V_{removed})}{V_{final}} = \frac{(2.914 \times 10^{-2} \text{ M})(5.00 \text{ mL})}{25.00 \text{ mL}} = 5.82 \times 10^{-3} \text{ M}$$

**7. Do the Results Make Sense?** To check for reasonableness, make sure you have used the correct volumes. The *stock solution* was prepared in a volume of 1.00 L. The *dilution* was from 5.00 to 25.00 mL, a factor of 5. We can see by inspection that the final concentration is lower than the stock concentration by a factor of about 5.

**Practice Exercise 1.10**

Calculate the concentration of ammonium nitrate in a solution prepared by dissolving 3.20 g of the salt in enough water to make 100 mL of solution, and then diluting 5.00 mL of this solution to a volume of 25.00 mL.

**Figure 1-16**

Flowchart showing the connections among measured quantities in dilution procedures.

**SECTION EXERCISES**

●**1.6.1** An urban gardener buys a 1.0-kg bag of ammonium nitrate to fertilize house plants. Calculate concentrations of the ionic species present in a stock solution prepared by dissolving 7.5 g of the salt in 0.300 L of water.

●**1.6.2** Plants can be badly damaged by solutions that are highly concentrated. A friend of the urban gardener recommends against using fertilizer solutions that are greater than 1.0 ($1.0 \times 10^{-3}$ M) in total nitrogen. What volume (in mL) of the stock solution from Section Exercise 1.6.1 will be needed to prepare 750 mL of 1.0 mM fertilizer solution?

●**1.6.3** Draw a molecular picture of a portion of the solution described in Section Exercise 1.6.2. Make sure the solution is electrically neutral (omit water molecules to simplify your drawing).

## 1.7 WRITING CHEMICAL EQUATIONS

### Balanced Equations

In chemical reactions, *the amount of each element is always conserved.* This is consistent with the statements of Dalton's atomic theory. In addition, *the total amount of electrical charge is always conserved.* This is the law of conservation of charge. A **balanced chemical equation** describes a chemical reaction in which the amounts of all elements and of electrical charge are conserved. We refer to the species that take part in a chemical reaction as **reagents. Reactants** are reagents that are consumed in a reaction, whereas **products** are reagents that are produced in a reaction. In addition to providing a list of reactants and products of the reaction, a balanced chemical equation tells us the relative amounts of all the reagents.

As an example, consider the industrial synthesis of ammonia ($NH_3$). Ammonia is made by the Haber process, a single chemical reaction between molecules of hydrogen ($H_2$) and nitrogen ($N_2$). Although it is simple, this synthesis has immense industrial importance. Canada produces some 4 billion kilograms of ammonia annually.

In this synthesis, the reactants are $N_2$ and $H_2$, which react to produce $NH_3$. Here is the balanced chemical equation:

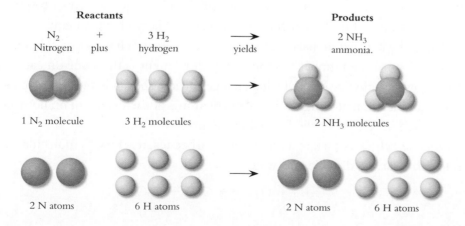

| Reactants | | | | Products |
|---|---|---|---|---|
| $N_2$ | + | $3 H_2$ | $\longrightarrow$ | $2 NH_3$ |
| Nitrogen | plus | hydrogen | yields | ammonia. |

1 $N_2$ molecule     3 $H_2$ molecules          2 $NH_3$ molecules

2 N atoms     6 H atoms          2 N atoms     6 H atoms

All balanced chemical equations have the following features:

1. The reactants appear on the left and the products appear on the right. The arrow joining them indicates the direction of reaction.
2. An integer precedes the formula of each substance. These numbers are the **stoichiometric coefficients.** When no number appears (as for $N_2$ in this equation), the stoichiometric coefficient is 1.
3. The stoichiometric coefficients in a chemical equation are the smallest integers that give a balanced equation.
4. Charge is conserved. In this equation, all participants are neutral species, so charge is conserved regardless of the stoichiometric coefficients.

Stoichiometric coefficients describe the relative numbers of molecules involved in the reaction. In any actual reaction, immense numbers of molecules are involved, but the relative numbers are always related through the stoichiometric coefficients. Furthermore, these coefficients describe both the relative numbers of *molecules* and the relative numbers of *moles* involved in the reaction. For example, the Haber reaction always involves immense numbers of molecules, but the equation describing the synthesis of ammonia tells us the following:

1. Each molecule of nitrogen reacts with three molecules of hydrogen to give two molecules of ammonia.
2. Each mole of molecular nitrogen reacts with 3 moles of molecular hydrogen to give 2 moles of ammonia.

Keep in mind that the key feature of balanced chemical equations is the conservation law:

*The number of atoms of each element is conserved in any chemical reaction.*

## Balancing Equations

The stoichiometric coefficients in a balanced chemical equation must be chosen so that the atoms of each element are conserved. Many chemical equations can be balanced by inspection. Balancing by inspection means changing stoichiometric coefficients until the number of atoms of each element is the same on each side of the arrow. Usually, we can tell what changes need to be made by looking closely at the reaction and matching the numbers of atoms of each element on both sides of the equation. Consider the following example.

Propane, which is used as a fuel for gas barbecues, reacts with molecular oxygen to form carbon dioxide and water:

$$C_3H_8 + O_2 \longrightarrow CO_2 + H_2O$$

To determine if this equation is balanced, make a list of the elements and numbers of atoms on each side:

$$C_3H_8 + O_2 \longrightarrow CO_2 + H_2O$$
$$3\,C + 8\,H + 2\,O \longrightarrow 1\,C + 2\,H + 3\,O$$

The equation is not balanced, because there are too many carbon and hydrogen atoms on the left and too many oxygen atoms on the right. We need to change the numbers of molecules by changing the stoichiometric coefficients until the numbers of atoms of each element are equal.

It is easiest to balance a chemical equation one element at a time, starting with the elements that appear in only one substance on each side. Notice that all of the carbon atoms in propane end up in carbon dioxide molecules, and all of propane's hydrogen atoms appear in water molecules. This feature allows us to balance carbon and hydrogen easily.

To take care of the three carbon atoms per propane molecule, we need three molecules of $CO_2$. Thus, the carbon atoms are balanced by changing the stoichiometric coefficient of $CO_2$ from 1 to 3. In this reaction the ratio of $CO_2$ to propane is 3:1. Similarly, we need four molecules of water for the eight hydrogen atoms in one molecule of propane. Using this information, we modify the equation as follows:

$$C_3H_8 + O_2 \longrightarrow 3\,CO_2 + 4\,H_2O$$
$$3\,C + 8\,H + 2\,O \longrightarrow 3\,C + 8\,H + 10\,O$$

The situation is looking better because atoms of carbon and hydrogen are now conserved. However, the equation is still not balanced because there are too few oxygen atoms on the left side. To balance oxygen, we must give $O_2$ a stoichiometric coefficient of 5:

$$C_3H_8 + 5\,O_2 \longrightarrow 3\,CO_2 + 4\,H_2O$$
$$3\,C + 8\,H + 10\,O \longrightarrow 3\,C + 8\,H + 10\,O$$

Now the equation is balanced. Combustion of one molecule of propane gas produces three molecules of $CO_2$ and four molecules of $H_2O$ and consumes five molecules of $O_2$. Also, combustion of 1 mole of propane produces 3 moles of $CO_2$ and 4 moles of $H_2O$ and consumes 5 moles of $O_2$. Example 1-11 shows how to balance another reaction.

## EXAMPLE 1-11 BALANCING CHEMICAL REACTIONS

Ammonium nitrate ($NH_4NO_3$) explodes when it is heated above 300 °C. The products are three gases: molecular nitrogen, molecular oxygen, and steam (water vapour). Write a balanced equation for the explosion of ammonium nitrate.

**Strategy:** The description in the problem tells us what happens to the starting material: $NH_4NO_3$ breaks apart into molecules of $N_2$, $O_2$, and $H_2O$. An unbalanced form of the equation can be written from this description. Then, we must balance each element in turn by inspection.

$$NH_4NO_3 \longrightarrow N_2 + H_2O + O_2$$

**Solution:** Count atoms of each element to see if the elements are in balance:

$$2\,N + 4\,H + 3\,O \longrightarrow 2\,N + 2\,H + 3\,O$$

Focus first on the elements that appear in only one reactant and one product, nitrogen and hydrogen in this case. Nitrogen is already balanced. To balance the H atoms, change the stoichiometric coefficient of water from 1 to 2:

$$NH_4NO_3 \longrightarrow N_2 + 2\,H_2O + O_2$$
$$2\,N + 4\,H + 3\,O \longrightarrow 2\,N + 4\,H + 4\,O$$

Now, atoms of nitrogen and hydrogen are conserved, but the equation is not balanced with respect to oxygen. We cannot change the $1:1$ ratio of $NH_4NO_3$ to $N_2$ and the $1:2$ ratio of $NH_4NO_3$ to $H_2O$, because these ratios must be retained to keep the hydrogen and nitrogen atoms conserved. Thus, we must adjust the coefficient of $O_2$ to balance with respect to oxygen. Numerically, we can take care of oxygen by changing the coefficient of $O_2$ to $^1/_2$:

$$NH_4NO_3 \longrightarrow N_2 + 2\,H_2O + {}^1\!/_2\,O_2$$
$$2\,N + 4\,H + 3\,O \longrightarrow 2\,N + 4\,H + 3\,O$$

This balances the equation, but it is unrealistic from a molecular perspective because there is no such thing as half a molecule. To get rid of the $^1/_2$ without unbalancing the reaction, we multiply all the coefficients by 2:

$$2\,NH_4NO_3 \longrightarrow 2\,N_2 + 4\,H_2O + O_2$$
$$4\,N + 8\,H + 6\,O \longrightarrow 4\,N + 8\,H + 6\,O$$

**Does the Result Make Sense?** Count the atoms of each element to be sure that the reaction is balanced and reasonable from a molecular perspective.

**Common Pitfalls:** Missing the subscripts or parentheses in a chemical formula will make it impossible to balance a reaction.

Later, we show that it can be convenient to use fractional coefficients such as $\frac{1}{2}$. When this is the case, the coefficient refers to $\frac{1}{2}$ mol, not $\frac{1}{2}$ molecule.

**Practice Exercise 1.11**

Some racing cars use methanol, $CH_3OH$, as their fuel. Balance the combustion reaction of methanol.

## SECTION EXERCISES

●**1.7.1** Although gasoline is a complex mixture of molecules, the chemical reaction that takes place in an automobile engine can be represented by combustion of one of its components, octane ($C_8H_{18}$). Such burning of fossil fuels releases millions of tonnes of carbon dioxide into the Earth's atmosphere each year. Write a balanced equation for the combustion of octane.

●**1.7.2** Additional reactions accompany the combustion of octane. For example, molecular nitrogen reacts with molecular oxygen in an automobile cylinder to give nitrogen oxide. After leaving the engine, nitrogen oxide reacts with atmospheric oxygen to give nitrogen dioxide, the red-brown gas seen in the air over many urban areas. Write balanced equations for these two reactions.

●**1.7.3** Acrylonitrile is used to make synthetic fibres such as Orlon. About 1.5 billion kg of acrylonitrile is produced each year. Balance the following chemical equation, which shows how acrylonitrile is made from propene, ammonia, and oxygen:

$$C_3H_6 + NH_3 + O_2 \longrightarrow C_3H_3N + H_2O$$

Propene                    Acrylonitrile

## 1.8 THE STOICHIOMETRY OF CHEMICAL REACTIONS

*Stoichiometry* is pronounced
"stoy-key-om-etry." It combines two Greek words, *sto-
icheion* ("element") and
*metron* ("measure").

Sections 1.4 and 1.5 describe how to use the relationships among atoms, moles, and masses to answer "how much?" questions about *individual substances.* Combining these ideas with the concept of a balanced chemical equation lets us answer "how much?" questions about *chemical reactions.* The study of the amounts of materials consumed and produced in chemical reactions is called **stoichiometry.**

A chemical synthesis requires the proper amounts of starting materials for a successful outcome. Just as a cake recipe provides the amounts of ingredients needed for successful baking, a balanced chemical equation is a chemical recipe for successful synthesis. For example, how many grams of hydrogen do we need to produce 68 g of ammonia? To find out, we begin with the balanced chemical equation:

$$N_2 + 3H_2 \longrightarrow 2NH_3$$

According to the balanced equation, the synthesis of 2.0 moles of ammonia requires 3.0 moles of hydrogen and 1.0 mole of nitrogen. However, the balanced equation does not give us any direct information about the *masses* involved in the synthesis. One gram of $N_2$ plus 3 g of $H_2$ *does not* make 2 g of $NH_3$. Remember that calculations in chemistry are based on the concept of the mole.

In other words, a chemical recipe for making ammonia requires amounts in moles, not masses. Thus, to prepare 68 g of ammonia, we need to know the number of moles of ammonia in 68 g. To do that, we must divide the mass of ammonia by the molar mass of ammonia, 17.0 g/mol, to convert grams to moles:

$$n = \frac{m}{M} = \frac{68 \text{ g } NH_3}{17.0 \text{ g mol}^{-1}} = 4.0 \text{ mol } NH_3$$

How many moles of $H_2$ are needed to make 4.0 moles of $NH_3$? According to the balanced equation, 2 moles of $NH_3$ requires 3 moles of $H_2$. The ratio of stoichiometric coefficients from a balanced equation is called the **stoichiometric ratio.** We now apply this ratio to the desired amount of $NH_3$:

$$4.0 \text{ mol } NH_3 \left( \frac{3 \text{ mol } H_2}{2 \text{ mol } NH_3} \right) = 6.0 \text{ mol } H_2$$

Note that the units "mol $NH_3$" cancel, leaving only "mol $H_2$." All that remains is to convert from moles of $H_2$ to mass: $m = nM = (6.0 \text{ mol } H_2)(2.02 \text{ g/mol}) = 12 \text{ g } H_2$.

The same procedure can be used to show that the synthesis of 68 g of $NH_3$ requires 56 g of $N_2$.

To summarize, the amounts of different reagents that participate in a chemical reaction are related through the stoichiometric coefficients in the balanced chemical equation. To convert from moles of one reagent to moles of any other reagent, multiply by the stoichiometric ratio that leads to proper cancellation of units:

$$\text{Moles}_B = \left( \frac{\text{Coefficient}_B}{\text{Coefficient}_A} \right) \text{Moles}_A \tag{1-6}$$

68 g
$NH_3$

↓ Divide by molar mass

4.0 mol
$NH_3$

↓ Multiply by $C_B/C_A$

6.0 mol
$H_2$

↓ Multiply by molar mass

12 g
$H_2$

A flowchart illustrates the logic of the stoichiometric ratio.

Examples 1-12 and 1-13 illustrate this procedure.

## EXAMPLE 1-12  HOW MUCH PRODUCT CAN BE MADE?

Geranyl formate is used as a synthetic rose essence in cosmetics. The compound is prepared from formic acid and geraniol:

$$HCO_2H + C_{10}H_{18}O \longrightarrow C_{11}H_{18}O_2 + H_2O$$

A perfumery needs some geranyl formate for a batch of perfume. How many grams of geranyl formate can a chemist make from 375 g of geraniol?

**Strategy:** The stoichiometric coefficients state that 1 mole of each reactant will produce 1 mole of each product. A flowchart summarizes the steps used to convert the mass of geraniol into the mass of geranyl formate.

**Solution:** First, convert the mass of geraniol into moles by dividing by the molar mass:

$$n = \frac{m}{M} = \frac{375 \text{ g geraniol}}{154.2 \text{ g/mol}} = 2.432 \text{ mol geraniol}$$

A 1 : 1 stoichiometric ratio links moles of geraniol to moles of geranyl formate:

$$\text{Moles geranyl formate} = \left(\frac{1 \text{ mol geranyl formate}}{1 \text{ mol geraniol}}\right)(2.432 \text{ mol geraniol})$$

Thus, 2.432 moles of geraniol can produce 2.432 moles of geranyl formate. To finish the problem, multiply by the molar mass of geranyl formate to determine the mass of geranyl formate produced in the reaction. The mole units are eliminated, leaving gram units for the answer:

$$(2.432 \text{ mol geranyl formate})\left(\frac{182.3 \text{ g geranyl formate}}{1 \text{ mol geranyl formate}}\right) = 443 \text{ g geranyl formate}$$

**Does the Result Make Sense?** The result has three significant figures, just like the initial mass of geraniol. Geranyl formate has a slightly larger molar mass than geraniol, and we can see that the mass of the product is slightly more than the mass of the starting material.

**Practice Exercise 1.12**
What mass of formic acid is required to produce 443 g of geranyl formate?

## EXAMPLE 1-13  AMOUNTS OF REACTANTS AND PRODUCTS

Poisonous hydrogen cyanide (HCN) is an important industrial chemical. It is produced from methane (CH₄), ammonia, and molecular oxygen. The reaction also produces water. An industrial manufacturer wants to convert 175 kg of methane into HCN. How much molecular oxygen will be required for this synthesis?

**Strategy:**

1. This is a stoichiometry problem, in which we are asked to find the mass of a reactant.

2. To visualize this kind of problem, we need a balanced chemical equation, as follows:

Write the unbalanced form of the equation from the information given; namely methane, ammonia, and molecular oxygen react to form hydrogen cyanide and water:

$$CH_4 + NH_3 + O_2 \longrightarrow HCN + H_2O$$

Balance the equation by inspection, one element at a time. Carbon and nitrogen are in balance when there are equal numbers of molecules of $CH_4$, $NH_3$, and HCN. Hydrogen and oxygen, however, are not in balance. There are seven hydrogen atoms on the left but only three on the right. The coefficient for HCN cannot be changed without unbalancing C and N. Thus, to balance H, we must multiply the coefficient for $H_2O$ by 3:

$$CH_4 + NH_3 + O_2 \longrightarrow HCN + 3\,H_2O$$

This gives three oxygen atoms on the right, which corresponds to $\frac{3}{2}$ $O_2$ molecules among the starting materials:

$$CH_4 + NH_3 + \frac{3}{2}\,O_2 \longrightarrow HCN + 3\,H_2O$$

To avoid the impossible $\frac{1}{2}$ molecule, multiply all of the stoichiometric coefficients by 2. This gives the balanced equation:

$$2\,CH_4 + 2\,NH_3 + 3\,O_2 \longrightarrow 2\,HCN + 6\,H_2O$$

**Solution:**

3. The problem gives only one piece of data: 175 kg of $CH_4$ will be converted into HCN. The problem asks about another starting material, so we will need two molar masses:

$$M_{CH_4} = 16.04 \text{ g/mol}, \quad M_{O_2} = 32.00 \text{ g/mol}$$

4. Knowing the mass of one reagent, we are asked to find the mass of another. We need to determine molar amounts and then convert to kilograms. The calculations require two equations for interconversion of moles and mass:

$$n = \frac{m}{M} \quad \text{and} \quad n_B = \left(\frac{\text{Coefficient}_B}{\text{Coefficient}_A}\right) n_A$$

5. After converting mass to moles and $n_A$ to $n_B$, we need to convert moles back to mass: $m = nM$

6. Begin by converting mass of methane to moles:

$$(175 \text{ kg CH}_4)\left(\frac{10^3 \text{ g}}{1 \text{ kg CH}_4}\right) = 1.75 \times 10^5 \text{ g CH}_4$$

$$\frac{1.75 \times 10^5 \text{ g CH}_4}{16.04 \text{ g/mol}} = 1.091 \times 10^4 \text{ mol CH}_4$$

Next, use the ratio of coefficients to find the number of moles of $O_2$ that are required for the synthesis:

$$(1.091 \times 10^4 \text{ mol CH}_4)\left(\frac{3 \text{ mol O}_2}{2 \text{ mol CH}_4}\right) = 1.637 \times 10^4 \text{ mol O}_2$$

Notice that including the formula in the unit, "mol $O_2$" instead of "mol," for example, clarifies the cancellation of units. Finally, use the molar mass of $O_2$ to convert from moles back to mass:

$$(1.637 \times 10^4 \text{ mol O}_2)\left(\frac{32.00 \text{ g O}_2}{1 \text{ mol O}_2}\right)\left(\frac{1 \text{ kg}}{10^3 \text{ g}}\right) = 524 \text{ kg O}_2$$

**7. Does the Result Make Sense?** The units of the result, kilograms, indicate that the calculations are in order, because the question asked for the mass of oxygen. The mass of $O_2$ required for the synthesis, 524 kg, is a large quantity, but it is reasonable because the reaction also involves a large quantity of methane (175 kg).

**Common Pitfalls:** Inadvertently inverting the ratio of coefficients will give an incorrect result. Inspect the balanced reaction and predict whether there will be a greater or smaller number of moles of product than reactant.

**Practice Exercise 1.13**
In this synthesis, what mass of ammonia will be required, and what mass of HCN could be prepared?

---

## SECTION EXERCISES

● **1.8.1** Phosphorus trichloride is produced from the reaction of solid phosphorus and chlorine gas:

$$P_4 + 6\ Cl_2 \longrightarrow 4\ PCl_3$$

(a) What mass of phosphorus trichloride can be prepared from 75.0 g of phosphorus?

(b) What mass of chlorine will be consumed in the reaction?

● **1.8.2** Tooth enamel consists, in part, of $Ca_5(PO_4)_3(OH)$. Tin(II) fluoride (also called *stannous fluoride*) is added to some toothpastes because it exchanges $F^-$ for $OH^-$ to give the more decay-resistant compound fluoroapatite, $Ca_5(PO_4)_3F$. In addition, this reaction produces tin(II) oxide and water. What mass of $Ca_5(PO_4)_3(OH)$ can be converted to $Ca_5(PO_4)_3F$ by reaction with 0.115 g of tin(II) fluoride?

● **1.8.3** Combustion reactions require molecular oxygen. In an automobile the fuel-injection system must be adjusted to provide the right "mix" of gasoline and air. Compute the number of grams of oxygen required to react completely with 1.00 L of octane ($C_8H_{18}$, which has a density $\rho = 0.80$ g/mL). What masses of water and carbon dioxide are produced in this reaction?

---

# 1.9 YIELDS OF CHEMICAL REACTIONS

Under practical conditions, chemical reactions almost always produce smaller amounts of products than the amounts predicted by stoichiometric analysis. There are three major reasons for this.

1. Many reactions stop before reaching completion. For example, solid sodium metal reacts with oxygen gas to give sodium peroxide ($Na_2O_2$). As the reaction proceeds, a solid crust of the product builds up on the surface of the metal, as Figure 1-17 shows. The crust prevents oxygen from reaching the sodium metal, and the reaction stops, even though both starting materials are still present.

   Other reactions do not go to completion because they reach dynamic equilibrium. While reactant molecules continue to form product molecules, product molecules also interact to re-form reactant molecules. The Haber reaction and many precipitation reactions, described later in this chapter, are examples

**Figure 1-17**

Pure sodium metal

Sodium metal with a crust of
sodium peroxide ($2Na^+ + O_2^{2-}$)

The reaction of sodium metal with oxygen gas produces a crust of sodium peroxide, which blocks additional reactant molecules from reaching the unreacted metal.

of reactions that reach dynamic equilibrium rather than going to completion. We treat chemical equilibria in detail in Chapters 13–15.

2. Competing reactions often consume some of the starting materials. For example, sodium metal reacts with water to produce sodium hydroxide. If a sample of oxygen is contaminated with water vapour, both $O_2$ and $H_2O$ will react with the sodium metal. The more water present in the gas mixture, the less $Na_2O_2$ will be formed.

3. When the product of a reaction is purified and isolated, some of it is inevitably lost during the collection process. Gases may escape while being pumped out of a reactor. Liquids adhere to glass surfaces, making it impossible to transfer every drop of a liquid product. Likewise, it is impossible to scrape every trace of a solid material from a reaction vessel.

We can compare the incomplete formation of a chemical product with the process of harvesting. When a farmer harvests a fruit crop, some of the fruit is left on the trees. Some may spill off the truck on the way to market. Some of the fruit may be unsuitable for sale. The result is that less than 100% of the potential crop eventually finds its way to market. Likewise, a synthetic chemist loses some of the potential product of a reaction at each step in the preparation.

The amount of a product obtained from a reaction is often reported as a **yield**. The amount of product predicted by stoichiometry is the **theoretical yield**, whereas the amount actually obtained is the **actual yield**. The **percent yield** is the percentage of the theoretical amount that is actually obtained:

$$\text{Percent Yield} = 100\% \left( \frac{\text{Actual amount}}{\text{Theoretical amount}} \right) \qquad (1\text{-}7)$$

When we calculate a percent yield, the amounts can be expressed in either moles or mass, provided both the actual and theoretical amounts are in the same units. Example 1-14 shows how to use Equation 1-7.

## EXAMPLE 1-14 CALCULATING PERCENT YIELD

According to Example 1-12, it is possible to make 443 g of geranyl formate from 375 g of geraniol. A chemist making geranyl formate for the preparation of a perfume uses 375 g of starting material and collects 417 g of purified product. What is the percent yield of this synthesis?

**Strategy:** To calculate a percent yield, we need to compare the actual amount obtained in the synthesis with the theoretical amount that could be produced, using Equation 1-7.

**Solution:** The theoretical yield of geranyl formate is 443 g. This is the amount of product that would result from complete conversion of geraniol into geranyl formate. The actual yield, 417 g, is the quantity of the desired product that the chemist collects. The percent yield is their ratio multiplied by 100%:

$$\text{Percent yield} = (100\%)\left(\frac{417 \text{ g}}{443 \text{ g}}\right) = 94.1\%$$

 **Does the Result Make Sense?** Yields can never exceed the theoretical amount, so a percent yield always must be less than 100%. A percent yield between 90% and 100% is quite reasonable

**Practice Exercise 1.14**

If the reaction described in Example 1-13 produces 195 kg of HCN from 175 kg of methane reactant, what is the percent yield?

If the percent yield of a reaction is already known, we can calculate how much of a product to expect from a synthesis that uses a known amount of starting material. For example, the Haber synthesis of ammonia stops when 13% of the starting materials have formed products. Knowing this, how much ammonia could an industrial producer expect to make from 2.0 tonnes of molecular hydrogen? First, calculate the theoretical yield:

$$\text{Theoretical yield} = \frac{(2.0 \text{ tonne } H_2)(10^6 \text{ g/tonne})}{(2.016 \text{ g/mol } H_2)}\left(\frac{2 \text{ mol } NH_3}{3 \text{ mol } H_2}\right)$$

$$= 6.61 \times 10^5 \text{ mol } NH_3$$

Rearrange Equation 1-7 and substitute known values:

$$\text{Actual yield} = (6.61 \times 10^5 \text{ mol})\left(\frac{13\%}{100\%}\right) = 8.60 \times 10^4 \text{ mol } NH_3$$

Finally, convert back to mass:

$$\text{Actual mass} = (8.60 \times 10^4 \text{ mol})(17.03 \text{ g/mol})(10^{-6} \text{ tonne/g}) = 1.5 \text{ tonne}$$

Chemists often perform yield calculations in reverse to determine the masses of reactants needed to obtain a desired amount of product. If the percent yield is known from previous experiments, we can calculate the amount of starting material needed to make a specific amount of product. Example 1-15 shows how to do this.

## EXAMPLE 1-15 CALCULATING REACTANT MASS FROM YIELD

The industrial production of hydrogen cyanide is described in Example 1-13. If the yield of this synthesis is 97.5%, how many kilograms of methane should be used to produce $1.50 \times 10^5$ kg of HCN?

**Strategy:** This is a two-stage problem that requires a yield calculation and a conversion among molar amounts of products and reactants. This is typical of yield calculations. Use the percent yield to determine the theoretical yield, and then use the stoichiometric ratios to calculate masses of starting materials. A flowchart summarizes the stepwise strategy.

**Solution:** First, rearrange the equation for percent yield to find the theoretical yield that will give an actual yield of $1.50 \times 10^5$ kg:

$$\text{Theoretical yield} = (100\%)\left(\frac{\text{Actual yield}}{\text{Percent yield}}\right)$$

$$\text{Theoretical yield} = (100\%)\left(\frac{1.50 \times 10^5 \text{ kg}}{97.5\%}\right) = 1.538 \times 10^8 \text{ kg HCN}$$

$$(1.538 \times 10^5 \text{ kg HCN})(10^3 \text{ g/kg}) = 1.538 \times 10^8 \text{ g HCN}$$

$$\frac{1.538 \times 10^8 \text{ g HCN}}{27.03 \text{ g/mol}} = 5.690 \times 10^6 \text{ mol HCN}$$

Next, to convert this mass to moles, apply the stoichiometric ratio and multiply by the molar mass of methane:

$$(5.690 \times 10^6 \text{ mol HCN})\left(\frac{2 \text{ mol CH}_4}{2 \text{ mol HCN}}\right) = 5.690 \times 10^6 \text{ mol CH}_4$$

$$(5.690 \times 10^6 \text{ mol CH}_4)(16.04 \text{ g/mol})(10^{-3} \text{ kg/g}) = 9.13 \times 10^4 \text{ kg CH}_4$$

**Does the Result Make Sense?** Compare the mass of the reactant with the mass of the product. Notice that both masses are the same order of magnitude, $10^5$ kg. This makes sense when the yield is close to 100%, the stoichiometric ratio is 1 : 1, and the molar masses are comparable.

**Common Pitfalls:** Doing too many of the steps at once may lead to errors. Follow the procedure outlined above for this type of problem.

**Practice Exercise 1.15**
What mass of $NH_3$ is required for this synthesis?

Suppose a chemical reaction gives a 50% yield. Is that a good yield or a poor yield? The answer depends on the circumstances. Many valuable chemicals are manufactured in processes that require several steps. The overall yield of a multistep synthesis is the product of the yields of the individual steps. For example, suppose that a reaction requires two steps. If the first reaction has a yield of 85% and the second a yield of 65%, then the overall yield is $(100\%)(0.85)(0.65) = 55\%$.

## FOCUS 1-2  CHEMISTRY AND LIFE: FEEDING THE WORLD

Fertilizers are immensely important to humanity. Agriculture requires fertilizers because growing plants remove various chemical elements from the soil. In a fully contained ecosystem, decaying organic matter replenishes the soil, but the elements contained in crops that are harvested and shipped elsewhere are not replenished. Thus, intensive agriculture inevitably depletes the soil of essential elements, which must be replaced by fertilization.

Lacking means of fertilization, primitive people farmed a plot of land until the nutrients in the soil were exhausted. Then, they moved to new ground, where they burned the natural vegetation and began farming again. This slash-and-burn method of growing food is still used extensively in South America, where this human activity is destroying vast tracts of rain forest.

Fertilization using animal products has been practiced since ancient times. Animal manure returns nutrients to the soil, replenishing elements that are depleted as crops are grown and harvested. It is likely that the use of animal fertilizers quickly followed the domestication of goats, sheep, and cattle.

Growing plants require a variety of chemical elements. Nitrogen, phosphorus, and potassium are required in the greatest amounts, but plants also need trace amounts of calcium, copper, iron, zinc, and other elements. By far, the most substantial need is for nitrogen. The Earth's atmosphere is 80% molecular nitrogen, but plants cannot use $N_2$. Instead, most plants absorb nitrogen from the soil in the form of nitrate ions.

The production of nitrogen fertilizers is a major activity of the chemical industry. Every year, the top 15 chemicals in industrial production include several nitrogen-containing compounds whose major use is in fertilizers. Molecular nitrogen serves as the primary source of nitrogen for chemical production. Gaseous ammonia ($NH_3$), which is synthesized from $N_2$ and $H_2$, can be injected directly into the ground, where it dissolves in moisture in the soil and serves as a fertilizer. Ammonia is more widely used in reactions with acids to produce other fertilizers: ammonia and nitric acid produce ammonium nitrate ($NH_4NO_3$), whereas ammonia and sulphuric acid produce ammonium sulphate. These chemicals and urea (($H_2N)_2CO$) are among the most important fertilizers.

$NH_4NO_3$　　　　$(NH_2)_2CO$　　　　$NH_3$

Crop yields can rise dramatically with the use of commercial fertilizers. For example, in 1800, a hectare of land in the United States produced about 70 bushels of corn. In the 1980s, the same acre of land produced 375 bushels. Worldwide, approximately 10 billion hectares of land are used to grow food crops. This would probably be enough land to feed the world's population if the entire area could be fertilized commercially. It has been estimated that world crop production would increase by about 50% if about $100 per hectare were spent to apply modern chemical fertilizers. However, it would cost about $160 trillion to produce this additional food. Furthermore, the use of chemical fertilizers can lead to the contamination of streams, lakes, and bays with phosphates and nitrates.

Contamination of the world's oceans by fertilizer runoff is an emerging environmental problem. Ironically, the very fertilizers that make the land bloom can cause the oceans to wither. When water containing excess nitrogen drains into the seas, the nitrogen triggers rapid growth of phytoplankton. These microscopic algae eventually die, and as they rot, they consume the oxygen dissolved in seawater. Without oxygen, animal life of all types, including shellfish, crustaceans, and fish, can no longer survive. Oxygen-starved areas have existed in the Gulf of Mexico and Chesapeake Bay for some time, but in 2003 the United Nations Environmental Program reported that the known dead zones have doubled in the last decade of the twentieth century, to nearly 150 dead zones around the world. Among these zones are portions of the Baltic and Adriatic seas.

In addition, the production of fertilizers from molecular nitrogen is very energy-intensive. In the United States alone, hundreds of millions of barrels of oil are used every year to produce fertilizers. Finding ways to produce non-contaminating and inexpensive fertilizers without huge energy investment is a major challenge to today's chemical industry.

As world supplies of petroleum are depleted, and as the Earth's population steadily increases, society will be forced to develop more efficient ways to make fertilizer. Genetic engineering offers a promising solution. There is a remarkable class of bacteria called *Rhizobium* that live in the roots of leguminous plants such as soybeans, peas, and

The left side of this field has been fertilized.

peanuts. This organism can convert molecular nitrogen into ammonia. The plant and the bacterium have a symbiotic relationship. Ammonia produced by the bacterium nourishes the plant, and the plant provides other nutrients to the bacterium. Exactly how the bacterium converts nitrogen to ammonia is the subject of vigorous research. Scientists hope eventually to transfer the bacterial gene responsible for the conversion of nitrogen to ammonia into the cells of non-leguminous plants.

Notice that the overall yield is lower than the yield of the least efficient step. For this reason, the more steps there are in a synthesis, the more important the yields of the individual reactions become. For example, the synthesis of a growth hormone might require a sequence of 16 different chemical reactions. At an average yield per step of 95%, the overall yield of this synthesis is 44%, but at an average yield of 75% the overall yield of the synthesis falls to 1.0%!

Multistep reactions that have only moderate yields at each step are wasteful and expensive. For this reason, chemists devote much time, effort, and ingenuity to devising reaction sequences and conditions that improve the yields of chemical syntheses. Our most important industrial chemicals are produced in billion-kilogram quantities on an annual basis. Here, improving the synthesis yield by even a few tenths of a percent can save a company millions of dollars each year. A good example is fertilizer production, discussed in Focus 1-2.

## SECTION EXERCISES

● **1.9.1** When heated, potassium chlorate decomposes to potassium chloride and gaseous molecular oxygen:

$$2\ KClO_3 \longrightarrow 2\ KCl + 3\ O_2$$

What is the theoretical yield of oxygen when 5.00 g of potassium chlorate decomposes? Calculate the percent yield if a 5.00 g sample gives 1.84 g $O_2$ on decomposition. Give possible reasons why the actual yield is less than the theoretical yield.

●**1.9.2** Sulphuric acid is made from elemental sulphur in a three-step industrial process. How many tonnes of sulphur are required to produce 15 tonnes of $H_2SO_4$ if the three steps have yields of 94%, 92.5%, and 97%?

●**1.9.3** Aspirin is produced by treating salicylic acid with acetic anhydride:

$$2\ C_7H_6O_3\ +\ C_4H_6O_3\ \longrightarrow\ 2\ C_9H_8O_4\ +\ H_2O$$

Salicylic acid    Acetic    Aspirin
anhydride

Aspirin

If the synthesis has an 87% yield, what mass of salicylic acid should be used to produce 75 g of aspirin?

## 1.10 THE LIMITING REACTANT

The reactions encountered so far in this chapter were set up so that all the starting materials were used up at the same time. In other words, there were no leftovers at the end of the reaction. Often, chemical reactions are run with an excess of one or more starting materials. This means that one reactant will "run out" before the others. The reactant that runs out is called the **limiting reactant** because it limits the amount of product that can be made. The other starting materials are said to be *in excess.*

We illustrate the limiting reactant concept with the bicycle analogy shown in Figure 1-18. It takes two wheels, a frame, and a chain to build a bicycle. Imagine

**Figure 1-18**

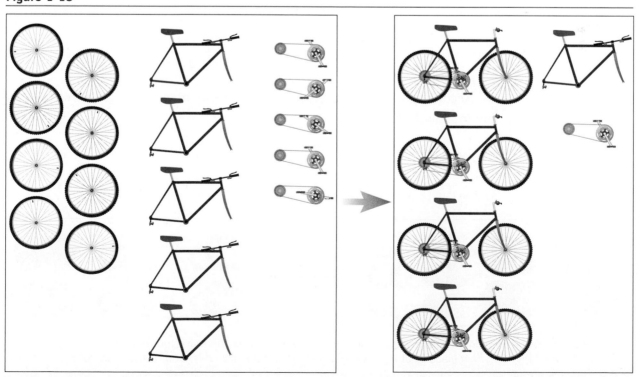

The number of bicycles that can be assembled is limited by whichever part runs out first. In the inventory shown in this figure, wheels are that part.

that a bicycle shop has the following parts on hand: eight wheels, five frames, and five chains. How many bicycles can be built with these parts? There are enough frames and chains to make five bicycles. However, because each bicycle needs two wheels, eight wheels are enough to make only four bicycles. Even though there are more wheels around the shop than any of the other parts, the shop will run out of wheels first. After four bicycles are built, the shop inventory will be four bicycles, no wheels, one frame, and one chain. There are not enough parts left to make a fifth bicycle because there is a shortage of two wheels. In this bicycle shop, wheels are the limiting reactant.

Chemical reactions must be analyzed in this same way. Instead of comparing numbers of bicycle parts, we compare the number of moles of each starting material on hand with the number of moles required to make the desired product. Consider the Haber synthesis again. Figure 1-19 shows a limiting reactant situation for this synthesis. If we start with six molecules of $H_2$ and four molecules of $N_2$, the six molecules of $H_2$ will combine with two molecules of $N_2$ to make four molecules of $NH_3$. When all the $H_2$ molecules are consumed, two molecules of $N_2$ will be left over. Here, $H_2$ is the limiting reactant. This molecular analysis applies equally well to numbers of moles. In molar terms, if six moles of $H_2$ and four moles of $N_2$ react completely, all the $H_2$ is consumed, four moles of $NH_3$ are produced, and two moles of $N_2$ are left over.

To solve a quantitative limiting reactant problem, we identify the limiting reactant by working with amounts in moles and the stoichiometric coefficients from the balanced equation. For the ammonia synthesis, if we start with 84.0 g of molecular nitrogen and 24.2 g of molecular hydrogen, what mass of ammonia can be prepared? First, convert from masses to moles:

$$\frac{84.0 \text{ g } N_2}{28.01 \text{ g/mol}} = 3.00 \text{ mol } N_2 \qquad \frac{24.2 \text{ g } H_2}{2.016 \text{ g/mol}} = 12.0 \text{ mol } H_2$$

According to the balanced equation, the stoichiometric ratio of nitrogen to ammonia is $1:2$.

$$N_2 + 3\,H_2 \longrightarrow 2\,NH_3$$

**Figure 1-19**

A molecular view of a limiting reactant situation for the ammonia synthesis. To make four molecules of $NH_3$ requires two molecules of $N_2$ and six molecules of $H_2$. If we start with four molecules of $N_2$ and six molecules of $H_2$, then $H_2$ is the limiting reactant.

There are 3.00 moles of $N_2$, which is enough to make 6.00 moles of $NH_3$.

$$(3.00 \text{ mol } N_2)\left(\frac{2 \text{ mol } NH_3}{1 \text{ mol } N_2}\right) = 6.00 \text{ mol } NH_3$$

The stoichiometric ratio of hydrogen to ammonia is $3 : 2$. There are 12.0 moles of $H_2$, which is enough to make 8.00 moles of $NH_3$.

$$(12.0 \text{ mol } H_2)\left(\frac{2 \text{ mol } NH_3}{3 \text{ mol } H_2}\right) = 8.00 \text{ mol } NH_3$$

There is enough $H_2$ on hand to make 8.00 moles of $NH_3$, but only enough $N_2$ to make 6.00 moles of $NH_3$. Nitrogen is the limiting reactant, and hydrogen is present in excess. Once 6.00 moles of ammonia have been produced, there will still be 3.00 moles of hydrogen left over. However, no more ammonia can be produced because there is no nitrogen left for the reaction.

We can identify the limiting reactant by dividing each amount in moles by the stoichiometric coefficient for that reactant. To see how this works, rearrange the ratios for $H_2$ and $N_2$:

$$\left(\frac{12.0 \text{ mol } H_2}{3 \text{ mol } H_2}\right)(2 \text{ mol } NH_3) = 4.00 \,(2 \text{ mol } NH_3) = 8.00 \text{ mol } NH_3$$

$$\left(\frac{6.00 \text{ mol } N_2}{2 \text{ mol } N_2}\right)(2 \text{ mol } NH_3) = 3.00 \,(2 \text{ mol } NH_3) = 6.00 \text{ mol } NH_3$$

The reactant with the smaller ratio of amount to coefficient can produce a smaller amount of product. That reactant ($N_2$ in this case) is limiting because it runs out first. This leads to the following generalization:

The limiting reactant is the one whose number of moles divided by its stoichiometric coefficient has the smallest value.

## Tables of Amounts

A table of amounts is a convenient way to organize the data and summarize the calculations of a stoichiometry problem. Such a table helps to identify the limiting reactant, shows how much product will form during the reaction, and indicates how much of the excess reactant will be left over. A table of amounts has the balanced chemical equation at the top. The table has one column for each substance involved in the reaction and three rows listing amounts. The first row lists the starting amounts for all the substances. The second row shows the changes that occur during the reaction, and the last row lists the amounts present at the end of the reaction. Here is a table of amounts for the ammonia example:

| Reaction | $N_2$ | + | $3\,H_2$ | $\longrightarrow$ | $2\,NH_3$ |
|---|---|---|---|---|---|
| Starting amount (mol) | 3.00 | | 12.0 | | 0 |
| Change in amount (mol) | −3.00 | | −9.00 | | +6.00 |
| Final amount (mol) | 0 | | 3.0 | | 6.00 |

A table of amounts has three important features, all involving *changes in amounts*:

1. *Changes are negative for reactants and positive for products.* This is because the amounts of reactants decrease during the reaction and the amounts of products increase during the reaction.
2. *All changes are related by stoichiometry.* Each ratio of changes in amount equals the ratio of stoichiometric coefficients in the balanced equation. In the example above, the changes in amounts for $H_2$ and $N_2$ are in the ratio $3:1$, the same as the ratio for the coefficients of $H_2$ and $N_2$ in the balanced equation.
3. In each column, *the starting amount plus the change in amount gives the final amount.* Each entry in the final amount row of the table is the sum of the previous entries in the same column.

The table of amounts for the ammonia example illustrates these features:

1. The changes are negative for $N_2$ and $H_2$, which are consumed, and positive for $NH_3$, which is produced.
2. The changes in amount for $H_2$ and $N_2$ are in the ratio of $3:1$, the same as the stoichiometric ratio for the coefficients of $H_2$ and $N_2$ in the balanced equation.
3. The change of $-9.00$ moles for $H_2$ leaves $(12.0 - 9.00) = 3.0$ moles as the final amount.

You can use these three key features to construct a table of amounts for any reaction.

When a reaction goes to completion, the limiting reactant is consumed completely, so its final amount must be zero. Two facts help identify the limiting reactant. First, final amounts can never be negative. If a negative amount appears in the bottom row of the table, a mistake has been made in identifying the limiting reactant. Second, the limiting reactant is always the one whose ratio of moles to stoichiometric coefficient is smallest. The smallest ratio indicates the limiting reactant, in this case $N_2$. Thus, dividing the starting amounts by coefficients is a quick way to identify the limiting reactant. Example 1-16 applies the features of limiting reactants and amounts tables.

## EXAMPLE 1-16  LIMITING REACTANT CALCULATION

Nitric acid, a leading industrial chemical, is used in the production of fertilizers and explosives. One step in the industrial production of nitric acid is the reaction of ammonia with molecular oxygen to form nitrogen oxide:

$$4\,NH_3 + 5\,O_2 \longrightarrow 4\,NO + 6\,H_2O$$

In a study of this reaction, a chemist mixed 125 g of ammonia with 256 g of oxygen and allowed them to react to completion. What masses of NO and $H_2O$ were produced, and which reactant was left over and what was its final mass?

**Strategy:** We apply the seven-step problem-solving method, but without numbering the steps. We are asked about masses of reagents in a chemical reaction. From the fact that starting amounts of both reactants are given, we identify this as a limiting reactant problem (one reactant will be consumed before the other is consumed). A table of amounts will help organize the information, and we need the balanced chemical equation, which is given in the statement of the problem.

**Solution:** The data available are starting masses of the reactants: 125 g of $NH_3$ and 256 g of $O_2$. We also need the molar masses, which are easily calculated:

$$M_{NH_3} = 17.03 \text{ g/mol}, \ M_{O_2} = 32.00 \text{ g/mol}, \ M_{NO} = 30.01 \text{ g/mol}, \ M_{H_2O} = 18.02 \text{ g/mol}$$

As always in stoichiometric calculations, we need to work with moles:

$$n_{NH_3} = \frac{125 \text{ g}}{17.03 \text{ g/mol}} = 7.34 \text{ mol}, \ n_{O_2} = \frac{256 \text{ g}}{32.00 \text{ g/mol}} = 8.00 \text{ mol}$$

Determine the limiting reactant by dividing each starting amount by the stoichiometric coefficient for that reactant:

$$\frac{7.34 \text{ mol NH}_3}{4 \text{ mol NH}_3} = 1.84, \ \frac{8.00 \text{ mol O}_2}{5 \text{ mol O}_2} = 1.60$$

ratio for $O_2$ is smaller, so oxygen is the limiting reactant.

Now set up and complete a table of amounts. Because $O_2$ is limiting, it will be completely consumed, so its change is 8.00 moles. Knowing this, we calculate all other changes in amounts using ratios of stoichiometric coefficients:

$$\text{Change in NH}_3 = (8.00 \text{ mol O}_2)\left(\frac{4 \text{ mol NH}_3}{5 \text{ mol O}_2}\right) = 6.40 \text{ mol NH}_3 \text{ used}$$

$$\text{Change in NO} = (8.00 \text{ mol O}_2)\left(\frac{4 \text{ mol NO}}{5 \text{ mol O}_2}\right) = 6.40 \text{ mol NO produced}$$

$$\text{Change in H}_2\text{O} = (8.00 \text{ mol O}_2)\left(\frac{6 \text{ mol H}_2\text{O}}{5 \text{ mol O}_2}\right) = 9.60 \text{ mol H}_2\text{O produced}$$

Remember that changes in amounts are negative for reactants and positive for products. Here is the complete amounts table:

| Reaction | 4 NH$_3$ | + | 5 O$_2$ | $\longrightarrow$ | 4 NO | + | 6 H$_2$O |
|---|---|---|---|---|---|---|---|
| Starting amount (mol) | 7.34 | | 8.00 | | 0 | | 0 |
| Change in amount (mol) | −6.40 | | −8.00 | | +6.40 | | +9.60 |
| Final amount (mol) | 0.94 | | 0.94 | | 6.40 | | 9.60 |

All that remains is to convert back to masses using $m = nM$. We leave it to you to verify that the masses present at the end of the reaction are 16 g of $NH_3$, 192 g of NO, and 173 g of $H_2O$.

**Does the Result Make Sense?** Are these masses reasonable? First, it is reasonable that none of the final amounts is negative. Second, we can check to see if total mass is conserved. We started with 125 g of ammonia and 256 g of oxygen, a total of 381 g. The masses present at the end also total 381 g, indicating that the calculations are consistent and the results are reasonable.

**Common Pitfalls:** Make sure the "change in amount" numbers are based on the stoichiometry of the balanced reaction.

**Practice Exercise 1.16**
Construct an amounts table for this reaction if the starting amounts are 85.15 g of $NH_3$ and 224 g of $O_2$.

A reaction that is carried out under limiting reactant conditions nevertheless has a yield that generally will be less than 100%. The reasons why reactions yield less than the theoretical amounts, given in Section 1.9, apply to all reactions. When a reaction operates under limiting reactant conditions, we calculate the theoretical yield assuming that the limiting reactant will be completely consumed. We then determine the percent yield as described in Section 1.9. Example 1-17 shows how to do this.

## EXAMPLE 1-17 YIELD UNDER LIMITING REACTANT CONDITIONS

The synthesis of aspirin appears in Section Exercise 1.9.3:

| Chemical formula: | $C_7H_6O_3$ | $C_4H_6O_3$ | $C_9H_8O_4$ |
|---|---|---|---|
| Name: | Salicylic acid | Acetic anhydride | Aspirin |
| M: | 138.1 g/mol | 102.1 g/mol | 180.2 g/mol |

Suppose a chemist started with 152 g of salicylic acid and 86.8 g of acetic anhydride and produced 133 g of aspirin. What is the yield of this reaction?

**Strategy:** The problem asks for a yield, so we identify this as a yield problem. In addition, we recognize this as a limiting reactant situation because we are given the masses of both starting materials. First, identify the limiting reactant by working with moles and stoichiometric coefficients. Then, carry out standard stoichiometry calculations to determine the theoretical amount that could form. A table of amounts helps organize these calculations. Calculate the percent yield from the theoretical amount and the actual amount formed.

**Solution:** Begin by converting mass to moles. The problem gives the molar masses.

$$\frac{152 \text{ g } C_7H_6O_3}{138.1 \text{ g/mol}} = 1.10 \text{ mol } C_7H_6O_3 \qquad \frac{86.8 \text{ g } C_4H_6O_3}{102.1 \text{ g/mol}} = 0.850 \text{ mol } C_4H_6O_3$$

Now identify the limiting reactant:

$$\frac{1.10 \text{ mol } C_7H_6O_3}{2 \text{ mol } C_7H_6O_3} = 0.550 \qquad \frac{0.850 \text{ mol } C_4H_6O_3}{1 \text{ mol } C_4H_6O_3} = 0.850$$

Because it has the smaller ratio, $C_7H_6O_3$ is the limiting reactant.

We determine the entries in the change row using the amount of the limiting reactant and stoichiometric ratios.

$$(1.10 \text{ mol } C_7H_6O_3)\left(\frac{1 \text{ mol } C_4H_6O_3}{2 \text{ mol } C_7H_6O_3}\right) = 0.550 \text{ mol } C_4H_6O_3 \text{ consumed}$$

$$(1.10 \text{ mol } C_7H_6O_3)\left(\frac{2 \text{ mol } C_9H_8O_4}{2 \text{ mol } C_7H_6O_3}\right) = 1.10 \text{ mol } C_9H_8O_4 \text{ produced}$$

$$(1.10 \text{ mol } C_7H_6O_3)\left(\frac{1 \text{ mol } H_2O}{2 \text{ mol } C_7H_6O_3}\right) = 0.550 \text{ mol } H_2O \text{ produced}$$

We can now complete the table of amounts:

| Reaction | $2\ C_7H_6O_3$ | + | $C_4H_6O_3$ | $\longrightarrow$ | $2\ C_9H_8O_4$ | + | $H_2O$ |
|---|---|---|---|---|---|---|---|
| Starting amount (mol) | 1.10 | | 0.850 | | 0 | | 0 |
| Change in amount (mol) | −1.10 | | −0.550 | | +1.10 | | +0.550 |
| Final amount (mol) | 0 | | 0.300 | | 1.10 | | 0.550 |

The final amount in this table represents the theoretical yield. To determine the percent yield, we need to have actual and theoretical amounts in the same units. In this example, we work with moles:

$$\text{Actual yield} = \frac{133\ \text{g}\ C_9H_8O_4}{180.2\ \text{g}\ C_9H_8O_4/\text{mol}} = 0.738\ \text{mol}$$

$$\text{Percent yield} = (100\%)\left(\frac{0.738\ \text{mol}}{1.10\ \text{mol}}\right) = 67.1\%$$

 **Does the Result Make Sense?** The entries in the amounts table are all reasonable, and the yield turns out to be less than 100%, as we know must be the case.

**Practice Exercise 1.17**

If 152 g of salicylic acid and 41.2 g of acetic anhydride react to form 114 g of aspirin, what is the percent yield?

Notice that neither the masses of the reactants nor the number of moles indicate directly which reactant is limiting. In this example, salicylic acid $(C_7H_6O_3)$ is the limiting reactant even though the mass of salicylic acid and the number of moles of salicylic acid are both larger than the corresponding amounts of acetic anhydride $(C_4H_6O_3)$.

It is often desirable to run practical syntheses under limiting reactant conditions, even though this may appear to be wasteful. For example, the aspirin synthesis typically is run with salicylic acid as the limiting reactant. Making salicylic acid the limiting reactant means that the product mixture will contain acetic anhydride and aspirin but a negligible amount of salicylic acid. This is advantageous for two reasons. First, salicylic acid and aspirin have similar molecular structures, making it difficult to separate these two substances. In contrast, aspirin is easy to separate from acetic anhydride. Second, salicylic acid is more expensive than acetic anhydride, and it is more economical to consume the more expensive reactant completely than to leave some of it unreacted.

Masses and number ratios cannot be used to determine the "limiting reactant" in bicycle manufacture either. Five chains are fewer in number than eight wheels and have less mass than eight wheels. Nevertheless, the wheels are used up before the chains are.

## SECTION EXERCISES

● **1.10.1** Suppose that the industrial synthesis of HCN described in Example 1-13 is carried out using $5.00 \times 10^2$ kg each of ammonia and methane in excess oxygen. What is the maximum mass of HCN that could be produced? Which reactant would be left over and what would be its final mass?

● **1.10.2** Lithium metal is one of the few substances that reacts directly with molecular nitrogen:

$$6\ Li + N_2 \longrightarrow 2\ Li_3N$$

What mass of the product, lithium nitride, can be prepared from 4.5 g of lithium metal and 9.5 g of molecular nitrogen?

● **1.10.3** Solid carbon reacts with chlorine gas to give carbon tetrachloride. If 150 g of carbon reacts with 250 g of $Cl_2$ with an 85% yield, what mass of $CCl_4$ is produced?

# CHAPTER 1 VISUAL SUMMARY

## 1.1 ATOMS, MOLECULES, AND COMPOUNDS

**LEARNING OBJECTIVES:** Our world is molecular

**SKILLS TO MASTER:** Recognizing elemental symbols and names of the elements. (p. 5)

Naming compounds from molecular pictures. (p. 5)

**KEY CONCEPTS:** Properties of molecules (the microscopic level) translate into properties of materials (the macroscopic level).

## 1.2 MEASUREMENTS IN CHEMISTRY

| Precision | Accuracy |
| --- | --- |
| Significant figures | |
| Time ($t$) | Volume ($V$) |
| Temperature ($T$) | Length |
| Scientific notation | Mass ($m$) |

$$T\,(K) = T\,(^oC) + 273.15$$
$$T\,(^oC) = T\,(K) - 273.15$$

**LEARNING OBJECTIVES:** Recognizing the SI units commonly used in chemistry, and performing some common unit conversions.

**SKILLS TO MASTER:** Working with unit conversions. (p. 10)

Determining the correct number of significant figures in the result of a calculation. (p. 12)

**KEY CONCEPTS:** The correct conversion ratio leads to cancellation of unwanted units.

Proper use of significant figures is important to tell the reader what the accuracy and precision of the measurement are.

## 1.3 CHEMICAL PROBLEM SOLVING

**SOLVING QUANTITATIVE PROBLEMS**
Step 1. Determine what is asked for.
Step 2. Visualize the problem.
Step 3. Organize the data.
Step 4. Identify a process to solve the problem.
Step 5. Manipulate the equations.
Step 6. Substitute and calculate.
Step 7. Does the result make sense?

**LEARNING OBJECTIVES:** Analyzing and solving problems in a consistent, organized fashion.

## 1.4 COUNTING ATOMS: THE MOLE

One mole is the number of atoms in exactly 12 g of pure carbon-12.

Molar mass ($M$)
Avogadro's number ($N_A$)

Elemental molar mass is the sum of the contribution of its isotopes.

$$M = \Sigma(\text{Fractional abundance})(\text{Isotopic } M)$$

Chemical calculations are built around the mole.

$$n = \frac{\#}{N_A} \qquad n = \frac{m}{M}$$

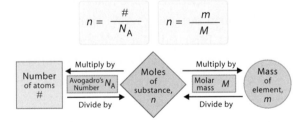

**LEARNING OBJECTIVES:** Solving mass-number-molecular weight type problems

**SKILLS TO MASTER:** Determining elemental molar masses. (p. 17)

**KEY CONCEPTS:** Chemical calculations are built around the mole.

One mole is the number of atoms in exactly 12 g of the pure isotope carbon-12.

The molar mass of any naturally occurring element is the sum of the contributions from its isotopes.

## 1.5 AMOUNT OF COMPOUNDS

$$M\ Compound = \Sigma(\#\ Atoms)(Element\ M)$$

| Element | # | M (g/mol) | Sum (g/mol) |
|---------|---|-----------|-------------|
| N | 2 | 14.007 | 28.014 |
| H | 4 | 1.008 | 4.032 |
| O | 3 | 15.999 | 47.997 |
| $NH_4NO_3$ | | M = 80.043 g/mol | |

Moles of substance → **Multiply by** *Molar Ratio* → Moles of each element

**Divide by**

**LEARNING OBJECTIVES:** Performing mole-mass-number conversions.

**SKILLS TO MASTER:** Drawing line structures of compounds. (p. 19)

Calculating molecular weights of compounds. (p. 22)

**KEY CONCEPTS:** A molecule is a combination of two or more atoms held together in a specific shape by attractive forces.

The molar mass of a compound is found by adding together the molar masses of all of its elements, taking into account the number of moles of each element present.

Molecular weights are used to convert between moles of substance and mass.

## 1.6 AQUEOUS SOLUTIONS

Solvent
Solute
Aqueous

*Polyatomic ions remain intact.*

$$c = \frac{n}{V} \qquad c_i V_i = c_f V_f$$

Concentration (c)
Molarity Dilution

**LEARNING OBJECTIVES:** Calculating concentrations of solutions and of diluted solutions.

**KEY CONCEPTS:** Polyatomic ions remain intact when a salt dissolves in water.

Concentration is the amount of substance per unit volume of solution.

Diluting a solution decreases the concentrations of the solutes.

## 1.7 WRITING CHEMICAL EQUATIONS

Stoichiometric coefficients    Balanced chemical equation

Reagents
Reactants    Products

$$N_2 + 3\ H_2 \longrightarrow 2\ NH_3$$

*Atoms are conserved.*

**LEARNING OBJECTIVES:** Balancing chemical reactions.

**KEY CONCEPTS:** Molecules of reagents react in whole number ratios to form products.

The number of atoms of each element is conserved in any chemical reaction.

## 1.8 STOICHIOMETRY OF CHEMICAL REACTIONS

$$n_B = \left(\frac{Coeff_A}{Coeff_B}\right) n_A$$

Stoichiometric ratio

*Calculations center on the mole.*

68 g $NH_3$ → 4.0 mol $NH_3$ → 6.0 mol $H_2$ → 12 g $H_2$

Stoichiometry

**LEARNING OBJECTIVES:** Calculating the amount of a product from the amounts of the reactants and a balanced chemical reaction.

**SKILLS TO MASTER:** Using stoichiometric ratios. (p. 42)

## 1.9 YIELDS OF CHEMICAL REACTIONS

Yield
   Theoretical yield
   Actual yield
   Percent yield

$$Percent\ yield = 100\% \left(\frac{Actual\ amount}{Theoretical\ amount}\right)$$

**LEARNING OBJECTIVES:** Calculating reaction yields.

## 1.10 THE LIMITING REACTANT

Limiting reactant    *Smallest value of moles divided by coefficient*

| Reaction: | $N_2$ | + $3\ H_2$ | $\longrightarrow 2\ NH_3$ |
|-----------|-------|------------|---------------------------|
| Starting amount (mol) | 3.00 | 12.0 | 0 |
| Change in amount (mol) | −3.00 | −9.00 | +6.00 |
| Final amount (mol) | 0 | 3.0 | 6.00 |

**LEARNING OBJECTIVES:** Solving limiting–reagent-type problems.

**SKILLS TO MASTER:** Constructing tables of amounts. (p. 53)

**KEY CONCEPTS:** The limiting reactant is the one whose number of moles divided by its stoichiometric coefficient has the smallest value.

The quantity of the limiting reactant determines how much products are formed.

## LEARNING EXERCISES

1.1   Make a list of all terms new to you in Chapter 1. Give a one-sentence definition for each. Consult the glossary if you need help.

1.2   Diagram the process for converting from the mass of a compound of a known chemical formula to the number of atoms of one of its constituent elements. Include all necessary equations and conversion factors.

1.3   Describe what a solution of magnesium nitrate looks like to a molecular-sized observer.

1.4   Update your list of memory bank equations. For each equation, add a phrase that describes the kinds of chemical problems for which the equation is used.

## PROBLEMS

**ilw** = interactive learning ware problem. Visit the website at www.wiley.com/canada/olmsted

### Atoms, Molecules, and Compounds

1.1   What is the elemental symbol for each of the following elements? (a) hydrogen; (b) helium; (c) hafnium; (d) nitrogen; (e) neon; and (f) niobium.

1.2   What is the elemental symbol for each of the following elements? (a) potassium; (b) platinum; (c) plutonium; (d) lead; (e) palladium; and (f) phosphorus.

1.3   What is the name of each of the following elements? (a) As; (b) Ar; (c) Al; (d) Am; (e) Ag; (f) Au; (g) At; and (h) Ac.

1.4   What is the name of each of the following elements? (a) Br; (b) Be; (c) B; (d) Bk; (e) Ba; and (f) Bi.

1.5   Examine the following molecular pictures and determine the corresponding molecular formulas.

(a)         (b)         (c)         (d)

1.6   Examine the following molecular pictures and determine the corresponding molecular formulas.

(a)         (b)         (c)

(d)         (e)

1.7   Write the chemical formula of the compound whose molecules contain (a) four atoms of chlorine and one atom of carbon; (b) two atoms of oxygen and two atoms of hydro-

gen; (c) four atoms of phosphorus and ten atoms of oxygen; (d) two atoms of iron and three atoms of sulphur.

1.8   Write the chemical formula of the compound whose molecules contain (a) five carbon atoms and twelve hydrogen atoms; (b) four atoms of fluorine and one atom of silicon; (c) two atoms of nitrogen and five atoms of oxygen; (d) one iron atom and three chlorine atoms.

### Measurements in Chemistry

1.9   Convert to scientific notation: (a) 100,000, precise to $\pm 1$; (b) ten thousand, precise to $\pm 1000$; (c) 0.000 400; (d) 0.0003; and (e) 275.3.

1.10  Convert to scientific notation: (a) 175,906; (b) 0.000 0605; (c) two and a half million, precise to $\pm 100$; and (d) two and a half billion, precise to $\pm$ one million.

1.11  Express each of the following in SI base units using scientific notation (e.g., 1.45 mm = $1.45 \times 10^{-3}$ m): (a) 432 kg; (b) 624 ps; (c) 1024 ng; (d) 93,000 km, precise to $\pm 10$; (e) 1 day; and (f) 0.0426 in.

1.12  Express each of the following in SI base units using scientific notation: (a) 1 week; (b) 1.35 mm; (c) 15 km; (d) 4.567 $\mu$s; (e) 6.45 mL; and (f) 47 kg.

1.13  The mass unit most commonly used for precious stones is the carat. 1 carat = 3.168 grains, and 1 g = 15.4 grains. Find the total mass in kilograms (kg) of a ring that contains a $5.0 \times 10^{-1}$ carat diamond and 7.00 g of gold.

1.14  What is the total mass in grams, expressed in scientific notation with the correct number of significant figures, of a solution containing 2.000 kg of water, 6.5 g of sodium chloride, and 47.546 g of sugar?

### Chemical Problem Solving

1.15  A chemist who wished to verify the density of water constructed a cylindrical container of aluminum 11.43 cm high, with an inside radius measuring 2.22 cm. The empty cylinder had a mass of 93.054 g. When filled with water, its mass was 270.064 g. Find the density of water from these data, expressing your result in SI units with the correct precision (cylinder volume, $V = \pi r^2 h$).

1.16  A chemist who prepared a new organic liquid wanted to determine its density. Having only a small sample to work with, the chemist had to use a small container. A tube with a volume of $8.00 \times 10^{-3}$ cm$^3$ weighed 0.4763 g when empty and 0.4827 g when filled with the liquid. Compute the density of the liquid in g/cm$^3$.

1.17  **ilw** Determine which possesses more mass: a sphere of gold with a diameter of 2.00 cm or a cube of silver measuring 2.00 cm on each side (sphere volume, $V = 4\pi r^3/3$).

1.18  Determine which possesses more mass: a sphere of zinc with a diameter of 2.00 cm or a cube of aluminum measuring 2.00 cm on each side (sphere volume, $V = 4\pi r^3/3$).

1.19  Bromine is one of two elements that are liquids at room temperature (mercury is the other). The density of bromine at room temperature is 3.12 g/mL. What volume of bromine is required if a chemist needs 36.5 g for an experiment?

**1.20** The density of gasoline at room temperature is 0.70 g/mL. If the gas tank of a car holds 12.0 gallons (45.4 L), what is the mass of a tankful of gasoline?

## Counting Atoms: The Mole

**1.21** (ilw) Use the data in the following table to calculate the molar mass of naturally occurring argon:

| Isotope | Isotopic Molar Mass (g/mol) | Abundance (%) |
|---------|-----------------------------|---------------|
| $^{36}Ar$ | 35.96755 | 0.337 |
| $^{38}Ar$ | 37.96272 | 0.063 |
| $^{40}Ar$ | 39.96238 | 99.600 |

**1.22** Use the data in the following table to calculate the molar mass of naturally occurring silicon:

| Isotope | Isotopic Molar Mass (g/mol) | Abundance (%) |
|---------|-----------------------------|---------------|
| $^{28}Si$ | 27.97693 | 92.23 |
| $^{29}Si$ | 28.97649 | 3.10 |
| $^{30}Si$ | 29.97376 | 4.67 |

**1.23** Calculate the number of moles in the following: (a) 7.85 g of Fe; (b) 65.5 $\mu$g of carbon; (c) 4.68 mg of Si; and (d) 1.46 tonnes of Al (1 tonne = $10^3$ kg).

**1.24** Calculate the number of moles in the following: (a) 3.67 kg of titanium; (b) 7.9 mg of calcium; (c) 1.56 g of ruthenium; and (d) 9.63 pg of technetium.

**1.25** Determine the number of atoms present in 5.86 mg of each of the following elements: (a) beryllium; (b) phosphorus; (c) zirconium; and (d) uranium.

**1.26** Determine the number of atoms present in 2.44 kg of each of the following elements: (a) argon; (b) silver; (c) iodine; and (d) potassium.

## Amounts of Compounds

**1.27** Calculate the molar mass of each of the following substances: (a) carbon tetrachloride; (b) potassium sulphide; (c) $O_3$ (ozone); (d) lithium bromide; (e) GaAs (a semiconductor); and (f) silver nitrate.

**1.28** Calculate the molar mass of each of the following compounds: (a) ammonium carbonate; (b) $N_2O$ (laughing gas); (c) calcium carbonate; (d) $NH_3$ (ammonia); (e) sodium sulphate; and (f) $C_4H_{10}$ (butane).

**1.29** Determine the molecular formula and calculate the molar mass of each of the following essential amino acids:

(a) Tyrosine    (b) Tryptophan

(c) Glutamic acid    (d) Lysine

**1.30** Determine the molecular formula and calculate the molar mass of each of the following B vitamins:

(a) Biotin    (b) Nicotinamide

(c) Pyridoxamine

(d) Pantothenic acid

**1.31** Calculate the number of molecules present in 5.86 mg of each of the following compounds: (a) methane ($CH_4$); (b) phosphorus trichloride; (c) dimethyl ether ($C_2H_6O$); and (d) uranium hexafluoride.

**1.32** The U.S. Recommended Daily Allowances (RDAs) of several vitamins follow. In each case, calculate how many molecules are in the U.S. RDA. (a) 60. mg vitamin C, $C_6H_8O_6$; (b) 400. mg folic acid, $C_{19}H_{19}N_7O_6$; (c) 1.5 mg vitamin A, $C_2OH_{30}O$; and (d) 1.70 mg vitamin $B_2$ (riboflavin), $C_{17}H_2ON_4O_6$.

**1.33** Calculate the mass of the following: (a) $3.75 \times 10^5$ molecules of methane; (b) $2.5 \times 10^9$ molecules of adrenaline, $C_9H_{13}NO_3$; and (c) one molecule of chlorophyll, $C_{55}H_{72}MgN_4O_5$.

**1.34** Calculate the mass of the following: (a) $1.0 \times 10^{15}$ molecules of ozone; (b) $5.000 \times 10^3$ molecules of cholesterol, $C_{27}H_{46}O$; and (c) one molecule of vitamin $B_{12}$, $C_{63}H_{88}CoN14O_{14}P$.

**1.35** (ilw) One carrot may contain 0.75 mg of vitamin A ($C_{20}H_{30}O$). How many moles of vitamin A is this? How many molecules? How many hydrogen atoms are in 0.75 mg of vitamin A? What mass of hydrogen is this?

**1.36** A particular oral contraceptive contains 0.035 mg of ethynyl estradiol in each pill. The formula of this compound is $C_{20}H_{24}O_2$. How many moles of ethynyl estradiol are there in one pill? How many molecules is this? How many carbon atoms are in a 0.035-mg sample of ethynyl estradiol? What mass of carbon is this?

## Aqueous Solutions

**1.37** (a) Calculate the molarities of the ionic species in $1.50 \times 10^2$ mL of aqueous solution that contains 4.68 g of magnesium chloride. (b) Draw a molecular picture that shows a portion of this solution, showing the relative proportions of the ions present.

1.38 A chemist places 3.25 g of sodium carbonate in a 250.-mL volumetric flask and fills it to the mark with water. (a) Calculate the molarities of the major ionic species. (b) Draw a molecular picture that shows a portion of this solution, making sure the portion is electrically neutral.

1.39 **ilw** A student prepares a solution by dissolving 4.75 g of solid KOH in enough water to make 275 mL of solution. (a) Calculate the molarities of the major ionic species present. (b) Calculate the molarities of the major ionic species present if 25.00 mL of this solution is added to a 100-mL volumetric flask, and water is added to the mark. (c) Draw molecular pictures of portions of the solutions in (a) and (b), showing how they differ.

1.40 A mildly antiseptic mouthwash can be prepared by dissolving sodium chloride in water. (a) Calculate the molarities of the ionic species present in 0.150 L of solution containing 27.0 g of sodium chloride. (b) Calculate the new molarities if 50.0 mL of this solution is diluted with water to give 450. mL of a new solution. (c) Draw molecular pictures of portions of the solutions described in (a) and (b), showing how they differ.

1.41 Concentrated hydrochloric acid is 12.1 M. What volume of this solution should be used to prepare 0.500 L of 0.125 M HCl?

1.42 Concentrated ammonia is 14.8 M. What volume of this solution should be used to prepare 1.25 L of 0.500 M $NH_3$?

1.43 Calculate the concentrations of the major ionic species present in each of the following solutions: (a) 4.55 g of $Na_2CO_3$ in 245 mL of solution; (b) 27.45 mg of $NH_4Cl$ in $1.55 \times 10^{-2}$ L of solution; and (c) 1.85 kg of potassium sulphate in $5.75 \times 10^3$ L of solution.

1.44 Calculate the concentrations of the major ionic species present in each of the following solutions: (a) 1.54 g of sodium hydrogen carbonate in 75.0 mL of solution; (b) 1.44 mg of $FeCl_3$ in 2.75 mL of solution; and (c) 8.75 kg of $KNO_3$ in 235 L of solution.

## Writing Chemical Equations

1.45 Balance the following chemical equations:
(a) $NH_4NO_3 \longrightarrow N_2O + H_2O$
(b) $P_4O_{10} + H_2O \longrightarrow H_3PO_4$
(c) $HIO_3 \longrightarrow I_2O_5 + H_2O$
(d) $As + Cl_2 \longrightarrow AsCl_5$

1.46 Balance the following chemical equations:
(a) $N_2O_5 + H_2O \longrightarrow HNO_3$
(b) $KClO_3 \longrightarrow KCl + O_2$
(c) $Fe + O_2 + H_2O \longrightarrow Fe(OH)_2$
(d) $P_4 + Cl_2 \longrightarrow PCl_3$

1.47 Draw a molecular picture that illustrates reaction 1.45 (d).

1.48 Draw a molecular picture that illustrates reaction 1.46 (d).

1.49 Balance the chemical equations for the following important industrial processes:
(a) Molecular hydrogen and carbon monoxide react to form methanol ($CH_3OH$).

(b) $CaO + C \longrightarrow CO + CaC_2$
(c) $C_2H_4 + O_2 + HCl \longrightarrow C_2H_4Cl_2 + H_2O$

1.50 The following reactions play roles in the manufacture of nitric acid. Write a balanced equation for each of them.
(a) $NH_3 + O_2 \longrightarrow NO + H_2O$
(b) Nitrogen oxide and molecular oxygen combine to give nitrogen dioxide.
(c) Nitrogen dioxide and water react to form nitric acid and nitrogen oxide.
(d) $NH_3 + O_2 \longrightarrow N_2 + H_2O$
(e) $NH_3 + NO \longrightarrow N_2 + H_2O$

1.51 Draw a molecular picture that illustrates reaction 1.49 (a).

1.52 Draw a molecular picture that illustrates reaction 1.50 (b).

1.53 Here are some reactions involving water. Balance the equations:
(a) $Ca(OH)_2 + H_3PO_4 \longrightarrow H_2O + Ca_3(PO_4)_2$
(b) $Na_2O_2 + H_2O \longrightarrow NaOH + H_2O_2$
(c) $BF_3 + H_2O \longrightarrow HF + H_3BO_3$
(d) $NH_3 + CuO \longrightarrow Cu + N_2 + H_2O$

1.54 Here are some reactions of phosphorus and its compounds. Balance the equations:
(a) $P_4 + Na \longrightarrow Na_3P$
(b) $Na_3P + H_2O \longrightarrow PH_3 + NaOH$
(c) $PH_3 + O_2 \longrightarrow P_4O_{10} + H_2O$
(d) $P_4O_{10} + H_2O \longrightarrow H_3PO_4$

## The Stoichiometry of Chemical Reactions

1.55 For each reaction in Problem 1.49, calculate the mass of the second reactant that is required to react completely with 5.00 g of the first reactant.

1.56 For each reaction in Problem 1.50, calculate the mass (in kg) of the first reactant that is required to react completely with 875 kg of the second reactant.

1.57 Iodine can be prepared by bubbling chlorine gas through an aqueous solution of sodium iodide:

$$2\,NaI(aq) + Cl_2(g) \longrightarrow I_2(s) + 2\,NaCl(aq)$$

What mass (in g) of NaI is required to produce 1.50 kg of $I_2$?

1.58 The fertilizer ammonium sulphate is prepared by the reaction between ammonia and sulphuric acid:

$$2\,NH_3(g) + H_2SO_4(aq) \longrightarrow (NH_4)_2SO_4(aq)$$

What mass (in kg) of $NH_3$ is needed to make 3.50 tonnes (1 tonne = 1000 kg) of ammonium sulphate?

1.59 The fermentation of sugar to produce ethyl alcohol occurs by the following reaction:

$$C_6H_{12}O_6(s) \xrightarrow{\text{yeast}} 2\,C_2H_5OH(l) + 2\,CO_2(g)$$

What mass of ethyl alcohol can be made from 1.00 kg of sugar?

1.60 One starting material for the preparation of nylon is adipic acid. Adipic acid is produced from the oxidation of cyclohexane:

Cyclohexane
$C_6H_{12}$

Adipic acid
$C_6H_{10}O_4$

If 375 kg of cyclohexane reacts with an unlimited supply of oxygen, how much adipic acid (in kg) can be formed?

**1.61** ⓘⓛⓦ The use of Freons is being phased out because of the damage these compounds do to the stratospheric ozone layer. One of the Freons, $CCl_2F_2$, is manufactured from carbon tetrachloride and hydrogen fluoride. The other reaction product is hydrogen chloride. If a manufacturer wishes to convert 175 kg of carbon tetrachloride into $CCl_2F_2$, what is the minimum mass of hydrogen fluoride required? What would be the masses of the two products obtained if this reaction is 100% efficient?

$CCl_2F_2$

**1.62** One way to clean up emissions of sulphur dioxide formed during the burning of coal is to pass the exhaust gases through a water slurry of powdered limestone. In the reaction, sulphur dioxide reacts with calcium carbonate and oxygen gas to produce calcium sulphate and carbon dioxide. A coal-burning plant that uses coal with a typical sulphur content produces $2.00 \times 10^4$ tonnes (1 tonne = 1000 kg) of sulphur dioxide in a month. How many tonnes of calcium carbonate will be consumed during the scrubbing process, assuming 100% efficiency? How many tonnes of calcium sulphate waste will the plant have to dispose of? (The process is actually about 90% efficient.)

## Yields of Chemical Reactions

**1.63** ⓘⓛⓦ Approximately 12 billion kilograms of phosphoric acid are produced annually for fertilizers, detergents, and agents for water treatment. Phosphoric acid can be prepared by heating the mineral fluoroapatite with sulphuric acid in the presence of water:

$Ca_5(PO_4)_3F + 5 H_2SO_4 + 10 H_2O \longrightarrow$

Fluoroapatite

$3 H_3PO_4 + 5 CaSO_4 2 H_2O + HF$

Phosphoric acid

If every kilogram of fluoroapatite yields $4.00 \times 10^2$ g of phosphoric acid, what is the percent yield?

**1.64** When HgO is heated, it decomposes into elemental mercury and molecular oxygen gas. If 60.0 g of Hg is obtained from 80.0 g of the oxide, what is the percent yield of the reaction?

**1.65** Phenobarbital is a sleep-inducing drug whose chemical formula is $C_{12}H_{12}N_2O_3$. It is manufactured in an eight-step process starting from toluene, $C_7H_8$. Theoretically,

each molecule of toluene yields one molecule of phenobarbital. If each of the steps has a yield of 88%, what mass of toluene is needed to manufacture 25 kg of phenobarbital?

**1.66** Most of the ammonia produced by the Haber process is used as fertilizer. A second important use of $NH_3$ is in the production of nitric acid, a top-15 industrial chemical. Nitric acid is produced by a three-step synthesis called the *Ostwald process:*

$$4 NH_3 + 5O_2 \longrightarrow 4 NO + 6 H_2O$$
$$2 NO + O_2 \longrightarrow 2 NO_2$$
$$3 NO_2 + H_2O \longrightarrow 2 HNO_3 + NO$$

The NO is recycled so that every mole of ammonia theoretically yields 1 mole of nitric acid. Starting with $7.50 \times 10^2$ kg of ammonia, what mass of nitric acid can be produced if the average yield of the three steps is 94.5% efficient?

**1.67** What is the yield of the reaction described in Problem 1.61 if the manufacturer obtains 105 kg of $CCl_2F_2$ from 175 kg of $CCl_4$? Given this reaction yield, what masses of $CCl_4$ and hydrogen fluoride should be used in order to make 155 kg of $CCl_2F_2$?

**1.68** If the oxidation reaction described in Problem 1.60 is 76.5% efficient, what mass of cyclohexane is required to produce 3.50 kg of adipic acid?

## The Limiting Reactant

**1.69** A fast-food restaurant makes double cheeseburgers using one hamburger roll, two quarter-pound patties of beef, one slice of cheese, a fourth of a tomato, and 15 g of shredded lettuce. At the start of the day, the restaurant manager has 12 dozen rolls, 40 lb of beef patties, 2 packages of sliced cheese containing 65 slices each, 40 tomatoes, and 1 kg of lettuce. Which ingredient will run out first, and how many cheeseburgers can be made?

**1.70** A hardware store sells a "handyman assortment" of wood screws containing the following:

| Type | Number of screws | Mass per screw (g) |
| --- | --- | --- |
| #8, $^1/_2$″ | 10 | 12.0 |
| #8, 1″ | 8 | 21.5 |
| #10, $^1/_2$″ | 8 | 14.5 |
| #10, $1^1/_2$″ | 5 | 31.0 |
| #12, 2″ | 4 | 45.0 |

The store buys these screws in 1-lb boxes (1 lb = 454 g). From two boxes of each type of screws, how many handyman assortments can the store prepare?

**1.71** ⓘⓛⓦ Ammonia is produced industrially using the Haber process:

$$N_2 + 3 H_2 \longrightarrow 2 NH_3$$

Suppose that an industrial reactor is charged with 75.0 kg each of $N_2$ and $H_2$. Use a table of amounts to determine what mass of ammonia could be produced if the reaction went to completion.

1.72 Acrylonitrile is an important building block for synthetic fibres and plastics. Over 1.4 billion kilograms of acrylonitrile are produced in the United States each year. The compound is synthesized from propene in the following reaction:

$$2 C_3H_6(g) + 2 NH_3(g) + 3 O_2(g) \longrightarrow$$
$$2 C_3H_3N(l) + 6 H_2O(g)$$

Propene                    Acrylonitrile

Use a table of amounts to determine how many kilograms of acrylonitrile can be prepared from $1.50 \times 10^3$ kg of propene, $6.80 \times 10^2$ kg of ammonia, and $1.92 \times 10^3$ kg of oxygen.

1.73 For each of the reactions given in Problem 1.49, consider an industrial process that starts with 1.00 tonne (1000 kg) of each reactant. Construct a table of amounts, identify the limiting reactant, and determine the maximum mass of each product that could be produced.

1.74 For each of the reactions given in Problem 1.50, consider an industrial process that starts with $7.50 \times 10^3$ kg of each reactant. Construct a table of amounts, identify the limiting reactant, and determine the maximum mass of each product that could be produced.

1.75 Elemental phosphorus, $P_4$, reacts vigorously with oxygen to give $P_4O_{10}$. Use a table of amounts to determine how much $P_4O_{10}$ can be prepared from 3.75 g of $P_4$ and 6.55 g $O_2$, and how much of the excess reactant will remain at the end of the reaction.

1.76 The following unbalanced reaction is called the *thermite reaction*. It releases tremendous amounts of energy and is sometimes used to generate heat for welding:

$$Al + Fe_3O_4 \longrightarrow Fe + Al_2O_3$$

Use a table of amounts to determine the masses of all substances present after the reaction if $2.00 \times 10^2$ g of Al and $7.00 \times 10^2$ g of $Fe_3O_4$ react to completion.

## ADDITIONAL PAIRED PROBLEMS

1.77 Draw a molecular picture that represents a homogeneous mixture of molecular fluorine and molecular chlorine in a 3:1 ratio of fluorine to chlorine. Both are gases, and both are diatomic molecules. The drawing should include at least 20 molecules. Use Figure 1-4 as a guide.

1.78 Gasoline does not dissolve in water. Instead, molecules of gasoline float on the surface of water. Draw a molecular picture that shows a heterogeneous mixture of gasoline and water. Draw enough molecules of each to show the structure of the mixture clearly. Use circles for water molecules and ovals for gasoline molecules.

1.79 Ethylene glycol is used as antifreeze in car radiators. The freezing temperature of ethylene glycol is $-11.5 \, °C$. Convert this freezing temperature to kelvins (K).

1.80 The melting point of silver metal is 1235 K. Convert this melting point to degrees Celsius (°C).

1.81 Perform the following calculations and report your answers with the correct number of significant figures:

(a) $\dfrac{(6.531 \times 10^{13})(6.02 \times 10^{23})}{(435)(2.000)}$

(b) $\dfrac{4.476 + (3.44)(5.6223) + 5.666}{(4.3)(7 \times 10^4)}$

1.82 Perform the following calculations, and report your answers with the correct number of significant figures:

(a) $\dfrac{(3.14159)(4.599 \times 10^6) - (1.12 \times 10^7)}{(4.756 \times 10^8) + (3.67 \times 10^4)}$

(b) $\dfrac{(6.577 \times 10^{-6}) + 0.00369 + (8.234 \times 10^{-4})}{(0.0002567) + (6.9377 \times 10^{-5})}$

1.83 Draw appropriately scaled and coloured molecular pictures of each of the following molecules: HCN (linear); $H_2O$ (bent, O in the middle); CO; and NNO (linear). Use Figure 1-4 as a guideline.

1.84 Draw appropriately scaled and coloured molecular pictures of each of the following molecules: $CO_2$ (linear, C in the middle); $ClO_2$ (bent, Cl in the middle); ClBr; and HOCl (linear, O in the middle). (Use Figure 1-4 as a guide.)

1.85 Calculate the mass of lithium that has the same number of atoms as 5.75 g of platinum.

1.86 Calculate the mass of neon that has the same number of atoms as 37.5 mg of xenon.

1.87 The chemical formula for aluminum sulphate is $Al_2(SO_4)_3$. (a) Compute its molar mass. (b) Compute the number of moles contained in 25.0 g of this compound. (c) Determine its percent composition. (d) Determine the mass of this compound that contains 1.00 mole of O.

1.88 The chemical formula for nickel sulphate hexahydrate is $NiSO_4 \cdot 6H_2O$. (a) Compute its molar mass. (b) Compute the number of moles contained in 25.0 g of this compound. (c) Determine its percent composition. (d) Determine the mass of this compound that contains 1.00 mole of O.

1.89 Sea water at 25 °C contains 8.3 mg/L of oxygen. What molarity is this?

1.90 The oceans contain 1.3 mg of fluoride ions per litre. What molarity is this?

1.91 Heart disease causes 31% of the deaths in Canada. However, the death rate from heart disease has dropped significantly in recent years partly because of the development of new drugs for heart therapy by chemists working in the pharmaceutical industry. One of these drugs is verapamil, used for the treatment of arrhythmia, angina, and hypertension. A tablet contains 120.0 mg of verapamil. Determine the following quantities: (a) the molar mass of verapamil; (b) the number of moles of verapamil in one tablet; and (c) the number of nitrogen atoms in one tablet.

Verapamil
$C_{27}H_{38}O_4N_2$

**1.92** Penicillin was discovered in 1928 by Alexander Fleming, who was a bacteriologist at the University of London. This molecule was originally isolated from a mould that contaminated some of Fleming's experiments. Penicillin destroys bacterial cells without harming animal cells, so it has been used as an antibiotic and has saved countless lives. The molecular formula of penicillin G is $C_{16}H_{18}N_2O_4S$. (a) Calculate the molar mass of penicillin G. (b) How many moles of penicillin are in a 50.0-mg sample? (c) How many carbon atoms are in a 50.0-mg sample? (d) What is the mass of sulphur in a 75.0-mg sample?

Penicillin G
$C_{16}H_{18}N_2O_4S$

**1.93** Although xenon is a "noble gas," it reacts with fluorine and oxygen. When xenon gas reacts with $F_2$ gas, $XeF_4$ is one of the products. If 5.00 g of xenon reacts with excess fluorine and generates 4.00 g of $XeF_4$, what is the percent yield and how much xenon remains unreacted, assuming no other product forms?

**1.94** Hydrogen fluoride is produced industrially by the action of sulphuric acid on $CaF_2$. If 365 kg of $CaF_2$ is treated with excess sulphuric acid and 155 kg of HF is produced, what is the percent yield and how much $CaF_2$ remains unreacted, assuming no other fluorine-containing product forms?

**1.95** The reaction of atom X with atom Y is represented in the following diagram. Write the balanced equation for this reaction.

● = atom X   ● = atom Y

**1.96** The reaction of element U with element V is represented in the diagram shown below. Write the balanced equation for this reaction.

● = Atom U   ● = Atom V

**1.97** A 4.6-g sample of ethanol ($C_2H_5OH$) is burned completely in air. (a) How many moles of $H_2O$ are formed? (b) How many molecules? (c) How many grams?

**1.98** An automobile engine completely burns 3.5 g of octane ($C_8H_{18}$). (a) How many moles of $CO_2$ are formed? (b) How many molecules? (c) How many grams?

**1.99** Propylene oxide is used primarily in the synthesis of poly(propylene glycol), a polymer used in the manufacture of automobile seats, bedding, and carpets. Around 2 billion kilograms of this compound are produced annually in the United States. Propylene oxide is produced by the following reaction:

*t*-Butyl
hydroperoxide
$C_4H_{10}O_2$

Propene
$C_3H_6$

Propylene
oxide
$C_3H_6O$

*t*-Butanol
$C_4H_{10}O$

(a) How many kilograms of propylene oxide can be prepared from 75 kg of *t*-butyl hydroperoxide? (b) What mass of propene will be required for the synthesis?

**1.100** Vinyl chloride, one of the top 20 industrial compounds, is used primarily to make the polymer poly(vinyl chloride), better known as PVC. This versatile polymer is used to make piping, siding, gutters, floor tile, clothing, and toys. Vinyl chloride is made by oxychlorination of ethylene ($C_2H_4$). The overall balanced equation follows:

Ethylene

Vinyl chloride

Annual production of vinyl chloride is about 6.8 billion kilograms. What minimum masses (in billions of kilograms) of $C_2H_4$ and HCl are required to produce this much vinyl chloride?

## ADDITIONAL PROBLEMS

**1.101** The diameter of a chlorine atom is $2.00 \times 10^2$ pm. How many chlorine atoms lined up end to end would form a line 2.54 cm long?

**1.102** The day is defined to contain exactly 24 hours of 60 minutes each, and each minute contains exactly 60 seconds. The year is 365.24 days long. How many seconds are there in a century? Express your answer in scientific notation with the correct number of significant figures.

**1.103** How many atoms of chromium-53 are there in a 0.123 g sample of chromium metal? (The isotopic abundance of $^{53}Cr$ is 9.5%.)

**1.104** Solution A is prepared by dissolving 90.0 g of $Na_3PO_4$ in enough water to make 1.5 L of solution. Solution B is 2.5 L of 0.705 M $Na_2SO_4$. (a) What is the molar concentration of $Na_3PO_4$ in Solution A? (b) How many millilitres of Solution A will give 2.50 g of $Na_3PO_4$? (c) A

50.0-mL sample of Solution B is mixed with a 75.00 mL sample of Solution A. Calculate the concentration of Na ions in the final solution.

**1.105** The seventh-ranked industrial chemical in U.S. production is phosphoric acid, whose chemical formula is $H_3PO_4$. One method of manufacture starts with elemental phosphorus, which is burned in air; the product of this reaction then reacts with water to give the final product. In 1995, the United States manufactured $2.619 \times 10^8$ pounds of phosphoric acid. How many moles is this? If 15% of this material was made by burning elemental phosphorus, how many moles and how many kilograms of phosphorus were consumed?

**1.106** Police officers confiscate a packet of white powder that they believe contains heroin. Purification by a forensic chemist yields a 38.70-mg sample for combustion analysis. This sample gives 97.46 mg of $CO_2$ and 20.81 mg of $H_2O$. A second sample is analyzed for its nitrogen content, which is 3.8%. Show by calculations whether these data are consistent with the formula for heroin, $C_{21}H_{22}NO_5$.

**1.107** In the Haber synthesis of ammonia, $N_2$ and $H_2$ react at high temperature, but they never react completely. In a typical reaction, 24.0 kg of $H_2$ and 84.0 kg of $N_2$ react to produce 68 kg of $NH_3$. Using a table of amounts, find the theoretical yield, the percent yield, and the masses of $H_2$ and $N_2$ that remain unreacted, assuming that no other products form.

**1.108** One set of reactants for rocket fuel is hydrazine and hydrogen peroxide, which react vigorously when mixed:

$$N_2H_4(l) + H_2O_2(l) \longrightarrow N_2(g) + H_2O(g) \text{(unbalanced)}$$

The density of liquid hydrazine is 1.44 g/mL, and that of hydrogen peroxide is 1.01 g/mL. What volume ratio of these two liquids should be used if both fuels are to be used up at the same time?

**1.109** The largest single use of sulphuric acid is for the production of phosphate fertilizers. The acid reacts with calcium phosphate in a 2 : 1 mole ratio to give calcium sulphate and calcium dihydrogen phosphate. The mixture is crushed and spread on fields, where the salts dissolve in rain water. (Calcium phosphate, commonly found in phosphate rock, is too insoluble to be a direct source of phosphate for plants.) (a) Write a balanced equation for the reaction of sulphuric acid with calcium phosphate. (b) How many kilograms each of sulphuric acid and calcium phosphate are required to produce 50.0 kg of the calcium sulphate–dihydrogen phosphate mixture? (c) How many moles of phosphate ion will this mixture provide?

**1.110** The $CO_2$ exhaled by astronauts must be removed from the spacecraft atmosphere. One way to do this is with solid LiOH:

$$CO_2(g) + 2 LiOH(s) \longrightarrow Li_2CO_3(s) + H_2O(l)$$

The $CO_2$ output of an astronaut is about 1.0 kg/day. What is the minimum mass of LiOH required for a six-day space shuttle flight involving five astronauts?

**GROUP EXERCISES**

**1.111** On average, a sample of human blood with dimensions of 0.1 mm × 0.1 mm × 0.1 mm contains $6.0 \times 10^3$ red blood cells. The volume of blood in a typical adult is about 5 L. How many red blood cells are there in an adult?

**1.112** An object will float on water, even though it is made of a material that is denser than water, if its net density is less than 1. Thus, a hollow sphere with a large enough interior volume will float, because air has a very low density (1.189 g/L at room temperature and pressure). Determine the volume percentage that the interior of a sphere must have to float if it is made of aluminum. If the sphere has an interior volume of 1.50 L, how thin must its metal shell be? (Hint: for a sphere,

$$V = \frac{4}{3}\pi r^3 \quad \text{and} \quad A = 4\pi r^2.$$ Repeat your calculations for copper.

**1.113** Consult the *CRC Handbook of Chemistry and Physics* and compile two lists, one of the $Z$ values of the elements that have just one stable isotope and the other of the $Z$ values of the elements that have four or more stable isotopes. Determine if there are any patterns in the atomic number values of the elements occurring in each of these lists.

**1.114** Aluminum sulphate is used in the manufacture of paper and in the water purification industry. In the solid state, aluminum sulphate is a hydrate. The formula is $Al_2(SO_4)_3 \cdot 18H_2O$. (a) How many grams of sulphur are there in 0.570 moles of solid aluminum sulphate? (b) How many water molecules are there in a 5.1-g sample of solid aluminum sulphate? (c) How many moles of sulphate ions are there in a sample of solid aluminum sulphate that contains 12.5 moles of oxygen atoms? (d) An aqueous solution of aluminum sulphate contains 1.25% by mass aluminum and has a density of 1.05 g/mL. What is the molarity of aluminum ions in the solution?

**1.115** $CF_3CH_2F$ (HFC-134a) has replaced chlorofluorocarbon compounds for use as refrigerants. HFC-134a is produced from trichloroethylene by the following reactions:

$$CCl_2{=}CHCl + 3\ HF \xrightarrow{\text{catalyst}} CF_3CH_2Cl + 2\ HCl$$

$$CF_3CH_2Cl + HF \xrightarrow{\text{catalyst}} CF_3CH_2F + HCl$$

In 1999, over 1.3 million tonnes of HFC-134a were produced, and production is increasing at a rate of nearly 20% annually. Manufacturers sell HFC-134a for $5.50/kg. Assume that you are a chemist working for a company that sold 65 million kilograms of HFC-134a last year. (a) If the synthesis had a reaction yield of 47%, how much trichloroethylene was used to produce the HFC-134a? (b) You have developed a new catalyst that improves the yield of the synthesis of HFC-134a from 47% to 53%. If the company uses the same mass of trichloroethylene for next year's production, how many kilograms of HFC-134a can the company produce if they use your new process? (c) Assuming that the selling price of HFC-134a does not change, how much more money will the company make next year because of your work?

## ANSWERS TO SECTION EXERCISES

**1.1.1** Ce, Cs, Cu, Ca, and C

**1.1.2** Zirconium, nickel, tin, tungsten, selenium, beryllium, and gold

**1.1.3** (a) $H_2O_2$; (b) $SO_2$; (c) $N_2O$; (d) $C_2H_4O_2$; and (e) $C_2H_4$

**1.2.1** $4.63 \times 10^{-7}$ $m^3$, $1.7935 \times 10^7$ m, and $9.36 \times 10^8$ s

**1.2.2** $3.0 \times 10^8$ m/s and $9.5 \times 10^{15}$ m

**1.2.3** 70.3 kg, 110. m, 312 K, and $3.155\,693 \times 10^7$ s

**1.3.1** $4.8 \times 10^{-3}$ $m^3$

**1.3.2** 1.3 g/cm$^3$

**1.4.1** 32.06 g/mol

**1.4.2** (a) $2.21 \times 10^{23}$ atoms; (b) $3.21 \times 10^{22}$ atoms; and (c) $7.80 \times 10^{21}$ atoms

**1.4.3** $4.0 \times 10^{-16}$ mol and $2.4 \times 10^8$ atoms

**1.5.1** (a) 129.91 g/mol; (b) 184.04 g/mol; (c) 142.98 g/mol; (d) 261.34 g/mol; and (e) 504.82 g/mol

**1.5.2** (a) 155.16 g/mol; (b) 121.16 g/mol; (c) 132.12 g/mol

**1.5.3** (a) $9.86 \times 10^{-5}$ mol; (b) $5.94 \times 10^{19}$ molecules; (c) $1.78 \times 10^{20}$ atoms; (d) $2.97 \times 10^{20}$ atoms; (e) 20.5 mg; and (f) $2.79 \times 10^{21}$ atoms

**1.6.1** $[NH_4^+] = 0.31$ M, $[NO_3^-] = 0.31$ M

**1.6.2** 1.2 mL

**1.6.3**

**1.7.1** $2\,C_8H_{18} + 25\,O_2 \rightarrow 16\,CO_2 + 18\,H_2O$

**1.7.2** $N_2 + O_2 \rightarrow 2\,NO$; $2\,NO + O_2 \rightarrow 2\,NO_2$

**1.7.3** $2\,C_3H_6 + 2\,NH_3 + 3\,O_2 \rightarrow 2\,C_3H_3N + 6\,H_2O$

**1.8.1** (a) 333 g $PCl_3$ can be prepared; (b) 258 g $Cl_2$ is consumed.

**1.8.2** 0.737 g

**1.8.3** $2.80 \times 10^3$ g $O_2$ is required; $1.14 \times 10^3$ g $H_2O$ and $2.47 \times 10^3$ g $CO_2$ are produced.

**1.9.1** Theoretical yield is 1.96 g, and percent yield is 94.0%. Possible reasons are that not all the $KClO_3$ decomposes or that some $O_2$ escapes and is not collected.

**1.9.2** Start with 5.8 tonnes of sulphur.

**1.9.3** 66 g salicylic acid is required.

**1.10.1** The maximum mass is 793 kg HCN, and 29 kg $CH_4$ is left.

**1.10.2** 7.5 g

**1.10.3** $2.3 \times 10^2$ g

## ANSWERS TO PRACTICE EXERCISES

**1.1** 0.703 g/cm$^3$

**1.2**

$C_3H_8$    $(CH_3)_2O$    $C_2H_5OH$    $C_2H_6$    $C_2H_4$    $C_2H_2$

**1.3**

Chemical formula: $C_8H_9NO_2$

**1.4** 330.36 g/mol

**1.5** 375.13 g/mol

**1.6** 13.9 moles and $8.36 \times 10^{24}$ atoms

**1.7** 3.20 g

**1.8** $[Ni^{2+}] = 0.255$ M; $[Cl^-] = 0.510$ M

**1.9** 11.8 mL

**1.10** $8.00 \times 10^{-2}$ M

**1.11** $2\,CH_3OH + 3\,O_2 \rightarrow 2\,CO_2 \rightarrow 4\,H_2O$

**1.12** 112 g formic acid is required.

**1.13** 186 kg of $NH_3$ is required as a starting material, and the reaction produces 295 kg of HCN.

**1.14** 66.1% yield

**1.15** $9.69 \times 10^4$ kg of $NH_3$ is required.

**1.16**

| Reaction | $4\,NH_3$ + | $5\,O_2$ | $\longrightarrow$ | $4\,NO$ + | $6\,H_2O$ |
|---|---|---|---|---|---|
| Start (mol) | 5.00 | 7.00 | | 0 | 0 |
| Change (mol) | −5.00 | −6.25 | | +5.00 | +7.50 |
| Final (mol) | 0 | 0.75 | | 5.00 | 7.50 |

**1.17** Yield = 78.4%

# THE BEHAVIOUR OF GASES

## EARTH'S ATMOSPHERE

The Earth's atmosphere is a shroud of gas that surrounds our planet. Most of the atmosphere lies within 100 km of the surface. The atmosphere contains two major components: a bit less than 80% $N_2$ and a bit more than 20% $O_2$. In addition to these two species, the atmosphere contains small amounts of water and traces of other species. Our molecular inset is a representation of the atmosphere.

In recent years scientists have discovered that some of the trace components of the atmosphere are changing in ways that may cause dramatic changes in the environment. All of us have heard the alarms: acid rain damages ecosystems. Burning fossil fuels may lead to global warming. Human activities have caused a "hole" in the ozone layer. More immediately visible are the spectacular sunsets—such as the one in our background photo—that result when the atmosphere contains a significant concentration of fine particles.

How could molecules present in mere trace amounts have such extensive effects on the environment? Acid rain provides a good example. Burning

gasoline and coal produces some gaseous $NO_2$ and $SO_2$. These oxides dissolve in water droplets and are oxidized to produce nitric acid ($HNO_3$) and sulphuric acid ($H_2SO_4$). Rain that forms from such acid-containing water vapour is highly acidic and can damage forests and aquatic life. For example, rain falling in Europe now regularly registers 10 times "normal" acidity. In 1994, it was reported that 25% of the trees in Europe suffer from leaf loss or leaf damage, most likely as a result of acid rain.

Sunlight acting on trace components in automobile exhaust gases causes the formation of smog. Cities worldwide, including Los Angeles, Mexico City, Shanghai, and Toronto (shown in our inset photo), suffer from smog. Smog contains nitrogen oxides, ozone, and larger molecules. The chemistry of smog is complex and not fully understood. Atmospheric scientists are studying how smog forms and how it can be prevented.

Acid rain and smog show that changes in the amounts of trace components of the atmosphere can

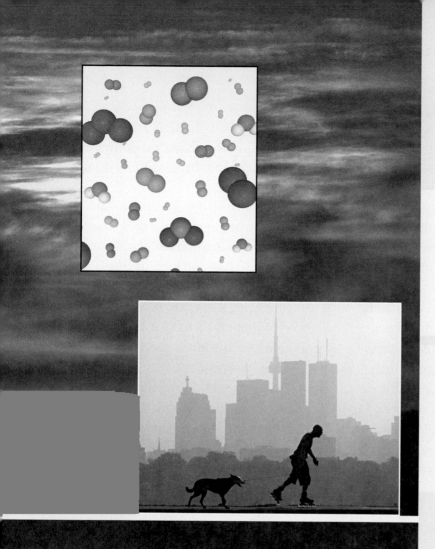

## LEARNING OBJECTIVES

*Upon completion of this chapter, you should be able to:*

- understand gas pressure and be able to express pressure in various units
- relate pressure, volume, temperature and amount of gas
- use the concept of partial pressures in gas mixtures
- use stoichiometry to solve problems involving gas-phase chemical reactions
- explain the basic concepts of kinetic molecular theory: molecular speed, energy, and the effects of temperature and volume on gas pressure
- calculate gas densities and molar masses from pressure–volume–temperature data
- calculate the pressure of a gas under non-ideal conditions, and explain the deviations from ideality
- do calculations involving water vapour pressure and relative humidity
- describe some of the basic chemistry of the troposphere

have dramatic consequences on the environment. Ozone depletion in the upper atmosphere (treated in Chapter 12) and gradual global warming (treated in Chapter 4) are long-term effects that may be equally or even more damaging. Atmospheric research scientists seek for ways to mitigate these effects.

We begin this chapter with a discussion of the variables that characterize gases. Then, we develop a molecular description that explains gas behaviour. Next, we explore additional gas properties and show how to do stoichiometric calculations for reactions involving gas-phase species. Finally, we return to the Earth's atmosphere and describe some aspects of its composition and chemical reactions.

## 2.1 PRESSURE

Like a liquid or a solid, a gas can be characterized by its volume ($V$), temperature ($T$), and amount ($n$). Unlike liquids and solids, however, the properties of gases are highly sensitive to pressure ($p$). We can understand this sensitivity using a molecular perspective.

An object that strikes a surface exerts a force against that surface. For example, Figure 2-1 shows that a basketball exerts force when it strikes a backboard. Gas molecules also exert forces. At the molecular level, atoms and molecules exert forces through never-ending collisions—among themselves and with the walls of their container.

Although pressure is caused by molecular collisions, it is a macroscopic property, the collective result of countless collisions. We can get a feel for the macroscopic characteristics of gas pressure by examining Earth's atmosphere.

The air around us is a huge reservoir of gas that exerts pressure on the Earth's surface. This pressure of the atmosphere can be measured with an instrument called a **barometer.** Figure 2-2 shows a schematic view of a simple mercury barometer Although mercury barometers are no longer in wide use owing to the toxicity of mercury, the way they work is very instructive to our understanding of pressure, and a widely used pressure unit is based on how they work.

A long glass tube, closed at one end, is filled with liquid mercury. The filled tube is inverted carefully into a dish that is partially filled with more mercury. The force of gravity pulls downward on the mercury in the tube. With no opposing force, all the mercury would run out of the tube and mix with the mercury in the dish. The mercury does fall, but the flow stops at a fixed height. The column of mercury stops falling because the atmosphere exerts pressure on the mercury in the dish, pushing the column up the tube. The column is in balance when the height of the mercury column generates a downward force on the inside of the tube that exactly balances the force exerted by the atmosphere on the outside of the tube. At sea level, atmospheric pressure supports a mercury column approximately 760 mm in height. Changes in altitude and weather cause fluctuations in atmospheric pressure. Nevertheless, at sea level the height of the mercury column seldom varies by more than 10 mm, except under extreme conditions, such as in the eye of a hurricane, when the mercury in a barometer may fall below 700 mm.

## Units of Pressure

The **standard atmosphere (atm)** is defined as the pressure that will support a column of mercury 760 mm in height, and pressures are therefore often expressed in units of mm Hg, also known as Torr. The accepted SI unit for pressure, however, is the **pascal (Pa)**. Pressure is defined as force per unit area, $F/A$, so the pascal can be expressed by combining the SI units for these two variables. The SI unit of force is the newton (N), and area is measured in square meters ($m^2$). Thus, the pascal is 1 $N/m^2$. By international agreement,

### Figure 2-1

When a basketball strikes a backboard, it exerts a force on the backboard. The gas molecules inside the basketball (*inset*) also exert forces when they strike the walls of the basketball. The net result of many collisions is gas pressure.

### Figure 2-2

Vacuum

Pressure due to weight of mercury

Atmospheric pressure

760 mm

Diagram of a mercury barometer.

1 atm is defined as $1.01325 \times 10^5$ Pa. A more convenient SI unit for pressures in the vicinity of 1 atm is the **bar**, defined to be exactly $10^5$ pascals. Thus, 1 atm = 1.01325 bar = 760 Torr, and 1 bar = 750.06 Torr. Chemists consider 1 bar to be the standard pressure for tabulating various properties of substances.

$$1 \text{ bar} = 100 \text{ kPa} = 100\,000 \text{ Pa} = 100\,000 \text{ N/m}^2 = 750.06 \text{ Torr}$$

$$1 \text{ atm} = 1.01325 \text{ bar} = 101\,325 \text{ Pa} = 760 \text{ Torr}$$

$$1 \text{ Torr} = 1 \text{ mm Hg} = 133.3 \text{ Pa}$$

A **manometer** is similar to a barometer, but in a manometer, gases exert pressure on *both* liquid surfaces. Consequently, a manometer measures the *difference* in pressures exerted by two gases. A simple manometer, shown in Figure 2-3, is a U-shaped glass tube containing mercury. One side of the tube is exposed to the atmosphere and the other to a gas of which the pressure we want to measure. In Figure 2-3, the pressure exerted by the atmosphere is less than the pressure exerted by the gas in the bulb. The difference in heights of mercury ($\Delta h$, in mm) between the two sides of the manometer depends on the difference in the pressures. Example 2-1 illustrates pressure measurement and unit conversions.

**Figure 2-3**

The difference in heights, $\Delta h$, of liquid on the two sides of a manometer, is a measure of the difference in gas pressures ($\Delta p$) applied to the two sides.

## EXAMPLE 2-1 MEASURING PRESSURE

A scientist collected an atmospheric gas sample. A manometer attached to the gas sample gave the reading shown in the figure (see also Figure 2-3), and the barometric pressure in the laboratory was 752 mm Hg. Calculate the pressure of the sample in atmospheres and in bars.

**Strategy:** We must combine the reading from the manometer with the barometric pressure to find the pressure of the gas sample. The manometer displays the pressure difference in mm Hg, so conversion factors are needed to express the pressure in atm and bar.

**Solution:** The difference in mercury levels in the manometer is 19 mm. The pressure difference between the sample and the atmosphere is therefore 19 mm Hg, or 19 Torr. The atmospheric pressure, as measured with a barometer, is 752 Torr. Shall we add or subtract the 19 Torr pressure difference? Notice that the mercury level in the manometer is lower on the side exposed to the atmosphere. Thus, the atmosphere pushes on the mercury harder than does the gas sample, meaning that the pressure of the gas sample is lower than the pressure of the atmosphere. Subtract the pressure difference:

$$p_{gas} = p_{atm} - \Delta p = 752 \text{ Torr} - 19 \text{ Torr} = 733 \text{ Torr}$$

Use the appropriate conversion factors to give the pressure in atm and bars.

$$1 \text{ atm} = 760 \text{ Torr and } 1 \text{ atm} = 1.01325 \text{ bar}$$
$$p \text{ (atm)} = 733 \text{ Torr } (1 \text{ atm}/760 \text{ Torr}) = 0.964 \text{ atm}$$
$$p \text{ (bar)} = 0.964 \text{ atm } (1.01325 \text{ bar/atm}) = 0.977 \text{ bar}$$

**Do the Results Make Sense?** The manometer shows that the pressure is lower on the inside than on the outside, so pressures that are smaller than 1 atm and 1 bar are reasonable.

**Common Pitfalls:** Adding $\Delta p$ to $p_{atm}$ instead of subtracting or vice versa. Remember that the higher pressure will push the mercury down the manometer further than the lower pressure.

**Practice Exercise 2.1**
In Figure 2-3, suppose that $p = 34.5$ mm Hg on a day when the barometer reads 765.3 Torr. Determine the gas pressure in the bulb in Torr, atm, and bar.

## SECTION EXERCISES

● **2.1.1** The average air pressure in Whitehorse, YT is 698 mm Hg. Express this pressure in Torr, atm and bar. What percentage change from standard atmospheric pressure is this?

● **2.1.2** Oxygen gas is sold in pressurized tanks. The gas in a tank exerts a pressure of 15.75 bar. Convert this pressure to pascals, kPa, and atm.

● **2.1.3** What pressure is exerted on the left side of the manometer shown in Figure 2-3 when the difference in column heights is 4.75 cm, the right side is open to the atmosphere, and barometric pressure is 752.8 Torr? Express your result in Torr, atm, kPa, and bar.

## 2.2 DESCRIBING GASES

The volume occupied by a gas changes dramatically in response to changes in conditions. This variability in volume is an obvious difference between gases and the other two phases, liquids and solids. Gas volumes change because the atoms or molecules of a gas are relatively far apart compared with those in a liquid or solid, and move freely about, as shown schematically in Figure 2-4. Gas molecules thus fill whatever volume is available to them.

### Variations in Gas Volume

Early scientists carried out experiments that determined how the volume of a gas changes with conditions. Starting around 1660, the Englishman Robert Boyle investigated the properties of gases confined in J-shaped glass tubes. Figure 2-5*a* illustrates Boyle's experiments. He added liquid mercury to the open end of a tube, trapping a fixed amount of air on the closed side. When additional mercury was added to the open end of the tube, the additional pressure compressed the trapped gas into a smaller volume. Boyle found that doubling the total pressure on the trapped air reduced its volume by a factor of 2. In more general terms, the volume of the trapped air was inversely proportional to the total pressure applied by the mercury plus the atmosphere, as the graph in Figure 2-5*b* shows:

$$V_{gas} \propto \frac{1}{p_{gas}} \qquad \text{(fixed temperature and amount)}$$

**Figure 2-4**

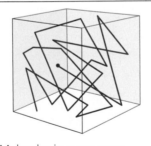

Molecules in a gas move freely throughout the entire volume of a container, changing direction whenever they collide with other molecules or with a wall. The line traces a possible path of a single molecule.

**Figure 2-5**

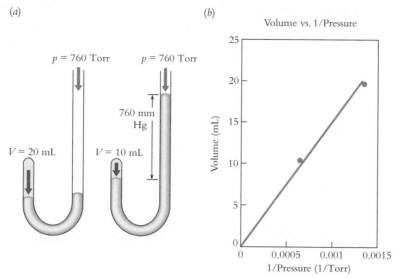

(a) Schematic illustration of Boyle's experiments on air trapped in J-shaped tubes. (b) Graph showing the linear variation of V vs. 1/p.

Boyle also observed that heating a gas causes it to expand in volume, but more than a century passed before Jacques-Alexandre-César Charles reported the first quantitative studies of gas volume as a function of temperature. Charles found that for a fixed amount of gas, a graph of gas volume vs. temperature gives a straight line, as shown in Figure 2-6. In other words, the volume of a gas is directly proportional to its temperature:

$$V_{gas} \propto T_{gas} \qquad \text{(fixed pressure and amount)}$$

In the early nineteenth century, Amedeo Avogadro found that the volume of a gas also changes when the amount of the gas changes. In this respect, gases are like liquids or solids. If we double the amount of any gas while keeping the temperature and pressure fixed, the gas volume doubles. In other words, gas volume is proportional to the amount of gas:

$$V_{gas} \propto n_{gas} \qquad \text{(fixed pressure and temperature)}$$

**Figure 2-6**

Plots of volume vs. temperature for 1 mole of air at three constant pressures.

## The Ideal Gas Equation

The work of Boyle, Charles, and Avogadro showed that the volume of any gas is proportional to the amount of gas, proportional to the temperature, and inversely proportional to the pressure of the gas. As the experiments show, p, T, and n can vary independently. This independence lets us link the four variables in a single proportionality:

$$V \propto \frac{nT}{p}$$

We can convert this proportionality into an equality by introducing a constant. This constant is known as the **universal gas constant** and is represented by the symbol **R**. Introducing R and multiplying through by p gives

$$pV = nRT \qquad\qquad (2\text{-}1)$$

We call Equation 2-1 the **ideal gas equation**, because it refers to the behaviour of a so-called ideal gas. As we describe in section 2-7, no gas is truly ideal, but under conditions close to 1 bar and room temperature, the ideal gas equation is adequate to describe most gases.

To use the ideal gas equation for quantitative calculations, we must express each quantity in appropriate units. The ideal gas equation holds only when temperature is expressed using an absolute scale, in other words in kelvins. If pressure is expressed in kPa, volume in litres, and $n$ in moles, then $R = 8.314\ \text{L kPa mol}^{-1}\ \text{K}^{-1}$. If pressure is expressed in bar, volume in litres, and $n$ in moles, then $R = 0.08314\ \text{L bar mol}^{-1}\ \text{K}^{-1}$. Both of these values are useful, since kPa and bar are both in common use.

## EXAMPLE 2-2 CALCULATION OF GAS PRESSURE

A 10 m$^3$ steel storage tank contains 8.85 kg of methane ($CH_4$). If the temperature is 25 °C, what is the pressure inside the tank?

**Strategy:** We use the seven-step strategy for problem solving.

1. The problem asks for the pressure exerted by a gas.

2. The gas is stored in a steel tank, so the volume of the gas cannot change. No chemical changes are described in the problem.

**Solution:**

3. We summarize the data given in the problem:

$$V = 10\ \text{m}^3,\ T = 25\ °C,\ m = 8.85\ \text{kg of methane},\ p = ?$$

4. We can calculate the pressure of the gas using the ideal gas equation, but we need to make sure all the variables are expressed in consistent units. Temperature must be in kelvins, amount of methane in moles, and volume in L. If $R = 8.314\ \text{L kPa mol}^{-1}\ \text{K}^{-1}$, then the calculated pressure will be in kPa.

5. We rearrange the ideal gas equation so that pressure is isolated on the left:

$$pV = nRT \quad \text{so} \quad p = \frac{nRT}{V}$$

Convert °C to kelvins:

$$T = 25 + 273.15 = 298.15\ \text{K}$$

Convert m$^3$ to L:

$$10\ \text{m}^3 \times 1000\ \text{L/m}^3 = 10\ 000\ \text{L}$$

Use the molar mass of methane to calculate the number of moles:

$$n = \frac{(8.85\ \text{kg})(10^3\ \text{g/kg})}{(16.04\ \text{g/mol})} = 5.517 \times 10^2\ \text{mol CH}_4$$

6. Now substitute and calculate the pressure:

$$p = \frac{(5.517 \times 10^2\ \text{mol})(8.314\ \text{L kPa mol}^{-1}\ \text{K}^{-1})(298.15\ \text{K})}{(1.00 \times 10^4\ \text{L})} = 137\ \text{kPa}$$

**7. Does the Result Make Sense?** You should verify that the units cancel properly, giving a result in pressure units. The problem describes a small amount of methane in a relatively large volume, so a low pressure is reasonable. This pressure is also low enough that the ideal gas equation applies. Note that 137 kPa is slightly more than 1 bar.

**Common Pitfalls:** Not using a consistent set of units. Make sure the units of $R$ are correct and that temperature is expressed in kelvins.

**Practice Exercise 2.2**
Determine the new pressure if the tank is stored in a hot location where the temperature reaches 42 °C.

## EXAMPLE 2-3 CALCULATION OF GAS AMOUNT

A pressure gauge on a tank of molecular oxygen gas reads 5.67 bar. If the tank holds 75.0 L and the temperature in the lab is 27.6 °C, how many grams of molecular oxygen are in the tank?

**Strategy:** The question asks for the mass of oxygen. We can use the ideal gas equation to calculate the number of moles of oxygen, and then molar mass leads us from moles to grams.

**Solution:** Begin by summarizing the data:

$$p = 5.67 \text{ bar } (= 567 \text{ kPa}), V = 75.0 \text{ L}, T = 27.6 + 273.15 = 300.8 \text{ K}$$

Rearrange the ideal gas equation to solve for the number of moles:

$$pV = nRT \quad \text{so} \quad n = \frac{pV}{RT}$$

Now substitute and do the calculation:

$$n = \frac{(567 \text{ kPa})(75.0 \text{ L})}{(8.314 \text{ L kPa K}^{-1} \text{ mol}^{-1})(301 \text{ K})} = 17.0 \text{ mol}$$

Complete the problem by converting moles to grams:

$$(17.0 \text{ mol } O_2)(32.00 \text{ g/mol}) = 544 \text{ g } O_2$$

**Does the Result Make Sense?** The units are correct and the numbers are all within the appropriate magnitudes, so the result is reasonable.

**Practice Exercise 2.3**
A 1.23-g sample of neon in a 1.50-L glass bulb has a pressure of 745 Torr. What is the temperature in the lab?

A useful quantity is the **molar volume** of a gas. As the name suggests, this is the volume of one mole of gas, and it must be expressed at some specified pressure and temperature. The molar volume can be easily calculated using the ideal gas equation. At 25.0 °C and 100 kPa,

$$\frac{V}{n} = \frac{RT}{p} = \frac{8.314 \text{ L kPa K}^{-1} \text{ mol}^{-1}(298.15 \text{ K})}{100.000 \text{ kPa}} = 24.79 \text{ L/mol}$$

This result is the molar volume of a gas under these conditions, and is useful for doing gas calculations at these conditions, and for estimating answers to gas problems at conditions close to these conditions. To visualize this volume, 24.79 L is the volume of a cube approximately 29 cm on a side.

## EXAMPLE 2-4  USING THE MOLAR VOLUME OF A GAS

A cement factory produces large volumes of $CO_2$ gas by the reaction $CaCO_3(s) \longrightarrow CaO(s) + CO_2(g)$. If the factory processes 10.0 tonnes of $CaCO_3(s)$ per hour, what volume of $CO_2(g)$ per hour is produced at $25.0\,^\circ C$ and 100.0 kPa?

**Strategy:** Since the question asks for a volume of gas measured at $25.0\,^\circ C$ and 100.0 kPa, we can use the molar volume of a gas calculated above to solve it.

**Solution:**

$$\frac{10.0 \text{ tonnes } CaCO_3}{h} \times \frac{1000 \text{ kg}}{\text{tonnes}} \times \frac{1000 \text{ g}}{\text{kg}} \times \frac{1 \text{ mol } CaCO_3}{100.1 \text{ g/mol}} = 9.99 \times 10^4 \text{ mol } CaCO_3/h$$

Since 1 mole of $CaCO_3$ yields 1 mole of $CO_2$, then $9.99 \times 10^4$ moles of $CO_2$ are produced per hour:

$$9.99 \times 10^4 \text{ mol/h} \times 24.79 \text{ L/mol} = 2.48 \times 10^6 \text{ L/h}$$

**Do the Results Make Sense?** This huge volume of $CO_2$ gas produced each hour is expected since we begin with a large mass of solid $CaCO_3$ and produce gaseous $CO_2$ from it. The factory would have to be designed to deal with so much $CO_2$ gas.

**Practice Exercise 2.4**

Pressing the button on a disposable cigarette lighter releases butane gas ($C_4H_{10}$). If 20 mL of gas is released at $25.0\,^\circ C$ and 100.0 kPa, what mass of butane is released?

## Variations on the Ideal Gas Equation

During chemical and physical transformations, any of the four variables in the ideal gas equation ($p$, $V$, $n$, $T$) may change, and any of them may remain constant. The experiments carried out by Robert Boyle are a good example. Boyle worked with a fixed amount of air trapped in a glass tube, so $n$ was held constant. Boyle also worked at only one temperature, so $T$ remained constant. Example 2-5 applies the ideal gas equation to this situation.

## EXAMPLE 2-5  PRESSURE–VOLUME VARIATIONS

A sample of helium gas is held at constant temperature inside a cylinder of volume 0.80 L when a piston exerts a pressure of 1.5 bar. If the external pressure on the piston is increased to 2.1 bar, what will be the new volume?

**Strategy:** We follow the seven-step strategy for problem solving.

1. The question asks for the new volume.

2 and 3. Visualize the conditions by drawing a schematic diagram of the initial and final conditions:

$p_i = 1.5$ bar

$p_f = 2.1$ bar

Increase
pressure →

$V_i = 0.80$ L

$V_f = ?$

## Solution:

4. Gas behaviour is involved, so the equation that applies is the ideal gas equation, $pV = nRT$.

5. Rearranging the gas equation to solve for $V$ will not help because we do not know $n$, the number of moles of helium present in the system, nor do we know $T$, the temperature of the gas. We do know that $n$ and $T$ remain unchanged as the pressure increases.

   To determine the final volume of the helium gas, apply the ideal gas equation to the initial (i) and final (f) conditions: $p_iV_i = n_iRT_i$ and $p_fV_f = n_fRT_f$

   In this problem, the quantity of helium inside the cylinder and the temperature of the gas are constant: $n_i = n_f$ and $T_i = T_f$ so $n_iRT_i = n_fRT_f$

   Therefore,

$$p_iV_i = p_fV_f \qquad \text{(constant } n \text{ and } T)$$

   Notice that this equality can be solved for $V_f$ without knowing the values for $n$ and $T$.

6. Now substitute and calculate the final volume:

$$V_f = \frac{p_iV_i}{p_f} = \frac{(1.5 \text{ bar})(0.80 \text{ L})}{(2.1 \text{ bar})} = 0.57 \text{ L}$$

**7. Does the Result Make Sense?** This answer is reasonable because a pressure increase has caused a volume decrease.

**Practice Exercise 2.5**

The piston in Example 2-5 is withdrawn until the gas volume is 2.55 L. Calculate the final pressure.

In Example 2-5, the quantities on the right side of the ideal gas equation are fixed, whereas the quantities on the left are changing. A good strategy for organizing gas calculations is to determine which variables do not change, and rearrange the gas equation to group them all on the right.

Examples 2-6 and 2-7 further demonstrate the analysis of gas problems.

## EXAMPLE 2-6 CHANGING GAS CONDITIONS

A sample of carbon dioxide in a 10.0-L gas cylinder at 25 °C and 1.00 bar pressure is compressed and heated. The final temperature and volume are 55 °C and 5.00 L. Compute the final pressure.

**Strategy:** We use the seven-step approach in compact form. The problem asks for the pressure of a gas after a change in conditions. A simple diagram helps us organize the information:

| Initial | | Final |
|---|---|---|
| $V_i = 10.0$ L<br>$T_i = 25 + 273 = 298$ K<br>$p_i = 1.00$ bar | Compress,<br>Heat $\longrightarrow$ | $V_f = 5.00$ L<br>$T_f = 55 + 273 = 328$ K<br>$p_f = ?$ |

The data tell us that $p$, $V$, and $T$ all change. The number of moles of $CO_2$ is not mentioned directly, but the physical transformation performed on the gas does not affect the number of molecules, so we conclude that $n_i = n_f$.

**Solution:** We rearrange $pV = nRT$, dividing both sides by $T$ to group the constants $n$ and $R$ on one side:

$$\frac{p_i V_i}{T_i} = nR = \frac{p_f V_f}{T_f}$$

Rearrange this expression to solve for the final pressure, insert the known information, and perform the calculations:

$$p_f = \left(\frac{p_i V_i}{T_i}\right)\left(\frac{T_f}{V_f}\right) = \frac{(1.00\ \text{bar})(10.0\ \text{L})(328\ \text{K})}{(298\ \text{K})(5.00\ \text{L})} = 2.20\ \text{bar}$$

**Does the Result Make Sense?** The result has the correct units and number of significant figures. The pressure has increased, which is what we expect, because compression increases the pressure, and a temperature increase also increases the pressure.

**Common Pitfalls:** Incorrectly rearranging the equation. Decide whether the final pressure should be higher or lower than the initial pressure before doing the calculation.

**Practice Exercise 2.6**

The sample of gas in Example 2-6 is placed in an ice bath and the pressure is increased to 1.35 bar. What is the new volume?

## EXAMPLE 2-7 GAS CALCULATIONS

Two natural gas storage tanks, with volumes of $1.5 \times 10^4$ and $2.2 \times 10^4$ L, are at the same temperature. The tanks are connected by pipes that equalize their pressures. What fraction of the stored natural gas is in the larger tank?

**Strategy:** The problem asks us to determine how the gas is distributed between the two tanks. This is a gas problem, so we use the ideal gas equation.

**Solution:** There are two pieces of data:

$$V_i = 1.5 \times 10^4\ \text{L} \quad \text{and} \quad V_2 = 2.2 \times 10^4\ \text{L}$$

Examine the conditions to determine which variables are the same for both tanks. The problem states that the tanks have the same pressure and temperature. Rearrange the gas equation to group the constants and find an equation that applies:

$$pV = nRT \quad \text{so} \quad \frac{n}{V} = \frac{p}{RT} = \text{constant}$$

The problem asks about the fraction of gas in the larger tank. This fraction is the ratio of the number of moles in the larger tank to the total number of moles in both tanks. Because $n/V$ is constant, we can write:

$$\frac{n_{\text{large tank}}}{V_{\text{large tank}}} = \frac{n_{\text{total}}}{V_{\text{total}}}$$

Now solve for the fraction of gas in the larger container:

$$\frac{n_{\text{large tank}}}{n_{\text{total}}} = \frac{V_{\text{large tank}}}{V_{\text{total}}} = \frac{2.2 \times 10^4 \text{ L}}{(2.2 \times 10^4 \text{ L}) + (1.5 \times 10^4 \text{ L})} = 0.59$$

 **Does the Result Make Sense?** A fraction of 0.59 means that 59% of the gas is in the larger tank. This seems reasonable because more than half the gas has to be in the larger tank.

**Practice Exercise 2.7**

Natural gas storage tanks have movable roofs that rise and fall as the volume of gas changes, keeping the pressure equal to that of the atmosphere. By what percentage does the volume of a tank change if the pressure changes from 768 Torr to 749 Torr?

## SECTION EXERCISES

●**2.2.1** According to the ideal gas equation, what will happen to the pressure of a gas if we make each of the following changes, holding all other conditions constant? (a) Double $T$ (in kelvins); (b) reduce the container's volume by a factor of two; (c) triple the amount of gas in the container; and (d) replace the gas with an equal number of moles of another gas whose molar mass is twice as great.

●**2.2.2** The molecular picture represents a small portion of an ideal gas exerting a pressure $p$. Redraw the picture in two ways, each of which gives a new pressure that is half as great.

●**2.2.3** A 1.54-L gas bulb in a chemistry laboratory contains oxygen gas at 21 °C and 758 Torr. The air conditioning in the laboratory breaks down, and the temperature rises to 31 °C. What pressure does the gauge show now?

## 2.3 GAS MIXTURES

Many gases are mixtures of two or more species. The atmosphere, with its mixture of nitrogen, oxygen, water, and various trace gases, is an obvious example. Another example is the gas used by deep-sea divers, which contains a mixture of helium and oxygen. The ideal gas model provides guidance as to how we describe mixtures of gases.

In an ideal gas, all molecules act independently. This notion applies to gas mixtures as well as to pure substances. Gas behaviour depends on the *number* of gas molecules but not on the *identity* of the gas molecules. The ideal gas equation applies to each gas in the mixture, as well as to the entire collection of molecules.

Suppose we pump 4.0 moles of helium into a deep-sea diver's tank. If we pump in another 4.0 moles of helium, the container now contains 8.0 moles of gas. The pressure can be calculated using the ideal gas equation, with $n = 4.0$ mol $+ 4.0$ mol $= 8.0$ mol. Now suppose that we pump in 4.0 moles of molecular oxygen. Now the container holds a total of 12.0 moles of gas. According

**Figure 2-7**

(a)          (b)

(c)          (d)

When 8.0 moles of He is added to a diver's tank (a), the atoms quickly become distributed uniformly throughout the tank (b). When 4.0 moles of $O_2$ is added to this tank (c), the molecules move about independently of the He atoms, causing the gases to mix uniformly (d).

The partial pressure of each gas is designated with a lower case *p with the gas formula in subscript* to distinguish it from the total pressure of the mixture, $p_{total}$, also written as *p* without subscripts.

to the ideal gas model, it does not matter whether we add the same gas or a different gas. Because all molecules in a sample of an ideal gas behave independently, the pressure increases in proportion to the increase in the total number of moles of gas. Thus, we can calculate the total pressure from the ideal gas equation, using $n = 8.0 \text{ mol} + 4.0 \text{ mol} = 12.0 \text{ mol}$.

How does a 1:2 mixture of $O_2$ and He appear on the molecular level? As $O_2$ is added to the container, its molecules move throughout the volume and become distributed uniformly. Diffusion causes gas mixtures to become homogeneous. Figure 2-7 shows this.

## Dalton's Law of Partial Pressures

The total pressure exerted by an ideal gas mixture is determined by the total number of moles of gas:

$$p = \frac{n_{total} \, RT}{V}$$

We can express the total number of moles of gas as the sum of the amounts of the individual gases. For the diver's He and $O_2$ mixture,

$$n_{total} = n_{He} + n_{O_2}$$

Substitution gives a two-term equation for the total pressure:

$$p_{total} = \frac{(n_{He} + n_{O_2})}{V} = \frac{n_{He}RT}{V} + \frac{n_{O_2}RT}{V}$$

Notice that each term on the right resembles the ideal gas equation rearranged to express pressure. Each term therefore represents the **partial pressure** (*p*) of one of the gases. As Figure 2-8 illustrates, the partial pressure of a gas is the pressure that would be present in a gas container if that gas were present by itself:

$$p_{He} = \frac{n_{He}RT}{V} \qquad \text{and} \qquad p_{O_2} = \frac{n_{O_2}RT}{V}$$

**Figure 2-8**

| 3.0 L at 273 K | 3.0 L at 273 K | 3.0 L at 273 K |
|---|---|---|
| 4.0 mol $O_2$ | 8.0 mol He | 12.0 mol gas |
| $p_{O_2}$ = 30 bar | $p_{He}$ = 60 bar | $p_{O_2}$ = 30 bar |
| | | $p_{He}$ = 60 bar |
| | | $p_{total}$ = 90 bar |

Molecular pictures of a sample of $O_2$, a sample of He, and a mixture of the two gases. Both components are distributed uniformly throughout the gas volume. Each gas behaves in the same way, whether it is pure or part of a mixture.

The total pressure in the container is the sum of the partial pressures:

$$p_{\text{total}} = p_{\text{He}} + p_{O_2}$$

We have used He and $O_2$ to illustrate the behaviour of a mixture of ideal gases, but the same result is obtained regardless of the number and identity of the gases.

This is **Dalton's law of partial pressures**. To obtain a total pressure of a gas containing n species, simply add the contributions from all gases present:

$$p_{\text{total}} = p_1 + p_2 + p_3 + \cdots + p_n$$

When doing calculations on a mixture of gases, we can apply the ideal gas equation to each component to find its partial pressure, $p_i$. Alternatively, we can treat the entire gas as a unit, using the total number of moles to determine the total pressure of the mixture ($p_{\text{total}}$).

## Describing Gas Mixtures

There are several ways to describe the chemical composition of a mixture of gases. The simplest method is merely to list each component with its partial pressure or number of moles. Two other descriptions, mole fractions and parts per million, are also used frequently.

Chemists often express chemical composition in fractional terms, stating the number of moles of a substance as a fraction of the number of moles of all substances in the mixture. This way of stating composition is the **mole fraction (X)**:

$$\text{Mole fraction of A} = X_A = \frac{n_A}{n_{\text{total}}}$$

Mole fractions provide a simple way to relate the partial pressure of one component to the total pressure of the gas mixture:

$$p_A = \frac{n_A RT}{V} \quad \text{and} \quad p_{\text{total}} = \frac{n_{\text{total}} RT}{V}$$

Dividing $p_A$ by $p_{\text{total}}$ gives

$$\frac{p_A}{p_{\text{total}}} = \frac{\left(\dfrac{n_A RT}{V}\right)}{\left(\dfrac{n_{\text{total}} RT}{V}\right)} = \frac{n_A}{n_{\text{total}}} = X_A$$

Rearranging gives Equation 2-2:

$$p_A = X_A p_{\text{total}} \tag{2-2}$$

The partial pressure of a component in a gas mixture is its mole fraction times the total pressure. Example 2-8 illustrates calculations with gas mixtures.

## EXAMPLE 2-8 GAS MIXTURES

Deep-sea divers use mixtures of helium and molecular oxygen in their breathing tanks.

The amount of gas introduced into a diving tank can be determined by weighing the tank before and after charging the tank with gas. A diving shop placed 80.0 g of $O_2$ and 20.0 g of He in a 5.00-L tank at 298 K. Determine the total pressure of the mixture, and find the partial pressures and mole fractions of the two gases.

**Strategy:** We have a mixture of two gases in a container with known volume and temperature. The problem asks for pressures and mole fractions. Because molecular interactions are negligible, each gas can be described independently by the ideal gas equation. As usual, we need molar amounts for the calculations.

**Solution:** Begin with the data provided:

$$V = 5.00 \text{ L}, T = 298 \text{ K}, m_{He} = 20.0 \text{ g}, m_{O_2} = 80.0 \text{ g}$$

Convert the mass of each gas into an amount in moles:

$$n = m/M, \text{ thus } n_{He} = 5.00 \text{ mothus l}, n_{O_2} = 2.50 \text{ mol}$$

Next, use the ideal gas equation to compute the partial pressure of each gas:

$$p_{He} = \frac{(5.00 \text{ mol})(8.314 \text{ L kPa mol}^{-1} \text{ K}^{-1})(298 \text{ K})}{5.00 \text{ L}} = 2480 \text{ kPa}$$

The same calculation for $O_2$ gives $p_{O_2} = 1240$ kPa.
    The total pressure is the sum of the partial pressures:

$$p_{total} = p_{He} + p_{O_2} = 2480 \text{ kPa} + 1240 \text{ kPa} = 3720 \text{ kPa}$$

Mole fractions can be calculated from moles or partial pressures:

$$X_{He} = n_{He}/n_{total} = 5.00 \text{ mol}/7.50 \text{ mol} = 0.667$$
$$X_{O_2} = p_{O_2}/p_{total} = 1240 \text{ kPa}/3720 \text{ kPa} = 0.333$$

**Do the Results Make Sense?** The partial pressures are large, but diving tanks are built to withstand high pressures. To check for consistency, add the mole fractions, which must add to 1.00.

**Practice Exercise 2.8**
Another 5.00-L diving tank is charged with 75.0 g of $O_2$ and 25.0 g of He. Calculate the mole fractions and the partial pressures at 25 °C.

When referring to sparse components of gas mixtures, scientists typically use **parts per million (ppm)** to designate the relative number of molecules of a substance present in a sample. Parts per million measures how many molecules of a substance are present in one million molecules of sample. As an example, concentrations of atmospheric pollutants often are given in ppm. If a pollutant is present at a concentration of 1 ppm, there is one molecule of the pollutant in every one million molecules of the atmosphere. In molar terms, 1 ppm means that there is one mole of pollutant for every one million moles of gas. Or, to put it another way, there is $10^{-6}$ mole of pollutant in every mole of air.

The Canadian objective for ground-level ozone is 0.082 ppm. This may not seem like much, but ozone is a very toxic substance that is particularly damaging to soft tissue such as the lungs.

Even lower concentrations in a gas mixture are expressed in **parts per billion (ppb)**. Parts per billion measures how many molecules of substance are present in one billion molecules of sample.

Air quality in Los Angeles, California, on Sept. 11, 2000

$$1 \text{ ppm} = 1 \text{ molecule out of every } 10^6 \text{ molecules}$$

$$1 \text{ ppb} = 1 \text{ molecule out of every } 10^9 \text{ molecules}$$

Mole fractions, parts per million, and parts per billion all are ratios of moles of a particular substance to total moles of sample. Mole fraction is moles per mole, ppm is moles per million moles, and ppb is moles per billion moles. These measures are related by scale factors: $\text{ppm} = 10^6 X$, and $\text{ppb} = 10^9 X$. In other words, a concentration of 1 ppm is a mole fraction of $10^{-6}$, and a concentration of 1 ppb is a mole fraction of $10^{-9}$. When the ozone concentration in the atmosphere reaches 0.5 ppm, the mole fraction of ozone is $0.5 \times 10^{-6}$, or $5 \times 10^{-7}$. Example 2-9 shows how to work with concentrations in parts per million.

## EXAMPLE 2-9 WORKING WITH CONCENTRATIONS IN PPM

The exhaust gas from an average automobile contains 206 ppm of the pollutant nitrogen oxide, NO. If an automobile emits 125 L of exhaust gas at 100.0 kPa and 350 K, what mass of NO has been added to the atmosphere? Assume the atmospheric pressure is 100 kPa.

**Strategy:** The question asks for mass of NO. Information about ppm tells us how many moles of NO are present in one mole of exhaust gas. We can use the ideal gas equation to determine the total number of moles of gas emitted, use the ppm information to find moles of NO, and do a mole–mass conversion to get the required mass.

**Solution:** Data: Concentration of NO = 206 ppm, $V = 125$ L, $p = 100$ kPa, $T = 350$ K

$$n_{gas} = \frac{pV}{RT} = \frac{(100 \text{ kPa})(125 \text{ L})}{(8.314 \text{ L kPa mol}^{-1} \text{ K}^{-1})(350 \text{ K})} = 4.30 \text{ mol gas}$$

$$X_{NO} = 10^{-6}(206 \text{ ppm}) = 2.06 \times 10^{-4}$$

$$n_{NO} = X_{NO} n_{gas} = (2.06 \times 10^{-4})(4.30 \text{ mol}) = 8.85 \times 10^{-4} \text{ mol}$$

$$m = nM = (8.85 \times 10^{-4} \text{ mol})(30.0 \text{ g/mol}) = 2.65 \times 10^{-2} \text{ g}$$

**Does the Result Make Sense?** The mass of NO is rather small, only 27 mg, but the mole fraction is also small, so this is a reasonable result.

**Common Pitfalls:** Incorrectly converting mole fraction to ppm. Remember that $X = 1$ means $10^6$ ppm.

**Practice Exercise 2.9**

The maximum NO emission allowed by the state of California is 762 ppm. What mass of NO is this per litre of exhaust emitted at a temperature of 50 °C?

The description presented in this section applies to a gas mixture that is not undergoing chemical reactions. As long as reactions do not occur, the number of moles of each gas is determined by the amount of that substance initially present. When reactions occur, the numbers of moles of reactants and products change as predicted by the principles of stoichiometry. Changes in composition must be taken into account before the properties of the gas mixture can be computed. Gas stoichiometry is described in the next section.

## SECTION EXERCISES

●**2.3.1** Find the partial pressures and mole fractions of a gas mixture that contains 1.00 g each of $H_2$ and $N_2$, if the total pressure of this mixture is 2.30 bar.

●**2.3.2** Draw a molecular picture that shows how a sample of the gas mixture in Section Exercise 2.3.1 appears at the molecular level.

●**2.3.3** An atmospheric chemist reported that the air in an urban area contained CO at a concentration of 35 ppm. The air temperature was 29 °C, and the atmospheric pressure was 745 Torr. (a) What was the mole fraction of CO? (b) What was the partial pressure of CO? (c) How many molecules of CO were in $1.0 \times 10^3$ L of this air?

## 2.4 GAS STOICHIOMETRY

The principles of stoichiometry apply equally to solids, liquids, and gases. That is, no matter what phase substances are in, their chemical behaviour can be described in molecular terms, and their transformations must be visualized and balanced using molecules and moles.

The ideal gas equation relates the number of moles of gas to the physical properties of that gas. When a chemical reaction involves a gas, the ideal gas equation provides the link between the molar amount of that gas and its partial pressure:

$$n_i = \frac{p_i V}{RT}$$

Stoichiometric calculations always require amounts in moles. Example 2-10 shows how to apply this idea.

## EXAMPLE 2-10 GAS STOICHIOMETRY

This example describes the synthesis of acetylene ($C_2H_2$) from calcium carbide ($CaC_2$). Modern industrial production of acetylene ($C_2H_2$) is based on a reaction of methane ($CH_4$) under carefully controlled conditions. At temperatures greater than 1,600 K, two methane molecules rearrange to give three molecules of hydrogen and one molecule of acetylene:

$$2\ CH_4(g) \xrightarrow{1,600\ K} C_2H_2(g) + 3\ H_2(g)$$

A 50.0-L steel vessel, filled with $CH_4$ to a pressure of 10.0 bar at 298 K, is heated to 1600 K to convert $CH_4$ into $C_2H_2$. What mass of $C_2H_2$ can be produced? What pressure does the reactor reach at 1600 K?

**Strategy:** We apply the seven-step approach to this problem.

1. This is a stoichiometry problem involving gases. The mass of a product and the final pressure must be calculated.

2 and 3. List the data:

$$CH_{4(g)}: \qquad V = 50.0 \text{ L}, T = 298 \text{ K}, p = 10.0 \text{ bar} = 1000 \text{ kPa}$$
$$C_2H_{2(g)} + H_{2(g)}: V = 50.0 \text{ L}, T = 1600 \text{ K}, p = \text{?}$$

**Solution:**

4. In any stoichiometry problem, work with moles. This problem involves gases, so use the ideal gas equation to convert $p$–$V$–$T$ information into moles.

5. Use the ideal gas equation to calculate the amount of $CH_4$ present initially:

$$n = \frac{p_i V}{RT_i} = \frac{(1000 \text{ kPa})(50.0 \text{ L})}{(8.314 \text{ L kPa mol}^{-1} \text{ K}^{-1})(298 \text{ K})} = 20.18 \text{ mol CH}_4$$

(Carry one additional significant figure to avoid errors caused by premature rounding.)
Now construct an amounts table to determine how much of each product forms:

| Reaction | 2 CH$_4$ | $\longrightarrow$ | C$_2$H$_2$ | + | 3 H$_2$ |
|---|---|---|---|---|---|
| Initial amount (mol) | 20.18 | | 0 | | 0 |
| Change in amount (mol) | −20.18 | | +10.09 | | +30.27 |
| Final amount (mol) | 0 | | 10.09 | | 30.27 |

Do a mole–mass conversion to determine the mass of acetylene formed:

$$m = nM = (10.09 \text{ mol})(26.04 \text{ g/mol}) = 263 \text{ g acetylene}$$

6. Use the ideal gas equation and the total number of moles present at the end of the reaction to calculate the final pressure:

$$n_{total} = 10.09 + 30.27 = 40.36 \text{ mol}$$

$$\text{Rearrange } pV = nRT \text{ to } p = \frac{nRT}{V}$$

$$p = \frac{(40.36 \text{ mol gas})(8.314 \text{ L kPa mol}^{-1})(1600 \text{ K})}{50.0 \text{ L}} = 10\,700 \text{ kPa} = 107 \text{ bar}$$

7. **Do the Results Make Sense?** From the initial amount of methane, we can calculate the initial mass of methane, which is 323 g. The mass of acetylene, 263 g, is less than this, a reasonable result given that $H_2$ is also produced. The final pressure, 107 bar, seems high, but the temperature of the reactor has increased substantially and the amount of products is twice the amount of reactants, so a large pressure increase is to be expected in this reaction.

**Common Pitfalls:** Incorrectly converting between kPa and bar. Remember that 1 bar = 100 kPa.

**Practice Exercise 2.10**
Calculate the volume of hydrogen gas generated when 1.52 g of Mg reacts with an excess of aqueous HCl, if the gas is collected at 755 Torr and 22.5 °C (1 mole of Mg produces 1 mole of $H_2$).

Many types of problems can involve gases. The strategy for doing stoichiometric calculations is the same whether the species involved are solids, liquids, or gases. In this chapter, we add the ideal gas equation to our equations for converting measured quantities into moles. Example 2-11 is a limiting reactant problem that involves a gas.

## EXAMPLE 2-11 LIMITING REACTANTS IN A GAS MIXTURE

Margarine can be made from natural oils, such as coconut oil, by hydrogenation:

$$C_{57}H_{104}O_6(l) + 3\,H_2(g) \xrightarrow{\text{200 °C, 7 atm, Ni catalyst}} C_{57}H_{110}O_6(s)$$
$$\text{Oil} \qquad\qquad\qquad\qquad\qquad\qquad\qquad\qquad\qquad\qquad \text{Margarine}$$

An industrial hydrogenator with a volume of $2.50 \times 10^2$ L is charged with 12.0 kg of oil and 7.09 bar (=709 kPa) of hydrogen ($H_2$) at 473 K (200 °C), and the reaction goes to completion. What is the final pressure of $H_2$ and how many kilograms of margarine will be produced?

**Strategy:** We have data for the amounts of both starting materials, so this is a limiting reactant problem. Given the chemical equation, the first step in a limiting reactant problem is to determine the number of moles of each starting material present at the beginning of the reaction. Next, compute ratios of moles to coefficients to identify the limiting reactant. After that, a table of amounts summarizes the stoichiometry.

**Solution:** The ideal gas equation is used for the gaseous starting material:

$$\text{moles } H_2 = \frac{pV}{RT} = \frac{(709\text{ kPa})(2.50 \times 10^2\text{ L})}{(8.314\text{ L kPa mol}^{-1}\text{ K}^{-1})(473\text{ K})} = 45.07\text{ mol}$$

For the other starting material, do a mole–mass conversion:

$$\text{moles oil} = \frac{m}{M} = \frac{1.20 \times 10^4\text{ g}}{885.4\text{ g/mol}} = 13.55\text{ mol}$$

Now divide each number of moles by the stoichiometric coefficient to identify the limiting reactant:

$$\frac{45.09\text{ mol } H_2}{3\text{ mol } H_2} = 15.03 \qquad \frac{13.55\text{ mol oil}}{1\text{ mol oil}} = 13.55$$

The reactant with the smaller ratio, the oil, is limiting.
The table of amounts follows:

| Reaction | $3\,H_2$ | + | $C_{37}H_{104}O_6$ | $\longrightarrow$ | $C_{37}H_{110}O_6$ |
|---|---|---|---|---|---|
| Initial amount (mol) | 45.09 | | 13.55 | | 0 |
| Change in amount (mol) | −40.65 | | −13.55 | | 13.55 |
| Final amount (mol) | 4.4 | | 0 | | 13.6 |

(The final amounts have been rounded off correctly.)
Use the ideal gas equation to calculate the pressure of hydrogen at the end of the reaction:

$$p = \frac{nRT}{V} = \frac{(4.4\text{ mol})(8.314\text{ L kPa mol}^{-1}\text{ K}^{-1})(473\text{ K})}{2.50 \times 10^2\text{ L}} = 69\text{ kPa} = 0.69\text{ bar}$$

Now calculate the mass of margarine:

$$m = nM = (13.6\text{ mol})(891.5\text{ g/mol}) = 1.21 \times 10^4\text{ g} = 12.1\text{ kg}$$

**Do the Results Make Sense?** In this example, oil is the limiting reactant. Excess hydrogen is easily recovered from a gas-phase reactor, so margarine manufacturers make the oil the limiting reactant to ensure complete conversion of oil into margarine. The excess $H_2$ gas is recovered and used again in a subsequent reaction.

**Practice Exercise 2.11**
In the industrial production of nitric acid, one step is oxidation of nitrogen oxide: $2\,NO + O_2 \longrightarrow 2\,NO_2$ (all gases). A reaction chamber is charged with 5.00 bar each of NO and $O_2$. If the reaction goes to completion without a change in temperature or volume, calculate the final pressures of each reagent.

A commercial hydrogenator

## Summary of Mole Conversions

Because moles are the currency of chemistry, all stoichiometric computations require amounts in moles. In the real world, we measure mass, volume, temperature, and pressure. With the ideal gas equation, our catalogue of relationships for mole conversion is complete. Table 2-1 lists three equations, each of which applies to a particular category of chemical substances.

All three of these equations should be firmly embedded in your memory, *along with the substances to which they apply.* Using $pV = nRT$ on an aqueous solution, for example, gives impossible results. Example 2-12 uses all three relationships. Viewed as a whole, the example may seem complicated. As the solution illustrates, however, breaking the problem into separate parts allows each part to be solved using simple chemical and stoichiometric principles. Complicated problems are often simplified considerably by looking at them one piece at a time.

## TABLE 2-1  SUMMARY OF MOLE RELATIONSHIPS

| Substance | Relationship | Equation |
|---|---|---|
| Pure liquid or solid | Moles = Mass/Molar mass | $n = \dfrac{m}{M}$ |
| Liquid solution | Moles = (Molarity)(Volume) | $n = cV$ |
| Gas | Moles = $\dfrac{\text{(Pressure)(Volume)}}{\text{(Constant)(Temperature)}}$ | $n = \dfrac{pV}{RT}$ |

## EXAMPLE 2-12 GENERAL STOICHIOMETRY

Mg reacting with aqueous HCl

The reaction of a metal with an acid generates hydrogen gas and an aqueous solution of ions. Suppose that 3.50 g of magnesium metal is dropped into 0.150 L of 6.00 $M$ HCl in a 5.00-L cylinder with an initial gas pressure of 1.00 bar at 25.0 °C, and the cylinder is immediately sealed. Find the final partial pressure of hydrogen, the total pressure in the container, and the concentrations of all ions in solution.

**Strategy:** Data are given for all reactants, so this is a limiting reactant problem. We must balance the chemical equation and then work with a table of molar amounts.

**Solution:** Begin by analyzing the chemistry. First, list all major species present in the system before reaction:

$$Mg(s), \; H_3O^+(aq), \; Cl^-(aq), \; \text{and } H_2O(l).$$

Magnesium, a Group 2 metal, reacts with acids to generate +2 cations:

$$Mg(s) + H_3O^+(aq) \longrightarrow Mg^{2+}(aq) + H_2(g) \quad \text{(unbalanced)}$$

This reaction can be balanced by inspection. Charge balance requires that the +2 charge on the $Mg^{2+}$ be matched by two hydronium ions among the starting materials. Two water molecules on the right side must then balance two oxygen atoms on the left:

$$Mg(s) + 2 \, H_3O^+(aq) \longrightarrow Mg^{2+}(aq) + H_2(g) + 2 \, H_2O(l)$$

The problem asks for pressures and ion concentrations. The final pressure can be determined from $p$–$V$–$T$ data and $n_{H_2}$. Moles of hydrogen can be found from the mass of magnesium, the stoichiometric ratio, and a table of amounts. Here is a summary of the data:

$$V_{container} = 5.00 \text{ L} \qquad T = 298 \text{ K} \qquad m_{Mg} = 3.50 \text{ g}$$
$$V_{solution} = 0.150 \text{ L} \qquad p_{air} = 1.00 \text{ bar} \qquad [H_3O^+] = [Cl^-] = 6.00 \text{ } M$$

Now we analyze the stoichiometry of the reaction. The starting amounts of $H_3O^+$ and Mg are given, but the data must be converted to moles before we can construct the table of amounts:

$$\text{Moles of Mg} = n_{Mg} = \frac{m}{M} = \frac{3.50 \text{ g Mg}}{24.31 \text{ g/mol}} = 0.1439 \text{ mol Mg}$$

$$\text{Moles of } H_3O^+ = n_{H_3O^+} = cV = (6.00 \text{ mol/L})(0.150 \text{ L}) = 0.900 \text{ mol } H_3O^+$$

You should be able to use the numbers of moles and the stoichiometric ratio to show that magnesium is the limiting reactant. The table of amounts follows. We omit the amounts of water because as the solvent, it is present in excess:

| Reaction | Mg | + | 2 H₃O⁺ | ⟶ | Mg²⁺ | + | H₂ | + | 2 H₂O |
|---|---|---|---|---|---|---|---|---|---|
| Initial amount (mol) | 0.1439 | | 0.900 | | 0 | | 0 | | |
| Change in amount (mol) | −0.1439 | | −0.2878 | | +0.1439 | | +0.1439 | | |
| Final amount (mol) | 0 | | 0.6122 | | 0.1439 | | 0.1439 | | |

From the final numbers of moles, we can calculate the final concentrations:

$$[Mg^{2+}] = \frac{0.1439 \text{ mol}}{0.150 \text{ L}} = 0.959 \text{ } M$$

$$[H_3O^+] = \frac{0.6122 \text{ mol}}{0.150 \text{ L}} = 4.08 \text{ } M$$

$$[Cl^-] = 6.00 \text{ M}$$

Remember, chloride is a spectator ion, so its concentration does not change. Before calculating the pressures, we must visualize the reaction vessel. The container's total volume is 5.00 L, but 0.150 L is occupied by the aqueous solution. This leaves 4.85 L for the gas mixture. The partial pressure of hydrogen is calculated using the ideal gas equation and assuming that no $H_2$ remains in solution; this is a good assumption because hydrogen gas is not very soluble in water:

$$p_{H_2} = \frac{nRT}{V} = \frac{(0.1439 \text{ mol})(8.314 \text{ L kPa K}^{-1} \text{ mol}^{-1})(298 \text{ K})}{4.85 \text{ L}} = 73.5 \text{ kPa} = 0.735 \text{ bar}$$

The amount of air originally present does not change in the reaction, so the pressure exerted by the air remains constant at 1.00 bar. The final total pressure is the sum of the partial pressures:

$$p_{total} = p_{H_2} + p_{air} = 0.735 \text{ bar} + 1.00 \text{ bar} = 1.73 \text{ bar}$$

 **Do the Results Make Sense?** The values of all the quantities calculated in this example are in reasonable ranges for a reaction that consumes somewhat less than 1 mole of reactants.

**Common Pitfalls:** Not remembering that partial pressures are additive. The total pressure is always the sum of the partial pressures.

**Practice Exercise 2.12**
Repeat the calculations if all conditions are the same, except that 14.0 g of Mg is added to the HCl solution.

## SECTION EXERCISES

●**2.4.1** In an oxyacetylene torch, acetylene ($C_2H_2$) burns in a stream of $O_2$. This combustion reaction produces a flame with a temperature exceeding 3000 K. A robotic welding machine welds the inside of a tank that has an air volume of $1.50 \times 10^2$ L, at a temperature of 30 °C. What mass of $C_2H_2$ must be burned to raise the $CO_2$ partial pressure to 0.10 bar? Assume that a negligible amount of the $CO_2$ escapes from the tank before all the $C_2H_2$ has been burned. What mass of water will be produced during the burn?

●**2.4.2** A reaction vessel contains $H_2$ gas at 2.40 bar. If just enough $O_2$ is added to react completely with the $H_2$ to form $H_2O$, what will the total pressure in the vessel be before any reaction occurs? ($H_2$ and $O_2$ can also be used as a torch. The flame produced in an oxyhydrogen torch is about 4000 K.)

●**2.4.3** When ammonium nitrate ($NH_4NO_3$) explodes, all the products are gases:

$$2 \text{ NH}_4\text{NO}_3(s) \longrightarrow 4 \text{ H}_2\text{O}(g) + \text{O}_2(g) + 2 \text{ N}_2(g)$$

If 5.00 g of $NH_4NO_3$ explodes in a closed 2.00-L container that originally contains air at $p = 1.00$ bar and $T = 25$ °C, and the temperature rises to 205 °C, what total pressure develops?

## 2.5 MOLECULAR VIEW OF GASES

To understand why all gases can be described by a single equation, we need to explore how gases behave at the molecular level. In this section, we examine the molecular properties of gases and how they result in the ideal gas equation.

### Molecular Speeds

How fast do gas molecules move? Molecular speeds can be measured using a molecular beam apparatus, shown schematically in Figure 2-9a. Gas molecules escape from an oven through a small hole into a chamber in which the molecular density is very low. A set of slits blocks the passage of all molecules except those moving in the forward direction. The result is a beam of molecules, all moving in the same direction. A rotating shutter blocks the beam path except for a small slit that allows a packet—"pulse"— of molecules to pass through. Each molecule moves down the beam axis at its own speed. The faster a molecule moves, the less time it takes to travel the length of the chamber. A detector at the end of the chamber measures the number of molecules arriving as a function of time, giving a profile of speeds.

When the speed profile of a gas is measured in this way, the results give a distribution like the one shown in Figure 2-9b. If all the molecules travelled at the same speed, they would reach the detector at the same time, in a single clump. Instead, faster molecules move ahead of the main packet, and slower molecules fall behind. This experiment shows that molecules in a gas have a distribution of speeds.

A pattern emerges when this molecular beam experiment is repeated for various gases at a common temperature: molecules with small masses move faster than those with large masses. Figure 2-10 shows this for $H_2$, $CH_4$, and $CO_2$. Of these molecules, $H_2$ has the smallest mass and $CO_2$ the largest. The vertical line drawn for each gas shows the speed at which the distribution reaches its maximum height. More molecules have this speed than any other, so this is the most probable speed for molecules of that gas. The most probable speed for a molecule of hydrogen at 300 K is $1.57 \times 10^3$ m/s, which is 5650 km/h.

Example 2-13 describes an experiment with a molecular beam apparatus.

**Figure 2-9**

(a) Diagram of a molecular beam apparatus designed to measure the speeds of gas molecules. (b) Distribution of molecules observed by the detector as a function of time after opening the shutter.

**Figure 2-10**

Molecular speed distributions for $CO_2$, $CH_4$, and $H_2$ at a temperature of 300 K.

## EXAMPLE 2-13  A MOLECULAR BEAM EXPERIMENT

The figures below represent mixtures of neon atoms and hydrogen molecules.

Sample A                           Sample B

One of the gas mixtures was used in a pulsed molecular beam experiment. The result of the experiment is shown below. Which of the two gas samples, A or B, was used for this experiment?

**Strategy:** Begin by taking an inventory of the two samples, and then decide how each of them would behave in a beam experiment.

**Solution:** Sample A contains eight $H_2$ molecules and four Ne atoms, and Sample B contains four $H_2$ molecules and eight Ne atoms. In a beam experiment, both samples would give two peaks in relative areas of $2:1$. Which sample was used for this experiment? Particles with small mass move faster than particles with large mass, so we expect $H_2$ ($M = 2.02$ g/mol) to reach the detector before Ne ($M = 20.2$ g/mol). The data show that the first substance to reach the detector is present in the larger amount. Consequently, the sample used in the beam experiment is the one with the larger amount of the hydrogen molecules, Sample A.

 **Does the Result Make Sense?** Figure 2-10 shows that $H_2$ moves much faster than other substances, so it makes sense that the $H_2$ molecules get to the detector faster.

**Practice Exercise 2.13**

Suppose that Sample A is used and that the second gas is helium instead of neon. Sketch a graph showing the number of molecules/atoms vs. time.

## Speed and Energy

The energy of a molecule is related to its speed. Any moving object has kinetic energy ($E_{kinetic}$) whose magnitude is given by Equation 2-3:

$$E_{kinetic} = \frac{1}{2}mv^2 \qquad (2\text{-}3)$$

In this equation, $m$ is the object's mass and $v$ its speed. The most probable speed of hydrogen molecules at 300 K is $1.57 \times 10^3$ m/s. We need the mass of one $H_2$ molecule to calculate its most probable kinetic energy. The SI unit of energy is the joule (J), which equals 1 kg m$^2$ s$^{-2}$. Thus, we need the mass ($m$) in kilograms:

$$m = \frac{M(1\ kg/10^3\ g)}{N_A} = \frac{(2.016\ g/mol)(1\ kg/10^3\ g)}{6.0223 \times 10^{23}\ molecules/mol}$$

$$= 3.348 \times 10^{-27}\ kg/H_2\ molecule$$

Now apply Equation 2-3:

$$E_{kinetic}\ (most\ probable) = \frac{1}{2}(3.348 \times 10^{-27}\ kg/H_2\ molecule)(1.57 \times 10^3\ m/s)^2$$

$$= 4.13 \times 10^{-21}\ kg\ m^2\ s^{-2}/H_2\ molecule$$

$$= 4.13 \times 10^{-21}\ J/H_2\ molecule$$

The most probable speeds of methane and carbon dioxide are lower than the most probable speed of hydrogen, but $CH_4$ and $CO_2$ molecules have larger masses than $H_2$. When kinetic energy calculations are repeated for these gases, they show that the most probable kinetic energy is the same for all three gases.

For $CH_4$:

$$\frac{(16.04\ g/mol)(10^{-3}\ kg/g)(5.57 \times 10^2\ m/s)^2}{2(6.022 \times 10^{23}\ molecules/mol)} = 4.13 \times 10^{-21}\ J/CH_4\ molecule$$

For $CO_2$:

$$\frac{(44.01\ g/mol)(10^{-3}\ kg/g)(3.37 \times 10^2\ m/s)^2}{2(6.022 \times 10^{23}\ molecules/mol)} = 4.13 \times 10^{-21}\ J/CO_2\ molecule$$

Even though the speed distributions for these three gases are different, the most probable kinetic energies are identical, since the temperature is the same in each case.

Molecular beam experiments show that molecules move faster as temperature increases. Molecules escaping from the oven at 900 K take less time to reach the detector than molecules escaping at 300 K. As molecular speed increases, so does kinetic energy (Equation 2-3). Figure 2-11 shows the molecular kinetic energy distributions at 300 K and 900 K. Comparing this figure with Figure 2-10, we see that molecular energies show a wide distribution that is similar to the distribution in molecular speeds. Unlike speeds, however, molecular energy distributions are the same for all gases at any particular temperature. The distributions shown in Figure 2-11 apply to any gas.

**Figure 2-11**

Distribution of molecular energies for a gas at 300 K and 900 K.

## Average Kinetic Energy

The most probable kinetic energy is not the same as the average kinetic energy. Figure 2-11 shows that the distribution of kinetic energies is not symmetrical. To find the average kinetic energy per molecule ($\overline{E}_{kinetic}$), we must add all the individual molecular energies and divide by the total number of molecules. The result is Equation 2-4, which describes how the average kinetic energy of gas molecules depends on the temperature of the gas:

A bar over a symbol means "average value."

$$\overline{E}_{kinetic} = \frac{3RT}{2N_A} \qquad (2\text{--}4)$$

In this equation, $T$ is the temperature in kelvins, $N_A$ is Avogadro's number (units of $mol^{-1}$), and $R$ is the gas constant. Kinetic energy is expressed in joules per molecule, and the appropriate value of $R$ is $8.314 \, J \, K^{-1} \, mol^{-1}$.

The average kinetic energy expressed by Equation 2-4 is kinetic energy *per molecule*. We find the total kinetic energy ($\overline{E}_{kinetic,molar}$) of 1 mole of gas molecules by multiplying Equation 2-4 by Avogadro's number:

$$\overline{E}_{kinetic,molar} = N_A \frac{3RT}{2N_A} = \frac{3}{2}RT$$

Thus, 1 mole of any gas has a total molecular kinetic energy of 3/2 *RT*. Example 2-14 applies these equations to sulphur hexafluoride.

### EXAMPLE 2-14 MOLECULAR KINETIC ENERGIES

Determine the average molecular kinetic energy and molar kinetic energy of gaseous sulphur hexafluoride, $SF_6$, at 150 °C.

**Strategy:** The problem asks for a calculation of kinetic energies and provides a temperature. We have two equations for the kinetic energy of a gas:

$$\overline{E}_{kinetic} = \frac{3RT}{2N_A} \qquad \text{and} \qquad \overline{E}_{kinetic,molar} = \frac{3}{2}RT$$

To carry out the calculations, we need the constants $R$ and $N_A$ as well as $T$ in kelvins. The constants are found in standard reference tables, but both are used so frequently in chemical calculations that it is wise to memorize them:

$$R = 8.314 \, J \, K^{-1} \, mol^{-1} \text{ and } N_A = 6.022 \times 10^{23} \text{ molecules/mol}$$

**Solution:** Convert the temperature from °C to K by adding 273.15:

$$T = 150 \,°C + 273.15 = 423 \, K$$

Now substitute and calculate:

$$\overline{E}_{kinetic} = \frac{3(8.314 \, J \, K^{-1} \, mol^{-1})(423 \, K)}{2(6.022 \times 10^{23} \text{ molecules/mol})} = 8.76 \times 10^{-21} \text{ J/molecule}$$

$$\overline{E}_{kinetic, molar} = \frac{3}{2}(8.314 \, J \, K^{-1} \, mol^{-1})(423 \, K) = 5.28 \times 10^3 \text{ J/mol}$$

**Do the Results Make Sense?** Does it bother you to find that neither the chemical formula nor the molar mass is needed for these calculations? Remember that not all data are necessarily required for any particular calculation. Because average kinetic energy depends on temperature but not on molar mass, we do not need mass information to do this problem.

**Practice Exercise 2.14**

The atmospheric temperature at altitudes where jetliners fly is approximately −35 °C. Calculate the average and molar kinetic energies of molecular nitrogen at this temperature.

## Ideal Gases

The behaviour of gases suggests that their molecules have little effect on one another. Gases are easy to compress, showing that there is lots of empty space between molecules. Gases also escape easily through any opening, indicating that their molecules are not strongly attracted to one another. An **ideal gas** is defined as one for which both the volume of molecules and the forces among the molecules are so small that they have no effect on the behaviour of the gas.

According to this model, a substance will be a gas when its molecules are small and do not attract one another strongly. As we describe in Chapter 8, small molecules usually have small attractive forces. Table 2-2 lists the substances that exist as gases under normal conditions. In accord with the model, they are mostly small molecules with molar masses less than 50 g/mol.

How does an ideal gas behave? We can answer this question by considering how changes in the other conditions affect the pressure. Each time a molecule strikes a wall, it exerts a force on the wall. During each second, many collisions exert many such forces. Pressure is the sum of all these forces per unit area. In an ideal gas, each molecule is independent of all others. This independence means that the total pressure is the sum of the pressure created by each individual molecule.

To see how pressure depends on the other properties, we consider the effect of changing one property at a time, holding the other properties constant. We analyze what happens to the molecular collisions as each property changes.

### TABLE 2-2 SOME GASEOUS SUBSTANCES*

| Elemental Gases | | | Binary Gases | | |
|---|---|---|---|---|---|
| Substance | Formula | $M$ (g/mol) | Substance | Formula | $M$ (g/mol) |
| Hydrogen | $H_2$ | 2.0 | Methane | $CH_4$ | 16.0 |
| Helium | He | 4.0 | Ammonia | $NH_3$ | 17.0 |
| Neon | Ne | 20.2 | Carbon monoxide | CO | 28.0 |
| Nitrogen | $N_2$ | 28.0 | Ethane | $C_2H_6$ | 30.0 |
| Oxygen | $O_2$ | 32.0 | Nitric oxide | NO | 30.0 |
| Fluorine | $F_2$ | 38.0 | Hydrogen sulphide | $H_2S$ | 34.1 |
| Argon | Ar | 39.9 | Hydrogen chloride | HCl | 36.5 |
| Ozone | $O_3$ | 48.0 | Carbon dioxide | $CO_2$ | 44.0 |
| Chlorine | $Cl_2$ | 71.0 | Nitrogen dioxide | $NO_2$ | 46.0 |

*At 298.15 K, 100.0 kPa.

**Figure 2-12**

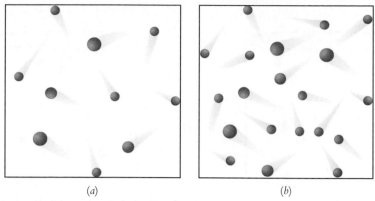

(a)                    (b)

Schematic view of the effect of doubling the number of gas molecules in a fixed volume. Container (b) has twice as many molecules as (a). Consequently, the molecular density is twice as large in (b), with twice as many collisions per second with the walls.

First, consider increasing the amount of the gas while keeping the temperature and volume fixed. Figure 2-12 shows that doubling the amount of gas in a fixed volume doubles the number of collisions with the walls. Thus, pressure is directly proportional to the amount of gas. This agrees with the ideal gas equation.

Next, consider keeping the temperature and amount fixed and changing the volume of the gas. Figure 2-13 shows that compressing a gas into a smaller volume has the same effect as adding more molecules. The result is more collisions with the walls. If the molecules act independently, cutting the volume in half will double the pressure. In other words, pressure is inversely proportional to volume, again in agreement with the ideal gas equation.

To complete our analysis, we must determine the effect of a change in temperature. According to Equation 2-3, kinetic energy is proportional to temperature, and according to Equation 2-2, kinetic energy is proportional to the square of the molecular speed. Thus, the square of the molecular speed is proportional to temperature.

**Figure 2-13**

(a)                    (b)

Volume decreases, molecular density increases

Pressure increases

Schematic view of the effect of compressing a fixed quantity of gas into a smaller volume at constant $T$. Decreasing the volume increases the molecular density, which increases the number of collisions per second with the walls.

**Figure 2-14**

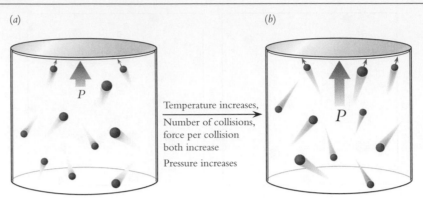

(a)        (b)

Temperature increases,

Number of collisions, force per collision both increase

Pressure increases

Schematic view of the effect of increasing the temperature of a gas. Molecular speeds increase, resulting in more wall collisions and more force per collision, so the pressure increases.

Molecular speed affects pressure in two ways that are illustrated in Figure 2-14. First, faster-moving molecules hit the walls more often than slower-moving molecules. The *number of collisions* each molecule makes with the wall is proportional to the molecule's speed. Second, the *force* exerted when a molecule strikes the wall depends on the molecule's speed. A fast-moving molecule exerts a larger force than the same molecule moving slower. Force per collision increases with speed, and number of collisions increases with speed, so the total effect of a single molecule on the pressure of a gas is proportional to the square of its speed.

Pressure is proportional to the square of molecular speed, which in turn is proportional to temperature. For an ideal gas, then, the pressure is directly proportional to temperature, and a plot of $p$ vs. $T$ yields a straight line. Again, this agrees with the ideal gas equation.

A gas will obey the ideal gas equation whenever it meets the conditions that define the ideal gas. Molecular sizes must be negligible compared with the volume of the container, and the energies generated by forces between molecules must be negligible compared with molecular kinetic energies. The behaviour of any real gas departs somewhat from ideality, because real molecules occupy volume and exert forces on one another. Nevertheless, departures from ideality are small enough to neglect under many circumstances. We consider departures from ideal gas behaviour in Section 2-7.

## SECTION EXERCISES

**●2.5.1** A molecular beam experiment of the type illustrated in Figure 2-9 is performed with an equimolar mixture of He and $CO_2$. Sketch the appearance of a graph of the number of molecules reaching the detector as a function of time.

**●2.5.2** Calculate each of the following energies: (a) total kinetic energy of 1.00 mole of He atoms at −100 °C; (b) total kinetic energy of 1.00 g of $N_2$ molecules at 0 °C; and (c) kinetic energy of a single molecule of $SF_6$ that has twice as much energy as the average molecular energy at 200 °C.

**●2.5.3** Is a gas more likely to behave ideally at high pressure or at low pressure? Explain your answer in terms of the two assumptions made for ideal gases.

## 2.6 ADDITIONAL GAS PROPERTIES

The ideal gas equation and the molecular view of gases lead to several useful applications. We have already described how to carry out calculations involving $p$–$V$–$n$–$T$ relationships. In this section, we examine the use of the gas equation to determine molar masses, gas density, and rates of gas movement.

### Determination of Molar Mass

The ideal gas equation can be combined with the mole–mass relation to find the molar mass of an unknown gas:

$$pV = nRT \text{(ideal gas equation)} \quad \text{and} \quad n = \frac{m}{M}$$

If we know the pressure, volume, and temperature of a gas sample, we can use this information to calculate how many moles are in the sample:

$$n = \frac{pV}{RT}$$

If we also know the mass of the gas sample, we can use that information to determine the molar mass of the gas:

$$M = \frac{m}{n}$$

Example 2-15 illustrates a molar mass calculation using gas data.

### EXAMPLE 2-15 MOLAR MASS DETERMINATION

Acetylene is a high-energy molecule used as a fuel for oxyacetylene welding.

Calcium carbide ($CaC_2$) is a hard, grey-black solid that has a melting point of 2000 °C. This compound reacts strongly with water to produce a gas and a solution containing $OH^-$ ions. A 12.8-g sample of $CaC_2$ was treated with excess water. The resulting gas was collected in an evacuated 5.00-L glass bulb with a mass of 1254.49 g. The filled bulb had a mass of 1259.70 g and a pressure of 98.8 kPa when its temperature was 26.8 °C. Calculate the molar mass and determine the formula of the gas.

**Strategy:** We can use the ideal gas equation to calculate the molar mass. Then, we can use the molar mass to identify the correct molecular formula among a group of possible candidates, knowing that the products must contain the same elements as the reactants. The problem involves a chemical reaction, so we must make a connection between the gas measurements and the chemistry that takes place. Because the reactants and one product are known, we can write a partial equation that describes the chemical reaction:

$$CaC_2(s) + H_2O(l) \longrightarrow Gas + OH^-(aq)$$

In any chemical reaction, atoms must be conserved, so the gas molecules can contain only H, O, C, and/or Ca atoms. To determine the chemical formula of the gas, we must find the combination of these elements that gives the observed molar mass.

**Solution:** Use the gas data to determine the molar mass. The problem gives the following data about the unknown gas:

$$V_{bulb} = V_{gas} = 5.00 \text{ L}, \, T = 26.8 \, °C = 300 \text{ K}, \, p = 98.8 \text{ kPa}$$

$$m_{bulb+gas} = 1259.70 \text{g}, \, m_{bulb} = 1254.49 \text{ g}$$

$$m_{gas} = m_{bulb+gas} - m_{bulb} = 5.21 \text{ g}$$

We use $V$, $T$, $p$, and the ideal gas equation to find the number of moles of gas. Then, with the mass of the gas sample, we can determine the molar mass:

$$\frac{pV}{RT} = n_{gas} = \frac{m_{gas}}{M}$$

We could determine $n$ and then solve for $M$, but we can also solve the equality for $M$ and then substitute the data directly:

$$M = \frac{RTm_{gas}}{pV}$$

$$M = \frac{(8.314 \text{ L kPa K}^{-1} \text{ mol}^{-1})(300.0 \text{ K})(5.21 \text{ g})}{(98.8 \text{ kPa})(5.00 \text{ L})} = 26.3 \text{ g/mol}$$

To identify the gas, we examine the formulas and molar masses of known compounds that contain H, O, C, and Ca:

| Formula | M (g/mol) | Comment |
| --- | --- | --- |
| Ca | 40 | A gas with $M$ = 26.0 g/mol cannot contain Ca |
| CO | 28 | This is close but too high |
| $O_2$ | 32 | This is also too high, and $H_2O_2$ is even higher |
| $H_2O$ | 18 | $H_{10}O$, $M$ = 26.0 g/mol, does not exist |
| $CH_4$ | 16 | $CH_{14}$, $M$ = 26.0 g/mol, does not exist |
| $C_2H_2$ | 26 | This substance has the observed molar mass |

A little trial and error leads to the molecular formula $C_2H_2$, commonly known as acetylene.

**Does the Result Make Sense?** Knowing the formula of the gaseous product and that hydroxide is another product, we can write a balanced equation that generates acetylene and a basic solution:

$$CaC_2(s) + 2 \, H_2O(l) \longrightarrow Ca^{2+}(aq) + 2 \, OH^-(aq) + C_2H_2(g)$$

The balanced equation suggests that the result is reasonable.

**Common Pitfalls:** Using an inconsistent set of units. Use the correct units for $R$ and make sure the temperature is in kelvins.

**Practice Exercise 2.15**
The glass bulb of Example 2-15 is filled with an unknown gas until the pressure is 774 Torr at a temperature of 24.5 °C. The mass of bulb plus contents is 1260.33 g. Determine the molar mass of the unknown gas.

## Gas Density

Whereas liquids' and solids' densities are not strong functions of temperature or pressure, the density of a gas varies strongly with these conditions. To see this, we combine the ideal gas equation and the mole–mass relation and rearrange to obtain an equation for density ($\rho = m/V$):

$$n = \frac{pV}{RT} \quad \text{and} \quad n = \frac{m}{M}$$

Set the two expressions for $n$ equal to each other:

$$\frac{m}{M} = \frac{pV}{RT}$$

Now multiply both sides of the equality by $M$ and divide both sides by $V$:

$$\rho_{gas} = \frac{m}{V} = \frac{pM}{RT}$$

This equation reveals three features of gas density:

1. For any given gas at fixed temperature, the density increases linearly with pressure. The reason is that increasing the pressure compresses the gas into a smaller volume without changing its mass.

2. For any given gas at fixed pressure, the density decreases linearly with temperature. The reason is that increasing the temperature causes the gas to expand without changing its mass.

3. For different gases at the same temperature and pressure, the density increases linearly with molar mass. The reason is that equal numbers of *moles* of different gases occupy equal volumes at a given temperature and pressure.

Hot-air balloons rise because hot air is less dense than cooler air.

There are practical applications of Features 2 and 3. Balloons inflated with helium rise in the atmosphere because the molar mass of helium is substantially lower than that of air. Consequently, the density of a helium-filled balloon is less than the density of air, and the balloon rises, just as a cork released under water rises to the surface. Hot-air balloons exploit Feature 3. When the air beneath a hot-air balloon is heated, its density decreases, becoming smaller than the density of the outside air. With sufficient heating, the balloon rises and floats over the landscape. In contrast, cold air is less dense than warm air, so cold air sinks. For this reason, valleys are often colder than the surrounding hillsides during winter.

When a gas is released into the atmosphere, whether it rises or sinks depends on its molar mass. If the molar mass of the gas is greater than the average molar mass of air, the gas remains near the ground. Carbon dioxide fire extinguishers are effective because of this feature. The molar mass of $CO_2$ is greater than that of $N_2$ or $O_2$, so a $CO_2$ fire extinguisher lays down a blanket of this gas that excludes oxygen from the fire, snuffing it out. Example 2-16 treats another practical example of gas density.

## EXAMPLE 2-16 GAS DENSITY

A hot-air balloon will rise when the density of its air is 15% lower than that of the atmospheric air. Calculate the density of air at 295 K, 100 kPa (assume that dry air is 78% $N_2$ and 22% $O_2$) and determine the minimum temperature of air that will cause a balloon to rise.

**Strategy:** The problem has two parts. First, we must calculate the density of atmospheric air. To do this, we need to determine the molar mass of dry air, which is the weighted average of the molar masses of its components. Then, we must calculate the temperature needed to reduce the density by 15%. For both calculations, we use the ideal gas equation as rearranged to give gas density:

$$\rho_{gas} = \frac{pM}{RT}$$

**Solution:** Begin by calculating the molar mass of dry air. Multiply the fraction of each component by its molar mass:

$$M_{air} = \left(\frac{78\%}{100\%}\right)(28.02 \text{ g/mol}) + \left(\frac{22\%}{100\%}\right)(32.00 \text{ g/mol}) = 28.9 \text{ g/mol}$$

The other conditions are stated in the problem:

$$T = 295 \text{ K}, \, p = 100 \text{ kPa}$$

Substitute into the equation for gas density and calculate:

$$\rho_{gas} = \frac{pM}{RT} = \frac{(100 \text{ kPa})(28.9 \text{ g/mol})}{(8.314 \text{ L kPa mol}^{-1} \text{ K}^{-1})(295 \text{ K})} = 1.18 \text{ g/L}$$

As a final result, we would round to two significant figures, but for purposes of the next calculation we retain the extra significant figure until the calculations are complete.

The balloon will rise when the density of its contents is 15% less than the density of the exterior air:

$$\rho_{gas} = 1.18 \text{ g/L} - \left(\frac{15\%}{100\%}\right)(1.18 \text{ g/L}) = 1.00 \text{ g/L}$$

Rearrange the density equation to solve for temperature, then substitute and calculate:

$$\rho_{gas} = \frac{pM}{RT} \quad \text{so} \quad T = \frac{pM}{R\rho_{gas}}$$

$$T = \frac{(100 \text{ kPa})(28.9 \text{ g/mol})}{(8.314 \text{ L kPa mol}^{-1} \text{ K}^{-1})(1.00 \text{ g/L})} = 348 \text{ K}$$

Because some of the data are stated only to two significant figures, round both results to two significant figures: $\rho_{gas} = 1.0$ g/L and T = 350 K.

**Do the Results Make Sense?** A temperature of 350 K is 77 °C, a reasonable result. Notice that the molar mass and the gas constant appear in both calculations, so we could find the temperature requirement by simple proportions:

$$T_{inside} = T_{outside} \frac{\rho_{outside}}{\rho_{inside}} = \frac{(295 \text{ K})(1.18 \text{ g/L})}{(1.00 \text{ g/L})} = 348 \text{ K, which rounds to 350 K}$$

**Practice Exercise 2.16**

Helium is used for lighter-than-air blimps, whereas argon is used to exclude air from flasks in which air-sensitive chemical syntheses are performed. Calculate the densities of these two noble gases at 295 K and 100 kPa, and explain why the two gases have different uses.

Gas density has a significant effect on the interactions among molecules of a gas. As molecules move about, they collide regularly with one another and with the walls of their container. Figure 2-15 shows that the frequency of collisions depends on the density of the gas. At low density, a molecule may move all the way across a container before it encounters another molecule. At high density, a molecule travels only a short distance before it collides with another molecule. As our Focus 2-1: Tools for Discovery describes, many scientific experiments require gas densities low enough to provide collision-free environments.

**Figure 2-15**

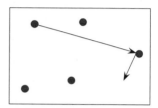

As molecular density decreases, the average distance travelled between molecular collisions increases.

## Rates of Gas Movement

The average speed of a gas molecule depends on the temperature of the gas. To state this dependence quantitatively, we start by applying Equation 2-3 to the average speed and energy:

$$\overline{E}_{\text{kinetic}} = \frac{1}{2} m \overline{v^2}$$

Equation 2-4 provides a second expression for the average kinetic energy:

$$\overline{E}_{\text{kinetic}} = \frac{3RT}{2N_A}$$

We set the two expressions for kinetic energy equal to each other:

$$\frac{1}{2} m \overline{v^2} = \frac{3RT}{2N_A}$$

Now solve this equality for $v$, noting that the product $mN_A$ is the mass, $M$, of one mole of gas molecules:

$$\overline{v} = \left( \frac{3RT}{M} \right)^{1/2} \tag{2-5}$$

This speed is called the **root-mean-square** speed. It is apparent that the average speed of gas molecules is directly proportional to the square root of the temperature and is inversely proportional to the square root of the molar mass.

Equation 2-5 can be applied directly to the movement of molecules escaping from a container into a vacuum. This process is **effusion**, and is shown diagrammatically in Figure 2-16.

A second type of gas movement is **diffusion**, the movement of one type of molecule through molecules of another type. Diffusion is exemplified by air escaping from a punctured tire. The escaping molecules must diffuse among the molecules already present in the atmosphere. Diffusing molecules undergo frequent collisions, so their paths are similar to that shown in Figure 2-4. Nevertheless, their average rate of movement depends on temperature and molar mass according to Equation 2-5. Figure 2-16 shows a molecular-level comparison of effusion and diffusion.

**Figure 2-16**

Effusion

Gas 1　　　　　Vacuum

Diffusion

Gas 1　　　　　Gas 2

Effusion is the movement of a gas into a vacuum. Diffusion is the gradual mixing of two or more gases.

## FOCUS 2-1 TOOLS FOR DISCOVERY: HIGH VACUUM

"Nature abhors a vacuum." True? On Earth, it is certainly difficult to remove the gas from a container, thereby generating and maintaining a vacuum—the absence of gas. In outer space, on the other hand, a vacuum is the rule rather than the exception. Most of the volume of the universe is nearly empty space, close to a perfect vacuum. If a spacecraft sprang a leak, its gaseous atmosphere would quickly escape into that vacuum. Perhaps it would be better to say that conditions on Earth are unfavourable for a vacuum.

Over the years, scientists have spent immense efforts to achieve high vacuums. Powerful and sophisticated pumps are required to remove the gas from a chamber. Once a high vacuum is established, the chamber must be leak-free at the molecular level to keep atmospheric gas from seeping back into an evacuated system. Moreover, the materials used to make a vacuum chamber cannot contain substances that give off gases. A modern high-vacuum instrument is an elaborate array of pumps, valves, seals, and windows, as the photograph at right shows.

Why bother to remove the gases from a small volume of space? High vacuum is used in chemistry and physics research to achieve any of the following conditions: the absence of molecular collisions, the maintenance of an ultra-clean environment, or a simulation of conditions in outer space.

A mass spectrometer generates a beam of ions that is sorted according to specifications of the particular instrument. Usually, the sorting depends on differences in speed, trajectory, and mass. For instance, one type of mass spectrometer measures how long it takes ions to travel from one end of a tube to another. Mass spectrometers require a vacuum to function, and so residual gas must be removed from the tube to eliminate collisions between gas molecules and the ions that are being analyzed. As the diagram shows, collisions with unwanted gas molecules deflect the ions from their paths and change the expected mass spectral pattern.

In a high-vacuum, collision-free tube, a beam of ions is sorted according to ion speed (mass).

If the vacuum is not high enough, residual gas molecules (black circle) collide with the ions, deflecting them from their paths and destroying the sorting process.

A mass spectrometer provides an example of a *molecular beam*, in this case a beam of molecular ions. Molecular beams are used in many studies of fundamental chemical interactions. In a high vacuum, a molecular beam allows chemists to study the reactions that take place through specifically designed types of collisions. For example, a crossed-beam experiment involves the intersection of two molecular beams of two different substances. The types of substances, molecular speeds, and orientations of the beams can be changed systematically to give detailed information about how chemical reactions occur at the molecular level. Chemists have also learned how to create molecular beams in which the molecules have very little energy of motion. These isolated, low-energy molecules are ideal for studies of fundamental molecular properties.

An ultra-clean environment is another major reason for generating high vacuum. At atmospheric pressure, every atom on a solid surface is bombarded with gas molecules at a rate of trillions per second. Even under a reasonably high vacuum, $10^{-9}$ bar, a gas molecule strikes every atom on a solid surface about once per second. If the surface is reactive, these collisions result in chemical reactions that contaminate the surface. The study of pure surfaces of metals or semiconductors requires ultra-high vacuum, with pressures on the order of $10^{-12}$ bar.

High vacuum on a large scale has long played a role in studies of nuclear reactions. To study fundamental nuclear processes, physicists have constructed very long tubes through which they can accelerate beams containing nuclei of various elements. The photo shows one of the longest of these, the high-energy linear accelerator at Stanford University. The interior of the accelerator is a one-mile-long pipe that is an obvious feature in this photo. The pipe must be maintained at high vacuum. Otherwise, the particles in the beam would be deflected by molecular collisions, ruining the experiments. This long pipe has massive vacuum pumps located at regular intervals along its entire length, pumping continually to keep the pressure inside the pipe at an acceptably low level.

An example of diffusion is shown in Figure 2-17, in which aqueous solutions of hydrochloric acid (HCl) and ammonia ($NH_3$) are placed at opposite ends of a glass tube. Molecules of HCl gas and $NH_3$ gas escape from the solutions and diffuse through the air in the tube. When the two gases meet, they undergo an acid–base reaction to make ammonium chloride, a white solid salt:

$$\underset{\text{Strong acid}}{HCl(g)} \; + \; \underset{\text{Weak base}}{NH_3(g)} \; \longrightarrow \; \underset{\text{Whitesolid}}{NH_4Cl(s)}$$

The lighter $NH_3$ molecules diffuse more rapidly than the heavier HCl molecules, so the white band of salt forms closer to the HCl end of the tube, as can be seen in Figure 2-17.

Rates of molecular motion are directly proportional to molecular speeds, so Equation 2-5 predicts that for any gas, rates of effusion or diffusion increase with the square root of the temperature in kelvins. Also, at any particular temperature, effusion and diffusion are faster for molecules with small molar masses. These facts are combined in Graham's law, which compares the rates of effusion or diffusion of two gases under identical conditions:

> The movement of gases through the atmosphere, such as the fragrance of a rose moving from the flower to our noses, generally involves air currents as well as diffusion. Convection, the flow of gas in a current, moves molecules much more rapidly than diffusion.

$$\frac{\text{rate}_1}{\text{rate}_2} = \sqrt{\frac{M_2}{M_1}} \quad \text{(Graham's law)} \tag{2-6}$$

Figure 2-17 shows a practical demonstration of Graham's law.

Uranium naturally contains 0.71% $^{235}U$, the isotope that undergoes fission in nuclear reactors to produce electricity. The balance is almost all $^{238}U$, which does not undergo fission. Unfortunately, this concentration of $^{235}U$ is too low for many reactors in the world and must be increased. This is done by chemically converting the uranium to $UF_6$, which is a gas above 57 °C, and taking advantage of Graham's

**Figure 2-17**

NH$_3$ diffuses through a glass tube faster than HCl. When the two gases meet, they form ammonium chloride (NH$_4$Cl), which appears as a white band closer to the end of the tube that contains HCl. (An aqueous solution of ammonia is also known as ammonium hydroxide.)

law to separate the $^{235}$UF$_6$ from the $^{238}$UF$_6$. Since $^{235}$UF$_6$ has a lower molecular weight than $^{238}$UF$_6$, it diffuses faster:

$$\frac{\text{rate}_{235_{UF_6}}}{\text{rate}_{238_{UF_6}}} = \sqrt{\frac{M_{238_{UF_6}}}{M_{235_{UF_6}}}} = \sqrt{\frac{352.04 \text{ g/mol}}{349.03 \text{ g/mol}}} = 1.0043$$

Thus, the separation of the two isotopes afforded by each diffusion step is very small, and many steps must be employed in series to obtain sufficient enrichment of the $^{235}$U. Large-scale "diffusion plants" increase the $^{235}$UF$_6$ content, and the UF$_6$ is then converted to UO$_2$ for use in power reactors. In Canada, this enrichment step is not necessary—the CANDU reactors used to generate power in Canada operate on UO$_2$ containing the natural abundance of $^{235}$U (0.71%).

## EXAMPLE 2-17  GRAHAM'S LAW: RELATIVE RATES OF DIFFUSION

Refer to Figure 2-17, and calculate how far from the HCl end of the tube the NH$_4$Cl($s$) forms. Assume the distance between the HCl and NH$_3$ is initially 60 cm.

**Strategy:** This is a diffusion problem, so Graham's law should be used. We expect the NH$_4$Cl($s$) to form closer to the compound with the higher molecular weight.

**Solution:** The only data required are the length of the tube (60 cm) and the molecular weights of the two gases:

$$M_{\text{HCl}} = 1.008 + 34.453 = 35.461 \text{ g/mol}$$

$$M_{\text{NH}_3} = 14.0067 + 3(1.0079) = 17.030 \text{ g/mol}$$

According to Graham's law,

$$\frac{rate_{NH_3}}{rate_{HCl}} = \sqrt{\frac{M_{HCl}}{M_{NH_3}}} = \sqrt{\frac{35.461 \text{ g/mol}}{17.030 \text{ g/mol}}} = 1.44$$

Since the $NH_3$ diffuses faster, it must diffuse further than HCl in the same time. Thus, the distance that the $NH_3$ diffuses is 1.44 times that of the HCl. If we let $x$ = distance diffused by the $NH_3$ and $y$ = distance diffused by the HCl, then $x + y = 60$ cm, and $x = 1.44\,y$. Solving these two equations simultaneously, $x = 35.4$ cm, and $y = 24.6$ cm. Thus, the ring of $NH_4Cl(s)$ forms 24.6 cm from the left (HCl) end of the tube.

 **Does the Result Make Sense?** The $NH_4Cl(s)$ forms closer to the HCl end of the tube, since HCl diffuses more slowly than $NH_3$. We can see from Figure 2-17 that this is qualitatively correct.

**Common Pitfalls:** Not inverting the ratio of molecular weights. Notice that the lighter species diffuses faster.

**Practice Exercise 2.17**
Calculate the relative rates of diffusion of the following pairs of gases: (a) CO and $CO_2$; (b) CO and $N_2$; and (c) Xe and Rn.

Graham's law can also be used to calculate the molecular weight of an unknown gas by comparing its rate of effusion to that of a known gas. Example 2-18 shows how to do this.

### EXAMPLE 2-18 GRAHAM'S LAW: MEASURING MOLECULAR WEIGHTS

A sample of krypton gas (Kr) escapes through a small hole in 64.4 s. The same amount of a gas whose identity is unknown escapes through the hole in 31.6 s, under identical conditions. Calculate the molecular weight of the unknown gas.

**Strategy:** The problem involves diffusion rates, so Graham's law should be used. Since we know the diffusion rates of both gases and the molecular weight of one of them, we can calculate the molecular mass of the other.

**Solution:** We have the effusion times of Kr and the unknown gas. The ratio of these is:

$$\frac{\text{effusion time}_{Kr}}{\text{effusion time}_{unknown}} = \frac{64.4 \text{ s}}{31.6 \text{ s}} = 2.037$$

But the rate is the inverse of time, so

$$\frac{\text{effusion rate}_{Kr}}{\text{effusion rate}_{unknown}} = \frac{1}{2.037} = 0.491$$

and so

$$\frac{\text{effusion rate}_{Kr}}{\text{effusion rate}_{unknown}} = 0.491 = \sqrt{\frac{M_{unknown}}{M_{Kr}}}$$

or,

$$M_{unknown} = 0.491^2(M_{Kr}) = 0.491^2(83.8 \text{ g/mol}) = 20.2 \text{ g/mol}$$

(The unknown gas could be Neon, Ne, which has this atomic weight.)

 **Does the Result Make Sense?** The unknown gas is lighter than Kr, and diffuses faster, which makes sense. Also, the unknown gas has a molecular weight which matches that of an actual gas.

**Common Pitfall:** Not realizing that rate is inversely related to diffusion time. Graham's law, as written in Equation 2-6, is used to calculate relative rates of diffusion, not times.

**Practice Exercise 2.18**

A quantity of Neon (Ne) effused through a small hole in 6.00 min. The same quantity of an unknown gas composed only of C and H took 6.82 min to effuse under the same conditions. If the ratio of C to H atoms in this compound is 1 : 1, deduce the chemical formula of the compound.

## SECTION EXERCISES

●**2.6.1** On a summer day, the temperature inside a house rises to 37 °C. In the evening, the house may cool to 21 °C. Does air enter or leave the house as it cools? If the volume of the house is $9.50 \times 10^5$ L and the barometric pressure has held steady at 768 mm Hg, what mass of air is transferred? (The average molar mass of air is 28.8 g/mol.)

●**2.6.2** A 2.96-g sample of a compound of mercury and chlorine is vapourized in a 1.000-L bulb at 307 °C, and the final pressure is found to be 394 Torr. What are the molar mass and chemical formula of the compound?

●**2.6.3** Calculate the average speeds of $H_2O$ and $D_2O$ molecules when $T = 500$ °C. (D represents deuterium, the isotope of hydrogen that contains one neutron.) Use these speeds to calculate the ratio of effusion rates for these two gases from an oven at 500 °C.

●**2.6.4** Calculate how long it will take 1.00 mole of $SO_2$ to effuse through a small hole if it takes 10.0 minutes for 1.00 mole of $SO_3$ to effuse under the same conditions.

**Figure 2-18**

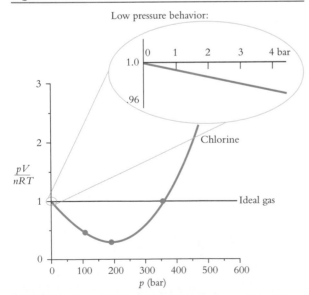

The variation in $pV/nRT$ with pressure shows that chlorine is not an ideal gas, except at very low pressure.

## 2.7 NON-IDEAL (REAL) GASES

There is plenty of evidence that gases behave non-ideally. Figure 2-18 is a plot of the ratio of $pV$ to $nRT$ (called the compressibility factor) versus $p$ for chlorine gas. For an ideal gas, $pV = nRT$, and so $pV/nRT = 1$, no matter what the pressure is. At low pressure, the compressibility factor is very close to 1 for all gases. However, Figure 2-18 shows that as the pressure is increased, the compressibility factor of chlorine initially decreases below 1, and at very high pressure increases to values above 1.

The ideal gas model is based on two assumptions: namely, that the molecules have no volume and that there are no forces among the molecules. When a gas is at low pressure, both of these assumptions are valid—the molecules are far enough apart that any intermolecular forces are negligible, and the volume occupied by the molecules is small in comparison with the volume of the container. However, when the pressure of a gas is increased, by decreasing the volume of the container for instance, two things happen that cause the ideal gas equation to become invalid. First, the molecules become close enough to one another that attractive intermolecular forces (discussed in detail in Chapter 8) cause a reduction in pressure below that which the ideal gas equation would predict. Second, as the volume is decreased, the molecules begin to take up a significant portion of the volume of the container. The volume in which any one molecule can move is therefore less than the volume of the container. This would cause the pressure to be higher than that predicted by the ideal gas equation. Whether the actual pressure is higher or lower than the ideal pressure depends on the relative magnitudes of these two effects.

**Figure 2-19**

Variations in $pV/nRT$ for He, $F_2$, $CH_4$, and $N_2$ at 300 K.

Several gases are compared on one plot of the compressibility factor versus pressure in Figure 2-19. The differences in the shapes of these curves tell us something about the relative molecular sizes and intermolecular forces present in these gases. For instance, the compressibility factor of helium gas is always greater than 1, indicating very small intermolecular forces. The compressibility factor of methane is less than that of nitrogen, indicating larger intermolecular forces in the larger $CH_4$ molecule than in $N_2$.

Johannes van der Waals realized the importance of molecular size and intermolecular forces as early as 1873, and developed an equation to accurately relate $p$, $V$, and $T$, even at high pressures. The van der Waals equation takes into account the size of the gas molecules, as well as the strength of the intermolecular forces:

$$p = \frac{nRT}{V - nb} - a\left(\frac{n^2}{V^2}\right) \tag{2-7}$$

where $b$ is a term proportional to molecular size, and $a$ is proportional to the magnitude of the intermolecular forces. For larger molecules, $b$ is larger, and the van der Waals equation predicts a larger pressure. For molecules with larger intermolecular forces, $a$ is larger, and the predicted pressure is smaller. Note that Equation 2-7 reduces to $p = nRT/V$ if $a$ and $b$ are zero; that is, if the gas behaves ideally.

Table 2-3 lists the van der Waals $a$ and $b$ terms for a number of common gases. Note that in general, as the size of the molecule increases, the $b$ term increases. (Compare the values for He and $Cl_2$, for instance.) Furthermore, note that the $a$ term also increases with the size of the molecule, since larger molecules experience larger intermolecular forces. There are other factors that determine these intermolecular forces, however, and these are discussed in more detail in Chapter 8.

## TABLE 2-3  VAN DER WAALS EQUATION CONSTANTS

| Gas | $a$ (kPa L$^2$ mol$^{-2}$) | $b$ (L/mol) | Gas | $a$ (kPa L$^2$ mol$^{-2}$) | $b$(L/mol) |
|---|---|---|---|---|---|
| He | 3.46 | 0.0237 | $Cl_2$ | 657.9 | 0.0562 |
| Ne | 21.4 | 0.0171 | $CO_2$ | 364.0 | 0.0427 |
| Ar | 136.3 | 0.0322 | $H_2O$ | 553.6 | 0.0305 |
| $H_2$ | 24.7 | 0.0266 | $NH_3$ | 422.5 | 0.0371 |
| $N_2$ | 140.8 | 0.0391 | $CH_4$ | 228.3 | 0.0428 |
| $O_2$ | 137.8 | 0.0318 | $C_2H_6$ | 556.2 | 0.0638 |
| CO | 147.2 | 0.0395 | $C_6H_6$ | 1824 | 0.0115 |
| $F_2$ | 117.1 | 0.0290 | HCl | 370.0 | 0.0406 |

## EXAMPLE 2-19  REAL GAS PRESSURES

Gases such as methane are supplied in steel cylinders. A 15.0-L cylinder contains 62.0 mol $CH_4$ when full, and 0.620 mol $CH_4$ after prolonged use. Compare the ideal and real (van der Waals) pressures before and after use if the cylinder is at 27 °C.

**Strategy:** For the ideal gas pressures, simply rearrange the ideal gas equation to solve for $p$. For the real pressure, look up the $a$ and $b$ terms and use the van der Waals equation.

**Solution:** The van der Waals constants for methane are $a$ = 228.2 kPa L$^2$ mol$^{-2}$ and $b$ = 0.0428 L/mol. $T$ = 27°C = 300 K. When the tank is full,

$$p_{real} = \frac{nRT}{V - nb} - a\left(\frac{n}{V}\right)^2$$

$$= \frac{62.0 \text{ mol}(8.314 \text{ L kPa K}^{-1} \text{ mol}^{-1})(300 \text{ K})}{15.0 \text{ L} - 62.0 \text{ mol}(0.0428 \text{ L/mol})} - 228.2 \text{ kPa L}^2\text{mol}^{-2}\left(\frac{62.0 \text{ mol}}{15.0 \text{ L}}\right)^2$$

$$= 12\,520 \text{ kPa} - 3900 \text{ kPa} = 8620 \text{ kPa}$$

$$p_{ideal} = \frac{nRT}{V} = \frac{62.0 \text{ mol}(8.314 \text{ L kPa K}^{-1} \text{ mol}^{-1})(300 \text{ K})}{15.0 \text{ L}} = 10\,300 \text{ kPa}$$

After use,

$$p_{real} = \frac{nRT}{V - nb} - a\left(\frac{n}{V}\right)^2$$

$$= \frac{0.620 \text{ mol}(8.314 \text{ L kPa K}^{-1} \text{ mol}^{-1})(300 \text{ K})}{15.0 \text{ L} - 0.620 \text{ mol}(0.0428 \text{ L/mol})} - 228.2 \text{ kPa L}^2 \text{ mol}^{-2}\left(\frac{0.620 \text{ mol}}{15.0 \text{ L}}\right)^2$$

$$= 103.3 \text{ kPa} - 0.4 \text{ kPa} = 103 \text{ kPa}$$

$$p_{ideal} = \frac{nRT}{V} = \frac{0.620 \text{ mol}(8.314 \text{ L kPa K}^{-1} \text{ mol}^{-1})(300 \text{ K})}{15.0 \text{ L}} = 103 \text{ kPa}$$

**Do the Results Make Sense?** Before use, the two calculated pressures are quite different, reflecting the failure of the ideal gas equation at such a high pressure. After use, when the pressure is low, either gas equation will accurately predict the actual pressure.

**Common Pitfalls:** Making mathematical errors. Break the van der Waals equation into two parts as above and calculate each separately. The second term should be smaller than the first.

**Practice Exercise 2.19**

Repeat the above example using helium in place of methane. Which gas, $CH_4$ or He, shows the greatest deviation from ideal gas behaviour? Why?

**SECTION EXERCISES**

● **2.7.1** Why does helium exert a greater pressure than the same amount of ammonia at the same temperature and volume?

● **2.7.2** Calculate the pressure exerted by 1 kg of $F_{2(g)}$ at 100 °C in a 200-L container. Use both the ideal gas equation and the van der Waals equation and comment on why the two calculated pressures are different.

● **2.7.3** By looking at the data in Table 2-3, predict which of $He_{(g)}$ or the same amount of $Ne_{(g)}$ should exert the higher pressure at the same temperature and volume.

## 2.8 CHEMISTRY OF THE EARTH'S ATMOSPHERE

The blanket of air that cloaks our planet behaves as an ideal gas, but the atmosphere is bound to the Earth by gravitational attraction, not by confining walls. The pressure exerted by the atmosphere can be thought of as the pressure of a column of air, just as the pressure exerted by mercury in a barometer is the pressure of the column of mercury. The higher we rise into the atmosphere, the less air there is above us. Less air above us means that the pressure exerted by the column of air is lower. Lower pressure, in turn, means lower molecular density, as indicated by the ideal gas equation:

$$\frac{n}{V} = \frac{p}{RT}$$

Figure 2-20 shows a molecular profile of a column of atmospheric air.

Pressure also varies with atmospheric conditions. For example, molecules in the atmosphere move faster when they are heated by the Sun. This in turn causes an increase in pressure that is described by the ideal gas equation:

$$p = \frac{nRT}{V}$$

Molecules in high-pressure areas flow into regions of lower pressure, generating wind. This continual interplay of temperature and pressure plays a major role in determining our planet's weather.

### Composition of the Lower Atmosphere

The atmosphere is a complex, dynamic mixture of gases. The composition and chemical reactivity patterns of the atmosphere change significantly as altitude above the surface of the Earth increases, as we explore in Chapter 4. Here, we focus on the layer closest to the surface, called the **troposphere**. Nitrogen and oxygen make up more than 99% of the troposphere. Only three other substances, $H_2O$, Ar, and $CO_2$, are present in amounts greater than 0.01%. Of these five principal constituents, all of them but water are present in nearly constant amounts. Because water can condense and evaporate readily, the water content of the atmosphere changes from day to day, depending on temperature and geography. The composition of dry air at sea level, expressed in mole fractions, is shown in Table 2-4.

**Figure 2-20**

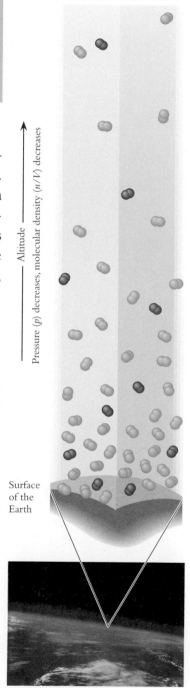

Altitude ⟶

Pressure ($p$) decreases, molecular density ($n/V$) decreases

Surface of the Earth

Molecular profile of the Earth's atmosphere, showing a column above some point on the Earth's surface. As altitude increases, both pressure ($p$) and molecular density ($n/V$) decrease.

## TABLE 2-4 COMPOSITION OF DRY AIR AT SEA LEVEL*

| Major Components | Mole Fraction |
|---|---|
| $N_2$ | 0.7808 |
| $O_2$ | 0.2095 |
| Ar | $9.34 \times 10^{-3}$ |
| $CO_2$ | $3.85 \times 10^{-4}$ |

| Trace Components | Abundance (ppm) |
|---|---|
| Ne | 18.2 |
| He | 5.24 |
| $CH_4$ | 1.4 |
| Kr | 1.14 |
| $H_2$ | 0.5 |
| NO | 0.5 |
| $N_2O$ | 0.25 |

*Note: Figures are for water-free air. The $H_2O$ content of air varies between $X = 0$ and $X = 0.04$.

## Vapour Pressure

After a rainfall, puddles of water slowly disappear. The higher the temperature, the faster the puddles vanish. Puddles disappear because water molecules move from the liquid phase to the gas phase through **evaporation**. Evaporation is common to all substances in condensed phases, not just to water. We use the term **vapour** to describe a gaseous substance that forms by evaporation. Evaporation from bodies of water guarantees that the Earth's atmosphere always contains water vapour.

Evaporation can be understood by examining the kinetic energies of molecules. All molecules—solid, liquid, and gas—have some kinetic energy of motion, and any sample of molecules has a distribution of kinetic energies determined by its temperature (see Figure 2-11). Attractive forces between molecules confine liquids and solids to a fixed volume, but some molecules at the surface of that volume have enough kinetic energy to overcome the forces holding them in the condensed phase. These molecules evaporate from a liquid or solid and move into the vapour phase. The rate of evaporation increases with temperature, because the fraction of molecules with sufficient energy to escape into the vapour phase increases as the sample warms up.

The water vapour that evaporates from a puddle moves away into the atmosphere. This process continues until the puddle has vapourized completely. The situation is quite different when a substance is confined to an enclosed space. If the vapourized molecules cannot escape, their numbers increase, and they exert pressure on the walls of the closed container. As evaporation proceeds, more and more molecules enter the vapour phase, and the partial pressure of the vapour rises accordingly. Figure 2-21 shows that as this happens, some gas molecules are recaptured when they collide with the surface of the liquid. Eventually, the partial pressure of the substance in the vapour phase reaches a level at which the number of molecules being recaptured (condensing) during any time interval equals the number of molecules escaping (evaporating) from the surface during that same time interval.

When the rate of evaporation equals the rate of condensation, the system is in a state of **dynamic equilibrium**. The system is dynamic because molecular transfers

**Figure 2-21**

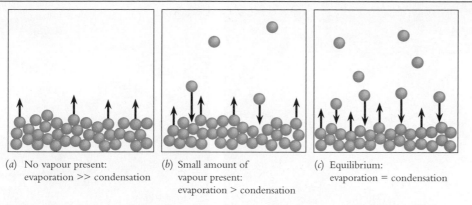

(a) No vapour present:
evaporation >> condensation

(b) Small amount of
vapour present:
evaporation > condensation

(c) Equilibrium:
evaporation = condensation

Schematic view of evaporation and condensation in a closed system. (a) Initial conditions.
(b) Intermediate conditions. (c) Equilibrium conditions: partial pressure = vapour pressure.

continue, and it has reached equilibrium because no further net change occurs. The pressure of the vapour at dynamic equilibrium is called the **vapour pressure ($p_{vap}$)** of the substance. The vapour pressure of any substance increases rapidly with temperature because the kinetic energies of the molecules increase as the temperature rises. Table 2-5 lists the vapour pressures for water at various temperatures. We describe intermolecular forces and vapour pressure in more detail in Chapter 8.

A vapour pressure is the pressure exerted by a gas in equilibrium with its condensed phase. When this equilibrium has been reached, the gas is saturated with that particular vapour. Notice in Table 2-5 that at 25 °C the atmosphere is saturated with water vapour when the partial pressure of $H_2O$ is 23.756 Torr. At this pressure, the molecular density of $H_2O$ in the gas phase is sufficient to make the rate of condensation equal to the rate of evaporation. Addition of more water molecules to the gas phase results in condensation to hold the partial pressure of $H_2O$ fixed at 23.756 Torr.

In most cases, the atmosphere contains less water vapour than the maximum amount it can hold; that is, $p_{H_2O} < p_{vap,H_2O}$. The amount of water vapour actually present in the atmosphere is described by the percent relative humidity, which is the partial pressure of water present in the atmosphere, divided by the vapour pressure of water at that temperature multiplied by 100:

$$\text{Relative humidity} = (100\%)\left(\frac{p_{H_2O}}{p_{vap,H_2O}}\right) \qquad (2\text{–}8)$$

## TABLE 2-5  VAPOUR PRESSURES ($p_{vap}$) OF WATER AT VARIOUS TEMPERATURES ($T$)

| $T$ (°C) | $p_{vap}$ (Torr) | $T$ (°C) | $p_{vap}$ (Torr) | $T$ (°C) | $p_{vap}$ (Torr) |
|---|---|---|---|---|---|
| 0 | 4.579 | 35 | 42.175 | 70 | 233.7 |
| 5 | 6.543 | 40 | 55.324 | 75 | 289.1 |
| 10 | 9.209 | 45 | 71.88 | 80 | 355.1 |
| 15 | 12.788 | 50 | 92.51 | 85 | 433.6 |
| 20 | 17.535 | 55 | 118.04 | 90 | 525.76 |
| 25 | 23.756 | 60 | 149.38 | 95 | 633.9 |
| 30 | 31.824 | 65 | 187.54 | 100 | 760 |

Dew and fog form when water vapour condenses from the atmosphere.

Relative humidity thus describes how close the atmosphere is to being saturated with water vapour.

Because the vapour pressure of water varies with temperature, a given amount of water in the atmosphere represents a higher relative humidity as the temperature falls. For example, a partial pressure of $H_2O$ of 6.54 Torr at 15 °C corresponds to a relative humidity of about 50% ($p_{vap}$ = 12.788 Torr, from Table 2-5):

$$\text{Relative humidity} = \left( \frac{6.54 \text{ Torr}}{12.788 \text{ Torr}} \right)(100) = 51.1\%$$

At 5 °C, however, this same partial pressure is 100% relative humidity.

The formation of dew and fog are consequences of this variation in relative humidity. Warm air at high relative humidity may cool below the temperature at which its partial pressure of $H_2O$ equals the vapour pressure. When air temperature falls below this temperature, called the *dew point*, some $H_2O$ must condense from the atmosphere. Example 2-20 shows how to work with vapour pressure variations with temperature, and the Focus 2-2: Chemistry and the Environment explores how variations in other trace gases affect climate.

### EXAMPLE 2-20  WATER VAPOUR PRESSURE

Fog forms when humid warm air from above a body of water moves inland and cools. What is the highest temperature at which fog could form from air that is at 65% relative humidity when its temperature is 27.5 °C?

**Strategy:** This problem asks about the partial pressure of water vapour in the atmosphere. Fog forms when that partial pressure exceeds the vapour pressure. Partial pressures are not given among the data, but relative humidity describes how close the partial pressure of water vapour is to its vapour pressure at the given temperature. Because vapour pressure varies strongly with temperature, we must use the information given in Table 2-5.

**Solution:** To convert relative humidity at 27.5 °C into a partial pressure, we need the vapour pressure of water at that temperature. Table 2-5 lists vapour pressures in 5 °C increments, but we can interpolate to find the correct value. Because 27.5 °C is halfway between 25 °C and 30 °C, the vapour pressure is **approximately** halfway between its values at 25 °C and 30 °C:

| $T$ (°C) | 25 | 30 | 27.5 |
|---|---|---|---|
| $p_{vap, H_2O}$ (Torr) | 23.756 | 31.824 | $\frac{1}{2}$ (23.756 + 31.824) = 27.790 |

Next, we find the partial pressure from a rearranged version of Equation 2-8:

$$p_{H_2O} = \frac{(\text{Relative humidity})(p_{vap,H_2O})}{100\%} = \frac{65\%(27.790 \text{ Torr})}{100\%} = 18 \text{ Torr}$$

This partial pressure remains constant as the temperature drops, meaning that fog can form below the temperature at which the vapour pressure of water is 18 Torr. Examining Table 2-5, we find that the first listed temperature at which $p_{vap,H_2O} < 18$ Torr is $T = 20$ °C, where $p_{vap,H_2O} = 17.535$ Torr. Thus, under these conditions, fog can form if the temperature drops below 20 °C.

 **Does the Result Make Sense?** If you live near a large body of water, you probably have encountered fog at temperatures similar to this.

**Practice Exercise 2.20**

If the relative percent humidity is 65% at 20 °C, at what temperature will dew form?

## Chemistry in the Troposphere

Every year, humanity releases millions of tons of gaseous and particulate pollutants into the Earth's atmosphere. The main source of atmospheric pollution is the burning of fossil fuels. For example, diesel fuel and gasoline provide energy for automobiles, trucks, and buses, but burning these mixtures also produces vast quantities of $CO_2$ and lesser amounts of CO. Carbon monoxide is highly toxic, and carbon dioxide is a principal agent in global warming, as we discuss in Chapter 4. In addition, internal combustion engines are rather inefficient, so significant amounts of unburned hydrocarbons are emitted in exhaust gases. Combustion also generates nitrogen oxides and sulphur oxides, both of which undergo reactions in the atmosphere that contribute to pollution, including acid rain.

## Oxides of Nitrogen

Normally, $N_2$ is a stable, unreactive molecule. However, under the extreme conditions found in an automobile engine cylinder, nitrogen molecules may react with oxygen molecules to produce nitrogen oxide:

$$N_2 + O_2 \xrightarrow{\text{High temperature}} 2\,NO$$

Nitrogen oxide reacts in the atmosphere with $O_2$ to form nitrogen dioxide:

$$2\,NO + O_2 \longrightarrow 2\,NO_2$$

$NO_2$ is a red-brown gas that can be seen in the atmosphere over many large cities, where concentrations may reach 0.9 ppm. Tolerable limits for this toxic, irritating gas are 3–5 ppm.

Nitrogen dioxide absorbs energy from sunlight and decomposes into NO molecules and oxygen atoms. Remember that oxygen is one of the elements that exist as a diatomic molecule in its natural state, so oxygen *atoms* are very reactive, since they contain unpaired electrons. Molecular nitrogen is too stable to react with oxygen atoms, but molecules of oxygen react with oxygen atoms to produce ozone ($O_3$), a toxic and highly reactive molecule:

Ozone irritates respiratory tissue to such an extent that the World Health Organization recommends a maximum concentration of 0.05 ppm for people breathing the air for an eight hour period.

## FOCUS 2-2 CHEMISTRY AND THE ENVIRONMENT: DOES HUMAN ACTIVITY CHANGE THE WEATHER?

In 1997–98, meteorologists predicted successfully that an observed increase in temperature of Pacific Ocean waters, known as El Niño, would lead to unusual and devastating global weather patterns. As predicted, rain and snowfall in California were more than double the normal values, and spring tornadoes in the South and Midwest were more devastating than usual. Weather elsewhere was also significantly affected. A relatively small change in the temperatur of one portion of the Pacific Ocean resulted in substantial changes in global weather patterns. The El Niño phenomenon highlights the fact that the Earth's weather patterns can be changed by variations in the environment.

Can human activities modify the environment in ways that cause changes in weather patterns? Growing evidence suggests that they can. The activities of modern society appear to be changing the average temperature of the atmosphere. That temperature is determined by the balance between the amount of sunlight reaching the Earth's surface and the amount of heat radiated back into space. Water, $CO_2$, and $CH_4$ in the atmosphere act as "greenhouse" gases that reduce the planet's heat loss. Increased amounts of these trace components thus are expected to increase the average temperature of our world.

Atmospheric concentrations of both $CO_2$ and $CH_4$ are increasing. The graph shows data for $CO_2$, the amount of which has increased by 30% as humans combust more and more fossil fuels. The graph also shows that the global mean temperature has risen by nearly 1 °C.

The three main variables that most influence the weather are air temperature, pressure, and water content. Their intricate interplay, energized by sunlight, creates weather patterns. Given this complexity, how can we determine if the increase in global temperature is due to the higher levels of greenhouse gases, and how can we predict the effects of further increases? One approach is computer modeling to simulate the atmosphere and project how the weather might change as a result of changes in conditions. These models predict that if the current output of greenhouse gases continues, average global temperature will rise by at least 1 °C over the next 50 to 100 years. The models also predict the consequences of global warming, including increases in extreme weather events such as heat waves and tornadoes.

A rise in average temperature of 1 °C may not seem enough to cause dramatic shifts in the weather, but the historical record shows otherwise. The period 1500–1850 is called the Little Ice Age, because there were extensive increases in the sizes of the glaciers in all alpine regions. During that period, the average global temperature was just 0.5 °C lower than in 1900.

Although current trends in climate lead experts to predict further global warming, natural events could change the balance. The geological record shows a series of Ice Ages alternating with eons of global warming. The cataclysmic eruption of the Indonesian volcano Krakatau in 1883 spewed huge quantities of volcanic dust and ash into the atmosphere. Enough global cooling was caused such that the following year was known in Europe, half a globe away, as "the year without a summer."

In the light of such uncertainties, calls to curb the pace of greenhouse emissions are controversial. Proponents point to the potentially devastating consequences of predicted global warming. Opponents cite the high current costs of reducing the use of fossil fuels, coupled with the uncertainties of how natural events affect the weather. Meanwhile, climate experts continue to collect data on current weather trends and refine their computer models, hoping to make more definitive projections about what lies ahead for our planet's weather.

Both ozone and oxygen atoms react with unburned hydrocarbons to produce many compounds that are harmful to the respiratory system. The mixture of all these pollutants is sometimes called **photochemical smog**. To control photochemical smog, governments have set limits on the levels of NO and unburned hydrocarbons allowed in automobile exhaust. Today, chemists and engineers are researching cleaner-burning fuels, more efficient engines, and fuel cells as alternatives to combustion engines.

Catalytic converters convert much of the NO and $NO_2$ from exhaust gases into $N_2$ and $O_2$ before they are released into the atmosphere. These have helped alleviate pollution from nitrogen oxides at only a small additional cost. We could reduce pollution emissions even further, but consumers and manufacturers are reluctant to pay the higher costs required to develop and produce cleaner fuels and engines.

## Oxides of Sulphur

About half the electricity produced in the United States is generated by burning coal. Like petroleum, coal was formed millions of years ago from decaying plants under conditions of high temperature and pressure. Coal is a network of carbon atoms that also contains small amounts of other elements, including hydrogen, oxygen, nitrogen, mercury, and sulphur. When coal burns, its sulphur combines with $O_2$ to produce sulphur dioxide:

$$S \; (s, \text{from coal}) \; + \; O_2 \, (g) \; \longrightarrow \; SO_2 \, (g)$$

In the presence of dust particles or ultraviolet (UV) light, atmospheric $SO_2$ reacts further with $O_2$ to form $SO_3$

The combustion products from sulphur impurities in coal are particularly damaging to the environment. In humans, prolonged exposure to sulphur dioxide diminishes lung capacity and aggravates respiratory problems such as asthma, bronchitis, and emphysema. Concentrations as low as 0.15 ppm can incapacitate persons with these diseases, and at about 5 ppm everyone experiences breathing difficulties. In 1952, a particularly serious episode of $SO_2$ pollution in London caused approximately 4000 deaths over several days.

One particularly troubling aspect of $SO_2$ and $SO_3$ pollution is acid rain, which occurs when these gases combine with water to produce acid mists:

$$SO_2 \, (g) \; + \; H_2O \, (g) \; \longrightarrow \; H_2SO_3 \, (\textit{mist})$$

$$SO_3 \, (g) \; + \; H_2O \, (g) \; \longrightarrow \; H_2SO_4 \, (\textit{mist})$$

Raindrops passing through such mists become acidic, increasing the acidity of rainwater as much as 1000-fold. Acid rain is common in the heavily industrialized areas of the United States, Canada, Europe, and Asia. In areas where acid rain is particularly severe, some lakes are so acidic that fish can no longer survive. In addition to the devastating effect on aquatic life, acid rain leaches nutrients from the soil, alters the metabolism of organisms in the soil, accelerates the corrosion of metals, and damages important building materials such as limestone and marble.

Sulphur dioxide can be removed from power plant exhaust gas by a scrubber system. One common method involves the reaction of $SO_2$ with calcium oxide (lime) to form calcium sulphite:

$$SO_2 \, (g) \; + \; CaO \, (s) \; \longrightarrow \; CaSO_3 \, (s)$$

Unfortunately, scrubber systems are expensive to operate, and the solid $CaSO_3$ is generated in large enough quantities to create significant disposal problems.

Sulphur pollution may become increasingly significant in the United States, hence in eastern Canada, if increasing amounts of energy production come from coal-fired power plants. Because different deposits of coal have different sulphur concentrations, the amount of pollution will depend on the type of coal used. The oldest coal, anthracite, is 90% carbon and has a sulphur content lower than 1%, whereas bituminous coal can contain as much as 5% sulphur by mass. Unfortunately, vast North American coal reserves are mostly high-sulphur bituminous coal. Thus, increased dependence on coal as a source of energy will come with a high cost to society. The choices are to find other energy sources; to spend more money to develop efficient scrubber systems; to reduce consumption through conservation; or to face potentially disastrous consequences for our environment.

Statues are slowly decomposed by acid rain.

## SECTION EXERCISES

●**2.8.1** What are the mole fractions of the three most abundant trace atmospheric constituents listed in Table 2-4?

●**2.8.2** Argon gas can be recovered from the atmosphere by appropriate cooling and liquefaction processes. How many litres of dry air at 1.00 bar and 27 °C must be processed to collect 10.0 mol of Ar?

●**2.8.3** A weather report gives the current temperature as 18 °C and sets the dew point at 10 °C. Using data from Table 2-5, determine the partial pressure of water vapour in the atmosphere and calculate the relative humidity.

# CHAPTER 2 VISUAL SUMMARY

## 2.1 PRESSURE

Standard atmosphere (atm)
Torr
Pascal (Pa)
Bar

760 mm

$P_{atm}$

$p_{gas}$

Barometer

Manometer

**LEARNING OBJECTIVES:** Understanding gas pressure and expressing pressure in various units

**SKILLS TO MASTER:** Converting among pressure units. (p. 71)

## 2.2 DESCRIBING GASES

$p = 0.33$ atm
$p = 0.75$ atm
$p = 2$ atm

The gas constant, $R$

Ideal gas equation     $pV = nRT$

*The only equation needed for solving quantitative problems about gases*

$$\frac{p_i V_i}{n_i T_i} = \frac{p_f V_f}{n_f T_f}$$

**LEARNING OBJECTIVES:** Relating pressure, volume, temperature, and amount of gas.

**SKILLS TO MASTER:** Using the ideal gas equation. (p. 73)
   Relating final conditions to initial conditions. (p. 76)

**KEY CONCEPTS:** Solving quantitative problems about gases at moderate temperatures and pressures requires only one equation, the ideal gas equation.

## 2.3 GAS MIXTURES

Dalton's law of partial pressures

$$p_{total} = p_1 + p_2 + p_3 \cdots$$

Partial pressure

Mole fraction ($x$)
Parts/million (ppm)
Parts/billion (ppb)

$$p_A = x_A p_{total}$$

**LEARNING OBJECTIVES:** Using the concept of partial pressures in gas mixtures.

**SKILLS TO MASTER:** Determining partial pressures. (p. 80)

**KEY CONCEPTS:** In a mixture of gases, each gas contributes to the total pressure the amount that it would exert if the gas were present in the container by itself.

## 2.4 GAS STOICHIOMETRY

|  | $N_2$ | + | $3 H_2$ | → | $2 NH_3$ |
|---|---|---|---|---|---|
| Initial $p$ (bar) | 1 | | 4 | | 0 |
| Change in $p$ (bar) | −1 | | −3 | | +2 |
| Final $p$ (bar) | 0 | | 1 | | 2 |

**Summary of Mole Relationships**

| Substance | Equation |
|---|---|
| Liquid or solid | $n = \dfrac{m}{M}$ |
| Aqueous solution | $n = cV$ |
| Gas | $n = \dfrac{pV}{RT}$ |

**LEARNING OBJECTIVES:** Using stoichiometry to solve problems involving gas-phase chemical reactions.

**SKILLS TO MASTER:** Using the ideal gas equation in gas stoichiometry problems. (p. 84)

## 2.5 MOLECULAR VIEW OF GASES

$$E_{kinetic} = \frac{1}{2}mv^2 \qquad \overline{E}_{kinetic} = \frac{3RT}{2N_A}$$

At a given T, all gases have the same kinetic energy distribution.

| Ideal gas | Negligible molecular sizes Negligible intermolecular forces |

**LEARNING OBJECTIVES:** Explaining the basic concepts of kinetic molecular theory: molecular speed, energy, and the effects of temperature and volume on gas pressure.

**KEY CONCEPTS:** At a given temperature, all gases have the same molecular kinetic energy distribution.

The volume occupied by the molecules of an ideal gas is negligible compared with the volume of its container. The energies generated by forces among ideal gas molecules are negligible compared with molecular kinetic energies.

## 2.6 ADDITIONAL GAS PROPERTIES

$$\overline{u} = \left(\frac{3RT}{M}\right)^{\frac{1}{2}}$$

Root-mean-square speed

**LEARNING OBJECTIVES:** Calculating gas densities and molar masses from pressure–volume–temperature data.

**SKILLS TO MASTER:** Using gas properties to determine molar mass. (p. 97)

Using the ideal gas equation to calculate gas densities. (p. 99)

Using Graham's law for effusion and diffusion problems. (p. 103)

## 2.7 NON-IDEAL (REAL) GASES

$$p = \frac{nRT}{V - nb} - a\left(\frac{n^2}{V^2}\right)$$

**LEARNING OBJECTIVES:** Calculating the pressure of a gas under non-ideal conditions, and explaining the deviations from ideality.

**SKILLS TO MASTER:** Working with the van der Waals equation. (p. 107)

## 2.8 CHEMISTRY OF THE EARTH'S ATMOSPHERE

$$H_2O\ (l) \rightleftharpoons H_2O\ (g)$$

Dynamic equilibrium

Troposphere

NO     SO₂
NO₂  O₃  SO₃

Photochemical smog

$$\text{Humidity} = (100\%)\left(\frac{p_{H_2O}}{p_{vap,H_2O}}\right)$$

Vapour
Vapour pressure ($p_{vap}$)
Evaporation

**LEARNING OBJECTIVES:** Doing calculations involving water vapour pressure and relative humidity and describing some of the basic chemistry of the troposphere.

## LEARNING EXERCISES

**2.1** List the macroscopic properties and the microscopic properties of gases.

**2.2** Update and reorganize your list of memory bank equations so that all equations that involve moles are grouped together.

**2.3** Summarize the strategy for calculating final pressures of gases in a limiting reactant problem.

**2.4** Define partial pressure, vapour pressure, and relative humidity. Explain how they are related.

**2.5** List all terms new to you that appear in Chapter 2. Write a one sentence definition of each in your own words. Consult the glossary if you need help.

## PROBLEMS

**ilw** = interactive learning ware problem. Visit the website at www.wiley.com/canada/olmsted

### Pressure

**2.1** Describe what would happen to the barometer in Figure 2-2 if the tube holding the mercury had a pinhole at its top.

**2.2** Redraw the U-tube shown in Figure 2-3 to show how the manometer appears if the pressure on the side exposed to the atmosphere is greater than the pressure on the side exposed to the gas by an amount $2\Delta h$.

**2.3** Describe how the difference between an inflated and a flat automobile tire shows that a gas exerts pressure.

**2.4** A sailboat moves across the water making use of the wind. How does the motion of a sailboat demonstrate that gas molecules exert pressure?

**2.5** Express the following in units of pascal and bar: (a) 455 Torr; (b) 2.45 atm; (c) 0.46 Torr; and (d) $1.33 \times 10^{-3}$ atm.

**2.6** Convert the following to Torr: (a) 1.00 Pa; (b) 125.6 bar; (c) 75.0 atm; and (d) $4.55 \times 10^{-10}$ atm.

### Describing Gases

**2.7** A sample of air was compressed to a volume of 20.0 L. The temperature was 298 K and the pressure was 5.00 bar. How many moles of gas were in the sample? If the sample had been collected from air at $p = 1.00$ bar, $T = 298$ K, what was the original volume of the gas?

**2.8** A bicycle pump inflates a tire of volume 565 mL until the internal pressure is 6.47 bar at a temperature of 21.7 °C. How many moles of air does the tire contain? What volume of air at 1.01 bar and 21.7 °C did the pump transfer?

**2.9** Rearrange the ideal gas equation to give the following expressions: (a) an equation that relates $p_i$, $T_i$, $p_f$, and $T_f$ when $n$ and $V$ are constant; (b) $V = ?$; and (c) an equation that relates $p_i$, $V_i$, $p_f$, and $V_f$ when $n$ and $T$ are constant.

**2.10** Rearrange the ideal gas equation to give the following expressions: (a) $n = ?$; (b) an equation that relates $V_i$, $n_i$, $V_f$, and $n_f$ when $p$ and $T$ are constant; and (c) $n/V = ?$.

**2.11** **ilw** It requires 0.255 L of air to fill a metal foil balloon to 1.000 bar pressure at 25 °C. The balloon is tied off and placed in a freezer at −15 °C. What is the new volume of air in the balloon?

**2.12** A pressurized can of whipping cream has an internal pressure of 1.075 bar at 25 °C. If it is placed in a freezer at −15 °C, what is the new value for its internal pressure?

**2.13** Under which of the following conditions could you use the equation $p_f V_f = p_i V_i$? (a) A gas is compressed at constant $T$. (b) A gas-phase chemical reaction occurs. (c) A container of gas is heated. (d) A container of liquid is compressed at constant $T$.

**2.14** Under which of the following conditions could you *not* use the equation $p_f V_f / T_f = p_i V_i / T_i$? (a) $p$ is expressed in Torr; (b) $T$ is expressed in °C; (c) $V$ is changing; and (d) $n$ is changing.

### Gas Mixtures

**2.15** **ilw** The amount of $NO_2$ in a smoggy atmosphere was measured to be 0.78 ppm. The barometric pressure was 758.4 Torr. Compute the partial pressure of $NO_2$ in atmospheres.

**2.16** California's automobile emission standards require that exhaust gases contain less than 220 ppm hydrocarbons and less than 1.2% CO (both of these values are in moles per mole of air). At standard atmospheric pressure, what are the partial pressures, in Torr and in bar, that correspond to these values?

**2.17** In dry atmospheric air, the four most abundant components are $N_2$, $X = 0.7808$; $O_2$, $X = 0.2095$; Ar, $X = 9.34 \times 10^{-3}$; and $CO_2$, $X = 3.25 \times 10^{-4}$. Calculate the partial pressures of these four gases, in Torr, under standard atmospheric conditions.

**2.18** Find the partial pressures, total pressure, and mole fractions of a gas mixture in a 4.00-L container at 375 °C if it contains 1.25 g each of Ar, CO, and $CH_4$.

**2.19** The following questions refer to Figure 2-7d: (a) Which gas exerts a higher partial pressure: $O_2$ or He? (b) Which has a higher mole fraction? (c) Compute the mole fraction of He.

**2.20** The figures shown below represent very small portions of three gas mixtures, all at the same volume and temperature.

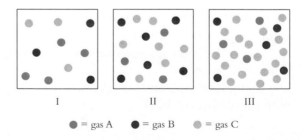

(a) Which sample has the highest partial pressure of gas A?

(b) Which sample has the highest mole fraction of gas B?

(c) In sample III, what is the concentration of gas A in ppm?

**2.21** The natural gas in a storage tank is analyzed. A 2.00-g sample of the gas contains 1.57 g $CH_4$, 0.41 g $C_2H_6$, and 0.020 g $C_3H_8$. Calculate the partial pressures of each gas in the storage tank if the total pressure in the tank is 2.35 bar.

**2.22** A sample of automobile exhaust gas is analyzed at an Ontario Drive Clean centre. The gas contains 487.4 ppm $CO_2$, 10.3 ppb NO, and 4.2 ppb CO. Calculate the partial pressures of these gases if the exhaust is emitted at 113.1 kPa pressure.

## Gas Stoichiometry

**2.23** **ilw** Humans metabolize glucose into products. The products of glucose consumption are $CO_2$ and $H_2O$:

$$C_6H_{12}O_6(s) + O_2(g) \longrightarrow$$
$$CO_2(g) + H_2O(l) \quad \text{(unbalanced)}$$

What volume of $CO_2$ is produced under body conditions (37 °C, 101.325 kPa) during the consumption of 4.65 g of glucose?

**2.24** Oxygen gas can be generated by heating $KClO_3$ in the presence of a catalyst:

$$KClO_3(s) \longrightarrow KCl(s) + O_2(g) \quad \text{(unbalanced)}$$

What volume of $O_2$ gas will be generated at $T = 25$ °C and $p = 765.1$ Torr from 1.57 g of $KClO_3$?

**2.25** Sodium metal reacts with molecular chlorine gas to form sodium chloride. A closed container of volume $3.00 \times 10^3$ mL contains chlorine gas at 27 °C and $1.25 \times 10^3$ Torr. Then, 6.90 g of solid sodium is introduced, and the reaction goes to completion. What is the final pressure if the temperature rises to 47 °C?

**2.26** Carbon monoxide and molecular oxygen react to form carbon dioxide. A 50.0-L reactor at 25.0 °C is charged with 1.00 bar of CO. The gas is then pressurized with $O_2$ to give a total pressure of 3.56 bar. The reactor is sealed, heated to 350 °C to drive the reaction to completion, and cooled back to 25.0 °C. Compute the final partial pressure of each gas.

**2.27** **ilw** Ammonia is produced industrially by reacting $N_2$ with $H_2$ at elevated pressure and temerature in the presence of a catalyst:

$$N_2(g) + H_2(g) \longrightarrow NH_3(g) \quad \text{(unbalanced)}$$

Assuming 100% yield, what mass of ammonia would be produced from a 1 : 1 mole ratio mixture in a reactor that has a volume of $8.75 \times 10^3$ L, under a total pressure of 275 bar at $T = 455$ °C?

**2.28** Ethylene oxide is produced industrially from the reaction of ethylene with oxygen at 1 bar and 280 °C, in the presence of a silver catalyst:

$$C_2H_4(g) + O_2(g) \longrightarrow C_2H_4O(g) \quad \text{(unbalanced)}$$

Assuming a 100% yield, how many kilograms of ethylene oxide would be produced from $5.00 \times 10^4$ L of a mixture containing ethylene and oxygen in a 1 : 1 mole ratio?

**2.29** In actual practice, the reaction of Problem 2.27 gives a yield of only 13%. Repeat the calculation of Problem 2.27 taking this into account.

**2.30** In actual practice, the ethylene oxide reaction of Problem 2.28 gives only a 65% yield under optimal conditions. Repeat the calculation of Problem 2.28 assuming a 65% yield.

## Molecular View of Gases

**2.31** Redraw Figure 2-9b to show the distribution of molecules if the temperature of the oven is doubled.

**2.32** Redraw Figure 2-10 to show a temperature of 200 K.

**2.33** Draw a single graph that shows the *speed* distributions of $N_2$ at 200 K, $N_2$ at 300 K, and He at 300 K.

**2.34** Draw a single graph that shows the *speed* distributions of $Cl_2$ at 500 K, $Br_2$ at 500 K, and $Cl_2$ at 300 K.

**2.35** Redraw the graph in Problem 2.33 as a graph of *energy* distributions for the three gases.

**2.36** Redraw the graph in Problem 2.34 as a graph of *energy* distributions for the three gases.

**2.37** Calculate the most probable speed and average kinetic energy per mole for each of the following: (a) He at 627 °C; (b) $O_2$ at 27 °C; and (c) $SF_6$ at 627 °C.

**2.38** Calculate the most probable speed and average kinetic energy per mole for each of the following: (a) Ar at 127 °C; (b) Xe at 127 °C; and (c) $C_3H_8$ at 327 °C.

**2.39** Explain in molecular terms why each of these statements is true: (a) At very high pressure, no gas behaves ideally. (b) At very low temperature, no gas behaves ideally.

**2.40** A cylinder with a movable piston contains a sample of gas. Describe in molecular terms the effect on pressure exerted by the gas for each of the following changes: (a) The piston is pushed in. (b) Some gas is removed while the piston is held in place. (c) The gas is heated while the piston is held in place.

**2.41** The figure represents an ideal gas in a container with a movable friction-free piston. (a) The external pressure on the piston exerted by the atmosphere is 1 bar. If the piston is not moving, what is the pressure inside the container? Explain in terms of molecular collisions. (b) Redraw the sketch to show what would happen if the temperature of the gas in the container is doubled. Explain in terms of molecular collisions.

**2.42** Consider the figure appearing in Problem 2.40. (a) The monatomic gas shown in the figure is replaced by an equal number of molecules of a diatomic gas, all other conditions remaining the same. What is the pressure inside the container? Explain in terms of molecular collisions. (b) The external pressure is reduced to 0.75 bar, and the piston moves as a result. Redraw the sketch to show the new situation.

2.43 Describe the molecular changes that account for the result in Problem 2.11.

2.44 Describe the molecular changes that account for the result in Problem 2.12.

## Additional Gas Properties

2.45 **ilw** Freons (CFCs) are compounds that contain carbon, chlorine, and fluorine in various proportions. They have been extensively used as foaming agents, propellants, and refrigeration fluids. Freons are controversial because of the damage they do to the ozone layer in the stratosphere. A 2.55-g sample of a particular Freon in a 1.50-L bulb at 25.0 °C has a pressure of 262 Torr. What is the molar mass and formula of the compound?

2.46 Gaseous hydrocarbons, which contain only carbon and hydrogen, are good fuels because they burn in air to generate large amounts of heat. A sample of hydrocarbon with $m = 1.63$ g exerts a pressure of 1.50 bar in a 945-mL bulb at 21.5 °C. Determine the molar mass and chemical formula of this hydrocarbon.

2.47 What is the density (g/L) of $SF_6$ gas at 755 Torr and 27 °C?

2.48 What is the density (g/L) of $H_2$ gas at 380 Torr and −23 °C?

2.49 Determine the root-mean-square speed of $SF_6$ molecules under the conditions of Problem 2.47.

2.50 Determine the root-mean-square speed of $H_2$ molecules under the conditions of Problem 2.48.

2.51 If a gas line springs a leak, which will diffuse faster through the atmosphere and why: $CH_4$ or $C_2H_6$?

2.52 A student proposes to separate CO from $N_2$ by an effusion process. Is this likely to work well? Why or why not?

2.53 Calculate the relative rates of effusion of $^{14}CO_2$ and $^{12}CO_2$. Repeat for $^{14}CO$ and $^{12}CO$. Is it easier to separate $^{14}C$ from $^{12}C$ as the monoxide or as the dioxide?

2.54 The percentage of $^{235}U$ in natural uranium is 0.71%. If each step in the enrichment of $^{235}U$ by gaseous diffusion of $UF_6$ results in an enrichment of 1.0043 times, what will the percentage of $^{235}U$ be after 100 steps? After 500 steps?

## Non-Ideal (Real) Gases

2.55 Use the data in Table 2-3 to calculate the pressure of 1.00 mole of $F_2(g)$ in a 30.0-L container at 25 °C using the ideal gas law and using the van der Waals equation. Why are these two values so similar?

2.56 Calculate the pressure of 100.0 mole of $F_2(g)$ in a 30.0-L container at 25 °C using the ideal gas law and using the van der Waals equation. Why are these two values so different?

2.57 Calculate the pressure of 100.0 mole of $H_2(g)$ in a 30.0-L container at 25 °C using the van der Waals equation. Why is this value different from the one calculated in Problem 2.56?

2.58 Use the data in Table 2-3 to predict which of the same amount of $F_2$ or $Cl_2$ at the same temperature and volume will have the higher pressure and explain your reasoning.

2.59 Oxygen gas ($O_2$) is stored in steel cylinders. Assume we have 75 mol $O_2$ stored in a 15.0-L cylinder at 25 °C. Use the van der Waals equation to calculate the pressure.

2.60 A fire in the laboratory causes the temperature of the oxygen cylinder in problem 2.59 to increase to 700 °C. The cylinder can withstand a pressure of only 200 bar before exploding. Does the cylinder explode?

## Chemistry of the Earth's Atmosphere

2.61 How much does the mole fraction of $O_2$ in the atmosphere change at 20 °C as the relative humidity varies from 100% to 0% at 1 bar total pressure?

2.62 How much does the partial pressure of $N_2$ gas in the atmosphere change at 30 °C and 1 bar as the relative humidity varies from zero to 100%?

2.63 Find the dew point for a relative humidity of 78% at 25 °C.

2.64 In the tropics, water will condense in human lungs when the temperature and relative humidity are too high. Using Table 2-5, estimate the vapour pressure of water at a body temperature of 37 °C. If atmospheric temperature is 40 °C, at what relative humidity does this life-threatening process occur?

2.65 **ilw** Neon was discovered by cooling dry air until it is liquefied and then boiling off the components one at a time. What volume of dry air at 298 K and 1.00 bar must be treated in this way to obtain 10 mg of neon? (See Table 2-4.)

2.66 Compute the total mass of krypton contained in 1.00 cubic kilometre of dry air, assuming constant atmospheric pressure of 1.00 bar (see Table 2-4).

## ADDITIONAL PAIRED PROBLEMS

2.67 What mass of krypton is present in a 725-mL container at 925 °C in which the pressure of krypton is 10.0 bar? How many atoms is this?

**2.68** Nitrogen gas is available commercially in pressurized 9.50-L steel cylinders. If a tank has a pressure of 145 bar at 298 K, how many moles of $N_2$ are in the tank? What is the mass of $N_2$?

**2.69** The figure shows three chambers with equal volumes, all at the same temperature, connected by closed valves. Each chamber contains helium gas, with amounts proportional to the number of atoms shown.

A        B        C

Answer each of the following questions, briefly stating your reasoning: (a) Which of the three has the highest pressure? (b) If the pressure of B is 1.0 bar, what is the pressure of A? (c) If the pressure of A starts at 1.0 bar, and then all of the atoms from B and C are added to A's container, what will be the new pressure? (d) If the pressure in B is 0.50 bar, what will the pressure be after the valves are opened?

**2.70** The following figure shows two tanks connected by a valve. Each tank contains a different gas, both at 0.0 °C. (a) Redraw the system to show how it appears after the valve has been opened. (b) If each molecule in the diagram represents 1.0 mole of gas, what is the system's total pressure after the valve is opened? (Assume that the valve has negligible volume.) (c) If all the gas is pumped into the smaller tank, what are the partial pressures of the two gases?

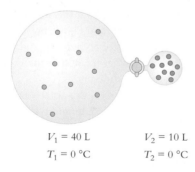

$V_1 = 40$ L        $V_2 = 10$ L
$T_1 = 0$ °C        $T_2 = 0$ °C

**2.71** At an altitude of 40 km above the Earth's surface, the temperature is about −25 °C, and the pressure is about 3.0 Torr. Calculate the average molecular speed of ozone ($O_3$) at this altitude.

**2.72** The average kinetic energy of a 1.55-g sample of argon gas in a 5.00-L bulb is $1.02 \times 10^{-20}$ J/atom. What is the pressure of the gas? What is the average speed of the argon atoms under these conditions?

**2.73** At an altitude of 150 km above the Earth's surface, atmospheric pressure is $10^{-10}$ bar and the temperature is 310 K. What is the molecular density at this altitude?

**2.74** Molecular clouds composed mostly of hydrogen molecules have been detected in interstellar space. The molecular density in these clouds is about $10^{10}$ molecules/m$^3$, and their temperature is around 25 K. What is the pressure in such a cloud?

**2.75** Use data from Table 2-5 to estimate the dew point for air that has (a) relative humidity of 80% at 35 °C; (b) relative humidity of 50% at 15 °C; and (c) relative humidity of 30% at 25 °C.

**2.76** A sample of warm moist air (100% relative humidity) is collected in a container at $p = 756$ Torr and $T = 30.0$ °C. Find (a) the new pressure after a drying agent is added to remove the water vapour and (b) the mole fraction and concentration in grams per litre of water vapour in the original air sample.

**2.77** Describe a gas experiment that would show that the element oxygen exists naturally as diatomic molecules.

**2.78** A student proposes that $SO_2$ gas contains a significant fraction of dimer with formula $S_2O_4$. Describe a gas experiment that would verify or disprove this proposal.

**2.79** A mixture of cyclopropane gas ($C_3H_6$) and oxygen ($O_2$) in 1.00 : 4.00 mole ratio is used as an anaesthetic. What mass of each of these gases is present in a 2.00-L bulb at 23.5 °C if the total pressure is 1.00 bar?

**2.80** A gas cylinder of volume 5.00 L contains 1.00 g of Ar and 0.0500 g of Ne. The temperature is 275 K. Find the partial pressures, total pressure, and mole fractions.

**2.81** Consider two gas bulbs of equal volume, one filled with $H_2$ gas at 0 °C and 2 bar, the other containing $O_2$ gas at 25 °C and 1 bar. Which bulb has (a) more molecules; (b) more mass; (c) higher average kinetic energy of molecules; and (d) higher average molecular speed?

**2.82** Consider two gas bulbs of equal volume, one filled with $Cl_2$ gas at 100 °C and 2 bar, the other containing $N_2$ gas at 25 °C and 1 bar. Which bulb has (a) more molecules; (b) more mass; (c) higher average kinetic energy of molecules; and (d) higher average molecular speed?

**2.83** Refer to Figure 2-11 to determine the following for $O_2$ gas at 300 K: (a) What is the most probable kinetic energy? (b) What is the most probable speed?

**2.84** Molecular beam experiments on ammonia at 425 K give the speed distribution shown in the figure:

(a) What is the most probable speed? (b) What is the most probable kinetic energy?

**2.85** Explain in your own words how sulphur impurities in coal can lead to acid rain. Use balanced equations where appropriate.

**2.86** Explain in your own words how the reactions occurring in the cylinders of automobile engines contribute to the production of photochemical smog. Use balanced equations where appropriate.

**2.87** A sample of gas is found to exert a pressure of 525 Torr when it is in a 3.00-L flask at 0.00 °C. Compute (a) the new volume if $p$ becomes 755 Torr and $T$ is unchanged; (b) the new pressure if $V$ becomes 2.00 L and $T$ is unchanged; and (c) the new pressure if the temperature is raised to 50.0 °C and $V$ is unchanged.

**2.88** A sample of gas is found to exert a pressure of 322 Torr when it is in a 2.00-L flask at 100.00 °C. Compute (a) the new volume if $p$ becomes 525 Torr and $T$ is unchanged; (b) the new pressure if the volume is reduced to 1.50 L and $T$ becomes 50.0 °C; and (c) the new pressure if half the gas is removed but $V$ and $T$ remain the same.

**2.89** Redraw the central figure in Figure 2-8 to show each of the following changes in conditions: (a) The temperature increases by 50%. (b) 25% of the helium atoms are removed. (c) Hydrogen molecules replace half the helium atoms.

**2.90** Redraw the central figure in Figure 2-8 to show each of the following changes in conditions: (a) Half of the helium atoms are removed. (b) The helium atoms are replaced with an equal number of nitrogen molecules. (c) The temperature is lowered by 50%.

**2.91** Determine whether each of the following statements is true or false. If false, rewrite the statement so that it is true. (a) At constant $T$ and $V$, $p$ is inversely proportional to the number of moles of gas. (b) At constant $V$, the pressure of a fixed amount of gas is directly proportional to $T$. (c) At fixed $n$ and $V$, the product of $p$ and $T$ is constant.

**2.92** Determine whether each of the following statements is true or false. If false, rewrite the statement so that it is true. (a) At fixed $n$ and $p$, $V$ is independent of $T$. (b) At fixed $n$ and $T$, the product of $p$ and $V$ is constant. (c) When gas is added to a chamber at fixed $V$ and $T$, the pressure increases as $n^2$.

**2.93** A 3.00-g sample of an ideal gas occupies 0.963 L at 22 °C and 0.969 bar. What will be its volume at 15 °C and 1.00 bar?

**2.94** A 96.0-g sample of $O_2$ gas at 0.0 °C and 380 Torr is compressed and heated until the volume is 3.00 L and the temperature is 27 °C. What is the final pressure in Torr?

**2.95** The gas $SF_6$ is used to trace air flows because it is non-toxic and can be detected selectively in air at a concentration of 1.0 ppb. What partial pressure is this? At this concentration, how many molecules of $SF_6$ are contained in 1.0 cm$^3$ of air at $T = 21$ °C?

**2.96** In 2008, carbon dioxide levels in the atmosphere reached 385 ppm by volume. (The 1958 reading was 314.6 ppm by volume.) Convert this reading to a partial pressure in bar. At this level, how many $CO_2$ molecules are there in 1.0 L of dry air at −45 °C if the total pressure is 1 bar?

## ADDITIONAL PROBLEMS

**2.97** A 15.00-g piece of dry ice (solid $CO_2$) is dropped into a bottle with a volume of 0.750 L, and the air is pumped out. The bottle is sealed and the $CO_2$ is allowed to evaporate. What will be the final pressure in the bottle if the final temperature is 0.0 °C?

**2.98** Two chambers are connected by a valve. One chamber has a volume of 15 L and contains $N_2$ gas at a pressure of 2.0 bar. The other has a volume of 1.5 L and contains $O_2$ gas at 3.0 bar. The valve is opened, and the two gases are allowed to mix thoroughly. The temperature is constant at 300 K throughout this process. (a) How many moles each of $N_2$ and $O_2$ are present? (b) What are the final pressures of $N_2$ and $O_2$, and what is the total pressure? (c) What fraction of the $O_2$ is in the smaller chamber after mixing?

**2.99** Calculate the density of (a) dry air at 1 bar and 298 K and (b) air of 100% relative humidity at 1 bar and 298 K.

**2.100** On a smoggy day, the ozone content of air over Toronto reaches 0.50 ppm. Compute the partial pressure of ozone and the number of molecules of ozone per cubic centimetre if atmospheric pressure is 762 Torr and the temperature is 28 °C.

**2.101** Liquid oxygen, used in some large rockets, is produced by cooling dry air to −183 °C. How many litres of dry air at 25 °C and 750 Torr would have to be processed to produce 150 L of liquid oxygen (density = 1.14 g/mL)?

**2.102** Benzaldehyde is a fragrant molecule used in artificial cherry flavouring. Combustion of 125 mg of benzaldehyde gives 363 mg of carbon dioxide and 63.7 mg of water. In another experiment, a 110-mg sample is vapourized at 150 °C in a 0.100-L bulb. The vapour gives a pressure of 274 Torr. Determine the molecular formula of benzaldehyde.

**2.103** The methanation reaction, $3\,H_2 + CO \longrightarrow CH_4 + H_2O$, is used commercially to prepare methane. A gas reactor with a volume of $1.00 \times 10^2$ L is pressurized at 575 K with 20.0 bar of $H_2$ gas and 10.0 bar of CO gas; 145 g of $CH_4$ is produced. What is the percent yield of the synthesis?

**2.104** In an explosion, a compound that is a solid or a liquid decomposes very rapidly, producing large volumes of gas. The force of the explosion results from the rapid expansion of the hot gases. For example, TNT (trinitrotoluene) explodes according to the following balanced equation:

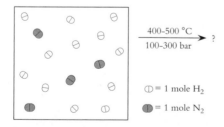

(a) How many moles of gas are produced in the explosion of 1.0 kg of TNT? (b) What volume will these gases occupy if they expand to a total pressure of 1.0 bar at 25 °C? (c) At 1.0 bar total pressure, what would be the partial pressure of each gas?

**2.105** The Haber synthesis of ammonia occurs in the gas phase at high temperature (400–500 °C) and pressure (100–300 bar). The starting materials for the Haber synthesis are placed inside a container, in proportions shown in the figure. Assuming 100% yield, draw a sketch that illustrates the system at the end of the reaction.

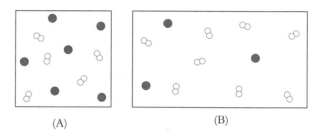

**2.106** When a sealed bulb containing a gas is immersed in an ice bath, it has a gas pressure of 345 Torr. When the same bulb is placed in an oven, the pressure of the gas rises to 745 Torr. What is the temperature of the oven in degrees Celsius?

**2.107** Many of the transition metals form complexes with CO; these complexes are called metal carbonyls and have the general formula $M(CO)_x$. A 0.500-g sample of gaseous nickel carbonyl in a 0.100-L bulb generates a pressure of 552 Torr at 30 °C. What is the formula of nickel carbonyl?

**2.108** At low temperature, nitrogen dioxide molecules join together to form dinitrogen tetroxide.

$$2\,NO_2 \longrightarrow N_2O_4 \text{ (low temperature)}$$

A sample of $NO_2$ sealed inside a glass bulb at 23 °C gave a pressure of 691 Torr. Lowering the temperature to −5 °C converted the $NO_2$ to $N_2O_4$. What was the final pressure inside the bulb?

**2.109** People often remark that "the air is thin" at higher elevation. Explain this comment in molecular terms using the fact that the atmospheric pressure atop Mount Everest is only about 250 Torr.

**2.110** The figures shown below represent mixtures of argon atoms and hydrogen molecules. The volume of container B is twice the volume of container A.

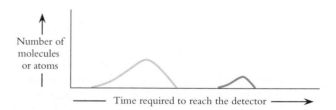

(A)          (B)

(a) Which container has a higher total gas pressure? Explain. (b) Which container has a higher partial pressure of molecular hydrogen? Explain. (c) One of the two gas mixtures shown above was used in a pulsed molecular beam experiment. The result of the experiment is shown below. Which of the two gas samples, A or B, was used for this experiment? Explain.

Number of molecules or atoms ↑

⟶ Time required to reach the detector ⟶

**2.111** Find the partial pressures, in bar, of the eight most abundant atmospheric components listed in Table 2-4 at 25 °C, 50% relative humidity, and $p = 765$ Torr.

**2.112** A 0.1054-g mixture of $KClO_3$ and a catalyst was placed in a quartz tube and heated vigorously to drive off all the oxygen as $O_2$. The $O_2$ was collected at 25.17 °C and a pressure of 759.2 Torr. The volume of gas collected was 22.96 mL. (a) How many moles of $O_2$ were produced? (b) How many moles of $KClO_3$ were in the original mixture? (c) What was the mass percent of $KClO_3$ in the original mixture?

2.113 Elemental analysis of an organic liquid with a fishy odour gives the following elemental mass percentages: H, 14.94%; C, 71.22%; and N, 13.84%. Vapourization of 250 mg of the compound in a 150-mL bulb at 150 °C gives a pressure of 435 Torr. What is the molecular formula of the compound?

2.114 When heated to 150 °C, $CuSO_4 \cdot 5H_2O$ loses its water of hydration as gaseous $H_2O$. A 2.50-g sample of the compound is placed in a sealed 4.00-L steel vessel containing dry air at 1.00 bar and 27 °C and the vessel is then heated to 227 °C. What are the final partial pressure of $H_2O$ and the final total pressure?

## GROUP STUDY PROBLEMS

2.115 A balloon filled with helium gas at 1.0 bar and 25 °C is to lift a 350-kg payload. What is the minimum volume required for the balloon? Assume that dry air at the same pressure and temperature makes up the atmosphere.

2.116 What temperature must the air be if the balloon described in Problem 2.115 is going to use hot air instead of helium?

2.117 A mouse is placed in a sealed chamber filled with air at 765 Torr and equipped with enough solid KOH to absorb any $CO_2$ and $H_2O$ produced. The gas volume in the chamber is 2.05 L, and its temperature is held at 298 K. After 2 hours, the pressure inside the chamber has fallen to 725 Torr. What mass of oxygen has the mouse consumed?

2.118 Liquid helium at 4.2 K has a density of 0.147 g/mL. A 2.00-L metal bottle at 95 K contains air at 1.0 bar pressure. We introduce 0.100 L of liquid helium, seal the bottle, and allow the entire system to warm to room temperature (25 °C). What is the pressure inside the bottle?

2.119 You are on vacation in Hawaii, where the temperature is 27 °C. You wish to go for a scuba dive. Your tank, with a volume of 12.5 L, is filled with dry air to a pressure of 165 bar. If your body requires you to consume 14.0 g $O_2$ per minute, how long can you dive at a depth of 70 feet if you must allow 6.0 minutes to return to the surface?

2.120 If 1.00 metric ton of soft coal containing 4.55% (by mass) sulphur is burned in an electric power plant, what volume of $SO_2$ is produced at 55 °C and 1.00 bar? What mass of CaO is required to "scrub" this $SO_2$? What mass of $CaSO_3$ must be disposed of?

2.121 Construct the following graphs for an ideal gas when $T = 298$ K: (a) $V$ vs. $p$; (b) $pV$ vs. $p$; and (c) $1/V$ vs. $p$. Describe in words the shape of each of these graphs.

## ANSWERS TO SECTION EXERCISES

2.1.1 698 Torr, 0.92 atm, 0.93 bar, and 8.2% change

2.1.2 $1.575 \times 10^6$ Pa, $1.575 \times 10^3$ kPa, and 15.54 atm

2.1.3 800.3 Torr, 1.053 atm, $1.067 \times 10^2$ kPa and 106.7 bar

2.2.1 (a) and (b) pressure doubles; (c) pressure triples; and (d) no change

2.2.2

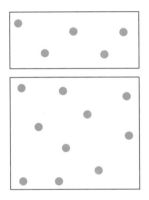

2.2.3 784 Torr

2.3.1 $H_2$, $X = 0.933$ and $p = 2.15$ bar; $N_2$, $X = 0.067$ and $p = 0.15$ bar

2.3.2

2.3.3 (a) $3.5 \times 10^{-5}$; (b) $2.6 \times 10^{-2}$ Torr; and (c) $8.3 \times 10^{20}$ molecules

2.4.1 7.8 g $C_2H_2$ and 5.4 g $H_2O$

2.4.2 3.60 bar

2.4.3 5.89 bar (Remember to calculate the new partial pressure of air at the higher temperature.)

2.5.1

2.5.2 (a) $2.16 \times 10^3$ J; (b) 122 J; and (c) $1.96 \times 10^{-20}$ J

**2.5.3** Low pressure, because fewer molecular interactions and there is more space between the molecules

**2.6.1** 59.1 kg of air enters as the house cools.

**2.6.2** 272 g/mol, $HgCl_2$

**2.6.3** Speeds are $1.03 \times 10^3$ m/s for $H_2O$ and $9.82 \times 10^2$ m/s for $D_2O$. Ratio of rates is 1.05.

**2.6.4** 8.94 min.

**2.7.1** Helium is a small, non-polar atom. Ammonia is a larger, polar molecule. The larger attractive forces among ammonia molecules cause its pressure to be smaller.

**2.7.2** $p_{ideal} = 408.1$ kPa; $p_{vdW} = 407.6$ kPa

**2.7.3** Ne has a larger "$a$" factor, and a smaller "$b$" factor than He. Both of these tend to make the pressure of Ne lower than that of He.

**2.8.1** Ne, $1.82 \times 10^{-5}$; He, $5.24 \times 10^{-6}$; and $CH_4$, $1.4 \times 10^{-6}$

**2.8.2** $2.64 \times 10^4$ L

**2.8.3** $p_{vap} = 9.209$ Torr, 59% relative humidity

## ANSWERS TO PRACTICE EXERCISES

**2.1** $p_{gas} = 799.8$ Torr $= 1.052$ atm $= 1.066$ bar

**2.2** $p = 145$ kPa

**2.3** $T = 294$ K $= 21\,°C$

**2.4** $m = 0.047$ g

**2.5** $p = 0.47$ bar

**2.6** $V = 6.79$ L

**2.7** Increase of 2.5%

**2.8** $p_{O_2} = 11.5$ atm; $p_{He} = 30.6$ atm; $X_{He} = 0.727$

**2.9** $m = 8.5 \times 10^{-4}$ g

**2.10** $V = 1.53$ L

**2.11** $p_{NO} = 0$ (Limiting Reagent), $p_{O_2} = 2.50$ bar, $p_{NO_2} = 5.00$ bar

**2.12** $[H_3O^+] = 0$ M (LR), $m_{Mg} = 3.06$ g, $[Mg^{2+}] = 3.00$ M, $[Cl^-] = 6.00$ M, $p_{H_2} = 2.30$ bar, $p = 3.30$ bar

**2.13** He atoms move slower than $H_2$ molecules

**2.14** $\overline{E}_{kinetic} = 4.93 \times 10^{-21}$ J/molecule; $\overline{E}_{kinetic}$, molar $= 2.97 \times 10^3$ J/mol

**2.15** $M = 28.0$ g/mol

**2.16** $\rho_{He} = 0.16$ g/L and $\rho_{Ar} = 1.67$ g/L. Ar is more dense than air, making it a good gas blanket, whereas He is considerably less dense than air, giving it good lifting ability.

**2.17** (a) 0.7978; (b) 0.9999; (c) 0.7690

**2.18** $C_2H_2$

**2.19** Before use, $p_{vdW} = 11\,371$ kPa and $p_{ideal} = 10\,300$ kPa. After use, $p_{vdW} = 103$ kPa and $p_{ideal} = 103$ kPa. Methane has the greatest deviation from ideality because of stronger intermolecular forces than in He.

**2.20** Dew forms at about 13 °C.

# ENERGY AND ITS CONSERVATION

## OUR ENERGY FUTURE

A plentiful supply of energy is essential for life. Human societies make a vast enterprise out of obtaining energy in useful forms. The Sun provides nearly all of the Earth's energy. Green plants use sunlight for photosynthesis, which produces energy-rich chemicals such as glucose $(C_6H_{12}O_6)$. Living organisms extract energy from glucose through a complex web of chemical reactions, the net result of which is the combustion of glucose with oxygen:

$$C_6H_{12}O_6\ (s) + 6\ O_2\ (g) \longrightarrow 6\ CO_2\ (g)$$
$$+ 6\ H_2O\ (l) + \text{Energy}$$

The fuels of modern society—natural gas, petroleum products, and coal—are called fossil fuels because they are formed from the decomposition of green plants over many millions of years. Like glucose, fossil fuels release energy when they react with oxygen to form carbon dioxide and water.

Fossil fuels are non-renewable, and combustion products contaminate the atmosphere. Consequently, scientists are searching for new sources of energy. One possibility is molecular hydrogen, which releases energy when it reacts with oxygen:

$$2\ H_2\ (g) + O_2\ (g) \longrightarrow 2\ H_2O\ (l) + \text{Energy}$$

Hydrogen powers the rockets of the space shuttle, but it is not a major fuel source at this time because there is no economical source of molecular hydrogen, and transporting this gas is difficult.

Water could become a source of molecular hydrogen, but the decomposition of $H_2O$ into $H_2$ and $O_2$ requires the input of just as much energy as is released during the combustion of molecular hydrogen:

$$2\ H_2O\ (l) + \text{Energy} \longrightarrow 2\ H_2\ (g) + O_2\ (g)$$

Sunlight may be the future source of energy for "splitting water," but this remains a major technological challenge.

Small engines may incorporate the fuel cell as a power source. A fuel cell (shown in our molecular inset) combines hydrogen and oxygen in a way that generates electricity, which drives a motor. Leading

## LEARNING OBJECTIVES

*Upon completion of this chapter you should be able to:*

- recognize the types of energy of interest to chemists
- understand the first law of thermodynamics and the concepts of heat and work
- understand the origins of energy changes in chemical reactions
- apply the principles of calorimetry to determine energy changes in a chemical reaction
- understand and calculate enthalpy and internal energy
- be familiar with our sources of energy

automobile manufacturers have already developed fuel-cell vehicles.

Iceland, an island nation with abundant hydro-electric and geothermal power sources close to where people live, is using natural power sources to convert water into hydrogen and oxygen. In April 2003, Iceland opened its first hydrogen filling station for fuel-cell vehicles. If this initiative proves successful, perhaps the rest of the industrialized world will eventually convert to a hydrogen-fuelled economy.

This chapter describes the properties and behaviour of energy. We discuss the flow of energy from place to place and the conversions of energy from one form to another. We describe the law of conservation of energy and how to measure energy changes. We introduce enthalpy, a useful energy-related quantity. To end the chapter, we describe current and future energy sources.

## 3.1 TYPES OF ENERGY

Personal experiences provide an intuitive sense of what energy is about. Backpackers climbing toward a mountain pass use energy to raise themselves and their packs up the mountain. When they reach the top, the hikers are tired after spending energy to make the climb. Scientifically, however, we need more than an intuitive sense of a concept as important as energy.

In science, **energy is defined as the ability to do work, and work is defined as the displacement of an object against an opposing force.** Here are some examples:

- Backpackers do work by climbing against the force of gravity.
- A baseball player does work by hitting the ball over the fence.
- Water does work by turning a turbine that generates electricity.
- A gasoline engine does work by moving an automobile along the highway.

These examples involve two categories of energy. One is **kinetic energy**, the energy of moving objects. The other is **potential energy**, which is energy that is stored. In addition to kinetic and potential energies, **thermal energy** refers to the energy content of a hot object, and **radiant energy** is the energy content of electromagnetic radiation (light). Kinetic, potential, thermal, and radiant energies all play important roles in chemistry.

## Kinetic and Potential Energies

Every moving object has kinetic energy. A speeding jet plane has a large amount of kinetic energy, and so do atoms and molecules. Whether we describe an airplane or a molecule, the amount of kinetic energy is always given by the kinetic energy equation, introduced in Chapter 2:

$$E_{kinetic} = \frac{1}{2}mv^2 \qquad (2\text{-}3)$$

The SI units for kinetic energy are $kg\ m^2\ s^{-2}$, commonly referred to as a joule (J):

$$1\ J = 1\ kg\ m^2\ s^{-2}$$

The kinetic energy of an object changes with its conditions. For example, on accelerating from a stoplight, an automobile increases its kinetic energy. Conversely, when that automobile brakes to a stop, its kinetic energy decreases.

There are several types of potential energy, but the most familiar form is **gravitational energy**. A rock teetering high on a ledge has gravitational energy that will change to kinetic energy as the rock falls to lower ground. Water behind a dam has stored gravitational energy that is converted into electrical energy when the water drives a turbine. Although gravitational energy is most familiar in everyday life, **electrical energy**, **chemical energy**, and **mass** are the forms of potential energy that are particularly important in chemistry.

## Electrical Energy

Matter is electrical by its very nature. Atoms are composed of nuclei and electrons that are subject to electrical force. The nucleus bears a positive electrical charge, whereas the electrons are negatively charged. Countless studies have shown that when two objects have opposite charges, electrical forces pull the objects together. Figure 3-1 demonstrates the interactions of charged objects by showing two small metal balls that bear electrical charges. The balls are pulled together when they have opposite charges. This attractive force between opposite charges is what holds electrons and nuclei together in atoms. It is also the reason that atoms combine into molecules, and it is why gases condense to form liquids and solids. Figure 3-1 also shows that objects possessing like charges push one another apart. Electrons repel one another, and so do nuclei.

**Figure 3-1**

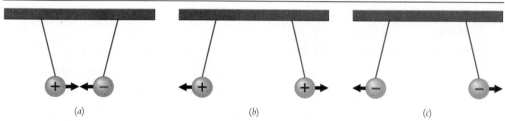

(a)          (b)          (c)

Charged objects attract one another if the charges have opposite signs (a). Charged objects with the same sign, either positive (b) or negative (c), repel one another.

Electrical forces lead to electrical potential energy. Energy is released when oppositely charged objects move close together, whereas energy must be supplied to pull oppositely charged objects apart. For example, separating an electron from its nucleus requires an energy input, just as lifting a backpack requires an energy input. Figure 3-2 shows this in diagrammatic form.

Just as kinetic energy can be calculated from an equation, the electrical potential energy of two charges, $q_1$ and $q_2$, separated by distance $r$ can be calculated using Equation 3-1:

$$E_{electrical} = k\frac{q_1 q_2}{r} \qquad (3\text{-}1)$$

When distance is expressed in picometres (1 pm = $10^{-12}$ m) and charges are in electronic units, the constant in the equation is $k = 2.31 \times 10^{-16}$ J pm. To remove electrons from atoms or molecules, energy must be supplied to overcome this electrical energy. We use Equation 3-1 in later chapters to analyze the energetics of electrons, ions, and nuclei.

**Figure 3-2**

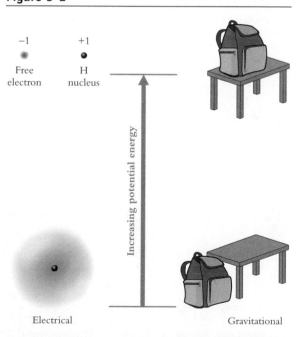

Removing an electron from a nucleus increases the electrical potential energy. Lifting a backpack from the floor to a tabletop increases the gravitational potential energy.

## Chemical Energy

Atoms bind together to form molecules through the influence of electrical forces. As we describe in Chapters 6 and 7, chemical bonds result from negative electrons shared between positive nuclei. Electrons and nuclei in a molecule are arranged in a way that minimizes electrical potential energy. Chemists give this form of potential energy a specific name, chemical energy or bond energy. Every different molecule has a specific stability that is related to its chemical energy. Consequently, energy is either absorbed or released during any particular chemical reaction. Figure 3-3 shows the chemical energy changes involved in the formation of molecular hydrogen, oxygen, and water from atoms of hydrogen and oxygen. The formation of chemical bonds makes $H_2$ and $O_2$ more stable than collections of separated atoms, but molecules of $H_2O$ are even more stable than these diatomic molecules, so energy is released when $H_2$ and $O_2$ react to form $H_2O$.

## Mass

We think of mass ($m$) and energy ($E$) as separate quantities, but, as Albert Einstein was the first to show, mass actually represents a form of potential energy. Einstein's most famous equation, $E = mc^2$, where $c$ is the speed of light, describes the mass–energy equivalence. Conversion of mass into other forms of energy is not important in standard chemical processes, but, as we describe in Chapter 20, nuclear energy is the result of transformations of large amounts of nuclear mass into useable energy.

## Thermal Energy

Consider a monatomic gas such as helium or argon. Recall from Chapter 2 that the atoms of this gas move continuously with a distribution of kinetic energies. The total energy of these random movements is called thermal energy. Equation 2-4 expresses the average kinetic energy of a gas:

$$\overline{E}_{\text{kinetic}} = \frac{3RT}{2N_A} \tag{2-4}$$

Thus, average molecular kinetic energy increases as temperature increases. For a monatomic gas such as He or Ar, the only contribution to thermal energy is the kinetic energy of the atoms.

Equation 2-4 also describes the kinetic energy of gaseous molecules, but molecules possess additional thermal energy. Figure 3-4 shows that molecules rotate in space, giving them rotational kinetic energy, and Figure 3-5 shows that a molecule also vibrates by internal movement of its atoms. The energies of molecular vibrations are combinations of kinetic and potential energies. Like translational kinetic energy, the rotational and vibrational energies of molecules increase with temperature and are part of the total thermal energy of the gas.

**Figure 3-3**

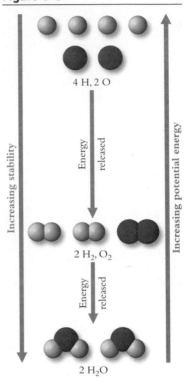

4 H, 2 O

2 H₂, O₂

2 H₂O

Increasing stability

Energy released

Energy released

Increasing potential energy

A collection of two $H_2$ and one $O_2$ molecules is more stable than a collection of four H and two O atoms, but a pair of $H_2O$ molecules is more stable than either of these.

**Figure 3-4**

A water molecule has rotational energy.

In the liquid phase, atoms and molecules still have energies of translation, rotation, and vibration, but the nearby presence of many other molecules significantly restricts their freedom of motion. In the solid phase, in contrast, atoms and molecules no longer translate freely. Each atom or molecule in a solid can vibrate back and forth but cannot easily change places with other atoms and molecules.

Whatever factors contribute to thermal energy, heating an object causes all its atoms and molecules to move faster, leading to an increase in thermal energy. A higher temperature indicates greater thermal energy, and a lower temperature indicates lesser thermal energy.

## Radiant Energy

When you bake in the sun, your body absorbs energy from sunlight. Infrared radiation from a heat lamp in a restaurant keeps food warm until the server delivers the meal to the customer. When a microwave oven cooks food, the food absorbs energy from microwave radiation. Sunlight, infrared light, and microwaves are examples of electromagnetic radiation. Much of what we know about the structure of atoms and molecules comes from studies of how matter interacts with radiation, as we discuss in Chapter 4.

## Energy Transfers and Transformations

Energy can be transferred from one object to another. Anyone who has played a game of pool has firsthand experience with energy transfers. When the cue ball strikes the pack, the balls rebound in all directions. The cue ball loses most of its speed, and it may even come to a complete stop. The collision transfers kinetic energy from the cue ball to the other balls. Whereas transfers of energy among pool balls occur only when initiated by a pool cue, transfers of energy are never-ending at the molecular level. When atoms or molecules of substances collide with one another, energy transfers cause some molecules to speed up and others to slow down.

Thermal energy can also be transferred between objects. When a hot object is placed in contact with a cold object, thermal energy flows from the hot object to the cold object until the two reach the same temperature. At the molecular level, there is a decrease in the average energy of the molecules in the hotter object, so its temperature decreases. There is a corresponding increase in the average energy of the molecules in the cooler object, so its temperature increases.

If you stir a cup of hot coffee with a cool spoon, energy flows from the coffee to the spoon. If the spoon is left in the coffee long enough, the two will reach the same temperature. You can detect the energy transfer easily by touching the spoon after it has been removed from the hot liquid.

**Figure 3-5**

A water molecule can vibrate by internal motion of its atoms.

**Figure 3-6**

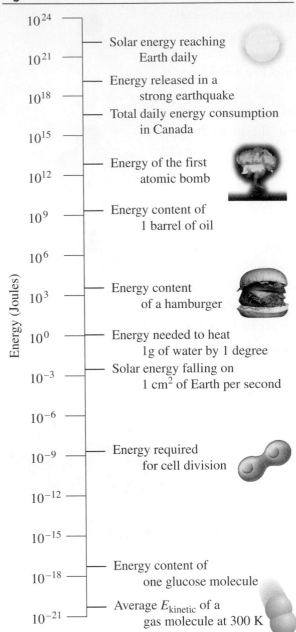

Energy (Joules)

$10^{24}$ —
$10^{21}$ — Solar energy reaching Earth daily
$10^{18}$ — Energy released in a strong earthquake
— Total daily energy consumption in Canada
$10^{15}$ —
$10^{12}$ — Energy of the first atomic bomb
$10^{9}$ — Energy content of 1 barrel of oil
$10^{6}$ —
$10^{3}$ — Energy content of a hamburger
$10^{0}$ — Energy needed to heat 1g of water by 1 degree
$10^{-3}$ — Solar energy falling on 1 cm$^2$ of Earth per second
$10^{-6}$ —
$10^{-9}$ — Energy required for cell division
$10^{-12}$ —
$10^{-15}$ —
$10^{-18}$ — Energy content of one glucose molecule
$10^{-21}$ — Average $E_{kinetic}$ of a gas molecule at 300 K

Chemists deal with energies that span over 40 orders of magnitude, from $10^{-21}$ J (the kinetic energy of a typical gas molecule at room temperature) to $10^{22}$ J (the energy reaching the Earth from the Sun daily).

Energy can also be transformed from one type to another. For example, when a rock falls from a ledge, gravitational potential energy is transformed into kinetic energy. When the rock hits the ground, its kinetic energy is transformed into thermal energy. As backpackers hike up a trail, their bodies oxidize glucose, and so chemical energy is transformed into gravitational potential energy.

Energy transformations accompany chemical reactions. If a reaction releases energy, chemical energy is converted into other forms of energy. A reaction that releases energy almost always produces thermal energy, but chemical energy can be transformed into a variety of other forms, depending on the conditions under which the reaction occurs. An example is the combustion of gasoline. When a gasoline engine powers an automobile, chemical energy is transformed into thermal energy of the hot combustion products, which expand against a piston, transforming the energy into kinetic energy. When a gasoline engine operates a water pump, chemical energy is transformed into gravitational potential energy, whereas if the engine operates an electrical generator, chemical energy is transformed into electrical potential energy.

Energy transformations and energy transfers often occur together. For instance, when a sidewalk absorbs sunlight, radiant energy of the sunlight is transferred to the sidewalk and transformed into thermal energy. When water courses through a turbine, gravitational potential energy of the water is transferred to the turbine blades and eventually transformed into electrical energy.

In addition to its many different types, energy comes in amounts that span a huge range. Figure 3-6 gives some indication of the magnitudes of energies encountered in chemistry.

## SECTION EXERCISES

●**3.1.1** Calculate the kinetic energy (in J) of an electron whose speed is 75 km/h (see inside back cover for electron properties).

●**3.1.2** What type of energy is each of the following? (a) A single molecule of methane moves at 100 km/s; (b) 1 mole of methane molecules has an average speed of 100 km/s; (c) the flame from burning methane emits blue light; and (d) methane is a fuel that releases energy when it burns.

●**3.1.3** What transformations of energy take place during the following processes? (a) Water is heated in a microwave oven; (b) snow breaks loose in an avalanche; and (c) a speeding automobile brakes to a stop.

## 3.2 THERMODYNAMICS

In this section, we develop the basic principles of **thermodynamics**, which is the quantitative study of energy transfers and transformations. Simply put, thermodynamics is the study of "how much energy goes where."

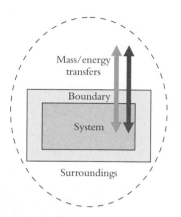

### Terms of Thermodynamics

A thermodynamic **system** is whatever we want to describe and study by itself. Once we select something as a system, everything else is the **surroundings**. A system is separated from its surroundings by a **boundary**, across which matter and/or energy can move. An aquarium is a good example of a system. The aquarium and its contents are separated from the surroundings by a boundary consisting of the glass walls and by the interface between air and water. Matter and energy move between the aquarium system and the surroundings as water evaporates, as food is added, and as the aquarium warms or cools during the course of a day.

An aquarium is an open system.

### Conservation of Energy

Energy can be transferred from one object to another, and energy can be transformed from one type to another. Nevertheless, all studies of thermodynamics indicate that the total amount of energy remains the same during transfers and transformations. That is, *energy is conserved*, as summarized in the **law of conservation of energy**:

*Energy is neither created nor destroyed in any process, although it may be transferred from one body to another or transformed from one form to another.*

Innumerable experiments confirm this fundamental law of science. Whenever the energy of one body increases, a compensating decrease occurs in the energy of some other body. Whenever the amount of one form of energy increases, the amount of some other form decreases by an equal amount.

A baseball helps to illustrate features of conservation of energy. A baseball that has been popped up has kinetic energy. As it rises in the air, the ball slows until it reaches its highest point. At that instant, the ball is not moving vertically, so it has no vertical kinetic energy. What has happened to the initial kinetic energy of the baseball? During its climb, the baseball slows because gravitational force acts to convert the ball's kinetic energy into potential energy. When the ball reaches its highest point, all the vertical kinetic energy of its initial upward motion has been stored as gravitational potential energy. If the ball were to be trapped at this point (e.g., if it landed on top of the screen behind home plate), the energy would remain stored until the ball fell back to the playing field. During that fall, the stored energy would be released again as kinetic energy.

When a catcher catches a fastball, the ball loses the kinetic energy it had while it was in flight. This energy has not been transformed into gravitational potential energy because the ball is no higher above the Earth than before the pitch. It seems that energy has not been conserved. However, careful measurements of the temperature of the ball and of the glove would reveal that as the ball came to rest, the temperature of both objects increased slightly. When a catcher catches a fastball, the kinetic energy lost by the ball shows up as thermal energy of the glove and the ball. This increase in thermal energy results in a higher temperature.

Extending the baseball analogy, a baseball pitcher is very tired after throwing fastballs for nine innings. That tired feeling is the body's signal that stored chemical energy was consumed in throwing all those pitches. Each time your body moves, it does so by transforming chemical energy. The body releases this stored energy by breaking down sugars and fats into carbon dioxide and water. To replenish our supply of stored chemical energy, we must eat.

Kinetic energy

Thermal energy

## Heat

A system can exchange thermal energy with its surroundings. When this happens, the amount of energy that is transferred is referred to as **heat** ($q$). Heat flows are changes in energy, so they are measured in joules (J). Chemical reactions usually result in heat flows. When a reaction releases energy, some of the released energy may flow from the chemical system to its surroundings. When a reaction absorbs energy, that energy is likely to be supplied by a heat flow from the surroundings. Figure 3-7 illustrates heat flows in schematic fashion.

Some heat flows result only in changes in temperature. Under these conditions, the amount of heat can be determined from the temperature change ($\Delta T = T_{final} - T_{initial}$). Experiments show that changes in temperature depend on four factors:

1. $\Delta T$ depends on $q$, the *amount of heat transferred.* As an example, the transfer of 50 J of heat to an object causes an increase in temperature that is twice as large as the increase caused by 25 J of heat.

2. $\Delta T$ depends on the *direction of the heat flow.* If a substance absorbs heat, $\Delta T$ will be positive; but if a substance releases heat, $\Delta T$ will be negative.

3. $\Delta T$ depends inversely on the *amount of material.* For example, $\Delta T$ for the transfer of 50 J of heat to 1 mole of a substance is twice as large as $\Delta T$ for the transfer of 50 J of heat to 2 moles of the same substance.

4. $\Delta T$ depends on the *identity of the material.* For instance, 50 J of heat increases the temperature of 1 mole of gold more than it increases the temperature of 1 mole of water. The dependence of $\Delta T$ on the identity of the material is expressed by the **molar heat capacity** ($C$, units J mol$^{-1}$ °C$^{-1}$). The molar heat capacity is the amount of heat needed to raise the temperature of 1 mole of substance by 1 °C. The molar heat capacities of several chemical substances are listed in Table 3-1.

**Figure 3-7**

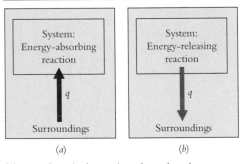

(a)          (b)

(a) In a chemical reaction that absorbs energy, heat flows from the surroundings to the system. (b) In a chemical reaction that releases energy, heat flows from the system to the surroundings.

## TABLE 3-1  MOLAR HEAT CAPACITIES (C) OF REPRESENTATIVE SUBSTANCES AT 25 °C

| Substance | $C$ (J mol$^{-1}$ °C$^{-1}$) | Substance | $C$ (J mol$^{-1}$ °C$^{-1}$) | Substance | $C$ (J mol$^{-1}$ °C$^{-1}$) |
|---|---|---|---|---|---|
| Ag (s) | 25.351 | $C_2H_5OH$ (l) | 111.46 | KCl (s) | 51.3 |
| Al (s) | 24.35 | CO (g) | 29.142 | $K_2Cr_2O_7$ (s) | 219.24 |
| $Al_2O_3$ (s) | 79.04 | $CO_2$ (g) | 37.11 | $MgSO_4$ (s) | 96.48 |
| Ar (g) | 20.786 | $Cl_2$ (g) | 33.907 | $N_2$ (g) | 29.125 |
| $BaCl_2$ (s) | 75.14 | Cu (s) | 24.435 | $NH_3$ (g) | 35.06 |
| $Br_2$ (l) | 75.689 | Fe (s) | 25.10 | $NH_4Cl$ (s) | 84.1 |
| C (s, graphite) | 8.527 | $H_2$ (g) | 28.824 | NO (g) | 29.844 |
| $CaCO_3$ (s) | 81.88 | HCl (g) | 29.12 | Na (s) | 28.24 |
| $CCl_4$ (l) | 131.75 | $H_2S$ (g) | 34.23 | NaOH (s) | 59.54 |
| $CH_3CN$ (l) | 91.5 | He (g) | 20.786 | $O_2$ (g) | 29.355 |
| $CH_3CO_2H$ (l) | 124.3 | Hg (l) | 27.983 | S (s) | 22.64 |
| $CH_3OH$ (l) | 81.6 | $H_2O$ (l) | 75.291 | Si (s) | 20.00 |

One equation incorporates all four factors and describes the temperature change resulting from a heat flow:

$$\Delta T = \frac{q}{nC}$$

In this equation, $q$ is the amount of heat transferred, $n$ is the number of moles of material involved, and $C$ is the molar heat capacity of the substance. Solving for $q$ gives Equation 3-2, the essential equation for calculating heat flows from temperature measurements:

$$q = nC\Delta T \tag{3-2}$$

Energy transfer is directional, and for this reason it is essential to keep track of the signs associated with heat flows. For example, if we drop a heated block of metal into a beaker of cool water, we know that heat will flow from the metal to the water: $q$ is negative for metal and positive for the water. The *amount* of heat flow is the same, however, for both objects. In other words, $q_{water} = -q_{metal}$. In more general terms, the heat is transferred from the system to the surroundings, and so:

$$q_{surr} = -q_{sys}$$

Note that neither temperature nor a change in temperature equates directly with the heat flow, $q$. We can relate $q$ to $\Delta T$ only if we know the identity and amount of the material that undergoes a change of state. Example 3-1 provides some practice.

## EXAMPLE 3-1  HEAT CAPACITY AND TEMPERATURE CHANGE

Calculate the temperature change that results from adding 250 J of thermal energy to each of the following: (a) 0.75 mole of Hg; (b) 0.35 mole of Hg; and (c) 0.35 mole of $H_2O$.

**Strategy:** We are asked to determine the change in temperature, $T$, that accompanies a heat flow. Thermal energy is added to each substance, so we expect an increase in temperature for each case. A diagram similar to Figure 3-7 summarizes the process:

**Solution:** The data provided are $q$ and $n$. These are related to $T$ through Equation 3-2, which requires values of $C$ from Table 3-1:

$$q = nC\Delta T \quad \text{from which} \quad \Delta T = \frac{q}{nC}$$

Substitute values of $q$, $n$, and $C$ and calculate:

(a) $q = 25$ J, $n = 0.75$ mol, $C = 27.983$ J mol$^{-1}$ °C$^{-1}$

$$\Delta T = \frac{q}{nC} = \frac{250 \text{ J}}{(0.75 \text{ mol})(27.983 \text{ J mol}^{-1} \text{ °C}^{-1})} = 12 \text{ °C}$$

(b) $q = 25$ J, $n = 0.35$ mol, $C = 27.983$ J mol$^{-1}$ °C$^{-1}$

$$\Delta T = \frac{q}{nC} = \frac{250 \text{ J}}{(0.35 \text{ mol})(27.983 \text{ J mol}^{-1} \text{ °C}^{-1})} = 26 \text{ °C}$$

(c) $q = 25$ J, $n = 0.35$ mol, $C = 75.291$ J mol$^{-1}$ °C$^{-1}$

$$\Delta T = \frac{q}{nC} = \frac{250 \text{ J}}{(0.35 \text{ mol})(75.291 \text{ J mol}^{-1} \text{ °C}^{-1})} = 9.5 \text{ °C}$$

**Do the Results Make Sense?** In each case, the sign of $T$ is positive because the substance absorbs heat from its surroundings. Notice that the magnitudes of $T$ make sense. The smaller quantity of Hg experiences a larger temperature change than the larger quantity, and the substance with the larger heat capacity, water, experiences a smaller temperature change than the substance with the smaller heat capacity, Hg.

**Common Pitfalls:** Not understanding which way the heat is flowing. Heat always flows from the warmer to the cooler body.

**Practice Exercise 3.1**

Determine $\Delta T$ for 27.5 g of liquid methanol ($CH_3OH$) that transfers 76.5 J of energy to its surroundings.

Equation 3-2 demonstrates that a change of temperature accompanies a heat flow. The equation is also used to compute the amount of heat transferred in a specific temperature change, as Example 3-2 shows.

## EXAMPLE 3-2  HEAT TRANSFER AND TEMPERATURE CHANGE

An aluminum frying pan that weighs 745 g is heated on a stove from 25 °C to 205 °C. What is $q$ for the frying pan?

**Strategy:** A temperature change signals a heat flow. In this example, an increase in the temperature of the aluminum pan means that heat flows from the surroundings (including the stove) to the pan, which represents the system. As in Example 3-1, a diagram summarizes the process:

**Solution:** Equation 3-2 is used to calculate how much heat flows, provided we know $n$, $C$, and $\Delta T$. Use the mass of the frying pan and the molar mass of aluminum to calculate $n$:

$$n = \frac{745 \text{ g Al}}{26.98 \text{ g/mol}} = 27.61 \text{ mol Al}$$

Consult Table 3-1 for the molar heat capacity of Al:

$$C_{Al} = 24.35 \text{ J mol}^{-1}\,{}^\circ\text{C}^{-1}$$

Temperatures are given in the problem:

$$T_i = 25\ {}^\circ\text{C}, \ T_f = 205\ {}^\circ\text{C}$$
$$\Delta T = T_f - T_i = 205\ {}^\circ\text{C} - 25\ {}^\circ\text{C} = 180\ {}^\circ\text{C}$$

Now substitute and calculate, converting to kilojoules to avoid large numbers:

$$q = nC\Delta T = (27.61 \text{ mol})(24.35 \text{ J mol}^{-1}\,{}^\circ\text{C}^{-1})(180\ {}^\circ\text{C})(10^{-3} \text{ kJ/J}) = 121 \text{ kJ}$$

As a frying pan is heated, it loses some energy as heat to the surrounding air at the same time as it gains energy as heat from the burner. The calculation in Example 3-2 gives the *net* heat gain for the pan.

**Does the Result Make Sense?** The result has the right units, and its significant figures are consistent with the data. The quantity is positive, which is what we expect for a temperature increase. The magnitude seems rather large, but so are the temperature change and the amount of Al being heated.

**Practice Exercise 3.2**
A bottle that contains 235 g of drinking water cools in a refrigerator from 26.5 °C to 4.9 °C. Calculate $q$ for the water.

# Work

Backpackers climbing a mountain trail convert chemical energy into gravitational potential energy. Their bodies undergo displacements (increases in altitude) against an external force (force of gravity). Energy used to move an object against an opposing force is called **work** ($w$). Like heat, work is a flow of energy between objects or between a chemical system and its surroundings. The amount of work depends on the magnitude of the force that must be overcome and the amount of movement or displacement. Work is the product of force ($F$) and displacement ($d$): $w = Fd$.

It is just as important to keep track of the signs associated with work as it is to keep track of the signs associated with heat flows. We view work from the perspective of the system. When a system does work on the surroundings, the sign of $w$ is *negative* because the system *loses* energy. When the surroundings do work on the system, $w$ is *positive* for the system because the system *gains* energy. Work is always energy transferred between the system and surroundings, so the sign of $w_{surr}$ is always opposite to the sign of $w_{sys}$. Moreover, the *amount* of work is the same for both the system and surroundings. In other words,

$$w_{surr} = -w_{sys}$$

**Figure 3-8**

All physical human activities involve doing thermodynamic work, which the body supplies from the chemical energy stored in foods.

Chemical energy fuels all human activity, whether it be work or play (Figure 3-8). In any physical activity, a person does work to accomplish movement, and the energy for this work comes from the chemical energy stored in food.

## First Law of Thermodynamics

Energy is exchanged in only two ways: as heat or as work. For example, when gasoline burns in an automobile engine, some of the energy released does work in moving a piston, and some of the energy heats the engine and is lost to the surroundings. A simple equation summarizes the relationship linking work and heat to the energy change of a system:

$$\Delta E = q + w \tag{3-3}$$

Equation 3-3 applies to any process. This equation is one version of the **first law of thermodynamics** and is a restatement of the law of conservation of energy. We can show this by substituting into Equation 3-3 the equalities for heat and work:

$$q_{sys} = -q_{surr} \quad \text{and} \quad w_{sys} = -w_{surr}$$

$$\Delta E_{sys} = q_{sys} + w_{sys} = (-q_{surr}) + (-w_{surr}) = -(q_{surr} + w_{surr}) = -\Delta E_{surr}$$

$$\Delta E_{sys} = -\Delta E_{surr} \tag{3-4}$$

Our primary interest is in a chemical system rather than its surroundings. For convenience, the subscript denoting the *system* usually is omitted in thermodynamic equations. Thus, an energy term without a subscript always refers to a system. To avoid ambiguity, we *never* omit the subscript indicating the surroundings.

Equation 3-4 is a quantitative statement of the law of conservation of energy. Whenever a system undergoes a change in energy, its surroundings undergo an equal and opposite change in energy, ensuring that the total energy remains unchanged. This leads to still another expression for the first law of thermodynamics:

$$\Delta E_{\text{total}} = \Delta E_{\text{sys}} + \Delta E_{\text{surr}} = 0$$

This relationship is profound, because it means that regardless of the conditions and circumstances, we can always completely account for energy changes. This powerful thermodynamic tool helps us analyze many different types of chemical processes.

## State and Path Functions

The properties that describe a system and its transformations can be grouped in two broad categories. Some properties depend only on the conditions that describe the system before and after the transformation. These properties are called **state functions**. Other properties depend on *how* the change occurs. Properties that depend on how a change takes place are called **path functions**.

An everyday example illustrates the difference between state functions and path functions. The Daltons, who live in Vancouver, decide to visit relatives in Iqaluit. They take different routes, as Figure 3-9 shows. Mr. Dalton takes a flight

**Figure 3-9**

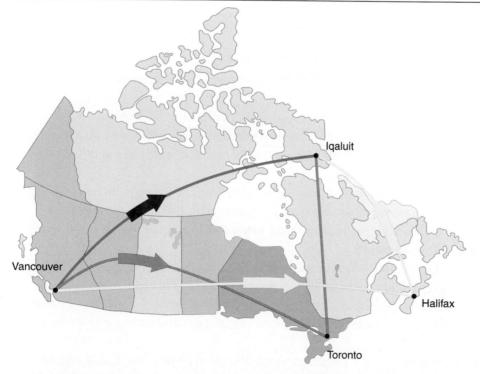

Distance between two cities is a state function, but distance travelled is not a state function. Each of the Daltons travels a different distance, but the distance from Vancouver to Iqaluit is a fixed amount.

## FOCUS 3-1 CHEMISTRY IN CANADA: A NEW ENERGY SOURCE

Plasco Energy Group, an Ottawa area company, is developing a technology to turn garbage into energy and other useful materials. In the Plasco process, any materials with high value are removed from the garbage and collected for recycling. The remaining materials are shredded and enter a high temperature conversion reactor. This step is a thermal process that uses the intense heat from an electrically generated plasma. A plasma is a gaseous state of matter so hot (3000–7000 °C) that molecules become atomized, and a significant portion of the atoms are ionized.

In contrast to incineration or combustion, which operate with excess oxygen to produce $CO_2$, $H_2O$, and oxides of sulphur and nitrogen, the plasma process produces a gas containing CO, $H_2$, $H_2S$, and $NH_3$. Depending on the origin of the waste material, the product gases may also include HCl, $CO_2$, $CH_4$, and other hydrocarbons, as well as carbon and ash. The gases of interest are $H_2$ and CO, a 1 : 1 mixture of which is referred to as synthesis gas. Thus, several chemical processes must be used to remove undesired species from the gas stream.

Acidic compounds such as HCl can be removed by reacting with an alkaline solution such as aqueous sodium hydroxide:

$$HCl\ (g) + NaOH\ (aq) \longrightarrow H_2O\ (l) + NaCl\ (aq)$$

The only product from this step is therefore a dilute salt solution. Hydrogen sulphide can be removed by first absorbing the gas in an alkaline solution:

$$H_2S\ (g) + OH^-\ (aq) \longrightarrow HS^-\ (aq) + H_2O\ (l)$$

followed by oxidation of the $HS^-$ to elemental sulphur (a valuable product) by bacteria in the presence of oxygen:

$$HS^-\ (aq) + \tfrac{1}{2}\,O_2\ (g) \xrightarrow{\text{Bacteria}} S\ (s) + OH^-\ (aq)$$

The hydroxide ion produced by this step is recycled to remove more $H_2S$. Such a procedure is typical in large-scale chemical processes—recycling as much as possible often makes the difference between an economic and an uneconomic process.

Mercury is a common contaminant in waste, but must be prevented from entering the environment because of its toxicity. This is done by contacting the gas stream with activated carbon, a form of carbon having a very high surface area. The mercury atoms adsorb onto the carbon surface, and the small volume of contaminated carbon can then be safely disposed of.

After removal of these contaminants, the gas stream contains mostly $H_2$ and CO. This can be burned in an internal combustion engine to drive a generator and produce electrical power. **The Plasco Conversion System is currently the only waste conversion technology that can generate more than a megawatt-hour of net power per tonne of waste processed.**

During the initial plasma processing step, some solid materials result. This residue is sent to a separate plasma process where the solids are melted, and any remaining volatile compounds and carbon are converted into more crude synthesis gas. This additional gas is fed back into the conversion reactor. Any remaining solids are melted into a liquid slag and cooled into small pellets. These pellets are sold as construction aggregate, to be used in concrete or road paving applications.

The overall process produces very little waste, and gas-phase emissions are extremely low. Plasco intends to install these waste-to-energy conversion plants in cities, which will supply the garbage and consume the electricity.

directly from Vancouver to Iqaluit, but Ms. Dalton goes to Toronto for a business meeting and then flies from Toronto to Iqaluit. The Daltons' daughter goes first to Halifax, where she catches a flight to Iqaluit. On arrival, each of the Daltons is asked two questions by their relatives: "How far is Iqaluit from Vancouver?" and "How far did you travel to get here?" Each answers "3579 km" to the first question, because the distance between the two cities is a difference in values of a state function. Each answers differently to the second question, however, because each Dalton travelled a different distance to reach Iqaluit. Distance travelled depends on the path and is a change in a path function.

The travels of the Dalton family illustrate two features of state functions:

1. The change in a state function can be determined without knowing the details of a process, because that change is independent of path. Iqaluit is 3579 km from Vancouver, no matter how the Daltons travel between the two cities. This means that state function values can be tabulated and looked up when needed. The distance from Vancouver to Iqaluit can be found in an atlas of Canada. Energy is a state function, and changes in energy that accompany standard chemical reactions are tabulated in reference books such as the *CRC Handbook of Chemistry and Physics.*

2. The change in a state function can be measured using the most convenient method available. To determine the distance between two cities, modern geographers make use of satellite mapping. We describe convenient methods for determining energy changes of chemical reactions later in this chapter.

These two features of state functions are more important for scientists than for cross-country travellers like the Daltons. Chemists determine values of state function changes by doing careful experiments using convenient paths. These values are collected in tables, just as distances between cities are collected in an atlas. Energy is one such state function, and chemists use tabulated energy values to analyze chemical processes from a thermodynamic point of view.

How do we know that energy is a state function? Consider the combustion of methane from a molecular perspective. Every molecule of methane that burns reacts with two molecules of oxygen to produce one molecule of carbon dioxide and two molecules of water. Figure 3-10 shows that the value of $\Delta E$ for this process is the difference between the energy required to break all the bonds in the reactant molecules and the energy released in forming all the bonds in the product molecules. This energy difference depends only on the strengths of the various chemical bonds, not on how the bonds are broken and formed. Whether this process occurs in a car fuelled by natural gas or in a home furnace, the combustion reaction has the same overall energy change. In summary, $\Delta E$ depends on the net change in the conditions of the system but not on *how* that change occurs.

**Figure 3-10**

Regardless of the conditions, the reaction of methane with molecular oxygen to form water and carbon dioxide involves breakage of two $O=O$ bonds and four $C-H$ bonds and subsequent formation of four $O-H$ bonds and two $C=O$ bonds for every molecule of methane that reacts.

## Thermodynamic Path Functions

Energy is a state function, but heat and work are path functions. To illustrate this, Figure 3-11 describes two different paths for the combustion of 1 mole of methane. Path 1 represents what happens in an automobile fuelled by natural gas: as methane burns, the system does work on its surroundings by driving back the piston. At the same time, some heat is transferred from the chemicals to the engine block:

$$\Delta E_{\text{engine}} = q_{\text{engine}} + w_{\text{engine}}$$

Path 2 represents what happens when natural gas burns in a furnace: the heat released from burning methane is transferred completely to the surroundings:

$$\Delta E_{\text{furnace}} = q_{\text{furnace}}$$

No work is done along Path 2, so $w_{\text{furnace}} = 0$. For Path 1, $w_{\text{engine}} > 0$, showing that the work done is different for these two paths. Work is a path function, not a state function.

**Figure 3-11**

A block diagram of two different paths for the reaction of methane and oxygen. Occurring inside an engine (Path 1), the process transfers heat and work. Occurring in a furnace (Path 2), the process does no work and transfers only heat.

Knowing that energy is a state function, we can state that $\Delta E$ is the same for both paths:

$$\Delta E_{\text{engine}} = \Delta E_{\text{furnace}}$$

Since in the engine, some of this energy accomplishes work, and in the furnace, all of this energy is transferred as heat, then $q_{\text{engine}} < q_{\text{furnace}}$. The heat transferred is different for the two paths, so $q$ is a path function, not a state function.

As with the distance each Dalton travelled in going from Vancouver to Iqaluit (see Figure 3-9), $q$ and $w$ are path functions. As with the distance between Vancouver and Iqaluit, $\Delta E$ is a state function. The fact that heat transfer depends on the path, whereas energy change is independent of path, has important consequences in chemistry, as we describe later in this chapter.

**SECTION EXERCISES**

●**3.2.1** How much heat is required to raise the temperature of 25.0 g of water from 25 °C to 65 °C on a methane-burning stove (see Table 3-1)?

●**3.2.2** Cyclists in the Tour de France are on the road an average of 4.0 h/day, and their average speed is approximately 40 km/h. Under these conditions, a cyclist uses approximately 3860 kJ/h. How much ground beef must he consume daily to maintain his weight? If the cyclist ate fruit rather than ground beef, how much fruit would he require? (The energy derived from ground beef is 16 kJ/g, and for fruit is 2.5 kJ/g.)

●**3.2.3** Which of the following are state functions? (a) Height of a mountain; (b) distance travelled in climbing that mountain; (c) energy consumed in climbing the mountain; and (d) gravitational potential energy of a climber on top of the mountain

## 3.3 ENERGY CHANGES IN CHEMICAL REACTIONS

Chemical reactions involve rearrangements of atoms: some chemical bonds break, and others form. Bond breakage always requires an input of energy, and bond formation always results in a release of energy. The sum of these energies determines the net energy change for the reaction.

Figure 3-12 summarizes the balance between gains and losses of energy in chemical reactions. If the bonds of the products are more stable than the bonds of the reactants, energy is released during the reaction. For example, the reactions of fuels with oxygen gas release large amounts of energy that can heat a home or drive an engine. For chemical fuels, the energy released by bond formation is greater than the energy consumed by bond breakage (Figure 3-12a).

Conversely, energy is absorbed during a reaction if the bonds of the products are less stable than the bonds of the reactants. The surroundings must transfer energy to the chemical system for the reaction to occur (Figure 3-12b). An example is the photosynthesis reaction, in which plants use radiant energy from sunlight to form energy-rich products from water and carbon dioxide. In photosynthesis, the energy consumed in bond breakage exceeds the energy released by bond formation, so energy must be provided by the surroundings.

Uphill: energy absorbed

Downhill: energy released

If a reaction releases energy when going in one direction, it must absorb an equal amount of energy to go in the opposite direction. Reversing the *direction* of a reaction changes the *sign* of the energy change, but it does not affect the *amount* of energy change. This applies to all energy transfers, not only those occurring in chemical reactions. If a change is "uphill" going in one direction, it must be "downhill" by an equal amount in the other direction.

## Features of Reaction Energies

The thermodynamic conventions for amounts and sign allow us to include the energy change when we write the equation for a balanced chemical reaction. For example, when two $NO_2$ molecules collide, they may form a chemical bond in a reaction that releases energy. The energy flow for $NO_2$ reacting to produce $N_2O_4$ corresponds to Figure 3-12a:

$$2NO_2 \longrightarrow N_2O_4 \qquad \Delta E = -52.8 \text{ kJ}$$

The balanced equation indicates that the reaction of 2 moles of $NO_2$ to form 1 mole of $N_2O_4$ releases 52.8 kJ of energy to the surroundings.

When a chemical reaction releases energy, $\Delta E$ has a negative sign. When a reaction absorbs energy, $\Delta E$ has a positive sign. If we reverse the direction of a chemical reaction, we also reverse the direction of energy flow. A reaction that releases energy in one direction must absorb energy to run in the opposite direction. For example, upon heating, $N_2O_4$ decomposes into a pair of $NO_2$ molecules:

$$N_2O_4 \longrightarrow 2 NO_2 \qquad \Delta E = +52.8 \text{ kJ}$$

The decomposition reaction absorbs energy, with an energy profile similar to the one in Figure 3-12b.

**Figure 3-12**

(a) Energy-releasing Reaction

(b) Energy-absorbing Reaction

Schematic diagrams showing energy flows that accompany reactions. (a) When a reaction releases energy, the change in energy ($\Delta E$) is negative for the chemical reaction and positive for the surroundings. (b) When a reaction absorbs energy, $\Delta E$ is positive for the chemical reaction and negative for the surroundings.

The amount of energy released or absorbed in a chemical reaction is proportional to the amounts of chemicals that react. For instance, when 10 moles of $NO_2$ react to form 5 moles of $N_2O_4$, five times as much energy is released as when 2 moles of $NO_2$ form 1 mole of $N_2O_4$:

$$2\ NO_2 \longrightarrow N_2O_4 \qquad \Delta E = -52.8\ kJ$$

$$10\ NO_2 \longrightarrow 5\ N_2O_4 \qquad \Delta E = -264\ kJ$$

We must take stoichiometric coefficients into account if we are interested in the energy change per mole of any specific reagent. The molar energy change for any component of a reaction can be found by dividing the overall energy change by the stoichiometric coefficient of the specific reagent. Here is the calculation for the energy change per mole of $NO_2$ consumed to produce $N_2O_4$:

$$2\ NO_2 \longrightarrow N_2O_4 \qquad \Delta E = -52.8\ kJ$$

$$\Delta E_{molar,\ NO_2} = \frac{-52.8\ kJ}{2\ mol\ NO_2} = -26.4\ kJ/mol\ NO_2$$

Knowing the overall energy change for a reaction, we can use stoichiometric principles to determine $\Delta E$ for a given amount of starting materials. Example 3-3 shows how to do this.

### EXAMPLE 3-3  ENERGY CHANGE IN A REACTION

The Haber reaction for the formation of ammonia releases energy:

$$N_2 + 3\ H_2 \longrightarrow 2\ NH_3 \qquad \Delta E = -40.9\ kJ$$

How much energy is released in the production of 1.00 kg of ammonia?

**Strategy:** The problem asks for the amount of energy released in a chemical reaction that forms a specified mass of product. The balanced equation is as much visualization as is needed for this problem. The data provided are $\Delta E$ for the balanced equation and the mass of product formed, 1.00 kg.

Our goal is to link the mass of the product with the energy released. We must determine the number of moles of ammonia and take account of the stoichiometric coefficients in the balanced equation. A flowchart summarizes the calculations:

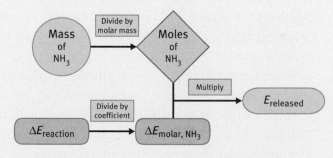

**Solution:** Use the mass and the molar mass of ammonia to calculate $n$. You should verify that the molar mass of ammonia is 17.03 g/mol.

$$n = \frac{m}{M} = \frac{1.00 \text{ kg } (1000 \text{ g/kg})}{17.03 \text{ g/mol}} = 58.72 \text{ mol NH}_3$$

Use the stoichiometric coefficient for $NH_3$ and $\Delta E$ for the balanced equation to determine $\Delta E_{molar}$ for $NH_3$:

$$\frac{-40.9 \text{ kJ}}{2 \text{ mol NH}_3} = -20.45 \text{ kJ/mol NH}_3$$

Now multiply $\Delta E_{molar}$ by the number of moles of $NH_3$ produced:

$$\Delta E = (-20.45 \text{ kJ/mol NH}_3)(58.72 \text{ mol NH}_3) = -1.20 \times 10^3 \text{ kJ}$$

 **Does the Result Make Sense?** The units and number of significant figures are correct. The sign is negative, as it should be for an energy-releasing reaction. A magnitude of 1200 kJ is reasonable for the amount produced, about 60 mol, with an energy change of about $-20$ kJ/mol.

**Common Pitfalls:** Putting the wrong sign on an energy change. Make sure you understand which direction the energy is flowing.

**Practice Exercise 3.3**

The acid–base neutralization reaction releases energy:

$$H_3O^+ \text{ (aq)} + OH^- \text{ (aq)} \longrightarrow 2H_2O \text{ (l)} \qquad \Delta E = -55.8 \text{ kJ}$$

Calculate the energy change when 25.0 mL of 0.100 M HCl reacts with excess aqueous NaOH.

## Path Independence

A change in any thermodynamic state function is independent of the path used to accomplish that change. This feature of state functions tells us that the energy change in a chemical reaction is independent of the manner in which the reaction takes place. In the real world, chemical reactions often follow very complicated paths. Even a relatively simple overall reaction such as the combustion of $CH_4$ and $O_2$ can be very complicated at the molecular level:

$$CH_4 \text{ (g)} + 2O_2 \text{ (g)} \longrightarrow CO_2 \text{ (g)} + 2H_2O \text{ (l)}$$

Nevertheless, path independence tells us that we can calculate the energy change for any reaction by choosing the path that makes the calculation easiest.

Figure 3-13 shows three different paths for the combustion reaction of methane. One path, indicated with the blue arrow, is the path that might occur when natural gas burns on a stove burner. As $CH_4$ and $O_2$ combine in a flame, many chemical species can form, including OH, $CH_3O$, and so on. This is not a convenient path for calculating the energy change for the net reaction, because the process involves many steps and several unstable chemical species.

We can imagine other paths for this reaction that are much simpler than the actual one. Figure 3-13 shows two examples. One path (green arrows) decomposes the reactants into individual atoms, and then combines the atoms to form the products. The other path (orange arrows) converts the reactants into elements in the form that they normally take at room temperature, and then reacts the elements to form the products.

Because energy is a state function, the change of energy in a chemical reaction is determined by the conditions of the reactants and products, as the figure shows. Thus, we can calculate this change by following whichever path we choose. Just as geographers can determine the distance between two cities using the most convenient method, chemists can select a simple path to determine the change in energy for a chemical process.

## Bond Energies

If we wish to use the path that breaks molecules apart into atoms, we must imagine breaking chemical bonds. A chemical bond is a stable arrangement of electrons shared between the bonded atoms. Therefore, a bond cannot be broken without an input of the **bond energy** (*BE*). Bond energies are positive quantities because a molecule must absorb energy to break its chemical bonds. Imagine grasping a molecule and "pulling it apart" at a junction between two particular atoms. The energy that you must supply to accomplish this separation is the bond energy.

Bond energies are expressed most often in units of kilojoules of energy per mole of bonds. To separate 1 mole of hydrogen molecules into hydrogen atoms, for example, requires an input of 435 kJ to break the chemical bonds:

**Figure 3-13**

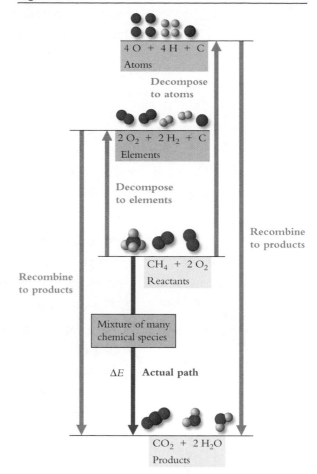

An energy-level diagram demonstrates that the net energy change for a chemical reaction ($\Delta E$) has the same value regardless of the path used to accomplish the reaction. Three different paths are shown here.

$$H_2 \, (g) \longrightarrow H \, (g) + H \, (g) \qquad \Delta E_{\text{bond breaking}} = BE = +435 \text{ kJ/mol}$$

In the opposite direction, energy is released when a bond forms. The formation of 1 mole of $H_2$ from 2 moles of hydrogen atoms releases 435 kJ of energy:

$$H \, (g) + H \, (g) \longrightarrow H_2 \, (g) \qquad \Delta E_{\text{bond forming}} = -BE = -435 \text{ kJ/mol}$$

Bond energies depend on the types of atoms that are bonded together. The bond energies of representative diatomic molecules show that bond energies span a wide range of values from substance to substance:

$$Br_2 \, (g) \longrightarrow Br \, (g) + Br \, (g) \qquad \Delta E = BE = 190 \text{ kJ/mol}$$
$$HCl \, (g) \longrightarrow H \, (g) + Cl \, (g) \qquad \Delta E = BE = 430 \text{ kJ/mol}$$
$$N_2 \, (g) \longrightarrow N \, (g) + N \, (g) \qquad \Delta E = BE = 945 \text{ kJ/mol}$$

In polyatomic molecules, the bond energies depend not only on the specific atoms in the bonds but also on the rest of the molecular structure. Figure 3-14 shows that an ethanol molecule has five C—H bonds in two different bonding

**Figure 3-14**

C—H 410 kJ/mol
C—H 393 kJ/mol

The two different types of C—H bonds in ethanol have slightly different bond energies, because the bonding environment about one C atom differs slightly from the bonding environment about the other C atom.

environments. The C—H bonds shown in red are part of a $CH_3$ group that is bonded to another carbon atom. The C—H bonds in blue belong to a $CH_2$ group that is bonded to both another carbon atom and to an oxygen atom. The different bonding patterns result in different C—H bond energies. To break a red C—H bond requires 410 kJ/mol, whereas to break a blue C—H bond requires slightly less energy, 393 kJ/mol. These variations in the C—H bond energies show that individual bond energies change with the structure of the entire molecule.

Despite the fact that any particular bond energy depends on the entire molecule, many bonds between specific atoms cluster around an average value. For example, C—H bond energies have an average value of 415 kJ/mol.

Table 3-2 gives average bond energies for a variety of chemical bonds. The table has the following features:

1. All values (other than those for diatomic molecules) are averages over different compounds. A bond in any given compound may have a value that differs from the average by between 5 and 20 kJ/mol.

2. The values are given to the nearest 5 kJ/mol.

3. Each value is the energy per mole required to break a gaseous molecule into a pair of neutral gaseous fragments. As we describe in Chapter 8, molecules in liquids and solids experience attractive forces between the molecules that modify the energies of individual bonds. Consequently, average bond energies are less reliable when they are applied to substances in the liquid phase and the solid phase.

Table 3-2 reveals that a few elements, most notably carbon, nitrogen, and oxygen, form more than one type of bond. The average energies of the different types of bonds show striking differences. As we describe in detail in Chapter 7, these differences arise from differences in the number of electrons involved in bond formation. Most of the bonds listed in the table are single bonds, which involve the sharing of two electrons. Double bonds, which involve the sharing of four electrons, are substantially stronger than single bonds. Triple bonds, which involve the sharing of six electrons, are stronger yet. The bond energies of carbon–carbon bonds illustrate this trend:

| C—C | Single bond | $BE$ = 345 kJ/mol |
| C=C | Double bond | $BE$ = 615 kJ/mol |
| C≡C | Triple bond | $BE$ = 835 kJ/mol |

## Reaction Energy

A chemist who runs a chemical reaction needs to know if that reaction releases or absorbs energy. A reaction that releases energy may go out of control or even cause an explosion unless provisions are made to transfer energy efficiently to the surroundings. Conversely, a reaction that absorbs energy is unlikely to proceed efficiently unless provisions are made to provide energy from the surroundings. Fortunately, tabulated data can be used to calculate the energy change for any chemical reaction, including one that has not been previously carried out.

## TABLE 3-2  AVERAGE BOND ENERGIES*

| Diatomic Molecules (Dissociation Energy of Gaseous Molecules) | | | | | | | |
|---|---|---|---|---|---|---|---|
| H—H | 435 | F—F | 155 | O—O | 495 | | |
| H—F | 565 | Cl—Cl | 240 | N=O | 605 | | |
| H—Cl | 430 | Br—Br | 190 | N≡N | 945 | | |
| H—Br | 365 | I—I | 150 | C≡O | 1070 | | |
| H—I | 295 | | | | | | |

| Single Covalent Bonds (Average Values) | | | | | | | | | |
|---|---|---|---|---|---|---|---|---|---|
| H—C | 415 | C—C | 345 | Si—Si | 225 | N—N | 160 | O—O | 145 |
| H—N | 390 | C—N | 305 | Si—F | 565 | N—O | 200 | O—Si | 450 |
| H—O | 460 | C—O | 360 | Si—Cl | 390 | N—F | 285 | O—P | 335 |
| H—Si | 320 | C—Si | 300 | Si—Br | 310 | N—Cl | 200 | O—F | 190 |
| H—P | 320 | C—P | 265 | Si—N | 320 | P—P | 210 | O—Cl | 220 |
| H—S | 365 | C—S | 270 | Sn—Sn | 145 | P—F | 490 | O—Br | 200 |
| H—Te | 240 | C—F | 485 | Sn—Cl | 315 | P—Cl | 320 | S—O | 265 |
| | | C—Cl | 330 | | | P—Br | 270 | S—F | 285 |
| | | C—Br | 275 | | | P—I | 185 | S—S | 240 |
| | | C—I | 215 | | | As—Cl | 295 | S—Cl | 255 |

| Multiple Covalent Bonds (Average Values) | | | | | | | |
|---|---|---|---|---|---|---|---|
| C=C | 615 | N=N | 420 | C≡C | 835 | | |
| C=N | 615 | N=O | 605 | C≡N | 890 | | |
| C=O | 750–800† | O=P | 545 | | | | |
| C=S | 575 | O=S | 515 | | | | |

*All values are in kilojoules per mole (kJ/mol) and are rounded to the nearest 5 kJ/mol.
†C=O bond energy in $CO_2$

Bond energies can be used to estimate the energy change that occurs in a chemical reaction. The reaction of molecular hydrogen with molecular oxygen to form gaseous water provides a simple example:

$$2H_2\,(g) + O_2\,(g) \longrightarrow 2\,H_2O\,(g)$$

Imagine that this reaction occurs along a two-step path. In the first step, the bonds in the reactants break, yielding four hydrogen atoms and two oxygen atoms. In the second step, the hydrogen and oxygen atoms form two water molecules that contain four O—H bonds.

We can calculate the energy change for this reaction, on a molar basis, using bond energies from Table 3-2. The energy required for the bond-breaking step ($\Delta E_{\text{bond breaking}}$) is the sum of the bond energies for all bonds in the reactant molecules:

Step 1
Break 1 mol O=O bonds
Break 2 mol H–H bonds

Step 2
Make 4 mol O–H bonds

$\Delta E_{\text{reaction}}$

$$\Delta E_{\text{bond breaking}} = (2\ \text{mol } H_2)(BE\ H_2) + (1\ \text{mol } O_2)(BE\ O_2)$$

$$\Delta E_{\text{bond breaking}} = (2\ \text{mol } H_2)(435\ \text{kJ/mol}) + (1\ \text{mol})(495\ \text{kJ/mol}) = 1365\ \text{kJ}$$

This energy change is *positive* because the chemical system *gains* the energy required to break the chemical bonds.

The energy change in the second step, $\Delta E_{\text{bond forming}}$, is the sum of all the energies released during bond formation of the products. The reaction produces

2 moles of water containing 4 moles of O—H bonds. Bond formation always releases energy, so the chemical system loses energy to the surroundings. Consequently, the energy change is negative:

$$\Delta E_{\text{bond forming}} = -(4 \text{ mol O—H bonds})(460 \text{ kJ/mol}) = -1840 \text{ kJ}$$

The overall energy change for the reaction is the sum of these two terms:

$$\Delta E_{\text{reaction}} = \Delta E_{\text{bond breaking}} + \Delta E_{\text{bond forming}} = (1365 \text{ kJ}) + (-1840 \text{ kJ})$$
$$= -475 \text{ kJ}$$

The reaction of $H_2$ with $O_2$ fuels the space shuttle.

The Greek letter $\Sigma$ means "the sum of." Thus, $\Sigma BE$ means "the sum of all the bond energies."

The large negative $\Delta E$ for this reaction is one of the reasons why molecular hydrogen is a candidate to replace fossil fuels, as described in our chapter opener.

We can apply the analysis of this reaction to any chemical reaction. The energy change for a chemical reaction can be estimated as the sum of all bond energies of the reactants minus the sum of all bond energies of the products, as summarized in Equation 3-5:

$$\Delta E_{\text{reaction}} = \Sigma BE_{\text{bonds broken}} - \Sigma BE_{\text{bonds formed}} \qquad (3\text{-}5)$$

To apply Equation 3-5, we need complete inventories of the bonds present in all the reagents that participate in the reaction. We build such inventories from structural formulas, from which we count the number of bonds of each type. The procedures for generating structural formulas are not introduced until Chapter 6, so we provide the correct structures for examples in this chapter.

As an example that uses structural formulas and Equation 3-5, consider the energy change that takes place during the combustion reaction of propane $(C_3H_8)$. Combustion is a reaction with molecular oxygen, and the products of propane combustion are carbon dioxide and water:

$$C_3H_8 \ (g) + 5 \ O_2 \ (g) \longrightarrow 3 \ CO_2 \ (g) + 4 \ H_2O \ (g)$$

To estimate the amount of energy absorbed or released in this reaction, we must compile an inventory of all the bonds that break and all the bonds that form. A ball-and-stick model shows that propane contains 8 C—H bonds and 2 C—C bonds. These bonds break in each propane molecule, and one O═O bond breaks in each oxygen molecule. Two C═O bonds form in each $CO_2$ molecule, and two O—H bonds form in each $H_2O$ molecule. In summary:

Now we apply Equation 3-5, using average bond energy values from Table 3-2:

$$\Delta E_{reaction} = \Sigma BE_{bonds\ broken} - \Sigma BE_{bonds\ formed}$$

A table helps us carry out the sums:

| Bonds Broken | Bonds Formed |
|---|---|
| (2 mol C—C) (345 kJ/mol) | (6 mol C═O) (800 kJ/mol) |
| (8 mol C—H) (415 kJ/mol) | (8 mol O—H) (460 kJ/mol) |
| (5 mol O═O) (495 kJ/mol) | |
| Sum: 6485 kJ | Sum: 8480 kJ |

$$\Delta E_{reaction} = 6485\ kJ - 8480\ kJ = -1995\ kJ$$

Equation 3-5 predicts that burning 1 mole of propane releases 1995 kJ of energy. The energy change accompanying this reaction has been measured accurately to be −2044 kJ/mol of propane. Thus, average bond energies predict a value that is off by about 50 kJ/mol, an error of about 2.5%. This error occurs because the bond energies in any given molecule usually differ somewhat from average bond energies. These variations are typically only a few percent, so calculations using average bond energies provide reliable estimates of reaction energies.

The energy released by combusting 1 mole of propane is large compared with the energy changes in typical chemical reactions. Propane is typical of compounds that come from fossil fuels: large negative molar energy values characterize the burning of all these compounds. Why do combustion reactions release so much energy? Table 3-3 summarizes the properties of the bonds involved in the combustion of fossil fuels. The most important features of the data are the bond energies *per pair of electrons*. Notice that the bond of molecular oxygen has a significantly smaller bond energy per electron pair than any of the other bonds in the table. In addition, the O—H and C═O bonds that make up the products of combustion are more stable than the C—H and C—C bonds found in fossil fuels. Thus, burning a compound that contains carbon and hydrogen releases a great amount of energy that can be used to heat homes, cook food, power vehicles, or produce electricity. In fact, most of the energy requirements of our society are provided by the energy released during the combustion of hydrocarbons derived from three fossil fuels: petroleum, natural gas, and coal.

Energy analyses can be extended to reactions other than combustion, as shown in Example 3-4.

## TABLE 3-3 COMPARISON OF COMBUSTION REACTION BOND ENERGIES

| Bond Type | Bond Energy (kJ/mol) | Electron Pairs per Bond | Energy per Electron Pair (kJ) |
|---|---|---|---|
| **Reactant Bonds** | | | |
| O═O | 495 | 2 | 248 |
| C—H | 415 | 1 | 415 |
| C—C | 345 | 1 | 345 |
| **Product Bonds** | | | |
| C═O | 800 | 2 | 400 |
| O—H | 460 | 1 | 460 |

## EXAMPLE 3-4 REACTION ENERGIES

The world produces tens of billions of kilograms of vinyl chloride annually. Most is converted to the polymer poly(vinyl chloride) (PVC), which is used to make piping, siding, gutters, floor tiles, clothing, and toys. Vinyl chloride is made in a two-step process. The balanced overall equation is as follows:

Ethylene　　　　　　　　Vinyl chloride

Based on average bond energies, what energy change accompanies the formation of 1 mole of vinyl chloride? Does the synthesis require an input of energy?

**Strategy:**

1. The question asks for an energy change.

2. The balanced equation presented in the problem is the visual aid we need to identify the bonds that break and form.

**Solution:**

3. The structures are given in the equation. Here is the inventory:

| Bonds Broken | Atom Inventory | Bonds Formed |
|---|---|---|
| 8 C—H bonds | 10 H atoms | 6 C—H bonds |
| 2 C=C bonds | 4 C atoms | 2 C=C bonds |
| 2 H—Cl bonds | 2 Cl atoms | 2 C—Cl bonds |
| 1 O=O bond | 2 O atoms | 4 O—H bonds |

Look up bond energies in Table 3-2:

　C—H: 415 kJ/mol　　C=C: 615 kJ/mol　　　　H—Cl: 430 kJ/mol
　O=O: 495 kJ/mol　　C—Cl: 330 kJ/mol　　　　O—H: 460 kJ/mol

4 and 5. We use Equation 3-5, which does not need to be manipulated:

$$\Delta E_{reaction} = \sum BE_{bonds\ broken} - \sum BE_{bonds\ formed}$$

6. Sum the bond energies for bonds broken and bonds formed and subtract the latter from the former:

| Bonds Broken | Bonds Formed |
|---|---|
| (8 mol C—H) (415 kJ/mol) | (6 mol C—H) (415 kJ/mol) |
| (2 mol C=C) (615 kJ/mol) | (2 mol C=C) (615 kJ/mol) |
| (2 mol H—Cl) (430 kJ/mol) | (2 mol C—Cl) (330 kJ/mol) |
| (1 mol O=O) (495 kJ/mol) | (4 mol O—H)(460 kJ/mol) |

$$\Delta E_{reaction} = (5905\ kJ) - (6220\ kJ) = -315\ kJ$$

Our calculation predicts that 315 kJ of energy is released for every 2 moles of vinyl chloride. However, the problem asks for the energy change associated with 1 mole of product, so divide by 2 to obtain an estimate of 157.5 kJ of energy released per mole of vinyl chloride. Remember, though, that this is an estimate and that tabulated values are given to the nearest 5 kJ, so round this value to 160 kJ. The negative sign for $\Delta E_{reaction}$ indicates that the chemicals release energy to the surroundings.

**Does the Result Make Sense?** Units and significant figures are correct. You should be developing a sense of the energies involved in chemical reactions. Although the sum of reactant bond energies and the sum of product bond energies are very large, the result is a typical reaction energy.

**Practice Exercise 3.4**

Vinyl chloride can also be formed by reacting ethylene with $Cl_2$, which produces 1,2-dichloroethane ($ClCH_2$—$CH_2Cl$). This compound is heated to eliminate HCl. Estimate $\Delta E$ for each of these reactions.

---

## SECTION EXERCISES

● **3.3.1** Methanol ($CH_3OH$) is used in the engines of some high-powered racing cars. Estimate how much energy is released when 1 mole of methanol is burned in air.

● **3.3.2** Draw a diagram like the one in Figure 3-13 to illustrate the different paths for the reaction in Section Exercise 3.3.1.

● **3.3.3** Methyl chloride ($CH_3Cl$) can be produced from methane in either of the following reactions:

$$CH_4 \ (g) + HCl \ (g) \longrightarrow CH_3Cl \ (g) + H_2 \ (g)$$

$$CH_4 \ (g) + Cl_2 \ (g) \longrightarrow CH_3Cl \ (g) + HCl \ (g)$$

Create a bond inventory and estimate the energy change for each reaction. Based on the values that you calculate, which reaction is more suitable for industrial production of methyl chloride? Explain.

methanol

methyl chloride

---

## 3.4 MEASURING ENERGY CHANGES: CALORIMETRY

The bond energies that appear in Table 3-2 were computed from the results of thousands of measurements of the energy changes accompanying chemical reactions. How are such energy changes measured? Recall that any energy change is the sum of two transfers: eat ($q$) and work ($w$). To determine the energy change that accompanies a chemical process, chemists run that process under controlled conditions that allow them to determine heat and work. This section describes those controlled conditions.

## Calorimeters

A **calorimeter** is a device used to measure heat flows that accompany chemical processes. The basic features of a calorimeter include an insulated container and a thermometer that monitors the temperature of the calorimeter. A block diagram of a calorimeter appears in Figure 3-15. In a calorimetry experiment, a chemical reaction takes place within the calorimeter, resulting in a heat flow between the chemicals and the calorimeter. The temperature of the calorimeter rises or falls in response to this heat flow.

Calorimeters are insulated to ensure that negligible heat is transferred between the calorimeter and its surroundings during the experiment. As a result, the only heat exchange is between the chemicals and the calorimeter:

$$q_{chemicals} = -q_{calorimeter}$$

*Exo* is Greek for "out," *endo* is Greek for "within," and *therm* is Greek for "heat."

If the chemicals release heat, $q_{chemicals}$ is negative and the reaction is termed **exothermic**. The calorimeter gains this heat ($q_{calorimeter}$ is positive), and the temperature of the calorimeter rises. Conversely, if the chemicals absorb heat, $q_{chemicals}$ is positive and the reaction is termed **endothermic**. The calorimeter loses this heat ($q_{calorimeter}$ is negative), and the temperature of the calorimeter falls.

We can determine $q_{calorimeter}$ from its temperature change using Equation 3-6, which is similar to Equation 3-2:

$$q_{calorimeter} = C_{cal}\Delta T \qquad (3\text{-}6)$$

Here, $C_{cal}$ is the total heat capacity of the calorimeter. That is, $C_{cal}$ is the amount of heat required to raise the temperature of the entire calorimeter (water bath, container, and thermometer) by 1 °C.

The total heat capacity, $C_{cal}$, can be found by determining the temperature change resulting from a known amount of heat. One way of doing this uses an electrical heater to supply heat ($q_{heater}$) that can be determined accurately by measuring current, voltage, and time. Then, $q_{heater}$ and the measured temperature increase can be used in Equation 3-6 to calculate $C_{cal}$. Example 3-5 illustrates this technique.

**Figure 3-15**

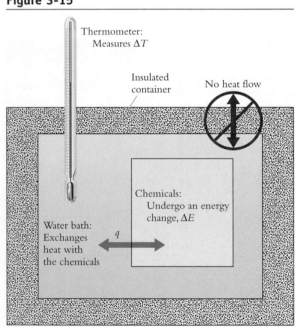

Block diagram of a typical calorimeter.

---

### EXAMPLE 3-5  DETERMINING TOTAL HEAT CAPACITY

A calorimeter is calibrated with an electrical heater. Before the heater is turned on, the calorimeter temperature is 23.6 °C. The addition of $2.02 \times 10^3$ J of electrical energy from the heater raises the temperature to 27.6 °C. Determine the total heat capacity of this calorimeter.

**Strategy:** We are asked to determine the total heat capacity of a calorimeter. A sketch of the process helps to identify what takes place. The electrical heater converts electrical energy into heat that flows into the calorimeter and raises the temperature of the water bath.

**Solution:** The problem statement gives us the temperature data and the heat flow. The heater releases energy to the calorimeter, so $q_{heater}$ has a negative value:

$$q_{heater} = -2.02 \times 10^3 \text{ J}, \quad \Delta T = T_f - T_i = (27.6\,°C - 23.6\,°C) = 4.0\,°C$$

Calorimetry experiments are designed so that the heat transfer is confined to the calorimeter. Equation 3-6 relates heat flow and temperature change:

$$q_{calorimeter} = -q_{heater} \quad \text{and} \quad q_{calorimeter} = C_{cal}\Delta T$$

Combine these equations and rearrange and solve for $C_{cal}$:

$$C_{cal} = \frac{-q_{heater}}{\Delta T}$$

Now, substitute and calculate:

$$C_{cal} = \frac{2.02 \times 10^3 \text{ J}}{4.0\,°C} = 5.1 \times 10^2 \text{ J/°C}$$

A *total* heat capacity has units of J/°C, whereas a *molar* heat capacity has units of J mol$^{-1}$ °C$^{-1}$. If we have $n$ moles of a pure substance, its total heat capacity is $C_{total} = nC_{molar}$.

**Does the Result Make Sense?** The result has the correct units, the value is positive, and is rather large; this is what we expect for a calorimeter, so this is a reasonable result. We round to two significant figures because, although each temperature is precise to three significant figures, the temperature *difference* is only known to two significant figures.

**Common Pitfalls:** Using the wrong value or units for the heat capacity. Make sure you understand the difference between the total heat capacity and the molar heat capacity.

**Practice Exercise 3.5**

What will be the final temperature of this calorimeter if the initial temperature is 23.7 °C and the heater supplies 4.65 kJ of electrical energy?

## Types of Calorimeters

Volume, pressure, temperature, and amounts of substances may change during a chemical reaction. When scientists make experimental measurements, however, they prefer to control as many variables as possible, to simplify the interpretation of their results. In general, it is possible to hold volume or pressure constant, but not both. In *constant-volume calorimetry*, the volume of the system is fixed, whereas in constant-pressure calorimetry the pressure of the system is fixed. Constant-volume calorimetry is most often used to study reactions that involve gases, whereas constant-pressure calorimetry is particularly convenient for studying reactions in liquid solutions. Whichever type of calorimetry is used, temperature changes are used to calculate $q$.

*Constant-pressure calorimetry* requires only a thermally insulated container and a thermometer. A simple, inexpensive constant-pressure calorimeter can be made using two nested Styrofoam cups. Figure 3-16 shows an example. The inner cup holds the water bath, a magnetic stir bar, and the reactants. The thermometer is inserted through the cover. The outer cup provides extra thermal insulation.

**Figure 3-16**

A constant-pressure calorimeter can be constructed from two Styrofoam cups, a cover, a stirrer, and a thermometer.

In one use of this calorimeter, aqueous solutions containing the reactants are mixed in the cup, and the thermometer registers the resulting temperature change. The heat capacity of the calorimeter must be determined independently. One way to do this would be through electrical heating, as described in Example 3-5. However, satisfactory accuracy is often obtained by assuming that the heat capacity of the calorimeter is the same as the heat capacity of the water contained in the calorimeter. This neglects the contribution from the nested cups and the thermometer, which may be only about 1% of $C_{cal}$.

Example 3-6 illustrates an application of constant-pressure calorimetry. Our Focus 3-2: Tools for Discovery describes uses of constant-pressure calorimetry in studies of biological systems.

## EXAMPLE 3-6  CONSTANT-PRESSURE CALORIMETRY

Ammonium nitrate ($NH_4NO_3$, $M = 80.05$ g/mol) is used in cold packs to "ice" injuries. When 20.0 g of this compound dissolves in 125 g of water in a coffee-cup calorimeter, the temperature falls from 23.5 °C to 13.4 °C. Determine $q$ for the dissolving of the compound. Is the process exothermic or endothermic?

### Strategy:

1. We are asked to find $q$ and determine its sign for ammonium nitrate dissolving in water.

2. A sketch helps to identify the process. The temperature of the calorimeter decreases as $NH_4NO_3$ dissolves.

23.5 °C

13.4 °C

125 g

20.0 g

**Solution:**

3. We are given temperature data and information about the calorimeter's contents:

$$\Delta T = T_f - T_i = 13.4\ °C - 23.5\ °C = -10.1\ °C$$

$$NH_4NO_3:\ 20.0\ g,\quad H_2O:\ 125\ g$$

4. Equation 3-6 lets us determine $q$ after heat capacity and temperature change are known.

$$q_{calorimeter} = C_{cal}\Delta T$$

5. Given no additional information, we make the approximation that $C_{cal}$ is the heat capacity of its water content:

$$C_{cal} \cong n_{water}C_{water}$$

$$n = \frac{m}{M} = \frac{125\ g}{18.016\ g/mol} = 6.938\ mol$$

$$C_{water} = (6.938\ mol)(75.291\ J\ mol^{-1}\ °C^{-1}) = 522.4\ J/°C$$

6. Now we are ready to solve for the quantities asked for. Begin by calculating the heat flow for the calorimeter

$$q_{calorimeter} = C_{cal}\Delta T$$

$$q_{calorimeter} = (522.4\ J/°C)(-10.1\ °C) = -5276\ J = -5.28\ kJ$$

The temperature of the calorimeter falls during the process, which means the calorimeter loses heat as the ammonium nitrate dissolves. The ions gain energy, so the dissolving process is endothermic and $q_{salt} = +5.28\ kJ$.

**7. Does the Result Make Sense?** The units are kJ, appropriate for an energy calculation. The drop in temperature of the calorimeter indicates a heat-absorbing process for the calorimeter, so the chemical system absorbs heat from the calorimeter as the compound dissolves. Thus, a positive $q$ is reasonable. Does the magnitude seem reasonable to you?

**Common Pitfalls:** Not understanding that the water in a problem like this one is part of the surroundings, not part of the system. Make sure you understand what the system and what the surroundings are.

**Practice Exercise 3.6**

When 10.00 mL of 1.00 M HCl solution is mixed with 115 mL of 0.100 M NaOH solution in a constant-pressure calorimeter, the temperature rises from 22.45 °C to 23.25 °C. Assuming that the heat capacity of the calorimeter is the same as that of 125 g of water, calculate $q$ for this reaction.

Figure 3-17 illustrates a constant-volume calorimeter of a type that is often used to measure $q$ for combustion reactions. A sample of the substance to be burned is placed inside the sealed calorimeter in the presence of excess oxygen gas. When the sample burns, energy flows from the chemicals to the calorimeter. As in a constant-pressure calorimeter, the calorimeter is well insulated from its surroundings, so all the heat released by the chemicals is absorbed by the calorimeter. The temperature change of the calorimeter, with the calorimeter's heat capacity, gives the amount of heat released in the reaction.

**Figure 3-17**

Control panel    Stirrer

Electronic thermometer

Oxygen bomb (made of steel)

Ignition wires

Water    Insulated walls    Stainless steel bucket    Sample pellet in sample dish

A commercial instrument for constant-volume calorimetry is called a *bomb calorimeter*, because the container in which the reaction occurs resembles a bomb.

## Calculating Energy Changes

In a calorimetry experiment, the heat flow resulting from a process is determined by measuring the temperature change of the calorimeter. Then, $q$ can be related to energy change through the first law of thermodynamics (Equation 3-3):

$$\Delta E = q + w$$

To determine $\Delta E$ using measured values of $q$, we also must know $w$. Because heat and work are path functions, however, we proceed differently for constant volume

### FOCUS 3-2  TOOLS FOR DISCOVERY: CALORIMETRY IN BIOLOGY

Calorimetry is an important technique in biology as well as in chemistry. The inventor of the calorimeter was Antoine Lavoisier, who is shown in the illustration. Lavoisier was a founder of modern chemistry, but he also carried out calorimetric measurements on biological materials. Lavoisier and Pierre Laplace reported in 1783 that respiration is a very slow form of combustion. Thus, calorimetry has been applied to biology virtually from its invention.

Calorimetry is valuable in biological studies because every living thing is an energy-processing system. An analysis of energy flows in biology can provide information about how organisms use energy for growth, reproduction, and many other processes.

Despite Lavoisier's early work on the link between energy and life, calorimetric measurements played a relatively minor role in biology until recent years, primarily because of practical obstacles. Every organism must take in and give off matter as part of its normal function, and it is very difficult to make accurate heat-flow measurements when matter is transferred. Moreover, the sizes of many organisms are poorly matched to the sizes of calorimeters. Although a chemist can adjust the amount of a substance on which to carry out calorimetry, a biologist often cannot.

Improvements in calorimeters have led to a blossoming of biology-related studies within the last two decades. One development is the differential scanning calorimeter, in which the temperature difference is measured between two matched chambers, only one of which contains the material being studied. A second development is improved sensitivity of temperature detection using differential amplifiers and microscopic-sized sensors. These and other advances in instrumentation make it possible to detect temperature changes of less than 0.001 °C and heat flows as small as 0.001 J. The ability to measure very small amounts of heat has been exploited by biologists, as the following examples illustrate.

Calorimetry shows that the rates of metabolism of plant tissues vary widely with species, with cell types, and with environmental conditions. This provides a means of exploring the mechanisms by which various agents influence the health of a plant community. Studies are being done on beneficial agents such as growth promoters, and detrimental ones such as atmospheric pollutants. For example, a correlation has been found between the metabolic heat rates and the extent of damage to pine needles by ozone.

Muscle activity is accompanied by cellular pumping of sodium ions. The energy requirements of the sodium pump have been studied on an individual cardiac muscle mounted inside a tiny differential calorimeter and stimulated by electrical impulses. The heat evolved was different in the presence and absence of a known inhibitor of the sodium pump.

The metabolism of microbes is relatively easy to study by calorimetry because microbial growth is accompanied by energy release in the form of heat. Pharmaceutical calorimetry uses the heat profiles of microbial cultures to assess how various drugs and drug combinations affect the growth of the culture. A graph of heat profiles shows that a drug acting as a growth inhibitor reduces the heat output of a culture, whereas a lethal drug kills the culture and eliminates its heat production. A flow calorimeter, in which a bathing solution is slowly passed through the cell culture, is particularly well suited for these studies, because it mimics a living system and provides information about how quickly a given drug acts.

The pharmaceutical industry has found that microcalorimetry can be a powerful tool both for quality control and in improving drug action. A drug generally releases energy when it degrades. Because of this, measurements of the heat released when a drug is subjected to temperature or humidity changes allow manufacturers to determine how rapidly a drug will lose its effectiveness under different conditions. Bacteria such as those that cause tooth decay continually release energy.

The heat flow from a bacterial colony is significantly reduced when a drug reduces bacterial action, and this feature makes microcalorimetry useful in probing the effectiveness of antibacterial agents.

Microcalorimetry also provides insight into how drugs act at the molecular level. Drug action generally requires the drug to form a chemical interaction with some specific biomolecule in a cell, and these interactions involve energy changes. Using computational techniques like those described in Section 3.3, pharmaceutical chemists can predict the heat profile that will accompany various pharmaceutical interactions. Microcalorimetric measurements can then verify whether a drug is acting effectively.

These are just some of the ways in which calorimetry is used in contemporary biological research. Our examples highlight studies at the cellular level, but ecologists also use calorimetry to explore the energy balances in ecosystems, and whole-organism biologists have found ways to carry out calorimetric measurements on fish, birds, reptiles, and mammals, including humans.

than for constant pressure. To distinguish between these different paths, we use a subscript $v$ for constant-volume calorimetry and a subscript $p$ for constant-pressure calorimetry. This gives different expressions for the two types of calorimeters:

$$\Delta E \text{ (constant volume)} = q_v + w_v \quad \text{and} \quad \Delta E \text{ (constant pressure)} = q_p + w_p$$

As described in Section 3-2, work is the product of force and displacement. In a constant-volume calorimeter, the chemical reaction is contained within the sealed calorimeter, so there is no displacement and $w_v = 0$. Thus:

$$\Delta E = \Delta q_v \quad \text{(constant–volume process)}$$

For a constant-pressure calorimeter, the volume of the reacting chemicals may change, so $w_p \neq 0$ and must be evaluated. We do this in Section 3.5.

## Molar Energy Change

Energy change is an extensive quantity, which means that the amount of energy released or absorbed depends on the amounts of substances that react. For example, as more methane burns, more energy is released. When we report an energy change, we must also report the amounts of the chemical substances that generate the energy change. For tabulation purposes, all changes in thermodynamic functions such as $\Delta E$ are given *per mole*, just as heat capacities are conveniently tabulated as molar values. We use $\Delta E_{molar}$ for the energy change that accompanies the reaction of 1 mole of a particular substance.

Experimental measurements by calorimetry usually involve amounts different from 1 mole. The molar energy change can be found from an experimental energy change by dividing by the number of moles that reacted, as expressed by Equation 3-7:

$$\Delta E_{molar} = \frac{\Delta E}{n} \tag{3-7}$$

Example 3-7 shows how to evaluate $\Delta E_{molar}$ for the combustion of octane.

## EXAMPLE 3-7  MOLAR ENERGY OF COMBUSTION

A 0.1250-g sample of octane ($C_8H_{18}$, $M = 114.2$ g/mol) is burned in excess $O_2$ in the constant-volume calorimeter described in Example 3-5. The temperature of the calorimeter rises from 21.1 °C to 32.9 °C. What is $\Delta E_{molar}$ for the combustion of octane?

### Strategy:

1. We are asked to find the molar energy change for combustion of octane.

2. A sketch helps to identify the process. The heat given off in the combustion reaction is absorbed by the calorimeter.

### Solution:

3. The value of $C_{cal}$ is calculated in Example 3-5, and $\Delta T$ can be found from the data:

$$C_{cal} = 5.1 \times 10^2 \text{ J/°C}, \quad \Delta T_{reaction} = 32.9 \text{ °C} - 21.1 \text{ °C} = 11.8 \text{ °C}$$

We also know the mass of octane (0.1250 g) and its molar mass (114.2 g/mol).

4. The calculation takes more than one step, so we need to identify a process. Use Equation 3-6 to find $q_{\text{calorimeter}}$ The heat gained by the calorimeter is supplied by the chemical reaction, so $q_{\text{reaction}} = -q_{\text{calorimeter}}$. Because the calorimeter operates at constant volume, $w = 0$, so $\Delta E = q_{\text{reaction}}$. This energy change is for 0.1250 g of octane. Use $n = m/M$ to determine $n$, then use Equation 3-7 to convert to the molar energy change: $\Delta E_{\text{molar}} = \Delta E/n$. A flowchart summarizes the process:

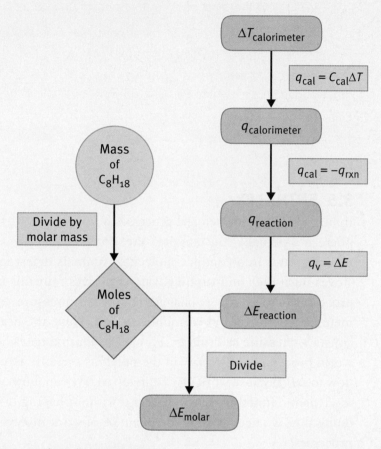

5. None of the equations needs to be manipulated.
6. Now substitute and calculate, step by step:

$$q_{\text{calorimeter}} = C_{\text{cal}}\Delta T = (5.1 \times 10^2 \text{ J/°C})(11.8 \text{ °C})(10^{-3} \text{ kJ/J}) = 6.0 \text{ kJ}$$

$$\Delta E = q_{\text{reaction}} = -q_{\text{calorimeter}} = -6.0 \text{ kJ}$$

$$n = \frac{m}{M} = \frac{0.1250 \text{ g}}{114.2 \text{ g/mol}} = 1.09 \times 10^{-3} \text{ mol}$$

$$\Delta E_{\text{molar}} = \frac{\Delta E}{n} = \frac{-6.0 \text{ kJ}}{1.1 \times 10^{-3} \text{ mol}} = -5.5 \times 10^3 \text{ kJ/mol octane}$$

 **Does the Result Make Sense?** The units are correct, and we know that octane is a good fuel, so the large negative value is reasonable.

## Practice Exercise 3.7

When 0.100 g of graphite (elemental carbon) burns in the calorimeter of Example 3-7, the temperature rises from 23.5 °C to 29.9 °C. Determine the molar energy of combustion of graphite.

## SECTION EXERCISES

● **3.4.1** A coffee-cup calorimeter is calibrated using a small electrical heater. The addition of 3.45 kJ of electrical energy raises the calorimeter temperature from 21.65 °C to 28.25 °C. Calculate the heat capacity of the calorimeter.

● **3.4.2** Adding $1.530 \times 10^3$ J of electrical energy to a constant-pressure calorimeter changes the water temperature from 20.50 °C to 21.85 °C. When 1.75 g of a solid salt is dissolved in the water, the temperature falls from 21.85 °C to 21.44 °C. Find the value of $q_p$ for the solution process.

● **3.4.3** In a constant-volume calorimeter, 3.56 g of solid sulphur is burned in excess oxygen gas:

$$S\ (s) + O_2\ (g) \longrightarrow SO_2\ (g)$$

The calorimeter has a total heat capacity of 4.32 kJ/°C. The combustion reaction causes the temperature of the calorimeter to increase from 25.93 °C to 33.56 °C. Calculate the molar energy for combustion of sulphur.

## 3.5 ENTHALPY

In our world, most chemical processes occur in contact with the Earth's atmosphere at a virtually constant pressure. For example, plants convert carbon dioxide and water into complex molecules; animals digest food; water heaters and stoves burn fuel; and running water dissolves minerals from the soil. All these processes involve energy changes at constant pressure. Most laboratory experiments are performed at constant pressure. Thus, the heat flow measured using constant-pressure calorimetry, $q_p$, closely approximates heat flows in many real-world processes. As we saw in the previous section, we cannot equate this heat flow to $\Delta E$, because work may be involved. We can, however, identify a new thermodynamic function that we can use without having to calculate work. Before doing this, we need to describe one type of work involved in constant-pressure processes.

### Expansion Work

Whenever a chemical process occurs at constant pressure, the volume can change, particularly when gases are involved. In a constant-pressure calorimeter, for instance, the chemical system may expand or contract. In this change of volume, the system moves against the force exerted by the constant pressure. Because work is force times displacement, $w = Fd$, this means that work is done whenever a volume change occurs at constant pressure.

The cylinder and piston in Figure 3-18 illustrate expansion work. If the chemical system expands, it pushes the piston through a displacement ($d$). The opposing force is related to the pressure ($p$) on the piston. As described in Chapter 2, pressure is force per unit area:

$$\text{Pressure} = \frac{\text{Force}}{\text{Area}} = \frac{F}{A} \quad \text{or} \quad F = pA$$

Substituting this into the equation for work, we see how the work done by the expanding system ($w$) is related to the constant pressure ($p$):

$$w = Fd = pAd$$

As Figure 3-18 shows, the volume change of the chemical system is the product of area multiplied by displacement:

$$\Delta V = (\text{Area})(\text{Displacement}) = Ad$$

This lets us write an expression giving the work done on the surroundings in terms of the opposing constant pressure and the change of volume:

$$w_{\text{surr}} = p\Delta V$$

Every energy transfer thus has a *direction*. When a system expands, it does work on the surroundings, transferring energy to the surroundings. This means that $w$ is positive for the surroundings and negative for the system in an expansion:

$$w_{\text{sys}} = -w_{\text{surr}}$$

$$w_{\text{sys}} = -p\Delta V_{\text{sys}} \quad \text{(constant pressure)} \tag{3-8}$$

Equation 3-8 is used to calculate the amount of work done during any process that occurs at constant pressure. When the process is an expansion, $\Delta V_{\text{sys}}$ is positive and $w_{\text{sys}}$ is negative. In a contraction, on the other hand, $\Delta V_{\text{sys}}$ is negative and $w_{\text{sys}}$ is positive.

Because of this expansion work, a process at constant pressure involves both heat and work:

$$\Delta E = q_p + w_p = q_p - p\Delta V_{\text{sys}}$$

Rearranging this equation gives an equation for $q_p$:

$$q_p = \Delta E + p\Delta V$$

We omit the subscript for $V$ because all quantities refer to the chemical system.

**Figure 3-18**

When a reaction occurs in a constant-pressure calorimeter, the external pressure is fixed but the volume of the chemical system can change, so work is done.

## Definition of Enthalpy

Equation 3-9 defines a new state function, called **enthalpy** ($H$), that we can relate to $q_p$:

$$H = E + pV \tag{3-9}$$

We can use Equation 3-9 to relate a change in enthalpy to changes in energy, pressure, and volume:

$$\Delta H = \Delta(E + pV) = E + \Delta(pV)$$

For a process occurring at constant pressure, $\Delta(pV) = p\Delta V$:

$$\Delta H = \Delta E + p\Delta V \quad \text{(constant pressure)}$$

As shown above, the quantity on the right is $q_p$, so a simple equality links $H$ and the heat flow in a constant-pressure process:

$$\Delta H = q \qquad (3\text{-}10)$$

Equation 3-10 lets us describe enthalpy in words:

> *Enthalpy is a thermodynamic function that describes heat flow at constant pressure.*

Just as the heat flow in a constant-volume process gives the energy change, the heat flow in a constant-pressure process gives the enthalpy change:

$$q_v = \Delta E \quad \text{and} \quad q_p = \Delta H$$

Whereas a constant-volume calorimeter provides direct measurements of energy changes, a constant-pressure calorimeter provides direct measurements of enthalpy changes. Table 3-4 summarizes the properties of the two different types of calorimetry.

The units of enthalpy are the same as the units of energy: joules or kilojoules. Tables give enthalpy changes per mole of substance, usually expressed in kilojoules per mole.

### TABLE 3-4 PROPERTIES OF CALORIMETERS

| Type | Reaction Medium | Measurement | Example |
|------|-----------------|-------------|---------|
| Constant pressure | Solution | $q_p = \Delta H$ | Coffee cup |
| Constant volume | Gas | $q_v = \Delta E$ | Bomb |

## Energy and Enthalpy

Energy is a *fundamental* thermodynamic property. In contrast, enthalpy is a *defined* thermodynamic property that is convenient when working at constant pressure. Enthalpy changes and energy changes are related through the *PV* product:

$$\Delta H = \Delta E + \Delta(pV)$$

For solids and liquids, volume changes during chemical reactions are small enough to neglect:

$$\Delta(pV)_{(\text{condensed phases})} \cong 0$$

Thus, enthalpy changes and energy changes are essentially equal for processes that involve only liquids and solids.

An example of such a process is the dissolving of ammonium nitrate in water to produce aqueous ions:

$$NH_4NO_3\,(s) \longrightarrow NH_4^+\,(aq) + NO_3^-\,(aq)$$

As Example 3-6 shows, 20.0 g of $NH_4NO_3$ absorbs 5.28 kJ of energy when the compound dissolves in a constant-pressure calorimeter. This information lets us calculate the molar enthalpy change for the dissolving process:

$$\Delta H = q_p = 5.28 \text{ kJ} \quad \text{and} \quad n = \frac{20.0 \text{ g}}{80.05 \text{ g/mol}} = 0.2498 \text{ mol}$$

$$\Delta H_{molar} = \frac{\Delta H}{n} = \frac{5.28 \text{ kJ}}{0.2498 \text{ mol}} = 21.1 \text{ kJ/mol}$$

Since none of the reactants or products is a gas, the energy and enthalpy changes for dissolving are essentially equal:

$$\Delta E_{molar} = \Delta H_{molar} = 21.1 \text{ kJ/mol}$$

Both pressure and volume may change significantly when a reaction produces or consumes gases. Consequently, the enthalpy change is likely to differ from the energy change for any chemical reaction involving gaseous reagents. To determine the difference, we can use the ideal gas equation to relate the change in the pressure–volume product to the change in number of moles:

$$pV = nRT \quad \text{so} \quad \Delta(pV)_{gases} = \Delta(nRT)_{gases}$$

Combining the equations for enthalpy and $(pV)$, we find that the enthalpy and energy changes are related through Equation 3-11:

$$\Delta H_{reaction} = \Delta E_{reaction} + \Delta(pV)_{reaction}$$

$$\Delta(pV)_{reaction} \cong (pV)_{gases} = \Delta(nRT)_{gases}$$

$$\Delta H_{reaction} \cong \Delta E_{reaction} + \Delta(nRT)_{gases}$$

Because $R$ and $T$ are constant, $RT$ can be brought outside the parentheses.

$$\Delta H_{reaction} \cong \Delta E_{reaction} + RT\Delta n_{gases} \quad \text{(at constant } T\text{)} \qquad (3\text{-}11)$$

In this equation, $\Delta n$ gases is the difference between the numbers of moles of gaseous products and gaseous reactants in the balanced chemical equation:

$$\Delta n_{gases} = n_{gas \ products} - n_{gas \ reactants}$$

Example 3-8 demonstrates a typical magnitude for the difference between $\Delta E_{reaction}$ and $\Delta H_{reaction}$.

## EXAMPLE 3-8  $\Delta E$ AND $\Delta H$ FOR COMBUSTION

Find the difference between the molar $\Delta H$ and $\Delta E$ for the combustion of octane at 298 K.

**Strategy:** In this example, we need to pay particular attention to relationships between different quantities.

1. The problem asks about the difference between a molar $\Delta H$ and $\Delta E$.

2 and 3. Because we are given no specific data, visualizing includes figuring out what information we need. Equation 3-11 links $\Delta H$ and $\Delta E$:

$$\Delta H_{reaction} \cong E_{reaction} + RT\Delta n_{gases}$$

To calculate the difference between $\Delta H_{reaction}$ and $\Delta E_{reaction}$, we need information about the $RT\Delta n$ term as it applies to the combustion reaction.

**Solution:**

4. The balanced chemical equation for combustion applies to 2 moles of octane:

$$2\ C_8H_{18}\ (l) + 25\ O_2\ (g) \longrightarrow 16\ CO_2\ (g) + 18\ H_2O\ (l)$$

The question asks about a molar amount of octane, so divide each coefficient by 2 to obtain a reaction in which 1 mole of octane burns at 298 K:

$$C_8H_{18}\ (l) + \frac{25}{2}O_2\ (g) \longrightarrow 8\ CO_2\ (g) + 9\ H_2O\ (l)$$

5. To find the difference between $H$ and $E$, rearrange Equation 3-11:

$$\Delta H - \Delta E = RT\Delta n_{gases}\ \text{(at constant } T)$$

6. Here are the values needed to complete the calculation:

$$n_f = \text{mol gas (products) mol } CO_2 = 8$$

$$n_i = \text{mol gas (reactants)} = \text{mol } O_2 = \frac{25}{2}$$

$$\Delta n_{gases} = 8\ \text{mol } CO_2 - \frac{25}{2}\text{mol } O_2 = \frac{9}{2}\text{mol gas}$$

$$T = 298\ K \quad \text{(Temperature must be in kelvins to match the units of } R)$$

$$\Delta H - \Delta E = (8.314\ \text{J mol}^{-1}\ K^{-1})(298\ K)\left(-\frac{9}{2}\text{mol gas/mol octane}\right)$$

$$\Delta H - \Delta E = -1.11 \times 10^4\ \text{J/mol octane} = -11.1\ \text{kJ/mol octane}$$

**Does the Result Make Sense?** The units are kJ/mol, as required for molar energy and enthalpy, and the number of significant figures is appropriate. The difference is negative because the number of moles of gas decreases during the reaction. This means the volume change is negative, $\Delta(pV)$ is negative, and so $\Delta H = \Delta E + \Delta(pV)$ is more negative than $\Delta E$. The magnitude of the difference is in the range of typical molar energy values.

**Practice Exercise 3.8**

Determine the difference between molar $\Delta H$ and $\Delta E$ for the combustion of graphite (C).

In Example 3-7, we used calorimetry data to determine that when 1 mole of octane burns, the energy change is −5500 kJ/mol. The result of Example 3-8 shows that the enthalpy change for 1 mole of octane differs from this by −11.1 kJ. To two significant figures, $\Delta H$ and $\Delta E$ are the same, even though the value of $\Delta n_{gases}$ is significant. The fractional difference between $\Delta H_{reaction}$ and $\Delta E_{reaction}$ is 11.1/5500, or about 0.2%.

This calculation shows that reaction energies and reaction enthalpies are usually about the same, even when reactions involve gases. For this reason, chemists often use $\Delta E_{reaction}$ and $\Delta H_{reaction}$ interchangeably. Because many everyday processes occur at constant pressure, thermodynamic tables usually give values for enthalpy changes. Nevertheless, bear in mind that these are different thermodynamic quantities. For processes with modest $\Delta E$ values and significant volume changes, $\Delta E$ and $\Delta H$ can differ substantially.

## Energy and Enthalpy of Vapourization

One process for which $\Delta H$ and $\Delta E$ differ significantly is vapourization, the conversion of a liquid or solid into a gas, or sublimation, the conversion of a solid into a gas. Two examples are:

$$H_2O \ (l) \longrightarrow H_2O \ (g) \qquad \Delta H_{vap}(373 \text{ K}) = 40.79 \text{ kJ/mol}$$

$$I_2 \ (s) \longrightarrow I_2 \ (g) \qquad \Delta H_{sub}(298 \text{ K}) = 62.4 \text{ kJ/mol}$$

A substance changes from a condensed phase (liquid or solid) to the gas phase, so $\Delta n$ (gas) = 1 mol. The change in volume for this type of transformation is almost equal to the volume of the resulting gas. For example, according to the ideal gas equation, 1 mole of water vapour at 373 K and at 1 bar pressure has a volume of 31.0 L:

$$\frac{V}{n} = \frac{RT}{p} = \frac{(0.08314 \text{ L bar mol}^{-1} \text{ K}^{-1})(373 \text{ K})}{(1.00 \text{ bar})} = 31.0 \text{ L/mol}$$

The same amount of liquid water at 373 K has a volume of only 18 mL, or 0.018 L. This small volume can be neglected compared with the larger volume of the gas. Thus, we can simplify Equation 3-11 for vapourizations:

$$\Delta H_{vap} \cong \Delta E_{vap} + RT_{vap} \quad \text{and} \quad \Delta E_{vap} \cong \Delta H_{vap} - RT_{vap}$$

Or for sublimations:

$$\Delta H_{sub} \cong \Delta E_{sub} + RT_{sub} \quad \text{and} \quad \Delta E_{sub} \cong \Delta H_{sub} - RT_{sub}$$

Applying this equation, we calculate $\Delta E_{vap}$ for water and $\Delta E_{sub}$ for iodine:

$$H_2O \ (l) \longrightarrow H_2O \ (g)$$

$$\Delta E_{vap}(373 \text{ K}) = 40.79 \text{ kJ/mol} - (8.314 \text{ J mol}^{-1} \text{ K}^{-1})(373 \text{ K})(10^{-3} \text{ kJ/J})$$

$$= 40.79 \text{ kJ/mol} - 3.10 \text{ kJ/mol} = 37.69 \text{ kJ/mol}$$

$$I_2 \ (s) \longrightarrow I_2(g)$$

$$\Delta E_{sub}(298 \text{ K}) = 62.4 \text{ kJ/mol} - (8.314 \text{ J mol}^{-1} \text{ K}^{-1})(298 \text{ K})(10^{-3} \text{ kJ/J})$$

$$= 62.4 \text{ kJ/mol} - 2.48 \text{ kJ/mol} = 59.9 \text{ kJ/mol}$$

Notice that the difference between $\Delta E$ and $\Delta H$ in each case depends only on temperature. At 373 K (100 °C), $\Delta H_{vap}$ for water is always 3.10 kJ/mol greater than $\Delta E_{vap}$. At 298 K, $\Delta H_{sub}$ for iodine is always 2.48 kJ/mol greater than $\Delta E_{vap}$. The difference between $\Delta E$ and $\Delta H$ arises because, in addition to overcoming intermolecular forces in the condensed phase ($\Delta E$), the escaping vapour must do work, $w = \Delta(pV) = RT$, as it expands against the constant external pressure of the atmosphere.

## Enthalpies of Formation

How do we determine the energy and enthalpy changes for a chemical reaction? We could perform calorimetry experiments and analyze the results, but to do this for every chemical reaction would be an insurmountable task. Fortunately, it turns out to be unnecessary. Using the first law of thermodynamics and the idea of a state function, we can calculate enthalpy changes for almost any reaction using experimental values for one set of reactions, the **formation reactions**.

To make use of these formation reactions, we must also understand the concept of a **standard state**. The standard state of a substance is simply the form that is most stable under specified conditions. Normally, these conditions are 1 bar, 298 K, and a concentration of 1 M if the substance is in solution. For example, bromine is most stable as $Br_2$, not Br, and as a liquid, not as a gas or a solid. Most metals are solids in their standard states. Oxygen and nitrogen are gases in their standard states.

Formation reactions have the following features:

1. There is a *single product* with a stoichiometric coefficient of *1*.

2. All the starting materials are *elements*, and each is in its standard state.

3. Enthalpies of reactions involving gases vary with pressure (because of the $pV$ term in the definition of $H$), so pressures must be specified.

4. Enthalpies of reactions occurring in solution vary with concentration, so concentrations must be specified.

When the partial pressure of each gaseous reagent is 1 bar and the concentration of each species in solution is 1 M, the enthalpy change in a formation reaction is the **standard enthalpy of formation** $(\Delta H_f^\circ)$.

$$\Delta H_f^\circ \quad \begin{matrix} \leftarrow \text{Standard} \\ \leftarrow \text{Formation} \end{matrix}$$
$$\underset{\text{Change}}{\uparrow} \ \underset{\text{Enthalpy}}{\uparrow}$$

Reaction enthalpies may also vary with temperature. Most tabulations of thermodynamic functions are for $T = 298$ K, so we assume that is the temperature unless otherwise specified.

Here are several examples of formation reactions:

$$Fe\ (s) + \frac{1}{2}O_2\ (g) \longrightarrow FeO\ (s) \qquad\qquad \Delta H_f^\circ = -272.0\ \text{kJ/mol}$$

$$Br_2\ (l) \longrightarrow Br_2\ (g) \qquad\qquad \Delta H_f^\circ = 30.9\ \text{kJ/mol}$$

$$\frac{1}{2}N_2\ (g) + \frac{1}{2}O_2\ (g) \longrightarrow NO\ (g) \qquad\qquad \Delta H_f^\circ = 91.3\ \text{kJ/mol}$$

$$C(\text{graphite}) + 2\ H_2\ (g) + \frac{1}{2}O_2\ (g) \longrightarrow CH_3OH\ (l) \qquad \Delta H_f^\circ = -239.2\ \text{kJ/mol}$$

$$C\ (\text{graphite}) \longrightarrow C\ (\text{graphite}) \qquad\qquad \Delta H_f^\circ = 0\ \text{kJ/mol}$$

These examples illustrate additional features of formation reactions. First, formation reactions can be either exothermic or endothermic. Second, an element in a state that is different from the standard state has a positive standard formation enthalpy, as the formation reaction for $Br_2$ (g) shows. Third, fractional stoichiometric coefficients are common in formation reactions because the reaction must

The formation reaction of FeO proceeds vigorously.

generate exactly 1 mole of product. Finally, an element that is already in its standard state has $\Delta H_f^\circ = 0$, because the formation "reaction" involves no change.

Some formation reactions, such as the oxidation of metals, proceed readily and directly. Others do not. Methanol, for example, cannot form by direct combination of elemental carbon, hydrogen, and oxygen. Whether or not a formation reaction occurs in nature, however, the enthalpy of formation is a well-defined quantity.

Over the years, scientists have measured standard enthalpies of formation of many chemical substances. Reference books such as the *CRC Handbook of Chemistry and Physics* contain tables of these values, and Appendix D of this text lists values for many common substances.

## Enthalpy Changes for Chemical Reactions

Standard enthalpies of formation are particularly useful because they can be used to calculate the enthalpy change for any reaction. As a first example, consider nitrogen dioxide reacting to form $N_2O_4$:

$$2\,NO_2\,(g) \longrightarrow N_2O_4\,(g)$$

Imagine this reaction following the two-step pathway that appears in Figure 3-19. In the first step, 2 moles of $NO_2$ molecules decompose into nitrogen molecules and oxygen molecules:

$$2\,NO_2\,(g) \longrightarrow N_2\,(g) + 2\,O_2\,(g)$$

In the second step, these nitrogen and oxygen molecules react to produce $N_2O_4$:

$$N_2\,(g) + 2\,O_2\,(g) \longrightarrow N_2O_4\,(g)$$

The pathway shown in Figure 3-19 is not how the reaction actually occurs, but enthalpy is a state function. Because the change of any state function is independent of the path of the reaction, we can use any convenient path for calculating the enthalpy change. **Hess' law** summarizes this feature.

In this example, the enthalpy change for the overall reaction can be determined by adding the enthalpy changes of the two steps of Figure 3-19:

$$\Delta H_{\text{reaction}}^\circ = \Delta H_{\text{decomposition}}^\circ + \Delta H_{\text{formation}}^\circ$$

$$2NO_2\,(g) \longrightarrow \cancel{N_2}\,(g) + 2\,\cancel{O_2}\,(g) \qquad \Delta H_{\text{decomposition}}^\circ$$

$$\cancel{N_2}\,(g) + 2\,\cancel{O_2}\,(g) \longrightarrow N_2O_4\,(g) \qquad \Delta H_{\text{formation}}^\circ$$

$$\overline{2NO_2\,(g) \longrightarrow N_2O_4\,(g) \qquad \Delta H_{\text{reaction}}^\circ}$$

The decomposition step is related to the formation reaction of $NO_2$ in two ways. First, decomposition is the opposite of formation. Second, this decomposition step involves 2 moles of $NO_2$, whereas the formation reaction involves 1 mole of $NO_2$. Using these features, we can express the enthalpy change of the decomposition reaction for $NO_2$ in terms of the standard enthalpy of formation of $NO_2$. The sign for the enthalpy change of the decomposition reaction is oppo-

**Figure 3-19**

$2\,NO_2\,(g)$

$\Delta H^\circ$ reaction

$\Delta H^\circ$ decomposition

$N_2O_4\,(g)$

Reactants

Products

$\Delta H^\circ$ formation

Elements

$2\,O_2\,(g) \quad + \quad N_2\,(g)$

The enthalpy change for the reaction of $NO_2$ to produce $N_2O_4$ can be calculated using a two-step path. In the first step, $NO_2$ decomposes to $N_2$ and $O_2$. In the second step, the elements react to form $N_2O_4$. Enthalpy is a state function, so $\Delta H$ for the overall reaction is the sum of the enthalpy changes for these two steps.

Hess' law states that the enthalpy change for any overall process is equal to the sum of enthalpy changes for any set of steps that leads from the starting materials to the products.

site that of the corresponding formation reaction, and the total enthalpy of decomposition requires that we multiply by the stoichiometric coefficient:

$$\Delta H^{\circ}_{decomposition} = -2\,\Delta H^{\circ}_f\,(NO_2)$$

The enthalpy change of the formation reaction is just $\Delta H^{\circ}_f$ of $N_2O_4$:

$$\Delta H^{\circ}_{formation} = H^{\circ}_f\,(N_2O_4)$$

Thus, the molar enthalpy change of the overall reaction can be expressed entirely in terms of standard formation reactions:

$$\Delta H^{\circ}_{reaction} = \Delta H^{\circ}_f(N_2O_4) - 2\Delta H^{\circ}_f(NO_2)$$

The enthalpy change of the overall reaction is the sum of the formation enthalpies of the products minus the sum of the formation enthalpies of the reactants.

We can calculate the enthalpy change for the overall reaction using standard enthalpies of formation from Appendix D:

$$\Delta H^{\circ}_{reaction} = (1\ mol\ N_2O_4)(11.1\ kJ/mol\ N_2O_4)$$
$$- (2\ mol\ NO_2)(33.2\ kJ/mol\ NO_2)$$

$$\Delta H^{\circ}_{reaction} = -55.3\ kJ$$

Qualitatively, the negative value for the enthalpy change is logical, because during the reaction, a chemical bond forms between the nitrogen atoms, but none of the bonds of the reactants break.

The same type of path can be visualized for any chemical reaction, as Figure 3-20 shows. The reaction enthalpy for any chemical reaction can be found from the standard enthalpies of formation for all the reactants and products. Multiply each standard enthalpy of formation by the appropriate stoichiometric coefficient, add the values for the products, add the values for the reactants, and subtract the sum for reactants from the sum for products. Equation 3-12 summarizes this procedure:

$$\Delta H^{\circ}_{reaction} = \Sigma\,coeff_{products}\,\Delta H^{\circ}_{f\,products} - \Sigma\,coeff_{reactants}\,\Delta H^{\circ}_{f\,reactants} \qquad (3\text{-}12)$$

Remember that the Greek symbol $\Sigma$ means "the sum of," and "coeff" refers to the stoichiometric coefficients in the balanced equation.

As can be seen from Equation 3-12, the value of $\Delta H^{\circ}_{reaction}$ depends on the coefficients used for the reagents. This means that the enthalpy change for 1

**Figure 3-20**

Any chemical reaction can be imagined to proceed in two stages. First, reactants decompose into their constituent elements. Second, these elements recombine to form products.

mole of any particular reactant may differ from $\Delta H^{\circ}_{\text{reaction}}$. Returning to our earlier example, the enthalpy change of $-55.3$ kJ is for 2 moles of $NO_2$ reacting to form 1 mole of $N_2O_4$. The enthalpy change per mole of $NO_2$ is $(-55.3 \text{ kJ/2 mol}) = -28.6$ kJ/mol. We can generalize this calculation: to find the molar enthalpy change for any reagent, divide $\Delta H^{\circ}_{\text{reaction}}$ by the appropriate stoichiometric coefficient.

As another illustration of the use of Equation 3-12, we calculate the enthalpy of combustion of methane, the principal component of natural gas. We begin with the balanced chemical equation:

$$CH_4 \text{ (g)} + 2 \, O_2 \text{ (g)} \longrightarrow CO_2 \text{ (g)} + 2 \, H_2O \text{ (l)}$$

From Equation 3-12, we have:

$$\Delta H^{\circ}_{\text{reaction}} = [\Delta H^{\circ}_{\text{f}}(CO_2) + 2 \, \Delta H^{\circ}_{\text{f}}(H_2O)] - [\Delta H^{\circ}_{\text{f}}(CH_4) + 2 \, H^{\circ}_{\text{f}}(O_2)]$$

The data required for the calculation come from Appendix D. Notice that the value for $O_2$ is zero, because this is an element in its standard state:

$$\Delta H^{\circ}_{\text{reaction}} = \left[ (1 \text{ mol } CO_2)\left(\frac{-393.5 \text{ kJ}}{\text{mol } CO_2}\right) + (2 \text{ mol } H_2O)\left(\frac{-285.8 \text{ kJ}}{\text{mol } H_2O}\right) \right]$$
$$- \left[ (1 \text{ mol } CH_4)\left(\frac{-74.81 \text{ kJ}}{\text{mol } CH_4}\right) + (2 \text{ mol } O_2)\left(\frac{0 \text{ kJ}}{\text{mol } O_2}\right) \right]$$

$$\Delta H^{\circ}_{\text{reaction}} = [-965.1 \text{ kJ}] - [-74.81 \text{ kJ}] = -890.3 \text{ kJ}$$

## Enthalpy Changes Under Non-Standard Conditions

Standard formation enthalpies refer to 1 bar pressure, 1 M concentration, and $T = 298$ K. What if a reaction occurs under other conditions? Energies and enthalpies of substances change as temperature, concentration, and pressure change, so we must expect that $\Delta H$ for a reaction depends on these variables, too. Calculations of $\Delta H$ under non-standard conditions are beyond the scope of our coverage. Fortunately, changing conditions have only a small effect on $\Delta H$, so calculations of reaction enthalpies using standard enthalpies of formation provide reasonable results, even when the temperature, pressure, and concentration depart somewhat from 298 K, 1 bar, and 1 M.

Example 3-9 applies tabulated enthalpies to a reaction occurring at elevated temperature and pressure.

### EXAMPLE 3-9  MOLAR ENTHALPY CHANGE

Nitric acid ($HNO_3$), which is produced in the gas phase at elevated temperature and pressure, is an important chemical in the fertilizer industry because it can be converted into ammonium nitrate. The following overall chemical equation summarizes the industrial reaction:

$$12 \, NH_3 \text{ (g)} + 21 \, O_2 \text{ (g)} \longrightarrow 8 \, HNO_3 \text{ (g)} + 4 \, NO \text{ (g)} + 14 \, H_2O \text{ (g)}$$

Determine whether this overall process is exothermic or endothermic and estimate the energy change per mole of $HNO_3$ formed.

**Strategy:** Again, we follow the seven-step procedure. The first question asks whether this reaction is exothermic (releases energy) or endothermic (absorbs energy). The second question asks us to estimate the energy change for the reaction per mole of $HNO_3$. Although the reaction conditions differ from standard conditions, $\Delta H$ varies slowly with temperature and pressure, so we can use Equation 3-12 and tabulated standard enthalpies of formation to estimate the reaction enthalpy. Because reaction enthalpy is almost the same as reaction energy, this calculation will give a satisfactory estimate of the reaction energy.

**Solution:** Appendix D lists the following standard heats of formation, all in kJ/mol:

$$NH_3 \, (g): -45.9; \quad HNO_3 \, (g): -133.9; \quad H_2O \, (g): -241.8; \quad NO \, (g): +91.3$$

Substitute these values into Equation 3-12 and use the stoichiometric coefficients from the balanced chemical equation:

$$\Delta H_{reaction} \cong [(8 \text{ mol})(-133.9 \text{ kJ/mol}) + (4 \text{ mol})(91.3 \text{ kJ/mol}) + (14 \text{ mol})(-241.8 \text{ kJ/mol})]$$
$$- [(12 \text{ mol})(-45.9 \text{ kJ/mol}) + (21 \text{ mol})(0 \text{ kJ/mol})]$$
$$= [-1071.2 \text{ kJ} + 365.2 \text{ kJ} - 3385.2 \text{ kJ}] - [-550.8 \text{ kJ} + 0 \text{ kJ}]$$
$$= [-4091.2 \text{ kJ}] - [-550.8 \text{ kJ}]$$
$$\Delta H_{reaction} \cong -3540.4 \text{ kJ} = -3.5404 \times 10^3 \text{ kJ}$$

It is important to pay careful attention to the signs, which we have done by carrying out the calculations in sequence. First, multiply each standard enthalpy, including its sign, by the appropriate coefficient. Then combine values for reactants and values for products, adding or subtracting according to the sign of each enthalpy. Finally, subtract the sum for reactants from the sum for products to obtain the result.

The negative sign for $\Delta H_{reaction}$ indicates that this reaction is exothermic. This is the value for a reaction that generates 8 moles of $HNO_3 \, (g)$. Therefore, the value *per mole* of nitric acid produced is

The heat of formation of a substance depends on its phase, because heat is required to melt or boil a substance. $\Delta H_f^\circ$ for liquid $H_2O$ is $-285.8$ kJ/mol, and $\Delta H_f^\circ$ for gaseous $H_2O$ is $-242$ kJ/mol (the difference is the enthalpy of vapourization of water). This illustrates the importance of keeping track of the phase of each species.

$$\Delta H_{molar} = \frac{-3.5404 \times 10^3 \text{ kJ}}{8 \text{ mol } HNO_3} = -442.55 \text{ kJ/mol } HNO_3$$

**Does the Result Make Sense?** The units are correct and the numerical value is typical of reaction energies and enthalpies. Moreover, it makes sense for the chemical industry to develop exothermic reactions for large-scale industrial production.

**Common Pitfalls:** Forgetting to multiply each standard heat of formation by the stoichiometric coefficient in the balanced reaction.

**Practice Exercise 3.9**
Determine the energy change per mole of $NH_3$ consumed for the first step in nitric acid formation, the reaction of ammonia with oxygen:

$$4 \, NH_3 \, (g) + 7 \, O_2 \, (g) \longrightarrow 4 \, NO_2 \, (g) + 6 \, H_2O \, (l)$$

Equation 3-12 can also be used to calculate the standard enthalpy of formation of a substance whose formation reaction does not proceed cleanly and rapidly. The enthalpy change for some other chemical reaction involving the substance can be determined by calorimetric measurements. Then, Equation 3-12 can be used to calculate the unknown standard enthalpy of formation. Example 3-10 shows how to do this using experimental data from a constant-volume calorimetry experiment combined with standard heats of formation.

**EXAMPLE 3-10  ENTHALPY OF FORMATION OF OCTANE**

The enthalpy of combustion of octane ($C_8H_{18}$) is $-5.5 \times 10^3$ kJ/mol. Using tabulated standard enthalpies of formation in Appendix D, determine the standard enthalpy of formation of octane.

**Strategy:** Because enthalpy is a state function, we can visualize the reaction as occurring through decomposition and formation reactions. Appendix D lists enthalpies of formation, and the experimental heat of combustion is provided. We can use Equation 3-12 to relate the enthalpy of combustion to the standard enthalpy of formation for octane.

**Solution:** Begin with the balanced chemical equation for the combustion of 1 mole of octane:

$$C_8H_{18}\,(l) + \frac{25}{2}O_2\,(g) \longrightarrow 8\,CO_2\,(g) + 9\,H_2O\,(l)$$

Use Equation 3-12 to find the heat of combustion, $\Delta H_{comb}$, for this reaction:

$$\Delta H_{comb} = \Sigma coeff_{products}\,\Delta H^\circ_{f\ products} - \Sigma coeff_{reactants}\,\Delta H^\circ_{f\ reactants}$$

$$= 8\,\Delta H^\circ_f\,(CO_2) + 9\,\Delta H^\circ_f\,(H_2O) - \Delta H^\circ_f\,(octane) - \left(\frac{25}{2}\right)\Delta H^\circ_f\,(O_2)$$

Now rearrange the equation to solve for the enthalpy of formation of octane:

$$\Delta H^\circ_f(octane) = 8\,\Delta H^\circ_f\,(CO_2) + 9\,\Delta H^\circ_f\,(H_2O) - \left(\frac{25}{2}\right)\Delta H^\circ_f\,(O_2) + \Delta H_{comb}$$

Substitute the heat of combustion and experimental values for the heats of formation from Appendix D:

$$\Delta H^\circ_f(octane) = (8\ mol\ O_2)(-393.5\ kJ/mol) + (9\ mol\ H_2O)(-285.8\ kJ/mol)$$

$$-\left(\frac{25}{2}mol\ O_2\right)(0\ kJ/mol) + 5.5 \times 10^3\ kJ/mol$$

$$\Delta H^\circ_f(octane) = -220\ kJ/mol$$

**Does the Result Make Sense?** The units are correct, and the negative value indicates that octane is somewhat more stable than the elements from which it forms. From Appendix D, we see that methane ($CH_4$) and ethane ($C_2H_6$), similar substances to octane, have $\Delta H^\circ_f$ values that are negative, so the result is reasonable.

**Practice Exercise 3.10**

Determine the enthalpy of formation of phosphoric acid from data in Appendix D and the following information:

$$P_4O_{10}\,(s) + 6\,H_2O\,(l) \longrightarrow 4\,H_3PO_4\,(s) \quad \Delta H^\circ_{reaction} = -417\ kJ$$

## Bond Energies and Enthalpies of Formation

As described earlier, we can estimate the energies of chemical reactions from bond energies rather than calculating them from formation reactions. Equation 3-5, like Equation 3-12, contains two sums:

$$\Delta E_{reaction} = \Sigma BE_{bonds\ broken} - \Sigma BE_{bonds\ formed} \qquad (3\text{-}5)$$

The similarity arises because both equations describe similar processes and are consequences of Hess' law. For Equation 3-12, we imagine a process in which reactants decompose into elements in their standard states, which then recombine to give products. For Equation 3-5, we imagine a process in which reactants break up entirely into gaseous atoms, which then recombine to give products.

It is preferable to use Equation 3-12 when standard enthalpies of formation are available for the reagents taking part in a chemical reaction, because this equation uses actual experimental values for standard enthalpies of formation. Consequently, Equation 3-12 is exact at $T = 298$ K, pressure $= 1$ bar, and concentration $= 1$ M, and provides very good estimates even under other conditions. Equation 3-5, on the other hand, uses tabulated bond energies, which are average values for each chemical bond rather than exact values for the bonds in any specific substance. Consequently, Equation 3-5 provides *estimates* of reaction energies rather than *exact* values. The advantage of average values is that we can make a reasonable estimate of the sign and magnitude of $\Delta E$ and $\Delta H$ for reactions that involve substances for which standard formation enthalpies are not available.

## SECTION EXERCISES

●**3.5.1** In a steam engine, water vapour drives a piston to accomplish work. What amount of work, in joules, is done when 25.0 g of water at 100.0 °C vapourizes to steam, pushing a piston against 1.00 bar of external pressure? (*Hint:* Use the ideal gas equation to find the volume change when liquid water vapourizes.)

●**3.5.2** Which of the following substances have zero values for their standard enthalpy of formation? (a) Ozone, $O_3$ (*g*); (b) solid mercury, Hg (*s*); (c) liquid bromine, $Br_2$ (*l*); (d) graphite, C (*s*); (e) atomic fluorine, F (*g*); and (f) solid sulphur, $S_8$ (*s*)

●**3.5.3** Hydrogen gas is prepared industrially from methane and steam:

$$CH_4\ (g) + H_2O\ (g) \longrightarrow CO\ (g) + 3\ H_2\ (g)$$

Use standard enthalpies of formation to determine $\Delta H_{reaction}$ and $\Delta E_{reaction}$ for the balanced reaction at 298 K.

## 3.6 ENERGY SOURCES

Our chapter introduction emphasizes the importance of energy to human civilizations. Civilization may have begun when humans learned how to use wood fires as energy sources. The mastery of fire was just the first step in the use of a variety of energy sources to accomplish increasingly energy-intense goals, such as the launching of spacecraft that escape from the gravitational pull of the Earth.

### Energy and Civilization

The advances of human civilization can be viewed as the results of people figuring out how to increase the availability of energy. As Figure 3-21 shows, the amount of energy used per person per day has multiplied by more than a factor of 100 as humans moved beyond the application of their own muscles, by making use of animals, water power, fossil fuels, and nuclear energy.

The taming of fire was undoubtedly the first advance in energy usage by humans. Burning wood provided heat, light, cooking, and eventually the mastery of chemical arts, such as the firing of pottery and the smelting of metals. Although these applications did not directly increase the amount of energy that humans could use for locomotion or construction, they made life considerably easier and more comfortable.

The domestication of animals occurred at least 6000 years ago. Besides providing a reliable food source, domestic animals supplied energy for many purposes. Hitched to a plough, oxen or water buffalo allowed humans to cultivate larger parcels of land. Land travel over long distances became much easier with horses, camels, and even elephants.

The next stage in exploitation of energy sources was the harnessing of moving air or water. This was the first method of creating mechanical energy to replace work by humans or animals. The Greek poet Antipater of Thessalonica, writing around 80 BC, describes how young women were freed from work by watermills. Wind energy was used to move sailboats from very early times: Egyptian drawings from around 3200 BC show sailboats. It took a long time, however, for humans to learn how to use wind energy for stationary applications. The first windmills, used for grinding grain and pumping water, appear to have been developed in Persia some time around 600 AD.

The Industrial Revolution came hand-in-hand with the use of fossil fuels. Although coal had been used for heating and in metallurgy since at least the thirteenth century, it was not until the invention and refinement of the steam engine that coal consumption increased greatly. By the middle of the nineteenth century, work done by machines exceeded the work done by animal power. Whereas steam engines were mainly fuelled by coal, the advent of the internal combustion engine required a volatile fuel, and petroleum distillates are perfectly suited for this purpose.

Figure 3-22 illustrates how important fossil fuel sources have been since the dawn of the Industrial Revolution. From about 1900 until the present, these

**Figure 3-21**

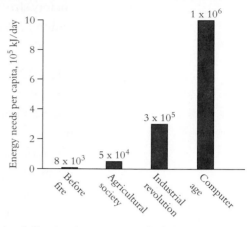

As civilization has progressed, the energy needs per capita have multiplied over 100-fold.

**Figure 3-22**

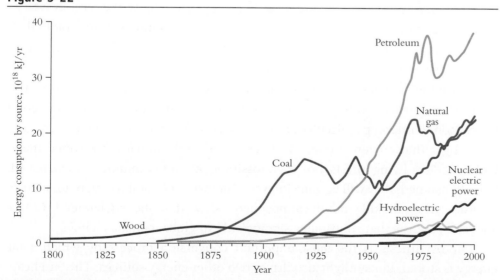

Since 1800, the main sources of energy have been fossil fuels, with other sources contributing less than 20% of the total.

non-renewable sources have supplied more than 80% of the energy needs of our society. In the year 2008, total world energy production was $4.73 \times 10^{20}$ J, of which 85% came from petroleum, coal, and natural gas.

## Ultimate Energy Sources

The vast majority of our energy sources originate in solar energy. The Earth is bathed in a flood of electromagnetic radiation from the Sun, about $8 \times 10^{18}$ J/min. About half of this energy reaches the Earth's surface, the rest being reflected by the atmosphere or radiated from the Earth. Almost none of this incoming energy is used by us directly, however, because solar energy is too diffuse to accomplish energy-intensive tasks like running an automobile.

Fortunately, natural processes convert some solar energy into more concentrated forms. The most obvious process is photosynthesis, by which green plants use the energy of visible light to drive energy-storing chemical reactions. Photosynthesis combines carbon dioxide and water to form compounds that store substantial amounts of chemical energy. One example is glucose:

$$6\ CO_2\ (g) + 6\ H_2O\ (l) + 2880\ kJ \longrightarrow C_6H_{12}O_6\ (s) + 6\ O_2\ (g)$$

Plants use photosynthesis to produce energy-rich compounds. Cattle and other herbivores obtain energy-rich compounds by eating plants. Carnivores eat herbivores, but the food chain ultimately leads back to energy storage via photosynthesis. When a compound such as glucose reacts with oxygen, the energy that was stored during photosynthesis is released and can be used to drive the various processes of life:

$$C_6H_{12}O_6\ (s) + 6\ O_2\ (g) \longrightarrow 6\ CO_2\ (g) + 6\ H_2O\ (l)\quad +2880\ kJ$$

From an energy perspective, forests are bank accounts that store the results of photosynthesis. When we cut and burn wood, we make withdrawals from these energy accounts. So long as we do not cut more rapidly than trees can replace themselves, forests represent a renewable energy resource. In the past, as well as in some parts of the world today, forests were harvested much more rapidly than they reproduced. Fortunately, attitudes about forest use are changing, and modern forestry practices make it possible to sustain this energy resource.

Fossil fuels can also be thought of as bank accounts, because these fuels originally were forests. Geological processes acting over millions of years converted the carbon-containing molecules of these ancient forests into coal, petroleum, and natural gas that remained trapped underground until humans discovered them. Unlike today's living forests, however, fossil fuel accounts cannot be replenished. Once consumed, they will be gone forever. Moreover, the combustion of fossil fuels releases carbon dioxide into the atmosphere. As we describe in Chapter 4, $CO_2$ is a greenhouse gas that contributes significantly to global climate change.

Photosynthesis is the most important conversion process by which solar energy is stored, but sunlight also drives two other energy sources. The Sun heats and evaporates water from the oceans and lakes, and this evaporated water eventually returns to Earth as precipitation. In addition to providing the fresh water that is essential to terrestrial life, the precipitation that falls at high altitudes has stored gravitational energy that can be harnessed through watermills, dams, and hydroelectric turbines. The Sun also heats air masses at different rates. This leads to pressure differences between different locations, as we describe in Chapter 2. Wind is air moving from high-pressure zones to low-pressure zones, and sails and windmills can harness this air movement.

The most prominent non-solar energy source is nuclear energy. Worldwide, nuclear energy supplies about 3.5% of human energy production. Nuclear power plants are used exclusively to produce electricity. Some countries, most notably Belgium (76%) and France (58%), rely heavily on nuclear-generated electricity. Canada's electricity comes mainly from hydroelectric power, but the Province of Ontario relies on nuclear power for approximately 50% of its electricity. The United States currently produces the largest quantity of nuclear energy, but the future of this energy source is uncertain. Many people have strong concerns over the safety of nuclear power plants and the long-term storage of the radioactive waste from these plants, as we describe in Chapter 20.

The Earth's hot interior is an additional non-solar energy source that has not been extensively exploited. Beneath the Earth's crust is a layer called the mantle, where the temperature reaches 4000 °C. For the most part, this energy source is too deep underground to be exploited. In certain areas, however, geological formations allow water to act as a carrier of energy that comes to the surface as superheated water or steam. Power plants have been installed in some of these geothermal areas, most notably in Iceland (look at the photo inset on the opening pages of this chapter), but these represent only a tiny fraction of the overall world energy usage.

## Future Resources

Industrial civilization was built by the consumption of fossil fuels. Currently, well over 80% of world energy comes from the three main fossil fuels: petroleum, natural gas, and coal. Because these are non-renewable resources, there will come a time when these energy bank accounts are exhausted. Experts disagree on how soon the depletion of fossil fuels will occur, but eventually it will be necessary to find other energy sources.

Even if fossil fuel supplies are sufficient for many years to come, there are environmental reasons to reduce our reliance on them. If world consumption of fossil fuels continues at the current pace, there are likely to be significant undesirable climate effects.

From an economic perspective, ideal energy sources have high intensity and are readily extracted and transported. From an environmental perspective, the ideal energy source is both renewable and environmentally benign. Fossil fuels have been nearly ideal from the economic standpoint, which is why they dominate world usage even though they have environmentally undesirable characteristics. It is a major challenge to find and develop other energy sources that are both economically feasible and environmentally friendly.

The solutions to world energy problems most likely will differ by sector of use. Currently in Canada, approximately 36% of energy demands are from industry, 29% from transportation, and 29% from commercial and residential needs. Industry gets approximately one third of its energy needs from electricity, whereas commercial and residential consumers get about two thirds of theirs from electricity.

The production of electricity traditionally requires massive installations such as nuclear power plants, coal-fired generators, or dams with huge turbines. Each of these carries environmental liabilities. Advocates of clean power propose a dual approach to reduce the environmental impacts of electricity generation. One part is large-scale power plants using renewable energy sources: biomass, wind, and geothermal energies. The economics of such power generators generally has not been favourable, but several governments have mandated increased use of so-called green power sources, and power companies increasingly are buying electricity from green power producers. The second part is a web of lower-intensity power generators such as photovoltaic cells and small windmills. Although such generators cannot provide reliable electricity continuously, they can augment central power sources, thereby reducing demand and the consumption of fossil fuels.

Industries that require large amounts of energy have the option of establishing dedicated power plants. Often these can reduce both the cost and the environmental impact of power generation. One example is cogeneration, in which electrical power generation is coupled with heating and cooling. The heat that would be dumped into the environment at a large central power plant is instead used to operate the heating and cooling system for the facility. Another example is the use of biomass, such as the waste from paper mills, wood products manufacturing, orchard pruning, and agricultural by-products. In the United States, more than 350 biomass power plants generate over 7500 megawatts of electricity—enough power to meet the energy needs of several million homes.

The transportation sector relies nearly totally on fossil fuels. This is because the best energy sources for transportation are fluids that deliver the largest possible amount of energy per unit of volume. Gasoline is nearly ideal in this regard, but because it is non-renewable and contributes to the build-up of carbon dioxide in the atmosphere, there is growing interest in replacing gasoline with a clean, renewable fluid fuel.

As described in our chapter introduction, hydrogen may be such a fuel. It can be produced by electrolysis of water and consumed in a fuel cell that generates electricity, with the only product being non-polluting water. As an energy source for transportation, however, hydrogen is far from ideal. One reason is that as a gas, the energy content per unit volume of hydrogen is much smaller than that of liquid fuels like gasoline or methanol. A second reason is that hydrogen must be produced by electrolysis, which requires substantial energy, or by chemical treatment of a fossil fuel, which releases carbon dioxide to the atmosphere. Hydrogen may be the clean transportation fuel of the future, but only if efficient means of production, distribution, and storage can be developed.

## Energy Conservation

Your weight depends on how much food you eat and how much exercise you get. Similarly, an economy needs to balance its energy demands with its energy supplies.

In addition to finding clean, renewable sources of energy, society can reduce the environmental impact of its energy uses by using energy more efficiently.

One good example of improved energy efficiency is cogeneration, described above. Generating electricity at a central power plant and dumping the heat by-products of that generation into the environment, and then using that electricity at another site to heat or cool a facility, is hugely wasteful. Cogeneration is much more energy-efficient. Electricity is generated for the other needs of the facility, such as running motors, lighting, and computers. The waste heat from the generation facility is used to operate heaters and refrigerators.

Improved energy efficiency in heating and cooling can also be accomplished through proper design. Throughout history, humans have designed their homes to maximize comfort. To stay cool in hot climates, they have used contact with the ground (cliff dwellings throughout the world) or air (wind towers in Iran), the evaporation of water (pools and fountains at Alhambra in Spain), or the reflection of the sun's rays (whitewash in Southern Europe and North Africa). To stay warm in cold climates, humans have used southern exposures that capture as much solar energy as possible and thick walls that absorb heat during the day and release it slowly at night.

Contemporary architects and builders have rediscovered and improved on these passive heating and cooling techniques. Passive heating relies on capturing incoming solar radiation and storing it in walls of concrete, brick, or rock. Use of high-quality insulation, including double-glazing of windows, reduces the need for additional heating. Passive cooling relies on shading, good ventilation, evaporators, and contact with cool sources such as the ground.

There is a major potential for energy conservation in transportation, by increasing the energy efficiency of automobiles. The recent commercialization of hybrid vehicles, which combine electric and gasoline motors, demonstrates how much more efficient automobile transport can be. Hybrid power systems deliver double the fuel efficiency of conventional engines. Moreover, as fuel cells are perfected, even greater energy efficiencies may be achieved.

No matter how efficient we become at our uses of energy, however, it will remain true that technological modern societies require copious energy supplies. Meeting those needs without overwhelming our planet with pollution and greenhouse gases is an ongoing challenge that many chemists are working to meet.

## SECTION EXERCISES

●**3.6.1** What was the most important energy source in each of the following years: 1850, 1900, 1950, and 2000?

●**3.6.2** In the nineteenth century, coal-fuelled steam engines were widespread. Today, petroleum-fuelled internal combustion engines have replaced them. Suggest reasons for this replacement.

# CHAPTER 3 VISUAL SUMMARY

## 3.1 TYPES OF ENERGY

Energy transfer

Kinetic energy
Potential energy
   Electrical energy
   Chemical energy
   Gravitational energy
   Mass
Thermal energy
Radiant energy

$$E_{kinetic} = \frac{1}{2}mv^2$$

$$E_{electrical} = k\,\frac{q_1 q_2}{r}$$

$$\overline{E}_{kinetic} = \frac{3RT}{2N_A}$$

Energy transformation

**LEARNING OBJECTIVES:** Recognizing the types of energy of interest to chemists.

**SKILLS TO MASTER:** Identifying forms of energy. (p. 130)

## 3.2 THERMODYNAMICS

Law of conservation of energy

*Energy is neither created nor desctroyed.*

Work (w)

Boundary    System    Vapourization

Heat (q)    $q = nC\Delta T$    Molar heat capacity (C)

Surroundings    State function

Path function

First law of thermodynamics

$$\Delta E = q + w \qquad \Delta E_{sys} = -\Delta E_{surr}$$

**LEARNING OBJECTIVES:** Understanding the first law of thermodynamics and the concepts of heat and work.

**SKILLS TO MASTER:** Calculating the heat that flows between the system and the surroundings. (p. 136)

Calculating the work done on or by the system. (p. 139)

**SKILLS TO MASTER:** Applying the first law of thermodynamics. (p. 144)

**KEY CONCEPTS:** Energy is neither created nor destroyed in any process, although it may be transferred from one body to another or transformed from one form into another.

The value of a state function does not depend on the path taken nor on the rate of the change.

## 3.3 ENERGY CHANGES IN CHEMICAL REACTIONS

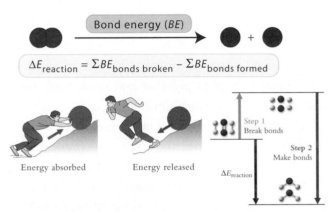

Bond energy (BE)

$$\Delta E_{reaction} = \Sigma BE_{bonds\ broken} - \Sigma BE_{bonds\ formed}$$

Energy absorbed    Energy released

Step 1 Break bonds

Step 2 Make bonds

$\Delta E_{reaction}$

**LEARNING OBJECTIVES:** Understanding the origins of energy changes in chemical reactions.

**SKILLS TO MASTER:** Calculating reaction energies from bond energies. (p. 150)

**KEY CONCEPTS:** Breaking bonds is endothermic. Making bonds is exothermic.

## 3.4 MEASURING ENERGY CHANGES: CALORIMETRY

Exothermic    Endothermic

Surroundings    System: Chemicals    Reactants    $+\Delta E$    $-\Delta E$    Products    Energy flow

Surroundings    System: Chemicals    Products    $-\Delta E$    $+\Delta E$    Reactants    Energy flow

Calorimeter

$$q_{calorimeter} = C_{cal}\Delta T \qquad \Delta E_{sys} = \frac{\Delta E}{n}$$

**LEARNING OBJECTIVES:** Applying the principles of calorimetry to determine energy changes in a chemical reaction.

**SKILLS TO MASTER:** Calculating reaction enthalpies from calorimetric data. (p. 155)

**KEY CONCEPTS:** The heat transferred between the system and the surroundings can be used to calculate the enthalpy change of the system.

The heat transferred in a constant-volume calorimeter is the enthalpy change.

The heat transferred in a constant-volume calorimeter is the energy change.

Calculating the enthalpy change of a chemical reaction from standard enthalpies of formation. (p. 171)

Calculating the enthalpy change of a chemical reaction under non-standard conditions. (p. 173)

**KEY CONCEPTS:** A formation reaction produces 1 mole of a chemical substance from the elements in their standard states.

The enthalpy change for any overall process is equal to the sum of enthalpy changes for any set of steps that leads from the starting materials to the products.

Enthalpy is a thermodynamic state function that describes heat flow at constant pressure.

## 3.5 ENTHALPY

**LEARNING OBJECTIVES:** Understanding and calculating enthalpy and internal energy.

**SKILLS TO MASTER:** Calculating energy and work. (p. 162)

## 3.6 ENERGY SOURCES

**LEARNING OBJECTIVES:** Being familiar with our sources of energy.

## LEARNING EXERCISES

3.1  List as many different types of energy as you can, and provide a specific example illustrating each energy type.

3.2  Write a paragraph that explains the relationships among energy, heat, work, and temperature.

3.3  Prepare a table listing all the bond energies involving halogen atoms (F, Cl, Br, I), organized so it matches the periodic table. What regularities can you find in this listing, and what exceptions are there?

3.4  Write a paragraph describing what happens to the energy released during a chemical reaction that occurs in a constant-pressure calorimeter.

3.5  Describe in your own words what enthalpy is and why it is preferred to energy in describing the thermochemistry of many chemical processes.

3.6  Make a list of energy sources that you personally use. For each item on your list, describe a way that you might be able to reduce your energy consumption.

3.7  Update your list of memory bank equations. Write a sentence that describes the restrictions on each equation. (For example, $H = E + pV$ is a definition that has no restrictions.)

3.8  Prepare a list of the terms in Chapter 3 that are new to you. Write a one-sentence definition for each, using your own words. If you need help, consult the glossary.

## PROBLEMS

**ilw** = interactivelearning ware problem. Visit the website at www.wiley.com/canada/olmsted

### Types of Energy

3.1  Explain how each of the following observations is consistent with conservation of energy: (a) An apple gains kinetic energy as it falls from a tree. (b) When that apple hits the ground, it loses its kinetic energy.

3.2  Explain how each of the following observations is consistent with conservation of energy: (a) Water running into a pond loses kinetic energy. (b) A rock lying in the sun increases in temperature.

3.3  Calculate the kinetic energy of an electron moving at a speed of $4.55 \times 10^5$ m/s (see inside back cover for electron properties).

3.4  Calculate the kinetic energy of a proton moving at a speed of $2.32 \times 10^3$ m/s (see inside back cover for proton properties).

3.5  In KCl solid, the distance between each $K^+$ cation and the nearest $Cl^-$ anion is 320 pm. Calculate the attractive electrical energy for this pair of ions.

3.6  In CaO solid, the distance between each $Ca^{2+}$ cation and the nearest $O^{2-}$ anion is 240 pm. Calculate the attractive electrical energy for this pair of ions.

3.7  In each of the following processes, energy is transformed from one type to another. Identify what type of energy is consumed and what type of energy is produced. (a) Sunlight heats the roof of a house. (b) A baseball pitcher throws a fastball. (c) Wax burns in a candle flame.

3.8  In each of the following processes, energy is transformed from one type to another. Identify what type of energy is consumed and what type is produced. (a) Methane burns on the element of a stove. (b) An elevator carries passengers from the ground floor to the fourth floor. (c) A firefly produces light on a summer evening.

3.9  Compute the speed of a neutron that has a kinetic energy of $3.75 \times 10^{-23}$ J (see inside back cover for neutron properties).

3.10  Compute the speed of a proton that has a kinetic energy of $2.75 \times 10^{-25}$ J (see inside back cover for proton properties).

### Thermodynamics

3.11  An electric heater supplies 25.0 joules of energy to each of the following samples. Compute the final temperature in each case: (a) 10.0-g block of Al originally at 15.0 °C; (b) 25.0-g block of Al originally at 29.5 °C; (c) 25.0-g block of Ag originally at 29.5 °C; and (d) 25.0-g sample of liquid $H_2O$ originally at 22.0 °C.

3.12  Each of the following is placed in an ice bath until it has lost 65.0 joules of energy. Compute the final temperature in each case: (a) 35.0-g block of Al originally at 65.0 °C; (b) 50.0-g block of Al originally at 65.0 °C; (c) 50.0-g block of Ag originally at 65.0 °C; and (d) 50.0-g sample of $H_2O$ originally at 32.5 °C.

3.13  An iron kettle weighing 1.35 kg contains 2.75 kg of water at 23.0 °C. The kettle and water are heated to 95.0 °C. How many joules of energy are absorbed by the water and by the kettle?

3.14  A piece of silver whose mass is 15.0 g is immersed in 25.0 g of water. This system is heated electrically from 24.0 °C to 37.6 °C. How many joules of energy are absorbed by the silver and how many by the water?

3.15  **ilw** A silver coin weighing 27.4 g is heated to 100.0 °C in boiling water. It is then dropped into 37.5 g of water initially at 20.5 °C. Find the final temperature of the coin + water.

3.16  A stainless steel spoon weighs 24.7 g and is at a temperature of 18.5 °C. It is immersed in 85.0 mL of hot coffee ($T = 84.0$ °C) in a Thermos flask. What is the final temperature of the spoon + coffee?
(Assume that the heat capacity of the spoon is the same as that of pure iron and that the heat capacity of the coffee is the same as that of pure water.)

3.17  A pot containing 475 mL of water at 21.5 °C is heated on a stove until its temperature is 87.6 °C. What is $q$ for the water?

**3.18** A dish containing 145 g of water at 54.0 °C is put in a refrigerator to cool. It is removed when its temperature is 5.50 °C. What is $q$ for the water?

**3.19** A 55 kg person walking 6.0 km/hr requires 1090 kJ of energy per hour. How much further does he have to walk to consume the additional energy contained in 250 g of ground beef relative to 250 g of broiled chicken? The energy content of ground beef is 16 kJ/g, while that of chicken is 6.0 kJ/g.

**3.20** A person weighing 85 kg running at 16 km/hr requires 4245 kJ of energy per hour. How far must this person run to consume the energy contained in 0.455 kg of sugar? The energy content of sugar is 16 kJ/g.

## Energy Changes in Chemical Reactions

**3.21** When 1.350 g of benzoic acid ($C_7H_6O_2$) burns completely in excess $O_2$ gas at constant volume and 298 K, it releases 35.61 kJ of energy. (a) What is the balanced chemical equation for this reaction? (b) What is the molar energy of combustion of benzoic acid? (c) How much energy is released per mole of $O_2$ consumed?

**3.22** Acetylene ($C_2H_2$) is used in welding torches because it has a high heat of combustion. When 1.00 g of acetylene burns completely in excess $O_2$ gas at constant volume, it releases 48.2 kJ of energy. (a) What is the balanced chemical equation for this reaction? (b) What is the molar energy of combustion of acetylene? (c) How much energy is released per mole of $O_2$ consumed?

**3.23** Consult Table 3-2 to find the bond energies for bonds between hydrogen and Group 16 elements. Is there a trend? If so, use the trend to predict the bond energies of H—Se and H—Po.

**3.24** Consult Table 3-2 to find the single-bond energies for bonds between identical row 3 elements (e.g., P—P). Is there a trend? If so, use the trend to predict the bond energy of an Al—Al bond.

**3.25** (ilw) Use average bond energies to estimate the energy change for the reaction of $N_2$ and $O_2$ to give $N_2O_4$.

$$N{\equiv}N + 2\,O{=}O \longrightarrow \underset{O}{\overset{O}{N}}{-}\underset{O}{\overset{O}{N}}$$

**3.26** Acrylonitrile is an important starting material for the manufacture of plastics and synthetic rubber. The compound is made from propene in a gas-phase reaction at elevated temperature:

$$2\,C_3H_6 + 2\,NH_3 + 3\,O_2 \longrightarrow$$
$$2\,H_2C{=}H{-}C{\equiv}N + 6\,H_2O$$

Use average bond energies to estimate the energy change for the acrylonitrile synthesis.

**3.27** Draw a diagram like the one in Figure 3-13 to illustrate the different paths for the reaction, $N_2O_4 \rightarrow 2\,NO_2$, which absorbs energy.

**3.28** Draw a diagram like the one in Figure 3-13 to illustrate the different paths for the reaction in Problem 3.26.

**3.29** Estimate the reaction energy for each of the following reactions (all bonds are single bonds except as noted):

$$2\,H_2 + O_2 \text{ (double bond)} \longrightarrow 2\,H_2O$$
$$3\,H_2 + N_2 \text{ (triple bond)} \longrightarrow 2\,NH_3$$
$$3\,H_2 + CO \text{ (triple bond)} \longrightarrow CH_4 + H_2O$$

**3.30** Estimate the energy change for the following reaction when $X$ is (a) F; (b) Cl; (c) Br; and (d) I (all bonds are single bonds):

$$2\,HX\,(g) \longrightarrow H_2\,(g) + X_2\,(g)$$

## Measuring Energy Changes: Calorimetry

**3.31** When 5.34 g of a salt dissolves in 155 mL of water (density = 1.00 g/mL) in a coffee-cup calorimeter, the temperature rises from 21.6 °C to 23.8 °C. Determine $q$ for the solution process, assuming that $C_{cal} \cong C_{water}$.

**3.32** When 10.00 mL of a solution of a strong acid is mixed with 100.0 mL of a solution of a weak base in a coffee-cup calorimeter, the temperature falls from 24.6 °C to 22.7 °C. Determine $q$ for the acid–base reaction, assuming that the liquids have densities of 1.00 g/mL and the same heat capacity as pure water.

**3.33** Constant-volume calorimeters are sometimes calibrated by running a combustion reaction of known $\Delta E$ and measuring the change in temperature. For example, the combustion energy of glucose is 15.57 kJ/g. When a 1.7500-g sample of glucose burns in a constant-volume calorimeter, the calorimeter temperature increases from 21.45 °C to 23.34 °C. Find the total heat capacity of the calorimeter.

**3.34** Constant-pressure calorimeters can be calibrated by electrical heating. When a calorimeter containing 125 mL of water is supplied with 1150 J of electrical energy, its temperature rises from 23.45 °C to 25.25 °C. What is the total heat capacity of the calorimeter, and what percentage of this is due to the water?

**3.35** (ilw) A 1.35-g sample of caffeine ($C_8H_{10}N_4O_2$) is burned in a constant-volume calorimeter that has a heat capacity of 7.85 kJ/°C. The temperature increases from 24.65 °C to 29.04 °C. Determine the amount of heat released and the molar energy of combustion of caffeine.

**3.36** An electric heater adds 19.75 kJ of heat to a constant-volume calorimeter. The temperature of the calorimeter increases by 4.22 °C. When 1.75 g of methanol is burned in the same calorimeter, the temperature increases by 8.47 °C. Calculate the molar energy of combustion of methanol.

## Enthalpy

**3.37** How much work is done in blowing up a balloon from zero volume to a volume of 2.5 L, assuming that $p = 1.00$ bar and no work is required to stretch the rubber? (In reality, the work that goes into stretching the rubber is substantial.)

**3.38** A typical hot-air balloon has a volume of 19.5 $m^3$ when inflated. How much work must be done to inflate such a balloon when atmospheric pressure is 755 Torr?

**3.39** Determine the standard enthalpy change for each of the following reactions:
  (a) $C_2H_4 (g) + 3 O_2 (g) \longrightarrow 2 CO_2 (g) + 2 H_2O (l)$
  (b) $2 NH_3 (g) \longrightarrow N_2 (g) + 3 H_2 (g)$
  (c) $5 PbO_2 (s) + 4 P(s, white) \longrightarrow P_4O_{10} (s) + 5 Pb (s)$
  (d) $SiCl_4 (l) + 2 H_2O (l) \longrightarrow SiO_2 (s) + 4 HCl (aq)$

**3.40** Determine the standard enthalpy change for each of the following reactions:
  (a) $2 NH_3 (g) + 3 O_2 (g) + 2 CH_4 (g) \longrightarrow$
  $2 HCN (g) + 6 H_2O (g)$
  (b) $2 Al (s) + 3 Cl_2 (g) \longrightarrow 2 AlCl_3 (s)$
  (c) $3 NO_2 (g) + H_2O (l) \longrightarrow 2 HNO_3 (g) + NO (g)$
  (d) $2 C_2H_2 (g) + 5 O_2 (g) \longrightarrow 4 CO_2 (g) + 2 H_2O (l)$

**3.41** Find $\Delta E°_{reaction}$ for each of the reactions in Problem 3.39.

**3.42** Find $\Delta E°_{reaction}$ for each of the reactions in Problem 3.40.

**3.43** Write a balanced equation for the formation reaction of each of the following substances: (a) $K_3PO_4 (s)$; (b) acetic acid, $CH_3CO_2H (l)$; (c) trimethylamine, $(CH_3)_3N (g)$; and (d) bauxite, $Al_2O_3 (s)$.

**3.44** Write the balanced equation for the formation reaction of each of the following substances: (a) butanol, $C_4H_9OH (l)$; (b) sodium carbonate, $Na_2CO_3 (s)$; (c) ozone, $O_3 (g)$; and (d) rust, $Fe_3O_4 (s)$.

**3.45** (ilw) Using standard heats of formation, determine $H$ for the following reactions:
  (a) $4 NH_3 (g) + 5 O_2 (g) \longrightarrow 4 NO (g) + 6 H_2O (l)$
  (b) $4 NH_3 (g) + 3 O_2 (g) \longrightarrow 2 N_2 (g) + 6 H_2O (l)$

**3.46** Using standard heats of formation, determine $\Delta H$ for the following reactions:
  (a) $Fe_2O_3 (s) + 3 H_2O (l) \longrightarrow 2 Fe(OH)_3 (s)$
  (b) $B_2O_3 (s) + 3 H_2O (l) \longrightarrow 2 H_3BO_3 (s)$

## Energy Sources

**3.47** The total population of Canada in July of 2008 was 33 212 696. From Figure 3-21, calculate the annual total energy usage of Canada.

**3.48** The total population of Canada in 1867 was 3 463 000. Using data in Figure 3-21 and assuming that the energy usage at that time was typical of the Industrial Revolution, calculate the annual total energy usage of Canada in 1867.

**3.49** Using information in Figure 3-22, calculate the total world energy consumption in 1975. Then construct a pie chart showing the percentages of that total contributed by each major source.

**3.50** Repeat the calculation described in Problem 3.49 for the years 1923 and 2000.

## ADDITIONAL PAIRED PROBLEMS

**3.51** Does more heat have to be removed from an automobile engine when it burns 1 g of gasoline while idling in a traffic jam or when it burns 1 g of gasoline while accelerating? Explain in terms of $\Delta E$, $q$, and $w$.

**3.52** One way to vapourize a liquid is to inject a droplet into a high vacuum. If this is done, is the heat absorbed by the droplet equal to $\Delta E$ or $\Delta H$ for the phase change? Explain in terms of $q$ and $w$.

**3.53** Estimate the difference between $\Delta H_{reaction}$ and $\Delta E_{reaction}$ for the reaction

$$C(graphite) + H_2O (l) \longrightarrow CO (g) + H_2 (g),$$

$$at\ T = 298\ K$$

**3.54** Estimate the difference between $\Delta H_{reaction}$ and $\Delta E_{reaction}$ for the combustion of liquid butanol ($C_4H_9OH$) in excess $O_2 (g)$, at $T = 298$ K.

**3.55** The amount of heat produced in an "ice calorimeter" is determined from the quantity of ice that melts, knowing that it takes 6.01 kJ of heat to melt exactly 1 mole of ice. Suppose that a 12.7 g copper block at 200.0 °C is dropped into an ice calorimeter. How many grams of ice will melt?

**3.56** A gold coin whose mass is 7.65 g is heated to 100.0 °C in a boiling water bath and then quickly dropped into an ice calorimeter. What mass of ice melts? (See Problem 3.55 for useful information.)

**3.57** The human body "burns" glucose ($C_6H_{12}O_6$) for energy. Burning 1.00 g of glucose produces 15.7 kJ of heat. (a) Write the balanced equation for the combustion (burning) of glucose. (b) Determine the molar heat of combustion of glucose. (c) Using appropriate thermodynamic data, determine the heat of formation of glucose.

3.58 Solid urea, $(NH_2)_2CO$, burns to give $CO_2$, $N_2$, and liquid $H_2O$. Its heat of combustion is $-632.2$ kJ/mol. (a) Write the balanced combustion equation. (b) Calculate the heat generated per mole of $H_2O$ formed. (c) Using this heat of combustion and the appropriate thermodynamic data, determine the heat of formation of urea.

3.59 Use average bond energies (Table 3-2) to compare the stabilities of ethanol, $C_2H_5OH$, and dimethyl ether, $(CH_3)_2O$, which have the same empirical formula, $C_2H_6O$ (all the bonds are single bonds).

3.60 Use average bond energies (Table 3-2) to compare the stabilities of allyl alcohol, $H_2C=CHCH_2OH$, and acetone, $(CH_3)_2C=O$, which have the same empirical formula, $C_3H_6O$. (Each compound has one double bond, as indicated in the formulas.)

3.61 What energy transformations take place when an airplane accelerates down a runway and takes off into the air?

3.62 What energy transformations take place when a moving automobile brakes and skids to a stop?

3.63 It takes 100.0 J of heat to raise the temperature of 52.5 g of lead from 28.0 °C to 47.6 °C. What is the molar heat capacity of lead?

3.64 It takes 87.7 J of heat to raise the temperature of 27.0 g of tin from 0.0 °C to 15.0 °C. What is the molar heat capacity of lead?

3.65 Use average bond energies (Table 3-2) to compare the combustion energies of ethane, ethylene, and acetylene. Calculate which of these hydrocarbons releases the most energy per gram.

3.66 Isooctane, $C_8H_{18}$, is the basis of the octane rating system for gasoline because it burns smoothly, with minimal engine "knocking." Use average bond energies to estimate the energy released during the combustion of 1 mole of pure isooctane.

Isooctane
$C_8H_{18}$

3.67 What is the kinetic energy (in J) of a 2250-pound automobile traveling at 57.5 km per hour?

3.68 What is the speed (in km per hour) of a 60-kg runner whose kinetic energy is 345 J?

3.69 Use standard enthalpies of formation to determine $\Delta H^\circ_{reaction}$ for the following reactions:
(a) $2 SO_2 (g) + O_2 (g) \longrightarrow 2 SO_3 (g)$
(b) $2 NO_2 (g) \longrightarrow N_2O_4 (g)$
(c) $Fe_2O_3 (s) + 2 Al (s) \longrightarrow Al_2O_3 (s) + 2 Fe (s)$

3.70 Using standard enthalpies of formation, calculate $\Delta H^\circ_{reaction}$ for the following reactions:
(a) $2 NH_3 (g) + 3 O_2 (g) + 2 CH_4 (g) \longrightarrow$
$$2 HCN (g) + 6 H_2O (g)$$
(b) $2 C_2H_2 (g) + 5 O_2 (g) \longrightarrow 4 CO_2 (g) + 2 H_2O (g)$
(c) $C_2H_4 (g) + O_3 (g) \longrightarrow CH_3CHO (g) + O_2 (g)$

3.71 Use tabulated bond energies in Table 3-2 to estimate the energy change when HCl adds to ethylene ($H_2C=CH_2$) to produce $CH_3CH_2Cl$ and when $Cl_2$ adds to ethylene to produce $ClCH_2CH_2Cl$.

3.72 Phosgene ($Cl_2C=O$) is a highly toxic gas that was used for chemical warfare during World War I. Use the bond energies in Table 3-2 to estimate the energy change that occurs when carbon monoxide and chlorine combine to make phosgene.

$$C=O (g) + Cl_2 (g) \longrightarrow Cl_2C=O (g)$$

3.73 "Strike anywhere" matches contain $P_4S_3$, a compound that ignites when heated by friction. It reacts vigorously with oxygen, as follows:

$$P_4S_3 (s) + 8 O_2 (g) \longrightarrow P_4O_{10} (s) + 3 SO_2 (g)$$
$$\Delta H^\circ_{reaction} = -3,677 \text{ kJ}$$

Use data from Appendix D to determine $\Delta H^\circ_f$ for $P_4S_3 (s)$.

3.74 When a corpse decomposes, much of the phosphorus in the body is converted to phosphine, $PH_3$, a colourless gas with the odour of rotting fish. Phosphine is a highly reactive molecule that ignites spontaneously in air. In the graveyard, phosphine that escapes from the ground ignites in air, giving small flashes of flame. These flashes are sometimes attributed to supernatural causes, such as a will-o'-the-wisp. Determine $\Delta H^\circ$ for the combustion of phosphine.

$$PH_3 (g) + O_2 (g) \longrightarrow P_4O_{10} (s) + H_2O (g)$$
$$\text{(unbalanced)}$$

3.75 A 44.0-g sample of an unknown metal at 100.0 °C is placed in a constant-pressure calorimeter containing 80.0 g of water at 24.8 °C. Assume that the heat capacity of the calorimeter equals the heat capacity of the water it contains. The final temperature is 28.4 °C. Calculate the heat capacity of the metal and use the result to identify the metal:

$$Al = 0.903 \text{ J/g K; Cr} = 0.616 \text{ J/g K;}$$
$$Co = 0.421 \text{ J/g K; or Cu} = 0.385 \text{ J/g K.}$$

3.76 A coin dealer, offered a rare silver coin, suspected that it might be a counterfeit nickel copy. The dealer heated the coin, which weighed 15.5 g, to 100.0 °C in boiling water and then dropped the hot coin into 21.5 g of water at $T = 15.5\,°C$ in a coffee-cup calorimeter. The temperature of the water rose to 21.5 °C. Was the coin made of silver or nickel?

3.77 In some liquid-fuel rockets, such as the lunar lander module of the Apollo Moon missions, the fuels are liquid hydrazine ($N_2H_4$) and dinitrogen tetroxide gas ($N_2O_4$). The two chemicals ignite on contact to release very large amounts of energy:

$$2\,N_2H_4\,(l) + N_2O_4\,(g) \longrightarrow 3\,N_2\,(g) + 4\,H_2O\,(g)$$

(a) Calculate the enthalpy change that takes place when 1 mole of hydrazine burns in a lunar lander.
(b) Calculate the enthalpy change if $O_2$ was used in the lander instead of $N_2O_4$.

3.78 For spacecraft fuels, the energy content *per gram* of fuel should be as large as possible. Which of the following has the largest energy content per gram and which has the smallest?
(a) dimethylhydrazine,
$(CH_3)_2NNH_2$, $\Delta H^{\circ}_{combustion} = -1694\ kJ/mol$;
(b) methanol, $CH_3OH$, $\Delta H^{\circ}_{combustion} = -726\ kJ/mol$; or
(c) octane, $C_8H_{18}$, $\Delta H^{\circ}_{combustion} = -5590\ kJ/mol$

3.79 Refer to Figure 3-22 to answer the following questions:
(a) Before 1850, how many main sources of energy were there, and what were they? (b) Before 1980, what was the highest value for natural gas production, and in what year? (c) What percentage of energy was provided by fossil fuels (petroleum, natural gas, and coal) in 1950?

3.80 Refer to Figure 3-22 to answer the following questions:
(a) Before 1985, what was the highest value for petroleum production, and in what year? (b) What percentage of energy was provided by coal in 1945? (c) When was the first year in which coal provided less energy than either natural gas or petroleum?

3.81 For the constant-temperature process that follows, give the sign (+, −, or 0) for each of the specified thermodynamic functions. In each case, give a brief account of your reasoning: (a) $\Delta H_{sys}$; (b) $\Delta E_{surr}$; and (c) $\Delta E_{univ}$.

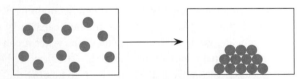

3.82 For the constant-temperature process that follows, give the sign (+, −, or 0) for each of the specified thermodynamic functions. In each case, give a brief account of your reasoning: (a) $w_{sys}$; (b) $q_{sys}$; and (c) $\Delta E_{surr}$.

**ADDITIONAL PROBLEMS**

3.83 A piece of rhodium metal whose mass is 4.35 g is heated to 100.0 °C and then dropped into an ice calorimeter. As the Rh metal cools to 0.0 °C, 0.316 g of ice melts (it takes 6.01 kJ of heat to melt exactly 1 mole of ice). What is the molar heat capacity of rhodium?

3.84 Gases are sold and shipped in metal tanks under high pressure. A typical tank of compressed air has a volume of 30.0 L and is pressurized to 15.0 bar at $T = 298\ K$. What work had to be done in filling this tank if the compressor operates at 15.0 bar? (*Hint:* What volume did the air occupy before it was compressed?)

3.85 A 70-kg person uses 220 kJ of energy to walk 1.0 km. This energy comes from "burning" glucose (see Problem 3.57), but only about 30% of the heat of combustion of glucose can be used for propulsion. The rest is used for other bodily functions or is "wasted" as heat. Assuming that a "sugar-coated" breakfast cereal contains 35% sugar (which can be considered glucose) and no other energy source, calculate how many grams of cereal provide enough energy to walk 1.0 km.

3.86 A home swimming pool contains 155 $m^3$ of water. At the beginning of swimming season, the water must be heated from 20 °C to 30 °C. (a) How much heat energy must be supplied? (b) If a natural gas heater supplies this energy with an 80% heat transfer efficiency, how many grams of methane must be burned? (The heat of combustion of methane is $-803\ kJ/mol$.)

3.87 One way to cool a hot beverage is with a cold spoon. A silver spoon weighing 99 g is placed in a Styrofoam cup containing 205 mL of hot coffee at 83.2 °C. Find the final temperature of the coffee, if the initial temperature of the spoon is 22.0 °C and assuming that coffee has the same heat capacity as water. Would an aluminum spoon of the same mass cool the coffee more or less effectively?

3.88 A room in a home measures 3.0 m by 5.0 m by 4.0 m. Assuming no heat or material losses, how many grams of natural gas (methane, $CH_4$) must be burned to heat the air in this room from 15 °C to 25 °C? Assume that air is 78% $N_2$ and 22% $O_2$, and use data from Table 3-1 and Appendix D.

3.89 Only one of the following expressions describes the heat of a chemical reaction under all possible conditions. Which is it? For each of the others, give an example for which the expression gives the wrong value for the heat. (a) $\Delta E$; (b) $\Delta H$; (c) $q_v$; (d) $q_p$; and (e) $\Delta E - w$.

3.90 For the following constant-temperature process, give the sign (+, −, or 0) for each of the specified thermodynamic functions. In each case, give a brief account of your reasoning: (a) $\Delta w_{sys}$; (b) $\Delta E_{surr}$; and (c) $q_{sys}$.

**3.91** The power ratings of microwave ovens are listed in watts (1 watt = 1 J/s). How many seconds will it take a microwave oven rated at 750 watts to warm 575 mL of soup from 25 °C to 85 °C? Assume that the heat capacity and the density of the soup are the same as that of pure water.

**3.92** In metric terms, a typical automobile averages 6.0 km/L of gasoline burned. Gasoline has a heat of combustion of 48 kJ/g and a density of 0.68 g/mL. How much energy is consumed in driving an automobile 1.0 km?

**3.93** A 9.50-g copper block, initially at 200.0 °C, is dropped into a Thermos flask containing 200 mL of water initially at 5.00 °C. What is the final temperature of the Thermos flask contents?

**3.94** Use average bond energies (see Table 3-2) to estimate the net energy change per mole of silicon for the conversion of a silicon chain into an Si—O—Si chain. Repeat this calculation to estimate the net energy change per mole of carbon for the conversion of a carbon chain into a C—O—C chain.

**3.95** The following figure represents a piston and cylinder containing a collection of gas molecules. The piston can move in either direction. Assume that the gas molecules are the system. (a) Redraw the figure to show what happens when some work is done on the system. (b) Redraw the figure to show what happens when some heat is added to the system.

**3.96** A sample of carbon monoxide gas with mass of 7.75 g is heated from 25 °C to 175 °C at constant pressure of 5.00 bar. Calculate $q$, $w$, and $E$ for the process.

**3.97** Determine the values of $\Delta H_1^{\circ}$ and $\Delta H_2^{\circ}$ and the identity and coefficient of the missing species in the diagram below. (*Hint:* Use Appendix D).

**3.98** The most common form of elemental sulphur is $S_8$, in which eight sulphur atoms link in a ring of single bonds. At high temperature, in the gas phase, $S_8$ can break apart to give $S_2$, the sulphur analogue of molecular oxygen:

$$S_8 (g) \longrightarrow 4 S_2 (g) \qquad \Delta H = +239 \text{ kJ at } T = 800. \text{ K}$$

Using the appropriate data in Table 3-2, calculate the $S{=}S$ double-bond energy in $S_2 (g)$.

**3.99** The reaction that forms calcium carbonate deposits in caves is as follows:

$$Ca^{2+} (aq) + 2 HCO_3^{-} (aq) \longrightarrow$$
$$CaCO_3 (s) + CO_2 (g) + H_2O (l)$$
$$\Delta H^{\circ} = +39 \text{ kJ}$$

(a) Assuming that the inside of the cave is at $T = 15 \,°C$, $p = 1.00$ bar, determine $\Delta E$ for this reaction.
(b) Using data from Appendix D, determine the heat of formation of $HCO_3^{-} (aq)$.

**3.100** The five stable oxides of nitrogen are $NO$, $NO_2$, $N_2O$, $N_2O_4$, and $N_2O_5$. Balance each of the following oxidation reactions, and then use standard formation enthalpies to calculate the heats of reaction per mole of *atomic* nitrogen for each reaction:
(a) $N_2 + O_2 \longrightarrow NO$
(b) $N_2O + O_2 \longrightarrow NO$
(c) $NO + O_2 \longrightarrow NO_2$
(d) $NO_2 + O_2 \longrightarrow N_2O_5$

## GROUP STUDY PROBLEMS

**3.101** The heat required to sustain animals that hibernate comes from the biochemical combustion of fatty acids, one of which is arachidonic acid. For this acid, (a) determine its structural formula; (b) write its balanced combustion reaction; (c) use average bond energies to estimate the energy released in the combustion reaction; and (d) calculate the mass of arachidonic acid needed to warm a 500-kg bear from 5 °C to 25 °C. (Assume that the average heat capacity of bear flesh is 4.18 J/g K.)

Arachidonic acid ($C_{20}H_{32}O_2$)

3.102 Photosynthesis on the Earth's land surfaces converts approximately $5.5 \times 10^{16}$ g/year of $CO_2$ into energy-rich compounds. Assume that all of the carbon is incorporated into glucose, and use information found in this chapter to determine the following: (a) How much energy is stored per day? (b) The land mass of the Earth covers 29% of its surface. What fraction of the solar energy that reaches the land surface is stored via photosynthesis? (c) If the biosphere is at steady state, consuming all the energy that it stores, at what rate is the biosphere using energy? Express your result in MW (1 W = 1 J/s). (d) If a typical nuclear power plant is rated at $10^3$ MW, how many nuclear power plants would it take to generate energy at the same rate as the biosphere does?

3.103 According to Table 3-1, molar heat capacities of monatomic gases (He, Ar) are significantly smaller than those of diatomic gases ($N_2$, $O_2$, $H_2$). Explain in molecular terms why more heat must be supplied to raise the temperature of 1 mole of diatomic gas by 1 K

than to raise the temperature of 1 mole of monatomic gas by 1 K.

3.104 Suppose 100.0 mL of 1.00 M HCl and 100.0 mL of 1.00 M NaOH, both initially at 25.0 °C, are mixed in a Thermos flask. When the reaction is complete, the temperature is 31.8 °C. Assuming that the solutions have the same heat capacity as pure water, compute the heat released. Use this value to evaluate the molar heat of the neutralization reaction:

$$H_3O^+ (aq) + OH^- (aq) \longrightarrow 2\,H_2O\ (l)$$

3.105 Ethanol, $CH_3CH_2OH$, is used as a gasoline additive because it boosts octane ratings. Gasoline that contains ethanol is known as gasohol. Calculate the amount of energy released by burning one gallon of ethanol. The density of ethanol is 0.787 g/mL. Use the data in Appendix D and average bond energies from Table 3-2.

## ANSWERS TO SECTION EXERCISES

3.1.1  $2.0 \times 10^{-28}$ J

3.1.2  (a) kinetic energy; (b) thermal energy; (c) radiant energy; and (d) chemical energy

3.1.3  (a) Radiant energy is transformed into thermal energy; (b) gravitational (potential) energy is transformed into kinetic energy; and (c) kinetic energy is transformed into thermal energy.

**3.2.1** $q_{sys} = 4.2 \times 10^3$ J

**3.2.2** 0.97 kg of ground beef or 6.2 kg of fruit

**3.2.3** (a) and (d) are state functions

**3.3.1** −630 kJ

**3.3.2** See figure below.

2 C + 8 H + 8 O

Atoms

Decompose to atoms

2 C + 4 H$_2$ + 4 O$_2$

Elements

Decompose to elements

Recombine to products

2 CH$_3$OH + 3 O$_2$

Reactants

Recombine to products

Mixture of many chemical species

$\Delta E$    **Actual path**

2 CO$_2$ + 4 H$_2$O

Products

**3.3.3** CH$_4$, four C—H bonds; CH$_3$Cl, three C—H bonds, one C—Cl bond; others have one bond each. The first reaction absorbs 80 kJ, and the second reaction releases 105 kJ. The second reaction is used in industry because it does not require an energy input.

**3.4.1** $C = 523$ J/K

**3.4.2** $q_p = +4.6 \times 10^2$ J

**3.4.3** $\Delta E_{molar} = -297$ kJ/mol

**3.5.1** $w_{sys} = -4.31 \times 10^3$ J

**3.5.2** (c), (d), and (f)

**3.5.3** $\Delta H_{reaction} = +206.1$ kJ and $\Delta E_{reaction} = +201.1$ kJ

**3.6.1** 1850, wood; 1900, coal; 1950 and 2000, petroleum

**3.6.2** Petroleum is easier to transport and store; internal combustion engines are more powerful and compact than steam engines.

## ANSWERS TO PRACTICE EXERCISES

**3.1** $\Delta T = -1.09$ K

**3.2** $q = -21.2$ kJ

**3.3** $\Delta E = -0.140$ kJ

**3.4** $\Delta E_1 = -150$ kJ; $\Delta E_2 = +45$ kJ

**3.5** Final temperature would be 32.8 °C.

**3.6** $q = -4.2 \times 10^2$ J

**3.7** $\Delta E_{molar} = -3.9 \times 10^2$ kJ/mol of graphite

**3.8** $\Delta n = 0$, so $\Delta H - \Delta E = 0$

**3.9** $\Delta E = -349$ kJ/mol NH$_3$

**3.10** $H_f^\circ(\text{H}_3\text{PO}_4) = -1279$ kJ/mol

# ATOMS AND LIGHT

## LASERS

In the right combination, atoms and light combine to create one of the most remarkable tools of modern technology, the laser. Lasers have many commercial applications, ranging from laser light shows to CD scanners, eye surgery, and fibre optics communications. Lasers have also become versatile tools for advanced scientific research. To give just one example, lasers can be finely tuned to deposit vapour-phase atoms in regular patterns. Our inset shows chromium atoms deposited in a dot array using laser focusing. Each peak is 13 nm high.

The word *laser* stands for *light amplification by stimulated emission of radiation*. In a laser, atoms and light interact to generate a beam of light that has precisely defined properties. In simple terms, when a sample of atoms (or molecules) is "pumped" uphill in energy, light of a specific wavelength (colour) may interact with the atoms or molecules to produce more light of that wavelength. Mirrors reflect the light back into the sample, so the process can repeat itself many times, amplifying the intensity of the light immensely.

Light from a laser differs substantially from conventional light. Laser light is monochromatic—has a single colour—whereas conventional light sources typically produce light of many colours. A laser is highly directional, whereas conventional sources send light in all directions. Consequently, laser light is more highly organized than light from normal sources. Many of the applications of lasers take advantage of this high degree of organization.

Bar code scanners, for example, exploit the directionality of a laser beam. The laser light reflects more strongly from white stripes than from black. Because of the directionality of the laser beam, the reflected laser beam faithfully mirrors the pattern of the bar code. A sensor reads these variations and converts the light pattern into an electronic representation of the bar code. The sensor transmits the electronic representation to a computer.

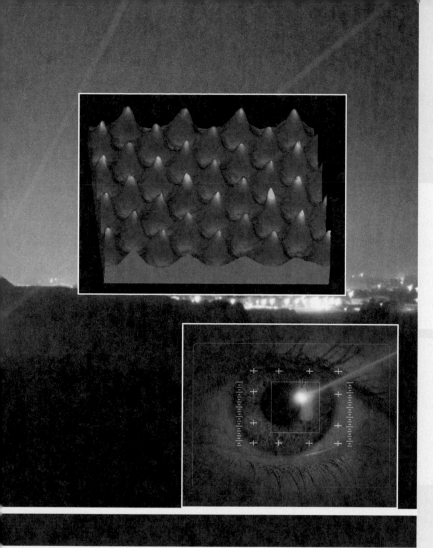

The combination of high intensity and directionality is the basis for laser surgery. Essentially, a laser acts as a "light knife." In eye surgery, a laser beam can be focused to a tiny spot at the back of the retina. This allows the surgeon to carry out delicate operations such as repairing a detached retina without physically invading the eyeball.

Regardless of the application, every laser is based on the interactions between light and atoms or molecules. We describe these interactions in this chapter. First, we summarize the properties of atoms and light. Next, we discuss the energy changes that accompany the interactions between electrons and light. Finally, we describe the properties of electrons bound to atoms and construct a picture of atomic structure. Differences in atomic structure distinguish the elements from one another and generate the rich array of chemical behaviour exhibited by different types of atoms.

## LEARNING OBJECTIVES
*Upon completion of this chapter you should be able to:*

- **understand some of the fundamental aspects of atoms**
- **understand some of the fundamental aspects of light**
- **explain the origins of atomic spectra and relate electron energies in the hydrogen atom to its emission spectrum**
- **describe properties of free electrons and those in atoms or molecules**
- **write valid sets of quantum numbers for a given set of orbitals**
- **recognize shapes of *s*, *p*, *d*, and *f* orbitals**
- **describe the major chemical reactions in the thermosphere and stratosphere and explain the chemistry of the greenhouse effect**

## 4.1 CHARACTERISTICS OF ATOMS

We begin with a review of the fundamental characteristics of atoms.

*Atoms possess mass.* Matter possesses mass, and matter is made up of atoms, so atoms possess mass. This property was already recognized in the time of John Dalton, who made it one of the postulates of his atomic theory.

Rutherford's first publication on the existence of particles smaller than atoms was based on work done at McGill University in Montreal.

*Atoms contain positive nuclei.* Rutherford's scattering experiment demonstrated that every atom contains a tiny central core where all the positive charge and most of the mass are concentrated. Subsequent experiments showed that whereas the masses of nuclei can have various values, the mass of every nucleus is at least 1800 times larger than the mass of the electron. Thus, more than 99.9% of the mass of an atom is contained in its nucleus.

*Atoms contain electrons.* For an atom to be neutral, the number of electrons that it contains must equal the total positive charge on its nucleus. Because each element has a characteristic positive charge associated with its nucleus, ranging from 1 for hydrogen to greater than 100 for the heaviest elements, atoms of different elements have different numbers of electrons.

*Atoms occupy volume.* Matter occupies space, and matter is made up of atoms, so atoms occupy space. It is extremely difficult to compress a solid such as copper or a liquid such as mercury, because the electron cloud of each atom occupies some volume that no other atom is able to penetrate because of electron–electron repulsion. Example 4-1 shows how to estimate the volume of an atom from the density of a sample, the molar mass of the substance, and Avogadro's number.

## EXAMPLE 4-1 ATOMIC VOLUMES

A cylindrical block of lithium with diameter and height of 2.4 cm contains approximately 1 mole of Li atoms.

The density of lithium, the lightest metal, is 0.534 g/cm$^3$. Estimate the volume occupied by a single atom in solid lithium.

**Strategy:** The mole is a convenient place to begin the calculations. Visualize a piece of lithium containing 1 mole of atoms. The molar mass, taken from the periodic table, tells us the number of grams of Li in 1 mole. The density equation can be used to convert from mass to volume. Once we have the volume of 1 mole of lithium, we divide by the number of atoms per mole to find the volume of a single atom.

**Solution:** First, assemble the data:

$$\rho = 0.534 \text{ g/cm}^3 \text{ (stated in the problem)}$$

$$M = 6.941 \text{ g/mol (from periodic table)}$$

$$N_A = 6.022 \times 10^{23} \text{ atom/mol}$$

Now rearrange the density equation and solve for the molar volume of lithium.

$$\rho = \frac{m}{V} \quad \text{so} \quad V_{molar} = \frac{M}{\rho} = \frac{6.941 \text{ g/mol}}{0.534 \text{ g/cm}^3} = 13.00 \text{ cm}^3/\text{mol}$$

Finally, divide by Avogadro's number to find volume per atom:

$$V_{atom} = \frac{V_{molar}}{N_A} = \frac{13.00 \text{ cm}^3/\text{mol}}{6.022 \times 10^{-23} \text{ atom/mol}} = 2.16 \times 10^{-23} \text{ cm}^3/\text{atom}$$

**Does the Result Make Sense?** The units are correct. This is a very tiny volume, as we would expect, knowing that atoms are very tiny. A cube with this volume has sides measuring $2.78 \times 10^{-10}$ m in length (0.278 nm).

**Practice Exercise 4.1**

Estimate the volume occupied by a single atom of iron with density of 7.874 g/cm$^3$.

The volume of an atom is determined by the size of its electron cloud. Example 4-1 demonstrates that atomic dimensions are a little over $10^{-10}$ m, whereas Rutherford's experiments showed that nuclear dimensions are only about $10^{-15}$ m. This is 100 000 times smaller than atomic dimensions, so the nucleus is buried deep within the electron cloud. If an atom were the size of a sports stadium, its nucleus would be the size of a pea. Figure 4-1 shows a schematic view of two atoms with their electron clouds in contact with each other.

*Atoms have various properties.* The periodic table is a catalogue of the elements, each with its own unique set of physical and chemical properties. Each element has a unique value for $Z$, the positive charge on its nucleus. The number of electrons possessed by a neutral atom of that element is also equal to $Z$. The different properties of elements arise from these variations in nuclear charges and numbers of electrons.

*Atoms attract one another.* Every gas changes into a liquid if the pressure is high enough and the temperature is low enough. The atoms or molecules of a liquid or solid stick together in a finite volume rather than expanding, as a gas does, to fill all available space. This cohesiveness comes from electrical forces of attraction between the negative electron cloud of each atom and the positive nuclei of other atoms. We describe intermolecular forces in Chapter 8.

*Atoms can combine with one another.* Atoms can form chemical bonds with one another to construct molecules. The details of chemical bonding appear in Chapters 6 and 7.

What does an atom experience in an encounter with another atom? The nucleus, which contains most of the mass of the atom, is confined to a tiny volume. Electrons, on the other hand, are spread out through space. Therefore, a collision between two atoms is a collision of their electron clouds. The electron clouds repel each other but are attracted by the nuclei. Chemists describe molecular structure, properties of materials, and chemical reactions in terms of how electrons respond to these electrical forces.

Our catalogue of atomic characteristics emphasizes electrons, because electrons determine the chemical properties of atoms. For the same reason, the next several chapters examine electrons and the way they influence chemical properties. First, however, we describe light and its interaction with atoms, because light is an essential tool for probing properties of electrons.

**Figure 4-1**

$\longleftarrow 10^{-10}$ m $\longrightarrow$

$10^{-15}$ m

Depiction of two atoms with their electron clouds in contact.

## SECTION EXERCISES

●4.1.1 The density of gold metal is 18.9 g/cm³. (a) What is the volume occupied by one gold atom? (b) If the diameter of a gold nucleus is 1/100 000 times that of a gold atom, what is the volume of one gold nucleus?

●4.1.2 (a) Use the volume occupied by a gold atom to estimate the thickness of one atomic layer of gold. (b) Estimate how many layers of gold atoms there are in a strip of gold foil that is 1.0 $\mu$m thick (1 $\mu$m = 10⁻⁶ m).

## 4.2 CHARACTERISTICS OF LIGHT

The most useful tool for studying the structure of atoms is **electromagnetic radiation**. What we call **light** is one form of this radiation. We need to know about the properties of light in order to understand what electromagnetic radiation reveals about atomic structure.

### Light Has Wave Aspects

Light has wave-like properties. A wave is a regular oscillation in some particular property, such as the up-and-down variation in position of water waves. Water waves vary with time. A surfer waiting for a "big one" bobs up and down as "small ones" pass by. Light waves vary with time, too. This variation is characterized by the **frequency ($\nu$)** of the wave, which is the number of wave crests passing a point in space in one second. The frequency unit is $s^{-1}$, also designated Hertz (Hz). Water waves also vary in space; that is, wave height differs from one place to another. Light waves vary in space in a manner illustrated in Figure 4-2. This variation in space is characterized by the **wavelength ($\lambda$)**, which is the distance between successive wave crests, measured in units of distance such as metres or nanometres.

Like electromagnetic radiation, water waves vary in time (frequency) and space (wavelength).

λ is the Greek letter *lambda*, and $\nu$ is the Greek letter *nu*.

The value of $c$ can be rounded to 2.998 × 10⁸ m/s for most calculations.

The height of a wave is called its **amplitude**. The amplitude of a light wave measures the **intensity** of the light. As Figure 4-3 shows, a bright light is more intense than a dim one.

Light waves always move through empty space at the same speed. The speed of light is a fundamental constant, denoted by the symbol $c$ ($c$ = 2.998 × 10⁸ m/s). For any wave, its wavelength (in units of m) multiplied by its frequency ($s^{-1}$) equals its speed (m/s). Thus, light obeys Equation 4-1:

$$\lambda\nu = c \qquad (4\text{-}1)$$

Example 4-2 applies Equation 4-1.

## EXAMPLE 4-2  WAVELENGTH–FREQUENCY CONVERSION

An FM radio station transmits its signal at 88.1 MHz. What is the wavelength of the radio signal?

**Strategy:** The link between wavelength ($\lambda$) and frequency ($\nu$) is given by Equation 4-1.

**Solution:** First, summarize the data:

$$c = 2.998 \times 10^8 \text{ m/s}, \qquad \nu = 88.1 \text{ MHz}$$

Next, rearrange Equation 4-1 to solve for wavelength.

$$\nu\lambda = c \quad \text{so} \quad \lambda = \frac{c}{\nu}$$

To obtain equivalent units, convert the frequency units from MHz to Hz. The prefix "M" stands for "mega," which is a factor of $10^6$. Remember that Hz is equivalent to $s^{-1}$.

$$\lambda = \frac{2.998 \times 10^8 \text{ m/s}}{88.1 \times 10^6 \text{ s}^{-1}} = 3.40 \text{ m}$$

 **Does the Result Make Sense?** A wavelength of 3.40 m is rather long, but radio waves are known as long-wavelength radiation. See Figure 4-4 for a sense of the wavelengths of electromagnetic radiation.

**Common Pitfalls:** Errors commonly occur when rearranging equations. If Equation 4-1 was rearranged incorrectly, the units on the final answer in this problem would not be length. Always check that the units on the answer are correct.

**Practice Exercise 4.2**
What is the frequency of light that has a wavelength of 1.40 cm?

**Figure 4-2**

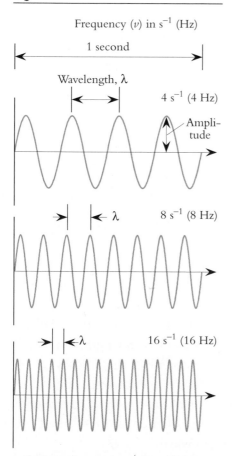

A light wave can be described by its wavelength and frequency, which are connected. As wavelength increases, frequency decreases, and vice versa.

**Figure 4-3**

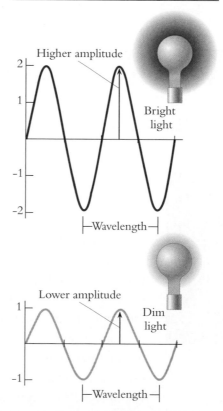

The amplitude of a light wave represents the intensity of the light. A bright light is more intense and has greater amplitude than a dim light.

The wavelengths and frequencies of electromagnetic radiation cover an immense range. What we call light is the tiny part of the spectrum that the human eye can detect. Figure 4-4 shows that the visible spectrum of light covers the wavelength range from about 400 nm (violet) to 700 nm (red). The centre of this range is yellow light, with a wavelength around 580 nm and a frequency around $5.2 \times 10^{14} \text{ s}^{-1}$. Although visible light is extremely important to living creatures for seeing, the gamma-ray, X-ray, ultraviolet, infrared, microwave, and radio frequency portions of the electromagnetic spectrum have diverse effects and applications in our lives.

**Figure 4-4**

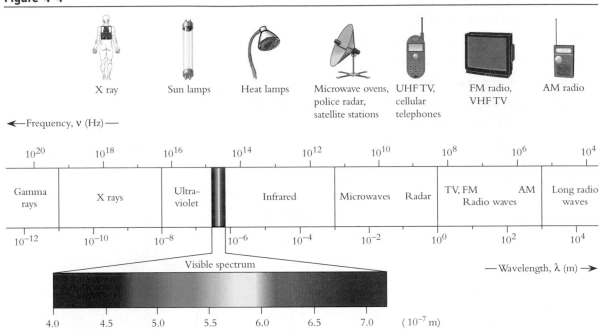

The electromagnetic spectrum, showing its various regions and the wavelengths and frequencies associated with each.

Light rays refract as they pass through a prism, causing white light to separate into its rainbow of colours.

Radiation with short wavelengths, in the X-ray and gamma-ray regions, can generate ions by removing electrons from atoms and molecules. These ions are highly reactive and can cause serious damage to the material that absorbs the light. However, under closely controlled conditions, X rays are used in medical imaging, and gamma rays are used to treat cancer. Ultraviolet (UV) radiation falls in between X rays and visible light in terms of wavelength. Ultraviolet radiation can also damage materials, especially in high doses.

Radiation with long wavelengths falls in the infrared, microwave, or radio frequency regions. Heat lamps make use of infrared radiation, microwave ovens cook with microwave radiation, and radio and television signals are transmitted by radio waves.

What we perceive as white light actually contains a range of wavelengths. These wavelengths are revealed when sunlight passes through a prism or through raindrops. These objects refract (bend) different wavelengths of light through different angles, so the light that passes through spreads out in space, with each wavelength appearing at its own characteristic angle.

## The Photoelectric Effect

Light carries energy. When our bodies absorb sunlight, for example, we feel warm because the energy of the sunlight has been transferred to our skin. Moreover, the total energy of a beam of light depends on its intensity (how bright the light is). Sunlight at midday is more intense than the rays of the setting Sun, because light from the setting Sun is reduced in intensity as it passes through a larger thickness of the atmosphere.

A phenomenon known as the **photoelectric effect** shows how the energy of light depends on its frequency and intensity. As Figure 4-5 illustrates, the photoelectric effect involves light and electrons. In a photoelectric experiment, a beam of light strikes the surface of a metal. Under the right conditions, the light causes electrons to be ejected from the metal surface. These electrons strike a detector that measures the number of electrons. The kinetic energies of the electrons can be determined by changing the voltage applied to the detector.

A detailed study of the photoelectric effect reveals how the behaviour of the electrons is related to the characteristics of the light:

1. Below a characteristic threshold frequency, $\nu_0$, no electrons are observed, regardless of the intensity of the light.

2. Above the threshold frequency, the maximum kinetic energy of ejected electrons increases linearly with the frequency of the light, as shown in Figure 4-6.

3. Above the threshold frequency, the *number* of emitted electrons increases with the intensity of the light, but the *kinetic energy* per electron does not depend on the intensity of the light.

4. All metals exhibit the same pattern, but as Figure 4-6 shows, each metal has a different threshold frequency.

In 1905, Albert Einstein provided an elegant explanation of the photoelectric effect. Einstein postulated that light comes in packets or bundles, called **photons**. Each photon has an energy that is directly proportional to the frequency:

$$E_{\text{photon}} = h\nu_{\text{photon}} \tag{4-2}$$

In this equation, $E$ is the energy of light and is its frequency. The proportionality constant between energy and frequency is known as **Planck's constant ($h$)** and has a value of $6.626 \times 10^{-34}$ J s. Example 4-3 illustrates the use of Equation 4-2.

**Figure 4-5**

A diagrammatic view of the photoelectric effect. When light of high enough frequency strikes a metal surface, electrons are ejected from the surface.

The photoelectric effect is the basis for many light-sensing devices, such as automatic door openers and camera exposure meters.

**Figure 4-6**

Variation in the maximum kinetic energy of electrons ejected from two different metal surfaces (a and b) by light of various frequencies.

Planck's constant ($h$) is named in honour of Max Planck, who first introduced this constant in 1900 to explain another phenomenon of light.

## EXAMPLE 4-3 THE ENERGY OF LIGHT

What is the energy of a photon of red light of wavelength 655 nm?

**Strategy:** This conversion problem requires two steps. Equations 4-1 and 4-2 relate the energy of a photon to its frequency and wavelength.

**Solution:** Summarize the data:

$$h = 6.626 \times 10^{-34} \text{ J s}, \quad \lambda = 655 \text{ nm}, \quad c = 2.998 \times 10^8 \text{ m/s}$$

Combine the equations into an equation that relates energy to wavelength:

$$E = h\nu \quad \text{and} \quad \nu = \frac{c}{\lambda} \quad \text{so} \quad E = \frac{hc}{\lambda}$$

Substitute and evaluate:

$$E_{\text{photon}} = \frac{(6.626 \times 10^{-34} \text{ J s})(2.998 \times 10^8 \text{ m/s})}{(655 \text{ nm})(10^{-9} \text{ m/nm})} = 3.03 \times 10^{-19} \text{ J}$$

**Does the Result Make Sense?** An energy of $10^{-19}$ J seems very small, but as we describe later in this section, a calculation of the energy of 1 mole of these photons gives $(3.03 \times 10^{-19} \text{ J})(6.022 \times 10^{23}/\text{mol})$ $(10^{-3} \text{ kJ/J}) = 182$ kJ/mol. This energy is comparable to the energies of chemical bonds.

**Common Pitfalls:** Knowing when to multiply or divide by Avogadro's number sometimes causes confusion. If you get into the habit of writing the units on all terms in a calculation and the final answer, such an error will be immediately obvious.

**Practice Exercise 4.3**

What is the energy of an UV photon whose wavelength is 254 nm?

Einstein applied the law of conservation of energy to the photoelectric effect, as shown schematically in Figure 4-7. When a metal surface absorbs a photon, the energy of the photon is transferred to an electron:

$$\Delta E_{\text{electron}} = E_{\text{photon}}$$

Some of this energy is used to overcome the forces that bind the electron to the metal, and the remainder shows up as kinetic energy of the ejected electron.

The energy of a photon that has the threshold frequency ($\nu_0$) corresponds to the binding energy of the electron. In other words, the energy of a photon at the threshold frequency equals the minimum energy needed to overcome the forces that bind the electron to the metal. Putting these ideas together, Einstein obtained a simple linear equation that matches the graph in Figure 4-6:

$$\text{Electron kinetic energy} = \text{Photon energy} - \text{Binding energy}$$

$$E_{\text{kinetic}} \text{ (electron)} = h\nu - h\nu_0 \qquad (4\text{-}3)$$

Einstein's explanation accounts for the observed properties of the photo-electric effect. First, when the energy of the photon is less than $h\nu_0$ (low-frequency light), there is not enough energy per photon to overcome the binding energy of the electron. Under these conditions, no electrons can escape from the metal surface, no matter how intense the light. Second, after the energy of the photon exceeds the threshold value ($hv > hv_0$), electrons are ejected. The extra energy of the photon is transferred to the ejected electron as kinetic energy; this extra kinetic energy increases linearly with $v$. Third, the intensity of a light beam is a measure of the number of photons: light with higher amplitude carries more photons than light of lower amplitude. The intensity of the light *does not* determine the amount of energy per photon. **"Higher intensity" means more photons but not more energy per photon.** More photons striking the metal result in more electrons being ejected, but the energy of each photon and each electron is unchanged. Finally, each metal has its own characteristic threshold frequency because electrons are bound more tightly to some metals than to others.

Example 4-4 shows how to apply Einstein's analysis of the photoelectric effect.

**Figure 4-7**

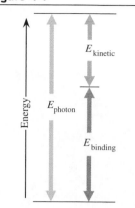

Diagram of the energy balance for the photoelectric effect.

## EXAMPLE 4-4  THE PHOTOELECTRIC EFFECT

The minimum energy needed to remove an electron from a potassium metal surface is $3.7 \times 10^{-19}$ J. Will photons of frequencies $4.3 \times 10^{-14}$ s$^{-1}$ (red light) and of $7.5 \times 10^{-14}$ s$^{-1}$ (blue light) trigger the photoelectric effect? If so, what is the maximum kinetic energy of the ejected electrons?

**Strategy:** We must analyze the energy requirements for ejection of an electron. No electrons will be ejected unless the energy of the photons exceeds some threshold value characteristic of the metal. If the photon energy exceeds this threshold value, electrons will be ejected with kinetic energy given by Equation 4-3. An important part of this problem is the conversion of photon frequency to photon energy.

**Solution:** The threshold energy of potassium metal, $3.7 \times 10^{-19}$ J, is given in the problem. To determine whether the red and blue photons will eject electrons, we convert the frequencies of these photons to their corresponding energies:

$$E_{\text{red photon}} = hv = (6.626 \times 10^{-34} \text{ J s})(4.3 \times 10^{14} \text{ s}^{-1}) = 2.8 \times 10^{-19} \text{ J}$$

$$E_{\text{blue photon}} = hv = (6.626 \times 10^{-34} \text{ J s})(7.5 \times 10^{14} \text{ s}^{-1}) = 5.0 \times 10^{-19} \text{ J}$$

According to these calculations, a photon of red light does not have enough energy to overcome the forces that bind electrons to the metal. A blue photon, however, will eject an electron from the surface of potassium metal because its energy exceeds the threshold value. What happens to the extra energy of a blue photon? Equation 4-3 indicates that it is transferred to the electron as kinetic energy:

$$E_{\text{kinetic}} \text{ (electron)} = hv - hv_0$$

$$E_{\text{kinetic}} \text{ (electron)} = (5.0 \times 10^{-19} \text{ J}) - (3.7 \times 10^{-19} \text{ J}) = 1.3 \times 10^{-19} \text{ J}$$

**Do the Results Make Sense?** Figure 4-4 shows that red light has a lower frequency than blue light, and Equation 4-2 shows that photon energy is directly proportional to frequency. It makes sense that the photon of higher energy can eject an electron, but the photon of lower energy cannot.

**Common Pitfalls:** Be sure that the photon energy is greater than the threshold energy, otherwise no electrons will be ejected.

### Practice Exercise 4-4

Will a green photon ($\lambda = 515$ nm) eject an electron from a potassium surface? If so, what is the maximum kinetic energy of ejected electrons?

## Light Has Particle Aspects

Albert Einstein is considered the greatest modern physicist because his theories influenced topics in physics ranging from the basis for lasers to the expanding universe.

Einstein's explanation of the photoelectric effect showed that light has some properties of particles. Light consists of photons, each of which has the discrete energy $E = h\nu$. Although simple, this particle-like explanation of light was revolutionary, because before 1905, the properties of light had been explained with a wave picture. A complete description of light includes wave-like and particle-like properties.

Neither the particle nor the wave view of light is wrong or entirely right. Light has some properties of waves and some properties of particles. When light interacts with a relatively large body such as a raindrop or a prism, its wave properties dominate the interaction. On the other hand, when light interacts with a small body such as an atom or an electron, particle properties dominate the interaction. Each view provides different information about the properties of light, and when we think about light, we must think of both types of features.

A photon of a particular frequency has a specific energy. As the photoelectric effect shows, when something absorbs that photon, the energy is transferred to whatever absorbs the photon. Sunlight striking your skin is absorbed and warms you up. When an individual atom absorbs a photon, the energy of the atom increases by an amount equal to the energy of the photon. We analyze these energy exchanges in the next section.

### SECTION EXERCISES

**4.2.1** A compact disc player uses laser light of frequency $3.85 \times 10^{14}$ s$^{-1}$ to read the information on the disc. (a) What is this light's wavelength? (b) In what portion of the electromagnetic spectrum (visible, ultraviolet, and so on) does this wavelength fall? (c) What is the energy of 1 mole of photons at this frequency?

**4.2.2** The light reaching us from distant stars is extremely dim, so astronomers use instruments capable of detecting small numbers of photons. An infrared photon detector registered a signal at 1250 nm from Alpha Centauri with a total energy of $1.20 \times 10^{-16}$ J. How many photons were detected?

**4.2.3** In a photoelectric effect experiment, a metal absorbs photons with $E = 6.00 \times 10^{-19}$ J. The maximum kinetic energy of the ejected electrons is $2.00 \times 10^{-19}$ J. Calculate the frequency of the light and the binding energy in joules per electron, and convert this energy into kilojoules per mole.

## 4.3 ABSORPTION AND EMISSION SPECTRA

In the photoelectric effect, energy absorbed from photons provides information about the binding energies of electrons to metal surfaces. When light interacts with free atoms, the interaction reveals information about electrons bound to individual atoms.

## Light and Atoms

The absorption of photons by free atoms has two different results, depending on the energy of the photon. When an atom absorbs a photon of sufficiently high energy, an electron is ejected, a process termed photoionization. Here, we focus on a second type of result, in which the atom gains energy but does not ionize. Instead, the atom is transformed to a higher energy state called an **excited state**. Atoms in excited states subsequently give up their excess energy to return to lower energy states. The lowest energy state of an atom, which is its most stable state, is called the **ground state**.

Attractive electrical force holds a bound electron within an atom, and energy must be supplied to remove a bound electron from an atom. The lower an energy state is, the more energy must be supplied to remove its electron. These are energy changes that are measured relative to the energy of a free electron. By convention, scientists define the energy of a free stationary electron to be zero. As Figure 4-8 indicates, the kinetic energy of a freely moving electron is positive relative to this conventional zero point. In contrast, bound electrons are lower in energy than free stationary electrons, so they have negative energy values relative to the zero point. Therefore, the term "most stable" in this context corresponds to "lowest" or "most negative." A bound electron has "negative energy" only in relation to a free electron, because energy is released when a free stationary electron binds to an atom.

Atomic energy transformations can be represented using an **energy-level diagram** such as the one shown in Figure 4-9. These diagrams show energy along the vertical axis. Each energy state of the atom is represented by a horizontal line. Absorption of a photon, shown by an upward arrow, causes a transition from the ground state to one of the excited states. Excited states are not stable. They lose their excess energy in collisions with other atoms or by emitting photons, shown by downward arrows.

Energy-level diagrams are compact ways to summarize large amounts of information about atomic energies.

The key feature in the exchange of energy between atoms and light is that energy is conserved. This requires that the change in energy of the atom exactly equals the energy of the photon. Equation 4-4 states this equality:

$$\Delta E_{\text{atom}} = \pm h\nu_{\text{photon}} \qquad (4\text{-}4)$$

When an atom *absorbs* a photon, the atom gains the photon's energy, so $\Delta E_{\text{atom}}$ is positive. When an atom *emits* a photon, the atom loses the photon's energy, so $\Delta E_{\text{atom}}$ is negative. As an atom returns from an excited state to the ground state, it must lose exactly the amount of energy that it originally gained. However, excited atoms usually lose excess energy in several steps involving small energy changes, so the frequencies of emitted photons often are lower than those of absorbed photons. Example 4-5 analyzes energy changes associated with the emission of light.

**Figure 4-8**

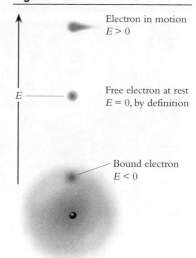

By convention, a free stationary electron has zero energy, so bound electrons have negative energies.

**Figure 4-9**

An atomic energy-level diagram showing the relationships among atomic energy levels and photon absorption and emission.

## EXAMPLE 4-5 EMISSION ENERGIES

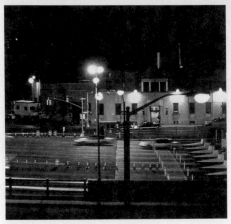

Sodium vapour lamps emit yellow light.

A sodium-vapour street lamp emits yellow light at wavelength $\lambda = 589$ nm. What is the energy change for a sodium atom involved in this emission? How much energy is released per mole of sodium atoms?

**Strategy:** This problem relates energies of photons to energy changes of atoms. The solution requires a conversion involving wavelength and energy.

**Solution:** Sodium atoms lose energy by emitting photons of light ($\lambda = 589$ nm). We can use Equations 4-1 and 4-2 to relate the wavelength of one of these photons to its energy:

$$E_{photon} = h\nu = \frac{hc}{\lambda}$$

Energy is conserved, so the energy of the emitted photon must exactly equal the energy lost by the atom:

$$\Delta E_{atom} = -E_{photon} = -\frac{hc}{\lambda}$$

The equality includes a negative sign because the atom *loses* energy.
To calculate the energy of a photon, we need the following data:

$$h = 6.626 \times 10^{-34} \text{ J s}, \quad c = 2.998 \times 10^8 \text{ m/s}, \quad \lambda = 589 \text{ nm} = 589 \times 10^{-9} \text{ m}$$

$$\Delta E_{atom} = -E_{photon} = \frac{(6.626 \times 10^{-34} \text{ J s})(2.998 \times 10^8 \text{ m/s})}{(589 \times 10^{-9} \text{ m})}$$

$$= -3.37 \times 10^{-19} \text{ J}$$

This calculation gives the energy change for one sodium atom emitting one photon. We use Avogadro's number to convert from energy per atom to energy per mole of atoms:

$$E_{molar} = (\Delta E_{atom})(N_A) = (-3.37 \times 10^{-19} \text{ J})(6.022 \times 10^{23} \text{ mol}^{-1})$$

$$= -2.03 \times 10^5 \text{ J/mol} = -203 \text{ kJ/mol}$$

 **Do the Results Make Sense?** The energy change is negative, as required by the fact that the emitted photon carries energy. The magnitude is reasonable, being the same order of magnitude as bond energies.

### Practice Exercise 4.5

Mercury lamps emit photons with wavelength of 436 nm. Calculate the energy change of the mercury atoms in J/atom and kJ/mol.

## Atomic Spectra

When a light beam passes through a tube containing a monatomic gas, the atoms absorb specific and characteristic frequencies of the light. As a result, the beam emerging from the sample tube contains fewer photons at these specific frequencies. We can see where the frequencies are missing in the visible portion of the spectrum by passing the emerging light through a prism. The prism deflects the light, with different frequencies deflecting through different angles. After leav-

**Figure 4-10**

Schematic representation of an apparatus that measures the absorption spectrum of a gaseous element. The gas in the tube absorbs light at specific wavelengths, called lines, so the intensity of transmitted light is low at these particular wavelengths.

ing the prism, the beam strikes a screen, where the missing frequencies appear as gaps or dark bands. These are the frequencies of light absorbed by the atoms in the sample tube. The resulting pattern, shown schematically in Figure 4-10, is called an **absorption spectrum**.

A similar experiment can be performed to measure the energies of the photons *emitted* by atoms in excited states. Figure 4-11 outlines the features of an apparatus that measures these emitted photons. An electrical discharge excites a collection of atoms from their ground state into higher-energy states. These excited atoms lose all or part of their excess energy by emitting photons. This emitted light can be analyzed by passing it through a prism to give an **emission spectrum**. This is a plot of the intensity of light emitted as a function of frequency. The emission spectrum for hydrogen, shown in Figure 4-11, shows several sharp emission lines of high intensity. The frequencies of these lines correspond to photons emitted by the hydrogen atoms as they return to their ground state.

**Figure 4-11**

Schematic representation of an apparatus that measures the emission spectrum of a gaseous element. Emission lines appear bright against a dark background. The spectrum shown is the emission spectrum for hydrogen atoms.

Sodium (Na)

Neon (Ne)

Mercury (Hg)

Each element has a unique emission pattern that provides valuable clues about atomic structure.

Each element has unique absorption and emission spectra. That is, each element has its own set of characteristic frequencies of light that it can absorb or emit. Also, Figures 4-10 and 4-11 show only the visible portions of absorption and emission spectra. Electron transitions also take place in regions of the electromagnetic spectrum that the human eye cannot detect. Instruments allow scientists to "see" into these regions.

Each frequency absorbed or emitted by an atom corresponds to a particular energy change for the atom. These characteristic patterns of energy gains and losses provide information about atomic structure.

## Quantization of Energy

When an atom absorbs light of frequency $v$, the light beam loses energy $hv$, and the atom gains that amount of energy. What happens to the energy that the atom gains? A clue is that when the frequency of the bombarding light is high enough, it produces cations and free electrons. In other words, a photon with high enough energy can cause an atom to lose one of its electrons. This implies that absorption of a photon results in an energy gain for an electron in the atom. Consequently, the energy change for the atom equals the energy change for an atomic electron:

$$\Delta E_{\text{atom}} = \Delta E_{\text{electron}} = hv$$

The atomic spectra of most elements are complex and show little regularity. However, the emission spectrum of the hydrogen atom is sufficiently simple to be described by a single formula:

$$v_{\text{emission}} = (3.29 \times 10^{15} \text{ s}^{-1})\left(\frac{1}{n_1^2} - \frac{1}{n_2^2}\right)$$

The Swiss mathematician and physicist Johann Balmer proposed a form of this equation in 1885. At that time, the link between frequency and energy was not known. Balmer did not understand the significance of $n_1$ and $n_2$, both of which are integers (1, 2, 3, and so on). Then, in 1913, Niels Bohr used the discovery that $E = hv$ to interpret Balmer's observations. Bohr realized that the emission

Niels Bohr (1885–1962) was a Danish physicist whose discovery of the quantization of atomic energy levels won him a Nobel Prize in 1922. Bohr headed a world renowned institute for atomic studies in Copenhagen in the 1920s and 1930s.

frequencies have specific values because the electron in a hydrogen atom is restricted to specific energies described by an equation containing an integer $n$:

$$E_n = -\frac{2.18 \times 10^{-18}\,\text{J}}{n^2} \qquad (4\text{-}5)$$

The negative sign in Equation 4-5 reflects the fact that bound electrons have negative energies relative to a stationary free electron.

Bohr's idea of restricted energy levels was revolutionary, because scientists at that time thought that the electron in a hydrogen atom could have any energy, not just the ones described by Equation 4-5. In contrast, Bohr interpreted the hydrogen emission spectrum to mean that electrons bound to atoms can have only certain specific energy values. A property that is restricted to specific values is said to be **quantized**. The atomic energy levels of hydrogen (and other elements) are quantized. In Equation 4-5, each integral value of $n$ describes one of the allowed energy levels of the hydrogen atom. For example, the energy of an electron in hydrogen's fourth level is

$$E_4 = \frac{2.18 \times 10^{-18}\,\text{J}}{4^2} = 1.36 \times 10^{-19}\,\text{J}$$

When an electron changes energy levels, the change is an electronic transition between quantum levels. When a hydrogen atom absorbs or emits a photon, its electron changes from one energy level to another. Thus, the change in energy of the atom is the difference between the two levels:

$$\Delta E_{\text{atom}} = E_{\text{final}} - E_{\text{initial}}$$

Photons always have positive energies, but energy changes ($\Delta E$) can be positive or negative. When absorption occurs, an atom gains energy, $\Delta E$ for the atom is positive, and a photon disappears:

$$E_{\text{absorbed photon}} = \Delta E_{\text{atom}}$$

When emission occurs, an atom loses energy, $\Delta E$ for the atom is negative, and a photon appears:

$$E_{\text{emitted photon}} = -\Delta E_{\text{atom}}$$

We can combine these two equations by using absolute values:

$$E_{\text{photon}} = |\Delta E_{\text{atom}}| \qquad (4\text{-}6)$$

Example 4-6 applies Equations 4-5 and 4-6 to the hydrogen atom.

The constants in Bohr's equation and Balmer's equation are related through $E = h\nu$.

## EXAMPLE 4-6 HYDROGEN ENERGY LEVELS

What is the energy change when the electron in a hydrogen atom changes from the fourth energy state to the second energy state? What is the wavelength of the photon emitted?

**Strategy:** The problem asks about energy and the wavelength of a photon emitted by a hydrogen atom. Wavelength is related to energy, and photon energy is determined by the difference in energy between the two levels involved in the transition. In this case, the electron moves from the fourth to the second energy level. The energy difference between these two states is given by

$$\Delta E_{atom} = E_{final} - E_{initial} = (E_2 - E_4)$$

**Solution:** As shown in the text, $E_4 = -1.36 \times 10^{-19}$ J. A similar calculation using Equation 4-5 gives the energy of the second level:

$$E_2 = -\frac{2.18 \times 10^{-18} \text{ J}}{2^2} = -5.45 \times 10^{-19} \text{ J}$$

This lower energy level has a more negative energy. The energy difference is

$$\Delta E_{atom} = E_{final} - E_{initial} = (-5.45 \times 10^{-19} \text{ J}) - (-1.36 \times 10^{-19} \text{ J})$$
$$= -4.09 \times 10^{-19} \text{ J}$$

This energy change is *negative* because the atom *loses* energy. This lost energy appears as a photon whose energy is given by Equation 4-6:

$$E_{photon} = |\Delta E_{atoms}| = |-4.09 \times 10^{-19} \text{ J}| = 4.09 \times 10^{-19} \text{ J}$$

To determine the wavelength of the photon, we use Equations 4-1 and 4-2:

$$E_{photon} = h\nu = \frac{hc}{\lambda}$$

Solving for $\lambda$ gives

$$\lambda_{photon} = \frac{hc}{E_{photon}} = \frac{(6.626 \times 10^{-34} \text{ J s})(2.998 \times 10^8 \text{ m/s})(10^9 \text{ nm/m})}{(4.09 \times 10^{-19} \text{ J})} = 486 \text{ nm}$$

**Do the Results Make Sense?** The signs of the energies are consistent—the atom loses energy, the photon has positive energy. The wavelength of light falls in the visible region. According to Figure 4-4, a photon with a wavelength of 486 nm has a blue colour.

**Common Pitfalls:** The wording of these problems must be interpreted correctly before the calculation is attempted. You must understand, for instance, that "ground state" means the electron is in it lowest energy state, $E_1$.

### Practice Exercise 4.6

What energy and wavelength must a photon have to excite a hydrogen atom from its ground state to the $n = 4$ level?

## Energy-Level Diagrams

The quantum levels of an electron bound to an atom are crudely analogous to the gravitational potential energies available to a ball on a staircase. As illustrated in Figure 4-12, a ball may sit on any of the steps. If we define the top of the steps to be $E = 0$, the ball has a negative potential energy when it is on any of the lower steps. To move a ball from the bottom of the staircase to step 5 requires the addition of a specific amount of energy, $\Delta E = E_5 - E_1$. If too little energy is supplied, the ball cannot reach this step. Conversely, if a ball moves down the staircase, it

**Figure 4-12**

A ball on a staircase shows some properties of quantized energy states.

releases specific amounts of energy. If a ball moves from step 5 to step 3, it loses energy, $\Delta E = E_3 - E_5$. Although a ball may rest squarely on any step, it cannot be suspended at some position between the stair steps. Electrons in atoms, like balls on steps, cannot exist "between steps" but must occupy one of the specific, quantized energy levels. (Remember that this is an analogy; atomic energy levels are not at all similar to staircases except in being quantized.)

According to Equation 4-5, a hydrogen atom has a regular progression of quantized energy levels. Figure 4-13 shows the energy-level diagram for hydrogen atoms; arrows represent some of the absorption and emission transitions. Notice that the energies of absorption from the lowest energy level are identical to the energies of emission to the lowest energy level. This means that the wavelengths of light absorbed in these upward transitions are identical to the wavelengths of light emitted in the corresponding downward transitions.

Elements other than hydrogen also have quantized energy levels, but they lack the regular spacing described by Equation 4-5. Scientists use experimental values for observed absorption and emission lines to calculate the allowed energy levels for each different element. As an example, Figure 4-14 shows the energy-level diagram for mercury. The absorption spectrum of mercury shows two dominant lines (upward arrows). Its emission spectrum, generated by an electrical discharge, contains light of many different wavelengths (downward arrows). Energy-level diagrams such as Figures 4-13 and 4-14 summarize the quantized energy levels of an atom as deduced from light measurements and

**Figure 4-13**

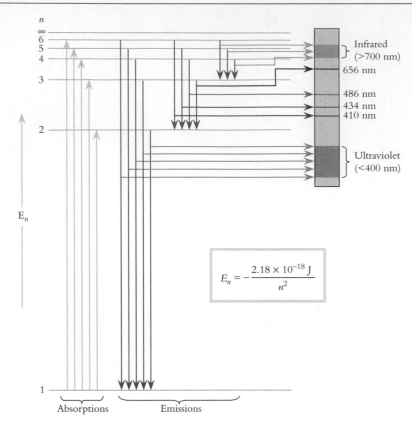

$$E_n = -\frac{2.18 \times 10^{-18}\,\text{J}}{n^2}$$

Energy levels for the hydrogen atom and some of the transitions that occur between levels. Upward arrows represent absorption transitions, and downward arrows represent emissions.

**Figure 4-14**

Energy-level diagram for mercury (Hg) atoms showing the most prominent absorption and emission lines. The numbers accompanying the arrows are the wavelengths in nanometres (nm) of the photons associated with these transitions.

energy conservation. Example 4-7 provides more practice in this type of reasoning, and our Focus 4-1: Tools for Discovery discusses how spectroscopy—the observation of spectra—can be used to study otherwise inaccessible objects.

## EXAMPLE 4-7 ENERGY-LEVEL DIAGRAMS

Light from a ruby laser is generated from $Cr^{3+}$ ions.

Ruby lasers use crystals of $Al_2O_3$. The crystals contain small amounts of $Cr^{3+}$ ions, which absorb light between 400 and 560 nm. These excited-state ions lose some energy as heat. After losing heat, the $Cr^{3+}$ ions return to the ground state by emitting red light of wavelength 694 nm. Calculate (a) the molar energy of the 500-nm radiation used to excite the $Cr^{3+}$ ions, (b) the molar energy of the emitted light, and (c) the fraction of the excitation energy emitted as red photons and the fraction lost as heat. (d) Draw an energy-level diagram, in kilojoules per mole, that summarizes these processes.

**Strategy:** This problem asks about energies, light, and atoms. With multipart problems, the best strategy is to work through the parts one at a time. Parts (a) and (b) concern the link between light and energy, discussed in Section 4.2. Once the transition energies have been determined, we can calculate the fraction of the excited-state energy lost as heat, part (c), and we can draw an energy-level diagram that shows how the levels are related, part (d).

**Solution:** We solve part (a) using Equations 4-1 and 4-2 and Avogadro's number:

$$E_{photon} = h\nu \quad \text{and} \quad \lambda\nu = c \quad \text{so} \quad E_{photon} = \frac{hc}{\lambda}$$

$$E_{photon\ absorbed} = \frac{(6.626 \times 10^{-34}\ J\ s)(2.998 \times 10^8\ m/s)}{(500\ nm)(10^{-9}\ m/nm)} = 3.97 \times 10^{-19}\ J$$

This is the energy change per atom. To calculate the change per mole, we multiply by Avogadro's number. We also need to convert from J to kJ:

$$E_{photon\ absorbed} = (3.97 \times 10^{-19}\ J)(6.022 \times 10^{23}\ mol^{-1})(10^{-3}\ kJ/J) = 239\ kJ/mol$$

Apply the same reasoning to part (b). The result gives the energy of the red photon.

$$E_{photon\ emitted} = 172\ kJ/mol$$

To solve part (c), remember that energy is conserved. The sum of the emitted heat and the emitted photon must equal the energy absorbed by the ion:

$$E_{photon\ absorbed} = E_{photon\ emitted} + E_{heat\ emitted}$$

Because 239 kJ/mol is absorbed and 172 kJ/mol is emitted, the fraction of the excitation energy re-emitted is $172/239 = 0.720$. The fraction converted to heat is the difference between this value and 1.000, or 0.280. In other words, 72.0% of the energy absorbed by the chromium ion is emitted as red light, and the other 28.0% is lost as heat.

Part (d) asks for an energy-level diagram for this process. The electron starts in the ground state. On absorption of a photon, the electron moves to an energy level that is higher by 239 kJ/mol. The chromium ion loses 28.0% of its excited-state energy as heat as the electron moves to a different level that is 172 kJ/mol above the ground state. Finally, emission of the red photon returns the $Cr^{3+}$ ion to the ground state. The numerical values allow us to construct an accurate diagram.

**Do the Results Make Sense?** When $Cr^{3+}$ ions absorb light, they are pumped to an energy level higher than the energy that they later emit. If calculations had shown that the emitted light had a larger energy than the absorbed light, the result would have been unreasonable because the system would violate the law of conservation of energy.

### Practice Exercise 4.7

After excitation in an electric discharge, an atom of Hg returns to the ground state by emitting two photons with wavelengths of 436 nm and 254 nm. Identify the excited state on Figure 4-14 and calculate its excitation energy in kJ/mol.

## FOCUS 4-1  TOOLS FOR DISCOVERY: SPECTROSCOPY, OBSERVING FROM AFAR

How do we know the composition of the Sun and other stars? How can we measure the temperature inside a flame so hot that any thermometer would melt? How can we explore chemical reactions among molecules that are much too tiny to see directly? Light allows us to do all these things. The study of matter with electromagnetic radiation is called *spectroscopy.*

Astronomers use spectroscopy to identify the composition of the Sun and other stars. A striking example is the discovery of the element helium. In 1868, astronomers viewing a solar eclipse observed emission lines that did not match any known element. The English astronomer Joseph Lockyer attributed these lines to a new element that he named *helium*, from *helios*, the Greek word for the Sun. For 25 years, the only evidence for the existence of helium was these solar spectral lines.

In 1894, the Scottish chemist William Ramsay removed nitrogen and oxygen from air through chemical reactions. From the residue, Ramsay isolated argon, the first noble gas to be discovered. A year after discovering argon, Ramsay obtained an unreactive gas from uranium-containing mineral samples. The gas exhibited the same spectral lines that had been observed in the solar eclipse of 1868. After helium was shown to exist on Earth, this new element was studied and characterized.

The tip of the flame of a gas stove is much too hot to measure using physical probes, but its temperature can be measured spectroscopically. The outer portion of a gas flame glows bright blue. Other hot objects

emit different colours: a candle flame is yellow, and the heating element of a hairdryer glows red. All hot objects lose energy by giving off light, and the hotter the object, the higher the average energy of the emitted light. An object at around 1000 K gives off red light. At around 2000 K, an object emits predominantly yellow light, and blue light indicates a temperature of around 3000 K.

Just as hot objects on Earth display different colours, the colours of stars range from distinctly red through yellow to quite blue, indicating different temperatures at their surfaces. The stars making up the Orion constellation illustrate the differences in star colours. Betelgeuse, at the upper left, is a relatively cool red star whose surface temperature is around 3000 K. The distinctly blue colour of Rigel, at lower right, results from its surface temperature of about 11000 K. The surface temperature of our own Sun is around 6000 K, resulting in its lovely yellow hue. The photo of Orion is a time exposure using different focal lengths, which results in the colour trails that highlight different star hues.

Dr. Gerhard Herzberg is one of Canada's most recognized scientists, especially in the area of spectroscopy. Dr. Herzberg was awarded the Nobel Prize in Chemistry in 1971 for his discovery of the spectrum of the methylene radical ($CH_2$), an important molecule in organic chemistry. Dr. Herzberg also developed and applied spectroscopic tools to investigate astrophysics, chemistry, biology and medicine.

The previous examples illustrate the use of light given off—emission spectra—to explore inaccessible objects. Absorption spectroscopy is a powerful complementary technique. In this application, light shines on the system to be studied. The researcher then examines how that light changes during the interaction with the system. Absorption spectroscopy has been enhanced greatly by the development of powerful lasers. For example, by using a combination of a pulsed laser (laser #1 in the illustration below) and a continuous laser (laser #2), chemists can observe chemical bonds in the process of breaking. The light from laser #1 is absorbed by a specific chemical bond, which stretches and breaks. The molecule absorbs light from laser #2 only while it is in the process of breaking. The graph shows the observed intensity of the second laser beam as a function of time.

If the pulse from laser #1 is ultrafast, the bond-energizing step occurs in a very short time, about 10 fs (1 fs = $10^{-15}$ s). As the bond stretches through the specific length at which it absorbs photons from laser #2, the molecule can absorb a photon from the second laser beam. This absorption causes the transmitted intensity of laser #2 to fall rapidly as the bond stretches. When the bond breaks, photons from laser #2 are no longer absorbed and the transmitted intensity returns to its original value. By measuring the time it takes for this to occur, chemists have determined that it takes about 200 fs for a chemical bond to break.

A. Before the pulse    B. During bond breaking    C. After the bond breaks

## SECTION EXERCISES

● **4.3.1** When minerals absorb invisible ultraviolet light from a mercury lamp, they emit visible light of a longer wavelength, converting the remaining energy into heat. How much energy per mole is converted to heat by a mineral that absorbs ultraviolet light at 366 nm and emits green light at 545 nm?

● **4.3.2** Calculate the wavelengths that hydrogen atoms in the fifth energy level can emit.

● **4.3.3** Gaseous helium atoms absorb UV light of wavelength 53.7 nm. After absorbing a photon of this wavelength, a helium atom may emit light of wavelength 501.6 nm. What is the net energy change for a helium atom that has gone through this absorption–emission sequence? Draw an energy-level diagram that shows the sequence.

## 4.4 PROPERTIES OF ELECTRONS

The energies of electrons in atoms play a central role in determining chemical behaviour. Several other properties of electrons also influence the physical and chemical characteristics of atoms and molecules. Some properties are characteristic of all electrons, but others arise only when electrons are bound to atoms or molecules. In this section, we describe the properties possessed by all electrons.

### Properties Shared by All Electrons

*Each electron has the same mass and charge.* Every electron has a mass of $9.109 \times 10^{-31}$ kg and a charge of $1.602 \times 10^{-19}$ C.

*Electrons behave like magnets.* Some types of atoms behave like tiny magnets. The best-known example is iron, the material used to make many permanent magnets. Experiments have shown that the magnetic behaviour of atoms is caused by magnetic properties of their component parts, especially electrons.

The magnetic properties of electrons arise from a property called **spin**, which we describe in more detail in Section 4.5. All electrons have spin of the same magnitude, but electron spin can respond to a magnet in two different ways. Most magnetic effects associated with atoms are caused by the spins of their electrons. Iron and nickel are permanent magnets because of the cooperative effect of many electrons.

*Electrons have wave properties.* We are used to thinking of electrons as particles. As it turns out, electrons display both particle properties and **wave properties**. The French physicist Louis de Broglie first suggested that electrons display wave–particle duality like that exhibited by photons. De Broglie reasoned, from nature's tendency toward symmetry: if things that behave like waves (light) have particle characteristics, then things that behave like particles (electrons) should also have wave characteristics.

Experiments had shown that a beam of light shining on an object exerts a pressure, and this, in turn, implies that a photon has momentum. Quantitative measurements of the pressure exerted by light showed that a simple equation relates the momentum of light ($p$) to its energy:

$$E = pc$$

As we have already described, light energy also is related to its wavelength:

$$E = h\nu = \frac{hc}{\lambda}$$

Setting these two energy expressions equal to each other gives an expression relating $p$ to $\lambda$:

$$pc = \frac{hc}{\lambda}$$

The speed of light cancels, leaving an equation that de Broglie suggested should apply to electrons and other particles as well as to photons:

$$p = \frac{h}{\lambda}$$

The momentum of a particle is the product of its mass and speed, $p = mv$. (**Do not confuse the symbol for speed, $v$, with the symbol for frequency, $\nu$.**)

**Figure 4-15**

Examples of wave patterns. (*a*) Floats produce standing water waves. (*b*) X rays generate wave interference patterns. (*c*) Protruding atoms on a metal surface generate standing electron waves.

Making this substitution and solving for λ gives a form of the de Broglie equation that links the wavelength of a particle with its mass and speed:

$$\lambda_{particle} = \frac{h}{mv} \tag{4-7}$$

De Broglie's theory predicted that electrons are wave-like. How might this be confirmed? Figure 4-15 shows examples of the characteristic intensity patterns displayed by waves. In Figure 4-15*a*, water waves radiate away from two bobbing floats and form a standing pattern. In Figure 4-15*b*, X rays form a similar wave pattern. Here, high-energy photons have passed through the regular array of atoms in a crystal, the nuclei of which scatter the photon waves. If electrons have wave properties, they should display regular wave patterns like these.

In 1927, American physicists Clinton Davisson and Lester Germer and British physicist George Thomson carried out experiments in which they exposed metal films to electron beams with well-defined kinetic energies. Both experiments generated patterns like those shown in Figure 4-15*b*, confirming the validity of the de Broglie equation for electron wavelengths. This established the wave nature of electrons.

In recent years, scanning tunnelling electron microscopes have produced pictures of electron waves, an example of which appears in Figure 4-15*c*. Here, two atoms on an otherwise smooth metal surface act like the floats in Figure 4-15*a*, and cause the electrons in the metal to set up a standing wave pattern.

Both photons and electrons are particle–waves, but different equations describe their properties. Table 4-1 summarizes the properties of photons and free electrons, and Example 4-8 shows how to use these equations.

## TABLE 4-1 EQUATIONS FOR PHOTONS AND FREE ELECTRONS

| Property | Photon Equation | Electron Equation |
|---|---|---|
| Energy | $E = h\nu$ | $E_{kinetic} = \dfrac{mv^2}{2}$ |
| Wavelength | $\lambda = \dfrac{hc}{E}$ | $\lambda = \dfrac{h}{mv}$ |
| Speed | $c = 3 \times 10^8$ m/s | $v = \sqrt{\dfrac{2E_{kinetic}}{m}}$ |

*h*, Planck's constant; *ν*, frequency; *m*, mass; *v*, speed.

## EXAMPLE 4-8  WAVELENGTHS

The structure of a crystal can be studied by observing the wave interference patterns that result from passing particle–waves through the crystal lattice. To generate well-defined patterns, the wavelength of the particle–wave must be about the same as the distance between atomic nuclei in the crystal. In a typical crystal, this distance is 0.25 nm. Determine the energy of a photon particle–wave beam with this wavelength and of an electron particle–wave beam with this wavelength.

**Strategy:** This problem has two parts, one dealing with photons and the other with electrons. We are asked to relate the wavelengths of the particle–waves to their corresponding energies. Table 4-1 emphasizes that photons and electrons have different relationships between energy and wavelength. Thus, we use different equations for the two calculations.

**Solution:** Photon energy is $E = h\nu = hc/\lambda$. We substitute and evaluate, being careful with units:

$$E_{photon} = \frac{(6.626 \times 10^{-34} \text{ J s})(2.998 \times 10^8 \text{ m/s})}{(0.25 \text{ nm})(10^{-9} \text{ m/nm})} = 7.9 \times 10^{-16} \text{ J}$$

For an electron, we need to work with two equations. The de Broglie equation links the speed of an electron with its wavelength:

$$\lambda_{particle} = \frac{h}{mv}$$

The kinetic energy equation links the speed of an electron with its kinetic energy:

$$E_{kinetic} = \frac{1}{2}mv^2$$

Begin by determining the speed of the electron:

$$v_{electron} = \frac{h}{m\lambda} = \frac{(6.626 \times 10^{-34} \text{ kg m}^2/\text{s})}{(9.109 \times 10^{-31} \text{ kg})(0.25 \times 10^{-9} \text{ m})} = 2.91 \times 10^6 \text{ m/s}$$

Next, use the speed to find the kinetic energy of the electron:

$$E_{kinetic} = \frac{1}{2}mv^2 = \frac{(9.109 \times 10^{-31} \text{ kg})(2.91 \times 10^6 \text{ m/s})^2}{2}$$

$$E_{kinetic} = 3.9 \times 10^{-18} \text{ kg m}^2 \text{ s}^{-2} = 3.9 \times 10^{-18} \text{ J}$$

**Do the Results Make Sense?** To see if these results are reasonable, we need to compare the values to other values of these quantities. The photon has a much higher energy than those associated with visible light, which is reasonable for X rays. The speed of the electron is about 1% of the speed of light—very fast but not out of the realm of possibility.

### Practice Exercise 4.8

In a photoelectric-effect experiment, a photon with energy of $1.25 \times 10^{-18}$ J is absorbed, causing ejection of an electron with kinetic energy of $2.5 \times 10^{-19}$ J. Calculate the wavelengths (in nm) associated with each.

The de Broglie equation predicts that every particle has wave characteristics. The wave properties of subatomic particles such as electrons and neutrons play important roles in their behaviour, but larger particles such as Ping-Pong balls or automobiles do not behave like waves. The reason is the scale of the waves. For all except subatomic particles, the wavelengths involved are so short that we are unable to detect the wave properties. Example 4-9 illustrates this.

De Broglie received the Nobel Prize in physics in 1929, only two years after experiments confirmed his theory. Davisson, a student of Nobel laureate Robert Millikan, and Thomson, the son and student of J. J. Thomson (who won the Nobel prize for discovering the electron), shared the Nobel Prize in physics in 1937.

## EXAMPLE 4-9 MATTER WAVES

Compare the wavelengths of an electron travelling at $1.00 \times 10^5$ m/s and a Ping-Pong ball of mass 11 g travelling at 2.5 m/s.

**Strategy:** This problem deals with particle–waves that have mass. Equation 4-7, the de Broglie equation, relates the mass and speed of an object to its wavelength.

**Solution:** For the electron: $m_e = 9.109 \times 10^{-31}$ kg, $v = 1.00 \times 10^5$ m/s

$$\lambda_{electron} = \frac{h}{mv} = \frac{6.626 \times 10^{-34} \text{ kg m}^2/\text{s}}{(9.109 \times 10^{-31} \text{ kg})(1.00 \times 10^5 \text{ m/s})} = 7.27 \times 10^{-9} \text{ m}$$

For the Ping-Pong ball: $m_{ball} = 11$ g, $v = 2.5$ m/s

$$\lambda_{ball} = \frac{h}{mv} = \frac{6.626 \times 10^{-34} \text{ kg m}^2/\text{s}}{(11 \text{ g})(1 \text{ kg}/10^3 \text{ g})(2.5 \text{ m/s})} = 2.4 \times 10^{-32} \text{ m}$$

**Do the Results Make Sense?** The wavelength of the electron is significant because it is about the same size as the radius of an atom, but the wavelength of the Ping-Pong ball is inconsequential compared with its size.

**Practice Exercise 4.9**

Calculate the wavelength associated with a proton that is moving at a speed of $2.85 \times 10^5$ m/s.

## Heisenberg's Uncertainty Principle

A particle occupies a particular location, but a wave has no exact position. A wave extends over some region of space. Because of their wave properties, electrons are always spread out rather than located in one particular place. As a result, *the position of a moving electron cannot be precisely defined*. We describe electrons as *delocalized* because their waves are spread out rather than pinpointed.

Mathematically, the position and momentum of a photon are linked. Werner Heisenberg, a German physicist, found in the 1920s that the momentum and position of a particle–wave cannot be simultaneously pinned down. If a particular particle–wave can be pinpointed in a specific location, its momentum cannot be known. Conversely, if the momentum of a particle–wave is known precisely, its location cannot be known. Heisenberg summarized this uncertainty in what has become known as the **uncertainty principle**: the more accurately we know position, the more uncertain we are about momentum, and vice versa. Uncertainty is a feature of all objects, but it becomes noticeable only for very small objects, and in this chapter is relevant especially to electrons. Mathematically, the uncertainty principle is

$$(\Delta x)(\Delta mv) \geq \frac{h}{4\pi}$$

where $\Delta x$ is the uncertainty in the position of the electron, $\Delta mv$ is the uncertainty in its momentum, and $h$ is Planck's constant. Suppose for instance that an electron ($m = 9.11 \times 10^{-31}$ kg) is moving at $v = 2.2 \times 10^6$ m/s in a hydrogen atom. If this velocity is known to within 5%, or to within $1.10 \times 10^5$ m/s, then we can calculate the uncertainty in the position of the electron:

$$(\Delta x) = \frac{h}{4\pi(\Delta mv)} = \frac{6.63 \times 10^{-34} \, \text{J s}}{4\pi(9.11 \times 10^{-31} \, \text{kg} \times 1.10 \times 10^5 \, \text{m/s})}$$

$$= 5.26 \times 10^{-10} \, \text{m}$$

However, note that the diameter of the hydrogen atom is only $2.40 \times 10^{-10}$ m, so the uncertainty in the position of the electron is comparatively very large! The physical significance of this will become clear in the next section, where the positions of electrons in atoms are described in terms of probabilities.

Heisenberg's uncertainty principle forced a change in thinking about how to describe the universe. In a universe subject to uncertainty, many things cannot be measured exactly, and it is never possible to predict with certainty exactly what will occur next. This uncertainty has become accepted as a fundamental feature of the universe at the scale of electrons, protons, and neutrons.

## SECTION EXERCISES

**●4.4.1** Calculate the wavelength associated with a photon whose energy is $1.00 \times 10^{-19}$ J and the wavelength associated with an electron having a kinetic energy of $1.00 \times 10^{-19}$ J.

**●4.4.2** Describe the differences between the properties of a free electron and those of a Ping-Pong ball.

**●4.4.3** The smallest distance that can be resolved by a microscope is typically 1.25 λ, where λ is the wavelength of light or electrons used. The microscope can be greatly improved if a beam of electrons is used instead of a beam of light. If an electron microscope is to successfully resolve objects 0.600 nm apart, what kinetic energy must the electrons have?

## 4.5 QUANTIZATION AND QUANTUM NUMBERS

The properties of electrons described so far (mass, charge, spin, and wave nature) apply to all electrons. Electrons travelling freely in space, electrons moving in a copper wire, and electrons bound to atoms all have these characteristics. Bound electrons, those held in a specific region in space by electrical forces, have additional important properties relating to their energies and the shapes of their waves. These additional properties can have only certain specific values, so they are said to be quantized.

*Energies of bound electrons are quantized.* As described in Section 4.3, atoms of each element have unique, quantized electronic energy levels (see Figures 4-12 and 4-13). This quantization of energy is a property of *bound* electrons. The absorption and emission spectra of atoms consist of discrete energies because electrons undergo transitions from one bound state to another. In contrast, if an atom absorbs enough energy to remove an electron completely, the electron is no longer bound and can take on any amount of kinetic energy. *Bound* electrons have quantized energies; *free* electrons can have any amount of energy.

Absorption and emission spectroscopies provide experimental values for the quantized energies of atomic electrons. The theory of quantum mechanics provides a mathematical explanation that links quantized energies to the wave characteristics of electrons. These wave properties of atomic electrons are described by the Schrödinger equation, a mathematical equation describing the kinetic and potential energies of the atom.

The Schrödinger equation has solutions only for specific energy values. In other words, the energy of an atom is quantized, restricted to certain values. For each quantized energy value, the Schrödinger equation generates a wave function that describes how the electrons are distributed in space.

To picture the spatial distribution of an electron around a nucleus, we must try to visualize a three-dimensional wave. Scientists have coined a name for these three-dimensional waves that characterize electrons: they are called **orbitals**. The word comes from *orbit*, which describes the path that a planet follows when it moves about the Sun. An orbit, however, consists of a specific path, typically a circle or an ellipse. In contrast, an orbital is a three-dimensional volume; for example, a sphere or an hourglass. The shape of a particular orbital shows how an atomic or a molecular electron fills three-dimensional space. Just as energy is quantized, orbitals have specific shapes and orientations. We describe the details of orbitals in Section 4.6.

Each quantized property can be identified, or indexed, using a **quantum number**. These are integers that specify the values of the electron's quantized properties, and are the direct result of solving the Schrödinger equation. Each electron in an atom has three quantum numbers that specify its three variable properties. A set of three quantum numbers is a shorthand notation that describes a particular energy, orbital shape, and orbital orientation. A fourth quantum number, with a value of $+\frac{1}{2}$ or $-\frac{1}{2}$ describes spin orientation. To describe an atomic electron completely, chemists specify a value for each of its four quantum numbers.

## Principal Quantum Number

The most important quantized property of an atomic electron is its energy. The quantum number that indexes energy is the **principal quantum number** (*n*). For the simplest atom, hydrogen, we can use Equation 4-5 to calculate the energy of the electron if we know *n*. However, that equation applies only to the hydrogen atom.

**Figure 4-16**

$n = 1$        $n = 2$

Sizes of atomic orbitals

$E_1$              $E_2 > E_1$

Average heights of bouncing balls

The orbital size of an atomic electron (*top*) is correlated with its orbital energy, just as the bounce height of a tennis ball is correlated with its kinetic energy (*bottom*).

No known equation provides the exact energies of an atom that has more than one electron. Nevertheless, each electron in an atom can be assigned a value of *n* that is a positive integer and that correlates with the energy of the electron. The most stable energy for an atomic electron corresponds to $n = 1$, and each successively higher value of *n* describes a less stable energy state.

Values such as $n = 0$, $n = -3$, and $n = 5/2$ are unacceptable because they do not represent solutions to the Schrödinger equation, meaning that they do not correspond to reality.

The principal quantum number also tells us something about the size of an atomic orbital, because the energy of an electron is correlated with its distribution in space. That is, the more energy the electron has, the more it is spread out in space. As illustrated by Figure 4-16, this situation is roughly analogous to a bouncing tennis ball: the more kinetic energy the ball has, the higher it bounces and the greater its average distance above the court. Similarly, the higher the principal quantum number, the more energy the electron has and the greater its average distance from the nucleus.

To summarize, the principal quantum number (*n*) can have any positive integral value. It indexes the energy of the electron and is correlated with orbital size. As *n* increases, the energy of the electron increases, its orbital gets bigger, and the electron is less tightly bound to the atom.

## Azimuthal Quantum Number

When the Schrödinger equation for a one-electron atom is solved mathematically, the restrictions on *n* and *l* emerge as quantization conditions that correlate with energy and the shape of the wave function.

In addition to size, an atomic orbital also has a specific shape. The solutions for the Schrödinger equation and experimental evidence show that orbitals have a variety of shapes. A second quantum number indexes the shapes of atomic orbitals. This quantum number is the **azimuthal quantum number (*l*)**.

We can categorize the shapes of objects—such as the basketball, football, and tire iron shown in Figure 4-17—according to their preferred axes. A basketball has no preferred axis because its mass is distributed equally in all directions about its centre. A football has one preferred axis, with more mass along this axis than in any other direction. A tire iron has two preferred axes, at right angles to each other. In analogous fashion, electron density in an orbital can be concentrated along preferred axes.

**Figure 4-17**

Some everyday objects have their masses concentrated along preferred axes. A basketball has no such axes, but a football has one, and a tire iron has two.

The value of *l* correlates with the number of preferred axes in a particular orbital and thereby identifies the orbital shape. According to quantum theory, orbital shapes are highly restricted. These restrictions are linked to energy, so the value of the principal quantum number (*n*) limits the possible values of *l*. The smaller *n* is, the more compact the orbital and the more restricted its possible shapes.

Historically, orbital shapes have been identified with letters rather than numbers. These letter designations correspond to the values of *l* as follows:

| Value of *l* | 0 | 1 | 2 | 3 |
|---|---|---|---|---|
| Orbital letter | *s* | *p* | *d* | *f* |

The letters *s*, *p*, *d*, and *f* are derived from early spectroscopic measurements. Scientists assigned the qualitative words *sharp*, *principle*, *diffuse*, and *fundamental* to the spectra they observed of alkali metals.

An orbital is named by listing the numerical value for *n*, followed by the letter that corresponds to the numerical value for *l*. Thus, a 3*s* orbital has quantum numbers $n = 3, l = 0$. A 5*f* orbital has $n = 5, l = 3$. Notice that the restrictions on *l* mean that many combinations of *n* and *l* do not correspond to orbitals that exist. For example, when $n = 1$, *l* can only be zero. In other words, 1*s* orbitals exist, but there are no 1*p*, 1*d*, 1*f*, or 1*g* orbitals. Similarly, there are 2*s* and 2*p* orbitals, but no 2*d*, 2*f*, or 2*g* orbitals. Remember that *n* restricts *l*, but *l* does not restrict *n*. Thus, a 10*d* orbital ($n = 10, l = 2$) is rather high in energy, and perfectly legitimate, but there is no orbital with $n = 2, l = 10$.

## Magnetic and Spin Orientation Quantum Numbers

A sphere has no preferred axis, so it has no directionality in space. When there is a preferred axis, as for a football, Figure 4-18 shows that the axis can point in many different directions relative to an *xyz* coordinate system. Thus, objects with preferred axes have directionality as well as shape.

Among atomic orbitals, *s* orbitals are spherical and have no directionality. Other orbitals are nonspherical, so, in addition to having shape, every orbital points in some direction. Like energy and orbital shape, orbital direction is quantized. Unlike footballs, *p*, *d*, and *f* orbitals have restricted numbers of possible orientations. The **magnetic quantum number** ($m_l$) indexes these restrictions.

Just as orbital size (*n*) limits the number of preferred axes (*l*), the number of preferred axes (*l*) limits the orientations of the preferred axes ($m_l$). When $l = 0$, there is no preferred axis and there is no orientation, so $m_l = 0$. One preferred axis ($l = 1$) can orient in any of three directions, giving three possible values for $m_l$: $+1, 0$, and $-1$. Two preferred axes ($l = 2$) can orient in any of five directions, giving five possible values for $m_l$: $+2, +1, 0, -1$, and $-2$. Each time *l* increases in value by one unit, two additional values of $m_l$ become possible, and the number of possible orientations increases by 2.

The magnetic quantum number derives its name from the fact that different orbital orientations generate different behaviours in the presence of magnetic fields.

**Figure 4-18**

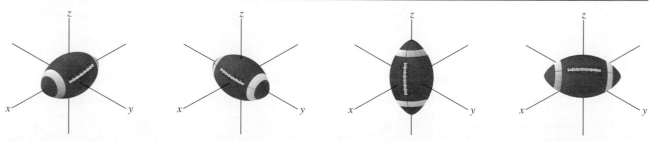

A football has directionality and shape. The figure shows four of the many ways in which a football can be oriented relative to a set of *x*-, *y*-, and *z*-axes.

Spin "up"　　　Spin "down"

As mentioned in Section 4-4, an electron has magnetism associated with a property called spin. Magnetism is directional, so the spin of an electron is directional, too. Like orbital orientation, spin orientation is quantized: electron spin has only two possible orientations, "up" or "down." The **spin orientation quantum number** ($m_s$) indexes this behaviour. The two possible values of $m_s$ are $+\frac{1}{2}$ and $-\frac{1}{2}$. This can also be thought of as an electron spinning either clockwise or counterclockwise. Electrons are also commonly referred to as being "spin up" or "spin down," respectively.

## Sets of Quantum Numbers

A complete description of an atomic electron requires a set of four quantum numbers, $n$, $l$, $m_l$, and $m_s$, which must meet all the restrictions summarized in Table 4-2. Any set of quantum numbers that does not obey these restrictions does not correspond to an orbital and cannot describe an electron.

### TABLE 4-2 RESTRICTIONS ON QUANTUM NUMBERS FOR ATOMS

| Quantum Number | Restrictions | Range |
|---|---|---|
| $n$ | Positive integers | $1, 2 \ldots, \infty$ |
| $l$ | Positive integers less than $n$ | $0, 1 \ldots, (n-1)$ |
| $m_l$ | Integers between $l$ and $-l$ | $-l \ldots, -1, 0, +1, \ldots, +l$ |
| $m_s$ | Half integers, $+\frac{1}{2}$ or $-\frac{1}{2}$ | $-\frac{1}{2}, +\frac{1}{2}$ |

An atomic orbital is designated by its $n$ and $l$ values, such as 1s, 4p, 3d, and so on. When $l > 0$, there is more than one orbital of each designation: three $n\,p$ orbitals, five $n\,d$ orbitals, and so on. When an electron occupies any orbital, its spin quantum number, $m_s$, can be either $+\frac{1}{2}$ or $-\frac{1}{2}$ Thus, there are many sets of valid quantum numbers. An electron in a 3p orbital, for example, may have any one of six valid sets of quantum numbers:

$$n = 3, l = 1, m_l = +1, m_s = +\frac{1}{2} \quad n = 3, l = 1, m_l = +1, m_s = -\frac{1}{2}$$
$$n = 3, l = 1, m_l = \phantom{+}0, m_s = +\frac{1}{2} \quad n = 3, l = 1, m_l = \phantom{+}0, m_s = -\frac{1}{2}$$
$$n = 3, l = 1, m_l = -1, m_s = +\frac{1}{2} \quad n = 3, l = 1, m_l = -1, m_s = -\frac{1}{2}$$

As $n$ increases, so does the number of valid sets of quantum numbers. Example 4-10 shows this.

### EXAMPLE 4-10 VALID QUANTUM NUMBERS

Determine how many valid sets of quantum numbers exist for 4d orbitals, and give two examples.

**Strategy:** The question asks for the sets of quantum numbers that have $n = 4$ and $l = 2$. Each set must meet all the restrictions listed in Table 4-2. The easiest way to see how many valid sets there are is to list all the valid quantum numbers.

**Solution:** Because this is a 4d orbital, $n$ and $l$ are specified and cannot vary. The other two quantum numbers, however, have several acceptable values. For each value of $m_l$, either value of $m_s$ is acceptable, so the total number of possibilities is the product of the number of possible values for each quantum number:

| Quantum number | $n$ | $l$ | $m_l$ | $m_s$ |
|---|---|---|---|---|
| Possible values | 4 | 2 | 2, 1, 0, −1, −2 | $+\frac{1}{2}, -\frac{1}{2}$ |
| Number of possible values | 1 | 1 | 5 | 2 |

Total of possible sets of values for a 4d electron: $(1)(1)(5)(2) = 10$

There are 10 sets of quantum numbers that describe a 4d orbital. Here are two of them, chosen randomly:

$$n = 4, l = 2, m_l = 1, m_s = +\tfrac{1}{2}$$
$$n = 4, l = 2, m_l = -2, m_s = -\tfrac{1}{2}$$

You should be able to list the other eight sets.

**Do the Results Make Sense?** If you write all the possible sets, you will find there are 10 of them. Notice that there are always $2l + 1$ possible values of $m_l$ and 2 possible values of $m_s$, so every nd set of d orbitals has 10 possible sets of quantum numbers.

**Common Pitfalls:** Errors in these types of problems usually result from not knowing the restrictions on the values of the quantum numbers. Make sure you know these restrictions; see Table 4-2.

**Practice Exercise 4.10**
Write all the valid sets of quantum numbers of the 5p orbitals.

## SECTION EXERCISES

**4.5.1** List all the valid sets of quantum numbers for $n = 3$. Give each its proper orbital name (e.g., the orbital with $n = 1, l = 0$ is 1s).

**4.5.2** Which of the following sets of quantum numbers describe actual orbitals, and which are non-existent? For each one that is non-existent, list the restriction that makes it forbidden.

|     | $n$ | $l$ | $m_l$ | $m_s$ |
|-----|-----|-----|-------|-------|
| (a) | 4   | 1   | 1     | 0     |
| (b) | 4   | 4   | 1     | $+\tfrac{1}{2}$ |
| (c) | 4   | 0   | 1     | $+\tfrac{1}{2}$ |
| (d) | 4   | 2   | 2     | $-\tfrac{1}{2}$ |

**4.5.3** Determine the number of different allowable sets of quantum numbers that have $n = 4$.

## 4.6 SHAPES OF ATOMIC ORBITALS

Wave-like properties cause electrons to be smeared out rather than localized at an exact position. This smeared-out distribution can be described using the notion of electron density: where electrons are most probably found, there is high electron density. Low electron density correlates with regions where electrons are least probably found. Each electron, rather than being a point charge, is a three-dimensional particle–wave that is distributed over space in an orbital. Orbitals describe the delocalization of electrons. Moreover, when the energy of an electron changes, the size and shape of its distribution in space change as well.

An atom that contains many electrons can be described by superimposing (adding together) the orbitals for all of its electrons to obtain the overall size and shape of the atom.

The chemical properties of atoms are determined by the behaviour of their electrons. Because atomic electrons are described by orbitals, the interactions of electrons can be described in terms of orbital interactions. The two characteristics of orbitals that determine how electrons interact are their shapes and their energies. Orbital shapes, the subject of this section, describe the distribution of electrons in three-dimensional space. Orbital energies, which we describe in Chapter 5, determine how easily electrons can be moved.

The quantum numbers $n$ and $l$ determine the size and shape of an orbital. As $n$ increases, the size of the orbital increases, and as $l$ increases, the shape of the orbital becomes more elaborate.

## Orbital Depictions

We need ways to visualize electrons as particle–waves delocalized in three-dimensional space. Orbital pictures provide maps of how an electron wave is distributed in space. There are several ways to represent these three-dimensional maps. Each one shows some important orbital features, but none shows all of them. We use three different representations: plots of electron density, pictures of electron density, and pictures of electron contour surfaces.

An **electron density plot** is useful because it represents the electron distribution in an orbital as a two-dimensional plot. These graphs show electron density (or probability) along the $y$-axis and distance from the nucleus, $r$, along the $x$-axis. Figure 4-19$a$ shows an electron density plot for the 2$s$ orbital.

Electron density plots are useful because those for several orbitals can be superimposed to indicate the relative sizes of various orbitals. The simplicity of such a graph is also a drawback, however, because it does not show the three-dimensionality of an orbital.

Orbital pictures have an advantage over electron density plots in that they can indicate the three-dimensional nature of orbitals. One type of orbital picture is a two-dimensional colour pattern in which the density of colour represents electron density. Figure 4-19$b$ shows such an **orbital density picture** of the 2$s$ orbital. This two-dimensional pattern of colour density shows a cross-sectional slice through the middle of the orbital.

Orbital density pictures are probably the most comprehensive views we can draw, but they require much time and care. An **electron contour drawing** provides a simplified orbital picture. In this representation, we draw a contour surface that encloses almost all the electron density. Commonly, "almost all" means 90%. Thus, the probability of finding the electron inside this surface is 90%, and there is a 10% probability of finding the electron outside this surface at any given time. Figure 4-19$c$ shows a contour drawing of the 2$s$ orbital.

A useful analogy for understanding the value of contour surfaces is a swarm of bees around a hive. At any one time, some bees will be off foraging for nectar, so a contour surface drawn around *all* the bees might cover several acres. This would not be a very useful map of bee density. A contour surface containing 90%

**Figure 4-19**

(a) Electron density — Distance from the nucleus, $r$ — Node (zero electron density)

(b)

(c)

Different depictions of the 2$s$ orbital. (a) A plot of electron density vs. distance from the nucleus. (b) An orbital density picture. (c) An electron contour drawing.

of the bees, on the other hand, would be just a bit bigger than the hive itself. This would be a very useful map of bee density, because anyone inside that contour surface would most probably interact with bees.

The drawback of contour drawings is that all details of electron density *inside* the surface are lost. Thus, if we want to convey the maximum information about orbitals, we must use combinations of the various types of depictions.

The advantages and disadvantages of the three types of plots are highlighted by how they show one characteristic feature of orbitals. Figure 4-19*a* shows clearly that there is a value for *r* where the electron density falls to zero. A place where electron density is zero is called a **node**. Figure 4-19*b* shows the node for the 2s orbital as a white ring. In three dimensions, this node is a spherical surface. Figure 4-19*c* does not show the node, because the spherical nodal surface is hidden inside the 90% contour surface. The graph shows the location of the node most clearly, the orbital density picture gives the best sense of the shape of the node, whereas the contour drawing fails to show it at all.

You might wonder how an electron in the 2s orbital, for instance, could exist on either side of the node, but not at the node itself. To answer this, you must think of the electron as having wave-like properties—this wave has zero amplitude at the node, but a non-zero amplitude on either side of the node. Thus, in effect, the electron can move across the node without actually existing at the node!

Figure 4-20 shows electron density plots for 1s, 2s, and 3s orbitals. Note that the size of the s orbitals increases as the value of *n* increases, which is discussed in the next section. However, the number of nodes increases as well. A 1s orbital has no nodes, a 2s orbital has one node, and a 3s orbital has two nodes (and so on). In the 2s orbital, for instance, the nodes would appear as concentric spheres; that is, one inside the other.

## Orbital Size

How large are orbitals? Experiments that measure atomic radii provide information about the size of an orbital. In addition, theoretical models of the atom predict how the electron density of a particular orbital changes with distance from

**Figure 4-20**

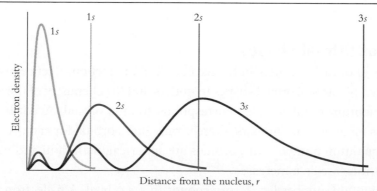

Electron density plots for the 1s, 2s, and 3s atomic orbitals of the hydrogen atom. The vertical lines indicate the values of *r* where the 90% contour surface would be located.

**Figure 4-21**

Electron density plots for the 3s (*maroon line*), 3p (*lavender line*), and 3d (*teal line*) orbitals for the copper atom. All three orbitals are nearly the same size.

the nucleus, *r*. When these sources of information are combined, they reveal several regular features about orbital size.

*In any particular atom, orbitals get larger as the value of n increases.* For any particular atom, the $n = 2$ orbitals are larger than the 1s orbital, the $n = 3$ orbitals are larger than the $n = 2$ orbitals, and so on. The electron density plots in Figure 4-20 show this trend for the first three s orbitals of the hydrogen atom. This plot also shows that the number of nodes increases as *n* increases.

*In any particular atom, all orbitals with the same principal quantum number are similar in size.* As an example, Figure 4-21 shows that the $n = 3$ orbitals of the copper atom have their maximum electron densities at similar distances from the nucleus. The same regularity holds for all other atoms. The quantum numbers other than *n* affect orbital size only slightly. We describe these small effects in the context of orbital energies in Chapter 4.

*Each orbital becomes smaller as nuclear charge increases.* As the positive charge of the nucleus increases, the electrical force exerted by the nucleus on the negatively charged electrons increases as well, and electrons become more tightly bound. This in turn reduces the radius of the orbital. As a result, each orbital shrinks in size as atomic number increases. For example, the 2s orbital steadily decreases in size across the second row of the periodic table from Li ($Z = 3$) to Ne ($Z = 10$). The atomic number, $Z$, is equal to the number of protons in the nucleus, so increasing $Z$ means increasing nuclear charge.

The radius of the 2s orbital decreases as $Z$ increases.

## Details of Orbital Shapes

The shapes of orbitals strongly influence chemical interactions. Hence, we need to have detailed pictures of orbital shapes to understand the chemistry of the elements.

The quantum number $l = 0$ corresponds to an *s* orbital. According to the restrictions on quantum numbers, there is only one s orbital for each value of the principal quantum number. All s orbitals are spherical, with radii and number of nodes that increase as *n* increases.

The quantum number $l = 1$ corresponds to a *p* orbital. A *p* electron can have any of three values for $m_l$, so for each value of *n* there are three different *p* orbitals.

**Figure 4-22**

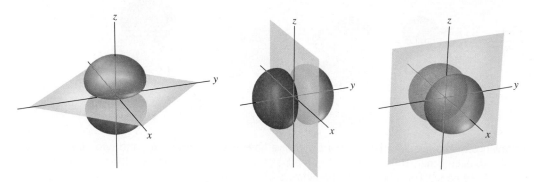

Contour drawings of the three *p* orbitals. The three orbitals have the same shape, but each is oriented perpendicular to the other two.

The *p* orbitals, which are not spherical, can be shown in various ways. The most convenient representation shows the three orbitals with identical shapes but pointing in three different directions. Figure 4-22 shows electron contour drawings of the 2*p* orbitals. Each *p* orbital has high electron density in one particular direction, perpendicular to the other two orbitals, with the nucleus at the centre of the system. The three different orbitals can be represented so that each has its electron density concentrated on both sides of the nucleus along a preferred axis. We can write subscripts on the orbitals to distinguish the three distinct orientations: $p_x$, $p_y$, and $p_z$. Each *p* orbital also has a nodal plane that passes through the nucleus. The nodal plane for the $p_x$ orbital is the *yz* plane, for the $p_y$ orbital is the *xz* plane, and for the $p_z$ orbital is the *xy* plane.

Note in Figure 4-22 that each *p* orbital has also been coloured differently on either side of the node. Since waves can have positive or negative amplitudes (phases), and since orbitals are wave-like, then each orbital has two phases. One colour therefore represents the positive phase of the wave function described by the Schrödinger equation, and the other represents the negative phase. The electron, however, occupies both phases of any orbital.

As *n* increases, the detailed shapes of the *p* orbitals become more complicated (the number of nodes increases, just as for *s* orbitals). Nevertheless, the *directionality* of the orbitals does not change. Each *p* orbital is perpendicular to the other two in its set, and each *p* orbital has its lobes along its preferred axis, where electron density is high. To an approaching atom, therefore, an electron in a 3*p* orbital presents the same characteristics as one in a 2*p* orbital, except that the 3*p* orbital is bigger. Consequently, the shapes and relative orientations of the 2*p* orbitals in Figure 4-22 represent the prominent spatial features of all *p* orbitals.

The quantum number *l* = 2 corresponds to a *d* orbital. A *d* electron can have any of five values for $m_l$ (−2, −1, 0, +1, and +2), so there are five different orbitals in each set. Each *d* orbital has two nodal planes. Consequently, the shapes of the *d* orbitals are more complicated than their *s* and *p* counterparts. The contour

**Figure 4-23**

$3d_{xy}$          $3d_{yz}$          $3d_{xz}$          $3d_{x^2-y^2}$          $3d_{z^2}$

Contour drawings of the five $3d$ orbitals. Four of the five have two preferred axes and two nodal planes at right angles to each other. The $d_{z^2}$ orbital has the $z$-axis as its only preferred axis. It has two nodes that are shaped like cones, one above and one below the $xy$ plane.

drawings in Figure 4-23 show these orbitals in the most convenient way. In these drawings, three orbitals look like three-dimensional cloverleaves, each lying in a plane with the lobes pointed between the axes. A subscript identifies the plane in which each lies: $d_{xy}$, $d_{xz}$, and $d_{yz}$. A fourth orbital is also a cloverleaf in the $xy$ plane, but its lobes point along the $x$- and $y$-axes. This orbital is designated $d_{x^2-y^2}$. The fifth orbital looks quite different. Its major lobes point along the $z$-axis, but there is also a "doughnut" of electron density in the $xy$ plane. This orbital is designated $d_{z^2}$.

If $l = 3$, $f$ orbitals result. An $f$ electron can have any one of seven values of $m_l$ (from $-3$ to $+3$), thus there are seven different $f$ orbitals in each set. The shapes and names of the $f$ orbitals are shown in Figure 4-24.

**Figure 4-24**

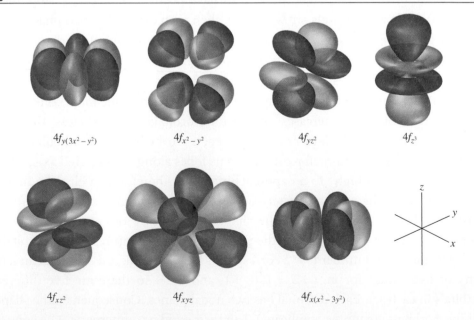

$4f_{y(3x^2-y^2)}$          $4f_{x^2-y^2}$          $4f_{yz^2}$          $4f_{z^3}$

$4f_{xz^2}$          $4f_{xyz}$          $4f_{x(x^2-3y^2)}$

Contour drawings of the seven $4f$ orbitals.

## 4.7 SUNLIGHT AND THE EARTH

The Sun bathes the Earth in a never-ending flux of electromagnetic radiation. This light interacts with atoms and molecules in the atmosphere according to the principles developed in this chapter. We can divide the atmosphere into the regions designated in Figure 4-25. Each region occupies a particular altitude range, and each region absorbs different wavelengths of incoming light. Although these regions blend into one another without distinct boundaries, there are distinctive interactions of light with the atoms and molecules of each region. We describe these interactions from the top down, starting with the first layer encountered by incoming light from the Sun.

### Reactions in the Thermosphere

The **thermosphere** is the outermost part of the atmosphere, above an altitude of about 85 km. Here, molecules of nitrogen and oxygen absorb X-ray radiation coming from the Sun. These photons have enough energy to ionize molecules and to break chemical bonds. Here are two examples:

$$N_2 + h\nu \longrightarrow N_2^+ + e^- \qquad O_2 + h\nu \longrightarrow O + O$$

The products of these reactions are unstable, so they eventually recombine, releasing energy in the form of heat:

$$N_2^+ + e^- \longrightarrow N_2 + \text{Heat} \qquad O + O \longrightarrow O_2 + \text{Heat}$$

The thermosphere, which gets its name from the heat released in these reactions, is a complex mixture of atoms, ions, and molecules, including a high mole fraction of oxygen atoms. At the same time, however, the total atmospheric pressure in the thermosphere is less than $10^{-7}$ bar, which means that the density of atoms and molecules in the thermosphere is quite low.

The aurora borealis, a spectacular atmospheric light show shown in Figure 4-26, originates in the thermosphere. In addition to electromagnetic radiation, the Sun emits a steady stream of protons and electrons. The Earth's magnetic field deflects most of these particles, but some reach the

**Figure 4-25**

The Earth's atmosphere is divided into several regions. Note that the scale for pressure is logarithmic.

thermosphere above the north and south poles of the planet, particularly during times of solar storms, when their emission from the Sun is at its highest. Upon entering the atmosphere, these high-energy particles collide with the atoms and molecules in the thermosphere, transferring energy and generating highly excited states. As these excited atoms and molecules relax back to their ground states, they emit visible light. The colour of the emission depends on the species involved. For example, nitrogen molecules emit red light during an aurora event. Oxygen atoms can emit either red or green light, and nitrogen atoms emit blue light. In other words, the aurora is based on the same principles as emission spectra: atoms and molecules that reach excited states by gaining energy from some external source return to their ground states by emitting photons.

As solar radiation passes through the thermosphere, absorption by atoms and molecules removes its highest-energy photons. The result is that the intensity of high-energy light decreases as sunlight moves down through the atmosphere toward the Earth's surface. The boundary between the thermosphere and the mesosphere is the altitude where almost all the ionizing radiation has been removed. This altitude is about 85 km above the Earth's surface.

Sunlight not absorbed in the thermosphere passes through the mesosphere, which is about 35 km thick, with little absorption. Consequently, there are no important light-induced chemical reactions in the mesosphere.

**Figure 4-26**

The aurora borealis is due to the emission of photons by excited-state atoms and molecules in the thermosphere. In the Northern Hemisphere, the aurora is called the Northern Lights.

## Reactions in the Ozone layer

Near the top of the **stratosphere**, solar radiation generates an abundance of ozone ($O_3$) molecules. The resulting **ozone layer** has important consequences for life on Earth. Ozone forms in two steps. First, a photon with a wavelength between 180 and 240 nm breaks an $O_2$ molecule into two atoms of oxygen:

$$O_2 + h\nu_{(\lambda=180-240 \text{ nm})} \longrightarrow O + O$$

Second, an oxygen molecule captures one of these oxygen atoms to form an ozone molecule:

$$O_2 + O \longrightarrow O_3 + \text{Heat}$$

The second step occurs twice for each $O_2$ fragmentation, giving the overall balanced process for ozone formation:

$$3\, O_2 + h\nu_{(\lambda=180-240 \text{ nm})} \longrightarrow 2\, O_3 + \text{Heat}$$

Why is the ozone layer confined to one region of the atmosphere? The production of ozone requires both a source of oxygen atoms and frequent collisions between the atoms and the molecules that make up the atmosphere. At altitudes lower than 20 km, all the light energetic enough to split oxygen molecules into oxygen atoms has already been absorbed. Consequently, below this altitude, the dissociation of $O_2$ does not occur and ozone does not form. At altitudes higher than 35 km, on the other hand, there is plenty of light to dissociate $O_2$, but the molecular density and the rate of molecular collisions are too low. Above 35 km, there are not enough atoms and molecules for the combination reaction to occur.

Another light-induced reaction limits the ozone concentration in the ozone layer. Ozone strongly absorbs UV light in the region between 200 and 340 nm. The energies of these photons are high enough to break ozone apart into $O_2$ molecules and oxygen atoms:

$$O_3 + hv_{(\lambda = 200-340 \text{ nm})} \longrightarrow O + O_2$$

The interactions of molecules and light in the ozone layer result in a delicate balance that holds ozone concentration at a relatively constant value. Three reactions maintain this balance:

1. Photons with wavelengths of 180–240 nm break apart $O_2$ molecules.
2. Photons with wavelengths of 200–340 nm break apart $O_3$ molecules.
3. Oxygen atoms combine with $O_2$ molecules to produce $O_3$ molecules and heat (which causes this region of the atmosphere to be relatively warm).

The absorption of UV light by ozone and oxygen molecules is critical for life on Earth. If this radiation reached the Earth's surface, it would cause severe biological damage, because light of wavelengths around 300 nm has enough energy to break biological molecules apart. If we were continually bathed in such light, death rates would increase dramatically for almost all living species. Our Focus 4-2: Chemistry and the Environment explores the effects of ultraviolet light, and in Chapter 12 we examine the ozone layer in more detail and discuss how this delicate balance has been endangered by human activities.

## The Greenhouse Effect

The energy balance between light absorbed and emitted determines the Earth's temperature. The sunlight that reaches the Earth's surface is in the ultraviolet, visible, and infrared regions. The Earth's surface absorbs much of this sunlight, which heats the Earth's surface to comfortable temperatures. This sunlight also heats and evaporates water from the oceans. This water later condenses and falls as rain or snow. Furthermore, sunlight drives photosynthesis, the chemical engine of life.

Warm bodies emit radiation, and the Earth is no exception. The Earth's temperature is determined by the balance between solar energy absorbed by the planet and the energy lost through infrared radiation. Most of the radiation emitted by the Earth lies in the infrared region of the spectrum. Nitrogen and oxygen, the major components of the atmosphere, are transparent to infrared radiation. However, some trace gases in the troposphere, the layer of atmosphere closest to the surface of the Earth, are strong infrared absorbers. Foremost among these are carbon dioxide, water vapour, and methane. These gases are known as **greenhouse gases**. Figure 4-27 shows how this works. Much as the glass panes of a greenhouse keep the greenhouse warm, greenhouse gases moderate temperature changes from day to night by absorbing some of the infrared photons emitted by the Earth. Following absorption, the gases re-emit still longer wavelength

**Figure 4-27**

The role of atmospheric $CO_2$ in the greenhouse effect. Carbon dioxide is transparent to incoming sunlight, but it absorbs and re-emits a significant amount of the infrared radiation emitted by the Earth. This alters Earth's energy balance, raising its average temperature.

photons, some of which return to the surface where they are reabsorbed by the planet. This process keeps the temperature from falling dramatically at night when the ground is no longer absorbing energy from the Sun.

The degree of temperature control depends on the amount of water and carbon dioxide in the atmosphere. The amount of $H_2O$ in the atmosphere is very low in desert regions, and the atmosphere is less dense at high altitude. Consequently, deserts and mountains experience more severe day-to-night temperature variations than regions that have higher humidity or are at lower altitude.

Over the past century, the atmospheric concentration of one greenhouse gas, $CO_2$, has increased dramatically. Figure 4-28 shows these concentrations over the last 1100 years. The data up to 1960 are from air trapped in ice cores collected in Antarctica, whereas data from 1958 to the present are from an air sampling station atop Mauna Loa volcano in Hawaii. Many atmospheric chemists have warned for years that this increase is causing a small but dramatic rise in the Earth's average temperature. In 1995, the report of the United Nations Intergovernmental Panel on Climate Change documented an increase in average global air temperature of between 0.3 and 0.6 °C over the past century. More dramatically, an increase of between 1.0 and 3.5 °C is predicted for the current century.

A temperature change of 3 °C may seem trivial, but data indicate that the average global temperature during the last major ice age was only a few degrees lower than it is today. Moreover, computer models predict that higher average temperatures lead to increased heat waves and more severe weather patterns such as hurricanes. There is evidence that such effects may already be occurring. The extreme heat suffered by much of Europe in the summer

**Figure 4-28**

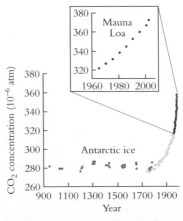

The concentration of $CO_2$ in the atmosphere has increased dramatically in the course of the last century, as humans have increased their consumption of fossil fuels.

of 2003 is a recent example. Additional predicted consequences include a rise in sea levels that could inundate coastal urban centres and disturbances of the ecological balance in many parts of the globe.

Among the components of our atmosphere, the concentration of carbon dioxide is a mere 385 parts per million (ppm). In other words, 999 615 of every million molecules in the air are *not* $CO_2$. (Almost all the molecules are $N_2$ or $O_2$.) At such a low concentration, how could $CO_2$ possibly cause a measurable change in the Earth's surface temperature? The answer lies in the role that minor atmospheric species play in the global energy balance.

Carbon dioxide does not affect the energy *input* to the planet because $CO_2$ is transparent to most of the incoming solar radiation. In contrast, $CO_2$ is extremely effective at absorbing infrared radiation, so the energy *output* from the planet decreases when the amount of carbon dioxide in the atmosphere increases by even a small amount. Thus, a little $CO_2$ goes a long way.

At present, the atmospheric $CO_2$ content is increasing by about 1.5 ppm/year. Moreover, concentrations of two other greenhouse gases, methane and $N_2O$, have also increased during the industrial age. Methane has more than doubled, $CO_2$ has increased by 30%, and $N_2O$ has increased by 15%. The increased presence of each of these gases in the atmosphere means that less energy escapes from the Earth's surface, causing the surface to heat up until the amount of radiant energy that escapes once again balances the incoming amount of solar energy.

## FOCUS 4-2 CHEMISTRY AND THE ENVIRONMENT: EFFECTS OF ULTRAVIOLET LIGHT

Some ultraviolet photons have enough energy to break some chemical bonds. When UV light breaks bonds in biochemical molecules, the products can undergo chemical reactions that lead to cell damage.

The sunlight that normally reaches the Earth's surface has very little UV light with wavelengths shorter than about 360 nm. As the diagram shows, the thermosphere and stratosphere absorb photons with shorter wavelengths.

Ultraviolet light with wavelengths between 360 and 400 nm, which is designated UVA, reaches the Earth's surface. This long-wavelength UV light, although not as damaging as shorter-wavelength radiation, nevertheless can damage skin cells. Our bodies have developed a mechanism to protect us from such damage. When exposed to sunlight, skin produces melanin molecules, which absorb UV light and convert its energy into heat. Every photon that is absorbed by a melanin molecule is prevented from being absorbed by some other component of skin.

Although melanin provides some protection from the damaging effects of ultraviolet light, studies have shown that prolonged exposure to sunlight increases the risk that skin cells will become cancerous. Skin cancer is much more prevalent among people who have a long history of exposure to the Sun. Fair-skinned people are more prone to skin cancer, because their bodies produce less melanin and therefore less protection from the damage caused by UV light.

Cinnamic acid
$C_9H_8O_2$

Benzophenone
$C_{13}H_{10}O$

Sun worshippers and those who work outside can protect themselves from ultraviolet exposure with skin creams that contain UV-absorbing molecules. The UV-absorbing molecule in the first sunscreens was para-aminobenzoic acid (PABA). However, this compound may have toxic effects of its own. The UV-absorbing components of current sunscreens are derivatives of cinnamic acid and benzophenone.

Ultraviolet light is also harmful to the eyes. Although our eyes do not detect UV photons, the lens of the eye is particularly susceptible to damage by UV light. The lens focuses visible photons, but it does not absorb them. In contrast, the lens does absorb ultraviolet photons, causing cell damage that eventually can result in the formation of cataracts. Optometrists and ophthalmologists recommend that eyeglasses contain UV absorbers to protect the eyes from exposure to ultraviolet light.

In recent years, the amount of ozone in the stratosphere has diminished through reactions with human-generated pollutants, as we describe in Chapter 12. As the concentration of ozone in the stratosphere fell, the amount of ultraviolet light that reached the Earth's surface increased. There is great concern about the biological effects of this higher-energy UV light, which includes light in the UVB range (290–320 nm). In the lower Southern Hemisphere, where ozone depletion has been greatest, the incidence of eye damage among sheep has shown a sharp increase in recent years, perhaps because of increased UV exposure.

Sunlight is not the only source of UV light. Ultraviolet light is emitted by mercury vapour black lights. These high-energy UV photons (UVC, between 200 and 290 nm) can be absorbed by nearby atoms and molecules. The electrons excited by this UV absorption often lose energy by emitting visible light. For example, metal ions in certain minerals absorb UV photons and emit visible light, as shown in the photos below.

The increase in greenhouse gases over the past century is a documented reality, but several uncertainties make it difficult to say how much the greenhouse effect will change Earth's climate. For example, the $CO_2$ build-up may not continue at the accelerating pace of recent times, because the biosphere and the oceans may be able to absorb much of the additional $CO_2$ that results from human activities. Although increased greenhouse gases inevitably result in global warming, some argue that these effects may be smaller than changes in the Earth's temperature caused by natural causes. The ice ages demonstrate, for example, that the Earth undergoes significant warming and cooling cycles without human intervention. Nevertheless, the human species is running a huge meteorological experiment on our planet, the long-term outcome of which might be very damaging. By the time we know the results of this experiment, it may be too late to do anything about it.

## SECTION EXERCISES

**4.7.1** Volcanic eruptions change the Earth's energy balance by adding large amounts of smoke particles to the troposphere and stratosphere. Spectacular red sunsets result, because these particles scatter shorter-wavelength light so that less of it reaches the Earth's surface. What effect does this have on the Earth's average temperature? Draw a graph similar to the one in the Focus 4-2 Chemistry and the Environment box to support your answer. On this graph, show the effect on incoming sunlight and the resulting shift, if any, in outgoing radiation.

**4.7.2** The ozone cycle is a delicate balance described by the three equations mentioned in the text. Combine these three equations in appropriate numbers with cancellation of common species to show that the overall ozone cycle converts UV light into heat:

$$h\nu_{(\lambda=180-340 \text{ nm})} \longrightarrow \text{Heat}$$

(*Hint:* You need to show one of the three reactions running backward.)

# CHAPTER 4 VISUAL SUMMARY

## 4.1 CHARACTERISTICS OF ATOMS

Mass
Volume
Nuclei
Electrons

**LEARNING OBJECTIVES:** Understanding some of the fundamental aspects of atoms.

**SKILLS TO MASTER:** Calculating the volume of a single atom. (p. 194)

**KEY CONCEPTS:** Atoms have mass, contain positive nuclei, contain electrons, and occupy volume.

## 4.2 CHARACTERISTICS OF LIGHT

$$E_{photon} = h\nu \qquad \lambda\nu = c$$

Frequency (ν)
Wavelength (λ)
Amplitude
Intensity
Photons

$$E_{electron} = h\nu - h\nu_0$$   Photoelectric effect

**LEARNING OBJECTIVES:** Understanding some of the fundamental aspects of light.

**SKILLS TO MASTER:** Interconverting frequency and wavelength of light. (p. 196)

Calculating photon energies from frequency or wavelength. (p. 199)

Calculating the energies of electrons ejected by the photoelectron effect. (p. 200)

**KEY CONCEPTS:** Light has wave-like and particle-like characteristics.

## 4.3 ABSORPTION AND EMISSION SPECTRA

$$E_{H\,atom} = \frac{2.18 \times 10^{-18}\ J}{n^2}$$

$$\Delta E_{atom} = h\nu$$

$$E_{photon} = |\Delta E_{atom}|$$

Energy level diagram

Quantized

Excited state

Heat loss

Photon emission

Photon emission

Absorption spectrum

Photon absorption

Emission spectrum

Ground state

**LEARNING OBJECTIVES:** Explaining the origins of atomic spectra and relating electron energies in the hydrogen atom to its emission spectrum.

**SKILLS TO MASTER:** Calculating energies of electrons in hydrogen atoms. (p. 207)

Working with energy-level diagrams. (p. 209)

**KEY CONCEPTS:** The energies of electrons in atoms are quantized.

## 4.4 PROPERTIES OF ELECTRONS

Wave properties    Spin

$$\lambda = \frac{h}{mv}$$    Mass Charge

Uncertainty principle

**LEARNING OBJECTIVES:** Describing properties of free electrons and those in atoms or molecules.

**SKILLS TO MASTER:** Using the de Broglie equation. (p. 215)

Using the uncertainty principle. (p. 217)

**KEY CONCEPTS:** Electrons have particle-like and wave-like properties

## 4.5 QUANTIZATION AND QUANTUM NUMBERS

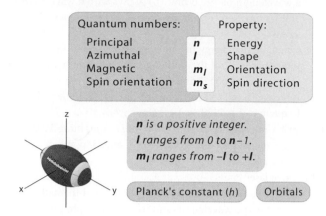

| Quantum numbers: | | Property: |
|---|---|---|
| Principal | $n$ | Energy |
| Azimuthal | $l$ | Shape |
| Magnetic | $m_l$ | Orientation |
| Spin orientation | $m_s$ | Spin direction |

$n$ is a positive integer.
$l$ ranges from 0 to $n-1$.
$m_l$ ranges from $-l$ to $+l$.

Planck's constant ($h$)    Orbitals

**LEARNING OBJECTIVES:** Writing valid sets of quantum numbers for a given set of orbitals

**SKILLS TO MASTER:** Determining sets of quantum numbers. (p. 222)

**KEY CONCEPTS:** Each atomic energy level is associated with a specific three-dimensional atomic orbital.

The principal quantum number must be a positive integer.

The azimuthal quantum number ($l$) can be zero or any positive integer smaller than $n$.

The magnetic quantum number ($m_l$) ranges from $-l$ to $+l$.

The spin quantum number can be $+\frac{1}{2}$ or $-\frac{1}{2}$.

Each atomic energy level is associated with a specific three-dimensional atomic orbital.

## 4.6 SHAPES OF ATOMIC ORBITALS

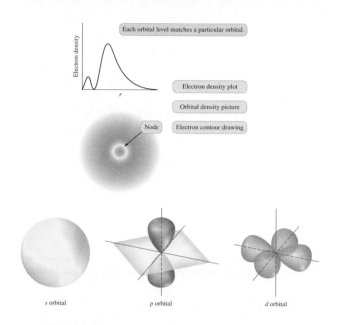

Each orbital level matches a particular orbital.

Electron density

Electron density plot

Orbital density picture

Node    Electron contour drawing

*s* orbital    *p* orbital    *d* orbital

**LEARNING OBJECTIVES:** Recognizing shapes of $s$, $p$, $d$, and $f$ orbitals

## 4.7 SUNLIGHT AND THE EARTH

Greenhouse gases

Incoming sunlight    Infrared radiation emitted by the Earth

Thermosphere

Mesosphere

Ozone layer

Stratosphere

Troposphere

**LEARNING OBJECTIVES:** Describing the major chemical reactions in the thermosphere and stratosphere and explaining the chemistry of the greenhouse effect.

## LEARNING EXERCISES

**4.1** List the properties of electrons and of photons, including the equations used to describe each.

**4.2** Write a short description of (a) the photoelectric effect; (b) wave–particle duality of light; (c) electron spin; and (d) the uncertainty principle.

**4.3** Describe an atomic energy-level diagram and the information it incorporates.

**4.4** Update your list of memory bank equations. For each new equation, specify the conditions under which it can be used.

**4.5** Make a list of all terms in this chapter that are new to you. Write a one-sentence definition of each in your own words. Consult the glossary if you need help.

## PROBLEMS

**ilw** = interactive learning ware problem. Visit the website at www.wiley.com/canada/olmsted

### Characteristics of Atoms

**4.1** The density of silver is $1.050 \times 10^4$ kg/m$^3$, and the density of lead is $1.134 \times 10^4$ kg/m$^3$. For each metal, (a) calculate the volume occupied per atom; (b) estimate the atomic diameter; and (c) using this estimate, calculate the thickness of a metal foil containing $6.5 \times 10^6$ atomic layers of the metal.

**4.2** The density of aluminum metal is $2.700 \times 10^3$ kg/m$^3$, and the density of copper is $8.960 \times 10^3$ kg/m$^3$. For each metal, (a) calculate the volume occupied per atom; (b) estimate the atomic diameter; and (c) using this estimate, calculate the thickness of a metal foil containing $6.5 \times 10^6$ atomic layers of the metal.

**4.3** Describe evidence that indicates that atoms have mass.

**4.4** Describe evidence that indicates that atoms have volume.

**4.5** Draw an atomic picture of a layer of aluminum metal atoms.

**4.6** Steel is iron containing a small amount of carbon, with the relatively small carbon atoms occupying "holes" between larger iron atoms. Draw a picture that shows what a layer of stainless steel looks like on the atomic level.

### Characteristics of Light

**4.7** Convert the following wavelengths into frequencies (Hz), using power-of-10 notation: (a) 4.33 nm; (b) $2.35 \times 10^{-10}$ m; (c) 735 mm; and (d) 4.57 $\mu$m.

**4.8** Convert the following wavelengths into frequencies (Hz), using power-of-10 notation: (a) 2.76 km; (b) 1.44 cm; (c) $3.77 \times 10^{-7}$ m; and (d) 348 nm.

**4.9** Convert the following frequencies into wavelengths, expressing the result in the indicated units: (a) 4.77 GHz (m); (b) 28.9 kHz (cm); (c) 60. Hz (mm); and (d) 2.88 MHz ($\mu$m).

**4.10** Convert the following frequencies into wavelengths, expressing the result in the indicated units: (a) 2.77 MHz (mm); (b) 90.1 kHz (m); (c) 50. Hz (km); and (d) 8.88 GHz ($\mu$m).

**4.11** Calculate the energy in joules per photon and in kilojoules per mole of the following: (a) blue-green light with a wavelength of 490.6 nm; (b) X rays with a wavelength of 25.5 nm; and (c) microwaves with a frequency of $2.5437 \times 10^{10}$ Hz.

**4.12** Calculate the energy in joules per photon and in kilojoules per mole of the following: (a) red light with a wavelength of 665.7 nm; (b) infrared radiation whose wavelength is 1255 nm; and (c) ultraviolet light with a frequency of $4.5528 \times 10^{15}$ Hz.

**4.13** **ilw** A nitrogen laser puts out a pulse containing 10.0 mJ of energy at a wavelength of 337.1 nm. How many photons is this?

**4.14** A dye laser emits a pulse at 450 nm that contains $2.75 \times 10^{15}$ photons. What is the energy content of this pulse?

**4.15** What are the wavelength and frequency of photons with the following energies: (a) 745 kJ/mol; (b) $3.55 \times 10^{-19}$ J/photon?

**4.16** What are the wavelength and frequency of photons with the following energies: (a) 355 J/mol; (b) $2.50 \times 10^{-18}$ J/photon?

**4.17** **ilw** When light of frequency $1.30 \times 10^{15}$ s$^{-1}$ shines on the surface of cesium metal, electrons are ejected with a maximum kinetic energy of $5.2 \times 10^{-19}$ J. Calculate (a) the wavelength of this light; (b) the binding energy of electrons to cesium metal; and (c) the longest wavelength of light that will eject electrons.

**4.18** The binding energy of electrons to a chromium metal surface is $7.21 \times 10^{-19}$ J. Calculate (a) the longest wavelength of light that will eject electrons from chromium metal; (b) the frequency required to give electrons with kinetic energy of $2.5 \times 10^{-19}$ J; and (c) the wavelength of the light in part (b).

**4.19** Draw energy-level diagrams that illustrate the difference in electron binding energy between cesium metal and chromium metal. Refer to Problems 4.17 and 4.18.

**4.20** A phototube delivers an electrical current when a beam of light strikes a metal surface inside the tube. Phototubes do not respond to infrared photons. Draw an energy-level diagram for electrons in the metal of a phototube and use it to explain why phototubes do not respond to infrared light.

**4.21** Refer to Figure 4-4 to answer the following questions: (a) What is the wavelength range for radio waves? (b) What colour is light whose wavelength is $5.8 \times 10^{-7}$ m? (c) In what region does radiation with frequency of $4.5 \times 10^8$ Hz lie?

**4.22** Refer to Figure 4-4 to answer the following questions: (a) What is the wavelength range for infrared radiation? (b) What colour is light of wavelength $4.85 \times 10^{-7}$ m? (c) In what region does radiation with frequency of $4.5 \times 10^{18}$ Hz lie?

### Absorption and Emission Spectra

**4.23** **ilw** From Figure 4-14, calculate the energy difference in kilojoules per mole between the excited state of mercury that emits 404-nm light and the ground state.

**4.24** From Figure 4-14, determine the wavelength of light needed to excite an electron from the ground state to the lowest excited state of the mercury atom.

**4.25** Determine the wavelengths that hydrogen atoms absorb to reach the $n = 8$ and $n = 9$ states from the ground state. In what region of the electromagnetic spectrum do these photons lie?

**4.26** Determine the frequencies that hydrogen atoms emit in transitions from the $n = 6$ and $n = 5$ levels to the $n = 3$ level. In what region of the electromagnetic spectrum do these photons lie?

**4.27** Using Figure 4-13, explain why more lines appear in emission spectra than in absorption spectra.

**4.28** Using Figure 4-14, explain why mercury emits photons at longer wavelengths than the wavelengths that mercury absorbs.

## Properties of Electrons

**4.29** What is the mass in grams of 1 mole of electrons?

**4.30** What is the charge in coulombs of 1 mole of electrons?

**4.31** Determine the wavelengths of electrons with the following kinetic energies: (a) $1.15 \times 10^{-19}$ J; (b) 3.55 kJ/mol; and (c) $7.45 \times 10^{-3}$ J/mol.

**4.32** Determine the wavelengths of electrons with the following kinetic energies: (a) 76.5 J/mol; (b) $4.77 \times 10^{-18}$ J; and (c) $3.21 \times 10^{-11}$ J.

**4.33** Determine the kinetic energies in joules of electrons with the following wavelengths: (a) 3.75 nm; (b) 4.66 m; and (c) 8.85 mm.

**4.34** Determine the kinetic energies in joules of electrons with the following wavelengths: (a) 3.75 m; (b) 4.66 $\mu$m; and (c) 2.85 mm.

## Quantization and Quantum Numbers

**4.35** List all the valid sets of quantum numbers for a $6p$ electron.

**4.36** List all the valid sets of quantum numbers for a $4f$ electron.

**4.37** (ilw) If you know that an electron has $n = 3$, what are the possible values for its other quantum numbers?

**4.38** If you know that an electron has $m_l = -2$, what are the possible values for its other quantum numbers?

**4.39** For the following sets of quantum numbers, determine which describe actual orbitals and which are nonexistent. For each one that is non-existent, list the restriction that forbids it:

|     | $n$ | $l$ | $m_l$ | $m_s$ |
| --- | --- | --- | --- | --- |
| (a) | 5 | 3 | $-2$ | $-1$ |
| (b) | 5 | 3 | $-3$ | $\frac{1}{2}$ |
| (c) | 3 | 3 | $-3$ | $\frac{1}{2}$ |
| (d) | 3 | 0 | 0 | $-\frac{1}{2}$ |

**4.40** For the following sets of quantum numbers, determine which describe actual orbitals and which are non-existent. For each one that is non-existent, list the restriction that forbids it:

|     | $n$ | $l$ | $m_l$ | $m_s$ |
| --- | --- | --- | --- | --- |
| (a) | 3 | $-1$ | $-1$ | $\frac{1}{2}$ |
| (b) | 3 | 1 | $-1$ | $-\frac{1}{2}$ |
| (c) | 3 | 1 | 2 | $\frac{1}{2}$ |
| (d) | 3 | 2 | 2 | $-\frac{1}{2}$ |

**4.41** List the values for the quantum numbers for a $3d$ electron that has spin up and the largest possible value for its magnetic quantum number.

**4.42** List the values for the quantum numbers for a $4p$ electron that has spin down and the most negative possible value for its magnetic quantum number.

## Shapes of Atomic Orbitals

**4.43** Refer to Figures 4-19 and 4-22. Draw the analogous set of three depictions for an orbital that has $n = 2, l = 1$.

**4.44** Refer to Figures 4-19 and 4-21. Draw the analogous set of three depictions for an orbital that has $n = 3, l = 0$.

**4.45** Draw pictures showing how the $p_y$ orbital looks when viewed along the $z$-axis, the $y$-axis, and the $x$-axis (some views may look the same).

**4.46** Draw pictures showing how the $d_{xz}$ orbital looks when viewed along the $z$-axis, the $y$-axis, and the $x$-axis (some views may look the same).

**4.47** Identify each of the following orbitals, and state the values for $n$ and $l$ for each. (*Hint:* Use the size of the $3s$ orbital to identify the other three.)

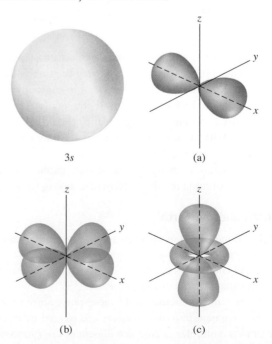

$3s$          (a)

(b)          (c)

4.48 Shown below are electron density pictures and electron density plots for the 1s, 2s, 2p, and 3p orbitals. Assign the various depictions to their respective orbitals.

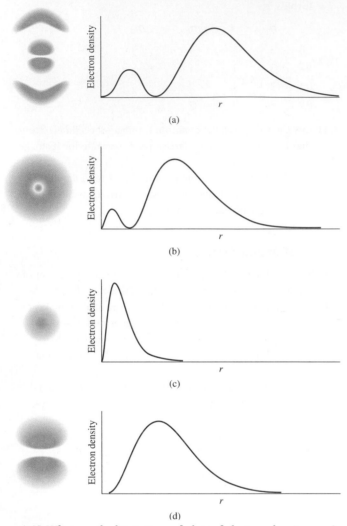

(a)

(b)

(c)

(d)

4.49 What are the limitations of plots of electron density vs. r?

4.50 The conventional method of showing the three-dimensional shape of an orbital is an electron contour surface. What are the limitations of this representation?

4.51 Construct contour drawings for the orbitals graphed in Figure 4-20, appropriately scaled to illustrate the size differences among these orbitals.

4.52 Draw contour drawings for the orbitals graphed in Figure 4-21, illustrating the shape differences among these orbitals.

## Sunlight and the Earth

4.53 When molecules of nitrogen and oxygen in the thermosphere absorb short-wavelength light, $N_2$ molecules ionize and $O_2$ molecules break into atoms. Draw molecular pictures that illustrate these processes.

4.54 When ozone molecules in the mesosphere absorb UV light, they fragment into oxygen atoms and oxygen molecules. Draw a molecular picture that illustrates this process.

4.55 **ilw** Light of wavelength 340 nm or shorter is required to fragment ozone molecules. What is the minimum energy

in kJ/mol for this process? If an ozone molecule absorbs a 250-nm photon, how much excess kinetic energy will the fragments possess?

4.56 It requires 496 kJ/mol to break $O_2$ molecules into atoms and 945 kJ/mol to break $N_2$ molecules into atoms. Calculate the maximum wavelengths of light that can break these molecules apart. What part of the electromagnetic spectrum contains these photons?

4.57 List the region of the atmosphere and the atmospheric gases that absorb light in each of the following spectral regions: (a) less than 200 nm; and (b) 240–310 nm.

4.58 List the region of the atmosphere and the atmospheric gases that absorb light in each of the following spectral regions: (a) 200–240 nm; and (b) 700–2000 nm.

4.59 A high-altitude balloon equipped with a transmitter and pressure sensor reports a pressure of $10^{-3}$ bar (see Figure 4-25). (a) What altitude has the balloon reached? (b) In what region of the atmosphere is the balloon? (c) What chemical processes take place in this region?

4.60 At one stage of the return to Earth of a space shuttle flight, its instruments report that the atmospheric pressure is $10^{-6}$ bar (see Figure 4-25). (a) What is the altitude? (b) In what region of the atmosphere is the shuttle? (c) What chemical processes take place in this region?

## ADDITIONAL PAIRED PROBLEMS

4.61 Calculate the following for a photon of frequency $4.5 \times 10^{13}$ Hz that is reflected off the moon: (a) its wavelength; (b) its energy; and (c) how long it takes to reach the Earth, which is $4.60 \times 10^5$ km from the moon.

4.62 Calculate the following for a photon ($\lambda = 525$ nm) emitted by the sun: (a) its frequency; (b) its energy; and (c) the time it takes to reach the Earth, which is $1.50 \times 10^8$ km from the Sun.

4.63 It requires 243 kJ/mol to fragment $Cl_2$ molecules into Cl atoms. What is the longest wavelength (in nm) of sunlight that could accomplish this? Will $Cl_2$ molecules in the troposphere fragment?

4.64 It requires 364 kJ/mol to break the chemical bond in HBr molecules. What is the longest wavelength (in nm) of light that has enough energy to cause this bond to break? Will HBr molecules in the troposphere be fragmented into atoms?

4.65 How many sets of quantum number values are there for a 4p electron? List them.

4.66 How many sets of quantum number values are there for a 3d electron? List them.

4.67 Redraw the first light wave in Figure 4-3 to show a wave whose frequency is twice as large as that shown in the figure, but whose amplitude is the same.

4.68 Redraw the first light wave in Figure 4-3 to show a wave whose wavelength and amplitude are each twice as large as that shown in the figure.

4.69 A minimum of 216.4 kJ/mol is required to remove an electron from a potassium metal surface. What is the longest wavelength of light that can do this?

4.70 A minimum of 216.4 kJ/mol is required to remove an electron from a potassium metal surface. If UV light at 255 nm strikes this surface, what is the maximum speed of the ejected electrons?

4.71 In a photoelectric effect experiment, photons whose energy is $6.00 \times 10^{-19}$ J are absorbed by a metal, and the maximum kinetic energy of the resulting electrons is $E_{kinetic} = 2.70 \times 10^{-19}$ J. Calculate (a) the binding energy of electrons in the metal; (b) the wavelength of the light; and (c) the wavelength of the electrons.

4.72 In a photoelectric effect experiment, the minimum frequency needed to eject electrons from a metal is $7.5 \times 10^{14}$ s$^{-1}$. Suppose that a 366-nm photon from a mercury discharge lamp strikes the metal. Calculate (a) the binding energy of the electrons in the metal; (b) the maximum kinetic energy of the ejected electrons; and (c) the wavelength associated with those electrons.

4.73 One frequency of a CB (Citizen's Band) radio is 27.3 MHz. Calculate the wavelength and energy of photons at this frequency.

4.74 Microwave ovens use radiation whose wavelength is 12.5 cm. What is the frequency and energy in kJ/mol of this radiation?

4.75 Refer to Figure 4-25 to answer the following questions: (a) What is the pressure at an altitude of 60 km? (b) What atomic and molecular species are present at that altitude? (c) At what altitude is the pressure 8 Torr? (d) What region of the atmosphere is this?

4.76 Refer to Figure 4-25 to answer the following questions: (a) What is the pressure at an altitude of 100 km? (b) Describe the chemistry that takes place at that altitude. (c) What are the atomic and molecular species present at that altitude?

4.77 Barium salts in fireworks generate a yellow-green colour. Ba$^{2+}$ ions emit light with $\lambda = 487, 514, 543, 553,$ and 578 nm. Convert these wavelengths into frequencies and into energies in kJ/mol.

4.78 The bright red colour of highway safety flares comes from strontium ions in salts such as $Sr(NO_3)_2$ and $SrCO_3$. Burning a flare produces strontium ions in excited states, which emit red photons at 606 nm and at several wavelengths between 636 and 688 nm. Calculate the frequency and energy (kJ/mol) of emissions at 606, 636, and 688 nm.

4.79 A hydrogen atom emits a photon as its electron changes from $n = 5$ to $n = 1$. What is the wavelength of the photon? In what region of the electromagnetic spectrum is this photon found?

4.80 A hydrogen atom emits a photon as its electron changes from $n = 7$ to $n = 3$. What is the wavelength of the photon? In what region of the electromagnetic spectrum is this photon found?

## ADDITIONAL PROBLEMS

4.81 The photoelectric effect for magnesium metal has a threshold frequency of $8.95 \times 10^{14}$ s$^{-1}$. Can Mg be used in photoelectric devices that sense visible light? Do a calculation in support of your answer.

4.82 Energetic free electrons can transfer their energy to bound electrons in atoms. In 1913, James Franck and Gustav Hertz passed electrons through mercury vapour at low pressure to determine the minimum kinetic energy required to produce the excited state that emits ultraviolet light at 253.7 nm. What is that minimum kinetic energy? What wavelength is associated with electrons of this energy?

4.83 The radius of a typical atom is $10^{-10}$ m, and the radius of a typical nucleus is $10^{-15}$ m. Compute typical atomic and nuclear volumes and determine what fraction of the volume of a typical atom is occupied by its nucleus.

4.84 Neutrons, like electrons and photons, are particle–waves whose diffraction patterns can be used to determine the structures of molecules. Calculate the kinetic energy of a neutron with a wavelength of 75 pm.

4.85 The human eye can detect as little as $2.35 \times 10^{-18}$ J of green light of wavelength 510 nm. Calculate the minimum number of photons that can be detected by the human eye.

4.86 Gaseous lithium atoms absorb light of wavelength 323 nm. The resulting excited lithium atoms lose some energy through collisions with other atoms. The atoms then return to their ground state by emitting two photons with $\lambda = 812.7$ and 670.8 nm. Draw an energy-level diagram that shows this process. What fraction of the energy of the absorbed photon is lost in collisions?

4.87 Calculate the wavelengths associated with an electron and a proton, each travelling at 5.0% of the speed of light.

4.88 One hydrogen emission line has a wavelength of 486 nm. Identify the values for $n_{final}$ and $n_{initial}$ for the transition giving rise to this line.

4.89 An atomic energy-level diagram, shown to scale, follows:

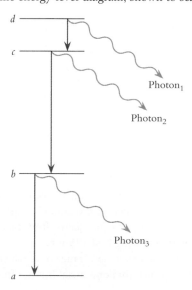

An excited-state atom emits photons when the electron moves in succession from level $d$ to level $c$, from level $c$ to level $b$, and from level $b$ to the ground state (level $a$). The wavelengths of the emitted photons are 565, 152, and 121 nm (not necessarily in the proper sequence). Match each emission with the appropriate wavelength and calculate the energies of levels $b$, $c$, and $d$ relative to level $a$.

4.90 Small helium–neon lasers emit 1.0 mJ/s of light at 634 nm. How many photons does such a laser emit in one minute?

4.91 The argon-ion laser has two major emission lines, at 488 and 514 nm. Each of these emissions leaves the $Ar^+$ ion in an energy level that is $2.76 \times 10^{-18}$ J above the ground state. (a) Calculate the energies of the two emission wavelengths in joules. (b) Draw an energy-level diagram (in joules per atom) that illustrates these facts. (c) What frequency and wavelength radiation is emitted when the $Ar^+$ ion returns to its lowest energy level?

4.92 The series of emission lines that results from excited hydrogen atoms undergoing transitions to the $n = 3$ level is called the "Paschen series." Calculate the energies of the first five lines in this series of transitions, and draw an energy-level diagram that shows them to scale.

4.93 It takes 486 kJ/mol to remove electrons completely from sodium atoms. Sodium atoms absorb and emit light of wavelengths 589.6 and 590.0 nm. (a) Calculate the energies of these two wavelengths in kJ/mol. (b) Draw an energy-level diagram for sodium atoms that shows the levels involved in these transitions and the ionization energy. (c) If a sodium atom has already absorbed a 590.0-nm photon, what is the wavelength of the second photon a sodium atom must absorb in order to remove an electron completely?

4.94 As a general rule, the temperature drops more on clear nights than on cloudy nights. What feature of clouds accounts for this?

4.95 The graph below shows the results of photoelectron experiments on two metals, using light of the same energy:

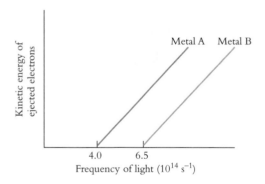

(a) Calculate the binding energy of each metal. Which has the higher binding energy? Explain. (b) Calculate the kinetic energies of electrons ejected from each metal by photons with wavelength of 125 nm.
(c) Calculate the wavelength range over which photons can eject electrons from one metal but not from the other.

## GROUP STUDY PROBLEMS

4.96 For the $3d_{yz}$ orbital, draw graphs of electron density vs. $r$: (a) for $r$ lying along the $z$-axis; (b) for $r$ lying along the $x$-axis; (c) for $r$ pointing halfway between the $y$- and $z$-axes in the $yz$ plane. (*Hint*: Consult Figure 4-23.)

4.97 A hydrogen atom undergoes an electronic transition from the $n = 4$ to the $n = 2$ state. In the process, the H atom emits a photon, which then strikes a cesium metal surface and ejects an electron. It takes $3.23 \times 10^{-19}$ J to remove an electron from Cs metal. Calculate (a) the energy of the $n = 4$ state of the H atom; (b) the wavelength of the emitted photon; (c) the energy of the ejected electron; and (d) the wavelength of the ejected electron.

4.98 The Sun's atmosphere contains vast quantities of $He^+$ cations. These ions absorb some of the Sun's thermal energy, promoting electrons from the $He^+$ ground state to various excited states. A $He^+$ ion in the fifth energy level may return to the ground state by emitting three successive photons: an infrared photon ($\lambda = 1014$ nm), a green photon ($\lambda = 469$ nm), and an X ray ($\lambda = 26$ nm). (a) Calculate the excitation energies of each of the levels occupied by the $He^+$ ion as it returns to the ground state. (b) Draw an energy-level diagram for $He^+$ cations that illustrates these processes.

4.99 Gaseous Ca atoms absorb light at 422.7, 272.2, and 239.9 nm. After the 272.2-nm absorption, an emission at 671.8 nm occurs. Absorption of light at 239.9 nm is followed by emission at 504.2 nm. Construct an energy-level diagram for Ca atoms and answer the following questions: (a) After the 504.2-nm emission, what wavelength would have to be emitted to return to the lowest energy state? (b) What wavelength of light corresponds to the energy difference between the state reached using 422.7-nm light and that reached using 272.2-nm light? (c) Do any sequences described by the data lead to a common energy level? If so, which ones?

4.100 Design a figure that summarizes the chemistry described in Section 4.7. Begin your figure with the regions of the atmosphere. Add the key atomic, molecular, and ionic species found in each region. Include three downward arrows that represent the penetration of sunlight into the atmosphere. Label two of these arrows "Ionizing radiation" and "High-energy UV light." Label the third arrow with the components of sunlight that reach the Earth's surface. Show on your figure the ozone layer, the region associated with the greenhouse effect, and the location of the aurora borealis. The text and figures of Section 4.7 contain all of the information you need.

## ANSWERS TO SECTION EXERCISES

4.1.1  (a) $1.73 \times 10^{-29}$ m$^3$; and (b) $1.73 \times 10^{-34}$ m$^3$

4.1.2  (a) $3.21 \times 10^{-10}$ m; and (b) $3.1 \times 10^3$ layers

4.2.1  (a) 779 nm; (b) infrared; and (c) 154 kJ

4.2.2  755 photons

4.2.3  $\nu = 9.06 \times 10^{14}$ s$^{-1}$; $E_{binding} = 4.00 \times 10^{-19}$ J = 241 kJ/mol

4.3.1  107 kJ/mol

4.3.2  $4.05 \times 10^3$ nm; $1.28 \times 10^3$ nm; 434 nm; and 95.0 nm

**4.3.3** $3.30 \times 10^{-18}$ J See energy-level diagram.

$$E = 3.70 \times 10^{-18} \text{ J}$$
$$\lambda = 501.6 \text{ nm}$$
$$E = 3.30 \times 10^{-18} \text{ J}$$

Energy, $E$

$$\lambda = 53.7 \text{ nm}$$

$$E = 0$$

**4.4.1** Photon: 1990 nm; electron: 1.55 nm

**4.4.2** Their masses are very different. A free electron carries electrical charge, is a small magnet, and sometimes displays wave behaviour; a Ping-Pong ball is electrically neutral and non-magnetic and displays particle behaviour.

**4.4.3** $1.05 \times 10^{-18}$ J

**4.5.1**

| | | | | |
|---|---|---|---|---|
| $n = 3$ | $l = 0$ | $m_l = 0$ | $m_s = +^1/_2$ | $3s$ |
| $n = 3$ | $l = 0$ | $m_l = 0$ | $m_s = -^1/_2$ | $3s$ |
| $n = 3$ | $l = 1$ | $m_l = 0$ | $m_s = +^1/_2$ | $3p$ |
| $n = 3$ | $l = 1$ | $m_l = 0$ | $m_s = -^1/_2$ | $3p$ |
| $n = 3$ | $l = 1$ | $m_l = +1$ | $m_s = +^1/_2$ | $3p$ |
| $n = 3$ | $l = 1$ | $m_l = +1$ | $m_s = -^1/_2$ | $3p$ |
| $n = 3$ | $l = 1$ | $m_l = -1$ | $m_s = +^1/_2$ | $3p$ |
| $n = 3$ | $l = 1$ | $m_l = -1$ | $m_s = -^1/_2$ | $3p$ |
| $n = 3$ | $l = 2$ | $m_l = 0$ | $m_s = +^1/_2$ | $3d$ |
| $n = 3$ | $l = 2$ | $m_l = 0$ | $m_s = -^1/_2$ | $3d$ |
| $n = 3$ | $l = 2$ | $m_l = +2$ | $m_s = +^1/_2$ | $3d$ |
| $n = 3$ | $l = 2$ | $m_l = +2$ | $m_s = -^1/_2$ | $3d$ |
| $n = 3$ | $l = 2$ | $m_l = -2$ | $m_s = +^1/_2$ | $3d$ |
| $n = 3$ | $l = 2$ | $m_l = -2$ | $m_s = -^1/_2$ | $3d$ |
| $n = 3$ | $l = 2$ | $m_l = +1$ | $m_s = +^1/_2$ | $3d$ |
| $n = 3$ | $l = 2$ | $m_l = +1$ | $m_s = -^1/_2$ | $3d$ |
| $n = 3$ | $l = 2$ | $m_l = -1$ | $m_s = +^1/_2$ | $3d$ |
| $n = 3$ | $l = 2$ | $m_l = -1$ | $m_s = -^1/_2$ | $3d$ |

**4.5.2** (a) Non-existent, $m_s$ must be $+^1/_2$ or $-^1/_2$; (b) non-existent, $l$ must be less than $n$; (c) non-existent, $m_l$ cannot exceed $l$; and (d) actual

**4.5.3** 32

**4.6.1** Refer to Figures 4-19, 4-22, and 4-23.

**4.6.2** (a)          (b)

**4.6.3**

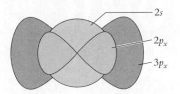

$$2s$$
$$2p_x$$
$$3p_x$$

The $n = 2$ orbitals are not important when $3p_x$ is occupied.

**4.7.1** Decreases the temperature, because amount of incoming blue light is reduced.

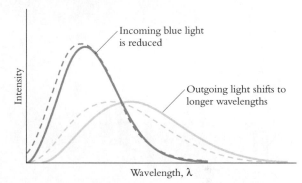

Incoming blue light is reduced

Outgoing light shifts to longer wavelengths

Intensity

Wavelength, $\lambda$

**4.7.2**
$$h\nu + O_2 \longrightarrow 2 \, O$$
$$O + O_2 \longrightarrow O_3 + \text{heat}$$
$$O + O_2 \longrightarrow O_3 + \text{heat}$$
$$h\nu + O_3 \longrightarrow O_2 + O$$
$$h\nu + O_3 \longrightarrow O_2 + O$$
$$\underline{2 \, O \longrightarrow O_2 + h\nu}$$
$$h\nu \longrightarrow \text{heat}$$

## ANSWERS TO PRACTICE EXERCISES

**4.1** $V_{atom} = 1.178 \times 10^{-23}$ cm$^3$/atom

**4.2** $\nu = 2.14 \times 10^{10}$ Hz

**4.3** $E_{photon} = 7.82 \times 10^{-19}$ J

**4.4** Yes; $E_{kinetic}$ (electron) $= 2 \times 10^{-20}$ J

**4.5** $\Delta E = 4.56 \times 10^{-19}$ J $= 274$ kJ/mol

**4.6** $E = 2.04 \times 10^{-18}$ J, $\lambda = 97.2$ nm

**4.7** The level is the sixth from the bottom of Figure 4-14; $E = 745$ kJ/mol

**4.8** Photon has $\lambda = 159$ nm, electron has $\lambda = 0.98$ nm.

**4.9** Proton wavelength is $1.39 \times 10^{-12}$ m.

**4.10**

$$n = 5, l = 1, m_l = +1, m_s = +^1/_2$$
$$n = 5, l = 1, m_l = +1, m_s = -^1/_2$$

$$n = 5, l = 0, m_l = +1, m_s = +^1/_2$$
$$n = 5, l = 0, m_l = +1, m_s = -^1/_2$$

$$n = 5, l = 1, m_l = -1, m_s = +^1/_2$$
$$n = 5, l = 1, m_l = -1, m_s = -^1/_2$$

# ATOMIC ENERGIES AND PERIODICITY

## THE CHEMISTRY OF FIREWORKS

People love the spectacle of fireworks. From Bastille Day in France to Guy Fawkes Day in Britain, from Chinese New Year to Canada Day, fireworks bring joy to celebrations all around the world. Fireworks date back more than 1000 years to the discovery of black powder in China. This first gunpowder was brought to Europe during the Middle Ages and was used widely in weapons, in construction, and for fireworks.

Aerial fireworks are launched using fast-burning fuses that ignite propellants of black powder. After launching, a delayed fuse sets off a "bursting charge" that contains pellets containing an oxidizing agent and a fuel. An exothermic redox reaction within the pellets causes the components of the fireworks to emit light. A random distribution of pellets gives an irregular spray of light, whereas arranging the pellets on a sphere gives a highly symmetrical explosion.

To produce white light in fireworks, a reactive metal such as magnesium produces particles of solid magnesium oxide that reach temperatures in excess of 3000 °C during the explosion of the fireworks.

These hot particles appear white by emitting light across the visible spectrum. Iron particles burn at a lower temperature and generate golden sparks. Larger particles stay hot longer than smaller ones and continue to burn using $O_2$ in the atmosphere, emitting long-lived sparks rather than short flashes.

The colours of fireworks displays are produced by emission from atomic ions as described in Chapter 4. The explosions of fireworks promote electrons to excited states. The energy-level scheme of every element is different, so fireworks manufacturers can change colours by incorporating different elements. Sodium ions emit yellow light, strontium ions produce red light, and green comes from barium ions. Excited copper ions emit blue light, but good blue fireworks require precisely optimized conditions, so a fireworks show can be judged by the quality of its blue explosions.

The colours of fireworks depend on the energies of the atomic orbitals of the various atomic ions, but orbital energy levels have consequences that are much

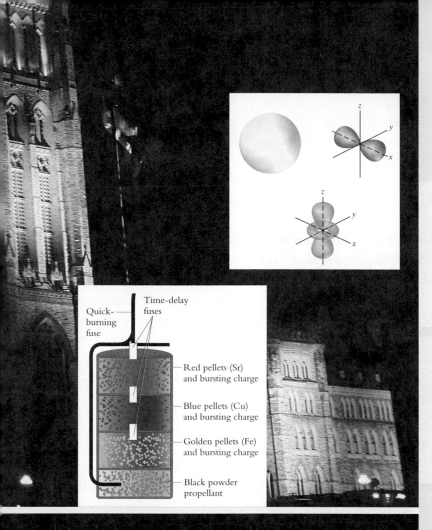

more far-reaching. Orbital energies determine the stabilities of atoms and how atoms react. The structure of the periodic table is based on orbital energy levels. In this chapter, we explore the details of orbital energies and relate them to the form and structure of the periodic table. This provides the foundation for interpreting chemical behaviour patterns.

# 5

## CHAPTER CONTENTS

## LEARNING OBJECTIVES

*Upon completion of this chapter you should be able to:*

- **explain the effects of nuclear charge and screening on the energies of electrons**
- **understand the relationships between the structure of the periodic table and electron configurations**
- **use the Pauli exclusion principle, Hund's rule, and the orbital filling order to predict electron configurations of atoms and ions**
- **understand the trends in atomic radius, ionization energy and electron affinity and their relationships to nuclear charge and electron configuration**

## 5.1 ORBITAL ENERGIES

A hydrogen atom can absorb a photon and change from its most stable state (ground state) to a less stable state (excited state), as described in Section 4.3. We can account for this process in terms of atomic orbitals. When a hydrogen atom absorbs a photon, its electron transfers to an orbital that is higher in energy and has a larger principal quantum number. Figure 5-1 illustrates this process.

The atomic orbital model explains perfectly the spectra and the energy levels of the hydrogen atom. Does this model apply to other atoms? Experiments show that although the details are different for each kind of atom, the underlying principles are the same. Variations in nuclear charge and in the number of electrons change the magnitudes of the electrical forces that hold electrons in their orbitals. Differences in forces cause changes in orbital energies that can be understood qualitatively using forces of electrical attraction and repulsion, as we describe later in this chapter.

### The Effect of Nuclear Charge

A helium +1 cation, like a hydrogen atom, has just one electron. Absorption and emission spectra show that $He^+$ has energy levels that depend on $n$, just like the hydrogen atom. Nevertheless, Figure 5-2 shows that the emission spectra of $He^+$ and H differ, which means that these two species must have different energy levels. We conclude that something besides $n$ influences orbital energy.

The difference between $He^+$ and H can be found in their nuclei. A hydrogen nucleus is a single proton with a +1 charge, whereas a helium nucleus contains two protons and two neutrons and has a charge of +2. The larger nuclear charge of $He^+$ attracts the single electron

**Figure 5-1**

$n = 4$  $E_4$ ———

$n = 3$  $E_3$ ——

$n = 2$  $E_2$ ——

Energy, $E_n$

*2p orbital*

+Photon →

$n = 1$  $E_1$ ——

*1s orbital*

Ground state          Excited state

When a hydrogen atom absorbs light, the energy of the photon converts it from the ground state to an excited state. In this process, its electron transfers to an orbital that has a higher energy and a larger principal quantum number.

**Figure 5-2**

The emission spectra of $He^+$ and H reveal transitions at characteristic energies. The emitted photons have different wavelengths and energies because $He^+$ has quantized energy levels that are different from those of H.

more strongly than does the smaller charge of H. As a result, He⁺ binds the electron with stronger force. Thus, any given energy level in the helium ion is more stable (lower in energy) than the corresponding level in the hydrogen atom.

The stability of an orbital can be determined by measuring the amount of energy required to remove an electron completely. This is the **ionization energy** (*IE*):

$$H \longrightarrow H^+ + e^- \qquad IE_H = 2.18 \ 10^{-18} \, J$$
$$He^+ \longrightarrow He^{2+} + e^- \qquad IE_{He+} = 8.72 \times 10^{-18} \, J$$

The ionization energy of He⁺ is four times as large as the *IE* of H. Thus, the ground state orbital for He⁺ must be four times as stable as the ground state orbital for H. Spectral analysis shows that each orbital of a helium cation is four times as stable as its counterpart orbital in a hydrogen atom, showing that orbital stability increases with the square of the nuclear charge, $Z^2$. Figure 5-3 shows the relationship among the energy levels of He⁺ and H. The diagram is in exact agreement with calculations based on the Schrödinger equation.

## Effects of Other Electrons

A hydrogen atom or a helium cation contains just one electron, but nearly all other atoms and ions contain many electrons. In a multielectron atom, each electron affects the properties of all the other electrons. These electron–electron interactions make the orbital energies of every element unique.

A given orbital is less stable in a multielectron atom than it is in the single-electron ion with the same nuclear charge. For instance, Table 5-1 shows that it takes more than twice as much energy to remove the electron from He⁺ (one electron) as it does to remove one of the electrons from a neutral He atom (two electrons). This demonstrates that the 1s orbital in He⁺ is more than twice as stable as the 1s orbital in neutral He. The nuclear charge of both species is +2, so the smaller ionization energy for He must result from the presence of the second electron. A negatively charged electron in a multielectron atom is attracted to the positively charged nucleus, but it is repelled by the other negatively charged electrons. This electron–electron repulsion accounts for the lower ionization energy of the helium atom.

**Figure 5-3**

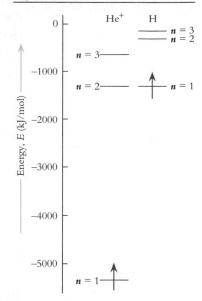

An energy-level diagram for He⁺ and H. Each species has just one electron, but the different nuclear charges make each He⁺ orbital four times more stable than the corresponding H orbital.

## TABLE 5-1 COMPARATIVE IONIZATION ENERGIES

| Orbital | H Atom | He Atom | He⁺ Ion |
|---------|--------|---------|---------|
| 1s | $2.18 \times 10^{-18}$ | $3.94 \times 10^{-18}$ | $8.72 \times 10^{-18}$ |
| 2p | $0.545 \times 10^{-18}$ | $0.585 \times 10^{-18}$ | $2.18 \times 10^{-18}$ |

*IE* values are in units of J/atom.

Figure 5-4

As a free electron approaches a He$^+$ cation, it is attracted to the +2 charge on the nucleus but repelled by the −1 charge on the 1s electron. When it is far from the cation, the electron experiences a net charge of +1.

**Figure 5-5**

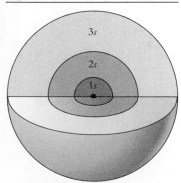

Cutaway view of the first three s orbitals. The 1s orbital screens the 2s and 3s orbitals. The 2s orbital screens the 3s orbital but not the 1s orbital. The 3s orbital is ineffective in screening the 1s and 2s orbitals.

## Screening

Figure 5-4 shows a free electron approaching a helium cation. The +2 charge of the nucleus attracts the incoming electron, but the negative charge on the He$^+$ 1s electron repels the incoming electron. This electron–electron repulsion cancels a portion of the attraction between the nucleus and the incoming electron. Chemists call this partial cancellation **screening**.

With its −1 charge, a bound electron could reduce the total charge by a maximum of one charge unit. Indeed, when an approaching electron is far enough away from a He$^+$ ion, it feels an attraction because of the net charge on the ion, +1. However, the 1s orbital is spread out all around the nucleus. This means that as an approaching electron gets close enough, the 1s electron screens only part of the total nuclear charge. Consequently, an approaching electron feels a net attraction resulting from some **effective nuclear charge** ($Z_{\text{eff}}$) greater than +1 but less than +2.

Incomplete screening can be seen in the ionization energies of hydrogen atoms, helium atoms, and helium ions (Table 5-1). Without *any* screening, the ionization energy of a helium atom would be the same as that of a helium ion; both would be $8.72 \times 10^{-18}$ J. With *complete* screening, one helium electron would compensate for one of the protons in the nucleus, making $Z_{\text{eff}} = +1$. The energy required to remove an electron from a helium atom would then be the same as the energy required to remove an electron from a hydrogen atom, $2.18 \times 10^{-18}$ J. The actual ionization energy of a helium atom is $3.94 \times 10^{-18}$ J, about twice the fully screened value and about half the totally unscreened value. Screening is incomplete because both helium electrons occupy an extended region of space, so neither is completely effective at shielding the other from the +2 charge of the nucleus.

Electrons in compact orbitals pack around the nucleus more tightly than do electrons in large, diffuse orbitals. As a result, the effectiveness in screening nuclear charge decreases as orbital size increases. Because the size of an orbital increases with *n*, an electron's ability to screen decreases as *n* increases. In a multielectron atom, lower-*n* electrons are concentrated between the nucleus and higher-*n* electrons. The negative charges of these inner electrons counteract most of the positive charge of the nucleus.

The efficient screening by electrons with small values of *n* can be appreciated by comparing the ionization energies of the 2p orbitals listed in Table 5-1. Consider an excited-state helium atom that has one of its electrons in the 1s orbital and its other electron in a 2p orbital. It takes $0.585 \times 10^{-18}$ J to remove the 2p electron from this excited state helium atom. This value is almost the same as that of an excited hydrogen atom with its lone electron in a 2p orbital, $0.545 \times 10^{-18}$ J. It is much less than the 2p orbital ionization energy of an excited He$^+$ ion, $2.18 \times 10^{-18}$ J. These data show that $Z_{\text{eff}}$ is quite close to +1 for the 2p orbital of an excited atom. In the excited He atom, the electron in the 1s orbital is very effective at screening the electron in the 2p orbital from the full +2 charge of the nucleus.

In multielectron atoms, electrons with any given value of $n$ provide effective screening for any orbital with a larger value of $n$. That is, $n = 1$ electrons screen the $n = 2$, $n = 3$, and larger orbitals, whereas $n = 2$ electrons provide effective screening for the $n = 3$, $n = 4$, and larger orbitals but provide little screening for the $n = 1$ orbital. Figure 5-5 illustrates this for the 1s, 2s, and 3s orbitals.

The amount of screening also depends on the shape of the orbital. The shaded area of the electron density plot shown in Figure 5-6 emphasizes that the 2s orbital has a region of significant electron density near the nucleus. A 2p orbital lacks this inner layer, so virtually all of its electron density lies outside the region occupied by the 1s orbital. Consequently, a 1s orbital screens the 2p orbital more effectively than the 2s orbital, even though both $n = 2$ orbitals are about the same size. Thus, a 2s electron feels a larger effective nuclear charge than a 2p electron. This results in stronger electrical attraction to the nucleus, which makes the 2s orbital more stable than the 2p orbitals. The 2p orbitals of any multielectron atom are always less stable (higher in energy) than the 2s orbital.

The screening differences experienced by the 2s and 2p orbitals also extend to larger values of $n$. The 3s orbital is more stable than the 3p orbital, the 4s is more stable than the 4p, and so on. Orbitals with higher $l$ values show similar effects. The 3d orbitals are always less stable than the 3p orbitals, and the 4d orbitals are less stable than the 4p orbitals. These effects can be summarized in a single general statement:

> *The higher the value of the l quantum number, the more that orbital is screened by electrons in smaller, more stable orbitals.*

In a one-electron system (H, $He^+$, $Li^{2+}$, and so on) the stability of the orbitals depends only on $Z$ and $n$. In multielectron systems, orbital stability depends primarily on $Z$ and $n$, but it also depends significantly on $l$. In a sense, $l$ fine-tunes orbital energies.

Electrons with the same $l$ value but different values of $m_l$ do not screen one another effectively. For example, when electrons occupy different $p$ orbitals that have the same $n$ value, and hence the same energies as one another (such orbitals are said to be *degenerate*), the amount of mutual screening is slight. This is because screening is effective only when much of the electron density of one orbital lies between the nucleus and the electron density of another. Recall from Chapter 4 (Figure 4-22) that the $p$ orbitals are perpendicular to one another, with high electron densities in different regions of space. The electron density of the $2p_x$ orbital does not lie between the $2p_y$ orbital and the nucleus, so there is little screening. The $d$ orbitals also occupy different regions of space from one another, so mutual screening among electrons in these orbitals is also small. Example 5-1 provides another look at screening.

**Figure 5-6**

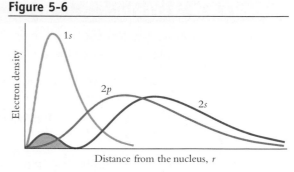

Electron density plots for the 1s, 2s, and 2p orbitals. Unlike the 2p orbital, the 2s orbital has significant electron density very near the nucleus (shaded region).

## EXAMPLE 5-1  SCREENING

Make an electron density plot showing the 1s, 2p, and 3d orbitals to scale. Label the plot in a way that summarizes the screening properties of these orbitals.

**Strategy:** This is a qualitative problem that asks us to combine information about three different orbitals on a single plot. We need to find electron density information and draw a single graph to scale.

**Solution:** Figures 4-20, 4-21, and 5-6 show electron density plots of the *n* = 1, *n* = 2, and *n* = 3 orbitals. We extract the shapes of the 1s, 2p, and 3d orbitals from these graphs. Then, we add labels that summarize the screening properties of these orbitals. Screening is provided by small orbitals, the electron density of which is concentrated inside larger orbitals. In this case, 1s screens both 2p and 3d; 2p screens 3d, but not 1s; and 3d screens neither 1s nor 2p. The screening patterns can be labelled as shown.

 **Does the Result Make Sense?** We know that the most important factor for orbital size is the value of *n* and that small orbitals screen better than large ones, so the screening sequence makes sense.

**Practice Exercise 5.1**

Construct and label a graph illustrating that the 2s orbital screens the 3s orbital more effectively than the 3s screens the 2s.

---

Quantitative information about energies of atomic orbitals is obtained using **photoelectron spectroscopy**, which applies the principles of the photoelectric effect to gaseous atoms. Our Focus 5-1: Tools for Discovery (on the next page) explores this powerful spectroscopic technique.

There are many different atomic orbitals, and each has a characteristic energy and shape. How the electrons of an atom distribute themselves among the atomic orbitals is the subject of the next two sections.

## SECTION EXERCISES

●**5.1.1** Figures 4-20, 4-21, and 5-6 show electron density plots of *n* = 1, *n* = 2, and *n* = 3 orbitals. Draw a plot that shows the *n* = 1 and *n* = 3 orbitals to scale. Use different colours to keep the figure as clear as possible. Shade the regions of the 3s and 3p plots where screening by 1s electrons is relatively ineffective.

●**5.1.2** The outer layers of the Sun contain He atoms in various excited states. One excited state contains one 1s electron and one 3p electron. Based on the effectiveness of screening, estimate the ionization energy of the 3p electron in this excited atom.

●**5.1.3** Redraw Figure 5-3 so that it includes the energy levels for $Li^{2+}$.

## FOCUS 5-1 TOOLS FOR DISCOVERY: PHOTOELECTRON SPECTROSCOPY

Absorption and emission spectra of atoms and ions yield information about energy differences *between* orbitals, but they do not give an orbital's *absolute* energy. The most direct measurements of orbital energies come from a technique called *photoelectron spectroscopy*.

Photoelectron spectroscopy works like the photoelectric effect described in Chapter 4, except that the sample often is a gas. Light shines on the sample, and when the sample absorbs a photon, the photon's energy is transferred to an electron. If the photon energy is high enough, the electron is ejected from the sample. The kinetic energy of the ejected electron is equal to the difference between the photon energy and the binding energy (ionization energy) of the electron, as the diagram shows.

A photoelectron spectrometer uses high-energy photons, typically 11 900 kJ/mol (emitted by excited Mg atoms). These photons have more than enough energy to eject an electron from an atom or molecule of the sample. The kinetic energy of the departing electron can be determined by measuring its speed. Knowing the photon energy and the electron's kinetic energy, we can find the binding energy of the orbital from which the electron came:

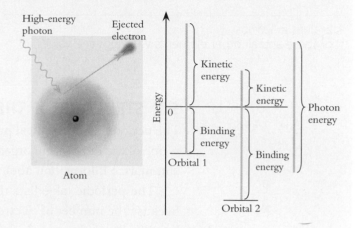

$$\text{Orbital binding energy} = E_{\text{photon}} - E_{\text{kinetic}} \text{ (electron)}$$

The photoelectron spectrum of a monatomic gas is a set of peaks representing the energies of orbitals. The figure shows the spectrum of neon. Two peaks correspond to orbital energies of 2080 and 4680 kJ/mol. The smaller value is the ionization energy of the 2p orbital, and the larger value is the ionization energy of the 2s orbital. The ionization energy of neon's 1s electrons is so great that they cannot be ejected by the photons used in photoelectron experiments. The ionization energy of an electron measures its stability when the electron is bound to the atom. In other words, a 2p electron of a neon atom is 2080 kJ/mol more stable than a free electron in a vacuum.

Photoelectron spectra of the first five elements, shown schematically in the graph, illustrate how atomic energy levels change with atomic number and with quantum numbers. The peaks in black are due to 1s electrons. As Z increases, the stability of the 1s orbital increases dramatically (for Be and B, these orbitals have ionization energies greater than 104 kJ/mol). Peaks in blue are due to 2s electrons, and the peak in magenta is attributed to a 2p electron. The table below gives values for orbital ionization energies (all in kJ/mol) for elements 11–21 obtained from photoelectron spectra.

| Orbital | \multicolumn{12}{c}{Element} |
|---|---|---|---|---|---|---|---|---|---|---|---|---|
|  | Na | Mg | Al | Si | P | S | Cl | Ar | K | Ca | Sc |
| 2s | 6840 | 9070 | * |  |  |  |  |  |  |  |  |
| 2p | 3670 | 5310 | 7790 | * |  |  |  |  |  |  |  |
| 3s | 500 | 740 | 1090 | 1460 | 1950 | 2050 | 2440 | 2440 | 3930 | 4650 | 5440 |
| 3p |  | 580 | 790 | 1010 | 1000 | 1250 | 1250 | 2380 | 2900 | 3240 |
| 3d |  |  |  |  |  |  |  |  |  |  | 770 |
| 4s |  |  |  |  |  |  |  |  | 420 | 590 | 630 |

All values in kJ/mol.
*For these and all other elements with higher $Z$ values, $IE > 11\,900$ kJ/mol.

## 5.2 STRUCTURE OF THE PERIODIC TABLE

The periodicity of chemical properties, which is summarized in the periodic table, is one of the most useful organizing principles in chemistry. Periodic patterns also summarize information about electron arrangements in atoms.

The periodic table lists the elements in order of increasing atomic number. Because the number of electrons in a neutral atom is the same as its atomic number, this list is also in order of increasing number of atomic electrons. Hydrogen, with $Z = 1$ and one electron, appears first, followed sequentially by helium (two electrons), lithium (three electrons), and so on.

A list of the elements in one long row is not a periodic table, because it does not reveal periodic patterns. To convert this single, long row into a periodic table, the "ribbon" of elements is cut at appropriate points to generate strips that are placed in rows. The rows are positioned so that each column contains elements with similar chemical properties. Figure 5-7 illustrates the positioning for the first 20 elements. The first cut comes between helium (two electrons) and lithium

**Figure 5-7**

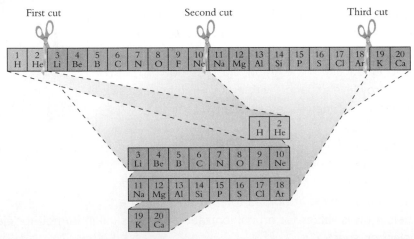

To generate the periodic table, the ribbon of elements is cut, and the segments are placed in rows so that each column contains elements with similar chemical properties.

(three electrons); the next comes between neon (10 electrons) and sodium; and the third is between argon (18 electrons) and potassium. With this arrangement, the first column contains Li, Na, and K—three highly reactive metals—whereas the last column contains He, Ne, and Ar—three unreactive gases.

Before establishing the connection between atomic orbitals and the periodic table, we must first describe two additional features of atomic structure: the Pauli exclusion principle and the aufbau principle.

## The Pauli Exclusion Principle

It is a general principle of nature that any system tends to find its most stable (lowest energy) arrangement. Atoms obey this principle, so hydrogen atoms are normally in their most stable state, the ground state. A hydrogen atom in its ground state has its one electron in the 1s orbital. We might expect a multielectron atom in its ground state to have all its electrons in the most stable, 1s orbital. However, studies show that this is not the case. Instead, bound electrons obey a fundamental law of quantum mechanics called the **Pauli exclusion principle**:

Wolfgang Pauli won the Nobel Prize for Physics in 1945 for discovering the exclusion principle.

> *Each electron in an atom has a unique set of quantum numbers.*

Named for the Austrian physicist Wolfgang Pauli (1900–1958), this principle can be derived from the mathematics of quantum mechanics, but it cannot be rationalized in a simple way. Nevertheless, all experimental evidence upholds the idea. When one electron in an atom has a particular set of quantum numbers, no other electron in the atom is described by that same set. There are no exceptions to the Pauli exclusion principle.

## The Aufbau Principle

The ground state of an atom is, by definition, the most stable arrangement of its electrons. *Most stable* means that the electrons occupy the lowest-energy orbitals available. We construct the **ground-state configuration** of an atom by placing electrons in the orbitals starting with the most stable in energy and moving progressively upward. In accordance with the Pauli principle, each successive electron is placed in the *most stable orbital* whose quantum numbers *are not already assigned* to another electron. This is the **aufbau principle**.

*Aufbau* is a German word meaning "construction."

In applying the aufbau principle, remember that a full description of an electron requires four quantum numbers: $n$, $l$, $m_l$, and $m_s$. Each combination of the quantum numbers $n$ and $l$ describes one quantized energy level. Moreover, each level includes one or more orbitals, each with a different value of $m_l$. Within a set, all orbitals with the same value of $l$ have the same energy. For example, the 2p energy level ($n = 2$, $l = 1$) consists of three distinct p orbitals ($m_l = -1$, 0, and +1), all with the same energy. In addition, different values of the spin orientation quantum number, $m_s$, describe the different spin orientations of an electron. When the two possible values of $m_s$ are taken into account, we find that six different sets of quantum numbers can be used to describe an electron in a 2p energy level:

$$n = 2 \qquad l = 1 \qquad m_l = +1 \qquad m_s = +\tfrac{1}{2}$$
$$n = 2 \qquad l = 1 \qquad m_l = +1 \qquad m_s = -\tfrac{1}{2}$$
$$n = 2 \qquad l = 1 \qquad m_l = 0 \qquad m_s = +\tfrac{1}{2}$$
$$n = 2 \qquad l = 1 \qquad m_l = 0 \qquad m_s = -\tfrac{1}{2}$$
$$n = 2 \qquad l = 1 \qquad m_l = -1 \qquad m_s = +\tfrac{1}{2}$$
$$n = 2 \qquad l = 1 \qquad m_l = -1 \qquad m_s = -\tfrac{1}{2}$$

In other words, the 2p energy level consists of three orbitals with different $m_l$ values that can hold as many as six electrons without violating the Pauli exclusion principle. The same is true of every set of p orbitals (3p, 4p, etc.). Each set has six equivalent descriptions and can hold six electrons. A similar analysis for other values of l shows that each s energy level contains a single orbital and can hold up to two electrons, each d energy level consists of five different orbitals than can hold up to 10 electrons, and each f energy level consists of seven different orbitals that can hold up to 14 electrons.

The Pauli and aufbau principles dictate where the cuts occur in the ribbon of elements. After two electrons have been placed in the 1s orbital (He), the next electron must go in a less stable, $n = 2$ orbital (Li). After eight additional electrons have been placed in the 2s and 2p orbitals (Ne), the next electron must go in a less stable, $n = 3$ orbital (Na). The ends of the rows in the periodic table are the points at which the next electron occupies an orbital of next higher principal quantum number.

Which $n = 2$ orbital does the third electron in a lithium atom occupy? Screening causes the orbitals with the same principal quantum number to decrease in stability as l increases. Consequently, the 2s orbital, being more stable than the 2p orbital, fills first. Similarly, 3s fills before 3p, which fills before 3d, and so on.

Here is a summary of the conditions for atomic ground states:

1. Each electron in an atom occupies the most stable available orbital.
2. No two electrons in an atom can have identical quantum numbers.
3. Orbital capacities are as follows: s, 2 electrons; p set, 6 electrons; d set, 10 electrons; f set, 14 electrons.
4. The higher the value of n, the less stable the orbital.
5. For equal n, the higher the value of l, the less stable the orbital.

Armed with these conditions, we can correlate the rows and columns of the periodic table with values of the quantum numbers n and l. This correlation appears in the periodic table shown in Figure 5-8. Remember that the elements are arranged so that Z increases one unit at a time from left to right across a row. At the end of each row, we move down one row, to the next higher value of n,

**Figure 5-8**

Group number

| 1 | 2 | 3 | 4 | 5 | 6 | 7 | 8 | 9 | 10 | 11 | 12 | 13 | 14 | 15 | 16 | 17 | 18 |
|---|---|---|---|---|---|---|---|---|---|---|---|---|---|---|---|---|---|
| 1A | 2A | 3B | 4B | 5B | 6B | 7B | 8B | 8B | 8B | 1B | 2B | 3A | 4A | 5A | 6A | 7A | 8A |

Row number

Row 1:
$1$ **H** $1s^1$ (Group 1)

(1–18 is IUPAC system; A, B designation is older U.S. system)

$2$ **He** $1s^2$ (Group 18)

Main group (s block) — Main group (p block) — Transition metals (d block)

Row 2: $3$ **Li** $2s^1$ | $4$ **Be** $2s^2$ | $5$ **B** $2s^2 2p^1$ | $6$ **C** $2s^2 2p^2$ | $7$ **N** $2s^2 2p^3$ | $8$ **O** $2s^2 2p^4$ | $9$ **F** $2s^2 2p^5$ | $10$ **Ne** $2s^2 2p^6$

Row 3: $11$ **Na** $3s^1$ | $12$ **Mg** $3s^2$ | $13$ **Al** $3s^2 3p^1$ | $14$ **Si** $3s^2 3p^2$ | $15$ **P** $3s^2 3p^3$ | $16$ **S** $3s^2 3p^4$ | $17$ **Cl** $3s^2 3p^5$ | $18$ **Ar** $3s^2 3p^6$

Row 4: $19$ **K** $4s^1$ | $20$ **Ca** $4s^2$ | $21$ **Sc** $4s^2 3d^1$ | $22$ **Ti** $4s^2 3d^2$ | $23$ **V** $4s^2 3d^3$ | $24$ **Cr** $4s^1 3d^5$ | $25$ **Mn** $4s^2 3d^5$ | $26$ **Fe** $4s^2 3d^6$ | $27$ **Co** $4s^2 3d^7$ | $28$ **Ni** $4s^2 3d^8$ | $29$ **Cu** $4s^1 3d^{10}$ | $30$ **Zn** $4s^2 3d^{10}$ | $31$ **Ga** $4s^2 4p^1$ | $32$ **Ge** $4s^2 4p^2$ | $33$ **As** $4s^2 4p^3$ | $34$ **Se** $4s^2 4p^4$ | $35$ **Br** $4s^2 4p^5$ | $36$ **Kr** $4s^2 4p^6$

Row 5: $37$ **Rb** $5s^1$ | $38$ **Sr** $5s^2$ | $39$ **Y** $5s^2 4d^1$ | $40$ **Zr** $5s^2 4d^2$ | $41$ **Nb** $5s^1 4d^4$ | $42$ **Mo** $5s^1 4d^5$ | $43$ **Tc** $5s^2 4d^5$ | $44$ **Ru** $5s^1 4d^7$ | $45$ **Rh** $5s^1 4d^8$ | $46$ **Pd** $4d^{10}$ | $47$ **Ag** $5s^1 4d^{10}$ | $48$ **Cd** $5s^2 4d^{10}$ | $49$ **In** $5s^2 5p^1$ | $50$ **Sn** $5s^2 5p^2$ | $51$ **Sb** $5s^2 5p^3$ | $52$ **Te** $5s^2 5p^4$ | $53$ **I** $5s^2 5p^5$ | $54$ **Xe** $5s^2 5p^6$

Row 6: $55$ **Cs** $6s^1$ | $56$ **Ba** $6s^2$ | $71$ **Lu** $6s^2 4f^{14} 5d^1$ | $72$ **Hf** $6s^2 5d^2$ | $73$ **Ta** $6s^2 5d^3$ | $74$ **W** $6s^2 5d^4$ | $75$ **Re** $6s^2 5d^5$ | $76$ **Os** $6s^2 5d^6$ | $77$ **Ir** $6s^2 5d^7$ | $78$ **Pt** $6s^1 5d^9$ | $79$ **Au** $6s^1 5d^{10}$ | $80$ **Hg** $6s^2 5d^{10}$ | $81$ **Tl** $6s^2 6p^1$ | $82$ **Pb** $6s^2 6p^2$ | $83$ **Bi** $6s^2 6p^3$ | $84$ **Po** $6s^2 6p^4$ | $85$ **At** $6s^2 6p^5$ | $86$ **Rn** $6s^2 6p^6$

Row 7: $87$ **Fr** $7s^1$ | $88$ **Ra** $7s^2$ | $103$ **Lr** $7s^2 5f^{14} 6d^1$ | $104$ **Rf** $7s^2 6d^2$ | $105$ **Db** $7s^2 6d^3$ | $106$ **Sg** $7s^2 6d^4$ | $107$ **Bh** $7s^2 6d^5$ | $108$ **Hs** $7s^2 6d^6$ | $109$ **Mt** $7s^2 6d^7$ | $110$ **Ds** $7s^1 6d^9$ | $111$ **Rg** $7s^1 6d^{10}$ | $112$ **Cn** $7s^2 6d^{10}$

Inner transition metals (f block)

Row 6 — Lanthanides: $57$ **La** $6s^2 5d^1$ | $58$ **Ce** $6s^2 4f^1 5d^1$ | $59$ **Pr** $6s^2 4f^3$ | $60$ **Nd** $6s^2 4f^4$ | $61$ **Pm** $6s^2 4f^5$ | $62$ **Sm** $6s^2 4f^6$ | $63$ **Eu** $6s^2 4f^7$ | $64$ **Gd** $6s^2 4f^7 5d^1$ | $65$ **Tb** $6s^2 4f^9$ | $66$ **Dy** $6s^2 4f^{10}$ | $67$ **Ho** $6s^2 4f^{11}$ | $68$ **Er** $6s^2 4f^{12}$ | $69$ **Tm** $6s^2 4f^{13}$ | $70$ **Yb** $6s^2 4f^{14}$

Row 7 — Actinides: $89$ **Ac** $7s^2 6d^1$ | $90$ **Th** $7s^2 6d^2$ | $91$ **Pa** $7s^2 5f^2 6d^1$ | $92$ **U** $7s^2 5f^3 6d^1$ | $93$ **Np** $7s^2 5f^4 6d^1$ | $94$ **Pu** $7s^2 5f^6$ | $95$ **Am** $7s^2 5f^7$ | $96$ **Cm** $7s^2 5f^7 6d^1$ | $97$ **Bk** $7s^2 5f^9$ | $98$ **Cf** $7s^2 5f^{10}$ | $99$ **Es** $7s^2 5f^{11}$ | $100$ **Fm** $7s^2 5f^{12}$ | $101$ **Md** $7s^2 5f^{13}$ | $102$ **No** $7s^2 5f^{14}$

The periodic table with its rows and blocks labelled to show the relationship between sectors of the table and ground-state configurations. Rows are labelled with the highest principal quantum number of the occupied orbitals, and each block is labelled with the letter (s, p, d, f) indicating the orbital set that is filling.

and return to the left side to the next higher $Z$ value. Inspection of Figure 5-8 reveals that the ribbon of elements is cut after elements 2, 10, 18, 36, 54, and 86.

As the atomic number increases, the length of ribbon between cuts increases as well. The first segment contains only hydrogen and helium. Then there are two 8-element segments, followed by two 18-element segments, and finally two 32-element pieces. The last segment stops before reaching its full length, however, because these elements have not been discovered in nature, nor have they been prepared in the laboratory. As we describe in Chapter 20, nuclei of elements with very high $Z$ are unstable.

In Figure 5-8, each *row* is labelled with the highest principal quantum number of its occupied orbitals. For example, elements of the third row (Na → Ar) have electrons in orbitals with $n = 3$ (in addition to electrons with $n = 1$ and 2). Each *column* is labelled with its group number, starting with Group 1 on the left and proceeding to Group 18 on the right (the $f$ block does not have group numbers). In general, elements in the same group of the periodic table have the same arrangement of electrons in their least stable occupied orbitals.

**Figure 5-9**

The calculated energy-level diagram for neutral atoms with $Z$ between 19 and 30 shows that the $3d$ and $4s$ atomic orbitals have nearly the same energy.

## Order of Orbital Filling

The variations in orbital stability with $n$ and $l$ ensure that $1s$ fills before $2s$ and $2s$ fills before $2p$. After $2p$, the next orbitals to fill are $3s$ and $3p$, but then what? Both $3d$ and $4s$ are less stable than $3p$, so either might be the next orbital to fill. Calculations of screening show that the $4s$ and $3d$ orbital energies for the elements from $Z = 19$ to $Z = 30$ are nearly the same, as shown schematically in Figure 5-9. We cannot predict which orbital fills next from this stability ladder.

The periodic table provides the answer. Each cut in the ribbon of the elements falls at the end of the $p$ block. This indicates that when the $np$ orbitals are full, the next orbital to accept electrons is the $(n + 1)s$ orbital. For example, after filling the $3p$ orbitals from Al $(Z = 13)$ to Ar $(Z = 18)$, the next element, potassium, has its final electron in the $4s$ orbital rather than in one of the $3d$ orbitals. According to the aufbau principle, this shows that the potassium atom is more stable with one electron in its $4s$ orbital than with one electron in one of its $3d$ orbitals. The $3d$ orbitals fill after the $4s$ orbital is full, starting with scandium $(Z = 21)$.

A similar situation exists at the end of the next row. When the $4p$ orbital is full $(Kr, Z = 36)$, the next element $(Rb, Z = 37)$ has an electron in the $5s$ orbital rather than either a $4d$ orbital or a $4f$ orbital. In fact, electrons are not added to the $4f$ orbitals until element 58, after the $5s$, $5p$, and $6s$ orbitals have filled.

The arrangement of the periodic table provides a simple way to determine the filling order of the elements, as shown in Figure 5-10 and applied in Example 5-2.

**Figure 5-10**

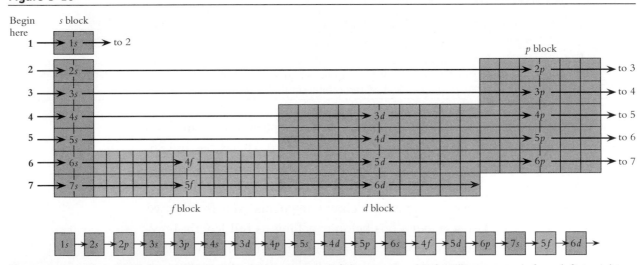

The periodic table in block form, showing the filling sequence of the atomic orbitals. Filling proceeds from left to right across each row and from the right end of each row to the left end of the succeeding row.

### EXAMPLE 5-2 ORBITAL FILLING SEQUENCE

Which orbitals are filled, and which set of orbitals is partially filled in a germanium atom?

**Strategy:** For this qualitative problem, use the periodic table to determine the order of orbital filling. Locate the element in a block and identify its row and column. Move along the ribbon of elements to establish the sequence of filled orbitals.

**Solution:** Germanium is element 32. Consult Figure 5-8 to determine that Ge is in Group 14, row 4 of the *p* block:

Starting from the top left of the periodic table and working left to right across the rows until we reach Ge, we identify the filled orbitals: 1*s*, 2*s*, 2*p*, 3*s*, 3*p*, 4*s*, 3*d*. Germanium is in row 4 of the *p* block, so the 4*p* set of orbitals is partially filled.

**Does the Result Make Sense?** Ge has $Z = 32$, meaning its neutral atoms contain 32 electrons. We can count how many each orbital can hold, 2 for each *s* orbital, 6 for each *p* orbital set, and 10 for each *d* orbital set. 2 (1*s*) + 2 (2*s*) + 6 (2*p*) + 2 (3*s*) + 6 (3*p*) + 2 (4*s*) + 10 (3*d*) = 30 electrons, leaving 2 in the partially filled 4*p* orbital set.

**Common Pitfalls:** Putting 3*d* electrons before 4*s* electrons in energy. Consult Figure 5-10 for the correct filling order.

**Practice Exercise 5.2**

Determine which orbitals are filled and which one is partially filled for the element Zr.

## Valence Electrons

The chemical behaviour of an atom is determined by the electrons that are accessible to an approaching chemical reagent. Accessibility, in turn, has a spatial component and an energetic component. An electron is accessible *spatially* when it occupies one of the largest orbitals of the atom. Electrons on the perimeter of the atom, farthest from the nucleus, are the first ones encountered by an incoming chemical reagent. An electron is accessible *energetically* when it occupies one of the least stable occupied orbitals of the atom. Electrons in less stable (higher energy) orbitals are thus more chemically active than electrons in more stable orbitals.

Similar electron accessibility generates similar chemical behaviour. For example, iodine has many more electrons than chlorine, but these two elements display similar chemical behaviour, as reflected by their placement in the same group of the periodic table. This is because the chemistry of chlorine and iodine is determined by the number of electrons in their largest and least stable occupied orbitals: 3*s* and 3*p* for chlorine and 5*s* and 5*p* for iodine. Each of these elements has seven accessible electrons, and this accounts for the chemical similarities.

Accessible electrons are called **valence electrons**, and inaccessible electrons are called **core electrons**. Valence electrons participate in chemical reactions, but core electrons do not. Orbital size increases and orbital stability decreases as the principal quantum number *n* gets larger. Therefore, the valence electrons for most atoms are the ones in orbitals with the largest value of *n*. Electrons in orbitals with lower *n* values are core electrons. In chlorine, valence electrons have $n = 3$, and core electrons have $n = 1$ and $n = 2$. In iodine, valence electrons have $n = 5$, and all others are core electrons.

The nearly equal energies of $ns$ and $(n-1)d$ orbitals creates some ambiguity about valence and core electrons for elements in the $d$ and $f$ blocks. For example, titanium forms a chloride and an oxide, the chemical formulas of which are consistent with four valence electrons: $TiCl_4$ and $TiO_2$. This shows that the two $3d$ electrons of titanium participate in its chemistry, even though they have a lower principal quantum number than its two $4s$ electrons. On the other hand, zinc forms compounds such as $ZnCl_2$ and $ZnO$, indicating that only two electrons are involved in its chemistry. This chemical behaviour indicates that zinc's ten $3d$ electrons, which completely fill the $3d$ orbitals, are inaccessible for chemical reactions. When the $d$ orbitals are partially filled, the $d$ electrons participate in chemical reactions, but when the $d$ orbitals are completely filled, the $d$ electrons do not participate in reactions. Analogous behaviour is observed for the $f$ orbitals, leading to a general rule for identifying valence electrons:

> *Valence electrons are all those of highest principal quantum number plus those in partially filled d and f orbitals.*

Applying this general rule, we find that the number of valence electrons can be determined easily from group numbers. For Groups 1–8, the number of valence electrons equals the group number. As examples, potassium and rubidium, members of Group 1, have just one valence electron each. Tungsten, in Group 6, has six valence electrons: two $6s$ electrons and four $5d$ electrons. For Groups 12–18, the number of valence electrons equals the group number minus 10 (the number of electrons it takes to fill the $d$ orbitals). Thus, antimony and nitrogen, in Group 15, have $15 - 10 = 5$ valence electrons each (two $s$ and three $p$). For Groups 9–11, the number of valence electrons cannot be stated unambiguously, because the $d$ electrons may or may not participate in bonding.

Numbering the groups in the periodic table as 1 through 18 is the modern standard, recommended by International Union of Pure and Applied Chemistry (IUPAC). Many periodic tables still have the older numbering scheme, which used the letter "A" for the main group elements (currently Groups 1, 2, and 13–18), and "B" for the transition metals (currently Groups 3–12), each in conjunction with a number. It is also common to find periodic tables with both numbering systems along the top.

We have described the layout of the periodic table in terms of the orbital descriptions of the various elements. As our Focus 5-2: Chemical Milestones describes, the periodic table was first proposed well before quantum theory was developed, when the only guidelines available were patterns of chemical and physical behaviour.

## FOCUS 5-2  CHEMICAL MILESTONES: HISTORY OF THE PERIODIC TABLE

Today we work confidently with the rows and columns of the periodic table. Yet less than 150 years ago, only about half of all elements known today had been discovered, and these presented a bewildering collection of chemical and physical properties. The discovery of the patterns that underlie this apparent randomness is a tale of inspired chemical detective work.

One early attempt to organize the elements clustered them into groups of three, called *triads*, whose members display similar chemical properties. Lithium, sodium, and potassium, for example, have many common properties and were considered to be a triad. This model was severely limited, for many elements could not be grouped into triads. The triad model is just one of nearly 150 different periodic arrangements of the elements that have been proposed.

Our modern periodic table was developed independently in the late 1860s by Dimitri Mendeleev (Russian) and Julius Lothar Meyer (German). At that time, about 60 elements had been discovered, but nothing was known about atomic structure. Lothar Meyer and Mendeleev had to work with elemental molar masses and other known elemental properties.

Lothar Meyer, a physicist, examined atomic volumes. He plotted atomic volume against molar mass and observed the pattern shown in the figure. There is a clear pattern of "waves," cresting successively at Li, Na, K, Rb, and Cs.

Mendeleev, a chemist, examined the relative numbers of the atoms of different elements that combine in chemical compounds. In $MgCl_2$, for example, each magnesium atom combines with two chlorine atoms. When he matched combining ability against molar mass, Mendeleev found the same sort of pattern as Lothar Meyer, with Li, Na, K, Rb, and Cs all combining 1 : 1 with Cl. Thus, each scientist was led to propose a table in which elements are arranged in rows of increasing mass, with breaks so that these five elements fall in the same column.

Mendeleev was bolder in his interpretation than Lothar Meyer, and for this reason we honour him as the primary discoverer of the modern periodic table. A few elements did not fit the pattern of variation in combining numbers with molar mass. Mendeleev proposed that these irregularities meant that the element's molar mass had been measured incorrectly. For example, Mendeleev predicted that the correct molar mass of indium is 113 g/mol, not 75 g/mol, the value assigned at that time on the assumption that the formula for indium oxide is InO. Later experiments showed that the correct formula is $In_2O_3$, and indium's true molar mass is 114.8 g/mol.

Mendeleev also predicted the existence of elements that had not yet been discovered. His arrangement of the then-known elements left some obvious holes in the periodic table. For instance, between zinc (combines with 2 Cl) and arsenic (combines with 5 Cl) were holes for one element that would combine with three chlorine atoms and another that would combine with four. Mendeleev assigned these holes to two new elements. He predicted that one element would have a molar mass of 68 g/mol and chemical properties like those of aluminum, whereas the other would have a molar mass of 72 g/mol and chemical properties similar to silicon. These elements, gallium (Z = 31, M = 69.7 g/mol) and germanium (Z = 32, M = 72.6 g/mol), were discovered within 15 years. Chemists soon verified that gallium resembles aluminum in its chemistry, whereas germanium resembles silicon, just as Mendeleev had predicted.

Ge (*left*) has many similarities with Si (*right*).

The predictions made by Mendeleev provide an excellent example of how a scientific theory allows far-reaching predictions of as-yet-undiscovered phenomena. Today's chemists still use the periodic table as a predictive tool. For example, modern semiconductor materials such as gallium arsenide were developed in part by predicting that elements in the appropriate rows and columns of the periodic table should have the desired properties. At present, scientists seeking to develop new superconducting materials rely on the periodic table to identify elements that are most likely to confer superconductivity.

●**5.2.1** What is the atomic number of the (as yet unproven) element that would occupy the position in Row 7, Column 17 of the periodic table?

●**5.2.2** Determine which orbitals are filled and which is partially filled for Br.

●**5.2.3** List all elements that have two valence electrons and specify the orbitals to which the valence electrons belong.

## 5.3 ELECTRON CONFIGURATIONS

A complete specification of how an atom's electrons are distributed in its orbitals is called an **electron configuration**. There are three common ways to represent electron configurations. One is a complete specification of quantum numbers. The second is a shorthand notation from which the quantum numbers can be inferred. The third is a diagrammatic representation of orbital energy levels and their occupancy.

A list of the values of all quantum numbers is easy for the single electron in a hydrogen atom: $n = 1, l = 0, m_l = 0$, and $m_s = +\frac{1}{2}$ or $n = 1, l = 0, m_l = +0$, and $m_s = -\frac{1}{2}$. Either designation is equally valid, because under normal conditions these two states are equal in energy. In a large collection of hydrogen atoms, half the atoms have one designation and the other half have the other designation.

As the number of electrons in an atom increases, a listing of all quantum numbers quickly becomes tedious. For example, iron, with 26 electrons, would require the specification of 26 sets of 4 quantum numbers. To save time and space, chemists have devised a shorthand notation to write electron configurations. The orbital symbols ($1s$, $2p$, $4d$, etc.) are followed by superscripts designating how many electrons are in each set of orbitals. The compact configuration for a hydrogen atom is $1s^1$, indicating one electron in the $1s$ orbital.

The third way to represent an atomic configuration uses an energy-level diagram similar to the one shown in Figure 5-9 to designate orbitals. Each electron is represented by an arrow and is placed in the appropriate orbital. The direction of the arrow indicates the value of $m_s$. The arrow points upward for $m_s = +\frac{1}{2}$ and downward for $m_s = -\frac{1}{2}$. The configuration of hydrogen can be represented by a single arrow in a $1s$ orbital.

$$1s \quad \uparrow$$

A neutral helium atom has two electrons. To write the ground-state electron configuration of He, we apply the aufbau principle. One unique set of quantum numbers is assigned to each electron, moving from the most stable orbital upward until all electrons have been assigned. The most stable orbital is always $1s$ ($n = 1, l = 0, m_l = 0$). Both helium electrons can occupy the $1s$ orbital, provided one of them has $m_s = +\frac{1}{2}$ and the other has $m_s = -\frac{1}{2}$. Here are the three representations of helium's ground-state electron configuration:

$$n = 1, \boldsymbol{l} = 0, \boldsymbol{m_l} = 0, \boldsymbol{m_s} = +\tfrac{1}{2}$$
$$n = 1, \boldsymbol{l} = 0, \boldsymbol{m_l} = 0, \boldsymbol{m_s} = -\tfrac{1}{2} \qquad 1s^2 \qquad 1s \; \text{}$$

The two electrons in this configuration are said to be *paired* electrons, meaning that they are in the same energy level, with opposing spins. Opposing spins cancel, so paired electrons have zero net spin.

A lithium atom has three electrons. The first two electrons fill lithium's lowest possible energy level, the 1s orbital, and the third electron occupies the 2s orbital. The three representations for the ground-state electron configuration of a lithium atom are as follows:

$$n = 1, l = 0, m_l = 0, m_s = +\tfrac{1}{2}$$
$$n = 1, l = 0, m_l = 0, m_s = -\tfrac{1}{2} \qquad 1s^2 2s^1$$
$$n = 2, l = 0, m_l = 0, m_s = +\tfrac{1}{2}$$

The set $n = 2, l = 0, m_l = 0, m_s = -\tfrac{1}{2}$ is equally valid for the third electron.

The next atoms of the periodic table are beryllium and boron. You should be able to write the three different representations for the ground-state configurations of these elements. The filling principles are the same as we move to higher atomic numbers. Example 5-3 shows how to apply these principles to aluminum.

## EXAMPLE 5-3  AN ELECTRON CONFIGURATION

Construct an energy-level diagram and the shorthand representation of the ground-state configuration of aluminum. Provide one set of valid quantum numbers for the highest-energy electron.

**Strategy:** First consult the periodic table to locate aluminum and determine how many electrons are present in a neutral atom. Then construct the electron configuration using the patterns of the periodic table.

**Solution:** Aluminum has $Z = 13$, so a neutral atom of Al has 13 electrons. Place the 13 electrons sequentially, using arrows, into the most stable orbitals available. Two electrons fill the $n = 1$ orbital, eight electrons fill the $n = 2$ orbitals, two electrons fill the 3s orbital, and one electron goes in a 3p orbital.

The last electron could be placed in any of the 3p orbitals, because these three orbitals are equal in energy. The final electron also could be given either spin orientation. By convention, we place electrons in unfilled orbitals starting with the left-hand side, with spins pointed up.

The shorthand configuration is $1s^2\ 2s^2\ 2p^6\ 3s^2\ 3p^1$.

The least stable electron is in a 3p orbital, meaning $n = 3$ and $l = 1$. The value of $m_l$ can be any of three values: $+1$, $+1$, or 0. The spin quantum number, $m_s$, can be $+\tfrac{1}{2}$ or $-\tfrac{1}{2}$. One valid set of quantum numbers is

$$n = 3, l = 1, m_l = 1, \text{ and } m_s = +\tfrac{1}{2}$$

You should be able to write the other five possible sets.

**Do the Results Make Sense?** Aluminum, in Group 13, has three valence electrons. The configurations show three electrons with $n = 3$, so the configuration is consistent with the valence electron count.

**Practice Exercise 5.3**

Determine the energy-level diagram and shorthand notation for the electron configuration of the fluorine atom.

Electron configurations become longer as the number of electrons increases. To make the writing of a configuration even more compact, chemists make use of the regular pattern for the electrons with lower principal quantum numbers. Compare the configurations of neon and aluminum:

| | |
|---|---|
| Ne (10 electrons) | $1s^2 \, 2s^2 \, 2p^6$ |
| Al (13 electrons) | $1s^2 \, 2s^2 \, 2p^6 \quad 3s^2 \, 3p^1$ |

The description of the first 10 electrons in the configuration of aluminum is identical to that of neon, so we can represent that portion as [Ne]. With this notation, the configuration of Al becomes [Ne] $3s^2 \, 3p^1$. The element at the end of each row of the periodic table has a **noble gas configuration**. These configurations can be written in the following shorthand notation:

| Notation | | Configuration | Element |
|---|---|---|---|
| [He] | = | $1s^2$ | He (2 electrons) |
| [Ne] | = | [He] $2s^2 \, 2p^6$ | Ne (10 electrons) |
| [Ar] | = | [Ne] $3s^2 \, 3p^6$ | Ar (18 electrons) |
| [Kr] | = | [Ar] $4s^2 \, 3d^{10} \, 4p^6$ | Kr (36 electrons) |
| [Xe] | = | [Kr] $5s^2 \, 4d^{10} \, 5p^6$ | Xe (54 electrons) |
| [Rn] | = | [Xe] $6s^2 \, 5d^{10} \, 4f^{14} \, 6p^6$ | Rn (86 electrons) |

To write the configuration of any other element, we first consult the periodic table to find its location relative to the noble gases. Then, we specify the noble gas configuration and build the remaining portion of the configuration according to the aufbau principle. Example 5-4 applies this procedure to indium.

## EXAMPLE 5-4  A SHORTHAND ELECTRON CONFIGURATION

Determine the configuration of indium, first in shorthand form and then in full form.

**Strategy:** Locate the element in the periodic table, and find the nearest noble gas with smaller atomic number. Start with the configuration of that noble gas, and add enough additional electrons to the next filling orbitals to give the neutral atom.

**Solution:** Indium (In, $Z = 49$) is in Row 5, Group 13. The nearest noble gas of smaller $Z$ is Kr ($Z = 36$). Thus, the configuration of In has 36 electrons in the Kr configuration and 13 additional electrons. The last orbital to fill in Kr is $4p$, and the periodic table shows that the next orbitals to fill are the $5s$, $4d$, and $5p$ orbitals:

$$\text{Configuration of indium: [Kr] } 5s^2 \, 4d^{10} \, 5p^1$$

To write the full configuration, decompose the krypton configuration:

$$[\text{Kr}] = [\text{Ar}] \, 4s^2 \, 3d^{10} \, 4p^6 = 1s^2 \, 2s^2 \, 2p^6 \, 3s^2 \, 3p^6 \, 4s^2 \, 3d^{10} \, 4p^6$$

Full configuration of In (49 electrons):

$$1s^2 \, 2s^2 \, 2p^6 \, 3s^2 \, 3p^6 \, 4s^2 \, 3d^{10} \, 4p^6 \, 5s^2 \, 4d^{10} \, 5p^1$$

**Does the Result Make Sense?** Indium, in Group 13, has three valence electrons. The configurations show three electrons with $n = 5$, so the configuration is consistent with the valence electron count.

**Common Pitfalls:** Not ordering the energy levels of the electrons correctly. Refer to Figure 5-10 for the correct filling order.

**Practice Exercise 5.4**
Determine the compact notation for the electron configuration of the cadmium atom.

## Electron–Electron Repulsion

The aufbau principle allows us to assign quantum numbers to aluminum's 13 electrons without ambiguity. The first 12 electrons fill the 1s, 2s, 2p, and 3s energy levels, and the last electron can occupy any 3p orbital with either spin orientation. But what happens when more than one electron must be placed in a p energy level? Carbon atoms, for example, have six electrons, two of which occupy 2p orbitals. How should these two electrons be arranged in the 2p orbitals? As Figure 5-11 shows, three different arrangements of these electrons obey the Pauli principle and appear to be consistent with the aufbau principle:

**Figure 5-11**

1. The electrons could be paired in the same 2p orbital (same $m_l$ value but different $m_s$ values).

2. The electrons could occupy different 2p orbitals with the same spin orientation (different $m_l$ values but the same $m_s$ value).

3. The electrons could occupy different 2p orbitals with opposite spin orientations (different $m_l$ values and different $m_s$ values).

Three different arrangements of two 2p electrons obey the Pauli and the aufbau principles. Only one (number 2) also obeys Hund's rule.

These three arrangements have different energies, because electrons that are close together repel each other more than electrons that are far apart. As a result, for two or more orbitals having nearly equal energies, greatest stability results when electrons occupy the orbitals that keep them farthest apart. Placing two electrons in different p orbitals keeps them relatively far apart, so an atom is more stable with the two electrons in different p orbitals. Thus, arrangements 2 and 3 are more stable than arrangement 1.

Arrangements 2 and 3 look spatially equivalent, but experiments show that a configuration that gives unpaired electrons the same spin orientation is always more stable than one that gives them opposite orientations. **Hund's rule** summarizes the way in which electrons occupy orbitals of equal energies:

> *The most stable configuration involving orbitals of equal energies is the one with the maximum number of electrons with the same spin orientation.*

According to Hund's rule, the ground-state configuration for carbon atoms is arrangement 2. Example 5-5 provides practice in the application of Hund's rule.

## EXAMPLE 5-5  APPLYING HUND'S RULE

Write the shorthand electron configuration and draw the ground-state orbital energy-level diagram for the valence electrons in a sulphur atom.

**Strategy:** From the periodic table, we see that sulphur has 16 electrons and is in the $p$ block, Group 16. To build the ground-state configuration, apply the normal filling rules and then apply Hund's rule if needed.

**Solution:** The first 12 electrons fill the four lowest-energy orbitals:

$$1s^2\ 2s^2\ 2p^6\ 3s^2$$

Sulphur's remaining 4 electrons occupy the three $3p$ orbitals. The complete configuration is $1s^2\ 2s^2\ 2p^6\ 3s^2\ 3p^4$, or $[\text{Ne}]\ 3s^2\ 3p^4$.

   To minimize electron–electron repulsion, put three of the $3p$ electrons in different orbitals, all with the same spin, and then place the fourth electron, with opposite spin, in the first orbital. In accord with Hund's rule, this gives the same value of $m_s$ to all electrons that are not paired. Here is the energy-level diagram for the valence electrons:

**Do the Results Make Sense?** Sulphur, in Group 16, has six valence electrons. The configurations show six electrons with $n = 3$, so the configuration is consistent with the valence electron count. The electrons are spread among the three $3p$ orbitals, which minimizes electron–electron repulsion.

**Common Pitfalls:** Not following Hund's rule. Make sure there is maximum unpairing of electrons in degenerate orbitals.

**Practice Exercise 5.5**

Determine the compact configuration for the nitrogen atom and write a valid set of quantum numbers for its valence electrons.

## Orbitals with Nearly Equal Energies

The filling order embodied in the periodic table predicts a regular progression of ground-state configurations. Experiments show, however, that some elements have ground-state configurations different from the predictions of the regular progression. Among the first 40 elements, there are only two exceptions: copper and chromium. Chromium ($Z = 24$) is in Group 6, four elements into the $d$ block. We would predict that chromium's valence configuration should be $4s^2\ 3d^4$. Instead, experiments show that the ground-state configuration of this element is $4s^1\ 3d^5$. Likewise, the configuration of copper ($Z = 29$) is $4s^1\ 3d^{10}$ rather than the predicted $4s^2\ 3d^{10}$.

   Look again at Figure 5-9, which shows that these two sets of orbitals are nearly the same in energy. Each $(n - 1)d$ orbital has nearly the same energy as its $ns$ counterpart. In addition, each $(n - 2)f$ orbital has nearly the same energy as its $(n - 1)d$ counterpart. Table 5-2 lists these orbitals and the atomic numbers of the elements for which the filling sequence differs from the expected pattern. These anomalous configurations are also indicated in Figure 5-8.

   Often, an $s$ orbital contains only one electron rather than two. Five of the exceptional ground-state configurations have a common pattern and are easy to

## TABLE 5-2 ATOMIC ORBITALS WITH NEARLY EQUAL ENERGIES

| Orbitals | Atomic Numbers Affected | Example |
|---|---|---|
| $4s, 3d$ | 24, 29 | Cr ([Ar] $4s^1\,3d^5$) |
| $5s, 4d$ | 41, 42, 44, 45, 46, 47 | Ru ([Kr] $5s^1\,4d^7$) |
| $6s, 5d, 4f$ | 57, 58, 64, 78, 79 | Au ([Xe] $6s^1\,4f^{14}\,5d^{10}$) |
| $6d, 5f$ | 89, 91, 92, 93, 96, 110, 111 | U ([Rn] $7s^2\,5f^3\,6d^1$) |

remember. For example, Cr and Mo are $s^1\,d^5$, not $s^2\,d^4$ as aufbau would predict. This occurs because of the **exceptional stability of filled or half-filled sets of degenerate orbitals**. In this case, moving an electron from the $s$ to the $d$ orbitals makes both half-filled. This removes electron–electron repulsion in the $s$ orbital, and in this case, the additional electron–electron repulsion in the $d$ orbitals is more than offset by the reduction of repulsion in the $s$ orbital. Thus, the $s^1\,d^5$ configuration is at a lower energy (i.e., is more stable) than $s^2\,d^4$. Note that this does not occur for all $s^2\,d^4$ elements, however. Tantalum (Ta), for instance, is more stable as $s^2\,d^4$.

Other examples of anomalous electron configurations are Cu, Ag, and Au, which are all $s^1\,d^{10}$. Moving an electron out of the $s$ orbital results in less electron–electron repulsion, and again the increase in repulsion among the $d$ electrons is more than offset by this. The other exceptional cases follow no recognizable patterns, because they are generated by subtle interactions among all the electrons. Among elements whose valence electrons are filling orbitals with nearly equal energies, several factors help to determine the ground-state configuration. The details are beyond the scope of general chemistry, except that you should recognize that even a subtle change can cause variations in the filling pattern predicted by the periodic table.

## Configurations of Ions

The electron configurations of atomic ions are written using the same procedure as for neutral atoms, taking into account the proper number of electrons. An anion has one *additional* electron for each unit of negative charge. A cation has one *fewer* electron for each unit of positive charge.

For most atomic ions, the filling order of orbitals is the same as that of neutral atoms. For example, $Na^+$, Ne, and $F^-$ all contain 10 electrons, and each has the configuration $1s^2\,2s^2\,2p^6$. Atoms and ions that have the same number of electrons are said to be **isoelectronic**.

The nearly-equal energies of $ns$ and $(n-1)d$ orbitals causes the configurations of some cations to differ from the configurations predicted by the filling pattern of the periodic table. This feature is particularly important for the transition metals. Experiments show that in transition metal cations, the $(n-1)d$ orbitals are *always* more stable than the $ns$ orbitals. For example, an $Fe^{3+}$ cation contains 23 electrons. The first 18 electrons fill the 1s, 2s, 2p, 3s, and 3p orbitals,

**Figure 5-12**

V atom
$4s < 3d$

$Fe^{3+}$ cation
$4s > 3d$

Transition metal atoms and cations have different valence configurations, even when the species contain the same number of electrons (23 in this example).

as predicted by the periodic table. However, the five remaining electrons populate the $3d$ set, leaving the $4s$ orbital empty. Thus, the configuration of the $Fe^{3+}$ cation is $[Ar] 3d^5$.

Vanadium atoms ($[Ar] 4s^2 3d^3$) and $Fe^{3+}$ cations ($[Ar] 3d^5$) have different configurations, even though each has 23 electrons. Remember that the energy ranking of orbitals such as $4s$ and $3d$ depends on a balance of several factors, and that even a small variation in that balance can change the filling order of the orbitals. That is the case for transition metal atoms and cations, as shown in Figure 5-12. In neutral transition metal atoms, a configuration that fills the $ns$ orbital is slightly more stable than one that places all valence electrons in the $(n - 1)d$ orbital. In transition metal cations, however, the stable configuration always places all valence electrons in the $(n - 1)d$ orbitals. Example 5-6 shows how to write the configuration of a transition metal cation.

## EXAMPLE 5-6　CONFIGURATION OF A CATION

What is the ground-state electron configuration of a $Cr^{3+}$ cation?

**Strategy:** Use the aufbau approach, remembering that because $Cr^{3+}$ is a transition metal cation, its $3d$ orbital is more stable than its $4s$ orbital.

**Solution:** A neutral chromium atom has 24 electrons, so the corresponding $Cr^{3+}$ cation has 21 electrons. The first 18 electrons follow the usual filling order to give the argon core configuration: $1s^2 2s^2 2p^6 3s^2 3p^6$, or $[Ar]$. Place the remaining three electrons in the $3d$ set of orbitals, following Hund's rule: $[Ar] 3d^3$

**Does the Result Make Sense?** For any cation, the empty $4s$ orbital is slightly higher in energy than the partially filled $3d$ orbital. Thus, the isoelectronic $V^{2+}$ and $Cr^{3+}$ cations both have the $[Ar] 3d^3$ configuration. On the other hand, the isoelectronic neutral atom scandium has the configuration $[Ar] 4s^2 3d^1$.

**Practice Exercise 5.6**

Determine the ground-state electron configuration (compact form) of a $Ru^{3+}$ cation.

Here are the guidelines for building atomic or ionic electron configurations:

1. Count the total number of electrons.
   (a) Add electrons for anions.
   (b) Subtract electrons for cations.
2. Fill orbitals to match the nearest noble gas of smaller atomic number.
3. Add remaining electrons to the next filling orbitals according to Hund's rule.
   (a) For neutral atoms and anions, place electrons in $ns$ before $(n - 1)d$.
   (b) For cations, place electrons in $(n - 1)d$ before $ns$.
4. Look for exceptions and correct the configuration, if necessary.

## Magnetic Properties of Atoms

How do we know that an $Fe^{3+}$ ion in its ground state has the configuration $[Ar]\ 3d^5$ rather than the $[Ar]\ 4s^2\ 3d^3$ configuration predicted by the periodic table? Remember from Chapter 4 that electron spin gives rise to magnetic properties. Consequently, any atom or ion with unpaired electrons has nonzero net spin and is attracted by a strong magnet. We can divide the electrons of an atom or ion into two categories with different spin characteristics. In the filled orbitals, all the electrons are paired. Each electron with spin orientation $+\frac{1}{2}$ has a partner with spin orientation $-\frac{1}{2}$ The spins of these electrons cancel each other, giving a net spin of zero. An atom or ion with all electrons paired is not attracted by strong magnets and is termed **diamagnetic**. In contrast, spins do not cancel when unpaired electrons are present. An atom or ion with unpaired electrons is attracted to strong magnets and is termed **paramagnetic**. Moreover, the spins of all the unpaired electrons are additive, so the amount of paramagnetism shown by an atom or ion is proportional to the number of unpaired spins.

**Figure 5-13**

Glass tube

Magnet    Sample

The number of unpaired electrons in a paramagnetic substance can be measured with a Gouy balance.

The magnetic properties of chemical species can be measured with an instrument known as a Gouy balance, as shown schematically in Figure 5-13. The paramagnetic sample is suspended from one pan of the balance and placed just above the poles of a magnet. The paramagnetic substance is pulled into the magnetic field, creating a downward force on the sample. Weights are added to the second pan until the force is balanced. The paramagnetism of the sample is proportional to the mass required to balance the pans.

In $Fe^{3+}$, Hund's rule dictates that the five $d$ electrons all have the same spin orientation. For these five electrons, the spins all act together, giving a net spin of $5 \times \frac{1}{2} = \frac{5}{2}$. The alternative configuration for $Fe^{3+}$, $[Ar]\ 4s^2\ 3d^3$, is paramagnetic as well, but its net spin is $3 \times \frac{1}{2} = \frac{3}{2}$. Experiments show that $Fe^{3+}$ has a net spin of $\frac{5}{2}$.

In fact, all magnetic measurements on a wide range of transition metal cations are consistent with the $(n - 1)d$ orbitals being occupied rather than the $ns$ orbital.

Example 5-7 shows how to use configurations to predict whether an atom or ion is diamagnetic or paramagnetic.

---

### EXAMPLE 5-7 UNPAIRED ELECTRONS

Which of these species is paramagnetic: $F^-$, $Zn^{2+}$, and Ti?

**Strategy:** Paramagnetism results from unpaired spins, which exist only in partially filled sets of orbitals. We need to build the configurations and then look for any orbitals that are partially filled.

**Solution:** $F^-$: A fluorine atom has 9 electrons, so $F^-$ has 10 electrons. The configuration is $1s^2\ 2s^2\ 2p^6$. There are no partially filled orbitals, so fluoride ion is diamagnetic.

$Zn^{2+}$: The parent zinc atom has 30 electrons, and the cation has 28, so the configuration for $Zn^{2+}$ is $[Ar]\ 3d^{10}$. Again, there are no partially filled orbitals, so this ion is also diamagnetic.

Ti: A neutral titanium atom has 22 electrons. The ground-state configuration is [Ar] $4s^2\ 3d^2$. The spins of the $4s$ electrons cancel, but the two electrons in $3d$ orbitals have the same spin orientation, so their effect is additive. This ion is paramagnetic, with net spin of $\frac{1}{2} + \frac{1}{2} = 1$.

 **Does the Result Make Sense?** Filled orbitals always have all electrons paired, and two of these three species have completely filled orbitals. Only Ti has a partially filled orbital set.

**Practice Exercise 5.7**

Most transition metal cations are paramagnetic. Which cations in the first transition metal series have net charges less than +4 and are exceptions to this generalization?

The ground-state configurations of most neutral atoms and many ions contain unpaired electrons, so we might expect most materials to be paramagnetic. On the contrary, most substances are diamagnetic. This is because stable substances seldom contain free atoms. Instead, atoms are bonded together in molecules, and as we show in Chapter 6, bonding usually (but not always!) results in the pairing of electrons and the cancellation of spin. As a result, paramagnetism is observed primarily among salts of the transition and rare-earth metals, whose cations have partially filled $d$ and $f$ orbitals.

## Excited States

A ground-state configuration is the most stable arrangement of electrons, so an atom or ion will usually have this configuration. When an atom absorbs energy, however, it can reach an excited state with a new electron configuration. For example, sodium atoms normally have the ground-state configuration [Ne] $3s^1$, but when sodium atoms are in the gas phase, an electrical discharge can induce transitions that transfer the $3s$ electron to a higher-energy orbital, such as $3p$ or $4s$. Excited atoms are unstable and spontaneously return to the ground-state configuration, giving up their excess energy in the process. This is the principle behind sodium vapour lamps, which are used for street lighting. The light from a sodium vapour lamp comes from photons emitted as excited sodium atoms return to their ground states.

Excited-state configurations are perfectly valid as long as they meet the restrictions given in Table 4-2. In the electrical discharge of a sodium vapour lamp, for instance, we find some sodium atoms in excited states with configurations such as $1s^2\ 2s^2\ 2p^6\ 3p^1$ or $1s^2\ 2s^2\ 2p^5\ 3s^2$. These configurations use valid orbitals and are in accord with the Pauli principle, but they describe atoms that are less stable than those in the ground state. Each atom or ion has only one state that is most stable, and that is the ground state.

Excited states play important roles in chemistry. Recall from Chapter 4 that the properties of atoms can be studied by observing excited states. In fact, chemists and physicists use the characteristics

The red light in highway flares comes from excited $Sr^{2+}$ ions.

of excited states extensively to probe the structure and reactivity of atoms, ions, and molecules. Excited states also have practical applications. For example, street lamps use the emissions from excited sodium atoms, the dazzling colours of a fireworks display come from photons emitted by metal ions in excited states, and the red light in highway flares often comes from excited $Sr^{2+}$ ions.

---

**SECTION EXERCISES**

● **5.3.1** Determine the ground-state electron configurations and predict the net spin of (a) $Au^+$; (b) the neutral element with $Z = 118$; and (c) $S^{2-}$.

● **5.3.2** The ground-state configuration of Np is [Rn] $7s^2\, 5f^4\, 6d^1$. Draw an energy-level diagram that shows how the $7s^2$, $5f$, and $6d$ orbitals are related. Include arrows representing the seven electrons that occupy these orbitals.

● **5.3.3** Determine the ground-state configurations of Mo, I, and $Hg^+$. Use shorthand notation.

---

## 5.4 PERIODICITY OF ATOMIC PROPERTIES

The physical and chemical properties of the elements show regular periodic trends that can be explained using electron configurations and nuclear charges. We focus on the physical properties of the elements in this section. A preliminary discussion of the chemical properties of some of the elements appears in Section 5.6. Other chemical properties are discussed after we introduce the principles of chemical bonding in Chapters 6 and 7.

### Underlying Patterns

Two fundamental features of orbitals form the basis of periodicity and are summarized on the periodic table in Figure 5-14:

*As principal quantum number n increases, atomic orbitals become larger and less stable. As atomic number Z increases, any given atomic orbital becomes smaller and more stable.*

**Figure 5-14**

Underlying features of the periodic table.

Moving from left to right across a row of the periodic table, the *n* value of the least-stable occupied orbital remains the same while Z increases. A larger nuclear charge exerts a stronger electrical attraction on the electron cloud, and this stronger attraction results in smaller orbitals. Furthermore, electrons closer to the nucleus are energetically more stable than those farther from the nucleus. Moving from *left to right* across a row, orbitals become *smaller and more stable*.

Proceeding down a column of the periodic table, *n* and Z both increase. As *n* increases, orbitals become larger and less stable, but as Z increases, orbitals become smaller and more stable. Which trend dominates here? Recall that the number of core electrons increases as we move down any column. For example, sodium (Z = 11) has 10 core electrons and 1 valence electron. In the next lower row, potassium (Z = 19) has 18 core electrons and 1 valence electron. The screening provided by potassium's additional eight core electrons largely cancels the effect of the additional eight protons in its nucleus. Consequently, increased screening largely offsets the increase in Z value from one row to the next. For this reason, *n* is the most important factor in determining orbital size and stability within a column. From *top to bottom* of a column, valence orbitals become *larger and less stable*.

Atomic radii

## Atomic Radii

Because most of the volume of an atom is occupied by its electron cloud, the size of an atom is determined by the sizes of its orbitals. Atomic size follows these periodic trends:

> *Atomic size decreases from left to right and increases from top to bottom of the periodic table.*

A convenient measure of atomic size is the radius of the atom. Figure 5-15 shows the trends in **atomic radii**. For example, the atomic radius decreases smoothly across row 3, from 186 pm for sodium to 100 pm for chlorine. The atomic radius increases smoothly down Group 1, from 152 pm for lithium to 265 pm for cesium. Notice, however, that the atomic radius changes very little across the *d* and *f* blocks of the table. This is due to screening. For these elements, the largest orbital is the filled *ns* orbital. Moving from left to right across a row, Z increases, but electrons add to the smaller $(n - 1)d$ or $(n - 2)f$ orbitals. An increase in Z by one unit is matched by the addition of one screening electron. From the perspective of the outlying *s* orbital, the increases in Z are balanced by increased screening from the added *d* or *f* electrons. Thus, the electron in the outermost occupied orbital, *ns*, feels an effective nuclear charge that changes very little across these blocks. As a consequence, atomic size remains nearly constant across each row of the *d* and *f* blocks.

**Figure 5-15**

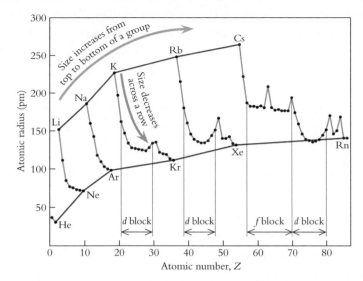

The radii of gaseous atoms vary in periodic fashion. Atomic radius decreases from left to right within any row (*blue lines*) and increases from top to bottom within any group (*red lines*).

It is important to be familiar with periodic trends in physical and chemical properties, but it is just as important to understand the principles that give rise to these trends. Example 5-8 shows how to analyze trends in terms of the underlying principles.

## EXAMPLE 5-8 TRENDS IN ATOMIC RADII

For each of the following pairs, predict which atom is larger and why: Si or Cl, S or Se, and Mo or Ag.

**Strategy:** Qualitative predictions about atomic size can be made on the basis of electron configurations and the effects of $Z$ and $n$ on size.

**Solution:** Silicon and chlorine are in the third row of the periodic table:

Chlorine's nuclear charge ($+17$) is larger than silicon's ($+14$), so chlorine's nucleus exerts a stronger pull on its electron cloud. Chlorine also has three more electrons than silicon, which raises the possibility that screening effects could counter the extra nuclear charge. Remember, however, that electrons in the same type of orbital do a poor job of screening one another from the nuclear charge. For Si and Cl, screening comes mainly from the core electrons, not from the electrons in the $3p$ orbitals. Because screening effects are similar for these elements, nuclear charge determines which of the two atoms is larger. Therefore, we conclude that chlorine, with its greater nuclear attraction for the electron cloud, is the smaller atom.

Sulphur and selenium are in Group 16 of the periodic table.

Although they both have the $s^2p^2$ valence configurations, selenium's least stable electrons are in orbitals with a larger $n$ value. Orbital size increases with $n$. Selenium also has a greater nuclear charge than sulphur, which raises the possibility that nuclear attraction could offset increased $n$. Remember, however, that much of this extra nuclear charge is offset by the screening influence of the core electrons. Selenium has 18 core electrons, and sulphur has 10. Thus, we conclude that selenium, with its larger $n$ value, is larger than sulphur.

Molybdenum and silver are in the same row of the $d$ block:

They have the following configurations:

$$Mo = [Kr]\ 5s^1\ 4d^5,\ Ag = [Kr]\ 5s^1\ 4d^{10}$$

In each case, $5s$ is the largest occupied orbital. The $4d$ orbitals are smaller, with their electron density located mostly inside the $5s$ orbital. Consequently, $4d$ is effective at screening $5s$. The nuclear charge of silver is five units larger than that of molybdenum, but silver also has five extra screening electrons. These offset the extra nuclear charge, making Mo and Ag nearly the same size.

**Do the Results Make Sense?** The trends in Figure 5-15 confirm the results. Chlorine lies to the right of silicon in the same row of the periodic table. Size decreases from left to right in any row; thus, chlorine is smaller than silicon. Selenium is immediately below sulphur in the same column of the periodic table. Size increases down a column; thus, selenium is larger than sulphur. Molybdenum and silver occupy the same row of the $d$ block of the periodic table, across which size changes very little; thus, molybdenum and silver are nearly the same size.

**Practice Exercise 5.8**

Use periodic trends to determine which of the following are smaller than As and which are larger than As: P, Ge, Se, and Sb.

## Ionization Energy

When an atom absorbs a photon, the gain in energy promotes an electron to a less stable orbital. As electrons move into less stable orbitals, they have less electrical attraction for the nucleus. If the absorbed photon has enough energy, an electron can be ejected from the atom, as occurs in photoelectron spectroscopy.

The minimum amount of energy needed to remove an electron from a neutral atom is the first ionization energy $(IE_1)$. Variations in ionization energy mirror variations in orbital stability, because an electron in a less stable orbital is easier to remove than one in a more stable orbital.

Figure 5-16 shows how the first ionization energies of gaseous atoms vary with atomic number. Notice the trends in ionization energy. Ionization energy increases regularly from left to right across each row (Row 3: 496 kJ/mol for Na to 1,520 kJ/mol for Ar) and decreases regularly from top to bottom of each column (Group 18: 2372 kJ/mol for He to 1037 kJ/mol for Rn). As with atomic radius, ionization energy does not change much for elements in the $d$ and $f$ blocks, because increased screening from the $d$ and $f$ orbitals offsets increases in $Z$.

1st *IE*

**Figure 5-16**

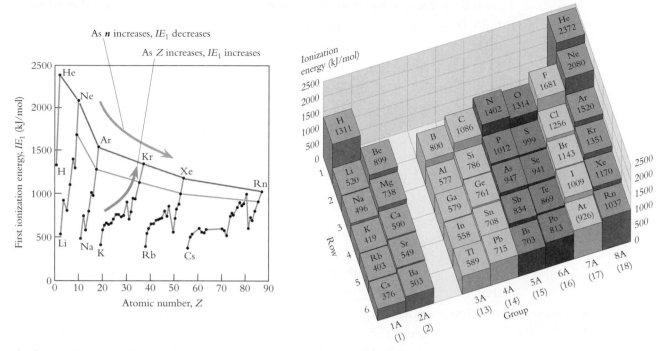

The first ionization energy increases from left to right (*orange arrow*) and decreases from top to bottom (*blue arrow*) of the periodic table.

## Higher Ionizations

A multielectron atom can lose more than one electron, but ionization becomes more difficult as cationic charge increases. The first three ionization energies for a magnesium atom in the gas phase provide an illustration. (Ionization energies are measured on gaseous elements to ensure that the atoms are isolated from one another.)

| **Process** | **Configurations** | ***IE*** |
|---|---|---|
| $Mg\ (g) \longrightarrow Mg^+\ (g) + e^-$ | $[Ne]\ 3s^2 \longrightarrow [Ne]\ 3s^1$ | 739 kJ/mol |
| $Mg^+\ (g) \longrightarrow Mg^{2+}\ (g) + e^-$ | $[Ne]\ 3s^1 \longrightarrow [Ne]$ | 1450 kJ/mol |
| $Mg^{2+}\ (g) \longrightarrow Mg^{3+}\ (g) + e^-$ | $[Ne] \longrightarrow [He]\ 2s^2\ 2p^5$ | 7730 kJ/mol |

Notice that the second ionization energy of magnesium is almost twice as large as the first, even though each electron is removed from a 3s orbital. This is because $Z_{eff}$ increases as the number of electrons decreases. That is, the positive charge on the magnesium nucleus remains the same throughout the ionization process, but the net charge of the electron cloud decreases with each successive ionization. As the number of electrons decreases, each electron feels a greater electrical attraction to the nucleus, resulting in a larger ionization energy.

The third ionization energy of magnesium is more than 10 times the first ionization energy. This large increase occurs because the third ionization removes a *core* electron (2p) rather than a *valence* electron (3s). Removing core electrons from any atom requires much more energy than removing valence electrons. The

Appendix C gives the first three ionization energies for the first 36 elements.

second ionization energy of any Group 1 metal is substantially larger than the first ionization energy; the third ionization energy of any Group 2 metal is substantially larger than the first or second ionization energy; and so on.

## Electron Affinity

A neutral atom can add an electron to form an anion. The energy change when an electron is added to an atom is called the **electron affinity** (*EA*). Both ionization energy (*IE*) and electron affinity measure the stability of a bound electron, but for different species. Here, for example, are the values for fluorine:

$$F \longrightarrow F^+ + e^- \qquad IE_1 = 1681 \text{ kJ/mol}$$

$$F + e^- \longrightarrow F^- \qquad EA = -322 \text{ kJ/mol}$$

Energy is released when an electron is added to a fluorine atom to form a fluoride anion. In other words, a fluoride anion is more stable than a fluorine atom plus a free electron. Another way of saying this is that fluorine atoms have an affinity for electrons.

Note that the energy associated with removing an electron to convert an anion to a neutral atom (i.e., the reverse of electron attachment) has the same magnitude as the electron affinity, but the opposite sign. Removing an electron from $F^-$, for example, requires energy, giving a positive energy change:

$$F^- \longrightarrow F + e^- \qquad \Delta E = 322 \text{ kJ/mol}$$

The aufbau principle must be obeyed when an electron is added to a neutral atom, so the electron goes into the most stable orbital available. Hence, we expect trends in electron affinity to parallel trends in orbital stability. However, electron–electron repulsion and screening are more important for negative ions than for neutral atoms, so there is no clear trend in electron affinities as *n* increases. Thus, there is only one general pattern:

> *Electron affinity tends to become more negative from left to right across a row of the periodic table.*

The plot in Figure 5-17 shows how electron affinity changes with atomic number. The blue lines reveal the trend: electron affinity increases in magnitude across each row of the periodic table. This trend is due to increasing effective nuclear charge, which binds the added electron more tightly to the nucleus. Notice that in contrast to the pattern for ionization energies (Figure 5-16), values for electron affinities remain nearly constant among elements occupying the same column of the periodic table.

The electron affinity values for many of the elements shown in Figure 5-17 appear to lie on the *x*-axis. In fact, these elements have positive electron affinities, meaning the resulting anion is less stable than the neutral atom. Moreover, the second electron affinity of every element is large and positive. Positive electron affinities cannot be measured directly. Instead, these values are estimated by other methods, as we show in Section 5.5.

Tables in reference sources often give electron affinities as positive values when the negative ion is more stable than the neutral atom. This convention is contrary to the sign convention for other energetic processes, which uses negative values when energy is released.

**Figure 5-17**

The electron affinity of atoms varies with atomic number. In moving across any main group (*blue lines*), the electron affinity becomes more negative, but this is the only clear trend.

Although electron affinity values show only one clear trend, there is a recognizable pattern in the values that are positive. When the electron that is added must occupy a new orbital, the resulting anion is unstable. Thus, all the elements of Group 2 have positive electron affinities, because their valence *ns* orbitals are filled. Similarly, all the noble gases have positive electron affinities, because their valence *np* orbitals are filled. Elements with half-filled orbitals also have lower electron affinities than their neighbours. As examples, N (half-filled $2p$ orbital set) has a positive electron affinity, and so does Mn (half-filled $3d$ orbital set).

## Irregularities in Ionization Energies

Ionization energies deviate somewhat from smooth periodic behaviour. These deviations can be attributed to screening effects and electron–electron repulsion. Aluminum, for example, has a smaller ionization energy than either of its neighbours in Row 3:

| Element | $Z$ | Atom Configuration | $IE_1$ | Cation Configuration |
|---------|-----|--------------------|--------|----------------------|
| Mg | 12 | [Ne] $3s^2$ | 738 kJ/mol | [Ne] $3s^1$ |
| Al | 13 | [Ne] $3s^2\,3p^1$ | 577 kJ/mol | [Ne] $3s^2$ |
| Si | 14 | [Ne] $3s^2\,3p^2$ | 786 kJ/mol | [Ne] $3s^2\,3p^1$ |

The configurations of these elements show that a $3s$ electron is removed to ionize magnesium, whereas a $3p$ electron is removed to ionize aluminum or silicon. Screening makes the $3p$ orbital significantly less stable than a $3s$ orbital, and this difference in stability more than offsets the increase in nuclear charge in going from magnesium to aluminum.

As another example, oxygen has a smaller ionization energy than either of its neighbours in Row 2:

| Element | Z | Atom Configuration | $IE_1$ | Cation Configuration |
|---|---|---|---|---|
| N | 7 | $1s^2\,2s^2\,2p^3$ | 1402 kJ/mol | $1s^2\,2s^2\,2p^2$ |
| O | 8 | $1s^2\,2s^2\,2p^4$ | 1314 kJ/mol | $1s^2\,2s^2\,2p^3$ |
| F | 9 | $1s^2\,2s^2\,2p^5$ | 1681 kJ/mol | $1s^2\,2s^2\,2p^4$ |

Remember that electron–electron repulsion has a destabilizing effect. The ionization energy of oxygen is less than that of nitrogen, despite the increased nuclear charge, because the $p^4$ configuration in the O atom has significantly greater electron–electron repulsion than the $p^3$ configuration in the N atom.

## Sizes of Ions

An atomic cation is always smaller than the corresponding neutral atom. Conversely, an atomic anion is always larger than the neutral atom. Figure 5-18 illustrates these trends, and electron–electron repulsion explains them. A cation has fewer electrons than its parent neutral atom. This reduction in the number of electrons means that the cation's remaining electrons experience less electron–electron repulsion. An anion has more electrons than its parent neutral atom. This increase in the number of electrons means that there is greater electron–electron repulsion in the anion than in the parent neutral atom.

Figure 5-18 also highlights the relationships among isoelectronic species, those possessing equal numbers of electrons. As noted in Section 5.3, the $F^-$ anion and the $Na^+$ cation are isoelectronic, each having 10 electrons and the configuration [He] $2s^2\,2p^6$. For isoelectronic species, properties change

**Figure 5-18**

Comparison of the sizes ($r$, in pm) of neutral atoms and their ions for some representative elements.

Group: 1, 2, 16, 17

Row 2: Li 1+ 152/76; O 2– 73/140; F 1– 72/133

Isoelectronic series

Row 3: Na 1+ 186/102; Mg 2+ 160/72; S 2– 103/184; Cl 1– 100/181

Row 4: K 1+ 227/138; Ca 2+ 197/99; Br 1– 114/196

Row 5: Rb 1+ 248/152; Sr 2+ 215/112; I 1– 133/220

## TABLE 5-3 TRENDS IN AN ISOELECTRONIC SEQUENCE

| Property | Species | | | | |
|---|---|---|---|---|---|
| | $O^{2-}$ | $F^-$ | Ne | $Na^+$ | $Mg^{2+}$ |
| $Z$ | 8 | 9 | 10 | 11 | |
| Radius (pm) | 140 | 133 | * | 102 | 72 |
| Ionization energy (kJ/mol) | <0 ($-EA_2$) | 322 ($-IEA$) | 2100 ($IE_1$) | 4560 ($IE_2$) | 7730 ($IE_3$) |

*Radii of ions are determined from dimensions of ionic crystals, and neutral atoms cannot be measured by this method.

regularly with $Z$. For example, Table 5-3 shows two properties of five isoelectronic species, each of which has 10 electrons. A progressive increase in nuclear charge results in a corresponding decrease in ionic radius, a result of stronger electrical force between the nucleus and the electron cloud. For the same reason, as $Z$ increases, it becomes progressively more difficult to remove an electron.

---

**SECTION EXERCISES**

●**5.4.1** Explain the following observations, using variations in $Z$, $Z_{eff}$, and $\textbf{\textit{n}}$: (a) Cl has a higher ionization energy than Al. (b) Cl has a higher ionization energy than Br. (c) Sr is larger than Mg. (d) Cl$^-$ is larger than K$^+$. (e) The first electron affinity of O is negative, but the second is positive.

●**5.4.2** Use periodic properties to explain the following observations: (a) Niobium and indium have nearly the same atomic size. (b) The first ionization energy of gallium is significantly lower than that of zinc. (c) Manganese has a positive electron affinity.

●**5.4.3** Explain why ionization energy decreases substantially as the configuration changes from $np^6$ to $np^6 (\textbf{\textit{n}} + 1)s^1$.

---

## 5.5 ENERGETICS OF IONIC COMPOUNDS

The electronic structures of atoms developed in the preceding sections allow us to determine why some elements form ionic compounds whereas others do not. Soft, lustrous sodium metal reacts vigorously with yellow-green chlorine gas to form sodium chloride, a white crystalline solid that we know as common table salt. Sodium chloride contains sodium and chlorine in a 1:1 elemental ratio. As shown in Figure 5-19, a crystal of sodium chloride is an ionic compound made up of equal numbers of Na$^+$ and Cl$^-$ ions. This contrasts starkly with the result of the reaction of elemental carbon with oxygen gas. Under appropriate conditions, these elements form CO, containing carbon and oxygen in a 1:1 elemental ratio, but CO does not contain ions. In fact, carbon and oxygen never react directly to form an ionic compound.

A limited number of elements form ionic compounds. As we describe in the next two chapters, most substances contain neutral molecules rather than charged ions. The trends in ionization energies and electron affinities indicate which elements tend to form ions. Ionic compounds form when the stabilization gained through ionic attraction exceeds the energy required to create ions from neutral atoms. In this section, we use ionization energy, electron affinity, and electrical forces to analyze the energetics of ionic compounds.

**Figure 5-19**

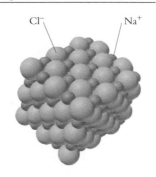
Cl$^-$　　Na$^+$

Sodium chloride contains Na$^+$ cations and Cl$^-$ anions organized in a three-dimensional cubic array.

Sodium and chlorine react rapidly to form sodium chloride, the white cloud rising from the flask.

Recall from Chapter 3 that energy changes for reactions do not depend on the path by which starting materials are converted to products. Thus, we can determine the energy change accompanying a reaction using any convenient path. To evaluate the energy change for formation of ionic NaCl, we make use of a reaction path that includes an ionization step and an electron attachment step. Here is one such path:

| | |
|---|---|
| Na $(s) \longrightarrow$ Na $(g)$ | Sublimation energy |
| Na $(g) \longrightarrow$ Na$^+$ $(g)$ | Ionization energy |
| $\frac{1}{2}$ Cl$_2$ $(g) \longrightarrow$ 2 Cl$(g)$ | Bond energy |
| Cl $(g)$ + e$^- \longrightarrow$ Cl$^-$ $(g)$ | Electron affinity |
| Na$^+$ $(g)$ + Cl$^-$ $(g) \longrightarrow$ NaCl $(s)$ | Condensation |
| Na $(s)$ + $\frac{1}{2}$ Cl$_2$ $(g) \longrightarrow$ NaCl $(s)$ | Formation of NaCl from the elements |

Notice that adding all the individual steps of this path gives a net reaction that matches the overall stoichiometry of the reaction. Consequently, the net energy change for the reaction is the sum of the energies of the individual steps.

**Step 1: Sublimation.** Sodium atoms must be removed from the solid to form sodium gas. Energy must be supplied to do this because, as we describe in Chapter 8, interatomic forces hold the atoms together in the solid metal. The tabulated value for the enthalpy of sublimation of Na is 107.5 kJ/mol. As described in Chapter 3, at 298 K, the energy of sublimation is 2.5 kJ/mol less than this:

$$\text{Na } (s) \longrightarrow \text{Na } (g) \qquad \Delta E_{sub} = 105 \text{ kJ/mol}$$

**Step 2: Bond Dissociation.** Chlorine molecules must be broken apart into chlorine atoms. Table 3-2 gives the bond energy ($BE$) of molecular chlorine, 240 kJ/mol. We use $\frac{1}{2}$ mole of Cl$_2$ to form 1 mole of NaCl, so the energy requirement is half this amount:

$$\frac{1}{2} \text{ Cl}_2 (g) \longrightarrow \text{Cl } (g) \qquad \Delta E = \frac{1}{2} BE = 120 \text{ kJ/mol}$$

**Step 3: Ionization of Na.** Ionizing sodium atoms requires that energy be supplied, the amount being the first ionization energy for sodium:

$$\text{Na } (g) \longrightarrow \text{Na}^+ (g) + \text{e}^- \qquad \Delta E = IE = 495.5 \text{ kJ/mol}$$

**Step 4: Electron Attachment to Cl.** The electron affinity of chlorine is negative, which means that energy is released when a chlorine atom gains an electron:

$$\text{Cl } (g) + \text{e}^- \longrightarrow \text{Cl}^- (g) \qquad \Delta E = EA = -348.5 \text{ kJ/mol}$$

**Step 5: Condensation.** Individual Na$^+$ and Cl$^-$ ions must condense into a three-dimensional array of ions. Energy is released in this condensation, because cations and anions attract each other. We need to examine this attraction in detail to determine how much energy is released in the condensation step.

Figure 5-20 reminds us that an ionic crystal contains many ions. All the ions exert electrical forces on one another. To calculate the energy resulting from these forces, we recall Equation 3-1, which describes the energy arising from the electrical forces between one pair of charged particles:

**Figure 5-20**

$$E_{electrical} = k\frac{q_1 q_2}{r} \tag{3-1}$$

In this equation, $q_1$ and $q_2$ are the electrical charges of the ions and $r$ is the distance between them. In this set of calculations, all the energies are in kJ/mol. When charges are expressed in units of electronic charge, and $r$ is in pm, the value of $k$ is $1.389 \times 10^5$ kJ pm mol$^{-1}$.

For sodium cations and chloride anions, $q_1 = +1$ and $q_2 = -1$. To complete the calculation, we need to know how closely the ions approach each other before their mutual attraction is balanced by electron cloud repulsion. In the sodium chloride crystal, this distance is 313 pm. Using this value for $r$, we can calculate the energy released in forming 1 mole of ion pairs:

In a sodium chloride crystal, each sodium cation has six chloride anions closest to it in an octahedral array, but all the other sodium cations and chloride anions also exert electrical forces.

$$Na^+ (g) + Cl^- (g) \longrightarrow NaCl (g, \text{ion pair})$$

$$E_{\text{ion pair}} = \frac{(1.389 \times 10^5 \text{ kJ pm/mol})(+1)(-1)}{(313 \text{ pm})} = -444 \text{ kJ/mol}$$

This is only part of the energy released in forming the crystal, because electrical interactions do not stop at individual ion pairs. In solid sodium chloride, each sodium cation is attracted to all the surrounding chloride anions. As Figure 5-20 shows, one $Na^+$ ion has six nearby $Cl^-$ ions. Moreover, there are many other chloride ions farther away, and all of them contribute to the total energy of the crystal in proportion to their distance from a particular sodium ion. At the same time, there are repulsive interactions from ions of the same charge that are close to one another. Equation 5-1 can be applied to all these ion–ion interactions in all three dimensions. When all these energies are added up for sodium chloride, the calculated value of the energy released is $-769$ kJ/mol:

$$Na^+ (g) + Cl^- (g) \longrightarrow NaCl (s) \qquad \Delta E_{\text{calculated}} = -769 \text{ kJ/mol}$$

This is the energy released when the solid forms from separated gaseous ions. The reverse process, in which an ionic solid decomposes into gaseous ions, is termed the **lattice energy** (**LE**) and is a positive quantity

$$NaCl (s) \longrightarrow Na^+ (g) + Cl^- (g) \qquad \Delta E = LE = +769 \text{ kJ/mol}$$

The overall energy change for the sodium chloride reaction is obtained by summing the energies of the five steps. Figure 5-21 summarizes the process.

$$\Delta E_{\text{calculated}} = \Delta E_{\text{sub}} + \tfrac{1}{2} BE + IE + EA - LE$$
$$\Delta E_{\text{calculated}} = 105 + 120 + 495.5 + (-348.5) - 769 = -397 \text{ kJ/mol}$$

This calculated energy is within 5% of the experimental value for the energy released in the actual reaction:

$$Na (s) + \tfrac{1}{2} Cl_2 (g) \longrightarrow NaCl (s) \qquad \Delta E_{\text{experimental}} = -411 \text{ kJ/mol}$$

**Figure 5-21**

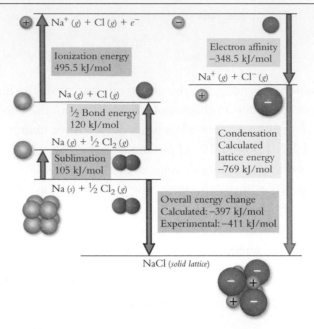

The reaction of sodium metal with molecular chlorine gas to produce solid sodium chloride can be analyzed by breaking the overall process into a series of steps involving ions in the gas phase.

This close agreement between the actual energy released in the reaction and the energy calculated by assuming that the final product is composed of ions shows that solid sodium chloride is best described as being made up of sodium cations and chloride anions. An energy cycle like the one in Figure 5-21 is called a *Born–Haber cycle*. Example 5-9 uses a Born–Haber cycle to estimate an electron affinity.

## EXAMPLE 5-9  BORN–HABER CYCLE

Magnesium metal burns in air to produce magnesium oxide, a white solid that contains $Mg^{2+}$ and $O^{2-}$. The oxide anion, $O^{2-}$, is not stable except in a crystalline solid such as MgO. This makes it impossible to measure directly the second electron affinity of oxygen. Use a Born–Haber cycle and the following data to calculate oxygen's second electron affinity. Sublimation enthalpy of Mg = $\Delta H_{sub}$ = 147.1 kJ/mol.

1st ionization energy of Mg = $IE_1$ = 738 kJ/mol

2nd ionization energy of Mg = $IE_2$ = 1451 kJ/mol

Bond energy of $O_2$ = $BE$ = 495 kJ/mol

1st electron affinity of O = $EA_1$ = 141 kJ/mol

Calculated lattice energy of MgO = $LE$ = 3795 kJ/mol

Experimental reaction energy = $\Delta E_{reaction}$ = −602 kJ/mol

**Strategy:** We are asked to find the second electron affinity of oxygen:

$$O^- (g) + e^- \longrightarrow O^{2-} (g) \qquad EA_2 = ?$$

As noted in the previous section, second electron affinities are all large and positive.

The different steps of a Born–Haber cycle can be connected together as shown in Figure 5-21. The overall energy change for the reaction described by the cycle is equal to the sum of the energy changes for the individual steps. A diagram of the steps, similar to Figure 5-21, helps sort out the calculation. Remember that the overall reaction consumes $1/2$ mole of $O_2$ for each mole of Mg. Therefore, our calculation requires $1/2$ the bond energy of molecular oxygen.

**Solution:** All the energy changes for the cycle are found among the data, except for the electron affinity of $O^-$, which is $EA_2$. We need to subtract 2.48 kJ/mol from the enthalpy of sublimation to convert it into an energy of sublimation. Then, we set up the sum of the individual energy changes and solve for $EA_2$.

$$\Delta E_{sub} = 147.1 \text{ kJ/mol} - 2.48 \text{ kJ/mol} = 144.6 \text{ kJ/mol, rounds to 145 kJ/mol}$$
$$\Delta E_{reaction} = \Delta E_{sub} + IE_1 + IE_2 + \tfrac{1}{2} BE + EA_1 + EA_2 - LE$$
$$EA_2 = \Delta E_{reaction} - \Delta E_{sub} - IE_1 - IE_2 - \tfrac{1}{2} BE - EA_1 + LE$$
$$EA_2 = [-602 - 145 - 738 - 1451 - 248 - (-141) + 3795] \text{ kJ/mol}$$
$$EA_2 = 752 \text{ kJ/mol}$$

   **Does the Result Make Sense?** The result is a large positive value, reflecting the fact that an isolated $O^{2-}$ anion is very unstable.

**Practice Exercise 5.9**
The formation reaction of $Na_2O$ has $\Delta E = -413$ kJ/mol. Use this value and energy values in Example 5-9 and the text to calculate the lattice energy of this compound.

## Why Not $Na^{2+}Cl^{2-}$?

The electrical attraction between doubly charged ions is significantly greater than the attraction between singly charged ions, because $q_1$ and $q_2$ are twice as large. This suggests that the transfer of a second electron might give an even more stable crystal composed of $Na^{2+}$ and $Cl^{2-}$ ions. We can extend our calculations to see if this is true.

The energy of attraction between $Na^{2+}$ and $Cl^{2-}$ ions in one $Na^{2+}Cl^{2-}$ ion pair would be about four times larger than that of one $Na^+Cl^-$ pair. If we assume that the dimensions and arrangement of the $Na^{2+}Cl^{2-}$ crystal are the same as those for the $Na^+Cl^-$ crystal, this gives a calculated lattice energy for $Na^{2+}Cl^{2-}$ of around 3075 kJ/mol. To collect this energy, however, the ion pair must pay the price of forming the doubly charged ions. Appendix C provides the first two ionization energies for sodium:

$$Na \longrightarrow Na^+ + e^- \quad Is^2 \, 2s^2 \, 2p^6 \, 3s^1 \longrightarrow 1s^2 \, 2s^2 \, 2p^6 \quad IE_1 = 495.5 \text{ kJ/mol}$$
$$Na^+ \longrightarrow Na^{2+} + e^- \quad Is^2 \, 2s^2 \, 2p^6 \longrightarrow 1s^2 \, 2s^2 \, 2p^5 \quad IE_2 = 4562 \text{ kJ/mol}$$

We do not know the second electron affinity for Cl, but the result of Example 5-9 suggests that the value will be large and positive, perhaps around 700 kJ/mol.

These energies demonstrate why sodium and chlorine do not form $Na^{2+}Cl^{2-}$. Forming $Na^{2+}$ and $Cl^{2-}$ from neutral gaseous atoms requires more than 5000 kJ/mol, much more energy than the 3075 kJ/mol that would be released in the formation of the $Na^{2+}Cl^{2-}$ lattice

The second ionization energy of sodium is much larger than its first ionization energy because a *core* 2p electron must be removed to create $Na^{2+}$ from $Na^+$. Removal of a core electron always requires a great deal of energy, so it is a general feature of ionic systems that ions formed by removing core electrons are not found in stable ionic compounds.

## Cation Stability

Knowing that the energy cost of removing core electrons is always excessive, we can predict that the ionization process will stop when all valence electrons have been removed. Thus, a knowledge of ground-state configurations is all that we need to make qualitative predictions about cation stability.

Each element in Group 1 of the periodic table has one valence electron. These elements form ionic compounds containing $A^+$ cations. Examples are KCl and $Na_2CO_3$. Each element in Group 2 of the periodic table has two valence electrons and forms ionic compounds containing $A^{2+}$ cations. Examples are $CaCO_3$ and $MgCl_2$.

Beyond these two columns, the removal of all valence electrons is usually not energetically possible. For example, iron has eight valence electrons but forms only two stable cations, $Fe^{2+}$ and $Fe^{3+}$. Compounds of iron containing these ions are abundant in the Earth's crust. Pyrite ($FeS_2$) and iron(II) carbonate ($FeCO_3$, or siderite) are examples of $Fe^{2+}$ salts. Iron(III) oxide ($Fe_2O_3$, or hematite) can be viewed as a network of $Fe^{3+}$ cations and $O^{2-}$ anions. One of the most abundant iron ores, magnetite, has the chemical formula $Fe_3O_4$ and contains a $2:1$ ratio of $Fe^{3+}$ and $Fe^{2+}$ cations. The formula of magnetite can also be written as $FeO \cdot Fe_2O_3$ to emphasize the presence of two different cations.

Other metallic elements form ionic compounds with cation charges ranging from $+1$ to $+3$. Aluminum nitrate nonahydrate, $Al(NO_3)_3 \cdot 9H_2O$, is composed of $Al^{3+}$ cations, $NO_3^-$ anions, and water molecules. Silver nitrate ($AgNO_3$), which contains $Ag^+$ cations, is a soluble silver salt that is used in silver plating.

## Anion Stability

Halogens, the elements in Group 17 of the periodic table, have the largest electron affinities of all the elements, so halogen atoms ($ns^2\,np^5$) readily accept electrons to produce halide anions ($ns^2\,np^6$). This allows halogens to react with many metals to form binary compounds, called **halides**, which contain metal cations and halide anions. Examples include NaCl (chloride anion), $CaF_2$ (fluoride anion), AgBr (bromide anion), and KI (iodide anion).

Isolated atomic anions with charges more negative than $-1$ are always unstable, but oxide ($O^{2-}$, $1s^2\,2s^2\,2p^6$) and sulphide ($S^{2-}$, [Ne] $3s^2\,3p^6$) are found in many ionic solids, such as CaO and $Na_2S$. The lattice energies of these solids are large enough to make the overall reaction energy-releasing despite the large positive second electron affinity of the anions. In addition, three-dimensional arrays of surrounding cations stabilize the $-2$ anions in these solids.

## Trends in Lattice Energies

The lattice energy is the sum of all ion interactions, each of which is described by Equation 3-1. Looking at this equation, we can predict that lattice energy will increase as ionic charge increases and that it will decrease as ionic size increases. A third trend occurs in the summing of all the ion contributions: lattice energy increases with the number of ions in the chemical formula of the salt.

## TABLE 5-4  LATTICE ENERGIES OF HALIDES AND OXIDES

| Cation | Anion | | | | |
|---|---|---|---|---|---|
| | $F^-$ | $Cl^-$ | $Br^-$ | $I^-$ | $O^{2-}$ |
| $Li^+$ | 1030 | 834 | 788 | 730 | 2799 |
| $Na^+$ | 910 | 769 | 732 | 682 | 2481 |
| $K^+$ | 808 | 701 | 671 | 632 | 2238 |
| $Rb^+$ | 774 | 680 | 651 | 617 | 2163 |
| $Cs^+$ | 744 | 657 | 632 | 600 | * |
| $Mg^{2+}$ | 2913 | 2326 | 2097 | 1944 | 3795 |
| $Ca^{2+}$ | 2609 | 2223 | 2132 | 1905 | 3414 |
| $Sr^{2+}$ | 2476 | 2127 | 2008 | 1937 | 3217 |
| $Ba^{2+}$ | 2341 | 2033 | 1950 | 1831 | 3029 |
| $Fe^{2+}$ | 2769 | 2525 | 2464 | 2382 | 3795 |
| $Fe^{3+}$ | 5870 | 5364 | 5268 | 5117 | 14 309 |

All values are in kJ/mol.
*Exists as superoxide rather than oxide.

These trends are apparent in the values of lattice energies that appear in Table 5-4. Notice, for example, that the lattice energies of the alkali metal chlorides decrease as the size of the cation increases, and the lattice energies of the sodium halides decrease as the size of the anion increases. Notice also that the lattice energy of MgO is almost four times the lattice energy of LiF. Finally, notice that the lattice energy of $Fe_2O_3$, which contains five ions in its chemical formula, is four times as large as that of FeO, which contains only two ions in its chemical formula.

The ionic model describes a number of metal halides, oxides, and sulphides, but it does not describe most other chemical substances adequately. Whereas substances such as CaO, NaCl, and $MgF_2$ behave like simple cations and anions held together by electrical attraction, substances such as CO, $Cl_2$, and HF do not. In a crystal of $MgF_2$, electrons have been *transferred* from magnesium atoms to fluorine atoms, but the stability of HF molecules arises from the *sharing* of electrons between hydrogen atoms and fluorine atoms. We describe electron sharing, which is central to molecular stability, in Chapters 6 and 7.

## SECTION EXERCISES

● **5.5.1** Magnesium fluoride forms from the elements as follows:

$$Mg\ (s) + F_2\ (g) \longrightarrow MgF_2\ (s) \qquad \Delta E_{reaction} = -1123\ kJ/mol$$

The energy of sublimation of Mg is 145 kJ/mol, and the bond energy of $F_2$ is 155 kJ/mol. Use this information and data from Appendix C to calculate the lattice energy of $MgF_2$. Compare your value with the value in Table 5-4.

● **5.5.2** Iron and cobalt form compounds that can be viewed as containing $A^{3+}$ cations, but nickel does not. Use the ionization energies in Appendix C to predict which other transition metal elements are unlikely to form stable cations with charges greater than +2.

● **5.5.3** From the location of each element in the periodic table, predict which ion of each of the following elements will be found in ionic compounds: Ca, Cs, Al, and Br.

## 5.6 IONS AND CHEMICAL PERIODICITY

The elements that form ionic compounds are found in specific places in the periodic table. Atomic anions are restricted to elements on the right side of the table: the halogens, oxygen, and sulphur. All the elements in the $s$, $d$, and $f$ blocks, on the other hand, form compounds containing atomic cations.

Ion formation is only one pattern of chemical behaviour. Many other chemical trends can be traced ultimately to valence electron configurations, but we need the description of chemical bonding that appears in Chapters 6 and 7 to explain such periodic properties. Nevertheless, we can relate important patterns in chemical behaviour to the ability of some elements to form ions.

The elements that can form cations relatively easily are metals. All metals have similar properties, in part because their outermost $s$ electrons are relatively easy to remove. All elements in the $s$ block have $ns^1$ or $ns^2$ valence configurations. The $d$-block elements have one or two $ns$ electrons and various numbers of $(n-1)d$ electrons. Examples are titanium ($4s^2\,3d^2$) and silver ($5s^1\,4d^{10}$). Elements in the $f$ block have two $ns$ electrons and a number of $(n-2)f$ electrons. Samarium, for example, has the valence configuration $6s^2\,4f^6$. As we describe in Chapter 7, the metallic behaviour of these elements occurs partly because the $s$ electrons are shared readily among all atoms. Metals form ionic salts because $s$ electrons (and some $d$, $p$, and $f$ electrons) can be readily removed from the metal atoms to form cations.

Whereas the other blocks contain only metals, elemental properties vary widely within the $p$ block. We have already noted that aluminum ($3s^2\,3p^1$) can lose its three valence electrons to form $Al^{3+}$ cations. Draw a right triangle in the lower left portion of this block whose hypotenuse runs from Ga to Bi. The six elements within this triangle also lose $p$ electrons easily and therefore have metallic properties. Examples are tin ($5s^2\,5p^2$) and bismuth ($6s^2\,6p^3$). In contrast, the halogens and noble gases on the right of this block are distinctly non-metallic. The noble gases, Group 18 of the periodic table, are monatomic gases that resist chemical attack because their electron configurations contain completely filled $s$ and $p$ orbitals.

Elements in any intermediate column of the $p$ block display a range of chemical properties even though they have the same valence configurations. Carbon, silicon, germanium, and tin all have $ns^2\,np^2$ valence configurations; yet carbon is a non-metal, silicon and germanium are metalloids, and tin is a metal.

Qualitatively, we can understand this variation by recalling that as the principal quantum number increases, the valence orbitals become less stable. In tin, the four $n = 5$ valence electrons are bound relatively loosely to the atom, resulting in the metallic properties associated with electrons that are easily removed. In carbon, the four $n = 2$ valence electrons are bound relatively tightly to the atom, resulting in non-metallic behaviour. Silicon ($n = 3$) and germanium ($n = 4$) fall in between these two extremes. Example 5-10 describes the elements with five valence electrons.

## EXAMPLE 5-10 CLASSIFYING ELEMENTS

Nitrogen is a colourless diatomic gas. Phosphorus has several elemental forms, but the most common is a red solid that is used for match tips. Arsenic and antimony are gray solids, and bismuth is a silvery solid. Classify these elements of Group 15 as metals, non-metals, or metalloids.

**Strategy:** All elements except those in the *p* block are metals. Group 15, however, is part of the *p* block, within which elements display all forms of elemental behaviour. To decide the classifications of these elements, we must examine this group relative to the diagonal arrangement of the metalloids:

**Solution:** We see that Group 15 passes through all three classes of elements. The elements with the lowest *Z* values, nitrogen and phosphorus, are non-metals. The element with highest *Z* value, bismuth, is a metal, and the two elements with intermediate *Z* values, arsenic and antimony, are metalloids.

**Do the Results Make Sense?** As the principal quantum number increases, valence electrons become progressively easier to remove, and metals are those elements with valence electrons that are easily removed.

**Practice Exercise 5.10**

Classify the 4*p* set of elements, from Ga to Kr.

## *s*-Block Elements

The electron configuration of any element in Groups 1 and 2 of the periodic table contains a core of tightly bound electrons and one or two *s* electrons that are loosely bound. The **alkali metals** (Group 1, $ns^1$ configuration) and the **alkaline earth metals** (Group 2, $ns^2$ configuration) form stable ionic salts because their valence electrons are easily removed. Nearly all salts of alkali metals and many salts of alkaline earth metals dissolve readily in water, so naturally occurring sources of water frequently contain these ions.

*s* block

The four most abundant *s*-block elements in the Earth's crust are sodium, potassium, magnesium, and calcium; their occurrence is summarized in Table 5-5. These elements are found in nature in salts such as $NaCl$, $KNO_3$, $MgCl_2$, $MgCO_3$, and $CaCO_3$. Portions of these solid salts dissolve in rainwater as it percolates through the Earth's crust. The resulting solution of anions and cations eventually finds its way to the oceans. When water evaporates from the oceans, the ions are left behind. Over many eons, the continual influx of river water containing these ions has built up the substantial salt concentrations found in the Earth's oceans.

Table 5-5 shows that each of the four common *s*-block ions is abundant not only in seawater but also in body fluids, where these ions play essential biochemical roles. Sodium is the most abundant cation in fluids that are outside of cells, and proper functioning of body cells requires that sodium concentrations be maintained within a narrow range. One of the main functions of the kidneys is to control the excretion of sodium. Whereas sodium cations are abundant in the fluids

## TABLE 5-5  ABUNDANCE OF s-BLOCK ELEMENTS

| Element | Abundance in Crust (% by Mass) | Abundance in Seawater (mol/L) | Abundance in Blood Plasma (mol/L) |
|---------|-------------------------------|-------------------------------|-----------------------------------|
| Na | 2.27 | 0.462 | 0.142 |
| K | 1.84 | 0.097 | 0.005 |
| Mg | 2.76 | 0.053 | 0.003 |
| Ca | 4.66 | 0.01 | 0.005 |

outside of cells, potassium cations are the most abundant ions in the fluids inside cells. The difference in ion concentration across cell walls is responsible for the generation of nerve impulses that drive muscle contraction. If the difference in potassium ion concentration across cell walls deteriorates, muscular activity, including the regular muscle contractions of the heart, can be seriously disrupted.

The cations $Mg^{2+}$ and $Ca^{2+}$ are major components of bones. Calcium occurs as hydroxyapatite, whose chemical formula is $Ca_5(PO_4)_3(OH)$. The structural form of magnesium in bones is not fully understood. In addition to being essential ingredients of bone, these two cations also play key roles in various biochemical reactions, including photosynthesis, the transmission of nerve impulses, and the formation of blood clots.

Beryllium behaves differently from the other s-block elements because the $n = 2$ orbitals are more compact than orbitals with higher principal quantum number. The first ionization energy of beryllium, 899 kJ/mol, is comparable with those of non-metals, so beryllium does not form compounds that are clearly ionic.

The calcium compound that makes up bones is very durable.

Some compounds of the s-block elements are important industrial chemicals as well. For example, Canada is the largest producer of $K_2O$ (potassium oxide, whose common name is "potash") in the world, and produces some 10 billion kilograms each year. This compound, which is obtained from mineral deposits, is the most common source of potassium for fertilizers. Fertilization with potassium is necessary because this element is essential for healthy plant growth. Moreover, potassium salts are highly soluble in water, so potassium quickly becomes depleted from the soil. Consequently, agricultural land requires frequent addition of potassium fertilizers.

Three other compounds of s-block elements—calcium oxide (CaO, known as "lime"), sodium hydroxide (NaOH), and sodium carbonate ($Na_2CO_3$)—are among the top 15 industrial chemicals in annual production. Lime is perennially in the top 10 because it is the key ingredient in construction materials such as concrete, cement, mortar, and plaster. Two other compounds, calcium chloride ($CaCl_2$) and sodium sulphate ($Na_2SO_4$), rank just below the top 50 in industrial importance.

Many industrial processes make use of chemically useful anions such as hydroxide ($OH^-$), carbonate ($CO_3^{2-}$), and chlorate ($ClO_3^-$). These anions must be supplied as chemical compounds that include cations. Sodium is most frequently used as this spectator cation because it is abundant, inexpensive, and non-toxic. Hydroxide ion is industrially important because it is a strong base. Sodium hydroxide is used to manufacture other chemicals, textiles, paper, soaps, and detergents. Sodium carbonate and sand are the major starting materials in the manufacture of glass. Glass contains sodium and other cations embedded in a matrix of silicate $(SiO_3^{2-})$ anions. About half the sodium carbonate produced in the world is used in glass making.

## *p*-Block Elements

The properties of elements in the *p* block vary across the entire spectrum of chemical possibilities. The elements in Group 13, with a single electron in a *p* orbital as well as two valence *s* electrons, display chemical reactivity characteristic of three valence electrons. Except for boron, the elements of this group are metals that form stable cations with $+3$ charge. Examples are $Al(OH)_3$ and $GaF_3$. Metallic character diminishes rapidly as additional *p* electrons are added. This change culminates in the elements in Group 18. With filled *p* orbitals, these elements are so unreactive that for many years they were thought to be completely inert. Xenon is now known to form compounds with the most reactive non-metals, oxygen, fluorine, and chlorine; krypton forms a few highly unstable compounds with these elements.

Although the non-metals do not readily form cations, many of them combine with oxygen to form polyatomic oxoanions. These anions have various stoichiometries, but there are some common patterns. Two second-row elements form oxoanions with three oxygen atoms: carbon (four valence electrons) forms carbonate $(CO_3^{2-})$, and nitrogen (five valence electrons) forms nitrate $(NO_3^-)$. In the third row, the most stable oxoanions contain four oxygen atoms: $SiO_4^{4-}$, $PO_4^{3-}$, $SO_4^{2-}$, and $ClO_4^-$.

*p* block

Many of the minerals that form the Earth's crust contain oxoanions. Examples of carbonates are $CaCO_3$ (limestone) and $MgCa(CO_3)_2$ (dolomite). Barite $(BaSO_4)$ is a sulphate mineral. An important phosphate is $Ca_5(PO_4)_3F$ (apatite). Two silicates are zircon $(ZrSiO_4)$ and olivine (a mixture of $MgSiO_4$ and $FeSiO_4$).

Several leading industrial chemicals contain these anions or are acids resulting from addition of $H^+$ to the anions. Sulphuric acid $(H_2SO_4)$ is one of the highest-production chemicals in the world. Two other industrially important acids are nitric acid $(HNO_3)$ and phosphoric acid $(H_3PO_4)$. Heating $CaCO_3$ drives off $CO_2$ and forms $CaO$, the essential ingredient of building materials mentioned earlier. Industrially important salts include ammonium sulphate, $(NH_4)_2SO_4$; aluminum sulphate, $Al_2(SO_4)_3$; sodium carbonate, $Na_2CO_3$; and ammonium nitrate, $NH_4NO_3$. The chemistry of *p*-block elements (also called "main group elements") is discussed in more detail in Chapter 19.

## SECTION EXERCISES

● **5.6.1** Consult the table of first ionization energies in Appendix C and calculate the average values for the non-metals, metalloids, and *s*-block elements. How does the trend in these averages relate to the ionic chemistry of these elements?

● **5.6.2** Classify each of the following elements as a metal, non-metal, or metalloid: S, Si, Sr, Se, Sc, Sg, and Sn.

# CHAPTER 5 VISUAL SUMMARY

## 5.1 ORBITAL ENERGIES

Ionization energy (IE)

Screening increases as *l* increases.

Effective nuclear charge ($Z_{eff}$)

**LEARNING OBJECTIVES:** Explaining the effects of nuclear charge and screening on the energies of electrons.

**SKILLS TO MASTER:** Predicting the effects of screening. (p. 247)

Drawing electron density plots. (p. 249)

**KEY CONCEPTS:** The higher the value of the *l* quantum number, the more that orbital is screened by electrons in smaller, more stable orbitals.

## 5.2 STRUCTURE OF THE PERIODIC TABLE

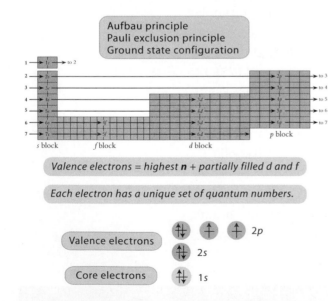

Aufbau principle
Pauli exclusion principle
Ground state configuration

Valence electrons = highest **n** + partially filled d and f

Each electron has a unique set of quantum numbers.

Valence electrons

Core electrons

**LEARNING OBJECTIVES:** Understanding the relationships between the structure of the periodic table and electronic configurations.

**SKILLS TO MASTER:** Counting valence electrons. (p. 257)

**KEY CONCEPTS:** The Pauli exclusion principle states that each electron in an atom has a unique set of quantum numbers.

The aufbau principle states that electrons are placed into atomic orbitals beginning with the lowest energy electrons, followed by successively higher energy electrons.

Valence electrons are all those of highest principal quantum number plus those in partially filled *d* and *f* orbitals.

## 5.3 ELECTRON CONFIGURATIONS

Electron configuration

Paramagnetic

Diamagnetic

Hund's rule

Noble gas configuration

As many electrons as possible have the same spin orientation.

Isoelectronic

**LEARNING OBJECTIVES:** Using the Pauli exclusion principle, Hund's rule, and the orbital filling order to predict electron configurations of atoms and ions.

**SKILLS TO MASTER:** Writing electron configurations. (p. 260)

Filling orbital with electrons in increasing order of energy. (p. 262)

Explaining anomalous electron configurations. (p. 264)

**KEY CONCEPTS:** The most stable configuration involving orbitals of equal energies is the one with the maximum number of electrons with the same spin orientation (Hund's rule).

## 5.4 PERIODICITY OF ATOMIC PROPERTIES

*Orbital energy and size increase with **n** and decrease with Z.*

*Electron affinity increases across each row of the periodic table.*

Atomic radii
Ionization energy
Electron affinity

Atomic radii          1st Ionization energy

**LEARNING OBJECTIVES:** Relating trends in atomic radius, ionization energy, and electron affinity to nuclear charge and electronic configuration.

**SKILLS TO MASTER:** Predicting periodic variations in atomic radius. (p. 270)

Predicting periodic variations in ionization energy. (p. 272)

Predicting periodic variations in electron affinity. (p. 274)

**KEY CONCEPTS:** As principal quantum number $n$ increases, atomic orbitals become larger and less stable. As atomic number $Z$ increases, any given atomic orbital becomes smaller and more stable.

Atomic size decreases from left to right and increases from top to bottom of the periodic table.

First ionization energy increases from left to right across each row and decreases from top to bottom of each column of the periodic table.

Electron affinity tends to become more negative from left to right across a row of the periodic table.

## 5.5 ENERGETICS OF IONIC COMPOUNDS

Step 1: Sublimation
Step 2: Bond breakage
Step 3: Ionization
Step 4: Electron attachment
Step 5: Condensation

Energy of reaction

$$E_{electrical} = k\frac{q_1 q_2}{r}$$

Lattice energy

**LEARNING OBJECTIVES:** Understanding why ionic compounds exist and the energetics of their formation.

**SKILLS TO MASTER:** Using a Born–Haber cycle to predict the energy change of a reaction, or of one of the steps of the reaction. (p. 278)

Explaining the effects of charge and ionic radius on lattice energy. (p. 279)

**KEY CONCEPTS:** The lattice energy is the energy required to separate an ionic compound into its ions.

The Born-Haber cycle is an example of the use of Hess' law.

## 5.6 IONS AND CHEMICAL PERIODICITY

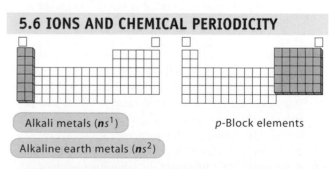

Alkali metals ($ns^1$)

Alkaline earth metals ($ns^2$)

$p$-Block elements

**LEARNING OBJECTIVES:** Understanding the trends in atomic radius, ionization energy and electron affinity and their relationships to nuclear charge and electronic configuration.

**SKILLS TO MASTER:** Predicting the charges on ions and knowing some common properties of $s$-, $p$-, and $d$-block elements. (p. 285)

## LEARNING EXERCISES

**5.1**  Draw appropriately scaled pictures of all the occupied orbitals in a krypton atom.

**5.2**  Write brief explanations of (a) screening; (b) the Pauli exclusion principle; (c) the aufbau principle; (d) Hund's rule; and (e) valence electrons.

**5.3**  Construct an orbital energy-level diagram for all orbitals with $n < 8$ and $l < 4$. Use the periodic table to help determine the correct order of the energy levels.

**5.4**  Describe periodic variations in electron configurations; explain how they affect ionization energy and electron affinity.

**5.5**  List the 18 elements classified as non-metals. Give the name, symbol, and atomic number for each.

**5.6**  Update your list of memory bank equations by adding principles for atomic configurations.

**5.7**  List all terms new to you that appear in Chapter 5. Write a one-sentence definition of each, using your own words. Consult the glossary if you need help.

## PROBLEMS

**ilw** = interactive learning ware problem. Visit the website at www.wiley.com/canada/olmsted

### Orbital Energies

**5.1**  For each pair of orbitals, determine which is more stable and explain why: (a) He 1s and He 2s; (b) Kr 5p and Kr 5s; and (c) He 2s and He$^+$ 2s.

**5.2**  For each pair of orbitals, determine which is more stable and explain why: (a) C 2s and C 2p; (b) Ar 5p and Ar$^+$ 5p; and (c) Ar 4s and Ar 5s.

**5.3**  In a hydrogen atom, the 3s, 3p, and 3d orbitals all have the same energy. In a helium atom, however, the 3s orbital is lower in energy than the 3p orbital, which in turn is lower in energy than the 3d orbital. Explain why the energy rankings of hydrogen and helium are different.

**5.4**  The energy of the $n = 2$ orbital of the He$^+$ ion is the same as the energy of the $n = 1$ orbital of the H atom. Explain this fact.

**5.5**  Refer to the table of ionization energies in Focus 5-1 on page 252 to answer the following questions. In each case, provide a brief explanation of your choice. (a) Which ionization energies show that an electron in a 1s orbital provides nearly complete screening of an electron in a 2p orbital? (b) Which ionization energies show that the stability of $n = 2$ orbitals increases with $Z^2$?

**5.6**  Refer to Table 5-1 to answer the following questions. In each case, provide a brief explanation of your choice. (a) Which ionization energies show that stability of $n = 1$ orbitals increases with $Z$? (b) Which ionization energies show that an electron in a 1s orbital incompletely screens another electron in the same orbital?

### Structure of the Periodic Table

**5.7**  Draw the periodic table in block form, and outline and label each of the following sets: (a) Elements that are one electron short of filled p orbitals; (b) elements for which $n = 3$ orbitals are filling; (c) elements with half-filled d orbitals; and (d) the first element that contains a 5s electron.

**5.8**  Draw the periodic table in block form, and outline and label each of the following sets: (a) Elements for which $n = 1$ orbitals are filling; (b) elements for which the 5f orbitals are filling; (c) elements with $s^2 p^4$ configurations; and (d) elements with filled valence s orbitals but empty valence p orbitals.

**5.9**  Determine the atomic number and valence orbital that is filling for the element that would appear below lead in the periodic table.

**5.10**  Determine the atomic number and valence orbital that is filling for the element that would appear below francium in the periodic table.

**5.11**  Predict the location in the periodic table (row and column) of element 117.

**5.12**  Predict the location in the periodic table (row and column) of element 113.

**5.13**  How many valence electrons does each of the following atoms have? O, V, Rb, Sn, and Cd.

**5.14**  How many valence electrons does each of the following elements have? P, Cr, Kr, I, and Ba.

### Electron Configurations

**5.15**  **ilw** Write the complete electron configuration, and list a correct set of values of the quantum numbers for each of the valence electrons in the ground-state configurations of (a) Be, (b) O, (c) Ne, and (d) P.

**5.16**  Write the complete electron configuration, and list a correct set of values of the quantum numbers for each of the valence electrons in the ground-state configurations of (a) Li, (b) C, (c) F, and (d) Mg.

**5.17**  Which of the atoms of Problem 5.15 are paramagnetic? Draw orbital energy-level diagrams to support your answer.

**5.18**  Which of the atoms of Problem 5.16 are paramagnetic? Draw orbital energy-level diagrams to support your answer.

**5.19**  The following are hypothetical configurations for a beryllium atom. Which use non-existent orbitals, which are forbidden by the Pauli principle, which are excited states, and which is the ground-state configuration? (a) $1s^3 2s^1$; (b) $1s^1 2s^3$; (c) $1s^1 2p^3$; (d) $1s^2 2s^1 2p^1$; (e) $1s^2 2s^2$; (f) $1s^2 1p^2$; and (g) $1s^2 2s^1 2d^1$.

**5.20**  None of the following hypothetical configurations describes the ground state of a fluorine atom. For each, state the reason why it is not correct: (a) $1s^2 2s^2 2p^4$; (b) $1s^2 2s^1 2p^6$; (c) $1s^3 2s^2 2p^4$; and (d) $1s^2 2s^2 1p^5$.

**5.21**  The ground state of Mo has higher spin than that of Tc. Construct energy-level diagrams for the valence electrons that show how electron configurations account for this difference.

**5.22**  The ground state of V has lower spin than that of Cr. Construct energy-level diagrams for the valence electrons that show how electron configurations account for this difference.

**5.23**  Write the correct ground-state electron configuration, in shorthand notation, for C, Cr, Sb, and Br.

**5.24** Write the correct ground-state electron configuration, in shorthand notation, for N, Ti, As, and Xe.

**5.25** For nitrogen, how many different excited-state configurations are there in which no electron has $n > 2$? Write all of them.

**5.26** For fluorine, how many different excited-state configurations are there in which no electron has $n > 2$? Write all of them.

## Periodicity of Atomic Properties

**5.27** Arrange the following atoms in order of decreasing first ionization energy: Ar, Cl, Cs, and K.

**5.28** Arrange the following atoms in order of increasing size: Cl, F, P, and S.

**5.29** One element has these ionization energies and electron affinity (all in kJ/mol): $IE_1 = 376$, $IE_2 = 2420$, $IE_3 = 3400$, 400, $EA = -45.5$. In what column of the periodic table is this element found? Give your reasoning. Refer to Appendix C if necessary.

**5.30** One element has these ionization energies and electron affinity (all in kJ/mol): $IE_1 = 503$, $IE_2 = 965$, $IE_3 = 3600$, $EA = 46$. In what group of the periodic table is this element found? Give your reasoning. Refer to Appendix C if necessary.

**5.31** According to Appendix C, each of the following elements has a positive electron affinity. For each one, construct its valence orbital energy-level diagram and use it to explain why the anion is unstable: N, Mg, and Zn.

**5.32** According to Appendix C, each of the following elements has a positive electron affinity. For each one, construct its valence orbital energy-level diagram and use it to explain why the anion is unstable: Be, Ar, and Mn.

**5.33** List the atomic ions that are isoelectronic with $Br^-$ and have net charges (absolute values) that are less than 4 units. Arrange these in order of increasing size.

**5.34** List the ionic species that are isoelectronic with Ar and have net charges (absolute values) that are less than 4 units. Arrange these in order of increasing size.

## Energetics of Ionic Compounds

**5.35** From the following list, select the elements that form ionic compounds: Ca, C, Cu, Cs, Cl, and Cr. Indicate whether each forms a stable cation or a stable anion.

**5.36** From the following list, select the elements that form ionic compounds: B, Ba, Be, Bi, and Br. Indicate whether each forms a stable cation or a stable anion.

**5.37** Given the following data, estimate the energy released when gaseous K and I atoms form gaseous $[K^+I^-]$.

| Element | EA (kJ/mol) | $IE_1$ (kJ/mol) | $IE_2$ (kJ/mol) | Ion Radius (pm) |
|---------|------------|----------------|----------------|-----------------|
| K | −48.4 | 418.8 | 3051 | 133 (cation) |
| I | −295.3 | 1008.4 | 1845.9 | 220 (anion) |

**5.38** Repeat the calculation of Problem 5.37 for $K^{2+}$ and $I^{2-}$, using 500 kJ/mol as the estimated second electron affinity

of iodine and assuming no change in distance of closest approach.

**5.39** Consider three possible ionic compounds formed by barium and oxygen: $Ba^+O^-$, $Ba^{2+}O^{2+}$, and $Ba^{3+}O^{3-}$.
(a) Which would have the greatest lattice energy?
(b) Which would require the least energy to form the ions? (c) Which compound actually exists, and why?

**5.40** Consider three possible ionic compounds formed by calcium and the chloride anion: $CaCl$, $CaCl_2$, and $CaCl_3$.
(a) Which would have the greatest lattice energy? (b) Which would require the least energy to form the ions? (c) Which compound actually exists, and why?

**5.41** **ilw** Calculate the overall energy change for the formation of lithium fluoride from lithium metal and fluorine gas. In addition to data found in Appendix C and Table 5-4, the following information is needed: the bond energy of $F_2$ is 155 kJ/mol, and lithium's enthalpy of sublimation is 159.3 kJ/mol.

**5.42** Calculate the overall energy change for the formation of calcium bromide from calcium metal and liquid bromine. In addition to data found in Appendix C and Table 5-4, the following information is needed: the bond energy of $Br_2$ is 224 kJ/mol; bromine's enthalpy of sublimation is 30.9 kJ/mol; calcium's enthalpy of sublimation is 177.8 kJ/mol.

**5.43** Draw a Born–Haber diagram similar to Figure 5-21 for the reaction described in Problem 5.41.

**5.44** Draw a Born–Haber diagram similar to Figure 5-21 for the reaction described in Problem 5.42.

**5.45** Refer to the lattice energy values in Table 5-4. If $Cs_2O$ existed, what lattice energy would you predict it would have? Explain your prediction.

**5.46** Refer to the lattice energy values in Table 5-4. Predict the lattice energy of MgS, and explain your prediction.

## Ions and Chemical Periodicity

**5.47** Classify each of the elements from Group 16 of the periodic table as a metal, a non-metal, or a metalloid.

**5.48** Classify each of the elements from Group 14 of the periodic table as a metal, a non-metal, or a metalloid.

**5.49** Classify each of the elements listed in Problem 5.35 as a metal, a non-metal, or a metalloid.

**5.50** Classify each of the elements listed in Problem 5.36 as a metal, a non-metal, or a metalloid.

**5.51** We list polonium as a metal, but some chemists classify it as a metalloid. List other metals that might be expected to show properties in between those of metals and metalloids.

**5.52** What is the maximum number of valence $p$ electrons possessed by a metallic element? Which metal(s) have this configuration?

## ADDITIONAL PAIRED PROBLEMS

**5.53** Use the periodic table to find and list (a) all elements with ground-state configurations indicating that the $4s$ and $3d$ orbitals are nearly equal in energy; (b) the elements in the column that has two elements with one valence configuration and two with another valence configuration; and (c) a set of elements with valence configurations indicating that the 6d and 5f orbitals are nearly equal in energy.

5.54 Use the periodic table to find and list (a) all elements with ground-state configurations indicating that the 5s and 4d orbitals are nearly equal in energy; (b) two elements with configurations indicating that the 5d and 4f orbitals are nearly equal in energy; and (c) three elements in the same group that have different valence configurations.

5.55 Write the correct electron configuration for the $Mn^{2+}$ ground state, and give a correct set of quantum numbers for all electrons in the least-stable occupied orbital.

5.56 Write the correct electron configuration for the $Co^{3+}$ ground state, and give a correct set of quantum numbers for all electrons in the least-stable occupied orbital.

5.57 Predict the total electron spin for P, $Br^-$, and $Cu^+$.

5.58 Predict the total electron spin for Gd, Sr, and $Ag^+$.

*For Problems 5.59–5.62, explain your rankings in terms of quantum numbers and electrical interactions.*

5.59 Arrange the following in order of increasing ionization energy: N, O, Ne, Na, and $Na^+$.

5.60 Arrange the following in order of decreasing ionization energy: Br, Ar, $Ar^+$, and Cl.

5.61 Arrange the following in order of decreasing size (radius): $Cl^-$, $K^+$, Cl, and $Br^-$.

5.62 Arrange the following in order of increasing size: K, $K^+$, Ar, and Ca.

5.63 Write correct ground-state electron configurations for the neutral atoms with atomic numbers 9, 20, and 33.

5.64 Write correct ground-state electron configurations for the neutral atoms and doubly positive ions with atomic numbers 14, 34, and 92.

5.65 Using information in Table 5-4, determine the average difference in lattice energy between alkali fluorides and alkali chlorides, and the average difference in lattice energy between alkali bromides and alkali iodides. Is there a trend in the individual values? If so, describe and explain it.

5.66 Using information in Table 5-4, determine the average difference in lattice energy between lithium halides and sodium halides, and the average difference in lattice energy between rubidium halides and cesium halides. Is there a trend in the individual values? If so, describe and explain it.

5.67 Draw energy-level diagrams that show the ground-state valence electron configurations for $Cu^+$, $Mn^{2+}$, and $Au^{3+}$.

5.68 Draw energy-level diagrams that show the ground-state valence electron figurations for $Ir^+$, $Cd^{2+}$, and $V^{2+}$.

5.69 Write a brief explanation for each of the following: (a) In a hydrogen atom, the 2s and 2p orbitals have identical energy. (b) In a helium atom, the 2s and 2p orbitals have different energies.

5.70 Write a brief explanation for each of the following: (a) All three 2p orbitals of a helium atom have identical energy. (b) The IE of the He excited state is nearly the same as the IE of the H $2p^1$ excited state.

5.71 Refer to Figure 5-16 to answer the following questions about first ionization energies: (a) Which element shows the greatest decrease from its neighbour of next lower Z? (b) What is the atomic number of the element with the lowest value? (c) Identify three ranges of Z across which the value changes the least. (d) List the atomic numbers of all elements whose values are between 925 and 1050 kJ/mol.

5.72 Refer to Figure 5-17 to answer the following questions about electron affinities:
(a) Which column of the periodic table has the second-most negative set of values?
(b) Which element has the highest value?
(c) Apart from the end of a row, which element shows the greatest decrease compared with its neighbour of immediately lower Z?
(d) What do elements 6, 14, and 32 have in common?
(e) List the atomic numbers of all elements whose values are within 10% of −50 kJ/mol.

5.73 Which has the most unpaired electrons: $S^+$, S, or $S^-$? Use electron configurations to support your answer.

5.74 Which has the most unpaired electrons: Si, P, or S? Use electron configurations to support your answer.

5.75 The figure below shows four proposed electron energy diagrams for a phosphorus atom. Which are forbidden by the Pauli principle, which are excited states, and which is the ground-state configuration?

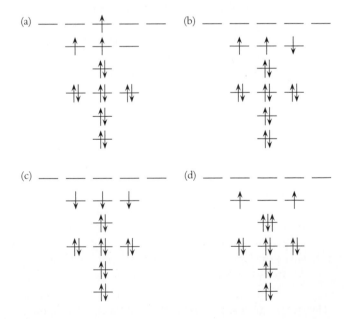

5.76 None of the four proposed electron energy diagrams shown below describes the ground state of a sulphur atom. For each, state the reason why it is not correct.

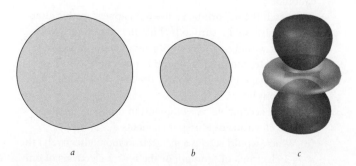

(a) Rank the orbitals in order of stability for a multielectron atom.
(b) Provide two sets of quantum numbers for an electron in orbital *a*.
(c) Give the atomic number of an element that has two electrons in orbital *a* but no electrons in orbital *c*.
(d) Give the name of a cation that has one electron in orbital *c*.
(e) How many other orbitals have the same principal quantum number as orbital *c*?
(f) An element has its two least-stable electrons in orbital *b*. If an atom of that element loses one electron, will orbital *b* become larger or smaller or remain the same size?

5.80 Consider the following three atomic orbitals. They are drawn to scale, and orbital *a* has $n = 2$.

*a*          *b*          *c*

(a) How many electrons can be placed in orbital *b*?
(b) Provide three sets of quantum numbers that describe an electron in orbital *c*.
(c) Arrange these orbitals in order of increasing stability for a multielectron atom.
(d) In a sodium atom, how many other orbitals have exactly the same energy as orbital *b*?
(e) Give the name of the element that has orbital *a* filled but orbitals *b* and *c* empty.
(f) Give the name of an element that has both an empty and a partially filled orbital *c*.
(g) Give the name of a common anion in which all orbitals of type *c* are filled.

5.77 Identify the ionic compounds that best fit the following descriptions:
(a) the alkaline earth cation with the second-smallest radius and a Group 16 anion that is isoelectronic with the noble gas from Row 3;
(b) a +1 ion that is isoelectronic with the noble gas from Row 3, combined with the anion formed from the Row 2 element with the highest electron affinity; and
(c) the alkaline earth metal with the highest second ionization energy that combines in a 1 : 2 ratio with an element from Row 3.

5.78 Identify the ionic compounds that best fit the following descriptions:
(a) the naturally occurring alkali metal with the lowest ionization energy combined with the Group 17 element with the most negative electron affinity;
(b) the anion from Group 16 that has the smallest radius and is isoelectronic with a noble gas, combined with an alkali cation that is isoelectronic with the same noble gas; and
(c) the compound formed from the elements of Row 4 that have the lowest ionization energy and highest electron affinity.

5.79 Consider the three atomic orbitals that follow. They are drawn to scale, and orbital *c* has $n = 3$.

*a*          *b*          *c*

## ADDITIONAL PROBLEMS

5.81 Write the electron configuration for the lowest-energy *excited* state of each of the following: Be, $O^{2-}$, $Br^-$, $Ca^{2+}$, and $Sb^{3+}$.

5.82 Write the ground-state configurations for the isoelectronic species $Ce^{2+}$, $La^+$, and Ba. Are they the same? What features of orbital energies account for this?

5.83 The ionization energy of lithium atoms in the gas phase is about half as large as the ionization energy of beryllium atoms in the gas phase. In contrast, the ionization energy of $Li^+$ is about four times larger than the ionization energy of $Be^+$. Explain the difference between the atoms and the ions.

**5.84** The figures show $Cl^-$, Ar, and $K^+$ drawn to scale. Decide which figure corresponds to which species and explain your reasoning.

       *a*                     *b*                     *c*

**5.85** Make an electron density plot that shows how the 3s and 3p orbitals are screened effectively by the 2p orbitals. Provide a brief explanation of your plot.

**5.86** Use the data in Appendix C to explain why the noble gases seldom take part in chemical reactions.

**5.87** From its location in the periodic table, predict some of the physical and chemical properties of francium. What element does it most closely resemble?

**5.88** What would be the next two orbitals to fill after the 7p orbital?

**5.89** Which has the more stable 2s orbital: a lithium atom or a $Li^{2+}$ cation? Explain your reasoning.

**5.90** The first four ionization energies of aluminum are as follows: $IE_1 = 579$ kJ/mol, $IE_2 = 1817$ kJ/mol $IE_3 = 2,745$ kJ/mol, and $IE_4 = 11\ 578$ kJ/mol.
(a) Explain the trend in ionization energies.
(b) Rank ions of aluminum in order of ionic radius, from largest to smallest.
(c) Which ion of aluminum has the largest electron affinity?

**5.91** The idea of an atomic radius is inherently ambiguous. Explain why this is so.

**5.92** Use electron–electron repulsion and orbital energies to explain the following irregularities in first ionization energies: (a) Boron has a lower ionization energy than beryllium. (b) Sulphur has a lower ionization energy than phosphorus.

## GROUP STUDY PROBLEMS

**5.93** Suppose that the effects of screening were always less than the effects of increasing principal quantum number; for example, making the 3d orbital always more stable than the 4s orbital. Construct the periodic table for the resulting filling order of the orbitals. What is the lowest value of Z that would have a different configuration, and how does it differ? Use this periodic table to identify the atomic numbers of elements that would be alkali metals. Which elements do you think would be halogens (adding one electron easily)? Which would be noble gases (chemically unreactive)? What similarities do you see, and what differences are there? Do you think the periodic table would have been easier to discover, harder to discover, or about the same had the filling order been like this? Why?

**5.94** No elements have ground-state configurations with electrons in g ($l = 4$) orbitals, but excited states can have

such electrons. (a) How many different g orbitals are there? (b) What are the possible values of $m_l$? (c) What is the lowest principal quantum number for which there are g orbitals? (d) Which orbitals may have nearly the same energy as the lowest-energy g orbitals? (e) What is the atomic number of the first element that has a g electron in its ground-state configuration, assuming the g orbital begins to fill after an s orbital?

**5.95** Use data in Appendix C, Table 5-4, and the following information to calculate the overall energy changes for the formation of CaCl and $CaCl_2$. Use your results to determine which compound is more likely to form in the reaction of solid calcium with chlorine gas. Sublimation enthalpy of Ca = 178 kJ/mol, $Cl_2$ bond energy = 240 kJ/mol, lattice energy for CaCl (estimated) = 720 kJ/mol.

**5.96** In the imaginary galaxy Topsumturvum, all of the rules for atomic configurations are the same as ours, except that orbital stability increases slightly as $l$ increases. As a result, the 3s orbital is more stable than the 4d orbitals, but the 4f orbitals are slightly more stable than the 3s orbital. (a) Draw an orbital energy ladder for Topsumturvum. (b) Determine the ground-state configuration for the Topsumturvum atom containing 27 electrons. (c) Draw a block diagram that shows the first 52 elements in the periodic table for Topsumturvum.

*Questions 5.97– 5.100 refer to a hypothetical universe named Morspin, where electrons have three possible spin orientations $(m_s = +^1/_2, -^1/_2, 0)$ rather than the two $(m_s = +^1/_2, -^1/_2,)$ in our own universe. All other physical laws in the two universes are the same. The periodic table of Morspin follows:*

| 1 | 2 | | | | | | | | | | | | | 3 |
|----|----|----|-------|----|----|----|----|----|----|----|----|----|----|----|
| 4 | 5 | 6 | | 7 | 8 | 9 | 10 | 11 | 12 | 13 | 14 | 15 | | |
| 16 | 17 | 18 | | 19 | 20 | 21 | 22 | 23 | 24 | 25 | 26 | 27 | | |
| 28 | 29 | 30 | 31-45 | 46 | 47 | 48 | 49 | 50 | 51 | 52 | 53 | 54 | | |
| 55 | 56 | 57 | 58-72 | 73 | 74 | 75 | 76 | 77 | 78 | 79 | 80 | 81 | | |

**5.97** (a) Which would have the larger first ionization energy: Morspin element 18 or Morspin element 30? Explain.
(b) Which would have the larger radius: Morspin element 15 or cation $17^{2+}$? Explain.
(c) Which would have the larger electron affinity: Morspin element 47 or 48? Explain.

**5.98** (a) What would be the electron configuration of Morspin cation $38^{3+}$?
(b) Sketch the $d_{xy}$ orbital as it would appear in this alternative universe. Be sure to include the coordinate axes.
(c) What would be the correct ground-state configurations for Morspin atoms of atomic numbers 3 and 14?

**5.99** (a) How many electrons would the first three noble gases in Morspin have?
(b) Write acceptable sets of quantum numbers for the valence electrons of Morspin elements 4, 7, and 32.
(c) Which would screen a 4s electron more effectively: the $n = 3$ electrons of Morspin or the $n = 3$ electrons of our own universe? Explain.

**5.100** (a) What would be the electron configuration of Morspin element 19?

(b) How many unpaired electrons would there be in element 28? Explain.

(c) What would be the highest positive charge of cations found in ionic salts that could form from Morspin element 30? Explain.

## ANSWERS TO SECTION EXERCISES

**5.1.1**

Distance from the nucleus, $r$

**5.1.2** $2.4 \times 10^{-19}$ J

**5.1.3**

**5.2.1** 117

**5.2.2** Br, with 35 electrons, is in the $p$ block. Filled orbitals are 1s, 2s, 2p, 3s, 3p, 4s, and 3d. The 4p orbital is partially filled.

**5.2.3** Be, Mg, Ca, Sr, Ba, Ra, Zn, Cd, and Hg; all have $s^2$.

**5.3.1** (a) [Xe] $4f^{14} 5d^{10}$, zero spin; (b) [Rn] $7s^2 5f^{14} 6d^{10} 7p^6$, zero spin; and (c) [Ar], zero spin

**5.3.2**

**5.3.3** Mo, [Kr] $5s^1 4d^5$; I, [Kr] $5s^2 4d^{10} 5p^5$; and $Hg^+$, [Xe] $6s^1 5d^{10} 4f^{14}$

**5.4.1** (a) Cl has larger $Z$; (b) Cl has lower $n$; (c) Sr has higher $n$; (d) $Cl^-$ has smaller $Z$; and (e) $O^-$ has smaller $Z_{eff}$.

**5.4.2** (a) Between Nb and In, a $d$ block fills, so $Z_{eff}$ remains nearly constant; (b) Ga loses a $4p$ electron, which is easier to remove than the $4s$ electrons of Zn; and (c) Mn has $d^5$ valence configuration, so the next electron must pair with an existing electron.

**5.4.3** Stability decreases markedly with $n$.

**5.5.1** $LE = 2956$ kJ/mol, compares closely with the value in Table 5-4, 2913 kJ/mol.

**5.5.2** Cu and Zn

**5.5.3** $Ca^{2+}$, $Cs^+$, $Al^{3+}$, and $Br^-$

**5.6.1** Non-metals, 1390; metalloids, 824; s block, 610 (all in kJ/mol); low ionization energy leads to formation of ions

**5.6.2** S, non-metal; Si, metalloid; Sr, metal; Se, non-metal; Sc, metal; Sg, metal; and Sn, metal

## ANSWERS TO PRACTICE EXERCISES

**5.1**

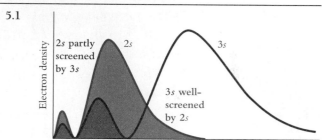

Distance from the nucleus, $r$

**5.2** Zr, with 40 electrons, is in the $d$ block. Filled orbitals are 1s, 2s, 2p, 3s, 3p, 4s, 3d, 4p, and 5s. The 4d orbital is partially filled.

**5.3** Configuration is $1s^2 2s^2 2p^5$

**5.4** Configuration is [Kr] $5s^2 4d^{10}$.

**5.5** Compact configuration is [He] $2s^2 2p^3$. Valid quantum numbers for the valence electrons are as follows:

$n = 2, l = 0, m_l = 0, \quad m_s = +\frac{1}{2}$

$n = 2, l = 0, m_l = 0, \quad m_s = -\frac{1}{2}$

$n = 2, l = 1, m_l = +1, m_s = +\frac{1}{2}$

$n = 2, l = 1, m_l = 0, \quad m_s = +\frac{1}{2}$

$n = 2, l = 1, m_l = -1, m_s = +\frac{1}{2}$

**5.6** $Ru^{3+}$ cation has $44 - 3 = 41$ electrons. Configuration is [Kr] $4d^5$.

**5.7** Exceptions will be the ions with the 3d orbital set completely filled or completely empty: $Zn^{2+}$, $Cu^+$, and $Sc^{3+}$.

**5.8** P and Se are smaller than As; Ge and Sb are larger than As.

**5.9** Lattice energy $= 2473$ kJ/mol.

**5.10** Ga is a metal; Ge and As are metalloids; and Se, Br, and Kr are non-metals.

# FUNDAMENTALS OF CHEMICAL BONDING

## NANOTECHNOLOGY

A gear-shaped molecule spins freely on a copper metal surface. A molecular "soccer ball" nestles snugly in a cavity on a protein. It sounds like science fiction, but these creations are real developments in the blossoming field of nanotechnology. The term "nanotechnology" refers to science carried out at the nanometre scale ($1 \text{ nm} = 10^{-9} \text{ m}$).

Our background image, taken by a scanning tunnelling microscope, shows a single layer of gear-shaped organic molecules lying on a copper metal surface. Most of the molecules are locked in place by nesting close to their neighbours. One of them, however, appears distinctly "fuzzy," which suggests that the molecule may be spinning rapidly. This particular molecule was "unlocked" by moving it a fraction of a nanometre with the tip of a scanning tunnelling microscope. (See Focus 1-1 in Chapter 1 for a description of scanning tunnelling microscopy.)

The inset image is a computer simulation of this system. The molecule on the right has been moved slightly. Computer-based calculations show that this small change in position allows the molecule in the centre to rotate freely, just as is observed experimentally.

Another example of nanotechnology research is an attempt to develop biological molecules that can interact with fullerene, $C_{60}$. By themselves, $C_{60}$ molecules are difficult to manipulate because they are quite chemically inert. Scientists envision using proteins bound to $C_{60}$, like the one illustrated here, as molecular machines that can deliver $C_{60}$ units to build larger carbon structures.

A spinning molecule on a copper surface and a soccer-ball molecule tethered to a protein may seem no more useful than a spinning ice-skater or a tetherball. Nonetheless, advocates of nanotechnology cite a wealth of potential applications for this new field, including tailored synthetic membranes that can collect specific toxins from industrial waste and computers that process data much faster than today's best models. The list of possible benefits from nanotechnology is limited only by our imaginations.

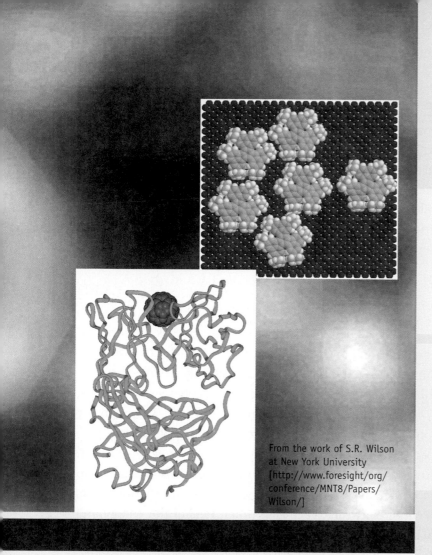

From the work of S.R. Wilson
at New York University
[http://www.foresight/org/
conference/MNT8/Papers/
Wilson/]

# 6

## CHAPTER CONTENTS

## LEARNING OBJECTIVES

*Upon completion of this chapter you should be able to:*

- use the concept of electronegativity to determine the polarity of a chemical bond
- draw optimized Lewis structures of covalent compounds, including resonance structures
- recognize the importance of the tetrahedral shape in molecules
- use the VSEPR model to predict the shapes of molecules with steric numbers 2, 3, 4, 5 and 6.
- understand the factors that influence bond angles, lengths and energies

Nanotechnology involves designing and constructing specific molecular structures, one atom at a time. This requires a thorough understanding of chemical bonding, because every molecule is held together by chemical bonds in which electrons are shared between atoms. The basic principles of bond formation are always the same, but different elements have different bond-forming abilities, and chemical bonds show a variety of forms and strengths.

This chapter and the next describe chemical bonding. First, we explore the interactions among electrons and nuclei that account for bond formation. Then we show how atoms are connected together in simple molecules such as water ($H_2O$). We show how these connections lead to a number of characteristic molecular geometries. In Chapter 7, we discuss more elaborate aspects of bonding that account for the properties of materials as diverse as deoxyribonucleic acid (DNA) and transistors.

## 6.1 OVERVIEW OF BONDING

A collection of electrons and nuclei responds to electrical forces, just as a body of water responds to the force of gravity. Without a barrier to block its path, water runs downhill, finding its most stable position. In a similar way, any group of charged particles arranges itself to maximize electrical attraction and give the most stable arrangement of charges.

Equation 3-1 describes electrical potential energy as a function of charges ($q_1$ and $q_2$) and their separation distance ($r$):

$$E_{\text{electrical}} = k\frac{q_1 q_2}{r} \qquad (3\text{-}1)$$

The equation states that electrical energy between charged species is proportional to the magnitudes of the charges and inversely proportional to the distance between them. Also, charges of opposite sign attract one another, but like charges repel.

Equation 3-1 describes the potential energy of one pair of charges. Molecules, however, contain two or more nuclei and two or more electrons. To obtain the total electrical energy of a molecule, Equation 3-1 must be applied to every possible pair of charged species. These pairwise interactions are of three types. First, electrons and nuclei attract one another. Attractive interactions release energy, so an electron attracted to a nucleus is at lower energy, and therefore more stable, than a free electron. Second, electrons repel each other, raising the energy and reducing the stability of a molecule. Third, nuclei repel each other, so these interactions also reduce the stability of a molecule.

The electrons and nuclei in a molecule balance these three interactions in a way that gives the molecule its greatest possible stability. This balance is achieved when the electrons are concentrated between the nuclei. We view the electrons as shared between the nuclei and call this sharing a **covalent bond**. In any covalent bond, the attractive energy between nuclei and electrons exceeds the repulsive energy arising from nuclear–nuclear and electron–electron interactions.

### The Hydrogen Molecule

We demonstrate these ideas for the simplest stable neutral molecule, molecular hydrogen. A hydrogen molecule contains just two nuclei and two electrons.

Consider what happens when two hydrogen atoms come together and form a covalent bond. As the atoms approach, each nucleus attracts the opposite electron, pulling the two atoms closer together. At the same time, the two nuclei repel each other, and so do the two electrons. These repulsive interactions drive the atoms apart.

For $H_2$ to be a stable molecule, the sum of the attractive energies must exceed the sum of the repulsive energies. Figure 6-1 shows a static arrangement of electrons and nuclei in which the electron–nucleus distances are shorter than the electron–electron and nucleus–nucleus distances. In this arrangement, attractive interactions exceed repulsive interactions, leading to a stable molecule. Notice that the two electrons occupy the region between the two nuclei, where they can interact with both nuclei at once. In other words, the atoms share the electrons in a covalent bond.

**Figure 6-1**

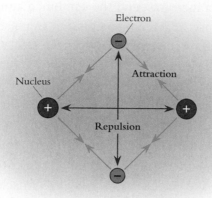

When electrons are in the region between two nuclei, attractive electrical forces exceed repulsive electrical forces, leading to the stable arrangement of a chemical bond. Remember that electrons are not point charges but are spread out over a relatively large volume.

An actual molecule is dynamic, not static. Electrons move continuously and can be thought of as being spread over the entire molecule. In a covalent bond, nevertheless, the distribution of electrons has the general characteristics shown by the static view in the figure. The most probable electron locations are between the nuclei, where they are best viewed as being shared between the bonded atoms.

## Bond Length and Bond Energy

As two hydrogen atoms come together to form a molecule, attractive forces between nuclei and electrons make the hydrogen molecule more stable than the individual hydrogen atoms. The amount of increased stability depends on the distance between the nuclei, as shown in Figure 6-2. At distances greater than

**Figure 6-2**

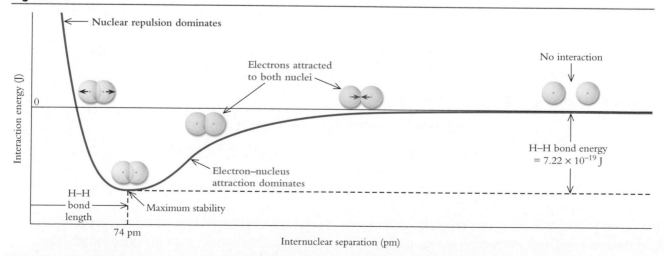

The interaction energy of two hydrogen atoms depends on the distance between the nuclei.

300 pm, the total interaction energy is almost zero because of the large separation. At closer distances, the attraction between the electron of one atom and the nucleus of the other atom increases, and the combined atoms become more stable. Moving the atoms closer together generates greater stability until the nuclei are 74 pm apart. At distances closer than 74 pm, the nucleus–nucleus repulsion increases more rapidly than the electron–nucleus attraction. Thus, at a separation distance of 74 pm, the hydrogen molecule is at the bottom of a potential energy "well." The atoms are combined in a molecule, sharing two electrons in a covalent bond.

Experimental studies of molecular motion reveal that nuclei vibrate continuously, oscillating about their optimum separation distance like two balls attached to opposite ends of a spring. Figure 6-3 shows this in schematic fashion for a hydrogen molecule vibrating about its optimum separation distance of 74 pm.

Figure 6-2 shows two characteristic features of chemical bonds. The separation distance where the molecule is most stable (74 pm for $H_2$) is known as the **bond length**. As described in Chapter 3, the amount of stability at this separation distance is the bond energy, or strength of the bond. Bond lengths and strengths are important properties of bonds that describe the characteristics of chemical bonding. The bond strength of the hydrogen molecule is 435 kJ/mol. Dividing this value by $N_A$ ($6.022 \times 10^{23}$ $mol^{-1}$) and converting to joules gives a bond energy of $7.22 \times 10^{-19}$ J/molecule.

Every bond has a characteristic length and strength. Example 6-1 compares molecular hydrogen with molecular fluorine.

**Figure 6-3**

Molecules vibrate continually about their bond length, like two balls attached to a spring.

---

## EXAMPLE 6-1 BOND LENGTH AND ENERGY

The bond length of molecular fluorine is 142 pm, and the bond energy is 155 kJ/mol. Draw a figure similar to Figure 6-2 that includes both $F_2$ and $H_2$. Write a caption for the figure that summarizes the comparison of these two diatomic molecules.

**Strategy:** We are asked to draw a graph that compares the bond properties of $F_2$ and $H_2$. The bond energy is the lowest point on the interaction energy curve, representing the energy minimum for the pair of atoms. Energy in joules is plotted on the $y$-axis. The bond distance represents optimal separation of the nuclei, plotted along the $x$-axis in units of picometre (pm). Scale the axes so that both sets of data can be seen clearly. The caption should be a succinct statement that summarizes the main point of the figure.

**Solution:** Begin by assembling the data.

$$\text{Bond length } (x\text{–axis): } H_2 = 74 \text{ pm, } F_2 = 142 \text{ pm}$$

$$\text{Interaction energy } (y\text{–axis): } H_2 = 7.22 \times 10^{-19} \text{ J/molecule}$$

$$F_2 = \frac{(155 \text{ kJ/mol})(10^3 \text{ J/kJ})}{6.022 \times 10^{23} \text{ molecules/mol}} = 2.57 \times 10^{-19} \text{ J/molecule}$$

The curves for the two molecules should have the same general appearance, with zero interaction energy at long distance, and with interaction energy rising very sharply at distances shorter than the bond length. The caption should summarize the content of the figure: interaction energy plots show that the bond in $F_2$ (red line) is both longer and weaker than the bond in $H_2$ (blue line).

**Do the Results Make Sense?** Fluorine is a second-row atom, so its atoms are much bigger than first-row hydrogen atoms. Thus, it makes sense that F atoms cannot get as close to one another as can H atoms.

### Practice Exercise 6.1

Make an interaction energy plot similar to Figure 6-2 for H—F (92 pm, 565 kJ/mol) and H—Br (141 pm, 365 kJ/mol). Write a brief caption for the figure that summarizes the data.

## Other Diatomic Molecules: $F_2$

Bond formation in $H_2$ is relatively easy to describe, because we need to place just two electrons. Bond formation in other molecules, however, requires us to consider the fates of many electrons. This is a complex task, but one simplified way to describe many bonds requires the sharing of only two electrons, as we illustrate using $F_2$.

As two fluorine atoms approach each other, the electrons of each atom feel the attraction of the nucleus of the other atom, just as happens in $H_2$. Remember from Chapter 5 that atomic electrons are described by an electron configuration, which assigns each electron to an atomic orbital. The electron configuration of the F atom is $1s^2\,2s^2\,2p^5$. Orbitals have different sizes and shapes, so when one fluorine atom approaches another, the electrons in each orbital experience different forces of electrical attraction. The largest of these orbitals is the $2p$ set, so $2p$ electrons are closest to the neighbouring nucleus and experience the strongest attraction. We can arrange the $2p$ orbitals so that one orbital of each atom points directly at the approaching atom, as shown in Figure 6-4. An electron in this orbital gets closer to the opposite nucleus than any of the other electrons do. Consequently, we can describe the F—F bond as the sharing of two electrons from the fluorine $2p$ orbitals that point directly at the approaching atom.

**Figure 6-4**

Strong $2p$–$2p$ overlap

The chemical bond in $F_2$ forms from strong electrical attraction of the electron in the fluorine $2p$ orbital that points directly at the approaching atom.

## Unequal Electron Sharing

Each nucleus of a hydrogen molecule has a charge of +1. Consequently, both nuclei attract electrons equally. The result is a symmetrical distribution of the electron density between the atoms. Each nucleus of a fluorine molecule has a charge of +9, and again the electrons experience the same net attraction toward both nuclei. In a chemical bond between identical atoms, the two nuclei share the bonding electrons equally.

In contrast to the symmetrical forces in $H_2$ and $F_2$, the bonding electrons in HF experience *unsymmetrical* attractive forces. The bonding electrons are attracted in one direction by the +1 charge of the hydrogen nucleus and in the other direction by the +9 charge of the fluorine atom. Recall from Chapter 5 that the electrons close to the nucleus effectively screen the nuclear charge of a multielectron atom. This means that the electron in an approaching H atom will not experience a charge of +9. Still, the effective nuclear charge of a fluorine atom is significantly greater than the +1 nuclear charge of a hydrogen atom, so the electrons shared between H and F feel stronger attraction to the F atom than to the H atom.

Unequal attractive forces lead to an unsymmetrical distribution of the bonding electrons. The HF molecule is most stable when its electron density is concentrated closer to the fluorine atom and away from the hydrogen atom. This unequal distribution of electron density gives the fluorine end of the molecule a partial negative charge and the hydrogen end a partial positive charge. These partial charges are less than one charge unit and are equal in magnitude. To indicate such partial charges, we use the symbols $\delta+$ and $\delta-$ or an arrow pointing from the positive end toward the negative end, as shown for HF in Figure 6-5. This unequal electron sharing results in a **polar covalent bond**. We describe details of bond polarity and its consequences in Section 6.5.

## Electronegativity and Polar Covalent Bonds

Any covalent bond between atoms of different elements is polar to some extent, because each element has a different effective nuclear charge. Each element has a characteristic ability to attract bonding electrons, which is called **electronegativity**, and is symbolized by the Greek letter chi ($\boldsymbol{\chi}$). When two elements have different electronegativity values, a bond between their atoms is polar, and the greater the difference ($\Delta\boldsymbol{\chi}$), the more polar the bond.

Electronegativity is related to but distinct from ionization energy and electron affinity. As described in Chapter 5, ionization energy measures how strongly an atom attracts one of its own electrons. Electron affinity specifies how strongly an atom attracts a free electron. Figure 6-6 provides a visual summary of these three different types of attraction.

Electronegativities, which have no units, are estimated by using combinations of atomic and molecular properties. The American chemist Linus Pauling developed one commonly used set of electronegativities, which are presented in Figure 6-7. Modern X-ray techniques can measure the electron density distributions of chemical bonds. The distributions obtained in this way agree with those predicted from estimated electronegativities.

**Figure 6-5**

Unequal sharing of electron density in HF results in a polar covalent bond. The colour gradient represents the variation in electron density shared between the atoms.

**Figure 6-6**

Electronegativity

Ionization energy

Electron affinity

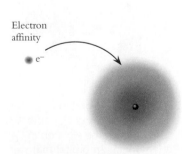

Electronegativity, ionization energy, and electron affinity are distinct properties.

**Figure 6-7**

Electronegativities of the elements vary periodically, increasing from left to right and decreasing from top to bottom of the periodic table. Values for Group 18 have not been determined.

Figure 6-7 shows that electronegativity is a periodic property. Electronegativities increase from the lower left to the upper right of the periodic table. Cesium ($\chi = 0.70$) has the lowest value, and fluorine ($\chi = 4.0$) has the highest value. Notice also that electronegativities decrease down most columns and increase from left to right across the $s$ and $p$ blocks. As with ionization energies and electron affinities, variations in nuclear charge and principal quantum number explain electronegativity trends. Metals generally have low electronegativities ($\chi = 0.7$ to $2.4$) and non-metals have high electronegativities, ranging from $\chi = 2.1$ to $4.0$. As we would expect from their intermediate character between metals and non-metals, metalloids have electronegativities that are larger than the values for most metals and smaller than those for most non-metals.

Linus Pauling (1901–1994) won the 1954 Nobel Prize in chemistry for his ideas about chemical bonds. Pauling was a leader in the movement to limit nuclear weapons, for which he was awarded the Nobel Peace Prize in 1963. He was also an advocate for the benefits of large daily doses of Vitamin C.

Electronegativity differences ($\Delta X$) between bonded atoms provide a measure of where any particular bond lies on the continuum of bond polarities. Three fluorine-containing substances, $F_2$, HF, and CsF, represent the range of variation. At one end of the continuum, the bonding electrons in $F_2$ are shared equally between the two fluorine atoms ($\Delta X = 4.0 - 4.0 = 0$), resulting in a covalent bond. At the other limit, CsF ($\Delta X = 4.0 - 0.7 = 3.3$) is an ionic compound in which electrons have been fully transferred from Cs to F to give $Cs^+$ cations and $F^-$ anions. Most bonds, including the bond in HF ($\Delta X = 4.0 - 2.1 = 1.9$), fall between these extremes. These are polar covalent bonds, in which two atoms share electrons unequally but do not fully transfer the electrons. Example 6-2 illustrates the periodicity of electronegativity and bond polarity variations.

## EXAMPLE 6-2 ELECTRONEGATIVITIES

Use the periodic table, without looking up electronegativity values, to rank each set of three bonds from least polar to most polar: (a) S—Cl, Te—Cl, Se—Cl; and (b) C—S, C—O, C—F.

**Strategy:** The larger the difference in electronegativity, the more polar the bond. Therefore, we can use periodic trends in electronegativities to arrange these bonds in order of polarity. Electronegativities decrease down most columns and increase from left to right across the s and p blocks. Use the periodic table to compare electronegativity values and rank the bond polarities.

**Solution:** (a) All these bonds contain chlorine, so the trend in bond polarity matches the trend in electronegativity of the bonding partners, S, Te, and Se. These elements are from Group 16 of the periodic table, so their electronegativities decrease going down the column: S > Se > Te. The trends also reveal that chlorine is more electronegative than its three partners are, so $\Delta X$ and bond polarity increases as $X$ of the bonding partner decreases:

$$\underset{\text{Least polar}}{\text{S—Cl}} < \text{Se—Cl} < \underset{\text{Most polar}}{\text{Te—Cl}}$$

(b) These three bonds contain carbon, so the trend in bond polarities depends on the trend in electronegativities of S, O, and F. Sulphur is below oxygen in Group 16 of the periodic table, so sulphur is less electronegative than oxygen. Fluorine is to the right of oxygen in the same row, so fluorine is more electronegative than oxygen, because electronegativity increases across a row. All three have larger electronegativities than carbon, so the electronegativity difference and bond polarity increases as electronegativity increases:

$$\underset{\text{Least polar}}{\text{C—S}} < \text{C—O} < \underset{\text{Most polar}}{\text{C—F}}$$

 **Do the Results Make Sense?** You can use actual electronegativity values to verify that these rankings are correct.

### Practice Exercise 6.2

List in order of increasing polarity all the possible bonds formed from the following elements: H, O, F. Consult Figure 6-7 for electronegativity values.

## 6.2 LEWIS STRUCTURES

In this section, we develop a process for making schematic drawings of molecules called **Lewis structures**. A Lewis structure shows how the atoms in a molecule are bonded together and reveals the distribution of bonding and non-bonding valence electrons in a molecule. In a sense, a Lewis structure is a "molecular blueprint" that shows how a molecule is laid out. From this perspective, writing a Lewis structure is the first step in developing a bonding description of a molecule. Lewis structures are named for their inventor, G. N. Lewis, an American chemist who was a professor of chemistry at the University of California, Berkeley, during the first half of the twentieth century. Drawing a correct Lewis structure is the first step in determining the shape of a covalent compound.

### The Conventions

Lewis structures are drawn according to the following conventions, which we illustrate in Figure 6-8 for the HF molecule:

1. *Each atom is represented by its elemental symbol.* In this respect, a Lewis structure is an extension of the chemical formula.

2. *Only the valence electrons appear in a Lewis structure.* Recall from Chapter 5 that valence electrons are accessible for bonding. Core electrons are too close to the nucleus and too tightly bound to be shared with other atoms.

3. *A line joining two elemental symbols represents one pair of electrons shared between two atoms.* Two atoms may share up to three pairs of electrons: in **single bonds** (two shared electrons, one line), **double bonds** (four shared electrons, two lines), or **triple bonds** (six shared electrons, three lines).

4. *Dots placed next to an elemental symbol represent non-bonding electrons on that atom.* Non-bonding electrons usually occur in pairs with opposing spins.

These conventions divide molecular electrons into three groups. Core electrons are purely atomic in nature and do not appear in Lewis structures. Bonding valence electrons are shared between atoms and appear as lines. Non-bonding valence electrons are localized on atoms and appear as dots.

**Figure 6-8**

| 1 | Elemental symbols represent atoms | H | F |
| 2 | Only valence electrons appear | | |
| 3 | Lines represent bonds | H—F | |
| 4 | Dots represent nonbonding electrons | H—F̤: | |

The Lewis structure conventions for hydrogen fluoride.

### The Bonding Framework

A central feature of a Lewis structure is the bonding framework, which shows all the atoms connected as they are in the actual molecule. There is no foolproof method

**Figure 6-9**

Outer atoms      Inner atoms

Outer atoms bond to only one other atom. Inner atoms bond to two or more other atoms.

for putting the atoms together in the correct arrangement, but the following guidelines frequently lead from the chemical formula to the correct arrangement of atoms.

1. An *outer atom* bonds to only one other atom. An *inner atom* bonds to more than one other atom. Figure 6-9 shows a ball-and-stick model of glycine, an amino acid, using coloured highlights to show the difference between inner atoms and outer atoms.

2. Hydrogen atoms are always outer atoms in the compounds discussed in this text. Notice that all five of glycine's hydrogen atoms are outer atoms. (Hydrogen does adopt an inner position in some molecules, but these examples are relatively uncommon and are not described in this textbook.)

3. In inorganic compounds, outer atoms other than hydrogen usually are the ones with the highest electronegativities. Here are some examples. Sulphur dioxide ($SO_2$) has two outer oxygen atoms ($\chi = 3.5$) bonded to an inner sulphur atom ($\chi = 2.5$). Silicon tetrachloride ($SiCl_4$) contains an inner silicon atom ($\chi = 1.8$) bonded to four outer chlorine atoms ($\chi = 3.0$). Chlorine trifluoride ($ClF_3$) has an inner chlorine atom that bonds to three outer fluorine atoms ($\chi = 4.0$):

These three frameworks and the framework for glycine in Figure 6-9 illustrate an important point about Lewis structures. Although Lewis structures show how atoms are connected to one another, a Lewis structure is not intended to show the actual shape of a molecule. Silicon tetrachloride is not flat and square, $SO_2$ is not linear, and the fluorine atoms in $ClF_3$ are not all equivalent. We describe how to use Lewis structures to determine the shapes of molecules later in this chapter.

4. The order in which atoms appear in the formula often indicates the bonding pattern. For example, in HCN the framework is H—C—N, and the atoms in the $OCN^-$ anion are in the order O—C—N. This guideline is particularly useful for organic substances. As an example, $CH_3NH_2$ has the framework shown in the margin:

Notice in this case that the hydrogen atoms are grouped in the formula with their bonding partners, three hydrogen atoms with carbon and two more with nitrogen.

$H_3PO_4$

The bonding framework of phosphoric acid

5. The hydrogen atoms appear first in the formula of an oxoacid. Nevertheless, in almost all cases, these acidic hydrogen atoms bond to oxygen atoms, not to the

central atom of the oxoacid. In phosphoric acid, $H_3PO_4$, for example, there are three O—H bonds but *no* H—P bond. This exception to Guideline 4 above arises because of the convention of listing acidic hydrogen atoms first in the formula of an oxoacid, as we describe in Chapter 14.

These five guidelines allow us to figure out the frameworks of many compounds from their chemical formulas. When the guidelines do not provide an unambiguous framework, however, we need additional structural information. For example, the chemical formula of benzene is $C_6H_6$. The guidelines tell us that the six H atoms are in outer positions but provide no help in deducing the rest of the connections. In fact, there are several possible frameworks for the chemical formula $C_6H_6$. We must be told that the six C atoms of benzene form a ring, with one H bonded to each C, before we can write the framework for this molecule.

The bonding framework of benzene

## Building Lewis Structures

We build Lewis structures according to the six-step procedure described in the box, Building Lewis Structures. Learning how to apply these steps is best done through examples. We begin with phosphorus trichloride, $PCl_3$.

### Building Lewis Structures

Step 1 Count the valence electrons.

Step 2 Assemble the bonding framework, placing two electrons per bond.

Step 3 Place three non-bonding pairs of electrons on each outer atom, except H.

Step 4 Assign the remaining electrons to inner atoms.

Step 5 Optimize electron configurations of the inner atoms.

Step 6 Identify equivalent or near-equivalent Lewis structures.

**Step 1. *Count the valence electrons.*** A Lewis structure shows *all* valence electrons and *only* valence electrons, so a correct count of valence electrons is essential. Recall from Section 5.2 that the number of valence electrons of an atom can be found from its position in the periodic table. Add the contributions from all atoms to obtain a total count of valence electrons. If the species is an anion, *add* one electron for each negative charge; if the species is a cation, *subtract* one electron for each positive charge.

In $PCl_3$, phosphorus contributes five valence electrons (Group 15, $s^2 p^3$). Each chlorine atom contributes seven electrons (Group 17, $s^2 p^5$), so the three chlorine atoms of phosphorus trichloride contribute 21 electrons, giving a total valence electron count of $5 + 21 = 26$.

**Step 2. *Assemble the bonding framework.*** To draw a Lewis structure correctly, we must know how the atoms are connected. The guidelines help us do this. The Lewis structure does not have to match the actual three-dimensional molecular shape, so we can arrange the framework in any convenient way.

The three chlorine atoms of $PCl_3$, being more electronegative than phosphorus, are outer atoms (Guideline 3). Each Cl atom forms a bond to the inner P atom. To build the bonding framework, draw lines to indicate the bonds between the atoms. In the following structures, we show the most recent modification in red:

$$Cl\!-\!P\!-\!Cl$$
$$|$$
$$Cl$$

It is convenient to show the framework of phosphorus trichloride as a T shape. Later in this chapter, we will see that this is not the actual geometry of the molecule.

**Step 3. *Place three non-bonding pairs of electrons on each outer atom, except H.*** Studies of the electron distributions around outer atoms consistently show that hydrogen is always associated with two electrons (one pair). All other outer atoms always have eight electrons (four pairs). The reason for this regularity is that each atom in a molecule is most stable when its valence shell of electrons is complete. For hydrogen, this requires a single pair of electrons, enough to make full use of the hydrogen $1s$ orbital. Any other atom needs four pairs of electrons, the maximum number that can be accommodated by an $ns\,np$ valence shell. Details of these features can be traced to the properties of atoms (Chapter 5) and are discussed further in Chapter 7.

One bonding electron pair plus three non-bonding pairs, also called **lone pairs**, give each outer atom four electron pairs. Such a set of four pairs of electrons associated with an atom is often called an **octet**. Working from the bonding framework for $PCl_3$, we add three lone pairs to each Cl atom in phosphorus trichloride, completing the octets for all three.

$$:\!\ddot{C}l\!-\!P\!-\!\ddot{C}l\!:$$
$$|$$
$$:\!\ddot{C}l\!:$$

**Step 4. *Assign the remaining valence electrons to inner atoms.*** This step is straightforward when there is only one inner atom. If the molecule has more than one inner atom, place non-bonding pairs around the most electronegative atom until it has an octet of electrons. If there are still unassigned electrons, do the same for the next most electronegative atom. Continue in this manner until all the electrons have been assigned.

Continuing with our example of $PCl_3$, we need to take an inventory of the valence electrons before completing Step 4. The framework contains three bonds that require a total of six electrons. Each of three Cl atoms has three lone pairs, for a total of 18 electrons. Thus, the first three steps of the procedure account for all but two of the 26 valence electrons. Step 4 directs us to place the remaining pair of electrons on the inner phosphorus atom.

$$:\!\ddot{C}l\!-\!\ddot{P}\!-\!\ddot{C}l\!:$$
$$|$$
$$:\!\ddot{C}l\!:$$

We have now accounted for all the valence electrons of $PCl_3$.

After completing Step 4, we have built a *provisional Lewis structure*, which accounts for all the valence electrons but may or may not represent the optimum arrangement of electrons. Steps 5 and 6 optimize a Lewis structure. Before we describe these steps in detail, Examples 6-3 and 6-4 reinforce Steps 1–4.

## EXAMPLE 6-3  LEWIS STRUCTURE OF AN ANION

Determine the provisional Lewis structure of the $BF_4^-$ anion.

**Strategy:** The strategy for a Lewis structure is always the same. Follow the six-step procedure.

**Solution:**

1. Use the periodic table to count valence electrons. Boron, in Group 13, has three valence electrons ($s^2 p^1$). Each fluorine atom contributes seven electrons ($s^2 p^5$). The species is an anion, so we *add* one extra electron to account for the negative charge:

$$1 \text{ B} = (1)(3e^-) = 3 \text{ e}^-$$

$$4 \text{ F} = (4)(7e^-) = 28 \text{ e}^-$$

$$-1 \text{ ion} = 1 \text{ e}^-$$

$$\overline{\text{Total valence e}^- = 32 \text{ e}^-}$$

2. Fluorine, having the highest electronegativity among the elements, is always an outer atom. Four F atoms bond to the inner B atom, accounting for eight valence electrons:

$$
\begin{array}{ll}
\begin{array}{c}
\phantom{F-}F \\
\phantom{F-}| \\
F-B-F \\
\phantom{F-}| \\
\phantom{F-}F
\end{array}
&
\begin{array}{l}
32 \text{ valence e}^- \\
\underline{-\ 8 \text{ framework e}^-} \\
24 \text{ e}^- \text{ left to place}
\end{array}
\end{array}
$$

3. Next, complete the octets of the four fluorine atoms:

$$
\begin{array}{ll}
\begin{array}{c}
\phantom{:}:\!\ddot{\text{F}}\!: \\
\phantom{:\ddot{F}}| \\
:\!\ddot{\text{F}}\!-\text{B}-\ddot{\text{F}}\!: \\
\phantom{:\ddot{F}}| \\
\phantom{:}:\!\ddot{\text{F}}\!:
\end{array}
&
\begin{array}{l}
32 \text{ valence e}^- \\
-\ 8 \text{ framework e}^- \\
\underline{-24 \text{ lone pair e}^-} \\
0 \text{ e}^- \text{ left to place}
\end{array}
\end{array}
$$

4. At the end of Step 3, we have placed all 32 valence electrons, so Step 4 is not needed. The result of Step 3 is the provisional Lewis structure of the $BF_4^-$ ion

 **Does the Result Make Sense?** Check a provisional Lewis structure by determining that all the valence electrons are assigned. Also check that identical atoms—the four F atoms in this example—all have the same electron distribution.

**Common Pitfalls:** Neglecting to add an electron for each negative charge or subtracting one for each positive charge.

**Practice Exercise 6.3**

Determine the provisional Lewis structure of the hydronium ion, $H_3O^+$. Don't forget to take account of the positive charge on this species!

## EXAMPLE 6-4  PARENTHESES AND LEWIS STRUCTURES

Determine the provisional Lewis structure of diethylamine, $(CH_3CH_2)_2NH$.

**Strategy:** Follow the six-step procedure, using the chemical formula to determine the bonding framework.

**Solution:**

1. The molecule has one nitrogen atom ($s^2 p^3$). Interpreting the parentheses in the formula reveals that there are 4 carbon atoms ($s^2 p^2$) and 11 hydrogen atoms ($s^1$). You should be able to verify that the valence electron count is 32 $e^-$.

2. The parentheses in the formula tell us that diethylamine has two $CH_3CH_2$ groups bonded to the nitrogen atom:

$$\underset{\underset{H}{|}}{H}{-}\underset{\underset{H}{|}}{\overset{\overset{H}{|}}{C}}{-}\underset{\underset{H}{|}}{\overset{\overset{H}{|}}{C}}{-}\overset{\overset{H}{|}}{N}{-}\underset{\underset{H}{|}}{\overset{\overset{H}{|}}{C}}{-}\underset{\underset{H}{|}}{\overset{\overset{H}{|}}{C}}{-}H$$

The framework contains 15 bonds, accounting for 30 of the 32 valence electrons

3. All of the outer atoms are hydrogen atoms, so step 3 is not required for diethylamine.

4. Nitrogen is more electronegative than carbon. Place the last two valence electrons on this atom, completing its octet and giving the provisional Lewis structure:

$$\underset{\underset{H}{|}}{H}{-}\underset{\underset{H}{|}}{\overset{\overset{H}{|}}{C}}{-}\underset{\underset{H}{|}}{\overset{\overset{H}{|}}{C}}{-}\overset{\overset{H}{|}}{\underset{\cdot\cdot}{N}}{-}\underset{\underset{H}{|}}{\overset{\overset{H}{|}}{C}}{-}\underset{\underset{H}{|}}{\overset{\overset{H}{|}}{C}}{-}H$$

 **Does the Result Make Sense?** Remember from line structure rules that carbon always has four bonds. Do all the carbon atoms in this structure meet that requirement?

**Practice Exercise 6.4**

Build the provisional Lewis structure of $(CH_3)_3COH$, whose proper name is 2-methyl-2-propanol.

## Optimizing the Structure

The first four steps in our procedure lead to a provisional Lewis structure that contains the correct bonding framework and the correct number of valence electrons. Although the provisional structure is the correct structure in some cases, many other molecules require additional reasoning to reach the optimum Lewis structure. This is because the *distribution* of electrons in the provisional structure may not be the one that makes the molecule most stable. Step 3 of the procedure places electrons preferentially on outer atoms, ensuring that each *outer* atom has its full complement of electrons. However, this step does not always give the optimal configuration for the *inner* atoms. Step 5 of the procedure addresses this need.

**Step 5.** *Optimize electron configurations of the inner atoms.* To optimize the electron distributions about inner atoms, we first check to see if any inner atom lacks an octet. Examine the provisional structures of $PCl_3$, $BF_4^-$, and $(CH_3CH_2)_2NH$. All the inner atoms have octets in these structures, indicating that the provisional structures are the optimum Lewis structures. No optimization is required for these three examples.

As the next examples show, the provisional structure may contain one or more inner atoms with less than octets of valence electrons. These provisional structures must be optimized in order to reach the most stable molecular configuration. To optimize the electron distribution about an inner atom, move electrons from adjacent outer atoms to make double or triple bonds until the octet is complete. Examples 6-5 and 6-6 illustrate this procedure.

## EXAMPLE 6-5 OPTIMIZING A PROVISIONAL STRUCTURE

Aqueous solutions of formaldehyde, $H_2CO$, are used to preserve biological specimens. Determine the Lewis structure of formaldehyde.

**Strategy:** Apply the six-step procedure.

**Solution:**

1–4. You should be able to follow the first four steps of the procedure to obtain the following provisional structure:

5. In the provisional structure, there are only six electrons (three single bonds) associated with the inner carbon atom. The carbon atom is more stable when it is associated with eight electrons. All the valence electrons have been placed, so the only source of additional electrons is the non-bonding electrons on the outer oxygen atom. We move one pair of electrons from the outer oxygen to make a second bond between carbon and oxygen (the three pairs are equivalent, so it doesn't matter which pair we move). In the completed Lewis structure, the double bond signifies four electrons shared between carbon and oxygen:

Sharing four electrons allows both carbon and oxygen to have octets of electrons. The right-hand figure above is the correct Lewis structure for formaldehyde.

 **Does the Result Make Sense?** Carbon has four bonds and oxygen has an octet, as expected.

**Common Pitfalls:** Not moving electrons inward to complete the octet on the inner atom.

**Practice Exercise 6.5**

Determine the Lewis structure of ethylene, $H_2CCH_2$.

## EXAMPLE 6-6 WORKING WITH MULTIPLE INNER ATOMS

Acrylonitrile, $H_2CCHCN$, is used to manufacture polymers for synthetic fibres. Draw the Lewis structure of acrylonitrile.

**Strategy:** Apply the six-step procedure.

**Solution:**

1. You should be able to show that the molecule has 20 valence electrons.

2. To assemble the framework, remember that hydrogen atoms are always outer atoms and that the atoms are usually bonded in the order listed in the formula:

$$
\begin{array}{cc}
\text{H} & \text{H} \\
| & | \\
\text{H—C—C—C—N}
\end{array}
\qquad
\begin{array}{l}
\text{20 valence } e^- \\
\underline{-12 \text{ framework } e^-} \\
8 \ e \text{ left to place}
\end{array}
$$

The framework has six bonds, which use 12 of the 20 valence electrons, leaving 8 left to place.

3. Use six of the remaining electrons to complete the octet of the outer nitrogen atom:

$$
\begin{array}{ll}
& 20 \text{ valence e}^- \\
& -12 \text{ framework e}^- \\
& \underline{-6 \text{ outer e}^-} \\
& 2 \text{ e}^- \text{ left to place}
\end{array}
$$

4. There are two electrons left to place in the molecule, and none of the carbon atoms has an octet. Which carbon atom gets that last pair of electrons? Anticipating Step 5, we see that we will be able to complete the octet on the carbon adjacent to the nitrogen atom by shifting lone pairs. This suggests that we place the electrons on either the left-most or the middle carbon atom:

5. In the provisional structure, two of the carbon atoms lack octets of electrons. Complete the Lewis structure by shifting lone pairs and making a double bond and a triple bond, as shown here:

Provisional structure          Final Lewis structure

**Does the Result Make Sense?** Because the inner atoms are electron-deficient in the provisional structure, we should pay particular attention to their electron count in the final structure. Each of the carbon atoms has an octet composed of four bonds, in accordance with line structure rules.

**Practice Exercise 6.6**
Acrolein, $H_2CCHCHO$, is used to manufacture polymers for plastics. Draw the Lewis structure of acrolein.

## Beyond the Octet

The concept of an octet of electrons is one of the foundations of chemical bonding. In fact, C, N, and O, the three elements that occur most frequently in organic and biological molecules, rarely stray from the pattern of octets. Nevertheless, an octet of electrons does not guarantee that an inner atom is in its most stable configuration. In particular, elements that occupy the third and higher rows of the periodic table and have more than four valence electrons may be most stable with more than an octet of electrons. Atoms of these elements have valence $d$ orbitals, which allow them to accommodate more than eight electrons. In the third row, phosphorus, with five valence electrons, can form as many as five bonds. Sulphur, with six valence electrons, can form six bonds, and chlorine, with seven valence electrons, can form as many as seven bonds.

A molecule that contains an inner atom from the third or higher row of the periodic table may have electrons left over after each inner atom has an octet of electrons. In such cases, we place any remaining electrons on those inner atoms. Example 6-7 describes one such molecule.

## EXAMPLE 6-7 LEWIS STRUCTURE OF ClF₃

Chlorine trifluoride is used to recover uranium from nuclear fuel rods in a high-temperature reaction that produces gaseous uranium hexafluoride:

$$2 \, ClF_3 \, (g) + U \, (s) \longrightarrow UF_6 \, (g) + Cl_2 \, (g)$$

Determine the Lewis structure of $ClF_3$.

**Strategy:** "Determine the Lewis structure" means "Use the six-step procedure."

**Solution:**

1. All four atoms are halogens (Group 17, $s^2 p^5$), so $ClF_3$ has 28 valence electrons.

2. Build the framework. Chlorine, with lower electronegativity than fluorine, is the inner atom. Make a single bond to each of the three fluorine atoms:

$$
\begin{array}{c}
F \\
| \\
F-Cl-F
\end{array}
$$

3. Add three non-bonding pairs to each fluorine outer atom.

$$
\begin{array}{c}
:\ddot{F}: \\
| \\
:\ddot{F}-Cl-\ddot{F}:
\end{array}
$$

4. Four electrons remain to be assigned. Place the last two electron pairs on the inner chlorine atom.

$$
\begin{array}{c}
:\ddot{F}: \\
| \\
:\ddot{F}-\ddot{Cl}-\ddot{F}:
\end{array}
$$

**Does the Result Make Sense?** Step 4 results in 10 electrons around the chlorine atom. However, chlorine, from Row 3 of the periodic table, has *d* orbitals available to participate in bonding, so this is a legitimate Lewis structure. Notice that shifting electrons to make double bonds would not reduce the electron count around chlorine. Instead, making a Cl—F double bond would give fluorine more than an octet, which is not possible because fluorine is a second-row element.

**Common Pitfalls:** Assuming that atoms cannot have more than an octet of electrons.

**Practice Exercise 6.7**

Determine the Lewis structure of sulphur tetrafluoride, $SF_4$.

When do we entertain the possibility of valence electron counts greater that eight? For some molecules, such as $ClF_3$, the first four steps of the procedure lead to an inner atom with an electron count that exceeds an octet. In other cases, the optimization carried out in Step 5 may result in electron counts higher than eight. Optimization leads to an inner atom with more than eight electrons when the provisional Lewis structure leaves an inner atom with a deficiency of electrons. To determine if there is a deficiency of electrons, we make use of the concept of **formal charge (FC)**. The formal charge of an atom is the difference between the number of valence electrons in the free atom and the number of electrons assigned to that atom in the Lewis structure:

Formal charge = (Valence electrons in the free atom) −

      (Valence electrons assigned to that atom in the Lewis structure)     (6-1)

When an inner atom has a positive formal charge, a better Lewis structure may be obtained by shifting electrons to make multiple bonds.

We determine the number of electrons assigned to an atom in the Lewis structure by making two assumptions. First, lone pairs are localized on the atoms to which they are assigned. Second, bonding electrons are shared equally between bonded atoms. In this way, an atom is assigned all of its lone-pair electrons and one-half of its bonding electrons. The other half of the bonding electrons are assigned to the atom's bonding partners. Thus, a quick way to determine electrons assigned to an atom is by counting dots and lines around that atom. Each dot represents a non-bonding electron, and each line represents a bonding pair, one of whose electrons is assigned to the atom.

For the molecule treated in Example 6-7, the formal charge on the inner chlorine atom is calculated as follows:

$$\begin{aligned}
FC_{Cl} &= \text{Valence electrons} - \text{Assigned electrons} \\
&= (7 \text{ valence electrons}) - [\tfrac{1}{2} (6 \text{ bonding electrons}) \\
&\quad + (4 \text{ lone-pair electrons})] \\
&= 0
\end{aligned}$$

Each of the three fluorine atoms also has a formal charge of zero:

$$\begin{aligned}
FC_{F} &= \text{Valence electrons} - \text{Assigned electrons} \\
&= (7 \text{ valence electrons}) - [\tfrac{1}{2} (2 \text{ bonding electrons}) \\
&\quad + (6 \text{ lone-pair electrons})] \\
&= 0
\end{aligned}$$

A useful feature of formal charges is that the sum of the formal charges on all atoms equals the charge of the species. For a neutral molecule, the sum of the formal charges must be zero. For a cation or anion, the sum of the formal charges equals the charge on the ion.

In general, the best Lewis structure is the one that results in minimum formal charges. Thus, the formal charge of each atom is zero in the Lewis structure of $ClF_3$. This leads us to a general strategy for optimizing electron configurations of inner atoms: *if Step 4 of the procedure leaves a positive formal charge on an inner atom beyond the second row, shift electrons to make double or triple bonds to minimize formal charge, even if this gives an inner atom with more than an octet of electrons.* Example 6-8 illustrates the use of formal charge in optimizing the configuration of an inner atom.

## EXAMPLE 6-8 LEWIS STRUCTURE OF SULPHUR DIOXIDE

As described in Chapter 2, sulphur dioxide, a by-product of burning fossil fuels, is the primary contributor to acid rain. Determine the Lewis structure of $SO_2$.

**Strategy:** Use the six-step procedure for drawing Lewis structures.

**Solution:**

1. All three atoms are from Group 16 of the periodic table, so each has six valence electrons, giving 18 for the molecule.

2. The more electronegative oxygen atoms are outer atoms:

$$O-S-O$$

3. The two bonds use 4 valence electrons, leaving 14 to place. Six go on each of the outer oxygen atoms, using 12:

$$:\ddot{O}-S-\ddot{O}:$$

4. Place the last two valence electrons on the sulphur atom:

$$:\ddot{O}-\ddot{S}-\ddot{O}:$$

5. Because sulphur is from Row 3, we determine how to optimize the structure by evaluating formal charge. Sulphur has six valence electrons (Group 16) and four assigned electrons (2 bonds + 2 lone–pair electrons):

$$FC_S = \text{Valence electrons} - \text{Assigned electrons}$$
$$= (6 \text{ valence electrons}) - [\tfrac{1}{2}\,(4 \text{ bonding electrons}) + (2 \text{ lone–pair electrons})]$$
$$= +2$$

It is helpful also to evaluate the formal charges on the O atoms:

$$FC_O = \text{Valence electrons} - \text{Assigned electrons}$$
$$= (6 \text{ valence electrons}) - [\tfrac{1}{2}\,(2 \text{ bonding electrons}) + (6 \text{ lone–pair electrons})]$$
$$= -1$$

All three formal charges can be reduced to zero if we move two electrons from each of the outer oxygen atoms to form two S=O double bonds.

$$FC = +2$$
$$:\ddot{O}-\ddot{S}-\ddot{O}: \implies :O=\ddot{S}=O:$$
$$FC = -1 \quad FC = -1 \quad\quad \text{All FC} = 0$$

**Does the Result Make Sense?** Sulphur has more than an octet, but that is permissible because of the *d* orbitals accessible to electrons in elements with valence orbitals of *n* ≥ 3. The two oxygen atoms have identical electron distributions. We describe additional evidence that this is a reasonable structure in Section 6.5.

**Common Pitfalls:** Not ensuring that the formal charges are as low as possible.

**Practice Exercise 6.8**
Determine the Lewis structure of phosphoric acid, $H_3PO_4$. Note that each of the H atoms are bound to an O atom in this molecule.

To summarize, the provisional Lewis structure reached after Step 4 may not allocate an optimum number of electrons to one or more of the inner atoms. The electron distribution must be optimized when any inner atom does not have at least eight electrons or when an inner atom from beyond the second row has a positive formal charge. In either of these situations, a more stable structure results from transferring non-bonding electrons from outer atoms to inner atoms to create double bonds (four shared electrons) or triple bonds (six shared electrons).

We do not need to calculate the formal charges of an inner atom from Row 2, because these species are limited to octets. Nevertheless, formal charges can indicate the best way to optimize a structure, as Example 6-9 shows.

## EXAMPLE 6-9 LEWIS STRUCTURE OF AN ORGANIC MOLECULE

Acetic acid ($CH_3CO_2H$, a carboxylic acid) is an important industrial chemical and is the sour ingredient in vinegar. Build its Lewis structure.

**Strategy:** Once again, follow the stepwise procedure.

**Solution:**

1. The valence electron count is 24 $e^-$.

2. The way the chemical formula is written helps us construct the correct framework. Three of the hydrogen atoms are connected to one of the carbon atoms, the two oxygen atoms are connected to the other carbon atom, and the fourth hydrogen atom is connected to one of the oxygen atoms:

3. Only one outer atom is not a hydrogen atom. This outer oxygen atom requires three pairs of non-bonding electrons:

4. Four valence electrons must be placed around the three inner atoms. The most electronegative inner atom is oxygen, so place two pairs of non-bonding electrons on that atom, completing its octet. All the valence electrons have now been placed:

(We label the different atoms of each element to help identify them in Step 5.)

5. Among the inner atoms, $C_a$ and $O_b$ have eight electrons each, but $C_b$ has only six. Both oxygen atoms have lone pairs that could be transferred, but $O_a$ has a formal charge of −1, so using one of its pairs of electrons to form a double bond gives $C_b$ its octet and reduces all formal charges in the molecule to zero.

Another way to complete the octet of $C_b$ would be to make a double bond to the inner oxygen, $O_b$. Notice, however, that doing so creates formal charge on both of the oxygen atoms:

This is not a realistic depiction, particularly when compared with a structure with no formal charges.

> 🔗 **Does the Result Make Sense?** Each carbon atom has four bonds, and each oxygen atom has zero formal charge. The —$CO_2H$ group, with one $C{=}O$ double bond and an acidic C—O—H linkage, is characteristic of carboxylic acids.

**Common Pitfalls:** Not remembering that the H atom is usually bound to an O atom in an oxoacid.

**Practice Exercise 6.9**
Build the Lewis structure of carbonic acid, $H_2CO_3$.

## Resonance Structures

In completing Step 5 of the Lewis structure procedure, we often find that there is more than one way to optimize the configurations of inner atoms. This leads to the sixth and final step of the Lewis structure procedure.

**Step 6. *Identify equivalent or near-equivalent Lewis structures.*** The nitrate ion, $NO_3^-$, is a good example for this step. The anion has 24 valence electrons, and nitrogen is the inner atom. Three N—O bonds use 6 electrons, and completing the octets of the three outer oxygen atoms places the remaining 18 electrons, giving the provisional Lewis structure shown in Figure 6-10. The nitrogen atom has only six electrons in the provisional structure, so optimization is required. We need to make one N$=$O double bond to complete nitrogen's octet. As Figure 6-10 shows, a lone pair from any of the three oxygen atoms can be moved to form this double bond.

Which of these options is the "best" Lewis structure? Actually, no single Lewis structure by itself is an accurate representation of $NO_3^-$. Any single structure of the anion shows nitrate with one N$=$O double bond and two N—O single bonds. In Section 6.5, we show that single and double bonds between the same types of atoms have different lengths and different energies. In contrast, experiments show that the three nitrate N—O bonds are identical. To show that the nitrate N—O bonds are all alike, we use a composite of the three equivalent Lewis structures. These are traditionally called **resonance structures**. Resonance structures are connected by double-headed arrows to emphasize that a complete depiction requires all of them.

**Figure 6-10**

To convert from the provisional structure of the nitrate anion to complete the nitrogen octet, any of the three oxygen atoms can supply a pair of electrons.

It is essential to realize that electrons in the nitrate anion *do not* "flip" back and forth among the three bonds, as implied by separate structures. The true character of the anion is a blend of the three, in which all three nitrogen–oxygen bonds are equivalent. The need to show several equivalent structures for such species reflects the fact that Lewis structures are approximate representations. They reveal much about how electrons are distributed in a molecule or ion, but they are imperfect instruments that cannot describe the entire story of chemical bonding. In Chapter 7, we show how to interpret these structures from a more detailed bonding perspective.

Example 6-10 examines another anion that displays resonance structures.

**EXAMPLE 6-10  LEWIS STRUCTURE OF AN OXOANION**

Determine the Lewis structure of dihydrogen phosphate, $H_2PO_4^-$.

**Strategy:** Follow the six-step procedure.

**Solution:**

1. The phosphorus atom contributes five valence electrons, the four oxygen atoms contribute six valence electrons each, and the hydrogen atoms each contribute one electron. The negative charge indicates that there is an extra electron, for a total of 32 valence electrons.

2. Guidelines for the framework indicate that the phosphorus atom is in the middle, bonded to the four oxygen atoms. By recognizing the link between dihydrogen phosphate and phosphoric acid, we know that the hydrogen atoms bond to two of the oxygen atoms.

3 and 4.  Twelve electrons are required for the framework, leaving 20 to complete the octets of the four oxygen atoms. Here is the provisional Lewis structure:

$$\begin{array}{c} :\ddot{O}: \\ | \\ H-\ddot{O}-P-\ddot{O}-H \\ | \\ :\ddot{O}: \end{array}$$

5. Although the inner phosphorus atom has its octet of electrons, a formal charge calculation indicates that we can shift electrons to optimize the provisional structure:

$$FC_P = \text{Valence electrons} - \text{Assigned electrons}$$
$$= (5 \text{ valence electrons}) - [\tfrac{1}{2} (8 \text{ bonding electrons}) + (0 \text{ lone–pair electrons})]$$
$$= +1$$

The +1 formal charge can be eliminated by making one P=O double bond. As in acetic acid, we do not shift electrons from either inner oxygen atom, because that would create a +1 formal charge on that atom. We can shift electrons from either of the outer O atoms, so there are two options for the double bond, and this signals the possibility of resonance structures.

6. The two candidates are equivalent Lewis structures, so the best depiction of $H_2PO_4^-$ shows two resonance structures connected by a double-headed arrow:

$$\begin{array}{c} :O: \\ \| \\ H-\ddot{O}-P-\ddot{O}-H \\ | \\ :\ddot{O}: \\ FC = -1 \end{array} \longleftrightarrow \begin{array}{c} \overset{\cdot\cdot}{\phantom{x}}FC = -1 \\ :\ddot{O}: \\ | \\ H-\ddot{O}-P-\ddot{O}-H \\ \| \\ :O: \end{array}$$

**Does the Result Make Sense?** Notice that the two outer O atoms have identical descriptions and so do the two inner O atoms. Evaluating formal charges on all the atoms gives a net −1 charge, matching the overall charge of the anion.

**Common Pitfalls:** Not remembering that the sum of the formal charges must equal the total charge on the species.

**Practice Exercise 6.10**

Draw the Lewis structure of the acetate anion, $CH_3CO_2^-$. Example 6-9 will give you an idea of where to start.

In the examples presented so far, all the resonance structures are equivalent, but resonance structures are not always equivalent. Resonance structures that are not equivalent occur when Step 5 requires shifting electrons from atoms of different elements. In such cases, different possible structures may have different formal charge distributions, and the optimal set of resonance structures includes those forms with the least amount of formal charge. Example 6-11 treats a molecule that has near-equivalent resonance structures.

## EXAMPLE 6-11 LEWIS STRUCTURE OF $N_2O$

Determine the Lewis structure of dinitrogen oxide (NNO), a gas used as an anaesthetic, a foaming agent, and a propellant for whipped cream.

**Strategy:** Once again, follow the six-step procedure. We condense the steps.

**Solution:**

1–4. By now the first four steps should be familiar to you. Here is the provisional Lewis structure:

$$:\ddot{N}-N-\ddot{O}:$$

5 and 6. In the provisional structure, the inner nitrogen atom has only four electrons, so we complete its octet by transferring two pairs of electrons from the outer atoms. We can choose any two pairs, which leads to three possible structures:

Evaluate the three potential resonance structures of dinitrogen oxide using their formal charges. The third structure shows more accumulation of formal charge than the first two. Thus, the optimal Lewis structure of the NNO molecule is a composite of the first two structures, but not the third:

$$:N{=}N{=}\ddot{O}: \quad \longleftrightarrow \quad :N{\equiv}N-\ddot{O}:$$

**Does the Result Make Sense?** Experimental studies support our conclusion about the resonance structures of NNO. As we describe in Section 6.5, double bonds are shorter than single bonds. The length of the nitrogen–oxygen bond in NNO is less than typical N—O single bonds but greater than typical N=O double bonds.

**Practice Exercise 6.11**

Draw the Lewis structure of ozone, $O_3$, in which the three oxygen atoms are linked in a row.

The Lewis structure of a molecule shows how its valence electrons are distributed. These structures present simple, yet information-filled views of the bonding in chemical species. In the remaining sections of this chapter, we build on Lewis structures to predict the shapes and some of the properties of molecules. In Chapter 7, we use Lewis structures as the starting point to develop orbital overlap models of chemical bonding.

## SECTION EXERCISES

●**6.2.1** Determine Lewis structures for the following molecules: (a) $Br_2$; (b) $NCl_3$; and (c) $BrF_5$.

●**6.2.2** Determine Lewis structures for the following molecules: (a) $H_2NCH_2CH_2NH_2$; (b) HCCH (acetylene, used in welding torches); and (c) $(H_3C)_2CO$ (acetone, an organic solvent).

●**6.2.3** Determine Lewis structures for the following ions: (a) $OH^-$; (b) $CO_3{}^{2-}$; and (c) $ClO_3{}^-$. Include any possible resonant structures.

## 6.3 MOLECULAR SHAPES: TETRAHEDRAL SYSTEMS

The Lewis structure of a molecule shows how the valence electrons are distributed among the atoms. This gives a useful qualitative picture, but a more thorough understanding of chemistry requires more detailed descriptions of molecular bonding and molecular shapes. In particular, the *three-dimensional* structure of a molecule, which plays an essential role in determining chemical reactivity, is not shown directly by a Lewis structure.

### The Shape of Methane

The Lewis structure of methane ($CH_4$) shows that the molecule contains four C—H single covalent bonds. How are these bonds arranged in three-dimensional space? Based on its chemical behaviour, chemists concluded more than a century ago that the methane molecule has a highly symmetrical shape. Modern experiments show that the carbon atom is at the centre of the molecule, with the four hydrogen atoms located at the four corners of a regular **tetrahedron**. Figure 6-11 shows four representations of the tetrahedral methane molecule. Each H—C—H set makes a bond angle of 109.5° (Figure 6-11b), and all C—H bond lengths are 109 pm. A regular tetrahedron has the shape of a pyramid with four identical faces and four identical corners (Figure 6-11c). Each face is an equilateral triangle.

Chemists use a variation on the ball-and-stick model to depict more clearly the three-dimensional character of molecules, as shown for methane in Figure 6-11d. The central carbon atom is placed in the plane of the paper. In these models, solid lines represent bonds lying in the plane of the paper, solid wedges represent bonds that protrude outward from the plane of the paper, and dashed wedges represent bonds extending backward, behind the plane.

Methane is the simplest molecule with a tetrahedral shape, but many molecules contain atoms with tetrahedral geometry. Because tetrahedral geometry is so prevalent in chemistry, it is important to be able to visualize the shape of a tetrahedron.

$$H—\overset{\displaystyle H}{\underset{\displaystyle H}{C}}—H$$

The Lewis structure of methane

**Figure 6-11**

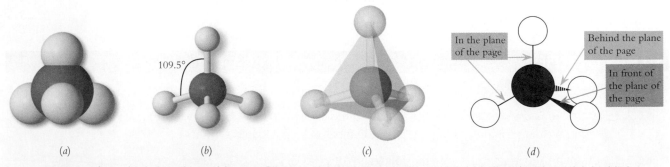

(a)          (b)          (c)          (d)

Representations of tetrahedral methane: (a) space-filling model; (b) ball-and-stick model; (c) ball-and-stick model, highlighting the tetrahedral faces; (d) ball-and-stick drawing using wedge representations for the out-of-plane bonds.

## Why a Tetrahedron?

The most stable shape for any molecule maximizes electron–nuclear attractive interactions while minimizing nuclear–nuclear and electron–electron repulsions. The distribution of electron density in each chemical bond is the result of attractions between the electrons and the nuclei. The distribution of chemical bonds relative to one another, on the other hand, is dictated by electrical repulsion between electrons in different bonds. The spatial arrangement of bonds must minimize electron–electron repulsion. This is accomplished by keeping chemical bonds as far apart as possible. The principle of minimizing electron–electron repulsion is called **valence shell electron pair repulsion**, usually abbreviated **VSEPR**.

Methane has four pairs of valence electrons, each shared in a chemical bond between the carbon atom and one of the hydrogen atoms. The electron density in each C—H bond is concentrated between the two nuclei. At the same time, methane's four pairs of bonding electrons repel one another. Electron–electron repulsion in methane is minimized by keeping the four C—H bonds as far apart as possible.

Consider building methane by sequential addition of H atoms to a C atom, as shown schematically in Figure 6-12. The first hydrogen atom can approach from any direction. To stay as far away from the first as possible, the second atom approaches from the opposite side of the carbon atom, generating a linear array. The third hydrogen atom approaches the $CH_2$ fragment from one side and repels the two existing C—H bonds to form a triangular shape, also called **trigonal planar**. The fourth atom approaches from above or below the plane of the existing bonds and repels the three existing C—H bonds. This converts the triangular shape into the tetrahedral geometry of methane.

VSEPR theory was first proposed in 1963 by Dr. Ronald Gillespie, a Canadian Chemist working at McMaster University.

**Figure 6-12**

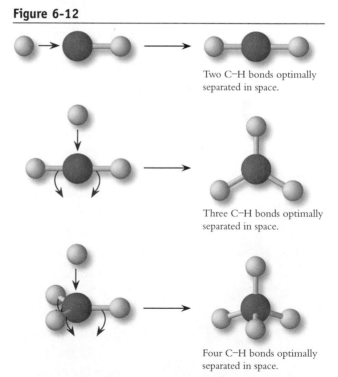

Two C–H bonds optimally separated in space.

Three C–H bonds optimally separated in space.

Four C–H bonds optimally separated in space.

The tetrahedron can be visualized as built by the sequential addition of hydrogen atoms to carbon, always keeping the C—H bonds as far apart as possible.

## Carbon and the Tetrahedron

Methane is the smallest member of a huge class of compounds called **hydrocarbons**, whose molecules contain only carbon and hydrogen. Hydrocarbons in which each carbon atom forms bonds to four other atoms are called **alkanes**. The three smallest alkanes are methane ($CH_4$), ethane ($C_2H_6$), and propane ($C_3H_8$). The structures of ethane and propane, shown in Figure 6-13, are unambiguous, because the carbon atoms can be placed only in a row. We can think of ethane as a methane molecule with one hydrogen atom replaced by a $CH_3$ (methyl) group. Similarly, propane is a methane molecule with two hydrogen atoms replaced by methyl groups. The Lewis structures of both these alkanes show each carbon atom surrounded by four pairs of bonding electrons. The

**Figure 6-13**

H—C—C—H
(*a*) Ethane

H—C—C—C—H
(*a*) Propane

Lewis structure and ball-and-stick models of ethane (*a*) and propane (*b*). All the carbon atoms have tetrahedral shapes, because each has four pairs of electrons to separate in three-dimensional space.

**Figure 6-14**

Propane

Replace H with $CH_3$    Replace H with $CH_3$

2-Methylpropane

Butane

The two structural isomers of $C_4H_{10}$ at left have the same chemical formula but different bonding arrangements and different properties.

bonds around each carbon are arranged in a tetrahedron to keep the four bonding pairs as far apart as possible.

Any hydrocarbon with four or more carbon atoms has more than one possible backbone. Figure 6-14 shows the possibilities for hydrocarbons with four carbon atoms. Building on propane, we can replace a hydrogen atom on a terminal carbon with a methyl group to form butane, an alkane with four carbon atoms in a row. Alternatively, we can replace either hydrogen atom on the inner carbon of propane to give a different compound, 2-methylpropane. In this molecule, three carbon atoms are in a row, but the fourth carbon atom is off to one side.

Two or more compounds that have the same molecular formula but different arrangements of atoms are called **structural isomers**. Butane and 2-methylpropane are structural isomers with the formula $C_4H_{10}$.

As the number of carbon atoms in the alkane increases, so does the number of possible structural isomers. Thousands of different alkanes exist, because there are no limits on the length of the carbon chain. Regardless of the number of the chain length, alkanes have tetrahedral geometry around all of their carbon atoms. The structure of decane, $C_{10}H_{22}$, shown in Figure 6-15, illustrates this feature. Notice that the carbon backbone of decane has a zigzag pattern because of the 109.5° bond angles that characterize the tetrahedron.

**Figure 6-15**

Decane and other hydrocarbons consist of chains of tetrahedral carbon groups, emphasized here by highlighting one of the groups.

## The VSEPR Model

Having introduced methane and the tetrahedron, we now begin a systematic coverage of the VSEPR model and molecular shapes. The VSEPR model assumes that electron–electron repulsion determines the arrangement of valence electrons around each inner atom. This is accomplished by positioning electron pairs as far apart as possible. Figure 6-12 shows the optimal arrangements for two electron pairs (linear), three electron pairs (trigonal planar), and four electron pairs (tetrahedral). Using Lewis structures, we can explore these and other shapes in detail.

We begin by introducing a few terms that are useful in describing molecular shapes. An **electron group** is a set of electrons that occupies a particular region around an atom. A group can be two electrons in a single bond, four electrons in a double bond, or six electrons in a triple bond. Non-bonding electrons almost always occur in groups of two electrons (pairs) with opposite spins, but when a species has an odd number of electrons, a single electron may be a "group." We use **ligand** to refer to an atom or a group of atoms bonded to an inner atom. A ligand can be as simple as a hydrogen atom or as complex as dozens of atoms held together as a unit by covalent bonds of their own. The **steric number (SN)** of an inner atom is the sum of the number of ligands plus the number of lone pairs; in other words, the total number of groups associated with that atom.

Figure 6-16 shows the molecular shapes of methane, ammonia, and water, all of which have hydrogen ligands bonded to an inner atom. These molecules have different numbers of ligands, but they all have the same steric number.

The steric number identifies how many groups of electrons must be widely separated in three-dimensional space. In ammonia, for example, the nitrogen atom bonds to three hydrogen atoms, and it has one lone pair of electrons. How are the three hydrogen atoms and the lone pair oriented in space? Just as in methane, the four groups of electrons are positioned as far apart as possible, thus minimizing electron–electron repulsion.

The word *ligand* is derived from the Latin verb *ligare,* which means "to bind."

**Figure 6-16**

Steric number = 4
Electron group geometry = Tetrahedral

| Tetrahedral | Trigonal pyramidal | Bent |
|:---:|:---:|:---:|
| $CH_4$ | $NH_3$ | $H_2O$ |
| Ligands = 4 | Ligands = 3 | Ligands = 2 |
| Lone pairs = 0 | Lone pairs = 1 | Lone pairs = 2 |

Electron group geometries and molecular shapes for a steric number of 4.

The **electron group geometry** of an atom refers to the three-dimensional arrangement of the valence shell electron groups. Since electrons repel one another regardless of whether they are bonding pairs or lone pairs, it is the steric number that identifies the electron group geometry of an atom. The tetrahedral arrangement leads to greatest stability for a steric number of 4.

The inner atoms of methane, ammonia, and water, shown in Figure 6-16, all have tetrahedral electron group geometry. Nevertheless, each molecule has a different shape. The steric number dictates electron group geometry, but the **molecular shape** describes how the *ligands*, not the *electron groups*, are arranged in space. For example, the shape of ammonia is found by ignoring the tetrahedral arm occupied by the lone pair. What remains is a nitrogen atom atop three hydrogen ligands. This shape is called a **trigonal pyramid**, consistent with the arrangement of the four atoms that make up the molecule. Nitrogen is at the apex of the pyramid, and the three hydrogen atoms make up the triangular base of the pyramid. To find the shape of a water molecule, we ignore the two tetrahedral arms that contain lone pairs. This leaves a planar H—O—H atomic system with a **bent shape**.

Our approach to these molecules illustrates the general strategy for determining the electron group geometry and the molecular shape of each inner atom in a molecule. The process has four steps, beginning with the Lewis structure and ending with the molecular shape.

### Determining Molecular Shapes

Step 1 Determine the Lewis structure.
Step 2 Use the Lewis structure to find steric numbers for inner atoms.
Step 3 Determine the electron group geometries from the steric numbers.
Step 4 Use the ligand count to derive molecular shapes from electron group geometries.

Examples 6-12 and 6-13 illustrate this strategy.

### EXAMPLE 6-12  SHAPE OF THE HYDRONIUM ION

Describe the shape of the hydronium ion ($H_3O^+$). Make a sketch of the ion that shows the three-dimensional shape, including any lone pairs that may be present.

**Strategy:** Follow the four-step process described in the flowchart. Begin with the Lewis structure. Use this structure to determine the steric number, which indicates the electron group geometry. Then take into account any lone pairs to deduce the molecular shape.

**Solution:**

1. Determine the Lewis structure. A hydronium ion has eight valence electrons. Six are used to make three O—H single bonds, and two are placed as a lone pair on the oxygen atom.

$$\begin{array}{c} \text{H} \\ | \quad + \\ \text{H}-\underset{\cdot\cdot}{\text{O}}-\text{H} \end{array}$$

Eight electrons surround the oxygen atom, a Row 2 element, so this is the correct Lewis structure.

2. From the Lewis structure, we see that oxygen bonds to three hydrogen ligands and has one lone pair. The sum of the lone pairs and the ligands yields a steric number of 4.

3. A steric number of 4 identifies four electron groups that must be separated in three-dimensional space. Four groups are as far apart as possible in tetrahedral electron group geometry.

4. The number of ligands determines the molecular shape. With three ligands, the hydronium ion has trigonal pyramidal geometry, just like ammonia. Sketch the structure using solid and dashed wedges to indicate hydrogen atoms in front of and behind the plane of the page. The lone pair can be at any of the four tetrahedral positions, but the structure is easiest to visualize and to draw if the lone pair occupies the top position of the tetrahedron. A ball-and-stick model helps us to visualize the molecular shape.

 **Does the Result Make Sense?** The hydronium ion has the same number of electrons and atoms as the ammonia molecule, so it is reasonable for these two species to have the same shape.

**Common Pitfalls:** Forgetting the distinction between ligands and electron groups.

**Practice Exercise 6.12**
Describe the shape of chloromethane, $ClCH_3$ a sketch of the molecule that shows its three-dimensional shape.

## EXAMPLE 6-13 SHAPES AROUND INNER ATOMS

Describe the shape of hydroxylamine, $HONH_2$.

**Strategy:** Apply the four-step process to each of the inner atoms.

Solution:

1. There are 14 valence electrons, and the atoms are bonded in the order listed. In the final Lewis structure, oxygen has two lone pairs and nitrogen has one.

$$H-\ddot{O}-\overset{\overset{\displaystyle H}{|}}{N}-H$$

2. The oxygen atom has two ligands; one is a hydrogen atom and the other is the $NH_2$ group. Combined with two lone pairs, this gives oxygen a steric number of 4. The nitrogen atom also has a steric number of 4, with one lone pair and three ligands (H, H, and OH).

3. Both atoms have tetrahedral electron group geometry.

4. Oxygen has two ligands, so the geometry around the oxygen atom is bent. The nitrogen atom has three ligands, giving it trigonal pyramidal geometry. A ball-and-stick model shows the overall shape of the molecule. We show the lone pairs in this model to emphasize the electron group geometries of the two inner atoms.

 **Does the Result Make Sense?** You can visualize hydroxylamine as ammonia (trigonal pyramid) with one H replaced by —OH, or as water (bent) with one H replaced by —$NH_2$.

**Practice Exercise 6.13**
Describe the shape of ethanol, $CH_3CH_2OH$.

## Silicon

Many elements of the periodic table, from titanium and tin to carbon and chlorine, exhibit tetrahedral electron group geometry and tetrahedral molecular shapes. In particular, silicon displays tetrahedral shapes in virtually all of its stable compounds.

Compounds of silicon with oxygen are prevalent in the Earth's crust. About 95% of crustal rock and its various decomposition products (sand, clay, soil) are composed of silicon oxides. In fact, oxygen is the most abundant element in the Earth's crust (45% by mass) and silicon is second (27%). In the Earth's surface layer, four of every five atoms are silicon or oxygen.

The principal oxide of silicon is **silica**. The empirical formula of silica is $SiO_2$, but the compound consists of a continuous network of Si—O bonds rather than individual $SiO_2$ molecules. Figure 6-17 shows part of this network. Each silicon atom is at the centre of a regular tetrahedron, bonded to four oxygen atoms. As in water molecules, each oxygen atom displays a bent shape, with two lone pairs and two O—Si bonds generating a steric number of 4.

More than 20 different forms of silica exist, because the bonds and lone pairs around the oxygen atoms can be arranged in various ways. Each arrangement creates a different structural form for the silica network. Quartz, the most common form of silica, is found in granite, sandstone, and beach sand.

Closely related to silica are the silicate minerals, all of which contain polyatomic anions made of silicon and oxygen. The simplest silicates, called *orthosilicates*, contain $SiO_4^{4-}$ anions. The $SiO_4^{4-}$ anion is tetrahedral, with a central silicon atom bonded to four outer oxygen atoms.

Different minerals contain different metal cations to balance the $-4$ charge on the orthosilicate ion. Examples include calcium silicate ($Ca_2SiO_4$), an important

**Figure 6-17**

Quartz, a common form of silica, is a network of Si—O bonds. Silicon and oxygen both have tetrahedral electron group geometry. All the silicon atoms have tetrahedral shapes and all the oxygen atoms have bent shapes.

ingredient in cement, and zircon ($ZrSiO_4$), which is often sold as "artificial diamond." One of the most prevalent minerals in the Earth's mantle is olivine, $M_2(SiO_4)$, in which $M$ is one or two of the abundant metal cations, $Fe^{2+}$, $Mg^{2+}$, and $Mn^{2+}$.

The orthosilicates have entirely different structures than silica. In contrast to the fully covalent bonding network in silica, orthosilicates are ionic, containing metal cations and discrete $SiO_4^{4-}$ anions. Orthosilicate minerals have $4:1$ ratios of oxygen atoms to silicon atoms, and every oxygen atom is a negatively charged outer atom. In contrast, every oxygen atom in silica is inner, bonded to two silicon atoms. Silica has no ionic units and a $2:1$ ratio of oxygen atoms to silicon atoms. Nevertheless, the silicon atoms in both structures have four Si—O bonds and tetrahedral shapes.

Many other silicate minerals contain inner *and* outer oxygen atoms. In *metasilicates*, for example, two oxygen atoms bonded to each silicon atom are inner and two are outer, with negative formal charges. Metasilicate networks can be linear chains or rings, examples of which appear in Figure 6-18. As the examples show, the ratio of silicon to oxygen in a metasilicate is $3:1$. The models also show that metasilicates consist of large networks of Si—O—Si linkages bearing negatively charged outer O atoms. These charges must be counterbalanced by metal cations. In jade ($NaAlSi_2O_6$), which has a linear chain structure, the metal cations are $Na^+$ and $Al^{3+}$. Beryl ($Be_3Al_2Si_6O_{18}$), the principal commercial source of beryllium metal, has six silicon tetrahedra linked in a ring and contains $Be^{2+}$ and $Al^{3+}$ cations. Pure beryl is colourless, but most beryl crystals contain small amounts of transition metal ions such as $Mn^2$ (red) or $Cr^3$ (green). Emerald gemstones are beryl with 2% of its $Al^{3+}$ ions replaced by $Cr^{3+}$ ions. Figure 6-19 shows some of these colourful silicate minerals.

Lewis structure

Ball-and stick model

Lewis structure and ball-and-stick model of the orthosilicate ion

**Figure 6-18**

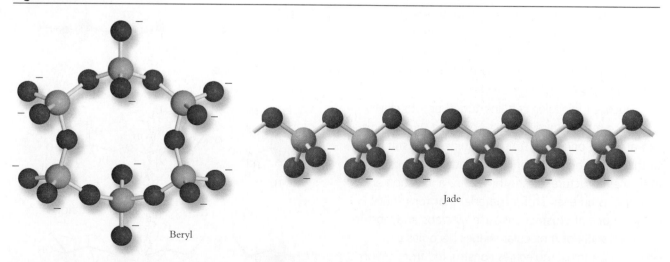

Beryl

Jade

Ball-and-stick representations of two forms of metasilicates, the six-Si ring of beryl (*left*) and a portion of the linear chain of jade (*right*).

**Figure 6-19**

Photos of four silicate minerals: red beryl, emerald, mica, and asbestos. Silicates take many forms, depending on the detailed structure of the Si—O bonding network.

## FOCUS 6-1  CHEMISTRY AND LIFE: THE IMPORTANCE OF SHAPE

The regular tetrahedron is a simple yet elegant geometric form. The ancient Greeks identified it as one of only five regular solids that can be placed inside a sphere so that every vertex touches the surface of the sphere. The Greeks had no idea, however, of the importance that tetrahedral shapes have for the chemical processes of life.

Start with water, which is essential for life as we know it. If the water molecule were linear rather than bent, it would lack the properties that life-forms require. Linear water would not be polar and would be a gas like carbon dioxide. Why is water bent? Its four electron pairs adopt tetrahedral geometry, putting lone pairs at two vertices of a tetrahedron and hydrogen atoms at the other two vertices.

Biological molecules contain carbon, nitrogen, and oxygen atoms in inner positions. These second-row atoms are frequently associated with four electron pairs, giving them a steric number of 4 and tetrahedral electron group geometry. When many tetrahedra are attached to one another, an immense variety of shapes become possible. One relatively simple example is glucose, $C_6H_{12}O_6$. This sugar molecule forms a puckered ring of tetrahedral atoms. Polar O—H groups extend off the ring, and this is why glucose can dissolve in water or blood. Glucose can fuel life processes because its shape makes it easy to transport in and out of biological cells.

Glucose molecules can link together into chains, with each ring tethered to the next by a bridging oxygen atom. In one form, this is cellulose, the stiff material that gives the stalks of plants and the trunks of trees their structural strength. Chitin, a variation on cellulose, is an even stiffer material that forms the exoskeletons of crustaceans such as crabs and lobsters.

The variety of molecular shapes becomes even larger for proteins, molecules constructed from amino acids. An amino acid contains a tetrahedral nitrogen atom bound to a tetrahedral carbon atom, followed by

Cellulose

Chitin

a planar carbon atom and a tetrahedral oxygen atom. Any number of amino acids can be linked together. Moreover, the tetrahedral geometry allows bending and folding that leads to an infinite variety of protein sizes and shapes, including the helix and the pleated sheet:

Helix                    Pleated sheet

From water to sugars to proteins, tetrahedral shapes are at the heart of the molecules of biology. Our sex drives are controlled by the release of testosterone, a steroid molecule with many tetrahedral carbon atoms. Soothing odours and salves such as those made with menthol (oil of mint) are the result of molecules with tetrahedral carbon atoms. The disgusting smell of rotting flesh is due to putrescine, another molecule with atoms forming tetrahedral shapes. As long as we are alive, we are awash in that elegant shape that owes its name to the Greeks.

Putrescine

Testosterone

Menthol

Silicates also exist in which each silicon atom bonds to one outer oxygen and to three inner oxygen atoms. The result is a linked network in which every silicon atom forms three Si—O—Si links, giving a planar, sheet-like structure. The empirical formula of this silicate is $Si_2O_5^{2-}$. In many minerals, aluminum atoms replace some of the silicon atoms to give aluminosilicates. The *micas*—one has the chemical formula $KMg_3(AlSi_3O_{10})(OH)_2$—contain sheets in which every fourth silicon atom is replaced by an aluminum atom. Because of the planar arrangement of its aluminosilicate network, mica is easily broken into flakes. Figure 6-19 includes a photograph of mica.

When asbestos is handled, microscopic fibres become suspended in the atmosphere and are breathed into the lungs. There, they lodge in lung tissue, where they remain for many years, causing irritation that eventually leads to loss of lung function and possibly cancer. Asbestos, which was once used extensively as insulation, is now recognized as a significant health hazard.

Even more complicated chemical formulas result when some silicon atoms have one outer oxygen atom, whereas others in the same mineral have two outer oxygen atoms. The asbestos minerals—one is crocidolite, $Na_2Fe_5(OH)_2(Si_4O_{11})_2$ —have this type of structure. The photograph in Figure 6-19 shows that this silicate contains long fibrous chains. At the molecular level, the chains can be visualized as linked $Si_4O_{11}^{6-}$ units.

In summary, despite the structural diversity exhibited by the silicates, their silicon atoms always have tetrahedral geometry. In addition, every outer oxygen atom contributes a net charge of $-1$ to the structure, whereas every inner oxygen atom is electrically neutral and has an $Si-O-Si$ bond angle close to $109.5°$.

Tetrahedral shapes dominate the structures of biological molecules, as our Focus 6-1: Chemistry and Life describes.

## SECTION EXERCISES

**●6.3.1** Tin tetrachloride, $SnCl_4$, is an important starting material for the preparation of a variety of tin compounds. Build the Lewis structure for $SnCl_4$ and determine its shape.

**●6.3.2** Hydrazine, occasionally used as a rocket fuel, has the chemical formula $N_2H_4$. Determine the Lewis structure of hydrazine and draw a picture that shows the shape of the molecule.

**●6.3.3** Draw a ball-and-stick model of the $Si_2O_5^{2-}$ unit found in mica. Your sketch should have one $Si-O-Si$ linkage and should show that it connects to four other $Si_2O_5^{2-}$ units to form an interlocking sheet.

## 6.4 OTHER MOLECULAR SHAPES

Tetrahedral geometry may be the most common shape in chemistry, but several other shapes also occur frequently. This section applies the VSEPR model to four additional electron group geometries and their associated molecular shapes.

## Steric Number 2: Linear Geometry

A molecule with two ligands and no lone pairs has a steric number of 2, meaning there are two electron groups that must be separated in three-dimensional space. This is accomplished by placing the groups on opposite sides of the inner atom, separated by $180°$. We illustrate this arrangement using a compound of zinc.

Zinc forms both ionic and covalent compounds. One important covalent example is dimethylzinc, $Zn(CH_3)_2$, a substance that has two $Zn-C$ bonds. Used in synthesis reactions since the mid-1800s, dimethylzinc finds uses today in the preparation of catalysts and semiconductors. Zinc is in Group 12 of the periodic table (configuration $[Ar]\, 3d^{10}\, 4s^2$), so it has only two valence electrons.

Each $CH_3$ group contributes seven electrons, giving the molecule 16 valence electrons. The bonding framework uses all 16:

Coordination number = 2
Steric number = 2

Linear molecular shape
Bond angle = 180°

Notice that the zinc atom is associated with only four valence electrons. Although this is less than an octet, the adjacent carbon atoms have no lone pairs available to form multiple bonds. In addition, the formal charge on the zinc atom is zero. Thus, Zn has only four electrons in the optimal Lewis structure of dimethylzinc. This Lewis structure shows two pairs of bonding electrons and no lone pairs on the inner atom, so Zn has a steric number of 2. Two pairs of electrons are kept farthest apart when they are arranged along a line. Thus, the $C-Zn-C$ bond angle is 180°, and **linear geometry** exists around the zinc atom.

A second example of linear geometry comes from the familiar gas, carbon dioxide:

Coordination number = 2
Steric number = 2

Linear molecular shape
Bond angle = 180°

The carbon atom in $CO_2$ has two groups of electrons. Recall from our definition of a group that a double bond counts as one group of four electrons. Although each double bond includes four electrons, all four are concentrated between the nuclei. Remember also that the VSEPR model applies to electron *groups*, not specifically to electron pairs (despite the name of the model). It is the number of ligands and lone pairs, not the number of shared electrons, that determines the steric number and hence the molecular shape of an inner atom.

## Steric Number 3: The Trigonal Plane

Triethylaluminum, $Al(CH_2CH_3)_3$, has long been used in the chemical industry in the production of alkenes—hydrocarbons that have $C=C$ double bonds. In the presence of triethylaluminum, two or more ethylene molecules link together to form straight-chain hydrocarbons that contain an even number of carbon atoms and one double bond. For example, four ethylene molecules form octene:

Ethylene                Octene

The other elements in Group 12, Cd and Hg, also form dimethyl compounds. Dimethylmercury, which can be synthesized by bacteria from industrial wastes, is fatal in very small quantities.

A description of the bonding in triethylaluminum begins with the Lewis structure. The chemical formula, $Al(CH_2CH_3)_3$, indicates that aluminum bonds to three $CH_2CH_3$ fragments. There are 42 valence electrons, all of which are used to complete the bonding framework. Each of the six carbon atoms in triethylaluminum has an octet of electrons and a steric number of 4. Thus, each ethyl group of $Al(CH_2CH_3)_3$ can be described exactly as in ethane (see Figure 6-13), except that one C—H bond is replaced by a C—Al bond. Although aluminum has less than an octet of electrons, the formal charge on the Al atom is zero. Thus, the Al atom has only six electrons in the optimal Lewis structure of triethylaluminum.

With three ligands and no lone pairs, the aluminum atom has a steric number of 3. The three pairs of bonding electrons must be as far apart as possible to minimize electron–electron repulsion. This is accomplished by placing the ligands in a triangular array called a trigonal plane, with bond angles of 120°. Figure 6-20a shows the molecule in a ball-and-stick model, and Figure 6-20b is another view, with the ethyl groups shown schematically to emphasize the trigonal planar geometry around the aluminum atom.

Trigonal planar geometry is common in carbon compounds that have double bonds. For example, ethylene has two trigonal planar carbon atoms. Each carbon atom has three ligands: two hydrogen atoms and one $CH_2$ group.

Each C atom has
Steric number = 3

Each C atom has
Trigonal planar geometry

**Figure 6-20**

(a) Ball-and-stick model of $Al(C_2H_5)_3$        (b) Top view of simplified ball-and-stick model

Ball-and-stick model of triethylaluminum (a) and a stylized model (b) that emphasizes the trigonal planar geometry around the aluminum atom.

## Steric Number 5: The Trigonal Bipyramid

The elements beyond Row 2 of the periodic table can accommodate more than four groups of electrons, and this results in steric numbers greater than 4.

Phosphorus pentachloride exemplifies molecules with a steric number of 5. The Lewis structure of $PCl_5$ shows the molecule with five P—Cl bonds. The five groups of bonding electrons must be arranged to keep the chlorine atoms as far apart as possible. The resulting geometry, shown in Figure 6-21, is called **trigonal bipyramidal** because it can be viewed as two pyramids that share a triangular base. As Figure 6-21*b* shows, a trigonal bipyramid has six triangular faces.

Unlike the geometries for other steric numbers, the five positions in a trigonal bipyramid are not all equivalent, as shown in Figure 6-21*a*. Three positions lie at the corners of an equilateral triangle around the phosphorus atom, separated by 120° bond angles. Atoms in the trigonal plane are in *equatorial positions*. The other two positions lie along an axis above and below the trigonal plane, separated from equatorial positions by 90° bond angles. Atoms in these sites are in *axial positions*.

The difference between equatorial and axial positions determines the arrangement of bonding pairs and lone pairs around an atom with a steric number of 5. An example is provided by sulphur tetrafluoride, a colourless gas that has industrial uses as a potent fluorinating agent. The Lewis structure of $SF_4$ shows four S—F bonds and one lone pair of electrons on the sulphur atom. These five pairs of electrons are distributed in a trigonal bipyramid around the sulphur atom.

Because equatorial and axial positions differ, two molecular geometries are possible for $SF_4$. As Figure 6-22 shows, placing the lone pair in an axial position gives a trigonal pyramid, whereas placing the lone pair in an equatorial position gives a **seesaw** shape.

Experiments show that $SF_4$ has the seesaw geometry, which means that this shape is more stable than the trigonal pyramid. This is explained by the fact that lone pairs are attracted to just one nucleus whereas bonding pairs are attracted to two nuclei. As a result, lone pairs are more spread out in space than are electrons in bonds. Consequently, a structure is more stable when lone pairs are placed as far as possible from other lone pairs and from bonding pairs. The greater stability resulting from placing the lone pair in an equatorial position indicates that less net electron–electron repulsion is associated with this arrangement. Studies

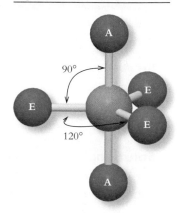

Lewis structure of $PCl_5$

**Figure 6-21**

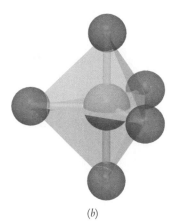

A = Axial position
E = Equatorial position

*(a)*

*(b)*

(a) Ball-and-stick model of $PCl_5$ with axial and equatorial sites labelled. (b) Model showing the equatorial plane and opposing pyramids that make up the trigonal bipyramid.

**Figure 6-22**

Lewis structure          Trigonal pyramid          Seesaw

Sulphur tetrafluoride has two possible molecular geometries. The stable isomer is the seesaw form.

of the geometries about other atoms with a steric number of 5 show that lone pairs *always* occupy equatorial positions.

The trigonal bipyramid ($PCl_5$) and the seesaw ($SF_4$) are two of the four geometries for an atom with a steric number of 5. Example 6-14 introduces the other two: **T-shaped** and linear.

## EXAMPLE 6-14 GEOMETRY OF $ClF_3$

The Lewis structure of chlorine trifluoride is treated in Example 6-7. Determine the molecular geometry and draw a three-dimensional picture of the molecule.

**Strategy:** Use the Lewis structure of $ClF_3$ to determine the steric number of the chlorine atom. Obtain the molecular shape from the orbital geometry after placing lone pairs in appropriate positions.

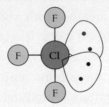

**Solution:** The steric number for chlorine is 5, leading to a trigonal bipyramidal electron group geometry.

When the steric number is 5, there are two distinct positions: equatorial and axial. Placing lone pairs in equatorial positions always leads to the greatest stability. Thus, $ClF_3$ is T-shaped with two equatorial lone pairs.

**Does the Result Make Sense?** Think of making $ClF_3$ from $SF_4$ by replacing one F atom with a lone pair. An axial lone pair would be at 90° to the existing equatorial lone pair, whereas a second equatorial lone pair is at 120° to the existing one. The larger angle means less repulsion between lone pairs.

**Practice Exercise 6.14**
The fourth molecular shape arising from a steric number of 5 is represented by the triiodide anion, $I_3^-$. Determine the molecular geometry and draw a three-dimensional picture of the triiodide ion.

Figure 6-23 summarizes the characteristics of atoms with a steric number of 5.

**Figure 6-23**

Steric number = 5
Electron group geometry = Trigonal bipyramidal

| Trigonal bipyramidal | Seesaw | T-shaped | Linear |
|---|---|---|---|

| $PCl_5$ | $SF_4$ | $ClF_3$ | $I_3^-$ |
|---|---|---|---|
| Ligands = 5 | Ligands = 4 | Ligands = 3 | Ligands = 2 |
| Lone pairs = 0 | Lone pairs = 1 | Lone pairs = 2 | Lone pairs = 3 |

Electron group geometries and molecular shapes for a steric number of 5.

## Steric Number 6: The Octahedron

Sulphur hexafluoride is a colourless, odourless, non-toxic, unreactive gas. It is prepared commercially by reacting sulphur with excess fluorine. Because of its unusual stability, sulphur hexafluoride is used as an insulating gas for high-voltage electrical devices.

The Lewis structure of $SF_6$, shown in Figure 6-24a, indicates that sulphur has six S—F bonds and no lone pairs. The molecular geometry that keeps the six fluorine atoms as far apart as possible is **octahedral** in shape, as Figure 6-24b shows. Figure 6-24c shows that an octahedron has eight triangular faces.

The six positions around an octahedron are equivalent, as Figure 6-25 demonstrates. Replacing one fluorine atom in $SF_6$ with a chlorine atom gives $SF_5Cl$. No matter which fluorine is replaced, the $SF_5Cl$ molecule has four fluorine atoms in a square, with the fifth fluorine and the chlorine atom on opposite sides, at right angles to the plane of the square.

Three common molecular shapes are associated with octahedral electron group geometry. Most often, an inner atom with a steric number of 6 has octahedral molecular shape with no lone pairs. Example 6-15 uses a compound of xenon, whose chemical behaviour is described in the Focus 6-2: Chemical Milestones, to show a second common molecular shape, **square planar**.

**Figure 6-24**

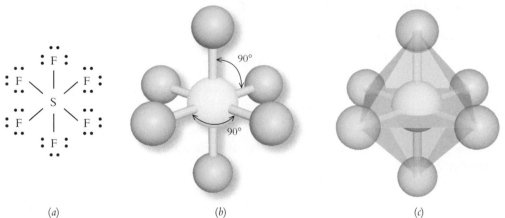

(a)                           (b)                           (c)

Views of sulphur hexafluoride: (a) Lewis structure; (b) ball-and-stick model; (c) ball-and-stick model showing the triangular faces of the octahedron.

**Figure 6-25**

Replacing any of the fluorine atoms of $SF_6$ by a chlorine atom gives a molecule with a square of fluorine ligands capped by one fluorine and one chlorine. This shows that all six positions of the octahedron are equivalent.

## FOCUS 6-2 CHEMICAL MILESTONES: DO NOBLE GASES REACT?

The noble gases are the only elements that exist naturally as individual atoms. Electron configurations make it clear why noble gas atoms prefer to remain as single atoms. Each noble gas has a filled shell configuration: $1s^2$ for He and $s^2 p^6$ for the others. All electrons are paired, and there are no vacant valence orbitals. Moreover, the ionization energies of these elements are extremely high—over 2000 kJ/mol for He and Ne, and over 1000 kJ/mol for the other noble gases.

Nevertheless, as early as 1933, it was suggested that xenon might form stable compounds with the most electronegative elements fluorine and oxygen. Early attempts to react xenon directly with fluorine were unsuccessful, and chemists ignored this possibility for the next 30 years.

In 1962, the English chemist Neil Bartlett, working at the University of British Columbia, overturned the conventional wisdom. Bartlett was exploring the reactions of platinum hexafluoride, an extremely reactive molecule. He found that $PtF_6$ reacted cleanly and rapidly with molecular oxygen:

$$O_2 + PtF_6 \longrightarrow O_2^+ + PtF_6^-$$

Bartlett knew that the ionization energies of $O_2$ and Xe are nearly identical (1180 and 1170 kJ/mol, respectively). He reasoned that if $PtF_6$ reacted with molecular oxygen, it might also react with xenon. Sure enough, mixing Xe gas with $PtF_6$ resulted in an immediate reaction that formed a yellow solid. With this simple experiment, the field of noble gas chemistry was inaugurated. Within a year of Bartlett's first experiment, eight different compounds of xenon had been made and studied. For example, when heated or illuminated, xenon reacts with fluorine to form a mixture of three different xenon fluorides: $XeF_2$, $XeF_4$, and $XeF_6$. The hexafluoride reacts with one mole of water to give an oxofluoride:

$$XeF_6 + H_2O \longrightarrow XeOF_4 + 2\ HF$$

Reaction with excess water produces xenon trioxide:

$$XeF_6 + 3H_2O \longrightarrow XeO_3 + 6\ HF$$

Most xenon compounds involve bonds with the most electronegative elements fluorine and oxygen. More exotic compounds containing Xe—S, Xe—H, and Xe—C bonds can be formed under carefully controlled conditions; for example, in solid matrices at liquid nitrogen temperature. The three Lewis structures below are examples of these compounds in which the xenon atom has a steric number of 5 and trigonal bipyramidal electron group geometry.

xenon difluorodioxide

hydrogen xenon
hydrogen sulfide

bis (2,6-difluorophenyl) xenon

Chlorine, the next most electronegative element, reacts with xenon to form transient species that decompose at room temperature. Krypton forms $KrF_2$ and a few other compounds, but the chemistry of krypton is much more restricted than that of xenon.

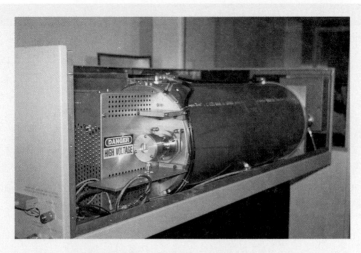

The ability of xenon to react is easily explained. Although it has a closed-shell configuration, [Kr] $5s^2 4d^{10} 5p^6$, xenon's outermost electrons can be promoted relatively easily into $5d$ orbitals, [Kr] $5s^2 4d^{10} 5p^{6-n} 5d^n$. This produces incompletely filled orbital sets that can form stable chemical bonds.

The Canadian Society for Chemistry designated the work of Neil Bartlett on the reactive noble gases an International Historic Chemical Landmark at the University of British Columbia on May 23, 2006.

At first, noble gas chemistry had almost no practical applications. Recently, however, lasers have been developed that are based on the chemical reactions of noble gases.

The gas in an excimer laser is a mixture containing about 2% of one noble gas and 0.2% halogen in a second noble gas, typically neon. The mixtures include $Ar-F_2$, $Kr-F_2$, and $Xe-Cl_2$. In a strong electric discharge, the noble gas atoms become excited and ionized and the halogen molecules fragment into atoms. Excited states of the halogen atoms and noble gas atoms are created. These species collide and form excited diatomic molecules: ArF*, KrF*, XeCl*. These excited molecules emit ultraviolet light and break up into atoms in the process:

$$KrF^* \longrightarrow Kr + F + hv$$

Excimer lasers emit ultraviolet light that can vapourize solid substances like semiconductor surfaces and polymers, making it possible for such a laser to serve as a cutting device. Laser light can be focused to very small areas, so an excimer laser can perform very delicate cutting tasks.

A recent application of the excimer laser is eye surgery to correct for nearsightedness or astigmatism. In this technique, known as photorefractive keratectomy or PRK, a surgeon wields a computer-controlled excimer laser to selectively vapourize a portion of the corneal lens. Thinning the lens in just the right way can improve the ability of the eye to focus light correctly without the need for glasses or contact lenses. This procedure can be performed in just a few minutes with only a local anaesthetic.

The capability to control an excimer laser beam also is exploited in the semiconductor industry, where these lasers are used to etch elaborate features during the fabrication of semiconductor chips. Neil Bartlett probably never dreamed that his explorations of the chemistry of xenon would lead to such exotic applications.

## EXAMPLE 6-15 STRUCTURE AND BONDING OF XeF₄

Describe the geometry and draw a ball-and-stick sketch of xenon tetrafluoride.

**Strategy:** Follow the usual procedure. Determine the Lewis structure, and then use it to find the steric number for xenon and to deduce electron group geometry. Next, use the number of ligands to identify the molecular shape.

**Solution:**

1. The xenon atom contributes eight valence electrons to the molecule ($5s^2 5p^6$). Four fluorine atoms add 28 more for a total of 36 valence electrons. Eight electrons are used to make the four Xe—F bonds, and 24 more fill the valence shells of the fluorine atoms. We are left with four electrons that must be placed as lone pairs on the xenon atom:

$$8 \; e^- \text{ used to make Xe—F bonds}$$
$$24 \; e^- \text{ used to fill the F valence shells}$$
$$\underline{4 \; e^- \text{ used in two Xe lone pairs}}$$
$$36 \text{ total valence electrons}$$

$$:\!\ddot{F}\!:$$
$$|$$
$$:\!\ddot{F}\!-\!\ddot{Xe}\!-\!\ddot{F}\!:$$
$$|$$
$$:\!\ddot{F}\!:$$

2. The steric number is the sum of the ligands (four) and lone pairs (two): 6.

3. The ligands must be located at the corners of an octahedron to minimize electron–electron repulsion between the electron pairs. To give the greatest stability, the two lone pairs must be as far apart as possible, because lone pairs take up more space than bonding pairs. Placing the lone pairs at opposite ends of one axis, 180° apart, minimizes their mutual repulsion. This leaves the four fluorine atoms in a square plane around xenon:

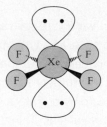

**Does the Result Make Sense?** Where we place the first lone pair makes no difference, inasmuch as all six positions are identical. If we place the second lone pair anywhere except opposite the first, there is a 90° angle between the two lone pairs, generating large lone-pair repulsion.

### Practice Exercise 6.15

The third molecular shape arising from an octahedron is exemplified by chlorine pentafluoride, ClF₅. Describe the geometry and draw a ball-and-stick sketch of chlorine pentafluoride.

The **square pyramidal** geometry of ClF₅ completes our inventory of molecular shapes. Figure 6-26 summarizes the characteristics of atoms with a steric number of 6.

A summary of molecular shapes appears in Table 6-2 at the end of Section 6.5.

**Figure 6-26**

Steric number = 6
Electron group geometry = Octahedral

| Octahedral | Square pyramidal | Square planar |
|---|---|---|
| $SF_6$ | $ClF_5$ | $XeF_4$ |
| Ligands = 6 | Ligands = 5 | Ligands = 4 |
| Lone pairs = 0 | Lone pairs = 1 | Lone pairs = 2 |

Electron group geometries and molecular shapes for a steric number of 6.

## SECTION EXERCISES

●**6.4.1** Describe the geometry and draw a ball-and-stick sketch of gallium triiodide.

●**6.4.2** Describe the geometry and draw a ball-and-stick sketch of the $AsCl_4^-$ anion.

●**6.4.3** Determine the Lewis structure and describe the geometry of $XeOF_4$.

## 6.5 PROPERTIES OF COVALENT BONDS

Having developed ideas about Lewis structures and shapes of molecules, we are now in a position to explore some of the important properties of covalent bonds. These properties provide revealing evidence about molecular shapes.

### Bond Angles

Each of the steric numbers described in Sections 6.3 and 6.4 results in electron groups separated by well-defined **bond angles**. If the VSEPR model is accurate, the actual bond angles found by experimental measurements on real molecules should match the optimal angles predicted by applying the model.

Experimental results agree with the predictions of the model. For instance, measurements show bond angles of 109.5° in $CH_4$, 120° in $Al(C_2H_5)_3$, and 90° in $SF_6$. Moreover, when the steric number of an atom changes, bond angles change exactly as the model predicts. Phosphorus pentachloride provides an example. Experiments show that gaseous $PCl_5$ is a trigonal bipyramid with two sets of bond angles, 120° and 90°. When this compound solidifies, however, it forms ionic crystals that contain $PCl_4^+$ cations and $PCl_6^-$ anions. Molecular structure determination reveals that the bond angles in the cation are 109.5°, whereas those in the anion are 90°. These are exactly what the VSEPR model predicts for tetrahedral and octahedral geometries.

Symmetrical molecules without lone pairs on their inner atoms, such as $CH_4$, $PCl_5$, and $SF_6$, have exactly the bond angles of regular geometric shapes. However, lone pairs alter bond angles by small but important amounts. In ammonia, for example, the experimental H—N—H angles are 107°, and in water the H—O—H bond angle is only 104.5°. These distortions from the ideal 109.5° angle of tetrahedral geometry are caused by greater electron–electron repulsion for lone pairs than for bonding pairs. A pair of electrons in a bonding orbital is attracted to both nuclei involved in the bond, but the lone-pair electrons in a non-bonding orbital are attracted to just one nucleus. As a result, lone-pair electron density is spread more widely over space than the electron density in a bonding orbital. This means that the electron–electron repulsion generated by non-bonding pairs is always greater than that generated by bonding pairs.

Distorting the tetrahedral geometry to increase the angles around the lone pair minimizes the overall electron–electron repulsion in an ammonia molecule. Thus, the three hydrogen atoms move slightly closer together to minimize the repulsion between their bonding pairs and the larger lone pair. Similarly, the angle between the two lone pairs in a water molecule increases by a few degrees, pushing the two hydrogen atoms closer together. The magnitude of distortion depends on the details of electron distribution in the molecule, but the qualitative effect is always the same.

Example 6-16 examines bond angles in another molecule.

## EXAMPLE 6-16  BOND ANGLES

Experiments show that sulphur tetrafluoride has bond angles of 86.8° and 101.5°. Give an interpretation of these bond angles.

**Strategy:** To interpret bond angles, we must construct a model of the molecule by using Lewis structures and steric numbers.

**Solution:** Refer to the previous section, where we describe sulphur tetrafluoride. The sulphur atom has a steric number of 5, with one lone pair in the equatorial plane. The model predicts that $SF_4$ is a seesaw-shaped molecule, with F—S—F bond angles of 90° and 120°. However, the equatorial lone pair extends over more space than a bonding pair. To reduce overall electron–electron repulsion, the two axial fluorine atoms move slightly away from the lone pair and toward the equatorial fluorine atoms to give a bond angle slightly smaller than the expected 90°. Likewise, the two equatorial fluorine atoms move closer together, resulting in a bond angle of 101.5°.

**Does the Result Make Sense?** Notice that the change of the equatorial–equatorial bond angle is much greater than the change in the angle involving the axial fluorine atoms. Starting from an ideal angle of 120°, the equatorial ligands have much more space for angle reduction than do the 90° axial ligands.

### Practice Exercise 6.16

Experiments show that chlorine trifluoride has bond angles of 87.5°. Explain these bond angles, and make a sketch of the molecule.

Three triatomic molecules, water, ozone, and carbon dioxide, provide particularly strong experimental evidence to support the VSEPR model:

The inner O atom in water has a steric number of 4, for which VSEPR predicts tetrahedral geometry and a bond angle of 109.5°. The H—O—H bond angle in water is reduced by 5°, to 104.5°, to minimize repulsion between the lone pairs on the oxygen atom. The inner oxygen atom of ozone has just one lone pair and a steric number of 3, giving a predicted bond angle of 120°. Reducing the bond angle by 3° minimizes electron–electron repulsion between the lone pair and the two bonding pairs. In contrast, the carbon atom in $CO_2$ has no lone pairs and a steric number of 2. In agreement with the observed bond angle of 180°, the model predicts that the two C=O bonds point in opposite directions along a line.

You may be wondering how we know what the bond angles actually are. The answer is that bond angles can be measured using various spectroscopic methods. For instance, the frequencies at which a molecule vibrates (see Figure 3-5 in Chapter 3) depend on the masses of the atoms and the bond lengths, but also on the bond angles. Molecules will typically absorb radiation in the infrared portion of the spectrum, and respond by vibrating at particular frequencies. These frequencies can be used to deduce bond angles.

Bond angles are summarized with other data in Table 6-2 at the end of Section 6.5.

## Dipole Moments

As described in Section 6.1, most chemical bonds are polar, meaning that one end is slightly negative, and the other is slightly positive. Bond polarities, in turn, tend to give a molecule a negative end and a positive end. A molecule with this type of asymmetrical distribution of electron density is said to have a **dipole moment**, symbolized by the Greek letter mu, **$\mu$**.

The dipole moment of a polar molecule can be measured by placing a sample in an electrical field. For example, Figure 6-27 shows a sample of hydrogen fluoride molecules between a pair of metal plates. In the absence of an applied field, the molecules are oriented randomly throughout the volume of the device. When an electrical potential is applied across the plates, the HF molecules align spontaneously according to the principles of electrical force. The positive ends of the molecules point toward the negative plate, and the negative ends point toward the positive plate. The extent of alignment depends on the magnitude of the dipole moment. The SI unit for the dipole moment is the coulomb metre (C m), but experimental values are usually recorded in units of debyes, D ($1 D = 3.34 \times 10^{-30}$ C m). Dipole moments span a range of about 2 debyes. Hydrogen fluoride has $\mu = 1.78$ D. A non-polar molecule like $F_2$ has no dipole moment ($\mu = 0$ D).

**Figure 6-27**

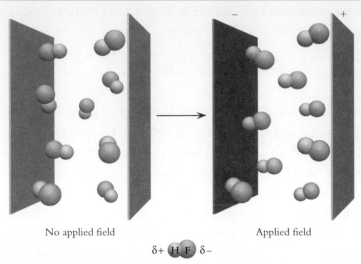

No applied field                    Applied field

δ+ H F δ–

An applied electrical field causes an alignment of polar HF molecules. The extent of alignment depends on the magnitude of the dipole moment.

The large dipole moment of the HF molecule indicates that electron density in the chemical bond of this molecule is strongly concentrated at the fluorine end. This simple molecule demonstrates that formal charge calculations, done to determine if Lewis structures are correct, do not indicate how charge is *actually* distributed in a molecule. The formal charges of the H atom and the F atom in HF are zero, yet HF has a large dipole moment. We determine formal charges by *assuming* that all bonding electrons are shared equally. In reality, most bonds, the H—F bond being just one example, involve unequal electron sharing. Consequently, formal charge calculations are extremely useful for assessing whether a valence electron distribution is reasonable, but they do not reliably predict bond polarity or the distribution of actual charge.

Dipole moments depend on bond polarities. For example, the trend in dipole moments for the hydrogen halides follows the trend in electronegativity differences; the more polar the bond (indicated by $\Delta \chi$), the larger the molecular polarity (indicated by the dipole moment, $\mu$):

**Figure 6-28**

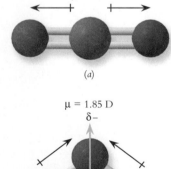

$\mu = 0$ D

(a)

$\mu = 1.85$ D

δ–

δ+

(b)

(a) When identical polar bonds point in opposite directions, their polarity effects cancel, giving zero net dipole moment. (b) When identical polar bonds do not point in opposite directions, there is a net dipole moment, symbolized by $\mu$.

HCl
$\Delta \chi = 1.9$
$\mu = 1.07$ D

HBr
$\Delta \chi = 0.7$
$\mu = 0.79$ D

HI
$\Delta \chi = 0.4$
$\mu = 0.38$ D

Dipole moments also depend on molecular shape. Any diatomic molecule with different atoms has a dipole moment. For more complex molecules, we must evaluate dipole moments using both bond polarity and molecular shape. A molecule with polar bonds has no dipole moment if a symmetrical shape causes polar bonds to cancel one another.

Figure 6-28 illustrates the dramatic effect of shape on the dipole moments of triatomic molecules. Recall from Section 6.1 that an arrow with a crossbar at one end can be used to represent a polar bond. The head of the arrow points to the partial negative end ($\delta-$) of the polar bond, and the "+" end of the arrow points toward the partial positive end ($\delta+$). Figure 6-28a shows that although both bonds in linear $CO_2$ are polar ($\Delta\chi = 1.0$), the two arrows indicating individual bond polarities point in opposite directions along a line. Thus, the effect of one polar bond exactly cancels the effect of the other. In bent $H_2O$, in contrast, Figure 6-28b shows that the effects of the two polar bonds do not cancel. Water has a partial negative charge on its oxygen atom and partial positive charges on its hydrogen atoms. The resulting dipole moment of the molecule is $\mu = 1.85$ D, shown by the coloured arrow in Figure 6-28b.

No fully symmetrical molecule has a non-zero dipole moment. Phosphorus pentachloride, for example, has $\mu = 0$ D. The two axial P—Cl bonds point in opposite directions, and although three P—Cl bonds arranged in a trigonal plane have no counterparts pointing in opposite directions, trigonometric analysis shows that the polar effects of three identical bonds in a trigonal plane cancel exactly. Likewise, the bonds in a tetrahedron are arranged so that their polarities cancel exactly, so neither $CH_4$ nor $CCl_4$ has a dipole moment.

The perfect symmetry of these geometric forms is disrupted when a lone pair replaces a bond, giving a molecule with a dipole moment. Examples include $SF_4$ (seesaw), $ClF_3$ (T shape), $NH_3$ (trigonal pyramid), and $H_2O$ (bent), all of which have dipole moments. Replacing one or more bonds with a bond to a different kind of atom also introduces a dipole moment. Thus, chloroform ($CHCl_3$) has a dipole moment, but $CCl_4$ does not. The carbon atom of chloroform has four bonds in a near-regular tetrahedron, but the four bonds are not identical. The C—Cl bonds are more polar than the C—H bond, so the polarities of the four bonds do not cancel. Figure 6-29 shows that chloroform has $\mu = 1.04$ D.

In a symmetrical octahedral system such as $SF_6$, each polar S—F bond has a counterpart pointing in the opposite direction. The bond polarities cancel in pairs, leaving this molecule without a dipole moment. Example 6-17 examines molecular variations on octahedral geometry.

Bond angles are summarized with other data in Table 6-2 at the end of Section 6.5.

**Figure 6-29**

CCl₄
Nonpolar

$\mu = 0$

CHCl₃
Polar

$\mu = 1.04$ D

Carbon tetrachloride is a symmetrical tetrahedral molecule, so the individual bond polarities cancel. Chloroform is also a tetrahedral molecule, but the four bonds are not identical, so the bond polarities do not cancel.

---

### EXAMPLE 6-17 PREDICTING DIPOLE MOMENTS

Does either $ClF_5$ or $XeF_4$ have a dipole moment?

**Strategy:** Molecules have dipole moments unless their symmetries are sufficient to cancel their bond polarities. Therefore, we must examine the structure and layout of bonds in each molecule.

**Solution:** Chlorine pentafluoride and xenon tetrafluoride appear in Figure 6-26. Each has an inner atom with a steric number of 6, but their electron group arrangements include lone pairs. As a result, $ClF_5$ has a square pyramidal shape, whereas $XeF_4$ has a square planar shape. Pictures can help us determine whether or not the bond polarities cancel:

XeF$_4$
nonpolar

ClF$_5$
polar

Each molecule has four fluorine atoms at the corners of a square. The Xe—F bond polarities cancel in pairs, leaving $XeF_4$ with no dipole moment. Four bond polarities also cancel in $ClF_5$, but the fifth Cl—F bond has no counterpart in the opposing direction, so $ClF_5$ has a dipole moment that points along the axis containing the lone pair and the fifth Cl—F bond.

**Do the Results Make Sense?** $XeF_4$, like $I_3^-$, is a species that has lone pairs but zero dipole moment. Symmetry explains this. We can see why by examining how the lone pairs are placed. In $I_3^-$, the three lone pairs form a symmetrical trigonal plane. In $XeF_4$, the two lone pairs oppose each other.

**Practice Exercise 6.17**

Use molecular symmetry to determine if ethane ($C_2H_6$) and ethanol ($C_2H_5OH$) have dipole moments.

The importance of polarity in chemical bonds cannot be understated. For example, as will be detailed in Chapter 9, the solubility of a substance depends on the polarity of the substance and of the substance in which it is dissolved (the "solvent"). In order to dissolve a polar substance such as a salt, a polar solvent such as water is required. Also, if two substances are both non-polar, one will dissolve in the other. This can have important consequences. Many toxic and persistent pollutants, for example, are essentially non-polar. For this reason, they tend to be insoluble in water, but very soluble in non-polar substances such as the fatty tissues of animals. These substances undergo a process called biomagnification, whereby their concentrations increase moving up the food chain. Animals at the top of the food chain (humans included) tend to have high concentrations of these substances in their tissues compared with those in water. Polar substances tend to dissolve in water and not move up the food chain in this manner.

## Bond Length

As described in Section 6.1, the bond length of a covalent bond is the nuclear separation distance where the molecule is most stable. The H—H bond length in molecular hydrogen is 74 pm. At this distance, attractive interactions are maximized relative to repulsive interactions (see Figure 6-2). Having developed ideas about Lewis structures and molecular shapes, we can now examine bond lengths in more detail.

Table 6-1 lists average bond lengths for the most common chemical bonds and the period number, *n*, for each atom. The table displays several trends. One

## TABLE 6-1  AVERAGE BOND LENGTHS*

### H—X Bonds

| $n_a$ | $n_b$ | | | | | | | | |
|---|---|---|---|---|---|---|---|---|---|
| 1 | 1 | H—H | 74 | | | | | | |
| 1 | 2 | H—C | 109 | H—N | 101 | H—O | 96 | H—F | 92 |
| 1 | 3 | H—Si | 148 | H—P | 144 | H—S | 134 | H—Cl | 127 |
| 1 | 4 | | | | | | | H—Br | 141 |

### Second-Row Elements

| $n_a$ | $n_b$ | | | | | | | | |
|---|---|---|---|---|---|---|---|---|---|
| 2 | 2 | C—C | 154 | C—N | 147 | C—O | 143 | C—F | 135 |
| 2 | 2 | C—Si | 185 | N—N | 145 | O—O | 148 | F—F | 142 |
| 2 | 3 | O—Si | 166 | C—P | 184 | C—S | 182 | C—Cl | 177 |
| 2 | 3 | F—Si | 157 | O—P | 163 | O—S | 158 | N—Cl | 175 |
| 2 | 3 | | | F—P | 157 | F—S | 156 | | |
| 2 | 4, 5 | | | F—Xe | 190 | C—Br | 194 | C—I | 214 |

### Larger Elements

| $n_a$ | $n_b$ | | | | | | | | |
|---|---|---|---|---|---|---|---|---|---|
| 3 | 3 | Si—Si | 235 | P—P | 221 | S—S | 205 | Cl—Cl | 199 |
| 3 | 3 | Si—Cl | 202 | P—Cl | 203 | S—Cl | 207 | | |
| 4 | 4 | | | | | | | Br—Br | 228 |
| 5 | 5 | | | | | | | I—I | 267 |

### Multiple Bonds

| | | | | | | | |
|---|---|---|---|---|---|---|---|
| C=C | 133 | C=N | 138 | C=O | 120 | O=O | 121 |
| P=O | 150 | S=O | 143 | | | | |
| C≡C | 120 | C≡N | 116 | C≡O | 113 | N≡N | 110 |

*All values are in picometres; 1 pm = $10^{-12}$ m. $n_a$ and $n_b$ are the period numbers of the atoms.

trend is that bonds become longer as the radii of the atoms become larger. Compare the bond lengths of the diatomic halogens:

| | $F_2$ | $Cl_2$ | $Br_2$ | $I_2$ |
|---|---|---|---|---|
| Bond length (pm) | 142 | 199 | 228 | 267 |
| Atomic radius (pm) | 72 | 100 | 114 | 133 |

The trend is consistent with our introduction to orbital overlap. Recall that bonding involves valence orbitals, and it is the filled valence orbitals that determine the size of an atom. Because atomic size increases with the principal quantum number, bond lengths vary predictably with the *n*-value of the valence orbitals.

Nuclear charge also affects bond length. Recall from Section 5.4 (Chapter 5) that atomic size decreases from left to right across a row of the periodic table because of the progressive rise in nuclear charge. Smaller orbitals form shorter bonds, so Cl—Cl bonds are shorter than S—S bonds, and S—S bonds are shorter than P—P bonds.

Bond polarity also contributes to bond length because partial charges generate electrical attraction that pulls the atoms closer together. For example, notice in Table 6-1 that C—O bonds are slightly shorter than either C—C or O—O bonds. This is a result of the polarity of the C—O bond.

The bond lengths in Table 6-1 show one final feature: a multiple bond is shorter than the corresponding single bond between the same two atoms. This is because placing additional electrons between atoms increases net attraction and pulls the atoms closer together. Thus, triple bonds are the shortest of all bonds among second-row elements.

Because bond lengths depend on the number of electrons involved in the bond, we can use bond lengths to decide which Lewis structure best represents the actual electron distribution in a molecule. Sulphur–oxygen bonds provide a good example. Figure 6-30 shows two possible sets of Lewis structures for species that contain sulphur–oxygen bonds. One set places octets of electrons on the inner sulphur atoms, whereas the other set reduces the formal charge on sulphur to zero at the expense of exceeding the octet on the S atom. As the figure shows, the two sets of structures predict different bond types.

Experimental bond length values clearly support the optimized Lewis structures. In sulphuric acid, there are two distinctly different bond types: the S—OH bond lengths are 157 pm, whereas the S—O bond lengths are 142 pm. This indicates that an S—O single bond has a length of 157 pm, and an S=O double bond has a length of 142 pm. In the sulphate anion, the bond lengths are 147 pm, indicating intermediate bond character consistent with the optimized Lewis structure. The bond lengths in $SO_2$ and $SO_3$ are 143 pm and 142 pm, respectively. These values also support the optimized Lewis structures, which predict double bonds rather than intermediate bonds.

**Figure 6-30**

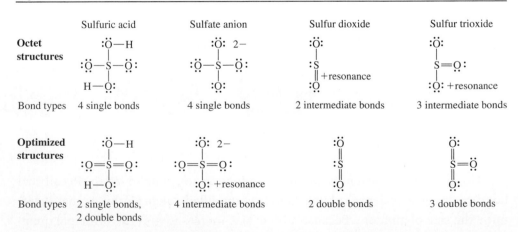

Octet structures and optimized structures predict different bond types for sulphur–oxygen bonds.

To summarize, the following factors influence bond lengths:

1. The smaller the atoms, the shorter the bond.
2. The higher the bond multiplicity, the shorter the bond.
3. The higher the effective nuclear charges of the bonded atoms, the shorter the bond.
4. The larger the electronegativity difference of the bonded atoms, the shorter the bond.

Example 6-18 provides practice in the use of these factors.

### EXAMPLE 6-18  BOND LENGTHS

What factor accounts for each of the following differences in bond length?

(a)  $I_2$ has a longer bond than $Br_2$.
(b)  C—N bonds are shorter than C—C bonds.
(c)  H—C bonds are shorter than the C≡O bond.
(d)  The carbon–oxygen bond in formaldehyde, $H_2C$=O, is longer than the bond in carbon monoxide, C≡O.

**Strategy:** Bond lengths are controlled by four factors, some of which are more influential than others. To explain a difference in bond length, we need to determine the way that the factors are balanced.

**Solution:**

(a)  I—I > Br—Br. Iodine is just below bromine in the periodic table, so the valence orbitals of iodine are larger than the valence orbitals of bromine. Thus, the $I_2$ bond is longer than the $Br_2$ bond because iodine has a larger atomic radius.

(b)  C—N < C—C. Carbon and nitrogen are second-row elements. Nitrogen has a higher nuclear charge than carbon, however, so nitrogen has the smaller radius. This makes C—N bonds shorter than C—C bonds. In addition, a C—N bond is polar, which contributes to the shortening of the C—N bond.

(c)  H—C < C≡O. Here, we are comparing bonds in which the atomic radii and the amount of multiple bonding influence bond length. The experimental fact that H—C bonds are shorter than the triple C≡O bond indicates that the size of the hydrogen orbital is a more important factor than the presence of multiple bonding.

(d)  C=O > C≡O. Both bonds are between carbon and oxygen, so $n$, atomic number ($Z$), and electronegativity difference ($\Delta\chi$) are the same. However, carbon monoxide contains a triple bond, whereas formaldehyde has a double bond. The triple bond in CO is shorter than the double bond in $H_2CO$ because more shared electrons means a shorter bond.

**Do the Results Make Sense?** When factors work in the same direction (effective nuclear charge and bond polarity in the case of C—N < C—C), we can make confident predictions. When factors work in the opposite direction ($n$ value and bond multiplicity in the case of H—C < C≡O), we can explain the observed values but could not have predicted them.

### Practice Exercise 6.18

What factor accounts for each of the following differences in bond length? (a) The C=C bond is longer than the C≡C bond. (b) The C—Cl bond is shorter than the Si—Cl bond. (c) The C—C bond is longer than the O—O bond.

## Bond Energy

As described in Section 3.3, energy must be supplied to break any chemical bond. Bond energies, like bond lengths, vary in ways that can be traced to atomic properties. There are three consistent trends in bond strengths:

1. *Bond strength increases as more electrons are shared between the atoms.* We described this trend in Section 3.3. Shared electrons are the "glue" of chemical bonding, so sharing more electrons strengthens the bond.

2. *Bond strength increases as the electronegativity difference ($\Delta\chi$) between bonded atoms increases.* Polar bonds gain stability from the electrical attraction between the negative and positive fractional charges around the bonded atoms. Bonds between oxygen and other second-row elements exemplify this trend:

| | | |
|---|---|---|
| O—O | $\Delta\chi = 0.0$ | $BE = 145$ kJ/mol |
| O—N | $\Delta\chi = 0.5$ | $BE = 200$ kJ/mol |
| O—C | $\Delta\chi = 1.0$ | $BE = 360$ kJ/mol |

3. *Bond strength decreases as bonds become longer.* For larger atoms, the electron density of a bond is spread over a wider region. This decreases the net attraction between the electrons and the nuclei. The following bond energies illustrate this effect:

| | | |
|---|---|---|
| H—F | 92 pm | $BE = 565$ kJ/mol |
| H—Cl | 127 pm | $BE = 430$ kJ/mol |
| H—Br | 141 pm | $BE = 360$ kJ/mol |
| H—I | 161 pm | $BE = 295$ kJ/mol |

Like bond lengths, bond energies result from the interplay of several factors, including nuclear charge, principal quantum number, electrical forces, and electronegativity. Thus, it should not be surprising that there are numerous exceptions to these three trends in bond energies. Although it is possible to *explain* many differences in bond energy, it frequently is not possible to *predict* differences with confidence.

## Summary of Molecular Shapes

Table 6-2 summarizes the relationships among steric number, electron group geometry, and molecular shape. If you remember the electron group geometry associated with each steric number, you can deduce molecular shapes, bond angles, and existence of dipole moments.

This relatively small catalogue of molecular shapes accounts for a remarkable number of molecules. Even complicated molecules such as proteins and other polymers have shapes that can be traced back to these relatively simple templates. The overall shape of a large molecule is a composite of the shapes associated with its inner atoms. The shape around each inner atom is determined by steric numbers and the number of lone pairs.

## TABLE 6-2 FEATURES OF MOLECULAR GEOMETRIES

| Steric Number | Ligands | Lone Pairs | Electron Group Geometry | Molecular Shape | Bond Angles | Dipole Moment* | Picture |
|---|---|---|---|---|---|---|---|
| 2 | 2 | 0 | Linear | Linear | 180° | No | |
| 3 | 3 | 0 | Trigonal planar | Trigonal planar | 120° | No | |
| | 2 | 1 | Trigonal planar | Bent | <120° | Yes | |
| 4 | 4 | 0 | Tetrahedral | Tetrahedral | 109.5° | No | |
| | 3 | 1 | Tetrahedral | Trigonal pyramidal | <109.5° | Yes | |
| | 2 | 2 | Tetrahedral | Bent | <109.5° | Yes | |
| 5 | 5 | 0 | Trigonal bipyramidal | Trigonal bipyramidal | 90°, 120° | No | |
| | 4 | 1 | Trigonal bipyramidal | Seesaw | <90°, <120° | Yes | |
| | 3 | 2 | Trigonal bipyramidal | T-shaped | <90°, <120° | Yes | |
| | 2 | 3 | Trigonal bipyramidal | Linear | 180° | No | |
| 6 | 6 | 0 | Octahedral | Octahedral | 90° | No | |
| | 5 | 1 | Octahedral | Square pyramidal | 90°, <120° | Yes | |
| | 4 | 2 | Octahedral | Square planar | 90° | No | |

*Applies only to molecules with identical ligands.

## SECTION EXERCISES

●6.5.1 Predict the bond angles for $BCl_3$, $SF_4$, and $SnCl_4$. Which of these molecules, if any, has a dipole moment?

●6.5.2 Experimental evidence shows that $PF_3Cl_2$ has a dipole moment, whereas $PCl_3F_2$ does not. Determine the structures of these two compounds and explain how the structures minimize total repulsion.

●6.5.3 Arrange the following bonds in order of increasing length (shortest first) and increasing strength (weakest first). State the factors responsible for the position of each bond in your sequences: $C-C$, $C=O$, $C=C$, $C-H$, and $C-Cl$.

# CHAPTER 6 VISUAL SUMMARY

## 6.1 OVERVIEW OF BONDING

$$E_{electrical} = k\,\frac{q_1 q_2}{r}$$

**LEARNING OBJECTIVES:** Use the concept of electronegativity to determine the polarity of a chemical bond.

**SKILLS TO MASTER:** Using electronegativity differences to assess bond polarity (p. 304).

**KEY CONCEPTS:** A covalent bond arises from the mutual attraction of the bonding electrons to the two nuclei.

A greater difference in electronegativity leads to a more polar bond.

## 6.2 LEWIS STRUCTURES

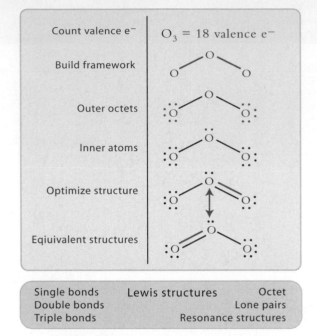

Outer atoms other than hydrogen have octets.

$FC$ = Valence e$^-$ (free atom) – Valence e$^-$ (Lewis)

**LEARNING OBJECTIVES:** Draw optimized Lewis structures of covalent compounds, including resonance structures.

**SKILLS TO MASTER:** Following the four steps for drawing a provisional Lewis structure (p. 307).

Optimizing the configuration of the inner atoms (p. 310).

Calculating formal charges (p. 313).

Identifying and drawing resonance structures (p. 317).

**KEY CONCEPTS:** Most atoms other than hydrogen are most stable when associated with an octet of electrons.

Only the valence electrons appear in a Lewis structure.

The most likely Lewis structure has the lowest formal charges.

## 6.3 MOLECULAR SHAPES: TETRAHEDRAL SYSTEMS

SN 4 = Tetrahedral electron group geometry

| | | |
|---|---|---|
| Lewis structure | | Tetrahedron |
| Steric number | | Trigonal pyramid |
| Electron groups | | Bent shape |
| Molecular shape | | |

VSEPR    Electron group geometry    Ligand

Steric number (SN)    Molecular shape

**LEARNING OBJECTIVES:** Recognize the importance of the tetrahedral shape in molecules.

**SKILLS TO MASTER:** Predicting the shapes of molecules that have a steric number of 4 (p. 320).

**KEY CONCEPTS:** The steric number of an inner atom is the sum of the number of ligands plus the number of lone pairs.

An electron group can be two electrons in a single bond, four electrons in a double bond, six electrons in a triple bond, a pair of nonbonding electrons or a single electron.

Molecular shape describes how the *ligands*, not the *electron groups*, are arranged in space.

Molecular shapes can be derived from the steric number and number of ligands bonded to a central atom.

## 6.4 OTHER MOLECULAR SHAPES

| SN | Electron group geometry |
|----|-------------------------|
| 2 | Linear |
| 3 | Trigonal planar |
| 5 | Trigonal bipyramidal |
| 6 | Octahedral |

**LEARNING OBJECTIVES:** Use the VSEPR model to predict the shapes of molecules with steric numbers 2, 3, 5 and 6.

**SKILLS TO MASTER:** Predicting the shapes of molecules that have a steric number of 2 (p. 330).

Predicting the shapes of molecules that have a steric number of 3 (p. 331).

Predicting the shapes of molecules that have a steric number of 5 (p. 333).

Predicting the shapes of molecules that have a steric number of 6. (p. 335).

## 6.5 PROPERTITES OF COVALENT BONDS

Bond strength increases as:
More electrons are shared
$\Delta\chi$ increases
Bonds become shorter

*Lone pairs decrease bond angles.*

**LEARNING OBJECTIVES:** Understand the factors that influence bond angles, lengths and energies.

**KEY CONCEPTS:** Electron – electron repulsion generated by nonbonding pairs is always greater than that generated by bonding pairs.

Dipole moments are affected by electronegativity differences and molecular geometry.

Bond lengths are affected by atomic radii, bond multiplicity, effective nuclear charges and electronegativity difference.

## LEARNING EXERCISES

6.1  Design a flowchart that shows how to determine the Lewis structure of a molecule.

6.2  Design a flowchart that shows how to determine the shape of a molecule.

6.3  Describe the role that electrical forces play in determining each of the following properties: (a) bond length; (b) bond polarity; (c) bond angle; and (d) molecular shape.

6.4  Write a paragraph that explains why formal charges do not match actual charges.

6.5  Draw as many different pictures as you can that illustrate geometric shapes of molecules that have no dipole moments.

6.6  Explain in your own words the meaning of each of the following terms: (a) electronegativity; (b) bonding framework; (c) inner atom; (d) ligand; (e) steric number; (f) multiple bond; (g) silicate; (h) alkane; and (i) dipole moment.

6.7  In your own words, write a one-sentence definition of each term in Chapter 6 that is new to you. Consult the glossary if you need help.

## PROBLEMS

**ilw** = interactive learning ware problem. Visit the website at www.wiley.com/canada/olmsted

### Overview of Bonding

6.1  For the following elements, write the complete electron configuration and identify which of the electrons will be involved in bond formation: (a) O; (b) P; (c) B; and (d) Br.

6.2  For the following elements, write the complete electron configuration and identify which of the electrons will be involved in bond formation: (a) Si; (b) N; (c) Cl; and (d) S.

6.3  Describe bond formation between a hydrogen atom and an iodine atom to form a molecule of HI, and include a picture of the overlapping orbitals.

6.4  Describe bond formation between two bromine atoms to form a molecule of $Br_2$, and include a picture of the overlapping orbitals.

6.5  Give the group number and the number of valence electrons for the following elements: (a) aluminum; (b) arsenic; (c) fluorine; and (d) tin.

6.6  Give the group number and the number of valence electrons for the following elements: (a) selenium; (b) iodine; (c) germanium; and (d) nitrogen.

6.7  Hydrogen forms diatomic molecules with elements from Group 1 of the periodic table. Describe the bonding in LiH and include a picture of the overlapping orbitals.

6.8  Molecules of dilithium can form in the gas phase at low pressure. Describe the bonding in $Li_2$ and include a picture of the overlapping orbitals.

6.9  For each of the following pairs, identify which element tends to attract electron density from the other in a covalent bond: (a) C and N; (b) S and H; (c) Zn and I; and (d) S and As.

6.10  For each of the following pairs, identify which element tends to attract electron density from the other in a covalent bond: (a) C and O; (b) O and H; (c) Hg and C; and (d) Si and Cl.

6.11  Show the direction of bond polarity for the following bonds using $\delta+/\delta-$ notation: (a) Si—O; (b) N—C; (c) Cl—F; and (d) Br—C.

6.12  Show the direction of bond polarity for the following bonds using $\delta+/\delta-$ notation: (a) C—P; (b) B—O; (c) C—Sn; and (d) N—H.

6.13  Arrange the following molecules in order of increasing bond polarity: $H_2O$, $NH_3$, $PH_3$, and $H_2S$.

6.14  Arrange the following molecules in order of increasing bond polarity: $SiCl_4$, $SCl_2$, $NCl_3$, and $PCl_3$.

### Lewis Structures

6.15  Count the total number of valence electrons in the following species: (a) $H_3PO_4$; (b) $(C_6H_5)_3C^+$; (c) $(NH_2)_2CO$; and (d) $SO_4^{2-}$.

6.16  Count the total number of valence electrons in the following species: (a) $(CH_3)_4N^+$; (b) $H_2SO_4$; (c) $PO_4^{3-}$; and (d) $(CH_3)_3SiCl$.

6.17  Convert the following formulas into molecular frameworks. For each molecule, tell how many valence electrons are required to construct the framework. (a) $(CH_3)_3CBr$; (b) $(CH_3CH_2CH_2)_2NH$; (c) $HClO_3$; and (d) $OP(OCH_3)_3$.

6.18  Convert the following formulas into molecular frameworks. For each molecule, tell how many valence electrons are required to construct the framework. (a) $(CH_3)_3SiCl$; (b) $(CH_3CH_2)_3N$; (c) $H_2CO_3$; and (d) $CH_3CH_2CH_2OH$.

6.19  Determine the Lewis structure of (a) $NH_3$; (b) $NH_4^+$; and (c) $H_2N^-$.

6.20  Determine the Lewis structure of (a) $H_3O^+$; (b) $H_2O$; and (c) $OH^-$.

6.21  **ilw** Determine the Lewis structure of (a) $PBr_3$; (b) $SiF_4$; and (c) $BF_4^-$.

6.22  Determine the Lewis structure of (a) $CBr_4$; (b) $AlCl_4^-$; and (c) $H_2S$.

6.23  Use the standard procedures to determine the Lewis structure of (a) $H_3CNH_2$; (b) $CF_2Cl_2$; and (c) $OF_2$.

6.24  Use the standard procedures to determine the Lewis structure of (a) $HCCl_3$; (b) $H_2NNH_2$; and (c) $H_2SiCl_2$.

6.25  Use the standard procedures to determine the Lewis structure of (a) $(CH_3)_2CO$; (b) $CH_3CN$; (c) $CH_2CHCH_3$; and (d) $CH_3CHNH$.

6.26  Use the standard procedures to determine the Lewis structure of (a) $CH_2CHCHO$; (b) $HCCCH_2OH$; (c) $(CH_3)_2CCH_2$; and (d) $H_2CCCH_2$.

6.27  Determine the Lewis structure of (a) $IF_5$; (b) $SO_3$; (c) $OPCl_3$; and (d) $XeF_2$.

6.28  Determine the Lewis structure of (a) $PBr_5$; (b) $O_2SCl_2$; (c) $AsCl_3$; and (d) $KrF_4$.

**6.29** Determine the Lewis structure of each of the following polyatomic ions. Include all resonance structures and formal charges, where appropriate: (a) $NO_3^-$; (b) $HSO_4^-$; (c) $CO_3^{2-}$; and (d) $ClO_2^-$.

**6.30** Determine the Lewis structure of each of the following polyatomic ions. Include all resonance structures and formal charges, where appropriate: (a) $SO_3^{2-}$; (b) $HPO_4^{2-}$; (c) $BrO^-$; and (d) $HCO_3^-$.

## Molecular Shapes: Tetrahedral Systems

**6.31** Sketch and name the shapes of the following molecules: (a) $CF_2Cl_2$; (b) $SiF_4$; and (c) $PBr_3$.

**6.32** Sketch and name the shapes of the following molecules: (a) $OF_2$; (b) $HCCl_3$; and (c) $NH_2Cl$.

**6.33** Draw a ball-and-stick model that shows the geometry of 1,2-dichloroethane, $ClH_2CCH_2Cl$.

**6.34** Draw a ball-and-stick model that shows the geometry of 1-bromopropane, $BrCH_2CH_2CH_3$.

**6.35** Draw Lewis structures of all possible structural isomers of the alkanes with formula $C_6H_{14}$.

**6.36** Draw Lewis structures of all possible structural isomers of the bromoalkanes with formula $C_4H_9Br$.

**6.37** Write the Lewis structure of dimethylamine, $(CH_3)_2NH$. Determine its geometry, and draw a ball-and-stick model of the molecule, showing it as an ammonia molecule with two hydrogen atoms replaced by $CH_3$ groups.

**6.38** Write the Lewis structure of dimethyl ether, $(CH_3)_2O$. Draw a ball-and-stick model of this molecule, showing it as a water molecule with each hydrogen atom replaced by a $CH_3$ group.

**6.39** Silicon forms a tetramethyl compound with the formula $(CH_3)_4Si$. Determine the Lewis structure, determine the shape about each inner atom, and draw a ball-and-stick model of the compound.

**6.40** The second simplest silicate is the disilicate anion, $Si_2O_7^{6-}$, in which one oxygen bridges between the two silicon atoms. Determine the Lewis structure, determine the shape about each inner atom, and draw a ball-and-stick model of the anion.

## Other Molecular Shapes

**6.41** Name the molecular shape of an inner atom that has the following characteristics: (a) two lone pairs and three ligands; (b) a steric number of 5 and one lone pair; (c) a steric number of 3 and no lone pairs; and (d) five ligands and a steric number of 6.

**6.42** Name the molecular shape of an inner atom that has the following characteristics: (a) two lone pairs and a steric number of 6; (b) two ligands and three lone pairs; (c) one lone pair and five ligands; and (d) a steric number of 5 and two lone pairs.

**6.43** Fluorine forms compounds of chemical formula $XF_4$ with elements from Groups 14, 16, and 18. Determine the

Lewis structure, describe the shape, and draw a ball-and-stick model of $GeF_4$, $SeF_4$, and $XeF_4$.

**6.44** Iodine forms three compounds with chlorine: $ICl$, $ICl_3$, and $ICl_5$. Determine the Lewis structure, describe the shape, and draw a ball-and-stick model of each compound.

**6.45** Determine the molecular shape and the ideal bond angles of each of the following: (a) $SO_2$; (b) $SbF_5$; (c) $ClF_4^+$; and (d) $ICl_4^-$.

**6.46** Determine the molecular shape and the ideal bond angles of each of the following: (a) $SO_3$; (b) $I_3^+$; (c) $SbCl_5^{2-}$; and (d) $XeO_4$.

## Properties of Covalent Bonds

**6.47** Determine the Lewis structures of the following compounds, and determine which ones have dipole moments. For each molecule that has a dipole moment, draw a ball-and-stick model and include an arrow to indicate the direction of the dipole moment: (a) $SiF_4$; (b) $H_2S$; (c) $XeF_2$; (d) $GaCl_3$; and (e) $NF_3$.

**6.48** Determine the Lewis structures of the following compounds, and determine which ones have dipole moments. For each molecule that has a dipole moment, draw a ball-and-stick model and include an arrow to indicate the direction of the dipole moment: (a) $CH_4$; (b) $CHCl_3$; (c) $CH_2Cl_2$; (d) $CH_3Cl$; and (e) $CCl_4$.

**6.49** **ilw** Carbon dioxide has no dipole moment, but sulphur dioxide has $\mu = 1.63$ D. Use Lewis structures to account for this difference in dipole moments.

**6.50** Sulphur trioxide has no dipole moment, but arsenic trifluoride has $\mu = 2.59$ D. Use Lewis structures to account for this difference in dipole moments.

**6.51** Which of the following molecules would you expect to have bond angles that deviate from the ideal VSEPR values? For the molecules that do, make sketches that illustrate the deviations. (a) $PF_5$; (b) $CH_3I$; and (c) $BrF_5$.

**6.52** Which of the following molecules would you expect to have bond angles that deviate from the ideal VSEPR values? For the molecules that do, make sketches that illustrate the deviations. (a) $TeCl_4$; (b) $XeF_4$; and (c) $SbCl_3$.

**6.53** Arrange the following bonds in order of increasing length (shortest first). List the factors responsible for each placement: $H{-}N$, $N{-}N$, $Cl{-}N$, $N{\equiv}N$, and $C{\equiv}N$.

**6.54** Arrange the following bonds in order of increasing bond length (shortest first). List the factors responsible for each placement: $Cl{-}Cl$, $Br{-}Br$, $O{-}Cl$, $P{=}O$, and $H{-}F$.

**6.55** Use Table 3-2 to arrange the following bonds in order of increasing bond strength (weakest first). List the single most important factor for each successive increase in strength: $C{=}C$, $H{-}N$, $C{=}O$, $N{\equiv}N$, and $C{-}C$.

**6.56** Use Table 3-2 to arrange the following bonds in order of increasing bond strength (weakest first). List the single most important factor for each successive increase in strength: $Si{-}Si$, $C{-}C$, $C{-}Si$, $H{-}C$, and $Sn{-}Sn$.

## ADDITIONAL PAIRED PROBLEMS

**6.57** Draw Lewis structures and ball-and-stick structures showing the correct geometries for molecules of the following substances: (a) dichlorine oxide, $Cl_2O$ (used for bleaching wood pulp and water treatment, about $10^5$ tonnes produced each year); (b) benzene, $C_6H_6$, which contains a ring of six carbon atoms, each bonded to one hydrogen atom (one of the top 20 industrial chemicals, used in production of polymers); and (c) ethylene oxide, $C_2H_4O$, which contains a C—C—O triangular ring (one of the top 50 industrial chemicals, used in polymer production).

**6.58** Draw Lewis structures and ball-and-stick structures showing the geometries of molecules of the following substances: (a) methyl isocyanate, $CH_3NCO$ (a toxic compound responsible for thousands of deaths in Bhopal, India, in 1984); and (b) tetrafluorohydrazine, $N_2F_4$, (a colourless liquid used as rocket fuel).

**6.59** Determine the molecular geometries of the following molecules: $SiCl_4$, $SeF_4$, and $CI_4$.

**6.60** Determine the molecular geometries of the following ions: $ClF_2^-$, $BF_4^-$, and $PF_4^-$.

**6.61** Shorter bonds are usually stronger bonds, but this is not always the case. Using Tables 6-1 and 3-2, find and list any X—Y bonds, for X and Y both Row 2 elements, that are (a) shorter but weaker and (b) longer but stronger than the corresponding X—X bond.

**6.62** Using Tables 6-1 and 3-2, prepare a list of bonds to fluorine that are both longer and stronger than F—F bonds.

**6.63** How many different structural isomers are there for octahedral molecules with the general formula $AX_3Y_3$? Draw three-dimensional structures of each.

**6.64** Identify which of the four octahedral molecules shown here are equivalent:

$$
\begin{array}{cccc}
\text{X} & \text{Y} & \text{Y} & \text{Y} \\
\text{Y}\cdots\!\overset{|}{\underset{|}{\text{A}}}\!\cdots\text{Y} & \text{Y}\cdots\!\overset{|}{\underset{|}{\text{A}}}\!\cdots\text{Y} & \text{X}\cdots\!\overset{|}{\underset{|}{\text{A}}}\!\cdots\text{Y} & \text{X}\cdots\!\overset{|}{\underset{|}{\text{A}}}\!\cdots\text{Y} \\
\text{Y}\quad\text{Y} & \text{Y}\quad\text{X} & \text{X}\quad\text{Y} & \text{Y}\quad\text{X} \\
\text{X} & \text{X} & \text{Y} & \text{Y} \\
(a) & (b) & (c) & (d)
\end{array}
$$

**6.65** Describe the bonding and determine the empirical chemical formula of the silicon–oxygen network of zircon.

**6.66** Describe the bonding and determine the empirical chemical formula of the silicon–oxygen network of asbestos.

**6.67** Write Lewis structures and calculate formal charges for the following polyatomic ions: (a) bromate; (b) nitrite; (c) phosphate; and (d) hydrogen carbonate.

**6.68** Carbon, nitrogen, and oxygen form two different polyatomic ions: cyanate ion $(NCO^-)$ and fulminate ion $(CNO^-)$. Write Lewis structures for each anion, including near-equivalent resonance structures and indicating formal charges.

**6.69** Species with chemical formula $XY_4$ can have the following shapes. For each, name the molecular geometry, identify the ideal VSEPR bond angles, tell how many lone pairs are present in the structure, and give a specific example.

(a)　　　　(b)　　　　(c)

**6.70** Species with chemical formula $XY_3$ can have the following shapes. For each, name the molecular geometry, identify the ideal VSEPR bond angles, tell how many lone pairs are present in the structure, and give a specific example.

(a)　　　　(b)　　　　(c)

**6.71** Four different bonds are found in $H_2O$, $CO_2$, and HCN. Arrange the four different bonds in order of increasing bond length. Give the reasons for your placements.

**6.72** Four different bonds are found in $NH_3$, CO, and HOCl. Arrange the four different bonds in order of increasing bond length. Give the reasons for your placements.

**6.73** The following partial structure is that of azodicarbonamide:

$$
\begin{array}{ccccccc}
 & \text{O} & & & & \text{H} & \\
 & \| & & & & | & \\
\text{H}\diagdown & \text{C} & \diagup\text{N}\diagdown & & \text{C} & \diagup\text{N}\diagdown & \\
\text{N} & & & \text{N} & & & \text{H} \\
| & & & & & | & \\
\text{H} & & & & & \text{O} & \\
\end{array}
$$

$$\underset{H}{\overset{O}{\underset{|}{\underset{N}{H-}}}}\,C\,N\,N\,\underset{O}{\overset{H}{C}}\,N-H$$

When heated, azodicarbonamide breaks apart into gaseous carbon monoxide, nitrogen, and ammonia. Azodicarbonamide is used as a foaming agent in the polymer industry. (a) Add non-bonding electron pairs and multiple bonds as required to complete the Lewis structure of this molecule. (b) Determine the geometry around each inner atom.

**6.74** In the lower atmosphere, $NO_2$ participates in a series of reactions in air that is also contaminated with unburned hydrocarbons. One product of these reactions is peroxyacetyl nitrate (PAN).

$$
\begin{array}{c}
\quad\text{H}\quad\text{O}\qquad\qquad\text{O} \\
\quad|\quad\ \ \|\qquad\qquad\ \ \| \\
\text{H}-\overset{*}{\text{C}}-\overset{*}{\text{C}}-\text{O}-\text{O}-\overset{*}{\text{N}}\leftarrow \\
\quad\nearrow|\qquad\qquad\quad\ \ | \\
\quad\text{H}\qquad\qquad\qquad\text{O} \\
\end{array}
$$

The skeletal arrangement of the atoms in PAN appears above. (a) Complete the Lewis structure of this compound. (b) Determine the shape around each atom marked with an asterisk. (c) Give the approximate values of the bond angles indicated with arrows.

**6.75** The H—O—H bond angle in a water molecule is 104.5°. The H—S—H bond angle in hydrogen sulphide is only 92.2°. Explain these variations in bond angles, using orbital sizes and electron–electron repulsion arguments. Draw space-filling models to illustrate your explanation.

6.76 The bond angles are 107.3° in $NH_3$, 100.3° in $PCl_3$, and 93.3° in $PH_3$. Explain these variations in bond angles, using orbital sizes and electron–electron repulsion arguments. Draw space-filling models to illustrate your explanation.

6.77 List the following $X$—$H$ bonds from smallest bond polarity to largest bond polarity: C—H, F—H, N—H, O—H, and Si—H.

6.78 List the following $X$—$O$ bonds from smallest bond polarity to largest bond polarity: C—O, N—O, S—O, O—H, and Br—O.

6.79 Do the following structures represent the same compound or are they different compounds? Explain.

H
|
C—Br     *and*     C—Br
H   Br         H   H
                Br

6.80 Do the following structures represent the same compound or are they different compounds? Explain.

F—S—Cl     *and*     F—S—F
F   F           F   Cl

6.81 Benzyne is an unstable molecule that can be generated as a short-lived species in solution. Suggest a reason why benzyne is very reactive. (*Hint:* Examine the bond angles.)

H
|
H   C
C       C
H    C       C
C
|
H
Benzyne

6.82 Cyclopropane, $C_3H_6$, which has three carbon atoms in a ring, is far more reactive than other alkanes. Determine the Lewis structure and the molecular shapes of the carbon atoms of this molecule. Suggest a reason for the reactivity of $C_3H_6$. (*Hint:* Examine the bond angles.)

## ADDITIONAL PROBLEMS

6.83 Determine the Lewis structures for the two possible arrangements of the $N_2O$ molecule, N—N—O and N—O—N Experiments show that the molecule is linear and has a dipole moment. What is the arrangement of atoms? Justify your choice.

6.84 The inner atom of a triatomic molecule can have any of four different electron group geometries. Identify the four, describe the shape associated with each, and give a specific example of each.

6.85 The molecule dichlorobenzene, $C_6H_4Cl_2$, which contains a ring of six carbon atoms, has three structural isomers. These differ in the relative positions of the Cl atoms around the ring. Determine the Lewis structures and geometries of these three isomers. Draw line structures of each. Which isomers have dipole moments? Which isomer has the largest dipole moment?

6.86 Determine the Lewis structure of borazine, $B_3N_3H_6$. The molecule contains a planar ring of alternating boron and nitrogen atoms, with a hydrogen atom attached to each ring atom.

6.87 Both $PF_3$ and $PF_5$ are known compounds. $NF_3$ also exists, but $NF_5$ does not. Why is there no molecule with the formula $NF_5$?

6.88 Phosphorous acid ($H_3PO_3$) is an exception to the rule that hydrogen always bonds to oxygen in oxoacids. In this compound, one of the hydrogen atoms bonds to phosphorus. Determine the Lewis structure for phosphorous acid and determine the geometry around the phosphorus atom. Draw a ball-and-stick sketch of the molecule.

6.89 Tellurium compounds, which are toxic and have a hideous stench, must be handled with extreme care. Predict the formula of the tellurium–fluorine molecule or ion that has the following molecular geometry: (a) bent; (b) T-shaped; (c) square pyramid; (d) trigonal bipyramid; (e) octahedron; and (f) seesaw.

6.90 Indium triiodide exists in the gas phase as individual $InI_3$ molecules. In the liquid phase, however, two $InI_3$ molecules combine to give $In_2I_6$, in which two iodine atoms bridge between indium atoms, and there are four outer iodine atoms. Describe the bonding in $InI_3$ and $In_2I_6$. Draw a ball-and-stick model of $In_2I_6$.

6.91 Imagine a square planar molecule, $XY_2Z_2$ in which $X$ is the central atom and $Z$ is more electronegative than $Y$. Two structural isomers are possible. (a) Draw a ball-and-stick model of each isomer. (b) Suppose you have made one isomer, but do not know which one. Explain how measuring the dipole moment of the substance would identify the isomer.

6.92 Sulphur and fluorine form seven different molecules: $SF_2$, $SSF_2$, $FSSF$, $F_3SSF$, $SF_4$, $F_5SSF_5$, and $SF_6$. Draw the Lewis structure of each molecule, and identify the geometry around each inner sulphur atom.

6.93 Among the halogens, only one known molecule has the formula $XY_7$. It has pentagonal bipyramidal geometry, with five $Y$ atoms in a pentagon around the central atom $X$. The other two $Y$ atoms are in axial positions. Draw a ball-and-stick model of this compound. Based on electron–electron repulsion and atomic size, determine the identities of atoms $X$ and $Y$. Explain your reasoning. (Astatine is not involved. This element is radioactive and highly unstable.)

6.94 Spiroalkanes are hydrocarbons in which two carbon-containing rings share one carbon atom. One of the simplest spiroalkanes is spiroheptane. Describe the bonding in spiroheptane. Draw a ball-and-stick model that shows the geometry of the molecule.

Spiroheptane
$C_7H_{12}$

## GROUP STUDY PROBLEMS

**6.95** Many transition metal cations form metal complexes with geometries that match those described in this chapter. In developing Lewis structures for these complexes, we assume that the metal cation has no valence electrons. Each ligand forms a metal–ligand bond using one of its lone electron pairs. Determine the Lewis structure and describe the geometry around each inner atom of (a) $[Sc(H_2O)_6]^{3+}$; (b) $[Ni(CN)_4]^{2-}$ (not tetrahedral); and (c) $[CuCl_5]^{3-}$.

**6.96** Sulphur forms neutral compounds with oxygen, fluorine, and chlorine that display a variety of steric numbers and molecular shapes. Find as many examples as you can, describe the geometry of each, and identify the steric numbers and shapes that are not found.

**6.97** Convert each of the following line structures into a complete Lewis structure. For each inner atom, identify the electron group geometry and shape. Draw ball-and-stick pictures that show the molecular geometries accurately.

**6.98** In the following reactions, phosphorus forms a bond to a Row 2 element. In one reaction, phosphorus donates two electrons to make the fourth bond, but in the other reaction, phosphorus accepts two electrons to make the fourth bond. Use Lewis structures of starting materials and products to determine in which reaction phosphorus is a donor and in which it acts as an acceptor.

$$PCl_3 + N(CH_3)_3 \longrightarrow Cl_3PN(CH_3)_3$$

$$PCl_3 + BBr_3 \longrightarrow Cl_3PBBr_3$$

**6.99** Describe the bonding differences among silica, silicate minerals, and metasilicate minerals.

**6.100** Determine the Lewis structures, electron group geometries, and molecular shapes of the following compounds, which contain odd numbers of valence electrons. (*Hint:* Treat the odd electron like a lone pair.) (a) Nitrogen dioxide, $NO_2$ (red-brown gas that pollutes the air over many cities); and (b) chlorine dioxide, $ClO_2$ (highly explosive gas used as an industrial bleach).

## ANSWERS TO SECTION EXERCISES

**6.1.1** The Cl nucleus attracts the electron in the H atom, and the H nucleus most strongly attracts an electron in the $3p$ orbital of Cl that points along the bond axis. Two electrons are shared in the resulting bond, which is polar, with greater electron density at the Cl end of the molecule.

**6.1.2** (a) O; (b) C; (c) Cl; and (d) Cl

**6.1.3** C—H < As—Cl < Sn—Cl < Si—O

**6.2.1**

(a) $:\!\ddot{B}r\!-\!\ddot{B}r\!:$    (b) $:\!\ddot{C}l\!-\!\ddot{N}\!-\!\ddot{C}l\!:$   with Cl below    (c) Br with F's arranged

**6.2.2**

(a) H—N—C—C—N—H (with H's and lone pairs as drawn)    (b) H—C≡C—H    (c) H—C—C—C—H (with O double bonded to central C)

**6.2.3**

(a) $H\!-\!\ddot{O}\!:^-$   (b) three resonance structures of $CO_3^{2-}$

(c) three resonance structures of $ClO_3^-$

**6.3.1** Sn is in the same group of the periodic table as C. It forms tetrahedral compounds, of which $SnCl_4$ is one.

$$:\!\ddot{C}l\!-\!\underset{\underset{:\ddot{C}l:}{|}}{\overset{\overset{:\ddot{C}l:}{|}}{Sn}}\!-\!\ddot{C}l\!:$$

**6.3.2** Each N atom has electron group number 4 and one lone pair, and the shape around each N atom is a trigonal pyramid.

$:\!N\!-\!N\!:$ with H atoms

**6.3.3** The atoms inside the circle make up one $Si_2O_5^{2-}$ unit:

**6.4.1**

$:\!\ddot{I}\!-\!Ga\!\begin{smallmatrix}\ddot{I}:\\ \ddot{I}:\end{smallmatrix}$

The bonding about Ga is similar to that around Al in $Al(C_2H_5)_3$. Gallium forms three bonds and has no lone pairs. The molecule is trigonal planar in shape.

**6.4.2**

$:\!\ddot{C}l\!\underset{:\ddot{C}l: \;\; :\ddot{C}l:}{\overset{}{\diagdown}}\!As\!\diagup\!\ddot{C}l\!:^-$

The bonding is similar to that in $SF_4$. The ion has a see-saw shape.

**6.4.3** Xe is the inner atom. It has a steric number of 6 and one lone pair, giving square pyramidal geometry with oxygen at the top of the pyramid

**6.5.1** $BCl_3$, 120°; $SF_4$, <90° and <120°; and $SnCl_4$, 109.5°; only $SF_4$ has a dipole moment.

**6.5.2**

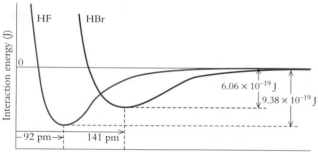

Asymmetric        Symmetric

The chlorine atoms, being larger than fluorine atoms, occupy the equatorial positions because this minimizes atom–atom repulsion.

**6.5.3** Increasing length: C—H ($n = 1$) < C=O (polar double) < C=C (double) < C—C (single) < C—Cl ($n = 3$); increasing strength: C—Cl ($n = 3$) < C—C < C—H ($n = 1$, polar) < C=C (double) < C=O (polar double)

## ANSWERS TO PRACTICE EXERCISES

**6.1** Your plots should show that the bond in HBr is both longer and weaker than the bond in HF.

**6.2** H—H, O—O, F—F < O—F < O—H < H—F

**6.3**

**6.4**

**6.5**

**6.6**

**6.7**

**6.8**

**6.9**

**6.10**

**6.11**

**6.12** Chloromethane is tetrahedral like methane, with a Cl atom replacing one H atom.

Chloromethane

**6.13** The three inner atoms in ethanol all have tetrahedral group geometry. Each of the carbon atoms has tetrahedral geometry, and so does the oxygen atom. With two lone pairs, the geometry about the O atom is bent.

**6.14** The inner I atom has a steric number of 5 and trigonal bipyramidal electron group geometry. Three lone pairs occupy equatorial positions, so the ion is linear.

**6.15** The inner Cl atom has a steric number of 6 and octahedral electron group geometry. One lone pair occupies one octahedral position. The resulting shape is a pyramid built on a square, or square pyramid.

**6.16** The electron pair geometry of $ClF_3$ is trigonal bipyramidal, with predicted bond angles of 90° and 120°. Lone pairs occupy two of the equatorial positions, however, so there is only one equatorial F atom and no 120° bond angles. Lone-pair repulsion distorts the axial positions, pushing them away and reducing the bond angle below 90°.

87.5°

**6.17** Ethane is symmetrical, so it has no dipole moment. Ethanol has a polar group at one end, so it has a dipole moment.

**6.18** (a) Bond lengths decrease as the number of electron pairs in the bond increases. (b) Bond lengths increase with the principal quantum number of the valence electrons. (c) Both bonds are between identical second-row atoms. The O—O bond is shorter because orbital size decreases across a row.

# THEORIES OF CHEMICAL BONDING

## PHOTOSYNTHESIS

Photosynthesis is the process used by green plants to harvest light energy from the Sun and store it in the form of chemical energy. Through a complex sequence of chemical reactions, plants use solar energy to convert water and carbon dioxide into energy-rich glucose and molecular oxygen. Plants then convert glucose into more complex sugars, starch, or cellulose. These compounds, in turn, are available to plant-eating animals as sources of energy and building blocks for biochemical synthesis.

Photosynthesis begins with chlorophyll, a ball-and-stick structure of which appears in our inset figure. This molecule is a pigment that gives plants their green colours. Chlorophyll molecules absorb photons of visible light, and the photon energy creates a molecular excited state. This process is analogous to the atomic absorption leading to excited states, as described in Chapter 4. The excited electron generates a cascade of reactions in which an electron moves from chlorophyll to plastoquinone to cytochrome. The structure of plastoquinone appears later in this chapter, and we discuss cytochromes briefly in Chapter 18. These three molecules are packed together in chloroplasts, subcellular components in the cells of higher plants and animals. Chloroplasts can be seen in our inset photo of a plant cell.

Chlorophyll, plastoquinone, and cytochrome are complex molecules, but each has an extended pattern of single bonds alternating with double bonds. Molecules that contain such networks are particularly good at absorbing light and at undergoing reversible oxidation–reduction reactions. These properties are at the heart of photosynthesis.

In this chapter, we develop a model of bonding that can be applied to molecules as simple as $H_2$ or as complex as chlorophyll. We begin with a description of bonding based on the idea of overlapping atomic orbitals. We then extend the model to include the molecular shapes described in Chapter 6. Next, we apply the model to molecules with double and triple bonds. Then we present variations on the orbital overlap model that encompass electrons distributed across three, four, or more atoms, including the extended systems of molecules such as chlorophyll. Finally, we show how to generalize the model to describe the electronic structures of metals and semiconductors.

## LEARNING OBJECTIVES

*Upon completion of this chapter you should be able to:*

- use the orbital overlap model to explain the bonding in simple molecules

- assign the correct hybrid orbitals used by each inner atom in a molecule, and the molecular geometry that results

- describe the σ and π bonding systems in multiple bonds

- use molecular orbital theory to calculate a bond order, predict magnetic properties of a molecule and explain trends in bond length and energy

- describe the bonding in three-atom π systems

- describe the bonding in extended π systems

- explain such properties as electrical conductivity and colour of metals, non-metals and metalloids in terms of band theory

## 7.1 LOCALIZED BONDS

Lewis structures are blueprints that show the distribution of valence electrons in molecules. However, the dots and lines of a Lewis structure do not show any details of how bonds form, how molecules react, or the shape of a molecule. In this respect, a Lewis structure is like the electron configuration of an atom: both tell us about electron distributions, but neither provides detailed descriptions. Just as we need atomic orbitals to understand how electrons are distributed in an atom, we need an orbital view to understand how electrons are distributed in a molecule.

The localized bond model is often called *valence bond theory*.

In this chapter, we describe two ways to think about bonding: **localized bonds** and **delocalized bonds**. The localized bond approach to molecules takes its cue from Lewis structures. According to Lewis structures, electrons are either localized in bonds between two atoms or localized on a single atom, usually in pairs. Accordingly, the localized bonding model develops orbitals of two types. One type is a bonding orbital that has high electron density between two atoms. The other type is an orbital located on a single atom. In other words, any electron is restricted to the region around a single atom, either in a bond to another atom or in a non-bonding orbital.

Localized bonds are easy to apply, even to very complex molecules, and they do an excellent job of explaining much chemical behaviour. In many instances, however, localized bonds are insufficient to explain molecular properties and chemical reactivity. In the second half of this chapter, we show how to construct delocalized bonds, which spread over several atoms. Delocalization requires a more complex analysis, but it explains chemical properties that localized bonds cannot.

## Orbital Overlap

As described in the opening pages of Chapter 6, the electrons in a hydrogen molecule are "smeared out" between the two nuclei in a way that maximizes electron/nucleus attraction. To understand chemical bonding, we must develop a new orbital model that accounts for shared electrons. In other words, we need to develop a set of **bonding orbitals**.

The model of bonding used by most chemists is an extension of the atomic theory that incorporates a number of important ideas from Chapters 4 and 5 of this textbook. Bonding orbitals are described by combining atomic orbitals. Remember that orbitals have wave properties. Waves interact by addition of their amplitudes. Two waves that occupy the same region of space become superimposed, generating a new wave that is a composite of the original waves. In regions where the amplitudes of the superimposed waves have the same sign, the waves add. This gives a new wave amplitude that is larger than either original wave, as Figure 7-1a shows. In regions where the amplitudes have opposite signs, the waves subtract, giving a new wave amplitude that is smaller than either of the original waves, as illustrated in Figure 7-1b.

Because electrons have wave-like properties, orbital interactions involve similar addition or subtraction of wave functions. When two orbitals are superimposed,

**Figure 7-1**

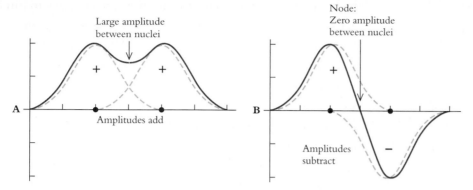

(a) When two wave amplitudes (*dashed lines*) add, the resulting new wave (*solid line*) has a large amplitude in the overlap region. (b) When two wave amplitudes (*dashed lines*) subtract, the resulting new wave (*solid line*) has a small amplitude in the overlap region.

**Figure 7-2**

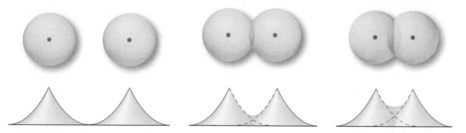

As two hydrogen atoms approach each other, the overlap of their 1*s* atomic orbitals increases. The wave amplitudes add, generating a new orbital with high electron density between the nuclei.

one result is a new orbital that is a composite of the originals, as shown for molecular hydrogen in Figure 7-2. This interaction is called **orbital overlap**, and it is the foundation of the bonding models described in this chapter.

The simplest bonding model requires only the addition of wave functions. Later in this chapter, we show that a complete description of bonding often requires both addition and subtraction of wave functions.

## Conventions of the Orbital Overlap Model

Orbital overlap models proceed from the following assumptions:

1. Each electron in a molecule is assigned to a specific orbital.
2. No two electrons in a molecule have identical descriptions, because the Pauli exclusion principle applies to electrons in molecules as well as in atoms.
3. The electrons in molecules obey the aufbau principle, meaning that they occupy the most stable orbitals available to them.
4. Even though every atom has an unlimited number of atomic orbitals, the valence orbitals are all that are needed to describe bonding.

**Figure 7-3**

High energy orbitals: too unstable for bonding

Valence orbitals: form bonding orbitals

Core orbitals: do not interact

The energy-level diagram for a fluorine atom ($1s^2\, 2s^2\, 2p^5$).

Bonding involves the valence orbitals almost exclusively, because these orbitals have the appropriate energies to interact strongly. Examine the electron energy-level diagram of a fluorine atom, shown in Figure 7-3. The core $1s$ orbitals on a fluorine atom are at much lower energies than any of the $n = 2$ orbitals and do not participate effectively in bonding with other atoms. On the other hand, orbitals with $n > 2$ lie at considerably higher energy than the $n = 2$ orbitals. These orbitals can interact, but the resulting orbitals are too unstable to form strong covalent bonds. Therefore, the only orbitals of fluorine that form chemical bonds are the valence orbitals, those with $n = 2$. Typically, the highest energy-occupied orbitals are the ones used to bond with other atoms.

## Diatomic Molecules: HF and $F_2$

Consider the orbital interactions of a hydrogen atom and a fluorine atom as they combine to form a molecule of hydrogen fluoride. The electron in the hydrogen atom occupies the $1s$ orbital. According to the orbital overlap model, this orbital interacts with the valence orbitals from the fluorine atom. The $2s$ orbital of fluorine is substantially more stable than the set of $2p$ orbitals. Moreover, each $p$ orbital has a strong directional component that is not present in the spherical $2s$ orbital. Both factors make a fluorine $2p$ orbital a better choice for orbital interaction than the $2s$ orbital. Consequently, the bonding in HF can be described as shown in Figure 7-4. One covalent bond forms from the overlap between the hydrogen $1s$ orbital and the fluorine $2p$ orbital that points along the bond axis.

H—F̈:

Lewis structure of hydrogen fluoride

The Lewis structure of hydrogen fluoride shows three lone pairs on the fluorine atom. These non-bonding electrons are localized in atomic orbitals that belong solely to fluorine. Remembering that one of the fluorine $2p$ orbitals is used to form the H—F bond, we conclude that the three lone pairs must occupy the remaining pair of $2p$ orbitals and the $2s$ orbital of the fluorine atom.

A similar situation arises when two fluorine atoms approach each other. The first valence orbitals to overlap are the $2p$ orbitals pointing along the axis that joins the atoms. Figure 7-5 shows that bond formation in molecular fluorine results from the strong directional overlap of these two atomic $2p$ orbitals.

**Figure 7-4**

Strong $1s$–$2p$ overlap

H        F        HF

The bond in H—F forms from orbital overlap between the hydrogen $1s$ orbital and the fluorine $2p$ orbital that points along the bond axis.

## Bonding in Hydrogen Sulphide

Hydrogen sulphide is a toxic gas with the foul odour of rotten eggs. The Lewis structure of $H_2S$ shows two bonds and two lone pairs on the S atom. Experiments show that hydrogen sulphide has a bond angle of 92.1°. We can describe the bonding of $H_2S$ by applying the orbital overlap model.

Sulphur has the valence configuration $3s^2\,3p^4$, with two of the valence $p$ orbitals occupied by single electrons. Recall from Section 4.6 that atomic $p$ orbitals are separated by 90° and thus are in just the right orientation to form a pair of H—S bonds at the appropriate bond angle. As Figure 7-6 shows, the $1s$ orbitals of the two hydrogen atoms overlap with two of the sulphur $3p$ orbitals to form a pair of bonding orbitals at right angles to one another, consistent with the experimental data for $H_2S$. The slight increase of the H—S—H bond angle from the ideal value of 90° is consistent with repulsion of the two hydrogen atoms. This minor deviation is not large enough to justify alteration of the model. The Lewis structure of $H_2S$ shows the molecule with two lone pairs localized on the sulphur atom. These non-bonding electrons occupy the $3s$ and the remaining $3p$ orbital of the sulphur atom.

Example 7-1 applies the orbital overlap model to phosphine, $PH_3$.

**Figure 7-5**

Strong $2p$–$2p$ overlap

The bond in $F_2$ forms from orbital overlap between a pair of fluorine $2p$ orbitals that point along the bond axis.

Lewis structure and ball-and-stick model of hydrogen sulphide

Valence electron configuration of sulphur

**Figure 7-6**

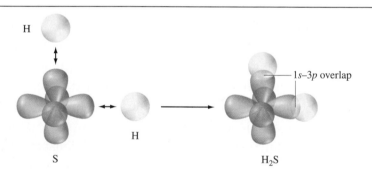

The bonding in hydrogen sulphide can be described as the overlap of two hydrogen $1s$ orbitals with a pair of sulphur $3p$ orbitals.

### EXAMPLE 7-1 ORBITAL OVERLAP IN PHOSPHINE

Phosphine is a colourless, highly toxic gas with bond angles of 93.6°. Describe the bonding in $PH_3$.

Valence configuration of phosphorus

**Strategy:** We begin with the Lewis structure of $PH_3$, and apply the principles of orbital overlap. The bonding description of the molecule must be consistent with the bond angles, so we must pay particular attention to the orientation of the valence orbitals.

**Solution:** The Lewis structure of phosphine shows three P—H bonds and one lone pair on the phosphorus atom.

$$H\!-\!\overset{\displaystyle ..}{P}\!-\!H$$
$$|$$
$$H$$

Ball-and-stick model of phosphine. The bond angles are 93.6°.

The valence electron configuration of phosphorus is $3s^2\ 3p^3$, with a single electron in each of the three $3p$ orbitals. The angles between these $p$ orbitals are 90°, so the phosphorus atom is in the proper configuration to form three P—H bonds by direct orbital overlap with three hydrogen $1s$ orbitals. The angles increase by 3.6° because of repulsion among the hydrogen atoms. Finally, the lone pair of electrons on the phosphorus atom occupies the $3s$ atomic orbital. A ball-and-stick model shows the shape of the molecule.

**Does the Result Make Sense?** When bond angles are close to 90°, as in phosphine, it is sensible to describe the bonding using atomic orbitals that point at right angles to one another; in other words, the valence $p$ orbitals.

**Practice Exercise 7.1**

Hydrogen selenide is yet another offensive-smelling toxic gas. The bond angle in $H_2Se$ is 92.1°. Describe the bonding of hydrogen selenide and draw a ball-and-stick model that shows the shape of the molecule.

## Methane: The Need for an Expanded Model

**Figure 7-7**

Tetrahedral molecular geometry, with 109.5° bond angles, minimizes repulsion among the bonding electron pairs of methane.

As described in Chapter 6, methane adopts tetrahedral molecular geometry because of electron–electron repulsions among the four pairs of bonding electrons. Figure 7-7 shows 109.5° angles between the four C—H bonds, as predicted by the valence shell electron pair repulsion (VSEPR) model. How can the carbon atom bond to four hydrogen atoms in this geometry? The valence $2p$ orbitals of the carbon atom are separated by 90° angles, and the $2s$ orbital has a spherical shape. Clearly, the four valence orbitals of carbon cannot overlap individually with four hydrogen $1s$ orbitals and give the known tetrahedral geometry of methane. Nevertheless, experiments verify the bond angles predicted by the VSEPR model, 109.5°. When theory and experiment disagree, the theory must be modified. In the next section, we present additional features of the orbital overlap model that allow for bond angles that are impossible to describe using traditional atomic orbitals.

## SECTION EXERCISES

● **7.1.1** Draw a figure similar to Figure 7-2 that shows changes in the wave functions as a hydrogen atom approaches a fluorine atom to form an H—F bond.

● **7.1.2** Describe the bond between a hydrogen atom and a bromine atom in a molecule of hydrogen bromide, HBr.

● **7.1.3** The gaseous compound stibine, $SbH_3$, has experimental bond angles of 91.3°. Give a complete description of the bonding in stibine, including a ball-and-stick sketch that shows the shape of the molecule.

## 7.2 HYBRIDIZATION OF ATOMIC ORBITALS

We cannot generate a tetrahedron by simple overlap of atomic orbitals, because atomic orbitals do not point toward the corners of a tetrahedron. In this section, we present a modification of the localized bond model that accounts for tetrahedral geometry and several other common molecular shapes.

## Methane: *sp³* Hybrid Orbitals

The most convenient way to visualize tetrahedral bonding in methane is to imagine the four carbon valence orbitals mixing together to generate a set of four new orbitals. This can be done in such a way that each new orbital points toward a different corner of a tetrahedron. Combining atomic orbitals on a particular atom to form a special set of directional orbitals is referred to as **hybridization**. Every hybrid orbital is directional, with a lobe of high electron density pointing in one specific direction.

Any hybrid orbital is named from the atomic valence orbitals from which it is constructed. To match the geometry of methane, we need four orbitals that point at the corners of a tetrahedron. We construct this set from one *s* orbital and three *p* orbitals, so the hybrids are called *sp³* **hybrid orbitals**. Figure 7-8a shows the detailed shape of an *sp³* hybrid orbital. For simplicity, the small lobe is often omitted from diagrams involving sp³ hybrid orbitals. Figure 7-8b is an orbital depiction of methane, showing four *sp³* hybrid orbitals interacting with 1s orbitals from four H atoms. Note that all four *sp³* hybrids have the same shape, but each points to a different corner of a regular tetrahedron.

Orbital mixing always generates changes in orbital energies as well as in orbital shapes. Hybrid orbitals always have different energies than do the pure atomic orbitals from which they form. The energy-level diagram shown in Figure 7-9 summarizes the hybridization process.

**Figure 7-8**

(a)

(b)

(a) The shape of an *sp³* hybrid orbital. The smaller lobe is omitted in subsequent diagrams for clarity. (b) Orbital depiction of methane, showing four *sp³* hybrid orbitals interacting with 1s orbitals from four H atoms.

**Figure 7-9**

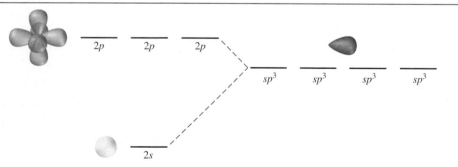

Hybrid orbitals form from combinations of atomic valence orbitals.

A complete orbital overlap view of methane appears in Figure 7-10. Hybridization gives each carbon orbital a strongly favoured direction for overlap with an atomic 1s orbital from an approaching hydrogen atom. Four such interactions generate four localized bonds that use all the valence electrons of the five atoms involved.

## Other Tetrahedral Systems

According to the VSEPR model developed in Chapter 6, an inner atom with a steric number of 4 adopts tetrahedral electron group geometry. This tetrahedral arrangement of four electron groups is very common, the only important exceptions being the hydrides of elements beyond the second row, such as $H_2S$ and $PH_3$.

**Figure 7-10**

Methane forms from orbital overlap between the hydrogen $1s$ orbitals and the $sp^3$ hybrid orbitals of the carbon atom.

Example 7-2 applies the orbital overlap molecule to the hydronium ion, and Example 7-3 treats a molecule with two inner atoms.

## EXAMPLE 7-2  BONDING IN THE HYDRONIUM ION

Describe the bonding of the hydronium ion, $H_3O^+$.

**Strategy:** Use the strategies from Chapter 6 to determine the Lewis structure, steric number of the inner atom, and electron group geometry. The steric number also determines the hybridization.

**Solution:** A hydronium ion has eight valence electrons. Six are used to make three O—H single bonds, and two are placed as a lone pair on the oxygen atom.

The Lewis structure shows the inner oxygen atom with three ligands and one lone pair, which gives a steric number of 4. A steric number of 4 indicates that the electron group geometry is tetrahedral and that the bonding can be described with $sp^3$ hybrids for the oxygen atom. Three hybrid orbitals overlap with hydrogen $1s$ atomic orbitals to form bonds. The fourth $sp^3$ hybrid contains a lone pair of electrons.

**Does the Result Make Sense?** Remember that the molecular shape ignores the lone pair. The hydronium ion has a trigonal pyramidal shape described by the three $sp^3$ hybrid orbitals that form bonds to hydrogen atoms.

**Common Pitfalls:** Not determining the correct Lewis structure. The Lewis structure is needed in order to determine the hybridization of the inner atom.

**Practice Exercise 7.2**

Describe the bonding of a water molecule.

## EXAMPLE 7-3 A MOLECULE WITH TWO INNER ATOMS

Describe the bonding of methanol, $CH_3OH$. Sketch an orbital overlap picture of the molecule.

**Strategy:** Begin with the Lewis structure, and then survey each inner atom for its steric number and hybridization. Here is a summary of the information:

$$H-\overset{\overset{\displaystyle H}{|}}{\underset{\underset{\displaystyle H}{|}}{C}}-\overset{..}{\underset{..}{O}}-H$$

Ligands = 4          Ligands = 2
Lone pairs = 0       Lone pairs = 2
Steric number = 4    Steric number = 4
Hybridization = $sp^3$   Hybridization = $sp^3$

**Solution:** Both inner atoms have steric numbers of 4 and tetrahedral electron group geometry, so both can be described using $sp^3$ hybrid orbitals. All four hydrogen atoms occupy outer positions, and these form bonds to the inner atoms through $1s-sp^3$ overlap. The oxygen atom has two lone pairs, one in each of the two hybrid orbitals not used to form O—H bonds.

Sketching an orbital overlap picture for a molecule like methanol takes practice. Because drawing is an important skill in chemistry, time spent practicing is a good investment. One way to begin is to draw the bond between the inner atoms, showing direct end-on overlap of the hybrid orbitals. Next, fill in the tetrahedral geometry around each inner atom, and then add the outer hydrogen atoms. Finish the sketch by adding the two lone pairs.

Remember that the four positions of a tetrahedron are equivalent, so it does not matter which orbitals hold the lone pairs and which ones form bonds. The use of colour helps to clarify the sketch. At the right is one possible sketch of the molecule, but other orientations would be equally valid.

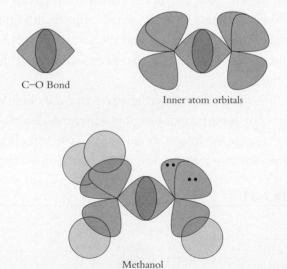

C—O Bond

Inner atom orbitals

Methanol

**Does the Result Make Sense?** You can visualize methanol either as methane with one —H replaced by —OH or as water with one —H replaced by —$CH_3$. Either way of visualizing the molecule reveals a tetrahedral inner atom, for which $sp^3$ hybridization is appropriate.

### Practice Exercise 7.3
Describe the bonding of methylamine, $CH_3NH_2$. Sketch an orbital overlap picture of the molecule.

## General Features of Hybridization

When direct overlap of atomic orbitals cannot predict bond angles correctly, we generate hybrid orbitals on the inner atoms. Consequently, each of the electron group geometries described in Chapter 6 is associated with its own specific set of hybrid orbitals. Each type of hybrid orbital scheme shares the characteristics described in our discussion of methane:

1. The number of hybrid orbitals generated by the hybridization process equals the number of valence atomic orbitals participating in hybridization.

2. The steric number of an inner atom uniquely determines the number and type of hybrid orbitals.

3. Hybrid orbitals form localized bonds by overlap with atomic orbitals or with other hybrid orbitals.

4. There is no need to hybridize orbitals on outer atoms, because these atoms do not have limiting geometries. Hydrogen always forms localized bonds with its 1s orbital. The bonds formed by all other outer atoms can be described using valence $p$ orbitals.

To visualize bond formation by an outer atom other than hydrogen, recall the bond formation in HF. One valence $p$ orbital from the fluorine atom overlaps strongly with the hydrogen 1s orbital to form the bond. We can describe bond formation for any outer atom except H through overlap of one of its valence $p$ orbitals with the appropriate hybrid orbital of the inner atom. An example is dichloromethane, $CH_2Cl_2$, which appears in Figure 7-11. We describe the C—H bonds by $sp^3$–1s overlap, and the C—Cl bonds by $sp^3$–3p overlap. Each C—Cl bond results from overlap between a carbon $sp^3$ hybrid and a chlorine 3p atomic orbital.

It is possible to describe outer atoms as hybridized, and some computer programs for predicting molecular structures describe bonding by hybridizing every atom except hydrogen. It is easier to use unhybridized orbitals to describe outer atoms, for two reasons. First, hybridization is invoked in order to produce sets of

**Figure 7-11**

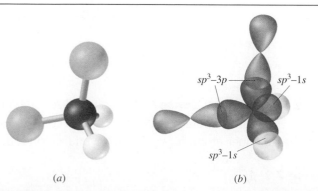

(a)    (b)

(a) Ball-and-stick model of $CH_2Cl_2$. (b) The bonding in $CH_2Cl_2$ involves two $sp^3$–1s C—H bonds and two $sp^3$–3p C—Cl bonds.

orbitals that point in directions that match observed bond angles. There are no bond angles centred on outer atoms, so there is no need for such directional orbitals. Second, hybridization of an outer atom places all non-bonding electrons in hybrid orbitals that have the same energy. In contradiction to this prediction, photoelectron spectroscopy reveals that the non-bonding electrons on outer atoms have two distinct energies that correspond more closely to the energies of $s$ and $p$ atomic orbitals. Remember, however, that all bonding descriptions are models that oversimplify the electron distribution of any actual molecule. All models of bonding describe some features well but fall short of describing all features accurately.

## $sp^2$ Hybrid Orbitals

Triethylaluminum appears in Chapter 6 as an example of a molecule with a trigonal planar inner atom, with 120° angles separating the three Al—C bonds (Figure 6-20). The valence orbitals of the Al atom have the wrong geometry to form a trigonal plane. Furthermore, a set of $sp^3$ hybrids is not appropriate, because these would give C—Al—C bond angles of 109.5° rather than 120°. We need a different set of hybrid orbitals to represent an atom with trigonal planar electron group geometry. Proper mixing of the $p$ orbital and two $p$ orbitals gives a set of three **$sp^2$ hybrid orbitals**. A set of $sp^2$ hybrid orbitals thus makes use of two of the valence $p$ orbitals, but not the third. An individual $sp^2$ hybrid orbital looks very much like its $sp^3$ counterpart, with high electron density along one axis. As Figure 7-12 shows, however, these hybrid orbitals differ from the $sp^3$ set in their orientations. The three $sp^2$ hybrid orbitals point to the three corners of an equilateral triangle. Figure 7-12 also shows that the unused $p$ orbital is perpendicular to the plane of the three hybrids.

**Figure 7-12**

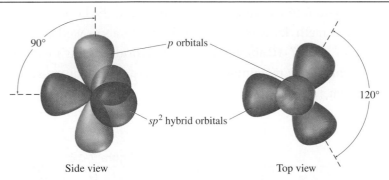

Side view          Top view

An $sp^2$–hybridized atom has three coplanar hybrid orbitals separated by 120° angles. One unchanged $p$ orbital is perpendicular to the plane of the hybrids.

In triethylaluminum, each Al—C bond is the result of an $sp^2$ hybrid on aluminum overlapping with an $sp^3$ hybrid on a carbon atom. Figure 7-13 shows this bonding representation, with three equivalent Al—C bonds and the unused 3$p$ orbital on the aluminum atom.

Remember from Chapter 3 that energy is released when a bond forms. Consequently, atoms that form covalent bonds tend to use all their valence $s$ and $p$ orbitals to make as many bonds as possible. We might expect the $sp^2$–hybridized

**Figure 7-13**

Side view

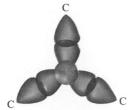

Top view

An orbital overlap picture for the Al atom in triethylaluminum, showing its three equivalent bonds and the remaining 3$p$ valence orbital. For clarity, we show only one hybrid orbital from each $sp^3$–hybridized carbon atom.

aluminum atom to form a fourth bond with its unused $3p$ orbital. A fourth bond does not form in $Al(C_2H_5)_3$ because the carbon atoms bonded to aluminum have neither orbitals nor electrons available for additional bond formation. However, the potential to form a fourth bond with some other atom makes triethylaluminum a very reactive molecule.

## *sp* Hybrid Orbitals

Dimethylzinc appears in Chapter 6 as an example of linear geometry. There are two bonds to the zinc atom and a C—Zn—C bond angle of 180°. To describe linear orbital geometry, we need a hybridization scheme that generates two orbitals pointing in opposite directions. This new scheme is a pair of **$sp$ hybrid orbitals**, formed from the zinc $4s$ orbital and one of its $4p$ orbitals. Figure 7-14 shows the shapes and orientations of the two $sp$ hybrid orbitals, and Figure 7-15 presents the orbital overlap picture of dimethylzinc. Each C—Zn bond in dimethylzinc can be described as a localized orbital resulting from the overlap of an $sp^3$ hybrid on the carbon atom and an $sp$ hybrid on the zinc atom. For clarity, the figure shows only one of the $sp^3$ hybrid orbitals for each methyl group.

Dimethylzinc

**Figure 7-14**

$s$     $p$     $sp$   $sp$

A pair of $sp$ hybrid orbitals forms by interaction of an $s$ atomic orbital and one $p$ atomic orbital. The two hybrids point in opposite directions.

As with the aluminum atom in triethylaluminum, the bonding description of the zinc atom in dimethylzinc includes vacant $p$ orbitals as well as hybrid orbitals for bonding. As Figure 7-16 shows, a set of $sp$ hybrid orbitals makes use of only one valence $p$ orbital. The remaining two $p$ orbitals are perpendicular to each other and perpendicular to the pair of hybrids.

**Figure 7-15**

The Zn—C bonds of dimethylzinc are formed by overlap of an $sp$ hybrid orbital from the metal with an $sp^3$ hybrid orbital from carbon.

**Figure 7-16**

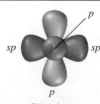

Side view

An $sp$-hybridized atom has two hybrid orbitals separated by 180°. The remaining two atomic $p$ orbitals are perpendicular to the hybrids and perpendicular to each other.

Compounds like triethylaluminum and dimethylzinc that have metal–carbon bonds are rather uncommon. Nevertheless, trigonal planar geometry ($sp^2$) and linear geometry ($sp$) occur frequently in organic chemistry. As we show in Section 7.3, these geometries and their corresponding hybridizations occur in molecules with double bonds and triple bonds, respectively.

## The Participation of *d* Orbitals

Elements beyond the second row of the periodic table can form bonds to more than four ligands and can be associated with more than an octet of electrons. These features are possible for two reasons. First, elements with $n > 2$ have atomic radii that are large enough to bond to 5, 6, or even more ligands. Second, elements with $n > 2$ have *d* orbitals with energies close to the energies of the valence *p* orbitals. An orbital overlap description of the bonding in these species relies on the participation of *d* orbitals of the inner atom.

Phosphorus pentachloride (Figure 6-21) exemplifies the electron group geometry associated with a steric number of 5 and trigonal bipyramidal geometry. Four valence 3s and 3p orbitals can form no more than four covalent bonds, but phosphorus can form additional bonds using the 3d orbitals in its valence shell. One way to involve a 3d orbital in bond formation is by combining the 3s orbital, the set of 3p orbitals, and one 3d orbital to form a set of five $sp^3d$ **hybrid orbitals**. In $PCl_5$, each P — Cl bond results from overlap between a chlorine 3p orbital and an $sp^3d$ hybrid on the phosphorus atom.

A molecule with a steric number of 6 requires six hybrid orbitals arranged in octahedral geometry. In Chapter 6, sulphur hexafluoride is the primary example of a molecule with a steric number of 6 (Figure 6-24). Six equivalent orbitals for sulphur can be constructed for the inner sulphur atom by combining the 3s, the three 3p, and two of the 3d orbitals to form a set of $sp^3d^2$ *hybrid orbitals*. Each S — F bond results from overlap of an $sp^3d^2$ hybrid orbital from sulphur with a 2p orbital from fluorine. Figure 7-17 shows ball-and-stick models and orbital overlap sketches of $PCl_5$ and $SF_6$.

**Figure 7-17**

(a)    (b)    (c)    (d)

Ball-and-stick models and orbital overlap pictures of $PCl_5$ (a and b) and $SF_6$ (c and d).

Example 7-4 examines the role of lone pairs in $sp^3d$ and $sp^3d^2$ hybrid systems.

## EXAMPLE 7-4 $sp^3d$ AND $sp^3d^2$ HYBRID SYSTEMS

Describe the bonding of chlorine trifluoride and xenon tetrafluoride.

**Strategy:** To describe bonding, start with a Lewis structure, determine the steric number, and use the steric number to assign the electron group geometry and hybridization of the inner atom. Treat each molecule separately.

**Solution:** These molecules appear in Examples 6-14 and 6-15. The results are reproduced here, but you should review those examples in Chapter 6 if the procedure is not clear to you:

With a steric number of 5, chlorine has trigonal bipyramidal electron group geometry. This means the inner atom requires five directional orbitals, which are provided by an $sp^3d$ hybrid set. Fluorine uses its valence $2p$ orbitals to form bonds by overlapping with the hybrid orbitals on the chlorine atom. Remember that the trigonal bipyramid has non-equivalent axial and equatorial sites. As we described in Chapter 6, lone pairs always occupy equatorial positions. See the orbital overlap view below.

With a steric number of 6, xenon has octahedral electron group geometry. This means the inner atom requires six directional orbitals, which are provided by an $sp^3d^2$ hybrid set. Fluorine uses its valence $2p$ orbitals to form bonds by overlapping with the hybrid orbitals on the xenon atom. The two lone pairs are on opposite sides of a square plane, to minimize electron–electron repulsion. See the orbital overlap view below.

$ClF_3$                    $XeF_4$

**Do the Results Make Sense?** The hybridization schemes match the shapes that we describe in Chapter 6. Notice that the steric numbers of inner atoms uniquely determine both the electron group geometry and the hybridization. This makes sense, because the steric number describes how many electron groups must be accommodated around an inner atom.

### Practice Exercise 7.4

Describe the bonding and draw orbital overlap pictures of $SF_4$ and $ClF_5$.

The $sp^3d$ and $sp^3d^2$ hybrid orbital sets complete our survey of the common chemical geometries. Table 7-1 summarizes the results. Example 7-5 completes this section of the chapter.

### TABLE 7-1  A SUMMARY OF VALENCE ORBITAL HYBRIDIZATION

| Steric Number | Electron Group Geometry | Hybridization | Number of Hybrid Orbitals | Number of Unused $p$ Orbitals | Picture |
|---|---|---|---|---|---|
| 2 | Linear | $sp$ | 2 | 2 | |
| 3 | Trigonal planar | $sp^2$ | 3 | 1 | |
| 4 | Tetrahedral | $sp^3$ | 4 | 0 | |
| 5 | Trigonal bipyramidal | $sp^3d$ | 5 | 0 | |
| 6 | Octahedral | $sp^3d^2$ | 6 | 0 | |

### EXAMPLE 7-5  ASSIGNING HYBRID ORBITALS

For each of the following Lewis structures, name the electron group geometry and the hybrid orbitals used by the inner atoms.

(a) $:\ddot{C}l-\underset{\textstyle |}{\overset{\textstyle :\ddot{C}l:}{N}}-H$   (b) $:\ddot{I}-\ddot{I}-\ddot{I}:^-$   (c) $H_3C-\underset{\textstyle |}{\overset{\textstyle CH_3}{Ga}}-CH_3$

**Strategy:** The steric number of an inner atom determines the electron group geometry, each of which is associated with one specific type of hybrid orbital.

**Solution:**

(a)  The inner nitrogen atom bonds to three ligands and has one lone pair. The steric number is 4, and the electron group geometry is tetrahedral. Four groups of electrons require four hybrid orbitals, so the nitrogen atom uses $sp^3$ hybrid orbitals.

(b) The inner iodine atom bonds to two ligands and has three lone pairs. The steric number is 5, and the electron group geometry is trigonal bipyramidal. The inner atom adopts $sp^3d$ hybridization.

(c) Gallium bonds to three methyl groups and has no lone pairs, thus giving a steric number of 3. The electron group geometry is trigonal planar, and the gallium atom uses $sp^2$ hybrid orbitals. (The carbon atoms of the methyl group ligands are tetrahedral and utilize $sp^3$ hybridization.)

**Do the Results Make Sense?** The hybridization choices for inner atoms are correct when they match the steric numbers as indicated by the Lewis structures.

**Practice Exercise 7.5**
For each of the following Lewis structures, name the electron group geometry and the hybrid orbitals used by the inner atoms.

(a)    (b) :F̈—Ö—F̈:    (c) :C̈l—S̈n—C̈l:

---

## SECTION EXERCISES

● **7.2.1** Draw an energy-level diagram that summarizes the $sp^2$ hybridization process. Use Figure 7-9 as a guide.

● **7.2.2** Give a complete bonding description of $SCl_2$. Your answer should include an orbital overlap sketch in which you label all of the orbitals with their names. Name the electron group geometry and the molecular shape.

● **7.2.3** The following species all have four ligands, but each has a different geometry and different hybridization: $SiF_4$, $TeF_4$, and $ClF_4^-$. Draw a complete Lewis structure for each of them, name each electron group geometry, and identify the hybrid orbitals used by each inner atom.

## 7.3 MULTIPLE BONDS

Many of the Lewis structures in Chapter 6 and elsewhere in this book represent molecules that contain double or triple bonds. From simple molecules such as ethylene and acetylene to complex biochemical compounds such as chlorophyll and plastoquinone, multiple bonds are abundant in chemistry. Double and triple bonds can be described by extending the orbital overlap model of bonding. We begin with ethylene, a simple hydrocarbon with the formula $C_2H_4$.

Ethylene is a colourless, flammable gas with a boiling point of $-104\ °C$. Approximately 120 billion kilograms of ethylene are produced annually in the world (4.6% of this in Canada), making it one of the top five industrial chemicals. The manufacture of plastics (polyethylene is the most common example) consumes 75% of this output, and much of the rest is used to make antifreeze (ethylene glycol). Because ethylene stimulates the breakdown of cell walls, it is used commercially to hasten the ripening of fruit, particularly bananas.

## Bonding in Ethylene

Every description of bonding starts with a Lewis structure. Ethylene has 12 valence electrons. The bond framework of the molecule has one C—C bond and four C—H bonds, requiring 10 of these electrons. We place the final two electrons as a lone pair on one of the carbon atoms, leaving the second carbon atom with only six electrons. Making a double bond between the carbon atoms gives both carbon atoms octets and completes the Lewis structure.

Lewis structure            Ball-and-stick Model

What orbital overlap picture best describes ethylene? Both carbon atoms have three ligands, and neither has a lone pair. Thus, each carbon atom has a steric number of 3, which predicts trigonal planar electron group geometry. Experiments show that $C_2H_4$ is a planar molecule with bond angles close to 120°. These features are consistent with $sp^2$ hybridization, so a bonding framework can be constructed using carbon $sp^2$ hybrid orbitals and hydrogen $1s$ orbitals. According to this model, each C—H bond involves overlap of a carbon $sp^2$ hybrid with a hydrogen $1s$ atomic orbital, and the C—C bond forms from the overlap of two $sp^2$ hybrid orbitals. Figure 7-18 shows the bonding framework for ethylene.

The orbital overlap view in Figure 7-18 shows one bond joining the carbon atoms in ethylene, but the Lewis structure indicates that a *double* bond exists between the carbon atoms. A look at the full set of valence orbitals lets us see how to construct a second bond between the carbon atoms of ethylene. The $sp^2$ hybrid orbitals require three of the valence orbitals, leaving one valence $p$ orbital that is not part of the hybrid set. The electron density of this $p$ orbital is concentrated above and below the $sp^2$ plane. As Figure 7-19 shows, these "leftover" $p$ orbitals on adjacent carbon atoms overlap in a side-by-side fashion. This is *one* bond that has two regions of high electron density, one on either side of the plane defined by the bonding framework. Figure 7-19 shows a π bond from four perspectives.

## σ Bonds and π Bonds

In the bond framework in Figure 7-18, all the bonds form from end-on overlap of orbitals directed toward each other. As illustrated by the three examples in Figure 7-20, this type of overlap gives high electron density distributed symmetrically *along* the internuclear axis. A bond of this type is called a **sigma (σ)** bond, and a bonding orbital that describes a σ bond is a σ orbital.

As Figure 7-19 shows, bonds that form from the side-by-side overlap of atomic *p* orbitals have different electron density profiles than σ bonds. A *p* orbital has zero electron density—a node—in a plane passing through the nucleus, so bonds that form from side-by-side overlap have no electron density directly on

**Figure 7-18**

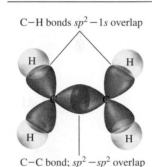

C—H bonds $sp^2 - 1s$ overlap

C—C bond; $sp^2 - sp^2$ overlap

A top view shows that the bonding framework of ethylene involves $sp^2$ hybrids on the carbon atoms and $1s$ orbitals on the hydrogen atoms.

**Figure 7-19**

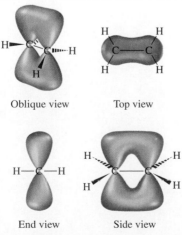

Oblique view            Top view

End view            Side view

Four views of a π bond.

**Figure 7-20**

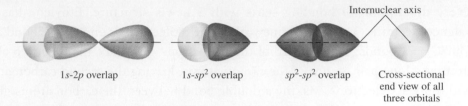

1s-2p overlap    1s-sp² overlap    sp²-sp² overlap    Cross-sectional end view of all three orbitals

Internuclear axis

All σ bonds have high electron density concentrated along the internuclear axis and axial symmetry, so their end-on profiles are circles.

the bond axis. High electron density exists between the bonded atoms, but it is concentrated *above and below* the bond axis. A bond of this type is called a **pi (π)** bond, and a bonding orbital that describes a π bond is a π orbital.

The double bond in ethylene contains one σ bond and one π bond. The σ bond forms from the end-on overlap of two hybrid orbitals, and the π bond forms from the side-by-side overlap of two atomic *p* orbitals. Figure 7-21 shows the complete orbital picture of the bonding in ethylene. Ethylene is the simplest of a class of molecules, the alkenes, all of which contain C═C double bonds. The alkenes are the subject of Focus 7-1: Chemistry and Life.

**Figure 7-21**

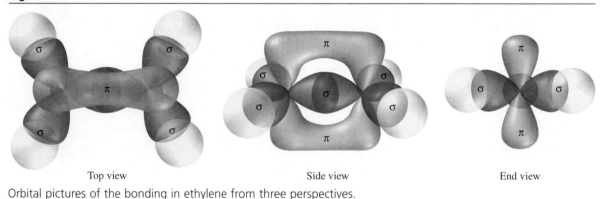

Top view                    Side view                    End view

Orbital pictures of the bonding in ethylene from three perspectives.

The availability of a leftover valence *p* orbital to form π bonds is characteristic of *sp²* hybridization and is not restricted to carbon atoms. Example 7-6 shows how the same bonding picture applies to a compound that has a C═N double bond. We construct this bonding picture using the following procedure:

1. Determine the Lewis structure.

2. Use the Lewis structure to determine steric numbers and hybridizations.

3. Construct the σ bond framework.

4. Add the π bonds.

## EXAMPLE 7-6  ORBITAL OVERLAP IN DOUBLE BONDS

An imine is a molecule that contains a carbon═nitrogen double bond. Describe the bonding of the simplest possible imine, $H_2CNH$, by sketching the $\sigma$ and $\pi$ bonding systems.

**Strategy:** Follow the four-step procedure. Use the Lewis structure to identify the appropriate hybridization. Then draw sketches of the various orbitals involved in the bonding.

**Solution:** Each molecule of $H_2CNH$ has 12 valence electrons. The bonding framework requires eight electrons, leaving four electrons. Placing these on nitrogen, the more electronegative inner atom, leaves the carbon atom with only three pairs of electrons. Give the carbon atom an octet by transferring one lone pair from the nitrogen atom to make a C═N bond:

$$
\begin{array}{c}
H \\
\phantom{H}\diagdown \\
C=\ddot{N} \\
\phantom{H}\diagup \phantom{aa} \diagdown \\
H \phantom{aaaa} H
\end{array}
$$

The carbon and nitrogen atoms each have steric numbers of 3, so their bonding frameworks can be represented using $sp^2$ hybrid orbitals. Now we can sketch the $\sigma$ bonding system of imine:

The carbon and nitrogen atoms each have a $p$ orbital left over after construction of the $sp^2$ hybrids. These two $p$ orbitals are perpendicular to the plane that contains the five nuclei. Side-by-side overlap of the $p$ orbitals gives a $\pi$ bond that completes the bonding description of this molecule:

$2p - 2p\ \pi$ bond

**Does the Result Make Sense?** Imine and ethylene have the same number of valence electrons and similar bonding frameworks. Compare these bonding pictures with those for ethylene. Notice that they are identical, except that a lone pair on the nitrogen atom replaces one of the C—H bonds.

**Common Pitfalls:** Neglecting the $p$ orbitals left over after hybridization. These are used to make $\pi$ bonds.

### Practice Exercise 7.6

Chloroethylene, $H_2CCHCl$ (also known as vinyl chloride), is the precursor to the common polymer "PVC." Describe the bonding of chloroethylene by sketching the $\sigma$ and $\pi$ bonding systems.

## π Bonds Involving Oxygen Atoms

In addition to carbon and nitrogen, oxygen is a common participant in the formation of π bonds. The bonding patterns for oxygen are illustrated by acetic acid ($CH_3CO_2H$), the Lewis structure of which appears in Example 6-9. The Lewis structure shows that acetic acid has a carbon atom and an inner oxygen with steric numbers of 4, indicating tetrahedral orbital geometry and bonding that can be described using $sp^3$ hybrids. The second carbon atom has a steric number of 3, indicating trigonal planar orbital geometry and bonding that can be described using $sp^2$ hybrids. Remember that outer atoms do not require hybrid orbitals, so the outer oxygen atom in acetic acid uses a $2p$ orbital to form a σ bond to carbon. The σ bond framework appears in Figure 7-22a.

Once the σ bonding framework is completed, the only atoms with unused valence orbitals are the $sp^2$–hybridized carbon atom and the outer oxygen atom. As shown in Figures 7-22b and 7-22c, each of these atoms has a $2p$ atomic orbital that is positioned perfectly for side-by-side overlap, just as in ethylene and imine.

The description of acetic acid is completed by identifying the orbitals that contain the two lone pairs on the outer oxygen atom. The σ bond and the π bond account for two valence $2p$ orbitals of the oxygen atom. This leaves the third $2p$ orbital and the $2s$ orbital for the lone pairs.

Lewis structure of acetic acid

### Figure 7-22

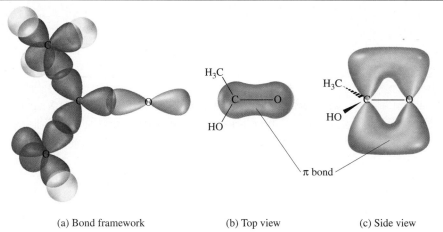

(a) Bond framework        (b) Top view        (c) Side view

(a) The bonding framework of acetic acid is composed of σ bonds. (b) A top view shows how the π bond forms by side-by-side overlap of $2p$ orbitals from oxygen and carbon. (c) Side view of the π bond.

## FOCUS 7-1 CHEMISTRY AND LIFE: THE ALKENES

Alkenes are hydrocarbons that contain one or more carbon=carbon double bonds. The simplest alkene, ethylene, is produced by plants to stimulate the ripening of fruit. Larger alkenes are abundant in the biological world and play a wide variety of roles.

Animals communicate by excreting minute amounts of chemical signalling agents known as *pheromones*. Some pheromones repel predators, others mark trails, and still others signal alarm. Pheromones have a wide variety of chemical structures, but many of them are alkenes. For example, the trail marker used by termites is neocembrene, an alkene with four carbon–carbon double bonds. Animals produce pheromones to attract members of

Neocembrene ($C_{20}H_{32}$)
(a termite trail marker)

the opposite sex. One such pheromone is released by female moths to signal that they are ready to mate. A male moth that senses this pheromone immediately seeks out the female. Remarkably, the same molecule has been isolated from the urine of female elephants that are about to ovulate. The presence of this pheromone sends a signal to bull elephants that the mating season has arrived. The similar function of this same compound in two different species is a surprising coincidence of evolution.

Plants produce a vast array of *terpenes,* alkenes built in multiples of five carbon atoms. Many terpenes have characteristic fragrances. For example, the fresh odour of a pine forest is due to pinene, a 10-carbon molecule with a ring structure and one double bond. The fragrances of terpenes make them important in the flavour and fragrance industry. Limonene, another 10-carbon molecule with a ring and two double bonds, is the principal component of lemon oil. Geraniol, a chain-like molecule with two double bonds, is one of the molecules responsible for the fragrance of roses and is used in many perfumes. Many other terpenes have important medicinal properties.

Moth/elephant sex attractant

Pinene ($C_{10}H_{16}$)     Limonene ($C_{10}H_{16}$)     Geraniol ($C_{10}H_{18}O$)

Alkenes with many double bonds in a row are coloured. Some plant pigments are alkenes of this kind. One example is β–carotene, which gives carrots their distinctive orange colour. Animals break down β–carotene into Vitamin A, which is essential for vision. Xanthin molecules, relatives of β–carotene that contain oxygen atoms, occur in corn, orange juice, and shellfish. The xanthin shown at right makes the flamingo pink.

Xanthin ($C_{38}H_{48}O_2$)

Isoprene may be the naturally occurring alkene with the greatest economic impact. This compound, a major component of the sap of the rubber tree, is used to make the long-chain molecules of natural rubber (polyisoprene). As we describe in Chapter 17, the synthetic rubbers that make up most of today's tires are made from other alkenes.

Long chain     Long chain

Polyisoprene

Used tires create serious environmental problems, because they do not degrade easily and they release noxious contaminants when they burn. Recently, a process has been developed that breaks tires down into a polyisoprene oil that can then be further decomposed into limonene. The yield of limonene is only a few percent, but if this process can be made more efficient, it may become possible to turn ugly, smelly old tires into fragrant oil of lemon.

Ozone, $O_3$, is the only exception to this generalization about inner oxygen atoms.

The bonding pattern of the oxygen atoms in acetic acid repeats in most other oxygen-containing compounds. Outer oxygen atoms frequently have some $\pi$ character in their bonding. Inner oxygen atoms, on the other hand, do not participate in $\pi$ bonding. Any inner oxygen atom has a steric number of 4 and can be described appropriately using $sp^3$ hybrid orbitals.

## To $\pi$ Bond or Not To $\pi$ Bond: Carbon vs. Silicon

Carbon forms $\pi$ bonds readily, but silicon, its Group 14 neighbour, forms $\sigma$ bonds with tetrahedral geometry in almost all its compounds. Compare the most stable oxide of carbon, $CO_2$, with the most stable oxide of silicon, $SiO_2$. Despite their similar chemical formulas, these two oxides could hardly be more different. Carbon dioxide exists as discrete linear triatomic molecules with its atoms strongly bonded together by a combination of $\sigma$ bonds and $\pi$ bonds. Silicon dioxide, on the other hand, exists as a network of silicon and oxygen atoms containing tetrahedral bond angles and no $\pi$ bonds. In fact, the most abundant silicon-containing materials are silica and the silicate minerals, in which silicon and oxygen form extensive networks of $\sigma$ bonds formed by overlap of $sp^3$ hybrid orbitals (see Figures 6-17 and 6-18). These differences in bond formation indicate that the $3p$ valence orbitals of silicon atoms form much more stable $\sigma$ bonds than $\pi$ bonds. In contrast, the $2p$ valence orbitals of carbon form stable $\pi$ and $\sigma$ bonds, and carbon atoms can have linear, trigonal planar, and tetrahedral geometries. The versatility in bonding of carbon atoms gives rise to the vast, intricate worlds of organic chemistry and biochemistry.

Why does silicon not form $\pi$ bonds as readily as carbon, its Group 14 neighbour? The answer can be found by examining the sizes of the valence orbitals of silicon and carbon. Figure 7-23 compares side-by-side $p$-orbital overlap for carbon atoms and silicon atoms. Notice that for the larger silicon atom, the $3p$ orbitals are too far apart for strong side-by-side overlap. The result is a very weak $\pi$ bond. Consequently, silicon makes more effective use of its valence orbitals to form four $\sigma$ bonds, which can be described using $sp^3$ hybrid orbitals.

In general, $\pi$ bonding requires the presence of a second-row element. Carbon, nitrogen, and oxygen readily form $\pi$ bonds with one another. Oxygen also readily forms $\pi$ bonds with third- and fourth-row elements. Phosphate ($PO_4^{3-}$), sulphate ($SO_4^{2-}$), perchlorate ($ClO_4^-$), and other similar anions contain $O—X$ $\pi$ bonds, which we describe in Section 7.6. Many other examples of $\pi$ bonding exist, but we confine our descriptions in this textbook to bonds between second-row elements and in oxoanions.

**Figure 7-23**

133 pm          210 pm

C—C $\pi$ bonds form from the overlap of $2p$ orbitals

Si—Si $\pi$ bonds form from the overlap of $3p$ orbitals

Comparison of $\pi$ bonding between two carbon atoms and between two silicon atoms.

**Figure 7-24**

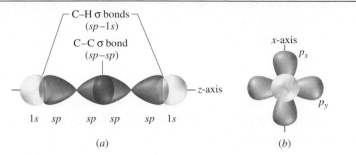

(a)    (b)

(a) A side view of the σ bond framework of acetylene. The unhybridized p orbitals of carbon are not shown in this view. (b) In an end view, the hydrogen atom obscures the bonding orbitals. This view shows the p orbitals of carbon, perpendicular to the σ bond axis.

**Figure 7-25**

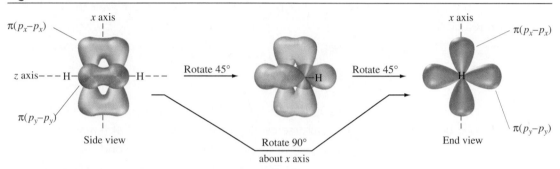

Three views of the bonding in acetylene. The two π bonds are perpendicular to each other and have electron density above and below the internuclear axis.

## Acetylene: Formation of a Triple Bond

The Lewis structure of acetylene $(C_2H_2)$ shows a triple bond between the carbon atoms:

$$H—C≡C—H$$

Each carbon atom has a steric number of 2, indicating that acetylene is a linear molecule and that sp hybrid orbitals can be used to construct the bonding orbital framework. Figure 7-24a shows the σ bonding system of acetylene.

The sp hybrid set requires just one of the valence p orbitals. Figure 7-24b shows that the remaining two p orbitals are oriented at right angles to the axis of the acetylene molecule and at right angles to each other. Each p orbital can overlap with its counterpart on the other carbon atom, just as in ethylene. Thus, acetylene has two π bonds. Each one forms from the side-by-side overlap of p orbitals on each carbon atom. Figure 7-25 illustrates this. According to this description, a triple bond is made up of one σ bond and two π bonds. Figure 7-26, which superimposes the views in Figures 7-24 and 7-25, depicts all the bonds in the acetylene molecule. Example 7-7 treats another molecule with bonding that can be described by sp hybrid orbitals.

**Figure 7-26**

Composite orbital overlap view of the σ and π bonds of acetylene.

## EXAMPLE 7-7  ORBITAL OVERLAP IN TRIPLE BONDS

Hydrogen cyanide, HCN, is an extremely poisonous gas with an odour resembling that of almonds. Approximately one million tonnes of HCN are produced each year in the world, most of which is used to prepare starting materials for polymers. Construct a complete bonding picture for HCN and sketch the various orbitals.

**Strategy:** Use the four-step procedure described earlier. Begin with the Lewis structure for the molecule, and then identify the appropriate hybrid orbitals. Construct a $\sigma$ bond framework, and complete the bonding picture by assembling the $\pi$ bonds from the unhybridized $p$ orbitals.

**Solution:** The six steps of our procedure for writing Lewis structures lead to a triple bond between carbon and nitrogen:

$$H—C≡N:$$

Hydrogen and nitrogen are outer atoms, so they use atomic orbitals to form covalent bonds. The carbon atom has a steric number of 2, so it can be described using $sp$ hybrids. With this information, we can construct the $\sigma$ bonding network for HCN:

The triple bond includes two $\pi$ bonds formed by the pairwise overlap of the $p$ orbitals that remain on the carbon and nitrogen atoms. Two figures from different perspectives help illustrate the triple bond:

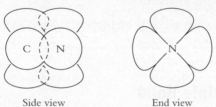

Side view                    End view

The non-bonding pair on the nitrogen atom occupies the $2s$ orbital, which is the only valence orbital of this atom not used for bonding.

The $\sigma$ and $\pi$ bonding networks are shown separately for clarity. Keep in mind, however, that the molecule is a *composite* of both networks, so visualize the HCN molecule with the two bonding networks superimposed.

**Does the Result Make Sense?** Just as imine resembles ethylene, hydrogen cyanide is similar to acetylene, with the same number of valence electrons. A lone pair on the N atom of HCN replaces one C—H bond of acetylene.

**Practice Exercise 7.7**
Describe and draw the bonding system of 2-butyne, $CH_3CCCH_3$.

## SECTION EXERCISES

●**7.3.1** Determine the Lewis structure and describe the bonding completely (including an orbital sketch) for the formaldehyde molecule, $H_2CO$. Formaldehyde is a pungent, colourless gas that is highly toxic and is a suspected carcinogen.

● **7.3.2** Propene is the three-carbon alkene, and propyne is the three-carbon hydrocarbon that contains a triple bond:

Propene          Propyne

Determine the chemical formulas and Lewis structures of these two substances. Describe their bonding completely, including the geometry and hybridization for each carbon atom.

● **7.3.3** Convert the line structure of limonene (see Focus 7-1: Chemistry and Life) into a Lewis structure. Determine how many carbon atoms participate in π bonding and draw orbital sketches for a π bond, a C—C σ bond, and a C—H σ bond in this molecule.

## 7.4 MOLECULAR ORBITAL THEORY: DIATOMIC MOLECULES

Hybrid orbitals and localized bonds provide a model of bonding that can be applied easily to a wide range of molecules. This model does an excellent job of rationalizing and predicting chemical structures, but localized bonds cannot predict or interpret other aspects of bonding and reactivity. For example, localized bonds cannot explain either the green colour of chlorophyll or the ability of this molecule to collect energy from sunlight. Another example is the electrical properties of metals and semiconductors, which cannot be accounted for by a localized bonding model. In the remaining sections of this chapter, we explore the properties of electrons that are said to be *delocalized* over three, four, or even dozens of atoms. Delocalized electrons occupy **molecular orbitals (MOs)**, so called because they may span entire molecules.

Molecular orbital theory is more complex than the hybrid orbital approach, but the foundations of the model are readily accessible. Molecular orbital theory opens the door to many fascinating aspects of modern chemistry. In this section, we introduce the molecular orbital approach through diatomic molecules.

### Molecular Orbitals of H₂

The opening pages of this chapter introduce the principles of orbital overlap by describing molecular hydrogen as a composite of overlapping spherical 1s orbitals. This interaction is the starting point for our discussion of molecular orbital theory.

Because electrons have wave-like properties, orbital interactions involve addition or subtraction of amplitudes, as described in Section 7-1. So far, we have described only additive orbital interactions. The wave amplitudes add in the overlap region, generating a new bonding orbital with larger amplitude between the nuclei. However, a complete mathematical treatment of orbital overlap requires that orbitals be conserved. In other words, whenever several orbitals interact, they must generate an equal number of new orbitals.

**Figure 7-27**

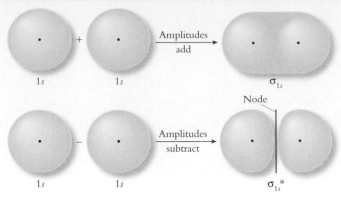

When two hydrogen 1$s$ orbitals interact, they generate two molecular orbitals, one bonding ($\sigma_{1s}$) and one antibonding ($\sigma_{1s}^{*}$).

When two hydrogen atoms interact, their 1$s$ orbitals generate two new molecular orbitals, as shown in Figure 7-27. One of these interactions is additive, leading to a molecular orbital with high electron density between the nuclei. High electron density between the nuclei results in more attraction than repulsion (see Figure 6-1 in Chapter 6), so this combination is called a **bonding molecular orbital**. Electrons in bonding orbitals are more stable than they would be in individual atoms. The bonding MO of molecular hydrogen is labelled $\sigma_{1s}$. It is $\sigma$ because the electron distribution is concentrated along the bond axis, and 1$s$ because the MO is constructed from 1$s$ atomic orbitals.

The second orbital interaction is subtractive, generating an orbital with low electron density between the nuclei. Low electron density between the nuclei results in more nucleus–nucleus repulsion than electron–nucleus attraction. This combination is destabilizing compared with individual atoms and is referred to as an **antibonding molecular orbital**. The antibonding combination is labelled as $\sigma_{1s}^{*}$. The MO is a $\sigma$ orbital because the electron density lies along the bond axis, although the electron density in this orbital is concentrated outside the nuclei rather than between the nuclei. The asterisk in the symbol signifies an antibonding orbital. Figure 7-27 shows that the $\sigma_{1s}^{*}$ orbital has a node between the nuclei. Nodes that pass between nuclei are one of the characteristics of antibonding molecular orbitals. Recall from Section 4.6 (Chapter 4) that nodes represent regions within an orbital where there is no electron density. It is important to realize that $\sigma_{1s}^{*}$ is one orbital with lobes of electron density on either side of the node. This is analogous to atomic $p$ orbitals, as described in Chapter 4.

Figure 7-28 is an energy-level diagram, called a **molecular orbital diagram**, summarizing the MO treatment of $H_2$. It shows that the bonding orbital is more stable than the orbitals from which it forms, whereas the antibonding orbital is less stable. When a hydrogen molecule forms, its two electrons obey the same principles for distributing electrons as was described in Chapter 4. Electrons must obey the aufbau process, the Pauli exclusion principle, and Hund's rule, no matter what types of orbitals they occupy. The two valence electrons of molecular hydrogen fill the most stable orbital, the $\sigma_{1s}$ orbital, with their spins oriented in opposite directions. The 1s orbitals of the hydrogen atoms have been used to form the MO pair, and the $\sigma_{1s}^*$ orbital, being higher in energy, is empty.

As their names suggest, molecular orbitals can span an entire molecule, whereas localized bonds cover just two nuclei. Because diatomic molecules contain just two nuclei, the localized view gives the same general result as molecular orbital theory. The importance of molecular orbitals and delocalized electrons becomes apparent as we move beyond diatomic molecules in the following sections of this chapter. Meanwhile, diatomic molecules offer the simplest way to develop the ideas of molecular orbital theory.

The MO diagram shown in Figure 7-28 applies to all diatomic species from the first row of the periodic table. We can, for instance, use molecular orbital theory to rationalize why helium atoms do not combine to form molecules of $He_2$. This hypothetical diatomic molecule would have four electrons. Two of these would populate the bonding $\sigma_{1s}$ orbital, imparting stability to the molecule. The remaining two electrons would have to occupy the antibonding $\sigma_{1s}^*$ orbital, and their destabilizing effect would cancel the influence of the bonding electrons, as shown by the MO diagram in Figure 7-29. A convenient way to summarize this argument is by calculating the **bond order**, which represents the net amount of bonding between two atoms:

Bond order = $\frac{1}{2}$(number of electrons in bonding MOs

− number of electrons in antibonding MOs)   (7-1)

For molecular hydrogen, with two bonding electrons and zero electrons in the antibonding orbitals, the bond order is 1. A bond order of 1 represents a single bond, consistent with the Lewis structure of $H_2$. For the hypothetical $He_2$ molecule, equal numbers of bonding and antibonding electrons give a bond order of 0. Helium atoms do not bond together into stable diatomic molecules.

Non-integer bond orders are possible, and will occur if the molecule has an odd number of electrons. The physical interpretation of a bond order of 1.5, for instance, is a bond that is intermediate between a single and a double bond. The bond energy and bond length will also be intermediate between those of a single and double bond. Example 7-8 looks at an intermediate case.

**Figure 7-28**

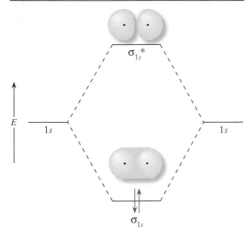

Molecular orbital diagram for molecular hydrogen.

**Figure 7-29**

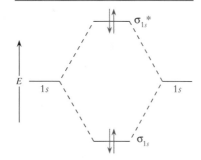

Diatomic $He_2$ does not exist because the stability imparted by the bonding electrons is offset by the destabilization due to the antibonding electrons.

## EXAMPLE 7-8  DOES He₂⁺ EXIST?

Use a molecular orbital diagram to predict if it is possible to form the $He_2^+$ cation.

**Strategy:** The MO diagram shown in Figure 7-28 can be applied to any of the possible diatomic molecules or ions formed from the first-row elements, hydrogen and helium. Count the electrons of $He_2^+$, place the electrons in the MO diagram, and calculate the bond order. If the bond order is greater than zero, the species can form, under the right conditions.

**Solution:** One He atom has two electrons, so a $He_2^+$ cation has three electrons. Following the aufbau process, two electrons fill the lower-energy $\sigma_{1s}$ orbital, so the third must be placed in the antibonding $\sigma_{1s}^*$ orbital in either spin orientation. A shorthand form of the MO diagram appears below. The bond order of the cation is $\frac{1}{2}$.

$$\text{Bond order} = \frac{1}{2}(2-1) = \frac{1}{2}$$

We predict that $He_2^+$, with a net bond order of $\frac{1}{2}$, can be prepared in the laboratory.

**Does the Result Make Sense?** Electrical discharges through samples of helium gas generate $He^+$ cations, some of which bond with He atoms to form $He_2^+$ cations. These fall apart as soon as they capture electrons, but they last long enough to be studied spectroscopically. The bond dissociation energy is 250 kJ/mol, approximately 60% as strong as the bond in the $H_2$ molecule, the bond order of which is 1.

**Practice Exercise 7.8**

Which species has the stronger bond, $H_2$ or $H_2^-$? Use a molecular orbital diagram to support your answer.

## Second-Row Diatomic Molecules

The next step in the development of molecular orbital theory is to consider the MOs formed by additive and subtractive overlap of atomic $p$ orbitals. We use molecular oxygen as a case study.

$$\ddot{\text{O}}\!=\!\ddot{\text{O}}$$

Lewis structure of $O_2$

The Lewis structure of molecular oxygen shows the two atoms connected by a double bond, with two non-bonding electron pairs on each oxygen atom. Both atoms in $O_2$ are outer atoms, so there are no constraining bond angles and no need for hybridization. Atomic valence 2s and 2p orbitals can be used to describe the bonding in this molecule.

The valence orbitals on the two oxygen atoms interact to form a set of molecular orbitals. Each oxygen atom has four valence orbitals, so the MO diagram for $O_2$ contains eight molecular orbitals. The first interaction involves the 2s atomic orbitals. This pair interacts in exactly the same way as the 1s orbitals described for $H_2$. The additive combination generates $\sigma_{2s}$ and the subtractive combination forms $\sigma_{2s}^*$. Because of the larger size of the atomic orbitals, these two MOs are larger than their hydrogen $\sigma_{1s}$ and $\sigma_{1s}^*$ counterparts, but their overall appearances are similar.

Figure 7-30 shows the construction of the 2p-based molecular orbitals. One pair of MOs forms from the $p$ orbitals that point toward each other along the bond axis. By convention, we label this as the $z$-axis. This end-on overlap gives $\sigma_p$ and $\sigma_p^*$ orbitals that concentrate electron density between the two oxygen nuclei, as shown in Figure 7-30a. The remaining four $p$ orbitals form pairs of $\pi$ and $\pi^*$ MOs through side-by-side overlap. One of these pairs comes from the $p_y$ orbitals, and

**Figure 7-30**

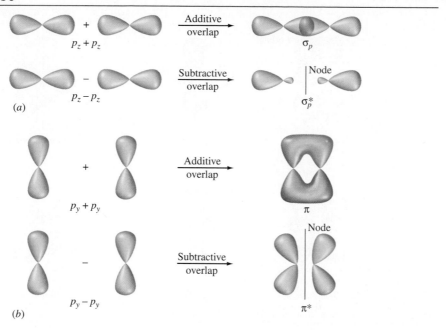

Additive and subtractive combinations of *p* orbitals lead to bonding and antibonding orbitals. (*a*) End-on overlap gives σ orbitals. (*b*) Side-by-side overlap gives π orbitals.

the other pair comes from the $p_x$. Figure 7-30*b* shows only the $p_y$ pair of π orbitals. The $p_x$ pair has the same appearance but is perpendicular to the one shown in the figure. Figure 7-31 shows complete sets of the π and π* orbitals from three perspectives. Notice that the π molecular orbitals closely resemble π bonds of acetylene (Figure 7-25).

Remember that electrons preferentially fill the lowest-energy (most stable) orbitals that are available. Before placing the valence electrons in these MOs, therefore, we need an orbital energy level diagram. The orbital overlap model and the properties of electrons allow us to draw some general conclusions about the relative energies of the eight molecular orbitals generated by $n = 2$ atomic orbitals:

1. The bonding and antibonding $\sigma_s$ orbitals are more stable than any of the six molecular orbitals derived from the $2p$ orbitals. This is because the $2s$ orbitals that generate the $\sigma_s$ and $\sigma_s$* orbitals are more stable than the $2p$ atomic orbitals.

2. The two π bonding orbitals are degenerate (i.e., they have identical energies), because the corresponding atomic *p* orbitals are degenerate. Likewise, the two π* orbitals are degenerate.

3. The antibonding orbitals formed from the atomic $2p$ orbitals are at the top of the energy-level diagram, with the $\sigma_p$* orbital being higher than π*.

After applying these features, we are left with the bonding $\sigma_p$ and π molecular orbitals. These must be placed between $\sigma_s$* and π*, but which of them is

**Figure 7-31**

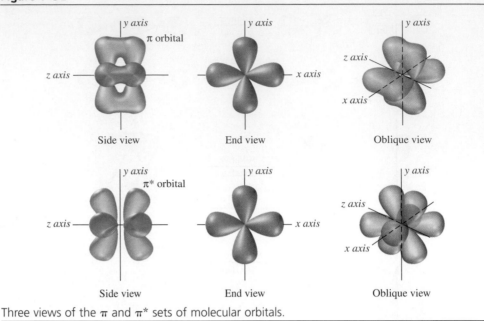

Three views of the π and π* sets of molecular orbitals.

lower in energy? Orbital overlap arguments suggest that $\sigma_p$ should be more stable than π, because end-on overlap is more effective than side-by-side overlap. Molecular oxygen displays this ranking. Figure 7-32 shows the complete diagram and places the valence electrons of $O_2$ in accordance with the Pauli and aufbau principles as well as Hund's rule.

**Figure 7-32**

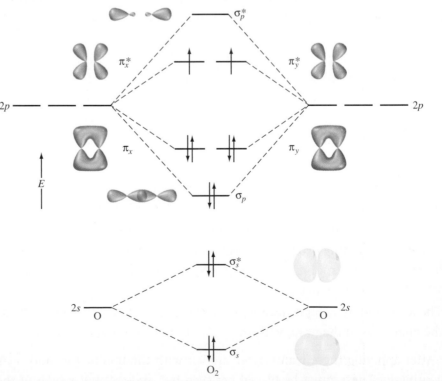

The molecular orbital diagram for $O_2$. The energy ranking of the MOs applies to second-row diatomic molecules with $Z > 7$.

We can write configurations for diatomic molecules by naming the orbitals ($\sigma_s$, $\pi_x$, etc.), using superscripts to show how many electrons each orbital contains. Here is the configuration for the ground state of $O_2$:

$$(\sigma_s)^2 \, (\sigma_s{}^*)^2 \, (\sigma_p)^2 \, (\pi_x)^2 \, (\pi_y)^2 \, (\pi_x{}^*)^1 \, (\pi_y{}^*)^1$$

## Evidence for Antibonding Orbitals

The molecular orbital model developed in this section is more elaborate than the localized bonds described earlier in this chapter. Is this more complex model necessary to give a thorough picture of chemical bonding? Experimental evidence for molecular oxygen suggests that the answer is yes.

Figure 7-33 shows that liquid oxygen adheres to the poles of a magnet. Attraction to a magnetic field shows that molecular oxygen is paramagnetic. Recall from Chapter 5 that paramagnetism arises when the electron configuration includes unpaired electrons. Neither the Lewis structure of $O_2$ nor the simple localized bond model reveals the presence of unpaired electrons. The molecular orbital description, however, shows that the least stable electrons of the oxygen atom occupy the two degenerate $\pi^*$ antibonding orbitals (Figure 7-32).

Bond length and bond energy measurements on oxygen and its cation also indicate that the highest-energy occupied orbital is antibonding. Removing a bonding electron from a molecule gives a cation with a weaker, longer bond than what the neutral molecule has. In contrast, removing an electron from $O_2$ to form $O_2{}^+$ increases the bond energy and decreases the bond length:

**Figure 7-33**

Molecular oxygen is paramagnetic, so it clings to the poles of a magnet.

| Species | Bond Length (pm) | Bond Energy (kJ/mol) | Configuration | Bond Order |
|---------|------------------|----------------------|---------------|------------|
| $O_2$ | 121 | 496 | $\ldots (\sigma_p)^2 \, (\pi)^4 \, (\pi^*)^2$ | 2 |
| $O_2{}^+$ | 112 | 643 | $\ldots (\sigma_p)^2 \, (\pi)^4 \, (\pi^*)^1$ | 2.5 |

These data show that the least stable occupied orbital of $O_2$ is antibonding in character: removing an electron reduces the amount of antibonding and strengthens the bond.

Our treatment of $O_2$ shows that the extra complexity of the molecular orbital approach explains features that a simpler description of bonding cannot explain. The Lewis structure of $O_2$ does not reveal its two unpaired electrons, but an MO approach does. The simple $\sigma-\pi$ description of the double bond in $O_2$ does not predict that the bond in $O_2{}^+$ is stronger than that in $O_2$, but the MO approach does. As we show in the following sections, the molecular orbital model has even greater advantages in explaining bonding when Lewis structures show the presence of resonance.

## Homonuclear Diatomic Molecules

The MO diagram for molecular oxygen can be generalized for any of the second-row diatomic molecules. For instance, we can apply it to $B_2$. Although elemental boron is a solid under normal conditions, gaseous $B_2$ molecules form when

**Figure 7-34**

Energies of the $n = 2$ valence orbitals as a function of $Z$.

**Figure 7-35**

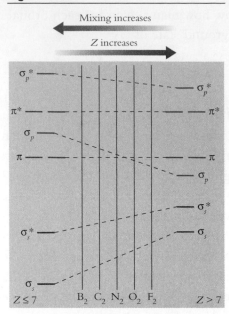

Mixing of the $2s$ and $2p_z$ orbitals causes the $\sigma_s$ MOs to become more stable and the $\sigma_p$ MOs to become less stable. The amount of mixing decreases across Row 2. For $B_2$, $C_2$, and $N_2$, the $\pi$ MOs are more stable than $\sigma_p$. For $O_2$ and $F_2$, $\sigma_p$ is more stable than $\pi$.

the solid is heated. The MO diagram for $O_2$ predicts that $B_2$, with six valence electrons, should have the configuration $(\sigma_s)^2 (\sigma_s{}^*)^2 (\sigma_p)^2$. With all its electrons paired, the species should be diamagnetic. In contradiction to this prediction, experiments show that $B_2$ molecules are paramagnetic, with two unpaired electrons. When theory and experiment conflict, theory must be revised. The diatomic molecular orbital diagram must be revised to account for the paramagnetism of $B_2$.

The view of molecular orbital theory that leads to the MO diagram in Figure 7-32 assumes that the $2s$ and $2p$ orbitals act independently. A more refined treatment considers interactions among the $2s$ and $2p$ sets of orbitals. Remember that the $2s$ and $2p$ orbitals have similar radii. Consequently, when two atoms approach each other, the $2s$ and $2p_z$ orbitals on one atom overlap with *both* the $2s$ and the $2p_z$ orbitals of the other atom. This mixed interaction of $2s$ and $2p_z$ stabilizes $\sigma_s$ and destabilizes $\sigma_p$. In other words, **orbital mixing** causes $\sigma_s$ and $\sigma_p$ to move farther apart in energy. The amount of mixing depends on the difference in energy between the $2s$ and $2p$ atomic orbitals. Mixing is largest when the energies of the orbitals are nearly the same. As Figure 7-34 shows, the energies of atomic orbitals diverge as $Z$ increases across the second row, so mixing is large for $B_2$ but small for $F_2$. Figure 7-35 shows how mixing affects the stability of the $\sigma_s$ and $\sigma_p$ orbitals. Notice that crossover of the energy levels takes place between molecular nitrogen and molecular oxygen. Because of orbital mixing, we require two generalized molecular orbital diagrams for diatomic molecules: one for $B_2$, $C_2$, and $N_2$, and the other for $O_2$ and $F_2$. Figure 7-35 shows both diagrams.

These molecular orbital diagrams help to rationalize experimental observations about molecules, as Example 7-9 shows.

### EXAMPLE 7-9   TRENDS IN BOND ENERGY

Use molecular orbital diagrams to explain the trend in the following bond energies: $B_2 = 290$ kJ/mol, $C_2 = 600$ kJ/mol, and $N_2 = 942$ kJ/mol.

**Strategy:** The data show that bond energies for these three diatomic molecules increase across the second row of the periodic table. We must construct molecular orbital diagrams for the three molecules and use the results to interpret the trend.

**Solution:** The crossover point for the $\sigma_p/\pi$ energy levels takes place just after $N_2$, so the general diagram for $Z \leq 7$ in Figure 7-35 applies to all three molecules. The valence electron counts are $B_2 = 6\ e^-$, $C_2 = 8\ e^-$, and $N_2 = 10\ e^-$. Place these electrons in the MO diagram following the aufbau process. Here are the results:

| $B_2$ | $C_2$ | $N_2$ |

The diagrams show that the increasing electron counts, as we progress from boron to carbon to nitrogen, result in the filling of bonding molecular orbitals. This is revealed most clearly by calculating the bond orders for the three molecules: $B_2 = 1$, $C_2 = 2$, and $N_2 = 3$. Increasing bond order corresponds to greater electron sharing between the nuclei, resulting in stronger bonding between the atoms.

**Does the Result Make Sense?** Of these three diatomic molecules, only $N_2$ exists under normal conditions. Boron and carbon form solid networks rather than isolated diatomic molecules. However, molecular orbital theory predicts that $B_2$ and $C_2$ are stable molecules under the right conditions, and in fact both molecules can be generated in the gas phase by vapourizing solid boron or solid carbon in the form of graphite.

**Common Pitfalls:** Not correctly ordering the molecular orbitals with respect to their energies. Refer to Figure 7-35 for the correct order.

**Practice Exercise 7.9**

Use molecular orbital diagrams to explain the trend in bond energies for the following diatomic molecules: $N_2 = 942$ kJ/mol, $O_2 = 495$ kJ/mol, and $F_2 = 155$ kJ/mol.

## Heteronuclear Diatomic Molecules

Nitrogen oxide, NO, is an example of heteronuclear diatomic molecules, those composed of different atoms. This interesting molecule has been in the news several times in recent years, because of important discoveries about the role of NO as a biological messenger, as we describe in our introduction to Chapter 19.

Because the qualitative features of orbital overlap do not depend on the identity of the atoms, the bonding in NO can be described by the same sets of orbitals that describe homonuclear diatomic molecules. Which of the two general MO diagrams for diatomic molecules applies to nitrogen oxide? The crossover point for the energy rankings of the $\sigma_p$ and $\pi$ molecular orbitals falls between N and O, so we expect the orbital energies to be nearly equal. Photoelectron spectroscopy confirms this but also shows that the $\sigma_p$ MO is a bit more stable than the $\pi$ MO. Consequently, the MO configuration for the 11 valence electrons of NO mirrors

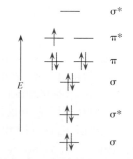

Molecular orbital diagram of NO

that for the 12 valence electrons of $O_2$, except that there is a single electron in the $\pi^*$ orbital:

$$(\sigma_s)^2 \, (\sigma_s{}^*)^2 \, (\sigma_p)^2 \, (\pi_x)^2 \, (\pi_y)^2 \, (\pi_x{}^*)^1$$

There are eight bonding electrons and three antibonding electrons, so the bond order of NO is 2.5. This bond is stronger and shorter than the double bond of $O_2$ but weaker and longer than the triple bond of $N_2$.

| Species | Bond Length (pm) | Bond Energy (kJ/mol) | Configuration | Bond Order |
|---|---|---|---|---|
| $O_2$ | 121 | 495 | $\ldots (\sigma_p)^2 \, (\pi)^4 \, (\pi^*)^2$ | 2 |
| NO | 115 | 605 | $\ldots (\sigma_p)^2 \, (\pi)^4 \, (\pi^*)^1$ | 2.5 |
| $N_2$ | 110 | 945 | $\ldots (\pi)^4 \, (\sigma_p)^2$ | 3 |

Electron configurations for these molecules lead to a guideline for molecular orbital configurations of second-row diatomic molecules and ions, based on average nuclear charge, $Z_{average}$:

> $\sigma_p$ is lower in energy than $\pi$ when $Z_{average} > 7$, and $\pi$ is lower in energy than $\sigma_p$ when $Z_{average} \leq 7$.

To conclude this section, Example 7-10 compares three diatomic molecules. Section 7.5 explores molecular orbitals in triatomic molecules.

## EXAMPLE 7-10 COMPARING IONIZATION ENERGIES

The first ionization energy of NO is 891 kJ/mol, that of $N_2$ is 1500 kJ/mol, and that of CO is 1350 kJ/mol. Use electron configurations to explain why NO ionizes more easily than either $N_2$ or CO.

**Strategy:** Recall from Chapter 5 that ionization energy refers to the removal of an electron from an atom, or, in this case, from a molecule. We must count the valence electrons, choose the correct MO diagram, follow the aufbau process in placing the electrons, and then use the configurations to explain the ionization energy data.

**Solution:** Begin with a summary of the important information for each molecule: NO has 11 valence electrons and $Z_{average} = 7.5$; $N_2$ has 10 valence electrons and $Z_{average} = 7$; CO has 10 valence electrons and $Z_{average} = 7$. Applying the general diatomic MO diagrams of Figure 7-35 gives the following configurations:

$$NO = (\sigma_s)^2 \, (\sigma_s{}^*)^2 \, (\sigma_p)^2 \, (\pi_x)^2 \, (\pi_y)^2 \, (\pi_x{}^*)^1$$

$$N_2 = (\sigma_s)^2 \, (\sigma_s{}^*)^2 \, (\sigma_p)^2 \, (\pi_x)^2 \, (\pi_y)^2$$

$$CO = (\sigma_s)^2 \, (\sigma_s{}^*)^2 \, (\sigma_p)^2 \, (\pi_x)^2 \, (\pi_y)^2$$

CO and $N_2$ are isoelectronic species, meaning they have the same electron configurations (see Section 5.3). Each has a bond order of 3, and the least stable electrons are in bonding orbitals. It makes sense that these two molecules have comparable ionization energies. NO has one electron more than $N_2$ and CO, and this electron occupies one of the antibonding $\pi^*$ orbitals. It is much easier to remove this antibonding electron from NO than it is to remove a bonding electron from either of the other two species.

**Does the Result Make Sense?** The difference in ionization energies between NO and the other two molecules is around 500 kJ/mol. We can say that an electron in a $\pi$ orbital strengthens a bond by about 250 kJ/mol, and an electron in a $\pi^*$ orbital weakens a bond by about 250 kJ/mol. Look at the bond energies for NO, $O_2$, and $N_2$ and decide if this conclusion is valid.

**Practice Exercise 7.10**

Use electron configurations to predict which of the following is the most stable diatomic combination of carbon and nitrogen: $CN^-$, $CN$, or $CN^+$.

---

**SECTION EXERCISES**

●**7.4.1** Dilithium molecules can be generated from lithium metal at very low pressure. Construct a molecular orbital diagram of dilithium. What is the bond order of this molecule?

●**7.4.2** Draw an orbital sketch of each of the occupied valence orbitals for dilithium.

●**7.4.3** Draw a complete molecular orbital diagram for the diatomic molecule $Ne_2$. Use the diagram to explain why $Ne_2$ is not a stable molecule. Could $Ne_2^+$ exist, in theory?

## 7.5 THREE-CENTRE π ORBITALS

The π molecular orbitals described so far involve two atoms, so the orbital pictures look the same for the localized bonding model applied to ethylene and the MO approach applied to molecular oxygen. In the organic molecules described in the introduction to this chapter, however, π orbitals spread over three or more atoms. Such **delocalized π orbitals** can form when *more than two p* orbitals overlap in the appropriate geometry. In this section, we develop a molecular orbital description for three-atom π systems. In the following sections, we apply the results to larger molecules.

### Ozone and the Nature of Resonance

We begin our exploration of delocalized π bonds with ozone, $O_3$. As described in Chapter 4, ozone in the upper stratosphere protects plants and animals from hazardous ultraviolet radiation. Ozone has 18 valence electrons and a Lewis structure that appears in Figure 7-36a. Experimental measurements show that ozone is a bent molecule with a bond angle of 118°.

The bond framework of ozone can be represented by the localized σ bonds shown in Figure 7-36b. The inner oxygen atom, with a steric number of 3, is trigonal planar and can be described using $sp^2$ hybrid orbitals. Two of these $sp^2$ orbitals form σ bonds by overlap with $2p$ orbitals from the outer oxygen atoms. The third $sp^2$ hybrid contains a lone pair of electrons. The Lewis structure also indicates the presence of a π bond and five non-bonding electron pairs on the two outer oxygen atoms.

The $sp^2$–hybridized inner oxygen atom has one unused $2p$ orbital, which is oriented perpendicular to the plane defined by the hybrid orbitals (review Figure 7-12). This $p$ orbital is in the perfect position to form a π bond through side-by-side overlap with a $p$ orbital on an outer oxygen atom. The resonance structures in the Lewis structure of ozone indicate that either outer oxygen is an equally good choice for this overlap. However, the two distinct localized π bonds shown in the ozone res-

**Figure 7-36**

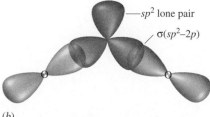

*(a)*

*(b)*

The structure of ozone. (a) The Lewis structure shows resonance that involves the bond (blue lines) and one lone pair (red dots). (b) The bonding framework.

onance structures are misleading, because all three of the perpendicular *p* orbitals overlap simultaneously. Neither Lewis structures nor the localized bonding model gives an adequate description of the π bonding in ozone. Molecular orbital theory, on the other hand, does an excellent job of treating the interactions of multiple *p* orbitals.

Molecular orbital interactions are possible only for atomic orbitals with overlapping electron densities. In particular, π molecular orbitals form when *p* atomic orbitals overlap side by side, on adjacent atoms. As always, the resulting number of molecular orbitals must equal the number of contributing atomic orbitals. Another general feature of orbital interactions is that there are equal numbers of bonding MOs and antibonding MOs. Thus, ozone, with three overlapping atomic *p* orbitals, has three π MOs, one of which is bonding and one of which is antibonding. What is the third molecular orbital?

A clue to the nature of the third π MO can be found in the placement of electrons in the two resonance structures for ozone, which are shown with colour highlights in Figure 7-36*a*. Notice that in one resonance structure, the left outer atom has three lone pairs and a single bond, whereas the right outer atom has two lone pairs and a double bond. In the other resonance structure, the third lone pair is on the *right* outer atom, with the double bond to the *left* outer atom. The double bond appears in different positions in the two structures, and one of the lone pairs also appears in different positions. These variations signal delocalized π orbitals.

The double bond shown in different locations in the two resonance structures represents a pair of electrons in a delocalized π bonding molecular orbital that spans all three of the oxygen atoms, as shown in Figure 7-37*a*. One lone pair also appears in different positions in the two resonance structures, again signalling a delocalized orbital. This lone pair is spread over both *outer* atoms but not across the *inner* atom, as shown in Figure 7-37*b*. This is a **non-bonding molecular orbital**, $\pi_{nb}$. The lone pair shown in different positions in the resonance structures occupies the delocalized $\pi_{nb}$ orbital.

To summarize, the interactions among the three *p* orbitals of ozone generate three molecular orbitals, of which the π and $\pi_{nb}$ are filled. The third orbital, a π* antibonding orbital (Figure 7-37*c*), is empty. Except in diatomic molecules,

**Figure 7-37**

*(a)* π          *(b)* $\pi_{nb}$          *(c)* π*
(Bonding)   (Nonbonding)   (Antibonding)

The three π MOs of ozone. (a) The π orbital spreads over all three atoms. (b) The $\pi_{nb}$ orbital has a node on the inner O atom. (c) The π* orbital has nodes in each of the bonding regions.

ground-state configurations seldom include occupied antibonding π orbitals. The bond lengths in ozone provide evidence that the π electrons extend over the entire molecule rather than being isolated between two atoms. Both bond lengths are 128 pm, intermediate between the length of the double bond in $O_2$ (121 pm) and the length of the single $O—O$ bond in $H_2O_2$ (148 pm).

## A Composite Model of Bonding

Notice that our description of ozone mixes the two approaches to bonding described in this chapter. We describe the framework of the molecule using localized bonds made from hybrid and atomic orbitals. We describe the π system of ozone using molecular orbitals. This mixed approach is useful, because although the π systems of many molecules require delocalized molecular orbitals, the bonding framework of any molecule is easy to describe using hybridization and localized bonds. A combination of the two models is a straightforward and powerful way to describe many molecules. The composite approach to bonding begins with the Lewis structure and relies on the following guidelines:

1. Construct the σ bonding framework using hybrid orbitals for inner atoms and atomic orbitals for outer atoms, as described in Section 7.2. Any hybrid orbital not used to form σ bonds contains lone pairs of electrons.

2. If the molecule contains multiple bonds, construct the π bonding system using molecular orbital theory, as described in this section and in the remaining pages of Chapter 7. Watch for resonance structures, which signal the presence of delocalized electrons.

3. Place one pair of valence electrons in any atomic orbital that is not used in hybridization or in the π system.

4. Sum the electrons allocated in Steps 1–3. The result must match the total number of valence electrons used in the Lewis structure. In addition, a complete bonding description must account for all the valence orbitals.

   A review of ozone illustrates the four-step procedure. The Lewis structure shows nine pairs of valence electrons. As Figure 7-36a highlights, two pairs form the bond framework, two pairs are in different locations in the two resonance structures, and there are five localized lone pairs. The σ bonding system and the $sp^2$ lone pair account for 6 of the 18 valence electrons (Step 1, see Figure 7-36b). The π system is constructed using MO theory, paying attention to the orbital overlap characteristics of the atomic p orbitals not used for hybridization. There are three π molecular orbitals, two of which are filled. As is often the case, the antibonding π* orbital is empty. The π bonds account for four valence electrons, leaving eight left to be placed (Step 2, see Figure 7-37).

**Figure 7-38**

The 2s orbital and one 2p orbital on each outer atom account for four lone pairs in ozone.

The remaining eight valence electrons are the four pairs shown in grey on the outer atoms of Figure 7-36a. Which orbitals hold these lone pairs? We answer this question by taking an inventory of the valence orbitals. For the inner atom, the $sp^2$ hybrid orbitals account for the 2s orbital and the two 2p orbitals that lie in the plane of the molecule, and the third p orbital, perpendicular to the molecular plane, is part of the π system. All of the valence orbitals of the inner oxygen atom are accounted for. For each outer atom, one p orbital in the plane of the page contributes to the σ bonds, and the p orbital perpendicular to the plane of the page is part of the π system. The remaining valence orbitals on each outer atom are the 2s orbital and the 2p orbital that lies in the molecular plane and is perpendicular to the σ bonds. These four atomic orbitals hold the final eight electrons (Step 3). Figure 7-38 shows the orientation of these four non-bonding orbitals. Our inventory accounts for all 18 valence electrons and the full set of the 12 valence orbitals (Step 4).

Example 7-11 applies the composite model of bonding to the acetate anion.

## EXAMPLE 7-11  BONDING IN ACETATE

The acetate anion ($CH_3CO_2^-$) forms when acetic acid, the acid present in vinegar, is treated with hydroxide ion:

$$CH_3CO_2H + OH^- \longrightarrow CH_3CO_2^- + H_2O$$

Use the four guidelines of the composite model of bonding to describe the bonding of this anion. Sketch the σ bonding system and the occupied π orbitals.

**Strategy:** As with any description of bonding, the procedure begins with the Lewis structure, from which the steric numbers and hybridizations can be determined. Then, the σ and π bonds in the species can be described using localized bonds for the framework and molecular orbitals for delocalized π systems.

**Solution:** The acetate anion has 24 valence electrons, leading to a Lewis structure with two resonance forms:

From the Lewis structure, we see that the $CH_3$ carbon has a steric number of 4, so its bonding can be described by $sp^3$ hybrid orbitals. Three of the $sp^3$ hybrid orbitals form σ bonds by overlap with 1s orbitals on the hydrogen atoms. The fourth $sp^3$ hybrid of the $CH_3$ group forms a σ bond with the neighbouring carbon atom. This second carbon has a steric number of 3, indicating trigonal planar geometry and $sp^2$ hybrid orbitals, one of which bonds to the $CH_3$ group. The two remaining $sp^2$ hybrids bond to 2p orbitals from the outer oxygen atoms. This gives a total of six σ bonds and uses 12 of the 24 valence electrons, as the sketch below shows:

The resonance structures indicate that acetate has a delocalized π system. The $sp^2$–hybridized carbon atom and the two outer oxygen atoms all have $2p$ orbitals available for side-by-side overlap. From the perspective of the orbital overlap drawing shown above, these three $p$ orbitals are perpendicular to the plane of the page. As in the ozone molecule, these three orbitals form a set of three delocalized π orbitals. The C=O double bond is in different positions in the two resonance structures, which indicates two delocalized π electrons. One of the oxygen lone pairs also exchanges position in the resonance structures, accounting for two more delocalized π electrons. The π system for acetate has the same orbital assignments as the π system for ozone. Below are sketches of the occupied orbitals, π and $\pi_{nb}$:

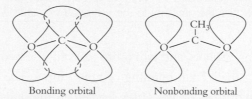

Bonding orbital          Nonbonding orbital

**Does the Result Make Sense?** In the acetate anion, the two oxygen atoms are equivalent, so we expect that the orbital descriptions about each O atom should be identical. You can verify that they are, and experiments further verify this: the two carbon–oxygen bond lengths in acetate are identical.

**Practice Exercise 7.11**
Describe the bonding in the nitrite anion, $NO_2^-$.

## Carbon Dioxide

Carbon dioxide is essential to photosynthesis because plants use the carbon atoms of $CO_2$ to make glucose. Furthermore, animals and plants use fats and carbohydrates for energy production and eliminate the carbon atoms from these molecules as $CO_2$.

The molecule has 16 valence electrons. Its Lewis structure shows that it has two double bonds, with a steric number of 2 for the carbon atom. Consistent with this, the molecule is linear. Figure 7-39 shows the two σ bonds formed by end-on overlap between $sp$ hybrid orbitals on the carbon atom and $2p_z$ atomic orbitals of oxygen.

The $2p_x$ and $2p_y$ orbitals of the carbon atom are not used in forming the hybrid set. Each of these orbitals is oriented perfectly to overlap side by side with $p$ orbitals of each oxygen neighbour. As in ozone, the three $2p_y$ orbitals interact to generate three delocalized π orbitals. One of these orbitals is bonding, one is non-bonding, and the third is antibonding, as shown in Figure 7-40. These MOs have the same forms as their counterparts in ozone but follow the linear shape of the bond framework. The $2p_x$ atomic orbitals combine in exactly the same way, except that the resulting three orbitals point at right angles to the $p_y$ set. These two sets of delocalized π orbitals have exactly the same energies, but they are independent of one another because they point in different directions.

The eight pairs of valence electrons in the $CO_2$ molecule occupy the eight most stable valence orbitals. Two pairs form the σ bonds shown in Figure 7-39. Another two pairs fill the two delocalized π

**Figure 7-39**

The σ bonding system of carbon dioxide. The bonds result from $sp$–$p_z$ overlap.

**Figure 7-40**

(a) π
(Bonding)

(b) $\pi_{nb}$
(Nonbonding)

(c) π*
(Antibonding)

One set of delocalized π orbitals of carbon dioxide (there is another set perpendicular to the page). Notice the similarity to the π orbitals of ozone.

$\ddot{N}{=}N{=}\ddot{O}$

Dinitrogen oxide
16 electrons

$\ddot{O}{=}C{=}\ddot{O}$

Carbon dioxide
16 electrons

bonding orbitals, and two more pairs fill the two delocalized π non-bonding MOs, leaving the π* antibonding orbitals empty. The remaining two pairs are in the 2s atomic orbitals on the outer oxygen atoms. This completes the valence configuration and accounts for all the valence orbitals.

## Other Second-Row Triatomics

Triatomic species can be linear, like $CO_2$, or bent, like $O_3$. The principles of orbital overlap do not depend on the identity of the atoms involved, so all second-row triatomic species with 16 valence electrons have the same bonding scheme as $CO_2$ and are linear. For example, dinitrogen oxide ($N_2O$) has 16 valence electrons, so it has an orbital configuration identical to that of $CO_2$. Each molecule is linear with an inner atom of which the steric number is 2. As in $CO_2$, the bonding framework of $N_2O$ can be represented with sp hybrid orbitals. Both molecules have two perpendicular sets of three π molecular orbitals. The resonance structures of $N_2O$, described in Example 6-11, reveal four pairs of delocalized π electrons, enough to fill the bonding and non-bonding MOs. Second-row triatomic ions with 16 valence electrons include azide ($N_3^-$), cyanate ($NCO^-$), and fulminate ($CNO^-$). These are also linear, with sp-hybridized inner atoms and two sets of delocalized π orbitals.

Ozone, which has 18 valence electrons, exemplifies bent molecules. Another example is the nitrite anion $NO_2^-$, the subject of Practice Exercise 7.11. The bonding of $NO_2^-$ can be represented using $sp^2$ hybrid orbitals for the inner nitrogen atom and one set of delocalized π orbitals.

## SECTION EXERCISES

**7.5.1** Construct the bonding description for the formate anion ($HCO_2^-$) and compare its π system with that of the acetate anion.

**7.5.2** Explain the difference between the nitride ion ($N^{3-}$) and the azide ion ($N_3^-$). Describe in detail the distribution of electrons in both species.

**7.5.3** Describe the bonding in $NO_2^+$ and contrast it with the bonding in $NO_2^-$. Your answer should include separate sketches of the σ and π bonding systems. Tell how many delocalized π electrons, if any, are present in each species.

## 7.6 EXTENDED π SYSTEMS

Ozone and carbon dioxide demonstrate that *p* orbitals can overlap side by side with more than one neighbour. This feature can lead to a system of π bonds that can extend over many atoms. Extended systems can be long chains, or they can be more compact clusters or rings. We begin by describing two examples of four-atom π systems, butadiene and the carbonate anion.

### Butadiene

Butadiene, used in the chemical industry as a precursor of synthetic rubber, is a hydrocarbon with the formula $C_4H_6$. (See Focus 17-1 in Chapter 17 to learn more about the rubber industry.) The Lewis structure of butadiene contains two double bonds on sequential pairs of carbon atoms. The chemistry of butadiene, including its ability to form rubber, can be traced to the delocalized electrons in the π system of the molecule.

Each of the four carbon atoms of butadiene has a steric number of 3 and can be described using $sp^2$ hybridization. There are no lone pairs in the molecule, so all of the hybrid orbitals are used to form σ bonds. There are three C—C bonds formed by $sp^2$–$sp^2$ overlap, and six C—H bonds formed by $sp^2$–$1s$ overlap. The σ bonding system, which Figure 7-41 shows, has nine bonds that account for 18 of the 22 valence electrons, leaving 4 for the π system.

All the atoms of butadiene lie in a plane defined by the $sp^2$ hybrid orbitals. Each carbon atom has one remaining *p* orbital that points perpendicular to the plane, in perfect position for side-by-side overlap. Figure 7-42 shows that all four *p* orbitals interact to form four delocalized π molecular orbitals; two are bonding MOs and two are antibonding. The four remaining valence electrons fill the π orbitals, leaving the two π* orbitals empty.

Whenever the Lewis structure of a molecule shows double or triple bonds alternating with single bonds, the bonding description includes delocalized π orbitals. Molecules that have alternating single and double bonds, like butadiene, are said to have **conjugated π systems**. Each double bond in a conjugated system adds two electrons to the corresponding set of delocalized MOs. Example 7-12 examines another molecule with conjugated double bonds.

**Figure 7-41**

Lewis structure and orbital overlap view of the bonding framework of butadiene. The framework contains three C—C bonds ($sp^2$–$sp^2$) and six C—H bonds ($sp^2$–$1s$).

**Figure 7-42**

Each of the carbon atoms in butadiene has a *p* orbital oriented to overlap with those on neighbouring carbon atoms, generating a set of four π MOs, of which one is shown.

## EXAMPLE 7-12 CONJUGATED DOUBLE BONDS

Describe the bonding of methyl methacrylate. This compound is an important industrial chemical, used mainly to make plastics such as poly(methyl methacrylate) (PMMA).

PMMA is a common polymer used to make plastics.

**Strategy:** Follow the four-step procedure for the composite model of bonding. Use localized bonds and hybrid orbitals to describe the bonding framework and the inner atom lone pairs. Next, analyze the $\pi$ system, paying particular attention to resonance structures or conjugated double bonds. Finally, make sure the bonding inventory accounts for all the valence electrons and all the valence orbitals.

**Solution:** The Lewis structure shows that methyl methacrylate has the formula $C_5H_8O_2$, with 40 valence electrons. You should be able to verify that the two $CH_3$ groups have $sp^3$–hybridized carbons, the inner oxygen atom is $sp^3$ hybridized, the outer oxygen atom uses $2p$ atomic orbitals, and the three double-bonded carbons are $sp^2$ hybridized. These assignments lead to the following inventory of $\sigma$ bonds and inner-atom lone pairs:

| | | |
|---|---|---|
| 6 $sp^3$–1s C—H $\sigma$ bonds | 2 $sp^2$–1s C—H $\sigma$ bonds | 2 $sp^2$–$sp^2$ C—C $\sigma$ bonds |
| 1 $sp^3$–$sp^2$ C—C $\sigma$ bond | 1 $sp^3$–$sp^3$ O—C $\sigma$ bond | 1 $sp^2$–$sp^3$ C—O $\sigma$ bond |
| 1 $sp^3$–$2p$ C—O $\sigma$ bond | 2 $sp^3$ O lone pairs | |

Summing these contributions, the 14 $\sigma$ bonds and 2 $sp^3$ lone pairs account for 32 of the 40 valence electrons.

As the Lewis structure shows, one single bond separates the two double bonds, which signals conjugation and delocalized $\pi$ orbitals. The $\pi$ system of methyl methacrylate forms from the leftover $p$ orbitals on the three $sp^2$–hybridized carbon atoms, plus one $p$ orbital from the outer oxygen atom. These four $p$ orbitals are located on adjacent atoms, positioned for side-by-side overlap. They form a set of four delocalized $\pi$ molecular orbitals, two bonding and two antibonding. Four electrons occupy the $\pi$ bonding orbitals. The $\pi^*$ antibonding MOs are empty. Finally, the two lone pairs on the outer oxygen atom complete the valence electron count, filling the only remaining valence orbitals, the $2s$ orbital and the final $2p$ orbital on the outer oxygen atom.

A ball-and-stick model of methyl methacrylate shows the planar part of the molecule covered by the delocalized $\pi$ orbitals. The atoms that contribute $p$ orbitals to the delocalized $\pi$ system are marked with stars, and the other atoms that lie in the same plane are marked with dots.

**Does the Result Make Sense?** Notice that the π system in methyl methacrylate is nearly the same as that in butadiene. The only difference is the replacement of one C atom by an O atom. Both systems have four overlapping *p* orbitals, giving rise to two bonding MOs and two antibonding MOs, so it is reasonable that the π systems are quite similar.

**Practice Exercise 7.12**
Describe the bonding in acryloyl chloride, $C_3H_3OCl$.

The effect of conjugation on bond stability is revealed by comparing 1,3-pentadiene and 1,4-pentadiene. Figure 7-43 shows that both have eight $C—H$ bonds, two $C—C$ bonds, and two $C=C$ bonds. The only significant difference between these molecules is that the π system of 1,3-pentadiene is conjugated, generating delocalized π orbitals. In contrast, 1,4-pentadiene has an $sp^3$–hybridized $CH_2$ group acting as a "spacer" to separate two localized π bonds.

Does the delocalized π system of 1,3-pentadiene stabilize the molecule relative to 1,4-pentadiene? Each molecule reacts with two molecules of $H_2$ to give a common product, pentane. Thus, a comparison of their energies of hydrogenation

**Figure 7-43**

Energy-level diagram for the hydrogenation of 1, 3- and 1,4-pentadiene. The difference in these energies, 28 kJ/mol, represents the additional stability that delocalization provides to 1,3-pentadiene.

allows us to determine the relative stabilities of two delocalized $\pi$ bonds compared with two isolated $\pi$ bonds. The hydrogenation of 1,4-pentadiene releases 252 kJ/mol, but the hydrogenation of 1,3-pentadiene releases only 224 kJ/mol. As Figure 7-43 demonstrates, these energies of hydrogenation show that 1,3-pentadiene is 28 kJ/mol *more stable* than is 1,4-pentadiene. Because the bonding in these two molecules is otherwise the same, we can conclude that the extra stability comes from the delocalized $\pi$ system.

## Carbonate Ion

In addition to four atoms in a row, a four-atom delocalized framework can have a central atom surrounded by three other atoms with which it interacts. An example is the carbonate anion, the Lewis structure of which displays resonance:

Carbonate anion ($CO_3^{2-}$)

The existence of resonance structures indicates that a complete description of the carbonate ion requires delocalized $\pi$ orbitals. The changing positions of the double bond and one of the lone pairs tell us there are four delocalized electrons in the $\pi$ system. The carbon atom has a steric number of 3, meaning that the $\sigma$ bonding framework for $CO_3^-$ can be described using $sp^2$ hybrid orbitals on the carbon atom. These hybrid orbitals overlap end-on with $2p$ orbitals of the three oxygen atoms, as shown in Figure 7-44a.

The carbon atom has one unused $2p$ orbital perpendicular to the molecular plane. As Figure 7-44b shows, all three oxygen atoms also have perpendicular $p$ orbitals, oriented properly for side-by-side overlap. These four $p$ orbitals give four $\pi$ molecular orbitals delocalized over the entire anion. Figure 7-44c shows that the bonding $\pi$ molecular orbital resembles a propeller when viewed from above.

### Figure 7-44

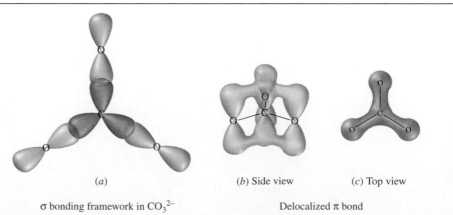

(a)

$\sigma$ bonding framework in $CO_3^{2-}$

(b) Side view          (c) Top view

Delocalized $\pi$ bond

(a) The framework of the carbonate anion consists of three $sp^2$–$2p$ $\sigma$ bonds. (b) The $\pi$ system forms from four overlapping $p$ orbitals on adjacent atoms. (c) Viewed from above, the bonding $\pi$ MO resembles a propeller. The other three $\pi$ MOs are not shown.

The remaining π orbitals, which consist of two non-bonding MOs and one $\pi^*$ antibonding MO, are not shown in the figure. Depictions of molecular orbitals become complicated as the number of contributing atomic orbitals increases. These details are beyond the scope of an introductory course. In agreement with the Lewis structure, a bonding orbital and one non-bonding orbital are filled in the carbonate π system.

## Larger Delocalized π Systems

The dienes are the first members of a set of chain compounds with conjugated double bonds. These chains can be quite long, as shown by the delocalized π systems of some biological molecules. Figure 7-45 shows line structures of three examples: Vitamin A, retinal, and β–carotene. β–Carotene is an orange compound that gives carrots their colour. Our bodies convert carotene into Vitamin A. This compound, in turn, is used to make retinal, a molecule that plays an essential role in the chemistry of vision. Notice that each of these molecules contains a sequence of conjugated double bonds. Vitamin A has a row of 10 carbon atoms with $sp^2$ hybridization. The leftover $p$ orbitals on these 10 atoms generate 10 π–type molecular orbitals, five bonding and five antibonding. The line structure shows five double bonds, consistent with the five bonding π orbitals being filled and the five $\pi^*$ orbitals being empty. β–Carotene has 22 atoms that participate in the delocalized π system. Retinal has 11 sequential $sp^2$–hybridized carbon atoms and an outer oxygen atom, resulting in six conjugated double bonds.

Plastoquinone, the line structure of which appears in the margin, has 10 double bonds. Unlike the π bonds in the tail of retinal, however, the π bonds in the long tail of plastoquinone are not delocalized because $sp^3$–hybridized carbon atoms separate them. The delocalized π system of plastoquinone is its planar ring of six carbon atoms, with two of the carbon atoms double-bonded to outer oxygen atoms.

Plastoquinone

Delocalized π system

**Figure 7-45**

Vitamin A ($C_{20}H_{30}O$)

Retinal ($C_{20}H_{28}O$)

β-Carotene ($C_{40}H_{56}$)
(an orange plant pigment, and the
precursor to vitamin A)

The line structures of three biologically important molecules that contain extended conjugated π systems. The conjugated systems are highlighted.

The orange colour of β–carotene and the pink of xanthin (see Focus 7-1) demonstrate that compounds with extended π systems often are coloured. A molecule is coloured when it absorbs photons in the visible region of the electromagnetic spectrum. Recall from Chapter 4 that substances absorb only photons whose energies match differences among electron energy levels. Visible photons have wavelengths between 700 and 400 nm and energies between 180 and 330 kJ/mol. Thus, in a coloured substance, the energy needed to promote an electron from the highest filled orbital to the lowest unoccupied orbital is between 180 and 330 kJ/mol.

Molecules with isolated π bonds, such as ethylene or 1,4-pentadiene, have energy gaps between the filled π and empty π* that are greater than 330 kJ/mol. These molecules absorb UV radiation but are colourless. However, it is a general feature of π orbital energies that the larger the number of atoms over which the orbital extends, the smaller the energy difference between orbitals. The long conjugated chains of β–carotene and xanthin reduce their energy gaps into the visible region and make these compounds coloured.

In addition to long conjugated chains, planar rings with delocalized π bonding are common in biological systems. Look at the structure of chlorophyll in the introduction to this chapter. Chlorophyll contains four five-atom rings (four C and one N) connected in a larger ring that forms a plane. The alternating double and single bonds in the ring portion of chlorophyll indicates a delocalized π system.

Benzene ($C_6H_6$) is the simplest example of a stable ring molecule with a delocalized π system. This compound has a ring of six carbon atoms, and the Lewis structure in Figure 7-46a shows alternating double and single bonds. The figure also shows that benzene has two resonance structures. The molecule has six adjacent $sp^2$–hybridized atoms, each with a p orbital perpendicular to the plane of the ring (Figure 7-46b). These six p orbitals are in perfect position for side-by-side overlap. Together they form a set of six molecular orbitals: three bonding π orbitals and three antibonding π* orbitals. There are six π electrons in benzene, just enough to fill the bonding molecular orbitals. The antibonding orbitals remain empty. The most stable π orbital of benzene consists of circular orbital rings above and below the plane of the carbon atoms (Figure 7-46c). Two other bonding π orbitals have more complicated shapes and are not shown. These delocalized molecular orbitals are highly resistant to chemical attack, making the benzene ring remarkably stable.

Benzene is highly stable to chemical attack. β–Carotene absorbs visible light and is orange. Both properties are direct consequences of delocalized π systems. Delocalized π systems also affect the oxidation–reduction characteristics of organic substances. Our chapter introduction noted that plastoquinone and cytochrome play important redox roles in photosynthesis. As another example, β–carotene, with its delocalized π system, is thought to act as an antioxidant, protecting the body from cancer-causing oxidation processes. Oxidation and reduction reactions involve the transfer of electrons from occupied orbitals in one substance to empty orbitals in another substance. Whether or not a redox

**Figure 7-46**

(a)

(b)

(c)

(a) The Lewis structure of benzene. (b) The p orbitals that interact to form delocalized MOs. (c) The most stable bonding π molecular orbital of benzene has two rings of electron density above and below the molecular plane.

reaction takes place depends on orbital energies, so the stabilization of π systems by delocalization has a direct effect on redox behaviour. An exciting field involving materials containing conjugated systems is that of photovoltaic substances, which convert sunlight directly into electricity. Parts of these devices are "electron donors," which are usually conjugated polymers containing delocalized π electrons resulting from *p*-orbital hybridization.

## π Bonding Beyond the Second Row

Several common polyatomic oxoanions, including sulphate, perchlorate, and phosphate, have inner atoms from the third row of the periodic table. In these anions, valence *d* orbitals are available to participate in π bonding. Figure 7-47 shows how a π orbital can form through side-by-side overlap of a *d* orbital on one atom with a 2*p* orbital on another atom. As with other π bonds, electron density is concentrated above and below the bond axis.

Several elements in Row 3 of the periodic table form π bonds to oxygen through side-by-side overlap of 3*d* and 2*p* orbitals. An example is the sulphate anion, the Lewis structure of which appears in Figure 7-48. The steric number of the sulphur atom in $SO_4^{2-}$ is 4, indicating that the anion has tetrahedral geometry. We can describe the bonding framework using $sp^3$ hybrid orbitals. In addition, however, the six resonance structures show that there are delocalized π bonds that involve all five atoms. These bonds are the result of delocalized π orbitals created from a 2*p* orbital on each oxygen atom and valence *d* orbitals on the sulphur atom. The details of these interactions, which also occur for other

**Figure 7-47**

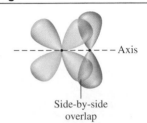

Side-by-side overlap

A π bond can form through side-by-side overlap between a *d* orbital and a *p* orbital.

**Figure 7-48**

:Ö:—S—Ö:   ⟷   Ö=S=Ö   ⟷   Ö=S—Ö:   ⟷   Ö=S—Ö:   ⟷   :Ö—S=Ö   ⟷   :Ö—S=Ö

The six resonance structures of the sulphate ion indicate that delocalized orbitals span the entire anion.

### SECTION EXERCISES

● **7.6.1** Describe the π bonding in the hydrogen carbonate ion, $HCO_3^-$.

● **7.6.2** Shown below is the line structure of azulene, a deep-purple solid, which forms a blue solution (see photo in margin). How many delocalized orbitals are in an azulene molecule? How many of these orbitals are occupied?

Azulene
$C_{10}H_8$

● **7.6.3** The perchlorate ion ($ClO_4^-$) has the same number of valence electrons as the sulphate anion. Describe the bonding in $ClO_4^-$.

A solution of azulene has a blue colour.

third-row oxoanions such as phosphate, are beyond the scope of this text. You need only recognize that they include $\pi$ orbitals formed through the side-by-side overlap of $d$ orbitals and $p$ orbitals.

## 7.7 BAND THEORY OF SOLIDS

Copper, iron, and aluminum are three common metals in modern society. Copper wires carry the electricity that powers most appliances, including the lamp by which you may be reading. The chair in which you are sitting may have an iron frame, and you may be sipping a soft drink from an aluminum can. The properties that allow metals to be used for such a wide range of products can be traced to the principles of bonding and electronic structure.

Iron and other metals have tremendous mechanical strength, which suggests that the bonds between their atoms must be strong. At the same time, most metals are malleable, which means they can be shaped into thin sheets to make objects such as aluminum cans. Metals are also ductile, which means they can be drawn into wires. The properties of malleability and ductility suggest that atoms in metals can be moved about without weakening the bonding. Finally, metals conduct electricity and heat, which shows that some of the electrons in a metal are free to move throughout the solid.

Metals are malleable and ductile.

All these properties of metals are consistent with a bonding description that places the valence electrons in delocalized orbitals. This section describes the **band theory of solids**, an extension of the delocalized orbital ideas that are the theme of this chapter. Band theory accounts for the properties of metals, and it also explains the properties of metalloids such as silicon.

### DELOCALIZED ORBITALS IN LITHIUM METAL

We use lithium, the lightest metal, to demonstrate the principles of band theory. Solid lithium contains atoms held together in a three-dimensional crystal lattice. Bonding interactions among these atoms can be described by orbital overlap. To see how this occurs, consider building an array of lithium atoms one at a time.

Figure 7-49 indicates the orbital energies resulting from such sequential construction of ever-larger atom clusters. We describe the bonding using overlapping valence orbitals. In $Li_2$, the two valence $2s$ orbitals interact to form a $\sigma$ bonding orbital and a $\sigma^*$ antibonding orbital. The bonding orbital is filled, and the antibonding orbital is empty. The single bond in $Li_2$ is relatively weak, 105 kJ/mol, because compact $2s$ orbitals do not overlap strongly.

Clusters of lithium atoms such as $Li_4$, $Li_6$, and $Li_{18}$ can be generated in the gas phase by bombarding a piece of lithium metal with gaseous metal atoms in a chamber at very low pressure. The $Li_4$ molecule is bonded together by four $2s$ orbitals that interact to give two filled delocalized bonding orbitals and two empty delocalized antibonding orbitals. Likewise, $Li_6$ has six delocalized orbitals, $Li_{18}$ has 18 delocalized orbitals and so on. As with all molecular orbitals, these interactions generate equal numbers of bonding and antibonding orbitals. In 1 cm$^3$ of Li metal, there are roughly $4.6 \times 10^{22}$ atoms, the $2s$ atomic orbitals of which generate approximately $4.6 \times 10^{22}$ delocalized orbitals. This is a huge number of bonding orbitals

**Figure 7-49**

The energy band describing the bonding in lithium metal can be constructed by adding atoms one at a time, allowing all the 2*s* valence orbitals to interact to form delocalized orbitals.

and antibonding orbitals, yet they all obey the principles developed in this chapter. Most important, each orbital extends over the entire lattice of atoms.

Figure 7-49 shows that the energy spacing between orbitals decreases as the number of delocalized orbitals increases. For the huge number of atoms in a piece of metal, the orbitals are spaced so closely that they behave as if they were merged into an energy band. This bonding model is called band theory because energy bands are one of its main features.

To keep our picture as simple as possible, we include only the 2*s* orbitals in Figure 7-49. For most metals, however, a comprehensive picture of the bonding includes interactions among all valence *s*, *p*, *d*, and even *f* orbitals. Each group of orbitals interacts to generate an energy band. Interactions among bands are complex, and often the bands overlap. Despite this complexity, however, the fundamental features illustrated in Figure 7-49 remain valid.

## Electrical Conductivity

The most important feature of metallic bonding is the energy bands that arise because of the close spacing between orbital energies. The countless orbitals within the band are separated by infinitesimally small energy gaps. The valence electrons of the metal atoms occupy these orbitals according to the aufbau and Pauli principles. In a metal, the occupied orbitals of highest energy are so close in energy to the unoccupied orbitals of lowest energy that it takes little energy to transfer an electron from an occupied to an unoccupied orbital.

When an electrical potential is applied to a metal, the negative pole repels electrons and the positive pole attracts them. In energy terms, the occupied orbitals near the negative pole are pushed higher in energy than unoccupied orbitals near the positive pole. This "tilts" the energy levels, as shown in Figure 7-50, and electrons can move downhill out of filled orbitals into empty orbitals at lower energy. As a result, electrons flow through the metal from the negative end to the positive end, generating an electrical current.

**Figure 7-50**

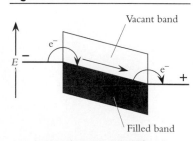

An electrical potential shifts the energy levels in a metal. The arrows show the direction of electron flow.

## Insulators and Conductors: Carbon vs. Lead

Metallic lead is dark in colour and is an electrical conductor. Diamond, the most valuable form of carbon, is transparent and is an electrical insulator. These properties are very different; yet both lead and carbon are in Group 14 of the periodic table and have the same valence configuration, $s^2p^2$. Why, then, are diamonds transparent insulators, whereas lead is a dark-coloured conductor?

Both diamond and lead can be described using delocalized orbitals and energy bands, but, as Figure 7-51 shows, the energy distributions of the bands are quite different. Lead is metallic because the energy separation between its filled and vacant orbitals is infinitesimally small. One continuous band of orbitals results in highly mobile electrons and electrical conductivity. The valence orbitals of carbon, on the other hand, form two distinct bands. The filled bonding orbitals are well separated in energy from the empty, antibonding orbitals. The energy difference between these two bands is called the **band gap** ($E_g$). The band gap for diamond is 580 kJ/mol.

Lead and diamond have distinctly different band structures for two reasons. First, the valence orbitals of carbon are smaller than those of lead, resulting in stronger overlap for diamond. This results in a large stabilization energy and a large band gap. Second, lead has valence $d$ and $f$ orbitals that are involved in band formation. These $d$ and $f$ bands overlap the $s$ and $p$ bands, contributing to the continuous distribution of orbital energies that gives lead its metallic properties.

Carbon in the form of diamond is an electrical insulator because of its huge band gap. In fact, its band gap of 580 kJ/mol substantially exceeds the C—C bond energy of 345 kJ/mol. In other words, it requires more energy to promote an electron from band to band in diamond than to break a covalent bond. Lead, in contrast, is a metallic conductor because it has $E_g = 0$.

The differences in their band structures also explain why diamond is transparent and lead is dark. Remember that substances can absorb only photons whose energies match differences between energy levels, and a coloured substance absorbs visible photons, whose energies are between 180 and 330 kJ/mol. For diamond, the energy difference between the top of the filled band and the bottom of the empty band is 580 kJ/mol. Diamond cannot absorb visible light because there are no energy levels between 180 and 330 kJ/mol above the filled band. Visible light passes straight through a diamond. Lead, on the other hand, has many vacant orbitals into which visible photons can promote electrons. Thus, lead can absorb all visible wavelengths, making it grey.

**Figure 7-51**

Lead and carbon have different properties because the energy differences between their filled bonding orbitals and their empty antibonding orbitals are different.

## Metalloids

Orbital energies are determined largely by the amount of spatial overlap. Because atomic orbitals become increasingly diffuse (spread out in space) as their principal quantum number ($n$) increases, the spatial overlap of the valence orbitals decreases as the valence shell $n$ value increases (see Figure 7-51). Less overlap means less difference between the energy of the bonding and antibonding levels, so the band gap shrinks from carbon to lead. Carbon ($n = 2$) has a large band gap and is a non-metal, but tin ($n = 5$) and lead ($n = 6$) have $E_g = 0$ and are metals. Silicon and germanium ($n = 3$ and $n = 4$, respectively) have intermediate band gap values, which cause them to behave as **metalloids**.

Like carbon, silicon and germanium have low-energy bands of bonding orbitals that are completely filled with electrons and higher-energy bands that are empty. For silicon, $E_g = 105$ kJ/mol, whereas $E_g = 64$ kJ/mol for germanium. Electrons cannot make the transition between these bands unless sufficient energy is supplied. Thus, silicon and germanium are non-conductors in the absence of an energy source. However, electrons can move between the bands at higher temperatures or in the presence of photons with energy that matches $E_g$. When electrons are transferred into the high-energy band, they can move freely among the many vacant orbitals of the band. In addition, electron transfer creates vacancies (holes) among the bonding orbitals in the low-energy band. The holes and high-energy electrons allow electricity to flow. Thus, silicon and germanium become conductors in the presence of an appropriate energy source. They are called **semiconductors** to indicate their ability to conduct electricity under special conditions. Figure 7-52 uses band-gap diagrams to summarize the behaviour of a semiconductor.

Transfer of electrons from one band to another requires energy equal to the band gap. This energy can be provided by heating (exploited in thermistors) or by absorption of light (used in photoconductors). Example 7-13 deals with photoconductors.

**Figure 7-52**

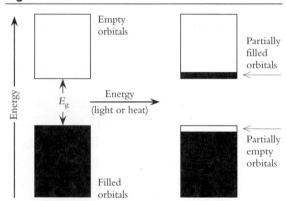

When a semiconductor absorbs heat or light, electrons are excited from the filled band to the empty band, and the semiconductor conducts electricity.

---

## EXAMPLE 7-13 PROPERTIES OF PHOTOCONDUCTORS

"Electric eye" door openers use photoconductors that respond to infrared light with a wavelength ($\lambda$) of 1.5 $\mu$m. Which is suitable for photoconductors operating at this wavelength: germanium ($E_g = 64$ kJ/mol) or silicon ($E_g = 105$ kJ/mol)?

**Strategy:** To conduct electricity, a semiconductor must be provided with energy that is *at least* equal to its band gap. In a photoconductor, this energy must come from photons. Thus, we need to compare the band gaps with the energy of the infrared photons.

**Solution:** First, calculate the energy of 1.5 $\mu$m light, recalling from Chapter 4 that photon energy is related to wavelength:

$$E_{photon} = h\nu = \frac{hc}{\lambda}$$

$$E_{photon} = \frac{(6.626 \times 10^{-34}\ \text{J s})(2.998 \times 10^8\ \text{m/s})}{(1.5\ \mu\text{m})(10^{-6}\ \text{m}/\mu\text{m})} = 1.32 \times 10^{-19}\ \text{J}$$

Multiply by Avodagro's number to convert the energy of one photon into the energy of 1 mole of photons in kilojoules per mole:

$$(1.32 \times 10^{-19} \text{ J/photon})(6.022 \times 10^{23} \text{ photons/mol})(10^{-3} \text{ kJ/J}) = 79 \text{ kJ/mol}$$

This energy is insufficient to overcome the band gap of silicon, so silicon could not be used for an infrared photoconductor. Germanium, however, with its smaller band gap ($E_g$ = 64 kJ/mol), becomes an electrical conductor when illuminated by infrared light with $\lambda$ = 1.5 $\mu$m.

**Does the Result Make Sense?** Silicon is closer to carbon, and germanium is closer to lead, in the periodic table. We expect the band gap in silicon to be larger than that in germanium, so it makes sense that lower-energy radiation can generate electrical conductivity in germanium than in silicon.

**Common Pitfalls:** Be careful with unit conversions. For instance, 1 $\mu$m = $10^{-6}$ m.

**Practice Exercise 7.13**
The band gap of silicon is 105 kJ/mol. What is the minimum wavelength of light that can promote an electron from the valence band to the conduction band? What region of the electromagnetic spectrum matches this wavelength?

## Doped Semiconductors

The relatively large band gaps of silicon and germanium limit their usefulness in electrical devices. Fortunately, adding tiny amounts of other elements that have different numbers of valence electrons alters the conductive properties of these solid elements. When a specific impurity is added deliberately to a pure substance, the resulting material is said to be *doped*. A **doped semiconductor** has almost the same band structure as the pure material, but it has different electron populations in its bands.

In pure silicon, all orbitals of the low-energy band are filled, and all orbitals of the high-energy band are empty. When pure silicon is doped with arsenic atoms, As atoms replace some Si atoms in the crystal lattice. Each arsenic atom contributes five valence electrons instead of the four contributed by a silicon atom. The fifth valence electron cannot enter the low-energy band of the solid because that band is already filled. The fifth valence electron from each arsenic atom must therefore occupy the higher-lying energy band. Silicon or germanium doped with atoms from Group 15 is called an *n*-type **semiconductor** because extra <u>n</u>egative charges exist in its high-energy band. These few electrons in a partially occupied band make the doped material an electrical conductor.

Silicon can also be doped with gallium, a Group 13 element. Gallium has three valence electrons, so each Ga atom in the Si lattice has one less electron than the atom that the Ga atom replaces. In gallium-doped silicon, there are not quite enough electrons to fill all the bonding orbitals. Silicon and germanium doped with atoms from Group 13 are known as *p*-type **semiconductors** because their low-energy bands have <u>p</u>ositive vacancies. Electrons move through the crystal by flowing from filled orbitals into these vacant ones.

**Figure 7-53**

Schematic views of an *n*-type and a *p*-type semiconductor.

Semiconductors such as silicon are extremely sensitive to impurities. Replacing just 0.000 01% of the Si atoms with a dopant can cause as much as a 100 000-fold increase in electrical conductivity.

Figure 7-53 shows band-gap diagrams of *n*-type and *p*-type semiconductors. Electrical current flows in a doped semiconductor in the same way as current flows in a metal (see Figure 7-50). Only a small energy difference exists between the top of the filled band and the next available orbital, so the slightest applied potential tilts the bands enough to allow electrons to move and current to flow.

Not all semiconductors are made from silicon or germanium. Compounds made from equimolar amounts of Group 13 and Group 15 elements also are semiconductors. Gallium arsenide (GaAs) is a typical example. One Ga atom and one As atom have a total of eight valence electrons, so gallium arsenide is isoelectronic with germanium. Gallium arsenide can be doped with zinc atoms to make a *p*-type semiconductor or with tellurium atoms to make an *n*-type semiconductor. Other semiconductors contain equimolar compositions of an element such as zinc that has two valence electrons and a Group 16 element (S, Se, or Te) with six valence electrons. Zinc sulphide is an example. Our Focus 7-2: Chemistry and Technology describes the light-emitting characteristics of this type of semiconductor.

## SECTION EXERCISES

**7.7.1** CdS, a semiconductor studied for use in solar cells, absorbs blue light at 470 nm. Calculate its band gap in kJ/mol.

**7.7.2** Draw a band-gap diagram that illustrates a semiconductor made of gallium arsenide doped with zinc atoms.

**7.7.3** You have been asked to develop a *p*-type semiconductor using Al, Si, and/or P. Which two elements would you choose, and what role would each element play in the semiconductor?

## FOCUS 7-2 CHEMISTRY AND TECHNOLOGY: LIGHT-EMITTING DIODES

One application of modern solid-state electronic devices is semiconductor materials that convert electrical energy into light. These *light-emitting diodes* (LEDs) are used for visual displays and solid-state lasers. Many indicator lights are LEDs, and diode lasers read compact discs in a CD player. The field of diode lasers is expanding particularly rapidly, driven by such applications as fibre optic telephone transmission.

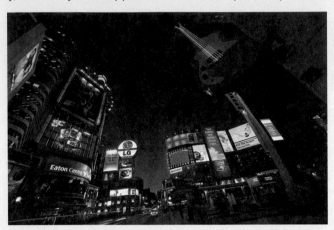

LED displays at Dundas Square, Toronto

The heart of a light-emitting diode is a junction between a *p*-type semiconductor and an *n*-type semiconductor. The different semiconductor types have different electron populations in their bands. The lower-energy band of a *p*-type semiconductor is deficient in electrons, whereas the upper-energy band of an *n*-type semiconductor has a small population of electrons. The band structure in the junction region is shown schematically in the figure below.

When an electrical potential is applied to an LED device, the *p* region is made positive relative to the *n* region. The positive potential attracts electrons, which can jump from the higher energy band of the *n* region to the lower-energy band of the *p* region. In doing so, these electrons emit their excess energy as photons of light.

The colour of the light depends on the energy loss for the electrons and increases with the size of the band gap of the semiconductor. Thus, red LEDs use semiconductors with relatively small band gaps, whereas green LEDs have larger band gaps.

Semiconductors containing arsenic are particularly versatile. A GaAs semiconductor can be made a *p* type by doping with a small excess of Ga, or an *n* type by doping with a small excess of As. The band gap of these semiconductors can be varied by replacing some Ga with Al, or some As with P. Since GaAs forms solid solutions with both AlAs and GaP, the composition can be varied continuously, from pure AlAs to pure GaAs and from pure GaAs to pure GaP. These materials have non-stoichiometric chemical formulas, $Al_xGa_{1-x}As$ and $GaP_xAs_{1-x}$.

The band gap of the LED varies with composition for both these solid solutions, as shown in the figure below. The cause of the variation is different for the two substances. Semiconductor band gaps increase when orbital overlap decreases. A decrease in orbital overlap can arise from increased spacing between atoms or increased ionic character of the bonds.

Phosphorus and arsenic have nearly identical electronegativities, so in $GaP_xAs_{1-x}$, the dominant effect is the smaller atomic radius of P relative to As. Substituting P atoms for As atoms shrinks the dimensions of the semiconductor lattice. This leads to greater overlap of the valence orbitals, increased stability of the bonding orbitals (valence band), and an increased band gap.

Substituting Al atoms for Ga atoms does not change the lattice dimensions, which are determined by the size of As, the largest atom of this set. Here, the lower electronegativity of Al relative to Ga leads to more ionic character and a smaller band gap.

# CHAPTER 7 VISUAL SUMMARY

## 7.1 LOCALIZED BONDS

Bonding orbitals
Orbital overlap
Localized bonds
Delocalized bonds

*Bonding orbitals result from combining atomic orbitals.*

H + F → HF

**LEARNING OBJECTIVES:** Use the orbital overlap model to explain the bonding in simple molecules.

**SKILLS TO MASTER:** Drawing sketches that show orbital overlap (p. 361).

Describing the bonding in molecules using the orbital overlap model (p. 362).

**KEY CONCEPTS:** Bonding orbitals are constructed by combining atomic orbitals from adjacent atoms.

Only the valence orbitals are needed to describe bonding.

## 7.2 HYBRIDIZATION OF ATOMIC ORBITALS

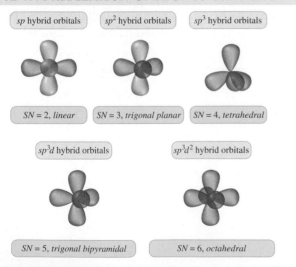

*sp* hybrid orbitals  $sp^2$ hybrid orbitals  $sp^3$ hybrid orbitals

*SN = 2, linear*  *SN = 3, trigonal planar*  *SN = 4, tetrahedral*

$sp^3d$ hybrid orbitals  $sp^3d^2$ hybrid orbitals

*SN = 5, trigonal bipyramidal*  *SN = 6, octahedral*

**LEARNING OBJECTIVES:** Assign the correct hybrid orbitals used by each inner atom in a molecule, and the molecular geometry that results.

**SKILLS TO MASTER:** Correlating the steric number with the electron group geometry, hybridization, number of hybrid orbitals and number of unused p-orbitals on an inner atom (p. 368).

**KEY CONCEPTS:** The number of hybrid orbitals generated by the hybridization process equals the number of valence atomic orbitals participating in hybridization.

The steric number of an inner atom can be used to infer the hybrid orbitals it is using.

Elements beyond the second row of the periodic table can form bonds to more than four ligands and can be associated with more than an octet of electrons.

## 7.3 MULTIPLE BONDS

Sigma bond ($\sigma$)
Pi bond ($\pi$)

$\sigma$ bond: high electron density on the bond axis
$\pi$ bond: high electron density off the bond axis

**LEARNING OBJECTIVES:** Describe the $\sigma$ and $\pi$ bonding systems in multiple bonds.

**SKILLS TO MASTER:** Constructing the $\sigma$ and $\pi$ bonding framework of a molecule (p. 375).

**KEY CONCEPTS:** A sigma ($\sigma$) bond has high electron density distributed symmetrically along the bond axis.

A pi ($\pi$) bond has high electron density concentrated above and below the bond axis.

## 7.4 MOLECULAR ORBITAL THEORY

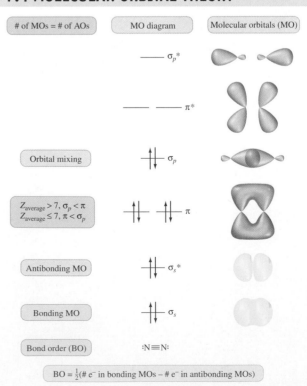

| # of MOs = # of AOs | MO diagram | Molecular orbitals (MO) |

$\sigma_p{}^*$

$\pi^*$

Orbital mixing  $\sigma_p$

$Z_{average} > 7, \sigma_p < \pi$
$Z_{average} \leq 7, \pi < \sigma_p$  $\pi$

Antibonding MO  $\sigma_s{}^*$

Bonding MO  $\sigma_s$

Bond order (BO)  :N≡N:

BO = $\frac{1}{2}$(# e⁻ in bonding MOs − # e⁻ in antibonding MOs)

**LEARNING OBJECTIVES:** Use molecular orbital theory to calculate a bond order, predict magnetic properties of a molecule and explain trends in bond length and energy.

**SKILLS TO MASTER:** Drawing MO diagrams for heteronuclear diatomic molecules (p. 383).

Calculating the bond order (p. 385).

Predicting the effects of ionization on bond length and energy (p. 389).

**KEY CONCEPTS:** The total number of molecular orbitals produced by a set of interacting atomic orbitals is equal to the number of interacting atomic orbitals.

Delocalized electrons occupy molecular orbitals (MOs), so called because they may span entire molecules.

A Bonding MO results from positive overlap of atomic orbitals wavefunctions and stabilizes the bond.

An antibonding MO results from negative overlap of atomic orbital wavefunctions and destabilizes the bond.

Paramagnetism arises from unpaired electrons.

## 7.5 THREE-CENTRE π ORBITALS

Delocalized π orbitals

π orbital     Nonbonding MO ($\pi_{nb}$)     π* orbital

Delocalized π orbitals result from side-by-side overlap of atomic π orbitals on more than two adjacent atoms.

**LEARNING OBJECTIVES:** Describe the bonding in three-atom π systems.

**SKILLS TO MASTER:** Using the composite model of bonding (p. 395).

**KEY CONCEPTS:** Hybrid orbitals can be used to construct the molecular framework. Molecular orbitals can then be used to construct the π bonding system.

## 7.6 EXTENDED π SYSTEMS

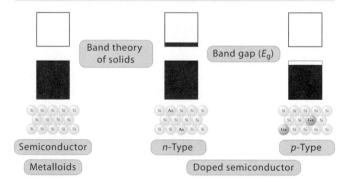

Vitamin A

Conjugated π systems

**LEARNING OBJECTIVES:** Describe the bonding in extended π systems.

**KEY CONCEPTS:** A delocalized π system is present whenever p orbitals on more than two adjacent atoms are in position for side-by-side overlap.

Molecules that have alternating single and double bonds are said to have conjugated π systems.

Delocalized π systems result in increased stability and often in coloured compounds.

## 7.7 BAND THEORY OF SOLIDS

Band theory of solids          Band gap ($E_g$)

Si Si Si Si      Si As Si Si      Si Si Si Si
Si Si Si Si      Si Si Si Si      Si Si Ga Si
Si Si Si Si      Si As Si Si      Ga Si Si Si

Semiconductor          n-Type          p-Type

Metalloids                Doped semiconductor

**LEARNING OBJECTIVES:** Explain such properties as electrical conductivity and colour of metals, nonmetals and metalloids in terms of band theory.

**KEY CONCEPTS:** Metals conduct heat and electricity because their band gaps are so small.

Semiconductors have intermediate band gaps.

A doped semiconductor has almost the same band structure as the pure material, but different electron populations in its bands.

## LEARNING EXERCISES

**7.1** Design a detailed flowchart that shows how to describe the bonding of a second-row diatomic molecule.

**7.2** Design a detailed flowchart that shows how to describe the bonding of a polyatomic species that has multiple bonds.

**7.3** Prepare a table of hybridizations and their various shapes.

**7.4** Prepare a list of molecules and ions mentioned in this chapter that contain electrons in delocalized orbitals. Organize your list by the number of atoms over which delocalization occurs.

**7.5** Design a flowchart that summarizes the composite model of bonding.

**7.6** Describe in your own words your understanding of each of the following terms: (a) $\sigma$ bond; (b) $\pi$ bond; (c) anti-bonding molecular orbital; (d) non-bonding electrons; (e) bond order; and (f) delocalized orbital.

**7.7** Summarize the types of experimental evidence that support the idea of delocalized orbitals.

**7.8** Create a single figure that explains the difference between a conductor, an insulator, and a semiconductor.

**7.9** Prepare a list of all terms new to you that appear in Chapter 7. Write a one-sentence definition of each. Consult the glossary if you need help.

## PROBLEMS

**ilw** = interactive learning ware problem. Visit the website at www.wiley.com/canada/olmsted

### Localized Bonds

**7.1** Describe the bonding between two bromine atoms in $Br_2$.

**7.2** Describe the bonding between fluorine and chlorine atoms in the mixed halogen FCl.

**7.3** Hydrogen forms polar diatomic molecules with elements of Group 1. Describe the bonding in LiH and include a picture of the overlapping orbitals.

**7.4** Describe the bonding between hydrogen and chlorine atoms in HCl and include a picture of the overlapping orbitals.

**7.5** The bond angles in antimony trifluoride are 87°. Describe the bonding in $SbF_3$, including a picture of the orbital overlap interaction that creates the Sb—F bonds.

**7.6** Hydrogen telluride, $H_2Te$, is stable only at low temperature. The compound forms a colourless gas, with bond angles of 90°. Describe the bonding of $H_2Te$, including an orbital overlap sketch.

### Hybridization of Atomic Orbitals

**7.7** Name the hybridization of an inner atom that has the following characteristics: (a) two lone pairs and two ligands; (b) three ligands and one lone pair; (c) three ligands and no lone pairs; and (d) five ligands and one lone pair.

**7.8** Name the hybridization of an inner atom that has the following characteristics: (a) two ligands and no lone pairs; (b) two ligands and three lone pairs; (c) two lone pairs and four ligands; and (d) one lone pair and two ligands.

**7.9** Name the hybrid orbitals formed by combining the following sets of atomic orbitals: (a) one 3s and three 3p orbitals; (b) one 2s and one 2p orbital; and (c) one 3s, three 3p, and two 3d orbitals.

**7.10** Name the hybrid orbitals formed by combining the following sets of atomic orbitals: (a) one 2s and two 2p orbitals; (b) one 3s, three 3p, and one 3d orbital; and (c) one 4s and three 4p orbitals.

**7.11** Identify the hybridization of the underlined atom in the following species: (a) $(CH_3)_2\underline{N}H$; (b) $\underline{S}bF_5$; (c) $\underline{S}O_2$; and (d) $\underline{C}S_2$.

**7.12** Identify the hybridization of the underlined atom in the following species: (a) $\underline{S}F_6$, (b) $(CH_3)_2\underline{C}O$; (c) $\underline{B}F_4^-$; and (d) $HC\underline{C}CH_3$.

**7.13** Identify the hybrid orbitals used by the inner atoms.

(a) :Cl—S—Cl: with =O above    (b) P with F, F, Cl, Cl    (c) Xe with F, O, F

(d) :Cl—C—Cl: with =O above

**7.14** Identify the hybrid orbitals used by the inner atoms.

(a) O=S=O with :O: above    (b) I with F, F, F, F and Br below    (c) Xe with F, O, F, O

(d) :Br—P—Br: with Br above

**7.15** Describe the bonding of chloroform, $CHCl_3$. Sketch an orbital overlap picture of the molecule.

**7.16** Describe the bonding of dichlorosilane, $SiH_2Cl_2$. Sketch an orbital overlap picture of the molecule.

**7.17** Describe the bonding of hydrazine, $H_2NNH_2$. Sketch an orbital overlap picture of the molecule.

**7.18** Describe the bonding of hydrogen peroxide, HOOH. Sketch an orbital overlap picture of the molecule.

### Multiple Bonds

**7.19** Describe the bonding in the common solvent acetone, $(CH_3)_2CO$. Include sketches of all the bonding orbitals.

**7.20** Describe the bonding in the solvent used in nail polish, ethyl acetate, $C_2H_5CO_2CH_3$. Include sketches of all the bonding orbitals.

**7.21** **ilw** Describe the steric number, geometry, and hybridization of each of the atoms of neocembrene, whose line structure appears in the Focus 7-1: Chemistry and Life.

**7.22** Describe the steric number, geometry, and hybridization of each of the atoms of geraniol, whose line structure appears in the Focus 7-1: Chemistry and Life.

7.23 The five-carbon compounds 1,4-pentadiene, 1-pentyne, and cyclopentene all have molecular formula $C_5H_8$. Use steric numbers and hybridization to develop bonding pictures of these three molecules.

1,4-Pentadiene    1-Pentyne    Cyclopentene

7.24 The four-carbon compounds butane, 2-butene, and 2-butyne differ in the bonding between the second and third carbon atoms in the chain. Use steric numbers and hybridization to develop bonding pictures of these three molecules.

Butane        2-Butene        2-Butyne

7.25 Decide if the following pairs of orbitals overlap to form a $\sigma$ bond, $\pi$ bond, or no bond at all. Explain your reasoning in each case, including a sketch of the orbitals. Assume the bond lies along the z-axis. (a) $2p_z$ and $2p_z$; (b) $2p_y$ and $2p_x$; (c) $sp^3$ and $2p_z$; and (d) $2p_y$ and $2p_y$.

7.26 Decide if the following pairs of orbitals overlap to form a $\sigma$ bond, $\pi$ bond, or no bond at all. Explain your reasoning in each case, including a sketch of the orbitals. Assume the bond lies along the z-axis. (a) $sp^2$ and $2p_x$; (b) $2p_x$ and $2p_x$; (c) $sp$ and $2p_z$; and (d) $2p_y$ and $2p_z$.

## Molecular Orbital Theory: Diatomic Molecules

7.27 Use molecular orbital diagrams to rank the bond energies of the following diatomic species, from weakest to strongest: $H_2$, $H_2^-$, and $H_2^{2-}$.

7.28 Use molecular orbital diagrams to rank the bond energies of the following diatomic species, from weakest to strongest: $He_2$, $He_2^+$, and $He_2^{2+}$.

7.29 In each of the following pairs, which has the stronger bond? Use orbital configurations to justify your selections: (a) CO or $CO^+$; (b) $N_2$ or $N_2^+$; and (c) $CN^-$ or CN.

7.30 In each of the following pairs, which has the stronger bond? Use orbital configurations to justify your selections: (a) $O_2$ or $O_2^+$; (b) CO or $CO^-$; and (c) $F_2$ or $F_2^+$.

7.31 For each of the following interactions between orbitals of two different atoms, sketch the resulting molecular orbitals. Assume that the nuclei lie along the z-axis, and include at least two coordinate axes in your drawing. Label each MO as bonding or antibonding and $\sigma$ or $\pi$. (a) 2s and $2p_z$; and (b) $2p_x$ and $2p_x$.

7.32 For each of the following interactions between orbitals of two different atoms, sketch the resulting molecular orbitals. Assume that the nuclei lie along the z-axis, and include at least two coordinate axes in your drawing. Label each MO as bonding or antibonding and $\sigma$ or $\pi$. (a) $2p_z$ and $2p_z$; and (b) $2p_y$ and $2p_y$.

7.33 The active ingredient in commercial laundry bleach is sodium hypochlorite (NaClO). Describe the bonding in the hypochlorite anion, $ClO^-$. (*Hint:* In this ion, the valence orbitals of chlorine behave like $n = 2$ orbitals.)

7.34 The superoxide ion, $O_2^-$, is a reactive species that may play a role in the chemistry of aging. Describe the bonding in this ion.

## Three-Centre $\pi$ Orbitals

7.35 Describe the bonding of carbon disulfide, $CS_2$, in terms of orbital overlap and delocalized electrons. How many valence electrons occupy delocalized $\pi$ orbitals in this molecule?

7.36 Describe the bonding of sulphur dioxide, $SO_2$, in terms of orbital overlap and delocalized electrons. How many valence electrons occupy delocalized $\pi$ orbitals in this molecule?

7.37 Calcium cyanamide (CaNCN) is used as a pesticide, herbicide, and plant growth regulator. Write a Lewis structure for the $NCN^{2-}$ anion and describe its bonding, including any delocalized orbitals.

7.38 Detonators for explosive charges of trinitrotoluene (TNT) contain mercury fulminate, $Hg(CNO)_2$, which explodes when struck. Write a Lewis structure for the $CNO^-$ anion and describe its bonding, including any delocalized orbitals.

7.39 Hydrazoic acid has the formula $HN_3$, with the three nitrogen atoms in a row. Determine the Lewis structure, hybridization, and bond angles of this compound. Describe its $\pi$ bonding network.

7.40 Two molecules that have been detected in interstellar space are HNCO and OCS. Describe the bonding in each molecule, identify the hybrid orbitals used to make $\sigma$ bonds, and identify their delocalized $\pi$ systems.

## Extended $\pi$ Systems

7.41 Which of the molecules whose line structures appear in the Focus 7-1: Chemistry and Life are stabilized by delocalized $\pi$ electrons? For those that are, identify the atoms that contribute to the delocalized orbitals.

7.42 Which of the following molecules are stabilized by delocalized $\pi$ orbitals? For those that are, identify the atoms over which delocalization extends.

7.43 Vitamin C has the structure shown below:
(a) Identify the hybridization of each carbon and oxygen atom. (b) How many $\pi$ bonds are there? (c) How many electrons are in delocalized $\pi$ orbitals? (d) Redraw the structure and circle the largest continuous plane of atoms.

HO
HO
HO    OH
Vitamin C
$(C_6H_8O_6)$

**7.44** Carvone, the line structure of which appears below, is the principal flavour and fragrance ingredient in spearmint.
(a) Identify the hybridization of each carbon atom.
(b) How many $\pi$ bonds are there? (c) How many electrons are in delocalized $\pi$ orbitals? (d) Redraw the structure and circle the largest continuous plane of atoms.

Carvone
($C_{10}H_{14}O$)

**7.45** The two common oxides of carbon are CO and $CO_2$, which are linear. A third oxide of carbon, which is rare, is carbon suboxide, OCCCO. Draw a Lewis structure for carbon suboxide. Determine its geometry and identify all orbitals involved in delocalized $\pi$ orbitals. Make separate sketches of the $\sigma$ and $\pi$ bonding systems.

**7.46** The oxalate anion, $C_2O_4^{2-}$, has a C—C bond and two outer oxygen atoms bonded to each carbon atom. Determine the Lewis structure of this anion, including all resonance structures. What is the geometry about the carbon atoms, and how many atoms contribute $p$ orbitals to the delocalized $\pi$ system?

**7.47** Determine the Lewis structure and describe the bonding of the perchlorate anion, $ClO_4^-$.

**7.48** The permanganate ion, $MnO_4^-$, gives potassium permanganate solutions a rich burgundy colour. Determine the Lewis structure, including all resonance structures, and describe the bonding in this ion.

## Band Theory of Solids

**7.49** Identify the following as $p$-type, $n$-type, or undoped semiconductors: (a) GaP; (b) InSb doped with Te; and (c) CdSe.

**7.50** Identify the following as $p$-type, $n$-type, or undoped semiconductors: (a) Ge; (b) Ge doped with P; and (c) AlAs doped with Zn.

**7.51** Use band theory to explain why iron is harder and melts at a higher temperature than potassium. (*Hint:* Consider the number of valence electrons in each element.)

**7.52** The following energies measure the strength of bonding in sodium, magnesium, and aluminum metals:

$$Na(s) \longrightarrow Na\,(g) \qquad \Delta E = 99 \text{ kJ/mol}$$
$$Mg(s) \longrightarrow Mg\,(g) \qquad \Delta E = 127 \text{ kJ/mol}$$
$$Al(s) \longrightarrow Al\,(g) \qquad \Delta E = 291 \text{ kJ/mol}$$

Use band theory to explain these data. (*Hint:* Consider the number of valence electrons in each element.)

**7.53** Draw a band-gap diagram for gallium arsenide doped with zinc.

**7.54** Draw a band-gap diagram for germanium doped with phosphorus.

## ADDITIONAL PAIRED PROBLEMS

**7.55** Sulphur forms two stable oxides, $SO_2$ and $SO_3$. Describe the bonding and geometry of these compounds.

**7.56** Nitrogen forms two stable oxoanions, $NO_2^-$ and $NO_3^-$. Describe the bonding and geometry of these anions.

**7.57** Indium arsenide is a semiconductor with a band gap of 34.7 kJ/mol. What is the minimum frequency of light that will cause InAs to conduct electricity? What region of the electromagnetic spectrum corresponds to this frequency?

**7.58** Gallium arsenide is a common semiconductor. This material becomes a conductor when it is irradiated with infrared light of frequency $3.43 \times 10^{14}\,s^{-1}$. Calculate the band gap of GaAs in kJ/mol.

**7.59** Acrolein has the Lewis structure shown below. Give a complete description of its bonding, including orbital overlap sketches of the $\sigma$ and $\pi$ bonding systems, and an accounting of its delocalized electrons.

**7.60** The allyl cation has the Lewis structure shown below. Give a complete description of its bonding, including orbital overlap sketches of the $\sigma$ and $\pi$ bonding systems, and an accounting of its delocalized electrons.

**7.61** Cinnamic acid is one of the natural components of cinnamon oil. (a) Redraw the structure of cinnamic acid, adding the lone pairs. (b) What orbitals are used for bonding by atoms $A$, $B$, $C$, $D$, and $E$? (c) What are the bond angles about atoms $A$, $C$, and $D$? (d) How many $\pi$ bonds does cinnamic acid have?

Cinnamic acid ($C_9H_8O_2$)

**7.62** Capsaicin is the molecule responsible for the hot spiciness of chili peppers:

Capsaicin ($C_{18}H_{27}O_3N$)

(a) How many $\pi$ bonds does capsaicin have? (b) What orbitals are used for bonding by each of the labelled atoms? (c) What are the bond angles about each of the labelled atoms? (d) Redraw the structure of capsaicin, adding the lone pairs.

**7.63** Use electron configurations to decide if the following species are paramagnetic or diamagnetic: (a) CO; (b) $N_2^+$; (c) $O_2^+$; and (d) $CN^-$.

**7.64** Use electron configurations to decide if the following species are paramagnetic or diamagnetic: (a) NO; (b) $C_2$; (c) $O_2^-$; and (d) $Cl_2$.

**7.65** Determine the type of orbitals (atomic, $sp^3$, or $sp^2$) used by each atom in the molecules shown below:

    $H_3CNH_2$

**7.66** Determine the type of orbitals (atomic, $sp^3$, or $sp^2$) used by each atom in the molecules shown below:

    $N_2H_4$

**7.67** Draw band-gap diagrams for a metallic conductor and an *n*-type semiconductor.

**7.68** Draw band-gap diagrams for an insulator and a *p*-type semiconductor.

**7.69** Carbocations are unstable high-energy compounds that contain positively charged carbon atoms. Carbocations form during the course of a reaction but are usually consumed rapidly. Describe the bonding and geometry of the carbocation $(CH_3)_3C^+$.

**7.70** The methylene fragment $CH_2$ has been identified as a reactive species in some gas-phase chemical reactions. Determine its Lewis structure, describe its bonding, and predict the H—C—H bond angle.

## ADDITIONAL PROBLEMS

**7.71** Combine the features of Figures 7-50 and 7-52 in a single drawing that shows how silicon and germanium conduct electricity.

**7.72** Write the electron configuration and determine the bond order of NO, $NO^+$, and $NO^-$. Which ones display paramagnetism?

**7.73** Oxygen forms three different ionic compounds with potassium: potassium oxide ($K_2O$), potassium superoxide ($KO_2$), and potassium peroxide ($K_2O_2$). Write electron configurations for the superoxide and peroxide anions. Compare these configurations with the configuration of the oxygen molecule. Rank the three species in order of increasing bond order, bond energy, and bond length. Which of the three are paramagnetic? Which has the largest paramagnetism?

**7.74** Nitrogen molecules can absorb photons to generate excited-state molecules. Construct an energy-level diagram and place the valence electrons so that it describes the most stable *excited* state of an $N_2$ molecule. Is the N—N bond in this excited-state $N_2$ molecule stronger or

weaker than the N—N bond in ground-state nitrogen? Explain your answer.

**7.75** Two oxoanions containing chlorine are $ClO_3^-$ (chlorate) and $ClO_2^-$ (chlorite). Determine Lewis structures and describe the bonding for each of these anions, including delocalized $\pi$ bonds.

**7.76** The ionization energy of molecular oxygen is smaller than that of atomic oxygen (1314 vs. 1503 kJ/mol). In contrast, the ionization energy for molecular nitrogen is larger than that of atomic nitrogen (1503 vs. 1402 kJ/mol). Explain these data, using the electron configurations of the diatomic molecules.

**7.77** What is the relationship among bond length, bond energy (see Chapter 3), and bond order? Which of them can be measured?

**7.78** Buckminsterfullerene contains 60 carbon atoms, each of which uses $sp^2$ hybrid orbitals and contributes one $2p$ orbital to make delocalized molecular orbitals. How many $\pi$ orbitals and how many $\pi$ bonding electrons does buckminsterfullerene have?

**7.79** Acrylonitrile, shown below, is used to manufacture polymers for synthetic fibres. Describe the bonding of acrylonitrile. (a) Identify the hybrid and/or atomic orbitals involved in bonding for each atom. (b) Draw sketches that show the $\sigma$ and $\pi$ bonding systems. (c) If there are delocalized $\pi$ orbitals, identify and describe them.

**7.80** Use orbital energy diagrams to decide whether NF would be stabilized or destabilized by adding one electron to make the corresponding anion and by removing an electron to form the corresponding cation. Sketch the orbital involved in these changes.

**7.81** The molecule CN is not stable in our atmosphere, but it has been detected in the interstellar regions. Describe its bonding, and sketch its least stable occupied molecular orbital.

**7.82** The concept of threshold energy appears in Chapter 4 as part of the discussion of the photoelectric effect. What feature of band theory would you associate with the threshold energy? Consider electrons ejected by photons with a frequency significantly higher than the threshold frequency. Would these electrons all have the same kinetic energy? Explain.

**7.83** Chlorine forms one neutral oxide, $ClO_2$. Describe the bonding in this unusual compound. Explain why it is considered unusual.

**7.84** Use orbital sketches to illustrate the bonding of allene, $H_2CCCH_2$. Identify all orbitals in the bonding scheme. Make sure your sketches show the three-dimensional arrangement of the hydrogen atoms. Make separate sketches of the $\sigma$ and $\pi$ bonding systems.

7.85 Dilithium molecules can be generated by vapourizing lithium metal at very low pressure. Do you think it is possible to prepare diberyllium? Explain your reasoning using MO diagrams for $Li_2$ and $Be_2$.

7.86 Silicon carbide (carborundum) is a hard crystalline material similar to diamond. It is widely used for cutting tools. Describe the bonding in silicon carbide, whose empirical formula is SiC. Draw an energy band picture showing the band gap for SiC compared with those of diamond and silicon.

7.87 The bond strengths in the H—X molecules ($X$ = F, Cl, Br, I) show that bonds become weaker as orbitals become more diffuse. Make a series of orbital overlap sketches for the H—X molecules to illustrate this. Shade your figures in a way that illustrates the changes in electron density.

7.88 When an oxalate anion, $C_2O_4{}^{2-}$, adds two protons to form oxalic acid, two C—O bonds become longer and two become shorter than the bonds in oxalate anions. Which bonds get longer and which shorter? Use bonding principles to explain these changes.

7.89 Elements in the same group of the periodic table usually have similar chemical properties. Nitrogen and oxygen are unique among elements in their groups, however, because they are stable as diatomic molecules. Phosphorus and sulphur, the next two members of these groups, can exist as long-chain molecules. Use the properties of orbitals to account for these differences in chemical behaviour.

7.90 Explain how the following bond lengths (in picometres) provide evidence for the existence of delocalized $\pi$ orbitals in the carbonate ion: C—O (average), 143; C=O (average), 122; and $CO_3{}^{2-}$, 129.

7.91 Describe the bonding in ketene, $H_2CCO$. Make separate sketches of the $\sigma$ and $\pi$ bonding systems.

7.92 Consider the bond lengths of the following diatomic molecules: $N_2$, 110 pm; $O_2$, 121 pm; and $F_2$, 143 pm. Explain the variation in length in terms of the molecular orbital descriptions of these molecules.

## GROUP STUDY PROBLEMS

7.93 Pyrrole, $C_4H_5N$, is a common organic molecule. Pyrrole rings are common substructures of many biological molecules. According to the localized bonding model presented in this chapter, what geometry and hybridization are predicted for the nitrogen atom of pyrrole? In fact, pyrrole is a planar molecule with a trigonal planar nitrogen atom. Apply the composite model of bonding developed in this chapter to explain why the bonding of pyrrole differs from the predictions of the localized bond model. Your answer should include a sketch of the most stable $\pi$ orbital of pyrrole.

Pyrrole

7.94 The pentadienyl cation ($C_5H_7{}^+$) is a high-energy species that can be generated in the laboratory under carefully controlled conditions. (a) Describe the bonding orbitals used by each of the carbon atoms. (b) How many $p$ orbitals contribute to the $\pi$ bonding system? How many molecular orbitals do they form? (c) How many delocalized $\pi$ electrons are there in $C_5H_7{}^+$? (d) Would you expect this ion to have a more or less stable $\pi$ bonding system than 1,3-pentadiene or 1,4-pentadiene? (Give reasons.)

7.95 One of the first compounds discovered for the treatment of AIDS was azidothymidine (AZT). Below is the line structure of AZT. (a) Identify the type of orbitals used for $\sigma$ bonding by each of the nitrogen atoms. (b) What is the bond angle around the oxygen atom in the five-membered ring? (c) How many carbon atoms can be described as using $sp^3$ hybrid orbitals? (d) How many $\pi$ bonds are there? What orbitals are used to form these bonds?

7.96 Metal carbonyls are compounds in which molecules of carbon monoxide form covalent bonds to transition metals. Examples include $Cr(CO)_6$, $Fe(CO)_5$, and $Ni(CO)_4$. The principal bonding interaction in a metal carbonyl is a $\sigma$ bond formed between the carbon atom of CO and one of the valence $d$ orbitals of the metal. (a) What is the hybridization of the C atoms in metal carbonyls? (b) Use the $d_{z^2}$ orbital to make an orbital overlap sketch of the M—CO $\sigma$ bond. (c) Draw structures showing the geometries of these carbonyls. (d) Metal carbonyls also have $\pi$ bonds that result from overlap of the $\pi^*$ orbital of CO with one of the metal $d$ orbitals. Use the $d_{xz}$ orbital to make an orbital overlap sketch of the $\pi$ bond in a metal carbonyl.

7.97 Use the principles of orbital overlap to explain why two different isomers of 1,2-dichloroethylene exist but only one isomer of 1,2-dichloroethane exists.

*trans*-1,2-Dichloro-ethylene        *cis*-1,2-Dichloro-ethylene        1,2-Dichloro-ethane

Use bond strength arguments to estimate the amount of energy required to interconvert the *cis* and *trans* isomers.

**7.98** Titanium(IV) oxide, $TiO_2$, is a white solid that absorbs ultraviolet light at 350 nm. Heating this solid in the presence of molecular hydrogen reduces some of the Ti(IV) ions to Ti(III), which act as dopants. The product shows semiconducting properties and is blue, indicating that it absorbs red light. Develop appropriate band-gap diagrams for $TiO_2$ and the doped product. Use the diagrams to explain what type of semiconductor the product is and illustrate the changes in energy gaps indicated by the light-absorbing properties of the materials. Suggest possible chemical formulas for the product. (*Hints:* There is more than one possibility; they are non-stoichiometric but must be electrically neutral.)

**7.99** Imagine making an $O_4$ molecule by attaching an oxygen atom to the non-bonding pair of electrons on the central oxygen of ozone ($O_3$). Describe the bonding in $O_4$, which would be isoelectronic with $NO_3^-$ and $CO_3^{2-}$. Compare the expected stability of the $O_4$ molecule with that of two $O_2$ molecules. Does the simple picture of bonding explain why $O_4$ does not exist?

## ANSWERS TO SECTION EXERCISES

**7.1.1**

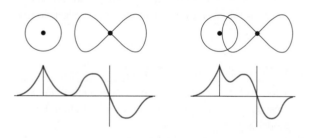

**7.1.2** The 1s orbital of the H atom overlaps with the 4p orbital of Br that points along the bond axis, giving a bond that looks similar to that in HF.

**7.1.3** Sb uses its 5p orbitals, which overlap with 1s orbitals of the H atoms, forming three bonds at right angles to one another.

**7.2.1**

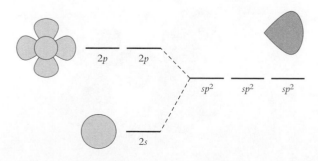

**7.2.2** S has SN = 4, uses $sp^3$ hybridization, has tetrahedral electron group geometry, and is bent, with bond angles somewhat less than 109.5°.

**7.2.3**

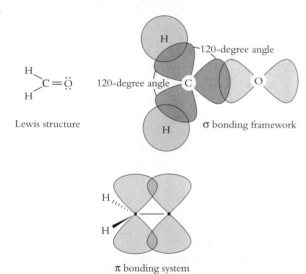

$SiF_4$ has tetrahedral electron group geometry and uses $sp^3$ hybridization; $TeF_4$ has trigonal bipyramidal electron group geometry and uses $sp^3d$ hybridization; $ClF_4^-$ has octahedral electron group geometry and uses $sp^3d^2$ hybridization.

**7.3.1**

The carbon atom uses $sp^2$ hybrids to form three σ bonds. Its unused *p* orbital overlaps with an oxygen *p* orbital to form a π bond.

**7.3.2**

Propene ($C_3H_6$)          Propyne ($C_3H_4$)

The $CH_3$ carbon in each compound uses $sp^3$ hybrids and has tetrahedral geometry. In propene, the other two carbon atoms have trigonal planar geometry, use $sp^2$ hybrids for three $\sigma$ bonds, and have a $\pi$ bond between them. In propyne, the other two carbon atoms have linear geometry, use $sp$ hybrids for two $\sigma$ bonds, and have two $\pi$ bonds between them.

**7.3.3**

Four carbon atoms participate in $\pi$ bonding. For appropriate orbital sketches, see Figures 7-20 and 7-21.

**7.4.1**

$$\underline{\hspace{1.2cm}}\ \sigma^*$$

$$\underset{\text{Li}_2}{\underline{\uparrow\downarrow}}\ \sigma$$

With two bonding electrons, $Li_2$ has a bond order of 1.

**7.4.2**  There is only one occupied valence orbital, $\sigma_s$.

**7.4.3**  $Ne_2$ has 16 valence electrons, which fill all the MOs formed from its valence orbitals. Your diagram should look like the right side of Figure 7-35, with a pair of electrons in each level. The calculated bond order is zero, which explains why this molecule cannot be prepared. In theory, $Ne_2^+$ could exist with a bond order of 0.5.

**7.5.1**  The formate anion is just like acetate, except it has a hydrogen atom in place of the methyl ($CH_3$) group of acetate. The bonding description is similar to that given for acetate in Example 7-11. In the bonding framework, replace the $sp^3$ hybrid–$sp^2$ hybrid $\sigma$ bond with a $1s$–$sp^2$ hybrid $\sigma$ bond. The $\pi$ bonding system is identical to that of acetate.

**7.5.2**  With 16 valence electrons, $N_3^-$ is isoelectronic with $CO_2$. Its bonding is identical to that of $CO_2$. It is linear, with the inner atom having a steric number of 2. There are two delocalized $\pi$ systems. The nitride ion, $N^{3-}$, contains a single atom with an atomic configuration identical to that of Ne. Electrons fill a spherical shell in this atomic anion.

**7.5.3**  $NO_2^+$ contains 16 valence electrons, making it isoelectronic with $CO_2$. It is linear, with the inner atom having a steric number of 2. There are two delocalized $\pi$ systems, each containing four electrons, for a total of eight delocalized electrons. $NO_2^-$ is treated in Practice Exercise 7-11. With 18 valence electrons, this anion is isoelectronic with ozone. Its inner atom has a steric number of 3, so it is bent. There is one delocalized system containing four electrons.

**7.6.1**  The Lewis structure shows two resonance forms, indicating three-atom $\pi$ delocalization that can be described exactly as in the acetate or formate anions.

$$\ddot{\underset{..}{O}}-\overset{\overset{\textstyle :\!O:}{\|}}{C}-\ddot{\underset{..}{O}}-H \longleftrightarrow \ddot{\underset{..}{O}}=\overset{\overset{\textstyle :\!\ddot{O}:}{\,}}{C}-\ddot{\underset{..}{O}}-H$$

**7.6.2**  Azulene has 10 delocalized $\pi$ orbitals; five of the orbitals are occupied.

**7.6.3**  The Lewis structure and bonding are the same as in the sulphate anion. The ion is tetrahedral. The Cl atom uses $sp^3$ hybrids to form four $\sigma$ bonds and uses $d$ orbitals to form delocalized $\pi$ bonds with the O atoms.

**7.7.1**  250 kJ/mol

**7.7.2**  This is a $p$-type semiconductor because Zn has one fewer valence electron than Ga. The band-gap diagram appears as follows:

$E$ —— Vacant orbitals

**7.7.3**  Si would be the semiconductor and Al would be the dopant. Aluminum has one less valence electron than silicon.

## ANSWERS TO PRACTICE EXERCISES

**7.1**  Two bonds form by overlap of H $1s$ orbitals with Se $4p$ orbitals. Your ball-and-stick model should look like the model of $H_2S$.

**7.2**  The oxygen atom has SN = 4, so it uses four $sp^3$ hybrid orbitals. Two overlap with $1s$ orbitals from the hydrogen atoms to form the bonds, and the other two hold the two lone pairs.

**7.3**  Each inner atom has SN = 4, so each uses four $sp^3$ hybrid orbitals. The carbon atom forms three bonds with H atoms, each described by $sp^3$–$1s$ overlap. The fourth orbital bonds to an $sp^3$ orbital of the nitrogen atom to form an $sp^3$–$sp^3$ bond. Two $sp^3$ hybrids on N overlap with $1s$ orbitals on H to form bonds, and the fourth hybrid holds the lone pair.

$CH_3NH_2$

**7.4** Bonds are between F $2p$ orbitals and hybrids on the inner atoms.

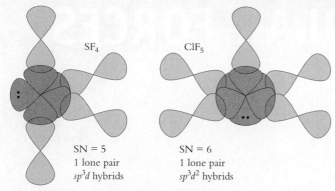

SF$_4$

SN = 5
1 lone pair
$sp^3d$ hybrids

ClF$_5$

SN = 6
1 lone pair
$sp^3d^2$ hybrids

**7.5** (a) The P atom in the $PF_6^-$ anion has SN = 6, octahedral electron group geometry, and $sp^3d^2$ hybridization. (b) The O atom in $OF_2$ has SN = 4, tetrahedral electron group geometry, and $sp^3$ hybridization. (c) The Sn atom in $SnCl_2$ has SN = 3, trigonal planar electron group geometry, and $sp^2$ hybridization.

**7.6** View chloroethylene as ethylene with one H atom replaced by a Cl atom. The bonding is identical to that of ethylene, except for a $\sigma$ bond formed by overlap of a Cl $3p$ orbital with an $sp^2$ hybrid.

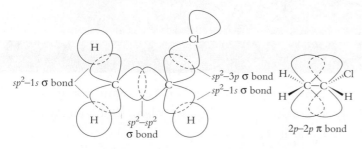

$sp^2$–$1s$ $\sigma$ bond

$sp^2$–$3p$ $\sigma$ bond
$sp^2$–$1s$ $\sigma$ bond

$sp^2$–$sp^2$
$\sigma$ bond

$2p$–$2p$ $\pi$ bond

**7.7** View 2-butyne as acetylene with both H atoms replaced by methyl groups. The two inner C atoms have $sp$ hybridization and a triple bond between them. The two outer C atoms use $sp^3$ hybrids to form three C—H bonds and one C—C bond.

$\sigma$ framework

$\pi$ bonds

**7.8** The third electron of $H_2^-$ must occupy an antibonding orbital, so $H_2$ has the stronger bond.

$H_2^-$

**7.9** The MO diagrams show that whereas $N_2$ has a bond order of $(8 - 2)/2 = 3$, $O_2$ has a bond order of $(8 - 4)/2 = 2$ and $F_2$ has a bond order of $(8 - 6)/2 = 1$. After $N_2$, the extra electrons occupy antibonding orbitals, causing bond energy to decrease in this set.

$N_2$      $O_2$      $F_2$

**7.10** The CN species have the same MO diagrams as CO, leading to these configurations: $CN^- = (\sigma_s)^2 (\sigma_s{}^*)^2 (\pi_p)^2 (\pi_x)^2 (\pi_y)^2$; $CN = (\sigma_s)^2 (\sigma_s{}^*)^2 (\sigma_p)^2 (\pi_x)^2 (\pi_y)^1$; and $CN^+ = (\sigma_s)^2 (\sigma_s{}^*)^2 (\sigma_p)^2 (\pi_x)^2$. $CN^-$ has the strongest bond (bond order of 3).

**7.11** The nitrite anion is isoelectronic with the ozone molecule. Describe its bonding exactly like that of ozone, but replace the inner O atom with N and include the $-1$ charge on the anion.

**7.12** The Lewis structure of acryloyl chloride shows a four-atom chain with alternating single and double bonds, so the $\pi$ system of this molecule is identical to that of methacrylate. Here is an inventory of the bond framework:

| | | |
|---|---|---|
| 3 $sp^2$–$1s$ C—H $\sigma$ bonds | 1 $sp^2$–$2p$ C—O $\sigma$ bond | 2 $sp^2$–$sp^2$ C—C $\sigma$ bonds |
| 1 $sp^2$–$3p$ C—Cl $\sigma$ bond | 2 $3p$ Cl lone pairs | 1 $3s$ Cl lone pair |
| 1 $2s$ O lone pair | 1 $2p$ O lone pair | |

**7.13** The wavelength is 1.14 μm, in the near-infrared spectral region.

# EFFECTS OF INTERMOLECULAR FORCES

## OPALS AND BUTTERFLY WINGS

Opals and butterfly wings seem very different, yet both shimmer with beautiful colours. Although they are made of very different chemical substances, both have highly regular structures containing hexagonal patterns.

Opals contain microscopic spheres of silica ($SiO_2$), packed into planar hexagonal layers. The shimmering colours of opals result from the way this layered structure interacts with light. Diffraction and interference of light passing through these layers depend on the wavelength, with the result that some colours are enhanced while other colours are diminished.

The surface of a butterfly wing has an "inverse opal" structure. Butterfly wings are composed of chitin, the same cellulose-like polymer that forms the shells of crabs. In a butterfly wing, however, the chitin forms a hexagonal network. Instead of tightly packed spheres, the result is a solid structure with a regular array of spherical holes. The shimmering colours of butterfly wings arise from interference of light scattered and refracted by this regular array of chitin at the wing surface. As with opals, colour variations arise from the way different wavelengths interact with molecular-level structure.

Materials scientists have learned how to make synthetic opals. The procedure requires solid spheres of uniform diameter around 250 nm. Silica spheres are made by reacting an organosilicon compound, $Si(OCH_3)_4$, with water. Hydrogen and methyl groups trade places, forming $Si(OH)_4$ and $CH_3OH$. The silicon-containing product molecules eliminate water and combine to form Si—O—Si bonds, and, if the reaction conditions are right, the network of bonds closes on itself to form spheres of uniform size. When these spheres are suspended in water, they slowly solidify into the regular array of opals.

Scientists also have learned how to mimic the surface of a butterfly wing. Polystyrene beads and smaller silica nanoparticles are suspended in water and mixed thoroughly using ultrasound. When a glass slide is dipped into the suspension and slowly withdrawn, a thin film forms on the glass surface. This film is a

# 8

## CHAPTER CONTENTS

regular array of beads encased in a matrix of nanoparticles. Heating the film destroys the polystyrene beads but leaves the silica web intact. The result is a silica "inverse opal" film.

The iridescent properties of opals and butterfly wings are due to their highly regular molecular structures. To understand how solid structures form ordered patterns, we need information about the forces of attraction that exist between molecules. This chapter begins with descriptions of intermolecular forces and their effects on gases. Then, we show how these forces, even though considerably weaker than the chemical bonds described in Chapters 6 and 7, are important in determining the behaviour of solids and liquids. We end the chapter with a discussion of phase changes.

## LEARNING OBJECTIVES

*Upon completion of this chapter you should be able to:*

- **predict the relative magnitudes of intermolecular forces and their effects on physical properties of substances (boiling and melting points, surface tension viscosity and vapour pressure).**

- **predict the relative magnitudes of intermolecular forces and their effects on physical properties of substances.**

- **explain trends in surface tension, capillary action, viscosity and vapour pressure in terms of intermolecular forces.**

- **explain the properties of solids in terms of the dominant intermolecular forces present.**

- **understand amorphous and crystalline solids at the molecular level.**

- **explain enthalpies of phase changes in terms of intermolecular forces.**

- **interpret a pressure-temperature phase diagram of a pure substance.**

## 8.1 EFFECTS OF INTERMOLECULAR FORCES

Forces of attraction between molecules are responsible for the existence of liquids and solids. In the absence of these **intermolecular forces**, all molecules would move independently, and all substances would be gases. The natural phases of the elements indicate the importance of intermolecular forces. At room temperature and pressure, only 11 elements are gases. Mercury and bromine are liquids, and all the rest of the elements are solids. For all but the 11 gaseous elements, intermolecular forces are too large to ignore under normal conditions.

### The Halogens

The halogens, the elements from Group 17 of the periodic table, provide an introduction to intermolecular forces. These elements exist as diatomic molecules: $F_2$, $Cl_2$, $Br_2$, and $I_2$. The bonding patterns of the four halogens are identical. Each molecule contains two atoms held together by a single covalent bond that can be described by end-on overlap of valence *p* orbitals.

Although they have the same bonding patterns, bromine and iodine differ from chlorine and fluorine in their macroscopic physical appearance and in their molecular behaviour. As Figure 8-1 shows, at room temperature and pressure, fluorine and chlorine are gases, bromine is a liquid, and iodine is a solid.

Gases and condensed phases look very different at the molecular level. Molecules of $F_2$ or $Cl_2$ move freely throughout their gaseous volume, travelling many molecular diameters before colliding with one another or with the walls of their container. Because much of the volume of a gas is empty space, samples of gaseous $F_2$ and $Cl_2$ readily expand or contract in response to changes in pressure. This freedom of motion exists because the intermolecular forces among these molecules are small. (Recall however that these forces are not zero; see Section 2.1 for a discussion of the van der Waals equation, which takes account of these forces.)

Molecules of liquid bromine also move about relatively freely, but there is not much empty space between molecules. A liquid cannot be compressed significantly by increasing the pressure, because its molecules are already in close contact with one another. Neither does a liquid expand significantly if the pressure is reduced, because the intermolecular forces in a liquid are large enough to prevent the molecules from breaking away from one another.

Solid iodine, like liquid bromine, has little empty space between molecules. Like liquids, solids have sufficiently strong intermolecular forces that they neither expand nor contract significantly when pressure changes. In solids, the intermolecular forces are strong enough to prevent molecules from moving freely past one another. Instead, the $I_2$ molecules in a sample of solid iodine are arranged in ordered arrays. Each molecule vibrates back and forth about a single most stable position, but it cannot slide easily past its neighbours.

**Figure 8-1**

Gaseous $Cl_2$

Liquid $Br_2$

Solid $I_2$

Under normal conditions, chlorine is a pale yellow-green gas, bromine is a dark red liquid, and iodine is a dark crystalline solid.

The interplay between molecular kinetic energies of motion and intermolecular attractive forces accounts for these striking differences, as illustrated in Figure 8-2. Recall from Chapter 2 that molecules are always moving. Intermolecular attractive forces tend to hold molecules together in a condensed phase, but molecules that are moving fast enough can overcome these forces and move freely in the gas phase. When the average kinetic energy of motion is large enough, molecules remain separated from one another and the substance is a gas. Conversely, when intermolecular attractive forces are large enough, molecules remain close to one another and the substance is a liquid or solid.

The bars in Figure 8-3 compare the attractive energies arising from intermolecular forces acting on the halogens. The figure also shows the average molecular kinetic energy at room temperature. For fluorine and chlorine, the attractive energy generated by intermolecular forces is smaller than the average molecular kinetic energy at room temperature. Thus, these elements are gases under normal conditions. In contrast, bromine molecules have enough kinetic energy to move freely about, but their energy of motion is insufficient to completely overcome the intermolecular forces, and a liquid results. Finally, attractive forces are strong enough in iodine to lock the $I_2$ molecules in position in the solid state.

## Melting and Boiling Points

Melting points and boiling points can be used as indicators of the strengths of intermolecular forces. Remember from Chapter 2 that the average kinetic energy of molecular motion increases with absolute temperature. The boiling point of a substance is the temperature at which the average kinetic energy of molecular motion balances the attractive energy of intermolecular attractions. When the pressure is 1 bar, that temperature is the **normal boiling point**. For example, the normal boiling point of bromine is 332 K (59 °C). Above this temperature, the average kinetic energy of motion exceeds the attractive energies created by intermolecular forces of attraction, and bromine exists as a gas. Under atmospheric pressure, bromine is a liquid at temperatures below 332 K.

The conversion of a liquid into a gas is called **vapourization**. A liquid vapourizes when molecules leave the liquid phase faster than they are captured from the gas. **Condensation** is the reverse process. A gas condenses when molecules leave the gas phase more rapidly than they escape from the liquid. We explore these processes in more detail in Section 8.6.

The molecules in a liquid are able to move about freely, even though they do not escape from the liquid. When a liquid is cooled, however, its molecular energies decrease. At temperatures below the melting point, the molecules become locked in place and the liquid solidifies. When the pressure is 1 bar, that temperature is the **normal melting point**. A liquid freezes when liquid molecules have too little energy of motion to slide past one another. Conversely, a solid melts when its molecules have enough energy of motion to move freely past one another. Just

Intermolecular force of attraction          Kinetic energy

A substance exists in a condensed phase when its molecules have too little average kinetic energy to overcome intermolecular forces of attraction.

Average kinetic energy at 298 K

Intermolecular attractive energy

$F_2$     $Cl_2$     $Br_2$     $I_2$

The balance of average kinetic energy (*red line*) and intermolecular forces of attraction favours the gas phase for $F_2$ and $Cl_2$ and the condensed phases for $Br_2$ and $I_2$.

**TABLE 8-1 MELTING AND BOILING POINTS**

| Substance | $T_m$ (K) | $T_b$ (K) | Substance | $T_m$ (K) | $T_b$ (K) |
|-----------|-----------|-----------|-----------|-----------|-----------|
| He | 0.95* | 4.2 | $Br_2$ | 266 | 332 |
| $H_2$ | 14.0 | 20.3 | $I_2$ | 387 | 458 |
| $N_2$ | 63.3 | 77.4 | $P_4$ | 317 | 553 |
| $F_2$ | 53.5 | 85.0 | Na | 371 | 1156 |
| Ar | 83.8 | 87.3 | Mg | 922 | 1363 |
| $O_2$ | 54.8 | 90.2 | Si | 1683 | 2628 |
| $Cl_2$ | 172 | 239 | Fe | 1808 | 3023 |

*Under high pressure.
He remains liquid at 0 K and 1 bar.

as intermolecular forces determine the normal boiling point, these forces also determine the melting point. The larger the intermolecular forces, the higher the melting point: fluorine freezes at 53.5 K, chlorine at 172 K, and bromine at 266 K.

Boiling points and melting points depend on the strengths of intermolecular forces because the rates of escape and capture depend on the balance between molecular kinetic energies and intermolecular forces of attraction. A substance with large intermolecular forces must be raised to a high temperature before its molecules have sufficient kinetic energies to overcome those forces. A substance with small intermolecular forces must be cooled to a low temperature before its molecules have small enough kinetic energies to coalesce into a condensed phase. Table 8-1 lists the boiling and melting points of some representative elemental substances.

Other chemical and physical properties are also the result of intermolecular forces, and are discussed later in this chapter, as well as in Chapter 9.

**SECTION EXERCISES**

●8.1.1 Based on the behaviour of the other elements of Group 17, predict whether $At_2$ will be a gas, liquid, or solid at room temperature. Redraw the bar graph of Figure 8-3 to include $At_2$.

●8.1.2 Refer back to Figure 2-19 and determine which of the four gases has the largest intermolecular forces and which has the smallest. State your reasoning.

●8.1.3 Use the data in Table 8-1 to answer the following: (a) Does Xe boil at a higher or lower temperature than Ar? (b) What is the order of increasing intermolecular forces for $F_2$, $N_2$, $H_2$, and $Cl_2$? (c) Which group in the periodic table has the smallest intermolecular forces?

## 8.2 TYPES OF INTERMOLECULAR FORCES

The magnitudes of intermolecular forces depend on whether each interacting particle is charged, polar, or non-polar. If both are charged, then the force is ionic, a form of chemical bonding that was dealt with in Chapter 5. In other cases, however, the forces are weaker because one or both particles are not charged. The

## TABLE 8-2 INTERMOLECULAR FORCES

| Name of Force | Approximate Potential Energy Range (kJ/mol) | Attraction | Examples |
|---|---|---|---|
| Ion–dipole | 40–600 | Charged ion is attracted to the oppositely charged end of the dipole | $Cl^-–H_2O$ $Na^+–H_2O$ |
| Hydrogen bond | 5–50 | Positively polarized H atom is attracted to a lone pair on an electronegative atom | $(CH_3)_2CO–H_2O$ |
| Dipole–dipole | 5–25 | Mutual attraction of two dipoles | $HCl–HCl$ |
| Ion-induced dipole | 3–15 | The ion induces a dipole in the other species, and is then attracted to it | $Ca^{2+}–O_2$ |
| Dipole-induced dipole | 2–10 | A dipole induces a dipole in the other species, and is then attracted to it | $HCl–Cl_2$ |
| Dispersion (London) | 0.1–5 | Momentary shifts in electron clouds produce momentary dipoles, which attract each other | $He–He$ |

possible interactions, their approximate energies, and some examples are presented in Table 8-2, and are detailed in the following sections. In general, the strongest force that exists between two molecules is the one we are interested in since it will dictate the magnitudes of properties such as the melting and boiling points.

## Ion–Dipole Forces

The strongest intermolecular force is that between an ion and a permanent dipole. For instance, when an ionic substance dissolves in water, strong forces of attraction between the ions in the solid material and the (numerous) polar water molecules overcome the ion–ion forces in the solid. This is discussed in more detail in Section 9.2.

## Dispersion Forces

The weakest forces between two molecules are **dispersion forces**. Dispersion forces exist because the electron cloud of any molecule distorts easily. For example, consider what happens when two halogen molecules approach each other. Each molecule contains positive nuclei surrounded by a cloud of negative electrons. As two molecules approach each other, the nuclei of one molecule attract the electron cloud of the other. Electrons are highly mobile, so their orbitals change shape in response to this attraction. At the same time, the two electron clouds repel each other, which leads to further distortion to minimize electron–electron repulsion. As Figure 8-4 indicates, this distortion of

**Figure 8-4**

Isolated molecule

Neighbouring molecules

Exaggerated view of how dispersion forces arise.

**Figure 8-5**

Boiling points of the halogens increase with the number of electrons.

**Figure 8-6**

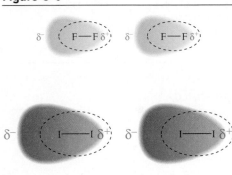

With many more electrons, the electron cloud of iodine is much larger and more polarizable than that of fluorine.

the electron cloud creates a charge imbalance, giving the molecule a slight positive charge at one end and a slight negative charge at the other. Dispersion forces are the net attractive forces among molecules generated by all these induced charge imbalances.

The magnitude of dispersion forces depends on how easy it is to distort the electron cloud of a molecule. This ease of distortion is called the **polarizability**, because distortion of an electron cloud generates a temporary polarity within the molecule. We can explore how polarizability varies by examining boiling points of the halogens, as shown by the graph in Figure 8-5. Since all four of these species are not charged and non-polar, the only intermolecular forces they experience are dispersion forces. The data reveal that boiling points increase with the total number of electrons. Fluorine, with 18 total electrons, has the lowest boiling point (85 K). Iodine, with 106 electrons, has the highest boiling point (458 K). The large electron cloud of $I_2$ distorts more readily than the small electron cloud of $F_2$.

Figure 8-6 shows schematically how a large electron cloud distorts more than a small one, generating larger dispersion forces and leading to a higher boiling point.

Molecular size increases with chain length as well as with the size of individual atoms. Figure 8-7 shows how the boiling points of alkanes increase as the carbon chain gets longer. As alkanes get longer, their electron clouds become larger and more polarizable, making dispersion forces larger and raising the boiling point. As examples, methane ($CH_4$, 10 electrons) is a gas at 298 K, pentane ($C_5H_{12}$, 45 electrons) is a low-boiling liquid, decane ($C_{10}H_{22}$, 115 electrons) is a high-boiling liquid, and eicosane ($C_{20}H_{42}$, 225 electrons) is a waxy solid.

**Figure 8-7**

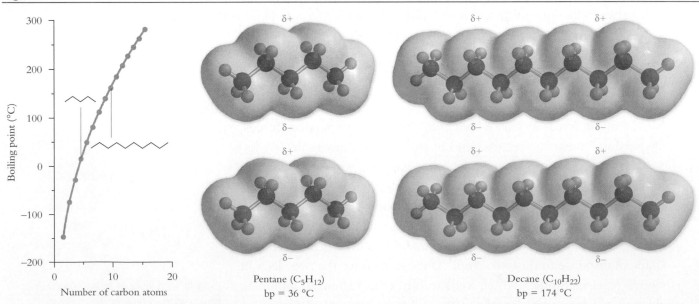

Pentane ($C_5H_{12}$)
bp = 36 °C

Decane ($C_{10}H_{22}$)
bp = 174 °C

The boiling points of alkanes increase with the length of the carbon chain, because a long electron cloud is more polarizable than a short one.

The boiling point increases progressively as molecules become larger and more polarizable. We apply this reasoning to the elemental noble gases in Example 8-1.

### EXAMPLE 8-1 BOILING POINT TRENDS

Neon and xenon are gases at room temperature, but both become liquids if the temperature is low enough. Draw a molecular picture showing the relative sizes and polarizabilities of atoms of neon and xenon, and use the picture to determine which substance has the lower boiling point.

**Strategy:** The boiling point of a substance depends on the magnitude of its intermolecular forces, which in turn depends on the polarizability of its electron cloud. Monatomic gases contain atoms rather than molecules, so we must assess interatomic forces for these substances.

**Solution:** The dispersion forces that act between atoms of the noble gases depend on the polarizabilities of their electron clouds. The total electron counts for these atoms are 10 for neon and 54 for xenon. When two atoms approach each other, the smaller electron cloud of neon distorts less than the larger electron cloud of xenon, as a molecular picture illustrates:

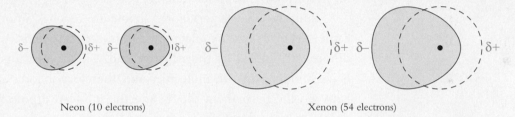

Neon (10 electrons)               Xenon (54 electrons)

Less polarizability means smaller partial charges and weaker intermolecular forces. Thus, neon has the lower boiling point.

 **Does the Result Make Sense?** The experimental values are 166.1 K for Xe and 27.1 K for Ne, in agreement with our analysis.

### Practice Exercise 8.1

Among $H_2S$, $H_2Se$, and $H_2Te$, use dispersion arguments to determine which has the highest boiling point and which has the lowest.

Dispersion forces increase in strength with the number of electrons, because larger electron clouds are more polarizable than smaller electron clouds. For molecules with comparable numbers of electrons, the shape of the molecule makes an important secondary contribution to the magnitude of dispersion forces. For example, Figure 8-8 shows the shapes of pentane and 2,2-dimethylpropane. Both of these molecules have the formula $C_5H_{12}$, with 72 total electrons. Notice that 2,2-dimethylpropane has a more compact structure than pentane. This compactness results in a less polarizable electron cloud and smaller dispersion forces. Accordingly, pentane has a boiling point of 36 °C, whereas 2,2-dimethylpropane boils at 10 °C.

Although dispersion forces are weak, many such forces are collectively strong. The bottom electron

**Figure 8-8**

Pentane ($C_5H_{12}$)          2,2-Dimethylpropane ($C_5H_{12}$)
bp = 36 °C                       bp = 10 °C

The boiling point of *n*-pentane is higher than the boiling point of 2,2-dimethylpropane, because an extended electron cloud is more polarizable than a compact one.

micrograph in the margin depicts tiny hairs on the bottom of the feet of a gecko (shown in the photo). The top electron micrograph depicts bundles of carbon nanotubes synthesized by researchers in an attempt to make a reusable adhesive. The huge number of dispersion forces between these tiny fibres and a surface results in substantial adhesion.

## Induced Dipoles

When an ion approaches a non-polar molecule, the electron cloud in the molecule is distorted by electrostatic attraction or repulsion to the ion. This results in a dipole in the otherwise non-polar molecule. We say that the ion has induced a dipole in the molecule. The ion and the **induced dipole** then experience a force of attraction. An example is the interaction of iodide ions ($I^-$) with molecular iodine. This interaction is sufficiently strong to create the triiodide ion, $I_3^-$, which is far more soluble in water than is $I_2$ alone.

Similarly, when a permanent dipole approaches a non-polar molecule, electrons in the molecule are attracted to the positive end of the dipole, or repelled by the negative end. This too results in a dipole being induced in the molecule, and therefore a force of attraction between the permanent dipole and the induced dipole is created. A dipole-induced dipole force is smaller than an ion-induced dipole force because the ion has a whole charge, whereas the dipole has only a partial charge at each end. Argon gas, for instance, has a small but measurable solubility in water, because the polar water molecule induces a dipole in the otherwise non-polar argon atom, and a net force of attraction results.

## Dipolar Forces

Dispersion forces exist among all molecules, but some substances remain liquid at much higher temperatures than can be accounted for by dispersion forces alone. Consider 2-methylpropane and acetone, whose structures are shown in Figure 8-9. These molecules have similar shapes and nearly the same number of electrons (34 vs. 32). They are so similar that we might expect the two to have nearly equal boiling points, but acetone is a liquid at room temperature, whereas 2-methylpropane is a gas. The reason is that acetone has a large dipole moment. Remember from Chapter 6 that chemical bonds are polarized toward the more electronegative atom. Thus, the C═O bond of acetone is highly polarized, with a partial negative charge on the O atom ($\chi = 3.5$) and a partial positive charge on the C atom ($\chi = 2.5$). The C─H bonds in these molecules, in contrast, are only slightly polar, because the electronegativity of hydrogen ($\chi = 2.1$) is only slightly smaller than that of carbon.

When two polar acetone molecules approach each other, they align so that the positive end of one molecule is close to the negative end of the other. In a liquid array, this repeating pattern of head-to-tail alignment gives rise to significant net attractive dipolar forces among the molecules.

The dispersion forces in acetone are nearly the same as those in 2-methylpropane, but the addition of dipolar forces makes the total amount of intermolecular attraction between acetone molecules substantially greater than the attraction between molecules of 2-methylpropane. Consequently, acetone boils at a considerably higher temperature than 2-methylpropane. Example 8-2 provides some additional comparisons of dispersion forces and dipole forces.

**Figure 8-9**

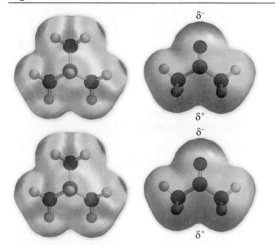

Acetone and 2-methylpropane have similar molecular shapes, but acetone has a large dipole moment resulting from its polar $C = O$ bond.

## EXAMPLE 8-2 BOILING POINTS AND STRUCTURE

The line structures of butane, methyl ethyl ether, and acetone follow. Explain the trend in boiling points: butane (0°C), methyl ethyl ether (8°C), and acetone (56°C).

Butane        Methyl ethyl ether        Acetone

**Strategy:** These boiling points can be explained in terms of dispersion forces and dipolar forces. First, assess the magnitudes of dispersion forces, which are present in all substances, and then look for molecular polarity.

**Solution:** A table helps to organize the available information:

| Substance | Boiling Point (°C) | Total Electrons |
|---|---|---|
| Butane | 0 | 34 |
| Methyl ethyl ether | 8 | 34 |
| Acetone | 56 | 32 |

The table shows that dispersion forces alone cannot account for the range in boiling temperatures. Methyl ethyl ether and butane have the same number of electrons and similar shapes; yet their boiling points are different. Acetone, which has fewer electrons than the other compounds, has slightly smaller dispersion forces; yet it boils at a higher temperature. The order of boiling points indicates that acetone is a more polar molecule than methyl ethyl ether, which in turn is more polar than butane.

We expect butane to be non-polar because of the small electronegativity difference between carbon and hydrogen. Acetone and methyl ethyl ether, on the other hand, contain polar carbon—oxygen bonds. Molecular geometry reveals why acetone is more polar than methyl ethyl ether. The full Lewis structures of these molecules show that the oxygen atom in the ether has a steric number of 4 and bent geometry. The two C—O bond dipoles in methyl ethyl ether partially cancel each other, leaving a relatively small molecular dipole moment. On the other hand, the polar $C = O$ bond in acetone is unopposed, so acetone has a larger dipole moment and is more polar than methyl ethyl ether.

**Does the Result Make Sense?** Drawings using arrows help us see why acetone has a larger dipole moment than dimethyl ether. The arrows show the charge displacement for each polar bond. Experimental values for the dipole moments are butane, 0 D; methyl ethyl ether, 1.12 D; and acetone, 2.88 D.

Methyl ethyl ether          Acetone

### Practice Exercise 8.2

Acetaldehyde, $CH_3COH$, has a structure like that of acetone but with one $CH_3$ group replaced by H. This substance boils at 21 °C. Explain its boiling point relative to the three compounds described in this Example.

## Hydrogen-Bonding Forces

Figure 8-10

The Lewis structure and ball-and-stick model of 1-propanol.

Methyl ethyl ether is a gas at room temperature (boiling point = 8 °C), but 1-propanol, shown in Figure 8-10, is a liquid (boiling point = 97 °C). The compounds have the same molecular formula, $C_3H_8O$, and each has a chain of four inner atoms, C—O—C—C and O—C—C—C. Consequently, the electron clouds of these two molecules are about the same size, and their dispersion forces are comparable. Each molecule has an $sp^3$–hybridized oxygen atom with two polar single bonds, so their dipolar forces should be similar. The very different boiling points of 1-propanol and methyl ethyl ether make it clear that dispersion and dipolar forces do not reveal the entire story of intermolecular attractions.

The forces of attraction between 1-propanol molecules are stronger than those between methyl ethyl ether molecules because of an intermolecular interaction called a **hydrogen bond**. A hydrogen bond occurs when a highly electronegative atom with a lone pair of electrons shares its non-bonding electrons with a positively polarized hydrogen atom on another molecule. Hydrogen bonds are intermolecular (or sometimes intramolecular) interactions, but are not chemical bonds. These interactions are comparable to and sometimes stronger than dipolar and dispersion interactions, but they are only 5–10% as strong as covalent bonds.

There are two requirements for hydrogen bond formation. First, there must be an electron-deficient hydrogen atom that can accept an electron pair. Hydrogen atoms in O—H, F—H, and N—H bonds meet this requirement. Second, there must be a highly electronegative atom that can donate an electron pair. Three sec-

### Figure 8-11

Hydrogen fluoride

Water–ethanol

Salicylic acid

Formic acid

Ammonia–water

Glycine (an amino acid)

Examples of hydrogen bonding.

ond-row elements, O, N, and F, meet this requirement, and there is recent evidence that S atoms in biological compounds also can donate electron pairs to form hydrogen bonds. Figure 8-11 shows representative examples of hydrogen bonding. The hydrogen bonds, which are highlighted, are shown as dashed lines to indicate the weakly bonding nature of these interactions.

Notice from the examples shown in Figure 8-11 that hydrogen bonds can form between different molecules (e.g., $H_3N-H_2O$) or between identical molecules (e.g., $HF-HF$). Also notice that molecules can form more than one hydrogen bond (e.g., glycine) and that hydrogen bonds can form within a molecule (e.g., salicylic acid) as well as between molecules. Example 8-3 explores the possibilities for hydrogen bond formation.

## EXAMPLE 8-3  FORMATION OF HYDROGEN BONDS

In which of the following systems will hydrogen bonding play an important role: $CH_3F$, $(CH_3)_2CO$ (acetone), $CH_3OH$, and $NH_3$ dissolved in $(CH_3)_2CO$?

**Strategy:** Hydrogen bonds require electron-deficient hydrogen atoms in polar $H-X$ bonds, and highly electronegative atoms with non-bonding pairs of electrons are present. Use Lewis structures to determine whether these requirements are met.

**Solution:** Here are the Lewis structures of the four molecules:

Acetone and $CH_3F$ contain electronegative atoms with non-bonding pairs, but neither has any highly polar $H-X$ bonds. Thus, there is no hydrogen bonding between molecules of these substances.

The $O-H$ bond in $CH_3OH$ meets both of the requirements for hydrogen bonding. The $O-H$ hydrogen atom on one molecule interacts with the oxygen atom of a neighbouring molecule:

For a solution of ammonia in acetone, we must examine both components. Acetone has an electronegative oxygen atom with non-bonding pairs, whereas $NH_3$ has a polar $N-H$ bond. Consequently, a mixture of these two compounds displays hydrogen bonding between the hydrogen atoms of ammonia and oxygen atoms of acetone:

**Do the Results Make Sense?** Methanol has a considerably higher boiling point than its alkane relatives, methane and ethane, consistent with significant intermolecular forces. Ammonia dissolves readily in acetone, also consistent with significant intermolecular forces.

### Practice Exercise 8.3
Draw a picture that shows the hydrogen-bonding interactions of an acetone molecule dissolved in water.

Hydrogen bonding is particularly important in biochemical systems, because biomolecules contain many oxygen and nitrogen atoms that participate in hydrogen bonding. For example, the amino acids from which proteins are made contain $NH_2$ (amino) and $CO_2H$ (carboxylic acid) groups. Four different types of hydrogen bonds exist in these systems: $O=H-N$, $N=H-O$, $O=H-O$, and $N=H-N$. When biological molecules contain S atoms, $S-H-O$ and $S=H-N$ hydrogen bonds also can form. Figure 8-11 includes a view of hydrogen bonding between glycine molecules, and we examine more details of hydrogen bonding in biomolecules in Chapter 17.

## Binary Hydrogen Compounds

Boiling points of the binary hydrogen compounds illustrate the interplay among different types of intermolecular forces. The graph in Figure 8-12 shows that there are periodic trends in the boiling points of these compounds. In general, the boiling points of the binary hydrogen compounds increase from top to bottom of each column of the periodic table. This trend is due to increasing dispersion forces: the more electrons the molecule has, the stronger the dispersion forces and the higher the boiling point. In Group 16, for example, $H_2S$ ($18\ e^-$) boils at $-60\ °C$, $H_2Se$ ($36\ e^-$) at $-41\ °C$, and $H_2Te$ ($54\ e^-$) at $-4\ °C$.

As Figure 8-12 shows, ammonia, water, and hydrogen fluoride depart dramatically from this periodic behaviour. This is because their molecules experience particularly large intermolecular forces resulting from hydrogen bonding. In hydrogen fluoride, for instance, partial donation of an electron pair from the highly electronegative fluorine atom of one HF molecule to the electron-deficient hydrogen atom of another HF molecule creates a hydrogen bond. Similar interactions among

**Figure 8-12**

Periodic trends in the boiling points of binary hydrogen compounds. Notice that $H_2O$, HF, and $NH_3$ are exceptions to the trends.

many HF molecules result in a network of hydrogen bonds that gives HF a boiling point much higher than those of HCl, HBr, and HI.

Fluorine has the highest electronegativity, so the strongest *individual* hydrogen bonds are those in HF. Every hydrogen atom in liquid HF is involved in a hydrogen bond, but there is only one polar hydrogen atom per molecule. Thus, each HF molecule participates in two hydrogen bonds with two other HF partners. There is one hydrogen bond involving the partially positive hydrogen atom and a second involving the partially negative fluorine atom.

Water has a substantially higher boiling point than hydrogen fluoride has, which indicates that the total amount of hydrogen bonding in $H_2O$ is greater than that in HF. The higher boiling point of water reflects the fact that it forms more hydrogen bonds *per molecule* than hydrogen fluoride. A water molecule has two hydrogen atoms that can form hydrogen bonds and two non-bonding electron pairs on each oxygen atom. This permits every water molecule to be involved in four hydrogen bonds to four other $H_2O$ partners, as shown in Figure 8-13(a).

Hydrogen bonding in solid ice creates a three-dimensional network that puts each oxygen atom at the centre of a distorted tetrahedron. Figure 8-13 shows that two arms of the tetrahedron are regular covalent O—H bonds, whereas the other two arms of the tetrahedron are hydrogen bonds to two different water molecules.

Dispersion forces, dipole interactions, and hydrogen bonds all are significantly weaker than covalent intramolecular bonds. For example, the average C—C bond energy is 345 kJ/mol, whereas dispersion forces are just 0.1–5 kJ/mol for small alkanes such as propane. Dipolar interactions between polar molecules such as acetone range between 5 and 25 kJ/mol, and hydrogen bonds range between 5 and 50 kJ/mol.

**Figure 8-13**

(a)    (b)

The structure of ice. (a) Each oxygen atom is at the centre of a distorted tetrahedron of hydrogen atoms. The tetrahedron is composed of two short covalent O—H bonds and two long H—O hydrogen bonds. (b) Water molecules are held in a network of these tetrahedra.

**SECTION EXERCISES**

●**8.2.1** Explain the following differences in boiling points: (a) Kr boils at −152 °C, and propane boils at −42 °C; (b) C(CH$_3$)$_4$ boils at 10 °C, and CCl$_4$ boils at 77 °C; and (c) N$_2$ boils at −196 °C, and CO boils at −191.5 °C.

●**8.2.2** Acetone and methanol have nearly equal boiling points. What types of intermolecular forces do each exhibit? What does the similarity in boiling points tell you about the relative magnitudes of each type of force in these two compounds?

●**8.2.3** There are nine important hydrogen-bonding interactions. One of them is O═HO. Draw the other eight. For each of the nine, draw a Lewis structure of a specific example using real molecules.

Mercury forms near-spherical drops.

**Figure 8-14**

Molecule is pulled away from surface

No net "pull" in any direction

In the interior of a liquid (*bottom*), each molecule experiences equal forces in all directions. A molecule at the surface of a liquid (*top*) is pulled back into the liquid by intermolecular forces.

## 8.3 LIQUIDS

A gas condenses to a liquid if it is cooled sufficiently. Condensation occurs when the average kinetic energy of molecules falls below the value needed for the molecules to move about independently. Thus, the molecules in a liquid are confined to a relatively small volume by intermolecular forces of attraction. Although they cannot readily escape, liquid molecules remain free to move about within the liquid phase. In this behaviour, liquid molecules behave like the molecules of a gas. The large-scale consequences of the molecular-level properties are apparent. Like gases, liquids are fluid, so they flow easily from place to place. Unlike gases, however, liquids are dense, and do not expand or contract significantly with changes in temperature or pressure.

## Properties of Liquids

In a liquid, intermolecular forces are strong enough to confine the molecules to a specific volume, but they are not strong enough to keep molecules from moving from place to place within the liquid. The relative freedom of motion of liquid molecules leads to three liquid properties arising from intermolecular forces: **surface tension**, **capillary action**, and **viscosity**.

A small amount of a liquid tends to take a spherical shape: for example, mercury drops are nearly spherical and water drips from a faucet in nearly spherical liquid droplets. Surface tension, which measures the resistance of a liquid to an increase in its surface area, is the physical property responsible for this behaviour.

Figure 8-14 illustrates at the molecular level why liquids exhibit surface tension. A molecule in the interior of a liquid is completely surrounded by other molecules. A molecule at a liquid surface, on the other hand, has other molecules beside it and beneath it but very few above it in the gas phase. This difference means that there is a net intermolecular force on molecules at the surface that pulls them toward the interior of the liquid.

Surface tension is therefore proportional to the magnitude of intermolecular forces among the molecules in the liquid, so it is not surprising that polar molecules such as water have high surface tensions, and non-polar molecules such as

## TABLE 8-3 SURFACE TENSION VALUES OF SOME COMMON LIQUIDS AT 20 °C

| Liquid | Dominant Intermolecular Force | Surface Tension $(J/m^2)$ |
|---|---|---|
| Diethyl ether | Dispersion | 0.0170 |
| *n*-Hexane | Dispersion | 0.0184 |
| Benzene | Dispersion | 0.0289 |
| Cyclohexanol | Hydrogen bonds | 0.0344 |
| Ethylene glycol | Hydrogen bonds | 0.0477 |
| Water | Hydrogen bonds | 0.0728 |
| Mercury | Metallic bonds | 0.425 |

hexane have very low surface tension. Table 8-3 list the surface tensions of several liquids in $J/m^2$, that is, the energy required to spread the surface of a liquid by 1 $m^2$.

The net attraction of surface molecules to the interior of the liquid indicates that molecules are most stable when attractive forces are maximized by as many neighbour molecules as possible. Consequently, a liquid is most stable when the fewest molecules are at its surface. This occurs when the liquid has minimal surface area. Spheres have less surface area per unit volume than any other shape, so small drops of a liquid tend to be spheres. Large drops are distorted from ideal spheres by the force of gravity.

A water strider is supported on water because of surface tension.

Other effects of surface tension include the fact that small objects whose density is greater than that of water can actually be supported by liquid water without sinking into it, and droplets of water tend to bead up and roll off a freshly waxed car.

Molecules in contact with the surface of their container experience two sets of intermolecular forces. **Cohesive forces** attract molecules in the liquid to one another. In addition, **adhesive forces** attract molecules in the liquid to the molecules of the container walls. One result of adhesive forces is the curved surface of a liquid, called a **meniscus**. As Figure 8-15 shows, water in a glass tube forms a concave meniscus that increases the number of water molecules in contact with the walls of the tube. This is because adhesive forces of water to glass are stronger than the cohesive forces among water molecules.

Figure 8-15 shows another result of adhesive forces. In a tube of small enough diameter, water actually climbs the walls, pulled upward by strong adhesive forces. This upward movement of water against the force of gravity is capillary action. Capillary action between sap and the cellulose walls of wood fibre plays a role in how trees transport sap from their roots to their highest branches.

Water can be poured very quickly from one container to another, salad oil pours more slowly, and honey sometimes seems to take forever. A liquid's resistance to flow is called its viscosity; the greater

**Figure 8-15**

Water in contact with glass takes on a concave meniscus, and water rises inside a small-diameter tube because of capillary action.

its viscosity, the more slowly the liquid pours. Viscosity measures how easily molecules slide by one another, and this depends strongly on molecular shapes. Liquids such as water, acetone, and benzene, whose molecules are small and compact, have low viscosity. In contrast, large molecules such as the sugars in honey and the hydrocarbons found in oils tend to be tangled up with one another. Tangling inhibits the flow of molecules and leads to high viscosity. In addition, strong intermolecular cohesive forces make it harder for molecules to move about.

Molecules move faster as temperature increases, and this allows them to slide by one another more easily. Thus, viscosity decreases as temperature increases. This dependence is quite noticeable for highly viscous substances such as honey and syrup, which are much easier to pour when hot than when cold.

Viscosity can also be decreased by dilution with less viscous solvents. This is the principle behind a process for extraction of useful oil from oil sands (also known as tar sands) in northern Alberta. Liquid hydrocarbon solvents are injected into the ground to dilute the very heavy oil. This reduces its viscosity, allowing it to flow and be pumped to the surface more easily.

## Vapour Pressure

Whenever a liquid has an exposed surface, some of its molecules will escape into the gas phase. Remember from Section 2.5 that any collection of molecules has a distribution of kinetic energies. A liquid has a fixed volume, because although its molecules have sufficient energy of motion to move about within the liquid, on average they do not have sufficient energy of motion to escape into the gas phase. Nevertheless, the distribution of molecular energies guarantees that some of the molecules in any liquid have enough kinetic energy to overcome the intermolecular forces that confine the liquid. These molecules may escape into the gas phase if they reach the surface of the liquid.

Our senses tell us that molecules escape from a liquid. The red gas phase above liquid bromine shows that bromine molecules are in both phases. The evaporation of a rain puddle in the sunshine shows that water molecules escape from the liquid into the gas phase. The smell of gasoline around an open tank informs our noses that gasoline molecules escape from the liquid phase into the gas phase.

How many of the molecules of a liquid have enough energy to escape into the gas phase? That depends on the temperature and the strength of intermolecular forces for the liquid. As Figure 8-16 shows, at any particular temperature, more molecules can escape from liquid bromine than from liquid water. This is because the intermolecular forces among water molecules are larger than those among bromine molecules. A water molecule needs more energy of motion to escape the liquid phase. The figure also illustrates that when the temperature increases, the fraction of molecules having enough energy to escape increases.

**Figure 8-16**

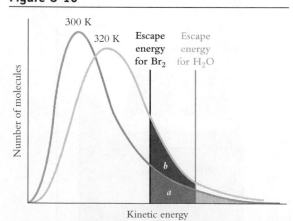

The fraction of molecules with enough kinetic energy to escape a liquid depends on the strength of intermolecular forces and temperature. (a) At 300 K, more bromine molecules can escape than water molecules. (b) More bromine (or water) molecules can escape at 320 K than at 300 K.

**Figure 8-17**

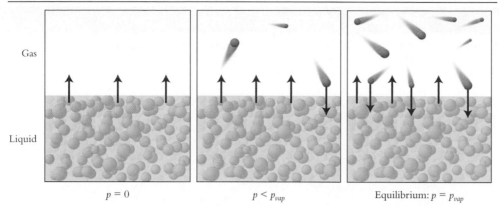

$p = 0$          $p < p_{vap}$          Equilibrium: $p = p_{vap}$

As the partial pressure of a substance in the gas above a liquid increases, the difference between molecules escaped and molecules recaptured decreases until, at the vapour pressure of the substance, equilibrium exists.

A liquid in an open container continually loses molecules until eventually it has evaporated completely. In a closed container, however, the partial pressure of the vapour increases as more and more molecules enter the gas phase. As this partial pressure builds, increasing numbers of molecules from the gas strike the liquid surface and are recaptured. Eventually, as Figure 8-17 shows, the number of molecules escaping from the liquid exactly matches the number of molecules being captured by the liquid. There is no net change, so this is a dynamic equilibrium (see Section 1.9). The pressure at which this equilibrium exists is the **vapour pressure** of the liquid. We can conclude from Figure 8-16 that the vapour pressure of any liquid increases with increasing temperature.

When molecular energies are nearly sufficient to overcome intermolecular forces, molecules of a substance move relatively freely between the liquid phase and the vapour phase. We describe these phase changes in Section 8.6.

## SECTION EXERCISES

●**8.3.1** The alcohol *n*-pentanol ($C_5H_{11}OH$) is more than 10 times as viscous as *n*-hexane ($C_6H_{14}$) at room temperature. Why is this?

●**8.3.2** The mercury inside the glass column of a barometer displays a convex meniscus rather than the concave meniscus shown by water inside a glass column. Explain in terms of intermolecular forces.

●**8.3.3** If the equilibrium view in Figure 8-17 represents bromine at 298 K, redraw the figure to show, qualitatively, how the liquid–vapour equilibrium for water at 298 K differs.

## 8.4 FORCES IN SOLIDS

Like liquids, solids are compact, with molecules nestled close to one another. In solids, though, intermolecular forces are so strong that the molecules cannot move past one another. Instead, they are locked in position. This rigidity gives solids stable shapes that we see in structures ranging from bones to airplane wings. One

## TABLE 8-4  FORCES IN DIFFERENT TYPES OF SOLIDS

| Solid Type | Molecular | Molecular | Molecular | Metallic | Network | Ionic |
|---|---|---|---|---|---|---|
| Attractive Forces | Dispersion | Dispersion + dipolar | Dispersion + dipolar + H bonding | Delocalized bonding | Covalent | Electrical |
| Energy (kJ/mol) | 0.05–40 | 5–25 | 10–40 | 75–1000 | 150–500 | 400–4000 |
| Example | Ar | HCl | $H_2O$ | Cu | $SiO_2$ | NaCl |
| Melting Point (K) | 84 | 158 | 273 | 1357 | 1983 | 1074 |
| Molecular Picture | | | | | | |

of the most active areas of research in modern chemistry, physics, and engineering is the development of new solid materials. Solids play an ever-larger role in society, from high-temperature superconductors, to heat-resistant tiles for the outer "skin" of the space shuttle, to new tissue-compatible solids for surgical implants. In this section, we describe the various forces that exist in solids.

Naphthalene
$C_{10}H_8$

## Magnitudes of Forces

The melting points of the elements presented in Table 8-1 span an immense range, from <15 K ($H_2$) to >1500 K (Si and Fe). These values indicate that the forces holding solids together range from very small to extremely large. This is because, in addition to the intermolecular forces described in Section 8.2, solids can be bound together by covalent bonds, metallic bonding (with delocalized electrons), and ionic interactions. There are four distinct types of solids—molecular, metallic, network, and ionic—each characterized by a different type of force. Table 8-4 contrasts the forces and energies in these four types of solids.

The molecules (or atoms, for noble gases) of a **molecular solid** are held in place by the types of forces already discussed in this chapter: dispersion forces, dipolar interactions, and/or hydrogen bonds. The atoms of a **metallic solid** are held in place by the delocalized bonding described in Section 7.7. A **network solid** contains an array of covalent bonds linking every atom to its neighbours. An **ionic solid** contains cations and anions, attracted to one another by electrical forces, as described in Section 5.5.

## Molecular Solids

Molecular solids are aggregates of molecules bound together by intermolecular forces. Substances that are gases under normal conditions form molecular solids when they condense at low temperature. Many larger molecules have sufficient dispersion forces to exist as solids at room temperature. One example is naphthalene ($C_{10}H_8$), a white solid that melts at 80°C. Naphthalene has a planar structure like that of benzene (see Section 7.6 ), with a cloud of 10 delocalized electrons that lie above and below the molecular plane. Naphthalene molecules are held in the solid state by strong dispersion forces among these highly polarizable electrons. The molecules in crystalline naphthalene stack face-to-face to minimize the total energy. This intermolecular molecular structure leads to plate-like crystals in the macroscopic solid state.

In addition to dispersion forces, molecular solids often involve dipolar interactions and hydrogen bonding. Benzoic acid ($C_6H_5CO_2H$) provides a good example, as illustrated in Figure 8-18. Molecules of benzoic acid are held in place by a combination of dispersion forces among the electrons, and hydrogen bonding among the —$CO_2H$ groups. With fewer electrons, benzoic acid has weaker dispersion forces than naphthalene, but its hydrogen bonding gives benzoic acid a higher melting point, 122 °C.

The effect of extensive hydrogen bonding is revealed by the relatively high melting point of glucose ($C_6H_{12}O_6$), the sugar found in blood and human tissue. Glucose melts at 155 °C, because each of its molecules has five —OH groups that form hydrogen bonds to neighbouring molecules. Although glucose lacks the highly polarizable electrons found in naphthalene and benzoic acid, its extensive hydrogen bonding gives this sugar the highest melting point of the three compounds.

The molecules of a molecular solid retain their individual properties. Solid $I_2$, for example, is dark in colour because individual $I_2$ molecules absorb visible light, as shown by the fact that $I_2$ vapour also is coloured. Glucose dissolved in coffee tastes just as sweet as solid glucose because the sweet taste of glucose comes from its shape, which is the same in the solid phase as it is in solution.

Glucose $C_6H_{12}O_6$

### Figure 8-18

Dispersion forces between planes

Hydrogen bonding between molecules in a plane

Crystals of benzoic acid contain pairs of molecules held together head to head by hydrogen bonds. These pairs then stack in planes that are held together by dispersion forces.

## Network Solids

In sharp contrast to molecular solids, network solids have very high melting points. Compare the behaviour of phosphorus and silicon, third-row neighbours in the periodic table. As listed in Table 8-1, white phosphorus ($P_4$) melts at 317 K, but silicon melts at 1683 K. Phosphorus is a molecular solid that contains individual $P_4$ molecules, but silicon is a network solid in which covalent bonds among Si atoms connect all the atoms. The vast array of covalent bonds in a network solid makes the entire structure behave as one giant "molecule."

Phosphorus is a molecular solid.

The cause of the great difference in the melting points of these two elements is evident from Table 8-4: bonding forces are much stronger than intermolecular forces. For solid silicon to melt, a significant fraction of its Si—Si covalent bonds must break. The average Si—Si bond energy is 225 kJ/mol, whereas attractive energies due to intermolecular forces are less than 40 kJ/mol, so it takes a much higher temperature to melt silicon than phosphorus.

Silicon is a network solid.

Bonding patterns determine the properties of network solids. Diamond and graphite, the two forms of elemental carbon that occur naturally on Earth, are both network solids, but they have very different physical and chemical properties. Diamond contains a three-dimensional array of bonds, with each tetrahedral ($sp^3$) carbon atom linked to all the others through a network of covalent bonds. This three-dimensional network of strong covalent bonds makes diamond extremely strong and abrasive. Graphite, in contrast, has trigonal planar ($sp^2$) carbon atoms in a two-dimensional array of bonds. The structure is supplemented by delocalized $\pi$ bonding above and below the plane

Diamond has three-dimensional bonding.

Graphite has a planar bond network.

of the bonds. Each two-dimensional layer is attracted to its neighbouring layers only by dispersion forces among the $\pi$ electrons. As a result, planes of carbon atoms easily slide past one another, making graphite a brittle lubricant.

Several oxides and sulphides display the characteristics of network solids. The bond network of silica appears in Section 6.3. Other examples are titania ($TiO_2$) and alumina ($Al_2O_3$). These two substances have extremely high melting points because their atoms are held together by networks of strong covalent bonds. Like graphite, $MoS_2$ is a two-dimensional network solid that serves as a solid lubricant.

Covalent bonds make network solids extremely durable. Geological examples include the "everlasting sands" and granite formations such as the Rock of Gibraltar. Like diamonds, other valuable gemstones are network solids as well. Rubies and sapphires are covalent crystals of aluminum oxide with small amounts of coloured transition metal ion impurities. Carborundum is a 1 : 1 network solid of silicon and carbon that has the same lattice structure as diamond. Carborundum is much less expensive than diamond but almost as strong and wear-resistant, so it is used for the edges of cutting tools. Example 8-4 compares the structures and properties of a network solid and a molecular solid.

## EXAMPLE 8-4  NETWORK AND MOLECULAR SOLIDS

Whereas $SiO_2$ melts at 1710 °C, other non-metal oxides melt at much lower temperatures. For example, $P_4O_6$ melts at 25 °C. Referring to the accompanying bonding pictures, describe the forces that hold these solids together.

$P_4O_6$

$SiO_2$ network

**Strategy:** Identify the type of solid. Solids may be network, ionic, metallic, or molecular. Different forces account for the stability of each type.

**Solution:** Because these are non-metal oxides, they cannot be described as metallic. Neither oxide contains ions, so they must be network or molecular. The melting points provide the information needed to categorize the oxides, and the molecular views support the identifications.

$SiO_2$: The high melting point of 1710 °C indicates that this is a network solid, in which strong covalent bonds must break to liquefy the substance. The bonding picture shows atoms linked together, each O atom bonded to two Si atoms and each Si atom bonded to four O atoms. This gives a three-dimensional array of strong covalent bonds, many of which must break for silica to melt.

$P_4O_6$: The relatively low melting point of 25 °C indicates a molecular solid. The molecular structure shows that $P_4O_6$ is a discrete molecule. Strong covalent bonding holds the atoms in each molecule together, but each molecule is attracted to the others only by dispersion forces. In this molecular solid, little energy is required to overcome dispersion forces and allow $P_4O_6$ solid to melt.

**Do the Results Make Sense?** Much of the Earth's silicon is found in durable rock formations based on silica. Granite is a common example. This is consistent with a network solid. Phosphorus, in contrast, occurs naturally in ionic solids containing the phosphate anion, indicating that phosphorus oxides readily undergo chemical reactions. The latter observation is consistent with a molecular solid.

### Practice Exercise 8.4

One neighbour of arsenic, germanium, melts at 1210 °C. Its other neighbour, selenium, melts at 490 °C. Based on these melting points, what type of solid is each of these elements?

## Metallic Solids

As described in Section 7.7, the bonding in solid metals comes from electrons in highly delocalized valence orbitals. There are so many such orbitals that they form energy bands, giving the valence electrons high mobility. Consequently, each metal

Sodium can be cut with a knife.

atom can be viewed as a cation embedded in a "sea" of mobile valence electrons. The properties of metals can be explained on the basis of this picture and the most obvious of these properties is electrical conductivity.

Metals display a range of melting points, indicating that the strength of metallic bonding is variable. The alkali metals of Group 1 are quite soft and melt near room temperature. Sodium, for example, melts at 98 °C and cesium at 28.5 °C. Bonding is weak in these metals, because each atom of a Group 1 metal contributes only one valence electron to the bond-forming energy band. Metals near the middle of the *d* block, on the other hand, are very hard and have some of the highest known melting points. Tungsten melts at 3407 °C, rhenium at 3180 °C, and chromium at 1857 °C. Atoms of these metals contribute two *s* electrons and several *d* electrons to the bond-forming energy band, leading to extremely strong metallic bonding.

Metals are ductile and malleable, meaning they can be drawn into wires or hammered into thin sheets. When a piece of metal forms a new shape, its atoms change position. Because the bonding electrons are fully delocalized, however, changing the positions of the atoms does not cause corresponding changes in the energy levels of the electrons. The "sea" of electrons is largely unaffected by the pattern of metal cations, as Figure 8-19 illustrates. Thus, metals can be forced into many shapes, including sheets and wires, without destroying the bonding nature of their filled bands of delocalized orbitals.

The transition metals display a range of properties. Copper and silver are much better electrical conductors than chromium. Tungsten has very low ductility. Mercury is a liquid at room temperature. All these differences result from variations in numbers of valence electrons. A transition metal such as vanadium or chromium has five or six valence electrons per atom, all of which occupy bond-

**Figure 8-19**

When a metal changes shape, its atoms shift position. However, because the valence electrons are fully delocalized, the energy of these electrons is unaffected.

ing orbitals in the metal lattice. As a result, there are strong attractive forces among the metal atoms, and vanadium and chromium are strong and hard. Beyond the middle of the transition metal series, the additional valence electrons occupy anti-bonding orbitals, which reduces the net bonding. This effect is most pronounced at the end of the $d$ block, where the number of antibonding electrons nearly matches the number of bonding electrons. Zinc, cadmium, and mercury, with $d^{10}$ configurations, have melting temperatures that are more than 600 °C lower than those of their immediate neighbours.

Tungsten can be heated to incandescence without melting.

## Ionic Solids

As described in Chapter 5, ionic solids contain cations and anions strongly attracted to each other by electrical forces. These forces act between *ions* rather than between *molecules*. Ionic solids must be electrically neutral, so their stoichiometries are determined by the charges carried by the positive and negative ions.

Many ionic solids contain metal cations and polyatomic anions. Here again, the stoichiometry of the solid is dictated by the charges on the ions and the need for the solid to maintain electrical neutrality. Some examples of 1 : 1 ionic solids containing polyatomic anions are $NaOH$, $KNO_3$, $CuSO_4$, $BaCO_3$, and $NaClO_3$. Some metallic ores have 1 : 1 stoichiometry, such as scheelite, $CaWO_4$ (contains $WO_4^{2-}$), zircon, $ZrSiO_4$ (contains $SiO_4^{4-}$), and ilmenite, $FeTiO_3$ (contains $TiO_3^{2-}$).

Minerals often contain more than one cation or anion. For example, apatite $(Ca_5(PO_4)_3F)$ contains both phosphate and fluoride anions. Beryl $(Be_3Al_2Si_6O_{18})$ contains beryllium and aluminum cations as well as the $Si_6O_{18}^{12-}$ polyatomic anion. An even more complex example is garnierite $[(Ni,Mg)_6Si_4O_{10}(OH)_2]$, which has a variable composition of $Ni^{2+}$ and $Mg^{2+}$ cations. Although the relative proportions of $Mg^{2+}$ and $Ni^{2+}$ vary, garnierite always has six cations and two $OH^-$ ions. for every anion of $Si_4O_{10}^{10-}$.

One of the most exciting developments in materials science in recent years involves mixed oxides containing rare earth metals. Some of these compounds are *superconductors*, as described in our Focus 8-1: Chemistry and Technology. Below a certain temperature, a superconductor can carry an immense electrical current without losses from resistance. Before 1986, it was thought that this property was limited to a few metals at temperatures below 25 K. Then, it was found that a mixed oxide of lanthanum, barium, and copper showed superconductivity at around 30 K, and since then the temperature threshold for superconductivity has been advanced to 138 K.

Ceramic superconductor materials display a characteristic that is unexpected in traditional chemistry but relatively common in solids: non-stoichiometric composition. When $YBa_2Cu_3O_{7-x}$ contains a stoichiometric amount of oxygen ($x = 0$ or 1), it has inferior superconducting properties. Optimal superconducting behaviour requires a slight deviation from stoichiometric composition. To maintain overall electrical neutrality, the average charge on the copper cations must deviate from integral values. Assigning characteristic charges to the other elements (+3 for Y, +2 for Ba, and −2 for O), we find that Cu must have an average charge of $+6.8/3 = 2.27$ when $x = 0.1$. Because fractional charges are impossible, this means that most of the Cu atoms in the substance bear $+2$ charges, but a few have $+3$ charges.

## SECTION EXERCISES

**8.4.1** Describe the forces that exist in (a) solid $CO_2$ (dry ice); (b) crystalline yellow elemental sulphur, $S_8$; (c) tin (soft, malleable, melts at 232 °C); and (d) $Li_2O$.

**8.4.2** Arrange the following in order of increasing melting point and explain the reasons for your rankings: C (diamond), $F_2$, K, Co.

**8.4.3** Determine the chemical formulas of the following ionic substances: (a) magnesium phosphate; (b) spodumene, a mineral containing lithium and aluminum cations and $Si_2O_6^{4-}$; and (c) the superconducting oxide that contains $Bi^{3+}$, $Sr^{2+}$, and $Cu^{2+}$ in $2:2:1$ atomic ratio.

## FOCUS 8-1  CHEMISTRY AND TECHNOLOGY: SUPERCONDUCTORS

Trains that run on frictionless tracks, and computer chips smaller than those of the present generation yet faster and with much larger capacities—these are potential applications of room-temperature superconductors. Research groups around the world are developing new materials in hopes of reaching this spectacular goal.

Metals conduct electricity because their valence electrons are highly mobile. Although charge flows easily through a metal wire, normal electrical conductors always have significant resistivity. Qualitatively, electrons are slowed by bumping into atoms along their paths. In contrast to normal conductivity, superconductivity is characterized by zero electrical resistance. Electrons move freely through a superconductor without any friction. A superconductor can carry immense amounts of current without losses resulting from heating.

A material that displays superconductivity does so only below a critical temperature, $T_c$. Above $T_c$, the material has normal resistivity, but as the temperature drops below $T_c$, its resistance abruptly disappears, as the graph shows.

The transition between normal conductivity and superconductivity occurs at different temperatures for different materials. In general, $T_c$ is near 4.2 K, the boiling point of liquid helium. For this reason, any device that makes use of superconductivity must be immersed in a bath of liquid helium. Little wonder that superconductivity was not discovered until early in this century and remained a laboratory curiosity until the mid-1980s.

Early work on superconductors concentrated on metals or metal mixtures (alloys). Niobium alloys are particularly good superconductors, and in 1973 a niobium alloy, $Nb_3Ge$, was found to have $T_c = 23$ K, the highest known value for a metal superconductor. In 1986, a ceramic oxide with formula $La_{2-x}Ba_xCuO_4$ was found to show superconductivity at 30 K. Through intense research efforts on ceramic oxides, $YBa_2Cu_3O_{7-x}$, with $T_c = 93$ K, was discovered in 1987.

A temperature above 77 K brought superconductivity to a temperature that can be achieved using inexpensive liquid nitrogen as a coolant. Moreover, the compound is easy to make: three oxides, $Y_2O_3$, BaO, and CuO, are ground together in the correct stoichiometric ratio and heated to 950 °C. The mixture is cooled and pressed into pellets, after which the pellets are heated to just below their melting point to bind the grains in the pellets tightly together. Finally, the pellets are heated in oxygen at around 550 °C.

Ceramic oxide superconductors have distinct atomic layers. The Cu-containing superconductors contain planes of copper and oxygen atoms, as the molecular view shows. These planes alternate with layers containing oxygen and the other metals that make up the superconductor. Superconductivity takes place in the Cu–O planes.

The record for the highest superconducting temperature as of 2009 is 138 K, held by a non-stoichiometric ceramic oxide, $Hg_{0.8}Tl_{0.2}Ba_2Ca_2Cu_3O_{8.33}$. This is still far below room temperature, but research continues.

Superconductivity has also been discovered in rather exotic materials, including the following: Buckminsterfullerene ($C_{60}$) doped with ICl; carbon nanotubes (superconductivity in just one direction); nickel borocarbides, which contain $Ni_2B_2$ layers alternating with $RC$ sheets, where $R$ is a rare earth element such as Er; and organic superconductors that contain planar organic cations and oxoanions. Chemists and physicists continue to study these and other families of superconductors.

For many applications, a superconductor must first be drawn into a wire. This has recently been accomplished. The photo shows a superconductor ribbon wrapped around the copper wires that it could replace.

Flexible superconducting tapes provide promise of uses for superconductors in motors, generators, and even electric transmission lines. Meanwhile, superconducting magnets cooled to the temperature of liquid helium are already in use. High-field nuclear magnetic resonance (NMR) spectrometers have become standard instruments in chemical research laboratories, and the same type of machine (called an MRI spectrometer) is used for medical diagnosis in hospitals worldwide.

O    Cu

La, Sr, or Ba

# 8.5 ORDER IN SOLIDS

In a solid, the atoms, molecules, or ions are in fixed positions. Their motions are restricted to vibrations about preferred locations. This contrasts with gases and liquids, in which molecules change their positions continuously. A solid that displays highly regular repeating patterns of organization is classified as a **crystalline**

**Figure 8-20**

Quartz (*left*) is a crystalline form of silicon dioxide containing a regular three-dimensional array of SiO₂ units. Silica glass (*right*) is an amorphous form containing irregular arrays of SiO₂ units.

solid. Examples include diamonds, sugar crystals, quartz, and table salt. Precious gemstones like rubies, sapphires, and emeralds are highly valued crystals of rare and richly coloured minerals. A solid having an irregular arrangement of particles with no observable pattern is classified as an **amorphous** solid. Amorphous solids include cotton candy, glass, and wax.

Whether a particular substance is crystalline or amorphous can depend on how the solid is prepared. Silicon dioxide exists in both forms. Quartz is a crystalline form of silicon dioxide found in minerals all over the world. Each tetrahedral silicon atom bonds to four oxygens in a highly symmetrical three-dimensional network. This strong bonding network gives silicon dioxide its high melting point of 1710 °C. Slow cooling of molten $SiO_2$ gives crystalline quartz, but rapid cooling gives an amorphous glass. Figure 8-20 shows molecular representations of quartz and glass. Both structures are three-dimensional networks, but the two-dimensional representations shown in the figure are sufficient to show the differences between crystalline and amorphous solids.

The photographs in Figure 8-21 illustrate the following characteristics of crystalline solids:

1. Parallel faces and edges characterize the shape of a crystal. The edges of a crystal usually intersect at characteristic angles.
2. When a crystal breaks into smaller pieces, fragmentation occurs along crystal edges. Small crystals have the same characteristic angles as larger crystals.
3. A crystal has a high degree of symmetry.

**Figure 8-21**

Minerals can form colourful regular crystals. Shown here are vanadinite, $Pb_5(VO_4)_3CCl$ (*left*); quartz, $SiO_2$ (*centre*); and stibnite, $Sb_2SS_3$ (*right*).

## The Crystal Lattice and the Unit Cell

Any crystal is a vast array of atoms, molecules, or ions arranged in some regular repeating pattern. In Section 5.5, we presented the regular arrangement of ions in an ionic crystal, known as the crystal lattice. In the crystal lattice, every ion of a given type has an identical environment. The notion of a crystal lattice extends to any crystalline substance. We place a point at the centre of each species (atom, molecule, or ion) contained in the crystal. The crystal lattice is the pattern formed throughout the crystal by these points. This pattern is regular and symmetrical, and the symmetry of a crystal reflects the symmetry of its lattice.

Because the pattern of the lattice repeats exactly, every crystal has one smallest unit from which the entire pattern can be assembled. This minimum unit is called a **unit cell**. The idea of a unit cell is illustrated in two dimensions by the art of M. C. Escher. Escher often used symmetrical patterns aligned together to create an overall design, one example of which appears in Figure 8-22. The repeating units can be visualized as tiles placed edge to edge. The faces of unit cells can be squares, rectangles, or parallelograms. In Figure 8-22, each tile is a parallelogram.

A unit cell is a three-dimensional block that can be stacked in orderly fashion to create an entire crystal. We show several different unit cells in the figures that follow.

**Figure 8-22**

A drawing by M. C. Escher that contains a repeating pattern. One two-dimensional unit cell is highlighted.

## Cubic Crystals

The easiest crystal lattice to visualize is the **simple cubic** structure. In a simple cubic crystal, layers of atoms stack one directly above another, so that all atoms lie along straight lines at right angles, as Figure 8-23 shows. Each atom in this structure touches six other atoms: four within the same plane, one above the plane, and one below the plane. Within one layer of the crystal, any set of four atoms forms a square. Adding four atoms directly above or below the first four forms a cube, for which the lattice is named. The unit cell of the simple cubic lattice, shown in the cutaway portion of Figure 8-23, contains $\frac{1}{8}$ of an atom at each of the eight corners of a cube. The total number of atoms in the unit cell is $8\left(\frac{1}{8}\right) = 1$.

When eight atoms form a cube, there is a cavity at the centre of the cube. The cavity is not large enough to hold an additional atom, but moving the eight corner atoms slightly away from each other does allow the structure to accommodate another atom. The result is the **body-centred cubic** structure.

**Figure 8-23**

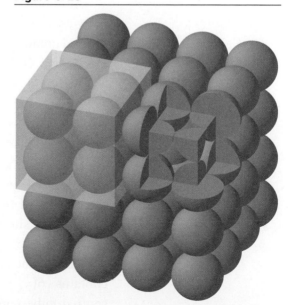

The simple cubic lattice is built from layers of spheres stacked one directly above another. The cutaway view shows one unit cell.

**Figure 8-24**

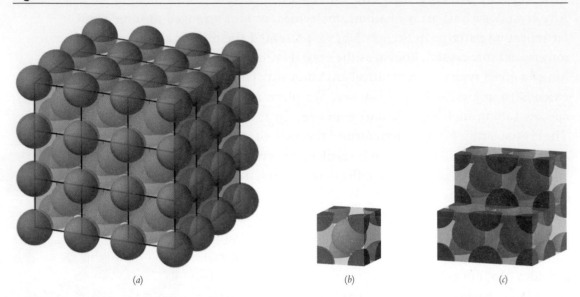

(a)     (b)     (c)

Views of the iron crystal, which has a body-centred cubic unit cell. (a) A molecular view, with the lattice representation; (b) the unit cell; (c) six unit cells stacked to form part of the crystal.

Each atom in a body-centred cubic lattice is at the centre of a cube and contacts eight neighbouring corner atoms. In other words, every atom in the structure can be viewed as the centre atom of one body-centred cube or as a corner atom in an alternate cube. The unit cell of a body-centred cube contains a total of $\left[1 + 8\left(\frac{1}{8}\right)\right] = 2$ atoms. The views of the iron crystal lattice in Figure 8-24 illustrate this arrangement.

Although the corner atoms must move apart to convert a simple cube into a body-centred cube, the extra atom in the centre of the structure makes the body-centred cubic lattice more compact than the simple cubic structure. All the alkali metals, as well as iron and the transition metals from Groups 5 and 6, form crystals with body-centred cubic structures.

Yet another common crystal lattice based on the simple cubic arrangement is known as the **face-centred cubic** structure. When four atoms form a square, there is open space at the centre of the square. A fifth atom can fit into this space by moving the other four atoms away from one another. Stacking together two of these five-atom sets creates a cube. When we do this, additional atoms can be placed in the centres of the four faces along the sides of the cube, as Figure 8-25 shows.

The unit cell of the face-centred cube consists of six $\frac{1}{2}$-atoms embedded in the faces of the cube and eight $\frac{1}{8}$-atoms at the corners of the cube. This unit cell contains $\left[6\left(\frac{1}{2}\right) + 8\left(\frac{1}{8}\right)\right] = 4$ atoms. As we describe below, each atom in a face-centred cube contacts 12 neighbouring atoms. Consequently, the face-centred cube is more compact than either of the other cubic structures.

**Figure 8-25**

| (a) | (b) | (c) | (d) |

The face-centred cubic lattice can be viewed as a simple cube with each face (a) expanded just enough to fit an additional atom in the centre of each face (b), giving eight atoms at the corners and six atoms embedded in the faces (c). The unit cell (d) has $\frac{1}{2}$ of an atom at the centre of each face and $\frac{1}{8}$ of an atom at each corner.

---

### EXAMPLE 8-5  CALCULATING THE MASS OF A UNIT CELL

Calculate the mass of a single unit cell of polonium (Po) metal, which crystallizes in a simple cubic structure.

**Strategy:** Knowing the number of atoms in a simple cubic unit cell and the atomic weight of the metal allows us to calculate the mass of the unit cell.

**Solution:** A simple cubic unit cell has 1/8 of an atom at each corner inside the unit cell, for a total of one atom. The molar mass of Po is 209 g/mol. The mass of the unit cell is therefore:

$$\text{mass} = \frac{209 \text{ g/mol}}{6.02 \times 10^{23} \text{ mol}^{-1}} = 3.47 \times 10^{-22} \text{ g}$$

 **Does the Result Make Sense?** We expect the mass of a unit cell to be the mass of only a small number of atoms (one in this case).

**Common Pitfall:** Not remembering how many atoms are in each type of cubic unit cell.

**Practice Exercise 8.5**

(a) Calculate the mass of a single unit cell of silver (Ag) metal, which crystallizes in a face-centred arrangement.

(b) Calculate the mass of a single unit cell of sodium (Na) metal, which crystallizes in a body-centred arrangement.

---

### EXAMPLE 8-6  CALCULATING THE DENSITY OF A CRYSTALLINE SOLID

Calculate the density of silver metal (in g/cm³), if the edge length of its face-centred cubic unit cell is 407 pm.

**Strategy:** Knowing the edge length of the unit cell, we can easily calculate its volume. Furthermore, we know that all face-centred unit cells contain the equivalent of four atoms, so we can calculate its mass. Density is mass per unit volume.

**Solution:** To calculate the volume of the unit cell, we need only cube the edge length:

$$V = l^3 = (407 \times 10^{-12} \text{ m})^3 = (407 \times 10^{-10} \text{ cm})^3 = 6.74 \times 10^{-23} \text{ cm}^3$$

The mass of the unit cell is the mass of four silver atoms:

$$\text{mass} = \frac{4(107.9 \text{ g/mol})}{6.02 \times 10^{23} \text{ mol}^{-1}} = 7.17 \times 10^{-22} \text{ g}$$

The density is therefore:

$$\rho = \frac{mass}{volume} = \frac{7.17 \times 10^{-22} \, g}{6.74 \times 10^{-23} \, cm^3} = 10.6 \, g/cm^3$$

 **Do the Results Make Sense?** Note that the density of a large piece of silver metal is the same as the density of one of its unit cells, since metals are in effect a large assembly of unit cells, and density is an intensive property (i.e., does not depend on the amount of substance). 10.6 g/cm$^3$ is a reasonable value for the density of a metal.

**Practice Exercise 8.6**

Calculate the density of iron metal, which crystallizes in a body-centred cubic structure having a unit cell edge length of 287 pm.

## EXAMPLE 8-7  CALCULATING THE RADIUS OF AN ATOM

Calculate the radius of a palladium (Pd) atom, given that the density of Pd metal is 12.02 g/cm$^3$ and that it has a face-centred cubic unit cell.

**Strategy:** Calculate the mass of the unit cell, and use this with the density to find the volume. Use the known arrangement of atoms on one face of the unit cell to find the radius of an atom.

**Solution:** A face-centred unit cell contains a total of four atoms: $\frac{1}{8}$ at each corner, and $\frac{1}{2}$ on each of the six faces. The mass of the unit cell is therefore:

$$mass = \frac{4(106.4 \, g/mol)}{6.02 \times 10^{23} \, mol^{-1}} = 7.07 \times 10^{-22} \, g$$

The volume of the unit cell can now be calculated:

$$V = \frac{m}{\rho} = \frac{7.07 \times 10^{-22} \, g}{12.02 \, g \, cm^{-3}} = 5.88 \times 10^{-23} \, cm^3$$

from which the edge length of the unit cell can be found:

$$l = \sqrt[3]{V} = \sqrt[3]{5.88 \times 10^{-23} \, cm^3} = 3.89 \times 10^{-8} \, cm$$

Referring to Figure 8-25d, it is apparent that the length of the diagonal across a face of the unit cell is four times the atomic radius. Since the face is a square, it must be true that $l^2 + l^2 = (4r)^2$, where $r$ is the radius of an atom. Rearranging this expression gives the radius:

$$2l^2 = 16r^2$$
$$l^2 = 8r^2$$
$$r^2 = \frac{l^2}{8}$$
$$r = \sqrt{\frac{l^2}{8}} = \frac{l}{\sqrt{8}} = \frac{l}{2.828} = \frac{3.89 \times 10^{-8} \, cm}{2.828} = 1.35 \times 10^{-8} \, cm$$

Atomic radii are commonly expressed in pm. In this case, $r$ = 135 pm.

 **Do the Results Make Sense?** Atomic radii are typically between 100 and 300 pm.

**Practice Exercise 8.7**

Calculate the radius of a chromium (Cr) atom, knowing that chromium crystallizes in a body-centred cubic structure and has a density of 7.14 g/cm$^3$.

## Close-Packed Crystals

A solid is most stable when each atom, molecule, or ion has as many close neighbours as possible, thus maximizing intermolecular attractions. An arrangement that accomplishes this is described as a **close-packed** structure. Close-packed structures are arranged so that the empty space around the atoms or molecules is minimized.

To visualize a close-packed atomic solid, think of the atoms as spheres placed as compactly as possible. Instead of square planar layers, a close-packed structure has hexagonal planar layers as shown in Figure 8-26a. Notice that this is a more compact planar arrangement than the square, because it places each sphere within a regular hexagon formed by six others. Each sphere has six nearest neighbours in the same plane. Now add a second layer of spheres. To achieve a most compact arrangement, each sphere sits in one of the "dimples" between a trio of spheres in the first layer, as shown in Figure 8-26b. This adds three more nearest neighbours for each sphere in the first layer, one in each of three neighbouring dimples. As additional spheres are added, the second layer eventually looks identical to the lower layer, except that it is offset slightly to allow the spheres to nestle in the dimples formed by the layer below.

Now consider adding a third layer of close-packed spheres. This new layer can be placed in two different ways because there are two sets of dimples in the second layer. Notice in Figure 8-26b that the view through one set of dimples reveals the maroon spheres of the first layer. If spheres in the third layer lie in these dimples, the third layer is directly above the first, and the structure is called **hexagonal close-packed**. The view through the second set of dimples passes unobstructed through both layers. If the spheres of the third layer lie in the unobstructed set of dimples, the third layer is offset from both of the lower layers. It turns out that this arrangement is the face-centred cubic lattice, also called **cubic close-packed**. Figure 8-27 compares the two close-packed structures.

**Figure 8-26**

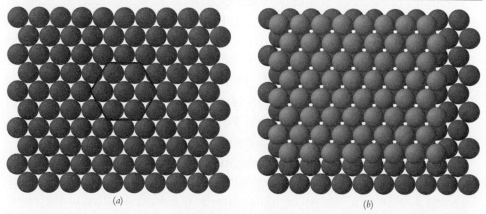

(a) Spheres close-packed in a layer generate a hexagonal pattern. (b) When a second layer is packed on top of the first, each sphere in the second layer nestles in the dimple created by three adjacent spheres in the lower layer. The second layer has two different sets of dimples, one directly above the spheres in the first layer (*maroon*) and the other offset from the spheres in the layer below it (*unshaded*).

**Figure 8-27**

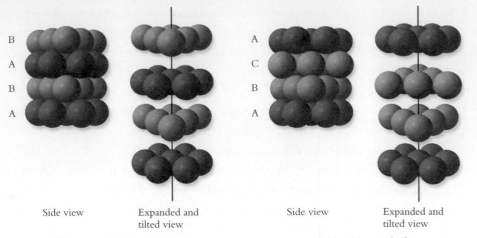

Side view     Expanded and tilted view        Side view     Expanded and tilted view

**Hexagonal close-packed**         **Cubic close-packed**

Side and expanded views of hexagonal and cubic close-packed crystal types. In the hexagonal close-packed structure, spheres on both sides of any plane are in the same positions, and the *third* layer is directly above the first. In the cubic close-packed structure, layers take up three different positions, and the *fourth* layer is directly above the first.

In either of these close-packed structures, each sphere has 12 nearest neighbours: 6 in the same plane, 3 in the dimples above, and 3 in the dimples below. The expanded views in Figure 8-27 show the different arrangements of the hexagonal and cubic close-packed crystalline types. In the hexagonal close-packed structure, notice that the third layer lies directly above the first, the fourth above the second, and so on. The layers can be labelled ABAB . . . . In the cubic close-packed structure, the third layer is offset from the other two, but the fourth layer is directly above the first. This arrangement can be labelled ABCABC . . . .

To see how a set of hexagonal layers can generate a cubic close-packed structure, we need to remove all but one of the atoms from the top layer, as shown in Figure 8-28*a*. The remaining atom from this top layer (shown in maroon in the figure) defines one corner of a cube. Each face of the cube contains an atom at each corner and one atom in the centre, as shown by Figure 8-28*b*. Three atoms from the second layer (shown in gold in the figure) lie at corners of the cube, and three more fall in the centres of three faces of the cube. Atoms from the third row (shown in teal in the figure) occupy the remaining three corners and three faces of the cube. Figure 8-28*c* shows that the atoms from two hexagonal layers fall along a diagonal plane through the cube. The perspectives of Figures 8-28*b* and 8-28*c* reveal the face-centred cubic arrangement.

Atoms often form crystals with hexagonal or cubic close-packed geometry. For instance, magnesium and zinc crystallize with their atoms in a hexagonal close-packed array, whereas silver, aluminum, and gold crystallize in the cubic close-packed arrangement. Argon solidifies at low temperature as a cubic close-packed crystal, and neon can solidify in either form.

Up to now, we have described the crystalline arrays favoured by spherical objects such as atoms, but most molecules are far from spherical. Stacks of pro-

**Figure 8-28**

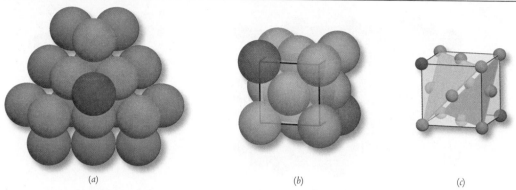

(a)                          (b)                          (c)

Three views of the hexagonal layers contained within a face-centred cubic array. (a) A view perpendicular to the hexagonal layers, with all but one atom removed from the top layer. (b) A side view showing an outline of the cube and atoms from three successive layers of the cubic array. (c) The same side view, with two hexagonal layers screened for emphasis.

duce illustrate that non-spherical objects require more elaborate arrays to achieve maximal stability. Compare a stack of bananas with a stack of oranges. Just as the stacking pattern for bananas is less symmetrical than that for oranges, the structural patterns for most molecular crystals are less symmetrical than those for crystals of spherical atoms, reflecting the lower symmetry of the molecules that make up molecular crystals.

Non-spherical bananas require more elaborate packing schemes than do spherical oranges.

## Ionic Solids

The packing in ionic crystals requires that ions of opposite charges alternate with one another to maximize attractions among ions. A second important feature of ionic crystals is that the cations are usually smaller than the anions. Consequently, ionic compounds adopt a variety of structures that depend on the charges and sizes of the ions. One way to discuss ionic structures is to identify a crystal lattice for one set of ions, and then describe how the other ions pack within the lattice of the first set.

For many $1:1$ ionic crystals such as NaCl, the most stable arrangement is a face-centred cubic array of anions with the cations packed into the holes between the anions. This structure appears in Figure 8-29. In addition to the face-centred cubic arrangement of chloride ions, the unit cell shows $\frac{1}{4}$ of a sodium cation along each edge of the cell, and there is an additional sodium cation in the centre of the unit cell. Summing the components reveals that each unit cell contains four complete NaCl formula units and is electrically neutral:

Chloride anions: 6 faces$\left(\frac{1}{2}\text{ anions/face}\right)$ + 8 corners$\left(\frac{1}{8}\text{ anion/corner}\right)$ = 4 Cl⁻ anions

Sodium cations: 12 edges$\left(\frac{1}{4}\text{ cation/edge}\right)$ + 1 in centre = 4 Na⁺ cations

**Figure 8-29**

(a)                                                                                              (b)

(a) A sodium chloride crystal can be viewed as two overlapping face-centred cubic lattices, one of $Na^+$ cations (*purple*) and the other of $Cl^-$ anions (*green*). (b) Because $Cl^-$ anions are considerably larger than $Na^+$ cations, the most convenient unit cell places anions at the corners. There is a cation at the centre of this unit cell.

Many ionic solids have the same unit cell structure as the sodium chloride lattice. Examples are all the halides of $Li^+$, $Na^+$, $K^+$, and $Rb^+$, as well as the oxides of $Mg^{2+}$, $Ca^{2+}$, $Sr^{2+}$, and $Ba^{2+}$.

**Figure 8-30**

The cesium chloride lattice consists of a simple cubic array of chloride anions with a cesium cation at the centre of each unit cell.

The cations are significantly smaller than the anions in most $1:1$ ionic salts, but cesium, which forms the largest monatomic cation, is an exception. Because its cations and anions are close to the same size, cesium chloride is most stable in a body-centred cubic lattice. There are $Cl^-$ anions at the corners of the cube, with a $Cs^+$ in the central body position, as Figure 8-30 shows. The unit cell contains $8\left(\frac{1}{8}\right) = 1$ chloride ion and 1 cesium ion. The cesium chloride lattice is found for CsCl, CsBr, CsI, and several other $1:1$ ionic crystals.

Many ionic compounds have stoichiometries that differ from $1:1$. To give just a single example, the fluorite lattice is named for the naturally occurring form of $CaF_2$ and is common for salts of the general formula $MX_2$. This lattice is easiest to describe by thinking of the $Ca^{2+}$ cations arranged in a face-centred cubic structure. The $F^-$ anions form a simple cube of eight anions in the interior of the face-centred cube.

Figure 8-31*a* shows the unit cell of the fluorite lattice, which contains a total of four $Ca^{2+}$ cations and eight $F^-$ anions:

Calcium anions: 6 faces$\left(\frac{1}{2}\text{cation/face}\right)$ + 8 corners$\left(\frac{1}{8}\text{cation/corner}\right)$ = 4 $Ca^{2+}$ cations

Fluoride anions: 8 complete anions within the face–centred cube = 8 $F^-$ anions

This gives the correct $1:2$ stoichiometry, resulting in electrical neutrality.

From the perspective of the fluoride ions, the fluorite structure can be seen as a body-centred cube with an $F^-$ anion in the centre. Instead of eight corner ions, however, the alternating corners are vacant, placing calcium ions at four of the corners. This structure appears in Figure 8-31*b*. If you look closely at this structure, you will see that the calcium ions are arranged in a tetrahedron about the fluoride ion.

**Figure 8-31**

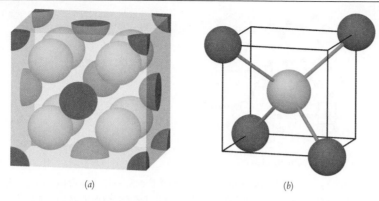

(a)                                    (b)

(a) The unit cell of the fluorite lattice has cations in a face-centred cubic arrangement, with a simple cube of anions within the unit cell. (b) The fluorite lattice can also be visualized as a body-centred cube with calcium ions at four of the corners, the other four corners vacant, and a fluoride ion in the body position.

We have touched briefly on three simple ionic lattices, but there are many others. Moreover, the structures of many crystalline network solids can also be described by the methods we have introduced here for NaCl, CsCl, and $CaF_2$.

## Amorphous Solids

Solid materials are most stable as crystals. When a liquid is cooled slowly, it generally solidifies as crystals. When solids form rapidly, on the other hand, their atoms or molecules may become locked into positions other than those of a regular crystal, giving materials that are amorphous, meaning "without form." Ordinary cane sugar (sucrose) is crystalline, and crystalline rock candy forms when melted sugar is cooled slowly. In contrast, rapid cooling of melted sugar gives an amorphous solid: cotton candy contains long threads of amorphous sugar.

What we call **glass** is an entire family of amorphous solids based on silicon dioxide, whose common name is silica ($SiO_2$). Pure silica is usually found as crystals containing the regular array of covalent bonds shown in Figure 8-20. Quartz is crystalline silica. When quartz is melted and then quickly cooled, however, it forms fused silica, an amorphous solid glass. Silica glass has many desirable properties. It resists corrosion, transmits light well, and withstands wide variations in temperature. Unfortunately, pure silica is very difficult to work with because of its high melting point (1710 °C). Consequently, silica glass is used only for special applications.

Mixing sodium oxide ($Na_2O$) with silica results in a glass that can be shaped at a lower temperature. Sodium oxide is ionic, so it breaks the Si—O—Si chain of covalent bonds, as shown in Figure 8-32. This weakens the lattice strength of the glass, lowers its melting point, and reduces the viscosity of the resulting liquid. However, the weakened lattice also means that glass made from mixed sodium and silicon oxides is vulnerable to chemical attack.

A desirable glass melts at a reasonable temperature, is easy to work with, and yet is chemically inert. Such a glass can be prepared by adding a third component

Cotton candy is amorphous. Rock candy is crystalline. Both are made of sucrose.

**Figure 8-32**

Quartz glass                    Sodium-containing glass

Adding sodium oxide to quartz glass breaks some of the Si—O covalent bonds (*left view*, red outlines) to create terminal O⁻ anions associated with Na cations (*right view*). This softens the glass, reducing its melting point.

that has bonding characteristics intermediate between those of purely ionic sodium oxide and those of purely covalent silicon dioxide. Several different components are used, depending on the properties desired in the glass.

The glass used for windowpanes and bottles is soda–lime–silica glass, a mixture of sodium oxide, calcium oxide, and silicon dioxide. The addition of CaO strengthens the lattice enough to make the glass chemically inert to most common substances. (Strong bases and HF, however, attack this glass.) Pyrex, the glass used in coffeepots and laboratory glassware, is a composite of $B_2O_3$, CaO, and $SiO_2$. This glass can withstand rapid temperature changes that would crack soda–lime–silica glass. Lenses and other optical components are made from glass that contains PbO. Light rays are strongly bent as they pass through lenses made of this glass. Coloured glasses contain small amounts of coloured metal oxides such as $Cr_2O_3$ (amber), NiO (green), or CoO (brown).

Many contemporary materials are amorphous solids composed of extremely large molecules called **polymers**. Polymeric solids are intermediate between molecular and covalent solids. They have discrete but extremely large molecules held together by dispersion forces and/or hydrogen bonds. Because polymer molecules are so large, their covalent bonding plays an important role in determining the properties of the solid. Plastic polymers can be shaped and moulded because the intermolecular forces between polymer molecules are weaker than chemical bonds, but a plastic polymer also has relatively high strength because its long-chain molecules are held together by strong covalent bonds. We discuss polymers in Chapter 17.

Amorphous solids resemble liquids in that their molecules are not organized in regular arrays. In fact, there is no clear distinction between amorphous solids and very viscous liquids. When asphalt gets hot, it becomes sticky and eventually melts, but whether sticky asphalt is a soft solid or a highly viscous liquid is a matter of perspective. Our Focus 8-2: Chemistry and Life explores another type of structure, the liquid crystal, that shares characteristics of liquids and solids.

## FOCUS 8-2 CHEMISTRY AND LIFE: LIQUID CRYSTALS

"Liquid crystal" sounds like a contradiction. Liquids are fluid, their molecules continually changing places in a manner that is not particularly well organized. Crystals are immobile, their molecules locked into fixed positions that form regular patterns. Yet, not only does this unusual combination of fluidity and regular patterns exist, it plays important roles in biological organisms.

We can visualize a liquid crystal as not quite liquid and not quite solid. The molecules in a liquid crystal can move about, but they do so in patterned ways. Molecules that form liquid crystals have cylindrical shapes. In the liquid phase, these "molecular rods" have random orientations, whereas in the solid phase, they are all aligned like boards of a fence. In the liquid crystal phase, there is partial alignment of molecules in layers. These crystalline layers rearrange easily, making it possible for small changes in forces to change the properties of the liquid crystal. The illustration at right shows the most common arrangements of liquid crystals. The nematic arrangement has the molecules aligned along their long axes, but the ends do not align. Smectic arrangements have greater alignment, with the ends of the molecules closely aligned. The molecules in a smectic arrangement can be upright (smectic A) or tilted (smectic C).

Liquid crystals were discovered by an Austrian biologist, Frederich Reinitzer, in 1888. Reinitzer found that cholesteryl benzoate, a biological chemical, melts to form a hazy liquid. At a higher temperature, the haziness disappears. This clear state is what we know now as a liquid crystal.

The development of high-magnification microscopy made it possible to create images of biological materials at the molecular level. Many of these images show structures that have liquid crystalline aspects. Shown here are aligned mosaic virus molecules and protein molecules in voluntary muscles. In addition, all cell walls are "picket fences" of rod-shaped molecules in regular yet fluid arrays.

Chemists learned how to design molecules that display liquid crystal behaviour, and physicists discovered how to use electrical force to make rapid changes in liquid crystal orientations. The result is LCD (liquid crystal display) technology, which brought us digital watches, laptop computer screens, and flat-screen television sets. This field continues to expand, with some predicting that "electronic paper" based on LCD technology will eventually make it possible to carry computer images in our pockets.

During the late twentieth century, most of the excitement about liquid crystals was in chemistry, physics, and technology. Now, however, the knowledge that made flat TV screens possible is being applied to problems in biology. These applications take a variety of forms. For example, DNA packs inside chromosomes in liquid-crystal-like structures. The physics of liquid crystals will give molecular biologists insights about chromosome structure. Furthermore, drugs and antibiotics that interact with DNA often have liquid crystal properties. Rather than being cylindrical, these liquid crystal molecules are planar, allowing them to slide into a DNA coil without destroying the organized pattern of the DNA. Do liquid crystal structures also exist in living cells, and, if so, what roles do liquid crystal properties play in the regulation of life processes?

Normal Liquid

Nematic

Smectic A

Smectic C

Line structure of
cholesteryl benzoate

Another direction in liquid crystal research is the fabrication of new molecules that mimic natural materials. The molecule shown below forms two different liquid crystals because of the cylindrical properties of the colour-shaded portions of the molecule. These two trios are tethered together by long hydrocarbon chains. In addition to forming liquid crystals, this molecule self-assembles into organized larger units, as do proteins and DNA. The techniques used to synthesize this particular molecule can be adapted to incorporate different cylindrical components that give a variety of properties, paving the way for the creation of new materials with potential applications in biotechnology and nanotechnology.

## Crystal Imperfections

The two extremes of ordering in solids are perfect crystals with complete regularity and amorphous solids that have little symmetry. Most solid materials are crystalline but contain defects. **Crystalline defects** can profoundly alter the properties of a solid material, often in ways that have useful applications. Doped semiconductors, described in Section 7.7, are solids into which impurity "defects" are introduced deliberately in order to modify electrical conductivity. Gemstones are crystals containing impurities that give them their colour. Sapphires and rubies are imperfect crystals of colourless $Al_2O_3$, but $Ti^{3+}$ or $Fe^{3+}$ makes sapphires blue, and $Cr^{3+}$ makes rubies red.

As noted earlier, yttrium–barium–copper superconductors have optimal properties when they have a slight deficiency of oxygen. This departure from stoichiometric composition is created by defects in the crystal structure. The superconductor crystal has oxygen anions missing from some positions in the crystal lattice, and the number of missing anions can vary sufficiently to give the material a non-stoichiometric overall composition. The solid remains electrically

Sapphires are blue; rubies are red.

neutral when its anion content varies, because some of the cations take on different charges. The copper cations in the superconductor can have a +2 or +3 charge. The relative number of $Cu^{3+}$ ions decreases as oxygen anions are removed from the structure.

Substitutional impurities replace one metal atom with another, whereas interstitial impurities occupy the spaces between metal atoms. Interstitial impurities create imperfections that play important roles in the properties of metals. For example, small amounts of impurities are deliberately added to iron to improve its mechanical properties. Pure iron is relatively soft and easily deformed, but the addition of a small amount of carbon creates steel, a much harder material. Carbon atoms fill some of the open spaces between iron atoms in the crystalline structure of steel. Although carbon atoms fit easily into these spaces, their presence reduces the ability of adjacent layers of iron atoms to slide past each other, as Figure 8-33 illustrates.

This section describes only a few of the organizing features of solid materials. The rapidly growing field of materials science addresses these and many other atomic and molecular aspects of solids that determine the technologically useful properties of materials.

**Figure 8-33**

The presence of even a few carbon atoms (*grey spheres*) in the interstitial holes in an iron lattice prevents adjacent layers of iron atoms from sliding past one another and hardens the iron into steel.

## SECTION EXERCISES

**8.5.1** Predict the angles found in the crystalline fragments broken from a hexagonal close-packed crystal. What additional angles would be found in fragments from cubic close-packed crystals? (*Hint:* Think about the angles found in hexagons and cubes.)

**8.5.2** Is the view of fluorite shown in Figure 8-31*b* a unit cell? If so, show how to assemble it to create the fluorite crystal. If not, explain why it is not a unit cell.

**8.5.3** Carbon atoms in steel occupy interstitial positions, as shown in Figure 8-33. Manganese atoms, which are also added to steel to increase its hardness, occupy substitutional positions, replacing iron atoms in the crystalline structure. Draw an atomic picture of a portion of manganese-containing steel.

**8.5.4** Calculate the atomic radius of a tungsten (W) atom, knowing that it crystallizes in a body-centred cubic lattice, and its density is 19.25 g/cm³.

## 8.6 PHASE CHANGES

Phase changes are characteristic of all substances. The normal phases displayed by the halogens appear in Section 8.1, where we also show that a gas liquefies or a liquid freezes at low enough temperatures. Vapour pressure, which results from molecules escaping from a liquid into the gas phase, is one of the liquid

**Figure 8-34**

When heat is supplied at a constant rate to a sample of liquid water, the temperature rises to 100 °C and then remains constant until all the water has evaporated.

properties described in Section 8.3. The temperatures and pressure at which phase changes occur depend on the magnitudes of intermolecular forces.

## Heats of Phase Changes

Any phase change that results in increased molecular mobility requires that intermolecular forces be overcome. This, in turn, requires an energy input. Consider a teakettle on a hot stove. The burner supplies energy that heats the water until the temperature reaches the boiling point of 100 °C. As Figure 8-34 shows, the temperature then stops rising even though heat is still transferred to the water. The added energy now causes molecules of water to move from the liquid phase into the gas phase, and the temperature remains constant as the water boils away. Steam escapes into the atmosphere as the burner transfers heat to the water molecules.

A molecular perspective reveals why energy must be supplied to boil water. A molecule of water cannot escape the liquid phase unless it has enough energy of motion to overcome the hydrogen bonding forces that hold liquid water together. About 40 kJ of heat must be supplied to transfer 1 mole of water molecules from the liquid phase into the vapour phase.

Phase changes require that energy must be either supplied or removed from the substance undergoing the phase change. This energy transfer process normally occurs in the form of heat transfers. Moreover, phase changes generally occur at constant pressure. Recall from Chapter 3 that for constant pressure changes, heat transfer equals the enthalpy change: $q_p = \Delta H$. A change of phase always involves changes in both the energy and the enthalpy of the substance. The magnitude of the changes depends on the strength of intermolecular forces in the substance undergoing the phase change.

The amount of heat required to vapourize a substance also depends on the size of the sample. Twice as much energy is required to vapourize 2 moles of water than 1 mole. The heat needed to vapourize 1 mole of a substance at its normal boiling point is called the **molar heat of vapourization, $\Delta H_{vap}$**.

Energy must also be provided to melt a solid substance. This energy is used to overcome the intermolecular forces that hold molecules or ions in fixed positions in the solid phase. Thus, the melting of a solid also has characteristic energy and enthalpy changes. The heat needed to melt 1 mole of a substance at its normal melting point is called the **molar heat of fusion, $\Delta H_{fus}$**.

Fusion and vapourization are the most familiar phase changes, but **sublimation** is also common. Sublimation is a phase change in which a solid converts directly to a vapour without passing through the liquid phase. Dry ice (solid $CO_2$)

sublimes at 195 K with $\Delta H_{subl} = 25.2$ kJ/mol. Mothballs contain naphthalene ($C_{10}H_8$, $\Delta H_{subl} = 73$ kJ/mol), a crystalline white solid that sublimes to produce a gas that repels moths. The purple colour of the gas above iodine crystals in a closed container provides visible evidence that this solid also sublimes at room temperature ($\Delta H_{subl} = 62.4$ kJ/mol).

Phase changes can go in either direction: steam condenses upon cooling, and liquid water freezes at low temperature. Each of these is *exothermic* because each is the reverse of an endothermic phase change. That is, heat is released as a gas condenses to a liquid and as a liquid freezes to a solid. To make ice cubes, for instance, water is placed in a freezer that absorbs the heat released during the formation of ice. A phase change that is exothermic has a negative enthalpy change:

The purple colour of the vapour above solid iodine is due to $I_2$ molecules in the gas phase.

$$\Delta H_{solidification} = -\Delta H_{fus} \quad \text{and} \quad \Delta H_{condensation} = -\Delta H_{vap}$$

By convention, tabulated values of heats of phase changes are always specified in the endothermic direction. Consequently, the reverse processes have the same magnitude but the opposite sign:

Solid $\longrightarrow$ Liquid $\Delta H_{fus}$          Liquid $\longrightarrow$ Solid $-\Delta H_{fus}$

Liquid $\longrightarrow$ Vapour $\Delta H_{vap}$          Vapour $\longrightarrow$ Liquid $-\Delta H_{vap}$

Solid $\longrightarrow$ Vapour $\Delta H_{subl}$          Vapour $\longrightarrow$ Solid $-\Delta H_{subl}$

Table 8-5 lists values of $H_{fus}$, $H_{vap}$, melting points, and boiling points for different chemical substances. Example 8-8 provides practice in using enthalpies of phase changes.

## TABLE 8-5 TEMPERATURES AND MOLAR ENTHALPIES OF PHASE CHANGES

| Substance | Formula | $T_m$ (K) | $H_{fus}$ (kJ/mol) | $T_b$ (K) | $H_{vap}^*$ (kJ/mol) |
|---|---|---|---|---|---|
| Argon | Ar | 83 | 1.3 | 87 | 6.3 |
| Oxygen | $O_2$ | 54 | 0.45 | 90 | 9.8 |
| Methane | $CH_4$ | 91 | 0.84 | 112 | 9.2 |
| Ethane | $C_2H_6$ | 90 | 2.85 | 184 | 15.5 |
| Diethyl ether | $(C_2H_5)_2O$ | 157 | 6.90 | 308 | 26.0 |
| Bromine | $Br_2$ | 266 | 10.8 | 332 | 30.5 |
| Ethanol | $C_2H_5OH$ | 156 | 7.61 | 351 | 39.3 |
| Benzene | $C_6H_6$ | 278.5 | 10.9 | 353 | 31.0 |
| Water | $H_2O$ | 273 | 6.01 | 373 | 40.79 |
| Mercury | Hg | 234 | 23.4 | 630 | 59.0 |

* at the normal boiling point, $T_b$

## EXAMPLE 8-8 HEAT OF PHASE CHANGE

A swimmer emerging from a pool is covered with a film containing about 75 g of water. How much heat must be supplied to evaporate this water?

**Strategy:** Energy in the form of heat is required to evaporate the water from the swimmer's skin. The energy needed to vapourize the water can be found using the molar heat of vapourization and the number of moles of water. The process can be shown with a simple block diagram:

**Solution:** The $\Delta H_{vap}$ of water is 40.79 kJ/mol at the normal boiling point (see Table 8-5). We will assume it has the same value at the swimmer's body temperature. The molar mass of water is 18.02 g/mol, so 75 g of water is 4.16 mol. Therefore, the heat that must be supplied is

$$q_p = n\,\Delta H_{vap} = (4.16\ \text{mol})(40.79\ \text{kJ/mol}) = 1.7 \times 10^2\ \text{kJ}.$$

**Does the Result Make Sense?** If the swimmer's body must supply all this heat, a substantial chilling effect occurs. Thus, swimmers usually towel off (to reduce the amount of water that must be evaporated) or lie in the sun (to let the sun provide most of the heat required).

**Practice Exercise 8.8**

Determine how much heat is involved in freezing 125 g of water in an ice-cube tray. What is the direction of heat flow in this process?

The graph in Figure 8-34 shows that adding heat to boiling water does not cause the temperature of the water to increase. Instead, the added energy is used to overcome intermolecular attractions as molecules leave the liquid phase and enter the gas phase. Other two-phase systems, such as an ice–water mixture, show similar behaviour. Heat added to an ice–water mixture melts some of the ice, but the mixture remains at 0 °C. Similarly, when an ice–water mixture in a freezer loses heat to the surroundings, the energy comes from some liquid water freezing, but the mixture remains at 0 °C until all the water has frozen. This behaviour can be used to hold a chemical system at a fixed temperature. A temperature of 100 °C can be maintained by a boiling water bath, and an ice bath holds a system at 0 °C. Lower temperatures can be achieved with other substances. Dry ice maintains a temperature of −78 °C; a bath of liquid nitrogen has a constant temperature of −196 °C (77 K); and liquid helium, which boils at 4.2 K, is used for research requiring ultracold temperatures.

## Phase Diagrams

The stable phase of a substance depends on temperature and pressure. A **phase diagram** is a map of the pressure–temperature world showing the phase behaviour of a substance. As Figure 8-35 shows, a phase diagram is a *P–T* graph that shows the ranges of temperature and pressure over which each phase is stable. Pressure is plotted along the *y*-axis, and temperature is plotted along the *x*-axis. In the upper left-hand region (low *T*, high *P*), the substance is stable as a solid. In the lower right-hand region (high *T*, low *P*), the substance is stable as a gas. In some intermediate range of *T* and *P*, the substance is stable as a liquid.

As illustrated by Figure 8-35, phase diagrams have many characteristic features:

1. Boundary lines between phases separate the regions where each phase is stable.

2. Movement across a boundary line corresponds to a phase change. The arrows on the figure show six different phase changes: sublimation and its reverse, deposition; melting and its reverse, freezing; and vapourization and its reverse, condensation.

3. At any point along a boundary line, the two phases on either side of the line coexist in a state of dynamic equilibrium. The normal melting point and normal boiling point of a substance (shown by red dots) are the points where the phase boundary lines intersect the horizontal line that represents *p* = 1 atm (1.01325 bar).

4. Three boundary lines meet in a single point (shown by a red dot), called a **triple point**. All three phases are stable simultaneously at this unique combination of temperature and pressure. Notice that, although *two* phases are stable under any of the conditions specified by the boundary lines, *three* phases can be simultaneously stable only at a triple point.

5. Above a sufficiently high temperature and pressure, called the **critical point** (shown by a red dot), the distinction between the gas phase and the liquid phase disappears. Instead, the substance is a supercritical fluid, with viscosity typical of a gas, density typical of a liquid, and able to expand or contract like a gas.

6. What happens to a substance as temperature changes at constant pressure can be determined by drawing a horizontal line at the appropriate pressure on the phase diagram.

The normal melting and boiling points are by convention defined at 1 atm pressure, which is 1.01325 bar.

**Figure 8-35**

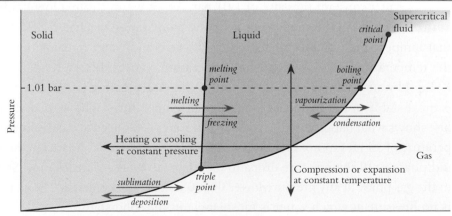

The general form of a phase diagram. Any point on the diagram corresponds to a specific temperature and pressure. Lines trace conditions under which phase changes occur, and the blue arrows show six types of phase transitions.

7. What happens to a substance as pressure changes at constant temperature can be determined by drawing a vertical line at the appropriate temperature on the phase diagram.

**Figure 8-36**

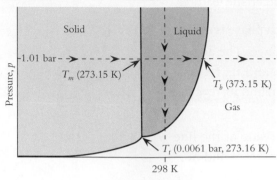

The phase diagram for water. The critical point is at 647 K, 218 bar, well off-scale on this graph.

**Figure 8-37**

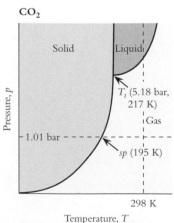

Phase diagrams for molecular nitrogen and carbon dioxide, two substances that are gases at room temperature and pressure.

8. The temperature for conversion between the vapour and a condensed phase depends strongly on pressure. Qualitatively, this is because compressing a gas increases the collision rate and makes condensation more favourable. We describe the quantitative details of this variation in Section 11.5.

9. The melting temperature is almost independent of pressure, making the boundary line between solid and liquid nearly vertical. Qualitatively, this is because pressure has hardly any effect on the compact liquid and solid phases.

10. The solid–vapour boundary line extrapolates to $p = 0$ and $T = 0$. This is a consequence of the direct link between temperature and molecular energy. At $T = 0$ K, molecules have minimum energy, so they cannot escape from the solid lattice. At 0 K, the vapour pressure of every substance would be zero.

The phase diagram for water, shown in Figure 8-36, illustrates these features for a familiar substance. The figure shows that liquid water and solid ice coexist at the normal melting point, $T_m = 273.15$ K and $p = 1.01$ bar. Liquid water and water vapour coexist at the normal boiling point, $T_b = 373.15$ K and $p = 1.01$ bar. The triple point of water occurs at $T_t = 273.16$ K and $p = 0.0061$ bar. The figure shows that when $p$ is lower than 0.0061 bar, there is no temperature at which water is stable as a liquid. At sufficiently low pressure, ice sublimes but does not melt.

The dashed lines on Figure 8-36 show two paths that involve phase changes for water. The *horizontal dashed line* shows what happens as the temperature increases at a constant pressure of 1.01 bar. As ice warms from a low temperature, it remains in the solid phase until the temperature reaches 273.15 K. At that temperature, solid ice melts to liquid water, and water remains liquid as the temperature increases until the temperature reaches 373.15 K. At 373.15 K, liquid water changes to water vapour. When the pressure is 1.01 bar, water is most stable in the gas phase at all higher temperatures. The *vertical dashed line* shows what happens as the pressure on water is reduced at a constant temperature of 298 K (approximately room temperature). Water remains in the liquid phase until the pressure drops to 0.03 bar. At 298 K, water is most stable in the gas phase at any pressure lower than 0.03 bar. Notice that at 298 K, there is no pressure at which water is most stable in the solid state.

All phase diagrams share the 10 common features listed above. However, the detailed appearance of a phase diagram is different for each substance, as determined by the strength of the intermolecular interactions for that substance. Figure 8-37 shows two examples, the phase diagrams for molecular nitrogen and for carbon dioxide. Both these substances are gases under normal conditions. Unlike $H_2O$, whose triple point lies close to 298 K, $N_2$ and $CO_2$ have triple points that are well below room temperature. Although both are gases at room temperature and pressure, they behave differently when cooled at $p = 1.01$ bar. Molecular nitrogen liquefies at 77.4 K and then solidifies at 63.3 K, whereas carbon dioxide condenses directly to the solid phase at 195 K. This difference in behaviour arises because the triple point of $CO_2$, unlike the triple points of $H_2O$ and $N_2$, occurs at a pressure greater than 1.01 bar. The phase diagram of $CO_2$ shows that at a pressure of 1.01 bar, there is no temperature at which the liquid phase is stable.

Phase diagrams are constructed by measuring the temperatures and pressures at which phase changes occur. Approximate phase diagrams such as those shown in Figures 8-36 and 8-37 can be constructed from the triple point, normal melting point, and normal boiling point of a substance. Example 8-9 illustrates this procedure.

## EXAMPLE 8-9 CONSTRUCTING A PHASE DIAGRAM

Ammonia is a gas at room temperature and pressure. Its normal boiling point is 239.8 K, and it freezes at 195.5 K. The triple point for $NH_3$ is $p = 0.0610$ bar and $T = 195.4$ K. Use this information to construct an approximate phase diagram for $NH_3$.

**Strategy:** The normal melting, boiling, and triple points give three points on the phase boundary curves. To construct the curves from knowledge of these three points, use the common features of phase diagrams: the vapour–liquid and vapour–solid boundaries of phase diagrams slope upward, the liquid–solid line is nearly vertical, and the vapour–solid line begins at $T = 0$ K and $p = 0$.

**Solution:** Begin by choosing appropriate scales, drawing the $p = 1$ bar line, and locating the given data points. An upper temperature limit of 300 K encompasses all the data:

Next, connect the points and label the domains:

**Does the Result Make Sense?** We know that ammonia is a gas under normal pressure and temperature, and the diagram shows that 298 K, 1.01 bar lies within the gas region of the phase diagram.

**Practice Exercise 8.9**
Molecular chlorine melts at 172 K, boils at 239 K, and has a triple point at 172 K, 0.014 bar. Sketch the phase diagram for this compound.

Phase diagrams can be used to determine what phase of a substance is stable at any particular pressure and temperature. They also summarize how phase changes occur as either condition is varied. Example 8-10 provides an illustration.

## EXAMPLE 8-10 INTERPRETING A PHASE DIAGRAM

A chemist wants to perform a synthesis in a vessel at $p = 0.50$ bar using liquid $NH_3$ as the solvent. What temperature range would be suitable? When the synthesis is complete, the chemist wants to boil off the solvent without raising $T$ above 220 K. Is this possible?

**Strategy:** The phase diagram for $NH_3$ shows the boundary lines for the liquid domain. These boundary lines can be used to determine the conditions under which phase changes occur.

**Solution:** Because the chemist wants to work at $p = 0.50$ bar, draw a horizontal line across the phase diagram at $p = 0.50$ bar. The top figure is an expanded view of the phase diagram between 150 and 300 K:

The horizontal line intersects the boundary lines at about 235 and 195 K. Liquid $NH_3$ is stable between these temperatures at this pressure.

At the completion of the synthesis, the chemist wants to remove the solvent without raising $T$ above 220 K. On the bottom figure, a vertical line at 220 K on the phase diagram represents this condition:

The line intersects the liquid–vapour boundary at about 0.25 bar. A vacuum pump capable of reducing $p$ below 0.25 bar can be used to vapourize and remove the $NH_3$ while keeping the temperature below 220 K.

**Do the Results Make Sense?** We know that ammonia is a gas at normal temperature and pressure, so it is sensible that the temperature must be lowered substantially to run a reaction in liquid ammonia. Knowing that ammonia is liquid at 0.50 bar and 220 K, it makes sense that a still lower pressure is required to boil off the liquid at this temperature.

**Practice Exercise 8.10**

A sample of $N_2$ gas is at $p = 0.1$ bar and $T = 63.1$ K. Use Figure 8-37 to determine what will happen to this sample if the temperature remains fixed while the pressure slowly increases to 1 bar.

## Variations on Phase Diagrams

We tend to think of three phases of matter: solid, liquid, and vapour. However, many substances have more than one solid phase, whereas others may display liquid crystal phases that are distinct from either the standard liquid phase or the solid phase. Phase diagrams of solids are particularly useful in geology, because many minerals undergo solid–solid phase transitions at the high temperatures and pressures found deep in the Earth's crust. Figure 8-38 is the phase diagram for silica, an important geological substance. Notice that the pressure scale ranges from 0 to 140 bar, and the temperature scale ranges from 0 to 2000 °C. These are geological pressures and temperatures. Notice also that there are six different forms of crystalline silica, each stable in a different temperature–pressure region.

A geologist who encounters a sample of stishovite, for example, can be confident that it solidified under extremely high pressure conditions.

Another complicating feature of phase diagrams at high pressure is the critical point, above which the liquid–vapour transition no longer occurs. As the pressure on a gas increases, the gas is compressed into an ever-smaller volume. If the temperature is low enough, this compression eventually results in liquefaction: the volume decreases dramatically as intermolecular forces hold the molecules close together in the condensed phase. At high enough temperature, however, no amount of compression is enough to cause liquefaction. The substance is a supercritical fluid, able to expand to fill the available space, but resistant to further compression.

The phase diagrams in Figures 8-36 and 8-37 do not show critical points, because the critical points of water, carbon dioxide, and nitrogen occur at higher pressures than those shown on these diagrams. The critical point of water is $p = 218$ bar, $T = 647$ K; that of $CO_2$ is $p = 72.9$ bar, $T = 304$ K; and that of $N_2$ is $p = 33.5$ bar, $T = 126$ K.

Although most critical pressures are many times greater than atmospheric pressure, supercritical fluids have important commercial applications. The most important of these is the use of supercritical $CO_2$ as a solvent. Supercritical $CO_2$ diffuses through a solid matrix rapidly, and it transports materials well because it has a lower viscosity than ordinary liquids. Because $CO_2$ is a natural product of biological processes, it is non-toxic, making it well suited for "Green Chemistry" applications. Supercritical $CO_2$ currently is used as a solvent for dry cleaning, for petroleum extraction, for decaffeination, and for polymer synthesis.

**Figure 8-38**

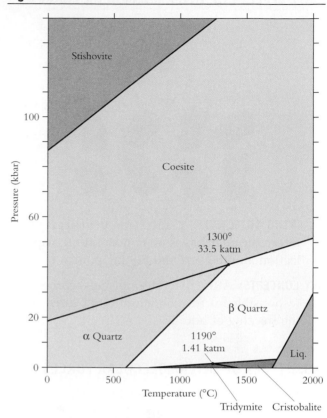

The phase diagram for silica shows six different forms of the solid, each stable under different temperature and pressure conditions.

---

**SECTION EXERCISES**

● **8.6.1** Use information in Table 8-5 to determine how much heat must be supplied to vapourize 15 g of the following substances: (a) methane, (b) ethane, and (c) ethanol.

● **8.6.2** If dry ice is heated under a pressure greater than 10 bar, it melts instead of subliming. If the pressure is then reduced to 1 bar, the liquid boils to give gaseous $CO_2$. Show these processes on a phase diagram for $CO_2$.

● **8.6.3** Refer to the phase diagrams in this chapter to answer the following: (a) What is the maximum pressure at which solid $N_2$ can sublime? (b) What happens if the pressure above ice at −10°C is reduced to 0.5 bar, to 0.05 bar, and finally to 0.005 bar? (c) What occurs if dry ice (solid $CO_2$) is heated from 180 K to room temperature in a container held at constant pressure of 7 bar?

# CHAPTER 8 VISUAL SUMMARY

## 8.1 EFFECTS OF INTERMOLECULAR FORCES

Intermolecular force of attraction

Kinetic energy

**LEARNING OBJECTIVES:** Understand the effects of intermolecular forces on condensation, vapourization and melting and boiling points.

**KEY CONCEPTS:** Attractive intermolecular forces are responsible for the condensation of gases, and the freezing of liquids.

## 8.2 TYPES OF INTERMOLECULAR FORCES

Dispersion forces

Dipolar forces

Hydrogen bonding

**LEARNING OBJECTIVES:** Predict the relative magnitudes of intermolecular forces and their effects on physical properties of substances.

Explain the origins of each type of intermolecular force.

**SKILLS TO MASTER:** Recognizing the dominant intermolecular force between two species (p. 429).

Predicting trends in boiling points based on molecular structures (p. 433).

**KEY CONCEPTS:** All intermolecular forces are electrostatic, and under most conditions are attractive.

Intermolecular forces exist among all ions, dipoles and uncharged non-polar species.

A dipole can be induced into a non-polar species by an ion or by a polar species.

Hydrogen bonds exist among species having a polar bond between H and O, N or F.

## 8.3 LIQUIDS

Vapour pressure

Capillary action
Meniscus
Viscosity
Surface tension

Equilibrium: $p = vp$

**LEARNING OBJECTIVES:** Explain trends in surface tension, capillary action, viscosity and vapour pressure in terms of intermolecular forces.

**KEY CONCEPTS:** Substances with higher intermolecular forces have higher surface tensions, higher viscosities and lower vapour pressures.

## 8.4 FORCES IN SOLIDS

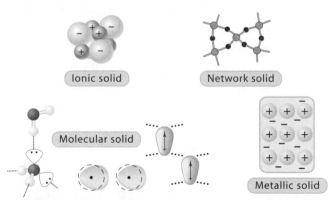

Ionic solid

Network solid

Molecular solid

Metallic solid

**LEARNING OBJECTIVES:** Explain the properties of solids in terms of the dominant intermolecular forces present.

**SKILLS TO MASTER:** Recognizing the dominant forces present in solids and their magnitudes (p. 442).

**KEY CONCEPTS:** Forces in solids may include dispersion, dipolar, H-bonding, metallic, covalent and ionic.

## 8.5 ORDER IN SOLIDS

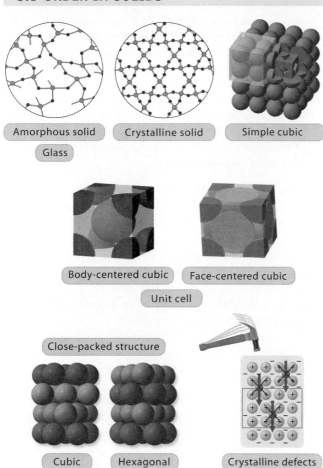

Amorphous solid

Glass

Crystalline solid

Simple cubic

Body-centered cubic

Face-centered cubic

Unit cell

Close-packed structure

Cubic

Hexagonal

Crystalline defects

**LEARNING OBJECTIVES:** Understand amorphous and crystalline solids at the molecular level.

**SKILLS TO MASTER:** Calculating the mass of a single unit cell (p. 453).

Calculating the density of a crystalline solid from unit cell data (p. 453).

Calculating the radius of an atom from unit cell data and the density of the solid (p. 454).

**KEY CONCEPTS:** Solid materials may be crystalline or amorphous.

The unit cell is the building block of crystalline solids.

One way to discuss ionic structures is to identify a crystal lattice for one set of ions, and then describe how the other ions pack within the lattice of the first set.

Crystalline defects can profoundly alter the properties of a solid.

## 8.6 PHASE CHANGES

Phase diagram

Triple point   Critical point

$\Delta H_{sub}$   Sublimation

$\Delta H_{vap}$

$\Delta H_{fus}$

Molar heat of fusion, $\Delta H_{fus}$

Molar heat of vapourization, $\Delta H_{vap}$

Solid   Liquid

$CO_2$   $tp$

Pressure, $p$

1.01 bar

$sp$

Gas   298 K

Temperature, $T$

**LEARNING OBJECTIVES:** Explain enthalpies of phase changes in terms of intermolecular forces.

Interpret a pressure-temperature phase diagram of a pure substance.

**SKILLS TO MASTER:** Calculate the total heat absorbed or evolved during a phase change (p. 464).

State which phases are present at any point on a pressure-temperature phase diagram (p. 467).

Predict what happens step-by-step to a substance as it is heated isobarically or pressurized isothermally (p. 468).

Construct a pressure-temperature phase diagram from physical (p. 469)

**KEY CONCEPTS:** A phase diagram shows which phases are the most stable at any conditions of temperature and pressure.

Two phases are in equilibrium on a line on a phase diagram and three phases coexist at the triple point.

## LEARNING EXERCISES

8.1 List all the types of interactions that can act to hold a solid together. Organize the list from strongest to weakest.

8.2 Draw molecular pictures that show every type of hydrogen bond that exists in a solution containing methanol, water, and ammonia.

8.3 Write a paragraph that describes the phenomenon of superconductivity.

8.4 Define and give an example of each of the following: (a) close-packed structure; (b) unit cell; (c) molecular solid; (d) covalent solid; (e) amorphous solid; and (f) lattice.

8.5 Write a paragraph describing what happens at the molecular level during each of the six phase changes shown in Figure 8-35.

8.6 Prepare a list of the terms in Chapter 8 that are new to you. Write a one-sentence definition for each, using your own words. If you need help, consult the glossary.

## PROBLEMS

**ilw** = interactive learning ware problem. Visit the website at www.wiley.com/canada/olmsted

### Effects of Intermolecular Forces

8.1 Xenon condenses at 166 K, krypton condenses at 121 K, and argon condenses at 87 K. Draw a bar graph similar to Figure 8-3 showing the relative magnitudes of interatomic attractive energies for these three noble gases. Include two horizontal lines representing the average kinetic energy at 140 and 100 K.

8.2 Methane condenses at 121 K, but carbon tetrachloride boils at 350 K. Draw a bar graph similar to that in Figure 8-3 showing the relative attractive energies between molecules for these two substances. Include a horizontal line representing the average kinetic energy at room temperature (298 K).

8.3 Predict whether the effects of molecular volume and of intermolecular attractions become more or less significant when the following changes are imposed: (a) A gas is expanded to a larger volume at constant temperature. (b) More gas is introduced into a container of constant volume at constant temperature. (c) The temperature of a gas increases at constant pressure.

8.4 Predict whether the effects of molecular volume and of intermolecular attractions become more or less significant when the following changes are imposed: (a) The temperature of a gas is lowered at constant volume. (b) A gas is compressed into a smaller volume at constant temperature. (c) Half the gas is removed from a sample at constant volume.

8.5 Draw pictures showing the atomic arrangements of samples of Ag $(s)$, Ar $(g)$, and Hg $(l)$.

8.6 Draw pictures showing the atomic arrangements of samples of Zn $(s)$, Kr $(g)$, and Ga $(l)$.

8.7 **ilw** From the following experimental data, calculate the percent deviation from ideal behaviour: 1.00 mol $CO_2$ in a 1.20-L container at 40.0 °C exerts 19.7 bar pressure.

8.8 From the following experimental data, calculate the percent deviation from ideal behaviour: 3.000 g $H_2$ at 0.00 °C and 193.5 bar occupies a volume of 189.18 cm$^3$.

8.9 Chlorine gas is commercially produced by electrolysis of seawater and then stored under pressure in metal tanks. A typical tank has a volume of 15.0 L and contains 1.25 kg of $Cl_2$. Use the van der Waals equation (see section 2.7) to calculate the pressure in this tank if the temperature is 295 K, and compare the result with the ideal gas value.

8.10 Chlorine gas used in chemical syntheses is transferred from tanks under high pressure to reactor flasks at considerably lower pressure. If 10.5 g of $Cl_2$ is introduced into a 5.00-L reaction flask at 145 °C, calculate the pressure of $Cl_2$ gas using the van der Waals equation (see section 2.7) and using the ideal gas equation.

### Types of Intermolecular Forces

8.11 Arrange the following in order of ease of liquefaction: $CCl_4$, $CH_4$, and $CF_4$. Explain your ranking.

8.12 Arrange the following in order of increasing boiling point: Ar, He, Ne, and Xe. Explain your ranking.

8.13 Draw molecular pictures of $CCl_4$ and $CH_4$ showing the relative amount of polarizability of each.

8.14 Draw molecular pictures of Ne and Xe showing the relative amount of polarizability of each.

8.15 List ethanol ($CH_3CH_2OH$), propane ($CH_3CH_2CH_3$), and n-pentane ($CH_3CH_2CH_2CH_2CH_3$) in order of increasing boiling point, and explain what features determine this order.

8.16 List n-propanol ($CH_3CH_2CH_2OH$), dimethyl ether ($CH_3OCH_3$), and diethyl ether ($CH_3CH_2OCH_2CH_3$) in order of increasing boiling point and explain what features determine this order.

8.17 Which of the following will form hydrogen bonds with another molecule of the same substance? Draw molecular pictures illustrating these hydrogen bonds. (a) $CH_2Cl_2$; (b) $H_2SO_4$; (c) $H_3COCH_3$; and (d) $H_2NCH_2CO_2H$.

8.18 Which of the following form hydrogen bonds with water? Draw molecular pictures illustrating these hydrogen bonds. (a) $CH_4$; (b) $I_2$; (c) HF; (d) $H_3COCH_3$; and (e) $(CH_3)_3COH$.

8.19 Draw Lewis structures showing all possible hydrogen-bonding interactions for (a) two $NH_3$ molecules; and (b) one $NH_3$ molecule and one $H_2O$ molecule.

8.20 Draw Lewis structures that show the hydrogen-bonding interactions for (a) two $CH_3OH$ molecules; and (b) one HF molecule and one acetone molecule [$(CH_3)_2C = O$].

### Liquids

8.21 Pentane is a $C_5$ hydrocarbon, gasoline contains mostly $C_8$ hydrocarbons, and fuel oil contains hydrocarbons in the $C_{12}$ range. List these three hydrocarbons in order of increasing viscosity, and explain what molecular feature accounts for the variation.

8.22 Given that a lubricant must flow easily to perform its function, which grade of motor oil is preferred for winter use: high- or low-viscosity? Why?

8.23 Aluminum tubing has a thin surface layer of aluminum oxide. What shape meniscus would you expect to find for water and for mercury inside aluminum tubing? Explain your answers in terms of intermolecular forces.

8.24 Platinum tubing has a clean metal surface. What shape meniscus would you expect to find for water and for mercury inside platinum tubing? Explain your answers in terms of intermolecular forces.

8.25 Paper towels are better at absorbing spilled wine than at absorbing spilled salad oil. Why is this?

8.26 Water forms "beads" on the surface of a freshly waxed automobile, but it forms a film on the clean windshield. Why is this?

8.27 For each of the following pairs of liquids, choose which has the higher vapour pressure at room temperature and state the reason why: (a) benzene ($C_6H_6$) or chlorobenzene ($C_6H_5Cl$); (b) hexane ($C_6H_{14}$) or 1-hexanol ($C_6H_{13}OH$); and (c) octane ($C_8H_{18}$) or heptane ($C_7H_{16}$).

8.28 For each of the following pairs of liquids, choose which has the lower vapour pressure at room temperature and state the reason why: (a) water ($H_2O$) or methanol ($CH_3OH$); (b) 1-pentanol ($C_5H_{12}OH$) or 1-hexanol ($C_6H_{13}OH$); and (c) methyl chloride ($CH_3Cl$) or chloroform ($CHCl_3$).

### Forces in Solids

8.29 Classify each of the following as an ionic, network, molecular, or metallic solid: Sn, $S_8$, Se, $SiO_2$, and $Na_2SO_4$.

8.30 Classify each of the following as an ionic, network, molecular, or metallic solid: Pt, $P_4$, Ge, $As_2O_3$, and $(NH_4)_3PO_4$.

8.31 Describe the differences in (a) bonding characteristics and (b) macroscopic properties between metals and network solids.

8.32 Describe the differences in (a) interparticle forces and (b) macroscopic properties between molecular and ionic solids.

8.33 Indicate what type of solid (ionic, network, metallic, or molecular/atomic) each of the following forms on solidification: (a) $Br_2$; (b) KBr; (c) Ba; (d) $SiO_2$; and (e) $CO_2$.

8.34 Indicate what type of solid (ionic, covalent, metallic, or molecular/atomic) each of the following forms on solidification: (a) HCl; (b) KCl; (c) $NH_4NO_3$; (d) Mn; and (e) Si.

### Order in Solids

8.35 The density of graphite is 2260 kg/m³, whereas that of diamond is 3513 kg/m³. Describe the bonding features that cause these two forms of carbon to have different densities.

8.36 Amorphous silica has a density of around 2.3 g/cm³, and crystalline quartz has a density of 2.65 g/cm³. Describe the bonding features that cause these two forms of the same substance to have different densities.

8.37 **ilw** The unit cell of the mineral perovskite follows. What is the formula of perovskite?

= Titanium (Ti)
= Calcium (Ca)
= Oxygen (O)

8.38 The unit cell of a compound of sodium (purple) and oxygen (red) appears at right. Determine the chemical formula of the compound.

8.39 Construct part of the Lewis structure of carborundum, the diamond-like compound of empirical formula SiC.

8.40 Construct part of the Lewis structure of quartz, the network solid of empirical formula $SiO_2$.

8.41 For the pattern shown below, draw a tile that represents a unit cell for the pattern and contains only complete fish. Draw a second tile that is a unit cell but contains no complete fish.

8.42 For the pattern shown at right, draw a tile that represents a unit cell for the pattern, but does not contain any complete stars. Draw a different tile that contains only complete stars.

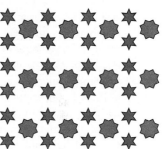

8.43 In the unit cell of lithium sulphide, the sulphide ions adopt a face-centred cubic lattice. The lithium ions are in the holes within the face-centred cube. If the unit cell is charge-neutral, how many complete lithium ions must be packed within each face-centred cube?

8.44 In the unit cell of calcium oxide, the oxide ions adopt a face-centred cubic lattice. The calcium ions are in the holes within the face-centred cube. If the unit cell is charge-neutral, how many complete calcium ions must be packed within each face-centred cube?

### Phase Changes

8.45 From Table 8-5, determine which of each pair has the higher heat value and use intermolecular forces to explain why: (a) heats of vapourization of methane and ethane; (b) heats of vapourization of ethanol and diethyl ether; and (c) heats of fusion of argon and methane.

8.46 From Table 8-5, determine which of each pair has the higher heat value and use intermolecular forces to explain why: (a) heats of vapourization of water and methane; (b) heats of fusion of benzene and ethane; and (c) heats of vapourization of molecular oxygen and argon.

8.47 Sketch the approximate phase diagram for $Br_2$ from the following information: normal melting point is 265.9 K, normal boiling point is 331.9 K, triple point at $p = 5.87 \times 10^{-2}$ bar, and $T = 265.7$ K. Label the axes and the area where each phase is stable.

8.48 Oxygen has a normal melting point of 55 K and a normal boiling point of 90.2 K, and its triple point occurs at $p = 0.0015$ bar and $T = 54$ K. Sketch the approximate phase diagram for oxygen, and label the axes and the area where each phase is stable.

8.49 Using the phase diagram of Problem 8.47, describe what happens to a sample of $Br_2$ as the following processes take place. Draw lines on the phase diagram showing each process. (a) A sample at $T = 400$ K is cooled to 250 K at constant $p = 1.00$ bar. (b) A sample is compressed at constant $T = 265.8$ K from $p = 1.00 \times 10^{-3}$ bar to $p = 1.00 \times 10^3$ bar. (c) A sample is heated from 250 to 400 K at constant $p = 2.00 \times 10^{-2}$ bar.

8.50 Using the phase diagram of Problem 8.48, describe what happens to a sample of $O_2$ as the following processes take place. Draw lines on the phase diagram showing each process. (a) A sample at $T = 125$ K is cooled to 25 K at constant $p = 1.00$ bar. (b) A sample is compressed at constant $T = 54.5$ K, starting at $p = 1.00 \times 10^{-2}$ bar, to $p = 1.00 \times 10^2$ bar. (c) A sample is heated from 25 to 100 K at constant $p = 1.00 \times 10^{-3}$ bar.

8.51 **ilw** The heat capacity of ethanol is 111 J °C$^{-1}$ mol$^{-1}$. Combine this information with the data in Table 8-5 to calculate the amount of heat required to convert 1.00 g of ethanol at 25 °C to ethanol vapour.

8.52 White phosphorus is a molecular solid made of molecules of $P_4$. A 1.00-g sample of white phosphorus starts at 25 °C. Calculate the amount of heat required to melt the sample. Use the following data: $T_m = 44$ °C; $C = 94$ J °C$^{-1}$ mol$^{-1}$); $\Delta H_{fusion} = 2.5$ kJ/mol.

## ADDITIONAL PAIRED PROBLEMS

8.53 List the different kinds of forces that must be overcome to convert each of the following from a liquid to a gas: (a) $NH_3$; (b) $CHCl_3$; (c) $CCl_4$; and (d) $CO_2$.

8.54 List all the intermolecular forces that stabilize the liquid phase of each of the following: (a) Xe; (b) $SF_4$; (c) $CF_4$; and (d) $CH_3CO_2H$ (acetic acid).

8.55 Manganese forms crystals in either body-centred cubic or face-centred cubic geometry. Which phase has the higher density? Explain your choice.

8.56 Beryllium forms crystals in either body-centred cubic or hexagonal close-packed geometry. Which phase has the higher density? Explain your choice.

8.57 For each of the following pairs, identify which has the higher boiling point, and identify the type of force that is responsible: (a) $H_3COCH_3$ and $CH_3OH$; (b) $SO_2$ and $SiO_2$; (c) HF and HCl; and (d) $Br_2$ and $I_2$.

8.58 The boiling points of the Group 16 binary hydrides are as follows: $H_2O$, 100 °C; $H_2S$, −60 °C; $H_2Se$, −41 °C. Explain in terms of intermolecular forces why $H_2S$ has a lower boiling point than either $H_2O$ or $H_2Se$.

8.59 Refer to Figure 8-37. Describe in detail what occurs when each of the following is carried out: (a) The pressure on a sample of liquid $N_2$ is reduced from 1.00 to 0.010 bar at a constant temperature of 70 K. (b) A sample of $N_2$ gas is compressed from 1.00 to 50.0 bar at a constant temperature of 298 K. (c) A sample of $N_2$ gas is cooled at a constant pressure of 1.00 bar from 298 to 50 K.

8.60 Refer to Figure 8-37. Describe in detail what occurs when each of the following is carried out: (a) A sample of $CO_2$ gas is compressed from 1.00 to 50.0 bar at $T = 298$ K. (b) Dry ice at 195 K is heated to 350 K at $p = 6.00$ bar. (c) A sample of $CO_2$ gas at $p = 1.00$ bar is cooled from 298 to 50 K.

8.61 The compound 1,2-dichloroethylene (ClCHCHCl) exists as *cis* and *trans* isomers. One isomer boils at 47 °C, the other at 60 °C. Draw Lewis structures of the two isomers. Use dipole moments and symmetry arguments to assign the boiling points to the isomers.

8.62 Why do the two isomers shown have very different melting points?

$C_8H_8O_3$
Methyl 2-hydroxybenzoate
(Oil of wintergreen)
mp = −8 °C

$C_8H_8O_3$
Methyl 4-hydroxybenzoate
mp = 127 °C

8.63 Which gas deviates more from ideal $pV/nRT$ behaviour: $F_2$ or $Cl_2$? Explain your choice.

8.64 Which gas deviates more from ideal $pV/nRT$ behaviour: $CH_4$ or $SnH_4$? Explain your choice.

8.65 Refer to Figure 8-38, and describe what happens to silica as it is slowly heated from room temperature to 2000 °C at atmospheric pressure.

8.66 Refer to Figure 8-38. How many triple points appear on this phase diagram? For each one, describe the conditions and state the three phases that coexist under these conditions.

8.67 Solid sodium metal can exist in two different crystalline forms: body-centred cubic and hexagonal. One is stable only below 5 K at relatively low pressure; the other is stable under all other conditions. Draw atomic pictures of both phases, identify the phase that is stable at low temperature and pressure, and explain your choice.

8.68 When neon solidifies, it can form either face-centred cubic or hexagonal close-packed crystals of virtually identical stability. Draw atomic pictures of both phases and explain why they are equally stable.

8.69 What type of solid is each of the following? Give reasons for your assignments. (a) $ZrO_2$ (mp 2677 °C); (b) Zr; and (c) $ZrSiO_4$.

8.70 What type of solid is each of the following? Give reasons for your assignments. (a) $SO_2$; (b) $ZnSO_4$; and (c) Zn.

8.71 The figure below shows the phase diagram for elemental sulphur. Use it to answer the following questions:

(a) Under what conditions does rhombic sulphur melt? (b) Under what conditions does rhombic sulphur convert to monoclinic sulphur? (c) Under what conditions does rhombic sulphur sublime?

8.72 Use the figure in Problem 8.71 to answer the following questions: (a) How many triple points are there in the diagram? (b) What phases coexist at each triple point? (c) If sulphur is at the triple point at 153 °C, what happens if the pressure is reduced?

8.73 The unit cell of rhenium oxide appears in the figure below. (a) Name the lattice occupied by the rhenium atoms. (b) Calculate the number of rhenium atoms and the number of oxygen atoms in the unit cell and identify the formula of rhenium oxide.

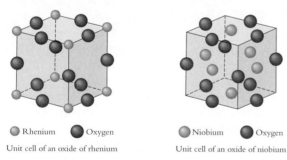

Unit cell of an oxide of rhenium        Unit cell of an oxide of niobium

8.74 One form of niobium oxide crystallizes in the unit cell shown above. Calculate the number of niobium and oxygen atoms in the unit cell and identify the formula of niobium oxide.

8.75 Mercury is a liquid at room temperature, and both cesium and gallium melt at body temperature. How do interatomic forces in these three metals compare with those in more typical metals such as copper or iron?

8.76 Tungsten does not melt even when heated to incandescence. How do interatomic forces in tungsten compare with those in more typical metals such as gold or nickel?

## ADDITIONAL PROBLEMS

8.77 Use the phase diagram of Problem 8.71 to answer the following questions: (a) Are there constant pressures at which each of the four phases becomes stable as the temperature is raised? If so, identify those pressures and describe what happens as the temperature increases. (b) Are there constant temperatures at which each of the

four phases becomes stable as the pressure is raised? If so, identify those temperatures and describe what happens as the pressure increases.

8.78 Does the boiling point of HCl (see the graph in Figure 8-12) suggest that it may form hydrogen bonds? Explain your answer and draw a molecular picture that shows the possible hydrogen bonds between HCl molecules.

8.79 Molecular hydrogen and atomic helium have two electrons, but He boils at 4.2 K, whereas $H_2$ boils at 20 K. Neon boils at 27.1 K, whereas methane, which has the same number of electrons, boils at 114 K. Explain why molecular substances boil at a higher temperature than atomic substances with the same number of electrons.

8.80 Quartz and glass are both forms of silicon dioxide. A piece of quartz breaks into a collection of smaller regular crystals with smooth faces. A piece of glass breaks into irregular shards. Use molecular structures to explain why the two solids break so differently.

8.81 Draw the unit cell of the NaCl crystal and determine the number of nearest neighbours of opposite charge for each ion in this unit cell.

8.82 In the two figures shown below, the green molecule is about to strike the wall of its container. Assuming all other conditions are identical, which collision will exert greater pressure on the wall? Explain in terms of intermolecular interactions.

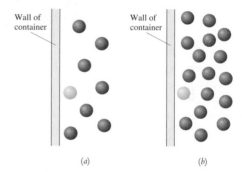

(a)                (b)

8.83 Aluminum silicate, $Al_2SiO_5$, is a prevalent mineral in the Earth's crust. The figure below shows a portion of the phase diagram for this mineral, which has three different crystalline forms. Use the phase diagram to answer the following questions: (a) What can you say about the melting point of $Al_2SiO_5$? (b) When liquid $Al_2SiO_5$ solidifies at atmospheric pressure, what crystalline form results? (c) Might there be a pressure above which the liquid solidifies to give a different crystalline form? If so, identify the form and estimate the minimum pressure needed. (d) Among the three forms of solid, which is the densest, and which is the least dense? State your reasoning.

**8.84** Arrange the following liquids in order of increasing viscosity, and state the factors that determine the ranking: *n*-butanol, $CH_3CH_2CH_2CH_2OH$; *n*-pentane, $CH_3CH_2CH_2CH_2CH_3$; propane-1,3-diol, $HOCH_2CH_2CH_2OH$; and 2,2-dimethylpropane, $(CH_3)_4C$.

**8.85** Construct a qualitative graph similar to the one in Figure 8-34 that summarizes the energy changes that accompany the following process. A sample of water at 35 °C is placed in a freezer until the temperature reaches 25 °C. Plot temperature along the *y*-axis and kilojoules of heat removed along the *x*-axis.

**8.86** Recently, a new group of solids was prepared that can act as superconductors at temperatures near the boiling point of liquid nitrogen. The unit cell of one of these new superconductors follows. Identify the formula of the compound.

● Barium  ● Yttrium  ● Oxygen  ● Copper

**8.87** As a vapour, acetic acid ($CH_3CO_2H$) exists as a mixture of individual molecules and pairs of molecules held together by hydrogen bonds. (a) Draw a Lewis structure that shows how two acetic molecules can form a pair by hydrogen bonding. (b) Does the fraction of paired acetic acid molecules increase or decrease as the temperature rises? Explain.

## GROUP STUDY PROBLEMS

**8.88** Classify each of the following solids as network, metallic, ionic, or molecular: (a) a solid that conducts electricity; (b) a solid that does not conduct electricity but dissolves in water to give a conducting solution; and (c) a solid that does not conduct electricity and melts below 100 °C to give a non-conducting liquid.

**8.89** Describe the similarities and differences between hexagonal close-packed and body-centred cubic structures.

**8.90** The packing efficiency of spheres in a crystal is the percentage of the space occupied by the spheres. Calculate the packing efficiencies of the simple cubic, body-centred cubic, and face-centred cubic cells. [*Hints:* The volume of a sphere is $4\pi r^3/3$. Use the number of spheres contained in each unit cell and determine the lengths of the sides of the unit cells from the spheres that are in contact with each other (edge of cube, body diagonal of body-centred, face diagonal of face-centred)].

**8.91** How many grams of methane must be burned to provide the heat needed to melt a 75.0-g piece of ice and bring the resulting water to a boil? Assume a constant pressure of 1 bar. Use the data in Tables 3-1 and 8-5 to develop your solution.

**8.92** Solid silver adopts a face-centred cubic lattice. The metallic radius of a silver atom is 144 pm. (a) How many silver atoms occupy one unit cell of solid silver? (b) Calculate the length of one side of a unit cell. (c) Calculate the volume of the unit cell. (d) What percentage of the volume of the unit cell is empty? Use the diagram at right to develop your answers.

**8.93** A chemist has stored pure bromine in a freezer at 265.9 K, the melting point of $Br_2$. A sample with a mass of 50.0 g is removed from the freezer and placed in a distillation flask, where it is heated until all the bromine has vapourized. The heater provides a constant flow of heat at 25.0 J/s. Prepare a heating curve for this process, showing temperature vs. time. Draw it as quantitatively as you can, using data from Table 8-5 and the following heat capacity information: $C_{liquid} = 76\,J\,°C^{-1}\,mol^{-1}$; $C_{gas} = 20.8\,J\,°C^{-1}\,mol^{-1}$.

## ANSWERS TO SECTION EXERCISES

**8.1.1** $At_2$ is a solid, with stronger intermolecular forces than $I_2$. Continuing the trends shown in Figure 8-3, the bar for $At_2$ should be higher than that for $I_2$.

**8.1.2** Intermolecular forces cause $pV/nRT$ to be smaller than 1.0. Fluorine reaches the smallest value, so it has the largest intermolecular forces. Helium never dips below 1.0, indicating very small intermolecular forces.

**8.1.3** (a) Xe boils at a higher temperature than Ar. (b) The order is $H_2$, $N_2$, $F_2$, $Cl_2$. (c) Group 18 has the smallest intermolecular forces.

**8.2.1** (a) Propane has larger polarizability because its electrons are spread over a larger volume than those in krypton. (b) $CCl_4$ has a larger polarizability because it has more electrons. (c) CO has a dipole moment, generating dipole–dipole attractions not present in $N_2$.

**8.2.2** Acetone has dispersion forces and dipole–dipole forces. Methanol has dispersion forces, dipole–dipole forces, and hydrogen-bonding forces. Their boiling points are similar because the dipolar and dispersion forces in acetone are larger than the analogous forces in methanol, counterbalancing the hydrogen bonding in methanol.

**8.2.3**

O—H---N     :Ö—H---:N⟨H/H⟩     O—H---F     :Ö—H---:F̈—H
                        \H

F—H---F     H—F̈:---H—F̈:     F—H---N     :F̈—H---:N⟨H/H⟩
                                                          \H

F—H---O     :F̈—H---:Ö⟨H⟩     N—H---F     :N⟨H/H⟩---:F̈—H
                      \H

N—H---N     :N⟨H/H⟩---:N⟨H/H⟩     N—H---O     :N⟨H/H⟩---:Ö⟨H⟩
                                                              \H

**8.3.1** Each *n*-pentanol molecule, with its highly polar O—H bond, forms hydrogen bonds with neighbouring *n*-pentanol molecules. These hydrogen bonds must break for *n*-pentanol molecules to move past one another, causing this liquid to have a higher viscosity than that of *n*-hexane.

**8.3.2** A meniscus forms because of differences between the forces holding molecules in the liquid and those attracting them to the walls of the container. For water in glass, water–wall forces are greater than water–water forces, so water is attracted to the wall and the meniscus is concave. The convex meniscus for mercury in glass signals that mercury–mercury forces are greater than mercury–wall forces, so the wall repels mercury.

**8.3.3** Figure 8-16 shows that fewer water molecules than bromine molecules have enough energy to escape the liquid at 298 K, so your figure should show fewer molecules in the gas phase and fewer arrows:

Equilibrium: $p = p_{vap}$

**8.4.1** (a) $CO_2$ is a molecular solid, with dispersion forces between the molecules. (b) $S_8$ is another molecular solid, with dispersion forces between the molecules. (c) Tin is a metal, with delocalized metallic bonding holding the atoms together. (d) $Li_2O$ is ionic, with electrical attractions between $Li^+$ cations and $O^{2-}$ anions.

**8.4.2** The order of melting points is $F_2$, K, Co, C (diamond). $F_2$ is molecular, with dispersion forces only. K and Co are metals, but Co has more valence electrons, resulting in stronger bonding. Diamond is a network solid with strong covalent bonds.

**8.4.3** (a) $Mg_3(PO_4)_2$; (b) $LiAlSi_2O_6$; and (c) $Bi_2Sr_2CuO_6$

**8.5.1** All angles in a hexagonal crystal are multiples of 60°, so we expect angles of 60°, 120°, and so on. A cubic close-packed crystal also has right angles, so we expect 90° angles as well.

**8.5.2** The view is not a unit cell, because if two of these views are placed side by side, the ion from one view occupies a vacancy from the other view.

**8.5.3**

**8.5.4** Radius of W atom = 137 pm.

**8.6.1** (a) Methane, 8.6 kJ; (b) ethane, 7.8 kJ; and (c) ethanol, 13 kJ

**8.6.2**

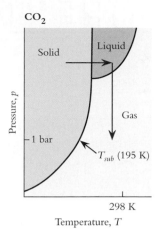

$CO_2$

**8.6.3** (a) 0.124 bar; (b) ice remains solid until the pressure drops to about 0.005 bar, when it sublimes; and (c) at 7 bar, dry ice melts at around 217 K and is liquid at room temperature.

## ANSWERS TO PRACTICE EXERCISES

**8.1** $H_2Te$, with the largest inner atom, has largest dispersion forces and highest boiling point. $H_2S$, with the smallest inner atom, has smallest dispersion forces and lowest boiling point.

**8.2** Acetaldehyde is more compact than acetone and has fewer electrons, so it boils at a lower temperature. Like acetone, acetaldehyde has a polar $C = O$ bond, giving it a higher boiling point than either butane or methyl ethyl ether.

**8.3** The O atom in an acetone molecule can form two H bonds to two different water molecules as shown at right.

**8.4** Selenium is a molecular solid; germanium is a network solid.

**8.5** (a) $m = 7.17 \times 10^{-22}$ g; (b) $m = 7.64 \times 10^{-23}$ g

**8.6** $\rho = 7.83$ g/cm$^3$

**8.7** $r = 125$ pm

**8.8** 41.7 kJ flows from the tray to the freezer compartment.

**8.9**

**8.10** The gas liquefies at $p = 0.126$ bar, and the liquid freezes when the pressure reaches about 0.5 bar. At 1 bar, 63.1 K, $N_2$ is a solid.

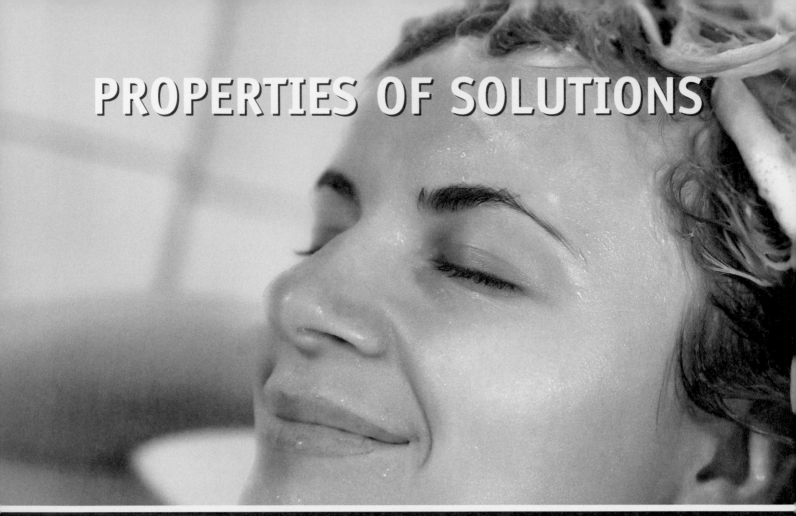

# PROPERTIES OF SOLUTIONS

## THE CHEMISTRY OF SHAMPOO

Products for personal care are big business: Canadians spend close to $1 billion annually on shampoos, conditioners, and styling gels. These products rely on intermolecular forces for their effectiveness. Shampoos, in particular, interact with water in ways that we can understand by knowing the properties of solutions.

The active ingredients in a shampoo play three fundamental roles. Some allow water to wash away the substances that make hair "dirty." Others adhere to hair to impart a desirable feel and texture. The rest are emulsifiers that keep the mixture from separating into its components. To accomplish these effects, ingredients combine two types of interactions: a strong attraction to water (hydrophilic) and an aversion to water (hydrophobic). It may seem that these properties are incompatible, but shampoos contain molecules that are designed to be simultaneously hydrophilic and hydrophobic. One example is sodium lauryl sulphate, our inset molecule. The ionic "head" of the molecule is hydrophilic, so it interacts attractively with water. The hydrocarbon "tail" is hydrophobic, so it interacts attractively with grease and dirt. Molecules of the shampoo associate with hydrophobic dirt particles to form hydrophilic clumps that dissolve in water and wash away.

Washing removes the natural coating of protective oil that gives hair its body and shine. To counter this, shampoos and conditioners contain hydrophobic oils that cling to the surface of hair and remain in place upon rinsing. Dimethicone, shown in our inset, is an artificial oil that contains hydrophobic silicon–oxygen chains and methyl groups.

To keep their various ingredients mixed together uniformly, shampoos also contain emulsifiers with structures that include hydrophobic and hydrophilic components. Our inset shows the structure of panthenol, one such emulsifier. Panthenol has hydrophobic hydrocarbon regions interspersed with hydrophilic polar groups such as N — H and O — H.

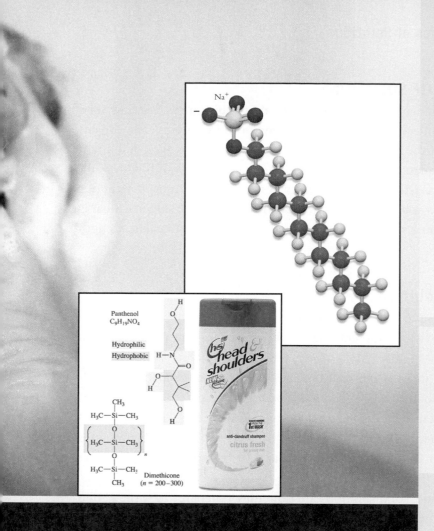

# 9

## LEARNING OBJECTIVES

*Upon completion of this chapter you should be able to:*

- work with and interconvert concentration units (mass %, molarity, molality, mole fraction, ppm and ppb).
- predict the relative solubilities of a solute in various solvents, and explain in terms of intermolecular forces.
- calculate and explain solubilities in water.
- understand the reason for and calculate the magnitudes of colligative properties.
- understand some aspects of colloidal suspensions and surfactant solutions.

In addition to cleaning and conditioning hair, a shampoo must be non-toxic, biodegradable, and pleasant-smelling. Therefore, chemists who formulate hair care products face challenging chemical design problems. That is why a typical list of ingredients contains a multitude of compounds.

This chapter explores solutions. Its main emphasis is on aqueous solutions, because water is the most important solvent for life. We first describe the nature of solutions. Then we examine the molecular features that determine whether one substance dissolves in another. We introduce some characteristic properties of solutions, including effects on phase behaviour. The chapter concludes by looking more closely at dual-nature molecules such as those present in shampoos.

## 9.1 THE NATURE OF SOLUTIONS

A solution is a homogeneous mixture of two or more substances. As described in Chapter 1, a solution contains a solvent and one or more solutes. The solvent determines the state of the solution, and normally the solvent is the component present in the greatest quantity. The most common solutions are liquids with water as solvent, but solutions exist in all three states of matter. The atmosphere of our planet, air, is a gaseous solution with molecular nitrogen as the solvent. Steel is a solid solution containing solutes such as chromium and carbon that add strength to the solvent, iron.

## Components of Solutions

The solvent determines the phase of a solution, but the solutes may be substances that would normally exist in different phases. Table 9-1 lists an example of each possibility.

Gaseous solutions are easy to prepare and easy to describe. The atoms or molecules of a gas move about freely. When additional gases are added to a gaseous solvent, each component behaves independently of the others. Unless a chemical reaction occurs, the ideal gas equation and Dalton's law of partial pressures describe the behaviour of gaseous solutions at and below atmospheric pressure (see Chapter 2).

At the other extreme, solid solutions are difficult to prepare. To form a solid solution, individual atoms or molecules must mix homogeneously within the solid network. Given that the atoms or molecules in a solid cannot move freely about, such mixing is difficult to achieve. Even if we grind two solids to fine powders and then mix them thoroughly, the resulting mixture will not be a solution, because the mixture will remain heterogeneous at the microscopic and molecular levels.

Solid solutions generally form in one of two ways, both of which involve forming the solid from the liquid phase. One way is to heat the solid solvent until it melts, add the solutes into the molten material, and then cool the melt until it solidifies. Solid solutions of one metal in another, such as brass and steel, are prepared in this way. A second method is to dissolve the solid solvent and solutes in an appropriate liquid, and then cool or evaporate the liquid until a solid precipitates. Solid solutions of organic substances can form in this manner.

## TABLE 9-1 PHASES OF SOLUTIONS

| Solvent Type | Solute Type | | |
|---|---|---|---|
| | **Gas** | **Liquid** | **Solid** |
| Gas | Diving gas (He, $O_2$) | Humid air ($N_2$, $O_2$, $H_2O$) | Air above $I_2$ ($N_2$, $O_2$, $I_2$) |
| Liquid | Carbonated water ($H_2O$, $CO_2$) | Vodka ($H_2O$, $C_2H_5OH$) | Saline solution ($H_2O$, NaCl) |
| Solid | $H_2$ storage alloy (La, Ni, $H_2$) | Plastic (PVC, dioctylphthalate) | Steel (Fe, C, Mn) |

Usually we think of solutions as liquids, and aqueous solutions are the most common liquid solutions in everyday life. Blood, sweat, and tears all are aqueous solutions, as are the beverages that we drink. The oceans are aqueous solutions of sodium chloride and other mineral salts. Even the fresh waters of streams and lakes are aqueous solutions containing salts and organic solutes. Most of this chapter discusses aqueous solutions.

Ocean water is an aqueous solution containing many solutes.

## Solution Concentration

The composition of a solution can vary, so we must specify the concentrations of solutes as well as their identities. There are several ways to express concentration, each having advantages as well as limitations. Any concentration value is a ratio of amounts. The amount of one component, usually a solute, appears in the numerator, and some other amount, describing either the solvent or the total solution, appears in the denominator.

In the laboratory, chemists often make solutions by weighing the solute. Thus, mass percent is a common measure of solution concentration:

$$\text{Mass percent} = \frac{\text{Mass solute}}{\text{Total mass of solution}} \times 100\%$$

For example, dissolving 12.5 g of NaCl in 100 g of water gives a solution that is 11.1 mass percent:

$$\text{Mass \% NaCl} = \frac{12.5 \text{ g}}{12.5 \text{ g} + 100.0 \text{ g}} \times 100\% = 11.1 \text{ mass \%}$$

Molarity, introduced in Chapter 1, is the most common concentration measure for aqueous solutions. Molarity is the number of moles of a solute divided by the volume of solution:

$$\text{Molarity} = \frac{\text{Moles of solute}}{\text{Total volume of solution}} \quad \text{or} \quad c = \frac{n_{\text{solute}}}{V_{\text{solution}}} \qquad (1\text{-}4)$$

Molarity is convenient for expressing concentrations of liquid solutions, because we often measure liquid solutions volumetrically.

Molarity is not a convenient concentration measure when the temperature of the solution changes. This is because a solution expands as it is heated. Consequently, the total volume of solution changes with temperature, causing molarity to vary even though the amounts of solutes and solvent remain fixed.

For applications where the solution temperature changes, chemists prefer to use the **molality** ($b$). Molality is defined as the number of moles of solute divided by the mass of solvent in kilograms.

$$\text{Molality} = \frac{\text{Moles of solute}}{\text{Kilograms of solvent}} \quad \text{or} \quad b = \frac{n_{\text{solute}}}{m_{\text{solvent}}} \qquad (9\text{-}1)$$

The IUPAC recommendation for the symbol for molality is either $m$ or $b$. We use $b$ in this book to avoid confusion with mass, $m$.

Neither moles of solute nor mass of solvent change with temperature, so molality is independent of temperature. Example 9-1 illustrates molality calculations.

## EXAMPLE 9-1 MOLALITY

Hydrogen peroxide disinfectant typically contains 3.0% $H_2O_2$ by mass. Assuming that the rest of the contents is water, what is the molality of this disinfectant?

**Strategy:** To calculate molality, we use Equation 9-1, for which we need moles of solute and mass of solvent in kilograms. To find these quantities, it is convenient to consider 100 g of solution, for which the mass percentage is the mass of one component present.

**Solution:** In 100 g of disinfectant, there are 3.0 g of hydrogen peroxide and 97 g of water. We need to convert the mass of solute into moles and the mass of solvent into kilograms:

$$M_{H_2O_2} = 2(16.00 \text{ g/mol}) + 2(1.008 \text{ g/mol}) = 34.02 \text{ g/mol}$$

$$n_{H_2O_2} = \frac{m}{M} = \frac{3.0 \text{ g}}{34.02 \text{ g/mol}} = 0.0882 \text{ mol}$$

$$m_{water} = (97 \text{ g})\left(\frac{1 \text{ kg}}{10^3 \text{ g}}\right) = 0.0970 \text{ kg}$$

$$b = \frac{n_{solute}}{m_{solvent}} = \frac{0.0882 \text{ mol}}{0.0970 \text{ kg}} = 0.91 \text{ mol/kg}$$

 **Does the Result Make Sense?** The units match those of molality, and the result is rounded to two significant figures to match the precision of the mass percentage information.

**Common Pitfalls:** Confusing molarity with molality.

**Practice Exercise 9.1**

Determine the molality of a sugar solution that contains 3.94 g of sucrose ($C_{12}H_{22}O_{11}$) dissolved in 285 g of water.

A convenient concentration measure for gaseous solutions is the mole fraction, introduced in Section 2.3. The mole fraction ($X$, dimensionless) is the number of moles of one component divided by the total moles of all components.

$$\text{Mole fraction of } A = \frac{\text{Moles of } A}{\text{Total number of moles}} \quad \text{or} \quad X_A = \frac{n_A}{n_{total}}$$

Mole fractions are convenient for expressing concentrations of gas mixtures, because we often measure gas pressures, and partial pressures are related to total pressure through the mole fraction. Furthermore, mole fractions are not a function of either temperature or pressure, whereas other concentration units (e.g., molarity) are because of a change in volume of the solution if either temperature or pressure is changed.

Parts per million (ppm) and parts per billion (ppb) are additional concentration measures related to mole fraction. As described in Section 2.3, parts per million is the number of items of one kind present in a sample containing a million $(10^6)$ total items. Parts per billion is the number of items of one kind present in a sample containing a billion $(10^9)$ items. When the items are molecules,

$$1 \text{ ppm} = 1 \text{ molecule out of every } 10^6 \text{ molecules}$$

$$1 \text{ ppb} = 1 \text{ molecule out of every } 10^9 \text{ molecules}$$

These measures are useful when the concentration of a solute is quite small. Krypton, for example, is present in the Earth's atmosphere at a mole fraction of $1.14 \times 10^{-6}$. This is more conveniently expressed as 1.14 ppm.

Each concentration measure is convenient for some types of calculations but inconvenient for others. Consequently, a chemist may need to convert a concentration from one measure to another. Example 9-1 illustrates the conversion from a mass-based concentration, percent by mass, to a mole-based concentration, molality. Mole fraction, molarity, and molality all are mole-based for the solute, but the denominators of these concentration measures are quite different. Mole fraction uses total number of moles, molarity uses total volume, and molality uses mass of *solvent*.

To convert between molarity and the other mole-based concentration measures, we must relate volume to mass and so the density of the solution is required: Density = Mass/Volume, or $\rho = m/V$.

Conversions between molarity and other concentration measures require two steps. First, we use density to convert volume of solution to mass of solution. Then, we find the mass of solvent by subtracting the mass of solute from the mass of solution. Example 9-2 provides practice in conversions among mole-based concentration measures.

## EXAMPLE 9-2 CONCENTRATION CONVERSIONS

Concentrated aqueous ammonia is known commercially as ammonium hydroxide ($NH_4OH$), because $H_2O$ molecules can transfer protons to $NH_3$ molecules to generate $NH_4^+$ cations and $OH^-$ anions.

Concentrated aqueous ammonia (also known as ammonium hydroxide) is 14.8 M and has a density of 0.898 g/mL. Determine the molality and mole fraction of ammonia in this solution.

**Strategy:** We have information about molarity (mol/L) and density (g/mL) and are asked to find molality (mol/kg) and mole fraction (mol/mol). A good way to approach conversions from molarity to another measure is to choose a convenient volume for the solution, determine its mass and the mass of solute, and find the mass of water by difference. Then convert mass of water to kilograms and to moles to complete the calculations.

**Solution:** Because we know molarity (mol/L), it is most convenient to work with 1.000 L of solution, which has a volume of 1000 mL and contains 14.8 mol $NH_3$. Use the density to calculate the mass of this solution:

$$\rho = \frac{m}{V} \quad \text{so} \quad m = \rho V$$

$$m_{\text{solution}} = (0.898 \text{ g/mL})(1000 \text{ mL}) = 898 \text{ g}$$

Calculate the mass of $NH_3$ from moles of $NH_3$:

$$m_{NH_3} = nM = (14.8 \text{ mol})(17.0 \text{ g/mol}) = 252 \text{ g } NH_3$$

Subtract this mass from the total mass to find mass of $H_2O$:

$$m_{solvent} = 898\ g - 252\ g = 646\ g$$

Convert solvent mass to kilograms and to moles:

$$m_{solvent} = (646\ g)\left(\frac{1\ kg}{10^3\ g}\right) = 0.646\ kg$$

$$n_{solvent} = \frac{m}{M} = \frac{646\ g}{18.0\ g/mol} = 35.9\ mol$$

Now calculate molality and mole fraction:

$$b = \frac{n_{solute}}{m_{solvent}} = \frac{14.8\ mol}{0.646\ kg} = 22.9\ mol/kg$$

$$X = \frac{n_{solute}}{n_{solute} + n_{solvent}} = \frac{14.8\ mol}{14.8\ mol + 35.9\ mol} = 0.292$$

 **Do the Results Make Sense?** Concentrated ammonia has a high molarity, so we expect that it will also have a high molality and mole fraction.

**Practice Exercise 9.2**
Determine the molality and mole fraction of concentrated aqueous HCl, which is 12.0 M and has a density of 1.19 g/mL.

## SECTION EXERCISES

● **9.1.1** Brass contains 63% copper and 37% zinc. Identify the solvent and solute, and describe what type of solution this is.

● **9.1.2** Tonic water contains 31 ppm *by mass* of quinine ($C_{20}H_{24}O_2N_2$). Calculate the mole fraction, molarity, and molality of quinine in tonic water, assuming the density of tonic water is 1.00 g/mL.

● **9.1.3** Calculate the mass of $Na_2SO_4$ to be added to 2.5 L of water to make a solution that has a total molality of ions of 4.2 mol/kg. Assume water has a density of 1.00 g/mL.

## 9.2 DETERMINANTS OF SOLUBILITY

Solubilities vary tremendously. At one extreme, some substances form solutions in all proportions and are said to be **miscible**. For example, acetone and water can be mixed in any proportion, from pure water to pure acetone. At the other extreme, a substance may be **insoluble** in another. One example is common salt, NaCl, of which the solubility in gasoline is virtually zero.

Many combinations display solubility that is between the two extremes of miscible and insoluble. In other words, the substance dissolves, but there is a limit to the amount of solute that will dissolve in a given amount of solvent. When that limit has been reached, the solution is **saturated**. The concentration of a saturated solution is the **solubility** of the substance in that particular solvent at a specified temperature. For instance, 350 g of NaCl will dissolve in 1 L of water at room temperature, but only 0.0015 g of AgCl will dissolve under the same conditions.

## Like Dissolves Like

Solubility is a complex phenomenon that depends on the balance of several properties. A complete understanding of solubility is beyond an introductory course, but the general features of solubility are summarized by the expression *like dissolves like*. This means that one substance dissolves in another if the forces of attraction between the solute and the solvent are similar to the solvent–solvent and solute–solute interactions. This generalization can be applied to a variety of solution types. When two liquids are mixed, three sets of intermolecular interactions must be assessed. Water ($H_2O$) and methanol ($CH_3OH$) provide a first example. When water and methanol mix, $H_2O \cdots H_2O$ attractive forces and $CH_3OH \cdots CH_3OH$ attractive forces must be overcome, but they are replaced by $H_2O \cdots CH_3OH$ attractive forces. The two liquids are alike in that both substances contain O—H groups that form hydrogen bonds readily. When these liquids are mixed, $H_2O \cdots H_2O$ hydrogen bonds and $CH_3OH \cdots CH_3OH$ hydrogen bonds break, but $H_2O \cdots CH_3OH$ hydrogen bonds form. The net result is that the degree of hydrogen bonding in the solution is about the same as in either of the pure liquids, making these two liquids miscible.

Octane and cyclohexane make up another liquid pair of which the intermolecular interactions are alike. Both have low polarities, so these molecules in the pure liquids are held together only by the dispersion forces caused by their polarizable electron clouds. Dispersion forces in solutions of octane and cyclohexane are about the same as in the pure liquids. Again, these two liquids are miscible.

In contrast, water and octane are not alike. As Figure 9-1 shows, an octane molecule disrupts the hydrogen-bonding network of water. Octane does not form hydrogen bonds, so the only forces of attraction between water molecules and octane molecules are dispersion forces. Because hydrogen bonds are stronger than dispersion forces, the cost of disrupting the hydrogen-bonding network in

Water
$H_2O$

Methanol
$CH_3OH$

Octane
$C_8H_{18}$

Cyclohexane
$C_6H_{12}$

**Figure 9-1**

Water and octane are not alike. Rather than dissolving, octane floats on water. The presence of an octane molecule in the water layer disrupts a portion of the hydrogen-bonding network in water.

**Figure 9-2**

Acetone is a versatile solvent that dissolves both polar and non-polar substances. The three-carbon chain is compatible with non-polar molecules such as octane and cyclohexane, whereas the polar C=O bond is compatible with polar molecules such as methanol and water.

water is far greater than the stability gained from octane–water dispersion forces. Thus, octane and water are nearly insoluble in each other. When mixed, octane and water partition into distinct layers, one of nearly pure water and the other of nearly pure octane. For the same reason, water and cyclohexane are also insoluble in each other and form two layers when mixed.

Some liquids can interact with other substances in multiple ways. Acetone, for instance, has a polar C=O bond and a three-carbon bonding framework (Figure 9-2). The bonding framework is similar to that of a hydrocarbon, so acetone mixes with cyclohexane and octane. The polar C=O group makes acetone miscible with other polar molecules such as acetonitrile ($CH_3CN$). Finally, the oxygen atom in acetone has lone pairs of electrons that can form hydrogen bonds with hydrogen atoms of ammonia or water. Because of its versatility, acetone is an important industrial solvent. It is also used in laboratories to clean and rinse glassware. Example 9-3 treats several alcohols that also display multiple types of interactions.

## EXAMPLE 9-3  SOLUBILITY TRENDS

Give a molecular explanation for the following trend in alcohol solubilities in water:

| | | |
|---|---|---|
| *n*-Propanol | $CH_3CH_2CH_2OH$ | Miscible |
| *n*-Butanol | $CH_3CH_2CH_2CH_2OH$ | 1.1 M |
| *n*-Pentanol | $CH_3CH_2CH_2CH_2CH_2OH$ | 0.30 M |
| *n*-Hexanol | $CH_3CH_2CH_2CH_2CH_2CH_2OH$ | 0.056 M |

**Strategy:** Solubility limits depend on the stabilization generated by solute–solvent interactions balanced against the destabilization that occurs when solvent–solvent interactions are disrupted by solute, or when solute–solute interactions are disrupted by solvent. Thus, we must examine intermolecular interactions involving water and alcohol molecules.

**Solution:** When an alcohol dissolves in water, the non-polar hydrocarbon part of the alcohol disrupts the hydrogen-bonding network of water. Counterbalancing this solvent disruption, the —OH groups of the alcohol form hydrogen-bonding interactions with water molecules.

As the non-polar region of an alcohol grows longer, each solute molecule disrupts more and more hydrogen bonds in the solvent. At the same time, each of these alcohols has only one —OH group, so the amount of compensating solute–solvent hydrogen bonding is the same for all the alcohols.

Longer-chain alcohols are progressively less soluble in water, because as the hydrocarbon chain gets longer, more destabilization is involved in inserting the alcohol into the water matrix.

**Does the Result Make Sense?** Another way to look at this solubility trend is to note that *n*-propanol ($CH_3CH_2CH_2OH$) is not much different from ethanol, which mixes completely with water. In contrast, *n*-hexanol ($CH_3CH_2CH_2CH_2CH_2CH_2OH$) is not much different from octane, which is insoluble in water.

**Practice Exercise 9.3**

Here are the room-temperature solubilities (mass percentages) of some liquids in cyclohexane ($C_6H_{12}$): water, 0.01%; methanol, 5%; ethanol, completely miscible. Give a molecular explanation for the trend.

## Solubility of Solids

As described in Chapter 8, there are four different kinds of solids: network, molecular, metallic, and ionic. Each is held together by a different kind of interaction, so each has its own solubility characteristics.

*Network solids* such as diamond, graphite, or silica cannot dissolve without breaking covalent chemical bonds. Because intermolecular forces of attraction are always much weaker than covalent bonds, solvent–solute interactions are never strong enough to offset the energy cost of breaking bonds. Although network solids may react with specific liquids or vapours, they will not dissolve in most solvents.

At the opposite extreme, *molecular solids* contain individual molecules bound together by various combinations of dispersion forces, dipole forces, and hydrogen bonds. Conforming to "like dissolves like," molecular solids dissolve readily in solvents with similar types of intermolecular forces. Non-polar $I_2$, for instance, is soluble in non-polar liquids such as carbon tetrachloride ($CCl_4$). Many organic compounds are molecular solids that dissolve in organic liquids such as cyclohexane and acetone.

Hydrogen bonding allows water to dissolve materials that form hydrogen bonds. Hydrogen bonding makes sugars such as sucrose ($C_{12}H_{22}O_{11}$) and glucose ($C_6H_{12}O_6$) highly soluble in water. Figure 9-3 shows that glucose is an organic molecule with five polar O—H groups, each of which forms hydrogen bonds with water molecules. When glucose dissolves, hydrogen bonds between water and glucose replace the hydrogen bonds lost by the water molecules of the solvent. This balance means that the energy requirements for solution formation are small, and glucose is quite soluble in water. On the other hand, naphthalene, a similarly sized solid hydrocarbon limited to dispersion forces, is nearly insoluble in water.

The best solvent for a molecular solid is one whose intermolecular forces match the forces holding the molecules in the crystal. For a solid held together by dispersion forces, good solvents are non-polar liquids such as carbon tetrachloride ($CCl_4$) and cyclohexane ($C_6H_{12}$). For polar solids, a polar solvent such as acetone works well. Example 9-4 provides some practice in recognizing solubility types.

**Figure 9-3**

All five polar O—H groups of glucose can form hydrogen bonds with water molecules (*red dotted lines*).

### EXAMPLE 9-4 SOLUBILITIES OF VITAMINS

Vitamins are organic molecules that are required for proper function but not synthesized by the human body. Thus, vitamins must be present in the foods people eat. Vitamins fall into two categories: fat-soluble, which dissolve in fatty hydrocarbon-like tissues, and water-soluble. The structures of several vitamins follow. Assign each one to the appropriate category.

Vitamin E

Vitamin A

Pantothenic acid
(Vitamin B₅)

Vitamin C

Pyridoxamine
(Vitamin B₆)

Vitamin D

**Strategy:** At first glance it may seem that "like dissolves like" does not apply here. Certainly, none of these complex molecules looks like water, and the resemblance to simple hydrocarbons such as cyclohexane also is remote. Keep in mind, however, that the basis for the principle is that similar compounds dissolve in each other because they have common patterns of intermolecular interactions. Example 9-3 indicates that alcohols containing large non-polar segments do not dissolve well in water. We can categorize vitamins similarly by the amounts of their structures that can be stabilized by hydrogen bonding to water molecules.

**Solution:** A hydrogen-bond donor must have a hydrogen atom bonded to F, O, or N, and a hydrogen-bond acceptor is an electronegative atom with a lone pair of electrons. By these criteria, all the vitamins shown are capable of some hydrogen bonding. However, Vitamins A, D, and E have large regions containing only non-polar C—C and C—H bonds. Like the longer alcohols in Example 9-3, the hydrogen-bonding sites in these molecules are insufficient to make them soluble in water. Their large hydrocarbon-like regions make these vitamins fat-soluble.

The remaining molecules, Vitamin C, pantothenic acid, and pyridoxamine, have comparatively large numbers of O—H and N—H groups. These groups allow each of these vitamin molecules to form many hydrogen bonds, so they are all water-soluble (all the B vitamins are soluble in water).

**Do the Results Make Sense?** The different solubilities of these two kinds of vitamins have important metabolic consequences. Aqueous body fluids do not dissolve fat-soluble vitamins, so these molecules can be stored in fatty body tissue for a long time. As a result, too much of a fat-soluble vitamin can overload the body's storage capabilities and lead to a toxic reaction. In contrast, the body cannot store water-soluble vitamins; instead, it excretes anything more than the amount it can use immediately. People must therefore have a steady supply of water-soluble vitamins in their diets to remain healthy.

**Practice Exercise 9.4**
Classify each of the following as water-soluble or fat-soluble:

Uric acid

Testosterone

*Metals* do not dissolve in water, because they contain extensive delocalized bonding networks that must be disrupted before the metal can dissolve. A few metals *react* with water, and several *react* with aqueous acids, but no metal will simply *dissolve* in water. Likewise, metals do not dissolve in non-polar liquid solvents.

When an alkali metal contacts water, metal atoms donate electrons to water molecules, producing hydrogen gas and a solution of the metal cation (e.g., Na). When a metal such as Ca, Zn, or Fe is treated with a strong aqueous acid, hydronium ions in the acid solution accept electrons from metal atoms, creating cations that then dissolve. Zinc metal, for example, reacts with hydrochloric acid to generate $H_2$ gas and $Zn^{2+}$ cations in solution:

$$Zn\ (s) + 2H_3O^+\ (aq) \longrightarrow Zn^{2+}\ (aq) + H_2\ (g) + 2\ H_2O\ (l)$$

These processes differ from other solution processes in that the solute undergoes a chemical transformation. The aqueous medium dissolves the metal by a chemical reaction that converts the insoluble metal into soluble cations. The solution produced when zinc reacts with aqueous HCl is an aqueous solution of $Zn^{2+}$ ions, not a solution of Zn metal in water. If this solution is boiled to dryness, the remaining solid is $ZnCl_2$, not Zn metal.

Zinc metal reacts with aqueous acids to produce hydrogen gas and $Zn^{2+}$ ions.

## Alloys

Metals are insoluble in common liquid solvents but can dissolve in one another (like dissolves like). A mixture of substances with metallic properties is called an **alloy**. Some alloys are true solutions, but microscopic views show that others are heterogeneous mixtures. Brass, for instance, is a homogeneous solution of copper (20–97%) and zinc (80–3%), but common plumber's solder is a heterogeneous alloy of lead (67%) and tin (33%). When solder is examined under a microscope, separate regions of solid lead and solid tin can be seen. When brass is examined, no such regions can be detected.

One way that a solid metal can accommodate another is by substitution. For example, sterling silver is a solid solution containing 92.5% silver and 7.5% copper. Copper and silver occupy the same column of the periodic table, so they share many properties, but copper atoms (radius of 128 pm) are smaller than silver atoms (radius of 144 pm). Consequently, copper atoms can readily replace silver atoms in the solid crystalline state, as shown schematically in Figure 9-4.

A second way for a solid to accommodate a solute is interstitially, with solute atoms fitting in between solute atoms in the crystal structure. An important alloy of this type is carbon steel, a solid solution of carbon in iron, also shown in Figure 9-4. Steels actually are both substitutional and interstitial alloys. Iron is the solvent and carbon is present as an interstitial solute, but varying amounts of manganese, chromium, and nickel are also present and can be in substitutional positions.

Mercury, the only metal that is a liquid at room temperature, dissolves a number of metals to give liquid solutions. Any solution of another metal dissolved in mercury is called an *amalgam*. Metals close to mercury in the periodic table, such

**Figure 9-4**

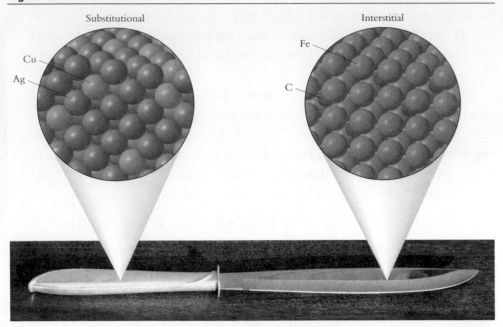

Metal alloys are solid solutions that can be either substitutional (copper in silver) or interstitial (carbon in iron).

as silver, gold, zinc, and tin, are particularly soluble in mercury. An amalgam of silver, tin, and mercury is used to make dental fillings. When the intermetallic compound $Ag_3Sn$ is ground with mercury, it forms a semisolid amalgam that can be shaped to fill a cavity. On standing, mercury binds with the other metals to form a hard solid mixture of $Ag_5Hg_8$ and $Sn_7Hg_8$. The mixture expands slightly during reaction, forming a tight fit within the cavity. The mercury atoms in dental fillings are chemically bound and do not dissolve, so they have been traditionally considered safe for the wearer, despite the fact that free mercury metal is highly toxic.

> Because of concerns about possible leaching of mercury from fillings, new resin materials have replaced mercury amalgams as the material of choice for dental fillings.

## Solubility of Salts

*Ionic solids,* also known as salts, contain cations and anions held in three-dimensional lattices by strong electrical attractions, as described in Section 5.5. Typical ionic salts have such strong attractive forces that these substances are solids at room temperature. Thus, we might think that there are no solvents that dissolve ionic solids at room temperature. Experience tells us differently, however, because many ionic solids dissolve in water. Moreover, chemists have discovered how to make ionic substances that combine inorganic ions with organic groups. These substances, although ionic, can be liquids at room temperature, and their potential as solvents is being extensively studied.

Water is highly polar, but it is not ionic. How, then, can water act as a solvent for ionic solids? A salt dissolves only if the interactions between the ions and the solvent are strong enough to overcome the attractive forces that hold ions in the crystal lattice. When an ionic solid forms an aqueous solution, the cations and anions are *solvated* by strong ion–dipole interactions with water molecules.

**Figure 9-5**

An ionic solid dissolves in water through the formation of ion–dipole interactions that overcome the forces of the crystal lattice. The dotted lines represent hydrogen bonds among water molecules. Water molecules cluster around ions in solution with the negatively charged oxygen atoms towards cations, and the positively charged hydrogen atoms towards anions.

Figure 9-5 illustrates the solvation of $Na^+$ and $Cl^-$ ions as NaCl dissolves in water. A cluster of water molecules surrounds each ion in solution. Notice how the water molecules are oriented so that their dipole moments align with charges of the ions. The partially negative oxygen atoms of water molecules point toward $Na^+$ cations, whereas the partially positive hydrogen atoms of water molecules point toward $Cl^-$ anions.

Remember that any liquid is dynamic, with molecules and ions moving about continuously. The solvent molecules do not remain fixed in their positions. In fact, the water molecules solvating the ions change places many millions of times per second with water molecules in the bulk solvent. Nevertheless, the solvation of ions is the primary driving force in the formation of a salt solution. We describe the energy changes that accompany the formation of aqueous solutions of salts in more detail in the following section.

## SECTION EXERCISES

**9.2.1** List the types of intermolecular interactions that stabilize a solution of acetone in methanol, and draw molecular pictures that illustrate any dipole–dipole and hydrogen-bonding interactions that exist between molecules of these substances.

**9.2.2** On the basis of their molecular structures, predict which of the following silicon-containing materials is soluble in water: elemental Si, $SiO_2$, $Na_4SiO_4$, and $Si(CH_3)_4$.

**9.2.3** Explain in terms of attractive forces why some salts containing $Zn^{2+}$ are soluble in water, but Zn metal is insoluble.

## 9.3 CHARACTERISTICS OF AQUEOUS SOLUTIONS

Although there are many different solvents, water is by far the most important. Water is the liquid responsible for much of the Earth's characteristics. Furthermore, water is the medium in which life begins and is sustained. In this section, we focus specifically on the characteristics of water and its solutions.

### The Uniqueness of Water

Why does water play central roles in geology and biology? One important reason is that water is the only small molecule that is not a gas under terrestrial conditions. Compare the boiling point of water (100 °C) with the boiling points of methane (−91 °C) and ammonia (−34 °C) or the sublimation temperature of carbon dioxide (−78.5 °C). Moreover, water is a liquid over a uniquely large temperature range, 0–100 °C. Compare this 100-degree range with the 44-degree range for ammonia. Because water is a liquid under the most prevalent terrestrial conditions, it is uniquely available to play the role of solvent.

The biochemical structures of even the simplest life forms require a liquid medium.

Life as we know it depends on the existence of water as a liquid. Biochemical processes require free movement of chemicals, which cannot occur in the solid phase. Biochemical structures contain many interlocking parts that would not be stable in the gas phase. Thus, the liquid phase is best suited for life. Moreover, water is an excellent solvent, particularly for molecules that can form hydrogen bonds. As we describe in Chapter 17, the molecular building blocks of living matter are rich in groups that form hydrogen bonds. This allows biological molecules to be synthesized, move about, and assemble into complex structures, all in aqueous solution.

The structure of the water molecule accounts for its unique properties. The O—H bonds are highly polar, and the geometry of the molecule is bent. These features give $H_2O$ a large dipole moment, generating strong dipole–dipole attractive forces. Equally important are the hydrogen bonds that water forms, both with other water molecules and with other substances.

The combination of a large dipole moment and hydrogen-bonding ability allows water molecules to establish strong attractive interactions of several types. Ionic salts that dissolve in water do so because of the strongly attractive forces between the dipoles of water molecules and the ions of the solute. Water-soluble organic and biological molecules contain —OH and —NH$_2$ groups that form strongly attractive hydrogen-bonding interactions with solvent water molecules. Even small non-polar molecules such as $O_2$ dissolve in water to a certain extent because of dipole-induced dipole forces.

### Solubility Equilibrium

Nearly every substance that dissolves in water has an upper limit to its solubility. Solids, liquids, and gases all display this characteristic. The room-temperature solubility of solid NaCl in water is about 6 M. Liquid n-hexanol forms a saturated aqueous solution at a concentration of $5.6 \times 10^{-2}$ M. Gaseous $O_2$

in the Earth's atmosphere dissolves in natural waters to give concentrations of about $2.5 \times 10^{-4}$ M.

If the concentration of a solute is lower than its solubility, additional solute can dissolve, but once the concentration of solute reaches the solubility of that substance, no further net changes occur. Individual solute molecules still enter the solution, but the solubility process is balanced by precipitation, as Figure 9-6 illustrates. A saturated solution in contact with excess solute is in a state of dynamic equilibrium. For every molecule or ion that enters the solution, another returns to the solid state. We represent dynamic equilibria by writing the equations using double arrows, showing that both processes occur simultaneously:

$$\text{Solute } (pure\ state) \rightleftarrows \text{Solute } (aq)$$

$$\text{NaCl } (s) \rightleftarrows \text{Na}^+ (aq) + \text{Cl}^- (aq)$$

$$\text{CH}_3\text{CH}_2\text{CH}_2\text{CH}_2\text{CH}_2\text{CH}_2\text{OH } (l) \rightleftarrows \text{CH}_3\text{CH}_2\text{CH}_2\text{CH}_2\text{CH}_2\text{CH}_2\text{OH } (aq)$$

$$\text{O}_2 (g) \rightleftarrows \text{O}_2 (aq)$$

Several observations show that saturated solutions are at dynamic equilibrium. For example, if $\text{O}_2$ gas enriched in the oxygen-18 isotope is introduced into the gas phase above water that is saturated with oxygen gas, the gas in the solution eventually also becomes enriched in the heavier isotope. As another example, if finely divided crystalline salt is in contact with a saturated solution of the salt, the small crystals slowly disappear and are replaced by larger crystals. Each of these observations shows that molecules are moving between the two phases, yet the concentrations of the saturated solutions remain constant.

**Figure 9-6**

Solution

Solute

When a solution is saturated, there is dynamic equilibrium. Solute molecules move back and forth between the pure solute phase and the solution phase at equal rates.

## Energetics of Salt Solubility

When a salt dissolves, energy must be supplied to separate the ions and to break hydrogen bonds in the solvent. At the same time, energy is released when solvent molecules form ion–dipole interactions with the ions. The **molar heat of solution** $(\Delta H_{soln})$ is the net energy flow resulting from these three factors. In any actual dissolving process, all these energy changes occur simultaneously, but we can visualize a path that treats each step separately. Figure 9-7 illustrates this path for NaCl. The first step is separating the ions, for which energy equal to the lattice energy (see Section 5.5) must be supplied. Because ionic attractions extend throughout the crystal, lattice energies are quite large. The lattice energy of NaCl is 787 kJ/mol, and that of MgO is 3791 kJ/mol.

Additional energy is needed to move solvent molecules apart to make room for the dissolving ions. In this step, hydrogen bonds break. Hydrogen bonds are substantially weaker than ion–ion interactions, so this step requires much less energy than breaking apart the lattice.

The third step is solvation of the ions by solvent molecules. Water molecules cluster around each ion, oriented to give attractive ion–dipole interactions. This

**Figure 9-7**

The dissolving of an ionic salt to produce an aqueous solution can be analyzed by breaking the overall process into three steps.

**Figure 9-8**

Whether the dissolving of a salt is exothermic or endothermic depends on the balance between the attractive forces of the crystal lattice and the ion–dipole forces that stabilize the ions in solution.

step releases energy. Although each individual ion–dipole interaction is weak, each ion forms from four to eight such interactions, depending on the size of the ion and the concentration and temperature of the solution. Taken together, the vast number of ion–dipole interactions results in a substantial release of energy.

Large positive energy changes accompany the breaking apart of a lattice, and large negative energy changes accompany ion solvation. Nevertheless, molar heats of solution are typically less than 100 kJ/mol. As Figure 9-7 suggests, the reason for this is that the energy absorbed in separating ions of opposite charge is largely, if not entirely, balanced by the energy released when water molecules form ion–dipole interactions with the ions.

The dissolving process can be either exothermic or endothermic. If the energy released through ion–dipole attractions is larger than the crystal lattice energy, energy is released as the salt dissolves, giving an exothermic process with negative $\Delta H_{soln}$. Some salts, including calcium chloride ($CaCl_2$, $\Delta H_{soln} = -83$ kJ/mol) and magnesium sulphate ($MgSO_4$, $\Delta H_{soln} = -91.2$ kJ/mol), release relatively large amounts of energy when they dissolve in water. Conversely, if the crystal lattice energy is larger than the ion–dipole attractive energy, the salt must absorb energy from the surroundings as it dissolves, giving an endothermic solution process with positive enthalpy change. One soluble salt with a relatively large positive $\Delta H_{soln}$ is ammonium nitrate ($NH_4NO_3$, $\Delta H_{soln} = 21.1$ kJ/mol). This balance between the lattice energy and the energies of ion–dipole interactions is summarized in Figure 9-8.

Heats of solution are the basis for "instant cold packs" and "instant hot packs" used for the first-aid treatment of minor sprains and pulled muscles. These packs have two separate compartments. One contains water, and the other contains a salt: $NH_4NO_3$ for cold packs and $MgSO_4$ or $CaCl_2$ for hot packs. Kneading the pack breaks the wall between the compartments, allowing the salt to mix with water. As the salt dissolves to form an aqueous solution, the temperature of the pack changes. Heat is absorbed or released only as the salt dissolves, however, so after all of the salt has dissolved, the pack gradually returns to room temperature. Further manipulation of the pack has no effect.

As mentioned above, complete solvation of ions involves from four to eight water molecules per ion. In highly concentrated aqueous solutions, there are not enough water molecules to fully solvate all the ions. Because of this, ion–solvent interactions may increase as water is added to a concentrated solution. When a concentrated solution is diluted by adding water, energy is released and the temperature of the solution increases. The molar energy change accompanying this process is called the **heat of dilution**.

Strong acids often have large heats of dilution, because proton transfer from a strong acid to water is highly exothermic. This can lead to problems when a concentrated acid is diluted with water. When water contacts the solution, the dilution process releases a large amount of energy. This energy heats the solution, and the heating can occur rapidly enough to form a "hot spot" in which the temperature is greater than the boiling temperature of water. If this occurs, droplets of acid will be ejected from the solution. Hot acid rapidly burns the skin and attacks many other materials, including clothing. Sulphuric acid is particularly likely to splatter, because concentrated $H_2SO_4$ has a high viscosity and does not mix readily with added water. Therefore, acids should always be diluted by slowly adding the concentrated acid to a larger volume of water. In this way the transformation of concentrated acid into a more dilute solution occurs relatively slowly, and no local "hot spots" of high temperature can develop.

Whether heat is absorbed or released when a solution is formed does not by itself allow us to predict the solubility of the solute. The entropy change (covered in Chapter 11) is also important, and in the case of making solutions, is usually more important than the enthalpy change in determining solubilities.

Instant hot and cold packs rely on heats of solution to generate instant heating or cooling.

When diluting a strong acid, the acid always should be added slowly to the water, with vigorous stirring.

## Effect of Temperature

In general, solids are more soluble in hot water than in cold water. This is true for nearly all organic solids and for most ionic salts. Figure 9-9 shows how the solubilities of several salts vary with temperature. The figure indicates that some ionic salts behave contrary to this general rule. Whereas the solubility of potassium nitrate increases dramatically with temperature, that of sodium chloride changes hardly at all, and the solubility of lithium sulphate decreases. Solubility in water decreases with increasing temperature for a number of other sulphate salts, including sodium sulphate and calcium sulphate.

A compound having a solubility that increases with temperature can be purified by recrystallization. The impure solid is dissolved in a minimum volume of hot water. The hot solution is filtered to remove insoluble impurities, and then the solution is cooled in an ice bath. The solubility of the compound decreases as the temperature drops, causing the substance to precipitate from solution. Soluble impurities usually remain in solution. Purification by recrystallization is not restricted to aqueous solutions, however. An organic solid can be purified by recrystallization from an appropriate organic solvent.

**Figure 9-9**

Solubilities of salts in water vary with temperature.

## Gas–Solution Equilibria

Non-polar gases are only slightly soluble in water. For example, water in contact with the Earth's atmosphere contains $O_2$ at a concentration of only about $2.5 \times 10^{-4}$ M and $CO_2$ at about $1 \times 10^{-5}$ M. Nevertheless, these small concentrations are essential for aquatic life. Fish and other aquatic animals use their gills to extract $O_2$ dissolved in water, and unless that oxygen is replenished, these species die. Submerged green plants carry out photosynthesis using dissolved carbon dioxide, which also must be replenished for these plants to survive.

Figure 9-10 is a molecular view showing that the equilibrium concentration of a dissolved gas varies with the partial pressure of that gas. An increase in the partial pressure of gas results in an increase in the rate at which gas molecules enter the solution. This increases the concentration of gas in solution, which, in turn, results in an increase in the rate at which gas molecules escape from the solution. Equilibrium is re-established when the solute concentration is high enough that the rate of escape equals the rate of capture.

**Figure 9-10**

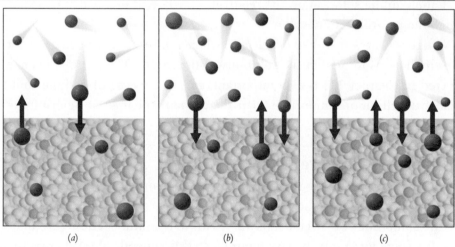

(a)          (b)          (c)

Molecular view of a gas–solution equilibrium. (a) At equilibrium, the rate of escape of gas molecules from the solution equals the rate of capture of gas molecules by the solution. (b) An increase in gas pressure causes more gas molecules to dissolve, throwing the system out of equilibrium. (c) The concentration of solute increases until the rates of escape and capture once again balance.

Fish and submerged plants depend on the equilibria between $O_2$ and $CO_2$ in the atmosphere and dissolved in the water. Home aquariums use air bubblers to maintain equilibrium concentrations by replenishing these gases.

Experiments show that the equilibrium concentration of dissolved gas increases linearly with the partial pressure of the gas. The equation describing this linear behaviour is **Henry's law**:

$$[\text{gas } (aq)]_{eq} = K_H(p_{gas})_{eq} \tag{9-2}$$

The proportionality constant for a gas–solution equilibrium is designated $K_H$ and is called the Henry's law constant. This constant has a different value for each combination of gas and solvent, and it also varies with temperature. Table 9-2 lists values for $K_H$ for representative gases dissolving in water. Note that gases are less soluble at higher temperatures.

## TABLE 9-2 HENRY'S LAW CONSTANTS ($K_H$) IN AQUEOUS SOLUTION (M/Bar)

| Gas | 0.0 °C | 25 °C | 30 °C |
|---|---|---|---|
| $N_2$ | $1.1 \times 10^{-3}$ | $6.6 \times 10^{-4}$ | $4.0 \times 10^{-4}$ |
| $O_2$ | $2.5 \times 10^{-3}$ | $1.3 \times 10^{-3}$ | $8.8 \times 10^{-4}$ |
| CO | $1.6 \times 10^{-3}$ | $9.5 \times 10^{-4}$ | $4.4 \times 10^{-4}$ |
| Ar | $2.5 \times 10^{-3}$ | $1.5 \times 10^{-3}$ | $1.0 \times 10^{-3}$ |
| He | $4.1 \times 10^{-4}$ | $3.8 \times 10^{-4}$ | $3.8 \times 10^{-4}$ |
| $CO_2$ | $7.7 \times 10^{-2}$ | $3.4 \times 10^{-2}$ | $1.6 \times 10^{-2}$ |

Example 9-5 uses Henry's law to determine the concentrations of atmospheric gases that dissolve in water, and our Focus 9-1: Chemistry and Life addresses an important consequence of Henry' s law.

## EXAMPLE 9-5 SOLUBILITIES OF ATMOSPHERIC GASES

The Earth's atmosphere contains 78% $N_2$, 21% $O_2$, and minor amounts of other gases, including $CO_2$ (0.0385%). Determine the concentrations of $N_2$, $O_2$, and $CO_2$ in water at equilibrium with the Earth's atmosphere at 25 °C.

**Strategy:** Each gas establishes its own dynamic equilibrium with water. The concentration depends on the partial pressure of the gas in the atmosphere and on the value of its Henry's law constant at 25 °C. Recall from Chapter 2 that the partial pressure of any gas in a mixture is given by the mole fraction ($X_i$) multiplied by total pressure.

**Solution:** Use 1.00 bar for the total pressure:

$$p_{O_2} = X_{O_2}p = \left(\frac{21\% \ O_2}{100\%}\right)(1.00 \text{ bar}) = 0.21 \text{ bar}$$

Analogous calculations show that the partial pressure of $N_2$ is 0.78 bar, and that of $CO_2$ is $3.85 \times 10^{-4}$ bar. Now use Equation 9-2 and values of $K_H$ from Table 9-2 to calculate the concentration of each dissolved gas:

$$[\text{gas } (aq)] = K_H(p_{gas})$$

$$[N_2 \ (aq)] = (6.6 \times 10^{-4} \text{ M/bar})(0.78 \text{ bar}) = 5.1 \times 10^{-4} \text{ M}$$

$$[O_2(aq)] = (1.3 \times 10^{-3} \text{ M/bar})(0.21 \text{ bar}) = 2.7 \times 10^{-4} \text{ M}$$

$$[CO_2 \ (aq)] = (3.4 \times 10^{-2} \text{ M/bar})(3.85 \times 10^{-4} \text{ bar}) = 1.3 \times 10^{-5} \text{ M}$$

 **Do the Results Make Sense?** All the molarities are rather small, in the $10^{-5}$–$10^{-4}$ M range. Given that these are non-polar gases, it does make sense that their solubilities are small in a polar solvent such as water.

**Common Pitfall:** Make sure to calculate the partial pressures correctly. In this example, the total pressure is 1.00 bar, which makes this calculation easy.

### Practice Exercise 9.5

What are the concentrations of these gases in water at equilibrium with the atmosphere at 30 °C?

A few gases interact with water to form concentrated aqueous solutions. For example, a 12 M HCl (*aq*) solution can be prepared by bubbling HCl gas through water, because proton transfer occurs to generate $H_3O^+$ and $Cl^-$ ions:

$$HCl \ (g) + H_2O \ (l) \longrightarrow Cl^- \ (aq) + H_3O^+ \ (aq)$$

Ammonia is another gas that is very soluble in water, giving solutions as concentrated as 14.8 M. Ammonia dissolves because it forms strong hydrogen bonds with water molecules (so strong that there is a proton transfer from water to ammonia). When an ammonia molecule displaces a water molecule, one hydrogen-bonding interaction is exchanged for another.

## SECTION EXERCISES

●**9.3.1** Draw a molecular picture illustrating the hydrogen bonds that form between a methanol ($CH_3OH$) molecule and water molecules in a dilute aqueous solution of methanol.

●**9.3.2** List the salts in Figure 9-9 in order of increasing solubility at 0 °C.

●**9.3.3** Some gases can be collected by liquid displacement. The gas is bubbled through water into an inverted container. If 0.18 mole of $CO_2$ at $p = 0.98$ bar is bubbled through 450 mL of water at 298 K, what fraction of the gas dissolves in the water?

## ▮ FOCUS 9-1 CHEMISTRY AND LIFE: SCUBA DIVING AND HENRY'S LAW

According to Henry's law, gases become more soluble as pressure increases. This solubility property has minimum effects on everyday life, because changes in altitude or weather cause only modest variations in atmospheric pressure. Scuba divers, however, must pay careful attention to the solubility equilibria of gases. The pressure exerted on a diver increases by 1 bar for every 10 m of descent. Increasing pressure causes more and more gas to dissolve in a diver's blood. When the diver returns to the surface, the amount of dissolved gas far exceeds its solubility. Consequently, ascending too rapidly can be disastrous.

Carbonated beverages illustrate what happens when a dissolved gas undergoes a rapid drop in pressure. Soft drinks, soda water, and champagne are bottled under several bars pressure of carbon dioxide. When a bottle is opened, the total pressure quickly falls to 1 bar. At this lower pressure, the concentration of $CO_2$ in the solution is much higher than its solubility, so the excess $CO_2$ forms gas bubbles and escapes from the liquid. As the photo shows, this process can be dramatic.

Scuba divers experience similar pressure changes. The amount of air dissolved in the blood increases significantly as the diver descends. If a diver returns to the surface too quickly, nitrogen gas dissolved in the blood forms bubbles in the same way as the $CO_2$ in a freshly opened carbonated drink. These bubbles interfere with the transmission of nerve impulses and restrict the flow of blood. The effect is extremely painful and can cause paralysis or death. The bubbles tend to collect in the joints, where they cause severe contractions. This is the source of the name of this dangerous condition—the "bends."

Divers avoid the bends by returning to the surface slowly, taking short "decompression stops" at intermediate depths to allow excess gas to escape from their blood without forming bubbles. Another way divers reduce the risk of the bends is by using helium–oxygen gas mixtures instead of compressed air. Helium is only half as soluble in water as nitrogen is, so less gas dissolves in blood.

Scuba divers face other hazards from the effects of Henry's law. At depths between 25 and 40 m, divers experience an intoxicating feeling from the high concentration of $N_2$ in the blood. This dissolved $N_2$ is thought to interfere with nerve transmission, giving rise to feelings of euphoria known as "rapture of the deep." Like other forms of intoxication, this causes a loss of judgment that can be deadly. The famous oceanographer Jacques-Yves Cousteau wrote: "I am personally quite receptive to nitrogen rapture. I like it and fear its doom. It destroys the instinct of life."

Still another hazard associated with Henry's law is oxygen toxicity. In high enough concentration, molecular oxygen is poisonous. Breathing pure oxygen at 1 bar pressure for just a few hours causes pain in the chest and coughing. Lung congestion and permanent tissue damage result after 24 hours of breathing pure $O_2$, and longer exposure causes death. Breathing compressed air (21% $O_2$) at a dive depth of 38 m is equivalent to breathing pure oxygen at sea level. At 61 m ($p_{O_2} = 1.3$–1.5 bar), oxygen toxicity causes muscle twitching, vomiting, and dizziness, and can result in deadly seizures. Although recreational divers seldom reach depths of 61 m, those who do venture this far into the sea must breathe mixtures of gas that contain less than 10% oxygen.

Despite these hazards, humans continue to venture beneath the surface of the seas. Adventurers seek to salvage materials from sunken ships. Archaeologists wish to explore ancient seaport sites that have since sunk beneath the waves. Biologists and lovers of nature study marine life in its natural habitat. In addition, hordes of amateur divers pursue this sport for its sheer enjoyment.

Scuba divers experience an undersea world filled with mystery and beauty, but diving can be a perilous hobby. Little wonder that novice divers must undergo rigorous training courses before they are free to explore the depths.

## 9.4 COLLIGATIVE PROPERTIES

Solute molecules alter many properties of a liquid. For instance, adding salt to water gives a solution that boils at a higher temperature than pure water, and adding ethylene glycol to the water in an automobile radiator gives a solution that protects against freezing. The presence of solute molecules causes changes in four common properties of solutions: vapour pressure, freezing point, boiling point, and osmotic pressure. These four are known as the **colligative properties**. Changes in each of these can be understood from a molecular perspective by examining how added solute molecules affect the dynamics of phase changes. These examinations reveal that the colligative properties generally are independent of the nature of the solute.

### Vapour Pressure Reduction

The easiest of the colligative properties to visualize is the effect of solute molecules on the vapour pressure exerted by a liquid. In a closed system, the solvent and its vapour reach dynamic equilibrium at a partial pressure of solvent equal to the vapour pressure. At this pressure, the rate of condensation of solvent vapour equals the rate of evaporation from the liquid.

**Figure 9-11**

Addition of solute molecules (*right view, orange balls*) reduces the rate of escape of solvent molecules compared with pure solvent (*left view*). As a result, the vapour pressure of a solution is lower than the vapour pressure of a pure solvent.

Figure 9-11 is a molecular view of how a solute changes this liquid–vapour equilibrium of the solvent. The presence of a solute means that there are fewer solvent molecules at the surface of the solution. As a result, the rate of solvent evaporation from a solution is slower than the rate of evaporation of pure solvent. At equilibrium, the rate of condensation must be correspondingly slower than the rate of condensation for the pure solvent at equilibrium with its vapour. In other words, the vapour pressure drops when a solute is added to a liquid. A solute decreases the concentration of solvent molecules in the gas phase by reducing the rates of both evaporation and condensation.

This molecular view of Figure 9-11 suggests that the extent of vapour pressure lowering will depend on the fraction of solvent molecules that have been replaced. In other words, the vapour pressure should be proportional to the mole fraction of the solvent. The molecular view also suggests that this effect does not depend on the nature of the solute, but only on its mole fraction. Experiments show that this is often the case, particularly for dilute solutions. A simple equation, **Raoult's law**, expresses this proportionality between vapour pressure and mole fraction. If we represent the solvent as "A" and the solute as "B," then

$$p_{\text{vap,solution}} = X_A p_{\text{vap,A}} \tag{9-3}$$

Raoult's law indicates that the vapour pressure decreases as the amount of solute in the solution is increased for the reasons described above.

Example 9-6 shows how to apply Raoult's law.

### EXAMPLE 9-6 RAOULT'S LAW

Calculate the vapour pressure of a 5% by mass benzoic acid $[C_7H_6O_{2(aq)}]$ in ethanol solution at 35 °C. The vapour pressure of pure ethanol at this temperature is 13.40 kPa.

**Strategy:** To calculate the vapour pressure of a solution, we need to first calculate the mole fraction of the solute, and then use Raoult's law.

**Solution:** To calculate the mole fraction of benzoic acid in this example, begin by assuming a certain mass of solution, 100 g:

$$m_{\text{benzoic acid}} = \left(\frac{5\%}{100\%}\right)(100 \text{ g}) = 5.00 \text{ g} \qquad m_{\text{water}} = 100.\ g - 5.00 \text{ g} = 95.00 \text{ g}$$

$$n_{\text{benzoic acid}} = \frac{5.00 \text{ g}}{122.1 \text{ g/mol}} = 0.0410 \text{ mol}$$

$$n_{\text{ethanol}} = \frac{95.00 \text{ g}}{46.0 \text{ g/mol}} = 2.07 \text{ mol}$$

The mole fraction of the solvent can now be calculated:

$$X_{\text{water}} = \frac{2.07 \text{ mol}}{2.07 \text{ mol} + 0.0410 \text{ mol}} = 0.981$$

We can then use Raoult's law (Equation 9-3) to calculate the vapour pressure of the solution:

$$p_{\text{vap,solution}} = X_{\text{ethanol}}p_{\text{vap,ethanol}} = 0.981(13.40 \text{ kPa}) = 13.14 \text{ kPa}$$

 **Does the Result Make Sense?** The vapour pressure is slightly lower than that of the pure solvent, which is expected in a dilute solution.

### Practice Exercise 9.6

How much is the vapour pressure reduced below 101.3 kPa by dissolving 12.5 g of sucrose ($C_{12}H_{22}O_{11}$) in 225 g of coffee at 100 °C? Assume coffee has the same vapour pressure as pure water.

Equation 9-3 describes the total vapour pressure above a solution only when the solute does not have a significant vapour pressure of its own. When both the solvent and solute are volatile, such as for a solution of benzene and toluene, two similar organic liquids, the total vapour pressure above the solution is a sum of contributions from both solvent and solute. Denoting the solvent as "A" and the solute as "B":

$$p_{\text{vap,solution}} = X_A p_{\text{vap,A}} + X_B p_{\text{vap,B}} \qquad (9\text{-}4)$$

Equation 9-4 predicts that the vapour pressure of a solution is between those of the pure solvent and pure solute. This is illustrated in Figure 9-12, for mixtures of benzene and toluene. The x-axis represents the composition of the liquid phase; that is, the mole fractions of each component. The red line is the vapour pressure of toluene, calculated by Raoult's law (Equation 9-3). Note that if $X_{\text{toluene}} = 0$, its vapour pressure is zero, and if $X_{\text{toluene}} = 1$, then the vapour pressure is that of pure toluene. The blue line is similarly the vapour pressure of benzene. If the solution contains some of each, then the total vapour pressure (the green line) can be predicted by Equation 9-4.

Example 9-7 illustrates the use of Equation 9-4 to calculate the total vapour pressure above a solution of a given composition.

**Figure 9-12**

Partial pressures and total pressure above liquid mixtures of benzene and toluene.

## EXAMPLE 9-7 VAPOUR PRESSURE OF A SOLUTION

A closed bottle contains a solution of 5.0% by mass acetone $[(CH_3)_2CO]$ in water. Calculate the vapour pressure of this solution at 25 °C. The vapour pressure of pure water at 25 °C is 3.17 kPa and that of pure acetone is 30.7 kPa.

**Strategy:** To calculate vapour pressure above a solution, we need to know the mole fractions of each of the solvent and solute, as well as the vapour pressures of both pure substances. The information about amounts of acetone acid and water must be converted into moles.

**Solution:** First, use the mass percentage to find the masses of acetone and water in a convenient amount of the solution, exactly 100 g:

$$m_{acetone} = \left(\frac{5.0\%}{100\%}\right)(100.\ g) = 5.00\ g \quad m_{water} = 100\ g - 5.0\ g = 95.0\ g$$

Convert from mass to moles using molar mass in the usual manner:

$$M_{acetone} = 58.0\ g/mol \quad so \quad n_{acetone} = \frac{m}{M} = \frac{5.00\ g}{58.0\ g/mol} = 0.0862\ mol$$

$$M_{water} = 18.0\ g/mol \quad so \quad n_{water} = \frac{m}{M} = \frac{95.0\ g}{18.0\ g/mol} = 5.28\ mol$$

Use these amounts to determine the mole fractions:

$$X_{water} = \frac{n_{water}}{n_{water} + n_{acetone}} = \frac{5.28\ mol}{5.28\ mol + 0.0862\ mol} = 0.984$$

thus, $X_{acetone} = 1 - X_{water} = 1 - 0.984 = 0.016$ and the vapour pressure of the solution can then be calculated:

$$p_{vap,solution} = X_{acetone}p_{vap,acetone} + X_{water}p_{vap,water}$$
$$= 0.016(30.7\ Pa) + 0.984(3.17\ Pa)$$
$$= 3.61\ kPa$$

**Does the Result Make Sense?** The vapour pressure of the solution is intermediate between those of the solvent and solute, which is expected. It is closer to that of water because the mole fraction of water is much larger than that of acetone.

**Practice Exercise 9.7**
A closed bottle contains a solution of 3.0% by mass ethyl acetate ($CH_3COOC_2H_5$) in water. Calculate the vapour pressure of this solution at 25 °C. The vapour pressure of pure water at 25 °C is 3.17 kPa and that of pure ethyl acetate is 12.6 kPa.

Solutions that obey Equation 9-4 are said to be ideal. Ideality in a solution is analogous with ideality in gases—the molecules are assumed to not interact with one another. However, many solutions do not strictly obey Equation 9-4. In such cases, we can infer something about the intermolecular forces present. If the solute–solvent intermolecular forces are strong (stronger than solute–solute or solvent–solvent), then more energy will be required to overcome these forces and the vapour pressure will be lower than what Equation 9-4 predicts. Conversely, relatively weak solute–solvent forces lead to higher than predicted solution vapour pressure. In either case, we refer to the solution as being non-ideal.

## Distillation

Distillation is an important separation process used in the laboratory as well as on a large industrial scale, and is based on Equation 9-4. The essence of separating a mixture of liquids by distillation is that the composition of the gas phase in equilibrium with the solution is different from the composition of the solution.

We can illustrate this by considering a mixture of benzene and toluene, two similar organic liquids. Benzene has the higher vapour pressure at 25 °C (12.7 kPa compared with 3.79 kPa for toluene). Suppose the mole fractions in the liquid mixture are $X_{benzene} = 0.400$ and $X_{toluene} = 0.600$. We can calculate the partial pressure of each compound in the gas phase using Raoult's law:

$$p_{vap,benzene} = X_{benzene}p_{vap,benzene}$$
$$= 0.400(12.7 \text{ kPa})$$
$$= 5.08 \text{ kPa}$$

$$p_{vap,toluene} = X_{toluene}p_{vap,toluene}$$
$$= 0.600(3.79 \text{ kPa})$$
$$= 2.27 \text{ kPa}$$

The composition of the gas phase can then be calculated using Equation 2-2 (Chapter 2):

$$X_{benzene} = \frac{p_{benzene}}{p_{total}} = \frac{5.08 \text{ kPa}}{5.08 \text{ kPa} + 2.27 \text{ kPa}} = 0.691$$

$$X_{toluene} = \frac{p_{toluene}}{p_{total}} = \frac{2.27 \text{ kPa}}{5.08 \text{ kPa} + 2.27 \text{ kPa}} = 0.309$$

Thus, the gas phase is more concentrated in the more volatile component (benzene) even though the liquid phase is more concentrated in toluene. If we now condense this gas phase, we will have a liquid phase consisting of 69.1% benzene and 30.9% toluene. We have therefore enriched the benzene from its original 40.0% to 69.1%.

**Fractional distillation** is the process of separating a mixture of volatile components by performing this evaporation–condensation sequence many times. At each step, the gas phase becomes more enriched in the more volatile component than in the previous step. The liquid phase therefore becomes more concentrated in the less volatile component, and thus the boiling point of the liquid mixture increases as the distillation proceeds. These ideas are conveniently displayed on a plot of boiling point versus solution composition. Figure 9-13 is such a plot for the benzene–toluene system. The two lines represent the compositions of the gas phase (upper line) and the liquid phase (lower line) in equilibrium with one another at a specific temperature.

Beginning with a liquid mixture having $X_{benzene} = 0.300$, we can see from Figure 9-13 that the composition of the gas phase will be approximately $X_{benzene} = 0.500$ in the first stage of the distillation. If this vapour phase is condensed and is allowed to equilibrate with its vapour, the vapour will have a composition of approximately $X_{benzene} = 0.700$ in Stage 2. Repeating this process eventually yields very pure benzene in the vapour phase.

In some cases where the liquid mixture is far from ideal, it is not possible to separate the components by distillation. These solutions are called **azeotropes**, and at some particular liquid-phase composition have the same vapour-phase composition. An example is the ethanol–water system. As seen in Figure 9-14, this system is called a "minimum boiling azeotrope" because the boiling temperature is a minimum where this happens (maximum boiling azeotropes also exist). Beginning with a dilute solution of ethanol in water, it is not possible to distill the mixture to more than 95.6% by mass ethanol. For this reason, ethanol is commonly sold as an "azeotropic mixture" that has this particular composition.

**Figure 9-13**

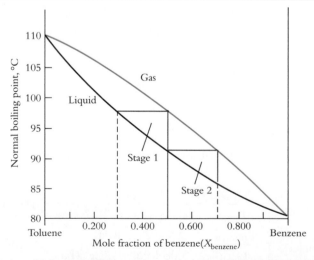

Normal boiling point of benzene–toluene mixtures as a function of the liquid-phase composition. The upper line is the composition of the vapour phase. The lower line is the composition of the liquid phase. The horizontal lines (called "tie-lines") connect these two compositions at particular temperatures.

**Figure 9-14**

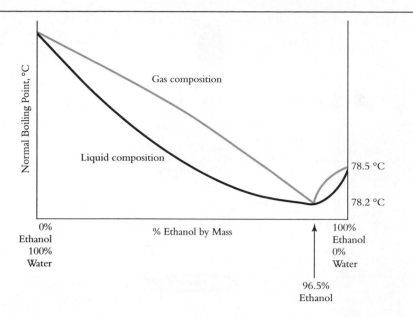

The ethanol–water system is a minimum boiling azeotrope. It is not possible to distill dilute solutions of ethanol to more than 95.6% by mass ethanol, because at this point, the compositions of the liquid and vapour phases are the same.

On a large scale, fractional distillation is used extensively in the oil industry to separate crude oil into components such as gasoline, diesel fuel, and lubricating oils. As shown in Figure 9-15, crude oil is introduced at the bottom of a distillation tower and is heated. The temperature is higher near the bottom than near the top. The more volatile components have higher mole fractions in the gas phase and are transported upward through the tower. The less volatile components condense at higher temperatures and are drawn off as liquids lower in the column.

On a chemistry laboratory scale, fractional distillation is also a very common procedure. Figure 9-16 shows a common distillation apparatus setup for separating liquid mixtures. In this setup, the vapour phase (called the distillate) is condensed and collected in a receiver. As the temperature is increased, components of lower vapour pressure become more volatile and are distilled out of the apparatus.

## Boiling and Freezing Points

The lower vapour pressure of a solution compared with pure solvent results in changes in the boiling point and freezing point of the solution. The molecular perspective is similar to that for vapour pressure reduction. In a solution, solute molecules replace some of the solvent molecules, so a given volume of a solution contains a smaller number of solvent molecules than does the same volume of pure solvent. Consequently, the presence of solute molecules reduces the rate at

**Figure 9-15**

Diagram of a fractional distillation column used to separate useful products from crude oil.

**Figure 9-16**

Diagram of a laboratory-scale fractional distillation setup. The mixture to be separated is placed in the round bottom flask and heated. The more volatile fraction (the distillate) is condensed and collected in the receiver.

which solvent molecules leave the liquid phase. Figure 9-17 shows how this affects a liquid–solid equilibrium: changing one rate without changing the other rate throws the dynamic equilibrium out of balance.

The addition of solutes *decreases* the freezing point of a solution. In the solution, solvent molecules collide with crystals of solid solvent less frequently than they do in the pure solvent. Consequently, fewer molecules are captured by the solid phase than escape from the solid to the liquid. Cooling the solution restores dynamic equilibrium because it simultaneously reduces the number of molecules that have sufficient energy to break away from the surface of the solid and increases the number of molecules in the liquid with small enough kinetic energy to be captured by the solid.

The effect of a solute on the boiling point of a solution is opposite to its effect on the freezing point. A non-volatile solute *increases* the boiling point of a solution. This is because the solute blocks some of the solvent molecules from reaching the surface of the solution and thus decreases the rate of escape into the gas phase.

**Figure 9-17**

Dynamic equilibrium:
Two solid molecules escape,
two liquid molecules are captured

Solute disrupts equilibrium:
Two solid molecules escape,
one liquid molecule is captured

Molecular views of the rates of solid–liquid phase transfer of a pure liquid and a solution at the normal freezing point. The addition of solute does not change the rate of escape from the solid, but it decreases the rate at which the solid captures solvent molecules from the solution. This disrupts the dynamic equilibrium between escape and capture.

To get back to dynamic equilibrium, the solution must be heated so that more molecules acquire sufficient energy to escape from the liquid phase.

These effects can be displayed on a phase diagram that compares the behaviour of pure solvent and a solution. Figure 9-18 shows such a phase diagram for water and aqueous solutions. Depression of the vapour pressure by adding a solute causes the freezing point to decrease (by $\Delta T_f$) and the boiling point to increase (by $\Delta T_b$).

Experiments show that at low solute concentration, the changes in freezing point and boiling point of a solution, $\Delta T_f$ and $\Delta T_b$, depend on the molality of the solute. Furthermore, we need to take account of any dissociation of the solute. For example, a 1 molal solution of NaCl is actually 2 molal in ions, since $NaCl_{(s)} \longrightarrow Na^+_{(aq)} + Cl^-_{(aq)}$ in water. The extent of dissociation is expressed as the **van't Hoff factor, $i$**:

$$i = \frac{\text{moles of particles in solution}}{\text{moles of solute dissolved}}$$

Taking these factors into account, the depression of the freezing point and elevation of the boiling points can be calculated by the equations:

$$\Delta T_f = iK_f b \tag{9-5}$$

$$\Delta T_b = iK_b b \tag{9-6}$$

Since these equations deal with temperature changes, we need to use molality, which is independent of temperature. The constant $K_f$ is called the **freezing point**

**Figure 9-18**

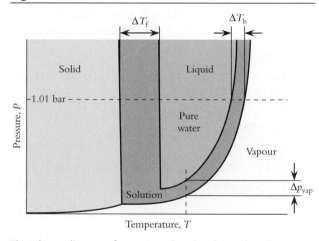

The phase diagram for water, showing how the phase boundaries change when a non-volatile solute dissolves in the liquid.

Equations 9-5 and 9-6 can be derived from our simple molecular picture and kinetic molecular theory. The derivation is independent of the nature of solute and solvent, so the equations are valid for other solvents besides water, except that $K_f$ and $K_b$ have different values for each solvent.

depression constant, and $K_b$ is called the **boiling point elevation constant**. These constants are different for different solvents but do not depend on the identity of the solutes. For water, $K_f$ is 1.858 °C kg/mol and $K_b$ is 0.512 °C kg/mol.

The experimental values of $\Delta T_f$ or $\Delta T_b$ for ionic solutes are often slightly lower than those predicted by Equations 9-5 and 9-6. This is because some of the cations and anions in aqueous solutions form ion pairs, reducing the total number of solute particles, thereby reducing the molality.

Equations 9-5 and 9-6 are both used in the same manner. Example 9-8 shows the procedure.

## EXAMPLE 9-8  FREEZING POINT DEPRESSION

Ethylene glycol (1,2-ethanediol) is added to automobile radiators to prevent cooling water from freezing. Estimate the freezing point of coolant that contains 2.00 kg of ethylene glycol in 5.00 L of water. Is this a high enough concentration to protect a radiator in Montreal, where the temperature may be as low as −40 °C?

**Strategy:** The question asks for the freezing point of a solution. The phrase *to prevent the water from freezing* reveals that we are dealing with depression of the freezing point of water. Equation 9-5 describes this process for a dilute solution: $\Delta T_f = iK_f b$. A coolant solution is quite concentrated, so this equation will not apply exactly, but we can use it to obtain an estimate of the freezing point.

**Solution:** The freezing point depression constant for water is known from experiments and can be found in tables: $K_f$ = 1.858 °C kg/mol. To calculate the freezing point, we must first determine the molality of the solute in this solution, using Equation 9-1:

$$b = \frac{n_{solute}}{m_{solvent}}$$

Determine the number of moles of solute from the mole–mass relationship. The molar mass of ethylene glycol is obtained from its chemical formula, $C_2H_6O_2$:

$$M = 62.07 \text{ g/mol}$$

$$n_{C_2H_6O_2} = \frac{(2.00 \text{ kg})(10^3 \text{ g/kg})}{62.07 \text{ g/mol}} = 32.22 \text{ mol}$$

Find the mass of water from its density, $\rho$ = 1.00 g/mL, and then convert to kilograms:

$$m_{water} = \rho V = (5.00 \text{ L})(10^3 \text{ mL/L})(1.00 \text{ g/mL})(10^{-3} \text{ kg/g}) = 5.00 \text{ kg}$$

Now calculate the molality and substitute into Equation 9-5 to estimate the difference between the freezing point of the solution and that of pure water:

$$b = \frac{32.22 \text{ mol}}{5.00 \text{ kg}} = 6.444 \text{ mol/kg}$$

Note that ethylene glycol does not dissociate in water, so in this case the van't Hoff factor $i$ = 1.

$$\Delta T_f = iK_f b = (1)(1.858 \text{ °C kg/mol})(6.444 \text{ mol/kg}) = 12 \text{ °C}$$

We round to two significant figures because we do not expect the equation to be accurate at this high concentration.

The freezing point of the solution differs from that of pure water by this amount. Because the freezing point of water is 0 °C and freezing points are depressed by adding solutes, the new freezing point is below 0 °C: $T_f = -12$ °C. Obviously, this is not a high enough concentration to protect the radiator on a cold day in Montreal.

 **Does the Result Make Sense?** If you read the label on a container of radiator antifreeze, you will see that to protect your car engine in a Canadian winter requires a 50:50 mixture of ethylene glycol and water.

**Common Pitfall:** Neglecting the dissociation of salts when calculating colligative properties. In this example, the solute does not dissociate, but be aware that salts do.

**Practice Exercise 9.8**

Calculate the boiling point of an 8% by mass aqueous solution of $Na_2SO_4$. Remember that this is an ionic solute.

## Osmosis

**Osmosis** is the movement of solvent molecules through a semipermeable membrane. The membrane is a thin, pliable sheet of material, perforated with molecular-scale holes. The holes are large enough to allow water molecules to pass back and forth through the membrane, but too small to allow the passage of solute molecules or hydrated ions; hence, the membrane is semipermeable.

If a semipermeable membrane separates two identical solutions, then solvent molecules move in both directions at the same rate and there is no net osmosis. The two sides of the membrane are at dynamic equilibrium. The situation changes when the solutions on the two sides of the membrane are different. Consider the membrane in Figure 9-19a, which has pure water on one side and a solution of sugar in water on the other. The sugar molecules reduce the concentration of water molecules in the solution. Consequently, fewer water molecules pass through the membrane from the solution side than from the pure solvent side. Water flows from the side containing pure solvent to the side containing solution, so there is a net rate of osmosis.

**Figure 9-19**

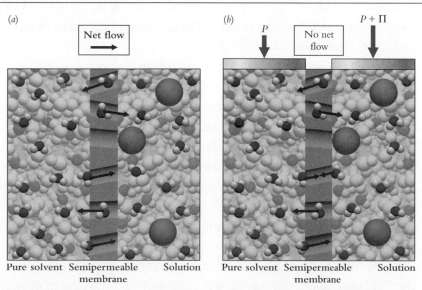

(a) The solvent concentration in a solution is lower than in the pure liquid solvent, leading to an imbalance in flow through a semipermeable membrane. (b) By increasing the pressure on the solution, we can increase the rate of solvent flow out of the solution until it matches the flow out of the pure solvent.

In the absence of other forces, osmosis continues until the concentration of solvent is the same on both sides of the membrane. However, pressure can be used to stop this process. An increase in pressure on the solution side pushes solvent molecules against the membrane and thereby increases the rate of transfer of water molecules from the solution side to the solvent side. Figure 9-19*b* shows that dynamic equilibrium can be established by increasing the pressure on the solution until the rate of solvent transfer is equal in both directions.

The pressure increase needed to equalize the transfer rates is called the **osmotic pressure (Π)**. Osmotic pressure is a pressure *difference*. Pressure is exerted on both sides of a semipermeable membrane, but Π is the extra pressure that must be exerted on the solution to maintain dynamic equilibrium.

Like freezing point depression and boiling point elevation, osmotic pressure is proportional to the concentration of solute molecules. Experiments show that osmotic pressure is proportional to both concentration, *c* (expressed as molarity) and temperature:

$$\Pi = cRT \tag{9-7}$$

In Equation 9-7, *c* is the total molarity of all solutes, *T* is the temperature in kelvins, and *R* is the gas constant, $8.314 \text{ L kPa mol}^{-1} \text{ K}^{-1}$.

Osmotic pressure effects can be substantial. For example, ocean waters contain dissolved salts at a total ionic concentration of about 1.13 M. The osmotic pressure of ocean water can be calculated:

$$\Pi = cRT = (1.13 \text{ mol/L})(8.314 \text{ L kPa mol}^{-1} \text{ K}^{-1})(298 \text{ K}) = 2800 \text{ kPa}$$

This means that if samples of pure water and ocean water were placed on opposite sides of a semipermeable membrane, an external pressure of 2800 kPa (28.0 bar) would have to be applied to the ocean water side to prevent osmosis. The osmotic pressure of ocean water is thus approximately 28 times atmospheric pressure. By comparison, the freezing point of ocean water is depressed by only about 1% from the freezing point of pure water, from 273 K to about 271 K (−2°C).

Osmotic pressure plays an important role in biological chemistry because the cells of the human body are encased in semipermeable membranes and bathed in body fluids. Under normal physiological conditions, the body fluid outside the cells has the same total solute molarity as the fluid inside the cells, and there is no net osmosis across cell membranes. Solutions with the same solute molarity are called *isotonic* solutions.

The situation changes when there is a concentration imbalance. Figure 9-20 shows red blood cells immersed in solutions of different concentrations. When the fluid outside the cell has a higher solute concentration, the result is slower movement of water through the membrane into the cell. The net result is that water leaves the cell, causing it to shrink. When the fluid outside the cell has a lower concentration, movement of water into the cell increases. The extra water in the cell causes an increase in internal pressure. Eventually, the internal pressure of the cell matches the osmotic pressure, and water transport reaches dynamic equilibrium. Unfortunately,

**Figure 9-20**

In isotonic solution, red blood cells are spherical (*top*). At higher ion concentration (*centre*), osmotic flow removes water from the cell interior, shrinking the cell. At lower ion concentration, osmotic flow pumps water into the cell and may cause it to burst (*bottom*).

osmotic pressures are so large that cells can burst under the increased pressure before they reach equilibrium.

Red blood cells are particularly susceptible to these potentially damaging concentration changes because they are suspended in the aqueous medium of the blood. Consequently, solutions used for intravenous feeding must be isotonic. Example 9-9 deals with isotonic solutions.

## EXAMPLE 9-9 ISOTONIC SOLUTIONS

Isotonic intravenous solutions contain 49 g/L of glucose ($C_6H_{12}O_6$). What is the osmotic pressure of blood?

**Strategy:** Isotonic solutions, by definition, exert equal osmotic pressure. Therefore, $\Pi$ for blood is the same as $\Pi$ for the glucose solution. We can calculate $\Pi$ from Equation 9-7 after converting the concentration into moles per litre:

$$\Pi = cRT \quad \text{and} \quad c = \frac{n}{V} = \frac{m/M}{V}$$

**Solution:** From the formula of glucose, $M = 180$ g/mol. Substitute to find the molarity of the glucose solution:

$$c = \frac{(49 \text{ g})/(180 \text{ g/mol})}{(1 \text{ L})} = 0.272 \text{ mol/L}$$

Because we are working with blood in the human body, $T$ is human body temperature, 37 °C.

$$T = 37 \text{ °C} = 37 + 273 = 310 \text{ K} \qquad R = 8.314 \text{ L kPa mol}^{-1} \text{ K}^{-1}$$

$$\Pi = (0.272 \text{ mol/L})(8.314 \text{ L kPa mol}^{-1} \text{ K}^{-1})(310 \text{ K}) = 700 \text{ kPa}$$

If a sample of red blood cells is added to pure water, osmosis carries water into the cells. This process would continue until the internal pressure of the cell was 700 kPa higher than the pressure on the outside of the cell, which is much more than the cell membrane can tolerate. Consequently, red blood cells burst when immersed in pure water.

**Does the Result Make Sense?** The high osmotic pressure of seawater, 2800 kPa, makes it clear that osmotic pressures can be quite large. Thus, an osmotic pressure of 700 kPa is a reasonable value for a solution such as blood.

**Common Pitfall:** Osmotic pressure is the one colligative property whose magnitude depends on molarity, not molality. Be sure to use the correct units for concentration.

### Practice Exercise 9.9

Calculate the osmotic pressure of the sodium sulphate solution described in Practice Exercise 9-8, assuming that it has cooled to 50°C and the density of the solution is 1.00 g/mL.

When the additional pressure applied to a solution is less than the osmotic pressure, solvent molecules flow from pure solvent into the solution. When the additional pressure equals the osmotic pressure, equilibrium is established, and there is no net flow of solvent molecules. What if we apply an additional pressure that is greater than the osmotic pressure? Now the pressure on the solution is sufficient to drive solvent molecules from the solution into the pure solvent. Osmosis now transfers solvent in the opposite direction; **reverse osmosis** occurs.

Reverse osmosis can be used to purify water, because the liquid passing through the semipermeable membrane is pure solvent. A water purifier that uses reverse osmosis requires semipermeable membranes that do not rupture under the high pressures required. Recall that seawater has an osmotic pressure of nearly

Reverse osmosis is carried out on both large and small scales.

28 bar and that red blood cells rupture at 7 bar. Nevertheless, membranes have been developed that make it feasible to purify water using this technique. Reverse osmosis currently supplies pure drinking water to individual households as well as entire municipalities.

## Determination of Molar Mass

The magnitude of osmotic pressure is large enough that measurements of $\Pi$ provide a convenient way to determine the molar mass of a compound. The osmotic pressure equation (Equation 9-7) can be solved for molar mass after molarity is expressed in terms of mass and molar mass:

$$\Pi = cRT \quad \text{so} \quad \Pi = \frac{mRT}{V(M)}$$

Rearrangement gives an equation for molar mass:

$$M = \frac{mRT}{\Pi V} \tag{9-8}$$

For determination of the molar mass of an unknown compound, a measured mass of material is dissolved to give a measured volume of solution. The system is held at constant temperature, and the osmotic pressure is determined. Osmotic pressure measurements are particularly useful for determining the molar masses of large molecules such as polymers and biological materials, as Example 9-10 illustrates.

### EXAMPLE 9-10 DETERMINING MOLAR MASS

A 25.00-mL aqueous solution containing 0.420 g of haemoglobin has an osmotic pressure of 0.613 kPa at 27 °C. What is the molar mass of haemoglobin?

**Strategy:** We have mass, volume, temperature, and osmotic pressure. We use Equation 9-8 to calculate the molar mass.

**Solution:**

$$m_{solute} = 0.420 \text{ g} \quad\quad R = 8.314 \text{ L kPa mol}^{-1}\text{K}^{-1} \quad\quad T = 27\,°C + 273 = 300.\text{ K}$$

$$V_{solution} = (25.00 \text{ mL})(1 \text{ L}/1000 \text{ mL}) = 2.500 \times 10^{-2} \text{ L}$$

$$M = \frac{mRT}{\Pi V} = \frac{(0.420 \text{ g})(8.314\, L \text{ kPa K}^{-1}\text{mol}^{-1})(300 \text{ K})}{(0.613 \text{ kPa})(2.500 \times 10^{-2} \text{ L})}$$

$$M = 6.8 \times 10^4 \text{ g/mol}$$

**Does the Result Make Sense?** This is a large value for a molar mass, but it is reasonable because biological molecules often are quite large. As in any calculation, be careful to express all data in appropriate units. The osmotic pressure was measured to two significant figures, so the result has two significant figures.

### Practice Exercise 9.10

When 7.50 mg of a protein is dissolved in water to give 10.00 mL of solution, the osmotic pressure is found to be 0.221 kPa at 21 °C. Determine the molar mass of the protein.

## 9.5 BETWEEN SOLUTIONS AND MIXTURES

Sugar dissolves in water to give a solution that contains individual sugar molecules distributed uniformly among the water molecules. The aqueous sugar solution is stable and remains uniform indefinitely. A solution is a homogeneous mixture—on the microscopic scale, one microscopic portion of a solution looks the same as every other microscopic portion.

Cornstarch granules are too large to dissolve or to remain suspended in water.

On the other hand, when cornstarch and water are stirred together and allowed to stand, the resulting mixture soon separates into its components. Because cornstarch is insoluble in water, a combination of cornstarch and water forms a heterogeneous mixture. That is, one microscopic portion of this mixture has a different composition than another microscopic portion. An insoluble granule of cornstarch has an entirely different composition than a microscopic portion of water.

One of the characteristic differences between a solution and a heterogeneous mixture is the sizes of the particles composing each of them. Molecule-sized particles such as sugar molecules, with dimensions in the nanometre range, tend to form solutions. Particles such as cornstarch granules, with dimensions larger than micrometres, tend to form heterogeneous mixtures.

### Colloidal Suspensions

Particles with dimensions between 1 nanometre and 1 micrometre, called **colloids**, are larger than the typical molecule but smaller than can be seen under an optical microscope. When a colloid is mixed with a second substance, the colloid can become uniformly spread out, or dispersed, throughout the dispersing medium. Such a dispersion is a **colloidal suspension** that has properties intermediate between those of a true solution and those of a heterogeneous mixture. As Table 9-3 demonstrates, colloidal suspensions can involve nearly any combination of the three phases of matter. Gas–gas mixtures are the exception, because any gas mixes uniformly with any other gas to form a true solution.

The examples in Table 9-3 illustrate the diversity of colloidal suspensions. Colloidal suspensions are commonly found in nature, and people have learned how to make useful suspensions. Some of these are foods such as butter, ice cream, and whipped cream. Others are personal care items like shampoos, shaving creams,

## TABLE 9-3 TYPES OF COLLOIDAL SUSPENSIONS

| Colloid Phase | Phase of Medium | Name | Natural Example | Technical Example |
|---|---|---|---|---|
| Liquid | Gas | Aerosol | Fog | Hair spray |
| Solid | Gas | Aerosol | Pollen | Fumigant |
| Gas | Liquid | Foam | Whipped cream | Fire extinguisher |
| Liquid | Liquid | Emulsion | Milk | Shampoo |
| Solid | Liquid | Sol | River water | Ink |
| Gas | Solid | Solid foam | Pumice | Styrofoam |
| Liquid | Solid | Solid emulsion | Opal | Butter |
| Solid | Solid | Solid sol | Bone | Plumber's solder |

and hair sprays. Ink, paint, and solder also fall in this category. Our Focus 9-2: Chemistry and the Environment explores aerosols in detail.

The stable condition for a liquid or solid is the bulk phase, so we would expect the colloidal particles making up aerosols and emulsions to coagulate into larger particles, destroying the suspension. Although this frequently happens, for example when rain condenses from clouds, many colloidal suspensions last indefinitely. Why is this? Intermolecular forces pull colloidal particles together, so a colloidal suspension must be stabilized by opposing repulsive forces. Electrical charges provide these repulsive forces. The particles of some colloids preferentially attract positive ions, and the particles of other colloids preferentially attract negative ions. When particles have charges of the same sign, the particles repel one another, and this electrical repulsion can be sufficient to prevent condensation into larger particles.

**Figure 9-21**

Sodium stearate
$NaC_{17}H_{35}CO_2$

Sodium stearate is a typical surfactant molecule. It has an ionic, hydrophilic head and a non-polar, hydrophobic tail.

## Surfactants

As described in our chapter introduction, substances that do not dissolve in water, such as organic fats and oils, are called **hydrophobic**. Substances that are miscible with water, including hydrogen-bonding molecules such as methanol and acetone, are called **hydrophilic**. Some molecules contain both hydrophilic and hydrophobic regions. Such a dual-nature molecule, which is known as a **surfactant**, may have a polar or ionic "head" that is hydrophilic and a long hydrocarbon "tail" that is hydrophobic. Figure 9-21 shows the structure of one surfactant, sodium stearate. The head of the stearate anion resembles the water-soluble acetate anion, and the tail is a hydrocarbon chain containing 17 carbon atoms.

The definition of a surfactant is a compound that reduces the surface tension of a liquid at low concentrations. They are able to do this because they can form three different types of structures in water. They may form a molecular **monolayer** on the surface. The polar head groups immerse themselves in the water whereas the non-polar tails aggregate together on the surface. Agitating the solution may cause the molecules to arrange into spherical aggregates called **micelles**, in which the hydrophobic tails point inward. The polar heads lie on the outside of the micelle, where they interact with the aqueous solvent. Surfactants may also form

**Figure 9-22**

Cross-sectional molecular views of the structures that can form when surfactant molecules are placed in water.

enclosed bilayers, called **vesicles**. Vesicles have two parallel rows of molecules oriented so that their hydrocarbon tails cluster together. Figure 9-22 illustrates these three types of structures.

## FOCUS 9-2 CHEMISTRY AND THE ENVIRONMENT: AEROSOLS

You have undoubtedly used one of the common products of modern society, the aerosol spray can. We use these handy containers to deliver fine mists of many useful products. Do you need to keep your hair in place? Use an aerosol spray can to apply mousse. Do you want to get rid of unpleasant room odours? Spray the air with an aerosol air freshener. Has the paint worn off your favourite outdoor chair? Touch it up using an aerosol paint can. These are just three of the many products that can be delivered using aerosol sprays.

Aerosols, like solutions, are mixtures. Unlike solutions, however, they are not single phases. Instead, an aerosol is a suspension of tiny particles of a condensed phase, either liquid or solid, in a gas. The particles that make up an aerosol can have a range of diameters between 10 nm and 10 μm.

Individual aerosol particles are too small to be seen with the naked eye, but we can nevertheless see an aerosol. Aerosol particles scatter light, and we can observe this scattered light from the side. Our atmosphere always contains some aerosol particles, and spectacular sky views sometimes result from such scattering, especially just before sunset. Closer to the ground, we can "see" a searchlight beam from the side because of light scattered by aerosol particles. Light scattering also reduces visibility, so aerosols generate haziness. The Great Smoky Mountains are named for the near perpetual haziness caused by aerosols created by the oxidation and polymerization of organic molecules emitted from their forested slopes.

People invented the aerosol spray can, but naturally occurring aerosols play critical roles in regulating the Earth's weather. Erupting volcanoes spew immense aerosol plumes into the stratosphere. These aerosols persist for years, scattering incoming sunlight and reducing the Earth's average temperature. The water in droplets of sea spray evaporates, leaving behind an aerosol suspension of minuscule salt crystals. These solid particles provide sites where water vapour in the atmosphere can condense, resulting in fog. Indeed, it is believed that the formation of clouds and fog always requires the presence of aerosol particles.

Even though aerosols occur naturally in the atmosphere, anthropogenic aerosols can have substantial effects on atmospheric processes. The burning of tropical forests releases significant aerosol smoke clouds that can change surface temperatures far from their point of origin. Desertification caused by human activities has been followed by the generation of huge dust clouds, whipped up by prevailing winds and carried around the world. Probably the best-known atmospheric effect caused by humans, the ozone hole, was contributed to by freons that were used as propellants in aerosol cans. This usage has now been banned, and recent measurements show that the ozone layer may be on its way to recovery.

Aerosol spray cans were invented in 1929, and perfection of a reliable valve and development of disposable cans took place in the 1940s. Shortly thereafter, "aerosol" became a household word. Like many other modern conveniences, however, the aerosol spray can has drawbacks as well as advantages. Because the particles in an aerosol are extremely tiny, they are quite mobile. They last for a long time in the atmosphere and can affect the climate, as already described. They can penetrate deep into our lungs and cause adverse health effects. Thus, anthropogenic aerosols have both global and local side effects. Despite increasing scientific studies, these effects are not yet fully understood.

All these arrangements obey the principle of like dissolving like. The hydrocarbon tails aggregate through dispersion forces because they are incompatible with the aqueous medium. Incorporating these tails into the solution would disrupt the hydrogen-bonding network of the solvent. The ionic heads, on the other hand, interact strongly with water to maximize ion–dipole and hydrogen-bonding interactions.

Surfactants are widely used in industrial chemistry to modify the behaviour of aqueous solutions. Common surfactant head groups include carboxylate ($-CO_2^-$), sulphonate ($-SO_3^-$), sulphate ($-OSO_3^-$), and ammonium ($-NR_3^+$), where R is a carbon-containing group of atoms). The negative charge of anionic head groups usually is neutralized by $Na^+$, and the positive charge of

$NR_3^+$ usually is neutralized by $Cl^-$. These ions are used because they are non-toxic and their salts are highly soluble.

The most familiar surfactants are soaps and detergents. Many types of substances soil clothing; some are water-soluble, but others are not. In a washing machine, a soap or detergent removes water-insoluble grease (e.g., butter, fat, and oil) from the clothing surface. Dispersion forces are responsible for the attraction of grease particles to the hydrocarbon tails of surfactant molecules, which form aggregates. Agitation removes these aggregates from the fabric, suspending them in solution as tiny micelles with grease particles trapped inside. The micelles do not redeposit on the fabric because the hydrophilic heads of the surfactants hold them in solution. As water drains from the washing machine, it carries away the grease-containing micelles, leaving clean clothes behind.

Soaps are derived from natural sources such as animal fats. Soap made by boiling animal fat in an alkaline solution obtained from ashes has been known since the time of the ancient Sumerians, 2500 BC. Such soaps contain stearic acid and other long-chain organic acids. These carboxylate surfactants form insoluble salts with $Ca^{2+}$ and $Mg^{2+}$. In regions where water is hard, these soaps precipitate calcium and magnesium stearate scum that inhibits cleansing action and is responsible for bathtub rings. Detergents such as sodium lauryl sulphate, on the other hand, are "synthetic" compounds, originally prepared in the laboratory, and contain sulphonate and sulphate head groups. These groups do not form precipitates with +2 cations. The cleaning actions of soaps and detergents are similar, but detergents have largely replaced soaps because of their superior behaviour in hard water.

Some surfactants are used as emulsifiers in processed foods such as bottled salad dressing. An emulsifier causes normally incompatible liquids such as the oil and water in salad dressing to disperse in each other, by forming molecular connections between the liquids. The hydrophobic tails of emulsifier molecules interact with oil molecules, whereas the hydrophilic heads on the emulsifier molecules interact with water molecules.

Surfactants are used in such a wide variety of ways that billions of dollars are spent on them every year. They appear in many household products, including cleansing agents and shampoos (recall the chapter introduction). Half of the surfactants produced in North America are used in household and industrial cleaning products, but the remaining half are used in a wide range of industries.

In agriculture, surfactants are used as wetting agents that assist in the uniform application of sprayed pesticides. Because surfactants reduce the surface tension of the solvent, this allows the solution of the pesticide to more effectively wet the surface it is being applied to. Surfactants also are used to prevent caking of fertilizers. Additives to agricultural products must not interfere with the active agents and must be biodegradable and environmentally benign. In the food industry, various surfactants are used as emulsifiers, cleaners, foaming agents, and antifoaming agents. Paints are dispersions of dyes, binding agents, and fillers. Most

Lecithin, a common phospholipid, has a hydrophobic tail and a hydrophilic head.

paints contain surfactants that improve their flow and mixing properties. Surfactants are used widely in the plastics industry as foaming agents to assist in the production of plastic foams and to improve moldability and extrudability of specially shaped products. In the manufacture of textiles, surfactants are used to clean natural fibres, as lubricants that reduce friction during the spinning and weaving processes, as emulsifiers that improve the application of dyes and finishes, and as antistatic agents.

## Cell Membranes

It may seem like a huge conceptual leap from industrial surfactants to biological cell membranes, but the same principles apply to both sets of substances. Every biological cell is surrounded by a membrane only a few molecules thick. Among the major components of membranes are **phospholipids**, which are surfactants. Although their chemical structures are more complex than simple surfactants such as sodium stearate, phospholipids nevertheless have hydrophilic heads and hydrophobic tails. One phospholipid, lecithin, has a cationic $N(CH_3)_3^+$ group and eight oxygen atoms with non-bonding pairs of electrons, all of which form hydrogen bonds with water molecules. The hydrophobic portion of lecithin consists of two hydrocarbon tails.

Phospholipids form bilayers in aqueous media. The molecules form two approximately parallel rows with tails on the inside and heads toward the outside, in contact with the solution. This arrangement, shown in Figure 9-23, is analogous to the vesicles in Figure 9-22. The bilayer forms a closed sac that contains the aqueous cytoplasm and all the cellular components. Thus, a cell can be viewed as a large and complex vesicle.

**Figure 9-23**

The lipid bilayer of a cell membrane contains two layers of a phospholipid such as lecithin, arranged tail-to-tail.

One purpose of a cellular lipid bilayer is to control the passage of molecules into and out of the cell. Uncharged small molecules such as water, ammonia, and oxygen diffuse readily through the membrane. Hydrophobic molecules such as hydrocarbons can also pass through, because they are soluble in the overlapping tails that make up the interior of the bilayer. Ions and water-soluble polar molecules such as glucose and urea, on the other hand, cannot pass through the membrane.

For cells to carry out their functions, glucose and other nutrients must be brought in, and urea and other waste products must be expelled. This would be an impossible task if cell membranes were composed only of phospholipids. Large protein molecules act as molecular "gates" through the membranes (see Chapter 17 for the structures of proteins). These proteins are embedded in the bilayers but protrude into the surrounding water and/or into the cell interiors, as Figure 9-24 indicates.

**Figure 9-24**

In addition to phospholipids, cell membranes contain protein molecules that carry out special functions such as transporting ions and molecules through the membrane.

## SECTION EXERCISES

**9.5.1** Line drawings of some molecules follow. Identify the hydrophilic and hydrophobic regions of each and determine which are surfactants.

Tetramethylammonium chloride

Benzalkonium chloride
(used as a disinfectant)

Dipentyl ether

**9.5.2** Explain why glucose and other large, water-soluble molecules cannot pass through a lipid bilayer.

# CHAPTER 9 VISUAL SUMMARY

## 9.1 THE NATURE OF SOLUTIONS

Solution concentration
  Molarity
  Mole fraction
  ppm
  ppb

Solute
Solvent

$$c_m = \frac{n_{solute}}{m_{solvent}}$$

Molality

**LEARNING OBJECTIVES:** Work with and interconvert concentration units (mass %, molarity, molality, mole fraction, ppm and ppb).

**SKILLS TO MASTER:** Calculating concentrations in mass %, molarity, molality, mole fraction, ppm and ppb (p. 483).

Converting from one concentration unit to another (p. 485).

**KEY CONCEPTS:** The solvent determines the phase of the solution.

Concentrations are expressed as amount of solute divided by the amount of solvent, or by the amount of solution.

## 9.2 DETERMINANTS OF SOLUBILITY

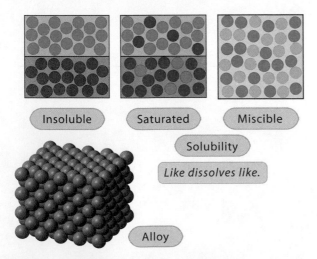

Insoluble    Saturated    Miscible

Solubility

*Like dissolves like.*

Alloy

**LEARNING OBJECTIVES:** Predict the relative solubilities of a solute in various solvents, and explain in terms of intermolecular forces.

**KEY CONCEPTS:** Substances that dissolve in each other usually have similar types of intermolecular interactions ("like dissolves like").

## 9.3 CHARACTERISTICS OF AQUEOUS SOLUTIONS

$H_2O$

Molar heat of solution ($\Delta H_{soln}$)

$$[\text{gas } (aq)]_{eq} = K_H (p_{gas})_{eq}$$

Henry's Law

Heat of dilution

**LEARNING OBJECTIVES:** Calculate and explain solubilities in water.

**SKILLS TO MASTER:** Using Henry's law to calculate gas solubility (p. 498).

**KEY CONCEPTS:** Enthalpies of solution and of dilution depend on the lattice energy of the salt as well as the hydration energies of its ions.

The solubility of a gas in a liquid depends on the partial pressure of the gas.

## 9.4 COLLIGATIVE PROPERTIES

Colligative properties
 Raoult's Law
 Boiling point elevation
 Freezing point depression

$$\Delta T_f = iK_f b$$

$$\Delta T_b = iK_b b$$

$$p_{vap,\,solution} = X_{solvent}\,p_{vap,\,solvent}$$

$$p_{vap,\,solution} = X_A\,p_{vap,A} + X_B\,p_{vap,B}$$

*Colligative properties depend on the amount of solute, but not on the nature of the solute.*

Osmotic pressure ($\Pi$)
Osmosis
Reverse osmosis

$$\Pi = cRT$$

$$M = \frac{mRT}{\Pi V}$$

Pure solvent    Semipermeable    Solution
                membrane

**LEARNING OBJECTIVES:** Understand the reason for and calculate the magnitudes of colligative properties.

**SKILLS TO MASTER:** Using Raoult's Law to calculate the vapour pressure of a solution (p. 502).

Using Raoult's law to calculate the composition of the products of distillation (p. 505)

Calculating the boiling point elevation and freezing point depression due to the addition of a solute (p. 509).

Calculating the osmotic pressure of a solution (p. 512).

Using osmotic pressure to calculate the molecular weight of a solute (p. 514).

**KEY CONCEPTS:** Differences in vapour pressure can be used to separate liquid mixtures by fractional distillation.

For a broad range of solutes, colligative properties depend on the amount of solute but not on the nature of the solute.

Boiling point elevation and freezing point depression result from a decrease in solution vapour pressure compared to the pure solvent.

## 9.5 BETWEEN SOLUTIONS AND MIXTURES

$Na^+$

Hydrophilic

Sodium stearate
$NaC_{17}H_{35}CO_2$

Hydrophobic

Colloid    Colloidal suspension

Monolayer

Micelle

Surfactant

Vesicle

**LEARNING OBJECTIVES:** Understand some aspects of colloidal suspensions and surfactant solutions.

**KEY CONCEPTS:** A colloidal suspension has properties intermediate between those of a true solution and those of a heterogeneous mixture.

A surfactant molecule has a polar head group and a non-polar tail.

Monolayers, micelles and vesicles can form when surfactants are placed in water.

## LEARNING EXERCISES

9.1   Prepare a table of the various solution concentration measures. Describe how to convert from each of them to each of the others.

9.2   Draw molecular pictures that show every type of hydrogen bond that exists in a solution containing methanol, water, and ammonia.

9.3   Write a paragraph that explains the reasons why "like dissolves like" and includes several examples involving different types of solvents and solutes.

9.4   Define and give an example of each of the following: (a) alloy; (b) amalgam; (c) aerosol; (d) colligative property; and (e) surfactant.

9.5   Update your list of memory-bank equations. Be sure to mention how the equations in this chapter are used.

9.6   Write a paragraph that describes the types of substances that form monolayers, micelles, and vesicles in water. Explain the differences among these structures.

9.7   Prepare a list of the terms in Chapter 9 that are new to you. Write a one-sentence definition for each, using your own words. If you need help, consult the glossary.

## PROBLEMS

**ilw** = interactive learning ware problem. Visit the website at www.wiley.com/canada/olmsted

### The Nature of Solutions

9.1   Identify the solvent, the primary solutes, and the normal phase of each in each of the following solutions: (a) carbonated water; (b) white wine; and (c) humid air.

9.2   Identify the solvent, the primary solutes, and the normal phase of each in each of the following solutions: (a) seawater; (b) steel; and (c) vinegar.

9.3   Calculate the mole fractions of a solution made by mixing 14.5 g of methanol ($CH_3OH$) with 101 g of water.

9.4   Calculate the mole fractions of a solution made by mixing 1.755 g of *p*-xylene ($C_8H_{10}$) with 125 g of cyclohexane ($C_6H_{12}$).

9.5   A solution contains 1.521 g of maleic acid ($HO_2CCH{=}CHCO_2H$) dissolved in 85.0 mL of acetone (solution density = 0.818 g/mL). Calculate the molality, mole fraction, and mass percent of maleic acid in the solution.

9.6   Camphor is a natural compound from a tree found in Java, Brazil, China, and other countries. A solution of camphor in ethanol is a mild topical antiseptic. One such solution contains 1.47 g of camphor dissolved in 125 mL of ethanol (solution density = 0.785 g/mL). Calculate the molality, mole fraction, and mass percent of camphor in the solution.

Camphor
($C_{10}H_{16}O$)

9.7   **ilw** A saturated solution of hydrogen peroxide in water contains 30.% by mass $H_2O_2$ and has a density of 1.11 g/mL. Calculate the mole fractions, molarity, and molality of this solution.

9.8   Vodka is 35% by mass ethanol ($C_2H_5OH$) and has a density of 0.94 g/mL. Assuming no other components (other than water) calculate the mole fractions, molarity, and molality of vodka.

9.9   Aqueous ammonia at a concentration of 2.30 M has a density of 0.9811 g/mL. Calculate the mole fractions, mass fraction, and molality of this solution.

9.10  Aqueous ammonia at a concentration of 6.69 M has a density of 0.9501 g/mL. Calculate the mole fractions, mass fraction, and molality of this solution.

9.11  You are asked to prepare an aqueous solution of methanol ($CH_3OH$) with a mole fraction of 0.105. (a) If you use 25.0 g of methanol, what mass of water should you use? (b) What is the molality of the resulting solution?

9.12  You are asked to prepare an aqueous solution of ethylene glycol ($HOCH_2CH_2OH$) with a mole fraction of 0.095. (a) If you use 545 g of water, what mass of ethylene glycol should you use? (b) What is the molality of the resulting solution?

### Determinants of Solubility

9.13  Salad dressing contains long-chain fatty acids such as $C_{17}H_{33}CO_2H$) and vinegar (a dilute aqueous solution of acetic acid, $CH_3CO_2H$), which are not miscible. Explain.

$C_{17}H_{35}CO_2H$

9.14  Acetonitrile ($CH_3CN$) is miscible with water and miscible with cyclohexane ($C_6H_{12}$), but water and cyclohexane are nearly insoluble in each other. Explain.

$H_3C{-}C{\equiv}N{:}$          Acetonitrile          Cyclohexane

9.15  Ammonia condenses to a liquid at low temperature. What kinds of solids would you expect to be soluble in liquid ammonia?

9.16  Mercury is the only metallic element that is a liquid at room temperature. What kinds of solids would you expect to be soluble in liquid mercury?

9.17  Select all the miscible pairs from among the following liquids, and explain your choices: $H_2O$, $CH_3OH$, $C_8H_{18}$, and $CCl_4$.

**9.18** Select all the miscible pairs from among the following liquids, and explain your choices: $H_2O$, $CCl_4$, $C_6H_{12}$, and $Br_2$.

**9.19** Match each solute to its most appropriate solvent, and explain your choices:

| Solute: | $I_2$ | NaCl | Au | paraffin |
|---------|-------|------|------|----------|
| Solvent: | Hg | $CCl_4$ | *n*-octane | water |

**9.20** Match each solute to its most appropriate solvent, and explain your choices:

| Solute: | benzene | $Br_2$ | glucose | Sn |
|---------|---------|--------|---------|------|
| Solvent: | Hg | $CCl_4$ | gasoline | water |

**9.21** Draw a molecular picture of an alloy that contains 90% copper and 10% nickel.

**9.22** Draw a molecular picture of an alloy that contains 95% iron and 5% boron.

## Characteristics of Aqueous Solutions

**9.23** Titan, one of the moons of Jupiter, appears to have oceans composed of liquid methane. Describe how this liquid differs from liquid water, and predict whether methane-based life-forms are likely.

**9.24** At high pressures, carbon dioxide can be liquefied at room temperature. Describe how this liquid differs from liquid water, and predict whether carbon–dioxide-based life-forms are likely.

**9.25** Draw a molecular picture illustrating the solubility equilibrium between KCl (*s*) and KCl (*aq*).

**9.26** Draw a molecular picture illustrating the solubility equilibrium between $NH_3$ (*g*) and $NH_3$ (*aq*).

**9.27** A drop in temperature accompanies the escape of $CO_2$ from a carbonated beverage. Explain this phenomenon in terms of intermolecular forces.

**9.28** A rise in temperature accompanies the addition of water to a concentrated solution of HCl. Explain this phenomenon in terms of intermolecular forces.

**9.29** One of the reasons that different aquatic life-forms thrive in water of different temperatures is the variation with temperature in the concentration of dissolved oxygen. Using data in Table 9-2, calculate the percentage change in the equilibrium oxygen concentration when water warms from 0.0 °C to 25.0 °C.

**9.30** One of the detrimental effects of the "thermal pollution" of water supplies is that a rise in temperature reduces the amount of dissolved oxygen available for fish. Using the information in Table 9-2, calculate how many litres of water a fish requires at 30.0 °C to obtain the same amount of oxygen that it could obtain from 1.00 L of water at 25.0 °C.

**9.31** ⓘⓁⓦ At 25 °C, the equilibrium pressure of ammonia vapour above a 0.500 M aqueous ammonia solution is 0.84 kPa. Calculate the Henry's law constant and determine the equilibrium pressure of ammonia vapour above a 2.5 M solution.

**9.32** At 25 °C, the equilibrium pressure of HCl vapour above a 0.100 M aqueous solution of HCl is 0.53 kPa. Calculate the Henry's law constant and determine the equilibrium pressure of HCl vapour above a 6.0 M solution.

**9.33** Calculate the total pressure above a solution of acetone in water having $X_{acetone} = 0.100$ at 25 °C. At this temperature, the vapour pressure of pure acetone is 30.17 kPa and that of water is 3.17 kPa.

**9.34** Calculate the composition (in mole fractions) of the vapour phase in Problem 9.33.

**9.35** An aqueous solution having $X_{ethanol} = 0.2$ is to be concentrated using fractional distillation. The solution is brought to its boiling point (82.6 °C), at which the vapour pressures of pure ethanol and water are 119.8 kPa and 52.6 kPa, respectively. The distillate is condensed and collected (see Figure 9-16). What is the mole fraction of ethanol in the distillate?

## Colligative Properties

**9.36** Urea, a fertilizer, has the chemical formula $(NH_2)_2CO$. Calculate the vapour pressure of water above a fertilizer solution containing 7.50 g of urea in 15.0 mL of water (solution density = 1.00 g/mL), at a temperature at which the vapour pressure of pure water is 4.400 kPa.

**9.37** Ethylene glycol, an automobile coolant, has the chemical formula $HOCH_2CH_2OH$. Calculate the vapour pressure of water above a coolant solution containing 65.0 g of ethylene glycol dissolved in 0.500 L of water (density 1.00 g/mL), at 100 °C, the boiling point of pure water.

**9.38** ⓘⓁⓦ Estimate the freezing point of a wine that is 12% by mass ethanol. (Ignore all other solutes.)

**9.39** Calculate the freezing point of a solution that contains 12.50 g of sucrose $(C_{12}H_{22}O_{11})$ in 155 mL of water.

**9.40** Do you have enough information to calculate the boiling point of the wine in Problem 9.38? If so, calculate it. If not, explain what feature of wine prevents you from doing this calculation.

**9.41** Do you have enough information to calculate the boiling point of the solution in Problem 9.39? If so, calculate it. If not, explain what feature of the solution prevents you from doing this calculation.

**9.42** List the following aqueous solutions in order of decreasing freezing point, and explain your listing: 0.30 M NaCl, 0.75 M $NH_3$, 0.25 M $MgCl_2$, 0.25 M $NH_3$.

**9.43** List the following aqueous solutions in order of increasing boiling point and explain your listing: 0.20 M NaCl, 0.20 M sucrose, 0.25 M $MgCl_2$, 0.40 M sucrose.

**9.44** An aqueous solution contains 1.00 g/L of a derivative of the detergent lauryl alcohol. The osmotic pressure of this solution at 25.0 °C is measured to be 8.64 kPa. (a) What is the molar mass of the detergent? (b) The hydrocarbon portion of the molecule is an 11-carbon chain. What is the molar mass of the polar portion?

9.45 An aqueous solution containing 1.00 g of a sugar in 100 mL of solution has an osmotic pressure of 1.38 bar at 25 °C. (a) What is the molar mass of this sugar? (b) Sugars have the empirical formula $C_x(H_2O)_y$, where $y = x, x - 1$, or $x - 2$. What is the molecular formula of this sugar?

9.46 **ilw** The boiling point of an aqueous solution containing sucrose ($C_{12}H_{22}O_{11}$) is 101.45 °C. Calculate the osmotic pressure of this solution at 35 °C, at which the solution density is 1.036 g/mL.

9.47 The freezing point of an aqueous solution containing pantothenic acid ($M = 205.3$ g/mol) is −0.65 °C. Calculate the osmotic pressure of this solution at 18 °C, at which the density of the solution is 1.015 g/mL.

## Between Solutions and Mixtures

9.48 Describe the similarities and differences between fog and smoke.

9.49 Describe the similarities and differences between milk and whipped cream.

9.50 Of the following compounds, which will be the best and which will be the worst surfactant? Support your choices with molecular pictures. (a) Propanoic acid, $H_3CCH_2CO_2H$; (b) sodium lauryl sulphate, $H_3C(CH_2)_{11}OSO_3{}^-Na^+$; and (c) lauryl alcohol, $H_3C(CH_2)_{11}OH$.

9.51 Of the following compounds, which will be the best and which will be the worst surfactant? Support your choices with molecular pictures. (a) Sodium alkylbenzene-sulphonate, $C_{18}H_{29}SO_3Na$; (b) octadecane, $C_{18}H_{38}$; and (c) decanoic acid, $H_3C(CH_2)_8CO_2H$.

9.52 The figure below represents a monolayer of the surfactant sodium stearate on a liquid surface. Decide if the solvent is polar or non-polar. Explain your reasoning.

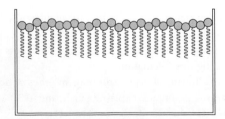

Sodium stearate      Simplified depiction
                     of the stearate anion

9.53 Stearic acid forms a monolayer on the surface of water. Draw a molecular picture that shows how stearic acid molecules are arranged in this monolayer.

## ADDITIONAL PAIRED PROBLEMS

9.54 Commercial phosphoric acid is 85% by mass $H_3PO_4$ and 15% water. The density of the acid solution is 1.685 g/mL.

Calculate the molality, the mole fraction, and the molarity of this solution.

9.55 A commercial solution contains 2.0 M ammonia in methanol. The density of the solution is 0.787 g/mL. Calculate the molality, the mole fraction, and the percent by mass of the ammonia solution.

9.56 Dichlorodiphenyltrichloroethane (DDT) has the following line structure:

Is this compound hydrophilic or hydrophobic? Will animals excrete DDT readily, or will the compound concentrate in fatty tissues? Does your answer explain why DDT has been banned as a pesticide?

9.57 Butylated hydroxytoluene (BHT) is used as a food preservative. It has the following line structure:

Would you expect to find this compound to be excreted in urine or stored in body fat? Explain your reasoning. BHT is non-toxic to humans.

9.58 What molality of ethylene glycol is required to protect the water in an automobile cooling system from freezing at −20 °C?

9.59 Compute the molar mass of Vitamin C, if a solution containing 22.0 g in 100 g of water freezes at −2.33 °C.

9.60 Rank the following substances in order of increasing solubility in water, and state the reasons for your rankings: $C_6H_6$ (benzene), $HOCH_2CH(OH)CH(OH)CH_2OH$ (erythritol), and $C_5H_{11}OH$ (pentanol).

9.61 Rank the following substances in order of increasing solubility in cyclohexane ($C_6H_{12}$) and state the reasons for your rankings: KCl, $C_2H_5OH$, and $C_3H_8$.

9.62 Water and carbon tetrachloride are not miscible. When mixed together, they form two layers, like water and oil. If an aqueous solution of $I_2$ is shaken with $CCl_4$, the iodine moves into the $CCl_4$ layer. Explain this behaviour based on your knowledge of intermolecular forces.

9.63 If some benzene is shaken with a mixture of water and carbon tetrachloride, the resulting mixture contains two layers (see Problem 9.62). Which layer contains the benzene? Explain this behaviour based on your knowledge of intermolecular forces.

9.64 To make a good solder joint, the liquid metal solder must adhere well to the metal surfaces being joined. "Flux" is used to clean the metal surfaces. What types of substances must flux remove?

9.65 A pipette is "dirty" when water forms beads on its walls rather than forming a thin film that drains well. Which of the following on the surface of a pipette wall will make it "dirty"? In each case, explain the intermolecular forces underlying your classification. (a) grease; (b) $Mg^{2+}$ ions; (c) acetone; and (d) $SiO_2$.

9.66 Carbon disulphide ($CS_2$) has a density of 1.261 g/mL and boils at 46.30 °C. When 0.125 mol of a solute is dissolved in 200.0 mL of $CS_2$, the solution boils at 47.46 °C. (a) Determine the molal boiling point constant for this solvent. (b) When 2.7 g of an unknown compound is dissolved in 25.0 mL of $CS_2$, the solution boils at 47.08 °C. Determine the molar mass of the unknown.

9.67 Benzene ($C_6H_6$) has a normal freezing point of 5.50 °C and a density of 0.88 g/mL. When 1.28 g of naphthalene ($C_{10}H_8$) is dissolved in 125 mL of benzene, the freezing point of the solution is 5.03 °C.
(a) Determine the molal freezing point constant for this solvent.
(b) When 0.125 g of an unknown compound is dissolved in 25.0 mL of benzene, the solution freezes at 5.24 °C. Determine the molar mass of the unknown.

## ADDITIONAL PROBLEMS

9.68 One of the earliest methods of preserving fish was by salting. Explain what happens when fish is placed in a concentrated salt solution.

9.69 Would water dissolve salts as well as it does if it had a linear structure (such as $CO_2$) instead of a bent one? Explain.

9.70 When an aqueous solution cools to low temperature, part of the water freezes as pure ice. What happens to the freezing point of the remaining solution when this occurs? A glass of wine placed in a freezer at −10 °C for a very long time forms some ice crystals but does not completely freeze. Compute the molality of ethanol in the remaining liquid phase.

9.71 Some surfactants form membranes that span small holes between two aqueous solutions. These membranes are liquid bilayers two molecules thick. Draw a molecular picture of one of these membranes.

9.72 Explain using intermolecular forces whether water can dissolve gasoline. Knowing that gasoline is less dense than water, would you use water to fight a gasoline fire? Explain.

9.73 Saltwater fish have blood that is isotonic with seawater, which freezes at −1.96 °C. What is the osmotic pressure of fish blood at 15 °C?

9.74 Brackish water, with a salt content around 0.5% by mass, is found in semi-arid regions such as the American southwest. Assuming that brackish water contains only sodium chloride and that the ions form no ion pairs, estimate the osmotic pressure of brackish water at 298 K.

9.75 Hard candy is made from very hot solutions of sugar and water with small amounts of added flavourings. In a typical preparation, the boiling point of the sugar–water mixture reaches 145 °C. What mass ratio of sugar ($C_{12}H_{22}O_{11}$) to water is required to raise the boiling point of a sugar–water solution to 145 °C?

9.76 Using the information in Table 9-2, calculate the following: (a) the mass of $CO_2$ that dissolves in 225 mL of a carbonated beverage at 1.11 bar pressure and 25 °C; and (b) the partial pressure of $CO_2$ in the gas space above the liquid if a bottle of this carbonated beverage is stored at 0.0 °C.

## GROUP STUDY PROBLEMS

9.77 Drinking seawater can be dangerous because it is more concentrated than a person's body fluids. Discuss the consequences, at both the cellular and the molecular levels, of drinking too much seawater.

9.78 Polystyrene is a plastic that dissolves in benzene. When a benzene solution (density = 0.88 g/mL) containing 8.5 g/L of polystyrene is placed in an osmometer at 25 °C, the solution reaches equilibrium when the height of liquid inside the osmometer is 12.9 cm higher than outside. Calculate the molar mass of the polystyrene.

9.79 Colloidal suspensions of gold are used to label biological molecules for imaging by electron microscopy. A typical colloidal suspension contains 0.10 g of gold per litre of suspension, and the particle size is 15 nm diameter. The density of gold is 19.3 g/cm³. The volume of a sphere is $4\pi r^3/3$, and the surface area of a sphere is $4\pi r^2$. (a) What is the volume in litres of one colloidal particle? (b) How many gold atoms are there in one colloidal particle? (c) How many colloidal particles are there in 1 L of this solution? (d) What is the total surface area of gold in part (c)? (e) What would be the surface area of a cube of gold of mass 0.10 g?

9.80 The table lists the parts per million by mass of the principal ions in seawater:

| Ion | $Cl^-$ | $Na^+$ | $Mg^{2+}$ | $SO_4^{2-}$ | $Ca^{2+}$ | K | $Br^-$ |
|---|---|---|---|---|---|---|---|
| $10^2$ ppm | 194 | 108 | 12.9 | 9.04 | 4.11 | 3.92 | 0.67 |

Assuming that each ion acts independently of all the others and that seawater has a density of 1.026 g/mL, calculate the freezing point and osmotic pressure of seawater. The actual freezing point of seawater is −1.96 °C. What conclusion can you reach about your assumptions? Take this into account and recalculate the osmotic pressure of seawater.

9.81 At body temperature (37 °C), the Henry's law constant for $N_2$ dissolving in water is $3.8 \times 10^{-4}$. Calculate the solubility of dinitrogen in blood that is in contact with air (78% $N_2$) at atmospheric pressure and at 4.05 bar, the pressure at a sea depth of 30.5 m. Determine the volume of $N_2$ that will be released from 1 L of blood if a deep-sea diver surfaces quickly from this depth. If this escape of $N_2$ occurs in the form of gas bubbles that are 1 mm in diameter, how many bubbles per litre is this?

9.82 Consider Raoult's law, as given in Equation 9-3. Raoult's law applies to the total vapour pressure only when the vapour pressure of the solute is negligible compared with the vapour pressure of the solvent. Explain this limitation in molecular terms. What equation from this chapter might you use to correct for the limitation? Will a solution containing a volatile solute have a higher or lower vapour pressure and boiling point than that predicted using Raoult's law? Use Figure 9-11 as a starting point.

## ANSWERS TO SECTION EXERCISES

9.1.1  Copper is the solvent, zinc is the solute. This is a substitutional solid solution.

9.1.2  $X = 1.7 \times 10^{-6}$, $[C_{20}H_{24}O_2N_2] = 9.6 \times 10^{-5}$M, $b = 9.6 \times 10^{-5}$ moles/kg

9.1.3  1492 g $Na_2SO_4$

9.2.1  Intermolecular interactions are dispersion, dipole–dipole, and hydrogen bonding.

Dipole-dipole          Hydrogen bonding

9.2.2  Only $Na_4SiO_4$, which is an ionic solid, is water-soluble.

9.2.3  Zinc salts are soluble because the solutions are stabilized by ion–dipole interactions between partially negative O atoms in water molecules and the $Zn^{2+}$ cations, which replace ion–ion interactions in a zinc salt. Zn metal is insoluble because the "sea" of electrons in delocalized orbitals strongly binds the cations in the metal.

9.3.1  The O atom and the H atom of the methanol —OH group can form hydrogen bonds to water molecules:

9.3.2  $KNO_3 < Li_2SO_4 < KCl < KBr < NaCl < NaNO_3$

9.3.3  The fraction that dissolves is $8.3 \times 10^{-2}$ (8.3%). This is not an efficient way to collect $CO_2$ because its solubility in water is significant.

9.4.1  Freezing point depression = $1.42 \times 10^{-3}$ °C; boiling point elevation = $3.90 \times 10^{-4}$ °C; and osmotic pressure = $1.89 \times 10^{-2}$ bar = 14.2 Torr

9.4.2  The figure should look the same, except there should be more arrows moving from right to left than from left to right.

9.4.3 Your description should include the features shown in your figure for 9.4.2 and the prevention of ion passage through the pores of the semipermeable membrane.

9.5.1

$$H_3C\!-\!\overset{\overset{\displaystyle CH_3}{|}}{\underset{\underset{\displaystyle CH_3}{|}}{N^+}}\!-\!CH_3 \qquad Cl^-$$  All hydrophilic (ions)

Hydrophilic            Hydrophobic

(This molecule is a surfactant.)

All hydrophobic

9.5.2 Because they are large, they cannot slip easily between the molecules of a lipid bilayer membrane. Because they are hydrophilic, they do not dissolve in the bilayer.

## ANSWERS TO PRACTICE EXERCISES

9.1   $b = 4.04 \times 10^{-2}$ mol/kg

9.2   $b = 15.9$ mol/kg; $X_{HCl} = 0.223$

9.3   Each liquid has extensive hydrogen bonding that must be overcome for it to dissolve in cyclohexane. Solubility increases as the hydrocarbon portion gets longer because the extent of hydrogen bonding decreases and the magnitude of cyclohexane–solute dispersion forces increases.

9.4   Uric acid forms several hydrogen bonds and is water-soluble. Testosterone contains mostly non-polar C — C and C — H bonds and is fat-soluble.

9.5   $3.1 \times 10^{-4}$ M $N_2$, $1.9 \times 10^{-4}$ M $O_2$, $5.2 \times 10^{-6}$ M $CO_2$

9.6   $\Delta p_{vap} = 0.296$ kPa

9.7   $p_{vap,\ solution} = 3.23$ kPa

9.8   $T_b = 100.94°C$

9.9   $\Pi = 45.4$ bar

9.10   $M = 8.28 \times 10^3$ g/mol

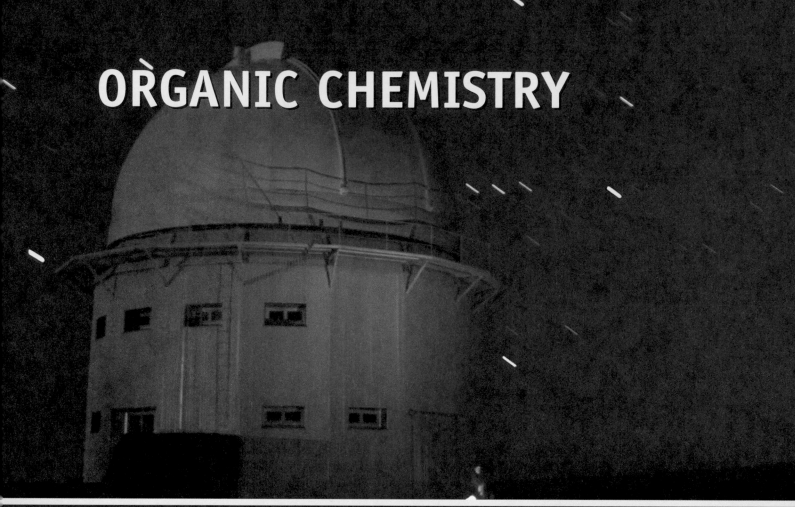

# ORGANIC CHEMISTRY

## THE HANDEDNESS OF LIFE

Molecules of the amino acids that make up our proteins have the property of being non-superimposable on their mirror image. Because of this, they are said to be chiral, or to possess "handedness." Although both mirror image forms are theoretically possible, such as those for the amino acid alanine above, life on Earth involves amino acids that are mainly of the mirror-image form said to be "left-handed." The reason that most amino acids are of the left-handed form is not known, however. In the absence of an influence that possesses handedness, such as a living system, chemical reactions produce an equal mixture of both mirror-image forms. Since almost all theories about the origin of life presume that amino acids and other molecules central to life were present before self-replicating organisms came into being, it was assumed that they were present in equal mirror-image forms in the primordial soup. But could the mirror-image forms of these molecules actually have been present in unequal amounts before life began, leading to some sort of preference as life evolved? A meteorite discovered in 1970, known as the Murchison meteorite, fuelled speculation about this topic. Analysis of the meteorite showed that amino acids and other complex molecules associated with life were present, proving that molecules required for life could arise outside the confines of Earth. Even more interesting, recent experiments have shown that a 7–9% excess of four left-handed amino acids is present in the Murchison meteorite. The origin of this unequal distribution is uncertain, but some scientists speculate that electromagnetic radiation, emitted in a corkscrew fashion from the poles of spinning neutron stars, could lead to a bias of one mirror-image isomer over another when molecules form in interstellar space.

In this chapter, you will learn how to recognize, name, and draw some common types of organic molecules. Families of organic molecules are characterized

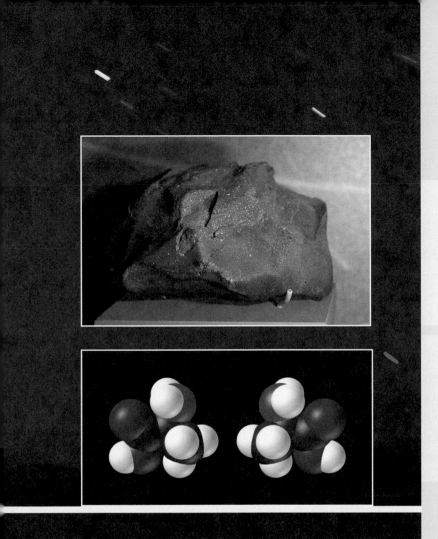

# 10

## CHAPTER CONTENTS

## LEARNING OBJECTIVES

*Upon completion of this chapter you should be able to:*

- draw and name hydrocarbons using the IUPAC system
- draw and name aromatic compounds using the IUPAC system
- draw and name compounds containing common functional groups
- recognize, name, and draw stereoisomers
- predict the products from $S_N2$ and $S_N1$ reactions
- write the mechanisms of $S_N2$ and $S_N1$ reactions
- predict the products from E2 and E1 reactions
- write the mechanisms of E2 and E1 reactions
- predict the products of addition reactions
- write the mechanisms of electrophilic addition of hydrogen halides, water, or halogens to alkenes

by groups of atoms called functional groups, which impart specific properties and reactivities to molecules. It is important to learn to recognize and name compounds based on their functional groups. The three-dimensional nature of molecules, or stereochemistry, is also introduced. Later in the chapter, we describe three important classes of organic reactions, and describe under what conditions they occur, as well as some aspects of the mechanisms by which they proceed.

## 10.1 HYDROCARBONS

Carbon's ability to form as many as four strong bonds to other carbon atoms, as well as to hydrogen, oxygen, sulphur, and nitrogen atoms, is its most important characteristic and accounts for the vast number of different molecules required for complex living organisms. The word "organism" is in fact the origin of "organic" chemistry.

## Hydrocarbons: Representative Alkanes, Alkenes, and Alkynes

Hydrocarbons, as the name implies, are compounds with molecules that contain only carbon and hydrogen atoms. Methane ($CH_4$, see Figure 10-1) and ethane ($C_2H_6$) are hydrocarbons. They also belong to a subgroup of hydrocarbons known as alkanes, which have only single bonds between carbon atoms. Hydrocarbons with molecules that have a carbon=carbon double bond are called alkenes, and those with a carbon≡carbon triple bond are called alkynes. Alkanes are also referred to as being **saturated**, which means they have the maximum number of hydrogen atoms. Compounds with multiple bonds, such as alkenes and alkynes, are called **unsaturated** compounds because they possess fewer than the maximum number of hydrogen atoms.

The generalized formula of an alkane is $C_nH_{2n+2}$. Owing to unsaturation, alkenes have fewer hydrogens than alkanes, and have the general formula $C_nH_{2n}$. Similarly, alkynes have still fewer hydrogens, and their general formula is $C_nH_{2n-2}$.

## Alkanes

The principal sources of alkanes are natural gas and petroleum. The smaller alkanes (having one to four carbon atoms) are gases under ambient conditions. The higher molecular weight alkanes are obtained largely by refining petroleum. Methane ($CH_4$) was a major component of the early atmosphere of this planet, but is no longer present in the atmosphere in appreciable amounts (probably a good thing since it is a very efficient greenhouse gas). It is, however, a major component of the atmospheres of Jupiter, Saturn, Uranus, and Neptune. Titan, Saturn's largest moon, contains a large amount of methane in its atmosphere. Scientists believe Titan's methane atmosphere is continuously being replenished from methane on or near the moon's surface.

On Earth, methane is the major component of natural gas. Some living organisms produce methane from carbon dioxide and hydrogen. These very primitive creatures, called *methanogens*, may be Earth's oldest organisms, and they may represent a separate form of evolutionary development. Methanogens can survive

Most organic chemicals are derived from oil and natural gas.

**Figure 10-1**

**Methane**        **Ethene**        **Ethyne**
              **(ethylene)**     **(acetylene)**

Structures of three simple hydrocarbons: methane (an alkane), ethene (an alkene), and ethyne (an alkyne).

only in an anaerobic (i.e., oxygen-free) environment. They have been found in ocean trenches, in mud, in sewage, and in cows' stomachs.

## Alkenes

Ethene ($C_2H_4$, see Figure 10-1) and propene ($C_3H_6$) are the two simplest alkenes and are among the most important industrial chemicals produced in the world. Each year, the chemical industry worldwide produces more than 75 billion kilograms of ethene and about half as much propene.

Ethene is used as a starting material for the synthesis of many industrial compounds, including ethanol, ethylene oxide, ethanal, and the polymer polyethylene (see Section 17.2). Ethene also occurs in nature as a plant hormone. It is produced naturally by fruits such as tomatoes and bananas and is involved in the ripening process of these fruits. Ethene is commonly used in the commercial fruit industry to bring about the ripening of tomatoes and bananas picked green because the green fruits are less susceptible to damage during shipping. Propene is used in making the polymer polypropylene, and is a starting material for the synthesis of acetone, a very useful solvent. There are many naturally occurring alkenes. The two line structures in Figure 10-2 are of limonene (found in the rind of many citrus fruits) and beta-carotene (the compound that makes carrots orange), both examples of alkenes.

**Figure 10-2**

Limonene                                        beta-Carotene

Limonene and beta-carotene are examples of naturally occurring alkenes.

Recall from Chapter 1 that the lines in line structures represent bonds, and it is assumed that carbon atoms are at each bend in the line and at the ends of lines. The number of hydrogen atoms is inferred by assuming that each carbon atom makes a total of four bonds with other atoms, hydrogen included. Double and triple bonds are shown by double and triple lines, respectively. Non-carbon atoms are shown explicitly, as are any hydrogen atoms bound to those atoms. Many examples of line structures appear in this chapter.

## Alkynes

The simplest **alkyne** is ethyne ($C_2H_2$; also called acetylene, see Figure 10-1). Alkynes occur in nature and can be synthesized in the laboratory. Two examples of alkynes (see Figure 10-3) among thousands that have a biosynthetic origin are

**Figure 10-3**

Capillin                        Dactylyne                       Ethinyl estradiol

Capillin and dactylyne are naturally occurring alkynes, whereas ethyinyl estradiol is a synthetic alkyne.

capillin, an antifungal agent, and dactylyne, a marine natural product that is an inhibitor of pentobarbital metabolism (pentobarbital is a potent anaesthetic). Ethinyl estradiol (Figure 10-3) is a synthetic alkyne with estrogen-like properties that is used in oral contraceptives.

## Naming Alkanes

The development of a formal system for naming organic compounds did not come about until near the end of the nineteenth century. Before that time, many organic compounds had already been discovered. The names given these compounds sometimes reflected a source of the compound. Acetic acid, for example, can be obtained from vinegar and takes its name from the Latin word for vinegar, *acetum*. Formic acid can be obtained from some ants; it is named from the Latin word for ants, *formicae*. Ethanol (or ethyl alcohol) was at one time called grain alcohol because it was obtained by fermentation of grains. These older names for organic compounds are now called "common" or "trivial" names, and are still widely used by chemists and biochemists and in commerce. You will learn some of these common names because we will use them occasionally, but it is more important to learn the systematic method for naming compounds.

The formal system of nomenclature used today is one proposed by the International Union of Pure and Applied Chemistry (IUPAC). Underlying the IUPAC system of nomenclature for organic compounds is a fundamental principle: *each different compound has an unambiguous name.* Thus, the IUPAC system provides different names for the tens of millions of known organic compounds, and names can be devised for any other compound yet to be synthesized or discovered. In the same way, we can derive the structure of a given compound from its IUPAC name. The IUPAC system for naming alkanes is not difficult to learn, and the principles involved are used in naming compounds in other families as well; thus, we begin our study of the IUPAC system with the rules for naming alkanes.

The simplest alkanes are ones in which the carbon atoms are connected in a straight chain. The names for several of the unbranched alkanes are listed in Table 10-1. The ending for all of the names of alkanes is *-ane*. Learning the names is like learning to count in organic chemistry. Thus, one, two, three, four, and five become meth-, eth-, prop-, but-, and pent-, respectively.

## TABLE 10-1 THE STRAIGHT-CHAIN ALKANES

| Name | Number of Carbon Atoms | Structure | Line Structure |
|---|---|---|---|
| Methane | 1 | $CH_4$ | |
| Ethane | 2 | $CH_3CH_3$ | — |
| Propane | 3 | $CH_3CH_2CH_3$ | |
| Butane | 4 | $CH_3(CH_2)_2CH_3$ | |
| Pentane | 5 | $CH_3(CH_2)_3CH_3$ | |
| Hexane | 6 | $CH_3(CH_2)_4CH_3$ | |
| Heptane | 7 | $CH_3(CH_2)_5CH_3$ | |
| Octane | 8 | $CH_3(CH_2)_6CH_3$ | |
| Nonane | 9 | $CH_3(CH_2)_7CH_3$ | |
| Decane | 10 | $CH_3(CH_2)_8CH_3$ | |

## Structural Isomers of Alkanes

In many cases, the carbon atoms of alkanes are not arranged in straight chains, but are instead "branched." This leads to structural isomers—structures having the same formula, but with the carbon atoms connected differently. For instance, there is a branched version of butane called *iso*butane in which three carbons are in a row, and a fourth is bonded to the middle of these:

Butane          *Iso*butane

Note that both of these structures are butane, since both have the formula $C_4H_{10}$. For pentane, there are two structural isomers other than the straight chain, named *iso*pentane and *neo*pentane:

Pentane          *Iso*pentane          *Neo*pentane

As the number of carbon atoms increases, the number of structural isomers increases quickly. There are 5 structural isomers of hexane, 9 of heptane, 18 of octane, 35 of nonane, and 75 of decane.

## Naming Straight-Chain Alkyl Groups

Many larger molecules contain groups of atoms that are almost identical to alkanes. If we remove one hydrogen atom from an alkane, we obtain an **alkyl group**. These alkyl groups have names that end in -*yl*. When the hydrogen atom that is removed is a **terminal** hydrogen atom (i.e., at one end of the chain), the names are straightforward. The first four examples are given in Table 10-2.

### TABLE 10-2 NAMES AND ABBREVIATIONS OF ALKYL GROUPS

| Alkane | | Alkyl Group | Abbreviation |
|---|---|---|---|
| $CH_3$–**H**<br>**Methane** | becomes | $CH_3$–<br>**Methyl** | Me– |
| $CH_3CH_2$–**H**<br>**Ethane** | becomes | $CH_3CH_2$–<br>**Ethyl** | Et– |
| $CH_3CH_2CH_2$–**H**<br>**Propane** | becomes | $CH_3CH_2CH_2$–<br>**Propyl** | Pr– |
| $CH_3CH_2CH_2CH_2$–**H**<br>**Butane** | becomes | $CH_3CH_2CH_2CH_2$–<br>**Butyl** | Bu– |

## Naming Branched-Chain Alkanes

In some cases, such as *iso*butane and *neo*pentane, the carbon atoms are not all arranged in a straight chain. Instead, smaller alkyl groups are attached to carbons along the longer chain, and a so-called "branched" alkane results. Branched-chain alkanes are named according to the following rules:

1. **Locate the longest continuous chain of carbon atoms; this chain determines the parent name for the alkane.** The following compound, for example, is a hexane because the longest continuous chain contains six carbon atoms:

$$CH_3CH_2CH_2CH_2CHCH_3$$
$$|$$
$$CH_3$$
or

The longest continuous chain may not always be obvious from the way the formula is written. For example, the following alkane is named as a heptane because the longest chain contains seven carbon atoms:

$$CH_3CH_2CH_2CH_2CHCH_3$$
$$|$$
$$CH_2$$
$$|$$
$$CH_3$$

2. **Number the longest chain beginning with the end of the chain nearer the substituent.** Applying this rule, we number the above two alkanes as follows:

$$\overset{6}{C}H_3\overset{5}{C}H_2\overset{4}{C}H_2\overset{3}{C}H_2\overset{2}{C}H\overset{1}{C}H_3$$
$$|$$
$$CH_3$$
**Substituent**

Substituent

$$\overset{7}{C}H_3\overset{6}{C}H_2\overset{5}{C}H_2\overset{4}{C}H_2\overset{3}{C}HCH_3$$
$$|$$
$$^2CH_2$$
$$|$$
$$^1CH_3$$

3. **Number the substituent group accordingly.** The parent name is placed last, and the substituent group, preceded by the number showing its location on the chain, is placed first. Numbers are separated from words by a hyphen. Our two examples are therefore 2-methylhexane and 3-methylheptane, respectively:

$$\overset{6}{C}H_3\overset{5}{C}H_2\overset{4}{C}H_2\overset{3}{C}H_2\overset{2}{C}H\overset{1}{C}H_3$$
$$|$$
$$CH_3$$

$$\overset{7}{C}H_3\overset{6}{C}H_2\overset{5}{C}H_2\overset{4}{C}H_2\overset{3}{C}HCH_3$$
$$|$$
$$^2CH_2$$
$$|$$
$$^1CH_3$$

**2-Methylhexane**          **3-Methylheptane**

4. **When two or more substituents are present, give each substituent a number corresponding to its location on the longest chain.** For example, the following compound is 4-ethyl-2-methylhexane:

$$CH_3CHCH_2CHCH_2CH_3$$

$$CH_3 \quad CH_2$$

$$CH_3$$

**4-Ethyl-2-methylhexane**

The substituent groups should be listed *alphabetically* (e.g., ethyl before methyl).

5. **When two substituents are present on the same carbon atom, use that number twice:**

$$CH_3$$

$$CH_3CH_2CCH_2CH_2CH_3$$

$$CH_2$$

$$CH_3$$

**3-Ethyl-3-methylhexane**

6. **When two or more substituents are identical, indicate this by the use of the prefixes di-, tri-, tetra-, and so on.** Then make certain that each and every substituent has a number. Commas are used to separate numbers from one another:

$$CH_3CH-CHCH_3$$
$$CH_3 \quad CH_3$$
**2,3-Dimethylbutane**

$$CH_3$$
$$CH_3CHCHCHCH_3$$
$$CH_3 \quad CH_3$$
**2,3,4-Trimethylpentane**

$$CH_3 \quad CH_3$$
$$CH_3CCH_2CCH_3$$
$$CH_3 \quad CH_3$$
**2,2,4,4-Tetramethylpentane**

2,2,4-Trimethylpentane is also known as *iso*octane. The octane rating of gasoline is a measure of its resistance to detonation (knocking), and does not reflect its octane content.

Application of these six rules allows us to name most of the alkanes encountered. Two other rules, however, may be required occasionally:

7. **When two chains of equal length compete for selection as the parent chain, choose the chain with the greater number of substituents:**

$$\overset{7}{CH_3}-\overset{6}{CH_2}-\overset{5}{CH}-\overset{4}{CH}-\overset{3}{CH}-\overset{2}{CH}-\overset{1}{CH_3}$$
$$CH_3 \quad CH_2 \quad CH_3 \quad CH_3$$
$$CH_2$$
$$CH_3$$

**2,3,5-Trimethyl-4-propylheptane**
**(four substituents)**

Note also from this example that the multiplying prefixes such as "di" and "tri" are ignored when alphabetizing the groups. *Trimethyl* therefore comes before *propyl*.

8. **When branching first occurs at an equal distance from either end of the longest chain, choose the name that gives the lower number at the first point of difference:**

$$\overset{6}{CH_3}-\overset{5}{CH}-\overset{4}{CH_2}-\overset{3}{CH}-\overset{2}{CH}-\overset{1}{CH_3}$$
$$CH_3 \quad\quad CH_3 \quad CH_3$$

**2,3,5-Trimethylhexane**
(*not* 2,4,5-trimethylhexane)

## EXAMPLE 10-1  NAMING ALKANES

Name the following alkanes:

(a)          (b)          (c)

**Strategy:** Follow the eight rules in order. Some of the last rules may not be necessary.

**Solution:**

(a)   This is a straight chain of eight carbon atoms. From Table 10-1, we see that its name is octane.

(b)   The longest chain in this structure has seven carbon atoms, hence it is a heptane. We can try numbering the molecule in two different ways:

1          7   or   7          1

The numbering on the right will give us lower numbers for placement of the substituents, so we choose it. We see that there are three methyl groups bonded to the chain, hence it is a trimethylheptane. They are located at positions 2, 3, and 6, hence the molecule is 2,3,6-trimethylheptane. (Note that if we had numbered the chain from the other end, we would conclude that the name is 2,5,6-trimethylheptane, which is incorrect according to rule number 8.)

(c)   The longest chain in this structure has five carbon atoms, hence it is a pentane. There are three methyl groups, hence trimethylpentane. Counting from the left side will give us the lower numbers: 2,2,4-trimethylpentane.

 **Do the Results Make Sense?** In each case, the parent chain is longer than the substituents, and the numbers are as low as possible.

**Common Pitfalls:** Not finding the longest continuous chain. Examine the structure carefully, considering several possibilities, before naming it.

**Practice Exercise 10.1**

Give the IUPAC systematic name for each of the following alkanes:

(a)          (b)                              (c)

## EXAMPLE 10-2  DRAWING ALKANES

Draw line structures of the following alkanes: (a) 2,3,6-trimethyloctane; (b) 5-ethyl-2-methyl-5-propylnonane; and (c) 2-methylbutane.

**Strategy:** Drawing a structure is done in the same order as naming a structure. Begin by identifying the longest chain from the name. Then number the carbon atoms in this chain and place the substituents at the appropriate places.

**Solution:**

(a) The chain is octane, which according to Table 10-1 has eight carbons:

Add the substituents at the appropriate places:

(b) The chain is nonane, which has nine carbon atoms:

Add the substituents at the appropriate places:

(c) The chain is butane, which has four carbon atoms:

Add the substituent at the appropriate place:

 **Do the Results Make Sense?** After drawing these structures, try to name them using the IUPAC system. If the name matches the original name, the structure is likely correct.

**Common Pitfalls:** Not using the lowest possible numbers. Try numbering from both ends to make sure your numbers are the lowest.

### Practice Exercise 10.2

Draw line structures of the following alkanes: (a) 2,2,7-trimethyldecane; (b) 4-ethyl-2-methylhexane; and (c) 3,3-diethyl-5,5-dimethyloctane.

## Other Alkyl Groups

It makes no difference which hydrogen atom is removed from methane ($CH_4$) to make a methyl group because all four hydrogen atoms are identical. Likewise, in ethane ($CH_3CH_3$), all six hydrogens are identical, and removing any one of them makes the ethyl group. However, in larger alkanes, not all hydrogen atoms are equivalent. In propane ($CH_3CH_2CH_3$), for instance, the six hydrogens on the terminal methyl groups are identical to one another, but different from the two on the middle carbon atom. Thus, two different propyl groups can be formed, depending on which hydrogen is removed:

$$-CH_2CH_2CH_3 \qquad CH_3\overset{|}{C}HCH_3$$

Propyl            *Iso*propyl

The prefixes *iso* and *neo* are used without a hyphen, as in *iso*butyl. The prefixes *sec-* and *tert-* are used with hyphens, as in *sec*-butyl and *tert*-butyl.

Similarly for the butyl group, there are four possibilities:

$$-CH_2CH_2CH_2CH_3 \qquad CH_3CH_2\overset{|}{C}HCH_3 \qquad \underset{CH_3}{\overset{\displaystyle -CH_2\overset{|}{C}HCH_3}{}} \qquad \underset{CH_3}{CH_3\overset{|}{\underset{|}{C}}CH_3}$$

Butyl       *sec*-Butyl       *Iso*butyl       *tert*-Butyl

## Cycloalkanes

Cycloalkanes, as the name implies, are alkanes that have a cyclic (ring) structure. Completing the ring structure requires making an extra C—C bond, so the number of hydrogens is less than in the corresponding alkane, and the formula for a cycloalkane is $C_nH_{2n}$. Compounds from 3 to 30 carbon atoms arranged in a ring have been prepared, and many cycloalkanes exist in nature. Line structures of the first four cycloalkanes are shown in Figure 10-4.

**Figure 10-4**

Cyclopropane    Cyclobutane    Cyclopentane    Cyclohexane
$C_3H_6$           $C_4H_8$           $C_5H_{10}$           $C_6H_{12}$

Line structures of the first four cycloalkanes.

The bond angles in cyclopropane and cyclobutane are much less than the ideal $109.5°$, even though the carbon atoms are $sp^3$ hybridized. As a result, these C—C bonds are fairly weak and these two compounds are quite reactive.

Naming cycloalkanes requires numbering the carbon atoms. For instance, all the carbon atoms in cyclopentane are equivalent, so the numbering serves only to indicate the relative positions of substituents. Some examples of cycloalkanes are shown in Figure 10-5.

**Figure 10-5**

Methylcyclopropane          1-Ethyl-3-methylcyclobutane          1,4-Dimethylcyclohexane

Numbering of the carbon atoms in cycloalkanes serves to designate the relative positions of substituents.

## Other Substituents

Alkyl groups are of course not the only substituents in organic chemistry. Some of the commonly encountered groups and their names as substituents are given below.

| | |
|---|---|
| $-Cl$ | chloro |
| $-Br$ | bromo |
| $-F$ | fluoro |
| $-I$ | iodo |
| $-NH_2$ | amino |
| $-NO_2$ | nitro |
| $-OH$ | hydroxy |

For naming purposes, these are used in the same manner as alkyl groups. Some examples are given in Figure 10-6.

**Figure 10-6**

2-Nitrobutane          3-Chloroaminocyclopentane          2-Bromo-*iso*propylcyclobutane

Examples of naming substituted cycloalkanes.

## Naming Alkenes and Alkynes

In ethene, the double bond can only be in one place—between the two carbon atoms. In propene, this is also the case, because no matter where we place the double bond, we obtain the same molecule; that is, there is only one structural isomer of propene:

These are the same molecule.

For butene, however, there are two structural isomers, which differ in the location of the double bond. "1-Butene" means butene with a double bond from carbon number 1 to the next highest number, 2. Similarly, "2-butene" means the double bond goes from carbon 2 to carbon 3:

1-Butene          2-Butene

Alkynes are named in a similar fashion. For example, there are three structural isomers of hexyne:

1-Hexyne    2-Hexyne    3-Hexyne

To name an alkene or alkyne, follow the same rules as for alkanes, but specify where the multiple bond is. Some examples are

2-Methyl-1-butene    1-Chloro-2-butene    4,5-Dimethyl-2-hexyne

Note that the third example is not 2,3-dimethyl-4-hexyne. The numbering is done such that the multiple bond is at as low a number as possible, and only then are the substituent positions numbered accordingly.

It is possible to have unsaturation in cyclic structures. Cyclohexene and 1,3-cyclopentadiene ("diene" meaning two double bonds) are common examples:

1,3-Cyclopentadiene    Cyclohexene

The familiar smell of new carpeting comes from 4-phenylcyclohexene. This compound forms as an undesirable by-product during the manufacture of rubber-backed carpets. Its toxicity is currently being investigated.

4-Phenylcyclohexene

## SECTION EXERCISES

**10.1.1** Draw and name all structural isomers of hexane, $C_6H_{14}$.

**10.1.2** Name the following compounds:

(a)    (b)    (c)    (d)    (e)    (f)

**10.1.3** Draw the following structures: (a) 2,2-dinitropropane; (b) *tert*-butylcyclohexane; (c) 4-amino-2,2,3-trimethylhexane; (d) 2,3-dichloro-4-ethylcyclohexene; (e) propadiene; and (f) 4,5-difluoro-2-heptyne.

## 10.2 AROMATIC COMPOUNDS

The compound known as **benzene** is the prototype of a group of unsaturated cyclic hydrocarbons known as aromatic compounds. Benzene can be written as a six-membered ring with alternating single and double bonds, called a Kekulé structure after August Kekulé, who first conceived of this arrangement of atoms (reportedly in a dream!):

**Kekulé structure for benzene**     **Bond-line representation of Kekulé structure**

Even though the Kekulé structure is frequently used for benzene compounds, there is much evidence that this representation is inadequate and incorrect. For example, if benzene had alternating single and double bonds as the Kekulé structure indicates, we would expect the lengths of the carbon–carbon bonds around the ring to be alternately longer and shorter, as we typically find with carbon–carbon single and double bonds. In fact, the carbon–carbon bonds of benzene are all the same length (1.39 Å), a value in between that of a carbon—carbon single bond and a carbon-carbon double bond. There are two ways of dealing with this problem: with resonance theory or with molecular orbital theory. If we use resonance theory, we visualize benzene as being represented by either of two equivalent Kekulé structures:

**Two contributing Kekulé structures for benzene**     **A representation of the resonance hybrid**

Based on the principles of resonance theory (see Section 6.2) we recognize that benzene cannot be represented adequately by either structure, but that, instead, it should be visualized as a hybrid of the two structures. We represent this hybrid by a hexagon with a circle in the middle.

In the molecular orbital explanation, we begin by recognizing that the carbon atoms of the benzene ring are $sp^2$ hybridized. Therefore, each carbon has a $p$ orbital that has one lobe above the plane of the ring and one lobe below, as shown in Figure 10-7.

**Figure 10-7**

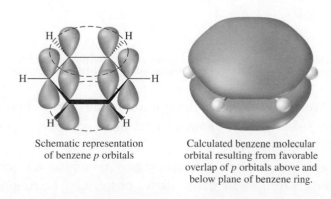

Schematic representation of benzene $p$ orbitals

Calculated benzene molecular orbital resulting from favorable overlap of $p$ orbitals above and below plane of benzene ring.

Schematic and calculated $p$ orbital representations for benzene.

The lobes of each *p*-orbital above and below the ring overlap with the lobes of neighbouring *p*-orbitals. This kind of overlap of *p*-orbitals leads to a set of bonding molecular orbitals that encompass all of the carbon atoms of the ring, as shown in the calculated molecular orbital. Therefore, the six electrons associated with these *p*-orbitals (one electron from each orbital) are **delocalized** about all six carbon atoms of the ring. This delocalization of electrons explains how all the carbon–carbon bonds are equivalent and have the same length.

## Naming Benzene Compounds

Some compounds are substituted benzenes, in which a hydrogen atom has been replaced by another atom or group of atoms. Naming these compounds consists of naming the atoms or groups of atoms, and specifying where they are attached to the ring structure relative to one another.

In many simple compounds, benzene is the parent name, and the substituent is written as a prefix. Some of these compounds also have trivial names (indicated in parentheses):

CH₃                Cl                NH₂               OH

Methylbenzene      Chlorobenzene     Aminobenzene      Hydroxybenzene
(Toluene)                            (Aniline)         (Phenol)

In the above examples, it makes no difference which carbon atom the substituent is bonded to since all six carbons in the ring are identical. When there are two atoms or groups bonded to the benzene ring, however, it is necessary to indicate where the substituents are relative to one another. This can be done by numbering the carbon atoms in the ring, or by using the prefixes *ortho*, *meta*, and *para* (*abbreviated o-, m-, and p-*). For instance, the dichlorobenzenes are

*o*-Dichlorobenzene      *m*-Dichlorobenzene      *p*-Dichlorobenzene
(1,2-Dichlorobenzene)    (1,3-Dichlorobenzene)    (1,4-Dichlorobenzene)

For 1,2-dichlorobenzene, we say the chlorine atoms are "ortho" to one another. Similarly, they are "meta" to one another in 1,3-dichlorobenzene and "para" to one another in 1,4-dichlorobenzene.

When there are more than two substituent groups bonded to a benzene ring, we must use numbers to indicate their relative positions. The ring carbons are numbered in such a way that the numbers are as low as possible. Some examples are

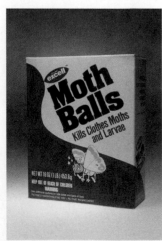

Moth balls are pieces of solid
*p*-dichlorobenzene.

## 1,2,4-Trichlorobenzene
(*not* 1,3,4-trichlorobenzene)

## 1-Butyl-2,4-difluorobenzene

## 1,2,3,5-Tetrabromobenzene
(*not* 1,3,4,5-tetrabromobenzene)

## Pentabromobenzene
(numbers not necessary)

## 1-Methyl-2,5-Dinitrobenzene
(2,5-dinitrotoluene)
(*not* 1-methyl-3,6-dinitrobenzene)

## Pentachlorohydroxybenzene
(pentachlorophenol)
(numbers not necessary)

In three of the above examples, there are two possible numbering schemes—in all cases, we choose the scheme that gives the lowest initial numbers, as for alkanes. Note also that in some of the examples shown above, numbers are not necessary. For instance, since all six carbons in benzene are equivalent, numbering in pentabromobenzene would be redundant. Similarly, in pentachlorophenol, all five carbons not bonded to the —OH group must be bonded to bromine atoms, so numbering is not necessary.

When a benzene ring is attached to some other group of atoms in a molecule, it is called a **phenyl group** (see Figure 10-8).

Example 10-3 gives some practice in naming benzene compounds.

**Figure 10-8**

2-Phenylpropane
(*iso*propylbenzene)

Phenylmethane
(methylbenzene)
(toluene)

Phenylcyclohexane

Examples of phenyl-containing compounds.

## EXAMPLE 10-3 NAMING BENZENE COMPOUNDS

Name the following compounds:

(a)   (b)   (c)   (d)   (e)

**Strategy:** Often there is more than one way to name simple aromatic compounds such as these. We can number the carbon atoms on the ring; we can use the ortho-, meta-, and para-designations of relative positions of substituents; and sometimes, we can use common names. All three approaches are used here.

**Solution:**

(a) The two fluorine atoms are in the 1 and 4 positions on the ring. An acceptable name is therefore 1,4-difluorobenzene. The 1 and 4 positions are *para* to one another, so *p*-difluorobenzene is also acceptable.

(b) Using numbers to name this compound, we must decide whether the –OH group or the –CH₃ group is at position number 1. Thus, 3-methylhydroxybenzene or 3-hydroxymethylbenzene are both correct IUPAC names for this compound. We could also note that the two substituents are *meta* to one another, so *m*-hydroxymethylbenzene is an acceptable name. Hydroxybenzene is also called phenol, so *m*-methylphenol and 3-methylphenol describe this compound. Likewise, methylbenzene is commonly called toluene, so this compound could be called *meta*-hydroxytoluene or 3-hydroxytoluene. Eight names all describing the same compound unambiguously!

(c) This aromatic compound must be named using numbers since there are more than two substituents. It is 1,3,5-trimethylbenzene.

(d) Naming it as a benzene compound, it is propylbenzene. Alternatively, it can be named 1-phenylpropane.

(e) This can be named bromobenzene or phenylbromide.

 **Do the Results Make Sense?** There are several ways to name these compounds. Make sure you can name them by numbering the ring, or by using *o*-, *m*-, and *p*- designations where possible.

**Common Pitfalls:** Incorrect numbering of the ring carbons. Count both ways and make sure your numbers are the lowest possible.

**Practice Exercise 10.3**

Name the following compounds using numbers and the *o*-, *m*-, and *p*- designations where possible:

(a)     (b)     (c)     (d)

**SECTION EXERCISES**

●**10.2.1** Draw and name all the possible structural isomers of tribromobenzene.

●**10.2.2** Name the following compounds:

(a)     (b)     (c)

●**10.2.3** Draw the following structures: (a) *p*-dichlorobenzene; (b) *m*-iodophenol; and
(c) 3-iodocyclopentylbenzene.

## 10.3 FUNCTIONAL GROUPS

A **functional group** is the part of a molecule where most of its chemical reactions occur. It is the part that effectively determines the compound's chemical properties (and many of its physical properties as well). Simple hydrocarbons are arranged in classes, as we saw in the first two sections of this chapter. In this section, we introduce other common functional groups and their properties. Table 10.3 at the end of this section summarizes the most important functional groups.

We can simplify our discussion if we introduce a symbol that is widely used in designating general structures of organic molecules: the symbol R. *R is used as a general symbol to represent any alkyl group.* For example, R might be a methyl group, an ethyl group, a propyl group, or an isopropyl group. Thus, the general formula for an alkane is R–H.

### Alkyl Halides

Alkyl halides are compounds in which a halogen atom (fluorine, chlorine, bromine, or iodine) replaces a hydrogen atom of an alkane. For example, $CH_3Cl$ and $CH_3CH_2Br$ are alkyl halides. Alkyl halides are also called **haloalkanes**. The generic formula for an alkyl halide is R–X, where X is fluorine, chlorine, bromine, or iodine.

Some examples of haloalkanes are given below. They are named in the same fashion as branched alkanes, by including the halogen atoms as substituents in the name:

| $CH_3CH_2Cl$ | $CH_3CH_2CH_2F$ | $CH_3CHBrCH_3$ |
|:---:|:---:|:---:|
| **Chloroethane** | **1-Fluoropropane** | **2-Bromopropane** |

When the parent chain has both a halo and an alkyl substituent attached to it, number the chain from the end nearer the first substituent, regardless of whether it is halo or alkyl. If two substituents are at equal distance from the end of the chain, then number the chain from the end nearer the substituent that has alphabetical precedence:

**2-Chloro-3-methylpentane**      **2-bromo-4-chloropentane**

Many simple haloalkanes are instead named as alkyl halides. This naming style is used in particular for derivatives of a few common small alkyl groups with accepted trivial names such as *iso*propyl, *iso*butyl, *tert*-butyl, and *neo*pentyl.

| $CH_3CH_2Cl$ | $CH_3CHCH_3$ | $(CH_3)_3CBr$ | $CH_3CHCH_2Cl$ | $CH_3CCH_2Br$ |
|:---:|:---:|:---:|:---:|:---:|
| | $\mid$ | | $\mid$ | $\mid$ |
| | **Br** | | $CH_3$ | $CH_3$ |
| **Ethyl chloride** | **Isopropyl bromide** | ***tert*-Butyl bromide** | **Isobutyl chloride** | **Neopentyl bromide** |

Alkyl halides are classified as being primary (1°), secondary (2°), or tertiary (3°), depending on the carbon atom to which the halogen is directly attached. If the carbon atom that bears the halogen is attached to only one other carbon, it is said to be a **primary carbon atom** and the alkyl halide is classified as a **primary alkyl halide**. If the carbon that bears the halogen is itself attached to two other carbon atoms, then the carbon is a **secondary carbon** and the alkyl halide is a **secondary alkyl halide**. If the carbon that bears the halogen is attached to three other carbon atoms, then the carbon is a **tertiary carbon** and the alkyl halide is a **tertiary alkyl halide**. Some examples of primary, secondary, and tertiary alkyl halides are given in Figure 10-9.

**Figure 10-9**

Examples of primary, secondary, and tertiary alkyl halides.

## EXAMPLE 10-4 ALKYL HALIDES

Draw line structures and give the IUPAC names for all of the structural isomers of $C_4H_9Br$. State whether each is a primary, secondary, or tertiary alkyl halide.

**Strategy:** Draw all possible structural isomers of the alkane $C_4H_{10}$. For each, place the Br atom at all non-redundant positions.

**Solution:** The simplest isomer of $C_4H_{10}$ is ⌃⌄⌃ . The Br atom could be placed at only two positions:

Br⌄⌃⌄ (1-bromobutane, a primary alkyl halide) or ⌃⌄⌃ (2-bromobutane, a secondary alkyl

halide). The other structural isomer of $C_4H_{10}$ is *iso*butane, ⅄ . This also has two possible placements of

the Br atom: ⅄ (2-bromo-2-methylpropane or *tert*-butyl bromide, a tertiary alkyl halide) and ⅄

(1-bromo-2-methylpropane or *iso*butyl bromide, a primary alkyl halide).

**Do the Results Make Sense?** Each structural isomer has the formula $C_4H_9Br$, so the results do make sense.

**Common Pitfalls:** Not considering all structural isomers of the hydrocarbon. Make sure you have drawn all of them, and that you have considered each possible placement of the other atom (Br in this case).

## Practice Exercise 10.4

(a)  Write line structures for any two isomeric alkyl bromides with the formula $C_5H_{11}Br$. Write

formulas for (b) ethyl fluoride and (c) *iso*propyl chloride. What are the names for (d) [structure with F] and (e) $C_6H_5I$?

## Alcohols

Methanol (also called methyl alcohol) has the structural formula $CH_3OH$ and is the simplest member of a family of organic compounds known as **alcohols**. The characteristic functional group of this family is the hydroxyl (−OH) group attached to an $sp^3$−hybridized carbon atom. Another example of an alcohol is ethanol, $CH_3CH_2OH$ (also called ethyl alcohol).

**This is the functional group of an alcohol.**

Alcohols may be viewed in two ways structurally: (1) as hydroxyl derivatives of alkanes or (2) as alkyl derivatives of water. Ethyl alcohol, for example, can be seen as an ethane molecule in which one hydrogen has been replaced by a hydroxyl group, or as a water molecule in which one hydrogen has been replaced by an ethyl group:

The above drawings show the two lone pairs of electrons on the oxygen atom, but these are often not shown explicitly.

As with alkyl halides, alcohols are classified into three groups: primary, secondary, and tertiary alcohols. Some examples of primary alcohols are

If the carbon atom that bears the hydroxyl group is a secondary carbon, then the alcohol is a secondary alcohol:

Isopropyl alcohol
(a 2° alcohol)

Menthol
(a 2° alcohol found
in peppermint oil)

If the carbon atom that bears the hydroxyl group is a tertiary carbon, then the alcohol is a tertiary alcohol:

*tert*-Butyl alcohol
(a 3° alcohol)

Norethindrone
(an oral contraceptive that contains a 3° alcohol
group as well as a ketone group and carbon–
carbon double and triple bonds)

The presence of a hydroxyl group in an alcohol has such important consequences to the chemistry of the molecule that we emphasize its importance by changing the base name from that of an alkane to an *alkanol*. Alcohols are therefore named following a procedure similar to that for branched chain alkanes. First, select the longest continuous carbon chain *to which the hydroxyl is directly attached*. Change the name of the alkane corresponding to this chain by dropping the final "e" and adding the suffix *-ol*. Then number the longest continuous carbon chain so as to give the carbon atom bearing the hydroxyl group the lower number. Indicate the positions of other substituents (as prefixes) by using the numbers corresponding to their positions along the carbon chain. Figure 10-10 shows how these rules are applied.

Alcohols containing two hydroxyl groups are commonly called **glycols**. In the IUPAC system they are named as **diols** (Figure 10-11).

**Figure 10-10**

Examples of naming alcohols.

**Figure 10-11**

| Substitutive | 1,2-Ethanediol or ethane-1,2-diol | 1,2-Propanediol or propane-1,2-diol | 1,3-Propanediol or propane-1,3-diol |
| --- | --- | --- | --- |
| *Common* | *Ethylene glycol* | *Propylene glycol* | *Trimethylene glycol* |

Examples of naming glycols.

Syntheses of alcohols are very important industrial processes. Tens of billions of litres of methanol are produced worldwide each year by reacting carbon monoxide with hydrogen gas over a catalyst:

$$CO + 2\,H_2 \xrightarrow[\text{ZnO/Cr}_2\text{O}_3 \text{ catalyst}]{400°\ C} CH_3OH$$

Methanol is a good solvent, and is also the starting material for the syntheses of formaldehyde ($H_2CO$) and acetic acid ($CH_3COOH$), as well as many other useful compounds.

Ethanol, the alcohol in all wines, beers, and distilled liquors, is commonly produced by fermentation of sugars or grains in the presence of yeast:

$$\underset{\text{glucose}}{C_6H_{12}O_6} \xrightarrow{\text{Yeast}} \underset{\text{ethanol}}{2\,C_2H_5OH} + 2\,CO_2$$

*Iso*propanol, also known as rubbing alcohol, is produced by reacting propene with water in the presence of an acid and a catalyst:

$$\underset{\text{Propene}}{CH_3CH = CH_3} + H_2O \xrightarrow[\text{Catalyst}]{\text{Acid}} \underset{\textit{Iso}\text{propanol}}{CH_3\overset{\displaystyle OH}{\underset{|}{C}}HCH_3}$$

**EXAMPLE 10-5 ALCOHOLS**

Write line structures for each of the following, and state whether each is a primary, secondary, or tertiary alcohol: (a) 3-pentanol; (b) 2-methyl-2-butanol; and (c) 1-hexanol.

**Strategy:** Draw the line structures and inspect the carbon to which the –OH group is attached.

**Solution**

(a)  The structure is [structure with OH and carbon number 3]. Carbon number 3 is bonded to two other carbons, so this is a secondary alcohol.

(b)  The structure is [structure with OH and carbon number 2]. Carbon number 2 is bonded to three other carbons, so this is a tertiary alcohol.

(c)  The structure is [structure with HO and carbon number 1]. Carbon number 1 is bonded to just one other carbon, so this is a primary alcohol.

**Do the Results Make Sense?** The –OH group in a primary alcohol must be at the end of a chain. For a secondary alcohol, it must be attached to a –CH group, and in a tertiary alcohol, it must be attached to a tertiary carbon, so all three answers make sense.

**Practice Exercise 10.5**

(a)  Write line structures for (i) two primary alcohols, (ii) a secondary alcohol, and (iii) a tertiary alcohol—all having the molecular formula $C_4H_{10}O$.

(b)  Another way of naming alcohols is to name the alkyl group that is attached to the –OH and add the word *alcohol*, just like we did for alkyl halides. Write line structures for (i) propyl alcohol and (ii) *iso*propyl alcohol.

Diethyl ether was first used as an aneasthetic in the mid-nineteenth century. In this device, sponges in the glass vessel were soaked in ether. As the patient breathed in through the mouthpiece, air was drawn into the chamber, over the sponges, and into the patient's lungs.

## Ethers

Ethers have the general formula R−O−R or R−O−R′, where R′ is an alkyl (or phenyl) group that may be different from R. Ethers can be thought of as derivatives of water, in which both hydrogen atoms have been replaced by alkyl groups. The bond angle at the oxygen atom of an ether is only slightly larger than that of water:

**General formula for an ether**          **Dimethyl ether**

One way of naming ethers is to name the two alkyl groups attached to the oxygen atom in alphabetical order and add the word *ether*. If the two alkyl groups are the same, we use the prefix *di-*, as in *dimethyl ether*, for example.

Ethers can be synthesized by the dehydration of alcohols in the presence of sulphuric acid. The term "dehydration" means removal of water:

$$R\text{—}OH + HO\text{—}R \xrightarrow{H_2SO_4} R\text{—}O\text{—}R + H_2O$$

Diethyl ether is made commercially, for instance, by the dehydration of ethanol:

$$CH_3CH_2OH \xrightarrow[140\,°C]{H_2SO_4} CH_3CH_2OCH_2CH_3$$

Diethyl ether is the most common ether found in an organic chemistry laboratory. It is a reasonably good solvent, and can be easily removed from a reaction mixture because its boiling point is so low (34.6 °C).

## EXAMPLE 10-6  ETHERS

Write line structures for (a) diethyl ether, (b) ethyl propyl ether, and (c) ethyl *isopropyl* ether.

What name would you give to the following:

(d)          (e)          (f)

**Strategy:** Ethers are named for the two groups bonded to the oxygen atom.

**Solution:**

(a)  Diethyl ether must have an ethyl group on either side of the O atom:

(b)  Ethyl propyl ether must have an ethyl group on one side and a propyl group on the other side of the O

   atom:

(c)  Ethyl *isopropyl* ether must have an ethyl group on one side and an *isopropyl* group on the other side of

   the O atom:

(d)  We see a propyl group on the left. Me is an abbreviation for a methyl group. This is therefore methyl propyl ether (the methyl and propyl groups are in alphabetical order in the name).

(e)  Two isopropyl groups gives us di*isopropyl* ether

(f)  $C_6H_5$ is a phenyl group. This is therefore methyl phenyl ether.

**Do the Results Make Sense?** Each name has two groups—one on either side of the O atom—and the groups are in alphabetical order, so the names make sense.

**Practice Exercise 10.6**

Write line structures for (a) methyl ethyl ether, (b) dibutyl ether, and (c) methyl *tert*-butyl ether. Name the following:

(d)          (e)          (f)

## Amines

Just as alcohols and ethers may be considered to be organic derivatives of water, amines may be considered to be organic derivatives of ammonia:

**Ammonia**     **An amine**     **Amphetamine**
(a dangerous stimulant)     **Putrescine**
(found in decaying meat)

Amines are classified as primary, secondary, or tertiary amines. This classification is based on the number of organic groups that are attached to the nitrogen atom:

**A primary (1°)**
**amine**     **A secondary (2°)**
**amine**     **A tertiary (3°)**
**amine**

Putrescine (1,4-diaminobutane) and cadaverine (1,5-diaminopentane) are both produced by the breakdown of amino acids in living and dead organisms. As their names suggest, both have foul odours.

Putrescine

Cadaverine

Notice that this is quite different from the way alcohols and alkyl halides are classified. *Iso*propylamine, for example, is a primary amine even though its $-NH_2$ group is attached to a secondary carbon atom. It is a primary amine because only one organic group is attached to the *nitrogen* atom:

**Isopropylamine**
(a 1° amine)     **Piperidine**
(a cyclic 2° amine)

Amines can be named by naming the groups attached to the nitrogen atom (in alphabetical order), and adding *amine*. Some examples are

Ethylamine     Ethylmethylpropylamine     Diethylphenylamine

Amines are like ammonia with a trigonal pyramidal shape around the nitrogen atom. The $C-N-C$ bond angles of trimethylamine are 108.7°, a value very close to the ideal tetrahedral $H-C-H$ bond angles of methane. The nitrogen atom of an amine is $sp^3$ hybridized, with the lone pair of electrons occupying one orbital. This means that the unshared pair is relatively exposed (red region in the electron density map in Figure 10-12), which is important because it is involved in almost all of the reactions of amines.

**Figure 10-12**

**Trimethylamine**

VSEPR structure and electron density map of trimethylamine.

---

## EXAMPLE 10-7 AMINES

Name the following amines and classify each as primary, secondary, or tertiary:

(a)          (b)          (c)

**Strategy:** Name the alkyl groups attached to the nitrogen atom in alphabetical order, using the prefixes *di*- and *tri*- if the groups are the same, and add *amine* to complete the name.

**Solution:** (a) dimethylamine (a secondary amine; the N atom is bonded to two carbon-containing groups); (b) pentylamine (a primary amine); and (c) triethylamine (a tertiary amine).

 **Do the Results Make Sense?** Each amine has one, two, or three alkyl groups bonded to it, and the group names are in alphabetical order, so the compound names make sense.

**Common Pitfalls:** Not putting the group names in alphabetical order. Also remember that the name of an amine is one word—no spaces.

**Practice Exercise 10.7**

Name the following amines:

(a)          (b)          (c)          (d)

Write line structures for (e) propylamine, (f) trimethylamine, and (g) ethylmethyl*iso*propylamine. Which of these amines (a–g) are primary, secondary, or tertiary amines?

## Aldehydes and Ketones

Aldehydes and ketones both contain the **carbonyl group**—a group in which a carbon atom has a double bond to oxygen (there are five important functional groups containing the carbonyl group):

The carbonyl group

The carbonyl group of an aldehyde is bonded to one hydrogen atom and one carbon atom (except for formaldehyde, which is the only aldehyde bearing two hydrogen atoms). The carbonyl group of a ketone is bonded to two carbon atoms. Using R, we can designate the general formulas for aldehydes and ketones as follows:

| ALDEHYDES | KETONES |
|:---:|:---:|
| or   **RCHO** | or   **RCOR′** |
| **(R=H in formaldehyde)** | **(where R′ is an alkyl group that may be the same or different from R)** |

Aldehydes are named by naming the corresponding alkane, dropping the "e" and adding *-al*. The first two aldehydes are therefore methanal and ethanal, but these are commonly called formaldehyde and acetaldehyde, respectively (see Figure 10-13). The carbonyl carbon is number 1, by convention.

### Figure 10-13

| Methanal | Ethanal | Phenylmethanal | 4-Methylpentanal |
| (formaldehyde) | (acetaldehyde) | (benzaldehyde) | |

Examples of simple aldehydes.

Formaldehyde is used as a disinfectant and as a preservative for tissues or cells, and is a primary ingredient in the manufacture of many adhesives. Benzaldehyde is an important component in the scent of almonds, and is used as an almond flavouring agent in many foods.

Ketones can be named in either of two ways. We can take the name of the alkane, drop the "e" and add *-one*. It may be necessary to specify the position of the carbonyl carbon in the chain. Alternatively, we can name the two groups on either side of the carbonyl group and add *ketone* (see Figure 10-14). Some examples are

Aldehydes and ketones have a trigonal planar arrangement of groups around the carbonyl carbon atom, which is $sp^2$ hybridized. In formaldehyde, for example, the bond angles are as shown below. The slight distortion from 120° bond angles is due to the extra electron density around the oxygen atom repelling the two C—H bonds.

Aldehydes can be prepared by oxidation of primary alcohols. Formaldehyde, for instance, is prepared by the oxidation of methanol:

Ketones can be prepared by the oxidation of secondary alcohols. Dimethyl ketone (acetone) can be prepared by oxidizing 2-propanol:

Example 10-8 gives some practice in naming aldehydes and ketones.

**Figure 10-14**

Propanone
(dimethyl ketone)
(acetone)

Butanone
(ethyl methyl ketone)

2-Pentanone
(methyl propyl ketone)

3-Pentanone
(diethyl ketone)

Examples of both methods of naming ketones.

---

## EXAMPLE 10-8 ALDEHYDES AND KETONES

Give IUPAC names for the following compounds:

(a)          (b)          (c)          (d)

**Strategy:** Aldehydes are named by naming the corresponding alkane, dropping the "e" and adding -*al*. Ketones can be named by dropping the "e" and adding -*one*, or by naming the two groups on either side of the carbonyl group and adding *ketone*.

**Solution:**

(a) This aldehyde has four carbon atoms, and is therefore butanal.

(b) This is also a butanal, but with a methyl group as a side chain. The carbonyl carbon is number 1 by convention, so this is 2-methylbutanal.

(c) This five-carbon ketone can be named 2-pentanone (signifying that the carbonyl carbon is number 2), or as methyl propyl ketone, signifying that there are methyl and propyl groups attached to the carbonyl group.

(d) Methyl phenyl ketone or phenyl propanone.

**Do the Results Make Sense?** The carbonyl group is at the end of the chain in the aldehydes, but not in the ketones. All carbon atoms are accounted for in naming the ketones by either convention, so the names are appropriate.

**Common Pitfalls:** Mixing up the two ways of naming ketones. Both are in common use, so make sure you understand the two conventions. Also remember to put spaces in the names of ketones when naming the groups on either side of the carbonyl group.

**Practice Exercise 10.8**

Give correct names for the following compounds:

(a)          (b)          (c)          (d)

## Carboxylic Acids, Esters, and Amides

**Carboxylic acids** have a carbonyl group bonded to a hydroxyl group, and they have the general formula $R-\overset{\overset{\displaystyle O}{\|}}{C}-O-H$ (see also Figure 10-15). The functional group $-\overset{\overset{\displaystyle O}{\|}}{C}-O-H$ is called the carboxyl group (carbonyl + hydroxyl).

Naming carboxylic acids consists of naming the corresponding alkane, dropping the -*ane* and adding -*oic acid*. The first two carboxylic acids (see Figure 10-16) are therefore methanoic acid and ethanoic acid, but are commonly referred to as formic acid and acetic acid, respectively. Phenyl methanoic acid (also shown in Figure 10-16) is commonly referred to as benzoic acid.

Formic acid is an irritating liquid produced by ants. (The sting of the ant is caused, in part, by formic acid being injected under the skin.) Acetic acid, the substance responsible for the sour taste of vinegar, is produced when certain

**Figure 10-15**

A carboxylic acid                    The carboxyl group

Carboxylic acids and the carboxyl group can be represented in different ways.

**Figure 10-16**

Formic acid

Acetic acid

Benzoic acid

Three common carboxylic acids.

Spoilage of wines occurs when the ethanol is oxidized to acetaldehyde, which is subsequently oxidized to acetic acid.

bacteria act on the ethyl alcohol of wine and cause it to be oxidized by air. The word vinegar is in fact derived from the French word *vinaigre*, which means "sour wine."

As with aldehydes, the carbonyl carbon in a carboxylic acid is assigned the number 1. Using this fact, the names of the following compounds can be easily understood:

3-Chlorobutanoic acid    Bromoethanoic acid    2,3-Diaminopropanoic acid
(bromoacetic acid)

Carboxylic acids can be prepared by oxidation of other functional groups; for instance, oxidation of primary alcohols or alkyl benzenes by $KMnO_4$, or oxidation of aldehydes by $Ag(NH_3)_2OH$:

**Figure 10-17**

General formula for an ester.

**Esters** have the general formula $RCO_2R'$ (or $RCOOR'$), where a carbonyl group is bonded to an alkoxyl ($-OR$) group. R and R' may be the same or different (Figure 10-17).

Esters are named by first naming the group that is part of the alkoxyl group, and then the carboxylic acid part of the molecule is named according to the number of carbons in it, and adding *-ate*. Some examples are shown in Figure 10-18.

Isoamyl acetate (banana oil)

**Figure 10-18**

Ethyl ethanoate is an important solvent and is commonly called ethyl acetate.

Methyl butanoate      Pentyl pentanoate

Examples of naming esters.

Octyl acetate
(artifical orange flavouring)

Esters can be prepared by reacting a carboxylic acid with an alcohol through the acid-catalyzed loss of a molecule of water. For example,

$$CH_3-C(=O)OH \ + \ HOCH_2CH_3 \ \xrightarrow{\text{acid-catalyzed}} \ CH_3-C(=O)OCH_2CH_3 \ + \ H_2O$$

**Acetic acid**     **Ethyl alcohol**     **Ethyl acetate**

**Amides** have the formulas $RCONH_2$, $RCONHR'$, or $RCONR'R''$, where a carbonyl group is bonded to a nitrogen atom bearing hydrogen and/or alkyl (R' or R'') groups. Amides are named in a fashion similar to esters, but *N*- and *N,N*- are used to indicate substituents that are attached to the nitrogen atom. General formulas and some specific examples are shown in Figure 10-19.

Ethyl methylphenylglycidate
(artifical strawberry flavouring)

**Figure 10-19**

**An unsubstituted amide**    **An *N*-substituted amide**    **An *N,N*-disubstituted amide**

**General formulas for amides**

**Ethanamide**    ***N*-Methylethanamide**    ***N,N*-Dimethylethanamide**
**(acetamide)**    **(*N*-Methylacetamide)**    **(*N,N*-Dimethylacetamide)**

**Specific examples of amides**

Many esters have pleasant odours and flavours.

General formulas and examples of amides.

## EXAMPLE 10-9 CARBOXYLIC ACIDS, ESTERS, AND AMIDES

Draw the following compounds: (a) hexanoic acid; (b) 2-ethyl butanoic acid; (c) methyl propanoate, (d) phenyl ethanoate; (e) *N*-ethylethanamide (*N*-ethylacetamide); and (f) *N*-methyl-*N*-phenylpropanamide.

**Strategy:** For carboxylic acids, draw the corresponding alkane and substitute a carboxylate group for the appropriate carbon atom. For esters, remember that the first group named is on the oxygen linkage, and the rest of the molecule is named for the number of carbons including the carbonyl carbon. For amides, the procedure is similar to that of esters.

**Solution:**

**Do the Results Make Sense?** The carboxylic acid structures have the correct numbers of carbons. The esters have the first-named groups on the ester linkages, and the amides have the first-named group or groups on the amide linkages, so the structures look reasonable.

**Common Pitfalls:** For esters, remember that the first-named group is attached to the oxygen atom; likewise for amides and the nitrogen atom.

**Practice Exercise 10.9**

Draw the following compounds: (a) *N*,*N*-dimethyl-3-methylbutanamide; (b) 3-phenylpentanal; and (c) phenyl ethanoate.

# Nitriles

A nitrile has the formula R−C≡N (or R−CN). The carbon and the nitrogen of a nitrile are *sp* hybridized. In the IUPAC system, nitriles are named by adding the suffix *nitrile* to the name of the corresponding hydrocarbon. The carbon atom of the −C≡N group is assigned number 1. The simplest nitrile is ethane nitrile, $CH_3CN$, which is a useful solvent for many chemical reactions. The name acetonitrile is an acceptable common name for $CH_3CN$, and acrylonitrile is an acceptable common name for $CH_2$=CHCN:

$$\overset{2}{CH_3}-\overset{1}{C}≡N:$$

**Ethanenitrile**
**(acetonitrile)**

$$\overset{4}{CH_3}\overset{3}{CH_2}\overset{2}{CH_2}-\overset{1}{C}≡N:$$

**Butanenitrile**

**Propenenitrile**
**(acrylonitrile)**

**4-Pentenenitrile**

Table 10-3 summarizes the structures and names for all of the functional groups introduced in this section. You should learn to identify these common functional groups because they appear in other more complex molecules.

## TABLE 10-3 IMPORTANT FUNCTIONAL GROUPS IN ORGANIC CHEMISTRY

| Compound Class | Functional Group | General Formula | Specific Example | IUPAC Name | Common Name |
|---|---|---|---|---|---|
| **Haloalkane** | | RX | $CH_3CH_2Cl$ | Chloroethane | Ethyl chloride |
| **Alcohol** | | ROH | $CH_3CH_2OH$ | Ethanol | Ethyl alcohol |
| **Ether** | | ROR | $CH_3OCH_3$ | Methoxymethane | Dimethyl ether |
| **Amine** | | $RNH_2$ $R_2NH$ $R_3N$ | $CH_3NH_2$ | Methanamine | Methylamine |
| **Aldehyde** | | $\overset{O}{\underset{}{RCH}}$ | $\overset{O}{\underset{}{CH_3CH}}$ | Ethanal | Acetaldehyde |
| **Ketone** | | $\overset{O}{\underset{}{RCR'}}$ | $\overset{O}{\underset{}{CH_3CCH_3}}$ | Propanone | Acetone |
| **Carboxylic acid** | | $\overset{O}{\underset{}{RCOH}}$ | $\overset{O}{\underset{}{CH_3COH}}$ | Ethanoic acid | Acetic acid |
| **Ester** | | $\overset{O}{\underset{}{RCOR'}}$ | $\overset{O}{\underset{}{CH_3COCH_3}}$ | Methyl ethanoate | Methyl acetate |
| **Amide** | | $\overset{O}{\underset{}{RCNH_2}}$ $\overset{O}{\underset{}{RCNHR'}}$ $\overset{O}{\underset{}{RCNR'R''}}$ | $\overset{O}{\underset{}{CH_3CNH_2CH_3}}$ | *N*-Methyl ethanamide | *N*-Methyl acetamide |
| **Nitrile** | $-C\equiv N:$ | RCN | $CH_3C\equiv N$ | Ethanenitrile | Acetonitrile |

**SECTION EXERCISES**

● **10.3.1** Identify all functional groups in each of the following molecules:

(a)          (b)          (c)

● **10.3.2** Identify all primary, secondary, tertiary, and quaternary carbon atoms in the following structures (a quaternary carbon is attached to four other carbon atoms):

(a)     (b)          (c)

● **10.3.3** Draw structures of the following compounds: (a) butanenitrile; (b) *N*-ethyl-*N*-phenyl acetamide; and (c) phenyl butanoate.

## 10.4 STEREOCHEMISTRY

So far in this chapter, we have represented organic molecules as flat, two-dimensional objects. In reality, this is not the case—organic molecules have three-dimensional shapes because of the hybridization of each central atom, as discussed in Chapters 6 and 7. **Stereochemistry** is the study of the relative spatial arrangements of atoms in molecules. Seemingly minor differences in stereochemistry can have enormous effects on the function and reactivity of molecules, so it is important to understand some aspects of this subject.

## Stereoisomers

In Section 10.1, the concept of structural isomers was introduced—structures having the same formula, but with the carbon atoms connected differently. **Stereoisomers** have their atoms connected in the same way as one another, but differ in the arrangement of their atoms in space.

A simple example of two stereoisomers is *cis*- and *trans*-dichloroethene. The Latin words *cis* and *trans* mean "next to" and "across," respectively, and in this case refer to the relative positions of the two chlorine atoms:

*cis*-1,2-Dichloroethene     *trans*-1,2-Dichloroethene

| Stereoisomer | $T_m$ (°C) | $T_b$ (°C) |
|---|---|---|
| *cis*-Dichloroethene | −81 | 60 |
| *trans*-Dichloroethene | −49 | 48 |

*cis*-Dichloroethene and *trans*-dichloroethene have very different physical properties.

These two compounds have the same molecular formula ($C_2H_2Cl_2$) and the same connectivity of atoms ($CHCl=CHCl$), but a different spatial arrangement of the chlorines (and the hydrogens), so they are called stereoisomers. Similarly for the two stereoisomers of 2-pentene:

*cis*-2-Pentene          *trans*-2-Pentene

In cases where there are more than two different groups attached to the double bond (other than hydrogen), the *cis–trans* system is not sufficient to unambiguously name a compound. An unambiguous system assigns a **Z** (*zusammen*, German for "together") or **E** (*entgegen*, German for "opposite") to designate the stereoisomers. For instance, 1-bromo-1-chloropropene can exist as a pair of stereoisomers:

I                    II

To name these two compounds, we need to first assign priorities to the substituents by looking at the atoms attached directly to the double-bond carbons. The rules for assigning priority are attributed to Cahn, Ingold, and Prelog:

Vladimir Prelog won the 1975 Nobel prize in chemistry for explaining ways in which the chirality of molecules influenced the mechanisms of chemical reactions.

1. The higher the atomic number of the atom bound to the carbon, the higher the priority. For example, priority increases in the order

$$-H < -C < -N < -O < -Cl.$$

2. If two substituents have the same first atom, move to the next atom (away from the double bond) until a difference is found. For example,

$$-CH_3 < -C_2H_5 < -CH_2Cl < -CH_2Br < -OCH_3.$$

In the 1-bromo-1-chloropropene example, the groups bound to the double-bond carbons in order of priority are $-Br$ higher than $-Cl$ on one carbon and $-CH_3$ higher than $-H$ on the other. The highest priority pair is therefore $-Br$ and $-CH_3$. In structure I above, these two groups are *trans* to each other, and so the correct name of the compound is (*E*)-1-bromo-1-chloropropene. In structure II, the high priority groups are *cis* to each other, and the correct name is therefore (*Z*)-1-bromo-1-chloropropene.

Example 10-10 provides some practice in naming *E/Z* stereoisomers.

**EXAMPLE 10-10 *E/Z* STEREOISOMERS**

Draw and name the two stereoisomers of 1-bromo-3-chloro-4-isopropyl-3-heptene.

**Strategy:** Draw the line structure of the compound and determine the priority of the four groups on the double-bond carbons.

**Solution:** The line structure for one stereoisomer of this compound is

The immediate substituent atoms on carbon number 3 are C and Cl. The Cl has a higher atomic number and therefore a higher priority. On carbon number 4, the immediate substituent atoms are both carbons, so we need to look at the next higher atomic number atoms in the substituent chain. These are also carbon, but the isopropyl group has two carbons whereas the propyl group has only one. The priority order is therefore isopropyl > propyl. Since the two higher priority groups (chlorine and isopropyl) are on the same side of the double bond, this configuration is (*Z*). The correct name for the compound is therefore (*Z*)-1-bromo-3-chloro-4-isopropyl-3-heptene. The other stereoisomer is

This stereoisomer has the two high priority groups (Cl and isopropyl) *trans* to each other, so the correct name is (*E*)-1-bromo-3-chloro-4-isopropyl-3-heptene.

 **Do the Results Make Sense?** At this stage, you must follow the steps to assign *E* or *Z* to a stereoisomer. It is difficult to do so by simply looking at the structure.

**Common Pitfalls:** Not following the steps. Do not guess!

**Practice Exercise 10.10**

Draw and name the two stereoisomers of (a) 1-chloro-1,2-difluoroethene
(b) 1,3-dibromo-1-chloro-2-methylpropene, and (c) 1-nitro-2-pentene.

## Drawing Molecules in Three Dimensions

Drawing molecules to show their three-dimensional structures is easily done using three different symbols for bonds. A straight line indicates a bond in the plane of the paper. As much of the carbon skeleton of a molecule as possible is drawn in the plane of the paper. A dashed line is then used to represent a bond to an atom that is behind this plane, and a solid wedge to represent a bond to an atom that is in front of this plane. The tetrahedral shape of methane is easily visible when drawn in this manner (see Figure 10-20).

**Figure 10-20**

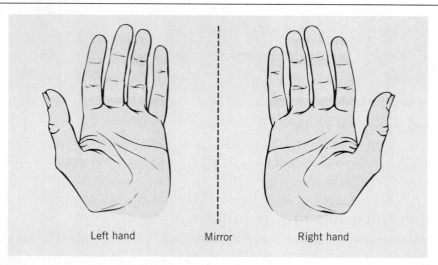

A three-dimensional representation of the methane molecule.

Some other examples of molecules drawn using dashes and wedges are shown below. Study them to make sure you understand which groups are behind the plane of the paper and which are in front.

Further classification of stereoisomers requires an understanding of the concept of mirror images. If you hold your left hand near a mirror, you will see an image that looks just like your right hand, so your two hands are said to be mirror images of one another (Figure 10-21).

**Figure 10-21**

The left hand is the mirror image of the right hand.

## Enantiomers and Diastereomers

Although your two hands are mirror images of one another, they are non-superimposable, as shown in Figure 10-22. In chemistry, molecules that are non-superimposable mirror images of one another are called **enantiomers**.

An example is 2-pentanol. The –OH group and the H-atom on the second carbon can have two different orientations:

**Figure 10-22**

To show that these two stereoisomers are enantiomers, one structure and its mirror image are drawn:

Your left and right hands are non-superimposable.

Look carefully at the mirror image of structure I to verify that it really is the same as structure II, only rotated 180°. Structures I and II are therefore mirror images of one another. However, it is not possible to superimpose the mirror images:

Because structures I and II are non-superimposable mirror images, they are enantiomers, much like your two hands.

Stereoisomers that are not mirror images of one another are called **diastereomers**. For example, *cis-* and *trans*-dichloroethene are diastereomers. As shown in Figure 10-23, the mirror image of *cis*-dichloroethene is not *trans*-dichloroethene.

**Figure 10-23**

| *cis*-Dichloroethene | Mirror image of *cis*-dichloroethene | *trans*-Dichloroethene |

*cis*-Dichloroethene, its mirror image, and *trans*-dichloroethene. *Cis-* and *trans*-dichloroethene are diastereoisomers.

## Chirality

A molecule is likely to have an enantiomer if it contains a single carbon atom with four different groups attached to it. This carbon atom is referred to as a **chirality centre**, and the molecule is said to be **chiral**. In the example of 2-pentanol, carbon number 2 is the chirality centre; it is bound to $-CH_3$, $-H$, $-OH$, and $-C_3H_7$. Molecules that can be superimposed on their mirror images are **achiral**. Some examples of chiral and achiral molecules are shown in Figure 10-24.

The term "chirality centre" is an IUPAC-recommended term. However, many chemists still use the terms "stereocentre," "stereogenic centre," or "chiral centre."

**Figure 10-24**

Examples of chiral and achiral molecules.

## EXAMPLE 10-11 CHIRAL COMPOUNDS

Two amino acids are glycine and leucine. Determine whether each of these compounds is chiral or achiral.

Glycine                    Leucine

**Strategy:** A compound is chiral if it contains a carbon attached to four different groups. Analyze each carbon in each molecule to see if this condition is met.

**Solution:** In glycine, carbon number 1 is attached to —H, —H, —NH$_2$, and —COOH. Carbon number 2 is attached to only three groups. Glycine is therefore achiral.

In leucine, carbon number 1 is attached to —H, —NH$_2$, —CH$_2$, and —COOH. This carbon is chiral, and so leucine is chiral.

**Do the Results Make Sense?** To be chiral, a compound must have at least one chirality centre. Leucine has one, but glycine does not have any.

**Common Pitfalls:** Forgetting to account for all groups attached to each carbon. Do not forget the implicit hydrogen atoms.

### Practice Exercise 10.11

Determine whether the amino acid arginine is chiral or achiral.

Arginine

## The R/S System for Naming Chiral Compounds

Enantiomers are different compounds, and therefore each must have a unique name. An example is the analgesic ibuprofin, which can exist as the following pair of enantiomers:

Inactive enantiomer of ibuprofen          Active enantiomer of ibuprofen

The active enantiomer has a pain-killing effect in the human body within minutes, but the inactive enantiomer does not, although it is slowly converted to the active one. Obviously, we need to unambiguously assign names to these two compounds. The system assigns an **R** (from the Latin word *rectus* meaning "straight" or "correct") or **S** (from the Latin word *sinister* meaning "left") to specify which enantiomer is which. To assign an *R* or *S* to a chiral molecule, we follow a three-step process:

Ibuprofen is sold as a **racemate**, an equimolar mixture of the *R* and *S* enantiomers. Only the *S* enantiomer is active in the human body, but it is less expensive to prepare the racemate than the pure *S* enantiomer.

1. Locate the chirality centre, identify the four groups attached to it, and assign priorities to them just as we did for the *E/Z* configurations of *cis–trans* stereoisomers.
2. Orient the molecule so that the group having the lowest priority is pointing away from you.
3. Read the other three groups (pointing toward you) from highest to lowest priority. If these are in a clockwise direction, the molecule is the *R* enantiomer. If they are in the counter-clockwise direction, it is the *S* enantiomer.

Take the inactive enantiomer of ibuprofen as an example. The groups attached to the chirality centre in increasing priority are $-H$, $-CH_3$, $-C_6H_4$, and $-COOH$. Orienting the molecule with the $-H$ away from our line of sight, we have

H atom is behind
the chirality centre.

CH₃

COOH

The other three groups from highest to lowest priority are $-COOH$, $-C_6H_4$, and $-CH_3$. These three groups are in a clockwise order, and so this is the (*R*)-ibuprofen, and the active enantiomer must be (*S*)-ibuprofen.

### EXAMPLE 10-12 IDENTIFYING *R* AND *S* ENANTIOMERS

Assign an *R* or *S* configuration to the chirality centre in the enantiomer of 3,3-dimethylcyclohexanol shown:

H          CH₃

HO    1    3    CH₃

**Strategy:** Follow the three steps for assigning an *R* or *S* configuration to this molecule.

**Solution:** Carbon number 1 is a stereocentre; the groups attached to it in increasing priority are —H, —CH$_2$—CH$_2$—CH$_2$—, —CH$_2$—C(CH$_3$)$_2$—, and —OH. Orienting the lowest priority group (—H) away from us, the other three groups from highest to lowest priority are —OH, –CH$_2$–C(CH$_3$)$_2$–, and —CH$_2$—CH$_2$—CH$_2$—.

These are oriented in a clockwise fashion, so this molecule is (*R*)-3,3-dimethylcyclohexanol.

**Does the Result Make Sense?** With practice, you will be able to "read" the chirality of a molecule. For now, follow the three steps and you will always arrive at the correct answer.

**Common Pitfalls:** When you orient the molecule with the lowest priority group facing away, make sure to reorient the other groups as well.

**Practice Exercise 10.12**
Determine whether the chirality centres in each of the following molecules are *R* or *S*:

(a)        (b)        (c)

---

If plane-polarized light is passed through a solution containing only one enantiomer of a chiral compound, the light is rotated in a specific direction. The other enantiomer will rotate light in the opposite direction.

Plane-polarized light

Rotation of the plane of plane-polarized light

α

Cell Containing One Enantiomer

Incident light

Transmitted light

This phenomenon allows chemists to determine which enantiomer is which. The effects of chirality, however, go far beyond this interesting behaviour. Many biologically important molecules are chiral, but often only one enantiomer has any biological activity. For example, the *R* and *S* enantiomers of carvone taste and smell very different from each other. (*S*)-Carvone is abundant in the oil from caraway seeds, and (*R*)-carvone in spearmint oil. These molecules taste and smell different because of specific interactions with chiral receptors in our olfactory system.

(*R*)-Carvone    (*S*)-Carvone

Some molecules have more than one chirality centre, and thus more than one pair of enantiomers. Galaxolide has two chirality centres (at carbon numbers 4 and 7), and hence two pairs of enantiomers.

A synthetic mixture of all four enantiomers is used to give a pleasant musky fragrance to detergents and cosmetics, although only the two stereoisomers shown here actually have an odour.

(4S,7R)-Galaxolide          (4S,7S)-Galaxolide

The (S,S) stereoisomer of ethambutol is used to treat tuberculosis, and apparently influences protein synthesis in mycobacteria (the causative agent of the disease) by inhibition of RNA synthesis. The (R,R) stereoisomer is inactive. Many chiral drugs must be made with high enantiomeric purity because of the potential side-effects of the other stereoisomers.

(S,S)-Ethambutol

Testosterone and epitestosterone are steroids produced in the human body. They differ only in the configuration of the −OH group, although there are several chirality centres in these molecules. Testosterone is an anabolic steroid, meaning it helps in the development of muscle mass, and can be used to enhance athletic performance. Most healthy adult males have a testosterone to epitestosterone ratio of about 1 : 1. Epitestosterone has no anabolic effect, but has been used by dishonest athletes to mask their use of testosterone by keeping their testosterone to epitestosterone ratio close to 1 : 1.

Testosterone          Epitestosterone

In some cases, one enantiomer of a chiral compound is beneficial whereas the other is dangerous. The drug thalidomide was prescribed in the late 1950s and early 1960s to pregnant women. The R enantiomer was effective at suppressing morning sickness, but the S enantiomer caused severe birth defects. The toxicity of the S enantiomer is due to its specific interactions with DNA at genes that control the development of limbs. Prescribing only the R enantiomer is not a safe alternative because it slowly converts to the S enantiomer in the body.

(S)-Thalidomide          (R)-Thalidomide

**SECTION EXERCISES**

● **10.4.1** Determine whether each of the following molecules is chiral or not:

(a)          (b)          (c)

● **10.4.2** Assign *E* or *Z* configurations to the following molecules:

(a)          (b)          (c)

● **10.4.3** Assign *R* or *S* configurations to each of the following molecules:

(a)          (b)          (c)

## 10.5 SUBSTITUTION REACTIONS

There are many types of organic compounds, and they all display their own characteristic types of reactivity. In this and the next two sections, we will cover a few of the most common reaction types. They all share the principle of having a *nucleophile* (a reagent searching for a positive charge) attacking an *electrophile* (a reagent searching for a negative charge). Any reaction of an electrophile also involves a nucleophile. Reactions differ depending on which part of a molecule acts as the nucleophile or electrophile, and whether a reactions happens in a single step or in several steps.

Nucleophiles are molecules or ions that have either a negative charge, a lone pair of electrons, or both. Electrophiles are molecules or ions that have a positive charge or at least a partial positive charge. This carbon atom in an alkyl halide, for example, has a partial positive charge because the electronegative halogen pulls the electrons of the carbon–halogen bond in its direction:

The word nucleophile comes from *nucleus*, the positive part of an atom, plus *-phile* from the Greek word *philos*, meaning "to love." Similarly, the term electrophile means "electron loving."

$$\delta+ \qquad \delta-$$
$$C \longrightarrow X$$

This is the positive centre that the nucleophile seeks.

The electronegative halogen polarizes the C–X bond.

## Nucleophilic Substitution Reactions

In a nucleophilic substitution reaction, one group (the nucleophile) substitutes for another (the **leaving group**). In the generalized reaction shown below, R–LG is called the **substrate** and contains the electrophilic centre that the nucleophile (Nu) seeks. LG is the leaving group, and the product of the reaction is Nu–R:

$$Nu \quad + \quad R\text{–}LG \quad \longrightarrow \quad Nu\text{–}R \quad + \quad LG$$

Nucleophile    Substrate          Product  Leaving group

If the substrate is an alkyl halide, the nucleophile replaces the halogen substituent. A substitution reaction takes place and the leaving group is a halide ion. Some products that can be prepared in this fashion are alcohols, ethers, and other alkyl halides. Three examples follow.

Preparation of methanol from hydroxide ion and methyl iodide:

$$HO^- \quad + \quad CH_3\text{–}I \quad \longrightarrow \quad CH_3\text{–}OH \quad + \quad I^-$$

Nucleophile    Substrate          Product          Leaving group

Preparation of ethyl methyl ether from methoxide ion and ethyl bromide:

$$CH_3O^- \quad + \quad CH_3CH_2\text{–}Br \quad \longrightarrow \quad CH_3CH_2\text{–}OCH_3 \quad + \quad Br^-$$

Nucleophile    Substrate          Product          Leaving group

Preparation of propyl iodide from iodide ion and propyl chloride:

I⁻ +          Cl          I  +  Cl⁻

Nucleophile    Substrate          Product          Leaving group

A neutral molecule such as water can also function as a nucleophile because of the region of negative charge on the oxygen atom. We can, for instance, prepare alcohols using water in a nucleophilic substitution reaction:

Nucleophile    Substrate          Product          Leaving group

Similarly, ammonia can be used as a nucleophile to prepare amines:

$$H_3N\text{:} \quad + \quad CH_3\text{–}Br \quad \longrightarrow \quad CH_3\text{–}NH_3^+ \quad + \quad Br^-$$

Nucleophile    Substrate          Product          Leaving group

---

### EXAMPLE 10-13 NUCLEOPHILIC SUBSTITUTION REACTIONS

Identify the nucleophile, substrate, product, and leaving group in the reaction
$CH_3I + CH_3CH_2O^- \longrightarrow CH_3OCH_2CH_3 + I^-$.

**Strategy:** The nucleophile is the reactant having a negative charge or at least one unshared electron pair. The substrate contains the positive centre (electrophile) created by the withdrawal of electrons from a carbon by an electronegative atom. Using these points, we can identify the nucleophile, substrate, product, and leaving group.

**Solution:** The ethoxide ion ($CH_3CH_2O^-$) is the nucleophile, which acts on the substrate $CH_3I$ to make the product ethyl methyl ether, and $I^-$ is the leaving group.

**Do the Results Make Sense?** The ethoxide ion functions as a nucleophile because of its negative charge and reacts with the electrophile in the substrate, which is the carbon attached to the iodine atom. All species are accounted for.

### Practice Exercise 10.13

Identify the nucleophile, substrate, product, and leaving group in each of the following reactions:

(a)   $I^- + CH_3CH_2Br \rightarrow CH_3CH_2I + Br^-$

(b)

(c)

There are two possible mechanisms for nucleophilic substitution reactions. The first is termed the **$S_N2$ mechanism**, in which the carbon–halogen bond breaks at the same time that the new bond between the nucleophile and the carbon forms. Using the example of an alkyl halide for the substrate, and using dotted lines to indicate the two bonds being formed and broken at the same time:

$$Nu + R\text{–}X \longrightarrow Nu \text{------} R \text{------} X \longrightarrow Nu\text{–}R + X^-$$

The second is termed the **$S_N1$ mechanism**, in which the carbon–halogen bond breaks first, generating an $R^+$ species (called a **carbocation**) and the leaving group:

$$R\text{–}X \longrightarrow R^+ + X^-$$

followed by reaction of the nucleophile with $R^+$ to form the product:

$$Nu + R^+ \longrightarrow Nu\text{–}R^+ \quad \text{(if the nucleophile is neutral)}$$

$$Nu^- + R^+ \longrightarrow Nu\text{–}R \quad \text{(if the nucleophile is charged)}$$

These two mechanisms are explored in more detail below.

## Mechanism of the $S_N2$ Reaction

In an $S_N2$ reaction, the nucleophile approaches the carbon bearing the leaving group from the back side; that is, from the side directly opposite the leaving group. As the reaction progresses, the bond between the nucleophile and the carbon atom strengthens, and the bond between the carbon atom and the leaving group weakens. The formation of the bond between the nucleophile and the carbon atom provides most of the energy necessary to break the bond between the carbon atom and the leaving group. For example, the mechanism for the reaction of hydroxide ion with chloromethane is shown in Figure 10-25.

Sir Christopher Ingold was responsible for introducing the concepts of nucleophile and electrophile, $S_N1$, $S_N2$, E1, and E2 mechanisms, and was a co-author of the Cahn-Ingold-Prelog priority rules introduced in Section 10.4.

**Figure 10-25**

The nucleophile is attracted to the positive end of the polar C–Cl bond and donates a pair of electrons.

The chlorine begins to move away with the pair of electrons that bonded it to the carbon.

In the transition state, a bond between the oxygen and carbon is partially formed, and the bond between the carbon and the chlorine is partially broken.

The bond between the carbon and oxygen is fully formed and the chloride ion leaves.

Mechanism of the S$_N$2 reaction.

The polarized C–Cl bond is indicated by the $\delta+$ and $\delta-$ symbols. Initially, the nucleophilic hydroxide ion is attracted to the electrophilic carbon atom (indicated by the red arrow). At the same time, the bonding electrons between the carbon and chlorine move toward the chlorine atom (indicated by the blue arrow). Eventually the chlorine atom leaves with both of these electrons.

Figure 10-25 indicates that the reaction proceeds through the formation of an unstable arrangement of atoms called the **transition state**. The transition state (denoted by the ‡ symbol) is a fleeting arrangement of the atoms in which the nucleophile and the leaving group are both partially bonded to the carbon atom undergoing substitution. The transition state has an extremely brief existence. It lasts only as long as the time required for one molecular vibration, about $10^{-13}$ s. The structure and energy of the transition state are highly important aspects of any chemical reaction. This is examined in more detail in Section 12.6.

## Mechanism of the S$_N$1 Reaction

The *overall* reaction of *tert*-butyl chloride with an hydroxide ion is analogous to the reaction of methyl chloride with an hydroxide ion:

The mechanism for this reaction, however, is quite different from the S$_N$2 mechanism discussed above, and involves two steps (Figure 10-26). In the first step, a molecule of *tert*-butyl chloride ionizes and becomes a *tert*-butyl cation and a chloride ion. Formation of this positively charged carbon species (a "carbocation") takes place slowly because it is usually a highly endothermic process. In the second step, the intermediate carbocation reacts rapidly with the nucleophilic hydroxide ion to produce *tert*-butyl alcohol.

**Figure 10-26**

Step 1

The leaving group leaves
with the pair of electrons
that bonded it to the carbon.

This slow step produces
the carbocation intermediate
and the chloride ion.

Step 2

This faster step produces the product
by reaction of the carbocation intermediate
from the first step with the nucleophile.

Mechanism of the $S_N1$ reaction.

## Rates of Substitution Reactions

In Chapter 12, we relate reaction mechanisms to reaction rates. A comparison of $S_N1$ and $S_N2$ reactions is incomplete, however, without looking at the factors that affect their rates. To study these factors, we need to keep in mind an important fact concerning reaction rates, which is detailed in Chapter 12: *a reaction proceeds at the rate of its slowest step*. For instance, as shown in Figure 10-26, formation of the carbocation in an $S_N1$ reaction is a slow process. As soon as the carbocation is formed, the nucleophile attacks it and produces the product. This second step is very fast, but the rate of production of the product can only be as fast as the slow first step in the mechanism because the second step cannot proceed without carbocations from the first step. Furthermore, since there is only one reactant in the slow step of the $S_N1$ reaction (the substrate), the rate of reaction depends only on how much of the substrate is present. (This is in fact the origin of the term $S_N1$—substitution, nucleophilic, one reactant in the slowest step.) The rate is not affected by the quantity of nucleophile, as long as some is present, or even by the strength of the nucleophile.

   In an $S_N2$ reaction, the rate depends on the concentrations of both the nucleophile and the substrate, since both are involved in the slowest (and only) step in the reaction. The identity of the nucleophile is also important in $S_N2$ reactions, since a stronger nucleophile will result in a faster reaction between it and the substrate. The best nucleophiles have the following characteristics:

1. A good nucleophile can easily donate electrons. For instance, the hydroxide ion is a better nucleophile than water because it is negatively charged. Ammonia is a better nucleophile than water because the nitrogen atom is less electronegative than the oxygen atom, and can therefore donate electrons more easily.

2. A nucleophile that is more polarizable is stronger. The iodide ion, I⁻, is a stronger nucleophile, because it is larger and more polarizable than the fluoride ion, F⁻. This large electron cloud is easily drawn toward the substrate to initiate the substitution reaction.

The large orbitals of the I⁻ ion overlap
well with those of the substrate.

The smaller orbitals of the F⁻ ion do not
overlap well with those of the substrate.

3. A molecule with less "steric hindrance" is a stronger nucleophile. Large alkyl groups get in the way ("sterically hinder") the nucleophile from interacting with the substrate:

The *tert*-butoxide ion is a
sterically hindered nucleophile.

The methoxide ion is smaller
and can attack the
substrate more easily.

Because the $S_N1$ mechanism involves a carbocation intermediate, only those substrates that ionize to produce stable carbocations will undergo $S_N1$ reactions. The reasons why some carbocations are more stable than others is beyond the scope of this book, but it can be summarized that tertiary alkyl carbocations are typically more stable than secondary ones, and the secondary are more stable than the primary ones. The $S_N1$ mechanism is therefore common for tertiary alkyl halides, is rare in the case of secondary alkyl halides, and is never followed for primary alkyl halides. For $S_N2$ reactions, however, the trend is the opposite. In an $S_N2$ reaction, a bulky substrate is not easily attacked by the nucleophile, and so secondary and tertiary substrates undergo $S_N2$ substitutions very slowly.

Faster $S_N2$ reactions

Substrate:  1°                    2°                    3°

Faster $S_N1$ reactions

Certain characteristics of the leaving group also dictate how easily it leaves, and thus how readily the reaction proceeds. In general, the weaker the group is as a base (i.e., the less likely it is to form a bond with a hydrogen ion in solution), the better it is as a leaving group. The reasons why something might be a weaker or stronger base are made clear in Chapters 14 and 15, but for now we can list some common groups in order of strength as leaving groups:

| $OH^-$, $RO^-$ | $F^-$ | $Cl^-$ | $Br^-$ | $I^-$, $H_2O$ |
|:---:|:---:|:---:|:---:|:---:|
| **Very poor** | **Poor** | **Fair** | **Good** | **Very good** |

## Stereochemistry of $S_N2$ Reactions

If you studied Section 10.4 about stereochemistry, you can now understand the change in stereochemistry that happens during an $S_N2$ reaction. Note in Figure 10-25 that the hydrogen atoms in the substrate, chloromethane, become inverted (pushed to the opposite side of the carbon atom) in the product, methanol. An inversion is easier to see if the substrate is chiral. For example, the $S_N2$ reaction of (R)-2-bromooctane with sodium hydroxide results in (S)-2-octanol:

(R)-2-Bromooctane    Transition State    (S)-2-Octanol

Note the inversion of configuration from R to S.

---

### EXAMPLE 10-14 $S_N1$ AND $S_N2$ REACTIONS

Predict the products of the following reactions and state whether they will occur via $S_N1$ or $S_N2$ mechanisms.

(a) [structure] Cl + $I^-$     (b) [structure] Cl + $^{\ominus}OCH_3$

**Strategy:** Identify the nucleophile and the substrate. The product and leaving group will then be obvious. Also remember that primary alkyl halides usually react via $S_N2$ and tertiary via $S_N1$ mechanisms.

**Solution:**

(a) The iodide ion is the nucleophile. The 1-chlorobutane is the substrate. The substitution results in

[structure] I (1-iodobutane), and the leaving group is the chloride ion. Since the substrate is a primary alkyl halide, the mechanism would be $S_N2$.

(b) The methoxide ion is the nucleophile. The 2-chloro-2-methylpropane is the substrate. The substitution

results in [structure] O (tert-butyl methyl ether), and the leaving group is the chloride ion. Since the substrate

is a tertiary alkyl halide, the mechanism would be $S_N1$.

**Do the Results Make Sense?** In each case, the leaving group is plausible, and the mechanisms match what we know about primary or tertiary alkyl halides.

**Practice Exercise 10.14**

Predict the products of the following reactions and state whether they will occur via $S_N1$ or $S_N2$ mechanisms.

(a) [structure: cyclohexane with a $CH_2CH_2Cl$ side chain] $\quad + \quad {}^{\ominus}OCH_3$

(b) [structure: tert-butyl bromide] $-Br \quad + \quad {}^{\ominus}OC_2H_5$

## Aromatic Substitution Reactions

In aromatic substitution reactions, a group substitutes for one of the hydrogen atoms of the aromatic ring, but the ring itself is unchanged. The mechanisms for these reactions are complex and are not dealt with here, but the reactions can be generalized as follows:

[benzene ring] $+ \ XY \ \longrightarrow \ $ [benzene ring with Y] $ + \ HX$

where XY can be any one of a number of strongly electrophilic species. For instance, XY can be a bromine molecule, and bromination will occur in the presence of $FeBr_3$ as a catalyst:

[benzene ring] $+ \ Br_2 \ \xrightarrow{FeBr_3} \ $ [benzene ring with Br] $ + \ HBr$

Bromine is electrophilic in this reaction because it is easily polarized, and the positive end will be strongly attracted to the $\pi-$electrons of the benzene ring. The analogous reaction will occur with $Cl_2$ and $AlCl_3$.

Substitution of a nitro group for a hydrogen will occur using nitric acid in the presence of sulphuric acid as a catalyst:

[benzene ring] $+ \ HNO_3 \ \xrightarrow{H_2SO_4} \ $ [benzene ring with $NO_2$] $ + \ H_2O$

Sulphonation of the ring can occur in the presence of fuming sulphuric acid (sulphuric acid to which high concentrations of $SO_3$ have been added):

[benzene ring] $+ \ H_2SO_4 \ \longrightarrow \ $ [benzene ring with $SO_3H$] $ + \ H_2O$

## SECTION EXERCISES

● **10.5.1** Define the terms nucleophile and electrophile.

● **10.5.2** In a nucleophilic reaction involving an alkyl halide, why is the nucleophile attracted to the carbon atom?

● **10.5.3** Describe the $S_N1$ and $S_N2$ mechanisms in your own words.

## 10.6 ELIMINATION REACTIONS

In an **elimination reaction**, the fragments of some molecule (YZ) are removed (eliminated) from adjacent atoms of the reactant. This elimination leads to the creation of a multiple bond:

$$\overset{\mid}{\underset{Y}{C}}-\overset{\mid}{\underset{Z}{C}} \quad \xrightarrow[\text{(--YZ)}]{\text{Elimination}} \quad \overset{}{\underset{}{C}}=\overset{}{\underset{}{C}}$$

In elimination reactions, as in $S_N1$ and $S_N2$ reactions, there is an attacking group that has a lone pair of electrons and a leaving group. The attacking group is called a Lewis base, because it readily accepts the hydrogen from the alkyl halide, forming HB:

$$-\overset{H}{\underset{\underset{X}{\mid}}{C}}-\overset{\mid}{C}- \; + \; :B^- \; \longrightarrow \; C{=}C \; + \; H:B \; + X^-$$

**A base**

Dehydrohalogenation

A widely used method for synthesizing alkenes is the elimination of HX from adjacent atoms of an alkyl halide, in which case the reaction is called a **dehydrohalogenation**. Reacting the alkyl halide with a strong base causes the reaction to take place. In the following two examples, the ethoxide ion acts as the base:

$$H_3C-\overset{H}{\underset{\underset{Br}{\mid}}{C}}-CH_3 \; + \; C_2H_5O^- \; \longrightarrow \; H_2C{=}\overset{}{\underset{H}{C}}{-}CH_3 \; + \; Br^- \; + \; C_2H_5OH$$

$$H_3C-\overset{CH_3}{\underset{\underset{Br}{\mid}}{C}}-CH_3 \; + \; C_2H_5O^- \; \longrightarrow \; \underset{H_3C}{\overset{CH_3}{\diagup}}C\underset{CH_2}{\diagdown} \; + \; Br^- \; + \; C_2H_5OH$$

Chloroalkanes also undergo the elimination of hydrogen chloride, iodoalkanes undergo the elimination of hydrogen iodide, and, in all cases, alkenes are produced.

Elimination reactions occur by a variety of mechanisms. With alkyl halides, two mechanisms (called E1 and E2) are especially important because they are closely related to the $S_N2$ and $S_N1$ reactions that we have just studied.

## Mechanism of the E2 Reaction

When isopropyl bromide is reacted with an ethoxide ion to form propene, experimental evidence indicates that the reaction takes place in the manner illustrated in Figure 10-27. In this mechanism, proton transfer to the base, formation of the carbon-carbon double bond, and elimination of the bromide ion occur

**Figure 10-27**

| The basic ethoxide ion is attracted to the H atom. The electron pair bonding the C to the H moves in to become the new pi-bond of a double bond and the Br atom begins to leave with the two electrons that bonded it to the carbon. | In the transition state, partial bonds extend throughout the molecule. | The product contains the newly formed double bond and the base has reacted with the H atom to form ethanol. Bromide ion is the other product. |

A mechanism for the E2 reaction.

simultaneously; in other words, all bond-breaking and bond-forming steps occur at the same time. It is referred to as E2, because the rate of reaction depends on the concentrations of two reagents—the base and the substrate, similar to the $S_N2$ reaction.

Whether the substrate is primary, secondary, or tertiary makes little difference in the E2 mechanism, because the base is removing a hydrogen atom, which is easily accessible. E2 reactions therefore occur in primary, secondary, and tertiary alkyl halides.

## The E1 Reaction

Eliminations may take a different pathway from E2 reactions. Reacting *tert*-butyl chloride with water, for example, gives 2-methylpropene:

The mechanism for this E1 reaction is detailed in Figure 10-28. E1 reactions happen in two distinct steps, much like $S_N1$ reactions. In the first step, a carbocation is formed. This is energetically unfavourable, so it is this step that limits the rate of the overall reaction. In the second step, the base (ethoxide ion in this example) removes a proton from the carbocation, and the electrons that bound the proton to the carbocation move to make the second carbon–carbon bond. Similar to the $S_N1$ mechanism, E1 reactions involve a carbocation intermediate and therefore rarely happen with secondary alkyl halides, and never with primary alkyl halides, but are more common with tertiary substrates. The mechanism is termed E1 because the rate of reaction depends on the rate of the slow first step, which in turn depends on the concentration of only one reagent—the substrate.

**Figure 10-28**

Step 1

A chlorine atom leaves
with the electron pair
that bonded it to the carbon.

A carbocation and chloride ion
are produced in this first step.
These ions are solvated and
therefore stabilized by
the solvent molecules.

Step 2

$^{\ominus}OC_2H_5$ + H—C—C$^+$ ⟶ C=C + $C_2H_5OH$

The negative charge on the ethoxide ion
is attracted to the H atom. The bonding
electrons of this H atom move in to
form a double bond between
the carbon atoms.

The alkene is produced
along with a molecule of ethanol.

A mechanism for the E1 reaction between *tert*-butyl chloride and the ethoxide ion.

## EXAMPLE 10-15  E1 AND E2 REACTIONS

Predict the products of the following elimination reactions and state whether they will occur via E1 or E2 mechanisms.

(a) ⟶ with Br, $^{\ominus}OC_2H_5$     (b) with Br, $^{\ominus}OH$

**Strategy:** E1 reactions involve loss of the leaving group to form a carbocation, removal of $H^+$ from the carbocation, and formation of a carbon=carbon double bond. E2 reactions involve the simultaneous $H^+$ removal, C=C bond formation, and loss of the leaving group. In either case, the result is a product with a C=C double bond. Primary alkyl halides can only undergo E2 eliminations. Secondary ones are more likely to undergo E1 eliminations.

**Solution:**

(a) The Br atom is the leaving group. The ethoxide anion removes the proton, forming ethanol. The double bond forms between the two carbon atoms, giving ethene. Since the substrate is a primary alkyl halide, the mechanism is E2:

$$\text{with Br} \xrightarrow{\ominus OC_2H_5} \text{=} + Br^- + C_2H_5OH$$

(b) The Br atom is the leaving group. The hydroxide anion removes the proton, forming water. The double bond forms between the two carbon atoms, giving propene. Since the substrate is a tertiary alkyl halide, the mechanism is E1:

$$CH_3CH{=}CH_2 + H_2 \xrightarrow[\substack{or\ Pt \\ 25°C}]{Ni,\ Pd,} CH_3CH_2{-}CH_3$$

**Do the Results Make Sense?** In each case, a double bond has been formed, a leaving group departs, and a proton is removed as well, all of which must happen in elimination reactions.

**Practice Exercise 10.15**

Predict the products of the following elimination reactions and state whether they will occur via E1 or E2 mechanisms.

(a)  [structure with Br]   $\xrightarrow{\ominus OCH_3}$       (b)  [structure with Cl]   $\xrightarrow{\ominus OC_2H_5}$

---

**SECTION EXERCISES**

● **10.6.1** Show the mechanism for the E2 reaction of 2-bromopropane with a propoxide ion.

● **10.6.2** Show the mechanism for the E1 reaction of 2-iodo-2-methylpropane with a hydroxide ion.

## 10.7 ADDITION REACTIONS

Addition reactions are, in a sense, the opposite of elimination reactions. In an addition reaction, two atoms are added to the carbon atoms involved in a double bond, and the double bond is converted into a single bond. Addition reactions are useful for producing alkanes, alcohols, alkyl halides, and other compounds. In this section, we look at the details of several addition reactions.

## Hydrogenation of Alkenes

Alkenes react with hydrogen in the presence of a variety of metal catalysts to add one hydrogen atom to each carbon atom of the double bond. Hydrogenation reactions that involve finely divided *insoluble* platinum, palladium, or nickel catalysts are said to proceed by **heterogeneous catalysis**, because the substrate is soluble in the reaction mixture but the catalyst is not. An example is the hydrogenation of ethene to produce ethane:

$$CH_3CH{=\!=}CH_2 + H_2 \xrightarrow[\substack{\text{or Pt} \\ 25°C}]{\text{Ni, Pd,}} CH_3CH_2{-\!-}CH_3$$

The product that results from the addition of hydrogen to an alkene is an alkane, and the term used to describe the addition of hydrogen is *catalytic hydrogenation*. Our Focus 10–2: Chemistry and Technology examines the use of hydrogenation in the food industry. There are many other examples of additions to the double bonds of alkenes. The additions of hydrogen halides, sulphuric acid, water (in the presence of an acid catalyst), and halogens are illustrated in Figure 10–29.

## FOCUS 10-2  CHEMISTRY AND TECHNOLOGY: HYDROGENATION IN THE FOOD INDUSTRY

The food industry makes use of catalytic hydrogenation to convert liquid vegetable oils to semisolid fats in making margarine and solid cooking fats. Examine the labels of many prepared foods and you will find that they contain "partially hydrogenated vegetable oils." There are several reasons why foods contain these oils, but one is that partially hydrogenated vegetable oils have a longer shelf life.

A product used in baking that contains oils and mono- and diacylglycerols that are partially hydrogenated.

Fats and oils are glyceryl esters of carboxylic acids with long carbon chains, called "fatty acids." Fatty acids are saturated (no double bonds), monounsaturated (one double bond), or polyunsaturated (more than one double bond). Oils typically contain a higher proportion of fatty acids with one or more double bonds than fats do. Hydrogenation of an oil converts some of its double bonds to single bonds, and this conversion has the effect of producing a fat with the consistency of margarine or a semisolid cooking fat.

Our bodies are incapable of making polyunsaturated fats, and therefore, such fats must be present in our diets in moderate amounts in order to maintain health. Saturated fats can be made in the cells of our body from other food sources; for example, from carbohydrates (sugars and starches). For this reason saturated fats in our diet are not absolutely necessary and, indeed, too much saturated fat has been implicated in the development of cardiovascular disease.

One potential problem that arises from using catalytic hydrogenation to produce partially hydrogenated vegetable oils is that the catalysts used for hydrogenation cause isomerization of some of the double bonds of the fatty acids (some of those that do not absorb hydrogen). In most natural fats and oils, the double bonds of the fatty acids have the *cis* configuration. The catalysts used for hydrogenation convert some of these *cis* double bonds to the unnatural *trans* configuration. The health effects of *trans* fatty acids are still under study, but experiments thus far indicate that they cause an increase in serum levels of cholesterol and triacylglycerols, which in turn increases the risk of cardiovascular disease.

No (or zero%) trans fatty acids.

## Hydrogenation: The Function of the Catalyst

Heterogeneous hydrogenation catalysts typically involve finely divided platinum, palladium, nickel, or rhodium deposited on the surface of powdered carbon (charcoal). Hydrogen gas introduced into the atmosphere of the reaction vessel adsorbs to the metal by a chemical reaction, where unpaired electrons on the surface of the metal *pair* with the electrons of hydrogen (Figure 10-30*a*) and bind the hydrogen to the surface. The collision of an alkene with the surface causes adsorption of the alkene as well (Figure 10-30*b*). A stepwise transfer of hydrogen atoms takes place, and this produces an alkane which desorbs from (leaves) the catalyst surface (Figures 10-30*c* and 10-30*d*).

**Figure 10-29**

Additions to alkenes.

**Figure 10-30**

The mechanism for the hydrogenation of an alkene as catalyzed by finely divided platinum metal: *(a)* hydrogen adsorption; *(b)* adsorption of the alkene; *(c,d )*, stepwise transfer of both hydrogen atoms to the alkene, followed by desorption of the alkane.

## Understanding Additions to Alkenes

Two characteristics of the double bond help us understand why these addition reactions occur:

1. An addition reaction results in the conversion of one $\pi$ bond and one $\sigma$ bond into two $\sigma$ bonds. The result of this change is usually energetically favourable. The energy evolved in making two $\sigma$ bonds exceeds that needed to break one $\sigma$ bond and one $\pi$ bond (because $\pi$ bonds are weaker), and, therefore, addition reactions usually release energy:

2. The electrons of the π bond are exposed. Because the π bond results from over-lapping *p* orbitals, the π electrons lie above and below the plane of the double bond:

An electrostatic potential map for ethene shows the higher density of negative charge in the region of the π bond.

The electron pair of the π bond is distributed throughout both lobes of the π molecular orbital.

The electrons in the π bond are therefore particularly susceptible to electron-seeking reagents (electrophiles). Electrophiles include positive reagents such as protons ($H^+$), and neutral reagents such as bromine (because it can be polarized so that one end is positive), as well as Lewis acids such as $BH_3$, $BF_3$, and $AlCl_3$. Metal ions that contain vacant orbitals—the silver ion ($Ag^+$), the mercuric ion ($Hg^{2+}$), and the platinum ion ($Pt^{2+}$), for example—also act as electrophiles.

Hydrogen halides react with alkenes by donating a proton to the π bond. The proton uses the two electrons of the π bond to form a σ bond to one of the carbon atoms. This leaves a vacant *p* orbital and a positive charge on the other carbon. The overall result is the formation of a carbocation and a halide ion from the alkene and HX:

$$X{-}H \quad + \quad \Large\diagup\hspace{-0.3em}\diagdown\hspace{-0.3em}C{=}C\hspace{-0.3em}\diagup\hspace{-0.3em}\diagdown \quad \longrightarrow \quad -\!\overset{\displaystyle |}{\underset{\displaystyle |}{C}}\!-\!\overset{\displaystyle H}{\overset{\displaystyle |}{C}}\!{}^+\diagup\diagdown \quad + \quad X^-$$

Electrophile          Nucleophile                                    Carbocation

In the next step, the carbocation is the electrophile and the halide ion is the nucleophile:

$$-\!\overset{\displaystyle H}{\underset{\displaystyle |}{\overset{\displaystyle |}{C}}}\!-\!\overset{\displaystyle |}{C}{}^+\diagup \quad + \quad X^- \quad \longrightarrow \quad -\!\overset{\displaystyle H}{\underset{\displaystyle |}{\overset{\displaystyle |}{C}}}\!-\!\overset{\displaystyle X}{\underset{\displaystyle |}{\overset{\displaystyle |}{C}}}\!-$$

Electrophile          Nucleophile

The order of reactivity of the hydrogen halides is HI > HBr > HCl > HF, because of the increasing H–X bond energy in this same order.

## Markovnikov's Rule

The addition of HX to an unsymmetrical alkene (an alkene with different groups on either side of the double bond) could conceivably occur in two ways. In practice, however, one product usually predominates. The addition of HBr to propene, for example, could conceivably lead to either 1-bromopropane or 2-bromopropane. The main product, however, is 2-bromopropane:

Similarly, when 2-methylpropene reacts with HBr, the main product is 2-bromo-2-methylpropane, not 1-bromo-2-methylpropane:

Consideration of many examples like this led the Russian chemist Vladimir Markovnikov in 1870 to formulate what is now known as **Markovnikov's rule**. One way to state this rule is to say that *in the addition of HX to an alkene, the hydrogen atom adds to the carbon atom of the double bond that already has the greater number of hydrogen atoms*. Reactions that obey Markovnikov's rule are said to be *Markovnikov additions*. The addition of HBr to propene is an illustration:

Markovnikov's rule also allows us to predict the outcome of the addition of a reagent such as ICl. Because of the greater electronegativity of chlorine, iodine is the positive portion of this molecule and will add to the carbon that already has the greater number of hydrogens. The addition of ICl to 2-methylpropene therefore produces 2-chloro-1-iodo-2-methylpropane:

## EXAMPLE 10-16 MARKOVNIKOV'S RULE

Give the structure and name of the product that would be obtained from the addition of IBr to propene.

**Strategy:** Use Markovnikov's rule and attach the more positive end if the IBr molecule to the carbon with the greater number of hydrogens bound to it.

**Solution:** First, we recognize that the more positive end of IBr is the I atom, since it is less electronegative than Br. Using Markovnikov's rule, the I atom will therefore substitute onto the carbon atom having more hydrogens:

This product is 2-bromo-1-iodopropane.

**Do the Results Make Sense?** The more positive part of the reagent ($I^-$) was added to the carbon that already had more hydrogens bonded to it according to Markovnikov' rule. The double bond was converted into a single bond. Both these actions make sense in an addition reaction.

**Common Pitfalls:** Not recognizing the more positive end of the addition reagent.

**Practice Exercise 10.16**

(a) Give the structure and name of the product that would be obtained from the addition of HCl to 1-butene.

(b) Give the structure and name of the product that would be obtained from for the addition of HBr to 2-cyclohexyl-1-butene.

## Addition of Water to Alkenes: Acid-Catalyzed Hydration

The acid-catalyzed addition of water to the double bond of an alkene (**hydration of an alkene**) is a method for the preparation of low-molecular-weight alcohols and has its greatest utility in large-scale industrial processes. The acids most commonly used to catalyze the hydration of alkenes are dilute aqueous solutions of sulphuric acid or phosphoric acid. In these reactions, the addition of water to the double bond follows Markovnikov's rule. The generalized reaction is as follows:

An example is the hydration of 2-methylpropene to prepare 2-methyl-2-propanol:

Because the reactions follow Markovnikov's rule, acid-catalyzed hydrations of alkenes do not yield primary alcohols except in the special case of the hydration of ethene:

## Electrophilic Addition of Bromine and Chlorine to Alkenes

Alkenes react rapidly with chlorine and bromine to form **vicinal** dihalides (abbreviated *vic*-dihalide, meaning that the two halogens are on adjacent carbon atoms). Some specific examples of this reaction are

When bromine is used for this reaction, it can serve as a test for the presence of carbon–carbon multiple bonds. If we add bromine to an alkene or alkyne, the red-brown colour of the bromine disappears almost instantly as long as the alkene (or alkyne) is present in excess:

Coloured                        Colourless

Rapid decolourization of the red-brown $Br_2$
solution is a positive test for unstauration.

This behaviour contrasts markedly with that of alkanes, which do not react appreciably with bromine or chlorine at room temperature under the same conditions.

## EXAMPLE 10-17 ADDITION OF HALOGENS TO A DOUBLE BOND

Predict all products of the addition reaction                + ICl ⟶

**Strategy:** We need to keep in mind that enantiomers are possible, and Markovnikov's rule.

**Solution:** In ICl, the Cl atom is more electronegative, so the I atom will bond to the carbon atom that has more hydrogen atoms bonded to it, according to Markovnikov's rule. Enantiomers are possible:

 **Do the Results Make Sense?** A vicinal dihalide is formed, and two enantiomers result, which makes sense.

### Practice Exercise 10.17
Draw and name the product of the addition reaction of cyclopentylethene with water.

## SECTION EXERCISES

● **10.7.1** Predict the product of addition of $Br_2$ to 2-methylpropene.

● **10.7.2** What is formed when HBr is added to 3-phenyl-1-butene?

● **10.7.3** Show an addition reaction that could be used to form 2-butanol.

# CHAPTER 10 VISUAL SUMMARY

## 10.1 HYDROCARBONS

Butane    2-Butene

Propyne    Cyclohexane

-Cl    chloro
-Br    bromo
-F    fluoro
-I    iodo
-NH₂    amino
-NO₂    nitro
-OH    hydroxy

4-Ethyl-2,3-dimethylheptane

**LEARNING OBJECTIVES:** Draw and name hydrocarbons using the IUPAC system.

**SKILLS TO MASTER:** Naming alkanes and alkyl groups (p. 534)

Naming branched-chain alkanes (p. 536)

Naming cycloalkanes, alkenes, and alkynes (p. 540).

Naming other substituents (p. 541)

**KEY CONCEPTS:** The IUPAC system of nomenclature gives each compound a unique name.

## 10.2 AROMATIC COMPOUNDS

1,2 positions are *ortho*
1,3 positions are *meta*
1,4 positions are *para*

Ethylbenzene
or phenylethane

CH₃    NH₂    OH

Toluene    Aniline    Phenol

**LEARNING OBJECTIVES:** Draw and name aromatic compounds using the IUPAC system.

**SKILLS TO MASTER:** Naming benzene compounds using numbers or the *o*-, *m*-, and *p*- designations to denote the locations of substituents (p. 544).

**KEY CONCEPTS:** All six carbon atoms in benzene are equivalent.

## 10.3 FUNCTIONAL GROUPS

Bromoethane or
ethyl bromide

*Tert*-butyl alcohol
or *tert*-butanol

Ethyl propyl ether

Ethyldimethylamine    3-Methylhexanal

2-Pentanone or
methyl propyl ketone

Chloroethanoic acid
or chloroacetic acid

Methyl ethanote
or methyl acetate

*N,N*-Dimethyl ethanamide
or *N,N*-dimethyl acetamide

**LEARNING OBJECTIVES:** Draw and name compounds containing common functional groups.

**SKILLS TO MASTER:** Naming and drawing alkyl halides (p. 547).

Naming and drawing alcohols (p. 549).

Naming and drawing ethers (p. 552).

Naming and drawing amines (p. 554).

Naming and drawing aldehydes and ketones (p. 556).

Naming and drawing carboxylic acids, esters, and amides (p. 558).

Naming and drawing nitriles (p. 561).

**KEY CONCEPTS:** The IUPAC system can be used to unambiguously assign a name to any compound.

## 10.4 STEREOCHEMISTRY

(Z)-1-bromopropene    (E)-1-bromopropene

Enantiomers are non-superimposable mirror images of one another

(R)-Ibuprofen          (S)-Ibuprofen

**LEARNING OBJECTIVES:** Recognize, name, and draw stereoisomers.

**SKILLS TO MASTER:** Naming *cis* and *trans* stereoisomers using the *E/Z* system (p. 563).

Drawing structures with wedged and dashed bonds (p. 565).

Recognizing enantiomers and diastereomers (p. 566).

Determining from a line structure whether a molecule is chiral or achiral (p. 567).

Identifying *R* and *S* enantiomers (p. 569).

**KEY CONCEPTS:** Enantiomers are non-superimposable mirror images of each other.

A compound is chiral if it contains a carbon atom bonded to four different groups.

## 10.5 SUBSTITUTION REACTIONS

$S_N2$ Mechanism:

$$Nu \;+\; R\text{–}X \longrightarrow Nu \text{-----} R \text{-----} X \longrightarrow Nu\text{–}R \;+\; X^-$$

R/S inversion

$S_N1$ Mechanism:

Step 1        $R\text{–}X \longrightarrow R^+ \;+\; X^-$

Step 2

$Nu \;+\; R^+ \longrightarrow Nu\text{–}R^+$   (if the nucleophile is neutral)

$Nu^- \;+\; R^+ \longrightarrow Nu\text{–}R$   (if the nucleophile is charged)

**LEARNING OBJECTIVES:** Predict the products from $S_N2$ and $S_N1$ reactions.

Write the mechanisms of $S_N2$ and $S_N1$ reactions.

**SKILLS TO MASTER:** Recognizing the nucleophile, substrate, product, and leaving group in a nucleophilic substitution reaction (p. 573).

Stating the factors that make a good nucleophile and a good leaving group (p. 576).

Predicting the stereochemistry of the product of an $S_N2$ reaction involving a chiral substrate (p. 578).

**KEY CONCEPTS:** $S_N2$ reactions occur in a single step.

$S_N1$ reactions occur via the formation of a carbocation, followed by its reaction with a nucleophile.

## 10.6 ELIMINATION REACTIONS

E2 Mechanism:

E1 Mechanism:

Step 1

Step 2

**LEARNING OBJECTIVES:** Predict the products from E2 and E1 reactions.

Write the mechanisms of E2 and E1 reactions.

**KEY CONCEPTS:** E2 reactions occur in a single step.

E1 reactions occur via the formation of a carbo-cation, followed by its reaction with a base.

Elimination reactions convert a single bond into a double bond.

Elimination reactions eliminate the leaving group as well a molecule made from a hydrogen atom from the substrate plus the base.

## 10.7 ADDITION REACTIONS

$$\text{C=C} + \text{X—Y} \longrightarrow -\overset{|}{\underset{X}{C}}-\overset{|}{\underset{Y}{C}}-$$

$$\underbrace{\underset{\pi \text{ bond}}{} \quad \underset{\sigma \text{ bond}}{}}_{\textbf{Bonds broken}} \longrightarrow \underbrace{\underset{2\sigma \text{ bonds}}{}}_{\textbf{Bonds formed}}$$

**Step 1**

$$\underset{\text{Electrophile}}{X–H} + \underset{\text{Nucleophile}}{\Large\diagup\!\!=\!\!\diagdown} \longrightarrow \underset{\text{Carbocation}}{-\overset{H}{\underset{|}{C}}-C^+} + X^-$$

**Step 2**

$$\underset{\text{Carbocation}}{-\overset{H}{\underset{|}{C}}-C^+} + \underset{\text{Nucleophile}}{X^-} \longrightarrow \underset{\text{Product}}{-\overset{H}{\underset{|}{C}}-\overset{X}{\underset{|}{C}}-}$$

Markovnikov's Rule: In the addition of HX to an alkene, the hydrogen atom adds to the carbon atom of the double bond that already has the greater number of hydrogen atoms.

**LEARNING OBJECTIVES:** Predict the products of elimination reactions.

Write the mechanisms of electrophilic addition of hydrogen halides, water, or halogens to alkenes.

**SKILLS TO MASTER:** Applying Markovnikov's rule (p. 586).

**KEY CONCEPTS:** Addition reactions convert a carbon-carbon double bond to a single bond and add an atom or group of atoms to each of the carbons.

## LEARNING EXERCISES

**10.1** Prepare a list of the functional groups described in this chapter. Describe at least one chemical property of each functional group.

**10.2** Prepare a list of the organic reactions described in this chapter. Write a one-sentence description of each reaction.

**10.3** Describe the conditions necessary to make a molecule chiral.

**10.4** Describe the factors that make a good nucleophile.

**10.5** Describe the factors that make a good leaving group.

**10.6** Describe the differences between the mechanism of an $S_N1$ and an $S_N2$ reaction.

**10.7** Describe the differences between the mechanism of an E1 and E2 reaction.

**10.8** List all of the terms new to you in Chapter 10. Write a one-sentence definition of each.

## PROBLEMS

**ilw** = interactive learning ware problem. Visit the website at www.wiley.com/canada/olmsted

## Hydrocarbons

**10.1** Shown below is the line drawing for the hydrocarbon-based natural product limonene. Re-draw limonene and indicate the number of $sp^3$–hybridized carbon atoms. How many $sp^2$–hybridized carbon atoms are there?

Limonene

**10.2** Draw all the structural isomers of the hydrocarbon with the molecular formula $C_6H_{14}$.

**10.3** Provide an unambiguous IUPAC name for each of the following hydrocarbons.
(a)  $CH_3(CH_2)_2CH(CH_3)_2$
(b)  $(CH_3)_2CHCH(CH_3)_2$

(c)

(d)

(e)

**10.4** Draw the hydrocarbons that correspond to the following IUPAC names: (a) 2-methyl-3-ethylhexane;
(b) 3-*iso*propylnonane; (c) 3-methylpentane;
(d) 1-methyl-4-ethylcylcohexane; and
(e) *tert*-butylcyclopentane.

**10.5** Draw line drawings for the alkenes that correspond to the following IUPAC names: (a) 2-methyl-2-butene;
(b) 2,3-dimethyl-2-butene; (c) 2-methyl-3-hexene;
(d) 3-methylcyclohexene; and (e) 1-*iso*propyl-3-methylcyclohexene.

**10.6** Draw line drawings for the alkynes that correspond to the following IUPAC names: (a) 2-methyl-3-hexyne;
(b) 3-ethyl-3-methyl-1-pentyne; (c) 4-methyl-2-pentyne;
(d) 3,3-dimethyl-1-butyne; and
(e) 4-chloro-5-methyl-2-hexyne.

## Aromatic Compounds

**10.7** Name the following compounds using *o*-, *m*- and *p*-designations:

(a)                    (b)                    (c)

**10.8** Name the following compounds using numbers:

(a)                    (b)                    (c)

**10.9** Draw the hydrocarbons that correspond to the following IUPAC names: (a) toluene;
(b) *m*-chlorotoluene;
(c) *m*-dimethylbenzene;
(d) 2,4,6-trinitrotoluene (the explosive in dynamite); and
(e) hexachlorobenzene.

## Functional Groups

**10.10** Describe the following alkyl halides as primary (1°), secondary (2°), or tertiary (3°).

(a)                    (b)                    (c)

(d)                    (e)

**10.11** Provide an unambiguous IUPAC name for the alkyl halides in Problem 10.10.

**10.12** Provide an unambiguous IUPAC name for each of the following alkyl halides.

(a)                    (b)

(c)                    (d)

(e)

**10.13** Describe the following alcohols as primary (1°), secondary (2°), or tertiary (3°).

(a)        (b)        (c)

(d)        (e)

**10.14** Provide an unambiguous IUPAC name for each of the alcohols in Problem 10.13.

**10.15** Draw the alcohols that correspond to the following IUPAC names: (a) 3-methyl-2-pentanol; (b) *iso*propanol; (c) 3,4-hexanediol; (d) cyclohexanol; and (e) 4-ethyl-2-hexanol.

**10.16** Draw the ethers that correspond to the following IUPAC names: (a) diethyl ether; (b) *tert*-butyl methyl ether; (c) di*iso*butyl ether; (d) dibutyl ether; and (e) methyl *iso*propyl ether.

**10.17** Describe the following amines as primary (1°), secondary (2°), or tertiary (3°).

(a)        (b)        (c)

(d)        (e)

**10.18** Identify the functional groups in each of the following molecules.

(a)                    (b)

(c)                    (d)

(e)

**10.19** Provide unambiguous IUPAC names for each of the compounds in Problem 10.18.

**10.20** Indicate the hybridization state and bond angles for each carbon atom in the molecule shown below.

**Stereochemistry**

**10.21** Indicate which of the following alkenes would show *E/Z* isomerism. For each that does draw both isomers.
(a) 1-pentene; (b) 3-ethyl-2-pentene; (c) 2-pentene; (d) 2,3-dimethyl-2-pentene; and (e) 2,4-dimethyl-2-hexene.

**10.22** Assign *E* or *Z* stereochemical descriptors to the following compounds.

(a)        (b)        (c)

(d)        (e)

**10.23** Identify the following molecules as chiral or achiral.

(a)        (b)        (c)

(d)        (e)

**10.24** Identify the chirality centres in each of the following molecules and assign the *R* or *S* stereochemical descriptors to them.

(a)

(b)

(c)

(d)

(e)

**10.25** Shown below is the antibiotic amoxicillin that is produced by fermentation cultures of the fungus *Penicillium chrysogenum*. Identify the chirality centres and assign the *R* or *S* stereochemical descriptors to each of them.

Amoxicillin

## Substitution Reactions

**10.26** Identify the nucleophile, substrate, and leaving group in each of the following reactions. Note that the sodium ion is a spectator in each case.

(a) $NaBr + CH_3I \longrightarrow NaI + CH_3Cl$

(b) $(CH_3)C-OH + HCl \longrightarrow (CH_3)_3C-Cl + H_2O$

(c) $CH_3-ONa + CHCH_2-I \longrightarrow$
$$CH_3-O-CH_2CH_3 + NaI$$

(d)

$+ NaOCH_3 \longrightarrow$ $+ NaBr$

(e)

$+ NaCN \longrightarrow$ $+ NaBr$

**10.27** Identify the leaving group on each of these molecules and decide if it is a good leaving group or poor leaving group.

$CH_3-Br$

(a)

(b)

(c)

(d)

(e)

**10.28** Predict whether the following alkyl chlorides would follow an $S_N1$ or an $S_N2$ mechanism.

$CH_3 - Cl$

(a)

(b)

(c)

(d)

(e)

**10.29** Predict the products from each of the following substitution reactions.

(a)  $-OH + HBr \longrightarrow$

(b)  $Br + HBr \longrightarrow$

(c)  $+ H_2O \longrightarrow$

(d)  $+ {}^{\ominus}OCH_3 \longrightarrow$

(e)  $+ NaCN \longrightarrow$

**10.30** Predict the product of the following aromatic substitution reactions.

(a) [benzene] + Br$_2$ $\xrightarrow{\text{FeBr}_3}$

(b) [benzene] + HNO$_3$ $\xrightarrow{\text{H}_2\text{SO}_4}$

(c) [benzene] + Cl$_2$ $\xrightarrow{\text{AlCl}_3}$

(d) [benzene] + Br–CH$_3$ $\xrightarrow{\text{FeBr}_3}$

(e) [benzene] + Br$-\!\!\prec$ $\xrightarrow{\text{FeBr}_2}$

**10.31** How many monochlorination products are possible when toluene is treated with Cl$_2$ and AlCl$_3$?

## Elimination Reactions

**10.32** Predict whether the following compounds would undergo elimination following the E1 or E2 mechanism.

(a) [structure with Br]

(b) [structure with Br]

(c) [structure with Br, ethyl]

(d) [structure with Br]

(e) [structure with Br]

**10.33** Draw the major elimination product for each of the alkyl halides shown in problem 10.32.

**10.34** Predict the major elimination product for each of the following alkyl bromides. Note that the sodium ion is a spectator in each case.

(a) [structure with Br] + NaOCH$_2$CH$_3$ $\longrightarrow$

(b) [structure with Br] + NaOCH$_2$CH$_3$ $\longrightarrow$

(c) [structure with Br] + NaOCH$_2$CH$_3$ $\longrightarrow$

(d) H$-$C$-$CH$_2-$Br with two CH$_3$ groups + NaOCH$_2$CH$_3$ $\longrightarrow$

(e) Br$-$C$-$CH$_3$ with three CH$_3$ groups + NaOCH$_2$CH$_3$ $\longrightarrow$

**10.35** Draw the E1 mechanism for the formation of phenylethene (styrene) from 1-bromo-1-phenylethane, as shown below. Clearly indicate the flow of all the electrons with curved arrows.

[structure with Br] $\xrightarrow{\text{H}_2\text{O}}$ [styrene structure]

## Addition Reactions

**10.36** The addition of H$_{2(g)}$ to alkenes is carried out in the presence of a heterogeneous catalyst such as Pt metal. Define what is meant by heterogeneous. Describe the role of the catalyst in a reaction.

**10.37** Predict the product of each of the following hydrogenation reactions.

(a) [alkene structure] + H$_{2(g)}$ $\xrightarrow{\text{Pt}}$

(b) [alkene structure] + H$_{2(g)}$ $\xrightarrow{\text{Pt}}$

(c) [cyclohexene structure] + H$_{2(g)}$ $\xrightarrow{\text{Pt}}$

(d) [methylenecyclohexane structure] + H$_{2(g)}$ $\xrightarrow{\text{Pt}}$

(e) [dimethylcyclohexene structure] + H$_{2(g)}$ $\xrightarrow{\text{Pt}}$

(HINT: Consider stereochemistry)

**10.38** Deuterium gas ($D_2$) is made up exclusively of the heavy isotope of hydrogen, which has one neutron in addition to a proton. Predict the product of each of the reactions in Problem 10.37 if the hydrogen gas was replaced with $D_{2(g)}$.

**10.39** Draw the mechanism for the electrophilic addition of HBr to 2,3-dimethyl-2-butene. Clearly indicate the flow of all the electrons with curved arrows.

**10.40** Predict the product for each of the following addition reactions:

(a)   + Br$_2$ ⟶

(b)   + H$_2$SO$_4$ ⟶

(c)   + HCl ⟶

(d)   + H$_2$O —cat. HCl→

(e)   + I–Cl —H$_2$O→

**10.41** Draw the mechanism of reaction for Problem 10.40(e). Clearly indicate the flow of all the electrons with curved arrows.

**10.42** Use Markovnikov's rule to predict the major products of the following addition reactions.

(a)   + HBr ⟶

(b)   + HBr ⟶

(c)   + H$_2$O —cat. HCl→

(d)   + H$_2$O —cat. HCl→

(e)   + I–Cl ⟶

**10.43** Provide the structure of the alkene that leads to the following addition products upon addition with HBr.

(a)   —HBr→

(b)   —HBr→

(c)   —HBr→

(d)   —HBr→

(e)   —HBr→

**10.44** Which of the addition products in Problem 10.43 are chiral?

**10.45** Find Markovnikov's rule in the textbook, restate this rule and draw an example that illustrates it.

**10.46** Use Markovnikov's rule to predict the major products from the addition of ICl (iodine chloride) to the alkenes listed below.

(a)   + ICl ⟶

(b)   + ICl ⟶

(c)   + ICl ⟶

(d)   + ICl ⟶

(e)   + ICl ⟶

**10.47** Draw the mechanism of reaction for Problem 10.46(e). Clearly indicate the flow of all the electrons with curved arrows.

**10.48** Which of the addition products in Problem 10.46 are chiral?

**10.49** Draw the structure for the missing intermediate compound in each of the following reactions.

(a)

(b)

(c)

(d)

(e)

**10.50** Provide the identity of the missing reagent(s) for each of the following transformations.

(a)

(b)

(c)

(d)

(e)

**10.51** Provide the identity of the missing reagent(s) for each of the following transformations.

**10.52** Fill in the missing reagents and intermediates in the following transformations.

## ADDITIONAL PAIRED PROBLEMS

**10.53** Hydrocarbons containing only carbon and hydrogen bonds are considered non-polar. Use an electronegativity argument to explain this.

**10.54** Use your knowledge of intermolecular forces to explain why hexane ($T_b = 69\ °C$) is a liquid at room temperature.

**10.55** Use an argument based upon intermolecular forces to explain why hexane is insoluble in water.

**10.56** Compare the structures of cyclopropane and cyclohexane. What are the internal bond angles for cyclopropane? Use your knowledge of the ideal bond angles for an $sp^3$–hybrid carbon atom to comment on the amount of strain that exists in the cyclopropane ring.

**10.57** Dichloromethane ($CH_2Cl_2$) is a polar molecule whereas $CCl_4$ is considered non-polar. Provide an explanation for this observation.

**10.58** Dimethylether is insoluble in water whereas ethanol (with the same molecular formula) is fully miscible. Provide an explanation for this observation.

**10.59** Consider the chemical structure of acetone (dimethyl ketone). Would you expect the carbonyl group of this molecule to have a dipole? Draw the structure of acetone and use an arrow to indicate the direction of the dipole. What would be the most likely product from attack of a nucleophile such as the hydroxide ion?

**10.60** Amines are considered bases, or proton acceptors. Draw a chemical equation that demonstrates this fact.

**10.61** Acetic acid is ~10 000 times more acidic (i.e., a better proton donor) than ethanol. Propose an explanation for this observation.

**10.62** There is a significant barrier at room temperature for the interconversion of $E$- and $Z$-1,2-dichloroethene. Use a model based upon the overlap of hybrid orbitals to explain this observation.

**10.63** The amino acids serine and cystine are shown below. The chirality centre for serine is assigned as having an $S$ configuration whereas the chirality centre in cystine is assigned $R$. Provide an explanation for why these chirality centres are assigned different configurations.

Serine                          Cystine

**10.64** In Focus 10-1 we state that a pure sample of a single enantiomer would rotate plane-polarized light in one direction. We also state that a pure sample of the opposite enantiomer would rotate plane-polarized light in the *opposite* direction by an equal amount. Predict what the observed rotation would be if you had an equal mixture of both enantiomers in the same sample.

**10.65** Draw all the possible stereoisomers for 2,3-dibromopentane. How many are related by a mirror-image relationship? Describe the relationship between the other isomers.

**10.66** The hydroxide ion ($OH^-$) is a much stronger nucleophile than water ($H_2O$). Provide some reasons for this observation. Based upon this, how do you think that hydronium ion ($H_3O^+$) compares as a nucleophile to water?

**10.67** The conversion of *tert*-butanol to *tert*-butyl chloride by the $S_N1$ requires a catalytic amount of $H^+$ (i.e., $H_3O^+$). Examine the mechanism for the $S_N1$ and provide an explanation for the role of the catalyst.

**10.68** The hydrogenation of fatty acids (see Focus 10-2) results in the production of a solid (i.e., margarine) from an oil. Explain how removing the double bonds from a long-chain fatty acid results in an increase in the melting point.

**10.69** The addition of HCl to tetrafluoroethene is considerably slower than the addition of HCl to ethene. Use the mechanism of electrophilic addition to provide an explanation for this observation.

**10.70** The $S_N2$ reaction of ($R$)-2-bromobutane with NaOH produces exclusively ($S$)-2-butanol. Use the mechanism of the $S_N2$ to explain why the configuration of the chirality centre changes during the course of this reaction.

**10.71** The reaction of 1-phenylethanol with HBr results in the production of 1-bromo-1-phenylethane. When pure ($R$)-1-phenylethanol is used in the reaction, the product is a $1:1$ mixture of ($R$)- and ($S$)-1-bromo-1-phenylethane. Use the mechanism of the $S_N1$ to explain this result.

**10.72** The E1 reaction of 1-phenyl-1-propanol is shown below. What conclusions can you draw about the relative stability of $E$ versus $Z$ alkenes? What factors do you think might contribute to this result?

95%                    5%

**10.73** Two reactions for *N*-hexyl bromide are shown below. Based upon this result, what conclusions can you make about the effect that the size of the base has on this reaction?

**10.74** Under identical $S_N2$ conditions, 2,2-dimethylpropane reacts at a rate that is ~$1 \times 10^{-5}$ times slower than ethyl bromide, even though both are primary alkyl halides. Provide an explanation for this observation.

**10.75** Under identical $S_N1$ conditions, *tert*-butyl chloride reacts ~30 000 times faster than *N*-propyl chloride. Provide an explanation that takes into account the role that the methyl groups might have in this result. Using the model of each CH bond in the adjacent methyl groups as a "cloud" of electrons, explain how this could stabilize the positive charge on the carbon atom.

**10.76** The compound below is essentially unreactive under all substitution conditions, both $S_N1$ and $S_N2$. Provide explanations that accounts for this.

$$Cl-\underset{\underset{CF_3}{|}}{\overset{\overset{CF_3}{|}}{C}}-CF_3$$

## ADDITIONAL PROBLEMS

**10.77** The heat released when 1 mole of (*E*)- and (*Z*)-2-butene is hydrogenated (i.e., heat of hydrogenation) is given below. Can you make any conclusions about the relative stability of alkenes from this data?

Heat of hydrogenation $= -115$ kJ/mol

Heat of hydrogenation $= -119$ kJ/mol

**10.78** The heat of hydrogenation for cyclohexene is $-119$ kJ/mol and for cyclohexadiene is $-230$ kJ/mol. Based on this trend, it would be a safe conclusion to say that a second double bond results in a doubling of the heat of hydrogenation. However, the heat of hydrogenation for cyclohexatriene (benzene) is only $-206$ kJ/mol. Provide an explanation that is consistent with this observation.

**10.79** The molecule cycloheptatriene is shown below along with the cycloheptatrienyl cation that can be prepared from it. The cycloheptatrienyl cation is quite stable and can be stored at room temperature. In addition, it has been shown that all seven carbon atoms are identical, the C–C bond lengths are all equivalent, and the molecule is completely planar. In cycloheptatriene the C–C bond lengths are a combination of short double bonds and longer single bonds, and one of the carbon atoms is bent out of the plane of the other atoms. Provide a model for

the bonding in the cycloheptatrienyl cation that accounts for these observations.

Cycloheptatriene      Cycloheptatrienyl cation

**10.80** It is about 10 000 times easier to remove the proton shown in red on pentane-2,4-dione than it is to remove a similar proton from acetone. Provide an explanation for this observation. (*Hint:* Comparing the structure of the anions formed may help guide your explanation.)

Pentane-2,4-dione      Acetone

**10.81** Develop a molecular orbital description of the $\pi$ bond in 1,2-dichloroethene that includes the overlap of the contributing *p* atomic orbitals. Be sure your model accounts for the restricted rotation about the C–C bond.

**10.82** Cyclohexane exists as two equivalent, rapidly interconverting "chair" conformations as shown below. However *tert*-butylcyclohexane exists locked in only one of two possible chair conformations, also shown below. Provide an explanation for this observation.

not observed

Cyclohexane          *tert*-Butylcyclohexane

**10.83** Shown below is the three-dimensional structure for D-glucose, a naturally occurring sugar produced by all photosynthetic plants. D-Glucose is the key building block for most carbohydrates and is a critical energy source for metabolism in all animals including humans. Identify the chirality centres in D-glucose and assign the *R* or *S* configuration to each of them (carbon atom 1 is identified). The enantiomer of D-glucose is L-glucose, which does not occur in nature and instead has to be synthesized chemically. L-Glucose has been investigated as an artificial sweetener because it retains the sweet taste of D-glucose but is not metabolized by our bodies. Draw the three-dimensional structure of L-glucose, identify all the chirality centres, and assign *R* or *S* configurations to each of them.

D-Glucose

**10.84** Enantiomers have identical physical properties including melting point, boiling point, polarity, and solubility (in water, for example). However diastereomers (stereoisomers not related by a mirror image) are expected to have completely different physical properties including melting point, boiling point, polarity, and solubility. Consider the reaction between a 1 : 1 mixture of (R)- and (S)-2-chloropropionic acid and a sample of pure (R)-2-butanol to produce a 1 : 1 mixture of stereoisomeric esters, as shown below. Analyze the structure of the esters and decide on the stereochemical relationship between them. Are they enantiomers or diastereomers? Do you think that it would be possible to separate this mixture into two pure samples of each stereoisomer?

(S)-2-Chloropropionic acid

(R)-2-Chloropropionic acid

Acid cat.
Heat

1:1 mixture of stereoisomers

**10.85** In the E2 reaction of isopropyl bromide with ethoxide ion (Figure 10-27), a competing reaction would be the $S_N2$ conversion that leads to isopropyl ethyl ether. Describe the general factors that would favour an E2 versus an $S_N2$ reaction.

**10.86** The version of Markovnikov's rule that we give in this chapter states, *the hydrogen atom adds to the carbon atom of the double bond that already has the greater number of hydrogen atoms.* Examine the mechanism of electrophilic addition of HCl to a double bond and review the discussion of carbocation stability. Restate Markovnikov's rule in a way that takes the relative stability of carbocations into account.

**10.87** It is sometimes said that the addition of $BH_3$ to an alkene gives the anti-Markovnikov product in a reaction that is shown below. Use the relative electronegativity numbers of hydrogen (2.1) and boron (2.0) and the description of Markovnikov's rule that you developed in Problem 10.86 to show that the addition of $BH_3$ to an alkene actually follows Markovnikov's rule.

## GROUP STUDY PROBLEMS

**10.88** Make a plot in which the y-axis is relative energy and the x-axis is progress of the reaction from starting material to product. Chart the changes in relative energy you would expect as the $S_N2$ conversion from $CH_3-Br$ to $CH_3-Cl$ takes place. (Note: the $CH_3-Br$ bond energy is 293 kJ/mol and the $CH_3-Cl$ bond energy is 356 kJ/mol.)

**10.89** Complete the same exercise as Problem 10.88 for the $S_N1$ conversion of *tert*-BuBr to *tert*-BuCl. (*Hint:* The formation of the carbocation requires the most energy.) Describe the difference between the two profiles.

**10.90** The bromination of benzene by $Br_2$ and $FeBr_3$ can be written as an attack of "$Br^+$" on the aromatic ring as shown below. Use the mechanism of this reaction to explain why the bromination of nitrobenzene proceeds considerably slower than bromination of benzene. Consider the resonance structures of the intermediate formed by attack of "$Br^+$" on the ring.

Fast

Very slow

**10.91** Explain the trends in the relative acidity of the following acetic acid derivatives.

| Acetic acid | Relative acidity |
|---|---|

H₃C—C(=O)—OH    1

H₂C(Cl)—C(=O)—OH    40

HC(Cl)(Cl)—C(=O)—OH    1600

Cl—C(Cl)(Cl)—C(=O)—OH    12 500

**10.92** Amides, such as *N*-methyl acetamide, have been shown to have significant double-bond character about the C–N bond. This is because the resonance structure on the right makes an important contribution to the overall resonance hybrid. What are the consequences of this double-bond character on the structure of an amide? Explain why in an ester, such as methyl acetate, the similar resonance structure does not make a significant contribution to the resonance hybrid. What consequence would this have for the structures of an ester?

*N*-Methyl acetamide

Methyl acetate     Not important

**10.93** Use the reactions that you learned in this chapter to propose a synthesis of the following compounds from the staring materials indicated.

from

from

from

**ANSWERS TO SECTION EXERCISES**

**10.1.1**

Hexane    2-Methylpentane    3-Methylpentane

2,2-Dimethylbutane    2,3-Dimethylbutane

**10.1.2** (a) 2-Methylbutane; (b) 1,2-dimethylcyclopentane; (c) 3-ethylheptane; (d) 2,3-dichloro-2-butene; (e) 2,3-dimethylcyclobutene; and (f) 3-methyl-1-butyne

**10.1.3**

(a)    (b)    (c)

(d)    (e)    (f)

**10.2.1**

1,2,3-Tribromobenzene    1,2,4-Tribromobenzene    1,3,5-Tribromobnzene
(a)      (b)      (c)

**10.2.2** (a) 2-Phenylbutane or *sec*-butylbenzene; (b) 2,5-dibromoaminobenzene; and (c) *p*-nitrocyclohexylbenzene or 4-cyclohexylnitrobenzene

**10.2.3**

(a)      (b)      (c)

**10.3.1**

Carboxylic acid

Aromatic

(a)

Amino — NH$_2$  O — Ketone

(b)

Ester

OH — Alcohol

Halogen

Br

Alkene

(c)

**10.3.2**

Primary — Quaternary

Primary

(a)

Secondary

Primary

Primary

NH$_2$

(b)

Secondary

Primary

O

Primary

OH

(c)

**10.3.3**

(a)          (b)          (c)

**10.4.1** (a) Achiral; (b) achiral; and (c) chiral

**10.4.2** (a) *Z*; (b) *Z*; and (c) *E*

**10.4.3** (a) *R*; (b) *S*; and (c) *S*

**10.5.1** A nucleophile is a species that is attracted to a positive centre such as a carbocation. An electrophile is a species that is attracted to a negative centre.

**10.5.2** In the alkyl halide, the electronegative halogen atom attracts electrons toward itself, leaving the carbon atom relatively positively charged. The nucleophile is attracted to this positive centre.

**10.5.3** In the S$_N$2 mechanism, the nucleophile is attracted to the positive centre, and a bond begins to form. At the same time, the bond between the leaving group and the positive centre lengthens. Eventually, this bond breaks at the same time that the new bond is formed. In the S$_N$1 mechanism, the leaving group leaves, forming a positive charge on the carbon atom acting as the electrophile. Then, the nucleophile attacks this positive centre, forming a new bond.

**10.6.1**

$+$ C$_3$H$_7$OH $+$ Br$^-$

**10.6.2**

---

**10.7.1**   1,2-Dibromo-2-methylpropane
**10.7.2**   2-Bromo-3-phenylbutane (by Markovnikov's rule)
**10.7.3**

## ANSWERS TO PRACTICE EXERCISES

**10.1** (a) 2-Methylpropane; (b) 4-ethyl-3-methyl-6-propylde-
cane; and (c) 4-ethyloctane

**10.2**

(a)                              (b)

(c)

**10.3** (a) *o*-Aminochlorobenzene (or 2-chloroaniline);
(b) *p*-bromomethylbenzene (or *p*-bromotoluene or
3-bromotoluene); (c) 2,5-dibromopropylbenzene; and
(d) butylbenzene or 1-phenylbutane

**10.4** (a) Any two of

(b) $C_2H_5F$, (c) $C_3H_7Cl$; (d) *iso*-propyl fluoride or
2-fluoropropane; and (e) phenyl iodide or iodobenzene

**10.5**

(a) (i)

(ii)                    (iii)

(b) (i)                  (ii)

**10.6** (a)            (b)

(c)

(d) Butyl propyl ether; (e) methyl *iso*pentyl ether; and
(f) ethyl phenyl ether

10.7 (a) Propyl*iso*propylamine; (b) tripropylamine;
  (c) methylphenylamine; and (d) dimethylphenylamine

(e)            (f)            (g)

  (e) is a primary amine
  (a) and (c) are secondary amines
  (b), (d), (f), and (g) are tertiary amines
10.8 (a) Diphenyl ketone; (b) methyl*tert*-butyl ketone;
  (c) 3-phenylpropanal; and (d) 4-ethylhexanal

10.9

(a)                          (b)

(c)

10.10

(a)        (*Z*)-1-Chloro-1,2-Difluoroethene        (*E*)-1-Chloro-1,2-Difluoroethene

(b) (*E*)-1,3-Dibromo-1-Chloro-2-Methypropene    (*Z*)-1,3-Dibromo-1-Chloro-2-Methypropene

(c)        (*Z*)-1-Nitro-2-Pentene        (*E*)-1-Nitro-2-Pentene

10.11

  The carbon labelled "1" is chiral, therefore arginine is
  chiral.

10.12 (a) *S*; (b) *S*; (c) *R*
10.13 (a) I⁻ is the nucleophile, CH₃CH₂Br is the substrate,
  CH₃CH₂I is the product, and Br⁻ is the leaving group.

  (b) CN⁻ is the nucleophile,                    is the
  substrate,                    is the product and, Br⁻
  is the leaving group.
  (c) OH⁻ is the nucleophile, *tert*-butyl chloride is the
  substrate, *tert*-butyl alcohol is the product, and Cl⁻ is
  the leaving group.
10.14 (a) methyl-3-cyclohexylpropyl ether and Cl⁻, $S_N2$
  mechanism; and (b) Ethyl-*tert*-butyl ether and Br⁻, $S_N1$
  mechanism

10.15

(a)                          +  Br⁻ +  CH₃OH (E2)

(b)                          +  C₂H₅OH + Cl⁻ (E1)

10.16        Cl

        2-Chlorobutane
(a)

        Br
        2-Bromo-2-Cyclohexylbutane
(b)

10.17

        +  H₂O  ⟶

        Cyclopentylmethanol

# SPONTANEITY OF CHEMICAL PROCESSES

## THINGS BECOME DISORGANIZED

Perhaps you have noticed that there is a natural tendency for structures to become disorganized. Death is followed by decay. The organized biological structures of the once-living creature decompose. The inscription on a gravestone erodes and becomes illegible. The gravestone tilts and eventually falls over.

Despite this natural tendency, living organisms are highly structured. At the molecular level, DNA is a regular spiral of repeating units carrying specific genetic messages. The order of a DNA sequence is very precise, and the slightest deviation can have disastrous consequences. Macroscopically, animals create regular structures, from the simple hexagonal pattern of a honeycomb to the elaborate network of a highway bridge. Living organisms expend large quantities of energy to produce and maintain organized structures such as these.

Chemists would like to invent molecules that assemble themselves into specific, well-organized arrays. Recently, there have been some exciting successes. Our inset is a molecular view of a self-assembled

"molecular wreath." This organized structure forms when six molecular chains weave themselves together in the presence of 24 copper cations. Self-assembling molecular systems could lead to new materials with useful properties. They may also shed light on how life originated.

Why do structures tend to disintegrate, and why must living organisms spend much time and effort creating organized structures? The first law of thermodynamics, presented in Chapter 3, requires that energy must be conserved but places no restrictions on what direction a process must take. The explanation for why processes have preferred directions lies in the second law of thermodynamics, which we describe in this chapter. The second law lets us determine whether or not a chemical process can occur. It also lets us determine whether a change of conditions will change the direction of a chemical process. These predictive features make the second law as important as the first law. Chemists need to know in advance if a particular product can form when particular

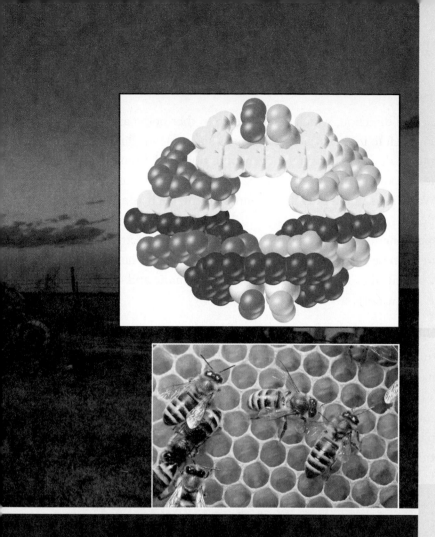

reactants are mixed, and if a different set of conditions would favour the production of products.

Two quantitative thermodynamic functions, energy and enthalpy, appear in Chapter 3. In the first section of this chapter, we show that we cannot predict the directions of processes using only energy and enthalpy changes. To extend thermodynamics to cover directionality, we need new thermodynamic functions. Sections 11.2 and 11.3 introduce and describe one of these functions, entropy. Section 11.4 describes a combination of energy and entropy, free energy, which is useful for treating chemical systems at constant temperature and pressure. We complete this chapter with two sections that address useful applications of thermodynamics.

# 11

## CHAPTER CONTENTS

## LEARNING OBJECTIVES

*Upon completion of this chapter you should be able to*

- identify the system and the surroundings in a process
- calculate the total entropy change of a process from the entropy changes of the system and the surroundings
- place substances in increasing order of absolute entropy
- calculate the entropy change of a reaction from standard absolute entropies
- calculate the free energy change of a reaction either from standard free energies of formation, or from standard enthalpies of formation and standard entropies
- calculate the temperature at which a reaction becomes spontaneous
- calculate vapour pressures from thermodynamic data
- calculate energy changes of coupled biochemical reactions

## 11.1 SPONTANEITY

Every process has a preferred direction, which is referred to in thermodynamics as the **spontaneous** direction. Left to itself, a process follows its spontaneous direction. For example, the spontaneous direction for water movement is downhill, from higher altitude to lower altitude. A spontaneous process can be reversed only by the action of some outside force. Water runs uphill only if an external agent, such as a pump, forces it to do so.

A process may be spontaneous, and yet the process may not occur. Extending our water example, water can be stored behind a dam for a very long time unless a spillway is opened. A chemical example is the reaction of methane and oxygen to form carbon dioxide and water:

$$CH_4\,(g) + 2\,O_2\,(g) \longrightarrow CO_2\,(g) + 2\,H_2O\,(l) \qquad \text{Spontaneous direction}$$

Methane–oxygen mixtures can be stored indefinitely, but a spark will cause the mixture to burst into flames.

A process that does not appear to occur may be spontaneous but very slow, or it may be non-spontaneous. Water does not run uphill, even when a spillway is present, without the action of a pump, because this is the non-spontaneous direction. Water and carbon dioxide do not rearrange into methane and molecular oxygen, even in the presence of a spark, because this is the non-spontaneous direction:

$$CO_2\,(g) + 2\,H_2O\,(l) \longrightarrow CH_4\,(g) + 2\,O_2\,(g) \qquad \text{Non-spontaneous direction}$$

Notice that the word *spontaneous* has a different meaning in thermodynamics than it does in everyday speech. Ordinarily, spontaneous refers to an event that takes place without any effort or premeditation. For example, a crowd cheers spontaneously for an outstanding performance. In thermodynamics, spontaneous refers only to the natural direction of a process, without regard to whether it occurs rapidly or slowly. This chapter deals with the spontaneity of chemical change, which refers to the direction that a process will take if left alone and given enough time. Chemical kinetics, which we introduce in Chapter 12, describes the factors that determine the speed of that change. Chapter 13 deals with equilibrium, which is the state that the chemical reaction (eventually) arrives at.

## Dispersal of Matter

To gain some insight into why processes have preferred directions, consider some large-scale events that go in one particular direction:

1. A bag of marbles dropped on the floor will scatter, but the marbles will not roll spontaneously back into the bag.

2. An untended wooden fence eventually falls apart, but piles of wood will not assemble spontaneously into fences.

3. A student's dormitory room naturally gets messy, but (unfortunately) does not spontaneously clean itself up!

Each of these processes has a spontaneous direction, in which matter becomes more spread out or dispersed. Another way to describe these processes states that

**Figure 11-1**

(a)               (b)

When the valve is opened between two bulbs, one of which is empty and the other filled with a gas, gas molecules move spontaneously from the filled bulb until the gas molecules are dispersed equally between the bulbs. The reverse process never occurs spontaneously.

each reduces the *constraints* on the objects. The marbles in a bag, the boards of a fence, and the books and clothing in a tidy dorm room are restricted, or constrained, in their positions. We can summarize these observations in a common-sense law: *things tend to become dispersed.*

Does this common-sense law also apply to events at the molecular level? Consider the two glass bulbs shown in Figure 11-1. One bulb contains nitrogen dioxide, a red-brown gas, and the second bulb is empty. When the valve that connects the bulbs is opened, the red-brown gas expands to fill both bulbs. The opposite process never occurs spontaneously. That is, if both bulbs contain $NO_2$ at the same pressure, opening the valve never causes the pressure to rise in one bulb and fall in the other.

The second part of Figure 11-1 shows a molecular view of what happens in the two bulbs. Recall from Chapter 2 that the molecules of a gas are in continual motion. The $NO_2$ molecules in the filled bulb are always moving, undergoing countless collisions with one another and with the walls of their container. When the valve between the two bulbs is opened, some molecules move into the empty bulb, and eventually the concentration of molecules in each bulb is the same. At this point, the gas molecules are in a state of dynamic equilibrium. Molecules still move back and forth between the two bulbs, but the concentration of molecules in each bulb remains the same. This state is discussed in detail in Chapter 13.

The gas molecules escaping from a bulb behave similarly to marbles rolling out of a bag. Molecules, like marbles, tend to become more dispersed, filling a larger volume. The opposite process, in which gas molecules become more constrained by all moving into one bulb, never occurs spontaneously. Thus, our common-sense law applies to this molecular example.

**Figure 11-2**

Sugar dissolving in coffee involves dispersal of molecules. Sugar molecules are distributed more widely in solution than they are in the solid crystal.

For a second example, consider Figure 11-2, which shows sugar dissolving in coffee from a molecular perspective. Before the solid sugar dissolves, the sugar molecules are organized in a crystal. As the molecules dissolve, they become dispersed uniformly throughout the liquid coffee. The opposite process, in which the molecules become more constrained by forming sugar cubes from sweet coffee, does not occur. Once again, the direction of spontaneous change at the molecular level conforms to the common-sense law. Sugar molecules are more dispersed when they are dissolved in coffee than when they are part of a solid crystal. A sugar cube is the molecular equivalent of a tidy dorm room. Sugar molecules disperse, destroying the compact molecular organization of a sugar cube as the cube dissolves in coffee.

**Figure 11-3**

When a hot object contacts a cold object, heat always flows from higher to lower temperature.

## Energy Dispersal

Processes that do not involve obvious dispersal of matter may nevertheless have preferred directions. For example, Figure 11-3 shows that when a hot block of metal is placed in a cold glass of water, the metal block cools and the water warms. This process continues until the two are at the same temperature. Whenever two objects at different temperatures contact each other, the object at higher temperature transfers energy to the object at lower temperature.

The molecular views shown in Figure 11-3 show that this is an example of energy dispersal. Recall from Chapter 3 that higher temperature means larger average thermal energy. The metal atoms in the hot block have higher thermal energies than the water molecules in the glass. When the two objects contact each other, the hotter atoms transfer energy to the colder molecules until the average thermal energy of each object is the same.

Figure 11-4 shows another example of energy dispersal. When a baseball pitcher throws a fastball, all the molecules of the ball move together through space. This represents an energy constraint. In addition to their random thermal energies of motion, all the molecules making up the ball have a portion of their kinetic energies that point in the same direction. As described in Chapter 3, when the ball comes to rest in the catcher's glove, the conversion of kinetic energy to thermal energy causes an increase in the temperature of the glove. The molecules in the glove move with larger average speeds. Moreover, these increased molecular speeds point randomly in all directions. In other words, the constrained, directional character of the kinetic energy of the ball has been dispersed into randomly directed energy of molecules in the glove.

## Energy and Spontaneity

Spontaneous processes result in the dispersal of matter and energy. In many cases, however, the spontaneous direction of a process may not be obvious. Can we use energy changes to predict spontaneity? To answer that question, consider two everyday events, the melting of ice at room temperature and the formation of ice in a freezer.

As diagrammed in Figure 11-5, ice melts spontaneously if a mixture of water and ice at 0 °C is placed on a table at 25 °C:

$$H_2O(s, 0\ °C) \xrightarrow{\text{Table top, 25 °C}} H_2O(l, 0\ °C)$$

Recall from Section 8.6 that a phase change from solid to liquid requires energy to overcome the intermolecular forces holding molecules in place in the solid. In this spontaneous process, the water–ice system gains energy as ice melts. The enthalpy change of the system also is positive.

Now consider what happens if the same mixture is placed in a freezer at −15 °C. Water freezes spontaneously, as shown schematically in Figure 11-6:

$$H_2O(l, 0\ °C) \xrightarrow{\text{Freezer, −15 °C}} H_2O(s, 0\ °C)$$

This process is the reverse of melting. To form a solid from a liquid, we must remove enough energy to allow the molecules to stick together in the solid phase. In this spontaneous process, the water–ice system releases energy as water solidifies, and the enthalpy of the system also decreases.

Both these everyday processes are spontaneous, but whereas one process is endothermic, the other is exothermic. The energy and enthalpy of the system increase in one process, but these quantities decrease in the

**Figure 11-4**

As a thrown baseball is caught, its kinetic energy is dispersed from the organized kinetic energy of the ball to randomly oriented kinetic energies of many individual molecules in the catcher's glove.

**Figure 11-5**

Schematic view of the spontaneous process for a water and ice mixture on a table. The energy-absorbing process, melting, is spontaneous under these conditions.

**Figure 11-6**

Schematic view of the spontaneous process for a water and ice mixture in a freezer. The energy-releasing process, freezing, is spontaneous under these conditions.

other process. This simple example demonstrates that analyzing energy and enthalpy changes alone is not enough to predict whether a process will occur spontaneously. We need a property other than energy and enthalpy if we hope to use thermodynamics to determine when a process will be spontaneous.

## Opposing Dispersal Trends

Liquid water contains molecules that are free to move about through their entire volume, whereas the water molecules in ice are held in a highly structured three-dimensional lattice. Thus, when water freezes, the spontaneous conversion leads to more constraint within the system. This seems like a violation of the common-sense law, but to complete our analysis we need to consider the energy change that accompanies freezing. As Figure 11-7 illustrates, water releases energy as it freezes. Energy is dispersed in the freezer as a result of the energy transfer. Freezing constrains matter but disperses energy.

The melting of ice is the reverse of the freezing of water. Energy becomes more constrained as it is transferred from the air in the room to the melting ice. At the same time, the molecules in the ice cube become less constrained, because they are free to move about in the liquid phase. Melting disperses matter but constrains energy.

If the common-sense law of dispersal is correct, then the freezing and melting processes both must have a net increase in dispersion. In freezing, the energy dispersal caused by the heat flow must be larger than the molecular constraints created by the fixed positions of the ice molecules. In melting, the dispersal of matter that results from releasing molecules from their fixed positions must be

**Figure 11-7**

When water freezes, its molecules become more constrained, but the energy released in this process becomes more dispersed in the surroundings.

larger than the energy constraints created by absorbing energy from the surroundings. The spontaneity of any process must therefore be evaluated in terms of the dispersal of both energy and matter. In Section 11.2, we define a new thermodynamic function that allows us to do such evaluations.

---

**SECTION EXERCISES**

● **11.1.1** Explain the following observations in terms of dispersal and constraints: (a) untended fences eventually fall down; (b) a plate is easy to break but difficult to mend; and (c) a wine cooler beverage does not spontaneously separate into alcohol, water, and fruit juice.

● **11.1.2** Solid sugar can be recovered from coffee by boiling off the water. (Coffee candy can be made in this way.) Sugar molecules become more constrained in this process. What else must occur for total dispersal to increase?

● **11.1.3** Draw molecular pictures of liquid water and water vapour that show what happens to the constraints among $H_2O$ molecules when water evaporates.

---

**Figure 11-8**

## 11.2 ENTROPY: THE MEASURE OF DISPERSAL

The spontaneous direction of any process is toward greater dispersal of matter and of energy. If we are to apply this criterion in a quantitative way, we need ways to measure amounts of dispersal. We analyze the constraints on a system to measure the dispersal of matter—the more the system is constrained, the less dispersed it is. We can also do calculations on the flow of heat to measure changes in the dispersal of energy.

### Entropy and Dispersal of Matter

The function that provides a quantitative measure of dispersal is called **entropy** and is symbolized $S$. In 1877, the Austrian physicist Ludwig Boltzmann derived Equation 11-1, which defines the entropy of a substance in terms of $W$, the number of ways of describing the system.

$$S = k \ln W \qquad (11\text{-}1)$$

The essence of this equation is that the more possible descriptions the system has, the greater is its entropy. The equation states that entropy increases in proportion to the natural logarithm of $W$, the proportionality being given by the Boltzmann constant, $k = 1.38 \times 10^{-23}$ J/K. Equation 11-1 also establishes a starting point for entropy. If there is only one way to describe the system, it is fully constrained and $W = 1$. Because ln (1) = 0, $S = 0$ when $W = 1$.

To get a feel for $W$, consider placing two marbles of different colours in a box containing nine compartments of equal size. Figure 11-8 shows two different ways to distribute the two marbles. We could place both marbles in any of the nine

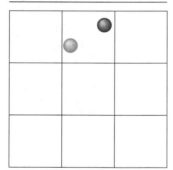

Two marbles confined
to one compartment
$W = 9 \times 1 = 9$

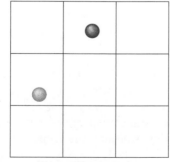

Two marbles occupying
different compartments
$W = 9 \times 8 = 72$

Two marbles distributed among nine compartments provides an example of $W$ and its relationship to the dispersal of matter.

**Figure 11-9**

Ice-like distribution of marbles
$W = 1$

Water-like distribution of marbles
$W = 9 \times 8 = 72$

Nine identical marbles distributed among nine compartments have distributions analogous to the distribution of water molecules in ice and water.

compartments. There are nine ways to place the first marble, but then we must place the second marble in the same compartment, so this distribution has $W = 9$. Another type of distribution places each marble in a *different* compartment. After placing the first marble in one of the nine compartments, the second can go into any of the other eight. The nine possible placements of the first marble multiplied by the eight possibilities for the second gives a total of 72 possible arrangements, so $W = 72$. The fact that $W$ is larger for the second distribution than for the first means that the second distribution has a larger entropy.

How does this relate to dispersal and constraint? When both marbles are confined to the same box, the marbles are more constrained (less dispersed) than when they occupy two different boxes.

A second example using marbles lets us tie this way of measuring dispersal to molecular structures. Consider placing nine identical marbles in the nine compartments, as shown in Figure 11-9. There is only one distribution that places each marble in a different compartment, so this arrangement has $W = 1$ and $S = 0$. Now consider removing one marble from a compartment and adding it to another one. We can remove the marble from any of the nine compartments and add it to any of the other eight, giving $W = 72$. There are fewer constraints on the second distribution, which has higher entropy. Compare this with ice melting to form water. In an ice crystal, all the molecules are constrained to fixed positions, analogous to one marble per compartment. In liquid water, molecules are free to move around, analogous to moving marbles between compartments.

Although the Boltzmann equation may appear simple, applying it to a molecular system is always challenging. The reason is that there are immense numbers of molecules in any realistic molecular system, so it would be necessary to count huge numbers of possibilities to determine the value of $W$. Instead, scientists have found ways to measure entropy by analyzing energy dispersal.

## Entropy and Dispersal of Energy

Any energy transfer results in either dispersal or constraint of energy, and thus generates a *change* in entropy ($\Delta S$). When a flow of heat occurs at constant temperature, Equation 11-2 provides a quantitative measure of the entropy change:

Equation 11-2 can be generalized to cover processes that are not at constant temperature, but this is beyond the scope of this text.

$$\Delta S = \frac{q_T}{T} \tag{11-2}$$

To put this in words, when heat flows at constant temperature, the entropy change is equal to the heat transferred ($q_T$) divided by the temperature in kelvins ($T$). The units of $\Delta S$ are energy/temperature, or J/K. The subscript $T$ in Equation 11-2 is a reminder that the quantity of heat transferred depends on the conditions. This equation is restricted to processes that occur at constant temperature.

The presence of $q$ in this equation is consistent with the earlier observation that heat flows change the dispersal of energy. The amount of change depends on the magnitude of the heat flow, but it depends on temperature as well.

Temperature tracks the thermal energy of motion of molecules in an object. At low temperature, the average molecular energy of motion is small, so the dispersal of a given amount of heat has a greater impact at low temperature than dispersing the same amount of heat at higher temperature. Dividing by $T$ takes this effect into account.

Heat flow at constant temperature may seem paradoxical, because we usually associate a flow of heat with a change in temperature. Nevertheless, three important types of chemical processes occur in which heat flows take place without a temperature change:

1. Phase changes take place at constant temperature, absorbing or releasing heat in the process.

2. A chemical reaction may occur under conditions in which temperature is held constant, such as in the human body or a thermostatted automobile engine. The exothermicity or endothermicity of the reaction generates a heat flow, even though the temperature is constant.

3. The surroundings may be so large that they can absorb or release significant amounts of heat before the temperature changes by a measurable amount. An experiment performed in a constant-temperature water bath is a common example of this category.

Example 11-1 illustrates this for ice forming in a freezer, the process diagrammed in Figure 11-6.

---

### EXAMPLE 11-1 ENTROPY CHANGE DURING FREEZING

What is the total entropy change when 2.00 moles of water ($\Delta H_{fus} = 6.01$ kJ/mol) freezes at 0.0 °C in a freezer compartment in which the temperature is held at −15 °C?

**Strategy:** The problem asks for the *total* entropy change, which includes $\Delta S$ for the water and $\Delta S$ for the freezer compartment. When water freezes, heat flows from the water to its surroundings, the freezer compartment (see Figure 11-6). Thus, $q$ is negative for the water, and its entropy decreases. At the same time, $q$ is positive for the freezer compartment, and its entropy increases.

**Solution:** We need a way of determining $q$, and we must identify $T$ for the water and for the freezer compartment. The water–ice mixture remains at a constant temperature of 0.0 °C (273.15 K) while the phase change occurs. (Eventually, the ice will cool to −15 °C, but the question asks only about the freezing process.) The freezer is at a constant temperature of −15 °C = 258 K. The problem states that $n = 2.00$ mol and $\Delta H_{fus} = 6.01 \times 10^3$ J/mol. Remember that $\Delta H$ for a phase change is the amount of heat required for the phase conversion of 1 mole of the substance. This gives us an equation to calculate $q$ for the phase change:

$$q = n\,\Delta H_{fus}$$

We calculate the entropy changes for these constant-temperature processes using Equation 11-2.

$$\Delta S = \frac{q_T}{T}$$

Remember that $\Delta H_{fus}$, which is positive, refers to ice melting. Remember also that the temperature must be in kelvins. The water–ice mixture releases heat as the water freezes, so $q$ for the water has a negative value:

$$q_{H_2O} = -(2.00 \text{ mol})(6.01 \times 10^3 \text{ J/mol}) = -1.202 \times 10^4 \text{ J}$$

$$\Delta S_{H_2O} = \frac{q_{H_2O}}{T_{H_2O}} = \frac{-1.202 \times 10^4 \text{ J}}{273.15 \text{ K}} = -44.0 \text{ J/K}$$

The freezer compartment absorbs the heat released by the water, but it does so at *its* temperature, which is $-15 \text{ °C} + 273 = 258$ K. Because the freezer absorbs heat, $q_{freezer}$ has a positive sign. Thus, for the freezer:

$$q_{freezer} = -q_{H_2O} = +1.202 \times 10^4 \text{ J}$$

$$\Delta S_{freezer} = \frac{q_{freezer}}{T_{freezer}} = \frac{1.202 \times 10^4 \text{ J}}{258 \text{ K}} = 46.6 \text{ J/K}$$

The total entropy change is the sum of these changes:

$$\Delta S_{total} = \Delta S_{H_2O} + \Delta S_{freezer} = (-44.0 \text{ J/K}) + (46.6 \text{ J/K}) = +2.6 \text{ J/K}$$

**Does the Result Make Sense?** The negative sign for the entropy change of the ice–water mixture is consistent with our qualitative view that matter is more dispersed in a liquid than in a solid. The positive sign for the entropy change of the freezer is consistent with heat being absorbed by the freezer, which increases the dispersal of energy. Note that $\Delta S_{total} > 0$, which must be the case for a spontaneous process.

**Common Pitfalls:** Not understanding the signs of the entropy changes of the system and of the surroundings.

**Practice Exercise 11.1**
The enthalpy of vapourization of water is 40.79 kJ/mol, and $T_b = 100$ °C. Calculate the entropy changes for water and for the hot plate, and the overall entropy change when a teakettle on a hot plate at 250 °C boils off 3.55 moles of water.

We found that the total entropy change is positive for the spontaneous freezing of water in a freezer. We can analyze the total entropy change for 2.00 moles of ice melting in a room at 25 °C (see Figure 11-5) in the same way. The heat required to melt the ice has the same magnitude but the opposite sign as the heat released in freezing the water in Example 11-1: $1.202 \times 10^4$ J/mol. Ice melts at the same temperature as water freezes, 273.15 K, so the entropy change for the water is also equal in magnitude but is opposite in sign: $\Delta S = 44.0$ J/K. To calculate the entropy change of the room, we must use its temperature, 25 °C + 273 = 298 K. The heat required to melt the ice comes from the air in the room, but the amount of heat is so small and the volume of the room is so large that the temperature remains essentially constant:

$$\Delta S_{room} = \frac{-1.202 \times 10^4 \text{ J}}{298 \text{ K}} = -40.3 \text{ J/K}$$

$$\Delta S_{total} = (44.0 \text{ J/K}) + (-40.3 \text{ J/K}) = +3.7 \text{ J/K}$$

We see that the total change in entropy is a positive quantity for both these spontaneous processes, even though one process is exothermic and the other is

**Figure 11-10**

Water freezes, releasing energy to the freezer

Ice melts, absorbing energy from the room

In both cases, $\Delta S_{total} = \Delta S_{sys} + \Delta S_{surr}$

When calculating entropy changes, be careful about the sign of $q$, use the appropriate temperatures, and sum the changes for system and surroundings.

endothermic. When this type of calculation is carried out for other processes, the same result is always obtained. For any spontaneous process, the total change of entropy is a positive quantity. Thus, this new state function of entropy provides a thermodynamic criterion for spontaneity, which is summarized in the **second law of thermodynamics**:

*In any spontaneous process, the total entropy increases:* $\Delta S_{total} > 0$

Figure 11-10 summarizes the ice and water example and illustrates potential pitfalls in evaluating entropy changes. First, be careful about the sign of $q$. Heat flows between a system (the water–ice mixture in this example) and its immediate surroundings (the freezer or air in the room). The sign for the system depends on whether the process is endothermic or exothermic, and the sign for the surroundings is always the opposite of the sign for the system. Table 11-1 summarizes the signs. Second, always use the absolute temperature of the component for which entropy change is being evaluated. In our examples, the freezer, the room, and the water–ice mixture all have different temperatures. Third, find the *total* entropy change by summing the entropy changes of the system and its surroundings, as described by Equation 11-3:

$$\Delta S_{total} = \Delta S_{sys} + \Delta S_{surr} \qquad (11\text{-}3)$$

Example 11-2 further illustrates these principles.

## TABLE 11-1 SIGNS ASSOCIATED WITH HEAT TRANSFERRED ($q$)

| Process | System | Surroundings | Example |
|---------|--------|--------------|---------|
| Exothermic | − | + | Water freezing |
| Endothermic | + | − | Ice melting |

## EXAMPLE 11-2 ENTROPY CHANGE OF A REFRIGERANT

In a refrigerator, a liquid refrigerant absorbs heat from the contents of the refrigerator. This heat vapourizes the refrigerant, and then a mechanical pump compresses the gas and recondenses it to a liquid. One common refrigerant is HFC-134a, $CH_2FCF_3$ ($M = 102.0$ g/mol). HFC-134a boils at $-27$ °C with $\Delta H_{vap} = 22.0$ kJ/mol. Calculate the total entropy change when 150 g of HFC-134a vapourizes at its boiling point, exchanging heat with contents of the refrigerator at 4 °C.

**Strategy:** The problem describes a relatively complex process, so we apply the seven-step problem-solving method. The problem asks for the total entropy change ($\Delta S_{total}$) caused by the refrigeration process. We must determine the entropy change of the liquid refrigerant, HFC-134a, and the entropy change of the contents of the refrigerator.

A block diagram helps us to visualize the thermodynamic processes. The liquid refrigerant evaporates at constant temperature as it absorbs heat from the contents of the refrigerator, which are at a different constant temperature.

**Solution:** Here are the data:

For $CH_2FCF_3$, $m = 150$ g, $M = 102.0$ g/mol, $T_b = -27$ °C $+ 273 = 246$ K, $\Delta H_{vap} = 22.0$ kJ/mol.

For the contents of the refrigerator, $T_{refrig} = 4$ °C $+ 273 = 277$ K.

Equation 11-3 gives the total entropy change:

$$\Delta S_{total} = \Delta S_{sys} + \Delta S_{surr}$$

We can define the HFC as the system and the contents of the refrigerator as the immediate surroundings:

$$\Delta S_{sys} = \Delta S_{HFC}$$

$$\Delta S_{surr} = \Delta S_{contents}$$

$$\Delta S_{total} = \Delta S_{HFC} + \Delta S_{contents}$$

Use the data to find $q$ for the HFC, and then use Equation 11-2 to calculate $\Delta S$ for the HFC and for the contents:

$$q = n \, \Delta H_{vap} \quad \text{and} \quad \Delta S = \frac{q_T}{T}$$

To evaluate $q$, we need the number of moles of HFC-134a:

$$n_{HFC} = \frac{150 \text{ g}}{102.0 \text{ g/mol}} = 1.471 \text{ mol}$$

$$q_{HFC} = (1.471 \text{ mol})(22.0 \text{ kJ/mol})(10^3 \text{ J/kJ}) = 3.235 \times 10^4 \text{ J}$$

$$\Delta S_{HFC} = \frac{q_{HFC}}{T_{HFC}} = \frac{3.235 \times 10^4 \text{ J}}{246 \text{ K}} = 132 \text{ J/K}$$

This is the entropy change of the refrigerant fluid. The contents of the refrigerator give up heat equal to the heat absorbed by the fluid:

$$q_{contents} = -q_{HFC}$$

$$\Delta S_{contents} = \frac{-q_{HFC}}{T_{contents}} = \frac{-3.235 \times 10^4 \text{ J}}{277 \text{ K}} = -117 \text{ J/K}$$

The total entropy change is therefore:

$$\Delta S_{total} = \underset{\uparrow \text{System}}{\Delta S_{HFC}} + \underset{\uparrow \text{Surroundings}}{\Delta S_{contents}} = 132 + (-117) = 15 \text{ J/K}$$

**Does the Result Make Sense?**

The refrigerant vapourizes, dispersing its matter, so a positive $\Delta S$ is reasonable. The contents release energy to the refrigerant, constraining their energy, so a negative $\Delta S$ is reasonable. The total entropy change is positive, as it must be for any spontaneous process.

**Common Pitfalls:** Not using the appropriate temperatures for the system or the surroundings. These are usually different from one another.

**Practice Exercise 11.2**

Calculate the total entropy change when 175 g of water at 0 °C freezes, transferring heat that causes HFC-134a to boil at −27 °C.

## Direction of Heat Flow

Equation 11-2 states that an entropy change at constant temperature is heat flow divided by temperature, and according to the second law of thermodynamics, the total entropy change is always positive. Taken together, these two requirements dictate that spontaneous heat-flow between two bodies at different temperatures always is from the warmer body to the colder body. For example, when a kettle of boiling water sits on the heating element of a stove, heat always flows from the hot burner to the relatively cooler boiling water.

To understand why this must occur, consider the entropy changes that would accompany heat transfer in the *opposite* direction. Suppose the burner is at 455 K and the water is at 373 K. We can calculate the entropy change that would occur if 100 J of heat flowed from the water to the burner. In this scenario, $q$ for the burner is positive, so the burner gains entropy. For the water, $q$ is negative, so the water loses entropy:

$$\Delta S = \frac{100 \text{ J}}{455 \text{ K}} = 0.220 \text{ J/K}; \quad \Delta S_{water} = \frac{-100 \text{ J}}{373 \text{ K}} = -0.268 \text{ J/K}$$

$$\Delta S_{total} = \Delta S_{burner} + \Delta S_{water} = 0.220 \text{ J/K} + (-0.268 \text{ J/K}) = -0.048 \text{ J/K}$$

A spontaneous "uphill" flow of heat, from the cooler water to the hotter burner, would result in a *decrease* in total entropy, which the second law of thermodynamics forbids. According to all observations, which the second law summarizes, heat never flows spontaneously from a cold to a hot body.

We can summarize the first and second laws of thermodynamics in two simple equations:

$$\text{First law:} \qquad \Delta E_{total} = 0 \text{ (always)}$$

$$\text{Second law:} \qquad \Delta S_{total} > 0 \text{ (always)}$$

● **11.2.1** Benzene ($C_6H_6$) has $\Delta H_{fus} = 10.9$ kJ/mol and a freezing point of 5.5 °C, whereas water has $\Delta H_{fus} = 6.01$ kJ/mol and freezes at 0.0 °C. Suppose that a sealed jar containing a mixture of solid and liquid benzene at 5.5 °C is immersed in a mixture of ice and water at 0.0 °C. What will happen?

● **11.2.2** Compute $\Delta S_{total}$ for 10.0 g of benzene changing phase as described in Section Exercise 11.2.1.

● **11.2.3** Suppose that a mixture of ice and water is placed in a refrigerator that is held at the freezing point of water, exactly 0.0 °C. Calculate $\Delta S$ for the water–ice mixture, $\Delta S$ for the refrigerator, and $\Delta S_{total}$ when 5.00 g of ice forms. Is this process spontaneous? Is the reverse process spontaneous?

## 11.3 ENTROPIES OF PURE SUBSTANCES

The Boltzmann equation, $S = k \ln W$, establishes the zero point for entropies. Because $\ln(1) = 0$, this equation predicts that the entropy will be zero for a system with only one possible description. Figure 11-11$a$ shows a system with $W = 1$, nine identical marbles placed in nine separate compartments. The figure also shows two types of conditions where $W > 0$. Figure 11-11$b$ shows a condition when the marbles are not identical. If one marble is a different colour than the other eight, there are nine different places to place the different marble. Figure 11-11$c$ shows a condition where there are more compartments than marbles. If there are more places to put the marbles than there are marbles, there are multiple ways to place the marbles in the compartments.

### Minimization of Entropy

We can apply the same reasoning to atoms or molecules as to marbles. First, any substance that is impure is like the depiction in Figure 11-11$b$. The impurity atoms or molecules can occupy any locations in the sample, increasing $W$ and giving the sample additional entropy. Second, if there is any extra space available for the atoms or molecules of a pure substance, the situation is like the depiction

**Figure 11-11**

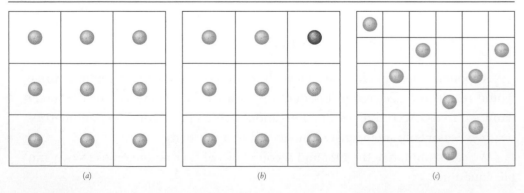

(a) When there is only one possible distribution, $W = 1$ and $S = 0$. (b) When items are not identical, there is more than one possible distribution, making $W > 1$ and $S > 0$. (c) When there are more places to put the items than there are items, $W > 1$ and $S > 0$.

in Figure 11-11c. A gas has much empty space between its atoms or molecules, so gases have very large values of $W$ and correspondingly high entropies. Liquids have much less available space, but the atoms and molecules still have considerable freedom of motion, meaning there are multiple locations available, so $W > 1$ and $S > 0$.

A pure crystalline solid comes closest to the depiction in Figure 11-11a. Nevertheless, each atom or molecule in a pure crystalline solid vibrates back and forth in its compartment, and this vibration can be thought of as similar to the depiction in Figure 11-11c. The vibrations move the atoms or molecules randomly about in the space available to them, making $W > 1$ and $S > 0$.

To reach $W = 1$ and $S = 0$, we must remove as much of this vibrational motion as possible. Recall that temperature is a measure of the amount of thermal energy in a sample, which for a solid is the energy of the atoms or molecules vibrating in their cages. Thermal energy reaches a minimum when $T = 0$ K. At this temperature, there is only one way to describe the system, so $W = 1$ and $S = 0$. This is formulated as the **third law of thermodynamics**, which states that a pure, perfect crystal at 0 K has zero entropy:

$$S_{(\text{pure, perfect crystal}; \, T = 0 \text{ K})} = 0 \qquad (11\text{-}4)$$

Each of the three specifications is essential in order to make $W = 1$. If the sample is not *pure*, the impurities can have multiple locations. If the sample is not a *perfect crystal*, the imperfections can have multiple locations. If the sample is not at $T = 0$ K, the vibrational energy gives the atoms or molecules multiple locations.

## Absolute Entropies

The third law of thermodynamics makes it possible to measure the **absolute entropy** of any substance at any temperature. Figure 11-12 shows the stages of a pure substance, in this case argon, as the temperature is raised from $T = 0$, $S = 0$ to $T = 298$ K. As temperature increases, entropy increases steadily.

When solid argon is heated, its atoms "rattle around" in their crystalline cages with increasing energy of motion. Equation 11-2 does not apply to this process because the temperature is changing, but the techniques of calculus can be used to determine the entropy change from the energy required for the heating.

When the solid melts at 83.8 K, entropy increases by an amount given by Equation 11-2:

$$q_{Ar} = n\Delta H_{fus} = (1.00 \text{ mol})(1.21 \text{ kJ/mol})(10^3 \text{ J/kJ}) = 1.21 \times 10^3 \text{ J}$$

$$\Delta S_{fus} = \frac{q_{Ar}}{T_{fus}} = \frac{1.21 \times 10^3 \text{ J}}{83.8 \text{ K}} = 14.4 \text{ J/K}$$

Raising the temperature of liquid argon from 83.8 K to its boiling point, 87.3 K, results in an additional increase in entropy as the argon atoms gain thermal energy.

**Figure 11-12**

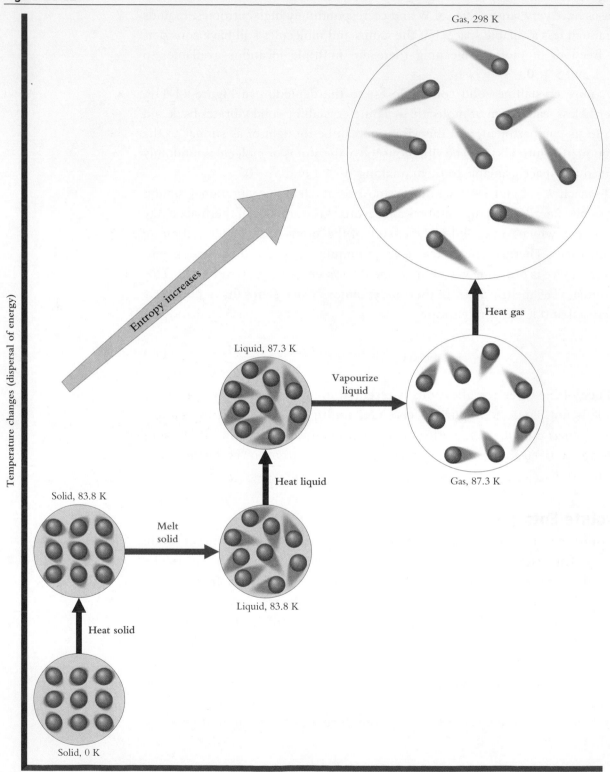

At $T = 0$ K, a pure perfect solid crystal has $S = 0$ J/K. The absolute entropy of the substance increases as the temperature increases and the phase changes. The temperatures shown here are for pure argon.

When argon vapourizes, there is another large increase in entropy that can be calculated using Equation 11-2:

$$q_{Ar} = n\Delta H_{vap} = (1.00 \text{ mol})(6.53 \text{ kJ/mol})(10^3 \text{ J/kJ}) = 6.53 \times 10^3 \text{ J}$$

$$\Delta S_{vap} = \frac{q_{Ar}}{T_{vap}} = \frac{6.53 \times 10^3 \text{ J}}{87.3 \text{ K}} = 74.8 \text{ J/K}$$

This large increase in entropy comes mainly from the increase in volume that accompanies vapourization. At 87.3 K, the molar volume of liquid argon is only 0.29 L, whereas the molar volume of argon gas is 7.17 L.

More energy is required to heat 1 mole of argon gas from 87.3 to 298 K, and this energy gets dispersed among the argon atoms, increasing the entropy still further.

Any substance can be subjected to the path outlined in Figure 11-12. Starting from the lowest possible temperature, the substance is heated to 298 K at a constant pressure of 1 bar. The amount of heat required to increase the temperature and the energies of any phase changes are measured using calorimetry. These calorimetry experiments may sound commonplace, but as our Focus 11-1: Chemistry and Technology describes, working at temperatures near 0 K is both difficult and intriguing.

The third law of thermodynamics establishes a "starting point" for entropies. At 0 K, any pure, perfect crystal is completely constrained and has $S = 0$ J/K. At any higher temperature, the substance has a positive entropy that depends on the conditions. The molar entropies of many pure substances have been measured at standard thermodynamic conditions, 1 bar and 298 K. The same thermodynamic tables that list standard enthalpies of formation usually also list standard molar entropies, designated $S°$, at these conditions. Table 11-2 lists representative values of $S°$ to give you an idea of the magnitudes of absolute entropies. Appendix D contains a more extensive list.

**TABLE 11-2  STANDARD MOLAR ENTROPIES ($S°$) OF SELECTED SUBSTANCES AT 298 K AND 1 BAR**

| Substance | Phase | $S°$ (J K$^{-1}$mol$^{-1}$) | Substance | Phase | $S°$ (J K$^{-1}$ mol$^{-1}$) |
|---|---|---|---|---|---|
| C | Solid (diamond) | 2.4 | He | Gas | 126.153 |
| C | Solid (graphite) | 5.74 | Ar | Gas | 154.846 |
| Si | Solid | 18.8 | Xe | Gas | 169.685 |
| Al | Solid | 28.3 | $H_2$ | Gas | 130.680 |
| Cu | Solid | 33.2 | CO | Gas | 197.660 |
| Ag | Solid | 42.6 | $F_2$ | Gas | 202.79 |
| $SiO_2$ | Solid | 41.5 | $Cl_2$ | Gas | 223.08 |
| NaCl | Solid | 72.1 | $CO_2$ | Gas | 213.78 |
| $I_2$ | Solid | 116.1 | $CH_4$ | Gas | 184.3 |
| $H_2O$ | Liquid | 69.95 | $C_2H_2$ | Gas | 200.9 |
| Hg | Liquid | 75.9 | $C_2H_4$ | Gas | 219.3 |
| $Br_2$ | Liquid | 152.2 | $C_2H_6$ | Gas | 229.2 |

## ■ FOCUS 11-1 CHEMISTRY AND TECHNOLOGY: SEEKING ABSOLUTE ZERO

Atoms and molecules at absolute zero temperature truly would be fully constrained. Because all excess energy would be removed, the inherent properties of atoms and molecules would be sharply defined. For this reason, chemists and physicists have worked extensively to develop techniques to reduce the temperature of a sample as close as possible to absolute zero. Along the way, a number of astonishing discoveries have been made, including superconductivity and superfluidity.

Reducing the temperature becomes ever more difficult, the nearer we approach absolute zero. If we wish to reduce the temperature of a sample below the temperature of its surroundings, we must find some way other than a heat flow to remove energy of motion, because "uphill" heat flows are never spontaneous. To cool a refrigerator below the temperature of its surroundings, we use a refrigerating fluid that condenses when compressed and absorbs energy from its surroundings as it vapourizes. Molecular nitrogen is liquefied by rapid expansion. As the gas expands, the molecules must overcome intermolecular forces of attraction. Their potential energy increases at the expense of their kinetic energy of motion. This slows them down, so the temperature drops. Nitrogen condenses at 77 K. Still lower temperatures require even more elaborate procedures.

Low-temperature research requires hard work and imagination, but successful advances are richly rewarded. Seven Nobel Prizes in physics and chemistry have been awarded for low-temperature research. The first, in 1913, went to the Dutch physicist Heike Kamerlingh Onnes, who discovered how to cool He gas to 4.2 K and convert it into a liquid. The American William Giauque received the 1949 prize in chemistry and the Russian Pyotr Kapitsa won the 1978 prize in physics. Each was honoured for a variety of discoveries resulting from low-temperature research, and each developed a new technique for achieving low temperature.

Giauque's technique is still used in contemporary low-temperature research. A sample containing paramagnetic ions (e.g., $Fe^{3+}$) is bathed in liquid helium under reduced pressure to chill it to a temperature below 4.2 K. The sample is held in a strong magnetic field, which aligns the magnetic directions of the paramagnetic ions. Then, the liquid helium is removed and the magnet turned off. The paramagnetic ions absorb energy in order to randomize their magnetic alignments. This absorption of energy causes the temperature to fall. Temperatures ranging from $10^{-3}$ to $10^{-6}$ K can be achieved using this technique.

The 1996 Nobel Prize in physics went to three researchers who studied liquid helium at a temperature of 0.002 K, discovering superfluid helium. A superfluid behaves completely unlike conventional liquids. Liquids are normally viscous because their molecules interact with one another to reduce fluid motion. Superfluid helium has zero viscosity, because all of its atoms move together like a single "superatom." This collective behaviour also causes superfluid liquid helium to conduct heat perfectly, so heating a sample at one particular spot results in an immediate and equal increase in temperature throughout the entire volume. A superfluid also flows extremely easily, so it can form a fountain (shown in the photo) in apparent defiance of gravity.

Superfluidity is just one of the surprising new properties discovered through low-temperature research. Another example is superconductivity, described in our Focus 8-1: Chemistry and Technology in Chapter 8. The 2003 Nobel Prize in physics went to three theoreticians who developed theories explaining these phenomena.

*A superfluid fountain of liquid helium*

In 1997 and 2001, the Nobel Prizes in physics were awarded for low-temperature research on gases. When an atomic gas expands through a small opening, it forms a beam of atoms travelling in one direction (recall the molecular beams described in Section 2.5). Normally, these atoms have speeds around 1 km/s. When the atoms are bombarded head-on with photons from an intense laser beam with just the right frequency, however, the interaction between photons and atoms slows down the atoms, much as a stiff headwind slows down a runner. A sophisticated combination of laser beams and magnetic fields makes it possible to cool a beam of helium atoms to temperatures around $2 \times 10^{-7}$ K. At this temperature, the atoms move at a leisurely 2 cm/s. The winners of the 2001 Nobel Prize studied beams of alkali metal atoms at a temperature around $1 \times 10^{-7}$ K. At this temperature, the atoms would instantaneously condense to form a metallic solid if they collided with one another, but the use of atomic beams ensures that the atoms do not collide, so they keep their atomic identities. These clusters of gaseous alkali metal atoms act like single "superatoms," just like the atoms in superfluid helium.

The existence of "superatoms" was first predicted in 1924 by an Indian physicist, S. N. Bose, and elaborated further by Albert Einstein. Over 70 years later, studies at ultralow temperature confirmed the predictions. Physicists and chemists continue to work at the limits of low temperature to test some of the most fundamental predictions of quantum mechanics. Undoubtedly, additional Nobel Prizes will reward such research.

One important feature not revealed directly by Table 11-2 is that binding atoms into molecules always constrains the atoms and reduces total entropy. As illustrated in Figure 11-13, a sample containing 2 moles of He has considerably more entropy (252 J/K) than 1 mole of $H_2$ (131 J/K), even though both samples contain the same total number of atoms.

The representative values of $S°$ listed in Table 11-2 reveal the following additional general trends in absolute entropies:

1. Unlike enthalpies of formation, $S°$ values are *never* zero at any temperature above 0 K. Absolute entropies can be zero only at 0 K.

2. The absolute entropy of a substance depends on its phase. For example, compare three substances that are quite similar except for their phase at 298 K, 1 bar (Figure 11-14):

$$I_2(s), S° = 116 \text{ J K}^{-1} \text{ mol}^{-1}$$
$$Br_2(l), S° = 152 \text{ J K}^{-1} \text{ mol}^{-1}$$
$$Cl_2(g), S° = 223 \text{ J K}^{-1} \text{ mol}^{-1}$$

**Figure 11-13**

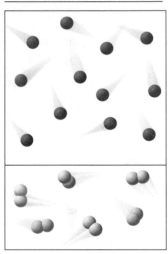

At the same temperature and pressure, the atoms in a 1-mole sample of $H_2$ (*bottom*) are more constrained than the atoms in a 2-mole sample of gaseous He (*top*).

**Figure 11-14**

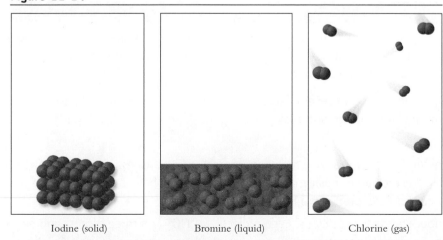

Iodine (solid)  Bromine (liquid)  Chlorine (gas)

The absolute entropies of otherwise similar substances depend on their phases:
$S° \; I_2(s) < S° \; Br_2(l) < S° \; Cl_2(g)$.

The substance in the more constrained phase has the lower entropy. For substances that are otherwise similar, $S_{solid} < S_{liquid} < S_{gas}$.

3. Among substances that share a phase and similar structure, $S°$ values increase with molar mass. For example:

| Substance | $M$ (g/mol) | $S°$ (J K$^{-1}$ mol$^{-1}$) | Substance | $M$ (g/mol) | $S°$ (J K$^{-1}$ mol$^{-1}$) |
|---|---|---|---|---|---|
| He ($g$) | 4.00 | 126 | Cu (s) | 63.6 | 33 |
| Ar ($g$) | 39.9 | 155 | Ag (s) | 108 | 43 |

The entropies of gases increase with molar mass because translational energy levels, which are quantized just like other forms of energy, are spaced more closely as mass increases, increasing the number of ways to distribute energy among the levels. The entropy of metallic solids increases with molar mass because atomic number also increases, meaning there are more electrons among which to disperse the available energy.

4. Molar entropies increase as the size and complexity of the molecule increases. Compare, for example, the standard molar entropies of the three two-carbon hydrocarbons:

| $C_2H_2$ | $C_2H_4$ | $C_2H_6$ |
|---|---|---|
| 200.9 J/(mol K) | 219.3 J/(mol K) | 229.2 J/(mol K) |

The reason behind this trend is that a molecule with many atoms has more variations in how it can vibrate than a molecule with few atoms. This means more ways to distribute energy—in other words, greater energy dispersal at any given temperature—and larger entropy.

Example 11-3 provides practice in identifying how absolute entropies vary with molecular properties.

## EXAMPLE 11-3  ABSOLUTE ENTROPIES

For each of the following pairs of substances under standard conditions, predict which has the larger standard entropy and give a reason why: (a) 1 mole each of Hg and Au; (b) 1 mole each of NO and $NO_2$; (c) 2 moles of $NO_2$ and 1 mole of $N_2O_4$; and (d) 1 mole each of Xe and Kr.

**Strategy:** Analyze each pair, keeping in mind that several features affect absolute entropy, including phase, number of atoms or molecules, amount of bonding, and molar mass.

**Solution:**

(a) At 298 K, mercury is a liquid metal, but gold is a solid metal. In general, a liquid is less constrained than a solid. The two elements have almost the same molar mass, so liquid Hg should have a larger $S°$ value than solid Au. Values in Appendix D confirm the prediction: Hg, 76 J K$^{-1}$ mol$^{-1}$; Au, 47 J K$^{-1}$ mol$^{-1}$.

(b) Both NO and $NO_2$ are gases under standard conditions. Each molecule of $NO_2$ has three atoms, and each molecule of NO has two atoms. Thus, $NO_2$ should have a larger standard molar entropy than NO, and experimental values confirm this: $NO_2$, 240 J K$^{-1}$ mol$^{-1}$; NO, 211 J K$^{-1}$ mol$^{-1}$.

(c) Two moles of $NO_2$ contain the same number of atoms as 1 mole of $N_2O_4$. In $N_2O_4$, however, pairs of $NO_2$ units are tied together by a bond that constrains them. Thus, 2 moles of $NO_2$ should have more entropy than 1 mole of $N_2O_4$, as confirmed by experimental values: $NO_2$, (240 J $K^{-1}$ $mol^{-1}$)(2 mol) = 480 J/K; $N_2O_4$, (304 J $K^{-1}$ $mol^{-1}$)(1 mol) = 304 J/K.

(d) Both Xe and Kr are monatomic gases from Group 18 of the periodic table. Because Xe has a higher molar mass than Kr, Xe should have greater entropy: Xe, 170 J $K^{-1}$ $mol^{-1}$; Kr, 164 J $K^{-1}$ $mol^{-1}$.

**Do the Results Make Sense?** In these examples, we apply atomic and molecular reasoning to predict trends in the absolute entropies. We verify that our predictions make sense by consulting tables of quantitative $S°$ values.

**Practice Exercise 11.3**

For each of the following pairs, select the one with smaller $S°$ at 298 K and give the reason: (a) $CH_4$ (g) or $SiH_4$; (b) $CH_4$ (g) or $H_2$ (g); and (c) $CH_4$ (g) or $H_2O$ (l).

## Entropy and Concentration

Entropy increases as a substance spreads out. The molar entropy of a gas is larger than the molar entropy of the same substance as a liquid or a solid, because molecules are more dispersed in a gas than in a condensed phase. As shown in Figure 11-15a, as a gas expands into a larger volume, it becomes more dispersed, so its entropy increases. There is a similar increase in dispersion when a solution is diluted by adding more solvent. Figure 11-15b illustrates that the dilution of a *solution* has the same dispersing effect as the expansion of a *gas* into a larger volume. Entropy increases when molecules are dispersed throughout a larger volume. This means that entropy increases as concentration decreases.

The mathematical machinery of thermodynamics allows this qualitative statement to be expressed quantitatively. Experiments and theory show that the molar entropy of a gas or solute varies logarithmically with concentration (c):

$$S = S° - R \ln\left(\frac{c}{c°}\right)$$

The ideal gas equation (Equation 2-1) can be rearranged to show that the concentration of a gas, which is moles of gas per unit volume, is conveniently measured by its partial pressure (p):

$$c = \frac{n}{V} = \frac{p}{RT}$$

Applying this equality to both c and c°, we find that for a gas, the concentration ratio equals the pressure ratio:

$$\frac{c}{c°} = \frac{p}{p°}$$

**Figure 11-15**

(a) When a gas expands, it becomes dispersed and entropy increases. (b) When a solution is diluted by a factor of 2, the increase in dispersion of the solute is the same as when the pressure of a gas is reduced by a factor of 2.

Since $c° = 1$ M and $p° = 1$ bar, we can simplify the equations for entropies of gases and solutes when their concentrations differ from standard conditions, obtaining Equation 11-5:

$$S_{(p \neq 1 \text{ bar})} = S° - R \ln p \qquad S_{(c \neq 1 \text{ M})} = S° - R \ln c \qquad (11\text{-}5)$$

Remember, however, that concentration must always be expressed using the same units as for the standard state of the substance: bar for gases and molarity for solutes.

Equation 11-5 refers to molar quantities. To obtain the total entropy of a sample, we must multiply its molar entropy by $n$, the number of moles. Example 11-4 illustrates the calculation of $\Delta S$ for a change in concentration.

---

## EXAMPLE 11-4 ENTROPY CHANGE ON EXPANSION

Oxygen gas has many applications, from welders' torches to respirators. The gas is sold commercially in pressurized steel tanks. One such tank contains $O_2$ at $p = 6.50$ bar and $T = 298$ K. Using standard thermodynamic data, compute the molar entropy of the gas in the tank at 6.50 bar and the change in entropy of a 0.155-mole sample of gas withdrawn from the tank at 1.10 bar and constant temperature.

**Strategy:** This is a quantitative thermodynamics problem that asks us to determine entropy under non-standard conditions using standard thermodynamic data. The problem asks for two quantities: the molar entropy of $O_2$ gas in the pressurized tank and the entropy change when a sample of that gas undergoes a pressure change.

Begin with a diagram of the process that includes a summary of all the information given in the problem: We also need the absolute entropy for $O_2$, which we find in Appendix D:

$$O_2 \ (g), \ S° = 205 \text{ J mol}^{-1} \text{ K}^{-1}$$

**Solution:** The gas stored in the tank is not at standard pressure, so apply Equation 11-5 to calculate its molar entropy. As the gas leaves the tank, it expands and its entropy increases. The final pressure is not standard pressure, so again use Equation 11-5 to calculate its molar entropy at the final pressure. Then calculate the entropy change for the expansion, taking the difference in molar entropies between initial and final conditions and multiplying by the number of moles undergoing the expansion.

$$S_{(p = 6.50 \text{ bar})} = S° - R \ln p = 205 \text{ J mol}^{-1}\text{K}^{-1} - (8.314 \text{ J mol}^{-1}\text{K}^{-1}) \ln 6.50$$

$$= 205 \text{ J mol}^{-1}\text{K}^{-1} - 15.562 \text{ J mol}^{-1}\text{K}^{-1} = 189 \text{ J mol}^{-1}\text{K}^{-1}$$

$$S_{(p = 1.0 \text{ bar})} = S° - R \ln p = 205 \text{ J mol}^{-1}\text{K}^{-1} - (8.314 \text{ J mol}^{-1}\text{K}^{-1}) \ln 1.10$$

$$= 205 \text{ J mol}^{-1}\text{K}^{-1} - 0.7924 \text{ J mol}^{-1}\text{K}^{-1} = 204 \text{ J mol}^{-1}\text{K}^{-1}$$

$$\Delta S = n \ (S_{\text{final}} - S_{\text{initial}}) = (0.155 \text{ mol})(204 \text{ J mol}^{-1}\text{K}^{-1} - 189 \text{ J mol}^{-1}\text{K}^{-1}) = 2.3 \text{ J/K}$$

**Do the Results Make Sense?** The positive value for this entropy change reflects the fact that the expansion is a spontaneous process.

**Practice Exercise 11.4**

Calculate the entropy change when 125 g of He gas, initially at 298 K and 1.00 bar, is compressed to a new pressure of 15.7 bar.

## Standard Reaction Entropies

Entropy changes are important in every process, but chemists are particularly interested in the effects of entropy on chemical reactions. If a reaction occurs under standard conditions, its entropy change can be calculated from absolute entropies using the same reasoning used to calculate reaction enthalpies from standard enthalpies of formation. The products of the reaction have molar entropies, and so do the reactants. The total entropy of the products is the sum of the molar entropies of the products multiplied by their stoichiometric coefficients in the balanced chemical equation. The total entropy of the reactants is a similar sum for the reactants. Equation 11-6 expresses the entropy change accompanying a reaction as the difference between these two quantities:

$$\Delta S^\circ_{reaction} = \sum coeff_{products}\, S^\circ_{products} - \sum coeff_{reactants}\, S^\circ_{reactants} \quad (11\text{-}6)$$

Notice that Equation 11-6 is similar to Equation 3-12, which is used to calculate reaction *enthalpies*.

$$\Delta H^\circ_{reaction} = \sum coeff_{products}\, \Delta H^\circ_{f,\,products} - \sum coeff_{reactants}\, \Delta H^\circ_{f,\,reactants} \quad (3\text{-}12)$$

As Example 11-5 shows, the applications of these equations follow parallel paths.

### EXAMPLE 11-5  ENTROPY AND ENTHALPY CHANGES

Polyacrylonitrile — Many repeat units

Acrylonitrile is an essential monomer in the polymer industry because it is used to make polyacrylonitrile for synthetic fibres.

Acrylonitrile is made from propene:

$$2\ C_3H_6\,(g) + 2\ NH_3\,(g) + 3\ O_2\,(g) \longrightarrow 2\ C_3H_3N\,(l) + 6\ H_2O\,(l)$$

Calculate $\Delta H^\circ_{reaction}$ and $\Delta S^\circ_{reaction}$ for the synthesis of acrylonitrile using standard thermodynamic data.

2 ⌐ + 2 NH₃ + 3 O₂ ⟶ 2 ⌐ₙ + 6 H₂O

Propene          Acrylonitrile

**Strategy:** We use a short version of the seven-step method. The problem asks for the entropy and enthalpy changes accompanying a chemical reaction, so we focus on the balanced chemical equation and the thermodynamic properties of the reactants and products.

**Solution:** Appendix D lists standard enthalpies of formation and standard entropies. Here are the values for the substances involved in this reaction:

| Substance | $C_3H_6$ (g) | $NH_3$ (g) | $O_2$ (g) | $C_3H_3N$ (l) | $H_2O$ (l) |
|---|---|---|---|---|---|
| $\Delta H_f^\circ$ (kJ/mol) | 20.0 | −45.9 | 0 | 172.9 | −285.83 |
| $S^\circ$ (J mol$^{-1}$K$^{-1}$) | 226.9 | 192.8 | 205.15 | 188 | 69.95 |
| Coefficient | 2 | 2 | 3 | 2 | 6 |

Standard thermodynamic changes for a reaction can be calculated using Equations 3-12 (from Chapter 3) and 11-6:

$$\Delta H_{reaction}^\circ = \sum coeff_{products}\,\Delta H_{f,products}^\circ - \sum coeff_{reactants}\,\Delta H_{f,reactants}^\circ$$

$$= [2(172.9) + 6(-285.83)] - [2(20.0) + 2(-45.9) + 3(0)] \text{ kJ/mol}$$

$$= -1317 \text{ kJ/mol}$$

$$\Delta S_{reaction}^\circ = \sum coeff_{products}\,S_{products}^\circ - \sum coeff_{reactants}\,S_{reactants}^\circ$$

$$= [2(188) + 6(69.95)] - [2(226.9) + 2(192.8) + 3(205.15)]$$

$$= -659 \text{ J/mol}$$

**Do the Results Make Sense?** According to these values, the reaction is exothermic, but the products are less dispersed than the reactants. Because the reaction produces liquids from gases, this decrease in dispersion is a reasonable result.

**Practice Exercise 11.5**

Calculate the entropy change accompanying the combustion of methane if the water is produced in the liquid phase.

## SECTION EXERCISES

●**11.3.1** Of the following pairs, which has the greater entropy? Explain each choice. (a) 1 g of dew or 1 g of frost; (b) 1 mole of gaseous hydrogen atoms or 0.5 mole of gaseous hydrogen molecules; (c) "perfect" diamond or flawed diamond, each ¼ carat; and (d) 5 mL of liquid ethanol at 0 °C or 5 mL of liquid ethanol at 50 °C.

●**11.3.2** Explain the following differences in entropies in molecular terms (substances are at standard conditions unless otherwise noted): (a) 1 mole of $O_2$ has less entropy than 1 mole of $O_3$; (b) 3 moles of $O_2$ has more entropy than 2 moles of $O_3$; (c) 1 mole of $I_2$ has less entropy than 1 mole of $O_2$; and (d) 1 mole of HCl (aq) in concentrated solution (12 M) has less entropy than 1 mole of HCl (aq) in dilute solution (0.100 M).

●**11.3.3** Compute the standard entropy change for the following reaction, and state why it is negative.

$$12\ NH_3\ (g) + 21\ O_2\ (g) \longrightarrow 8\ HNO_2\ (g) + 4\ NO\ (g) + 12\ H_2O\ (l)$$

## 11.4 SPONTANEITY AND FREE ENERGY

The second law of thermodynamics states that total entropy always increases during a spontaneous process. Hence, a calculation of $\Delta S_{total}$ should determine whether a particular chemical process is spontaneous. Unfortunately, this is not practical for most processes. It is usually possible to calculate the entropy change for a *system*. However, the surroundings are often more complex than the system, making it difficult to calculate $\Delta S$ of the *surroundings*. It would be much more convenient to have a way to determine the direction of spontaneous change using *just the system*, not the surroundings.

There is no single criterion for the system alone that applies to *all* processes. However, if we restrict the conditions to constant temperature and pressure, there is a state function whose change for the system predicts spontaneity. This new state function is the **free energy** (*G*), which was introduced by the American J. Willard Gibbs and is defined by Equation 11-7:

$$G = H - TS \qquad (11\text{-}7)$$

How does free energy allow us to predict spontaneity while ignoring the surroundings? To answer this question, we examine the change in free energy of a system,

$$\Delta G_{sys} = \Delta H_{sys} - \Delta(TS)_{sys}$$

If we hope to use $\Delta G_{sys}$ to predict spontaneity, we must relate $\Delta G_{sys}$ to $\Delta S_{total}$, which must be positive if a process is spontaneous. To do that, we must restrict the conditions.

First, the process must occur at *constant temperature*. This lets us relate $\Delta(TS)_{sys}$ to $\Delta S_{sys}$, resulting in Equation 11-8:

$$\Delta(TS)_{sys} = (TS)_{final} - (TS)_{initial} = T(S_{final} - S_{initial}) = T\Delta S_{sys}$$

$$\Delta G_{sys} = \Delta H_{sys} - T\Delta S_{sys} \text{ (constant } T) \qquad (11\text{-}8)$$

Second, the process must occur at *constant pressure*. Then we can relate the *enthalpy* change for the system to the *entropy* change for the surroundings. Recall that $\Delta H$ equals $q$ when $p$ is constant:

$$\Delta H_{sys} = q_{sys} \qquad (\text{constant } p)$$

Remember also that the heat flow for a system is always equal in magnitude but opposite in sign to the heat flow of the surroundings:

$$q_{sys} = -q_{surr} \qquad \text{so} \qquad \Delta H_{sys} = -q_{surr}$$

Furthermore, we have already required constant temperature, so the heat flow of the surroundings measures the entropy change of the surroundings. That is,

$$\frac{q_{surr}}{T} = \Delta S_{surr} \qquad \text{and} \qquad q_{surr} = T\Delta S_{surr}$$

We combine these equalities into an equation that relates $\Delta H_{sys}$ to $S_{surr}$:

$$\Delta H_{sys} = -T\Delta S_{surr} \qquad (\text{constant } p \text{ and } T)$$

Substituting this result into Equation 11-8 gives

$$\Delta G_{sys} = -T\,\Delta S_{surr} - T\,\Delta S_{sys} = T(\Delta S_{surr} + \Delta S_{sys}) \qquad \text{(constant } p \text{ and } T)$$

Finally, because $\Delta S_{surr} + \Delta S_{sys} = \Delta S_{total}$:

$$\Delta G_{sys} = -T\,\Delta S_{total} \qquad \text{(constant } p \text{ and } T)$$

In words, in any process that occurs at constant $T$ and $p$, the free energy change for the *system* is negative whenever the *total* entropy change is positive; that is, whenever the overall process is spontaneous. Defining a new function and imposing some restrictions provides a way to use properties of a system to determine whether a process is spontaneous.

Although the restrictions of constant $T$ and $p$ are stringent, they apply to many important chemical processes, including reactions that occur in the human body, which has a nearly constant temperature of 37 °C and nearly constant pressure of 1 bar. Any biochemical reaction that occurs in the body occurs under conditions in which the immediate surroundings are at constant $T$ and $P$. Figure 11-16 shows three examples of systems at constant $T$ and $p$.

Because free energy is a state function, its values can be tabulated for use in chemical calculations. As with standard enthalpies of formation, the **standard molar free energy of formation** ($\Delta G_f^\circ$) for any substance is defined to be the change of free energy when 1 mole of that substance forms from its elements in their standard states. The same reasoning that is used for enthalpy changes leads to Equation 11-9, from which we can calculate the standard free energy change for any chemical reaction:

$$\Delta G_{reaction}^\circ = \sum \text{coeff}_{products}\Delta G_{f,\ products}^\circ - \sum \text{coeff}_{reactants}\Delta G_{f,\ reactants}^\circ \qquad (11\text{-}9)$$

**Figure 11-16**

Many real-life situations operate under nearly constant temperature and pressure and thus are subject to $\Delta G_{sys} < 0$ as the condition for spontaneity. Examples shown here are a laboratory synthesis, a biological system, and an industrial process.

The form of Equation 11-9 should look familiar to you, because it is analogous to Equation 3-12 for reaction enthalpies and Equation 11-6 for reaction entropies.

As Equation 11-10 describes, the standard free energy change for a reaction can also be calculated from $\Delta H°$ and $\Delta S°$ for the reaction by making use of Equation 11-8 under standard conditions:

$$\Delta G° = \Delta H° - T\,\Delta S° \qquad (11\text{-}10)$$

Either of these equations can be used to find standard free energy changes. Which equation to use depends on the available data. Example 11-6 illustrates both types of calculations.

Pay close attention to units when using Equation 11-10. The values of $\Delta H°$ and $\Delta G°$ are usually given in kJ/mol, but $S°$ is usually expressed in J $K^{-1}$ or J $mol^{-1}$ $K^{-1}$. Thus, entropies must be multiplied by $10^{-3}$ kJ/J before adding the two terms.

### EXAMPLE 11-6 FREE ENERGY OF REACTION

Find the standard free energy change for the acrylonitrile synthesis discussed in Example 11-5.

**Strategy:** There are two ways to calculate $\Delta G°_{reaction}$. The first method uses standard free energies of formation and Equation 11-9. The second method uses Equation 11-10 and the values of $\Delta H°$ and $\Delta S°$ calculated earlier. Either method requires the balanced chemical equation:

$$2\ C_3H_6\ (g) + 2\ NH_3\ (g) + 3\ O_2\ (g) \longrightarrow 2\ C_3H_3N\ (l) + 6\ H_2O\ (l)$$

We perform both calculations to show that they give the same result.

**Solution:** (a) Using Equation 11-9 and $\Delta G°_f$ values from Appendix D:

$$\Delta G°_{reaction} = \sum coeff_{products}\Delta G°_{f,\ products} - \sum coeff_{reactants}\Delta G°_{f,\ reactants}$$

$$= [2(\Delta G°_f(C_3H_3N(l))) + 6(\Delta G°_f(H_2O(l)))]$$

$$- [2(\Delta G°_f(C_3H_6(g))) + 2(\Delta G°_f(NH_3(g))) + 3(\Delta G°_f(O_2(g)))]$$

Substitute values from Appendix D:

$$\Delta G°_{reaction} = [(2\ mol)(208.6\ kJ/mol) + (6\ mol)(-237.1\ kJ/mol)]$$

$$- [(2\ mol)(74.62\ kJ/mol) + (2\ mol)(-16.4\ kJ/mol) + (3\ mol)(0\ kJ/mol)]$$

$$\Delta G°_{reaction} = -1122\,kJ$$

(b) Using Equation 11-10 and the enthalpy and entropy results of Example 11-5:

$$\Delta H°_{reaction} = -1317\ kJ \quad\text{and}\quad \Delta S°_{reaction} = -659\ J/K$$

$$\Delta G°_{reaction} = \Delta H°_{reaction} - T\Delta S°_{reaction}$$

$$\Delta G°_{reaction} = [-1317\ kJ] - [(298\ K)(-659\ J/K)(10^{-3}\ kJ/J)] = -1121\ kJ$$

**Does the Result Make Sense?** In the second calculation, the entropy term $T\Delta S$ is positive, making $\Delta G°$ less negative than $H°$. This is due to the formation of liquid water, which makes the products of this reaction more constrained than the reactants. Nevertheless, the large negative enthalpy change accompanying the reaction means that much heat becomes dispersed in the surroundings, making the overall process highly favourable. The large negative value for $\Delta G°_{reaction}$ indicates that the production of acrylonitrile is highly spontaneous under standard conditions.

**Common Pitfalls:** Neglecting to use either kJ or J consistently. Remember that the usual units for $\Delta G°$ and $H°$ are kJ/mol, whereas those for $\Delta S°$ are J K$^{-1}$ mol$^{-1}$.

**Practice Exercise 11.6**

Calculate $\Delta G°_{reaction}$ for the combustion of methane, producing liquid water, under standard conditions (a) by using $\Delta G°_f$ values from Appendix D and (b) by using $\Delta H°_f$ and $\Delta S°$ values from Appendix D, and the fact that $\Delta G° = \Delta H° - T \Delta S°$.

## Free Energy and Temperature

The tabulated values for thermodynamic functions refer to unit concentrations and 298 K, but chemical reactions occur at many different concentrations and temperatures. To use $\Delta G$ as a measure of spontaneity under all conditions, we must know how $\Delta G$ depends on temperature and concentration.

To see how free energy varies with temperature, we turn to Equation 11-10. This equation is valid at *any* temperature as long as the temperature is *constant*. However, application of the equation at a temperature different from 298 K requires knowledge of $\Delta H_{reaction}$ and $\Delta S_{reaction}$ at the new temperature.

The entropy of any chemical substance increases as temperature increases. These changes in entropy as a function of temperature can be calculated, but the absolute entropy of every substance increases with temperature, and the entropy of the reactants often changes with temperature by almost the same amount as the entropy of the products. This means that the temperature effect on the entropy *change* for a reaction is usually small enough that we can consider $\Delta S_{reaction}$ to be independent of temperature.

As mentioned in Section 3.5, $\Delta H_{reaction}$ also does not change rapidly with temperature. This is because the enthalpy change is closely related to the energy change, and the energy change is determined by the strengths of the chemical bonds in reactants and products. These, in turn, are nearly temperature-independent. As a result, changes in free energy at temperatures reasonably close to 298 K can be estimated by assuming that standard enthalpies and entropies at 298 K also apply at any other temperature. This is described by Equation 11-11:

$$\Delta G°_{reaction,\ T} \cong \Delta H°_{reaction,\ 298\ K} - T\Delta S°_{reaction,\ 298\ K} \tag{11-11}$$

An immediate consequence of Equation 11-11 is that the spontaneity of any reaction with a large $\Delta S°$ is very sensitive to temperature. Example 11-7 shows this.

## EXAMPLE 11-7 TEMPERATURE AND SPONTANEITY

Dinitrogen tetroxide can decompose into two molecules of nitrogen dioxide:

$$N_2O_4\ (g) \longrightarrow 2\ NO_2\ (g)$$

(a) Show that this reaction is not spontaneous at 298 K under standard conditions.
(b) Find the temperature at which the reaction becomes spontaneous at standard pressure.

**Strategy:** The word *spontaneous* suggests that free energies are involved in this problem. The criterion for spontaneity at constant temperature and pressure is $\Delta G_{reaction} < 0$, so the problem asks us to show that $\Delta G^\circ_{reaction} > 0$ at $T = 298$ K. Then we can use Equation 11-11 to calculate the temperature at which $\Delta G^\circ_{reaction} < 0$. As temperature changes, $\Delta G^\circ$ must become zero before it becomes negative. The reaction will be spontaneous at any temperature farther from 298 K than the temperature at which $\Delta G^\circ = 0$.

**Solution:** (a) Because standard free energies of formation at 298 K are available in Appendix D, we can use Equation 11-9 to determine whether the decomposition reaction is spontaneous at 298 K, 1 bar:

$$\Delta G^\circ = (2 \text{ mol})(51.3 \text{ kJ/mol}) - (1 \text{ mol})(99.8 \text{ kJ/mol}) = 2.8 \text{ kJ}$$

The positive value for $\Delta G^\circ$ indicates that this reaction is *not* spontaneous at 298 K and standard pressures. In fact, the calculation tells us that the reaction will be spontaneous in the opposite direction. Under standard conditions at 298 K, $NO_2$ reacts to form $N_2O_4$:

$$2 \text{ NO}_2 (g) \longrightarrow \text{N}_2\text{O}_4 (g) \qquad \Delta G^\circ = -2.8 \text{ kJ}$$

(b) For temperatures different from 298 K, we use Equation 11-11 to estimate $\Delta G^\circ_{reaction}$. This equation lets us determine the temperature at which $N_2O_4$ decomposition becomes spontaneous when the partial pressure of each gas is 1 bar:

$$\Delta G^\circ_{reaction, \, T} \cong \Delta H^\circ_{reaction, \, 298 \text{ K}} - T\Delta S^\circ_{reaction, \, 298 \text{ K}}$$

To find the temperature at which $\Delta G^\circ = 0$, begin by calculating $\Delta H^\circ$ and $\Delta S^\circ$ using tabulated data:

$$\Delta H^\circ = (2 \text{ mol})(33.2 \text{ kJ/mol}) - (1 \text{ mol})(11.1 \text{ kJ/mol}) = 55.3 \text{ kJ}$$

$$\Delta S^\circ = (2 \text{ mol})(240.1 \text{ J mol}^{-1} \text{ K}^{-1}) - (1 \text{ mol})(304.4 \text{ J mol}^{-1} \text{ K}^{-1}) = 175.8 \text{ J/K}$$

Now set $\Delta G^\circ$ equal to zero and rearrange the equation to solve for temperature:

$$\Delta G^\circ = 0 = \Delta H^\circ - T\Delta S^\circ \quad \text{or} \quad T\Delta S^\circ = \Delta H^\circ$$

$$T = \frac{\Delta H^\circ}{\Delta S^\circ} = \frac{55.3 \text{ kJ}}{(175.8 \text{ J/K})(10^{-3} \text{ kJ/J})} = 315 \text{ K}$$

At 315 K, $\Delta G^\circ = 0$, so at all temperatures greater than 315 K, $\Delta G^\circ$ is negative, and decomposition of $N_2O_4$ is spontaneous at partial pressures of 1 bar.

**Do the Results Make Sense?** The decomposition of $N_2O_4$ requires a bond to break. This is the reason why the decomposition has a positive $\Delta H^\circ$. At the same time, the number of molecules doubles during decomposition, which is the reason $\Delta S^\circ$ has a positive value. The positive enthalpy change means that energy is removed from the surroundings and constrained, whereas the positive entropy change means that matter is dispersed. At temperatures below 315 K, the enthalpy term dominates and decomposition is not spontaneous, but at temperatures above 315 K, the entropy term dominates and decomposition is spontaneous.

**Practice Exercise 11.7**

Estimate $\Delta G^\circ$ for the combustion reaction of methane at 1920 °C, the temperature of a gas flame. (*Reminder*: $H_2O$ is a gas at this temperature!)

## Free Energy and Concentration

Gases that participate in chemical reactions typically are at pressures different from 1 bar. Substances in solution are likely to be at concentrations different from one molar. For example, a biochemist who wants to know what processes are spontaneous under physiological conditions will find that the substances dissolved in biological fluids are rarely at one molar concentration. How does $\Delta G$ vary with changes in molarity and pressure? Recall that enthalpy is virtually independent of concentration but that entropy obeys Equation 11-5:

$$S_{(p \neq 1 \text{ bar})} = S° - R \ln p \qquad\qquad S_{(c \neq 1 \text{ M})} = S° - R \ln c$$

Equation 11-5 applies to individual substances. To see how these equations affect $\Delta S$ for a chemical reaction, consider the decomposition of limestone ($CaCO_3$) to produce lime (CaO), carried out industrially in kilns in which the partial pressure of $CO_2$ is different from 1 bar:

$$CaCO_3 \text{ (s)} \longrightarrow CaO \text{ (s)} + CO_2 \text{ (g)}$$

The entropy change for the reaction is the difference in entropy between products and reactants:

$$\Delta S_{\text{reaction}} = S \text{ (CaO)} + S \text{ (CO}_2) - S \text{ (CaCO}_3)$$

The entropy of each solid is its standard entropy, $S°$, but the entropy of the gas must be corrected for the deviation of pressure from standard conditions. Equation 11-5 gives the molar entropy of carbon dioxide as a function of its partial pressure:

$$S(CO_2) = S° \text{ (CO}_2) - R \ln(p_{CO_2})$$

Now substitute the corrected entropy of $CO_2$ into the equation for $\Delta S_{\text{reaction}}$:

$$\Delta S_{\text{reaction}} = S° \text{ (CaO)} + [S° \text{ (CO}_2) - R \ln(p_{CO_2})] - S° \text{ (CaCO}_3)$$

Older limestone kilns have vents through which $CO_2$ can escape.

Next, rearrange the equation so that all the standard entropies are together:

$$\Delta S_{\text{reaction}} = S° \text{ (CaO)} + S° \text{ (CO}_2) - S° \text{ (CaCO}_3) - R \ln(p_{CO_2})$$

The first three terms are the standard entropy change for the reaction, allowing simplification:

$$\Delta S_{\text{reaction}} = \Delta S°_{\text{reaction}} - R \ln(p_{CO_2})$$

In this example, only one of the reagents has a concentration that can vary, and each stoichiometric coefficient is 1. What happens for a more complicated reaction? Consider the synthesis of ammonia carried out in a pressurized reactor containing $N_2$, $H_2$, and $NH_3$ at partial pressures different from 1 bar:

$$N_2 \text{ (g)} + 3 H_2 \text{ (g)} \longrightarrow 2 NH_3 \text{ (g)}$$

The entropy change for the reaction is the difference in entropy between products and reactants, obtained by multiplying each corrected entropy by the appropriate stoichiometric coefficient:

$$\Delta S_{\text{reaction}} = 2 S \text{ (NH}_3) - 3 S \text{ (H}_2) - S \text{ (N}_2)$$

Applying Equation 11-5 gives the molar entropy of each gas:

$$S \text{ (N}_2) = S° \text{ (N}_2) - R \ln (p_{N_2})$$

$$S \text{ (H}_2) = S° \text{ (H}_2) - R \ln (p_{H_2})$$

Modern limestone kilns use blowers as well as vents and are computer-controlled.

$$S \text{ (NH}_3) = S° \text{ (NH}_3) - R \ln (p_{NH_3})$$

Now substitute the corrected entropies:

$$\Delta S_{\text{reaction}} = 2\,[S°\,(NH_3) - R\ln\,(p_{NH_3})] - 3\,[S°\,(H_2) - R\ln\,(p_{H_2})]$$
$$- [S°\,(N_2) - R\ln\,(p_{N_2})]$$

Next, rearrange the equation so that all the logarithmic terms are together:

$$\Delta S_{\text{reaction}} = 2\,S°\,(NH_3) - 3\,S°\,(H_2) - S°\,(N_2) - 2\,R\ln\,(p_{NH_3})$$
$$+ 3\,R\ln\,(p_{H_2}) + R\,\ln\,(p_{N_2})$$

As before, the first three terms are the standard entropy change for the reaction:

$$\Delta S_{\text{reaction}} = \Delta S°_{\text{reaction}} - 2\,R\ln\,(p_{NH_3}) + 3\,R\ln\,(p_{H_2}) + R\ln\,(p_{N_2})$$

We use the fact that $a\ln x = \ln x^a$ to combine the logarithmic terms:

$$2\,R\ln\,(p_{NH_3}) = R\ln\,(p_{NH_3})^2 \quad\text{and}\quad 3\,R\ln\,(p_{H_2}) = R\ln\,(p_{H_2})^3$$

and the equation becomes

$$\Delta S_{\text{reaction}} = \Delta S°_{\text{reaction}} - R\,[\,\ln\,(p_{NH_3})^2 - \ln\,(p_{H_2})^3 - \ln\,(p_{N_2})]$$

Additional logarithmic properties, $\ln u + \ln v = \ln(uv)$ and $\ln u - \ln v = \ln(u/v)$, allow us to put all the logarithmic terms into a single ratio:

$$\Delta S_{\text{reaction}} = \Delta S°_{\text{reaction}} - R\ln\left[\frac{(p_{NH_3})^2}{(p_{N_2})(p_{H_2})^3}\right]$$

Note that the pressure ratio has product concentrations raised to their stoichiometric coefficients in the numerator, and reactant concentrations raised to their stoichiometric coefficients in the denominator. The form of this equation applies to all reactions, not just the synthesis of ammonia. A general reaction can be written as

$$aA + bB \longrightarrow dD + eE$$

Here, A and B represent any reactants, D and E represent any products, and the lower-case letters represent the stoichiometric coefficients. The expression for the entropy change accompanying the reaction is

$$\Delta S_{\text{reaction}} = \Delta S°_{\text{reaction}} - R\ln\left[\frac{(c_D)^d(c_E)^e}{(c_A)^a(c_B)^b}\right]$$

The concentrations must, as always, be expressed in the units consistent with the standard state: pressure in bar for any gas, molarity for any solute. The ratio in the logarithmic term is called the **reaction quotient (Q)** (Equation 11-12).

$$\left[\frac{(c_D)^d(c_E)^e}{(c_A)^a(c_B)^b}\right] = Q \tag{11-12}$$

The entropy change for a reaction under non-standard concentrations can be expressed in terms of the standard entropy change and Q:

$$\Delta S_{\text{reaction}} = \Delta S°_{\text{reaction}} - R\ln Q$$

This general expression describing how $\Delta S_{reaction}$ varies with concentrations leads to an equation for the free energy change when concentrations are non-standard:

$$\Delta G_{reaction} = \Delta H^\circ_{reaction} - T(\Delta S^\circ_{reaction} - R \ln Q)$$

Because $\Delta H^\circ - T\,\Delta S^\circ = G^\circ$, this equation reduces to

$$\Delta G_{reaction} = \Delta G^\circ_{reaction} + RT \ln Q \qquad (11\text{-}13)$$

**Where did the units go?** We can only take the logarithm of a dimensionless quantity, so what happened to the concentrations in the expression for $Q$? The answer is detailed in Section 13.2, but for now it is enough to know that Equation 11-13 will give the correct result even if $Q$ is not dimensionless.

An immediate consequence of Equation 11-13 is that the direction of spontaneity of a reaction depends on the concentrations of reactants and products. Product concentrations appear in the numerator of $Q$, so when the concentration of a product increases, $\ln Q$ increases, making $\Delta G$ less negative, so a reaction becomes less spontaneous as product concentrations increase. Conversely, increasing the concentration of a reactant decreases $\ln Q$, which in turn makes $\Delta G$ more negative. Thus, a reaction becomes more spontaneous as reactant concentrations increase.

If a reaction is not spontaneous at unit concentrations, we may be able to make it spontaneous by increasing the concentrations of reactants or by removing products as they form. If a reaction has a very positive $G^\circ_{reaction}$, however, it may not be possible to change the value of $Q$ sufficiently to make $\Delta G_{reaction} < 0$. Remember that $\ln Q$ changes more slowly than $Q$ itself. A 10-fold change in $Q$, for instance, changes $\ln Q$ only by a factor of 2.3. Nonetheless, reactions that are almost spontaneous under standard conditions can be driven forward by making appropriate changes in concentration. Example 11-8 provides an illustration.

---

## EXAMPLE 11-8  EFFECT OF CONCENTRATION ON SPONTANEITY

The decomposition of dinitrogen tetroxide produces nitrogen dioxide:

$$N_2O_4\,(g) \longrightarrow 2\,NO_2\,(g)$$

Find the minimum partial pressure of $N_2O_4$ at which the reaction is spontaneous if $p_{NO_2} = 1$ bar and $T = 298$ K.

**Strategy:** We need to calculate a pressure that is just sufficient to make the reaction spontaneous. That means solving for the pressure at which $\Delta G = 0$, using Equation 11-13:

$$\Delta G = \Delta G^\circ + RT \ln Q = \Delta G^\circ + RT \ln\left[\frac{(p_{NO_2})^2}{p_{N_2O_4}}\right]$$

**Solution:** The standard free energy change was determined in Example 11-7:

$$\Delta G^\circ = 2.8 \text{ kJ/mol}$$

A visual diagram summarizes the information:

| $N_2O_4 \longrightarrow$ | $2\,NO_2$ | $\Delta G^\circ = 2.8$ kJ |
|---|---|---|
| $p = ?$ | $p = 1$ bar | $\Delta G^\circ \leq 0$ |

The problem asks for the partial pressure of $N_2O_4$ that will make the decomposition spontaneous when $T = 298$ K and $p$ of $NO_2 = 1$ bar. The value of $\Delta G$ must be zero before it can become negative. Therefore, to find the threshold pressure of $N_2O_4$ that makes the decomposition spontaneous, set $\Delta G = 0$ and $p_{NO_2} = 1$ bar and then rearrange to solve for the partial pressure of $N_2O_4$:

$$0 = \Delta G° + RT \ln\left[\frac{(1 \text{ bar})^2}{p_{N_2O_4}}\right]$$

$$\ln\left[\frac{1 \text{ bar}}{p_{N_2O_4}}\right] = -\frac{\Delta G°}{RT} = -\frac{(2.8 \text{ kJ/mol})(10^3 \text{ J/kJ})}{(8.314 \text{ J mol}^{-1} \text{ K}^{-1})(298 \text{ K})} = -1.13$$

$$\frac{1 \text{ bar}}{p_{N_2O_4}} = e^{-1.13} = 0.323 \quad \text{and} \quad p_{N_2O_4} = \frac{1 \text{ bar}}{0.323} = 3.1 \text{ bar}$$

This decomposition is spontaneous as long as the partial pressure of $N_2O_4$ is greater than 3.1 bar.

**Does the Result Make Sense?** Under standard conditions (both pressures 1 bar), the reaction is not spontaneous. To force the reaction to go, we need more reactants, so a reactant pressure that is greater than 1 bar makes sense.

**Common Pitfalls:** Not using a consistent set of units. $\Delta G°$ is converted to Joules to use the appropriate value of the gas constant ($R$) in J mol$^{-1}$ K$^{-1}$.

**Practice Exercise 11.8**
Find the maximum partial pressure of $NO_2$ at which the reaction is spontaneous if $p_{N_2O_4} = 1$ bar and $T = 298$ K.

This example shows that a reaction with a small positive $\Delta G°$ can be made spontaneous by relatively small changes in concentrations.

If neither temperature nor concentrations are at standard values, free energy calculations must be done in two steps. First, correct for temperature to obtain $\Delta G_T°$ using Equation 11-11:

$$\Delta G°_{\text{reaction}, T} \cong \Delta H°_{\text{reaction}, 298 \text{ K}} - T\Delta S°_{\text{reaction}, 298 \text{ K}}$$

Second, use this value of $\Delta G_T°$ in Equation 11-13 to complete the calculation of $\Delta G$:

$$\Delta G_{\text{reaction}} = \Delta G°_{\text{reaction}, T} + RT \ln Q$$

Section exercise 11.4.3 can be solved in this manner.

Recall that $y = \ln x$ implies that $x = e^y$, where "e" is the base for natural logarithms.

The mol unit in the value of $\Delta G°$ refers to "per mole of reaction" and is included in the calculation to cancel the mole unit in $R$.

## Influencing Spontaneity

Suppose a chemist wants to carry out a particular chemical synthesis, but the reaction has a positive value for $\Delta G°$. The thermodynamic calculation indicates that the reaction is spontaneous in the wrong direction at 298 K and *standard* concentrations, but this does not prevent it from occurring under *all* conditions. What can be done to make the reaction go in the desired direction?

Example 11-8 illustrates that changing the reaction quotient changes $\Delta G$ for a reaction. Changing Q alters the entropy change of the chemical system.

In particular, increasing the pressure of $N_2O_4$ above 3.1 bar would cause spontaneous decomposition of $N_2O_4$, even though this reaction is not spontaneous under standard conditions. Reactions in liquid solutions likewise may proceed spontaneously if the concentrations of reactants are high enough or the concentrations of products are low enough.

Changing the temperature of the system is another way to influence the spontaneity of a reaction. The equation for $\Delta G$ has two parts, $\Delta H$ and $T\Delta S$, which can work together or in opposition:

$$\Delta G_T^\circ = \Delta H^\circ - T\Delta S^\circ$$

A *positive* $\Delta S^\circ$ promotes spontaneity because it makes $\Delta G^\circ$ more negative. This reflects the fact that a positive $\Delta S^\circ$ means the system becomes more dispersed during the reaction. A *negative* $\Delta H^\circ$ promotes spontaneity as well, because it also makes $\Delta G^\circ$ more negative. This reflects the fact that energy becomes more dispersed in the surroundings when a reaction releases energy. Thus, a reaction that has a positive $\Delta S^\circ$ and a negative $\Delta H^\circ$ is spontaneous at all temperatures. The combustion of propane to form gaseous products is an example of such a reaction:

$$C_3H_8\ (g) + 5\ O_2\ (g) \longrightarrow 3\ CO_2\ (g) + 4\ H_2O\ (g)$$

$$\Delta H^\circ = -897\ \text{kJ} \qquad \Delta S^\circ = +145\ \text{J/K}$$

The products of this reaction are more dispersed than the reactants, and the reaction releases energy. Consequently, $\Delta G^\circ$ is negative at all temperatures, and the reverse reaction cannot be made spontaneous by changing the temperature.

By the same reasoning, a *negative* $\Delta S^\circ$ and a *positive* $\Delta H^\circ$ oppose spontaneity, so a reaction in which the system becomes constrained and energy is absorbed is non-spontaneous regardless of temperature. The system and its surroundings both would experience decreases in entropy if such a process were to occur, and this would violate the second law of thermodynamics.

A reaction that has the same sign for $\Delta S^\circ$ and $\Delta H^\circ$ will be spontaneous at some temperatures but non-spontaneous at others. At low temperature, $\Delta S^\circ$ is multiplied by a small value for $T$, so at sufficiently low temperature, $\Delta H^\circ$ contributes more to $\Delta G^\circ$ than $T\Delta S^\circ$ does. At high temperature, $\Delta S^\circ$ is multiplied by a large value for $T$, so at sufficiently high temperature, $\Delta S^\circ$ contributes more to $\Delta G^\circ$ than does $\Delta H^\circ$.

Reactions that have positive $\Delta H^\circ$ and positive $\Delta S^\circ$ are favoured by entropy but disfavoured by enthalpy. Such reactions are spontaneous at high temperature, where the $T\Delta S^\circ$ term dominates $\Delta G^\circ$, because matter becomes dispersed during the reaction. A reaction is entropy-driven under these conditions. These reactions are non-spontaneous at low temperature, where the $\Delta H^\circ$ term dominates $\Delta G^\circ$.

The opposite situation holds for reactions that have negative values for $\Delta H^\circ$ and $\Delta S^\circ$. These reactions are spontaneous at low temperature because their release of heat disperses energy into the surroundings. The favourable $\Delta H^\circ$ dominates $\Delta G^\circ$ as long as $T$ does not become too large, and the reaction is enthalpy-driven. At high temperature, however, the unfavourable $\Delta S^\circ$ dominates $\Delta G^\circ$, and the reaction is no longer spontaneous. The effects of temperature on spontaneity are summarized in Table 11-3.

*At 298 K, 1 bar, the water product would be in its condensed phase, making the value of $\Delta S^\circ$ negative. Nevertheless, the highly negative $\Delta H^\circ$ value keeps $\Delta G^\circ$ highly negative.*

## TABLE 11-3 INFLUENCE OF TEMPERATURE ON SPONTANEITY

| $\Delta H°$ | $\Delta S°$ | $\Delta G°$ (high $T$) | $\Delta G°$ (low $T$) | Spontaneous |
|---|---|---|---|---|
| − | + | − | − | At all temperatures |
| + | − | + | + | Not at any temperature |
| + | + | − | + | At high temperature (entropy-driven) |
| − | − | + | − | At low temperature (enthalpy-driven) |

Calcium sulphate, the substance used to absorb water in desiccators, provides an example of this temperature sensitivity. Anhydrous calcium sulphate absorbs water vapour from the atmosphere to give the hydrated salt. The reaction has a negative $\Delta S°$ because water molecules become more constrained when gaseous water molecules move into the solid state. The reaction also has a negative $\Delta H°$ because of the electrical forces of attraction between the ions of the salt and the polar water molecules:

$$CaSO_4 \ (s) + 2 \ H_2O \ (g) \longrightarrow CaSO_4 \cdot 2H_2O \ (s)$$

$$\Delta H° = -104.9 \ kJ \qquad \Delta S° = -290.2 \ J/K$$

At 298 K, the favourable $\Delta H°$ contributes more to $\Delta G°$ than the unfavourable $\Delta S°$:

$$\Delta G°_{298K} = (-104.9 \ kJ) - (298 \ K)(-290.2 \ J/K)(10^{-3} \ kJ/J) = -18.4 \ kJ$$

Thus, anhydrous calcium sulphate acts as a "chemical sponge" at room temperature, trapping water vapour spontaneously to form calcium sulphate dihydrate.

The calcium sulphate in a desiccator is effective at removing water vapour only as long as some anhydrous salt remains. When all the anhydrous salt has been converted to the dihydrate, the desiccator can no longer maintain a dry atmosphere. Fortunately, the thermodynamics of this reaction makes it possible to regenerate the drying agent. At 100 °C (373 K), $\Delta S°$ contributes more to $\Delta G°$ than does $\Delta H°$:

$$\Delta G°_{373K} = (-104.9 \ kJ) - (373 \ K)(-290.2 \ J/K)(10^{-3} \ kJ/J) = +3.3 \ kJ$$

At this temperature, the reverse reaction is spontaneous. Calcium sulphate dihydrate can be converted to anhydrous calcium sulphate in a drying oven at 100 °C. Then it can be cooled and returned to a desiccator, ready once more to act as a chemical sponge for water.

Desiccators are used to store chemicals that react with water. Calcium sulphate chips in the bottom of the desiccator absorb water vapour from the atmosphere. The blue chips contain an indicator that turns pink when the calcium sulphate is saturated with water.

## SECTION EXERCISES

● **11.4.1** Estimate $\Delta G°$ for the formation of gaseous water from the elements at $T = 373$ K.

● **11.4.2** Does a temperature exist at which the water formation reaction is non-spontaneous under standard pressure? If so, compute this temperature. If not, explain why in molecular terms.

● **11.4.3** Using Examples 11-7 and 11-8, find the minimum partial pressure of $N_2O_4$ at which decomposition of $N_2O_4$ occurs, if $T = 127$ °C and $p_{NO_2} = 0.50$ bar.

## 11.5 SOME APPLICATIONS OF THERMODYNAMICS

### Nitrogen Fixation

Nitrogen is unevenly distributed between the Earth's crust and the atmosphere. In the crust, nitrogen is present at the level of 19 parts per million (ppm) by mass, four orders of magnitude less than oxygen ($4.55 \times 10^5$ ppm) and silicon ($2.72 \times 10^5$ ppm). In contrast, 80% of the atmosphere is molecular nitrogen. Paradoxically, nitrogen is essential for all life, but the sea of atmospheric nitrogen is virtually inaccessible to higher life-forms. Most biochemical systems lack the ability to break the strong triple bond between the nitrogen atoms in $N_2$. Molecular nitrogen must be converted to some other species, usually ammonia ($NH_3$) or nitrate anions ($NO_3^-$), before most life-forms can incorporate nitrogen atoms into their biochemical molecules. This process is known as **nitrogen fixation**.

In nature, nitrogen fixation is accomplished by nitrogenase, an enzyme that binds $N_2$ and weakens its bonding sufficiently to break the triple bond. Only a few algae and bacteria contain nitrogenase. Our Focus 11-2: Chemistry and Life describes what is known about this enzyme.

The thermodynamics of nitrogen chemistry helps explain why $N_2$ is so abundant in the atmosphere, and yet the element remains inaccessible to most life-forms. Table 11-4 shows that most of the abundant elements react with $O_2$ spontaneously under standard conditions. This is why many of the elements occur in the Earth's crust as their oxides. However, $N_2$ is resistant to oxidation, as shown by the positive $\Delta G_f^\circ$ for $NO_2$.

Because of their resistance to chemical attack, nitrogen atoms are not "locked up" in solid or liquid substances as are other elements such as Si, Al, Fe, and H. The most stable form of the element nitrogen is a gaseous diatomic molecule with a very strong triple bond. Therefore, the element nitrogen is concentrated in the Earth's gaseous atmosphere even though it is only a trace element in overall abundance.

$:N\equiv N:$

Bond energy = 945 kJ/mol.

According to Table 11-4, chlorine is also resistant to oxidation. Unlike nitrogen, however, chlorine reacts spontaneously with metals to generate salts such as NaCl and $MgCl_2$. Thus, among abundant elements on Earth, nitrogen is uniquely stable in its elemental form.

## TABLE 11-4 SURFACE-ABUNDANT ELEMENTS AND THEIR OXIDES

| Element | % by Mass | Oxide | $\Delta G_f^\circ$ (kJ/mol) |
|---|---|---|---|
| O | 49.1 | $O_2$ | 0 |
| Si | 26.1 | $SiO_2$ | −856 |
| Al | 7.5 | $Al_2O_3$ | −1582 |
| Fe | 4.7 | $Fe_3O_4$ | −1015 |
| Ca | 3.4 | CaO | −603 |
| Na | 2.6 | $Na_2O$ | −376 |
| K | 2.4 | $KO_2$ | −239 |
| Mg | 1.9 | MgO | −569 |
| H | 0.88 | $H_2O$ | −237 |
| Ti | 0.58 | $TiO_2$ | −885 |
| Cl | 0.19 | $Cl_2O$ | +98 |
| C | 0.09 | $CO_2$ | −394 |
| N | <0.1 | $NO_2$* | +51 |

*Several other oxides of nitrogen exist. All have even more positive free energies of formation than does $NO_2$.

Every breath of air we take is 80% nitrogen, but our bodies must obtain the nitrogen required for biosynthesis from nitrogen contained in the proteins we eat. In the plant kingdom, the most important sources of nitrogen are $NH_3$ and the ammonium cation ($NH_4^+$).

Why is the nitrogen atom in $NH_3$ accessible to living organisms? Consider an organism synthesizing the amino acid glycine from its nitrogen-deficient precursor, acetic acid. In elemental terms, the net reaction is

Acetic acid

$$CH_3CO_2H + N + H \longrightarrow H_2NCH_2CO_2H$$

Neither nitrogen nor hydrogen exists as free atoms, so the synthesis of glycine from acetic acid requires molecular sources of nitrogen and hydrogen. Atmospheric nitrogen and water are the most abundant sources of these two elements, but the production of glycine from $N_2$ and $H_2O$ is significantly non-spontaneous under standard conditions:

Glycine

$$4\ CH_3CO_2H + 2\ N_2 + 2\ H_2O \longrightarrow 4\ H_2NCH_2CO_2H + O_2$$

$$\Delta G^{\circ}_{reaction} = (1\ mol)(0\ kJ/mol) + (4\ mol)(-367\ kJ/mol)$$

$$- (4\ mol)(-389\ kJ/mol) - (2\ mol)(-237\ kJ/mol)$$

$$- (2\ mol)(0\ kJ/mol) = +564\ kJ$$

Per mole of glycine, this reaction is "uphill" by 141 kJ.

### FOCUS 11-2  CHEMISTRY AND LIFE: HOW DOES NITROGENASE WORK?

In the epic poem, *Rime of the Ancient Mariner*, a becalmed sailor laments, "Water, water, everywhere, nor any drop to drink." With regard to the element nitrogen, we might say "$N_2$, $N_2$, everywhere, nor any N to eat." Even though the atmosphere of our planet is 80% molecular nitrogen, plants are unable to break the extremely strong $N \equiv N$ triple bond. Instead, plants must rely on other sources for the nitrogen atoms that are essential for the synthesis of biomolecules such as amino acids and DNA.

For nitrogen to be available to life-forms, it must be either oxidized to nitrate ($NO_3^-$) or reduced to ammonia ($NH_3$). Either process is known as *nitrogen fixation*. Lightning oxidizes atmospheric nitrogen to nitrate, but lightning accounts for only about 1% of the fixed nitrogen required by the biosphere. The fertilizer industry reduces $N_2$ to $NH_3$ by reaction with $H_2$, but fertilizers supply only about 30% of fixed nitrogen. The bulk—about 70%—comes from the activity of one group of bacteria that can convert $N_2$ to $NH_3$. These bacteria contain nitrogenase, an enzyme that carries out this difficult chemical transformation.

Researchers are working to understand how this enzyme works. Research on nitrogenase takes two main forms. One is an examination of the structure and operation of the enzyme to determine the details of the reactions by which $N_2$ is converted to ammonia. The other form is the synthesis of artificial catalysts that mimic the operation of nitrogenase.

Nitrogenase is a complex protein with two metal-containing parts that "dock" together during the fixation of nitrogen. Part 1 contains an interconnected group of seven iron atoms, nine sulphur atoms, and a single molybdenum atom. Molecular nitrogen interacts with this cluster, where it gains the six electrons and six protons needed to become two ammonia molecules. However, exactly how $N_2$ binds to the [7Fe–9S–Mo] group and how it is reduced to $NH_3$ remain a mystery. Recent structural determinations indicate that a relatively light atom, perhaps nitrogen, sits in the centre of the cluster, interacting simultaneously with six of the iron atoms.

[4Fe-4S] cluster       [8Fe-7S] cluster       7Fe-9S-Mo binding site

Part 1 of the nitrogenase protein contains another interconnected group of Fe–S atoms, this one with eight iron atoms and seven sulphur atoms. This [8Fe–7S] group collects electrons and transmits them to the binding centre. Part 2 of nitrogenase contains a third Fe–S group, this one made up of four iron atoms and four sulphur atoms. This part of the enzyme also binds two molecules of ATP.

When the two parts of the enzyme join, the [4Fe–4S] group of Part 2 ends up close to the [8Fe–7S] group of Part 1. It is thought that when the parts of the enzyme join, two electrons are pumped from the ATP molecules to the [4Fe–4S] site, then to the

[8Fe–7S] site, and finally to the [7Fe–9S–Mo] binding site. Next, the parts separate, and two new ATP molecules recharge Part 2. This process must occur four times to provide the eight electrons needed for the fixation reaction. The entire fixation process requires eight electrons and eight protons, six of each to form the ammonia molecule and two more that are "wasted" by forming an $H_2$ molecule.

Although the mechanistic details of $N_2$ reduction have not been determined, chemists have been able to make a molybdenum-containing complex, shown below, that can bind $N_2$ and convert it to $NH_3$. In this complex, $N_2$ binds to Mo and is locked in place by three bulky groups. The complex enables $N_2$ to gain 6 electrons and 6 $H^+$, in steps, from other reagents. Further exploration of Mo–$N_2$ chemistry could lead to success in mimicking nitrogenase, or at least to a better understanding of how this remarkable enzyme works.

Perhaps chemists will be able to mimic nature without duplicating the iron–sulphur–molybdenum structure. For example, a zirconium complex with tetramethyl cyclopentadiene can bind dinitrogen in a manner that breaks the N≡N bond, as shown at right. Treatment with hydrogen gas results in formation of small amounts of ammonia. Although the yields are too low to make this a viable commercial process, researchers hope to make the process more efficient through chemical modifications and changes in conditions.

On the other hand, $NH_3$ can provide both nitrogen and hydrogen atoms. Each $NH_3$ molecule also contains two extra hydrogen atoms that can be oxidized to produce water:

$$CH_3CO_2H + NH_3 + \tfrac{1}{2}O_2 \longrightarrow H_2NCH_2CO_2H + H_2O$$

$$\Delta G^\circ_{\text{reaction}} = [(1\ \text{mol})(-237\ \text{kJ/mol}) + (1\ \text{mol})(-367\ \text{kJ/mol})]$$

$$- [(1\ \text{mol})(-398\ \text{kJ/mol}) + (1\ \text{mol})(-17\ \text{kJ/mol})$$

$$+ (0.5\ \text{mol})(0\ \text{kJ/mol})]$$

$$\Delta G^\circ_{\text{reaction}} = -198\ \text{kJ}$$

This reaction is significantly spontaneous under standard conditions because it couples the production of water with the formation of the N—C bond in glycine.

The intensive agriculture that is characteristic of industrialized countries requires much more fixed nitrogen than is readily available from natural sources. Consequently, one of the major products of the chemical industry is nitrogen-containing fertilizers. Among the top 50 industrial chemicals, nitrogen gas (separated from air by cooling and liquefaction) perennially ranks number 2, with $NH_3$ (ammonia), $HNO_3$ (nitric acid), $(NH_2)_2CO$ (urea), and $NH_2NO_3$ (ammonium nitrate) all in the top 15. All

Commercial agriculture uses substantial amounts of nitrogen-containing fertilizers.

these chemicals are produced in huge amounts because of their roles as fertilizers (urea and ammonium nitrate), fertilizer precursors (nitrogen gas and nitric acid), or both (ammonia).

Free energies of formation suggest that ammonia and ammonium nitrate could be produced spontaneously in nature:

$$N_2(g) + 3\ H_2(g) \longrightarrow 2\ NH_3(g) \qquad \Delta G_f^\circ = -32.8\ kJ$$
$$2\ N_2(g) + 3\ O_2(g) + 4\ H_2(g) \longrightarrow 2\ NH_4NO_3(s) \quad \Delta G_f^\circ = -32.8\ kJ$$

These reactions might occur in an atmosphere containing significant amounts of molecular hydrogen, but on Earth, water is the only readily available source of hydrogen. The free energies of the reactions that yield $NH_3$ and $NH_4NO_3$ from gaseous nitrogen and water are highly unfavourable:

$$2\ N_2 + 6\ H_2O \longrightarrow 4\ NH_3 + 3\ O_2 \quad \Delta G_{reaction}^\circ = +1358\ kJ$$
$$2\ N_2 + 4\ H_2O + O_2 \longrightarrow 2\ NH_4NO_3 \qquad \Delta G_{reaction}^\circ = +580\ kJ$$

Thus, large energy costs are involved in generating fixed nitrogen.

The strategy used in fertilizer production is to synthesize molecular hydrogen from methane:

$$CH_4 + H_2O \longrightarrow CO + 3\ H_2$$

$$\Delta G^\circ = +142\ kJ \qquad \Delta H^\circ = +205.9\ kJ \qquad \Delta S^\circ = +214.6\ J/K$$

Although this reaction is not spontaneous at room temperature, it is favourable at a temperature of 1000 K:

$$\Delta G_{1000\ K}^\circ = \Delta H^\circ - T\Delta S^\circ = +205.9\ kJ - (1000\ K)(214.6\ J/K)(10^{-3} kJ/J) = -8.7\ kJ$$

Heat energy must be supplied continuously to keep the temperature from falling while the endothermic reaction proceeds. Otherwise, the temperature would quickly fall below the minimum value at which $H_2$ synthesis is spontaneous. For this reason, the production of $H_2$ (and ultimately of fertilizer) requires considerable amounts of energy. The energy cost is even greater if we consider that methane is obtained from non-renewable reservoirs of natural gas, created over countless eons through the photosynthetic storage of solar energy.

**Figure 11-17**

Block diagram of the industrial routes from nitrogen gas to fertilizers. Methane, water, and energy in the form of heat are key ingredients in the first step.

With an ample supply of hydrogen, and at high pressure, production of ammonia from $N_2$ becomes feasible:

$$N_2 + 3\,H_2 \longrightarrow 2\,NH_3 \qquad \Delta G^\circ_{reaction} = -32.8\ kJ$$

**Figure 11-18**

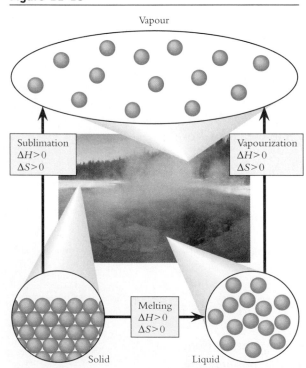

Schematic view of the three phase changes leading from more constrained to less constrained phases, illustrated by the phase changes for water. Each is accompanied by positive enthalpy and entropy changes for the substance.

Figure 11-17 shows other spontaneous reactions that lead from ammonia to nitric acid, urea, and ammonium nitrate.

## Phase Changes

Phase changes, which convert a substance from one phase to another, have characteristic thermodynamic properties: any change from a more constrained phase to a less constrained phase increases both the enthalpy and the entropy of the substance. Recall from our description of phase changes in Chapter 8 that enthalpy increases because energy must be provided to overcome the intermolecular forces that hold the molecules in the more constrained phase. Entropy increases because the molecules are more dispersed in the less constrained phase. Thus, when a solid melts or sublimes or a liquid vapourizes, both $\Delta H$ and $\Delta S$ are positive. Figure 11-18 summarizes these features.

As described in Section 11.4, when $\Delta H$ and $\Delta S$ have the same sign, the spontaneous direction of a process depends on $T$. For a phase change, enthalpy dominates $\Delta G$ at low temperature, and the formation of the more constrained phase is spontaneous. In contrast, entropy dominates $\Delta G$ at high temperature, and the formation of the less constrained phase is spontaneous. At one characteristic temperature, $\Delta G = 0$, and the phase change proceeds in both directions at the same rate. The two phases coexist, and the system is in a state of dynamic equilibrium.

The spontaneous direction of a phase change also depends on pressure, primarily because the molar entropy of a gas exhibits strong pressure-dependence:

$$S(\text{gas}) = S° - R \ln p_{\text{vap}}$$

$$\Delta G_{\text{vap}} = \Delta H_{\text{vap}} - T\Delta S_{\text{vap}} = \Delta H_{\text{vap}} - T\Delta S°_{\text{vap}} + RT \ln p_{\text{vap}}$$

The result of this pressure dependence is that at any temperature, there is some pressure at which $\Delta G_{\text{vap}} = 0$. This is the vapour pressure ($p_{\text{vap}}$) of the substance at that temperature. We can relate $p_{\text{vap}}$ to thermodynamic properties:

$$\Delta G_{\text{vap}} = 0 = \Delta H_{\text{vap}} - T\Delta S°_{\text{vap}} + RT \ln p_{\text{vap}}$$

$$RT \ln p_{\text{vap}} = -\Delta H_{\text{vap}} + T\Delta S°_{\text{vap}}$$

Rearranging gives Equation 11-14:

$$\ln p_{\text{vap}} = -\frac{\Delta H_{\text{vap}}}{RT} + \frac{\Delta S°_{\text{vap}}}{R} \qquad (11\text{-}14)$$

Equation 11-14 is one form of the Clausius–Clapeyron equation, which was first developed by Rudolf Clausius and Emile Clapeyron in the nineteenth century. Example 11-9 shows how to apply this equation.

> See Chapter 2 for information about vapour pressure. The vapour pressure of water at various temperatures appears in Table 2-5.

## EXAMPLE 11-9 VAPOUR PRESSURE

Liquid bromine has a significant vapour pressure.

Molecular bromine is a liquid at 298 K, 1 bar, and has a significant vapour pressure at this temperature. Calculate this vapour pressure in Torr.

**Strategy:** We apply the condensed form of the seven-step strategy. To calculate a vapour pressure, we need the enthalpy and entropy of vapourization. Then we can apply Equation 11-14.

**Solution:** Appendix D contains thermodynamic information for molecular bromine:

$$\text{Br}_2 \, (l): \quad \Delta H°_f = 0 \text{ kJ/mol} \qquad S° = 152.2 \text{ J mol}^{-1}\text{K}^{-1}$$

$$\text{Br}_2 \, (g): \quad \Delta H°_f = 30.9 \text{ kJ/mol} \quad S° = 245.5 \text{ J mol}^{-1}\text{K}^{-1}$$

We apply Equations 3-12 and 11-6 to the vapourization reaction:

$$\text{Br}_2 \, (l) \longrightarrow \text{Br}_2 \, (g)$$

$$\Delta H°_{\text{reaction}} = \sum \text{coeff}_{\text{products}} \, \Delta H°_{\text{f, products}} - \sum \text{coeff}_{\text{reactants}} \, \Delta H°_{\text{f, reactants}}$$

$$\Delta H°_{\text{reaction}} = 30.9 \text{ kJ} - 0 \text{ kJ} = 30.9 \text{ kJ}$$

$$\Delta S^{\circ}_{reaction} = \sum coeff_{products}\, S^{\circ}_{products} - \sum coeff_{reactants}\, S^{\circ}_{products}$$

$$\Delta S^{\circ}_{reaction} = 245.5 \text{ J/K} - 152.2 \text{ J/K} = 93.3 \text{ J/K}$$

Now substitute these values into Equation 11-14 and calculate the result:

$$\ln p_{vap} = -\frac{\Delta H_{vap}}{RT} + \frac{\Delta S^{\circ}_{vap}}{R}$$

$$= -\frac{(30.9 \text{ kJ/mol})(10^3 \text{ J/kJ})}{(8.314 \text{ J mol}^{-1}\text{ K}^{-1})(298 \text{ K})} + \frac{(93.3 \text{ J mol}^{-1}\text{K}^{-1})}{(8.314 \text{ J mol}^{-1}\text{ K}^{-1})}$$

$$\ln p_{vap} = -12.47 + 11.22 = -1.25 \quad \text{so} \quad p_{vap} = e^{-1.25} = 0.287 \text{ bar}$$

Remember that the standard unit of pressure is the bar. To complete the calculation, we need to convert from bar to Torr (1 bar = 750 Torr):

$$p_{vap} = 0.287 \text{ bar} \left(\frac{750 \text{ Torr}}{1 \text{ bar}}\right) = 215 \text{ Torr}$$

**Does the Result Make Sense?** We know that $Br_2$ is a liquid under standard conditions, but the margin photograph shows a distinct red colour in the vapour phase, indicating a significant vapour pressure. The calculated result is significant but well below atmospheric pressure of 760 Torr.

**Common Pitfalls:** As always, keeping a consistent set of units is important.

**Practice Exercise 11.9**

Use the values of standard thermodynamic functions to determine the vapour pressure of water, in Torr, at 50°C.

We can further modify Equation 11-14 to calculate the difference in the vapour pressure at two temperatures. This is useful if we know the vapour pressure at one temperature and want to calculate it at another. If the vapour pressure is $p_1$ at $T_1$ and is $p_2$ at $T_2$, then according to Equation 11-14:

$$\ln (p_2) - \ln (p_1) = \left[\left(\frac{-\Delta H_{vap}}{RT_2} + \frac{\Delta S^{\circ}_{vap}}{R}\right) - \left(\frac{-\Delta H_{vap}}{RT_1} + \frac{\Delta S^{\circ}_{vap}}{R}\right)\right]$$

$$= \frac{-\Delta H_{vap}}{RT_2} + \frac{\Delta H_{vap}}{RT_1}$$

$$\ln(p_2) - \ln(p_1) = \frac{\Delta H_{vap}}{R}\left[\frac{1}{T_1} - \frac{1}{T_2}\right] \tag{11-15}$$

Equation 11-15 can be used in several ways. For instance, the vapour pressure of a liquid is by definition 101.325 kPa (1 atm) at the normal boiling point (normal boiling points have not been corrected to the standard of 1 bar yet). Knowing this and $\Delta H_{vap}$ allows us to calculate the vapour pressure at any other temperature. Alternatively, if we have the vapour pressure at two different temperatures, we can calculate the value of $\Delta H_{vap}$. And finally, knowing the vapour pressure at one temperature and $\Delta H_{vap}$ allows us to calculate the temperature at which the vapour pressure attains some other value. The advantage of using Equation 11-15 is that entropy values are not required. Examples 11-10, 11-11, and 11-12 demonstrate the use of Equation 11-15.

## EXAMPLE 11-10 VAPOUR PRESSURE AT ANOTHER TEMPERATURE

Water boils at 100 °C. Calculate the vapour pressure of water at 90 °C. Assume that $\Delta H_{vap} = 40.79$ kJ/mol and that $\Delta H_{vap}$ is constant over this temperature range.

**Strategy:** We can use Equation 11-15 to solve for the vapour pressure at 90 °C. We know that water boils at 100 °C, so we know that by definition, its vapour pressure is 101.325 kPa at this temperature.

**Solution:** The data we have are

$$p_1 = 101.325 \text{ kPa} \qquad T_1 = 100 \text{ °C} = 373.15 \text{ K}$$

$$p_2 = ? \qquad T_2 = 90 \text{ °C} = 363.15 \text{ K}$$

$$\Delta H_{vap} = 40.79 \text{ kJ/mol}$$

Using Equation 11-15, we have

$$\ln(p_2) = \ln(p_1) + \frac{\Delta H_{vap}}{R}\left[\frac{1}{T_1} - \frac{1}{T_2}\right]$$

$$= \ln(101.325) + \frac{40\,790 \text{ J/mol}}{8.314\,\text{J}\,\text{K}^{-1}\text{mol}^{-1}}\left[\left(\frac{1}{373.15 \text{ K}} - \frac{1}{363.15 \text{ K}}\right)\right]$$

$$= 4.618 + (-0.362)$$

$$= 4.256$$

$$p_2 = e^{4.256} = 70.5 \text{ kPa}$$

Note that the units for $p_2$ will be the same as those we used for $p_1$, kPa in this example.

**Does the Result Make Sense?** The calculated vapour pressure at 90 °C is less than that at 100 °C, as expected. There is a considerable difference in the vapour pressures at temperatures only 10 °C apart because of the logarithmic relationship between vapour pressure and temperature.

**Common Pitfalls:** Not using a consistent set of units for $\Delta H_{vap}$ and $R$. In this example, we converted the units of $\Delta H_{vap}$ to J/mol to match the usual units for $R$.

**Practice Exercise 11.10**

Benzene ($C_6H_6$) has a vapour pressure of 100 Torr at 26.1 °C and $\Delta H_{vap} = 33.9$ kJ/mol. Calculate the vapour pressure of benzene at 40 °C.

## EXAMPLE 11-11 CALCULATING A TEMPERATURE FROM VAPOUR PRESSURE DATA

Acetone (($CH_3$)$_2$CO) has a vapour pressure of 53.3 kPa at 39.5 °C, and $\Delta H_{vap} = 31.3$ kJ/mol. Calculate the normal boiling point of acetone.

**Strategy:** We can again use Equation 11-15, but we will need to rearrange it to solve for a temperature. Again, the normal boiling point is the temperature at which the vapour pressure is 101.325 kPa.

**Solution:** The data we have are

$$p_1 = 53.3 \text{ kPa} \qquad\qquad T_1 = 39.5 \,°\text{C} = 312.6 \text{ K}$$

$$p_2 = 101.325 \text{ kPa} \qquad\qquad T_2 = ?$$

$$\Delta H_{vap} = 31.3 \text{ kJ/mol}$$

Equation 11-15 is $\ln(p_2) - \ln(p_1) = \dfrac{\Delta H_{vap}}{R}\left[\dfrac{1}{T_1} - \dfrac{1}{T_2}\right]$. This can be rearranged to solve for $T_2$ as follows:

$$\frac{R}{\Delta H_{vap}}\left(\ln(p_2) - \ln(p_1)\right) = \left[\frac{1}{T_1} - \frac{1}{T_2}\right]$$

$$\frac{1}{T_2} = \frac{1}{T_1} - \frac{R}{\Delta H_{vap}}\left(\ln(p_2) - \ln(p_1)\right)$$

$$T_2 = \left[\frac{1}{T_1} - \frac{R}{\Delta H_{vap}}\left(\ln(p_2) - \ln(p_1)\right)\right]^{-1}$$

Then the values can be substituted and the equation solved. Note the conversion of the units of $\Delta H_{vap}$ to match those of $R$:

$$T_2 = \left[\frac{1}{312.6 \text{ K}} - \frac{8.314\,\text{J}\,\text{K}^{-1}\text{mol}^{-1}}{31\,300 \text{ J/mol}}\left(\ln(101.325) - \ln(53.3)\right)\right]^{-1}$$

$$= [0.003198 \text{ K}^{-1} - 0.000170 \text{ K}^{-1}]^{-1}$$

$$= 330.2 \text{ K}$$

$$= 57.0 \,°\text{C}$$

**Does the Result Make Sense?** Again, we see a dramatic increase of the vapour pressure with temperature, but this is expected because of the logarithmic relationship. 57.0 °C is not an unreasonable boiling point for a small organic molecule.

**Common Pitfalls:** Improper rearrangement of Equation 11-15. Make sure you can do this rearrangement yourself.

**Practice Exercise 11.11**

Toluene ($C_6H_5CH_3$) boils at 110.6 °C and has $\Delta H_{vap} = 38.1$ kJ/mol. At what temperature is the vapour pressure of toluene 400 Torr?

## EXAMPLE 11-12 CALCULATING THE ENTHALPY OF VAPOURIZATION

The vapour pressure of methanol is 55.6 Torr at 10.0 °C and is 97.5 Torr at 20.0 °C. Calculate the enthalpy of vapourization of methanol.

**Strategy:** Equation 11-15 can be rearranged to solve for $\Delta H_{vap}$. As long as we have two vapour pressures at two temperatures, the calculation is simple.

**Solution:** The data we have are

$$p_1 = 55.6 \text{ Torr} \qquad\qquad T_1 = 10.0 \,°\text{C} = 283.15 \text{ K}$$

$$p_2 = 97.5 \text{ Torr} \qquad\qquad T_2 = 20.0 \,°\text{C} = 293.15 \text{ K}$$

$$\Delta H_{vap} = ?$$

Rearranging Equation 11-15 to solve for $\Delta H_{vap}$:

$$\Delta H_{vap} = \frac{R\left(\ln(p_2) - \ln(p_1)\right)}{\left[\dfrac{1}{T_1} - \dfrac{1}{T_2}\right]}$$

$$= \frac{8.314\,\text{J}\,\text{K}^{-1}\text{mol}^{-1}\left(\ln(97.5) - \ln(55.6)\right)}{\left[\dfrac{1}{283.15\ \text{K}} - \dfrac{1}{293.15\ \text{K}}\right]}$$

$$= 38\,760\ \text{J/mol}$$

$$= 38.8\ \text{kJ/mol}$$

**Does the Result Make Sense?** This magnitude of enthalpy of vapourization is correct. Note that it is intermediate between those of water and acetone (see previous two examples) because the extent of hydrogen bonding in methanol is less than that of water, and acetone does not H-bond.

**Common Pitfalls:** Improper rearrangement of Equation 11-15. Make sure you can do this yourself.

**Practice Exercise 11.12**
Calculate the enthalpy of vapourization of hexane ($C_6H_{14}$), given that its normal boiling point is 68.7 °C, and that its vapour pressure at 15.8 °C is 0.133 bar.

## Thermal Pollution

One of the most important practical consequences of the second law of thermodynamics is that heat must be transferred from the system to its surroundings whenever a process is driven in the direction that increases constraints on the system. This heat raises the temperature and increases the entropy of the surroundings, thus offsetting the entropy decrease of the system. Refrigerators, computers, automobile engines, nuclear power plants, and humans are subject to this requirement. All these objects increase the temperature of their surroundings as they operate. For industrial-scale operations such as nuclear power plants, the amount of heat generated can have serious consequences. This type of heating is called **thermal pollution**.

Any thermally powered electrical power plant uses heat flows to convert stored energy into work. The stored energy may be chemical (coal, oil, or natural gas) or nuclear (uranium). When this potential energy is released, it creates a heat flow that is used to generate steam, which drives turbines to generate electricity. Electricity is the organized motion of electrons, all going in the same direction in a wire. The electrons become constrained, so there is a decrease in entropy that must be paid for by dispersing energy somewhere else. Consequently, a significant portion of the energy used for thermal generation of electrical power has to be "wasted" by dumping it into the surroundings as heat.

Power plants require large cooling towers to dissipate heat.

The increase in temperature that results from thermal power generation creates biological stresses on local ecosystems. Fish, being cold-blooded, are particularly sensitive to changes in water temperature. As water temperature increases, the blood temperature of the fish increases, and this in turn increases its rate of metabolism. At the same time, however, the amount of oxygen available to supply the metabolic demands of the fish decreases, because oxygen is less soluble in warm water than in cold water. Species that thrive in cold water, such as trout, die if the water temperature rises by more than a few degrees. Simpler forms of aquatic life also are highly sensitive to temperature. Plankton, which provide the food base for many aquatic ecosystems, thrive at temperatures between 14 °C and 24 °C. When water temperatures exceed 24 °C, blue-green algae crowd out these essential plankton, leading to serious disruption of the ecological balance. Furthermore, algal blooms consume the oxygen dissolved in water and can cause fish to suffocate.

Hydroelectric generation of electricity need not lead to thermal pollution, because flowing water can drive turbines without a large increase in the energy constraints. Unfortunately, there are other detrimental ecological consequences of hydroelectric power plants, such as the flooding of vast natural habitats, disruption of rivers and streams, and changes in local weather patterns caused by the creation of huge bodies of water. Although these effects may be less destructive than thermal pollution, no truly "clean" power source seems to exist. The second law of thermodynamics states that a price must always be paid for concentrating energy. Inevitably, human existence does this, so we pay this price continually. Conservation measures minimize the thermodynamic costs, but these costs can never be eliminated completely.

## SECTION EXERCISES

●**11.5.1** Atmospheric nitrogen can be transformed to ammonia by various bacteria that fix nitrogen. The overall process is complex, but the net reaction can be written as

$$2\,N_2 + 6\,H_2O \longrightarrow 4\,NH_3 + 3\,O_2$$

Assuming standard conditions (not quite true but close enough for approximations), how much free energy must the bacteria consume to fix 1 mole of N atoms?

●**11.5.2** All automobile engines operate at relatively high temperature and require continual cooling. Using our description of thermal pollution, explain how the second law of thermodynamics requires automobile engines to "dump" heat to their surroundings.

## 11.6 BIOENERGETICS

Life creates order out of chaos, but thermodynamics shows that any spontaneous process must increase the total amount of dispersion in the universe. To constrain matter and energy in a living organism, it is necessary to release stored energy as heat, which, in turn, disperses energy in the surroundings. Consequently, living things must use large amounts of energy to survive. In this section, we describe representative energetic processes that operate in living organisms.

As this infrared photo shows, a person exercising emits heat to the surroundings.

## Biochemical Energy Production

Living organisms use carbohydrates as their source of energy. Plants make their own carbohydrates through photosynthesis. Animals, on the other hand, obtain carbohydrates by eating plants or other animals. Plants and animals use the oxidation of carbohydrates by $O_2$ as a source of free energy. They also transform carbohydrates into fats, which react spontaneously with $O_2$. The extraction of chemical energy from these reactions is an important component of metabolism. The highly spontaneous nature of these metabolic oxidation reactions is illustrated by glucose (a carbohydrate) and palmitic acid (a fatty acid, one component of fat):

$$\text{Glucose: } C_6H_{12}O_6 + 6\ O_2 \longrightarrow 6\ CO_2 + 6\ H_2O \quad \Delta G° = -2870 \text{ kJ}$$

$$\text{Palmitic acid: } C_{15}H_{31}CO_2H + 23\ O_2 \longrightarrow 16\ CO_2 + 16\ H_2O \quad \Delta G° = -9790 \text{ kJ}$$

The negative standard free energy changes of these reactions are due to two features. There is a large negative enthalpy change because the $O{=}O$ bond in molecular oxygen is converted into stronger $C{=}O$ bonds in $CO_2$. Entropy also favours these reactions because there are many more molecules on the product side of the equation than on the reactant side (recall from Section 11.3 that entropy always increases when large molecules break apart into smaller ones).

The large negative free energy change that accompanies the oxidation of glucose and palmitic acid means that a little fat or carbohydrate goes a long way as a fuel for life processes. However, a living cell would be destroyed quickly if all this change in free energy occurred in a single burst. To harness the free energy released in these reactions without being destroyed, cells use elaborate chains of sequential reactions that harvest free energy a little at a time.

The free energy changes in metabolic oxidation reactions serve several purposes. Part of this energy appears as heat flows that maintain body temperature and disperse energy to the surroundings. Another portion of the energy is used to synthesize high-energy molecules that the body uses to drive the many reactions that occur within cells. In addition to storing part of the energy produced in the oxidation reactions, these high-energy species serve as energy transport molecules, moving to different regions of the cell where energy is required for various metabolic processes.

**Adenosine triphosphate (ATP)** is the most important of these energy transport molecules. Some of the energy released during the oxidation of glucose is used to drive a condensation reaction in which adenosine diphosphate (ADP) and phosphoric acid link together and eliminate water. This reaction stores chemical energy, as indicated by its positive standard free energy change:

$$\text{ADP} + H_3PO_4 \longrightarrow \text{ATP} + H_2O \quad \Delta G° = 30.6 \text{ kJ}$$

Glucose

Palmitic acid

**Figure 11-19**

Phosphoric acid reacts with ADP to produce water and ATP.

Figure 11-19 illustrates the molecular details of this reaction. Although ATP is a complex molecule, the adenosine portion does not change during the condensation reaction. The condensation reaction adds a third phosphate group to the end of an existing chain. This reaction absorbs energy because the P—O—P and H—O—H bonds that form are somewhat weaker than the two polarized P—O—H linkages that break.

The exact processes by which carbohydrates and fats are converted to $CO_2$ and $H_2O$ depend on the conditions and the particular needs of the cell. Each possible route involves a complex series of chemical reactions, many of which are accompanied by the conversion of ADP to ATP. One molecule of glucose, for example, is oxidized to $CO_2$ and $H_2O$ in a sequence of many individual reactions that can convert as many as 36 ADP molecules into ATP molecules:

$$C_6H_{12}O_6 + 6\ O_2 + 36\ ADP + 36\ H_3PO_4 \longrightarrow 6\ CO_2 + 36\ ATP + 42\ H_2O$$

## Coupled Reactions

Cells obtain energy by reversing the ADP–ATP reaction, converting ATP back to ADP and phosphoric acid:

$$ATP + H_2O \longrightarrow ADP + H_3PO_4 \qquad \Delta G° = -30.6\ kJ$$

The energy released in converting ATP to ADP is used to drive reactions that are non-spontaneous under physiological conditions. This is accomplished by coupling the two reactions in a single system.

For example, the amino acid glutamine is synthesized in cells by the reaction of ammonia with another amino acid, glutamic acid:

This reaction is thermodynamically unfavourable, $\Delta G° = +14\ kJ$, but it can be driven by coupling it with the conversion of ATP into ADP. The net free energy change for the coupled process is the sum of the $G°$ values for the individual reactions:

$$\text{Glutamic acid} + NH_3 \longrightarrow \text{Glutamine} + H_2O \quad \Delta G^\circ_{\text{glutamine}} = +14 \text{ kJ}$$
$$\text{ATP} + H_2O \longrightarrow \text{ADP} + H_3PO_4 \quad \Delta G^\circ_{\text{ATP}} = -30.6 \text{ kJ}$$
$$\textit{Net}: \text{Glutamic acid} + \text{ATP} + NH_3 \longrightarrow \text{Glutamine} + \text{ADP} + H_3PO_4$$
$$\Delta G^\circ_{\text{reaction}} = \Delta G^\circ_{\text{Glutamine}} + \Delta G^\circ_{\text{ATP}} = 14 \text{ kJ} + (-30.6 \text{ kJ}) = -17 \text{ kJ}$$

The negative value of $\Delta G^\circ_{\text{reaction}}$ indicates that the free energy released in the ATP reaction is more than enough to drive the conversion of glutamic acid into glutamine.

For a spontaneous reaction to drive one that is non-spontaneous, the two sets of reactants must interact chemically. A coupled biochemical reaction occurs on the surface of a protein. The protein has a particular shape that accommodates the molecule that participates in the coupled reaction. In this way, the protein acts as an enabler for a particular biochemical reaction.

The net reaction between glutamic acid and ATP actually requires two steps, each of which is spontaneous. In the first step, an ATP molecule transfers a phosphate group to glutamic acid:

Glutamic acid

Next, an ammonia molecule reacts with the phosphorylated form of glutamic acid, producing phosphoric acid and glutamine:

Phosphorylated glutamic acid

Overall, one molecule of ATP is converted to ADP and phosphoric acid for each molecule of glutamine produced from glutamic acid.

The synthesis of ATP from ADP and phosphoric acid is non-spontaneous. Consequently, ATP synthesis must be coupled to some more spontaneous reaction. Example 11-13 describes one of these reactions.

## EXAMPLE 11-13 ATP-FORMING REACTIONS

One biochemical reaction that produces ATP involves the conversion of acetyl phosphate to acetic acid and phosphoric acid:

Acetyl phosphate        Acetic acid

$\Delta G^\circ = -46.9 \text{ kJ/mol}$

Couple this reaction with the ADP–ATP reaction, balance the net equation, and show that the coupled reaction is spontaneous.

**Strategy:** A coupled process links a spontaneous reaction with a non-spontaneous one. In this case, the negative free energy change of the acetyl phosphate reaction drives the conversion of ADP to ATP.

**Solution:** Combining the two reactions gives the overall balanced equation:

$$CH_3CO_2PO_3H_2 + H_2O \longrightarrow CH_3CO_2H + H_3PO_4$$
$$ADP + H_3PO_4 \longrightarrow ATP + H_2O$$
$$\overline{\text{Net: } CH_3CO_2PO_3H_2 + ADP \longrightarrow CH_3CO_2H + ATP}$$

The net energy change for the coupled process is the sum of the $G°$ values for the individual reactions:

$$\Delta G°_{net} = \Delta G°_{acetyl\ phosphate} + \Delta G°_{ATP} = -46.9\ kJ + 30.6\ kJ = -16.3\ kJ$$

 **Does the Result Make Sense?** The negative value of $\Delta G°_{net}$ shows that the free energy released in the acetyl phosphate reaction is more than enough to drive the conversion of ADP to ATP.

**Common Pitfalls:** Not properly adding the reactions to be coupled. The net energy change will only be correct if the coupled reactions add up to the net reaction.

**Practice Exercise 11.13**

The coupled reaction that adds phosphate to glucose is spontaneous:

$$\text{Glucose + ATP} \longrightarrow \text{Glucose phosphate + ADP} \qquad \Delta G°_{net} = -16.7\ kJ$$

Write the separate reactions involved in this process and determine $\Delta G°$ for each.

## Energy Efficiency

Cells store the energy released during the oxidation of glucose by converting ADP into ATP. The storage process cannot be perfectly efficient, however, because each step in the reaction sequence must have a negative free energy change. In practical terms, this requires that some energy be released to the surroundings as heat.

The oxidation of 1 mole of glucose can produce as many as 36 molecules of ATP under normal physiological conditions:

$$C_6H_{12}O_6 + 6\ O_2 + 36\ ADP + 36\ H_3PO_4 \longrightarrow 6\ CO_2 + 36\ ATP + 42\ H_2O$$

The overall free energy change for this process can be determined from the values for its uncoupled parts:

$$C_6H_{12}O_6 + 6\ O_2 \longrightarrow 6\ CO_2 + 6\ H_2O \qquad \Delta G° = -2870\ kJ$$

$$36\ (ADP + H_3PO_4 \longrightarrow ATP + H_2O)$$

$$\Delta G° = (36\ mol)(+30.6\ kJ/mol) = +1100\ kJ$$

$$\Delta G°_{overall} = -2870\ kJ + 1100\ kJ = -1770\ kJ$$

Although this coupled process stores 1100 kJ of free energy under standard conditions, 1770 kJ of free energy is "wasted." Thus, cells convert less than 40% of the chemical energy stored in glucose into chemical energy stored in ATP and later use it to drive the biochemical machinery of metabolism. Well over half of the energy is dispersed as heat, increasing the entropy of the surroundings.

A functioning cell does not operate under standard conditions, so the above calculation is only an estimate. The standard free energy change for conversion of ADP to ATP is 30.6 kJ/mol. Because the concentrations of phosphate species in cells are far from 1 M, however, the free energy change under cellular conditions can be as low as about 28 kJ/mol or as high as about 50 kJ/mol, depending on the exact concentrations of the various species involved in the reaction. Nevertheless, the qualitative result holds under cellular conditions: over half the free energy released in the oxidation of glucose is dispersed as heat.

Fats such as palmitic acid are metabolized through pathways similar to the ones for the oxidation of glucose. The complete oxidation of palmitic acid has a standard free energy change of $-9790$ kJ/mol and produces 130 ATP molecules per molecule of palmitic acid consumed. You should be able to verify that this metabolic process has about the same efficiency as the oxidation of glucose.

## SECTION EXERCISES

●**11.6.1** Nitrogen-fixing bacteria react $N_2$ with $H_2O$ to produce $NH_3$ and $O_2$ using ATP as their energy source. About 24 molecules of ATP are consumed per molecule of $N_2$ fixed. What percentage of the free energy derived from ATP is stored in $NH_3$?

●**11.6.2** For the conversion of ATP to ADP, $\Delta H° = -21.0$ kJ/mol and $\Delta G° = -30.6$ kJ/mol at 298 K. Calculate $\Delta S°$ for this reaction. What happens to the spontaneity of this reaction as the temperature increases to 37 °C?

●**11.6.3** In running a mile, an average person consumes about 500 kJ of free energy. (a) How many moles of ATP does this represent? (b) Assuming 38% conversion efficiency, how many grams of glucose must be oxidized?

# CHAPTER 11 VISUAL SUMMARY

## 11.1 SPONTANEITY

Spontaneous

Dispersal of energy

Dispersal of matter

*Things tend to become dispersed.*

**LEARNING OBJECTIVES:** Recognize the driving force behind all chemical change: dispersal of energy and matter.

**KEY CONCEPTS:** A process that is spontaneous in one direction is non-spontaneous in the opposite direction.

Energy in the form of heat always flows from a warmer object to a colder one.

The spontaneity of any process must be evaluated in terms of the dispersal of both energy and matter.

## 11.2 ENTROPY: THE MEASURE OF DISPERSAL

Ice

$-q$ $+q$

System
$S = -$
$T = 273$ K

Surroundings
$S = +$
$T = 258$ K

Dispersal of energy

Dispersal of matter

$$\Delta S = \frac{q_T}{T}$$

$W = 9 \times 8 = 72$

$$S = k \ln W$$

$$\Delta S_{total} = \Delta S_{system} + \Delta S_{surroundings}$$

*For a spontaneous process:* $\Delta S_{total} > 0$  |  Second law of thermodynamics

**LEARNING OBJECTIVES:** Predict the direction of change based on the entropy changes in the system and in the surroundings.

**SKILLS TO MASTER:** Using Boltzmann's equation for entropy changes in both the system and the surroundings (p. 615)

Calculating entropy changes from the flow of heat at constant temperature (p. 617)

**KEY CONCEPTS:** The second law of thermodynamics states that in any spontaneous process, the direction of change is such that total entropy increases: $\Delta S_{total} > 0$.

## 11.3 ENTROPIES OF PURE SUBSTANCES

Third law of thermodynamics  |  $S_{pure,\ perfect\ crystal;\ 0\ K} = 0$

$S_{solid} < S_{liquid} < S_{gas}$

$$S_{(p \neq 1\ bar)} = S^\circ - R \ln p$$

Expansion

$$S_{(c \neq 1\ M)} = S^\circ - R \ln c$$

Dilution

$$\Delta S^\circ_{reaction} = \Sigma\, coeff_{products}\, S^\circ_{products} - \Sigma\, coeff_{reactants}\, S^\circ_{reactants}$$

**LEARNING OBJECTIVES:** Understand and calculate entropies of pure substances.

**SKILLS TO MASTER:** Calculating absolute entropies (p. 622)

Calculating entropy changes with concentration or pressure (p. 629)

Calculating standard reaction entropies (p. 631)

**KEY CONCEPTS:** The third law of thermodynamics states that a pure, perfect crystal at 0 K has zero entropy.

Entropy increases as the temperature increases, when a solid melts, and when a liquid evaporates.

## 11.4 SPONTANEITY AND FREE ENERGY

$\Delta G_{sys} < 0$ *for any spontaneous process at constant T and P.*

Free energy ($G$)  $\quad G = H - TS$

$$\Delta G_{sys} = \Delta H_{sys} - T\Delta S_{sys}$$

$$\Delta G^{\circ}_{sys} = \Delta H^{\circ}_{sys} - T\Delta S^{\circ}_{sys}$$

$$\Delta G^{\circ}_{reaction,\,T} = \Delta H^{\circ}_{reaction,\,298\,K} - T\Delta S^{\circ}_{reaction,\,298\,K}$$

Standard molar free energy of formation ($\Delta G^{\circ}_f$)

$$\Delta G^{\circ}_{reaction} = \Sigma\, coeff_{products}\, \Delta G^{\circ}_{products} - \Sigma\, coeff_{reactants}\, \Delta G^{\circ}_{reactants}$$

Reaction quotient ($Q$)

$$\Delta G_{reaction} = \Delta G^{\circ}_{reaction} + RT \ln Q \qquad \frac{(c_D)^d(c_E)^e}{(c_A)^a(c_B)^b} = Q$$

**LEARNING OBJECTIVES:** Predict the direction of spontaneous change using the reaction free energy.

**SKILLS TO MASTER:** Calculating the standard free energy change for a reaction from standard free energies of formation (p. 634)

Calculating the standard free energy change for a reaction from standard enthalpies and entropies of reaction (p. 635)

Calculating a reaction quotient (p. 639)

Calculating the free energy of reaction under non-standard conditions (p. 640)

Predicting the influence of temperature on spontaneity (p. 643)

**KEY CONCEPTS:** The reaction free energy is negative for any spontaneous process at constant temperature and pressure.

A reaction free energy is a function of concentrations (or partial pressures) and of the absolute temperature.

## 11.5 SOME APPLICATIONS OF THERMODYNAMICS

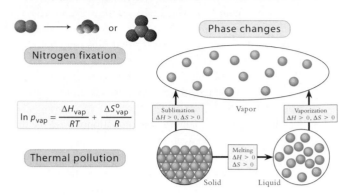

Nitrogen fixation

$$\ln p_{vap} = \frac{\Delta H_{vap}}{RT} + \frac{\Delta S^{\circ}_{vap}}{R}$$

Thermal pollution

Phase changes

Sublimation $\Delta H > 0, \Delta S > 0$   Vapor   Vaporization $\Delta H > 0, \Delta S > 0$

Melting $\Delta H > 0$ $\Delta S > 0$

Solid    Liquid

**LEARNING OBJECTIVES:** Apply thermodynamics to chemical reactions and phase changes.

**SKILLS TO MASTER:** Calculating vapour pressure using the Clausius–Clapeyron equation (p. 649)

Calculating the enthalpy of vapourization from vapour pressure data (p. 652)

## 11.6 BIOENERGETICS

Adenosine triphosphate (ATP)

Coupled reactions

| | | | |
|---|---|---|---|
| ATP + H₂O | → (Enzyme) | ADP + H₃PO₄ | $\Delta G^{\circ} = -$ |
| Reactant + H₃PO₄ | → | Product + H₂O | $\Delta G^{\circ} = +$ |
| Reactant + ATP | → | Product + ADP | $\Delta G^{\circ} = -$ |

**LEARNING OBJECTIVES:** Describe thermodynamically some representative energetic processes that operate in living organisms.

**KEY CONCEPTS:** Adenosine triphosphate is used in the body to store energy.

Cells obtain energy to drive otherwise non-spontaneous reactions via the dephosphorylation of ATP.

## LEARNING EXERCISES

**11.1**  Describe the thermodynamic criteria for spontaneity. Describe in your own words what entropy is and why it is not a conserved quantity.

**11.2**  Make a list of everyday processes that illustrate that matter naturally becomes less constrained.

**11.3**  Write a paragraph that explains why living creatures must convert some stored energy into heat in the course of their activities.

**11.4**  Update your list of memory bank equations. Describe the restrictions on each equation.

**11.5**  Make a list of all terms new to you that appear in Chapter 11. Using your own words, write a one-sentence definition for each. Consult the Glossary if you need help.

## PROBLEMS

**ilw** = interactive learning ware problem. Visit the website at www.wiley.com/canada/olmsted

### Spontaneity

**11.1**  Describe the dispersals and constraints that accompany the following processes: (a) Ocean waves wash away a sand castle. (b) Water and acetone, two liquids, mix to form a homogeneous liquid solution. (c) In the child's game "pick up sticks," a bundle of sticks is dropped to the floor. (d) Water evaporates from a puddle after a summer shower.

**11.2**  Describe the dispersals and constraints accompanying the following processes: (a) A secretary "straightens up" the boss' desk. (b) Wood burns, producing $CO_2$ and $H_2O$ vapours. (c) $I_2$ crystals form as a hot solution of $I_2$ in $CCl_4$ cools. (d) A skilled mechanic reassembles a torn-down engine.

**11.3**  The figure below shows what happens when a drop of ink is added to a beaker of water. Using ideas of dispersal and constraints, explain what is occurring.

●= Water  ●= Ink

**11.4**  "All the king's horses and all the king's men couldn't put Humpty together again." Describe Humpty-Dumpty's fate in terms of constraints and dispersals.

**11.5**  Explain each of the following observations using constraints and dispersals: (a) A puncture causes a tire to deflate. (b) An open bottle of perfume on a table eventually fills the room with the fragrance of the perfume.

**11.6**  Explain each of the following observations using constraints and dispersals: (a) A glass dropped on the floor shatters into many pieces. (b) The wind scatters raked leaves.

**11.7**  Use dispersal notions to explain why the following do not occur without the action of some outside force: (a) The handle of a spoon placed in ice water becomes too hot to handle. (b) Sand on a beach forms a sand castle.

**11.8**  Use dispersal notions to explain why the following do not occur without the action of some outside force: (a) Water runs uphill from the bottom of a dam to its top. (b) Air inflates a flat tire.

### Entropy: The Measure of Dispersal

**11.9**  A tic-tac-toe game contains nine compartments. What is $W$ for all possible first moves by each player (one X and one O placed in two different compartments)?

**11.10**  Consider a box of 16 equal-sized compartments into which two different-coloured marbles are to be placed. Calculate $W$ for placing each in a different compartment and for placing both in the same compartment.

**11.11**  Ice melts at 273.15 K with $\Delta H_{fus} = 6.01$ kJ/mol. An ice cube of mass 13.8 g is dropped into a swimming pool, the temperature of which is held at 27.5 °C. (a) What is the entropy change ($\Delta S$) for melting the ice? (b) What is $\Delta S$ of the pool? (c) What is the overall $\Delta S$?

**11.12**  Solid $CO_2$ (dry ice) sublimes with $\Delta H_{subl} = 25.2$ kJ/mol at 195 K. A block of dry ice of mass 27.5 g sublimes in a room of temperature 26.5 °C. (a) What is the entropy change ($\Delta S$) of the $CO_2$? (b) What is $\Delta S$ of the room? (c) What is the overall $\Delta S$?

**11.13**  Without doing any calculations, determine the signs of $\Delta S$ for the system, for the surroundings, and for the overall process in each of the following: (a) Water boils in a teakettle on a hot stove. (b) Ice in an ice-cube tray, left on a counter top, melts. (c) A cup of coffee is reheated in a microwave oven.

**11.14**  Without doing any calculations, determine the signs of $\Delta S$ for the system, for the surroundings, and for the overall process in each of the following: (a) Ice forms on the surface of a birdbath in winter. (b) A hot cup of coffee cools when left to stand. (c) A Popsicle melts when left on a table.

**11.15**  **ilw**  Calculate the entropy change of 15.5 g of steam that condenses to liquid water at 373.15 K. Without doing additional calculations, what can you say about the entropy change of the surroundings?

**11.16**  Calculate the entropy change of 75.4 g of water that freezes to ice at 273.15 K. Without doing additional calculations, what can you say about the entropy change of the surroundings?

**11.17**  Table 8-5 lists molar enthalpies of fusion of several substances. Calculate the molar entropy of fusion at its normal melting point for each of the following: (a) argon; (b) methane; (c) ethanol; and (d) mercury.

**11.18** Table 8-5 lists molar enthalpies of vapourization of several substances. Calculate the molar entropy of vapourization at its normal boiling point for each of the following: (a) molecular oxygen; (b) ethane; (c) benzene; and (d) mercury.

## Entropies of Pure Substances

**11.19** For each of the following pairs of substances, determine which has the larger molar entropy at 298 K and state the main reason for the difference: (a) NaCl ($aq$) and $MgCl_2$ ($aq$); (b) HgO ($s$) and HgS ($s$); and (c) $Br_2$ ($l$) and $I_2$ ($s$).

**11.20** For each of the following pairs of substances, determine which has the larger molar entropy at 298 K and state the main reason for the difference: (a) $Br_2$ and $Cl_2$; (b) Ni and Pt; (c) $C_5H_{12}$ (pentane, liquid) and $C_8H_{18}$ (octane, liquid); and (d) $SiF_4$ and $CH_4$ (both gases).

**11.21** Oxygen, ozone, and hydrogen are gases at $T = 298$ K, $p = 1$ bar. Their molar entropies are in the sequence $H_2 < O_2 < O_3$. Using molecular properties, explain why ozone has more entropy than oxygen but hydrogen has less entropy than either.

**11.22** Mercury, water, and bromine are liquids at $T = 298$ K, $p = 1$ bar. Their molar entropies are in the sequence $H_2O < Hg < Br_2$. Using molecular properties, explain why bromine has more entropy than mercury but water has the least entropy of these three.

**11.23** Using tabulated values of $S°$, calculate the standard entropy per mole of atoms for He, $H_2$, $CH_4$, and $C_3H_6$. Explain the trend that you find in terms of what bond formation does to atomic constraints.

**11.24** Using tabulated values of $S°$, calculate the standard entropy per mole of atoms for Ar, $O_2$, $O_3$, and $C_2H_6$. Explain the trend that you find in terms of what bond formation does to atomic constraints.

**11.25** Compute the absolute entropy of the following: (a) 2.50 moles of Ar gas at $p = 0.25$ bar; (b) 0.75 mole of $O_3$ gas at $p = 2.75$ bar; and (c) 0.45 mole of a mixture of $N_2$ ($X = 0.78$) and $O_2$ ($X = 0.22$) at $p_{total} = 1.00$ bar.

**11.26** Compute the absolute entropy of the following: (a) 1.00 mole of molecular hydrogen gas at $p = 5.0$ bar; (b) 0.25 mole of ethane gas at $p = 0.10$ bar; and (c) 1.00 mole of a mixture of $N_2$ ($p = 125$ bar) and $H_2$ ($p = 375$ bar).

**11.27** **ilw** Compute the standard entropy change for the following reactions:
(a) $N_2$ ($g$) + 3 $H_2$ ($g$) $\longrightarrow$ 2 $NH_3$ ($g$)
(b) 3 $O_2$ ($g$) $\longrightarrow$ 2 $O_3$ ($g$)
(c) $PbO_2$ ($s$) + 2 Ni ($s$) $\longrightarrow$ Pb ($s$) + 2 NiO ($s$)
(d) $C_2H_4$ ($g$) + 3 $O_2$ ($g$) $\longrightarrow$ 2 $CO_2$ ($g$) + 2 $H_2O$ ($l$)

**11.28** Compute the standard entropy change for the following reactions:
(a) 2 $H_2$ ($g$) + $O_2$ ($g$) $\longrightarrow$ 2 $H_2O$ ($l$)
(b) C ($s$) + 2 $H_2$ ($g$) $\longrightarrow$ $CH_4$ ($g$)

(c) $CH_3CH_2OH$ ($l$) + 3 $O_2$ ($g$) $\longrightarrow$ 2 $CO_2$ ($g$) + 3 $H_2O$ ($l$)
(d) $Fe_2O_3$ ($s$) + 2 Al ($s$) $\longrightarrow$ $Al_2O_3$ ($s$) + 2 Fe ($s$)

**11.29** For each reaction in Problem 11.27, explain what features of the atomic constraints for the reactants and products account for the magnitude and sign of $\Delta S°$.

**11.30** For each reaction in Problem 11.28, explain what features of the atomic constraints for the reactants and products account for the magnitude and sign of $\Delta S°$.

**11.31** Without doing calculations or looking up absolute entropy values, determine the sign of $\Delta S°$ for the following processes:
(a) $CO_2$ ($s$) $\longrightarrow$ $CO_2$ ($g$)
(b) 2 Na ($s$) + $F_2$ ($g$) $\longrightarrow$ 2 NaF ($s$)
(c) 2 CO ($g$) + $O_2$ ($g$) $\longrightarrow$ 2 $CO_2$ ($g$)

**11.32** Without doing calculations or looking up absolute entropy values, determine the sign of $\Delta S°$ for the following processes:
(a) $MgCl_2$ ($s$) $\longrightarrow$ $Mg^{2+}$ ($aq$) + 2 $Cl^-$ ($aq$)
(b) $SF_6$ ($g$) $\longrightarrow$ $SF_6$ ($l$)
(c) 2 $CH_3OH$ ($l$) + 3 $O_2$ ($g$) $\longrightarrow$ 2 $CO_2$ ($g$) + 4 $H_2O$ ($g$)

## Spontaneity and Free Energy

**11.33** Each of the following statements is false. Rewrite each so that it makes a correct statement about free energy:
(a) $\Delta G_{total} > 0$ for any spontaneous process.
(b) $\Delta G_{system}$ increases in any process at constant $T$ and $p$. (c) $\Delta H = \Delta G - T \Delta S$.

**11.34** Each of the following statements is false. Rewrite each so that it makes a correct statement about free energy:
(a) In any process at constant $T$ and $p$, the overall free energy decreases. (b) $\Delta G_{sys} = 0$ for any spontaneous process. (c) $\Delta G = \Delta H + \Delta S$.

**11.35** Compute the standard free energy change for each reaction in Problem 11.27.

**11.36** Compute the standard free energy change for each reaction in Problem 11.28.

**11.37** Estimate $\Delta G°$ at $-85°C$ for the reactions in Problem 11.27(b) and 11.27(c).

**11.38** Estimate $\Delta G°$ at $825°C$ for the reactions in Problem 11.28(b) and 11.28(d).

**11.39** Calculate $\Delta G_{reaction}$ at 298 K for the reactions in Problem 11.27(a) and 11.27(b) if $p = 15$ bar for each gaseous substance.

**11.40** Calculate $\Delta G_{reaction}$ at 298 K for the reactions in Problem 11.28(a) and 11.28(b) if $p = 0.25$ bar for each gaseous substance.

## Some Applications of Thermodynamics

**11.41** **ilw** Compute $\Delta H°$, $\Delta S°$, and $\Delta G°$ for the production of $NH_4NO_3$ from ammonia and oxygen:

2 $NH_3$ ($g$) + 2 $O_2$ ($g$) $\longrightarrow$ $NH_4NO_3$ ($s$) + $H_2O$ ($l$)

(This reaction is not feasible industrially, because $NH_3$ combustion cannot be controlled to give $NH_4NO_3$ as a product.)

**11.42** Compute $\Delta H°$, $\Delta S°$, and $\Delta G°$ for the production of urea from ammonia (see Figure 11-17).

**11.43** Without doing any calculations, answer the following for the formation of $NO_2$ from $N_2$ and $O_2$, and state your reasons: (a) Is there any temperature at which the reaction is spontaneous when all reagents are at 1 bar pressure? (b) At 298 K, is there any combination of partial pressures of the reagents that would make the reaction spontaneous?

**11.44** Without doing any calculations, answer the following for the formation of $ClO_2$ from $Cl_2$ and $O_2$, and state your reasons: (a) Is there any temperature at which the reaction is spontaneous when all reagents are at 1 bar pressure? (b) At 298 K, is there any combination of partial pressures of the reagents that would make the reaction spontaneous?

**11.45** **ilw** Using data from Appendix D, calculate the vapour pressure of $P_4$ (g) over solid white phosphorus at 298 K.

**11.46** Using data from Appendix D, calculate the vapour pressure of Hg (g) over liquid mercury at 298 K.

**11.47** Octane has a vapour pressure of 0.468 bar at 100 °C, and an enthalpy of vapourization of 50.1 kJ/mol. Calculate the normal boiling point of octane.

**11.48** Dichloromethane has a vapour pressure of 10 Torr at −43.3 °C and 40 Torr at −22.3 °C. Calculate its enthalpy of vapourization in kJ/mol.

**11.49** The enthalpy of vapourization is not actually constant, but rather is slightly dependent on temperature. Use the following data to calculate the enthalpy of vapourization of water between 10 °C and 20 °C. The value at the boiling point is 40.7 kJ/mol. Why is the value you calculated higher?

**11.50** Liquid hydrogen has a vapour pressure of 0.216 bar at 16 K and an enthalpy of vapourization of 0.46 kJ/mol. Calculate its vapour pressure at 17 K.

## Bioenergetics

**11.51** Two children on opposite ends of a seesaw can be used as an analogy for a coupled reaction. Describe the coupling of spontaneous and non-spontaneous processes during the actions of a seesaw.

**11.52** A pair of weights and a pulley can be used as an analogy for a coupled reaction. Use the figure to describe a coupled process. Identify spontaneous and non-spontaneous portions of the process.

**11.53** Glucose and fructose combine to make sucrose in a condensation reaction for which $\Delta G° = 23.0$ kJ/mol:

$$C_6H_{12}O_6 + C_6H_{12}O_6 \longrightarrow C_{12}H_{22}O_{11} + H_2O$$
$$\text{Glucose} \qquad \text{Fructose} \qquad\qquad \text{Sucrose}$$

Write the balanced equation for this reaction coupled with the ADP–ATP reaction and operating in the direction of overall spontaneity. Calculate $\Delta G°$ for the overall process.

**11.54** Although the ATP–ADP reaction is the principal energy shuttle in metabolic pathways, many other examples of coupled reactions exist. For example, the glutamic acid–glutamine reaction discussed in the text can couple with the acetyl phosphate reaction shown in Example 11-13. Write the balanced equation for the coupled reaction operating in the direction of overall spontaneity and calculate $\Delta G°$ for the overall process.

## ADDITIONAL PAIRED PROBLEMS

**11.55** For the chemical reaction

$$Al_2O_3 \text{ (s)} + 3 H_2 \text{ (g, } p = 1 \text{ bar)} \longrightarrow 2 \text{ Al (s)} + 3 H_2O \text{ (l)}$$

at $T = 298$ K, answer each of the following by doing a quantitative calculation of the appropriate thermodynamic function: (a) Is this a spontaneous reaction? (b) Does the reaction absorb or release heat? (c) Are the products more or less dispersed than the reactants?

**11.56** For the chemical reaction

$$3 \text{ Fe (s)} + 4 H_2O \text{ (l)} \longrightarrow Fe_3O_4 \text{ (s)} + 4 H_2 \text{ (g, } p = 1 \text{ bar)}$$

at $T = 298$ K, answer each of the following by doing a quantitative calculation of the appropriate thermodynamic function: (a) Is this a spontaneous reaction? (b) Does the reaction absorb or release heat? (c) Are the products more or less dispersed than the reactants?

**11.57** Use data from Appendix D to compute $\Delta S°$ for each of the following reactions:
(a) $2 ClO_2^- \text{ (aq)} + O_2 \text{ (g)} \longrightarrow 2 ClO_3^- \text{ (aq)}$
(b) $4 FeCl_3 \text{ (s)} + 3 O_2 \text{ (g)} \longrightarrow 2 Fe_2O_3 \text{ (s)} + 6 Cl_2 \text{ (g)}$
(c) $3 N_2H_4 \text{ (l)} + 4 O_3 \text{ (g)} \longrightarrow 6 \text{ NO (g)} + 6 H_2O \text{ (l)}$

**11.58** Calculate the standard entropy change at 298 K of each of the following reactions, which are important in the chemistry of coal. Assume that coal has the same thermodynamic properties as graphite.
(a) C (s, coal) + $H_2O$ (g) $\longrightarrow$ CO (g) + $H_2$ (g)
(b) C (s, coal) + $O_2$ (g) $\longrightarrow$ $CO_2$ (g)
(c) C (s, coal) + $\frac{1}{2} O_2$ (g) $\longrightarrow$ CO (g)
(d) CO (g) + $H_2O$ (g) $\longrightarrow$ $CO_2$ (g) + $H_2$ (g)

**11.59** Use data from Appendix D to compute $\Delta G°$ for each reaction in Problem 11.57.

**11.60** Use data from Appendix D to compute $\Delta G°$ for each reaction in Problem 11.58.

**11.61** Use data from Appendix D to compute $\Delta G°$ at 85.0°C for each reaction in Problem 11.57.

**11.62** Use data from Appendix D to compute $\Delta G°$ at 200.0°C for each reaction in Problem 11.58.

**11.63** For the following constant-temperature process, give the sign (+, −, or 0) for each of the specified thermodynamic functions. In each case, give a brief account of your reasoning: (a) $w_{sys}$; (b) $q_{sys}$; and (c) $\Delta S_{surr}$. (*Hint:* Review Chapter 3.)

**11.64** For the following constant-temperature process, give the sign (+, −, or 0) for each of the specified thermodynamic functions. In each case, give a brief account of your reasoning: (a) $q_{sys}$; (b) $\Delta S_{sys}$; and (c) $\Delta E_{total}$. (*Hint:* Review Chapter 3.)

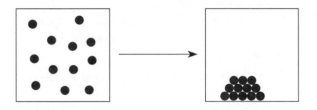

**11.65** One possible source of acid rain is the reaction between $NO_2$, a pollutant from automobile exhausts, and water:

$$3 \, NO_2 \, (g) + H_2O \, (l) \longrightarrow 2 \, HNO_3 \, (g) + NO \, (g)$$

Determine whether this is thermodynamically feasible (a) under standard conditions at 298 K; (b) under standard conditions at 277 K; and (c) at 298 K, with each product gas present at $p = 1.00 \times 10^{-6}$ bar.

**11.66** Filaments of elemental boron might be made by reducing boron trichloride with molecular hydrogen:

$$2 \, BCl_3 \, (g) + 3 \, H_2 \, (g) \longrightarrow 2 \, B \, (s) + 6 \, HCl \, (g)$$

Determine whether this is thermodynamically feasible (a) under standard conditions at 298 K; (b) under standard conditions at 995 K; and (c) at 995 K, if each reactant gas is present at $p = 5.5$ bar and HCl pressure is 0.1 bar.

**11.67** ATP and ADP have been referred to as the "energy currency" of the cell. Explain this analogy.

**11.68** Explain why a nuclear power plant must contribute thermal pollution to the environment.

**11.69** Use data from Appendix D to answer quantitatively the following questions about the dissolving of table salt in water under standard conditions:

$$NaCl \, (s) \longrightarrow Na^+ \, (aq) + Cl^- \, (aq)$$

(a) Is this a spontaneous reaction? (b) Does it release energy? (c) Does the chemical system become more constrained?

**11.70** Repeat the calculations of Problem 11.69 for silver chloride, AgCl, dissolving in water. What do the results reveal about the difference in solubility of these two salts and the reasons for the difference?

**11.71** An ice-cube tray containing 155 g of liquid $H_2O$ at 0.0 °C is placed in a freezer of temperature −20.0 °C. As soon as all the $H_2O$ has frozen, the tray is removed from the freezer. (a) Find $\Delta S$ for the $H_2O$. (b) Find $S$ for the overall process. (c) If the tray were left in the freezer until its temperature reached −20°C, would there be an additional net entropy change? Explain.

**11.72** Find $\Delta S$ for the system, surroundings, and overall when 25.0 g of liquid $H_2O$ is evaporated at 100. °C, if the heat required is provided by a hot plate of temperature 315 °C.

**11.73** Although ammonia can be used directly as a fertilizer, much of it is converted to urea or ammonium nitrate before being applied in the field. Considering the physical properties of these substances, suggest why ammonia is not preferred as a fertilizer.

**11.74** What are the four most important nitrogen-containing chemicals, and for what are they used?

**11.75** Determine $\Delta G_{combustion}$ per gram of palmitic acid ($C_{16}H_{32}O_2$, $\Delta G°_{combustion} = -9790$ kJ mol$^{-1}$) and of glucose ($C_6H_{12}O_6$, $\Delta G°_{combustion} = -2870$ kJ mol$^{-1}$). On a per-gram basis, which is the better energy source?

**11.76** The most important commercial process for generating hydrogen gas is the water − gas shift reaction:

$$CH_4 \, (g) + H_2O \, (g) \longrightarrow CO \, (g) + 3 \, H_2 \, (g)$$

Use tabulated thermodynamic data to find $\Delta G°$ and $\Delta G°_{1300K}$.

**11.77** Without doing any calculations, determine the sign of $\Delta S$ for the system for each of the following changes: (a) A soft drink is chilled in an ice chest. (b) The air in a bicycle pump is compressed. (c) A carton of juice concentrate is mixed with water.

**11.78** Without doing any calculations, determine the sign of $\Delta S$ for the system for each of the following changes: (a) A cold soft drink warms as it sits in the sun. (b) Air leaks out of a punctured automobile tire. (c) Solid AgCl precipitates from solution when silver nitrate solution is mixed with sodium chloride solution.

**11.79** A teaspoon from a freezer is placed in a glass of water, and the two equilibrate to the same temperature. In this process, what can you deduce about each of the following? (a) $\Delta E_{total}$; (b) $\Delta E_{teaspoon}$; (c) $\Delta S_{total}$; (d) $\Delta S_{water}$; and (e) $q_{teaspoon}$.

**11.80** A pie is removed from a hot oven and allowed to cool to room temperature. In this process, what can you deduce about the changes in each of the following? (a) $\Delta E_{total}$; (b) $\Delta E_{pie}$; (c) $\Delta S_{total}$; (d) $\Delta S_{pie}$; and (e) $q_{pie}$.

**11.81** Arrange the following in order of increasing entropy, from smallest to largest value: 1.0 mol $H_2O$ (liquid, 373 K); 0.50 mol $H_2O$ (liquid, 298 K); 1.0 mol $H_2O$ (liquid, 298 K); 1.0 mol $H_2O$ (gas, 373 K, 1.0 bar); and 1.0 mol $H_2O$ (gas, 373 K, 0.1 bar).

**11.82** Arrange the following in order of increasing entropy, from smallest to largest value: 1.00 g $Br_2$ (gas, 331.9 K, 0.10 bar); 1.00 g $Br_2$ (gas, 331.9 K, 1.00 bar); 1.00 g $Br_2$ (liquid, 331.9 K, 1.00 bar); and 1.00 g Br atoms (gas, 331.9 K, 0.10 bar).

**11.83** One reaction that generates pollutants in the internal combustion engine is the oxidation of nitrogen by oxygen:

$$N_2 \, (g) + O_2 \, (g) \longrightarrow 2 \, NO \, (g)$$

Determine $\Delta H°$, $\Delta S°$, and $\Delta G°$ for this reaction at room temperature, and estimate the temperature above which it becomes spontaneous at standard pressure. Given this result, explain how NO can form in automobile engines.

**11.84** Advertisements that state "diamonds are forever" assume that the reaction, C (diamond) $\longrightarrow$ C (graphite) is not spontaneous. Determine $\Delta H°$, $\Delta S°$, and $\Delta G°$ for this reaction at room temperature. Is there a temperature at which this reaction is not spontaneous? If so, determine that temperature.

**11.85** The normal boiling point of benzene is 80.1 °C, at which $\Delta H°_{vap} = 30.77$ kJ/mol. (a) Determine $\Delta S°_{vap}$ and $\Delta G°_{vap}$ for benzene at its normal boiling point. (b) Estimate $\Delta G°_{vap}$ for benzene at 21 °C. (c) Calculate the vapour pressure of benzene at 25 °C.

**11.86** The normal boiling point of ethane is 184 K, at which $\Delta H°_{vap} = 15.5$ kJ/mol. (a) Determine $\Delta S°_{vap}$ and $\Delta G°_{vap}$ for ethane at its normal boiling point. (b) Estimate $\Delta G°_{vap}$ for ethane at 21 °C. (c) Calculate the vapour pressure of ethane at −65 °C.

## ADDITIONAL PROBLEMS

**11.87** A sample of liquid $Br_2$ (0.080 mol) at 332 K is placed in a cylinder equipped with a piston at $P_{ext} = 1.0$ bar. The cylinder is immersed in a constant-temperature bath at 332 K, the boiling point of $Br_2$ ($\Delta H_{vap} = 30.9$ kJ/mol). The system undergoes a constant-temperature, constant-pressure expansion to a final volume of 2.2 L. The sketch shows the apparatus immediately after the cylinder is placed in the bath. (a) Draw a sketch that shows how the system looks when it reaches its final state. (b) Determine the total S for this process.

$P_{ext} = 1.0$ atm

Constant temperature bath at 332 K

**11.88** The notion of thermodynamic coupling of a non-spontaneous process with a spontaneous process is not restricted to chemical reactions. Identify the spontaneous and non-spontaneous portions of the following coupled processes: (a) Water behind a dam passes through a turbine and generates electricity. (b) A gasoline engine pumps water from a valley to the top of a hill.

**11.89** Crystalline KCl has $S° = 83$ J mol$^{-1}$ K$^{-1}$, and crystalline CaO has $S° = 55$ J mol$^{-1}$ K$^{-1}$. What accounts for the larger entropy of KCl crystals?

**11.90** Both $CCl_4$ (carbon tetrachloride) and $CS_2$ (carbon disulphide) are liquids used as solvents in special industrial applications. (a) Using data from Appendix D, calculate $\Delta H°$ and $\Delta G°$ for combustion of these liquids:

$$CCl_4 \, (l) + 5 \, O_2 \, (g) \longrightarrow CO_2 \, (g) + 4 \, ClO_2 \, (g)$$

$$CS_2 \, (l) + 3 \, O_2 \, (g) \longrightarrow CO_2 \, (g) + 2 \, SO_2 \, (g)$$

(b) What features of the reactants and products account for the differences? (c) Based on your results, would you recommend special precautions against fires for industrial plants using either solvent? Explain your recommendations.

**11.91** In a system operating without any restrictions, heat is not a state function. However, heat flow describes a state function change under certain restricted conditions. For each of the following conditions, identify the state function that corresponds to heat flow: (a) $q_V$; (b) $q_P$; and (c) $q_T$.

**11.92** A piece of dry ice at $T = 195$ K is dropped into a beaker containing water at $T = 273.15$ K. You may assume that all of the resulting $CO_2$ vapour escapes into the atmosphere. (a) Describe the process that takes place at the molecular level. (b) Calculate the overall entropy change if 12.5 g of dry ice undergoes this process.

**11.93** At its triple point, a dynamic equilibrium can exist among all three phases of matter. Draw a molecular picture of argon that shows what happens at the triple point. What is $\Delta G$ for each of the processes under these conditions? Describe the matter and energy dispersal taking place for each of the processes.

**11.94** The following diagram represents two flasks connected by a valve. Each flask contains a different gas. (a) If the valve is opened, the two gases can move back and forth between the two flasks. Redraw the figure to show the system at maximum entropy. (b) Redraw the figure in a way that shows the state of the system at its lowest possible entropy. (The substances need not remain gaseous.)

Valve

**11.95** Without doing any calculations, predict the signs (+, −, or 0) of $\Delta H$ and $\Delta S$ for the following processes, all occurring at constant $T$ and $p$. Explain your predictions.
(a) $2\, NO_2\,(g) \longrightarrow N_2O_4\,(g)$
(b) A block of gold melts in a jeweller's crucible
(c) $CH_4\,(g) + 2\, O_2\,(g) \longrightarrow CO_2\,(g) + 2\, H_2O\,(g)$

**11.96** Heat pumps extract heat from a cold reservoir (usually the ground) and use it to maintain the temperature of a warmer reservoir (e.g., a home). Explain why heat pumps cannot operate without converting some additional energy into heat.

**11.97** For the ATP $\longrightarrow$ ADP reaction, $\Delta G° = -30.6$ kJ/mol. Under cellular conditions, this reaction can release more than 50 kJ/mol. List the ways in which cellular conditions differ from standard conditions and describe the effect that each difference has on the free energy change.

**11.98** On a hot day, one of your friends suggests opening the door of your refrigerator to cool your kitchen. Will this strategy work? Explain your answer using principles of thermodynamics.

**11.99** What is the efficiency of the metabolic conversion of palmitic acid to ATP? Compute the number of grams of palmitic acid that would have to be metabolized to provide the heat to warm a swimmer from whose skin 75 g of water evaporates.

**11.100** Few of the elements exist on the Earth in their pure form. Most elements exist as oxides or sulphides. Explain why this is so, and explain why nitrogen exists as a pure element.

**11.101** A firefly produces light by converting the compound luciferin to oxyluciferin:

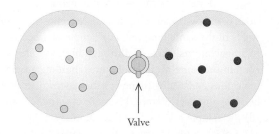

Luciferin + ATP ⟶

Luciferyl adenylate + Diphosphate + $H_2O$

+ $O_2$ ⟶

Oxyluciferin + AMP + $CO_2$ + light

Answer the following for each of these processes: (a) Is energy absorbed or released? (b) Does system entropy increase, decrease, or remain nearly the same? (c) Is $\Delta G$ negative or positive? Explain your reasoning.

**11.102** Humans perspire as a way of keeping their bodies from overheating during strenuous exercise. The evaporation of perspiration transfers heat from the body to the surrounding atmosphere. Calculate the total $\Delta S$ for evaporation of 1.0 g of water if the skin is at 37.5°C and air temperature is 23.5°C.

**11.103** Two moles of HCl gas have a larger standard entropy, 374 J mol$^{-1}$ K$^{-1}$, than 1 mole each of $H_2$ gas and $Cl_2$ gas, 354 J mol$^{-1}$ K$^{-1}$. What feature of the molecular structures of these substances accounts for this difference?

## GROUP STUDY PROBLEMS

**11.104** The enthalpy of sublimation of ice at 273.15 K is not the simple sum of the enthalpies of fusion and vapourization of water, but it can be calculated using Hess' law and an appropriate path that includes fusion and vapourization. Devise such a path, show it on a phase diagram for water, and carry out the calculation, making reasonable assumptions if necessary ($C$(liquid water) = 75.3 J mol$^{-1}$ K$^{-1}$, and $C$(water vapour) = 33.6 J mol$^{-1}$ K$^{-1}$).

**11.105** In the upper atmosphere, ozone is produced from oxygen:

$$3\, O_2\,(g) \longrightarrow 2\, O_3\,(g)$$

(a) Compute $\Delta H°$, $\Delta S°$, and $\Delta G°$ for this reaction. (b) Is there a temperature at which this reaction becomes spontaneous at 1 bar pressure? If so, find it. If not, explain why one does not exist. (c) Assume an atmosphere with $p_{O2}$ = 0.20 bar and $T$ = 298 K. Below what pressure of $O_3$ is $O_3$ production spontaneous? (d) In view of your answers to parts (a) through (c), how can the ozone layer form? (You may need to review Section 4.7).

**11.106** $S°$ of graphite is 3 times larger than $S°$ of diamond. Explain why this is so. (You may need to review the structures and properties of graphite and diamond in Section 8.4.) Buckminsterfullerene is a solid that consists of individual molecules with formula $C_{60}$. Is the molar entropy of buckminsterfullerene larger or smaller than that of graphite? How about the entropy per gram? Explain.

**11.107** Here are thermodynamic data for the fusion of $NH_3$:

$$NH_3 \ (s) \longrightarrow NH_3 \ (l) \quad \Delta H° = 5.65 \text{ kJ/mol,}$$
$$\Delta S° = 28.9 \text{ J mol}^{-1} \text{ K}^{-1}$$

(a) Calculate $\Delta G°$ for the melting of 1.00 mole of $NH_3$ at 298 K.

(b) Calculate the freezing point of $NH_3$.

**11.108** Construct a table of the absolute entropies of all the gaseous substances listed in Appendix D, arranged in order of increasing $S°$. Use the general trends described in the text to explain the order of entries in this table. Are there substances that do not appear to fit the general trends? If so, suggest reasons why their $S°$ values are higher or lower than expected.

**11.109** Phosphorus forms white crystals made up of $P_4$ molecules. There are two forms of white crystalline phosphorus, called α and β. The difference between $P_{4\alpha}$ and $P_{4\beta}$ is determined by the way the $P_4$ molecules pack together in the crystal lattice. The form is always obtained when liquid phosphorus freezes. However, at temperatures below $-77°C$, the $P_{4\alpha}$ crystals change spontaneously to $P_{4\beta}$:

$$P_{4\alpha} \xrightarrow{-77°C} P_{4\beta}$$

Answer the following, and explain your reasoning: (a) Which form of phosphorus has a more constrained crystalline structure? (b) What are the signs of $\Delta H$ and $\Delta S$ for this process? (c) Would the entropy of $P_{4\alpha}$ crystals be 0 at 0 K?

## ANSWERS TO SECTION EXERCISES

**11.1.1** (a) A fence is constrained; when it falls down, it becomes more dispersed; (b) a plate has a relatively constrained structure, which becomes dispersed when the plate breaks; and (c) forming a solution from pure substances involves dispersal of the components throughout the solution.

**11.1.2** Constraining one substance (formation of sugar crystals) must be accompanied by even greater

dispersal elsewhere. In this example, the water molecules that boil away are much more dispersed than water molecules in the liquid.

**11.1.3**

**11.2.1** Heat will flow from the warmer mixture to the cooler mixture, so benzene will freeze and ice will melt.

**11.2.2** 5.01 J/K

**11.2.3** $\Delta S_{\text{mixture}} = -6.11$ J/K, $\Delta S_{\text{refrigerator}} = 6.11$ J/K, $\Delta S_{\text{total}} = 0$; neither process is spontaneous.

**11.3.1** (a) 1 g of dew because liquids are less constrained than solids of the same material; (b) 1 mole of gaseous hydrogen atoms, because although both samples contain the same number of atoms, binding atoms into molecules constrains them; (c) the flawed diamond, because a perfect diamond is more constrained; and (d) the sample at 50°C, because increasing the temperature of a sample increases its molecular motion and makes it less constrained.

**11.3.2** (a) There are more atoms in the sample of $O_3$, so there is more dispersal of matter in the molecular arrangement; (b) both samples have the same number of atoms, but there are more molecules in 3 moles of $O_2$, so there is more dispersal of matter in its molecular arrangement; (c) $I_2$ is a solid under standard conditions, so it is more constrained than gaseous $O_2$; and (d) diluting a solution spreads its solute particles (ions in this case) over a larger region, giving it greater dispersal.

**11.3.3** $-2906.4$ J/K. Entropy is negative as 33 moles of gas are converted to 12 moles of gas and 12 moles of liquid, less dispersal of energy as number of moles is reduced and gas is converted to liquid.

**11.4.1** $-225$ kJ/mol

**11.4.2** Yes, $T = 5.3 \times 10^3$ K

**11.4.3** $2.7 \times 10^{-3}$ bar

**11.5.1** 340 kJ

**11.5.2** When an automobile engine moves an automobile, it does work and constrains energy (all parts of the automobile move together in the same direction). To do this without violating the second law of thermodynamics, enough heat must be transferred to the surroundings to disperse a greater amount of energy.

**11.6.1** 92%

**11.6.2** $+32.2$ J mol$^{-1}$ K$^{-1}$; the spontaneity increases as temperature increases.

**11.6.3** (a) about 16 moles of ATP; and (b) about 80 g of glucose

## ANSWERS TO PRACTICE EXERCISES

**11.1**   $\Delta S_{water} = 388$ J/K, $\Delta S_{hot\ plate} = -277$ J/K, and
$\Delta S_{total} = 111$ J/K

**11.2**   $\Delta S_{total} = 23$ J/K

**11.3**   (a) $CH_4$, lower $M$; (b) $H_2$, fewer atoms; (c) $H_2O$, liquid

**11.4**   $-715$ J/K

**11.5**   $-242.9$ J/K

**11.6**   $-818$ kJ; (b) $-818$ kJ

**11.7**   $-791$ kJ (This estimate is only approximate, inasmuch as
the temperature is much higher than 298 K.)

**11.8**   0.57 bar

**11.9**   96.4 Torr

**11.10**   $p_{vap} = 175$ Torr

**11.11**   $T = 91.0\ °C$

**11.12**   $\Delta H_{vap} = 31.6$ kJ/mol

**11.13**   $ATP + H_2O \longrightarrow ADP + H_3PO_4$     $\Delta G° = -30.6$ kJ

$$Glucose + H_3PO_4 \longrightarrow Glucose\ phosphate + H_2O$$
$$\Delta G° = 13.9\ kJ$$

# KINETICS: MECHANISMS AND RATES OF REACTIONS

## THE OZONE PROBLEM

Events that take place on a grand scale often can be traced to the molecular level. An excellent example is the depletion of the ozone layer in the Earth's stratosphere. The so-called ozone hole was first observed above the Antarctic in the 1980s and is now being observed above both the Arctic and Antarctic poles. The destruction of ozone in the stratosphere is caused primarily by reactions between chlorine atoms and ozone molecules, as depicted in our molecular inset view.

The chlorine atoms in the upper atmosphere come from the breakdown of $CF_2Cl_2$ and other similar chlorofluorocarbons (CFCs), known commercially as Freons. Production of these compounds was more than one million tonnes in 1988, largely for use in refrigerators and air conditioners. Once released into the atmosphere, CFCs diffuse slowly upward until they reach the ozone layer. There, ultraviolet light from the Sun splits off chlorine atoms. These react with ozone, with dramatic results. Annual ozone decreases have exceeded 50% above Antarctica. The background

photo shows the Antarctic hole (blue-violet) on September 24, 2003. This area of low ozone concentration almost disappears later in the year, only to reappear the following year. A complex set of physical and chemical mechanisms account for this behaviour, but there is no doubt about chlorine's central role.

Ozone in the stratosphere removes high-energy ultraviolet radiation coming from the Sun. Because this radiation is lethal, loss of ozone in the stratosphere imperils life on Earth. This danger led world nations to agree to ban CFC production by 1996 according to the "Montreal Protocol." Despite dramatic reductions in the use of CFCs, these and other ozone-depleting pollutants remain in the atmosphere. The Antarctic ozone hole was so large in the year 2000 that populated regions of Chile and Australia were exposed to significantly increased amounts of ultraviolet radiation.

The ozone hole would almost certainly be much worse if chemists had not studied the reactions of CFCs with atmospheric gases before ozone depletion was dis-

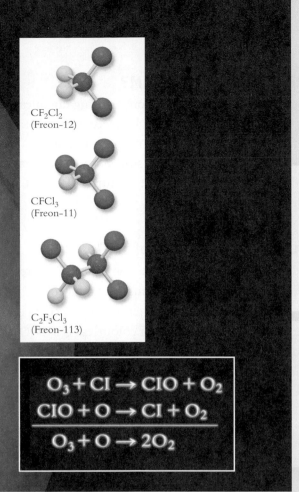

CF$_2$Cl$_2$
(Freon–12)

CFCl$_3$
(Freon–11)

C$_2$F$_3$Cl$_3$
(Freon–113)

$$O_3 + Cl \rightarrow ClO + O_2$$
$$ClO + O \rightarrow Cl + O_2$$
$$\overline{O_3 + O \rightarrow 2O_2}$$

# 12

covered. The 1995 Nobel Prize in chemistry was awarded to the three pioneers in this effort. A Dutch chemist, Paul Crutzen, discovered how ozone concentration is regulated in a normal stratosphere, and two Americans, F. Sherwood Rowland and Mario Molina, showed that CFCs can destroy ozone. These studies of molecular reactions allowed quick determination that CFCs are a likely cause of ozone depletion and led to the international restrictions described above.

The story of the ozone hole illustrates how important it is to learn the molecular details of chemical reactions. Some chemists use information about how reactions occur, to design and synthesize useful new compounds. Others explore how to modify reaction conditions to minimize the cost of producing industrial chemicals. This chapter explores how chemical reactions occur at the molecular level. We show how to describe a reaction from the molecular perspective, introduce the basic principles that govern these processes, and describe some experimental methods used to study chemical reactions.

## CHAPTER CONTENTS

## LEARNING OBJECTIVES

*Upon completion of this chapter you should be able to*

- explain the concepts of a mechanism and a rate-determining step in a chemical reaction
- determine the rate of a reaction based on the rate of change of concentration of a reactant or product
- determine the rate law, given the mechanism and knowledge of the relative rates of steps of a reaction
- determine the rate of a reaction based on the rate of change of concentration of a reactant or product
- show that the mechanism and rate law are closely related
- explain and quantify the effects of temperature on a reaction rate
- explain the mechanisms by which catalysts function

**Figure 12-1**

When a flask containing reddish-brown $NO_2$ gas at room temperature (*top*) is cooled in a low-temperature bath (*bottom*), $NO_2$ molecules combine to form colourless $N_2O_4$ molecules.

## 12.1  WHAT IS A REACTION MECHANISM?

Atoms, ions, and molecules rearrange and recombine during chemical reactions. These processes usually do not occur all at once. Instead, each reaction consists of a sequence of molecular events called a **reaction mechanism**.

We introduce the principles of mechanisms using simple chemical reactions. After presenting the principles, we will be able to return to the more complex mechanisms involved in the ozone problem.

## Example of a Mechanism: Formation of $N_2O_4$

Nitrogen dioxide, a red-brown gas, is stable at room temperature. However, Figure 12-1 shows that reducing the temperature causes a sample of $NO_2$ to become colourless. The colour change that happens at low temperature occurs because two $NO_2$ molecules combine to form one molecule of a colourless gas, $N_2O_4$.

The mechanism of this reaction describes what happens at the molecular level. Two $NO_2$ molecules collide, forming a collision complex. In this collision complex, a bond may form between the two nitrogen atoms, producing an $N_2O_4$ molecule.

The formation of $N_2O_4$ requires more than a random collision between two $NO_2$ molecules. The product contains a bond between the nitrogen atoms, so the collision must form a collision complex that brings the two nitrogen atoms into contact. A collision in which the nitrogen atom of one $NO_2$ molecule strikes an oxygen atom of a second will not lead to $N_2O_4$.

The Lewis structure of $NO_2$ indicates that the nitrogen atom has an unpaired electron. Two $NO_2$ molecules combine by using their unpaired electrons to form an N—N bond.

## Elementary Reactions

A mechanism is a description of the actual molecular events that occur during a chemical reaction. Each such event is an **elementary reaction**. Elementary reactions involve one, two, or occasionally three reactant molecules or atoms. In other words, elementary reactions can be unimolecular, bimolecular, or termolecular. A typical mechanism consists of a sequence of elementary reactions. Although an *overall* reaction describes the starting materials and final products, it is usually not *elementary* because it does not represent the individual steps by which the reaction occurs.

The most common type of elementary process is a **bimolecular reaction** that results from the collision of two molecules, atoms, or ions. The collision of two $NO_2$ molecules to give $N_2O_4$ is a bimolecular reaction. Here is another example:

$$OH^- \quad + \quad H_3O^+ \qquad\qquad Collision \qquad\qquad H_2O \quad + \quad H_2O$$

Notice that the characteristic feature of a bimolecular elementary reaction is a collision between two species, giving a collision complex that results in a rearrangement of chemical bonds. The two reaction partners react to form a new bond ($N_2O_4$ forming from two $NO_2$ molecules), or they form two new species by transferring one or more atoms or ions from one partner to another (2 $H_2O$ forming from $OH^-$ and $H_3O^+$).

In a **unimolecular reaction**, a single molecule fragments into two pieces or rearranges to a new isomer. A simple example of a unimolecular reaction is the decomposition of $N_2O_4$. An $N_2O_4$ molecule continually vibrates, and these vibrations cause its N—N bond to stretch. If the molecule has sufficient energy, the bond breaks and the molecule separates into two molecules of $NO_2$, much like a spring that breaks if it is stretched too far.

$$N_2O_4 \qquad\qquad\qquad\qquad\qquad NO_2 \quad NO_2$$

A stable molecule such as $N_2O_4$ does not decompose unless it first acquires the energy needed to break one of its bonds. Recall from Section 3-1 that any collection of molecules has a distribution of kinetic energies. Molecular collisions can convert kinetic energy of one molecule into vibrational energy of another molecule, giving some molecules enough energy to break apart.

In a **termolecular reaction**, three chemical species collide simultaneously. Termolecular reactions are rare because they require a collision of three species at the *same time* and in exactly the right orientation to form products. The odds against such a simultaneous three-body collision are high. Instead, processes involving three species usually occur in two-step sequences. In the first step, two molecules collide and form a collision complex. In a second step, a third molecule collides with the complex before it breaks apart. Most chemical reactions, including all those introduced in this book, can be described at the molecular level as sequences of bimolecular and unimolecular elementary reactions.

To summarize, the mechanism of a reaction converts starting materials to products through a specific sequence of bimolecular collisions and unimolecular rearrangements.

## Alternative Mechanisms

Very few chemical reactions are as simple as the low-temperature reaction of $NO_2$ to form $N_2O_4$. For example, at high temperature, $NO_2$ molecules have enough energy to decompose into nitrogen oxide and molecular oxygen, rather than forming $N_2O_4$:

$$2\,NO_2 \longrightarrow 2\,NO + O_2$$

In this reaction, one N—O bond breaks and a new O==O bond forms. We can visualize these changes occurring in more than one sequence.

One possible sequence that describes the decomposition of $NO_2$ molecules starts with a unimolecular reaction. At high temperature, collisions may transfer enough energy to allow some $NO_2$ molecules to break apart. That is, an N—O bond in $NO_2$ may break to produce a molecule of NO and an oxygen atom. Because oxygen atoms are highly reactive, a subsequent bimolecular collision between an oxygen atom and an $NO_2$ molecule would then generate NO and $O_2$. This reaction sequence can be summarized as Mechanism I for $NO_2$ decomposition:

*Mechanism I*

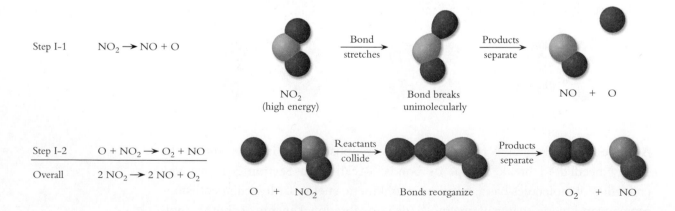

Step I-1     $NO_2 \longrightarrow NO + O$

Step I-2     $O + NO_2 \longrightarrow O_2 + NO$

Overall     $2\,NO_2 \longrightarrow 2\,NO + O_2$

Mechanism I illustrates an important requirement for reaction mechanisms. Because a mechanism is a summary of events at the molecular level, a mechanism *must* lead to the correct stoichiometry to be an accurate description of the chemical reaction. The sum of the steps of a mechanism must give the balanced stoichiometric equation for the overall chemical reaction. If it does not, the proposed mechanism must be discarded. In Mechanism I, the net result of two sequential elementary reactions is the observed reaction stoichiometry.

A second possible mechanism for $NO_2$ decomposition starts with a bimolecular reaction. When two fast-moving $NO_2$ molecules collide, an oxygen atom may be transferred between them to form molecules of $NO_3$ and $NO$. Molecules of $NO_3$ are unstable and readily break apart into $NO$ and $O_2$. This reaction sequence can be summarized as Mechanism II for $NO_2$ decomposition:

*Mechanism II*

| | |
|---|---|
| Step II-1 | $2\ NO_2 \longrightarrow NO_3 + NO$ |
| Step II-2 | $NO_3 \longrightarrow NO + O_2$ |
| Overall | $2\ NO_2 \longrightarrow 2\ NO + O_2$ |

Mechanism II gives the observed overall stoichiometry because the $NO_3$ molecule produced in the first step is consumed in the second step.

Either Mechanism I or Mechanism II can account for the decomposition of $NO_2$. Each consists of elementary steps, each generates a highly reactive species that undergoes further chemistry, and each accounts for the observed reaction stoichiometry. But does either mechanism represent what really happens at the molecular level? Before describing how to test whether a mechanism is realistic, we must introduce some additional features of mechanisms.

## Intermediates

Both proposed mechanisms for $NO_2$ decomposition contain chemical species produced in the first step and consumed in the second step. This is the defining characteristic of an **intermediate**. An intermediate is a chemical species *produced* in an early step of a mechanism and *consumed* in a later step. Intermediates never appear in the overall chemical equation. Notice for example that neither the O atoms of Mechanism I nor the $NO_3$ molecules of Mechanism II appear in the balanced chemical equation for $NO_2$ decomposition.

Intermediates are reactive chemical species that usually exist only briefly. They are consumed rapidly by bimolecular collisions with other chemical species or by unimolecular decomposition. The intermediate in Mechanism I is an oxygen atom that reacts rapidly with $NO_2$ molecules. The intermediate in Mechanism II is an unstable $NO_3$ molecule that rapidly decomposes.

The most direct way to test the validity of a mechanism is to determine what intermediates are present during the reaction. If oxygen atoms were detected, we would know that Mechanism I is a reasonable description of $NO_2$ decomposition. Likewise, the observation of $NO_3$ molecules would suggest that Mechanism II is reasonable. In practice, the detection of intermediates is quite difficult because they are usually reactive enough to be consumed as rapidly as they are produced. As a result, the concentration of an intermediate in a reaction mixture is very low. Highly sensitive measuring techniques are therefore required for the direct detection of chemical intermediates.

## The Rate-Determining Step

How fast do chemical reactions occur? The speed of a reaction is described by its **rate**. Rate is the number of events per unit time, such as the number of molecules reacting per second. Every elementary reaction has a characteristic rate. Some reactions are so fast that they are complete in the smallest measurable fraction of a second, whereas others are so slow that they require almost an eternity to reach completion. The observed rate of an *overall* chemical reaction is determined by the rates of the elementary reactions that make up the mechanism.

If one would study the individual steps of a mechanism in isolation, one would find that they each have their own unique rate, and one of them will be the slowest. This is called the **rate-determining step** because the reaction as a whole cannot go faster than the slowest step in its mechanism. The idea of the rate-determining step is central to the study of reaction mechanisms.

Anyone who has flown on a commercial airline is familiar with the consequences of rate-determining steps, as illustrated by Figure 12-2. At a busy airport, for example, airplanes queue up while waiting for a runway to become available. After the pilot receives permission to take off, the plane becomes airborne very quickly. No matter how quickly an airplane is loaded and ready to leave the gate, the rate of takeoffs is determined by the availability of runways. That is, the rate-determining step for airplane takeoff is the rate of takeoffs from the runways. Baggage claim is often the rate-determining step at the end of a flight. No matter how quickly the passengers leap from their seats, push through the aisle, and race through the terminal, passengers who checked their baggage cannot leave the airport any faster than the baggage is delivered.

Chemical knowledge often can be used to assess which step of a mechanism is likely to be rate-determining. For example, in both proposed mechanisms for the decomposition of $NO_2$, chemical knowledge suggests that the first step is much slower than the second step. In Mechanism I, an O atom is produced in the first step, $NO_2 \longrightarrow NO + O$. Oxygen atoms are known to be highly reactive, so it is reasonable to predict that this intermediate reacts rapidly with $NO_2$ molecules. Compared with the fast second step of this mechanism, the step that forms the oxygen atoms is expected to be slow and rate-determining. Similarly, the first step of Mechanism II, $2 NO_2 \longrightarrow NO_3 + NO$, produces $NO_3$. This species is known to be unstable, so it will decompose in the second step of

**Figure 12-2**

The slowest step in any sequential process is rate-determining. Airplanes take off at the rate that runways become available, and arriving passengers leave the terminal at the rate their baggage is delivered.

Mechanism II almost as soon as it forms. Again, the second step of the mechanism is expected to be fast, so the step that forms the reactive intermediate is slow and rate-determining. Later in this chapter, we discuss experiments that make it possible to distinguish between Mechanisms I and II.

---

**SECTION EXERCISES**

● **12.1.1** Nitric oxide (NO) in polluted air can react with ozone ($O_3$) to form $NO_2$ in a single bimolecular reaction. Identify the second molecular product of this reaction, and draw a molecular picture similar to the ones in this section that shows how this chemical reaction occurs.

● **12.1.2** One common analogy for the rate-determining step involves automobiles on a highway passing through a tollbooth. In your own words, explain rate-determining steps using the tollbooth analogy.

● **12.1.3** The following mechanism has been proposed for the reaction of NO with $H_2$:

$$2\ NO \longrightarrow N_2O_2$$

$$N_2O_2 + H_2 \longrightarrow H_2O + N_2O$$

$$N_2O + H_2 \longrightarrow N_2 + H_2O$$

Determine the balanced chemical equation for the reaction that occurs by this mechanism and identify any chemical intermediates.

---

## 12.2 RATES OF CHEMICAL REACTIONS

Some of the most important information about a mechanism comes from experiments that determine how fast a chemical reaction occurs under various conditions. In chemical reactions, amounts of reactants and products change, so reaction rates are given in units of amount per unit time; for example, molecules per second. Amounts can also be expressed as concentrations, so rates can be measured in units of concentration per unit time; for example, molar per minute.

Recall that concentrations are commonly expressed as molarities, mol/L.

The study of the rates of chemical reactions is called **kinetics**. Chemists study reaction rates for many reasons. To give just one example, Rowland and Molina used kinetic studies to show the destructive potential of CFCs. Kinetic studies are essential to the explorations of reaction mechanisms, because a mechanism can never be determined by calculations alone. Kinetic studies are important in many areas of science, including biochemistry, synthetic chemistry, biology, environmental science, engineering, and geology. The usefulness of chemical kinetics in elucidating mechanisms can be understood by examining the differences in rate behaviour of unimolecular and bimolecular elementary reactions.

### A Molecular View

To obtain a molecular perspective of reaction rates, consider the unimolecular reaction shown in Figure 12-3. At elevated temperature, the compound *cis*-2-butene can rearrange to form its isomer, *trans*-2-butene. The reaction occurs

**Figure 12-3**

*cis*-2-butene    Rotation around    *trans*-2-butene
                  the C–C σ bond

*cis*-2-Butene can rearrange to form *trans*-2-butene in a unimolecular reaction if collisions give the molecule enough extra energy to break the C — C π-bond.

after collisions transfer enough energy to a *cis*-2-butene molecule to break the C — C π bond. Once the bond breaks, rotation around the C — C σ bond takes place rapidly until a π bond forms again.

For a molecular view of rates, consider a sample containing 12 molecules of *cis*-2-butene, which represents a tiny portion of an immense reaction mixture. Figure 12-4 illustrates schematically how this reaction progresses with time. During the first minute, six *cis*-2-butene molecules isomerize to *trans*-2-butene. The average rate for this time period is 6 molecules/min. As the reaction proceeds, measurements show that the rate of reaction decreases. During the second minute, only three molecules isomerize, giving an average rate of 3 molecules/min for that period.

Why does the rate become slower as this reaction proceeds? Dividing the number of *cis*-2-butene molecules that react by the total number of *cis*-2-butene molecules present at the beginning of the interval reveals the answer:

First minute:    Fraction reaction $= \dfrac{6 \; cis\text{−}2\text{−butene molecules react}}{12 \; cis\text{−}2\text{−butene molecules present}} = 0.5$

Second minute:   Fraction reacting $= \dfrac{3 \; cis\text{−}2\text{−butene molecules react}}{6 \; cis\text{−}2\text{−butene molecules present}} = 0.5$

**Figure 12-4**

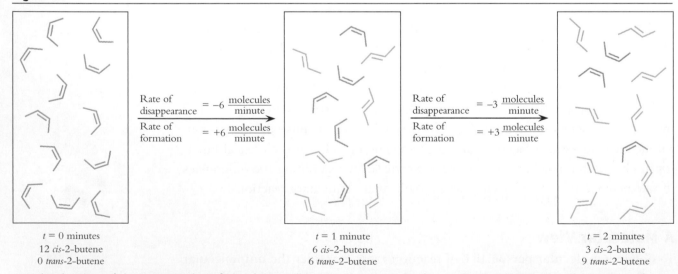

Rate of disappearance $= -6 \dfrac{\text{molecules}}{\text{minute}}$

Rate of formation $= +6 \dfrac{\text{molecules}}{\text{minute}}$

Rate of disappearance $= -3 \dfrac{\text{molecules}}{\text{minute}}$

Rate of formation $= +3 \dfrac{\text{molecules}}{\text{minute}}$

*t* = 0 minutes
12 *cis*-2-butene
0 *trans*-2-butene

*t* = 1 minute
6 *cis*-2-butene
6 *trans*-2-butene

*t* = 2 minutes
3 *cis*-2-butene
9 *trans*-2-butene

A molecular view of the isomerization of *cis*-2-butene illustrates that the rate of isomerization decreases as the number of *cis*-2-butene molecules decreases.

**Figure 12-5**

$t = 0$ minutes
12 molecules $NO_2$
0 molecules $N_2O_4$

$t = 1$ minute
8 molecules $NO_2$
2 molecules $N_2O_4$

$t = 2$ minutes
6 molecules $NO_2$
3 molecules $N_2O_4$

A molecular view of the formation of $N_2O_4$ from $NO_2$. The rate of reaction decreases as the number of molecules decreases because fewer $NO_2$ molecules are present to undergo collisions.

The fraction of the *cis*-2-butene molecules present that react during each one-minute interval is the same. That is, the rate of reaction is constant on a *per molecule* basis. As the reaction proceeds, however, fewer *cis*-2-butene molecules remain, causing the overall rate of reaction to decrease. This is characteristic of unimolecular elementary reactions. The rate per molecule is constant, but if the number of reactant molecules is cut in half, the rate of reaction is cut in half as well.

Bimolecular elementary reactions also slow down as the number of reactant molecules decreases, as can be seen for the formation of $N_2O_4$ from $NO_2$. Figure 12-5 shows 12 molecules of starting material, again representing a tiny fraction of a much larger sample. Four $NO_2$ molecules combine during the first minute, so the average rate during this period is 4 molecules/min. As the reaction proceeds, however, the rate decreases. During the second minute, only two molecules react, giving an average rate for that interval of 2 molecules/min.

Dividing the number of molecules reacting during each time interval by the number of reactant molecules present shows that this reaction behaves differently from a unimolecular reaction:

$$\text{First minute: Fraction reaction} = \frac{4 \; NO_2 \text{ molecules react}}{12 \; NO_2 \text{ molecules present}} = 0.33$$

$$\text{Second minute: Fraction reacting} = \frac{2 \; NO_2 \text{ molecules react}}{8 \; NO_2 \text{ molecules present}} = 0.25$$

In this case, not only do fewer molecules react, but the rate of reaction per molecule also decreases with time. If it were possible to count the number of collisions, we would find that the reaction slows down not only because fewer reactant molecules are present, but also because fewer collisions occur. For all bimolecular reactions, the rate *per collision* is constant, but as the reaction proceeds, fewer reactant molecules remain to collide with one another, so the rate per molecule decreases.

Figure 12-5 also shows that each species has its own rate, but the individual rates are linked by the stoichiometric coefficients. The reaction generates one molecule of $N_2O_4$ for every two molecules of $NO_2$ that are destroyed. That is, the 1:2 stoichiometry of this reaction results in a 1:2 relationship between the rate of disappearance of $NO_2$ and the rate of appearance of $N_2O_4$. The ratio of rates for different species is always equal to the ratio of their stoichiometric coefficients.

## A Macroscopic View: Concentration Changes

Molecules are too numerous to follow on an individual basis. Therefore, a chemist interested in measuring the rate of a reaction monitors the *concentration* of a particular compound as a function of time. The concentrations of reactants, products, or both may be monitored. For example, Figure 12-6 shows some experimental data obtained from a series of concentration measurements on the decomposition of $NO_2$:

$$2\ NO_2 \longrightarrow 2\ NO + O_2$$

The rate of the reaction at any specific time is given by how fast the concentration changes. The plots in Figure 12-6 show the same features that the molecular pictures of Figures 12-4 and 12-5 show: rates of reaction decrease as starting materials are consumed, and rates for different species are linked by stoichiometry.

The average rate of a reaction can be expressed as the change in concentration ($\Delta c$) over some time interval ($\Delta t$). For oxygen formation during the decomposition of $NO_2$:

$$\text{Rate of change of } O_2 \text{ concentration} = \frac{\Delta [O_2]}{\Delta t}$$

**Figure 12-6**

| Time | (s) | 0 | 5.0 | 10 | 15 | 20 | 30 | 50 | 80 | 120 | 240 |
|---|---|---|---|---|---|---|---|---|---|---|---|
| $[NO_2]$ | $(10^{-2}\ M)$ | 4.1 | 3.1 | 2.5 | 2.1 | 1.8 | 1.4 | 1.0 | 0.70 | 0.50 | 0.30 |
| $[NO]$ | $(10^{-2}\ M)$ | 0 | 1.0 | 1.6 | 2.0 | 2.3 | 2.7 | 3.1 | 3.4 | 3.6 | 3.8 |
| $[O_2]$ | $(10^{-2}\ M)$ | 0 | 0.50 | 0.80 | 1.0 | 1.1 | 1.3 | 1.6 | 1.7 | 1.8 | 1.9 |

A plot of $[NO_2]$, $[O_2]$, and $[NO]$ as a function of time (seconds) for the decomposition reaction of $NO_2$. The concentration data are shown in the table.

Because rates change continuously with time, accurate rate determinations must use small time intervals. At the outset of the $NO_2$ decomposition experiment shown in Figure 12-6, the time interval between measurements is 5 seconds. For the first 5 seconds, the average rate of change of $O_2$ concentration is as follows:

$$\text{Rate of change of } O_2 \text{ concentration} = \frac{\Delta[O_2]}{\Delta t}$$

$$= \frac{0.50 \times 10^{-2}\,\text{M} - 0\,\text{M}}{5.0\,\text{s} - 0\,\text{s}}$$

$$= 1.0 \times 10^{-3}\,\text{M/s}$$

At later times, the rate is different. Verify this for yourself by calculating the average rates for the second and third 5-second intervals.

We can also write a rate expression for the rate of change of the concentration of $NO_2$. Over the first 5 seconds of this reaction, this is as follows:

$$\text{Rate of change of } NO_2 \text{ concentration} = \frac{\Delta[NO_2]}{\Delta t}$$

$$= \frac{3.1 \times 10^{-2}\,\text{M} - 4.1 \times 10^{-2}\,\text{M}}{5.0\,\text{s} - 0\,\text{s}}$$

$$= -2.0 \times 10^{-3}\,\text{M/s}$$

The value is negative because $NO_2$ is being consumed in the reaction—its concentration decreases with time. Notice also that, over the same period, the rate of $O_2$ formation is only half the rate of $NO_2$ consumption. This follows from the molecular view of the mechanism and from the stoichiometry of the reaction. The rate relationship among the three species involved in $NO_2$ decomposition is given by the following expression:

$$\text{Reaction rate} = \frac{\Delta[O_2]}{\Delta t} = \left(\frac{1}{2}\right)\left(\frac{\Delta[NO]}{\Delta t}\right) = \left(-\frac{1}{2}\right)\left(\frac{\Delta[NO_2]}{\Delta t}\right)$$

The relationship among reaction rates and stoichiometric coefficients can be applied to any reaction. Equation 12-1 gives the relationships among the rate expressions for a reaction of the form $aA + bB \rightarrow dD + eE$:

$$\text{Reaction rate} = \left(-\frac{1}{a}\right)\left(\frac{\Delta[A]}{\Delta t}\right) = \left(-\frac{1}{b}\right)\left(\frac{\Delta[B]}{\Delta t}\right)$$

$$= \left(\frac{1}{d}\right)\left(\frac{\Delta[D]}{\Delta t}\right) = \left(\frac{1}{e}\right)\left(\frac{\Delta[E]}{\Delta t}\right) \qquad (12\text{-}1)$$

Here, lowercase letters represent stoichiometric coefficients and uppercase letters represent chemical substances.

Equation 12-1 is one of the few rate statements that follow directly from knowledge of the reaction stoichiometry. Almost all other rate information must be determined by carrying out experiments on how concentrations change with time. Example 12-1 illustrates the application of Equation 12-1.

## EXAMPLE 12-1 RELATIVE RATES OF REACTION

Acrylonitrile is produced from propene, ammonia, and oxygen by the following balanced equation (see Example 11-5 in Chapter 11):

$$2\ C_3H_6 + 2\ NH_3 + 3\ O_2 \longrightarrow 2\ CH_2CHCN + 6\ H_2O$$

Relate the rates of reaction of starting materials and products.

**Strategy:** The rates for different species participating in a chemical reaction are related by the stoichiometric coefficients of the balanced chemical equation. Equation 12-1 provides the exact relationship.

**Solution:** The balanced equation shows that three molecules of oxygen are consumed for every two molecules of propene and two molecules of ammonia. Thus, the rate of $C_3H_6$ and $NH_3$ consumption is only two-thirds the rate of $O_2$ consumption. Those seven molecules of starting materials produce two molecules of $CH_2CHCN$ and six molecules of $H_2O$. Thus, $CH_2CHCN$ is produced at the same rate as $C_3H_6$ is consumed, whereas $H_2O$ is produced three times as fast as $CH_2CHCN$ is. The link between relative reaction rates and reaction stoichiometry is Equation 12-1. Therefore,

$$\text{Reaction rate} = \left(-\frac{1}{2}\right)\left(\frac{\Delta[C_3H_6]}{\Delta t}\right) = \left(-\frac{1}{3}\right)\left(\frac{\Delta[O_2]}{\Delta t}\right) = \left(-\frac{1}{2}\right)\left(\frac{\Delta[NH_3]}{\Delta t}\right)$$

$$= \left(\frac{1}{2}\right)\left(\frac{\Delta[CH_2CHCN]}{\Delta t}\right) = \left(\frac{1}{6}\right)\left(\frac{\Delta[H_2O]}{\Delta t}\right)$$

**Does the Result Make Sense?** Each of the reactants disappears as the reaction proceeds, so negative signs are appropriate for these three species. Three molecules of $O_2$ are consumed for every six molecules of $H_2O$ formed, so it makes sense that the rate of disappearance of $O_2$ is half as large as the rate of disappearance of $H_2O$; in other words

$$\left(-\frac{1}{3}\right)\left(\frac{\Delta[O_2]}{\Delta t}\right) = \left(\frac{1}{6}\right)\left(\frac{\Delta[H_2O]}{\Delta}\right) \quad \text{or} \quad -\left(\frac{\Delta[O_2]}{\Delta t}\right) = \left(\frac{3}{6}\right)\left(\frac{\Delta[H_2O]}{\Delta t}\right) = \left(\frac{1}{2}\right)\left(\frac{\Delta[H_2O]}{\Delta t}\right)$$

**Common Pitfalls:** Not putting negative signs for the reactants in such an expression. Remember that the reaction rate is always positive, but the rates of change of the concentrations of reactants and products are negative and positive, respectively.

**Practice Exercise 12.1**
Relate the rate of formation of $N_2O_4$ to the rate of disappearance of $NO_2$ for the reaction $2\ NO_2 \longrightarrow N_2O_4$.

## SECTION EXERCISES

●**12.2.1** Ammonia is produced by the following reaction:

$$N_2 + 3\ H_2 \longrightarrow 2\ NH_3$$

Consider a portion of a flask that contains five molecules of $N_2$ and nine molecules of $H_2$. Draw a molecular picture showing the contents of this portion of the flask after two molecules of $NH_3$ have formed. Represent nitrogen atoms as dark spheres and hydrogen atoms as light spheres.

●**12.2.2** In the ammonia synthesis from the previous problem, which of the starting materials is consumed at a faster rate?

●**12.2.3** State the relative rates for the consumption of starting materials and the formation of products for the ammonia synthesis.

## 12.3 CONCENTRATION AND REACTION RATES

Both concentration and temperature affect the rate of a chemical reaction. This section examines how changes in the concentrations of starting materials and products affect the rate of a chemical reaction. We describe temperature effects in Section 12.6.

## Concentration Effects

A reaction of ozone provides an example of concentration effects. Ozone in the atmosphere near the Earth's surface is a serious pollutant that damages soft tissues such as the lungs. In major urban areas, smog alerts are issued whenever there are elevated concentrations of ozone in the lower atmosphere. Nitrogen oxide, another component of photochemical smog, is a colourless gas produced in a side reaction in automobile engines. One reaction that links these species is the reaction of NO and $O_3$ to produce $O_2$ and $NO_2$:

$$NO + O_3 \longrightarrow NO_2 + O_2$$

Experiments indicate that this reaction occurs through a single bimolecular collision, as shown in Figure 12-7.

**Figure 12-7**

$$O_3 \quad + \quad NO \qquad\qquad \text{Bonds rearrange as an} \qquad\qquad O_2 \quad + \quad NO_2$$
$$\text{oxygen atom is transferred}$$

The reaction between $O_3$ and NO is believed to occur by a mechanism that consists of the single bimolecular step shown here in a molecular view.

For a better understanding of the effect of changing concentrations on the rate of a chemical reaction, it helps to visualize the reaction at the molecular level. In this one-step bimolecular reaction, a collision between molecules that are in the proper orientation leads to the transfer of an oxygen atom from $O_3$ to NO. As with the formation of $N_2O_4$, the rate of this bimolecular reaction is proportional to the number of collisions between $O_3$ and NO. The more such collisions there are, the faster the reaction occurs.

To relate collision rates with concentrations, consider two containers of equal volume, each containing NO and $O_3$. Figure 12-8 shows molecular views of portions of two such containers. The container in Figure 12-8*b* has the same number of $O_3$ molecules as the container in Figure 12-8*a*, but the container in Figure 12-8*b* holds twice as many NO molecules as the container in Figure 12-8*a*. The larger amount of NO in the container in Figure 12-8*b* gives an $O_3$

**Figure 12-8**

(a)                                                                                    (b)

A schematic illustration of two containers of equal volume containing different amounts of $O_3$ and NO. Container *b* has the same amount of $O_3$ but twice as much NO as container *a*. With twice as many NO molecules, twice as many NO—$O_3$ collisions occur, so the reaction in container *b* proceeds twice as fast as the reaction in container *a*.

molecule in this container twice as many opportunities to collide with an NO molecule. Thus, the rate of reaction in the container in Figure 12-8*b* is twice the rate in the container in Figure 12-8*a*. To generalize, collision rates and reaction rates of bimolecular reactions increase linearly with the concentration of each participant. Example 12-2 extends this reasoning to $O_3$ molecules.

## EXAMPLE 12-2  RATES AND NUMBER OF MOLECULES

The container shown at right contains $O_3$ and NO. Compared with the container in Figure 12-8*a* (which has the same volume), how fast will the reaction proceed?

**Strategy:** The mechanism of this reaction requires a collision between an NO molecule and an $O_3$ molecule, as shown in Figure 12-7. The rate of the reaction therefore depends on the number of collisions that occur, and the collision frequency is proportional to the number of molecules of each type.

**Solution:** In Figure 12-8*a* there are 12 molecules of $O_3$ and 10 of NO. The container shown in this example contains 4 molecules of $O_3$ and 10 molecules of NO. The concentration of $O_3$ is only one-third as great, so any one molecule of NO will encounter an $O_3$ molecule three times less often than it would in the container in Figure 12-8*a*. This new mixture will react one-third as fast as the one in Figure 12-8*a*.

**Does the Result Make Sense?** If we reduce the amount of one reactant in a reaction flask, it makes sense that there are fewer collisions with the other reactant, and if there are fewer collisions, it makes sense that the rate of reaction will decrease.

**Practice Exercise 12.2**

Suppose the container contains 10 molecules of $O_3$ and 12 molecules of NO. Compare the rate of reaction with the rates of the two views in Figure 12-8 and the view in this example.

## Rate Laws

The effect of concentration on the rate of a particular chemical reaction can be summarized in an algebraic expression known as a **rate law**. A rate law links the rate of a reaction with the concentrations of the reactants through a **rate constant** ($k$). In addition, as we show later in this chapter, the rate law may contain concentrations of chemical species that are not part of the balanced overall reaction.

Although every reaction has its own unique rate law, many rate laws have a general form:

$$\text{Rate} = k[A]^y[B]^z \tag{12-2}$$

In general, each concentration has some exponent (here, $y$ and $z$). Each exponent is called the *order* of the reaction with respect to that particular species. In Equation 12-2, $y$ is the **order of reaction** with respect to species A, and $z$ is the order with respect to species B. If the value of $y$ is 1, the reaction is called **first order** in A; if the value of $z$ is 2, the reaction is called **second order** in B, and so on. Orders of reaction are usually small integers or simple fractions. The most common orders are 1 and 2. The sum of the exponents is known as the **overall order** of the reaction.

The exponents in a rate law depend on the reaction *mechanism* rather than on the *stoichiometry* of the overall reaction. *The order of a reactant often differs from its stoichiometric coefficient.* Consequently, a rate law must always be determined by conducting experiments; it can never be derived from the stoichiometry of the overall chemical reaction. For example, experimental studies show that the rate law for the reaction of $O_3$ with $NO_2$ to give $N_2O_5$ and $O_2$ is first order in each reactant:

$$2\,NO_2 + O_3 \longrightarrow N_2O_5 + O_2 \qquad \text{Experimental rate} = k[NO_2][O_3]$$

Notice that for this reaction, the order of reaction with respect to $NO_2$ is 1, whereas the stoichiometric coefficient is 2. This shows that the order of a reaction for a particular species cannot be predicted by looking at the overall balanced equation.

When a reaction proceeds in a single elementary step, however, its rate law will mirror its stoichiometry. An example is the rate law for $O_3$ reacting with NO. Experiments show that this reaction is first order in each of the starting materials and second order overall:

$$NO + O_3 \longrightarrow NO_2 + O_2 \qquad \text{Experimental rate} = k[NO][O_3]$$

This rate law is fully consistent with the molecular view of the mechanism shown in Figure 12-7. If the concentration of either $O_3$ or NO is doubled, the number of collisions between starting material molecules doubles too, and so does the rate of reaction. If the concentrations of *both* starting materials are doubled, the collision rate and the reaction rate increase by a factor of 4.

## Mechanisms and Rate Laws

The relationship between a mechanism and its rate law can be illustrated for the decomposition of $NO_2$. Experiments that we describe in Section 12.4 reveal that the rate of this reaction is proportional to the square of the $NO_2$ concentration:

$$2 NO_2 \longrightarrow 2 NO + O_2 \qquad \text{Experimental rate} = k[NO_2]^2$$

As described earlier, either Mechanism I or Mechanism II might describe this reaction:

**Mechanism I**

$$NO_2 \longrightarrow NO + O \qquad \text{Step I–1 (rate–determining)}$$
$$O + NO_2 \longrightarrow O_2 + NO \qquad \text{Step I–2 (fast)}$$

**Mechanism II**

$$2 NO_2 \longrightarrow NO_3 + NO \qquad \text{Step II–1 (rate–determining)}$$
$$NO_3 \longrightarrow NO + O_2 \qquad \text{Step II–2 (fast)}$$

Each mechanism *predicts* a rate behaviour that can be compared with the experimental rate law. If the prediction differs from the experimental observation, the mechanism is incorrect.

Remember that the overall rate of a reaction is determined by the rate of the slowest step. In other words, no reaction can proceed faster than the rate-determining step. Any step that comes *after* the rate-determining step cannot influence the overall rate of reaction. In both proposed mechanisms, the first step of each mechanism is rate-determining. That is, each proposed mechanism predicts an overall rate of $NO_2$ decomposition that is the same as the rate of the first step in the mechanism.

The first step in Mechanism I is the unimolecular decomposition of $NO_2$. Our molecular analysis shows that the rate of a unimolecular reaction is constant on a *per molecule* basis. Thus, if the concentration of $NO_2$ is doubled, twice as many molecules decompose in any given time. In quantitative terms, if $NO_2$ decomposes by Mechanism I, the rate law will be the following:

$$\text{Predicted rate (Mechanism I)} = k[NO_2]$$

Once an $NO_2$ molecule decomposes, the O atom that results from decomposition very quickly reacts with another $NO_2$ molecule.

The rate-determining step of Mechanism II is a bimolecular collision between two identical molecules. A bimolecular reaction has a constant rate on a *per collision* basis. Thus, if the number of collisions between $NO_2$ molecules increases, the rate of decomposition increases accordingly. Doubling the concentration of

$NO_2$ doubles the number of molecules present, and it also doubles the number of collisions for *each* molecule. Each of these factors doubles the rate of reaction, so doubling the concentration of $NO_2$ increases the rate for this mechanism by a factor of 4. Consequently, if $NO_2$ decomposes by Mechanism II, the rate law will be the following:

$$\text{Predicted rate (Mechanism II)} = k[NO_2][NO_2] = k[NO_2]^2$$

The two proposed mechanisms for this reaction predict different rate laws. Whereas Mechanism I predicts that the rate is proportional to $NO_2$ concentration, Mechanism II predicts that the rate is proportional to the *square* of $NO_2$ concentration. Experiments agree with the prediction of Mechanism II, so Mechanism II is consistent with the experimental behaviour of the $NO_2$ decomposition reaction, and therefore *could* be the correct mechanism. Mechanism I predicts rate behaviour contrary to what is observed experimentally, so Mechanism I *cannot* be correct.

We return to the relationship between rate laws and mechanisms in Section 12.5, after discussing experimental methods for determining the rate law of a reaction.

## Rate Constants

Every reaction has its own characteristic rate constant that depends on the intrinsic speed of that particular reaction. For example, the value of $k$ in the rate law for $NO_2$ decomposition is different from the value of $k$ for the reaction of $O_3$ with NO. Rate constants are independent of concentration and time (which is why we call them rate *constants*), but as we discuss in Section 12.6, rate constants are sensitive to temperature.

Rates of reaction have units of $(\text{concentration})(\text{time})^{-1}$. Because time does not appear in any other term on the right-hand side of the rate law (see Equation 12-2), the units of $k$ must always include time in the denominator. The concentration units of the rate constant depend on the overall order of the rate law, however, because the units of the rate constant must cancel concentration units to give the proper units for the rate of reaction. As an example, consider the rate law for the reaction of NO with $O_3$:

$$\text{Rate} = k[NO][O_3]$$

To agree with the units of $(\text{concentration})(\text{time})^{-1}$ for the rate of reaction, the rate constant must have units of $(\text{concentration})^{-1}(\text{time})^{-1}$ (such as $M^{-1} s^{-1}$). These units fit into the rate law as follows:

| Rate law: | Rate | = | $k$ | [NO] | $[O_3]$ |
|---|---|---|---|---|---|
| Units: | M/s | = | $M^{-1} s^{-1}$ | M | M |

Example 12-3 provides practice in deducing the units of a rate constant.

## EXAMPLE 12-3  UNITS OF THE RATE CONSTANT

Reactions in aqueous solution can have complex kinetics. An example is the reaction between arsenic acid and iodide ions:

$$H_3AsO_4\,(aq) + 3\,I^-\,(aq) + 2\,H_3O^+\,(aq) \longrightarrow H_3AsO_3\,(aq) + I_3^-\,(aq) + 3\,H_2O\,(l)$$

The rate law for this reaction has been found experimentally to be as follows:

$$Rate = k[H_3AsO_4][I^-][H_3O^+]$$

What are the units of the rate constant when the time is expressed in minutes?

**Strategy:** Units on the left must be the same as units on the right. The rate constant must have units that achieve this. Analyze the units for each component of the rate equation to find the units for $k$.

**Solution:** On the left side of the rate law, the rate of reaction is expressed in terms of changes in concentration and time:

$$Rate\ of\ reaction\ = \frac{\Delta c}{\Delta t}$$

For species in solution, concentration ($c$) is in molarity (M). According to the conditions of the problem, time ($t$) is in minutes (min). Therefore, the units of the rate must be:

$$M/min$$

The right-hand side of the rate law contains a product of three concentrations, so it shows third-order behaviour overall. This gives concentration units of $M^3$. This requires units of $M^{-2}\,min^{-1}$ for the rate constant:

| Term in equation: | Rate | $k$ | $[H_3AsO_4]$ | $[I^-]$ | $[H_3O^+]$ |
|---|---|---|---|---|---|
| Units: | M/min | $M^{-2}\,min^{-1}$ | M | M | M |

 **Does the Result Make Sense?** The units are the same on both sides of the equality. A rate is always in (concentration)/(time), so we have to adjust the units of the rate constant according to the overall order of the reaction.

**Practice Exercise 12.3**

What are the units of the rate constant for the reaction of hydroxide anions with hydronium cations, which proceeds by a single collision: $H_3O^+ + OH^- \longrightarrow H_2O$

Once again, we emphasize that the order of reaction and the value of the rate constant must be determined by doing experiments. Knowing the order of reaction then makes it possible to write the specific rate law for the chemical process. In the next three sections, we discuss how chemists determine orders of reactions and further explore how rate laws are related to chemical mechanisms.

## SECTION EXERCISES

● **12.3.1** Draw a container with numbers of NO and $O_3$ molecules that would react at half the rate of the one in Figure 12-8a.

● **12.3.2** The isomerization reaction of *cis*-2-butene, shown in Figure 12-3, is found experimentally to follow first-order kinetics. What is the rate law, and what are the units of the rate constant?

● **12.3.3** The reaction between hydrogen and bromine is as follows: $H_2\,(g) + Br_2\,(g) \longrightarrow 2\,HBr\,(g)$. Experiments show that this reaction is first order in $H_2$ and one-half order in $Br_2$. What is the rate law, what is the overall order of the reaction, and what are the units of the rate constant?

## 12.4 EXPERIMENTAL KINETICS

Every rate law must be determined experimentally. A chemist may imagine a reasonable mechanism for a reaction, but that mechanism must be tested by comparing the *actual* rate law for the reaction with the rate law *predicted* by the mechanism. To determine a rate law, chemists observe how the rate of a reaction changes with concentration. The graph of the data for the $NO_2$ decomposition reaction shown in Figure 12-6 is an example of such observations.

Regardless of the rate law, the rate of a reaction generally decreases with time because the concentrations of reactants decrease. The form of a rate law is determined by exploring the details of how the rate decreases with time. Because rate laws describe how rates vary with *concentration*, it is necessary to do mathematical analysis to convert rate laws into equations that describe how concentrations vary with *time*. These equations are different for different reaction orders.

## First-Order Reactions

Recall that Mechanism I for the decomposition of $NO_2$ and the mechanism for the isomerization reaction of *cis*-2-butene both predict a simple first-order rate law:

$$\text{rate} = k[NO_2] \qquad \text{and} \qquad \text{rate} = k[cis\text{--}2\text{--butene}]$$

The general form of a first-order rate law is rate $= k[A]$, where A is a reactant in the overall reaction. It is a simple matter to convert this into an expression relating the concentration of the reactant to time. We begin by noting that since the reactant is disappearing with time, the rate of the reaction can also be expressed as follows:

$$\text{Rate} = \frac{-d[A]}{dt}$$

Thus we have two expressions for the rate, and can set them equal to one another:

$$\frac{-d[A]}{dt} = k[A]$$

which can be solved by collecting the variables involving [A] on one side:

$$\frac{d[A]}{[A]} = -kdt$$

followed by integration:

$$\int \frac{d[A]}{[A]} = \int -kdt$$

The result is $\ln[A] = -kt + \text{constant}$. (In the physical sciences, there are many times that we need to integrate a term of the form $\dfrac{dx}{x}$, with the result that many solutions contain a $\ln(x)$ term.) The constant is easily evaluated by noting that at time zero, $[A] = [A]_0$, the starting concentration of A, which is a known quantity. Thus,

$$\ln[A] = -kt + \ln[A]_0$$

**Figure 12-9**

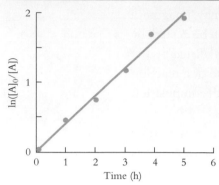

| Time (h) | [C$_5$HBr] (M)$_{11}$ | [C$_5$H$_{11}$Br]$_0$/[C$_5$H$_{11}$Br] | ln([A]$_0$/[A]) |
|----------|----------------------|----------------------------------------|-----------------|
| 0 | 0.150 | 1.00 | 0 |
| 1.00 | 0.099 | 1.52 | 0.42 |
| 2.00 | 0.073 | 2.05 | 0.72 |
| 3.00 | 0.047 | 3.19 | 1.16 |
| 4.00 | 0.028 | 5.36 | 1.68 |
| 5.00 | 0.022 | 6.82 | 1.92 |

A plot of ln([A]$_0$/[A]) vs. time in hours for the conversion of an alkyl bromide to an alkene. The experimental data are shown in the table. The linearity of this plot verifies that this reaction obeys a first-order rate law.

which can be rearranged to give Equation 12-3:

$$\ln\left(\frac{[A]_0}{[A]}\right) = kt \qquad (12\text{-}3)$$

An important feature of Equation 12-3 is that a graph of $\ln\left(\dfrac{[A]_0}{[A]}\right)$ vs. $t$ gives a straight line of which the slope is $k$. Consequently, experimental data can be tested for first-order behaviour by preparing such a graph and observing whether or not it is linear.

Figure 12-9 illustrates this procedure for measurements done on the conversion of an alkyl bromide to an alkene:

$$\text{Alkyl bromide} \ (\text{C}_5\text{H}_{11}\text{Br}) + \text{H}_2\text{O} \longrightarrow \text{Alkene} \ (\text{C}_5\text{H}_{10}) + \text{H}_3\text{O}^+ + \text{Br}^-$$

The linear appearance of the plot shows that this reaction obeys a first-order rate law. Additional mechanistic studies suggest that alkene formation proceeds in a two-step sequence. In the first step, which is rate-determining, the C—Br bond breaks to generate a bromide anion and an unstable cationic intermediate. In the second step, the intermediate transfers a proton (H$^+$) to a water molecule, forming the alkene and H$_3$O$^+$:

$$\text{C}_5\text{H}_{11}\text{Br} \xrightarrow{\ \text{Slow}\ } \text{Intermediate} \ (\text{C}_5\text{H}_{11}{}^+) + \text{Br}^-$$

$$\text{Intermediate} \ (\text{C}_5\text{H}_{11}{}^+) + \text{H}_2\text{O} \xrightarrow{\ \text{Fast}\ } \text{C}_5\text{H}_{10} + \text{H}_3\text{O}^+$$

This is known as an $S_N1$ reaction mechanism, and is explored in detail in Section 10.5. This example shows that first-order kinetics are not restricted to reactions with just one starting material, such as the decomposition of nitrogen dioxide or the isomerization of *cis*-2-butene.

The rate constant for the alkyl bromide reaction is equal to the slope of the line. The best way to determine a slope is by doing a linear curve fit using a spreadsheet or graphing calculator. Somewhat less accurately, any two points on the line determine the slope:

$$\text{Rate constant} = k = \text{Slope} = \frac{\Delta y}{\Delta x} = \frac{1.16 - 0}{3.00\ \text{h} - 0\ \text{h}} = \frac{1.16}{3.00\ \text{h}} = 0.387\ \text{h}^{-1}$$

Equation 12-3 is used to test whether a reaction is first order overall. The concentration of reactant A is monitored as a function of time. If a plot of $\ln\left(\dfrac{[A]_0}{[A]}\right)$ vs. $t$ is linear, then the reaction is first order in A. In Example 12-4, we apply this test to the data for the decomposition reaction of $NO_2$ presented in Figure 12-6.

## EXAMPLE 12-4 FIRST-ORDER KINETIC ANALYSIS

According to proposed Mechanism I, the decomposition of $NO_2$ should follow first-order kinetics. Do the experimental data of Figure 12-6 support this mechanism?

**Strategy:** According to the concentration–time form of first-order rate laws, a plot of $\ln\left(\dfrac{[A]_0}{[A]}\right)$ vs. $t$ must be a straight line if Mechanism I were correct.

**Solution:** Convert the concentration data into ratios, take the logarithms, and then prepare a logarithmic plot to determine whether the reaction is first order:

| Time (s) | 0 | 5 | 10 | 15 | 20 | 30 | 50 | 80 | 120 | 240 |
|---|---|---|---|---|---|---|---|---|---|---|
| $[NO_2]\ (10^{-2}\,M)$ | 4.1 | 3.1 | 2.5 | 2.1 | 1.8 | 1.4 | 1.0 | 0.70 | 0.50 | 0.30 |
| $\dfrac{[NO_2]_0}{[NO_2]}$ | 1.00 | 1.32 | 1.64 | 1.95 | 2.28 | 2.93 | 4.10 | 5.86 | 8.20 | 13.7 |
| $\ln\left(\dfrac{[NO_2]_0}{[NO_2]}\right)$ | 0.00 | 0.28 | 0.50 | 0.67 | 0.82 | 1.08 | 1.41 | 1.77 | 2.10 | 2.62 |

**Nitrogen Dioxide Decomposition First-Order Plot**

The graph is not linear, so we conclude that the decomposition of $NO_2$ does not follow first-order kinetics. Consequently, Mechanism I, which predicts first-order behaviour, cannot be correct.

**Does the Result Make Sense?** Remember that reaction order always must be determined by analyzing experiments. The decomposition of $NO_2$ does not follow first-order kinetics, so a mechanism that predicts first-order behaviour cannot be correct, no matter how "reasonable" that mechanism might appear on paper.

**Practice Exercise 12.4**

Here are experimental data for the reaction $SO_2Cl_2 (g) \longrightarrow SO_2 (g) + Cl_2 (g)$

| Time ($10^3$ s) | 0 | 2.5 | 5.0 | 7.0 | 10.0 | 15.0 | 20.0 | 30.0 | 50.0 |
|---|---|---|---|---|---|---|---|---|---|
| $[SO_2Cl_2]$ (Torr) | 760 | 720 | 680 | 645 | 610 | 545 | 490 | 395 | 255 |

Construct the appropriate graph to determine if these data are consistent with first-order kinetics. Note that pressure units can be used in the same manner as molar concentrations.

Data must be collected over a relatively long period to determine whether a first-order plot is linear. Notice that the first five points on the plot in Example 12-4 fall reasonably close to the dashed straight line. Only after more than 50% of the reactant has been consumed does this plot deviate substantially from linearity.

Another characteristic of first-order reactions is that the time it takes for half the reactant to disappear is the same, no matter what the concentration. This time is called the **half-life** ($t_{1/2}$). Applying Equation 12-3 to a time interval equal to the half-life results in an equation for $t_{1/2}$. When half the original concentration has been consumed, $[A] = 0.5[A]_0$. The $[A]_0$ terms cancel:

$$kt_{1/2} = \ln\left(\frac{[A]_0}{0.5[A]_0}\right) = \ln 2$$

Rearranging gives an equation for $t_{1/2}$:

$$t_{1/2} = \frac{\ln 2}{k} \tag{12-4}$$

Equation 12-4 does not contain the concentration of A, so the half-life of a first-order reaction is a constant that is independent of how much A is present. The decomposition reactions of radioactive isotopes provide excellent examples of first-order processes, as Example 12-5 illustrates. We describe the use of radioactive isotopes and their half-lives to determine the age of an object in more detail in Chapter 20.

## EXAMPLE 12-5 HALF-LIVES

Carbon-14 ($^{14}C$) is a radioactive isotope with a half-life of $5.73 \times 10^3$ years. The fractional amount of $^{14}C$ present in an object can be used to determine its age. Calculate the rate constant for decay of $^{14}C$ and determine how long it takes for 90% of the $^{14}C$ in a sample to decompose.

**Strategy:** We are asked to calculate two quantities, so we can apply the seven-step strategy.

1. We are asked to determine a rate constant and the time required for material to be consumed.

2. Visualize the process: atoms of $^{14}C$ decompose as time passes.

**Solution:**

3. The only piece of data is the half-life: $t_{1/2} = 5.73 \times 10^3$ years.

The ages of ancient pots can be determined by $^{14}C$ dating of charcoal found with the pots.

4. Equation 12-3 links time and concentration through the rate constant, and Equation 12-4 links the rate constant to the half-life. Knowing a half-life, we can calculate the rate constant using Equation 12-4. Then, the value of the rate constant and Equation 12-3 can be used to determine the time required to reach a certain concentration.

5. We need to break the problem into its two parts. To find the value of the rate constant, rearrange Equation 12-4 to isolate $k$:

$$t_{1/2} = \frac{\ln 2}{k} \quad \text{so} \quad k = \frac{\ln 2}{t_{1/2}}$$

To find the time, rearrange Equation 12-3 to isolate $t$:

$$\ln\left(\frac{[A]_0}{[A]}\right) = kt \quad \text{so} \quad t = \frac{\ln\left(\frac{[A]_0}{[A]}\right)}{k}$$

6. Now substitute and calculate the two quantities:

$$k = \frac{0.6931}{5.73 \times 10^3 \, \text{yr}} = 1.21 \times 10^{-4} \, \text{yr}^{-1}$$

When 90% has decayed, 10% remains, thus $[A] = 0.10[A]_0$:

$$\ln\left(\frac{[A]_0}{[A]}\right) = \ln\left(\frac{[A]_0}{0.10[A]_0}\right) = \ln(10)$$

$$t = \frac{\ln(10)}{k} = \frac{2.303}{1.21 \times 10^{-4} \, \text{yr}^{-1}} = 1.90 \times 10^4 \, \text{yr}$$

**Do the Results Make Sense?** It requires 19 000 years for 90% of a sample of $^{14}C$ to decompose. This is a long time, but it is reasonable because it takes 5730 years for half of the isotope to decompose.

**Common Pitfalls:** Using the wrong expression for the half-life. Make sure that if the reaction is first order, you use Equation 12-4. An expression for the half-life of a second-order reaction is given in Equation 12-5.

**Practice Exercise 12.5**

Determine the half-life for the decomposition reaction for which data are given in Practice Exercise 12.4.

## Second-Order Reactions

Mechanism II for the decomposition of $NO_2$ predicts that the reaction should have a second-order rate law:

$$\text{rate} = \frac{d[NO_2]}{[dt]} = -k[NO_2]^2$$

This can be mathematically converted into an expression relating concentration and time, as was done for the first-order process. Using [A] to denote the concentration of $NO_2$ in this example, we have

$$\frac{d[A]}{[dt]} = -k[A]^2$$

Collecting the variables gives

$$\frac{d[A]}{[A]^2} = -kdt$$

and so

$$\int \frac{d[A]}{[A]^2} = \int -kdt$$

$$\frac{-1}{[A]} = -kt + \text{constant}$$

Again, $[A] = [A]_0$ at $t = 0$. Thus, the constant $= \dfrac{-1}{[A]_0}$. Inserting this into the above expression and rearranging gives Equation 12-5:

$$\frac{1}{[A]} - \frac{1}{[A]_0} = kt \tag{12-5}$$

The most important feature of Equation 12-5 is that a graph of $\left(\dfrac{1}{[A]} - \dfrac{1}{[A]_0}\right)$ vs. $t$ gives a straight line of which the slope is $k$. Consequently, experimental data can be tested for second-order behaviour by preparing such a graph and observing whether or not it is linear. Figure 12-10 shows the decomposition data of Figure 12-6 plotted in this way. The graph is linear, showing that the reaction is in fact second order in $NO_2$. Thus, the kinetic behaviour of the decomposition of $NO_2$ matches the predicted rate law for Mechanism II (see earlier discussion on page 686). Chemists have accepted Mechanism II as the correct mechanism for this decomposition because it predicts the correct rate law and consists of reasonable elementary reactions, and because the intermediate species $NO_3$ has been detected when this reaction proceeds.

Example 12-6 provides another example of the analysis of rate data.

**Figure 12-10**

The decomposition data for $NO_2$ plotted as the reciprocal of concentration vs. time. This graph is linear, with a slope equal to the second-order rate constant.

## EXAMPLE 12-6 ANALYSIS OF RATE DATA

The Diels–Alder reaction, in which two alkenes combine to give a new product, is one of the most frequently used reactions for the synthesis of organic compounds. Thousands of examples are found in the chemical literature. The reaction of butadiene is a simple example.

$$\underbrace{\text{\LARGE $\diagdown\!\!\diagup$} + \text{\LARGE $\diagup\!\!\diagdown$}}_{\substack{2\ C_4H_6 \\ \text{(Butadiene)}}} \longrightarrow \quad C_8H_{12}$$

Use the data in the table to determine the rate law and the rate constant for the Diels–Alder reaction of butadiene:

| Time (min) | 0 | 4.0 | 8.0 | 12.0 | 16.0 | 20.0 | 30.0 |
|---|---|---|---|---|---|---|---|
| $C_4H_6$: (M) | 0.130 | 0.0872 | 0.0650 | 0.0535 | 0.0453 | 0.0370 | 0.0281 |

**Strategy:** When asked to determine a rate law and rate constant from concentration–time data, we must determine the order of the reaction. The rate law for this reaction has some order x with respect to butadiene ($C_4H_6$ = A), which we must determine:

$$\text{Rate} = k[A]^x$$

We can use the data provided in the problem to test graphically whether the reaction is first or second order in butadiene. If the reaction is first order, a plot of $\ln\left(\dfrac{[A]_0}{[A]}\right)$ vs. $t$ is linear. If the reaction is second order, a plot of $\left(\dfrac{1}{[A]} - \dfrac{1}{[A]_0}\right)$ vs. $t$ is linear. If neither plot is a straight line, the reaction is something other than first or second order.

**Solution:** Begin by calculating the values for the logarithms and the reciprocals of the concentrations:

| Time (min) | 0 | 4.0 | 8.0 | 12.0 | 16.0 | 20.0 | 30.0 |
|---|---|---|---|---|---|---|---|
| $\ln\left(\dfrac{[A]_0}{[A]}\right)$ | 0 | 0.400 | 0.693 | 0.888 | 1.05 | 1.26 | 1.53 |
| $\dfrac{1}{[A]} - \dfrac{1}{[A]_0}$ | 0 | 3.8 | 7.7 | 11.0 | 14.4 | 19.3 | 27.9 |

Next, prepare first- and second-order plots to see whether either gives a straight line. The graphs are shown below.

The first-order plot is not linear, so the reaction cannot be first order. The straight line in the $\left(\dfrac{1}{[A]} - \dfrac{1}{[A]_0}\right)$ vs. $t$ plot shows that the reaction is second order in butadiene. This is the rate law for this Diels–Alder reaction:

$$\text{Rate} = k[C_4H_6]^2$$

We could determine the slope using a spreadsheet or graphing calculator. Here, we use the first and last data points to evaluate the slope and rate constant:

$$k = \text{Slope} = \frac{\Delta y}{\Delta x} = \frac{27.9\ M^{-1} - 0\ M^{-1}}{30.0\ \text{min} - 0\ \text{min}} = \frac{27.9\ M^{-1}}{30.0\ \text{min}} = 0.930\ M^{-1}\ \text{min}^{-1}$$

 **Do the Results Make Sense?** A second-order dependence is consistent with a mechanism in which the first, rate-determining step is the collisional interaction of two molecules of butadiene.

**Practice Exercise 12.6**

Construct the appropriate graph to determine if the data in Practice Exercise 12.4 are consistent with a second-order reaction.

Equations 12-3 and 12-5 describe how concentration changes with time when only a *single* reactant is involved. However, most reactions involve concentration changes for more than one species. Although it is possible to develop equations relating concentration and time for such reactions, such equations are more complex to derive than the equations that involve just one reactant. Fortunately, it is often possible to simplify the experimental behaviour of a reaction. We describe two experimental methods that accomplish this: the isolation method and the method of initial rates.

## Isolation Experiments

One way to simplify the behaviour of a reaction is to adjust the conditions so that the initial concentration of one starting material is much smaller than the initial concentrations of the others. This establishes experimental conditions under which the concentration of only one of the starting materials changes significantly during the reaction. This concentration is then said to be *isolated*. We use another reaction that is important in atmospheric chemistry to illustrate this **isolation method** in detail.

Trees and shrubs contain a group of fragrant compounds called *terpenes*. The simplest terpene is isoprene. All other terpenes are built around carbon skeletons constructed from one or more isoprene units. Plants emit terpenes into the atmosphere, as anyone who has walked in a pine or eucalyptus forest will have noticed. The possible effect of terpenes on the concentration of ozone in the troposphere has been the subject of much debate and has led to careful measurements of rates of reaction with ozone.

The reactions of terpenes with ozone lead to a complicated array of products, but the rate behaviour for ozone reacting with a terpene can be studied by measuring the concentration of the terpene as a function of time:

$$O_3 + \text{Terpene} \longrightarrow \text{products} \qquad \text{Rate} = k[O_3]^y[\text{Terpene}]^z$$

We expect the reaction rate to depend on two concentrations rather than one, but we can isolate one concentration variable by making the initial concentration of one reactant much smaller than the initial concentration of the other. Data collected under these conditions can then be analyzed using Equations 12-3 and 12-5, which relate concentration to time. For example, an experiment could be performed on the reaction of ozone with isoprene with the following initial concentrations:

$$[O_3]_0 = 5.40 \times 10^{-4} \text{ M} \qquad [\text{Isoprene}]_0 = 3.0 \times 10^{-6} \text{ M}$$

Isoprene
$C_5H_8$

The familiar odours of a pine forest are caused by terpenes, the simplest of which is isoprene.

Under these initial conditions, the isoprene is consumed entirely before there is an appreciable change in the concentration of ozone, as Figure 12-11 illustrates. For 1:1 reaction stoichiometry and an initial ozone concentration 200 times larger than the initial isoprene concentration, the ozone concentration changes by less than 1%:

$$(5.40 \times 10^{-4} \text{ M}) - (3.0 \times 10^{-6} \text{ M}) = 5.37 \times 10^{-4} \text{ M}$$

| Initial ozone concentration | Concentration decrease from reaction with isoprene | Final ozone concentration |

In other words, the reaction is flooded with a large excess of ozone. Because the concentration of ozone does not change appreciably, the near-constant concentration of ozone can be grouped with the rate constant:

$$\text{Rate} = \underbrace{k[O_3]^y}[\text{isoprene}]^z$$

This quantity remains constant over the course of the reaction because $[O_3] >> [\text{isoprene}]$.

**Figure 12-11**

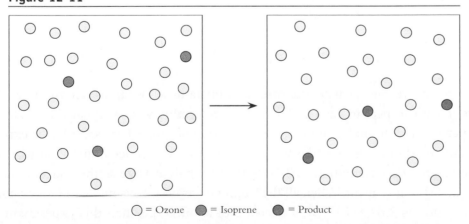

○ = Ozone    ● = Isoprene    ● = Product

When the initial concentration of ozone (*green circles*) is much greater than that of isoprene (*lavender circles*), isoprene reacts completely to give products (*orange circles*) without changing the concentration of ozone appreciably.

**Figure 12-12**

Data for the reaction of isoprene with ozone under conditions when the concentration of isoprene is isolated. The experimental behaviour is first order under these conditions.

We can identify this combination of constant quantities as a new constant, $k_{obs}$:

$$k[O_3]^y = k_{obs}$$

Here, "obs" is an abbreviation for "observed." Using $k_{obs}$, we can simplify the rate law to rate $= k_{obs}[\text{isoprene}]^z$. It is important to remember that $k_{obs}$ is an observed, *experimental* rate constant, but it is not the *true* rate constant ($k$) that appears in the general rate law.

Isolating isoprene as the only reactant with a concentration change simplifies the general rate law to a form that can be tested against the single-concentration rate laws:

$$\text{First-order case: } \ln\left(\frac{[\text{isoprene}]_0}{[\text{isoprene}]}\right) = k_{obs}t$$

$$\text{Second-order case: } \frac{1}{[\text{isoprene}]} - \frac{1}{[\text{isoprene}]_0} = k_{obs}t$$

Experimental data collected under these conditions are plotted in Figure 12-12. Notice that the first-order plot is linear but the second-order plot is not, demonstrating that the reaction is first order in isoprene. The slope of the first-order plot gives the observed rate constant:

$$\text{Slope} = \frac{4.216 - 0.00}{0.955\,\text{s} - 0.00\,\text{s}} = 4.4\,\text{s}^{-1} = k_{obs} = k[O_3]^y$$

Before we can calculate the true rate constant ($k$), we must determine $y$, the order with respect to ozone. One way to determine $y$ is to repeat the isolation experiment with a different initial concentration of ozone. For example, the experiment could be repeated with the same initial isoprene concentration but with an initial ozone concentration of $2.70 \times 10^{-4}$ M, half as large as the concentration in the first experiment. When all the isoprene has been consumed, the ozone concentration is $2.67 \times 10^{-4}$ M, a decrease of about 1%. When this experiment is done, the result is $k_{obs} = 2.2\,\text{s}^{-1}$, half as large as $k_{obs}$ in the first experiment:

$$k_{obs,I} = k[O_3]_I^y = 4.4\,\text{s}^{-1} \qquad k_{obs,II} = k[O_3]_{II}^y = 2.2\,\text{s}^{-1}$$

We need to determine two unknowns, $k$ and $y$, but we can eliminate $k$ by taking the ratio of the two expressions for $k_{obs}$:

$$\frac{k_{obs,I}}{k_{obs,II}} = \frac{k[O_3]_I^y}{k[O_3]_{II}^y} = \frac{[O_3]_I^y}{[O_3]_{II}^y}$$

Now substitute the appropriate values to obtain a numerical expression for $y$:

$$\frac{k_{obs,I}}{k_{obs,II}} = \frac{4.4\,s^{-1}}{2.2\,s^{-1}} = 2.0$$

$$\frac{[O_3]_I^y}{[O_3]_{II}^y} = \left(\frac{5.4 \times 10^{-4}\,M}{2.7 \times 10^{-4}\,M}\right)^y = (2.0)^y$$

The ratio of experimental rate constants is 2.0, and the ratio of the ozone concentration terms is $(2.0)^y$. To evaluate $y$, we set these ratios equal to each other and take the logarithm of each side of the equality:

$$\ln (2.0) = y \ln (2.0)$$

Solving for $y$ gives $y = \ln (2.0)/\ln (2.0) = 1$, so the reaction is first order in ozone. Thus, the complete experimental rate law for the reaction is as follows:

$$\text{Rate} = k[O_3]\,[\text{Isoprene}]$$

Finally, the true rate constant can be calculated from $k_{obs}$ using the data from either experiment:

$$k_{obs} = k[O_3] \quad \text{so} \quad k = \frac{k_{obs}}{[O_3]} = \frac{4.4\,s^{-1}}{5.4 \times 10^{-4}\,M} = 8.1 \times 10^3\,M^{-1}s^{-1}$$

Example 12-7 provides another illustration of the isolation technique.

## EXAMPLE 12-7 RATE LAW FROM ISOLATION EXPERIMENTS

The reaction of hydrogen and bromine produces hydrogen bromide, a highly corrosive gas:

$$H_2\,(g) + Br_2\,(g) \longrightarrow 2\,HBr\,(g)$$

To determine the rate law for this reaction, a chemist performed two isolation experiments using different initial concentrations. Both experiments gave linear graphs of $\ln\left(\frac{[H_2]_0}{[H_2]}\right)$ vs. $t$, but with different slopes. Here are the details:

| $[Br_2]_0$ (M) | $[H_2]_0$ (M) | Slope of Graph ($s^{-1}$) |
|---|---|---|
| $3.50 \times 10^{-5}$ | $2.50 \times 10^{-7}$ | $8.87 \times 10^{-4}$ |
| $2.00 \times 10^{-5}$ | $2.50 \times 10^{-7}$ | $6.71 \times 10^{-4}$ |

Determine the rate law and the rate constant for the reaction.

**Strategy:** When asked to determine a rate law and rate constant, we must determine the order of the reaction. The rate law for this reaction can be expected to contain the concentrations of $H_2$ and $Br_2$ raised to powers $y$ and $z$ that must be determined:

$$\text{Rate} = k\,[H_2]^y[Br_2]^z$$

Because we expect the rate law to contain the concentrations of more than one species, isolation experiments are required to determine the orders of reaction. In the experiments for which data are shown, the initial concentrations of $Br_2$ are about 100 times the initial concentrations of $H_2$, so the $Br_2$ concentration remains essentially constant over the course of the reaction. Thus, the rate law can be rewritten with $[Br_2]$ included as part of the observed rate constant:

$$\text{Rate} = k_{obs}[H_2]^y \quad \text{where} \quad k_{obs} = k[Br_2]^z$$

**Solution:** In each experiment, a first-order graph gives a straight line, which means that the reaction is first order in hydrogen ($y = 1$). For a first-order reaction, the slope gives $k_{obs}$ as follows:

$$k_{obs,I} = \text{Slope}_I = 8.87 \times 10^{-4}\ s^{-1} \quad \text{and} \quad k_{obs,II} = \text{Slope}_{II} = 6.71 \times 10^{-4}\ s^{-1}$$

To find the order of reaction with respect to bromine, we must compare the ratio of the observed rate constants to the ratio of the initial concentrations of $Br_2$:

$$\frac{k_{obs,I}}{k_{obs,II}} = \frac{k[Br_2]_I^z}{k[Br_2]_{II}^z}$$

We calculate the ratio of observed rate constants from the slopes of the graphs:

$$\frac{k_{obs,I}}{k_{obs,II}} = \frac{8.87 \times 10^{-4}\ s^{-1}}{6.71 \times 10^{-4}\ s^{-1}} = 1.32$$

We calculate the ratio of concentrations from the different initial concentrations of $Br_2$:

$$\frac{[Br^2]_I}{[Br^2]_{II}} = \frac{3.50 \times 10^{-5}\ M}{2.00 \times 10^{-5}\ M} = 1.75$$

Putting these two together yields an equality where $z$ is the only unknown:

$$1.32 = \frac{k_{obs,I}}{k_{obs,II}} = \frac{[Br_2]_I^z}{[Br_2]_{II}^z} = (1.75)^z$$

The exponent $z$ can be isolated by taking the logarithm of both sides:

$$\ln(1.32) = z \ln(1.75) \quad 0.278 = z(0.560) \quad \text{from which} \quad z = 0.496$$

Reaction orders are usually integers or simple fractions, so we round $z$ to 0.5. The reaction is half-order in bromine, and the rate law is as follows:

$$\text{Rate} = k[H_2][Br_2]^{1/2}$$

To calculate the true rate constant ($k$), use the relationship between the observed rate constant and $k$, substituting data from either of the two experiments:

$$k_{obs} = k[Br_2]^{1/2} \quad \text{so} \quad k = \frac{k_{obs}}{[Br_2]^{1/2}}$$

$$k = \frac{8.87 \times 10^{-4}\ s^{-1}}{(3.50 \times 10^{-5}\ M)^{1/2}} = \frac{8.87 \times 10^{-4}\ s^{-1}}{5.92 \times 10^{-3}\ M^{1/2}} = 0.150\ M^{-1/2}s^{-1}$$

 **Do the Results Make Sense?** A reaction order of one-half is not obvious. In Section 12.5, we describe a mechanism that explains the half-order dependence on bromine concentration.

**Practice Exercise 12.7**

When the reaction of methyl bromide with aqueous base was studied under isolation conditions, the following data were collected:

| $[CH_3Br]_0$ (M) | $[OH^-]_0$ (M) | Experimental Order | Slope of Graph ($s^{-1}$) |
|---|---|---|---|
| $5.0 \times 10^{-4}$ | 0.1 | First order | 17.2 |
| $5.0 \times 10^{-4}$ | 0.025 | First order | 4.3 |

Determine the rate law and find the value of $k$.

The half-life of a second-order reaction can also be calculated. Unlike for first-order reactions, however, it depends on the initial concentration. Beginning with Equation 12-5, we have

$$\frac{1}{[A]} - \frac{1}{[A]_0} = kt$$

After one half-life has elapsed, $[A] = 1/2\,[A]_0$ and $t = t_{1/2}$. Thus,

$$\frac{1}{1/2[A]_0} - \frac{1}{[A]_0} = kt_{1/2}$$

$$1 = [A]_0 kt_{1/2}$$

$$t_{1/2} = \frac{1}{k[A]_0} \quad \text{(second order only)}$$

This tells us that the half-life of a second-order reaction increases as the reaction progresses, since $[A]_0$, the concentration at the beginning of each half-life period, is decreasing.

## Initial Rates

A second way to simplify the behaviour of a reaction is the **method of initial rates**. In this method, we measure the rate at the very beginning of the reaction for different concentrations. A set of experiments is done, changing only one initial concentration each time. Instead of measuring the concentration at many different times during the reaction, we make just one measurement for each set of concentrations. The reaction orders can be evaluated from the relationships between the changes in concentration and the changes in initial rates. We illustrate how this works using a gas-phase reaction of $H_2$ with NO:

$$2\,H_2\,(g) + 2\,NO\,(g) \longrightarrow N_2\,(g) + 2\,H_2O\,(g)$$

At an elevated temperature of 1000 K, a sequence of initial-rate experiments gives the following information:

| Experiment | $[H_2\,(g)]$ (M) | $[NO\,(g)]$ (M) | Initial Rate (M/s) |
|---|---|---|---|
| 1 | 0.0010 | 0.0020 | $1.2 \times 10^{-4}$ |
| 2 | 0.0010 | 0.0030 | $1.8 \times 10^{-4}$ |
| 3 | 0.0020 | 0.0020 | $4.8 \times 10^{-4}$ |

For each of these experiments, we can write the same general equation for the initial rate:

$$\text{Initial rate} = k[H_2]_{\text{initial}}^x [NO]_{\text{initial}}^y$$

This equation contains three unknowns that we have to evaluate: $x$, $y$, and $k$. The essential feature of the initial-rate method is that we can take ratios of initial rates under different conditions. First, apply this technique to Experiments 1 and 2, which have the same initial concentration of $H_2$:

$$\text{Initial rate}_1 = 1.2 \times 10^{-4} \text{ M/s} = k (0.0010 \text{ M})^x (0.0020 \text{ M})^y$$

$$\text{Initial rate}_2 = 1.8 \times 10^{-4} \text{ M/s} = k (0.0010 \text{ M})^x (0.0030 \text{ M})^y$$

When we take the ratio of the second initial rate to the first, the rate constant and the initial concentration term for $H_2$ cancel:

$$\frac{\text{Initial rate}_2}{\text{Initial rate}_1} = \frac{1.8 \times 10^{-4} \text{ M/s}}{1.2 \times 10^{-4} \text{ M/s}} = \frac{(0.0030 \text{ M})^y}{(0.0020 \text{ M})^y} = \frac{(0.0030 \text{ M})^y}{(0.0020 \text{ M})^y}$$

Evaluating the ratios, we find $1.5 = (1.5)^y$, from which $y = 1$.

Now repeat this analysis for the third experiment and the first experiment, for which the initial concentrations of NO are the same:

$$\frac{\text{Initial rate}_3}{\text{Initial rate}_1} = \frac{4.8 \times 10^{-4} \text{ M/s}}{1.2 \times 10^{-4} \text{ M/s}} = \frac{(0.0020 \text{ M})^{xy}}{(0.0010 \text{ M})^{xy}} = \frac{(0.0020 \text{ M})^x}{(0.0010 \text{ M})^x}$$

This gives $4 = (2)^x$, from which $x = 2$. Thus, the rate law is rate $= k[H_2][NO]^2$.

Once we know $x$ and $y$, we can use any of the experiments to calculate the rate constant $k$:

$$1.2 \times 10^{-4} \text{ M/s} = k(0.0010 \text{ M})^2(0.0020 \text{ M})$$

$$k = \frac{1.2 \times 10^{-4} \text{ M/s}}{(0.0010 \text{ M})^2(0.0020 \text{ M})} = 6.0 \times 10^4 \text{ M}^{-2}\text{s}^{-1}$$

It is important to note that a single pair of measurements is usually not sufficient to establish the rate law. Care must be taken and extensive data must be collected before the mechanism is "proven."

Example 12-8 provides practice in applying the method of initial rates.

## EXAMPLE 12-8 RATE LAW FROM INITIAL RATES

The reaction of nitrogen dioxide with fluorine generates nitryl fluoride:

$$2 \text{ NO}_2 (g) + \text{F}_2 (g) \longrightarrow 2 \text{ NO}_2\text{F} (g)$$

To determine the rate law for this reaction, a chemist performed several initial-rate experiments using different initial concentrations. Here are the details:

| Experiment | [NO$_2$ (g)] (mM) | [F$_2$ (g)] (mM) | [NO$_2$F (g)] (mM) | Initial Rate (M/s) |
|---|---|---|---|---|
| 1 | 1.0 | 5.0 | 0.10 | $2.0 \times 10^{-4}$ |
| 2 | 2.0 | 5.0 | 0.10 | $4.0 \times 10^{-4}$ |
| 3 | 2.0 | 5.0 | 1.00 | $4.0 \times 10^{-4}$ |
| 4 | 2.0 | 7.5 | 0.10 | $6.0 \times 10^{-4}$ |

Determine the rate law and the rate constant for the reaction.

**Strategy:** When asked to determine a rate law and rate constant, we must determine the order of the reaction. The rate law for this reaction may contain the concentrations of $NO_2$, $F_2$, and $NO_2F$ raised to powers $x$, $y$, and $z$ that must be determined:

$$\text{Rate} = k[NO_2]^x\,[F_2]^y\,[NO_2F]^z$$

Because the rate law contains more than one species, we need to use either isolation or initial rates to determine the orders of reaction. In the experiments of which data are shown, initial-rate data are obtained for various combinations of initial concentrations. We apply the ratios of these initial rates to evaluate the orders.

**Solution:** First determine which pairs of initial rates to compare. Only $[NO_2]$ changes from Experiment 1 to Experiment 2, so we can use that ratio to evaluate $x$. Only $[F_2]$ changes from Experiment 2 to Experiment 4, so we can use that ratio to evaluate $y$. Finally, Experiments 2 and 3 are identical except for $[NO_2F]$, so we can use that pair to evaluate $z$.

To find $x$:

$$\frac{\text{Initial rate}_2}{\text{Initial rate}_1} = \frac{4.0 \times 10^{-4}\text{ M/s}}{2.0 \times 10^{-4}\text{ M/s}} = \frac{k(2.0\text{ mM})^x (5.0\text{ mM})^y (0.10\text{ mM})^z}{k(1.0\text{ mM})^x (5.0\text{ mM})^y (0.10\text{ mM})^z} = \frac{(2.0\text{ mM})^x}{(1.0\text{ mM})^x}$$

Simplifying, $2.0 = (2.0)^x$, so $x = 1$. The reaction is first order in $NO_2$.

To find $y$:

$$\frac{\text{Initial rate}_4}{\text{Initial rate}_2} = \frac{6.0 \times 10^{-4}\text{ M/s}}{4.0 \times 10^{-4}\text{ M/s}} = \frac{k(2.0\text{ mM})^x (7.5\text{ mM})^y (0.10\text{ mM})^z}{k(2.0\text{ mM})^x (5.0\text{ mM})^y (0.10\text{ mM})^z} = \frac{(7.5\text{ mM})^y}{(5.0\text{ mM})^y}$$

Simplifying, $1.5 = (1.5)^y$, so $y = 1$. The reaction is first order in $F_2$.

To find $z$:

$$\frac{\text{Initial rate}_3}{\text{Initial rate}_2} = \frac{4.0 \times 10^{-4}\text{ M/s}}{4.0 \times 10^{-4}\text{ M/s}} = \frac{k(2.0\text{ mM})^x (5.0\text{ mM})^y (1.0\text{ mM})^z}{k(2.0\text{ mM})^x (5.0\text{ mM})^y (0.10\text{ mM})^z} = \frac{(1.0\text{ mM})^y}{(0.10\text{ mM})^y}$$

Simplifying, $1.0 = (10)^z$, so $z = 0$. In other words, there is no effect of $[NO_2F]$ on the rate of the reaction. This means that $[NO_2F]$ does not appear in the rate law, and we say that the reaction is zeroth order in $[NO_2F]$. Although $NO_2F$ may be involved in the reaction, it is not a reactant in the rate-determining step. Thus, the rate law is as follows:

$$\text{Rate} = k[NO_2]\,[F_2]$$

To calculate the true rate constant ($k$), use the expression for the initial rate, substituting data from any of the experiments:

$$\text{Initial rate} = k[NO_2][F_2] \quad \text{so} \quad k = \frac{\text{Initial rate}}{[NO_2][F_2]}$$

$$k = \frac{2.0 \times 10^{-4}\text{ M/s}}{(1.0\text{ mM})(5.0\text{ mM})(10^{-3}\text{ M/mM})^2} = 40\text{ M}^{-1}\text{s}^{-1}$$

 **Do the Results Make Sense?** A rate law that is first order in each reactant is consistent with a mechanism of which the first step is a collision between one of each of these reactant molecules.

**Common Pitfall:** Not comparing the data from the right experiments. Make sure the concentration of only one reactant is changing at a time.

**Practice Exercise 12.8**

In the manufacture of cement, $CaCO_3$ (s) is heated to produce $CaO$ (s) and $CO_2$ (g). The temperature is high enough that the carbon dioxide can undergo a reaction with itself to produce carbon monoxide:

$$2\ CO_2\ (g) \longrightarrow 2\ CO\ (g) + O_2\ (g)$$

Neither CO nor $O_2$ affects the rate. Initial rate studies on this reaction give the following data:

$$[CO_2]_{initial} = 0.025\ M,\ \text{initial rate} = 4.7 \times 10^{-3}\ M/s$$

$$[CO_2]_{initial} = 0.037\ M,\ \text{initial rate} = 7.0 \times 10^{-3}\ M/s$$

Determine the rate law for the reaction and find the value of $k$.

This is the first example where we have tested to see if the concentration of a product affects the rate of a reaction. It may seem intuitive that products should not be involved in rate laws, but as we show in Section 12.5, a product may influence the rate of a reaction. In careful rate studies, this possibility must be considered. It is common for some reagents in a chemical system to have no effect on the rate of chemical reaction. Although it is unusual for *none* of the reagents to affect the rate, there are some reactions that are zero order overall. Such reactions have a particularly simple rate law: Rate $= k$.

Table 12-1 summarizes the parameters for the experimental kinetics of the three simplest types of rate laws.

## TABLE 12-1  RATE PARAMETERS FOR SIMPLE REACTION ORDERS

| Order | Rate Law | Concentration Dependence | Linear Plot | Units of $k$ (concentrations in M, time in s) |
|---|---|---|---|---|
| 0 | rate $= k$ | $[A]_0 - [A] = kt$ | $([A]_0 - [A])$ vs. $t$ | $M\ s^{-1}$ |
| 1 | rate $= k\,[A]$ | $\ln\left(\dfrac{[A]_0}{[A]}\right) = kt$ | $\ln\left(\dfrac{[A]_0}{[A]}\right)$ vs. $t$ | $s^{-1}$ |
| 2 | rate $= k\,[A]^2$ | $\dfrac{1}{[A]} - \dfrac{1}{[A]_0} = kt$ | $\dfrac{1}{[A]} - \dfrac{1}{[A]_0}$ vs. $t$ | $M^{-1}\ s^{-1}$ |

## SECTION EXERCISES

● **12.4.1** $N_2O_5$ can decompose to nitrogen dioxide and oxygen:

$$2\ N_2O_5\ (g) \longrightarrow 4\ NO_2\ (g) + O_2\ (g)$$

Here are some data for this decomposition reaction:

| Time (min) | 0 | 20.0 | 40.0 | 60.0 | 80.0 |
|---|---|---|---|---|---|
| $[N_2O_5]\ (10^{-2}\ M)$ | 0.92 | 0.50 | 0.28 | 0.15 | 0.08 |

Determine the reaction order and the rate constant by constructing appropriate graphs using these data.

●**12.4.2** $AB_2$ reacts with CB as follows:

$$AB_2 (g) + CB (g) \longrightarrow AB (g) + CB_2 (g)$$

Use the following information to determine the rate law: (1) When an experiment is performed with $[AB_2]_0 = 1.75 \times 10^{-3}$ M and $[CB]_0 = 0.15$ M, a plot of $\left( \dfrac{1}{[AB_2]} - \dfrac{1}{[AB_2]_0} \right)$ vs. $t$ gives a straight line. (2) When this experiment is repeated with $[AB_2]_0 = 1.75 \times 10^{-3}$ M and $[CB]_0 = 0.25$ M, the slope of a plot of $\left( \dfrac{1}{[AB_2]} - \dfrac{1}{[AB_2]_0} \right)$ vs. $t$ is identical to the slope of the plot for Experiment (1).

●**12.4.3** The ammonium ion can react with the nitrite ion in aqueous solution as follows:

$$NH_4^+ (aq) + NO_2^- (aq) \longrightarrow N_2 (g) + 2H_2O (l)$$

Use the following information about initial rates to determine the rate law and evaluate the rate constant for this reaction:

| Experiment | $[NH_4^+(aq)]$ (M) | $[NO_2^-(aq)]$ (M) | Initial Rate (M/s) |
|---|---|---|---|
| 1 | 0.040 | 0.200 | $2.15 \times 10^{-6}$ |
| 2 | 0.060 | 0.200 | $3.23 \times 10^{-6}$ |
| 3 | 0.060 | 0.100 | $1.66 \times 10^{-6}$ |

## 12.5 LINKING MECHANISMS AND RATE LAWS

The mechanism of any chemical reaction has the following characteristics:

1. The mechanism is one or more elementary reactions describing how the chemical reaction occurs. These elementary reactions may be unimolecular, bimolecular, or (rarely) termolecular.

2. The sum of the individual steps in the mechanism must give the overall balanced chemical equation. Sometimes a step may occur more than once in the mechanism.

3. The reaction mechanism must be consistent with the experimental rate law.

In this section, we show that every mechanism predicts a rate law. If the rate law predicted by a proposed mechanism matches the experimental rate law, the mechanism is a *possible* description of how the reaction proceeds (there may be more mechanisms leading to the same rate law). On the other hand, if the rate law predicted by the proposed mechanism differs from experimental rate law, the proposed mechanism *must* be wrong.

Remember that a bimolecular elementary reaction is a collision between two molecules; its rate law contains the concentrations of both reactants. For a bimolecular elementary reaction (A + B $\longrightarrow$ products), the elementary rate $= k[A][B]$.

A unimolecular elementary reaction is a fragmentation or rearrangement of one chemical species, so its rate law contains the concentration of only that species. For a unimolecular elementary reaction (C $\longrightarrow$ products), the elementary rate $= k[C]$.

## TABLE 12-2  SOME MECHANISMS AND RATE LAWS*

| Stoichiometry | Mechanism | Rate Law |
|---|---|---|
| $2\,NO_2 \longrightarrow N_2O_4$ | $NO_2 + NO_2 \longrightarrow N_2O_4$ | Rate $= k[NO_2]^2$ |
| $2\,NO_2 \longrightarrow 2\,NO + O_2$ | $2\,NO_2 \longrightarrow NO_3 + NO$ (slow) <br> $NO_3 + O_3 \longrightarrow NO_2 + O_2$ | Rate $= k[NO_2]^2$ |
| $NO_2 + O_3 \longrightarrow NO_2 + O_2$ | $NO + O_3 \longrightarrow NO_2 + O_2$ | Rate $= k[NO][O_3]$ |
| *cis*-2-butene $\longrightarrow$ *trans*-2-butene | *cis*-2-butene $\longrightarrow$ *trans*-2-butene | Rate $= k[$*cis*-2-butene$]$ |
| $C_5H_{11}Br + H_2O \longrightarrow C_5H_{10} + H_3O^+ + Br^-$ | $C_5H_{11}Br \longrightarrow C_5H_{11}^+ + Br^-$ (slow) <br> $C_5H_{11}^+ + H_2O \longrightarrow C_5H_{10} + H_3O^+$ | Rate $= k[C_5H_{11}Br]$ |

*Examples from Sections 12.1 through 12.4.

These rate laws for elementary steps are related to the experimental rate law for the overall reaction in a way that depends on which step in the mechanism is rate-determining.

## Rate-Determining First Step

Each example we have used to introduce the concepts of chemical mechanisms has a first step that is rate-determining. These mechanisms and their rate laws are summarized in Table 12-2.

In each of these mechanisms, the first elementary reaction is the rate-determining step. Because the overall reaction can go no faster than its rate-determining step, no elementary reaction that occurs after the rate-determining step affects the overall rate of reaction. The predicted rate law is first order for a reaction having a first step that is unimolecular and rate-determining. The predicted rate law is second order overall for a reaction having a first step that is bimolecular and rate-determining. For example, the first step of the mechanism for the $C_5H_{11}Br$ reaction is unimolecular and slow, so the rate law predicted by this mechanism is first order: Rate $= k[C_5H_{11}Br]$. Example 12-9 treats the rate law of another mechanism with a first step that is rate-determining.

### EXAMPLE 12-9  PREDICTED RATE LAWS

At elevated temperature, $NO_2$ reacts with CO to produce $CO_2$ and NO:

$$CO + NO_2 \longrightarrow CO_2 + NO$$

In one possible mechanism for this reaction, products form directly in a one-step bimolecular collision that transfers an oxygen atom from $NO_2$ to CO:

In another possible mechanism, two $NO_2$ molecules collide in the rate-determining step to form NO and $NO_3$. In a second and faster step, the highly reactive $NO_3$ intermediate transfers an oxygen atom to CO in a bimolecular collision:

What is the predicted rate law for each of these mechanisms?

**Strategy:** It is best to work with each mechanism separately. When the first step is rate-determining, the predicted rate law matches the rate expression for that first elementary step.

**Solution:** First, analyze the one-step mechanism:

$$CO + NO_2 \longrightarrow CO_2 + NO$$

This process is analogous to the reaction of NO and $O_3$ discussed in Section 12.3. In a simple one-step atom transfer, the reaction is first order in each of the starting materials and second order overall. Thus, rate = $k[CO][NO_2]$. The second mechanism occurs in two steps:

$$2\ NO_2 \longrightarrow NO_3 + NO \quad \text{(slow, rate determining)}$$
$$NO_3 + CO \longrightarrow NO_2 + CO_2 \quad \text{(fast)}$$

In this mechanism, the first step is slow and rate-determining. Because the collision partners are identical, this reaction is second order in $NO_2$ and second order overall. Thus, rate = $k[NO_2]^2$

**Do the Results Make Sense?** Carbon monoxide does not appear in the second rate law because it participates in the mechanism *after* the rate-determining step. Remember, any reaction after the rate-determining step does not affect the overall rate of reaction.

Experiments show that this reaction is second order in $NO_2$, as predicted by the second proposed mechanism. The one-step mechanism can be ruled out because it is not consistent with the experimental rate law. Agreement with the rate law does not prove that the second mechanism is the correct one, however, because other mechanisms may predict the same rate law. It is one strong piece of evidence that supports this particular two-step process.

**Practice Exercise 12.9**

Here is a proposed mechanism for a Diels–Alder reaction between two butadiene molecules:

What rate law does this mechanism predict?

## Rate-Determining Later Step

For many reaction mechanisms, the rate-determining step occurs after one or more faster steps. In such cases, the reactants in the early steps may or may not appear in the rate law. Furthermore, the rate law is likely to depart from simple first- or second-order behaviour. Fractional orders, negative orders, and overall orders greater than 2, all are signals that a fast first step is followed by a slow subsequent step.

Example 12-7 treats a reaction of this kind. The experimental rate law for the reaction of $H_2$ gas with $Br_2$ gas depends on the square root of the $Br_2$ concentration, and the reaction is also first order in $H_2$:

$$H_2 + Br_2 \longrightarrow 2\,HBr \qquad Rate = k[H_2][Br_2]^{1/2}$$

Despite the simple 1:1 stoichiometry of the overall reaction, this experimental rate law cannot be explained by a simple mechanism. For the first step of the mechanism for this reaction to be rate-determining, it would have to include a half-molecule of $Br_2$. There is no such species, so the first step cannot be rate-determining. Instead, some later step in the mechanism must be rate-determining.

Figure 12-13 uses molecular pictures to show the accepted mechanism for this reaction. The reaction begins with the dissociation of $Br_2$ molecules into Br atoms:

$$Br_2 \xrightarrow{\;k_1\;} 2\,Br$$

Almost all the Br atoms produced in the first step simply recombine to regenerate $Br_2$ molecules:

$$2\,Br \xrightarrow{\;k_{-1}\;} Br_2$$

**Figure 12-13**

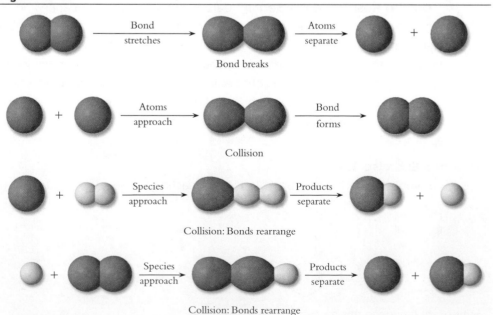

A molecular view of the accepted mechanism for the reaction of $H_2$ and $Br_2$ to produce 2 HBr.

This is an example of a **reversible reaction**, one that occurs rapidly in both directions. For simplicity, we combine these two elementary reactions in a single expression that uses double arrows:

$$Br_2 \underset{k_{-1}}{\overset{k_1}{\rightleftharpoons}} 2\,Br \quad \text{(Fast, reversible)}$$

Most bromine atoms produced in the first step recombine, but occasionally, a Br atom reacts with a molecule of $H_2$ to form one molecule of HBr and a hydrogen atom. This step is slow and rate-determining:

$$Br + H_2 \xrightarrow{k_2} HBr + H \quad \text{(Slow, rate–determining)}$$

To complete the mechanism, we need a step that consumes H atoms. The hydrogen atom produced in the rate-determining step undergoes a third elementary reaction, a bimolecular collision with a $Br_2$ molecule. The H atom rapidly forms a bond with one Br atom to form one molecule of HBr and a Br atom:

$$H + Br_2 \xrightarrow{k_3} HBr + Br \quad \text{(fast)}$$

Typically, elementary reactions in a mechanism are written with their rate constants above or below the reaction arrow. A reaction that is the reverse of another is labelled accordingly.

Is this mechanism satisfactory? A satisfactory mechanism must meet three criteria. It must be made up entirely of reasonable elementary steps, it must give the correct stoichiometry of the reaction, and it must predict the experimental rate law.

This mechanism has four steps. The first step is the unimolecular decomposition of $Br_2$. The remaining steps all are simple bimolecular collisions. We can readily visualize each step as a reaction that the molecules can undergo, so the mechanism meets the first criterion.

The easiest way to determine whether the steps lead to the correct stoichiometry is to write all four reactions in the same direction. Then cancel those species that appear both on the left and on the right:

$$\cancel{Br_2} \longrightarrow 2\,\cancel{Br}$$
$$2\,\cancel{Br} \longrightarrow \cancel{Br_2}$$
$$\cancel{Br} + H_2 \longrightarrow HBr + \cancel{H}$$
$$\cancel{H} + Br_2 \longrightarrow HBr + \cancel{Br}$$

Net reaction $\overline{\quad H_2 + Br_2 \longrightarrow 2\,HBr \quad}$

Combining the steps of the mechanism leads to the balanced equation for the overall reaction, so the second criterion also is satisfied.

To see if the proposed mechanism predicts the correct rate law, we start with the rate-determining step. The second step in this mechanism is rate-determining, so the overall rate of the reaction is governed by the rate of this step:

$$\text{Rate} = k_2[Br][H_2]$$

This rate law accurately describes the rate behaviour predicted by the proposed mechanism, but the law cannot be tested against experiments because it contains the concentration of Br atoms, which are intermediates in the reaction. As mentioned earlier, an intermediate has a short lifetime and is hard to detect, so it is difficult to make accurate measurements of its concentration. Furthermore, it is not possible to adjust the experimental conditions in a way that changes the concentration of an intermediate by a known amount. Therefore, if this proposed rate law is to be tested against experimental behaviour, the concentration of the intermediate must be expressed in terms of the concentrations of reactants and products.

## Equality of Rates

One way to relate the concentration of an intermediate to other concentrations is by assuming that earlier reversible steps have equal forward and reverse rates. The proposed mechanism begins with the decomposition of $Br_2$ molecules into Br atoms, most of which recombine rapidly to give $Br_2$ molecules. At first, $Br_2$ molecules decompose faster than Br atoms recombine, but the Br atom concentration quickly becomes large enough for recombination to occur at the same rate as decomposition. When the two rates are equal, so are their rate expressions:

$$\text{Rate of decomposition} = \text{Rate of recombination}$$

$$k_1[Br_2] = k_{-1}[Br]^2$$

A reaction that has equal forward and reverse rates is said to be in *equilibrium*. Equilibrium is the subject of the next chapter.

This equality can be used to express the concentration of the intermediate in terms of concentrations of the reactants. First solve for [Br] by rearranging the rate expressions and taking the square root of both sides:

$$[Br]^2 = \frac{k_1}{k_{-1}}[Br_2] \quad \text{from which} \quad [Br] = \left(\frac{k_1}{k_{-1}}\right)^{1/2}[Br_2]^{1/2}$$

Now substitute this equation for [Br] into the expression for the rate-determining step:

$$\text{Rate} = k_2[Br][H_2] = k_2\left(\frac{k_1}{k_{-1}}\right)^{1/2}[H_2][Br_2]^{1/2} = k[H_2][Br_2]^{1/2}$$

In this expression, $k$, the rate constant for the overall reaction, is related to the elementary rate constants:

$$k = k_2\left(\frac{k_1}{k_{-1}}\right)^{1/2}$$

The predicted rate law is first order in hydrogen and one-half order in bromine, in agreement with the experimental rate law determined in Example 12-7. Thus, the proposed mechanism satisfies the three requirements for a satisfactory mechanism.

As the mechanism for the reaction between $H_2$ and $Br_2$ shows, the observation of an order other than first or second indicates that some step beyond the first one in the mechanism is rate-determining. The one-half-order rate dependence for $Br_2$ comes about because of the initial rapid cleavage reaction in which each $Br_2$ molecule produces two Br atoms.

Our Focus 12-1: Chemistry and the Environment discusses reactions in the stratosphere, and the connection between mechanisms and rate laws is illustrated further by Example 12-10.

**EXAMPLE 12-10  REACTION BETWEEN NO AND $O_3$**

Nitrogen oxide converts ozone into molecular oxygen, as follows:

$$O_3 + NO \longrightarrow O_2 + NO_2$$

The experimental rate law is rate $= k[O_3][NO]$. Which of the following mechanisms are consistent with the experimental rate law?

**Mechanism I**

$$O_3 + NO \xrightarrow{k_1} O + NO_3 \quad \text{(slow)}$$

$$O + O_3 \xrightarrow{k_2} 2\,O_2 \quad \text{(fast)}$$

$$NO_3 + NO \xrightarrow{k_3} 2\,NO_2 \quad \text{(fast)}$$

**Mechanism II**

$$O_3 \underset{k_{-1}}{\overset{k_1}{\rightleftharpoons}} O_2 + O \quad \text{(fast, reversible)}$$

$$NO + O \xrightarrow{k_2} NO_2 \quad \text{(slow)}$$

**Mechanism III**

$$O_3 + NO \xrightarrow{k_3} O_2 + NO_2 \quad \text{(slow)}$$

**Strategy:** We need to determine whether the rate law predicted by each mechanism matches the experimental rate law, by calculating the rate law predicted by each.

**Solution:** Mechanism I is a three-step process in which the first step is rate-determining. When the first step of a mechanism is rate-determining, the predicted rate law is the same as the rate expression for that first step. Here, the rate-determining step is a bimolecular collision. The rate expression for a bimolecular collision is first order in each collision partner:

$$\text{Rate} = k_1[O_3][NO]$$

Mechanism I is consistent with the experimental rate law. If we add the elementary reactions, we find that it also gives the correct overall stoichiometry, so this mechanism meets all the requirements for a satisfactory one.

Mechanism II begins with fast reversible ozone decomposition followed by a rate-determining bimolecular collision of an oxygen atom with a molecule of NO. The rate of the slow step is as follows:

$$\text{Rate} = k_2[NO][O]$$

This rate expression contains the concentration of an intermediate, atomic oxygen. To convert the rate expression into a form that can be compared with the experimental rate law, assume that the rate of the first step is equal to the rate of its reverse process. Then solve the equality for the concentration of the intermediate:

$$k_1[O_3] = k_{-1}[O][O_2] \quad \text{from which} \quad [O] = \frac{k_1[O_3]}{k_{-1}[O_2]}$$

$$\text{Rate} = \frac{k_1 k_2 [O_3][NO]}{k_{-1}[O_2]}$$

Mechanism II is inconsistent with the experimental rate law. Notice that this mechanism predicts a rate that is inversely proportional to the concentration of molecular oxygen. That means that oxygen would slow down the reaction: as the concentration of oxygen *increased*, the rate of the reaction would *decrease*.

Mechanism III is a simple one-step bimolecular collision. Its predicted rate law is as follows:

$$\text{Rate} = k_1[O_3][NO]$$

Mechanism III is consistent with the experimental rate law.
Therefore, the $O_3 + NO$ reaction might go by Mechanism I or Mechanism III, but not by Mechanism II.

**Do the Results Make Sense?** The fact that two different mechanisms give the same predicted rate law illustrates one feature of mechanisms. Chemists can imagine more than one possible mechanism for almost any chemical reaction. We have to use additional chemical knowledge to distinguish among possible mechanisms. Here, Mechanism I is less likely than Mechanism III because the products of the first step, O and $NO_3$, are less stable than the $O_2$ and $NO_2$ formed in Mechanism III. Other factors being equal, molecules generally follow the path of lower energy, which is Mechanism III in this case.

### Practice Exercise 12.10

Here is a proposed mechanism for the formation of $N_2O_5$ from ozone:

$$NO_2\ (g) + O_3\ (g) \underset{k_{-1}}{\overset{k_1}{\rightleftharpoons}} NO_3\ (g) + O_2\ (g) \quad \text{(fast, reversible)}$$

$$NO_3\ (g) + NO_2\ (g) \overset{k_2}{\longrightarrow} N_2O_5\ (g) \quad \text{(slow)}$$

What rate law does this mechanism predict?

## ■ FOCUS 12-1  CHEMISTRY AND THE ENVIRONMENT: REACTIONS OF OZONE

The ozone layer is found 30–45 km above the surface of the Earth. Nowhere else in the atmosphere is ozone found in significant amounts. As described in Section 4.7, ozone is limited to this narrow band because only at this altitude can it form at an appreciable rate through sunlight interacting with $O_2$.

Ozone is formed in the stratosphere by the action of ultraviolet light on oxygen molecules. The overall reaction is:

$$3\ O_2 + h\nu_{(\lambda\ <\ 240\ nm)} \longrightarrow 2\ O_3$$

An oxygen molecule absorbs a photon ($h\nu$) of high-energy ultraviolet light ($\lambda < 240$ nm). The energy of the photon breaks $O_2$ into two O atoms:

$$O_2 + h\nu_{(\lambda\ <\ 240\ nm)} \longrightarrow O + O \quad \text{Step 1}$$

Oxygen atoms are reactive and add to $O_2$ to form $O_3$:

$$O + O_2 \longrightarrow O_3 \quad \text{Step 2}$$

This ozone molecule contains excess energy and can decompose again. The stratosphere contains enough $N_2$, however, that an ozone molecule usually collides with a nitrogen molecule and gives up its excess energy before it can break apart. Step 1 of the mechanism generates two oxygen atoms, each of which reacts with an $O_2$ molecule. Thus, each time the first step occurs, the second step occurs twice:

$$O_2 + h\nu_{(\lambda \,<\, 240 \text{ nm})} \longrightarrow O + O$$

$$O + O_2 \longrightarrow O_3$$

$$O + O_2 \longrightarrow O_3$$

Net reaction: $3\,O_2 + h\nu_{(\lambda \,<\, 240 \text{ nm})} \longrightarrow 2\,O_3$

The first step is the rate-determining step, because as soon as an oxygen atom forms, it is snapped up by the nearest available $O_2$ molecule.

The formation of $O_3$ goes on continuously, but $O_3$ itself is also photolyzed by ultraviolet light between 240 and 340 nm:

$$O_3 + h\nu_{(\lambda \,=\, 240\,-\,350 \text{ nm})} \longrightarrow O + O_2$$

The O atom can react with a second $O_3$ molecule:

$$O + O_3 \longrightarrow 2\,O_2$$

The result is a reduction in the amount of ozone:

Net: $2\,O_3 + h\nu_{(\lambda \,=\, 240\,-\,350 \text{ nm})} \longrightarrow 3\,O_2$

Short-wavelength ultraviolet light forms $O_3$, whereas long-wavelength ultraviolet light decomposes $O_3$. The reactions create a delicate balance in which the rate of $O_3$ decomposition matches the rate of $O_3$ production. The resulting ozone layer absorbs nearly all the solar photons in the 240–340 nm range. Photons in this wavelength range have enough energy to damage and even destroy living cells, so the ozone layer protects life on Earth.

Unfortunately, chemical species produced by humans can react in ways that reduce the $O_3$ concentration. For example, NO changes the oxygen balance through the following reactions:

$$NO + O_3 \longrightarrow NO_2 + O_2$$

$$NO_2 + h\nu_{(\lambda \,<\, 400 \text{ nm})} \longrightarrow NO + O$$

$$O + NO_2 \longrightarrow O_2 + NO$$

The first step in this process occurs twice to generate the two molecules of $NO_2$ needed for the second and third steps. The result of this sequence is that NO in the stratosphere increases the rate of $O_3$ decomposition:

$$\text{Net: } 2\,O_3 + hv_{(\lambda\,<\,400\text{ nm})} \longrightarrow 3\,O_2$$

As mentioned in the introduction to this chapter, CFCs also have potentially devastating effects on the ozone layer. In Section 12.7, we describe how CFCs destroy stratospheric ozone.

## SECTION EXERCISES

**12.5.1** A student proposes the following mechanism for the gas-phase decomposition of $N_2O_5$ (all substances are gases):

| | | |
|---|---|---|
| Overall reaction: | $2\,N_2O_5$ $\longrightarrow$ | $4\,NO_2 + O_2$ |
| Step 1: | $N_2O_5$ $\longrightarrow$ | $N_2O_4 + O$ |
| Step 2: | $O + N_2O_5$ $\longrightarrow$ | $N_2O_4 + O_2$ |

(a) Propose a third step to complete this mechanism.

(b) Experiments show that this decomposition reaction is first order. Which step in the mechanism must be rate-determining?

**12.5.2** What order for $N_2O_5$ does the mechanism in Section Exercise 12.5.1 predict if the first step is fast and reversible and the second step is rate-determining?

**12.5.3** The oxidation of NO by $O_2$ is an example of a third-order reaction:

$$2\,NO\,(g) + O_2\,(g) \longrightarrow 2\,NO_2\,(g)$$

$$\text{Rate} = k[NO]^2\,[O_2]$$

(Although a single elementary termolecular mechanism is consistent with the rate law, three-body collisions are rare. It is more likely that the first step is fast and reversible:

$$NO + NO \underset{k_{-1}}{\overset{k_1}{\rightleftharpoons}} N_2O_2$$

(a) Propose a slow second step to complete this mechanism.

(b) Show that this mechanism is consistent with the observed rate law.

## 12.6 REACTION RATES AND TEMPERATURE

Common experience tells us that chemical reactions proceed faster at higher temperature. Food is stored in refrigerators because food spoils more quickly at high temperature than at low temperature. Wood does not react with oxygen at an appreciable rate at room temperature, but it burns vigorously at high temperature. At the macroscopic level, higher temperature means faster reactions, but how does temperature affect the rate of a reaction at the molecular level? Because temperature is a measure of the energy of motion of molecules, we need to explore the relationship between reaction rates and molecular energy to answer this question.

**Figure 12-14**

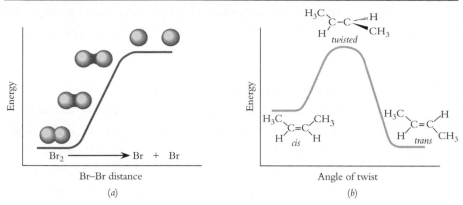

Energy profiles for two unimolecular processes: (a) the unimolecular decomposition of a bromine molecule; (b) the unimolecular isomerization of *cis*-2-butene.

## Energy Changes in a Unimolecular Reaction

In a unimolecular reaction, a molecule fragments into two pieces or rearranges to a different isomer. In either case, a chemical bond breaks. For example, in the fragmentation of bromine molecules, breaking a $\sigma$ bond gives a pair of bromine atoms:

$$Br_2 \longrightarrow 2\ Br$$

Recall that this unimolecular process is the first step of the reaction between molecular hydrogen and molecular bromine to give HBr.

Figure 12-14*a* shows the energy changes that occur during the fragmentation of a bromine molecule. The reaction begins with a stretching of the Br—Br bond. As the bond becomes longer, the system moves to higher energy. This decreased stability results from the loss of orbital overlap as the bromine atoms move apart. If the bond is stretched far enough, the molecule fragments into two bromine atoms. When a bond breaks, the fragment products are always at higher energy than the starting materials.

Like fragmentations, unimolecular rearrangements are always "uphill" at the beginning of the process, because a bond breaks. Unlike fragmentations, rearrangements are "downhill" at the end as a new bond forms. An example of this kind of energy profile for the isomerization reaction of *cis*-2-butene appears in Figure 12-14*b*.

## Energy Changes During Bimolecular Reactions

In a bimolecular reaction, some bonds break and other bonds form. For example, in the reaction between NO and $O_3$, an O atom is transferred from $O_3$ to NO:

$$O_3 + NO \longrightarrow O_2 + NO_2$$

During the collision, the ozone molecule distorts as the O—O bond stretches and weakens, and NO must also undergo distortion as it bonds with the incoming oxygen atom from ozone. Distorting a molecule from its most favoured configuration always requires the addition of energy. Thus, at the outset of the reaction, the system is destabilized by the molecular distortions required for atom transfer. At the

**Figure 12-15**

A schematic representation of the energy profile for the bimolecular reaction between $O_3$ and NO molecules.

end of the process, on the other hand, the system becomes stabilized as the products adopt their lowest-energy configuration. Figure 12-15 shows a schematic representation of the energy profile for this reaction.

## Activation Energy

The energy profiles for these elementary reactions share several common features, one of which is particularly important for a discussion of reaction rates: in each example, the system must move "uphill" from the energy of the reactants to an energy maximum. In almost all chemical reactions, the molecules must overcome an energy barrier before starting materials can become products. This energy barrier, the minimum energy that must be supplied before reaction can take place, is the **activation energy** ($E_a$) of the chemical reaction. Activation energies arise because chemical bonds in reactant molecules must distort or break before new bonds can form in product molecules. The activation energy for any reaction mechanism is independent of both *reactant concentrations* and *temperature*.

**Figure 12-16**

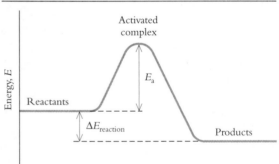

A generalized activation energy diagram for an elementary reaction. The figure shows a reaction that releases energy, resulting in a negative overall energy change ($\Delta E$).

Figures 12-14 and 12-15 are **activation energy diagrams** for elementary reactions. They are plots of the change in energy as reactants are transformed into products. Figure 12-16 shows a generalized activation energy diagram labelled with its characteristics. The *x*-axis represents the course of the reaction as starting materials are transformed into products. Because these transformations may involve complicated combinations of rotations, vibrations, and atom transfers, accepted practice is to label this axis as the **reaction coordinate** without specifying further details about what changes occur.

The *y*-axis of an activation energy diagram is the energy of the molecular system. The graph shows how chemical energy changes during the course of the reaction. Low values of energy represent high molecular stability, and high values represent low molecular stability. The reaction shown in the activation energy diagram in Figure 12-16 releases energy, so its products are lower in energy than its reactants are. For a reaction that absorbs energy, in contrast, the activation energy diagram shows the energy of the products to be higher than the energy of the reactants. Whether the net reaction releases or absorbs energy, however, there almost always is a positive activation energy because of the requirement to break or distort one chemical bond before another can form.

The molecular arrangement at the point of highest energy along the diagram is known as the **activated complex**. The difference in energy between the reactants and this activated complex is the activation energy, $E_a$. An activated complex is *not* considered to be an intermediate. An intermediate is stable for a short time, perhaps milliseconds, and may be observed with highly sensitive instruments. An activated complex forms and disappears in the time required for a molecular collision, which is about $10^{-13}$ second. For this reason, an activated complex is also known as a *transition state*. In other words, the activated complex represents the instantaneous configuration of atoms, at the top of the activation energy barrier, during the conversion from reactants with one set of bonds into products with a different set of bonds.

The magnitude of an activation energy depends on the details of the bonding changes that occur during the formation of the activated complex. As described in Chapter 3, the energy change for a reaction ($\Delta E$) can be estimated from average bond energies. In contrast, the activation energy ($E_a$) is not so easily estimated; it depends on how far bonds must be stretched before others can start to form, and this is different for every reaction. However, activation energies can be determined by measuring how rate constants vary with temperature.

## Activation Energy and the Rate Constant

A chemical reaction cannot occur unless the starting materials have enough energy to overcome the activation energy barrier. Where do molecules obtain this energy? Recall from Section 3.1 that every molecule has thermal energy associated with translation, rotation, and vibration. During a chemical reaction, some of this thermal energy is transformed into chemical energy as a molecule distorts or as two molecules collide.

Any collection of molecules includes molecules with a wide range of energies. Some molecules have much energy, but others have very little. This distribution of molecular energies, described in Section 3.1, depends on the temperature of the sample, as shown in Figure 12-17. Heating the system increases the translational, rotational, and vibrational energies of the molecules. Thus, as the temperature increases, the distribution broadens and shifts toward higher energy.

The energy distributions in Figure 12-17 show that only some of the molecules in any reaction system have energies greater than that required for reaction. How large this fraction is depends on $E_a$ and temperature:

**Figure 12-17**

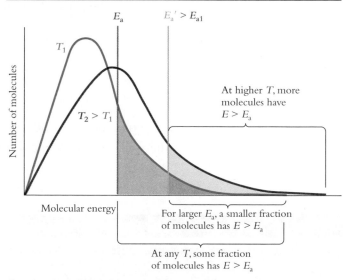

The distribution of molecular energies, shown at two different temperatures. Notice that the energy needed for reaction to occur is a fixed value that is independent of $T$.

1. At any $T$, a larger fraction of molecules can react if $E_a$ is small than if $E_a$ is large.

2. For any reaction, a larger fraction of molecules can react if $T$ is high than if $T$ is low.

The larger the fraction of molecules that can react, the faster a reaction will proceed. Thus, a reaction with a low activation energy is faster than one with a high activation energy, and the rate of any reaction increases with increasing temperature.

## The Arrhenius Equation

The value of the rate constant for a particular reaction depends on the activation energy of the reaction, the temperature of the system, and how often a collision occurs in which the atoms are in the required orientation. All these factors can be summarized in a single equation, called the **Arrhenius equation:**

$$k = Ae^{-E_a/RT} \qquad (12\text{-}6)$$

In this equation, $A$ is the value that the rate constant would have if all molecules had enough energy to react. The exponential term contains the dependence on activation energy ($E_a$ in J/mol) and temperature ($T$ in K). The gas constant ($R$ in J $mol^{-1}$ $K^{-1}$) serves as a unit conversion factor. The negative sign of the exponential means that the larger the value of the exponent, the smaller the rate constant. Qualitatively, this means that as $E_a$ gets larger, the rate constant gets smaller, and that as $T$ gets larger, the rate constant gets larger. The Arrhenius equation is an approximation, and in practice, $A$ varies somewhat with temperature, but in this chapter we will assume it is a true constant.

Equation 12-6 can be converted from exponential form to a form that is easier to treat graphically, by taking the natural logarithm of both sides and making use of the fact that the natural logarithm of an exponential is the exponent:

$$\ln k = \ln\left(Ae^{-E_a/RT}\right) = \ln A + \ln\left(e^{-E_a/RT}\right)$$

$$\ln k = \ln A - \frac{E_a}{RT} \qquad (12\text{-}7)$$

Equation 12-7 leads to a linear graph: plotting $\ln k$ along the $y$-axis vs. $1/T$ along the $x$-axis gives a straight line with a slope of $-E_a/R$ and an intercept of $\ln A$. Thus, $A$ and $E_a$ can be evaluated after $k$ has been measured at several different temperatures. Example 12-11 shows how this is done.

## EXAMPLE 12-11 GRAPHING TO DETERMINE $E_a$

At high temperature, cyclopropane isomerizes to propene:

Cyclopropane
($C_3H_6$)

Propene
($C_3H_6$)

When this reaction is studied at different temperatures, the following rate constants are obtained:

| $T$ (°C) | 477 | 523 | 577 | 623 |
|---|---|---|---|---|
| $k$ ($s^{-1}$) | $1.8 \times 10^{-4}$ | $2.0 \times 10^{-3}$ | $2.7 \times 10^{-2}$ | $2.3 \times 10^{-1}$ |

What is the activation energy for the isomerization of cyclopropane?

**Strategy:** The variation of a rate constant with temperature is described by the Arrhenius equation. According to its logarithmic form (Equation 12-7), a plot of ln k vs. 1/T, with temperature expressed in Kelvins, should be a straight line.

**Solution:** We need to convert the data into the appropriate form and then prepare a graph. Here are the values that should be graphed:

| T (K) | 750 | 796 | 850 | 896 |
|---|---|---|---|---|
| $1/T\,(10^{-3}\,K^{-1})$ | 1.33 | 1.26 | 1.18 | 1.12 |
| ln K | −8.62 | −6.21 | −3.61 | −1.43 |

A graph of these data has the expected linear form, with the following slope:

$$\text{Slope} = -3.27 \times 10^4 \text{ K}$$

According to Equation 12-7

$$\text{Slope} = -\frac{E_a}{R} \quad \text{from which} \quad E_a = -R\,(\text{Slope})$$

$$E_a = -(8.314 \text{ J/mol K})(-3.27 \times 10^4 \text{ K})(10^{-3} \text{ kJ/J}) = 2.7 \times 10^2 \text{ kJ/mol}$$

**Does the Result Make Sense?** A C—C bond must break for isomerization to occur, and typical C—C bond energies are around 350 kJ/mol (Table 6–2). The activation energy is considerably smaller than this because the cyclopropane ring is highly strained away from the ideal $sp^3$ hybridization angle, so the C—C bond is abnormally weak in this compound.

**Practice Exercise 12.11**

Here are some data for the rate constant of the reaction $O_3 + NO \longrightarrow O_2 + NO_2$:

| T (°C) | 125 | 175 | 225 | 275 |
|---|---|---|---|---|
| $k\,(10^7\,M^{-1}\,s^{-1})$ | 3.46 | 5.07 | 6.90 | 8.87 |

Prepare the appropriate graph and determine the activation energy for this reaction.

Graphing, using rate constants measured at several temperatures, is the most accurate method to find the value of $E_a$. However, a good estimate of $E_a$ can be calculated from rate constants measured at just two temperatures:

$$\ln k_1 = \ln A - \frac{E_a}{RT_1} \quad \text{and} \quad \ln k_2 = \ln A - \frac{E_a}{RT_2}$$

Thus, the $A$ term can be eliminated by subtracting one equation from the other:

$$\ln k_2 - \ln k_1 = \left(\ln A - \frac{E_a}{RT_2}\right) - \left(\ln A - \frac{E_a}{RT_1}\right) = \frac{E_a}{R}\left(\frac{1}{T_1} - \frac{1}{T_2}\right)$$

$$\ln\left(\frac{k_2}{k_1}\right) = \frac{E_a}{R}\left(\frac{1}{T_1} - \frac{1}{T_2}\right) \qquad (12\text{-}8)$$

Example 12-12 shows how to use Equation 12-8.

## EXAMPLE 12-12 CALCULATING AN ACTIVATION ENERGY

The reactions of $NO_2$ have been studied as a function of temperature. For the following decomposition reaction, the rate constant is $2.7 \times 10^{-2}$ $M^{-1}$ $s^{-1}$ at 227 °C and $2.4 \times 10^{-1}$ $M^{-1}$ $s^{-1}$ at 277 °C:

$$2\ NO_2 \longrightarrow 2\ NO + O_2$$

Studies of the conversion of $NO_2$ to $N_2O_4$ give $k = 5.2 \times 10^8$ $M^{-1}$ $s^{-1}$ at both 25 °C and 87 °C:

$$2\ NO_2 \longrightarrow N_2O_4$$

Calculate the activation energies of these two reactions.

**Strategy:** The graphical treatment of the variation of rate constants with temperature is appropriate when a set of values is available. When only two values of the rate constant are available, $E_a$ is calculated using Equation 12-8, which can be rearranged to solve for $E_a$:

$$E_a = \frac{R \ln\left(\dfrac{k_2}{k_1}\right)}{\left(\dfrac{1}{T_1} - \dfrac{1}{T_2}\right)}$$

Remember that the units must be consistent, so $T$ must be in K and $R$ in J mol$^{-1}$ K$^{-1}$.

**Solution:** For the decomposition of $NO_2$:
Convert the temperatures to kelvins:

$$227\ °C + 273 = 500\ K$$
$$227\ °C + 273 = 500\ K$$

$$E_a = \frac{(8.314\ \text{J mol}^{-1}\ \text{K}^{-1}) \ln\left(\dfrac{2.4 \times 10^{-1}\ M^{-1}\ s^{-1}}{2.7 \times 10^{-2}\ M^{-1}\ s^{-1}}\right)}{\left(\dfrac{1}{500\ K} - \dfrac{1}{550\ K}\right)}$$

$$E_a = \frac{(8.314\ \text{J mol}^{-1}\ \text{K}^{-1}) \ln(8.89)}{(2.000 \times 10^{-3}\ K^{-1} - 1.818 \times 10^{-3}\ K^{-1})} = \frac{(8.314\ \text{J mol}^{-1}\ \text{K}^{-1})(2.18)}{1.82 \times 10^{-4}\ K^{-1}}$$

As usual, we carry one additional significant figure until we round the final answer to two significant figures:

$$E_a = (9.96 \times 10^4\ \text{J/mol})(10^{-3}\ \text{kJ/J}) = 1.0 \times 10^2\ \text{kJ/mol}$$

For the formation of $N_2O_4$:

$$k_2 = k_1 \quad \text{so} \quad \ln\left(\frac{k_2}{k_1}\right) = \ln 1 = 0 \quad \text{giving} \quad E_a = 0\ \text{kJ/mol}$$

**Do the Results Make Sense?** Because bonds must be broken in most reactions, we expect activation energies to be comparable to but smaller than bond energies, so an activation energy of 100 kJ/mol is reasonable. An activation energy of 0 kJ/mol indicates that no bonds need to be distorted or broken in this reaction. We discuss this in more detail shortly.

**Practice Exercise 12.12**
Cyclobutane decomposes to form ethylene:

$$C_4H_8 \longrightarrow 2\, C_2H_4.$$

Experimental rate constants for this reaction are $k = 6.1 \times 10^{-8}\,s^{-1}$ at 327 °C and $k = 1.5 \times 10^{-11}\,s^{-1}$ at 245 °C. Determine the activation energy for the reaction.

After the activation energy of a reaction has been determined, we can use the Arrhenius equation to estimate values of the rate constant for the reaction at temperatures where experiments have not been carried out. This is particularly useful for temperatures at which a reaction is too slow or too fast to be studied conveniently. Example 12-13 illustrates this application.

## EXAMPLE 12-13 CALCULATING $k$ FROM $E_a$

When two $NO_2$ molecules collide and react, they can form a bond or exchange an oxygen atom. For the oxygen exchange, $E_a = 1.0 \times 10^2$ kJ/mol, and the rate constant at 250 °C is $8.6 \times 10^{-2}\,M^{-1}\,s^{-1}$. Estimate the rate constant for oxygen exchange at room temperature (25 °C).

**Strategy:** We can calculate a rate constant at any temperature using the Arrhenius equation provided that we know $E_a$ and the rate constant for the reaction at some other temperature.

**Solution:** Equation 12-8 relates rate constants at different temperatures to the activation energy:

$$\ln\left(\frac{k_2}{k_1}\right) = \frac{E_a}{R}\left(\frac{1}{T_1} - \frac{1}{T_2}\right)$$

The values that we need to substitute into this equation are given in the problem

$$k_1 = 8.6 \times 10^{-2}\,M^{-1}\,s^{-1} \quad T_1 = 250\,°C = 523\,K \quad T_2 = 25\,°C = 298\,K \quad E_a = 1.0 \times 10^2\,kJ/mol$$

After substituting, we solve for $k_2$, the rate constant of the decomposition reaction at 298 K:

$$\ln\left(\frac{k_2}{8.2 \times 10^{-2}\,M^{-1}\,s^{-1}}\right) = \frac{(1.0 \times 10^2\,kJ/mol)(10^3\,J/kJ)}{8.314\,J\,mol^{-1}\,K^{-1}}\left(\frac{1}{524\,K} - \frac{1}{298\,K}\right)$$

$$= -17.4$$

$$\frac{k_2}{8.6 \times 10^{-1}\,M^{-1}\,s^{-1}} = e^{-17.4} = 2.8 \times 10^{-8}$$

$$k_2 = 2.4 \times 10^{-9}\,M^{-1}s^{-1}$$

**Does the Result Make Sense?** Notice that at room temperature, the rate constant for exchange of oxygen atoms, $2.4 \times 10^{-9}\,M^{-1}\,s^{-1}$, is 17 powers of 10 smaller than the rate constant for bond formation between two N atoms, $5.2 \times 10^8\,M^{-1}\,s^{-1}$. This immense difference reflects the fact that exchange of atoms involves a large energy input to break an NO bond, whereas bond formation between two $NO_2$ molecules requires no energy input.

**Practice Exercise 12.13**
Estimate the rate constant for the decomposition of cyclobutane at 25 °C (see Practice Exercise 12.12).

## Values of Activation Energy

Most reactions between stable molecules have activation energies of 100 kJ/mol or greater, even when the overall reaction is exothermic. As an example, consider the reaction of hydrogen and oxygen:

$$2 \, H_2 \, (g) + O_2 \, (g) \longrightarrow 2 \, H_2O \, (g) \quad \Delta H° = -484 \text{ kJ/mol}$$
$$\text{and} \quad \Delta G° = -457 \text{ kJ/mol}$$

This combustion reaction is so spontaneous and exothermic that it is used to drive the main engines of the space shuttle. Nevertheless, mixtures of hydrogen and oxygen are stable indefinitely at room temperature. This reaction does not occur at room temperature because strong bonds must be broken to transform $H_2$ and $O_2$ into water molecules. In other words, the strong bonds in the reactant molecules result in a large activation energy for the reaction. As Figure 12-18 shows, a violent explosion occurs when a spark is applied to a mixture of $H_2$ and $O_2$. The spark gives some of the molecules enough energy to overcome the activation barrier, and after the reaction starts, the energy released in the formation of water molecules raises the temperature sufficiently that the reaction goes to completion very quickly.

**Figure 12-18**

A balloon filled with a mixture of hydrogen and oxygen is stable until a source of energy, such as a spark or flame, initiates the reaction. After the reaction begins, the formation of water releases enough energy to cause an explosion.

The combustion reactions of hydrocarbons such as natural gas and gasoline, which also release large amounts of energy, have negligible rates at room temperature because they have large activation energy barriers, typically 140–200 kJ/mol. For a hydrocarbon molecule to combine with oxygen to form $CO_2$ and $H_2O$, one of its C—H bonds first must be weakened substantially. This accounts for the large activation energy barrier for combustion.

Many elementary reactions have large activation energies. For example, the isomerization of *cis*-2-butene to *trans*-2-butene is a unimolecular rotation with an activation energy of 284 kJ/mol. This high value arises because a C—C $\pi$ bond must be broken during the course of the isomerization process (see Figure 12-3).

Example 12-12 shows that no energy barrier exists for the combination of two $NO_2$ molecules to form $N_2O_4$. The activation energy for this reaction is zero because $NO_2$ is an odd-electron molecule with a lone electron readily available for bond formation:

The recombination of two bromine atoms to form a $Br_2$ molecule has zero activation energy for the same reason.

## SECTION EXERCISES

● **12.6.1** Use bonding arguments to predict whether the ammonia synthesis reaction has a high or a low activation energy:

$$N_2\,(g) + 3\,H_2\,(g) \longrightarrow 2\,NH_3\,(g) \quad \Delta H° = -91.8\text{ kJ}$$

● **12.6.2** The reaction of $H_2$ with $Br_2$ discussed in Section 12.5 has been studied extensively as a function of temperature. Here are some experimental results:

| $T\,(10^2\text{ K})$ | 5.00 | 5.25 | 5.50 | 5.75 |
|---|---|---|---|---|
| $k\,(M^{-1/2}\,s^{-1})$ | $8.1 \times 10^{-7}$ | $6.5 \times 10^{-6}$ | $4.0 \times 10^{-5}$ | $2.1 \times 10^{-4}$ |

(a) Determine $E_a$ by graphical analysis; and (b) estimate $k$ at 425 K.

● **12.6.3** Use standard thermodynamic data to determine $\Delta E$ and construct the activation energy diagram for the following reaction. (Hint: $\Delta(pV) = 0$ for this reaction, so enthalpies and energies can be used interchangeably.)

$$CO + NO_2 \longrightarrow CO_2 + NO \quad E_a = 133\text{ kJ}$$

## 12.7 CATALYSIS

Many potentially useful chemical reactions proceed slowly at room temperature. To take advantage of these reactions, we must find ways to make them go faster. According to the discussion in Section 12.6, one way to make a reaction go faster is to run it at higher temperature. At higher temperature, however, undesirable reactions also go faster, so increasing the temperature may lead to lower yields because of competing reactions. In addition, the extra energy required to run a reaction at higher temperature adds to production costs. Furthermore, high-temperature reactions introduce safety concerns, and many chemical species are not stable at high temperature. Thus, although some industrial reactions are carried out at high temperature to maximize the amount of product that can be synthesized in a given time, the chemical industry prefers to find other ways to increase reaction rates.

Another way to make a reaction go faster is to add a substance called a **catalyst**. A catalyst functions by changing the mechanism of a reaction in a manner that lowers activation energy barriers. Although the catalyst changes the mechanism of a reaction, it is not part of the overall stoichiometry of the reaction. A catalyst always participates in an early step of a reaction mechanism, but when the reaction is over, the catalyst has been regenerated. When we write a net reaction that is influenced by a catalyst, we write the formula of the catalyst above or below the reaction arrow.

It is important to recognize the difference between a *catalyst* and an *intermediate*. A catalyst is a reactant in an early step of a mechanism and a product of a later step. The opposite is true of an intermediate. An intermediate is a transient species that is a product in an early step of a mechanism but is consumed in a later step:

$$\text{Catalyst} + \text{Reactant}_1 \longrightarrow \text{Intermediate}$$

$$\text{Intermediate} + \text{Reactant}_2 \longrightarrow \text{Products} + \text{Catalyst}$$

Catalysts are vital in the chemical industry. The market for catalysts in the United States alone exceeds $2.0 billion, including more than $600 million for petroleum refining and more than $750 million for chemical production. Although these are large sums of money, the products made available by catalysts are far more valuable than the catalysts themselves. The total value of fuels and chemicals produced by catalysts exceeds $900 billion.

## Catalysis and the Ozone Problem

Catalysts are immensely beneficial in industry, but accidental catalysis in the atmosphere can be disastrous. Recall from Focus 12-1 that the chemistry of ozone in the stratosphere involves a delicate balance of reactions that maintain a stable concentration of ozone. Chlorofluorocarbons (CFCs) shift that balance by acting as catalysts for the destruction of $O_3$ molecules.

Because CFCs are highly resistant to chemical attack, they are stable in the lower atmosphere, where they can exist for up to 100 years. This stability gives CFCs time to diffuse up through the troposphere and into the stratosphere. There, CFCs absorb short-wavelength ultraviolet light from the Sun that breaks carbon — chlorine bonds and produces chlorine atoms. For example:

$$CF_2Cl_2 \xrightarrow{h\nu} CF_2Cl + Cl$$

Chlorine atoms react with $O_3$ molecules to produce $O_2$ and ClO, as shown by the molecular pictures in Figure 12-19. This is a catalytic process, because chlorine monoxide reacts with an oxygen atom to produce a second $O_2$ molecule and *regenerate the chlorine atom*:

The first step of destruction of ozone by Cl atoms is generally accepted to be $Cl + O_3 \longrightarrow ClO + O_2$. The second step, however, is still the topic of active research. Other possible steps that regenerate the Cl atom are $ClO + O_3 \longrightarrow Cl + 2\, O_2$, or $2\, ClO \longrightarrow Cl_2O_2$ followed by $Cl_2O_2 \longrightarrow 2\, Cl + O_2$.

$$Cl + O_3 \longrightarrow ClO + O_2$$
$$ClO + O \longrightarrow Cl + O_2$$
$$\overline{\text{Net: } O_3 + O \xrightarrow{\quad Cl \quad} 2\, O_2}$$

The net reaction for this two-step mechanism is the conversion of an $O_3$ molecule and an oxygen atom into two $O_2$ molecules. In this mechanism, chlorine atoms catalyze ozone decomposition. They participate in the mechanism, but they do not appear in the overall stoichiometry. Although chlorine atoms are consumed in the first step, they are regenerated in the second. The cyclical nature of this process means that each chlorine atom can catalyze the destruction of many $O_3$

**Figure 12-19**

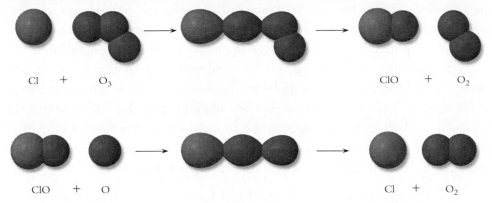

Cl   +   O₃                                    ClO   +   O₂

ClO   +   O                                    Cl   +   O₂

Chlorine atoms catalyze the reaction of ozone with oxygen atoms by forming an unstable ClO intermediate that readily reacts with an O atom, forming another $O_2$ molecule and regenerating the Cl atom.

molecules. It has been estimated that each chlorine atom produced by a CFC molecule in the upper stratosphere destroys about 100 000 molecules of ozone before it is removed by other reactions such as recombination:

$$CF_2Cl + Cl \longrightarrow CF_2Cl_2$$

Figure 12-20, an activation energy diagram for the ozone–oxygen atom reaction, illustrates how a catalyst modifies the energy profile of a reaction. The figure shows that the direct reaction between $O_3$ and $O$ has a substantially higher

**Figure 12-20**

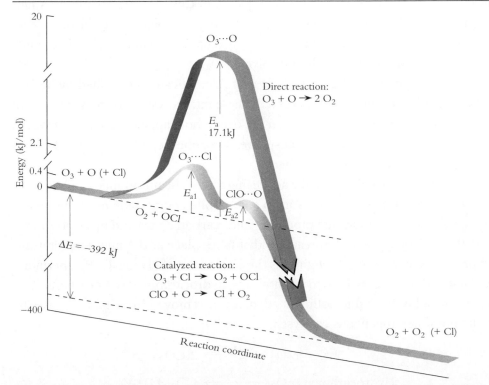

The activation energy diagram for the reaction between $O_3$ and $O$. Notice the break in the vertical scale: the overall reaction is exothermic by 392 kJ.

activation energy than the chlorine-catalyzed reaction sequence. In other words, the activated complex between Cl atoms and $O_3$ molecules lies at a lower energy than the activated complex between O atoms and $O_3$ molecules, making chlorine atoms much more effective than oxygen atoms at destroying $O_3$ molecules.

Notice that both steps of the chlorine-catalyzed reaction appear on the activation energy diagram. Each step of a mechanism has its own activation energy, so the diagram has two activation barriers. Experimental data indicate that the first barrier is higher than the second. Because the uncatalyzed reaction involves just one step, it has only one activation barrier. At the temperature of the ozone layer, 220 K, the rate data for the two pathways are as follows:

$$O_3 + O \longrightarrow 2\,O_2 \qquad\qquad E_a = 17.1 \text{ kJ/mol} \quad k = 4.1 \times 10^5 \text{ M}^{-1}\text{ s}^{-1}$$

$$O_3 + Cl \longrightarrow O_2 + OCl \qquad E_{a1} = 2.1 \text{ kJ/mol} \quad k_1 = 5.2 \times 10^9 \text{ M}^{-1}\text{ s}^{-1}$$

$$ClO + O \longrightarrow Cl + O_2 \qquad E_{a2} = 0.4 \text{ kJ/mol} \quad k_2 = 2.2 \times 10^{10} \text{ M}^{-1}\text{ s}^{-1}$$

We see that chlorine atoms provide an alternative mechanism for the reaction of ozone with oxygen atoms. The lower-energy pathway breaks down ozone in the stratosphere at a significantly faster rate than in the absence of the catalyst. This disturbs the delicate balance among ozone, oxygen atoms, and oxygen molecules in a way that poses a serious threat to the life-protecting ozone layer.

## Homogeneous and Heterogeneous Catalysts

The catalysis of ozone decomposition by chlorine atoms occurs entirely in the gas phase. Chlorine is classified as a **homogeneous catalyst** because the catalyst and the reactants are present in the *same phase*, in this case the gas phase. A **heterogeneous catalyst**, on the other hand, is in a *different phase* than the one where the reaction occurs. A heterogeneous catalyst is usually a solid, and the reactants are gases or are dissolved in a liquid solvent.

The development of the ozone hole over Antarctica is accelerated by heterogeneous catalysis on microcrystals of ice. These microcrystals form in abundance in the Antarctic spring, which is when the ozone hole appears. Ice microcrystals are less common in the Arctic atmosphere, so ozone depletion has not been as extensive in the Northern Hemisphere.

Heterogeneous catalysts are the active ingredients in automobile catalytic converters. When combustion occurs in an automobile engine, side reactions generate small amounts of undesired products. Some carbon atoms end up as poisonous CO rather than $CO_2$. Another reaction that takes place at the high temperatures and pressures in automobile engines is the conversion of $N_2$ to NO. Furthermore, the combustion process fails to burn all the hydrocarbons. Hydrocarbons, CO, and NO are all undesirable pollutants that can be removed from exhaust gases by the action of heterogeneous catalysts:

$$2\,CO + O_2 \xrightarrow{\text{Catalyst}} 2\,CO_2$$

$$\text{Hydrocarbons} + O_2 \xrightarrow{\text{Catalyst}} CO_2 + H_2O$$

$$2\,NO \xrightarrow{\text{Catalyst}} N_2 + O_2$$

Catalyst Cushion Mat
3-Way Monolithic Catalysts
Stainless Steel Body
Aluminized Heat Shield

**Figure 12-21**

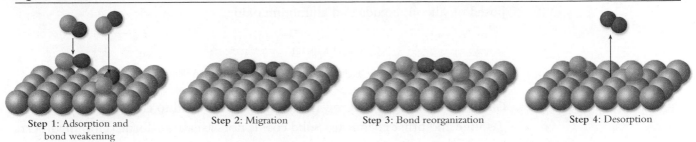

Step 1: Adsorption and bond weakening          Step 2: Migration          Step 3: Bond reorganization          Step 4: Desorption

Schematic molecular pictures of the four steps for the conversion of NO to $N_2$ and $O_2$ on a platinum metal surface.

A mixture of catalysts is needed to catalyze the variety of reactions that must be carried out in the converter. The most important catalyst is platinum metal, but palladium and rhodium are used as well, as are transition metal oxides such as CuO and $Cr_2O_3$.

Heterogeneous reactions are complicated because the reacting species must be transferred from one phase to another before the reaction can occur. Despite much research, chemists still have limited knowledge about the mechanisms of reactions that involve heterogeneous catalysts. However, it is known that heterogeneous catalysis generally proceeds in four steps, as illustrated in Figure 12-21 for the conversion of NO into $N_2$ and $O_2$.

1. The starting materials bind to the surface of the catalyst. This process is known as **adsorption**. When a substance is adsorbed, its internal bonds are weakened or broken in favour of bonds to the catalyst.

2. Bound reactants or fragments of them migrate over the surface of the catalyst.

3. These bound species react to form products.

4. Products escape from the surface of the catalyst. This step is **desorption**.

## Catalysis in Industry

The business of the chemical industry is to transform inexpensive substances into more valuable ones. In many cases, catalysts play important roles in these processes. Here, we describe the roles of catalysts in some important industrial reactions. Other catalyzed industrial reactions are considered in Chapter 13, where we describe properties of chemical equilibria.

*Petroleum refining.* Petroleum is the source not only of fuels but also of carbon compounds used for synthesis in the chemical industry. Crude petroleum is a complex mixture containing thousands of compounds, most of which are alkanes, alkenes, and derivatives of benzene. Petroleum refining converts this viscous black liquid into a variety of useful products, from gasoline to lubricants (see Figure 12-22). Refining begins with distillation, which separates the various components into fractions with different boiling point ranges. Large high-boiling hydrocarbons are less versatile than smaller ones, so a further step in refining is the catalytic cracking of large hydrocarbons into smaller ones. Cracking is

**Figure 12-22**

Hexadecane

Dodecane

Crude petroleum is refined to produce a variety of useful products. Shown are motor oil and lubricant jelly.

performed at high temperature in the presence of a heterogeneous catalyst composed of silicon dioxide and aluminum oxide:

$$C_{12} \text{ and higher} \xrightarrow[300°C]{\text{SiO}_2/\text{Al}_2\text{O}_3} C_5 \text{ to } C_{10}$$

$$\text{Kerosene} \qquad\qquad\qquad \text{Alkanes, alkenes}$$

*Chemicals from coal.* Heterogeneous catalysts are used to convert solid coal into gasoline and other chemicals. Solid coal is not easily transformed into hydrocarbon chains, so the conversion requires two general steps: gasification followed by catalytic hydrocarbon-forming reactions. Coal is first gasified (converted into a mixture of carbon monoxide and hydrogen) by reaction with steam:

$$C \text{ (coal)} + H_2O \text{ (g)} \rightarrow CO \text{ (g)} + H_2 \text{ (g)} \qquad \Delta H = 1131 \text{ kJ/mol}$$

This reaction is significantly endothermic, so it must be driven by added energy. Burning some of the coal provides the energy required to drive the gasification and produce the steam:

$$C \text{ (coal)} + O_2 \text{ (g)} \rightarrow CO_2 \text{ (g)} \qquad \Delta H = -407 \text{ kJ/mol}$$

The reaction of steam with coal generates a 1:1 gas mixture of $H_2$ and $CO$. Hydrocarbon formation requires a 2:1 mixture, so additional $H_2$ is produced by the reaction of steam with some of the $CO$:

$$CO(g) + H_2O \text{ (g)} \rightarrow CO \text{ (g)} + H_2 \text{ (g)} \qquad \Delta H = -42 \text{ kJ/mol}$$

A variety of solid metallic catalysts containing oxides of iron, chromium, or copper can be used for hydrocarbon formation. High gas pressure (20 bar) and elevated temperature (250 – 350 °C) are also required. Several hydrocarbon-forming reactions occur; the most important is the stepwise production of alkanes. Here are two simple examples:

$$CO + 3H_2 \xrightarrow[\text{High } p, T]{\text{Catalyst}} CH_4 + H_2O$$

$$CO + 2H_2 + CH_4 \xrightarrow[\text{High } p, T]{\text{Catalyst}} C_2H_6 + H_2O$$

Another important chemical produced from $CO$ and $H_2$ is methanol ($CH_3OH$). The synthesis is carried out at high temperature (200–300 °C) and pressure (50–100 bar) in the presence of copper(II) oxide or some other metal oxide catalyst:

$$CO + 2H_2 \xrightarrow[\text{High } p, T]{\text{Catalyst}} CH_3OH$$

Billions of litres of methanol are produced annually worldwide.

Mixtures of $CO$ and $H_2$ have immense potential as a feedstock for the chemical industry. The variety of products available from $CO$ and $H_2$ is virtually unlimited, and the Earth has huge coal reserves. Unfortunately, the direct conversion of coal is energy intensive and highly polluting. For all of these reasons, coal conversion has been studied intensively. Despite considerable research, however, coal conversion is currently not economical compared with petroleum refining. The

Most catalysts used in the chemical industry are solid metal halides, metal oxides, or pure metals.

catalytic process is not the main stumbling block. Instead, the problem is the high cost of gasification, both in energy and in capital investment. The development of a catalyst that is effective at low temperature would solve this problem.

*Acetic acid production.* Although most industrial catalysts are heterogeneous, a growing number of industrial reactions use homogeneous catalysts. One example is the production of acetic acid. Most of the 2.1 billion kilograms of acetic acid produced annually is used in the polymer industry. The reaction of methanol and carbon monoxide to form acetic acid is catalyzed by a rhodium compound that dissolves in methanol:

$$CH_3OH \ + \ CO \ \xrightarrow[\text{173 °C, 700 bar}]{[Rh(CO)_2I_3]} \ CH_3CO_2H$$

These are just a few examples of catalytic reactions used industrially. Catalyst research is one of the most active areas of chemistry and chemical engineering.

## Biocatalysis: Enzymes

Living organisms carry out an astonishing variety of chemical processes. Organisms organize small molecules into complex biopolymers such as deoxyribonucleic acid (DNA) and proteins (see Chapter 17). Conversely, organisms break down large energy-rich molecules in many steps to extract chemical energy in small portions to drive their many activities. Furthermore, organisms produce antibodies, medium-sized molecules that combat bacterial invaders. In many of these processes, organisms break chemical bonds selectively without resorting to high temperature.

### FOCUS 12-2  CHEMICAL MILESTONES: THE WORK OF DR. JOHN POLANYI

This chapter presents some of the more fundamental aspects of chemical reactions—the mechanisms by which they proceed and the rates at which they form products. Measuring concentrations as a function of time, deducing orders and mechanisms, and calculating activation energies all give us some information about how a reaction occurs at the molecular level. Transition states, such as those pictured in Figure 12-19, are inferred from the data, but until the work of Dr. John Polanyi, there was little experimental evidence of the dynamics of reactions. Exactly how do the atoms and molecules rearrange themselves during a reaction, and how does the energy flow among these species?

Dr. John Polanyi is a chemist at the University of Toronto whose work continues to bring new insights into the mechanisms of chemical reactions. Polanyi was a pioneer in measuring the energy states and movements of species at the very moment of reaction. The transition state is brief, perhaps only picoseconds long, and so only spectroscopic techniques can be used to study it.

Polanyi's first graduate student, Ken Cashion, examined the seemingly simple gas-phase reaction $H + Cl_2 \rightarrow HCl + Cl$. This reaction releases energy, but the distribution of this energy among rotation, vibration, and translation of the products was unknown. Cashion was able to measure the vibrational energy in the product HCl molecules as this energy was transformed into infrared radiation.

$$H + Cl_2 \longrightarrow HCl + Cl$$

Hydrogen

Transition State

Infrared light

Rotation

Vibration

Chlorine gas

The experiment was ingeniously simple and inexpensive. Atomic hydrogen was formed by flowing molecular hydrogen through an electric discharge, the power for which came from a neon-sign transformer. The H atoms were mixed with chlorine gas at low pressure in a glass vessel equipped with sodium chloride windows, which are transparent to infrared radiation. An infrared spectrometer was positioned in front of the sodium chloride window to collect and analyze the light produced by the reaction. The infrared spectrum indicated HCl molecules at a temperature of thousands of degrees. This molecular excitation is termed "infrared chemiluminescence," and originated from the vibrational de-excitation of newly formed HCl molecules. These experiments began a new field of chemistry called *reaction dynamics*.

Further work was required to tease out the rates at which particular states of vibration, rotation, and translation of the products were excited during the reaction. One major problem was "vibrational relaxation," by which the initially highly excited product molecules lost their vibrational energy to other molecules, vastly complicating the chemiluminescence spectrum. This was solved by measuring the infrared radiation at the instant that the products were formed and then immediately removing them from the chamber at a liquid-nitrogen-cooled wall before they had time to relax. Doing so allowed the vibrational (and rotational) energies at the transition state to be accurately quantified.

Quantifying the energies associated with translational motion was the last piece of the puzzle. Experiments by D.R. Herschbach and Y.T. Lee employed crossed molecular beam techniques, in which reactants are collided with one another and the products are analyzed according to their kinetic energies. Polanyi shared the 1986 Nobel Prize in chemistry with Herschbach and Lee. The citation read in part, "John C. Polanyi has developed the method of infrared chemiluminescence, in which the extremely weak infrared emission from a newly formed molecule is measured and analysed. He has used this method to elucidate the detailed energy disposal during chemical reactions."

A biochemical catalyst is called an **enzyme**. Enzymes are specialized proteins that catalyze specific biochemical reactions. Some enzymes are found in extracellular fluids such as saliva and gastric juices, but most are found inside cells. Each type of cell has a different array of enzymes that act together to determine what role the cell plays in the overall biochemistry of the organism. Enzymes are complex molecules. Biochemists have determined the molecular structures of some enzymes, but the structures of many enzymes are not yet known.

Despite the wide diversity of enzyme structures, most enzyme activity follows a general mechanism that has several reversible steps. In the first step, a reactant molecule known as a *substrate* (S) binds to a specific location on the enzyme (E), usually a groove or a pocket on the surface of the protein:

$$E + S \rightleftharpoons ES$$

The substrate binds to the active site through intermolecular interactions that usually include significant amounts of hydrogen bonding.

Binding causes the shape of the enzyme to change, and this change distorts the structure of the substrate:

$$ES \leftrightharpoons E \cdots S \qquad \text{(fast)}$$

Distortion allows the substrate to react more easily with another reactant (R) to form the desired product (P). Once the product forms, it no longer binds strongly to the enzyme:

$$E \cdots S + R \longrightarrow E \cdots P \longrightarrow E + P$$

The enzyme is regenerated at the end of this sequence, making it available to bind another substrate molecule. Note that the steps in this enzyme-catalyzed biochemical mechanism are similar to the steps in chemical heterogeneous catalysis: binding with bond weakening, reaction at the bound site, and release of products.

The names of enzymes usually end in *-ase*; for example, hexokinase and nitrogenase.

A specific illustration of enzymatic catalysis is the addition of a phosphate group to glucose. This reaction is the first step in glycolysis, the fragmentation of glucose. Virtually all cells use glycolysis to produce adenosine triphosphate (ATP), the energy-rich compound that supplies energy for many biological processes. This addition of phosphate is catalyzed by an enzyme, hexokinase. The first step is binding of glucose by the enzyme.

When glucose binds to hexokinase, chemical interactions between the two molecules cause the enzyme–glucose complex to change its shape, as Figure 12-23 shows.

The enzyme folds in on the substrate, isolating the bound glucose molecule from its aqueous environment. This allows an ATP molecule to transfer a phosphate group to the hydroxyl group in position 6 of glucose, as shown in Figure

**Figure 12-23**

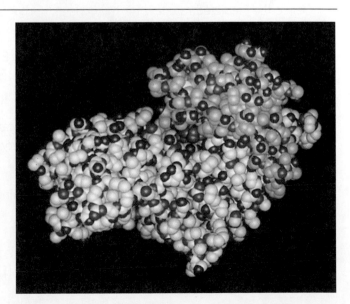

Computer models showing the shape of hexokinase (a) without and (b) with bound glucose. The enzyme folds around the substrate to bind it and isolate it from its aqueous environment.

**Figure 12-24**

Hexokinase catalyzes the phosphorylation of $\alpha$-glucose by adenosine triphosphate (ATP). "AD" represents the adenosine portion of ATP and ADP. The complete structure of ATP appears in Figure 11-19.

12-24. The product, glucose-6-phosphate, is too big to bind tightly to hexokinase. Consequently, the enzyme opens up, releases the product molecule, and is ready to bind another glucose molecule.

Although the molecular details of enzyme mechanisms are complex, the kinetic behaviour of many enzymatic processes is first order in both the substrate and the enzyme. Example 12-14 shows that the mechanism just outlined is consistent with this kinetic behaviour.

## EXAMPLE 12-14 ENZYME KINETICS

Derive the predicted rate law for the general mechanism for enzyme catalysis, assuming that the distortion step is rate-determining:

$$E + S \underset{k_{-1}}{\overset{k_1}{\rightleftharpoons}} ES \xrightarrow{k_2, \text{ slow}} E \cdots S$$

$$E \cdots S + R \xrightarrow{k_3} E + P$$

**Strategy:** The rate law predicted by a mechanism can be derived by setting the overall rate of reaction equal to the rate of the slowest step. As described in Section 12.5, any earlier steps are assumed to have equal forward and reverse rates.

**Solution:** The mechanism includes a rate-determining distortion step, for which the rate expression is

$$\text{rate} = k_2[ES]$$

This rate equation contains the concentration of an intermediate, so use the forward and reverse rates of the first step to derive an expression for [ES] in terms of [E] and [S]:

$$k_1[E][S] = k_{-1}[ES] \quad \text{so} \quad [ES] = \frac{k_1}{k_{-1}}[E][S]$$

Now substitute into the original rate expression:

$$\text{Rate} = \frac{k_1 k_2}{k_{-1}} [E][S] = k_{obs}[E][S] \quad \text{where} \quad k_{obs} = \frac{k_1 k_2}{k_{-1}}$$

**Does the Result Make Sense?** The mechanism predicts that the rate of this reaction depends on the concentrations of enzyme (E) and the reactant that binds to it (S) but does not depend on additional reactants (R). This makes sense from a mechanistic point of view, provided there is enough of R present so that the third step proceeds rapidly once distortion is complete. Experimentally, many enzyme-catalyzed reactions obey this form of rate law.

**Practice Exercise 12.14**

If the third reaction of this sequence is rate-determining and the distorted complex is at equilibrium with the undistorted complex, what is the resulting rate law?

Enzyme reactions are almost always first order in enzyme, as one would expect for a reaction in which one molecule of enzyme participates in the rate-determining step. As Figure 12-25 illustrates, the rate dependence on substrate concentration generally shows two distinct regions of kinetic behaviour. At low substrate concentration, the binding or distortion step is rate-determining, and the reaction is first order in substrate. However, when substrate concentration is high, all the enzyme molecules are bound either to substrate or to product. Under these conditions, the enzyme is catalyzing the reaction as fast as it can, so adding more substrate cannot make the enzyme work any faster. At high substrate concentrations, the reaction is zero order in substrate. A reaction under these conditions is obeying what are called "Michaelis–Menton" kinetics.

This has been a brief overview of a rich field. Details of enzyme structure and catalytic activity are studied in laboratories worldwide. Moreover, genetic engineering makes it possible to "manufacture" key enzymes in large quantities, so enzymes may become industrial catalysts that accomplish reactions rapidly and selectively.

**Figure 12-25**

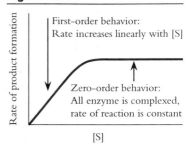

The typical profile of how the rate of an enzyme-catalyzed reaction varies with substrate concentration

## SECTION EXERCISES

● **12.7.1** When metals such as platinum, palladium, and rhodium are used as catalysts, they are usually deposited as thin layers over a highly porous material such as charcoal. Explain why these materials are less expensive and more effective catalysts than pieces of pure metal.

● **12.7.2** Which of the following statements are true? Correct the untrue statements so that they are true. (a) The concentration of a homogeneous catalyst appears in the rate law. (b) A catalyst changes an endothermic reaction into an exothermic reaction. (c) A catalyst lowers the activation energy of the rate-determining step. (d) A catalyst is a reactant in an early step of a mechanism and a product of a later step.

● **12.7.3** Why would it be advantageous to develop a catalyst for the following reaction?

$$4\,CO + 2\,NO_2 \rightarrow N_2 + 4\,CO_2$$

# CHAPTER 12 VISUAL SUMMARY

## 12.1 WHAT IS A REACTION MECHANISM?

A mechanism is the exact molecular pathway *that starting materials follow on their way to becoming products.*

Reaction mechanism

Rate-determining step — Elementary reaction

Bimolecular reaction

Unimolecular reaction

Intermediate — Rate — Termolecular reaction

**LEARNING OBJECTIVES:** Explain the concepts of a mechanism and a rate-determining step in a chemical reaction.

**SKILLS TO MASTER:** Visualizing elementary reactions (p. 623)

**KEY CONCEPTS:** The overall reaction is the sum of the elementary steps in the mechanism.

A reaction mechanism is the exact molecular pathway that starting materials follow on their way to becoming products.

The rate-determining step is the slowest step in the reaction mechanism. The overall reaction cannot go faster than the rate-determining step.

## 12.2 RATES OF CHEMICAL REACTIONS

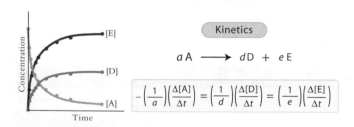

Kinetics

$$a\,A \longrightarrow d\,D + e\,E$$

$$-\left(\frac{1}{a}\right)\left(\frac{\Delta[A]}{\Delta t}\right) = \left(\frac{1}{d}\right)\left(\frac{\Delta[D]}{\Delta t}\right) = \left(\frac{1}{e}\right)\left(\frac{\Delta[E]}{\Delta t}\right)$$

**LEARNING OBJECTIVES:** Determine the rate of a reaction based on the rate of change of concentration of a reactant or product.

**SKILLS TO MASTER:** Calculating reaction rates from concentration versus time data (p. 679)

Calculating relative rates of reaction (p. 681)

**KEY CONCEPTS:** The reaction rate can be expressed using the rate of change of concentration of any of the reagents.

## 12.3 CONCENTRATION AND REACTION RATES

The order of a reaction must be determined by experiments.

Rate law — $\text{Rate} = k[A]^y[B]^z$ — Rate constant ($k$)

$\text{Rate} = k[NO_2]$ — First order

$\text{Rate} = k[NO_2]^2$ — Second order

**LEARNING OBJECTIVES:** Determine the rate law, given the mechanism and knowledge of the relative rates of steps of a reaction.

**SKILLS TO MASTER:** Predicting the rate law from the reaction mechanism (p. 685)

**KEY CONCEPTS:** A reaction rate depends on the concentrations of the reactants as expressed by the rate law for that reaction.

The order of a reactant often differs from its stoichiometric coefficient.

## 12.4 EXPERIMENTAL KINETICS

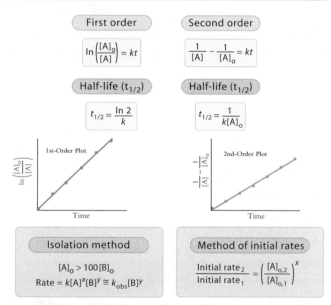

First order

$$\ln\left(\frac{[A]_o}{[A]}\right) = kt$$

Half-life ($t_{1/2}$)

$$t_{1/2} = \frac{\ln 2}{k}$$

1st-Order Plot

$\ln\left(\frac{[A]_o}{[A]}\right)$ vs Time

Second order

$$\frac{1}{[A]} - \frac{1}{[A]_o} = kt$$

Half-life ($t_{1/2}$)

$$t_{1/2} = \frac{1}{k[A]_o}$$

2nd-Order Plot

$\frac{1}{[A]} - \frac{1}{[A]_o}$ vs Time

Isolation method

$[A]_o > 100[B]_o$

$\text{Rate} = k[A]^x[B]^y \cong k_{obs}[B]^y$

Method of initial rates

$$\frac{\text{Initial rate}_2}{\text{Initial rate}_1} = \left(\frac{[A]_{o,2}}{[A]_{o,1}}\right)^x$$

**LEARNING OBJECTIVES:** Determine rate laws from concentration versus time data.

**SKILLS TO MASTER:** Determining the rate law from isolation experiments (p. 696)

Determining the rate law from the method of initial rates (p. 701)

**KEY CONCEPTS:** Rate laws must be determined from experimental data.

The half-life of a first-order reaction is a constant. The half-life of a second-order reaction depends on the initial concentration.

## 12.5 LINKING MECHANISMS AND RATE LAWS

*When the 1st step is rate-determining, the predicted rate law for the overall reaction is the rate expression for that 1st step.*

Reversible reaction

$$Br_2 \rightleftharpoons Br + Br$$

$$k_1[Br_2] = k_{-1}[Br]^2$$

Bimolecular elementary reaction: $A + B \longrightarrow products$
Elementary rate $= k[A][B]$

Unimolecular elementary reaction: $C \longrightarrow products$
Elementary rate $= k[C]$

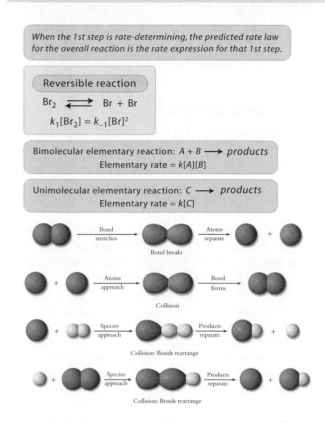

Bond stretches — Bond breaks — Atoms separate

Atoms approach — Collision — Bond forms

Species approach — Collision: Bonds rearrange — Products separate

Species approach — Collision: Bonds rearrange — Products separate

**LEARNING OBJECTIVES:** Show that the mechanism and rate law are closely related.

**SKILLS TO MASTER:** Determining the rate law when the first step is rate-determining (p. 706)

Determining the rate law when a later step is rate-determining (p. 708)

Using the equilibrium (equality of rates) assumption (p. 710)

**KEY CONCEPTS:** The mechanism must be the sum of the elementary reactions, and must explain the experimentally observed rate law.

## 12.6 REACTION RATES AND TEMPERATURE

$$k = Ae^{-E_a/RT}$$

Arrhenius equation

$$\ln k = \ln A - \frac{E_a}{RT}$$

$$\ln \frac{k_2}{k_1} = \frac{E_a}{R}\left(\frac{1}{T_1} - \frac{1}{T_2}\right)$$

Activation energy diagram

Activated complex

Activation energy ($E_a$)

Reactants

$\Delta E_{reaction}$

Products

Reaction coordinate

*The rate of a reaction increases with temperature.*

**LEARNING OBJECTIVES:** Explain and quantify the effects of temperature on a reaction rate.

**SKILLS TO MASTER:** Drawing and labelling an activation energy diagram for an elementary reaction (p. 715)

Using the Arrhenius equation to calculate a rate constant as a function of temperature (p. 718)

Using the Arrhenius equation to calculate an activation energy (p. 720)

**KEY CONCEPTS:** The activation energy is the minimum energy that must be supplied before a reaction can take place.

## 12.7 CATALYSIS

Catalyst

Adsorption     Desorption

Heterogeneous catalyst
Homogeneous catalyst
Enzyme

**LEARNING OBJECTIVES:** Explain the mechanisms by which catalysts function.

**SKILLS TO MASTER:** Recognizing the difference between a catalyst and an intermediate (p. 724)

Describing the mechanism of heterogeneous catalysis (p. 726)

Explaining the mechanism of catalysis by an enzyme (p. 730)

**KEY CONCEPTS:** A catalyst functions by providing a mechanism for which the rate-determining step has a lower activation energy.

## LEARNING EXERCISES

12.1   List the requirements of a satisfactory mechanism, and describe the characteristics of the steps of a mechanism.

12.2   Write one-sentence explanations for the following kinetic terms: (a) first order; (b) second order; (c) isolation method; (d) elementary step; (e) catalyst; (f) intermediate; and (g) enzyme.

12.3   Describe the differences among the rate, the rate law, and the rate constant for a chemical reaction.

12.4   Write a paragraph that describes what the activation energy is and how it affects the kinetic behaviour of a reaction.

12.5   Update your list of memory bank equations. Include information about how to apply each new equation.

12.6   Make a list of terms new to you that appear in Chapter 12. In your own words, write a one-sentence definition of each. Consult the Glossary if you need help.

## PROBLEMS

**ilw** = interactive learning ware problem. Visit the website at www.wiley.com/canada/olmsted

### What Is a Reaction Mechanism?

12.1   What is the rate-determining step in each of the following processes? (a) A line of people get coffee from a large coffee urn. (b) You go through the express line (10 items or less, cash only) at a supermarket. (c) A squad of parachutists makes a "jump" from an airplane cargo door.

12.2   What is the rate-determining step in each of the following processes? (a) People buy their lunches at a cafeteria. (b) Music enthusiasts enter an amphitheatre for a concert. (c) You leave a pay-as-you-exit parking lot.

12.3   The reaction of $H_2$ with $I_2$ is $H_2 + I_2 \longrightarrow 2\ HI$. Use reactant molecules to write appropriate elementary reactions that satisfy the following criteria: (a) a unimolecular decomposition that generates I; (b) a bimolecular collision that forms a square $H_2I_2$ complex; and (c) a bimolecular collision in which a hydrogen atom is transferred between reactants.

12.4   The reaction of NO with $Cl_2$ is

$$2\ NO + Cl_2 \longrightarrow 2\ NOCl$$

Use reactant molecules to write appropriate elementary reactions that satisfy the following criteria: (a) a unimolecular decomposition that generates Cl; (b) a bimolecular collision in which a Cl atom is transferred between reactants; and (c) a termolecular collision leading to the observed products.

12.5   Draw molecular pictures illustrating each part of Problem 12.3.

12.6   Draw molecular pictures illustrating each part of Problem 12.4.

12.7   Write three different satisfactory mechanisms for the reaction in Problem 12.3: one having your elementary reaction (a) as its first step; one having your elementary reaction (b) as its first step; and one having your elementary reaction (c) as its first step.

12.8   Write three different satisfactory mechanisms for the reaction in Problem 12.4: one having your elementary reaction (a) as its first step; one having your elementary reaction (b) as its first step; and one having your elementary reaction (c) as its first step.

### Rates of Chemical Reactions

12.9   For the reaction $2\ NO + Cl_2 \longrightarrow 2\ NOCl$, do the following: (a) Express the rate in terms of the disappearance of $Cl_2$. (b) Relate the rate of NOCl formation to the rate of $Cl_2$ disappearance. (c) If $Cl_2$ reacts at a rate of 47 M/s, state how fast NOCl will form.

12.10  Do the following for the ozone decomposition mechanism $2\ O_3 \longrightarrow 3\ O_2$: (a) Express the rate in terms of $O_2$ formation. (b) Relate the rate of $O_3$ consumption to the rate of $O_2$ production. (c) If $O_2$ forms at a rate of $2.7 \times 10^{-6}$ M/s, state how fast ozone disappears.

12.11  For the reaction $2\ NO + Cl_2 \longrightarrow 2\ NOCl$, do the following: (a) Draw a molecular picture showing a sample that contains 12 NO molecules and 5 $Cl_2$ molecules. (b) Redraw the picture to show the result after four molecules of NO have reacted. (c) Redraw the picture to show the result when one reactant has been consumed completely.

12.12  For the reaction $O_3 + NO \longrightarrow O_2 + NO_2$, do the following: (a) Draw a molecular picture showing a sample that contains 6 NO molecules and 8 $O_3$ molecules. (b) Redraw the picture to show the result after three molecules of NO have reacted. (c) Redraw the picture to show the result when one reactant has been consumed completely.

12.13  Calcium oxide, an important ingredient in cement, is produced by decomposing calcium carbonate at high temperature:

$$CaCO_3\ (s) \longrightarrow CaO\ (s) + CO_2\ (g)$$

In one reaction, 3.5 kg of calcium carbonate is heated at 550 °C in a 5.0-L vessel. The pressure of $CO_2$ is 0.15 bar after 5.0 minutes. (a) What is the average rate of $CO_2$ production in mol/min during the 5-minute interval? (b) How many moles of $CaCO_3$ decompose in the 5-minute interval?

12.14  A chemist is studying the rate of the Haber synthesis:

$$N_2 + 3\ H_2 \longrightarrow 2\ NH_3$$

Starting with a closed reactor containing 1.25 mol/L of $N_2$ and 0.50 mol/L of $H_2$, the chemist finds that the $H_2$ concentration has fallen to 0.25 mol/L after 30 seconds. (a) What is the average rate of reaction over this time? (b) What is the average rate of $NH_3$ production? (c) What is the $N_2$ concentration after 30 seconds?

**12.15** At high temperature, cyclopropane isomerizes to propene:

Cyclopropane          Propene
($C_3H_6$)            ($C_3H_6$)

In a small sample containing 10 molecules of cyclopropane, the reaction proceeds at an average rate of 0.25 molecules/min for 20 minutes. (a) Use line drawings to illustrate the small sample before the reaction begins. (b) Redraw the picture after 20 minutes of reaction.

**12.16** Under the appropriate conditions, *cis*-dichloroethylene isomerizes to *trans*-dichloroethylene:

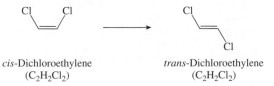

*cis*-Dichloroethylene        *trans*-Dichloroethylene
($C_2H_2Cl_2$)                ($C_2H_2Cl_2$)

In a small sample of gas containing 12 molecules of *cis*-dichloroethylene, the reaction proceeds at an average rate of 1.5 molecules/s for 2.0 seconds. (a) Use line drawings to illustrate the small sample before the reaction begins. (b) Redraw the picture after 2.0 seconds of reaction.

## Concentration and Reaction Rates

**12.17** Radioactive isotopes decay according to first-order kinetics. For one particular isotope, 1.00 nmole registers $1.2 \times 10^6$ decays in 1.00 min. (a) How many decays will occur in 1.00 min if 5.00 nmoles of this isotope are present? (b) What fraction of the isotope decays per minute in each case? (c) Explain the relationship between your answers to (a) and (b).

**12.18** Popcorn kernels pop independently (i.e., "unimolecularly"). At constant temperature, 6 kernels pop in 5 seconds when 150 kernels are present. (a) After 50 kernels have popped, how many kernels pop in 5 seconds? (b) Is there a change in the fraction of kernels popping per second? If so, by how much? (c) Explain the relationship between your answers to (a) and (b).

**12.19** For the net reaction $2\,AB + 2C \longrightarrow A_2 + 2\,BC$, the following slow first step has been proposed:

$$C + AB \longrightarrow BC + A$$

(a) What rate law is predicted by this step? (b) What units are associated with the rate constant for this rate law? (c) Write additional steps that complete the mechanism.

**12.20** For the net reaction in Problem 12.19, another possible slow first step is as follows:

$$2\,AB \longrightarrow A_2 + 2\,B$$

(a) What rate law is predicted by this step? (b) What units are associated with the rate constant for this rate law? (c) Write additional steps that complete the mechanism.

**12.21** For the net reaction in Problem 12.19, another possible slow first step is as follows:

$$2\,C + AB \longrightarrow AC + BC$$

(a) What rate law is predicted by this step? (b) What units are associated with the rate constant for this rate law? (c) Write additional steps that complete the mechanism.

**12.22** For the net reaction in Problem 12.19, another possible slow first step is as follows:

$$AB \longrightarrow A + B$$

(a) What rate law is predicted by this step? (b) What units are associated with the rate constant for this rate law? (c) Write additional steps that complete the mechanism.

**12.23** The two pictures shown below represent starting conditions for the following reaction:

$$O_3 + NO \longrightarrow O_2 + NO_2 \qquad \text{Rate} = k[O_3][NO]$$

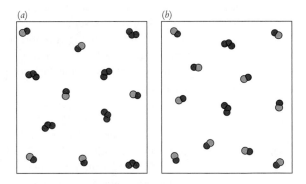

Will Flask *b* react faster or slower or at the same rate as Flask *a*? By how much? Explain your reasoning in terms of molecular collisions.

**12.24** The two pictures shown below represent starting conditions for the following reaction:

$$O_2 + NO_2 \rightarrow O_3 + NO \qquad \text{Rate} = k[O_2][NO_2]$$

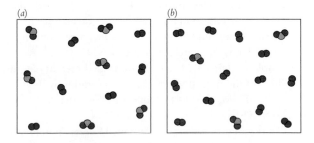

Will Flask *b* react faster or slower or at the same rate as Flask *a*? By how much? Explain your reasoning in terms of molecular collisions.

## Experimental Kinetics

**12.25** (ilw) The following data are obtained for the decomposition of $N_2O_5$ at 45 °C:

$$2 N_2O_5 (g) \longrightarrow 4 NO_2 (g) + O_2 (g)$$

| $t\,(10^2\text{ s})$ | 0 | 2.00 | 4.00 | 6.00 | 8.00 |
|---|---|---|---|---|---|
| $pN_2O_5$ (bar) | 2.50 | 2.22 | 1.96 | 1.73 | 1.53 |

(a) Use these data to determine the order of the decomposition process. Show your reasoning.
(b) Determine the rate constant, including appropriate units. (c) What is the pressure of $N_2O_5$ after 1600 s? (d) How long will it take for the pressure to decrease to 0.500 bar?

**12.26** Radioactive iodine ($^{131}I$) is used frequently in biological studies. A radiation biologist studies the rate of decomposition of this substance and obtains the following data:

| Time (days) | 0 | 4.0 | 8.0 | 12.0 | 16.0 |
|---|---|---|---|---|---|
| Mass $^{131}I$ (µg) | 12.0 | 8.48 | 6.0 | 4.24 | 3.0 |

(a) Use these data to determine the order of the decomposition process. Show your reasoning.
(b) Determine the rate constant, including appropriate units. (c) How many micrograms will be left after 32 days? (d) How many days will it take for the sample size to decrease to 1.2 µg?

**12.27** (ilw) The decomposition reaction of NOBr is second order in NOBr, with a rate constant at 20 °C of 25 $M^{-1}\,min^{-1}$. If the initial concentration of NOBr is 0.025 M, find (a) the time at which the concentration will be 0.010 M; and (b) the concentration after 125 minutes of reaction.

**12.28** The condensation reaction of butadiene ($C_4H_6$) is second order in $C_4H_6$, with a rate constant of 0.93 $M^{-1}\,min^{-1}$ (see Example 12-6). If the initial concentration of $C_4H_6$ is 0.240 M, find (a) the time at which the concentration will be 0.100 M; and (b) the concentration after 25 minutes of reaction.

**12.29** The conversion of $C_5H_{11}Br$ into $C_5H_{10}$ follows first-order kinetics, with a rate constant of 0.385 $h^{-1}$. If the initial concentration of $C_5H_{11}Br$ is 0.125 M, find (a) the time at which the concentration will be $1.25 \times 10^{-3}$ M; and (b) the concentration after 3.5 hours of reaction.

**12.30** The light-emitting decay of excited mercury atoms is first order, with a rate constant of $1.65 \times 10^6\,s^{-1}$. A sample contains $4.5 \times 10^{-6}$ M of excited mercury atoms. Find (a) the time at which the concentration will be $4.5 \times 10^{-7}$ M; and (b) the concentration after $2.5 \times 10^{-6}$ s.

**12.31** The reaction of $N_2O_5$ with gaseous $H_2O$ is a potentially important reaction in the chemistry of acid rain:

$$N_2O_5 (g) + H_2O (g) \longrightarrow 2 HNO_3 (g)$$

It is possible to monitor the concentration of $N_2O_5$ with time. Describe a set of isolation experiments from which the rate law and rate constant for this reaction could be determined.

**12.32** The reaction of $NO_2$ with CO is a reaction that may occur in automobile exhaust:

$$NO_2 (g) + CO (g) \longrightarrow NO (g) + CO_2 (g)$$

Because of its colour, it is possible to monitor the concentration of $NO_2$ with time. Describe a set of isolation experiments from which the rate law and rate constant for this reaction could be determined.

**12.33** The following initial rate information was collected at 25 °C for this aqueous reaction:

$$2 I^- (aq) + S_2O_8^{2-} (aq) \longrightarrow I_2 (aq) + 2 SO_4^{2-} (aq)$$

| $[I^-]_0$ | $[S_2O_8^{2-}]_0$ (M) | Initial Rate (M/min) |
|---|---|---|
| 0.125 | 0.150 | $4.4 \times 10^{-2}$ |
| 0.375 | 0.150 | $1.3 \times 10^{-1}$ |
| 0.125 | 0.050 | $1.5 \times 10^{-2}$ |

Determine the rate law and evaluate the rate constant for this reaction.

**12.34** The following initial rate information was collected at 25 °C for this aqueous reaction:

$$2 ClO_2 (aq) + 2 OH (aq) \longrightarrow ClO_3^- (aq) + ClO_2^- (aq) + 2 H_2O$$

| $[ClO_2]_0$ (M) | $[OH^-]_0$ (M) | Initial Rate (M/s) |
|---|---|---|
| 0.050 | 0.050 | $2.9 \times 10^{-2}$ |
| 0.075 | 0.050 | $6.5 \times 10^{-2}$ |
| 0.075 | 0.075 | $9.8 \times 10^{-2}$ |

Determine the rate law and evaluate the rate constant for this reaction.

## Linking Mechanisms and Rate Laws

**12.35** Problem 12.19 asks about a mechanism for the hypothetical reaction of AB with C. Assume that the first step described in Problem 12.19 is fast and reversible and your second step is rate-determining. Derive the rate law under these assumptions.

**12.36** Problem 12.8 asks for mechanisms for the reaction of NO with $Cl_2$. For parts (a) and (b) of that problem, assume that the first step in your mechanism is fast and reversible and the second step is rate-determining. Derive the rate laws under these assumptions.

**12.37** (ilw) A student proposes the following mechanism for the atmospheric decomposition of ozone to molecular oxygen:

$$O_3 + O_2 \rightleftharpoons O_5 \qquad \text{(fast, reversible)}$$
$$O_5 \longrightarrow 2 O_2 + O \qquad \text{(slow, rate-determining)}$$

(a) Propose a third step that completes this mechanism.
(b) Determine the rate law predicted by this mechanism.
(c) Atmospheric chemists would consider this mechanism to be molecularly unreasonable. Explain why.

**12.38** The reaction of CO with $Cl_2$ gives phosgene ($COCl_2$), a nerve gas used in World War I. Even though the stoichiometry is simple, the mechanism has several steps:

$$Cl_2 \rightleftharpoons 2\,Cl \qquad \text{(fast, reversible)}$$
$$Cl + Cl_2 \longrightarrow COCl \qquad \text{(slow, rate-determining)}$$
$$COCl + Cl_2 \longrightarrow COCl_2 + Cl \quad \text{(fast)}$$

(a) Show that this mechanism gives the correct overall stoichiometry. (b) What rate law does this mechanism predict? (c) Identify any reactive intermediates in the mechanism.

**12.39** The reaction of NO with $O_2$ to give $NO_2$ is an important process in the formation of smog in any large city:

$$2\,NO + O_2 \longrightarrow 2\,NO_2$$

Experiments show that this reaction is third order overall. The following mechanism has been proposed:

$$NO + NO \underset{k_{-1}}{\overset{k_1}{\rightleftharpoons}} N_2O_2$$
$$N_2O_2 + O_2 \overset{k_2}{\longrightarrow} NO_2 + NO_2$$

(a) If the second step is rate-determining, what is the rate law? (b) Is this rate law consistent with the overall third-order behaviour? Explain. (c) Draw molecular pictures that show different ways the intermediate species might bind together, and identify the one that is most reasonable with respect to the second step of the mechanism.

**12.40** Gaseous $N_2O_5$ decomposes according to the following equation:

$$2\,N_2O_5 \longrightarrow 4\,NO_2 + O_2$$

Much evidence suggests that the mechanism is as follows:

$$N_2O_5 \rightleftharpoons NO_2 + NO_3 \qquad \text{(fast decomposition)}$$
$$NO_2 + NO_3 \longrightarrow NO + NO_2 + O_2 \qquad \text{(slow)}$$
$$NO + NO_3 \longrightarrow 2\,NO_2 \qquad \text{(fast)}$$

(a) Show that this mechanism gives the correct overall stoichiometry.
(b) Determine the rate law predicted by this mechanism.
(c) This decomposition is first order experimentally. Does this information prove that Step 2, rather than Step 1, is rate-determining? Explain.

## Reaction Rates and Temperature

**12.41** If a reaction has an activation energy of zero, how will its rate constant change with temperature? Explain in molecular terms what $E_a = 0$ means.

**12.42** If a reaction has an activation energy of zero, how is $\Delta E$ for the forward reaction related to $E_a$ for the reverse reaction? Draw an activation energy diagram illustrating your answer.

**12.43** Consider the exothermic reaction $AC + B \longrightarrow AB + C$.
(a) Draw an activation energy diagram for this reaction.
(b) Label the energies of reactants and products.
(c) Show $\Delta E_{reaction}$ by a double-headed arrow. (d) Show $E_a$ for the forward reaction by a single-headed arrow.
(e) Label and draw a molecular picture of the activated complex.

**12.44** Consider the endothermic reaction $AB + C \longrightarrow AC + B$.
(a) Draw an activation energy diagram for this reaction.
(b) Label the energies of reactants and products.
(c) Show $\Delta E_{reaction}$ by a double headed arrow. (d) Show $E_a$ for the forward reaction by a single-headed arrow.
(e) Label and draw a molecular picture of the activated complex.

**12.45** For the isomerization that converts cyclopropane to propene, $\Delta H_{reaction} = -33$ kJ and $E_a = 272$ kJ/mol.
(a) Draw a molecular picture of cyclopropane converting into propene. (b) Draw the activation energy diagram for the reaction, and label it completely.

**12.46** Nitrogen dioxide in smog can combine in an elementary reaction to form $N_2O_4$ molecules. The combination reaction is exothermic by 57 kJ/mol. For the reverse reaction, the dissociation of $N_2O_4$, $E_a = 70$ kJ/mol.
(a) Draw a molecular picture of $NO_2$ combining to form $N_2O_4$. (b) Draw an activation energy diagram that shows the energy relationships as quantitatively as possible. Label the diagram completely.

**12.47** ⓘⓛⓦ Fireflies flash at a rate that depends on the temperature. At 29 °C, the average rate is 3.3 flashes every 10 seconds, whereas at 23 °C, the average rate falls to 2.7 flashes every 10 seconds. Calculate the "energy of activation" for the flashing process.

**12.48** The rate at which tree crickets chirp is 190/min at 28 °C but only 39.6/min at 5 °C. From these data, calculate the "energy of activation" for the chirping process.

## Catalysis

**12.49** The industrial process for forming methanol involves a catalyst:

$$CO + 2\,H_2 \xrightarrow[\text{High } p,\,T]{\text{Catalyst}} CH_3OH$$

Draw molecular pictures showing the bond breakage and formation that must occur in the course of this reaction. Suggest how the catalyst might make it easier for these reactions to occur.

**12.50** The industrial process for forming acetic acid involves a catalyst:

$$CH_3OH + CO \xrightarrow[\text{173 °C, 700 atm}]{[Rh(CO)_2I_3]} CH_3OH$$

Draw molecular pictures showing the bond breakage and formation that must occur in the course of this reaction. Suggest how the catalyst might make it easier for these reactions to occur.

**12.51** The addition of molecular hydrogen to ethylene is extremely slow unless Pd metal is present. In the presence of the metal, however, $H_2$ is adsorbed on the metal surface as H atoms, which then add to $C_2H_4$ when it strikes the surface:

(a) Draw an activation energy diagram for this mechanism that illustrates the effect of the catalyst. (b) Identify any catalysts or intermediates in the mechanism. Explain. (c) Draw a molecular picture that illustrates how this addition takes place.

**12.52** NO is an atmospheric pollutant that destroys ozone in the stratosphere. Here is the accepted mechanism:

$$O_3 + NO \longrightarrow NO_2 + O_2 \quad \text{(slow)}$$
$$NO_2 + O \longrightarrow NO + O_2 \quad \text{(fast)}$$
$$O_3 + O \longrightarrow 2\,O_2 \quad \text{(overall)}$$

(a) Draw an activation energy diagram for this mechanism. Include as many details as possible. (b) Identify any catalysts or intermediates in the mechanism. Explain. (c) The activation energy for decomposition of ozone promoted by NO is 11.9 kJ/mol. The activation energy for decomposition of ozone promoted by Cl is 2.1 kJ/mol. Which pollutant is a more serious threat to the ozone layer: Cl or NO? Explain.

## ADDITIONAL PAIRED PROBLEMS

**12.53** Oxygen reacts with NO to form $NO_2$ as the only product. Here is a proposed first step of the mechanism:

$$O_2 + NO \longrightarrow NO_2 + O$$

(a) What second step is required to give a satisfactory mechanism?
(b) Draw molecular pictures to illustrate each step.

**12.54** Oxygen reacts with CO to form $CO_2$ as the only product. Here is a proposed first step of the mechanism:

$$O_2 + CO \longrightarrow CO_2 + O$$

(a) What second step is required to give a satisfactory mechanism?
(b) Draw molecular pictures to illustrate each step.

**12.55** With appropriate catalysts, it is possible to make benzene from acetylene. The reaction has simple overall stoichiometry and can be studied in kinetics experiments:

$$3\,C_2H_2\,(g) \longrightarrow C_6H_6\,(g)$$

(a) Write the rate *expression* (see Equation 12-1) for this reaction in terms of the disappearance of $C_2H_2$ and in terms of the appearance of $C_6H_6$. (b) If enough

information is given to find the rate law, find it. If not, describe the data that you would need and how you would analyze the data to determine the rate law.

**12.56** The first step in the Ostwald process for the synthesis of nitric acid is the combustion of ammonia:

$$4\,NH_3 + 5\,O_2 \xrightarrow{\text{Pt gauze}} 4\,NO + 6\,H_2O$$

In a catalytic experiment, the rate of reaction is found to be

$$\text{Rate} = \frac{\Delta[NO]}{\Delta t} = 1.5 \times 10^{-3}\,\text{M/s}$$

What is the rate expressed in terms of $NH_3$, $O_2$, and $H_2O$?

**12.57** Use the following information to construct an activation energy diagram for the reaction, $S\,(s) + O_2\,(g) \longrightarrow SO_2\,(g)$

$$\Delta H_f^\circ(SO_2) = -296.1\,\text{kJ/mol},\ E_a = 150\,\text{kJ/mol}$$

**12.58** The reaction of ethylene ($C_2H_4$) with $H_2$ to form ethane ($C_2H_6$) (see Problem 12.51) is exothermic by 137 kJ/mol, has $E_a = 200$ kJ/mol, and is catalyzed by Pd metal, which reduces $E_a$ by 60 kJ/mol. Draw an activation energy diagram for this reaction, and label it clearly. Be sure all the given facts are illustrated in the diagram.

**12.59** What happens to the rate of a reaction involving $H_2$ if the concentration of $H_2$ is tripled and the reaction is (a) second order in $H_2$; (b) zero order in $H_2$; and (c) $\frac{3}{2}$ order in $H_2$?

**12.60** What happens to the rate of a reaction involving CO if the concentration is doubled and the reaction is (a) first order in CO; (b) half order in CO; and (c) inverse first order in CO?

**12.61** Azomethane decomposes into nitrogen and ethane at high temperature:

$$H_3C-N{=}N-CH_3 \longrightarrow N_2 + C_2H_6$$

A chemist studying this reaction at 300°C obtains the following data:

| Time ($10^2$ s) | 0 | 1.00 | 1.50 | 2.00 | 2.50 | 3.00 |
|---|---|---|---|---|---|---|
| [Azomethane] (mM) | 7.94 | 6.15 | 5.40 | 4.75 | 4.20 | 3.69 |

Prepare first- and second-order graphic plots of these data, determine the order of the reaction, and calculate its rate constant.

**12.62** Gas A decomposes by the reaction $3\,A \longrightarrow B + C$ at 45 °C, and its concentration changes as follows:

| Time ($10^2$ s) | 0 | 2.00 | 4.00 | 6.00 | 8.00 |
|---|---|---|---|---|---|
| [A] (M) | 2.50 | 2.22 | 1.96 | 1.73 | 1.53 |

Find the rate law and determine the rate constant.

**12.63** The hypothetical reaction $X_2 + 2\,Y \longrightarrow 2\,XY$ goes by the following mechanism:

$$X_2 \longrightarrow 2X \qquad \text{(slow)}$$
$$X + Y \longrightarrow XY \qquad \text{(fast)}$$

The pictures that follow represent starting conditions for the reaction. What would the rate of reaction be for Flask *b* compared with that for Flask *a*? Explain.

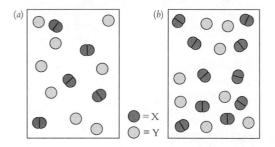

**12.64** Here is another set of starting conditions for the reaction described in Problem 12.63. What would the rate of reaction be for Flask *b* compared with that for Flask *a*? Explain.

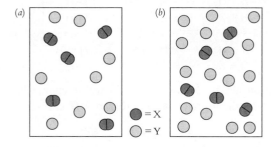

**12.65** A chemist carried out kinetic studies on the reaction described in Problem 12.31, obtaining the following data:

Experiment A: $[N_2O_5]_0 = 1.5 \times 10^{-4}$ M, $[H_2O]_0 = 0.025$ M

| Time (s) | 0 | 60 | 120 | 180 | 240 |
|---|---|---|---|---|---|
| $[N_2O_5]\,(10^{-5}\text{M})$ | 15 | 12 | 9.6 | 7.6 | 6.1 |

Experiment B: $[N_2O_5]_0 = 1.5 \times 10^{-4}$ M, $[H_2O]_0 = 0.050$ M

| Time (s) | 0 | 60 | 120 | 180 | 240 |
|---|---|---|---|---|---|
| $[N_2O_5]\,(10^{-5}\text{ M})$ | 15 | 9.6 | 6.1 | 3.9 | 2.5 |

Use graphs to determine the rate law for this reaction, and evaluate the rate constant.

**12.66** A chemist carried out kinetic studies on the reaction described in Problem 12.32, obtaining the following data:

Experiment A: $[NO_2]_0 = 3.5 \times 10^{-3}$ M, $[CO]_0 = 0.40$ M

| Time (s) | 0 | 60 | 120 | 180 | 240 |
|---|---|---|---|---|---|
| $[N_2O_5]\,(10^{-3}\text{ M})$ | 3.5 | 2.5 | 1.9 | 1.6 | 1.3 |

Use graphs to determine the rate law for this reaction, and evaluate the rate constant.

**12.67** Cyclopropane converts to propene by first-order kinetics with a rate constant at 500 °C of $5.5 \times 10^{-4}$ s$^{-1}$. Calculate how long it takes (a) for 10.0% of a sample to decompose; (b) for 50.0% to decompose; and (c) for 99.9% to decompose.

**12.68** The reaction of $Cl_2$ with $H_2S$ in aqueous solution is first order in each reactant, with $k = 3.5 \times 10^{-2}$ M$^{-1}$ s$^{-1}$ at 28 °C. A solution has $[Cl_2] = 0.035$ M and $[H_2S] = 5.0 \times 10^{-5}$ M. (a) Find the $H_2S$ concentration after 225 seconds of reaction. (b) Find the time at which the $H_2S$ concentration has fallen to $1.0 \times 10^{-5}$ M.

**12.69** Phosphine ($PH_3$) decomposes into phosphorus and molecular hydrogen:

$$4\,PH_3\,(g) \longrightarrow P_4\,(g) + 6\,H_2\,(g)$$

Experiments show that this is a first-order reaction, with a rate constant of $1.73 \times 10^{-2}$ s$^{-1}$ at 650 °C. Which of the following statements are true? (a) The overall reaction is elementary. (b) The data allow $E_a$ to be evaluated. (c) The rate constant will be smaller than $1.73 \times 10^{-2}$ s$^{-1}$ at 500 °C. (d) The reaction $PH_3 \longrightarrow PH_2 + H$ might be the first, rate-determining step in the mechanism of this reaction.

**12.70** The Haber reaction for the manufacture of ammonia is

$$N_2 + 3\,H_2 \longrightarrow 2\,NH_3$$

Without doing any experiments, which of the following can you say *must* be true? (a) Reaction rate $= -\Delta[N_2]/\Delta t$. (b) The reaction is first order in $N_2$. (c) The rate of disappearance of $H_2$ is three times the rate of disappearance of $N_2$. (d) The rate of disappearance of $N_2$ is three times the rate of disappearance of $H_2$. (e) $\Delta[H_2]/\Delta t$ has a positive value. (f) The reaction is not an elementary reaction. (g) The activation energy is positive.

**12.71** For each of the following reactions, what is the order with respect to each starting material and what is the overall order of the reaction?

$$2\,N_2O_5 \longrightarrow 4\,NO_2 + O_2 \qquad \text{Rate} = k[N_2O_5]$$

$$2\,NO + 2\,H_2 \longrightarrow N_2 + 2H_2O \qquad \text{Rate} = k[NO]^2[H_2]$$

$$\text{Glucose} + \text{ATP} \xrightarrow{\text{Enzyme}} \text{Glucose phosphate} + \text{ADP}$$
$$\text{Rate} = k[\text{enzyme}]$$

**12.72** For each of the following reactions, what is the order with respect to each starting material, and what is the overall order of the reaction?

$$N_2 + 3\,H_2 \xrightarrow{\text{Catalyst}} 2\,NH_3 \qquad \text{Rate} = k$$

$$\text{Sucrose} + H_2O \longrightarrow \text{Glucose} + \text{Fructose}$$
$$\text{Rate} = k[\text{Sucrose}][H_2O][H_3O^+]$$

$$CHCl_3 + Cl_2 \longrightarrow CCl_4 + HCl \quad \text{Rate} = k[CHCl_3][Cl]^{1/2}$$

**12.73** What is the difference between speed and spontaneity of a chemical reaction?

**12.74** Can the assumption that the forward and reverse rates are equal ever be exact for a reacting system? Explain your answer.

**12.75** Write chemical equations that show how $CF_2Cl_2$ contributes to the destruction of ozone in the stratosphere.

**12.76** Write the chemical equations for the reactions that maintain the balance among $O_3$, $O_2$, and $O$ in the unpolluted stratosphere.

**12.77** One possible mechanism for the reaction between $H_2$ and NO follows:

$$2NO \rightleftharpoons N_2O_2 \qquad \text{(fast)}$$
$$N_2O_2 + H_2 \longrightarrow N_2O + H_2O \qquad \text{(slow)}$$
$$N_2O + H_2 \longrightarrow N_2 + H_2O \qquad \text{(fast)}$$

(a) What is the overall stoichiometry of this reaction?
(b) What rate law is predicted by this mechanism?

**12.78** A chemical reaction is thought to proceed by the following mechanism:

$$A + B \underset{k_{-1}}{\overset{k_1}{\rightleftharpoons}} C$$

$$C + D \xrightarrow{k_2} A + E$$

$$E + B \xrightarrow{k_3} 2F$$

(a) What is the net stoichiometry of the reaction? (b) If the second step is rate-determining, what is the rate law? (c) Identify any catalysts and intermediates.

**12.79** Use molecular arguments to explain why each of the following factors speeds up a chemical reaction: (a) a catalyst; (b) an increase in temperature; and (c) an increase in concentration.

**12.80** Explain in molecular terms why the Haber synthesis cannot proceed in a single-step elementary reaction.

**12.81** Molecule A decomposes to give B and C. The following figures represent two experiments conducted to study this decomposition reaction.

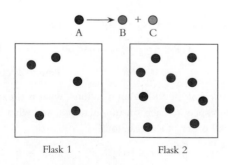

Flask 1            Flask 2

Flask 2 is found to react four times faster than Flask 1. What is the rate law for the decomposition of A? Explain your reasoning in terms of molecular collisions.

**12.82** Phosphorus-32 is a radioactive isotope that decomposes in a unimolecular first-order process. The following figures represent portions of two flasks, each containing six atoms of $^{32}P$.

Flask 1            Flask 2

Will the time required for the six atoms in Flask 2 to decompose to three atoms be faster, slower, or the same as that in Flask 1? Explain.

## ADDITIONAL PROBLEMS

**12.83** At least three possible reaction mechanisms exist for the reaction of hydrogen with halogens, $H_2 + X_2 \longrightarrow 2\,HX$. Determine the rate law predicted by each of them:

(a) $H_2 + X_2 \longrightarrow 2\,HX$     (bimolecular, direct)
(b) $X_2 \longrightarrow X + X$     (slow)
    $X + H_2 \longrightarrow HX + H$     (fast)
    $H + X \longrightarrow HX$     (fast)
(c) $X_2 \rightleftharpoons X + X$     (fast reversible)
    $X + H_2 \longrightarrow HX + H$     (slow)
    $H + X \longrightarrow HX$     (fast)

**12.84** Consider the following hypothetical reaction, which is 3/2 order overall and half order in

$$2\,\bigcirc + \,\bigcirc\bigcirc \longrightarrow 2\,\bigcirc\bigcirc$$

The following figure represents one set of initial conditions for the reaction. Draw a similar figure that represents a set of initial conditions for which the rate would be twice as fast as in the first case:

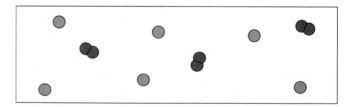

**12.85** The following mechanism has been proposed for the gas-phase reaction of $CHCl_3$ and $Cl_2$:

$$Cl_2 \underset{k_{-1}}{\overset{k_1}{\rightleftharpoons}} 2\,Cl \qquad \text{(fast, reversible)}$$

$$Cl + CHCl_3 \xrightarrow{k_2} HCl + CCl_3 \qquad \text{(slow)}$$

$$CCl_3 + Cl \xrightarrow{k_3} CCl_4 \qquad \text{(fast)}$$

(a) Write the net balanced equation. (b) What are the intermediates in the reaction? (c) What rate law is predicted by this mechanism?

**12.86** Ammonium cyanate ($NH_4NCO$) rearranges in water to urea ($NH_2CONH_2$). The following data are obtained at 50 °C:

| Time (h) | 0 | 1.0 | 2.0 | 3.0 | 5.0 | 7.0 | 9.0 |
|---|---|---|---|---|---|---|---|
| [NH₄NCO] (M) | 0.500 | 0.375 | 0.300 | 0.250 | 0.188 | 0.150 | 0.125 |

At 25 °C, the concentration falls from 0.500 M to 0.300 M in 6.0 hours.

(a) Determine the rate law. (b) Determine the rate constant at 50 °C.

(c) Determine the activation energy.

**12.87** Photographers use a rule of thumb, that a 10 °C rise in temperature cuts development time in half, to determine how to modify film development time as the temperature varies. (a) Calculate the activation energy for the chemistry of film developing, assuming that "normal" temperature is 20 °C. (b) If a certain film takes 10 minutes to develop at 20 °C, how long will it take at 25 °C?

**12.88** If chlorofluorocarbons catalyze $O_3$ decomposition, which is exothermic by 392 kJ/mol, they must also catalyze $O_3$ production from $O_2$. Using the energies shown in Figure 12-20, explain why catalysis results in net $O_3$ destruction even though reactions in both directions are accelerated.

**12.89** Acetaldehyde decomposes to methane and carbon monoxide:

$$H_3C-\overset{\overset{\textstyle O}{\|}}{C}-H \longrightarrow CH_4 + CO$$

Kinetic data for the decomposition of acetaldehyde follow:

| Time ($10^3$ s) | 0 | 1.000 | 2.000 | 3.000 | 4.000 |
|---|---|---|---|---|---|
| [CH₃CHO] (mol/L) | 0.250 | 0.118 | 0.0770 | 0.0572 | 0.0455 |

(a) Determine the rate law for the decomposition of acetaldehyde. (b) What is the value of the rate constant? (c) How long does it take for 75% of the acetaldehyde to decompose?

**12.90** A chemist interested in a reaction having overall stoichiometry

$$3A + 2B \longrightarrow \text{Products}$$

carries out two sets of experiments, both at 25 °C. In an experiment in which [B] is 1.00 M and [A] is 0.050 M at the start of the reaction, the chemist obtains these data:

| Time ($10^2$ s) | 0 | 1.0 | 2.0 | 3.0 | 4.0 | 5.0 |
|---|---|---|---|---|---|---|
| [A] (M) | 0.050 | 0.040 | 0.032 | 0.025 | 0.020 | 0.016 |

Then, the chemist performs another experiment with [B] = 1.50 M and [A] = 0.050 M. Again, it takes

$3.0 \times 10^2$ seconds for the concentration of A to fall to 0.025 M. (a) What is the rate law of the reaction? Show your reasoning. (b) Calculate the rate constant.

**12.91** The enzyme urease catalyzes the conversion of urea to ammonia and carbon dioxide:

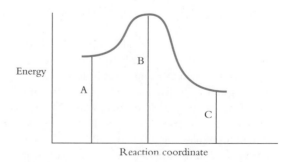

The uncatalyzed reaction has an activation energy of 125 kJ/mol. The enzyme catalyzes a mechanism that has $E_a = 46$ kJ/mol. By what factor does urease increase the rate of urea conversion at 21 °C? (*Hint:* Use the Arrhenius equation.)

**12.92** An atmospheric scientist interested in how NO is converted to $NO_2$ in urban atmospheres carries out two experiments to measure the rate of this reaction. The data are tabulated below. Find the rate law and rate constant for this reaction.

A: $[NO]_0 = 9.63 \times 10^{-3}$ M, $[O_2]_0 = 4.1 \times 10^{-4}$ M:

| t (s) | 0 | 3.0 | 6.0 | 9.0 | 12.0 |
|---|---|---|---|---|---|
| [O₂] ($10^{-4}$ M) | 4.1 | 2.05 | 1.02 | 0.51 | 0.25 |

B: $[NO]_0 = 4.1 \times 10^{-4}$ M, $[O_2]_0 = 9.75 \times 10^{-3}$ M:

| t ($10^2$ s) | 0 | 1.00 | 2.00 | 3.00 | 4.00 |
|---|---|---|---|---|---|
| [NO] ($10^{-4}$ M) | 4.1 | 2.05 | 1.43 | 1.02 | 0.82 |

**12.93** How can you identify intermediates and catalysts in a proposed mechanism for a reaction?

**12.94** For the reaction $NO + O_3 \longrightarrow NO_2 + O_2$, the experimental rate law is rate = $k$ [NO][O₃]. Which of the following sets of conditions will give the fastest rate? Explain your choice. (a) 0.5 mole of NO and 0.5 mole of $O_3$ in a 2.0-L vessel; (b) 0.5 mole of NO and 0.5 mole of $O_3$ in a 1.0-L vessel; and (c) 2.0 moles of NO and 0.1 mole of $O_3$ in a 1.0-L vessel.

**12.95** For the activation energy diagram shown, which of the following statements are true?

(a) $\Delta E_{\text{reaction}} = A - C$; (b) $\Delta E_{\text{reaction}} = B - C$;
(c) $E_{a \text{ (forward)}} = E_{a \text{ (reverse)}}$; (d) A represents the energy of the starting materials; (e) $E_{a \text{ (forward)}} = B - C$; and
(f) $E_{a \text{ (forward)}} < E_{a \text{ (reverse)}}$.

12.96 Below are some data for the decomposition of molecule A. Determine the rate law and rate constant.

| $t$ (s) | 0 | 10.0 | 20.0 | 30.0 |
|---|---|---|---|---|
| [A] (M) | 0.64 | 0.52 | 0.40 | 0.28 |

## GROUP STUDY PROBLEMS

12.97 According to the induced-fit model of enzyme activity, binding a reactant to the enzyme causes a distortion in the conformation of the reactant itself. This distortion decreases the activation energy of the catalyzed reaction. For the hypothetical unimolecular reaction A $\longrightarrow$ B + C, draw a reaction coordinate diagram for the uncatalyzed reaction and for the same reaction catalyzed by an enzyme. Use the enzyme model to explain the differences in the two diagrams.

12.98 In the preparation of cobalt complexes, one reaction involves displacement of $H_2O$ by $Cl^-$:

$$[Co(NH_3)_5H_2O]^{3+} + Cl^- \longrightarrow [Co(NH_3)_5Cl]^{2+} + H_2O$$

A proposed mechanism for this displacement starts with rapid reversible dissociation of water from the complex:

$$[Co(NH_3)_5H_2O]^{3+} \rightleftharpoons [Co(NH_3)_5]^{3+} + H_2O$$

(a) Propose a slow second step that completes the mechanism and gives the correct overall stoichiometry. (b) Derive the rate law that this mechanism predicts. (c) When the rate is studied in 1 M aqueous HCl solution that is 1 mM in $[Co(NH_3)_5H_2O]^{3+}$, first-order experimental kinetics are observed. Is this observation consistent with the proposed mechanism? State your reasoning clearly and in detail.

12.99 In your own words, describe the induced-fit model of enzyme specificity. Illustrate with diagrams, using a hypothetical enzyme that catalyzes the decomposition of a square but cannot catalyze the decomposition of a triangle:

12.100 Red and white Ping-Pong balls with small Velcro patches are placed in an air-blowing machine like the ones used to scramble numbered balls in lottery drawings on TV. If red and white balls collide at the Velcro points, they stick together. When 20 red and 10 white balls are put in the machine, 4 pairs form after 2 minutes of blowing. (a) If 10 balls of each colour had been placed in the machine, how many pairs would form under the same conditions? (b) What if there were 20 red and 15 white balls? (c) What if there were 10 red and 20 white balls? (d) What if the machine had been crammed with 40 red and 20 white balls? (e) Explain why the rates change with the numbers of balls.

12.101 Bromide ions catalyze the decomposition of aqueous $H_2O_2$:

$$2 H_2O_2 (aq) \longrightarrow 2 H_2O (l) + O_2 (g)$$

The first step is formation of $BrO^-$:

$$H_2O_2 (aq) + Br^- (aq) \longrightarrow H_2O (l) + BrO^- (aq)$$

The $BrO^-$ ion then quickly reacts with another $H_2O_2$ molecule:

$$BrO^- (aq) + H_2O_2 (aq) \longrightarrow H_2O (l) + A + B$$

(a) Identify A and B. (b) Construct an activation energy diagram for this reaction. Consult thermodynamic tables to get an estimate of $\Delta E$. Clearly show the effect of bromide ion catalysis and locate on your diagram the energy of the $BrO^-$ intermediate.

12.102 The reaction between ozone and nitrogen dioxide follows:

$$2 NO_2 (g) + O_3 (g) \longrightarrow N_2O_5 (g) + O_2 (g)$$
$$\Delta H = -200 \text{ kJ}$$

The reaction proceeds according to the experimental rate law, rate = $k$ [$NO_2$][$O_3$], and $NO_3$ has been identified as an intermediate in the reaction. The activation energy is 50 kJ/mol. (a) Devise a two-step mechanism for the reaction that is consistent with the experimental observations. Identify the rate-determining step. (b) Draw a molecular picture that illustrates your rate-determining step. (c) Sketch the activation energy diagram for this reaction and label it as completely as you can.

12.103 Hydroxide ion has an effect on the rate of the redox reaction between $I^-$ and $OCl^-$ as the following data show:

$$OCl^- (aq) + I^- (aq) \longrightarrow Cl^- (aq) + IO^- (aq)$$

| Experiment | $[OCl^-]_0$ [M] | $[I^-]_0$ [M] | $[OH^-]_0$ [M] | Initial Rate ($10^{-2}$ M/s) |
|---|---|---|---|---|
| 1 | 0.025 | 0.0025 | 0.10 | 3.8 |
| 2 | 0.025 | 0.0060 | 0.10 | 9.1 |
| 3 | 0.037 | 0.0025 | 0.10 | 5.6 |
| 4 | 0.037 | 0.0025 | 0.20 | 2.8 |

(a) Determine the rate law. (b) Evaluate the rate constant. (c) Devise a mechanism that is consistent with the rate law. The first step should be a rapid equilibrium involving $H_2O$. (d) Design additional experiments that you could use to test your mechanism.

## ANSWERS TO SECTION EXERCISES

**12.1.1** Second product is $O_2$: $NO + O_3 \longrightarrow NO_2 + O_2$.

**12.1.2** Your description should include highway speeds and the rate at which toll-takers take tolls.

**12.1.3** Chemical equation is $2\,NO + 2\,H_2 \longrightarrow 2\,H_2O + N_2$, and the intermediates are $N_2O_2$ and $N_2O$.

**12.2.1**

**12.2.2** Hydrogen

**12.2.3** Relative rate $= -\dfrac{\Delta[N_2]}{\Delta t} = -\dfrac{1}{3}\dfrac{\Delta[H_2]}{\Delta t} = +\dfrac{1}{2}\dfrac{\Delta[NH_3]}{\Delta t}$

**12.3.1**

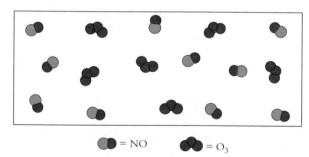

$\bullet\!\!\bullet$ = NO　　$\bullet\!\!\bullet\!\!\bullet$ = $O_3$

**12.3.2** The rate law is rate $= k\,[\textit{cis}$-2-Butene$]$, and the rate constant has units of time$^{-1}$.

**12.3.3** The rate law is rate $= k\,[H_2][Br_2]^{1/2}$, the overall order is $\dfrac{3}{2}$, and the rate constant has units of $M^{-1/2}$/time.

**12.4.1** A plot of $\ln([N_2O_5]_0/[N_2O_5])$ vs. $t$ is linear, so the reaction is first order; $k = 3.04 \times 10^{-2}$ min$^{-1}$.

**12.4.2** The rate law is rate $= k\,[AB_2]^2$.

**12.4.3** Rate $= k\,[NH_4^+][NO_2^-]$; $k = 2.7 \times 10^{-4}$ $M^{-1}$ s$^{-1}$

**12.5.1** (a) $N_2O_4 \longrightarrow 2\,NO_2$ (occurs twice); and (b) the first step must be rate-determining.

**12.5.2** If the second step is rate-determining, the rate law is second order in $N_2O_5$.

**12.5.3** (a) $N_2O_2 + O_2 \longrightarrow 2\,NO_2$; and
(b) rate $= k_2[N_2O_2][O_2]$, but using equality of rates, $[N_2O_2] = k_1[NO]^2/k_{-1}$, resulting in rate $= k\,[NO]^2[O_2]$.

**12.6.1** The $N_2$ triple bond is extremely strong. The $H_2$ bond, 435 kJ/mol, is a strong single bond. From these bond strengths, we predict a high activation energy for the reaction.

**12.6.2** (a) Graph $\ln k$ vs. $1/T$:

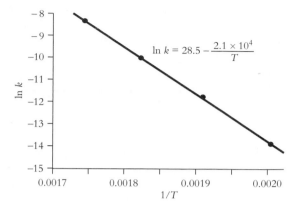

$$\ln k = 28.5 - \frac{2.1 \times 10^4}{T}$$

$E_a = 1.8 \times 10^2$ kJ/mol; and (b) $k = 4.4 \times 10^{-10}$ s$^{-1}$

**12.6.3**

**12.7.1** They are less expensive because only a thin layer of the expensive metal, rather than solid pieces, is required. They are more effective because a layer on porous charcoal has a larger surface area (where the catalyzed reaction occurs) than the relatively smooth surface of a pure piece of metal.

**12.7.2** (a) True; (b) a catalyst does not affect the overall energy change of a reaction; (c) a catalyst lowers the activation energy of an overall reaction by changing its mechanism; and (d) true.

**12.7.3** The reaction converts pollutants (CO and $NO_2$) into benign constituents of the atmosphere ($CO_2$ and $N_2$), so catalyzing this reaction could reduce atmospheric pollution levels.

## ANSWERS TO PRACTICE EXERCISES

**12.1** Reaction rate $= \left(-\dfrac{1}{2}\right)\left(\dfrac{\Delta[NO_2]}{\Delta t}\right) = \left(\dfrac{\Delta[N_2O_4]}{\Delta t}\right)$

**12.2** The rate will be the same as the rate for Figure 12-8a, so it will be half as fast as the rate for Figure 12-8b and three times as fast as the rate for the view shown in the example.

**12.3** The rate law is first order in each reactant and second order overall, so the rate constant has units of $M^{-1}$ s$^{-1}$.

**12.4**    A plot of ln ($[SO_2Cl_2]_0/[SO_2Cl_2]$) vs $t$ will be linear if the reaction is first order:

The plot is linear, so the reaction is first order.

**12.5**    $t_{1/2} = 3.18 \times 10^4$ s

**12.6**    A plot of $\left(\dfrac{1}{[SO_2Cl_2]} - \dfrac{1}{[SO_2Cl_2]_0}\right)$ will be linear if the reaction is second order:

The plot is not linear, so the data are not consistent with a second-order reaction.

**12.7**    Rate $= k\,[CH_3Br]\,[OH^-]$ and $k = 172\ M^{-1}\ s^{-1}$

**12.8**    Rate $= k\,[CO_2]$ and $k = 0.19\ s^{-1}$

**12.9**    Rate $= k_1[\text{Butadiene}]^2$

**12.10**   Rate $= \dfrac{k_2 k_1 [O_3][NO_2]^2}{k_{-1}[O_2]}$

**12.11**   $E_a = 11.4\ kJ/mol$

$y = -1370x + 20.8$

**12.12**   $E_a = 262\ kJ/mol$

**12.13**   At 25 °C (298 K), $k = 4.7 \times 10^{-31}\ s^{-1}$

**12.14**   Rate $= \dfrac{k_1 k_2 k_3}{k_{-1}k_{-2}}\,[E][S][R] = k_{obs}[E][S][R]$, with

$$k_{obs} = \dfrac{k_1 k_2 k_3}{k_{-1}k_{-2}}$$

# PRINCIPLES OF CHEMICAL EQUILIBRIUM

## AGRICULTURE AND NITROGEN FIXATION

Growing plants require a regular supply of carbon, hydrogen, and nitrogen to synthesize living tissue. Water, atmospheric carbon dioxide, and molecular oxygen supply carbon, hydrogen, and oxygen, but nitrogen is not readily available. Even though molecular nitrogen is the main component of the atmosphere, the nitrogen atoms in $N_2$ are held together by an exceptionally strong triple bond. Consequently, few organisms make direct use of $N_2$. Instead, plants extract nitrogen from nitrate and ammonium ions in the soil. The availability of usable nitrogen, known as "fixed" nitrogen, often limits the growth rate of vegetation.

The main natural source of fixed nitrogen is decaying animal and plant matter. In addition, lightning strikes convert atmospheric nitrogen into compounds containing fixed nitrogen, and a few types of bacteria are capable of using $N_2$ directly. These natural sources fall well short of providing enough nitrogen for intensive agriculture, thus farmers must supply nitrogen-containing fertilizers. Before World War I, the main source of fertilizer was natural deposits of sodium nitrate, but these supplies were quite limited. In 1898,

the English chemist Sir William Ramsay predicted that the limited supply of fixed nitrogen would lead to world famine by the middle of the twentieth century.

In the absence of a new source of fixed nitrogen, Ramsay's prediction might have come true, but just before the start of World War I, German chemists developed the **Haber process** for producing ammonia from molecular nitrogen and hydrogen:

$$N_2 + 3 H_2 \longrightarrow 2 NH_3$$

An astonishing 1–2% of the world energy demand is used just to make ammonia by this process. Approximately 4–5% of the world natural gas supply is also required (which is the source of the hydrogen gas). The Haber process now thoroughly dominates the modern fertilizer industry, because it provides a plentiful and relatively inexpensive industrial source of fixed nitrogen.

It took humanity a long time to discover how to fix nitrogen because of the extreme conditions of temperature and pressure required. The reasons for this are explained by chemical equilibrium. This is thus another example of how chemistry has strongly impacted our daily lives!

# 13

## CHAPTER CONTENTS

**13.1 Describing Chemical Equilibria**
**13.2 Properties of Equilibrium Constants**
**13.3 Thermodynamics and Equilibrium**
**13.4 Shifts in Equilibrium**
**13.5 Working with Equilibria**
**13.6 Equilibria in Aqueous Solutions**

## LEARNING OBJECTIVES

*Upon completion of this chapter you should be able to*

- **explain the dynamic nature of equilibrium in terms of reversibility**
- **understand some of the properties of equilibrium constants**
- **relate the equilibrium position to thermodynamic quantities**
- **predict the effects on the equilibrium position of changing concentrations or temperature**
- **solve quantitative equilibrium problems**
- **perform equilibrium calculations on reactions in aqueous solution**

As an indispensable source of fertilizer, the Haber process is one of the most important reactions in industrial chemistry. Nevertheless, even under optimal conditions, the yield of the ammonia synthesis in industrial reactors is only about 13%. This is because the Haber process does not go to completion; the net rate of producing ammonia reaches zero when substantial amounts of $N_2$ and $H_2$ are still present. At balance, the concentrations no longer change even though some of each starting material is still present. This balance point represents dynamic chemical equilibrium.

In this chapter, we present basic features of chemical equilibrium. We explain why reactions such as the Haber process cannot go to completion. We also show why using catalysts and elevated temperatures can accelerate the rate of this reaction but cannot shift its equilibrium position in favour of ammonia and why elevated temperature in fact shifts the equilibrium in the wrong direction. In Chapters 14 and 15, we turn our attention specifically to applications of equilibria, including acid–base chemistry.

## 13.1  DESCRIBING CHEMICAL EQUILIBRIA

The detailed chemistry describing the synthesis of ammonia is complex, so we introduce the principles of equilibrium using the chemistry of nitrogen dioxide. Molecules in a sample of nitrogen dioxide are always colliding with one another. As described in Chapter 12, a collision in the correct orientation can result in bond formation, producing an $N_2O_4$ molecule:

$$2\ NO_2 \longrightarrow N_2O_4$$

In a vessel that contains only $NO_2$ molecules, the production of $N_2O_4$ is the only reaction that takes place. However, once $N_2O_4$ molecules are present, the reverse reaction can also occur. An $N_2O_4$ molecule can fragment after collisions give it sufficient energy to break the N—N bond. These fragmentations regenerate $NO_2$:

$$N_2O_4 \longrightarrow 2\ NO_2$$

Figure 13-1 depicts these two processes from the molecular perspective.

## Dynamic Equilibrium

Collisions between $NO_2$ molecules produce $N_2O_4$ and consume $NO_2$. At the same time, decomposition of $N_2O_4$ produces $NO_2$ and consumes $N_2O_4$. When the concentration of $N_2O_4$ is very low, the decomposition occurs more often than the formation reaction. As the $N_2O_4$ concentration increases, however, the rate of decomposition increases. Eventually, the rate of $N_2O_4$ production equals the rate of its decomposition. Even though individual molecules continue to combine and decompose, the rate of one reaction exactly balances the rate of the other. This is a **dynamic equilibrium**. At dynamic equilibrium, the rates of the forward and reverse reactions are equal. The system is *dynamic* because individual molecules react continuously. It is *at equilibrium* because there is no *net* change in the system.

**Figure 13-1**

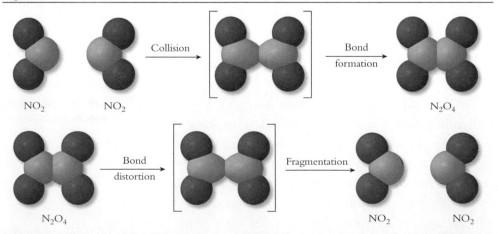

When two $NO_2$ molecules collide in the proper orientation, an N—N bond forms to produce a molecule of $N_2O_4$. When a molecule of $N_2O_4$ acquires sufficient energy through molecular collisions, its N—N bond distorts and eventually fragments to produce two molecules of $NO_2$.

The reactions that occur in a mixture of $NO_2$ and $N_2O_4$ lead to changes in concentrations. The concentration of $NO_2$ can be determined experimentally, because $NO_2$ is orange and $N_2O_4$ is colourless (see Figure 12-1). The colour intensity of the gas mixture is proportional to the concentration of $NO_2$. Figure 13-2 summarizes the results of quantitative experiments on the $NO_2/N_2O_4$ system. The data show that the concentrations of both gases level off to constant values as the reaction reaches equilibrium. If the reaction to form $N_2O_4$ went to completion, the concentration of $NO_2$ would drop to zero. If the reaction to decompose $N_2O_4$ went to completion, the concentration of $N_2O_4$ would drop to zero. Instead, these reactions reach equilibrium when substantial amounts of both gases are present.

## The Equilibrium Constant

When a mixture contains both $NO_2$ and $N_2O_4$, both the production reaction and the decomposition reaction occur. Each reaction has a characteristic rate constant, $k_p$ for the production reaction and $k_d$ for the decomposition reaction:

Production: $$NO_2 + NO_2 \xrightarrow{k_p} N_2O_4$$

Decomposition: $$N_2O_4 \xrightarrow{k_d} NO_2 + NO_2$$

Decomposition is the reverse of production, so we can combine the two reactions in a single equation with a double arrow to show that the reaction proceeds in both directions:

$$NO_2 + NO_2 \underset{k_d}{\overset{k_p}{\rightleftarrows}} N_2O_4$$

As this designation suggests, the reactions continue at equilibrium, even though the concentrations of $NO_2$ and $N_2O_4$ no longer change. That is, the rate of decomposition of $N_2O_4$ exactly balances the rate of its production, so no net change occurs. The rates of these elementary reactions can be expressed as described in Chapter 12:

$$\text{Rate of production of } N_2O_4 = k_p \left[ NO_2 \right]^2$$
$$\text{Rate of decomposition of } N_2O_4 = k_d \left[ N_2O_4 \right]$$

At equilibrium, the rate at which $N_2O_4$ decomposes equals the rate at which $N_2O_4$ forms. We designate equilibrium concentrations with subscript "eq":

$$k_p \left[ NO_2 \right]_{eq}^2 = k_d \left[ N_2O_4 \right]_{eq}$$

We rearrange this equality to group the rate constants on one side and the concentrations on the other, with product concentrations in the numerator. The concentration ratio is the **equilibrium constant** ($K_{eq}$) for the reaction:

$$K_{eq} = \frac{k_p}{k_d} = \frac{\left[ N_2O_4 \right]_{eq}}{\left[ NO_2 \right]_{eq}^2}$$

The equilibrium constant expresses the relationship between the concentrations of products and reactants at equilibrium.

**Figure 13-2**

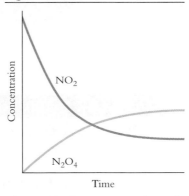

Changes in the concentrations of $NO_2$ and $N_2O_4$ with time. When the concentrations become constant with time, the system is said to be in equilibrium.

The double arrow, $\rightleftarrows$, is called an equilibrium sign. It must be written in this manner. Other arrows are used in chemistry but have different meanings. The resonance symbol, for instance, introduced in Chapter 6, is a double-headed arrow: $\leftrightarrow$.

Reactions in aqueous solution are treated in the same way as those in the gas phase. As the concentration of products increases, the products react to regenerate reactants, and eventually the reaction reaches equilibrium. Example 13-1 treats an equilibrium in aqueous solution.

## EXAMPLE 13-1 AQUEOUS EQUILIBRIUM

Molecular iodine dissolves in a solution containing iodide anions to form triiodide anions:

$$I_2\ (aq) + I^-\ (aq) \longrightarrow I_3^-\ (aq)$$

Describe the additional reaction that occurs as this system approaches equilibrium, draw molecular pictures illustrating the reactions that occur at equilibrium, and write the equilibrium constant expression for this reaction.

**Strategy:** The problem asks for a qualitative analysis of a chemical equilibrium. We must visualize what takes place at the molecular level, describe the system in words, draw pictures that summarize the reactions, and then use the ideas developed for the $NO_2/N_2O_4$ reaction to write an expression for the equilibrium constant.

**Solution:** Initially, the forward reaction to form triiodide anions is the only process that occurs. As the concentration of the product increases, however, the reverse reaction also takes place:

$$I_3^-\ (aq) \longrightarrow I_2\ (aq) + I^-\ (aq)$$

We need two molecular pictures to illustrate the chemistry. One shows an $I_2$ molecule colliding with an $I^-$ anion to form $I_3^-$. A second picture illustrates the reverse process:

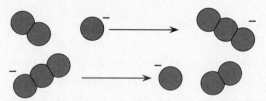

The equilibrium constant expression is determined from the rates for production of $I_3^-$ (the forward reaction) and for its decomposition (the reverse reaction). These are of the form of Equation 12-2:

$$\text{Rate of production of } I_3^- = k_p[I_2][I^-]$$
$$\text{Rate of decomposition of } I_3^- = k_d[I_3^-]$$

Setting these rates equal to each other gives the equilibrium condition, which can be solved for the equilibrium constant expression:

$$I_2\ (aq) + I^-\ (aq) \rightleftarrows I_3^-\ (aq)$$

$$k_p[I_2]_{eq}[I^-]_{eq} = k_d[I_3^-]_{eq} \quad \text{so} \quad K_{eq} = \frac{k_p}{k_d} = \frac{[I_3^-]_{eq}}{[I_2]_{eq}[I^-]_{eq}}$$

 **Do the Results Make Sense?** From a molecular perspective, just as $I_2$ and $I^-$ can combine to form $[I_3^-]$, it is reasonable that triiodide can break apart into the substances from which it originally formed.

### Practice Exercise 13.1

In the gas phase, formic acid molecules ($HCO_2H$) form molecular pairs, called dimers, through hydrogen bonding. Describe the resulting equilibrium, including the reactions, a molecular picture, and the equilibrium constant expression.

## Reversibility

In the $NO_2/N_2O_4$ system, molecules of $NO_2$ combine to give $N_2O_4$ molecules, and $N_2O_4$ molecules decompose to give $NO_2$ molecules. This is an example of the **reversibility** of molecular reactions. Look again at Figure 13-1: if two $NO_2$ molecules can form a bond when they collide, then that bond can also break apart when an $N_2O_4$ molecule distorts. The concept of reversibility is a general principle that applies to all molecular processes. Every *elementary* reaction that goes in the forward direction can also go in the reverse direction. As a consequence of reversibility, we can write each step in a chemical mechanism using a double arrow to describe what happens at chemical equilibrium.

To illustrate the generality of reversibility and the equilibrium expression, we extend our kinetic analysis to a different chemical reaction—one that has a **two-step mechanism**. At elevated temperature, $NO_2$ decomposes into $NO$ and $O_2$ instead of forming $N_2O_4$. The mechanism for the decomposition reaction, which appears in Chapter 12, contains two steps:

$$NO_2 + NO_2 \xrightarrow{k_1} NO + NO_3$$
$$NO_3 \xrightarrow{k_2} NO + O_2$$

Just as the decomposition of $N_2O_4$ is reversible, so is the decomposition of $NO_3$. As Figure 13-3 shows, $NO_3$ and $NO$ can collide and exchange an O atom. Furthermore, if an $O_2$ molecule and an $NO$ molecule collide in the proper orientation, they can bond to form $NO_3$:

$$NO + O_2 \xrightarrow{k_{-2}} NO_3$$

These reactions play negligible roles early in the decomposition of $NO_2$, when the concentration of $NO$ is low. As the products accumulate, however, collisions involving $NO$ become more likely. Eventually, product concentrations become large enough that the rates of the reverse reactions match the rates of the forward

**Figure 13-3**

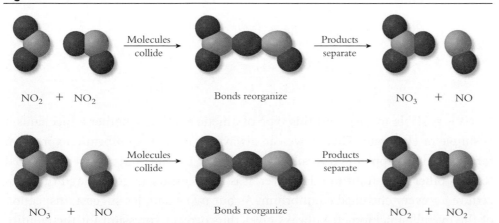

$$NO_2 \quad + \quad NO_2 \qquad\qquad \text{Bonds reorganize} \qquad\qquad NO_3 \quad + \quad NO$$

$$NO_3 \quad + \quad NO \qquad\qquad \text{Bonds reorganize} \qquad\qquad NO_2 \quad + \quad NO_2$$

If two molecules of $NO_2$ can exchange an O atom when they collide, forming $NO_3$ and $NO$, then a collision of $NO_3$ and $NO$ can also result in exchange of an O atom, forming two molecules of $NO_2$.

reactions. At this time, the system has reached equilibrium, and each step in the mechanism has a forward rate that equals its reverse rate:

$$NO_2 + NO_2 \underset{k_{-1}}{\overset{k_1}{\rightleftharpoons}} NO + NO_3$$

$$NO_3 \underset{k_{-2}}{\overset{k_2}{\rightleftharpoons}} NO + O_2$$

$$k_1[NO_2]_{eq}^2 = k_{-1}[NO]_{eq}[NO_3]_{eq} \quad \text{and} \quad k_2[NO_3]_{eq} = k_{-2}[NO]_{eq}[O_2]_{eq}$$

Rearranging each of these rate equalities gives equilibrium expressions for both steps of the mechanism:

$$\frac{k_1}{k_{-1}} = \frac{[NO]_{eq}[NO_3]_{eq}}{[NO_2]_{eq}^2} \quad \text{and} \quad \frac{k_2}{k_{-2}} = \frac{[NO]_{eq}[O_2]_{eq}}{[NO_3]_{eq}}$$

If we multiply one of these equations by the other, the concentration of the intermediate, $[NO_3]_{eq}$, cancels to produce an equilibrium expression entirely in terms of concentrations of reactants and products:

$$\frac{k_1}{k_{-1}} \frac{k_2}{k_{-2}} = \frac{[NO]_{eq}[NO_3]_{eq}}{[NO_2]_{eq}^2} \frac{[NO]_{eq}[O_2]_{eq}}{[NO_3]_{eq}}$$

The two concentration terms containing NO can be combined, and the ratio of rate constants can be replaced by a single constant, the equilibrium constant for the reaction:

$$K_{eq} = \frac{[NO]_{eq}^2[O_2]_{eq}}{[NO_2]_{eq}^2}$$

Each equilibrium expression described so far contains a ratio of concentrations of products and reactants. Moreover, each concentration is raised to a power equal to its stoichiometric coefficient in the balanced equation for the overall reaction. Concentration ratios always have products in the numerator and reactants in the denominator:

$$2\,NO_2 \rightleftharpoons 2\,NO + O_2 \qquad 2\,NO_2 \rightleftharpoons N_2O_4$$

$$K_{eq} = \frac{[NO]_{eq}^2[O_2]_{eq}}{[NO_2]_{eq}^2} \qquad K_{eq} = \frac{[N_2O_4]_{eq}}{[NO_2]_{eq}^2}$$

It is possible to carry out this type of kinetic analysis whether a mechanism is simple or elaborate. That is, we can always derive the equilibrium expression for a reaction by applying reversibility and setting forward and reverse rates equal to each other at equilibrium. However, it is unnecessary to go through this procedure for every chemical equilibrium. As our two examples suggest, inspection of the overall stoichiometry *always* gives the correct expression for the equilibrium constant. That is, a reaction of the form

$$a\mathrm{A} + b\mathrm{B} \rightleftharpoons d\mathrm{D} + e\mathrm{E}$$

has an equilibrium constant expression given by Equation 13-1:

$$K_{eq} = \frac{[D]_{eq}^{d}[E]_{eq}^{e}}{[A]_{eq}^{a}[B]_{eq}^{b}} \qquad (13\text{-}1)$$

Equation 13-1 expresses the fact that the equilibrium expression for any reaction can be written from its overall stoichiometry. Example 13-2 applies Equation 13-1 to the Haber process.

## EXAMPLE 13-2 EQUILIBRIUM CONSTANT EXPRESSION

What is the equilibrium expression for the Haber synthesis of ammonia?

$$N_2\,(g) + 3\,H_2\,(g) \rightleftharpoons 2\,NH_3\,(g)$$

**Strategy:** If the mechanism for this equilibrium were given, we could derive an equilibrium expression using the principle of reversibility, but Equation 13-1 shows that we can write the equilibrium expression directly without knowing details of the mechanism.

**Solution:** The product concentration, raised to a power equal to its stoichiometric coefficient, appears in the numerator. The concentrations of the two starting materials, each raised to a power equal to its stoichiometric coefficient, appear in the denominator:

$$K_{eq} = \frac{[NH_3]_{eq}^{2}}{[N_2]_{eq}[H_2]_{eq}^{3}}$$

 **Does the Result Make Sense?** Compare the equilibrium constant expression with the equilibrium reaction to verify that products and reactants are in the right places, with the correct powers.

**Common Pitfalls:** Neglecting the exponents in the equilibrium expression.

**Practice Exercise 13.2**
Write the balanced reaction and the equilibrium constant expression for the conversion of molecular oxygen into ozone, $O_2 \longrightarrow O_3$.

Equilibrium constant expressions have some similarities with the rate laws described in Chapter 12. For example, compare the equilibrium constant expression and the rate law for the $NO_2/NO$ reaction:

$$2\,NO_2 \rightleftharpoons 2\,NO + O_2 \qquad K_{eq} = \frac{[NO]_{eq}^{2}[O_2]_{eq}}{[NO_2]_{eq}^{2}} \qquad Rate = k[NO_2]^2$$

Both relationships include a constant and both involve concentrations raised to exponential powers. However, a rate law and an equilibrium expression describe fundamentally different aspects of a chemical reaction. A *rate law* describes how the *rate of a reaction changes* with concentration, and kinetics in general describes the rate and path of approach to equilibrium. As we describe in this chapter, an *equilibrium expression* describes the concentrations of reactants and products when *the net rate of the reaction is zero.*

**SECTION EXERCISES**

● **13.1.1** One possible mechanism for the $O_2/O_3$ equilibrium in the stratosphere is relatively simple. Use this mechanism to derive the expression for the equilibrium constant.

$$O_3 \underset{k_{-1}}{\overset{k_1}{\rightleftharpoons}} O + O_2$$

$$O + O_3 \underset{k_2}{\overset{k_1}{\rightleftharpoons}} 2\,O_2$$

Use this mechanism to derive the expression for the equilibrium constant.

● **13.1.2** Draw molecular pictures like the ones shown in Figure 13-1 to illustrate all the reactions that occur when the $O_3/O_2$ system is at equilibrium.

● **13.1.3** Write the equilibrium expression for the combustion of $NH_3$ during the synthesis of nitric acid:

$$4\,NH_3\,(g) + 5\,O_2\,(g) \rightleftharpoons 4\,NO\,(g) + 6\,H_2O\,(g)$$

## 13.2 PROPERTIES OF EQUILIBRIUM CONSTANTS

The equilibrium constant expression given by Equation 13-1 is completely general and can be applied to any chemical reaction.

Three features of equilibrium constants are especially important:

1. *$K_{eq}$ is related to the stoichiometry of the balanced net reaction.*

   The numerator contains only concentrations of products, and the denominator contains only concentrations of reactants. Each concentration is raised to a power equal to its stoichiometric coefficient.

2. *$K_{eq}$ applies only at equilibrium.*

   We use subscripts "eq" to emphasize that the concentrations of reactants and products used in the ratio must be concentrations at equilibrium.

3. *$K_{eq}$ is independent of initial conditions.*

   The equilibrium constant is a constant for any particular reaction at a given temperature. Whether initial conditions include pure reactants, pure products, or any composition in between, the system reaches a state of equilibrium that is determined by the value of $K_{eq}$.

## Concentration Units and Activities

Equilibrium constants are dimensionless numbers, yet the concentrations used in an equilibrium constant expression have units. To understand this, we need to explore the reaction quotient Q, introduced in Chapter 11. (In Section 13.3 we explore in detail the link between Q and $K_{eq}$.) Here, we use Q to address the issue of concentration units and the equilibrium constant.

Recall that Equation 11-12 defines Q for the reaction $aA + bB \rightleftharpoons dD + eE$:

$$\left[\frac{(c_D)^d (c_E)^e}{(c_A)^a (c_B)^b}\right] = Q \tag{11-12}$$

In this equation, concentrations are represented as $c_X$, whereas in Equation 13-1, concentrations are expressed as [X]. Either usage describes concentrations, so comparing Equations 13-1 and 11-12, we can see that both contain the same ratio. By either notation and in either equation, concentrations refer to molarity for species in aqueous solution and to bar for gases. Henceforth, to remind you of these conventions, we represent aqueous concentrations (molarities) as [X] and gas concentrations (bar) as $p_X$.

Recall as well that from the perspective of thermodynamics, every concentration is measured *relative to a defined standard concentration*. For any gas, the defined standard is 1 bar pressure; for any solute in aqueous solution, the defined standard is 1 M. To make Q dimensionless, we need to introduce the concept of **activity**, *a*. The activity of a solute in solution is usually very close to the ratio of its concentration to the standard concentration. For example, the activity of iodide ion in aqueous solution can be approximated very closely as

$$a_{I^-(aq)} \approx \frac{[I^-_{(aq)}]}{1\ \text{M}}$$

The activity thus has the same numerical value as the concentration, but has no units.

Similarly, the activity of $NO_2$ in the gas phase would be

$$a_{NO_2(g)} \approx \frac{p_{NO_2(g)}}{1\ \text{bar}}$$

and so the activity of any gas also has no units.

Strictly speaking, equilibrium constants and reaction quotients should be formulated using activities instead of concentrations or pressures. Consider for example, the $2\ NO_2 \rightleftharpoons N_2O_4$ equilibrium reaction. Using activities, the expression for the equilibrium constant is

$$K_{eq} = \frac{a_{N_2O_4}}{(a_{NO_2})^2} \approx \frac{\left(\dfrac{p_{N_2O_4}}{1\ \text{bar}}\right)_{eq}}{\left(\dfrac{p_{NO_2}}{1\ \text{bar}}\right)^2_{eq}}$$

Since the partial pressures of each gas are expressed in bar, neither activity has any units, and so the equilibrium constant is also dimensionless.

The same is true for a reaction occurring in solution, where each concentration is divided by the standard concentration (1 M) to obtain a unitless activity. Taking account of activities in the iodide–triiodide equilibrium shown in Example 13-1 for instance, the equilibrium constant should be written as follows:

$$K_{eq} = \frac{a_{I^-_3}}{a_{I_2}a_{I^-}} \approx \frac{\left(\dfrac{[I_3^-]_{eq}}{1\text{ M}}\right)}{\left(\dfrac{[I_2]_{eq}}{1\text{ M}}\right)\left(\dfrac{[I^-]_{eq}}{1\text{ M}}\right)}$$

In either of the above examples, the standard state concentrations or pressures cancel out, and we arrive at the more familiar expressions for the equilibrium constants:

$$K_{eq} = \frac{a_{N_2O_4}}{(a_{NO_2})^2} = \frac{(p_{N_2O_4})_{eq}}{(p_{NO_2})^2_{eq}} \quad\text{and}\quad K_{eq} = \frac{a_{I_3^-}}{a_{I_2}a_{I^-}} = \frac{([I_3^-]_{eq})}{([I_2]_{eq})([I^-]_{eq})}$$

Here (and henceforth in this book), we assume that the activity of a gas is numerically equal to its partial pressure, and that of a solute is numerically equal to its molarity. Why then do we bother with the notion of activities if we get the same values for equilibrium constants by simply using concentrations and partial pressures? The answer is that the numerical value of the activity of a substance can be significantly different from the concentration or partial pressure, especially at high concentrations in solution or at high pressure in the gas phase. Under these conditions, intermolecular forces become significant, and species behave as though their concentrations or partial pressures were quite different from their actual values. However, at low concentration and moderate pressures, the approximations that activity = concentration for solution phase species, and that activity = partial pressure for gas phase species, are reasonable.

An equilibrium constant expression thus contains concentrations or partial pressures, each of which has been divided by the reference concentration. For convenience, we omit these reference concentrations when we write the expressions for Q or for $K_{eq}$. You should remember, however, that the implicit presence of reference concentrations means that K values are dimensionless but that concentrations must be in mol/L for solutes and partial pressures must be in bar for gases.

By definition, the activity of any pure solid or liquid is unity (1.000). This is because the standard states for these substances are the pure solid or liquid themselves. The important result of this is that pure liquids or solids do not appear in equilibrium constant expressions. This is explored more in detail below, and illustrated in Example 13-3.

## Pure Liquids, Pure Solids, and Solvents

Chemical equilibria often involve pure liquids and solids in addition to gases and solutes. The concentration of a pure liquid or solid does not vary significantly. Figure 13-4 shows that although the amount of a solid or liquid can vary, the number of moles per unit volume remains fixed. In other words, the concentrations of pure liquids or solids are always equal to their standard concentrations. Thus, division by standard concentration results in a value of 1 for any pure liquid or

**Figure 13-4**

Although the amount of a liquid or solid can vary substantially, the concentration of substance in mol/L does not vary, as the highlighted regions illustrate.

solid. This allows us to omit pure liquids and solids from equilibrium constant expressions. For a general reaction:

$$aA + bB \leftrightarrows dD + sS$$

where S is a pure solid or liquid:

$$K_{eq} = \frac{[D]_{eq}^{d}}{[A]_{eq}^{d}[B]_{eq}^{b}}$$

Water often is a reagent in an aqueous equilibrium. For example, when carbon dioxide dissolves in water, it reacts with a water molecule to form carbonic acid:

$$CO_2\,(g) + H_2O\,(l) \rightleftharpoons H_2CO_3\,(aq) \qquad K_{eq} = \frac{[H_2CO_3]_{eq}}{(p_{CO_2})_{eq}[H_2O]_{eq}}$$

Here, carbonic acid concentration is expressed in molarity and carbon dioxide concentration in bar. What units are appropriate for the concentration of water, which is neither a solute nor a pure liquid? The thermodynamic convention is to express the concentration of solvent as its mole fraction, $X$.

$$\text{Mole fraction of solvent} = X_{solvent} = \frac{n_{solvent}}{n_{total}}$$

The mole fraction of water varies slightly as solutes are added, but in an aqueous solution, the concentration of water is much greater than the concentration of any solute. For instance, when $CO_2$ is bubbled through water at room temperature, the equilibrium concentration of $H_2CO_3$ is only $3.4 \times 10^{-2}$ M. Water has a density of 1.000 g/mL, so l L of liquid contains 55.5 moles of $H_2O$. Compared with 55.5 moles, the $3.4 \times 10^{-2}$ mole of carbonic acid in 1 L of this solution is negligible, so $n_{total} \cong n_{solvent}$, and the mole fraction of water is negligibly different from 1.00.

Even in relatively concentrated solutions, the mole fraction of water remains close to 1.00. At a solute concentration of 0.50 M, for example, $X_{H_2O} = 0.99$, only 1% different from 1.00.

Equilibrium calculations are seldom accurate to better than 5%, so this small deviation from 1.00 can be neglected. Consequently, we treat solvent water just like a pure substance: its concentration is essentially invariant, so it is omitted from the equilibrium constant expression.

Example 13-3 provides practice in manipulating equilibrium constant expressions.

## EXAMPLE 13-3 WRITING AN EQUILIBRIUM CONSTANT EXPRESSION

Write the equilibrium constant expression for the reaction of iron metal with strong aqueous acid, and indicate the concentration units for each reagent:

$$2\ Fe\ (s) + 6\ H_3O^+\ (aq) \leftrightarrows 2\ Fe^{3+}\ (aq) + 6\ H_2O\ (l) + 3\ H_2\ (g)$$

**Strategy:** The stoichiometry of the reaction determines the form of the equilibrium constant expression. Pure solids, liquids, or solvents do not appear in the expression, since their concentrations are constant.

**Solution:** In this reaction, Fe is a pure solid and $H_2O$ is the solvent, so they do not appear in the equilibrium expression. The other reagents have exponents equal to their stoichiometric coefficients:

$$K_{eq} = \frac{[Fe^{3+}]_{eq}^2 (p_{H_2})_{eq}^3}{[H_3O^+]_{eq}^6}$$

The concentrations of the two ionic solutes are expressed as molarities, and the concentration of $H_2$ is expressed as a partial pressure in bar.

**Does the Result Make Sense?** The products of which concentrations can vary appear in the numerator, and the reactant of which concentration can vary appears in the denominator, consistent with Equation 13-1.

**Practice Exercise 13.3**

Write the equilibrium constant expression for HCl gas dissolving in water to produce hydronium ions, and indicate the appropriate concentration units for each:

$$HCl\ (g) + H_2O\ (l) \leftrightarrows H_3O^+\ (aq) + Cl^-\ (aq)$$

## Direction of a Reaction at Equilibrium

For a reaction at equilibrium, the rate of the forward reaction is balanced exactly by the rate of the reverse reaction. For this reason, any equilibrium reaction can be written in either direction. The equilibrium constant for the Haber synthesis of ammonia, for example, can be expressed in two ways:

$$N_2 \,(g) + 3\,H_2 \,(g) \rightleftharpoons 2\,NH_3 \,(g) \qquad K_{eq,\,f} = \frac{(p_{NH_3})^2_{eq}}{(p_{N_2})_{eq}(p_{H_2})^3_{eq}}$$

$$2\,NH_3 \,(g) \rightleftharpoons N_2 \,(g) + 3\,H_2 \,(g) \qquad K_{eq,\,r} = \frac{(p_{N_2})_{eq}(p_{H_2})^3_{eq}}{(p_{NH_3})^2_{eq}}$$

The two equilibrium constants are reciprocals of each other:

$$K_{eq,\,f} = \frac{1}{K_{eq,\,r}}$$

$\uparrow$        $\uparrow$

Equilibrium constant     Equilibrium constant
for forward reaction      for reverse reaction

     The direction chosen for the equilibrium reaction is determined by convenience. A scientist interested in producing ammonia from $N_2$ and $H_2$ would use $K_{eq,\,f}$. On the other hand, someone studying the decomposition of ammonia on a metal surface would use $K_{eq,\,r}$. Either choice works as long as the *products* of the net reaction appear in the *numerator* of the equilibrium constant expression and the *reactants* appear in the *denominator*. Example 13-4 applies this reasoning to the iodine–triiodide reaction.

## EXAMPLE 13-4 FORMATION AND DECOMPOSITION

Write the equilibrium constant expression for the decomposition equilibrium for the triiodide anion, $I_3^- \,(aq) \rightleftharpoons I_2 \,(aq) + I^- \,(aq)$, and relate this expression to the equilibrium constant expression for production of triiodide anions.

**Strategy:** Equilibrium constant expressions are always written with the concentrations of products in the numerator and the concentrations of reactants in the denominator, and when a reaction is reversed, its equilibrium constant expression is inverted.

**Solution:** We can write the equilibrium constant expression by inspection of the decomposition equilibrium:

$$K_{eq} \,(\text{decomposition}) = \frac{[I_2]_{eq}[I^-]_{eq}}{[I_3^-]_{eq}}$$

Inverting this expression gives the equilibrium constant expression for the production reaction (Example 13-1):

$$I_2 \,(aq) + I^- \,(aq) \rightleftharpoons I_3^- \,(aq)$$

$$K_{eq} \,(\text{production}) = \frac{1}{K_{eq} \,(\text{decompostion})} = \frac{[I_3^-]_{eq}}{[I_2]_{eq}[I^-]_{eq}}$$

**Does the Result Make Sense?** When we reverse the direction of a reaction, products become reactants and vice versa. Thus, the concentrations that were in the numerator of $Q$ now are in the denominator, and vice versa.

### Practice Exercise 13.4

At 25 °C, experiments give $K_{eq} = 3.10$ for the $2\,NO_2 \rightleftharpoons N_2O_4$ equilibrium. Write the equilibrium constant expression for this equilibrium and determine the value of this equilibrium constant at 25 °C.

## Magnitudes of Equilibrium Constants

The magnitudes of equilibrium constants vary over a tremendous range and depend on the nature of the reaction as well as on the temperature of the system. Many reactions have very large equilibrium constants. For example, the reaction between $H_2$ and $Br_2$ to form HBr has a huge equilibrium constant:

$$H_2\ (g)\ +\ Br_2\ (g) \rightleftharpoons 2\ HBr\ (g) \qquad K_{eq} = \frac{(p_{HBr})^2_{eq}}{(p_{H_2})_{eq}(p_{Br_2})_{eq}} = 5.4 \times 10^{18}$$

The large value for this equilibrium constant indicates that the reaction goes virtually to completion. If the initial pressures are 1 bar each for $H_2$ and $Br_2$, then the pressure of HBr will be 2 bar when the system reaches equilibrium. The partial pressures of $H_2$ and $Br_2$ at equilibrium will be about $10^{-9}$ bar, which are negligible compared with the initial pressures.

Other reactions have extremely small equilibrium constants. For example, elemental fluorine, a diatomic molecule under standard conditions, is nevertheless at equilibrium with fluorine atoms:

$$F_2\ (g) \rightleftharpoons 2\ F\ (g) \qquad K_{eq} = \frac{(p_F)^2_{eq}}{(p_{F_2})_{eq}} = 2.1 \times 10^{-22}$$

The tiny value of this equilibrium constant indicates that a sample of fluorine at 25 °C consists almost entirely of $F_2$ molecules. If the partial pressure of $F_2$ is 1.0 bar at equilibrium, the partial pressure of fluorine atoms is $2.5 \times 10^{-11}$ bar, which is negligible compared with 1.0 bar. Nevertheless, the equilibrium constant is not zero, indicating that some fluorine atoms are present in the gas.

Some equilibrium constants are neither large nor small. As already mentioned, the equilibrium constant for the formation of $N_2O_4$ from $NO_2$ at 25 °C is 3.10. This moderate value indicates that an equilibrium mixture of $NO_2$ and $N_2O_4$ at 25 °C contains measurable amounts of each molecule.

## SECTION EXERCISES

●**13.2.1** Write expressions for the equilibrium constants for the following reactions:

(a) $PCl_5\ (s) \rightleftharpoons PCl_3\ (l) + Cl_2\ (g)$

(b) $CaCO_3\ (s) \rightleftharpoons CaO\ (s) + CO_2\ (g)$

(c) $Ca_3(PO_4)_2\ (s) \rightleftharpoons 3\ Ca^{2+}\ (aq) + 2\ PO_4^{3-}\ (aq)$

(d) $HCN\ (g) + H_2O\ (l) \rightleftharpoons H_3O^+\ (aq) + CN^-\ (aq)$

(e) $PCl_3\ (l) + Cl_2\ (g) \rightleftharpoons PCl_5\ (s)$

(f) $3\ Ca^{2+}\ (aq) + 2\ PO_4^{3-}\ (aq) \rightleftharpoons Ca_3(PO_4)_2\ (s)$

●**13.2.2** Draw a molecular picture illustrating the equilibrium in Section Exercise 13.2.1(d).

●**13.2.3** Write the equilibrium constant expression and determine the value of $K_{eq}$ for each of these reactions, which are related to reactions described in this section:

$$2\ F\ (g) \rightleftharpoons F_2\ (g)$$

$$2\ HBr\ (g) \rightleftharpoons H_2\ (g) + Br_2\ (g)$$

## 13.3 THERMODYNAMICS AND EQUILIBRIUM

Why do some reactions go virtually to completion, whereas others reach equilibrium when hardly any of the starting materials have been consumed? At the molecular level, bond energies and molecular organization are the determining factors. These features correlate with the thermodynamic state functions of enthalpy and entropy. As discussed in Chapter 11, free energy ($G$) is the state function that combines these properties. This section establishes the connection between thermodynamics and equilibrium.

### Free Energy and the Equilibrium Constant

Recall from Chapter 11 that the free energy change for a chemical process, $\Delta G$, is a signpost for spontaneity. Equation 11-13 relates $\Delta G$ to concentrations through the reaction quotient Q:

$$\Delta G_{reaction} = \Delta G^{\circ}_{reaction} + RT \ln Q \qquad (11\text{-}13)$$

A negative value for $\Delta G$ indicates that a process is spontaneous in the direction written, provided the system remains at constant temperature and pressure.

Consider as an example the reaction of $NO_2$ to produce $N_2O_4$:

$$2\,NO_2 \longrightarrow N_2O_4 \quad Q_{production} = \frac{p_{N_2O_4}}{(p_{NO_2})^2}$$

In a flask that contains only $NO_2$, $p_{N_2O_4} = 0$ and $Q_{production} = 0$. Because $\ln 0 = -\infty$, under these conditions the reaction that produces $N_2O_4$ has a negative value for $\Delta G$. In other words, the spontaneous reaction is the production of $N_2O_4$:

$$2\,NO_2 \longrightarrow N_2O_4 \quad \Delta G_{production} < 0 \quad \text{when} \quad p_{N_2O_4} = 0$$

Now consider what happens to Q and $\Delta G$ as the reaction proceeds. An increase in the amount of $N_2O_4$ causes an increase in the numerator of Q, whereas a decrease in the amount of $NO_2$ causes a decrease in the denominator of Q. Both of these changes increase the overall value of Q, making $\Delta G$ progressively less negative. Eventually, Q becomes large enough to make $\Delta G_{production} = 0$. At this point, the reaction is at equilibrium, and $Q = K_{eq}$, because each concentration in the expression for Q is an equilibrium concentration.

Starting from the other end of the reaction gives an analogous result. If Q is large and positive, so is ln Q, and $\Delta G$ has a positive value. In molecular terms, the reaction that forms $N_2O_4$ cannot proceed when there is no $NO_2$ present. Instead, the decomposition reaction of $N_2O_4$ has a negative value for $\Delta G$, so the decomposition reaction proceeds to form $NO_2$:

$$N_2O_4 \longrightarrow 2\,NO_2 \quad Q_{decomposition} = \frac{(p_{NO_2})^2}{p_{N_2O_4}}$$

$$\Delta G_{decomposition} < 0 \quad \text{when} \quad p_{NO_2} = 0$$

Decomposition of $N_2O_4$ causes Q to increase; $\Delta G$ becomes less negative, and eventually the reaction quotient reaches a value that makes $\Delta G = 0$.

**Figure 13-5**

Variation in $\Delta G$ with $Q$. At high $NO_2$ concentration (*left*), the spontaneous reaction is $2\,NO_2 \longrightarrow N_2O_4$; at high $N_2O_4$ concentration (*right*), the spontaneous reaction is $N_2O_4 \longrightarrow 2\,NO_2$. As the spontaneous reaction proceeds, $Q$ and $\Delta G$ change until the equilibrium condition is reached and $\Delta G = 0$.

Figure 13-5 shows schematically how $\Delta G$ for the $NO_2/N_2O_4$ system varies with Q. The formation of $N_2O_4$ is spontaneous, and the reaction proceeds to the right, if the relative amount of $NO_2$ is large. This causes $\Delta G_{\text{production}}$ to become less negative until $\Delta G_{\text{production}} = 0$ and the equilibrium position is reached. Decomposition of $N_2O_4$ is spontaneous, and the reaction proceeds to the left, if the relative amount of $N_2O_4$ is large. This causes $\Delta G_{\text{decomposition}}$ to become less negative until $\Delta G_{\text{decomposition}} = 0$, when the equilibrium position is reached. To summarize, regardless of the initial concentrations in a reaction system, the system reacts in the direction for which $G$ is negative. This makes $\Delta G$ less negative, and eventually $\Delta G = 0$ when the system has reached chemical equilibrium.

As this example indicates, the direction in which a reaction proceeds depends on the relationship between the magnitudes of Q and $K_{eq}$. Notice that if the concentration of any *product* is zero, then Q = 0, and the direction of the reaction must be to the *right*, toward products. Conversely, if any *reactant* concentration is zero, Q = ∞, and the direction of the reaction must be to the *left*, toward reactants.

The relationship between Q and $K_{eq}$ signals the direction of a chemical reaction. The free energy change, $\Delta G$, also signals the direction of a chemical reaction. These two criteria can be compared:

| Reaction goes right when | Equilibrium when | Reaction goes left when |
|:---:|:---:|:---:|
| $Q < K_{eq}$ | $Q = K_{eq}$ | $Q > K_{eq}$ |
| $\Delta G_{\text{reaction}} < 0$ | $\Delta G_{\text{reaction}} = 0$ | $\Delta G_{\text{reaction}} > 0$ |

## Calculating $K_{eq}$ from $\Delta G°$

The relationship between the equilibrium constant and free energy provides a connection between thermodynamics and equilibrium constants. Equation 11-13 is

$$\Delta G = \Delta G° + RT \ln Q$$

However, at equilibrium, $\Delta G = 0$ and $Q = K_{eq}$. Substituting, we have at equilibrium

$$0 = \Delta G° + RT \ln K_{eq}$$

Rearranging, we have a relationship between the equilibrium constant and the standard free energy change:

$$\Delta G° = -RT \ln K_{eq} \qquad (13\text{-}3)$$

Equation 13-3 is extremely important because it links thermodynamic data with equilibrium constants. Recall that Equation 11-9 makes it possible to use tabulated values to calculate the value for $\Delta G°$ for many reactions:

$$\Delta G°_{reaction} = \sum coeff_{products} \Delta G°_{f_{products}} - \sum coeff_{reactants} \Delta G°_{f_{reactants}} \qquad (11\text{-}9)$$

Then $K_{eq}$ can be calculated from Equation 13-3. Example 13-5 shows this for the Haber process.

### EXAMPLE 13-5 $K_{eq}$ FROM THERMODYNAMICS

Use standard thermodynamic data to calculate the value of $K_{eq}$ at 298 K for the Haber reaction:

$$N_2\,(g) + 3\,H_2\,(g) \rightleftharpoons 2\,NH_3\,(g)$$

**Strategy:** To show the logic of this kind of calculation, we apply the seven-step approach to problem-solving. We are asked to calculate $K_{eq}$ at 298 K for the Haber reaction. We visualize the process by remembering that there is a connection between free energy, $\Delta G°_{reaction}$, and equilibrium, $K_{eq}$. The problem provides only the balanced equation and the temperature. Any other necessary data will be found in tables and appendices.

**Solution:** Appendix D contains the appropriate values for $\Delta G°_f$, which we need for the calculation of $\Delta G°_{reaction}$

$$\Delta G°_f\ (kJ/mol):\quad N_2\,(g):0 \quad H_2\,(g):0 \quad NH_3\,(g):-16.4\ kJ/mol$$

We need two equations. We calculate the free energy change for the reaction using Equation 11-9, and Equation 13-3 links free energy with the equilibrium constant.

$$\Delta G°_{reaction} = \sum coeff_{products} \Delta G°_{f_{products}} - \sum coeff_{reactants} \Delta G°_{f_{reactants}} \qquad \Delta G°_{reaction} = -RT \ln K_{eq}$$

Rearrange Equation 13-3 to isolate $\ln K_{eq}$:

$$\ln K_{eq} = \frac{-\Delta G°_{reaction}}{RT}$$

Now substitute and calculate $\Delta G°_{reaction}$ and $K_{eq}$:

$$\Delta G°_{reaction} = (2 \text{ mol NH}_3)(-16.4 \text{ kJ/mol NH}_3) - 3(0) - 1(0) = -32.8 \text{ kJ}$$

$$\ln K_{eq} = \frac{-(-32.8 \text{ kJ})(10^3 \text{ J/kJ})}{(8.314 \text{ J/K})(298 \text{ K})} = 13.24 \quad K_{eq} = e^{13.24} = 5.6 \times 10^5$$

**Does the Result Make Sense?** When evaluating exponents of "e," the number of significant figures generally equals the number of decimal places in the ln value. In this case, whereas $e^{13.24} = 5.62 \times 10^5$, $e^{13.25} = 5.68 \times 10^5$, and $e^{13.23} = 5.57 \times 10^5$. The value of $K_{eq}$ is relatively large but reasonable, indicating that at room temperature, the equilibrium position for the Haber reaction favours the product, $NH_3$.

**Common Pitfalls:** Neglecting the coefficients in the balanced reaction. Make sure the reaction is in fact balanced before doing this type of calculation.

**Practice Exercise 13.5**
Use data from Appendix D to determine the equilibrium constant for formation of ozone from oxygen at 298 K.

## Equilibrium Constants and Temperature

Experimental studies on how temperature affects equilibria reveal a consistent pattern. The equilibrium constant of any *exothermic* reaction *decreases* with increasing temperature, whereas the equilibrium constant of any *endothermic* reaction *increases* with increasing temperature. We can use two equations for $\Delta G°$, Equations 13-3 and 11-10, to provide a thermodynamic explanation for this behaviour:

$$\Delta G° = -RT \ln K_{eq}$$

$$\Delta G° = \Delta H° - T\Delta S° \tag{11-10}$$

We set these two expressions for $\Delta G°$ equal to each other:

$$\Delta H° - T\Delta S° = -RT \ln K_{eq}$$

Solving for $\ln K_{eq}$ gives an equation relating $K_{eq}$ to standard enthalpy and entropy changes:

$$\ln K_{eq} = -\frac{\Delta H°}{RT} + \frac{\Delta S°}{R} \tag{13-4}$$

An exothermic reaction has a negative $\Delta H°$, making the first term on the right in Equation 13-4 positive. As $T$ increases, this term decreases, causing $K_{eq}$ to decrease. An endothermic reaction, in contrast, has a positive $\Delta H°$, making the first term on the right in Equation 13-4 negative. As $T$ increases, this term becomes less negative, causing $K_{eq}$ to increase. These variations in $K_{eq}$ with temperature, which often are substantial, can be estimated using Equation 13-4. Example 13-6 applies Equation 13-4 to the Haber synthesis.

## EXAMPLE 13-6 $K_{EQ}$ AND TEMPERATURE

Use tabulated thermodynamic data (see Appendix D) to estimate $K_{eq}$ for the Haber reaction at 500 °C.

**Strategy:** Again, we follow the seven-step procedure. This example is similar to Example 13-5, but note we are asked to find $K_{eq}$ at a temperature that is different from 298 K. Because the temperature is not 298 K, we cannot use the value of $\Delta G^\circ_{reaction}$ at 298 K. Instead, the process is as follows:

(a) Calculate values for $\Delta H^\circ_{reaction}$ and $\Delta S^\circ_{reaction}$ at 298 K using tabulated thermodynamic values and equations from earlier chapters.

$$\Delta H^\circ_{reaction} = \Sigma coeff_{products} \, \Delta H^\circ_{products} - \Sigma coeff_{reactants} \, \Delta H^\circ_{reactants} \qquad (3\text{-}12)$$

$$\Delta S^\circ_{reaction} = \Sigma coeff_{products} \, \Delta S^\circ_{products} - \Sigma coeff_{reactants} \, \Delta S^\circ_{reactants} \qquad (11\text{-}6)$$

(b) Use Equation 13-4 to determine the value of the equilibrium constant at 500 °C.

$$\ln K_{eq} = -\frac{\Delta H^\circ}{RT} + \frac{\Delta S^\circ}{R}$$

**Solution:** Appendix D contains the appropriate values:

| Substance | $N_2$ (g) | $H_2$ (g) | $NH_3$ (g) |
|---|---|---|---|
| $\Delta H^\circ_f$ (kJ/mol) | 0 | 0 | −45.9 |
| $S^\circ$ (J mol$^{-1}$ K$^{-1}$) | 191.61 | 130.680 | 192.8 |

First, determine $\Delta H^\circ_{reaction}$ and $\Delta S^\circ_{reaction}$:

$$\Delta H^\circ_{reaction} = (2)(-45.9 \text{ kJ/mol}) - 3(0) - 0 = -91.8 \text{ kJ mol}$$

$$\Delta S^\circ_{reaction} = (2)(192.8 \text{ J mol}^{-1} \text{ K}^{-1}) - (3)(130.680 \text{ J mol}^{-1} \text{ K}^{-1}) - 191.61 \text{ J mol}^{-1} \text{ K}^{-1}$$

$$\Delta S^\circ_{reaction} = -198.1 \text{ J mol}^{-1} \text{ K}^{-1}$$

Now calculate $K_{eq}$ at 500 °C:

$$\ln K_{eq} = -\frac{(-91.8 \text{ kJ mol})(10^3 \text{ J/kJ})}{(8.314 \text{ J mol}^{-1} \text{ K}^{-1})(500 + 273 \text{ K})} + \frac{-198.1 \text{ J mol}^{-1} \text{ K}^{-1}}{8.314 \text{ J mol}^{-1} \text{ K}^{-1}}$$

$$\ln K_{eq} = 14.28 - 23.83 = -9.55$$

Taking the antilogarithm, or $e^x$, of −9.55 gives the estimated equilibrium constant at 500 °C:

$$K_{eq} = 7.1 \times 10^{-5}$$

**Does the Result Make Sense?** As in Example 13-5, the result has only two significant figures because of the sensitivity of powers of "e" to small variations. We see that at this temperature, the equilibrium constant has a small value, indicating that the reactants are favoured. This is consistent with the observation that the Haber reaction has only a 13% yield at elevated temperatures.

**Practice Exercise 13.6**

Estimate the equilibrium constant for the ozone formation reaction at 825 °C.

Examples 13-5 and 13-6 underscore a dilemma faced by industrial chemists and engineers. Example 13-5 shows that at 298 K, the equilibrium position of the Haber reaction strongly favours the formation of ammonia. Why then is the Haber synthesis not carried out at 298 K? The reason is that even with a catalyst, the reaction is much too slow to be useful at this temperature. Thermodynamics favours the product, but the kinetics of the reaction are unfavourable. For the reaction to proceed at a practical rate, the temperature must be increased. This improves the kinetics of the process, but, unfortunately, the equilibrium constant falls dramatically as temperature increases. Example 13-6 shows that at 773 K, a realistic temperature for the Haber reaction, the equilibrium position does not favour $NH_3$:

$$N_2 + 3\,H_2 \rightleftharpoons 2\,NH_3$$

$$K_{eq,298\ K} = 5.6 \times 10^5 \qquad K_{eq,773\ K} = 7.1 \times 10^{-5}$$

As we describe in the Focus 11-2: Chemistry and Life in Chapter 11, chemists are searching for a catalyst that will allow his essential reaction to be carried out close to room temperature.

The Haber reaction is a practical example of the effect of temperature on an *exothermic* reaction. A practical example of an *endothermic* reaction is the use of methane and steam to produce the molecular hydrogen needed for the Haber reaction. The production of hydrogen in this way is highly endothermic, so the reaction is carried out at temperatures much greater than 1000 K to force the equilibrium toward the products:

Ammonia for making fertilizer is produced in reaction towers at elevated temperature and pressure.

$$CH_4\,(g) + H_2O\,(g) \rightleftharpoons CO\,(g) + 3\,H_2\,(g)$$

$$\Delta H^\circ = +206 \text{ kJ/mol} \qquad K_{eq,298\ K} = 1.3 \times 10^{-25} \qquad K_{eq,1500\ K} = 1.1 \times 10^4$$

You should be able to do the calculations to verify the value of the equilibrium constant at 1500 K.

Our Focus 13-1: Chemistry and Technology on the next page gives more information about the role of equilibrium in the chemical industry.

## SECTION EXERCISES

● **13.3.1** Use thermodynamic data (see Appendix D) to determine $K_{eq}$ for the oxidation of ClO at 298 K:

$$2\,ClO\,(g) + O_2\,(g) \rightleftharpoons 2\,ClO_2\,(g)$$

● **13.3.2** Does $K_{eq}$ for the reaction described in Section Exercise 13.3.1 increase or decrease when the temperature is raised above room temperature? State your reasoning.

● **13.3.3** Use thermodynamic data to estimate $K_{eq}$ at 1000 °C for the reaction in Section Exercise 13.3.1.

## FOCUS 13-1 CHEMISTRY AND TECHNOLOGY: INDUSTRIAL EQUILIBRIA

Reactions that reach equilibrium well before completion are not very attractive to the industrial chemist, because unreacted starting materials must be separated from the products, an expensive process. Nevertheless, gas-phase equilibria are involved in the industrial syntheses of $H_2SO_4$, $NH_3$, and $HNO_3$.

Sulphuric acid is produced beginning with elemental sulphur, which is burned in air to give $SO_2$. This gas reacts with additional $O_2$ to produce $SO_3$. The reaction requires a catalyst such as $V_2O_5$ and a temperature around 700 K. Even so, this reaction reaches equilibrium well before completion.

$$2\,SO_2\,(g) + O_2(g) \xrightleftharpoons[]{V_2O_5,\ 700\ K} 2\,SO_3\,(g)$$

To form sulphuric acid, $SO_3$ is added to $H_2SO_4$, and the resulting solution is treated with water:

$$SO_3\,(g) + H_2SO_4\,(l) \longrightarrow H_2S_2O_7\,(l)$$

$$H_2S_2O_7\,(l) + H_2O\,(l) \longrightarrow 2\,H_2SO_4\,(l)$$

The Haber synthesis for the conversion of $N_2$ into $NH_3$ is outlined below. The process uses $N_2$ from the atmosphere, but $H_2$ must be generated from natural gas (methane) and steam:

$$CH_4\,(g) + H_2O\,(g) \xrightleftharpoons[]{Ni,\ 800\ K} CO\,(g) + 3\,H_2\,(g)$$

The equilibrium constant for this reaction is small at 800 K, but the CO is used to make more hydrogen:

$$CO\,(g) + H_2O\,(g) \xrightleftharpoons[]{Fe_2O_3,Cr_2O_3,\ 525\ K} CO_2\,(g) + H_2\,(g)$$

This reaction is run at 525 K, a temperature at which it has a large equilibrium constant.

The Haber reaction must be catalyzed, and even then it does not approach completion:

$$N_2\,(g) + 3\,H_2\,(g) \xrightleftharpoons[]{Fe,\ K_2O,\ Al_2O_3,\ 725\ K} 2\,NH_3\,(g)$$

Recycling increases the yield. The product gas is chilled below 10 °C, causing ammonia to liquefy. The remaining gases are recycled back into the reactor to make more ammonia.

Much of the ammonia synthesized by the Haber process is used to make $HNO_3$. Ammonia and air are heated and passed over Pt gauze:

$$4NH_3\,(g) + 5\,O_2\,(g) \xrightleftharpoons[]{Pt\ gauze,\ 1200\ K} 4\,NO\,(g) + 6\,H_2O\,(g)$$

The combustion reaction goes nearly to completion, but the further reaction of NO and $O_2$ does not. As it leaves the reactor, NO is at equilibrium with $NO_2$:

$$2\ NO\ (g) + O_2\ (g) \rightleftharpoons 2\ NO_2\ (g)$$

Because this reaction is exothermic, lowering the temperature favours nitrogen dioxide, so as the products are cooled, more NO is converted into $NO_2$.

Finally, $NO_2$ is bubbled through water to give a concentrated aqueous solution of nitric acid, $HNO_3$:

$$3\ NO_2\ (g) + H_2O\ (l) \longrightarrow 2\ HNO_3\ (aq) + NO\ (g)$$

Because of the expense that accompanies inefficient syntheses, most industrial processes have been designed using reactions that go virtually to completion.

## 13.4  SHIFTS IN EQUILIBRIUM

What happens when the conditions change for a system that is already at equilibrium? Suppose additional amounts of one or more substances are introduced or the temperature of the system changes. How does a system that had been at equilibrium respond to these changes?

## Le Châtelier's Principle

A quick *qualitative* indication of how a system at equilibrium responds to a change in conditions can be obtained using **Le Châtelier's principle**, which was first formulated in 1884 by Henri-Louis Le Châtelier, a French industrial chemist:

> *When a change is imposed on a system at equilibrium, the system will react in the direction that reduces the amount of change.*

Le Châtelier's principle is a compact summary of how different factors influence equilibrium. Introducing a reagent causes a reaction to proceed in the direction that consumes the reagent. Reducing the temperature removes heat from the system and causes the reaction to produce heat by proceeding in the exothermic direction.

## Changes in Amounts of Reagents

A change in the amount of any substance that appears in the reaction quotient displaces the system from its equilibrium position. As an example, consider an industrial reactor containing a mixture of methane, hydrogen, steam, and carbon monoxide at equilibrium:

$$CH_4\ (g) + H_2O\ (g) \rightleftharpoons CO\ (g) + 3\ H_2\ (g) \quad Q = \frac{(p_{CO})(p_{H_2})^3}{(p_{CH_4})(p_{H_2O})}$$

How does this system respond if more steam is injected into the reactor? Adding one of the starting materials decreases the value of Q, making $Q < K_{eq}$. Because the system responds in a way that restores equilibrium, adding steam causes the reaction to proceed to the right, consuming $CH_4$ and $H_2O$ and producing $CO$ and $H_2$ until equilibrium is restored. Figure 13-6 provides a graphical illustration of how concentrations change when steam is added.

Removing a product from a system at equilibrium also makes $Q < K_{eq}$ and leads to the formation of additional products. This behaviour is commonly used to advantage in chemical synthesis. For example, calcium oxide (lime), an important material in the construction industry, is made by heating calcium carbonate in a furnace to about 1100 K:

$$CaCO_3\ (s) \rightleftarrows CaO\ (s) + CO_2\ (g) \qquad K_{eq} = (p_{CO_2})_{eq} = 1.0 \text{ at } 1100 \text{ K}$$

If the pressure of $CO_2$ in the furnace were to reach 1.0 bar, the system would attain equilibrium, and no additional products would form. If the $CO_2$ is allowed to escape from the reactor as it forms, on the other hand, the continuous removal of $CO_2$ drives the reaction to completion. Figure 13-7 illustrates this from a molecular perspective.

Continuous removal of a product maintains the pressure of $CO_2$ below 1.0 bar, so $Q$ has a smaller value than $K_{eq}$, and the reaction continues until all the $CaCO_3$ has been converted to $CaO$.

A change in the amount of a chemical species that has no impact on the reaction quotient will not disturb a chemical equilibrium. Thus, adding or removing air has no effect on the equilibrium conditions of the $CaO/CaCO_3/CO_2$ system, because the equilibrium constant depends only on the partial pressure of $CO_2$, not on the total pressure of gas in the system. Likewise, adding more $CaCO_3$ to a limestone reactor does not result in a higher

**Figure 13-6**

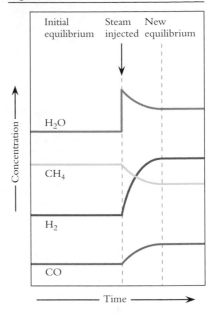

Adding some steam to a reactor that is at chemical equilibrium changes the value of $Q$, so the reaction is no longer at equilibrium. The reaction proceeds in the direction that consumes some of the added reagent in order to re-establish equilibrium.

**Figure 13-7**

(a) If $CO_2$ cannot escape, the decomposition reaction of $CaCO_3$ quickly establishes equilibrium. (b) Removing the $CO_2$ gas from a limestone kiln maintains $Q$ at a value below the equilibrium value, allowing the reaction to continue. (c) As long as $CO_2$ gas can escape, the reaction will proceed to completion.

**Figure 13-8**

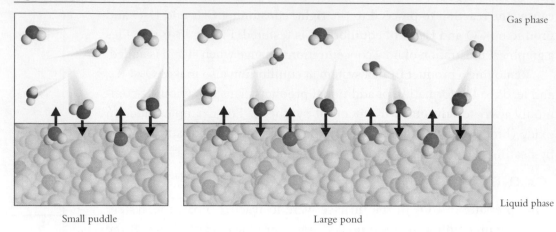

Small puddle                      Large pond

The equilibrium pressure (vapour pressure) above a liquid is independent of the amount of liquid, because the ratio of evaporating to condensing molecules is independent of amount.

equilibrium pressure of $CO_2$, because the concentration of pure solid $CaCO_3$ has a fixed value that is independent of how much solid is present.

Vapour pressure provides a simple illustration of why adding a pure liquid or solid does not change equilibrium concentrations. Recall from Chapter 8 that any liquid establishes a dynamic equilibrium with its vapour, and the partial pressure of the vapour at equilibrium is the vapour pressure. The vapour pressure is independent of the amount of liquid present. Figure 13-8 illustrates that the vapour pressure of water above a small puddle is the same as the vapour pressure above a large pond at the same temperature. More molecules escape from the larger surface of the pond, but more molecules are captured, too. The balance between captures and escapes is the same for both puddle and pond.

The effects of changes in amounts on a system at equilibrium can be summarized in accordance with Le Châtelier's principle:

1. Any change in conditions that *increases* the value of Q causes the reaction to consume products and produce reactants until equilibrium is re-established.

2. Any change in conditions that *decreases* the value of Q causes the reaction to form products and consume reactants until equilibrium is re-established.

3. Any change in amounts that has no effect on the value of Q has no effect on the equilibrium position.

Example 13-7 provides practice in applying Le Châtelier's principle at the molecular level.

## EXAMPLE 13-7  EFFECTS OF CONCENTRATION CHANGES

The left-hand view in the accompanying figure is a molecular picture showing a very small portion of an iodine–triiodide solution at equilibrium. The right-hand view shows this same solution, no longer at equilibrium because additional $I_2$ has been dissolved in the solution. Redraw this molecular picture to show qualitatively how the system responds to this change.

Initial equilibrium

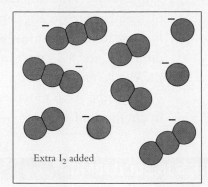
Extra I$_2$ added

**Strategy:** Because this problem asks for a qualitative answer, we do not need to do calculations. It is sufficient to apply Le Châtelier's principle to determine the direction of change and draw the new picture that shows the result of the change.

**Solution:** The change that has been imposed is addition of a reactant, I$_2$. The system responds by moving in the direction that reduces the concentration of this added reactant. We show this by combining one I$_2$ molecule with an I$^-$ anion to make one more I$_3^-$ anion.

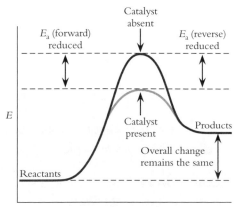
New equilibrium

**Does the Result Make Sense?** Because this is a qualitative question, we need only make sure that the changes are stoichiometrically correct and in the right direction, consuming added I$_2$. The new view contains one less I$_2$ molecule, one less I$^-$ anion, and one more molecule of the I$_3^-$ product.

**Practice Exercise 13.7**

Suppose that figure at the top right represents equilibrium and the top left view represents a displacement from equilibrium. Draw a molecular picture showing how the left-hand view would react to re-establish equilibrium.

## Effect of Catalysts

Catalysts *do not* affect the equilibrium constant. Figure 13-9 provides a reminder that a catalyst changes the mechanism of the reaction in a way that reduces the net activation energy barrier, but it does not alter the thermodynamic changes that accompany the reaction. In other words, a catalyst reduces the forward activation energy and the reverse activation energy, so the rate of reaction is increased in *both* directions. However, the catalyst does not affect the nature of the reactants or the products, so the standard free energy change, $\Delta G°$, does not change, and neither does $K_{eq}$. A catalyst changes the kinetics of a reaction but not its thermodynamics. Catalysis allows a reaction to reach equilibrium *more rapidly*, but it does not alter the equilibrium *position*.

## Effect of Temperature

As shown in Section 13.3, $K_{eq}$ varies with temperature in a way that can be understood using the principles of thermodynamics.

**Figure 13-9**

$E_a$ (forward) reduced

Catalyst absent

$E_a$ (reverse) reduced

$E$

Catalyst present

Products

Overall change remains the same

Reactants

Reaction coordinate

A catalyst changes the mechanism of a reaction and lowers the net activation energy barrier in both directions. It has no effect on the overall $\Delta G°$ and $K_{eq}$ for the reaction.

Temperature is the *only* variable that causes a change in the value of $K_{eq}$. The effect of temperature on $K_{eq}$ depends on the enthalpy change of the reaction, $H$. An *increase* in temperature always shifts the equilibrium position in the *endothermic* direction, and a *decrease* in temperature always shifts the equilibrium position in the *exothermic* direction.

Example 13-8 provides practice in analyzing changes that may lead to shifts in equilibrium.

## EXAMPLE 13-8  SHIFTS IN EQUILIBRIUM

Consider a saturated solution of $CaCl_2$ at equilibrium with excess $CaCl_2$ (s):

$$CaCl_2 \text{ (s)} \rightleftharpoons Ca^{2+} \text{ (aq)} + 2\,Cl^- \text{ (aq)} \qquad \Delta H° = +585 \text{ kJ}$$

How do the following changes affect the amount of dissolved $CaCl_2$?

(a) More $CaCl_2$ (s) is added.
(b) Some NaCl is dissolved in the solution.
(c) Some $NaNO_3$ is dissolved in the solution.
(d) Some pure water is added.
(e) The solution is heated.

Solid $CaCl_2$

**Strategy:** According to Le Châtelier's principle, the system will respond in the direction that reduces the amount of change. It will do so, however, only if an appropriate response exists. Changes in quantities of substances must be analyzed for their effect, if any, on the value of $Q$.

**Solution:**

(a) More $CaCl_2$ (s) is added. Because $Q = [Ca^{2+}][Cl^-]^2$, adding $CaCl_2$ (s) does *not* change $Q$. There is no effect on the amount of dissolved $CaCl_2$.

(b) Some NaCl is dissolved in the solution. This change increases $[Cl^-]$, making $Q > K_{eq}$. The reaction proceeds to the left, and some $CaCl_2$ (s) precipitates.

(c) Some $NaNO_3$ is dissolved in the solution. There is no effect on $Q$, so there is no effect on the amount of dissolved $CaCl_2$.

(d) Some pure water is added. Adding water dilutes the solution and lowers $[Ca^{2+}]$ and $[Cl^-]$. Because $Q$ is now less than $K_{eq}$, the reaction proceeds to the right. More $CaCl_2$ (s) dissolves.

(e) The solution is heated. Because the solubility reaction is endothermic ($\Delta H°$ is positive), $K_{eq}$ increases as $T$ increases. More $CaCl_2$ (s) dissolves.

 **Do the Results Make Sense?** The reasoning behind each part of the answer matches Le Châtelier's principle.

**Common Pitfalls:** Diluting a solution is often expected not to affect the equilibrium position. Part (d) of this question shows otherwise.

### Practice Exercise 13.8

The formation of triiodide from iodide and molecular iodine (Example 13-1) is exothermic. Prepare a list of changes that will lead to an increase in the concentration of triiodide present at equilibrium.

**13.4.1** The following solubility reaction is exothermic by about 70 kJ/mol:

$$Ba(OH)_2 \ (s) \rightleftarrows Ba^{2+} \ (aq) + 2 \ OH^- \ (aq)$$

Will more $Ba(OH)_2$ dissolve, or will some precipitate, or will no change occur, when each of the following changes is made on a saturated solution of $Ba(OH)_2$? (a) More $Ba(OH)_2 \ (s)$ is added to the solution; (b) more water is added to the solution; (c) some HCl is added to the solution; and (d) the solution is cooled on an ice bath.

**13.4.2** Refer to Examples 13-5 and 13-6. List four changes in conditions that might be used to increase the yield of ammonia in the Haber process.

**13.4.3** Chemists are optimistic that a catalyst will be found for the production of ammonia from hydrogen and nitrogen under standard conditions. In contrast, no hope exists of developing a catalyst for the production of hydrogen from methane and steam under standard conditions. Explain.

## 13.5 WORKING WITH EQUILIBRIA

The quantitative treatment of a reaction equilibrium usually involves one of two things. Either the equilibrium constant must be computed from a knowledge of concentrations, or equilibrium concentrations must be determined from a knowledge of initial conditions and $K_{eq}$. In this section, we describe the basic reasoning and techniques needed to solve equilibrium problems. Stoichiometry plays a major role in equilibrium calculations, so you may want to review the techniques described in Chapter 1, particularly the section on limiting reactants.

---

**SOLVING EQUILIBRIUM PROBLEMS**

Step 1   Determine what is asked for.

Step 2   Identify the major chemical species.

Step 3   Determine what chemical equilibria exist.

Step 4   Write the $K_{eq}$ expressions.

Step 5   Organize the data and unknowns.

Step 6   Carry out the calculations.

Step 7   Does the result make sense?

---

## Chemistry of Equilibria

Equilibrium conditions are determined by the chemical reactions that occur in a system. Consequently, it is necessary to analyze the chemistry of the system before doing *any* calculations. After the chemistry is known, a mathematical solution to the problem can be developed. We can modify the seven-step approach to problem-solving so that it applies specifically to equilibrium problems, proceeding from the chemistry to the equilibrium constant expression to the mathematical solution.

**Figure 13-10**

(a)                                    (b)

(a) When solid LiF is added to water, initially there are no ions in solution. (b) At equilibrium, Li$^+$ cations and F$^-$ anions are present.

We illustrate this approach using the equilibrium shown in Figure 13-10. When solid LiF is added to water, a small amount of the salt dissolves, leading to equilibrium between the solid and a solution. Chemical analysis reveals that the equilibrium concentration of F ions in the solution is $6.16 \times 10^{-2}$ M. We want to determine the equilibrium constant for this process.

1. **Determine what is asked for.**
   This is the usual first step. We need to know our destination before we can map out a route. We are asked for the value of the equilibrium constant for a solid dissolving to form an aqueous solution.

2. **Identify the major chemical species.**
   This step is important because chemical equilibria are dynamic interactions among *molecules and ions*. The key to success in working with equilibria is to "think molecules" (and ions). When solid LiF is added to water (Figure 13-10a), the species present before any reaction occurs are LiF (s) and H$_2$O (l).

3. **Determine what chemical equilibria exist.**
   Sometimes the net chemical reaction is provided, but in other cases, we have to examine the species present and determine what reactions can occur among them. The statement of the problem indicates that fluoride anions are present in solution when LiF dissolves in water. To maintain electrical neutrality, Li$^+$ ions must also be present in equal number. Here is the net reaction:

   $$\text{LiF (s)} \rightleftarrows \text{Li}^+ \text{ (aq)} + \text{F}^- \text{ (aq)}$$

4. **Write the $K_{eq}$ expressions.**
   This is the equation (or equations) that we must use to complete the calculations. The equilibrium constant expression for LiF dissolving in water is

   $$K_{eq} = [\text{Li}^+]_{eq} [\text{F}^-]_{eq}$$

5. **Organize the data and unknowns.**

   In some problems, concentrations at equilibrium are provided. In other problems, concentrations at equilibrium must be calculated, usually by using amounts tables. In this example, we are told that a solution of LiF at chemical equilibrium has $[F^-]_{eq} = 6.16 \times 10^{-2}$ M. The stoichiometric ratio of LiF is 1:1, so an equal amount of Li dissolves: $[Li^+]_{eq} = 6.16 \times 10^{-2}$ M.

6. **Substitute and calculate.**

   Refer to Step 1 to find out what must be calculated, the equilibrium constant or the equilibrium concentrations. In this example, we are after an equilibrium constant, which we calculate using the equilibrium concentrations:

   $$K_{eq} = [Li^+]_{eq}[F^-]_{eq} = (6.16 \times 10^{-2})(6.16 \times 10^{-2}) = 3.79 \times 10^{-3}$$

7. **Does the result make sense?**

   The concentration of ions in the saturated solution is in the range of $10^{-2}$ M, so an equilibrium constant in the $10^{-3}$ range is reasonable.

   As the LiF example illustrates, the most direct way to determine the value of an equilibrium constant is to mix substances that can undergo a chemical reaction, wait until the system reaches equilibrium, and measure the concentrations of the species present once equilibrium is established. Although the calculation of an equilibrium constant requires knowledge of the equilibrium concentrations of *all* species for which concentrations appear in the equilibrium constant expression, stoichiometric analysis often can be used to deduce the concentration of one species from the known concentration of another species. Example 13-9 shows how to approach an equilibrium constant problem from a molecular perspective.

---

**EXAMPLE 13-9  $K_{eq}$ FROM A MOLECULAR VIEW**

The figure represents a molecular view of a gas-phase reaction that has reached equilibrium. Assuming that each molecule in the molecular view represents a partial pressure of 1.0 bar, determine $K_{eq}$ for this reaction.

**Strategy:** Our seven-step approach to equilibrium problems will lead to the correct result. The first four steps are part of the strategy:

1. **Determine what is asked for.** The problem asks us to calculate an equilibrium constant, $K_{eq}$.

2. **Identify the major chemical species.** The chemical species present are the hypothetical gas-phase molecules, $AB_2$, A, and AB.

3. **Determine what chemical equilibria exist.** The forward reaction is provided. At equilibrium the reverse reaction proceeds at an equal rate:

$$2\ AB \rightleftarrows AB_2 + A$$

4. **Write the equilibrium constant expression.** We use the chemical reaction to determine the equilibrium constant expression:

$$K_{eq} = \frac{(p_{AB_2})_{eq}(p_A)_{eq}}{(p_{AB})^2_{eq}}$$

**Solution:**

5. **Organize the data and unknowns.** The problem states that each molecule represents a partial pressure of 1.0 bar, so we can determine the equilibrium concentrations of each reagent by counting molecules in the molecular picture:

$$(p_{AB_2})_{eq} = 4.0\ bar \quad (p_A)_{eq} = 1.0\ bar \quad (p_{AB})_{eq} = 4.0\ bar$$

6. **Substitute and calculate.** We substitute each concentration into the equilibrium constant expression and evaluate:

$$K_{eq} = \frac{(p_{AB_2})_{eq}\ (p_A)_{eq}}{(p_{AB})^2_{eq}} = \frac{(4.0\ bar)(1.0\ bar)}{(4.0\ bar)^2} = 0.25$$

 **7. Does the Result Make Sense?** There are comparable numbers of molecules of reactants and products present in this system at equilibrium, so an equilibrium constant close to 1 is reasonable.

**Practice Exercise 13.9**

Refer to the equilibrium view of the triiodide reaction shown in Example 13-7. If each circle represents a concentration of 1.0 mM, what is the value of the equilibrium constant?

## Initial Conditions and Concentration Tables

When we do equilibrium calculations, we are usually interested in the concentrations of species present *at equilibrium.* In many cases, however, we have information about what we call **initial concentrations**, *before* any net change has occurred. Initial concentrations are the concentrations that would be present if it were possible to mix all the reactants but block the reactions that lead to equilibrium. These concentrations are easy to calculate from the initial conditions, but they seldom exist in reality because substances begin to react as soon as they are mixed.

A chemical system reacts, often rapidly, from initial conditions to equilibrium. As this occurs, concentrations of starting materials decrease, and concentrations of products increase. These concentration changes are related in two ways, as described in Chapter 1. To review, the concentration of each reagent at equilibrium is its initial concentration plus the change that has occurred:

$$\begin{bmatrix} \text{Equilibrium} \\ \text{concentration} \end{bmatrix} = \begin{bmatrix} \text{Initial} \\ \text{concentration} \end{bmatrix} + \begin{bmatrix} \text{Change in concentration} \\ \text{during reaction} \end{bmatrix}$$

**Figure 13-11**

Schematic view of a concentration table. The arrows indicate how the relationships expressed in Equations 13-5 and 13-6 are used to complete the table.

$$[A]_{eq} = [A]_i + \Delta[A] \qquad\qquad (13\text{-}5)$$

Furthermore, stoichiometric ratios link the changes in concentration of the various reagents:

[Change in concentration of B] = [Stoichiometric ratio of B to A][Change in concentration of A]

$$\Delta[B] = \frac{\text{Coeff B}}{\text{Coeff A}}\Delta[A] \qquad\qquad (13\text{-}6)$$

As we show in upcoming examples, some changes in concentration are positive whereas others are negative, depending on whether a reagent is produced or consumed.

These relationships provide complete stoichiometric information regarding the equilibrium. Just as amounts tables are useful in doing stoichiometric calculations, a **concentration table** that provides initial concentrations, changes in concentrations, and equilibrium concentrations is an excellent way to organize Step 5 of the problem-solving procedure for equilibrium problems. Figure 13-11 diagrams how to complete a concentration table: Equation 13-5 applies to every column, but Equation 13-6 applies only across the "change" row.

Example 13-10 shows how to use stoichiometric reasoning and a concentration table to calculate an equilibrium constant.

## EXAMPLE 13-10 CALCULATING AN EQUILIBRIUM CONSTANT

Benzoic acid is a weak carboxylic acid that undergoes proton transfer with water. When a 0.125 M aqueous solution of benzoic acid ($C_6H_5CO_2H$) reaches equilibrium, $[H_3O^+] = 0.0028$ M. What is $K_{eq}$ for the proton transfer reaction?

**Strategy:** Again, the seven-step approach to equilibrium problems will lead to the correct result. This is a more complicated example than Example 13-9, so a concentration table as part of Step 5 helps keep track of the stoichiometric relationships.

Benzoic acid
($C_6H_5CO_2H$)

1. The problem asks us to calculate an equilibrium constant, $K_{eq}$.

2. The chemical species present initially are benzoic acid and water ($C_6H_5CO_2H$ and $H_2O$). The $H_3O^+$ mentioned in the problem is formed as the solution comes to equilibrium.

3. The equilibrium concentration of hydronium ions is provided, which suggests that one of the products is $H_3O^+$. Benzoic acid is a weak acid. Thus, the correct chemical reaction is proton transfer from benzoic acid to water:

$$C_6H_5CO_2H \ (aq) + H_2O \ (l) \rightleftharpoons C_6H_5CO_2^- \ (aq) + H_3O^+ \ (aq)$$

4. Use the chemical reaction to determine the equilibrium constant expression. Water is the solvent, so it is omitted from the equilibrium expression:

$$K_{eq} = \frac{[H_3O^+]_{eq}[C_6H_5CO_2^-]_{eq}}{[C_6H_5CO_2H]_{eq}}$$

**Solution:**

5. Construct a table of initial concentrations, changes in concentration, and equilibrium concentrations for each species that appears in the equilibrium constant expression. The equilibrium concentrations from the last row of the table are needed to find $K_{eq}$. Start by entering the data given in the problem. The initial concentration of benzoic acid is 0.125 M. Pure water contains no benzoate ions and a negligible concentration of hydronium ions. The problem also states the equilibrium concentration of hydronium ions, 0.0028 M.

| Reaction | $H_2O$ (l) + | $C_6H_5CO_2H$ (aq) $\rightleftharpoons$ | $C_6H_5CO_2^-$ (aq) + | $H_3O^+$ (aq) |
|---|---|---|---|---|
| Initial Concentration (M) | | 0.125 | 0 | 0 |
| Change in Concentration (M) | | | | |
| Equilibrium Concentration (M) | | *Need to find* | *Need to find* | 0.0028 |

We see in the last column of the table that the initial and equilibrium concentrations of $H_3O^+$ are known, so we can use Equation 13-5 to determine the change in concentration for this species:

$$[H_3O^+]_{eq} = [H_3O^+]_i + \Delta[H_3O^+]$$

$$0.0028\,M = 0\,M + \Delta[H_3O^+] \quad \text{so} \quad \Delta[H_3O^+] = 0.0028\,M$$

Once we know one change, Equation 13-6 allows us to complete the change row by calculating the changes in the other concentrations:

$$\Delta[B] = \frac{\text{Coeff B}}{\text{Coeff A}}\Delta[A]$$

All of the stoichiometric coefficients are equal to 1, so all three changes have the same magnitude, but they have different *signs*. As the reaction comes to equilibrium, concentrations of products *increase,* so the sign for changes in products is *positive.* The concentrations of starting materials *decrease* in the reaction, so the sign for the change in benzoic acid is *negative.*

| Reaction | $H_2O$ (l) + | $C_6H_5CO_2H$ (aq) ⇌ | $C_6H_5CO_2^-$ (aq) + | $H_3O^+$ (aq) |
|---|---|---|---|---|
| Initial Concentration (M) | | 0.125 | 0 | 0 |
| Change in Concentration (M) | | −0.0028 | +0.0028 | +0.0028 |
| Equilibrium Concentration (M) | | *Need to find* | *Need to find* | 0.0028 |

Now we know two of the three entries in each of the other columns. To complete the table, we apply Equation 13-5 to each column, obtaining the concentrations at equilibrium for these species:

$$[C_6H_5CO_2^-]_{eq} = 0 + 0.0028\ M = 0.0028\ M$$

$$[C_6H_5CO_2H]_{eq} = 0.125\ M - 0.0028\ M = 0.122\ M$$

6. Substitute the values from the last row of the concentration table into the equilibrium constant expression and evaluate the result:

$$K_{eq} = \frac{(2.8 \times 10^{-3})(2.8 \times 10^{-3})}{0.122} = 6.4 \times 10^{-5}$$

 **7. Does the Result Make Sense?** The concentration of hydronium ion at equilibrium is in the $10^{-3}$ range, so a value of $K_{eq}$ in the $10^{-5}$ range is reasonable.

**Common Pitfalls:** The most common error in making these tables is deciding what to enter for the change in concentration. Make sure, using Le Châtelier's principle, that you know which direction the equilibrium is shifting.

**Practice Exercise 13.10**

A 0.335 M aqueous solution of $NH_3$ (a weak base) has $[OH^-]_{eq} = 2.46 \times 10^{-3}$ M. Construct an amounts table for this equilibrium and use it to determine the value of $K_{eq}$.

## Calculating Equilibrium Concentrations

The second main type of equilibrium problem asks for values of equilibrium concentrations. We also use concentration tables for this type of problem, with one additional feature. In such problems, we need to assign a variable $x$ to one unknown concentration, and then we use the equilibrium constant to find the value of $x$ by standard algebraic techniques. Examples 13-11 and 13-12 illustrate this use and manipulation of unknowns.

### EXAMPLE 13-11  GAS EQUILIBRIUM CONCENTRATIONS

The molecular view represents a set of initial conditions for the reaction described in Example 13-9. Each molecule represents a partial pressure of 1.0 bar. Determine the equilibrium conditions and redraw the picture to illustrate those conditions.

View at equilibrium

**Strategy:** Apply the seven-step strategy.

1. The problem asks for the equilibrium pressures and a molecular view illustrating the equilibrium conditions.

2. The only species present initially is the diatomic gas AB. From Example 13-9, we know that this gas reacts to form A and $AB_2$.

3. The chemical equilibrium is the same one that we identified in Example 13-9.

4. The equilibrium constant expression is the same as in Example 13-9:

$$2AB \rightleftarrows AB_2 + A \qquad K_{eq} = \frac{(p_{AB_2})(p_A)_{eq}}{(p_{AB})^2_{eq}} = 0.25$$

**Solution:**

5.  From the molecular picture, we can determine the initial pressure of AB:

$$p(\text{initial, AB}) = 8.0 \text{ bar}$$

To relate this to equilibrium pressures, we need a concentration table. The only information available to us is the initial pressures:

| Reaction | 2 AB | ⇌ | AB$_2$ | + | A |
|---|---|---|---|---|---|
| Initial Pressure (bar) | 8.0 | | 0 | | 0 |
| Change in Pressure (bar) | | | | | |
| Equilibrium Pressure (bar) | | | | | |

We assign the unknown $x$ to represent the *change* in pressure of A during the reaction. Then we can find the changes for the other reagents by applying Equation 13-6 across the change row. The stoichiometric ratio indicates that two molecules of AB must be consumed for each molecule of A produced. Because AB molecules are *consumed* as the reaction comes to equilibrium, the change in pressure of AB has a *negative* value. Thus, if the change in pressure of A is $+x$ bar, the change in pressure of AB is $-2x$ bar. By stoichiometry, the change for AB$_2$ is also $+x$:

| Reaction | 2 AB | ⇌ | AB$_2$ | + | A |
|---|---|---|---|---|---|
| Initial Pressure (bar) | 8.0 | | 0 | | 0 |
| Change in Pressure (bar) | $-2x$ | | $+x$ | | $+x$ |
| Equilibrium Pressure (bar) | | | | | |

Now we apply Equation 13-5 to obtain expressions for the equilibrium pressures:

| Reaction | 2 AB | ⇌ | AB$_2$ | + | A |
|---|---|---|---|---|---|
| Initial Pressure (bar) | 8.0 | | 0 | | 0 |
| Change in Pressure (bar) | $-2x$ | | $+x$ | | $+x$ |
| Equilibrium Pressure (bar) | $8.0 - 2x$ | | $x$ | | $x$ |

6. To evaluate $x$, we substitute the equilibrium pressures into the equilibrium constant expression:

$$0.25 = \frac{(p_{AB_2})_{eq}(p_A)_{eq}}{(p_{AB})^2_{eq}} = \frac{x^2}{(8.0 - 2x)^2}$$

In general, we would have to solve for $x$ using the quadratic equation, but this particular expression can be simplified by taking the square root of each side:

$$0.50 = \frac{x}{8.0 - 2x}$$

Clear the fraction and then multiply through:

$$0.50 \, (8.0 - 2x) = x \qquad \text{so} \qquad 4.0 - x = x$$

Combine terms in $x$ and solve for $x$:

$$4.0 = 2x \quad \text{and} \quad x = 2.0$$

This lets us calculate the equilibrium pressures:

$$(p_{AB})_{eq} = 8.0 - 2(2.0) = 4.0 \text{ bar} \quad (p_{AB_2})_{eq} = (p_A)_{eq} = 2.0 \text{ bar}$$

This result can be checked by substituting the partial pressures into the equilibrium constant expression and verifying that the correct value for $K_{eq}$ is obtained:
[Designer: subscript "eq" is non-ital]

$$K_{eq} = \frac{(p_{AB_2})_{eq}(p_A)_{eq}}{(p_{AB})^2_{eq}} = \frac{2.0(2.0)}{4.0^2} = 0.25$$

To complete the problem, we draw a molecular view that shows four molecules of AB, two molecules of $AB_2$, and two atoms of A:

Initial conditions

 **Does the Result Make Sense?** There is less of the starting material and more of the products than were present at the beginning, and atoms of A and B have been conserved. This is a reasonable result.

**Practice Exercise 13.11**

If the initial conditions for the reaction described in this example are adjusted so that $p_{AB_2} = p_A = 2.0$ bar and $p_{AB} = 0$ bar, determine the equilibrium pressures and draw a molecular picture showing the equilibrium.

## EXAMPLE 13-12 EQUILIBRIUM CONCENTRATIONS

The equilibrium constant for the proton transfer reaction of benzoic acid, determined in Example 13-10, is $6.4 \times 10^{-5}$. Calculate the equilibrium concentration of benzoic acid and benzoate anions in a $5.0 \times 10^{-2}$ M aqueous solution of the acid.

**Strategy:** Apply the seven-step procedure.

1. This problem asks about equilibrium concentrations and provides the value of an equilibrium constant. Under these circumstances, we expect to have to define an appropriate unknown to represent the change in concentrations of one reagent.

2. Initially, benzoic acid and water are the only species present.

3. The proton transfer reaction is the same as in Example 13-10:

$$C_6H_5CO_2H \text{ (aq) } H_2O \text{ (l) } \rightleftharpoons C_6H_5CO_2^- \text{ (aq) } + H_3O^+ \text{ (aq)}$$

4. The equilibrium constant expression is the same as in Example 13-10:

$$K_{eq} = \frac{[H_3O^+]_{eq}[C_6H_5CO_2^-]_{eq}}{[C_6H_5CO_2H]_{eq}} = 6.4 \times 10^{-5}$$

## Solution:

5. Construct a concentration table. The only quantitative information provided in the problem is the value for the initial concentration of benzoic acid ($5.0 \times 10^{-2}$ mol/L).

| Reaction | $H_2O$ (*l*) | + | $C_6H_5CO_2H$ (*aq*) | $\rightleftharpoons$ | $C_6H_5CO_2^-$ (*aq*) | + | $H_3O^+$ (*aq*) |
|---|---|---|---|---|---|---|---|
| Initial Concentration (M) | | | 0.050 | | 0 | | 0 |
| Change in Concentration (M) | | | | | | | |
| Equilibrium Concentration (M) | | | *Need to find* | | *Need to find* | | *Need to find* |

Completing the table requires an appropriate unknown. If we let $x$ represent the *change* in concentration of $H_3O^+$ during the reaction, the changes in the other concentrations are $-x$ for benzoic acid and $+x$ for benzoate anion. Then, we apply Equation 13-5 to each column and complete the table:

| Reaction | $H_2O$ (*l*) | + | $C_6H_5CO_2H$ (*aq*) | $\rightleftharpoons$ | $C_6H_5CO_2^-$ (*aq*) | + | $H_3O^+$ (*aq*) |
|---|---|---|---|---|---|---|---|
| Initial Concentration (M) | | | 0.050 | | 0 | | 0 |
| Change in Concentration (M) | | | $-x$ | | $x$ | | $x$ |
| Equilibrium Concentration (M) | | | $0.050 - x$ | | $x$ | | $x$ |

6. Next, we substitute the values from the last row of the concentration table into the equilibrium constant expression:

$$K_{eq} = \frac{[H_3O^+]_{eq}[C_6H_5CO_2^-]_{eq}}{[C_6H_5CO_2H]_{eq}} = 6.4 \times 10^{-5} = \frac{x^2}{0.050 - x}$$

Multiplying through to clear the fraction gives

$$x^2 = (3.2 \times 10^{-6}) - (6.4 \times 10^{-5})x \quad \text{or} \quad x^2 + (6.4 \times 10^{-5})x - (3.2 \times 10^{-6}) = 0$$

This is a quadratic equation in the form $ax^2 + bx + c = 0$. In this case,

$$a = 1, \quad b = 6.4 \times 10^{-5}, \quad \text{and} \quad c = -3.2 \times 10^{-6}$$

Use the quadratic formula to find the value of $x$:

$$x = \frac{-b \pm \sqrt{b^2 - 4ac}}{2a} = \frac{-(6.4 \times 10^{-5}) \pm \sqrt{(6.4 \times 10^{-5})^2 - 4(1)(-3.2 \times 10^{-6})}}{2(1)}$$

$$x = \frac{-(6.4 \times 10^{-5}) \pm \sqrt{(4.096 \times 10^{-9}) + (1.28 \times 10^{-5})}}{2(1)}$$

There are two solutions to this equation, but one gives a negative value for $x$. We know that $x$ must be positive, because it represents the hydronium ion concentration. Hence, we evaluate the positive solution:

$$x = \frac{-(6.4 \times 10^{-5}) + (3.58 \times 10^{-3})}{2} = 1.76 \times 10^{-3}$$

Now we can solve for the concentrations of the ions at equilibrium:

$$[C_6H_5CO_2^-]_{eq} = x \text{ M} = 1.76 \times 10^{-3} \text{ M}$$

$$[C_6H_5CO_2H]_{eq} = (5.0 \times 10^{-2} \text{ M} - 1.76 \times 10^{-3} \text{ M}) = 4.82 \times 10^{-2} \text{ M}$$

To check the answer, substitute the values into the equilibrium constant expression:

$$K_{eq} = \frac{[H_3O^+]_{eq}[C_6H_5CO_2^-]_{eq}}{[C_6H_5CO_2H]_{eq}} = \frac{(1.76 \times 10^{-3})(1.76 \times 10^{-3})}{4.82 \times 10^{-2}} = 6.4 \times 10^{-5}$$

**Do the Results Make Sense?** Both concentrations are reasonable values: they are positive and smaller than the initial concentration of benzoic acid. The equilibrium constant expression using the calculated concentrations returns the correct value of $K_{eq}$.

**Common Pitfalls:** Not solving the quadratic equation correctly. It is best to write it out in parts as shown above and not to try to do it all in one step.

**Practice Exercise 13.12**
Calculate the concentrations of the species present in an aqueous solution of ammonia (see Practice Exercise 13-10) that is 0.0125 M $NH_3$.

Examples 13-11 and 13-12 involve equilibrium constants with moderate values. However, many chemical reactions have equilibrium constants that are either very small ($K_{eq} < 10^{-8}$) or very large ($K_{eq} > 10^8$). For these equilibria, we usually can use approximations to simplify the calculations.

A *very small* value of $K_{eq}$ means that concentrations of *products* at equilibrium are *very low* compared with reactant concentrations. As Figure 13-12a shows, when $K_{eq}$ is small, reactions barely get started before they reach equilibrium. One example is the equilibrium that results when solid AgBr is placed in water:

$$AgBr\ (s) \rightleftharpoons Ag^+\ (aq) + Br^-\ (aq) \qquad K_{eq} = 5.35 \times 10^{-13}$$

The equilibrium constant for this reaction is small, indicating that only a small amount of AgBr will dissolve in water.

At the other extreme, a *very large* value of $K_{eq}$ means that concentrations of *products* at equilibrium are *very high* compared with reactant concentrations. As Figure 13-12b shows, when $K_{eq}$ is large, the reaction proceeds almost to completion before equilibrium is reached. One example is the equilibrium that results from the mixing of aqueous solutions of $AgNO_3$ and KBr (the $K^+$ (aq) and $NO_3^-$ (aq) ions are spectators):

$$Ag + (aq) + Br^-\ (aq) \rightleftharpoons AgBr\ (s)$$
$$K_{eq} = 1/(5.35 \times 10^{-13}) = 1.87 \times 10^{12}$$

The equilibrium constant for this reaction is large, indicating that nearly all of the aqueous ions end up as solid precipitate.

**Figure 13-12**

Schematic representation of the relationship between $K_{eq}$ and the extent of reaction. (a) When $K_{eq} < 1$, the difference from initial conditions, $x$, is small. (b) When $K_{eq} > 1$, the difference from completion, $y$, is small.

## Working with Small Equilibrium Constants

When $K_{eq}$ is small, the changes in concentration are likely to be small compared with initial concentrations. Example 13-13 illustrates the use of approximations to simplify the mathematical calculations in such cases.

## Example 13-13 SMALL EQUILIBRIUM CONSTANT

Sodium benzoate is a common food preservative. This salt dissolves in water to produce benzoate anions, which accept protons from water:

$$H_2O \, (l) + C_6H_5CO_2^- \, (aq) \rightleftarrows OH^- \, (aq) + C_6H_5CO_2H \, (aq)$$

$$K_{eq} = \frac{[OH^-]_{eq}[C_6H_5CO_2H]_{eq}}{[C_6H_5CO_2^-]_{eq}} = 1.5 \times 10^{-10}$$

Benzoate anion
$(C_6H_5CO_2^-)$

What is the equilibrium concentration of hydroxide ions in a 0.135 M solution of sodium benzoate?

**Strategy:** We follow the step-by-step procedure for working with equilibria.

1. The problem asks for the equilibrium concentration of $OH^-$.

2. Under initial conditions, three chemical species are present: benzoate anions, $Na^+$, and $H_2O$.

3. The net reaction is given. A proton is transferred from a water molecule to a benzoate anion. A small amount of benzoic acid forms as the reaction reaches equilibrium. Notice that $Na^+$ ions do not participate in the net reaction.

4. The equilibrium constant expression is given in the problem.

## Solution:

5. Set up a concentration table. The concentration of water is not required because it does not appear in the equilibrium constant expression. A convenient choice for the unknown $x$ is the amount of benzoic acid formed in the reaction. This choice makes several other entries in the table equal to $x$.

| Reaction | $H_2O$ | $+$ | $C_6H_5CO_2^-$ | $\leftrightarrows$ | $C_6H_5CO_2H$ | $+$ | $OH^-$ |
|---|---|---|---|---|---|---|---|
| Initial Concentration (M) | | | 0.135 | | 0 | | 0 |
| Change to Equilibrium (M) | | | $-x$ | | $+x$ | | $+x$ |
| Equilibrium Concentration (M) | | | $0.135 - x$ | | $x$ | | $x$ |

6. Substitute the equilibrium concentrations in the expression for $K_{eq}$ and solve for the unknown:

$$K_{eq} = \frac{[OH^-]_{eq}[C_6H_5CO_2H]_{eq}}{[C_6H_5CO_2^-]_{eq}} = 1.5 \times 10^{-10} = \frac{(x)(x)}{0.135 - x}$$

The equilibrium constant has a small value ($10^{-10}$). This suggests that we can make an approximation by assuming that the change required to reach equilibrium ($x$) is very small. This situation is like that shown in Figure 13-12a. As shown in the concentration table, the concentration of benzoate ion is $(0.135 - x)$ M at equilibrium. Knowing that $x$ is small, we make the approximation that $x$ can be neglected relative to 0.135:

$$x \ll 0.135 \quad \text{so} \quad 0.135 - x \cong 0.135$$

This simplifies the equilibrium expression:

$$1.5 \times 10^{-10} = \frac{x^2}{0.135 - x} \cong \frac{x^2}{0.135}$$

$$x^2 \cong (0.135)(1.5 \times 10^{-10}) = 0.203 \times 10^{-10}$$

$$x \cong 0.45 \times 10^{-5} = 4.5 \times 10^{-6}$$

We use this value of $x$ to calculate the concentrations at equilibrium:

$$[C_6H_5CO_2^-] = (0.135 - x) \text{ M} = (0.135 - 4.5 \times 10^{-6}) \text{ M} = 0.135 \text{ M}$$

$$[C_6H_5CO_2H] = [OH^-] = x \text{ M} = 4.5 \times 10^{-6} \text{ M}$$

Check the result:

$$K_{eq} = \frac{[OH^-]_{eq}[C_6H_5CO_2H]_{eq}}{[C_6H_5CO_2^-]_{eq}} = \frac{(4.5 \times 10^{-6})(4.5 \times 10^{-6})}{0.135} = 1.5 \times 10^{-10}$$

**7. Do the Results Make Sense?** The approximation in this example is $x \ll 0.135$ M. We find $x = 4.5 \times 10^{-6}$ M, which is more than four orders of magnitude less than 0.135, so the approximation is a good one. The calculated concentrations return the correct equilibrium constant.

**Practice Exercise 13.13**
Use the approximation method to calculate the concentrations of the species present in concentrated ammonia, which is 14.8 M (see Practice Exercise 13.10 for $K$).

Whenever we make an approximation, we must verify that it is valid by comparing the value calculated using the approximation with the approximation itself. Most equilibrium constants are uncertain by about 5%, so $x$ can be neglected whenever its value is two or more orders of magnitude smaller than the value from which it is subtracted or added.

Notice, however, that we do not neglect the lone $x$ in determining the equilibrium concentrations of hydroxide ions and benzoic acid. We can neglect $x$ *only* when it appears in a sum or difference and *never when it stands alone.*

Usually, $x$ in a sum or difference is small enough to neglect if the equilibrium constant is smaller than $10^{-4}$ and the initial concentrations of starting materials are equal to or greater than 0.1 M. Nevertheless, the validity of an approximation must always be verified by comparing the result with the approximation. In other words, after calculating the value of $x$ using an approximation such as $(A - x) \cong A$, check to see whether the calculated value for $x$ indeed is less than 5% of $A$.

## Working with Large Equilibrium Constants

When $K_{eq}$ is very large, the situation is that represented by Figure 13-12b. In such a situation, the reaction proceeds nearly to completion, so the difference between completion and equilibrium, represented by $y$ in the figure, is very small. We choose $y$ as the variable rather than $x$ to highlight the fact that the system is coming back to equilibrium from completion as opposed to working forward from initial conditions. Because $y$ is small, working from completion *backward* to equilibrium allows us to neglect $y$ in many cases. When $K_{eq}$ is large, we could never neglect $x$ when working from initial concentrations *forward* to equilibrium.

When an equilibrium constant is very large, we use stoichiometry to determine what the concentrations would be at completion. Then we define $y$ to be the amount of back-reaction leading from completion to equilibrium. Solving for $y$ leads to the true equilibrium concentrations. The equilibrium position of the system lies very near the completion point of the reaction, so $y$ will be very small, allowing us to make approximations.

This is a two-step process, so for a reaction with a large $K_{eq}$, we can construct two concentration tables. The first table helps us determine the concentrations *at completion*, and the second table expresses the concentrations *at equilibrium*. Example 13-14 illustrates an equilibrium position close to completion.

## EXAMPLE 13-14 LARGE EQUILIBRIUM CONSTANT

One of the steps in the industrial production of nitric acid is the combustion of NO at room temperature:

$$2\ NO\ (g) + O_2\ (g) \rightleftarrows 2\ NO_2\ (g) \qquad K_{eq} = 4.2 \times 10^{12}$$

Suppose a reactor is charged with 10.0 bar each of NO and $O_2$. Find the partial pressure of each gas at equilibrium.

**Strategy:** Use the step-by-step approach for an equilibrium problem.

1. The problem asks us to calculate equilibrium pressures of all reagents.

2. The species present before reaction are NO gas and $O_2$ gas.

3. The reaction is given in the problem:

$$2\ NO\ (g) + O_2\ (g) \rightleftarrows 2\ NO_2\ (g)$$

4. We obtain the equilibrium constant expression from the reaction:

$$K_{eq} = \frac{(p_{NO_2})^2_{eq}}{(p_{NO})^2_{eq}(p_{O_2})_{eq}} = 4.2 \times 10^{12}$$

Because the magnitude of the equilibrium constant is very large, *almost all* of the starting materials will be converted to products. Because of this, the problem is easier to solve by taking the reaction to completion and then returning to equilibrium.

**Solution:**

5. Start the solution by setting up a concentration table. Because $K_{eq}$ is very large, we first take the reaction to completion. Finding the concentrations at completion is a limiting reactant problem. Because we have equal amounts of the two reactants and the coefficient for NO is 2, NO would be used up first, so it is the limiting reactant, and its partial pressure would be zero at completion. This requires a change in pressure of $-10.0$ bar for NO. By stoichiometry, the change of pressure for $NO_2$ is $+10.0$ bar and the change for $O_2$ is half the change for NO, $-5.0$ bar.

| Reaction | 2 NO (g) | + | O₂ (g) | ⇆ | 2 NO₂ (g) |
|---|---|---|---|---|---|
| Initial Pressure (bar) | 10.0 | | 10.0 | | 0 |
| Change to Completion (bar) | −10.0 | | −5.0 | | +10.0 |
| Pressure at Completion (bar) | 0 | | 5.0 | | 10.0 |

Now we work from completion to equilibrium. The equilibrium constant for this reaction is very large, but the partial pressure of NO cannot be zero at equilibrium. We define $y$ to be the change in NO pressure on going from completion to equilibrium. Then, the stoichiometric coefficients and Equation 13-6 give $+0.5y$ for the change in pressure of $O_2$ and $-y$ for the change in $NO_2$.

| Reaction | 2 NO (g) | + | O₂ (g) | ⇆ | 2 NO₂ (g) |
|---|---|---|---|---|---|
| Pressure at Completion (bar) | 0 | | 5.0 | | 10.0 |
| Change to Equilibrium (bar) | $+y$ | | $+0.5y$ | | $-y$ |
| Equilibrium Pressure (bar) | $y$ | | $5.0 + 0.5y$ | | $10.0 - y$ |

6. Having obtained expressions for the equilibrium pressures, we substitute these into the expression for $K_{eq}$:

$$K_{eq} = 4.2 \times 10^{12} = \frac{(10.0 - y)^2}{(y)^2(5.0 + 0.5y)}$$

Because $K_{eq}$ is very large, the amount of the limiting reactant present at equilibrium, $y$, will be very small compared with the amounts of materials present at completion. That is the point of taking the reaction to completion and then working back toward equilibrium. This leads to two approximations:

$$10.0 - y \cong 10.0 \quad \text{and} \quad 5.0 + 0.5y \cong 5.0$$

Now solve for $y$:

$$4.2 \times 10^{12} \cong \frac{(10.0)^2}{(y)^2(5.0)} \quad \text{giving} \quad y^2 \cong \frac{(10.0)^2}{(5.0)(4.2 \times 10^{12})}$$

$$y^2 \cong 4.76 \times 10^{-12} \quad \text{from which} \quad y \cong 2.2 \times 10^{-6}$$

Use the value of $y$ to determine the equilibrium pressures: $(p_{NO})_{eq} = y = 2.2 \times 10^{-6}$ bar $(p_{NO_2})_{eq} = (5.0 + 0.5y)$ bar $= 5.0$ bar $(p_{NO_2})_{eq} = (10.0 - y)$ bar $= 10.0$ bar
Check the result:

$$K_{eq} = \frac{(p_{NO_2})^2_{eq}}{(p_{NO})^2_{eq}(p_{O_2})_{eq}} = \frac{10.0^2}{(2.2 \times 10^{-6})^2(5.0)} = 4.2 \times 10^{12}$$

 **Do the Results Make Sense?** The calculations show that $y$ is more than five orders of magnitude smaller than 5.0 and 10.0 bar, so the approximations are valid.

### Practice Exercise 13.14
Figure 13-12b shows the precipitation reaction to form AgBr. Calculate the amount of AgBr solid that forms and the concentrations of $Ag^+$ and $Br^-$ ions remaining in solution if 0.500 L each of 1.00 M solutions of $AgNO_3$ and KBr are mixed together. $K_{eq}$ for the formation of AgBr from $Ag^+$ and Br is $1.87 \times 10^{12}$.

The examples of this section illustrate the general approach to equilibrium problems. Notice that these examples include gas-phase, precipitation, and acid–base chemistry. We use a variety of equilibrium examples to emphasize that the general strategy for working with equilibria is always the same, no matter what type of equilibrium is involved. In Chapters 14 and 15, we apply these ideas in more detail to important types of equilibria.

**SECTION EXERCISES**

●**13.5.1** Barium sulphate is a relatively insoluble salt used for medical radiographs of the gastrointestinal tract:

$$BaSO_4 \ (s) \rightleftharpoons Ba^{2+} \ (aq) + SO_4^{2-} \ (aq)$$

The concentration of $Ba^{2+}$ ions in a saturated aqueous solution is $1.05 \times 10^{-5}$ M. Determine $K_{eq}$ for barium sulphate dissolving in water.

●**13.5.2** A solution of acetic acid in water is described by the following equilibrium:

$$CH_3CO_2H \ (aq) + H_2O \ (l) \rightleftharpoons CH_3CO_2^- \ (aq) + H_3O^+ \ (aq) \qquad K_{eq} = 1.8 \times 10^{-5}$$

Calculate the equilibrium concentrations of acetic acid, acetate ion, and hydronium ion in a 2.5 M solution of acetic acid.

●**13.5.3** Silver bromide, a solid used in photographic film, can be prepared by mixing solutions of silver nitrate and potassium bromide:

$$Ag^+ \ (aq) + Br^- \ (aq) \rightleftharpoons AgBr \ (s) \qquad K_{eq} = 1.87 \times 10^{12}$$

Calculate the concentrations of Ag and Br ions remaining in solution after mixing 2.50 L of 0.100 M $AgNO_3$ solution with 2.50 L of 0.500 M KBr solution.

## 13.6 EQUILIBRIA IN AQUEOUS SOLUTIONS

Equilibria that occur in aqueous solution are of particular interest, because water is the medium of life and a major influence on the geography of our planet. Many substances dissolve in water, and the solutes in an aqueous solution may participate in a number of different types of equilibria. Solubility itself is one important type of equilibrium, as we describe in Chapter 15. Acid–base reactions, considered in detail in Chapter 14, are another. To conclude, we describe how to determine which equilibria are most important in any particular aqueous solution, and summarize many features of equilibrium reactions presented in this chapter.

### Species in Solution

The species in an aqueous solution can be categorized broadly into two groups present at different relative concentrations. We designate those present in relatively high concentrations as **major species**. We refer to those present in relatively low concentrations as **minor species**. Minor species in aqueous solutions generally have concentrations at least three orders of magnitude lower than the concentrations of the major solute species. However, minor species can be just as important as major species. In most aqueous solutions, one equilibrium plays the most important role. We call this the **dominant equilibrium**. The dominant equilibrium always has major species as its reactants, but the products of that equilibrium can be major or minor species.

The first step in analyzing an aqueous equilibrium is to identify the major species. Pure water contains $H_2O$ molecules at a concentration of 55.5 M, so an aqueous solution always contains $H_2O$ as a major species. In pure water, $H_2O$ is the only major species, but an aqueous solution contains two or more major species: $H_2O$ and those solute species that are present at highest concentration.

As discussed in Chapter 1, whenever an ionic solid dissolves in water, the salt dissociates to give a solution of cations and anions. Thus, in any aqueous salt solution, the major species are water molecules and the cations and anions generated by the salt. For example, a solution of potassium chloride contains $K^+$ and $Cl^-$ ions and $H_2O$ molecules as major species. Likewise, the major species in a solution of ammonium nitrate are $NH_4^+$, $NO_3^-$, and $H_2O$.

As will be described in Section 14.1, a strong acid is a substance that reacts virtually completely with water to produce hydronium ions. For example, when nitric acid ($HNO_3$) dissolves in water, its molecules react with water to produce hydronium ions:

$$HNO_3\ (aq) + H_2O\ (l) \longrightarrow H_3O^+\ (aq) + NO_3^-\ (aq)$$

This reaction is written with a single arrow to designate that it goes virtually to completion: the equilibrium constant for this proton transfer reaction is so large that the concentration of unreacted acid is very small. The major species in an aqueous solution of any strong acid are $H_2O$, $H_3O^+$, and the anion produced by removing a proton from the acid. This situation is illustrated in Figure 13-13, which shows a molecular picture of aqueous nitric acid. Table 13-1 lists the most important strong acids.

A strong base generates hydroxide ions when it dissolves in water. The most common examples of strong bases (see Table 13-1) are soluble metal hydroxides such as NaOH and KOH. These ionic substances dissociate into aqueous ions when they dissolve in water:

$$KOH\ (s) \xrightarrow{\ H_2O\ (l)\ } K^+\ (aq) + OH^-\ (aq)$$

The major species in any aqueous solution of a strong base are $H_2O$, $OH^-$, and the cation generated by the base.

In addition to salts, strong acids, and strong bases, many molecular substances dissolve in water. As Figure 13-14 shows, these solutes retain their structures,

**Figure 13-13**

The major species present in an aqueous solution of nitric acid are water molecules, hydronium ions, and nitrate anions. The concentration of $HNO_3$ molecules is negligible.

### TABLE 13-1 COMMON STRONG ACIDS AND BASES

| Acids of Industrial Importance | |
|---|---|
| $H_2SO_4$ | Sulphuric acid |
| $HNO_3$ | Nitric acid |
| HCl | Hydrochloric acid |

| Acids with Laboratory Applications | |
|---|---|
| $HClO_4$ | Perchloric acid |
| HBr | Hydrobromic acid |
| HI | Hydroiodic acid |

| Bases of Industrial Importance | |
|---|---|
| NaOH | Sodium hydroxide |
| $Ca(OH)_2$ | Calcium hydroxide |
| KOH | Potassium hydroxide |

| Bases with Laboratory Applications | |
|---|---|
| LiOH | Lithium hydroxide |
| $Ba(OH)_2$ | Barium hydroxide |

dissolving in aqueous solution to give intact molecules. The figure shows a molecular picture of vinegar, which is a dilute aqueous solution of acetic acid ($CH_3CO_2H$). The major species present in vinegar are $H_2O$ and $CH_3CO_2H$. (In Chapter 14, however, we will see that the minor species in many acidic or basic solutions are actually the more important ones.)

To identify the major species in any aqueous solution, first categorize the solutes. A soluble salt, strong acid, or strong base generates the appropriate cations and anions as major species. Every other solute generates its molecular species in solution. In addition, $H_2O$ is *always* a major species in aqueous solutions. Example 13-15 provides practice in identifying the major species in solution.

**Figure 13-14**

A molecular view of the major species present in an aqueous solution of acetic acid (vinegar).

## EXAMPLE 13-15 MAJOR SPECIES IN SOLUTION

Identify the major species in each of the following aqueous solutions: (a) NaCH$_3$CO$_2$ (sodium acetate); (b) HClO$_4$ (perchloric acid); (c) C$_6$H$_{12}$O$_6$ (glucose, used for intravenous feeding); and (d) NH$_3$ (ammonia, used for household cleaning).

**Strategy:** Identify the nature of each solute. If it is a salt, strong acid, or strong base, it generates ions in aqueous solution. All other solutes give aqueous solutions that contain neutral molecules as the major species.

**Solution:**

(a) From its formula, you should recognize that NaCH$_3$CO$_2$ is a salt. It contains the alkali metal cation Na$^+$ and the polyatomic acetate anion, CH$_3$CO$_2^-$. Thus, this salt dissolves, generating an aqueous solution in which the major species are H$_2$O, Na$^+$, and CH$_3$CO$_2^-$:

(b) HClO$_4$ is a strong acid, so it transfers a proton to water quantitatively. (See the list of strong acids in Table 13-1.) The major species in aqueous perchloric acid are H$_2$O, H$_3$O$^+$, and ClO$_4^-$:

(c) Glucose is not a strong acid, a strong base, or a salt. It dissolves in water without reacting, so the major species in intravenous feeding solutions are molecules of C$_6$H$_{12}$O$_6$ and H$_2$O:

Glucose (C$_6$H$_{12}$O$_6$)

(d) NH$_3$ is not a strong acid, a strong base, or a salt. The major species in household ammonia are molecules of NH$_3$ and H$_2$O:

 **Do the Results Make Sense?** Remember that major species are molecules unless the solute is a salt, strong acid, or strong base.

**Practice Exercise 13.15**

Identify the major species present in aqueous solutions of hydrochloric acid, sodium chlorate (NaClO$_3$), and chloroform (CHCl$_3$).

## Types of Aqueous Equilibria

Most aqueous equilibria fall into three broad categories: proton transfer, solubility, or complexation. The nature of the major species in the solution determines which category of equilibrium we need to consider.

One of the most important types of aqueous equilibrium involves proton transfer from an acid to a base. In aqueous solutions, water can act as an acid or a base. In the presence of an acid, symbolized H$A$, water acts as a base by accepting a proton. The equilibrium constant for transfer of a proton from an acid to a water molecule is called the **acid hydrolysis constant** ($K_a$):

$$HA\,(aq) + H_2O\,(l) \rightleftharpoons A^-\,(aq) + H_3O^+(aq) \qquad K_{eq} = \frac{[A^-]_{eq}[H_3O^+]_{eq}}{[HA]_{eq}} = K_a$$

For instance, when acetic acid, an example of a weak acid, dissolves in water, some acetic acid molecules transfer protons to water molecules to produce acetate anions and hydronium ions:

$$CH_3CO_2H\,(aq) + H_2O \rightleftharpoons CH_3CO_2^-\,(aq) + H_3O^+\,(aq) \qquad K_a = \frac{[CH_3CO_2^-]_{eq}[H_3O^+]_{eq}}{[CH_3CO_2H]_{eq}}$$

In the presence of a base $B$, water acts as an acid by donating a proton, and the equilibrium constant for the transfer of a proton from water to a base is called the **base hydrolysis constant** ($K_b$):

$$B\,(aq) + H_2O\,(l) \rightleftharpoons BH^+\,(aq) + OH^-\,(aq) \qquad K_{eq} = \frac{[BH^+]_{eq}[OH^-]_{eq}}{[B]_{eq}} = K_b$$

When ammonia, an example of a weak base, dissolves in water, some ammonia molecules accept protons from water to produce ammonium cations and hydroxide ions:

$$NH_3\,(aq) + H_2O\,(l) \rightleftharpoons NH_4^+(aq) + OH^-\,(aq) \qquad K_{eq} = \frac{[NH_4^+]_{eq}[OH^-]_{eq}}{[NH_3]_{eq}}$$

Notice that the expressions for $K_a$ and $K_b$ do not include the water molecules that act as starting materials for the proton transfer reactions. Water, as the solvent, is always present in huge excess. Thus, as described in Section 13.2, the concentration of water does not change significantly during an acid–base reaction and is omitted from $K_a$ and $K_b$.

Many substances that participate in aqueous reactions are soluble salts. These ionic solids dissolve in water to give solutions of cations and anions. For almost all salts, there is an upper limit to the amount that will dissolve in water. A salt solution is *saturated* when the amount dissolved has reached this upper limit of solubility. Any additional salt added to a saturated solution remains undissolved at the bottom of the vessel. When excess solid salt and a saturated solution are present, a solubility equilibrium exists. At equilibrium, ions dissolve continually, but other ions precipitate out at exactly the same rate. This equilibrium is illustrated for sodium chloride by the molecular picture in Figure 13-15.

By convention, a solubility equilibrium is written in the direction of a solid dissolving to give aqueous ions, and the equilibrium constant for this reaction is called the **solubility product** ($K_{sp}$). Here, for example, is the reaction describing the solubility equilibrium of copper(II) chloride:

**Figure 13-15**

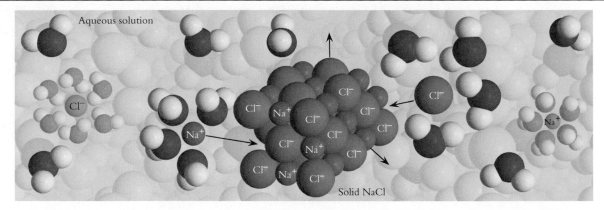

A molecular view of the solubility equilibrium for a solution of sodium chloride in water. At equilibrium, ions dissolve from the crystal surface at the same rate they are captured, so the concentration of ions in the solution remains constant.

$$CuCl_2\ (s) \rightleftharpoons Cu^{2+}\ (aq) + 2\ Cl^-\ (aq) \quad K_{eq} = [Cu^{2+}]_{eq}\,[Cl^-]^2_{eq} = K_{sp}$$

The evaporation of water from a saturated solution leaves a solution in which the ion concentrations exceed the solubility limit. To return to equilibrium, the salt must precipitate from the solution. Evaporation is used to "mine" sodium chloride and other salts from the highly salty waters of inland seas such as Great Salt Lake in Utah and Israel's Dead Sea.

Metal cations in aqueous solution often form chemical bonds to anions or neutral molecules that have lone pairs of electrons. A silver cation, for example, can associate with two ammonia molecules to form a silver–ammonia complex:

$$H{-}\!\!\underset{H}{\overset{H}{N}}\!\!: + Ag^+ + :\!\!\underset{H}{\overset{H}{N}}\!\!{-}H \rightleftharpoons H{-}\!\!\underset{H}{\overset{H}{N}}\!\!{-}Ag^+\!\!{-}\underset{H}{\overset{H}{N}}\!\!{-}H$$

The colour of an aqueous solution of copper(II) chloride shows that copper(II) cations are present in the aqueous phase at equilibrium with the solid salt.

The resulting species is called a **complex ion**. The equilibrium constant for the formation of a complex ion is called its **formation constant** ($K_f$). Tabulated values of $K_f$ always refer to the equilibrium constant for the complex forming from the metal cation. Here, for example, is the reaction describing the complexation equilibrium between $Ag^+$ and $NH_3$:

$$Ag^+\ (aq) + 2\ NH_3\ (aq) \rightleftharpoons [Ag(NH_3)_2]^+\ (aq) \qquad K_f = \frac{[Ag(NH_3)_2{}^+]_{eq}}{[Ag^+]_{eq}[NH_3]^2_{eq}}$$

Notice that square brackets are used in a new way for the silver–ammonia complex in the chemical reaction. Chemists use square brackets to identify a complex ion such as $[Ag(NH_3)_2]^+$, because the species involved in complexation, $NH_3$ in this instance, are set off by parentheses. When square brackets appear in an equilibrium constant expression, they always designate concentration; when they appear in a chemical reaction, they designate a complex. We consider the formation of metal complexes in more depth in Chapter 18.

## Identifying Types of Equilibria

The major species in an aqueous solution determine which categories of equilibria are important for that solution. Each major species present in the solution must be examined in light of these general categories. Are any of the major species weak acids or weak bases? Are there ions present that combine to form an insoluble salt? Do any of the major species participate in more than one equilibrium? Any chemical reaction can approach equilibrium from either direction. Consequently, there are six different types of aqueous equilibria in which major species are reactants:

1. An acid may donate a proton to water or some other base.

2. A base may accept a proton from water or some other acid.

3. A salt may dissolve in water.

4. Cations and anions in a solution may precipitate to form a solid.

5. A cation may associate with electron-donating species to form a complex.

6. A complex may dissociate.

We consider each of these in more detail in subsequent chapters, but being able to identify types of equilibria helps greatly in solving equilibrium problems. The equilibrium constants for many of these characteristic types of equilibria have been measured and tabulated. Representative $K_a$, $K_b$, and $K_{sp}$ values appear in Appendix E, and tables that are more extensive can be found in the *CRC Handbook of Chemistry and Physics*. Example 13-16 provides practice in identifying equilibria.

## EXAMPLE 13-16   TYPES OF AQUEOUS EQUILIBRIA

Write expressions for the equilibrium constants and decide which general category applies to each of these equilibria:

(a) $H_2CO_3 \, (aq) + H_2O \, (l) \rightleftharpoons HCO_3^- \, (aq) + H_3O^+ \, (aq)$

(b) $Fe^{3+} \, (aq) + 3 \, OH^- \, (aq) \rightleftharpoons Fe(OH)_3 \, (s)$

(c) $NH_4^+ \, (aq) + OH^- \, (aq) \rightleftharpoons NH_3 \, (aq) + H_2O \, (l)$

**Strategy:** Examine each reaction closely, and look for the characteristic features of the general equilibria. The equilibrium constant expression can be written by inspecting the overall stoichiometry.

**Solution:**

(a) In this reaction, water accepts a proton from $H_2CO_3$. The equilibrium constant for a proton transfer reaction that consumes $H_2O$ and produces $H_3O^+$ is called $K_a$:

$$\frac{[H_3O^+]_{eq}[HCO_3^-]_{eq}}{[H_2CO_3]_{eq}} = K_a$$

(b) This is a precipitation reaction. Solid $Fe(OH)_3$ forms in the forward direction. In the reverse direction, solid $Fe(OH)_3$ dissolves. This reverse reaction is a solubility reaction for which $K_{eq} = K_{sp}$. According to Equation 13-2, $K_{eq}$ must be $1/K_{sp}$ for the reaction in the direction written:

$$\frac{1}{[Fe^{3+}]_{eq}[OH^-]_{eq}^3} = \frac{1}{K_{sp}}$$

(c) This reaction involves proton transfer. In the *forward* direction, a hydroxide ion removes a proton from an ammonium ion. In the *reverse* direction, a proton is transferred from water to ammonia. This reverse reaction is included among the general categories; it is the transfer of a proton from a water molecule to a base. The equilibrium constant for this reaction is $K_b$. In the direction the reaction is written, the equilibrium constant is $1/K_b$:

$$\frac{[NH_3]_{eq}}{[NH_4^+]_{eq}[OH^-]_{eq}} = \frac{1}{K_b}$$

✏ **Do the Results Make Sense?** First, verify that each concentration quotient has products in the numerator and reactants in the denominator. Then examine the concentration ratio and relate it to the general types of equilibrium constants.

**Practice Exercise 13.16**
Relate each of the following equilibrium reactions to equilibrium constants of standard types:

(a) $HCO_3^- (aq) + H_3O^+ (aq) \rightleftharpoons H_2CO_3 (aq) + H_2O (l)$

(b) $Al(OH)_3 (s) \rightleftharpoons Al^{3+} (aq) + 3\ OH^- (aq)$

(c) $Co^{3+} (aq) + 6\ NH_3 (aq) \rightleftharpoons [Co(NH_3)_6]^{3+} (aq)$

(d) $CN^- (aq) + H_2O (l) \rightleftharpoons HCN (aq) + OH^- (aq)$

## Spectator Ions

With all the possible equilibria in aqueous systems, most species might be expected to participate in at least one equilibrium. Nonetheless, in many solutions, some of the ionic species undergo no significant reactions. These species are classified as **spectator ions**.

Consider what happens when a solution of potassium iodide is mixed with a solution of lead(II) nitrate. The major species present are $K^+ (aq)$, $I^- (aq)$, $Pb^{2+} (aq)$, $NO_3^- (aq)$, and $H_2O$. As shown in Figure 13-16, a bright yellow solid precipitates from the mixture. This solid is insoluble $PbI_2$. The $Pb^{2+}$ and I ions that remain in solution are at equilibrium with the precipitate:

$$PbI_2 (s) \rightleftharpoons Pb^{2+} (aq) + 2\ I^- (aq) \quad K_{sp} = 9.8 \times 10^{-9}$$

Two of the major species, $K^+$ and $NO_3^-$, are not involved in this equilibrium. They are spectator ions. Example 13-17 provides practice in identifying spectator ions.

**Figure 13-16**

Mixing aqueous solutions of KI and $Pb(NO_3)_2$ results in the formation of a yellow solid, $PbI_2$.

---

**EXAMPLE 13-17 SPECTATOR IONS**

When equal volumes of 0.100 M solutions of sodium bromide and silver nitrate are mixed, a white solid precipitates from the solution. Identify the precipitate, write the net ionic reaction for the solubility equilibrium, and identify any spectator ions.

**Strategy:** Analyze the problem from the molecular perspective. Begin by identifying the major species present. Then identify the equilibria in which these ions participate.

**Solution:** Sodium bromide and silver nitrate are both soluble ionic compounds, so the major species are $Na^+$, $Ag^+$, $Br^-$, $NO_3^-$, and $H_2O$ molecules. The solubility guidelines presented in Chapter 1 identify the precipitate as silver bromide:

$$AgBr (s) \rightleftharpoons Ag^+ (aq) + Br^- (aq) \quad K_{sp} = 5.35 \times 10^{-13}$$

Major ionic species that do not participate in aqueous equilibria are classified as spectator ions. In this case, the spectator ions are $Na^+$ and $NO_3^-$.

**Do the Results Make Sense?** All sodium salts are soluble, and so are all nitrate salts, so it makes sense that neither of these ions participates in a solubility equilibrium. Furthermore, nitrate and sodium cations are neither acidic nor basic, so it makes sense that neither participates in an acid–base equilibrium.

**Practice Exercise 13.17**

Identify the spectator ions present when a solution of sodium hydroxide is mixed with a solution of nitric acid.

The types of aqueous equilibria described in this section have been given special names, and it is essential that you be able to recognize them. Keep in mind, however, that the principles described in the previous sections apply to all chemical equilibria. Chemists categorize equilibria for convenience, but they treat all equilibria the same way. Our Focus 13-2: Chemistry and the Environment explores the roles of these equilibria in a spectacular natural process, the formation of limestone caverns.

## SECTION EXERCISES

●**13.6.1** What are the major species present in each of the following solutions? (a) 1.00 M perchloric acid; (b) 0.25 M ammonia; (c) 0.50 M potassium hydrogen carbonate; and (d) 0.010 M hypochlorous acid, HClO.

●**13.6.2** A green precipitate forms when solutions of sodium hydroxide and nickel(II) sulphate are mixed. Identify the precipitate, write the net ionic equation for the solubility equilibrium, and identify the spectator species.

●**13.6.3** Relate each of the following equilibrium reactions to equilibrium constants of standard types: $K_{sp}$, $K_a$, and $K_b$:

(a) $Al^{3+} (aq) + 3\ OH^- (aq) \leftrightarrows Al(OH)_3 (s)$

(b) $HClO (aq) + H_2O (l) \leftrightarrows ClO^- (aq) + H_3O^+ (aq)$

(c) $NH_3 (aq) + H_2O (l) \leftrightarrows NH_4^+ (aq) + OH^- (aq)$

## FOCUS 13-2   CHEMISTRY AND THE ENVIRONMENT: LIMESTONE CAVERNS

Limestone caverns are among nature's most spectacular displays. These caves occur in many parts of the world. Examples are Carlsbad Caverns in New Mexico, Jeita Caves in Lebanon, the Blue Grotto in Italy, and the Jenolan Caves in Australia. Wherever they occur, the chemistry of their formation involves the aqueous equilibria of limestone, which is calcium carbonate. Three such equilibria, linked to one another by Le Châtelier's principle, play essential roles in cave dynamics.

The starting point for the aqueous chemistry of calcium carbonate is its solubility product:

$$CaCO_3 (s) \rightleftarrows Ca^{2+} (aq) + CO_3^- (aq) \quad K_{sp} = 3.36 \times 10^{-9}$$

As the small value of $K_{sp}$ indicates, this substance is almost insoluble. The material from which caverns form was deposited over 250 million years ago, during an age when inland seas covered many regions of the continents. In these seas, corals, mollusks, and other marine organisms flourished, incorporating calcium ions dissolved in seawater into solid calcium carbonate shells and skeletons. As these sea creatures died, their

A spectacular gallery of Carlsbad Caverns.

remains accumulated on the ocean floor because of the insolubility of calcium carbonate. Over many eons, the shells and skeletons were compressed into limestone rock, eventually forming deposits hundreds of miles long.

When the climate became drier, the ancient seas evaporated, and these limestone deposits were buried beneath beds of silt. In time, uplift and erosion exposed the limestone deposits to nature's elements–primarily the action of running water. Water would not dissolve limestone to any great extent were it not for additional equilibria involving carbon dioxide. First is the solubility equilibrium between gaseous and dissolved carbon dioxide:

$$CO_2 (g) \rightleftarrows CO_2 (aq)$$

Dissolved $CO_2$ reacts with one molecule of water to form carbonic acid, which undergoes proton transfer with a second water molecule. These two reactions establish a second equilibrium that makes rainwater slightly acidic:

$$CO_2 (aq) + 2 H_2O (l) \rightleftarrows HCO_3^- (aq) + H_3O^+ (aq)$$

When exposed to acidic water, limestone slowly dissolves by forming hydrogen carbonate anions:

$$CaCO_3 (s) + H_3O^+ (aq) \rightleftarrows Ca^{2+} (aq) + HCO_3^- (aq) + H_2O (l)$$

Over thousands of years, water percolating through limestone dissolved some of the rock, forming underground chambers.

The story does not end with cavern formation. Many limestone caves, such as Carlsbad Caverns, contain spectacular formations that include stalagmites, stalactites, and limestone columns. We need to examine the equilibria more closely to understand how these structures form.

Inside newly formed caverns, calcium carbonate precipitates from water dripping from the ceilings of the chambers. This happens when water that is saturated with carbon dioxide and calcium hydrogen carbonate comes into contact with air. Some of the dissolved $CO_2$ escapes into the gas phase. This shifts the two equilibria to the left, and solid calcium carbonate precipitates:

$$CaCO_3 (s) + CO_2 (g) \uparrow^{Escape} + H_2O (l) \xleftarrow{\text{Shifts to left}} Ca^{2+} (aq) + 2HCO_3^- (aq)$$

The amount of calcium carbonate precipitating from any particular drop is imperceptibly small. Nevertheless, over the years, these deposits grow into translucent hollow tubes of $CaCO_3$ called "soda straws" (see photo inset). Soda straws lengthen as water drops fall from their tips. These delicate structures can reach lengths of more than a metre. In time, water flowing over the outside of the tube adds width to the growing formation, and the soda straw matures into the familiar stalactite.

Stalactites (*left*), "soda straws" (*inset*), and stalagmites.

Over the eons, the flow and evaporation of water inside a cavern creates a stunning array of rock sculptures. Stalagmites grow upward from the floor, sometimes joining stalactites to form massive columns. Limestone dams create beautiful pools of water. Limestone draperies fall like curtains from water flowing around overhanging rock. Delicate mineral flowers sprout from the walls. All these features result from the aqueous solubility equilibrium of calcium carbonate.

As water slowly dissolves the limestone roof of a cave, the roof becomes weak and may eventually collapse. The result is a sinkhole. When a source of water remains, the sinkhole fills to form a lake that can be very scenic.

Montezuma's Well, a sinkhole in Arizona.

# CHAPTER 13 VISUAL SUMMARY

## 13.1 DESCRIBING CHEMICAL EQUILIBRIA

Dynamic equilibrium

Reversibility

$$K_{eq} = \frac{[I_3^-]_{eq}}{[I_2]_{eq}[I^-]_{eq}}$$

Equilibrium constant

$$a A + b B \rightleftharpoons d D + e E \qquad K_{eq} = \frac{[D]_{eq}^d [E]_{eq}^e}{[A]_{eq}^a [B]_{eq}^b}$$

**LEARNING OBJECTIVES:** Explain the dynamic nature of equilibrium in terms of reversibility.

**SKILLS TO MASTER:** Formulating the equilibrium expression for a given reaction (p. 755)

**KEY CONCEPTS:** At equilibrium, the rates of the forward and reverse reactions are equal.

An equilibrium expression relates the concentrations of reactants and products at equilibrium.

Every elementary reaction that goes in the forward direction can also go in the reverse direction.

## 13.2 PROPERTIES OF EQUILIBRIUM CONSTANTS

Properties of equilibrium constants

$K_{eq}$ is related to the stoichiometry of the net reaction.

$K_{eq}$ applies only at equilibrium.

$K_{eq}$ is independent of initial conditions.

Pure solids, liquids, or solvents do not appear in the expression for $K_{eq}$.

$$K_{eq, fwd} = \frac{1}{K_{eq, rev}}$$

Magnitudes of $K_{eq}$ vary over a wide range.

**LEARNING OBJECTIVES:** Understand some of the properties of equilibrium constants.

**SKILLS TO MASTER:** Relating concentrations to activities (p. 757)

Understanding the magnitudes of equilibrium constants (p. 762)

**KEY CONCEPTS:** Concentrations or partial pressures can usually be used in place of activities in equilibrium constant expressions.

## 13.3 THERMODYNAMICS AND EQUILIBRIUM

Reaction goes left: $Q > K_{eq}$ & $\Delta G_{reaction} > 0$

Equilibrium when: $Q = K_{eq}$ & $\Delta G_{reaction} = 0$

Reaction goes right: $Q < K_{eq}$ & $\Delta G_{reaction} < 0$

$2\ NO_2 \rightleftharpoons N_2O_4$

$K_{eq} > Q$: $NO_2$ formation favored

$K_{eq} < Q$: $N_2O_4$ formation favored

$K_{eq} = Q$: No net change

$$\ln K_{eq} = -\frac{\Delta H^o}{RT} + \frac{\Delta S^o}{R}$$

$$\Delta G^o = RT \ln K_{eq}$$

**LEARNING OBJECTIVES:** Relate the equilibrium position to thermodynamic quantities.

**SKILLS TO MASTER:** Predicting the direction of change by comparing Q with $K_{eq}$ (p. 764)

Calculating the equilibrium constant from the reaction free energy change (p. 765)

Calculating the equilibrium constant at non-standard temperatures (p. 766)

**KEY CONCEPTS:** Reactions always move in the direction that minimizes the total free energy.

## 13.4 SHIFTS IN EQUILIBRIUM

Le Châtelier's principle

A system at equilibrium responds to offset changes.

No catalyst

Catalysts do not affect the value of $K_{eq}$.

Catalyst

Products

Overall change remains the same

Reactants

Reaction coordinate

Initial equilibrium / Steam injected / New equilibrium

$H_2O$

$CH_4$

$H_2$

$CO$

**LEARNING OBJECTIVES:** Predict the effects on the equilibrium position of changing concentrations or temperature.

**SKILLS TO MASTER:** Determining the effects of increasing or decreasing the concentration of a reactant or product (p. 770)

Determining the effects of changing the temperature (p. 773)

**KEY CONCEPTS:** When a change is imposed on a system at equilibrium, the system will react in the direction that reduces the amount of change (Le Châtelier's principle).

## 13.5 WORKING WITH EQUILIBRIA

Substance      A    B

Initial concentration    $[A]_i$   $[B]_i$

Change in concentration   $\Delta[A]$   $\Delta[B]$

Final concentration   $[A]_{eq}$   $[B]_{eq}$

Use Equation 13–6 across the change row:

$$\Delta[B] = \frac{\text{Coeff. B}}{\text{Coeff. A}}\,\Delta[A]$$

Use Equation 13–5 down each column:

$$[A]_{eq} = [A]_i + \Delta[A]$$

Equilibrium position

$$AgBr\,(s) \rightleftharpoons Ag^+\,(aq) + Br^-\,(aq)$$
$$K_{eq} \ll 1$$

Extent of reaction

$$[A]_i - x \cong [A]_i$$

**SOLVING EQUILIBRIUM PROBLEMS**

1. Determine what is asked for.
2. Identify the major species.
3. Determine what equilibria exist.
4. Write the expression for $K_{eq}$.
5. Organize the data and unknowns.
6. Carry out the calculations.
7. Check for reasonableness.

**LEARNING OBJECTIVES:** Solve quantitative equilibrium problems.

**SKILLS TO MASTER:** Constructing concentration tables (p. 779)

Using concentration tables to calculate equilibrium concentrations (p. 779)

Doing equilibrium calculations with very small equilibrium constants (p. 785)

Doing equilibrium calculations with very large equilibrium constants (p. 787)

**KEY CONCEPTS:** If initial concentrations and the equilibrium constant are known, equilibrium concentrations can be calculated.

## 13.6 EQUILIBRIA IN AQUEOUS SOLUTIONS

Major species
Minor species
Spectator ions
Dominant equilibrium    Water    Ions    Molecules

Aqueous solution

Solid NaCl

Solubility product ($K_{sp}$)    $AB\,(s) + H_2O \rightleftharpoons A^+\,(aq) + B^-\,(aq)$

Acid ionization constant ($K_a$)    $HA + H_2O \rightleftharpoons A^- + H_3O^+$

Base ionization constant ($K_b$)    $B + H_2O \rightleftharpoons HB^+ + OH^-$

Formation constant ($K_f$)    $M^{n+} + xL \rightleftharpoons [M^{n+}L_x]$

**LEARNING OBJECTIVES:** Perform equilibrium calculations on reactions in aqueous solution.

**SKILLS TO MASTER:** Identifying the major and minor species in solution (p. 792)

Writing equilibrium expressions for acid and base hydrolysis reactions (p. 794)

Writing solubility product and formation constant expressions (p. 795)

**KEY CONCEPTS:** Water is frequently a reagent in aqueous phase reactions.

## LEARNING EXERCISES

13.1   List the types of chemical equilibrium introduced in this chapter and give a specific example of each.

13.2   Draw a molecular picture that illustrates each of the examples that you gave in Learning Exercise 13-1.

13.3   Describe in your own words how to set up and solve a problem that asks for concentrations at equilibrium.

13.4   Describe how to calculate $K_{eq}$ from equilibrium concentrations and from thermodynamic functions.

13.5   Write a paragraph that describes the logic, process, and justification for using approximations in solving equilibrium problems.

13.6   Use your own words to define (a) initial concentrations, (b) change to equilibrium, (c) change to completion, and (d) equilibrium concentrations.

13.7   Summarize in writing the connections between equilibrium constants and thermodynamic functions.

13.8   Update your list of memory bank equations and their uses.

13.9   List all terms new to you that appear in Chapter 13. Write a one-sentence definition for each in your own words. Consult the Glossary if you need help.

## PROBLEMS

 =interactive learning ware problem. Visit the website at www.wiley.com/canada/olmsted

## Describing Chemical Equilibria

13.1   The alkene 2-butene has two isomers, *cis*-butene and *trans*-butene, which interconvert at high temperature or in the presence of a catalyst:

*cis*-Butene (*g*)          *trans*-Butene (*g*)

Suppose that the equilibrium constant for this reaction is 3.0. Draw a qualitative graph that shows how the pressure of each gas changes with time if the system initially contains pure *cis*-butene at a pressure of 1.0 bar.

13.2   Redraw the graph for the reaction in Problem 13.1, showing what happens if the system initially contains pure *trans*-butene at a pressure of 1.0 bar.

13.3   Aqueous solutions of sodium hypochlorite undergo decomposition according to the following two-step mechanism:

$$ClO^- \ (aq) + ClO^- \ (aq) \xrightarrow{k_1} Cl^- \ (aq) + ClO_2^- (aq)$$

$$ClO^- \ (aq) + ClO_2^- (aq) \xrightarrow{k_2} Cl^- \ (aq) + ClO_3^- (aq)$$

(a)   What additional reactions will be important as the reaction approaches equilibrium?

(b)   What is the equilibrium constant expression for the overall reaction?

(c)   Express the equilibrium constant in terms of the rate constants of the elementary reactions.

13.4   The reaction of $H_2$ gas with CO gas to give formaldehyde has been described by this mechanism (all species in the gas phase):

$$H_2 \xrightleftharpoons[k_{-1}]{k_1} 2 \, H$$

$$H + CO \xrightarrow{k_2} HCO$$

$$HCO + H \xrightarrow{k_3} H_2CO$$

(a)   What reactions describe the decomposition of $H_2CO$ under these same conditions?

(b)   What is the equilibrium constant expression for this reaction?

(c)   Express the equilibrium constant in terms of rate constants for the elementary reactions.

13.5   Draw molecular pictures that illustrate the reversibility of the reactions involved in the hypochlorite decomposition given in Problem 13.3. (See Figure 13-1 for examples.)

13.6   Draw molecular pictures that illustrate the reversibility of the reactions involved in the reaction of $H_2$ with CO to form $H_2CO$, given in Problem 13.4. (See Figure 13-1 for examples.)

13.7   Describe experiments that could be done to demonstrate that the decomposition reaction of $ClO^-$ anions (Problem 13.3) is reversible.

13.8   Describe experiments that could be done to demonstrate that the reaction of $H_2$ with CO (Problem 13.4) is reversible.

## Properties of Equilibrium Constants

13.9   Write the equilibrium constant expression for each of the following reactions:
(a)   $I_2 \ (s) + 5 \, F_2 \ (g) \rightleftharpoons 2 \, IF_5 \ (g)$
(b)   $P_4 \ (s) + 5 \, O_2 \ (g) \rightleftharpoons P_4O_{10} \ (s)$
(c)   $BaCO_3 \ (s) + C \ (s) \rightleftharpoons BaO \ (s) + 2 \, CO \ (g)$
(d)   $CO \ (g) + 2 \, H_2 \ (g) \rightleftharpoons CH_3OH \ (l)$
(e)   $H_3PO_4 \ (aq) + 3 \, H_2O \ (l) \rightleftharpoons PO_4^{3-} \ (aq) + 3 \, H_3O^+ \ (aq)$

13.10  Write the equilibrium constant expression for each of the following reactions:
(a)   $2 \, H_2S \ (g) + 3 \, O_2 \ (g) \rightleftharpoons 2 \, H_2O \ (l) + 2 \, SO_2 \ (g)$
(b)   $Fe_2O_3 \ (s) + 3 \, CO \ (g) \rightleftharpoons 2 \, Fe \ (s) + 3 \, CO_2 \ (g)$
(c)   $Cl_2 \ (g) + 3 \, F_2 \ (g) \rightleftharpoons 2 \, ClF_3 \ (g)$
(d)   $NH_3 \ (g) + H_3O^+ \ (aq) \rightleftharpoons NH_4^+ \ (aq) + H_2O \ (l)$
(e)   $SnO_2 \ (s) + 2 \, H_2 \ (g) \rightleftharpoons Sn \ (s) + 2 \, H_2O \ (l)$

13.11  Write each of the chemical reactions of Problem 13.9 in the opposite direction and determine the equilibrium constant expressions for these reactions.

13.12  Write each of the chemical reactions of Problem 13.10 in the opposite direction and determine the equilibrium constant expressions for these reactions.

13.13  State the standard (reference) concentration for each substance appearing in the equilibria of Problem 13.9.

13.14  State the standard (reference) concentration for each substance appearing in the equilibria of Problem 13.10.

13.15  Using copper as an example, draw a molecular picture illustrating that the concentration of atoms in a pure metal is independent of the volume of the sample.

**13.16** Using mercury as an example, draw a molecular picture illustrating that the concentration of atoms in a pure liquid is independent of the volume of the sample.

## Thermodynamics and Equilibrium

**13.17** (ilw) Using standard thermodynamic data from Appendix D, calculate $K_{sp}$ at 298 K for LiCl.

**13.18** Using standard thermodynamic data from Appendix D, calculate $K_{sp}$ at 298 K for $MgCl_2$.

**13.19** Using standard thermodynamic data from Appendix D, calculate the equilibrium constant at 298 K for each of the following chemical equilibria:
(a) $CO\ (g) + H_2O\ (l) \leftrightarrows CO_2\ (g) + H_2\ (g)$
(b) $2\ CO\ (g) + O_2\ (g) \leftrightarrows 2\ CO_2\ (g)$
(c) $BaCO_3\ (s) + C\ (s, graphite) \leftrightarrows BaO\ (s) + 2\ CO\ (g)$
(d) $3\ CO\ (g) + 6\ H_2\ (g) \leftrightarrows C_3H_6\ (g) + 3\ H_2O\ (l)$

**13.20** Using standard thermodynamic data from Appendix D, calculate the equilibrium constant at 298 K for each of the following chemical equilibria:
(a) $CH_4\ (g) + H_2O\ (l) \leftrightarrows CO\ (g) + 3\ H_2\ (g)$
(b) $4\ NH_3\ (g) + 5\ O_2\ (g) \leftrightarrows 4\ NO\ (g) + 6\ H_2O\ (l)$
(c) $SnO_2\ (s) + 2\ H_2\ (g) \leftrightarrows Sn\ (s, white) + 2\ H_2O\ (l)$
(d) $3\ Fe\ (s) + 4\ H_2O\ (l) \leftrightarrows Fe_3O_4\ (s) + 4\ H_2\ (g)$

**13.21** Estimate $K_{eq}$ for the equilibria in Problems 13.19 (b) and 13.19 (c) at 250 K.

**13.21** Estimate $K_{eq}$ for the equilibria in Problems 13.20 (c) and 13.20 (d) at 350 K.

**13.23** Determine $K_{eq}$ for the equilibria in Problems 13.19 (a) and 13.19 (d) at 395 K, if $H_2O$ is present as a gas rather than as a liquid.

**13.24** Estimate $K_{eq}$ for the equilibria in Problems 13.20 (c) and 13.20 (d) at 425 K, if $H_2O$ is present as a gas rather than as a liquid.

## Shifts in Equilibrium

**13.25** For each equilibrium in Problem 13.19, predict the effect of injecting additional $CO\ (g)$ into the system.

**13.26** For each equilibrium in Problem 13.20, predict the effect of injecting additional $H_2O\ (l)$ into the system.

**13.27** Consider the following reaction at equilibrium in water:

$$PbCl_2\ (s) \leftrightarrows Pb^{2+}\ (aq) + 2\ Cl^-\ (aq)$$

Predict whether $PbCl_2$ will dissolve, precipitate, or do neither after each of the following changes: (a) more $PbCl_2\ (s)$ is added; (b) more $H_2O$ is added; (c) solid NaCl is added; and (d) solid $KNO_3$ is added.

**13.28** The following exothermic gas-phase reaction is at equilibrium:

$$2\ SO_2\ (g) + O_2\ (g) \leftrightarrows 2\ SO_3\ (g)$$

Predict what happens to the amount of $SO_3$ in the system when each of the following changes is made: (a) the temperature is raised; (b) more $O_2$ is added; and (c) some Ar gas is introduced.

**13.29** Consider the following gas-phase reaction:

$$SO_2\ (g) + Cl_2\ (g) \leftrightarrows SO_2Cl_2\ (g)\ (exothermic)$$

Describe four changes that would drive the equilibrium to the left.

**13.30** Consider the following gas-phase reaction:

$$PCl_5\ (g) \leftrightarrows PCl_3\ (g) + Cl_2\ (g)\ (endothermic)$$

Describe four changes that would drive the equilibrium to the left.

## Working with Equilibria

**13.31** (ilw) An industrial chemist puts 1.00 mole each of $H_2\ (g)$ and $CO_2\ (g)$ in a 1.00-L container at constant temperature of 800°C. This reaction occurs:

$$H_2\ (g) + CO_2\ (g) \leftrightarrows H_2O\ (g) + CO\ (g)$$

When equilibrium is reached, 0.49 mole of $CO\ (g)$ is in the container. Find the value of $K_{eq}$ for the reaction.

**13.32** At high temperature, HCl and $O_2$ react to give $Cl_2$ gas:

$$4\ HCl\ (g) + O_2\ (g) \leftrightarrows 2\ Cl_2\ (g) + 2\ H_2O\ (g)$$

If HCl at 2.30 bar and $O_2$ at 1.00 bar react at 750 K, the equilibrium pressure of $Cl_2$ is measured to be 0.93 bar. Determine the value of $K_{eq}$ at 750 K.

**13.33** When 0.0500 mole of propanoic acid ($C_2H_5CO_2H$) is dissolved in 0.500 L of water, proton transfer occurs:

$$C_2H_5CO_2H + H_2O \leftrightarrows C_2H_5CO_2^- + H_3O^+$$

The equilibrium concentration of $H_3O^+$ ions is $1.15 \times 10^{-3}$ M. Evaluate $K_{eq}$.

**13.34** Cyanic acid, HCNO, is a weak acid:

$$HCNO + H_2O \leftrightarrows CNO^- + H_3O^+$$

In a 0.20 M aqueous solution of $HCNO^+$, the concentration of $H_3O^+$ cations is $6.5 \times 10^{-3}$ M. Evaluate $K_{eq}$.

**13.35** (ilw) The equilibrium constant for the following reaction is $1.6 \times 10^5$ at 1024 K:

$$H_2\ (g) + Br_2\ (g) \leftrightarrows 2\ HBr\ (g)$$

Find the equilibrium pressures of all gases if 10.0 bar of HBr is introduced into a sealed container at 1024 K.

**13.36** At 100 °C, $K_{eq} = 1.5 \times 10^8$ for the following reaction:

$$CO\ (g) + Cl_2\ (g) \leftrightarrows COCl_2\ (g)$$

Using appropriate approximations, calculate the partial pressure of CO at 100 °C at equilibrium in a chamber that initially contains $COCl_2$ at a pressure of 0.250 bar.

**13.37** At 1000 °C, $K_{eq} = 0.403$ for the following reaction:

$$FeO\ (s) + CO\ (g) \leftrightarrows Fe\ (s) + CO_2\ (g)$$

If CO gas at 5.0 bar is injected into a container at 1000 °C that contains excess FeO, what are the partial pressures of all gases present at equilibrium?

**13.38** $K_{eq}$ for the Haber reaction is $2.81 \times 10^{-5}$ at 472 °C. If a reaction starts with 3.0 bar of $H_2$ and 5.0 bar of $N_2$ at 472 °C, what is the equilibrium pressure of $NH_3$?

## Equilibria in Aqueous Solutions

**13.39** Identify the major species present in an aqueous solution of each of the following substances: (a) $CH_3CO_2H$; (b) $NH_4Cl$; (c) KCl; (d) $NaCH_3CO_2$; and (e) NaOH.

**13.40** Identify the major species present in an aqueous solution of each of the following substances: (a) HClO; (b) $CaBr_2$; (c) KClO; (d) $HNO_3$; and (e) HCN.

**13.41** Identify the acid–base equilibria, if any, for the major species in Problem 13.39.

**13.42** Identify the acid–base equilibria, if any, for the major species in Problem 13.40.

**13.43** When the following substances dissolve in water, what major species are present? (a) acetone; (b) potassium bromide; (c) lithium hydroxide; and (d) sulphuric acid.

**13.44** When the following substances dissolve in water, what major species are present? (a) sodium hydrogen carbonate; (b) methanol; (c) hydrogen bromide; and (d) benzoic acid.

**13.45** Write the equilibrium constant expression for each of the following reactions and relate it to equilibrium constants of standard types ($K_{sp}$, $K_a$, and $K_b$):
(a) $HClO_2\ (aq) + H_2O\ (l) \rightleftharpoons ClO_2^-\ (aq) + H_3O^+\ (aq)$
(b) $Fe^{3+}\ (aq) + 3\ OH^-\ (aq) \rightleftharpoons Fe(OH)_3\ (s)$
(c) $CN^-\ (aq) + H_3O^+\ (aq) \rightleftharpoons HCN\ (aq) + H_2O\ (l)$

**13.46** Write the equilibrium constant expression for each of the following reactions and relate it to equilibrium constants of standard types ($K_{sp}$, $K_a$, and $K_b$):
(a) $Ag_2SO_4\ (s) \rightleftharpoons 2\ Ag^+\ (aq) + SO_4^{2-}\ (aq)$
(b) $H_3PO_4\ (aq) + H_2O\ (l) \rightleftharpoons H_2PO_4^-\ (aq) + H_3O^+\ (aq)$
(c) $HCO_2^-\ (aq) + H_3O^+\ (aq) \rightleftharpoons HCO_2H\ (aq) + H_2O\ (l)$

**13.47** Identify the spectator ions present when the following are mixed:
(a) A solution of $CH_3CO_2H$ and a solution of NaOH.
(b) A solution of $CaCl_2$ and a solution of $K_3PO_4$.
(c) A solution of $HNO_3$ and a solution of KOH.

**13.48** Identify the spectator ions present when the following are mixed:
(a) A solution of $FeCl_3$ and a solution of NaOH.
(b) A solution of $NH_3$ and a solution of $HNO_3$.
(c) A solution of $KCH_3CO_2$ and a solution of $HClO_4$.

## ADDITIONAL PAIRED PROBLEMS

**13.49** Write the equilibrium constant expressions for the following:
(a) $2\ NaHCO_3\ (s) \rightleftharpoons Na_2CO_3\ (s) + CO_2\ (g) + H_2O\ (g)$
(b) $2\ N_2\ (g) + 6\ H_2O\ (l) \rightleftharpoons 4\ NH_3\ (g) + 3\ O_2\ (g)$
(c) $2\ C_2H_4\ (g) + O_2\ (g) \rightleftharpoons 2\ CH_3CHO\ (g)$
(d) $Ag_2SO_4\ (s) \rightleftharpoons 2\ Ag^+\ (aq) + SO_4^{2-}\ (aq)$
(e) $NH_4HS\ (s) \rightleftharpoons H_2S\ (g) + NH_3\ (g)$

**13.50** Write the equilibrium constant expressions for the following:
(a) $4\ NH_3\ (g) + 5\ O_2\ (g) \rightleftharpoons 4\ NO\ (g) + 6\ H_2O\ (l)$
(b) $HCN\ (aq) + H_2O\ (l) \rightleftharpoons CN^-\ (aq) + H_3O^+\ (aq)$
(c) $2\ NH_4^+\ (aq) + SO_4^{2-}\ (aq) \rightleftharpoons (NH_4)_2SO_4\ (s)$
(d) $SO_4^{2-}\ (aq) + H_2O\ (l) \rightleftharpoons OH^-\ (aq) + HSO_4^-\ (aq)$
(e) $2\ O_3\ (g) \rightleftharpoons 3\ O_2\ (g)$

**13.51** Phosgene is a deadly gas and its use in warfare has been outlawed. The gas decomposes at elevated temperature:

$$COCl_2\ (g) \rightleftharpoons CO\ (g) + Cl_2\ (g) \quad K_{eq} = 8.3 \times 10^{-4} \text{ at } 352\ °C$$

What are the pressures at equilibrium if a 2.55-L metal container is charged with 3.00 bar of phosgene at 25 °C and then heated to 352 °C?

**13.52** The equilibrium constant for the dissociation of $Cl_2$ into atomic chlorine at 1200 K is $2.5 \times 10^{-5}$. If $Cl_2$ gas at 298 K, 0.57 bar is placed in a sealed container that is then heated to 1200 K, what is the equilibrium pressure of Cl atoms?

**13.53** The figure represents two chambers containing gases that react as follows:

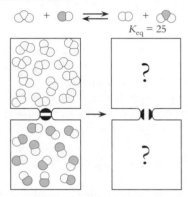

Draw a molecular picture that shows what the system looks like if the stopcock is opened and the reaction proceeds to equilibrium.

**13.54** The figure represents two chambers containing gases that react as follows:

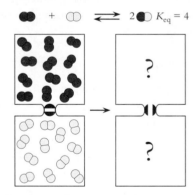

Draw a molecular picture that shows what the system looks like if the stopcock is opened and the reaction proceeds to equilibrium.

**13.55** A saturated solution of chloroform ($CHCl_3$) in water contains one drop of excess chloroform. If more liquid chloroform is added to this mixture, does the concentration of chloroform in the aqueous solution change? Explain your answer in molecular terms.

**13.56** A student mixes sodium chloride and water until the solid and the aqueous solution are at equilibrium. If more solid sodium chloride is added to this mixture, does the concentration of sodium chloride in the aqueous solution change? Explain your answer in molecular terms.

**13.57** At 350 K, $K_{eq}$ for the following reaction is 322:

$$Br_2 (g) + I_2 (g) \leftrightharpoons 2 \, IBr (g)$$

Suppose the equilibrium partial pressure of bromine is 0.512 bar and that of iodine is 0.327 bar. What is the pressure of IBr?

**13.58** Coal (solid carbon plus a collection of impurities) can be converted to a mixture of carbon dioxide and hydrogen gas by the following process:

$$C (s) + 2 \, H_2O (g) \leftrightharpoons CO_2 (g) + 2 \, H_2 (g) \quad K_{eq} \, 0.38 \text{ at } 1300 \text{ K}$$

At an industrial plant, a mixture of coal and water is placed inside a steel reaction vessel. When the mixture is heated to 1300 K, the equilibrium partial pressure of water is $2.80 \times 10^2$ bar. Calculate the partial pressures of $H_2$ and $CO_2$.

**13.59** Using tabulated standard thermodynamic data from Appendix D, calculate $K_{eq}$ for the reaction of $NO_2$ to form $N_2O_4$ at 298 K and at 525 K:

$$2 \, NO_2 (g) \leftrightharpoons N_2O_4 (g)$$

**13.60** Using tabulated standard thermodynamic data from Appendix D, calculate $K_{eq}$ for the following reaction at 298 K and at 825 K:

$$2 \, N_2O (g) + O_2 (g) \leftrightharpoons 4 \, NO (g)$$

**13.61** Write the equilibrium reaction and equilibrium constant expression for each of the following processes: (a) Trimethylamine $((CH_3)_3N)$, a weak base, is added to water. (b) Hydrofluoric acid (HF), a weak acid, is added to water. (c) Solid calcium sulphate $(CaSO_4)$, a sparingly soluble salt, is added to water.

**13.62** Write the equilibrium reaction and equilibrium constant expression for each of the following processes: (a) Trichloroacetic acid $(Cl_3CCO_2H)$ is added to water. (b) Aniline $((C_6H_5)NH_2)$, a weak base, is dissolved in water. (c) Solid calcium hydroxide $(Ca(OH)_2)$, a sparingly soluble salt, is added to water.

**13.63** Predict the effect of each of the following changes on the equilibrium position of the reaction in Example 13-14: (a) The partial pressure of $NO_2$ is cut in half. (b) The volume of the reactor is doubled. (c) A total of 10.0 bar of Ar gas is added to the reactor.

**13.64** Predict the effect each of the following changes will have on the equilibrium position of the reaction in Example 13-12: (a) Some sodium benzoate is dissolved in the solution. (b) An additional 1.0 L of water is added to the solution. (c) Some NaCl is dissolved in the solution.

**13.65** Use the appropriate data from Appendix D to calculate $K_{sp}$ for LiCl at 100 °C. Is LiCl more or less soluble at high temperature than at low temperature? (See Problem 13.17.)

**13.66** Use the appropriate data from Appendix D to calculate $K_{sp}$ for $MgCl_2$ at 100 °C. Is $MgCl_2$ more or less soluble at high temperature than at low temperature? (See Problem 13.18.)

## MORE CHALLENGING PROBLEMS

**13.67** At elevated temperature, carbon tetrachloride decomposes to its elements:

$$CCl_4 (g) \leftrightharpoons C (s) + 2 \, Cl_2 (g)$$

At 700 K, if the initial pressure of $CCl_4$ is 1.00 bar, the total pressure at equilibrium is 1.35 bar. Use these pressures to calculate $K_{eq}$ at 700 K.

**13.68** Suppose that the equilibrium system described in Problem 13.67 is expanded to twice its initial volume. Find the new equilibrium pressure.

**13.69** Using tabulated thermodynamic data from Appendix D, compute (a) $K_{eq}$ at 298 K; (b) the temperature at which the equilibrium pressure is 1.00 bar; and (c) $K_{eq}$ at 1050 K for the following reaction:

$$Hg (g) + HgCl_2 (s) \leftrightharpoons Hg_2Cl_2 (s)$$

**13.70** For the following reaction, $K_{eq}$ is $1.83 \times 10^3$ at 390. K:

$$PCl_5 (g) \leftrightharpoons PCl_3 (g) + Cl_2 (g)$$

If 2.00 g of $PCl_5$ is placed in a 3.00-L bulb at 390 K, what is the equilibrium pressure of $Cl_2$?

**13.71** Consider the following reaction:

$$CO_2 (g) + 2 \, OH^- (aq) \leftrightharpoons CO_3^{2-} (aq) + H_2O (l)$$

(a) Write the equilibrium constant expression for this reaction. (b) What will happen to the pressure of $CO_2$ in this equilibrium system if some $Na_2CO_3$ solid is dissolved in the solution? (c) What will happen to the pressure of $CO_2$ in this equilibrium system if some HCl gas is bubbled through the solution?

**13.72** Benzene can be sulphonated by concentrated sulphuric acid:

$$C_6H_6 + H_2SO_4 \leftrightharpoons C_6H_5SO_3^- + H_3O^+$$

This reaction occurs by the following three-step mechanism:

$$H_2SO_4 \leftrightharpoons SO_3 + H_2O$$

$$SO_3 + C_6H_6 \leftrightharpoons C_6H_6SO_3$$

$$C_6H_6SO_3 + H_2O \leftrightharpoons C_6H_5SO_3^- + H_3O^+$$

Determine the relationship between the equilibrium constant for the net reaction and the rate constants for the various elementary steps.

**13.73** At 1020 °C, $K_{eq} = 167.5$ for the conversion of $CO_2 (g)$ to CO (g) by solid graphite, C (s). A 1.00-L, high-pressure chamber containing excess graphite powder is charged with 0.494 mole each of $CO_2$ and CO and then is heated to 1020 °C. What is the equilibrium total pressure?

**13.74** A chemist claims to have discovered a new gaseous element, effluvium (Ef), which reacts with atmospheric nitrogen to form effluvium nitride:

$$2 \, Ef (g) + 3 \, N_2 (g) \leftrightharpoons 2 \, EfN_3 (g)$$

In a container initially containing $N_2$ at 1.00 bar pressure and Ef gas at 0.75 bar, the total gas pressure at equilibrium, according to the chemist's measurements, is 0.85 bar. Compute $K_{eq}$ for this reaction.

**13.75** $K_{eq} = 1.1 \times 10^{-33}$ for the following reaction at 298 K:

$$Sn\ (s) + 2\ H_2\ (g) \leftrightarrows SnH_4\ (g)$$

Find the equilibrium pressure of $SnH_4$ (g) in a container at 298 K containing 10.0 g of Sn (s) and $2.00 \times 10^2$ bar of $H_2$. What volume of container is expected to contain a single molecule of $SnH_4$?

**13.76** In the gas phase, acetic acid is at equilibrium with a dimer held together by a pair of hydrogen bonds. (a) If the total pressure of acetic acid gas in a glass bulb is 0.75 bar, what is the partial pressure of the dimer? (b) Is the equilibrium constant for this reaction higher or lower at 200 °C? (*Hint:* Hydrogen bonds must be broken for the dimer to decompose.)

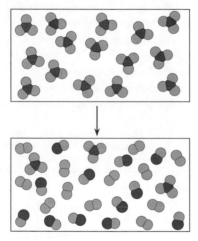

Dimer

**13.77** The following figure represents a system coming to equilibrium:

(a) Use molecular pictures similar to those in Problem 13.55 to write a balanced equation for the equilibrium reaction. (b) What is $K_{eq}$ for the reaction?

**13.78** Hydrogen fluoride is a highly reactive gas. It has many industrial uses, but the most familiar property of HF is its ability to react with glass. As a result, HF is used to etch glass and frost the inner surfaces of light bulbs. Hydrogen fluoride gas must be stored in stainless steel containers, and aqueous solutions must be stored in plastic bottles. Hydrogen fluoride can be produced from $H_2$ and $F_2$:

$$H_2\ (g) + F_2\ (g) \leftrightarrows 2\ HF\ (g) \qquad K_{eq} = 115$$

In a particular experiment, 3.00 bar each of $H_2$ and $F_2$ are added to a 1.50-L flask. Calculate the equilibrium partial pressures of all species.

## GROUP STUDY PROBLEMS

**13.79** (a) A container contains $CO_2$ at $p = 0.464$ bar. When graphite is added to the container, some $CO_2$ is converted to CO, and at equilibrium the total pressure is 0.746 bar. Compute $K_{eq}$ for the reaction:

$$CO_2\ (g) + C\ (s) \leftrightarrows 2\ CO\ (g)$$

(b) Suppose that the equilibrium system is compressed to one-third its initial volume. Find the new equilibrium pressure.

**13.80** When 1.00 mole of gaseous HI is sealed in a 1.00-L flask at 225 °C, it decomposes until the equilibrium amount of $I_2$ present is 0.182 moles:

$$2\ HI\ (g) \leftrightarrows H_2\ (g) + I_2\ (g)$$

(a) Use these data to calculate $K_{eq}$ for this reaction at 225 °C.

(b) Using standard thermodynamic data from Appendix D, estimate $K_{eq}$ at 625 °C.

(c) Using the result of (b), calculate the equilibrium partial pressures of all reagents under these conditions.

**13.81** One reaction that plays a role in photochemical smog is the following:

$$O_3\ (g) + NO\ (g) \leftrightarrows O_2\ (g) + NO_2\ (g) \qquad K_{eq} = 6.0 \times 10^{34}$$

Automobile engines produce both $O_3$ and NO. If the morning rush hour results in an atmosphere containing 1.5 ppm NO and 6.5 ppb $O_3$, calculate the equilibrium partial pressures of the three pollutants ($O_2$ is 21% of the atmosphere).

**13.82** The key step in the synthesis of sulphuric acid is the combustion of $SO_2$ to give sulphur trioxide:

$$2\ SO_2\ (g) + O_2\ (g) \leftrightarrows 2\ SO_3\ (g) \qquad K_{eq} = 5.60 \times 10^4 \text{ at } 350\ °C$$

(a) Sulphur dioxide and oxygen are mixed initially at 0.350 and 0.762 bar, respectively, at 350 °C. What are the partial pressures of the three gases when the mixture reaches equilibrium?

(b) Use thermodynamic data (see Appendix D) to calculate $K_{eq}$ for the sulphur trioxide reaction at 298 K. Based on this result, should this industrial synthesis be run at a higher or lower temperature than 350 °C to improve the yield?

(c) Suggest reasons why this reaction is run industrially at 700 K.

**13.83** The combustion of sulphur dioxide is one of the steps in the synthesis of sulphuric acid:

$$2\ SO_2\ (g) + O_2\ (g) \leftrightarrows 2\ SO_3\ (g)$$

The following diagram represents a mixture of $SO_2$, $O_2$, and $SO_3$ at equilibrium at $T = 1100$ K:

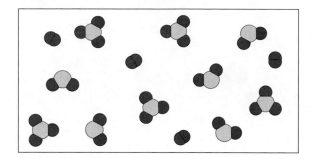

(a) Some gas is removed, resulting in the following conditions.

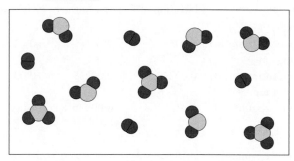

What happens to the position of the equilibrium? Explain.

(b) When the temperature of the system is increased to 1300 K, reaction occurs, leading to the following equilibrium conditions:

Is the reaction exothermic or endothermic? Use equilibrium arguments to explain your answer.

(c) Some gas is added to the original equilibrium mixture (at 1100 K), giving the following conditions:

What happens to the position of the equilibrium? Explain.

## ANSWERS TO SECTION EXERCISES

**13.1.1** $K_{eq} = \dfrac{[O_2]^3}{[O_3]^2}$

**13.1.2**

**13.1.3** $K_{eq} = \dfrac{[H_2O]^6[NO]^4}{[NH_3]^4[O_2]^5}$

**13.2.1** (a) $K_{eq} = (p_{Cl_2})_{eq}$

(b) $K_{eq} = (p_{CO_2})_{eq}$

(c) $K_{eq} = [PO_4^{3-}]^2_{eq}[Ca^{2+}]^3_{eq}$

(d) $K_{eq} = \dfrac{[H_3O^+]_{eq}[CN^-]_{eq}}{(p_{HCN})_{eq}}$

(e) $K_{eq} = \dfrac{1}{(p_{Cl_2})_{eq}}$

(f) $K_{eq} = \dfrac{1}{[PO_4^{3-}]^2_{eq}[Ca^{2+}]^3_{eq}}$

**13.2.2**

**13.2.3** $K_{eq} = \dfrac{(p_{F_2})_{eq}}{(p_F)^2_{eq}} = 4.8 \times 10^{21}$

$K_{eq} = \dfrac{(p_{H_2})_{eq}(p_{Br_2})_{eq}}{(p_{HBr})^2_{eq}} = 1.9 \times 10^{-19}$

**13.3.1** $K_{eq} = 1.4 \times 10^{-8}$

**13.3.2** Increases because the reaction is endothermic

**13.3.3** $K_{eq} = 2.4 \times 10^{-8}$

**13.4.1** (a) No change in Q, thus no change; (b), (c), and (d) more $Ba(OH)_2$ dissolves

**13.4.2** Lower $T$, remove $NH_3$, and increase pressures of $H_2$ and/or $N_2$

**13.4.3** The $H_2$–$N_2$ reaction has a large $K_{eq}$ at room temperature, whereas the $CH_4$–$H_2O$ reaction has a small $K_{eq}$ at room temperature.

**13.5.1** $K_{eq} = 1.10 \times 10^{-10}$

**13.5.2** $[CH_3CO_2^-]_{eq} = [H_3O^+]_{eq} = 6.7 \times 10^{-3}$ M and $[CH_3CO_2H]_{eq} = 2.5$ M

**13.5.3** $[Ag^+]_{eq} = 2.7 \times 10^{-12}$ M; $[Br^-]_{eq} = 0.200$ M

**13.6.1** (a) $H_3O^+$, $ClO_4^-$, and $H_2O$; (b) $NH_3$ and $H_2O$; (c) $K^+$, $HCO_3^-$, and $H_2O$; and (d) $HClO$ and $H_2O$

**13.6.2** $Ni(OH)_2$; net reaction is $Ni^{2+}$ $(aq)$ + 2 $OH^-$ $(aq) \longrightarrow Ni(OH)_2$ $(s)$; spectators are $SO_4^{2-}$ and $Na^+$.

**13.6.3** (a) $1/K_{sp}$; (b) $K_a$; and (c) $K_b$

## ANSWERS TO PRACTICE EXERCISES

**13.1**  $2\ HCO_2H\ (g) \rightleftarrows (HCO_2H)_2\ (g); \quad K_{eq} = \dfrac{[(HCO_2H_2)]_{eq}}{[HCO_2H]_{eq}}$

**13.2**  $3\ O_2(g) \rightleftarrows 2\ O_3(g); \quad K_{eq} = \dfrac{[O_3]^2_{eq}}{[O_2]^3_{eq}}$

**13.3**  $K_{eq} = \dfrac{[Cl^-]_{eq}[H_3O^+]_{eq}}{(p_{HCl})_{eq}};$

pressure of HCl should be in bar, concentration of the two ions in molarity

**13.4**  $K_{eq} = \dfrac{(p_{NO_2})^2_{eq}}{(p_{N_2O_4})^2_{eq}} = \dfrac{1}{3.10} = 0.323$

**13.5**  $K_{eq} = 6.1 \times 10^{-58}$

**13.6**  $K_{eq} = 1.7 \times 10^{-21}$

**13.7**  The shift is removal of some $I_2$, so your new picture should show one less triiodide and one more each of $I_2$ and iodide.

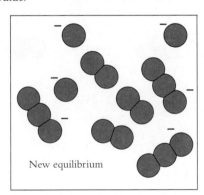

New equilibrium

**13.8**  Add $I_2$, add a soluble iodide salt such as KI, cool the solution, and boil off some solvent.

**13.9**  $K_{eq} = 1.0 \times 10^3$

**13.10**

| Reaction | $H_2O\ (l)$ $+NH_3\ (aq)$ | $\leftrightarrows$ $OH^-\ (aq)$ | $+$ $NH_4^+\ (aq)$ |
|---|---|---|---|
| Initial Concentration (M) | 0.335 | 0 | 0 |
| Change in Concentration (M) | −0.00246 | +0.00246 | +0.00246 |
| Equilibrium Concentration (M) | 0.333 | 0.00246 | 0.00246 |

$K_{eq} = 1.8 \times 10^{-5}$

**13.11**  At equilibrium, $p(AB_2) = p(A) = 1.0$ bar and $p(AB) = 2$ bar

View at equilibrium

**13.12**  $[NH_4^+] = [OH^-] = 4.7 \times 10^{-4}\ M; [NH_3] = 1.20 \times 10^{-2}\ M$

**13.13**  $[NH_4^+] = [OH^-] = 1.6 \times 10^{-2}\ M; [NH_3] = 14.8\ M$

**13.14**  0.50 mol AgBr forms; $[Ag^+] = [Br^-] = 7.3 \times 10^{-7}\ M$

**13.15**  Major species in aqueous hydrochloric acid are $H_3O^+$, $Cl^-$, and $H_2O$; in aqueous sodium chlorate are $Na^+$, $ClO_3^-$, and $H_2O$; and in aqueous chloroform are $CHCl_3$ and $H_2O$

**13.16**  (a) $1/K_a$; (b) $K_{sp}$; (c) $K_f$; and (d) $K_b$

**13.17**  $Na^+$ and $NO_3^-$

# AQUEOUS ACID–BASE EQUILIBRIA

## ACIDS AND TASTE

The tangy flavour of a fresh orange, the sour bite of lemon juice, and the tart pleasure of a Granny Smith apple are familiar flavours. Tangy, sour, and tart are taste sensations created by carboxylic acids, which contain the $—CO_2H$ functional group. Although it is not yet known exactly how tastes arise, a reasonable speculation is that the sensing process involves proton transfer reactions from this functional group to protein molecules in our taste buds.

Our molecular inset shows citric acid, which has three carboxylic acid functional groups. Citric acid is present in all citrus fruits as well as many other tart-tasting foods, including berries, pineapples, pears, and tomatoes. Lemons are acidic because they contain as much as 3% citric acid by mass.

The grape is another fruit containing carboxylic acids that contribute to its taste. Grapes contain several carboxylic acids as well as sugar molecules. When grapes are fermented to make wine, most of their sugar is converted into ethanol, but the carboxylic acids remain unchanged. A wine is "sweet" if

its residual sugar content masks the sour taste of these residual acids, and "dry" if its residual sugar content is less than about 1%.

Vegetables also contain carboxylic acids that contribute to their flavours. One example is oxalic acid, prevalent in spinach and rhubarb. Raw rhubarb leaves are mildly poisonous, and folklore holds that the toxic substance is oxalic acid. However, raw spinach can be eaten safely despite its equally high content of this acid.

Bacteria in milk feed on lactose (milk sugar) to produce lactic acid, another carboxylic acid. Under controlled conditions, this process makes yogurt. Similarly, the proper combination of yeast and bacteria acts on flour to form just enough lactic acid and acetic acid ($CH_3CO_2H$) to give sourdough bread its distinctive tang.

In addition to conveying acidity, the carboxylic acid group is highly polar, so smaller carboxylic acids such as acetic acid are water-soluble. Vinegar is a solution of acetic acid in water, typically about 5% by

# 14

## LEARNING OBJECTIVES
*Upon completion of this chapter you should be able to*

- use the Brønsted–Lowry theory of acids and bases
- relate pH to concentrations of ions in solution
- calculate concentrations and pH in weak acid and base solutions
- recognize and name some common acids and bases
- calculate the pH of solutions of salts of weak acids or bases
- explain the factors that contribute to the strength of an acid
- calculate concentrations in solutions involving multiple equilibria

mass of the acid. Acetic acid can be produced from sugar-containing foods by a combination of fermentation and oxidation, which breaks down glucose molecules into ethanol and then converts ethanol to acetic acid. The word *vinegar* comes from the French term for "sour wine"; indeed, poor wine tastes sour because some of its ethanol has become acetic acid.

This chapter describes the details of acid–base chemistry. We begin with a molecular view and a way of measuring acidity. Then we look at acid–base equilibrium calculations, making use of the general approaches to equilibrium developed in Chapter 13. We describe the important applications of buffers and titrations in Chapter 15.

## 14.1 PROTON TRANSFERS IN WATER

A simple but useful theory of acid–base reactions was developed independently in 1923 by the Danish chemist Johannes Brønsted and English chemist Thomas Lowry. The **Brønsted–Lowry** theory of acids and bases identifies an acid as a species that donates a proton (an $H^+$ ion), and a base is a species that accepts a proton. In an acid–base reaction, both things happen, and a proton is transferred from the acid to the base. Thus, every acid–base reaction has two reactants: an acid and a base.

A simple example is the reaction of hydrochloric acid with water. HCl is a strong acid, and as discussed in Section 13.6, hydrolyzes fully in water. In fact, the proton of the HCl is transferred to a water molecule:

$$H^+$$

$$HCl\ (aq) + H_2O\ (l) \longrightarrow H_3O^+\ (aq) + Cl^-\ (aq)$$

Thus, the water acts as a base in this reaction, accepting the proton from the acid HCl.

Among several peculiar traits of the water molecule is the fact that it can act as a base or as an acid. Dissolving a salt containing the phosphate ion in water, for instance, results in a proton being transferred from water to the phosphate ion, so in this case the water must be acting as an acid:

$$H^+$$

$$PO_4^{3-}\ (aq) + H_2O\ (l) \longrightarrow HPO_4^{2-}\ (aq) + OH^-\ (aq)$$

A molecular view of the above two reactions is given in Figure 14-1.

**Figure 14-1**

Every acid–base reaction involves transfer of a proton from an acid to a base.

Most examples in this chapter and the next take place in aqueous solution, but water is not always a reactant. For example, water is a product in the acid–base reaction:

$$\text{H}^+$$

$$\text{CH}_3\text{CO}_2\text{H} \ (aq) + \text{OH}^- \ (aq) \longrightarrow \text{CH}_3\text{CO}_2^- \ (aq) + \text{H}_2\text{O} \ (l)$$

## Conjugate Acid–Base Pairs

In a solution of hydrofluoric acid, HF molecules donate protons to water molecules, generating $\text{H}_3\text{O}^+$ and $\text{F}^-$ ions. As discussed in Chapter 13, the forward reaction and the reverse reaction must occur at equal rates in a solution that is at equilibrium. In an aqueous solution of HF that is at equilibrium, hydronium ions donate protons to fluoride ions to generate molecules of $\text{H}_2\text{O}$ and HF:

The reaction involves proton transfer in both directions, so there are two proton donors, HF and $\text{H}_3\text{O}^+$. There are also two proton acceptors, $\text{H}_2\text{O}$ and $\text{F}^-$. In other words, HF is an acid, and $\text{F}^-$ is a base. When a $\text{H}_2\text{O}$ molecule accepts a proton, the resulting cation is a proton donor. In other words, $\text{H}_3\text{O}^+$ is an acid, and $\text{H}_2\text{O}$ is a base. This is a general feature of proton transfer equilibria. Any proton donor and the species generated by removing one of its protons are called a **conjugate acid–base pair**. The hydronium ion is the **conjugate acid** of $\text{H}_2\text{O}$, and the fluoride ion is the **conjugate base** of HF. Thus, HF and $\text{F}^-$ are a conjugate acid–base pair; $\text{H}_3\text{O}^+$ and $\text{H}_2\text{O}$ are another conjugate acid–base pair.

Water can act as an acid or a base, so there are two conjugate acid–base pairs for water: $\text{H}_3\text{O}^+$ and $\text{H}_2\text{O}$ are a conjugate acid–base pair, and $\text{H}_2\text{O}$ and $\text{OH}^-$ are another conjugate acid–base pair. Example 14-1 reinforces the structural relationships of conjugate acid–base pairs.

### EXAMPLE 14-1  CONJUGATE ACID–BASE PAIRS

Write the chemical formula and the Lewis structure and draw a molecular picture of each of the following: (a) the conjugate acid of ammonia, $\text{NH}_3$; (b) the conjugate base of formic acid, $\text{HCO}_2\text{H}$; and (c) the conjugate acid of the hydrogen sulphate ion, $\text{HSO}_4^-$.

**Strategy:** Conjugate acid–base pairs are related through proton transfer. Removing a proton from an acid generates its conjugate base. Adding a proton to a base generates its conjugate acid. We use the procedures of Chapter 6 to construct the Lewis structures.

**Solution:**

(a) Add $H^+$ to $NH_3$ to obtain its conjugate acid, $NH_4^+$. With a steric number of 4, the ammonium ion has tetrahedral geometry:

(b) Remove $H^+$ from $HCO_2H$ to obtain its conjugate base, $HCO_2^-$:

(c) Add $H^+$ to $HSO_4^-$ to obtain its conjugate acid, $H_2SO_4$:

**Do the Results Make Sense?** An acid always differs from its conjugate base partner by the addition of $H^+$, and a base always differs from its conjugate acid partner by the removal of $H^+$. Notice that $HSO_4^-$, like $H_2O$, has a conjugate base ($SO_4^{2-}$) as well as a conjugate acid ($H_2SO_4$).

**Common Pitfalls:** Common Pitfalls: Not recognizing that many species have both a conjugate base and a conjugate acid ($HSO_4^-$, for instance).

**Practice Exercise 14.1**

Identify the conjugate acid and the conjugate base of the dihydrogen phosphate anion ($H_2PO_4^-$) and the hydrogen phosphate anion ($HPO_4^{2-}$).

## Autohydrolysis of Water

These examples of acid–base reactions show that water can act as either an acid or a base: water accepts a proton from an HCl molecule, but it donates a proton to a $PO_4^{3-}$ anion. As an acid, water donates a proton to a base and becomes a hydroxide anion. As a base, water accepts a proton from an acid and becomes a hydronium cation. A chemical species that can both donate and accept protons is said to be **amphiprotic**. This ability to serve as both a proton donor and a proton acceptor suggests that a proton transfer equilibrium must also exist for pure water, as Figure 14-2 illustrates.

$$2\,H_2O\,(l) \rightleftharpoons OH^-\,(aq) + H_3O^+\,(aq) \qquad K_{eq}\,(= K_w) = [H_3O^+]_{eq}\,[OH^-]_{eq}$$

**Figure 14-2**

| $H_2O$ | $H_2O$ | Collision and transfer | $OH^-$ | $H_3O^+$ |
| (Acid: A hydrogen ion donor) | (Base: A hydrogen ion acceptor) | | | |

The water equilibrium illustrates the amphiprotic nature of $H_2O$. In this reaction, one water molecule acts as a proton donor (acid), and another acts as a proton acceptor (base).

The equilibrium constant for this reaction is the **water autohydrolysis equilibrium constant**, and although it is formulated like any other equilibrium constant, it is particularly important in aqueous chemistry, and so is given its own symbol, $K_w$. (As usual, since the water is a pure liquid, its activity is equal to 1, and so $[H_2O\ (l)]$ does not enter into the equilibrium expression.) The value of $K_w$ is $1.0 \times 10^{-14}$ at 25 °C, which turns out to be a particularly convenient value for calculations in acid–base chemistry.

In pure water, the hydronium and hydroxide ion concentrations are equal: we find these concentrations by taking the square root of $K_w$:

$$[H_3O^+]_{eq} = [OH^-]_{eq} = 1.0 \times 10^{-7}\ M \quad \text{(pure water at 25 °C)}$$

Equal concentrations of these two ions mean that pure water is neither acidic nor basic. We refer to such a solution as being **neutral**.

The water equilibrium also describes an inverse relationship between $[H_3O^+]_{eq}$ and $[OH^-]_{eq}$. When an acid dissolves in water, the hydronium ion concentration increases, so the hydroxide ion concentration must decrease to maintain the product of the concentrations at $1.0 \times 10^{-14}$. Similarly, the hydroxide ion concentration increases when a base dissolves in water, so the hydronium ion concentration must decrease.

## Strong Acids

An acid that quantitatively donates protons to water molecules (i.e., essentially all protons are donated to water molecules, and no intact acid molecules remain) is called a **strong acid**. Table 13-1 lists the six most common strong acids: $HNO_3$, $HClO_4$, $H_2SO_4$, $HCl$, $HBr$, and $HI$. In an aqueous solution of a strong acid, the hydronium ion concentration is equal to the concentration of the acid solution. The concentration of hydroxide ions in a solution of a strong acid can be calculated from the concentration of $H_3O^+$ and $K_w$. Example 14-2 provides an illustration.

### EXAMPLE 14-2 ION CONCENTRATIONS IN A SOLUTION OF STRONG ACID

Determine the ion concentrations in a $5.0 \times 10^{-2}$ M aqueous solution of $HClO_4$, a strong acid.

**Strategy:** The seven-step method described in Chapter 13 provides the strategy for working any quantitative problem that involves equilibria. As a reminder, we detail each step in this first example of acid–base equilibria.

1. **Determine what is asked for.** This problem asks for concentrations of all ions in an acid solution.

2. **Identify the major chemical species.** Because $HClO_4$ is a strong acid, it transfers protons to water molecules quantitatively:

$$H^+$$

$$HClO_4\ (aq) + H_2O\ (l) \longrightarrow H_3O^+\ (aq) + ClO_4^-\ (aq)$$

Thus, the major species in solution are $H_3O^+$, $ClO_4^-$, and $H_2O$.

3. **Determine what chemical equilibria exist.** The water equilibrium always exists in an aqueous solution:

$$2\ H_2O\ (l) \rightleftharpoons OH^-\ (aq) + H_3O^+\ (aq)$$

4. **Write the $K_{eq}$ expressions.** The strong acid hydrolyzes completely, so the water equilibrium is the only one that applies to this problem:

$$K_{eq} = K_w = [H_3O^+]_{eq}\,[OH^-]_{eq} = 1.0 \times 10^{-14}$$

**Solution:**

5. **Organize the data and unknowns.** This aqueous solution is $5.0 \times 10^{-2}$ M $HClO_4$, so $[H_3O^+] = [ClO_4^-] = 5.0 \times 10^{-2}$ M. The final concentrations are found using an equilibrium analysis. Set up a concentration table for the water autohydrolysis equilibrium, and define the change in hydronium ion concentration as $x$:

| Reaction | $2\ H_2O\ (l)$ | $\rightleftharpoons$ | $H_3O^+\ (aq)$ | $+$ | $OH^-\ (aq)$ |
|---|---|---|---|---|---|
| Initial Concentration (M) | | | $5.0 \times 10^{-2}$ | | 0 |
| Change in Concentration (M) | | | $+x$ | | $+x$ |
| Equilibrium Concentration (M) | | | $5.0 \times 10^{-2} + x$ | | $x$ |

6. **Substitute and calculate.** The equilibrium constant is much smaller than the initial concentration, $5.0 \times 10^{-2}$, so it is reasonable to assume that $x \ll 5.0 \times 10^{-2}$. This leads to an approximation:

$$(5.0 \times 10^{-2} + x) \cong 5.0 \times 10^{-2}$$

Solving for $x$ gives the equilibrium concentration of hydroxide ions:

$$K_w = [H_3O+]_{eq}\,[OH^-]_{eq} = (5.0 \times 10^{-2})x = 1.0 \times 10^{-14}$$

$$x = \frac{1.0 \times 10^{-14}}{5.0 \times 10^{-2}} = 2.0 \times 10^{-13} \quad \text{so} \quad [OH^-]_{eq} = 2.0 \times 10^{-13}\ M$$

**Do the Results Make Sense?** Each of the ion concentrations is equal to or smaller than the starting concentration of the acid, which is reasonable. Also, the approximation $x \ll 5.0 \times 10^{-2}$ is valid, because $x = 2.0 \times 10^{-13}$. Note also that the question includes the phrase "$5.0 \times 10^{-2}$ M aqueous solution of $HClO_4$," but the calculation shows that there is essentially no $HClO_4$ left in solution. The phrase refers to how such a solution was made (i.e., "$5.0 \times 10^{-2}$ M moles of $HClO_4$ were dissolved into 1 L of water"), not what is actually in solution when equilibrium is reached.

**Common Pitfalls:** As in any equilibrium calculation, it is important to first write down the reactions occurring in the solution.

**Practice Exercise 14.2**

Determine the ion concentrations in a solution of HCl of concentration 6.5 mM.

In any solution of an acid, the total hydronium and hydroxide ion concentrations include the $10^{-7}$ M contribution from the water reaction. This example illustrates, however, that the change in hydronium ion concentration due specifically to the water equilibrium is negligibly small in an aqueous solution of a strong acid. This is true for any strong acid at a concentration greater than $10^{-5}$ M. Consequently, the hydronium ion concentration equals the concentration of the strong acid, and the hydroxide ion concentration will be small, but it can be calculated from

$$[OH^-]_{eq} = \frac{1.0 \times 10^{-14}}{[H_3O^+]_{eq}}$$

## Strong Bases

A substance that generates hydroxide ions quantitatively in aqueous solution is a **strong base**. The most common strong bases are the soluble metal hydroxides, among which NaOH perennially ranks among the top 10 industrial chemicals (tens of millions of tonnes are produced annually worldwide). When a soluble metal hydroxide dissolves in water, it generates metal cations and hydroxide anions:

$$NaOH\ (s) \xrightarrow{H_2O} Na^+\ (aq) + OH^-\ (aq)$$

For any aqueous strong base, the hydroxide ion concentration can be calculated directly from the overall solution molarity. As is the case for aqueous strong acids, the hydronium and hydroxide ion concentrations are linked through the water equilibrium, as shown by Example 14-3.

### EXAMPLE 14-3 ION CONCENTRATIONS IN A STRONG BASE SOLUTION

What are the ion concentrations in 0.500 L of an aqueous solution that contains 5.0 g of NaOH?

**Strategy:** For this example, we summarize the first four steps of the method. The problem asks for the concentration of ions. Sodium hydroxide is a strong base that dissolves in water to generate $Na^+$ cations and $OH^-$ anions quantitatively. The concentration of hydroxide ion equals the concentration of the base. The water equilibrium links the concentrations of $OH^-$ and $H_3O^+$, so an equilibrium calculation is required to determine the concentration of hydronium ion. What remains is to organize the data, carry out the calculations, and check for reasonableness.

**Solution:** The overall molarity of NaOH can be calculated from mass, molar mass, and volume:

$$M_{NaOH} = 40.00 \text{ g/mol}$$

$$n_{NaOH} = \frac{m}{M} = \frac{5.0 \text{ g}}{40.00 \text{ g/mol}}\ 0.125 \text{ mol}$$

$$c_{NaOH} = \frac{n}{V} = \frac{0.125 \text{ mol}}{0.500 \text{ L}} = 0.25 \text{ mol/L}$$

The solution is 0.25 M in NaOH, so the initial concentrations of the two ions are $[Na^+] = [OH^-] = 0.25$ M. Use the water equilibrium to determine the concentration of $H_3O^+$.

$$2 H_2O\ (l) \rightleftharpoons OH^-\ (aq) + H_3O^+\ (aq) \qquad K_w = [H_3O^+]_{eq}[OH^-]_{eq} = 1.0 \times 10^{-14}$$

Set up a concentration table, defining the change in hydronium ion concentration as $x$:

| Reaction | 2 $H_2O$ (l) | $\rightleftharpoons$ | $H_3O^+$ (aq) | + | $OH^-$ (aq) |
|---|---|---|---|---|---|
| Initial Concentration (M) | | | 0 | | 0.25 |
| Change in Concentration (M) | | | $+x$ | | $+x$ |
| Equilibrium Concentration (M) | | | $x$ | | $0.25 + x$ |

The equilibrium constant is much smaller than 0.25, so it is reasonable to assume that $x \ll 0.25$ and make the approximation that $0.25 + x \cong 0.25$. Solving for $x$ gives the equilibrium concentration of hydronium ions:

$$K_w = 1.0 \times 10^{-14} = x(0.25 + x) \cong 0.25\,x$$

$$x \cong \frac{1.0 \times 10^{-14}}{0.25} = 4.0 \times 10^{-14} \quad \text{so}$$

$$[H_3O^+]_{eq} = 4.0 \times 10^{-14}\,M \quad \text{and} \quad [OH^-]_{eq} = 0.25\,M$$

 **Do the Results Make Sense?** Since 5.0 g of NaOH is about 0.1 mol, and the solution volume is 0.5 L, concentrations less than 1 M are reasonable. Note that $x$ is indeed much smaller than 0.250.

**Practice Exercise 14.3**

Determine the ion concentrations in a solution prepared by dissolving 68 mg of LiOH in enough water to make 125 mL of solution.

Just as for solutions of strong acids, the water equilibrium contributes negligibly to the total hydroxide ion concentration in any solution of strong base having a concentration greater than $10^{-5}$ M. Consequently, the hydroxide ion concentration equals the concentration of the strong base, and

$$[H_3O^+]_{eq} = \frac{1.0 \times 10^{-14}}{[OH^-]_{eq}}$$

## SECTION EXERCISES

● **14.1.1**  Write the net proton transfer reaction and draw a molecular picture to illustrate the reaction between $H_3O^+$ and $SO_4^{2-}$.

● **14.1.2**  What is $[OH^-]$ in a sample of acid rain that has $[H_3O^+] = 2.5 \times 10^{-4}$ M?

● **14.1.3**  A student prepares 500.0 mL of aqueous solution containing 5.61 g of KOH. Determine the concentrations of all ions in this solution.

● **14.1.4**  Identify the conjugate acid and base of (a) $HCO_3^-$ (aq) and (b) $HSO_4^-$ (aq).

## 14.2 THE pH SCALE

The hydronium ion concentration in aqueous solution ranges from extremely high to extremely low. Here are three examples:

$$6.0\,M\,HCl,\,[H_3O^+] = 6.0\,M$$

$$\text{Pure } H_2O,\,[H_3O^+] = 1.0 \times 10^{-7}\,M$$

$$0.25\,M\,NaOH,\,[H_3O^+] = 4.0 \times 10^{-14}\,M$$

We use logarithmic notation to express this huge range of concentrations. As one example, $\log (4.0 \times 10^{-14}) = \log (4.0) + \log (10^{-14}) = 0.60 + (-14) = -13.40$. Notice that a logarithm has two parts. The numbers preceding the decimal point determine the power of 10. The numbers after the decimal point determine the numerical value. Therefore, when two digits follow the decimal place of the logarithm, the corresponding numerical value has two significant figures.

Because hydronium concentrations usually involve negative powers of 10, chemists use a negative log scale in expressing these concentrations. Equation 14-1 defines the **pH** scale of acid concentration:

$$pH = -\log [H_3O^+] \tag{14-1}$$

The pH of a solution is obtained by taking the logarithm of the hydronium ion concentration and then changing the sign. For example, the pH of pure water is

$$pH = -\log [H_3O^+] = -\log (1.0 \times 10^{-7}) = -(-7.00) = 7.00$$

The conversion from pH to $[H_3O^+]$ uses powers of 10. For example, lemon juice has a pH of about 2.0:

$$[H_3O^+] = 10^{-pH} = 10^{-2.0} = 1 \times 10^{-2}\,M$$

A logarithm carries no units, but the result of this calculation is a concentration in mol/L (M).

### Features of the pH Scale

1. pH 7.00 defines a "neutral" solution, $[H_3O^+] = [OH^-] = 10^{-7}$ M, neither acidic nor basic. Acidic solutions have pH $< 7$, and basic solutions have pH $> 7$.

2. The *more acidic* the solution, the *lower* its pH: lemon juice, at pH 2, is much more acidic than acid rain, pH $\sim 4$.

3. A change in pH of one unit reflects a 10-fold change in hydronium ion concentration: normal rainfall has pH $\sim 5$, but acid rain has a 10 times larger hydronium ion concentration, pH $\sim 4$.

4. The immense range of concentrations from $>1$ M to $<10^{-14}$ M is compressed into a more convenient range, from $\sim 1$ to $\sim +14$. (Note that pH values can be less than zero, or greater than 14.)

Figure 14-3 shows the range of pH and hydronium ion concentrations. The measurement of pH is a routine operation in most laboratories. Litmus paper, which turns red when dipped in acidic solution and blue when dipped in basic solution, gives a quick, qualitative indication of acidity. As Figure 14-4 shows, approximate measures of pH can be made using pH paper. Universal pH paper displays a range of colours in response to different pH values and is accurate to about 0.5 pH unit.

For quantitative pH determinations, a pH meter is commonly used. A pH meter uses a probe to make voltage measurements between a reference solution

**Figure 14-3**

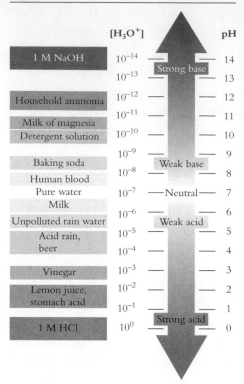

The pH values of commonly encountered aqueous solutions range from 0 to 14.

**Figure 14-4**

pH measurements can be made using universal pH paper or a pH meter.

containing acid at known concentration and the sample being tested. (The mechanism by which the hydronium ion concentration is converted into a voltage is described in section 16.5.) The pH meter is calibrated by immersing the probe in a solution of known pH and adjusting the meter to read the correct value. Then the probe is then immersed into the sample solution, and the pH appears on a digital display. Example 14-4 illustrates how to convert from pH measurements to ion concentrations.

## EXAMPLE 14-4 pH AND ION CONCENTRATIONS

What are the concentrations of hydronium and hydroxide ions in a beverage with a pH of 3.05?

**Strategy:** To convert from pH to ion concentrations, first apply Equation 14-1 to calculate $[H_3O^+]$.

$$pH = -\log [H_3O^+]$$

Then make use of the water equilibrium to calculate $[OH^-]$.

**Solution:** We must rearrange Equation 14-1, solving for the concentration:

$$\log [H_3O^+] = -pH \quad \text{and} \quad [H_3O^+] = 10^{-pH}$$

Now substitute and evaluate:

$$[H_3O^+] = 10^{-3.05} = 8.9 \times 10^{-4} \text{ M}$$

To determine the hydroxide ion concentration, we need to rearrange the water equilibrium expression:

$$K_w = 1.0 \times 10^{-14} = [H_3O^+]_{eq}[OH^-]_{eq} \quad \text{and} \quad [OH^-]_{eq} = \frac{1.0 \times 10^{-14}}{[H_3O^+]_{eq}}$$

Now substitute and evaluate:

$$[OH^-]_{eq} = \frac{1.0 \times 10^{-14}}{8.9 \times 10^{-4}} = 1.1 \times 10^{-11}\,M$$

**Do the Results Make Sense?** A pH of close to 3 represents an acidic solution, so we expect the hydronium ion concentration to be much larger than the hydroxide ion concentration. Remember that although pH and $K_w$ are dimensionless, concentrations of solutes are always expressed in mol/L. Our results have two significant figures because the logarithm has two decimal places.

**Common Pitfalls:** Not remembering that in a aqueous solution, $K_w = 1.0 \times 10^{-14} = [H_3O^+]_{eq}[OH^-]_{eq}$, whether the solution is acidic, basic, or neutral.

### Practice Exercise 14.4

According to Figure 14-3, milk of magnesia, an aqueous suspension of $Mg(OH)_2$, has a pH of about 10.5. Determine the corresponding concentrations of hydronium ions and hydroxide ions.

A logarithmic scale is useful not only for expressing hydronium ion concentrations, but also for expressing hydroxide ion concentrations. pOH is defined in the same manner as pH:

$$pOH = -\log[OH^-]$$

Use of this definition and the properties of logarithms leads to a statement of the water equilibrium constant in terms of pH:

$$K_w = [H_3O^+][OH^-] = 1.0 \times 10^{-14}$$

$$\log K_w = \log(1.0 \times 10^{-14}) = -14.00$$

$$\log\{[H_3O^+][OH^-]\} = \log[H_3O^+] + \log[OH^-] = -14.00$$

$$\text{so} \; -\log[H_3O^+] - \log[OH^-] = 14.00$$

Substituting pH = $-\log[H_3O^+]$ and pOH = $-\log[OH^-]$ gives Equation 14-2, which connects pH and pOH for any aqueous solution:

$$pH + pOH = 14.00 \qquad (14\text{-}2)$$

Example 14-5 demonstrates the usefulness of this equation.

### EXAMPLE 14-5  pH AND pOH

What is the pH of a 0.25 M solution of NaOH?

**Strategy:** To find the pH of a solution, first compute $[H_3O^+]$ or $[OH^-]$ and then apply Equation 14-1 or 14-2. Sodium hydroxide is a strong base, and the water equilibrium provides the link between hydroxide and hydronium ion concentrations.

**Solution:** In this solution of strong base, $[OH^-]$ is equal to the concentration of sodium hydroxide:

$$[OH^-] = 0.25 \text{ M}$$
$$pOH = -\log(0.25) = -(-0.60) = 0.60$$
$$pH = 14.00 - pOH = 14.00 - 0.60 = 13.40$$

Notice how much easier this calculation is than the analysis in Example 14-3. The detailed reasoning of the concentration table, $K_w$, and an approximation are compressed into two steps.

 **Does the Result Make Sense?** Verify that this result is the same as the one obtained in Example 14-3 by converting from pH to $[H_3O^+]$:

$$13.40 = pH = -\log[H_3O^+]$$
$$\log[H_3O^+] = -13.40 \quad \text{so} \quad [H_3O^+] = 10^{-13.40} = 4.0 \times 10^{-14} \text{ M}$$

**Practice Exercise 14.5**

Determine the pH of the solution of LiOH that appears in Practice Exercise 14.3.

## SECTION EXERCISES

● **14.2.1**  Determine the pH values of $1.5 \times 10^{-3}$ M HCl and $2.5 \times 10^{-2}$ M NaOH.

● **14.2.2**  The pH of rainwater in the Toronto area is approximately 4.5. Calculate the hydronium and hydroxide concentrations in this water.

**Figure 14-5**

In a solution of the weak acid HF, $H_2O$ and HF are present as major species, and $H_3O^+$ and $F^-$ are present in much lower concentrations.

## 14.3  WEAK ACIDS AND BASES

Most acids and bases are weak. A solution of a weak acid contains the acid and water as major species, and a solution of a weak base contains the base and water as major species. Proton transfer equilibria determine the concentrations of hydronium ions and hydroxide ions in these solutions. To determine the concentrations at equilibrium, we must apply the general equilibrium strategy developed in Chapter 13 to these types of solutions.

### Weak Acids: Proton Transfer to Water

Measurements of the pH of solutions of acids show that in solutions of many acids, the hydronium ion concentration is smaller than would be expected if proton transfer were quantitative. An acid that reaches equilibrium when only a small fraction of its molecules has transferred protons to water is called a **weak acid**. For example, the pH of a 0.25 M solution of HF is 1.92, giving $[H_3O^+] = 1.2 \times 10^{-2}$ M. This is substantially lower than 0.25 M (pH 0.60), the expected result from quantitative proton transfer. In other words, when HF dissolves in water, equilibrium is established when only a fraction of the HF molecules have transferred protons to water molecules:

$$2 \text{ H}_2\text{O } (l) + \text{HF } (aq) \rightleftharpoons \text{H}_3\text{O}^+ (aq) + \text{F}^- (aq)$$

In a solution of a weak acid, the *major* species are water molecules and the acid, HA. The products of the proton transfer reaction, $H_3O^+$ and $A^-$, are present in smaller concentrations as *minor* species. Figure 14-5 provides a molecular view.

The strength of a weak acid is measured by its acid hydrolysis constant, $K_a$. This equilibrium constant can be calculated from the measured pH of the solution, as illustrated in Example 14-6.

## EXAMPLE 14-6 CALCULATING $K_a$

The pH of a 0.25 M aqueous HF solution is 1.92. Calculate $K_a$ for this weak acid.

**Strategy:** Use the seven-step method. We are asked to evaluate an equilibrium constant. Visualize the molecular situation: what are the species involved in this equilibrium? After identifying the chemistry, set up and complete the appropriate concentration table and solve for the desired quantity—in this case, $K_a$.

**Solution:** The problem states that hydrogen fluoride is a weak acid, so the major species in solution are $H_2O$ and HF molecules. In aqueous solution, HF transfers protons to $H_2O$:

$$H_2O \ (l) + HF \ (aq) \rightleftharpoons H_3O^+ \ (aq) + F^- \ (aq)$$

From the pH, calculate the concentration of hydronium ions at equilibrium:

$$[H_3O^+] = 10^{-pH} = 10^{-1.92} = 1.2 \times 10^{-2} \text{ M}$$

Now construct a concentration table. The key is to use the initial and equilibrium concentrations of $H_3O^+$ to complete the "change" row. Shading indicates the concentrations that provide the starting points.

| Reaction | $H_2O \ (l)$ | + | HF (aq) | $\leftrightarrows$ | $H_3O^+$ (aq) | + | $F^-$ (aq) |
|---|---|---|---|---|---|---|---|
| Initial Concentration (M) | | | 0.25 | | 0 | | 0 |
| Change in Concentration (M) | | | $-1.2 \times 10^{-2}$ | | $+1.2 \times 10^{-2}$ | | $+1.2 \times 10^{-2}$ |
| Equilibrium Concentration (M) | | | 0.24 | | $1.2 \times 10^{-2}$ | | $1.2 \times 10^{-2}$ |

Now substitute numerical values of the concentrations and solve for $K_a$:

$$K_a = \frac{(1.2 \times 10^{-2})^2}{0.24} = 6.0 \times 10^{-4}$$

**Does the Result Make Sense?** The equilibrium concentration of hydronium ions is in the $10^{-2}$ M range, so this value for $K_a$ appears reasonable. Notice that the problem identified this solution as 0.25 M HF, but at equilibrium, [HF] = 0.24 M. **Conventionally, a solution's concentration is stated as its initial concentration, even though the equilibrium concentrations may differ from this initial concentration.**

### Practice Exercise 14.6

An aqueous solution of propanoic acid ($C_2H_5CO_2H$) with a concentration of 85 mM has pH = 2.97. Calculate $K_a$ for this weak acid.

Although it is not strongly reactive as an acid, HF has other unique reactive properties. It is almost the only common substance that reacts with glass. For this reason, artists use aqueous solutions of HF to etch glass, and it is used industrially to frost light bulbs. Because it destroys glass bottles, aqueous HF must be stored in plastic containers. Besides attacking glass, HF has a devastating effect on human nerve tissues. It is rapidly absorbed through the skin, so people who work with it must always wear HF-resistant gloves.

Aqueous hydrofluoric acid is used to etch glass.

**Figure 14-6**

The concentrations of molecular and ionic species are dramatically different in a solution of a strong acid (HCl) and a solution of a weak acid (HF).

Although proton transfer is never complete for a weak acid dissolved in water, some proton transfer always occurs, so the hydronium ion concentration in any solution of a weak acid is greater than that in pure water. Consequently, the pH of an aqueous solution of a weak acid is always less than 7.00.

Aqueous solutions of strong acids and weak acids both have pH <7, but their moleuular compositions are quite different, as Figure 14-6 displays. Both solutions are acidic, because both contain hyronium ion concentrations greater than $10^{-7}$ M. The strong acid transfers protons quantitatively, generating a high concentration of hydronium ions and leaving a negligible number of unhydrolyzed acid molecules. In contrast, the weak acid undergoes only a small amount of proton transfer, generating a much lower concentration of hydronium ions and remaining almost entirely unhydrolyzed.

As Figure 14-6 indicates, only a fraction of weak acid molecules undergoes proton transfer to form hydronium ions. Equation 14-3 defines the percent hydrolysis of any acid solution to be the percentage of the initial concentration of acid that has generated hydronium ions:

$$\%HA \text{ hydrolyzed} = 100\% \frac{[H_3O^+]_{eq}}{[HA]_{initial}} \qquad (14\text{-}3)$$

Notice that the concentration ratio in Equation 14-3 is the *equilibrium* concentration of hydronium ions divided by the *initial* concentration of the acid. The reason for this is that we are interested in how much of the starting material has undergone proton transfer.

When we know the pH and initial concentration of a weak acid solution, we can calculate the percent hydrolysis quickly. For the solution of HF in Example 14-6, we have pH 1.92, from which $[H_3O^+]_{eq} = 1.2 \times 10^{-2}$ M; and we have $[HF]_{initial} = 0.25$ M. Substituting, we find

$$\%HA \text{ hydrolyzed} = 100\% \frac{[H_3O^+]_{eq}}{[HA]_{initial}} = 100\% \left( \frac{1.2 \times 10^{-2}\,M}{0.25\,M} \right) = 4.8\%$$

Example 14-7 shows how to determine percent hydrolysis from initial concentration and an acid hydrolysis constant, $K_a$.

## EXAMPLE 14-7 DETERMINING PERCENT HYDROLYSIS

Determine the percent hydrolysis for an aqueous solution of HF that is 25 mM.

**Strategy:** To determine percent hydrolysis, we need to know the equilibrium concentration of hydronium ions. This requires an equilibrium calculation, for which we follow the seven-step method. We need to set up the appropriate equilibrium expression and solve for $[H_3O^+]_{eq}$, after which we can use Equation 14-3 to obtain the final result.

**Solution:** As in Example 14-6, the major species in solution are $H_2O$ and HF molecules. In aqueous solution, HF transfers protons to $H_2O$:

$$H_2O \ (l) + HF \ (aq) \rightleftharpoons H_3O^+ \ (aq) + F^- \ (aq)$$

The result of Example 14-6 provides the value for the equilibrium constant:

$$K_a = \frac{[H_3O^+]_{eq}[F^-]_{eq}}{[HF]_{eq}} = 6.0 \times 10^{-4}$$

Now construct a concentration table. All we know is that the initial concentration of HF is 25 mM (0.025 M):

| Reaction | $H_2O \ (l)$ + | HF (aq) | $\rightleftharpoons$ | $H_3O^+$ (aq) + | $F^-$ (aq) |
|---|---|---|---|---|---|
| Initial Concentration (M) | | 0.025 | | 0 | 0 |
| Change in Concentration (M) | | $-x$ | | $+x$ | $+x$ |
| Equilibrium Concentration (M) | | $0.025 - x$ | | $x$ | $x$ |

We substitute these values into the equilibrium constant expression, and then solve for $x$:

$$K_a = \frac{[H_3O^+]_{eq}[F^-]_{eq}}{[HF]_{eq}} = \frac{x^2}{0.025 - x} = 6.0 \times 10^{-4}$$

We need to decide if we can make the approximation that 0.025 M $- x \cong$ 0.025 M. Recall from Chapter 13 that the difference will be small enough to neglect if $K_{eq} < 10^{-4}$ and initial concentrations are greater than 0.10 M. In this case, $K_a$ is marginally larger than $10^{-4}$, but the initial HF concentration is *less than* 0.10 M. It is not clear that $x$ will be much smaller than 0.025, so we solve for $x$ using the quadratic formula:

$$x^2 = (0.025 - x)(6.0 \times 10^{-4}) \quad \text{or} \quad x^2 + (6.0 \times 10^{-4})x - 1.5 \times 10^{-5} = 0$$

$$x = \frac{-b \pm \sqrt{b^2 - 4ac}}{2a} = \frac{-(6.0 \times 10^{-4}) \pm \sqrt{(6.0 \times 10^{-4})^2 - 4(1)(-1.5 \times 10^{-5})}}{2(1)}$$

$$x = \frac{-(6.0 \times 10^{-4}) \pm \sqrt{(3.6 \times 10^{-7}) + (6.0 \times 10^{-5})}}{2(1)}$$

There are two solutions to this equation, but one gives a negative value for $x$. We know that $x$ must be positive, because it represents the equilibrium hydronium ion concentration. Hence, we solve for the positive solution:

$$x = \frac{-(6.0 \times 10^{-4}) + (7.7 \times 10^{-3})}{2} = 3.6 \times 10^{-3}$$

Thus,

$$[H_3O^+]_{eq} = x \ M = 3.6 \times 10^{-3} \ M$$

Now substitute this value and the initial concentration of HF into Equation 14-3 to determine the percent hydrolysis:

$$\%HA \text{ hydrolysis} = 100\% \frac{[H_3O^+]_{eq}}{[HA]_{initial}} = 100\% \left( \frac{3.6 \times 10^{-3} \ M}{0.025 \ M} \right) = 14\%$$

**Does the Result Make Sense?** The equilibrium concentration of hydronium ions is smaller than the starting concentration of HF, as it must be. We use two significant figures to match the precision in the $K_a$ value.

**Common Pitfalls:** Making the simplifying assumption when it is not warranted. If you are not sure, use the quadratic equation—it will always give the correct result.

**Practice Exercise 14.7**
Determine the percent hydrolysis for a solution of HF that is 1.0 M.

The results of Example 14-7 and Practice Exercise 14.7 illustrate an important feature of percent hydrolysis. As the concentration of a weak acid increases, the percent hydrolysis decreases. The concentration of hydronium ion decreases as a weak acid is diluted, but it decreases by a smaller fraction than the dilution factor. This behaviour of weak acids played an important role in the development of the ionic view of aqueous solutions, as described in our Focus 14-1: Chemical Milestones.

## Weak Bases: Proton Transfer from Water

When ammonia dissolves in water, the resulting solution has a pH greater than 7.00, indicating that the solution is basic (Figure 14-3). In other words, the concentration of hydroxide ions in aqueous ammonia is greater than that in pure water. Ammonia molecules accept protons from water molecules, generating ammonium cations and hydroxide anions:

$$NH_3 \ (aq) + H_2O \ (aq) \rightleftarrows NH_4^+ \ (aq) + OH^- \ (aq)$$

Ammonia is an example of a **weak base**. A weak base generates hydroxide ions by accepting protons from water but reaches equilibrium when only a fraction of its molecules have done so. The equilibrium constant for this type of equilibrium is designated $K_b$:

$$K_b = \frac{[NH_4^+]_{eq}[OH^-]_{eq}}{[NH_3]_{eq}}$$

Example 14-8 explores the ammonia equilibrium in more detail.

## FOCUS 14-1  CHEMICAL MILESTONES: ARRHENIUS AND THE IONIC THEORY

The existence of ions in aqueous solutions was first proposed by Svante Arrhenius, a young Swedish chemist, during the 1880s, well before the electronic structure of atoms had been discovered. This insight came while Arrhenius was pursuing his Ph.D. in chemistry, exploring why aqueous solutions conduct electricity.

The types of conductivity measurements carried out by Arrhenius were much like the demonstrations of electrical conductivity used today. Using a voltage source and a current meter, one can determine how much

current a solution carries, and this is proportional to its conductivity. A light bulb can serve as a qualitative current meter. As the photos show, solutes fall into three categories. A solution of sugar (left) does not conduct, a solution of NaCl (right) conducts extremely well, and a solution of acetic acid (center) conducts only a little. When Arrhenius began his work, ions were called "electrolytic molecules," so the categories were termed non-electrolytes, strong electrolytes, and weak electrolytes.

While Arrhenius was studying conductivity, others were characterizing colligative properties of solutions. The Dutch chemist J. T. van't Hoff studied osmotic pressure and derived the law of osmotic pressure, $\Pi V = nRT$. van't Hoff noted the parallel between this law and the ideal gas equation, and he proposed that solute molecules in solution act independently of one another. van't Hoff's law worked for solutions of non-electrolytes and weak electrolytes, but for strong electrolytes, van't Hoff had to multiply $n$ by a coefficient, $i$. For HCl and NaCl, the value of $i$ was close to 2, and $i$ was close to 3 for $CaCl_2$. For this reason, strong electrolytes were considered to be exceptions to van't Hoff's law.

From his conductivity measurements on solutions, Arrhenius concluded that strong electrolytes are not exceptions. Instead, they dissociate into ions. When $i = 2$, it meant that each solute species dissociated to give two ions. A compound with $i = 3$ dissociated to give three ions. Moreover, interpreting the results of his experiments at varying levels of concentration, Arrhenius concluded that at sufficiently high dilution, every electrolyte becomes fully dissociated.

Arrhenius devised demonstrations to support his ideas. The diagrams illustrate two of these. Layers of acetic acid solution and pure water, separated by a thin layer of concentrated sugar solution, show low conductivity. When the layers are mixed, the conductivity *increases*. To Arrhenius, this demonstrated that diluting acetic acid leads to more hydrolysis into ions. Layers of solutions of silver sulphate and barium chloride conduct well. When these layers are mixed, precipitates of silver chloride and barium sulphate form, and the light goes out. To Arrhenius, this showed that ions were present in solutions of salts, and that these ions were removed by precipitation.

Arrhenius knew that these ideas were revolutionary. He avoided mentioning "ions" in his thesis, for fear his examiners would reject the thesis. Even so, his committee gave him the lowest possible passing grade. In desperation, Arrhenius sent copies of his work to famous professors in several countries. Wilhelm Ostwald, a brilliant chemist, recognized the genius of the ion hypothesis and travelled to Sweden to meet Arrhenius.

Still, the chemical establishment remained opposed to the notion of ions in solution. In an attempt to convince "the wild army of Ionians" of how wrong their ideas were, the British Association scheduled a discussion titled "Theories of Solution," and invited van't Hoff, Arrhenius, and Ostwald to present their views. The rest of the discussion was packed with conservative older chemists, the idea being that reason would prevail and the Ionians would give up their views. Instead, most of the younger chemists sought out the Ionians for spirited exchanges, while the old chemists delivered their lectures to nearly empty rooms.

By 1895, nearly all chemists accepted Arrhenius's views, and Arrhenius finally was awarded a professorship in Stockholm. In 1903, the triumph of Arrhenius became complete, when he was awarded the Nobel Prize for chemistry. The Nobel Prize committee wrote that Arrhenius' award was given "in recognition of the extraordinary services he has rendered to the advancement of chemistry by his electrolytic theory of dissociation."

## EXAMPLE 14-8 pH OF A WEAK BASE

Ammonia has $K_b = 1.8 \times 10^{-5}$. What is the pH of 0.25 M aqueous ammonia?

**Strategy:** Use the seven-step procedure. We want to determine pH, for which we need to know the equilibrium concentration of either $H_3O^+$ or $OH^-$. The major species present in aqueous ammonia are molecules of $NH_3$ and $H_2O$. Both of these compounds produce hydroxide ions as minor species in solution:

$$NH_3\,(aq) + H_2O\,(l) \rightleftarrows NH_4^+\,(aq) + OH^-\,(aq) \qquad K_b = 1.8 \times 10^{-5}$$

$$2\,H_2O\,(l) \rightleftarrows H_3O^+\,(aq) + OH^-\,(aq) \qquad K_w = 1.0 \times 10^{-14}$$

These two equilibria are linked by the fact that the solution can have only one hydroxide ion concentration. Both equilibria must be satisfied, but which reaction should we use to calculate $[OH^-]_{eq}$? Notice that both equilibrium constants are much less than 1, but $K_b$ for ammonia is more than a billion times larger than $K_w$. This indicates that almost all the hydroxide ions in solution come from the ammonia equilibrium. Thus, the ammonia reaction is the appropriate choice for the calculation.

**Solution:** Determine $[OH^-]_{eq}$ by setting up a concentration table, solving the equilibrium expression for the unknown, and finding $[OH^-]_{eq}$. We know the initial concentrations but must identify $x$ as the change in concentration needed to reach equilibrium.

| Reaction | $H_2O\,(l)$ + | $NH_3\,(aq)$ | $\rightleftarrows$ | $NH_4^+\,(aq)$ + | $OH^-\,(aq)$ |
|---|---|---|---|---|---|
| Initial Concentration (M) | | 0.25 | | 0 | 0 |
| Change in Concentration (M) | | $-x$ | | $+x$ | $+x$ |
| Equilibrium Concentration (M) | | $0.25 - x$ | | $x$ | $x$ |

$$K_b = 1.8 \times 10^{-5} = \frac{x^2}{0.25 - x} \cong \frac{x^2}{0.25}$$

$$x^2 = 4.5 \times 10^{-6} \quad \text{so} \quad x = 2.1 \times 10^{-3}$$

The table shows that $[OH^-]_{eq} = x$, so $[OH^-]_{eq} = 2.1 \times 10^{-3}$ M.

Complete the problem using Equation 14-2 to calculate the pH of the solution:

$$pOH = -\log[OH^-]_{eq} = -\log(2.1 \times 10^{-3}) = 2.68$$

$$pH = 14.00 - pOH = 14.00 - 2.68 = 11.32$$

 **Does the Result Make Sense?** Note that $x \ll 0.25$, validating the approximation $(0.25 - x) \cong 0.25$. A pH greater than 7 but less than 14 indicates a basic solution, so this is a reasonable result.

**Common Pitfalls:** Neglecting the autohydrolysis of water. In this example, this was not significant, but you should always check.

**Practice Exercise 14.8**

Determine the pH of 85 mM aqueous ammonia.

The results of Examples 14-5 and 14-8 illustrate the differences between strong and weak bases:

| Strong Base | Weak Base |
|---|---|
| 0.25 M NaOH (aq) | 0.25 M $NH_3$ (aq) |
| $[OH^-]_{eq} = 0.25$ M | $[OH^-]_{eq} \times 2.1 \times 10^{-3}$ M |
| pH = 13.40 | pH = 11.32 |

The strong base is a soluble hydroxide that ionizes completely in water, so the concentration of OH⁻ matches the 0.25 M concentration of the base. For the weak base, in contrast, the equilibrium concentration of OH⁻ is substantially smaller than the 0.25 M concentration of the base. At any instant, only 0.8% of the ammonia molecules have accepted protons from water molecules, producing a much less basic solution in which OH⁻ is a minor species. The equilibrium concentration of unprotonated ammonia is nearly equal to the initial concentration. Figure 14-7 summarizes these differences.

Examples 14-6 through 14-8 illustrate the two main types of equilibrium calculations as they apply to solutions of acids and bases. Notice that the techniques are the same as those introduced in Chapter 13 and applied to weak acids in Examples 13-10 and 13-12. We can calculate values of equilibrium constants from a knowledge of concentrations at equilibrium (Examples 13-10 and 14-6), and we can calculate equilibrium concentrations from a knowledge of equilibrium constants and initial concentrations (Examples 13-12, 14-7, and 14-8).

**Figure 14-7**

The concentrations of molecular and ionic species are dramatically different in a solution of a strong base (NaOH) and a solution of a weak base ($NH_3$).

## SECTION EXERCISES

●**14.3.1** A $4.8 \times 10^{-2}$ M solution of hypochlorous acid (HClO) has a pH of 4.36. Calculate the percent hydrolysis and $K_a$ for this acid.

●**14.3.2** Draw a molecular picture showing the proton transfer reaction that takes place in an aqueous solution of HClO.

●**14.3.3** Construct a table comparing the concentrations of species in solutions of HCl and HClO, each of which is $4.8 \times 10^{-2}$ M.

## 14.4 RECOGNIZING ACIDS AND BASES

Learning to recognize the properties of a substance by examining a chemical formula is an important part of mastering chemistry. Fortunately, common acids and bases fall into a small number of structural categories. In this section, we describe how to recognize acids and bases. In Section 14.6, we explain how chemical structure influences acidity.

### Oxoacids

One general type of acid contains an inner atom bonded to a variable number of oxygen atoms and acidic OH groups. Five common examples of these **oxoacids** are shown in Figure 14-8. An oxoacid can usually be recognized by its formula. The general formula is $H_xEO_y$, where $x = 1$–3 and $y = 1$–4. The inner atom, $E$, can

**Figure 14-8**

| HNO$_3$ | H$_2$CO$_3$ | H$_3$PO$_4$ | H$_2$SO$_4$ | HClO$_2$ |
| Nitric acid | Carbonic acid | Phosphoric acid | Sulfuric acid | Chlorous acid |

Oxoacids have the general formula H$_x$EO$_y$ ($x = 1$–3, $y = 1$–4).

be a number of different elements, including B, C, N, P, S, and Cl. Hydrogen atoms of an oxoacid are listed first in the formula to emphasize that these compounds are acids. In almost all cases, the acidic hydrogen atoms are bonded to oxygen atoms, not to the inner atom $E$. Three of the strong acids, HNO$_3$, H$_2$SO$_4$, and HClO$_4$, are oxoacids. Most other oxoacids are weak, as shown by the examples in Table 14-1.

## TABLE 14-1 SOME WEAK OXOACIDS

| Name | Formula | $K_a$ | Conjugate Base |
|------|---------|-------|----------------|
| Chlorous acid | HClO$_2$ | $1.1 \times 10^{-2}$ | Chlorite, ClO$_2^-$ |
| Nitrous acid | HNO$_2$ | $5.6 \times 10^{-4}$ | Nitrite, NO$_2^-$ |
| Hypochlorous acid | HClO | $4.0 \times 10^{-8}$ | Hypochlorite, ClO$^-$ |
| Hypoiodous acid | HIO | $3.2 \times 10^{-11}$ | Hypoiodite, IO$^-$ |
| Iodic acid | HIO$_3$ | $1.7 \times 10^{-1}$ | Iodate, IO$_3^-$ |

## RULES FOR NAMING OXOACIDS

1. An oxoacid that forms from a polyatomic anion having a name ending in *-ate* has a name ending in *-ic*. For example, HNO$_3$ forms by adding a proton to the nitrate polyatomic anion, so HNO$_3$ is nitric acid. Likewise, HClO$_4$ is perchloric acid from the perchlorate anion.
2. An acid that forms from a polyatomic anion having a name ending in *-ite* has a name ending in *-ous*. For example, HNO$_2$ forms by adding a proton to the nitrite polyatomic anion, so HNO$_2$ is nitrous acid. Likewise, HClO is hypochlorous acid from the hypochlorite anion.

## Carboxylic Acids

A **carboxylic acid** can be represented as $RCO_2H$, in which the symbol $R$ can be as simple as a hydrogen atom or a methyl group (CH$_3$) or as elaborate as a structure containing dozens of atoms. The $R$ group of a carboxylic acid almost always contains C—H bonds, but hydrogen atoms bonded to carbon are *not* acidic. These atoms do not participate in aqueous proton transfer reactions. All carboxylic acids are weak. In an aqueous solution at equilibrium, a small fraction of the carboxylic acid molecules have undergone proton transfer to water molecules, generating hydronium ions and anions that contain the $-CO_2^-$ *carboxylate* group. An example is acetic acid:

$$CH_3CO_2H \ (aq) + H_2O \ (l) \rightleftarrows CH_3CO_2^- \ (aq) + H_3O^+ \ (aq)$$

Acetic acid        Acetate anion

The name of a carboxylic acid always ends in *-ic*, and the name of the anion ends in *-ate*. Figure 14-9 shows three simple carboxylic acids: formic acid ($K_a = 1.8 \times 10^{-4}$), acetic acid ($K_a = 1.8 \times 10^{-5}$), and benzoic acid ($K_a = 6.4 \times 10^{-5}$). $HCO_2^-$ is formate, $CH_3CO_2^-$ is acetate, and $C_6H_5CO_2^-$ is benzoate. Examples 13-10 and 13-12 treat benzoic acid quantitatively.

**Figure 14-9**

Formic acid
$HCO_2H$

Acetic acid
$CH_3CO_2H$

Benzoic acid
$C_6H_5CO_2H$

Ball-and-stick models of three common carboxylic acids, with acidic carboxyl groups highlighted.

## Other Acids

A number of small molecules that are acids do not fall into the categories mentioned above. These acids have no clear patterns in their structures, so it is best simply to learn their names and structures. Three hydrogen halides, HCl, HBr, and HI, are strong acids, but the fourth, HF, is a weak acid. Hydrogen sulphide ($H_2S$) and hydrocyanic acid (HCN) are weak acids that also happen to be quite poisonous. Table 14-2 summarizes the acid–base properties of these and other relatively

## TABLE 14-2 SOME WEAK ACIDS

| Name | Formula | $K_a$ | Lewis Structure |
|---|---|---|---|
| Hydrofluoric acid | HF | $6.3 \times 10^{-4}$ | :F̈—H |
| Hydrazoic acid | $HN_3$ | $2.5 \times 10^{-5}$ | :N̈=N=N̈—H |
| Hydrogen sulphide | $H_2S$ | $8.9 \times 10^{-8}$ | H—S̈—H |
| Hydrocyanic acid | HCN | $6.2 \times 10^{-10}$ | :N≡C—H |
| Hydrogen peroxide | $H_2O_2$ | $2.4 \times 10^{-12}$ | H—Ö—Ö—H |

small weak acids, and Example 14-9 provides some practice in identifying acids. **Polyprotic acids** (acids having more than one acidic hydrogen) are dealt with in Section 14.7.

## EXAMPLE 14-9 IDENTIFYING ACIDS

Examine the following formulas. Decide if each represents a strong acid, a weak acid, or neither. Justify your conclusions. (a) $Cl_3CCO_2H$; (b) $CH_3CH_2CH_2OH$; (c) HCN; and (d) $HClO_4$.

**Strategy:** We identify acids by recognizing the formulas of oxoacids, carboxylic acids, or one of the few other acids listed in Table 14-2.

**Solution:**

(a) The formula ends with a carboxyl group ($—CO_2H$). The $Cl_3C$ fragment fills the role of the $R$ group in the general formula of a carboxylic acid, $RCO_2H$. All carboxylic acids are weak.

(b) There is an $—OH$ fragment at the end of the formula, but it is not part of a carboxyl group. This compound is an alcohol, not a carboxylic acid.

(c) The hydrogen atom listed first in the formula suggests this compound is an acid. This is hydrocyanic acid, listed in Table 14-2. Since it is not one of the six common strong acids, we can conclude that HCN is a weak acid.

(d) $HClO_4$ matches the general formula of an oxoacid, $H_xEO_y$, where $x = 1$, $y = 4$, and $E = Cl$. This compound is perchloric acid, one of the six common strong acids.

   **Do the Results Make Sense?** The Lewis structures of these molecules confirm the assignments. The acidic hydrogen atoms are highlighted.

**Practice Exercise 14.9**

Decide if each of the following formulas represents a strong acid, a weak acid, or neither. Justify your conclusions. (a) HF; (b) $HCO_2H$; (c) $(CH_3)_2O$; and (d) $H_2CO_3$.

## Weak Bases

Besides water, the most common weak base is ammonia ($NH_3$). Many other weak bases are derivatives of ammonia called **amines**. In these organic compounds, one, two, or three of the N—H bonds in ammonia have been replaced with N—C bonds. The nitrogen atom in an amine, like its counterpart in ammonia, has a lone pair of electrons that can form a bond to a proton. Water does not protonate an amine to an appreciable extent, so all amines are weak bases. Table 14-3 lists several examples of bases derived from ammonia.

Methylamine    Dimethylamine    Trimethylamine

Pyridine

Aniline    Hydrazine    Hydroxylamine

## TABLE 14-3 REPRESENTATIVE ORGANIC BASES

| Name | Formula | $K_b$ |
|---|---|---|
| Methylamine | $CH_3NH_2$ | $4.6 \times 10^{-4}$ |
| Dimethylamine | $(CH_3)_2NH$ | $5.4 \times 10^{-4}$ |
| Trimethylamine | $(CH_3)_3N$ | $6.5 \times 10^{-5}$ |
| Hydrazine | $H_2NNH_2$ | $1.3 \times 10^{-6}$ |
| Hydroxylamine | $HONH_2$ | $8.7 \times 10^{-9}$ |
| Pyridine | $C_5H_5N$ | $1.7 \times 10^{-9}$ |
| Aniline | $C_6H_5NH_2$ | $7.4 \times 10^{-10}$ |

Example 14-10 shows how the acid–base properties of a molecule can be determined from formulas and structures.

## EXAMPLE 14-10 ACIDIC AND BASIC PROPERTIES OF AMINO ACIDS

Lysine and aspartic acids are two of the amino acids found in proteins. Describe the acid–base characteristics of these compounds.

Lysine                    Aspartic acid

**Strategy:** The structures of the amino acids reveal any functional groups that result in acidic or basic properties. To determine the acid–base properties of these compounds, look for carboxyl groups and amine groups.

**Solution:** Being amino acids, lysine and aspartic acid each has a carboxyl group ($CO_2H$) at one end and an amine group ($NH_2$) at the other. Lysine has a second amine group in the side chain and aspartic acid has a second carboxyl group.

**Do the Results Make Sense?** Remember that each functional group retains its characteristics, even if other functional groups are present in a molecule. The amine groups of these molecules can accept protons, and the carboxyl groups of the molecules can donate protons.

### Practice Exercise 14.10

Describe the acidic and basic characteristics of the amino acids histidine and threonine.

Histidine                    Threonine

The amino acids, basic building blocks of proteins, all share this dual acid–base character. See Chapter 17 for a description of the amino acids and their biological chemistry. Organic bases also have a long and varied history as painkillers and narcotics, as our following Focus 14-2: Chemistry and Life describes.

## FOCUS 14-2  CHEMISTRY AND LIFE: DRUGS AND THE BRAIN

For thousands of years, humans have used plants to make medicines and drugs. Many of the biologically active ingredients in plant extracts contain at least one nitrogen atom with a lone pair of electrons, making them weak bases. Such naturally occurring weak bases are classified as alkaloids. Modern pharmaceutical chemists have built upon nature's chemistry by synthesizing new compounds that are important medicines.

Coniine
$(C_8H_{17}N)$

Morphine
$(C_{17}H_{19}NO_3)$

Quinine
$(C_{20}H_{24}N_2O_2)$

Atropine
$(C_{17}H_{23}NO_3)$

The profound physiological effects of alkaloids have been known for centuries. For example, Socrates was put to death with an extract of hemlock, which contains a poisonous alkaloid, coniine. Other alkaloids have long been valued for their beneficial medical effects. Examples include morphine (a painkiller), quinine (used to treat malaria), and atropine (used to treat Parkinson's disease and in eye drops that dilate the pupils).

Because of their basic properties, alkaloids were among the first natural substances that early chemists extracted and purified. Morphine was isolated from poppies in 1805 and was the first alkaloid to be characterized. When treated with aqueous strong acid, alkaloids accept protons to produce water-soluble cations. The protonated alkaloids dissolve, leaving the rest of the plant materials behind. Adding a strong base to the aqueous extract reverses the proton transfer reaction, converts the alkaloid back to its neutral base form, and causes pure alkaloid to precipitate from the solution:

$$\text{Alkaloid } (s) + H_3O^+ (aq) \rightleftarrows \text{Alkaloid } H^+ (aq) + H_2O (l)$$

$$\text{Alkaloid } H^+ (aq) + OH^- (aq) \rightleftarrows \text{Alkaloid } (s) + H_2O (l)$$

Most of us have had to endure the discomfort of dental work, be it repairing cavities or more elaborate oral surgery. Dentists reduce the discomfort immensely by the use of a local painkiller, Novocain, to block the nerves in the mouth. Novocaine is a weak base containing two nitrogen atoms. A related analgesic, lidocaine (also known as Xylocaine), is sufficiently strong that applying it to the skin causes the nerves in the immediate area to shut down temporarily.

Procaine
(Novocain)
$(C_{13}H_{20}N_2O_2)$

A variety of drugs have been developed that act as sedatives, antidepressants, or stimulants; some of these are effective in treating psychiatric disorders. Many of these drugs are weak bases. Examples are barbiturates such as phenobarbital, tranquilizers like diazepam (Valium), and amphetamines derived from phenylethylamine.

Although nitrogen-containing drugs can be highly beneficial, others—both natural and synthetic—are notorious narcotics. Among alkaloids, morphine is responsible for the narcotic effects of opium and is also the starting material for the synthesis of codeine and heroin. Three other well-known narcotic alkaloids are nicotine, mescaline, and cocaine. Synthetic drugs as well as alkaloids can be abused. Overuse of barbiturates, tranquilizers, or amphetamines can lead to addiction or even death. The darkest side of these substances may be the so-called "designer drugs" such as Ecstasy, which alter brain chemistry in unpredictable and sometimes devastating ways.

Pentobarbital
(Nembutal)
$(NaC_{11}H_{17}N_2O_3)$

Diazepam
(Valium)
$(C_{16}H_{13}ClN_2O)$

Nicotine
$(C_{10}H_{14}N_2)$

Mescaline
$(C_{11}H_{17}NO_3)$

Cocaine
$(C_{17}H_{21}NO_4)$

Ecstasy or MDMA
$(C_{11}H_{15}NO_2)$

## SECTION EXERCISES

● **14.4.1** Classify each of the following substances as a weak acid, strong acid, weak base, strong base, or neither an acid nor a base: (a) $HClO_4$; (b) $NaOH$; (c) $CH_3OH$; (d) $C_2H_5CO_2H$; and (e) $C_2H_5NH_2$.

● **14.4.2** Draw ball-and-stick models for the conjugate acid–base pair of each weak acid and weak base in Section Exercise 14.4.1.

● **14.4.3** Identify the acid–base properties of nicotine and atropine, the structures of which appear below:

Nicotine
$(C_{10}H_{14}N_2)$

Atropine
$(C_{17}H_{23}NO_3)$

## 14.5  ACIDIC AND BASIC SALTS

Some compounds that are not immediately recognizable as acids and bases nevertheless display acid–base properties. In this section, we describe the acid–base chemistry of salts.

### Salts of Weak Acids

An aqueous solution of a soluble salt contains cations and anions, which may have acid–base properties. Anions that are conjugate bases of weak acids make a solution basic. For example, sodium fluoride dissolves in water to give $Na^+$, $F^-$, and $H_2O$ as major species. The fluoride anion is the conjugate base of the weak acid HF. This anion establishes a proton transfer equilibrium with water:

$$F^-(aq) + H_2O\,(l) \rightleftarrows HF(aq) + OH^-(aq) \quad K_{eq} = K_{b,HF} = \frac{[OH^-]_{eq}[HF]_{eq}}{[F^-]_{eq}}$$

Collision and proton transfer

The reaction generates hydroxide anions, so the solution is basic. Fluoride acts as a base, so the equilibrium constant is the base hydrolysis constant of HF, $K_{b,HF}$.

The fluoride ion equilibrium is linked to two other proton transfer equilibria. Combining the proton transfer equilibrium for HF with the proton transfer equilibrium for F reveals the relationship:

$$HF(aq) + H_2O\,(l) \rightleftarrows F^-(aq) + H_3O^+(aq) \quad K_a$$
$$H_2O\,(l) + F^-(aq) \rightleftarrows OH^-(aq) + HF(aq) \quad K_b$$
$$\overline{\phantom{xxxx}2H_2O\,(l) \rightleftarrows OH^-(aq) + H_3O^+(aq) \quad K_w}$$

Combining these two equilibria leads to cancellation of HF and $F^-$, so the sum of the two is the water equilibrium. How are the equilibrium constants for these three equilibria related? Whenever two or more equilibria are added, the equilibrium constant for the net reaction is the product of the individual equilibrium constants of the summed reactions. The expressions for the three equilibrium constants are

$$K_a = \frac{[H_3O^+]_{eq}[F^-]_{eq}}{[HF]_{eq}} \quad K_b = \frac{[OH^-]_{eq}[HF]_{eq}}{[F^-]_{eq}} \quad K_w = [H_3O^+]_{eq}[OH^-]_{eq}$$

Multiplying the expressions for $K_a$ and $K_b$ leads to cancellation of the concentrations of $[F^-]$ and $[HF]$, just as concentrations cancel when the individual reactions are added:

$$\frac{[H_3O^+]_{eq}[F^-]_{eq}}{[HF]_{eq}}\frac{[OH^-]_{eq}[HF]_{eq}}{[F^-]_{eq}} = [H_3O^+]_{eq}[OH^-]_{eq}$$

This can be generalized to any system:

$$K_a K_b = K_w \qquad (14\text{-}4)$$

Equation 14-4 applies to any acid and its conjugate base. The equation can also be expressed in logarithmic form using $pK$ notation and $pK_w = 14.00$. Note that, in analogy with pH, we can express the magnitudes of $K_a$ or $K_b$ as logarithmic quantities:

$$pK_a = -\log_{10}(K_a) \quad \text{and} \quad pK_b = -\log_{10}(K_b)$$

Since $K_a K_b = K_w$, it is easy to show that

$$pK_a + pK_b = 14.00 \qquad (14\text{-}5)$$

Because $K_a$ for HF and $K_w$ are known, Equation 14-4 can be used to determine $K_b$ for the fluoride anion. From Example 14-6, $K_a = 6.0 \times 10^{-4}$ for HF, and as always, $K_w = 1.0 \times 10^{-14}$. Thus,

$$K_b = \frac{K_w}{K_a} = \frac{1.0 \times 10^{-14}}{6.0 \times 10^{-14}} = 1.7 \times 10^{-11}$$

Notice that both $K_b$ and $K_a$ are much smaller than 1. This reflects the fact that the conjugate base of a weak acid is itself weak: HF is a weak acid, and $F^-$ is a weak base. Nevertheless, the inverse relationships in Equations 14-4 and 14-5 indicate that the stronger an acid is, the weaker is its conjugate base, and the weaker an acid is, the stronger is its conjugate base.

Example 14-11 shows how to deal quantitatively with the basicity of a salt solution.

## EXAMPLE 14-11 SALT OF A WEAK ACID

Sodium hypochlorite (NaOCl) is the active ingredient in laundry bleach. Typically, bleach contains 5.0% of this salt by mass, which is a 0.67 M solution. Determine the concentrations of all species and calculate the pH of laundry bleach.

**Strategy:** This is a quantitative acid–base equilibrium problem, so we use the seven-step method.

1. We are asked to determine pH, which requires that we find the hydronium ion concentration.

2. Sodium hypochlorite is a salt, so its aqueous solution contains the ionic species $Na^+$ and $OCl^-$. Thus, the major species in solution are $Na^+$, $OCl^-$, and $H_2O$.

3. Water is a proton donor and a proton acceptor. The sodium ion is neither an acid nor a base, so it is a spectator ion in this solution. Hypochlorite is the conjugate base of the weak oxoacid, HOCl. The $OCl^-$ anion accepts a proton from a water molecule:

$$OCl^- (aq) + H_2O\ (l) + \;\rightleftharpoons\; HOCl\ (aq) + OH^-\ (aq) \qquad K_{eq} = K_b(OCl^-)$$

**Solution:**

4. Determine $K_b$ for $OCl^-$ from the information given in Table 14-1:

$$K_a(HOCl) = 4.0 \times 10^{-8}$$

$$K_b(OCl^-) = \frac{K_w}{K_a} = \frac{1.0 \times 10^{-14}}{4.0 \times 10^{-8}} = 2.5 \times 10^{-7}$$

5. Set up a concentration table for this equilibrium. The initial concentration of $OCl^-$ is given, and $x$ can represent the change in concentration of $OCl^-$:

| Reaction | $H_2O$ | + | $OCl^-$ | $\leftrightarrows$ | $HOCl$ | + | $OH^-$ |
|---|---|---|---|---|---|---|---|
| Initial Concentration (M) | | | 0.67 | | 0 | | 0 |
| Change in Concentration (M) | | | $-x$ | | $+x$ | | $+x$ |
| Equilibrium Concentration (M) | | | $0.67 - x$ | | $x$ | | $x$ |

6. Solve the equilibrium constant expression to find the desired concentrations.

$$K_b = \frac{[HOCl]_{eq}[OH^-]_{eq}}{[OCl^-]_{eq}} = \frac{(x)(x)}{0.67 - x} = 2.5 \times 10^{-7}$$

Now make the approximation that $x \ll 0.67$:

$$0.67 - x \cong 0.67 \quad \text{and} \quad 2.5 \times 10^{-7} \cong \frac{x^2}{0.67}$$

$$x^2 \cong 1.68 \times 10^{-7} \quad \text{from which} \quad x \cong 4.1 \times 10^{-4}$$

The value of $x$ provides three of the required concentrations:

$$[OCl^-] = (0.67 - x)\,M = (0.67 - 4.1 \times 10^{-4})\,M = 0.67\,M$$

$$[OH^-] = [HOCl] = x\,M = 4.1 \times 10^{-4}\,M$$

The sodium ion concentration is the initial concentration of the salt:

$$[Na^+] = 0.67\,M$$

To find $[H_3O^+]$, rearrange the expression for $K_w$:

$$[H_3O^+] = \frac{K_w}{[OH^-]} = \frac{1.0 \times 1.0^{-14}}{4.1 \times 10^{-14}} = 2.4 \times 10^{-11}\,M$$

Finally, use $[H_3O^+]$ to determine the pH of the bleach solution:

$$pH = -\log[H_3O^+] = -\log(2.4 \times 10^{-11}) = 10.62$$

 **Does the Result Make Sense?** The value of $x$ is three orders of magnitude smaller than 0.67, so the approximation is valid. We know that the hypochlorite anion is the conjugate base of hypochlorous acid, so a pH > 7 is a reasonable result.

**Common Pitfalls:** Not recognizing that a species such as $OCl^-$ (aq) is a base. Keep the concept of conjugates in your mind when doing acid–base problems.

**Practice Exercise 14.11**

Potassium nitrite is used as a food preservative. Determine the pH and concentrations of all species in an aqueous solution of $KNO_2$ that has a concentration of 0.22 M (see $K_a$ values in Table 14-1).

## Salts of Weak Bases

Salts that contain cations of weak bases are acidic. For example, the ammonium cation is the conjugate *acid* of ammonia. When ammonium salts dissolve in water, $NH_4^+$ ions transfer protons to $H_2O$ molecules, generating $H_3O^+$ and making the solution slightly acidic:

$$NH_4^+ \ (aq) + H_2O \ (l) \rightleftharpoons H_3O^+ \ (aq) + NH_3 \ (aq)$$

The equilibrium constant for this reaction can be calculated from Equation 14-4 and $K_b$ for ammonia (Example 14-8):

$$K_a = \frac{K_w}{K_b} = \frac{1.0 \times 10^{-14}}{1.8 \times 10^{-5}} = 5.6 \times 10^{-10}$$

The salt of a weak base, pyridinium chloride, is the subject of Example 14-12.

---

### EXAMPLE 14-12 SALT OF A WEAK BASE

What are the important acid–base equilibria in an aqueous solution of pyridinium chloride ($C_5H_5NHCl$)? What are the values of their equilibrium constants?

Pyridine
($C_5H_5N$)

Pyridinium ion
($C_5H_5NH^+$)

**Strategy:** This is a qualitative problem. First identify the major species, and then list their equilibria. Values for equilibrium constants can be found in tables in this chapter and in Appendix E.

**Solution:** Pyridinium chloride is a salt that generates ions in solution. The major species are the pyridinium cation ($C_5H_5NH^+$), $Cl^-$, and $H_2O$. The formula identifies the pyridinium cation as the conjugate acid of the weak base, pyridine (see Table 14-3). There are two acid–base equilibria with major species as reactants:

$$C_5H_5NH^+ \ (aq) + H_2O \ (l) \rightleftharpoons H_3O^+ \ (aq) + C_5H_5N \ (aq) \qquad K_a$$

$$H_2O \ (l) + H_2O \ (l) \rightleftharpoons H_3O^+ \ (aq) + OH^- \ (aq) \qquad K_w = 1.0 \times 10^{-14}$$

$K_b$ for pyridine, listed in Table 14-3, is used to calculate $K_a$:

$$K_b(C_5H_5N) = 1.7 \times 10^{-9}$$

$$K_a = \frac{K_w}{K_b} = \frac{1.0 \times 10^{-14}}{1.7 \times 10^{-9}} = 5.9 \times 10^{-6}$$

This $K_a$ value is much greater than the value of $K_w$, so the weak acid, not water, determines the pH of a pyridinium chloride solution.

 **Does the Result Make Sense?** The water equilibrium seldom is the dominant one when a solution contains any solute with acid–base properties, so it is reasonable that the pH is determined by $K_a$.

**Practice Exercise 14.12**

Identify the major species and the dominant equilibrium for an aqueous solution of anilinium chloride (see Table 14-3).

In contrast to the acid–base character of ions arising from weak acids or bases, cations and anions resulting from strong acids or bases do not affect solution pH. The sodium ions in a solution of NaF do not participate in proton transfer reactions, nor do the chloride ions in a solution of pyridinium chloride. Consequently, the water equilibrium determines the pH of a solution of a salt containing neither weak acid nor weak base components. Consider as an example an aqueous solution of sodium chloride. The major ionic species present are $Na^+$ and $Cl^-$. Neither of these ions contains a hydrogen atom, so neither is a proton donor. The sodium cation has no lone pairs of electrons, so it is not a base. The chloride anion does have lone pairs, but $Cl^-$ is the conjugate base of a *strong* acid, HCl. Remember that a strong acid donates its protons to water quantitatively. This means that the chloride anion has a negligible tendency to accept protons from hydronium ions in aqueous solution. Because water is a much weaker acid than the hydronium ion, $Cl^-$ will not accept protons from water. Both $Na^+$ and $Cl^-$ are spectator ions, and the only important proton transfer equilibrium in an aqueous solution of NaCl is the water equilibrium.

## Summarizing Acids and Bases

In the last two sections we have described several different categories of strong and weak acids and bases. Table 14-4 summarizes the characteristics of these species.

## TABLE 14-4 RECOGNIZING ACIDS AND BASES

| Recognizing Acids |
| --- |
| **1. Strong Acids.** Memorize the formulas and names of the six common strong acids. |

| | | |
| --- | --- | --- |
| HCl (hydrochloric acid) | HBr (hydrobromic acid) | HI (hydriodic acid) |
| $HNO_3$ (nitric acid) | $HClO_4$ (perchloric acid) | $H_2SO_4$ (sulphuric acid) |

**2. Weak Acids.** Recognize these from general formulas.

  **A.** Oxoacids: $H_xEO_y$, where $x = 1-3$, $y = 1-4$, $E =$ B, C, N, P, S, Cl, Br, I, others

| | | |
| --- | --- | --- |
| HClO (hypochlorous acid) | HBrO (hypobromous acid) | $HSO_4^-$ (hydrogen sulphate) |
| $HNO_2$ (nitrous acid) | $H_2CO_3$ (carbonic acid) | $H_2SO_3$ (sulphurous acid) |

  **B.** Carboxylic acids: $RCO_2H$, where $R =$ H or any organic group

| | | |
| --- | --- | --- |
| $HCO_2H$ (formic acid) | $CH_3CO_2H$ (acetic acid) | $HOCH_2CO_2H$ (glycolic acid) |
| $C_6H_5CO_2H$ (benzoic acid) | $(CO_2H)_2$ (oxalic acid) | $HO_2C(CH_2)_4CO_2H$ (adipic acid) |

  **C.** Conjugate acid of a weak base

| | | |
| --- | --- | --- |
| $NH_4^+$ (ammonium ion) | $C_5H_5NH^+$ (pyridinium ion) | $C_6H_5NH_3^+$ (anilinium ion) |

  **D.** Miscellaneous examples

| | | |
| --- | --- | --- |
| HF (hydrofluoric acid) | HCN (hydrocyanic acid) | $HN_3$ (hydrazoic acid) |
| $H_2S$ (hydrogen sulphide) | $H_2O_2$ (hydrogen peroxide) | |

| Recognizing Bases | | |
|---|---|---|
| **1. Strong Bases.** Memorize the Group 1 hydroxides ($MOH$) and the soluble Group 2 hydroxides ($M(OH)_2$). | | |
| LiOH | NaOH | KOH |
| $Ca(OH)_2$ | $Sr(OH)_2$ | $Ba(OH)_2$ |
| **2. Weak Bases.** Recognize these from general formulas. | | |
| **A.** Ammonia ($NH_3$) and amines $RNH_2$, $R_2NH$, $R_3N$ | | |
| $CH_3NH_2$ (methylamine) | $(CH_3CH_2)_3 N$ (triethylamine) | $H_2NCH_2CH_2NH_2$ (ethylenediamine) |
| $(C_6H_5)_2 NH$ (diphenylamine) | $(C_6H_5)NH_2$ (aniline) | $C_5H_5N$ (pyridine) |
| **B.** Conjugate base of a weak acid | | |
| $F^-$ (fluoride) | $OCl^-$ (hypochlorite) | $HCO_3^-$ (hydrogen carbonate) |
| $CN^-$ (cyanide) | $SO_4^{2-}$ (sulphate) | $NO_2^-$ (nitrite) |
| $CH_3CO_2^-$ (acetate) | $PO_4^{3-}$ (phosphate) | $C_6H_5CO_2^-$ (benzoate) |

## SECTION EXERCISES

● **14.5.1** Determine the concentrations of all species present in a 0.35 M solution of KCN. (Potassium cyanide is a deadly poison.)

● **14.5.2** Write the acid–base equilibrium that determines the pH of aqueous solutions of each of the following salts, and state whether the resulting solution is acidic, basic, or neither: (a) $NH_4I$; (b) $NaClO_4$; and (c) $NaCH_3CO_2$.

● **14.5.3** Calculate the pH of a 0.75 M solution of $NH_4NO_3$ ($K_b$ for $NH_3 = 1.8 \times 10^{-5}$).

## 14.6 FACTORS AFFECTING ACID STRENGTH

Why is NaOH a strong base, $HNO_3$ a strong acid, and $CH_3OH$ neither? Each of these compounds contains O—H bonds, yet their proton transfer properties are strikingly different. In this section, we examine the effect of molecular structure on acid strength.

### Effect of Charge

The charge on a species has a major effect on its ability to donate or accept protons. Remember that opposite electrical charges attract, and like charges repel. An anion is both a better proton acceptor and a poorer proton donor than is a neutral molecule. Likewise, a cation is a poorer proton acceptor and a better proton donor.

The effect of charge is shown clearly by the acid–base properties of water and its ions (see Figure 14-10). The water molecule is both a weak acid and a weak base.

Hydroxide anion is a strong base because its negative charge strongly attracts protons. Despite having an O—H bond, the hydroxide ion is not acidic because the negative charge of $OH^-$ makes it exceedingly difficult to remove the remaining

**Figure 14-10**

Hydronium cations, neutral water molecules, and hydroxide anions illustrate the effect of charge on acid–base behaviour.

Acid strength decreases as negative charge increases.

proton. The hydronium ion, $H_3O^+$, is a strong acid because its positive charge enhances the removal of one of its protons. The hydronium ion shows negligible basicity even though it has a lone pair of electrons on its oxygen atom. The positive charge of $H_3O^+$ makes it exceedingly difficult to bind another proton.

Polyprotic acids also show clearly the effect of charge on acidity. As mentioned earlier, the successive $K_a$ values for a polyprotic oxoacid decrease by several orders of magnitude. The neutral parent acid is always a stronger acid than is the anion produced by removing one proton.

## Structural Factors

A proton transfer reaction involves breaking a covalent bond. For an acid, an H—$X$ bond breaks as the acid transfers a proton to the base, and the bonding electrons are converted to a lone pair on $X$. Breaking the H—$X$ bond becomes easier to accomplish as the bond energy becomes smaller or as the bonding electrons become less polarized toward $X$. Bond strengths and bond polarities help explain trends in acidity among neutral molecules.

The acidities of the binary hydrides increase from left to right across a row of the periodic table. This trend follows bond polarity: as the electronegativity of $X$ increases, the H—$X$ bond polarity increases and so does acidity. The C—H bonds of methane are essentially non-polar, so the hydrogen atoms are not at all acidic. On the other hand, with an electronegativity difference of 1.9, the electrons of the H—F bond are highly polarized toward the fluorine atom. The resulting partial positive charge on the hydrogen atom assists proton transfer.

Acidity increases moving down a column of the periodic table. The difference between HF (weak acid) and the other hydrogen halides (strong acids) exemplifies the vertical trend. Recall that moving down a column of the periodic table, the principal quantum number of the valence orbitals of the halogen increases, orbital overlap in bonding decreases, and bond strength decreases. Thus, the vertical trend follows bond strength. The H—F bond is substantially more polar, and hence stronger, than the H—Cl bond, so HF is a weaker acid than HCl.

The acidity of binary hydrides shows periodic variations.

**Figure 14-11**

The strength of an oxoacid increases as more and more electron density is withdrawn from the O—H bond.

Among oxoacids, the strengths of O—H bonds depend on the amount of electron density in the bond. This, in turn, depends on how strongly the rest of the atoms in the molecule attract electrons. The halogen oxoacids show this effect in two ways, as shown in Figure 14-11. The acidity of the hypohalous acids increases as the electronegativity of the halogen increases. This is because the more electronegative the atom is, the more it attracts electrons. The acidity of the chlorine oxoacids increases as the number of oxygen atoms increases. This is because each oxygen atom attracts electron density from the rest of the molecule, so the more oxygen atoms there are, the less electron density resides in the O—H bond, thus weakening the bond and increasing the acid strength.

The presence of a second oxygen atom also explains why alcohols like methanol ($CH_3OH$) are not acidic in aqueous solution, whereas carboxylic acids like formic acid ($HCO_2H$) are acidic:

The second oxygen atom in a carboxyl group withdraws electron density from the O—H bond, weakening the bond and enhancing its acidity. The non-polar C—H bonds in an alcohol do not have the same effect. At least equally important, the carboxylate conjugate base is stabilized by delocalization of two types: both the $\pi$ electron cloud and the negative charge are spread over both oxygen atoms rather than being localized on one atom.

Example 14-13 explores another feature of relative acid strengths.

## EXAMPLE 14-13 ACIDITIES OF SIMPLE CARBOXYLIC ACIDS

Oxalic acid ($HO_2C—CO_2H$) has $K_{a1} = 1.3 \times 10^{-2}$ and $K_{a2} = 1.5 \times 10^{-4}$. Formic acid ($HCO_2H$) has $K_a = 1.8 \times 10^{-4}$. Explain why the first proton of oxalic acid is substantially more acidic than the proton of formic acid, but the second proton is less acidic.

**Strategy:** To explain features of acid strength, we need to look at the effects of bond strength, polarity, and charge.

**Solution:** Examine the structures of these three species to gain insight into their relative acid strengths:

As the left-hand structure indicates, the second —$CO_2H$ group in oxalic acid is better at withdrawing electron density than is the —H in formic acid. This makes the O—H bonds in oxalic acid more polar than the O—H bond in formic acid, so the first proton of oxalic acid is more easily transferred, making oxalic acid stronger than formic acid. The resulting anion (right-hand structure) has a negative charge that hinders removal of the second proton and makes hydrogen oxalate a weaker acid than formic acid.

 **Does the Result Make Sense?** Notice that the trends and electron shifts are similar to those shown in Figure 14-11.

**Practice Exercise 14.13**

Iodic acid, $HIO_3$, has $K_a = 0.17$. Explain why this acid is stronger than four of the acids shown in Figure 14-11 but weaker than the other two acids in that figure.

## SECTION EXERCISES

**14.6.1** Sodium metal reacts with ammonia to form an ionic salt, $NaNH_2$, which contains $NH_2^-$ anions. Is this salt acidic or basic and why? What reaction occurs if this salt is placed in water?

**14.6.2** Among the following pairs of acids, which is stronger and why? (a) $H_3PO_4$ and $H_3PO_3$; (b) $H_2S$ and $H_2Se$; (c) $H_2SO_3$ and $HSO_3^-$.

**14.6.3** Rank the following acids from weakest to strongest: HBr, HBrO, HClO, and $HClO_2$.

## 14.7 MULTIPLE EQUILIBRIA

Most of the examples in previous sections appear to involve a single acid–base equilibrium. A closer look reveals that nearly any solution that displays acid–base properties has at least two acid–base equilibria. Look again at Example 14-8, where we list two equilibria as potentially important:

$$NH_3\ (aq) + H_2O\ (l) \rightleftharpoons NH_4^+\ (aq) + OH^-\ (aq) \qquad K_b = 1.8 \times 10^{-5}$$

$$2\ H_2O\ (l) \rightleftharpoons H_3O^+\ (aq) + OH^-\ (aq) \qquad K_w = 1.0 \times 10^{-14}$$

The water equilibrium always exists in aqueous solution. In general, we can focus our initial attention on the equilibria involving other major species ($NH_3$ in this example). Nevertheless, the water equilibrium does exert its effect on the concentrations of $OH^-$ and $H_3O^+$. In this example, the concentration of hydroxide anion is established by the ammonia equilibrium, but the concentration of hydronium cations must be found by applying the water equilibrium. We use this feature in several of our examples in this chapter.

Because many species exhibit acid–base properties, it is often possible to write several proton transfer equilibrium expressions for an aqueous solution. Each such expression is valid if the reactants are species that are actually present in the solution. We have already seen how to focus on the dominant equilibrium: consider only those expressions that have major species as reactants, and look for the one with the largest equilibrium constant.

After completing our analysis of the effects of the dominant equilibrium, we may need to consider the effects of other equilibria. The calculation of $[H_3O^+]$ in a solution of weak base illustrates circumstances where this secondary consideration is necessary. Here, the dominant equilibrium does not include the species $H_3O^+$, the concentration of which we wish to know. In such cases, we must turn to an equilibrium expression that has the species of interest as a product. The reactants should be species that are involved in the dominant equilibrium, because the concentrations of these species are determined by the dominant equilibrium. We can use these concentrations as the initial concentrations for our calculations based on secondary equilibria. Look again at Example 14-11 for another application of this idea. In that example, the dominant equilibrium is the reaction between hypochlorite anions and water molecules:

$$OCl^- (aq) + H_2O (l) \rightleftarrows HOCl (aq) + OH^- (aq)$$

Working with this equilibrium, we can determine the concentrations of $OCl^-$, HOCl, and $OH^-$. To find the concentration of hydronium ions, however, we must invoke a second equilibrium, the water equilibrium:

$$2 H_2O (l) \rightleftarrows H_3O^+ (aq) + OH^- (aq)$$

## Polyprotic Acids

An acid that contains more than one acidic hydrogen atom is called a **polyprotic acid**. Figure 14-8 shows the structures of two polyprotic oxoacids, sulphuric acid ($H_2SO_4$, diprotic) and phosphoric acid ($H_3PO_4$, triprotic). Many carboxylic acids are also polyprotic. Table 14-5 lists several examples of polyprotic acids. Other examples are listed in Appendix E.

## TABLE 14-5 REPRESENTATIVE POLYPROTIC ACIDS.

| Name | Formula | $K_{a1}$ | Monoanion | $K_{a2}$ |
|---|---|---|---|---|
| Sulphuric acid | $H_2SO_4$ | Strong | $HSO_4^-$ | $1.0 \times 10^{-2}$ |
| Oxalic acid | $HO_2CCO_2H$ | $5.6 \times 10^{-2}$ | $HO_2CCO_2^-$ | $1.5 \times 10^{-4}$ |
| Sulphurous acid | $H_2SO_3$ | $1.4 \times 10^{-2}$ | $HSO_3^-$ | $6.3 \times 10^{-8}$ |
| Phosphoric acid ($K_{a3} = 4.8 \times 10^{-13}$) | $H_3PO_4$ | $6.9 \times 10^{-3}$ | $H_2PO_4^-$ | $6.3 \times 10^{-8}$ |
| Phthalic acid | $HO_2CC_6H_4CO_2H$ | $1.1 \times 10^{-3}$ | $HO_2CC_6H_4CO_2^-$ | $3.7 \times 10^{-6}$ |
| Carbonic acid | $H_2CO_3$ | $4.5 \times 10^{-7}$ | $HCO_3^-$ | $4.7 \times 10^{-11}$ |

Each successive proton transfer reaction of a polyprotic acid has its own value for $K_a$, and each successive value is a few orders of magnitude smaller than its predecessor, as explained in Section 14.6.

Other than the water equilibrium, polyprotic acids provide the most common examples of secondary equilibria that play roles in determining concentrations of minor species in aqueous solution. When $K_{a1}$ is three or more orders of magnitude larger than $K_{a2}$, we can base our initial analysis completely on $K_{a1}$. To find the concentrations of minor species, however, we must use successive $K_a$ values and carry out more than one calculation, using additional concentration tables. When multiple equilibria must be taken into account, it is particularly important to identify the dominant equilibrium, which will have major species as reactants. After solving for the concentrations of ions resulting from the dominant equilibrium, we then consider how other equilibria affect the concentrations of the minor species. Example 14-14 shows how this works.

## EXAMPLE 14-14 ION CONCENTRATIONS IN A POLYPROTIC ACID SOLUTION

Carbonated water contains carbonic acid, a diprotic acid that forms when carbon dioxide dissolves in water:

$$CO_2 \ (g) + H_2O \ (l) \rightleftarrows H_2CO_3 \ (aq)$$

A typical carbonated beverage contains 0.050 M $H_2CO_3$. Determine the concentrations of the ions present and the pH in this solution.

**Strategy:** The problem describes a weak diprotic acid and asks for ion concentrations. To determine concentrations of all ions, we need to consider more than one equilibrium. This is done in stages, starting with the dominant equilibrium. We apply the seven-step strategy. In carbonated water, the major species are $H_2CO_3$ and $H_2O$.

There are two chemical equilibria involving these major species. The acid undergoes proton transfer with equilibrium constant $K_{a1}$:

$$H_2CO_3 \ (aq) + H_2O \ (l) \rightleftarrows H_3O^+ \ (aq) + HCO_3^- \ (aq) \qquad K_{a1} = 4.5 \times 10^{-7}$$

As always, water undergoes proton transfer:

$$H_2O \ (l) + H_2O \ (l) \rightleftarrows H_3O^+ \ (aq) + OH^- \ (aq) \qquad K_w = 1.0 \times 10^{-14}$$

Because carbonic acid is diprotic, a second proton transfer equilibrium has an effect on the ion concentrations:

$$HCO_3^- \ (aq) + H_2O \ (l) \rightleftarrows H_3O^+ \ (aq) + CO_3^{2-} \ (aq) \qquad K_{a2} = 4.7 \times 10^{-11}$$

**Solution:** Because $K_{a1}$ is several orders of magnitude larger than $K_{a2}$ or $K_w$, we identify $K_{a1}$ as dominant. Notice, however, that the water equilibrium generates some hydroxide ions in the solution, so this equilibrium must be used to find the concentration of hydroxide ions. The third reaction involves a *minor* species, $HCO_3^-$, as a reactant, so it cannot be the dominant equilibrium. However, just as the water equilibrium generates some hydroxide ions, the hydrogen carbonate equilibrium generates some carbonate anions of which the concentration must be determined.

We can write an equilibrium constant expression for each equilibrium. Because the initial calculations involve the dominant equilibrium, we start with the expression for $K_{a1}$:

$$K_{a1} = \frac{[H_3O^+]_{eq}[HCO_3^-]_{eq}}{[H_2CO_3]_{eq}}$$

Now we are ready to organize the data and the unknowns and do the calculations. There are multiple equilibria affecting ion concentrations, so we must work with more than one concentration table, starting with the dominant equilibrium. Set up a concentration table to determine concentrations of the ions generated by this reaction:

| Reaction | $H_2O$ (*l*) + $H_2CO_3$ (*aq*) $\rightleftharpoons$ $H_3O^+$ (*aq*) + $HCO_3^-$ (*aq*) | | |
|---|---|---|---|
| Initial Concentration (M) | 0.050 | 0 | 0 |
| Change in Concentration (M) | $-x$ | $+x$ | $+x$ |
| Equilibrium Concentration (M) | $0.050 - x$ | $x$ | $x$ |

Substitute the equilibrium concentrations into the equilibrium constant expression and solve for $x$, making the approximation that $x \ll 0.050$:

$$K_{a1} = 4.5 \times 10^{-7} = \frac{(x)(x)}{0.050 - x} = \frac{x^2}{0.050 - x} \cong \frac{x^2}{0.050}$$

$$x^2 = (0.050)(4.5 \times 10^{-7}) = 2.25 \times 10^{-8} \quad \text{from which} \quad x = 1.5 \times 10^{-4}$$

We round to two significant figures because the $K$ value has two significant figures. The concentrations are

$$[H_3O^+] = [HCO_3^-] = 1.5 \times 10^{-4} \, M$$

$$[H_2CO_3] = 0.050 - 1.5 \times 10^{-4} = 0.050 - 0.00015 = 0.050 \, M$$

Note that $x$ is about 0.3% of 0.050, so the approximation is valid. Furthermore, since $[H_3O^+] = x$, then $pH = -\log_{10}[H_3O^+] = -\log_{10}(1.5 \times 10^{-4}) = 3.82$.

Next, we take into account the proton transfer equilibrium involving hydrogen carbonate anion. To do this, we complete a second concentration table, using as "initial" concentrations those calculated for the first equilibrium:

| Reaction | $H_2O$ (*l*) + $HCO_3^-$ (*aq*) $\rightleftharpoons$ $H_3O^+$ (*aq*) + $CO_3^{2-}$ (*aq*) | | |
|---|---|---|---|
| Initial Concentration (M) | $1.5 \times 10^{-4}$ | $1.5 \times 10^{-4}$ | 0 |
| Change in Concentration (M) | $-x$ | $+x$ | $+x$ |
| Equilibrium Concentration (M) | $1.5 \times 10^{-4} - x$ | $1.5 \times 10^{-4} + x$ | $x$ |

Substitute equilibrium concentrations into the equilibrium constant expression and solve for $x$, making the approximation that $x \ll 1.5 \times 10^{-4}$:

$$K_{a2} = 4.7 \times 10^{-11} = \frac{[H_3O^+]_{eq}[CO_3^{2-}]_{eq}}{[HCO_3^-]_{eq}}$$

$$4.7 \times 10^{-11} = \frac{(x)(1.5 \times 10^{-4} + x)}{1.5 \times 10^{-4} - x} \cong \frac{x(1.5 \times 10^{-4})}{1.5 \times 10^{-4}} = x$$

This value is too small to cause a measurable change in the concentrations already calculated, but it does tell us the concentration of carbonate anions in the solution:

$$[CO_3^{2-}] = 4.7 \times 10^{-11} \text{ M}$$

One more ion remains: $OH^-$, generated from the water equilibrium. It is possible to set up a concentration table, but it is easier to apply the water equilibrium expression directly:

$$K_w = 1.0 \times 10^{-14} = [H_3O^+][OH^-] \qquad \text{so}$$

$$[OH^-] = \frac{1.0 \times 10^{-14}}{[H_3O^+]} = \frac{1.0 \times 10^{-14}}{1.5 \times 10^{-4}} = 6.7 \times 10^{-11} \text{ M}$$

**Do the Results Make Sense?** Compare the product concentrations with the concentration of the acid that is responsible for the various ions in the solution, $[H_2CO_3] = 0.050$ M. The ions generated when this acid undergoes proton transfer with water have concentrations that are two orders of magnitude smaller than the concentration of the parent acid.

Products of dominant equilibria: $[H_3O^+] = [HCO_3^-] = 1.5 \times 10^{-4}$ M

The products generated by other equilibria ($[CO_3^{2-}] = 4.7 \times 10^{-11}$ M and $[OH^-] = 6.7 \times 10^{-11}$ M) have concentrations that are more than six orders of magnitude smaller than the concentrations of ions produced in the reaction corresponding to the dominant equilibrium.

The major species have the highest concentrations; the species generated directly by reactions of major species have the next highest concentrations; and species generated by secondary reactions have very low concentrations. This is a reasonable outcome.

**Common Pitfalls::** Thinking that the hydronium ion from the second hydrolysis is significant. It rarely is. The value of $[H_3O^+]$ from the first hydrolysis can be used in subsequent calculations and will not change.

### Practice Exercise 14.14

Determine the concentrations of the ionic species present in a 1.0 M solution of phosphoric acid, $H_3PO_4$ (see Table 14-5 for $K_a$ values).

There are several acidic species present in any solution of a polyprotic acid. The solution of carbonic acid in Example 14-14 contains the following concentrations of acidic species:

$$[H_2CO_3] = 0.050 \text{ M} \qquad [H_3O^+] = 1.5 \times 10^{-4} \text{ M} \qquad [HCO_3^-] = 1.5 \times 10^{-4} \text{ M}$$

What happens if we add a base to this solution? The strongest acid donates protons preferentially, but the concentration of hydronium ions is so small that this ion is rapidly consumed. Then, the next strongest acid, $H_2CO_3$, reacts with added base. Generalizing, when a base is added to a solution that contains both a polyprotic acid and its anion, the base accepts protons preferentially from the neutral acid. Only after the neutral acid has been consumed completely does the anion participate significantly in proton transfer. Example 14-15 provides molecular pictures of this fact.

## EXAMPLE 14-15 MOLECULAR VIEW OF A POLYPROTIC ACID

The drawing shows a molecular view of a very small region of an aqueous solution of oxalic acid. For clarity, water molecules are not shown. Redraw this molecular picture to show the solution (a) after two hydroxide ions react with these molecules, and (b) after four hydroxide ions react with these molecules. Include in your drawings the water molecules that form as products.

**Strategy:** To draw molecular pictures illustrating a proton transfer process, we must visualize the chemical reactions that occur, see what products result, and then draw the resulting solution. When a strong base is added to a weak acid, hydroxide ions remove protons from the molecules of the weak acid. When more than one acidic species is present, the stronger acid loses protons preferentially.

**Solution:** The oxalic acid solution contains water molecules and $H_2C_2O_4$ molecules as major species. Added hydroxide ions remove protons from the strongest acid, $H_2C_2O_4$:

$$H_2C_2O_4\ (aq) + OH^-\ (aq) \rightleftarrows HC_2O_4^-\ (aq) + H_2O(l)$$

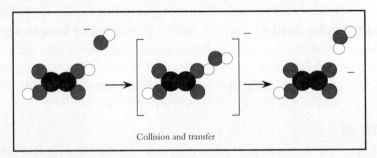

Collision and transfer

Each proton transfer reaction creates $HC_2O_4^-$, which is itself a weak acid. The $K_a$ values show that $H_2C_2O_4$ is substantially more acidic than $HC_2O_4^-$:

$$H_2C_2O_4 \qquad K_{a1} = 5.6 \times 10^{-2}$$

$$H_2C_2O_4^- \qquad K_{a1} = 5.6 \times 10^{-4}$$

Consequently, hydroxide ions react preferentially with $H_2C_2O_4$. Any added $OH^-$ ions react with $H_2C_2O_4$ until every oxalic acid molecule has lost one proton.

(a) The first two hydroxide ions convert two $H_2C_2O_4$ molecules into $HC_2O_4^-$ ions. Thus, the molecular picture should contain two $HC_2O_4^-$ anions, two water molecules, and one unreacted $H_2C_2O_4$ molecule:

(b) When four OH⁻ ions are added, the first three consume all the $H_2C_2O_4$ molecules. The remaining OH⁻ ion reacts with the weaker acid, $HC_2O_4^-$:

$$HC_2O_4^- \, (aq) + OH^- \, (aq) \rightleftharpoons C_2O_4^{2-} \, (aq) + H_2O \, (l)$$

Collision and transfer

This reaction goes virtually to completion, so at equilibrium this system contains two $HC_2O_4^-$, one $C_2O_4^{2-}$, and four water molecules.

**Do the Results Make Sense?** Keep in mind that added base accepts protons from acids that are present but leaves the rest of the acid molecule unchanged. Notice that each of the molecular pictures includes the same number of oxalate species. In the first view, each oxalate has two protons attached to it. In the second view, two of them have released one proton each to the added base. In the third view, all three have lost one proton; additionally, one has lost a second proton.

**Practice Exercise 14.15**

Draw a molecular picture of the carbonic acid solution of Example 14-14, with one molecule for each 0.01 M unit of concentration, showing how the solution looks after the addition of three hydroxide ions.

## Salts of Polyprotic Acids

Any anion of a weak acid, including the anions of polyprotic acids, is a weak base. The acid–base properties of monoanions of polyprotic acids are complicated, however, because the monoanion is simultaneously the conjugate base of the parent acid and an acid in its own right. For example, hydrogen carbonate anions undergo two proton transfer reactions with water:

$$HCO_3^- \, (aq) + H_2O \, (l) \rightleftharpoons H_2CO_3(l) + OH^-(aq)$$

and

$$K_{b1} = \frac{K_w}{K_{a1}} = \frac{1.0 \times 10^{-14}}{4.5 \times 10^{-7}} = 2.2 \times 10^{-8}$$

The larger of these two equilibrium constants is $K_{b1}$, so a solution of $NaHCO_3$ will be basic. Nevertheless, these two equilibrium constants have values that are sufficiently close together that both equilibria must be taken into account in careful quantitative work, but such calculations are beyond the scope of this text.

In contrast to the intermediate anions, an anion without hydrogen atoms no longer has acidic possibilities. Common examples are phosphate ($PO_4^{3-}$), sulphate ($SO_4^{2-}$), carbonate ($CO_3^{2-}$), and sulphite ($SO_3^{2-}$). Example 14-16 shows how to treat a solution of one of these salts.

## EXAMPLE 14-16  ION CONCENTRATIONS IN A POLYPROTIC ANION SOLUTION

Potassium sulphite is commonly used as a food preservative, because the sulphite anion undergoes reactions that release sulphur dioxide, an effective preservative. Determine the concentrations of the ionic species present and the pH in a 0.075 M solution of potassium sulphite, $K_2SO_3$ (aq).

**Strategy:** The problem describes a salt of a polyprotic acid and asks for ion concentrations. To determine concentrations of all ions, we need to consider more than one equilibrium. This is done in stages, starting with the dominant equilibrium. We apply the seven-step strategy. The problem asks for the concentrations of the ions in aqueous potassium sulphite, in which the major species are $K^+$, $SO_3^{2-}$, and $H_2O$. The potassium ion is a spectator ion. The sulphite anion undergoes proton transfer with equilibrium constant $K_{b2}$:

$$SO_3^{2-}\ (aq) + H_2O(l) \rightleftarrows HSO_3^-\ (aq) + OH^-\ (aq) \quad K_{b2} = \frac{[HSO_3^-]_{eq}[OH^-]_{eq}}{[SO_3^{2-}]_{eq}}$$

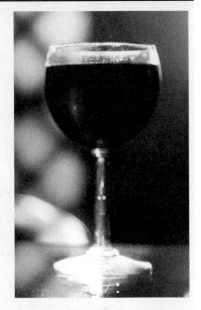

Wines are preserved using sulphites.

Table 14-5 provides the value of the acid equilibrium constant, $K_{a2} = 6.3 \times 10^{-8}$:

$$K_{b2} = \frac{K_w}{K_{a2}} = \frac{1.0 \times 10^{-14}}{6.3 \times 10^{-8}} = 1.6 \times 10^{-7}$$

This is much larger than $K_w$, so this is the dominant equilibrium, which we use to begin our calculations. Because sulphurous acid is diprotic, a second proton transfer equilibrium has an effect on the ion concentrations, and the water equilibrium also plays a secondary role:

$$HSO_3^-\ (aq) + H_2O\ (l) \rightleftarrows H_2SO_3\ (aq) + OH^-\ (aq) \quad K_{b1}$$

$$H_2O\ (l) + H_2O(l) \rightleftarrows H_3O^+\ (aq) + OH^-\ (aq) \quad K_w$$

**Solution:** Now we are ready to organize the data and the unknowns and do the calculations. The spectator ion is easiest to deal with:

$$[K^+] = 2[SO_3^{2-}] = 2(0.075\ M) = 0.15\ M$$

There are multiple equilibria affecting ion concentrations, so we must work with more than one concentration table, starting with the dominant equilibrium. Set up a concentration table to determine concentrations of the ions generated by this reaction:

| Reaction | $H_2O$ (l) + | $SO_3^{2-}$ (aq) $\leftrightarrows$ | $OH^-$ (aq) + | $HSO_3^-$ (aq) |
|---|---|---|---|---|
| Initial Concentration (M) | | 0.075 | 0 | 0 |
| Change in Concentration (M) | | $-x$ | $+x$ | $+x$ |
| Equilibrium Concentration (M) | | $0.075 - x$ | $x$ | $x$ |

Substitute the equilibrium concentrations into the equilibrium constant expression and solve for $x$, making the approximation that $x \ll 0.075$:

$$K_{b2} = 1.6 \times 10^{-7} = \frac{(x)(x)}{0.75 - x} \cong \frac{x^2}{0.075}$$

$$x^2 = (0.075)(1.6 \times 10^{-7}) = 1.2 \times 10^{-8} \quad \text{from which} \quad x = 1.1 \times 10^{-4} \, \text{M}$$

We round to two significant figures because the $K$ value has two significant figures. The concentrations are

$$[OH^-] = [HSO_3^-] = 1.1 \times 10^{-4} \, \text{M}$$

$$[SO_3^{2-}] = 0.075 - 1.1 \times 10^{-4} = 0.075 - 0.00011 = 0.075 \, \text{M}$$

Note that $x$ is about 0.1% of 0.075, so the approximation is valid.

Next, we take into account the proton transfer equilibrium involving the hydrogen sulphite anion. To do this, we complete a second concentration table, using as "initial" concentrations those calculated for the first equilibrium:

| Reaction | $H_2O \, (l)$ | $+$ | $HSO_3^- \, (aq)$ | $\rightleftharpoons$ | $OH^- \, (aq)$ | $+$ | $H_2SO_3 \, (aq)$ |
|---|---|---|---|---|---|---|---|
| Initial Concentration (M) | | | $1.1 \times 10^{-4}$ | | $1.1 \times 10^{-4}$ | | 0 |
| Change in Concentration (M) | | | $-x$ | | $+x$ | | $+x$ |
| Equilibrium Concentration (M) | | | $1.1 \times 10^{-4} - x$ | | $1.1 \times 10^{-4} + x$ | | $x$ |

Substitute equilibrium concentrations into the equilibrium constant expression and solve for $x$, making the approximation that $x \ll 1.1 \times 10^{-4}$:

$$K_{b1} = \frac{K_w}{K_{a1}} = \frac{1.0 \times 10^{-14}}{1.4 \times 10^{-2}} = 7.1 \times 10^{-13} = \frac{[OH^-]_{eq}[H_2SO_3]_{eq}}{[HSO_3^-]_{eq}}$$

$$7.1 \times 10^{-13} = \frac{(x)(1.1 \times 10^{-4} + x)}{1.1 \times 10^{-4} - x} \cong \frac{x(1.1 \times 10^{-4})}{1.1 \times 10^{-4}} \cong x$$

This value is too small to cause a measurable change in the concentrations already calculated, but it does tell us the concentration of sulphurous acid in the solution:

$$[H_2SO_3] = 7.1 \times 10^{-13} \, \text{M}$$

One more ion remains, $H_3O^+$, generated from the water equilibrium. It is possible to set up a concentration table, but it is easier to apply the water equilibrium expression directly:

$$K_w = 1.0 \times 10^{-14} = [H_3O^+][OH^-]$$

so $[H_3O^+] = \dfrac{1.0 \times 10^{-14}}{[OH^-]} = \dfrac{1.0 \times 10^{-14}}{1.1 \times 10^{-4}} = 9.1 \times 10^{-11} \, \text{M}$ and $\text{pH} = -\log[H_3O^+] = -\log(9.1 \times 10^{-11}) = 10.0$

**Do the Results Make Sense?** Compare the product concentrations with the concentration of the salt in the solution, 0.075 M. The ions generated when the anion undergoes proton transfer with water have concentrations that are over two orders of magnitude smaller than the concentration of the original solution.

Products of dominant equilibria:     $[OH^-] = [HSO_3^-] = 1.1 \times 10^{-4} \, \text{M}$

The products generated by other equilibria have concentrations that are more than six orders of magnitude smaller than the concentrations of ions produced in the reaction corresponding to the dominant equilibrium:

Products of other equilibria:     $[H_2SO_3] = 7.1 \times 10^{-13} \, \text{M}$     and     $[H_3O^+] = 9.1 \times 10^{-11} \, \text{M}$

The major species have the highest concentrations; the species generated directly by reactions of major species have the next highest concentrations, and species generated by secondary reactions have very low concentrations. This is a reasonable outcome.

**Practice Exercise 14.16**

Determine the concentrations of the ionic species present and the pH in a 0.36 M solution of sodium carbonate, $Na_2CO_3$ (see Table 14-5 for $K_a$ values).

Review the quantitative examples in this chapter and compare the techniques used to solve them. You should recognize a common approach. Although the species differ depending on the substances present, identifying the dominant equilibrium is a key step. Once this is done, the approach is always the same: set up and complete concentration tables, use their results to write algebraic expressions linking concentrations to equilibrium constants, and do the algebra to get the results.

## SECTION EXERCISES

**14.7.1** Calculate the hydroxide ion concentration in a 0.333 M aqueous solution of $NH_4Cl$.

**14.7.2** Write all the acid–base equilibrium reactions that have major species as reactants for an aqueous solution of sodium dihydrogen phosphate ($NaH_2PO_4$).

**14.7.3** Determine which of the equilibria in Section Exercise 14.7.2 will be the most important and predict whether this solution is acidic or basic.

# CHAPTER 14 VISUAL SUMMARY

## 14.1 PROTON TRANSFERS IN WATER

Strong acid + Base

NaOH
Strong base

$$HA + B \rightleftharpoons A^- + HB^+$$
Acid, Base

Brønsted-Lowry definition

Acid Base
Amphiprotic

$$K_W = [H_3O^+][OH^-]$$

Water equilibrium constant ($K_W$)

**LEARNING OBJECTIVES:** Use the Bronsted–Lowry theory of acids and bases.

**SKILLS TO MASTER:** Calculating concentrations in solutions of strong acids and bases (p. 815)

**KEY CONCEPTS:** Water can act as an acid or a base.

Any proton donor and the species generated by removing one of its protons are called a conjugate acid–base pair.

$K_W = [H_3O^+]_{eq} [OH^-]_{eq} = 1.0 \times 10^{-14}$ at 25 °C in any aqueous solution.

## 14.2 THE pH SCALE

pH

$$pH = -\log[H_3O^+]$$

$$pH + pOH = 14.00$$

**LEARNING OBJECTIVES:** Relate pH to concentrations of ions in solution.

**SKILLS TO MASTER:** Calculating pH (p. 819)
Relating pH with pOH (p. 821)

**KEY CONCEPTS:** pH is a logarithmic scale. A change of 1 pH unit is a 10-fold change of $[H_3O^+_{(aq)}]$.

## 14.3 WEAK ACIDS AND BASES

Concentration table

| Reaction | $H_2O + HA$ | $\rightleftharpoons$ | $H_3O^+$ | $+ A^-$ |
|---|---|---|---|---|
| Initial concentration | $[HA]_0$ | | 0 | 0 |
| Changes in concentration | $-x$ | | $+x$ | $+x$ |
| Final concentrations | $[HA]_0 - x$ | | $x$ | $x$ |

$$\% \, HA \text{ ionized} = 100\% \frac{[H_3O^+]_{eq}}{[HA]_{initial}}$$

Weak acid

$$K_a = \frac{[H_3O^+]_{eq}[A^-]_{eq}}{[HA]_{eq}}$$

$$K_b = \frac{[HB^+]_{eq}[OH^-]_{eq}}{[B]_{eq}}$$

Weak base

**LEARNING OBJECTIVES:** Calculate concentrations and pH in weak acid and base solutions.

**SKILLS TO MASTER:** Calculating $K_a$ from pH (p. 823)

Calculating % hydrolysis and concentrations in solutions of weak acids (p. 824)

Calculating concentrations in solutions of weak bases (p. 826)

**KEY CONCEPTS:** Using a concentration table simplifies calculations of this type.

## 14.4 RECOGNIZING ACIDS AND BASES

Recognizing acids

Polyprotic acid

Conjugate acid

Carboxylic acid    Oxoacid

An oxoacid can be recognized by the general formula $H_xEO_y$.

A carboxylic acid can be recognized by the general formula $RCO_2H$.

Recognizing bases

Amine    Conjugate base

Conjugate acid-base pair

$K_a$

Conjugate acid   Conjugate base        Conjugate base   Conjugate acid

**LEARNING OBJECTIVES:** Recognize and name some common acids and bases.

**SKILLS TO MASTER:** Naming oxoacids (p. 830)

Writing the reaction for hydrolysis of a carboxylic acid (p. 831)

Naming other weak acids (p. 831)

Naming some weak bases (p. 832)

**KEY CONCEPTS:** Oxoacids contain an inner atom bonded to a variable number of oxygen atoms and acidic OH groups.

Carboxylic acids have the general formula $RCO_2H$.

## 14.5 ACIDIC AND BASIC SALTS

Conjugate base    Conjugate acid         Conjugate acid    Conjugate base

$$K_a K_b = K_w \qquad pK_a + pK_b = 14.00$$

**LEARNING OBJECTIVES:** Calculate the pH of solutions of salts of weak acids or bases.

**SKILLS TO MASTER:** Calculating the pH of a solution of the salt of a weak acid (p. 837)

Calculating the pH of a solution of the salt of a weak base (p. 839)

**KEY CONCEPTS:** In any aqueous system, $K_a K_b = K_w$ and $pK_a + pK_b = pK_w$.

Salts of weak acids are basic in aqueous solution; salts of weak bases are acidic in aqueous solution.

## 14.6 FACTORS AFFECTING ACID STRENGTH

**LEARNING OBJECTIVES:** Explain the factors that contribute to the strength of an acid.

**SKILLS TO MASTER:** Understanding how structural factors influence the acidity of binary hydrides and oxoacids (p. 842)

Understanding how structural factors influence the acidity of carboxylic acids (p. 843)

**KEY CONCEPTS:** The strength of an oxoacid or carboxylic acid increases as electron density is withdrawn from the O—H bond.

Successive $K_a$ values of polyprotic acids are smaller than previous ones because of increasing negative charge on the anion.

## 14.7 MULTIPLE EQUILIBRIA

**LEARNING OBJECTIVES:** Calculate concentrations in solutions involving multiple equilibria.

**SKILLS TO MASTER:** Calculating the pH and concentrations in solutions of polyprotic acids (p. 846)

Calculating the pH and concentrations in solutions of salts of polyprotic acids (p. 851)

**KEY CONCEPTS:** The water equilibrium exists in any aqueous solution.

## LEARNING EXERCISES

14.1  Outline the procedure for working an equilibrium problem for a weak acid–base system.

14.2  Prepare a table listing the various types of acids and bases and the identifying features of each.

14.3  Write a paragraph describing the conjugate acid–base pair and explaining how each interacts with water.

14.4  Update your list of memory bank equations.

14.5  Prepare a list of all terms in Chapter 14 that are new to you. In your own words, write a one-sentence definition of each. Consult the Glossary if you need help.

## PROBLEMS

**ilw** = interactive learning ware problem. Visit the website at www.wiley.com/canada/olmsted

### Proton Transfers in Water

14.1  Draw a set of molecular pictures that show the proton transfer reaction occurring when HBr dissolves in water.

14.2  Draw a set of molecular pictures that show the proton transfer reaction occurring when $HNO_3$ dissolves in water.

14.3  Determine the concentrations of hydronium and hydroxide ions in $1.25 \times 10^{-3}$ M aqueous perchloric acid.

14.4  Determine the concentrations of hydronium and hydroxide ions in $3.65 \times 10^{-2}$ M aqueous sodium hydroxide.

14.5  **ilw** Concentrated aqueous HCl has a concentration of 12.1 M. Calculate the concentrations of all ions present in a solution prepared by pipetting 1.00 mL of concentrated HCl into a 100-mL volumetric flask and filling to the mark.

14.6  Concentrated aqueous $HClO_4$ has a concentration of 14.8 M. Calculate the concentrations of all ions present in a solution prepared by pipetting 5.00 mL of concentrated HCl into a 1000-mL volumetric flask and filling to the mark.

14.7  Calculate the concentrations of hydronium and hydroxide ions in a solution prepared by dissolving 0.488 g of HCl gas in enough water to make 325 mL of solution.

14.8  Calculate the concentrations of hydronium and hydroxide ions in a solution prepared by dissolving 0.345 g of solid NaOH in enough water to make 225 mL of solution.

### The pH Scale

14.9  Convert each of the following $H_3O^+$ concentrations to pH: (a) 4.0 M; (b) $3.75 \times 10^{-6}$ M; (c) 4.8 mM; and (d) 0.000 255 M.

14.10  Convert each of the following $H_3O^+$ concentrations to pH: (a) 1.25 M; (b) 2.95 μM; (c) 0.0366 M; and (d) $7.45 \times 10^{-4}$ M.

14.11  Convert each of the following $OH^-$ concentrations to pH: (a) 4.0 M; (b) $3.75 \times 10^{-6}$ M; (c) 4.8 mM; and (d) 0.000 255 M.

14.12  Convert each of the following $OH^-$ concentrations to pH: (a) 1.25 M; (b) 2.95 μM; (c) 0.0366 M; and (d) $7.45 \times 10^{-4}$ M.

14.13  Convert each of the following pH values into a hydronium ion concentration: (a) 0.66; (b) 7.85; (c) 3.68; and (d) 14.33.

14.14  Convert each of the following pH values into a hydronium ion concentration: (a) 1.56; (b) 3.85; (c) 9.75; and (d) 11.22.

14.15  Convert each of the following pH values into a hydroxide ion concentration: (a) 0.66; (b) 7.85; (c) 3.68; and (d) 14.33.

14.16  Convert each of the following pH values into a hydroxide ion concentration: (a) 1.56; (b) 3.85; (c) 9.75; and (d) 11.22.

### Weak Acids and Bases

14.17  Calculate the pH of a 1.5 M solution of each of the following compounds (see Appendix E for $K$ values): (a) $C_5H_5N$; (b) $HONH_2$; and (c) $HCO_2H$.

14.18  Calculate the pH of a $2.5 \times 10^{-2}$ M solution of the following compounds (see Appendix E for $K$ values): (a) $NH_3$; (b) HClO; and (c) HCN.

14.19  Draw molecular pictures illustrating the proton transfer reactions that determine the pH of the solutions in Problems 14.14 (b) and 14.14 (c).

14.20  Draw molecular pictures illustrating the proton transfer reactions that determine the pH of the solutions in Problems 14.18 (a) and 14.18 (c).

14.21  **ilw** For a 1.50 M aqueous solution of hydrazoic acid ($HN_3$), do the following: (a) Identify the major and minor species. (b) Compute concentrations of all species present. (c) Find the pH. (d) Draw a molecular picture illustrating the equilibrium reaction that determines the pH.

14.22  For a $2.75 \times 10^{-2}$ M aqueous solution of cyanic acid (HCNO), do the following: (a) Identify the major and minor species. (b) Compute concentrations of all species present. (c) Find the pH. (d) Draw a molecular picture illustrating the equilibrium reaction that determines the pH.

14.23  Determine the percent hydrolysis of the solution of hydrazoic acid in Problem 14.21.

14.24  Determine the percent hydrolysis of the solution of cyanic acid in Problem 14.22.

14.25  For a 0.350 M aqueous solution of trimethylamine ($N(CH_3)_3$), do the following: (a) Identify major and minor species. (b) Compute concentrations of all species. (c) Find the pH. (d) Draw a molecular picture illustrating the equilibrium reaction that determines the pH.

14.26  For a $1.85 \times 10^{-3}$ M aqueous solution of aniline ($C_6H_5NH_2$), do the following: (a) Identify major and minor species. (b) Compute concentrations of all species. (c) Find the pH. (d) Draw a molecular picture illustrating the equilibrium reaction that determines the pH.

14.27  Determine the percent hydrolysis of a solution of acetic acid that is 0.75 M (see Appendix E for $K$ values).

**14.28** Determine the percent hydrolysis of a solution of boric acid that is 75 mM (see Appendix E for $K$ values).

## Recognizing Acids and Bases

**14.29** Identify each of the following substances as a weak acid, strong acid, weak base, strong base, both weak acid and weak base, or neither an acid nor a base: (a) $NH_3$; (b) HCNO; (c) HClO; and (d) $Ba(OH)_2$.

**14.30** Identify each of the following substances as a weak acid, strong acid, weak base, strong base, both weak acid and weak base, or neither an acid nor a base: (a) $HNO_3$; (b) $CH_3CO_2H$; (c) HOH; (d) HOCl; and (e) $NH_2OH$.

**14.31** For each molecular picture shown in the figure, determine the chemical formula, identify it as an acid or base, and draw a picture of its conjugate partner.

**14.32** For each molecular picture shown in the figure, determine the chemical formula, identify it as an acid or base, and draw a picture of its conjugate partner.

**14.33** For each weak acid in Problem 14.17, identify the conjugate base. For each weak base, identify the conjugate acid.

**14.34** For each weak acid in Problem 14.18, identify the conjugate base. For each weak base, identify the conjugate acid.

**14.35** Write all the conjugate acid–base equilibrium expressions that apply to an aqueous solution of each of the substances in Problem 14.29, and identify each conjugate acid and base.

**14.36** Write all the conjugate acid–base equilibrium expressions that apply to an aqueous solution of each of the substances in Problem 14.30, and identify each conjugate acid and base.

## Acidic and Basic Salts

**14.37** **ilw** For a 0.45 M solution of $Na_2SO_3$, do the following: (a) Identify the major species. (b) Identify the equilibrium that determines the pH. (c) Compute the pH.

**14.38** For a $6.75 \times 10^{-3}$ M solution of sodium benzoate ($NaC_6H_5CO_2$), do the following: (a) Identify the major species. (b) Identify the equilibrium that determines the pH. (c) Compute the pH.

**14.39** For a solution that is 0.0100 M in $NH_4NO_3$, do the following: (a) Identify the major species. (b) Identify the equilibrium that determines the pH. (c) Calculate the pH.

**14.40** For a solution that is $4.75 \times 10^{-2}$ M in $NH_4Br$, do the following: (a) Identify the major species. (b) Identify the equilibrium that determines the pH. (c) Calculate the pH.

**14.41** Decide if aqueous solutions of the following salts are acidic, basic, or neutral. For each, write the balanced equation that determines the pH of the solution. (a) NaHS; (b) NaOI; (c) $LiClO_4$; and (d) $HC_5H_5NCl$ (pyridinium chloride). (See Appendix E for $K$ values.)

**14.42** Decide if aqueous solutions of the following salts are acidic, basic, or neutral. For each, write the balanced equation that determines the pH of the solution. (a) $Mg(NO_3)_2$; (b) $Li_2CO_3$; (c) $(CH_3)_3NHBr$; and (d) $NaCH_3CO_2$. (See Appendix E for $K$ values.)

**14.43** List the following sodium salts in order of increasing pH of their 0.25 M solutions: NaI, NaOH, NaF, $Na_3PO_4$, and $NaC_6H_5CO_2$. (See Appendix E for $K$ values.)

**14.44** List the following substances in order of increasing pH of their 0.25 M solutions: $NH_4Cl$, $MgCl_2$, HCl, NaOCl, $C_6H_5NH_3Cl$, and $NaClO_2$. (See Appendix E for $K$ values.)

## Factors Affecting Acid Strength

**14.45** Among the following pairs of acids, which is stronger and why? (a) $H_2SO_4$ and $HSO_4^-$; (b) HClO and HIO; and (c) HClO and $HClO_2$.

**14.46** Among the following pairs of acids, which is stronger and why? (a) $HBrO_3$ and $HBrO_2$; (b) $H_2S$ and $H_2O$; and (c) $H_2S$ and $HS^-$.

**14.47** Draw Lewis structures of the acids in Problems 14.45 (b) and 14.45 (c), and use arrows to show electron density shifts that account for their different acid strengths.

**14.48** Draw Lewis structures of the acids in Problems 14.46 (a) and 14.46 (b), and use arrows to show electron density shifts that account for their different acid strengths.

**14.49** Alcohols generally are not acidic, yet phenol ($C_6H_5OH$) is a weak acid. Draw the Lewis structure of the phenolate anion (the conjugate base of phenol), with a $C=O$ double bond, and use it to explain the acidity of the —OH group in phenol relative to the —OH group in ethanol.

Phenol

**14.50** Acetic acid ($CH_3CO_2H$) is weaker than chloroacetic acid ($ClCH_2CO_2H$). Draw Lewis structures of these two acids. Draw an arrow indicating the effect of the Cl atom on the electron density in the rest of the molecule, and use this drawing to explain why chloroacetic acid is the stronger acid.

## Multiple Equilibria

**14.51** (ilw) Use two concentration tables to calculate the concentrations of all species present in a 0.250 M solution of sodium acetate ($NaCH_3CO_2$). (See Appendix E for $K$ values.)

**14.52** Use two concentration tables to calculate the concentrations of all species present in a $3.45 \times 10^{-2}$ M solution of KBrO. (See Appendix E for $K$ values.)

**14.53** Determine the concentrations of the ionic species present in a $1.55 \times 10^{-2}$ M solution of the diprotic acid $H_2CO_3$. (See Table 14-5 for $K$ values.)

**14.54** Determine the concentrations of the ionic species present in a 0.355 M solution of the diprotic acid $H_2SO_3$. (See Table 14-5 for $K$ values.)

**14.55** Determine the concentrations of the ionic species present in a 55 mM solution of sodium carbonate ($Na_2CO_3$). (See Table 14-5 for $K$ values.)

**14.56** Determine the concentrations of the ionic species present in a 0.35 M solution of sodium sulphite ($Na_2SO_3$). (See Table 14-5 for $K$ values.)

## ADDITIONAL PAIRED PROBLEMS

**14.57** Boric acid ($H_3BO_3$), $K_a = 5.4 \times 10^{-10}$, is frequently used as an eyewash. (a) Use Lewis structures to illustrate the equilibrium reaction of $K_a$. (b) Calculate the pH of 0.050 M boric acid solution.

**14.58** Hydrazine ($N_2H_4$) has $K_b = 1.3 \times 10^{-6}$. (a) Use Lewis structures to illustrate the equilibrium reaction of $K_b$. (b) Calculate the pH of a $2.00 \times 10^{-1}$ M solution of $N_2H_4$.

**14.59** In the mid-1930s a substance was isolated from a fungus that is a parasite of ryes and other grasses. This alkaloid, lysergic acid, has been of great interest to chemists because of its strange, dramatic action on the human mind. Many derivatives of lysergic acid are known, some with medicinal applications. Perhaps the best known derivative of lysergic acid is the potent hallucinogen lysergic acid diethylamide (LSD):

LSD
($C_{20}H_{25}N_3O$)

Like other alkaloids, LSD is a weak base, with $K_b = 7.6 \times 10^{-7}$. What is the pH of a 0.55 M solution of LSD?

**14.60** The addictive painkiller morphine ($C_{17}H_{19}NO_3$), is the principal molecule in the milky juice that exudes from unripe poppy seed capsules. Calculate the pH of a 0.015 M solution of morphine, given that $K_b = 7.9 \times 10^{-7}$.

Morphine
($C_{17}H_{19}NO_3$)

**14.61** The pH of a 0.0100 M solution of the sodium salt of a weak acid is 11.00. What is the $K_a$ of the weak acid?

**14.62** The pH of a 0.060 M solution of a weak acid is 2.72. Calculate $K_a$ and identify the acid from among those listed in Appendix E.

**14.63** Determine the percent hydrolysis of the weak acid solution in Problem 14.57.

**14.64** Determine the percent hydrolysis of the weak acid solution in Problem 14.62.

**14.65** Putrescine, a substance with a vile odour, has two basic amino functional groups, as shown in the line structure above. Use Lewis structures to show the two proton-transfer reactions of putrescine and water.

$H_2N$ ⌇⌇⌇ $NH_2$     Putrescine
($C_4H_{12}N_2$)

**14.66** Draw Lewis structure sketches showing the aqueous-phase reaction between acetic acid and ammonia.

**14.67** Calculate the pH of a 0.250 M solution of aqueous NaF.

**14.68** Calculate the pH of a 0.025 M aqueous solution of $NH_4Cl$.

**14.69** Determine the concentrations of all species present in 2.00 M $H_2SO_4$.

**14.70** Determine the concentrations of all species present in 0.200 M aqueous $NaNO_2$.

**14.71** For each of the following compounds, identify the major species in solution and write a balanced equation for the equilibrium reaction that determines the pH: (a) $H_2SO_4$; (b) $Na_2SO_4$; (c) $NaHSO_4$; and (d) $NH_4Cl$.

**14.72** Identify major species and proton transfer equilibria for aqueous solutions of the following: (a) $NH_4NO_3$; (b) $KH_2PO_4$; (c) $H_3PO_4$; and (d) $HCO_2H$.

**14.73** Use tabulated $K_{eq}$ values to find the value of $K_{eq}$ for the following reaction:

$$HPO_4^{2-} (aq) + OH^- (aq) \leftrightarrows PO_4^{3-} (aq) + H_2O (l)$$

**14.74** Use tabulated $K_{eq}$ values to find the value of $K_{eq}$ for the following reaction:

$$HPO_4^{2-} (aq) + H_3O^+ (aq) \leftrightarrows H_2PO_4^- (aq) + H_2O (l)$$

## ADDITIONAL PROBLEMS

**14.75** Oxalic acid, a diprotic carboxylic acid found in many plants, including rhubarb, is an effective stain remover. Consider the following diagram to be a small section of an aqueous solution of oxalic acid:

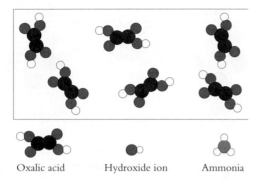

Oxalic acid      Hydroxide ion      Ammonia

(a) Draw a new picture that shows the appearance of the solution at equilibrium after four hydroxide ions enter the region. (b) Draw a new picture that shows the appearance of the solution at equilibrium after eight hydroxide ions enter the region. (c) Draw a new picture that shows the appearance of the solution at equilibrium after four ammonia molecules enter the region ($K_b = 1.8 \times 10^{-5}$).

**14.76** By coincidence, the $K_a$ of $CH_3CO_2H$ and the $K_b$ of $NH_3$ both are $1.8 \times 10^{-5}$. Write the appropriate acid–base equilibrium reactions and use them to determine whether each of the following 1.00 M solutions is acidic, basic, or neutral: (a) $CH_3CO_2H$; (b) $NH_3$; (c) $NH_4Cl$; (d) $NaCH_3CO_2$; and (e) $NH_4CH_3CO_2$.

**14.77** A solution is prepared by dissolving 3.5 g of $P_4O_{10}$ in 1.50 L of water. The oxide reacts with water quantitatively to form phosphoric acid. (a) Identify the major species in the solution. (b) Identify the minor species in the solution and rank them in order of concentration, highest first. (c) Calculate the pH of the solution.

**14.78** Aqueous solutions of $Na_2SO_3$ and $CH_3CO_2H$ are mixed. (a) List the major species in each solution. (b) Write the net ionic reaction that occurs on mixing. (c) Identify the acid, base, conjugate acid, and conjugate base.

**14.79** In aqueous solution, amino acids exist as zwitterions (German for "double ions"), compounds in which internal proton transfer gives a molecule with two (oppositely charged) functional groups. Use Lewis structures to illustrate the proton-transfer equilibrium between the uncharged form of glycine ($NH_2CH_2CO_2H$) and its zwitterion form.

**14.80** Vinegar is a dilute aqueous solution of acetic acid. A sample of vinegar has a pH of 2.39 and a density of 1.07 g/mL. What is the mass percentage of acetic acid in the vinegar?

**14.81** For each of the following reactions, write a balanced net ionic equation. Use different sizes of arrows to indicate whether the reaction goes nearly to completion or proceeds to only a small extent. (*Hint:* You may need to compare $K_a$ values.)
(a) $NaOH\ (aq) + C_6H_5CO_2H\ (s) \leftrightharpoons\ ?$
(b) $(CH_3)_3N\ (aq) + HNO_3\ (aq) \leftrightharpoons\ ?$
(c) $Na_2SO_4\ (aq) + CH_3CO_2H\ (aq) \leftrightharpoons\ ?$
(d) $NH_4Cl\ (aq) + Ca(OH)_2\ (aq) \leftrightharpoons\ ?$
(e) $K_2HPO_4\ (aq) + NH_3\ (aq) \leftrightharpoons\ ?$

**14.82** Identify each of the following as an acid, base, both, or neither and then write balanced reactions that illustrate the acidic or basic properties of the compound: (a) $H_2CO_3$; (b) $KHCO_3$; (c) $NH_3$; (d) $NaCl$; (e) $Na_2SO_4$; (f) $NH_4HSO_4$; and (g) $NaO_2CCO_2H$.

## GROUP STUDY PROBLEMS

**14.83** According to its label, each tablet of Alka-Seltzer contains 1.916 g (0.0228 mol) of sodium hydrogen carbonate. Consider an Alka-Seltzer tablet dissolved in 150 mL of water: (a) What equilibrium determines the pH? (b) What is the pH of the solution?

**14.84** When ammonium acetate dissolves in water, both the resulting ions undergo proton transfer reactions with water, but the net reaction can be written without using water:

$$NH_4^+\ (aq) + CH_3CO_2^-\ (aq) \leftrightharpoons NH_3\ (aq) + CH_3CO_2H\ (aq)$$

(a) Combine proton transfer equilibrium reactions involving water so that the net result is the above reaction, and use this sequence of reactions to derive an expression for $K_{eq}$ for this reaction in terms of $K_a$, $K_b$, and $K_w$. (b) Make a list of all the proton transfer reactions that occur in this solution. (c) Use $K_a$ and $K_b$ to calculate the concentrations of $H_3O^+$ and $OH^-$ that result from the reactions of $NH_4^+$ and $CH_3CO_2^-$ with water in a 0.25 M solution of ammonium acetate. (d) Based on the results of part (c), what pH would you expect to find for this solution?

**14.85** In each of the following situations, one reaction goes essentially to completion. In each case, identify the major species in solution under initial conditions and write a balanced net ionic equation for the reaction that goes essentially to completion. (a) Gaseous HBr is bubbled through a 0.015 M solution of $Ca(OH)_2$; (b) 4.0 g of NaOH is added to 0.75 L of 0.055 M $NaHSO_4$; and (c) 25.0 mL of saturated $NH_4I$ is mixed with 50.0 mL of 0.95 M $Pb(NO_3)_2$.

**14.86** For each of the following reactions, write a balanced net ionic equation. Use different sizes of arrows to indicate whether the reaction goes nearly to completion or proceeds to only a small extent. (*Hint:* You may need to compare $K_a$ values.)
(a) $H_2S\ (aq) + NH_3\ (aq) \leftrightharpoons\ ?$
(b) $C_2H_5NH_2\ (aq) + KHSO_4\ (aq) \leftrightharpoons\ ?$
(c) $C_5H_5N\ (aq) + HCN\ (aq) \leftrightharpoons\ ?$
(d) $NH_4Br\ (aq) + Na_3PO_4\ (aq) \leftrightharpoons\ ?$
(e) $HClO\ (aq) + HONH_2\ (aq) \leftrightharpoons\ ?$

14.87 Pure sulphuric acid ($H_2SO_4$) is a viscous liquid that causes severe burns when it contacts the skin. Like water, sulphuric acid is amphiprotic, so a proton-transfer equilibrium exists in pure sulphuric acid. (a) Write this proton-transfer equilibrium reaction. (*Hint*: $H_2O$ is NOT involved.) (b) Construct the Lewis structure of sulphuric acid and identify the features that allow this compound to function as a base. (c) Perchloric acid ($HClO_4$) is a stronger acid than sulphuric acid. Write the proton transfer reaction that takes place when perchloric acid dissolves in pure sulphuric acid.

14.88 Prepare a graph of the percent hydrolysis of aqueous acetic acid as a function of the logarithm of its concentration. To do this, calculate the percent hydrolysis at a variety of molarities ranging from 1.0 M to $10^{-7}$ M. Repeat for aqueous hypoiodous acid. What similarities are there between the two graphs? What differences are there? (*Note*: This problem is suitable for spreadsheet analysis.)

## ANSWERS TO SECTION EXERCISES

**14.1.1**  $H_3O^+ (aq) + SO_4^{2-} (aq) \leftrightarrows H_2O (l) + HSO_4^- (aq)$

**14.1.2**  $4.0 \times 10^{-11}$ M

**14.1.3**  $[K^+] = [OH^-] = 0.200$ M; $[H_3O^+] = 5.0 \times 10^{-14}$ M

**14.1.4**  The conjugate base of $HCO_3^-$ is $CO_3^{2-}$; the conjugate acid is $H_2CO_3$. The conjugate base of $HSO_4^-$ is $SO_4^{2-}$; the conjugate acid is $H_2SO_4$.

**14.2.1**  $1.5 \times 10^{-3}$ M HCl, pH = 2.82; $2.5 \times 10^{-2}$ M NaOH, pH = 12.40

**14.2.2**  $[H_3O^+] = 3.16 \times 10^{-5}$ M; $[OH^-] = 3.16 \times 10^{-10}$ M

**14.3.1**  % Hydrolysis = 0.091%; $K_a = 4.0 \times 10^{-8}$

**14.3.2**

**14.3.3**

|            | **HCl**              | **HClO**             |
|------------|----------------------|----------------------|
| **[HA]**   | 0                    | $4.8 \times 10^{-2}$ M |
| **[H₃O⁺]** | $4.8 \times 10^{-2}$ M | $3.7 \times 10^{-5}$ M |
| **[A⁻]**   | $4.8 \times 10^{-2}$ M | $3.7 \times 10^{-5}$ M |

**14.4.1**  (a) strong acid; (b) strong base; (c) neither; (d) weak acid; and (e) weak base

**14.4.2**

(d)

Acid          Conjugate base

(e)

Base          Conjugate acid

**14.4.3**  Nicotine is a weak base, having two N atoms with lone pairs of electrons. Atropine is also a weak base, having one N atom with a lone pair of electrons. Atropine also contains O atoms, but it lacks the $CO_2H$ group, so atropine is not acidic.

**14.5.1**  $[HCN] = [OH^-] = 2.4 \times 10^{-3}$ M; $[CN^-] = [K^+] = 0.35$ M; $[H_3O^+] = 4.2 \times 10^{-12}$ M

**14.5.2**  (a) $NH_4^+ (aq) + H_2O (l) \leftrightarrows NH_3 (aq) + H_3O^+ (aq)$, acidic

(b) $2 H_2O (l) \leftrightarrows OH^- (aq) + H_3O^+ (aq)$, neither

(c) $CH_3CO_2^- (aq) + H_2O (aq) \leftrightarrows CH_3CO_2H (aq) + OH^- (aq)$, basic

**14.5.3**  pH = 4.69

**14.6.1**  The salt is basic, because the anion of $NH_3$ is more basic than $NH_3$. The anion accepts a proton from water:

$$NH_2^- + H_2O \longrightarrow NH_3 + OH^-$$

**14.6.2**  (a) $H_3PO_4$ is stronger, more O atoms to withdraw electrons; (b) $H_2Se$ is stronger, higher principal quantum number means weaker H—X bond; and (c) $H_2SO_3$ is stronger, neutral acid is stronger than anion.

**14.6.3**  HBrO is weakest, then HClO, $HClO_2$, and HBr (a strong acid) is strongest.

**14.7.1**  $[OH^-] = 7.4 \times 10^{-10}$ M

**14.7.2**  The major species are $H_2PO_4^-$, $H_2O$, and $Na^+$. There are three acid–base equilibria with these species as reactants:

$$H_2PO_4^- (aq) + H_2O (aq) \leftrightarrows HPO_4^{2-} (aq) + H_3O^+ (aq)$$

$$H_2PO_4^- (aq) + H_2O (aq) \leftrightarrows H_3PO_4 (aq) + OH^- (aq)$$

$$2 H_2O (aq) \leftrightarrows H_3O^+ (aq) + OH^- (aq)$$

**14.7.3** The most important equilibrium is the one with the largest $K_{eq}$. Here are the values:

$$H_2PO_4^-\ (aq) + H_2O\ (aq) \rightleftharpoons HPO_4^{2-}\ (aq) + H_3O^+\ (aq)$$

$$K_{a2} = 6.3 \times 10^{-8}$$

$$H_2PO_4^-\ (aq) + H_2O\ (aq) \rightleftharpoons H_3PO_4\ (aq) + OH^-\ (aq)$$

$$K_b = K_w/K_{a1} = 1.4 \times 10^{-12}$$

$$2H_2O\ (aq) \rightleftharpoons H_3O^+\ (aq) + OH^-\ (aq)$$

$$K_w = 1.0 \times 10^{-14}$$

The value of $K_{a2}$ is larger than the others, so this reaction is the most important. The solution is acidic, since this reaction produces $H_3O^+$ ions.

## ANSWERS TO PRACTICE EXERCISES

**14.1** $H_2PO_4^-$: conjugate acid is $H_3PO_4$, conjugate base is $HPO_4^{2-}$; $HPO_4^{2-}$: conjugate acid is $H_2PO_4^-$, conjugate base is $PO_4^{3-}$.

**14.2** $[H_3O^+] = [Cl^-] = 6.5 \times 10^{-3}$ M; $[OH^-] = 1.5 \times 10^{-12}$ M

**14.3** $[Li^+] = [OH^-] = 0.023$ M; $[H_3O^+] = 4.4 \times 10^{-13}$ M

**14.4** $[H_3O^+] = 3 \times 10^{-11}$ M; $[OH^-] = 3 \times 10^{-4}$ M

**14.5** pOH = 1.64, pH = 12.36

**14.6** $K_a = 1.4 \times 10^{-5}$

**14.7** 2.4% ionized

**14.8** pH = 11.09

**14.9** (a), (b), and (d) weak acid; (c) neither

**14.10** Both histidine and threonine have a basic amino ($NH_2$) group and an acidic carboxyl group. The histidine side chain has two amine nitrogen atoms in the ring. The side chain of threonine has no significant acid–base properties under normal conditions.

**14.11** $[K^+] = [NO_2^-] = 0.22$ M; $[OH^-] = [HNO_2] = 2.0 \times 10^{-6}$ M; pH = 8.30

**14.12** Major species: $C_6H_5NH_3^+$, $Cl^-$, and $H_2O$; dominant equilibrium: $H_2O + C_6H_5NH_3^+ \rightleftharpoons H_3O^+ + C_6H_5NH_2$

**14.13** $HIO_3$ is stronger than HIO, HBrO, HClO, and $HClO_2$ because it has more O atoms to withdraw more electron density from the O—H bond. $HIO_3$ is weaker than $HClO_3$ because I, being less electronegative than Cl, is less effective at withdrawing electron density. $HIO_3$ is weaker than $HClO_4$ both because of electronegativity differences and because four O atoms withdraw more electron density than three O atoms do.

**14.14** $[H_3PO_4] = 0.92$ M; $[H_3O^+] = [H_2PO_4^-] = 0.080$ M; $[HPO_4^{2-}] = 6.3 \times 10^{-8}$ M; $[PO_4^{-3}] = 3.78 \times 10^{-19}$ M; $[OH^-] = 1.3 \times 10^{-13}$ M.

**14.15** Your picture should contain two carbonic acid molecules, three hydrogen carbonate anions, and three "new" water molecules.

**14.16** $[CO_3^{2-}] = 0.35$ M; $[HCO_3^-] = [OH^-] = 8.8 \times 10^{-3}$ M; $[H_2CO_3] = 2.2 \times 10^{-8}$ M; $[H_3O^+] = 1.1 \times 10^{-12}$ M; $[Na^+] = 0.72$ M; pH = 11.9

# APPLICATIONS OF AQUEOUS EQUILIBRIA

## THE pH OF BLOOD

The human body generates a steady flow of acidic by-products during its normal metabolic processes. Foremost among these is carbon dioxide, which is a major product of the reactions the body uses to produce energy (see Section 11.6). An average person produces from 10 to 20 moles (440–880 g) of $CO_2$ every day. Blood carries $CO_2$ from the cells to the lungs to be exhaled. In aqueous solution, dissolved $CO_2$ is in equilibrium with carbonic acid:

$$H_2O\ (l)\ +\ CO_2\ (aq) \rightleftarrows H_2CO_3\ (aq)$$

In the absence of some regulatory mechanism, the continuous production of $CO_2$ would give blood a pH of about 5.6. The proper functioning of blood chemistry, however, requires a pH of 7.40. Severe illness or death can result from changes of blood pH by only a few tenths of a unit.

When blood pH falls below normal limits, the condition is termed acidosis. A number of body malfunctions can lead to acidosis, including diabetes, kidney failure, and persistent diarrhoea. Temporary acidosis

can result from prolonged vigorous exercise. These situations stimulate production of $CO_2$, increasing the acidity of the blood.

The opposite condition, in which blood pH rises above tolerable limits, is called alkalosis. Alkalosis can occur as a result of hyperventilation or severe vomiting. Mountain climbers put themselves at risk of alkalosis by climbing too high without allowing proper time to become acclimated to high elevation. At high altitude there is less oxygen in the air, and climbers must compensate by breathing faster. Under these conditions, the body may lose $CO_2$ too rapidly, causing a decrease in blood acidity and a corresponding rise in pH. The result can be severe altitude sickness, which results in death if the climber is not given supplemental oxygen or moved quickly to lower elevation.

Healthy humans do not suffer these ailments, because blood pH is tightly regulated by a pair of reactions that involve $H_2CO_3$ and $HCO_3^-$. Hydronium ions that enter the bloodstream react with hydrogen carbonate, and hydroxide anions react with carbonic

# 15

## LEARNING OBJECTIVES

*Upon completion of this chapter you should be able to*

- calculate the pH of a buffered solution
- explain how to prepare a buffered solution of known pH and capacity
- calculate an acid or base concentration from titration data
- use the concepts of $K_{sp}$ and the common-ion effect to calculate solution concentrations
- calculate the concentrations of species involved in complex-ion formation

acid. These reactions work together to maintain the pH of blood at $7.40 \pm 0.05$:

$$HCO_3^- + H_3O^+ \rightleftarrows H_2CO_3 + H_2O$$

$$H_2CO_3 + OH^- \rightleftarrows HCO_3^- + H_2O$$

This process for the chemical control of pH in an aqueous solution is known as buffering.

Despite buffering, blood pH would quickly become acidic were it not for the rapid elimination of $CO_2$ from the body. In the lungs, $CO_2$ is transferred from blood cells to the gas phase and then exhaled from our bodies.

This chapter describes several important applications of aqueous equilibria. We begin with a discussion of buffer chemistry, followed by a description of acid and base titration reactions. Then we change our focus to examine the solubility equilibria of inorganic salts. The chapter concludes with a discussion of the equilibria of complex ions.

## 15.1  BUFFER SOLUTIONS

Human blood contains a variety of acids and bases that maintain the pH very close to 7.4 at all times. Close control of blood pH is critical because death results if the pH of human blood drops below 7.0 or rises above 7.8. This narrow pH range corresponds to only a five-fold change in the concentration of hydronium ions. Chemical equilibria work in the blood to hold the pH within this narrow window. Close control of pH is achieved by a **buffer solution**, so called because it protects, or buffers, the solution against pH variations.

## The Composition of Buffer Solutions

To protect a solution against pH variations, a major species in the solution must react with added hydronium ions, and another major species must react with added hydroxide ions. The conjugate base of a weak acid will react readily with hydronium ions, and the weak acid itself will react readily with hydroxide ions. This means that a buffer solution can be defined in terms of its composition: *a buffer solution contains both a weak acid and its conjugate base as major species.* As one example, dissolving sodium acetate in a solution of acetic acid produces a buffer solution in which both acetic acid and acetate anions are major species. Example 15-1 describes an acetic acid–acetate buffer solution.

### EXAMPLE 15-1  CONCENTRATIONS IN A BUFFER SOLUTION

A solution contains 0.125 mole of solid sodium acetate dissolved in 1.00 L of 0.250 M acetic acid. Determine the concentrations of hydronium ions, acetate ions, and acetic acid.

**Strategy:** The seven-step procedure described in Chapter 13 can be applied to quantitative equilibrium problems.

1. **Determine what is asked for.** This problem asks about concentrations of species in an aqueous solution.

2. **Identify the major chemical species.** The original solution contains water and acetic acid molecules. Adding sodium acetate introduces two new major species, acetate anions and sodium cations. Thus, the resulting buffer solution has four major species: $H_2O$, $Na^+$, $CH_3CO_2^-$, and $CH_3CO_2H$.

3. **Determine what chemical equilibria exist.** Acetic acid is a weak acid, acetate anion is a weak base, water can act as an acid or a base, and $Na^+$ is a spectator ion. These species are reactants in three acid–base equilibria:

$$CH_3CO_2H\ (aq) + H_2O\ (l) \rightleftarrows CH_3CO_2^-\ (aq) + H_3O^+\ (aq) \qquad K_a = 1.8 \times 10^{-5}$$

$$CH_3CO_2^-\ (aq) + H_2O\ (l) \rightleftarrows CH_3CO_2H\ (aq) + OH^-\ (aq) \qquad K_{b,\ CH_3CO_2^-} = \frac{K_w}{K_{a,\ CH_3CO_2H}} = 5.6 \times 10^{-10}$$

$$H_2O\ (l) + H_2O\ (l) \rightleftarrows H_3O^+\ (aq) + OH^-\ (aq) \qquad K_w = 1.0 \times 10^{-14}$$

4. **Write the $K_{eq}$ expressions.** Among these reactions, the first one has the largest equilibrium constant, so the acetic acid equilibrium will generate the largest changes from initial concentrations:

$$K_a = 1.8 \times 10^{-5} = \frac{[H_3O^+]_{eq}\ [CH_3CO_2^-]_{eq}}{[CH_3CO_2H]_{eq}}$$

**Solution:**

5. *Organize the data and unknowns.* Set up a concentration table for this equilibrium:

| Reaction | $H_2O$ $(l)$ | + | $CH_3CO_2H$ $(aq)$ | $\rightleftarrows$ | $CH_3CO_2^-$ $(aq)$ | + | $H_3O^+$ $(aq)$ |
|---|---|---|---|---|---|---|---|
| Initial Concentration (M) | | | 0.250 | | 0.125 | | 0 |
| Change in Concentration (M) | | | $-x$ | | $+x$ | | $+x$ |
| Equilibrium Concentration (M) | | | $0.250 - x$ | | $0.125 + x$ | | $x$ |

6. *Substitute and calculate.* Because the initial concentrations are much larger than $K_{eq}$, we make the approximation that $x$ is negligible compared with the initial concentrations:

$$0.250 - x \cong 0.250 \qquad\qquad 0.125 + x \cong 0.125$$

To determine the concentration of hydronium ions, rearrange the expression for $K_a$ and substitute these concentrations:

$$K_a = \frac{[H_3O^+]_{eq}[CH_3CO_2^-]_{eq}}{[CH_3CO_2H]_{eq}} \quad \text{so} \quad [H_3O^+]_{eq} = \frac{K_a[CH_3CO_2H]_{eq}}{[CH_3CO_2^-]_{eq}}$$

$$x = \frac{(1.8 \times 10^{-5})(0.250)}{0.125} = 3.6 \times 10^{-5}$$

$$[H_3O^+]_{eq} = 3.6 \times 10^{-5} \text{ M}$$

$$[CH_3CO_2H]_{eq} = 0.250 \text{ M} \qquad\qquad [CH_3CO_2^-]_{eq} = 0.125 \text{ M}$$

**Do the Results Make Sense?** Notice that the $H_3O^+$ concentration is much smaller than the concentrations of acetic acid and acetate, so the approximation that $x$ is negligibly small is valid. The concentrations of acetic acid and acetate are essentially unchanged, reflecting the small equilibrium constant.

**Common Pitfalls:** Not writing down the chemical equilibria that apply. Such problems cannot be solved without first doing this.

**Practice Exercise 15.1**
A solution contains 0.360 mole of ammonia dissolved in 1.5 L of 0.125 M ammonium chloride. Determine the concentrations of hydroxide ions, ammonium ions, and ammonia molecules.

The analysis carried out in Example 15-1 reveals one of the key features of buffer solutions: the equilibrium concentrations of both the weak acid and its conjugate base are essentially the same as their initial concentrations.

In the laboratory, chemists prepare buffer solutions in three different ways. Each results in a solution containing a weak acid and its conjugate base as major species. The most straightforward way to produce a buffer solution is by dissolving a salt of a weak acid in a solution of the same weak acid, as described in Example 15-1.

A second way to prepare a buffer is by adding a strong base to a solution of a weak acid. This produces a buffer solution if the number of moles of strong base is about half the number of moles of weak acid. As a simple example, if 1 L of

0.5 M NaOH is mixed with 1 L of 1.0 M $CH_3CO_2H$, hydroxide anions react quantitatively with acetic acid molecules:

$$OH^- + CH_3CO_2H \rightleftarrows H_2O + CH_3CO_2^-$$

$$K_{eq} = \frac{1}{K_{b,\,CH_3CO_2^-}} = \frac{K_{a,CH_3CO_2H}}{K_w} = 1.8 \times 10^9$$

The 0.5 mole of added hydroxide converts 0.5 mole of acetic acid into acetate, producing a buffer solution containing 0.5 mole of acetate and 0.5 mole of acetic acid.

A third approach to buffer solutions is to add strong acid to a solution of a weak base. This produces a buffer solution if the amount of strong acid is about half the amount of weak base. Continuing with our examples of acetic acid–acetate buffers, if a solution of hydrochloric acid is added to a solution of sodium acetate, then hydronium ions react quantitatively with acetate anions:

$$H_3O^+ + CH_3CO_2^- \rightleftarrows H_2O + CH_3CO_2H \qquad K_{eq} = \frac{1}{K_{a,\,CH_3CO_2H}} = 5.6 \times 10^4$$

Example 15-2 treats a buffer solution made in this way.

## EXAMPLE 15-2 STRONG ACID–WEAK BASE BUFFER

Determine if a solution prepared by mixing 5.0 mL of concentrated HCl (12 M) with 75 mL of 1.0 M sodium acetate is a buffer solution.

**Strategy:** A buffer solution contains both a weak acid and its conjugate base as major species. Thus, to answer the question, we must calculate the concentrations of acetate anions and acetic acid in the solution. We use a compressed version of the seven-step method to obtain these concentrations.

**Solution:** Begin by identifying the major species in the two starting solutions:

$$HCl:\ H_3O^+ \text{ and } Cl^- \qquad \text{Sodium acetate: } CH_3CO_2^- \text{ and } Na^+$$

Because the hydronium ion is a strong acid and the acetate anion is a weak base, mixing the solutions results in near-quantitative reaction between these two ions:

$$H_3O^+ + CH_3CO_2^- \rightleftarrows H_2O + CH_3CO_2H \qquad K_{eq} = \frac{1}{K_{a,\,CH_3CO_2H}} = 5.6 \times 10^4$$

We calculate the initial amounts of the species from the volumes and concentrations, using $n = cV$:

$$n(H_3O^+) = (12\text{ M})(0.0050\text{ L}) = 0.060\text{ mol}$$

$$n(CH_3CO_2^-) = (1.9\text{ M})(0.075\text{ L}) = 0.075\text{ mol}$$

When the solutions are mixed, the reaction between $H_3O^+$ and $CH_3CO_2^-$ consumes all the hydronium ions and generates acetic acid molecules. After mixing the starting solutions, we have the following amounts:

$$n(CH_3CO_2^-) = 0.075\text{ mol} - 0.060\text{ mol} = 0.015\text{ mol}$$

$$n(CH_3CO_2H) = 0.060\text{ mol}$$

Both the weak acid and its conjugate base are major species in the mixture, so this is indeed a buffer solution.

**Does the Result Make Sense?** The sum of the moles of the products is equal to the number of initial moles of acetate, and the final amounts of acid and base are comparable. Thus, the calculations support the conclusion that the final mixture is a buffer solution.

**Practice Exercise 15.2**
Calculate the concentrations of the major species present in a buffer solution prepared by adding 1.40 g NaOH to 140 mL of 0.750 M NH₄Cl.

## Molecular View of a Buffer Solution

The purpose of a buffer solution is to maintain the pH within a narrow range. The reactions that occur when $H_3O^+$ or $OH^-$ is added to a buffer solution show how this is accomplished. Consider what happens when a small amount of hydroxide ion is added to the acetic acid–acetate buffer solution described in Example 15-1. Hydroxide ion is a strong base and acetic acid is a weak acid, so proton transfer from $CH_3CO_2H$ to $OH^-$ goes essentially to completion:

Collision and proton transfer

$$OH^- + CH_3CO_2H \rightleftharpoons H_2O + CH_3CO_2^- \qquad K_{eq} = \frac{1}{K_{b,\,CH_3CO_2^-}} = 1.8 \times 10^9$$

As long as the buffer solution contains acetic acid as a major species, a small amount of hydroxide ion added to the solution will be neutralized completely. Figure 15-1 shows two hydroxide ions added to a portion of a buffer solution. When a hydroxide ion collides with a molecule of weak acid, proton transfer forms a water molecule and the conjugate base of the weak acid. As long as there are more weak acid molecules in the solution than the number of added hydroxide ions, the proton transfer reaction goes virtually to completion. Weak acid molecules change into conjugate base anions as they react with added hydroxide.

**Figure 15-1**

When a strong base is added to a buffer solution, the weak acid $HA$ donates protons to hydroxide ions to form the conjugate base $A^-$, preventing a large increase in hydroxide ion concentration. (All water molecules except those produced in the proton transfer process are omitted for clarity.)

The same molecular reasoning shows that a buffer solution can absorb added hydronium ions. Consider what happens when some hydronium ions are added to the acetic acid–acetate buffer solution described in Example 15-1. The hydronium ion is a strong acid, and the acetate anion is a weak base, so proton transfer from $CH_3CO_2^-$ to $H_3O^+$ goes essentially to completion:

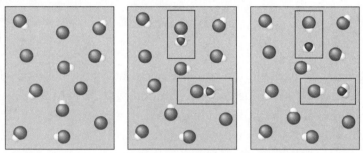

Collision and proton transfer

$$H_3O^+ + CH_3CO_2^- \rightleftharpoons H_2O + CH_3CO_2H \qquad K_{eq} = \frac{1}{K_{a,\ CH_3CO_2H}} = 5.6 \times 10^4$$

As long as the buffer solution contains a weak base as a major species, a small amount of hydronium ion added to the solution will be neutralized completely. Figure 15-2 shows two hydronium ions added to a portion of a buffer solution. When a hydronium ion collides with a weak base ion, proton transfer occurs, forming a water molecule and the conjugate acid of the weak base. Example 15-3 examines another buffer solution at the molecular level.

**Figure 15-2**

When a strong acid is added to a buffer solution, the conjugate base $A^-$ accepts protons from hydronium ions to form the weak acid $HA$, preventing a large increase in hydronium ion concentration. (All water molecules except those produced in the proton transfer process are omitted for clarity.)

---

## EXAMPLE 15-3 MOLECULAR VIEW OF A BUFFER SOLUTION

The molecular picture at right represents a small portion of a buffer system. Solvent water molecules are omitted for clarity.
Redraw the original figure to show the equilibrium condition that is established when (a) three hydroxide ions enter the region, and (b) seven hydronium ions enter the region. Include any water molecules that are part of the buffer chemistry.

**Strategy:** The problem asks for a drawing that represents equilibrium conditions. We need a stoichiometric analysis of the reaction components. A table of amounts helps organize the information. The problem has two parts, and it is best to treat them individually.

**Solution:** For part (a), hydroxide ions are neutralized by the weak acid component of the buffer. Set up a table of amounts, using a molecular picture to represent the balanced equation.

| Reaction |  | | | |
|---|---|---|---|---|
| Start (molecules/ions) | 6 | 3 | 5 | "0" |
| Change | −3 | −3 | +3 | +3 |
| Final | 3 | 0 | 2 | 3 |

Of course, the true amount of water is not "0." Water is the solvent, so there is an abundance of water molecules in the solution. This column represents only the water molecules produced by the buffer chemistry. We now redraw the system to show the molecular species present at the end of the reaction:

In part (b), the weak base of the buffer system reacts with hydronium ions:

For this example, the weak base is the limiting reactant, leaving excess $H_3O^+$ in the final solution. The buffer has been destroyed. We leave it to you to construct a table of amounts that leads to the final drawing, which appears at right:

**Do the Results Make Sense?** Inventories of the components of the figures match the stoichiometry of the reactions, so the final pictures make sense.

**Practice Exercise 15.3**
Redraw the original figure of Example 15-3 to show the final condition that is established when (a) nine $OH^-$ ions enter the region, and (b) four $H_3O^+$ ions enter the region.

# The Buffer Equation

The pH of a buffer solution depends on the weak acid equilibrium constant and the concentrations of the weak acid and its conjugate base. To show this, we begin by taking the logarithm of the acid equilibrium constant:

$$K_a = \frac{[H_3O^+]_{eq}[A^-]_{eq}}{[HA]_{eq}} \quad \text{so} \quad \log K_a = \log\left(\frac{[H_3O^+]_{eq}[A^-]_{eq}}{[HA]_{eq}}\right)$$

The logarithm of the concentration quotient can be separated by using the fact that $\log xy = \log x + \log y$:

$$\log K_a = \log[H_3O^+]_{eq} + \log\left(\frac{[A^-]_{eq}}{[HA]_{eq}}\right)$$

Multiplying both sides of the equation by $-1$ allows the use of $pK_a$ and pH instead of $\log K_a$ and $\log [H_3O^+]_{eq}$:

$$-\log K_a = -\log[H_3O^+]_{eq} - \log\left(\frac{[A^-]_{eq}}{[HA]_{eq}}\right) \quad \text{and} \quad pK_a = pH - \log\left(\frac{[A^-]_{eq}}{[HA]_{eq}}\right)$$

Now we rearrange this equation to solve for pH:

$$pH = pK_a + \log\left(\frac{[A^-]_{eq}}{[HA]_{eq}}\right)$$

This equation is exact, but it can be simplified by applying one of the key features of buffer solutions. Any buffer solution contains both members of a conjugate acid–base pair as major species. In other words, both the weak acid and its conjugate base are present in relatively large amounts. As a result, the change to equilibrium, $x$, is small relative to each initial concentration, and the equilibrium concentrations are virtually the same as the initial concentrations:

$$[A^-]_{eq} = [A^-]_{initial} - x \cong [A^-]_{initial}$$

$$[HA]_{eq} = [HA]_{initial} + x \cong [HA]_{initial}$$

Thus, instead of using equilibrium concentrations in this equation, we substitute the initial concentrations to give the **buffer equation:**

$$pH = pK_a + \log\left(\frac{[A^-]_{initial}}{[HA]_{initial}}\right) \qquad (15\text{-}1)$$

This is the Henderson–Hasselbalch equation, commonly known as the buffer equation, and is used to calculate the equilibrium pH of a buffer solution directly from initial concentrations. The approximation is valid as long as the difference between initial concentrations and equilibrium concentrations is negligibly small. As a rule of thumb, the buffer equation can be applied when initial concentrations of $HA$ and $A^-$ differ by less than a factor of 10. Example 15-4 provides an illustration of the use of the buffer equation.

## EXAMPLE 15-4 THE BUFFER EQUATION

Buffer solutions with pH values around 10 are prepared using sodium carbonate ($Na_2CO_3$) and sodium hydrogen carbonate ($NaHCO_3$). What is the pH of a solution prepared by dissolving 10.0 g each of these two salts in enough water to make 0.250 L of solution?

**Strategy:** We use the seven-step strategy for equilibrium problems, except that we identify this as a buffer solution. This allows us to use the buffer equation in place of an equilibrium constant expression.

1. We are asked to calculate the pH of a buffer solution.

2. Both compounds are salts that dissolve in water to give their constituent ions, so the major species in this buffer solution are $H_2O$, $Na^+$, $HCO_3^-$, and $CO_3^{2-}$.

3. A buffer solution must contain a weak acid and its conjugate base as major species. In this solution, $HCO_3^-$ is the weak acid, and $CO_3^{2-}$ is the conjugate base:

$$H_2O\ (l) + HCO_3^-\ (aq) \rightleftarrows CO_3^{2-}\ (aq) + H_3O^+\ (aq)$$

4. This proton transfer reaction involves the second acidic hydrogen atom of carbonic acid, so the appropriate equilibrium constant is $K_{a2}$, the p$K$ of which is found in Appendix E: p$K_{a2}$ = 10.33. Because this is a buffer solution, we apply the buffer equation:

$$pH = pK_a + \log\left(\frac{[A^-]_{initial}}{[HA]_{initial}}\right)$$

**Solution:**

5. We find the initial concentrations from the masses of the *salts* and the volume of the solution:

$$[HCO_3^-]_{initial} = \frac{m}{(M)V} = \frac{10.0\ g}{(84.01\ g/mol)(0.250\ L)} = 0.476\ M$$

$$[CO_3^{2-}]_{initial} = \frac{m}{(M)V} = \frac{10.0\ g}{(106.0\ g/mol)(0.250\ L)} = 0.377\ M$$

6. Now substitute the appropriate values into the buffer equation and evaluate:

$$pH = pK_a + \log\left(\frac{[A^-]_{initial}}{[HA]_{initial}}\right) = 10.33 + \log\left(\frac{0.377\ M}{0.476\ M}\right)$$

$$pH = 10.33 + (-0.101) = 10.23$$

 **Does the Result Make Sense?** The pH is close to the p$K_a$ of the conjugate acid–base pair, so this is a reasonable result.

**Practice Exercise 15.4**

An acidic buffer solution can be prepared from phosphoric acid and dihydrogen phosphate. What is the pH of solution prepared by mixing 23.5 g of $NaH_2PO_4$ and 15.0 mL of concentrated phosphoric acid (14.7 M) in enough water to give 1.25 L of solution?

# Buffer Action

When a strong acid is added to a buffer solution, hydronium ions react with the conjugate base $A^-$, lowering the concentration of $A^-$ and increasing the concentration of HA. This reaction lowers the pH of the buffer solution, but the increase in $H_3O^+$ concentration is much smaller than would be generated by the same amount of strong acid in an unbuffered solution. Similarly, when a strong base is added to a buffer solution, hydroxide ions react with the acid HA, lowering the concentration of HA and increasing the concentration of $A^-$. In this case, the pH of the buffer solution rises, but the increase in hydroxide ion concentration is much smaller than would be generated by the same amount of strong base added to an unbuffered solution. Example 15-5 shows a quantitative calculation.

## EXAMPLE 15-5 CHANGE IN BUFFER pH

By how much does the pH of the buffer solution of Example 15-4 change on the addition of 3.50 mL of 6.0 M HCl?

**Strategy:** The first four steps of the seven-step strategy are identical to the ones in Example 15-4. In this example, addition of a strong acid or base modifies the concentrations that go into the buffer equation. We need to determine the new concentrations (Step 5) and then apply the buffer equation (Step 6). In dealing with changes in amounts of acid and base, it is often convenient to work with moles rather than molarities. The units cancel in the concentration term of the buffer equation, so the ratio of concentrations can be expressed as a ratio of moles as well as a ratio of molarities:

$$\frac{[A^-]}{[HA]} = \frac{n_{A^-}/V}{n_{HA}/V} = \frac{n_{A^-}}{n_{HA}}$$

**Solution:** First, determine the amounts present in the solution before addition of the HCl:

$$n_{HA} = \frac{10.0 \text{ g}}{84.01 \text{ g/mol}} = 1.19 \times 10^{-1} \text{ mol}$$

$$n_{A^-} = \frac{10.0 \text{ g}}{106.0 \text{ g/mol}} = 9.43 \times 10^{-2} \text{ mol}$$

Next, calculate the amount of hydronium ions added:

$$n_{H_3O^+} = cV = (6.0 \text{ mol/L})(3.5 \text{ mL})(10^{-3} \text{ L/mL}) = 2.1 \times 10^{-2} \text{ mol}$$

The hydronium ions react completely with conjugate base anions, increasing the amount of $HCO_3^-$ and reducing the amount of $CO_3^{2-}$:

$$n_{HA} = 1.19 \times 10^{-1} \text{ mol} + 2.1 \times 10^{-2} \text{ mol} = 1.40 \times 10^{-1} \text{ mol}$$

$$n_{A^-} = 9.43 \times 10^{-2} \text{ mol} - 2.1 \times 10^{-2} \text{ mol} = 7.3 \times 10^{-2} \text{ mol}$$

Finally, substitute these new amounts into the buffer equation to compute the new pH:

$$pH = pK_a + \log\left(\frac{n_{A^-}}{n_{HA}}\right) = 10.33 + \log\left(\frac{0.073 \text{ mol}}{0.140 \text{ mol}}\right) = 10.33 - 0.28 = 10.05$$

**Does the Result Make Sense?** From Example 15-4, the pH of the buffer before adding the acid was 10.23, so the addition of $2.1 \times 10^{-2}$ mole of acid reduces the pH by 0.18 pH units. This small change is a reasonable outcome for a buffer solution.

**Common Pitfalls:** Inverting the quantity inside the parentheses. Make sure that if acid is added, the pH *decreases*.

### Practice Exercise 15.5

By how much does the pH of the buffer solution of Example 15-4 change on the addition of 0.92 g of NaOH?

The change in pH experienced by the buffer solution in Example 15-5 is tiny compared with the change that would result from adding $2.1 \times 10^{-2}$ mol of acid to 250 mL of water. The pH of water would change from 7.00 to 1.08, a decrease of 5.92 pH units. Figure 15-3 illustrates and contrasts the two changes.

**Figure 15-3**

Buffer solution $\xrightarrow[\text{6.0 M HCl}]{\text{Add 3.50 mL}}$ $\Delta(\text{pH}) = -0.18$          Pure water $\xrightarrow[\text{6.0 M HCl}]{\text{Add 3.50 mL}}$ $\Delta(\text{pH}) = -6$

Addition of strong acid to a buffer solution changes the pH by much less than does addition of strong acid to water.

## SECTION EXERCISES

●**15.1.1** Which of the following sets of chemicals can be used to prepare buffer solutions? For each one that can, specify the weak acid and its conjugate base that are major species in the buffer solution: (a) HCl and KCl; (b) HCl and $KNO_2$; (c) HCl and $NH_4Cl$; (d) NaOH and $Na_2HPO_4$; and (e) NaCl and $NaC_2H_3O_2$.

●**15.1.2** A buffer solution made from $NH_4Cl$ and $NH_3$ is used to control pH from pH $= 8$ to 10. Write balanced equations that show how this buffer system neutralizes $H_3O^+$ and $OH^-$.

●**15.1.3** A 1.50-L buffer solution is prepared from 0.200 mol $NH_4Cl$ and 0.112 mol $NH_3$. (a) What is the pH of the buffer solution? (b) What would be the pH after adding 1.70 g of NaOH to the buffer described in Part (a)? (c) What would be the pH of a solution made by adding 1.70 g of NaOH to 1.50 L of water?

## 15.2 CAPACITY AND PREPARATION OF BUFFER SOLUTIONS

Buffer solutions are practical and commonplace. In fact, many chemists and biologists use buffer solutions on a daily basis. Thus, it is important to know the limits of a buffer solutioní̈s capacity to control pH, as well as how to make buffer solutions.

## Buffer Capacity

When small amounts of hydronium or hydroxide ions are added to a buffer solution, the pH changes are very small. There is a limit, however, to the amount of protection that a buffer solution can provide. After either buffering agent is consumed, the solution loses its ability to maintain near-constant pH. The **buffer capacity** of a solution is the amount of added $H_3O^+$ or $OH^-$ that the buffer solution can tolerate without exceeding a specified pH range.

Buffer capacity is determined by the amounts of weak acid and conjugate base present in the solution. If enough $H_3O^+$ is added to react completely with the conjugate base, the buffer is destroyed. Likewise, the buffer is destroyed if enough $OH^-$ is added to consume all of the weak acid. Consequently, buffer capacity depends on the overall concentration as well as the volume of the buffer solution. A buffer solution of overall concentration 0.50 M has five times the capacity as an equal volume of a buffer solution of overall concentration 0.10 M. Two litres of 0.10 M buffer solution has twice the capacity as 1 L of the same buffer solution. Example 15-6 includes a calculation involving buffer capacity.

## EXAMPLE 15-6 BUFFER CAPACITY

Biochemists and molecular biologists use phosphate buffers to match physiological conditions. A buffer solution that contains $H_2PO_4^-$ as the weak acid and $HPO_4^{2-}$ as the weak base has a pH value very close to 7.0. A biochemist prepares 0.250 L of a buffer solution that contains 0.225 M $HPO_4^{2-}$ and 0.330 M $H_2PO_4^-$. What is the pH of this buffer solution? Is the buffering action of this solution destroyed by addition of 0.40 g of NaOH?

**Strategy:** Use the seven-step strategy to calculate the pH of the buffer solution using the buffer equation. Then compare the amount of acid in the solution with the amount of added base. Buffer action is destroyed if the amount of added base is sufficient to react with all the acid. The buffering action of this solution is created by the weak acid $H_2PO_4^-$ and its conjugate base $HPO_4^{2-}$. The equilibrium constant for this pair is $K_{a2}$ of phosphoric acid:

$$H_2O\ (l) + H_2PO_4^-\ (aq) \rightleftharpoons HPO_4^{2-}\ (aq) + H_3O^+\ (aq) \qquad K_{eq} = K_{a2}$$

**Solution:** Substitute the tabulated value of $pK_{a2}$ and the given concentrations into the buffer equation to calculate pH:

$$pH = pK_a + \log\left(\frac{[A^-]_{initial}}{[HA]_{initial}}\right) = 7.21 + \log\left(\frac{0.225\ M}{0.330\ M}\right) = 7.04$$

The question asks whether addition of base destroys the buffering capacity of the solution, so compare the amount of acid present with the amount of base added. As solid NaOH is added, each hydroxide ion that enters the buffer solution consumes one $H_2PO_4^-$ ion and produces one ion of $HPO_4^{2-}$ and one water molecule:

$$OH^-\ (aq) + H_2PO_4^-\ (aq) \longrightarrow H_2O\ (l) + HPO_4^{2-}\ (aq)$$

$$n_{OH^-\,added} = m/M = 0.40\ g/40.00\ g/mol = 0.010\ mol$$

$$n_{H_2PO_4^-\,present} = cV = (0.330\ M)(0.250\ L) = 0.0825\ mol$$

The amount of base added is considerably less than the amount of acid initially present in the buffer solution, so the solution will still act as a buffer. The added base raises the pH, however, because reaction of $OH^-$ with the buffer increases the amount of $HPO_4^{2-}$ and decreases the amount of $H_2PO_4^-$. We leave it to you to use the procedure illustrated in Example 15-5 to show that the pH after addition of this solid is 7.17, an increase of 0.13 pH units.

**Do the Results Make Sense?** The pH of the buffer solution, both before and after adding the solid NaOH, is close to the $pK_a$ of the conjugate acid–base pair. Moreover, the pH increases when NaOH is added. The solution becomes more basic as a consequence of the added hydroxide anions.

### Practice Exercise 15.6

Addition of 5.25 g of NaOH to the buffer solution described in Example 15-4 would exceed the capacity of the buffer. Calculate the concentration of excess hydroxide ion and the pH of the solution.

## Buffer Preparation

The buffer equation indicates that the pH of a buffer solution is close to the $pK_a$ of the acid used to prepare the buffer if the concentration of the acid and its conjugate base are roughly equal:

$$pH = pK_a + \log\left(\frac{[A^-]_{initial}}{[HA]_{initial}}\right)$$

Every weak acid has a specific $pK_a$ that determines the pH range over which it can serve as a buffering agent. Remember that a buffer solution must contain a weak acid and its conjugate weak base as major species. This condition is met when the ratio of weak base to weak acid is between 0.1 and 10. The buffer equation translates this restriction into a pH range over which the acid and its conjugate base can serve as an effective buffer:

$$pH_{low} = pK_a + \log 0.1 = pK_a - 1 \qquad pH_{high} = pK_a + \log 10 = pK_a + 1$$

$$\text{or, pH range} = pK_a \pm 1$$

With a given weak acid, a buffer solution can be prepared at any pH within about one unit of its $pK_a$ value. Suppose, for example, that a biochemist needs a buffer system to maintain the pH of a solution close to 5.0. What reagents should be used? According to the previous analysis, the weak acid can have a $pK_a$ between 4.0 and 6.0. As the $pK_a$ deviates from the desired pH, however, the solution has a reduced buffer capacity. Thus, a buffer has maximum capacity when its acid has its $pK_a$ as close as possible to the target pH. Table 15-1 lists some acid–base pairs often used as buffer solutions. For a pH = 5.0 buffer, acetic acid ($pK_a = 4.75$) and its conjugate base, acetate, would be a good choice.

A buffer solution must contain both the acid and its conjugate base, so at least two reagents must be added to water to prepare a buffer solution. An acetate buffer can be prepared, for example, from pure water, concentrated acetic acid, and an acetate salt. The cation contained in the salt should not have acid–base properties of its own, so sodium acetate would be an appropriate choice, but ammonium acetate would not.

## TABLE 15-1 COMMON BUFFER SYSTEMS

| Acid | Conjugate Base | $pK_a$ | pH Range |
|------|----------------|--------|----------|
| $H_3PO_4$ | $H_2PO_4^-$ | 2.16 | 1–3 |
| $HCO_2H$ | $HCO_2^-$ | 3.75 | 3–5 |
| $CH_3CO_2H$ | $CH_3CO_2^-$ | 4.75 | 4–6 |
| $H_2PO_4^-$ | $HPO_4^{2-}$ | 7.21 | 6–8 |
| $NH_4^+$ | $NH_3$ | 9.25 | 8–10 |
| $HCO_3^-$ | $CO_3^{2-}$ | 10.33 | 9–11 |
| $HPO_4^{2-}$ | $PO_4^{3-}$ | 12.32 | 11–13 |

The chemist must prepare a solution that contains acetic acid and acetate in amounts that generate a buffer with pH = 5.00. The buffer equation is used to calculate the desired molar ratio:

$$5.00 = 4.75 + \log\frac{[\text{Acetate}]}{[\text{Acetic acid}]} \quad \text{so} \quad \log\frac{[\text{Acetate}]}{[\text{Acetic acid}]} = 0.25$$

$$\frac{[\text{Acetate}]}{[\text{Acetic acid}]} = 10^{0.25} = 1.8$$

For the preparation of a buffer solution of pH = 5.00, sodium acetate and acetic acid should be added to pure water in a molar ratio of 1.8:1.0. The exact amounts must be calculated using the volume and concentration of the solution, as Example 15-7 illustrates.

## EXAMPLE 15-7 BUFFER PREPARATION

What mass of sodium acetate ($NaCH_3CO_2$, $M = 82.04$ g/mol) and what volume of concentrated acetic acid (17.45 M) should be used to prepare 1.5 L of a buffer solution at pH = 5.00 that is 0.150 M overall?

**Strategy:** Because we know we are dealing with a buffer solution made from a specific conjugate acid–base pair, we can work directly with the buffer equation. We need to calculate the ratio of concentrations of conjugate base and acid that will produce a buffer solution of the desired pH. Then we use mole–mass–volume relationships to translate the ratio into actual quantities.

**Solution:** The problem specifies a solution volume of 1.5 L with a total molarity of 0.50 mol/L. The total molarity is the combined concentration of the two buffer components:

$$M_{\text{acetate}} + M_{\text{acetic acid}} = 0.150 \text{ M}$$

Use the total volume of the solution, 1.5 L, to determine the total number of moles in the system:

$$n = cV = (0.150 \text{ mol/L})(1.5 \text{ L}) = 0.225 \text{ mol}$$

$$n_{\text{acetate}} + n_{\text{acetic acid}} = 0.225 \text{ mol}$$

The calculation involving the buffer equation appears in the text: a buffer solution with pH = 5.00 requires an acetate–acetic acid molar ratio of 1.8. This ratio can be rewritten as a molar equality:

$$\frac{n_{\text{acetate}}}{n_{\text{acetic acid}}} = 1.8 \quad \text{so} \quad n_{\text{acetate}} = 1.8 \, n_{\text{acetic acid}}$$

Now substitute and calculate the required moles of acetic acid:

$$1.8 \, n_{\text{acetic acid}} + n_{\text{acetic acid}} = 0.225 \text{ mol}$$

$$n_{\text{acetic acid}} = \frac{0.225 \text{ mol}}{2.8} = 0.0804 \text{ mol}$$

The rest of the 0.225 mol must be acetate:

$$n_{\text{acetate}} = 0.225 \text{ mol} - 0.0804 \text{ mol} = 0.145 \text{ mol}$$

Finally, use molarity and molar mass to convert from moles to measurable amounts:

$$m_{\text{sodium acetate}} = (0.145 \text{ mol})(82.04 \text{ g/mol}) = 12 \text{ g}$$

$$V_{\text{acetic acid}} = \frac{(8.04 \times 10^{-2} \text{ mol})(10^3 \text{ mL/L})}{17.45 \text{ mol/L}} = 4.6 \text{ mL}$$

The final values are rounded to two significant figures to match the precision of the mole ratio and the total volume of the solution.

 **Does the Result Make Sense?** The units are correct and the quantities are appropriate for approximately tenth-molar concentrations, so the answer appears to be reasonable.

**Practice Exercise 15.7**

What masses of $Na_2HPO_4$ and $NaH_2PO_4$ are required to prepare 2.0 L of a buffer solution that is at pH = 7.40 and a total concentration of 0.20 M?

As described in Section 15.1, there are two other ways to prepare the buffer solution described in Example 15-7. Concentrated acetic acid could be added to pure water, followed by enough sodium hydroxide to generate the required 1.8:1.0 ratio of acetate to acetic acid:

$$CH_3CO_2H\ (aq) + OH^-\ (aq) \longrightarrow CH_3CO_2^-\ (aq) + H_2O\ (l)$$

Alternatively, a solution of sodium acetate could be prepared and a strong acid added to reach the proper acetate:acetic acid ratio:

$$CH_3CO_2^-\ (aq) + H_3O^+\ (aq) \longrightarrow CH_3CO_2H\ (aq) + H_2O\ (l)$$

Buffer preparation requires detailed, step-by-step calculations. Example 15-8 illustrates the complete procedure.

## EXAMPLE 15-8 PREPARING A BUFFER

A chemist wants to prepare a buffer solution at pH = 9.00 with an overall concentration of 0.125 mol/L. She has solutions of 1.00 M HCl and NaOH and bottles of all common salts. What reagents should be used, and in what quantities, to prepare 1.00 L of a suitable buffer?

**Strategy:** A practical problem in solution preparation usually requires a different strategy than our standard seven-step procedure. The chemist must first identify a suitable conjugate acid–base pair and decide what reagents to use. Then the concentrations must be calculated, using pH and total concentration. Finally, she must determine the amounts of starting materials. Of the buffer systems listed in Table 15-1, the combination of $NH_3$ and $NH_4^+$ has the proper pH range for the required buffer solution.

Apparently, no bottles of aqueous ammonia are present in the laboratory, so the components of the buffer solution must come from the salts. The chemist needs an ammonium salt with a counter anion that has no acid–base properties. Ammonium chloride ($NH_4Cl$) would be an appropriate choice. This salt contains the conjugate acid, $NH_4^+$, and the technician can generate $NH_3$ by adding a strong base to the ammonium chloride solution:

$$NH_4^+\ (aq) + OH^-\ (aq) \longrightarrow NH_3\ (aq) + H_2O\ (l)$$

**Solution:** What concentrations of $NH_4^+$ and $NH_3$ are required? First use the buffer equation to find the proper ratio of base to acid, then use the total molarity of the solution to determine the concentrations of $NH_3$ and $NH_4^+$. The ratio of ammonia to ammonium ion must produce a pH of 9.00:

$$pH = pK_a + \log\frac{[Base]}{[Acid]} \qquad 9.00 = 9.25 + \log\frac{[NH_3]}{[NH_4^+]}$$

$$\log\frac{[NH_3]}{[NH_4^+]} = -0.25 \quad \text{from which} \quad \frac{[NH_3]}{[NH_4^+]} = 10^{-0.25} = 0.56 \quad \text{or} \quad [NH_3] = 0.56[NH_4^+]$$

Use the total molarity of the buffer solution to find the concentrations:

$$[NH_3] + [NH_4^+] = 0.125 \text{ M}$$

Substitute for $[NH_3]$:    $0.56[NH_4^+] + [NH_4^+] = 0.125 \text{ M}$

$$[NH_4^+] = \frac{0.125 \text{ M}}{1.56} = 0.0801 \text{ M}$$

$$[NH_3] = 0.125 \text{ M} - 0.0801 \text{ M} = 0.045 \text{ M}$$

Finally, use stoichiometry to calculate actual amounts. The buffer solution contains both ammonia and ammonium ions, but both species are derived from ammonium chloride. For the preparation of the solution, some of the $NH_4^+$ ions in an aqueous solution of ammonium chloride must be converted into $NH_3$ molecules. Thus, the amount of salt required is found from the total molarity of the buffer solution:

$$n_{NH_4Cl} = cV = (0.125 \text{ mol/L})(1.00 \text{ L}) = 0.125 \text{ mol}$$

$$m_{NH_4Cl} = nM = (0.125 \text{ mol})(53.49 \text{ g/mol}) = 6.69 \text{ g}$$

Ammonia is generated from $NH_4^+$ by adding sodium hydroxide solution:

$$n_{OH^-} = n_{NH_3} = (0.045 \text{ M})(1.00 \text{ L}) = 0.045 \text{ mol}$$

$$\text{Volume of 1.00 mol NaOH } (aq) = \frac{0.045 \text{ mol}}{1.00 \text{ mol/L}} = 0.045 \text{ L}$$

To make the buffer, the chemist should therefore mix together 6.69 g of $NH_4Cl$, 45 mL of 1.00 M NaOH, and enough water to make 1.00 L of solution.

 **Does the Result Make Sense?** The units are correct and the quantities seem reasonable for the target volume and concentrations, so the answer does make sense.

**Practice Exercise 15.8**
A chemist needs a buffer solution at pH = 5.25 with an overall concentration of 0.175 M. The chemist has bottles of all common salts and solutions of 1.50 M HCl and NaOH. What reagents should be used, and in what quantities, to prepare 2.50 L of a suitable buffer solution?

## SECTION EXERCISES

●**15.2.1**  A student adds 30 mL of 5.00 M HCl to the buffer solution described in Section Exercise 15.1.3. Is the buffering capacity of the solution destroyed? What is the final pH of the solution?

●**15.2.2**  Determine the mass of solid sodium formate ($NaHCO_2$) and the volume of 0.500 M HCl solution required to generate 250 mL of buffer solution with pH = 3.50 and a total concentration (conjugate acid plus base) of 0.225 M.

●**15.2.3**  Describe how you would prepare 2.5 L of a pH = 10.50 buffer solution with a total concentration of 0.15 M, using an appropriate salt and solid NaOH. Use Table 15-1 to choose the buffer system.

## 15.3 ACID–BASE TITRATIONS

The amount of an acid in a sample can be determined by adding a solution of base of known concentration until the amount of base added exactly matches the amount of acid in the sample. Similarly, the amount of a base in a sample can be determined by adding a solution of acid of known concentration until the amount of acid added exactly matches the amount of base in the sample. This procedure is called **titration**. The solution to be analyzed is placed in a beaker or flask. A second solution of known concentration, called the **titrant**, is added slowly by means of a calibrated measuring vessel called a buret.

In one type of titration, a solution of a strong base such as sodium hydroxide is added slowly to a solution that contains an unknown amount of an acid. Each hydroxide ion added to the acid solution accepts one proton from a molecule of acid. As the titration proceeds, fewer and fewer acid molecules remain in the acid solution, but the solution is still acidic. At the **stoichiometric point**, just enough hydroxide ions have been added to react with every acidic proton present in the acid solution before the titration was started. The hydroxide ions in the next drop of titrant do not react because acid molecules are no longer present in the solution. Before the stoichiometric point, the solution contains excess acid. After the stoichiometric point, the solution contains excess $OH^-$ ions. Figure 15-4 shows a titration setup and molecular views illustrating titration of a strong acid by a strong base.

If we know the molarity of the titrant and measure the volume of titrant required to reach the stoichiometric point, we can calculate the number of moles of hydroxide required to react with all the acid. This allows us to determine the concentration of the unknown acid solution. Example 15-9 illustrates how this is done.

**Figure 15-4**

(a)     (b)     (c)

Titration of HCl by NaOH. Spectator ions have been omitted for clarity. (a) Initially, the HCl solution contains many $H_3O^+$ ions. (b) At the stoichiometric point, the solution contains only $H_2O$ molecules and spectator ions. (c) Past the stoichiometric point, the solution contains many $OH^-$ ions.

## EXAMPLE 15-9 ACID–BASE TITRATION

Industrial wastewater often is contaminated with strong acids. Environmental regulations require that such wastewater be neutralized before it is returned to the environment. A 150-mL sample of wastewater was titrated with 0.1250 M sodium hydroxide, and 38.65 mL of the base was required to reach the stoichiometric point. What was the molarity of hydronium ions in this sample of wastewater?

**Strategy:** We are asked for the molarity of an acid. The analysis is a titration. Knowing that the wastewater contains a strong acid, we can write the general acid–base neutralization reaction:

$$H_3O^+ (aq) + OH^- (aq) \longrightarrow 2\, H_2O\, (l)$$

The key feature of a titration is that at the stoichiometric point, the number of moles of $OH^-$ added equals the number of moles of acid originally present. This lets us use the data for the base to calculate the number of moles of $H_3O^+$ in the sample.

**Solution:** Use the titration to calculate the number of moles of $OH^-$ required to reach the stoichiometric point:

$$\text{mol } OH^- \text{ added} = cV$$

$$\text{mol } OH^- \text{ added} = (0.1250 \text{ mol/L})(38.65 \text{ mL})(10^{-3} \text{ L/mL}) = 4.831 \times 10^{-3}\text{mol}$$

This is also therefore the number of moles of acid in the original sample. Because the original sample had a volume of $1.50 \times 10^2$ mL, its molarity is as follows:

$$c = \frac{n}{V} = \frac{4.831 \times 10^{-3}\text{mol}}{150\,\text{mL}(10^{-3}\,\text{L/mL})} = 3.22 \times 10^{-2}\text{mol/L}$$

 **Does the Result Make Sense?** An acid concentration around $10^{-2}$ M may not appear high, but it is more than 1000 times the concentration of hydronium ion in unpolluted streams.

### Practice Exercise 15.9

The acidic component of vinegar is acetic acid. When 5.00 mL of vinegar was added to water and the solution titrated with the 0.1250 M NaOH, it took 33.8 mL to reach the stoichiometric point. Calculate the molarity of acetic acid in vinegar.

How do we know when the stoichiometric point of a titration has been reached? At the stoichiometric point, the nature of the acid–base species in the solution changes. The simplest method to obtain a visual measure of what is going on is to place a tiny amount of a substance called an **indicator** in the solution to be titrated. An indicator is a molecule whose colour depends on the concentration of hydronium ions. A properly chosen indicator shows a distinct colour change when the titration reaches the stoichiometric point and the acid in the solution has all reacted. In Figure 15-4, the indicator is phenolphthalein, a molecule that is colourless when hydronium ions are present but pink when hydroxide ions are present. This colour change signals an end to the titration. As soon as we see the colour change, we stop adding base and measure the volume that has been delivered from the buret.

A second method to follow the progress of a titration and to accurately find the stoichiometric point is to measure the solution pH as a function of the volume of titrant added. If we titrate a strong acid such as HCl (*aq*) using a strong base such as NaOH (*aq*), the resultant **titration curve** is like the one shown in Figure 15-5.

The pH is initially that of the acid being titrated. As strong base is added, the pH rises slowly, until almost all of the acid is consumed. At that point, small amounts of added titrant cause a large change in the pH because the hydronium ion concentration is so low. The stoichiometric point is reached when the moles of added OH⁻ equals the moles of $H_3O^+$ originally present. The pH is exactly 7 at the stoichiometric point for the titration of a strong acid by a strong base. Beyond the stoichiometric point, the pH rises as more titrant is added.

**Figure 15-5**

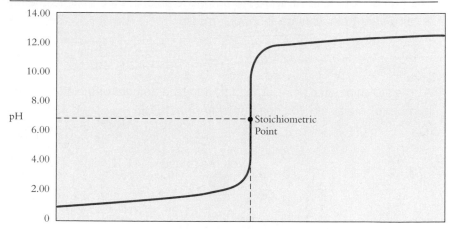

A titration curve for the titration of a strong acid with a strong base.

A quantitative titration requires a titrant solution of known concentration. In Example 15-9, the NaOH solution used as the titrant was known to be 0.1250 M. A titrant of known concentration is called a standard solution, and the concentration of such a solution is determined by a **standardization** titration. In a standardization titration, the solution being titrated contains a known amount of acid or base. An excellent acid for standardization is potassium hydrogen phthalate ($KHC_8H_4O_4$). This substance, a carboxylic acid that contains one weakly acidic hydrogen atom per molecule, is easily obtained as a highly pure solid. A known number of moles can be weighed on an analytical balance, dissolved in pure water, and then titrated with the base solution to be standardized. Example 15-10 illustrates the standardization procedure.

$KHC_5H_4O_4$

---

### EXAMPLE 15-10 STANDARDIZATION

A biochemist needed to standardize a solution of KOH. A sample of potassium hydrogen phthalate weighing 0.6745 g was dissolved in 100.0 mL of water and a drop of indicator was added. The solution was then titrated with the KOH solution. The titration required 41.75 mL of base to reach the stoichiometric point. Find the molarity of the KOH solution.

**Strategy:** First, we must identify the chemistry. This is an acid–base titration in which hydrogen phthalate anions (the acid) react with OH⁻ (the base). We use the molar equality of acid and base at the stoichiometric point together with the equations that link moles with mass and volume.

**Solution:** Hydrogen phthalate has one acidic hydrogen atom. As the net reaction shows, there is a 1:1 molar ratio between hydrogen phthalate and hydroxide:

$$HC_8H_4O_4^- \ (aq) + OH^- \ (aq) \longrightarrow C_8H_4O_4^{2-} \ (aq) + H_2O \ (l)$$

The number of moles of hydrogen phthalate anions is the same as the number of moles of $KHC_8H_4O_4$, which is calculated from its mass and molar mass:

$$n_{HC_8H_4O_4^-} = n_{KHC_8H_4O_4} = \frac{0.6745\,g}{204.2\,g/mol} = 3.303 \times 10^{-3}\,mol$$

At the stoichiometric point of the titration, the number of moles of hydroxide added from the buret equals the number of moles of hydrogen phthalate:

$$n_{OH^-}\ added = 3.303 \times 10^{-3}\,mol$$

$$Volume\ added = 41.75\ mL = 41.75 \times 10^{-3}\,L$$

$$c = \frac{n}{V} = \frac{3.303 \times 10^{-3}\,mol}{41.75 \times 10^{-3}\,L} = 7.911 \times 10^{-2}\,mol/L$$

**Does the Result Make Sense?** As we have already seen, base concentrations around $10^{-1}$ M are common. This concentration is a bit lower but still in the range that we would expect for laboratory solutions.

### Practice Exercise 15.10

To check the results of the first standardization, the biochemist repeated the titration. This time, a mass of 0.7015 g of potassium hydrogen phthalate dissolved in 135 mL of water required 43.36 mL of the base. Find the molarity and decide if the two results are consistent.

The technique of titration is equally useful for the titration of an unknown base by a solution of strong acid. The calculations proceed exactly as described previously. For the titration of a base, the stoichiometric point is reached when the number of moles of added acid in the titrant equals the number of moles of base in the unknown solution. Acid–base reactions that go to completion can be exploited in chemical analysis using the method of titration. Titrations can be understood in greater detail from the perspective of acid–base equilibria. Protonation of a weak base (B) by a strong acid is a reaction that goes virtually to completion because of its large equilibrium constant:

$$B + H_3O^+ \longrightarrow BH^+ + H_2O \qquad K_{eq} \gg 1$$

Likewise, when strong base is added to a solution containing an acid (HA), the reaction goes essentially to completion, again because of a large equilibrium constant:

$$OH^- + HA \longrightarrow H_2O + A^- \qquad K_{eq} \gg 1$$

Even though these reactions go essentially to completion, equilibria determine the pH of the solution before and during the titration.

Titrations are treated like any other equilibrium analysis, but we must pay special attention to the major species present in the solution, because these change during the titration. The most common titrations are analysis of a weak acid using a solution of strong base and analysis of a weak base using a solution of strong acid.

**Figure 15-6**

Schematic profile of the titration curve for a weak acid HA titrated with hydroxide ions. The titration can be divided into four regions that differ in the major species present in solution. The pH values are those for titration of 0.500 M acetic acid.

## Titration of a Weak Acid by OH⁻ Ions

The titration of a solution of a weak acid with a solution of a strong base is shown graphically in Figure 15-6. This titration curve has four distinct regions, each characterized by different major species:

1. Near the beginning of the titration, $HA$ and $H_2O$ are the only major species. The pH of the solution can be calculated using $K_a$ and the initial concentration of $HA$.

2. During most of the titration, both $HA$ and its conjugate base, $A^-$, are major species. As a result, the solution is buffered, and the pH changes relatively slowly with added hydroxide. In this region, the pH of the solution can be calculated using the buffer equation.

3. When nearly all of the $HA$ molecules have reacted with added $OH^-$, the only major species are $A^-$ and $H_2O$. Here, the pH is determined by $K_b$ and the concentration of $A^-$. Addition of $OH^-$ ions causes a sharp increase in pH.

4. After all the $HA$ molecules have reacted, the solution contains excess $A^-$ and $OH^-$ ions as the major species in solution. The pH is determined by the excess concentration of hydroxide ion.

These features of the weak acid titration curve are summarized in the flow chart shown in Figure 15-7.

As an example, consider the titration of 0.150 L of 0.500 M acetic acid solution with 2.50 M potassium hydroxide. Very early in the titration, the major species are water and acetic acid, and the equilibrium that determines solution pH is proton transfer between these major species:

**Figure 15-7**

A flow chart summarizes the major species in solution and the pH calculations for the four key regions of a weak acid titration curve.

$$H_2O\ (l) + CH_3CO_2H\ (aq) \rightleftharpoons H_3O^+\ (aq) + CH_3CO_2^-\ (aq) \quad K_a = 1.8 \times 10^{-5}$$

Using the standard method for solving equilibrium problems, you should be able to show that the initial solution contains 0.0750 mole of acetic acid, has a hydronium ion concentration of $3.0 \times 10^{-3}$ M, and has a pH of 2.52.

As the titration proceeds, the hydroxide ions in each added volume of titrant convert acetic acid molecules to acetate ions:

As a result, both acetate ions and acetic acid molecules are present as major species in solution. The presence of an acid and its conjugate base means that in this region of the titration, the solution is buffered, so the pH changes slowly as hydroxide ions are added to the solution.

This buffer region contains the **midpoint** of the titration, the point at which the amount of added $OH^-$ is equal to exactly half the weak acid originally present. In the current example, the solution at the midpoint contains 0.0375 mole each of acetic acid and acetate. Applying the buffer equation reveals the key feature of the midpoint:

> At the midpoint of a titration of a weak acid by a strong base, the pH of the solution equals the pKa of the weak acid.

$$\text{pH} = \text{p}K_a + \log\frac{n_{\text{acetate ion}}}{n_{\text{acetic acid}}} = 4.75 + \log(1) = 4.75$$

Beyond the buffer region, when nearly all of the acetic acid has been consumed, the pH increases sharply with each added drop of hydroxide solution. The titration curve passes through an almost vertical region before levelling off again. The **stoichiometric point** of an acid titration (also called the equivalence point) is the point at which the number of moles of added base is exactly equal to the number of moles of acid present in the original solution. At the stoichiometric point of a weak acid titration, the conjugate base is a major species in solution, but the weak acid is not.

Example 15-11 illustrates that the pH at the stoichiometric point in a titration of a weak acid differs from 7.0.

## EXAMPLE 15-11 pH AT THE STOICHIOMETRIC POINT

What is the pH at the stoichiometric point of the titration of 0.150 L of 0.500 M acetic acid with 2.50 M KOH solution?

**Strategy:** This is another equilibrium calculation to which the standard seven-step procedure applies. Special attention must be given, however, to analyzing the initial conditions *at the stoichiometric point*, bearing in mind that the reaction between hydroxide ions and a weak acid goes essentially to completion.

1. We are asked to calculate the pH at the stoichiometric point.

2. At the stoichiometric point, the amount of added hydroxide ions equals the amount of acetic acid that was originally present:

$$\text{mol } OH^- \text{ added} = \text{mol } CH_3CO_2H \text{ originally present}$$

This added hydroxide reacts essentially to completion with acetic acid:

$$CH_3CO_2H\ (aq) + OH^-\ (aq) \longrightarrow CH_3CO_2^-\ (aq) + H_2O\ (l)$$

Thus, the major acid–base species present at the stoichiometric point are $CH_3CO_2^-$ and $H_2O$.

3. The pH is determined by the proton transfer from water to acetate ions:

$$CH_3CO_2^-\ (aq) + H_2O\ (l) \rightleftarrows CH_3CO_3H\ (aq) + OH^-\ (aq)$$

4. The equilibrium constant expression for this equilibrium is related to $K_a$ for acetic acid:

$$K_{eq} = K_{b,\text{acetate}} = \frac{K_w}{K_{a,\text{ acetic acid}}} = 5.6 \times 10^{-10}$$

### Solution:

5. We need to calculate "initial" conditions for the stoichiometric point. Recall that the added base has reacted to form acetate anions:

$$n_{OH^-\text{ added}} = c_{\text{acid}}\ V_{\text{acid}} = (0.500\text{ mol/L})(0.150\text{ L}) = 0.0750\text{ mol}$$

$$n_{CH_3CO_2H} = 0,\ n_{OH^-} = 0,\text{ and } n_{CH_3CO_2^-} = 0.0750\text{ mol}$$

Before constructing a concentration table, convert moles of acetate to molarity. The volume at the stoichiometric point is the original volume plus the volume of added titrant:

$$V_{\text{initial}} = 0.150\text{ L}$$

$$V_{\text{titrant}} = \frac{n}{c} = \frac{0.0750\text{ mol}}{2.50\text{ mol/L}} = 0.0300\text{ L}$$

$$V_{\text{total}} = 0.150\text{ L} + 0.0300\text{ L} = 0.180\text{ L}$$

$$[CH_3CO_2^-]_{\text{initial}} = \frac{0.0750\text{ mol}}{0.180\text{ L}} = 0.417\text{ M}$$

| Reaction | $H_2O\ (l)$ + | $CH_3CO_2^-\ (aq)$ | $\leftrightarrows$ | $CH_3CO_2H\ (aq)$ + | $OH^-\ (aq)$ |
|---|---|---|---|---|---|
| Initial Concentration (M) | | 0.417 | | 0 | 0 |
| Change in Concentration (M) | | $-x$ | | $+x$ | $+x$ |
| Equilibrium Concentration (M) | | $0.417 - x$ | | $x$ | $x$ |

6. We have done this type of calculation many times, so you should be able to show that the pH of the solution at the stoichiometric point is 9.18.

**Does the Result Make Sense?** It may surprise you that the pH at the stoichiometric point is not 7.0, but this is a reasonable outcome. The major species present at the stoichiometric point are water and a conjugate base, so the resulting solution is basic in nature.

### Practice Exercise 15.11

What is the pH at the stoichiometric point of the titration of 0.250 L of 0.0350 M hypochlorous acid (HClO) with 1.00 M NaOH?

Although the exact pH at the stoichiometric point depends on what weak acid is being titrated, the qualitative result of Example 15-11 is reproduced for every titration of a weak acid with a strong base. At the stoichiometric point of a weak acid titration, the exact value of the pH is determined by $K_b$ for the conjugate base, and it is always *greater than* 7.0.

Beyond the stoichiometric point, in the final region of the titration curve, the concentration of acetic acid is very close to zero. There are no acid molecules to react with any further hydroxide ions, so excess hydroxide ions are present in solution as a major species. Beyond the stoichiometric point, the pH of the solution is determined by the amount of excess hydroxide ion.

## Titration of a Weak Base with $H_3O^+$ Ions

The principles that describe the titration of a weak acid also describe the titration of a weak base with hydronium ions. The titration curve for a weak base is shown in Figure 15-8.

Notice that the titration curve for a weak base has the same four regions seen in the titration curve of a weak acid:

1. At the beginning of the titration, the base $B$ and water are the only major species.

2. In the long, flat buffer region, both $B$ and its conjugate acid $BH^+$ are major species. The midpoint of the titration occurs in the buffer region. At this point, the pH of the solution is equal to the $pK_a$ of the conjugate acid of the base.

3. When nearly all the $B$ molecules have been protonated, the solution is no longer buffered. Further addition of hydronium ions causes a sharp drop in pH.

4. Beyond the stoichiometric point, the solution contains excess $H_3O^+$ ions, which determine the pH.

Example 15-12 demonstrates that at the stoichiometric point in any weak base titration, the pH is less than 7.0.

**Figure 15-8**

Schematic profile of the titration curve for a weak base $B$ titrated with hydronium ions. The pH values are those for titration of ephedrine, a weak base that is the active ingredient in many decongestants.

## EXAMPLE 15-12 TITRATION OF A WEAK BASE

Ephedrine, a weak base, is the active ingredient in many commercial decongestants. To analyze a sample of ephedrine dissolved in 0.200 L of water, a chemist carries out a titration with 0.900 M HCl, monitoring the pH continuously. The data obtained in this titration are shown in Figure 15-8. Calculate $K_b$ for ephedrine and determine the pH of the solution at the stoichiometric point.

**Strategy:** This problem asks for two different quantities, requiring two separate analyses. Thus, the problem should be solved in two stages. The reaction that takes place during the titration is proton transfer from hydronium ions (added from the buret) to ephedrine molecules (in the solution). For simplicity, we designate ephedrine as Ep and its conjugate acid as $EpH^+$. The problem provides a titration curve (Figure 15-8) and asks about two different equilibrium results, $K_b$ and pH at the stoichiometric point. Ephedrine is not listed in Appendix E, so $K_b$ must be determined by analyzing the titration curve. The pH calculation requires the standard procedure using $K_b$.

**Solution:**

**Determination of $K_b$.** The titration curve does not give $K_b$ directly. However, at the midpoint of the titration, the concentration of Ep and $EpH^+$ are identical. Use this information in the Henderson–Hasselbalch equation to show that at the midpoint of the titration, the pH of the solution equals the $pK_a$ for $EpH^+$:

$$EpH^+ \ (aq) + H_2O \ (l) \rightleftharpoons Ep \ (aq) + H_3O^+ \ (aq)$$

$$pH = pK_a + \log\frac{[Base]}{[Acid]} = pK_a + \log(1) = pK_a$$

We are asked to determine $K_b$, which is related to $pK_a$ through Equation 14-5:

$$pK_a + pK_b = 14.00$$

To locate the midpoint of the titration, use the volume needed to reach the stoichiometric point. At the stoichiometric point, all the ephedrine has been converted to $EpH^+$ ions, whereas at the midpoint, exactly half the ephedrine has been converted to $EpH^+$ ions. Thus, the volume of the HCl solution at the midpoint of the titration is exactly half the volume needed to reach the stoichiometric point. According to the titration curve, the volume needed to reach the stoichiometric point is 12.0 mL. The midpoint, therefore, corresponds to the addition of 6.0 mL, and the pH at the midpoint is the pH reading when 6.0 mL of solution has been added. Reading from Figure 15-8:

$$pH_{midpoint} = pK_a = 9.96$$

Now apply Equation 14-5 to find $K_b$ for ephedrine:

$$pK_a + pK_b = 14.00$$

$$pK_b = 14.00 - pK_a = 14.00 - 9.96 = 4.04$$

$$K_b = 10^{-4.04} = 9.1 \times 10^{-5}$$

**pH at the Stoichiometric Point.** At the stoichiometric point, virtually all the ephedrine molecules have been converted to $EpH^+$ ions by proton transfer from $H_3O^+$:

$$Ep \ (aq) + H_3O^+ \ (aq) \longrightarrow EpH^+ \ (aq) + H_2O \ (l)$$

At this point, the solution contains $EpH^+$, $H_2O$, and $Cl^-$ as major species. The chloride ion is the conjugate base of a *strong* acid, HCl, so the pH of the solution is determined by the acid–base equilibrium of water and the acid, $EpH^+$:

$$EpH^+ (aq) + H_2O (l) \rightleftharpoons Ep (aq) + H_3O^+ (aq) \quad pK_a = 9.96$$

$$K_a = \frac{[Ep]_{eq} [H_3O^+]_{eq}}{[EpH^+]_{eq}} = 10^{-9.96} = 1.1 \times 10^{-10}$$

The amount of weak acid present at the stoichiometric point is equal to the amount of added hydronium ion:

$$n(EpH^+) = n(H_3O^+) \text{ needed to reach the stoichiometric point}$$

$$n(H_3O^+) \text{ needed} = (0.0120 \text{ L})(0.900 \text{ mol/L}) = 0.0108 \text{ mol}$$

$$n(EpH^+) = 0.0108 \text{ mol}$$

The initial concentration of $EpH^+$ is found by dividing the number of moles by the total volume of the original solution plus the volume of added HCl solution:

$$[EpH^+]_{initial} = \frac{0.0180 \text{ mol}}{0.200 \text{ L} + 0.0120 \text{ L}} = 0.0509 \text{ M}$$

You should be able to set up a concentration table and use the $K_a$ expression to solve for $[H_3O^+]_{eq}$. The pH of the solution at the stoichiometric point is 5.63.

**Do the Results Make Sense?** Ephedrine is a weak base, so $K_b = 9.1 \times 10^{-5}$ is a reasonable value. At the stoichiometric point, the only major species present are the conjugate acid and water, so the acidic pH of 5.63 is also reasonable.

**Practice Exercise 15.12**

Triethylamine ($N(CH_3)_3$, $pK_b = 4.19$) is a weak base with an odour of fish. A chemist analyzed a 1.36-g sample of $N(CH_3)_3$ dissolved in 0.250 L of water. Titration with 0.974 M HCl required 23.6 mL to reach the stoichiometric point. Determine (a) the pH at the beginning of the titration, (b) the pH at the midpoint, (c) the pH at the stoichiometric point, and (d) the pH after 27.5 mL of the HCl solution had been added.

You may wonder why we did not use the titration curve to determine the pH at the stoichiometric point of the ephedrine titration, as we did to find the pH at the midpoint. Notice from Figure 15-8 that near the stoichiometric point, the pH changes very rapidly with added $H_3O^+$. At the stoichiometric point, the curve is nearly vertical. Thus, there is much uncertainty in reading a graph to determine the pH at the stoichiometric point. In contrast, a titration curve is nearly flat in the vicinity of the midpoint, minimizing uncertainty caused by errors in graph reading.

The key to understanding the titrations of weak acids and bases is to be familiar with the species in solution and the dominant equilibrium at each point along the titration curve. Example 15-13 reinforces these qualitative features.

## EXAMPLE 15-13 QUALITATIVE FEATURES OF A TITRATION

The figures below represent a small portion of a weak acid solution and a titration curve for the acid. Redraw the molecular picture to show how the figure should look for each of the points A–C along the titration curve. Include in your drawings any water molecules formed as part of the titration process.

**Strategy:** Identify the major species in solution by assigning each of the points to one of the four characteristic regions of the titration curve. In the titration reaction, hydroxide ions react with molecules of weak acid:

The figure shows six molecules of H$A$ at the start of the titration. Apply stoichiometric reasoning to determine the numbers of H$A$, $A^-$, and $H_2O$ molecules to include in the figures.

**Solution:** Point A lies along the section of the titration curve known as the buffer region. Buffering action comes from the presence of a weak acid and its conjugate base as major species in solution. Moreover, Point A lies beyond the midpoint of the titration, which tells us that more than half of the weak acid has been consumed. We represent this solution with two molecules of H$A$, four ions of $A^-$, and four $H_2O$ molecules:

Point B represents the stoichiometric point of the titration. All of the weak acid has been converted to the conjugate base, so the figure should show six $A^-$ ions and the six $H_2O$ molecules formed in the titration:

Beyond the equivalence point, there is no weak acid present to neutralize added hydroxide ions. The figure for Point C should show six $A^-$ ions, six molecules of $H_2O$, and several $OH^-$ ions:

**Do the Results Make Sense?** The sequence of figures, from the initial state to Point C, shows a progressive loss of H$A$ with matching increases in $A^-$ and $H_2O$, as we would expect for a weak acid titration. The figures make sense.

**Practice Exercise 15.13**

The molecular view below represents a small portion of a solution that matches Point A on the corresponding titration curve. Redraw the molecular picture to show how the figure should look for each of the points B–D along the titration curve. Your drawings should show any water molecules formed as part of the titration process.

## Titration of Polyprotic Acids

In the titration of a polyprotic acid, the added base reacts first with the more acidic hydrogen atoms of the neutral acid. For example, the titration of maleic acid takes place in two steps. For removal of one acidic hydrogen atom of maleic acid, $pK_a = 1.82$ ($K_{a1} = 1.5 \times 10^{-2}$):

The second acidic hydrogen atom, with a $pK_a$ of 6.59 ($K_{a2} = 2.6 \times 10^{-7}$), is dramatically less acidic than the first. Consequently, the second proton transfer reaction does not occur until all of the neutral diprotic acid has been consumed:

$pK_{a2} = 6.59$

The titration curve for maleic acid, which appears in Figure 15-9, shows this two-step process clearly. These data represent the titration of 50.0 mL of 0.10 M maleic acid with 1.0 M NaOH. The titration curve has six distinct regions.

**Region 1.** At the beginning of the titration, the diprotic acid (represented by $H_2A$) and $H_2O$ are the only major species in the solution. As we describe in Chapter 14, the hydronium ion concentration can be calculated from the familiar equilibrium expression:

$$K_{a1} = 1.5 \times 10^{-2} = \frac{[HA^-]_{eq}[H_3O^+]_{eq}}{[H_2A]_{eq}} = \frac{x^2}{0.10 - x}$$

**Figure 15-9**

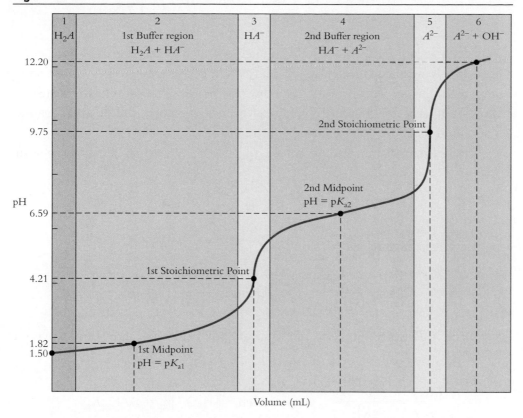

The titration of a diprotic acid, $H_2A$, can be divided into six regions that differ in the major species present in solution. Shown here is the titration curve for maleic acid (50.0 mL, 0.10 M) titrated with 1.0 M NaOH.

Solving the quadratic equation gives $x = [H_3O^+]_{eq} = 3.19 \times 10^{-2}$ M and an initial pH of 1.50.

**Region 2.** Within the first buffer region, both $H_2A$ and $HA^-$ are major species in solution, and we can apply the buffer equation to calculate the pH. Halfway to the first equivalence point of the titration, $[H_2A] = [HA^-]$ and $pH = pK_{a1} = 1.82$.

**Region 3.** At the first stoichiometric point of the titration, all the diprotic acid has been converted to its conjugate base, $HA^-$. This amphiprotic anion can react with itself, analogous to the self-ionization of water:

$$2\ HA^-\,(aq) \rightleftharpoons H_2A\,(aq) + A^{2-}\,(aq) \qquad K_{eq} = \frac{[H_2A]_{eq}[A^{2-}]_{eq}}{[HA^-]^2_{eq}}$$

We can relate this equilibrium constant to $K_{a1}$ and $K_{a2}$:

$$H_2O\,(l) + HA^-\,(aq) \rightleftharpoons H_3O^+\,(aq) + A^{2-}\,(aq) \qquad K_{eq} = K_{a2}$$

$$H_3O^+\,(aq) + HA^-\,(aq) \rightleftharpoons H_2O\,(l) + H_2A\,(aq) \qquad K_{eq} = 1/K_{a1}$$

Adding these two equations gives the reaction of the anion with itself, and as we show in Chapter 13, when equilibria are added, their equilibrium constants are multiplied. Thus:

$$K_{eq} = \frac{[H_2A]_{eq}[A^{2-}]_{eq}}{[HA^-]^2_{eq}} = \frac{K_{a2}}{K_{a1}} = \frac{2.6 \times 10^{-7}}{1.5 \times 10^{-2}} = 1.7 \times 10^{-5}$$

This value is larger than the equilibrium constants for the other acid–base equilibria involving major species, so we can treat the self-reaction of $HA^-$ as the dominant equilibrium at the first stoichiometric point. The equilibrium constant expression does not contain $[H_3O^+]$, however, so we need to do further manipulations to relate the dominant equilibrium to $[H_3O^+]$.

At the first stoichiometric point $[H_2A]_{eq} = [HA^-]_{eq}$, which allows us to write

$$\frac{K_{a2}}{K_{a1}} = \frac{[H_2A]^2_{eq}}{[HA^-]^2_{eq}}$$

Next, we rearrange the expression for $K_{a1}$:

$$K_{a1} = \frac{[HA^-]_{eq}[H_3O^+]_{eq}}{[H_2A]_{eq}} \quad \text{giving} \quad \frac{[H_2A]_{eq}}{[HA^-]_{eq}} = \frac{[H_3O^+]_{eq}}{K_{a1}}$$

Substitute this expression into the ratio of equilibrium constants and solve for $[H_3O^+]_{eq}$:

$$\frac{K_{a2}}{K_{a1}} = \frac{[H_3O^+]^2_{eq}}{K^2_{a1}} \quad \text{from which} \quad [H_3O^+]_{eq} = \sqrt{K_{a1}K_{a2}}$$

Now evaluate $[H_3O^+]_{eq}$:

$$[H_3O^+]_{eq} = \sqrt{(2.6 \times 10^{-7})(1.5 \times 10^{-2})} = 6.2 \times 10^{-5}\,M$$

$$pH = -\log(6.2 \times 10^{-5}) = 4.21$$

The pH at the first stoichiometric point is 4.21. This derivation is completely general, so Equation 15-2 can be applied at the first stoichiometric point of any polyprotic acid titration:

$$[H_3O^+]_{eq,\,\text{1st stoichiometric point}} = \sqrt{K_{a1}K_{a2}} \qquad (15\text{-}2)$$

**Region 4.** Moving beyond the first stoichiometric point, the titration enters the second buffer region. Here, the major species are $HA^-$ and its conjugate base, $A^{2-}$. The pH in this region is given by the buffer equation, using the $pK_a$ of $HA^-$. At the second midpoint, $[HA^-] = [A^{2-}]$, and $pH = pK_{a2}$. For the titration of maleic acid, pH = 6.59 at the second midpoint.

**Region 5.** At the second stoichiometric point, all of the diprotic acid has been converted to $A^{2-}$. The dominant equilibrium for maleic acid is proton transfer from water:

$$A^{2-}\,(aq) + H_2O\,(l) \rightleftharpoons HA^-\,(aq) + OH^-\,(aq) \qquad K_b = \frac{K_w}{K_{a2}}$$

The titration of maleic acid requires 10.0 mL to reach the second equivalence point, which gives a concentration of 0.833 M for $A^{2-}$. The equilibrium constant expression is straightforward:

$$\frac{[HA^-]_{eq}[OH^-]_{eq}}{[A^{2-}]_{eq}} = \frac{x^2}{0.833 - x} \cong \frac{x^2}{0.833} = \frac{1.00 \times 10^{-4}}{2.6 \times 10^{-7}} = 3.85 \times 10^{-8}$$

Solving for $x$ gives $[OH^-] = 5.66 \times 10^{-5}$, a value that justifies the usual assumption. The pH at the second stoichiometric point is 9.75.

**Region 6.** Finally, beyond the second stoichiometric point, the major species are $H_2O$, $A^{2-}$, and $OH^-$. The pH of the solution is determined by the concentration of excess hydroxide. For example, 11.0 mL of NaOH solution added in the maleic acid titration gives 0.016 M $OH^-$ and pH = 12.20. You should be able to verify these calculations.

By recognizing species in solution and their dominant equilibrium, we can construct titration curves for other diprotic acids. Example 15-14 shows how this is done for sulphurous acid.

### EXAMPLE 15-14 DRAWING A TITRATION CURVE

Sulphurous acid ($H_2SO_3$) has two acidic hydrogen atoms, with $pK_a$ values of 1.85 ($K_{a1} = 1.40 \times 10^{-2}$) and 7.20 ($K_{a2} = 6.3 \times 10^{-8}$). Construct a titration curve for the titration of 125 mL of 0.150 M sulphurous acid with 0.800 M NaOH.

**Strategy:** Calculate a pH value for each of the six regions of the titration, and then extrapolate to complete the titration curve.

**Solution:**

*Region 1:* The starting pH of the solution is calculated using $K_{a1}$ and the initial molarity of the diprotic acid. We use the standard approach to a weak acid equilibrium:

$$K_{a1} = \frac{[HSO_3^-]_{eq}[H_3O^+]_{eq}}{[H_2SO_3]_{eq}} = \frac{x^2}{0.150 - x} = 1.4 \times 10^{-2}$$

Using the quadratic equation lets use determine that $x = 0.0664$:

$$[H_3O^+]_{eq} = 6.64 \times 10^{-2} \text{ M} \quad \text{giving} \quad pH = 1.18$$

*Region 2:* At the first midpoint of the titration, the pH of the solution equals $pK_{a1}$:

$$pK_{a1} = 1.85$$

*Region 3:* We solve Equation 15-2 to determine the concentration of $[H_3O^+]_{eq}$ at the first stoichiometric point:

$$[H_3O^+]_{eq} = \sqrt{K_{a1}K_{a2}} = \sqrt{(1.4 \times 10^{-2})(6.3 \times 10^{-8})} = 2.75 \times 10^{-5} \text{ M}$$

$$pH = 4.56$$

*Region 4:* The pH at the second midpoint is given by $pK_{a2}$:

$$pK_{a2} = 7.20$$

*Region 5:* At the second stoichiometric point, all of the sulphurous acid has been converted to sulphite, and the pH is determined using $K_{b2}$. For this calculation, we need the concentration of $SO_3^{2-}$, for which we need to know the volume of NaOH solution required to reach the second stoichiometric point. Start by finding the number of moles of sulphurous acid:

$$n = cV = (0.150 \text{ mol/L})(0.125 \text{ L}) = 0.01575 \text{ mol } H_2SO_3$$

To calculate the required volume of NaOH solution, remember that the acid has two acidic hydrogen atoms, so 2 moles of $OH^-$ are required for each mole of $H_2SO_3$:

$$V_{base} = 0.01875 \text{ mol } H_2SO_3 \left( \frac{2 \text{ mol NaOH}}{1 \text{ mol } H_2SO_3} \right) \left( \frac{1 \text{ L}}{0.800 \text{ mol NaOH}} \right) = 0.0469 \text{ L}$$

Now calculate the sulphite concentration using the total volume of the solution:

$$c_{SO_3^{2-}} = \frac{n_{SO_3^{2-}}}{V_{total}} = \frac{0.01875 \text{ mol}}{0.125 \text{ L} + 0.0469 \text{ L}} = 0.1091 \text{ M}$$

Next, we need $K_{b2}$ for the sulphite anion:

$$K_{b2} = \frac{K_w}{K_{a2}} = \frac{1.00 \times 10^{-14}}{6.3 \times 10^{-8}} = 1.59 \times 10^{-7}$$

We are now ready to set up the base equilibrium expression for $SO_3^{2-}$:

$$K_{b2} = \frac{[HSO_3^-]_{eq}[OH^-]_{eq}}{[SO_3^{2-}]_{eq}} = \frac{x^2}{0.1091 - x} \cong \frac{x^2}{0.1091} = 1.59 \times 10^{-7}$$

This gives $x = 1.32 \times 10^{-4}$, from which we find $[OH^-]_{eq}$:

$$[OH^-]_{eq} = 1.32 \times 10^{-4} \text{ M} \qquad \text{giving} \quad pH = 10.12$$

*Region 6:* Finally, beyond the second stoichiometric point, the pH is determined by the amount of excess hydroxide ion. Calculate the amount of hydroxide ion in 1.00 mL of the NaOH solution:

$$n = cV = (0.800 \text{ mol/L})(0.00100) = 8.00 \times 10^{-4} \text{ mol}$$

The total volume is now $0.125 \text{ L} + 0.0469 \text{ L} + 0.00100 \text{ L} = 0.173 \text{ L}$:

$$[OH^-]_{eq} = \frac{8.00 \times 10^{-4} \text{ mol}}{0.173 \text{ L}} = 4.62 \times 10^{-3} \text{ M} \qquad \text{giving} \qquad pH = 11.67$$

We now have six pH values for key regions along the titration curve. A table helps to organize these pH values. You should be able to verify that the volumes are correct:

| Region | 1 | 2 | 3 | 4 | 5 | 6 |
|---|---|---|---|---|---|---|
| pH | 1.18 | 1.85 | 4.56 | 7.20 | 10.12 | 11.67 |
| Volume (mL) | 0 | 11.7 | 23.4 | 35.1 | 46.9 | 47.9 |

The titration curve plots pH against volume:

Caclulated Titration Curve of $H_2SO_3$

When plotted on a graph of pH vs. volume of NaOH solution, these six points reveal the gross features of the titration curve. Adding additional calculated points helps define the pH curve. On the curve shown here, the red points A–D were calculated using the buffer equation with base:acid ratios of 1:3 and 3:1. Point E was generated from excess hydroxide ion concentration, 2.00 mL beyond the second stoichiometric point. You should verify these additional five calculations.

**Does the Result Make Sense?** The calculated curve shows the general features of the pH titration curve for a diprotic acid. The pH of the solution is acidic at the first stoichiometric point (major species = weak acid $HA^-$) and basic at the second (major species = conjugate base $A^{2-}$). These features tell us the answer does make sense.

### Practice Exercise 15.14

Amino acids form diprotic cations. For example, the diprotic form of alanine is shown here:

$$pK_{a2} = 9.87$$
$$^+H_3N-CH-C-OH$$
$$CH_3 \quad pK_{a1} = 2.35$$

Diprotic alanine

Estimate the titration curve for 0.200 L of 0.120 M diprotic alanine with 3.00 M NaOH.

## Indicators

An acid–base titration can be performed without using a pH meter if an **indicator** is present that changes colour as the titration passes the stoichiometric point. An indicator is a weak organic acid that contains a highly delocalized $\pi$ bonding network. We use $pK_{In}$ to designate the $pK_a$ of an indicator. At pH $<$ $pK_{In}$, the indicator is in its protonated form, which we designate H*In*. At pH $>$ $pK_{In}$, the indicator gives up its proton and is converted to its conjugate base, designated *In*$^-$.

Deprotonation of H*In* changes the structure of the indicator, which alters the pattern of electron delocalization in the $\pi$ system. This change in the delocalized $\pi$ system causes a change in the colour of the indicator. Figure 15-10 shows these changes for one acid–base indicator, phenol red. The $pK_{In}$ of phenol red is 7.9. When the pH is less than 7.9, the protonated acid dominates, giving a yellow

**Figure 15-10**

The line structures of phenol red in its acidic and basic forms.

## TABLE 15-2 INDICATORS AND THEIR pH RANGES

| Indicator | $pK_{In*}$ | pH Range | Colours Acid | Base |
|---|---|---|---|---|
| Thymol blue** | 1.75 | 1.2–2.8 | Red | Yellow |
| Methyl orange | 3.40 | 3.1–4.4 | Red | Yellow |
| Bromocresol green | 4.68 | 4.0–5.6 | Yellow | Blue |
| Methyl red | 4.95 | 4.4–6.2 | Red | Yellow |
| Bromocresol purple | 6.3 | 5.2–6.8 | Yellow | Purple |
| Phenol red | 7.9 | 6.4–8.0 | Yellow | Red |
| Thymol blue** | 8.9 | 8.0–9.6 | Yellow | Blue |
| Phenolphthalein | 9.4 | 8.0–10.0 | Colourless | Red |
| Thymolphthalein | 10.0 | 9.4–10.6 | Colourless | Blue |

\* The acid dissociation constant of an indicator is designated $K_{In}$.

\*\*Thymol blue has two acidic hydrogen atoms, so it can be used as an indicator for two different pH regions.

solution. When the pH is higher than 7.9, the conjugate base dominates, giving a red solution:

$$\underset{\text{Yellow}}{HIn} + H_2O \rightleftharpoons H_3O^+ + \underset{\text{Red}}{In^-} \qquad pK_{In} = 7.9$$

During a titration, as the pH passes the value of $pK_{In}$, the indicator changes colour. A good indicator changes colour very near the pH of the stoichiometric point. Because this pH depends on the $pK_a$ of the substance being titrated, different titrations require different indicators. The best indicator for a titration has a $pK_{In}$ that is as close as possible to the pH of the solution at the stoichiometric point:

$$pK_{In} \cong pH_{\text{stoichiometric point}}$$

Because the pH changes rapidly near the stoichiometric point, an indicator is suitable for a titration if the pH at the stoichiometric point is within one unit of $pK_{In}$:

$$pH_{\text{stoichiometric point}} = pK_{In} \pm 1 \qquad (15\text{-}3)$$

Table 15-2 lists a selection of acid–base indicators, and Example 15-15 shows how to select an appropriate indicator.

## EXAMPLE 15-15 SELECTING AN INDICATOR

A student wants to titrate a solution of ammonia that has an approximate concentration of $10^{-2}$ M. What indicator would be appropriate?

**Strategy:** The student needs an indicator whose colour changes as close to the stoichiometric point as possible:

$$pH_{\text{stoichiometric point}} = pK_{In} \pm 1$$

Thus, the student must calculate the pH at the stoichiometric point and then consult Table 15-2.

**Solution:** The titration reaction is

$$NH_3 \, (aq) + H_3O^+ \, (aq) \longrightarrow NH_4^+ \, (aq) + H_2O \, (l)$$

At the stoichiometric point, all the ammonia molecules have been converted to ammonium ions, so the major species present are $NH_4^+$ and $H_2O$. The pH of the solution is thus determined by the acid–base equilibrium of $NH_4^+$ ions:

$$NH_4^+ \, (aq) + H_2O \, (l) \rightleftharpoons NH_3 \, (aq) + H_3O^+ \, (aq)$$

$$K_{eq} = \frac{[NH_3]_{eq}[H_3O^+]_{eq}}{[NH_4^+]_{eq}} = \frac{K_w}{K_b} = 5.6 \times 10^{-10}$$

Set up a concentration table at the stoichiometric point:

| Reaction | $H_2O^+ \, (l)$ | $NH_4^+ \, (aq)$ | $\rightleftharpoons$ | $NH_3 \, (aq)$ | $+$ | $H_3O^+ \, (aq)$ |
|---|---|---|---|---|---|---|
| Initial Concentration (M) | | $10^{-2}$ | | 0 | | 0 |
| Change in Concentration (M) | | $-x$ | | $+x$ | | $+x$ |
| Equilibrium Concentration (M) | | $10^{-2} - x$ | | $x$ | | $x$ |

Now substitute and solve:

$$5.6 \times 10^{-10} = \frac{(x)(x)}{10^{-2} - x} \cong \frac{x^2}{10^{-2}}$$

$$x^2 = 5.6 \times 10^{-12} \quad \text{giving} \quad x = 2.4 \times 10^{-6}$$

$$pH = 5.62$$

Table 15-2 shows that bromocresol purple ($pK_{In} = 6.3$) or methyl red ($pK_{In} = 4.95$) would be suitable indicators for the titration of ammonia. Methyl red is typically used for titrations of weak bases by strong acids. The figure shows its colour behaviour.

 **Does the Result Make Sense?** We see that $x$ is small relative to $10^{-2}$, and an acidic pH at the stoichiometric point is what we expect for this titration of weak base by strong acid.

**Practice Exercise 15.15**

A chemist needs to titrate a solution of hydrofluoric acid that is approximately 0.15 M. What indicator would be appropriate for the titration?

The objective of an acid–base titration is to determine the amount of an acid or base in a solution. Because an indicator is itself a weak acid, it may appear that adding an indicator would alter equilibrium concentrations and influence the titration. However, a useful indicator gives a noticeable colour to a solution at a concentration of $10^{-6}$ M. This is negligible compared with the concentration of the solution being titrated, which is usually in the range of $10^{-2}$ to $10^{-3}$ M.

The low concentration of an indicator also explains why the presence of this weak acid does not change the pH of the solution. Indicators are always present as minor species in solution, never as major species. Thus, the dominant equilibrium that determines the pH of a solution never involves the indicator. The $K_a$ of the substance being titrated establishes the equilibrium concentration of hydronium ions. This, in turn, establishes whether the indicator is in its acidic or basic form.

**SECTION EXERCISES**

● **15.3.1** Glycolic acid ($HOCH_2CO_2H$), a constituent of sugar cane juice, has a p$K_a$ of 3.9. Sketch the titration curve for the titration of 60.0 mL of 0.010 M glycolic acid with 0.050 M KOH. Indicate the stoichiometric point, the buffer region, and the point of the titration where pH = p$K_a$. Sketch the curve qualitatively without doing any quantitative calculations.

● **15.3.2** What is the pH of the glycolic acid solution at the stoichiometric point? What is the pH of the solution when 3.0 mL of the KOH solution has been added?

● **15.3.3** Choose an indicator suitable for the titration of glycolic acid. What is the colour of the solution at the stoichiometric point?

## 15.4 SOLUBILITY EQUILIBRIA

The limestone deposits that decorate Carlsbad and other caverns are the result of the solubility equilibrium of calcium carbonate in groundwater, as described in Chapter 13:

$$CaCO_3\,(s) \rightleftharpoons Ca^{2+}\,(aq) + CO_3^{2-}\,(aq) \quad K_{sp} = [Ca^{2+}]_{eq}[CO_3^{2-}]_{eq}$$

Solubility equilibria are important in the natural environment and in our daily lives. Coral reefs are built up over many years through deposits of calcium carbonate from tiny sea organisms. Thermal vents on the oceanís floor spew out insoluble metal sulphides and other minerals. Deposits from hard water can wreak havoc with plumbing.

In an aqueous solubility reaction, a salt dissolves to yield ions in solution. The amount of a salt that dissolves in water varies over a large range, as the following examples show:

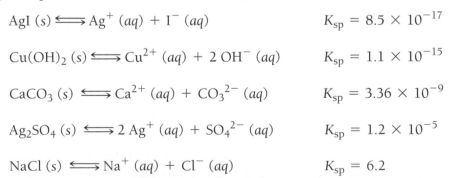

$$AgI\,(s) \rightleftharpoons Ag^+\,(aq) + I^-\,(aq) \qquad\qquad K_{sp} = 8.5 \times 10^{-17}$$

$$Cu(OH)_2\,(s) \rightleftharpoons Cu^{2+}\,(aq) + 2\,OH^-\,(aq) \qquad K_{sp} = 1.1 \times 10^{-15}$$

$$CaCO_3\,(s) \rightleftharpoons Ca^{2+}\,(aq) + CO_3^{2-}\,(aq) \qquad K_{sp} = 3.36 \times 10^{-9}$$

$$Ag_2SO_4\,(s) \rightleftharpoons 2\,Ag^+\,(aq) + SO_4^{2-}\,(aq) \qquad K_{sp} = 1.2 \times 10^{-5}$$

$$NaCl\,(s) \rightleftharpoons Na^+\,(aq) + Cl^-\,(aq) \qquad\qquad K_{sp} = 6.2$$

Substances that have $K_{sp} < 10^{-5}$, such as calcium carbonate, copper(II) hydroxide, and silver iodide, are classified as **insoluble** (even though they are not totally insoluble). Salts that have $K_{sp}$ between $10^{-2}$ and $10^{-5}$, such as silver sulphate, are **slightly soluble**, and solids such as NaCl, which have $K_{sp} > 10^{-2}$, are **soluble**. This classification provides a more specific description of soluble and insoluble than other, more qualitative solubility guidelines.

Solubility equilibria can be treated using our standard approach to equilibrium, as demonstrated in Examples 15-16, 15-17, and 15-18.

## EXAMPLE 15-16 SOLUBILITY PRODUCTS

Gypsum is a relatively soft rock made of calcium sulphate. Rainwater percolates through gypsum, dissolves some of the rock, and eventually becomes saturated with $Ca^{2+}$ ions and $SO_4^{2-}$ ions. A geochemist takes a sample of groundwater from a cave and finds that it contains $8.4 \times 10^{-3}$ M $SO_4^{2-}$ and $5.8 \times 10^{-3}$ M $Ca^{2+}$. (The ratio is not 1:1 because other sulphate rock contributes some of the $SO_4^{2-}$ ions to the solution.) Use these data to determine the solubility product of calcium sulphate.

**Strategy:** This problem is straightforward, so a truncated version of the seven-step strategy is appropriate. We are asked to evaluate $K_{sp}$ of $CaSO_4$:

$$CaSO_4 \ (s) \rightleftharpoons Ca^{2+} \ (aq) + SO_4^{2-} \ (aq) \qquad K_{sp} = [Ca^{2+}]_{eq}[SO_4^{2-}]_{eq}$$

**Solution:** The problem gives the required equilibrium concentrations, since it is stated that the groundwater sample is saturated (i.e., at equilibrium) with respect to these ions.

$$K_{sp} = (5.8 \times 10^{-3})(8.4 \times 10^{-3}) = 4.9 \times 10^{-5}$$

 **Does the Result Make Sense?** The equilibrium constant is small, indicating a slightly soluble salt, as we would expect for a naturally occurring mineral like gypsum.

**Practice Exercise 15.16**

Calcium hydroxide, known as slaked lime, is an ingredient of mortar and plaster. A saturated solution of $Ca(OH)_2$ has $[Ca^{2+}]_{eq} = 0.011$ M and $[OH^-]_{eq} = 0.0216$ M. Calculate the solubility product of calcium hydroxide.

Example 15-17 shows a calculation in which the stoichiometry differs from 1:1.

## EXAMPLE 15-17 CALCULATING $K_{SP}$

When solid $PbI_2$ is added to pure water at 25 °C, the salt dissolves until the concentration of $Pb^{2+}$ reaches $1.35 \times 10^{-3}$ M. After this concentration is reached, excess solid remains undissolved. What is $K_{sp}$ for this salt?

**Strategy:** The problem asks for an equilibrium constant, which means we need to find equilibrium concentrations of the species involved in the solubility reaction. Use the seven-step strategy, which we present here without step numbers.

The starting materials are solid $PbI_2$ and $H_2O$. These are the only major species present before any solid dissolves. The only equilibrium involving these reactants is the solubility reaction. Solid $PbI_2$ dissolves in water to produce its ions in solution, $Pb^{2+}$ and $I^-$:

$$PbI_2 \ (s) \rightleftharpoons Pb^{2+} \ (aq) + 2 \ I^- \ (aq) \qquad K_{eq} = K_{sq} = [Pb^{2+}]_{eq} \ [I^-]_{eq}^2$$

**Solution:** The equilibrium concentration of $Pb^{2+}$ ions is stated to be $1.35 \times 10^{-3}$ M, but the equilibrium concentration of iodide ions is not stated:

$$[Pb^{2+}]_{eq} = 1.35 \times 10^{-3} \ M \qquad [I^-]_{eq} = ?$$

The equilibrium concentration of $I^-$ is determined by stoichiometric analysis. Initially, the system contains only pure water and solid lead(II) iodide. Enough $PbI_2$ dissolves to make $[Pb^{2+}]_{eq} = 1.35 \times 10^{-3}$ M. One formula unit of $PbI_2$ contains one $Pb^{2+}$ cation and *two* $I^-$ anions. Thus, twice as many iodide ions as lead ions enter the solution. The concentration of $I^-$ at equilibrium is double that of $Pb^{2+}$ cations:

$$[I^-]_{eq} = 2 \ [Pb^{2+}]_{eq} = 2.70 \times 10^{-3} \ M$$

Substitute the values of the concentrations at equilibrium into the equilibrium expression and calculate the result:

$$K_{sp} = [Pb^{2+}]_{eq}[I^-]_{eq}^2 = (1.35 \times 10^{-3})(2.70 \times 10^{-3})^2 = 9.84 \times 10^{-9}$$

**Does the Result Make Sense?** Remember that although equilibrium calculations require concentrations in molarities for solutes, the equilibrium constant expression is dimensionless. The solubility product has a small value, which is reasonable given the small concentrations, $10^{-3}$ M, of the ions at equilibrium.

**Common Pitfalls:** Neglecting to square the concentrations of ions that have a stoichiometric coefficient of 2, for example.

**Practice Exercise 15.17**
A saturated solution of magnesium hydroxide has a pH of 10.35. What is the solubility product of $Mg(OH)_2$?

Example 15-18 deals with the second type of calculation, determining a concentration at equilibrium when the value of the solubility product is known.

## EXAMPLE 15-18 SOLUBILITY CALCULATIONS

Cadmium is an extremely toxic metal that finds its way into the aqueous environment as a result of some human activities. A major cause of cadmium pollution is zinc mining and processing, because natural deposits of ZnS ores usually also contain CdS. During the processing of these ores, highly insoluble cadmium sulphide ($K_{sp} = 7.9 \times 10^{-27}$) may be converted into considerably less insoluble cadmium hydroxide ($K_{sp} = 7.2 \times 10^{-15}$). What mass of $Cd(OH)_2$ will dissolve in $1.00 \times 10^2$ L of an aqueous solution?

**Strategy:** Determining what mass will dissolve requires that we find equilibrium concentrations, so we apply our standard approach:

Major species present: $Cd(OH)_2$ (s) and $H_2O$ (l)

Equilibrium: $Cd(OH)_2$ (s) $\rightleftharpoons Cd^{2+}$ (aq) + 2 OH$^-$ (aq)

$$K_{sp} = [Cd^{2+}]_{eq}[OH^-]_{eq}^2 = 7.2 \times 10^{-15}$$

**Solution:** Set up a concentration table:

| Reaction | Cd(OH)₂ (s) | $\rightleftharpoons$ | Cd²⁺ (aq) | + | 2 OH⁻ (aq) |
|---|---|---|---|---|---|
| Initial Concentration (M) | (solid) | | 0 | | 0 |
| Change in Concentration (M) | (solid) | | +x | | +2x |
| Equilibrium Concentration (M) | (solid) | | x | | 2x |

Evaluation:

$$7.2 \times 10^{-15} = (x)(2x)^2 = 4x^3 \quad \text{so} \quad x^3 = 1.8 \times 10^{-15} \quad \text{and} \quad x = 1.2 \times 10^{-5}$$

The concentration of $Cd^{2+}$ ions in a saturated solution is $1.2 \times 10^{-5}$ M. Therefore, in $1.00 \times 10^2$ L of solution, there will be

$$(1.2 \times 10^{-3} \text{ mol/L})(1.00 \times 10^2 \text{ L}) = 1.2 \times 10^{-3} \text{ mol } Cd^{2+}$$

Cadmium hydroxide generates $Cd^{2+}$ in a 1:1 stoichiometric ratio, so the number of moles of $Cd^{2+}$ in solution equals the number of moles of $Cd(OH)_2$ that dissolves. Find the amount in grams using the molar mass of $Cd(OH)_2$:

$$m = nM = (1.2 \times 10^{-3} \text{ mol})(146 \text{ g/mol}) = 0.18 \text{ g}$$

In contrast, only $1.3 \times 10^{-9}$ g of CdS dissolves in the same volume of water.

**Does the Result Make Sense?** The units are correct, and the overall solubility is low, as we expect for a hydroxide salt.

**Common Pitfalls:** Neglecting that the stoichiometric coefficient multiplies the concentration *and* is the exponent. For instance, in the above example, $[Cd^{2+}][OH^-] = x(2x)^2 = 4x^3$.

**Practice Exercise 15.18**

What mass of lead(II) bromide ($pK_{sp} = 5.18$) will dissolve in 1.25 L of water?

## Precipitation Equilibria

The solubility product ($K_{sp}$) describes the equilibrium of a salt dissolving in water. In the laboratory and in industry, solubility equilibria are often exploited in the opposite direction. Two solutions are mixed to form a new solution in which the solubility product of one substance is exceeded. This salt precipitates and can be collected by filtration. Example 15-19 illustrates how precipitation techniques can be used to remove toxic heavy metals from aqueous solutions.

### EXAMPLE 15-19 PRECIPITATION REACTIONS

As illustrated in Example 15-18, wastewater resulting from metal processing often contains significant amounts of toxic heavy metal ions that must be removed before the water can be returned to the environment. One method uses sodium hydroxide solution to precipitate insoluble metal hydroxides. Suppose that $1.00 \times 10^2$ L of wastewater containing $1.2 \times 10^{-5}$ M $Cd^{2+}$ is treated with 1.0 L of 6.0 M NaOH solution. What is the residual concentration of $Cd^{2+}$ after treatment, and what mass of $Cd(OH)_2$ precipitates?

**Strategy:** Again, we use the standard approach to an equilibrium calculation. In this case, the reaction is a precipitation, for which the equilibrium constant is quite large. Thus, taking the reaction to completion by applying limiting reactant stoichiometry is the appropriate approach to solving the problem.

The wastewater contains $Cd^{2+}$, so an anion must also be present in the solution to balance the charge of the cadmium ions. Other species may exist as well. The problem asks only about the cadmium in the wastewater, so assume that any other ions are spectators. The sodium hydroxide solution contains $Na^+$ and $OH^-$, so the major species in the treated wastewater include $H_2O$, $Cd^{2+}$, $OH^-$, and $Na^+$. The equilibrium constant for the precipitation reaction is the inverse of $K_{sp}$ for $Cd(OH)_2$ :

$$Cd^{2+} (aq) + 2\, OH^- (aq) \rightleftarrows Cd(OH)_2 (s)$$

$$K_{eq} = \frac{1}{[Cd^{2+}]_{eq}[OH^-]^2_{eq}} = \frac{1}{K_{sp}} = \frac{1}{7.2 \times 10^{-15}} = 1.4 \times 10^{14}$$

**Solution:** The problem asks for the residual concentration of cadmium ions. In other words, what is the $Cd^{2+}$ ion concentration in the solution after the NaOH is added? Before a concentration table can be completed, initial ion concentrations must be determined. The volume of the mixed solutions is $1.01 \times 10^2$ L, which is less than 5% different from $1.00 \times 10^2$ L. This means the initial $Cd^{2+}$ concentration is $1.2 \times 10^{-5}$ M. Do a dilution calculation to find the "initial" concentration (after mixing but before reaction) of hydroxide ions:

$$c_2 = \frac{c_1 V_1}{V_2} = \frac{(6.0\ M)(1.0\ L)}{1.00 \times 10^2\ L} = 6.0 \times 10^{-2}\ M = [OH^-]_{initial}$$

Now set up a concentration table. The equilibrium constant for precipitation is very large, so imagine the precipitation in two steps (see Example 13-14). First, take the reaction to completion by applying limiting reactant stoichiometry. Then "switch on" the solubility equilibrium:

| Reaction | $Cd^{2+}$ (aq) | + | 2 OH$^-$ (aq) | $\leftrightarrows$ | Cd(OH)$_2$ (s) |
|---|---|---|---|---|---|
| Initial Concentration (M) | $1.2 \times 10^{-5}$ | | $6.0 \times 10^{-2}$ | | (solid) |
| Change to Completion (M) | $-1.2 \times 10^{-5}$ | | $-2.4 \times 10^{-5}$ | | (solid) |
| Concentration at Completion (M) | 0 | | $6.0 \times 10^{-2}$ | | (solid) |
| Change to Equilibrium (M) | $+y$ | | $+2y$ | | (solid) |
| Equilibrium Concentration (M) | $y$ | | $6.0 \times 10^{-2} + 2y$ | | (solid) |

A tiny amount of Cd(OH)$_2$ dissolves in this approach to equilibrium, so we use the solubility equilibrium and $K_{sp}$:

$$K_{sp} = [Cd^{2+}]_{eq}[OH^-]^2_{eq} = 7.2 \times 10^{-15}$$

The variable $y$ represents the change to equilibrium in a reaction with a small equilibrium constant, so make the approximation that $y$ will be small compared with $6.0 \times 10^{-2}$:

$$K_{sp} = (y)(6.0 \times 10^{-2} + 2y)^2 \cong (y)(6.0 \times 10^{-2})^2 = 7.2 \times 10^{-15}$$

$$y = \frac{7.2 \times 10^{-15}}{(6.0 \times 10^{-2})^2} = 2.0 \times 10^{-12} \quad \text{and} \quad [Cd^{2+}]_{eq} = 2.0 \times 10^{-12} \text{ M}$$

The problem also asks for the mass of Cd(OH)$_2$ that precipitates. The number of moles of precipitate equals the number of moles of Cd$^{2+}$ ions removed by hydroxide treatment. The necessary information is contained in the concentration table in the row that gives the completion conditions. The precipitation equilibrium constant is so large ($>10^{14}$) that equilibrium considerations can be neglected for this calculation:

$$n_{Cd(OH)_2} = n_{Cd^{2+}} = cV$$

$$= (1.2 \times 10^{-5} \text{ M})(1.00 \times 10^2 \text{ L}) = 1.2 \times 10^{-3} \text{ mol}$$

$$m_{Cd(OH)_2} = nM = (1.2 \times 10^{-3} \text{ mol})(146 \text{ g/mol}) = 0.18 \text{ g}$$

**Does the Result Make Sense?** Notice that virtually all the mass that dissolved originally (Example 15-18) has been recovered, and the residual concentration of Cd$^{2+}$ is very small. Both these observations are consistent with a water treatment process.

**Practice Exercise 15.19**
A student mixes 0.175 L of 0.500 M AgNO$_3$ with 0.225 L of 0.900 M Na$_2$CO$_3$. A white precipitate of silver carbonate forms. Calculate the mass of the precipitate and the residual concentration of Ag$^+$ ion in the final solution. (See Appendix E for $K_{sp}$ values.)

## The Common-Ion Effect

The concentrations of Cd$^{2+}$ ions in Example 15-18 reveal an important feature of aqueous equilibria. Notice in the calculations that the amount of dissolved Cd$^{2+}$ ions decreased by a factor of $10^{10}$ upon addition of excess OH$^-$. This reduction in concentration is an example of the **common-ion effect**.

**Figure 15-11**

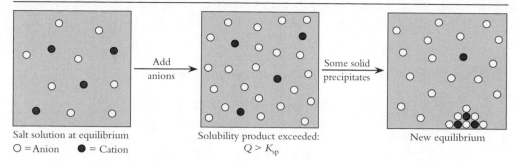

Salt solution at equilibrium
O = Anion   ● = Cation

Solubility product exceeded:
$Q > K_{sp}$

New equilibrium

When additional anions (*yellow circles*) are added to a solution of a salt at equilibrium (*left*), the solubility product is exceeded (*centre*), and some solid precipitates, reducing the concentration of cations (*blue circles*) in the solution. At the new equilibrium (*right*), there is a higher concentration of anions, but a lower concentration of cations. Water molecules have been omitted for clarity.

The solubility product of a salt contains the concentrations of both cations and anions, as shown here for cadmium hydroxide:

$$Cd(OH)_2 \ (s) \rightleftarrows Cd^{2+} \ (aq) + 2 \ OH^- \ (aq) \qquad K_{sp} = [Cd^{2+}]_{eq} \, [OH^-]^2_{eq} = 7.2 \times 10^{-15}$$

By the common-ion effect, increasing the amount of hydroxide ion in a saturated solution of $Cd(OH)_2$ causes a drop in the concentration of $Cd^{2+}$ ions and a precipitation of solid $Cd(OH)_2$. The common-ion effect is an application of Le Châtelier's principle (see Section 13.4). Added $OH^-$ shifts the equilibrium position in the direction that reduces this added hydroxide concentration. The drop in $[Cd^{2+}]$ is a response to the increase in $[OH^-]$, moving the system to a new equilibrium position.

Figure 15-11 is a molecular illustration of the common-ion effect, and Example 15-20 shows a quantitative application.

## EXAMPLE 15-20 THE COMMON-ION EFFECT

The concentration of chloride ion in seawater is around 0.55 M. To compare the solubility of $Pb^{2+}$ in freshwater vs. seawater, calculate the solubility in g/L of $PbCl_2$ in pure water and in 0.55 M NaCl.

**Strategy:** The problem asks for the g/L solubility of $PbCl_2$ in pure $H_2O$ and in 0.55 M NaCl. The same solubility equilibrium applies to each solution.

$$PbCl_2 \ (s) \rightleftarrows Pb^{2+} \ (aq) + 2 \ Cl^- \ (aq) \qquad K_{sp} = [Pb^{2+}]_{eq}[Cl^-]^2_{eq} = 1.2 \times 10^{-5}$$

The standard approach to equilibrium problems works well for both parts of this example. Solve the parts independently.

**Solution:**

**Pure water.** We have shown this type of calculation several times, so we summarize the results here. Review Example 15-18, if necessary.

$$K_{sp} = [Pb^{2+}]_{eq} \, [Cl^-]^2_{eq} = 4x^3 = 1.2 \times 10^{-5} \qquad \text{and} \qquad x = 0.014$$

$$[Pb^{2+}]_{eq} = 0.014 \text{ M}$$

Knowing that $PbCl_2$ produces $Pb^{2+}$ cations in 1:1 ratio tells us that 0.014 mole of $PbCl_2$ dissolves in 1 L of pure water. Convert to grams to finish the calculation. The molar mass of $PbCl_2$ is 278.1 g/mol:

$$(0.014 \text{ mol/L})(278.1 \text{ g/mol}) = 3.9 \text{ g/L } PbCl_2$$

**Salt water.** Set up a table of amounts that incorporates the starting concentration of chloride ion:

| Reaction | $PbCl_2$ (s) | $\leftrightarrows$ | $Pb^{2+}$ (aq) | + | 2 $Cl^-$ (aq) |
|---|---|---|---|---|---|
| Initial Concentration (M) | (solid) | | 0 | | 0.55 |
| Change in Concentration (M) | (solid) | | +x | | +2x |
| Final Concentration (M) | (solid) | | x | | 0.55 + 2x |

Substitute the results in the $K_{sp}$ expression and solve for $x$:

$$K_{sp} = [Pb^{2+}]_{eq}[Cl^-]^2_{eq} = (x)(0.55 + 2x)^2 \cong 0.303x = 1.2 \times 10^{-5}$$

$$x = 4.0 \times 10^{-5} \quad \text{and} \quad [Pb^{2+}]_{eq} = 4.0 \times 10^{-5} \text{ M}$$

The value of $x$ is small enough to justify the approximation. Now calculate the mass solubility. Again, the concentration of $Pb^{2+}$ in solution equals the number of moles of solid that dissolves:

$$(4.0 \times 10^{-5} \text{ mol/L})(278.1 \text{ g/mol}) = 0.011 \text{ g/L } PbCl_2.$$

**Do the Results Make Sense?** Both solubilities are low, as we would expect for a salt with a small value of $K_{sp}$. Notice that $PbCl_2$ is about 350 times less soluble in the NaCl solution. This makes sense in terms of the common-ion effect. The excess chloride ion suppresses the solubility of $Pb^{2+}$ by Le Châtelier's principle. The actual concentration of lead in seawater is much less than $4.0 \times 10^{-5}$ M. This is because other lead salts are much less soluble than lead(II) chloride. The ocean contains carbonate, for example, and $K_{sp}$ for lead(II) carbonate is quite small ($7.4 \times 10^{-14}$).

**Common Pitfalls:** Not remembering that the solid salt does not enter into the equilibrium expression. Its activity is 1.00 by convention.

**Practice Exercise 15.20**
(a) Calculate the hydronium ion concentration in 1.25 M $NH_4Cl$. (b) Calculate the hydronium ion concentration in a solution that is 1.25 M in $NH_4Cl$ and 0.755 M $NH_3$. (c) Explain the difference in concentrations.

The common-ion effect is quite general. The chemistry of buffer solutions is another important application of this principle. A buffer solution relies on the common-ion effect to suppress the concentration of hydronium ions and maintain a steady pH:

$$CH_3CO_2H \ (aq) + H_2O \ (l) \rightleftharpoons CH_3CO_2^- \ (aq) + H_3O^+ \ (aq) \quad K_a = 1.8 \times 10^{-5}$$

The graph of Figure 15-12 shows that the pH of an acetic acid solution is attenuated by added acetate anion. As acetate concentration increases, the equilibrium position shifts in the direction of more acetic acid and less $H_3O^+$.

## Effects of pH

Calcium carbonate is insoluble in pure water but dissolves in weakly acidic water. The role of this solubility phenomenon in the geochemistry of caverns is described in the Focus 13-2: Chemistry and the Environment. We can understand this depen-

**Figure 15-12**

The pH of an acetic acid solution is increased by the presence of added acetate ion. Added acetate ion decreases the concentration of hydronium ion, thereby increasing the pH of the solution.

dence on pH by examining the acid–base properties of the species involved in the solubility equilibrium.

$$CaCO_3 \ (s) \rightleftharpoons Ca^{2+} \ (aq) + CO_3{}^{2-} \ (aq)$$

Carbonate is the conjugate base of hydrogen carbonate, so we can write an acid–base equilibrium for a saturated solution of calcium carbonate:

$$CO_3{}^{2-} \ (aq) + H_2O \ (l) \rightleftharpoons HCO_3{}^{-} \ (aq) + OH^{-} \ (aq)$$

As a result of this equilibrium, a saturated solution of $CaCO_3$ is slightly basic.

Now consider what happens if some acid is added to this saturated solution. Hydronium ions react with hydroxide ions to form water:

$$OH^{-} \ (aq) + H_3O^{+} \ (aq) \rightleftharpoons 2 \ H_2O \ (l)$$

By Le Châtelier's principle, the system responds in the direction that will reduce the effect of this added acid, meaning that additional carbonate ions accept protons from water molecules:

$$CO_3{}^{2-} \ (aq) + H_2O \ (l) \rightleftharpoons HCO_3{}^{-} \ (aq) + OH^{-} \ (aq)$$

Responding to the reduction in carbonate ion concentration, additional calcium carbonate dissolves:

$$CaCO_3 \ (s) \rightleftharpoons Ca^{2+} \ (aq) + CO_3{}^{2-} \ (aq)$$

Qualitatively, making the solution more acidic increases the amount of calcium carbonate that dissolves in water.

The analysis can be made quantitative by writing the various equilibria and their $K_{eq}$ values. The reactions can be added to obtain the net reaction that occurs when calcium carbonate is exposed to acidic water, and the equilibrium constant for the net reaction is the product of the individual $K_{eq}$ values:

$$CaCO_3(s) \rightleftharpoons Ca^{2+}(aq) + CO_3^{2-}(aq) \qquad K_{eq} = K_{sp}$$

$$CO_3^{2-}(aq) + H_2O(l) \rightleftharpoons HCO_3^-(aq) + OH^-(aq) \qquad K_{eq} = K_b = \frac{K_w}{K_a}$$

$$OH^-(aq) + H_3O^+(aq) \rightleftharpoons 2 H_2O(l) \qquad K_{eq} = \frac{1}{K_w}$$

Net: $CaCO_3(s) + H_3O^+(aq) \rightleftharpoons Ca^{2+}(aq) + HCO_3^-(aq) + H_2O(l)$

$$K_{acidic} = K_{sp}\left(\frac{K_w}{K_a}\right)\left(\frac{1}{K_w}\right) = \frac{K_{sp}}{K_a}$$

Substituting the numerical values gives the equilibrium constant in acidic solution:

$$K_{acidic} = \frac{K_{sp}}{K_a} = \frac{3.36 \times 10^{-9}}{4.7 \times 10^{-11}} = 71$$

Taking the ratio of the equilibrium constants shows that calcium carbonate is 10 orders of magnitude more soluble in acidic solution than in water:

$$\frac{K_{acidic}}{K_{neutral}} = \frac{71}{K_{sp}} = \frac{71}{3.36 \times 10^{-9}} = 2.1 \times 10^{10}$$

Our Focus 15-1: Chemistry and the Environment in this chapter explores the effects of equilibria on $CO_2$ in the atmosphere, and Example 15-21 shows that the acidity of the solvent places an upper limit on the amount of salt that will dissolve.

### EXAMPLE 15-21 SALT DISSOLVING IN ACID

The most acid rain on record had pH = 1.87, recorded at Inverpolley Forest, Scotland in 1983. Calculate the concentration of $Ca^{2+}$ cations in a solution formed when this rain becomes saturated with calcium carbonate.

**Strategy:** We apply the seven-step method. The key feature is to identify the dominant equilibrium in this acidic solution. We do that by paying close attention to the major species present initially.

1. We are asked to calculate the equilibrium concentration of calcium ions, $[Ca^{2+}]_{eq}$.

2. The major species present initially, other than water, are solid calcium carbonate and aqueous hydronium ions.

3. These major species participate in the equilibrium described in the text:

$$CaCO_3(s) + H_3O^+(aq) \rightleftharpoons Ca^{2+}(aq) + HCO_3^-(aq) + H_2O(l)$$

4. The equilibrium constant expression is taken from the balanced equation. The value of the equilibrium constant was determined in the text:

$$K_{acidic} = \frac{[Ca^{2+}]_{eq}[HCO_3^-]_{eq}}{[H_3O^+]_{eq}} = 71$$

Since the pH is initially 1.87, the initial $[H_3O^+] = 10^{-1.87} = 0.0135$ M

**Solution:**

5. Organize the data using a concentration table. The large value for $K_{acidic}$ indicates that the reaction should be taken to completion by applying limiting reactant stoichiometry and then brought back to equilibrium. We can omit $H_2O$ and $CaCO_3$ from the table, because their concentrations do not appear in the equilibrium constant expression.

| Reaction | $CaCO_3$ (s) + | $H_3O^+$ (aq) | $\leftrightarrows$ | $Ca^{2+}$ (aq) | + | $HCO_3^-$ (aq) | + | $H_2O$ (l) |
|---|---|---|---|---|---|---|---|---|
| Initial Concentration (M) | | 0.0135 | | 0 | | 0 | | |
| Change to Completion (M) | | −0.0135 | | +0.0135 | | +0.0135 | | |
| Concentration at Completion (M) | | 0 | | 0.0135 | | 0.0135 | | |
| Change to Equilibrium (M) | | +y | | −y | | −y | | |
| Equilibrium Concentration (M) | | y | | 0.0135 − y | | 0.0135 − y | | |

6. We expect $y$ to be small, so we make the following approximation:

$$(0.0135 - y) \cong 0.0135$$

Next, substitute and solve for $y$:

$$\frac{[Ca^{2+}]_{eq}[HCO_3^-]_{eq}}{[H_3O^+]_{eq}} = \frac{(0.0135)^2}{y} = 71$$

$$y = \frac{(0.0135)^2}{71} = 2.56 \times 10^{-6}$$

This is quite small relative to 0.010, so

$$[Ca^{2+}]_{eq} = (0.0135\ M - y) = 0.0135\ M$$

**Does the Result Make Sense?** To determine if the overall calculation is reasonable, notice that the hydronium concentration in acid rain is five orders of magnitude higher than in pure water. This increased hydronium ion concentration can convert a substantial amount of carbonate anions into hydrogen carbonate anions, promoting the solubility of calcium carbonate. The resulting equilibrium concentration of calcium cations is about five orders of magnitude larger than the concentration in a saturated solution in pure water.

**Practice Exercise 15.21**

Calculate the g/L solubility of $CaCO_3$ in (a) pure water and in acidic solutions with (b) pH = 3.00 and (c) pH = 1.00.

The equilibrium constant expression derived for calcium carbonate dissolving in acidic solution is a general one that applies to any salt containing the conjugate base of a weak acid:

$$K_{acidic} = \frac{K_{sp}}{K_a} \qquad (15\text{-}4)$$

Thus, many salts are more soluble in acidic solution than in pure water. In addition to the carbonates, important examples include the phosphates ($PO_4^{3-}$) and the sulphides ($S^{2-}$). Insoluble hydroxides also dissolve in acidic solution, as illustrated by magnesium hydroxide:

$$Mg(OH)_2\ (s) \leftrightarrows Mg^{2+}\ (aq) + 2\ OH^-\ (aq) \qquad K_{sp} = 5.6 \times 19^{-12}$$

$$2\ [OH^-\ (aq) + H_3O^+\ (aq) \leftrightarrows 2\ H_2O\ (l)] \qquad K = \left(\frac{1}{K_w}\right)^2$$

Net: $Mg(OH)_2\ (s) + 2\ H_3O^+\ (aq) \leftrightarrows Mg^{2+}\ (aq) + 2\ H_2O\ (l) \qquad K = \dfrac{K_{sp}}{K_w^{\ 2}}$

Knowing that $K_w = 1.0 \times 10^{-14}$, you should be able to show that the hydroxide is 28 orders of magnitude more soluble in acid solution than in pure water.

## FOCUS 15-1 CHEMISTRY AND THE ENVIRONMENT: THE CARBON CYCLE

Carbon accounts for only 0.08% of our planet's mass. Nevertheless, life on Earth is based on carbon. Living organisms take in carbon, process it through biochemical reactions, and expel it as waste.

Carbon dioxide is especially important in the exchange of carbon between the biosphere and its environment. Green plants store energy from sunlight by converting $CO_2$ and $H_2O$ into carbohydrates. Most carbohydrates have the general formula $(CH_2O)_n$, so the chemical equation for carbohydrate formation appears simple, even though carbohydrates are complex (as described in Chapter 17):

$$n\,CO_2 + n\,H_2O \xrightarrow{\;h\nu\;} (CH_2O)_n + n\,O_2$$

Plants and animals use enzymes to catalyze the conversion of carbohydrates back to $CO_2$ and $H_2O$, a process that releases chemical energy and powers life processes:

$$(CH_2O)_n + n\,O_2 \xrightarrow{\;Enzymes\;} n\,CO_2 + n\,H_2O + energy$$

These two processes cycle carbon between the atmosphere, where it is found primarily as $CO_2$, and the biosphere, where it is found primarily as carbohydrates.

The carbon cycle is complicated by several reactions that involve $CO_2$. These reactions transfer carbon between the atmosphere, the hydrosphere (Earth's surface waters), and the lithosphere (Earth's crustal solids). The processes that move carbon from one sphere to another are illustrated schematically in the figure below.

Carbon dioxide is divided between the atmosphere and the hydrosphere. This division strongly favours the hydrosphere, because the large volume of water in the oceans has an immense capacity to dissolve $CO_2$. The dissolution of $CO_2$ into water causes the formation of carbonic acid ($H_2CO_3$):

$$CO_2\,(g) + H_2O\,(l) \leftrightharpoons H_2CO_3\,(aq)$$

However, almost all the Earth's carbon is found in the lithosphere as carbonate sediments that have precipitated from the oceans. Shells of aquatic animals also contribute $CaCO_3$ to the lithosphere. Carbon returns to the hydrosphere when carbonate minerals dissolve in water percolating through the Earth's crust. This process is limited by the solubility products for carbonate salts, so lithospheric carbonates represent a relatively inaccessible storehouse of carbon.

Until recently, the distribution of carbon among the different terrestrial spheres was stable. When humans began burning fossil fuels, however, such burning transferred carbon into the atmosphere as $CO_2$. This has become a rapidly changing feature of the overall carbon cycle. Over the last quarter century, the atmospheric concentration of $CO_2$ has grown by more than 10%.

Why should there be concern about an increase in atmospheric $CO_2$ of which the concentration in air is only 385 parts per million? Even a doubling of $CO_2$ concentration still leaves it well below 0.1% of the atmosphere. Despite its low concentration, $CO_2$ exerts a major effect on the average temperature of Earth's surface. Recall from Chapter 4 that $CO_2$ is a "greenhouse" gas that traps outgoing radiation. Increasing $CO_2$ in the atmosphere may contribute to global warming that could have catastrophic consequences. In a warmer climate, for example, much of the polar ice caps would melt, and the oceans would expand as their temperatures increased. Sea level could rise by several metres and inundate many of the world's most populous regions, such as Bangladesh, the Netherlands, and the East and Gulf Coasts of the United States. In addition, even small temperature fluctuations may alter global weather patterns and perhaps lead to serious droughts.

Our planet is such a complicated, dynamic set of interconnected systems that scientists do not know the result of doubling the concentration of atmospheric $CO_2$. Scientists are studying the carbon cycle in hopes of learning what lies ahead for the planet. There is some urgency to such studies, because recent global weather patterns show some of the characteristics predicted from computer models of global warming: average temperatures higher than those in previous years and an increased incidence of extreme weather such as hurricanes and heavy rainfall.

## SECTION EXERCISES

● **15.4.1** What is the concentration of $Mg^{2+}$ ions in a saturated aqueous solution of $Mg(OH)_2$?

● **15.4.2** Determine the concentration of $Mg^{2+}$ ions remaining in solution after 0.150 L of 0.125 M $MgCl_2$ solution is treated with 0.100 L of 0.65 M NaOH solution. Does NaCl also precipitate from this solution?

● **15.4.3** Determine the net reaction and the value of $K_{eq}$ for $Al(OH)_3$ ($K_{sp} = 4.6 \times 10^{-33}$) dissolving in acid solution.

## 15.5 COMPLEXATION EQUILIBRIA

In Section 13-6, we describe how metal cations in aqueous solution can form bonds to anions or neutral molecules that have lone pairs of electrons. This leads to formation of complex ions and to chemical equilibria involving complexation. The complexation equilibrium between $Ag^+$ and $NH_3$ is an example:

$$Ag^+ (aq) + 2 NH_3 (aq) \rightleftharpoons [Ag(NH_3)_2]^+ (aq) \qquad K_f = \frac{[Ag(NH_3)_2^+]_{eq}}{[Ag^+]_{eq}[NH_3]_{eq}^2}$$

The treatment of complexation equilibria is more complicated than treatments for acid–base or solubility equilibria, both because the stoichiometries of complexes are not obvious and because complex formation and dissociation frequently involve multiple steps.

## Stoichiometry of Complexes

A species that bonds to a metal to form a complex is known as a **ligand**. Any species that has a lone pair of electrons has the potential to be a ligand, but in this section, we confine our description to a few of the most common ligands: ammonia, compounds derived from ammonia, cyanide, and halides. These ligands are termed **Lewis bases** because they can donate the pair of electrons. The metal acts as a **Lewis acid** because it accepts the electrons. In these examples, the lone pair of electrons becomes the bonding pair. This bond is no different than a covalent bond in which the electrons come from both atoms, but it has its own name—a **coordinate covalent bond**. We describe additional examples in Chapter 18, which addresses the chemistry of the transition metals.

The stoichiometry of a metal complex is described by its chemical formula. For example, each cation of the silver–ammonia complex contains one $Ag^+$ cation bound to two neutral $NH_3$ ligands and carries a net charge of $+1$, as shown in Figure 15-13. The formula of a complex ion is enclosed in square brackets, as in $[Ag(NH_3)_2]^+$. The coordination number, which is the number of ligand atoms bonded directly to the metal, provides a shorthand notation for the stoichiometry of a metal complex. Coordination numbers of 6 and 4 are most common, and 2 is not unusual. There are no simple and reliable guidelines for predicting coordination numbers. Thus, coordination numbers must be determined by experiment. In this section, we provide information about the coordination number for each complex that we describe.

**Figure 15-13**

$[Ag(NH_3)_2]^{2+}$

Silver forms a linear complex ion with two neutral ammonia molecules as ligands.

## Complexation Calculations

In principle, the calculation of concentrations of species of a complexation equilibrium is no different from any other calculation involving equilibrium constant expressions. In practice, we have to consider multiple equilibria whenever a complex is present. This is because each ligand associates with the complex in a separate process with its own equilibrium expression. For instance, the silver–ammonia equilibrium is composed of two steps:

$$Ag^+ (aq) + NH_3 (aq) \rightleftarrows [Ag(NH_3)]^+ (aq)$$

$$K_1 = \frac{[Ag(NH_3)^+]_{eq}}{[Ag^+]_{eq}[NH_3]_{eq}} = 2.1 \times 10^3$$

$$[Ag(NH_3)]^+ (aq) + NH_3 (aq) \rightleftarrows [Ag(NH_3)_2]^+ (aq)$$

$$K_2 = \frac{[Ag(NH_3)_2{}^+]_{eq}}{[Ag(NH_3)^+]_{eq}[NH_3]_{eq}} = 8.2 \times 10^3$$

Because the two equilibrium constant expressions have similar magnitudes, a solution of the silver–ammonia complex generally has a significant concentration of each of the species that participates in the equilibria. The details of such calculations are beyond the scope of general chemistry. When the solution contains a large excess of ligand, however, each step in the complexation process proceeds nearly to completion. Under these conditions, we can apply the standard seven-step approach to a single expression that describes the formation reaction of the complete complex.

One of the characteristics of complexation equilibria is that, in the presence of excess concentration of a complexing ligand, formation of a complex often reduces the concentration of a free metal cation essentially to zero. This is another application of the common-ion effect discussed in the previous section. Example 15-22 treats a situation of this sort.

### EXAMPLE 15-22 FORMATION OF A GOLD COMPLEX

The small amounts of gold contained in low-grade ores can be extracted using a combination of oxidation and complexation. Gold is oxidized to $Au^+$, which forms a very strong complex with cyanide anions:

$$Au^+ (aq) + 2\ CN^- (aq) \rightleftharpoons [Au(CN)_2]^- (aq) \qquad K_f = 2 \times 10^{38}$$

Suppose that a sample of ore containing $2.5 \times 10^{-3}$ mole of gold is extracted with 1.0 L of $4.0 \times 10^{-2}$ M aqueous KCN solution. Calculate the concentrations of the three species involved in the complexation equilibrium.

$[Au(CN)_2]^{2-}$

**Strategy:** Determining concentrations requires a quantitative equilibrium calculation, so we apply the seven-step strategy. The complexation equilibrium is given in the problem.

1. The problem asks for the equilibrium concentrations of the three ionic species: $[Au^+]_{eq}$, $[CN^-]_{eq}$, and $[Au(CN)_2^-]_{eq}$.

2. Potassium cyanide is a salt that dissociates completely into ions, and the problem states that gold is oxidized, so the major species present are $Au^+$, $CN^-$, $K^+$, and $H_2O$.

3. Both $CN^-$ and $H_2O$ have acid–base properties, but the problem asks only about the species involved in the complexation equilibrium, so the important equilibrium is the complexation reaction:

$$Au^+ (aq) + 2\ CN^- (aq) \rightleftharpoons [Au(CN)_2]^- (aq)$$

4. Now we write the equilibrium constant expression for formation of the complex ion:

$$K_f = \frac{[Au(CN)_2^-]_{eq}}{[Au^+]_{eq}[CN^-]_{eq}^2} = 2 \times 10^{38}$$

**Solution:**

5. Organize the data using a concentration table. Initial concentrations can be found from the data stated in the problem, and because the equilibrium constant is quite large, we take the reaction to completion and then allow back-reaction to equilibrium:

| Reaction | $Au^+ (aq)$ | + | $2\ CN^- (aq)$ | $\rightleftharpoons$ | $[Au(CN)_2]^- (aq)$ |
|---|---|---|---|---|---|
| Initial Concentration (M) | $2.5 \times 10^{-3}$ | | $4.0 \times 10^{-2}$ | | 0 |
| Change to Completion (M) | $-2.5 \times 10^{-3}$ | | $-5.0 \times 10^{-3}$ | | $+2.5 \times 10^{-3}$ |
| Concentration at Completion (M) | 0 | | $3.5 \times 10^{-2}$ | | $2.5 \times 10^{-3}$ |
| Change to Equilibrium (M) | $+y$ | | $+2y$ | | $-y$ |
| Equilibrium Concentration (M) | $y$ | | $3.5 \times 10^{-2} + 2y$ | | $2.5 \times 10^{-3} - y$ |

6. We expect $y$ to have a small value, so we make two approximations:

$$3.5 \times 10^{-2} + 2y \cong 3.5 \times 10^{-2} \qquad \text{and} \qquad 2.5 \times 10^{-3} - y \cong 2.5 \times 10^{-3}$$

$$K_f = 2 \times 10^{38} = \frac{2.5 \times 10^{-3}}{y(3.5 \times 10^{-2})^2} \qquad \text{so} \qquad y = \frac{2.5 \times 10^{-3}}{K_f(3.5 \times 10^{-2})^2}$$

$$y = \frac{2.5 \times 10^{-3}}{(2 \times 10^{38})(3.5 \times 10^{-2})^2} = 1.0 \times 10^{-38}$$

$$[Au^+]_{eq} = 1 \times 10^{-38} \text{ M}; \quad [CN^-]_{eq} = 4 \times 10^{-2} \text{ M}; \quad [Au(CN)_2^-]_{eq} = 3 \times 10^{-3} \text{ M}$$

**Do the Results Make Sense?** Each equilibrium concentration has only one significant figure, matching the single significant figure in the $K_f$ value. The calculated concentration for $Au^+$ is too small to measure, but this is reasonable in light of the very large value of $K_f$. The tiny value of $y$ shows that the approximations are reasonable.

### Practice Exercise 15.22

Suppose that 0.275 g of silver nitrate dissolves in 0.85 L of 0.250 M ammonia:

$$Ag^+ (aq) + 2\, NH_3\, (aq) \rightleftharpoons [Ag(NH_3)_2^+]\, (aq) \qquad K_f = 1.6 \times 10^7$$

Calculate the equilibrium concentrations of all species involved in the complexation equilibrium.

## The Chelate Effect

Ethylenediamine
(en)

Ligands that have two or more donor atoms are **chelating** ligands. One of the most common chelating ligands is ethylenediamine (often abbreviated "en"), $H_2NCH_2CH_2NH_2$. Each nitrogen atom has a lone pair of electrons, so both can be donor atoms at the same time. Thus, ethylenediamine is said to be a **bidentate** ligand that can coordinate to a metal at two sites simultaneously. The word *dentate* is derived from the Latin *dentis*, meaning "tooth." A monodentate ligand has "one tooth" for binding to a metal, a bidentate ligand has "two teeth," and so on.

Chelating ligands bind much more tightly to their metal cations than do ligands that possess only one donor atom. A good example is the ethylenediamine complex with $Ni^{2+}$, which is much more stable than the analogous ammonia complex:

$$Ni^{2+} (aq) + 6\, NH_3\, (aq) \rightleftharpoons [Ni(NH_3)_6]^{2+}\, (aq) \qquad K_f = 2.0 \times 10^8$$

$$Ni^{2+} (aq) + 3\, en\, (aq) \rightleftharpoons [Ni(en)_3]^{2+}\, (aq) \qquad K_f = 4.1 \times 10^{17}$$

Even though both ligands form Ni—N bonds of similar strength, ethylenediamine binds to $Ni^{2+}$ many orders of magnitude more strongly than does $NH_3$. Why is this? Think about complexation at the molecular level. One of the Ni—N bonds in either complex can be broken fairly easily. When this happens to $[Ni(NH_3)_6]^{2+}$, the ammonia molecule drifts away and is replaced by another ligand, typically a water molecule from the solvent:

$$[Ni(NH_3)_6]^{2+} \longrightarrow [Ni(NH_3)_5]^{2+} + NH_3$$

$$Ni(NH_3)_5]^{2+} + H_2O \longrightarrow [Ni(NH_3)_5H_2O]^{2+}$$

The result is an exchange of ligands.

This simple ligand exchange reaction is less likely for $[Ni(en)_3]^{2+}$. As Figure 15-14 shows, when one Ni—N bond breaks and one $NH_2$ group of an ethylenediamine ligand dissociates from the metal complex, the ligand is still bound to the metal complex by the second $NH_2$ group. Consequently, the en ligand is not free to float away in solution, so the first $NH_2$ group binds again to the metal rather than being replaced by a competing ligand. For an ethylenediamine ligand to be replaced,

**Figure 15-14**

Breaking one bond in an ethylenediamine complex leaves the en ligand dangling but "tethered" to the metal by the second nitrogen atom. (Each carbon atom bonds to two hydrogen atoms that are not shown.)

both ends of the ligand must be released at almost the same time. This stabilization of a metal complex by a ligand with more than one donor atom is known as the **chelate effect**. (The word *chelate* comes from the Greek *chele*, meaning "claw.")

The chelate effect has important practical applications. Heavy metal ions such as $Hg^{2+}$ and $Pb^{2+}$ are extremely poisonous, and their effects are cumulative because the body has no mechanism for excreting heavy metal ions. Fortunately, chelating ligands can be used to treat poisoning by mercury and lead via so-called chelation therapy. One important example of a ligand used to treat metal ion poisoning is ethylenediamine tetraacetate ion, abbreviated $[EDTA]^{4-}$ or simply EDTA. With six donor atoms that can bond to a single metal ion, EDTA is a hexadentate ligand.

Dosages of EDTA are delivered as the calcium disodium salt, $Na_2[Ca(EDTA)]$. The calcium complex prevents EDTA from extracting iron from the blood. Unlike iron, heavy metal ions such as $Hg^{2+}$ preferentially displace calcium cations from $[Ca(EDTA)]^{2-}$:

$$[Ca(EDTA)]^{2-}\,(aq) + Hg^{2+}\,(aq) \longrightarrow [Hg(EDTA)]^{2-}\,(aq) + Ca^{2+}\,(aq)$$

The kidneys are able to pass chelated heavy metal ions into the urine, so treatment with $[Ca(EDTA)]^{2-}$ cleanses mercury from the blood, counteracting its poisonous effects.

$[Ca(EDTA)]^{2-}$, with the hydrogen atoms removed for clarity

## Complex Formation and Solubility

The effect of complex formation on the solubility of a solid can be observed in the home. Silver dinnerware eventually becomes discoloured by an unsightly black tarnish of $Ag_2S$, formed from the reaction of the silver surface with small amounts of $H_2S$ present in the atmosphere. Silver sulphide is highly insoluble in water. Commercial silver polishes contain ligands that form strong soluble complexes with $Ag^+$ ions. If a tarnished silver item is rubbed with a polish, the black tarnish dissolves, returning the silver to its brilliant shine.

Silver salts demonstrate how complexation enhances solubility. Figure 15-15 shows that silver will precipitate or dissolve in aqueous solution depending on the species that are present. Starting with a solution of silver nitrate, addition of aqueous NaCl causes AgCl to precipitate:

$$AgCl\,(s) \rightleftarrows Ag^+\,(aq) + Cl^-\,(aq) \qquad K_{sp} = 1.8 \times 10^{-10}$$

**Figure 15-15**

(a)    AgCl(s),
$K_{sp} = 1.8 \times 10^{-10}$

(b)    $[Ag(NH_3)_2]^+(aq)$

(c)    AgBr(s),
$K_{sp} = 5.35 \times 10^{-13}$

(d)    $[Ag(S_2O_3)_2]^{3-}(aq)$

Silver precipitates from solution when a solution is added that contains an anion that forms an insoluble silver compound (a and c). The precipitates redissolve when a solution is added that contains a species that can form a strong complex with $Ag^+$ (b and d).

If aqueous ammonia is added to the mixture, the precipitate dissolves by forming the silver–ammonia complex. The formation equilibrium combines with the solubility equilibrium to give the overall process:

$$Ag^+(aq) + 2\,NH_3(aq) \rightleftharpoons [Ag(NH_3)_2]^+(aq) \qquad K_f = 1.6 \times 10^7$$
$$AgCl(s) \rightleftharpoons Ag^+(aq) + Cl^-(aq) \qquad K_{sp} = 1.8 \times 10^{-10}$$

Net: $AgCl(s) + 2\,NH_3(aq) \rightleftharpoons [Ag(NH_3)_2]^+(aq) + Cl^-(aq) \quad K_{eq} = K_{sp}K_f = 2.9 \times 10^{-3}$

Although still less than 1, this equilibrium constant is large enough to allow solid AgCl to dissolve in strong aqueous ammonia. In fact, if you open a bottle of commercial silver polish, you will surely notice the sharp odour of ammonia. These polishes contain a variety of ligands, including ammonia.

Addition of a solution of NaBr results in a new precipitate, because AgBr is significantly less soluble than AgCl:

$$[Ag(NH_3)_2]^+(aq) \rightleftharpoons Ag^+(aq) + 2\,NH_3(aq) \qquad \frac{1}{K_f} = \frac{1}{1.6 \times 10^7}$$

$$Ag^+(aq) + Br^-(aq) \rightleftharpoons AgBr(s) \qquad \frac{1}{K_{sp}} = \frac{1}{5.35 \times 10^{-13}}$$

Net: $[Ag(NH_3)_2]^+(aq) + Br^-(aq) \rightleftharpoons AgBr(s) + 2\,NH_3(aq)$

$$K_{eq} = \frac{1}{K_f K_{sp}} = \frac{1}{(1.6 \times 10^7)(5.35 \times 10^{-13})} = 1.2 \times 10^5$$

This large positive value indicates that the equilibrium position lies well to the right, favouring formation of solid AgBr.

Upon addition of sodium thiosulphate solution, the AgBr precipitate dissolves, because $S_2O_3{}^{2-}$ forms a very strong complex with $Ag^+$. This ligand is capable of dissolving AgBr:

$$AgBr(s) \rightleftharpoons Ag^+ (aq) + Br^- (aq) \qquad K_{sp} = 5.35 \times 10^{-13}$$

$$Ag^+ (aq) + 2\, S_2O_3^{2-} (aq) \rightleftharpoons [Ag(S_2O_3)_2]^{3-} (aq) \qquad K_f = 2.0 \times 10^{13}$$

Net: $\quad AgBr\ (s) + 2\, S_2O_3^{2-} (aq) \rightleftharpoons [Ag(S_2O_3)_2]^{3-} (aq) + Br^- (aq) \qquad K_{eq} = K_{sp}K_f = 10.7$

These reactions find practical applications in film photography. Most photographic films contain silver halides, primarily silver bromide. Exposure to light causes some $Ag^+$ cations to gain electrons, forming Ag granules that, after developing, generate the images captured on the film. But what about the unreacted silver bromide? If it were left in the film, further exposure to light would cause further reaction and eventual blackening of the image. To prevent this, film must be "fixed" by removing all unreacted AgBr. This is done using thiosulphate, which forms a complex with otherwise insoluble AgBr and allows it to be washed from the film. As Example 15-23 shows, fixer solution has a limited capacity to accept AgBr, so after repeated use, a solution must be discarded and a new solution prepared.

## EXAMPLE 15-23 CAPACITY OF THIOSULPHATE FIXER

What mass of AgBr will dissolve in 1.5 L of fixer solution that contains 0.50 M of thiosulphate anions?

**Strategy:** We can calculate the mass that dissolves from the total concentration of $Ag^+$ ions at equilibrium. We apply the seven-step method.

1. The problem asks for mass of AgBr that will dissolve. To find this, we must calculate the equilibrium concentration of silver-containing ions in the solution.

2. In addition to water, the initial species are AgBr (s), $S_2O_3^{2-}$ (aq), and an unspecified spectator cation.

3. The equilibrium is the one described in the text, by which thiosulphate dissolves AgBr through complexation:

Thiosulfate, $S_2O_3^{2-}$
(one resonance form)

$$AgBr\ (s) + 2\, S_2O_3^{2-} (aq) \rightleftharpoons [Ag(S_2O_3)_2]^{3-} (aq) + Br^- (aq)$$

4. This reaction is the sum of the solubility reaction of AgBr and the complexation reaction between Ag and thiosulphate, so its $K_{eq}$ is the product of the two equilibrium constants as described in the text:

$$AgBr\ (s) + 2\, S_2O_3^{2-} (aq) \rightleftharpoons [Ag(S_2O_3)_2]^{3-} (aq) + Br^- (aq) \quad K_{eq} = 10.7$$

**Solution:**

5. Set up a concentration table for the aqueous species, using the initial concentration of thiosulphate, 0.50 M:

| Reaction | AgBr (s) | + | 2 $S_2O_3^{2-}$ (aq) | ⇆ | $[Ag(S_2O_3)_2]^{3-}$ (aq) | + | Br (aq) |
|---|---|---|---|---|---|---|---|
| Initial Concentration (M) | | | 0.50 | | 0 | | 0 |
| Change to Equilibrium (M) | | | $-2x$ | | $+x$ | | $+x$ |
| Equilibrium Concentration (M) | | | $0.50 - 2x$ | | $x$ | | $x$ |

6. Substitute the equilibrium concentrations into the equilibrium constant expression and solve for $x$:

$$\frac{[Ag(S_2O_3)_2^{3-}]_{eq}[Br^-]_{eq}}{[S_2O_3^{2-}]_{eq}^2} = \frac{x^2}{(0.50 - 2x)^2} = 10.7$$

We could solve by the quadratic equation, but notice that we can simplify this expression by taking the square root of each side:

$$\frac{x}{0.50 - 2x} = \sqrt{10.7} = 3.27 \quad \text{from which} \quad x = 1.64 - 6.54\,x$$
$$7.54x = 1.64 \quad \text{so} \quad x = 0.217 = 0.22$$

Thus, the equilibrium concentration of silver complex and bromide ion in the solution is 0.22 M. Multiplying by the volume of the solution gives the amounts in moles:

$$n_{Br^-} = cV = (0.22\ \text{M})(1.5\ \text{L}) = 0.33\ \text{mol}$$

The stoichiometric ratio between AgBr and the ions in solution is 1:1, so this amount is also the amount of AgBr that dissolves. Convert to mass by multiplying by the molar mass of AgBr:

$$m = nM = (0.33\ \text{mol})(187.8\ \text{g/mol}) = 62\ \text{g}$$

**Does the Result Make Sense?** Notice that all the variables for the solution are somewhat less than 1 M, and the solution volume is somewhat greater than 1 L, so a result that is somewhat less than the molar mass of AgBr appears reasonable.

### Practice Exercise 15.23

Zinc oxalate ($Zn(C_2O_4)$) is sparingly soluble in water ($K_{sp} = 1.4 \times 10^{-8}$). The $Zn^{2+}$ ion forms a tetrahedral-shaped complex with ammonia. The formation constant for the complex is $4.1 \times 10^8$. How many moles of zinc oxalate will dissolve in 1.0 L of 0.200 M aqueous ammonia?

$[Zn(NH_3)_4]^{2+}$

Buffer action and complexation occur together in many natural settings. We conclude this chapter with the Focus 15-2: Chemistry and Life that describes one important example, the pH of soil.

## SECTION EXERCISES

●**15.5.1** Determine the concentration of the species present at equilibrium if 0.10 mole of ethylenediamine is dissolved in 2.00 L of a solution that contains $6.5 \times 10^{-3}$ M $Ni^{2+}$. ($K_f = 4.1 \times 10^{17}$ for $[Ni(en)_3]^{2+}$.)

●**15.5.2** When $Zn^{2+}$ is added to aqueous ammonia, four ammonia ligands complex in stepwise fashion. Write all the equilibria involving $Zn^{2+}$ and $NH_3$.

●**15.5.3** Calculate the mass of AgCl that will dissolve in 250 mL of 1.0 M aqueous ammonia.

## FOCUS 15-2  CHEMISTRY AND LIFE: THE pH OF SOIL

In 1914, two European immigrants, Karen Blixen and her husband, established a coffee plantation outside of Nairobi in what is now Kenya. Conditions seemed perfect for coffee, as both the altitude and latitude closely matched that of Colombia, a highly successful coffee-growing country. Nevertheless, the coffee plants withered and eventually died, because the soil in that part of Kenya is too acidic for coffee plants to flourish. Blixen eventually became famous as the author of *Out of Africa*, which was made into an Oscar-winning movie.

Every type of plant requires soil with a pH that falls within a particular range. No crops like strongly acid soils. If soil pH falls much below 5, only grasses grow well. Blixen and her husband might have been successful had they tried cattle ranching rather than growing coffee.

Ironically, coffee does need relatively acidic soil, with pH between 5 and 6. Conifers and shrubs such as azaleas and rhododendrons thrive on soils with this acidity, as do tea, potatoes, rice, and rye. The vast majority of crop plants, including most vegetables, need soils just on the acidic side of neutral, pH between 6 and 7. Only a few crops—barley, sugar beets, cotton, and sugarcane—like soils on the mildly basic side, between pH 7 and 8, and only desert plants can cope with soils having pH greater than 8.

Plants are highly sensitive to soil acidity because many equilibria involving plant nutrients are affected by pH. Phosphorus is a primary example. This essential element for plant growth occurs in soils mainly as phosphates, which are subject to phosphate—hydrogen phosphate equilibria. Consequently, phosphorus is most readily available to plants when the soil pH is near neutral, between 6 and 8. Conversely, aluminum, which is toxic to most plants, is soluble in acidic soil but occurs as insoluble hydroxides at pH values above 5. In highly acidic soils, plants are deprived of phosphorus and subjected to toxic aluminum.

Delphinidin-3-glucoside
Chlorogenic acid
**Low pH: blue**
**High pH: pink**

The acidity of soils also strongly affects the availability of metal cations such as $K^+$, $Mg^{2+}$, and $Ca^{2+}$. Most soils contain significant amounts of clay, the chemical composition of which is dominated by aluminosilicates. Silicates are anionic, and the anions can be neutralized either by accepting protons or by associating with metal cations. When soil is too acidic, protons replace these metal cations, and the soil becomes depleted in these essential nutrients.

The colours of flowering plants such as hydrangeas are highly sensitive to soil acidity. At pH > 6.5, these showy flowers are deep pink, but at pH < 5, the blossoms are vivid blue. The chemistry of these changes involves complexation of aluminum by pigments that have acidic groups, as the structures above show.

Several processes act to increase soil acidity. As organic matter decays, it forms many organic carboxylic acids. Peat moss, for example, is entirely organic and has quite low pH, around 4. Acid rain, the result of industrial pollution, can seriously acidify soils. Chemical fertilizers that contain ammonium cations generate hydronium ions. Intensive agriculture increases soil acidity, because growing crops remove basic metal cations from the soil, replacing them with hydronium ions.

To reduce soil acidity and keep cropland productive, farmers amend the soil by "liming," which is the application of limestone. Limestone is calcium carbonate, and the carbonate anion is the conjugate base of hydrogen carbonate. Liming therefore increases the concentration of hydroxide in the soil, thereby increasing the pH:

$$CO_3^{2-} (aq) + H_2O (l) \rightleftharpoons HCO_3^- (aq) + OH^- (aq)$$

# CHAPTER 15 VISUAL SUMMARY

## 15.1 BUFFER SOLUTIONS

> A buffer soution contains both a weak acid and its conjugate base as major species in solution.

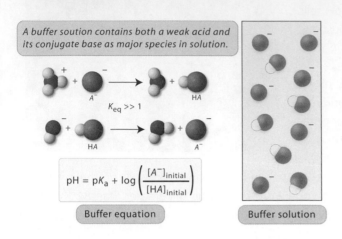

$$pH = pK_a + \log\left(\frac{[A^-]_{initial}}{[HA]_{initial}}\right)$$

Buffer equation

Buffer solution

**LEARNING OBJECTIVES:** Calculate the pH of a buffered solution.

**SKILLS TO MASTER:** Using the buffer equation to calculate pH of a buffer solution (p. 870)

Calculating the pH of a buffered solution after adding a small amount of acid or base (p. 872)

**KEY CONCEPTS:** A buffer solution contains both a weak acid and its conjugate base, or a weak base and its conjugate acid, as major species in solution.

## 15.2 CAPACITY AND PREPARATION OF BUFFER SOLUTIONS

pH range = $pK_a \pm 1$    Buffer capacity

**Table 15-1 Common Buffer Systems**

| Acid | Base | $pK_a$ | pH Range |
|------|------|--------|----------|
| $H_3PO_4$ | $H_2PO_4^-$ | 2.16 | 1–3 |
| $HCO_2H$ | $HCO_2^-$ | 3.75 | 3–5 |
| $CH_3CO_2H$ | $CH_3CO_2^-$ | 4.75 | 4–6 |
| $H_2PO_4^-$ | $HPO_4^{2-}$ | 7.21 | 6–8 |
| $NH_4^+$ | $NH_3$ | 9.25 | 8–10 |
| $HCO_3^-$ | $CO_3^{2-}$ | 10.33 | 9–11 |
| $HPO_4^{2-}$ | $PO_4^{3-}$ | 12.32 | 11–13 |

**LEARNING OBJECTIVES:** Explain how to prepare a buffered solution of known pH and capacity.

**SKILLS TO MASTER:** Calculating the capacity of a buffered solution (p. 873)

**SKILLS TO MASTER:** Preparing a buffered solution (p. 875)

## 15.3 ACID–BASE TITRATIONS

**LEARNING OBJECTIVES:** Calculate an acid or base concentration from titration data.

**SKILLS TO MASTER:** Calculating the concentration of a standard base solution (p. 881)

Calculating the pH at the stoichiometric point of a titration (p. 884)

Recognizing qualitative features of a titration curve (p. 889)

Constructing a titration curve for a polyprotic acid (p. 890)

Choosing an appropriate indicator for a particular acid–base titration (p. 895)

**KEY CONCEPTS:** At the stoichiometric point, just enough hydroxide ions have been added to react with every acidic proton present in the acid solution before the titration was started.

## 15.4 SOLUBILITY EQUILIBRIA

$$Cd(OH)_2(s) \rightleftharpoons Cd^{2+}(aq) + 2\,OH^-(aq)$$

| Insoluble | $K_{sp} < 10^{-5}$ |
| Slightly soluble | $10^{-5} < K_{sp} < 10^{-2}$ |
| Soluble | $K_{sp} > 10^{-2}$ |

$$K_{acidic} = \frac{K_{sp}}{K_a}$$

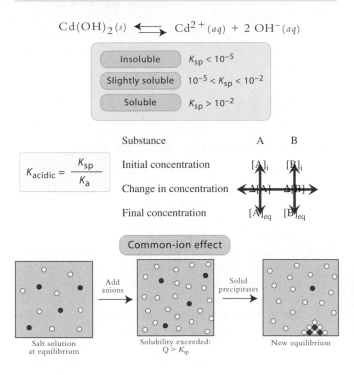

| Substance | A | B |
| Initial concentration | $[A]_i$ | $[B]_i$ |
| Change in concentration | $\Delta[A]$ | $\Delta[B]$ |
| Final concentration | $[A]_{eq}$ | $[B]_{eq}$ |

Common-ion effect

Salt solution at equilibrium → Add anions → Solubility exceeded: $Q > K_{sp}$ → Solid precipitates → New equilibrium

**LEARNING OBJECTIVES:** Use the concepts of $K_{sp}$ and the common-ion effect to calculate solution concentrations.

**SKILLS TO MASTER:** Using $K_{sp}$ to determine solution concentrations and precipitate masses (p. 898)

Using the common-ion effect to calculate solution concentrations (p. 902)

Calculating the effect of pH on the solubility of a salt (p. 904)

**KEY CONCEPTS:** Addition of a common ion decreases the solubility of a salt.

Lowering the pH increases the solubility of a salt that contains a basic anion.

## 15.5 COMPLEXATION EQUILIBRIA

$$2\,NH_3 + Ag^+ \rightleftharpoons [Ag(NH_3)_2]^+$$

Complex ion

Ligand

Ethylenediamine (en)

Bidentate ligand

Chelating ligand    EDTA

Chelate effect

(a) AgCl(s), $K_{sp}=1.8 \times 10^{-10}$ → $NH_3(aq)$ → (b) $[Ag(NH_3)_2]^+(aq)$ → $NaBr(aq)$ → (c) AgBr(s), $K_{sp}=5.35 \times 10^{-13}$ → $Na_2S_2O_3(aq)$ → (d) $[Ag(S_2O_3)_2]^{3-}(aq)$

**LEARNING OBJECTIVES:** Calculate the concentrations of species involved in complex-ion formation.

**SKILLS TO MASTER:** Understanding the role of complex formation on the solubility of a solid (p. 909)

**KEY CONCEPTS:** Electron-donating ligands are Lewis bases. Electron-accepting metals are Lewis acids.

The chelate effect results in very high formation constants of complexes.

## LEARNING EXERCISES

15.1   Write a paragraph that explains the chemical principles of buffer action.

15.2   Explain why the pH is not necessarily 7.0 at the stoichiometric point in an acid–base titration.

15.3   List all the types of calculations described in Chapter 15 in which acid–base equilibrium expressions play a role.

15.4   Update your list of memory bank equations and their uses.

15.5   Use your own words to define (a) buffer solution, (b) stoichiometric point, (c) common-ion effect, and (d) chelate.

## PROBLEMS

**ilw** = interactive learning ware problem. Visit the website at www.wiley.com/canada/olmsted

## Buffer Solutions

15.1   Which of the following solutions show buffer properties? Calculate the pH of each solution that is buffered.
(a) 0.100 L of 0.25 M $NH_4Cl$ + 0.150 L of 0.25 M NaOH
(b) 0.100 L of 0.25 M $NH_4Cl$ + 0.050 L of 0.25 M NaOH
(c) 0.100 L of 0.25 M $NH_4Cl$ + 0.050 L of 0.25 M HCl
(d) 0.100 L of 0.25 M $NH_3$ + 0.050 L of 0.25 M HCl

15.2   Which of the following solutions show buffer properties? Calculate the pH of each solution that is buffered.
(a) 0.100 L of 0.25 M $NaCH_3CO_2$ + 0.150 L of 0.25 M HCl
(b) 0.100 L of 0.25 M $NaCH_3CO_2$ + 0.050 L of 0.25 M HCl
(c) 0.100 L of 0.25 M $NaCH_3CO_2$ + 0.050 L of 0.25 M NaOH
(d) 0.100 L of 0.25 M $CH_3CO_2H$ + 0.050 L of 0.25 M NaOH

15.3   Calculate the change in pH resulting from the addition of 5.0 mmoles of strong acid to each solution in Problem 15.1 that is buffered.

15.4   Calculate the change in pH resulting from the addition of 2.5 mmoles of strong base to each solution in Problem 15.2 that is buffered.

15.5   Calculate how many moles of base it takes to change the pH of each buffered solution in Problem 15.1 by 0.10 pH unit.

15.6   Calculate how many moles of acid it takes to change the pH of each buffered solution in Problem 15.2 by 0.20 pH unit.

15.7   A buffer solution made from $NaHCO_3$ and $Na_2CO_3$ has a pH in the range of 9–11. Write balanced equations that show how this buffer system neutralizes $H_3O^+$ and $OH^-$.

15.8   A buffer solution made from formic acid ($HCO_2H$) and sodium formate ($NaHCO_2$) has a pH in the range of 3–5. Write balanced equations that show how this buffer system neutralizes $H_3O^+$ and $OH^-$.

15.9   Which of the following sets of chemicals can be used to prepare buffer solutions? For each one that can, specify the weak acid and the conjugate base that make up the major species of the buffer solution: (a) $NaHCO_3$ + NaOH; (b) NaOH + $NH_3$; (c) $H_3PO_4$ + HCl; and (d) HCl + $Na_2CO_3$.

15.10  Which of the following sets of chemicals can be used to prepare buffer solutions? For each one that can, specify the weak acid and the conjugate base that make up the major species of the buffer solution: (a) $Na_2HPO_4$ + HCl; (b) $NH_4Cl$ + $NH_3$; (c) $Na_3PO_4$ + NaOH; and (d) $NH_4Cl$ + $NaNO_3$.

## Capacity and Preparation of Buffer Solutions

15.11  From Table 15-1, select the best conjugate acid–base pairs for buffer solutions at pH 3.50 and 12.60. If you were going to add HCl solution as part of the buffer preparation, what other substance should you use in each case?

15.12  From Table 15-1, select the best conjugate acid–base pairs for buffer solutions at pH 6.85 and 10.80. If you were going to add NaOH solution as part of the buffer preparation, what other substance should you use in each case?

15.13  **ilw** What mass of ammonium chloride must be added to 1.25 L of 0.25 M ammonia to make a buffer solution of pH = 8.90?

15.14  What mass of sodium acetate must be added to 2.50 L of 0.55 M acetic acid to make a buffer solution of pH = 5.75?

15.15  What is the maximum volume of 2.0 M HCl (*aq*) that the buffer described in Problem 15.13 can tolerate without showing a pH change greater than 0.25 units?

15.16  What is the maximum volume of 2.5 M NaOH (*aq*) that the buffer described in Problem 15.14 can tolerate without showing a pH change greater than 0.25 units?

15.17  Determine the mass of $Na_2CO_3$ and the volume of 0.500 M HCl solution required to make 1.5 L of a buffer solution at pH = 9.85 and at a total concentration of 0.350 M.

15.18  Determine the mass of NaOH and the volume of 3.50 M $H_3PO_4$ solution required to make 2.5 L of a buffer solution at pH = 2.25 and at a total concentration of 0.175 M.

15.19  How many grams of solid NaOH must be added to the buffer solution described in Problem 15.17 to change the pH by 0.15 units? After this NaOH has dissolved, is the solution still a buffer? Explain.

15.20  How many millilitres of 2.00 M HCl must be added to the buffer solution described in Problem 15.18 to change the pH by 1.25 units? After this HCl has dissolved, is the solution still a buffer? Explain.

## Acid–Base Titrations

15.21  For each of the following, decide whether the pH at the stoichiometric point is greater than, less than, or equal to 7. In each case, identify the equilibrium that determines the pH. (a) $NH_3$ (*aq*) titrated with $HClO_4$ (*aq*);

(b) $HClO_4$ (*aq*) titrated with KOH (*aq*); and (c) $NaCH_3CO_2$ (*aq*) titrated with HCl (*aq*).

**15.22** For each of the following, decide whether the pH at the stoichiometric point is greater than, less than, or equal to 7. In each case, identify the equilibrium that determines the pH. (a) NaClO (*aq*) titrated with HCl (*aq*); (b) $HNO_3$ (*aq*) titrated with KOH (*aq*); and (c) $NH_4Cl$ (*aq*) titrated with NaOH (*aq*).

**15.23** A laboratory technician wants to determine the aspirin content of a headache pill by acid–base titration. Aspirin has a $K_a$ of $3.0 \times 10^{-4}$. The pill is dissolved in water to give a solution that is about $10^{-2}$ M and is then titrated with KOH solution. Find the pH at each of the following points, neglecting dilution effects: (a) before titration begins; (b) at the stoichiometric point; and (c) at the midpoint of the titration.

**15.24** Sleeping pills often contain barbital, which is weakly acidic ($pK_a = 8.0$). For analysis of the barbital content of a sleeping pill, a titration is carried out with strong base. It takes 12.00 mL of 0.200 M base to reach the stoichiometric point. If the initial acid concentration is 0.0120 M and the solution volume is 200 mL, what is the pH of the solution after adding the following volumes? (a) 0.0 mL; (b) 5.0 mL; (c) 12.00 mL; and (d) 13.0 mL

**15.25** Sketch the titration curve for the titration described in Problem 15.23, and choose an appropriate indicator for this titration.

**15.26** Sketch the titration curve for the titration described in Problem 15.24, and choose an appropriate indicator for this titration.

**15.27** **ilw** Calculate the pH at the second stoichiometric point when 150 mL of a 0.015 M solution of phthalic acid ($K_{a1} = 1.1 \times 10^{-3}$, $K_{a2} = 3.9 \times 10^{-6}$) is titrated with 1.00 M NaOH. What is a suitable indicator for this titration?

**15.28** Calculate the pH at the second stoichiometric point when 250 mL of a 0.025 M solution of tartaric acid ($K_{a1} = 9.2 \times 10^{-4}$, $K_{a2} = 4.3 \times 10^{-5}$) is titrated with 1.00 M NaOH. What is a suitable indicator for this titration?

**15.29** The dihydrogen phosphate ion is a diprotic acid ($pK_{a1} = 7.21$, $pK_{a2} = 12.32$). A titration sample was prepared by dissolving 1.51 g of $NaH_2PO_4$ in enough water to make 0.100 L. The solution was titrated with 1.452 M NaOH. Calculate the pH at the following points: (a) before the titration; (b) the first midpoint; (c) the first stoichiometric point; (d) the second midpoint; and (e) the second stoichiometric point.

**15.30** A sample of carbonic acid (0.125 L, 0.120 M, $pK_{a1} = 6.35$, $pK_{a2} = 10.33$) was titrated with 1.504 M NaOH. Calculate the pH at the following points: (a) before the titration; (b) the first midpoint; (c) the first stoichiometric point; (d) the second midpoint; and (e) the second stoichiometric point.

## Solubility Equilibria

**15.31** For the following salts, write a balanced equation showing the solubility equilibrium and write the solubility product expression for each: (a) silver chloride; (b) barium sulphate; (c) iron(II) hydroxide; and (d) calcium phosphate.

**15.32** For the following salts, write a balanced equation showing the solubility equilibrium and write the solubility product expression for each: (a) lead(II) chloride; (b) magnesium carbonate; (c) nickel(II) hydroxide; and (d) silver acetate.

**15.33** For each of the salts in Problem 15.31, determine the mass that dissolves in 475 mL of water at 25 °C. (See Appendix E for $K_{sp}$ values.)

**15.34** For each of the salts in Problem 15.32, determine the mass that dissolves in 375 mL of water at 25 °C. (See Appendix E for $K_{sp}$ values; for $AgC_2H_3O_2$, $K_{sp} = 1.94 \times 10^{-3}$.)

**15.35** Only 6.1 mg of calcium oxalate ($CaC_2O_4$) dissolves in 1.0 L of water at 25 °C. What is the solubility product of calcium oxalate?

**15.36** The solubility of sodium sulphate ($Na_2SO_4$) in water is 9.5 g/100 mL. What is $K_{sp}$ for sodium sulphate?

**15.37** **ilw** How many grams of $PbCl_2$ ($K_{sp} = 1.7 \times 10^{-5}$) will dissolve in 0.750 L of 0.650 M $Pb(NO_3)_2$ solution?

**15.38** How many grams of $BaF_2$ ($K_{sp} = 1.8 \times 10^{-7}$) will dissolve in 0.500 L of 0.100 M NaF solution?

**15.39** Determine the value of $K_{eq}$ for $CaSO_3$ ($K_{sp} = 1.0 \times 10^{-4}$) dissolving in acid solution, given that $K_{a2} = 6.3 \times 10^{-8}$ for $H_2SO_3$.

**15.40** Determine the value of $K_{eq}$ for $FeCO_3$ ($K_{sp} = 3.1 \times 10^{-11}$) dissolving in acid solution, given that $K_{a2} = 4.7 \times 10^{-11}$ for $H_2CO_3$.

**15.41** Determine the concentration of $Ca^{2+}$ in a solution obtained by treating solid $CaSO_3$ with 0.125 M HCl.

**15.42** Determine the concentration of $Fe^{2+}$ in a solution obtained by treating solid $FeCO_3$ with 0.255 M $HNO_3$.

## Complexation Equilibria

**15.43** Write the chemical formulas, including charge, of the complexes that form between the following metal cations and ligands: (a) $Fe^{3+}$ and $CN^-$, coordination number = 6; (b) $Zn^{2+}$ and $NH_3$, coordination number = 4; and (c) $V^{3+}$ and ethylenediamine (en), coordination number = 6.

**15.44** Write the chemical formulas, including charge, of the complexes that form between the following metal cations and ligands: (a) $Co^{2+}$ and $SCN^-$, coordination number = 4; (b) $Hg^{2+}$ and $NH_3$, coordination number = 2; and (c) $Mg^{2+}$ and $EDTA^{4-}$, coordination number = 6.

**15.45** Zinc forms a tetrahedral complex ion with four molecules of ammonia ($K_f = 4.1 \times 10^8$). A solution was prepared by dissolving 0.275 g of $ZnCl_2$ in 375 mL of 0.250 M ammonia. Calculate the equilibrium concentrations of $Zn^{2+}$, $[Zn(NH_3)_4]^{2+}$, and $NH_3$.

15.46 Copper forms a trigonal planar complex ion with three chloride ions ($K_f = 5.0 \times 10^5$). A solution was prepared by dissolving 0.325 g of $CuCl_2$ in 425 mL of 0.250 M NaCl. Calculate $[Cu^{2+}]_{eq}$, $[CuCl_3^-]_{eq}$, and $[Cl^-]_{eq}$.

15.47 **ilw** Determine the concentration of the species present at equilibrium after 0.25 mole of NaCl is dissolved in 1.50 L of a solution that contains $7.5 \times 10^{-3}$ M $Pb^{2+}$. ($K_f = 2.5 \times 10^{15}$ for $[PbCl_4]^{2-}$.)

15.48 Determine the concentration of the species present after 0.15 mole of bipyridyl (bipy, a bidentate ligand) is dissolved in 2.50 L of a solution that contains $4.5 \times 10^{-2}$ M $Fe^{2+}$. ($K_f = 1.6 \times 10^{17}$ for $[Fe(bipy)_3]^{2+}$.)

15.49 When a solution of ethylenediamine (en) is added to a solution containing $Mn^{2+}$, three en ligands bond to the metal in stepwise fashion. Write all the equilibria involving $Mn^{2+}$ and en.

15.50 When a solution containing $Co^{2+}$ is mixed with a solution containing oxalate anions ($C_2O_4^{2-}$), three bidentate oxalate ligands bond to the metal in stepwise fashion. Write all the equilibria involving $Co^{2+}$ and $C_2O_4^{2-}$.

15.51 Calculate the mass of $CaSO_3$ that will dissolve in 0.50 L of 0.25 M solution of the tridentate ligand nitrilotriacetate ($NTA^{3-}$):

$$Ca^{2+} + 2\,NTA^{3-} \rightleftharpoons [Ca(NTA)_2]^{4-} \quad K_f = 3.2 \times 10^{11}$$

15.52 Calculate the mass of $FeCO_3$ that will dissolve in 0.33 L of a solution that is 0.20 M in the oxalate anion ($C_2O_4^{2-}$):

$$Fe^{2+} + 3\,C_2O_4^{2-} \rightleftharpoons [Fe(C_2O_4)_3]^{4-} \quad K_f = 3.3 \times 10^{20}$$

## ADDITIONAL PAIRED PROBLEMS

15.53 The pH of a formic acid–formate buffer solution is 4.04. Calculate the acid:conjugate base ratio for this solution. Draw a molecular picture that shows a small region of the buffer solution. (You may omit spectator ions and water molecules.) Use the following symbols:

 = Formic acid     = Formate

15.54 The pH of an $NH_4^+/NH_3$ buffer solution is 8.77. Calculate the acid/conjugate base ratio for this solution. Draw a molecular picture that shows a small region of the buffer solution. (You may omit spectator ions and water molecules.) Use the following symbols:

= $NH_4^+$     = $NH_3$

15.55 The solubility product for $PbCl_2$ is $2 \times 10^{-5}$. If 0.50 g of solid $PbCl_2$ is added to 300 mL of water, will all the solid dissolve? (*Hint:* Calculate Q if all the solid dissolves.)

15.56 The solubility product for $MgCO_3$ is $7 \times 10^{-6}$. If 1.55 g of solid $MgCO_3$ is added to 255 mL of water, will all the solid dissolve? (*Hint:* Calculate Q if all the solid dissolves.)

15.57 Consider a buffer solution that contains 0.50 M $NaH_2PO_4$ and 0.20 M $Na_2HPO_4$. (a) Calculate its pH. (b) Calculate the change in pH if 0.120 g of solid NaOH is added to 150 mL of this solution. (c) If the acceptable buffer range of the solution is $\pm 0.10$ pH units, calculate how many moles of $H_3O^+$ can be neutralized by 250 mL of the buffer.

15.58 Consider a buffer solution that contains 0.45 M $H_3PO_4$ and 0.55 M $NaH_2PO_4$. (a) Calculate its pH. (b) Calculate the change in pH if 0.260 g of solid KOH is added to 250 mL of this buffer solution. (c) If the acceptable buffer range of the solution is $\pm 0.15$ pH units, calculate how many moles of $OH^-$ can be neutralized by 150 mL of the buffer.

15.59 Bromocresol purple is an indicator that changes colour from yellow to purple in the pH range from 5.2 to 6.8. Without doing a calculation, determine the colour of the indicator in each of the following solutions: (a) 0.15 M HCl; (b) 0.25 M NaOH; (c) 1.0 M KCl; and (d) 0.55 M $NH_3$.

15.60 A solution has a pH of 8.5. What would be the colour of the solution if the following indicators were present in the solution? (a) Methyl orange; (b) phenol red; (c) bromocresol green; and (d) thymol blue (see Table 15-2.)

15.61 When a solution of leucine (an amino acid) in water is titrated with strong base, the pH before titration is 1.85, the pH at the midpoint of the titration is 2.36, and the pH at the stoichiometric point is 6.00. Determine the value of $K_a$ for leucine.

15.62 The following figure shows the data obtained in the pH titration of a biochemical substance that is a weak acid. From the information provided, determine the $pK_a$ of this compound.

**15.63** If 150 mL of $1.50 \times 10^{-2}$ M $MgSO_4$ solution is mixed with 350 mL of $2.00 \times 10^{-2}$ M $CaCl_2$ solution, will solid $CaSO_4$ form? ($K_{sp} = 4.9 \times 10^{-5}$)

**15.64** If 200 mL of $2.50 \times 10^{-2}$ M NaCl solution is mixed with 300 mL of $4.00 \times 10^{-2}$ M $Pb(NO_3)_2$ solution, will solid $PbCl_2$ form? ($K_{sp} = 1.7 \times 10^{-5}$)

**15.65** When lead(II) fluoride dissolves in water at 25 °C, the equilibrium concentration of $Pb^{2+}$ is $2.0 \times 10^{-3}$ M. Use this information to calculate $K_{sp}$ for $PbF_2$.

**15.66** When silver sulphate dissolves in water at 25 °C, the equilibrium concentration of $Ag^+$ is $3.0 \times 10^{-2}$ mol/L. Use this information to calculate $K_{sp}$ for $Ag_2SO_4$.

**15.67** Ammonia is a convenient buffer system in the slightly basic range. (a) What is the pH of a buffer solution containing 35.0 g of $NH_4Cl$ dissolved in 1.00 L of 1.00 M $NH_3$? (b) How many moles of acid are required to change the pH of this solution by 0.05 pH units? (c) Suppose 5.0 mL of 12.0 M HCl solution is added to 250 mL of the solution of Part (a). Calculate the new pH.

**15.68** Acetic acid is a convenient buffer system in the slightly acidic range. (a) How many grams of sodium acetate must be added to 0.750 L of a 0.100 M solution of acetic acid to make a buffer of pH = 5.00? (b) How many moles of acid are required to change the pH of this solution by 0.05 pH units? (c) Suppose 5.0 mL of a 12.0 M HCl solution is added to 250 mL of the solution of Part (a). Calculate the new pH.

**15.69** The solubility of calcium phosphate ($Ca_3(PO_4)_2$) in water is $3.5 \times 10^{-5}$ g/L. Use this information to calculate $K_{sp}$ for this salt.

**15.70** The solubility of calcium arsenate ($Ca_3(AsO_4)_2$) in water is 0.036 g/L. Use this information to calculate $K_{sp}$ for this salt.

**15.71** Suppose you titrate 0.300 L of a 0.200 M solution of sodium formate with 6.0 M HCl. ($K_a$ for formic acid is $1.8 \times 10^{-4}$.) (a) What is the pH of the solution before beginning the titration? (b) What is the pH of the solution halfway through the titration? (c) What is the pH at the stoichiometric point? (d) What is a suitable indicator for this titration?

**15.72** Suppose you titrate 0.200 L of a 0.150 M solution of $NaNO_2$ with 6.0 M HCl. ($K_a$ for $HNO_2$ is $5.6 \times 10^{-4}$.) (a) What is the pH of the solution before beginning the titration? (b) What is the pH of the solution halfway through the titration? (c) What is the pH at the stoichiometric point? (d) What is a suitable indicator for this titration?

**15.73** One way to remove heavy metal ions from water is by treatment with sodium carbonate. Calculate the mass of the precipitate that will form and the concentration of $Mn^{2+}$ remaining in solution after 0.750 L of a solution that contains $Mn^{2+}$ at $2.50 \times 10^{-2}$ M is mixed with 0.150 L of 0.500 M $Na_2CO_3$ solution. ($K_{sp}$ for $MnCO_3$ is $2.2 \times 10^{-11}$.)

**15.74** Calculate the mass of the precipitate that will form and the concentration of $Ag^+$ remaining in solution after 0.250 L of a 0.200 M $AgNO_3$ solution is mixed with 0.350 L of a 0.300 M $Na_2CO_3$ solution. ($K_{sp}$ for $Ag_2CO_3$ is $8.46 \times 10^{-12}$.)

**15.75** If 0.10 g of $Zn(OH)_2$ is added to 1.00 L of water, what mass will dissolve? (See Appendix E for $K_{sp}$ values.)

**15.76** If 0.025 g of $Fe(OH)_3$ is added to 1.50 L of water, what mass will dissolve? (See Appendix E for $K_{sp}$ values.)

**15.77** In a biochemistry laboratory, you are asked to prepare a buffer solution to be used as a solvent for isolation of an enzyme. On the shelf are the following solutions, all 1.00 M: formic acid ($K_a = 1.8 \times 10^{-4}$), acetic acid ($K_a = 1.8 \times 10^{-5}$), sodium formate ($NaHCO_2$), and sodium acetate ($NaCH_3CO_2$). Describe how you would prepare 1.0 L of a pH = 4.80 buffer solution with total molarity of 0.35 M.

**15.78** You are doing undergraduate research for a biochemistry professor. Your first assignment is to prepare a pH = 7.50 phosphate buffer solution to be used in the isolation of DNA from a cell culture. The buffer must have a total concentration of 0.500 M. On the shelf you find the following chemicals: solid NaOH, concentrated HCl (12.0 M), concentrated $H_3PO_4$ (14.7 M), $KH_2PO_4$, and $K_2HPO_4$. Write a detailed set of quantitative instructions that describe how you would prepare 1.5 L of the buffer solution.

**15.79** Zinc forms an octahedral complex ion with three bidentate oxalate ions:

$$Zn^{2+} (aq) + 3\ C_2O_4{}^{2-} (aq) \rightleftarrows [Zn(C_2O_4)_3]^{4-} (aq)$$
$$K_f = 1.4 \times 10^8$$

A solution was prepared by dissolving 0.275 g of $ZnCl_2$ in 0.450 L of 0.250 M $Na_2C_2O_4$. Calculate the equilibrium concentrations of $Zn^{2+}$, $[Zn(C_2O_4)_3]^{4-}$, and $C_2O_4{}^{2-}$.

**15.80** Cadmium forms a tetrahedral complex ion with cyanide ions:

$$Cd^{2+} (aq) + 4\ CN^- (aq) \rightleftarrows [Cd(CN)_4]^{2-} (aq)$$
$$K_f = 6.0 \times 10^{18}$$

A solution was prepared by dissolving 0.125 g of $CdCl_2$ in 0.225 L of 0.210 M NaCN. Calculate the equilibrium concentrations of $Cd^{2+}$, $[Cd(CN)_4]^{2-}$, and $CN^-$.

## ADDITIONAL PROBLEMS

**15.81** At 25 °C, $[Mg^{2+}] = 1.14 \times 10^{-3}$ M in a saturated aqueous solution of $MgF_2$. (a) Write the equilibrium reaction and the equilibrium constant expression. (b) Use the concentration of the saturated solution to compute $K_{sp}$ for $MgF_2$ at 25 °C. (c) Using standard thermodynamic data from Appendix D, calculate $K_{sp}$ at 100.0 °C.

**15.82** Using the appropriate $K_{sp}$ values from Appendix E, find the concentrations of all ions in the solution at equilibrium after 0.500 L of a 0.300 M aqueous $Cu(NO_3)_2$ solution is mixed with 0.500 L of a 0.400 M aqueous KOH solution.

**15.83** Calculate the concentration of $Hg^{2+}$ cations in a saturated solution of HgS, $K_{sp} = 4.0 \times 10^{-53}$. What volume of saturated solution can be expected to contain a single $Hg^{2+}$ cation?

**15.84** A biochemist wants to use X-ray diffraction to determine the structure of a protein. The biochemist must isolate crystals of the protein from buffered solutions. Write instructions for preparing 1.0 L of a buffer that holds the pH of the protein solution at 5.20. Make the solution 0.50 M in an appropriate acid, and decide what else must be added to complete the preparation.

**15.85** Phosphate ions are a major pollutant of water supplies. They can be removed by precipitation using solutions of $Ca^{2+}$ ions because the $K_{sp}$ of calcium phosphate is $2.0 \times 10^{-33}$. Suppose that $3.00 \times 10^3$ L of wastewater containing $PO_4^{3-}$ at $2.2 \times 10^{-3}$ M is treated by adding 120 moles of solid $CaCl_2$ (which dissolves completely). (a) What is the concentration of phosphate ions after treatment? (b) What mass of calcium phosphate precipitates?

**15.86** The pH of an acetic acid–acetate buffer solution is 4.27. (a) Calculate the H$A$: $A^-$ concentration ratio. (b) Draw a molecular picture that shows a small region of the buffer solution. (You may omit spectator ions and water molecules.) Use the following symbols:

**15.87** Strontium iodate ($Sr(IO_3)_2$) is considerably more soluble in hot than in cold water. At 25 °C, 0.030 g of this compound dissolves in 0.100 L of water, but 0.80 g dissolves at 100.0 °C. Calculate $K_{sp}$ and $\Delta G°$ at each temperature.

**15.88** The titration of 25.0 mL of a mixture of hydrochloric acid and formic acid with 0.578 M NaOH gave the titration curve below. What are the molarities of HCl and $HCO_2H$ in the solution?

**15.89** Limestone caverns are formed by the reaction of $H_2O$ and $CO_2$ with natural deposits of calcium carbonate:

$$CaCO_3\ (s) + CO_2\ (g) + H_2O\ (l) \rightleftharpoons Ca^{2+}\ (aq) + 2\ HCO_3^-\ (aq) \qquad K_{eq} = 1.56 \times 10^{-8}$$

The partial pressure of $CO_2$ in the atmosphere is $3.2 \times 10^{-4}$ bar. What is the equilibrium concentration of calcium ions in groundwater?

**15.90** The solubility of Ni(II) hydroxide can be increased by adding either $NH_3\ (aq)$ or HCl $(aq)$. Explain, including balanced equations. Will $NH_3\ (aq)$ or HCl $(aq)$ increase, decrease, or have no effect on the solubility of Ni(II) chloride? Explain your answer, including balanced equations.

**15.91** Divide the silver salts listed in Appendix E into two sets: those that are more soluble in acidic solution than in pure water, and those of which solubility is independent of pH.

**15.92** One of the most common buffers used in protein chemistry is a weak base called TRIS (the nitrogen atom of the amino group is the basic portion of the molecule):

$$HOCH_2 \quad —NH_2 = (HOCH_2)_3CNH_2 = TRIS$$

(a) Write a balanced equation that shows how TRIS acts as a weak base in water. (b) Buffer solutions are prepared from TRIS by adding enough 12 M HCl to produce concentrations appropriate for the desired pH. Write a balanced equation that shows what happens when 12 M HCl is added to an aqueous solution of TRIS. (c) Write a balanced equation that shows what happens when hydroxide ions are added to a TRIS buffer solution. (d) Write a balanced equation that shows what happens when hydronium ions are added to a TRIS buffer solution.

**15.93** A biochemist prepares a buffer solution by adding enough TRIS (see Problem 15.92) and 12 M HCl to give 1.0 L of a buffer solution that has concentrations [TRIS] = 0.30 M and [TRISH$^+$] = 0.60 M. (p$K_b$ of TRIS is 5.91.) (a) Calculate the pH of the buffer solution. (b) Suppose that 5.0 mL of 12 M HCl is added to 1.0 L of the buffer solution; calculate the new pH of the solution.

**15.94** A technician accidentally pours 35 mL of 12 M HCl into the 1.0 L of buffer solution freshly prepared as described in Problem 15.93. (a) Do a calculation to determine whether the buffer has been ruined. (b) Is it possible to bring the buffer solution back to the original pH calculated in Problem 15.93? If so, what reagent, and how much, must be added to restore the buffer?

**15.95** The graph shows the titration curve for a solution of formic acid, which is a principal component in the venom of stinging ants. Identify the major species in the solution at equilibrium, identify the dominant equilibrium, and write the equilibrium constant expression for points A, B, C, and D on the curve.

Formic acid
$MM = 46.1$ g/mol
$pK_a = 3.75$

pH

Moles OH⁻

**15.96** Quinine, an alkaloid derived from a tree that grows in tropical rain forests, is used in the treatment of malaria. Like all alkaloids, quinine is a sparingly soluble weak base: 1.00 g of quinine will dissolve in $1.90 \times 10^2$ L of water. (a) What is the pH of a saturated solution of quinine? (b) A 100.0-mL sample of saturated quinine is titrated with 0.0100 M HCl solution. What is the pH at the stoichiometric point of the titration?

HO

Quinine
$(C_{20}H_{24}N_2O_2)$
$pK_b = 5.1$
$M = 324$ g/mol

**15.97** The following figure represents a small portion of an acetic acid–acetate buffer. Solvent molecules have been omitted for clarity ($pK_a$ of acetic acid $= 4.75$). (a) What is the pH of the buffer solution? (b) Redraw the original figure to show the equilibrium condition that is established when one hydroxide ion enters the region. (c) Redraw the original figure to show the equilibrium condition that is established when one ion of $HSO_4^-$ enters the region. ($pK_a$ of $HSO_4^- = 2.0$)

Acetic Acid          Acetate
$CH_3CO_2H$          $CH_3CO_2^-$

**15.98** Determine the solubility (g/L) of silver carbonate ($Ag_2CO_3$, $K_{sp} = 8.46 \times 10^{-12}$) in a pH $= 10.75$ buffer solution containing $CO_3^{2-}$ and $HCO_3^-$ at a total concentration of 0.860 M.

**15.99** Nickel(II) cations form a strong complex with cyanide:

$$Ni^{2+}(aq) + 4\,CN^-(aq) \rightleftharpoons [Ni(CN)_4]^{2-}(aq) \quad K_f = 1 \times 10^{22}$$

The complexation process allows aqueous cyanide solutions to dissolve insoluble nickel hydroxide:

$$Ni(OH)_2(s) + 4\,CN^-(aq)\,[Ni(CN)_4]^{2-}(aq) \rightleftharpoons 2\,OH^-(aq)$$

Use appropriate information from Appendix E to determine the equilibrium constant for this reaction, and calculate the mass of $Ni(OH)_2$ that will dissolve in 225 mL of 0.500 M aqueous cyanide solution.

### GROUP STUDY PROBLEMS

**15.100** The amine group of an amino acid readily accepts a proton, and the protonated form of an amino acid can be viewed as a diprotic acid. The $pK_a$ values for serine ($H_2NCHRCO_2H$, $R = CH_2OH$) are $pK_a$ ($H_3N^+$) $= 9.2$ and $pK_a$ ($CO_2H$) $= 2.2$. (a) What is the chemical formula of the species that forms when serine dissolves in pure water? (b) If this species is titrated with strong acid, what reaction occurs? (c) A chemist adds 10.00 mL of 1.00 M HCl to 200. mL of a 0.0500 M serine solution. This mixture is then titrated with 0.500 M NaOH. Draw the titration curve, indicating the pH at various stages of this titration.

**15.101** Virtually all investigations in cell biology and biochemistry must be carried out in buffered aqueous solutions. Imagine that you are studying an enzyme that is active only between pH $= 7.1$ and 7.4 and that you need to prepare 1.5 L of a phosphate buffer at pH $= 7.25$ that has a total phosphate concentration of 0.085 M. On the laboratory shelves, you find the following reagents: concentrated $H_3PO_4$ (14.75 M), solid $KH_2PO_4$, and solid $K_2HPO_4$. (a) Which of these will you use to prepare the buffer solution? (b) What quantities of each will you use? (c) The enzyme generates $H_3O^+$ as it functions. If you are running an experiment in 250 mL of the buffer, how many moles of $H_3O^+$ can the enzyme generate before it loses its activity?

**15.102** Seawater is approximately 0.5 M each in $Na^+$ and $Cl^-$ ions. By evaporation, NaCl ($K_{sp} = 6.2$) can be precipitated from this solution. If $1.00 \times 10^2$ L of seawater is evaporated, at what volume will the first solid NaCl appear?

**15.103** The $K_{sp}$ of $Fe(OH)_3$ is $2.8 \times 10^{-39}$, and the $K_{sp}$ of $Zn(OH)_2$ is $3 \times 10^{-17}$. (a) Can $Fe^{3+}$ be separated from $Zn^{2+}$ by the addition of an NaOH solution to an acidic solution that contains 0.300 M $Zn^2$ and 0.100 M $Fe^{3+}$? (b) At what hydroxide ion concentration will the second cation precipitate? (c) What is the concentration of the other cation at that hydroxide ion concentration?

**15.104** The figure below shows the titration curves for 50.0-mL samples of weak bases A, B, and C. The titrant was 0.10 M HCl. (a) Which is the strongest base?

(b) Which base has the largest $pK_b$? (c) What are the initial concentrations of the three bases? (d) What are the approximate $pK_a$ values of the conjugate acids of the three bases? (e) Which of the bases can be titrated quantitatively using indicators? Explain your answers, and identify an appropriate indicator for each base that can be titrated successfully.

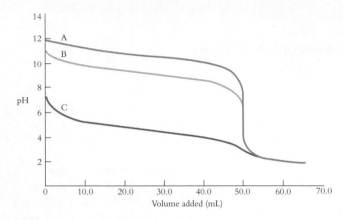

## ANSWERS TO SECTION EXERCISES

**15.1.1** (a) not suitable; (b) $HNO_2$ and $NO_2^-$; (c) not suitable; (d) $HPO_4^{2-}$ and $PO_4^{3-}$; and (e) not suitable

**15.1.2** $NH_4^+ (aq) + OH^- (aq) \rightarrow NH_3 (aq) + H_2O (l)$ and $NH_3 (aq) + H_3O^+ (aq) \rightarrow NH_4^+ (aq) + H_2O (l)$

**15.1.3** (a) 9.00; (b) 9.24; and (c) 12.45

**15.2.1** Buffering capacity destroyed, final pH = 1.6

**15.2.2** 3.8 g sodium formate and 72 mL of 0.500 M HCl solution

**15.2.3** Dissolve 9.0 g NaOH and 35 g NaHCO₃ in enough water to make 2.5 L of solution.

**15.3.1**

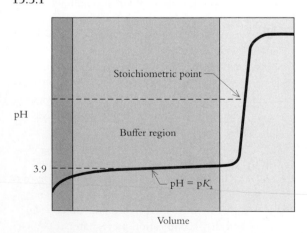

**15.3.2** pH = 7.91 at stoichiometric point and 3.0 after 3.0 mL of KOH is added

**15.3.3** Phenol red would be suitable; it would be red at the stoichiometric point.

**15.4.1** $1.7 \times 10^{-4}$ M

**15.4.2** $1.5 \times 10^{-9}$ M; NaCl does not precipitate

**15.4.3** $Al(OH)_3 (s) + 3 H_3O^+ (aq) \rightleftarrows Al^{3+} (aq) + 6 H_2O (l)$; $K_{eq} = 4.6 \times 10^9$

**15.5.1** $[Ni^{2+}] = 5.6 \times 10^{-16}$ M; $= 3.0 \times 10^{-2}$ M; $[Ni(en)_3^{2+}] = 6.5 \times 10^{-3}$ M

**15.5.2** $Zn^{2+} (aq) + NH_3 (aq) \rightleftarrows [Zn(NH_3)]^{2+} (aq)$
$[Zn(NH_3)]^{2+} (aq) + NH_3 (aq) \rightleftarrows [Zn(NH_3)_2]^{2+} (aq)$
$[Zn(NH_3)_2]^{2+} (aq) + NH_3 (aq) \rightleftarrows [Zn(NH_3)_3]^{2+} (aq)$
$[Zn(NH_3)_3]^{2+} (aq) + NH_3 (aq) \rightleftarrows [Zn(NH_3)_4]^{2+} (aq)$

**15.5.3** 1.8 g of AgCl dissolves.

## ANSWERS TO PRACTICE EXERCISES:

**15.1** $[OH^-]_{eq} = 3.36 \times 10^{-5}$ M; $[NH_4^+]_{eq} = 0.125$ M; $[NH_3]_{eq} = 0.240$ M

**15.2** $[NH_3]_{eq} = 0.25$ M; $[NH_4^+]_{eq} = 0.50$ M; $[Cl^-]_{eq} = 0.75$ M

**15.3**

**15.4** pH = 2.10

**15.5** Increases by 0.18, to pH = 10.41

**15.6** $[OH^-] = 0.049$ M, pH = 12.69

**15.7** 34 g Na₂HPO₄ and 18 g NaH₂PO₄

**15.8** The buffer should contain $CH_3CO_2H$ and $CH_3CO_2^-$, made from sodium acetate solid and hydrochloric acid. Use 35.8 g NaCH₃CO₂ and 0.070 L HCl (aq).

**15.9** The concentration of acetic acid is 0.845 M.

**15.10** $7.923 \times 10^{-2}$ M, which is consistent with the previous result to within 1 part in 800

**15.11** pH = 9.96

**15.12** (a) 11.39; (b) 9.81; (c) 5.44; (d) 1.85

**15.13**

**15.14**

Calculated titration curve:
0.200 L of 0.120 M

**15.15** Thymol blue

**15.16** $K_{sp} = 5.1 \times 10^{-6}$

**15.17** $K_{sp} = 5.6 \times 10^{-12}$

**15.18** 5.4 g

**15.19** $m_{Ag_2CO_3} = 12.1$ g; $[CO_3^{2-}]_{eq} = 0.40$ M; $[Ag^+]_{eq} = 2.31 \times 10^{-6}$ M

**15.20** (a) $[H_3O^+]_{eq} = 2.67 \times 10^{-5}$ M; (b) $[H_3O^+]_{eq} = 9.45 \times 10^{-10}$ M; (c) $NH_3$, being a product of the reaction that generates $H_3O^+$, drives the equilibrium to the left, reducing $[H_3O^+]_{eq}$.

**15.21** (a) $H_2O$: $5.8 \times 10^{-3}$ g; (b) pH = 3.00: 0.10 g/L; (c) pH = 1.00: 10. g/L

**15.22** $[Ag^+]_{eq} = 2.0 \times 10^{-9}$ M; $[NH_3]_{eq} = 0.246$ M; $[Ag(NH_3)^{2+}]_{eq} = 0.0019$ M

**15.23** 0.025 mol

# ELECTRON TRANSFER REACTIONS

## CORROSION

Reactions that involve the transfer of electrons are immensely beneficial in a vast range of chemical processes—from the operation of a battery to the transmission of nerve impulses. However, in the form of corrosion, electron-transfer reactions can also be highly destructive. The rusting of iron objects is a prominent example of corrosion.

Metallic iron is made up of neutral iron atoms held together by shared electrons (see Section 7.7). The formation of rust involves electron-transfer reactions. Iron atoms lose three electrons each, forming $Fe^{3+}$ cations. At the same time, molecular oxygen gains electrons from the metal, each molecule adding four electrons to form a pair of oxide anions. As our inset figure shows, the $Fe^{3+}$ cations combine with $O^{2-}$ anions to form insoluble $Fe_2O_3$, rust. Over time, the surface of an iron object becomes covered with flaky iron(III) oxide and pitted from loss of iron atoms.

Corrosion of iron has serious economic impacts. Approximately 20% of the iron and steel produced in North America, for instance, is used to replace rusted metal, and the cost of replacing corroded materials is several billion dollars per year. Consequently, the prevention of corrosion is a major business.

Corrosion can be prevented by covering the metal surface with a protective coating that blocks oxygen and water from reaching the metal surface. Structures such as Vancouver's Lion's Gate Bridge are continually being painted, ensuring unbroken layers that protect steel cables and towers from corrosion.

Electron-transfer reactions occur all around us. Objects made of iron become coated with rust when they are exposed to moist air. Animals obtain energy from the reaction of carbohydrates with oxygen to form carbon dioxide and water. Turning on a flashlight generates an electric current from a chemical reaction in the batteries. In an aluminum refinery, huge quantities of electricity drive the conversion of aluminum oxide into aluminum metal. These different chemical processes share one common feature:

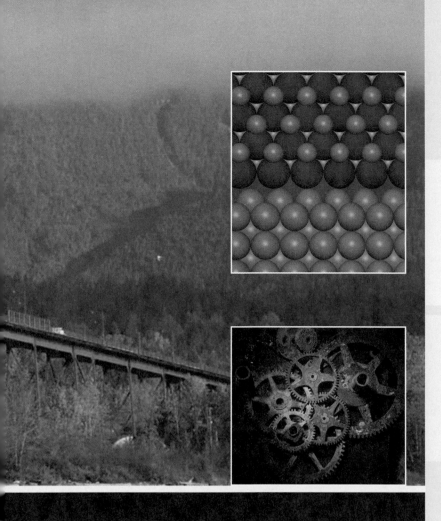

each is an oxidation–reduction reaction, commonly called a redox reaction, in which electrons are transferred from one chemical species to another.

Redox reactions can proceed by direct transfer of electrons between chemical species. Examples include the rusting of iron and the metabolic breakdown of carbohydrates. Redox processes can also take place by indirect electron transfer from one chemical species to another via an electrical circuit. When a chemical reaction is coupled with electron flow through a circuit, the process is called electrochemical. Flashlight batteries and aluminum smelters involve electrochemical processes.

We begin this chapter with a discussion of the principles of redox reactions, including redox stoichiometry. Then, we introduce the principles of electrochemistry. Practical examples of redox chemistry, including corrosion, batteries, and metallurgy, appear throughout the chapter.

# 16

## CHAPTER CONTENTS

## LEARNING OBJECTIVES

*Upon completion of this chapter you should be able to*

- **use oxidation numbers to show what is being oxidized and what is being reduced in a redox reaction**
- **balance redox reactions using the half-reaction method**
- **describe galvanic cells**
- **calculate standard cell potentials**
- **relate cell potential to the reaction conditions**
- **explain the chemistry of everyday redox reactions**
- **explain electrolytic reactions and cells**

## 16.1 RECOGNIZING REDOX REACTIONS

In a redox reaction, one species loses electrons while another species gains electrons. A chemical species that loses electrons is said to become oxidized, and a chemical species that gains electrons is said to become reduced.

Remember also that electrons are conserved in all chemical processes. That is, electrons can be *transferred* from one species to another, but they are neither created nor destroyed. Thus, it is impossible for oxidation to occur without reduction, because when one species gains electrons, another species must lose electrons. A simple example of a redox reaction is the violent reaction of magnesium metal with molecular oxygen to generate magnesium oxide, shown in Figure 16-1:

$$2 \, \text{Mg} \, (s) \, + \, \text{O}_2 \, (g) \longrightarrow 2 \, \text{MgO} \, (s)$$

In magnesium metal, each atom is neutral; the electrons exactly counterbalance the nuclear charge. When a magnesium atom reacts with an oxygen molecule, the metal atom loses its valence electrons to produce a +2 cation:

$$\text{Mg} \longrightarrow \text{Mg}^{2+} + 2e^-$$

Magnesium atoms are oxidized in this reaction, so some other species must be reduced. Oxygen molecules accept the electrons lost by the magnesium atoms. Each oxygen *atom* gains two electrons from a magnesium atom, generating two oxide anions:

$$\text{O}_2 + 4 \, e^- \longrightarrow 2 \, \text{O}^{2-}$$

Magnesium cations and oxide anions attract each other strongly, forming the ionic solid MgO. Notice that in the balanced redox reaction, there is no net change in the number of electrons; two Mg atoms lose four electrons, and one $\text{O}_2$ molecule gains four electrons.

**Figure 16-1**

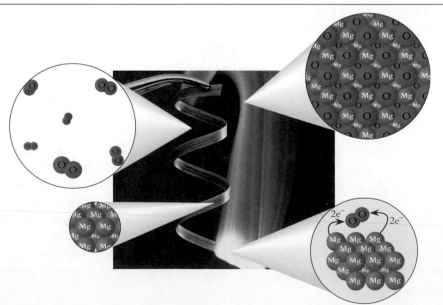

When set ablaze with a match, magnesium burns with an intense flame. The reaction involves electron transfer from the metal to oxygen.

The species that loses electrons in a redox process causes the reduction of some other species. Consequently, the species that loses electrons is a **reducing agent**. Magnesium metal acts as a reducing agent in the presence of atmospheric oxygen. Similarly, the species that gains electrons causes the oxidation of some other chemical species and is an **oxidizing agent** (also commonly referred to as an **oxidant**). Molecular oxygen acts as an oxidizing agent in the presence of magnesium metal. Every redox reaction has both an oxidizing agent and a reducing agent. Figure 16-2 summarizes these features using the magnesium–oxygen reaction as an example.

Electron transfer is often difficult to recognize, because electrons normally do not appear in chemical formulas and equations. The formula of a molecule or ion specifies the atomic composition and the net charge but does not account for the electrons. Likewise, electrons never appear in a balanced chemical equation. The electrons are "hidden" in the chemical formulas of the species involved in the reaction. For example, the formulas $Mg$, $O_2$, and $MgO$ do not suggest immediately that oxygen is reduced to oxide anions, and magnesium oxidized to magnesium cations, when magnesium burns. To recognize when electron transfer occurs, we need a method to keep track of electrons in chemical species.

**Figure 16-2**

Schematic view of the processes involved in redox reactions.

## Oxidation Numbers

The difficulty in recognizing redox reactions is illustrated by two of the reactions that occur during the extraction of iron from iron ores, a process that we describe in detail in Chapter 18:

$$FeO\ (s) + CO\ (g) \xrightarrow{\text{Heat}} Fe\ (l) + CO_2\ (g)$$

$$FeCO_3\ (s) \xrightarrow{\text{Heat}} FeO\ (s) + CO_2\ (g)$$

One of these is a redox reaction, but the other is not; how can we determine which one involves electron transfer?

To determine whether electrons are transferred in chemical reactions, chemists use a procedure that assigns an **oxidation number** (also known as an **oxidation state**) to each atom in each chemical species. In a redox reaction, electron transfer causes some of the atoms to change their oxidation numbers. Thus, we can identify redox reactions by noting changes in oxidation numbers.

To determine oxidation numbers, we assign each valence electron to a specific atom in a compound. This means that the oxidation number of an atom is the charge it would have if the compound were composed of ions. Ionic $FeO$, for example, would contain $Fe^{2+}$ cations and $O^{2-}$ anions. Thus, in $FeO$ we assign iron an oxidation number of $+2$ and oxygen an oxidation number of $-2$.

Most substances contain covalent bonds rather than ions. Nevertheless, the electrons in a bond between atoms of two different elements, such as in $FeO$ or $CO$, are polarized toward the more electronegative atom (see Figure 16-3 for electronegativities). For oxidation number purposes, we imagine that these electrons are transferred *completely* to the more electronegative atom.

**Figure 16-3**

| H 2.1 | | | | | | | | | | | | | | | | | |
|---|---|---|---|---|---|---|---|---|---|---|---|---|---|---|---|---|---|
| Li 1.0 | Be 1.5 | $(\chi) < 2.0$ | | $2.0 < \chi < 2.4$ | | $\chi > 2.4$ | | | | | | B 2.0 | C 2.5 | N 3.0 | O 3.5 | F 4.0 | |
| Na 0.9 | Mg 1.2 | | | | | | | | | | | Al 1.5 | Si 1.8 | P 2.1 | S 2.5 | Cl 3.0 | |
| K 0.8 | Ca 1.0 | Sc 1.3 | Ti 1.5 | V 1.6 | Cr 1.6 | Mn 1.5 | Fe 1.8 | Co 1.9 | Ni 1.9 | Cu 1.9 | Zn 1.6 | Ga 1.6 | Ge 1.8 | As 2.0 | Se 2.4 | Br 2.8 |
| Rb 0.8 | Sr 1.0 | Y 1.2 | Zr 1.4 | Nb 1.6 | Mo 1.8 | Tc 1.9 | Ru 2.2 | Rh 2.2 | Pd 2.2 | Ag 1.9 | Cd 1.7 | In 1.7 | Sn 1.8 | Sb 1.9 | Te 2.1 | I 2.5 |
| Cs 0.7 | Ba 0.9 | Lu 1.2 | Hf 1.3 | Ta 1.5 | W 1.7 | Re 1.9 | Os 2.2 | Ir 2.2 | Pt 2.2 | Au 2.4 | Hg 1.9 | Tl 1.8 | Pb 1.9 | Bi 1.9 | Po 2.0 | At 2.2 |

Electronegativities of the elements.

The bonding electrons in CO and $CO_2$ are polarized in the direction of the O atoms, as shown by the electronegativities of C (2.5) and O (3.5). Thus, if these substances were ionic, CO would consist of $C^{2+}$ cations and $O^{2-}$ anions, and $CO_2$ would contain $C^{4+}$ cations and $O^{2-}$ anions. Accordingly, we assign O an oxidation number of −2 in both these compounds, whereas C has an oxidation number of +2 in CO and +4 in $CO_2$.

It is always possible to determine oxidation numbers starting from electronegativity differences. A more systematic method for determining oxidation numbers uses the following four guidelines:

1. Treat polyatomic ions separately.

2. The sum of all the oxidation numbers must equal the charge of the species.

3. Hydrogen usually has an oxidation number of +1.

4. The most electronegative atom in a species has a negative oxidation number equal to the number of electrons needed to complete its valence octet.

Some examples illustrate these guidelines.

1. In an ionic compound, each ion is an individual chemical species with its own set of oxidation numbers. For example, we treat ammonium nitrate ($NH_4NO_3$) as $NH_4^+$ cations and $NO_3^-$ anions.

2. The sum of the oxidation numbers for $NO_3^-$ must be −1, and the sum for $NH_4^+$ must be +1. For a neutral molecule such as $NH_3$, the sum of the oxidation numbers must be zero. By this guideline, the oxidation number of every atom in pure neutral elements is zero. As examples, each atom in $H_2$, Fe, $O_2$, $O_3$, $P_4$, and $C_{60}$ has an oxidation number of zero. In addition, the oxidation number of any monatomic ion equals its charge: the oxidation number of $K^+$ is +1, $Al^{3+}$ is +3, and $Cl^-$ is −1.

3. Hydrogen has a +1 oxidation number in most of its compounds. In $NH_3$, for example, each hydrogen atom is assigned an oxidation number of +1, leaving a −3 value for nitrogen. There is one common exception to this guideline: in the metal hydrides such as NaH and $CaH_2$, hydrogen is assigned an oxidation number of −1. These assignments are consistent with electronegativities:

$$\text{Na (0.9) and Ca (1.0)} < \text{H (2.1)} < \text{N (3.0) and Cl (3.0)}$$

4. Nitrogen, which is in Group 15 of the periodic table, has five valence electrons and needs three more to complete its valence octet. In the cyanide anion, $CN^-$, nitrogen is the more electronegative atom, so N is assigned an oxidation number of $-3$. In contrast, oxygen is the more electronegative atom in the nitrate ion, $NO_3^-$, so each oxygen atom is assigned an oxidation number of $-2$ (Group 16, six valence electrons). With this assignment, the nitrogen atom in a nitrate anion must have a $+5$ oxidation number to satisfy Guideline 2. The only important exceptions to Guideline 4 are the peroxides, species that contain an O—O bond: oxygen has a $-1$ oxidation number in the peroxides. Examples include hydrogen peroxide, $H_2O_2$ (H is $+1$ by Guideline 3, so O is $-1$ by Guideline 2), and sodium peroxide, $Na_2O_2$ (treated as $Na^+$ and $O_2^{2-}$ by Guideline 1, giving O the oxidation number of $-1$ by Guideline 2.

Hydrogen peroxide

Oxidation numbers do not indicate how charge is *actually* distributed in a molecule. Remember that most bonds are polar, meaning that the electrons in a bond are skewed toward the more electronegative atom. In contrast, we determine oxidation numbers by *assuming* that all bonds are ionic. Consequently, oxidation numbers are useful for assessing whether electron exchange occurs during a reaction, but they do not represent actual charge distributions. A good example is $CN^-$, where we assign oxidation numbers of $-3$ to N and $+2$ to C. In reality, the ion carries a $-1$ net charge, and both atoms have slightly negative charges.

Examples 16-1 and 16-2 illustrate the application of these guidelines.

## EXAMPLE 16-1 ASSIGNING OXIDATION NUMBERS

Assign oxidation numbers to all atoms in these two reactions involving iron:

$$FeO\ (s) + CO\ (g) \xrightarrow{\text{Heat}} Fe\ (l) + CO_2\ (g)$$

$$FeCO_3\ (s) \xrightarrow{\text{Heat}} FeO\ (s) + CO_2\ (g)$$

**Strategy:** Consider each substance separately, and follow the four guidelines.

**Solution:** FeO: Oxygen has an oxidation number of $-2$ (Guideline 4). For the sum of the oxidation numbers to be zero (Guideline 2), iron must be $+2$.

CO: Oxygen has a value of $-2$ (Guideline 4). For the sum of the oxidation numbers to be zero (Guideline 2), carbon must have an oxidation number of $+2$.

Fe: This is a neutral element, so iron has an oxidation number of zero (Guideline 2).

$CO_2$: Each oxygen is $-2$, so carbon must be $+4$.

$FeCO_3$: This ionic compound contains the carbonate polyatomic anion, $CO_3^{2-}$. The $-2$ charge on carbonate requires that iron have $+2$ oxidation number (Guideline 2). In the carbonate anion, each oxygen atom is $-2$, for a total of $-6$. The oxidation numbers must add up to the net charge on the anion ($-2$), so the carbon atom must have an oxidation number of $+4$.

Here are the reactions again, showing all oxidation numbers:

$$\overset{+2}{\text{FeO}}(s) + \overset{+2}{\text{CO}}(g) \xrightarrow{\text{Heat}} \overset{0}{\text{Fe}}(l) + \overset{+4}{\text{CO}_2}(g)$$
$$\underset{-2}{} \quad \underset{-2}{} \qquad \qquad \underset{-2}{}$$

$$\overset{+2+4}{\text{FeCO}_3}(s) \xrightarrow{\text{Heat}} \overset{+2}{\text{FeO}}(s) + \overset{+4}{\text{CO}_2}(g)$$
$$\underset{-2}{} \qquad \qquad \underset{-2}{} \quad \underset{-2}{}$$

The first reaction is a redox reaction because the oxidation number of carbon increases (oxidation), and that of iron decreases (reduction).

Reduction: +2 → 0
(Fe gains 2 electrons)

FeO (s) + CO (g) —Heat→ Fe (l) + CO₂ (g)

Oxidation: +2 → +4
(C loses 2 electrons)

Electrons gained by iron matched by electrons lost by carbon

In the second reaction, none of the atoms changes its oxidation number, so this is not a redox process.

**Do the Results Make Sense?** Although these reactions look similar, the redox reaction includes a pure element. Nearly all reactions involving pure elements are redox, because the oxidation number changes from zero to either a positive or a negative value.

**Common Pitfalls:** Not following the guidelines for assigning oxidation numbers. Eventually you will be able to assign them by inspection. Until then, follow the guidelines.

**Practice Exercise 16.1**
Here are two reactions involved in the metallurgy of nickel. Assign oxidation numbers and determine if either is a redox reaction.

$$\text{NiO}(s) + \text{C}(s) \xrightarrow{\text{Heat}} \text{Ni}(l) + \text{CO}(g)$$

$$\text{Ni}(s) + 4\,\text{CO}(s) \xrightarrow{\text{Heat}} \text{Ni(CO)}_4(g)$$

## EXAMPLE 16-2  EXAMPLES OF OXIDATION NUMBERS

Assign oxidation numbers to each element in the following substances: $H_3PO_4$, $O_3$, NaOCl, $ClF_3$, and KH.

**Strategy:** Apply the guidelines for oxidation numbers to each substance.

**Solution:** $H_3PO_4$: Phosphoric acid is a covalent compound with a net charge of zero. Each hydrogen atom has an oxidation number of +1 (Guideline 3), and each oxygen has an oxidation number of −2 (Guideline 4). Now add the contributions from these atoms: $3(+1) + 4(-2) = -5$. For the oxidation numbers to sum to zero (Guideline 2), the phosphorus atom of phosphoric acid must have an oxidation number of +5.

$O_3$: Ozone is a neutral form of the element, so the oxidation number of its oxygen atoms is zero (Guideline 2).

NaOCl: Sodium hypochlorite is ionic, containing $Na^+$ cations and $OCl^-$ anions. The sodium cation has an oxidation number equal to its charge, +1. In the anion, oxygen is −2 (Guideline 4), so chlorine must be +1 for the sum of the oxidation numbers to match the −1 charge of the hypochlorite anion (Guideline 2).

$ClF_3$: Fluorine is the more electronegative atom, so each fluorine atom has an oxidation number of $-1$ (Guideline 4). For the sum of the oxidation numbers to be zero (Guideline 2), chlorine must be $+3$.

KH: In most cases hydrogen has an oxidation number of $+1$. In compounds with metals, however, the metal has a positive oxidation number and therefore the oxidation number of hydrogen is $-1$. This requires potassium to be $+1$.

**Do the Results Make Sense?** Use chemical behaviour and electronegativities to assess the reasonableness of the assignments. Phosphoric acid can be viewed as the phosphate anion ($-3$ charge) associated with three $H^+$ ions. Sodium, an alkali metal, readily forms cations with $+1$ charge. In KH, the assignments are consistent with the electronegativities ($\chi$) of the two elements: $\chi_K = 0.8$, $\chi_H = 2.1$.

**Practice Exercise 16.2**
Assign oxidation numbers for all atoms in $H_2SO_4$ and $KClO_3$.

## SECTION EXERCISES

**16.1.1** Determine the oxidation numbers of all atoms in each of the following substances: (a) $KMnO_4$ (potassium permanganate, a powerful oxidizing agent found in many general chemistry laboratories); (b) $SiCl_4$ (an intermediate in the production of ultrapure silicon for computer chips); (c) $Na_2SO_3$ (sodium sulphite, a food preservative); (d) $S_8$, the most stable form of sulphur; and (e) $LiAlH_4$ (a reducing agent in organic chemistry. *Hint:* This is an ionic compound with a lithium cation and a polyatomic anion.)

**16.1.2** Determine which of the following are redox reactions. For each redox reaction, state which species is oxidized and which is reduced.
(a) $CH_4 (g) + 2 O_2 (g) \longrightarrow CO_2 (g) + 2 H_2O (l)$
(b) $HCl (g) + NH_3 (aq) \longrightarrow NH_4Cl (aq)$
(c) $SiCl_4 (l) + 2 Mg (s) \longrightarrow Si (s) + 2 MgCl_2 (s)$
(d) $PbS (s) + 4 H_2O_2 (aq) \longrightarrow PbSO_4 (s) + 4 H_2O (l)$
(e) $3 O_2 (g) \longrightarrow 2 O_3 (g)$

**16.1.3** Nitrogen can have many different oxidation states, as illustrated by the following examples. Assign oxidation states to the nitrogen atoms in each species. (a) $NH_3$; (b) $N_2$; (c) $N_2H_4$; (d) $HNO_3$; (e) $NaNO_2$; (f) $NO$; and (g) $NO_2$.

## 16.2 BALANCING REDOX REACTIONS

Some redox reactions have relatively simple stoichiometry and can be balanced by inspection. Others are much more complex. Because redox reactions involve the transfer of electrons from one species to another, electrical charges must be considered explicitly when balancing complex redox equations.

The key to balancing complicated redox equations is to balance electrons as well as atoms. Because electrons do not appear in chemical formulas or balanced net reactions, however, the number of electrons transferred in a redox reaction often is not obvious. To balance these redox reactions, therefore, we need a procedure that shows the electrons involved in the oxidation and the reduction. One such procedure separates redox reactions into two parts, an oxidation and a reduction. Each part is a **half-reaction** that describes half of the overall redox process.

## Half-Reactions

In any redox reaction, some species are oxidized and others are reduced. To reveal the number of electrons transferred, we must separate the oxidation and reduction parts. This is most easily accomplished by dividing the redox reaction into two half-reactions, one describing the oxidation and the other describing the reduction. In half-reactions, electrons appear as reactants or products.

A starting material loses electrons in an oxidation, so electrons appear among the *products* of the oxidation half-reaction. A starting material gains electrons in a reduction, so electrons appear among the *reactants* of the reduction half-reaction. The reaction of magnesium metal with hydronium ions to produce hydrogen gas provides an example:

$$Mg\ (s) + 2\ H_3O^+\ (aq) \longrightarrow Mg^{2+}\ (aq) + H_2\ (g) + 2\ H_2O\ (l)$$

Here are the half-reactions for this redox reaction:

Oxidation:  $Mg\ (s) \longrightarrow Mg^{2+}\ (aq) + 2\ e^-$

Reduction:  $2\ H_3O^+\ (aq) + 2\ e^- \longrightarrow H_2\ (g) + 2\ H_2O\ (l)$

The half-reactions show that two electrons are required to produce each $H_2$ molecule in the reduction, and that the oxidation of one magnesium atom releases two electrons. In other words, electron transfer is balanced when one magnesium atom reacts for every $H_2$ molecule that forms. This example reveals an essential feature of redox reactions: the number of electrons lost in oxidation must equal the number of electrons gained in reduction.

Mg undergoes a redox reaction with strong aqueous acid.

The first step in balancing a redox reaction is to divide the unbalanced equation into half-reactions. Identify the participants in each half-reaction by noting that each half-reaction must be balanced. That is, each element in each half-reaction must be conserved. Consequently, any element that appears as a reactant in a half-reaction must also appear among the products. Hydrogen and oxygen frequently appear in both half-reactions, but other elements usually appear in just one of the half-reactions. Water, hydronium ions, and hydroxide ions often play roles in the overall stoichiometry of redox reactions occurring in aqueous solution. Chemists frequently omit these species in preliminary descriptions of such redox reactions.

When reactions do not divide clearly into an oxidation and a reduction, changes in oxidation numbers reveal how to divide the starting materials into half-reactions. Example 16-3 illustrates the separation process.

### EXAMPLE 16-3   IDENTIFYING HALF-REACTIONS

Separate the following unbalanced redox processes into half-reactions. All occur in aqueous solution:

(a) $Cr_2O_7^{2-}\ (aq) + Fe^{2+}\ (aq) \longrightarrow Cr^{3+}\ (aq) + Fe^{3+}\ (aq)$

(b) $H_2O_2\ (aq) + SO_3^{2-}\ (aq) \longrightarrow SO_4^{2-}\ (aq)$

(c) $BrO_2^-\ (aq) \longrightarrow BrO_3^-\ (aq) + Br^-\ (aq)$

**Strategy:** A redox equation can usually be separated by placing elements other than hydrogen and oxygen in separate half-reactions. If this does not work, use oxidation numbers to identify the oxidized and reduced species. Occasionally, an element other than O or H may be involved in both half-reactions.

**Solution:**

(a) This reaction can be divided by inspection. Iron must appear in one half-reaction and chromium in the other. Oxygen, which is present only in $Cr_2O_7^{2-}$, will be balanced in a later step.

$$Fe^{2+} \longrightarrow Fe^{3+} \qquad Cr_2O_7^{2-} \longrightarrow Cr^{3+}$$

(b) In this reaction, sulphur is the only element other than hydrogen and oxygen. Thus, sulphur must appear on both sides of one half-reaction. Furthermore, since sulphur appears in only one of the starting materials and in only one of the products, the second half-reaction must contain only hydrogen and oxygen. No product containing hydrogen is given, but water is always present in aqueous solution. The second half-reaction is the conversion of hydrogen peroxide to water:

*It is convenient to omit the phase designation when doing the steps in balancing redox reactions.*

$$SO_3^{2-} \longrightarrow SO_4^{2-} \qquad H_2O_2 \longrightarrow H_2O$$

(c) How can a reaction that contains just one reactant be divided into half-reactions? The only way is for this reactant to act as *both* the oxidizing agent and the reducing agent. Such a reaction is called a *disproportionation* reaction. Anions of $BrO_2^-$ react with one another; some are reduced and others are oxidized. Therefore, $BrO_2^-$ must appear in *both* half-reactions:

$$BrO_2^- \longrightarrow BrO_3^- \qquad \text{and} \qquad BrO_2^- \longrightarrow Br^-$$

**Do the Results Make Sense?** The separation in Part (a) is obvious. Oxidation numbers confirm the other assignments. In Part (b), recall that whereas the oxygen atom in $H_2O$ has an oxidation number of −2, the oxygen atoms in $H_2O_2$ have an oxidation number of −1. In Part (c), the oxidation number of the bromine atoms in $BrO_2^-$ is +3. Among the products, bromine is +5 in $BrO_3^-$ and −1 in $Br^-$. Thus, bromite ions are oxidized to bromate ions in one half-reaction, and other bromite ions are reduced to bromide ions in the other half-reaction.

**Practice Exercise 16.3**

Divide the following reactions into half-reactions:

(a) $Cu\ (s) + NO_3^-\ (aq) \longrightarrow Cu^{2+}\ (aq) + NO\ (g)$

(b) $ClO^-\ (aq) \longrightarrow ClO_3^-\ (aq) + Cl^-\ (aq)$

## Balancing Half-Reactions

After a redox reaction is divided into half-reactions, each half-reaction can be balanced independently. This is done in four steps, which *must be done in the order listed:*

a. Balance all elements, other than oxygen and hydrogen, by adjusting stoichiometric coefficients.

b. Balance oxygen by adding $H_2O$ to the side that is deficient in oxygen.

c. Balance hydrogen without unbalancing oxygen, as follows: unless the solution is basic, add $H_3O^+$ cations to the side that is deficient in hydrogen and an equal number of $H_2O$ molecules to the other side. If the solution is basic, add $H_2O$ to the side that is deficient in hydrogen and an equal number of $OH^-$ anions to the other side.

d. Balance net charge by adding electrons to the side that has the more positive (or less negative) net charge.

These four steps result in a balanced half-reaction in which elements, electrons, and total charge all are conserved. Example 16-4 illustrates the procedure.

## EXAMPLE 16-4  BALANCING HALF-REACTIONS

Balance the half-reactions in the following redox equation occurring in an acidic solution:

$$Cr_2O_7^{2-} + Fe^{2+} \longrightarrow Cr^{3+} + Fe^{3+}$$

**Strategy:** The half-reactions were identified in Example 16-3. Balance each half-reaction using the four-step procedure.

**Solution:**
Oxidation half-reaction: $Fe^{2+} \longrightarrow Fe^{3+}$

Steps a, b, c: Iron is already balanced, and neither oxygen nor hydrogen is present. Step d: The net charge on the left is +2, and on the right it is +3. The charge on the right needs to be made less positive: To balance the charges, add one electron on the right:

$$Fe^{2+} \longrightarrow Fe^{3+} + e^-$$

Now for the reduction half-reaction: $Cr_2O_7^{2-} \longrightarrow Cr^{3+}$

Step a: To balance chromium, give $Cr^{3+}$ a coefficient of 2:

$$Cr_2O_7^{2-} \longrightarrow 2\ Cr^{3+}$$

Step b: There are seven oxygen atoms on the left, so add seven water molecules on the right:

$$Cr_2O_7^{2-} \longrightarrow 2\ Cr^{3+} + 7\ H_2O$$

Step c: There are 14 hydrogen atoms on the right, requiring 14 $H_3O^+$ ions on the left and 14 additional $H_2O$ molecules on the right:

$$14\ H_3O^+ + Cr_2O_7^{2-} \longrightarrow 2\ Cr^{3+} + 21\ H_2O$$

Step d: The left-hand side has $14(+1) + 1(-2) = +12$ charge. The right-hand side has $2(+3) = +6$ charge. The charge balance thus requires six electrons on the left:

$$14\ H_3O^+ + Cr_2O_7^{2-} + 6\ e^- \longrightarrow 2\ Cr^{3+} + 21\ H_2O$$

**Do the Results Make Sense?** We can verify the electron counts by looking at oxidation numbers. The oxidation number of Fe increases by one, meaning that one electron must be released. The oxidation number of Cr changes from +6 to +3, meaning that each Cr atom must gain three electrons.

**Practice Exercise 16.4**
Balance the half-reactions from Practice Exercise 16.3.

## Balancing Redox Equations

After oxidation and reduction half-reactions are balanced, they can be combined to give the balanced chemical equation for the overall redox process. Although electrons are reactants in reduction half-reactions and products in oxidation half-reactions, they must cancel in the overall redox equation. To accomplish this, multiply each half-reaction by an appropriate integer that makes the number of electrons in the reduction half-reaction equal to the number of electrons in the oxidation half-reaction. The entire half-reaction must be multiplied by the integer to maintain charge balance. Example 16-5 illustrates this procedure.

### EXAMPLE 16-5  COMBINING HALF-REACTIONS

Complete the balancing of the reaction presented in Example 16-3(a):

$$Cr_2O_7^{2-} + Fe^{2+} \longrightarrow Cr^{3+} + Fe^{3+}$$

**Strategy:** The balanced half-reactions appear in Example 16-4. Balance the overall equation by combining the half-reactions in such a way that electrons cancel.

**Solution:** The half-reaction for iron contains one electron on the right, whereas the half-reaction for chromium contains six electrons on the left. To combine these half-reactions so that the electrons cancel, multiply the iron half-reaction by 6 and add it to the chromium half-reaction:

Oxidation:          $6\,Fe^{2+} \longrightarrow 6\,Fe^{3+} + 6\,e^-$

Reduction:     $14\,H_3O^+ + Cr_2O_7^{2-} + 6\,e^- \longrightarrow 2\,Cr^{3+} + 21\,H_2O$

Overall:     $14\,H_3O^+ + Cr_2O_7^{2-} + 6\,Fe^{2+} \longrightarrow 6\,Fe^{3+} + 2\,Cr^{3+} + 21\,H_2O$

 **Does the Result Make Sense?** An inventory of the reaction verifies that atoms and charge are balanced:

| Reactants | Products |
|---|---|
| 42 H, 2 Cr, 21 O, 6 Fe | 42 H, 2 Cr, 21 O, 6 Fe |
| $14(+1) + (-2) + 6(+2) = +24$ charge | $6(+3) + 2(+3) = +24$ charge |

All the elements are balanced, as is the total charge, so the overall equation is balanced.

**Practice Exercise 16.5**

Complete the balancing of the reactions from Practice Exercise 16.3.

Our Focus 16-1: Chemistry and the Environment describes a detoxification problem that was solved by applying redox chemistry.

### FOCUS 16-1  CHEMISTRY AND THE ENVIRONMENT: PURIFYING GROUNDWATER

Industrial operations use large quantities of chemicals that pose contamination threats to groundwater. In the recent past, industrial wastes such as chlorinated hydrocarbons (used as solvents and for cleaning) and chromium-containing solutions (used in metal plating) were routinely dumped into storage tanks from which they eventually leaked, leading to groundwater contamination. Heightened public awareness of the health and environmental hazards of such practices resulted in laws restricting the dumping of chemical waste, but preventing groundwater contamination remains a major problem. Many former dump sites continue to leak hazardous wastes, and industrial processes continue to generate chemical wastes, the safe disposal of which is costly and difficult.

The ideal disposal method is a chemical treatment that can convert hazardous waste into environmentally benign materials. For example, trichloroethylene ($Cl_2C$=$CHCl$) is highly toxic to aquatic life. The maximum allowable concentration in Canadian drinking water is thus very low: 5 $\mu$g/L. However, this compound can be destroyed by chemical treatment that reduces its chlorine atoms to chloride anions. Similarly, the chromium-containing waste from electroplating operations contains highly toxic $CrO_4^{2-}$ (chromate) anions, but a chemical treatment that reduces $CrO_4^{2-}$ to $Cr^{3+}$ causes the chromium to precipitate from the solution as insoluble $Cr(OH)_3$. This removal of chromium detoxifies the water.

Both these conversion processes involve the addition of electrons to the toxic substances. The trichloroethylene molecule is electrically neutral and must gain electrons in reactions that generate chloride anions. In addition, water or hydronium ions must supply hydrogen atoms that replace the chlorine atoms in the organic substance. The reaction mechanism is complex, but the net reaction is relatively simple, and the products are ethylene and chloride ions, both of which are non-toxic:

$$Cl_2C = CHCl + 3\ H_3O^+ + 6\ e^- \longrightarrow H_2C=CH_2 + 3\ Cl^- + 3\ H_2O$$

The chromium atom in $CrO_4^{2-}$ has an oxidation number of +6, so it must gain three electrons to be reduced to $Cr^{3+}$. Again, water or hydronium ions must supply hydrogen atoms that combine with the O atoms in chromate to form water:

$$CrO_4^{2-} + 8\ H_3O^+ + 3\ e^- \longrightarrow Cr^{3+} + 12\ H_2O$$

Because electrons are conserved, chemical detoxification of water containing either of these contaminants requires a substance that easily donates electrons. A suitable electron donor for groundwater cleanup must also be relatively inexpensive and must not itself be a source of toxicity. These are stringent requirements; yet, a very common material does this job spectacularly well: iron filings. Metallic iron readily donates electrons to become $Fe^{2+}$ or $Fe^{3+}$ cations, neither of which is significantly toxic.

$$Fe \longrightarrow Fe^{2+} + 2\ e^-$$

Thus, when iron reacts with trichloroethylene or chromate in a redox process, all the products are relatively non-toxic.

Turning this relatively simple chemistry into a workable solution for existing contaminated sites was the invention of Dr. R.W. Gillham at the University of Waterloo. As the figure indicates, the "iron wall" process takes advantage of the same natural water flow that leads to contamination. First, the source of contamination and direction in which groundwater flows are determined. A pit is then dug downstream from the contamination source. This pit is filled with a mixture of sand and iron filings. The contaminants react with iron filings as the groundwater flows through the pit, and the water emerging downstream from the pit contains only trace amounts of the contaminants.

This technique has proven effective for a number of other contaminants and continues to be the focus of research in Dr. Gillham's laboratory. For instance, RDX (hexahydro-1,3,5-trinitro-1,3,5-triazine) is an explosive, and as a result is a common groundwater and soil contaminant at military facilities. Reaction with iron has been shown to be an effective method for degrading this toxic substance into harmless smaller chemical species. For some contaminants, reduction by iron is only part of the solution to the problem. Nitrobenzene, for instance, is reduced by iron to aniline, which itself is still toxic. The aniline, however, can then be degraded by aerobic bacteria to below-detection limits.

Dr. Gillham's work is an interesting example of how a real-life problem prompts work at the laboratory scale, which is then scaled up as required to solve the problem. In the laboratory, questions about reaction mechanisms, reaction rates, and potential chemical interferences can be solved. Once a large-scale process such as an iron wall has been installed, the data from long-term monitoring of the decontaminated groundwater will suggest other laboratory-scale experiments that are needed. The result is an ever-improving understanding of the chemistry of such a process, and more effective remediation of contaminated sites.

When half-reactions are combined, there is often a duplication of some chemical species, particularly $H_2O$ and $H_3O^+$ or $OH^-$. The overall equation is "cleaned up" by combining species that appear twice on the same side. Also, when a species appears on both sides of the balanced equation, equal numbers of the species are subtracted from each side.

To summarize, the balancing of a redox equation occurs in several steps that are carried out in sequence.

---

**BALANCING REDOX REACTIONS**

Step 1    Break the unbalanced reaction into half-reactions by inspection.

Step 2    Balance each half-reaction, following the four-step procedure:

    a. Balance all elements other than oxygen and hydrogen by adjusting the stoichiometric coefficients.

    b. Balance oxygen by adding $H_2O$ to the side deficient in oxygen.

    c. Balance hydrogen. If the solution is neutral or acidic, add $H_3O^+$ to the side deficient in hydrogen, and the same amount of $H_2O$ to the other side. If the solution is basic, add $H_2O$ to the side deficient in oxygen, and the same amount of $OH^-$ to the other side.

    d. Balance the charge by adding electrons to the side with the more positive charge.

Step 3    Multiply the half-reactions by integers to make each have the same number of electrons.

Step 4    Add the two half-reactions and cancel any species duplicated on both sides of the total reaction.

---

Example 16-6 illustrates the entire balancing process, and Example 16-7 illustrates the balancing procedure in basic solution.

## EXAMPLE 16-6  BALANCING A REDOX EQUATION

Concentrated aqueous sulphuric acid ($H_2SO_4$) is a strong oxidizing agent that can react with elemental carbon:

$$H_2SO_4 + C \longrightarrow CO_2 + SO_2 \quad \text{(unbalanced)}$$

What is the balanced equation for this process?

**Strategy:** Divide the reaction into half-reactions, balance each using the stepwise procedure, combine the half-reactions, and then clean up the result to eliminate duplicated species.

**Solution:**

1. The half-reactions can be identified by inspection:

$$\text{Oxidation:} \quad C \longrightarrow CO_2$$

$$\text{Reduction: } H_2SO_4 \longrightarrow SO_2$$

2. Each half-reaction is balanced separately using the four-step process. Concentrated sulphuric acid contains some water, so $H_3O^+$ and $H_2O$ can be used as needed. The new feature in each step is highlighted by a yellow background.

   Step a: C is already balanced                S is already balanced

   Step b: $C + 2\,H_2O \longrightarrow CO_2$         $H_2SO_4 \longrightarrow SO_2 + 2\,H_2O$

   Step c: $C + 6\,H_2O \longrightarrow CO_2 + 4\,H_3O^+$     $H_2SO_4 + 2\,H_3O^+ \longrightarrow SO_2 + 4\,H_2O$

   Step d: $C + 6\,H_2O \longrightarrow CO_2 + 4\,H_3O^+ + 4\,e^-$    $H_2SO_4 + 2\,H_3O^+ + 2\,e^- \longrightarrow SO_2 + 4\,H_2O$

3. Multiply the reduction equation by 2, so the electrons will cancel, and add the half-reactions:

$$C + 6\,H_2O + 4\,H_3O^+ + 2\,H_2SO_4 + 4\,e^- \longrightarrow CO_2 + 4\,H_3O^+ + 4\,e^- + 2\,SO_2 + 8\,H_2O$$

4. Cancel common terms: $4\,H_3O^+$, four electrons, and $6\,H_2O$ on each side:

$$C + 2\,H_2SO_4 \longrightarrow CO_2 + 2\,SO_2 + 2\,H_2O$$

**Does the Result Make Sense?** An inventory verifies that this equation is balanced:

| Reactants | Products |
|---|---|
| 1 C, 4 H, 2 S, 8 O | 1 C, 4 H, 2 S, 8 O |
| 0 charge | 0 charge |

**Practice Exercise 16.6**

Balance this reaction in aqueous solution: $MnO_4^- + ClO_3^- \longrightarrow MnO_2 + ClO_4^-$.

---

## EXAMPLE 16-7   REDOX IN BASIC SOLUTION

Household ammonia should never be mixed with chlorine bleach, because a redox reaction occurs that generates toxic chlorine gas and hydrazine:

$$NH_3 + OCl^- \longrightarrow Cl_2 + N_2H_4 \quad \text{(unbalanced)}$$

Balance this equation in basic solution.

**Strategy:** Follow the step-by-step procedure.

Never mix household ammonia and chlorine bleach!

**Solution:**

1. Break into half-reactions, one containing Cl and the other containing N:

$$\text{Oxidation:} \quad NH_3 \longrightarrow N_2H_4$$

$$\text{Reduction:} \quad OCl^- \longrightarrow Cl_2$$

2. Balance the half-reactions individually.
   Step a. Balance N and Cl by inspection:

$$2\,NH_3 \longrightarrow N_2H_4 \qquad\qquad 2\,OCl^- \longrightarrow Cl_2$$

   Step b. Add $H_2O$ to balance oxygen atoms:

$$2\,NH_3 \longrightarrow N_2H_4 \qquad\qquad 2\,OCl^- \longrightarrow Cl_2 + 2\,H_2O$$

   Step c. Add $H_2O$ and $OH^-$ to balance hydrogen atoms:

$$2\,NH_3 + 2\,OH^- \longrightarrow N_2H_4 + 2\,H_2O \qquad\qquad 2\,OCl^- + 4\,H_2O \longrightarrow Cl_2 + 2\,H_2O + 4\,OH^-$$

   Two water molecules can be cancelled from each side of the reduction half-reaction:

$$2\,OCl^- + 2\,H_2O \longrightarrow Cl_2 + 4\,OH^-$$

   Step d. Add electrons to balance the charges. The nitrogen half-reaction needs two electrons on the right, and the chlorine reaction needs two electrons on the left:

$$2\,NH_3 + 2\,OH^- \longrightarrow N_2H_4 + 2\,H_2O + 2\,e^- \qquad\qquad 2\,OCl^- + 2\,H_2O + 2\,e^- \longrightarrow Cl_2 + 4\,OH^-$$

3. Because there are two electrons in each half-reaction, adding the two half-reactions leads to cancellation of the electrons.

4. Now recombine the half-reactions and cancel common terms:

$$2\,NH_3 + 2\,OH^- + \cancel{2\,H_2O} + 2\,OCl^- + \cancel{2\,e^-} \longrightarrow N_2H_4 + \cancel{2\,e^-} + \cancel{2\,H_2O} + Cl_2 + 4\,OH^-$$

   Cancel two $OH^-$ ions:

$$2\,NH_3 + 2\,OCl^- \longrightarrow N_2H_4 + Cl_2 + 2\,OH^-$$

 **Does the Result Make Sense?** Make sure the equation is balanced by taking an inventory:

| Reactants | Products |
|---|---|
| 2 N, 6 H, 2 Cl, 2 O | 2 N, 6 H, 2 Cl, 2 O |
| −2 charge | −2 charge |

The equation is balanced. Although we might have balanced this equation by inspection, the detailed procedure reveals the number of electrons transferred, which we will see is useful for other calculations.

**Practice Exercise 16.7**

Balance this reaction in basic aqueous solution: $MnO_4^- + Cd \longrightarrow MnO_2 + Cd(OH)_2$.

**SECTION EXERCISES**

●**16.2.1** Balance the following redox reactions from Example 16-3:

(a) $H_2O_2 + SO_3^{2-} \longrightarrow SO_4^{2-}$ (basic)
(b) $BrO_2^- \longrightarrow BrO_3^- + Br^-$ (acidic)

●**16.2.2** The tarnish that collects on objects made of silver is silver sulphide, a black solid. Tarnish can be removed by heating the tarnished object in an aluminum pan containing mildly basic water. Balance this equation:

$$Al\ (s) + Ag_2S\ (s) \longrightarrow Ag\ (s) + Al(OH)_3\ (s) + HS^-\ (aq)$$

●**16.2.3** Copper refining traditionally involves "roasting" sulphide ores with oxygen, which produces CuO and $SO_2$. An alternative process uses aqueous nitric acid to convert the CuS into soluble copper(II) hydrogen sulphate. (The $Cu^{2+}$ is reduced to Cu metal in a subsequent step.) One reaction which occurs is shown below. Balance this redox equation in an acidic solution:

$$CuS\ (s) + NO_3^-\ (aq) \longrightarrow NO\ (g) + Cu^{2+}\ (aq) + HSO_4^-\ (aq)$$

## 16.3 GALVANIC CELLS

The reaction of potassium metal with water to generate hydrogen gas, shown in Figure 16-4, is highly spontaneous:

$$2\,K\,(s) + 2\,H_2O\,(l) \xrightarrow{\text{Rapid}} 2\,K^+\,(aq) + 2\,OH^-\,(aq) + H_2\,(g)$$

In contrast, no reaction occurs when a silver spoon is dipped into water. The balanced redox equation for the possible oxidation of silver by water looks similar to that for potassium:

$$2\,Ag\,(s) + 2\,H_2O\,(l) \xrightarrow{\text{No reaction}} 2\,Ag^+\,(aq) + 2\,OH^-\,(aq) + H_2\,(g)$$

One metal reacts with water so vigorously that a fire or explosion may result, whereas the other does not react with water, even at high temperature. Despite their apparent similarities, the reaction of potassium is spontaneous, but that of silver is not. Thermodynamic calculations verify these observations. Remember

**Figure 16-4**

Potassium reacts spontaneously and violently with water, but silver does not react with water at all.

from Chapter 11 that spontaneous reactions have negative values for $\Delta G$. We can calculate the standard free energy changes for these reactions using tabulated free energies of formation. The $\Delta G°$ for the reaction of potassium has a large negative value, but the value for the reaction of silver is large and positive:

$$2\ K\ (s)\ +\ 2\ H_2O\ (l)\ \longrightarrow\ 2\ K^+(aq)\ +\ 2\ OH^-(aq)\ +\ H_2(g)\quad \Delta G°_{reaction} = -407\ kJ$$

$$2\ Ag(s)\ +\ 2\ H_2O\ (l)\ \longrightarrow\ 2\ Ag^+(aq)\ +\ 2\ OH^-(aq)\ +\ H_2(g)\quad \Delta G°_{reaction} = +314\ kJ$$

Experiments and calculations both indicate that electron transfer from potassium to water is spontaneous and rapid, whereas electron transfer from silver to water does not occur. In redox terms, potassium oxidizes easily, but silver resists oxidation. Because oxidation involves the loss of electrons, these differences in reactivity of silver and potassium can be traced to how easily each metal loses electrons to become an aqueous cation. One obvious factor is their first ionization energies, which show that it takes much more energy to remove an electron from silver than from potassium: 731 kJ/mol for Ag and 419 kJ/mol for K. The other alkali metals with low first ionization energies, Na, Rb, Cs, and Fr, all react violently with water.

## Direct and Indirect Electron Transfer

The reactivities of potassium and silver with water represent extremes in the spontaneity of electron-transfer reactions. The redox reaction between two other metals illustrates less drastic differences in reactivity. Figure 16-5 shows the reaction

Figure 16-5

When a strip of zinc metal is dipped in a solution of copper(II) sulphate, zinc is oxidized to $Zn^{2+}$ (aq), and $Cu^{2+}$ (aq) is reduced to copper metal. The insoluble metal precipitates from the solution. In the molecular views, water molecules and spectator anions have been omitted for clarity.

that occurs between zinc metal and an aqueous solution of copper(II) sulphate: zinc slowly dissolves, and copper metal precipitates. This spontaneous reaction has a negative standard free energy change:

$$Cu^{2+} (aq) + Zn (s) \longrightarrow Cu (s) + Zn^{2+} (aq) \qquad \Delta G^{\circ}_{reaction} = -213 \text{ kJ}$$

Each zinc atom transfers two electrons to a copper cation. Balanced half-reactions show the electron gains and losses more clearly:

Oxidation: $\qquad Zn (s) \longrightarrow Zn^{2+} (aq) + 2 e^-$

Reduction: $\qquad Cu^{2+} (aq) + 2 e^- \longrightarrow Cu (s)$

The oxidation of Zn metal by $Cu^{2+}$ ions is an example of *direct* electron transfer. A copper ion accepts two electrons when it collides with the surface of the zinc strip. Direct electron transfer occurs when electrons are transferred during a collision between the species being oxidized and the species being reduced, as shown by the molecular views in Figure 16-5.

Spontaneous redox reactions can also occur by *indirect* electron transfer, whereby species involved in the redox chemistry are not allowed to come into direct contact with one another. Instead, the oxidation occurs at one end of a wire and transfers electrons to the wire. Reduction occurs at the other end of the wire and removes electrons from the wire. The wire conducts electrons between the oxidation site and the reduction site.

Figure 16-6 is a schematic view of two beakers set up for indirect electron transfer. The beaker on the left contains an aqueous solution of zinc sulphate and a strip of zinc metal. The beaker on the right contains an aqueous solution of copper(II) sulphate and a strip of copper metal. A wire connects the two metal strips to allow indirect electron transfer. The oxidation half-reaction transfers electrons to the wire and releases $Zn^{2+}$ ions into the solution containing the zinc electrode:

**Figure 16-6**

If the zinc- and copper-containing portions of this redox system are physically separated, electron transfer can occur only through an external wire.

$$Zn (s) \longrightarrow Zn^{2+} (aq) + 2 e^- \text{ (wire)}$$

Because metals conduct charge, electrons can move from the zinc strip through the wire into the copper strip. Copper cations can collect electrons when they collide with the surface of the copper strip:

$$Cu^{2+} (aq) + 2 e^- \text{ (wire)} \longrightarrow Cu (s)$$

The arrangement shown in Figure 16-6 does not generate a sustained flow of electrons, however, because as electrons are transferred, the solutions in both beakers quickly become unbalanced in electrical charge. Oxidation of the zinc strip releases $Zn^{2+}$ ions into the solution, generating excess positive charge. On the other side, reduction of $Cu^{2+}$ ions to copper metal produces a solution that is deficient in positive charge. This charge imbalance generates an electrical force that stops the further flow of electrons.

# Ion Transport

Redox by way of indirect electron transfer cannot continue unless there is a way to remove the charge imbalance created by electron flow. Ion movement can remove this charge imbalance, if we provide a way for anions to move from the region of excess negative charge to the region of excess positive charge. (Cations will move in the opposite direction.) Such ion migration must not allow the oxidation–reduction pair to come into direct contact, however, because then direct electron transfer would replace indirect electron transfer.

In our zinc–copper example, electrons flowing through the wire from left to right generate surplus positive charge on the left and surplus negative charge on the right. Metal cations moving from left to right and sulphate anions moving from right to left can remove the surplus charges, but this flow of ions must occur in a way that prevents the two solutions from mixing freely. Figure 16-7 shows an arrangement that meets these needs. A porous glass plate connects the two vessels.

**Figure 16-7**

Schematic view of an arrangement that allows a sustained redox reaction accompanied by an external flow of electrons.

This plate allows ions to migrate between the two compartments, but the mixing process is too slow to lead to direct electron transfer. Electrons flow through the wire, and ions diffuse through the plate. Adding the porous plate completes a circuit that can be used to generate a sustained flow of electrons through the external wire.

Salt bridges and membranes can also serve as separators that allow ion migration without free mixing of solutions. A porous plate contains many tiny channels that allow water molecules and ions to diffuse slowly from one side to the other. A salt bridge contains gelatin impregnated with cations and anions that are similar in size and mobility, usually $K^+$ and $Cl^-$ or $NH_4^+$ and $NO_3^-$. The gelatin prevents solvent from moving freely but allows the ions to diffuse slowly. A membrane functions like a porous plate but is very thin, often only micrometres in width.

## Electrodes

The spontaneous redox reaction shown in Figure 16-7 takes place at the surfaces of metal plates, where electrons are gained and lost by metal atoms and ions. These metal plates are examples of **electrodes**. At an electrode, redox reactions transfer electrons between the aqueous phase and the external circuit. An oxidation half-reaction releases electrons to the external circuit at one electrode. A reduction half-reaction withdraws electrons from the external circuit at the other electrode. The electrode where oxidation occurs is the **anode**, and the electrode where reduction occurs is the **cathode**.

Some electrodes are made of substances that participate in the redox reactions that transfer electrons. These are **active electrodes**. Other electrodes serve only to supply or accept electrons but are not part of the redox chemistry; these are **passive electrodes**. In Figure 16-7, both metal strips are active electrodes. During the redox reaction, zinc metal dissolves from the anode while copper metal precipitates at the cathode. The reactions that take place at these active electrodes are conversions between the metals contained in the electrodes and their aqueous cations.

At another type of active electrode, found in many batteries, the reaction is the conversion between a metal and an insoluble salt. At the surface of this type of electrode, metal cations combine with anions from the solution to form the salt. One example is the cadmium anode of a rechargeable nickel–cadmium battery, the surface at which cadmium metal loses electrons and forms $Cd^{2+}$ cations. These cations combine immediately with hydroxide ions in solution to form insoluble cadmium hydroxide:

$$Cd \ (s) + 2 \ OH^- \ (aq) \longrightarrow Cd(OH)_2 \ (s) + 2 \ e^-$$

A redox half-reaction at an active electrode also may convert one metal salt into another. For example, the cathode in a nickel-cadmium battery is $NiO(OH)$, which is reduced to nickel(II) hydroxide. The half-reaction reduces Ni from a +3 to a +2 oxidation state:

$$NiO(OH) \ (s) + H_2O \ (l) + e^- \longrightarrow Ni(OH)_2 \ (s) + OH^- \ (aq)$$

In contrast with these active electrodes, a passive electrode conducts electrons to and from the external circuit but does not participate chemically in the half-reactions. Figure 16-8 shows a redox setup that contains passive electrodes. One compartment contains an aqueous solution of iron(III) chloride in contact with a platinum electrode. Electron transfer at this electrode reduces $Fe^{3+}$ $(aq)$ to $Fe^{2+}$ $(aq)$:

Reduction: $\qquad Fe^{3+}\,(aq) + e^{-} \longrightarrow Fe^{2+}\,(aq)$

In the second compartment, hydrogen gas bubbles over the surface of a platinum electrode. On the surface of this passive electrode, hydrogen molecules are oxidized, giving up two electrons to form pairs of protons, which bond to water molecules to give $H_3O^{+}$ ions:

Oxidation: $H_2\,(g) + 2\,H_2O\,(l) \longrightarrow 2\,H_3O^{+}\,(aq) + 2\,e^{-}$

Each half-reaction occurs at a platinum metal electrode, but notice that Pt is not a participant in either half-reaction. Neither electrode takes part in the redox chemistry.

Platinum metal is often used as a passive electrode because platinum is one of the least reactive elements. Platinum has a large ionization energy, so it can act as an "electron shuttle" without participating in redox chemistry.

The combination of hydrogen gas, $H_3O^{+}$ ions, and a platinum electrode is referred to as a *hydrogen electrode*. This electrode appears in the right-hand portion of Figure 16-8. When a hydrogen electrode operates under standard conditions, $p_{H_2} = 1.00$ bar and $[H_3O^{+}] = 1.00$ M, it is called a **standard hydrogen electrode** (SHE). The standard hydrogen electrode is particularly important in electrochemistry, as we describe in Section 16.4.

**Figure 16-8**

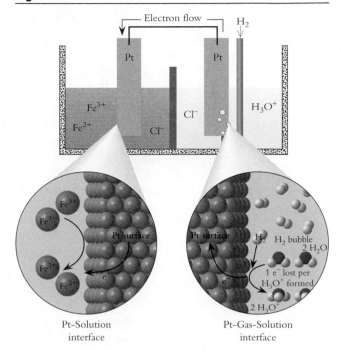

Passive electrodes serve only to conduct electrons to and from the interfaces. They do not take part in the redox reactions.

## Components of Galvanic Cells

Electrochemistry is the coupling of a chemical redox process with electron flow through a wire. The process represented in Figure 16-7 is electrochemical because the redox reaction releases electrons that flow through an external wire as an electrical current. On the other hand, Figure 16-5 shows a redox process that is not electrochemical, because direct electron transfer cannot generate an electrical current through a wire.

A **galvanic cell** uses redox reactions to generate electrical current. Galvanic cells find wide usage in the form commonly called *batteries*. To give just one example, flashlight batteries make use of electrochemical reactions to generate electrical current. When a flashlight is turned on, chemical redox reactions occur in the battery and generate an electron flow through the light bulb.

**Figure 16-9**

The essential components of a galvanic cell.

The cell shown diagrammatically in Figure 16-9 uses the $Cu^{2+}/Zn$ reaction to illustrate the essential components of a galvanic cell:

1. *A spontaneous redox reaction.* In Figure 16-9, the spontaneous reaction is electron transfer between zinc and copper:

$$Zn\ (s) + Cu^{2+}\ (aq) \longrightarrow Zn^{2+}\ (aq) + Cu\ (s)$$

2. *A physical barrier blocking direct electron transfer.* Here, placing the solutions in two separate compartments provides the barrier.

3. *Physical contact between chemical and electrical parts of the cell.* This is the role of the electrodes. The electrodes in Figure 16-9 are strips of copper and zinc.

4. *An external electrical circuit.* This circuit may simply transfer electrons, as in the zinc–copper example. In useful applications, the flow of electrons in the external circuit provides power to operate a flashlight, radio, or other device.

5. *Some means of completing the circuit.* In Figure 16-9, the porous glass plate allows ions to diffuse slowly between the two solutions, completing the circuit.

6. *Ions present in the solutions.* All parts of the circuit must allow the passage of charge. When an aqueous phase is present, it must contain ions, which are supplied by a soluble salt or strong acid.

To apply the features that characterize galvanic cells, Example 16-8 describes the lead storage battery.

---

## EXAMPLE 16-8   DESCRIBING A GALVANIC CELL

The electrical current needed to start an automobile engine is provided by a lead storage battery. This battery contains aqueous sulphuric acid in contact with two electrodes. One electrode is metallic lead, and the other is solid $PbO_2$. (The $PbO_2$ is packed into the pores of a lead electrode for mechanical strength, but the $PbO_2$ itself is the active substance of this electrode.) Each electrode becomes coated with solid $PbSO_4$ as the battery operates. Determine the balanced half-reactions, the overall redox reaction, and the anode and cathode in this galvanic cell.

**Strategy:** We are asked to identify the redox chemistry occurring in this battery. The problem provides a description of the chemical composition of a galvanic cell. To determine what redox reactions take place, examine the species present at each electrode. Then use the standard procedure to balance the half-reactions.

**Solution:** The description of the battery states that one electrode is Pb in contact with $PbSO_4$, and the other electrode is $PbO_2$ in contact with $PbSO_4$. This information identifies the two half-reactions:

Oxidation:     $Pb\ (s) \longrightarrow PbSO_4\ (s)$

Reduction:   $PbO_2\ (s) \longrightarrow PbSO_4\ (s)$

Lead is already balanced. To balance sulphur, we need to work with the major species in solution. The aqueous solution contains a strong acid, $H_2SO_4$, which generates $H_3O^+$ and $HSO_4^-$. Use the sulphur-containing species, $HSO_4^-$, to balance sulphur:

$$Pb + HSO_4^- \longrightarrow PbSO_4 \qquad PbO_2 + HSO_4^- \longrightarrow PbSO_4$$

Now all elements other than hydrogen and oxygen are balanced.

Here are the additional steps for the left half-reaction:

Oxygen is already balanced. Add $H_3O^+$ and $H_2O$ to balance hydrogen:

$$Pb + HSO_4^- + H_2O \longrightarrow PbSO_4 + H_3O^+$$

Add electrons to balance charge:

$$Pb + HSO_4^- + H_2O \longrightarrow PbSO_4 + H_3O^+ + 2\,e^-$$

Electrons appear as products, so this half-reaction is the oxidation, which takes place at the anode. The lead electrode is an active anode in a lead storage battery.

Now balance the second half-reaction. Add water to balance oxygen:

$$PbO_2 + HSO_4^- \longrightarrow PbSO_4 + 2\,H_2O$$

Add $H_3O^+$ and $H_2O$ to balance hydrogen:

$$PbO_2 + HSO_4^- + 3\,H_3O^+ \longrightarrow PbSO_4 + 5\,H_2O$$

Add $e^-$ to balance charge:

$$PbO_2 + HSO_4^- + 3\,H_3O^+ + 2\,e^- \longrightarrow PbSO_4 + 5\,H_2O$$

Electrons appear as starting materials, so this half-reaction is the reduction, which takes place at the cathode. Lead(IV) oxide is an active cathode in a lead storage battery.

To remember how electrodes are named, notice that *o*xidation and *a*node both begin with vowels; *r*eduction and *c*athode both begin with consonants.

Because two electrons are transferred in each half-reaction, the overall redox reaction can be obtained by adding the half-reactions and cleaning up:

$$PbO_2\,(s) + Pb\,(s) + 2\,HSO_4^-\,(aq) + 2\,H_3O^+\,(aq) \longrightarrow 2\,PbSO_4\,(s) + 4\,H_2O\,(l)$$

 **Do the Results Make Sense?** You should verify that each of the elements is balanced in this reaction. In addition, the net charge is zero on each side.

### Practice Exercise 16.8

The mercury battery, use of which is being discontinued because of the toxicity of mercury, contains HgO and Zn in contact with basic aqueous solution. The redox products are Hg and ZnO. Determine the oxidation and reduction half-reactions and the overall reaction for these batteries.

## Shorthand Cell Notations

It can be cumbersome to draw an electrochemical cell, or even to write out the full redox reaction taking place in the cell. Because of this, chemists have developed a shorthand notation for electrochemical cells, which simplifies the task of communicating this information. As an example, consider the oxidation of zinc metal by copper ions under standard conditions, which has the following oxidation and reduction half-reactions, and the balanced overall redox reaction:

Oxidation:    $Zn\ (s) \longrightarrow Zn^{2+}\ (aq) + 2\ e^-$

Reduction:    $Cu^{2+}\ (aq) + 2\ e^- \longrightarrow Cu\ (s)$

Overall:    $Zn\ (s) + Cu^{2+}\ (aq) \longrightarrow Zn^{2+}\ (aq) + Cu\ (s)$

Writing a shorthand notation requires us to specify

- what the anode and cathode are made of.
- what is in each of the oxidation and reduction compartments of the cell, and their concentrations.
- any boundaries between different phases.

The shorthand notation for this cell is

$$Zn\ (s)\,|\,Zn^{2+}\ (aq,\ 1\ M)\,\|\,Cu^{2+}\ (aq,\ 1\ M)\,|\,Cu\ (s)$$

By convention, the anode is placed at the left and the cathode at the right. The vertical solid line ($|$) represents the phase boundary and the double solid line ($\|$) represents the membrane or salt bridge separating the two half-cells. Note that only the species participating in the reaction are listed, so the spectator ions (anions in this case) are not shown.

In cases where there is more than one species in a compartment, reactants are listed first followed by the products. For example, consider the oxidation of $Fe^{2+}$ to $Fe^{3+}$ by permanganate ion, $MnO_4^-$, at standard state conditions:

Oxidation:    $Fe^{2+}\ (aq) \longrightarrow Fe^{3+}\ (aq) + e^-$

Reduction:    $MnO_4^-\ (aq) + 8\ H^+\ (aq) + 5\ e^- \longrightarrow Mn^{2+}(aq) + 4\ H_2O\ (l)$

Overall:    $MnO_4^-\ (aq) + 8\ H^+\ (aq) + 5\ Fe^{2+}\ (aq)$
$\longrightarrow Mn^{2+}\ (aq) + 5\ Fe^{3+}\ (aq) + 4\ H_2O\ (l)$

Since there are no solid metals in this reaction, both the anode and cathode are passive electrodes, and as usual we use platinum metal:

$Pt\ (s)\ |\ Fe^{2+}\ (aq,\ 1\ M),\ Fe^{3+}\ (aq,\ 1\ M)\ \|\ MnO_4^-\ (aq,\ 1\ M),$
$$H^+\ (aq,\ 1\ M),\ Mn^{2+}\ (aq,\ 1\ M)\ |\ Pt\ (s)$$

The reasons for being specific about concentrations are explored in the next two sections of this chapter. Example 16-9 illustrates further the idea of shorthand notations.

## EXAMPLE 16-9  WRITING SHORTHAND CELL NOTATIONS

(a) Write the shorthand cell notation for the redox reaction at standard state conditions:

Oxidation:    $Fe\ (s) \longrightarrow Fe^{2+}\ (aq) + 2\ e^-$

Reduction:    $Cr_2O_7^{2-}\ (aq) + 14\ H^+\ (aq) \longrightarrow 2\ Cr^{3+}\ (aq) + 7\ H_2O\ (l)$

**Strategy:** First, determine what the anode and cathode are, and then divide the reaction into oxidation and reduction species. Write the shorthand notation following the steps outlined above.

**Solution:** Iron metal is being oxidized, so the anode will be made of iron metal. There are no solid metals in the reduction reaction, so we use Pt ($s$) as a passive electrode. The outline of the shorthand notation will therefore look like

$$\text{Fe } (s) \,|\, \text{oxidation species} \,\|\, \text{reduction species} \,|\, \text{Pt } (s)$$

The oxidation species consist only of $Fe^{2+}$. (Remember that although there must be anions in this solution, they do not participate in this particular redox reaction, and are therefore not listed in the shorthand notation.) On the reduction side, $Cr_2O_7^{2-}$ and $H^+$ are the reactants and $Cr^{3+}$ is the only product. $H_2O$ is not explicitly listed in the shorthand notation, but is implicit since several species are aqueous. All aqueous species are 1 M, since the reaction takes place at standard state conditions. The final answer is therefore

$$\text{Fe } (s) \,|\, Fe^{2+} (aq, \text{ 1 M}) \,\|\, Cr_2O_7^{2-} (aq, \text{ 1 M}), H^+ (aq, \text{ 1 M}), Cr^{3+} (aq, \text{ 1 M}) \,|\, \text{Pt } (s)$$

(b) Write a balanced redox reaction for the cell described by the following shorthand notation:

$$\text{Al } (s) \,|\, Al^{3+} (aq, \text{ 1 M}) \,\|\, Br_2 (l) \,|\, Br^- (aq, \text{ 1 M}) \,|\, \text{Pt } (s)$$

**Strategy:** Separate the shorthand notation into the oxidation and reduction species. Balance the half-reactions separately and add together.

**Solution:** Since the anode always appears at the left side of a shorthand notation, aluminum metal must oxidized in this redox reaction. The half-reaction is thus

$$\text{Oxidation: Al } (s) \longrightarrow Al^{3+} (aq) + 3e^-$$

On the reduction side, the cathode is platinum metal, and so is not involved in the redox reaction itself. The only other species are liquid bromine and aqueous bromide ions, so the reduction reaction must be

$$\text{Reduction: } Br_2 (l) + 2\,e^- \longrightarrow 2\,Br^- (aq)$$

Multiplying the oxidation reaction by 3 and the reduction reaction by 2 and adding gives the final result:

$$\text{Overall: } 2\,\text{Al } (s) + 3\,Br_2 (l) \longrightarrow 2\,Al^{3+} (aq) + 6\,Br^- (aq)$$

 **Do the Results Make Sense?** In each case, all species appearing in the balanced redox reaction also appear in the shorthand notation, except for water, which is implied by the ($aq$) species.

**Practice Exercise 16.9**
(a) Write the shorthand cell notation for the lead-acid battery described in Example 16-8.
(b) Write a balanced redox reaction for the cell described by the following shorthand notation:

$$\text{Pt } (s) \,|\, Br^- (aq), Br_2 (l) \,\|\, Cr_2O_7^{2-} (aq), H^+ (aq), Cr^{3+} (aq) \,|\, \text{Pt } (s)$$

## SECTION EXERCISES

● **16.3.1** The thermite reaction between aluminum metal and iron oxide is so rapid and exothermic that it generates a fountain of sparks (see photo) and can melt the container in which it takes place. The spontaneity of this reaction suggests the possibility of a galvanic cell involving aluminum and iron:

$$2\,Al + 6\,OH^- \longrightarrow Al_2O_3 + 3\,H_2O + 6\,e^-$$

$$Fe_2O_3 + 3\,H_2O + 6\,e^- \longrightarrow 2\,Fe + 6\,OH^-$$

Draw a schematic diagram, similar to Figure 16-7, showing the operation of this galvanic cell. Include the directions of all charge flows.

● **16.3.2** Use tabulated thermodynamic data to verify that the galvanic cell of Section Exercise 16.3.1 is spontaneous in the direction written.

● **16.3.3** Draw molecular pictures similar to the one shown in Figure 16-5 to illustrate the difference between an active electrode and a passive electrode. Use an iron anode in contact with a solution containing $Fe^{2+}$ ions to illustrate an active electrode. Use a platinum cathode immersed in a solution containing $Fe^{2+}$ and $Fe^{3+}$ ions to illustrate a passive electrode.

The thermite reaction

## 16.4 CELL POTENTIALS

Water at high altitude has gravitational potential energy that can be converted to other forms. As water flows, the gravitational potential energy that it loses is converted into kinetic energy in a waterfall, kinetic energy in a waterwheel, or electrical energy in a hydroelectric turbine. Similarly, as electrons flow, electrical potential energy can be converted to other forms: light energy in a flashlight, mechanical energy in the starter motor of an automobile, or thermal energy in an electrical heater.

Batteries supply electrical current that consists of electrons flowing from one place to another. In some ways, flowing electrons are like flowing water. Because of gravitational force, water always flows downhill, from higher altitude to lower altitude. Similarly, electrons flow under the influence of electrical force. Electrons flow from regions of more negative electrical potential energy to regions of more positive electrical potential energy.

A potential energy difference is referred to simply as a *potential*. Continuing our analogy with water, differences in gravitational potential are measured by height differences. The top and bottom of the spillway of a dam are at different heights and have different gravitational potentials: water falling down the spillway converts gravitational potential energy into kinetic energy of motion. Analogously, one electrode generally is at higher electrical potential than another.

**Figure 16-10**

Water flows down a spillway from higher to lower gravitational potential. Electrons flow "downhill" through a wire from higher to lower negative electrical potential.

Whereas gravitational potentials are measured in heights, electrical potential differences are measured in volts. For electrons, a point at a more negative electrical potential is "uphill" from a point at a more positive electrical potential. The parallel between gravitational potential and electrical potential is summarized in Figure 16-10.

## Electrode Equilibrium

To visualize how electrochemical cells generate electrical potential differences, consider a zinc electrode dipped into a solution of zinc sulphate. From the macroscopic perspective, nothing happens. At the molecular level, however, some of the zinc atoms of the electrode are oxidized to $Zn^{2+}$ ions:

$$Zn \ (s) \longrightarrow Zn^{2+} \ (aq) + 2 \ e^- \ (metal)$$

The ions move off into the solution, leaving an excess of electrons in the metal strip. The reverse process also occurs. A $Zn^{2+}$ ion that collides with the metal surface may capture two electrons from the metal and be reduced to a neutral zinc atom:

$$Zn^{2+} \ (aq) + 2 \ e^- \ (metal) \longrightarrow Zn \ (s)$$

When a zinc strip is dipped into the solution, the initial rates of these two processes are different. The different rates of reaction lead to a charge imbalance across the metal–solution interface. If the concentration of zinc ions in solution is low enough, the initial rate of oxidation is more rapid than the initial rate of reduction. Under these conditions, excess electrons accumulate in the metal, and excess cationic charges accumulate in the solution. As excess charge builds, however, the rates of reaction change until the rate of reduction is balanced by the rate of oxidation. When this balance is reached, the system is at dynamic equilibrium. Oxidation and reduction continue, but the net rate of exchange is zero:

$$Zn \ (s) \Longleftrightarrow Zn^{2+} \ (aq) + 2 \ e^- \ (metal)$$

**Figure 16-11**

A strip of zinc metal immersed in a solution of zinc sulphate reaches a dynamic equilibrium when a few excess charges occur in each phase and the rates of oxidation and reduction are equal.

Figure 16-11 illustrates the zinc equilibrium at the molecular level. At equilibrium, the charge imbalance in the zinc strip is about one excess electron for every $10^{14}$ zinc atoms. This is negligible from the macroscopic perspective but significant at the molecular level.

An analogous dynamic equilibrium is established when a strip of copper is dipped in a dilute solution of copper sulphate:

$$Cu\ (s) \Longleftrightarrow Cu^{2+}\ (aq)\ +\ 2\ e^-\ (metal)$$

Once again, nothing appears to happen at the macroscopic level. At the molecular level, however, some of the copper atoms lose electrons and enter the solution as $Cu^{2+}$ ions, and some of the $Cu^{2+}$ ions capture electrons from the metal and deposit on the metal as Cu atoms. As with zinc, dipping a strip of copper metal in the solution generates a small charge imbalance.

The charge imbalances for copper and zinc have different values, because zinc is easier to oxidize than copper. Consequently, zinc creates a greater charge imbalance than copper. The concentration of excess electrons in the zinc electrode is greater than the concentration of excess electrons in the copper electrode, giving the zinc electrode more excess charge than the copper electrode.

When two electrodes contain different amounts of excess charge, there is a difference in electrical potential between them. Because it has more excess electrons, the zinc electrode is at a higher electrical potential than the copper electrode. In a galvanic cell, the difference in electrical potential causes electrons to flow from a region where the concentration of electrons is higher to a region where the concentration of electrons is lower. In this case, electrons flow from the zinc electrode toward the copper electrode, as shown at the molecular level in Figure 16-12.

The difference in electrical potential between two electrodes is the **cell potential**, designated $E$ and measured in volts (V). The magnitude of $E$ increases as the amount of charge imbalance between the two electrodes increases. For any galvanic cell, the value of $E$ and the direction of electron flow can be determined experimentally by inserting a voltmeter in the external circuit.

**Figure 16-12**

In a galvanic cell, electrons flow "downhill" from a region of higher electrical potential to a region of lower electrical potential.

Electrons always flow spontaneously "downhill" from higher electrical potential to lower electrical potential. In a galvanic cell, the electrode with the higher potential is designated by convention as the *negative electrode*. The electrode with the lower potential is designated as the *positive electrode*.

## Standard Electrical Potential

Electrochemical cells can be constructed using an almost limitless combination of electrodes and solutions, and each combination generates a specific potential. Keeping track of the electrical potentials of all cells under all possible situations would be extremely tedious without a set of standard reference conditions. By definition, the standard electrical potential is the potential developed by a cell in which all chemical species are present under standard thermodynamic conditions. Recall that standard conditions for thermodynamic properties include concentrations of 1 M for solutes in solution and pressures of 1 bar for gases. Chemists use the same standard conditions for electrochemical properties. As in thermodynamics, standard conditions are designated with a superscript °. A standard electrical potential is thus designated $E°$.

The zinc–copper galvanic cell is under standard conditions when the concentration of each ion is 1.00 M, as shown in Figure 16-13. The cell potential under these conditions can be determined by connecting the electrodes to a voltmeter. The measured potential is 1.10 V, with the Zn electrode at the higher (more negative) potential, so Zn gives up electrons and $E°_{cell} = 1.10$ V:

The cathode of a battery is at lower potential and is labelled with a "+" sign, designating that electrons flow to this terminal from the negative (anode) terminal.

**Figure 16-13**

Diagram of a zinc–copper electrochemical cell operating under standard conditions.

$$Zn\ (s) + Cu^{2+}\ (aq, 1\ M) \longrightarrow Zn^{2+}\ (aq, 1\ M) + Cu\ (s) \qquad E°_{cell} = 1.10\ V$$

A table giving the cell potentials of all possible redox reactions would be immense. Instead, chemists use the fact that any redox reaction can be broken into two distinct half-reactions, an oxidation and a reduction. They assign a potential to every half-reaction and tabulate $E°$ values for all half-reactions. The standard cell potential for any redox reaction can then be obtained by combining the potentials for its two half-reactions.

To do this, one particular half-reaction has to be selected as a reference reaction with zero potential. Once a reference half-reaction has been selected, all other half-reactions can then be assigned values relative to this reference value of 0 V. This is necessary because an experiment always measures the *difference* between two potentials rather than an absolute potential. The standard potential of 1.10 V for the zinc–copper cell, for example, is the difference between the $E°$ values of its two half-reactions.

Chemists have chosen the reduction of hydronium ions to hydrogen gas as the reference half-reaction:

$$2\,H_3O^+\,(aq,\,1\,M) + 2\,e^- \longrightarrow H_2\,(g,\,1\,bar) + 2\,H_2O\,(l) \quad E^\circ = 0\,V \text{ by definition}$$

This standard hydrogen electrode (SHE) is shown schematically in Figure 16-14.

**Figure 16-14**

Schematic diagram of a cell for measuring $E^\circ$ of the $Cu^{2+}/Cu$ half-reaction relative to the standard hydrogen electrode (SHE).

In addition to defined standard conditions and a reference potential, tabulated half-reactions have a defined reference *direction*. Just as the value of an equilibrium constant depends on the direction in which the equilibrium reaction is written, the values of $E^\circ$ depend on whether electrons are reactants or products. For half-reactions, the conventional *reference* direction is *reduction*, with electrons always appearing as *reactants*. Thus, each tabulated $E^\circ$ value for a half-reaction is a **standard reduction potential**.

It is important to remember that every redox reaction includes both an oxidation and a reduction. Half-reactions are tabulated as reductions by convention, but in any real redox process, one half-reaction occurs as a reduction and another occurs as an oxidation.

Defining a reference value for the SHE makes it possible to determine $E^\circ$ values of all other redox half-reactions. As an example, Figure 16-14 shows a cell in which an SHE is connected to a copper electrode in contact with a 1.00 M solution of $Cu^{2+}$. Measurements on this cell show that the SHE is at higher electrical potential than the copper electrode, indicating that electrons flow from the SHE to the Cu electrode. Reduction of $Cu^{2+}$ and oxidation of $H_2$ occur spontaneously to generate this electron flow:

$$Cu^{2+}\,(aq,\,1\,M) + H_2\,(g,\,1\,bar) + 2\,H_2O\,(l) \longrightarrow Cu\,(s) + 2\,H_3O^+\,(aq,\,1\,M)$$

In other words, oxidation takes place at the SHE (anode), and reduction occurs at the copper electrode (cathode).

Measurements show that the copper–hydrogen cell has a voltage of 0.34 V. Because the potential of the SHE is defined to be 0 V, this means that the reduction half-reaction $Cu^{2+}/Cu$ has $E^\circ = +0.34$ V *relative to the SHE*:

$$Cu^{2+}\,(aq,\,1\,M) + 2\,e^- \longrightarrow Cu\,(s) \qquad E^\circ = 0.34\,V \text{ (relative to SHE)}$$

Over the years, chemists have carried out many measurements similar to the one depicted in Figure 16-14. The resulting values for standard reduction potentials are tabulated in reference sources such as the *CRC Handbook of Chemistry and Physics*. Many values appear in Appendix F, and some representative values appear in Table 16-1.

Every standard reduction potential has a specific sign. When a substance is easier to reduce than hydronium ions under standard conditions, its $E^\circ$ is positive. When a substance is more difficult to reduce than hydronium ions under standard conditions, its $E^\circ$ is negative.

## TABLE 16-1 REPRESENTATIVE STANDARD REDUCTION POTENTIALS (AT 25 °C, 1 BAR)

| Reduction Reaction | Standard Reduction Potential, $E°$ (V) |
|---|---|
| $F_2$ (g) + 2 e$^-$ ⟶ 2 F$^-$ (aq) | 2.866 |
| $PbO_2$ + $HSO_4^-$ (aq) + 3 $H_3O^+$ (aq) + 2 e$^-$ ⟶ $PbSO_4$ (s) + 5 $H_2O$ (l) | 1.691 |
| $MnO_4^-$ (aq) + 8 H$^+$ (aq) + 5 e$^-$ ⟶ $Mn^{2+}$ (aq) + 4 $H_2O$ (l) | 1.507 |
| $Au^{3+}$ (aq) + 3 e$^-$ ⟶ Au (s) | 1.498 |
| $Cl_2$ (g) + 2 e$^-$ ⟶ 2 Cl$^-$ (aq) | 1.358 |
| $Cr_2O_7^{2-}$ (aq) + 14 $H_3O^+$ (aq) + 6 e$^-$ ⟶ 2 $Cr^{3+}$ (aq) + 21 $H_2O$ (l) | 1.232 |
| $O_2$ (g) + 4 $H_3O^+$ (aq) + 4 e$^-$ ⟶ 6 $H_2O$ (l) | 1.229 |
| $Pt^{2+}$ (aq) + 2 e$^-$ ⟶ Pt (s) | 1.18 |
| $Br_2$ (l) + 2 e$^-$ ⟶ 2 Br$^-$ (aq) | 1.066 |
| $Hg^{2+}$ (aq) + 2 e$^-$ ⟶ Hg (l) | 0.851 |
| $Ag^+$ (aq) + e$^-$ ⟶ Ag (s) | 0.800 |
| $I_2$ (s) + 2 e$^-$ ⟶ 2 I$^-$ (aq) | 0.536 |
| $Cu^+$ (aq) + e$^-$ ⟶ Cu (s) | 0.521 |
| $O_2$ (g) + 2 $H_2O$ (l) + 4 e$^-$ ⟶ 4 OH$^-$ (aq) | 0.401 |
| $Cu^{2+}$ (aq) + 2 e$^-$ ⟶ Cu (s) | 0.342 |
| AgCl (s) + e$^-$ ⟶ Ag (s) + Cl$^-$ (aq) | 0.222 |
| **2 $H_3O^+$ (aq) + 2 e$^-$ ⟶ $H_2$ (g) + 2 $H_2O$ (l)** | **0.000** |
| $Fe^{3+}$ (aq) + 3 e$^-$ ⟶ Fe (s) | −0.037 |
| $Pb^{2+}$ (aq) + 2 e$^-$ ⟶ Pb (s) | −0.126 |
| $Sn^{2+}$ (aq) + 2 e$^-$ ⟶ Sn (s) | −0.137 |
| $PbSO_4$ (s) + $H_3O^+$ (aq) + 2 e$^-$ ⟶ Pb (s) + $HSO_4^-$ (aq) + $H_2O$ (l) | −0.359 |
| $Fe^{2+}$ (aq) + 2 e$^-$ ⟶ Fe (s) | −0.447 |
| $Ni(OH)_2$ (s) + 2 e$^-$ ⟶ Ni (s) + 2 OH$^-$ (aq) | −0.72 |
| $Cr^{3+}$ (aq) + 3 e$^-$ ⟶ Cr (s) | −0.744 |
| $Zn^{2+}$ (aq) + 2 e$^-$ ⟶ Zn (s) | −0.762 |
| $Cd(OH)_2$ (s) + 2 e$^-$ ⟶ Cd (s) + 2 OH$^-$ (aq) | −0.860 |
| $Al^{3+}$ (aq) + 3 e$^-$ ⟶ Al (s) | −1.662 |
| $Mg^{2+}$ (aq) + 2 e$^-$ ⟶ Mg (s) | −2.37 |
| $Na^+$ (aq) + e$^-$ ⟶ Na (s) | −2.71 |
| $Ca^{2+}$ (aq) + 2 e$^-$ ⟶ Ca (s) | −2.868 |
| $Ca(OH)_2$ (s) + 2 e$^-$ ⟶ Ca (s) + 2 OH$^-$ (aq) | −3.02 |
| $Li^+$ (aq) + e$^-$ ⟶ Li (s) | −3.040 |

## Standard Cell Voltages

In any galvanic cell that is under standard conditions, electrons are produced by the half-reaction with the more negative standard reduction potential and consumed by the half-reaction with the more positive standard reduction potential. In other

**Figure 16-15**

The half-reaction with the more negative $E°$ produces electrons at higher electrical potential through oxidation. The half-reaction with the more positive $E°$ consumes electrons at lower electrical potential through reduction.

words, the half-reaction with the more negative $E°$ value occurs as the oxidation, and the half-reaction with the more positive $E°$ value occurs as the reduction. Figure 16-15 summarizes the conventions used to describe galvanic cells.

Tabulated standard reduction potentials allow us to determine the potential of any cell under standard conditions. This net standard cell potential for a galvanic cell is obtained by *subtracting* the more negative standard reduction potential from the more positive standard reduction potential, giving a positive overall potential. Combining these features gives an equation that summarizes the calculation of the standard potential for a galvanic cell:

$$E°_{cell} = E°_{cathode} - E°_{anode} \qquad (16\text{-}1)$$

Here is the procedure for the copper–hydrogen cell illustrated in Figure 16-14. We write the two half-reactions as reductions:

$$Cu^{2+}\,(aq, 1\,M) + 2\,e^- \longrightarrow Cu\,(s) \qquad\qquad E° = +0.34\,V$$

$$2\,H_3O^+\,(aq, 1\,M) + 2\,e^- \longrightarrow H_2\,(g, 1\,bar) + 2\,H_2O\,(l) \qquad E° = 0.00\,V$$

To obtain the overall reaction, we reverse the half-reaction that has the more negative potential, and we find $E°_{cell}$ by subtracting that $E°$ from the more positive $E°$:

$$H_2\,(g, 1\,bar) + 2\,H_2O\,(l) + Cu^{2+}\,(aq, 1\,M) \longrightarrow 2\,H_3O^+\,(aq, 1\,M) + Cu\,(s)$$

$$E°_{cell} = (E°_{Cu^{2+}/Cu}) - (E°_{H_3O^+/H_2}) = (0.34\,V) - (0.00\,V) = +0.34\,V$$

In this spontaneous redox reaction, oxidation of $H_2$ releases electrons at the platinum anode. Excess electrons in the anode create a negative electrical potential. In the other compartment of the cell, $Cu^2$ ions are reduced to copper metal by capturing electrons from the copper cathode. In the wire connecting the electrodes, electrons flow "downhill" from the region of more negative electrical potential (the anode, $E° = 0$ V) to the region of less negative electrical potential (the cathode, $E° = 0.34$ V).

Consider another galvanic cell, one in which a standard $Zn^{2+}/Zn$ half-reaction combines with the standard hydrogen electrode. According to Table 16-1, $E°$ for the standard $Zn^{2+}/Zn$ half-reaction is $-0.76$ V:

$$Zn^{2+}\,(aq, 1\,M) + 2\,e^- \longrightarrow Zn\,(s) \qquad\qquad E° = -0.762\,V$$

$$2\,H_3O + (aq, 1\,bar) + 2\,e^- \longrightarrow H_2\,(g, 1\,bar) + 2\,H_2O \qquad E° = 0.00\,V$$

The $Zn^{2+}/Zn$ reduction potential is more negative than the $H_3O^+/H_2$ reduction potential ($-0.762$ V vs. 0 V), so zinc is the anode in this cell. Zinc is oxidized and hydronium ions are reduced, causing electrons to flow from the more negative zinc electrode to the less negative SHE. Again, we reverse the direction of the half-reaction with the more negative potential and find $E°_{cell}$ by subtracting the half-cell potentials:

$$2\,H_3O^+\,(aq, 1\,M) + Zn\,(s) \longrightarrow H_2\,(g, 1\,bar) + 2\,H_2O\,(l) + Zn^{2+}\,(aq, 1\,M)$$

$$E°_{cell} = (E°_{H_3O^+/H_2}) - (E°_{Zn^{2+}/Zn}) = (0.00\,V) - (-0.762\,V) = +0.762\,V$$

The overall voltage generated by a standard galvanic cell is always obtained by *subtracting* one standard reduction potential from the other in the way that gives a *positive* value for $E°_{cell}$. Example 16-10 applies this reasoning to zinc and iron.

## EXAMPLE 16-10 STANDARD CELL POTENTIAL

A galvanic cell can be constructed from a zinc electrode immersed in a solution of zinc sulphate and an iron electrode immersed in a solution of iron(II) sulphate. What is the standard potential of this cell, and what is its spontaneous direction under standard conditions?

**Strategy:** First, identify the half-reactions. Then look up the standard reduction potentials in a table, and subtract the more negative value from the more positive value.

**Solution:** Write both half-reactions as reductions, and then look up the $E°$ values in Table 16-1:

$$Zn^{2+} (aq) + 2 e^- \longrightarrow Zn (s) \qquad E°_{Zn^{2+}/Zn} = -0.762 \text{ V}$$

$$Fe^{2+} (aq) + 2 e^- \longrightarrow Fe (s) \qquad E°_{Fe^{2+}/Fe} = -0.447 \text{ V}$$

Now, subtract the more negative value (−0.762 V) from the more positive value (−0.447 V) to obtain the standard cell potential:

$$E°_{cell} = E°_{Fe^{2+}/Fe} - E°_{Zn^{2+}/Zn} = (-0.447 \text{ V}) - (-0.762 \text{ V}) = 0.315 \text{ V}$$

Under standard conditions, the iron–zinc cell operates spontaneously (i.e., is a galvanic cell) in the direction that reduces $Fe^{2+}$ and oxidizes Zn:

$$Fe^{2+} (aq) + Zn (s) \longrightarrow Fe (s) + Zn^{2+} (aq) \qquad E°_{cell} = 0.315 \text{ V}$$

 **Does the Result Make Sense?** The calculation shows that zinc is oxidized preferentially over iron. Later in this chapter, we describe the use of zinc as a sacrificial anode to prevent corrosion of iron.

**Practice Exercise 16.10**

If $Cu/Cu^{2+}$ is used in place of $Zn/Zn^{2+}$ in the above example, what is the standard potential and spontaneous direction of reaction under standard conditions?

## Conventions for Standard Reduction Potentials

Here is a summary of the definitions and conventions for working with electrochemical potentials:

1. Standard conditions for electrochemical half-reactions are the same as those in thermodynamics: 25 °C (298.15 K), 1 M for solutes, and 1 bar for gases.

2. Reduction is the reference direction for electrochemical half-reactions: electrons are reactants.

3. The standard potential for reducing $H_3O^+ (aq)$ to $H_2 (g)$ and $H_2O (l)$ is defined to be 0.00 V.

4. The standard potential for any galvanic cell is determined by subtracting the more negative standard reduction potential from the more positive standard reduction potential. A positive $E°$ indicates spontaneity under standard conditions.

5. In any galvanic cell, the half-reaction with the more negative reduction potential occurs as oxidation at the anode, and the half-reaction with the more positive reduction potential occurs as reduction at the cathode.

Example 16-11 provides another illustration of standard cell potentials.

## EXAMPLE 16-11 HALF-CELL POTENTIALS

The following half-reactions occur during use of the rechargeable nickel–cadmium battery:

$$Cd(OH)_2 (s) + 2 e^- \longrightarrow Cd (s) + 2 OH^- (aq)$$

$$NiO(OH) (s) + H_2O (l) + e^- \longrightarrow Ni(OH)_2 (s) + OH^- (aq)$$

This battery has a potential of 1.35 V under standard conditions, with nickel as the cathode. Determine the net reaction, and use the tabulated standard reduction potential for the cadmium half-reaction to find $E°$ for the nickel half-reaction.

**Strategy:** We need to determine which reaction occurs as reduction and which occurs as oxidation when this battery is operating in its spontaneous direction. Then we can use Equation 16-1 to determine $E°$ for the nickel half-reaction.

**Solution:** The problem states that the nickel electrode is the cathode. Because reduction takes place at the cathode, we know that the nickel half-reaction occurs as reduction. In the oxidation half-reaction, cadmium is oxidized at the anode.

The problem also states that $E°_{cell} = 1.35$ V. The $E°$ value for the cadmium reaction appears in Table 16-1, but we must determine the value for the nickel half-reaction:

$$E°_{cell} = E°_{cathode} - E°_{anode}$$

$$E°_{cell} = E°_{NiO(OH)/Ni(OH)_2} - E°_{Cd(OH)_2/Cd}$$

$$1.35 \text{ V} = E°_{NiO(OH)/Ni(OH)_2} - (-0.860 \text{ V})$$

$$E°_{NiO(OH)/Ni(OH)_2} = (1.35 \text{ V}) + (-0.860 \text{ V}) = +0.49 \text{ V}$$

To balance the net reaction, multiply the nickel reduction half-reaction by 2 to balance the electrons and combine the result with the cadmium oxidation half-reaction. Electrons and hydroxide ions cancel:

$$2 NiO(OH) (s) + 2 H_2O (l) + 2e^- \longrightarrow 2 Ni (OH)_2 (s) + 2OH^- (aq)$$
$$\underline{Cd (s) + 2 OH^- (aq) \longrightarrow Cd(OH)_2 (s) + 2 e^-}$$
$$\text{Net: } 2 NiO(OH) (s) + 2 H_2O (l) \longrightarrow 2 Ni(OH)_2 (s) + Cd(OH)_2 (s)$$

**Does the Result Make Sense?** The relatively simple chemistry of this redox reaction is one reason why nickel–cadmium batteries are rechargeable. As we show later in this chapter, applying an external voltage can reverse this reaction.

### Practice Exercise 16.11

A cell is constructed with indium metal and a 1.00 M solution of $InCl_3$ in one compartment and Cu metal in a 1.00 M solution of $CuSO_4$ in the other compartment. The cell has a voltage of 0.680 V, with the copper electrode being the cathode. Determine the net reaction and the standard potential for the $In/In^{3+}$ half-reaction.

The calculation of $E°$ for this cell illustrates an important feature of cell potentials. A standard cell potential is the difference between two standard reduction potentials. This difference does not change when one half-reaction is multiplied by 2 to cancel electrons in the overall redox reaction. To understand why this is so, recall that cell potentials are analogous to altitude differences for water. Whether 10 000 or 20 000 L of water flows down a spillway, the altitude difference between the top and bottom of the spillway is the same. In the same way, multiplying a reaction by some integer changes the total number of moles of electrons transferred, but it does not change the potential difference through which the electrons are transferred. We return to this point in Section 16.5.

## SECTION EXERCISES

●**16.4.1** Use standard reduction potentials to determine the net reaction and standard cell potential for cells of two compartments, each containing a 1.00 M solution of the indicated cation in contact with an electrode of that neutral metal (i.e., cells similar to the one shown in Figure 16-13): (a) $Fe^{2+}$ and $Cr^{3+}$; (b) $Cu^{2+}$ and $Pb^{2+}$; and (c) $Au^{3+}$ and $Ag^+$.

●**16.4.2** Draw molecular pictures that illustrate the charge transfer process that takes place in the cell in Section Exercise 16.4.1(b).

●**16.4.3** Using the appropriate values from Table 16-1, calculate $E°$ for one cell of a lead storage battery. (Six of these cells are connected in series in an automobile storage battery.)

$$PbO_2\,(s) + Pb\,(s) + 2\,HSO_4^-\,(aq) + 2\,H_3O^+\,(aq) \longrightarrow 2\,PbSO_4\,(s) + 4\,H_2O\,(l)$$

## 16.5 FREE ENERGY AND ELECTROCHEMISTRY

Electrochemically, a spontaneous reaction generates a positive cell potential, $E_{cell}$. Thermodynamically, a spontaneous reaction has a negative change in free energy, $\Delta G$. Thus, a redox reaction that has a negative change in free energy generates a positive cell potential:

$$\Delta G_{reaction} < 0:\qquad \text{Spontaneous redox reaction}$$
$$E_{cell} > 0:\qquad \text{Spontaneous redox reaction}$$

## Cell Potential and Free Energy

The linkage between free energy and cell potentials can be made quantitative. The more negative the value of $\Delta G°$ for a reaction, the more positive its standard cell potential, as the following two examples illustrate:

$$Cu^{2+}\,(aq) + Pb\,(s) \longrightarrow Cu\,(s) + Pb^{2+}\,(aq)\quad \Delta G° = -71.0\ \text{kJ/mol}\quad E°_{cell} = +0.368\ \text{V}$$

$$Pb^{2+}\,(aq) + Sn\,(s) \longrightarrow Pb\,(s) + Sn^{2+}\,(aq)\quad \Delta G° = -2.1\ \text{kJ/mol}\quad E°_{cell} = +0.011\ \text{V}$$

For further insight into how these two quantities are related, consider again the analogy between gravitational and electrical potentials. When water tumbles down a spillway, the total energy change depends on the height through which the water falls and how much water flows. Similarly, when electrons flow in an electrochemical cell, the change in free energy ($\Delta G$) depends on the potential difference ($E$) and on the amount of charge that flows through the cell ($n$, the number of moles of electrons).

To complete the connection between electrical potential and free energy, electrical units must be converted to energy units. The fundamental unit of electrical charge ($e$) is the *coulomb* (C), defined as the quantity of electricity transferred by a current of 1 ampere in 1 second (the ampere and second are fundamental SI units). The charge on a single electron has been measured to a very high degree of accuracy: $1.6021 \times 10^{-19}$ C. The amount of charge provided by 1 mole of electrons is obtained by multiplying the absolute charge on a single electron by Avogadro's number ($N_A$). This quantity, the **Faraday constant ($F$)**, is the link between chemical amounts (moles) and electrical amounts (coulombs):

$$F = e\,N_A = (1.6021 \times 10^{-19}\ \text{C})(6.0221 \times 10^{23}\ \text{mol}^{-1}) = 96\,485\ \text{C/mol}$$

Michael Faraday, shown here delivering a popular lecture to the Royal Institution in 1855, developed many of the fundamental principles of electricity.

Putting these factors together, we find that the molar free energy change is the product of the number of electrons transferred, the Faraday constant, and the electrical potential:

$$\Delta G = -nFE \tag{16-2}$$

The units of $\Delta G$ are J/mol. On the right side of Equation 16-2, the Faraday constant has units of C/mol. Potential differences are in volts, and $1\ \text{J} = 1\ \text{V C}$, so $1\ \text{V} = 1\ \text{J/C}$ and the product $FE$ has units of J/mol. In this equation, $n$ is dimensionless because it is a *ratio*, the number of electrons transferred per atom reacting. Equation 16-2 has a negative sign because a spontaneous reaction has a negative value for $\Delta G$ but a positive value for $E$.

Remember that the number of electrons transferred is not explicitly stated in a net redox equation. This means that any overall redox reaction must be broken down into its balanced half-reactions to determine the value of $n$.

Equation 16-2 applies under all conditions, including standard conditions. At standard conditions, $\Delta G = \Delta G°$ and $E = E°$:

$$\Delta G° = -nFE° \tag{16-3}$$

**Figure 16-16**

A galvanic cell can be constructed from a silver chloride–silver electrode in contact with a solution containing chloride anions, and an iron electrode in contact with a solution containing iron(III) cations.

Equation 16-3 expresses an important link between two standard quantities. The equation lets us calculate standard electrical potentials from tabulated values for standard free energies. Equally important, accurate potential measurements on galvanic cells yield experimental values for standard potentials that can be used to calculate standard free energy changes for reactions.

The cell shown in Figure 16-16 can serve as an example for calculations using Equation 16-3. One cell contains aqueous 1.00 M iron(III) chloride in contact with an iron metal electrode, and the other cell contains 1.00 M KCl in contact with a silver chloride—silver (AgCl/Ag) electrode. The half-reactions for these electrodes follow:

$$AgCl\ (s) + e^- \longrightarrow Ag\ (s) + Cl^-\ (aq) \qquad E° = 0.2223\ V$$
$$Fe^{3+}\ (aq) + 3\ e^- \longrightarrow Fe\ (s) \qquad E° = -0.037\ V$$

In the spontaneous redox reaction under standard conditions, AgCl is reduced and Fe is oxidized:

$$Fe\ (s) + 3\ AgCl\ (s) \longrightarrow Fe^{3+}\ (aq) + 3\ Ag\ (s) + 3\ Cl^-\ (aq)$$
$$E° = (0.222\ V) - (-0.037\ V) = 0.259\ V = 0.259\ J/C$$

Equation 16-3 can be used to calculate the standard free energy change for this net reaction, applying the appropriate conversion factors:

$$\Delta G° = -nFE° = -(3)(9.6485 \times 10^4\ C/mol)(0.259\ J/C)$$
$$= -7.50 \times 10^4\ J/mol$$
$$= -75.0\ kJ/mol$$

## Cell Potentials and Chemical Equilibrium

Equation 16-3 links $\Delta G°$ for a reaction with $E°$. From Chapter 13, Equation 13-3 links $\Delta G°$ for a reaction to $K_{eq}$:

$$\Delta G° = -nFE°$$

$$\Delta G° = -RT \ln K_{eq} \qquad (13\text{-}3)$$

Combining these two equalities and grouping the constants gives a new equation that links standard potentials directly to equilibrium constants:

$$-RT \ln K_{eq} = -nFE°$$

$$E° = \frac{RT}{nF} \ln K_{eq} \qquad (16\text{-}4)$$

Many calculations using Equation 16-4 refer to standard temperature, 298.15 K. Furthermore, $K_{eq}$ often has a very large or very small value that is expressed using power-of-ten notation. For such values, calculations using the logarithm to base 10 (log) rather than ln are more convenient: $\log x = 2.303 \ln x$. We can substitute these values and the value for $F$ and then evaluate the multiplier for the log term at standard temperature:

$$\frac{2.303RT}{F} = \frac{(2.303)(8.314\ J/mol\ K)(298.15\ K)(1\ V\ C/1\ J)}{96\ 485\ C/mol} = 0.0592\ V$$

$$E° = \frac{0.0592\ V}{n} \log K_{eq} \qquad (16\text{-}5)$$

Equations 16-4 and 16-5 provide a method for calculating equilibrium constants from tables of standard reduction potentials. Example 16-12 illustrates the technique.

**EXAMPLE 16-12 CELL POTENTIALS AND EQUILIBRIUM**

Use tabulated standard reduction potentials to determine $K_{eq}$ at standard conditions for the spontaneous redox reaction that has the shorthand notation Cu $(s)$ | $Cu^{2+}$ $(aq)$ || $Br^-$ $(aq)$, $Br_2$ $(l)$ | Pt $(s)$

**Strategy:** This is a quantitative calculation, so it is appropriate to use the seven-step problem-solving strategy. We are asked to determine an equilibrium constant from standard reduction potentials. Visualizing the problem involves breaking the redox reaction into its two half-reactions:

$$Br_2 (l) + 2\ e^- \longrightarrow 2\ Br^- (aq) \quad \text{and} \quad Cu\ (s) \longrightarrow Cu^{2+} (aq) + 2\ e^-$$

Then, we use tabulated values and Equation 16-5 (since the temperature is 25 °C) to calculate $K_{eq}$.

**Solution:**

Table 16-1 provides standard reduction potentials for the two reactions:

$$Br_2 (l) + 2\ e^- \longrightarrow 2\ Br^- (aq) \quad E° = 1.066\ V$$

$$Cu^{2+} (aq) + 2\ e^- \longrightarrow Cu\ (s) \quad\quad E° = 0.342\ V$$

The link between standard reduction potentials and the equilibrium constant is Equation 16-5.

$$E° = \frac{0.0592\ V}{n} \log K_{eq}$$

To obtain the cell potential, subtract the less positive standard reduction potential from the more positive standard reduction potential. The spontaneous direction is reduction of $Br_2$ and oxidation of Cu.

$$E° = E°_{cathode} - E°_{anode} = (1.066\ V) - (0.342\ V) = 0.724\ V$$

Now use Equation 16-5 with the appropriate value of $n$, found from the half-reactions:

$$E° = \frac{0.0592\ V}{n} \log K_{eq}$$

$$\log K_{eq} = \frac{nE°}{0.0592\ V} = \frac{(2)(0.724\ V)}{0.0592\ V} = 24.5$$

$$K_{eq} = 10^{24.5} = 3 \times 10^{24}$$

**Does the Result Make Sense?** The magnitude of this equilibrium constant indicates that the redox reaction goes essentially to completion. This reflects the fact that bromine is a potent oxidizing agent and copper is relatively easy to oxidize. Note that we could have used Equation 16-4 instead of 16-5, but Equation 16-4 is simpler and can be used as long as the temperature is 25 °C.

**Practice Exercise 16.12**

Use values in Table 16-1 to calculate the equilibrium constant for the redox reaction
Pb $(s)$ + $Sn^{2+}$ $(aq)$ $\longrightarrow$ $Pb^{2+}$ $(aq)$ + Sn$(s)$.

## The Nernst Equation

In most laboratories, electrochemistry is practiced under non-standard conditions. That is, concentrations of dissolved solutes often are not 1 M, and gases are not necessarily at 1 bar. Recall from Chapter 11 that $\Delta G$ changes with concentration and pressure. The equation that links $\Delta G°$ with free energy changes under non-standard conditions is Equation 11-13:

$$\Delta G_{reaction} = \Delta G°_{reaction} + RT \ln Q \qquad (11\text{-}13)$$

Here, Q is the reaction quotient.

We can see how concentration and pressure affect cell potentials by substituting for $\Delta G$ and $\Delta G°$ using Equations 16-2 and 16-3.

$$\Delta G = -nFE \qquad \text{and} \qquad \Delta G° = -nFE°$$

$$-nFE = -nFE° + RT \ln Q$$

Dividing both sides by $-nF$ gives the **Nernst equation**:

$$E = E° - \frac{RT}{nF} \ln Q \qquad (16\text{-}6)$$

The Nernst equation is used to convert between standard cell potentials and potentials of electrochemical cells operating under non-standard concentration conditions.

When measurements are made at standard temperature, 298.15 K, the numerical value calculated earlier can replace $RT/F$:

$$E = E° - \frac{0.0592\ \text{V}}{n} \log Q \quad \text{when} \quad T = 298.15\ \text{K}$$

Note that the logarithmic term in this equation is $\log_{10}$ instead of $\ln$. Recall that the numerator of Q contains concentrations of products, and the denominator contains concentrations of reactants, all raised to powers equal to their stoichiometric coefficients. Solvents and pure solids and liquids do not appear in the concentration quotient because their activities are equal to 1. Thus, only solutes and gases appearing in the cell reaction affect its cell potential. Nevertheless, most cell reactions involve solutes, so a typical cell potential differs from $E°$. Examples 16-13 and 16-14 provide illustrations.

## EXAMPLE 16-13 NON-STANDARD CELL CONDITIONS

The permanganate ion is a powerful oxidizing agent that oxidizes water to oxygen under standard conditions. The shorthand notation for this cell is

$$\text{Pt } (s) \,|\, O_2 \,(g),\ H_3O^+ \,(aq)\ ||\ MnO_4^- \,(aq),\ H_3O^+ \,(aq),\ Mn^{2+} \,(aq) \,|\, \text{Pt } (s)$$

What is the potential of a permanganate–oxygen cell operating at $T = 25\ °C$, pH = 7.00, oxygen at $p = 0.200$ bar, 0.100 M $MnO_4^-$, and 0.100 M $Mn^{2+}$?

**Strategy:** This is a quantitative problem, so we follow the standard strategy. The problem asks about an actual potential under non-standard conditions. Before we determine the potential, we must visualize the electrochemical cell and determine the balanced chemical reaction. The half-reactions are deduced from the shorthand notation and balanced in the usual manner. Standard reduction potentials are found in Appendix F:

$$MnO_4^- \,(aq) + 8\ H_3O^+ \,(aq) + 5\ e^- \longrightarrow Mn^{2+} \,(aq) + 12\ H_2O \,(l) \qquad E° = 1.507\ \text{V}$$

$$O_2 \,(g) + 4\ H_3O^+ \,(aq) + 4\ e^- \longrightarrow 6\ H_2O \,(l) \qquad E° = 1.229\ \text{V}$$

To obtain the balanced equation, combine the two half-reactions and multiply each by an integer in order to make the electrons cancel:

$$4\ MnO_4^- \,(aq) + 32\ H_3O^+ \,(aq) + 20\ e^- \longrightarrow Mn^{2+} \,(aq) + 48\ H_2O \,(l)$$

$$\underline{30\ H_2O \,(l) \longrightarrow 5\ O_2 \,(g) + 20\ H_3O^+ \,(aq) + 20\ e^-}$$

$$4\ MnO_4^- \,(aq) + 12\ H_3O^+ \,(aq) \longrightarrow 4\ Mn^{2+} \,(aq) + 5\ O_2 \,(g) + 18\ H_2O \,(l)$$

**Solution:**

Under standard conditions, permanganate is reduced to $Mn^{2+}$ at the cathode, and water is oxidized at the anode:

$$E°_{cell} = E°_{cathode} - E°_{anode} = 1.507\ V - 1.229\ V = 0.278\ V$$

The cell operates at standard temperature, but non-standard concentrations, so use the Nernst equation:

$$E = E° - \frac{0.0592\ V}{n} \log Q$$

The form of the reaction quotient comes from the balanced overall redox equation.

$$Q = \frac{[p_{O_2}]^5 [Mn^{2+}]^4}{[MnO_4^-]^4 [H_3O^+]^{12}}$$

Use the concentrations to evaluate $Q$:

$$Q = \frac{[p_{O_2}]^5 [Mn^{2+}]^4}{[MnO_4^-]^4 [H_3O^+]^{12}} = \frac{(0.200)^5 (0.100)^5}{(0.100)^4 (1.00 \times 10^{-7})^{12}} = 3.20 \times 10^{80}$$

$$\log Q = \log (3.20 \times 10^{80}) = 80.505$$

Substitute into the Nernst equation to find the cell potential. The balanced half-reactions show that $n = 20$:

$$E = E° - \frac{0.0592\ V}{n} \log Q = 0.278\ V - \left(\frac{0.0592\ V}{20}\right)(80.505) = 0.040\ V$$

**Does the Result Make Sense?** The result has three decimal places because the $E°$ value is accurate to three decimal places, and the correction is a subtraction. The low concentration of $H_3O^+$ in pure water reduces this cell potential nearly to zero, but the reaction is still spontaneous, so permanganate can oxidize water. Solutions of potassium permanganate slowly deteriorate and cannot be stored for long times. The reaction is slow enough, however, that significant oxidation does not occur over days or weeks.

**Practice Exercise 16.13**

Determine $E$ for the reaction of Pb with $Sn^{2+}$ described in Practice Exercise 16.12, if $[Sn^{2+}] = 0.255\ M$ and $[Pb^{2+}] = 4.75 \times 10^{-4}\ M$.

## EXAMPLE 16-14 CALCULATING A CONCENTRATION IN AN ELECTROCHEMICAL CELL

In Example 16-13, we considered the cell reaction

$$4\ MnO_4^-\ (aq) + 12\ H_3O^+\ (aq) \longrightarrow 4\ Mn^{2+}\ (aq) + 5\ O_2\ (g) + 18\ H_2O\ (l)$$

for which the standard cell potential is +0.278 V. If this cell is operated at 25 °C, pH = 3.00, 0.200 M $MnO_4^-$, and 0.0500 M $Mn^{2+}$, and the cell potential is found to be 0.200 V, what must be $pO_2$?

**Strategy:** This problem requires calculating a cell potential under non-standard conditions, so we use the Nernst equation. In particular, concentrations and temperature are not standard, so we must use Equation 16-6:

$$E = E° - \frac{RT}{nF} \ln Q$$

and solve this for the partial pressure of oxygen, which is one term in the reaction quotient $Q$.

**Solution:** Rearranging Equation 16-6 to solve for the reaction quotient, we have

$$\ln Q = \frac{-nF(E - E^\circ)}{RT}$$

$$= \frac{-20(96\ 485\ \text{C mol}^{-1})(0.200 - 0.278)\ \text{J C}^{-1}}{8.314\ \text{J K}^{-1}\ \text{mol}^{-1}\ (25 + 273.15)\ \text{K}}$$

$$= 60.72$$

$$Q = e^{60.72} = 2.35 \times 10^{26}$$

$$\text{but, } Q = \frac{[p_{O_2}]^5\ [\text{Mn}^{2+}]^4}{[\text{MnO}_4^-]^4\ [\text{H}_3\text{O}^+]^{12}}$$

$$\text{thus, } [p_{O_2}] = \left(\frac{Q[\text{MnO}_4^-]^4\ [\text{H}_3\text{O}^+]^{12}}{[\text{Mn}^{2+}]^4}\right)^{1/5} = \left(\frac{2.35 \times 10^{26}(0.200)^4(10^{-3.00})^{12}}{(0.0500)^4}\right)^{1/5} = 0.036\ \text{bar}$$

**Does the Result Make Sense?** The temperature and concentrations of all species are different from those in Example 16-13, and each will have either a positive or negative effect on the cell potential. However, the values have not changed substantially, so a partial pressure of oxygen of 0.036 bar seems reasonable.

**Practice Exercise 16.14**

For the reaction $S_4O_6^{2-}\ (aq) + Cr^{2+}\ (aq) \longrightarrow S_2O_3^{2-}\ (aq) + Cr^{3+}\ (aq)$, the cell potential is +0.479 V when everything (including the temperature) is at standard state conditions except $[S_4O_6^{2-}\ (aq)]$. Calculate $[S_4O_6^{2-}\ (aq)]$. Refer to Appendix F for standard reduction potentials.

## The pH Meter

As the Nernst equation describes, cell potentials are linked quantitatively to concentrations. One practical consequence of this relationship is that potential measurements can be used to determine concentrations of ions in solution. The most common example is the **pH meter**, which scientists use to measure the acidity of aqueous solutions. Figure 16-17a illustrates this device: a probe is dipped into the sample, and the pH appears on a digital display. A pH meter measures the electrical potential difference between the sample and a reference solution containing acid at known concentration.

Contemporary pH meters use single probes that contain two electrodes, shown diagrammatically in Figure 16-17b. One electrode contains a buffer solution of known pH. A glass membrane separates this buffer solution from the solution whose pH is to be measured, so this electrode is called a glass electrode. The overall cell potential depends on the hydronium ion concentration in the solution whose pH is being measured, because the difference in hydronium ion concentrations causes a potential difference across this thin glass membrane. The second electrode serves as a reference electrode, and has a precisely known potential. This electrode contains $Hg_2Cl_2$ in contact with liquid mercury and aqueous KCl. The reactions occurring in this electrode are as follows:

Oxidation: $2 \, Ag \, (s) + 2 \, Cl^- \, (aq) \longrightarrow 2 \, AgCl \, (s) + 2 \, e^-$  $\quad E° = -0.22 \, V$

Reduction: $Hg_2Cl_2 \, (s) + 2 \, e^- \quad \longrightarrow 2 \, Hg \, (l) + 2 \, Cl^- \, (aq)$  $\quad E° = +0.28 \, V$

Overall: $\quad 2 \, Ag \, (s) + Hg_2Cl_2 \, (s) \longrightarrow 2 \, AgCl \, (s) + 2 \, Hg \, (l)$  $\quad E° = +0.06 \, V$

These pH electrodes are simple to use and maintain. They respond selectively to hydronium ion concentration and provide accurate measurements of pH values between about 0 and 10. They can be small enough to be implanted into blood vessels or even inserted into individual living cells. In precision work, these electrodes are calibrated before each use, because their characteristics change somewhat with time and exposure to solutions. The electrode is dipped into a buffer solution of known pH, and the meter is electronically adjusted until it reads the correct value.

**Figure 16-17**

(a)

Ag/AgCl electrode

Silver wire coated with AgCl

HCl (aq)

Glass membrane

Hg₂Cl₂/Hg electrode

Platinum wire

Hg₂Cl₂, Hg, and KCl

Saturated KCl (aq)

Porous membrane

(b)

(a) Modern pH meters are compact digital electronic devices. (b) The probe consists of two electrodes, the compositions of which are shown in the diagram.

## Electrochemical Stoichiometry

The coefficients of any balanced redox equation describe the stoichiometric ratios between chemical species, just as for other balanced chemical equations. Additionally, in redox reactions, we can relate moles of chemical change to moles of electrons. Because electrons always cancel in a balanced redox equation, however, we need to look at half-reactions to determine the stoichiometric coefficients for the electrons. A balanced half-reaction provides the stoichiometric coefficients needed to compute the number of moles of electrons transferred for every mole of reagent.

Electricity is normally measured in units of charge, the coulomb (C), or as rate of electrical current flow, the ampere (A; $1 \, A = 1 \, C/s$). The total amount of charge is the product of the current, symbolized by $I$, and the time ($t$) for which this current flows:

$$\text{Charge} = It$$

Just as molar mass provides the link between mass and moles, the Faraday constant provides the link between charge and moles. The number of moles of electrons transferred in a specific amount of time is the charge in coulombs divided by the charge per mole, $F$:

$$\text{Moles of electrons} = n = \frac{It}{F} \tag{16-7}$$

$$I = \text{Current (C/s)}, \, t = \text{Time (s)}, \, \text{and} \, F = 96 \, 485 \, C/mol$$

Here, $n$ refers to moles of electrons reacting and has units, whereas in Equations 16-3 through 16-6, $n$ is a dimensionless ratio that expresses the coefficient for the electrons involved in a balanced chemical reaction.

Equation 16-7 links the stoichiometry of a redox reaction with the characteristics of an electrochemical cell, and Example 16-15 shows how to apply this equation.

## EXAMPLE 16-15 ELECTROCHEMICAL STOICHIOMETRY

Automobile headlights typically draw 5.9 A of current. The galvanic cell of a lead storage battery, described in Example 16-8, consumes Pb and $PbO_2$ as it operates. A typical battery contains about 250 g of $PbO_2$. Assuming that the battery can supply 5.9 A of current until all the $PbO_2$ has been consumed, how long will it take for a battery to run down if the lights are left on after the engine is turned off?

**Strategy:** This is an electrochemical stoichiometry problem, in which an amount of a chemical substance is consumed as electrical current flows. Current flows as long as there is lead(IV) oxide present to accept electrons, and the battery dies when all the lead(IV) oxide is consumed. We need to have a balanced half-reaction to provide the stoichiometric relationship between moles of electrons and moles of $PbO_2$.

**Solution:** There is 250 g of $PbO_2$, and the headlights draw 5.9 A of current. Equation 16-8 links current with moles of electrons. Moles of electrons and moles of $PbO_2$ are related as described by the balanced half-reaction, determined in Example 16-8:

$$PbO_2\,(s) + HSO_4^-\,(aq) + 3\,H_3O^+\,(aq) + 2\,e^- \longrightarrow PbSO_4\,(s) + 5\,H_2O\,(l)$$

This gives us the stoichiometric ratio needed to relate moles of electrons to moles of chemical substance:

$$n_{e^-} = n_{PbO_2}\left(\frac{2\ mol\ e^-}{1\ mol\ PbO_2}\right)$$

Use the molar mass of $PbO_2$ to convert mass of $PbO_2$ to moles:

$$n_{PbO_2} = \frac{m_{PbO_2}}{M_{PbO_2}} = \frac{250\ g}{239.2\ g/mol} = 1.05\ mol\ PbO_2$$

$$n_{e^-} = \left(\frac{2\ mol\ e^-}{1\ mol\ PbO_2}\right)(1.05\ mol\ PbO_2) = 2.10\ mol\ electrons$$

We rearrange Equation 16-8, which provides the link between moles of electrons and the time it takes for the battery to become discharged:

$$n_{e^-} = \frac{It}{F} \quad or \quad t = \frac{n_{e^-}F}{I}$$

$$t = \frac{(2.10\ mol)(96\ 485\ C/mol)}{(5.9\ C/s)(60\ s/min)(60\ min/h)} = 9.5\ hours$$

This is the maximum time that the battery could supply current. In practice, the battery runs down before all the $PbO_2$ is consumed, because the voltage drops as the battery operates. This can be understood from the Nernst equation, which predicts that the potential is a function of the concentrations of reactants and products.

**Does the Result Make Sense?** If you leave your automobile headlights on for several hours when the engine is off, you will have a dead battery, as this calculation predicts. When an automobile engine is running, however, the engine drives an electrical alternator that recharges the battery and provides the current required for the headlights and other needs.

**Practice Exercise 16.15**

The reaction in a nickel–cadmium battery is

$$2\ NiO(OH)\ (s) + 2\ H_2O\ (l) + Cd\ (s) \longrightarrow 2\ Ni(OH)_2\ (s) + Cd(OH)_2\ (s)$$

If a battery contains 0.355 g of Cd and excess NiO(OH), how long can it operate if it draws 1.25 mA of current?

Redox reactions may involve solids, solutes, gases, or charge flows. Consequently, you must be prepared for all the various conversions from molar amounts to measurable variables. As a reminder, Table 16-2 lists the four relationships used for mole calculations.

Galvanic cells use redox reactions to generate electrical current. Electrical current can also drive redox reactions, and the same stoichiometric relationships apply to such processes, as we describe in Section 16.7.

## TABLE 16-2 RELATIONSHIPS USED TO CALCULATE MOLES

| Type of Material | Equation |
| --- | --- |
| Pure substance | $n = m/M$ |
| Solutes | $n = cV$ |
| Gases | $n = pV/RT$ |
| Electrons | $n = It/F$ |

## SECTION EXERCISES

●**16.5.1** Using standard reduction potentials, determine $\Delta G°$ and $K_{eq}$ at 25 °C for each of the following reactions:

(a) $2\ Cu\ (s) + Hg^{2+}\ (aq) \longrightarrow 2\ Cu^+\ (aq) + Hg\ (l)$

(b) $4\ Cr^{3+}\ (aq) + 21\ H_2O\ (l) \longrightarrow 2\ Cr\ (s) + Cr_2O_7^{2-}\ (aq) + 14\ H_3O^+\ (aq)$

●**16.5.2** Use the Nernst equation to determine the potential at 298 K for each of the cells described in Section Exercise 16.4.1, if all cation concentrations are $2.5 \times 10^{-2}$ M rather than 1.00 M: (a) $Fe^{2+}$ and $Cr^{3+}$; (b) $Cu^{2+}$ and $Pb^{2+}$; and (c) $Au^{3+}$ and $Ag^+$.

●**16.5.3** In one form of lithium battery, the spontaneous cell reaction is

$$4\ Li\ (s) + FeS_2\ (s) \longrightarrow Fe\ (s) + 2\ Li_2S\ (s)$$

Suppose that a lithium battery contains 250 mg each of Li and $FeS_2$. (a) Which reagent will be consumed first? (b) How many coulombs of charge can the battery deliver before this reagent is consumed? (c) What is the lifetime of the battery (in days) if it powers a cell phone requiring 1.50 mA of current?

## 16.6 REDOX IN ACTION

Electrochemical reactions have many practical applications. Some are spontaneous, and others are driven "uphill" by applying an external potential. In this section, we present practical examples of spontaneous redox processes. We describe externally driven redox reactions in Section 16.7.

## Batteries

A battery is a galvanic cell that generates electrical current to power a practical device. Batteries can be as small as the "buttons" that power cameras and hearing aids or large charge storage banks like those of electric automobiles. A battery must use cell reactions that generate and maintain a large electrical potential difference. This requires two half-reactions with substantially different standard reduction potentials. The ideal battery would be compact, lightweight, inexpensive, indefinitely rechargeable, and environmentally safe. This is a stringent set of requirements. No battery meets all of them, and only a few come close.

Batteries power hearing aids as well as automobiles.

A flashlight battery usually is an **alkaline dry cell**, so called because it is basic (alkaline) but contains a minimum amount of water (dry). In these batteries, zinc and manganese dioxide are the working materials (see the schematic diagram in Figure 16-18). The dry cell has a zinc anode in contact with a moist paste of $MnO_2$, KOH, and graphite powder. The paste contacts a passive graphite cathode. Both half-reactions involve multiple steps, but the chemistry can be approximated by single half-reactions:

Cathode: $\quad 2\,MnO_2\,(s)\,+\,H_2O\,(l)\,+\,2\,e^- \longrightarrow\, Mn_2O_3\,(s)\,+\,2\,OH^-$

Anode: $\quad\quad\quad Zn\,(s)\,+\,2\,OH^-\,(paste) \longrightarrow\, Zn(OH)_2\,(s)\,+\,2\,e^-$

Net: $\quad 2\,MnO_2\,(s)\,+\,H_2O\,(l)\,+\,Zn\,(s) \longrightarrow\, Mn_2O_3\,(s)\,+\,Zn(OH)_2\,(s)$

**Figure 16-18**

Graphite cathode

Paste of $MnO_2$, KOH, and graphite powder

Porous separator

Zinc anode

Schematic diagram and examples of alkaline dry cells. These batteries provide electrical current for many portable devices.

A dry cell generates a potential of about 1.5 V. These cells run irreversibly, and their cell potential slowly decreases with time. Nevertheless, dry cells have many uses because they are compact and made of inexpensive materials with low toxicities. Approximately 4 billion of these batteries are produced annually worldwide.

The zinc–air battery is more expensive than the dry cell and deteriorates relatively quickly once it is exposed to air. High capacity and a cell potential that does not vary with use offset these disadvantages. Like the dry cell, a zinc–air battery uses zinc for the anode reaction. Uniquely among batteries in common use, this battery relies on molecular oxygen from the atmosphere for its cathode reaction.

Cathode: $O_2 (g) + 2 H_2O (l) + 4 e^- \longrightarrow 4 OH^- (aq)$

Anode:　　　$2 Zn (s) + 4 OH^-(aq) \longrightarrow 2 ZnO (s) + 2 H_2O (l) + 4 e^-$

Net:　　　　$O_2(g) + 2 Zn (s) \longrightarrow 2 ZnO (s)$

Figure 16-19 shows the molecular process that occur in a zinc–air battery. In this cell, none of the components of the redox reaction changes concentration as the battery operates. Both zinc-containing substances are solids, and atmospheric air provides a constant supply of $O_2$ gas through a small hole in the battery casing. As the balanced redox reaction indicates, there is no net change in concentration of hydroxide ions, which are consumed at the cathode but regenerated at the anode. The aqueous zinc paste allows free diffusion of hydroxide ions from the anode to the cathode. Because none of the concentrations change as the battery functions, this battery delivers a constant voltage of approximately 1.4 V throughout its lifetime. Zinc–air batteries are used widely in devices that require constant voltage, including watches, calculators, hearing aids and cameras.

The **lead storage battery** provides electrical power in automobiles. It is well suited for this use because it supplies the large current needed to drive starter motors

**Figure 16-19**

In a zinc–air battery, zinc is oxidized and molecular oxygen is reduced, but no net change occurs in the activities of any species in solution. The migration of OH⁻ **through the zinc paste carries current and maintains a uniform concentration.**

**Figure 16-20**

Schematic view of a 6-V lead storage battery. Lead is oxidized and lead(IV)oxide is reduced during the operation of this battery, which contains three cells connected in series.

and headlights and can be recharged easily. Figure 16-20 shows the lead storage cell in a schematic view. The half-reactions are the subject of Example 16-8:

$$PbO_2\,(s) + HSO_4^-(aq) + 3\,H_3O^+\,(aq) + 2\,e^- \longrightarrow PbSO_4\,(s) + 5\,H_2O\,(l)$$
$$E° = +1.6913\text{ V}$$

$$Pb\,(s) + HSO_4^-\,(aq) + H_2O\,(l) \longrightarrow PbSO_4\,(s) + H_3O^+\,(aq) + 2\,e^-$$
$$E° = -0.3588\text{ V}$$

The net reaction is

$$Pb\,(s) + PbO_2\,(s) + 2\,HSO_4^-\,(aq) + 2\,H_3O^+(aq) \longrightarrow 2\,PbSO_4\,(s) + 4\,H_2O\,(l)$$

The standard cell potential for the lead acid battery is
$$E° = (1.6913) - (-0.3588) = 2.0501\text{ V}.$$

The anodes are lead plates that become coated with $PbSO_4$ as the battery discharges. The cathodes are lead containing $PbO_2$, which also become coated with $PbSO_4$ as the battery discharges. Both sets of electrodes are immersed in aqueous sulphuric acid that is about 5 M. Although the concentrations are not standard, the working potential of the cell is close to its $E°$ value of about 2 V. Automobile batteries use six such cells, connected in series to generate a total electrical potential of 12 V. This battery does not require a porous separator between its electrodes because the starting materials and products for both half-reactions are insoluble in sulphuric acid. The lead-containing materials never contact each other, so there is no possibility of direct electron transfer. Note also that the only species that have concentration changes as the battery is discharged are $HSO_4^-$ $(aq)$ and $H_3O^+$ $(aq)$.

Any battery has a finite life. When its chemicals are consumed, it no longer generates electricity and the battery is "dead." The lead storage battery can be recharged by using electricity to run its reactions in the opposite direction. Lead sulphate produced by the redox reaction adheres to the electrodes of a lead storage battery, so the reactions are easily reversed. While the auto is running, its motor drives an electrical generator, known as an alternator, which meets the vehicle's electrical needs. The alternator also recharges the battery by supplying the energy needed to drive its redox reactions in the reverse direction (see Section 16.7). When the electrical system is functioning properly, the battery is maintained at its optimum charge level, giving these batteries long lifetimes. This process does not work for dry cells or zinc–air batteries, because applying a reverse voltage to these batteries causes other reactions to occur.

The **nickel–cadmium battery** is another common battery that can be recharged. As described in Example 16-11, nickel and cadmium are the working substances in this battery:

$$\text{Cathode: } NiO(OH)(s) + H_2O(l) + e^- \longrightarrow Ni(OH)_2(s) + OH^-(aq)$$

$$\underline{\text{Anode: } Cd(s) + 2\,OH^-(aq) \longrightarrow Cd(OH)_2(s) + 2\,e^-}$$

$$\text{Net: } 2\,NiO(OH)(s) + Cd(s) + 2\,H_2O(l) \longrightarrow 2\,Ni(OH)_2(s) + Cd(OH)_2(s)$$

$$E^\circ = 1.35\,V$$

Nickel–cadmium batteries are rechargeable because the nickel and cadmium hydroxides products adhere tightly to the electrodes. More than 1.5 billion Ni–Cd batteries are produced every year. These batteries are not without drawbacks, however. Cadmium is a toxic heavy metal that adds weight to the battery. Moreover, even rechargeable batteries degrade eventually and must be recycled or disposed of in a benign way. Sales of Ni–Cd batteries are therefore declining.

**Lithium-ion batteries** are gradually replacing nickel–cadmium batteries in most portable applications. They have a very high power to weight ratio, partly because they are lightweight and partly because their potential is so high (approximately 3 V per cell). In a lithium-ion battery, the anode is typically made of graphite. The cathode may be made from a variety of materials, but a popular choice is lithium manganese oxide. Lithium ions are *intercalated* (dissolved) into both the anode and cathode materials. During discharge, lithium ions move from the anode to the cathode, and during recharging they move in the opposite direction. The chemistry during discharge is

$$\text{Oxidation: } \qquad\qquad\qquad Li_xC_6 \longrightarrow x\,Li^+ + x\,e^- + 6\,C$$
$$\text{Reduction: } Li_{1-x}MnO_2(s) + x\,Li^+ + x\,e^- \longrightarrow LiMnO_2$$

**Nickel metal hyride** batteries employ the same cathode reaction as in nickel-cadmium batteries. The anode reaction, however, involves the oxidation of hydrogen atoms that are initially dissolved in the anode metal, M:

$$MH + OH^- \longrightarrow M + H_2O + e^-$$

The metal used is actually an intermetallic compound composed of two different metals. The first is either nickel, cobalt, aluminum or maganese, and the second is one of lanthanum, cerium, neodymium or praseodymium. Nickel metal hyride batteries have 2 or 3 times the power generation capacity of nickel-cadmium batteries, but the capacity of the lithium ion battrey is higher still.

Another development in batteries is the fuel cell, which is described in our Focus 16-2: Chemistry and Technology.

## Corrosion

**Corrosion** is a natural redox process that returns refined metals to their more stable metal oxides. The oxidizing agent in corrosion chemistry is atmospheric oxygen. Oxygen is a potent oxidizing agent, particularly in the presence of aqueous acids:

$$O_2\,(g) + 2\,H_2O\,(l) + 4\,e^- \longrightarrow 4\,OH^-\,(aq) \qquad E° = +0.401\,V$$

$$O_2\,(g) + 4\,H_3O^+\,(aq) + 4\,e^- \longrightarrow 6\,H_2O\,(l) \qquad E° = +1.229\,V$$

Atmospheric $O_2$ has a partial pressure of 0.21 bar, and atmospheric water vapour is saturated with carbon dioxide. This dissolved $CO_2$ forms carbonic acid, which generates a hydronium ion concentration of about $2.0 \times 10^{-6}$ M. The Nernst equation allows calculation of the half-cell potential for the reduction of $O_2\,(g)$ at 25 °C under these conditions:

$$E = E° - \frac{0.0592}{n}\log\left(\frac{1}{pO_2[H_3O^+_{(aq)}]^4}\right)$$

$$E = 1.229\,V - \frac{0.0592\,V}{4}\log\left(\frac{1}{(0.20)(2.0 \times 10^{-6})^4}\right) = 0.88\,V$$

Since this half-reaction potential is positive, common terrestrial conditions promote corrosion.

In damp air, therefore, materials with standard reduction potentials of less than 0.88 V oxidize spontaneously. Atmospheric $O_2$ easily oxidizes iron and aluminum, the most important structural metals:

$$4\,Fe\,(s) + 3\,O_2\,(g) + 12\,H_3O^+\,(aq) \longrightarrow 4\,Fe^{3+}\,(aq) + 18\,H_2O\,(l)$$
$$E°_{cell} = (0.88\,V) - (-0.037\,V) = +0.92\,V$$

$$4\,Al\,(s) + 3\,O_2\,(g) + 12\,H_3O^+\,(aq) \longrightarrow 4\,Al^{3+}\,(aq) + 18\,H_2O\,(l)$$
$$E°_{cell} = (0.88\,V) - (-1.662\,V) = +2.54\,V$$

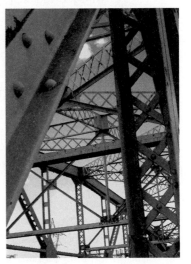

Preventing or repairing corrosion costs billions of dollars per year.

### FOCUS 16-2 CHEMISTRY AND TECHNOLOGY: FUEL CELLS

Imagine an automobile that runs in silence and without polluting emissions. Such an automobile, long a dream of the environmentally conscious, has recently become a reality. The power source is a fuel cell, an electrochemical cell that uses a redox reaction to produce electricity. Hydrocarbons such as natural gas and propane can be used in fuel cells, but the "cleanest" fuel is molecular hydrogen.

The potential advantage of a fuel cell over conventional power sources is the use of hydrogen as a fuel, with the result that the only waste product is water. Just as we now pull into a service station to refill the gas tank, owners of automobiles powered by fuel cells will likely refill their fuel tanks with hydrogen.

Although the principle was first proposed in 1839, making a practical fuel cell eluded scientists for a century and a half. The concept is simple, but the chemistry is difficult. A hydrogen fuel cell must oxidize $H_2$ at the anode and cleanly reduce $O_2$ at the cathode. Ballard Power Systems in Burnaby, British Columbia is a world leader in the development of fuel cell technology for both stationary and mobile systems. In a Ballard hydrogen fuel cell, oxidation of $H_2$ at the anode releases electrons into the device being powered (such as an electric motor) and produces protons. Reduction of $O_2$ at the cathode consumes these protons, and the electrons returning from the motor, and generates water, the only waste product. The reactions are thus very simple:

Anode:                                    $2 H_2(g) \longrightarrow 4 H^+ + 4 e^-$

Cathode:      $O_2(g) + 4 H^+ + 4 e^- \longrightarrow 2 H_2O$

Net:                       $2 H_2 + O_2(g) \longrightarrow 2 H_2O$        $E° = 1.229$ V

As the diagram at right shows, the heart of the Ballard fuel cell is the proton exchange membrane. This is a polymer material that is permeable to protons. Protons generated at the anode diffuse through the membrane to the cathode, where they are reduced to water. The cathode and anode are integral parts of the membrane, each is a layer of porous platinum metal, which serves as an electrode as well as a catalytic surface for the chemical reactions taking place.

A hydrogen fuel cell is environmentally friendly, but $H_2$ is much more difficult to store than liquid fuels. The production, distribution, and storage of hydrogen present major difficulties, so researchers are also working on fuel cells that use liquid hydrocarbon fuels. One such fuel cell is composed of layers of yttria-stabilized zirconia (YSZ), which is solid $ZrO_2$ containing around 5% $Y_2O_3$. This cell uses the combustion of a hydrocarbon as its energy source:

Anode:      $2(C_4H_{10} + 13 O^{2-} \longrightarrow 4 CO_2 + 5 H_2O + 26 e^-)$

Cathode:           $13(O_2 + 4 e^- \longrightarrow 2 O^{2-})$

Net:      $2 C_4H_{10} + 13 O_2 \longrightarrow 8 CO_2 + 10 H_2O$

As the reaction shows, a hydrocarbon fuel cell produces carbon dioxide, a "greenhouse" gas. These fuel cells would nevertheless be less polluting than internal combustion engines, because fuel cells are more efficient and do not generate polluting by-products such as CO and $NO_x$.

Alkali fuel cells containing KOH and platinum- and gold-coated electrodes were developed for the space program, but these have been too expensive for down-to-Earth vehicles. In addition, these cells require pure oxygen rather than $CO_2$-containing air. However, two new technologies have reduced the cost of alkali fuel cells to the point where a European company markets taxis that use them. One is

the use of $CO_2$ "scrubbers" to purify the air supply, making it possible to use atmospheric $O_2$ rather than purified oxygen. The other is the development of ultrathin films of platinum so that a tiny mass of this expensive metal can provide the catalytic surface area needed for efficient fuel cell operation.

Automobile manufacturers have invested heavily in fuel cells. Buses powered by fuel cells are on the road, and prototype cars are being tested. The chances are good that there is a fuel cell car in your future.

**Figure 16-21**

Oxidation: $Fe \longrightarrow Fe^{2+} + 2\,e^-$

Reduction: $O_2 + 4\,H_3O^+(aq) + 4\,e^- \longrightarrow 6\,H_2O$

Further Redox: $2\,Fe^{2+} + 6\,H_2O + \frac{1}{2}O_2 \longrightarrow Fe_2O_3 + 4\,H_3O^+$

A water droplet on an iron surface is a miniature electrochemical cell that corrodes iron metal and generates iron oxide (rust).

Both $Fe^{3+}$ and $Al^{3+}$ form insoluble oxides that precipitate from solution. Overall, $O_2$ oxidizes each metal to give its metal oxide:

$$4\,Fe\ (s) + 3\,O_2\ (g) \longrightarrow 2\,Fe_2O_3\ (s)$$

$$4\,Al\ (s) + 3\,O_2\ (g) \longrightarrow 2\,Al_2O_3\ (s)$$

The formation of rust on a steel surface is an obvious manifestation of corrosion. Even more damaging effects of corrosion take place beneath the surface of the metal, where tiny cracks weaken the metal.

The corrosion of iron occurs particularly rapidly when an aqueous solution is present. This is because water that contains ions provides an oxidation pathway with an activation energy that is much lower than the activation energy for the direct reaction of iron with oxygen gas. As illustrated schematically in Figure 16-21, oxidation and reduction occur at different locations on the metal surface. In the absence of dissolved ions to act as charge carriers, a complete electrical circuit is missing, so the redox reaction is slow. In contrast, when dissolved ions are present, such as in salt water and acidic water, corrosion can be quite rapid.

Various strategies are employed to prevent corrosion. The use of paint as a protective coating is described in our chapter introduction. A metal surface can also be protected by coating it with a thin film of a second metal. When the second metal is easier to oxidize than the first, the process is *galvanization*. Objects made of iron, including automobile bodies and steel girders, are dipped in molten zinc to provide sacrificial coatings. If a scratch penetrates the zinc film, the iron is still protected because zinc oxidizes preferentially:

$$Fe \longrightarrow Fe^{2+} + 2\,e^- \qquad E° = +0.447\ \text{V}$$

$$Zn \longrightarrow Zn^{2+} + 2\,e^- \qquad E° = +0.762\ \text{V}$$

Many iron objects, such as the hulls of ships or oil drilling platforms, are too large to dip in pots of molten zinc. Instead, these objects are protected from corrosion by connecting them to blocks of some more easily oxidized metal. Magnesium and zinc are frequently used for this purpose. As in galvanization, the more active metal is oxidized preferentially. The sacrificial metal must be replaced as it is consumed, but the cost of replacing a block of zinc or magnesium is minuscule compared with the cost of replacing a sophisticated iron structure.

Another way to protect a metal uses an impervious metal oxide layer. This process is known as *passivation*. In some cases, passivation is a natural process. Aluminum oxidizes readily in air, but the result of oxidation is a thin protective layer of $Al_2O_3$ through which $O_2$ cannot readily penetrate. Aluminum oxide adheres to the surface of unoxidized aluminum, protecting the metal from further reaction with $O_2$. Passivation is not effective for iron, because iron oxide is porous and does not adhere well to the metal. Rust continually flakes off the surface of the metal, exposing fresh iron to the atmosphere. Alloying iron with nickel or chromium, the oxides of which adhere well to, and therefore protect, metal surfaces, can be used to prevent corrosion. For example, stainless steel contains as much as 17% chromium and 10% nickel, the oxides of which adhere to the metal surface and prevent corrosion.

## SECTION EXERCISES

⬢**16.6.1** Mercury batteries have the following cathode reaction:

$$HgO\ (s) + H_2O\ (l) + 2\ e^- \longrightarrow Hg\ (l) + 2\ OH^-\ (aq)$$

Draw molecular pictures to illustrate this process.

⬢**16.6.2** In Section Exercise 16.3.1, an aluminum–iron galvanic cell was proposed. Using information presented in this section, explain why this cell is unlikely to make a good battery.

⬢**16.6.3** "Tin" cans are made of iron, coated with a thin film of tin. After a break occurs in the film, a tin can corrodes much more rapidly than zinc-coated iron. Using standard reduction potentials, explain this behaviour.

## 16.7 ELECTROLYSIS

In a galvanic cell, a spontaneous chemical reaction generates an electrical current. It is also possible to use an electrical current to drive a non-spontaneous chemical reaction. The recharging of a dead battery uses an external electrical current to drive the battery reaction in the reverse, or "uphill," direction.

An electrochemical reaction can be driven in the non-spontaneous direction only if the external electrical power source generates a larger potential than the potential developed by the cell. Compare this relationship with the flow of water under the influence of gravity. The spontaneous direction for the flow of water is "downhill," from higher to lower gravitational potential. Nevertheless, we can pump water uphill–in the "non-spontaneous" direction–if we exert an opposing force that is greater than the force of gravity. Similarly, we can reverse the direction of a galvanic cell by imposing an external potential to drive electrons "uphill."

In the presence of an opposing voltage that is greater than its cell potential, the cell reaction is reversed. This is an **electrolytic cell** rather than a galvanic cell. The process of using electrical current to drive redox reactions is **electrolysis**.

Electrolysis can be illustrated by the recharging of the lead storage battery. This electrolytic process is shown in Figure 16-22. As a galvanic cell, this battery oxidizes lead metal and reduces lead(IV) oxide spontaneously, generating an electrical potential of 2.05 V. Electrons flow from the Pb electrode to the $PbO_2$/Pb electrode. The reaction can be driven in the opposite direction by an external potential of *opposite sign and greater than 2.05 V.* Under these conditions, the external potential drives electrons through the external wire from $PbO_2$ toward Pb. At the $PbSO_4$–$PbO_2$/Pb interface, lead sulphate is oxidized to lead oxide:

**Figure 16-22**

A lead storage battery is recharged by forcing electrons to flow in the opposite direction from galvanic operation.

$$PbSO_4\,(s) + 5\,H_2O\,(l) \longrightarrow PbO_2\,(s) + HSO_4^-\,(aq) + 3\,H_3O^+\,(aq) + 2\,e^-$$

The electrons released in this reaction are pushed into the lead electrode, and at the $PbSO_4$–Pb interface, lead sulphate captures electrons and is reduced to lead:

$$PbSO_4\,(s) + H_3O^+\,(aq) + 2\,e^- \longrightarrow Pb\,(s) + HSO_4^-\,(aq) + H_2O\,(l)$$

Whether the cell operates galvanically or electrolytically, the electrode where oxidation occurs is always called the anode. Thus, under recharging conditions, the $PbO_2$/Pb electrode is the anode, and the Pb electrode is the cathode.

## Electrolysis of Water

As world deposits of petroleum and coal are exhausted, new sources of hydrogen will have to be developed for use as a fuel and in the production of ammonia for fertilizer. At present, most hydrogen gas is produced from hydrocarbons, but hydrogen gas can also be generated by the electrolysis of water. Figure 16-23 shows an electrolytic cell setup to decompose water. Two platinum electrodes are dipped in a dilute solution of sulphuric acid. The cell requires just one compartment, because hydrogen and oxygen escape from the cell much more rapidly than they react with each other.

**Figure 16-23**

A photograph and a schematic diagram of a cell for the electrolysis of water. The cell generates $H_2$ gas and $O_2$ gas in a 2:1 mole ratio.

During the electrolysis of water, hydronium ions capture electrons from the cathode, producing hydrogen gas. Water molecules lose electrons to the anode, producing oxygen gas:

Oxidation: $\qquad$ $6 \, H_2O \, (l) \longrightarrow O_2 \, (g) + 4 \, H_3O^+ \, (aq) + 4 \, e^-$

Reduction: $\quad 4 \, H_3O^+ \, (aq) + 4 \, e^- \longrightarrow 2 \, H_2 \, (g) + 4 \, H_2O \, (l)$

Overall: $\qquad\qquad$ $2 \, H_2O \, (l) \longrightarrow 2 \, H_2 \, (g) + O_2 \, (g)$

The potential supplied to an electrolytic cell determines whether or not electrolysis can occur, but the current flow and the time of electrolysis determine the amount of material electrolyzed. Recall Equation 16-7 from Section 16.5:

$$\text{Moles of electrons} = n = \frac{It}{F}$$

This equation provides the link between electrical measurements and amount of electrons, and the balanced half-reactions for the electrolytic process provide the link between the amount of electrons and amounts of chemical substances. Example 16-16 shows a calculation regarding electrolytic stoichiometry.

## EXAMPLE 16-16 ELECTROLYSIS AND STOICHIOMETRY

If the electrolytic cell shown in Figure 16-23 draws a current of 0.775 A for 45.0 min, calculate the volumes of $H_2$ and $O_2$ produced if each gas is collected at 25° C and $p = 1.00$ bar.

**Strategy:** "Volumes of $H_2$ and $O_2$" identifies this as a stoichiometry problem, for which our seven-step approach, presented here in compact form, is appropriate. The problem asks about amounts of chemicals produced in electrolysis and provides data about current and time. Visualize the electrolysis shown in Figure 16-23. Calculating amounts in electrochemistry requires the use of Equation 16-7 and knowledge of the numbers of electrons transferred in the half-reactions.

**Solution:** We calculate moles of electrons using Equation 16-7. Recall that 1 A = 1 C/s:

$$n_{\text{electrons}} = \frac{It}{F} = \frac{(0.775 \text{ C/s})(45.0 \text{ min})(60 \text{ s/min})}{96\,485 \text{ C/mol}} = 2.169 \times 10^{-2} \text{ mol}$$

To convert from moles of electrons to moles of chemical species, we need the molar ratios between electrons and chemical species, which we find by examining the half-reactions:

$$6 \, H_2O \, (l) \longrightarrow O_2 \, (g) + 4 \, H_3O^+ \, (aq) + 4 \, e^-$$

$$4 \, H_3O^+ \, (aq) + 4 \, e^- \longrightarrow 2 \, H_2 \, (g) + 4 \, H_2O \, (l)$$

The transfer of 4 moles of $e^-$ produces 2 moles of $H_2$ and 1 mole of $O_2$ :

$$n_{H_2} = (2.169 \times 10^{-2} \text{ mol } e^-)\left(\frac{2 \text{ mol } H_2}{4 \text{ mol } e^-}\right) = 1.084 \times 10^{-2} \text{ mol } H_2$$

$$n_{O_2} = (2.169 \times 10^{-2} \text{ mol } e^-)\left(\frac{1 \text{ mol } O_2}{4 \text{ mol } e^-}\right) = 0.5423 \times 10^{-2} \text{ mol } O_2$$

To complete the calculation, convert moles of gas into volumes using the ideal gas equation:

$$V = \frac{nRT}{p}$$

We leave this stoichiometric calculation to you. The volume of $H_2$ is 0.265 L, and that of $O_2$ is 0.133 L.

**Do the Results Make Sense?** Notice that the volume of $H_2$ gas is twice the volume of $O_2$ gas, as required by the stoichiometry of the overall reaction.

**Practice Exercise 16.16**
Calculate the mass (in mg) of Li transferred from the cathode to the anode when a lithium-ion battery is recharged for 35 min using a constant current of 0.155 A.

## Competitive Electrolysis

When an electrolytic cell is designed, care must be taken in the selection of the cell components. For example, consider what happens when an aqueous solution of sodium chloride is electrolyzed using platinum electrodes. Platinum is used for passive electrodes, because this metal is resistant to oxidation and does not participate in the redox chemistry of the cell. There are three major species in the solution: $H_2O$, $Na^+$, and $Cl^-$. Chloride ions cannot be reduced further, so there are just two candidates for reduction: water molecules and sodium ions.

$$Na^+ (aq) + e^- \longrightarrow Na (s) \qquad E° = -2.17 \text{ V}$$

$$2 H_2O (l) + 2e^- \longrightarrow H_2 (g) + 2 OH^- (aq) \qquad E° = -0.828 \text{ V}$$

In an electrolytic cell, the most easily oxidized species gets oxidized, and the most easily reduced species gets reduced. The standard potentials show that water is much easier to reduce than sodium ions, so the electrolysis will produce hydrogen and hydroxide ions rather than sodium metal.

The possibilities for oxidation are more varied in this solution. Water and chloride ions can be oxidized, but sodium ions cannot. The various possibilities for oxidation can be found in Appendix F. Because oxidation is the reverse of reduction, the half-reactions chosen must have *products* that include $Cl^-$ or $H_2O$. Although several possibilities exist, only the two with the lowest reduction potentials need to be considered:

$$O_2 (g) + 4 H_3O^+ (aq) + 4 e^- \longrightarrow 6 H_2O (l) \qquad E° = 1.229 \text{ V}$$

$$Cl_2 (g) + 2 e^- \longrightarrow 2 Cl^- (aq) \qquad E° = 1.358 \text{ V}$$

In general, we expect that an external potential will drive the pair of reactions of which the spontaneous reaction has the least positive cell potential. In this case, combining the reduction of water to hydrogen and hydroxide with the oxidation of water to $H_3O^+$ and oxygen gas gives the least positive $E°$:

$$2 H_2 (g) + O_2 (g) \longrightarrow 2 H_2O (l)$$

$$E°_{cell} = (1.229 \text{ V}) - (-0.828 \text{ V}) = +2.057 \text{ V}$$

Notice, however, that the cell potential for the reaction involving $Cl^-$ is only slightly larger:

$$H_2 (g) + 2 OH^- (aq) + Cl_2 (g) \longrightarrow 2 H_2O (l) + 2 Cl^- (aq)$$

$$E°_{cell} = (1.358 \text{ V}) - (-0.828 \text{ V}) = 2.186 \text{ V}$$

Experimentally, when an opposing potential sufficient to cause electrolysis is applied to a cell containing aqueous NaCl, the electrolytic reaction produces chlorine gas rather than oxygen gas at the anode:

$$\text{Cathode:} \quad 2\,H_2O(l) + 2\,e^- \longrightarrow H_2(g)\ 2\,OH^-(aq)$$

$$\text{Anode:} \quad\quad 2\,Cl^-(aq) \longrightarrow Cl_2(g) + 2\,e^-$$

**Figure 16-24**

The electroplating of silver.

The reason is that electrochemical reactions are determined by kinetic considerations as well as thermodynamics. The oxidation of water is favoured by thermodynamics, but the oxidation of chloride is a much faster reaction. Notice that the oxidation of water involves four electrons per water molecule, but the oxidation of chloride anions involves only two electrons per molecule of chlorine gas. When reactions have similar potentials, the one that requires fewer electrons is often favoured because it proceeds faster than the one that requires more electrons.

## Electroplating

Electrolysis can be used to deposit one metal on top of another, a process known as **electroplating**. Electroplating is done for cosmetic reasons as well as for protection against corrosion and wear. A thin layer of gold or silver is often plated on jewellery and tableware made from inexpensive metals such as copper. Figure 16-24 shows a schematic diagram of an electroplating apparatus. In this example, the spoon acts as a cathode where silver cations are reduced. The anode is a silver rod. An external power source removes electrons from the silver rod, causing oxidation to $Ag^+$ cations, which enter the solution. Electrons flow into the spoon, where silver ions in solution capture the electrons to become neutral silver atoms. These neutral atoms adhere to the copper surface as a layer of silver metal.

The quantity of metal deposited during a plating process is linked stoichiometrically to the current flow through Equation 16-7, as Example 16-17 illustrates.

## EXAMPLE 16-17 ELECTROPLATING

An electroplating apparatus is used to coat jewellery with gold. What mass of gold can be deposited from a solution that contains $Au(CN)_4^-$ ions if a current of 5.0 A flows for 30.0 min?

**Strategy:** This, like Example 16-16, is an electrochemical stoichiometry problem. It asks about the amount of gold deposited in an electrolytic reaction. The method is the same as that of Example 16-16.

**Solution:**

$$n_{electrons} = \frac{It}{F} = \frac{(5.0\ C/s)(30.0\ min)(60\ s/min)}{96\ 485\ C/mol} = 0.0933\ mol\ e^-$$

To determine moles of gold, we need a balanced half-reaction. The reactant is $Au(CN)_4^-$, and the product is Au. Apply the usual procedure for balancing a half-reaction, recognizing the presence of cyanide ions:

$$Au(CN)_4^-(aq) + 3\,e^- \longrightarrow Au(s) + 4\,CN^-(aq)$$

The stoichiometric coefficients, 1 and 3, indicate that 1 mole of gold will be plated for every 3 moles of electrons:

$$(0.0933 \text{ mol e}^-)\left(\frac{1 \text{ mol Au}}{3 \text{ mol e}^-}\right) = 0.0311 \text{ mol Au}$$

Finally, the molar mass of gold gives the mass of gold in grams:

$$(0.0311 \text{ mol Au})(197 \text{ g/mol}) = 6.1 \text{ g Au}$$

 **Does the Result Make Sense?** The final result is rounded to two significant figures to match the precision of the current measurement. The amount of gold is significantly less than 1 mole, consistent with the total charge.

**Practice Exercise 16.17**
Calculate the mass of silver that will be plated onto a spoon from a plating solution containing $[Ag(NH_3)_2]^+$ if a current of 3.75 A flows for 27.5 min.

Steel objects are often protected from corrosion by electroplating with chromium. The most straightforward process would be to electrolyze a solution of $Cr^{3+}$ cations. This fails because aqueous $Cr^{3+}$ ions are not reduced at a useful rate. Instead, solutions containing chromate anions are used:

$$8 \text{ H}_3\text{O}^+ (aq) + \text{CrO}_4^{2-} (aq) + 6 \text{ e}^- \longrightarrow \text{Cr} (s) + 12 \text{ H}_2\text{O} (l)$$

This reaction proceeds much faster than the direct reduction of $Cr^{3+}$. It requires twice as much electrical current, however, since 6 moles of electrons must be supplied to deposit 1 mole of Cr rather than 3 moles for direct electrolysis of $Cr^{3+}$.

Chromium is plated on the surface of iron to prevent corrosion and to improve appearance.

## SECTION EXERCISES

●**16.7.1** Fluorine is manufactured by the electrolysis of hydrogen fluoride dissolved in molten KF:

$$2 \text{ HF } (KF \text{ melt}) \longrightarrow \text{H}_2 (g) + \text{F}_2 (g)$$

(a) Identify the half-reactions. (The only species present in the molten phase are HF, $K^+$, and $F^-$.)

(b) $E° = -0.187$ V for reducing HF to $H_2$. What is the minimum external potential required to drive the electrolysis of HF at standard concentrations?

●**16.7.2** If a commercial reactor for the electrolysis of HF operates continuously at a current of 1500 A, how many kilograms of $F_2$ are produced in 24 hours of operation? What is the volume of the tank required to store the $F_2$ at $T = 298$ K, $p = 60$ bar?

●**16.7.3** Explain why neither aqueous KF nor pure liquid HF can be used for the electrolytic production of fluorine, even though both liquids are easier to handle than molten potassium fluoride.

# CHAPTER 16 VISUAL SUMMARY

## 16.1 RECOGNIZING REDOX REACTIONS

$$2 \text{ Mg} + \text{O}_2 \longrightarrow 2 \text{ MgO}$$
$$0 \quad\quad 0 \quad\quad\quad\quad +2 \; -2$$

Reducing agent

Oxidizing agent

Oxidation numbers

*Oxidation is the loss of electrons from a substance.*
*Reduction is the gain of electrons by a substance.*

*Oxidation and reduction always occur together.*

**LEARNING OBJECTIVES:** Use oxidation numbers to show what is being oxidized and what is being reduced in a redox reaction.

**SKILLS TO MASTER:** Recognizing the oxidant and the reducing agent (p. 931)

Assigning oxidation numbers (p. 931)

**KEY CONCEPTS:** Oxidation is the loss of electrons from a substance. Reduction is the gain of electrons by a substance. Oxidation and reduction always occur together.

## 16.2 BALANCING REDOX REACTIONS

*Electrons lost in oxidation = Electrons gained in reduction*

1. Break into half-reactions
2. Balance each half-reaction
   a. Balance all elements except O and H.
   b. Balance O by adding $H_2O$.
   c. Balance H by adding $H_3O^+$ and $H_2O$ (acid), or
      Balance H by adding $OH^-$ and $H_2O$ (base).
   d. Balance charge by adding electrons.
3. Multiply and cancel electrons
4. Recombine and simplify

Half-reaction

**LEARNING OBJECTIVES:** Balance redox reactions using the half-reaction method.

**SKILLS TO MASTER:** Identifying half-reactions (p. 936)

Balancing half-reactions (p. 937)

Combining half-reactions (p. 939)

**KEY CONCEPTS:** Redox reactions can be separated into two half-reactions, one for the oxidation and one for the reduction.

## 16.3 GALVANIC CELLS

**LEARNING OBJECTIVES:** Describe galvanic cells.

**SKILLS TO MASTER:** Describing a galvanic cell from the balanced redox reaction (p. 946)

Writing the shorthand notation of an electrochemical cell from the balanced redox reaction and vice-versa (p. 951)

**KEY CONCEPTS:** Oxidation always occurs at the anode and reduction always occurs at the cathode.

A galvanic cell has a spontaneous redox reaction and a positive standard cell potential.

## 16.4 CELL POTENTIALS

*The half-reaction with the most negative reduction potential occurs at the anode as oxidation.*

*The half-reaction with the least negative reduction potential occurs at the cathode as reduction.*

$$\text{Zn } (s) + \text{Cu}^{2+}(aq) \longrightarrow \text{Zn}^{2+}(aq) + \text{Cu } (s)$$
Anode            Cathode

Cell potential

$$E^{\circ}_{\text{cell}} = E^{\circ}_{\text{cathode}} - E^{\circ}_{\text{anode}}$$

$$2 \text{ H}_3\text{O}^+ (aq, 1.00 \text{ M}) + 2 \text{ e}^- \rightleftharpoons \text{H}_2 (g) + 2 \text{ H}_2\text{O} (l)$$

Standard reduction potential     $E^{\circ} = 0$ V (defined)

*When a reaction is multiplied by an integer, its cell potential remains unchanged.*

**LEARNING OBJECTIVES:** Calculate standard cell potentials.

**SKILLS TO MASTER:** Calculating standard cell potentials from standard reduction potentials (p. 954)

**KEY CONCEPTS:** The half-reaction with the more negative reduction potential occurs at the anode as an oxidation. The half-reaction with the more positive reduction potential occurs at the cathode as a reduction.

All reduction potentials are relative to one another. The standard potential for reducing $H_3O^+$ (aq) to $H_2$ (g) and $H_2O$ (l) is defined to be 0.00 V.

When a half-reaction is multiplied by any integer, its potential remains unchanged.

## 16.5 FREE ENERGY AND ELECTROCHEMISTRY

$$\Delta G = -nFE$$
$$\Delta G^{\circ} = -nFE^{\circ}$$

Faraday constant (F)

$$n_{electrons} = \frac{It}{F}$$

Charge-mole equivalence

$$E^{\circ} = \frac{RT}{nF} \ln K_{eq}$$

$$E^{\circ} = \frac{0.0592\,V}{n} \log K_{eq}$$

Nernst equation

$$E = E^{\circ} - \frac{RT}{nF} \ln Q$$

pH meter

**LEARNING OBJECTIVES:** Relate cell potential to the reaction conditions.

**SKILLS TO MASTER:** Calculating the free energy change from the cell potential (p. 964)

Calculating equilibrium constants of redox reactions (p. 965)

Calculating the cell potential under non-standard conditions using the Nernst equation (p. 966)

Performing stoichiometric electrochemical calculations (p. 970)

**KEY CONCEPTS:** A cell potential depends on the concentrations of all species in the reaction quotient and on the temperature.

## 16.6 REDOX IN ACTION

Alkaline dry cell

Nickel–cadmium battery

Lead storage battery

Zinc–air battery

Corrosion

**LEARNING OBJECTIVES:** Explain the chemistry of everyday redox reactions.

**SKILLS TO MASTER:** Explaining the chemistries of several types of batteries (p. 973)

Explaining under what conditions corrosion happens (p. 976)

**KEY CONCEPTS:** Batteries are galvanic cells.

A fuel cell is a galvanic cell requiring a continuous supply of fuel and oxidant.

## 16.7 ELECTROLYSIS

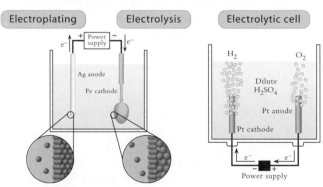

Electroplating

Electrolysis

Electrolytic cell

**LEARNING OBJECTIVES:** Explain electrolytic reactions and cells.

**SKILLS TO MASTER:** Predicting the reactions that occur at each electrode in an electrolytic cell (p. 983)

Doing quantitative electroplating calculations (p. 984)

**KEY CONCEPTS:** An electrolytic cell has a non-spontaneous redox reaction and a negative standard cell potential.

## LEARNING EXERCISES

**16.1** Describe the steps that must be followed in (a) balancing a redox equation; (b) drawing a molecular diagram of a galvanic cell; (c) calculating the potential of a galvanic cell operating under non-standard conditions; and (d) determining what reactions occur in an electrolytic cell.

**16.2** Describe the mechanism for charge movement in each of the following components of an electrochemical cell: (a) the external wire; (b) the electrode-solution interface; (c) the solution; and (d) the porous barrier.

**16.3** Write a paragraph explaining the linkages among cell potential, free energy, and the equilibrium constant.

**16.4** List at least six practical examples of redox chemistry.

**16.5** Update your list of memory bank equations. For each new equation, specify the type of calculations for which it is useful.

**16.6** List all terms new to you that appear in Chapter 16, and write a one-sentence definition of each in your own words. Consult the Glossary if you need help.

## PROBLEMS

**ilw** =interactive learning ware problem. Visit the website at www.wiley.com/canada/olmsted

## Recognizing Redox Reactions

**16.1** Determine the oxidation numbers of all atoms in the following: (a) $Fe(OH)_3$; (b) $NF_3$; (c) $CH_3OH$; (d) $K_2CO_3$; (e) $NH_4NO_3$; (f) $TiCl_4$; (g) $PbSO_4$; and (h) $P_4$.

**16.2** Determine the oxidation numbers of all atoms in the following: (a) $H_2CO$; (b) $AlCl_3$; (c) $XeF_4$; (d) $F_2O$; (e) $K_2Cr_2O_7$; (f) $NaIO_3$; (g) $P_2O_5$; and (h) $Na_2O_2$.

**16.3** Which of the following are redox reactions?
(a) $HBr\,(g) + H_2O\,(l) \longrightarrow H_3O^+\,(aq) + Br^-\,(aq)$
(b) $2\,Fe^{2+}\,(aq) + H_2O_2\,(aq) \longrightarrow 2\,Fe^{3+}\,(aq) + 2\,OH^-\,(aq)$
(c) $Fe^{2+}\,(aq) + 2\,OH^-\,(aq) \longrightarrow Fe(OH)_2\,(s)$
(d) $2\,CH_3OH\,(l) + 3\,O_2\,(g) \longrightarrow 2\,CO_2\,(g) + 4\,H_2O\,(l)$
(e) $N_2\,(g) + 3\,H_2\,(g) \longrightarrow 2\,NH_3\,(g)$

**16.4** Which of the following are redox reactions?
(a) $CO\,(g) + H_2O\,(g) \longrightarrow CO_2\,(g) + H_2\,(g)$
(b) $CO_2\,(g) + 2\,H_2O\,(l) \longrightarrow H_3O^+\,(aq) + HCO_3^-\,(aq)$
(c) $2\,CuS\,(s) + O_2\,(g) \longrightarrow 2\,Cu\,(s) + SO_2\,(g)$
(d) $2\,AgNO_3\,(aq) + Cu\,(s) \longrightarrow Cu(NO_3)_2\,(aq) + 2\,Ag\,(s)$
(e) $2\,AgNO_3\,(aq) + Na_2SO_4\,(aq) \longrightarrow 2\,NaNO_3\,(aq) + Ag_2SO_4\,(s)$

**16.5** Chlorine displays a wide range of oxidation numbers. Determine the oxidation number of chlorine in each of the following species: (a) $ClF_5$; (b) $MgCl_2$; (c) $NaClO_4$; (d) $Cl_2$; (e) $KClO_2$; and (f) $NaClO$.

**16.6** Sulphur displays a wide range of oxidation numbers. Determine the oxidation number of sulphur in each of the following species: (a) $H_2S$; (b) $S_2Cl_2$; (c) $Li_2SO_4$; (d) $S_8$; (e) $Na_2SO_3$; and (f) $SF_4$.

## Balancing Redox Reactions

**16.7** Determine the half-reactions for the following redox processes: (a) Sodium metal reacts with water to give hydrogen gas. (b) Gold metal dissolves in "aqua regia" (a mixture of HCl and $HNO_3$) to give $[AuCl_4]^-$ and NO. (c) Acidic potassium permanganate reacts with aqueous $K_2C_2O_4$, producing carbon dioxide and $Mn^{2+}$. (d) Coal (solid carbon) reacts with steam to produce molecular hydrogen and carbon monoxide.

**16.8** What are the half-reactions for these redox processes? (a) Aqueous hydrogen peroxide acts on $Co^{2+}$, and the products are hydroxide ion and $Co^{3+}$, in basic solution. (b) Methane reacts with oxygen gas and produces water and carbon dioxide. (c) To recharge a lead storage battery, lead(II) sulphate is converted to lead metal and to lead(IV) oxide. (d) Zinc metal dissolves in aqueous hydrochloric acid to give $Zn^{2+}$ ions and hydrogen gas.

**16.9** Balance the following half-reactions:
(a) $Cu^+ \longrightarrow CuO$ (acidic solution)
(b) $S \longrightarrow H_2S$ (acidic solution)
(c) $AgCl \longrightarrow Ag$ (basic solution)
(d) $I^- \longrightarrow IO_3^-$ (basic solution)
(e) $IO_3^- \longrightarrow IO^-$ (basic solution)
(f) $H_2CO \longrightarrow CO_2$ (acidic solution)

**16.10** Balance the following half-reactions:
(a) $SbH_3 \longrightarrow Sb$ (acidic solution)
(b) $AsO_2^- \longrightarrow As$ (basic solution)
(c) $BrO_3^- \longrightarrow Br_2$ (acidic solution)
(d) $Cl^- \longrightarrow ClO_2^-$ (basic solution)
(e) $Sb_2O_5 \longrightarrow Sb_2O_3$ (acidic solution)
(f) $H_2O_2 \longrightarrow O_2$ (basic solution)

**16.11** Balance the net redox reaction resulting from combining the following half-reactions in Problem 16.9 (the first listed in each case is the reduction): (a) 9a and 9b; (b) 9d and 9c; (c) 9d and 9e; and (d) 9f and 9b.

**16.12** Balance the net redox reaction resulting from combining the following half-reactions in Problem 16.10 (the first listed in each case is the reduction): (a) 10a and 10c; (b) 10a and 10e; (c) 10b and 10d; and (d) 10b and 10f.

**16.13** Consider the following redox reaction:

$$2\,MnO_4^- + 10\,Cl^- + 16\,H_3O^+ \longrightarrow 2\,Mn^{2+} + 5\,Cl_2 + 24\,H_2O$$

(a) Which species gets oxidized? (b) Which species gets reduced? (c) Which species is the oxidizing agent? (d) Which species is the reducing agent? (e) Which species gains electrons? (f) Which species loses electrons?

**16.14** Consider the following redox reaction:

$$4\,NO + 3\,O_2 + 6\,H_2O \longrightarrow 4\,NO_3^- + 4\,H_3O^+$$

(a) Which species gets oxidized? (b) Which species gets reduced? (c) Which species is the oxidizing agent? (d) Which species is the reducing agent? (e) Which species gains electrons? (f) Which species loses electrons?

**16.15** Balance the following redox equations:
(a) $CN^- + MnO_4^- \longrightarrow CNO^- + MnO_2$ (basic solution)
(b) $O_2 + As \longrightarrow HAsO_2 + H_2O$ (acidic solution)
(c) $Br^- + MnO_4^- \longrightarrow MnO_2 + BrO_3^-$ (basic solution)
(d) $NO_2 \longrightarrow NO_3^- + NO$ (acidic solution)
(e) $ClO_4^- + Cl^- \longrightarrow ClO^- + Cl_2$ (acidic solution)
(f) $AlH_4^- + H_2CO \longrightarrow Al^{3+} + CH_3OH$ (basic solution)

**16.16** Balance the following redox equations:
(a) $H_5IO_6 + Cr \longrightarrow IO_3^- + Cr^{3+}$ (acidic)
(b) $Se + Cr(OH)_3 \longrightarrow Cr + SeO_3^{2-}$ (basic)
(c) $HClO_2 + Co \longrightarrow Cl_2 + Co^{2+}$ (acidic)
(d) $CH_3COH + Cu^{2+} \longrightarrow CH_3CO_2^- + Cu_2O$ (basic)
(e) $NO_3^- + H_2O_2 \longrightarrow NO + O_2$ (basic)
(f) $BrO_3^- + Fe^{2+} \longrightarrow Br^- + Fe^{3+}$ (acidic)

## Galvanic Cells

**16.17** Use standard thermodynamic values to determine whether or not each of the following redox reactions is spontaneous under standard conditions:
(a) $O_2 + 2 Cu \longrightarrow 2 CuO$
(b) $O_2 + 2 Hg \longrightarrow 2 HgO$
(c) $CuS + O_2 \longrightarrow Cu + SO_2$
(d) $FeS + O_2 \longrightarrow Fe + SO_2$

**16.18** Use standard thermodynamic values to determine whether or not each of the following redox reactions is spontaneous under standard conditions:
(a) $H_2O + CO \longrightarrow CO_2 + H_2$
(b) $2 Al + 3 MgO \longrightarrow 3 Mg + Al_2O_3$
(c) $PbS + Cu \longrightarrow CuS + Pb$
(d) $N_2 + 2 O_2 \longrightarrow 2 NO_2$

**16.19** Draw a sketch that shows a molecular view of the charge transfer processes that take place at a silver chloride—silver electrode in contact with aqueous HCl, undergoing reduction:

$$AgCl\ (s) + e^- \longrightarrow Ag\ (s) + Cl^-\ (aq)$$

**16.20** Draw a sketch that shows a molecular view of the charge transfer processes that take place at a nickel electrode in contact with aqueous nickel(II) chloride, undergoing oxidation:

$$Ni^{2+}\ (aq) + 2e^- \longrightarrow Ni\ (s)$$

**16.21** Draw a sketch of a cell that could be used to study the following redox reaction:

$$H_2\ (g) + 2 H_2O\ (l) + Cl_2\ (g) \longrightarrow 2 H_3O^+\ (aq) + 2 Cl^-\ (aq)$$

**16.22** Draw a sketch of a cell that could be used to study the following redox reaction:

$$3 Fe^{2+}\ (aq) \longrightarrow Fe\ (s) + 2 Fe^{3+}\ (aq)$$

**16.23** Which of the electrodes in Problems 16.19 and 16.21 are active, and which are passive?

**16.24** Which of the electrodes in Problems 16.20 and 16.22 are active, and which are passive?

**16.25** Write the shorthand notations for each of the following reactions:
(a) $2 Ag\ (s) + Fe^{2+}\ (aq) \longrightarrow 2 Ag^+\ (aq) + Fe\ (s)$
(b) $NO_3^-\ (aq) + H_2O_2\ (aq) \longrightarrow NO\ (g) + O_2\ (g)$
(c) $H_2\ (g) + 2 H_2O\ (l) + Cl_2\ (g) \longrightarrow$
$\quad 2 H_3O^+\ (aq) + 2 Cl^-\ (aq)$

**16.26** Write balanced redox reactions for the electrochemical cells denoted by the following shorthand notations:
(a) $Cu\ (s)\ |\ Cu^{2+}\ (aq)\ \|\ Cr_2O_7^{2-}\ (aq),\ Cr^{3+}\ (aq),$
$\quad H_3O^+\ (aq)\ |\ Pt\ (s)$
(b) $Pt\ (s)\ |\ Cl^-\ (aq)\ |\ Cl_2\ (g)\ \|\ MnO_4^-\ (aq),\ Mn^{2+}\ (aq),$
$\quad H_3O^+\ (aq)\ |\ Pt\ (s)$
(c) $Pt\ (s)\ |\ C_2O_4^{2-}\ (aq)\ |\ CO_2\ (g)\ \|\ Ag^+\ (aq)\ |\ Ag\ (s)$

## Cell Potentials

**16.27** Use standard reduction potentials in Appendix F to calculate $E°$ for the reactions in Problems 16.15(a), 16.15(b), and 16.15(e).

**16.28** Use standard reduction potentials in Appendix F to calculate $E°$ for the reactions in Problems 16.16(a), 16.16(b), and 16.16(e).

**16.29** **ilw** Balance and calculate the standard potential for the following reaction in acidic solution:

$$NO \longrightarrow N_2O + NO_3^-$$

**16.30** Balance and calculate the standard potential for the following reaction in acidic solution:

$$ClO_3^- \longrightarrow ClO_4^- + Cl^-$$

**16.31** Describe a cell that could be used to measure the $E°$ of the $F_2/F^-$ reduction reaction. Include a sketch similar to that shown in Figure 16-8. Which electrode would be the anode?

**16.32** A cell is set up with two Fe wire electrodes, one immersed in a 1.0 M solution of $Fe(NO_3)_2$, the other in a 1.0 M solution of $Fe(NO_3)_3$. Determine $E°$ of this cell, identify the anode, and draw a picture that shows the direction of electron flow at each electrode and in the external circuit.

**16.33** Consult Appendix F and list the metals that can reduce $Be^{2+}$ to Be ($E° = -1.97$ V) under standard conditions. What characteristics do these metals have in common?

**16.34** Consult Appendix F and list the elements that can oxidize $H_2O$ to $O_2$ under standard conditions. What characteristics do these elements have in common?

## Free Energy and Electrochemistry

**16.35** Use standard reduction potentials in Appendix F to calculate $\Delta G°$ for the reactions in Problems 16.15(a), 16.15(b), and 16.15(e).

**16.36** Use standard reduction potentials in Appendix F to calculate $\Delta G°$ for the reactions in Problems 16.16(a), 16.16(b), and 16.16(e).

**16.37** ⓘⓛⓦ Example 16-11 describes the nickel–cadmium battery. What potential does this battery produce if its hydroxide ion concentration is $1.50 \times 10^{-2}$ M?

**16.38** If the cell illustrated in Figure 16-14 contains $1.00 \times 10^{-3}$ M concentrations of HCl and $CuSO_4$, what potential does it produce?

**16.39** If it takes 15 sec to start your car engine and the battery provides 5.9 A of current to the starter motor, what masses of Pb and $PbO_2$ are used in each battery cell?

**16.40** Suppose that automobiles were equipped with "thermite" batteries as described in Section Exercise 16.3.1. What masses of Al and $Fe_2O_3$ would be consumed in the process described in Problem 16.39?

**16.41** An alternator recharges an automobile battery when the engine is running. If the alternator produces 1.750 A of current and the operation of the engine requires 1.350 A of this current, how long will it take to convert 0.850 g of $PbSO_4$ back into Pb metal?

**16.42** A digital watch draws 0.20 mA of current provided by a zinc–air battery, the net reaction of which is $O_2\ (g)\ +$ $2\ Zn\ (s) \longrightarrow 2\ ZnO\ (s)$. If a partially used battery contains 1.00 g of zinc, for how many more hours will the watch run?

**16.43** Dichromate ions ($Cr_2O_7^{2-}$) oxidize acetaldehyde ($CH_3CHO$) to acetic acid ($CH_3CO_2H$), and are reduced to $Cr^{3+}$. The reaction takes place in acidic solution. Balance the redox reaction and determine how many moles of electrons are required to oxidize 1.00 g of acetaldehyde. What mass of sodium dichromate would be required to deliver this many electrons?

**16.44** Permanganate ions ($MnO_4^-$) oxidize acetaldehyde ($CH_3CHO$) to acetic acid ($CH_3CO_2H$), and are reduced to $MnO_2$. The reaction takes place in acidic solution. Balance the redox reaction and determine how many moles of electrons are required to oxidize 1.00 g of acetaldehyde. What mass of $KMnO_4$ would be required to deliver this many electrons?

**16.45** For a galvanic cell using the reaction $Ni^{2+}\ (aq)\ +$ $2\ Cr^{2+}\ (aq) \rightarrow Ni\ (s) + 2\ Cr^{3+}\ (aq)$, calculate the cell potential at 25 °C when $[Ni^{2+}] = 1.0 \times 10^{-4}$ M, $[Cr^{2+}] = 2.0 \times 10^{-3}$ M and $[Cr^{3+}] = 1.5 \times 10^{-3}$ M.

**16.46** For the reaction $Cu\ (s) + 2\ Ag^+\ (aq) \longrightarrow Cu^{2+}\ (aq)\ +$ $2\ Ag\ (s)$ at 25 °C, the cell potential is found to be 0.62 V. Everything is at standard state except the $Cu^{2+}\ (aq)$. Calculate $[Cu^{2+}]$.

**16.47** The reaction of tin metal with acid is $Sn\ (s) + 2\ H^+\ (aq)$ $\longrightarrow Sn^{2+}\ (aq) + H_2\ (g)$. Calculate the cell potential (a) with standard state conditions; (b) with standard state conditions, except that the pH is 2.00; and (c) with standard state conditions, except that the pH is 5.00.

### Redox in Action

**16.48** ⓘⓛⓦ Set up the Nernst equation for the standard dry cell. Using this equation, show that the voltage of a dry cell must decrease with use.

**16.49** Set up the Nernst equation for the lead storage cell, and use it to show that the voltage of this cell must decrease with use.

**16.50** If a chromium-plated steel bicycle handlebar is scratched, breaking the protective film, will the chromium or the steel corrode? Use standard potentials to support your prediction.

**16.51** Using standard potentials, explain why the steel propeller of an ocean-going yacht has a zinc collar.

**16.52** Explain why zinc–air batteries find extensive use for cameras and pacemakers but are not used to start automobiles.

**16.53** Explain why the lead storage battery, despite being the battery of choice for automobiles, is not suitable for space flights.

**16.54** Calculate the equilibrium constant for the lead storage battery at 25 °C.

### Electrolysis

**16.55** In the electrolysis of aqueous NaCl, what mass of $Cl_2$ is generated by a current of 4.50 A flowing for 200.0 min?

**16.56** Ni–Cd rechargeable batteries power a portable radio that draws 150 mA of current. Compute the masses of Cd and $NiO_2$ consumed when the radio is operated for 65 min.

**16.57** The Zn–Cu galvanic cell shown in Figure 16-7 is not rechargeable because of competitive electrolysis. Describe the reactions that take place if an opposing potential sufficient to reverse the reactions is applied to this galvanic cell.

**16.58** The conventional alkaline dry cell is not rechargeable. The solid products separate from the electrodes, so the reverse reactions cannot occur. What reactions may take place if an opposing potential sufficient to reverse the reactions is applied to a dry cell?

**16.59** ⓘⓛⓦ After use, a nickel–cadmium battery has 1.55 g of $Cd(OH)_2$ deposited on its anode. It is inserted in a recharger that supplies 125 mA of current at a voltage of 1.45 V. (a) To which electrode, Ni or Cd, should the negative wire from the charger be connected? Write the half-reaction occurring at this electrode during charging. (b) Compute the time in hours needed to convert all 1.55 g of $Cd(OH)_2$ back to Cd metal.

**16.60** A "dead" 12-V lead storage battery has 4.80 g of $PbSO_4$ deposited on each of its anodes. It is connected to a "trickle charger" that supplies 0.120 A of current at a voltage of 13 V. (a) To which electrode, Pb or $PbO_2$, should the negative wire from the charger be connected? Write the half-reaction occurring at this electrode during charging. (b) Compute the time in hours needed to convert all 4.80 g of $PbSO_4$ back to $HSO_4^-$ ions in solution.

**16.61** Electrochemistry can be used to measure electrical current in a silver coulometer, in which a silver cathode is immersed in a solution containing $Ag^+$ ions. The cathode is weighed before and after passage of current. A silver cathode initially has a mass of 10.77 g, and its mass increases to 12.89 g after current has flowed for 15.0 minutes. Compute the quantity of charge in coulombs and the current in amperes.

16.62 When Thomas Edison first sold electricity, he used zinc coulometers to measure charge consumption (see Problem 16.61). If the zinc plate in one of Edison's coulometers increased in mass by 7.55 g, how much charge had been consumed?

16.63 In a silver coulometer (see Problem 16.61), both electrodes are silver metal. Draw a molecular picture that illustrates the reactions that occur during operation of this coulometer.

16.64 In Edison's zinc coulometer (see Problem 16.62), both electrodes were zinc metal. Draw a molecular picture that illustrates the reactions that occur during operation of this coulometer.

## ADDITIONAL PAIRED PROBLEMS

16.65 Here are two standard reduction potentials:

$$O_2 (g) + 2 H_2O + 4 e^- \longrightarrow 4 OH^- (aq) \quad E° = 0.401 \text{ V}$$

$$O_2 (g) + 4 H_3O^+ (aq) + 4 e^- \longrightarrow 6 H_2O \quad E° = 1.229 \text{ V}$$

Are these data sufficient to allow calculation of $K_w$ at 298 K? If so, do the calculation. If not, explain in detail what additional data (thermodynamic or electrochemical) you would need to do the calculation.

16.66 Here are two standard reduction potentials:

$$SO_4^{2-} (aq) + 4 H_3O^+ (aq) + 2 e^- \longrightarrow SO_2(g) + 6 H_2O (l)$$
$$E° = 0.20 \text{ V}$$

$$SO_4^{2-} (aq) + 4 H_3O^+ (aq) + 2 e^- \longrightarrow H_2SO_3 (aq) + 5 H_2O (l)$$
$$E° = 0.18 \text{ V}$$

Are these data sufficient to allow calculation of the equilibrium vapour pressure of $SO_2$ (g) over a 1.00 M solution of $H_2SO_3$ (aq) at 298 K? If so, do the calculation. If not, explain in detail what additional data (thermodynamic or electrochemical) you would need to do the calculation.

16.67 Use data in Appendix F to calculate the equilibrium constant of the following reaction:

$$Zn^{2+} (aq) + 4 NH_3 (aq) \longrightarrow Zn(NH_3)_4^{2+} (aq)$$

16.68 Use data in Appendix F to calculate $K_{sp}$ for AgI.

16.69 For the reduction of $Cr(OH)_3$ by $H_2$ in basic solution to give Cr and $H_2O$, do the following: (a) Write the balanced net equation. (b) Compute $E°$. (c) Compute $G°$.

16.70 Consider the following redox reaction:

$$MnO_2 \longrightarrow MnO_4^- + Mn^{2+} \quad \text{(unbalanced)}$$

(a) Balance this reaction in acidic solution. (b) Write the expression for Q. (c) Determine $n$, $E°$, $\Delta G°$, and $K_{eq}$ at 298 K.

16.71 Potassium chromate ($K_2CrO_4$) dissolves in acidic solution to generate strongly oxidizing $Cr_2O_7^{2-}$ ions:

$$K_2CrO_4 (s) + H_3O^+ (aq)$$
$$\longrightarrow 2 K^+ (aq) + HCrO_4^- (aq) + H_2O (l)$$

$$2 HCrO_4^- (aq) \longrightarrow Cr_2O_7^{2-} (aq) + H_2O (l)$$

This oxidizing agent is reduced to $Cr^{3+}$ as it oxidizes other substances. (a) Balance the reduction half-reaction. (b) What mass of $K_2CrO_4$ produces 0.250 mole of electrons?

16.72 Breathalyzers determine the alcohol content in a person's breath by a redox reaction using dichromate ions:

$$C_2H_5OH + Cr_2O_7^{2-} \longrightarrow CH_3CO_2H + Cr^{3+} \quad \text{(unbalanced)}$$

(a) Balance this reaction. (b) If a breath sample generates a concentration of $4.5 \times 10^{-4}$ M $Cr^{3+}$ in 50.0 mL of solution, how many milligrams of alcohol were in the sample?

16.73 The figure below shows a schematic sketch of a galvanic cell.

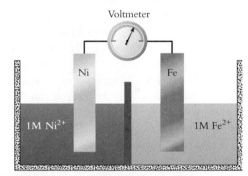

(a) Identify the two half-reactions. (b) Determine the potential of this cell. (c) Identify the anode and cathode. (d) Redraw the sketch to show the direction of electron flow and the molecular processes occurring at each electrode.

16.74 The figure below shows a schematic sketch of a galvanic cell.

(a) Identify the two half-reactions. (b) Determine the potential of this cell. (c) Identify the anode and cathode.

(d) Redraw the sketch to show the direction of electron flow and the molecular processes occurring at each electrode.

16.75 For the galvanic cell in Problem 16.73, which solution concentration would have to be reduced, and to what concentration, to reduce the cell potential to 0.0 V?

16.76 For the galvanic cell in Problem 16.74, which solution concentration would have to be reduced, and to what concentration, to reduce the cell potential to 0.0 V?

16.77 Balance the redox reactions between $MnO_4^-$ and each of the following sulphur-containing species. The final products are $Mn^{2+}$ and $HSO_4^-$ and the solution is acidic: (a) $H_2SO_3$; (b) $SO_2$; (c) $H_2S$; and (d) $H_2S_2O_3$.

16.78 Balance the redox reactions between $Cr_2O_7^{2-}$ and each of the following nitrogen-containing species. The final products are $Cr^{3+}$ and $NO_3^-$ and the solution is acidic: (a) NO; (b) $NO_2$; (c) $HNO_2$; and (d) $NH_4^+$.

16.79 How long would it take to electroplate all the $Cu^{2+}$ in 0.250 L of 0.245 M $CuSO_4$ solution with a current of 2.45 A?

16.80 Calcium metal is obtained by the direct electrolysis of molten $CaCl_2$. If a metallurgical electrolysis apparatus operates at 27.6 A, what mass of calcium metal will it produce in 24 hours of operation?

16.81 Draw a sketch of the electroplating apparatus that illustrates the process occurring in Problem 16.79. Include arrows showing the direction of electron flow, label the anode and cathode, and draw molecular pictures showing the processes occurring at each electrode.

16.82 Draw a sketch of an electroplating apparatus that illustrates the process occurring in Problem 16.80. Include arrows showing the direction of electron flow, label the anode and cathode, and draw molecular pictures showing the processes occurring at each electrode.

16.83 For each of the following pairs of species, select the one that is the better oxidizing agent under standard conditions (see Appendix F): (a) in acidic solution, $Cr_2O_7^{2-}$ or $MnO_4^-$; (b) in basic solution, $O_2$ or $NO_3^-$; and (c) $Fe^{2+}$ or $Sn^{2+}$.

16.84 For each of the following pairs of species, select the one that is the better reducing agent under standard conditions (see Appendix F): (a) Cu or Ag; (b) $Fe^{2+}$ or $Co^{2+}$; and (c) $H_2$ or $I^-$ (acid solution).

## ADDITIONAL PROBLEMS

16.85 Determine whether $O_2$ is capable of oxidizing each of the substances in Problem 16.84, under standard conditions (basic solution).

16.86 The following are spontaneous reactions. List the species in these reactions in order of oxidizing strength, from strongest oxidizing agent to weakest oxidizing agent:

$$2\,Cr^{2+} + Sn^{2+} \longrightarrow Sn + 2\,Cr^{3+}$$
$$Fe + Sn^{2+} \longrightarrow Sn + Fe^{2+}$$
$$Fe + U^{4+} \longrightarrow \text{No reaction}$$

$$U^{3+} + Cr^{3+} \longrightarrow U^{4+} + Cr^{2+}$$
$$Fe + 2\,Cr^{3+} \longrightarrow 2\,Cr^{2+} + Fe^{2+}$$

16.87 A cell is set up using two zinc wires and two solutions, one containing 0.250 M $ZnCl_2$ solution and the other containing 1.25 M $Zn(NO_3)_2$ solution. (a) What electrochemical reaction occurs at each electrode? (b) Draw a molecular picture showing spontaneous electron transfer processes at the two zinc electrodes. (c) Compute the potential of this cell.

16.88 Using standard reduction potentials, determine $K_{eq}$ for the decomposition reaction of hydrogen peroxide:

$$2\,H_2O_2 \longrightarrow 2\,H_2O + O_2$$

What does this value tell you about the stability of $H_2O_2$?

16.89 The first battery to find widespread commercial use was the carbon–zinc dry cell, in which the cathode reaction is

$$2\,MnO_2\,(s) + Zn^{2+}\,(aq) + 2\,e^- \longrightarrow ZnMn_2O_4\,(s)$$

In a flashlight, one of these batteries provides 0.0048 A. If the battery contains 4.0 g of $MnO_2$ and fails after 90% of its $MnO_2$ is consumed, calculate the operating life of the flashlight.

16.90 Electrolytic reactions, like other chemical reactions, are not 100% efficient. In a copper purification apparatus depositing Cu from $CuSO_4$ solution, operation for 5.0 hours at constant current of 5.8 A deposits 32 g of Cu metal. What is the efficiency?

16.91 A galvanic cell consists of a Pt electrode immersed in a solution containing $Fe^{2+}$ at 1.00 M and $Fe^{3+}$ at unknown concentration, as well as a Cu electrode immersed in a 1.00 M solution of $Cu^{2+}$. The cell voltage is 0.00 V. What is the concentration of $Fe^{3+}$ ions?

16.92 A galvanic cell is constructed using a silver wire coated with silver chloride and a nickel wire immersed in a beaker containing $1.50 \times 10^{-2}$ M $NiCl_2$. (a) Determine the balanced cell reaction. (b) Calculate the potential of the cell. (c) Draw a sketch showing the electron transfer reaction occurring at each electrode.

16.93 The sketch below shows a cell set up to electrolyze molten NaCl.

(a) What reactions occur? (b) Identify the anode and cathode. (c) Redraw the sketch showing the direction of electron flow, and include molecular pictures showing the processes at each electrode.

**16.94** Given that $E° = -0.34$ V for the reduction of $Tl^+$ to Tl, find the voltage developed by a cell consisting of Tl metal dipping in an aqueous solution that is 0.050 M in $Tl^+$, connected by a porous bridge to a 0.50 M aqueous solution of HCl in contact with a Pt electrode over which $H_2$ gas is bubbling at $p = 0.90$ bar.

**16.95** Draw a sketch of the cell in Problem 16.94. Include molecular views of the processes taking place at the electrodes.

**16.96** List all the metals that could be used as sacrificial anodes for iron. Which of these could also be sacrificial anodes for aluminum?

**16.97** For the reaction between strontium metal and magnesium ions $K_{eq} = 2.7 \times 10^{12}$:

$$Sr\ (s)\ +\ Mg^{2+}\ (aq) \rightleftharpoons Sr^{2+}\ (aq)\ +\ Mg\ (s)$$

Calculate $E°$ for a strontium–magnesium battery.

## GROUP STUDY PROBLEMS

**16.98** Use standard reduction potentials from Table 16-1 and Appendix F to determine $K_{sp}$ for as many metal hydroxides as the table allows. Compare your values with those in Appendix E. If there are $K_{sp}$ values for hydroxides in Appendix E that cannot be calculated from standard reduction potentials in Appendix F, use the $K_{sp}$ values to calculate the appropriate standard reduction potentials.

**16.99** Consider an electrochemical cell consisting of two vessels connected by a porous separator. One vessel contains a 0.500 M HCl solution and an Ag wire electrode coated with AgCl solid. The other vessel contains a 1.00 M $MgCl_2$ solution and an Mg wire electrode. (a) Determine the net reaction. (b) Calculate $E$ for the cell (see Appendix F). (c) Draw a molecular picture showing the reactions at each electrode.

**16.100** Use data from Appendix F for the following reaction:

$$5\ I^-\ (aq)\ +\ IO_3^-\ (aq)\ +\ 6\ H_3O^+\ (aq) \longrightarrow 3\ I_2\ (s)\ +\ 9\ H_2O\ (l)$$

(a) Determine the spontaneous direction at pH = 2.00 and $[I^-] = [IO_3^-] = 0.100$ M. (b) Repeat the calculation at pH = 11.00. (c) At what pH is this redox reaction at equilibrium at these concentrations of $I^-$ and $IO_3^-$?

**16.101** From the standard reduction potentials appearing in Table 16-1 and Appendix F, identify reaction pairs that are candidates for batteries that would produce more than 5 V of electrical potential under standard conditions. Suggest chemical reasons why no such battery has been commercially developed.

**16.102** A chemist wanted to determine $E°$ for the $Ru^{3+}/Ru$ reduction reaction. The chemist had all the equipment needed to make potential measurements, but the only chemicals available were $RuCl_3$, a piece of ruthenium wire, $CuSO_4$, copper wire, and water. Describe and

sketch a cell that the chemist could set up to determine this $E°$. Show how the measured voltage would be related to $E°$ of the half-reaction. If the cell has a measured voltage of 1.44 V, with the ruthenium wire being negative, determine $E°$ for $Ru^{3+}/Ru$.

## ANSWERS TO SECTION EXERCISES

**16.1.1** (a) K = +1, O = −2, Mn = +7; (b) Cl = −1, Si = +4; (c) Na = +1, O = −2, S = +4; (d) S = 0; and (e) Li = +1, Al = +3, H = −1

**16.1.2** (a) redox, C oxidized, $O_2$ reduced; (b) not redox; (c) redox, Mg oxidized, Si reduced; (d) redox, S oxidized, O reduced; and (e) not redox

**16.1.3** (a) −3; (b) 0; (c) −2; (d) +5; (e) +3; (f) +2; and (g) +4

**16.2.1** (a) $SO_3^{2-} + H_2O_2 \longrightarrow SO_4^{2-} + H_2O$; (b) $3\ BrO_2^- \longrightarrow Br^- + 2\ BrO_3^-$

**16.2.2** $2\ Al + 3\ Ag_2S + 3\ H_2O + 3\ OH^- \longrightarrow 6\ Ag + 2\ Al(OH)_3 + 3\ HS^-$

**16.2.3** $3\ CuS + 11\ H_3O^+ + 8\ NO_3^- \longrightarrow 8\ NO + 15\ H_2O + 3\ Cu^{2+} + 3\ HSO_4^-$

**16.3.1**

OH⁻ carries charge from solution to electrodes

Al loses electrons to become $Al_2O_3$

$Fe_2O_3$ gains electrons to become Fe

**16.3.2** $\Delta G° = -840.1$ kJ; negative value verifies spontaneity.

**16.3.3**

**Active electrode:** When an $Fe^{2+}$ ion collides with the surface of the active electrode, $Fe^{2+}$ gains two electrons and becomes part of the solid Fe electrode.

**Passive electrode:** When an $Fe^{2+}$ ion collides with the surface of the passive electrode, $Fe^{2+}$ loses one electron and becomes an $Fe^{3+}$ ion, which remains dissoved in solution.

**16.4.1** (a) $2\ Cr + 3\ Fe^{2+} \longrightarrow 2\ Cr^{3+} + 3\ Fe$, $E° = 0.297$ V

(b) $Pb + Cu^{2+} \longrightarrow Cu + Pb^{2+}$, $E° = 0.4681$ V

(c) $Au^{3+} + 3\ Ag \longrightarrow Au + 3\ Ag^{+}$, $E° = 0.698$ V

**16.4.2**

When a $Cu^{2+}$ ion collides with the Pb surface, $Cu^{2+}$ is reduced to Cu (elemental copper), and a Pb atom is oxidized to $Pb^{2+}$. Cu is insoluble in water, so it sticks to the Pb surface; $Pb^{2+}$ is soluble in water, so it dissolves.

**16.4.3** 2.0501 V

**16.5.1** (a) $\Delta G° = -63.7$ kJ, $K_{eq} = 1.4 \times 10^{11}$

(b) $\Delta G° = 1.14 \times 10^{3}$ kJ, $K_{eq} = 10^{-200}$

**16.5.2** (a) 0.281 V; (b) 0.4681 V; and (c) 0.761 V

**16.5.3** (a) $FeS_2$; (b) 804 C; and (c) 6.21 days

**16.6.1**

HgO captures 2 electrons to form Hg and $O^{2-}$

$O^{2-}$ reacts with $H_2O$ to form 2 OH$^-$

**16.6.2** At the aluminum electrode, the deposit of aluminum oxide is impervious to penetration by ions, so it blocks the passage of current.

**16.6.3** $E°$ for tin is less negative than that for iron, whereas $E°$ for zinc is more negative than that for iron. Thus, iron oxidizes instead of tin, but zinc oxidizes instead of iron.

**16.7.1** (a) $2\ HF + 2\ e^- \longrightarrow H_2 + 2\ F^-$ and $2\ F^- \longrightarrow F_2 + 2\ e^-$; and (b) 3.053 V

**16.7.2** 25.5 kg, requiring a tank of volume 278 L

**16.7.3** In aqueous KF, water would be electrolyzed rather than KF, generating $O_2$ instead of $F_2$; liquid HF does not conduct electricity and therefore does not support electrolysis.

## ANSWERS TO PRACTICE EXERCISES

**16.1** NiO, O is −2, Ni is +2; C, C is 0; Ni, Ni is 0; CO, O is −2, C is +2; redox reaction. Ni, Ni is 0; CO, O is −2, C is +2; Ni(CO)$_4$, CO has not changed, so O is −2, C is +2, and Ni is 0; not redox

**16.2** $H_2SO_4$, O is −2, H is +1, S is +6; $KClO_3$, O is −2, K is +1, Cl is +5

**16.3** (a) $Cu \longrightarrow Cu^{2+}$ and $NO_3^- \longrightarrow NO$;
(b) $ClO^- \longrightarrow ClO_3^-$ and $ClO^- \longrightarrow Cl^-$

**16.4** (a) $Cu \longrightarrow Cu^{2+} + 2\,e^-$ and $NO_3^- + 4\,H_3O^+ + 3\,e^- \longrightarrow NO + 6\,H_2O$
(b) $ClO^- + 6\,H_2O \longrightarrow ClO_3^- + 4\,H_3O^+ + 4\,e^-$ and $ClO^- + 2\,H_3O^+ + 2\,e^- \longrightarrow Cl^- + 3\,H_2O$

**16.5** (a) $3\,Cu + 2\,NO_3^- + 8\,H_3O^+ \longrightarrow 3\,Cu^{2+} + 2\,NO + 12\,H_2O$; (b) $3\,ClO^- \longrightarrow ClO_3^- + 2\,Cl^-$

**16.6** $2\,MnO_4^- + 3\,ClO_3^- + 2\,H_3O^+ \longrightarrow 2\,MnO_2 + 3\,ClO_4^- + 3\,H_2O$

**16.7** $2\,MnO_4^- + 3\,Cd + 4\,H_2O \longrightarrow 2\,MnO_2 + 3\,Cd(OH)_2 + 2\,OH^-$

**16.8** Oxidation half-reaction is
$Zn + 2\,OH^- \longrightarrow ZnO + H_2O + 2\,e^-$;
reduction half-reaction is
$HgO + H_2O + 2\,e^- \longrightarrow Hg + 2\,OH^-$; overall reaction is
$Zn + HgO \longrightarrow ZnO + Hg$.

**16.9** (a) Pb (s) | PbSO$_4$ (s) | HSO$_4^-$ (aq), H$_3$O$^+$ (aq) || HSO$_4^-$ (aq), H$_3$O$^+$ (aq) | PbO$_2$ (s) | PbSO$_4$ (s) | Pb (s)
(b) $6\,Br^-$ (aq) $+ Cr_2O_7^{2-}$ (aq) $+ 14\,H^+$ (aq) $\longrightarrow$ $3\,Br_2$ (l) $+ 2\,Cr^{3+}$ (aq) $+ 7\,H_2O$ (l)

**16.10** $E° = 0.789$ V; net reaction is
$Cu^{2+}$ (aq) $+ Fe$ (s) $\longrightarrow Cu$ (s) $+ Fe^{2+}$ (aq)

**16.11** $In^{3+}$ (aq) $+ 3\,e^- \rightleftarrows In$ (s), $E° = -0.338$ V; net reaction is
$3\,Cu^{2+}$ (aq) $+ 2\,In$ (s) $\longrightarrow 3\,Cu$ (s) $+ 2\,In^{3+}$ (aq).

**16.12** $K_{eq} = 0.43$

**16.13** $E = 0.070$ V

**16.14** $[S_4O_6^{2-}] = 0.55$M

**16.15** 135 hours

**16.16** 23 mg of Li is transferred.

**16.17** 6.92 g of Ag plates onto the spoon.

# MACROMOLECULES

## SILK, A NATURAL MACROMOLECULE

Stumble into a spider web and you will experience one of the strongest substances known. Spiders spin their webs from silk, a huge molecule made up of a vast chain of atoms, as shown in our molecular inset. A spider's web is at least five times stronger than steel. What's more, spider silk is waterproof and highly flexible. These properties come from the chemical bonding of the silk structure, as well as the intermolecular forces between the chains.

The feeling of a spider web may be unsettling, but a similar natural material has been used for centuries to make silk fabric that is prized for its smooth texture. Silkworms produce the silk fibres used to make clothing. They feast on mulberry leaves and convert the molecules from these leaves into silk, from which they spin cocoons.

The ancient Chinese discovered how to harvest silkworm cocoons, boil them to loosen the tangle, and unravel the silk into a fibre from which elegant clothing could be produced. A single silkworm cocoon can yield nearly a kilometre-long filament of silk, but the filament is so fine that it takes around 30 mulberry trees to yield enough cocoons to make one kilogram of silk.

The discovery of silk in China occurred many centuries BC, and by the time of the Roman Empire, silk fabric was a prized trade commodity. The caravan routes across Asia became known as the Silk Road. It is estimated that nearly 90% of the imports into the Roman Empire consisted of silk goods.

As we describe in this chapter, silk is a macromolecule, an extremely long molecule built from relatively simple molecular building blocks. Silk is made up of proteins, which are long sequences of amino acids linked together into very long chains. Spiders and silkworms secrete liquid proteins that solidify into strands of silk.

Chemists learned the basic composition of silk many years ago, but the reasons why this macromolecule is so strong, and yet flexible, are still not fully understood. Recent studies indicate that the secret lies in the way the chains of this protein nestle together.

Current research efforts focus on using techniques of genetic engineering to replicate natural spider silk on a useful scale.

Silk is just one example of macromolecules, also known as polymers, which are the subject of this chapter. In this chapter, we outline the principles of the structure and synthesis of the major classes of macromolecules and describe the properties that give these chemical substances central roles in industrial chemistry and biochemistry. We describe the components from which macromolecules are constructed, some important industrial polymers, and the macromolecules found in living systems.

# 17

## CHAPTER CONTENTS

## LEARNING OBJECTIVES

*Upon completion of this chapter you should be able to*

- describe functional groups and linkage groups in polymers
- describe polymers made by free radical polymerization
- describe polymers made by condensation polymerization
- recognize and describe some properties of plastics, fibres, and elastomers
- recognize and draw structures of monosaccharides and polysaccharides
- draw primary and secondary structures of DNA and RNA
- explain primary, secondary, and tertiary structures of proteins

## 17.1 STARTING MATERIALS FOR POLYMERS

The prefix "poly" means "many," and **polymers** are very large molecules constructed by linking together many copies of much smaller molecules called **monomers**. You can think of a polymer as the molecular equivalent of a very long chain of paper clips, of which a single paper clip is the monomer. Polymers are one class of **macromolecules**, which are very large molecules. Whereas all polymers are macromolecules, not all macromolecules are polymers. Some macromolecules that are not polymers appear in earlier chapters. A diamond, for example, is a macromolecule made entirely of carbon atoms. A grain of sand is a macromolecule made of silicon and oxygen atoms, in a 1:2 ratio, bonded together in a three-dimensional array.

In this chapter, we focus on polymers—from the plastics that are essential in our everyday lives to the proteins and DNA that are the materials of life. To understand the chemistry of polymers, we need to understand how monomers are able to link together into long chains. We begin with a description of the various monomers that are the most important starting materials for industrial and natural polymers.

For a polymer to form, monomers must react with one another to form links. Most monomers are organic molecules containing particular functional groups. The most important functional groups that participate in polymerization reactions are listed in Table 17-1. All these groups have electron pairs that can be incorporated

## TABLE 17-1 POLYMERIZABLE FUNCTIONAL GROUPS

| Formula | Name | Class | Example | | Reaction Partners |
|---|---|---|---|---|---|
| C=C | Double bond | Alkene | Ethylene | | Alkenes |
| —Ö—H | Hydroxyl | Alcohol | Ethanol | | Acids, amines, aldehydes, alcohols |
| Aldehyde | Aldehyde | Aldehyde | Formaldehyde | | Alcohols, amines |
| Carboxyl | Carboxyl | Acid | Acetic acid | | Alcohols, amines |
| Amine | Amine | Amine | Ethylenediamine | | Acids, alcohols |
| Phosphate | Phosphate | Phosphate | Phosphoric acid | | Phosphates, alcohols |
| —S̈—H | Sulphydryl | Thiol | Cysteine | | Thiols |

into new chemical bonds relatively easily. The $\pi$ electrons in a double bond are more reactive than electrons in $\sigma$ bonds, and the lone pairs of electrons on O, N, and S atoms are available for bond formation.

## C=C Double Bonds

A carbon–carbon double bond is a reactive functional group because of its $\pi$ electrons. Remember from Chapter 7 that ethylene has a C=C bond made up of one $\sigma$ bond plus one $\pi$ bond. As shown in Figure 17-1, the electrons in the $\pi$ bond are located off the $\sigma$ bond axis, making them more readily available for chemical reactions. Moreover, $\pi$ electrons are less tightly bound than $\sigma$ electrons. Consequently, the reactivity patterns of ethylene are dominated by the chemistry of its $\pi$ electrons. Polyethylene is one familiar polymer of which the monomer is ethylene. We describe the polymerization reaction of ethylene and other monomers containing C=C bonds in Section 17.2.

**Figure 17-1**

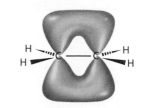

The $\pi$ electrons are readily available for chemical reactions that form bonds to other molecules.

## —OH and —SH

As described in Chapter 10, an **alcohol** is a molecule that contains a hydroxyl group (—OH) covalently bonded to carbon. Sugar molecules, which are described in Section 17.5, contain several hydroxyl groups bonded to carbon. Starch and cellulose are biological polymers containing sugar monomer units. Some chemical reactions break the O—H bond, whereas others break the C—O bond of an alcohol or sugar. In addition to forming linkages, the O—H bond is highly polar, so alcohols and sugars form hydrogen bonds readily. We describe later how hydrogen bonds play important roles in the properties of polymers, especially those that have biochemical importance.

CH₃OH
Methanol

CH₃CH₂OH
Ethanol

The hydroxyl group of an alcohol, which is covalently bonded to a carbon atom, has completely different properties than the hydroxide ion, OH⁻, which is present in salts such as NaOH and Mg(OH)₂.

A **thiol** contains an —SH group covalently bonded to carbon. Sulphur is just below oxygen in the periodic table, so a thiol is somewhat similar to an alcohol. Still, the chemical and physical properties of thiols differ significantly from those of alcohols. For example, whereas alcohols have inoffensive odours, thiols smell bad. The stench of skunk scent is due to thiols, including 3-methylbutanethiol. Thiols are important in proteins because of their abilities to form S—S linkages, which are described in Section 17-7.

## —NH₂

**Amines** are important compounds in macromolecular science. The —NH₂ group is polar and forms hydrogen bonds. Moreover, the lone pair of electrons on the nitrogen atom can accept a proton from a hydronium ion:

Skunk scent contains 3-methylbutanethiol.

When they are protonated, amines are compatible with water, a property critical in biochemical processes because it helps biological macromolecules dissolve in water. The N—H bond in an amine is fairly easy to break, so amines participate in the formation of several important classes of polymers, including nylon and proteins.

## The Carbonyl and Carboxyl Groups

The chemistry of carbonyl-containing functional groups was discussed in Chapter 10. In polymer chemistry, carboxylic acids, aldehydes, and, to a lesser extent, ketones, are all important. The chemistry of aldehydes and ketones is dominated by the polarity of the carbonyl group. The difference in electronegativity between carbon and oxygen polarizes the double bond, creating a partial positive charge on the carbon atom and a partial negative charge on the oxygen atom.

Formaldehyde

Although carboxylic acids react in many different ways, breaking the C—OH bond is the only reaction that is important in polymer formation. A carboxylic acid is highly polar and can give up H to form a carboxylate anion, $R$—$CO_2^-$. The carboxyl group (—$CO_2H$) also forms hydrogen bonds readily. These properties enhance the solubility of carboxylic acids in water, a particularly important property for biochemical macromolecules.

Carboxylic acid

## Phosphates

Phosphate

The **phosphate** group is derived from phosphoric acid ($H_3PO_4$) by replacing an O—H bond by an O—C or O—P bond. Phosphate is an important functional group in biochemistry, being involved in cellular energy production as well as acting as an important monomer in biopolymers, particularly in DNA. Bonds to phosphate groups form or break in the course of a number of important biochemical reactions.

To summarize from the perspective of polymer formation, the most important role of functional groups in polymerization is to provide bonds that are relatively easy to break. Because C—H and C—C $\sigma$ bonds are relatively strong and do not break easily, polymerization requires monomers that contain reactive functional groups. To form polymers, bonds in these groups must break, and new bonds that link monomers into macromolecules must form.

## Linkage Groups

When two monomers react by combining their functional groups, the result is a new functional group called a *linkage group*. The three linkage groups listed in Table 17-2 are particularly important in polymerization reactions. As a linkage forms, a small molecule is produced, often but not always water. A reaction that produces a linkage group and eliminates a small molecule is

## TABLE 17-2 IMPORTANT POLYMER LINKAGE GROUPS

| Linkage | Name | Precursors | Polymer Type |
|---|---|---|---|
| (ester structure) | Ester | Acid + alcohol | Polyesters |
| (amide structure) | Amide | Acid + amine | Polyamides, proteins |
| C—O—C | Ether | Alcohol + alcohol | Cellulose, starch |

called a **condensation reaction**. An ester is a simple example. A water molecule is eliminated as the oxygen atom of the alcohol links to the carbon atom of the carbonyl group. For example, ethanol reacts with acetic acid to give ethyl acetate:

$$CH_3CH_2-O-H \quad HO-\underset{\underset{CH_3}{}}{\overset{\overset{O}{||}}{C}} \quad \longrightarrow \quad CH_3CH_2-O-\underset{\underset{CH_3}{}}{\overset{\overset{O}{||}}{C}} \quad + \quad H_2O$$

Ethanol    Acetic acid       Ethyl acetate
(an alcohol) (a carboxylic acid)     (an ester)

A carboxylic acid can condense with an amine to form an amide, with the elimination of water. The amide linkage occurs in nature in proteins, biological macromolecules with amazing structural and functional diversity. Silk, described in the introduction to this chapter, is one such molecule. The monomers from which proteins form are *amino acids*, each of which contains both an amine group and a carboxyl group. The amino acids are strung together through a series of condensation reactions that give amide linkages.

The condensation reaction of two alcohol molecules to eliminate water and form a C—O—C bond (an ether) is also the basis of some common polymerizations. The direct condensation of alcohols has limited use as a route to polymers, however, because many alcohols undergo other reactions under the conditions required for condensation. Nature forges ether linkages in a more roundabout way than by eliminating water directly. The most important polymers containing ether linkages are two biochemical macromolecules, starch and cellulose.

Two molecules of phosphoric acid can undergo a condensation reaction, eliminating a water molecule and forming a P—O—P phosphate linkage:

Amine group    Carboxyl group

$$HO-\underset{\underset{OH}{|}}{\overset{\overset{O}{||}}{P}}-OH \quad H-O-\underset{\underset{OH}{|}}{\overset{\overset{O}{||}}{P}}-OH \quad \longrightarrow \quad HO-\underset{\underset{OH}{|}}{\overset{\overset{O}{||}}{P}}-O-\underset{\underset{OH}{|}}{\overset{\overset{O}{||}}{P}}-OH \quad + \quad H_2O$$

Phosphoric acid or phosphate can also condense with an alcohol to generate a phosphate linkage. As described later, such linkages are found in the structure of DNA.

$$HO-\overset{\overset{\displaystyle O}{\|}}{\underset{\underset{\displaystyle OH}{|}}{P}}-OH \quad H-O-\overset{\overset{\displaystyle H}{|}}{\underset{\underset{\displaystyle H}{|}}{C}}-H \longrightarrow HO-\overset{\overset{\displaystyle O}{\|}}{\underset{\underset{\displaystyle OH}{|}}{P}}-O-\overset{\overset{\displaystyle H}{|}}{\underset{\underset{\displaystyle H}{|}}{C}}-H + H_2O$$

## SECTION EXERCISES

● **17.1.1** Molecular pictures of several molecules appear below. Identify the functional groups or linkage groups present in each. (Some contain more than one functional group.)

Ethylene glycol          Alanine          Methyl acetate

● **17.1.2** Amino acids are the molecular building blocks of proteins. Any two amino acids can condense in two ways; each creates an amide linkage. Draw the structures of the two condensation products that link these two amino acids:

Alanine
(C_3H_7NO_2)          Phenylalanine
(C_9H_{11}NO_2)

● **17.1.3** Draw one complete specific line structure for each of the following: (a) an amine with the formula $C_6H_{15}N$; (b) an ester with at least seven carbon atoms; (c) an alcohol with a molar mass of 74 g/mol; and (d) a molecule with at least 10 carbons, which contains an aldehyde group and a phenyl ring ($C_6H_5$).

## 17.2 FREE RADICAL POLYMERIZATION

The simplest of all polymers is **polyethylene**, formed by bonding many ethylene molecules together in long chains containing 500 or more repeating $CH_2$ groups.

The linkage between two ethylene molecules involves reactions of their $\pi$ electrons. An electron from the $\pi$ bond of one ethylene molecule pairs with an electron from the $\pi$ bond of another ethylene molecule to form a new $\sigma$ bond between the two molecules. The second electron from each $\pi$ bond pairs with a $\pi$ electron from another ethylene molecule to continue the chain.

Writing the structural formula of a macromolecule such as polyethylene with thousands of atoms would be very time-consuming and tedious. Fortunately, the entire structure of a polyethylene molecule can be represented by simply specifying its repeat unit, as shown in Figure 17-2.

## Synthesis of Polyethylene

Ethylene forms a polymer through a three-step sequence. First is an **initiation** step, in which a reactive chemical substance attacks the $\pi$ bond of a single ethylene molecule. The product of the initiation step is able to react readily with the $\pi$ bond of another ethylene molecule in a **propagation** step. Many propagation steps occur, each one lengthening the growing polymer chain by adding one monomer unit. Finally, chain growth comes to an end through a **termination** step. Initiation, propagation, and termination are important steps in the synthesis of many polymers.

The polymerization of ethylene starts with the thermal decomposition of an **initiator** molecule, the general formula for which is $R$—O—O—$R$. Heating breaks the weak O—O single bond to form a pair of $R$—O· **free radicals**. Free radicals are highly reactive molecules that contain unpaired electrons. A free radical will attack any bond that has exposed electron density. In this case, a free radical attacks the $\pi$ bond of an ethylene molecule:

**Figure 17-2**

Polyethylene has a repeating pattern of $[-CH_2-CH_2-]$ units, each one derived from an ethylene molecule. Shown here are the Lewis structure of the repeat unit and a ball-and-stick model of a segment of a polyethylene chain.

In the initiation reaction, one $\pi$ electron of the ethylene molecule pairs with the single electron on the oxygen atom of the free radical. Together, these two electrons create a new C—O $\sigma$ bond (shown highlighted in purple). The second electron from the ethylene bond remains as an unpaired electron on the outermost carbon atom of the product.

The carbon atom with the unpaired electron is another free radical, so the stage is set for propagation. The unpaired electron on the carbon atom attacks the $\pi$ bond of another molecule of ethylene, making a new C—C $\sigma$ bond and leaving yet another carbon free radical:

This addition reaction can be repeated over and over in the propagation process, and each step adds two $CH_2$ groups to the growing polymer chain. The propagation can be written in the following general form:

Because chain growth involves free radicals, this type of polymerization is called **free radical polymerization**.

Free radical polymerization does not go on indefinitely, because sometimes two free radicals collide and react. This destroys the two free radicals and terminates the chain. For example, another initiator fragment can react with the carbon atom at the end of the chain:

$$RO\left[CH_2\!-\!CH_2\right]_n\!CH_2\!-\!\dot{C}H_2 + RO\cdot \longrightarrow RO\left[CH_2\!-\!CH_2\right]_n\!CH_2\!-\!CH_2\!-\!OR$$

The product of the last reaction may not look like polyethylene, but remember that $n$ can be a very large number, (hundreds, thousands or even larger). Thus, the OR ends of the macromolecule are insignificant compared with the many $CH_2-CH_2$ repeat units that make up the bulk of the polymer.

## Other Polyalkenes

A variety of polymers can be made using derivatives of ethylene in which one or more of the hydrogen atoms is replaced with other groups of atoms. Changing the structure of the monomer makes it possible to adjust the properties of the polymer. Table 17-3 lists important polymers that are made from alkene monomers. You may recognize some of their common names. Figure 17-3 shows the molecular structures of five of these monomers. Notice the variety of groups that can replace hydrogen atoms: $CH_3$, Cl, F, CN, and $C_6H_5$. Teflon® is a polymer of tetrafluoroethylene, and Example 17-1 shows how to write the structure of another polyethylene derivative.

Polyalkenes are used to make many of the items used in everyday life.

### TABLE 17-3  IMPORTANT POLYMERS MADE FROM ALKENES

| Polymer | Uses | North American Production ($10^9$ kg)* |
|---|---|---|
| Polyethylene | Piping, bottles, toys | 18.0 |
| Polypropylene | Carpets, labware, toys | 8.8 |
| Polyvinylchloride (PVC) | Piping, floor tile, clothing | 6.6 |
| Polystyrene | Containers, heat insulators, styrofoam | 2.7 |
| Styrene–butadiene | Tires, fibres | 2.0 |
| *2007 figures. | | |

**Figure 17-3**

Vinyl chloride    Propylene    Tetrafluoroethylene    Styrene    Acrylonitrile

Structures of alkene monomers used in the manufacture of important polymers.

## EXAMPLE 17-1 DRAWING THE STRUCTURE OF A POLYMER

Polyacrylonitrile, known commercially as Orlon, is made by polymerizing acrylonitrile (see Figure 17-3). Orlon is used to make fibres for carpeting and clothing. Draw the Lewis structure of polyacrylonitrile, showing at least three repeat units.

**Strategy:** Polyalkenes form by linking carbon atoms in a free radical polymerization. The polymer structure is constructed by connecting monomer units. The polymerization process converts the bonds of the monomers to bonds between polymer repeat units.

**Solution:** Draw three monomers connected in a symmetrical way that shows the polymer chain. A CN group is bound to every other carbon:

**Does the Result Make Sense?** Acrylonitrile polymerizes in the same way as ethylene. Notice that this polymer has the same structure as polyethylene, except that a CN group is bound to every second carbon atom, so the structure is a reasonable one. A line structure of polyacrylonitrile eliminates the clutter caused by the H atoms. A ball-and-stick model of the same polymer segment is included for comparison.

**Practice Exercise 17.1**

The polymer known as PVC is used to make plastic pipes. The monomer used to make PVC is vinyl chloride. Draw a segment of the PVC polymer containing four repeat units. See Figure 17-3 for the structure of vinyl chloride.

## Rubber

The first alkene polymer to be used in society was polyisoprene, a natural product extracted from the sap of rubber trees. See our Focus 17-1: Chemical Milestones for a description of the history of rubber. The monomer from which this polymer is constructed is isoprene, an alkene with two double bonds. The line structure of isoprene and the repeat unit of polyisoprene are shown in Figure 17-4. The figure also shows polybutadiene, a synthetic rubber, and its monomer, butadiene. When isoprene or butadiene polymerize, attack on one $\pi$ bond causes the second $\pi$ bond to shift, giving a macromolecule containing a four-carbon repeat unit with a double bond in its centre.

Natural rubber and polybutadiene each contain just one monomer, but the versatility of rubber materials is greatly increased by polymerizing mixtures of two different monomers to give **copolymers**. Because of their diverse properties,

**Figure 17-4**

Isoprene

Polyisoprene (natural rubber)

Butadiene

Polybutadiene (synthetic rubber)

Structures of isoprene, butadiene, and their polymers.

copolymers account for a large portion of the rubber produced each year. The largest production is a polymer of three-parts butadiene to one-part styrene and has the common name SBR (styrene butadiene rubber). During polymerization of butadiene and styrene, the chain grows by adding whichever monomer happens to collide with the chain end. The propagation process is random, so the exact sequence varies, but the resulting polymer contains three butadiene units for every styrene unit. Here is the line structure of a representative portion of such a polymer, with styrene units highlighted in light brown and butadiene units in gold:

Butadiene–styrene copolymer

At any point in the chain, there is a 75% probability of finding a butadiene unit and a 25% probability of finding a styrene unit.

## FOCUS 17-1 CHEMICAL MILESTONES: THE NATURAL RUBBER INDUSTRY

The rubber industry has a long and colourful history. Natural rubber is produced from latex, a milky fluid found in cells that lie between the bark and the wood of many plants. You may have seen latex flow from the broken stalks of milkweed plants, but the source of commercial rubber is the Hevea tree, a native of Brazil. When the bark of this tree is slashed, its milky white sap oozes out and can be collected in cups mounted on the tree's trunk. The people of the Amazon jungle made bouncing balls, shoes, and water jars out of rubber, and Portuguese explorers sent waterproof boots and a rubber-coated coat back to their king. The first commercial exports included some rubber shoes shipped to Boston in 1823.

Entrepreneurs in Europe and the United States experimented with rubber hats, coats, erasers, life preservers, and a number of other products. Through the first half of the nineteenth century, however, rubber items were impractical because natural rubber becomes sticky in hot weather and brittle in cold weather. In 1830, the world consumed just 156 tonnes of rubber, all from Brazil.

Nine years later, Charles Goodyear discovered how to stabilize rubber through vulcanization, and rubber quickly became an essential part of the world economy. Goodyear's discovery set off a rubber boom that rivalled the California gold rush. Adventurers swarmed to the jungles to amass fortunes as the price of rubber soared. The Amazon people were captured and forced to work at harvesting the latex fluid. Many of them died under the harsh conditions. New towns sprang to life along the river as paddle-wheel steamers carried luxury goods to rubber barons, who built grand mansions up the tributaries of the Amazon. The pinnacle of the rubber boom was the town of Manaus, which is located 1200 km up the Amazon system on the Negro river. At its height, Manaus teemed with mansions, cafes, and fine hotels. It had a streetcar system and even an enormous ornate opera house.

The Amazon rubber industry collapsed almost overnight. In 1876, the English botanist Henry Wickham shipped 70 000 Hevea seeds to the Royal Botanic Gardens in London. New strains of Hevea were developed that produced three to four times as much rubber and were more disease-resistant than their wild Amazonian cousins. Soon, seedlings were sent to Malaya, Java, and other islands of the East Indies. Thirty-five years later, rubber plantations on these islands took control of the industry.

In 1910, the price of rubber was $6300 (U.S.) per tonne, world production was 94 000 tonnes, and wild rubber from Brazil accounted for 83 000 tonnes. In 1912, the price collapsed, and by 1932, a tonne of rubber cost $55 (U.S.) per tonne. In 1937, the world bought more than 1.1 million tonnes of rubber, less than 2% of which came from Brazil. The rubber cities of the Amazon became near ghost towns as the market for wild rubber vanished. The opulent opera house in Manaus had barely opened its doors when the curtain came down on the Brazilian rubber economy.

The rubber industry changed again when the Japanese captured the East Indian rubber plantations during World War II. The resulting shortage of rubber prompted an intensive research program to produce synthetic rubber. Natural rubber is still produced in the tropics, but its importance pales compared with the glory days of the Brazilian rubber plantations.

The Manaus opera house

## Cross-Linking

The long-chain molecules described so far do not have the durability and strength associated with rubber products such as rubber bands and automobile tires. To achieve these properties, rubber must be treated chemically to create chemical bonds between long-chain molecules. This process is called **cross-linking**, because links are formed *across* the chains in addition to bonds *along* the chains. Vulcanization, the first way to form cross-links in rubber, was discovered (apparently by accident) in 1839 by Charles Goodyear, founder of the first U.S. rubber company.

The basic chemistry of vulcanization is simple. Sulphur is added to the polymer, and the mixture is heated under controlled conditions. Some of the polymer C—H bonds break and are replaced by C—S bonds. The cross-links consist of C—$(S_n)$—C chains, with *n* being 2 or more. Figure 17-5 represents the overall scheme as it applies to polybutadiene.

Cross-linking increases the tensile strength of rubber while also increasing its rigidity. A small amount of cross-linking gives rubber bands the ability to stretch and rebound, whereas a larger amount gives truck tires their toughness. Too much cross-linking is detrimental, however. If the sulphur content exceeds 10% by mass, vulcanized rubber becomes hard and brittle. Thus, the amount of cross-linking must be controlled to optimize flexibility and tensile strength without generating a brittle product.

**Figure 17-5**

When heated in the presence of sulphur, rubber forms cross-links between polymer chains.

Sulphur cross-links also play an important role in determining the structures of proteins, as discussed in Section 17.7.

---

## SECTION EXERCISES

●**17.2.1** Draw the structure of the polymer formed from each of these monomers:

(a) Propene ($C_3H_6$)    (b) Tetrafluoroethylene ($C_2F_4$)    (c) Methyl methacrylate ($C_5H_8O_2$)

●**17.2.2** In a 3:1 copolymer of butadiene and styrene, the placement of butadiene and styrene fragments along the chain is random. Draw a line structure for a portion of the copolymer that has this sequence:

–butadiene–styrene–styrene–butadiene–.

●**17.2.3** Neoprene is a synthetic rubber used to make gaskets. A section of neoprene follows. Draw the structure of the monomer used to make neoprene.

## 17.3 CONDENSATION POLYMERIZATION

In a condensation reaction, two molecules form bonds between their functional groups by eliminating water or some other small molecule. If no additional functional groups are present, that is as far as the linkage process goes. Polymer formation via condensation requires that each monomer contain *two* linkage-forming functional groups. After the two molecules link together, there are still two functional groups that can link to two more monomers, allowing the chain to grow into a polymer.

Figure 17-6 shows that there are two arrangements of functional groups that make it possible for condensation polymers to form. A single monomer may contain two *different* functional groups capable of linking, or one monomer may contain two of one particular functional group while a second monomer contains two of another functional group that can link with the first.

Amino acids belong to the first category. As its name implies, an amino acid contains an amine and a carboxyl group. Amino acids join through amide linkages:

The product is a new amino acid with an amine group at one end and a carboxyl group at the other. Consequently, each end can form another amide linkage, extending the chain.

The condensation of a diacid with a diol (dialcohol) belongs to the second category. Here, a monomer with two carboxyl groups links to a monomer with two hydroxyl groups to give an ester:

**Figure 17-6**

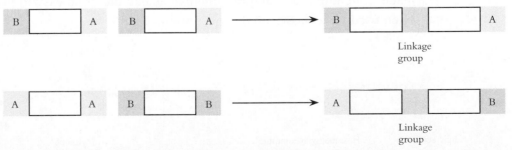

Two different arrangements of functional groups can give a condensation polymer. Notice that both condensation reactions generate a product that has an A end and a B end, allowing condensation to continue.

**Figure 17-7**

Nylon 6,6 can be drawn from a dish containing two starting materials. The bottom layer is an aqueous solution of 1,6-hexanediamine, and the top layer is adipoyl chloride (a more reactive derivative of adipic acid). Nylon 6,6 forms at the interface between the two liquids.

The product is a bifunctional molecule with a carboxyl group at one end and a hydroxyl group at the other. Chain growth can continue through repeated condensation reactions at both ends of the molecule.

Most polymers in production today are made from two different monomers because this synthetic route offers advantages over the use of bifunctional monomers. First, monomers with two identical functional groups are easier and less expensive to produce than monomers with two different groups. Second, the properties of the polymer can be varied easily by changing the structure of one of the monomers.

## Polyamides

Polymers that contain amide linkage groups are called **polyamides**. Proteins, which are biological polyamides, are described in Section 17.7. Here, we focus on two commercially important polyamides: Nylon 6,6 and Kevlar®.

Nylon 6,6 was the first polyamide to be produced commercially. Developed by Wallace Carothers at the DuPont Chemical Company in 1935, it still leads the polymer industry in annual production. Figure 17-7 illustrates that Nylon 6,6, made from adipic acid and hexamethylenediamine, is so easy to make that it is often used for a classroom demonstration:

Adipic acid · · · · · · · · · · · Hexamethylenediamine → Nylon 66 + $n\,H_2O$

Many other useful polymers can be made by combining different diamines and diacids. One example is Kevlar®, a polyamide made by condensing terephthalic acid with phenylenediamine:

Terephthalic acid　　　　Phenylenediamine

Figure 17-8 shows a portion of the structure of Kevlar®, which is so strong that it is used to make bulletproof vests. The tremendous strength of Kevlar® comes from a combination of intramolecular and intermolecular forces. The stiff phenyl rings make each polymer chain strong and rigid, while the chains are held together by hydrogen bonds. Interchain hydrogen bonds enhance the strength and durability of all polyamides, but Kevlar® has a particularly large number of hydrogen bonds per unit volume because of its compact, highly regular molecular structure.

We can identify the monomers from which a condensation polymer is made by visualizing the condensation reaction operating in reverse. The linkage bond breaks, and a small molecule is inserted. Example 17-2 illustrates this type of reasoning for a polyamide that was developed for use in fabrics.

**Figure 17-8**

A portion of the structure of Kevlar®. Pound for pound, this polymer is stronger than steel, and it has excellent flame resistance.

## EXAMPLE 17-2 THE STRUCTURE OF A POLYAMIDE

Qiana, a polyamide that feels much like silk, has the following structure:

Qiana cloth looks and feels much like silk.

Identify the monomers used to make Qiana.

**Strategy:** Visualize the condensation reaction that forms a polymer. Two monomers with appropriate functional groups combine, forming a new bond and eliminating water. Reverse this process to see what monomers make Qiana.

**Solution:** A polyamide is made from the condensation reaction of a diamine and a dicarboxylic acid. To identify the monomers, separate the amide linkage group and add water across the C—N bond:

Now construct the monomers:

**Does the Result Make Sense?** Deconstruction of the polyamide gives a pair of monomers, each one containing two of a different functional group. One monomer contains a pair of carboxyl groups, and the other contains a pair of amine groups. These are the components used to build amide linkages, so this monomer pair is reasonable.

### Practice Exercise 17.2

Nylon 6,12 forms stiff fibres that are used to make bristles for brushes. The repeat unit of Nylon 6,12 follows. Identify the monomers used to make this polyamide.

## Polyesters

Nylon was the first commercial polymer to make a substantial impact on the textile industry, but **polyesters** now comprise the largest segment of the market for synthetic fibres. In fact, polyesters account for 40% of the more than

4 billion kilograms of synthetic fibres produced in North America each year. The leading polyester, by far, is poly(ethylene terephthalate), or PET. This polymer is made from terephthalic acid and ethylene glycol in an acid–alcohol condensation reaction:

Terephthalic acid        Ethylene glycol        PET

A typical polyester molecule has 50 to 100 repeat units and a molar mass between 10 000 and 20 000 g/mol. PET can be formed into fibres (such as Dacron) or films (such as Mylar). Mylar films, which can be rolled into sheets 30 times thinner than a human hair, are used to make magnetic recording tape and packaging for frozen food. Dacron is best known for its use in clothing, but it has many other applications. For example, tubes of Dacron are used as synthetic blood vessels in heart bypass operations because Dacron is inert, non-allergenic, and non-inflammatory.

More than 2 billion kilograms of PET are produced in North America each year. This polymer is used to make tire cord, beverage bottles, home furnishings, small appliances, and many other common items.

## SECTION EXERCISES

● **17.3.1** The following monomers can be used to synthesize polymers:

Draw structures that show the repeat units of the polyamide and the polyester arising from these monomers.

● **17.3.2** Polycarbonates are colourless polymers nearly as tough as steel. One of the most common polycarbonates, Lexan, is used in bulletproof windows and as face plates in the helmets worn by astronauts. Lexan is made by condensing phosgene and a compound that contains two —OH functional groups, with the elimination of HCl:

Phosgene

Write the structure of Lexan. Show at least two complete repeat units of the polymer.

●**17.3.3** The following polymer is used to make carpets because its fibres have a single three-dimensional structure. Footprints quickly disappear from a carpet made of these fibres because compressed fibres quickly return to their most stable orientation. Identify the monomer used to make the polymer and write a balanced equation that shows the condensation of two of these monomers.

## 17.4 TYPES OF POLYMERS

To a chemist, a polymer is described by its chemical structure. To a manufacturer or a consumer, the important features of a polymer are its macroscopic properties. A tire manufacturer needs a flexible, tough, and durable polymer. A shirt-maker wants to spin a polymer into fibres that wear well, hold dyes, and are free of wrinkles. Makers of lenses look for transparent polymers that can be moulded into precise shapes. Each of these polymer applications has somewhat different requirements.

Although they have an endless variety of properties, polymers can be divided into three general categories, based on their form and resistance to stretching. These are **plastics**, **fibres**, and **elastomers**. Plastics differ in form from fibres: whereas plastics exist as blocks or sheets, fibres have been drawn into long threads. Unlike plastics or fibres, elastomers can be stretched without breaking. Polyethylene packaging films and polyvinylchloride (PVC) pipes are examples of plastics. Orlon carpets are made from polymer fibres, and rubber bands are elastomers. Some polymers, such as Nylon, can be formed into both plastics and fibres.

## Plastics

To the general public, the term *plastic* has become synonymous with *polymer*. More precisely, a plastic is the type of polymer that hardens on cooling or on evaporation of solvent, allowing it to be moulded or extruded into specific shapes or spread into thin films.

Plastics fall into two groups based on their response to heating. Those that melt or deform on heating are classed as **thermoplastic** materials, whereas plastics that retain their structural integrity are said to be **thermosetting**. Cross-linking makes it harder for a polymer to deform, so highly cross-linked polymers are thermosetting. Thermoplastic polymers have small amounts of cross-linking. For example, polyethylene consists of huge alkane molecules without cross-linking, held together only by dispersion forces between the chains. Polyethylene is a thermoplastic polymer that melts when heated because individual molecules acquire enough kinetic energy to overcome dispersion forces and slide past one another. In contrast, a rubber truck tire may be so extensively cross-linked that it can be viewed as an immense single molecule. To melt this structure would require the breakage of covalent bonds. This is a thermosetting polymer, which

**Figure 17-9**

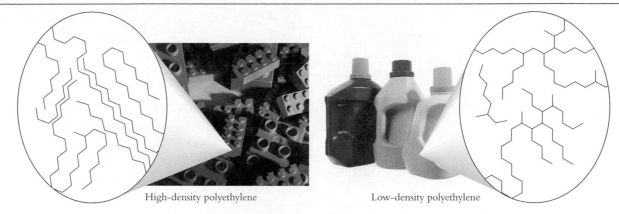

High-density polyethylene                    Low-density polyethylene

High-density polyethylene has aligned chains of $CH_2$ units, giving a tough, rigid polymer with a high melting point. Low-density polyethylene has branching chains that give a soft polymer with a low melting point.

may decompose irreversibly if it is heated to a high enough temperature but will not reversibly melt or deform.

Polyethylene, the best known thermoplastic material, exists in two general forms with different properties. High-density polyethylene is a rigid, strong polymer used to make bottle caps, toys, pipes, and cabinets for electronic devices such as computers and televisions. Low-density polyethylene is a soft, semirigid polymer used to make plastic bags, squeeze bottles, food packaging films, and other common items. Both kinds of polyethylene have the same repeat unit, $CH_2$, but different structures of individual polyethylene molecules, as shown in Figure 17-9.

High-density polyethylene forms under conditions that produce polymers made of straight chains of $CH_2$ units. These linear molecules maximize attraction resulting from dispersion forces by lining up in rows that create crystalline regions within the polymer. Maximizing attraction between the chains imparts strength and rigidity to the polymer. In contrast, low-density polyethylene has chains of $CH_2$ groups that branch off the main backbone of the polymer. These branches prevent the polymer molecules from packing closely together, thus decreasing dispersion forces and weakening the attraction between the chains. The result is an amorphous polymer that is flexible and melts at a relatively low temperature.

The flexibility of some plastics can be improved by the addition of small molecules called plasticizers. For example, pure PVC turns brittle and cracks too easily to make useful flexible plastic products. With an added plasticizer, however, PVC can be used to make seat covers for automobiles, raincoats, garden hoses, and other flexible plastic objects. Plasticizers must be liquids that mix readily with the polymer. In addition, they must have low volatility so that they do not escape rapidly from the plastic. Dioctylphthalate is a liquid plasticizer that is formed by condensing two molecules of 2-ethyl-hexan-1-ol with one molecule of phthalic acid, as illustrated in Figure 17-10. There is currently some concern over the toxicity of dioctylphthalate, especially to very young children.

**Figure 17-10**

Phthalic acid
(C₈H₆O₄)

2-ethyl-hexan-1-ol

Dioctylphthalate
(C₂₄H₃₈O₄)

+ 2 H₂O

The leading plasticizer dioctylphthalate is made by condensing two molecules of an eight-carbon alcohol with phthalic acid.

## Fibres

For centuries, humans have woven and spun cotton, wool, and silk into fabrics for clothing and other purposes. Like silk, cotton and wool are composed of polymer molecules. Today, synthetic fibres are equally important to the fabrics industry. Nylon 6, a polyamide with six carbon atoms in its repeat unit, makes resilient fibres that are used in carpeting. Part of the reason for the strength of these fibres is hydrogen bonding between neighbouring strands of the nylon polymer, as Figure 17-11 illustrates.

Synthetic fibres are thin threads of polymer made by forcing a fluid thermoplastic material through a set of tiny pores. This process requires that the polymer be in the liquid phase, so a fibre-forming polymer must melt at low temperature or dissolve in a convenient volatile solvent. A good fibre has to have high tensile strength, so most synthetic fibres are polyesters, polyamides, or polyacrylonitrile, all of which contain polar functional groups. These groups produce strong intermolecular forces that add significant tensile strength to the material.

**Figure 17-11**

One of the polymers used for carpet fibres is Nylon 6, in which neighbouring polymer strands are held together by hydrogen bonds.

## Elastomers

An elastomer is a polymer that is flexible, allowing it to be distorted from one shape to another. Polyisoprene (natural rubber), polybutadiene, and styrene-butadiene rubber are the most important commercial elastomers. All contain some C=C bonds, and their bulk properties are affected by the varying geometries about the carbon atoms that make up the polymer backbone.

The geometry about the carbon atoms of the CH₂ groups is tetrahedral, whereas that about the carbon atoms involved in double bonds is trigonal. Because of these varied shapes, the molecules lack the structural regularity required to form a polymer with crystalline properties. Consequently, elastomers such as polyisoprene are amorphous solids with individual polymer strands tangled together. This lack of structural regularity keeps the molecules from approaching too closely, thus minimizing attraction resulting from dispersion forces. This in turn allows individual polymer molecules to slide past one another and makes the polymer flexible.

Table 17-4 shows that the flexibility and strength of an elastomer are highly dependent on its amount of cross-linking. The linking of chains increases the strength of the elastomer but reduces its flexibility. Cross-linking also increases the ability of the elastomer to return to its original shape after distortion, as Figure 17-12 illustrates. For uses such as automobile tires that require both strength and flexibility, the amount of cross-linking must be carefully controlled.

## TABLE 17-4 EFFECTS OF CROSS-LINKING ON RUBBER

| Monomer Units Between Cross-Links | Degree of Flexibility | Strength | Product |
|---|---|---|---|
| 5–10 | Small | Very high | Casing for calculators |
| 10–20 | Restricted | High | Tires |
| 20–30 | Moderate | Moderate | Tire tubes |
| 30–40 | Moderate | Moderate | Artificial heart membrane |
| 50–80 | Moderate | Low | Kitchen gloves |
| 100–150 | High | Low | Surgical gloves |

## Polymer Stability

Any polymer that has commercial value must be stable under a variety of conditions. This means that it must not degrade when exposed to light, heat, or a variety of chemicals, including acids, bases, and oxidizing agents. Over time, polymers with superior stability have replaced less stable materials. For example, the first synthetic plastic, celluloid, is so highly flammable that it is no longer an important commercial polymer.

Many early motion pictures were recorded on film made from celluloid. Ping-Pong balls are among the few products still made from celluloid.

**Figure 17-12**

Stretching

The amount of cross-linking in an elastomer determines its strength and flexibility. A modest number of cross-links allows the material to stretch but then pulls it back to its original shape.

Among elastomers, artificial rubbers have replaced natural rubber for many uses because of their high resistance to chemical attack by ozone, an atmospheric pollutant. When ozone reacts with polymer chains, it breaks $C=C$ $\pi$ bonds and introduces additional cross-linking. Breaking $\pi$ bonds causes the rubber to soften, and cross-linking makes it more brittle. Both changes eventually lead to rupture of the polymer structure.

Among fibres, nylon and polyester have captured large segments of the clothing and floor-covering markets, in part because they last longer than natural fibres such as silk and wool. Moth holes in a wool sweater demonstrate that living creatures degrade natural fibres. Artificial fibres, by contrast, have no such natural enemies.

The stability of polymers is advantageous while they are being used, but it becomes a liability when these materials need to be discarded. Up to 15% of the volume of municipal waste dumps is polymeric material. Unlike natural materials, polymeric trash does not readily decompose after it is discarded. Thus, polymer disposal has become a cumulative problem.

Recycling polymers is one way to minimize the disposal problem. Approximately 87% of Canadian households regularly recycle plastic materials. A major obstacle to recycling plastics, however, is the great variation in the composition of polymeric material. Polyethylene and polystyrene have different properties, and a mixture of the two is inferior to either. Recyclers must either separate different types of plastics or process the recycled material for less specialized uses. Manufacturers label plastic containers with numbers that indicate their polymer type and make it easier to recycle these materials. Table 17-5 shows the recycling number scheme.

Thermoplastics are more suitable for recycling than elastomers or thermosetting polymers. Thermoplastics can be heated above their melting temperatures and then recast into new shapes. Elastomers and thermosets, on the other hand, have extensive cross-linking networks that must be destroyed and then reformed in the process of recycling. Processes that destroy cross-linking, however, generally break down the polymer beyond the point at which it can be easily reconstituted.

### TABLE 17-5 RECYCLING CATEGORIES FOR PLASTICS

| Number | Polymer | Abbreviation |
|:------:|---------|:------------:|
| 1 | Poly(ethylene terephthalate) | PET |
| 2 | High-density polyethylene | HDPE |
| 3 | Polyvinylchloride (PVC) | P |
| 4 | Low-density polyethylene | LDPE |
| 5 | Polypropylene | PP |
| 6 | Polystyrene | PS |
| 7 | Other | — |

One way to reduce disposal problems is to make polymers degradable. This poses a substantial challenge to polymer chemists—to fine-tune polymer properties so that the materials are stable while in use but degrade readily to innocuous materials when their useful life is over. A promising approach is to use polymers structurally similar to those found in nature. The goal is to develop polymers that can be degraded by biological organisms. Starch, which is described in the next section, is a natural polymer with properties that can be tailored to various commercial requirements.

---

## SECTION EXERCISES

● **17.4.1** Both polystyrene (see Figure 17-3) and Kevlar® (Figure 17-8) contain phenyl rings; yet polystyrene is a thermoplastic material like polyethylene, whereas Kevlar® is tough and rigid. Explain why the presence of phenyl rings in polystyrene does not lead to the rigidity characteristic of Kevlar®.

● **17.4.2** Describe what category of polymer is used to make the following items: (a) toothbrush bristles; (b) toothbrush handles; (c) chewing gum.

● **17.4.3** Make a drawing similar to Figure 17-12 that illustrates how cross-linking reduces the stretchability of a polymer.

---

## 17.5 CARBOHYDRATES

Cellulose and starch are macromolecules with empirical formulas that resemble "hydrated carbon," $C_x(H_2O)_y$, where $x$ and $y$ are integers. The monomers from which these macromolecules are constructed are sugars such as glucose and fructose. These monomers and macromolecules are the **carbohydrates**. Structurally, carbohydrates are very different from simple combinations of carbon and water. Even the smallest carbohydrates contain carbon chains with hydrogen atoms, OH groups, and occasional ether linkages.

Carbohydrates are an important food source for most organisms. Glucose, fructose, and sucrose are small carbohydrate molecules that can be broken down rapidly to provide quick energy for cells. Large amounts of energy are stored in carbohydrate macromolecules called **polysaccharides**. For example, glycogen is a polysaccharide used by humans for long-term energy storage. Other polysaccharides, such as cellulose and chitin, are building materials for plants and animals.

**Monosaccharides**, the molecular units of the saccharides, are carbohydrate molecules containing between three and six carbon atoms. Oligosaccharides contain small chains of 2–10 monosaccharide units, and polysaccharides contain long-chain polymers of monosaccharides.

### Monosaccharides

Monosaccharides have the formula $(CH_2O)_n$, where $n$ is between 3 and 6. Of the 70 or so monosaccharides that are known, 20 occur in nature. The most important naturally occurring monosaccharides contain five carbons (pentoses) or six

**Figure 17-13**

OH
⁵H₂C   H       H   OH
    ⁴C   O   C¹
   H   C₃   C₂   H
    HO       OH

Ribose
(C₅H₁₀O₅)

⁶CH₂OH
    ⁵C   O   H
 H   H
⁴C   OH   H   C¹
HO   C₃   C₂   OH
   H       OH

α-Glucose
(C₆H₁₂O₆)

Ribose and α-glucose are monosaccharides that occur in nature. The carbon atoms are numbered for identification purposes.

**Figure 17-14**

⁶CH₂OH
    ⁵C   O   OH
 H   H
⁴C   OH   H   C¹
HO   C₃   C₂   H
   H       OH

β-Glucose

The structure of β-glucose differs from that of α-glucose in the placement of the H atom and the —OH group on carbon atom number 1.

carbons (hexoses). Structures of ribose, an important pentose, and α-glucose, a hexose that is the most common monosaccharide, are shown in Figure 17-13. As shown in the figure, it is customary to number the carbon atoms in a monosaccharide, beginning with the HCOH group adjacent to the ether linkage.

The structures of ribose and α-glucose exemplify the characteristics of most monosaccharides. Each is a cyclic compound with an oxygen atom forming an ether linkage in one of the ring positions. Monosaccharides exist in several structural forms, but their ring forms are the basic building blocks of polysaccharides. Ribose has four hydroxyl groups bonded to the ring, and α-glucose has five. Monosaccharides are soluble in water because their several hydroxyl groups form hydrogen bonds with water.

Because the six-membered ring of α-glucose is an important structural component in many polysaccharides, we examine its structure in more detail. As Figure 17-13 shows, each carbon atom in the glucose ring is bonded to one hydrogen atom. Four of the ring carbons are bonded to hydroxyl groups (—OH), and the fifth ring carbon atom is bonded to a —CH₂OH fragment.

Glucose exists as two different isomers, α- and β-glucose. The β-glucose isomer, which is slightly more stable, is shown in Figure 17-14. The two isomers differ only in the placement of —H and —OH on carbon atom number 1. Whereas the —OH on carbon 1 of α-glucose lies below the ring in the view shown in Figure 17-13, that same —OH lies above the ring in β-glucose. This difference may not appear significant, but it has profound importance, especially for glucose polymers. The shape of a polysaccharide formed from α-glucose is quite different from the shape of a polysaccharide formed from β-glucose.

Every atom in the glucose ring has a steric number of 4 and approximately tetrahedral geometry, so the glucose ring is puckered rather than planar. This gives glucose a "lawn chair" shape. Although the ring is not planar, we can define a molecular plane that passes through the midpoints of the ring bonds, as shown in Figure 17-15. Each carbon atom of the ring is bonded to two non-ring atoms that lie in different positions relative to this molecular plane. The bond to one of the two atoms is perpendicular to the plane, whereas the other lies roughly parallel to the plane.

**Figure 17-15**

Oblique view          Side view

The six-membered ring of a hexose such as α-glucose takes on a puckered structure. A molecular plane can be defined that passes through the midpoints of all the bonds of the ring. Bonds to non-ring atoms are either perpendicular to this plane (shown in green) or roughly parallel to the plane (shown in brown).

Hexose monosaccharides can form both five- and six-membered rings. In most cases, the six-membered ring structure is more stable, but fructose is an important example of a hexose that is more stable as a five-membered ring. The structure of β-fructose is shown in Figure 17-16. Notice that there are —CH₂OH fragments bonded to two positions of this five-membered ring. Examples 17-3 and 17-4 explore the structures of monosaccharides in more detail.

**Figure 17-16**

Fructose is a hexose monosaccharide that forms a five-membered ring structure.

---

## EXAMPLE 17-3 MONOSACCHARIDE STRUCTURES

Describe the differences in the structures of ribose and fructose.

**Strategy:** Monosaccharides can differ in their formulas, their ring sizes, and the spatial orientations of their hydroxyl groups. To analyze the differences between two monosaccharides, begin with structural drawings of the molecules, oriented so the ether linkages are in comparable positions. Then examine the structures to locate differences in constituents and bond orientations.

**Solution:** Here are the structures of ribose (Figure 17-13) and β-fructose (Figure 17-16):

Ribose
($C_5H_{10}O_5$)

β-Fructose
($C_6H_{12}O_6$)

Although the chemical formulas indicate that ribose is a pentose and fructose is a hexose, the ring portions of the structures are the same size. Proceeding clockwise around the rings from the oxygen atom, we see that the structures differ at the first two positions. In the first position, ribose has a carbon atom bonded to —H and —OH, whereas β-fructose has a carbon bonded to —OH and —CH₂OH. In the second position the molecules have the same two bonds but in different orientations. The OH group points "up" in β-fructose and "down" in ribose. The two molecules have the same structures at the other positions.

 **Does the Result Make Sense?** The answer makes sense if the molecules have been described clearly, and if all the structural features have been accounted for.

**Common Pitfalls:** Not orienting the two molecules in the same manner before trying to make comparisons.

### Practice Exercise 17.3

Describe the differences in the structures of α-glucose and α-idose, a rare naturally occurring monosaccharide.

α-Glucose

α-Idose

## EXAMPLE 17-4 DRAWING MONOSACCHARIDES

The six-carbon sugar α-galactose is identical to α-glucose except at carbon atom number 4, where the orientations are different. Draw the molecular structure of α-galactose. Simplify the structure by using flat rings rather than the true three-dimensional forms.

**Strategy:** When a carbohydrate with a six-membered ring is drawn, it is best to start with the ring itself. Next, use α-glucose (see Figure 17-13) as a convenient template to obtain the proper orientations for the groups that galactose and glucose have in common. Finally, switch positions of groups as needed to obtain the correct final structure for galactose.

**Solution:** Here is the flat-ring structure of α-glucose:

6-membered ring          α-Glucose

The problem states that α-galactose differs from α-glucose at carbon atom number 4. This position is highlighted in the structure. Starting from α-glucose, exchange the hydroxyl group with the hydrogen on this carbon atom:

α-Galactose

**Does the Result Make Sense?** The two structures differ in only one place, the orientation at position 4. The numbering system is the same for both structures, so the proposed structure of galactose meets the specifications of the problem.

### Practice Exercise 17.4

The six-carbon sugar β-mannose is identical to α-glucose except at carbon atoms in positions 1 and 2, where the orientations are different. Draw the molecular structure of β-mannose.

## Disaccharides

Two monosaccharides can combine by a condensation reaction between two hydroxyl groups. This reaction forms a linkage in which an oxygen atom connects two saccharide rings:

A C—O—C linkage between two sugar molecules is termed a *glycosidic bond*. Monosaccharides contain several hydroxyl groups, so many glycosidic bonds are

**Figure 17-17**

α-Glucose
$C_6H_{12}O_6$

α-Glucose
$C_6H_{12}O_6$

Maltose
$C_{12}H_{22}O_{11}$

+ $H_2O$

The structure of maltose (*far right*). A glycosidic bond links different ring carbon atoms of two glucose molecules.

possible. Nevertheless, in all natural glycosidic bonds, the linkage uses a hydroxyl group on a carbon atom next to a ring oxygen atom of one sugar. The second sugar can link to the first through any of its hydroxyl groups. For example, α-maltose forms from two molecules of α-glucose. The glycosidic bond occurs at the position adjacent to the ring oxygen for one glucose molecule and a different position on the second glucose, as shown in Figure 17-17. Maltose is formed from starch and decomposes in the presence of yeast, first to give glucose and then ethanol and water.

Some disaccharides serve as soluble energy sources for animals and plants, whereas others are important because they are intermediates in the decomposition of polysaccharides. A major energy source for humans is sucrose, which is common table sugar. Sucrose contains α-glucose linked to β-fructose. About 80 million tonnes of sucrose are produced each year. Of that, 60% comes from sugarcane and 40% comes from sugar beets. Example 17-5 treats a disaccharide that is an energy source for insects.

Sucrose
($C_{12}H_{22}O_{11}$)

Sucrose, a disaccharide containing α-glucose and β-fructose, is extracted from sugarcane.

### EXAMPLE 17-5 DECOMPOSING A SUGAR

Whereas humans can obtain energy from sucrose, insects obtain energy from trehalose, the line structure of which follows. Identify the monosaccharides from which trehalose is constructed.

Trehalose
($C_{12}H_{22}O_{11}$)

**Strategy:** Identifying monomer building blocks requires us to visualize a condensation reaction operating in reverse. The linkage bond breaks, and water is added.

Insects extract energy from trehalose.

**Solution:** Break the glycosidic bond of trehalose and add water to generate these two monosaccharides:

α-Glucose
($C_6H_{12}O_6$)

α-Glucose
($C_6H_{12}O_6$)

The monosaccharide on the left is oriented with its ring oxygen atom in the back right position. Compare the locations of its —OH groups with those in the structures in Figure 17-14 to see that this sugar is α-glucose. The sugar on the right is also α-glucose.

**Do the Results Make Sense?** The monosaccharide on the right must be rotated clockwise about the plane of the ring to bring its ring oxygen atom to the back right position. When that is done, you should recognize that this sugar is identical to the other. Both the monosaccharides in trehalose are α-glucose.

α-Glucose

Rotate 180°

α-Glucose

**Practice Exercise 17.5**

The structure of the disaccharide melibiose is shown below. Identify the monosaccharides from which melibiose is constructed.

## Polysaccharides

Polysaccharides, macromolecules made up of linked monosaccharides, play two major roles in biological organisms. Some, such as cellulose, are structural materials. Others, including glycogen, act as reservoirs for energy storage. All carbohydrates are good sources of chemical energy, because they release energy upon reacting with oxygen to produce water and carbon dioxide.

**Figure 17-18**

Amylopectin

Stylized view of amylopectin. The polymer is made up entirely of $\alpha$-glucose units.

Polysaccharides formed from $\alpha$-glucose are called **starches**. A starch stores sugar until it is needed for energy production. Three important starches are glycogen, which animals produce in their livers, and amylose and amylopectin, produced by plants through photosynthesis. On average, plant starch is about 20% amylose and 80% amylopectin. Each of these polysaccharides contains glucose as its monomer, but they differ in how the monosaccharide units are linked.

Starch collects in granules in plant cells and in animal liver cells.

The simplest starch is amylose, which consists of long chains of $\alpha$-glucose linked end to end by glycosidic bonds. One molecule of amylose contains about 200 glucose units. Amylopectin and glycogen are predominantly linear chains, but in these polysaccharides, some of the rings form a second glycosidic linkage, generating branches along the main chain. Figure 17-18 shows a portion of the structure of amylopectin. Amylopectin contains about 1000 glucose monomers. Branches that are about 30 units long occur at intervals of 20–25 glucose units.

Starch is the principal carbohydrate reservoir in plants and is the major component of rice, grains, corn, and potatoes. The long chains are sparingly soluble in water, so they collect in cells as granules. Starch is digested by a series of enzymes into glucose molecules, which are metabolized for energy production. Humans, other animals, and plants have the enzymes necessary to digest plant starch.

Animals store glucose in the form of glycogen. This form of starch is structured like amylopectin but with more frequent, shorter branches. Glycogen is synthesized and stored as granules in the liver and in muscle tissue, where the sugar is readily available for rapid energy production. The liver has a limited capacity to store glycogen. When that capacity is exceeded, the body activates alternative biochemical pathways that convert sugar into fat.

**Figure 17-19**

Cellulose

Starch

Starch and cellulose have different shapes, even though both polysaccharides are made from glucose monomers.

The most abundant organic molecule in the biosphere is **cellulose**, a polysaccharide that is the principal building material for plants. Like amylose, cellulose is a linear polymer of glucose. Unlike amylose, however, the glucose monomers in cellulose are in the $\beta$ configuration (see Figure 17-14).

Figure 17-19 compares the glycosidic linkages of starch and cellulose. These polysaccharides are made from different forms of glucose, and as a result they have distinctly different shapes. Starch is a chain of $\alpha$-glucose units linked head to tail. The glycosidic linkage of two $\alpha$-glucose units imparts a distinct kink to the structure of the polysaccharide, resulting in a chain that coils on itself to produce the granular shape of starch deposits. In contrast, cellulose is made entirely from $\beta$-glucose units, the glycosidic linkages of which lead to a flat structure.

Cellulose is a long, ribbon-like chain of $\beta$-glucose units. As shown in Figure 17-20, the planar arrangement of cellulose makes it possible for hydrogen bonds to form between polysaccharide chains, generating extended packages of cellulose ribbons. A typical cellulose molecule contains 2000–3000 glucose units in an unbranched linear chain. Individual chains form hydrogen bonds to other chains, giving organized bundles. The cell walls of plants are made up of these bundles laid down in a cross-hatched pattern that gives cellulose strength in all directions.

**Figure 17-20**

Plant cell walls are made of bundles of cellulose chains laid down in a cross-hatched pattern that gives cellulose strength in all directions. Hydrogen bonding between the chains gives cellulose a sheet-like structure.

Chitin

Chitin, found in the exoskeletons of arthropods, is a derivative of cellulose in which an amide replaces one hydroxyl group.

The enzymes that cleave $\alpha$-glucose units from starches cannot attack the $\beta$-glucose linkages of cellulose because the geometry of the glycosidic linkage is different. As a result, cows and other ruminants must rely on bacteria in their digestive tracts to break down cellulose. These microorganisms, which are also present in termites, use a different group of enzymes to cleave β-glucose units.

Cellulose is the most abundant structural polysaccharide, but it is by no means the only one. Plants and animals use a variety of polysaccharides for a wide range of structural applications. Some polysaccharides contain sugars other than glucose, and others are derivatives of cellulose in which some of the hydroxyl groups on the ring have been converted into other functional groups. One important example is *chitin*, a derivative of cellulose present in many animals. Chitin makes up the exoskeletons of crustaceans, spiders, insects, and other arthropods. The structure of chitin is the same as that of cellulose, except that one hydroxyl group is replaced by an amide group.

---

**SECTION EXERCISES**

● **17.5.1** Gulose is a six-carbon sugar that differs from glucose in two positions. In gulose, the hydroxyl orientations at positions 2 and 3 are reversed from their orientations in glucose. Draw the structures of α- and β-gulose.

● **17.5.2** Approximately 5% of milk is lactose, which is made from β-galactose and β-glucose. The hydroxyl group adjacent to the ring oxygen atom in galactose links to the hydroxyl group in glucose that is in position 4. Draw the structure of lactose using the flattened views of the monosaccharide rings. (The structure of galactose appears in Example 17-4.)

● **17.5.3** Most animals cannot digest cellulose. Even ruminant animals such as cattle and deer rely on bacteria in the gut for the enzyme to decompose cellulose. Enzymes that break down a polysaccharide recognize the specific shapes of the various pieces formed in the decomposition process; thus, the enzyme that breaks down starch cannot break down cellulose because the pieces of the two polysaccharides have different shapes. Examine the structures of starch and cellulose (Figure 17-19) and identify the differences in their shapes that might account for this.

---

## 17.6 NUCLEIC ACIDS

All biological organisms have the ability to reproduce themselves. The instructions for self-replication are stored and transmitted by macromolecules called **nucleic acids**. There are two types of nucleic acids, one that stores genetic information and one that transmits the information. Genetic information is *stored* in molecules of **deoxyribonucleic acid (DNA)**, which are located in cell nuclei. These huge molecules have molar masses as large as $10^9$ g/mol.

The information stored in DNA is *transmitted* by **ribonucleic acid (RNA)**. There are several kinds of RNA; each has its own role in the operation of a cell. Molecules of RNA have molar masses of 20 000–40 000 g/mol, so they are much smaller than their DNA counterparts. In this section, we examine the structures of nucleic acids and survey their biochemical functions.

**Figure 17-21**

| Pyrimidines | | Purines | | |

Guanine, G
(found in RNA and DNA)

Adenine, A
(found in RNA and DNA)

Uracil, U
(found in RNA)

Thymine, T
(found in DNA)

Cytosine, C
(found in RNA and DNA)

The organic bases found in DNA are of two types. Guanine and adenine are purines, with two rings. Uracil, thymine, and cytosine are pyrimidines, with single rings. The hydrogen atoms that are eliminated during condensation are highlighted.

**Figure 17-22**

Ribose
$C_5H_{10}O_5$

Deoxyribose
$C_5H_{10}O_4$

The structures of ribose and deoxyribose. The atoms that are eliminated during condensation are highlighted.

## The Building Blocks

Nucleic acids are macromolecules made of three component parts:

1. A nitrogen-containing organic base. There are five such bases, the line structures of which appear in Figure 17-21. **Adenine** and **guanine** are *purines*, two-ring structures. **Thymine**, **cytosine**, and **uracil** are *pyrimidines*, one-ring structures. Thymine occurs only in DNA, and uracil occurs only in RNA.

2. A pentose sugar. In RNA the sugar is ribose, and in DNA the sugar is deoxyribose, a ribose in which one OH group has been replaced with one H atom (Figure 17-22).

3. A phosphate linkage derived from phosphoric acid.

These three components are linked through condensation reactions. As Figure 17-23 shows, the —OH group on position 1 in the ribose ring condenses with an N—H from the base, thus connecting the sugar to the base with a C—N bond. The combination of a base and a sugar is named for its base: cytidine, uridine, thymidine, guanosine, and adenosine. Phosphoric acid condenses with the $CH_2OH$ hydroxyl group of the sugar to complete the linkage and form a nucleotide. Adenosine monophosphate (AMP), the nucleotide that contains adenine, is shown in Figure 17-24. Example 17-6 shows the formation of another nucleotide.

The carbon atoms of DNA and RNA are numbered with primes to distinguish these numbers from the numbers used for positions around the rings of the nitrogen bases.

**Figure 17-23**

Ribose
(sugar)

Adenine
(base)

Adenosine

$+ H_2O$

Adenosine is formed by a condensation reaction between adenine and ribose.

**Figure 17-24**

The formation of adenosine monophosphate (AMP) by condensation of adenosine and phosphoric acid. The three linked units form the nucleotide building blocks required for nucleic acid synthesis.

## EXAMPLE 17-6 DRAWING NUCLEOTIDES

Draw the structure of uridine monophosphate (UMP).

**Strategy:** To construct a molecule from component parts, a good approach is to draw each piece separately and then combine them in the proper order. The pieces are linked by condensation reactions.

**Solution:** The UMP nucleotide contains uracil, ribose, and one phosphate group. The structure of uracil and the hydrogen eliminated during the condensation appear in Figure 17-21. Here are the component parts, drawn in position to eliminate water molecules and link:

After the components have been placed next to one another, drawing the final structure requires removal of the $H_2O$ units. From right to left in the drawing, the three components of UMP are as follows: uracil, a single-ring base; ribose, the sugar; and one phosphate group.

 **Does the Result Make Sense?** The best way to verify that this structure is reasonable is to compare the parts of the molecule with the components shown in Figures 17-21 and 17-22.

**Practice Exercise 17.6**

Draw the structure of cytosine monophosphate.

## The Primary Structure of Nucleic Acids

A nucleic acid polymer contains nucleotide chains in which the phosphate group of one nucleotide links to the sugar ring of a second. The resulting backbone is an alternating sequence of sugars and phosphates, as shown in

**Figure 17-25**

The backbone of a nucleic acid is formed by condensation reactions between nucleotides.

Figure 17-25. As the figure indicates, the backbone has directionality. Each phosphate unit forms a bridge from the 5′ position of one ribose to the 3′ position of another ribose.

Each position along a nucleic acid sequence is identical except for the identity of its base. The sequence of bases is called the **primary structure** of a polynucleotide chain. By convention, the listing always begins with the nucleotide that has the terminal phosphate group (5′ end), and it continues to the opposite end of the chain, where the sugar has an unreacted hydroxyl group in the linkage position (3′ end). The nucleotides are listed as their one-letter abbreviations, A, C, G, and T or U. For example, ACGT stands for the following sequence:

## Secondary Structure of DNA: The Double Helix

Although DNA was first isolated in 1868, the nature of nucleic acids remained a mystery for more than 50 years. Not until the 1920s were the structures of nucleotides

The structure of DNA was discovered by (shown left to right) Francis Crick, James Watson, Maurice Wilkins, and Rosalind Franklin.

determined. By then, scientists suspected that DNA was the genetic material, but no one could fathom how sequences of just four different nucleotides could store immense quantities of genetic information. During the 1940s, the British chemist Alexander Todd performed research on nucleic acids that eventually won him a Nobel Prize. Todd discovered the basic composition of the polynucleotide chain, with sugars, phosphates, and bases linked as shown in Figure 17-25. Another crucial step was made early in the 1950s by American chemist Edwin Chargaff, who studied the composition of DNA from a variety of plants and animals. Chargaff found that the relative amounts of different bases changed from one species to another. In every species he examined, however, the molar ratios of guanine to cytosine and of adenine to thymine were always very close to 1.0. Chargaff concluded that these constant ratios could not be coincidence. Somehow, adenine and thymine are paired in DNA, and so too are guanine and cytosine.

By now the stage was set for the discovery of the three-dimensional structure of DNA, the so-called **secondary structure** of the molecule. Some of the best minds in science were working on the problem. In 1953, James D. Watson and Francis Crick of Cambridge University announced that they had discovered the structure of DNA. Watson and Crick relied on X-ray diffraction patterns of crystalline DNA. The diffraction photographs were taken by Rosalind Franklin, a researcher in the laboratory of Maurice Wilkins at King's College in London. Combining Franklin's data with the earlier insights of Chargaff, Todd, and many others, Watson and Crick concluded that DNA must consist of two helices wound around each other in a **double helix**. In the double helix, shown schematically in Figure 17-26, the hydrophilic sugars and the phosphate groups lie on the outside of the molecule, with the hydrophobic bases tucked inside the structure. For their brilliant insight, Watson, Crick, and Wilkins shared the 1963 Nobel Prize in medicine and physiology.

According to the Watson–Crick model, hydrogen bonding holds the double helix together. The bases of one strand of DNA form hydrogen bonds to the bases of the second strand. Because of their matching structures, adenine pairs with thymine through two hydrogen bonds, and guanine pairs with cytosine through three hydrogen bonds. These sets are said to be **complementary base pairs**. As shown in Figure 17-27, complementary base pairs fit together like matching gears. The matching of bases in this way keeps the distance between the two strands constant through the entire length of the DNA molecule.

The nucleotide bases are flat molecules. Each base pair is parallel to the one below it, with 340 picometres separating the two. There is a rotation of 36° between pairs, giving 10 base pairs per complete turn of the helix. The two sugar–phosphate backbone strands wind around these stacked pairs, as shown in Figure 17-28. The two strands of DNA run in opposite directions, with the terminal phosphate end of one polynucleotide matched with the free hydroxyl end of the other.

**Figure 17-26**

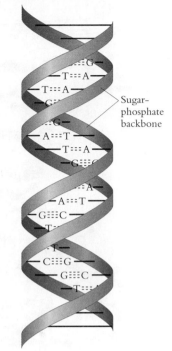

DNA consists of two strands of sugar–phosphate backbones wound around each other in a double helix. The two helices are connected by hydrogen bonds between the bases.

**Figure 17-27**

**Thymine · · · · · Adenine**

**Cytosine · · · · · Guanine**

The complementary base pairs of DNA. Thymine fits with adenine, and cytosine fits with guanine.

**Figure 17-28**

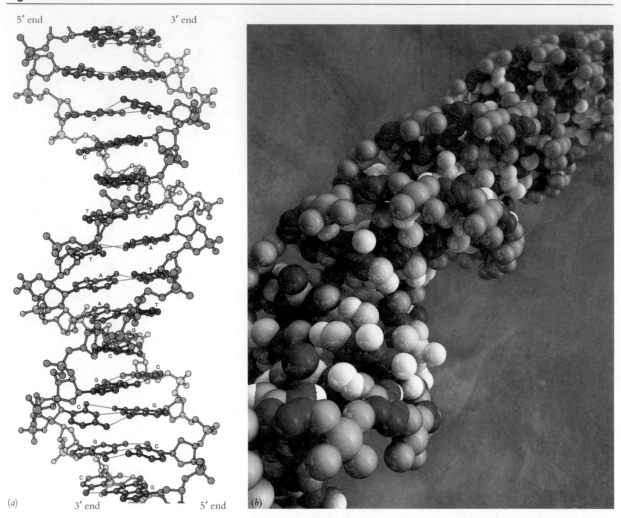

5′ end     3′ end

3′ end     5′ end

(a)     (b)

The structure of DNA. (a) A ball-and-stick model, with the sugar–phosphate backbone coloured blue and the bases coloured red. (b) A space-filling model, showing C atoms in blue, N atoms in dark blue, H atoms in white, O atoms in red, and P atoms in yellow.

In recent years, our detailed understanding of the structure of DNA and how it functions has led to many new methods of genetic manipulation. Our Focus 17-2: Chemistry and Technology explores one aspect of this subject.

## FOCUS 17-2 CHEMISTRY AND TECHNOLOGY: DNA "CHIPS"

Every gene contains DNA with a unique sequence of bases forming a genetic code containing the information an organism uses to live and replicate itself. Many years of research have resulted in an understanding of how the information content of DNA is translated into particular biochemical substances and how DNA replicates. The processes include unwinding of the DNA double helix so that its code can be read or duplicated.

In an unwound helix, the bases are exposed and can form new hydrogen-bonding interactions. In protein synthesis, the unwound DNA sequence serves as a template to build a molecule of RNA, the base sequence of which is complementary to that of the DNA sequence. The RNA molecule, in turn, serves as a blueprint for protein manufacture. In replication, as the illustration on the next page shows, unwinding and duplication generates two identical DNA helices from a single helix.

In both RNA synthesis and DNA replication, the bases of the original DNA strand are matched to a set of complementary bases through strong hydrogen-bonding interactions. This process is highly specific because each base interacts only with its complementary base. In DNA replication, the growing chain incorporates the correct complementary base by binding that base, then forming the backbone linkages.

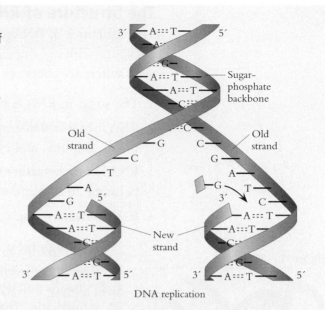

DNA replication

The promise of being able to predict and modify the genetic characteristics of an organism has fuelled massive efforts to determine the base sequences of human genes. The human genome project reached the goal of sequencing all the important DNA carried by humans in the year 2003.

A developing application of DNA technology uses a DNA "chip" that contains many small segments of bases of known sequence. Such DNA chips are being used to better understand the biochemistry of cancers. Cancers occur when defective genes cause cells to divide uncontrollably, but usually the process of protein synthesis is also modified. Different types of cancer result in different modifications to protein synthesis, depending on how the DNA sequence of the cancerous cell differs from that of a normal cell. The DNA chips allow clinicians to identify the proteins modified by a particular cancer.

This technology combines microfabrication techniques with the ability of DNA to recognize and bind to its complementary sequence. Light-sensitive chemical reactions are used to construct a DNA chip containing up to 10 000 distinct short DNA sequences, each one occupying a specific position on the chip. In one use of such chips, DNA is extracted from a cancerous cell, fragmented into segments, and chemically reacted with small fluorescent molecules that bind preferentially to specific sites along the DNA sequence. A solution of these fluorescent DNA segments is then applied to the chip.

As the schematic illustration shows, a segment containing a particular base sequence molecule will bind to an area of the chip that contains its complementary sequence, but it will not bind to any area that has a mismatched sequence.

When the chip is rinsed, fluorescent DNA remains bound to all those areas of the chip that have complementary sequences. Areas of the chip of which sequences are not complementary do not bind to the DNA. When the chip is rinsed and subsequently exposed to ultraviolet light, the areas with bound DNA fluoresce, generating light from each area to which the DNA has bound. In contrast, areas of the chip that contain non-complementary sequences remain dark. The fluorescence pattern reveals which protein synthesis processes have been modified in the cancerous cell, and this in turn helps physicians determine what treatment is most likely to be effective against the cancer.

The power of this new technique comes from the large number of DNA sequences that can be placed on a single chip—several tens of thousands on a single $1.3 \times 1.3$ cm chip. This gives a richly detailed fluorescence pattern that can be interpreted using a high-resolution optical scanner and computer analysis. Such detailed information makes it possible to distinguish among a large number of different types of cancerous tissues.

## The Structure of RNA

The structures of DNA and RNA are similar in that each has a sugar–phosphate backbone with one organic base bound to each sugar. However, there are four distinct differences between RNA and DNA:

1. The sugar in RNA is ribose, not deoxyribose.
2. RNA uses uracil instead of thymine. The component bases in RNA are adenine, uracil, guanine, and cytosine.
3. RNA is much smaller than DNA. The molecules of RNA range in molar mass between 20 000 and 40 000 g/mol instead of as much as $10^9$ g/mol.
4. RNA is usually single-stranded, not double-stranded.

**Figure 17-29**

```
---- Hydrogen
     bonding
□    Base
■    Backbone
```

The structure of an RNA molecule. Notice the folding caused by the intrastrand base pairing.

Although most of an RNA molecule is single-stranded, there often are some double-stranded regions. Intramolecular base pairing between guanine and cytosine and between adenine and uracil creates loops and kinks in the RNA molecule. The structure of one kind of RNA molecule is shown in Figure 17-29.

Whereas DNA has a single role as the storehouse of genetic information, RNA plays many roles in the operation of a cell. There are several different types of RNA, each having its own function. The principal job of RNA is to provide the information needed to synthesize proteins. Protein synthesis requires several steps, each assisted by RNA. One type of RNA copies the genetic information from DNA and carries this blueprint out of the nucleus and into the cytoplasm, where construction of the protein takes place. The protein is assembled on the surface of a ribosome, a cell component that contains a second type of RNA. The protein is constructed by sequential addition of amino acids in the order specified by the DNA. The individual amino acids are carried to the growing protein chain by yet a third type of RNA. The details of protein synthesis are well understood, but the process is too complex to be described in an introductory course in chemistry.

## SECTION EXERCISES

● **17.6.1**  Draw the structure of the RNA nucleotide that contains guanine.

● **17.6.2**  Part of a DNA sequence is G-C-C-A-T-A-G-G-T. What is its complementary sequence?

● **17.6.3**  Nucleotides can contain more than one phosphate group. One example is energy-storing adenosine triphosphate (ATP), which is discussed in Chapter 11. An ATP molecule is formed in two sequential condensation reactions between phosphoric acid and the phosphate group of AMP:

$$AMP + 2\ H_3PO_4 \longrightarrow ATP + 2\ H_2O$$

Draw the structure of ATP.

## 17.7 PROTEINS

Life is organized around the cell, the smallest functioning unit of an organism. The most important biochemicals in cells are proteins, including enzymes, antibodies, hormones, transport molecules, and the structural materials that make up the cell itself. Proteins protect organisms from disease, extract energy from food, move essential cellular components from place to place, and are responsible for vision, taste, and smell. Proteins even synthesize the genetic material contained in all cells. In other words, proteins are the molecular machinery of the cell. Structural materials in most animals, including skin, hooves, and feathers, are also made of proteins. How can this single group of molecules play so many different roles in the chemistry of life? To answer that question, we need to look at the structure of proteins. Remarkably, all proteins have amino acids as their common structural components.

### Amino Acids

As the name implies, an **amino acid** is a bifunctional molecule with a carboxylic acid group at one end and an amine group at the other. All proteins are polyamides made from condensation reactions of amino acids. Every amino acid in proteins has a central carbon atom bonded to one hydrogen atom and to a second group, symbolized in Figure 17-30 as $R$.

Twenty different amino acids are used to build proteins in living cells. The simplest is glycine, where $R = H$. In each of the 19 amino acids other than glycine, the side chain begins with a carbon atom. The side chains in naturally occurring amino acids are shown in Figure 17-31.

Among the common amino acids, 11 have side chains that contain polar functional groups that can form hydrogen bonds, such as $-OH$, $-NH_2$, and $-CO_2H$. These hydrophilic amino acids are commonly found on the outside of a protein, where their interactions with water molecules increase the solubility of the protein. The other nine amino acids have non-polar hydrophobic side chains containing mostly carbon and hydrogen atoms. These amino acids are often tucked into the inside of a protein, away from the aqueous environment of the cell.

Biochemists represent each amino acid with a three-letter abbreviation. These abbreviations appear under the names of the amino acids in Figure 17-31. For example, the abbreviation for glycine is Gly.

### Polypeptides

Proteins form in a sequence of condensation reactions in which the amine end of one amino acid combines with the carboxyl end of another, eliminating a water molecule to create an amide linkage. The amide group that connects two amino acids is called a **peptide linkage**, and the resulting molecule is known as a **peptide**. When two amino acids are linked, the product is a *dipeptide*. A dipeptide formed from alanine and glycine is shown in Figure 17-32.

**Figure 17-30**

Ball-and-stick representation and structural formula common to all of the naturally occurring amino acids.

**Figure 17-31**

Of the 20 amino acids, 11 have polar side chains (shown in colour), and 9 have nonpolar side chains. One, proline, has a unique ring structure, shown above in its entirety rather than as a side chain. Under the name of each amino acid is its three-letter abbreviation.

**Figure 17-32**

Alanine and glycine form a dipeptide by a condensation reaction.

Notice that this dipeptide is also an amino acid because the molecule retains an amine group at one end and a carboxyl group at the other end. Consequently, an additional amino acid can add to either end of a dipeptide to form a new peptide that also has an amine terminal group at one end and a carboxyl terminal group at its other end. Figure 17-33 shows the peptide that results from addition of another alanine molecule at the amine end and a cysteine molecule at the carboxyl end of the Ala-Gly dipeptide.

Protein synthesis in cells occurs by sequential condensations, always at the carboxyl end of the growing chain, eventually leading to a macromolecule called a **polypeptide**. All proteins are macromolecular polypeptides, but the lengths of proteins vary tremendously. Some small hormone proteins contain as few as eight or nine amino acids. Myosin, a very large muscle protein, has approximately 1750 amino acid units. Proteins exist with amino acid chain lengths that range between these extremes.

The **primary structure** of a polypeptide is its sequence of amino acids. It is customary to write primary structures of polypeptides using the three-letter abbreviation for each amino acid. By convention, the structure is written so that the amino acid on the left bears the terminal amino group of the polypeptide, and the amino acid on the right bears the terminal carboxyl group. Figure 17-34 shows the two dipeptides that can be made from glycine and serine. Although they contain the same amino acids, they are different molecules with chemical and physical properties that differ. Example 17-7 shows how to draw the primary structure of a peptide.

**Figure 17-33**

Ala-Ala-Gly-Cys tetrapeptide

Condensation reactions can take place at both ends of a dipeptide. The product can continue to grow into a polypeptide or a protein. The figure shows alanine and cysteine (Cys) adding to opposite ends of the Ala–Gly dipeptide.

**Figure 17-34**

Glycine–Serine
Gly–Ser

Serine–Glycine
Ser–Gly

Glycine and serine form two dipeptides, Gly–Ser and Ser–Gly. As the ball-and-stick models show, even dipeptides have distinctive shapes that depend on their primary structure.

---

## EXAMPLE 17-7 THE PRIMARY STRUCTURE OF A PEPTIDE

Draw the line structure of the peptide Asp–Met–Val–Tyr.

**Strategy:** We are asked to translate a shorthand designation into a line drawing. First, construct a backbone containing three amide linkages, putting the terminal $NH_2$ group on the left end and the terminal $CO_2H$ group on the right end of the peptide. Then attach the appropriate side groups, as determined from the molecular structures of the amino acids in Figure 17-31.

**Solution:** The four amino acids are joined with peptide linkages in the order given. Put the backbone in place as a line structure. Remember, however, that carbon atoms are not shown in line structures and hydrogens are shown only on atoms other than carbon.

Next, add the four side chains in the positions marked $R_1$, $R_2$, $R_3$, and $R_4$ to give the final structure of the peptide. The leftmost amino acid, aspartic acid, is at the amino end. Aspartic acid is followed by methionine, valine, and tyrosine. Tyrosine is at the carboxyl end. Here are the individual amino acids, with their $R$ groups highlighted:

To finish the structure, we replace the $R$ groups on the peptide backbone with their appropriate line structures:

 **Does the Result Make Sense?** The structure is reasonable if the peptide backbone is correct and if the four amino acid side chains match the structures shown in Figure 17-31.

**Practice Exercise 17.7**
Draw the line structure of the peptide Cys–Lys–Leu–Asn.

Proteins with similar primary structures can serve very different functions. For example, Figure 17-35 shows the primary structures of two pituitary hormones, vasopressin and oxytocin. These structures differ by just two amino acids (those highlighted in colour), but changing these two amino acids has a profound effect on the biochemistry of these molecules. Vasopressin regulates the rate at which water is reabsorbed by the kidneys and intestine. Alcohol suppresses the release of vasopressin, which is why consumption of alcoholic beverages leads to excessive urine production and dehydration. Oxytocin, on the other hand, induces contractions of the uterus during childbirth, and it triggers the release of milk by contracting muscles around the ducts that come from the mammary glands.

The enormous diversity of protein structure and function comes from the many ways in which 20 amino acids can combine into polypeptide chains. Consider how many tetrapeptide chains can be made using just two amino acids, cysteine and aspartic acid:

Cys–Cys–Cys–Cys   Cys–Cys–Cys–Asp   Cys–Cys–Asp–Cys   Cys–Cys–Asp–Asp
Cys–Asp–Cys–Cys   Cys–Asp–Cys–Asp   Cys–Asp–Asp–Cys   Cys–Asp–Asp–Asp
Asp–Cys–Cys–Cys   Asp–Cys–Cys–Asp   Asp–Cys–Asp–Cys   Asp–Cys–Asp–Asp
Asp–Asp–Cys–Cys   Asp–Asp–Cys–Asp   Asp–Asp–Asp–Cys   Asp–Asp–Asp–Asp

In all, there are $2^4 = 16$ different ways to combine two amino acids in chains of four units. The 20 common amino acids combine to give $20^4$ tetrapeptides, which gives $1.6 \times 10^5$ different molecules. Each time the chain adds an amino acid, the number of possibilities is multiplied by 20, so a set of $n$ amino acids can form $20^n$ different polypeptides. Because proteins can contain hundreds of amino acids, each of which can be any of the 20 possibilities, the number of possible structures becomes immense. In fact, $20^{100}$, the number of possible ways to construct a protein of 100 units, is greater than the estimated number of atoms in the universe.

## Secondary Protein Structure

A primary structure represents a polypeptide as a simple linear string of amino acids. Actually, within long polypeptides, certain sections fold into sheets or twist into coils. These regions with specific structural characteristics constitute the **secondary structure** of the protein. Figures 17-36 and 17-37 show the two most common secondary structures.

The secondary structure of a protein is determined by hydrogen bonding between C=O and N—H groups of the peptide linkages that make up the backbone of the protein. Hydrogen bonds can exist within the same protein chain or between different chains. Hydrogen bonding within a single protein chain gives a **helix** (Figure 17-36). Notice that the helix is held together by hydrogen bonds between the O atom of the carbonyl group in one amino acid and the H atom from an amide nitrogen three amino acids away. One complete turn of the helix contains 3.6 amino acids.

Another common form of secondary protein structure is a **pleated sheet** (Figure 17-37). The chains in a sheet are fully extended rather than coiled, and hydrogen bonds exist between different portions of the protein chains, with one part of

**Figure 17-35**

Vasopressin     Oxytocin

The primary structures of the pituitary hormones vasopressin and oxytocin differ by just two amino acids, but they have remarkably different properties. (The sulphur cross-links between Cys residues are described later in this section.)

**Figure 17-36**

A helical peptide backbone. The side chains are omitted to emphasize the shape of the helix. Notice the hydrogen bonding between N—H and C=O groups.

**Figure 17-37**

(*a*) Ball-and-stick model                    (*b*) Structural formula view

Two views of a pleated sheet. (*a*) The ball-and-stick view emphasizes the hydrogen bonding between the strands of the sheet. (*b*) The structural formula view emphasizes the folds of the sheet.

the chain bonded to another. In pleated regions, the oxygen and the amide hydrogen protrude at 90° angles to the axis of the extended protein backbone. This allows row on row of polypeptides to form hydrogen bonds and make a sheet of protein. The pleats in the sheet are caused by the bond angles of the peptide linkages.

Although other secondary protein structures play roles in determining the shapes of proteins, the helix and pleated sheet occur most frequently. A discussion of less common secondary structures is beyond the scope of this text.

## Tertiary Protein Structure

Each protein has a unique three-dimensional shape called its **tertiary structure**. The tertiary structure is the result of the bends and folds that a polypeptide chain adopts to achieve the most stable structure for the protein. As an analogy, consider the cord in Figure 17-38 that connects a computer to its keyboard. The cord can be pulled out so that it is long and straight; this corresponds to its primary structure. The cord has a helical region in its centre; this is its secondary structure. In addition, the helix may be twisted and folded on top of itself. This three-dimensional character of the cord is its tertiary structure.

The tertiary structure of a protein is determined primarily by the way in which water interacts with the side chains on the amino acids in the polypeptide chain. When the polypeptide folds into a three-dimensional shape that arranges its hydrophobic regions inside the overall structure, polar interactions with water molecules are maximized, making the system most stable. The folding of the protein is further directed and strengthened by a large number of hydrogen bonds. Hydrophilic side chains form hydrogen bonds with other side chains and with the protein backbone. In addition, water forms hydrogen bonds with the protein backbone and with hydrophilic side chains such as those in arginine, aspartic acid, and tyrosine, as Example 17-8 shows.

**Figure 17-38**

The cord that connects this computer to its keyboard has structures analogous to the primary, secondary, and tertiary levels of protein structure.

## EXAMPLE 17-8 HYDROGEN BONDING IN PROTEINS

Draw a line structure that shows the various ways in which water molecules form hydrogen bonds with a protein backbone.

**Strategy:** Remember from Chapter 8 that the hydrogen atoms of water molecules form hydrogen bonds with electronegative O and N atoms, whereas the oxygen atom of a water molecule forms hydrogen bonds with hydrogen atoms in highly polar N—H and O—H bonds.

**Solution:** Begin by drawing a section of protein backbone. Because the problem asks only about hydrogen bonds to the backbone, the side chains are not involved in this problem and may be designated simply as *R*.

Now add water molecules to illustrate the hydrogen-bonding interactions. Each hydrogen bond is shown as a dotted line. Because the problem does not ask that all hydrogen bonds be shown, it is sufficient to show one or two examples of each type. Lone pairs are shown only for atoms involved in hydrogen bonding.

**Does the Result Make Sense?** The structure involves water molecules acting as hydrogen bond acceptors to N—H and O—H groups or as hydrogen bond acceptors to electronegative oxygen atoms.

These are all legitimate hydrogen-bonding interactions, so the result is reasonable.

### Practice Exercise 17.8

Amino acid side chains on the outside of a protein can form hydrogen bonds to water molecules. Draw a line structure that shows how water molecules can form hydrogen bonds to the side chains in the protein segment shown below.

The amino acid cysteine plays a unique role in tertiary protein structure. The —SH groups of two cysteine side chains can cross-link through an S—S bond called a **disulphide bridge**, as shown below.

**Figure 17-39**

Disulfide bridge

The three-dimensional structure of oxytocin. The sulphur atoms of the disulphide bridge are in yellow. The side chains have been simplified as green spheres to emphasize the shape of the polypeptide chain.

The primary structures of oxytocin and vasopressin (Figure 17-35) have two cysteine residues connected by a disulphide bridge. The three-dimensional structure of the oxytocin backbone is shown in Figure 17-39. Notice that the disulphide bridge locks the peptide chain into a compact cyclic structure.

## Globular Proteins

A huge and diverse group of molecules called **globular proteins** carries out most of the work done by cells, including synthesis, transport, and energy production. Globular proteins have compact, roughly spherical tertiary structures containing folds and grooves. **Enzymes** are globular proteins that speed up biochemical reactions. The reactions of metabolism proceed too slowly to be of use to living organisms unless enzymes make them go faster. Antibodies, the agents that protect humans from disease, are also globular proteins. Other globular proteins transport smaller molecules through the blood. Haemoglobin, the macromolecule that carries oxygen in the blood, is a transport protein. Globular proteins, including oxytocin and vasopressin, also act as hormones. Others are bound in cell membranes, facilitating passage of nutrients and ions into and out of the cell. Many globular proteins have hydrophilic side chains over the outer surfaces of their tertiary structures, making them soluble in the aqueous environment of the cell.

The secondary structures of globular proteins include helices and pleated sheets in varying proportions. The unique primary structure of each globular protein leads to a unique distribution of secondary structures and to a specific tertiary structure. The stylized ribbon views of globular proteins in Figure 17-40 illustrate the diversity and intricate structures of these molecules. Regions that are helical are shown in blue, regions that are pleated are magenta, and sections that have no specific secondary structure are orange.

## Fibrous Proteins

Structural components of cells and tissues are made of proteins that form fibres. These **fibrous proteins** are the cables, girders, bricks, and mortar of organisms.

The helix is the prevalent secondary structure in keratins, which include wool, hair, skin, fingernails, and fur. These molecules are long strands of helical protein that lie with their axes parallel to the axis of the fibre. In hair, individual keratin molecules are wound like the strands of a rope. Because of its compact helical chains, keratin has a certain amount of elasticity. Stretching the fibres stretches and breaks the relatively weak hydrogen bonds of the helix, but as long as none of the stronger covalent bonds is broken, the protein relaxes to its original length when released. Hydrogen bonds can easily break and form again, but covalent bonds cannot.

A "permanent wave" changes the shape of a person's hair by changing the tertiary structure of hair protein. A solution that breaks the disulphide bridges between the protein chains is applied to the hair. The hair is then set in curlers to give the desired shape. A second solution that recreates S—S bonds is applied to the hair. When the curlers are removed, the new disulphide bridges hold the hair in its new configuration.

**Figure 17-40**

Immunoglobulin           Myoglobin           G–Actin

Ribbon views of proteins with varying amounts of helices and pleated sheets. Immunoglobulin, an antibody, is made up almost entirely of pleated sheets (magenta). Myoglobin, which stores oxygen in muscle tissue, is composed of about 70% helix (blue). G-Actin, a component of muscle protein fibres, is a complex mixture of helices and pleated sheets. Regions with no specific secondary structure are shown in orange.

Other fibrous proteins contain extensive regions of pleated sheets. The fibres spun by a silkworm, for example, are made almost entirely of fibroin, a protein composed primarily of just three amino acids: glycine (45%), alanine (30%), and serine (12%). Each chain of fibroin contains extensive regions where a sequence of six amino acids occurs repeatedly: . . . –Gly–Ser–Gly–Ala–Gly–Ala– . . . Notice that every other amino acid is glycine, which is the smallest amino acid. This alternating arrangement is an important feature in the packing of the strands that make up the pleated sheet.

## SECTION EXERCISES

**17.7.1** Write the three-letter shorthand form of this peptide sequence:

**17.7.2** Draw a line structure of the peptide sequence Ser–Gly–Lys–Asp and show how the side chains form hydrogen bonds to water molecules.

**17.7.3** Write a few sentences that explain the difference among primary, secondary, and tertiary protein structures. Your description should identify which types of intramolecular and intermolecular forces are responsible for each.

# CHAPTER 17 VISUAL SUMMARY

## 17.1 STARTING MATERIALS FOR POLYMERS

**LEARNING OBJECTIVES:** Describe functional groups and linkage groups in polymers.

**SKILLS TO MASTER:** Recognizing polymerizable functional groups (p. 999)

Recognizing linkage groups (p. 1000)

**KEY CONCEPTS:** Polymerization requires monomers that contain reactive functional groups.

A condensation reaction produces a linkage group and eliminates a small molecule.

## 17.2 FREE RADICAL POLYMERIZATION

**LEARNING OBJECTIVES:** Describe polymers made by free radical polymerization.

**SKILLS TO MASTER:** Explaining the mechanism of free radical polymerization (p. 1002)

Drawing the structure of a polymer (p. 1005)

**KEY CONCEPTS:** Free radical polymerization includes initiation, propagation, and termination steps.

Polymerizing mixtures of two different monomers to give copolymers increases the versatility of rubber materials.

Cross-linking strengthens rubber material by creating chemical links between long-chain molecules.

## 17.3 CONDENSATION POLYMERIZATION

**LEARNING OBJECTIVES:** Describe polymers made by condensation polymerization.

**KEY CONCEPTS:** Each monomer in a condensation polymerization has two functional groups.

Polyamides include nylons, Kevlar®, and proteins.

Polyesters include poly(ethylene terephthalate), Dacron, and Mylar.

## 17.4 TYPES OF POLYMERS

**LEARNING OBJECTIVES:** Recognize and describe some properties of plastics, fibres, and elastomers.

**KEY CONCEPTS:** Plastics are typically made in the form of blocks or sheets.

Fibres can be drawn into long threads.

Elastomers can be stretched without breaking.

## 17.5 CARBOHYDRATES

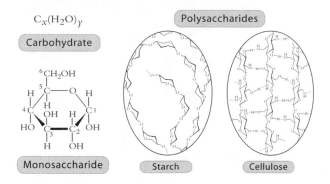

$C_x(H_2O)_y$

Carbohydrate

Polysaccharides

Monosaccharide

Starch

Cellulose

**LEARNING OBJECTIVES:** Recognize and draw structures of monosaccharides and polysaccharides.

**KEY CONCEPTS:** Monosaccharides have the formula $(CH_2O)_n$, where $n$ is between 3 and 6.

A C—O—C linkage between two sugar molecules is termed a glycosidic bond.

Polysaccharides in biological organisms act as structural materials or as reservoirs for energy storage.

## 17.6 NUCLEIC ACIDS

Primary structure

Sugar — Phosphate — Sugar — Phosphate — Sugar — Phosphate

A          G          C

Adenine   Guanine   Cytosine   Thymine   Uracil

DNA       RNA

Nucleic acids

To chain

Complementary base pairs

To chain

Double helix

Secondary structure

**LEARNING OBJECTIVES:** Draw primary and secondary structures of DNA and RNA.

**SKILLS TO MASTER:** Writing the primary structure of a nucleic acid (p. 1029)

Writing the complementary structure of a DNA sequence (p. 1031)

**KEY CONCEPTS:** The backbone of a nucleic acid is formed by condensation reactions between nucleotides.

Adenine pairs with thymine, and guanine pairs with cytosine, between two strands of DNA.

## 17.7 PROTEINS

Amino acid

Ala — Ser — Asn — Gly — Gly — Ser — Ala — His — Val

Polypeptide       Primary structure

Peptide linkage

Secondary structure

Tertiary structure

Helix     Pleated sheet

Globular proteins
Enzymes

Fibrous proteins

**LEARNING OBJECTIVES:** Explain primary, secondary, and tertiary structures of proteins.

**SKILLS TO MASTER:** Drawing the primary structure of a polypeptide (p. 1035)

**KEY CONCEPTS:** All proteins are polypeptides made from condensation reactions of amino acids.

Twenty different amino acids are used to build proteins in living cells.

Two common polypeptide secondary structures are a helix and a pleated sheet.

Tertiary structures of proteins may be globular or fibrous.

## LEARNING EXERCISES

**17.1** Prepare a list of the types of polymer linkage reactions described in this chapter. List at least one important polymer that forms from each type of reaction.

**17.2** Define primary, secondary, and tertiary structures. Give examples of each type of structure for a protein and for DNA.

**17.3** Draw molecular pictures that illustrate the linkages in each of the following polymer types: polyethylene, polyester, polyamide, polyether, and silicone.

**17.4** Draw pictures that illustrate the essential features of the helix, the pleated structure, and the double helix.

**17.5** Write a paragraph that describes the role of hydrogen bonding in the structures of biological macromolecules.

**17.6** List all terms new to you that appear in Chapter 17. Write a one-sentence definition of each, using your own words. Consult the index and glossary if you need help.

## PROBLEMS

**ilw** =interactive learning ware problem. Visit the website at www.wiley.com/canada/olmsted

### Starting Materials for Polymers

**17.1** Draw structural formulas and circle and identify the functional groups in the following molecules:

(a) (b) (c)

(d)

**17.2** Draw structural formulas and circle and identify the functional groups in the following molecules:

(a) (b)

(c) (d)

**17.3** Convert the ball-and-stick structures to line drawings and identify the functional groups in the following molecules:

(a)

(b)

(c)

(d)

**17.4** Convert the ball-and-stick structures to line drawings and identify the functional groups in the following molecules:

(a)

(b)

(c)

(d)

**17.5** Draw structural formulas and circle and identify the linkage groups in the following molecules:

**17.6** Draw structural formulas and circle and identify the linkage groups in the following molecules:

**17.7** Draw one example of (a) an amine with the formula $C_5H_{17}N$; (b) an ester with eight carbon atoms; (c) an aldehyde with a molar mass of at least 80 g/mol; and (d) an ether that contains a phenyl ring ($C_6H_5$).

**17.8** Draw one example of (a) an alcohol with the formula $C_4H_{10}O$; (b) a carboxylic acid with a molar mass of at least 60 g/mol; (c) a thiol with six carbon atoms; and (d) an amide that contains at least 10 carbons, including a phenyl ring ($C_6H_5$).

**17.9** Write balanced equations that show how the following molecules are produced in condensation reactions:

**17.10** Write balanced equations that show how the following molecules are produced in condensation reactions:

**17.11** Draw the structures of all possible products resulting from condensation reactions between aspartic acid and isoleucine:

Aspartic acid
Asp

Isoleucine
Ile

**17.12** Draw the structures of the condensation products of the following reactions:

## Free Radical Polymerization

**17.13** **ilw** Saran is a copolymer made from vinyl chloride and vinylidene chloride ($H_2C=CCl_2$). Draw the structure of this polymer, showing at least four repeat units in the polymer.

**17.14** Draw the structure of polypropylene, showing at least five repeat units.

**17.15** Identify the monomers used to make the following polymers:

(a)    $+CH_2CH-CH_2CH+$ (with $NO_2$ groups)

(b)    $+CFCl-CF_2-CFCl-CF_2+$

**17.16** Nitrile rubber, the structure of which follows, is a copolymer made from two monomers. The polymer is used to make automotive hoses and gaskets. Draw the structures of the monomers and name them.

**17.17** Draw a section of the polymer chain for polybutadiene and describe how it differs from polyethylene.

**17.18** Draw a section of the polymer chain showing at least four repeat units for each of the following polymers: (a) Teflon®; (b) PVC; and (c) Styrofoam.

## Condensation Polymerization

**17.19** The structure of Nylon 11 follows. Draw a line structure of each monomer used to make this polymer.

**17.20** Polybutylene terephthalate, used to make countertops and sinks, has the following structure. Draw the structural formulas of the monomers from which this polymer is made.

**17.21** Draw the structure of the repeat unit of the polymer that forms from ethylene glycol and *p*-phenylene diamine.

Ethylene glycol            *p*-Phenylene diamine

**17.22** Kodel is a polyester fibre made from terephthalic acid and cyclohexanedimethanol. Draw a segment of Kodel that contains at least four repeat units.

Cyclohexanedimethanol    $HOCH_2$—⬡—$CH_2OH$

## Types of Polymers

**17.23** Which is more extensively cross-linked: the rubber in an automobile tire or the rubber in a pair of surgical gloves? Explain.

**17.24** Plastic wrap can be stretched slightly to fit snugly over a food container. Is plastic wrap a thermoplastic or a thermosetting polymer? Explain.

**17.25** Describe what category of polymer to use to make each of the following items: (a) a balloon; (b) rope; and (c) a camera case.

**17.26** Describe what category of polymer to use to make each of the following items: (a) a countertop; (b) artificial turf; and (c) a bungee cord.

**17.27** Describe the changes in polymer properties that occur on adding dioctylphthalate to the reaction mixture.

**17.28** Describe the changes in polymer properties that occur on cross-linking.

## Carbohydrates

**17.29** The six-carbon sugar talose differs from glucose in the orientations of the hydroxyl groups at the 2 and 4 positions. Draw the structure of α-talose.

α-Glucose

**17.30** The five-carbon sugar ribose can form a six-membered ring, ribopyranose, that differs from glucose in the orientation of the —OH at position 3 and in having —H instead of —$CH_2OH$ in position 6. Draw the ring structure of α-ribopyranose.

**17.31** Draw the structure of β-talose (refer to Problem 17.29).

**17.32** Draw the structure of β-ribopyranose (refer to Problem 17.30).

**17.33** Glycogen is a glucose polymer that collects in granules in the liver. Cellulose is also a polymer of glucose, but cellulose forms sheet-like arrangements that are used in the cell walls of plants. What are the distinguishing structural features that allow glycogen to form granules whereas cellulose forms sheets? Illustrate your answer with structural drawings.

**17.34** Glycogen and cellulose are both polymers of glucose. Explain why humans can use glycogen but not cellulose as an energy source. Why can cows digest cellulose, but humans cannot?

## Nucleic Acids

**17.35** Part of a DNA sequence is A-A-T-G-C-A-C-T-G. What is its complementary sequence?

**17.36** Part of a DNA sequence is C-G-T-A-G-G-A-A. What is its complementary sequence?

**17.37** Draw the complete structure of the following segment of DNA: A-T-C-G.

**17.38** Draw the complete structure of the following segment of RNA: G-U-A-C.

**17.39** One strand of DNA contains the base sequence A-T-C. What is the base sequence of the complementary strand? Draw a structure of this section of DNA that shows the hydrogen bonding between the base pairs.

**17.40** Although RNA is a single-stranded molecule, it can have extensive regions of double-stranded structure resulting from intramolecular hydrogen bonding between guanine and cytosine and between adenine and uracil. Draw the hydrogen-bonding interaction between adenine and uracil.

## Proteins

**17.41** Draw the structures of the following amino acids: (a) Tyr; (b) Phe; (c) Glu; and (d) Met.

**17.42** Draw the structures of the following amino acids:
(a) Cys; (b) His; (c) Leu; and (d) Pro.

**17.43** Assign each amino acid in Problem 17.41 as possessing hydrophobic or hydrophilic side chains. Explain each assignment.

**17.44** Assign each amino acid in Problem 17.42 as possessing hydrophobic or hydrophilic side chains. Explain each assignment.

**17.45** Identify the following amino acids and characterize them as possessing hydrophilic or hydrophobic side chains:

**17.46** Identify the following amino acids and characterize them as possessing hydrophilic or hydrophobic side chains:

**17.47** Draw the line structures of all possible dipeptides that can form in condensation reactions between alanine, glutamic acid, and methionine.

**17.48** Draw the line structures of all possible dipeptides that can form from Cys, Gly, and Asn.

## ADDITIONAL PAIRED PROBLEMS

**17.49** Compute the molar mass of a polyethylene molecule that has 744 monomer units.

**17.50** Compute the molar mass of a polystyrene molecule that has 452 monomer units.

**17.51** There are four different bases in DNA strands. How many different 12-base combinations are there?

**17.52** The smallest proteins contain about 50 amino acids. How many different proteins containing 50 amino acids can be formed from the 20 common amino acids? Express your answer in power-of-ten notation.

**17.53** Hair spray is a solution of a polymer in a volatile solvent. When the solvent evaporates, a thin film of polymer that holds the hair in place is left behind. Many hair sprays contain a copolymer made of the following monomers:

(a) Assume that the polymer is made from a 1:1 mixture of the two monomers. Will the two monomers alternate in the polymer or will the arrangement be random? Explain. (b) Draw a section of the copolymer containing at least six repeat units. (c) What structural features of the copolymer allow it to hold hair in place?

**17.54** Chewing gum is mostly polyvinylacetate. The monomer used to make chewing gum is vinyl acetate:

Vinyl acetate
$(C_4H_6O_2)$

(a) Draw a portion of the polyvinylacetate polymer. Show at least three repeat units. (b) Is polyvinylacetate a plastic, a fibre, or an elastomer? Use the behaviour of chewing gum to justify your choice. (c) Chewing gum can be removed from clothing by cooling the polymer with a piece of ice. At low temperature, the gum crumbles easily and can be removed from the fabric. Explain this procedure at the molecular level.

**17.55** One nucleotide and its hydrogen-bonded partner in double-stranded DNA are called a *duplex*. Identify the duplex formed by guanine and draw its structure.

**17.56** Identify the duplex (see Problem 17.55) formed by thymine and draw its structure.

**17.57** Use the polymerization of acrylonitrile to describe each of the three steps of free radical polymerization. Write structures that illustrate the steps.

**17.58** Condensation polymers are usually made from two different monomers, each of which has two of the same functional group. Give three reasons that this strategy is used so often in polymer synthesis.

**17.59** Almost 1000 g of α-glucose will dissolve in 1 L of water. Draw the structure of α-glucose and include enough hydrogen-bonded water molecules to account for its tremendous solubility.

**17.60** Nylon 6 has the following repeat structure:

Draw two strands of Nylon 6, each with four repeat units. Show the two strands connected by hydrogen bonds.

**17.61** What features do nylon and proteins have in common? In what ways are they distinctly different?

**17.62** Complete the following table:

|  | RNA | DNA |
|---|---|---|
| Sugar |  |  |
| One-ring bases |  |  |
| Two-ring bases |  |  |

**17.63** Identify the monomers used to make the following polymers:

(a) Kevlar®; (b) PET; and (c) Styrofoam.

**17.64** Identify the monomers used to make the following polymers:

(a) Dacron; (b) PVC; and (c) Teflon®.

**17.65** High-density polyethylene has more $CH_2$ groups per unit volume than low-density polyethylene. Explain why this is so in terms of the structures of the two forms of the polymer.

**17.66** Explain the major differences between fibrous proteins and globular proteins. What role does each type of protein play in biological organisms?

**17.67** In the 1950s, Edwin Chargaff of Columbia University studied the composition of DNA from a variety of plants and animals. He found that the relative amounts of different bases changed from one species to another. However, in every species studied, the molar ratios of guanine to cytosine and of adenine to thymine were found to be very close to 1.0. Explain Chargaff's observations in terms of the Watson–Crick model of DNA structure.

**17.68** The melting point of DNA, which is the temperature at which the double helix unwinds, increases as the amount of guanine and cytosine increases relative to the amount of adenine and thymine. Explain this observation.

**17.69** For a cell to synthesize a particular protein, genetic information must be transmitted from the cell nucleus to the cytoplasm, where protein synthesis takes place. This shuttling of genetic information is accomplished by a type of RNA, messenger RNA (mRNA). The bases of mRNA are complementary to those in DNA. Assuming that mRNA is made from strand A of the DNA that follows, identify bases 1, 2, 3, and 4 in the mRNA.

Strand A    Strand B    mRNA

**17.70** Transfer-RNA molecules (tRNA) have a set of three bases at their tip that are exposed and can bind their complementary bases. What sequence of bases will be recognized by the following tRNA sequences: (a) GAU; (b) AGG; and (c) CCU?

**17.71** Describe how interactions with solvents affect the tertiary structure of a protein. Include explanations of the roles of regions containing mostly hydrophobic side groups and of regions containing mostly hydrophilic side groups. State where hydrophobic and hydrophilic side groups would be located on a water-soluble globular protein.

**17.72** Globular proteins adopt three-dimensional structures that place some of the amino acids on the inside of the molecule, out of contact with the aqueous environment, and others on the outer surface of the molecule. Which of the following amino acids would be most likely to be found on the inside of a globular protein? (a) Arg; (b) Val; (c) Met; (d) Thr; and (e) Asp. Explain your choices.

## ADDITIONAL PROBLEMS

**17.73** The first step of glucose metabolism is an enzyme-catalyzed condensation reaction between phosphoric acid and the $CH_2OH$ hydroxyl on glucose. Draw the structure of this glucose phosphate.

**17.74** Proteins are synthesized in the cell by adding one amino acid at a time to the growing polypeptide chain. Each amino acid is carried to the protein in a form in which the amino acid is linked to adenosine monophosphate. The amino acid is joined to AMP by a condensation of its carboxylic acid with the phosphate group. Draw the structure of Ala–AMP.

**17.75** Copolymerization of styrene with a small amount of divinylbenzene gives a cross-linked polymer that is hard and insoluble. Draw a picture of this polymer that shows at least two cross-links. (*Hint:* The cross-linking starts when divinylbenzene is incorporated into the growing polystyrene chain.)

Divinylbenzene

**17.76** Fungal laccase is an enzyme found in fungi that live on rotting wood. The enzyme is blue and contains 0.40% by mass copper. The molar mass of the enzyme is approximately 64 000 g/mol. How many copper atoms are there in one molecule of fungal laccase?

**17.77** One of the problems encountered in the polymerization of monomers that contain two different functional groups is that the molecules tend to cyclize rather than polymerize. Draw the structure of the cyclized product that would be produced from the following monomer:

**17.78** In the synthesis of glycogen, an enzyme catalyzes the transfer of a glucose molecule from glucose–UDP to the growing end of a glycogen polysaccharide. Glucose–UDP is a uridine diphosphate molecule linked to the hydroxyl group on the carbon adjacent to the ring oxygen atom of α-glucose. The glucose is at the end of UDPís phosphate chain. Draw the structure of glucose–UDP.

**17.79** The artificial sweetener aspartame (NutraSweet®) is the methyl ester of the following dipeptide:

What two amino acids are used to make aspartame?

**17.80** The nitrogen atom of a peptide linkage has trigonal planar geometry. What is the hybridization of the nitrogen atom in a peptide linkage? Explain why nitrogen adopts this form of hybridization.

**17.81** The development of artificial substances compatible with human tissue is an important area of research. One example is a polymer of lactic acid:

This polymer is used to make body implants needed for only a short time. Eventually, the polymer is converted back to lactic acid, which is metabolized to $CO_2$ and water in the same manner as natural lactic acid. Thus, the body absorbs the polymer without leaving any permanent residue. Draw at least four repeat units of the structure of the polymer made from lactic acid.

**17.82** Gentobiose is a disaccharide found bonded to a number of biological molecules. Gentobiose contains two linked β-glucose molecules. One molecule links through the hydroxyl group in position 1, and the other links through the $CH_2OH$ hydroxyl group. Draw the structure of gentobiose.

**17.83** Suppose that a polypeptide is constructed with alanine as the only monomer. What is the empirical formula of this polypeptide? If the polypeptide has a molar mass of $1.20 \times 10^3$ g/mol, how many repeat units of alanine does it contain?

**17.84** Draw the structure of the nucleotide formed from cytosine, ribose, and a phosphate.

## GROUP STUDY PROBLEMS

**17.85** Automobiles and major appliances such as refrigerators and washing machines require very tough, long-lasting paints that are baked onto the surface of the object. One

group of such paints is known as *alkyds*, which stands for *alcohol* and *acid*. These polyesters have extensive cross-linking that characterizes a tough coating material. One of the simplest alkyds is formed from glycerol and phthalic acid. Heating at 130 °C for about an hour maximizes the amount of cross-linking. Draw the structure of the condensation polymer that forms from these two monomers:

**17.86** Glyptal is a highly cross-linked polymer made by heating glycerol and phthalic anhydride. Show the structure of glyptal. (*Hint:* The highlighted section reacts with water in a reverse condensation reaction to give the monomer that polymerizes with glycerol.)

Phthalic anhydride

**17.87** Draw the structures of polyethylene and the copolymer of butadiene and styrene, showing at least six repeat units for each polymer. On the basis of their molecular structures, explain why polyethylene is more rigid than the butadiene–styrene copolymer.

**17.88** Design a protein containing 10 amino acids having a tertiary structure that would be roughly spherical with a hydrophobic interior and a hydrophilic exterior. Include one S—S bridge that would help stabilize the structure.

## ANSWERS TO SECTION EXERCISES

**17.1.1** (a) hydroxyl; (b) amine and carboxyl; and (c) ester

**17.1.2**

**17.1.3** (In each case, there are several correct answers.)

**17.2.1**

(a) 
Two repeat units

(b) 
Three repeat units

(c) 
Three repeat units

**17.2.2**

**17.2.3**

**17.3.1**

Polyamide

Polyester

**17.3.2**

**17.3.3**

$$HO-C(O)-OH + HO-C(O)-OH \longrightarrow HO-C(O)-O-C(O)-OH + H_2O$$

**17.4.1** In Kevlar®, hydrogen bonding between the amide linkage groups holds the polymer chains in a sheet-like arrangement. This gives the phenyl rings a spe-

cific ordered orientation and makes the polymer highly crystalline. On the other hand, polystyrene chains are held together by weaker dispersion forces. The polymer is less ordered and more flexible than Kevlar®.

**17.4.2** (a) fibre; (b) plastic; and (c) elastomer

**17.4.3**

**17.5.1**

α-Glucose        β-Glucose

**17.5.2**

β-Galactose

β-Glucose

**17.5.3** Refer to Figure 17.19. The β-linkages in cellulose result in a flat structure, whereas the linkages in starch result in a curved structure. Thus, an enzyme shaped to bind starch cannot bind cellulose.

**17.6.1**

**17.6.2** C-G-G-T-A-T-C-C-A

**17.6.3**

**17.7.1** Ala–Thr–Val–Gly–His

**17.7.2** (Lone pairs are shown only on hydrogen-bonded atoms.)

**17.7.3** The primary structure of a protein is the linear sequence of amino acids, which are held together by covalent bonds. Secondary structure involves regions of the protein chain that twist into coils or fold into sheets. Hydrogen bonding between the amide residues of the protein backbone accounts for secondary structure. The overall shape of a protein is its tertiary structure. Dispersion forces, hydrogen bonding, and covalent cross-links between cysteine side chains contribute to the tertiary structure.

## ANSWERS TO PRACTICE EXERCISES

**17.1**

**17.2**

**17.3** The structures differ at positions 2, 3, and 4. At each of these, the molecules have the same types of bonds but in different orientations. The OH group points "down–up–down" in $\alpha$-glucose and "up–down–up" in $\alpha$-idose. The two molecules have the same structures at the other positions.

**17.4**

β-Mannose

**17.5** Melibiose is composed of $\beta$-glucose and alpha-galactose.

**17.6**

**17.7**

**17.8**

# THE TRANSITION METALS

## METALS ESSENTIAL FOR LIFE

We tend to think of metals as lustrous solids: copper, tin, gold, silver, iron. We are less likely to think of individual metal atoms in biological "machines." Nevertheless, plants and animals require the presence of tiny amounts of transition metals. Humans require most of the elements in the first transition metal series and at least one element from the second transition metal series, molybdenum.

Among the transition metals, the biochemistry of iron is known in the most detail. Iron is essential for the synthesis of chlorophyll, the green pigment that drives photosynthesis. Our background photograph shows citrus trees suffering from the yellowing of leaves that is characteristic of iron deficiency. To combat iron deficiency, gardeners use fertilizers that contain iron chelates, octahedral complexes that slowly release their iron into the soil.

Iron is the transition metal present in largest quantities in the human body. The best-known biological iron-containing compound is the protein haemoglobin, the red component of blood that is responsible for the transport of oxygen. The central feature of haemoglobin is the porphyrin structure, a plane of carbon and nitrogen atoms that forms multiple rings with an extensive delocalized $\pi$-electron system. In haemoglobin, an $Fe^{2+}$ cation bonds to four nitrogen atoms that are part of the planar porphyrin ring; this structure is known as *haem*. An oxygen molecule can bind to the iron at a site perpendicular to the plane of the porphyrin ring. Haemoglobin exploits one of the special bonding properties of transition metal ions: their ability to form bonds by accepting electron pairs from atoms, such as nitrogen and oxygen. The structure of haem appears in our molecular inset. Variations of the haem structure are found in all life-forms.

Other iron-containing proteins feature clusters of iron and sulphur atoms. The iron atoms in these clusters have tetrahedral bonding geometry. Iron–sulphur proteins are essential for electron transfer processes, in which the clusters of iron and sulphur atoms gain and lose electrons. Here, another special property of

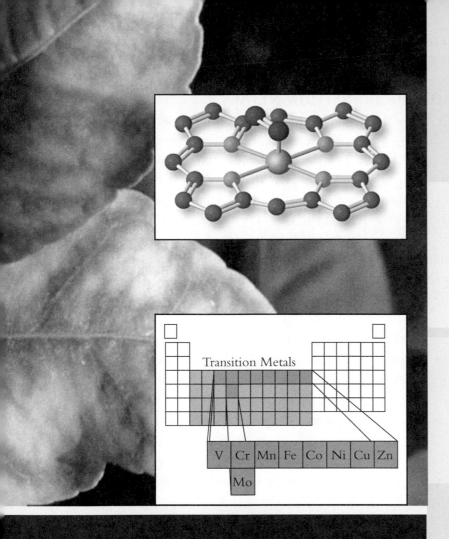

# 18

transition metal ions is exploited: their ability to take on several different oxidation states.

Metals display a remarkable range of chemical properties. The individual atoms or cations in biological macromolecules readily form and break bonds to other species as well as gain or lose electrons. Geologically, metals typically occur in minerals, where metal cations are associated with oxygen, sulphur, or polyatomic anions such as carbonate and silicate. In pure metals and alloys, metals participate in delocalized bonding that generates high strength and electrical conductivity.

In this chapter, we survey the diversity of transition metals, beginning with an overview. Then we describe the structure and bonding in transition metal complexes. We describe some biological roles of transition metals. The chapter ends with a discussion of metallurgy, the processes by which pure metals are extracted from mineral ores, and some applications of transition metals.

## CHAPTER CONTENTS
18.1 Overview of the Transition Metals
18.2 Coordination Complexes
18.3 Bonding in Coordination Complexes
18.4 Transition Metals in Biology
18.5 Metallurgy
18.6 Applications of Transition Metals

## LEARNING OBJECTIVES
*Upon completion of this chapter you should be able to*

- **predict periodic properties of transition metals**
- **recognize and name transition metal coordination complexes**
- **use crystal field theory to explain the colour and magnetic properties of complexes**
- **explain the importance of transition metal complexes in biological processes**
- **explain the chemistry of essential steps in the production of pure metals from ores**
- **recognize the importance of transition metals in everyday life**

## 18.1 OVERVIEW OF THE TRANSITION METALS

The **transition metals** lie in the $d$ block, at the centre of the periodic table, between the $s$-block metals and the elements in the $p$ block, as Figure 18-1 shows. As we describe in Chapter 5, most transition metal atoms in the gas phase have valence electron configurations of $s^2 d^{(x-2)}$, where $x$ is the group number of the metal. Titanium, for example, is in Group 4 and has valence configuration $4s^2 3d^2$. Recall, however, that the $(n + 1)s$ orbital and the set of $nd$ orbitals have nearly the same energy. As a result, several of the transition metals have valence configurations that differ from the general pattern, as listed in Table 5-2. Remember also that for transition metal *cations*, the $nd$ orbital *always* is more stable than the $(n + 1)s$ orbital, so transition metal cations have vacant $(n + 1)s$ orbitals.

A pure transition metal is best described by the band theory of solids, as introduced in Chapter 7. In this model, the valence $s$ and $d$ electrons form extended bands of orbitals that are delocalized over the entire network of metal atoms. These valence electrons are easily removed, so most elements in the $d$ block react readily to form compounds: oxides such as $Fe_2O_3$, sulphides such as $ZnS$, and mineral salts such as zircon, $ZrSiO_4$.

Between barium (Group 2, element 56) and lutetium (Group 3, element 71), the $4f$ orbitals fill with electrons, giving rise to the **lanthanides**, a set of 14 metals named for lanthanum, the first member of the series. The lanthanides are also called the *rare earths*, although except for promethium they are not particularly rare. Between radium (Group 2, element 88) and lawrencium (Group 3, element 103) are the 14 **actinides**, named for the first member of the set, actinium. The lanthanides and actinides are also known as the **inner transition metals**.

**Figure 18-1**

The transition metals are the elements in the $d$ block of the periodic table. The inner transition metals are those in the $f$ block.

**Figure 18-2**

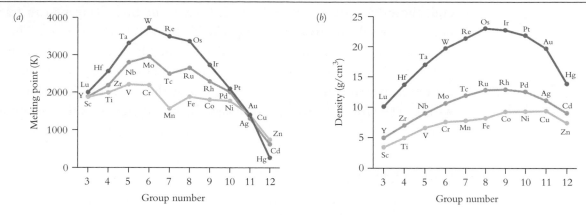

The melting points of the transition metals increase from Group 3 to Group 6 and then decrease to the end of the block. The densities of transition metals increase with atomic mass except at the end of the block.

## Physical Properties

As pure elements, almost all the transition metals are solids that conduct heat and electricity and are malleable and ductile. Although they share these general properties, transition metals display variations in other properties that can be traced to their different numbers of valence electrons.

Most pure transition metals have the shiny grey appearance that is termed "silvery" because of the appearance of silver metal. However, some transition metals have other colours; for example, the orange colour of copper and the yellow hue of gold.

The melting points of the transition metals show a periodic variation that is displayed in Figure 18-2a. Up to Group 6, the melting point increases with each added valence electron. This is because the first six valence electrons occupy bonding orbital bands that increase the overall bonding strength of the metal. Beyond Group 6, in contrast, the melting point decreases, falling precipitously for the elements of Group 12. This decrease is sufficient to make mercury unique among metals. This element has a melting point of −39 °C (234 K), making Hg a liquid at room temperature. Melting points decrease with added electrons beyond Group 6, because the added electrons occupy antibonding bands, weakening the overall bonding strength of the metal.

Samples of metals from the first transition series. The metals are Cu (wire), Fe (block), Ni (balls), and Zn (sticks).

The densities of transition metals also display regular periodic trends, as Figure 18-2b shows. Density increases moving down each column of the periodic table and increases smoothly across the first part of each row. Increasing atomic mass accounts for both these trends. The volume occupied by an individual atom in the metallic lattice varies slowly within the d block, so the more massive the nucleus, the greater the density of the metal. Toward the end of each row, density decreases for the same reason that melting point decreases. The added electrons occupy antibonding orbitals, and this leads to a looser array of atoms, larger atomic volume, and decreased density.

Transition metals are good conductors of both electricity and heat. In general, these two properties are linked, because mobile electrons can transport both charge (electricity) and energy (heat) through bulk materials. The metals of Group 11 (copper, silver, and gold) have the highest electrical conductivity of any of the elements. Silver is the best conductor, but copper is more abundant and consequently less expensive, so it is the metal of choice for electrical wiring. The Group 11 elements are also the best conductors of heat among the metals.

## Redox Behaviour

The chemistry of the transition metals is determined in part by their atomic ionization energies. Metals of the $3d$ and $4d$ series show a gradual increase in ionization energy with atomic number ($Z$), whereas the trend for the $5d$ series is more pronounced (Figure 18-3).

First ionization energies for transition metals in the $3d$ and $4d$ series are between 650 and 750 kJ/mol, somewhat higher than the values for Group 2 alkaline earth metals but lower than the typical values for non-metals in the $p$ block.

Except for the elements at the ends of the rows, each transition metal can exist in several different oxidation states. The oxidation states displayed by the $3d$ transition metals are shown in Table 18-1. The most important oxidation states are highlighted in the table. The most common oxidation state for the $3d$ transition metals is +2, known for all the elements except Sc. Chromium, iron, and cobalt are also stable in the +3 oxidation state, and for vanadium and manganese the +4 oxidation state is stable. Elements from scandium to manganese have a particularly stable oxidation state corresponding to the loss of all the valence electrons ($s^0 d^0$ configuration).

Cations with charges greater than +3 are generally unstable even in an ionic environment, because the energy required to remove an electron from an ion with +3 charge is very large. To convert $Ti^{3+}$ to $Ti^{4+}$, for instance, requires an energy

**Figure 18-3**

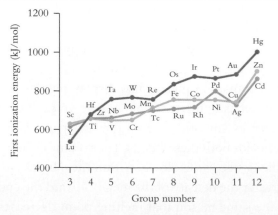

The first ionization energies of transition metals show gradual upward trends across each row of the periodic table.

## TABLE 18-1  OXIDATION STATES DISPLAYED BY THE 3d TRANSITION METALS*

| Element Group / Oxidation State | Sc 3 | Ti 4 | V 5 | Cr 6 | Mn 7 | Fe 8 | Co 9 | Ni 10 | Cu 11 | Zn 12 |
|---|---|---|---|---|---|---|---|---|---|---|
| +1 | | | $d^4$ | $d^5$ | $d^6$ | | $d^8$ | $d^9$ | $d^{10}$ | |
| +2 | | $d^2$ | $d^3$ | $d^4$ | $d^5$ | $d^6$ | $d^7$ | $d^8$ | $d^9$ | $d^{10}$ |
| +3 | $d^0$ | $d^1$ | $d^2$ | $d^3$ | $d^4$ | $d^5$ | $d^6$ | $d^7$ | $d^8$ | |
| +4 | | $d^0$ | $d^1$ | $d^2$ | $d^3$ | $d^4$ | $d^5$ | $d^6$ | | |
| +5 | | | $d^0$ | $d^1$ | $d^2$ | | $d^4$ | | | |
| +6 | | | | $d^0$ | $d^1$ | $d^2$ | | | | |
| +7 | | | | | $d^0$ | | | | | |

(Row labelled "Valence Configuration" spans the element columns.)

□ = +2    ■ = +3, +4    ▨ = $d^0$

*The table lists the configuration of the ion corresponding to each observed oxidation state. The most important oxidation states of each element are colour screened.

Zn $+^1$ ZnS, ZnCO₃

input of 4175 kJ/mol. Consequently, oxidation states greater than +3 are found in covalently bonded oxides, oxoanions, and halides, but not in ionic compounds such as sulphates and phosphates. Table 18-2 lists representative examples of compounds displaying the important oxidation states of the 3d transition metals.

## Transition Metal Compounds

A few of the transition metals, including gold, platinum, and iridium, are found in nature as pure elements, but most of the others are found associated with either

## TABLE 18-2  REPRESENTATIVE COMPOUNDS OF 3d TRANSITION METALS IN THEIR COMMONLY OCCURRING OXIDATION STATES

| Element | Oxidation State | Compound | Oxidation State | Compound | Oxidation State | Compound |
|---|---|---|---|---|---|---|
| Sc | +3 | Sc(NO₃)₃ · 6H₂O | | | | |
| Ti | | | +4 | TiBr₄, FeTiO₃ | | |
| V | | | +4 | VCl₄, VO₂ | +5 | Pb₂(VO₄)Cl |
| Cr | +2 | Cr(CH₃CO₂)₂ | +3 | Cr(OH)₃ | +6 | CrF₆, K₂Cr₂O₇ |
| Mn | +2 | MnCO₃, MnCl₂ | +4 | MnO₂, MnF₄ | +7 | KMnO₄ |
| Fe | +2 | FeSO₄ · H₂O, FeS₂ | +3 | FeCl₃, FePO₄ · 2H₂O | | |
| Co | +2 | CoS, Co(ClO₄)₂ | +3 | Co(OH)₃, CoF₃ | | |
| Ni | +2 | NiSO₄ · 7H₂O | | | | |
| Cu | +2 | Cu(NO₃)₂ · 6H₂O | | | | |
| Zn | +2 | ZnS, ZnCO₃ | | | | |

sulphur or oxygen. Iron, manganese, and the metals of Groups 3–6 (except for Mo) are most often found as oxides; less often, they occur as sulphates or carbonates. Molybdenum and the metals of Groups 7–12 (except for Mn and Fe) are most often found as sulphides.

Although naturally occurring compounds of transition metals are restricted in scope, a wide variety of compounds can be synthesized in the laboratory. Representative compounds appear in Table 18-2. These compounds fall into three general categories. There are many binary halides and oxides in a range of oxidation numbers. Ionic compounds containing transition metal cations and poly-atomic oxoanions are also common; these include nitrates, carbonates, sulphates, phosphates, and perchlorates. Finally, there are numerous ionic compounds in which the transition metal is part of an oxoanion.

The synthesis of transition metal compounds must be tailored to a particu-lar product. The details are complex, and existing patterns of synthesis contain so many exceptions that an overall view is beyond the scope of an introductory course in chemistry.

## SECTION EXERCISES

● **18.1.1** Predict which element of the following pairs will have the higher melting point and support your prediction using periodic trends: (a) Cr or W; (b) W or Os; (c) Pd or Ag; and (d) Y or Nb.

● **18.1.2** Predict which element of the following pairs will have the higher density and support your prediction using periodic trends: (a) Cr or W; (b) W or Os; (c) Pd or Ag; and (d) Y or Nb.

● **18.1.3** The following compounds can be purchased from chemical companies. Determine the oxidation state of the transition metal in each of them.
(a) $TaCl_5$;
(b) $Fe(NO_3)_3 \cdot 9H_2O$;
(c) $Rh_2O_3$;
(d) $CrO_2Cl_2$; and
(e) $Cu_2(OH)PO_4$.

## 18.2 COORDINATION COMPLEXES

Earlier chapters described how metal salts dissolve in water to give solutions of aqueous ions. Here are three examples:

$$Cu(ClO_4)_2 \, (s) \xrightarrow{H_2O} Cu^{2+} \, (aq) + 2 \, ClO_4^- \, (aq)$$

$$NiSO_4 \, (s) \xrightarrow{H_2O} Ni^{2+} \, (aq) + SO_4^{2-} \, (aq)$$

$$FeCl_3 \, (s) \xrightarrow{H_2O} Fe^{3+} \, (aq) + 3 \, Cl^- \, (aq)$$

Section 15.5 addresses the formation of complexes between metal ions and ligands. For example, $Ni^{2+} \, (aq)$ forms a complex with ammonia:

$$Ni^{2+} \, (aq) + 6 \, NH_3 \, (aq) \rightleftharpoons [Ni(NH_3)_6]^{2+} \, (aq)$$

**Figure 18-4**

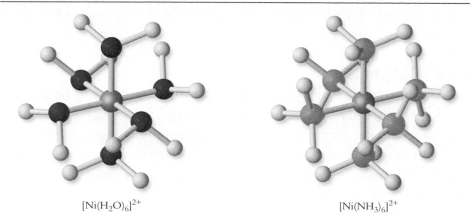

$[Ni(H_2O)_6]^{2+}$                                                    $[Ni(NH_3)_6]^{2+}$

The $Ni^{2+}$ cations in $[Ni(H_2O)_6]^{2+}$ and $[Ni(NH_3)_6]^{2+}$ bond to six ligands, one at each vertex of an octahedron.

For complexation purposes, a **ligand** is a species that has lone pairs of electrons available to donate to a metal atom or cation. Water molecules possess lone pairs of electrons, so water is a ligand that readily forms complex ions with metal cations. Although solubility and complexation equations show uncomplexed metal ions in solution, dissolved cations actually form chemical bonds to water molecules of the solvent. For example, $Ni^{2+}$ (aq) bonds to six water molecules in octahedral geometry, as illustrated in Figure 18-4. The aqueous $[Ni(H_2O)_6]^{2+}$ cation contains $Ni^{2+}$ bonded to six water ligands.

Colour changes often provide evidence for the interaction of ligands and metal cations, particularly for the transition metals. The $Ni^{2+}$ cation provides an example. Figure 18-5 shows that nickel(II) sulphate, a white crystalline solid, dissolves in water to give a green solution. The green colour cannot be due to $Ni^{2+}$ or $SO_4^{-2}$, which the white solid shows to be colourless. Rather, the colour comes from the octahedral complex that forms when each nickel ion binds to six water molecules:

$$NiSO_4\ (s) + 6\ H_2O\ (l) \longrightarrow [Ni(H_2O)_6]^{2+}\ (aq) + SO_4^{2-}\ (aq)$$

Replacing the water ligands with other ligands can also result in colour changes. Figure 18-5 shows the colour change that occurs when ammonia is added to a solution of hydrated $Ni^{2+}$. Ammonia molecules replace water ligands to give the blue species $[Ni(NH_3)_6]^{2+}$, the ball-and-stick structure of which appears in Figure 18-4:

$$\underset{\text{green}}{[Ni(H_2O_6)]^{2+}\ (aq)} + 6\ NH_3\ (aq) \longrightarrow \underset{\text{blue}}{[Ni(NH_3)_6]^{2+}\ (aq)} + 6\ H_2O\ (l)$$

Complex ions, also called **coordination complexes**, have well-defined stoichiometries and structural arrangements. Usually, the formula of a coordination complex is enclosed in brackets to show that the metal and all its ligands form a single structural entity. When an ionic coordination complex is isolated from aqueous solution, the product is composed of the complex ion and enough counter-ions to give a neutral salt. In the chemical formula,

**Figure 18-5**

Solid nickel(II) sulphate, nearly colourless, dissolves in water to give a green solution containing $[Ni(H_2O)_6]^{2+}$ cations. The addition of ammonia produces a blue solution of $[Ni(NH_3)_6]^{2+}$ cations. Solvent evaporation gives a blue–violet precipitate of $[Ni(NH_3)_6]SO_4$.

the counter-ions are shown outside the brackets. Examples include $[Ni(NH_3)_6]SO_4$ (the sulphate salt of $[Ni(NH_3)_6]^{2+}$) and $K_3[Fe(CN)_6]$ (the potassium salt of $[Fe(CN)_6]^{3-}$).

## Nature of Ligands

A ligand must have a lone pair of electrons that it donates to form a bond to the metal. For example, the Ni—N bonds in $[Ni(NH_3)_6]^{2+}$ form by overlap of an empty valence orbital on the metal (shown in green on the figure below) with the lone pair $sp^3$ orbital on the nitrogen atom. The water ligands in $[Ni(H_2O)_6]^{2+}$ coordinate to the metal in a similar manner, with the oxygen atoms donating lone pairs to form Ni—O bonds.

Donor                Metal                Bond formation

Hundreds of different ligands are known. Some of the most common are listed in Table 18-3. The simplest ligands coordinate to the metal through a single donor atom. The most common donor atoms in coordination chemistry are nitrogen, oxygen, and the halogens, but other important ligands have carbon, phosphorus, or sulphur donor atoms. Recall from Chapter 15 that a ligand with one donor atom is called **monodentate**. Ammonia is a monodentate ligand with a nitrogen donor atom. Other examples are water, halide ions, cyanide ion, hydroxide ion, and carbon monoxide.

A chelating ligand contains two or more donor atoms in a structure that allows the ligand to wrap around the metal. Examples featured in Chapter 15 are the bidentate ligand ethylenediamine (en, chemical formula $H_2NCH_2CH_2NH_2$) and the hexadentate ethylenediaminetetraacetate (EDTA).

## Structures of Coordination Complexes

Coordination complexes are a remarkably diverse group of molecules that form from virtually all transition metals in a variety of oxidation states. These compounds involve an extensive array of ligands, and they adopt several molecular geometries.

The molecular geometry of a complex depends on the **coordination number**, which is the number of ligand atoms bonded to the metal. The most common coordination number is 6, and almost all metal complexes with coordination number 6 adopt octahedral geometry. This preferred geometry can be traced to the valence shell electron pair repulsion (VSEPR) model introduced in Chapter 6. The ligands space themselves around the metal as far apart as possible, so as to minimize electron–electron repulsion.

Although a coordination number of 6 is most prevalent, 4 is also common, and several important complexes have a coordination number of 2. In addition,

## TABLE 18-3 COMMON LIGANDS

| | Ligand Name | Formula or Structure (Abbreviation) | Donor Atom(s) |
|---|---|---|---|
| **Monodentate** | Bromo | $Br^-$ | Br |
| | Chloro | $Cl^-$ | Cl |
| | Fluoro | $F^-$ | F |
| | Iodo | $I^-$ | I |
| | Cyano | $CN^-$ | C |
| | Carbonyl | CO | C |
| | Ammine* | $NH_3$ | N |
| | N-Nitrito | $NO_2^-$ | N |
| | Pyridine | (or pyr) | N |
| | N-Thiocyanato | $NCS^-$ | N |
| | Aqua | $H_2O$ | O |
| | Oxo | $O_2^-$ | O |
| | Hydroxo | $OH^-$ | O |
| | O-Nitrito | $NO_2^-$ | O |
| | Carbonato | $CO_3^{2-}$ | O |
| | Hydrido | $H^-$ | H |
| | S-Thiocyanato | $NCS^-$ | S |
| | Trimethylphosphine† | $P(CH_3)_3$ | P |
| **Bidentate** | Ethylenediamine | (en) | N, N |
| | Bipyridine | (bipy) | N, N |
| | Glycinato‡ | | N, O |
| | Oxalato | (o) | O, O |
| **Tetradentate** | Porphinato | See Figure 18-25 | N, N, N, N |
| **Hexadentate** | Ethylenediaminetetraacetato | (EDTA) | N, N, O, O, O, O |

*Amines (derivatives of $NH_3$), such as methylamine ($CH_3NH_2$), also are common ligands.
†This is the simplest of a group of ligands called *phosphines*, which have three organic groups bonded to phosphorus.
‡All of the common amino acids form bidentate ligands.

**Figure 18-6**

Complexes with coordination number 2 always adopt linear geometry about the metal cation.

a few complexes display coordination numbers of 3, 5, and 7. Examples of coordination number 2 include the silver–ammonia complex and the gold–cyanide complex, both described in Chapter 15. To minimize ligand–ligand repulsions, a complex with a coordination number of 2 is invariably linear, as Figure 18-6 shows.

Four-coordinate complexes may be either square planar or tetrahedral, as Figure 18-7 shows. Tetrahedral geometry is most common among the first-row transition metals. Examples include $[Zn(NH_3)_4]^{2+}$, $[Cd(en)_2]^{2+}$, $[FeCl_4]^-$, and $[Ni(CO)_4]$. Square planar geometry is characteristic of transition metal ions such as palladium(II), gold(III), and others that have eight $d$ electrons in the valence shell. We describe the differences between square planar and tetrahedral geometries in more detail in Section 18.3.

**Figure 18-7**

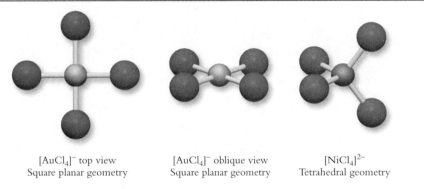

$[AuCl_4]^-$ top view
Square planar geometry

$[AuCl_4]^-$ oblique view
Square planar geometry

$[NiCl_4]^{2-}$
Tetrahedral geometry

Four-coordinate complexes adopt either square planar geometry, as in $[AuCl_4]^-$, or tetrahedral geometry, as in $[NiCl_4]^-$.

## Isomers

Isomerism of organic compounds was introduced in Chapter 10. The same concepts apply to transition metal complexes. Although they have the same formula, isomers usually have different chemical and physical properties. Isomers may have different colours, different melting or boiling points, different solubilities, and different rates of reactions with common reagents.

In $[Co(NH_3)_6]^{3+}$, one ammonia ligand occupies each apex of an octahedron; this complex has no isomers. Replacement of one of the six $NH_3$ ligands with a chloride ion generates $[Co(NH_3)_5Cl]^{2+}$. Regardless of which ligand is replaced, the geometry of the resulting complex is the same, because the six positions around an octahedron are equivalent by symmetry, as illustrated in Figure 18-8. Any octahedral coordination complex of the general formula $ML_5X$ has no isomers.

Now suppose a second ammonia ligand is replaced with a chloride ion. If an ammonia ligand *adjacent to* the first chloride ion is replaced, the two chloride ions are separated by a bond angle of 90°, and the result is called the *cis* isomer (Figure 18-9). If the ammonia ligand *opposite* the chloride ion is replaced, the two chloride ions are separated by a bond angle of 180°, and the complex is

**Figure 18-8**

There is only one isomer of $[Co(NH_3)_5Cl]^{2+}$. The top structure is identical to the bottom structure because the two can be superimposed on each other after a 90° rotation.

**Figure 18-9**

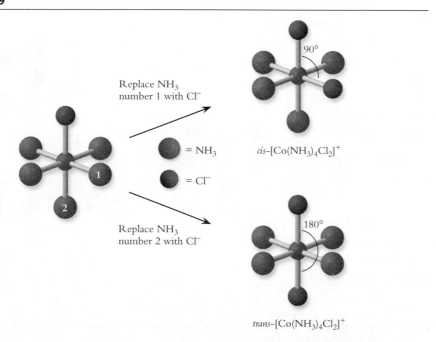

There are two isomers of $[Co(NH_3)_4Cl_2]^+$. The *cis* isomer is violet, and the *trans* isomer is green.

called the *trans* isomer. These two isomers of $[Co(NH_3)_4Cl_2]^+$ have the same chemical bonds but display different geometries, so they are **geometric isomers**. Geometric isomers have different properties. For example, whereas $cis\text{-}[Co(NH_3)_4Cl_2]^+$ is violet, the *trans* isomer is green. Example 18-1 introduces two more geometric isomers of coordination complexes.

*cis* means "next to," and *trans* means "across from."

**EXAMPLE 18-1 ISOMERS OF COORDINATION COMPLEXES**

Draw ball-and-stick models of all possible isomers of the octahedral compound $[Cr(NH_3)_3Cl_3]$.

**Strategy:** It is best to approach this problem by starting with $[Cr(NH_3)_6]^{3+}$ and replacing one ligand at a time, considering all possible structures. The first substitution gives only one compound. Introducing a second chloride ion creates *cis* and *trans* isomers of $[Cr(NH_3)_4Cl_2]^+$. Choose each of these isomers in turn, and consider all possible structures that result from introducing a third chloride ligand. This is best accomplished by drawing pictures.

In this isomer, three of the same ligands lie on an equatorial circle about the complex. Such a circle is called a *meridian*, and this isomer is the meridional isomer, abbreviated *mer*.

**Solution:** For the *trans* isomer, replacing any of the four remaining $NH_3$ ligands gives the same product, in which three of the same ligand are arranged in a T shape about the metal ion. This shape is designated *mer*:

*trans*-$[Cr(NH_3)_4Cl_2]^+$          *mer*-$[Cr(NH_3)_3Cl_3]$

The three like ligands lie in a plane that forms one face of the octahedron, shown in outline in the drawing. This is called the *facial* isomer, abbreviated *fac*.

Starting with *cis*-$[Cr(NH_3)_4Cl_2]^+$, it is possible to make two isomers of the final product. This can be seen most clearly by labelling the four $NH_3$ ligands. Replacing an $NH_3$ ligand at site *a* or *b* with $Cl^-$ generates the T-shaped arrangement. However, replacement at position *c* or *d* gives a new isomer in which the three chloride ligands occupy the corners of a triangle. This shape is designated *fac*:

This exhausts all possible structures of $[Cr(NH_3)_3Cl_3]$. Octahedral coordination complexes of general formula $ML_3X_3$ have two isomers.

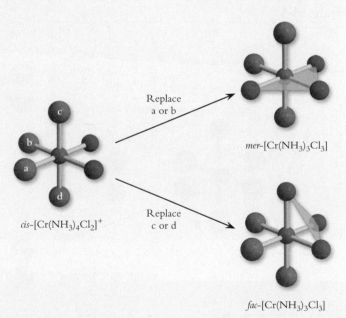

*cis*-$[Cr(NH_3)_4Cl_2]^+$

Replace
a or b

Replace
c or d

*mer*-$[Cr(NH_3)_3Cl_3]$

*fac*-$[Cr(NH_3)_3Cl_3]$

**Do the Results Make Sense?** No matter how many times the relative positions of $NH_3$ or Cl are exchanged around the metal centre, only two possible structures result: *fac* and *mer*.

**Practice Exercise 18.1**

Platinum forms a large collection of square planar complexes. Draw ball-and-stick models similar to those in Example 18-1 for the *cis* and *trans* isomers of $[Pt(NH_3)_2Cl_2]$.

Another type of isomerism displayed by coordination complexes is based on the bonding of the ligand. **Linkage isomers** occur when a ligand can bond to a metal using either of two donor atoms. Figure 18-10 shows the two linkage isomers of $[Co(NH_3)_5NO_2]^{2+}$, in which either the nitrogen atom or the oxygen atom of the nitrite ion can bind to cobalt. Linkage isomers are distinct compounds that have different properties. For example, the linkage isomers of $[Co(NH_3)_5NO_2]^{2+}$ have slightly different colours.

**Figure 18-10**

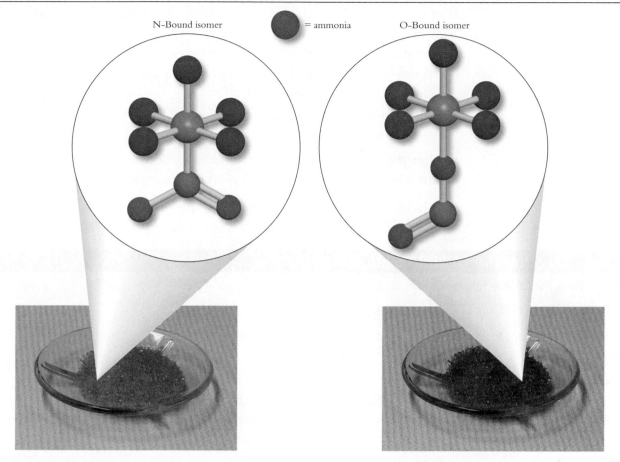

Linkage isomers occur when a ligand is capable of bonding to a metal through either of two donors. In $[Co(NH_3)_5NO_2]^{2+}$, the nitro ligand can bond to cobalt through the N atom or the O atom. These isomers have slightly different colours, as the photos show.

## Naming Coordination Compounds

Originally, compounds containing coordination complexes were given common names such as Prussian blue ($KFe[Fe(CN)_6]$), which is deep blue, or Reinecke's salt ($NH_4[Cr(NH_3)_2(NCS)_4]$), named for its first maker. Eventually, coordination compounds became too numerous for chemists to keep track of all the common names. To solve the nomenclature problem, the International Union of Pure and Applied Chemistry (IUPAC) created a systematic procedure for naming coordination compounds. The following guidelines are used to determine the name of a coordination compound from its formula, or vice versa:

1. As with all salts, name the cation before the anion.

2. Within the complex, first name the ligands in alphabetical order, and then name the metal.

3. If the ligand is an anion, add the suffix *-o* to the stem name: *bromo* ($Br^-$), *cyano* ($CN^-$), and *hydroxo* ($OH^-$). The simplest neutral ligands have special names: *aqua* ($H_2O$), *ammine* ($NH_3$), and *carbonyl* ($CO$). Other neutral ligands retain their usual names. Some common ligands and their names appear in Table 18-3.

4. Use a Greek prefix (*di-, tri-, tetra-, penta-, hexa-*) to indicate the number of identical ligands. If the name of the ligand already incorporates one of these prefixes (as in ethylenediamine), enclose the ligand name in parentheses and use the alternative prefixes *bis-* (two), *tris-* (three), and *tetrakis-* (four). These numerical prefixes are ignored in determining the alphabetical order of the ligands.

5. If the coordination complex is an anion, add the suffix *-ate* to the stem name of the metal.

6. After the name of the metal, give the oxidation number of the metal in parentheses as a Roman numeral or as 0 if the oxidation number is zero.

Example 18-2 provides practice in applying these guidelines.

## EXAMPLE 18-2 NAMING COORDINATION COMPOUNDS

What is the IUPAC name for each of the following coordination compounds?

(a) $[Ni(H_2O)_6]SO_4$; (b) $[Cr(en)_2Cl_2]Cl$; and (c) $K_2[CoCl_4]$.

**Strategy:** Follow the guidelines for naming coordination compounds. Break the complex ion down one piece at a time.

**Solution:**

(a) The cation is the complex ion. Name the ligand and include the appropriate prefix: hexaaqua. The formula lists one sulphate anion outside the brackets. Ions outside the brackets are not part of the complex, so the net charge on the nickel-containing cation is +2. Aqua is neutral, requiring that Ni be in the +2 oxidation state: hexaaquanickel(II) sulphate.

(b) The ligands are chloro and ethylenediamine. The coordination compound has a +1 charge and contains two anionic ligands, so chromium must be in the +3 oxidation state. The en ligand contains *di* as part of its formal name, so we use the prefix *bis* to specify two en ligands: dichlorobis(ethylenediamine)-chromium(III) chloride.

(c) The cation, potassium, is named first. The complex anion has four negative chloro ligands. The complex is an anion with −2 charge, requiring that Co be in the +2 oxidation state. Since the complex is an anion, the metal is given a suffix of *-ate*: potassium tetrachlorocobaltate(II).

 **Do the Results Make Sense?** Verify that the numerical prefixes give the correct total number of ligands, that the ligands are named alphabetically, and that the overall species is charge-neutral.

**Common Pitfalls:** Not following the guidelines for naming these compounds. If you learn the guidelines, you will always get the correct answer.

**Practice Exercise 18.2**

Name the following coordination compounds: (a) $[Co(NH_3)_5CO_3]NO_3$; (b) $[Mo(CO)_3(NH_3)_3]$; and (c) $[Al(H_2O)_5OH]Cl_2$.

The following examples illustrate further applications of the IUPAC guidelines:

| | |
|---|---|
| $Na_2[FeEDTA]$ | Sodium ethylenediaminetetraacetatoferrate(II) |
| $[Zn(NH_3)_4](NO_3)_2$ | Tetraamminezinc nitrate |
| $K[Ag(CN)_2]$ | Potassium dicyanoargentate(II) |
| $[Rh(NH_3)_5Br]Br_2$ | Pentaamminebromorhodium(III) bromide |

## TABLE 18-4 LATIN NAMES OF METALS IN ANIONIC COMPLEXES

| Metal (Symbol) | Latin Name |
|---|---|
| Iron (Fe) | Ferrate |
| Copper (Cu) | Cuprate |
| Silver (Ag) | Argentate |
| Tin (Sn) | Stannate |
| Gold (Au) | Aurate |
| Lead (Pb) | Plumbate |

These examples show that when information is not needed to identify the compound, it is omitted from the name. In the first name, for instance, it is not necessary to tell how many sodium ions are present, because we can deduce the number from the name of the complex anion. In the second name, the oxidation state of zinc is omitted because it is always +2. In the fourth name, the single bromo ligand is not preceded by the prefix mono.

The first and third examples illustrate a nuance of the naming rules. Iron and silver in anionic complexes are named by their Latin roots, *ferr-* and *argent-*, from which their symbols (Fe and Ag) are derived. Metals taking their Latin names in anionic coordination complexes are listed in Table 18-4.

Example 18-3 provides more practice in working with the names of coordination compounds. Our Focus 18-1: Chemical Milestones describes the detective work that led to the birth of coordination chemistry.

## FOCUS 18-1 CHEMICAL MILESTONES: THE BIRTH OF COORDINATION CHEMISTRY

Late in the nineteenth century, just as the principles of chemical bonding were being discovered, chemists carried out many studies of the interactions of ammonia with cations such as $Cr^{3+}$, $Co^{3+}$, $Pt^{4+}$, and $Pd^{2+}$. The most intriguing results were obtained for cobalt(III) chloride. By 1890, several ammonia compounds of $CoCl_3$ had been isolated. These coordination compounds differed in several of their properties, the most striking of which were their beautiful colours. At the time, the formulas of these cobalt complexes were written as follows:

| | |
|---|---|
| $CoCl_3 \cdot 6NH_3$ | Yellow-orange |
| $CoCl_3 \cdot 5NH_3$ | Rose |
| $CoCl_3 \cdot 4NH_3$ | Green |
| $CoCl_3 \cdot 3NH_3$ | Green |

Chemists were convinced that all the chlorine atoms had to be bonded to the cobalt in some way, and since ammonia is a gas at room temperature, they could not understand why the ammonia molecules did not evaporate away. One of the first proposed explanations was the chain theory, in which the ammonia molecules were assumed to form chains between the metal and the chloride. For example, a prominent coordination chemist of the time proposed the following structure for $CoCl_3 \cdot 6NH_3$:

$$
\begin{array}{c}
NH_3-Cl \\
| \\
Co-NH_3-NH_3-NH_3-NH_3-Cl \\
| \\
NH_3-Cl
\end{array}
$$

During the 1890s, these cobalt complexes attracted the attention of a Swiss chemist, Alfred Werner (1866–1919). Only in his early 20s, Werner had just earned his Ph.D. in organic chemistry. He studied the cobalt complexes in detail and developed the basis for our understanding of coordination chemistry.

Werner found that adding aqueous $AgNO_3$ to solutions of the various cobalt ammine complexes gave different amounts of silver chloride precipitate. For $CoCl_3 \cdot 6NH_3$, all three chloride ions precipitated as AgCl. Only two chloride ions precipitated for $CoCl_3 \cdot 5NH_3$, just one precipitated for $CoCl_3 \cdot 4NH_3$, and there was no precipitate for $CoCl_3 \cdot 3NH_3$. Werner explained these results by proposing that the cobalt ion in all four compounds had a coordination number of 6. He reformulated the compounds as follows:

$$CoCl_3 \cdot 6NH_3 = [Co(NH_3)_6]Cl_3$$
$$CoCl_3 \cdot 5NH_3 = [Co(NH_3)_5Cl]Cl_2$$
$$CoCl_3 \cdot 4NH_3 = [Co(NH_3)_4Cl_2]Cl$$
$$CoCl_3 \cdot 3NH_3 = [Co(NH_3)_3Cl_3]$$

The chloride ions that appear outside the brackets represent chloride anions that balance the positive charge on the coordination compound. When a coordination compound dissolves in water, the ligands (inside the brackets) remain bound to the metal cation, but the non-ligands (outside the brackets) exist as individual ions. These chloride ions precipitate in the presence of silver ions. The chloride ions inside the brackets, which are ligands bonded to the cobalt centre, do not precipitate as AgCl.

This elegant insight into chemical structure, based on simple stoichiometric relations, was one reason why Werner was awarded the Nobel Prize in chemistry in 1913. Since then, a long path of discovery has uncovered many remarkable properties of coordination complexes. To give just one example, there are two isomers of the square planar platinum complex $[Pt(NH_3)_2Cl_2]$. The *cis* isomer, known as cisplatin, is an effective anticancer drug, but the *trans* isomer shows no anticancer activity at all. The mechanism by which the *cis* isomer destroys cancer cells is not fully understood, but research indicates that cisplatin disrupts cell duplication by inserting into the DNA double helix. It appears that DNA binds this platinum complex by replacing the two chloride ions. In a sense, the DNA molecule becomes a huge ligand for the platinum atom. Apparently, the coordination complex can form bonds to DNA only when the chloride ions are adjacent to each other in the *cis* configuration.

*cis*-$[Pt(NH_3)_2Cl_2]$

## EXAMPLE 18-3  FORMULA OF A COORDINATION COMPOUND

Determine the formulas of the following coordination compounds:

(a) *fac*-Triamminetriiodoruthenium(III);

(b) *cis*-Chlorohydridobis(trimethylphosphine)platinum(II); and

(c) Sodium hexacyanoferrate(II).

**Strategy:** To obtain the formula, break the name down, one piece at a time.

**Solution:**

(a) Determine the number of each type of ligand: triammine = 3 $NH_3$; triiodo = 3 $I^-$. Identify the metal (ruthenium) and its oxidation state (III). Retain the italicized prefix to identify a particular isomer: [*fac*-$Ru(NH_3)_3I_3$].

(b) Chloro = $Cl^-$, Hydrido = $H^-$, and bis(trimethylphosphine) = 2 $P(CH_3)_3$. The metal is $Pt^{2+}$: *cis*-[$PtClH(P(CH_3)_3)_2$].

(c) Sodium is the cation. The metal is $Fe^{2+}$, and there are six cyano anions. Overall, then, the complex anion has a net charge of −4. Four $Na^+$ cations are required to balance the charge: $Na_4[Fe(CN)_6]$.

**Do the Results Make Sense?** Verify that the number of ligands matches the numerical prefixes in the name, that the ligands are listed alphabetically, and that the overall species is charge-neutral.

**Common Pitfalls:** Not correctly assigning charges to the ligands. Consult Table 18-3.

**Practice Exercise 18.3**

Determine the formulas of the following coordination compounds: (a) tris(bipyridine)ruthenium(II) chloride; (b) potassium hexachloroplatinate(IV); (c) pentaaquaoxoiron(III) ion.

## SECTION EXERCISES

**18.2.1** For each of the following, determine the charge on the complex ion, the oxidation state of the metal, and the coordination number of the metal: (a) $K_3[Fe(CN)_6]$; (b) [$V(NH_3)_4Cl_2$]; and (c) [$Ni(en)_2$]$SO_4$.

**18.2.2** Draw all of the isomers of [$Cr(en)(NH_3)_2I_2$]$^+$.

**18.2.3** Name the following coordination compounds: (a) [$Cr(NH_3)_5I$]$SO_4$; (b) $K_4[PtCl_6]$; (c) *cis*-[$Fe(CO)_4Cl_2$]; and (d) [$Fe(H_2O)_6$]$Cl_2$.

## 18.3 BONDING IN COORDINATION COMPLEXES

In the simplest view, a metal–ligand $\sigma$ bond can be described as the overlap of a filled donor orbital on the ligand with an empty acceptor orbital on the metal. There is one such interaction for each ligand donor atom, so an octahedral complex of the general formula [$ML_6$]$^{n+}$ has six $\sigma$ bonds. Although simple, this view of metal–ligand bonding does not explain the colours and other properties of coordination complexes. Molecular orbital theory provides the most complete description of the bonding in coordination complexes, but this approach is beyond the scope of introductory chemistry. Instead, we use a model called **crystal field theory** to explain the colours and magnetic properties of coordination compounds.

Crystal field theory focuses on electrical interactions between a transition metal ion and its ligands. In this view, the complex is held together by attractive forces between the negatively charged electrons of the ligand lone pairs and the positive charge of the metal ion. Consequently, bonding can be described much like the bonding in an ionic crystal. Like all models, crystal field theory has benefits and limitations. The model is easy to apply and can explain several important properties of transition metal complexes. Nevertheless, all metal–ligand bonds have covalent character, which limits the usefulness of an ionic model.

**Figure 18-11**

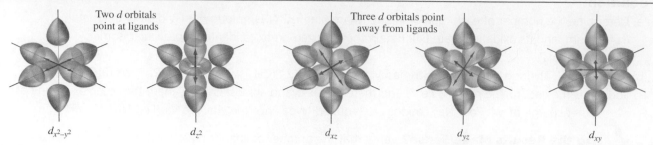

Two $d$ orbitals point at ligands

Three $d$ orbitals point away from ligands

$d_{x^2-y^2}$    $d_{z^2}$    $d_{xz}$    $d_{yz}$    $d_{xy}$

The five $d$ orbitals and their relationship to an octahedral set of ligands. Whereas two orbitals point directly at the ligands, the other three orbitals point between the ligands.

## Orbital Stability in Octahedral Complexes

In a free metal ion without any ligands, all five $d$ orbitals have identical energies, but what happens to the $d$ orbitals when six ligands are placed around a metal in octahedral geometry? The complex is stabilized by attractions between the positive charge of the metal ion and negative electrons of the ligands. At the same time, the electrons in the metal $d$ orbitals repel the electrons on the ligands. Electron–electron repulsion affects some of the $d$ orbitals more than others. Figure 18-11 shows that two orbitals, $d_{x^2-y^2}$ and $d_{z^2}$, point directly toward the ligands. As a consequence, electrons in these orbitals experience greater electron–electron repulsion than do electrons in $d_{xz}$, $d_{yz}$, and $d_{xy}$, which point between the ligands. This means that metal valence electrons are more stable when they occupy the $d$ orbitals that point away from the ligands. Figure 18-12 summarizes the electrical interactions of an octahedral complex ion. The three orbitals that are more stable are called $t_{2g}$ orbitals, and the two less stable orbitals are called $e_g$ orbitals. (The names $t_{2g}$ and $e_g$ are derived from symmetry properties of the orbitals that are beyond the scope of general chemistry.) The difference in energy between the two sets is known as the **crystal field splitting energy**, symbolized by the Greek letter $\Delta$.

**Figure 18-12**

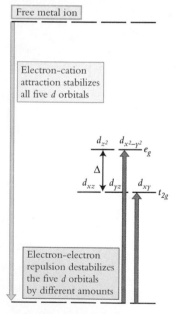

Free metal ion

Electron–cation attraction stabilizes all five $d$ orbitals

$d_{z^2}$  $d_{x^2-y^2}$  $e_g$

$\Delta$

$d_{xz}$  $d_{yz}$  $d_{xy}$  $t_{2g}$

Electron–electron repulsion destabilizes the five $d$ orbitals by different amounts

The crystal field energy-level diagram for octahedral coordination complexes. The energies of the $d$ orbitals differ because of differing amounts of electron–electron repulsion. The $t_{2g}$ orbitals point between the ligands, have less electron repulsion, and are more stable than the $e_g$ orbitals, which point directly at the ligands.

## Populating the $d$ Orbitals

Electron configurations of transition metal complexes are governed by the principles described in Chapter 5. The Pauli exclusion principle states that no two electrons can have identical descriptions, and Hund's rule requires that all unpaired electrons have the same spin orientation. These concepts are used in Chapter 5 for atomic configurations and in Chapters 6 and 7 to describe the electron configurations of molecules. They also determine the electron configurations of transition metal complexes.

The number of electrons in the $d$ orbitals depends on the electron configuration of the metal. That configuration can be found from the oxidation state of the metal and its atomic number. As an example, consider $[Cr(NH_3)_6]^{3+}$. The ionic charge on the complex is +3, and because ammonia is a neutral ligand, all the charge must arise from the metal. In other words, chromium has an oxidation state of +3, which means that three of its valence electrons have been removed. Chromium is in Group 6 of the periodic table, and Group 6 atoms have six valence electrons. Thus, $Cr^{3+}$ has three valence electrons, all in $3d$ orbitals:

[Ar] $3d^3$. One $d$ electron occupies each of the three $t_{2g}$ orbitals, and all three have the same spin:

The crystal field electron configuration of $[Cr(NH_3)_6]^{3+}$ can be summarized as $t_{2g}^3 e_g^0$. Example 18-4 describes the electron configuration of another coordination complex.

### EXAMPLE 18-4 ELECTRON CONFIGURATIONS

Draw an energy-level diagram and write the $d$ electron configuration of $[Pt(en)_3]Cl_2$.

**Strategy:** Identify the ligands and the geometry of the coordination complex, construct the crystal field energy-level diagram, and count $d$ electrons from the metal and place them according to the Pauli principle and Hund's rule.

**Solution:** Ethylenediamine is a bidentate ligand, so there are six donor atoms, giving the complex ion octahedral geometry. The two chlorides outside the bracket indicate that the complex ion is a +2 cation. Since en is a neutral ligand, Pt is in its +2 oxidation state. Platinum (Group 10) has 10 valence electrons, two of which are removed to give the +2 oxidation state, leaving eight $d$ electrons to be placed in the energy-level diagram. The Pauli principle and Hund's rule dictate the result:

The electron configuration is $t_{2g}^6 e_g^2$.

> **Do the Results Make Sense?** The $d$ electron count is correct (8), and the filling order follows both the Pauli principle and Hund's rule.

**Common Pitfalls:** Not following Hund's rule. The configuration in this case ($t_{2g}^6 e_g^2$) would be the same, but magnetic properties (see next page) depend on whether or not there are unpaired electrons.

**Practice Exercise 18.4**

Draw an energy-level diagram and write the $d$ electron configuration for $[V(H_2O)_6]^{3+}$.

The examples given so far lead to unambiguous electron configurations, but not all configurations are this straightforward. Consider any octahedral complex containing a metal atom with four $d$ electrons. Following the standard filling procedure, the first three electrons are placed in the $t_{2g}$ orbitals. The fourth electron might also be placed in a $t_{2g}$ orbital, but there is a price to pay in energy: two electrons in the same orbital are destabilized because two negative charges are confined to the same region of space (electron–electron repulsion). This destabilization is called the **pairing energy (P)**. There is no pairing energy if the fourth electron is placed in an $e_g$ orbital, but these orbitals are higher in energy than the $t_{2g}$ orbitals. The energy difference between the two sets is the crystal field splitting energy, $\Delta$.

Whether $t_{2g}^4 e_g^0$ or $t_{2g}^3 e_g^1$ is more stable depends on the relative magnitudes of $P$ and $\Delta$. If the energy required to pair electrons in a $t_{2g}$ orbital is less than the energy required to populate an $e_g$ orbital ($P < \Delta$), the ground-state configuration is $t_{2g}^4 e_g^0$. On the other hand, if $P > \Delta$, the most stable configuration is $t_{2g}^3 e_g^1$.

**Figure 18-13**

$P < \Delta$   $t_{2g}^4 e_g^0$

$P > \Delta$   $t_{2g}^3 e_g^1$

The two possible electron configurations for a $d^4$ metal complex. Notice that whereas $\Delta$ can be shown on this diagram, $P$ cannot, because $P$ measures electron–electron repulsion, not an orbital energy.

Both possibilities are shown in Figure 18-13 for the four-electron, octahedral case. In conformity with Hund's rule, a metal complex with the configuration $t_{2g}^3 e_g^1$ has four unpaired electrons. This configuration maximizes the number of electrons with unpaired spins, so it is described as **high-spin**. The alternative configuration, $t_{2g}^4 e_g^0$, has just two unpaired electrons and is described as **low-spin**.

## Magnetic Properties of Coordination Complexes

What experiments can be done to reveal whether a complex with four $d$ electrons is $t_{2g}^4 e_g^0$ or $t_{2g}^3 e_g^1$? One of the most common ways to determine the electron configuration of a coordination complex is to measure its magnetic properties.

Recall from Chapters 4 and 5 that the spin of an electron generates magnetism. When electrons are paired, their spins point in opposite directions, causing their magnetism to cancel. A molecule with all electrons paired has no magnetism from its electron spins and is diamagnetic, but a molecule that has unpaired electrons is paramagnetic. The amount of magnetism in a paramagnetic molecule depends on the number of unpaired electrons: the more unpaired electrons in the $d$ orbitals of a complex, the greater the magnetism. The magnetic properties of a complex ion can be measured with an instrument known as a magnetic susceptibility balance, as shown schematically in Figure 18-14. The sample is positioned between the poles of one pair of magnets to produce a vertical deflection of the beam, which is measured by optical sensors. A current is passed through a coil mounted between the poles of the other pair of magnets to produce a force equal and opposite to that exerted by the sample, returning the system to a position of equilibrium. The current through the coil is proportional to the force exerted by the sample. This provides an experimental method for determining the ground-state electron configuration of a coordination complex.

Example 18-5 illustrates the magnetic properties of another configuration.

**Figure 18-14**

The number of unpaired electrons in a paramagnetic complex can be determined with a magnetic susceptibility balance.

## EXAMPLE 18-5 HIGH- AND LOW-SPIN COMPLEXES

$[Fe(NH_3)_6]^{2+}$ is paramagnetic, but $[Co(NH_3)_6]^{3+}$ is not. Write the electron configuration for each of these metal complexes and draw energy-level diagrams showing which has the higher $\Delta$.

**Strategy:** Use the formula of a complex to determine its geometry, the form of its energy-level diagram, and the number of $d$ electrons. The spin information indicates how many of these electrons are paired.

**Solution:** Both complexes have six monodentate ligands, so both are octahedral.

Because ammine ligands are neutral molecules, the oxidation state of each metal is the same as the charge on the complex. Iron loses two of its eight valence electrons to reach the +2 oxidation state, leaving six electrons for the $d$ orbitals. Likewise, cobalt in its +3 oxidation state has six $d$ electrons.

The iron complex is paramagnetic, which means that it has unpaired electrons. This happens when the crystal field splitting energy is less than the pairing energy, so the energy-level diagram should show a relatively small $\Delta$. By Hund's rule, we place one electron in each of the five $d$ orbitals before pairing any. The sixth $d$ electron pairs up with an electron in any one of the $t_{2g}$ orbitals.

The cobalt complex is not paramagnetic, which means that it has all of its valence electrons paired. This happens when $\Delta$ is larger than $P$, so the energy-level diagram should show a relatively large $\Delta$. In a low-spin complex, the lower-energy $t_{2g}$ orbitals are filled completely before putting electrons in $e_g$. Here are the correct diagrams:

The paramagnetic configuration is $t_{2g}^4 e_g^2$, with four unpaired electrons. The other configuration is $t_{2g}^6 e_g^0$, with no unpaired electrons and zero magnetism.

$$[Fe(NH_3)_6]^{2+}$$
High-spin
$\Delta < P$
$t_{2g}^4 e_g^2$

$$[Co(NH_3)_6]^{3+}$$
Low-spin
$\Delta > P$
$t_{2g}^6 e_g^0$

 **Do the Results Make Sense?** Both configurations are $d^6$, and both follow Hund's rule, so the answer is reasonable.

**Practice Exercise 18.5**

$[Mn(H_2O)_6]^{2+}$ is more paramagnetic than $[Re(H_2O)_6]^{2+}$. Write the $d$ electron configuration for each of these metal complexes and draw energy-level diagrams showing which has the higher $\Delta$.

## Contributions to Crystal Field Splitting Energy

Whether a complex is low-spin or high-spin depends on the balance between pairing energy and crystal field splitting energy. Pairing energy changes very little from one coordination complex to the next. Consequently, electron configurations of coordination complexes are governed by the crystal field splitting energy, $\Delta$.

The most important factor that influences the value of $\Delta$ is the identity of the ligands. Compare $[Fe(CN)_6]^{4-}$ and $[FeCl_6]^{4-}$, two coordination complexes similar in every respect except for their ligands. Both contain $Fe^{2+}$ ions with six $d$ electrons, both have six ligands arranged in an octahedral geometry, and both have the same net charge. Despite these similarities, the cyano complex is low-spin and diamagnetic, whereas the chloro complex is high-spin and paramagnetic. For $[Fe(CN)_6]^{4-}$, the crystal field splitting energy is larger than the pairing energy, so the molecule is low-spin. In contrast, $P > \Delta$ for $[FeCl_6]^{4-}$. The chloro ligand generates a small energy gap between the $t_{2g}$ and $e_g$ orbitals, giving a high-spin coordination complex.

Studies of many coordination complexes reveal a common pattern in the energetic effects of the various ligands. This pattern is described by the **spectrochemical series**, in which ligands are listed in order of increasing energy level splitting:

$$I^- < Br^- < Cl^- < F^- < OH^- < H_2O < NH_3 < en < NO_2^- < CN^- < CO$$

Smaller splitting ⟶ Larger splitting

The spectrochemical series was established from experimental measurements. The ranking of ligands cannot be fully rationalized using crystal field theory, but more advanced bonding theories are beyond the scope of general chemistry.

The oxidation state of the metal also contributes to the crystal field splitting energy. For a given ligand, $\Delta$ increases as the oxidation state of the metal increases. This trend is easy to interpret using electrical forces. As the charge on the metal increases, so does the attraction between the metal and its ligands. The ligands approach the metal more closely, resulting in stronger orbital interactions and a larger crystal field splitting (Figure 18-15). The complexes of Example 18-5, $[Fe(NH_3)_6]^{2+}$ (high-spin) and $[Co(NH_3)_6]^{3+}$ (low-spin) illustrate this feature. These two species have the same ligands and the same molecular geometry, and both are $d^6$ metal ions, but $Co^{3+}$ generates a stronger crystal field than $Fe^{2+}$.

Another influence on the magnitude of the crystal field splitting is the position of the metal in the periodic table. Crystal field splitting energy increases substantially from $3d$ to $4d$ to $5d$ valence orbitals. Again, orbital shapes explain this trend. Orbital size increases as $n$ increases, and this means that the $d$ orbital set becomes more exposed to orbital interactions with approaching ligands. For example, the value of $\Delta$ for complexes of rhodium is about 50% greater than that

**Figure 18-15**

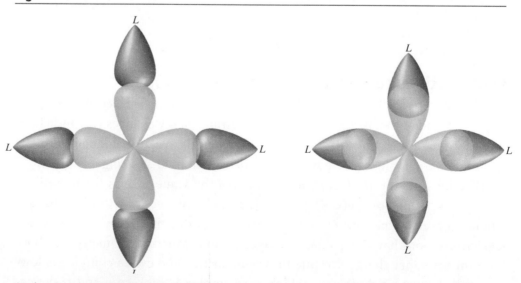

As the charge on the metal ion increases, electrical attraction pulls the ligands closer to the cation. This leads to greater repulsive interactions between valence $d$ electrons and ligand electrons.

for analogous cobalt complexes, and there is another 25% increase on moving from rhodium to iridium. As a result of this trend, coordination complexes of 4d and 5d elements are almost always low-spin. In contrast, the 3d transition metals have both low- and high-spin complexes. Example 18-6 applies the principles of crystal field splitting to a series of coordination complexes.

### EXAMPLE 18-6 CRYSTAL FIELD SPLITTING ENERGY

Arrange the following complexes in order of increasing crystal field splitting: $[Fe(H_2O)_6]^{2+}$, $[Fe(H_2O)_6]^{3+}$, $[FeCl_6]^{4-}$, and $[Ru(H_2O)_6]^{3+}$.

**Strategy:** The strength of splitting depends on the ligand, the charge on the metal, and the position of the metal in the periodic table. Examine these factors independently.

**Solution:**

*Ligand*: $Cl^-$ is lower in the spectrochemical series than $H_2O$. Thus, $[FeCl_6]^{4-}$ has low splitting relative to the others.

*Cation charge*: In $[FeCl_6]^{4-}$ and $[Fe(H_2O)_6]^{2+}$, iron is in its +2 oxidation state, but the metal is +3 in the other two complexes. Complexes with +3 oxidation states have larger splitting:

$$[FeCl_6]^{4-} < [Fe(H_2O)_6]^{2+} < [Fe(H_2O)_6]^{3+} \text{ and } [Ru(H_2O)_6]^{3+}$$

*Valence orbitals*: Three complexes contain iron, a metal with 3d valence electrons, but the fourth contains ruthenium, with 4d valence electrons. All other things being equal, the crystal field splitting for a 4d metal is about 50% larger than that for a 3d metal. Applying all three trends gives an unambiguous arrangement of these complexes:

$$[FeCl_6]^{4-} < [Fe(H_2O)_6]^{2+} < [Fe(H_2O)_6]^{3+} < [Ru(H_2O)_6]^{3+}$$

Lowest splitting ⟶ Highest splitting

 **Do the Results Make Sense?** The ranking is consistent with the three trends for determining $\Delta$: nature of the ligands, oxidation state of the metal, and position of the metal in the periodic table.

**Practice Exercise 18.6**

Arrange the following complexes in order of increasing crystal field splitting: $[Cr(NH_3)_6]^{3+}$, $[Cr(H_2O)_6]^{2+}$, $[Mo(NH_3)_6]^{3+}$, and $[Mo(H_2O)_6]^{3+}$.

## Colour in Coordination Complexes

Colour is a spectacular property of coordination complexes. For example, the hexaaqua cations of 3d transition metals display colours ranging from orange through violet (see photo at right). The origin of these colours lies in the *d* orbital energy differences and can be understood using crystal field theory.

Colour is caused by absorption of some of the light from the visible spectrum (Figure 18-16). The wavelengths of light absorbed by a collection of molecules are lost from the rest of the spectrum. Light that is not absorbed is reflected by an opaque sample and is transmitted through a transparent sample. In either case, an observer sees all wavelengths except those

**Figure 18-16**

The visible spectrum, showing the colours as a function of wavelength.

absorbed by the molecules. If a sample absorbs all wavelengths of visible light except blue, that sample will appear blue to the observer. A substance that appears black absorbs all wavelengths of visible light. A substance that absorbs no visible wavelengths appears white if light is reflected by the surface, and colourless if the light is transmitted through the sample.

Molecules that absorb a small wavelength region have a colour that is different from the colour absorbed. For example, a molecule that absorbs only orange light appears blue. Likewise, a molecule that absorbs only blue appears orange. Orange and blue are said to be **complementary colours**. Pairs of complementary colours are listed in Table 18-5.

When a molecule absorbs light, it gains energy from the absorbed photons. Energy conservation requires that the energy change for the molecule equals the energy of the absorbed photon:

$$\Delta E_{molecule} = E_{photon} = h\nu$$

This requirement means that when a coordination complex absorbs light, the crystal field splitting energy, $\Delta$, must match the energy of the absorbed light. Thus, the crystal field splitting energy of a complex can be determined from the wavelength of visible light that the complex absorbs:

$$\Delta E_{molecule} = \Delta = h\nu = \frac{hc}{\lambda}$$

## TABLE 18-5 RELATIONSHIPS AMONG WAVELENGTH, COLOUR, AND CRYSTAL FIELD SPLITTING ENERGY ($\Delta$)

| Wavelength (nm) | Colour Absorbed | Complementary Colour | $\Delta$ (kJ/mol) |
|---|---|---|---|
| >720 | Infrared | Colourless | <165 |
| 720 | Red | Green | 166 |
| 680 | Red-orange | Blue-green | 176 |
| 610 | Orange | Blue | 196 |
| 580 | Yellow | Indigo | 206 |
| 560 | Yellow-green | Violet | 214 |
| 530 | Green | Purple | 226 |
| 500 | Blue-green | Red | 239 |
| 480 | Blue | Orange | 249 |
| 430 | Indigo | Yellow | 279 |
| 410 | Violet | Lemon-yellow | 292 |
| <400 | Ultraviolet | Colourless | >299 |

**Figure 18-17**

The colours of $Cr^{3+}$ coordination complexes depend on the positions of the ligands in the spectrochemical series.

Spectroscopic properties of coordination compounds are highlighted by a series of $d^3$ $Cr^{3+}$ complexes, as shown in Figure 18-17. Each molecule has a ground-state configuration with three unpaired electrons in the $t_{2g}$ orbitals. Absorption of a photon excites one electron into an $e_g$ orbital, giving an excited-state configuration, $t_{2g}^2 e_g^1$. These three complexes have different values of $\Delta$, so they absorb light of different wavelengths. The cyano ligand is high in the spectrochemical series, generating a large crystal field splitting for $[Cr(CN)_6]^{3-}$. A photon of high-energy violet light is required to promote an electron from the $t_{2g}$ set to the $e_g$ set. Absorption of violet light gives $[Cr(CN)_6]^{3-}$ a yellow colour. The fluoro ligand is near the bottom of the spectrochemical series, generating a weak crystal field. The fluoro complex absorbs red light, generating a green colour for $[CrF_6]^{3-}$. When $H_2O$ is the ligand, there is an intermediate energy gap. Yellow-green light is absorbed, so $[Cr(H_2O)_6]^{3+}$ is violet.

Example 18-7 shows how to determine the value of $\Delta$ from an absorption spectrum.

## EXAMPLE 18-7 DETERMINING THE VALUE OF $\Delta$

Titanium(III) chloride dissolves in water to give $[Ti(H_2O)_6]^{3+}$. This complex ion has the absorption spectrum shown. From the wavelength at which maximum absorption occurs, predict the colour of the solution and calculate $\Delta$ in kilojoules per mole.

**Strategy:** The spectrum shows wavelengths of light absorbed by the metal complex. The wavelength absorbed most strongly corresponds to $\Delta$, but appropriate conversions are needed to find the molar energy associated with this wavelength. The colour of the solution is the complementary colour of the most strongly absorbed wavelength.

**Solution:** Interpolate on the graph to find that the maximum intensity occurs at 514 nm. Use this wavelength to find the energy of one photon:

$$E_{photon} = \frac{hc}{\lambda} = \frac{(6.626 \times 10^{-34}\text{J s})(2.988 \times 10^{8}\text{ m s}^{-1})}{(514 \text{ mn})(10^{-9}\text{m/nm})} = 3.86 \times 10^{-19}\text{J/photon}$$

Avogadro's number is used to convert to kilojoules per mole:

$$(3.86 \times 10^{-19} \text{ J/photon})(10^{-3} \text{ kJ/J})(6.022 \times 10^{23} \text{ photons/mol}) = 232 \text{ kJ/mol} = \Delta$$

The $[Ti(H_2O)_6]^{3+}$ ion absorbs at 514 nm, which is in the green to blue-green region of the visible spectrum. According to Table 18-5, the solution should be reddish purple.

 **Do the Results Make Sense?** The calculated value of $\Delta$ has the right units, and the numerical value matches the values found in Table 18-5. The colour is right, as shown by the photo.

**Common Pitfalls:** Not finding the complementary colour of the colour that is absorbed.

**Practice Exercise 18.7**

The complex ion $[Fe(ox)_3]^{3-}$ has $\Delta = 169$ kJ/mol. Calculate the wavelength absorbed by the complex, and determine the colour of the complex.

## Square Planar and Tetrahedral Complexes

In a square planar complex, the four ligands lie along the $x$- and $y$-axes. As shown in Figure 18-18, $d_{x^2-y^2}$ is the only $d$ orbital that points directly at the four ligands. Thus, in square planar geometry $d_{x^2-y^2}$ is significantly higher in energy than the other four $d$ orbitals. Although the $d_{xy}$ orbital also lies in the metal–ligand plane, it points *between* the ligands and is subject to less electron–electron repulsion than $d_{x^2-y^2}$. Because the remaining three $d$ orbitals point out of the metal–ligand plane, these orbitals are repelled less by the

**Figure 18-18**

$d_{x^2-y^2}$          Square planar geometry

In square planar geometry, the $d_{x^2-y^2}$ orbital is most strongly destabilized by interactions with the ligands. The dxy orbital also lies in the ligand plane, but it points between the ligands.

ligands and are lower in energy than $d_{xy}$. The $d_{z^2}$ orbital is slightly less stable than $d_{xz}$ and $d_{yz}$ because of the small band of electron density that circles the metal in the $xy$ plane. The energy-level diagram shown in Figure 18-18 shows the splitting pattern for a square planar complex. Square planar geometry is most common for $d^8$ metal ions with large crystal field splitting energies. The $d^8$ configuration fills the four lower-energy $d$ orbitals, but it leaves the high-energy $d_{x^2-y^2}$ empty.

The ligands of a tetrahedral complex occupy the corners of a tetrahedron rather than the corners of a square. The symmetry relationships between the $d$ orbitals and these ligands are not easy to visualize, but the splitting pattern of the $d$ orbitals can be determined using geometry. The result is the opposite of the pattern found in octahedral complexes: The $d_{xz}$, $d_{xy}$, and $d_{yz}$ orbitals are higher in energy than $d_{x^2-y^2}$ and $d_{z^2}$ . Figure 18-19 shows the splitting pattern for tetrahedral coordination. Because there are only four ligands rather than six, and because none of the $d$ orbitals point directly at the ligands, the crystal field splitting energy for tetrahedral geometry is only about half that for octahedral geometry. As a result, the pairing energy is almost always greater than the splitting energy ($P > \Delta$), and therefore almost all tetrahedral complexes are high-spin.

**Figure 18-19**

The crystal field energy-level diagram for tetrahedral complexes. The $d$ orbitals are split into two sets, with three orbitals destabilized relative to the two others.

**SECTION EXERCISES**

● **18.3.1** Draw crystal field splitting diagrams that show the electron configurations for the following complex ions: (a) $[Cr(H_2O)_6]^{2+}$; (b) $[IrCl_6]^{3-}$; (c) $[V(en)_3]^{3+}$; and (d) $[NiCl_4]^{2-}$ (tetrahedral).

● **18.3.2** Explain why hexacyano complexes of metals in their +2 oxidation state are usually yellow, but the corresponding hexaaqua compounds are often blue or green.

● **18.3.3** The value of $\Delta$ for $[RhCl_6]^{3-}$ is 243 kJ/mol. What wavelength of light will promote an electron from the $t_{2g}$ set to the $e_g$ set? What colour is the complex?

## 18.4 TRANSITION METALS IN BIOLOGY

Almost 90% of the atoms that make up a human body are either hydrogen or oxygen. Most of these are in water, which constitutes around 70% of a human by weight. The organic structures that make up the body, as well as the molecules involved in biosynthesis and energy production, are made almost entirely of C, H, N, and O. These four elements account for 99% of all the atoms present in a human. Another seven elements—Na, K, Ca, Mg, P, S, and Cl—are essential for all known life-forms. These seven add another 0.9% to the total atom count of a human being. The remaining 0.1%, the so-called "trace elements," are required by most biological organisms. Although these elements are present in only minute amounts, they are essential for healthy function. Figure 18-20 shows a periodic table summarizing the elemental composition of living organisms.

## Metalloproteins

The trace elements include nine transition metals: all members of the first row from vanadium to zinc, and molybdenum from the second row. Most transition metals in the body are natural constituents of proteins, biological macromolecules made of long chains of amino acids that are described in Chapter 17. These **metalloproteins** play three essential roles in biochemistry. Some act as transport and storage agents, moving small molecules from place to place within an organism. Others are enzymes, catalysts for a diverse group of biochemical reactions. Both transport and catalysis depend on the ability of transition metals to bind and release ligands. The third role of metalloproteins is to serve as redox reagents, adding or removing electrons in many different reactions. Transition metals are ideal for this purpose because of their abilities to shuttle between two or more oxidation states.

Metalloproteins typically are macromolecules containing many thousands of atoms. Determining the bonding, structure, and geometries of these macromolecules is a challenging task. Nevertheless, it is known that transition metals are bound to proteins through ligand—metal interactions of the sort described in this chapter. Among the 20 amino acids found in proteins, several have side chains that act as ligands: histidine, tyrosine, cysteine, glutamic acid, aspartic acid, and

**Figure 18-20**

= Bulk biological elements

= Trace elements believed to be essential for plants or animals

= Trace elements that may be essential for plants or animals

Elements essential for life.

methionine bind to metals through lone pairs on nitrogen, oxygen, or sulphur. Structures of the amino acid side chains appear in Figure 17-32.

## Transport and Storage Proteins

Organisms extract energy from food by using molecular oxygen to oxidize fats and carbohydrates. For most animals, the movement and storage of $O_2$ is accomplished by the iron-containing proteins *haemoglobin* and *myoglobin*. The iron atom in haemoglobin binds $O_2$ and transports this vital molecule from the lungs or the gills to various parts of the body where oxidation takes place. Haemoglobin makes $O_2$ about 70 times more soluble in blood than in water. Whereas haemoglobin transports $O_2$, myoglobin stores $O_2$ in tissues such as muscle that require large amounts of oxygen.

Haemoglobin and myoglobin have closely related structures. Both contain the haem structure shown in the introduction to this chapter. In myoglobin, the haem is bound to a polypeptide chain of 153 amino acids arranged in helical arrays. The ribbon structure of myoglobin is shown in Figure 18-21. The polypeptide chain folds in a manner that creates a "pocket" in the protein for a haem group. Haemoglobin is made up of four polypeptide chains, each of which is similar in shape and structure to a myoglobin molecule.

**Figure 18-21**

Myoglobin is a globular protein with a haem unit embedded in a pocket created by the folding of the protein chain.

Each haem unit in myoglobin and haemoglobin contains one $Fe^{2+}$ ion bound to four nitrogen donor atoms in a square planar arrangement. This leaves the metal with two axial coordination sites to bind other ligands. One of these sites is bound to a histidine side chain that holds the haem in the pocket of the protein. The other axial position is where reversible binding of molecular oxygen takes place.

The binding and release of oxygen by haemoglobin can be represented as a ligand-exchange equilibrium at the sixth coordination site on the $Fe^{2+}$ ion. Each of the four polypeptide chains of haemoglobin contains one haem unit, so a molecule of haemoglobin can bind as many as four oxygen molecules:

$$[\text{Haem}]_4 + 4O_2 \rightleftharpoons [\text{Haem}(O_2)]_4$$

Deoxyhaemoglobin                    Oxyhaemoglobin

In the absence of oxygen, the iron centre in each haem remains five-coordinate, with an approximately square pyramidal geometry, as shown in Figure 18-22a. In this form of haemoglobin, the $d^6$ metal ion is in a high-spin environment with four unpaired electrons in a $t_{2g}^4 e_g^2$ arrangement. Deoxyhaemoglobin has a bluish colour because the energy separation between the valence $d$ orbitals of the iron cations is small, and the molecule absorbs red light at the low-energy end of the visible spectrum. When oxygen moves into the pocket of the protein, it binds to the sixth coordination site on the metal, as shown in Figure 18-22b. One of the effects of the $O_2$ ligand is to increase the crystal field splitting energy. As a result, in oxyhaemoglobin, the metal is in a low-spin $t_{2g}^6 e_g^0$ environment and is diamagnetic. In this form, haemoglobin absorbs light at the blue end of the visible spectrum and appears bright red.

In the lungs, haemoglobin "loads" its four oxygen molecules and then moves through the bloodstream. In tissues, the concentration of $O_2$ is very low, but there is plenty of carbon dioxide, the end product of metabolism. The concentration of $CO_2$ has an important effect on haemoglobin–oxygen binding. Like oxygen, carbon dioxide can bind to haemoglobin. However, carbon dioxide binds to specific amino acid side chains of the protein, not to the haem group.

**Figure 18-22**

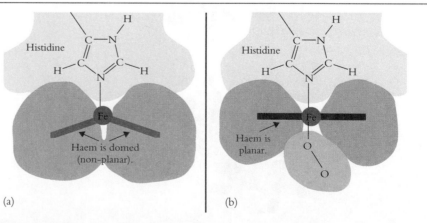

(a)                    (b)

(a) In deoxyhaemoglobin, iron is five-coordinate in a square pyramidal shape. (b) On coordination of molecular oxygen, the metal adopts octahedral geometry.

Binding carbon dioxide to the protein causes the shape of the haemoglobin molecule to change in ways that reduce the equilibrium constant for $O_2$ binding. The reduced binding constant allows haemoglobin to unload its $O_2$ molecules in oxygen-deficient, $CO_2$-rich tissue. The bloodstream carries this deoxygenated haemoglobin back to the lungs, where it releases carbon dioxide and loads four more molecules of oxygen. This $CO_2$ effect does not operate in myoglobin, which binds and stores the oxygen released by haemoglobin.

Carbon monoxide seriously impedes transport of oxygen. The deadly effect of inhaled CO results from its reaction with haemoglobin. A CO molecule is almost the same size and shape as $O_2$, so it fits into the binding pocket of the haemoglobin molecule. In addition, the carbon atom of CO forms a stronger bond to $Fe^{2+}$ than does $O_2$. Under typical conditions in the lungs, haemoglobin binds carbon monoxide 230 times more strongly than it binds $O_2$. Haemoglobin complexed to CO cannot transport oxygen, so when a significant fraction of haemoglobin contains CO, oxygen "starvation" occurs at the cellular level, leading to loss of consciousness and then to death.

Many deaths occur through accidental carbon monoxide poisoning. Burning fossil fuels generates some CO, particularly in an oxygen-depleted environment. Automobile engines, gas heaters, and charcoal braziers all produce some CO. The presence of colourless, odourless CO goes undetected, and in a poorly ventilated room carbon monoxide may build up to lethal concentrations without the occupants being aware of its presence.

An adult human contains about 4 g of iron, most of it in the form of haem-containing proteins. Yet the daily requirement of iron in the diet is only about 1 mg, indicating that the body recycles iron rather than excreting it. The recycling of iron requires a transport system and a storage mechanism. Transport of iron is accomplished by a protein called *transferrin*. Transferrin collects iron in the spleen and liver, where haemoglobin is degraded, and carries it to the bone marrow where fresh red blood cells are synthesized.

The protein that stores iron in the body is called *ferritin*. A ferritin molecule consists of a protein coat and an iron-containing core. The outer coat is made up of 24 polypeptide chains, each with about 175 amino acids. As Figure 18-23 shows, the polypeptides pack together to form a sphere. The sphere is hollow, and channels through the protein coat allow movement of iron in and out of the molecule. The core of the protein contains hydrated iron(III) oxide, $Fe_2O_3 \cdot H_2O$. The protein retains its shape whether or not iron is stored on the inside. When filled to capacity, one ferritin molecule holds as many as 4500 iron atoms, but the core is only partially filled under normal conditions. In this way, the protein has the capacity to provide iron as needed for haemoglobin synthesis or to store iron if an excess is absorbed by the body.

**Figure 18-23**

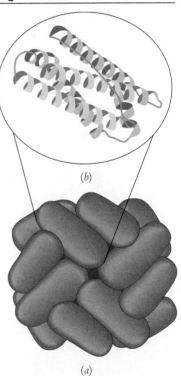

(b)

(a)

Schematic representation of ferritin, the iron storage protein. (a) The protein contains 24 nearly identical polypeptides. (b) A ribbon structure of one of the polypeptide chains.

## Enzymes

As described in Section 12.7, enzymes are the catalysts of biological reactions. Without enzymes, most of the reactions that occur in a cell would be imperceptibly

**Figure 18-24**

The structure of the enzymatic site of superoxide dismutase. The $Zn^{2+}$ and $Cu^{2+}$ cations lie in close proximity, with a histidine side chain (colour screened) acting as a bridge between the metals.

slow. Cations of transition metals play essential roles in the mechanisms of many enzyme-catalyzed reactions. Here, we introduce just one representative example, superoxide dismutase.

Molecular oxygen is essential to life, but reduced oxygen species such as superoxide, $O_2^-$, damage cells and are thought to play a role in the aging process. Cells contain enzymes that destroy these contaminants. One such enzyme, superoxide dismutase (SOD), is abundant in virtually every type of aerobic organism. The catalytic portion of the SOD enzyme contains one copper atom and one zinc atom, each bound in a tetrahedral arrangement. As Figure 18-24 shows, the two metals are held close together by a histidine ligand that forms a bridge between Cu and Zn.

The role of SOD is to convert superoxide ion to hydrogen peroxide:

$$2\,O_2^- + 2\,H_3O^+ \xrightarrow{\text{SOD}} O_2 + H_2O_2 + 2\,H_2O$$

The mechanism is believed to be a two-step process involving reduction and oxidation of the copper centre:

$$O_2^- + Cu^{2+}(SOD) \longrightarrow O_2 + Cu^+(SOD)$$

$$O_2^- + Cu^+(SOD) + 2\,H_3O^+ \longrightarrow Cu^{2+}(SOD) + H_2O_2 + 2\,H_2O$$

First, a superoxide ion transfers an electron to $Cu^{2+}$, giving molecular oxygen and $Cu^+$. In the second step, another superoxide anion reoxidizes the copper centre back to $Cu^{2+}$. The resulting $O_2^{2-}$ anion is protonated rapidly to give hydrogen peroxide. The source of the two protons is still unclear. It appears that the role of the $Zn^{2+}$ ion is to provide structural stability to the protein, because if zinc is removed, the protein degrades quite easily.

Interest in superoxide dismutase has increased in recent years with the discovery that a mutation in the gene coding for SOD is linked to certain types of the neurodegenerative disease amyotrophic lateral sclerosis (ALS), commonly known as Lou Gehrig's disease. Exactly how mutant forms of SOD are involved in ALS is a subject of intense research.

## Electron Transfer Proteins

From biochemical synthesis to bioenergetics, redox reactions are fundamental parts of the life process. Molecular oxygen is the oxidizing agent in most of these redox reactions. It has been estimated that the reduction of oxygen to water accounts for 90% of all the $O_2$ consumed in the biosphere:

$$O_2 + 4\,H_3O^+ + 4\,e^- \leftrightharpoons 6\,H_2O \qquad E° = 1.23\ V$$

The many redox reactions that take place within a cell make use of metalloproteins with a wide range of electron transfer potentials. To name just a few of their functions, these proteins play key roles in respiration, photosynthesis, and nitrogen fixation. Some of them simply shuttle electrons to or from enzymes that

**Figure 18-25**

(a) **Cytochrome** *c*                    (b) **Plastocyanin**

Ribbon structures of two redox proteins, cytochrome c (a) and plastocyanin (b). The blowups show the active sites where transition metal atoms are located.

require electron transfer as part of their catalytic activity. In many other cases, a complex enzyme may incorporate its own electron transfer centres. There are three general categories of transition metal redox centres: cytochromes, blue copper proteins, and iron–sulphur proteins.

*Cytochromes.* A cytochrome is a protein containing a haem with an iron cation bonded to four donor nitrogen atoms in a square planar array. Figure 18-25*a* shows the structure of cytochrome *c*, in which a histidine nitrogen atom and a cysteine sulphur atom occupy the fifth and sixth coordination sites of the octahedral iron centre.

The electron transfer properties of the cytochromes involve cycling of the iron between the +2 and +3 oxidation states:

$$(\text{Cytochrome})\text{Fe}^{3+} + e^- \rightleftharpoons (\text{Cytochrome})\text{Fe}^{2+} \qquad E° = -0.3 \text{ V to} + 0.4 \text{ V}$$

**Figure 18-26**

Cysteine    Bridging
             sulphides

*(a)*

Cysteine    *(b)*

Bridging sulphide

Bridging
sulphide

Bridging sulphide

*(c)*

There are three general types of iron–sulphur redox centres: (*a*) a single iron atom (brown) surrounded by four cysteine sulphur atoms (yellow); (*b*) two iron atoms bound to two cysteines and a pair of bridging sulphide ligands; and (*c*) a cube-like structure consisting of four irons and four sulphurs.

Different cytochromes have different side groups attached to the porphyrin ring. These side groups modify the electron density in the delocalized $\pi$-electron system of the porphyrin, which in turn changes the redox potential of the iron cation in the haem.

*Blue copper proteins.* A typical blue copper redox protein contains a single copper atom in a distorted tetrahedral environment. Copper performs the redox function of the protein by cycling between $Cu^+$ and $Cu^{2+}$. Usually, the metal binds to two N atoms and two S atoms through a methionine, a cysteine, and two histidines. An example is plastocyanin, shown in Figure 18-25b. As their name implies, these molecules have a beautiful deep-blue colour that is attributed to photon-induced charge transfer from the sulphur atom of cysteine to the copper cation centre.

*Iron–sulphur proteins.* In an iron–sulphur protein, the metal centre is surrounded by a group of sulphur donor atoms in a tetrahedral environment. The Focus 11-2: Chemistry and Life describes the roles that iron–sulphur proteins play in nitrogenase, and Figure 18-26 shows the structures about the metal in three different types of iron–sulphur redox centres. One type (Figure 18-26*a*) contains a single iron atom bound to four cysteine ligands. The electron transfer reactions at these centres involve cycling of the metal between $Fe^{2+}$ and $Fe^{3+}$.

A second type (Figure 18-26*b*) contains two iron atoms, each bound to two cysteine ligands. The metals are connected through a pair of sulphide ligands. The most complicated redox centre of the iron–sulphur proteins contains four iron atoms and four sulphur atoms arranged in a distorted cube (Figure 18-26*c*). Each iron is bonded to one cysteine and three bridging sulphides. Individual iron cations in iron–sulphur clusters do not undergo s imple changes in oxidation state. Instead, it appears that the electron transfer reactions involve orbitals that are delocalized over the entire cluster.

## SECTION EXERCISES

● **18.4.1** Make a table that summarizes the biological chemistry of iron discussed in this section. Your table should include the names of the metalloproteins, their functions, and the coordination environment around the metal.

● **18.4.2** Most iron-containing proteins have a reddish colour, and most copper proteins are blue. In contrast, zinc proteins are colourless. Why are zinc proteins colourless?

## 18.5 METALLURGY

Metallurgy is the production and purification of metals from naturally occurring deposits called *ores*. It has an ancient history and may represent the earliest useful application of chemistry. Metallurgical advances have had profound influences on the course of human civilization, so much so that historians speak of the

Bronze Age (ca. 3000–1000 BC) and the Iron Age (starting ca. 1000 BC). Except for aluminum and tin, the story of metallurgy is primarily about the extraction and purification of transition metals from their ores. In this section, we discuss some of the techniques of metallurgy.

Nearly all transition metals are oxidized readily, so most ores are compounds in which the metals have positive oxidation numbers. Examples include oxides ($TiO_2$, rutile; $Fe_2O_3$, hematite; $Cu_2O$, cuprite), sulphides (ZnS, sphalerite; $MoS_2$, molybdenite), phosphates ($CePO_4$, monazite; $YPO_4$, xenotime; both found mixed with other rare earth metal phosphates), and carbonates ($FeCO_3$, siderite). Other minerals contain oxoanions ($MnWO_4$, wolframite) and even more complex structures such as carnotite, $K_2(UO_2)_2(VO_4)_2 \cdot 3H_2O$.

Figure 18-27 shows in schematic fashion some of the alternative paths leading from ores to pure metals. These paths include four general processes of which the essential chemical process is **reduction** to yield the neutral metal. First is *separation*. Generally, a metal ore obtained from a mine contains a particular compound of some desired metal mixed with various other materials. The mineral must be separated from these other contaminants. Separation is often followed by *conversion*, in which the mineral is treated chemically to convert it into a form that can be easily reduced. The third step is *reduction*. After a suitable compound has been obtained, it is reduced to free metal by chemical reaction with a reducing agent or by electrolysis. The metal obtained by reduction often contains small amounts of impurities, so the final step is *refining* to purify the metal.

**Figure 18-27**

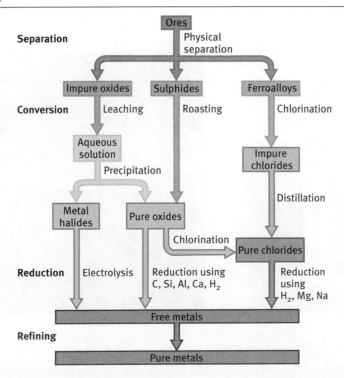

Metallurgy includes separation, conversion, reduction, and refining steps. The starting material is an impure ore, and the end product is pure metal.

## Overview of Metallurgical Processes

Ore obtained from a mining operation contains a desired mineral contaminated with other components, which may include sand, clay, and organic matter. This economically valueless portion of the ore, which is called gangue (pronounced "gang"), must be removed before the metal can be extracted and refined. Ores can be separated into components by physical or chemical methods.

**Flotation** is a common physical separation process in which the ore is crushed and mixed with water to form a thick slurry. As shown in Figure 18-28, the slurry is transferred to a flotation vessel and mixed with oil and a surfactant (see Chapter 9). The polar head groups of the surfactant coat the surface of the mineral particles, but the non-polar tails point outward, making the surfactant-coated mineral particles hydrophobic. Air is blown vigorously through the mixture, carrying the oil and the coated mineral to the surface, where they become trapped in the froth. Because the gangue has a much lower affinity for the surfactant, it absorbs water and sinks to the bottom of the flotation vessel. The froth is removed at the top, and the gangue is removed at the bottom.

A second separation technique is **leaching**, which uses solubility properties to separate the components of an ore. For example, modern gold production depends on the extraction of tiny particles of gold from gold-bearing rock deposits. After the rock is crushed, it is treated with an aerated aqueous basic solution of sodium cyanide. Oxygen gas oxidizes the metal, which forms a soluble coordination complex with the cyanide anion:

$$4\,Au\,(s) + 8\,CN^-\,(aq) + O_2\,(g) + 2\,H_2O\,(l)$$

$$\longrightarrow 4\,[Au(CN)_2]^-\,(aq) + 4\,OH^-\,(aq)$$

**Figure 18-28**

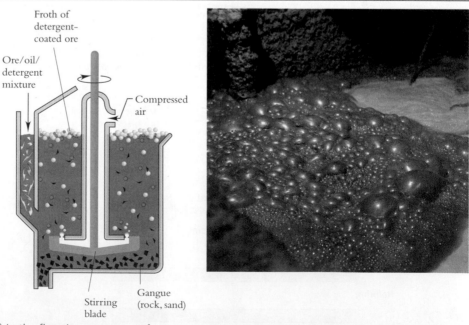

(a) In the flotation process, surfactant-coated mineral particles float to the surface of the mixture, where they become trapped in the froth. The gangue settles to the bottom.
(b) A photograph of froth in the flotation of a gold ore.

The aqueous gold-containing solution is then treated with zinc dust, which reduces gold back to the free metal:

$$Zn \ (s) + 2[Au(CN)_2]^- \ (aq) \longrightarrow 2 \ Au \ (s) + [Zn(CN)_4]^{2-} \ (aq)$$

Sulphide ores may be treated chemically to convert them to oxides before extraction can occur. The process of oxidizing an ore by heating to a high temperature in the presence of air is known as **roasting**. In the roasting of a sulphide ore, sulphide anions are oxidized and molecular oxygen is reduced. The conversion of galena (ZnS) is a typical example:

$$2 \ ZnS \ (s) + 3 \ O_2 \ (g) \longrightarrow 2 \ ZnO \ (s) + 2 \ SO_2 \ (g)$$

Unfortunately, roasting produces copious amounts of highly polluting $SO_2$ gas that has seriously damaged the environment around sulphide ore smelters.

Today, zinc and other metals can be extracted from sulphides by aqueous conversion processes that avoid the generation of $SO_2$. Aqueous acid reacts with the sulphides to generate free sulphur or sulphate ions rather than $SO_2$. Here are two examples:

$$2 \ ZnS \ (s) + 4 \ H_3O^+ \ (aq) + O_2 \ (g) \longrightarrow 2 \ Zn^{2+} \ (aq) + 2 \ S \ (s) + 6 \ H_2O \ (l)$$

$$3 \ CuS \ (s) + 8 \ NO_3^- \ (aq) + 8 \ H_3O^+ \ (aq) \longrightarrow$$
$$8 \ NO \ (g) + 3 \ Cu^{2+} \ (aq) + 3 \ SO_4^{2-} \ (aq) + 12 \ H_2O \ (l)$$

Taking into account the cost of $SO_2$ pollution of the atmosphere, these more elaborate aqueous separation procedures are economically competitive with conversion by roasting.

Once an ore is in suitably pure form, it can be reduced to the free metal. This is accomplished either chemically or electrolytically. Electrolysis is costly because it requires huge amounts of electrical energy. For this reason, chemical reduction is used unless the metal is too reactive for chemical reducing agents to be effective.

Mercury and lead are sufficiently easily reduced that roasting the sulphide ore frees the metal. Sulphide ion is the reducing agent, and both $O_2$ and the metal ion gain electrons:

$$HgS \ (s) + O_2 \ (g) \longrightarrow Hg \ (l) + SO_2 \ (g)$$

This reduction also produces $SO_2$, which must be removed from the exhaust gases.

One of the most common chemical reducing agents for metallurgy is coke, a form of carbon made by heating coal at high temperature until all of the volatile impurities have been removed. Metals that have cations with moderately negative reduction potentials—Co, Ni, Fe, and Zn—are reduced by coke. For example, direct reaction with coke in a furnace frees nickel from its oxide:

$$NiO \ (s) + C \ (s) \longrightarrow Ni \ (l) + CO \ (g)$$

Chemical reduction of an ore usually produces metal that is not pure enough for its intended use. Further **refining** of the metal removes undesirable impurities. Several important metals, including Cu, Ni, Zn, and Cr, are refined by electrolysis,

## TABLE 18-6  METALLURGY OF TRANSITION METALS

| Metal | Ore | Separation Method | Intermediate* | Reducing Agent |
|-------|-----|-------------------|---------------|----------------|
| Ti | $TiO_2$ | Chlorination | $TiCl_4$ | Mg |
| Cr | $FeCr_2O_4$ | Roasting | $Cr_2O_3$ | Al |
| Mn | $MnO_2$ | | $Mn_2O_3$ | Al |
| Fe | $Fe_3O_4$ | Slag formation | $(CaSiO_3)$ | C |
| Co | CoAsS | Roasting | CoO | C |
| Ni | $Ni_9S_8$ | Complexation | $Ni(CO)_4$ | $H_2$ |
| Cu | $CuFeS_2$ | Leaching | $Cu^{2+}$, $(SO_4^{2-})$ | |
| Zn | ZnS | Roasting | ZnO | C |
| Mo | $MoS_2$ | Roasting | $MoO_3$ | $H_2$ |
| W | $CaWO_4$ | Leaching | $WO_4^{2-}$, $WO_3$ | $H_2$ |
| Au | Au | Leaching | $[Au(CN)_2]^-$ | Zn |
| Hg | HgS | Roasting | $(SO_2)$ | $S^{2-}$ |

*Intermediates that represent impurities are shown in parentheses.

either from an aqueous solution of the metal salt or from anodes prepared from the impure metal. To give one example, $Zn^{2+}$ ions, obtained by dissolving ZnS or ZnO in acidic solution, can be reduced while water is oxidized:

$$Zn^{2+}\ (aq) + 2\ e^- \longrightarrow Zn\ (s)$$

$$3\ H_2O\ (l) \longrightarrow 1/2\ O_2\ (g) +\ 2\ H_3O^+\ (aq) + 2\ e^-$$

$$\overline{Zn^{2+}\ (aq) + 3\ H_2O\ (l) \longrightarrow Zn\ (s) + 1/2\ O_2\ (g) + 2\ H_3O^+\ (aq)}$$

Table 18-6 provides a summary of the chemical species and processes involved in the metallurgy of many transition metals.

## Iron and Steel

Iron has been a dominant structural material of modern times, and despite the growth in importance of aluminum and plastics, iron still ranks first in total use. Worldwide production of steel (iron with properties changed by additives) is on the order of 700 million tonnes per year. The most important iron ores are two oxides, hematite ($Fe_2O_3$) and magnetite ($Fe_3O_4$). The production of iron from its ores involves several chemical processes that take place in a blast furnace. As shown in Figure 18-29, this is an enormous chemical reactor where heating, reduction, and purification all occur together.

The raw materials placed in a blast furnace include the ore (usually hematite) and coke, which serves as the reducing agent. The ores always contain various amounts of silicon dioxide ($SiO_2$), which is removed chemically by reaction with limestone ($CaCO_3$). To begin the conversion process, pellets of ore, coke, and limestone are mixed and fed into the top of the furnace, and a blast of hot air is blown in at the bottom. As the starting materials fall through the furnace, the burning coke generates intense heat:

$$2\ C\ (s) + O_2\ (g) \longrightarrow CO\ (g) \qquad \Delta H° = -221\ kJ$$

**Figure 18-29**

A diagrammatic view showing the chemical reactions occurring within a blast furnace, which operates continuously at high temperatures (see photo).

The result is a temperature gradient ranging from about 800 °C at the top of the furnace to 1900 °C at the bottom.

The reduction of iron oxide takes place in several stages in different temperature zones within the furnace. The reducing agent is CO produced from burning coke. Here are the key reactions:

$$3\ Fe_2O_3\ (s)\ +\ CO\ (g)\ \longrightarrow\ 2\ Fe_3O_4\ (s)\ +\ CO_2\ (g)$$

$$Fe_3O_4\ (s)\ +\ CO\ (g)\ \longrightarrow\ 3\ FeO\ (s)\ +\ CO_2\ (g)$$

$$FeO\ (s)\ +\ CO\ (g)\ \longrightarrow\ Fe\ (l)\ +\ CO_2\ (g)$$

Once liberated from its oxides, the iron melts when the temperature reaches 1500 °C. Molten iron collects in a pool at the bottom of the furnace.

At the same time that heating and reduction occur, limestone decomposes into calcium oxide and $CO_2$. The CaO then reacts with $SiO_2$ impurities in the ore to generate calcium silicate:

$$CaCO_3\ (s)\ \xrightarrow{\ Heat\ }\ CaO\ (s)\ +\ CO_2\ (g)$$

$$CaO\ (s)\ +\ SiO_2\ (s)\ \xrightarrow{\ Heat\ }\ CaSiO_3\ (l)$$

At blast furnace temperatures, calcium silicate is a liquid, called *slag*. Being less dense than iron, slag pools on the surface of the molten metal. Both products are drained periodically through openings in the bottom of the furnace.

Although this chemistry is complex, the basic process is reduction of iron oxide by carbon in an atmosphere depleted of oxygen. Archaeologists have found ancient smelters in Africa (in what is now Tanzania) that exploited this chemistry to produce iron in prehistoric times. Early African peoples lined a hole with a fuel of termite residues and added iron ore. Charred reeds and charcoal provided the reducing substance. Finally, a chimney of mud was added. When this furnace was "fired," a pool of iron collected in the bottom.

The iron formed in a blast furnace, called *pig iron*, contains impurities that make the metal brittle. These include phosphorus and silicon from silicate and phosphate minerals that contaminated the original ore, as well as excess carbon and sulphur from the coke. This iron is refined in a converter furnace. Here, a stream of $O_2$ gas blows through molten impure iron. Oxygen reacts with the non-metal impurities, converting them to oxides. As in the blast furnace, CaO is added to convert $SiO_2$ into liquid calcium silicate, in which the other oxides dissolve. The molten iron is analyzed at intervals until its impurities have been reduced to satisfactory levels. Then, the liquid metal, now in the form called *steel*, is poured from the converter and allowed to solidify.

Most steels contain various amounts of other elements that are added deliberately to give the metal particular properties. These additives may be introduced during the converter process or when the molten metal is poured off. One of the most important additives is manganese, which adds strength and hardness to steel. Manganese is added to nearly every form of steel in amounts ranging from less than 1% to higher than 10%. More than 80% of manganese production ends up incorporated into steel.

The concentration of carbon in steel is intentionally varied to obtain steels having very different properties. Carbon and iron make very strong directional bonds, so the addition of more carbon generally makes the steel harder and stronger, but also makes the steel less ductile and more difficult to weld. Low-carbon steel, containing less than 0.29% carbon, is most often used as structural material. Medium-carbon steel (0.30–0.59% C) has good wear resistance and is used for making parts via forging (shaping using rolling and compressing forces). High-carbon steel (0.60–0.99% C) is very strong and is typically used for springs and high-strength wires. Ultrahigh carbon steel (1.0–2.0% C) is used to make cutting tools and axles.

## Other Metals

*Titanium.* The metallurgy of titanium illustrates the purification of one metal by another. The major titanium ores are rutile ($TiO_2$) and ilmenite ($FeTiO_3$). Either is converted to titanium(IV) chloride by a redox reaction with chlorine gas and coke. For rutile:

$$TiO_2\ (s) + C\ (s) + 2\ Cl_2\ (g) \xrightarrow{500\ °C} TiCl_4\ (g) + CO_2\ (g)$$

In this reaction, carbon is oxidized and chlorine is reduced. When the hot gas cools, titanium tetrachloride ($T_b = 140\ °C$) condenses to a liquid that is purified by distillation.

Titanium metal is obtained by reduction of $TiCl_4$ with molten magnesium metal at high temperature. The reaction gives solid titanium metal ($T_m = 1660\,°C$) and liquid magnesium chloride ($T_m = 714\,°C$):

$$TiCl_4\,(g) + 2\,Mg\,(l) \xrightarrow{\;850\,°C\;} Ti\,(s) + 2\,MgCl_2\,(l)$$

*Copper.* Copper is found mainly in the sulphide ore chalcopyrite ($FeCuS_2$), but chalcocite ($Cu_2S$), cuprite ($Cu_2O$), and malachite ($Cu_2CO_3(OH)_2$) are also important. Copper ores often have concentrations of copper less than 1% by mass, so achieving economic viability requires mining operations on a huge scale. The extraction and purification of copper are complicated by the need to remove iron from chalcopyrite. The first step in the process is flotation, which concentrates the ore to around 15% Cu by mass. In the next step, the concentrated ore is roasted to convert $FeCuS_2$ to CuS and FeO. Copper(II) sulphide is unaffected if the temperature is kept below 800 °C:

Copper mining is carried out on a huge scale.

$$2\,FeCuS_2\,(s) + 3\,O_2\,(g) \longrightarrow 2\,CuS\,(s) + 2\,FeO\,(s) + 2\,SO_2\,(g)$$

Heating the mixture of CuS and FeO to 1400 °C in the presence of silica ($SiO_2$) causes the material to melt and separate into two layers. The top layer is molten $FeSiO_3$ formed from the reaction of $SiO_2$ with FeO. As this takes place, the copper in the bottom layer is reduced from CuS to $Cu_2S$. This bottom layer consists of molten $Cu_2S$ contaminated with FeS. Reduction of the $Cu_2S$ takes place in a converter furnace following the same principle that converts impure iron into steel. Silica is added, and oxygen gas is blown through the molten mixture. Iron impurities are converted first to FeO and then to $FeSiO_3$, which is a liquid that floats to the surface. At the same time, $Cu_2S$ is converted to $Cu_2O$, which reacts with more $Cu_2S$ to give copper metal and $SO_2$:

$$2\,Cu_2S\,(l) + 3\,O_2\,(g) \longrightarrow 2\,Cu_2O\,(l) + 2\,SO_2\,(g)$$

$$2\,Cu_2O\,(l) + Cu_2S\,(l) \longrightarrow 6\,Cu\,(l) + SO_2\,(g)$$

Copper metal obtained from the converter furnace must be refined to better than 99.95% purity before it can be used to make electrical wiring. This is accomplished by electrolysis, as illustrated in Figure 18-30.

**Figure 18-30**

Diagram of an electrolytic cell for purification of copper.

Copper is used to make electrical wire.

The impure copper is formed into slabs that serve as anodes in electrolysis cells. The cathodes are constructed from thin sheets of very pure copper. These electrodes are immersed in a solution of $CuSO_4$ dissolved in dilute sulphuric acid. Application of a controlled voltage causes oxidation in which copper, along with iron and nickel impurities, is oxidized to its cations. Less reactive metal contaminants, including silver, gold, and platinum, are not oxidized. As the electrolysis proceeds and the anode dissolves, insoluble impurities fall to the bottom of the cell as sludge. This sludge is a valuable source of precious metals.

The metal cations released from the anode migrate through the solution to the cathode. Because $Cu^{2+}$ is easier to reduce than $Fe^{2+}$ and $Ni^{2+}$, careful control of the applied voltage makes it possible to reduce $Cu^{2+}$ to Cu metal, leaving $Fe^{2+}$ and $Ni^{2+}$ dissolved in solution.

## FOCUS 18-2  CHEMISTRY IN CANADA: THE CANADIAN METAL MINING INDUSTRY

The metal mining industry is one of Canada's economic engines. Every part of the country is involved in mining, smelting, and refining or fabrication of metal products. The Canadian shield, which spans eastern, northeastern, and central Canada, is a rich source of base metals (nickel, copper, zinc, lead, cobalt, and molybdenum), precious metals such as gold, silver, and platinum, and ferrous metals such as iron, niobium, and tantalum.

The scale of the industry in Canada is enormous. Sixty-six metal-producing mines operate from Newfoundland and Labrador to British Columbia. Half of these are precious-metal mines, most of the other half are base-metal mines, and a handful are ferrous-metal mines. Another 59 metal production facilities, including

Metal mining sites in Canada

aluminum smelters, steel makers, uranium processors, and producers and recyclers of base metals, operate from British Columbia to Newfoundland and Labrador. The total value of metals production in 2007 was approximately $26 billion, with iron ore, nickel, copper, and zinc leading in terms of tonnage and value. Two Canadian companies, Barrick Gold Corporation and Cameco Corporation, are the world's largest producers of gold and uranium, respectively.

Any industry having such large benefits generally has some problems as well. A major issue in the mining industry is acid rock drainage, which refers to the flow of highly acidic water from (usually abandoned) mine sites. Processing metal-bearing ores results in the generation of huge quantities of waste rock known as tailings. Water in contact with the tailings, or water that floods abandoned mines, can become acidified and can be very harmful to the environment if released. The greatest contributor to acidity is the oxidation of metal sulphides. Using pyrite (iron sulphide) as an example:

$$2 \text{ FeS}_2 \, (s) + 7 \text{ O}_2 \, (g) + 2 \text{ H}_2\text{O} \, (l) \longrightarrow \text{Fe}^{2+} \, (aq) + 4 \text{ SO}_4^{2-} \, (aq) + 4 \text{ H}^+ \, (aq)$$

Some of the $\text{Fe}^{2+}$ gets oxidized to $\text{Fe}^{3+}$, which then catalyzes oxidation of more pyrite, releasing more $\text{H}^+$:

$$\text{FeS}_2 \, (s) + 14 \text{ Fe}^{3+} \, (aq) + 8 \text{ H}_2\text{O} \, (l) \longrightarrow 15 \text{ Fe}^{2+} \, (aq) + 2 \text{ SO}_4^{2-} \, (aq) + 16 \text{ H}^+ \, (aq)$$

The net effect is high acidities and high concentrations of metal ions, both of which are toxic to plants and animals. The low pH oxidizes and solubilizes other metals that may also be toxic.

Mitigating the effects of acid rock drainage is the subject of much research. Common strategies include the addition of alkaline compounds ($\text{CaCO}_3$, $\text{CaO}$, $\text{Na}_2\text{CO}_3$, $\text{NaOH}$), which raise the pH to the point where metals precipitate out of solution as insoluble hydroxides, and coagulants and flocculants such as alum ($\text{Al}_2(\text{SO}_4)_3$) that cause the formation of solids that settle out of solution, carrying toxic metals with them. Enpar Technologies of Guelph, Ontario has developed a process to electrochemically reduce dissolved oxygen in the water on top of mine tailings. This has the effect of slowing the oxidation of sulphide minerals and the production of acidity. (The process makes use of a sacrificial anode, as described in Section 16.6.) Some bacteria are capable of reducing sulphate ions. Treatment of acid rock drainage using such bacteria results in precipitation of the metals as sulphides. These and other remediation strategies are in use at abandoned and operating mining sites across the country.

Nickel tailings at Sudbury, Ontario

● **18.5.1** Construct a flowchart that summarizes the chemistry that takes place in a blast furnace.

● **18.5.2** Examine Appendix F and explain why $ZnO$ can be reduced with coke, but $Cr_2O_3$ requires a more reactive metal such as aluminum.

● **18.5.3** Which of the reactions listed in Figure 18-29 are redox reactions? Identify the reducing agent in each case.

## 18.6 APPLICATIONS OF TRANSITION METALS

A complete discussion of all the transition metals is beyond the scope of an introductory course in chemistry. Instead, we provide a brief survey of several metals that highlights the diversity and utility of this group of elements.

### Titanium

Titanium, the ninth most abundant element in the Earth's crust, is characterized by its high strength, its low density (~57% that of steel), and its stability at very high temperature. When alloyed with small amounts of aluminum or tin, titanium has the highest strength-to-weight ratio of all the engineering metals. Its major use is in the construction of aircraft frames and jet engines. Because titanium is also highly resistant to corrosion, it is used in the construction of pipes, pumps, and vessels for the chemical industry.

Because it is difficult to purify and fabricate, titanium is an expensive metal. For example, although titanium bicycles are highly prized by avid riders, they are quite expensive, averaging well over $1000 for just the frame.

Titanium finds uses as diverse as aircraft turbine blades, lightweight bicycle frames, and $TiO_2$ "smoke" for skywriting.

The most important compound of titanium is titanium(IV) oxide, $TiO_2$. More than 2 million tonnes of $TiO_2$ are produced every year, much of it by the controlled combustion of $TiCl_4$:

$$TiCl_4 (g) + O_2 (g) \xrightarrow{1200\,°C} TiO_2 (s) + 2\,Cl_2 (g)$$

The $Cl_2$ produced in the combustion reaction is recycled to produce more $TiCl_4$ from rutile ore.

Titanium dioxide is brilliant white, highly opaque, chemically inert, and non-toxic. Consequently, it finds wide uses as a pigment in paints and other coatings, in paper, sunscreens, cosmetics, and toothpaste. Almost all white-coloured commercial products contain $TiO_2$.

## Chromium

Chromium makes up just 0.012% of the Earth's crust, yet it is an important industrial metal. The main use of chromium is in metal alloys. Stainless steel, for example, contains as much as 20% chromium. Nichrome, a 60:40 alloy of nickel and chromium, is used to make heat-radiating wires in electrical devices such as toasters and hair dryers. Another important application of chromium metal is as a protective and decorative coating for the surface of metal objects, as described in Chapter 16.

The only important ore of chromium is chromite, $FeCr_2O_4$. Reduction of chromite with coke gives ferrochrome, an iron–chromium compound:

$$FeCr_2O_4 (s) + 4\,C (s) \longrightarrow FeCr_2 (s) + 4\,CO (g)$$

Ferrochrome is mixed directly with molten iron to form chromium-containing stainless steel.

Chromium compounds of high purity can be produced from chromite ore without reduction to the free metal. The first step is the roasting of chromite ore in the presence of sodium carbonate:

$$4\,FeCr_2O_4 (s) + 8\,Na_2CO_3 (s) + 7\,O_2 (g) \xrightarrow{1100\,°C}$$
$$8\,Na_2CrO_4 (s) + 2\,Fe_2O_3 (s) + 8\,CO_2 (g)$$

The product is converted to sodium dichromate by reaction with sulphuric acid:

$$2\,Na_2CrO_4 (s) + H_2SO_4 (aq) \longrightarrow Na_2Cr_2O_7 (aq) + Na_2SO_4 (aq) + H_2O (l)$$

When the resulting solution is concentrated by evaporation, $Na_2Cr_2O_7 \cdot 2H_2O$ precipitates as a solid. This compound is the most important source of chromium compounds for the chemical industry. It is the starting material for most other chromium-containing compounds of commercial importance, including $(NH_4)_2Cr_2O_7$ (ammonium dichromate), $Cr_2O_3$ (chromium(III) oxide), and $CrO_3$ (chromium(VI) oxide).

Chromium metal is used as a protective and decorative coating.

Pure chromium metal is made by a two-step reduction sequence. First, sodium dichromate is reduced to chromium(III) oxide by heating in the presence of charcoal:

$$Na_2Cr_2O_7 (s) + C (s) \xrightarrow{\text{Heat}} Cr_2O_3 (s) + Na_2CO_3 (s) + CO (g)$$

Dissolving $Cr_2O_3$ in sulphuric acid gives an aqueous solution of $Cr^{3+}$ cations:

$$Cr_2O_3 (s) + 6 H_3O^+ (aq) \longrightarrow 2 Cr^{3+} (aq) + 9 H_2O (l)$$

Chromium compounds display a striking range of beautiful colours. Shown here are $Na_2CrO_4$ (yellow), $K_2Cr_2O_7$ (orange), $CrCl_3$ (green), and $CrO_3$ (dark purple).

Electrolysis of this solution reduces the cations to pure Cr metal, which forms a hard, durable film on the surface of the object serving as the cathode.

The name "chromium" is derived from the Greek word for colour, *chroma*, because this metal forms a wide variety of compounds with beautiful colours. Chromium compounds have been used for many years as pigments in paints and other coatings: $Na_2Cr_2O_7$ is bright orange, $Cr_2O_3$ is green, and the Zn and Pb salts of $CrO_4^{2+}$ are bright yellow. However, in recent years, the use of chromium pigments has diminished because chromium in the +6 oxidation state is highly toxic.

Chromium is also important in converting animal hides into leather. In the tanning process, hides are treated with basic solutions of Cr(III) salts, which causes cross-linking of collagen proteins. The hides toughen and become pliable and resistant to biological decay.

## Copper, Silver, and Gold

The first three pure metals known to humanity probably were copper, silver, and gold, known as the coinage metals because they found early use as coins. All three are found in nature in their pure elemental form, and all have been highly valued throughout civilization. The oldest known gold coins were used in Egypt around 3400 BC. At about the same time in the Middle East, copper was obtained from charcoal reduction of its ores. The first metallurgy of silver was developed in Asia Minor (Turkey) about 500 years later. All three of these metals are excellent electrical conductors and are highly resistant to corrosion.

The coffin of Tutankhamun contained over 100 kg of gold. Tutankhamun was only a minor pharaoh, who died at age 18.

These properties, coupled with its relatively low cost, make copper one of the most useful metals in modern society. About half of all copper produced is for electrical wiring, and the metal is also widely used for plumbing pipes. Copper is used to make several important alloys, the most important of which are bronze and brass. Both alloys contain copper mixed with lesser amounts of tin and zinc in various proportions. In bronze, the amount of tin exceeds that of zinc, whereas the opposite is true for brass. The discovery of bronze sometime around 3000 BC launched the advance of civilization known today as the Bronze Age. Because bronze is harder and stronger than other metals known in antiquity, it became a mainstay of the civilizations of India and the Mediterranean, used for tools, cookware, weapons, coins, and objects of art. Today, the principal use of bronze is for bearings, fittings, and machine parts.

Copper is resistant to oxidation, but over the course of time the metal acquires a coating of green corrosion called *patina*. The green compound is a mixed salt of $Cu^{2+}$ with hydroxide, sulphate, and carbonate ions that is formed by air oxidation in the presence of carbon dioxide and small amounts of sulphur dioxide:

$$3\ Cu\ (s) + 2\ H_2O\ (l) + SO_2\ (g) + 2\ O_2\ (g) \longrightarrow Cu_3(OH)_4SO_4\ (s)$$

$$2\ Cu\ (s) + H_2O\ (l) + CO_2\ (g) + O_2\ (g) \longrightarrow Cu_2(OH)_2CO_3\ (s)$$

Although trace amounts of copper are essential for all forms of life, the metal is toxic in large amounts. Thus, copper(II) salts, particularly $CuSO_4 \cdot 5H_2O$, are used as pesticides and wood preservatives. Wood soaked in solutions of $Cu^{2+}$ or coated with paints containing $Cu^{2+}$ resist degradation resulting from bacteria, algae, and fungi.

Silver is usually found as a minor component of ores of more abundant metals such as copper and zinc. Most commercial silver is produced as a by-product of the production of these common metals. For example, electrolytic refining of copper generates a solid anodic residue that is rich in silver and other precious metals. The silver from this residue can be isolated by oxidizing the metal into nitrate-containing solutions, silver nitrate being one of the few soluble silver salts. The pure metal is then deposited electrolytically. Silver is used for tableware in the form of sterling silver, an alloy containing small amounts of copper to make the metal harder. Silver is also used in jewellery, mirrors, and batteries, but the single most important use of the metal, accounting for about a third of all production, is in photography. This application is decreasing rapidly because digital cameras are now more popular than film cameras.

Silver does not form a simple oxide by direct oxidation in air, but the metal does form a black tarnish with oxygen and trace amounts of hydrogen sulphide in the atmosphere:

$$4\ Ag\ (s) + 2\ H_2S\ (s) + O_2\ (g) \longrightarrow \underset{\text{black}}{2\ Ag_2S\ (s)} + 2\ H_2O\ (l)$$

The extraction of gold by leaching is described in the previous section. Gold is used extensively in the manufacture of jewellery. Interestingly, Au(I) compounds are very effective in the treatment of rheumatoid arthritis, and there is recent evidence that certain gold-containing compounds have anticancer properties.

## Zinc and Mercury

Zinc and mercury are found in the Earth's crust as sulphide ores, the most common of which are sphalerite (ZnS) and cinnabar (HgS). (See the previous section for descriptions of the extraction and purification of these metals.)

The name *copper* and the symbol Cu are derived from the Latin *cuprum*, after the island of Cyprus, where the Romans first obtained copper metal. The symbols Ag and Au for silver and gold come from the Latin names for these elements: *argentum* and *aurum*.

The green patina on copper objects such as the roof of the Peace Tower of the Parliament Buildings in Ottawa is a mixture of copper salts, $Cu_3(OH)_4SO_4$ and $Cu_2(OH)_2CO_3$.

Most of the world's zinc output is used to prevent the corrosion of steel. Zinc is easier to oxidize than iron, as shown by the more negative reduction potential of $Zn^{2+}$:

$$Fe^{2+} + 2\,e^- \longrightarrow Fe \qquad E° = -0.447 \text{ V}$$
$$Zn^{2+} + 2\,e^- \longrightarrow Zn \qquad E° = -0.762 \text{ V}$$

Zinc oxide is an effective sunscreen.

Consequently, a zinc coating oxidizes preferentially and protects steel from corrosion. Zinc coatings are applied in several ways: by immersion in molten zinc, by paint containing powdered zinc, or by electroplating.

Zinc is also combined with copper and tin to make brass and bronze. Moreover, large amounts of zinc are used to make several types of batteries, as discussed in Chapter 16.

Zinc oxide is the most important zinc compound. The principal industrial use of zinc oxide is as a catalyst to shorten the time of vulcanization in the production of rubber. The compound is also used as a white pigment in paints, cosmetics, and photocopy paper. In everyday life, ZnO is also a common sunscreen.

The use of mercury for extracting silver and gold from their ores has been known for many centuries. Gold and silver form amalgams with liquid mercury, which is then distilled away to leave the pure precious metal. The Romans mined the mineral cinnabar (HgS) from deposits in Spain 2000 years ago, and in the sixteenth century the Spanish shipped mercury obtained from the same ore deposits to the Americas for the extraction of silver. Mercury is an important component of street lamps and fluorescent lights. It is used in thermometers and barometers and in gas-pressure regulators, electrical switches, and electrodes.

## The Platinum Metals

Six of the transition metals—Ru, Os, Rh, Ir, Pd, and Pt—are known as the *platinum metals*. The group is named for the most familiar and most abundant of the six. These elements are usually found mingled together in ore deposits, and they share many common features. Although they are rare (total annual production is only about 200 tonnes, most of which is platinum itself), the platinum metals play important roles in modern society.

The platinum metals are valuable by-products from the extraction of common metals such as copper and nickel. The anodic residue that results from copper refining is a particularly important source. The chemistry involved in their purification is too complex to describe here, except to note that the final reduction step involves reaction of molecular hydrogen with metal halide complexes.

By far the most important use of the platinum metals is for catalysis. The largest single use is in automobile catalytic converters, which ensure complete oxidation of CO to $CO_2$. Platinum is the principal catalyst, but catalytic converters also contain rhodium and palladium. These elements also catalyze a wide variety of reactions in the chemical and petroleum industries. For example, platinum metal is the catalyst for ammonia oxidation in the production of nitric acid, as described in Chapter 13:

$$4\,NH_3 + 5\,O_2 \xrightarrow{\text{Pt gauze, 1200 K}} 4\,NO + 6\,H_2O$$

Palladium is used as a catalyst for hydrogenation reactions in the food industry, and a rhodium catalyst is used in the production of acetic acid from methanol:

$$CH_3OH + CO \xrightarrow[\text{175 °C, 1 bar}]{\text{Rh--containing catalyst}} CH_3CO_2H$$

The combustion of ammonia is catalyzed by a gauze of platinum metal.

## SECTION EXERCISES

**18.6.1** In recent years, copper has replaced galvanized iron as the material of choice for plumbing pipes. Explain why copper is a better material for pipes than zinc-coated iron. Justify your answer with balanced equations.

**18.6.2** The chemistry of chromium and zinc can be used to remove traces of oxygen from bottled gases such as nitrogen and argon. The gas is bubbled through a solution of $Cr^{2+}$ in the presence of zinc metal. Trace oxygen in the gas reacts quickly with $Cr^{2+}$ to give $Cr^{3+}$ and water. The zinc in turn reduces $Cr^{3+}$ back to $Cr^{2+}$. Write balanced equations for these reactions.

**18.6.3** Refer to the standard reduction potentials for various metal cations (see Appendix F) and use these to predict which metals will be found in the anodic residue that results from the electrolytic refining of copper.

# CHAPTER 18 VISUAL SUMMARY

## 18.1 OVERVIEW OF THE TRANSITION METALS

**LEARNING OBJECTIVES:** Predict periodic properties of transition metals.

**SKILLS TO MASTER:** Naming groups of transition metals (p. 1056)

Knowing trends in melting point, density, and oxidation states (p. 1056)

Determining oxidation states in compounds of transition metals (p. 1058)

## 18.2 COORDINATION COMPLEXES

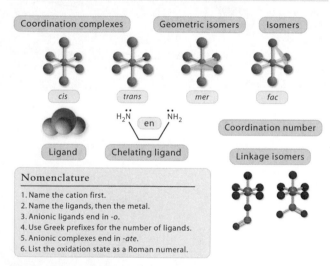

Nomenclature

1. Name the cation first.
2. Name the ligands, then the metal.
3. Anionic ligands end in -o.
4. Use Greek prefixes for the number of ligands.
5. Anionic complexes end in -ate.
6. List the oxidation state as a Roman numeral.

**LEARNING OBJECTIVES:** Recognize and name transition metal coordination complexes.

**SKILLS TO MASTER:** Recognizing and naming some common ligands (p. 1063)

Drawing *cis* and *trans* as well as *fac* and *mer* isomers (p. 1065)

Naming coordination compounds (p. 1067)

**KEY CONCEPTS:** A ligand bonds to a metal or metal cation using its lone pairs of electrons.

Linkage isomers occur when a ligand can bond to a metal using either of two donor atoms.

## 18.3 BONDING IN COORDINATION COMPLEXES

**LEARNING OBJECTIVES:** Use crystal field theory to explain the colour and magnetic properties of complexes.

**SKILLS TO MASTER:** Constructing a crystal field energy-level diagram for an octahedral complex (p. 1072)

Determining the electron configuration of a complex from magnetic properties (p. 1074)

Determining the value of $\Delta$ from the absorption spectrum of a complex (p. 1078)

**KEY CONCEPTS:** The spectrochemical series lists ligands in order of increasing energy-level splitting.

For a given ligand, $\Delta$ increases as the oxidation state of the metal increases.

If $P > \Delta$, a high-spin complex will likely result.

## 18.4 TRANSITION METALS IN BIOLOGY

Metalloproteins

Enzyme

Deoxyhaemoglobin:

Haem

Histidine

Porphyrin ring

Oxyhaemoglobin:

Molecular oxygen

Transport and storage

Electron transfer

**LEARNING OBJECTIVES:** Explain the importance of transition metal complexes in biological processes.

**KEY CONCEPTS:** Many transition metals (V, Cr, Mn, Fe, Co, Ni, Cu, Zn, and Mo) are essential for life.

Metalloproteins may act as storage or transport agents, enzymes or redox reagents.

## 18.5 METALLURGY

**LEARNING OBJECTIVES:** Explain the chemistry of essential steps in the production of pure metals from ores.

**KEY CONCEPTS:** Metallurgy includes separation, conversion, reduction, and refining steps. The starting material is an impure ore, and the end product is pure metal.

Separation processes include flotation and leaching.

Many sulphide ores are converted to oxides before reduction.

Chemical reducing agents include $H_2$, C, Al, Zn, Mg, and $S^{2-}$.

## 18.6 APPLICATIONS OF TRANSITION METALS

Ti Cr Cu Ag Au Zn Hg Pt metals

**LEARNING OBJECTIVES:** Recognize the importance of transition metals in everyday life.

**KEY CONCEPTS:** Titanium is prized for its high strength, low density, and resistance to high temperatures.

Chromium is used as a protective and decorative coating on other metals.

Copper is used primarily in electrical wire.

Gold and silver are prized for their resistance to oxidation.

Platinum is used as a catalyst in automobile catalytic converters.

## LEARNING EXERCISES

18.1   Write a description of the features that transition metals have in common.

18.2   Describe the features of the transition metals that are exploited in biology. Use an example to illustrate each feature.

18.3   Draw ball-and-stick models of all possible isomers of linear, tetrahedral, square planar, and octahedral complexes containing two different ligands.

18.4   Make a list of the transition metals discussed in this chapter and summarize their applications.

18.5   Write a summary of the biological chemistry of iron and copper described in this chapter.

18.6   List all terms new to you that appear in Chapter 18, and write a one-sentence definition of each in your own words. Consult the Glossary if you need help.

## PROBLEMS

**ilw** = interactive learning ware problem. Visit the website at www.wiley.com/college/olmsted

### Overview of the Transition Metals

18.1   The following compounds can be purchased from chemical supply companies. Determine the oxidation states of the transition metals in each: (a) $MnCO_3$; (b) $MoCl_5$; (c) $Na_3VO_4$; (d) $Au_2O_3$; and (e) $Fe_2(SO_4)_3 \cdot 5H_2O$.

18.2   The following compounds can be purchased from chemical supply companies. Determine the oxidation state of the transition metal in each: (a) $NiSO_4$; (b) $KMnO_4$; (c) $(NH_4)_2WO_4$; (d) $PbCrO_4$; and (e) $ZrOCl_2 \cdot 8H_2O$.

18.3   Give the names and symbols for the elements that have the following valence configurations: (a) $4s^1\, 3d^5$; (b) $5s^2\, 4d^{10}$; and (c) $4s^1\, 3d^{10}$.

18.4   Give the names and symbols for the elements that have the following valence configurations: (a) $5s^1\, 4d^8$; (b) $4s^2\, 3d^2$; and (c) $5s^1\, 4d^4$.

18.5   Write valence electron configurations for the following transition metal cations: (a) $Mn^{2+}$; (b) $Ir^{3+}$; (c) $Ni^{2+}$; and (d) $Mo^{2+}$.

18.6   Write valence electron configurations for the following transition metal cations: (a) $Ti^{2+}$; (b) $Fe^{3+}$; (c) $Pt^{2+}$; and (d) $Nb^{3+}$.

18.7   In each of the following pairs of transition metals, select the one with the higher value for the indicated property and give the reason: (a) melting points of Pd and Cd; (b) densities of Cu and Au; and (c) first ionization energies of Cr and Co.

18.8   In each of the following pairs of transition metals, select the one with the higher value for the indicated property and give the reason: (a) melting points of Zr and Mo; (b) densities of Ti and Cr; and (c) first ionization energies of Zr and Ag.

### Coordination Complexes

18.9   Determine the oxidation states and $d$ electron counts for the metal ions in the following coordination complexes: (a) $[Ru(NH_3)_6]Cl_2$; (b) *trans*-$[Cr(en)_2I_2]I$; (c) *cis*-$[PdCl_2(P(CH_3)_3)_2]$; (d) *fac*-$[Ir(NH_3)_3Cl_3]$; and (e) $[Ni(CO)_4]$.

18.10  Determine the oxidation states and $d$ electron counts for the metal ions in the following coordination complexes: (a) $[Rh(en)_3]Cl_3$; (b) *cis*-$[Mo(CO)_4Br_2]$; (c) $Na_3[IrCl_6]$; (d) *mer*-$[Ir(NH_3)_3Cl_3]$; and (e) $[Mn(CO)_5Cl]$.

18.11  Name the compounds in Problem 18.9.

18.12  Name the compounds in Problem 18.10.

18.13  Draw structures for the metal complexes in Problem 18.9.

18.14  Draw structures for the metal complexes in Problem 18.10.

18.15  Write the formulas of the following complex ions:
(a) *cis*-tetraamminechloronitrocobalt(III);
(b) amminetrichloroplatinate(II);
(c) *trans*-diaquabis (ethylenediamine)copper(II); and
(d) tetrachloroferrate(III).

18.16  Write the formulas of the following compounds:
(a) potassium tetrachloroplatinate(II);
(b) pentaammineaquachromium(III) iodide;
(c) tris (ethylenediamine)manganese(II) chloride; and
(d) pentaammineiodocobalt(III) nitrate.

18.17  Draw the structure for each complex ion in Problem 18.15.

18.18  Draw the structure for each complex ion in Problem 18.16.

### Bonding in Coordination Complexes

18.19  For an octahedral complex of each of the following metal ions, draw a crystal field energy-level diagram that shows the electron populations of the various $d$ orbitals. Where appropriate show both the high-spin and low-spin configurations: (a) $Ti^{2+}$; (b) $Cr^{3+}$; (c) $Mn^{2+}$; and (d) $Fe^{3+}$.

18.20  For an octahedral complex of each of the following metal ions, draw a crystal field energy-level diagram that shows the electron populations of the various $d$ orbitals. Where appropriate show both the high-spin and low-spin configurations: (a) $Zn^{2+}$; (b) $Cr^{2+}$; (c) $Co^{2+}$; and (d) $Rh^{3+}$.

18.21  Predict whether each of the following complexes is diamagnetic or paramagnetic. If a complex is paramagnetic, state its number of unpaired electrons. (a) $[Ir(NH_3)_6]^{3+}$; (b) $[Cr(H_2O)_6]^{2+}$; (c) $[PtCl_4]^{2-}$; and (d) $[Pd(P(CH_3)_3)_4]$.

18.22  Predict whether each of the following complexes is diamagnetic or paramagnetic. If a complex is paramagnetic, state its number of unpaired electrons. (a) $[Ru(CN)_6]^{4-}$; (b) $[Co(NH_3)_6]^{3+}$; (c) $[CoBr_4]^{2-}$ (tetrahedral); and (d) $[Pt(en)Cl_2]$.

18.23  Compounds of $Zr^{2+}$ are dark purple, but most $Zr^{4+}$ compounds are colourless. Explain.

**18.24** Of the coordination complexes $[Cr(H_2O)_6]^{3+}$ and $[Cr(NH_3)_6]^{3+}$, one is violet and the other is orange. Decide which is which and explain your reasoning.

**18.25** The complex $[Fe(en)_3]Cl_3$ is low-spin. Provide the following information: (a) the coordination number of the metal; (b) the oxidation number and $d$ electron count of the metal; (c) the geometry of the complex; (d) whether the complex is diamagnetic or paramagnetic; and (e) the number of unpaired electrons.

**18.26** The complex $[Mn(H_2O)_6]SO_4$ is high-spin. Provide the following information: (a) the coordination number of the metal; (b) the oxidation number and $d$ electron count of the metal; (c) the geometry of the complex; (d) whether the complex is diamagnetic or paramagnetic; and (e) the number of unpaired electrons.

**18.27** ⓘⓛⓦ An aqueous solution of $Cr^{2+}$ $(aq)$ is paramagnetic. Addition of sodium cyanide decreases the paramagnetism of the solution. Use crystal field energy-level diagrams to explain why the magnetic properties of the solutions are different.

**18.28** An aqueous solution of $CoCl_2$ is paramagnetic. Addition of hydrogen peroxide (an oxidizing agent) causes the solution to become diamagnetic. Use crystal field energy-level diagrams to explain why the magnetic properties of the solutions are different.

**18.29** ⓘⓛⓦ The complex $[Cr(NH_3)_6]^{3+}$ has its maximum absorbance at 465 nm. Calculate the crystal field splitting energy for the compound and predict its colour.

**18.30** The complex $[Fe(H_2O)_6]^{3+}$ has its maximum absorbance at 724 nm. Calculate the crystal field splitting energy for the compound and predict its colour.

### Transition Metals in Biology

**18.31** Summarize the differences between haemoglobin and myoglobin.

**18.32** What features do myoglobin and the cytochromes have in common?

**18.33** Draw a crystal field splitting diagram that illustrates the electron transfer reaction of the simple iron redox protein shown in Figure 18-25a.

**18.34** Draw a crystal field splitting diagram that illustrates the electron transfer reaction of a cytochrome.

### Metallurgy

**18.35** Write balanced chemical equations for the following metallurgical processes: (a) roasting of $CuFeS_2$; (b) removal of silicon from steel in a converter; and (c) reduction of titanium tetrachloride using sodium metal.

**18.36** Write balanced chemical equations for these metallurgical processes: (a) heating NiS in air; (b) reducing $Co_3O_4$ using Al metal; and (c) reduction of $MnO_2$ by coke.

**18.37** A copper ore contains 2.37% $Cu_2S$ by mass. If $5.60 \times 10^4$ kg of this ore is heated in air, calculate the mass of copper metal that is obtained and the volume of $SO_2$ gas produced at ambient conditions, 755 Torr and 23.5 °C.

**18.38** What mass of limestone, in kilograms, should be added for every kilogram of iron ore processed in a blast furnace if the limestone is 95.5% $CaCO_3$ and the iron ore contains 9.75% $SiO_2$?

**18.39** Calculate the standard free energy change at 25°C for reduction of ZnO to Zn using carbon and using carbon monoxide.

**18.40** Determine $\Delta G°$ for each oxidation reaction that occurs in a steel-making converter. Compare your values with $\Delta G°$ for the reaction of iron with $O_2$ to give $Fe_2O_3$.

### Applications of Transition Metals

**18.41** Identify the coinage metals and describe some of their applications.

**18.42** Identify the platinum metals and describe some of their applications.

**18.43** What features of titanium account for its use as an engineering metal?

**18.44** Explain why titanium(IV) oxide is used extensively as a white pigment.

### ADDITIONAL PAIRED PROBLEMS

**18.45** Draw all possible isomers of the following compounds: (a) $[Ir(NH_3)_3Cl_3]$; (b) $[Pd(P(CH_3)_3)_2Cl_2]$; (c) $[Cr(CO)_4Br_2]$; and (d) $[Cr(en)(NH_3)_2I_2]$.

**18.46** Draw the structures of all possible isomers of the following coordination compounds: (a) tetraamminedibromocobalt(III) bromide; (b) triamminetrichlorochromium(III); and (c) dicarbonylbis-(trimethylphosphine) platinum(0).

**18.47** Several commercial rust removers contain the bidentate ligand oxalate, $(O_2CCO_2)^{2-}$. Explain how these household products remove rust. Include a structural drawing of the species that forms.

**18.48** The carbonate ion can be either a monodentate or a bidentate ligand. Make sketches that show the ligand binding to a metal cation in both modes.

**18.49** Write electron configurations for the following: (a) Cr, $Cr^{2+}$, and $Cr^{3+}$; (b) V, $V^{2+}$, $V^{3+}$, $V^{4+}$, and $V^{5+}$; and (c) Ti, $Ti^{2+}$, and $Ti^{4+}$.

**18.50** Write electron configurations for the following: (a) Au, $Au^+$ and $Au^{3+}$; (b) Ni, $Ni^{2+}$, and $Ni^{3+}$; and (c) Mn, $Mn^{2+}$, $Mn^{4+}$, and $Mn^{7+}$.

**18.51** Name the following coordination compounds:

**18.52** Name the following coordination compounds:

(a)

(b) $K_3$

(c)

**18.53** Write a balanced equation for the reaction catalyzed by the enzyme superoxide dismutase. Make a sketch of the coordination environment around the two metal atoms.

**18.54** In superoxide dismutase, it is the $Cu^{2+}$ centre that oxidizes $O^{2-}$. Why is copper more suitable than the $Zn^{2+}$ centre for the role of the oxidizing agent in SOD?

**18.55** The $Cu^{2+}$ ion forms tetrahedral complexes with some anionic ligands. When $CuSO_4 \cdot 5H_2O$ dissolves in water, a blue solution results. The addition of aqueous KF solution results in a green precipitate, but the addition of aqueous KCl results in a bright green solution. Identify each green species and write chemical reactions for these processes.

**18.56** The $Cu^{2+}$ ion forms tetrahedral complexes with some anionic ligands. When $CuSO_4 \cdot 5H_2O$ dissolves in water, a blue solution results. The addition of aqueous KCN solution at first results in a white precipitate, but the addition of more KCN causes the precipitate to dissolve. Identify the precipitate and the dissolved species and write chemical reactions for these processes.

**18.57** Both vanadium and silver are lustrous silvery metals. Suggest why silver is widely used for jewellery, but vanadium is not.

**18.58** Titanium is nearly 100 times more abundant in the Earth's crust than copper; yet copper was exploited as a metal in antiquity, and titanium has found applications only in recent times. Explain.

**18.59** Draw a ball-and-stick model of the *mer* isomer of $[NiCl_3F_3]^{4-}$, oriented so that the fluoride ions are in the top, bottom, and left forward positions.

**18.60** Draw a ball-and-stick model of the *fac* isomer of $[NiCl_3F_3]^{4-}$, oriented so that the fluoride ions are in the top and two rear positions.

**18.61** What is the chemical name and the formula of the black tarnish that accumulates on objects made of silver? Write a balanced equation that shows how this black tarnish forms.

**18.62** What is the chemical name and the composition of the green tarnish that accumulates on objects made of copper? Write balanced equations that show how this green tarnish forms.

**18.63** As ligands, chloride and cyanide are at opposite ends of the spectrochemical series. Nevertheless, experiments show that $[CrCl_6]^{3-}$ and $[Cr(CN)_6]^{3-}$ have about the same amount of magnetism. Explain how this can be so.

**18.64** In octahedral geometry, the amount of magnetism in $d^4-d^7$ coordination compounds of the $3d$ transition metals depends on the ligands and the oxidation state of the metal. On the other hand, $d^1-d^3$ and $d^8-d^9$ compounds always have approximately the same amount of magnetism. Explain.

**18.65** The $d_{z^2}$ and $d_{x^2-y^2}$ orbitals have the same stability in an octahedral complex. However, in a square planar complex, the $d_{x^2-y^2}$ orbital is much less stable than the $d_{z^2}$ orbital. Use orbital sketches to explain the difference.

**18.66** Use orbital sketches to explain why the $d_{xz}$ and $d_{x^2-y^2}$ orbitals have different stabilities in octahedral coordination complexes.

## ADDITIONAL PROBLEMS

**18.67** Consider the chromium (III) complexes $[Cr(NH_3)_5L]^{n+}$, where $L$ is $Cl^-$, $H_2O$, or $NH_3$. (a) What is the value of $n^+$ for each of them? (b) The compounds absorb at 515 nm, 480 nm, and 465 nm, respectively. Predict the colours of the three complexes; and (c) calculate the crystal field splitting energy in kJ/mol for each of them and explain the trend in values.

**18.68** The complex ion $[Ag(NH_3)_2]^+$ has linear geometry. Predict the crystal field splitting diagram for this complex. Place the ligands along the $z$-axis.

**18.69** Tetracarbonylnickel(0) is $[Ni(CO)_4]$, and tetracyanozinc(II) is $[Zn(CN)_4]^{2-}$. Predict the geometry and colour of each complex.

**18.70** Carbon monoxide and tungsten form an octahedral complex. The molecule is colourless, even though W(0) is $d^6$. Sketch the ligand field diagram for hexacarbonyltungsten(0), and suggest why the complex is colourless.

**18.71** Some researchers use the term "the brass enzyme" to describe superoxide dismutase. Can you suggest a reason for this nickname?

**18.72** Blue copper proteins are blue when they contain $Cu^{2+}$ but colourless as $Cu^+$ compounds. The colour comes from an interaction in which a photon causes an electron to transfer from a sulphur lone pair on a cysteine ligand to the copper centre. Why does this charge transfer interaction occur for $Cu^{2+}$ but not $Cu^+$?

**18.73** The complex $[Ni(CN)_4]^{2-}$ is diamagnetic, but $[NiCl_4]^{2-}$ is paramagnetic. Propose structures for the two complexes and explain why they have different magnetic properties.

**18.74** Predict whether each of the following complexes is high-spin or low-spin: (a) $[Fe(CN)_6]^{4-}$; (b) $[MnCl_4]^{2-}$; (c) $[Rh(NH_3)_6]^{3+}$; and (d) $[Co(H_2O)_6]^{2+}$.

**18.75** The iron storage protein ferritin usually is neither empty of iron nor filled to capacity. Why is this situation advantageous for an organism?

**18.76** Explain how liquid mercury can be used to purify metals such as gold and silver.

**18.77** One of the most common approaches to the investigation of metalloproteins is to replace the naturally occurring metal ion with a different one that has a property advantageous for chemical studies. For example, zinc proteins are often studied by visible spectroscopy after $Co^{2+}$ has been substituted for $Zn^{2+}$. Explain, using crystal field energy-level diagrams, why $Co^{2+}$ is a better metal than $Zn^{2+}$ for visible spectroscopy.

**18.78** Draw a molecular picture of a surfactant coating a mineral particle in a flotation process. (See Chapter 9 for a review of surfactants.)

**18.79** In Zn purification by electrochemistry, ZnO is dissolved in sulphuric acid. Before deposition, zinc powder is added to displace less active metals, such as cadmium. Use $E°$ values and balanced equations to show how this is accomplished.

**18.80** Determine the Lewis structure and draw a ball-and-stick model showing the geometry of the dichromate anion, which contains one bridging oxygen atom.

## GROUP STUDY PROBLEMS

**18.81** Oxyhaemoglobin is bright red, but deoxyhaemoglobin has more of a blue colour. In both cases, the iron is in the +2 oxidation state. Give a detailed explanation for the difference in colour. How would you test your hypothesis? Based on your explanation, what colour would you predict for a sample of blood that is saturated with carbon monoxide?

**18.82** A portion of the absorption spectrum of a complex ion, $[Cr(H_2O)_4Cl_2]^+$, is represented by the following graph:

(a) Estimate the crystal field splitting energy $\Delta$ (in kilojoules per mole).

(b) What colour is the complex?

(c) Name the complex cation.

(d) Draw all possible isomers of the complex. (e) Draw the crystal field energy-level diagram and show the electronic transition that gives the complex its colour.

**18.83** On Earth, two main-group metals, Al and Pb, are used when low- and high-density metals are desired. Suppose you are transported to a planet elsewhere in our galaxy, where all transition metals are readily available but post-transition metals are rare. Where among the transition metals would you seek a replacement for Al for low-density uses, and where would you seek a replacement for Pb for high-density uses? Explain your reasoning.

**18.84** Often, a detailed figure can give an informative summary of a complex concept. The figure in the Focus 15-1: Chemistry and the Environment summarizing the carbon cycle is one example. Design a figure that summarizes the transport and storage of oxygen by haemoglobin and myoglobin.

**18.85** In the 1890s, Alfred Werner prepared several platinum complexes that contained both ammonia and chlorine. He determined the formulas of these species by precipitating the chloride ions with $Ag^+$. The empirical formulas and number of chloride ions that precipitate per formula unit follow:

| Empirical Formula | No. of Cl⁻ ions |
|---|---|
| $PtCl_4 \cdot 2NH_3$ | 0 |
| $PtCl_4 \cdot 3NH_3$ | 1 |
| $PtCl_4 \cdot 4NH_3$ | 2 |
| $PtCl_4 \cdot 5NH_3$ | 3 |
| $PtCl_4 \cdot 6NH_3$ | 4 |

Determine the molecular formulas, name these compounds, and draw the structures of the platinum complexes.

**18.86** Design a flowchart that summarizes the metallurgy of copper from chalcopyrite.

## ANSWERS TO SECTION EXERCISES

**18.1.1** (a) W, $T_m$ increases down a column; (b) W, $T_m$ decreases after Column 6; (c) Pd, $T_m$ decreases after Column 6; and (d) Nb, $T_m$ increases up to Column 6.

**18.1.2** (a) W, density increases with Z; (b) Os, density increases with Z; (c) Pd, density decreases at the end of a row; and (d) Nb, density increases with Z.

**18.1.3** (a) Ta(V); (b) Fe(III); (c) Rh(III); (d) Cr(VI); and (e) Cu(II).

**18.2.1** (a) −3, Fe(III), coordination number = 6; (b) 0, V(II), coordination number = 6; (c) +2, Ni(II), coordination number = 4.

**18.2.2**

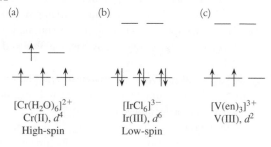

**18.2.3** (a) pentaammineiodochromium(III) sulphate;
  (b) potassium hexachloroplatinate(II);
  (c) *cis*-tetracarbonyldichloroiron(II); and
  (d) hexaaquairon(II) chloride

**18.3.1**

(a)    (b)    (c)

[Cr(H₂O)₆]²⁺    [IrCl₆]³⁻    [V(en)₃]³⁺
Cr(II), $d^4$    Ir(III), $d^6$    V(III), $d^2$
High-spin    Low-spin

(d)

[NiCl₄]²⁻
Ni(II), $d^8$

**18.3.2** Because cyanide is near the top of the spectrochemical series, it generates a relatively large energy gap between the two sets of $d$ orbitals. The hexacyano complexes are yellow because they absorb high-energy indigo light. The corresponding aqua complexes have a much smaller crystal field splitting energy. They absorb orange or red light, thus appearing blue or green.

**18.3.3** 492 nm, orange

**18.4.1**

**18.4.2** Colours in coordination compounds come from electronic transitions, either from a lower energy $d$ orbital to a higher energy $d$ orbital, or from a ligand to an empty $d$ orbital (charge transfer). In zinc proteins, $Zn^{2+}$ ion has a $d^{10}$ configuration with all the $d$ orbitals filled. Thus, electronic transitions in the visible portion of the spectrum are impossible.

**18.5.1**

**18.5.2** The standard reduction potentials are $-0.744$ V for $Cr^{3+}$ and $-0.7618$ V for $Zn^{2+}$, suggesting that these oxides should be equally easily reduced. However, $Cr^{2+}$ has a potential of $-0.913$ V, so reduction of chromium(III) oxide using coke stops at $Cr^{2+}$.

**18.5.3**

$$3\,Fe_2O_3 + CO \longrightarrow 2\,Fe_3O_4 + CO_2; \text{ reducing agent is CO}$$

$$Fe_3O_4 + CO \longrightarrow 3\,FeO + CO_2; \text{ reducing agent is CO}$$

$$FeO + CO \longrightarrow Fe + CO_2; \text{ reducing agent is CO}$$

$$CO_2 + C \longrightarrow 2\,CO; \text{ reducing agent is C}$$

$$C + O_2 \longrightarrow CO_2; \text{ reducing agent is C}$$

## TABLE: BIOLOGICAL ROLES OF IRON

| Protein | Function | Geometry | Coordination |
|---|---|---|---|
| Haemoglobin | O₂ transport | Octahedral and square pyramid | 4 N (porphyrin), I N (His), O₂ |
| Myoglobin | O₂ storage | Octahedral and square pyramid | 4 N (porphyrin), I N (His), O₂ |
| Transferrin | Fe transport | Octahedral | Not specified in text |
| Ferritin | Fe storage | Not specified in text | Fe₂O₃·H₂O |
| Cytochromes | Electron transfer | Octahedral | 4 N (porphyrin), |
| Cytochromes | Electron transfer | Octahedral | 2 N (His) or I N (His), I S (Met) |
| Fe–S Proteins | Electron transfer | Tetrahedral | 4 S (Cys) or 2 S (Cys), |
| Fe–S Proteins | Electron transfer | Tetrahedral | 2 S (sulphide) |

**18.6.1** Iron pipes are coated with zinc to prevent corrosion, zinc being easier to oxidize than iron. Copper is much more difficult to oxidize than either zinc or iron. Thus, copper makes better piping than galvanized iron because it is much less susceptible to corrosion. Standard reduction potentials show this trend:

$$Cu^{2+} + 2\,e^- \longrightarrow Cu \qquad E^o = +0.3419\ V$$

$$Fe^{2+} + 2\,e^- \longrightarrow Fe \qquad E^o = -0.447\ V$$

$$Zn^{2+} + 2\,e^- \longrightarrow Zn \qquad E^o = -0.7618\ V$$

**18.6.2** Trace oxygen in the gas reacts with $Cr^{2+}$:

$$4\,Cr^{2+}\,(aq) + O_2\,(g) + 4\,H_3O^+\,(aq)$$
$$\longrightarrow 4\,Cr^{3+}\,(aq) + 6\,H_2O\,(l)$$

Zinc reduces $Cr^{3+}$ back to $Cr^{2+}$:

$$Zn\,(s) + 2\,Cr^{3+}\,(aq) \longrightarrow Zn^{2+}\,(aq) + 2\,Cr^{2+}\,(aq)$$

**18.6.3** Metals with large positive standard reduction potentials will not electrolyze and will remain in the anodic residue: Au, Hg, Pt, Ag.

## ANSWERS TO PRACTICE EXERCISES

**18.1**

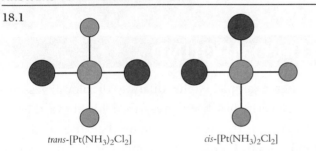

*trans*-[Pt(NH$_3$)$_2$Cl$_2$]      *cis*-[Pt(NH$_3$)$_2$Cl$_2$]

**18.2** (a) pentaamminecarbonatocobalt(III) nitrate;
(b) triamminetricarbonylmolybdenum(0); and
(c) pentaaquahydroxoaluminum chloride

**18.3** (a) [Ru(bipy)$_3$]Cl$_2$;
(b) K$_2$[PtCl$_6$];
(c) [Fe(H$_2$O)$_5$O]$^+$

**18.4**

$$\underline{\quad}\ \underline{\quad}\ e_g$$
$$\uparrow\ \uparrow\ \underline{\quad}\ t_{2g}$$

**18.5**

$$\underline{\quad}\ \underline{\quad}\ e_g$$

$\uparrow\quad\uparrow\ e_g$

$\uparrow\ \uparrow\ \uparrow\ t_{2g}$     $\uparrow\!\downarrow\ \uparrow\!\downarrow\ \uparrow\ t_{2g}$

[Mn(H$_2$O)$_6$]$^{2+}$     [Re(H$_2$O)$_6$]$^{2+}$
$\Delta < P$     $\Delta > P$
High-spin     Low-spin
$t_{2g}^3 e_g^2$     $t_{2g}^5 e_g^0$

**18.6** [Cr(H$_2$O)$_6$]$^{2+}$ < [Cr(NH$_3$)$_6$]$^{3+}$ < [Mo(H$_2$O)$_6$]$^{3+}$ < [Mo(NH$_3$)$_6$]$^{3+}$

**18.7** 708 nm; green

# THE MAIN GROUP ELEMENTS

## NEW DISCOVERIES ABOUT AN OLD COMPOUND

One might think that everything had long since been learned about a molecule as simple as nitrogen oxide (NO). After all, Joseph Priestley made the compound for the first time more than 200 years ago. Yet, recent research reveals that NO plays intriguing roles in a number of biochemical processes. Many animals, including humans and fruit flies, barnacles, and trout, synthesize NO. This little molecule plays roles in nerve function, regulation of blood pressure, blood clotting, and immune system responses. In many of these processes, NO is produced in one part of an organism and moves to another part, where its chemical properties trigger a biochemical reaction.

Nitrogen oxide helps to adjust the interactions between blood and blood vessels. One example is blood clotting. When the wall of a blood vessel is cut, an enzyme in the cell wall sends chemical signals that interact with blood platelets, causing clotting. Studies have shown that NO inhibits this process, an effect exploited by mosquitoes, the saliva of which contains NO-generating chemicals that prevent their victims' blood from clotting while the mosquito feeds.

Another example is the dilation of blood vessels. Nitroglycerin has been prescribed for more than a century as a treatment of angina because it dilates blood vessels, thereby decreasing blood pressure and increasing blood flow. A century later, it was discovered that NO is involved in the mechanism of this process. More recently, relief for men suffering from impotence has arrived in the form of Viagra®, a drug that also results in dilated blood vessels. Nitric oxide binds to the receptors of the enzyme guanylate cyclase, resulting in elevated levels of cyclic guanosine monophosphate (cGMP), which leads to smooth muscle relaxation. Viagra's molecular structure is similar to that of cGMP, and effectively competes with the enzyme cGMP-specific phosphodiesterase type 5 (PDE5), which is responsible for the degradation of cGMP. Less degradation of cGMP results in more muscle relaxation and more blood flow. Our background shows the Viagra® molecule conjugated with the PDE5 enzyme.

There is evidence that the neural processes leading to memory include the production and diffusion of NO. An enzyme in the brain produces NO from an amino acid. How does NO interact with more complex molecules as the brain functions? One possibility is that once a neural trigger fires, an adjacent neuron produces NO, which diffuses back to the original neuron and stimulates further activity. This could be how long-term memory develops.

Chemical reactions display a near-infinite variety, and the story of nitrogen oxide illustrates that there are still exciting discoveries to be made in chemistry. Many of these are likely to involve the elements in the *p* block of the periodic table, because the *p*-block elements are remarkably diverse. Aluminum, one of the most reactive metals, is a member of this group. So, too, are neon and argon, noble gases that are chemically inert. The elements of Groups 14–16 include non-metals (C, N, O, P, S), metals (Sn, Pb, Bi), and metalloids (Si, Ge, Te). In this chapter, we describe selected features of this diverse chemistry. The chapter opens with sections describing a system for organizing chemical reactions, including those of *p*-block elements. Then, we survey metals in the *p* block, metalloids, carbon, phosphorus, sulphur, and the halogens.

## CHAPTER CONTENTS

## LEARNING OBJECTIVES

*Upon completion of this chapter you should be able to*

- explain the chemistry of formation of adducts
- apply the concepts of hardness and softness to reactions
- explain the production, reactions, and uses of the main group metals
- explain the production, reactions, and uses of the metalloids
- explain the production, reactions, and uses of phosphorus
- explain the production, reactions, and uses of the other non-metals

## 19.1 LEWIS ACIDS AND BASES

Because chemical behaviour can be bewildering in its complexity, chemists search for general ways to organize chemical reactivity patterns. Two familiar patterns are Brønsted–Lowry acid–base (proton transfer) and redox (electron transfer) reactions. A related pattern of reactivity can be viewed as the donation of a pair of electrons to form a new bond. One example is the reaction between gaseous ammonia and trimethylboron, in which the ammonia molecule uses its non-bonding pair of electrons to form a bond between nitrogen and boron. The structures are drawn using the dash and wedge bonds introduced in Section 10.4:

$$NH_3 + B(CH_3)_3 \longrightarrow H_3NB(CH_3)_3$$

Viewed from the perspective of electrons, this reaction is similar to the transfer of a proton to an ammonia molecule:

In both cases, the nitrogen atom uses its pair of non-bonding electrons to make a new covalent bond. This similarity led G. N. Lewis to classify ammonia as a base in its reaction with $B(CH_3)_3$ as well as in its reaction with $H_3O^+$. Whereas the Brønsted–Lowry definition focuses on proton transfer, the Lewis definition of acids and bases focuses on electron pairs:

> *Any chemical species that acts as an electron-pair donor is a Lewis base.*
> *Any chemical species that acts as an electron-pair acceptor is a Lewis acid.*

Ammonia is a prime example of a **Lewis base**. In addition to its three N—H bonds, this molecule has a lone pair of electrons on its nitrogen atom, as Figure 19-1 shows. Although all of the valence orbitals of the nitrogen atom in $NH_3$ are occupied, the non-bonding pair can form a fourth covalent bond with a bonding partner that has a vacant valence orbital available.

Trimethylboron is an example of one type of **Lewis acid**. This molecule has trigonal planar geometry in which the boron atom is $sp^2$ hybridized with a vacant $2p$ orbital perpendicular to the plane of the molecule (Figure 19-1). Recall from Chapter 6 that atoms tend to use all their valence $s$ and $p$ orbitals to form covalent bonds. Second-row elements such as boron and nitrogen are most stable when surrounded by eight valence electrons divided among covalent bonds and lone pairs. The boron atom in $B(CH_3)_3$ can use its vacant $2p$ orbital to form a fourth covalent bond to a new partner, provided that the new partner supplies both

**Figure 19-1**

Nonbonding $sp^3$ pair

Vacant valence $2p$ orbital

Ammonia, which has a pair of non-bonding valence electrons, is a typical Lewis base. Trimethylboron, which has a vacant valence orbital, represents one type of Lewis acid.

electrons. Trimethylboron is a Lewis acid because it forms an additional bond by accepting a pair of electrons from some other chemical species.

## Formation of Lewis Acid–Base Adducts

The simplest type of Lewis acid–base reaction is the combination of a Lewis acid and a Lewis base to form a compound called an **adduct**. The reaction of ammonia and trimethylboron is an example. A new bond forms between boron and nitrogen, with both electrons supplied by the lone pair of ammonia (see Figure 19-2). Forming an adduct with ammonia allows boron to use all of its valence orbitals to form covalent bonds. As this occurs, the geometry about the boron atom changes from trigonal planar to tetrahedral, and the hybrid description of the boron valence orbitals changes from $sp^2$ to $sp^3$.

A general equation summarizes the formation of Lewis acid–base adducts:

$$A \ + \ :B \longrightarrow A-B$$

| Lewis | Lewis | Adduct |
| acid | base | |

Here are two more examples of adduct formation:

$$SiF_4 + :F^- \longrightarrow SiF_5^- \qquad AlCl_3 + :PCl_3 \longrightarrow Cl_3Al-PCl_3$$

Acid　Base　Adduct　　　Acid　Base　Adduct

The reaction between two $PCl_5$ molecules demonstrates two additional features of Lewis acid–base reactions. First, substances such as $PCl_5$ can act as both Lewis acids and Lewis bases. The chlorine atoms in $PCl_5$ have electron pairs to donate, imparting Lewis base properties to the molecule. The phosphorus atom in $PCl_5$, the bonding of which can be described using $sp^3 d$ hybrid orbitals, has additional vacant valence $d$ orbitals. One of these orbitals can form an additional bond by accepting an electron pair, resulting in $sp^3 d^2$ hybridization. This gives $PCl_5$ Lewis acid properties. Second, in addition to formation of a new P—Cl bond, the reaction between two $PCl_5$ molecules involves P—Cl bond breaking. A more detailed view of the process uses curved arrows to indicate shifts of electrons.

**Figure 19-2**

The N—B bond between $NH_3$ and $B(CH_3)_3$ can be described by the overlap of $sp^3$ orbitals on each atom, with the two electrons supplied by the lone pair of the N atom.

PCl₅ + PCl₅ ⟶ PCl₆⁻ + PCl₄⁺

## Recognizing Lewis Acids and Bases

A Lewis base must have valence electrons available for bond formation. Any molecule that has a Lewis structure showing non-bonding electrons can act as a Lewis base. Ammonia, phosphorus trichloride, and dimethyl ether, each of which contains lone pairs, are Lewis bases. Anions can also act as Lewis bases. In the first example of adduct formation above, the fluoride ion, with eight valence electrons in its $2s$ and $2p$ orbitals, acts as a Lewis base.

A Lewis acid must be able to accept electrons to form a new bond. Because bond formation can occur in several ways, compounds with several different structural characteristics can act as Lewis acids. Nevertheless, most Lewis acids fall into the following categories:

1. *A molecule that has vacant valence orbitals.* A good example is $B(CH_3)_3$, which uses a vacant $2p$ orbital to form an adduct with ammonia. The elements in the $p$ block beyond the second row of the periodic table have empty valence $d$ orbitals that allow them to act as Lewis acids. The silicon atom in $SiF_4$ is an example.

2. *A molecule with delocalized $\pi$ bonds involving oxygen.* Examples are $CO_2$, $SO_2$, and $SO_3$. Each of these molecules can form a $\sigma$ bond between its central atom and a Lewis base, at the expense of a $\pi$ bond. For example, the hydroxide anion, a good Lewis base, attacks the carbon atom of $CO_2$ to form hydrogen carbonate:

Lewis acid    Lewis base        Hydrogen carbonate adduct

In this reaction, the oxygen atom of the hydroxide ion donates a pair of electrons to make a new C—O bond. Because all the valence orbitals of the carbon atom in $CO_2$ are involved in bonding to oxygen, one of the C—O $\pi$ bonds must be broken to make an orbital available to overlap with the occupied orbital of the hydroxide anion.

3. *A metal cation.* Removing electrons from a metal atom always generates vacant valence orbitals. As described in Chapter 18, many transition metal cations form complexes with ligands in aqueous solution. In these complexes, the ligands act as Lewis bases, donating pairs of electrons to form metal–ligand bonds. The metal cation accepts these electrons, so it acts as a Lewis acid. Metal cations from the $p$ block also act as Lewis acids. For example, $Pb^{2+}$ $(aq)$ forms a Lewis acid–base adduct with four $CN^-$ anions, each of which donates a pair of electrons:

$$Pb^{2+}\,(aq) + 4\,CN^-\,(aq) \longrightarrow [Pb(CN)_4]^{2-}\,(aq)$$

Each of the four lead–carbon bonds forms by the overlap of an empty valence orbital on the metal ion with a lone pair on a carbon atom.

Example 19-1 provides practice in recognizing Lewis acids and bases.

## EXAMPLE 19-1 LEWIS ACIDS AND BASES

Identify the Lewis acids and bases in each of the following reactions and draw structures of the resulting adducts:

(a) $AlCl_3 + Cl^- \longrightarrow AlCl_4^-$

(b) $Co^{3+} + 6\ NH_3 \longrightarrow [Co(NH_3)_6]^{3+}$

(c) $SO_2 + OH^- \longrightarrow HSO_3^-$

**Strategy:** Every Lewis base has one or more lone pairs of valence electrons. A Lewis acid can have vacancies in its valence shell, or it can sacrifice a $\pi$ bond to make a valence orbital available for adduct formation. To decide whether a molecule or ion acts as a Lewis acid or base, examine its Lewis structure for these features. Review Section 6.2 for procedures used to determine Lewis structures.

**Solution:** On the Lewis structures, we show possible transfers of electron pairs with curved arrows.

(a) Both species of this pair contain chlorine atoms with lone pairs, so either might act as a Lewis base if a suitable Lewis acid is present. The aluminum atom of $AlCl_3$ has a vacant $3p$ orbital perpendicular to the molecular plane. The empty $p$ orbital accepts a pair of electrons from the $Cl^-$ anion to form the fourth Al—Cl bond. The Lewis acid is $AlCl_3$, and the Lewis base is $Cl^-$.

Acid          Base          Adduct

(b) As already noted, ammonia is a Lewis base because it has a donor pair of electrons on the nitrogen atom. Like other transition metal cations, $Co^{3+}$ is a Lewis acid. The cation uses vacant $3d$ orbitals to form bonds to $NH_3$:

Acid          Base          6 times          Adduct

(c) Sulphur dioxide has delocalized $\pi$ bonds, indicating Lewis acidity. The sulphur atom of $SO_2$ has a set of $3d$ orbitals that can be used to form an adduct. In this case, the hydroxide ion acts as a Lewis base. The anion uses one lone pair of electrons to form a new bond to sulphur:

Acid          Base          Adduct

 **Does the Result Make Sense?** The Lewis structures verify that each species identified as a Lewis base possesses lone pairs of electrons and that the Lewis acids have orbitals available to accept electrons.

**Common Pitfalls:** Drawing incorrect Lewis structures. Review the steps in Section 6.2 if necessary.

### Practice Exercise 19.1

Identify the Lewis acid and Lewis base, and draw the reaction showing electron transfers, in the formation of the aquo complex $[Ni(H_2O)_6]^{2+}$.

**SECTION EXERCISES**

● **19.1.1** Draw the Lewis structures of each of the reactants in the following reactions:

(a) $SO_3 + OH^- \longrightarrow HSO_4^-$

(b) $SnCl_2 + Cl^- \longrightarrow [SnCl_3]^-$

(c) $AsF_3 + SbF_5 \longrightarrow [AsF_2]^+ + [SbF_6]^-$

● **19.1.2** Identify the Lewis acid and the Lewis base in each of the reactions that appears in Section Exercise 19.1.1.

● **19.1.3** Draw structures that show the donation of electrons that takes place in each reaction in Section Exercise 19.1.1 and draw the Lewis structures of the products.

## 19.2 HARD AND SOFT LEWIS ACIDS AND BASES

Many years ago, geochemists recognized that whereas some metallic elements are found as sulphides in the Earth's crust, others are usually encountered as oxides, chlorides, or carbonates. Copper, lead, and mercury are most often found as sulphide ores; Na and K are found as their chloride salts; Mg and Ca exist as carbonates; and Al, Ti, and Fe are all found as oxides. Today chemists understand the causes of this differentiation among metal compounds. The underlying principle is how tightly an atom binds its valence electrons. The strength with which an atom holds its valence electrons also determines the ability of that atom to act as a Lewis base, so we can use the Lewis acid–base model to describe many affinities that exist among elements. This notion not only explains the natural distribution of minerals, but also can be used to predict patterns of chemical reactivity.

## Polarizability

Polarizability, described in Chapter 8, is a measure of how tightly electrons are bound to an atom or molecule. The polarizability of an atom, ion, or molecule is the ease with which its electron cloud can be distorted by an electrical charge. An electron cloud is polarized toward a positive charge and away from a negative charge, as shown in Figure 19-3. Pushing the electron cloud to one side of an atom causes a polarization of charge. The side with the concentrated electron density builds up a small negative charge; the protons in the nucleus give the opposite side a small positive charge.

Polarizability shows periodic variations that correlate with periodic trends in how tightly valence electrons are bound to the nucleus:

1. Polarizability decreases from left to right in any row of the periodic table. As the effective nuclear charge ($Z_{eff}$) increases, the nucleus holds the valence electrons more tightly.

**Figure 19-3**

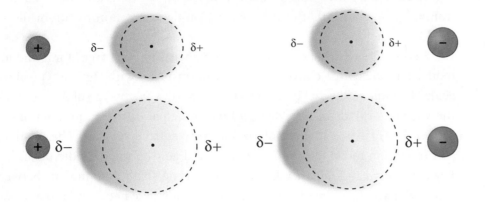

The electron cloud of an atom is polarized toward a positive charge and away from a negative charge. A smaller atom (top figures) binds its valence electrons more tightly and is less polarizable than a larger atom (bottom figures), which binds its electrons loosely.

2. Polarizability increases from top to bottom in any column of the periodic table. As the principal quantum number ($n$) increases, the valence orbitals become larger. This reduces the net attraction between valence electrons and the nucleus.

## The Hard–Soft Concept

Lewis acids and bases can be organized according to their polarizability. If polarizability is low, the species is categorized as "hard." If polarizability is high, the species is "soft."

A **hard Lewis base** has electron pairs of low polarizability. This characteristic correlates with high electronegativity. Fluoride, the anion of the most electronegative element, is the hardest Lewis base because it contains a small, dense sphere of negative charge. Molecules and ions that contain oxygen or nitrogen atoms are also hard bases, although not as hard as fluorine. Examples include $H_2O$, $CH_3OH$, $OH^-$, $NH_3$, and $H_2NCH_3$.

A **soft Lewis base** has a large donor atom of high polarizability and low electronegativity. Iodide ion has its valence electrons in large $n = 5$ orbitals, making this anion highly polarizable and a very soft base. Other molecules and polyatomic anions with donor atoms from Rows 3 to 6 are also soft bases. To summarize, the donor atom becomes softer from top to bottom of a column of the periodic table.

A **hard Lewis acid** has an acceptor atom with low polarizability. Most metal atoms and ions are hard acids. In general, the smaller the ionic radius and the larger the charge, the harder the acid. The $Al^{3+}$ ion, with an ionic radius of only 67 pm, is a prime example of a hard Lewis acid. The nucleus exerts a strong pull on the compact electron cloud, giving the ion very low polarizability.

Group 17

| F | Hardest |
| Cl | |
| Br | |
| I | |
| At | Softest |

Hard          Soft

The designation of hard acids is not restricted to metal cations. For example, in $BF_3$, the small boron atom in its +3 oxidation state is bonded to three highly electronegative fluorine atoms. All the B—F bonds are polarized away from a boron centre that is already electron-deficient. Boron trifluoride is a hard Lewis acid.

A **soft Lewis acid** has a relatively high polarizability. Large atoms and low oxidation states often convey softness. Contrast $Al^{3+}$ with $Hg^{2+}$, a typical soft acid. The ionic radius of $Hg^{2+}$ is 116 pm, almost twice the size of $Al^{3+}$, because the valence orbitals of $Hg^{2+}$ have a high principal quantum number, $n = 6$. Consequently, $Hg^{2+}$ is a highly polarizable, very soft Lewis acid. The relatively few soft transition metal ions are located around gold in the periodic table.

The terms *hard* and *soft* are relative, so there is no sharp dividing line between the two, and many Lewis acids and bases are intermediate between hard and soft. Example 19-2 shows how to categorize Lewis acids and bases according to their hard–soft properties.

---

## EXAMPLE 19-2 RANKING HARDNESS AND SOFTNESS

Rank the following groups of Lewis acids and bases from softest to hardest: (a) $H_2S$, $H_2O$, and $H_2Se$; (b) $Fe^0$, $Fe^{3+}$, and $Fe^{2+}$; and (c) $BCl_3$, $GaCl_3$, and $AlCl_3$.

| 5 B | | 7 N | 8 O | |
|---|---|---|---|---|
| 13 Al | | 15 P | 16 S | |
| 31 Ga | | | 34 Se | |
| | | | | |
| | | | | |

**Strategy:** The first task is to decide whether the members of a given group are Lewis acids or bases. Then evaluate the relative softness and hardness based on polarizability, taking into account correlations with electronegativity, size, and charge. Refer to the periodic table in assessing the trends.

**Solution:**

(a)

These three molecules have lone pairs, so they are Lewis bases. The central atoms are in the same column of the periodic table, so their polarizability and the softness of the molecules increase moving down the column. Thus, $H_2Se$ is softer than $H_2S$, which is softer than $H_2O$.

(b) $Fe^0$  $Fe^{2+}$  $Fe^{3+}$

Metal atoms and cations are Lewis acids. As valence electrons are removed from a metal atom, the remaining electron cloud undergoes an ever-larger pull from the nuclear charge. This decreases the size of the ion as well as its polarizability. Thus, $Fe^0$ is softer than $Fe^{2+}$, which is softer than $Fe^{3+}$.

(c)

These three molecules have trigonal planar geometry with $sp^2$-hybridized central atoms. Each has a vacant valence $p$ orbital perpendicular to the molecular plane, making the molecules Lewis acids. The size, polarizability, and softness of the central acceptor atom increase going down the column. A gallium atom is larger and more polarizable than an aluminum atom. Thus, $GaCl_3$ is softer than $AlCl_3$, which is softer than $BCl_3$. We have already noted that aluminum is hard. Thus, gallium trichloride is a soft Lewis acid, whereas $AlCl_3$ and $BCl_3$ are both hard.

**Does the Result Make Sense?** Hardness and softness are determined by atomic size, and the order of hardness in each of these sets is consistent with the size trends among the species.

**Common Pitfalls:** Not knowing the periodic trends in atomic sizes. Review Section 5.4 if necessary.

**Practice Exercise 19.2**

Identify $H_2O$, $NH_3$, and $PH_3$ as Lewis acids or Lewis bases, and rank in order of increasing hardness.

## The Hard–Soft Acid–Base Principle

The concept of hard and soft acids and bases can be used to interpret many trends in chemical reactivity. These trends are summarized in the hard–soft acid–base principle (HSAB principle), an empirical summary of results collected from many chemical reactions studied through decades of research:

> *Hard Lewis acids tend to combine with hard Lewis bases.*
> *Soft Lewis acids tend to combine with soft Lewis bases.*

The geochemical distribution of metals conforms to the HSAB principle. Metals that form hard acid cations have strong affinities for hard bases such as oxide, fluoride, and chloride. Most metal ions that are hard acids are found bonded to the oxygen atoms of various silicate anions. These elements are concentrated in the Earth's mantle. Hard acid metals also occur in combinations with other hard bases, including oxides or, less often, halides, sulphates, and carbonates. Examples include rutile ($TiO_2$), limestone ($CaCO_3$), gypsum ($CaSO_4$), and sylvite (KCl).

Metals that are soft acids, such as gold and platinum, have low affinities for hard oxygen atoms, so they are not affected by $O_2$ in the atmosphere. Consequently, these metals, including Ru, Rh, Pd, Os, Ir, Pt, and Au, are found in the crust of the Earth in their elemental form.

Other soft metals occur in nature as sulphides. The sulphide anion is a soft base, so this category includes some soft acids and many intermediate cases. Soft acids may also occur either in elemental form or as arsenide or telluride minerals. Examples include galena (PbS), cinnabar (HgS), chalcopyrite ($CuFeS_2$), argentite ($Ag_2S$), calaverite ($AuTe_2$), and sperrylite ($PtAs_2$).

The Group 13 elements illustrate the trend in hardness and softness among Lewis acids. At the top of the group, boron is a hard Lewis acid, and so is aluminum. Moving down the column, the valence orbitals increase in size and polarizability. Thus, gallium is a borderline acid, and indium is soft. The order of reactivity for the trihalides of these elements depends on the Lewis base. For a hard base such as ammonia, the reactivity trend for adduct formation is $BCl_3 > AlCl_3 > GaCl_3 > InCl_3$, reflecting the preference that the hard base has for the harder acid. The order is reversed when the trihalides form adducts with dimethyl-sulphide (($CH_3)_2S$), a soft base.

## Metathesis Reactions

A **metathesis reaction** is an exchange of bonding partners. Lewis acids and bases often undergo such exchanges, as exemplified by the following reaction:

$$BI_3 + GaF_3 \longrightarrow BF_3 + GaI_3$$

In this reaction, boron and gallium trade their halogen partners. According to the HSAB principle, metathesis reactions proceed in the direction that couples the harder acid with the harder base. Here, the sizes of the atoms reflect the fact that boron is harder than gallium, and fluoride is harder than iodide. The iodine atoms are significantly crowded in $BI_3$ but much more easily accommodated in $GaI_3$, because gallium is a larger atom than boron.

Metathesis is an important method for making molecules that have boron–carbon and aluminum–carbon bonds. These organoboron and organoaluminum species are valuable reagents in organic chemistry. Such compounds are often synthesized using alkyllithium reagents, powerful Lewis bases that are used widely in chemistry as sources of carbon atoms. Although highly reactive (they ignite spontaneously in air), alkyllithium reagents are available commercially, and they are easy to prepare in the laboratory by reacting alkyl halides with lithium metal:

$$ClCH_2CH_3 + 2\,Li \longrightarrow LiCH_2CH_3 + LiCl$$

Aluminum halides and boron halides undergo metathesis reactions with alkyllithium reagents according to the HSAB principle. Both boron and aluminum are softer acids than lithium, and carbon in these compounds is a very soft base. Metathesis generates lithium chloride and an organoaluminum or organoboron compound. To give a specific example, triethylaluminum, a colourless liquid that burns spontaneously in air, can be prepared by treating 1 mole of $AlCl_3$ with 3 moles of ethyllithium ($LiCH_2CH_3$):

$$3\,LiCH_2CH_3 + AlCl_3 \longrightarrow \underset{\text{Triethylaluminum}}{Al(CH_3CH_3)_3} + 3\,LiCl$$

Many other alkyl groups can be used in place of $—CH_2CH_3$. Organoaluminum compounds, including triethylaluminum, are used to make catalysts for the polymer industry.

Group 14 metal halides also undergo metathesis reactions. For instance, organotin compounds are prepared on an industrial scale using organoaluminum reagents. These reactions take place because tin is a softer Lewis acid than aluminum, and carbon (in an alkyl group, represented by $R$) is a softer Lewis base than chlorine:

$$3\,SnCl_4 + 4\,AlR_3 \longrightarrow 3\,SnR_4 + 4\,AlCl_3$$

Organotin compounds are important industrial chemicals. One major use is as stabilizers for poly(vinyl chloride) (PVC) plastics. These additives, one example of which is dioctyltinmaleate, inhibit degradation of the polymer by heat, light, and oxygen. In the absence of these tin compounds, PVC yellows and becomes brittle.

Organotin compounds are also used extensively in agriculture. More than a third of the world's food crops are lost to fungi, bacteria, insects, or weeds. Organotin compounds such as tributylhydroxytin $((CH_3CH_2CH_2CH_2)_3SnOH)$

Dioctyltinmaleate

inhibit growth of fungi among crops such as potatoes, peanuts, sugar beets, and rice. Marine paints for wooden boats also contain organotin compounds that inhibit the attachment of barnacles. Organotins are added to cellulose and wool to inhibit attack by moths. The tin compounds used in these applications are specific in their toxicity, so they present little danger to mammalian life.

## SECTION EXERCISES

● **19.2.1** Rank the following from softest to hardest: (a) $NCl_3$, $NH_3$, and $NF_3$; (b) $Pb^{2+}$, $Pb^{4+}$, and $Zn^{2+}$; (c) $ClO_4^-$, $ClO_2^-$, and $ClO_3^-$; and (d) $PCl_3$, $SbCl_3$, and $PF_3$.

● **19.2.2** Iron is always found in nature in compounds, often with oxygen. The other members of Column 8 in the periodic table, ruthenium and osmium, occur in elemental form. Explain these observations using the HSAB principle.

● **19.2.3** Sulphur tetrafluoride fluorinates boron trichloride according to the following unbalanced equation:

$$BCl_3 + SF_4 \longrightarrow BF_3 + SCl_2 + Cl_2$$

This is both a redox reaction and a metathesis reaction. (a) Balance the equation. (b) Identify the elements that change oxidation state. (c) Explain the metathesis portion of the reaction using hard–soft acid–base arguments.

## 19.3 THE MAIN GROUP METALS

The elements in the lower left portion of the $p$ block of the periodic table are the **main group metals**. Although the most important metals of technological society are transition metals from the $d$ block, three main group metals, aluminum, lead, and tin, have considerable technological importance.

### Production of Aluminum

Aluminum is unique among the main group metals. All other $p$-block metals have filled valence $d$ orbitals. As a consequence, these metals have much in common with their transition metal neighbours. They tend to be soft Lewis bases. Aluminum, on the other hand, lacks a filled $d$-orbital set and is a hard Lewis acid that has more in common with its nearest neighbour, magnesium. Highly reactive, aluminum is found naturally in the +3 oxidation state and is difficult to reduce to the pure metal. Thus, although tin and lead have been known since antiquity, aluminum was not discovered until 1825 and did not become a common commodity until more than 60 years later.

Bauxite, the main aluminum ore, is a mixed oxide–hydroxide, Al(O)OH, contaminated with $SiO_2$, $Fe_2O_3$, clay, and other hydroxide salts. To isolate the aluminum-containing material, the ore is treated with a strongly basic solution, and the high hydroxide concentration causes the solid to dissolve as a soluble complex ion, $[Al(OH)_4]^-$:

$$Al(O)OH\ (s) + OH^-\ (aq) + H_2O \longrightarrow [Al(OH)_4]^-\ (aq)$$

The main group metals (highlighted in yellow) are found in the corner of the $p$ block closest to the transition metals.

The impurities in bauxite are not soluble in strong base, and remain behind when the solution is separated from the undissolved solids. Although $[Al(OH)_4]^-$ is soluble, $Al(OH)_3$ is not, so diluting the strongly basic solution causes solid aluminum hydroxide to precipitate:

$$[Al(OH)_4]^- \, (aq) \xrightarrow{\text{Added water}} Al(OH)_3 \, (s) + OH^- \, (aq)$$

After this separation is complete, $Al(OH)_3$ is heated to drive off water, which converts the hydroxide into pure aluminum oxide:

$$Al(OH)_3 \, (s) \xrightarrow{125 \, °C} Al_2O_3 \, (s) + 3 \, H_2O \, (g)$$

Aluminum metal is produced from aluminum oxide by electrolysis using the **Hall–Héroult process**; this story is detailed in our Focus 19-1: Chemical Milestones. The melting point of $Al_2O_3$ is too high (2015 °C) and its electrical conductivity too low to make direct electrolysis commercially viable. Instead, $Al_2O_3$ is mixed with cryolite ($Na_3AlF_6$) containing about 10% $CaF_2$. This mixture has a melting point of 1000 °C, still a high temperature but not prohibitively so. Aluminum forms several complex ions with fluoride and oxide, so the molten mixture contains a variety of species, including $AlF_4^-$, $AlF_6^{3-}$, and $AlOF_3^{2-}$. These and other ions move freely through the molten mixture as electrolysis occurs.

Figure 19-4 shows a schematic representation of an electrolysis cell for aluminum production. An external electrical potential drives electrons into a graphite

**Figure 19-4**

Aluminum metal is produced by electrolysis of aluminum oxide dissolved in molten cryolite. $Al^{3+}$ is reduced to Al at the cathode, and the carbon anode is oxidized to $CO_2$.

cathode, where $Al^{3+}$ ions are reduced to Al metal:

$$\text{Cathode: } Al^{3+} \text{ (melt) } + 3\,e^- \longrightarrow Al\,(l)$$

The anode, which is also made of graphite, is oxidized during electrolysis. Carbon from the anode combines with oxide ions to form $CO_2$ gas:

$$\text{Anode: } 2\,O^{2-} \text{ (melt) } + C\,(s) \longrightarrow CO_2\,(g) + 4\,e^-$$

## FOCUS 19-1 CHEMICAL MILESTONES: THE STORY OF THE HALL–HÉROULT PROCESS

Aluminum is the third most abundant element in the Earth's crust and the most abundant metal. Nevertheless, aluminum was not discovered until 1825. The reason for this elusiveness is the high stability of $Al^{3+}$. The reduction of aluminum compounds to the free metal requires stronger reducing power than common chemical reducing agents can provide. The discovery of aluminum had to await the birth of electrochemistry and the development of electrolysis.

Twenty-five years after its discovery, aluminum was a precious metal. Then, a French chemist developed procedures for reducing aluminum compounds using sodium metal. The price of the metal dropped 100-fold. Even so, in 1885, aluminum was a semiprecious metal used for esoteric purposes such as a prince's baby rattle (commissioned by the French emperor to commemorate the birth of the prince imperial, Eugene Louis Jean Joseph in 1856) and the cap for the Washington Monument.

To convert aluminum from the stuff of princes' toys into recyclable kitchen foil required an inexpensive electrolytic reduction process. Two 22-year-old scientists, the American chemist Charles Hall and the French metallurgist Paul Héroult, discovered the same process independently in 1886. Both became famous as founders of the aluminum industry, Hall in the United States and Héroult in Europe.

Charles Hall was inspired by his chemistry professor at Oberlin College, who observed that whoever perfected an inexpensive way of producing aluminum would become rich and famous. After his graduation, Hall set to work in his home laboratory, trying to electrolyze various compounds of aluminum. He was aided by his sister Julia, who had studied chemistry and shared Charles' interests. Julia helped to prepare chemicals and witnessed many of the electrolysis experiments. After only eight months of work, Hall had successfully produced globules of the metal. Meanwhile, Héroult was developing the identical process in France.

Hall capitalized on his discovery by founding a company for the manufacture

Charles Hall        Julia Hall

of aluminum. That company became immensely successful, eventually growing into Alcoa, the world's largest aluminum-producing company.

Successful electrolysis of aluminum requires a liquid medium other than water that can conduct electricity. The key to the Hall–Héroult process is the use of molten cryolite, $Na_3AlF_6$, as a solvent. Cryolite melts at an accessible temperature, it dissolves $Al_2O_3$, and it is available in high purity. A second important feature is the choice of graphite to serve as the anode, which provides an easy oxidation process, the oxidation of carbon to $CO_2$.

How did two persons working independently on two different continents come up with an identical process at the same time? The reasoning that led Hall and Héroult to the identical process was probably similar. Electrolysis was recognized as a powerful reducing method. All attempts to reduce aqueous aluminum cations failed, making it clear that some molten salt would have to be used. Experimenting with various salts, no doubt guided by the principle that "like dissolves like," the two young men eventually tried $Na_3AlF_6$, a mineral with constituent elements that should not interfere with the reduction of aluminum. Graphite electrodes were already in use, so experimenting with them would have been a natural choice.

The ingredients for this invention may have all been in place, but that does not detract from its brilliance. Hall and Héroult had the courage to explore new procedures in homemade laboratories without the support of research grants. They explored the possibilities systematically to find a process that was a spectacular success. In more than 100 years of growth in the aluminum industry, the only significant change to the Hall–Héroult process has been the addition of $CaF_2$ to the melt to lower the operating temperature.

Canada ranks fourth in the world in terms of aluminum production, currently producing 3 million tonnes of the metal per year. The abundance of hydroelectric power in Québec and British Columbia is one reason that aluminum production facilities are located in these two areas. Another reason is that bauxite mines are located primarily in tropical countries, and so aluminum smelters must be located near shipping routes.

Because of the variety of ionic species present in the melt, the reactions that take place at the electrodes are considerably more complex than the simple representations given here. Nevertheless, the net reaction is the one given by these simplified reactions:

$$4\, Al^{3+} \,(\text{melt}) + 6\, O^{2-} \,(\text{melt}) + 3\, C \,(s) \longrightarrow 3\, CO_2 \,(g) + 4\, Al \,(l)$$

The electrolysis apparatus operates well above the melting point of aluminum (660 °C), and liquid aluminum has a higher density than the molten salt mixture, so pure liquid metal settles to the bottom of the reactor. The pure metal is drained through a plug and cast into ingots.

Aluminum refining consumes huge amounts of electricity. Approximately 4.6% of all electricity generated in Canada, for instance, is used to produce aluminum.

## Uses of Aluminum

Aluminum is one of society's most important structural metals. It has one of the lowest densities among the metals, yet it is very strong, and it resists corrosion by forming a thin layer of aluminum oxide on the surface of the metal. Alloys of aluminum are used for aircraft bodies, trailers, cooking utensils, highway signs, storage tanks, beverage cans, and many other objects. Because of its light weight, aluminum is often used for long-distance electricity transmission wires.

Although the major use of aluminum by far is as a metal, some aluminum compounds are also economically important. Foremost among these is aluminum chloride, which is an important industrial catalyst.

The aluminum atom in aluminum chloride behaves as a Lewis acid, and the chlorine atoms are Lewis bases. As a consequence, two $AlCl_3$ molecules form a Lewis acid–base adduct, $Al_2Cl_6$. As Figure 19-5 shows, the compound has two

**Figure 19-5**

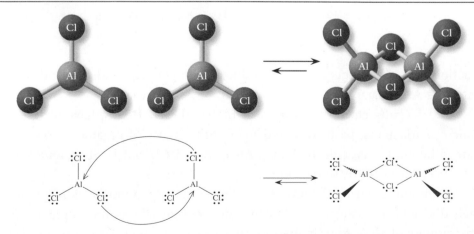

The Al atom in $AlCl_3$ has Lewis acid character, whereas the Cl atoms are Lewis bases. The compound forms a self-adduct, in which two Cl atoms donate electron pairs to form additional Al—Cl bonds.

tetrahedral aluminum atoms held together by a pair of chlorine atoms that bridge the two metal atoms. The 91° Al—Cl—Al bond angle suggests that the chlorine atom uses two valence $p$ orbitals to form the bridge. There is an equilibrium between $AlCl_3$ and $Al_2Cl_6$ molecules, but except in the gas phase above 750 °C, the formation of $Al_2Cl_6$ goes essentially to completion.

In the solid state, aluminum chloride exists in a crystalline lattice. Each aluminum atom is surrounded by six chlorine atoms arranged around the metal atoms at the corners of an octahedron. Aluminum bromide and aluminum iodide form $Al_2X_6$ molecules in all three phases.

The aluminum trihalides are particularly important Lewis acids in the chemical industry. They promote or catalyze a large variety of reactions. One of the most important applications is the Friedel–Crafts reaction, in which two molecules combine, forming a new C—C bond. For example, aluminum chloride or some other Lewis acid catalyzes the reaction between an acid chloride and benzene to form acetophenone:

An acid chloride is a derivative of a carboxylic acid, in which the hydroxyl group is replaced by a chlorine atom.

Benzene     Acetyl chloride          Acetophenone

The Lewis acid activates the acid chloride by forming an adduct with a chlorine atom bridge between carbon and aluminum. This chlorine bridge is similar to the one found in $Al_2Cl_6$:

Lewis base     Lewis acid          Adduct

The adduct fragments into an $[AlCl_4]^-$ anion and a very reactive cation:

Adduct                              Reactive cation

Reaction with a benzene molecule produces acetophenone and HCl and regenerates the $AlCl_3$ catalyst.

Friedel–Crafts chemistry is used in the synthesis of dyes, flavourings, fragrances, surfactants, pesticides, and many other types of organic compounds. Many different Lewis acids, including $AlCl_3$, $BF_3$, $TiCl_4$, $SnCl_4$, $SbF_5$, and $ZnCl_2$, serve as catalysts for these reactions.

Aluminum sulphate ($Al_2(SO_4)_3$) is widely used in water purification to remove finely divided particulate matter. When added to water, aluminum sulphate forms a precipitate of aluminum hydroxide that has a very open structure and large surface area. This precipitate, called a *gel*, traps dispersed particulate matter as it settles out of the liquid phase.

## Tin and Lead

Tin and lead are neighbours in Group 14 of the periodic table. Each metal displays two common oxidation states, +2 and +4, and each can be reduced to the free metal with relative ease. The major ore of lead is galena (PbS), in which lead is in the +2 state. In contrast, the major ore of tin is cassiterite ($SnO_2$), where tin is in the +4 state. The $Pb^{2+}$ state is substantially softer than the $Sn^{4+}$ state, not only because of lower charge but also because of a larger value for $n$. This accounts for lead being found associated with the softer base $S^{2-}$, whereas tin combines with the harder base $O^{2-}$.

Carbon in the form of charcoal is a sufficiently powerful reducing agent to convert lead and tin ores to the free metals. In the case of galena, the first step is roasting in air to form PbO:

$$2\ PbS + 3\ O_2 \xrightarrow{\text{Heat}} PbO + 2\ SO_2$$

$$PbO + C \xrightarrow{\text{Heat}} Pb + CO$$

$$SnO_2 + 2\ C \xrightarrow{\text{Heat}} Sn + 2\ CO$$

A bed of hot charcoal can supply both the heat and the reducing agent, and the simplicity of the process accounts for the early metallurgical development of these metals.

In former times, tin was used widely as a constituent of metal alloys, of which bronze, solder, and pewter are common examples. Bronze is an alloy of copper containing approximately 20% tin and smaller amounts of zinc. Pewter is another Cu—Sn alloy that contains tin as the major component ($\approx$85%), with roughly equal portions of copper, bismuth, and antimony. Solder consists of 67% lead and 33% tin, although lead-free solders (alloys of tin, antimony, bismuth, and sometimes silver) are now in common use because of toxicity concerns related to lead.

Because of its relatively low melting point (232 °C) and good resistance to oxidation, tin is used to provide protective coatings on metals such as iron that oxidize more readily. "Tin cans" are iron cans dipped in molten tin to provide a thin surface film of tin. Traditional metalsmiths use a similar process, coating copperware with a thin film of tin.

Metals that are soft Lewis acids (e.g., cadmium, mercury, and lead) are extremely hazardous to living organisms. Tin, in contrast, is not. One reason is that tin oxide is highly insoluble, so tin seldom is found at measurable levels in aqueous solution. Perhaps more important, the toxic metals generally act by binding to sulphur in essential enzymes. Tin is a harder Lewis acid than the other heavy metals, so it has a lower affinity for sulphur, a relatively soft Lewis base.

The history of lead is almost as ancient as that of tin. In Roman times, lead was formed into pipes that were used for water supplies (hence our word *plumbing*, derived from the same Latin word, *plumbum*, gives us the symbol Pb). Lead is a component of pewter and also was used as a glaze on drinking vessels. "White lead" ($Pb_3(OH)_2(CO_3)_2$) is an enduring white substance that was used for many years as a paint pigment and even as a component of cosmetics.

Bronze

More recently, the major use for lead has been in the automobile industry. The lead storage battery, described in Chapter 16, generates a high electrical potential, can deliver large currents, and is easily recharged. Tetraethyllead $((C_2H_5)_4Pb)$ is excellent at preventing the pre-ignition of gasoline in automobile cylinders. Once used widely as an "antiknock" fuel additive, $(C_2H_5)_4Pb$ is no longer added to gasoline because lead is highly toxic.

Unfortunately, lead easily enters the biosphere. Lead poisoning in humans causes learning impairment and behavioural disorders in children, and at higher levels it triggers mood swings, irritability, and loss of coordination. Historians think that the widespread use of lead in piping and drinking vessels caused the Romans to suffer a relatively high incidence of lead poisoning, which may even have contributed to the downfall of their empire. The toxicity of lead is also linked to the fact that lead salts are sweet—the Romans used lead barrels to sweeten their wine. Lead-based paints may have caused a similarly high incidence of lead poisoning among Europeans, and it is known that present-day North American children living in low-income housing are at high risk because of the presence of lead-based paints that have not been removed.

Pewter

Because of these damaging effects, most uses of lead that involve direct exposure for humans are being phased out. Unleaded gasoline and lead-free paints have replaced two former major commercial uses of lead. Lead has proved to be indispensable, however, in the lead storage battery, which now provides the major use of this metal. Although leakage from damaged batteries is still a potential hazard, contemporary batteries are manufactured in such a way that human exposure to battery contents is minimized.

## SECTION EXERCISES

●**19.3.1** Draw molecular pictures similar to those in Figure 19-5 that illustrate the electron pair donation that occurs when acetyl chloride ($CH_3COCl$, formed from acetic acid by replacing —OH with —Cl) forms an adduct with aluminum chloride.

●**19.3.2** Indium is a relatively soft Lewis acid. Use this fact to predict properties of indium compared with its horizontal neighbour (tin) and its diagonal neighbour (lead) in the periodic table.

●**19.3.3** Lead poisoning can be treated using the dianion of 2,3-dimercaptopropanol. Two ions bind one $Pb^{2+}$ ion in a soluble tetrahedral complex that can be excreted from the body. Using hard–soft acid–base concepts, draw the expected structure for this complex.

2,3-Dimercaptopropanol dianion

## 19.4 THE METALLOIDS

Six elements are **metalloids**: B, Si, Ge, As, Sb, and Te. Of these, silicon is by far the most abundant, making up over 27% of the Earth's crust, more than any other element except oxygen. In fact, $SiO_2$ and silicate minerals account for 80% of the atoms near the Earth's surface. Despite its great abundance, silicon was not discovered until 1824, probably because the strong bonds it forms with oxygen make silicon difficult to isolate. Two much rarer metalloids, antimony (known to the ancients) and arsenic (discovered ca. 1250 AD) were isolated and identified long before silicon.

Until quite recently, chemical interest in the metalloids consisted mainly of isolated curiosities, such as the poisonous nature of arsenic and the mildly therapeutic value of borax. The development of metalloid semiconductors, which we describe in Chapter 7, focused intense study on these elements.

## Boron

As is typical of second-row elements, boron has properties that distinguish it from the other elements in Group 13 as well as from the rest of the metalloids. The unique features of boron chemistry can be attributed to characteristics of its electron configuration. As a second-row element, boron has no valence *d* orbitals that can participate in bonding. Its ionization energies are considerably higher than those of aluminum and the other Group 13 elements, all of which are metals. However, the three valence electrons of the boron atom are insufficient to form the extensive bonding networks that are characteristic of carbon, its second-row neighbour.

Boric acid ($H_3BO_3$) and boron trifluoride ($BF_3$) exemplify the bonding patterns of boron compounds. As Figure 19-6 shows, both these compounds contain delocalized $\pi$ bonds. Each acts as a Lewis acid, readily adding an additional anion to form $[B(OH)_4]^-$ and $[BF_4]^-$ adducts. Both adducts are tetrahedral species in which the bonding can be described using four bonds formed from $sp^3$ hybrid orbitals on the boron atom.

**Figure 19-6**

The Lewis structures of $BF_3$ and $H_3BO_3$ indicate that these species contain delocalized bonds. The adducts $[BF_4]^-$ and $[B(OH)_4]^-$, in contrast, contain only single bonds.

The changes in bonding that accompany adduct formation are reflected in differences in bond lengths. The B—F bond lengths are 131 pm in $BF_3$ and 139 pm in $[BF_4]^-$; B—O bond lengths are 137 pm in $H_3BO_3$ and 147 pm in $[B(OH)_4]^-$. The shorter bonds in the neutral species provide clear evidence for multiple bonding. The unusual strength of the B—F bond in $BF_3$ also indicates the presence of $\pi$ bonding: 645 kJ/mol is an energy more consistent with double bonds than single bonds.

Because of their low polarizability, $BF_3$ and $BCl_3$ are gases. The larger, more polarizable electron clouds on bromine and iodine, and hence larger dispersion forces, make $BBr_3$ a volatile liquid and $BI_3$ a solid at room temperature.

Boron trihalides are strong Lewis acids that react with a wide collection of Lewis bases. Many adducts form with donor atoms from Group 15 (N, P, As) or Group 16 (O, S). Metal fluorides transfer $F^-$ ion to $BF_3$ to give tetrafluoroborate salts:

$$LiF + BF_3 \longrightarrow LiBF_4$$

Tetrafluoroborate anion is an important derivative of $BF_3$ because it is non-reactive. With four $\sigma$ bonds, the $[BF_4]^-$ anion has no tendency to coordinate further ligands. Tetrafluoroborate salts are used in synthesis when a bulky inert anion is necessary.

The observed order of reactivity for the boron halides is

$$BF_3 < BCl_3 < BBr_3 < BI_3$$

From electronegativity considerations, we might expect the opposite trend. The electronegativity difference between boron ($X = 2.0$) and fluorine ($X = 4.0$) is 2, whereas boron and iodine ($X = 2.5$) differ by only 0.5. Thus, fluorine atoms withdraw more electron density from boron than iodine atoms do, resulting in a more positive boron atom that should be a stronger Lewis acid. The observed reactivity trend indicates that some mechanism returns electron density to boron in the lighter boron halides. Such a mechanism is provided by the delocalized $\pi$ bonding resulting from the overlap between filled halogen $p$ orbitals and the empty $p$ orbital on the boron atom. The system shifts electron density from the halogen atoms into the bonding region between the atoms. Fluorine, being the smallest halogen, forms the strongest $\pi$ bond with boron, since the side-by-side overlap needed for bonding is greater when $n = 2$ than for larger, more diffuse $p$ orbitals.

## Silicon

The most abundant compounds of silicon are $SiO_2$ and the related silicate anions, all of which contain Si—O bonds. See Sections 6.3 and 7.3 for descriptions of the structure and bonding of these compounds, which involve $\sigma$ bond networks and tetrahedral geometry. As already mentioned, many minerals are combinations of hard silicate anions and hard metal cations.

Even though silicon is extremely abundant, only one silicon-containing compound appears in the list of top 50 industrial chemicals; that is sodium silicate ($Na_2SiO_3$), used for the manufacture of silica gel (a granular porous form of $SiO_2$) and glass. Nevertheless, with the advent of the electronic age, silicon has become an extremely important substance that is the primary ingredient of most semi-conductors. Because these are microscale devices, the quantity of production of silicon remains small compared with that of fertilizers and construction materials. Although relatively small in quantity, the value of silicon products is quite high.

The main source of silicon for semiconductor chips is silicon dioxide (silica). Silica can be reduced directly to elemental form by intense heating with coke in an electric arc furnace. At the temperature of the furnace, silicon is a gas:

$$SiO_2 + C \xrightarrow{\ 3000\ °C\ } Si\ (g) + 2\ CO$$

Figure 19-7

The product of this process has a purity of around 98%, which is sufficient for uses such as alloy formation but not nearly high enough for semiconductors.

To prepare pure silicon, silica is first converted to $SiCl_4$. A redox reaction between coke and chlorine gas is coupled with a metathesis reaction to give $SiCl_4$, which liquefies on cooling:

$$SiCl_4 + 2\ C + 2\ Cl_2 \xrightarrow{\ High\ T\ } SiCl_4\ (g) + 2\ CO$$

Distillation yields $SiCl_4$ of very high purity, which is then reduced with magnesium:

$$SiCl_4 + 2\ Mg \longrightarrow Si + 2\ MgCl_2$$

Silicon rod with impurities

Zone of molten silicon

Heating coil

Purer silicon

Inert gas

After the removal of $MgCl_2$ by washing with water, the silicon is purified still further by the technique of zone refining, shown schematically in Figure 19-7. A rod of impure silicon is melted and resolidified many times by a heating coil that passes back and forth along the rod. During this process, impurities remain preferentially in the liquid phase, yielding solid silicon containing less than 1 part per billion of impurities. Other elements are intentionally added to the silicon to make semiconductor materials, as discussed in Section 7.7.

In zone refining, a rod of impure silicon passes along a heating coil. The impurities concentrate in the molten zone, leaving solid material behind that is of higher purity. Repeated passage through the coil moves the impurities to the end of the rod.

## Silicones

In addition to its uses for electronic devices, silicon is a major component of silicone polymers. The silicone backbone consists of alternating silicon and oxygen atoms. The synthesis of these polymers begins with an organic chloride such as methyl chloride and an alloy of silicon and copper:

$$2\ CH_3Cl + Si(Cu) \longrightarrow (CH_3)_2SiCl_2 + Cu$$

Treatment of the resulting chlorosilane with water replaces the chlorine atoms with hydroxyl groups:

$$(CH_3)_2SiCl_2 + 2\ H_2O \longrightarrow (CH_3)_2Si(OH)_2 + 2\ HCl$$

Finally, the dihydroxysilane eliminates water in a condensation reaction to give a polymer with an Si—O—Si linkage, as Figure 19-8 shows.

Approximately 70 000 tonnes of silicone polymers are produced each year in the United States alone. Silicones are used as greases, caulking, gaskets, biomedical devices, cosmetics, surfactants, antifoaming agents, hydraulic fluids, and water repellents.

Figure 19-8

Structural formula and ball-and-stick model of a portion of a silicone polymer chain.

## Other Metalloids

The remaining metalloids are Ge, As, Sb, and Te. Each has an abundance of less than 2 ppm in the Earth's crust. As already mentioned, arsenic is highly poisonous. Besides being the stuff of many murder mysteries, arsenic is used as a pesticide. The major current use of antimony is in lead–acid batteries. Battery electrodes are much less likely to electrolyze water during recharging when they are made from a lead–antimony alloy containing up to 5% Sb. This virtually eliminates the danger of gas buildup and subsequent rupturing of the battery and has allowed the production of sealed batteries, which have much improved battery lifetimes compared with unsealed batteries that use pure lead electrodes.

A major and growing use of the minor metalloids is in semiconductor fabrication. Germanium, like silicon, exhibits semiconductor properties. Binary compounds between elements of Groups 13 and 15 also act as semiconductors. These "13–15" compounds, such as GaAs and InSb, have the same number of valence electrons as Si or Ge. The energy gap between the valence band and the conduction band of a 13–15 semiconductor can be varied by changing the relative amounts of the two components. This allows the properties of 13–15 semiconductors to be fine-tuned.

### SECTION EXERCISES

●**19.4.1** The anion present in aqueous solutions of boric acid is $[B(OH)_4]^-$ rather than $H_2BO_3^-$. This is because $H_3BO_3$ undergoes Lewis acid–base adduct formation with one water molecule, and the adduct then transfers a proton to a second water molecule to generate $[B(OH)_4]^-$ and a hydronium ion. Write Lewis structural diagrams illustrating these two transfer reactions. Show all formal charges, and include arrows that show the movement of electrons.

●**19.4.2** Draw band-gap diagrams (review Section 7.7 if necessary) illustrating that silicon is a semiconductor but carbon (diamond) is not. What feature of the valence atomic orbitals accounts for this difference?

## 19.5 PHOSPHORUS

Of all the elements, phosphorus is the only one that was first isolated from a human source. The element was extracted from human urine in 1669 by Hennig Brand, a German alchemist, using an unsavoury process: Brand boiled thousands of litres of urine until only a paste remained. Further heating of the paste at high temperature produced a gas that condensed to a waxy white solid when the vapour was bubbled into water. It was not until 1779 that phosphorus was discovered in mineral form, as a component of phosphate minerals.

Despite its relatively late discovery, phosphorus is the eleventh most abundant element in Earth's crustal rock. It has been estimated that world reserves of "phosphate rock" are sufficient to last for several hundred years. Virtually all phosphorus deposits contain apatite, with the general formula $Ca_5(PO_4)_3X$, where $X = F$, OH, or Cl. Fluoroapatite is the least soluble, and hence most abundant, of the three apatite minerals. Phosphorus is found in aqueous systems as $HPO_4^{2-}$ and $H_2PO_4^-$ ions. In biological organisms, phosphorus is a component of nucleic acids and energy-shuttling molecules such as adenosine triphosphate (ATP).

## Elemental Phosphorus

Modern production of elemental phosphorus uses a technique similar to the metallurgical processes described in Chapter 18. Apatite is mixed with silica and coke and then heated strongly in the absence of oxygen. Under these conditions, coke reduces phosphate to elemental phosphorus, the silica forms liquid calcium silicate, and the fluoride ions in apatite dissolve in the liquid calcium silicate. The reactions are not fully understood, but the stoichiometry for the calcium phosphate part of apatite is as follows:

$$2\ Ca_3(PO_4)_2 + 6\ SiO_2 + 10\ C \xrightarrow{1450\ °C} P_4 + 6\ CaSiO_3 + 10\ CO$$

Elemental phosphorus is a gas at the temperature of the reaction, so the product distills off along with carbon monoxide. When cooled to room temperature, $P_4$ condenses as a waxy white solid.

White phosphorus consists of individual $P_4$ molecules with the four atoms at the corners of a tetrahedron. Each atom bonds to three others and has one lone pair of electrons, for a steric number of 4. However, the triangular geometry of the faces of the tetrahedron constrains the bond angles in the $P_4$ tetrahedron to 60°, far from the optimal four-coordinate geometry of 109.5°. As a result, $P_4$ is highly reactive. As shown in Figure 19-9, samples of white phosphorus are stored under water because $P_4$ burns spontaneously in the presence of oxygen. The glow that emanates from $P_4$ as it burns in the dark led to its name, from the Greek words

**Figure 19-9**

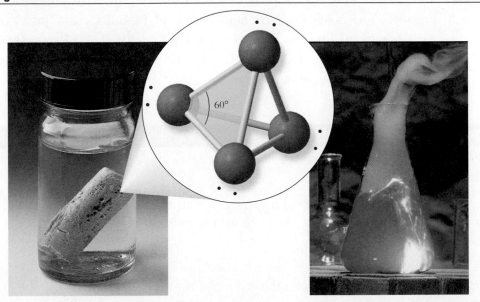

The strained 60° bond angles in tetrahedral $P_4$ make white phosphorus highly reactive. This form of the element must be kept out of contact with air, in which it spontaneously burns.

**Figure 19-10**

In red phosphorus, one of the P—P bonds of white phosphorus has been broken and used to link the tetrahedra together in a long chain.

*The Alchemist in Search of the Philosopher's Stone*, by Joseph Wright, 1771, depicts Hennig Brand's discovery of the element phosphorus.

*phos* ("light") and *phoros* ("bringing"). This form of phosphorus is very toxic—as little as 50 mg can cause death. As described later, most of the white phosphorus produced by the chemical industry is used to manufacture phosphoric acid.

Heating white phosphorus in the absence of oxygen causes the discrete $P_4$ units to link together, giving a chemically distinct elemental form, red phosphorus. As Figure 19-10 shows, one P—P bond of each tetrahedron breaks to allow formation of the bonds that link the $P_4$ fragments. The bond angles that involve the links are much closer to 109.5°, making red phosphorus less strained and less reactive than white phosphorus. Red phosphorus undergoes the same chemical reactions as the white form, but higher temperatures are required. In addition, red phosphorus is essentially non-toxic, and it is therefore easier and safer to handle.

The most thermodynamically stable form of the element is black phosphorus, which can be prepared by heating red phosphorus under high pressure. The black form contains chains of $P_4$ units cross-linked by P—P bonds, making this form even more polymerized and less strained than red phosphorus. Example 19-3 explores another difference between the elemental forms of phosphorus.

## EXAMPLE 19-3 MELTING POINTS OF PHOSPHORUS

The melting point of white phosphorus is 44.1 °C. In contrast, red phosphorus remains a solid up to 400 °C. Account for this large difference in melting point.

**Strategy:** As described in Chapter 8, melting points of solids may depend on both covalent bonds and intermolecular forces, so we must explain the melting point difference with reference to the bonding differences between the two forms.

**Solution:** White phosphorus consists of individual $P_4$ molecules. Because there are no polar bonds, the molecules are held in the solid state only by dispersion forces. This is a molecular solid, and such solids have relatively low melting temperatures. Red phosphorus is made of long chains of $P_4$ groups, each of which is held together by covalent P—P bonds, as shown in Figure 19-10. Thus, red phosphorus consists of macromolecules held together by dispersion forces, much like polyethylene, as described in Chapter 17. To melt red phosphorus, some chemical bonds must break. Bond breakage requires much more energy than the energy needed to overcome dispersion forces, giving red phosphorus the higher melting temperature.

 **Does the Result Make Sense?** The molecular models in Figures 19-9 and 19-10 show that white phosphorus is a molecular solid, whereas red phosphorus is intermediate between molecular and network solids. This correlates well with the melting points.

**Practice Exercise 19.3**
Predict whether black phosphorus melts at a higher or lower temperature than the other two forms, and rank the three forms of elemental phosphorus in order of increasing density, giving a reason for your ranking.

## Phosphoric Acid

The most important commercial product of phosphorus is phosphoric acid, $H_3PO_4$. Phosphoric acid consistently ranks among the top 10 industrial chemicals in the world. Canada alone produces 345 000 tonnes per year.

Almost all phosphoric acid is produced directly from apatite. The ore is partially purified, crushed, and then slurried with aqueous sulphuric acid:

$$Ca_5(PO_4)_3F\ (s) + 5\ H_2SO_4\ (aq) \longrightarrow 3\ H_3PO_4\ (aq) + 5\ CaSO_4\ (s) + HF\ (aq)$$

The dilute phosphoric acid obtained from this process is concentrated by evaporation. It is usually dark green or brown because of the presence of many metal ion impurities in the phosphate rock. However, this impure acid is suitable for the manufacture of phosphate fertilizers, which consumes almost 90% of phosphoric acid production.

High-purity $H_3PO_4$ is obtained using a more expensive redox process that starts from the pure element. Controlled combustion of white phosphorus gives phosphorus(V) oxide ($P_4O_{10}$), the structure of which is shown in Figure 19-11:

$$P_4 + 5\ O_2 \longrightarrow P_4O_{10}$$

Addition of water to $P_4O_{10}$ generates highly pure phosphoric acid:

$$P_4O_{10}\ (s) + 6\ H_2O\ (l) \longrightarrow 4\ H_3PO_4\ (aq)$$

More than 80% of the elemental phosphorus produced is converted to phosphoric acid. This pure product, which constitutes about 10% of the total industrial output of phosphoric acid, is the starting material for making food additives, pharmaceuticals, and detergents.

The elemental phosphorus that is not converted into phosphoric acid is used mainly to produce phosphorus chlorides ($PCl_3$ and $PCl_5$) and phosphorus oxychloride ($POCl_3$). These are important reagents for production of agrochemicals, drugs, and other specialty products.

**Figure 19-11**

Phosphorus(V) oxide ($P_4O_{10}$) contains the same triangular arrangement of phosphorus atoms as in $P_4$, but an oxygen atom is inserted into each P—P bond. An additional terminal O atom is double-bonded to each P atom.

## Phosphorus Fertilizers

Plants typically contain about 0.2% phosphorus by weight, but the element is easily depleted from soils. For this reason, the number one commercial application of phosphorus is in fertilizers. There are several common phosphorus fertilizers, but the single most important one is ammonium hydrogen phosphate, $(NH_4)_2HPO_4$. This phosphate compound is particularly valuable because it is highly soluble and provides both phosphorus and nitrogen. The compound can be made by treating phosphoric acid with ammonia:

$$H_3PO_4 + 2\,NH_3 \longrightarrow (NH_4)_2HPO_4$$

This is the predominant industrial route to $(NH_4)_2HPO_4$, accounting for the single largest use of phosphoric acid.

## Phosphate Condensations

Phosphoric acid can undergo phosphate condensation, a reaction that is similar to the condensation reactions of other species that contain O—H bonds (see Chapter 17):

The product, $H_4P_2O_7$, is pyrophosphoric acid. A second condensation leads to triphosphoric acid:

$$H_4P_2O_7 + H_3PO_4 \longrightarrow H_5P_3O_{10} + H_2O$$

The sodium salts of these acids, sodium pyrophosphate $(Na_4P_2O_7)$ and sodium triphosphate $(Na_5P_3O_{10})$, have been used widely in detergents. The polyphosphate anions are good additives for cleansing agents because they form complexes with metal ions, including those that make water "hard" $(Ca^{2+}, Mg^{2+})$ and those that cause colour stains $(Fe^{3+}, Mn^{2+})$. Moreover, polyphosphates are non-toxic, non-flammable, and non-corrosive; they do not attack dyes or fabrics, and they are readily decomposed during wastewater treatment.

These advantages appear to make polyphosphates ideal for use in cleaning agents. Unfortunately, adding phosphates to water leads to an imbalance in aquatic biosystems, particularly in lakes. Too much phosphate leads to runaway algae growth, which can overwhelm the lake with decaying organic matter. Organic decay consumes oxygen, depleting the lake of this life-sustaining substance and leading to the death of fish and other aquatic life-forms. This causes more organic decay and still more oxygen depletion, until eventually the lake supports no animal life.

Phosphate condensation reactions play an essential role in metabolism. Recall from Section 11.6 that the conversion of adenosine diphosphate (ADP) to adenosine triphosphate (ATP) requires an input of free energy:

$$ADP + H_3PO_4 \longrightarrow ATP + H_2O \qquad \Delta G^\circ = +30.6 \text{ kJ}$$

As also described in that section, ATP serves as a major biochemical energy source, releasing energy in the reverse, hydrolysis, reaction. The ease of interchanging O—H and O—P bonds probably accounts for the fact that a phosphate condensation/hydrolysis reaction is used for energy storage and transport.

In ADP and ATP, one end of the polyphosphate chain links to adenosine through an O—C bond. These bonds form in condensation reactions between hydrogen phosphate and alcohols. Other biochemical substances contain P—O—C linkages. An example is the thickening agent used in instant puddings and pie fillings, which is produced by the reaction of starch with sodium dihydrogen phosphate:

Starch molecules have many exposed O—H bonds, so this phosphorylation reaction results in multiple phosphate groups attached to each starch molecule. The remaining —OH group on each phosphate can condense with an O—H bond on another starch molecule. This cross-linking of starch chains gives the desired thick consistency of puddings and pies.

## Organophosphorus Compounds

Strictly speaking, the biochemical substances just mentioned fall into the class of organophosphorus compounds, which are substances containing both phosphorus and carbon. However, the term is usually used more specifically to describe smaller molecules of this type. Organophosphorus compounds have varied uses, ranging from gasoline additives to herbicides.

The phosphate structure plays a central role in many familiar insecticides and herbicides. These compounds often have other elements substituted in place of one or more O atoms in the phosphate group, giving structures that can penetrate living cells and disrupt their normal activities. Figure 19-12 shows the line

**Figure 19-12**

Parathion
($C_{10}H_{14}NO_5PS$)

Malathion
($C_{10}H_{19}O_6PS_2$)

Amiprophos
($C_{12}H_{19}N_2O_4PS$)

Sarin
($C_4H_{10}FO_2P$)

Line structures of four toxic organophosphorus compounds. Each has a modified phosphate group with atoms of other elements replacing one or more O atoms.

structures of some representatives of these species. Organophosphorus compounds such as Parathion and Malathion are potent insecticides. They act by phosphorylating an essential enzyme. Herbicides are represented by amiprophos, a compound that has a highly modified phosphate group. One oxygen atom is replaced with sulphur and another with nitrogen.

Not all toxic organophosphorus compounds have uses beneficial to humans. Sarin is an extremely toxic nerve gas that is lethal to humans. In March 1995, this substance was released in a terrorist attack on a Japanese subway, resulting in several deaths and many serious injuries. Sarin and related nerve gases bind an amino acid in the enzyme responsible for muscle action. When this enzyme is deactivated, muscles contract but cannot relax. Even a small dose can be lethal if the nerve gas reaches the muscles of the heart. The antidote for organophosphate poisoning is the drug Pralidoxime, which removes the phosphoryl group bound to the enzyme. Patients are usually also given atropine, which increases their heart rate.

## SECTION EXERCISES

**19.5.1** Figure 19-11 highlights the triangular arrangement of phosphorus atoms in $P_4O_{10}$. As every P atom bonds to four O atoms, however, the geometry about each P atom is tetrahedral, just as in the phosphate anion. Redraw Figure 19-11 in a way that highlights one of the tetrahedral $PO_4$ units.

**19.5.2** Casein is a polypeptide found in milk, in which the side chains of serine amino acids of the polypeptide have condensed with phosphoric acid to form a phosphoprotein that binds calcium phosphate. Draw the line structure of a serine–phosphate group. (See Section 17.7 for the structures of serine and polypeptides.)

**19.5.3** Shown below is a ball-and-stick model of glyphosate, the active ingredient in a commercial herbicide named Roundup. Draw the Lewis structure of glyphosate and describe its functional groups.

## 19.6 OTHER NON-METALS

Among the non-metals, the second-row elements nitrogen and oxygen display a rich variety of chemical properties that provide many of the examples used throughout this textbook. Because the structure and chemistry of compounds of these elements are described in other chapters, we do not present further details here. The noble gases of Group 18 are unreactive, for the most part. See the Focus 6-2: Chemical Milestones for a description of their limited chemical behaviour. The remaining reactive non-metals, all members of the *p* block, are carbon and sulphur and the halogens of Group 17 (F, Cl, Br, and I). Two of these elements are components of leading industrial chemicals: sulphuric acid perennially ranks as number one by a wide margin, and molecular chlorine ranked in the top 10 until very recently. Our survey of sulphur and the halogens emphasizes commercially important uses. We begin with a discussion of some of the more interesting aspects of the elemental forms of carbon.

## Carbon

Carbon is best known for its central role in organic chemistry, a subject introduced in Chapter 10 of this book. Outside of organic chemistry, carbon has a rich and interesting chemistry ranging from its many elemental forms to biologically significant molecules to industrially important compounds. Other aspects of the chemistry of carbon have been covered in this book—the chemistry of carbon dioxide has been discussed in terms of its roles in the atmosphere and in blood, and properties of solid and supercritical $CO_2$ were introduced. The role of carbon monoxide in impeding oxygen transport was detailed. Carbonic acid, bicarbonate, and carbonate ions have been used to illustrate a number of chemical principles. This section therefore concentrates on the properties of elemental carbon, in particular the more recently discovered forms of the element.

Carbon exists as different **allotropes**, forms of the element that are structurally different from one another (Figure 19-13). The two allotropes of carbon we are most familiar with are diamond and graphite. Both are elemental carbon, but their physical properties could hardly be more different. Diamond is transparent, very hard, and not electrically conductive. Graphite is black, soft, and electrically conductive. These differences arise because of how the carbon atoms are bonded to one another, as explained in Section 8.4. Until recently, these were believed to be the only allotropes of solid carbon, but many others have been discovered.

**Figure 19-13**

Graphite          Diamond

Structures of diamond and graphite.

Buckminsterfullerene has the molecular formula $C_{60}$, and was discovered by Sir Harold Kroto in the United Kingdom and Richard E. Smalley and Robert F. Curl, Jr. in the United States. These three scientists shared the 1996 Nobel Prize in chemistry for their discovery. Buckminsterfullerene is one of a group of molecules having the general formula $C_n$, with $n$ having several possible integral values, collectively called "fullerenes." Fullerenes can be made by heating graphite intensely in a helium atmosphere. Under these conditions, graphite vapourizes as molecular fragments, some of which recondense as fullerenes.

The $sp^2$-hybridized atoms in Buckminsterfullerene are bonded together into a hollow structure with 60 vertices and 32 faces, 12 of which are pentagons and 20 of which are hexagons, giving it the same geometry as a soccer ball. The allotrope

**Figure 19-14**

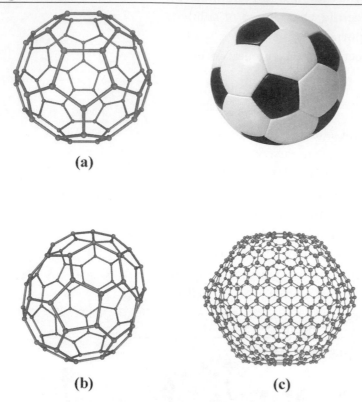

(a)

(b)                    (c)

Buckminsterfullerene (a) has the same shape as a soccer ball. Fullerenes $C_{70}$ (b) and $C_{540}$ (c) are also shown.

is named after the architect Buckminster Fuller, whose geodesic domes resemble these unusual chemical compounds. Buckminsterfullerene and two other fullerenes, $C_{70}$ and $C_{540}$, are shown in Figure 19-14.

Carbon nanotubes (Figure 19-15) were discovered in 1991, and consist of long cylinders of carbon atoms. The atoms are $sp^2$ hybridized as in fullerenes, but only in six-membered rings. Diameters as small as a few nanometres and lengths up to 1 mm have been reported. Some types of carbon nanotubes are as electrically conductive as copper metal, whereas others are semiconductors. Under tension, carbon nanotubes are the strongest material known so far. Theory predicts that they should also be excellent heat conductors along the long axis, but poor in the perpendicular (radial) direction. This opens the intriguing possibility of thermal insulating materials that could be turned on or off at will. A seemingly related allotrope is the carbon megatube, also shown in Figure 19-15. Note however that the

**Figure 19-15**

Carbon nanotubes may be single-walled, or as shown in the diagram at the left, multi-walled. At right is an electron micrograph of carbon megatubes, which have diameters of several micrometres.

**Figure 19-16**

A carbon nanobud is a composite of a nanotube and a fullerene.

megatubes have diameters in the micrometre range, approximately 1000 times greater than those of carbon nanotubes.

In 2009, scientists successfully reacted carbon nanotubes with fullerenes, and the result was carbon nanobuds, shown in Figure 19-16. This hybrid material may allow other chemical groups to be attached, since the carbon atoms in the fullerene are at higher energies than those in the nanotube because of more strained bond angles. In addition, the attached fullerene molecules would weaken the dispersion forces among neighbouring nanotubes, allowing the material to be dispersed in solvents more easily than carbon nanotubes themselves. Research into combining nanotubes with polymers is intense, and the fullerene groups in carbon nanobuds may function as molecular anchors to prevent the nanotubes from moving in these materials, thus improving the composite's mechanical properties. Carbon nano-onions, consisting of concentric spheres of carbon atoms, have been shown to be effective lubricants at the molecular scale, presumably because they act like tiny ball-bearings.

Carbon nanotubes, nanobuds, and nano-onions are such new materials that their potential applications are only starting to become apparent. For example, a carbon nanotube laid down on a silicon chip forms a molecular transistor. Such devices may eventually lead to further miniaturization of the chips that are at the heart of modern computers. Carbon nanotubes are also being investigated for their potential to deliver drugs, peptides, proteins, or nucleic acids in the human body. In the chemistry laboratory, carbon nanotubes have been used as filter material, adsorbents to remove impurities from water, and as components of electrochemical sensors. As an example of a sensor, Figure 19-17 shows that a small strand of DNA wound around a carbon nanotube emits light in the yellow region of the spectrum. In the presence of the chemotherapy drug Melphalan, the emission spectrum moves into the red portion of the spectrum.

**Figure 19-17**

A carbon nanotube with a strand of DNA wound around it emits light in the yellow portion of the spectrum (*left*). The emission moves to the red portion of the spectrum in the presence of the drug Melphalan.

## Sulphur

Sulphur displays rich and varied chemical behaviour, but from the commercial standpoint, sulphuric acid dominates the chemistry of this element. Sulphuric acid is used in every major chemical-related industry: fertilizers (60% of annual production), chemical manufacture (6%), petroleum refining (5%), metallurgy (5%), detergents, plastics, fibres, paints and pigments, and paper-making. In many of these cases, sulphur itself is not of interest and is eventually discarded as sulphate waste. Instead, $H_2SO_4$ is exploited for its Brønsted–Lowry acidity, its oxidizing ability, its affinity for water, and its ability to form sulphate precipitates.

Sulphur is encountered most often in sulphide minerals such as pyrite ($FeS_2$), molybdenite ($MoS_2$), chalcocite ($Cu_2S$), cinnabar (HgS), and galena (PbS). It is also present in huge amounts as $H_2S$ in natural gas and in sulphur-containing organic compounds in crude oil and coal. Moreover, sulphur is found in abundance as the pure element, particularly around hot springs and volcanoes and in capping layers over natural salt deposits. References to sulphur occur throughout recorded history, dating back as far as the sixteenth century BC. As shown in Figure 19-18, elemental sulphur is a yellow crystalline solid that consists of individual $S_8$ molecules. The eight atoms in each molecule form a ring that puckers in such a way that four atoms lie in one plane and the other four atoms lie in a second plane.

Elemental sulphur can be obtained from underground deposits. A well is drilled into the deposit, and superheated pressurized water (165 °C) is forced into the hole. The hot water melts the sulphur ($T_m = 119$ °C), and the liquid sulphur is forced to the surface when pressurized air is injected into the well. This method supplies about one-third of the world's sulphur. However, the major present-day source of the element is from hydrogen sulphide produced as a by-product of oil

**Figure 19-18**

Under normal conditions, sulphur forms yellow crystals. The crystals consist of individual $S_8$ molecules, with the eight sulphur atoms of each molecule arranged in a puckered ring.

and gas refining. Many petroleum and natural gas supplies contain some sulphur—up to 25% in some cases. Besides being undesirable in the final products, sulphur poisons many of the catalysts used in oil refining; hence, it must be removed from crude petroleum as a first step in refining. Sulphur is produced in a gaseous redox reaction between hydrogen sulphide and sulphur dioxide:

$$16\ H_2S\ +\ 8\ SO_2 \xrightarrow{Fe_2O_3} 3\ S_8\ +\ 16\ H_2O$$

If a supply of $SO_2$ is not already available, this gas is generated by reacting $H_2S$ with oxygen:

$$2\ H_2S\ +\ 3\ O_2 \longrightarrow 2\ SO_2\ +\ 2\ H_2O$$

This process and the subsequent oxidation of sulphur to sulphuric acid are relatively inexpensive because the reactions are exothermic. Thus, these reactions produce energy rather than requiring energy expenditure.

Sulphuric acid is manufactured from elemental sulphur by the process described in the Focus 13-1: Chemistry and Technology. About 90% of world output of sulphur is converted to sulphuric acid, and more than 60% of that sulphuric acid is used to extract phosphoric acid from phosphate minerals. This major use exploits the fact that sulphuric acid is the least expensive strong Brønsted acid. In addition to protonating the phosphate anions, the sulphate group of sulphuric acid sequesters $Ca^{2+}$ as $CaSO_4$, a waste by-product of the process. Another industrial use of sulphuric acid also exploits its acidity—the treatment of calcium fluoride ($CaF_2$) with sulphuric acid produces hydrogen fluoride (HF):

$$CaF_2\ (s)\ +\ H_2SO_4\ (l) \longrightarrow 2\ HF\ (g)\ +\ CaSO_4\ (s)$$

This source of hydrogen fluoride provides about 70% of the fluorine used industrially.

Sulphate solubility properties are the basis for other industrial uses of sulphuric acid. One example is in the metallurgy of titanium. One of the major ores of titanium is $FeTiO_3$, which is treated with sulphuric acid to separate titanium from iron:

$$FeTiO_3\ (s)\ +\ 2\ H_2SO_4\ (aq)\ +\ 5\ H_2O\ (l) \xrightarrow{150-180\ °C} 2\ TiSO_4F\ (aq)\ +\ FeSO_4 \cdot 7\ H_2O\ (s)$$

$$TiSO_4\ (aq)\ +\ H_2O\ (l) \xrightarrow{90\ °C} TiO(OH)_2\ (aq)\ +\ H_2SO_4\ (aq)$$

The $TiO(OH)_2$ precipitate is washed with dilute sulphuric acid to remove impurities and then converted to $TiO_2$ by heating to 1000 °C. As described in Section 18.4, $TiO_2$ is subsequently converted to $TiCl_4$, purified by distillation, and then reduced to titanium metal.

Another use for sulphuric acid is in water treatment. Aluminum sulphate, which is produced by treating sulphuric acid with aluminum oxide, is a top 50 industrial chemical because of its widespread use as a coagulant. Addition of $Al_2(SO_4)_3$ and $Ca(OH)_2$ generates a gelatinous precipitate of aluminum hydroxide mixed with calcium sulphate:

$$Al_2(SO_4)_3\ (aq)\ +\ Ca(OH)_2\ (aq) \longrightarrow Al(OH)_3\ (s)\ +\ CaSO_4\ (s)$$

The gel traps finely suspended particulate material as it settles slowly out of the solution. This process is used in the primary stage of wastewater treatment, which is the removal of materials that make the water cloudy.

Although not as important industrially as processes involving sulphate, sulphur displays a rich chemistry with the halogens. For example, sulphur forms seven different binary fluorides. One of these, $SF_4$, has significant Lewis acid–base properties. Sulphur tetrafluoride can act as either an electron pair donor or an electron pair acceptor. With a steric number of 5, $SF_4$ adopts a trigonal bipyramidal arrangement of valence orbitals with the lone pair of electrons in one of the equatorial positions. The molecule acts as a Lewis base but not as a sulphur lone-pair donor. Instead, $SF_4$ is a fluoride donor, as shown in the following reactions. The donor atom in $SF_4$ is fluoride, which is a harder base than sulphur. After forming an adduct with the Lewis acid, the S—F bond breaks to give the ionic products:

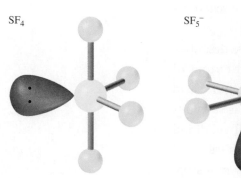

$SF_4$    $SF_5^-$

$$SF_4 + BF_3 \longrightarrow [SF_3]^+ [BF_4]^-$$

$$SF_4 + PF_5 \longrightarrow [SF_3]^+ [PF_6]^-$$

Sulphur tetrafluoride is also an important industrial fluorinating agent for both organic and inorganic compounds.

With its empty $3d$ orbitals, $SF_4$ also acts as a Lewis *acid.* This molecule forms adducts with pyridine and with fluoride ion. Each adduct has square pyramidal geometry, with a lone pair of electrons completing an octahedral arrangement of valence orbitals:

$$SF_4 + Pyr \longrightarrow SF_4 \cdot Pyr$$

$$SF_4 + CsF \longrightarrow Cs^+ + [SF_5]^-$$

Another key feature of sulphur chemistry is the Lewis acidity of sulphur dioxide. Sulphur dioxide is a common atmospheric pollutant that results from burning coal to produce electricity. Most coal reserves in North America include significant amounts of sulphur-containing impurities. When coal is burned, sulphur combines with $O_2$ to form $SO_2$, a hard Lewis acid.

In $SO_2$, the two S=O bonds are polarized toward the electronegative oxygen atoms, creating a partial positive charge on sulphur, as shown in Figure 19-19. The sulphur atom can use its empty $d$ orbitals to accept a pair of electrons from a water molecule to form an adduct. A second water molecule subsequently acts as a Brønsted–Lowry base and transfers a proton from one oxygen atom to another to give $H_2SO_3$:

**Figure 19-19**

| Highest electron density | Lowest electron density |

An electron density model of sulphur dioxide shows low electron density (*blue*) around the sulphur atom and high electron density (*red*) on the oxygen atoms.

Oxidation by molecular oxygen or $NO_2$ converts this weak acid into $H_2SO_4$, a major contributor to acid rain.

## TABLE 19-1 MAJOR USES OF CHLORINE

| Reactant | Intermediate | Final Products |
|---|---|---|
| **Organic Reagents** | | |
| Alkenes + $SO_2$ | Sulphonated alkenes | Surfactants, detergents |
| Benzene | Chlorobenzenes | Plastics, dyestuffs |
| Butadiene | Chloroprene | Neoprene (chlorinated rubber) |
| Ethylene | Chloroethenes | Plastics (PVC), solvents |
| Methane | Chloromethanes | Silicones |
| **Inorganic reagents** | | |
| CO | $COCl_2$ | Plastics (polycarbonate) |
| NaOH | NaOCl | Bleaches, disinfectants |
| Phosphorus | $PCl_3$, $POCl_3$ | Pesticides, flame retardants |
| Rutile ($TiO_2$ ore) | $TiCl_4$ | Paint pigment (pure $TiO_2$) |

## Chlorine

As mentioned earlier, molecular chlorine has long been one of the leading industrial chemicals. Table 19-1 provides a summary of the industrial importance of chlorine.

The starting material for all industrial chlorine chemistry is sodium chloride, obtained primarily by evaporation of seawater. The chloride ion is highly stable and must be oxidized electrolytically to produce chlorine gas. This is carried out on an industrial scale using the chlor–alkali process, which is shown schematically in Figure 19-20. The electrochemistry involved in the chlor–alkali process is discussed in Section 16.7. As with all electrolytic processes, the energy costs are very high, but the process is economically feasible because it generates three commercially valuable products: $H_2$ gas, aqueous NaOH, and $Cl_2$ gas.

Although chlorine displays a range of oxidation states from −1 to +7, the most stable by far is $Cl^-$, as the positive standard reduction potentials in Table 19-2 indicate. Thus, $Cl_2$ and all oxochloro anions are good oxidizing agents. This property accounts for two of the major uses of chlorine compounds. Molecular chlorine is used extensively as a purifying agent for water supplies because it destroys harmful bacteria, producing harmless $Cl^-$ in the process. Sodium hypochlorite (NaOCl) is the active ingredient in most bleaches. Hypochlorite ions oxidize many organic materials that are coloured, breaking them into smaller, colourless substances that are easily removed by detergents.

**Figure 19-20**

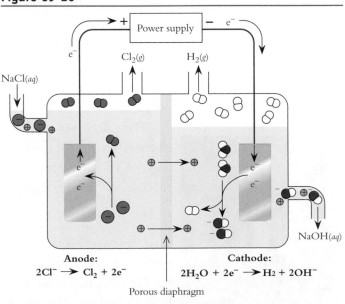

Anode:
$$2Cl^- \longrightarrow Cl_2 + 2e^-$$

Cathode:
$$2H_2O + 2e^- \longrightarrow H_2 + 2OH^-$$

Porous diaphragm

Schematic view of the electrolytic chlor–alkali process showing the molecular species involved in the redox reactions.

### TABLE 19-2 STANDARD REDUCTION POTENTIALS FOR CL SPECIES

| Reduction Half-Reaction | $E°$ (V) |
|---|---|
| $Cl_2 + 2\ e^- \longrightarrow 2\ Cl^-$ | +1.358 |
| $HClO + H_3O^+ + 2\ e^- \longrightarrow Cl^- + 2\ H_2O$ | +1.482 |
| $HClO_2 + 3\ H_3O^+ + 4\ e^- \longrightarrow Cl^- + 5\ H_2O$ | +1.570 |
| $ClO_3^- + 6\ H_3O^+ + 6\ e^- \longrightarrow Cl^- + 9\ H_2O$ | +1.451 |
| $ClO_4^- + 8\ H_3O^+ + 8\ e^- \longrightarrow Cl^- + 12\ H_2O$ | +1.389 |

A second major use for chlorine derives from its reactivity with organic materials, in particular with hydrocarbons. Two chlorinated hydrocarbons, ethylene dichloride (IUPAC name: 1,2-dichloroethane) and vinyl chloride (IUPAC name: chloroethene), rank among the leading industrial chemicals. Both substances are manufactured by reactions of $Cl_2$ with ethylene.

The first step in this process is a direct reaction carried out in liquid 1,2-dichloroethane with a metal chloride catalyst:

$$\underset{\substack{H\\ \\H}}{\overset{\substack{H\\ \\H}}{C=C}} + Cl_2 \xrightarrow{FeCl_3} \underset{\substack{H\\ \\H}}{\overset{\substack{Cl\\ \\Cl}}{\underset{}{C-C}}}\substack{H\\ \\H}$$

Thermal decomposition of this product yields vinyl chloride and HCl:

$$\underset{\substack{H\\ \\H}}{\overset{\substack{Cl\\ \\Cl}}{C-C}}\substack{H\\ \\H} \xrightarrow{Heat} \underset{\substack{H\\ \\H}}{\overset{\substack{Cl\\ \\H}}{C=C}} + HCl$$

The HCl produced in this reaction is used in an oxychlorination reaction to chlorinate additional ethylene:

$$2\ \underset{\substack{H\\ \\H}}{\overset{\substack{H\\ \\H}}{C=C}} + 2\ HCl + O_2 \longrightarrow 2\ \underset{\substack{H\\ \\H}}{\overset{\substack{Cl\\ \\H}}{C=C}} + 2\ H_2O$$

More than 95% of vinyl chloride is converted to PVC (polyvinylchloride) polymer.

The reaction of $Cl_2$ with ethylene to form vinyl chloride is just one example of a general process for the chlorination of hydrocarbons. One chlorine atom reacts with a hydrogen atom from the hydrocarbon to form HCl while a second chlorine atom replaces the hydrogen atom in the hydrocarbon. The generalized reaction is

$$RH + Cl_2 \longrightarrow RCl + HCl$$

Organic chlorides produced in this way are key starting materials for the synthesis of more complex organic compounds. The HCl by-product in these reactions may be exploited in an oxychlorination reaction as previously described, or it may be collected in water and marketed as hydrochloric acid (HCl). Sufficient HCl is produced through various industrial chlorination reactions as a by-product to account for over 90% of the production of hydrochloric acid.

PVC is used for a variety of common products.

In addition to making organic chlorine compounds, a significant fraction of $Cl_2$ production is used to make inorganic halides. One important use, described in Chapter 18, is in the metallurgy of titanium, in which molecular chlorine is used to convert $TiO_2$ into $TiCl_4$, which is easy to purify by distillation.

$$TiO_2 + C + 2\,Cl_2 \longrightarrow TiCl_4 + CO_2$$

In similar fashion that is described earlier in this chapter, easily purified $SiCl_4$ is produced from molecular chlorine and $SiO_2$ in the presence of coke:

$$SiO_2 + 2\,C + 2\,Cl_2 \longrightarrow SiCl_4 + 2\,CO$$

Another use of $Cl_2$ is the halogenation of phosphorus as an early step in the synthesis of organophosphorus compounds:

$$P_4 + 6\,Cl_2 \longrightarrow 4\,PCl_3$$

As Table 19-1 suggests, molecular chlorine is a tremendously versatile industrial chemical. This element is a leading industrial chemical because of this versatility rather than any single application, although polymers account for about one-third of its uses. In recent years, however, the industrial use of chlorine has come under strong attack from many environmentally conscious groups. One major reason is that dioxins and other chlorinated organics have detrimental effects on biosystems. The controversy over industrial chlorine is described in our Focus 19-2: Chemistry and the Environment.

## Other Halogens

No compound containing a halogen other than chlorine appears among the top 50 industrial chemicals, but fluorine nevertheless has considerable commercial value. Fluorine occurs as the mineral fluorite ($CaF_2$) and is prevalent in phosphate-bearing rock. As already mentioned, HF produced from sulphuric acid treatment of fluorite supplies some 70% of industrial HF; the remainder comes as a by-product of phosphoric acid production.

One major use of HF is in the manufacture of fluorinated hydrocarbons. Fluorinated ethylene is used for several specialty polymers. For example, the Teflon® coating on non-stick cookware is made from polytetrafluoroethylene (PTFE). This polymer is made from tetrafluoroethylene.

The starting material here is chloroform ($CHCl_3$), which is treated with HF to form the chlorofluorocarbon ($CHClF_2$):

$$\begin{array}{c} Cl \\ | \\ H-C-Cl \\ | \\ Cl \end{array} + 2\,HF \longrightarrow \begin{array}{c} Cl \\ | \\ H-C-F \\ | \\ F \end{array} + 2\,HCl$$

Subsequent heating in the presence of a platinum catalyst results in partial conversion to tetrafluoroethylene:

$$2\,\begin{array}{c} Cl \\ | \\ H-C-F \\ | \\ F \end{array} \xrightarrow[700\,°C]{Pt} \begin{array}{c} F \\ \diagdown \\ F \end{array}C=C\begin{array}{c} F \\ \diagup \\ F \end{array} + 2\,HCl$$

A coating of polytetrafluoroethylene gives Teflon® cookware its non-stick surface.

The tetrafluoroethylene is then polymerized to PTFE by addition polymerization as described in Chapter 17.

Until recently, chlorofluorocarbons (CFCs) for refrigeration were major end-products of HF chemistry, but these compounds are being phased out in accord with the Montreal Protocols because of their effect on the ozone layer (see Chapter 12).

Some industrial uses of fluorine require molecular fluorine, $F_2$, which is produced by electrolysis of HF. As Figure 19-21 shows, the electrolytic cell uses liquid HF (not aqueous) to which KF is added as an electrolyte. The redox chemistry is straightforward:

$$\text{Anode:} \quad 2F^- \longrightarrow F_2 + 2\,e^-$$
$$\text{Cathode:} \quad HF + 2e^- \longrightarrow H_2 + 2\,F^-$$
$$\overline{\text{Net:} \quad 2\,HF \longrightarrow F_2 + H_2}$$

Molecular fluorine reacts readily with a variety of organic and inorganic substances. It oxidizes metals to give metal fluorides. One useful metal fluoride is NaF, added to toothpaste to inhibit tooth decay. Uranium hexafluoride ($UF_6$) is used to enrich the $^{235}U$ content of uranium for nuclear power plants, and is made by reacting $UF_4$ (s) with gaseous fluorine. Molecular fluorine also combines with non-metals to give binary fluorides. The most important of these is sulphur hexafluoride, an inert and non-toxic gas that is used in electrical transformers as a dielectric to prevent arcing from one electrode to another:

$$S_8 + 24\,F_2 \longrightarrow 8\,SF_6$$

In addition, $F_2$ reacts with the other halogens to form interhalogen compounds, in which an atom of the second halogen is bonded to an odd number of F atoms. Examples are $ClF_3$, $BrF_5$, and $IF_7$.

The remaining halogens, bromine and iodine, are isolated from brines by treatment with $Cl_2$, which oxidizes the halide ions to their elemental forms:

$$2\,Br^-\,(aq) + Cl_2\,(g) \longrightarrow 2\,Cl^-\,(aq) + Br_2\,(l)$$

$$2\,I^-\,(aq) + Cl_2\,(g) \longrightarrow 2\,Cl^-\,(aq) + I_2\,(s)$$

Iodine is an essential element for humans because it is present in thyroxine, a hormone that regulates the rate of cellular use of oxygen:

**Figure 19-21**

Schematic view of an electrolytic cell used for the production of molecular fluorine, showing the molecular species involved in the redox reactions.

People whose diets are sparse in seafoods are susceptible to iodine-deficiency diseases, which are easily prevented by providing iodine in the diet. Iodized salt, which contains 0.1% KI, performs this function.

Molecular bromine is highly toxic, as is methyl bromide ($CH_3Br$), a dense gas used as an insecticide. Methyl bromide is produced by bromination of methane:

$$CH_4 + Br_2 \longrightarrow CH_3Br + HBr$$

Molecular bromine is also used in the synthesis of dyes and pharmaceuticals.

The most valuable bromine compound is probably silver bromide, which is the light-absorbing species on which most film photography is based. Brominated compounds are also used extensively as fire retardants, particularly in carpets, rugs, and clothing.

## FOCUS 19-2 CHEMISTRY AND THE ENVIRONMENT: THE CASE AGAINST THE INDUSTRIAL USE OF CHLORINE

Molecular chlorine has been a leading industrial chemical because it is the precursor for an immense variety of useful products—some 15 000 in all—that contain chlorine. Even a short list of the uses of chlorine indicates its versatility and commercial value: sugar refining, flame retardants, photography, deodorants, leather finishing, automobile bumpers, magnetic tape, and cosmetics. Nevertheless, groups ranging from environmental activists to government agencies have in recent years called for the reduction or even total elimination of chlorine as an industrial chemical.

The chloride anion is a major species in the oceans and plays an essential role in biochemistry. Compounds containing carbon–chlorine bonds occur much less frequently in nature. Volcanoes emit some halocarbons, and marine algae generate chloromethane. Other marine species produce toxic organohalogen molecules that protect them from predators. Nevertheless, organic chlorine compounds are uncommon, and consequently there are few mechanisms that degrade them.

The ecological problems caused by chlorine-containing compounds were first recognized years ago and called to public attention by Rachel Carson's best-selling book, *Silent Spring*. The chlorinated insecticide DDT accumulates in the tissues of animals. Birds of prey such as the peregrine falcon produced fragile, thin-shelled eggs that broke during incubation, and these species became endangered. Fortunately, the cause was determined and the use of DDT in many parts of the world was discontinued in time to allow gradual recovery of bird populations.

DDT

More recently, another class of organic chlorine compounds has emerged as an environmental hazard. These are the dioxins, which like DDT contain ring compounds with chlorine substituents. A relatively simple example is 2,3,7,8-tetrachlorodibenzo-*p*-dioxin:

Dioxin

This structure looks nothing like the structures of chlorine-containing compounds used in industrial processes. In fact, no dioxin is deliberately manufactured anywhere in the diverse chlorine industry. Nevertheless, dioxins are of concern for two reasons: first, dioxins appear to be inevitable trace by-products of some reactions involving chlorine, particularly combustion (e.g., in incinerators); and second, dioxins accumulate in the biosphere, where they have highly deleterious effects.

Dioxins have effects similar to and potentially even more far-reaching than those of DDT, because they apparently affect a wide variety of species. Predatory birds are especially susceptible, and there is growing evidence that humans may be at risk. Tests have shown that when the concentration of dioxins in the blood of laboratory animals reaches a critical level, reproductive and immune systems defects result. Moreover, recent data indicate that the concentration of dioxins in the blood of the average North American resident has nearly reached that level. A major reason is that dioxins are hydrophobic, so they accumulate in fatty tissue rather than being readily metabolized and excreted from the body.

Unfortunately, there is no easy way to eliminate production of dioxins without severely curtailing the use of all chlorinated compounds. This is because dioxins are formed when useful chlorine-containing compounds degrade, as for example when industrial wastes are incinerated. Consequently, several groups, including the International Joint Commission overseeing the quality of the Great Lakes, as well as environmental groups such as the Natural Resources Defense Council and the Sierra Club, have issued calls for the complete phase-out of the use of industrial chlorine.

Complete phase-out of chlorinated compounds is being resisted not only by chlorine producers, but also by the many industries that use chlorine compounds in the manufacture of products from paper to pharmaceuticals. Meanwhile, chemists seek ways to degrade chlorinated compounds to non-toxic substances.

## SECTION EXERCISES

●**19.6.1** Determine the changes in oxidation states for the reactions that convert $H_2S$ into $S_8$.

$$2\ H_2S + 3\ O_2 \longrightarrow 2\ SO_2 + 2\ H_2O$$

$$16\ H_2S + 8\ SO_2 \xrightarrow{Fe_2O_3} 3\ S_8 + 16\ H_2O$$

●**19.6.2** Determine the Lewis structures and describe the bonding and geometry of the following compounds that appear in this section: (a) $SF_4$; (b) $BrF_5$; and (c) $F_2C\!=\!CF_2$.

●**19.6.3** Although $Cl_2$ can be manufactured by electrolysis of aqueous brine, the analogous reaction cannot be used to make $F_2$. Explain why anhydrous HF must be used for production of $F_2$.

# CHAPTER 19 VISUAL SUMMARY

## 19.1 LEWIS ACIDS AND BASES

*A Lewis base acts as an electron pair donor.*

*A Lewis acid acts as an electron pair acceptor.*

Lewis base     Lewis acid     Adduct

**LEARNING OBJECTIVES:** Explain the chemistry of formation of adducts.

**SKILLS TO MASTER:** Identifying the Lewis acid and base in a reaction (p. 1114)

Drawing the Lewis structure of an adduct (p. 1115)

**KEY CONCEPTS:** Any chemical species that acts as an electron-pair donor is a Lewis base.

Any chemical species that acts as an electron-pair acceptor is a Lewis acid.

An adduct is formed by the combination of a Lewis base with a Lewis acid.

## 19.2 HARD AND SOFT LEWIS ACIDS AND BASES

Polarizability increases

Polarizability

Hard Lewis acid   Soft Lewis acid     Hard Lewis base   Soft Lewis base

Metathesis reaction

*Soft Lewis acids tend to combine with soft Lewis acids.*

*Hard Lewis acids tend to combine with hard Lewis bases.*

**LEARNING OBJECTIVES:** Apply the concepts of hardness and softness to reactions.

**SKILLS TO MASTER:** Writing metathesis reactions (p. 1122)

**KEY CONCEPTS:** A hard Lewis base has electron pairs of low polarizability.

A soft Lewis base has a large donor atom of high polarizability and low electronegativity.

A hard Lewis acid has an acceptor atom with low polarizability.

A soft Lewis acid has a relatively high polarizability.

Hard Lewis acids tend to combine with hard Lewis bases. Soft Lewis acids tend to combine with soft Lewis bases.

## 19.3 THE MAIN GROUP METALS

Hall-Héroult process

Main group metals

LEARNING OBJECTIVES: Explain the production, reactions, and uses of the main group metals.

SKILLS TO MASTER: Explaining the Hall–Héroult process for making aluminum metal (p. 1124)

## 19.4 THE METALLOIDS

Metalloids

Zone refining

Silicone polymer

LEARNING OBJECTIVES: Explain the production, reactions, and uses of the metalloids.

SKILLS TO MASTER: Explaining the bonding in boron compounds (p. 1131)

## 19.5 PHOSPHORUS

Red phosphorus

$H_3PO_4$    White phosphorus

Condensation reaction

LEARNING OBJECTIVES: Explain the production, reactions, and uses of phosphorus.

SKILLS TO MASTER: Describing the bonding in red and white phosphorus (p. 1135)

Explaining the processes for making phosphoric acid and phosphorus fertilizers (p. 1137)

Explaining phosphate condensation reactions (p. 1138)

## 19.6 OTHER NON-METALS

Chlor–alkali process

$S_8$

$SO_2$

LEARNING OBJECTIVES: Explain the production, reactions, and uses of the other non-metals.

SKILLS TO MASTER: Describing the various allotropes of carbon (p. 1141)

Explaining the electrolytic chlor–alkali process (p. 1147)

Explaining the electrolytic production of fluorine gas (p. 1150)

## LEARNING EXERCISES

**19.1** There are no new "memory bank" equations in this chapter, but several concepts from earlier chapters play major roles in the chemistry of main group elements. List the key concepts from earlier chapters that you are asked to apply in this chapter.

**19.2** Draw molecular pictures showing a typical Lewis acid–base reaction and a typical Brønsted acid–base reaction. Describe in words the differences and similarities of these two reactions.

**19.3** State the hard–soft acid–base (HSAB) principle. Define and give examples of hard and soft acids and bases.

**19.4** List the major industrial compounds containing main group elements and write a one-sentence description of the industrial importance of each.

**19.5** Write the structural formula for an example of each of the following: (a) Lewis acid–base adduct; (b) polyphosphate; (c) fluorine-containing polymer; and (d) alkyllithium reagent.

## PROBLEMS

**ilw** =interactive learning ware problem. Visit the website at www.wiley.com/canada/olmsted

### Lewis Acids and Bases

**19.1** In each of the following reactions, identify the Lewis acid and the Lewis base:
(a) $Ni + 4 CO \longrightarrow [Ni(CO)_4]$
(b) $SbCl_3 + 2 Cl^- \longrightarrow [SbCl_5]^{2-}$
(c) $(CH_3)_3P + AlBr_3 \longrightarrow (CH_3)_3P—AlBr_3$
(d) $BF_3 + ClF_3 \longrightarrow [ClF_2]^+[BF_4]^-$

**19.2** In each of the following reactions, identify the Lewis acid and the Lewis base:
(a) $AlCl_3 + CH_3Cl \longrightarrow CH_3^+ + [AlCl_4]^-$
(b) $Zn^{2+} + 4 CN^- \longrightarrow [Zn(CN)_4]^{2-}$
(c) $I_2 + I^- \longrightarrow I_3^-$
(d) $CaO + CO_2 \longrightarrow CaCO_3$

**19.3** Make a sketch that shows the three-dimensional structure of each of the following Lewis acids. Identify the specific orbital that acts as the acceptor during adduct formation: (a) $AlF_3$; (b) $SbF_5$; and (c) $SO_2$.

**19.4** Make a sketch that shows the three-dimensional structure of each of the following Lewis acids. Identify the specific orbital that acts as the acceptor during adduct formation: (a) $SnCl_4$; (b) $PF_5$; and (c) $CO_2$.

**19.5** **ilw** The reaction between $CO_2$ and $H_2O$ to form carbonic acid ($H_2CO_3$) can be described in two steps: formation of a Lewis acid–base adduct followed by Brønsted proton transfer. Draw Lewis structures illustrating these two steps, showing electron and proton movement by curved arrows.

**19.6** The reaction between $SO_3$ and $H_2O$ to form sulphuric acid ($H_2SO_4$) can be described in two steps: formation of a Lewis acid–base adduct followed by Brønsted proton transfer. Draw Lewis structures illustrating these two steps, showing electron and proton movement by curved arrows.

### Hard and Soft Lewis Acids and Bases

**19.7** Rank the following ions in order of increasing polarizability, and explain your reasoning: $Fe^{3+}$, $Fe^{2+}$, $Pb^{2+}$, and $V^{3+}$.

**19.8** Rank the following ions in order of increasing polarizability, and explain your reasoning: $SO_3^{2-}$, $NO_3^-$, $CO_3^{2-}$, and $ClO_3^-$.

**19.9** Rank the following Lewis acids from hardest to softest and explain your reasoning: (a) $BCl_3$, $BF_3$, and $AlCl_3$; (b) $Al^{3+}$, $Tl^{3+}$, and $Tl^+$; and (c) $AlCl_3$, $AlI_3$, and $AlBr_3$.

**19.10** Rank the following Lewis bases from hardest to softest and explain your rankings: (a) $NH_3$, $SbH_3$, and $PH_3$; (b) $PO_4^{3-}$, $ClO_4^-$, and $SO_4^{2-}$; and (c) $O^{2-}$, $Se^{2-}$, and $S^{2-}$.

**19.11** Explain why $SO_3$ is a harder Lewis acid than $SO_2$.

**19.12** Explain why iodide is a soft base but chloride is a hard base.

**19.13** For each of the following pairs of substances, determine whether metathesis will occur, and if so, identify the products: (a) $AlI_3 + NaCl$; (b) $TiCl_4 + TiI_4$; (c) $CaO + H_2S$; (d) $CH_3Li + PCl_3$; and (e) $AgI + SiCl_4$.

**19.14** For each of the following, state whether a reaction occurs and write a balanced equation for any reaction:
(a) $NBr_3 + GaCl_3 \longrightarrow$ ?
(b) $Al(CH_3)_3 + LiCH_3 \longrightarrow$ ?
(c) $SiF_4 + LiF \longrightarrow$ ?
(d) $LiCH_2CH_2CH_2CH_3 + SnCl_4 \longrightarrow$ ?

### The Main Group Metals

**19.15** Describe in detail the bonding in $Al_2Cl_6$. Explain the formation of the molecule in terms of Lewis acid–base chemistry.

**19.16** Write Lewis structures and describe the bonding in these three species found in the solution that is electrolyzed to form aluminum metal: $[AlF_4]^-$, $[AlF_6]^{3-}$, and $[OAlF_3]^{2-}$.

**19.17** From its position in the periodic table, predict the properties of thallium (Element 81).

**19.18** From its position in the periodic table, predict the properties of gallium (Element 31).

**19.19** Determine the Lewis structure of $SnCl_4$ and explain how it functions as a Lewis acid.

**19.20** Determine the Lewis structure of $SbF_5$ and explain how it functions as a Lewis acid.

### The Metalloids

**19.21** Borazine ($B_3N_3H_6$) is a planar molecule analogous to benzene ($C_6H_6$). Write the Lewis structure and describe the bonding of borazine.

**19.22** Boron nitride (BN) is a planar covalent solid analogous to graphite. Write a portion of the Lewis structure and describe the bonding of boron nitride, which has alternating B and N atoms.

19.23 Reasoning from periodic trends, determine whether Ge or Si has a larger band gap. Use orbital overlap arguments to support your choice.

19.24 Explain why, among the 13–15 semiconductors, Ga pairs with As whereas In pairs with Sb. Predict which of these has the smaller band gap.

19.25 **ilw** Write the reactions that generate a silicone polymer starting from ethyl chloride ($C_2H_5Cl$), and draw a portion of the structure of the resulting polymer. Show at least four repeat units of the polymer.

19.26 Write the reactions that generate a silicone polymer starting from a 1:1 mixture of $(F_3CCH_2CH_2)(CH_3)Si(OH)_2$ and $(CH_3)_2Si(OH)_2$, and draw a portion of the structure of the resulting polymer. Show at least four repeat units of the polymer.

## Phosphorus

19.27 Pyrophosphate ($P_2O_7^{4-}$) and triphosphate ($P_3O_{10}^{5-}$) are the first two polyphosphate anions. What is the chemical formula of the next largest polyphosphate anion? Draw a ball-and-stick model of this anion.

19.28 Three phosphate anions can condense to form a ring with a chemical formula of $P_3O_9^{3-}$. Determine the Lewis structure and draw a ball-and-stick model of this anion.

19.29 Describe the reactions that convert apatite to phosphoric acid. Identify Brønsted acid–base reactions and redox reactions, if any.

19.30 Describe the reactions that convert calcium phosphate to phosphoric acid. Identify Brønsted acid–base reactions and redox reactions, if any.

19.31 Write structural formulas showing the conversion of white phosphorus into red phosphorus. Use curved arrows to show how electrons move during this conversion.

19.32 Write structural formulas showing the reaction of ATP with water to form ADP. What is the other product?

## Other Non-metals

19.33 Prepare a list of the industrial reactions described in Section 19.6 that exploit the Brønsted acid character of sulphuric acid.

19.34 Prepare a list of the industrial reactions described in Section 19.6 that use sulphuric acid because of sulphate solubility characteristics.

19.35 Draw a portion of the repeating structure of polyvinylchloride.

19.36 Draw a portion of the repeating structure of polytetrafluoroethylene.

19.37 Identify the oxidizing agent, reducing agent, and changes of oxidation state that occur in the reaction forming $TiCl_4$ from $TiO_2$.

19.38 Identify the oxidizing agent, reducing agent, and changes of oxidation state that occur in the reaction forming $SiCl_4$ from $SiO_2$.

## ADDITIONAL PAIRED PROBLEMS

19.39 What mass of bauxite rock (Al(O)OH) must be processed to produce 2500 kg of pure aluminum if the bauxite rock is 85% pure and the processing steps have a net yield of 75%?

19.40 What mass of calcium phosphate rock ($Ca_3(PO_4)_2$) must be processed to produce 1500 kg of pure $P_4$ if the phosphate rock is 87% pure and the processing steps have a net yield of 68%?

19.41 Explain why $SF_6$ forms and is quite stable, whereas neither $OF_6$ nor $SBr_6$ is known.

19.42 Explain why the known forms of elemental carbon include forms with single, double, and triple bonds, whereas the known forms of elemental phosphorus all have single bonds.

19.43 Arsenic trichloride can act both as a Lewis acid and as a Lewis base. Explain why this is so and write a balanced equation for each using $BF_3$ and $Cl^-$ as reaction partners.

19.44 Boron trifluoride is a very strong Lewis acid, but trimethylboron ($B(CH_3)_3$) is a mild Lewis acid. Given the $\pi$ bonding character in $BF_3$, does this order of reactivity surprise you? Why or why not? Use bonding arguments to explain the trend in reactivity.

19.45 In basic aqueous solution, Al acts as a strong reducing agent, being oxidized to $AlO_2^-$. Balance this half-reaction, and determine balanced net reactions for Al reduction of the following: (a) $NO_3^-$ to $NH_3$; (b) $H_2O$ to $H_2$; and (c) $SnO_3^{2-}$ to Sn.

19.46 In acidic aqueous solution, $SnCl_2$ acts as a mild reducing agent, with tin being oxidized to $Sn^{4+}$ in the process. Balance this half-reaction, and determine balanced net reactions for this reagent causing the following reductions: (a) $MnO_4^-$ to $Mn^{2+}$; (b) $Cr_2O_7^{2-}$ to $Cr^{3+}$; and (c) $Hg_2^{2+}$ to Hg.

19.47 Briefly describe the changes in chemical properties that take place from top to bottom of Group 15 of the periodic table.

19.48 Briefly describe the changes in chemical properties that take place from top to bottom of Group 13 of the periodic table.

19.49 Describe the chemical reactions by which Al metal is obtained from its ore.

19.50 Describe the chemical reactions by which Pb metal is obtained from its ore.

19.51 Describe the industrial preparation of pure Si, starting from impure $SiO_2$. Include balanced chemical reactions.

19.52 Describe the industrial preparation of pure $H_3PO_4$ starting from phosphate rock. Include balanced chemical reactions.

**19.53** The only important ore of mercury is cinnabar, HgS. In contrast, zinc is found in several ores, including sulphides, carbonates, silicates, and oxides. Explain these observations in terms of hard and soft acids and bases.

**19.54** The only important ore of lead is galena, PbS. In contrast, tin is found primarily as cassiterite, $SnO_2$. Explain this difference in terms of hard and soft acids and bases.

## ADDITIONAL PROBLEMS

**19.55** The pressure of gaseous $Al_2Cl_6$ increases more rapidly with temperature than would be predicted by the ideal gas equation. Explain this behaviour.

**19.56** Water in thermal hot springs is often unpalatable owing to dissolved $H_2S$. Treatment with $Cl_2$ oxidizes the sulphur to $S_8$, which precipitates (the other product is HCl). Balance this reaction and calculate the mass of $Cl_2$ required to purify $6.0 \times 10^3$ L of water containing 25 ppm (by mass) dissolved $H_2S$.

**19.57** Calcium dihydrogen phosphate is a common phosphorus fertilizer that is made by treating fluoroapatite with phosphoric acid. Hydrogen fluoride is a by-product of the synthesis. Write a balanced equation for the production of this fertilizer and calculate the mass percent of phosphorus in the fertilizer.

**19.58** The black tarnish that forms on pure silver metal is the sulphide $Ag_2S$, formed by reaction with $H_2S$ in the atmosphere:

$$4\,Ag + 2\,H_2S + O_2 \longrightarrow 2\,Ag_2S + 2\,H_2O$$

The sulphide forms in preference to $Ag_2O$, even though the atmosphere is 20% $O_2$ with just a slight trace of $H_2S$. Use Lewis acid–base arguments to explain this behaviour.

**19.59** Thionyl chloride ($SOCl_2$) is used to remove water of hydration from metal halide hydrates. Besides the anhydrous metal halide, the products are $SO_2$ and HCl. (a) Draw the Lewis structure of $SOCl_2$. (b) Balance the reaction of iron(III) chloride hexahydrate with $SOCl_2$.

**19.60** Ammonium dihydrogen phosphate and ammonium hydrogen phosphate are common fertilizers that provide both nitrogen and phosphorus to growing plants. In contrast, ammonium phosphate is rarely used as a fertilizer because this compound has a high vapour pressure of toxic ammonia gas. Write balanced equations showing how each of these solid phosphates could generate ammonia gas. Why does ammonium phosphate have the highest vapour pressure of $NH_3$ among these three compounds?

**19.61** Boron trichloride is a gas, boron tribromide is a liquid, and boron triiodide is a solid. Explain this trend in terms of intermolecular forces and polarizability.

**19.62** Aluminum refining requires large amounts of electricity. Calculate the masses of Al and Na that are produced per mole of charge by electrolytic refining of $Al_2O_3$ and NaCl.

**19.63** Trisodium phosphate forms strongly basic solutions that are used as cleansers. Write balanced equations that show why $Na_3PO_4$ solutions are strongly basic. Include $pK_a$ values to support your equations.

**19.64** The fluorides $BF_3$, $AlF_3$, $SiF_4$, and $PF_5$ are Lewis acids. They all form very stable fluoroanions when treated with lithium fluoride. In contrast, three other fluorides, $CF_4$, $NF_3$, and $SF_6$, do not react with lithium fluoride. Use Lewis acid–base concepts to explain this behaviour.

**19.65** The first commercially successful method for the production of aluminum metal was developed in 1854 by H. Deville. The process relied on earlier work by the Danish scientist H. Oersted, who discovered that aluminum chloride is produced when chlorine gas is passed over hot aluminum oxide. Deville found that aluminum chloride reacts with sodium metal to give aluminum metal. Write balanced equations for these two reactions.

**19.66** Metal oxides and sulphide ores are usually contaminated with silica, $SiO_2$. This impurity must be removed when the ore is reduced to the pure element. Silica can be removed by adding calcium oxide to the reactor. Silica reacts with CaO to give $CaSiO_3$. Write a balanced equation for this reaction, and describe the reaction in terms of Lewis acids and bases.

**19.67** Complete the following reactions:
(a) $AlCl_3 + LiCH_3 \longrightarrow$ ?
(b) $SO_3 + \text{Excess } H_2O \longrightarrow$ ?
(c) $SbF_5 + LiF \longrightarrow$ ?
(d) $SF_4 + AsCl_5 \longrightarrow$ ?

**19.68** Some pure liquid interhalogen compounds are good electrical conductors, indicating that they contain cations and anions. Show a Lewis acid–base reaction between two bromine trifluoride molecules that would generate ionic species.

**19.69** Phosphorus(V) oxide has a very strong affinity for water; hence, it is often used as a drying agent in laboratory desiccators. One mole of $P_4O_{10}$ reacts with 6 moles of water. Based on this stoichiometry, identify the product of the reaction and balance the equation.

**19.70** A company that manufactures photographic film generates 2550 L/day of aqueous waste containing 0.125 g/L of $Br^-$ ions. To recover the bromine in the form of $Br_2$, the company bubbles $Cl_2$ gas through this waste. Calculate the volume of gas that is consumed daily if the gas is delivered at 1.06 bar and 21 °C.

## GROUP STUDY PROBLEMS

**19.71** Summarize the arguments for and against using phosphate-based detergents. Do you think these detergents should be used? Explain your position.

**19.72** Construct a table of bond lengths that supports the existence of bonding in boron trifluoride and boric acid.

The relevant data can be found in the text. Label your table thoroughly.

**19.73** Describe the ways in which the structures of DDT and dioxin (see Focus 19-2: Chemistry and the Environment) are similar. Describe the ways in which these structures differ.

**19.74** Summarize the arguments for and against using industrial chlorine. Do you think the industrial use of chlorine should be phased out? Explain your position.

**19.75** One of the factors that control the interaction of a Lewis base with a Lewis acid is the size of the molecules involved. For example, boron trifluoride forms a stronger adduct with tetrahydrofuran than with dimethyltetrahydrofuran. Use this example to explain size effects in adduct formation. (*Hint:* Think about electron–electron repulsion.)

Tetrahydrofuran
($C_4H_8O$)

Dimethyltetrahydrofuran
($C_6H_{12}O$)

## ANSWERS TO SECTION EXERCISES

### 19.1.1

(a)

(b)

(c)

**19.1.2** (a) Lewis acid: $SO_3$; Lewis base: $OH^-$; (b) Lewis acid: $SnCl_2$; Lewis base: $Cl^-$; and (c) Lewis acid: $SbF_5$; Lewis base: $AsF_3$

### 19.1.3

(a)

(One of three resonance structures)

(b)

(c)

**19.2.1** (a) $NH_3$, $NCl_3$, $NF_3$; (b) $Pb^{2+}$, $Pb^{4+}$, $Zn^{2+}$; (c) $ClO_4^-$, $ClO_3^-$, $ClO_2^-$; and (d) $SbCl_3$; $PCl_3$, $PF_3$

**19.2.2** Only the softest metals, those clustered around gold in the periodic table, exist in nature in elemental form. Ru and Os are in this category. Iron forms harder cations with an affinity for the hard anions abundant in the environment, including carbonate, oxide, and the silicates.

**19.2.3** (a) $4\,BCl_3 + 3\,SF_4 \longrightarrow 4\,BF_3 + 3\,SCl_2 + 3\,Cl_2$
(b) Chlorine is oxidized, and sulphur is reduced.
(c) Sulphur is softer than boron, and chlorine is softer than fluorine. According to the HSAB principle, sulphur prefers to bond with the softer chlorine, and boron prefers to bond with the harder fluorine.

**19.3.1**

**19.3.2** Indium, being relatively soft, has properties more like lead, which is softer than tin. Thus, indium is toxic, occurs naturally as a sulphide, and so on.

**19.3.3** Bonding is between the relatively soft Lewis-base atoms, S, of 2,3-dimercaptopropanol and the soft $Pb^{2+}$ cation:

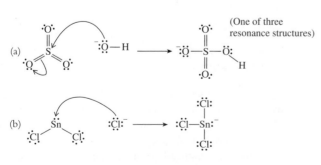

### 19.4.1

**19.4.2** The more diffuse $n = 3$ valence orbitals of Si do not overlap with one another as effectively as the compact $n = 2$ valence orbitals of C, so the difference in energy between bonding and antibonding orbitals (valence and conduction bands) is smaller for Si than for C:

Carbon: Large band gap          Silicon: Small band gap

**19.5.1**

**19.5.2**

Serine residue of the polypeptide

Phosphate group

**19.5.3** The structure includes a modified phosphate (P—C bond in place of one P—O bond), a carboxylic acid, and an amine:

**19.6.1** Oxidation states are shown beneath the atoms:

$$2 H_2\underset{-2}{S} + 3\underset{0}{O_2} \longrightarrow 2\underset{+4}{S}O_2 + 2 H_2\underset{-2}{O}$$

Sulphur is oxidized and oxygen is reduced.

$$16 H_2\underset{-2}{S} + 8\underset{+4}{S}O_2 \longrightarrow 3\underset{0}{S_8} + 16 H_2O$$

The sulphur in $H_2S$ is oxidized (loses electrons), and the sulphur in $SO_2$ is reduced (gains electrons).

**19.6.2** (a)

The S atom has SN = 5 and trigonal bipyramidal electron pair geometry. The lone pair is equatorial, giving a seesaw shape. Bonds can be described as formed from $2p$ atomic orbitals on F overlapping with $sp^3 d$ hybrids on S.

(b)

The Br atom has SN = 6 and octahedral electron geometry. The one lone pair results in a molecular shape that is square pyramidal. Bonds can be described as formed from $2p$ atomic orbitals on F overlapping with $sp^3 d^2$ hybrids on Br.

(c)

Each C atom has SN = 3, giving trigonal planar geometry and a planar molecule. There is a $\sigma$ bond network that can be described using $sp^2$ hybrids from C and $2p$ atomic orbitals from F. Side-by-side overlap of $2p$ orbitals from C gives a $\pi$ bond.

**19.6.3** The oxidation reactions and their standard potentials are as follows:

$$6 H_2 \longrightarrow O_2 + 4 H_3O^+ \ (aq) + 4e^- \quad -1.229 \ V$$
$$2 Cl^- \longrightarrow Cl_2 + 2 e^- \quad\quad\quad\quad\quad -1.358 \ V$$
$$2F^- \longrightarrow F_2 + 2 e^- \quad\quad\quad\quad\quad\quad -2.866 \ V$$

As described in Chapter 16, even though $C^-$ is thermodynamically harder to oxidize than $H_2O$, the reaction kinetically is much faster, so $Cl^-$ is preferentially oxidized in aqueous solution containing $Cl^-$. Oxidation of $F^-$ is thermodynamically so much more difficult that water is preferentially oxidized in aqueous solutions containing $F^-$.

## ANSWERS TO PRACTICE EXERCISES

**19.1** $Ni^{2+}$ is the Lewis acid, $H_2O$ is the Lewis base:

Acid      Base          Adduct

**19.2** Lewis bases: $PH_3$ is softer than $NH_3$, which is softer than $H_2O$.

**19.3** Because of its covalent cross-linking, black phosphorus melts at a higher temperature than either red or white phosphorus. White phosphorus is the least dense, red is intermediate in density, and black is most dense, because added bonds decrease the interatomic distances.

# NUCLEAR CHEMISTRY AND RADIOCHEMISTRY

## THE NUCLEAR DILEMMA

Life flourishes on Earth because of energy radiated by the Sun. That energy comes from transformations of atomic nuclei. At the extreme temperature of the Sun, small nuclei fuse together to form larger nuclei. The vast amounts of energy released by these solar fusion reactions continuously churn the matter making up the Sun, generating the hot spots and flares that appear in the X-ray image in our opening photo.

To unlock the energy secrets of the Sun, it was necessary to study matter at the subatomic level. The inset figure shows results from an experiment carried out in an instrument called a bubble chamber. Nuclear reactions generate high-energy particles that create ions as they pass through matter. In a bubble chamber, bubbles form around these ions, creating a track of bubbles along the path of the particle. Each type of particle creates a track with a different shape and length. The tracks tell scientists about the nature of nuclear reactions and have helped us understand what fuels the stars.

Unlocking the secrets of the nucleus was a mixed blessing, for in addition to our understanding of the

Sun and our ability to generate vast amounts of electrical energy, we also acquired nuclear weapons of immense destructive potential. The bombing of Hiroshima and Nagasaki with nuclear weapons was one of the last acts of the Second World War but the beginning of the nuclear dilemma: more than 50 years later, controversies still rage over how society should use the fruits of nuclear science.

Equally troublesome are the health hazards associated with the radioactive products of nuclear reactions. These hazards include genetic effects, cancer, and other illnesses that can be fatal. Unfortunately, there is no known way to make radioactive elements non-radioactive. Additionally, once radioactive substances are created, some of them will last for tens of thousands of years.

The benefits of nuclear reactions lie primarily in power generation and medical diagnosis and treatment. At present, society is divided as to whether these benefits outweigh their accompanying hazards. Nuclear medicine is firmly established. Many hospitals use radioactive materials for diagnosis and treatment; for example, to generate the images of diseased thyroid glands shown in our inset photo. Still, society has not solved the problem of what to do with the resulting radioactive waste.

One hundred years after the discovery of radioactivity and 50 years after the dawn of the "nuclear age," society continues to debate the benefits and costs of nuclear technology. Understanding nuclear transformations and the properties of radioactivity is necessary for intelligent discussions of the nuclear dilemma. In this chapter, we explore the nucleus and the nuclear processes that it undergoes. We describe the factors that make nuclei stable or unstable, the various types of nuclear reactions that can occur, and the effects and applications of radioactivity.

# CHAPTER CONTENTS

# LEARNING OBJECTIVES

*Upon completion of this chapter you should be able to*

- explain the forces that hold nuclei together
- describe nuclear decay processes and calculate rates of decay
- describe how products are synthesized in induced nuclear reactions
- describe the fission process and the essential features of a fission reactor
- describe the fusion process and the essential features of a fusion reactor
- explain the effects of ionizing radiation
- explain some practical applications of radioactivity

## 20.1 NUCLEAR STABILITY

In earlier chapters, we treated the nucleus as a structure that never changes. Although this is true for normal chemical processes, under the right conditions nuclei undergo transformations that change their structures. These processes depend strongly on energy as well as nuclear structure. Our discussion of nuclear chemistry begins with the structure and energetics of the nucleus.

### Nuclear Composition

The radius of a nucleus is about $10^{-14}$ m, 10 thousand times smaller than the radius of an atom. An atom the size of a football stadium would have a nucleus the size of a pea, but the density of nuclei is so great that such a pea would have a mass of more than 250 million tonnes.

Nuclei are composed of two different fundamental particles, protons and neutrons, which are called **nucleons**. A proton has a charge of $+1.60218 \times 10^{-19}$ C (commonly referred to as 1 electrostatic unit, esu) and a mass of $1.672\ 622 \times 10^{-27}$ kg. A neutron has almost the same mass, $1.674\ 927 \times 10^{-27}$ kg, but is electrically neutral. Ionized H atoms, which form from $H_2$ in electrical discharges, are free protons. A free proton does not last long, because it quickly captures an electron to become a hydrogen atom. Free neutrons are often generated in the course of nuclear reactions, but they do not last long either, because they are easily captured again when they collide with nuclei. Table 20-1 summarizes the properties of these nuclear building blocks. The other fundamental atomic particle, the electron, is included for comparison.

Each particular type of nucleus is called a **nuclide**. Nuclides are characterized by the number of protons $(Z)$ and neutrons $(N)$ that they possess. The number of protons in a nuclide is always the same as the atomic number of the element, $Z$. Recall however, that the number of neutrons can vary, and that isotopes are nuclides with the same number of protons but different numbers of neutrons. Copper, for example, has two stable isotopes. Each isotope has 29 protons, but one isotope has 34 neutrons, whereas the other has 36 neutrons.

The **mass number** of a nuclide, $A$, is its total number of protons and neutrons: $A = Z + N$. Because protons and neutrons each have molar mass near 1 g/mol, $A$ is always close to the numerical value of the molar mass of that isotope. For example, fluorine has $A = 19$ and a molar mass of 18.998 g/mol. A particular nuclide can be described by its elemental symbol, $X$, preceded by the value of $A$ as a superscript and that of $Z$ as a subscript: Examples include $^{63}_{29}Cu$, $^{4}_{2}He$, and $^{12}_{6}C$, the nuclei of which are shown schematically in Figure 20-1. The elemental symbol

**Figure 20-1**

Helium-4

Carbon-12

Copper-63

● = Neutron    ○ = Proton

Each nucleus contains $Z$ protons and $N$ neutrons. Shown here in schematic fashion are the nuclei of $^{4}_{2}He$, $^{12}_{6}C$, and $^{63}_{29}Cu$.

### TABLE 20-1 PROPERTIES OF FUNDAMENTAL PARTICLES

| Particle | Symbol | Charge ($10^{-19}$ C) | Mass ($10^{-27}$ kg) | Molar Mass (g/mol) |
|----------|--------|----------------------|----------------------|--------------------|
| Proton   | p      | +1.60218             | 1.672 622            | 1.007 276          |
| Neutron  | n      | 0                    | 1.674 927            | 1.008 665          |
| Electron | e      | −1.60218             | 0.000 911            | $5.486 \times 10^{-4}$ |

identifies $Z$, so the subscript is often omitted (for example, $^{63}$Cu). Alternatively, the name of the element is followed by its mass number, as in copper-63. Example 20-1 provides some practice in writing the symbols of nuclides.

---

### EXAMPLE 20-1 NUCLIDE SYMBOLS

Write the nuclear symbols for the following nuclides: (a) the one that contains 92 protons and 143 neutrons; and (b) the carbon isotope that has 8 neutrons.

**Strategy:** When determining symbols for nuclides, the key is to remember that the atomic number and number of protons are the same and that the mass number is the sum of the number of protons plus the number of neutrons.

**Solution:**

(a) The nuclide with 92 protons and 143 neutrons has an atomic number of 92 and a mass number that is the sum of the number of protons and neutrons: $A = 92 + 143 = 235$. Atomic number 92 corresponds to uranium, so the symbol for this nuclide is $^{235}_{92}$U, or just $^{235}$U.

(b) All carbon isotopes have the same atomic number, $Z = 6$. From the definition of the nuclide mass number, $A = Z + N = 8 + 6 = 14$. The symbol for this nuclide is $^{14}_{6}$C, or just $^{14}$C.

**Do the Results Make Sense?** The symbol of an element must match its atomic number. Verify this for uranium and carbon by consulting a periodic table. The superscript must be larger than the subscript, since the superscript is the sum of the number of protons (the subscript) and the number of neutrons.

**Common Pitfalls:** Omitting the subscript from an element symbol. Although $^{235}$U means the same as $^{235}_{92}$U, writing the subscript avoids confusion.

**Practice Exercise 20.1**

Write the nuclear symbols for the following nuclides: (a) the nuclide that contains 15 protons and 16 neutrons; (b) neon with the same number of neutrons as protons.

---

## Nuclear Binding Energy

Nuclei with more than one nucleon are held together by the **strong nuclear force**. Energy must be provided to overcome this force and remove a nucleon from a nucleus. This energy is called the **nuclear binding energy**.

Every nucleus contains $Z$ protons and $(A - Z)$ neutrons, so we can visualize a nuclear "formation reaction" for any nuclide, in which protons and neutrons combine to form the product nucleus:

$$Z \, p + (A - Z) \, n \longrightarrow {}^{A}_{Z}X$$

Protons and neutrons are symbolized formally as $^{1}_{1}$p and $^{1}_{0}$n, but they are often simplified to p and n.

Figure 20-2 illustrates this process schematically for fluorine:

$$9 \, p + 10 \, n \longrightarrow {}^{19}_{9}F$$

Because any stable nucleus is more stable than its separated nucleons, nuclear formation reactions of all stable nuclides are exothermic.

**Figure 20-2**

9 protons          10 neutrons          Fluorine nucleus

The nuclear formation reaction for $^{19}_9F$: $Z = 9$, $A - Z = 10$.

Formation reactions for nuclides are easy to visualize but impossible to carry out, so scientists must measure energies of nuclide formation indirectly. Accurate mass measurements are the best way to do this. Einstein recognized that mass is a form of stored energy. His famous equation links energy to mass and the speed of light, $c$:

$$E = mc^2 \qquad (20\text{-}1)$$

According to this equation, formation of a nucleus releases an amount of energy that is related to the change in mass accompanying the formation reaction:

$$\Delta E = (\Delta m)c^2 \qquad (20\text{-}2)$$

Equation 20-2 is the fundamental equation of nuclear energetics. It allows us to calculate the change in energy that accompanies any nuclear reaction from the masses of the particles before and after the transformation. When mass decreases, the energy change is negative, meaning that energy is released to the surroundings. In other words, nuclear reactions in which mass decreases are exothermic.

$c = 2.998 \times 10^8$ m/s

$1\,J = 1\,kg\,m^2/s^2$

Nuclide masses are ordinarily tabulated using units of grams per mole. Therefore, to compute $\Delta E$, it is useful to rewrite Equation 20-2 with energy in units of kilojoules per mole and the change in mass in grams per mole:

$$\Delta E = (\Delta m)c^2 = \Delta m(2.998 \times 10^8\,\text{m/s})^2\,(10^{-3}\,\text{kg/g})(10^{-3}\,\text{kJ/J})$$

$$\Delta E = (\Delta m)(8.988 \times 10^{10})\left(\frac{\text{kg m}^2\,\text{kJ}}{\text{s}^2\,\text{J g}}\right)$$

$$\Delta E = (\Delta m)(8.988 \times 10^{10}\,\text{kJ/g}) \qquad (20\text{-}3)$$

The large positive exponent in Equation 20-3 indicates that a tiny change in mass results in a huge change in energy. Note that when mass is destroyed ($\Delta m < 0$), energy is released ($\Delta E < 0$).

The change in mass that accompanies the formation of a nucleus is called the **mass defect**, and can be found from the difference between the mass of the product nucleus and the masses of its component nucleons:

$$\Delta m = m_{\text{nucleus}} - \left[Z(m_{\text{proton}}) + (Z - A)(m_{\text{neutron}})\right]$$

Once this change in mass is known, Equation 20-3 can be used to calculate the binding energy of any particular nuclide.

The tabulated molar mass of an element divided by Avogadro's number is the *average* mass per atom of that element, but it is not the *exact* mass of an individual nucleus. There are two reasons for this. First, molar masses refer to neutral atoms. The tabulated molar mass of an element includes the mass of its electrons in addition to the mass of its nucleus. Consequently, the mass of $Z$ electrons must be subtracted from the isotopic molar mass in computing the energy of formation of a nuclide. Second, molar masses of the elements are weighted averages of all naturally occurring isotopes of that element. As an example, the most abundant isotope of hydrogen, $^1$H, has an isotopic molar mass of 1.007 825 g/mol, but the presence of small amounts of the isotope $^2$H makes the elemental molar mass of naturally occurring hydrogen slightly larger, 1.007 94 g/mol. In making mass–energy conversions, we must work with isotopic molar masses rather than elemental molar masses.

The isotopic molar masses of all stable and many unstable isotopes have been determined using mass spectrometry. These masses can be found in standard data tables. We provide values as needed for calculations. Example 20-2 illustrates the calculation of nuclear binding energies from isotopic molar masses.

## EXAMPLE 20-2 NUCLEAR BINDING ENERGIES

The most abundant isotope of helium has two neutrons and an isotopic molar mass of 4.002 60 g/mol. Calculate the nuclear binding energy of this nuclide.

**Strategy:** A particular nuclide is made from the combination of $Z$ protons and $(A - Z)$ neutrons. Thus, a neutral atom of a specific isotope contains $Z$ protons, $Z$ electrons, and $(A - Z)$ neutrons. When these particles are brought together, a small amount of mass is converted to energy. To calculate that energy, first count protons, neutrons, and electrons, and then do a mass–energy calculation using Equation 20-3.

**Solution:** The helium isotope has two neutrons, two electrons, and two protons:

| | | |
|---|---|---|
| Protons: | 2(1.007 276 g/mol) | = 2.014 552 g/mol |
| Neutrons: | 2(1.008 665 g/mol) | = 2.017 330 g/mol |
| Electrons: | 2(0.000 5486 g/mol) | = 0.001 0972 g/mol |
| | Total molar mass of components | = 4.032 979 g/mol |

The actual mass of the helium isotope is slightly smaller than the mass of the component particles. The difference represents the amount of mass that is converted to energy when the nuclide forms:

$$\Delta m = (4.002\ 60\text{ g/mol}) - (4.032\ 979\text{ g/mol}) = -0.030\ 38\text{ g/mol}$$

Equation 20-3 lets us calculate an energy change from a mass change:

$$\Delta E = (-0.030\ 38\text{ g/mol})(8.988 \times 10^{10}\text{ kJ/g}) = -2.731 \times 10^9\text{ kJ/mol}$$

Notice that we carry all the significant figures for the masses. It is essential to do this, because a very small mass difference translates into an extremely large amount of energy. Even though the mass difference in this example is less than 1%, the energy difference is more than $10^9$ kJ/mol.

 **Does the Result Make Sense?** The energy change is negative, indicating that this helium nuclide is more stable than its separate component particles. We expect this for any stable nuclide.

**Common Pitfalls:** Neglecting the mass of the electrons. Make sure you account for all particles on both sides of the reaction.

**Practice Exercise 20.2**

The most abundant isotope of uranium is $^{238}U$, with an isotopic molar mass of 238.0508 g/mol. Calculate the nuclear binding energy of this nuclide.

Calculations show that as the mass number increases, so does the total binding energy of the nuclide. More important for the overall stability of the nucleus, however, is the binding energy *per nucleon.* This quantity is the total binding energy divided by the mass number, $A$. It describes how tightly each nucleon is bound to the nucleus. As the binding energy per nucleon becomes more negative, nuclides become more stable. Figure 20-3 shows the binding energy per nucleon plotted vs. mass number. Notice the broad minimum around $A = 60$, $Z = 26$, at a binding energy of $-8.3 \times 10^8$ kJ/mol of nucleons. The most stable nuclide of all is $^{56}_{26}Fe$.

The variations in binding energy shown in Figure 20-3 indicate that there are two types of nuclear reactions that release energy. When heavy nuclides fragment in a process called **fission**, energy is released. When light nuclides combine in a process called **fusion**, energy is also released. We describe these two energy-releasing processes in Sections 20.4 and 20.5.

When the energy change accompanying a process is known, Equation 20-3 can be used to calculate how much the mass changes during the process. Example 20-3 shows how to do this for a chemical reaction.

**Figure 20-3**

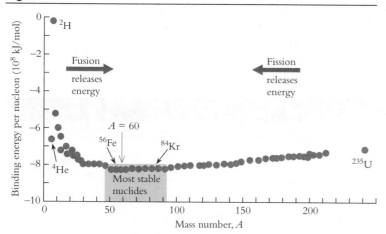

Plot of the binding energy per nucleon vs. mass number, $A$. The most stable nuclides lie in the region near $A = 60$.

---

## EXAMPLE 20-3  MASS EQUIVALENCE OF CHEMICAL ENERGY

Hydrogen, a fuel that releases a large amount of chemical energy when it burns, is used as an energy source in fuel cells. Use standard enthalpies of formation to calculate the change in mass that occurs when 1.00 mole of $H_2$ is burned.

**Strategy:** Whereas Example 20-2 addresses the energy change that accompanies a specific change in mass, Example 20-3 asks about the change in mass when a specific amount of energy is released in an exothermic chemical reaction. First determine $\Delta E$ for the chemical reaction and then use Equation 20-3.

Fuel cell used in the space shuttle

**Solution:** The combustion reaction per mole of hydrogen is as follows:

$$H_2\,(g) + 1/2\,O_2\,(g) \longrightarrow H_2O\,(g)$$

This reaction is also the formation reaction of gaseous $H_2O$. Its $\Delta H_f^{\circ}$ is listed in Appendix D. Enthalpy changes and energy changes are nearly the same, so $\Delta E \cong \Delta H$:

$$\Delta E \cong \Delta H = \Delta H_f^{\circ} = -242 \text{ kJ/mol}$$

To obtain the mass change, substitute this value into Equation 20-3:

$$\Delta E = (8.988 \times 10^{10} \text{ kJ/g})\,\Delta m$$

$$-242 \text{ kJ/mol} = (8.988 \times 10^{10} \text{ kJ/g})\,\Delta m$$

$$\Delta m = -2.69 \times 10^{-9} \text{ g/mol}$$

**Does the Result Make Sense?** When 1.00 mole of hydrogen reacts with oxygen, a few nanograms are converted to energy. This amount, which is typical of the mass consumed in conventional chemical reactions, is too small to weigh. Thus, within the precision of measurements, mass is conserved in ordinary chemical reactions.

**Common Pitfalls:** Assuming that there is no mass change in a chemical reaction. The mass changes are much smaller than in nuclear reactions, but are not zero.

**Practice Exercise 20.3**
A large natural gas-fired electrical generating station may burn 0.5 kg of methane per second. Use standard enthalpies of formation to calculate the change in mass that occurs if the power station operates for 24 hours.

The binding energy of a typical nuclide is about $8 \times 10^8$ kJ per mol of nucleons. To obtain a better feel for just how immense this quantity is, compare the energy released in a nuclear reaction with the energy released in a chemical reaction. As noted in Example 20-3, the chemical reaction of $H_2$ with $O_2$ is highly exothermic. When 1.00 mole of $H_2$ reacts chemically with $O_2$, 242 kJ of energy is released. If that same 1.00 mole of $H_2$ could be fused with neutrons to form He nuclei, the calculation of Example 20-2 indicates that $2.73 \times 10^9$ kJ of energy would be released. There is a difference of seven orders of magnitude between these two quantities; in other words, *fusion* of hydrogen nuclei releases 10 000 000 times as much energy as does *combustion* of molecular hydrogen.

## Energy Barriers

If fission of large nuclides and fusion of small nuclides are vastly exothermic, why have these processes not converted all elements into the most stable one, iron-56? The reason is that nuclear reactions have huge energy barriers. These barriers are analogous to the activation energy barriers for conventional combustion reactions. The combustion of gasoline, for example, is highly exothermic ($\Delta H < -5000$ kJ/mol), but the rate of the reaction is negligible at room temperature because an activation energy barrier prevents combustion from occurring without an external boost.

Massive nuclides do not fragment into lighter, more stable nuclides because an immense amount of energy must be supplied to overcome strong attractions among nucleons and pull a nucleus apart. Light nuclides do not fuse to give more massive, more stable nuclides because an immense amount of energy must be supplied to overcome electrical repulsion and bring nuclei together. The following analysis shows the magnitudes of typical activation energies for nuclear reactions.

Nuclei cannot fuse without first overcoming the repulsive electrical forces between them. Recall that this repulsion prevents the nuclei in molecules from approaching closer than the lengths of chemical bonds, which are about 100 pm. For nuclei to fuse, they must be brought within about $10^{-3}$ pm of each other. Equation 3-1, which describes electrical energy, allows calculation of the magnitude of this barrier.

$$E_{electrical} = k\frac{q_1 q_2}{r} \tag{3-1}$$

For a pair of nuclei, the charges $q_1$ and $q_2$ are equal to the nuclear charges $Z_1$ and $Z_2$, and the value of $k$ is $1.389 \times 10^5$ kJ pm/mol when charges are expressed in units of nuclear charge:

$$E_{electrical} = \frac{(1.389 \times 10^5 \text{ kJ pm/mol})(Z_1)(Z_2)}{d}$$

Because isotopes of hydrogen have the smallest possible nuclear charge (+1), they also have the minimal energy barrier to fusion. Consider, for example, the fusion of two deuterium nuclei:

$$^2_1H + {}^2_1H \longrightarrow {}^4_2He$$

In this example, $Z_1$ and $Z_2$ are each +1, and $d$ is the sum of the two nuclear radii, $2.8 \times 10^{-3}$ pm. Substituting these values gives $E = 5.0 \times 10^7$ kJ/mol. This is small relative to the energy released by fusion but very large relative to typical chemical activation energies of about 100 kJ/mol. The graph in Figure 20-4 is a schematic representation of the energy barriers for fusion of deuterium.

Activation energy barriers for nuclear reactions are more than a million times larger than the activation energies of conventional chemical reactions, so nuclear processes require immense energy inputs before they can occur. As one example, deuterium nuclei cannot fuse unless they first acquire $5 \times 10^7$ kJ/mol of energy. At room temperature, the average kinetic energy of deuterium nuclei is only 3.7 kJ/mol. A sample of deuterium needs only the energy boost from a spark or flame to react chemically with oxygen, but the temperature must be raised to $5 \times 10^6$ K before deuterium nuclei acquire enough kinetic energy to overcome the barrier to fusion.

The energy profile in Figure 20-4 also indicates that nuclei cannot eject nucleons without overcoming the strong nuclear

**Figure 20-4**

The energy profile for nuclear fusion of two deuterium nuclei.

forces that hold nucleons together, which are on the order of $10^8$ kJ/mol. To summarize, electrical repulsion keeps nuclei from fusing, and the strong nuclear attractive force keeps nuclei from disintegrating.

## Stable Nuclides

Although nuclides with mass numbers around 60 are the most stable, the balance of electrical repulsion and strong nuclear attraction makes many combinations of protons and neutrons stable for indefinite times. Nevertheless, many other combinations decompose spontaneously. For example, all hydrogen nuclides with $A > 2$ are so unstable that only one of them—tritium, $^3_1H$—exists even briefly. If other nuclides of hydrogen could be made, they would decompose rapidly by expelling neutrons. Hydrogen has two stable nuclides, and fluorine has just one stable nuclide, $^{19}_9F$. In contrast, tin has 10, with mass numbers 112, 114, 115, 116, 117, 118, 119, 120, 122, and 124. All stable nuclides, as well as a few that are unstable, are found in the Earth's crust and atmosphere. Those unstable nuclides that occur naturally have natural lifetimes that are longer than the age of the Earth or are produced in naturally occurring nuclear reactions.

For each different element, there are thus a few specific values of $A$ that result in stable nuclei. Figure 20-5 shows all the stable nuclei on a plot of the number of neutrons $(N)$ versus the number of protons $(Z)$. These data show a striking pattern: all stable nuclei fall within a "belt of stability." Any nucleus with a ratio of neutrons to protons that falls outside the belt of stability is unstable and decomposes spontaneously. Lighter nuclei lie along the $N = Z$ line, but the $N{:}Z$ ratio of stable nuclei rises slowly until it reaches 1.54. The trend is illustrated by the $N{:}Z$ ratios of the following stable nuclei: $^4_2He(1.00)$, $^{31}_{15}P(1.07)$, $^{56}_{26}Fe(1.15)$, $^{81}_{35}Br(1.31)$, $^{197}_{79}Au$ (1.49), and $^{208}_{82}Pb(1.54)$. Any nuclide with a ratio of neutrons to protons that falls outside the belt of stability is unstable and decomposes sponta-

Figure 20-5

Plot of the $Z$ and $N$ values of stable nuclei. All stable nuclei fall within a belt that lies between $N = Z$ and $N = 1.54\,Z$.

neously. Nuclides within the belt of stability contain just the right number of neutrons. These neutrons generate enough strong nuclear force to offset the electrical repulsions among protons. When there are too many protons, this becomes impossible: no nuclide with $Z > 83$ is stable. Some elements with higher $Z$ are found on Earth, notably radium $(Z = 88)$, thorium $(Z = 90)$, and uranium $(Z = 92)$, but all such elements are unstable and eventually disintegrate into nuclei with $Z < 83$. Consequently, the set of stable nuclei, those that make up the world of "normal" chemistry and provide the material for all terrestrial chemical reactions, is a small subset of all possible nuclei.

Electrical forces cause positively charged protons to repel one another, so an additional attractive force is needed to stabilize nuclei containing more than one proton. This attractive force, called the *strong nuclear force*, acts between protons and neutrons. The strong nuclear force operates only at very small distances within the nucleus, but at those distances, it is about 100 times stronger than proton–proton repulsion. This strong force is what binds neutrons and protons tightly in the nucleus. As $Z$ increases, the effect of electrical repulsion increases. This requires more neutrons to generate enough nuclear force to stabilize the nucleus.

The main features of nuclear stability can be summarized qualitatively. The balance between electrical repulsion and strong nuclear attraction changes with the number of protons and neutrons in the nucleus. Compressing more than one positively charged proton into a volume as small as the nucleus leads to strongly repulsive electrical forces that must be offset by the presence of specific numbers of neutrons. Neutrons provide strong nuclear binding forces without adding electrical repulsive forces. The extra binding force of the neutrons holds the protons within the nucleus.

Within the belt of stability, Table 20-2 shows that nuclides with even numbers of protons and neutrons are more prevalent than those with odd numbers of protons or neutrons. Almost 60% of all stable nuclides have even numbers of both protons and neutrons, whereas fewer than 2% have odd numbers of both. Moreover, of the five stable odd–odd nuclides, four are the lightest odd-$Z$ elements: $^{2}_{1}H$, $^{6}_{3}Li$, $^{10}_{5}B$, $^{14}_{7}N$. Above $Z = 7$, there are 152 stable even–even nuclides and just one, $^{158}_{57}La$, that is odd–odd.

## TABLE 20-2  DISTRIBUTION OF STABLE NUCLIDES

| Protons | Neutrons | Stable Nuclides | % |
|---------|----------|-----------------|-----|
| Even | Even | 154 | 58.8 |
| Even | Odd | 53 | 20.2 |
| Odd | Even | 50 | 19.1 |
| Odd | Odd | 5 | 1.9 |

## SECTION EXERCISES

●20.1.1  Write the symbols and determine $Z$, $N$, and $A$ for the following nuclides: (a) an oxygen nucleus with the same number of neutrons and protons; (b) element 43 with 55 neutrons; and (c) the unstable nuclide of hydrogen with the lowest mass.

●20.1.2  Predict whether each of the following is stable or unstable. If you predict that it is unstable, give your reason: (a) the nuclide with 94 protons and 150 neutrons; (b) the iodine nuclide with 73 neutrons; (c) $^{154}_{64}Gd$; and (d) the oxygen nuclide with 6 neutrons.

●20.1.3  Fluorine has only one stable isotope, $^{19}_{9}F$. Compute the total binding energy and the binding energy per nucleon for this nuclide.

## 20.2 NUCLEAR DECAY

Unstable nuclides decompose spontaneously into other, more stable nuclides. These decompositions are called **nuclear decay**, and unstable nuclides are called **radioactive**. Three features characterize nuclear decays: the products, the energy released, and the rate of decay. The products of nuclear decomposition include electrons, helium nuclei, and high-energy photons. Nuclear decay releases energy because the decay products are more stable than the starting materials. Furthermore, each unstable nuclide decomposes at some particular rate, which can take place on a timescale from less than a second to many millions of years. A key concept is that both mass number and electrical charge are conserved during a nuclear decay. That is, the sum of the number of protons and neutrons is the same after transformation as before, and the sum of the charges is also the same after transformation as before. Any description of a nuclear reaction must take these conservation requirements into account.

As an example, consider the decay of free neutrons. The stable particle with $A = 1$ is a proton, and neutron decay results in a proton. A neutron has $A = 1$, so its decay products must also have $A = 1$. The neutron has zero charge, so the sum of the charges of its decay products must also be zero. Because the proton carries a charge of $+1$, another particle with a $-1$ charge is required. This particle must have $A = 0$ to ensure that the mass number is conserved. The only particle with these properties is the electron. Thus, a neutron decays into a proton and an electron:

$$\mathrm{{}^{1}_{0}n \longrightarrow {}^{1}_{1}p + {}^{0}_{-1}e}$$

Although mass *number* is conserved, total mass is *not* conserved in nuclear reactions. Instead, some mass is converted into energy, or some energy is converted into mass. The notion of conservation of mass is valid within experimental error for all chemical transformations, as Example 20-3 illustrates. For nuclear processes, however, the amounts of energy produced or consumed are large enough to generate measurable mass changes. In nuclear decay, some mass is always converted into energy. We can use Equation 20-3, the mass of the decaying nucleus, and the masses of its products to calculate how much energy is released.

This reasoning is easily applied to the decay of a free neutron. The masses of the three participants are given in Table 20-1:

Neutron: 1.008665 g/mol    Proton: 1.007 276 g/mol    Electron: 0.000 5486 g/mol

$$\Delta m = 1.007\ 276 + 0.000\ 5486 - 1.008\ 665 = -0.000\ 840 \text{ g/mol}$$

$$\Delta E = (-0.000\ 840 \text{ g/mol})(8.988 \times 10^{10} \text{ kJ/g}) = -7.55 \times 10^{10} \text{ kJ/mol}$$

To summarize, the equation for a nuclear reaction is balanced when the total *charge* and total *mass number* of the products equal the total charge and total mass number of the reactants. This conservation requirement is one reason why the symbol for any nuclide includes its charge number ($Z$) as a subscript and its mass number ($A$) as a superscript. These features provide a convenient way to keep

Ernest Rutherford, working at McGill University in Montreal, observed that the paths taken by energetic particles emitted by radioactive uranium and thorium responded in three ways to magnetic fields: slightly bent, strongly bent, and unaffected. He gave them the designations $\alpha$, $\beta$, and $\gamma$. Even though scientists soon identified the particles, they still use these names to emphasize that they are nuclear decay products.

track of charge and mass balances. Notice that in the equation for neutron decay, the sum of the subscripts for reactants equals the sum of the subscripts for products. Likewise, the sum of the superscripts for reactants equals the sum of the superscripts for products. We demonstrate how to balance equations for other reactions as they are introduced.

**Figure 20-6**

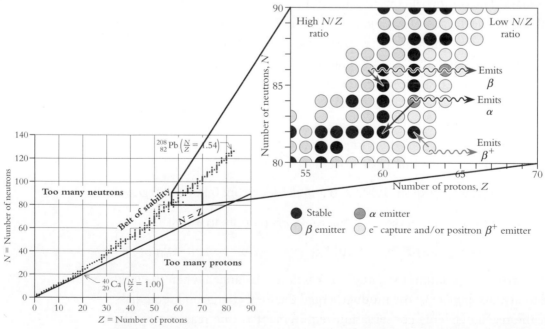

A detailed view of one portion of the $N$ vs. $Z$ plot of nuclides, illustrating the modes of nuclear decay for nuclides on either side of the belt of stability.

## TABLE 20-3 TYPES OF NUCLEAR DECAY

| Type of Decay | Cause of Instability | Emission | $\Delta Z$ of Emitter | $\Delta A$ of Emitter |
|---|---|---|---|---|
| $\alpha$ emission | Too massive ($Z > 83$) or $N:Z$ too large | $\alpha \left(^{4}_{2}\text{He}\right)$ | $-2$ | $-4$ |
| $\beta$ emission | $N:Z$ too large | $\beta \left(^{0}_{-1}\text{e}\right)$ | $+1$ | $0$ |
| $\beta^{+}$ emission | $N:Z$ too small | $\beta \left(^{0}_{+1}\text{e}\right)$ | $-1$ | $0$ |
| Electron capture | $N:Z$ too small | X-ray photon | $-1$ | $0$ |
| $\gamma$ emission | Excited nucleus | $\gamma$-ray photon | $0$ | $0$ |

## Decay Processes

There are five fundamental types of nuclear decay process, as listed in Table 20-3. Figure 20-6 diagrams how nuclear decays affect $N$ and $Z$. As the figure suggests, the decay process of any particular unstable nuclide depends on the reason for its instability.

Many nuclides that are too massive to be stable lose mass by emitting energetic helium nuclei, reducing $A$ by 4 units and $Z$ by 2 units. Energetic helium nuclei are called ($\alpha$) *particles*, so these nuclides are said to undergo **$\alpha$ emission**. Here are some examples:

$$^{146}_{62}\text{Sm} \longrightarrow {}^{142}_{60}\text{Nd} + {}^{4}_{2}\alpha$$

$$^{222}_{86}\text{Rn} \longrightarrow {}^{218}_{84}\text{Po} + {}^{4}_{2}\alpha \qquad \text{(see Figure 20-6)}$$

$$^{218}_{84}\text{Po} \longrightarrow {}^{214}_{82}\text{Pb} + {}^{4}_{2}\alpha$$

Nuclides that lie *above* the belt of stability have ratios of neutrons to protons that are too high. To become stable, these nuclides need to increase their nuclear charges. Such a nuclide converts a neutron into a proton plus an energetic electron. The nucleus expels the electron, thereby increasing its charge by 1 unit while leaving its mass number unchanged. Energetic electrons are called **$\beta$ particles**, so these nuclides are said to undergo $\beta$ emission. Here are some examples:

$$^{3}_{1}\text{H} \longrightarrow {}^{3}_{2}\text{He} + {}^{0}_{-1}\beta$$

$$^{145}_{59}\text{Pr} \longrightarrow {}^{145}_{60}\text{Nd} + {}^{0}_{-1}\beta \qquad \text{(see Figure 20-6)}$$

$$^{208}_{79}\text{Au} \longrightarrow {}^{208}_{80}\text{Hg} + {}^{0}_{-1}\beta$$

Nuclides that lie *below* the belt of stability have low neutron:proton ratios and must reduce their nuclear charges to become stable. These nuclides can convert protons into neutrons by **positron emission**. Positrons (symbolized $\beta^{+}$) are particles with the same mass as electrons but with a charge of $+1$ instead of $-1$. Here are three examples:

$$^{18}_{9}\text{F} \longrightarrow {}^{18}_{8}\text{O} + {}^{0}_{+1}\beta^{+}$$

$$^{52}_{26}\text{Fe} \longrightarrow {}^{52}_{25}\text{Mn} + {}^{0}_{+1}\beta^{+} \qquad \text{(see Figure 20-6)}$$

$$^{144}_{63}\text{Nd} \longrightarrow {}^{144}_{62}\text{Eu} + {}^{0}_{+1}\beta^{+}$$

**Figure 20-7**

Positron emission (a) and electron capture (b) are observed indirectly from the high-energy radiation that each produces.

The electron captured by a nucleus during electron capture is symbolized e, not β, because β is reserved for electrons ejected during nuclear decay.

Positrons cannot be observed directly because, as Figure 20-7a illustrates, when a positron encounters an electron, the two particles annihilate each other, converting their entire mass into a pair of photons. The occurrence of positron emission can be inferred from the observation of such a pair of photons. Each photon produced in this process has a specific energy: $E_{\text{photon}} = 9.87 \times 10^7$ kJ/mol. Photons with such high energy are called **γ rays**.

A nuclide with a low neutron: proton ratio can also reduce its nuclear charge by capturing one of its 1s orbital electrons in a process called **electron capture**. The captured electron combines with a proton to give a neutron, so $Z$ drops by one unit, whereas $A$ remains fixed. For example:

$$^{26}_{13}\text{Al} + ^{0}_{-1}\text{e} \longrightarrow ^{26}_{12}\text{Mg}$$

Like positron emission, electron capture is never observed directly. However, after electron capture, the product atom is missing one of its 1s electrons, as shown schematically in Figure 20-7b. When an electron from an outer orbital occupies this vacancy in the 1s orbital, a photon is emitted with energy that falls in the X-ray region of the spectrum ($E \cong 10^6$ kJ/mol).

Nuclides that are unstable because they have odd–odd composition can be converted into stable nuclides with even–even composition through any of three processes: electron emission, positron emission, or electron capture. Each process changes $Z$ and $N$ by one unit. The $^{64}_{29}\text{Cu}$ nuclide provides a convenient illustration of all three processes because it decays by all three modes:

$$^{64}_{29}\text{Cu} \longrightarrow ^{64}_{30}\text{Zn} + ^{0}_{-1}\beta$$

$$^{64}_{29}\text{Cu} \longrightarrow ^{64}_{28}\text{Ni} + ^{0}_{1}\beta^+$$

$$^{64}_{29}\text{Cu} + ^{0}_{-1}\text{e} \longrightarrow ^{64}_{28}\text{Ni}$$

It is unusual for a nuclide to decay by three different paths. Generally, an odd–odd nuclide decays preferentially by only one mode. However, predicting which mode of decay predominates is beyond the scope of this book.

In addition to the γ rays that result from the annihilation of a positron and an electron, γ rays are also emitted directly by many nuclides. The emission of a γ ray from an unstable nuclide changes neither $Z$ nor $A$, because γ rays are photons with zero mass and zero charge. Many nuclear reactions generate γ rays because of the release of energy that accompanies these reactions. Although most of this energy appears as kinetic energy of the products, some energy may be retained by the product nuclide as excitation energy. The excited nuclide can get rid of this excess energy by emitting a $\gamma$ ray, just as an excited atom can lose excess energy by emitting a photon. Excited nuclides (also called metastable nuclides) are identified with a superscript m after the value of $A$. For example, emission from radium-226 gives an excited radon nuclide, $^{222m}$Rn, which loses its excess energy by emitting a $\gamma$ ray:

$$^{226}\text{Ra} \longrightarrow {}^{222m}\text{Rn} + \alpha$$
$$^{222m}\text{Rn} \longrightarrow {}^{222}\text{Rn} + \gamma$$

This decay sequence illustrates a common feature of nuclear decays of very heavy elements: several decays often occur in sequence, because the product of the initial decay is also unstable. As another example, $^{222}$Rn is unstable, undergoing $\alpha$ decay:

$$^{222}\text{Ra} \longrightarrow {}^{218m}\text{Po} + \alpha$$

Example 20-4 traces this example of nuclear decay through its entire sequence of nuclear decompositions.

## EXAMPLE 20-4 DECAY SEQUENCES

Radon-222 is an unstable nuclide that has been detected in the air of some homes. Its presence is a concern because of high health hazards associated with exposure to its radioactivity. Gaseous radon easily enters the lungs, and once it decays, the products are solids that remain embedded in lung tissue. Radon-222 transmutes to a stable nuclide by emitting $\alpha$ and $\beta$ particles. The first four steps are $\alpha$, $\alpha$, $\beta$, $\beta$. Write this sequence of nuclear reactions and identify each product.

**Strategy:** Charge number and mass number must be conserved in each reaction. Thus, each $\alpha$ particle decreases the nuclear charge by two units and the mass number by four units. Similarly, each $\beta$ emission increases the nuclear charge by one unit but leaves the mass number unchanged. Consult a periodic table to identify the elemental symbol of each product nuclide.

**Solution:** The starting nuclide is $^{222}$Rn, which, according to the decay sequence, emits an $\alpha$ particle:

$$^{222}_{86}\text{Rn} \longrightarrow {}^{A}_{Z}\text{X} + {}^{4}_{2}\alpha$$

The product has $A = 222 - 4 = 218$ and $Z = 86 - 2 = 84$. Element 84 is Po, so radon-222 decomposes to give an $\alpha$ particle and polonium-218:

$$^{222}_{86}\text{Rn} \longrightarrow {}^{218}_{84}\text{Po} + {}^{4}_{2}\alpha$$

The polonium nucleus is unstable, and the sequence indicates that it decays via $\alpha$ emission:

$$^{218}_{84}\text{Po} \longrightarrow {}^{214}_{82}\text{Pb} + {}^{4}_{2}\alpha$$

The decay sequence indicates that lead-214 is unstable as well. This nuclide converts a neutron to a proton by $\beta$ emission:

$$^{214}_{82}\text{Pb} \longrightarrow {}^{A}_{Z}\text{X} + {}^{0}_{-1}\beta$$

For this product, $A = 214 - 0 = 214$ and $Z = 82 - (-1) = 83$. The product nucleus is Element 83, bismuth:

$$^{214}_{82}\text{Pb} \longrightarrow {}^{214}_{83}\text{Bi} + {}^{0}_{-1}\beta$$

The fourth decay in the sequence is also a β emission, which produces another isotope of Element 84, polonium-214:

$$^{214}_{83}\text{Bi} \longrightarrow {}^{214}_{84}\text{Po} + {}^{0}_{-1}\beta$$

**Do the Results Make Sense?** You can confirm that both *A* and *Z* are conserved in these processes. Extra-heavy nuclides like $^{222}$Rn frequently decay through sequences of α and β decays, because to reach the belt of stability, they must not only shed mass (α decays) but also decrease their *N:Z* ratios (β decays).

**Practice Exercise 20.4**

This decay sequence continues with four more steps: α, β, β, and α. Determine the correct equations for these reactions.

## Rates of Nuclear Decay

A sample of any unstable nuclide undergoes nuclear decay continuously as its individual nuclei undergo reaction. All nuclear decays obey first-order kinetics. For a chemical reaction, rate $= kc$. This rate law was integrated to give the form similar to Equation 12-3, which relates concentration, *c*, to time, *t*, for a first-order process ($c_0$ is the concentration present at $t = 0$):

$$\ln\left(\frac{c_0}{c}\right) = kt \qquad (12\text{-}3)$$

Recall from our discussion of kinetics in Chapter 12 that elementary unimolecular reactions are first order. The observed first-order kinetics of nuclear decays indicates that each nuclear decay mode is elementary and "uninuclear."

Recall also that for first-order reactions, the time required for exactly half of the substance to react is independent of how much material is present. This constant time interval is the half-life, $t_{1/2}$. Equation 12-4 relates the half-life to the reaction rate constant:

$$t_{1/2} = \frac{\ln 2}{k} = \frac{0.693}{k} \qquad (12\text{-}4)$$

The rate constant in a nuclear decay process is referred to as a **decay constant**. By convention, nuclear decay rates are expressed using half-lives rather than decay constants. Every unstable nuclide has its own characteristic half-life. Nuclide half-lives range from shorter than a second to longer than a billion years. The half-lives of some representative nuclides appear in Table 20-4.

The particles emitted during nuclear decay are so energetic that it is possible to count them individually. For this reason, the rate equations for nuclear decay are often expressed using the number of nuclei present, *N*, rather than molar concentrations:

$$\text{Rate} = \frac{-dN}{dt} = kN = \frac{0.693\,N}{t_{1/2}} \qquad (20\text{-}4)$$

## TABLE 20-4 HALF-LIVES OF REPRESENTATIVE NUCLIDES

| Nuclide | Mode | Half-Life |
|---------|------|-----------|
| $^{214}Po$ | $\alpha$ | $1.6 \times 10^{-4}$ s |
| $^{210}Tl$ | $\beta$ | 1.32 min |
| $^{239}U$ | $\beta$ | 24 min |
| $^{60}Co$ | $\beta, \gamma$ | 5.26 years |
| $^{3}H$ | $\beta$ | 12.3 years |
| $^{90}Sr$ | $\beta$ | 28.1 years |
| $^{14}C$ | $\beta$ | $5.73 \times 10^{3}$ years |
| $^{235}U$ | $\alpha$ | $7.0 \times 10^{8}$ years |
| $^{40}K$ | $\beta$ | $1.28 \times 10^{9}$ years |
| $^{238}U$ | $\alpha$ | $4.51 \times 10^{9}$ years |

In these equations, $N$ is the number of nuclei present at time $t$, and $N_0$ is the number of nuclei present at $t = 0$. When Equation 20-4 is integrated (see Chapter 12, Section 12.4 on first-order reactions to know how to do this), we obtain an expression for the number of nuclei remaining after time $t$:

$$\ln\left(\frac{N_0}{N}\right) = \frac{0.693\ t}{t_{1/2}} \tag{20-5}$$

As shown in Example 20-5, Equation 20-4 is used to find a nuclear half-life from measurements of nuclear decays. Equation 20-5 is used to find how much of a radioactive substance will remain after a certain time, or how long it will take for the amount of substance to fall by a given amount. Example 20-6 provides an illustration of this type of calculation. In Section 20.7, we show that Equation 20-5 also provides a way to determine the age of a material that contains radioactive nuclides.

## EXAMPLE 20-5 NUCLEAR HALF-LIFE

Plutonium is a synthetic element that is used in nuclear weapons. It is also proposed for use as a nuclear fuel. A sample of $^{239}Pu$ with mass of 1.00 mg decays at a rate of $2.3 \times 10^{6}$ counts/s. (Each count corresponds to the decay of one nucleus.) What is the half-life of this isotope?

**Strategy:** We are asked to determine the half-life of an isotope from the information about its decay rate. The emission rate is $-\Delta N/\Delta t$, so rearrange Equation 20-4 to calculate $t_{1/2}$:

$$\frac{dN}{dt} = \frac{-0.693\ N}{t_{1/2}} \quad \text{or} \quad t_{1/2} = \frac{0.693\ N}{-dN/dt}$$

**Solution:** The number of nuclei can be calculated from the mass of the sample and the molar mass of the isotope. Because the molar mass of an isotope is nearly equal to its mass number, we can use the mass number without introducing significant error:

$$N = \text{Number of nuclei} = \frac{mN_A}{M}$$

$$N = \frac{(1.00 \times 10^{-3}\,\text{g})(6.022 \times 10^{23}\,\text{nuclei/mol})}{239\,\text{g/mol}} = 2.52 \times 10^{18}\,\text{nuclei}$$

Now we can substitute to determine the half-life:

$$t_{1/2} = \frac{(2.52 \times 10^{18}\,\text{nuclei})(0.693)}{2.3 \times 10^{6}\,\text{nuclei/s}} = 7.6 \times 10^{11}\,\text{s}$$

This half-life is easier to interpret if expressed in years rather than seconds:

$$t_{1/2} = (7.6 \times 10^{11}\,\text{s})\left(\frac{1\,\text{min}}{60\,\text{s}}\right)\left(\frac{1\,\text{h}}{60\,\text{min}}\right)\left(\frac{1\,\text{day}}{24\,\text{h}}\right)\left(\frac{1\,\text{year}}{365\,\text{day}}\right) = 2.4 \times 10^{4}\,\text{years}$$

**Does the Result Make Sense?** A half-life of many years is reasonable, given that isotopic half-lives vary from less than a second to millions of years. It takes 24 000 years for half of the plutonium nuclei in any sample of $^{239}$Pu to decay.

**Common Pitfalls:** Assuming that a large decay rate implies a short half-life. The decay rate depends on both the half-life and the number of radioactive nuclei present.

**Practice Exercise 20.5**

There are no stable isotopes of technetium, but several radioactive isotopes can be prepared. A sample of $^{99}$Tc with mass of 2.35 μg emits $6.65 \times 10^{6}$ β particles in a time interval of 75 min. What is the half-life (in years) of this nuclide?

---

### EXAMPLE 20-6  RADIOACTIVE DECAY

One of the problems with radioactive nuclides such as $^{239}$Pu is that their decay cannot be stopped, so any sample of the nuclide continues to emit dangerous amounts of radiation until most of it has decayed. Use the result of Example 20-5 and assume that the radiation level from $^{239}$Pu is no longer a hazard when 99% of it has decayed. How long will this take?

**Strategy:** The question asks for the time it takes for 99% of a sample of plutonium to decay. The half-life is known from Example 20-5. Equation 20-5 relates the ratio $N_0/N$ to time and the half-life for decay. This equation can be solved for $t$, the time at which the ratio reaches the desired value:

$$\ln\left(\frac{N_0}{N}\right) = \frac{0.693\,t}{t_{1/2}} \quad \text{or} \quad t = \frac{t_{1/2}\ln(N_0/N)}{0.693}$$

**Solution:** When 99% of the Pu has decayed, 1% of the original amount, in other words $0.01N_0$, remains. Use this information to evaluate the ratio in the ln term:

$$\frac{N_0}{N} = \frac{N_0}{0.01\,N_0} = 100$$

Now substitute values and calculate $t$:

$$t = \frac{(2.4 \times 10^4 \text{ years})(\ln 100)}{0.693} = 1.6 \times 10^5 \text{ years}$$

**Does the Result Make Sense?** 160 000 years is a very long time, but the half-life of this nuclide is 24 000 years, and it makes sense that it will take many half-lives before the amount of Pu is 1% of its original amount. It will be a very long time before the plutonium that is stockpiled decays sufficiently that it no longer is a potential hazard.

**Practice Exercise 20.6**
How long will it take for two-thirds of the $^{99}$Tc in Practice Exercise 20.5 to decay?

## SECTION EXERCISES

●**20.2.1** Write the modes of decay that describe the following reactions: (a) $^{213}$Bi to $^{213}$Po; (b) $^{213}$Bi to $^{209}$Tl; (c) $^{213m}$Bi to $^{213}$Bi; and (d) $^{207}$Bi to $^{207}$Pb.

●**20.2.2** Identify the nuclides that decay in the following manner: (a) a nuclide undergoes $\beta$ and $\gamma$ decays to give $Z = 58$ and $A = 140$; (b) a nuclide undergoes $\alpha$ decay to give polonium-218; and (c) a nuclide captures an orbital electron to give tellurium with 73 neutrons.

●**20.2.3** Radioisotopes are used in many research applications. Because they decay continuously, the shelf life of a radioisotope is limited. An isotope used for bone marrow scanning is $^{111}$In, $t_{1/2} = 2.8$ days. Within what time must this isotope be used if it is effective down to 5% of its initial activity?

## 20.3 INDUCED NUCLEAR REACTIONS

Stable nuclides remain the same indefinitely, whereas unstable nuclides disintegrate continuously. In time, therefore, every element should be composed entirely of stable isotopes. On Earth, most elements have no naturally occurring unstable isotopes. With two exceptions, the unstable nuclides that are present either have half-lives longer than the age of the Earth (U, Ra, Th) or are products of the decays of these long-lived nuclides (Rn, Po). The exceptions are $^{3}$H ($t_{1/2} = 12.3$ years) and $^{14}$C ($t_{1/2} = 5730$ years). These half-lives are very short compared with the age of the Earth, so any $^{3}$H and $^{14}$C present during the Earth's formation decayed long ago. Thus, the $^{3}$H and $^{14}$C on Earth today must come from ongoing nuclear reactions. These nuclides are not formed in any nuclear-decay schemes. Instead, each forms through a binuclear reaction in which a nuclear projectile collides and reacts with another nucleus. Such reactions, which are called **induced nuclear reactions**, are categorized according to the nature of the nuclear projectile that induces the reaction.

### Neutron-Capture Reactions

One way to create unstable nuclides is by neutron capture. Because neutrons have no electrical charge, they readily penetrate any nucleus and may be captured by the nucleus. The Sun emits neutrons, so a continuous stream of solar neutrons bathes the Earth's atmosphere. The most abundant nuclide in the atmosphere, nitrogen-14, can capture a neutron to form the unstable nuclide $^{15m}N$. This nucleus rapidly ejects a proton, producing carbon-14:

$$^{14}_{7}N + ^{1}_{0}n \longrightarrow \left(^{15m}_{7}N\right) \longrightarrow ^{14}_{6}C + ^{1}_{1}p$$

Although carbon-14 decays via β emission with $t_{1/2} = 5730$ years, it is replenished continuously by this neutron-capture reaction. We show in Section 20.7 how this isotope is used to estimate the age of carbon-containing artefacts.

The other short-lived nuclide in the atmosphere, $^{3}H$, is also produced by disintegration of metastable $^{15}N$ nuclei. When nitrogen-14 captures a very energetic neutron, the metastable nucleus has sufficient excess energy that it fragments into $^{12}C$ and $^{3}H$ rather than $^{14}C$ and a proton:

$$^{14}_{7}N + ^{1}_{0}n \longrightarrow \left(^{15m}_{7}N\right) \longrightarrow ^{12}_{6}C + ^{3}_{1}H$$

Almost every nuclide undergoes neutron capture if a source of neutrons is available. Unstable nuclides used in radiochemical applications are manufactured by neutron bombardment. A sample containing a suitable target nucleus is exposed to neutrons coming from a nuclear reactor (see Section 20.4). When a target nucleus captures a neutron, its mass number increases by 1:

$$^{A}_{Z}X + ^{1}_{0}n \longrightarrow ^{(A+1)m}_{Z}X$$

Neutron capture is always exothermic, because the neutron is attracted to the nucleus by the strong nuclear force. Consequently, neutron capture generates a product nuclide in a metastable, excited state. These excited nuclei typically lose energy by emitting either γ rays or protons:

$$^{(A+1)m}_{Z}X \longrightarrow ^{(A+1)}_{Z}X + \gamma \qquad ^{(A+1)m}_{Z}X \longrightarrow ^{A}_{(Z-1)}Y + ^{1}_{1}p$$

The following are specific examples of each type of decay:

$$^{98}_{42}Mo + ^{1}_{0}n \longrightarrow ^{99}_{42}Mo + \gamma \qquad ^{14}_{7}N + ^{1}_{0}n \longrightarrow ^{14}_{6}C + ^{1}_{1}p$$
$$^{207}_{79}Au + ^{1}_{0}n \longrightarrow ^{208}_{79}Au + \gamma \qquad ^{3}_{2}H + ^{1}_{0}n \longrightarrow ^{3}_{1}H + ^{1}_{1}p$$

One of the most important isotopes in nuclear medicine, $^{99m}Tc$, is produced by bombarding molybdenum with neutrons. The initial product, $^{99}Mo$, has a high ratio of neutrons to protons, so it decomposes by releasing a β particle. The product is an excited technetium nuclide that emits a γ ray:

$$^{98}_{42}Mo + ^{1}_{0}n \longrightarrow ^{99}_{42}Mo + \gamma$$

$$^{99}_{42}Mo \longrightarrow ^{99m}_{43}Tc + ^{0}_{-1}\beta$$

$$^{99m}_{43}Tc \longrightarrow ^{99}_{43}Tc + \gamma$$

The product of this three-step nuclear process, $^{99}Tc$, is also unstable but lasts much longer than its predecessor. It decays by β emission with $t_{1/2} = 2.12 \times 10^{5}$ years.

## Other Binuclear Reactions

Neutrons readily induce nuclear reactions, but they always produce nuclides on the *high* neutron:proton side of the belt of stability. Protons must be added to the nucleus to produce an unstable nuclide with a *low* neutron:proton ratio. Because protons have positive charges, this means that the bombarding particle must have a positive charge. Nuclear reactions with positively charged particles require projectile particles that possess enough kinetic energy to overcome the electrical repulsion between two positive particles.

Ernest Rutherford was the first person to observe a binuclear reaction. In 1919, he exposed a sample of nitrogen to $\alpha$ particles from a naturally radioactive source. He observed the production of protons and deduced from the requirements of charge and mass balance that the other product was oxygen-17:

Compound Nucleus

The immediate product of a reaction between two nuclei is a **compound nucleus**. It has a charge equal to the sum of the charges of the reactants and a mass number equal to the sum of the mass numbers of the reactants. Every compound nucleus has excess energy that must be released after the two reactants bind together. Compound nuclei lose this excess energy by emitting one or more neutrons, protons, $\alpha$ particles, or deuterons ($^2H$ nuclei). As Example 20-7 shows, the conservation of charge and mass number permits identification of the participants in binuclear reactions.

### EXAMPLE 20-7  BALANCING BINUCLEAR REACTIONS

Identify $X$ and $Y$ in each of the following nuclear reactions:

(a) $^3He + n \longrightarrow (^mX) \longrightarrow Y + p$

(b) $^{112}Sn + X \longrightarrow (^{113m}Sn) \longrightarrow ^{113}Sn + Y$

(c) $X + \alpha \longrightarrow (^{113m}In) \longrightarrow Y + 2\,n$

**Strategy:** Charge number and mass number are conserved in nuclear reactions, so the missing components can be identified from the atomic numbers of the elements and the charge and mass numbers of elementary particles.

**Solution:**

(a) Both reactants are given: $^3He$ and n. Remember that a neutron has $Z = 0$ and $A = 1$. Therefore, the compound nucleus must have $A = 4$ and $Z = 2$ (He). It then loses a proton, leaving $A = 3$ and $Z = 1$ (H). To emphasize the conservation of charge and mass number, we write the reaction showing all charge and mass numbers:

$$^3_2He + ^1_0n \longrightarrow (^{4m}_2He) \longrightarrow ^3_1H + ^1_1p$$

(b) Because all the nuclides are isotopes of tin, $Z = 50$ for all of them. Therefore, $X$ and $Y$ must each have a charge number of zero. The possibilities are a neutron and a $\gamma$ ray. To balance mass numbers, $X$ must have a mass number of 1 (neutron), and $Y$ must have a mass number of 0 ($\gamma$ ray):

$$^{112}_{50}Sn + ^1_0n \longrightarrow (^{113m}_{50}Sn) \longrightarrow ^{113}_{50}Sn + \gamma$$

(c) The compound nucleus is $^{113m}$In, so $X$ must have this configuration minus an $\alpha$ particle, $^{4}_{2}\alpha$. Thus, $X$ must have $A = 113 - 4 = 109$ and $Z = 49 - 2 = 47$, and $X$ is $^{109}_{47}$Ag. Similarly, $Y$ must have the configuration $^{113m}_{49}$In minus two neutrons. Therefore, $Z$ does not change, but $A = 113 - 2 = 111$, so $Y$ is $^{111}_{49}$In:

$$^{109}_{47}\text{Ag} + {}^{4}_{2}\alpha \longrightarrow \left(^{113m}_{49}\text{In}\right) + {}^{111}_{49}\text{In} + 2\,{}^{1}_{0}\text{n}$$

 **Do the Results Make Sense?** Confirm that each nuclear reaction is correctly balanced by verifying that both $A$ and $Z$ are conserved.

**Practice Exercise 20.7**

Write the complete reaction for two $^{12}$C nuclei fusing and emitting an $\alpha$ particle.

## Making Synthetic Elements

Elements 43 (technetium), 61 (promethium), 85 (astatine), and all elements with $Z > 92$ do not exist naturally on the Earth, because no isotopes of these elements are stable. After the discovery of nuclear reactions early in the twentieth century, scientists set out to make these "missing" elements. Between 1937 and 1945, the gaps were filled and three actinides, neptunium ($Z = 93$), plutonium ($Z = 94$), and americium ($Z = 95$) were also made.

The production of synthetic elements requires binuclear reactions between two positive nuclei that must be forced together against the force of electrical repulsion. This necessitates the use of nuclear accelerators to give extremely high kinetic energies to the positive projectile nuclei. The first instrument applied to this task was the **cyclotron**, a particle accelerator developed by E. O. Lawrence at the University of California, Berkeley. The operation of the cyclotron is shown schematically in Figure 20-8. A combination of pulsed electrical and magnetic fields is used to accelerate a beam of particles to high kinetic energy. Lawrence used high-energy deuterons to increase the atomic number of target nuclei by one unit:

$$^{98}_{42}\text{Mo} + {}^{2}_{1}\text{H} \longrightarrow {}^{99}_{43}\text{Tc} + {}^{1}_{0}\text{n}$$
$$^{238}_{92}\text{U} + {}^{2}_{1}\text{H} \longrightarrow {}^{238}_{93}\text{Np} + 2\,{}^{1}_{0}\text{n}$$

**Figure 20-8**

A cyclotron accelerates particles in spiral paths. The photograph shows a modern cyclotron. The magnets (orange) are partially visible.

The most efficient way to make Elements 93 and 94 uses neutrons produced during fission in nuclear reactors instead of accelerated positive nuclei. Neutron capture by $^{238}$U followed by β emission gives isotopes with a mass number of 239:

$$^{238}_{92}\text{U} + \text{n} \longrightarrow {}^{239}_{92}\text{U} + \gamma$$

$$^{239}_{92}\text{U} \longrightarrow {}^{239}_{93}\text{Np} + \beta$$

$$^{239}_{93}\text{Np} \longrightarrow {}^{239}_{94}\text{Pu} + \beta$$

As $Z$ increases, the overall yield of the target element falls sharply, because many steps are required. Plutonium (Element 94) has been produced in tonnage quantities by neutron bombardment of uranium-238. Up to curium (Element 96), production in kilogram quantities is possible, but the yields fall by about one order of magnitude for each successive element beyond $Z = 96$.

Beyond $Z = 100$, synthesis by neutron bombardment of uranium is no longer effective. Instead, nuclides in the $Z = 95-99$ range are bombarded with beams of light nuclei. For example, mendelevium ($Z = 101$) was first made in 1955 by a team of Berkeley University chemists led by Glenn Seaborg. They bombarded Element 99 with helium nuclei that had been accelerated in the cyclotron:

$$^{253}_{99}\text{Es} + \alpha \longrightarrow {}^{256}_{101}\text{Md} + \text{n}$$

Beyond Element 101, increasingly heavier nuclei must be used as the projectiles, as the following examples illustrate:

$$^{246}_{96}\text{Cm} + {}^{12}_{6}\text{C} \longrightarrow {}^{(258-x)}_{102}\text{No} + x\,\text{n}$$

$$^{252}_{98}\text{Cf} + {}^{11}_{5}\text{B} \longrightarrow {}^{(263-x)}_{103}\text{Lw} + x\,\text{n}$$

$$^{249}_{98}\text{Cf} + {}^{16}_{8}\text{O} \longrightarrow {}^{263}_{106}\text{Sg} + 2\,\text{n}$$

This requires a different type of accelerator that is linear rather than circular. The schematic diagram in Figure 20-9 shows how a heavy-ion **linear accelerator** works. A packet of ions is accelerated down the centre of the accelerator by a series of electrically charged cylindrical tubes. The electrical potentials are varied so that

**Figure 20-9**

The beam in a linear accelerator follows a straight line. The photograph shows the interior of the Fermilab linear accelerator.

the tube just ahead of the ion packet is always negatively charged, and this pulls the packet ahead at ever-increasing kinetic energy.

As the charge of the projectile increases, it takes ever greater acceleration for the projectile to penetrate the target nucleus. Even when successful, bombardment generates only a few nuclei of a new element, and these are so unstable that they decay in a matter of seconds or less. Nevertheless, nuclear scientists continue to work at extending the list of known elements, in part because theory predicts that the element with $Z = 114$ will be more stable than its lighter neighbours. Recently, formation of the $Z = 111$ element has been confirmed, and this element was officially named Roentgenium (Rg) at a meeting of the International Union of Pure and Applied Chemistry in Ottawa in 2003. Element 112 was officially named Copernicium in 2009. Nuclides with $Z$ ranging from 112 to 118 have been reported, but more experiments are required before these are confirmed and given official names.

## SECTION EXERCISES

●**20.3.1** The only stable isotope of cobalt is $^{59}$Co. What induced reaction would generate each of the following nuclides from $^{59}$Co? Identify the compound nucleus in each case. (a) $^{59}$Fe; (b) $^{60}$Co; (c) $^{62}$Ni; and (d) $^{58}$Ni.

●**20.3.2** The two best-characterized isotopes of promethium ($Z = 61$) are $^{145}$Pm and $^{147}$Pm. The elements next to Pm in the periodic table have many naturally occurring isotopes. Neodymium ($Z = 60$) has $A = 142$, 143, 144, 145, 146, 148, and 150; and samarium ($Z = 62$) has A = 144, 147, 148, 149, 150, 152, and 154. Write nuclear reactions describing neutron capture by naturally occurring isotopes of Nd and Sm, followed by decay to produce $^{145}$Pm or $^{147}$Pm.

●**20.3.3** The following partial nuclear reactions show one reactant, the compound nucleus, and one product. Identify the other reactant and any additional products:

(a) $^{112}$In $\longrightarrow$ ($^{113m}$In) $\longrightarrow$ $^{113}$In

(b) $\alpha$ $\longrightarrow$ ($^{58m}$Ni) $\longrightarrow$ n

(c) $^{60}$Ni $\longrightarrow$ ($^{61m}$Cu) $\longrightarrow$ $^{57}$Co

## 20.4 NUCLEAR FISSION

The graph of binding energy in Figure 20-3 shows that large amounts of energy are released when heavy nuclei split into lighter ones (fission) and when light nuclei combine into heavier ones (fusion). Thus, both fission reactions and fusion reactions can serve as sources of energy. Fission is the subject of this section, and fusion is the subject of Section 20.5.

## Characteristics of Fission

Fission splits a nucleus into two fragments, each with a much lower $Z$ value. Several free neutrons are released during each fission event. When uranium-235 captures a neutron, many fission pathways are possible, two of which are

$$^{235}_{92}\text{U} + ^{1}_{0}\text{n} \longrightarrow (^{236m}_{92}\text{U}) \begin{cases} \nearrow ^{81}_{32}\text{Ge} + ^{152}_{60}\text{Nd} + 3\,^{1}_{0}\text{n} \\ \searrow ^{103}_{42}\text{Mo} + ^{131}_{50}\text{Sn} + 2\,^{1}_{0}\text{n} \end{cases}$$

The compound nucleus formed on neutron capture, $^{236m}$U, is highly unstable. As Figure 20-10 shows, it quickly splits into fragment nuclei plus several neutrons. Fission results in a wide range of product nuclides, of which we showed only two examples above. There are many modes of fragmentation, but free neutrons are always generated, and charge and mass number are always conserved.

Among naturally occurring nuclides, only $^{235}$U undergoes fission, but neutron capture followed by β decay converts two other naturally occurring nuclides, $^{238}$U and $^{232}$Th, into nuclides $^{239}$Pu and $^{233}$U that undergo fission.

The characteristics of nuclear fission can be summarized as follows:

1. Fission follows neutron capture by a small number of the heaviest nuclides, notably $^{235}$U, $^{239}$Pu, and $^{233}$U.

2. Fission gives a range of product nuclides. Neutron-induced fission of $^{235}$U yields the distribution shown in Figure 20-11. This distribution includes nuclides from $A = 77$ to $A = 157$. The most likely products are $A = 95$ and $A = 138$, but no single nuclide makes up more than 7% of the product fragments.

3. Each fission reaction generates from one to four free neutrons, with two or three being most common.

4. The process of fission releases large amounts of energy. The energy released for one set of fission fragments is calculated in Example 20-8. Although each set of fission fragments has a slightly different total mass, the average mass loss is about 0.2 g/mol. This translates into an average energy released in $^{235}$U fission of $1.8 \times 10^{10}$ kJ/mol.

5. Fission products often are radioactive. This is because the fissioning nucleus has an $N{:}Z$ ratio of 1.54, so its products have a similar $N{:}Z$ ratio. In contrast, stable nuclides in the $A = 77{-}157$ range have ratios of around 1.3, so the products of fission have excess neutrons, making them unstable.

**Figure 20-10**

| 1. Neutron capture destabilizes the nucleus | 2. The unstable nucleus distorts | 3. The nucleus splits into smaller fragment nuclei and free neutrons |

Schematic view of fission. Neutron capture produces a highly unstable nucleus that distorts and then splits into smaller nuclei and a few free neutrons.

**Figure 20-11**

The isotopic fission yield curve of $^{235}$U. Different mass numbers are produced in widely different percentages.

---

## EXAMPLE 20-8 FISSION ENERGY

Use isotopic molar masses (given below) to calculate the energy released per mole, per nucleus, and per gram when $^{235}$U undergoes fission in the following manner:

$$^{235}\text{U} + \text{n} \longrightarrow {}^{138}\text{Xe} + {}^{95}\text{Sr} + 3\,\text{n}$$

| Nuclide | n | $^{235}$U | $^{138}$Xe | $^{95}$Sr |
|---|---|---|---|---|
| M (g/mol) | 1.0087 | 235.0439 | 137.908 | 94.913 |

**Strategy:** The nuclear reaction, which is provided, lets us visualize what is occurring. Equation 20-3 relates the energy released in a nuclear transformation to the mass defect, $\Delta m$, which is the loss of mass per mole of reaction. The usual mole–mass conversion factors give energy released per nucleus and per gram.

**Solution:** The mass defect is the difference between the total mass of all products and the total mass of all reactants:

$$\Delta m = [(1\ ^{138}Xe)(137.908\ g/mol) + (1\ ^{95}Sr)(94.913\ g/mol) + (3\ n)(1.0087\ g/mol)]$$
$$- [(1\ ^{235}U)(235.0439\ g/mol) + (1\ n)(1.0087\ g/mol)] = -0.2055\ g/mol$$

Substitute this mass change into Equation 20-3 to find the energy released:

$$\Delta E = (\Delta m)(8.988 \times 10^{10}\ kJ/g)$$

$$\Delta E = (-0.2055\ g/mol)(8.988 \times 10^{10}\ kJ/g) = -1.847 \times 10^{10}\ kJ/mol$$

Finally, use Avogadro's number and the molar mass to convert to energy per nucleus and energy per gram:

$$\Delta E_{per\ nucleus} = \frac{-1.847 \times 10^{10}\ kJ/mol}{6.022 \times 10^{23}\ nuclei/mol} = -3.07 \times 10^{-14}\ kJ/nucleus$$

$$\Delta E_{per\ gram} = \frac{-1.847 \times 10^{10}\ kJ/mol}{235.0439\ g/mol} = -7.86 \times 10^{7}\ kJ/g$$

**Do the Results Make Sense?** The isotopic molar masses are precise to five or more significant figures, so we are tempted to express the result with five significant figures. The mass defect is determined by addition and subtraction, however, and two of the isotopic molar masses are known to just three decimal places, so the mass defect is precise to three decimal places, and the results are precise to only three significant figures. Fission of 1 g of $^{235}U$ releases enough energy to raise the temperature of about 250 million litres of water from 25 to 100 °C. For comparison, about 1.65 million grams of octane must be burned to release the same amount of energy. This immense energy explains why nuclear energy has been developed for power generation.

**Common Pitfalls:** Not accounting for all nuclides in the reaction. Make sure the reaction is balanced before calculating a mass defect.

**Practice Exercise 20.8**
Calculate the energy that is released per gram of $^{235}U$ when induced fission gives $^{144}Ba$ ($M = 143.92294$ g/mol), $^{90}Kr$ ($M = 89.91953$ g/mol), and two neutrons.

Every fission reaction releases some neutrons, and these neutrons can be recaptured by other nuclei, causing more fission reactions. When the amount of fissionable material is small, most neutrons escape from the sample, and only a few neutrons are recaptured. Increasing the amount of material increases the likelihood that neutrons will be recaptured and causes additional fission reactions. The **critical mass** is defined as the amount of material that is just large enough to recapture one neutron, on average, for every fission reaction.

As long as the amount of fissionable material is less than the critical mass, the rate of fission events does not grow, and the rate of energy release remains low. In contrast, a sample behaves quite differently when the amount of fissionable material is larger than the critical mass. Above the critical mass, more than

**Figure 20-12**

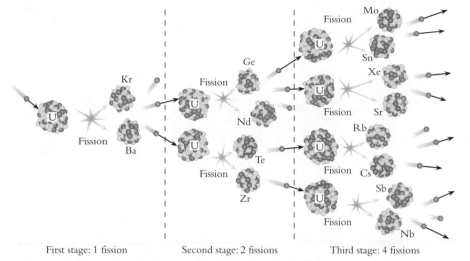

First stage: 1 fission          Second stage: 2 fissions          Third stage: 4 fissions

Schematic view of the start of a fission chain reaction. The chain continues to grow if more than one neutron, on average, is captured for every fission event.

one neutron, on average, is recaptured for every fission that occurs. Now the number of fission reactions grows rapidly. As an illustration, consider what happens when two neutrons are recaptured from each fission reaction. As shown in Figure 20-12, the neutrons produced by the first fission reaction trigger fragmentation of two more nuclei. Neutrons from these two fission events are recaptured by four additional nuclei, and this fission cascade goes on, doubling in each successive round. The result is a "chain" reaction that grows quickly to explosive proportions.

The recapture ratio does not have to be 2 for this effect to occur. Any recapture value larger than 1.0 results in explosive growth of the fission chain. The critical mass is called "critical" because any mass greater than this value sustains a chain reaction and may explode.

## Nuclear Reactors

A nuclear bomb is a terrifying example of the enormous amount of energy released by nuclear fission. However, nuclear fission can be used to generate electrical power if the rate of fission is controlled by adjusting the number of recaptured neutrons. The fission of just 1 g of $^{235}U$ releases as much energy as the combustion of 2200 L of gasoline or 6 tonnes of coal. Consequently, nuclear power is attractive, especially for countries that lack supplies of petroleum or coal.

Nuclear power plants, such as the one diagrammed in Figure 20-13, use the energy released in fission to generate electricity. The fission reaction takes place in a core that is heated by the released energy. A circulating fluid transfers this thermal energy to a heat exchanger, where the energy is used to convert water into steam. The steam drives a turbine connected to an electrical generator. Cooling water must be supplied to recondense the steam.

The turbine and generator components of a nuclear power plant have exact counterparts in power plants fuelled by fossil fuels. The uniqueness of the nuclear

**Figure 20-13**

1. **Nuclear reactor:** Water under high pressure carries heat generated in the nuclear reactor core to the steam generator.

2. **Steam generator:** Heat from the reactor vapourizes water in the steam generator, creating steam.

3. **Turbine and condenser:** Steam from the steam generator powers a turbine, producing useable electricity. The condenser uses cooling water from a river or ocean to recondense the steam from the turbine.

Schematic view of a nuclear power generating plant. The energy source is the core, in which a fission reaction occurs. The rest of the plant is designed to transfer the energy released during fission and convert it into electricity.

**Figure 20-14**

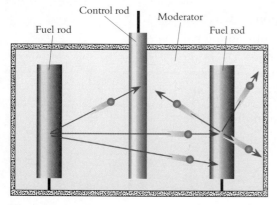

The neutron processes occurring in the core of a fission reactor. Fission of $^{235}U$ gives from one to four fast neutrons, which are slowed down by the moderator. Some neutrons are captured by other nuclides.

power plant lies in its core. The core is a nuclear reactor where fission takes place under conditions that keep the reactor operating just below the critical level. The core contains three parts: fuel rods, moderators, and control rods. These components act on the flow of neutrons within the core, as shown in Figure 20-14. The fate of neutrons must be controlled carefully. Fission must be sustained at a steady rate that produces sufficient energy to run a generator, but the rate must not be allowed to increase and destroy the reactor.

A nuclear reactor runs on its nuclear fuel rods, which contain fissionable material such as uranium, which for most reactor designs must first be enriched in $^{235}U$ as described in Section 2.6. The fuel rods contain more than the critical mass of fuel, but the rate of fission is kept under control by movable control rods. These rods contain $^{112}Cd$, which absorbs neutrons but does not undergo fission. The control rods capture neutrons that would otherwise trigger additional fission reactions. When pushed into the reactor, the control rods capture more neutrons, and the rate of fission decreases. When pulled out, the control rods capture fewer neutrons, and the rate of fission increases.

For the start-up of a reactor, its core is allowed to heat up by withdrawal of the control rods until the recapture ratio is slightly greater than 1.0. When the optimum operating temperature is reached, the control rods are inserted until the capture ratio is exactly 1, and the reaction proceeds at a steady rate. For the shutdown of a reactor, its control rods are fully inserted, reducing the recapture ratio to nearly zero.

The **moderator** component of a reactor slows neutrons without capturing them. Moderators are used because the neutrons released in fission have such high kinetic energies that they are difficult to capture. The critical mass of a nuclear fuel is much smaller for slow neutrons than for fast neutrons, so considerably less fuel is needed in a moderated reactor. Graphite ($^{12}$C), "heavy" water (deuterium oxide, $D_2O$), and ordinary water are used as moderators. In a nuclear power plant, heat must be transferred from the core to the turbines without any transfer of matter. This is because fission and neutron capture generate lethal radioactive products that cannot be allowed to escape from the core. A heat-transfer fluid such as liquid sodium metal flows around the core, absorbing the heat produced by nuclear fission. This hot fluid then flows through a steam generator, where its heat energy is used to vapourize water. After steam has been produced, its energy is used to produce electricity in a conventional steam turbine that has the same design as the steam turbines in fossil fuel power plants.

A Canadian-designed nuclear power generating system is in use in many countries including Korea, China, Romania, Argentina, and of course Canada. The CANDU (Canadian Deuterium Uranium) system differs from the pressurized water reactors described above in several important ways. First, the fuel in a CANDU reactor is made from natural grade (not enriched) uranium. This reduces the cost of the fuel considerably. Second, the primary coolant flowing through the reactor core is deuterium oxide ("heavy water"), which acts as both coolant and moderator, further simplifying the design. Third, a CANDU reactor can be fuelled on-line, whereas a pressurized water reactor must be shut down to do so.

The radioactive materials that are an unavoidable by-product of nuclear fission are life-threatening. To operate safely, therefore, a nuclear power plant must confine all its radioactive products until they can be disposed of safely.

The main danger in the operation of a nuclear power plant is potential loss of control over the nuclear reaction. If the core overheats, it may either explode or "melt down." In either event, radioactive materials escape from the reactor to contaminate the environment. Designers attempt to make nuclear reactors "fail-safe" by providing mechanisms that automatically shut the core down on overheating. One way this has been done is to design the control rods to fall into the core if their control mechanism fails.

Despite such safeguards, nuclear accidents have occurred. The worst occurred in 1986 at Chernobyl in Ukraine, shown in Figure 20-15a. During a test of the reactor, most of the control rods were removed. The reactor surged out of control, probably because the safety systems had been disabled in the course of the test. The excess heat generated steam at high pressure, which blew off the

**Figure 20-15**

(a)

(b)

(c)

Aerial views of three nuclear power plants.
(a) The Chernobyl nuclear power plant, site of a major nuclear accident in 1986.
(b) The Three Mile Island power plant, site of a minor nuclear accident in 1979.
(c) A CANDU plant in Canada, which has operated nuclear power plants safely for over 30 years.

roof of the reactor facility, spewing substantial amounts of radioactive material into the atmosphere. A fire released more radioactive material. Contamination was most severe within 30 km of the plant, where radioactivity levels became so high that the entire population had to be relocated. Significant contamination was also detected 1000 km away in Germany, and increased levels of radioactivity attributable to Chernobyl appeared throughout the Northern Hemisphere.

Nuclear power plants in the United States are supposed to be designed well enough to prevent accidents as serious as the one at Chernobyl. Nevertheless, the Three Mile Island plant in Pennsylvania, an aerial view of which is shown in Figure 20-15b, experienced a partial meltdown in 1979. This accident was caused by a malfunctioning coolant system. A small amount of radioactivity was released into the environment, but because there was no explosion, the extent of contamination was minimal.

A few nations rely heavily on nuclear power despite the possibility of accidents. In France and Japan, fission power from nuclear reactors provides two-thirds or more of overall energy needs. In Canada, all nuclear generating plants are in the Province of Ontario, where 17 reactors supply about half of the province's electricity requirements. Even when operated safely, nuclear power plants produce long-lived radioactive wastes, which must be sequestered from the biosphere until their radioactivity diminishes to acceptable levels. Plutonium-239, formed in nuclear fuel rods when $^{238}U$ captures neutrons and decays, is a particularly dangerous nuclear waste that is extremely toxic and has a half-life of nearly 25 000 years. In Europe, nuclear waste is placed far underground in abandoned salt mines. Deep-sea burial and ejection into outer space have also been proposed. Antinuclear activists contend that no current technology is acceptable, because there is no proof that radioactive material can be contained for the thousands of years necessary for these dangerous materials to decompose.

Modern nuclear reactors are highly technological, carefully engineered creations of advanced human societies, so it may seem impossible that a nuclear reactor could result from natural conditions. Our Focus 20-1: Chemistry and the Environment, however, describes evidence indicating that such a natural nuclear reactor did exist.

## FOCUS 20-1  CHEMISTRY AND THE ENVIRONMENT: A NATURAL NUCLEAR REACTOR?

Naturally occurring uranium mined today contains 99.28% $^{238}U$ and 0.72% $^{235}U$. Other than in a CANDU reactor, this is too low a percentage to sustain a fission reaction. In the past, however, the percentage of $^{235}U$ was higher than it is today, because $^{235}U$ has a shorter half-life than $^{238}U$ ($7.0 \times 10^8$ years compared with $4.5 \times 10^9$ years). Calculations using these half-lives show that 1.8 billion years ago, $^{235}U$ made up 3% of natural uranium. This is comparable to the enrichment levels used in most nuclear fuels. Long ago, then, the composition of uranium ores was such that a self-sustaining nuclear reaction could have occurred.

But did this ever happen? A sustained reaction requires a neutron recapture ratio of at least 1. Besides enrichment in $^{235}U$, there must be a high degree of purity that minimizes the fraction of neutrons captured by other nuclides. A sustained reaction also requires a moderator to slow the neutrons before they escape from the uranium mass. Because water is a relatively good moderator, the need for a moderator could be met by groundwater seeping through the deposit. These unique conditions are illustrated in our molecular view.

It may seem unlikely that all these conditions could have been met, but at least one deposit of uranium ore has characteristics indicating that, long ago, it operated as a natural nuclear reactor. At Oklo in the Gabon Republic near the western coast of equatorial Africa (see photo), there are uranium deposits of high purity that are about 1.8 billion years old. Thus, this uranium originally contained the 3% abundance of $^{235}U$ needed for a sustained reaction. Currently, the deposits in this mine differ from other uranium deposits in two ways that indicate that Oklo experienced sustained fission at one time. First, the $^{235}U$ content is slightly depleted, and second, there are unusual amounts of elements with mass numbers between 80 and 150.

The usual percentage of $^{235}U$ is 0.7207%, but the Oklo deposits contain 0.7071% of this isotope. This difference is small, but it is significant because it indicates that, at some time in the past, $^{235}U$ was consumed faster than its spontaneous decay rate. The question is whether that faster rate was caused by sustained fission.

What convinces scientists that sustained fission once occurred at Oklo is the presence of characteristic fission products in the ore. Elements of mass numbers between 75 and 160 occur in the ore in larger amounts than elsewhere. Furthermore, mass analysis of the elements in Oklo ore shows that they are distributed in the characteristic pattern shown in Figure 20-11. This isotopic "signature," which is not found in any other naturally occurring materials, is so characteristic that it has convinced most scientists that the ore deposits at Oklo once formed a huge nuclear reactor.

From the size of the ore deposit and the extent of $^{235}U$ depletion, scientists estimate that the Oklo deposits functioned as a natural nuclear reactor for about 100 000 years, generating $5 \times 10^{14}$ kJ of energy. This was an average power output of about 160 kilowatts, which is about enough to meet the needs of 100 present-day Canadian households (for 100 000 years!)

## SECTION EXERCISES

●20.4.1 Fission products have molar masses that average 0.09 g/mol less than their mass number: for $^{106}$Ru, $M = 105.91$ g/mol. Using this value, calculate mass losses and energy releases for $^{235}$U fissions that release 1 neutron, 2 neutrons, and 3 neutrons.

●20.4.2 Calculate the isotopic abundances of naturally occurring uranium $10^8$ years ago. (*Hint:* Start with a convenient amount of uranium of present-day abundances, and use half-lives to calculate the amounts that were present $10^8$ years ago.)

## 20.5 NUCLEAR FUSION

Fusion is a more attractive energy source than fission. Whereas fission occurs for only a few rare, extremely heavy nuclides, fusion is possible for abundant light nuclides such as $^1$H. Moreover, some fusion reactions release more energy per unit mass than fission reactions do. For example, the fusion of two hydrogen isotopes, deuterium and tritium, releases $3.4 \times 10^8$ kJ/g, compared with $7.9 \times 10^7$ kJ/g released in the fission of $^{235}$U. Another attractive feature of fusion reactions is that the product nuclides are usually stable, so smaller amounts of radioactive by-products result from fusion than from fission.

### The Threshold for Fusion

The major impediment to fusion reactions is that the reacting nuclei must have very high kinetic energies to overcome the electrical repulsion between positive particles. The fusion of two hydrogen nuclei has the lowest possible repulsion barrier because it involves two $Z = 1$ nuclei. Even so, this reaction requires kinetic energies equivalent to temperatures of $10^7$ K or greater. As $Z$ increases, so does this energy requirement: the fusion of two $^{12}$C nuclei requires kinetic energies equivalent to a temperature of $10^9$ K.

There are several ways to produce nuclei with enough kinetic energy to fuse. One method uses particle accelerators to generate small quantities of fast-moving nuclei. The characteristics of fusion reactions are studied with such accelerators, but the scale of fusion events in these experiments is too small to release useful amounts of energy. A second way to induce fusion uses the energy released in gravitational attraction to generate a temperature hot enough to start the reaction. The Sun and other stars operate in this way. A third way is the fusion bomb, which uses a fission bomb to generate a temperature high enough for fusion. None of these provides a method for harnessing fusion as a practical source of power on Earth. However, nuclear scientists have proposed that radiation might be able to initiate a useful fusion reaction by heating a gaseous plasma confined in a magnetic field. If successful, this method may harness fusion as a source of power.

A plasma is a high-temperature ionized gas consisting of free electrons and nuclei.

### Controlled Fusion

A long-standing goal of nuclear science has been to harness the energy of nuclear fusion in a sustained process. The technological problems are immense, however, because the fusing material must be confined to a volume small enough to generate high nuclear densities and many nuclear collisions. Because all materials vapourize at temperatures well below that required to sustain fusion reactions, containers that would withstand sustained fusion cannot be built.

The approach most often taken to address these problems is to use a hot, ionized gaseous plasma that contains cations of $^2$H, $^3$H, and $^6$Li, along with enough electrons to maintain charge neutrality. This plasma is heated intensely, usually by powerful laser beams. To keep the plasma from flying apart, it is confined by a donut-shaped magnetic field. Figure 20-16 shows a cutaway view of the ITER tokamak currently under construction in France. When completed in 2018, the demonstration reactor should be able to produce a peak power of 500 MW for approximately 1000 seconds.

Fusion cannot become a practical source of energy unless the energy released by fusion exceeds the energy used to heat and confine the plasma. Unfortunately, the energy input requirement is enormous, even in small-scale experimental studies. The best experiments conducted so far have managed to achieve fusion but not sustain it long enough to achieve any net production of energy.

Assuming that a sustained fusion reaction is achieved, some way must be found to harness its energy output. Because the field confining the plasma prevents the escape of charged particles, most of the energy output of the reactor is expected to be photons, ranging in energy from infrared rays to $\gamma$ rays. These photons will have to be absorbed by a suitable material that converts their energy, directly or indirectly, into electricity.

**Figure 20-16**

A version of a fusion reactor that has been the subject of intense development efforts is the tokamak. A cutaway view of the world's largest tokamak is shown here.

## Stellar Nuclear Reactions

The Sun and all other stars produce energy at a huge rate from sustained nuclear fusion. Over time, stars evolve through several stages, including stellar explosions. The products of a stellar explosion can form stars of more complex composition. Three distinct generations of stars have been identified, each fuelled by a different set of fusion reactions.

The formation of a star begins with a cloud of matter. Gravitational attraction pulls the matter together, and during this collapse, gravitational potential energy is converted into kinetic energy of motion: atoms move faster and faster, and the temperature rises. The temperature attained in the collapsing cloud increases with the total mass of the cloud, and if the mass is larger than about $2 \times 10^{29}$ kg, the temperature rises high enough to initiate the fusion of hydrogen nuclei.

## First-Generation Stars

Astrophysicists believe that the early universe was composed mostly of hydrogen. As a cloud of hydrogen collapses, heating breaks its hydrogen atoms into a plasma of protons and electrons. A **first-generation star** forms when the interior temperature of this plasma reaches $4 \times 10^7$ K. Above this temperature, protons combine in a reaction sequence that yields helium nuclei:

$$^1_1H + {}^1_1H \longrightarrow \left({}^2_2He\right) \longrightarrow {}^2_1H + {}^0_1\beta^+$$

$$^2_1H + {}^1_1H \longrightarrow {}^3_2He + \gamma$$

$$^0_1\beta^+ + {}^0_{-1}e \longrightarrow 2\gamma$$

$$^3_2He + {}^3_2He \longrightarrow \left({}^6_4Be\right) \longrightarrow {}^4_2He + 2{}^1_1H$$

The net reaction converts protons and electrons into helium nuclei and radiation, releasing about $2.5 \times 10^9$ kJ/mol of energy:

$$4{}^1_1H + 2{}^0_{-1}e \longrightarrow {}^4_2He + 6\gamma \qquad \Delta E \cong -2.5 \times 10^9 \text{ kJ/mol}$$

The pressure exerted by the radiation escaping from inner portions of the star counteracts the force of gravity, and the balance of the two opposing forces keeps the volume of the star constant as long as hydrogen fusion continues.

In time (about $10^{10}$ years for a star the size of our Sun), a star consumes most of its protons. Then, the preceding sequence of nuclear reactions ceases, gravitational forces take over once more, and the star collapses until its temperature reaches about $10^8$ K. At this temperature, a new fusion reaction begins in which helium nuclei fuse to form beryllium-8, which can fuse with a third helium nucleus:

The colour of a hydrogen-burning star depends on its mass; the higher the mass, the higher the temperature of the star and the more blue it appears.

$$^4_2He + {}^4_2He \longrightarrow \left({}^8_4Be\right)$$

$$\left({}^8_4Be\right) + {}^4_2He \longrightarrow {}^{12}_6C + \gamma$$

$$\text{Net: } 3{}^4_2He \longrightarrow {}^{12}_6C + \gamma \qquad \Delta E \cong -7.7 \times 10^8 \text{ kJ/mol}$$

The $^{12}$C that is produced also reacts with $^4$He:

$$^{12}_6C + {}^4_2He \longrightarrow {}^{16}_8O + \gamma$$

A star undergoing helium fusion has a dense, hot core and a large, cooler outer mantle. These stars appear large and red, so they are called red giants.

About $10^8$ years after fusion of helium begins, a star runs out of $^4$He fuel. When this happens, the star enters a new stage of gravitational collapse until the temperature increases to about $10^9$ K. This triggers an entirely new set of nuclear reactions between nuclides of carbon and oxygen. Here is one example:

$$^{12}_6C + {}^{16}_8O \longrightarrow {}^{24}_{12}Mg + {}^4_2He$$

As a result of these and other fusion reactions, the star eventually contains nuclides with $Z$ and $A$ values all the way up to iron-56, which is the most stable of all nuclides.

At this stage of their evolution, some stars generate energy more rapidly than it can be dissipated. These stars explode like giant hydrogen bombs, ejecting their various nuclides, ranging from hydrogen to iron, into space. The remaining core collapses on itself. Such explosions, called **supernovae**, have been observed by astronomers over the centuries, and the debris from supernovae explosions has been detected at numerous locations in the heavens. Figure 20-17 contains before-and-after photographs of a supernova that appeared in 1987.

The drawing in Figure 20-18 shows the evolutionary stages of a first-generation star.

**Figure 20-17**

Before-and-after photos of a supernova that appeared in 1987. The arrow in the left photo indicates the star that exploded. The photo on the right shows the aftermath of the explosion.

## Second-Generation Stars

Explosions of first-generation stars spew nuclides from $Z = 1$ to $Z = 26$ into interstellar space. There, this matter mixes with interstellar hydrogen, and eventually enough matter clumps together under the force of gravity to collapse and form a **second-generation star**. Young second-generation stars, like young first-generation stars, contain large amounts of hydrogen. Unlike first-generation stars, however, they also contain higher-$Z$ nuclides. One of these, $^{12}C$, catalyzes the fusion of H to form He. As shown schematically in Figure 20-19, protons add to $^{12}C$ in a sequence that ultimately produces $^4He$. Here are the reactions:

$$\,^1_1H + \,^{12}_6C \longrightarrow \left(\,^{13}_7N\right) \longrightarrow \,^{13}_6C + \,^0_1\beta^+$$

$$\,^1_1H + \,^{13}_6C \longrightarrow \,^{14}_7H$$

$$\,^1_1H + \,^{14}_7N \longrightarrow \left(\,^{15}_8O\right) \longrightarrow \,^{15}_7N + \,^0_1\beta^+$$

$$\,^1_1H + \,^{15}_7N \longrightarrow \,^{12}_6C + \,^4_2He$$

$$2\left[\,^0_1\beta^+ + \,^0_{-1}e \longrightarrow 2\,\gamma\right]$$

$$\text{Net: } 4\,^1_1H + 2\,^0_{-1}e \longrightarrow \,^4_2He + 4\,\gamma \qquad \Delta E \cong -2.5 \times 10^9 \text{ kJ/mol}$$

**Figure 20-18**

Schematic depiction of the evolutionary stages of a first-generation star.

**Figure 20-19**

In a second-generation star, fusion of protons to produce helium occurs in a catalytic cycle by the sequential addition of protons to $^{12}$C.

When most of the hydrogen has been consumed, a second-generation star collapses until it is hot enough for helium fusion to occur. Now, a larger range of reactions takes place because nuclides such as $^{13}$C generate neutrons when they fuse with $^4$He:

$$^{13}_{6}C + {}^{4}_{2}He \longrightarrow {}^{16}_{8}O + {}^{1}_{0}n$$

These neutrons are captured by iron to yield Co after β emission:

$$^{56}_{26}Fe + {}^{1}_{0}n \longrightarrow {}^{57}_{27}Co + {}^{0}_{-1}\beta$$

Neutron capture and β emission form nuclei of ever-higher atomic number. Neutron capture and β emission by Co ($Z = 27$) produce Ni ($Z = 28$), Ni produces Cu ($Z = 29$), and so on up the atomic-number ladder: neutron capture and β emission form all possible stable nuclides during the lifetime of a second-generation star.

Like first-generation stars, second-generation stars often become unstable when they reach the carbon-burning stage. If they explode in supernovae, their nuclear debris includes their content of heavier nuclides. In addition, supernova explosions generate large numbers of neutrons that can be captured by lighter nuclides in the exploding star to form still more heavy nuclides.

Our planet Earth contains significant amounts of elements all the way up to $Z = 92$. This indicates that our solar system resulted from the gravitational collapse of a cloud of matter that included debris from second-generation stellar supernovae. Thus, our Sun most likely is a **third-generation star**. The composition of a third-generation star includes high-$Z$ nuclides, but the nuclear reactions are the same as those in a second-generation star.

The nuclear reactions described in this section can account for the elements found on Earth, but much about the composition of the universe remains to be discovered.

## SECTION EXERCISES

●**20.5.1** One proposal for controlled fusion involves using an accelerator to propel deuterons into a lithium target, inducing the following reactions:

$$^{7}_{3}Li + {}^{2}_{1}H \longrightarrow {}^{8}_{3}Li + {}^{1}_{1}H \qquad {}^{8}_{3}Li \longrightarrow {}^{0}_{-1}\beta + 2{}^{4}_{2}He$$

(a) The first reaction can be viewed as the transfer of a neutron. Draw a nuclear picture illustrating this reaction.

(b) The reaction has an energy barrier that is 50% of the calculated barrier due to electrical force. What minimum kinetic energy must the deuterons have, if the radius of $^2$H is $1.4 \times 10^{-3}$ pm and that of $^7$Li is $2.4 \times 10^{-3}$ pm? (Refer to Section 20.1 to calculate the electrical energy barrier.)

(c) Use the following isotopic masses, all in grams per mole, to calculate the energy change for each step of this fusion reaction:

| $^7$Li | $^8$Li | $^1$H | $^2$H | $^4$He |
|--------|--------|-------|-------|--------|
| 7.016 005 | 8.022 488 | 1.007 822 | 2.014 1022 | 4.002 6036 |

(d) If 5.0 kg of $^2$H is completely consumed in the reaction, how much energy can be produced?

●**20.5.2** To show how zinc could form in a second-generation star, write a sequence of nuclear reactions that starts at $^{56}$Fe, ends at $^{66}$Zn, and is composed entirely of neutron capture and β emission. Your sequence should stay within the belt of stability depicted in Figures 20-5 and 20-6.

●**20.5.3** Describe how the compositions of planets around a second-generation star differ from that of the Earth. Could life as we know it exist on such a planet? Why or why not?

## 20.6 EFFECTS OF RADIATION

The nuclear explosions that devastated Hiroshima and Nagasaki killed 100 000–200 000 people instantaneously. Probably an equal number died later, victims of the radiation released in those explosions. Millions of people were exposed to the radioactivity released by the accident at the Chernobyl nuclear power plant. The full health effects of that accident may never be known, but 31 people died of radiation sickness within a few weeks of the accident, and more than 2000 people have developed thyroid cancer through exposure to radioactive iodine released in the accident. Even low levels of radiation can cause health problems. For this reason, workers in facilities that use radioisotopes monitor their exposure to radiation continually, and they must be rotated to other duties if their total exposure exceeds prescribed levels.

### Radiation Damage

Nuclear radiation causes damage because of its high energy content. Radiation passing through matter transfers energy to atoms and molecules in its path. The major result of this energy transfer is ionization. Electrons are torn from molecules, creating positive ions. A single nuclear emission has enough energy to generate many positive ions and free electrons. A typical α particle, for example, carries an energy of about $10^{-12}$ J, whereas ionization energies are typically $2 \times 10^{-18}$ J. Only about one-third of the energy of an α particle goes into generating ions, the rest being converted into heat. Still, one α particle generates about 150 000 cations. Beta and gamma rays have similar energies, and they also generate ions as they pass through matter.

As a high-energy particle passes through matter, it creates an ionization "track" that contains positive ions. These ions are chemically reactive because their bonds are weakened by the loss of bonding electrons. Even though each cation eventually recaptures an electron to return to electrical neutrality, many ions first undergo chemical reactions that are the source of the damage done by nuclear radiation.

### Immediate Health Effects

Living cells are delicately balanced chemical machines. The ionization track generated by a nuclear particle upsets this balance, almost always destroying the cell in the process. Although the body has a remarkable ability to repair and replace damaged cells, exposure to radiation can overload these control mechanisms, causing weakness, illness, and even death.

### TABLE 20-5 HEALTH EFFECTS OF RADIATION DOSES

| Dose (Sv) | Effect |
|---|---|
| 0–0.25 | Increased susceptibility to cancer; possible DNA damage |
| 0.25–0.50 | Reduced amounts of white blood cells |
| 0.50–1.0 | Fatigue and nausea in half of persons exposed |
| 1.0–2.0 | Nausea and vomiting; hair loss |
| 2.0–4.0 | Damage to bone marrow and spleen; 50% fatality rate |
| >6.0 | Usually death, even with treatment |

Because nuclear radiation varies considerably in energy, the potential to cause damage cannot be assessed simply by counting the number of emissions. The energy of emissions must also be taken into account. Furthermore, the three different types of nuclear radiation affect human cells to different extents. When the amount, energy content, and type of radiation are taken into account, the result is a measure of the effect of radiation on the human body. This is expressed using a unit called the **Sievert (Sv)**.

Different types of body cells show different sensitivities to nuclear radiation. Cells that divide most rapidly tend to be most easily damaged. These include bone marrow, white blood cells, blood platelets, the lining of the gastrointestinal tract, and cells in the gonads. Consequently, the symptoms of radiation sickness include loss of blood functions and gastrointestinal distress.

Individual ability to tolerate radiation damage varies, so a statistical variation exists in the relationship between dose level and health effects. Also, there are effective treatments, such as blood transfusions, for some radiation effects. The statistical patterns of human response to radiation are summarized in Table 20-5. Doses of over 6.0 Sv are almost always fatal.

Marie Curie, shown in this photograph with her daughter Irène Joliot-Curie, was the first woman to win a Nobel Prize and the first person to win two Nobel Prizes. The Nobel Prize in physics in 1903 was awarded for her explanation of radioactivity. The 1911 Nobel Prize in chemistry came for her discovery of the radioactive elements radium and polonium. Curie's husband Pierre Curie was also a Nobel laureate, as were her daughter Irene and her son-in-law Frédéric Joliot-Curie.

## Long-Term Effects

Exposure to low doses of radiation causes no short-term damage but makes the body more susceptible to cancers. In particular, people who have been exposed to increased radiation levels have a much higher incidence of leukemia than the general population has. Marie Curie, the discoverer of radium, eventually died of leukemia brought on by exposure to radiation in the course of her experiments. Medical researchers estimate that about 10% of all cancers are caused by exposure to high-energy radiation.

Another cumulative effect of radiation can be an irreversible alteration of DNA sequences. If part of a DNA molecule is ionized, its molecular chain may be broken. Chain breaks are repaired in the body, but after a serious rupture, the repaired unit may have a different sequence. This type of changed sequence is a genetic mutation. Altered DNA sequences in the reproductive organs are transmitted faithfully, thus passing on the genetic mutations to future generations. Because these effects are cumulative, individuals of childbearing age need to be especially careful about radiation exposure.

All humans are exposed to some level of radiation. Cosmic radiation continually bathes the Earth. This radiation and the naturally occurring radioactivity from nuclides such as $^{14}C$, $^{40}K$, and $^{222}Rn$ expose the average individual to about 0.001 Sv per year. This exposure is increased by radiation from human activities. Medical procedures, notably the use of X rays for imaging teeth and bones, contribute the largest amount. A typical medical X-ray procedure exposes the patient to 0.002 Sv. Although this is well below the level at which immediate effects occur, the possibility of cumulative effects makes it prudent to have X rays taken no more often than necessary for maintaining good health.

Total exposures vary considerably with human activities as well. "Frequent flyers," for example, receive higher doses of radiation because the intensity of cosmic radiation is significantly greater at high altitude than it is at ground level. Residents in locations where there are uranium deposits receive higher doses of radiation from radon, one of the radioactive decay products of uranium.

## Radiation Shielding

Radiation exposure can be reduced by placing the radiation source or the potential target behind a shield that captures the radiation. During exposure to X rays for dental imaging, the patient wears a lead-lined pad, because X rays are absorbed more effectively by lead than by any other material. A lead shield a few millimetres thick is sufficient to stop X rays.

Radiation of different types penetrates matter to different extents, as Figure 20-20 illustrates. Even though α particles are the most damaging radiation, they are the most easily stopped by shielding. In fact, α particles travel

**Figure 20-20**

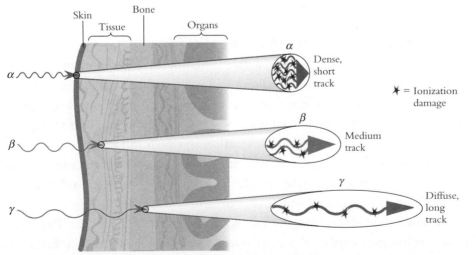

Different types of radiation penetrate matter to different degrees and produce damage at different rates along their paths.

less than 1 mm in solid materials before losing all their excess energy. Beta particles travel much farther, requiring shields 10–100 mm thick. Plastic shields that are several centimetres thick provide excellent shielding from both types of radiation.

The γ rays that accompany many nuclear decay processes can be more dangerous than α or β particles, because γ rays travel long distances before losing their energy. The radiation from $^{222}$Rn, which is responsible for as much as 40% of the background radiation to which humans are exposed, is typical. This nuclide decays by α emission accompanied by γ rays. The α particle is stopped within 10 cm in air or about 1 mm in a solid. The accompanying γ ray, on the other hand, may travel many metres and penetrate walls before finally losing its destructive power. Consequently, radioactive materials that emit γ rays must be shielded using lead blocks that are many centimetres thick.

## SECTION EXERCISES

●**20.6.1** Potassium-40 is a radioactive nuclide that occurs naturally. This nuclide has a half-life of $1.28 \times 10^9$ years and emits β particles with energies of $1.32 \times 10^8$ kJ/mol. Our bodies contain about 2.4 mg of potassium per kilogram of body weight, and $^{40}$K makes up 0.0118% of the element. To how many β particles, and with how much total energy, is a person weighing 70 kg exposed daily from this nuclide?

●**20.6.2** Although X-ray exposures pose little threat to patients, medical technicians must be careful to avoid exposure to the beam. Suppose that an X-ray machine is "leaking" radiation that exposes the technician to 1% of the dose received by a patient. After how many exposures would the technician begin to show reduced numbers of white blood cells? If the technician administers 40 X-ray studies daily and works 250 days per year, would this damage show up?

## 20.7 APPLICATIONS OF RADIOACTIVITY

Considering the frighteningly destructive potential of nuclear weapons and the health risks posed by exposure to too much nuclear radiation, radioactivity may seem like a curse. Nevertheless, scientists and medical practitioners use radioactivity routinely in many beneficial ways, such as positron emission tomography, a medical imaging technique that uses radioactive positron emitters and medical uses of radioactivity for cancer treatment. In this section, we describe two techniques that have broad scientific applications: radioactive dating and radioactive tracing. Dating techniques, which rely on the constant half-life of radioactive decay, allow scientists to estimate the ages of artefacts and of the Earth itself. Tracing techniques exploit the high sensitivity of radiation detectors to study processes such as the mechanisms of chemical reactions and the percolation of water through geological formations.

## Dating Using Radioactivity

Each radioactive nuclide has a characteristic constant half-life. This means that it acts as a clock, "ticking" (decaying) at a predictable rate. Suppose that a nuclear reactor generates 20 mg of $^{32}$P, an isotope with a half-life of 14.3 days. As Figure 20-21 illustrates, after 14.3 days, 10 mg will remain, and there will be 10 mg of the decay product, $^{32}$S. After 28.6 days, 5 mg of $^{32}$P will remain, and there will be 15 mg of $^{32}$S. A chemist who analyzes a sample of $^{32}$P and finds that it contains 15 mg of $^{32}$S and only 5 mg of $^{32}$P can therefore calculate that the sample was generated 28.6 days ago.

Radioactive dating is most valuable in estimating the age of materials that predate human records, such as the Earth itself. Calculating the age of a sample from the amounts of its radioactive nuclides and their decay products forms the basis for radioactive dating techniques. Equation 20-5 is used for such calculations:

$$\ln\left(\frac{N_0}{N}\right) = \frac{0.693\,t}{t_{1/2}}$$

In this equation, $N_0$ and $N$ are amounts of the nuclide present initially and at time $t$, respectively.

The Earth's age can be estimated by using the half-lives of unstable nuclides that are still present and the half-lives of those that are missing. Among those listed in Table 20-4, the shortest-lived naturally occurring nuclide is $^{235}$U ($t_{1/2} = 7.0 \times 10^8$ years). Nuclides with half-lives shorter than $10^8$ years are missing from the Earth's crust. Examples are $^{146}$Sm ($t_{1/2} = 7 \times 10^7$ years), $^{205}$Pb ($t_{1/2} = 3 \times 10^7$ yr), and $^{129}$I ($t_{1/2} = 1.7 \times 10^7$ years). If the Earth were less than $10^9$ years old, its crust would contain some of these nuclides. If it were older than $10^{10}$ years, $^{235}$U would no longer be present. Thus, isotopic abundances indicate that the Earth is older than $10^9$ years but younger than $10^{10}$ years.

More precise estimates come from accurate measurements of isotope ratios. Three pairs of radioactive isotopes and their products are abundant enough for such ratios to be measured:

| | | |
|---|---|---|
| $^{238}\text{U} \longrightarrow$ (Decay chain) $\longrightarrow$ $^{206}\text{Pb}$ | | $t_{1/2} = 4.5 \times 10^9$ years |
| $^{40}\text{K} \longrightarrow {}^{40}\text{Ar} + \beta$ | | $t_{1/2} = 1.28 \times 10^9$ years |
| $^{87}\text{Rb} \longrightarrow {}^{87}\text{Sr} + \beta$ | | $t_{1/2} = 4.9 \times 10^{11}$ years |
| $^{146}\text{Sm} \longrightarrow {}^{142}\text{Nd} + \alpha$ | | $t_{1/2} = 7 \times 10^7$ years |

If the amount of a radioactive nuclide in a rock sample is $N$, the sum of this amount plus the amount of its product nuclide is $N_0$. For argon dating, $N_0$ is the sum of potassium-40 and argon-40 present in a sample of rock. Assuming that Ar gas escapes from molten rock but is trapped when the rock cools and solidifies, the lifetime obtained by substituting these values into Equation 20-5 is the time since the rock solidified. In 2008, researchers from McGill University and l'Université du Québec à Montréal announced that they had found the oldest rock

**Figure 20-21**

Decay of $^{32}$P, a radioactive nuclide with a half-life of 14.3 days. The amount of the nuclide present decreases by half every 14.3 days.

ever recorded on Earth. They used the ratio of $^{146}$Sm to $^{142}$Nd to show that rocks collected in Northern Québec were 4.28 billion years old.

Analyses of this type are correct only if all of the product nuclide comes from radioactive decay. This is not known with certainty, but when age estimates using different pairs of nuclides give the same age and samples from different locations also agree, the age estimate is likely to be accurate. Note also that $4.28 \times 10^9$ years agrees with the qualitative limits derived from naturally occurring radioactive nuclides.

Samples that are $4.6 \times 10^9$ years old have been found in meteorites. This is the best present estimate for the age of the solar system. Example 20-9 illustrates this type of calculation for rock from the Earth's Moon.

## EXAMPLE 20-9 ARGON DATING

When a sample of Moon rock was analyzed by mass spectroscopy, the ratio of $^{40}$K to $^{40}$Ar was found to be 0.1295. Based on this ratio, how old is the Moon?

A sample of rock from the Moon

**Strategy:** This is a question about radioactive dating. Convert the isotopic ratio of the radioactive nuclide and its product into an amount ratio, and then use Equation 20-5.

**Solution:** Equation 20-5 requires a value for $N_0/N$, the ratio of potassium-40 present when the Moon was formed to the amount present today. We can determine $N_0/N$ from the isotopic ratio $N_K/N_{Ar}$ given in the problem. First note that $N = N_K$. Then relate $N_0$ to $N_K/N_{Ar}$. Each $^{40}$K nuclide that decays generates a $^{40}$Ar nuclide, so the initial amount, $N_0$, is the sum of the amounts currently present:

$$N_0 = N_K + N_{Ar}$$

Divide both sides of this equality by $N_K$:

$$\frac{N_K}{N_{Ar}} = 0.1295 \qquad \text{so} \qquad \frac{N_{Ar}}{N_K} = \frac{1}{0.1295} = 7.722$$

$$\frac{N_0}{N_K} = 1 + 7.722 = 8.722$$

Now rearrange Equation 20-5 to solve for $t$:

$$t = \frac{t_{1/2}}{\ln 2} \ln\left(\frac{N_0}{N_K}\right)$$

Find the half-life of $^{40}$K from Table 20-4: $t_{1/2} = 1.28 \times 10^9$ years:

$$t = \frac{1.28 \times 10^9 \,\text{yr}}{0.693} \ln(8.722) = 4.00 \times 10^9 \,\text{yr}$$

According to this analysis, the Moon solidified about 4 billion years ago.

**Does the Result Make Sense?** The age estimate for the Moon is similar to age estimates for the Earth based on several methods of radioactive dating. Thus, this is a sensible result.

**Practice Exercise 20.9**

A meteor that is believed to have come from Mars has a $^{40}K{:}^{40}Ar$ ratio of 0.1660. What age does this indicate?

Radioactivity serves as a useful clock only for times that are the same order of magnitude as the decay half-life. At times much longer than $t_{1/2}$, the amount of radioactive nuclide is too small to measure accurately. At times much shorter than $t_{1/2}$, the amount of the product nuclide is too small to measure. By an accident of nature, one naturally occurring radioisotope, $^{14}C$, has a half-life that is close to the age of human civilization. As a result, this isotope is used to determine the age of human artefacts.

The logic underlying carbon-14 dating differs from that for potassium–argon dating. Recall that bombardment of the upper atmosphere by cosmic rays generates small but measurable levels of $^{14}C$. This rate of production of $^{14}C$ by cosmic rays equals its rate of decay by β emission, so the percentage of $^{14}C$ in the atmosphere remains nearly constant. Living plants incorporate $CO_2$ from the atmosphere into carbohydrates, and animals use these carbohydrates as food. As a result, the percentage of carbon-14 in the tissues of living organisms is the same as that in the atmosphere. On death, however, the uptake of carbon stops. The carbon-14 in an object that was once alive slowly disappears with $t_{1/2} = 5.73 \times 10^3$ years. This makes it possible to date objects such as bone, wood, and cloth by assaying their levels of $^{14}C$.

The product of $^{14}C$ decay is $^{14}N$, but there is so much $^{14}N$ already present in the atmosphere that this isotope is not an accurate measure of how much $^{14}C$ has decayed. Instead, the ratio of $^{14}C$ to $^{12}C$ is used for radiocarbon dating. In the atmosphere and in living objects, the fraction of $^{14}C$ is $1.33 \times 10^{-12}$. That is, there is 1 atom of $^{14}C$ for every $7.54 \times 10^{11}$ atoms of $^{12}C$. Although this is a tiny amount, it nevertheless results in 15.3 decays of $^{14}C$ per gram of carbon per minute. Using modern mass spectrometers, the fractional abundance of $^{14}C$ can also be measured directly.

After a living organism dies, its $^{14}C$ content decreases according to the first-order decay law, whereas its $^{12}C$ content remains fixed. After $5.73 \times 10^3$ years, there is half as much $^{14}C$ present, for a fractional content of $6.63 \times 10^{-13}$ or one atom for every $1.508 \times 10^{12}$ atoms of $^{12}C$. The decay rate from a sample that is 5730 years old will be 7.65 decays $g^{-1}$ $min^{-1}$.

The usual procedure for radiocarbon dating is to burn a tiny sample of the object to be dated, collect the $CO_2$ that is produced, and compare its rate of radioactive decay with that of a fresh $CO_2$ sample. The ratio of counts gives $N_0/N$, which can then be substituted into Equation 20-5 to calculate $t$. Mass spectroscopic isotope analysis can also be used to obtain the $N_0/N$ value, as Example 20-10 illustrates.

## EXAMPLE 20-10 ISOTOPIC CARBON DATING

In 1988, the Shroud of Turin, claimed by some to have been used to bury Christ, was age-dated by isotopic analysis of its carbon content. The fractional content of $^{14}C$ from a sample from the shroud was $1.22 \times 10^{-12}$. How old is the shroud? If it were 2000 years old, what fractional content would it have?

**Strategy:** Calculation of an age, $t$, requires the value of $N_0/N$, which is used in Equation 20-5. The ratio can be found from the isotopic analyses of a fresh carbon sample and the old sample.

**Solution:** The data include the fractional content of $^{14}C$ in the sample from the shroud, $1.22 \times 10^{-12}$. The text states that the fractional content for a fresh sample is $1.33 \times 10^{-12}$. To obtain $N_0/N$, divide the ratio for a fresh sample by the ratio for the old sample:

$$\frac{N_0}{N} = \frac{1.33 \times 10^{-12}}{1.22 \times 10^{-12}} = 1.09$$

$$t = \frac{t_{1/2}}{\ln 2} \ln \left( \frac{N_0}{N} \right)$$

$$t = \frac{5730 \text{ years}}{0.693} \ln (1.09) = 714 \text{ years}$$

The Shroud of Turin is imprinted with an image that resembles a male human, thought by some to be a representation of Christ.

Thus, the cotton or flax from which the shroud was woven was harvested around 1300 AD.

Use the same equation to compute the isotopic ratio expected for an artefact that is 2000 years old:

$$\ln \left( \frac{N_0}{N} \right) = \frac{(2000 \text{ years})(0.693)}{5730 \text{ years}} = 0.242 \qquad \frac{N_0}{N} = e^{0.242} = 1.274$$

$$N = \frac{N_0}{1.274} = \frac{1.33 \times 10^{-12}}{1.274} = 1.04 \times 10^{-12}$$

**Does the Result Make Sense?** The ratio $N_0/N$ must be a number greater than 1, because the amount of radioactive carbon always decreases with time. The fraction for the shroud, 1.09, is significantly smaller than the ratio expected for a 2000-year-old object, 1.274. The smaller ratio means that less radioactive carbon has disintegrated, indicating that the object is not as old as 2000 years.

### Practice Exercise 20.10

Mass spectrometric analysis of charcoal from an archaeological dig gave a carbon-14 fractional abundance of $6.5 \times 10^{-13}$. What is the age of this dig?

Calculations of ages, as illustrated in Example 20-10, make the assumption that the level of $^{14}C$ in the atmosphere at the time of death was the same as it is today. To check the validity of $^{14}C$ dating, its age estimates have been compared with those obtained by tree-ring dating, which involves fewer assumptions than radiocarbon dating. Wood from a Roman shipwreck, for example, was dated at 40 BC ± 3 years using tree-ring dating and 80 BC ± 200 years using radiocarbon dating. A large number of such comparisons show small but consistent differences, indicating that the fractional amount of $^{14}C$ in the atmosphere has changed slowly but steadily with time. Archaeologists use a small multiplying factor to correct for this small drift when they use radiocarbon dating on objects that do not retain tree rings, such as charcoal and cloth.

## Radioactive Tracers

Radioactive isotopes are chemically identical to their natural, non-radioactive counterparts, but their high-energy decays allow them to be detected even though they may compose only a tiny fraction of the overall isotopic composition. As already described, radiocarbon dating can detect $^{14}C$ at a concentration of one part in $10^{12}$ in samples that contain carbon. If an isotopically enriched sample is introduced into some dynamic process, the course of that process can be followed by tracing the whereabouts of the radioactive isotope.

Tracer techniques are applied in fields that range from geology to medicine. For example, a holding tank at an industrial plant might be suspected of leaking contaminated water into wells in its vicinity, as shown diagrammatically in Figure 20-22. The path of water draining from the holding tank could be traced by adding water enriched in radioactive tritium to the holding tank. The water from the wells could then be sampled for radioactivity at regular intervals. If, after a few days, water in some wells showed significant radioactivity while the remaining wells were free of contamination even after several months, that would show which of the wells were receiving water from the holding tank.

Tracer techniques are used widely in biology. Botanists, for example, work to develop new plant hybrids that grow more rapidly. One common way to determine how fast plants grow is to measure how quickly they take up elemental phosphorus from the soil. New hybrids can be planted in a plot and fertilized with phosphorus enriched in radioactive $^{32}P$. When leaves from the plants are assayed for $^{32}P$ at regular intervals, the rate of increase in radioactivity measures the rate of phosphorus uptake.

**Figure 20-22**

Radioactive tracers can show flow pathways that cannot be detected using other methods of detection. Tritiated water added to a holding tank results in radioactivity in the water from wells that draw from the groundwater supply into which the holding tank drains.

## ▮ FOCUS 20-2 CHEMISTRY AND LIFE: PRESERVING FOOD WITH RADIATION

In a society where canned and frozen foods are so prevalent, we tend to take food preservation for granted. It is easy to forget that prevention of spoilage is a major undertaking when food is harvested at one time and place and potentially eaten much later and far away. A contemporary example is the International Space Station, where the astronauts must be supplied with edible food that is preserved without refrigeration for long times.

Spoilage is a biological process. Moulds, bacteria, and vermin eat foodstuffs, rendering them unfit for human consumption. To stop spoilage, processors treat food to kill microorganisms, chill it to slow the metabolism of destruction, and keep it sealed to ward off pests. Even in the developed world, these procedures are only partially successful, and in economically developing countries, up to 50% of crops may be lost to spoilage.

The bactericidal effects of ionizing radiation have been known for a century, and for over 80 years it has been known that ionizing radiation kills the *Trichinella spiralis* parasite, which infects raw pork. By the mid-1970s, international experts had concluded that irradiating foods preserves them without creating any toxicological hazards. Only in the last decade, however, has preservation by irradiation been applied commercially.

Irradiation of foods can be used for a variety of purposes. Low doses help preserve potatoes, onions, and garlic by inhibiting sprouting. Such doses also kill the parasite in pork that results in trichinosis, and they inactivate insects that feed on stored grain. Medium doses inhibit the growth of mould on fruits such as strawberries, and help to preserve meat, poultry, and fish by killing microorganisms that cause spoilage, and reducing the levels of pathogens such as salmonella. High doses of radiation are sufficiently lethal to microorganisms to provide long-term protection for meat as well as for spices and seasonings.

Irradiation preserves strawberries and other fruit. The green Radura logo is required in Canada to show a food has been treated with ionizing radiation.

Irradiation is a relatively simple process. The food to be irradiated is placed inside a chamber where it is exposed to γ rays from a radioactive source such as $^{60}$Co or $^{137}$Cs. Alternatively, an accelerated electron beam can be used, mimicking radiation.

Although γ irradiation of food is used widely in Europe, usage has been more restricted in Canada. To date, potatoes, onions, wheat, flour, whole wheat flour, whole and ground spices, and dehydrated seasoning preparations have been approved for irradiation. In the United States, irradiation has also been approved for fresh and frozen poultry and red meat to control pathogens, and for fresh eggs to kill salmonella bacteria. Opposition to irradiated foods arises from two main concerns. In the process of killing microorganisms, irradiation might also destroy important nutrients such as vitamins. Secondly, irradiation produces ions and free radicals such as OH that may react with foodstuffs to generate harmful compounds such as carcinogens. Although it is difficult to prove beyond a doubt that these effects do not occur, all the scientific evidence gathered thus far indicates that properly irradiated food is both nutritious and safe.

Furthermore, it is important to keep in mind that conventional methods of preservation also carry risks. For example, ethylene dibromide, which was used for years to kill pests in fruit, has been banned because it is implicated as a carcinogen. Nitrites, which are used widely in preserving bacon, bologna, and other processed meats, are now known to be unhealthy at high exposures. Inadequate treatment of foodstuffs, particularly raw meat and poultry, has led to several outbreaks of illness, some of which have resulted in deaths from pathogens such as *E. coli*. Irradiation is highly effective in eliminating the possibility of trichinosis from undercooked pork, salmonella from raw eggs, or gastrointestinal distress from medium-rare hamburgers. It is likely that the use of food irradiation in North America will increase in the future.

One of the first chemical applications of radioactive tracers was a set of elegant experiments on photosynthesis performed in the 1950s by Melvin Calvin. His goal was to determine the set of reactions used by plants to transform atmospheric $CO_2$ into carbohydrates. Calvin supplied growing plant cells with $CO_2$ enriched with $^{14}C$. By harvesting cells and using radioactive counting to determine which molecules gained enhanced $^{14}C$ radioactivity and in what sequence, Calvin was able to unravel the elaborate chain of reactions that is now known as the Calvin cycle. Dr. Calvin received the Nobel Prize in chemistry in 1961 in recognition of this work.

## SECTION EXERCISES

●**20.7.1** An archaeologist discovers a new site where small charcoal bits are mixed with human remains. Burning a small sample of this charcoal gives gaseous $CO_2$ that, when placed in a counter, registers 1.75 decays per second. When an equal mass of $CO_2$ from fresh charcoal is placed in this counter, the count rate is 3.85 decays per second. How old is the archaeological site?

●**20.7.2** Describe how radioactive iodine ($^{131}I$) could be used to determine how rapidly each of the following proposed chemical exchange reactions occurs: (a) $I_2$ (s) exchanging with $I^-$ in aqueous solution; and (b) $I_2$ (s) exchanging with $I_2$ (g) in a closed container.

# CHAPTER 20 VISUAL SUMMARY

## 20.1 NUCLEAR STABILITY

9 protons    10 neutrons    Fluorine-19, $^{19}_{9}F$

Nuclear binding energy

$$E = mc^2 \qquad \Delta E = (\Delta m)c^2$$

$$\Delta E = (\Delta m)(8.988 \times 10^{10} \text{ kJ/g})$$

Fission    Fusion

**LEARNING OBJECTIVES:** Explain the forces that hold nuclei together.

**SKILLS TO MASTER:** Writing nuclide symbols (p. 1164)

Calculating binding energies (p. 1165)

Predicting stabilities of nuclides (p. 1171)

**KEY CONCEPTS:** The mass number is the sum of the atomic number and the number of neutrons.

Energy and mass are interconvertible.

Most stable nuclides have even numbers of protons and neutrons.

## 20.2 NUCLEAR DECAY

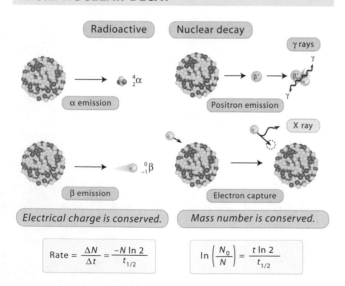

Electrical charge is conserved.    Mass number is conserved.

$$\text{Rate} = \frac{\Delta N}{\Delta t} = \frac{-N \ln 2}{t_{1/2}} \qquad \ln\left(\frac{N_0}{N}\right) = \frac{t \ln 2}{t_{1/2}}$$

**LEARNING OBJECTIVES:** Describe nuclear decay processes and calculate rates of decay.

**SKILLS TO MASTER:** Balancing nuclear reactions (p. 1173)

Calculating the decay rate of a nuclide (p. 1178)

Calculating the half-life of a nuclide (p. 1178)

**KEY CONCEPTS:** Mass number and electrical charge are conserved during a nuclear reaction.

The products of a nuclear reaction have higher binding energies than the reactants.

All nuclear decay processes obey first-order kinetics.

## 20.3 INDUCED NUCLEAR REACTIONS

Induced nuclear reaction

$^{14}_{7}N$    $^{4}_{2}He$    $[^{18m}_{9}F]$    $^{17}_{8}O$    $^{1}_{1}H$

Compound nucleus

Cyclotron    Linear accelerator

**LEARNING OBJECTIVES:** Describe how products are synthesized in induced nuclear reactions.

**SKILLS TO MASTER:** Balancing neutron-capture reactions (p. 1181)

Balancing binuclear reactions (p. 1183)

**KEY CONCEPTS:** In an induced nuclear reaction, a nuclear projectile collides and reacts with another nucleus.

## 20.4 NUCLEAR FISSION

**LEARNING OBJECTIVES:** Describe the fission process and the essential features of a fission reactor.

**SKILLS TO MASTER:** Calculating the energy released by a fission reaction (p. 1186)

**KEY CONCEPTS:** Neutron capture by a fissionable nucleus produces a highly unstable nucleus that distorts and then splits into smaller nuclei and a few free neutrons.

Fission gives a wide range of product nuclides from $A = 77$ to $A = 157$.

The critical mass is the amount of material that is just large enough to recapture one neutron, on average, for every fission reaction.

## 20.5 NUCLEAR FUSION

**LEARNING OBJECTIVES:** Describe the fusion process and the essential features of a fusion reactor.

**SKILLS TO MASTER:** Explaining the origins of heavy nuclides in the universe (p. 1195)

**KEY CONCEPTS:** Initiating a fusion reaction requires enormous inputs of energy to overcome electrical repulsion of protons in the nuclei.

## 20.6 EFFECTS OF RADIATION

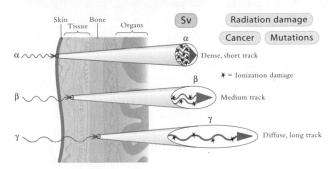

**LEARNING OBJECTIVES:** Explain the effects of ionizing radiation.

**SKILLS TO MASTER:** Knowing the health effects of radiation doses (p. 1200)

**KEY CONCEPTS:** Radiation damage is caused by the ionizing effects of high-energy particles.

Different types of radiation penetrate matter to different depths and produce different damage.

## 20.7 APPLICATIONS OF RADIOACTIVITY

**LEARNING OBJECTIVES:** Explain some practical applications of radioactivity.

**SKILLS TO MASTER:** Calculating the age of an object using radioactive dating (p. 1203)

**KEY CONCEPTS:** Radioactive nuclides can be used as tracers in biological and environmental studies.

Radiation can be used preserve food by killing bacteria.

## LEARNING EXERCISES

20.1 Write a paragraph summarizing the important features of each of the following topics: (a) nuclear stability; (b) nuclear decay; (c) fission; (d) fusion; and (e) binding energy.

20.2 Outline the requirements for a balanced nuclear equation, and explain how they differ from those for a balanced chemical equation.

20.3 Prepare a list of the beneficial and detrimental effects that have resulted from the discovery of nuclear reactions.

20.4 Make a list of all terms introduced in Chapter 20 that are new to you. Write a one-sentence definition for each, using your own words. If you need help, consult the Glossary.

20.5 Update your list of memory bank equations to include those that apply to nuclear chemistry and radiochemistry.

## PROBLEMS

**ilw** = interactive learning ware problem. Visit the website at www.wiley.com/canada/olmsted

## Nuclear Stability

20.1 Determine $Z$, $A$, and $N$ for each of the following nuclides: (a) the nuclide of neon that contains the same number of protons and neutrons; (b) the nuclide of lead that contains 1.5 times as many neutrons as protons; and (c) the nuclide of zirconium with a neutron:proton ratio of 1.25.

20.2 Determine $Z$, $A$, and $N$ for each of the following nuclides: (a) the helium nuclide with one less neutron than proton; (b) the nuclide of barium with a neutron:proton ratio of 1.25; and (c) the nuclide of thorium that contains 1.5 times as many neutrons as protons.

20.3 Manganese has only one stable isotope. Calculate the $N{:}Z$ ratio for each of the following four nuclides and determine which is the stable one: $^{53}$Mn, $^{54}$Mn, $^{55}$Mn, and $^{56}$Mn.

20.4 Rhodium has only one stable isotope. Calculate the $N{:}Z$ ratio for each of the following four nuclides and determine which is the stable one: $^{101}$Rh, $^{102}$Rh, $^{103}$Rh, and $^{104}$Rh.

20.5 Compute the energy released in kilojoules when the Sun converts 1.00 metric tonne of matter into energy.

20.6 Compute the energy released in joules per event and in kilojoules per mole when antiprotons (the antimatter corresponding to protons) annihilate with protons.

20.7 **ilw** Use atomic masses to compute the total binding energy and the binding energy per nucleon for elemental cesium, which has just one stable nuclide.

20.8 Use atomic masses to compute the total binding energy and the binding energy per nucleon for elemental cobalt, which has just one stable nuclide.

20.9 Is the repulsive barrier for fusion of $^1$H and $^6$Li larger, smaller, or about the same as that for the fusion of two $^4$He nuclei? Which gives a more stable product? Explain.

20.10 Is the repulsive barrier for fusion of two protons larger, smaller, or about the same as that for the fusion of two deuterium nuclei? Which of the resulting helium nuclei is less stable? Explain.

## Nuclear Decay

20.11 Write the correct symbol and give the name for each of the following products of nuclear decay: (a) a photon; (b) a positive particle with mass number 4; and (c) a positive particle with the same mass as an electron.

20.12 What are the charge number, mass number, and symbol for each of the following? (a) an alpha particle; (b) a beta particle; and (c) a neutron.

20.13 Identify the product of each of the following decay processes: (a) tellurium with 73 neutrons emits a $\gamma$ ray; (b) tellurium with 71 neutrons captures an electron; and (c) tellurium with 75 neutrons undergoes $\beta$ and $\gamma$ decays.

20.14 Identify the product of each of the following decay processes: (a) $^{52}$Fe emits a positron and a $\gamma$ ray; (b) $^{55}$Fe captures an electron; and (c) $^{59}$Fe undergoes $\beta$ and $\gamma$ decays.

20.15 Predict the most likely mode or modes of decay for each of the isotopes of Mn listed in Problem 20.3.

20.16 Predict the most likely mode or modes of decay for each of the isotopes of Rh listed in Problem 20.4.

20.17 **ilw** The iron isotope with 33 neutrons is used in medical applications. It is a $\beta$ emitter. A sample of iron containing 1.33 picograms (pico = $10^{-12}$) of this isotope registers 242 decays per second. Calculate the half-life of the iron isotope.

20.18 A radioactive nuclide of mass number 94 has been prepared by neutron bombardment. If 4.7 $\mu$g of this nuclide registers 20 counts per minute on a radioactivity counter, what is the half-life of this nuclide?

20.19 Thorium-232 decays by the following sequence of emissions: $\alpha$, $\beta$, $\beta$, $\alpha$, $\alpha$, $\alpha$, $\beta$, $\alpha$, $\beta$, $\alpha$. Identify the final product and all of the unstable intermediates.

20.20 Radon-222 decays by the following sequence of emissions: $\alpha$, $\alpha$, $\beta$, $\beta$, $\alpha$, $\beta$, $\beta$, $\alpha$. Identify the final product and all of the unstable intermediates.

## Induced Nuclear Reactions

20.21 Identify the compound nucleus and final product resulting from each of the following nuclear reactions: (a) carbon-12 captures a neutron and then emits a proton; (b) the nuclide with eight protons and eight neutrons captures an $\alpha$ particle and emits a $\gamma$ ray; and (c) curium-247 is bombarded with boron-11, and the product loses three neutrons.

20.22 Identify the compound nucleus and final product resulting from each of the following nuclear reactions: (a) lithium-7 captures a proton, and the product emits a neutron; (b) the nuclide with six protons and six neutrons captures an $\alpha$ particle and emits a $\gamma$ ray; and (c) plutonium-244 is bombarded with carbon-12, and the product loses two neutrons.

20.23 Draw a nuclear picture (see Figure 20-1 for pictures of nuclei) that illustrates the capture of an $\alpha$ particle by $^{14}$N, followed by the emission of a positron.

20.24 Draw a nuclear picture (see Figure 20-1 for pictures of nuclei) that illustrates the nuclear reaction responsible for the production of carbon-14. Include a picture of the compound nucleus.

$$^{14}N + n \longrightarrow {}^{14}C + p$$

## Nuclear Fission

20.25 When uranium-235 undergoes fission, the product with higher mass is most likely to have a mass number between 132 and 142. Use Figures 20-5 and 20-6 to identify which elements are most likely to result.

20.26 When uranium-235 undergoes fission, the product with lower mass is most likely to have a mass number between 88 and 100. Use Figures 20-5 and 20-6 to identify which elements are most likely to result.

20.27 (ilw) Compute the energy released in the following fission reaction:

$$^{235}U + n \longrightarrow {}^{152}Nd + {}^{81}Ge + 3\,n$$

The nuclide masses in grams per mole are U, 235.0439; Nd, 151.9233; and Ge, 80.9199. Compare your result with the calculation in Section Exercise 20.4.1.

20.28 Compute the energy released in the following fission reaction:

$$^{235}U + n \longrightarrow {}^{100}Mo + {}^{134}Sn + 2\,n$$

The nuclide masses in grams per mole are U, 235.0439; Mo, 99.9076; and Sn, 133.9125. Compare your result with the calculation in Section Exercise 20.4.1.

20.29 A coal-fired power plant requires just one heat exchanger, in which water is converted to the steam needed to drive the generator. A nuclear power plant requires two. Describe the features of nuclear reactions that require this second heat exchanger.

20.30 Describe the features of radioactivity that make an accident in a nuclear power plant more devastating than an accident in a coal-burning power plant.

## Nuclear Fusion

20.31 How much energy is released if 2.50 g of deuterons fuses with tritium and 50% of the neutrons are captured by $^6Li$? The relevant reactions and energies are:

$$^3_1H + {}^2_1H \longrightarrow ({}^5_2He) \longrightarrow {}^4_2He + {}^1_0n \quad \Delta E = -1.7 \times 10^9\,kJ/mol$$

$$^6_3Li + {}^1_0n \longrightarrow ({}^7_3Li) \longrightarrow {}^4_2He + {}^3_1H \quad \Delta E = -4.6 \times 10^8\,kJ/mol$$

20.32 How much energy will be released if 1.50 g of deuterons and 1.50 g of lithium-6 undergo fusion in a hydrogen bomb? The relevant reactions and energies are:

$$^3_1H + {}^2_1H \longrightarrow ({}^5_2He) \longrightarrow {}^4_2He + {}^1_0n \quad \Delta E = -1.7 \times 10^9\,kJ/mol$$

$$^6_3Li + {}^1_0n \longrightarrow ({}^7_3Li) \longrightarrow {}^4_2He + {}^3_1H \quad \Delta E = -4.6 \times 10^8\,kJ/mol$$

20.33 (ilw) Compute the speed of a tritium nucleus with enough kinetic energy to fuse with a deuteron. Calculate nuclear radii using $r = 1.2\,A^{1/3}$ fm ($1\,fm = 10^{-15}$ m).

20.34 Compute the speed of an $\alpha$ particle with enough kinetic energy to fuse with lithium-6. Calculate nuclear radii using $r = 1.2\,A^{1/3}$ fm ($1\,fm = 10^{-15}$ m).

20.35 Describe the features of fusion that have prevented fusion power from being commercialized as fission power has been.

20.36 Describe the features of fusion that make it more attractive as a power source than fission.

20.37 Tabulate the differences in temperature and composition of a first-generation star at the hydrogen-, helium-, and carbon-burning stages of its evolution.

20.38 Repeat the tabulation in Problem 20.37 for a second-generation star. List the differences between the two tabulations.

20.39 Using the characteristics of nuclear reactions, explain why no elements with $Z > 26$ form during the evolution and eventual explosion of first-generation stars.

20.40 Second- and third-generation stars may contain significant amounts of carbon; yet carbon burning does not begin until the third stage of their evolution. Explain this.

## Effects of Radiation

20.41 (ilw) The isotope $^{131}I$ has a half-life of 8.07 days and undergoes decay with an energy of $9.36 \times 10^7$ kJ/mol. If a patient with a thyroid disorder is given this isotope and 7.45 pg is incorporated into the thyroid gland, how much energy will the gland receive in one day?

20.42 A patient weighing 65 kg is given an intravenous dose of $^{99m}Tc$, a radioactive isotope that decays by $\gamma$ emission with energy of $1.35 \times 10^7$ kJ/mol and with a 6.0-hour half-life. If 15.0 mL of a 2.50-nanomolar solution is administered and all of the radioactive nuclide decays within the patient's body, what total energy does the patient receive?

20.43 How many $\beta$ particles are emitted per second by the dose of $^{131}I$ described in Problem 20.41?

20.44 How many $\gamma$ rays are emitted per second by the dose of $^{99m}Tc$ described in Problem 20.42?

20.45 Why does exposure to radiation in cancer therapy result in nausea and reduced resistance to infection?

20.46 Why is it more important to protect the reproductive organs from radiation than other organs?

## Applications of Radioactivity

20.47 (ilw) Extremely old rock samples can be dated using the ratio of $^{87}Sr$ to $^{87}Rb$, because $t_{1/2}$ for $^{87}Rb$ is $4.9 \times 10^{11}$ years. What is the approximate age of a meteorite that has a $^{87}Sr : {}^{87}Rb$ mass ratio of 0.0050?

20.48 Uranium deposits are dated by determining the ratio of $^{238}U$ to its final decay product, $^{206}Pb$. The half-life of $^{238}U$ is $4.5 \times 10^9$ years. If the ratio of $^{238}U$ to $^{206}Pb$ in an ore sample is 1.21, what is the age of the ore?

20.49 A river contains a high concentration of iron cations, and environmental activists contend that an industrial manufacturing plant is the source. The manufacturer says that although the plant generates aqueous iron waste, it is processed on-site and does not contaminate the river. How might this disagreement be resolved using radioactive tracers?

**20.50** Many places have underground water, called aquifers, that can be tapped using wells. Depending on the types of geological formations above these aquifers, water from the surface may or may not reach these underground reservoirs. How could radioactive tracers be used to determine if rainwater falling on a particular location makes its way down into an aquifer beneath that location?

**20.51** In the hydrolysis of an ester, one oxygen atom in the products comes from the ester and the other comes from a water molecule:

$$RCO-O-R' + H_2O \longrightarrow RCO-OH + HO-R'$$

Describe experiments that would determine if the oxygen atom in the alcohol, HO—R′, comes from water or from the ester.

**20.52** In aqueous solution, the hydrogen atom in the —$CO_2H$ group of a weak organic acid rapidly exchanges with a hydrogen atom in a water molecule through proton transfer:

$$RCO_2H + H_2O \rightleftharpoons RCO_2^- + H_3O^+$$

In contrast, the hydrogen atom in the —OH group of an alcohol is not thought to undergo this reaction. A research chemist claims that this is incorrect, and that rapid exchange of H atoms occurs between alcoholic —OH groups and water molecules. Describe experiments that would test this claim.

## ADDITIONAL PAIRED PROBLEMS

**20.53** Naturally occurring bismuth contains only one isotope, $^{209}Bi$. Compute the total molar binding energy and molar binding energy per nucleon of this element.

**20.54** Naturally occurring gold contains only one isotope, $^{197}Au$. Compute the total molar binding energy and molar binding energy per nucleon of this element.

**20.55** Polonium-210 has a half-life of 138.4 days. If a sample contains 5.0 mg of $^{210}Po$, how many milligrams will remain after 365 days? How many emissions per second will this residue emit?

**20.56** What fraction of $^{209}Po$, which has a half-life of 103 years, will remain intact after 365 days? If a sample of polonium contains equimolar amounts of $^{210}Po$ and $^{209}Po$, what will the isotope ratio be after 10 years (see Problem 20.55)?

**20.57** Neutrons decay into protons. What is the other product of this decay? If all of the decay energy is converted into kinetic energy of this other product, how much kinetic energy does it have?

**20.58** A positron has the same mass as an electron. When a positron and an electron annihilate, both masses are converted entirely into the energy of a pair of γ rays. Calculate the energy per γ ray and the energy of 1 mole of γ rays.

**20.59** The long-lived isotope of radium, $^{226}Ra$, decays by α emission with a half-life of 1622 years. Calculate how long it will take for 1% of this nuclide to disappear and how long until 1% of it remains.

**20.60** Phosphorus-30, which has a half-life of 150 seconds, decays by positron emission. How long will it take for (a) 2% and (b) 99.5% of this nuclide to decay?

**20.61** Two isotopes used in positron-emission imaging are $^{11}C$ and $^{15}O$. On which side of the belt of stability are these nuclides located? Write the nuclear reactions for their disintegrations.

**20.62** Neutron bombardment is an elegant way of "doping" silicon with phosphorus to convert it into an $n$-type semiconductor. Bombardment generates stable $^{31}P$ from $^{30}Si$. Write the nuclear reactions for this transformation. (Doped semiconductors are described in Chapter 7.)

**20.63** Sodium-24, $t_{1/2} = 15.0$ h, can be used to study the sodium balance in animals. If a saline solution containing 25 $\mu$g of $^{24}Na$ is injected into an animal, how long can the study continue before only 1 $\mu$g is left?

**20.64** What is the shelf life of a radioactive nuclide with a half-life of 14.6 days, if it loses its usefulness when less than 15% of the radioactivity remains?

**20.65** Complete the following nuclear reactions: (a) α emission from $^{238}U$; (b) $n + ^{60}Ni \longrightarrow ? + p$; (c) $^{239}Np + ^{12}C \longrightarrow ? + 3 n$; (d) $p + ^{35}Cl \longrightarrow \alpha + ?$; and (e) β emission from $^{60}Co$.

**20.66** Complete the following nuclear reactions: (a) positron emission from $^{26}Si$; (b) electron capture by $^{82}Sr$; (c) $^{210}Po \longrightarrow \alpha + ?$; (d) $\alpha + ^9Be \longrightarrow n + ?$; and (e) $^{99m}Tc \longrightarrow \gamma + ?$

**20.67** The heaviest transuranium elements are formed by bombardment with relatively heavy nuclides such as $^{48}Ti$. What nuclide could be formed by bombarding lead-208 with this nuclide?

**20.68** The heaviest transuranium elements are formed by bombardment with relatively heavy nuclides such as $^{58}Fe$. What nuclide could be formed by bombarding the stable bismuth nuclide with this nuclide?

**20.69** Gadolinium-152 ($M = 151.9205$ g/mol) emits an α particle ($E_{kinetic} = 3.59 \times 10^{-13}$ J) to form $^{148}Sm$ ($M = 147.9146$ g/mol). How much energy (in J/nucleus) is released in this transformation? What fraction of the energy is carried off by the α particle?

**20.70** Samarium-146 ($M = 145.9129$ g/mol) emits an α particle ($E_{kinetic} = 3.94 \times 10^{-13}$ J) and becomes $^{142}Nd$ ($M = 141.9075$ g/mol). How much energy (in J/nucleus) is released in this transformation? What fraction of the energy is carried off by the α particle?

**20.71** State why each of the following nuclides is unstable: (a) carbon-14; (b) plutonium-244; (c) $^{56}Mn$; and (d) a lithium nucleus with five neutrons.

**20.72** State why each of the following nuclides is unstable: (a) tritium (one proton, two neutrons); (b) uranium-238; (c) $^{40}K$; and (d) a beryllium nucleus with four neutrons.

**20.73** Describe the roles of the moderator and the control rods in the operation of a nuclear power plant.

**20.74** Describe how irradiation protects foods against spoilage, and describe how irradiation of foods might also produce undesirable by-products.

## ADDITIONAL PROBLEMS

**20.75** Free neutrons are unstable. Since they cannot be collected and weighed, it is difficult to measure their half-life accurately. Early estimates gave $t_{1/2} = 1.1 \times 10^3$ s, but more refined experiments give $t_{1/2} = 876 \pm 21$ s. Suppose a neutron source generates $10^5$ neutrons per second for 30 s. Assuming no neutrons are captured by nuclei, how many will be left after 5 h, according to each of these half-lives?

**20.76** Radioactive $^{64}$Cu ($M = 63.92976$ g/mol) decays either by emission of β particles with $9.3 \times 10^{-14}$ J of energy or by emission of positrons with $1.04 \times 10^{-13}$ J of energy. (a) Write the two decay reactions. (b) Calculate the molar masses of the two elemental products using mass–energy equivalence.

**20.77** Nuclear power plants often use boron control rods. Boron-10 absorbs neutrons efficiently, emitting α particles as a consequence. Write the nuclear reaction. Does this reaction pose a significant health hazard? Explain your reasoning.

**20.78** The radius of a $^{12}$C nucleus is $3.0 \times 10^{-15}$ m. Compute the energy barrier in J and in kJ/mol for the fusion of two $^{12}$C nuclei.

**20.79** A 250-mg sample of $CO_2$ collected from a small piece of wood at an archaeological site gave 1020 counts over a 24-h period. In the same counting apparatus, 1.00 g of $CO_2$ from freshly cut wood gave 18 400 counts in 20 h. What age does this give for the site?

**20.80** One isotope of nitrogen and one isotope of fluorine are positron emitters with relatively long half-lives. Identify them and write their decay reactions.

**20.81** The amount of radioactive carbon in any once-living sample eventually drops too low for accurate dating. This detection limit is about 0.03 $g^{-1}$ $min^{-1}$, whereas fresh samples exhibit a count rate of 15.3 $g^{-1}$ $min^{-1}$. What is the upper limit for age determinations using carbon dating?

**20.82** Carbon-14 dating gives 3250 years as the age of a charcoal sample, assuming a constant level of cosmic radiation. If the cosmic ray level in the atmosphere was 20% higher at the time the tree grew, what is the correct age of the sample?

**20.83** Strontium-90 is a dangerous fission product because it passes into human bodies and lodges in bones. It decays in two steps to give $^{90}$Zr. The molar masses of these two nuclides are 89.9073 and 89.9043 g/mol respectively. (a) Write the decay reactions that lead from $^{90}$Sr to $^{90}$Zr. (b) Which nuclide has the higher molar mass? (c) Calculate the total energy released in the decay scheme. (*Reminder:* Include the masses of the other decay products.)

**20.84** Suppose that fusion of tritium and deuterium involves two nuclei with equal kinetic energies totalling 75% of the electrical repulsion barrier. Calculate their speeds.

**20.85** Iodine-123, $t_{1/2} = 13.2$ h, is used in medical imaging of the thyroid gland. If 0.5 mg of this isotope is injected into a patient's bloodstream and 45% of it binds to the thyroid gland, how long will it take for the amount of iodine in the thyroid gland to fall to less than 0.1 μg?

**20.86** In 1934, Irene and Frédéric Joliot-Curie produced the first artificial radioisotope by bombarding $^{27}$Al with particles. The resulting compound nucleus decayed by neutron emission. Write the reaction and predict whether or not the final product is stable. If it is not stable, predict its mode of decay.

**20.87** Technetium has no stable isotopes. Which nuclide of this element would you predict to be nearly stable? (Make use of Figures 20-5 and 20-6.)

**20.88** Calculate the mass of $^{235}$U that reacts in a 20-kiloton bomb, given that one nucleus releases $2.9 \times 10^{-11}$ J. The designation "20-kiloton" means the same energy as $20 \times 10^3$ tonnes of TNT, which releases 2500 kJ of energy per kilogram.

**20.89** As discussed in Chapter 17, a condensation reaction between a carboxylic acid and an alcohol produces an ester with the elimination of water:

Explain how the unstable isotope $^{18}$O can be used to show whether the oxygen atom in the water molecule comes from the carboxylic acid or from the alcohol.

**20.90** A radioactivity counter gave a reading of 350 $min^{-1}$ for a 12.5-mg sample of cobalt(II) chloride partially enriched with cobalt-60 ($t_{1/2} = 5.2$ years). What percentage of the cobalt atoms is cobalt-60?

**20.91** A small amount of NaBr containing the radioactive isotope sodium-24 is dissolved in a hot solution of sodium nitrate containing the naturally occurring non-radioactive isotope sodium-23. The solution is cooled, and sodium nitrate precipitates from the solution. Will the precipitate be radioactive? Explain your answer.

## GROUP STUDY PROBLEMS

**20.92** From the mass number distribution given in Figure 20-11 and the belt of stability shown in Figures 20-5 and 20-6, predict which elements should be found in greatest abundance in the remains of a fission event like that of the Oklo reactor (Focus 20-1).

**20.93** The Earth captures $3.4 \times 10^{17}$ J/s of radiant energy, which is one part in $4.5 \times 10^{-10}$ of the Sun's total energy output. (a) By how much does the mass of the Sun change every second? (b) The Sun is in the hydrogen-burning stage of its evolution. Calculate how many moles of $^1$H are converted into $^4$He per second.

**20.94** For how many years could $^{235}$U supply the world's energy needs, given that uranium contains 0.72% of $^{235}$U, the world reserves of uranium are $1.0 \times 10^7$ tonnes, world energy consumption is $2.0 \times 10^{17}$ kJ/year, and each $^{235}$U nucleus releases $2.9 \times 10^{-11}$ J?

**20.95** Smoke detectors contain small quantities of $^{241}Am$, which decays by emitting an $\alpha$ particle with $5.44 \times 10^8$ kJ/mol energy and a 458-year half-life. It also emits $\gamma$ rays. (a) What is the neutron:proton ratio for $^{241}Am$? (b) If a smoke detector operates at 5.0 decays per second, what mass of $^{241}Am$ does it contain? (c) If this smoke detector malfunctions when its decay rate falls below 3.5 decays per second, for how many years could it be used? (d) If your bed were directly under this smoke detector, would you be exposed to its nuclear radiation? Explain.

**20.96** Unlike most other elements, different samples of lead have different molar masses. This is because lead is the stable end-product of two different radioactive decay schemes, one starting from $^{238}U$ and the other from $^{232}Th$. Both decay schemes consist entirely of $\alpha$, $\beta$, and $\gamma$ emissions. The most abundant stable lead isotopes have mass numbers 206, 207, and 208. Will a sample of lead, some of which came from $^{238}U$, have a higher or lower molar mass than another sample of lead, some of which came from $^{232}Th$? Draw a decay scheme that illustrates your reasoning.

**20.97** List the advantages and disadvantages of fission power and fusion power. Based on your list, do you think that Canada should continue to develop fission power plants? What about fusion power plants?

**20.98** Uranium-238 undergoes eight consecutive $\alpha$ emissions to give stable lead. In a uranium ore sample, all the $\alpha$ particles are quickly stopped, becoming trapped $^4He$ atoms. If analysis of a rock sample shows that it contains $6.0 \times 10^{-5}$ cm$^3$ of helium-4 (1 bar, 298 K) and $1.3 \times 10^{-7}$ g of $^{238}U$ per gram of rock, estimate the age of the rock.

## ANSWERS TO SECTION EXERCISES

**20.1.1** (a) $^{16}_{8}O$, $Z = 8$, $N = 8$, $A = 16$;
(b) $^{98}_{43}Tc$, $Z = 43$, $N = 55$, $A = 98$; and
(c) $^{3}_{1}H$, $Z = 1$, $N = 2$, $A = 3$

**20.1.2** (a) unstable, $Z > 83$; (b) unstable, odd–odd; (c) stable; and (d) unstable, $N < Z$

**20.1.3** Total binding energy is $-1.43 \times 10^{10}$ kJ/mol; binding energy per nucleon is $-7.52 \times 10^8$ kJ/mol

**20.2.1** (a) $\beta$ decay; (b) $\alpha$ decay; (c) $\gamma$ emission; and (d) positron emission or electron capture

**20.2.2** (a) $^{140}_{57}La$; (b) $^{222}_{86}Rn$; and (c) $^{125}_{53}I$

**20.2.3** 12 days

**20.3.1** (a) $^{59}_{27}Co + ^{1}_{0}n \longrightarrow \left(^{60m}_{27}Co\right) \longrightarrow ^{59}_{26}Fe + ^{1}_{1}p$;

(b) $^{59}_{27}Co + ^{1}_{0}n \longrightarrow \left(^{60m}_{27}Co\right) \longrightarrow ^{60}_{27}Co + \gamma$;

(c) $^{59}_{27}Co + ^{4}_{2}\alpha \longrightarrow \left(^{63m}_{29}Cu\right) \longrightarrow ^{62}_{28}Ni + ^{1}_{1}p$; and

(d) $^{59}_{27}Co + ^{1}_{1}p \longrightarrow \left(^{60m}_{28}Ni\right) \longrightarrow ^{58}_{28}Ni + 2^{1}_{0}n$

**20.3.2** There are several possibilities. Here are two:

$^{144}_{60}Nd + ^{1}_{0}n \longrightarrow \left(^{145m}_{60}Nd\right) \longrightarrow ^{145}_{61}Pm + ^{0}_{-1}\beta$

$^{147}_{62}Sm + ^{1}_{0}n \longrightarrow \left(^{148m}_{62}Sm\right) \longrightarrow ^{147}_{61}Pm + ^{1}_{1}p$

**20.3.3** (Answers are other reactant, then other product) (a) neutron, $\gamma$ ray; (b) $^{54}Fe$, $^{57}Ni$; and (c) proton, $\alpha$ particle

**20.4.1** One neutron: $-0.22$ g/mol, $-2.0 \times 10^{10}$ kJ/mol
Two neutrons: $-0.22$ g/mol, $-1.9 \times 10^{10}$ kJ/mol
Three neutrons: $-0.21$ g/mol, $-1.9 \times 10^{10}$ kJ/mol

**20.4.2** 99.22% $^{238}U$ and 0.78% $^{235}U$

**20.5.1** (a)

(b) $5.5 \times 10^7$ kJ/mol; (c) the first step is endothermic by $1.82 \times 10^7$ kJ/mol (but this energy could be provided by the kinetic energy of the deuterons); the second step is exothermic by $-1.50 \times 10^9$ kJ/mol; and (d) $4 \times 10^{12}$ kJ

**20.5.2** More than one sequence is possible. Here is one:

$^{56}Fe \xrightarrow{+n} ^{57}Fe \xrightarrow{+n} ^{58}Fe \xrightarrow{+n} ^{59}Fe \xrightarrow{-\beta} ^{59}Co$

$^{59}Co \xrightarrow{+n} ^{60}Co \xrightarrow{-\beta} ^{60}Ni$

$^{60}Ni \xrightarrow{+n} ^{61}Ni \xrightarrow{+n} ^{62}Ni \xrightarrow{+n} ^{63}Ni \xrightarrow{-\beta} ^{63}Cu$

$^{63}Cu \xrightarrow{+n} ^{64}Cu \xrightarrow{+n} ^{65}Cu \xrightarrow{+n} ^{66}Cu \xrightarrow{-\beta} ^{66}Zn$

**20.5.3** Planets around a second-generation star contain no elements with $Z > 26$. Simple life as we know it requires no elements with $Z > 26$, but mammalian life depends on enzymes containing copper, zinc, and cobalt.

**20.6.1** $4.4 \times 10^5$ $\beta$ particles, which deposit $9.6 \times 10^{-8}$ J of energy

**20.6.2** 12 500 X rays; the damage would not show up in a year.

**20.7.1** $6.5 \times 10^3$ years

**20.7.2** Prepare a sample of solid $I_2$ containing some of the radioactive isotope. (a) Put some of this solid in an aqueous solution containing a salt such as KI. Withdraw samples at various times, precipitate the $I^-$ as AgI, and measure the radioactivity of the solid. (b) Put some of the solid in a closed container equipped with a gas sampling device. Collect gas samples at various times and measure the radioactivity of the gas.

## ANSWERS TO PRACTICE EXERCISES

**20.1** (a) $^{31}_{15}P$ and (b) $^{20}_{10}Ne$

**20.2** $\Delta E = -1.738 \times 10^{11}$ kJ/mol

**20.3** $\Delta m = -0.0267$ g

**20.4** $^{214}_{84}Po \longrightarrow ^{4}_{2}\alpha + ^{210}_{82}Pb$; $^{210}_{82}Pb \longrightarrow ^{210}_{83}Bi + ^{0}_{-1}\beta$;
$^{210}_{83}Bi \longrightarrow ^{210}_{84}Po + ^{0}_{-1}\beta$; and $^{210}_{84}Po \longrightarrow ^{4}_{2}\alpha + ^{206}_{82}Pb$

**20.5** $t_{1/2} = 2.1 \times 10^5$ years

**20.6** $t = 3.4 \times 10^5$ years

**20.7** $^{12}_{6}C + ^{12}_{6}C \longrightarrow \left(^{24}_{12}Mg\right) \longrightarrow ^{4}_{2}\alpha + ^{20}_{10}Ne$

**20.8** $\Delta E_{per\ gram} = -7.37 \times 10^7$ kJ/g

**20.9** 3.6 billion years

**20.10** $5.9 \times 10^3$ years

# APPENDIX A: SCIENTIFIC NOTATION

Special notations are useful for expressing the precisions of very large and very small quantities. The diameter of a helium atom, for instance, is 0.000 000 000 24 metres, and the average distance from the Earth to its moon is 384,000,000 metres. Writing numbers with this many zeros is cumbersome, and it is easy to make a mistake. To shorten the writing of small and large numbers, scientists commonly use **scientific notation.**

Scientific notation is based on the fact that any number can be expressed as a number between 1 and 10 multiplied or divided by ten an appropriate number of times. For example, 384,000,000 metres can be written as follows:

$(3.84)(10)(10)(10)(10)(10)(10)(10)(10)$

$$= 3.84 \times 10^8 \text{ metres}$$

Similarly, 0.000 000 000 24 metres is written as follows:

$2.4/((10)(10)(10)(10)(10)(10)(10)(10)(10)(10))$

$$= 2.4 \times 10^{-10} \text{ metres}$$

The numeral "8" in $10^8$ and the numeral "$-10$" in $10^{-10}$ are examples of **exponents** or **powers**. The scientific notation, $3.84 \times 10^8$ m, contains an exponent (or power of ten) of eight.

## Converting to Scientific Notation

When a number is *divided* by ten, the decimal point moves one place to the *left;* when a number is *multiplied* by ten, the decimal point moves one place to the *right:*

$$\frac{384}{10} = 38.4 \qquad and \qquad (0.0024)(10) = 0.024$$

Multiplying *or* dividing a number by ten changes its value, but multiplying *and* dividing a number by ten leaves its value unchanged. Thus, to convert a number larger than ten to scientific notation, first divide by ten the number of times that gives a number between one and ten, then multiply by ten that same number of times. Here, for example, is the conversion of the number of seconds in a day into scientific notation:

$$\frac{86,400 \text{ s/day}}{(10)(10)(10)(10)} \times (10)(10)(10)(10)$$
$$\downarrow \qquad\qquad \downarrow$$
$$8.6400 \qquad\quad 10^4 \qquad = 8.6400 \times 10^4 \text{ s/day}$$

The four divisions by ten move the decimal point four places to the left, and the four multiplications by ten are written as $10^4$.

A number larger than ten can be quickly converted into scientific notation by moving the decimal point to the left enough places to give a number between one and ten, and multiplying by ten raised to the positive power that equals the number of places moved. For example, the number 96,485 can be expressed as 9.6485 multiplied by $10^4$:

$$96,485 = 9.6485 \times 10^4$$

To convert a number smaller than one into scientific notation, first multiply by ten as many times as needed to give a number between one and ten, then divide by ten that same number of times. Here, for example, is the conversion into scientific notation of the fraction of a day represented by one second:

$$0.000\ 011\ 574 \text{ day/s } (10)(10)(10)(10)(10) \times \frac{1}{(10)(10)(10)(10)(10)}$$
$$\downarrow \qquad\qquad\qquad\qquad \downarrow$$
$$1.1574 \qquad\qquad\qquad\qquad 10^{-5}$$
$$= 1.1574 \times 10^{-5} \text{ day/s}$$

To convert a number smaller than ten into scientific notation, move the decimal point to the right enough places to give a number between one and ten and multiply by ten raised to the negative power equal to the number of places moved.

With practice, these conversions can be carried out quite quickly. The key is to count the number of places the decimal point must be moved and write the exponent accordingly. Here are some additional examples:

$$1\ 000\ 000\ mm/km = 1.000\ 000 \times 10^6\ mm/km$$

$$299\ 792\ 458\ m/s = 2.997\ 924\ 58 \times 10^8\ m/s$$

$$0.08314\ \text{L } bar\ mol^{-1}K^{-1}$$
$$= 8.314 \times 10^{-2}\ \text{L } bar\ mol^{-1}K^{-1}$$

## Working with Powers of Ten

Chemistry students must be able to add, subtract, multiply, and divide numbers that are expressed in scientific notation. This is easily accomplished on a calculator by entering the numbers using power-of-ten notation, but it is also useful to understand how these operations are carried out. Multiplication and division are handled in one fashion, but addition and subtraction are handled in another fashion.

To *multiply* two numbers expressed in scientific notation, multiply the values and *add* the exponents:

$$(2.450 \times 10^2)(1.680 \times 10^3) = (2.450)(1.680) \times 10^{(2+3)}$$
$$= 4.116 \times 10^5$$

Retain signs when adding exponents:

$$(2.450 \times 10^2)(1.680 \times 10^{-3})(2.450)(1.680) \times 10^{(2-3)}$$
$$= 4.116 \times 10^{-1}$$

To *divide* numbers expressed in power of ten notation, divide the values and *subtract* the exponent of the divisor from that of the number divided:

$$\frac{2.450 \times 10^2}{1.680 \times 10^3} = \frac{2.450}{1.680} \times 10^{(2-3)} = 1.458 \times 10^{-1}$$

As with multiplication, retain signs when subtracting exponents:

$$\frac{2.450 \times 10^2}{1.680 \times 10^{-3}} = \frac{2.450}{1.680} \times 10^{(2-(-3))} = 1.458 \times 10^5$$

Addition and subtraction require a different approach. Consider, for example, the addition of these two numbers:

$$2.450 \times 10^2 + 1.680 \times 10^3 = ?$$

Using standard notation, this addition is as follows:

$$
\begin{array}{r}
1680 \\
\underline{245.0} \\
1925
\end{array}
$$

To obtain this result in scientific notation, express both numbers using the *same* power of ten, most conveniently the largest such power, then do the addition:

$$2.450 \times 10^2 = 0.2450 \times 10^3$$
$$0.2450 \times 10^3 + 1.680 \times 10^3 = 1.925 \times 10^3$$

To add or subtract numbers expressed in power of ten notation, first express all numbers using the *same power of ten*. Then add or subtract the values and retain the same power of ten. Calculators automatically take care of these power-of-ten manipulations.

Subtraction may give a result that is smaller than one. When this happens, the convention is to change the exponent to make the numerical value between one and ten. Here, for example, is the difference between the atomic radii of oxygen atoms and fluorine atoms:

O atom F atom Difference
$$1.40 \times 10^{-10}\text{m} - 1.35 \times 10^{-10}\text{m} = 0.05 \times 10^{-10}\text{m}$$

Proper notation
$$= 5 \times 10^{-12}\text{m}$$

To express this result using a number between one and ten, we moved the decimal point two places to the right and increase the negative power of ten by two.

This last example helps to show how the use of scientific notation simplifies the operations with the very small and very large numbers encountered in chemistry. Carried out in standard notation, this subtraction requires a whole lot of zeros:

$$
\begin{array}{r}
0.000\,000\,000\,140 \text{ m} \\
- \underline{0.000\,000\,000\,135} \text{ m} \\
0.000\,000\,000\,005 \text{ m}
\end{array}
$$

# APPENDIX B: QUANTITATIVE OBSERVATIONS

Chemists usually want to know not only *what* is happening but also to what extent (*how much*). Quantitative measurements have three equally important parts: a *numerical value*, appropriate *units*, and a *precision*. Any experimentally measured quantity *always includes all these parts*. Numerical value refers to the numbers. Units are what allow us to scale a numerical value appropriately. For example, 20 *kilometres* is a very different length than 20 *centimetres*. Precision requires further description.

## Precision and Significant Figures

The precision of a quantitative value is the degree of certainty with which it is known. For example, "about 20 km" is a less precise statement than "21.5 km." The basic rule for precision is that the number of digits in the numerical value expresses the precision of the measurement.

As an example of this rule, consider measuring the length of a table. After a quick measurement, you might estimate, "This table is 1.8 metres long." You have expressed the length as a two-digit number, meaning that you know the table to be longer than 1.7 metres but shorter than 1.9 metres. A more careful measurement with a tape measure might yield a result containing four digits, 1.826 metres. This means that the table is longer than 1.825 metres but shorter than 1.827 metres.

The number of digits in a numerical result is called its number of **significant figures.** Unless otherwise stated, the measurement is precise to within one unit in the last significant figure. Thus, a result reported as 1.826 metres has four significant figures. Unless additional information is given, the last digit (6) is uncertain by one unit: the length is $1.826 \pm 0.001$ metres. That is, the length falls somewhere between 1.825 metres and 1.827 metres.

When quantities are expressed in scientific notation, extra zeros always indicate extra precision. For instance, $3.840 \times 10^8$ metres means "not less than $3.839 \times 10^8$, nor more than $3.841 \times 10^8$ metres." On the other hand, $3.84 \times 10^8$ metres means "not less than $3.83 \times 10^8$, nor more than $3.85 \times 10^8$ metres."

The precision of a measurement depends on the quality of the measuring device used to obtain the measurement. Whereas very sensitive instruments yield measurements of high precision, less sensitive instruments yield results of lower precision. An electronic balance can determine the mass of 24 pennies to the nearest 0.0001 g, giving a measurement with six significant figures: 63.5465 g. A pan balance, on the other hand, can determine this same mass only to the nearest 0.01 g, giving a measurement with four significant figures: 63.55 g.

When a quantity fluctuates in value, the precision in its measurement is also limited by these fluctuations. For example, the distance between the Earth and its moon fluctuates with time, because the moon's orbit is elliptical rather than perfectly circular. The average distance is $3.844 \times 10^5$ km, but the actual distance varies between 3.564 and $4.067 \times 10^5$ km. Thus, the instantaneous distance between the Earth and its moon varies by $4.97 \times 10^4$ km in the course of a month.

## Relative and Absolute Precision

The degree of precision in a given experimental value can be expressed either as an **absolute** precision or as a **relative** precision. Absolute precision is the numerical uncertainty in the experimental value, and relative precision is the absolute precision divided by the experimental value. Absolute precisions have units, but relative precisions are always dimensionless ratios. Each of these ways of looking at precision has useful applications.

To illustrate these concepts, return to the example of table length. If we measure a table to be 1.826 metres long, the absolute precision of the measurement is 0.001 metres. The relative precision is the absolute precision divided by the length, (0.001 m)/(1.826 m), or $5 \times 10^{-4}$. This relative precision can be expressed as 1/1826, or $5 \times 10^{-4}$, or 0.05%, or 5 parts per 10,000. Each of these expressions is a ratio of two values that have the same units, so each ratio is dimensionless. Notice that because the absolute precision contains only one significant figure, this relative precision also contains only one significant figure.

When measurements are added or subtracted to compute a result, the *absolute* precision is more useful. When measurements are multiplied or divided to compute a result, the *relative* precision is more useful. As illustrations, return to the table example and consider the area and perimeter of the table top. Suppose that measurements show its width to be $3.20 \times 10^{-1}$ metres. The absolute precision of this measurement is 0.001 metres. The relative precision of the width measurement is the appropriate ratio: (0.001 m)/($3.20 \times 10^{-1}$ m), which is 1/320, or $3 \times 10^{-3}$.

Now combine these measurements to determine the perimeter $P$ and the area $A$. Perimeter is $P = 2L + 2W = 2(1.826 \text{ m}) + 2(3.20 \times 10^{-1} \text{ m}) = 4.292 \text{ m}$, and area is $A = (L)(W) = (1.826)(3.20 \times 10^{-1} \text{ m}) = 5.84 \times 10^{-1}$ m$^2$. How precisely is each of these computed values known? We find the precision of a computed result from the precisions of individual measurements by considering the largest possible variation in each individual measurement.

First, consider the table's perimeter. According to the precision of the measurements, the length might be as large as 1.827 m or as small as 1.825 m. The width might be as large as 0.321 m or as small as 0.319 m. Thus, the perimeter could be as large as $2(1.827) + 2(.321) = 4.296$ m, and it might be as small as $2(1.825) + 2(.319) = 4.288$ m. The precision in the perimeter is $\pm 0.004$ m. This is the *sum of the absolute precisions* of the individual values: $0.001 + 0.001 + 0.001 + 0.001 = 0.004$.

Applying the same logic to the area reveals that it might be as large as $1.827 \times 0.321 = 5.87 \times 10^{-1}$ m$^2$ or as small as $1.825 \times 0.319 = 5.82 \times 10^{-1}$ m$^2$. The area is $5.84 \times 10^{-1} \pm 0.025 \times 10^{-1}$ m$^2$, giving an imprecision of $(0.025 \times 10^{-1})/(5.84 \times 10^{-1}) = 4 \times 10^{-3}$. This is the *sum of the relative precisions* of length ($5 \times 10^{-4}$) and width ($3 \times 10^{-3}$).

## Rules for Determining Precision

The precision of a composite result can always be determined by the type of analysis described above, but this procedure is tedious. Fortunately the outcome is always the same. For *addition and subtraction*, the *absolute* precision of the result is the sum of the *absolute* precisions of the individual values. For *multiplication and division*, the *relative* precision of the result is the sum of the *relative* precisions of the individual values.

Another example illustrates the application of these rules. To determine the density of a liquid, a student filled a 25-mL graduated cylinder until it contained 25.0 mL of liquid. The mass of the full cylinder was 47.5764 g. The student removed liquid until the cylinder contained 20.0 mL of liquid, whereupon its mass was 43.0464 g. Table B-1 on the next page summarizes the measurements, computations, and precisions involved in this example.

The absolute precision of each *measured quantity* is found directly from the measurement. The absolute precision of the mass of the cylinder is the precision of the balance, 0.0001 g. Each relative precision is the absolute precision divided by the measured value; thus, the relative precision of the mass of the cylinder is 1/475 000. The precision of each *computed quantity* is determined by adding the precisions of quantities used in the calculation. The values in bold face are found directly. The volume transferred (5.0 mL) is obtained by *subtracting* two volumes, so its *absolute* precision is the sum of the *absolute* precisions of the volumes (0.1

## TABLE B-1 MEASURED AND DERIVED QUANTITIES AND PRECISIONS

| Quantity | Value | Absolute Precision | Relative Precision |
|---|---|---|---|
| **Measured Quantities** | | | |
| Initial cylinder volume | 25.0 mL | 0.1 mL | 1/250 |
| Initial cylinder mass | 47.5764 g | 0.0001 g | 1/475 000 |
| Final cylinder volume | 20.0 mL | 0.1 mL | 1/200 |
| Final cylinder mass | 43.0464 g | 0.0001 g | 1/430 000 |
| **Computed Quantities** | | | |
| Volume transferred | 5.0 mL | **0.2 mL** | 2/50 = 1/25 |
| Mass transferred | 4.5300 g | **0.0002 g** | 1/20 000 |
| Liquid density | 0.9060 g/mL | 0.04 | **1/25** |

+ 0.1 = 0.2 mL). Similar reasoning applies to the mass transferred. The density (0.9060 g/mL) is obtained by *dividing* mass by volume, so its *relative* precision is the sum of the *relative* precisions of transferred mass and transferred volume (1/25 + 1/20 000 = 1/25). Once the relative precision of the density has been calculated, it can be used to compute the absolute precision of the density: (1/25)(0.9060 g/mL) = ±0.04 g/mL.

## Rounding Off

In the above example, the density might be as large as 0.94 g/mL or as small as 0.87 g/mL. It would be incorrect to write it as 0.9060 g/mL, because the use of four significant figures implies that the value is known to 0.0001 g/mL. All nonsignificant figures must be eliminated, a process that is called **rounding off.** Because the uncertainty in the density is in the *second* decimal place, this result is rounded off to two decimals: 0.91 g/mL. This still overstates the precision, because 0.91 g/mL implies a precision of 0.01 g/mL, but this result is actually known only to 0.04 g/mL. However, rounding off one more place to give 0.9 g/mL would imply an uncertainty of 0.1 g/mL, which is larger than the actual uncertainty. The convention is to round off until dropping one more digit would result in an uncertainty larger than the actual uncertainty.

When the digit following the last significant digit is 5 or greater, the remaining digit is increased by 1 unit. For example, 0.9060 becomes 0.91. If the digit following the last significant digit is less than 5, the remaining digit remains unchanged. For example, 0.9045 rounded to 2 significant figures is 0.90.

## Shortcuts to Precision

You can spend much effort figuring out the appropriate number of significant figures in a result. Every time a scientist makes a measurement, precision and significant figures are a concern, but chemists are usually more interested in what experiments reveal than they are in significant figures. Precision may be extremely important, for example when determining the level of a particular carcinogen in a sample of ground water. Because the amount of carcinogen may be critical to human health, the precision of this measurement would have to be carefully stated. Precision is not so important to a chemist whose goal is to prepare new compounds. The chemist wishes to achieve a high yield but is not concerned about whether a yield is 90% or 90.05%.

The following simple guidelines are sufficient to determine the appropriate number of significant figures in General Chemistry.

To determine the number of significant figures in an individual measurement, count the number of digits, from left to right, beginning with the first one that is non-zero.

| | |
|---|---|
| 305 | 3 significant figures |
| 1.00 | 3 significant figures |
| 0.020 | 2 significant figures |

When adding or subtracting, set the number of *decimal places* in the answer equal to the number of *decimal places* in the number with the fewest places. The number of significant figures is irrelevant:

| | | |
|---|---|---|
| | 0.12 | 2 decimal places |
| | 1.6 | **1 decimal place** |
| | 11.490 | 3 decimal places |
| Sum: | 13.2 | **1 decimal place** |

When multiplying or dividing, set the number of *significant figures* in the answer equal to that of the quantity with the fewest *significant figures*. The number of decimal places is irrelevant:

$$1.365 \times 2.63 \times 0.33 = 1.2$$

| Significant figures: | 4 | 3 | 2 | 2 |
|---|---|---|---|---|

The last example is one where the simplified procedure gives a different result than the more elaborate one. One number (0.33) has a relative precision of 1/33, so the result, 1.1847, could be rounded to 1.18 (1 part in 118) rather than to 1.2 (1 part in 12). If the importance of the result is its quantitative value, 1.18 would be the more appropriate number to report; if its importance is in its qualitative significance, report 1.2. (But don't spend a lot of time worrying about it: either way of reporting is legitimate! What is not legitimate is to report this result as 1.1847.)

Calculators do not necessarily give results with the correct number of significant figures. They automatically drop trailing zeroes even when they are significant (try multiplying or adding 3.00 and 5.00 on your calculator) and they carry extra decimal places even when they are insignificant (try dividing 5.00 by 3.00 on your calculator). Never believe the number of significant figures on your calculator!

# APPENDIX C: IONIZATION ENERGIES AND ELECTRON AFFINITIES OF THE FIRST 36 ELEMENTS

| Z | Symbol | EA | $IE_1$ | $IE_2$ | $IE_3$ | $n^\dagger$ |
|---|--------|------|--------|--------|--------|-----|
| 1 | H | −72.8 | 1312 | − | − | |
| 2 | He | >0* | 2372 | 5250 | − | 1 |
| 3 | Li | −59.7 | 520.2 | 7298 | 11,815 | |
| 4 | Be | >0 | 899.4 | 1757 | 14,848 | |
| 5 | B | −26.8 | 800.6 | 2427 | 3660 | |
| 6 | C | −121.9 | 1086 | 2353 | 4620 | |
| 7 | N | >0 | 1402 | 2856 | 4578 | |
| 8 | O | −141.1 | 1314 | 3388 | 5300 | |
| 9 | F | −328.0 | 1681 | 3374 | 6050 | 2 |
| 10 | Ne | >0 | 2081 | 3952 | 6122 | |
| 11 | Na | −52.9 | 495.6 | 4562 | 6912 | |
| 12 | Mg | >0 | 737.7 | 1451 | 7733 | |
| 13 | Al | −42.7 | 577.6 | 1817 | 2745 | |
| 14 | Si | −133.6 | 786.4 | 1577 | 3232 | |
| 15 | P | −72.0 | 1012 | 1908 | 2912 | |
| 16 | S | −200.4 | 999.6 | 2251 | 3357 | 3 |
| 17 | Cl | −348.8 | 1251 | 2297 | 3822 | |
| 18 | Ar | >0 | 1520 | 2666 | 3931 | |
| 19 | K | −48.4 | 418.8 | 3051 | 4420 | |
| 20 | Ca | −2.4 | 589.8 | 1145 | 4912 | |
| 21 | Sc | −18.2 | 633 | 1235 | 2389 | |
| 22 | Ti | −7.7 | 658 | 1310 | 2653 | |
| 23 | V | −50.8 | 650 | 1414 | 2828 | |
| 24 | Cr | −64.4 | 652.8 | 1591 | 2987 | |
| 25 | Mn | >0 | 717.4 | 1509 | 3248 | |
| 26 | Fe | −14.6 | 763 | 1561 | 2957 | |
| 27 | Co | −63.9 | 758 | 1646 | 3232 | |
| 28 | Ni | −111.6 | 736.7 | 1753 | 3396 | 4 |
| 29 | Cu | −119.2 | 745.4 | 1958 | 3554 | |
| 30 | Zn | >0 | 906.4 | 1733 | 3833 | |
| 31 | Ga | −28.9 | 578.8 | 1979 | 2963 | |
| 32 | Ge | −119 | 762.1 | 1537 | 3302 | |
| 33 | As | −78 | 947 | 1798 | 2736 | |
| 34 | Se | −195.0 | 940.9 | 2045 | 2974 | |
| 35 | Br | −324.6 | 1140 | 2103 | 3500 | |
| 36 | Kr | >0 | 1351 | 2350 | 3565 | |

All values are in kJ/mol.

* A value >0 means that the anion is unstable, so its electron affinity cannot be experimentally determined.

$^\dagger$ $n$ is the principal quantum number of the electron whose ionization energy is listed.

Source: Copyright 2001 From *The CRC Handbook of Chemistry and Physics*, 82nd edition, by David R. Lide, ed. Reproduced by permission of Taylor and Grancis Group, LLC, a division of Informa plc.

# APPENDIX D: STANDARD THERMODYNAMIC FUNCTIONS

$T = 298.15$ K; $p = 1.000$ bar; Aqueous species at 1.000 M

| Substance | $\Delta H_f^0$ (kJ/mol) | $\Delta G_f^0$ (kJ/mol) | $S^0$ (J/mol K) | Substance | $\Delta H_f^0$ (kJ/mol) | $\Delta G_f^0$ (kJ/mol) | $S^0$ (J/mol K) |
|---|---|---|---|---|---|---|---|
| **Aluminum** | | | | **Carbon** | | | |
| Al (s) | 0 | 0 | 28.3 | C (s, graphite) | 0 | 0 | 5.7 |
| AlCl$_3$ (s) | −704.2 | −628.8 | 109.3 | C (s, diamond) | 1.9 | 2.9 | 2.4 |
| Al$_2$O$_3$ (s) | −1675.7 | −1582.3 | 50.9 | CH$_4$ (g) | −74.6 | −50.5 | 186.3 |
| **Antimony** | | | | C$_2$H$_2$ (g) | 227.4 | 209.9 | 200.9 |
| Sb (s) | 0 | 0 | 45.7 | C$_2$H$_4$ (g) | 52.4 | 68.4 | 219.3 |
| Sb$_4$O$_6$ (s) | −1417.1 | −1253.0 | 246.0 | C$_2$H$_6$ (g) | −84.0 | −32.0 | 229.2 |
| **Argon** | | | | C$_3$H$_6$ (g) | 20.0 | 74.62 | 226.9 |
| Ar (g) | 0 | 0 | 154.843 | C$_3$H$_8$ (g) | −103.8 | −23.4 | 270.3 |
| **Arsenic** | | | | C$_6$H$_6$ (l) | 49.1 | 124.5 | 173.4 |
| As (s) | 0 | 0 | 35.1 | CO (g) | −110.5 | −137.2 | 197.7 |
| As$_2$O$_5$ (s) | −924.9 | −782.3 | 105.4 | CO$_2$ (g) | −393.5 | −394.4 | 213.8 |
| AsCl$_3$ (l) | −305.0 | −259.4 | 216.3 | HCN (g) | 135.1 | 124.7 | 201.8 |
| **Barium** | | | | CS$_2$ (l) | 89.0 | 64.6 | 151.3 |
| Ba (s) | 0 | 0 | 62.5 | CCl$_4$ (l) | −128.2 | −65.21 | 216.40 |
| BaO (s) | −548.0 | −520.3 | 72.1 | CHCl$_3$ (l) | −134.1 | −73.7 | 201.7 |
| BaCO$_3$ (s) | −1213.0 | −1134.4 | 112.1 | CH$_3$CHO (g) | −166.2 | −127.6 | 263.8 |
| BaCl$_2$ (s) | −855.0 | −806.7 | 123.7 | CH$_3$CO$_2$H (l) | −484.3 | −389.9 | 159.8 |
| BaSO$_4$ (s) | −1473.2 | −1362.2 | 132.2 | CH$_3$OH (l) | −239.2 | −166.6 | 126.8 |
| **Beryllium** | | | | CH$_3$CH$_2$OH (l) | −277.6 | −174.8 | 160.7 |
| Be (s) | 0 | 0 | 9.5 | CH$_3$CN (l) | 40.6 | 86.5 | 149.6 |
| Be(OH)$_2$ (s) | −902.5 | −815.0 | 51.9 | **Chlorine** | | | |
| **Boron** | | | | Cl (g) | 121.3 | 105.3 | 165.2 |
| B (s) | 0 | 0 | 5.9 | Cl$_2$ (g) | 0 | 0 | 223.1 |
| B$_2$O$_3$ (s) | −1273.5 | −1194.3 | 54.0 | Cl$^-$ (aq) | −167.1 | −131.0 | 56.5 |
| H$_3$BO$_3$ (s) | −1094.3 | −968.9 | 90.0 | HCl (g) | −92.3 | −95.3 | 186.9 |
| BCl$_3$ (g) | −403.8 | −388.7 | 290.1 | ClO$^-$ (aq) | −107.1 | −36.8 | 42 |
| **Bromine** | | | | ClO$_2$$^-$ (aq) | −67 | 17 | 101 |
| Br (g) | 111.9 | 82.4 | 175.0 | ClO$_3$$^-$ (aq) | −104 | −3 | 162 |
| Br$_2$ (l) | 0 | 0 | 152.2 | ClO$_4$$^-$ (aq) | −128.1 | −8.52 | 184.0 |
| Br$_2$ (g) | 30.9 | 3.1 | 245.5 | ClO (g) | 101.8 | 98.1 | 226.6 |
| Br$^-$ (aq) | −121.4 | −104.0 | 82.4 | ClO$_2$ (g) | 102.5 | 120.5 | 256.8 |
| BrO$^-$ (aq) | −94.1 | −33.4 | 42 | **Chromium** | | | |
| HBr (g) | −36.3 | −53.4 | 198.7 | Cr (s) | 0 | 0 | 23.8 |
| **Cadmium** | | | | Cr$_2$O$_3$ (s) | −1139.7 | −1058.1 | 81.2 |
| Cd (s) | 0 | 0 | 51.8 | CrCl$_3$ (s) | −556.5 | −486.1 | 123.0 |
| CdO (s) | −258.4 | −228.7 | 54.8 | **Cobalt** | | | |
| CdCl$_2$ (s) | −391.5 | −343.9 | 115.3 | Co (s) | 0 | 0 | 30.0 |
| **Calcium** | | | | CoO (s) | −237.9 | −214.2 | 53.0 |
| Ca (s) | 0 | 0 | 41.6 | CoCl$_2$ (s) | −312.5 | −269.8 | 109.2 |
| Ca$^{2+}$ (aq) | −543.0 | −553.6 | −56.2 | **Copper** | | | |
| CaO (s) | −634.9 | −603.3 | 38.1 | Cu (s) | 0 | 0 | 33.2 |
| Ca(OH)$_2$ (s) | −985.2 | −897.5 | 83.4 | CuO (s) | −157.3 | −129.7 | 42.6 |
| CaCO$_3$ (s) | −1207.6 | −1129.1 | 91.7 | CuS (s) | −53.1 | −53.6 | 66.5 |
| CaCl$_2$ (s) | −795.4 | −748.8 | 108.4 | CuCl$_2$ (s) | −220.1 | −175.7 | 108.1 |
| CaF$_2$ (s) | −1228 | −1176 | 68.5 | CuCl (s) | −137.2 | −119.9 | 86.2 |
| CaH$_2$ (s) | −185.1 | −142.5 | 41.4 | CuBr (s) | −104.6 | −100.8 | 96.1 |
| CaS (s) | −482.4 | −477.4 | 56.5 | CuI (s) | −67.8 | −69.5 | 96.7 |
| CaSO$_4$ (s) | −1434.5 | −1322.0 | 106.5 | CuSO$_4$ (s) | −771.4 | −662.2 | 109.2 |

| Substance | $\Delta H_f^0$ (kJ/mol) | $\Delta G_f^0$ (kJ/mol) | $S^0$ (J/mol K) |
|---|---|---|---|
| **Fluorine** | | | |
| $F(g)$ | 79.4 | 62.3 | 158.8 |
| $F_2(g)$ | 0 | 0 | 202.8 |
| $F^-(aq)$ | $-335.4$ | $-278.79$ | $-13.8$ |
| $HF(g)$ | $-273.3$ | $-275.4$ | 173.8 |
| **Germanium** | | | |
| $Ge(s)$ | 0 | 0 | 31.1 |
| $GeCl_4(g)$ | $-495.8$ | $-457.3$ | 347.7 |
| $GeO_2(s)$ | $-580.0$ | $-521.4$ | 39.7 |
| **Gold** | | | |
| $Au(s)$ | 0 | 0 | 47.4 |
| **Helium** | | | |
| $He(g)$ | 0 | 0 | 126.153 |
| **Hydrogen** | | | |
| $H_2(g)$ | 0 | 0 | 130.680 |
| $H_2O(l)$ | $-285.83$ | $-237.1$ | 69.95 |
| $H_2O(g)$ | $-241.83$ | $-228.72$ | 188.835 |
| $H_2O_2(l)$ | $-187.8$ | $-120.4$ | 109.6 |
| $H^+(aq)$ | 0 | 0 | 0 |
| $H_3O^+(aq)$ | $-285.83$ | $-237.1$ | 69.95 |
| **Iodine** | | | |
| $I_2(s)$ | 0 | 0 | 116.1 |
| $I_2(g)$ | 62.4 | 19.3 | 260.7 |
| $I^-(aq)$ | $-56.78$ | $-51.57$ | 106.5 |
| $HI(g)$ | 26.5 | 1.7 | 206.6 |
| **Iron** | | | |
| $Fe(s)$ | 0 | 0 | 27.3 |
| $Fe_2O_3(s)$ | $-824.2$ | $-742.2$ | 87.4 |
| $FeO(s)$ | $-272.0$ | $-251.4$ | 60.8 |
| $Fe_3O_4(s)$ | $-1118.4$ | $-1015.4$ | 146.4 |
| $Fe(OH)_3(s)$ | $-823.0$ | $-696.5$ | 106.7 |
| $FeCl_2(s)$ | $-341.8$ | $-302.3$ | 118.0 |
| $FeCl_3(s)$ | $-399.5$ | $-334.0$ | 142.3 |
| $FeSO_4(s)$ | $-928.4$ | $-820.8$ | 107.5 |
| $FeS(s)$ | $-100.0$ | $-100.4$ | 60.3 |
| $FeS_2(s)$ | $-178.2$ | $-166.9$ | 52.9 |
| **Krypton** | | | |
| $Kr(g)$ | 0 | 0 | 164.085 |
| **Lead** | | | |
| $Pb(s)$ | 0 | 0 | 64.8 |
| $PbCl_2(s)$ | $-359.4$ | $-314.1$ | 136.0 |
| $PbO(s)$ | $-217.3$ | $-187.9$ | 68.7 |
| $PbO_2(s)$ | $-277.4$ | $-217.3$ | 68.6 |
| $PbS(s)$ | $-100.4$ | $-98.7$ | 91.2 |
| **Lithium** | | | |
| $Li(s)$ | 0 | 0 | 29.1 |
| $Li^+(aq)$ | $-278.47$ | $-293.31$ | 12.2 |
| $Li_2O(s)$ | $-597.9$ | $-561.2$ | 37.6 |
| $LiOH(s)$ | $-484.9$ | $-439.0$ | 42.8 |
| $LiCl(s)$ | $-408.6$ | $-384.4$ | 59.3 |
| **Magnesium** | | | |
| $Mg(s)$ | 0 | 0 | 32.67 |
| $Mg^{2+}(aq)$ | $-467.0$ | $-454.8$ | $-137$ |
| $MgO(s)$ | $-601.6$ | $-569.3$ | 27.0 |
| $Mg(OH)_2(s)$ | $-924.5$ | $-833.5$ | 63.2 |
| $MgCl_2(s)$ | $-641.3$ | $-591.8$ | 89.6 |
| $MgF_2(s)$ | $-1124.2$ | $-1071.1$ | 57.2 |
| $MgCO_3(s)$ | $-1095.8$ | $-1012.1$ | 65.7 |
| $MgSO_4(s)$ | $-1284.9$ | $-1170.6$ | 91.6 |

| Substance | $\Delta H_f^0$ (kJ/mol) | $\Delta G_f^0$ (kJ/mol) | $S^0$ (J/mol K) |
|---|---|---|---|
| **Manganese** | | | |
| $Mn(s)$ | 0 | 0 | 32.0 |
| $MnO(s)$ | $-385.2$ | $-362.9$ | 59.7 |
| $MnO_2(s)$ | $-520.0$ | $-465.1$ | 53.1 |
| $MnCl_2(s)$ | $-481.3$ | $-440.5$ | 118.2 |
| $MnCO_3(s)$ | $-894.1$ | $-816.7$ | 85.8 |
| $MnSO_4(s)$ | $-1065.25$ | $-957.36$ | 112.1 |
| **Mercury** | | | |
| $Hg(l)$ | 0 | 0 | 75.9 |
| $Hg(g)$ | 61.4 | 31.8 | 175.0 |
| $HgO(s)$ | $-90.8$ | $-58.5$ | 70.3 |
| $HgCl_2(s)$ | $-224.3$ | $-178.6$ | 146.0 |
| $Hg_2Cl_2(s)$ | $-265.4$ | $-210.7$ | 191.6 |
| $HgS(s)$ | $-58.2$ | $-50.6$ | 82.4 |
| **Neon** | | | |
| $Ne(g)$ | 0 | 0 | 146.328 |
| **Nickel** | | | |
| $Ni(s)$ | 0 | 0 | 29.9 |
| $NiO(s)$ | $-239.7$ | $-211.7$ | 37.99 |
| $NiCl_2(s)$ | $-305.3$ | $-259.0$ | 97.7 |
| **Nitrogen** | | | |
| $N_2(g)$ | 0 | 0 | 191.61 |
| $NH_3(g)$ | $-45.9$ | $-16.4$ | 192.8 |
| $NH_3(aq)$ | $-80.29$ | $-26.50$ | 111.3 |
| $NH_4^+(aq)$ | $-133.3$ | $-79.31$ | 111.2 |
| $N_2H_4(l)$ | 50.6 | 149.3 | 121.2 |
| $NO(g)$ | 91.3 | 87.6 | 210.8 |
| $NO_2(g)$ | 33.2 | 51.3 | 240.1 |
| $N_2O(g)$ | 81.6 | 103.7 | 220.0 |
| $N_2O_4(g)$ | 11.1 | 99.8 | 304.4 |
| $N_2O_5(g)$ | 13.3 | 117.1 | 355.7 |
| $NH_4Cl(s)$ | $-314.4$ | $-202.9$ | 94.6 |
| $NH_4NO_3(s)$ | $-365.6$ | $-183.9$ | 151.1 |
| $(NH_2)_2CO(s)$ | $-333.1$ | $-198$ | 105 |
| $HNO_3(g)$ | $-133.9$ | $-73.5$ | 266.9 |
| **Oxygen** | | | |
| $O_2(g)$ | 0 | 0 | 205.152 |
| $O_3(g)$ | 142.7 | 163.2 | 238.9 |
| $OH^-(aq)$ | $-230.0$ | $-157.244$ | $-10.9$ |
| **Phosphorus** | | | |
| $P(s, white)$ | 0 | 0 | 41.1 |
| $P(s, red)$ | $-17.6$ | $-12.1$ | 22.8 |
| $P_4(g)$ | 58.9 | 24.4 | 280.0 |
| $PH_3(g)$ | 5.4 | 13.4 | 210.2 |
| $P_4O_{10}(s)$ | $-2984.0$ | $-2697.7$ | 228.86 |
| $H_3PO_4$ | $-1271.7$ | $-1123.6$ | 150.8 |
| $PCl_5(g)$ | $-374.9$ | $-305.0$ | 364.6 |
| $PCl_3(g)$ | $-287.0$ | $-267.8$ | 311.8 |
| **Potassium** | | | |
| $K(s)$ | 0 | 0 | 64.7 |
| $K^+(aq)$ | $-252.1$ | $-283.27$ | 101.2 |
| $KO_2(s)$ | $-284.9$ | $-239.4$ | 116.7 |
| $KOH(s)$ | $-424.6$ | $-378.9$ | 78.9 |
| $KCl(s)$ | $-436.5$ | $-408.5$ | 82.6 |
| $KClO_3(s)$ | $-397.7$ | $-296.3$ | 143.1 |
| $KI(s)$ | $-327.9$ | $-324.9$ | 106.3 |
| **Selenium** | | | |
| $Se(s)$ | 0 | 0 | 42.4 |
| $H_2Se(g)$ | 29.7 | 15.9 | 219.0 |

| Substance | $\Delta H_f^0$ (kJ/mol) | $\Delta G_f^0$ (kJ/mol) | $S^0$ (J/mol K) | Substance | $\Delta H_f^0$ (kJ/mol) | $\Delta G_f^0$ (kJ/mol) | $S^0$ (J/mol K) |
|---|---|---|---|---|---|---|---|
| **Silicon** | | | | **Sulfur (cont.)** | | | |
| Si (s) | 0 | 0 | 18.8 | $SO_3$ (g) | −395.7 | −371.1 | 256.8 |
| $SiH_4$ (g) | 34.3 | 56.9 | 204.6 | $H_2SO_4$ (l) | −814 | −690.0 | 156.9 |
| $SiO_2$ (s) | −910.7 | −856.3 | 41.5 | $SO_4^{2-}$ (aq) | −909.3 | −774.53 | 18.5 |
| $SiCl_4$ (l) | −687.0 | −619.8 | 239.7 | $SF_6$ (g) | −1220.5 | −1116.5 | 291.5 |
| SiC (s) | −65.3 | −62.8 | 16.6 | **Tellurium** | | | |
| **Silver** | | | | Te (s) | 0 | 0 | 49.7 |
| Ag (s) | 0 | 0 | 42.6 | $TeO_2$ (s) | −322.6 | −270.3 | 79.5 |
| $Ag^+$ (aq) | 105.8 | 77.107 | 73.4 | **Tin** | | | |
| $Ag_2O$ (s) | −31.1 | −11.2 | 121.3 | Sn (s, white) | 0 | 0 | 51.2 |
| AgCl (s) | −127.0 | −109.8 | 96.3 | Sn (s, gray) | −2.1 | 0.13 | 44.1 |
| AgBr (s) | −100.4 | −96.9 | 107.1 | SnO (s) | −280.7 | −251.9 | 57.2 |
| $AgNO_3$ (s) | −124.4 | −33.4 | 140.9 | $SnO_2$ (s) | −577.6 | −515.8 | 49.0 |
| **Sodium** | | | | $SnCl_4$ (s) | −511.3 | −440.1 | 258.6 |
| Na (s) | 0 | 0 | 51.3 | **Titanium** | | | |
| $Na^+$ (aq) | −240.3 | −261.905 | 58.5 | Ti (s) | 0 | 0 | 30.7 |
| $Na_2O$ (s) | −414.2 | −375.5 | 75.1 | $TiCl_4$ (l) | −804.2 | −737.2 | 252.3 |
| NaOH (s) | −425.6 | −379.5 | 64.5 | $TiCl_4$ (g) | −763.2 | −726.3 | 353.2 |
| NaF (s) | −576.6 | −546.3 | 51.1 | $TiO_2$ (s) | −944.0 | −888.8 | 50.6 |
| NaCl (s) | −411.2 | −384.1 | 72.1 | **Uranium** | | | |
| NaBr (s) | −361.1 | −349.0 | 86.8 | U (s) | 0 | 0 | 50.2 |
| NaI (s) | −287.8 | −286.1 | 98.5 | $UO_2$ (s) | −1085.0 | −1031.8 | 77.0 |
| $Na_2CO_3$ (s) | −1130.7 | −1044.4 | 135.0 | $UF_6$ (g) | −2147.4 | −2063.7 | 377.9 |
| **Strontium** | | | | **Xenon** | | | |
| Sr (s) | 0 | 0 | 55.0 | Xe (g) | 0 | 0 | 169.685 |
| SrO (s) | −592.0 | −561.9 | 54.4 | $XeF_4$ (s) | −261.5 | −138 | 316 |
| $SrCl_2$ (s) | −828.9 | −781.1 | 114.9 | **Zinc** | | | |
| **Sulfur** | | | | Zn (s) | 0 | 0 | 41.63 |
| $S_8$ (s) | 0 | 0 | 31.80 | ZnO (s) | −350.5 | −320.5 | 43.7 |
| $S_8$ (g) | 102.30 | 49.63 | 430.23 | $ZnCl_2$ (s) | −415.1 | −369.4 | 111.5 |
| $H_2S$ (g) | −20.6 | −33.4 | 205.8 | ZnS (s) | −206.0 | −201.3 | 57.7 |
| $SO_2$ (g) | −296.8 | −300.1 | 248.2 | | | | |

Source: Copyright 2001 From *The CRC Handbook of Chemistry and Physics*, 82nd edition, by David R. Lide, ed. Reproduced by permission of Taylor and Grancis Group, LLC, a division of Informa plc.

# APPENDIX E: EQUILIBRIUM CONSTANTS

**$K_b$ VALUES**

| Name | Formula | $pK_b$ | $K_b$ |
|---|---|---|---|
| Ammonia | $NH_3$ | 4.75 | $1.8 \times 10^{-5}$ |
| Aniline | $C_6H_5NH_2$ | 9.13 | $7.4 \times 10^{-10}$ |
| Diethylamine | $(C_2H_5)_2NH$ | 3.16 | $6.9 \times 10^{-4}$ |
| Dimethylamine | $(CH_3)_2NH$ | 3.27 | $5.4 \times 10^{-4}$ |
| Ethylamine | $C_2H_5NH_2$ | 3.35 | $4.5 \times 10^{-4}$ |
| Ethylenediamine | $(CH_2NH_2)_2$ | 4.08 | $8.3 \times 10^{-5}$ |
| Hydrazine | $H_2NNH_2$ | 5.89 | $1.3 \times 10^{-6}$ |
| Hydroxylamine | $HONH_2$ | 8.06 | $8.7 \times 10^{-9}$ |
| Methylamine | $CH_3NH_2$ | 3.34 | $4.6 \times 10^{-4}$ |
| Pyridine | $C_5H_5N$ | 8.77 | $1.7 \times 10^{-9}$ |
| Triethylamine | $(C_2H_5)_3N$ | 3.25 | $5.6 \times 10^{-4}$ |
| Trimethylamine | $(CH_3)_3N$ | 4.19 | $6.5 \times 10^{-5}$ |
| Urea | $H_2NCONH_2$ | 13.82 | $1.5 \times 10^{-14}$ |

## $K_a$ VALUES

| Name | Formula | $pK_a$ | $K_a$ |
|------|---------|--------|-------|
| **Oxoacids** | | | |
| Arsenic | $H_3AsO_4$ | 2.26 | $5.5 \times 10^{-3}$ |
| Boric | $H_3BO_3$ | 9.27 | $5.4 \times 10^{-10}$ |
| Carbonic | $H_2CO_3$ | 6.35 | $4.5 \times 10^{-7}$ |
| $K_{a2}$ | $HCO_3^-$ | 10.33 | $4.7 \times 10^{-11}$ |
| Chlorous | $HClO_2$ | 1.94 | $1.1 \times 10^{-2}$ |
| Chromic | $H_2CrO_4$ | 0.74 | $1.8 \times 10^{-1}$ |
| $K_{a2}$ | $HCrO_4^-$ | 6.49 | $3.2 \times 10^{-7}$ |
| Hypobromous | HBrO | 8.55 | $2.8 \times 10^{-9}$ |
| Hypochlorous | HClO | 7.40 | $4.0 \times 10^{-8}$ |
| Hypoiodous | HIO | 10.5 | $3.2 \times 10^{-11}$ |
| Iodic | $HIO_3$ | 0.78 | $1.7 \times 10^{-1}$ |
| Nitrous | $HNO_2$ | 3.25 | $5.6 \times 10^{-4}$ |
| Paraperiodic | $H_5IO_6$ | 1.55 | $2.8 \times 10^{-2}$ |
| Periodic | $HIO_4$ | 1.64 | $7.3 \times 10^{-2}$ |
| Phosphoric | $H_3PO_4$ | 2.16 | $6.9 \times 10^{-3}$ |
| $K_{a2}$ | $H_2PO_4^-$ | 7.21 | $6.2 \times 10^{-8}$ |
| $K_{a3}$ | $HPO_4^{2-}$ | 12.32 | $4.8 \times 10^{-13}$ |
| Phosphorous | $H_3PO_3$ | 1.3 | $5.0 \times 10^{-2}$ |
| $K_{a2}$ | $H_2PO_3^-$ | 6.70 | $2.0 \times 10^{-7}$ |
| Sulfuric | $H_2SO_4$ | strong | |
| $K_{a2}$ | $HSO_4^-$ | 1.99 | $1.0 \times 10^{-2}$ |
| Sulfurous | $H_2SO_3$ | 1.85 | $1.4 \times 10^{-2}$ |
| $K_{a2}$ | $HSO_3^-$ | 7.20 | $6.3 \times 10^{-8}$ |

| Name | Formula | $pK_a$ | $K_a$ |
|------|---------|--------|-------|
| **Carboxylic acids** | | | |
| Acetic | $CH_3CO_2H$ | 4.75 | $1.8 \times 10^{-5}$ |
| Benzoic | $C_6H_5CO_2H$ | 4.20 | $6.3 \times 10^{-5}$ |
| Chloroacetic | $ClCH_2CO_2H$ | 2.87 | $1.4 \times 10^{-3}$ |
| Formic | $HCO_2H$ | 3.75 | $1.8 \times 10^{-4}$ |
| Oxalic | $(CO_2H)_2$ | 1.25 | $5.6 \times 10^{-2}$ |
| $K_{a2}$ | $HO_2CCO_2^-$ | 3.81 | $1.5 \times 10^{-4}$ |
| Propanoic | $C_2H_5CO_2H$ | 4.87 | $1.3 \times 10^{-5}$ |
| Trichloroacetic | $Cl_3CCO_2H$ | 0.66 | $2.2 \times 10^{-1}$ |
| **Other acids** | | | |
| Cyanic | HCNO | 3.46 | $3.5 \times 10^{-4}$ |
| Hydrazoic | $HN_3$ | 4.6 | $2.5 \times 10^{-5}$ |
| Hydrocyanic | HCN | 9.21 | $6.2 \times 10^{-10}$ |
| Hydrofluoric | HF | 3.20 | $6.3 \times 10^{-4}$ |
| Hydrogen peroxide | $H_2O_2$ | 11.62 | $2.4 \times 10^{-12}$ |
| Hydrogen sulfide | $H_2S$ | 7.05 | $8.9 \times 10^{-8}$ |
| $K_{a2}$ | $HS^-$ | 19 | $1 \times 10^{-19}$ |
| Phenol | $C_6H_5OH$ | 9.99 | $1.0 \times 10^{-10}$ |

## $K_{sp}$ VALUES, SALTS AT 25 °C

| Formula | $pK_{sp}$ | $K_{sp}$ |
|---------|-----------|----------|
| **Bromides** | | |
| AgBr | 12.27 | $5.35 \times 10^{-13}$ |
| CuBr | 8.20 | $6.3 \times 10^{-9}$ |
| **Carbonates** | | |
| $Ag_2CO_3$ | 11.07 | $8.46 \times 10^{-12}$ |
| $BaCO_3$ | 8.59 | $2.6 \times 10^{-9}$ |
| $CaCO_3$ | 8.47 | $3.36 \times 10^{-9}$ |
| $CdCO_3$ | 12.00 | $1.0 \times 10^{-12}$ |
| $CoCO_3$ | 12.85 | $1.4 \times 10^{-13}$ |
| $CuCO_3$ | 9.85 | $1.4 \times 10^{-10}$ |
| $FeCO_3$ | 10.51 | $3.1 \times 10^{-11}$ |
| $MgCO_3$ | 5.17 | $6.8 \times 10^{-6}$ |
| $MnCO_3$ | 10.66 | $2.2 \times 10^{-11}$ |
| $NiCO_3$ | 6.84 | $1.4 \times 10^{-7}$ |
| $PbCO_3$ | 13.13 | $7.4 \times 10^{-14}$ |
| $SrCO_3$ | 9.25 | $5.6 \times 10^{-10}$ |
| $ZnCO_3$ | 9.82 | $1.5 \times 10^{-10}$ |
| **Chlorides** | | |
| AgCl | 9.74 | $1.8 \times 10^{-10}$ |
| CuCl | 6.76 | $1.7 \times 10^{-7}$ |
| $PbCl_2$ | 4.77 | $1.7 \times 10^{-5}$ |
| **Fluorides** | | |
| $BaF_2$ | 6.74 | $1.8 \times 10^{-7}$ |
| $CaF_2$ | 10.46 | $3.45 \times 10^{-11}$ |
| $PbF_2$ | 7.48 | $3.3 \times 10^{-8}$ |
| **Hydroxides** | | |
| $Al(OH)_3$ | 32.34 | $4.6 \times 10^{-33}$ |
| $Ba(OH)_2$ | 3.59 | $2.55 \times 10^{-4}$ |
| $Ca(OH)_2$ | 5.30 | $5.0 \times 10^{-6}$ |
| $Cd(OH)_2$ | 14.14 | $7.2 \times 10^{-15}$ |
| $Co(OH)_2$ | 14.23 | $5.9 \times 10^{-15}$ |

| Formula | $pK_{sp}$ | $K_{sp}$ |
|---------|-----------|----------|
| **Hydroxides (cont.)** | | |
| $Co(OH)_3$ | 43.8 | $1.6 \times 10^{-44}$ |
| $Cu(OH)_2$ | 14.96 | $1.1 \times 10^{-15}$ |
| $Fe(OH)_2$ | 16.32 | $4.8 \times 10^{-17}$ |
| $Fe(OH)_3$ | 38.55 | $2.8 \times 10^{-39}$ |
| $Mg(OH)_2$ | 11.25 | $5.6 \times 10^{-12}$ |
| $Ni(OH)_2$ | 15.26 | $5.5 \times 10^{-16}$ |
| $Pb(OH)_2$ | 19.85 | $1.4 \times 10^{-20}$ |
| $Sn(OH)_2$ | 26.26 | $5.5 \times 10^{-27}$ |
| $Zn(OH)_2$ | 16.52 | $3 \times 10^{-17}$ |
| **Phosphates** | | |
| $Ba_3(PO_4)_2$ | 22.47 | $3.4 \times 10^{-23}$ |
| $Ca_3(PO_4)_2$ | 32.70 | $2.0 \times 10^{-33}$ |
| **Sulfates** | | |
| $Ag_2SO_4$ | 4.92 | $1.2 \times 10^{-5}$ |
| $BaSO_4$ | 9.96 | $1.1 \times 10^{-10}$ |
| $CaSO_4$ | 4.31 | $4.9 \times 10^{-5}$ |
| $PbSO_4$ | 7.60 | $2.5 \times 10^{-8}$ |
| **Sulfides** | | |
| $Ag_2S$ | 49.20 | $6.3 \times 10^{-50}$ |
| CoS | 20.40 | $4.0 \times 10^{-21}$ |
| $Cu_2S$ | 47.60 | $2.5 \times 10^{-48}$ |
| CuS | 25.20 | $6.3 \times 10^{-26}$ |
| FeS | 18.80 | $1.6 \times 10^{-19}$ |
| $Hg_2S$ | 47.00 | $1.0 \times 10^{-47}$ |
| HgS | 52.40 | $4.0 \times 10^{-53}$ |
| MnS | 13.34 | $4.6 \times 10^{-14}$ |
| NiS | 20.97 | $1.1 \times 10^{-21}$ |
| PbS | 28.05 | $8.9 \times 10^{-29}$ |
| SnS | 27.49 | $3.2 \times 10^{-28}$ |
| ZnS | 24.53 | $3.0 \times 10^{-25}$ |

# APPENDIX F:
# STANDARD REDUCTION POTENTIALS, $E°$

$T = 298.15$ K, and $p = 1.00$ bar.
Listed alphabetically by element; all values are in volts.

All ionic species are aqueous. $H_2O$ is liquid.
All other neutrals are solids unless otherwise specified.

**Aluminum**

| | |
|---|---|
| $Al^{3+} + 3\,e^- \rightarrow Al$ | $-1.662$ |
| $Al(OH)_3 + 3\,e^- \rightarrow Al + 3\,OH^-$ | $-2.31$ |

**Arsenic**

| | |
|---|---|
| $HAsO_2\,(aq) + 3\,H_3O^+ + 3\,e^- \rightarrow As + 5\,H_2O$ | $+0.248$ |

**Barium**

| | |
|---|---|
| $Ba^{2+} + 2\,e^- \rightarrow Ba$ | $-2.912$ |
| $Ba(OH)_2 + 2\,e^- \rightarrow Ba + 2\,OH^-$ | $-2.99$ |

**Boron**

| | |
|---|---|
| $H_3BO_3\,(aq) + 3\,H_3O^+ + 3\,e^- \rightarrow B + 6\,H_2O$ | $-0.8698$ |

**Bromine**

| | |
|---|---|
| $Br_2\,(aq) + 2\,e^- \rightarrow 2\,Br^-$ | $+1.0873$ |
| $Br_2\,(l) + 2\,e^- \rightarrow 2\,Br^-$ | $+1.066$ |
| $HBrO\,(aq) + H_3O^+ + 2\,e^- \rightarrow Br^- + 2\,H_2O$ | $+1.331$ |
| $BrO_3^- + 6\,H_3O^+ + 5\,e^- \rightarrow Br_2\,(l) + 3\,H_2O$ | $+1.423$ |

**Cadmium**

| | |
|---|---|
| $Cd^{2+} + 2\,e^- \rightarrow Cd$ | $-0.403$ |
| $Cd(OH)_2 + 2\,e^- \rightarrow Cd + 2\,OH^-$ | $-0.809$ |

**Calcium**

| | |
|---|---|
| $Ca^{2+} + 2\,e^- \rightarrow Ca$ | $-2.868$ |
| $Ca(OH)_2 + 2\,e^- \rightarrow Ca + 2\,OH^-$ | $-3.02$ |

**Carbon**

| | |
|---|---|
| $CO_2\,(g) + 2\,H_3O^+ + 2\,e^- \rightarrow HCO_2H + 2\,H_2O$ | $-0.199$ |
| $2\,CO_2\,(g) + 2\,H_3O^+ + 2\,e^- \rightarrow$ | |
| $\qquad\qquad\qquad H_2C_2O_4 + 2\,H_2O$ | $-0.49$ |
| $CNO^- + H_2O + 2\,e^- \rightarrow CN^- + 2\,OH^-$ | $-0.970$ |
| $p$-Benzoquinone $+ H_3O^+ + 2\,e^- \rightarrow$ | |
| $\qquad\qquad\qquad$ Hydroquinone $+ H_2O$ | $+0.6992$ |

**Cesium**

| | |
|---|---|
| $Cs^+ + e^- \rightarrow Cs$ | $-3.026$ |

**Chlorine**

| | |
|---|---|
| $Cl_2\,(g) + 2\,e^- \rightarrow 2\,Cl^-$ | $+1.35827$ |
| $2\,HClO\,(aq) + 2\,H_3O^+ + 2\,e^- \rightarrow Cl_2\,(g) + 2\,H_2O$ | $+1.611$ |
| $ClO^- + H_2O + 2\,e^- \rightarrow Cl^- + 2\,OH^-$ | $+0.81$ |
| $ClO_3^- + 6\,H_3O^+ + 6\,e^- \rightarrow Cl^- + 9\,H_2O$ | $+1.451$ |
| $ClO_4^- + 2\,H_3O^+ + 2\,e^- \rightarrow ClO_3^- + 3\,H_2O$ | $+1.189$ |
| $2\,ClO_4^- + 16\,H_3O^+ + 14\,e^- \rightarrow$ | |
| $\qquad\qquad\qquad Cl_2\,(g) + 24\,H_2O$ | $+1.39$ |
| $ClO_4^- + 6\,H_3O^+ + 6\,e^- \rightarrow ClO^- + 9\,H_2O$ | $+1.36$ |
| $ClO_4^- + 8\,H_3O^+ + 8\,e^- \rightarrow Cl^- + 12\,H_2O$ | $+1.389$ |

**Chromium**

| | |
|---|---|
| $Cr^{3+} + 3\,e^- \rightarrow Cr$ | $-0.744$ |
| $Cr^{3+} + e^- \rightarrow Cr^{2+}$ | $-0.407$ |
| $Cr^{2+} + 2\,e^- \rightarrow Cr$ | $-0.913$ |
| $Cr_2O_7^{2-} + 14\,H_3O^+ + 6\,e^- \rightarrow 2\,Cr^{3+} + 21\,H_2O$ | $+1.232$ |
| $Cr(OH)_3 + 3\,e^- \rightarrow Cr + 3\,OH^-$ | $-1.48$ |

**Cobalt**

| | |
|---|---|
| $Co^{3+} + e^- \rightarrow Co^{2+}$ | $+1.92$ |
| $Co^{2+} + 2\,e^- \rightarrow Co$ | $-0.28$ |
| $[Co(NH_3)_6]^{3+} + e^- \rightarrow [Co(NH_3)_6]^{2+}$ | $-0.108$ |

**Copper**

| | |
|---|---|
| $Cu^{2+} + 2\,e^- \rightarrow Cu$ | $+0.3419$ |
| $Cu^{2+} + e^- \rightarrow Cu^+$ | $+0.153$ |
| $Cu^+ + e^- \rightarrow Cu$ | $+0.521$ |
| $Cu(OH)_2 + 2\,e^- \rightarrow Cu + 2\,OH^-$ | $-0.222$ |

**Fluorine**

| | |
|---|---|
| $F_2\,(g) + 2\,e^- \rightarrow 2\,F^-$ | $+2.866$ |

**Gold**

| | |
|---|---|
| $Au^{3+} + 3\,e^- \rightarrow Au$ | $+1.498$ |
| $Au(CN)_2^- + e^- \rightarrow Au + 2\,CN^-$ | $-0.60$ |

**Hydrogen**

| | |
|---|---|
| $2\,H_3O^+ + 2\,e^- \rightarrow H_2\,(g) + 2\,H_2O$ | 0 (by definition) |
| $2\,H_2O + 2\,e^- \rightarrow H_2\,(g) + 2\,OH^-$ | $-0.828$ |
| $H_2O_2\,(aq) + 2\,H_3O^+ + 2\,e^- \rightarrow 2\,H_2O$ | $+1.776$ |

**Iodine**

| | |
|---|---|
| $I_2 + 2\,e^- \rightarrow 2\,I^-$ | $+0.5355$ |
| $I_3^- + 2\,e^- \rightarrow 3\,I^-$ | $+0.536$ |
| $2\,IO_3^- + 12\,H_3O^+ + 10\,e^- \rightarrow I_2 + 18\,H_2O$ | $+1.195$ |
| $H_5IO_6 + H_3O^+ + 2\,e^- \rightarrow IO_3^- + 4\,H_2O$ | $+1.601$ |

**Iron**

| | |
|---|---|
| $Fe^{3+} + 3\,e^- \rightarrow Fe$ | $-0.037$ |
| $Fe^{2+} + 2\,e^- \rightarrow Fe$ | $-0.447$ |
| $Fe^{3+} + e^- \rightarrow Fe^{2+}$ | $+0.771$ |
| $Fe(CN)_6^{3-} + e^- \rightarrow Fe(CN)_6^{4-}$ | $+0.358$ |

**Lead**

| | |
|---|---|
| $Pb^{2+} + 2\,e^- \rightarrow Pb$ | $-0.1262$ |
| $PbO_2 + HSO_4^-\,(aq) + 3\,H_3O^+\,(aq) + 2\,e^- \rightarrow$ | |
| $\qquad\qquad\qquad PbSO_4\,(s) + 5\,H_2O\,(l)$ | $+1.6913$ |
| $PbO_2 + 4\,H_3O^+ + 2\,e^- \rightarrow Pb^{2+} + 6\,H_2O$ | $+1.455$ |
| $PbSO_4 + H_3O^+ + 2\,e^- \rightarrow Pb + HSO_4^- + H_2O$ | $-0.3588$ |

**Lithium**

| | |
|---|---|
| $Li^+ + e^- \rightarrow Li$ | $-3.0401$ |

**Magnesium**

| | |
|---|---|
| $Mg^{2+} + 2\,e^- \rightarrow Mg$ | $-2.37$ |
| $Mg(OH)_2 + 2\,e^- \rightarrow Mg + 2\,OH^-$ | $-2.69$ |

**Manganese**

| | |
|---|---|
| $Mn^{2+} + 2\,e^- \rightarrow Mn$ | $-1.185$ |
| $MnO_4^- + 4\,H_3O^+ + 3\,e^- \rightarrow MnO_2 + 6\,H_2O$ | $+1.679$ |
| $MnO_4^- + 2\,H_2O + 3\,e^- \rightarrow MnO_2 + 4\,OH^-$ | $+0.595$ |
| $MnO_4^- + 8\,H_3O^+ + 5\,e^- \rightarrow Mn^{2+} + 12\,H_2O$ | $+1.507$ |
| $MnO_2 + 4\,H_3O^+ + 2\,e^- \rightarrow Mn^{2+} + 6\,H_2O$ | $+1.224$ |

**Mercury**

| | |
|---|---|
| $Hg^{2+} + 2\,e^- \rightarrow Hg\,(l)$ | $+0.851$ |
| $Hg_2Cl_2 + 2\,e^- \rightarrow 2\,Hg\,(l) + 2\,Cl^-$ | $+0.26808$ |
| $2\,Hg^{2+} + 2\,e^- \rightarrow Hg_2^{2+}$ | $+0.920$ |
| $Hg_2^{2+} + 2\,e^- \rightarrow 2\,Hg\,(l)$ | $+0.7973$ |

**Nickel**

| | |
|---|---|
| $Ni^{2+} + 2\,e^- \rightarrow Ni$ | $-0.257$ |
| $Ni(OH)_2 + 2\,e^- \rightarrow Ni + 2\,OH^-$ | $+0.72$ |
| $NiO_2 + 4\,H_3O^+ + 2\,e^- \rightarrow Ni^{2+} + 6\,H_2O$ | $+1.678$ |

**Nitrogen**

| | |
|---|---|
| $N_2(g) + 8\,H_3O^+ + 6\,e^- \rightarrow 2\,NH_4^+ + 8\,H_2O$ | +0.092 |
| $NO_2(g) + 2\,H_3O^+ + 2\,e^- \rightarrow NO(g) + 3\,H_2O$ | +1.03 |
| $N_2O(g) + 2\,H_3O^+ + 2\,e^- \rightarrow N_2(g) + 3\,H_2O(l)$ | +1.766 |
| $NO_3^- + 4\,H_3O^+ + 3\,e^- \rightarrow NO(g) + 6\,H_2O$ | +0.957 |
| $2\,NO(g) + 2\,H_3O^+ + 2\,e^- \rightarrow N_2O(g) + 3\,H_2O$ | +1.591 |
| $NO_2^-(g) + H_2O + 3\,e^- \rightarrow NO(g) + 2\,OH^-$ | −0.46 |
| $NO_3^- + 2\,H_2O + 3\,e^- \rightarrow NO(g) + 4\,OH^-$ | +0.109 |

**Oxygen**

| | |
|---|---|
| $O_2(g) + 2\,H_3O^+ + 2\,e^- \rightarrow H_2O_2(l) + 2\,H_2O$ | +0.695 |
| $O_2(g) + 2\,H_2O + 2\,e^- \rightarrow H_2O_2(l) + 2\,OH^-$ | −0.146 |
| $O_2(g) + 4\,H_3O^+ + 4\,e^- \rightarrow 6\,H_2O$ | +1.229 |
| $O_2(g) + 2\,H_2O + 4\,e^- \rightarrow 4\,OH^-$ | +0.401 |
| $O_3(g) + 2\,H_3O^+ + 2\,e^- \rightarrow O_2(g) + 3\,H_2O$ | +2.076 |

**Platinum**

| | |
|---|---|
| $Pt^{2+} + 2\,e^- \rightarrow Pt$ | +1.18 |

**Potassium**

| | |
|---|---|
| $K^+ + e^- \rightarrow K$ | −2.931 |

**Selenium**

| | |
|---|---|
| $SeO_3^{2-} + 3\,H_2O + 4\,e^- \rightarrow Se + 6\,OH^-$ | −0.366 |

**Silver**

| | |
|---|---|
| $Ag^+ + e^- \rightarrow Ag$ | +0.7996 |
| $AgCl + e^- \rightarrow Ag + Cl^-$ | +0.22233 |
| $AgBr + e^- \rightarrow Ag + Br^-$ | +0.07133 |
| $AgI + e^- \rightarrow Ag + I^-$ | −0.15224 |

**Sodium**

| | |
|---|---|
| $Na^+ + e^- \rightarrow Na$ | −2.71 |

**Sulfur**

| | |
|---|---|
| $S + 2\,H_3O^+ + 2\,e^- \rightarrow H_2S + 2\,H_2O$ | +0.14 |
| $S_4O_6^{2-} + 2\,e^- \rightarrow 2\,S_2O_3^{2-}$ | +0.08 |

**Tin**

| | |
|---|---|
| $Sn^{2+} + 2\,e^- \rightarrow Sn$ | −0.137 |
| $Sn^{4+} + 2\,e^- \rightarrow Sn^{2+}$ | +0.151 |

**Titanium**

| | |
|---|---|
| $Ti^{2+} + 2\,e^- \rightarrow Ti$ | −1.630 |
| $Ti^{3+} + e^- \rightarrow Ti^{2+}$ | −0.85 |
| $TiO_2 + 4\,H_3O^+ + 2\,e^- \rightarrow Ti^{2+} + 6\,H_2O$ | −0.502 |

**Zinc**

| | |
|---|---|
| $Zn^{2+} + 2\,e^- \rightarrow Zn$ | −0.7618 |
| $[Zn(NH_3)_4]^{2+} + 2\,e^- \rightarrow Zn + 4\,NH_3(aq)$ | −1.04 |

Source: Copyright 2001 From *The CRC Handbook of Chemistry and Physics*, 82nd edition, by David R. Lide, ed. Reproduced by permission of Taylor and Grancis Group, LLC, a division of Informa plc.

# SOLUTIONS TO ODD-NUMBERED PROBLEMS

## Chapter 1

**1.1** (a) H; (b) He; (c) Hf; (d) N; (e) Ne; (f) Nb

**1.3** (a) arsenic; (b) argon; (c) aluminum; (d) americium; (e) silver; (f) gold; (g) astatine; (h) actinium

**1.5** (a) $Br_2$; (b) HCl; (c) $C_2H_5I$; (d) $C_3H_6O$

**1.7** (a) $CCl_4$; (b) $H_2O_2$; (c) $P_4O_{10}$; (d) $Fe_2S_3$

**1.9** (a) $1.00000 \times 10^5$; (b) $1.0 \times 10^4$; (c) $4.00 \times 10^{-4}$; (d) $3 \times 10^{-4}$; (e) $2.753 \times 10^2$

**1.11** (a) $4.32 \times 10^2$ kg; (b) $6.24 \times 10^{-10}$ s; (c) $1.024 \times 10^{-9}$ kg; (d) $9.300 \times 10^7$ m; (e) $8.64 \times 10^4$ s; (f) $1.08 \times 10^{-3}$ m

**1.13** $7.10 \times 10^{-3}$ kg

**1.15** $9.98 \times 10^2$ kg/m$^3$

**1.17** Silver cube

**1.19** 11.7 mL

**1.21** 39.948 g/mol

**1.23** (a) 0.141 mol; (b) $5.45 \times 10^{-6}$ mol; (c) $1.67 \times 10^{-4}$ mol; (d) $5.41 \times 10^4$ mol

**1.25** (a) $3.92 \times 10^{20}$ atoms; (b) $1.14 \times 10^{20}$ atoms; (c) $3.87 \times 10^{19}$ atoms; (d) $1.48 \times 10^{19}$ atoms

**1.27** (a) 153.81 g/mol; (b) 110.27 g/mol; (c) 48.00 g/mol; (d) 86.84 g/mol; (e) 144.64 g/mol; (f) 169.88 g/mol

**1.29** (a) $C_9H_{11}NO_3$; 181.19 g/mol; (b) $C_{11}H_{12}N_2O_2$, 204.23 g/mol; (c) $C_5H_9NO_4$, 147.13 g/mol; (d) $C_6H_{14}N_2O_2$, 146.19 g/mol

**1.31** (a) $2.20 \times 10^{20}$ molecules; (b) $2.57 \times 10^{19}$ molecules; (c) $7.66 \times 10^{19}$ molecules; (d) $1.00 \times 10^{19}$ molecules

**1.33** (a) $9.99 \times 10^{-18}$ g; (b) $7.6 \times 10^{-13}$ g; (c) $1.484 \times 10^{-21}$ g

**1.35** $2.6 \times 10^{-6}$ mol; $1.6 \times 10^{18}$ molecules; $\#_H = 4.8 \times 10^{19}$ atoms; $8.0 \times 10^{-5}$ mg

**1.37** (a) $M_{Mg^{2+}} = 0.328$ M; $M_{Cl^-} = 0.655$ M;

(b)

**1.39** (a) $[K^+] = [OH^-] = 0.308$ M; (b) $[K^+] = [OH^-] = 0.077$ M; (c)

**1.41** 5.17 mL

**1.43** (a) $[CO_3^{2-}] = 0.175$ M; $[Na^+] = 0.350$ M; (b) $[NH_4^+] = [Cl^-] = 0.033$ M; (c) $[SO_4^{2-}] = 1.84 \times 10^{-3}$ M; $[K^+] = 3.69 \times 10^{-3}$ M

**1.45** (a) $NH_4NO_3 \rightarrow N_2O + 2 H_2O$; (b) $P_4O_{10} + 6 H_2O \rightarrow 4 H_3PO_4$; (c) $2 HIO_3 \rightarrow I_2O_5 + H_2O$; (d) $2 As + 5 Cl_2 \rightarrow 2 AsCl_5$

**1.47**

**1.49** (a) $2 H_2 + CO \rightarrow CH_3OH$; (b) $CaO + 3 C \rightarrow CO + CaC_2$; (c) $2 C_2H_4 + O_2 + 4 HCl \rightarrow 2 C_2H_4Cl_2 + 2 H_2O$

**1.51**

**1.53** (a) $3 Ca(OH)_2 + 2 H_3PO_4 \rightarrow 6 H_2O + Ca_3(PO_4)_2$; (b) $Na_2O_2 + 2 H_2O \rightarrow 2 NaOH + H_2O_2$; (c) $BF_3 + 3 H_2O \rightarrow 3 HF + H_3BO_3$; (d) $2 NH_3 + 3 CuO \rightarrow 3 Cu + N_2 + 3 H_2O$

**1.55** (a) 34.7 g CO; (b) 3.21 g C; (c) 2.85 g $O_2$

**1.57** $1.77 \times 10^3$ g

**1.59** 511 g

**1.61** 45.5 kg HF; 138 kg $CCl_2F_2$; 83.0 kg HCl

**1.63** 68.6%

**1.65** 28 kg

**1.67** 76.3%; 258 kg $CCl_4$; 67.2 kg HF

**1.69** lettuce; 66

**1.71** 91.3 kg

**1.73** (a) Limiting reactant = CO, 1.14 metric ton; (b) Limiting reactant = CaO, 0.499 metric ton CO and 1.14 metric ton $CaC_2$; (c) Limiting reactant = HCl, 1.36 metric ton $C_2H_4Cl_2$ and 0.247 metric ton $H_2O$

**1.75** 8.60 g; 1.7 g $O_2$

**1.77**

**1.79** 261.7 K

**1.81** (a) $4.52 \times 10^{34}$; (b) $1 \times 10^{-4}$

**1.83** HCN    $H_2O$    CO    $N_2O$

**1.85** 0.205 g

**1.87**  (a) 342.17 g/mol; (b) $7.31 \times 10^{-2}$ mol;
(c) Al: 15.77%, S: 28.12%, O: 56.11%; (d) 28.5 g

**1.89**  $2.6 \times 10^{-4}$ mol/L

**1.91**  (a) 454.59 g/mol; (b) $2.640 \times 10^{-4}$ mol;
(c) $3.179 \times 10^{20}$ atoms N

**1.93**  50.7%; 2.47 g

**1.95**  $Y + 2X \rightarrow YX_2$

**1.97**  (a) 0.30 mol; (b) $1.8 \times 10^{23}$ molecules; (c) 5.4 g

**1.99**  (a) 48 kg; (b) 35 kg

**1.101**  $1.38 \times 10^8$ atoms

**1.103**  $1.35 \times 10^{20}$ atoms of $^{53}Cr$

**1.105**  $1.212 \times 10^9$ mol $H_3PO_4$; $1.8 \times 10^8$ mol;
$5.6 \times 10^6$ kg of P consumed

**1.107**  102 kg; 67%; $N_2 = 28$ kg; $H_2 = 12$ kg

**1.109**  (a) $2 H_2SO_4 + Ca_3(PO_4)_2 \rightarrow 2 CaSO_4 + Ca(H_2PO_4)_2$;
(b) 30.6 kg; (c) 198 mol

# Chapter 2

**2.1**  A pinhole in the top of the tube would let air leak into the space until the internal pressure matched the external, atmospheric pressure. The barometer would then indicate zero pressure.

**2.3**  When air is pumped into a flat tire, the tire expands and resists compression. Both these observations are the result of the pressure exerted by the gas molecules making up air.

**2.5**  (a) $6.07 \times 10^4$ Pa, 0.607 bar; (b) $2.48 \times 10^5$ Pa, 2.48 bar; (c) 61 Pa, $6.1 \times 10^{-4}$ bar;
(d) $1.35 \times 10^2$ Pa, $1.35 \times 10^{-3}$ bar

**2.7**  4.04 mol; 100 L

**2.9**  (a) $p_i/T_i = p_f/T_f$; (b) $V = nRT/p$; (c) $p_iV_i = p_fV_f$

**2.11**  0.221 L

**2.13**  (a) equation is valid; (b), (c), and (d) not valid

**2.15**  $7.8 \times 10^{-7}$ atm

**2.17**  $p(N_2) = 593.4$ Torr; $p(O_2) = 159.2$ Torr;
$p(Ar) = 7.10$ Torr; and $p(CO_2) = 2.47 \times 10^{-1}$ Torr

**2.19**  (a) He; (b) He; (c) 0.67

**2.21**  $p(CH_4) = 2.05$ bar; $p(C_2H_6) = 0.28$ bar;
and $p(C_3H_8) = 9.5 \times 10^{-3}$ bar

**2.23**  3.94 L

**2.25**  0.443 bar

**2.27**  $2.26 \times 10^5$ g

**2.29**  $2.9 \times 10^4$ g

**2.31**

**2.33**

**2.35**

**2.37**  (a) $1.93 \times 10^3$ m/s, 11.2 kJ/mol;
(b) 394 m/s, 3.74 kJ/mol; (c) 319 m/s, 11.2 kJ/mol

**2.39**  (a) At very high pressure, molecules are very close together, so their volumes are significant compared to the volume of their container the first condition is not met and the gas is not ideal. (b) At very low temperature, molecules move very slowly, so the forces between molecules, even though small, are sufficient to influence molecular motion the second condition is not met and the gas is not ideal.

**2.41**  (a) 1 bar, generated by gas molecules colliding with the face of the piston; (b) increase in average molecular speed causes an increase in pressure, so the piston will move outward, leading to a lower frequency of collisions with the wall, reducing the pressure until the internal pressure is once again 1 bar.

**2.43**  As a gas cools, its molecules move more slowly, so they impart smaller impulses on the walls of their container. This reduces the internal pressure, so the balloon collapses until the increase in gas density inside the balloon brings the internal pressure back up to 1.000 bar.

**2.45**  121 g/mol; $CF_2Cl_2$

**2.47**  5.89 g/L

**2.49**  226 m/s

**2.51**  $CH_4$ diffuses faster, because it has the smaller $M$.

**2.53**  0.98; 0.97; easier to separate as a monoxide

**2.55**  0.826 bar; 0.826 bar; The two values are similar because the pressure is low enough that the molecules are far apart and intermolecular forces are small. Also, the temperature is high enough that the kinetic energy of the molecules is high enough to overcome what intermolecular forces do exist.

**2.57** 87.8 bar; this value is higher because the intermolecular forces in hydrogen are smaller and because hydrogen is a much smaller molecule than fluorine.

**2.59** 112.8 bar

**2.61** $+0.0049$

**2.63** 20.8 °C

**2.65** $6.8 \times 10^2$ L

**2.67** 6.10 g; $4.38 \times 10^{22}$ atoms

**2.69** (a) Chamber B, because it contains the most atoms; (b) 0.50 bar, because the pressures in the chambers are proportional to the number of atoms; (c) 4.5 bar, because the pressure will increase in proportion to the number of atoms; (d) When the valves are opened, the number of atoms in each chamber equalize: 27/3 = 9, giving 0.38 bar.

**2.71** 359 m/s

**2.73** $2 \times 10^{12}$ molecules/L

**2.75** (a) 30.9 °C; (b) 4.6 °C; (c) 6.1 °C

**2.77** Pump out a bulb of known volume, weigh it, fill with oxygen at a measured pressure, and weigh again. If gaseous oxygen were monatomic, this experiment would give $M = 16.00$ g/mol, while the diatomic gas gives $M = 32.00$ g/mol.

**2.79** $m(C_3H_6) = 0.682$ g; $m(O_2) = 2.08$ g

**2.81** (a) $H_2$; (b) $O_2$; (c) $O_2$; and (d) $H_2$

**2.83** (a) $4 \times 10^{-21}$ J; and (b) $4 \times 10^2$ m/s

**2.85** Sulphur in coal combines with oxygen from the air to form $SO_2$, and $SO_2$ reacts with water to form $H_2SO_4$: $S + O_2 \rightarrow SO_2$; $2\,SO_2 + O_2 \rightarrow 2\,SO_3$; $SO_3 + H_2O \rightarrow H_2SO_4$.

**2.87** (a) 2.09 L; (b) 788 Torr; and (c) 621 Torr

**2.89** (a) Your drawing should show the same atomic density as Figure 2-8 but with longer "tails" on the atoms, indicating that they are moving at higher speeds. (b) Your drawing should show the same lengths of "tails" as Figure 2-8, but there should be two fewer atoms. (c) Your drawing should be the same as Figure 2-8 except that half the atoms should be replaced with diatomic molecules.

**2.91** (a) $p$ is proportional to number of moles of gas; (b) true; and (c) $p/T$ is constant

**2.93** 0.911 L

**2.95** $1.0 \times 10^{-9}$ bar; $2.5 \times 10^{10}$ molecules/cm³

**2.97** 10.3 bar

**2.99** (a) 1.17 g/L; and (b) 1.15 g/L

**2.101** $6.32 \times 10^5$ L

**2.103** 64.4%

**2.105**

**2.107** $Ni(CO)_4$

**2.109** At higher elevations, Mount Everest, the atmospheric pressure, 250 Torr, is much lower than at ground level, 760 Torr. A lower pressure corresponds to a smaller molecular density. "The air is thin" refers to the fact that air is less dense at higher elevations.

**2.111** $p(H_2O) = 1.6 \times 10^{-2}$ atm; $p(\text{dry air}) = 1.004$ bar; $p(N_2) = 0.784$ bar; $p(O_2) = 0.210$ bar; $p(Ar) = 9.28 \times 10^{-3}$ bar; $p(CO_2) = 3.26 \times 10^{-4}$ bar; $p(Ne) = 1.83 \times 10^{-5}$ bar; $p(He) = 5.26 \times 10^{-6}$ bar; $p(CH_4) = 1.4 \times 10^{-6}$ bar; $p_{Kr} = 1.14 \times 10^{-6}$ bar

**2.113** $C_6H_{15}N$

## Chapter 3

**3.1** (a) On the tree, an apple possesses some gravitational potential energy. As an apple falls, that potential energy is converted into kinetic energy of motion. (b) When an apple hits the ground, the impact transfers energy to molecules in the earth and in the apple. As a result, there is a slight increase in temperature; the kinetic energy of the apple has been converted into thermal energy.

**3.3** $9.43 \times 10^{-20}$ J

**3.5** $-7.22 \times 10^{-19}$ J

**3.7** (a) Radiant energy is consumed and thermal energy is produced. (b) Chemical potential energy is consumed and kinetic energy is produced. (c) Chemical potential energy is consumed and thermal energy is produced.

**3.9** $2.12 \times 10^2$ m/s

**3.11** (a) 17.8 °C; (b) 30.6 °C; (c) 33.8 °C; and (d) 22.2 °C

**3.13** water: $8.29 \times 10^5$ J; kettle: $4.37 \times 10^4$ J

**3.15** 23.6 °C

**3.17** 131 kJ

**3.19** 14 km

**3.21** (a) $2\,C_7H_6O_2 + 15\,O_2 \rightarrow 14\,CO_2 + 6\,H_2O$; (b) $-3.221 \times 10^3$ kJ/mol; and (c) $-4.295 \times 10^2$ kJ/mol

**3.23** Moving down the column, the bond energy decreases. We predict that the H—Se bond has an energy between H—S and H—Te, and the H—Po bond is weaker than H—Te.

**3.25** 165 kJ

**3.27**

**3.29**   $H_2O$, $-475$ kJ; $NH_3$, $-90$ kJ; and $CH_4 + H_2O$, $-205$ kJ

**3.31**   $-1.4 \times 10^3$ J

**3.33**   14.4 kJ/°C

**3.35**   34.46 kJ; $-4.96 \times 10^3$ kJ/mol

**3.37**   $-2.5 \times 10^2$ J

**3.39**   (a) $-1411.1$ kJ; (b) 91.8 kJ; (c) $-1597.0$ kJ; and (d) $-21.2$ kJ

**3.41**   (a) $-1406.1$ kJ; (b) 86.8 kJ; (c) $-1597.0$ kJ; and (d) $-21.2$ kJ

**3.43**   (a) $3 K (s) + P (s) + 2 O_2 (g) \rightarrow K_3PO_4 (s)$; (b) $2 C (graphite) + 2 H_2 (g) + O_2 (g) \rightarrow CH_3CO_2H (l)$; (c) $3 C (graphite) + \frac{9}{2} H_2 (g) + \frac{1}{2} N_2 (g) \rightarrow (CH_3)_3N (g)$; and (d) $2 Al (s) + \frac{3}{2} O_2 (g) \rightarrow Al_2O_3 (s)$

**3.45**   (a) $-1166.2$ kJ; and (b) $-1531.4$ kJ

**3.47**   $1 \times 10^{16}$ J

**3.49**   Total is $7.89 \times 10^{19}$ kJ

**3.51**   Under idling conditions

**3.53**   $4.96 \times 10^3$ J

**3.55**   2.93 g

**3.57**   (a) $C_6H_{12}O_6 + 6 O_2 \rightarrow 6 CO_2 + 6 H_2O$; (b) $-2.83 \times 10^3$ kJ/mol; and (c) $-1.25 \times 10^3$ kJ/mol

**3.59**   Ethanol is more stable by 30 kJ/mol.

**3.61**   Conversion of stored chemical energy of the airplane's fuel into heat, kinetic energy, and gravitational potential energy of the airplane

**3.63**   20.1 J/mol K

**3.65**   $\Delta E_{ethane} = -46.3$ kJ/g, $\Delta E_{ethylene} = -45.6$ kJ/g, and $\Delta E_{acetylene} = -46.7$ kJ/g. Acetylene releases slightly more energy per unit mass than ethane or ethylene.

**3.67**   $8.68 \times 10^4$ J

**3.69**   (a) $-197.8$ kJ; (b) $-55.3$ kJ; and (c) $-851.5$ kJ

**3.71**   $-45$ kJ for HCl reaction; $-150$ kJ for $Cl_2$ reaction

**3.73**   $-197$ kJ

**3.75**   $C = 0.382$ J g$^{-1}$ °C$^{-1}$, Cu

**3.77**   (a) $-539.8$ kJ/mol; and (b) $-534.3$ kJ/mol

**3.79**   (a) wood; (b) natural gas production peaked in about 1972, at $23 \times 10^{18}$ kJ; (c) 91%

**3.81**   (a) negative; (b) positive; and (c) 0

**3.83**   25.2 J mol °C

**3.85**   133 g cereal

**3.87**   81.3 °C; Al spoon more effective

**3.89**   (a) $\Delta E \neq q$ when volume changes; (b) $\Delta H \neq q$ when pressure changes; (c) $q_v \neq q$ when volume changes; (d) $q_p \neq q$ when pressure changes; (e) always true

**3.91**   $1.9 \times 10^2$ s

**3.93**   5.85 °C

**3.95**

**3.97**   $\Delta H_1^\circ = 163.2$ kJ; $\Delta H_2^\circ = 85.7$ kJ; $2 NO_2 + N_2O_4$

**3.99**   (a) 37 kJ; (b) $-692$ kJ/mol

## Chapter 4

**4.1**   (a) $V_{Ag} = 1.706 \times 10^{-29}$ m³/atom, $V_{Pb} = 3.034 \times 10^{-29}$ m³/atom; (b) $d_{Ag} = 2.57 \times 10^{-10}$ m, $d_{Pb} = 3.12 \times 10^{-10}$ m; and (c) thickness$_{Ag} = 1.7 \times 10^{-3}$ m, thickness$_{Pb} = 2.0 \times 10^{-3}$ m

**4.3**   Gas pressure results from transfer of momentum between atoms and container walls, indicating that atoms have momentum and mass; mass spectrometers sort atomic and molecular ions according to their masses; all matter is made up of atoms and has mass.

**4.5**

**4.7**   (a) $6.92 \times 10^{16}$ Hz; (b) $1.28 \times 10^{18}$ Hz; (c) $4.08 \times 10^8$ Hz; and (d) $6.56 \times 10^{13}$ Hz

**4.9**   (a) $6.29 \times 10^{-2}$ m; (b) $1.04 \times 10^6$ cm; (c) $5.0 \times 10^9$ mm; and (d) $1.04 \times 10^8$ $\mu$m

**4.11**   (a) $4.049 \times 10^{-19}$ J/photon, $2.438 \times 10^2$ kJ/mol; (b) $7.79 \times 10^{-18}$ J/photon, $4.69 \times 10^3$ kJ/mol; and (c) $1.6855 \times 10^{-23}$ J/photon, $1.0150 \times 10^{-2}$ kJ/mol

**4.13** $1.7 \times 10^{16}$ photons

**4.15** (a) $1.61 \times 10^{-7}$ m, $1.86 \times 10^{15}$ Hz; and (b) $5.60 \times 10^{-7}$ m, $5.35 \times 10^{14}$ Hz

**4.17** (a) $2.31 \times 10^{-7}$ m; (b) $3.4 \times 10^{-19}$ J; and (c) $5.8 \times 10^{-7}$ m

**4.19**

**4.21** (a) 1 to 1000 m; (b) yellow; and (c) radar region

**4.23** 745 kJ/mol

**4.25** $\lambda_{8-1} = 92.57$ nm, $\lambda_{9-1} = 92.26$ nm; both are in the ultraviolet region

**4.27** Under normal conditions, absorptions occur between the lowest state and any higher-energy state. Emission originates from any higher state, to any lower state.

**4.29** $5.485\,7990 \times 10^{-4}$ g/mol

**4.31** (a) 1.45 nm; (b) 6.38 nm; and (c) 4.41 $\mu$m

**4.33** (a) $1.71 \times 10^{-20}$ J; (b) $1.11 \times 10^{-38}$ J; and (c) $3.08 \times 10^{-33}$ J

**4.35**

| $n$ | 6 | 6 | 6 | 6 | 6 | 6 |
|---|---|---|---|---|---|---|
| $l$ | 1 | 1 | 1 | 1 | 1 | 1 |
| $m_l$ | +1 | +1 | 0 | 0 | −1 | −1 |
| $m_s$ | $+\frac{1}{2}$ | $-\frac{1}{2}$ | $+\frac{1}{2}$ | $-\frac{1}{2}$ | $+\frac{1}{2}$ | $-\frac{1}{2}$ |

**4.37** $l = 0$, $m_l = 0$; $l = 1$, $m_l = 1, 0,$ or $−1$; and $l = 2$, $m_l = +2, +1, 0, −1,$ or $−2$; in all cases, $m_s = +\frac{1}{2}$ or $-\frac{1}{2}$

**4.39** (a) nonexistent: $m_s$ must be $+\frac{1}{2}$ or $-\frac{1}{2}$; (b) actual; (c) nonexistent: $l$ must be less than $n$; and (d) actual

**4.41** $3, 2, 2, +\frac{1}{2}$

**4.43**

**4.45** $x$ and $z$ axes     $y$ axis

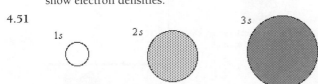

**4.47** (a) $2p$, $n = 2$, $l = 1$; (b) and (c) $3d$, $n = 3$, $l = 2$

**4.49** Electron density plots fail to show whether or not an orbital has spherical symmetry, and they do not directly show electron densities.

**4.51**

1s     2s     3s

**4.53**

**4.55** $E = 352$ kJ/mol; $E_{kinetic} = 2.10 \times 10^{-19}$ J

**4.57** (a) $O_2$, $N_2$, and $O_3$ (thermosphere); and (b) $O_3$ (stratosphere)

**4.59** (a) ~50 km; (b) upper stratosphere; and (c) formation of ozone

**4.61** (a) $6.7 \times 10^{-6}$ m; (b) $3.0 \times 10^{-20}$ J; and (c) 1.53 s

**4.63** 492 nm; yes

**4.65**

| Six: $n$ | 4 | 4 | 4 | 4 | 4 | 4 |
|---|---|---|---|---|---|---|
| $l$ | 1 | 1 | 1 | 1 | 1 | 1 |
| $m_l$ | +1 | +1 | 0 | 0 | −1 | −1 |
| $m_s$ | $+\frac{1}{2}$ | $-\frac{1}{2}$ | $+\frac{1}{2}$ | $-\frac{1}{2}$ | $+\frac{1}{2}$ | $-\frac{1}{2}$ |

**4.67** Original     2x frequency

**4.69** 552.7 nm

**4.71** (a) $3.30 \times 10^{-19}$ J; (b) $3.31 \times 10^{-7}$ m; and (c) $9.45 \times 10^{-10}$ m

**4.73** 11.0 m; $1.81 \times 10^{-26}$ J

**4.75** (a) $2 \times 10^{-4}$ bar; (b) $N O_2$, $O_3$; (c) ~35 km; and (d) stratosphere

**4.77**

| $\lambda$ (nm) | 487 | 514 | 543 | 553 | 578 |
|---|---|---|---|---|---|
| $\nu$ ($10^{14}$ s$^{-1}$) | 6.16 | 5.83 | 5.52 | 5.42 | 5.19 |
| $E$ (kJ/mol) | 246 | 233 | 220 | 216 | 207 |

**4.79** 95.0 nm; UV

**4.81** No

**4.83** $V_{atom} = 4 \times 10^{-30}$ m$^3$; $V_{nucleus} = 4 \times 10^{-45}$ m$^3$; $1/10^{15}$

**4.85** 6 photons

**4.87** Electron: $4.853 \times 10^{-11}$ m; proton: $2.642 \times 10^{-14}$ m

**4.89** Photon$_1$ = 565 nm ($d$ to $c$); Photon$_2$ = 121 nm ($c$ to $b$); Photon$_3$ = 152 nm ($b$ to $a$). The energies relative to $a$ are: $b$, $1.31 \times 10^{-18}$ J; $c$, $2.95 \times 10^{-18}$ J; $d$, $3.30 \times 10^{-18}$ J.

**4.91** (a) $E_{488} = 4.07 \times 10^{-19}$ J; $E_{514} = 3.86 \times 10^{-19}$ J; (b)

and (c) $4.17 \times 10^{15}$ s$^{-1}$ and 71.9 nm

**4.93** (a) $E_{589.6} = 202.8$ kJ/mol; $E_{590.0} = 202.7$ kJ/mol;

(b)

and (c) 423 nm

**4.95**   $E_A = 2.7 \times 10^{-19}$ J, $E_B = 4.3 \times 10^{-19}$ J; Metal B has the higher binding energy. (b) $1.59 \times 10^{-18}$ J; and (c) 460 nm to 750 nm

## Chapter 5

**5.1**   (a) He $1s$, lower value of $n$; (b) Kr $5s$, lower value of $l$; and (c) He$^+$ $2s$, less screened by $1s$ electrons

**5.3**   H atom, no screening effect, so orbital energy depends only on $n$ and $Z$, and all $n = 3$ orbitals have identical energy. He atom, the second electron screens the first, and the amount of screening decreases as $l$ increases.

**5.5**   (a) $IE$ of the He $2p$ orbital is not much larger than that of the H $2p$ orbital. (b) $IE$ of the He$^+$ $2p$ orbital is four times larger than that of the H $2p$ orbital.

**5.7**

**5.9**   Atomic number = 114; valence orbital = $7p$

**5.11**   Row 7, Column 17 (directly below astatine)

**5.13**   O, 6 valence electrons; V, 5; Rb, 1; Sn, 4; Cd, 2

**5.15**   (a) Be, 4 electrons, $1s^2\, 2s^2$

| $n$ | 2 | 2 |
|---|---|---|
| $l$ | 0 | 0 |
| $m_l$ | 0 | 0 |
| $m_s$ | $+\frac{1}{2}$ | $-\frac{1}{2}$ |

(b) O, 8, $1s^1\, 2s^2\, 2p^4$;

| $n$ | 2 | 2 | 2 | 2 | 2 | 2 |
|---|---|---|---|---|---|---|
| $l$ | 0 | 0 | 1 | 1 | 1 | 1 |
| $m_l$ | 0 | 0 | 1 | 0 | $-1$ | 1 |
| $m_s$ | $+\frac{1}{2}$ | $-\frac{1}{2}$ | $+\frac{1}{2}$ | $+\frac{1}{2}$ | $+\frac{1}{2}$ | $-\frac{1}{2}$ |

(c) Ne, 10 electrons, $1s^1\, 2s^2\, 2p^6$; first six valence electrons the same as for O; remaining two:

| $n$ | 2 | 2 |
|---|---|---|
| $l$ | 1 | 1 |
| $m_l$ | 0 | $-1$ |
| $m_s$ | $-\frac{1}{2}$ | $-\frac{1}{2}$ |

(d) P, 15 electrons, $1s^2\, 2s^2\, 2p^6\, 3s^2\, 3p^3$

| $n$ | 3 | 3 | 3 | 3 | 3 |
|---|---|---|---|---|---|
| $l$ | 0 | 0 | 1 | 1 | 1 |
| $m_l$ | 0 | 0 | $-1$ | 1 | 0 |
| $m_s$ | $+\frac{1}{2}$ | $-\frac{1}{2}$ | $+\frac{1}{2}$ | $+\frac{1}{2}$ | $+\frac{1}{2}$ |

**5.17**   O    $2p$ ⥮ ↑ ↑     P    $3p$ ↑ ↑ ↑

**5.19**   (a) Pauli-forbidden; (b) Pauli-forbidden; (c) excited-state configuration; (d) excited-state configuration; (e) ground-state configuration; (f) non-existent orbital; and (g) non-existent orbital

**5.21**   Mo    $5s$ ↑   $4d$ ↑ ↑ ↑ ↑ ↑     Tc    $5s$ ⥮   $4d$ ↑ ↑ ↑ ↑ ↑

**5.23**   C: [He] $2s^2\, 2p^2$; Cr: [Ar] $4s^1\, 3d^5$; Sb: [Kr] $5s^2\, 4d^{10}\, 5p^3$; Br: [Ar] $4s^2\, 3d^{10}\, 4p^5$

**5.25**   Seven: $1s^1\, 2s^2\, 2p^4$; $2s^2\, 2p^5$; $1s^1\, 2s^1\, 2p^5$; $1s^2\, 2s^1\, 2p^4$; $1s^2\, 2p^5$; $2s^1\, 2p^6$; $1s^1\, 2p^6$

**5.27**   Ar > Cl > K > Cs

**5.29**   Column 1 (Cs)

**5.31**   N, electron adds to an already-occupied orbital; electron–electron repulsion makes this an unfavourable process. Mg, electron adds to the next higher orbital, which is significantly higher in energy. Zn, electron adds to the next higher orbital, which is significantly higher in energy.

**5.33**   Y$^{3+}$ < Sr$^{2+}$ < Rb$^+$ < Br$^-$ < Se$^{2-}$ < As$^{3-}$

**5.35**   Stable anion: Cl; stable cations: Ca, Cu, Cs, Cr

**5.37**   $-270$ kJ/mol

**5.39**   (a) Ba$^{3+}$ O$^{3-}$; (b) Ba$^+$O$^-$; (c) Ba$^{2+}$ O$^{2-}$. The gain in lattice energy more than offsets the additional energy required to form the ions, but the energy required to remove the third electron from Ba is prohibitive.

**5.41**   $-610$ kJ/mol

**5.43**

**5.45** The decrease from $Rb_2O$ to $Cs_2O$ will be about the same as from $K_2O$ to $Rb_2O$, 75 kJ/mol. The predicted value is $2163 - 75 = 2088$ kJ/mol.

**5.47** Metal: Po; nonmetal: O, S, Se; metalloid: Te

**5.49** Metal: Ca, Cu, Cs, Cr; nonmetal: C, Cl

**5.51** Al, Ga, Sn, Bi

**5.53** (a) Cr, $4s^1 3d^5$; and Cu, $4s^1 3d^{10}$. (b) In column 6, Cr and Mo have $s^1 d^5$ configurations, while W and Sg have $s^2 d^4$ configurations. (c) The $6d$ and $5f$ orbitals fill in the second $f$ block, where Pa, U, Np, and Cm have both these orbitals partly filled, indicating nearly equal energies.

**5.55** $Mn^{2+}$: $1s^2 2s^2 2p^6 3s^2 3p^6 3d^5$;

| $n$ | 3 | 3 | 3 | 3 | 3 |
|---|---|---|---|---|---|
| $l$ | 2 | 2 | 2 | 2 | 2 |
| $m_l$ | 2 | 1 | 0 | $-1$ | $-2$ |
| $m_s$ | $+\frac{1}{2}$ | $+\frac{1}{2}$ | $+\frac{1}{2}$ | $+\frac{1}{2}$ | $+\frac{1}{2}$ |

**5.57** P: spin = $\frac{3}{2}$; $Br^-$: spin = 0; $Cu^+$: spin = 0

**5.59** $Na < O < N < Ne < Na^+$

**5.61** $Br^- > Cl^- > Cl > K^+$

**5.63** $Z = 9$ is F: [He] $2s^2 2p^5$; $Z = 20$ is Ca: [Ar] $4s^2$; and $Z = 33$ is As: [Ar] $4s^2 3d^{10} 4p^3$

**5.65** Average difference between alkali fluorides and alkali chlorides is 125 kJ/mol; average difference between alkali bromides and alkali iodides is 42.6 kJ/mol. Values decrease with the size of the alkali cation, because a larger cation cannot get as close to the anion as can a smaller cation.

**5.67**

**5.69** (a) In all one-electron atoms, orbital energy depends only on $n$ and $Z$, both of which are the same for the hydrogen atom $2s$ and $2p$ orbitals. (b) In multi-electron atoms, $n$ orbitals with different $l$ values are screened to different extents. The orbital with the lower $l$ value is less screened, hence more stable.

**5.71** (a) from element 80 (Hg) to 81 (Tl); (b) Cs, element 55; (c) $Z = 56$ to 71, $Z = 39$ to 46, and $Z = 20$ to 29; and (d) elements 15 (P), 16 (S), 33 (As), 34 (Se), 53 (I), 80 (Hg), 85 (At), and 86 (Rn)

**5.73** $S^+$

**5.75** Ground state is (c), (a) and (b) are excited states, but (d) is non-existent.

**5.77** (a) MgS; (b) KF; and (c) $BeCl_2$

**5.79** (a) $b$ is most stable (smallest $n$ value), and $a$ is more stable than $c$; (b) $n = 3, l = 0, m_l = 0, m_s = +\frac{1}{2}$; $n = 3$,

$l = 0, m_l = 0, m_s = -\frac{1}{2}$; (c) Any element from $Z = 12$ to $Z = 20$; (d) Several $3d$ transition metal cations, for example, $Fe^{3+}$; (e) eight; and (f) smaller

**5.81** Be: [He] $2s^1 2p^1$; $O^{2-}$: [He] $2s^2 2p^5 3s^1$; $Br^-$: [Ar] $4s^2 3d^{10} 4p^5 5s^1$; $Ca^{2+}$: [Ne] $3s^2 3p^5 4s^1$; $Sb^{3+}$: [Kr] $4d^{10} 5s^1 5p^1$

**5.83** The first ionization energies involve the $2s$ orbital for both atoms, so Be, with larger $Z$, has a larger $IE$. The second electron removed from Li is a core electron, so the $IE$ is much greater than the second $IE$ for Be.

**5.85**

**5.87** Francium is an alkali metal (Column 1) whose properties should closely resemble those of cesium: low $IE$, highly reactive, forms +1 cation, soft metal, melts at a low temperature.

**5.89** Removing electrons from any atom or ion reduces screening and electron–electron repulsion and therefore stabilizes the orbitals. The $Li^{2+}$ cation has a more stable $2s$ orbital.

**5.91** At what distance from the nucleus has the electron density decreased sufficiently that we consider it to be zero? That question has no unambiguous answer.

# Chapter 6

**6.1** (a) O: $1s^2 2s^2 2p^4$, six $n = 2$ electrons; (b) P: $1s^2 2s^2 2p^6 3s^2 3p^3$, five $n = 3$ electrons; (c) B: $1s^2 2s^2 2p^1$, three $n = 2$ electrons; and (d) Br: $1s^2 2s^2 2p^6 3s^2 3p^6 3d^{10} 4s^2 4p^5$, seven $n = 4$ electrons

**6.3** One valence $5p$ orbital of the iodine atom overlaps with the $1s$ orbital of the hydrogen atom to form a bond:

**6.5** (a) Al, Group 13, 3 valence electrons; (b) As, Group 15, 3 valence electrons; (c) F, Group 17, 3 valence electrons; and (d) Sn, Group 14, 3 valence electrons.

**6.7** The $2s$ orbital of a lithium atom can overlap with the $1s$ orbital of a hydrogen atom to produce an orbital with high electron density between the nuclei:

**6.9** (a) N (3.0) will attract electrons more than C (2.5); (b) S (2.5) will attract electrons more than H (2.1); (c) I (2.5) will attract electrons more than Zn (1.6); and (d) S (2.5) will attract electrons more than As (2.0).

**6.11**  (a) $\delta^+$ Si—O $\delta^-$; (b) $\delta^-$ N—C $\delta^+$; (c) $\delta^+$ Cl—F $\delta^-$; and (d) $\delta^-$ Br—C $\delta^+$.

**6.13**  $PH_3 < H_2S < NH_3 < H_2O$

**6.15**  (a) 32; (b) 90; (c) 24; and (d) 32

**6.17**

(a), (b), (c), (d) [molecular structures]

**6.19**  (a), (b), (c) [Lewis structures of $NH_3$-related species]

**6.21**  (a), (b), (c) [Lewis structures]

**6.23**  (a), (b), (c) [Lewis structures]

**6.25**  (a), (b), (c), (d) [Lewis structures]

**6.27**  (a), (b), (c), (d) [Lewis structures]

**6.29**  (a), (b) [resonance structures]

(c) [resonance structures, $O^{2-}$ carbonate-type]

(d) :O=Cl=O:⁻

**6.31**  (a) tetrahedral; (b) tetrahedral; and (c) trigonal pyramidal

(a) $CF_3Cl$ (b) $SiF_4$ (c) $PBr_3$ [structures]

**6.33**  [ball-and-stick structure with Cl atoms]

**6.35**  "Straight chain": C—C—C—C—C—C

[branched isomers]

Putting the branched carbon in Position 4 gives the same isomer as putting it in Position 2

**6.37**  Shape about the N atom is trigonal pyramidal; shape about C atoms is tetrahedral

[structures]

**6.39**  [structure of Si compound]

Hybridization = $sp^3$; geometry = tetrahedral

**6.41**  (a) T-shaped; (b) seesaw; (c) trigonal planar; and (d) square pyramidal

**6.43**  $GeF_4$, tetrahedron, $sp^3$; $SeF_4$, seesaw, $dsp^3$; $XeF_4$, square plane, $d^2sp^3$

[Lewis structures of $GeF_4$, $SeF_4$, $XeF_4$]

**6.45** (a) bent, 120°; (b) trigonal bipyramidal, 90° and 120°; (c) seesaw, 90° and 120°; and (d) square planar, 90°

**6.47** $H_2S$ and $NF_3$ have dipole moments.

**6.49** $CO_2$ has no lone pairs on the inner C atom, giving SN = 2, linear shape, and bond polarities that cancel. $SO_2$ has a lone pair on the inner S atom, bent shape, and polar bonds whose moments do not cancel.

**6.51** (a) ideal; (b) and (c) deviate

**6.53** $H-N < C\equiv N < N\equiv N < N-N < Cl-N$

**6.55** $C-C < H-N < C=C < C=O < N\equiv N$

**6.57**

(a)

(b)

resonance structures

(c)

**6.59** $SiCl_4$ and $CI_4$, tetrahedral; $SeF_4$, seesaw

**6.61** (a) C—N is shorter but weaker than C—C; (b) C—N is longer but stronger than N—N.

**6.63** Two isomers

**6.65** Empirical formula is $SiO_4^{4-}$. Orthosilicate networks are ionic, containing $SiO_4^{4-}$ anions and metal cations. Anions are tetrahedral.

**6.67** (a) $FC_{Br} = 0$

$\left[ :\ddot{O}-Br\overset{\ddot{O}:}{\underset{\ddot{O}:}{}} \right]^{-}$ + two others

(b) $FC_N = 0$

$\left[ :\ddot{O} \overset{\ddot{N}}{\diagup} \ddot{O}: \right]^{-} \longleftrightarrow \left[ :\ddot{O} \overset{\ddot{N}}{\diagdown} \ddot{O}: \right]^{-}$

(c) $FC_P = 0$

$\left[ \begin{array}{c} :\ddot{O}: \quad :\ddot{O}: \\ P \\ :\ddot{O}: \quad :\ddot{O}: \end{array} \right]^{3-} \longleftrightarrow \left[ \begin{array}{c} :\ddot{O}: \quad :\ddot{O}: \\ P \\ :\ddot{O}: \quad :\ddot{O}: \end{array} \right]^{3-}$ + two others

(d) $FC_C = 0$; $FC_O = 0$

$\left[ \overset{:\ddot{O}:}{\underset{:\ddot{O}:}{}} C - O \overset{H}{} \right]^{-} \longleftrightarrow \left[ \overset{:\ddot{O}:}{\underset{:\ddot{O}:}{}} C - O \overset{H}{} \right]^{-}$

**6.69** (a) seesaw, 90° and 120°, one lone pair, $SF_4$; (b) square planar, 90°, two lone pairs, $XeF_4$; and (c) tetrahedron, 109.5°, no lone pairs, $CH_4$

**6.71** $H-O < H-C < C\equiv N < C=O$

**6.73** (a)

(b) end N atoms, SN = 4, with one lone pair, trigonal pyramidal; C atoms and the centre N atoms, SN = 3, trigonal planar

**6.75**

**6.77** $Si-H < C-H < N-H < O-H < F-H$

**6.79** All positions around a tetrahedral center are equivalent, so there is only one isomer of $CH_2Br_2$. The two structures show different views of the same compound.

**6.81** The six-membered ring of benzyne requires bond angles of 120°. The C atoms in the triple bond have SN = 2, for which the optimum VSEPR angle is 180°. Thus, these atoms react readily.

**6.83** $N_2O$ must have the structure N—N—O if it is to have a dipole moment.

$\ddot{N}=N=\ddot{O} \qquad \ddot{N}=O=\ddot{N}$

**6.85** The isomer with Cl atoms at opposite ends is nonpolar, and the isomer with Cl atoms adjacent to each other has the largest dipole moment.

**6.87** Compounds with formula $XF_5$ have five electron pairs associated with the inner atom. This is possible for phosphorus, a third-row element that has $d$ orbitals available for bonding. It is not possible for nitrogen, a second-row element that lacks valence $d$ orbitals.

**6.89** (a) $TeF_2$; (b) $TeF_3^-$; (c) $TeF_5^-$; (d) $TeF_5^+$; (e) $TeF_6$; and (f) $TeF_4$

**6.91** (a)

(b) Like atoms opposite each other, no dipole moment; like atoms adjacent, dipole moment

**6.93** $IF_7$

## Chapter 7

**7.1** The bond forms by overlap of two $4p$ orbitals.

**7.3** The bond in LiH forms by overlap of the hydrogen $1s$ orbital and the lithium $2s$ orbital.

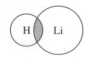

**7.5** Each bond results from overlap between a $5p$ orbital from Sb and a $2p$ orbital from F. There are three identical bonds that point at near-right angles to one another.

**7.7** (a) $sp^3$; (b) $sp^3$; (c) $sp^2$; and (d) $sp^3d^2$

**7.9** (a) $sp^3$; (b) $sp$; and (c) $sp^3d^2$

**7.11** (a) $sp^3$; (b) $sp^3d$; (c) $sp^2$; and (d) $sp$

**7.13** (a) $sp^2$; (b) $sp^3d$; (c) $sp^3d^2$; and (d) $sp^3$

**7.15** C has SN = 4, is tetrahedral, and uses $sp^3$ hybrids to form three C—Cl bonds and one C—H bond. Cl uses a $3p$ orbital for bond formation and has lone pairs in the $3s$ and two other $3p$ orbitals. H atom uses its 1s orbital to form the C—H bond.

C—H bond         C—Cl bond

**7.17** 4 $sp^3$ (N) − 1s (h) $\sigma$ bonds, 1 $sp^3$ (N) − $sp^3$ (N) $\sigma$ bond, 2 lone pairs in $sp^3$ hybrids

N—N bond       N—H bond

**7.19** Acetone has three inner C atoms. Two have only single bonds (one C—C and three C—H); these have SN = 4. The atom bonded to O has one $\pi$ bond and SN = 3.

**7.21** Some of the inner atoms have all single bonds, SN = 4, $sp^3$ hybridization, and tetrahedral geometry; others have one double bond, SN = 3, $sp^2$ hybridization, and trigonal planar geometry. The SN values for the 20 carbon atoms are shown in the line structure at right.

**7.23**

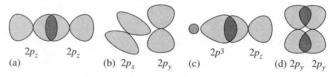

1,4-Pentadiene     1-Pentyne     Cyclopentene

**7.25** (a) $\sigma$ bond; (b) no bond; (c) $\sigma$ bond; and (d) $\pi$ bond

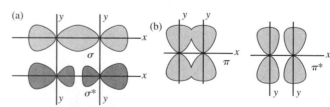

(a) $2p_z$ $2p_z$    (b) $2p_x$ $2p_y$   (c) $2p^3$ $2p_z$   (d) $2p_y$ $2p_y$

**7.27** From weakest to strongest bond, the order is $H_2^{2-}$, $H_2^-$, $H_2$

**7.29** (a) CO; (b) $N_2$; (c) $CN^-$

**7.31**

**7.33** There are eight bonding electrons and six antibonding electrons, for a net of two bonding electrons and a single bond for this ionic species.

**7.35** C atom has $sp$ hybrids that overlap with $3p$ orbitals from the S atoms to form two $\sigma$ bonds. There are two delocalized $\pi$ systems at right angles to each other, each made up of a $2p$ orbital on the C atoms overlapping side-by-side with a $3p$ orbital on each S atom. As in $CO_2$, eight electrons occupy the delocalized $\pi$ orbitals.

**7.37** Isoelectronic with $CO_2$; Lewis structure is

$$:\ddot{N}=C=\ddot{N}:^{2-}$$

C atom has SN = 2, $sp$ hybridization, and linear geometry. There are two sets of three-center delocalized $\pi$ orbitals.

**7.39** N bonded to H is bent, $sp^2$ hybridized, angle between 109–120°; outer N is not hybridized, middle N atom is linear, $sp$ hybridized.

**7.41** Xanthin has a delocalized $\pi$ system in which all 26 double-bonded C atoms contribute.

**7.43** (a) Hybridization is as shown on the line drawing ($3 = sp^3$, $2 = sp^2$).

(b-c) 2 $\pi$ bonds, on adjacent atoms, four electrons in the extended $\pi$ system; and (d) eight atoms screened in blue

**7.45** Lewis structure:

$$:\ddot{O}=C=C=C=\ddot{O}:$$

Linear, $\sigma$ bond network can be described using $sp$ hybrids; two sets of delocalized $\pi$ orbitals, each extending over all five atoms:

**7.47** Tetrahedral geometry, $sp^3$ hybrids on Cl overlap with $2p$ orbitals from O; $\pi$ system extends over all five atoms, incorporating $d$ orbitals from the inner Cl atom; three bonding $\pi$ orbitals occupied by six electrons.

**7.49** (a) undoped; (b) $n$-type; and (c) undoped

**7.51** Each K atom contributes one valence electron to the valence (bonding) band of the solid, each Fe atom contributes eight electrons. Thus, iron has much greater bonding, making it harder and giving it a higher melting point than potassium.

**7.53** $p$-type semiconductor

Vacant orbitals

**7.55** The S atom in each compound has SN = 3, so $sp^2$ hybrids overlapping with oxygen $2p$ orbitals describe the $\sigma$ bonds: two in $SO_2$ and three in $SO_3$. $SO_2$ is bent, and $SO_3$ is trigonal planar. There are two $\pi$ bonds in $SO_2$ and three $\pi$ bonds in $SO_3$. All the $\pi$ orbitals extend over the entire molecule.

**7.57** $\nu = 8.70 \times 10^{13}$/s; infrared

**7.59** 4 $sp^2$-1$s$ C—H $\sigma$ bonds, 2 $sp^2$-$sp^2$ C—C $\sigma$ bonds, 1 $sp^2$-2$p$ C–O $\sigma$ bond, delocalized $\pi$ network extending over four atoms; two bonding $\pi$ orbitals; one 2$s$ and one 2$p$ lone pair on the outer O atom.

$\sigma$ framework

$\pi$ system

**7.61** (a)

Cinnamic acid
$C_9H_8O_2$

; (b, c) A: $sp^2$ and $p$, 120°; B: $2p$, no angle (outer); C: $sp^3$, 109.5°; D: $sp^2$ and $p$, 120°; E: atomic 1$s$, no angle (outer); and (d) five $\pi$ bonds

**7.63** (a) and (d), diamagnetic; (b) and (c), paramagnetic

**7.65**

All C's use $sp^2$    C and N use $sp^3$

**7.67**

metal    $n$-semiconductor

**7.69** Central C, SN = 3, trigonal planar geometry, $sp^2$ hybrid orbitals; methyl C atoms, SN = 4, tetrahedral geometry, $sp^3$ hybrid orbitals

**7.71**

**7.73**  $O_2$: $(\sigma_s)^2(\sigma_s{}^*)^2(\sigma_p)^2(\pi_x)^2(\pi_y)^2(\pi_x{}^*)^1(\pi_y{}^*)^1$
Bond order = 2
$O_2{}^-$: $(\sigma_s)^2(\sigma_s{}^*)^2(\sigma_p)^2(\pi_x)^2(\pi_y)^2(\pi_x{}^*)^2(\pi_y{}^*)^1$
Bond order = 1.5
$O_2{}^{2-}$: $(\sigma_s)^2(\sigma_s{}^*)^2(\sigma_p)^2(\pi_x)^2(\pi_y)^2(\pi_x{}^*)^2(\pi_y{}^*)^2$
Bond order = 1

$O_2$ has the strongest, shortest bond and $O_2{}^{2-}$ the longest, weakest bond. $O_2$ and $O_2{}^-$ are both magnetic, with $O_2$ showing the largest magnetism.

**7.75**  Cl uses $sp^3$ hybrids that overlap with $2p$ orbitals from O. The $\pi$ system of the chlorate ion is extended over all four atoms, and there are two bonding $\pi$ orbitals occupied by four electrons. The $\pi$ system of the chlorite ion is extended over all three atoms, with one bonding $\pi$ orbital occupied by two electrons.

**7.77**  Higher bond order should be accompanied by increased bond strength and decreased bond length. Bond length and bond strength can be measured.

**7.79**  (a) The carbon atoms with SN = 3 have $\sigma$ bonding that can be described using $sp^2$ hybrid orbitals, and the $\sigma$ bonding of the carbon atom with SN = 2 can be described using $sp$ hybrids.

(b)

(c) There is an extended $\pi$ network that includes the three C atoms and the N atom.

**7.81**  There are 7 bonding electrons and 2 antibonding electrons, for a net of 5, or bond order 2.5. The least stable occupied orbital is a $\sigma_p$ orbital:

**7.83**  Its geometry is tetrahedral, the $\sigma$ bonds can be described using $sp^3$ hybrids from Cl, and the molecule has a bond angle near 109°. There is an extended $\pi$ system formed from $d$ orbitals on Cl overlapping side-by-side

with $2p$ orbitals from the two O atoms. It is unusual because it has an odd number of electrons.

**7.85**  MO diagram shows a bond order of zero for $Be_2$, which cannot be prepared.

**7.87**

**7.89**  Multiple bonds involving $\pi$ overlap are strong for N and O, but $\pi$ overlap is very slight for P and S.

**7.91**

# Chapter 8

**8.1**

**8.3**  (a) less significant; (b) more significant; and (c) less significant

**8.5**

**8.7**  $-8.9\%$

**8.9**  21.5 bar; $-25\%$ deviation from ideal gas behavior

**8.11**  $CH_4$ (hardest to liquefy) $<CF_4 < CCl_4$ (easiest to liquefy)

**8.13**

**8.15**  Propane (lowest bp) $< n$-pentane $<$ ethanol (highest bp)

**8.17**  (a) and (c) no hydrogen bonding; (b) and (d) hydrogen bonding

**8.19**

**8.21**  Pentane ($C_5$) < gasoline ($C_8$) < fuel oil ($C_{12}$)

**8.23**  Water will be concave, mercury will be convex.

**8.25**  Paper towels contain cellulose, which forms many hydrogen bonds with water. There are no strong intermolecular interactions between an oil and the fibers of paper towels.

**8.27**  (a) benzene, no dipole moment; (b) hexane, no H-bonding; and; (c) heptane, smaller dispersion forces

**8.29**  Sn: metallic solid; $S_8$: molecular solid; Se: network solid; $SiO_2$: network solid; $Na_2SO_4$: ionic solid

**8.31**  (a) The bonding in metals comes from extended networks of delocalized electrons, while the bonding in network solids includes many individual covalent bonds. (b) Metals conduct electricity, are malleable and ductile, and are shiny in appearance. Network solids are nonconductors or semiconductors, are brittle, and often have dull appearances.

**8.33**  (a) molecular solid; (b) ionic solid; (c) metallic solid; (d) network solid; and (e) molecular solid

**8.35**  The density of a solid depends not only on the nature of the elements it contains but also on the tightness of bonding. Graphite has a more open bonding pattern than diamond.

**8.37**  $CaTiO_3$

**8.39**

**8.41**

**8.43**  8

**8.45**  (a) methane, smaller dispersion forces; (b) ethanol, hydrogen bonding; and (c) argon, higher polarizability

**8.47**

**8.49**  (a) Gas cools, liquefies at 331.9 K, and solidifies at 265.9 K. (b) Compressing the gas causes it to liquefy at about $P = 6 \times 10^{-2}$ bar and solidify at about 0.5 bar.; (c) Solid sublimes to the vapour at about 265 K.

**8.51**  980. J

**8.53**  (a) dispersion interactions and hydrogen bonds; (b) dispersion interactions and polarity; (c) dispersion interactions; and (d) dispersion interactions

**8.55**  Face-centered cubic structure contains more atoms in a given volume and is denser.

**8.57**  (a) $CH_3OH$ (hydrogen bond); (b) $SiO_2$ (covalent bonds); (c) HF (hydrogen bond); and (d) $I_2$ (larger dispersion forces)

**8.59**  (a) At 70 K, 1.01 bar, nitrogen is liquid. When the pressure falls below about 0.5 bar, the liquid boils and forms a gas. (b) The sample remains a gas.; (c) The gas liquefies at 77 K, and the liquid solidifies at 63 K.

**8.61**  *trans*, no dipole moment, 47 °C; *cis*, dipole moment, 60 °C

*cis*
dipole moment
bp = 60 °C

*trans*
no dipole moment
bp = 47 °C

**8.63**  $Cl_2$, greater dispersion forces

**8.65**  At ~ 600 °C, α quartz converts to β quartz. At ~ 850 °C, the solid becomes tridymite, and just below 1500 °C, this converts to cristobalite. At ~ 1700 °C, cristobalite melts to give liquid silica.

**8.67**  More closely packed is stable at high pressure: body-centered cubic

**8.69**  (a) network; (b) metal; and (c) ionic

**8.71**  (a) $T > 153$ °C, $P > 1439$ bar; (b) $T$, $P$ combinations ranging from 95.3 °C, $5.2 \times 10^{-6}$ bar to 153 °C, 1439 bar; (c) $P < 5.2 \times 10^{-6}$ bar

**8.73**  (a) simple cubic, and (b) 3 O atoms per unit cell; $ReO_3$

**8.75**  Forces are smaller.

**8.77**  (a) At any pressure between $3.2 \times 10^{-5}$ bar and 1439 bar, there is a temperature at which each phase is stable. Below 95.3 °C, the rhombic phase is stable. At a temperature between 95.3 and 153 °C, the monoclinic phase becomes stable. At a temperature between 115.2 and 153 °C, monoclinic sulphur melts and the liquid becomes stable. At a still higher temperature that depends strongly on the pressure, the vapour phase is stable. (b) At temperatures between 115.2 and 153 °C, there is a pressure at which each phase is stable. Below $3.2 \times 10^{-5}$ bar, the vapour is stable. Above this pressure, the liquid is stable up to a pressure that depends on the exact temperature, but at higher pressure, the monoclinic solid phase is stable. At extremely high pressure, the rhombic phase becomes stable.

**8.79** The magnitudes of dispersion forces depend on how extended the valence electrons are. In a molecule, valence electrons are spread over a larger volume because they are shared between atoms.

**8.81** Each ion has six nearest neighbours of opposite charge:

**8.83** (a) mp > 900 °C; (b) sillimanite; (c) yes, kyanite; and (d) the stable phase at highest pressure, kyanite, is densest. The stable phase at lowest pressure, andalusite, is the least dense.

**8.85**

**8.87** (a)

H H O:---H—O H
H—C—C C—C—H
H O—H---:O H

(b) Fraction decreases as H-bonds break.

# Chapter 9

**9.1** (a) solvent, liquid water; primary solute, carbon dioxide (normally a gas); (b) solvent, liquid water; primary solute, ethanol (normally a liquid); (c) solvent, gaseous $N_2$; primary solutes, $O_2$ (normally a gas) and water (normally a liquid)

**9.3** $X_{methanol} = 0.0747$, $X_{water} = 0.925$

**9.5** $c_m = 1.88 \times 10^{-1}$ mol/kg; $X = 1.08 \times 10^{-2}$; % = 2.14%

**9.7** $X_{hydrogen\ peroxide} = 0.18$, $X_{water} = 0.82$; $M = 9.8$ M; $c_m = 13$ mol/kg

**9.9** $X_{ammonia} = 0.042$; $X_{water} = 0.958$; Mass fraction ammonia = 0.0399; $b = 2.44$ mol/kg

**9.11** (a) 120. g; (b) 6.50 mol/kg

**9.13** The hydrocarbon molecules in salad dressing have dispersion-type intermolecular forces but low polarity

and no hydrogen-bonding capability. A solution of acetic acid in water has high polarity and a large hydrogen-bonding capability. Hydrogen bonds would have to be broken for the two liquids to mix, making mixing energetically unfavourable.

**9.15** Ionic salts and polar organic materials such as alcohols

**9.17** $H_2O$ has strong hydrogen bonding, and so does $CH_3OH$, so this pair is miscible; $C_8H_{18}$ and $CCl_4$ both have only dispersion forces, so this pair is miscible.

**9.19** $CCl_4$ is most like $I_2$; NaCl is a salt, so it only dissolves in water; Au, a metal, requires another metal, Hg; and paraffin, a hydrocarbon, is best matched by $n$-octane, another hydrocarbon.

**9.21**

**9.23** Methane does not form H-bonds, so life is unlikely.

**9.25**

**9.27** Energy must be supplied to the molecules that escape.

**9.29** 48%

**9.31** $K_H = 59.53$ M/bar; $p = 4.2$ kPa

**9.33** $p_{total} = 5.87$ kPa

**9.35** $X_{ethanol} = 0.363$

**9.37** $p_{vap,\ solution} = (0.964)(1.01\ bar) = 0.974$ bar

**9.39** $T_f = -0.438$ °C.

**9.41** $T_b = 100.121$ °C

**9.43** 0.20 M sucrose (lowest) < 0.20 M NaCl < 0.40 M sucrose < 0.25 M $MgCl_2$ (highest)

**9.45** (a) M = 180 g/mol; (b) $C_6H_{12}O_6$.

**9.47** 8.00 bar

**9.49** Similarities: colloidal suspensions, similar compositions; difference: Milk is a suspension of solids (fats) in water, whereas whipped cream is a suspension of a gas (air) in a solid (coagulated fat) that also contains liquid.

**9.51** Best surfactant: Sodium alkylbenzenesulfonate; worst surfactant: octadecane

**9.53**

Water

**9.55** $b = 2.7$ mol/kg, $X = 4.6 \times 10^{-2}$, M = 4.3%

**9.57** BHT contains only one OH group, so it is highly hydrophobic. Thus, it is stored in fatty tissues (which are nonpolar) rather than being excreted from the body. BHT would be dangerous to organisms if the molecule was toxic.

**9.59** M = 175 g/mol

**9.61** KCl (lowest solubility) < $C_2H_5OH$ < $C_3H_8$ (highest solubility)

**9.63** Benzene is a nonpolar liquid; water is highly polar; $CCl_4$ is nonpolar. These two liquids are immiscible and benzene preferentially dissolves in $CCl_4$.

**9.65** (a) Grease = dirty; (b) $Mg^{2+}$ ions not dirty; (c) Acetone = not dirty; (d) $SiO_2$ = not dirty

**9.67** (a) 5.2°C kg/mol, (b) M = $1.1 \times 10^2$ g/mol.

**9.69** If water were linear rather than bent, the two polar O—H bonds would oppose each other and the molecule would be nonpolar like $CO_2$. Nonpolar water molecules could not associate with ions through ion-dipole interactions, so linear water would not be a good solvent for salts.

**9.71**

Water

Membrane          Membrane

Water

**9.73** 25.3 bar

**9.75** 8.3 kg sugar : 1 kg water

## Chapter 10

**10.1**

All other carbon atoms sp³

**10.3** (a) (2-methylpentane); (b) (2,3-dimethylbutane); (c) (2-methylpentane); (d) (2,4-dimethylpentane); (e) (3-methylpentane)

**10.5**

(a)          (b)          (c)

**10.7** (a) *o*-bromonitrobenzene, (b) *p*-dicyclohexylbenzene, (c) *m*-dinitrobenzene

**10.9** (a) $CH_3$  (b) $CH_3$  (c) $CH_3$

(d) $O_2N$ $CH_3$ $NO_2$ ... $NO_2$   (e) Cl Cl Cl Cl Cl Cl

**10.11** (a) 1-chloropropane (*or* n-propyl chloride);
(b) 2-chlorobutane;
(c) 2-chloro-3,3-dimethylbutane;
(d) 2-chloro-3-methylbutane;
(e) 3-chloro-3-methylheptane

**10.13** (a) Secondary 2°, (b) Primary 1°, (c) Secondary 2°, (d) Tertiary 3°, (e) Secondary 2°

**10.15**

(a) OH  (b) OH  (c) OH ... OH

(d) OH  (e) OH

**10.17** (a) Tertiary 3°, (b) Primary 1°, (c) Primary 1°, (d) Secondary 2°, (e) Tertiary 3°

**10.19** (a) Cyclohexanone, (b) Butanal, (c) Ethyl butanoate, (d) Butanoic acid, (e) Butanamide

**10.21** (a) No E/Z isomerization; (b) No E/Z isomerization

(c)

*E*-2-pentene     *Z*-2-pentene

(d) No E/Z isomerization; (e) No E/Z isomerization

**10.23** (a) Achiral, (b) Achiral, (c) Chiral, (d) Chiral, (e) Chiral

**10.25**

(R) ... $NH_2$ ... H ... (S) ... (R) ... S ... O ... N ... (R) ... $CO_2H$

Amoxicillin

**10.27** Good leaving groups are conjugate bases of strong acids (like $Cl^-$ and $Br^-$) ie weak bases. Poor leaving groups are conjugate bases of weak acids (like $OH^-$) ie strong bases.

(a) $Br^-$ = Good Leaving Group
(b) $CH_3—O^-$ = Poor Leaving Group
(c) $NH_2^-$ = Poor Leaving Group
(d) $Cl^-$ = Good Leaving Group
(e) $CH_3COO^-$ (acetate anion) = Good Leaving Group

**10.29**

(a)

(b)

(c)

(d)

(e)

**10.31**

There are three unique mono-chlorination products for toluene

**10.33** (a) (b) (c)

(d) (e)

**10.35**

**10.37** (a) + $H_{2(g)}$ $\xrightarrow{Pt}$

(b) + $H_{2(g)}$ $\xrightarrow{Pt}$

(c) + $H_{2(g)}$ $\xrightarrow{Pt}$

(d) + $H_{2(g)}$ $\xrightarrow{Pt}$

(e) + $H_{2(g)}$ $\xrightarrow{Pt}$

**10.39**

**10.41**

**10.43**

(a)

(b)

(c)

(d)

(e)

**10.45** Markovnikov's rule states that *in the addition of HX to an alkene, the hydrogen atom adds to the carbon atom of the double bond that already has the greater number of hydrogen atoms.* An example is:

**10.47**

**10.49**

**10.51**

**10.53** The Pauling electronegativity number for Carbon is 2.5 and for Hydrogen is 2.1. This small difference (0.4) means that the two electrons in the C—H bond are essentially equally shared between the two atoms. The result is a covalent bond that is essentially non-polar.

**10.55** Hexane is an essentially non-polar molecule with relatively weak dispersion forces as the main intermolecular forces. Water is a highly polar molecule with the comparatively stronger hydrogen bonding as the main intermolecular force. Non-polar hexane has no groups that can act as hydrogen bond donors or acceptors. As a result the intermolecular forces holding water molecules together (hydrogen bonds) are much stronger than any forces between water molecules and hexane molecules, thus the liquids are immiscible.

**10.57** In CCl₄, even thought there is an electronegative chlorine ($\chi = 3.0$) covalently bonded to a less electronegative carbon ($\chi = 2.5$) the four dipoles cancel each other out and the net result is a non-polar molecule. In $CH_2Cl_2$ there are two non-polar bonds (C—H) and two polar (C—Cl) bonds. These dipoles do **not** cancel each other out and the result is a polar molecule.

**10.59** The carbonyl group of acetone would have a dipole as shown below. The $\delta+$ end of the dipole is centered on the carbon atom while the $\delta-$ end is centered on the oxygen atom. Attack by a nucleophile (such as the OH⁻ ion) would occur at the positive end of the dipole as shown below.

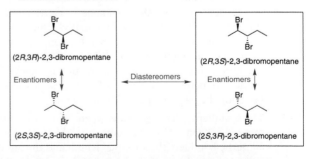

**10.61** There are two factors that make acetic acid more acidic than ethanol. First is the presence of a second electronegative oxygen atom. This helps make the negative charge that exists on the acetate anion more favorable overall. Second is the fact that the negative charge on the acetate anion can be delocalized over two oxygen atoms through a process known as resonance, which helps to stabilize the negative charge. There is neither an extra electronegative atom nor resonance stabilization to help make the ethoxide anion more stable.

**10.63** The configuration changes only because the atomic number for sulphur is higher than the atomic number for oxygen. This makes the CH₂—SH substituent higher priority than the COOH substituent on cystine. On serine, the COOH substituent has a higher priority (because of two oxygen atoms) than the CH₂—OH group. This change in priority results in a change in the configuration of the two different chirality centres.

**10.65** The two stereoisomers (2R,3R)-2,3-dibromopentane and (2S,3S)-2,3-dibromopentane are enantiomers of each other. The two stereoisomers (2R,3S)-2,3-dibromopentane and (2S,3R)-2,3-dibromopentane are also enantiomers of each other. The (2S,3S)-2,3-dibromopentane and the (2R,3R)-2,3-dibromopentane steroisomers are diasteromers of the (2R,3S)-2,3-dibromopentane and the (2S,3R)-2,3-dibromopentane. This is illustrated below.

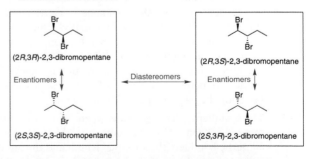

**10.67** In the absence of acid, the leaving group on *t*-butanol would be a hydroxide ion, which is the conjugate base of a weak acid ($H_2O$). Hydroxide ion is, therefore, a very poor leaving group. By adding acid, the first step in the mechanism is to protonate the oxygen atom of the alcohol. In this case then the leaving group is a molecule of water, the conjugate base of a strong acid ($H_3O^+$). This makes water a very good leaving group.

**10.69** In the addition of HCl to a double bond the first step is protonation of the double bond and the formation of a carbocation on one of the carbon atoms. The presence of the highly electronegative fluorine atoms strongly disfavors formation of this carbocation.

**10.71** The critical step in the $S_N1$ reaction is the formation of the carbocation intermediate. This carbocation is formally an $sp^2$-hybridized carbon atom that is planar with 120° bond angles and an empty p-orbital. The incoming nucleophile has no preference as to which side of the carbocation to attack to form the new covalent bond. In this example the result is an equimolar mixture of the ($S$)-1-bromo-1-phenylethane, the result of attack from the 'top' of the carbocation and ($R$)-1-bromo-1-phenylethane from attack at the 'bottom' of the carbocation.

**10.73** The reactions show that with a small base (i.e., $NaOCH_3$), substitution is the result via an $S_N2$ mechanism. The larger base (NaO-$t$—Bu) results in only the elimination product being formed. The conclusion is that elimination is favoured by larger bases.

**10.75** The $t$-butyl chloride reacts faster because under $S_N1$ conditions the rate determining step is formation of the carbocation. Tertiary carbocations are more stable than primary carbocations and this is demonstrated by this example. It is much easier to form a tertiary than a primary carbocation. If you picture each C—H bond on the adjacent methyl groups as a 'cloud' of electrons (and this is a pretty accurate picture) then each bond is able to donate a little of this cloud into the empty p-orbital on the carbocation through a process called hyperconjugation. Branched alkyl groups have a stronger electron donation effect than straight chain alkyl groups. It is this slight donation of each of the CH bonds in the alkyl branches that accounts for the extra stability of the tertiary carbocation.

**10.77** The fact that more energy is released when Z-2-butene is hydrogenated must mean that there was more energy stored in this double bond than in E-2-butene. This leads to the conclusion that E-alkenes are more stable (i.e., lower in energy) than Z-alkenes. The reason for this is that in Z-2-butene the two methyl groups that are on the same side of the double bond can interact with each other raising the energy of the molecule. In E-2-butene this unfavourable interaction is removed as the two methyl groups are as far apart as possible, lowering the overall energy of the system.

**10.79** The cycloheptatrienyl cation is an aromatic system the same as benzene. This is demonstrated by the fact that all of the bond lengths are equal length and the molecule is planar due to the presence of the $sp^2$ hybridized positively charged carbon and the $sp^2$ hybridized carbons of the double bonds.

**10.81** The molecular orbital for the π-bond would be formed by combination of the two atomic p-orbitals. The 'in-phase' overlap would then result in a region of greatest probability for finding the two electrons in the bonding molecular orbital between the two carbon atoms. This overlap would exist in a region above and below the plane of the carbon-carbon sigma bond preventing the rotation about this bond. Any rotation out of the plane would result in loss of the favorable atomic orbital overlap that generates the bonding molecular orbital.

**10.83** The enantiomer of D-glucose is L-glucose and is the complete mirror image of the molecule as shown below. All of the chiral centres in D-glucose have the (S) configuration, whereas all of the chiral centres in L-glucose have the (R) configuration.

D-Glucose          L-Glucose

**10.85** In general an E2 reaction can take place on any substrate with a hydrogen atom on a carbon atom adjacent to the leaving group. For the E2 reaction it makes little difference if the leaving group is on a primary, secondary or tertiary carbon atom. However, the $S_N2$ mechanism is favored by primary and some secondary but never by a tertiary substrate. Also the size of the base has an effect. Large strong and bulky bases, such as $t$-butoxide favour the elimination pathway. Smaller bases that are derived from weaker acids favor the substitution pathway.

**10.87** The electronegativity numbers for hydrogen (2.1) and boron (2.0) mean that the hydrogen is more electronegative than the boron atom and that the electrons in the B—H bond are polarized towards the hydrogen atom. This means that in an electrophillic addition the mechanism occurs as below where the boron actually adds first to the double bond. The second step is the attack of an $H^-$ (a hydride ion) on the carbocation. This mechanism agrees with the definition of Markovnikov's rule in **10.86** since the most stable carbocation is still the intermediate in the reaction.

## Chapter 11

**11.1** (a) A sand castle is organized in a particular way. Waves disperse the sand. (b) Mixing liquids disperses their molecules. (c) Sticks in a bundle are aligned in one direction. When dropped, the sticks lose their alignment and become more dispersed. (d) Water molecules in the gas phase are more dispersed than those in the liquid phase.

**11.3** Initially the ink is concentrated in a droplet, but the molecules gradually become dispersed throughout the liquid.

**11.5** (a) The air molecules in a tire are relatively constrained. A puncture allows gas molecules to escape from the tire and become more dispersed. (b) The fragrant molecules in a perfume bottle are relatively constrained. When the bottle is open, molecules escape from the confined volume and become more dispersed.

**11.7** (a) Energy dispersal requires that heat flows from high to low temperature, never in the opposite direction. (b) Matter dispersal requires that the spontaneous direction is toward greater dispersal, and a "sand castle" represents constrained matter.

**11.9** 72

**11.11** (a) 16.8 J/K; (b) −15.3 J/K; and (c) 1.5 J/K

**11.13** (a) $\Delta S_{sys} > 0$, $\Delta S_{surr} < 0$; (b) $\Delta S_{sys} > 0$, $\Delta S_{surr} < 0$; and (c) $\Delta S_{sys} > 0$, $\Delta S_{surr} < 0$. $\Delta S_{total} > 0$ always.

**11.15** −94.0 J/K; $\Delta S_{surr} > 94.0$ J/K

**11.17** (a) 16 J/mol K; (b) 9.3 J/mol K; (c) 48.8 J/mol K; and (d) $1.00 \times 10^2$ J/mol K

**11.19** (a) $MgCl_2$, because it produces more moles of ions per mole of substance; (b) HgS, because it has a higher molar mass; and (c) $Br_2(l)$, because it is a liquid

**11.21** Ozone has three atoms per molecule, whereas $O_2$ and $H_2$ have only two. $H_2$ has lower molar mass than $O_2$.

**11.23** He: $S° = 126.153$ J/mol K; $H_2$: $S° = 65.340$ J/mol K; $CH_4$: $S° = 37.26$ J/mol K; $C_3H_6$: $S° = 25.21$ J/mol K. Entropy per mole of atoms decreases as the number of atoms in a molecule increase, because tying together atoms into a molecule increases the amount of order among those atoms.

**11.25** (a) 416 J/K; (b) $1.7 \times 10^2$ J/K; and (c) $9.0 \times 10^1$ J/K

**11.27** (a) −198.1 J/K; (b) −137.7 J/K; (c) 12.3 J/K; and (d) −267.3 J/K.

**11.29** (a) reduction in number of moles of gaseous substances; (b) reduction in number of moles of gaseous substances; (c) all reagents are relatively constrained solid substances; and (d) reduction in number of moles of gaseous substances

**11.31** (a) positive; (b) negative; and (c) negative

**11.33** (a) $\Delta G_{system} < 0$ for any spontaneous process at constant $T$ and $P$; (b) the free energy of a system decreases in any process at constant $T$ and $P$; and (c) $\Delta G = \Delta H - T\Delta S$ at constant $T$

**11.35** (a) −32.8 kJ; (b) 326.4 kJ; (c) −206.1 kJ; and (d) −1331.4 kJ

**11.37** (11.27b) 311.3 kJ; and (11.27c) −204.3 kJ

**11.39** (a) −46.2 kJ; and (b) 319.7 kJ

**11.41** $\Delta H°_{reaction} = -559.6$ kJ; $\Delta S°_{reaction} = -574.9$ J/K; $\Delta G°_{reaction} = -388.2$ kJ

**11.43** (a) no temperature; and (b) zero concentration of products

**11.45** $5.2 \times 10^{-5}$ bar

**11.47** 119 °C

**11.49** 44.3 kJ/mol; This value is higher than that at the boiling point because at the boiling point, the water molecules have more kinetic energy, and therefore need less additional energy to escape into the gas phase.

**11.51** The first "reaction" would be child 1 starting on the ground and rising into the air. The second "reaction" would be child 2 starting in the air and falling to the ground. In this case child 2 would undergo the spontaneous reaction. The motion of child 1 would be nonspontaneous. The common intermediate is the seesaw board, which acts as a lever transferring energy from one child to the other.

**11.53** ATP + $C_6H_{12}O_6$ (fructose) + $C_6H_{12}O_6$ (glucose) → ADP + $C_{12}H_{22}O_{11}$ + $H_3PO_4$; $\Delta G° = -7.6$ kJ

**11.55** (a) not spontaneous; (b) absorbs heat; and (c) less dispersed

**11.57** (a) −83 J/K; (b) 328.7 J/K; and (c) 365.3 J/K

**11.59** (a) −40 kJ; (b) −148.4 kJ; and (c) −1997.7 kJ

**11.61** (a) −42 kJ; (b) −168.1 kJ; and (c) −2020.7 J/K

**11.63** (a) $w_{sys} = 0$ (no volume change); (b) $q_{sys} > 0$ (energy must be supplied to break chemical bonds); and (c) $\Delta S_{surr} < 0$ ($q_{sys} > 0$, $q_{surr} < 0$)

**11.65** (a) and (b) not thermodynamically feasible; and (c) thermodynamically feasible

**11.67** When a cell needs energy, it "spends" ATP; when fat or carbohydrates ("capital") are consumed, the energy is stored by converting ADP to ATP ("buying" ATP).

**11.69** (a) spontaneous; (b) absorbs energy; and (c) yes

**11.71** (a) −189 J/K; (b) 15 J/K; and (c) yes

**11.73** Ammonia is a toxic gas, while urea and ammonium nitrate both are relatively nontoxic solids. Thus, the transport and application of ammonia entail significant risks to humans. Even in aqueous solution, ammonia is highly irritating, as anyone knows who has used strong ammonia as a cleanser.

**11.75** Glucose, −15.9 kJ/g; palmitic acid, −38.2 kJ/g

**11.77** (a) negative; (b) negative; and (c) positive

**11.79** (a) $\Delta E_{total} = 0$; (b) $\Delta E_{teaspoon} > 0$; (c) $\Delta S_{total} > 0$; (d) $\Delta S_{water} < 0$; and (e) $q_{teaspoon} > 0$

**11.81** (0.5 mol, $l$, 298 K) < (1 mol, $l$, 298 K) < (1 mol, $l$, 373 K) < (1 mol, $g$, 1 bar, 373 K) < (1 mol, $g$, 0.1 bar, 373 K)

**11.83** $\Delta H° = 182.6$ kJ; $\Delta S° = 24.83$ J/K; $\Delta G° = 175.2$ kJ; spontaneous above 7350 K; although a large and positive free energy change means little products at equilibrium, it does not mean none.

**11.85** (a) $\Delta S°_{vap} = 87.1$ J/K; $\Delta G°_{vap} = 0$; (b) 5.16 kJ/mol; and (c) 0.14 bar

**11.87** (a)

(b) $\Delta S_{total} = 0$

**11.89** Attractive forces are larger for CaO, restricting dispersion by way of vibration.

**11.91** (a) $q_v = \Delta E$; (b) $q_p = \Delta H$; and (c) $q_T = T\Delta S$

**11.93**

$\Delta G = 0$ for all three processes; when matter becomes more constrained in the system, energy is dispersed into the surroundings, and vice versa.

**11.95** (a) $\Delta H$ is negative, $\Delta S$ is negative; (b) $\Delta H$ is positive, $\Delta S$ is positive; and (c) $\Delta H$ is negative, $\Delta S$ is small

**11.97** Temperature is 37 °C (310 K) and no solutes are present at 1 M. In particular, hydronium ion concentration is around $10^{-7}$ M.

**11.99** 40.6 %; 7.8 g

**11.101** Both processes release energy (consume ATP, generate light), have positive $\Delta S$ (increase in number of molecules) and have negative $\Delta G$ (spontaneous reactions).

**11.103** The HCl molecule is heteronuclear, while $H_2$ and $Cl_2$ are homonuclear. The homonuclear molecules are more symmetrical than the heteronuclear, so they have fewer possible rotational orientations in space. The heteronuclear molecules have more possible orientations and larger entropy.

# Chapter 12

**12.1** (a) pouring the coffee from the urn into the cup; (b) entering the items on the cash register (if the market has a good laser scanner, paying and receiving change may be rate-determining); and (c) preparing for the jump and passing through the door

**12.3** (a) $I_2 \longrightarrow I + I$; (b) $H_2 + I_2 \longrightarrow H_2I_2$; and (c) $H_2 + I_2 \longrightarrow H + HI_2$

**12.5**

**12.7** (a) $I_2 \longrightarrow I + I$
$I + H_2 \longrightarrow HI + H$
$H + I \longrightarrow HI$

(b) $H_2 + I_2 \longrightarrow H_2I_2$
$H_2I_2 \longrightarrow HI + HI$

(c) $H_2 + I_2 \longrightarrow H + HI_2$
$H + HI_2 \longrightarrow HI + HI$

**12.9** (a) Rate $= -\dfrac{\Delta[Cl_2]}{\Delta t}$

(b) $-\dfrac{\Delta[Cl_2]}{\Delta t} = \dfrac{1}{2}\dfrac{\Delta[NOCl]}{\Delta t}$; and (c) 94 M s$^{-1}$

**12.11**

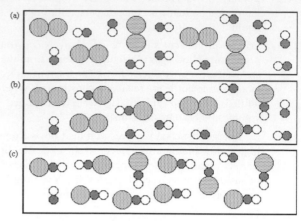

**12.13** (a) $2.2 \times 10^{-3}$ mol/min; and (b) $1.1 \times 10^{-2}$ mol

**12.15**

**12.17** (a) $6.0 \times 10^6$; (b) $2.0 \times 10^{-9}$; and (c) in first-order reactions, the fraction reacting is independent of concentration.

**12.19** (a) Rate $= k[C][AB]$; (b) units $=$ (conc.)$^{-1}$ (time)$^{-1}$; and (c) $C + AB \longrightarrow BC + A$
$A + AB \longrightarrow B + A_2$
$B + C \longrightarrow BC$
Net: $2 C + 2 AB \longrightarrow 2 BC + A_2$

**12.21** (a) Rate $= k[C]^2[AB]$; (b) units $=$ (conc.)$^{-2}$ (time)$^{-1}$; and (c) $2 C + AB \longrightarrow BC + AC$
$AC + AB \longrightarrow BC + A_2$
Net: $2 C + 2 AB \longrightarrow 2 BC + A_2$

**12.23** Slower by a factor of 0.56

**12.25** (a) This reaction is first-order; (b) $k = 6.14 \times 10^{-4}$ s$^{-1}$; (c) 0.936 bar; and (d) $2.62 \times 10^3$ s

**12.27** (a) 2.4 min; and (b) $3.2 \times 10^{-4}$ M

**12.29** (a) 12.0 hr; and (b) 0.0324 M

**12.31** Do two experiments; in each $[H_2O]_0 > 100 [N_2O_5]_0$ but with two different values for $[H_2O]_0$. Plot ln $\left(\dfrac{[N_2O_5]_0}{[N_2O_5]}\right)$ and $\dfrac{1}{[N_2O_5]} - \dfrac{1}{[N_2O_5]_0}$ vs. $t$ to determine order with respect to $N_2O_5$, and use the ratio of slope values and $H_2O$ concentrations to determine order with respect to $H_2O$. After determining the reaction order, rate constant can be determined from the slope of the linear plot (as described in Example 12-7 in the text).

**12.33** Rate $= k[S_2O_8^{2-}][I^-]$; $k = 2.3$ M$^{-1}$ min$^{-1}$

**12.35** Rate $= k_2\left(\dfrac{k_1}{k_{-1}}\right)\dfrac{[C][AB]^2}{[BC]}$

**12.37** (a) $O_3 + O \rightarrow 2 O_2$; (b) Rate $= k_2\dfrac{k_1}{k_{-1}}[O_3][O_2]$; and (c) it is improbable to have two simultaneous fragmentations of $O_5$.

**12.39** (a) Rate $= k_2 \left(\dfrac{k_1}{k_{-1}}\right)[NO]^2[O_2]$; (b) yes; and (c) the intermediate with an N—N bond is the most likely.

**12.41** When $E_a = 0$, $k$ is independent of temperature. This happens when a reaction can occur without first breaking any chemical bonds.

**12.43**

Activated complex:

**12.45**

**12.47** 25 kJ/mol

**12.49**

Bond breakage (wavy lines):     Bond formation (dotted lines):

Hydrogen gas could adsorb on the catalyst's surface as H atoms which will then react readily with CO molecules.

**12.51** (a)

(b) catalyst: Pd metal; intermediate: H atoms on the Pd metal surface; and (c)

H$_2$ adding to Pd metal and dissociating     C taking one H     2nd C taking other H     Ethane molecule

**12.53** (a) $O + NO \longrightarrow NO_2$;

and (b)

**12.55** (a) Rate $\dfrac{\Delta[C_6H_6]}{\Delta t} = -\dfrac{1}{3}\left(\dfrac{\Delta[C_2H_2]}{\Delta t}\right)$

(b) Experiments would have to be carried out measuring the rate as a function of $[C_2H_2]$.

**12.57**

**12.59** (a) nine-fold increase; (b) no change; and (c) increases by 5.2 times

**12.61** First order, with $k = 2.55 \times 10^{-3}\,\text{s}^{-1}$

**12.63** The rate for $b$ will be twice that for $a$, because the concentration of $X_2$ is twice as great.

**12.65** First-order plot is linear, with slope $= k_{obs} = 0.0038\,\text{s}^{-1}$; Rate $= k[N_2O_5][H_2O]$, $k = 0.15\,\text{M}^{-1}\,\text{s}^{-1}$

**12.67** (a) $1.9 \times 10^2\,\text{s}$; (b) $1.3 \times ^3\,\text{s}$; and (c) $1.3 \times 10^4\,\text{s}$

**12.69** (a) false; (b) false; (c) true; and (d) true

**12.71** (a) first order in $N_2O_5$ and first order overall; (b) second order in NO, first order in $H_2$, and third order overall; and (c) first order in enzyme and first order overall.

**12.73** The speed of a chemical reaction refers to how fast it proceeds. The spontaneity of a chemical reaction refers to whether or not the reaction can go in the direction written without outside intervention.

**12.75** $CF_2Cl_2 \xrightarrow{h\nu} CF_2Cl + Cl$

$O_3 \xrightarrow{h\nu} O_2 + O$

$Cl + O_3 \longrightarrow ClO + O_2$

$ClO + O \longrightarrow Cl + O_2$

**12.77** (a) $2\,NO + 2\,H_2 \longrightarrow N_2 + 2\,H_2O$; and

(b) Rate $= k_2 \dfrac{k_1}{k_{-1}}[NO]^2[H_2]$

**12.79** (a) A catalyst binds to reactant species in a way that weakens chemical bonds. (b) The average kinetic energy of the molecules increases, so a larger fraction of molecular collisions leads to reaction. (c) There are more molecules to react and a higher rate of molecular collisions.

**12.81** Rate $= k[A]^2$

**12.83** (a) Rate $= k[H_2][X_2]$; (b) Rate $= k[X_2]$; and
(c) Rate $= k_2\left(\dfrac{k_1}{k_{-1}}\right)^{1/2}[H_2][X_2]^{1/2}$

**12.85** (a) $Cl_2 + CHCl_3 \longrightarrow HCl + CCl_4$; (b) Cl and $CCl_3$;
and (c) Rate $= k_2\left(\dfrac{k_1}{k_{-1}}\right)^{1/2}[CHCl_3][Cl_2]^{1/2}$

**12.87** (a) 51 kJ/mol; (b) 7.0 min

**12.89** (a) Rate $= k[CH_3CHO]^2$; (b) $4.5 \times 10^{-3}\ M^{-1}\ s^{-1}$; and
(c) $2.7 \times 10^3\ s$

**12.91** $1.1 \times 10^{14}$

**12.93** Neither appears in the overall stoichiometry. Catalysts are reactants in early steps but are products of later steps, while intermediates are products of early steps and are consumed in later steps.

**12.95** Statements (d) and (f) are true.

## Chapter 13

**13.1**

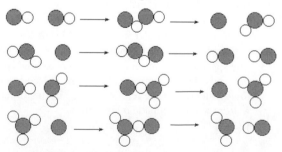

**13.3** (a) $Cl^-(aq) + ClO_2^-(aq) \xrightarrow{k_{-1}} ClO^-(aq) + ClO^-(aq)$
$Cl^-(aq) + ClO_3^-(aq) \xrightarrow{k_{-2}} ClO^-(aq) + ClO_2(aq)$
(b) $K_{eq} = \dfrac{[ClO_3^-]_{eq}[Cl^-]_{eq}^2}{[ClO^-]_{eq}^3}$; and (c) $K_{eq} = \dfrac{k_1 k_2}{k_{-1}k_{-2}}$

**13.5** Your molecular pictures of elementary reactions should show the reactants, the products, and (if applicable) the intermediate collision complex:

**13.7** To test the reversibility of a reaction, set up a system containing the products and observe whether or not reactants form. Here, a solution containing $Cl^-$ and $ClO_3^-$ ions should react to form some $ClO^-$ ions.

**13.9** (a) $K_{eq} = \dfrac{(p_{IF_5})_{eq}^2}{(p_{F_2})_{eq}^5}$; (b) $K_{eq} = \dfrac{1}{(p_{O_2})_{eq}^5}$;
(c) $K_{eq} = (p_{CO})_{eq}^2$; (d) $K_{eq} = \dfrac{1}{(p_{CO})_{eq}(p_{H_2})_{eq}^2}$; and

(e) $K_{eq} = \dfrac{[H_3O^+]_{eq}^3[PO_4^{3-}]_{eq}}{[H_3PO_4]_{eq}}$

**13.11** (a) $2\ IF_5(g) \rightleftharpoons I_2(s) + 5\ F_2(g)$, $K_{eq} = \dfrac{(p_{F_2})_{eq}^5}{(p_{IF_5})_{eq}^2}$;
(b) $P_4O_{10}(s) \rightleftharpoons P_4(s) + 5\ O_2(g)$, $K_{eq} = (p_{O_2})_{eq}^5$;
(c) $BaO(s) + 2\ CO(g) \rightleftharpoons$
$BaCO_3(s) + C(s)$, $K_{eq} = \dfrac{1}{(p_{CO})_{eq}^2}$;
(d) $CH_3OH(l) \rightleftharpoons CO(g) + 2\ H_2(g)$,
$K_{eq} = (p_{CO_2})_{eq}(p_{H_2})_{eq}^2$; and
(e) $PO_4^{3-}(aq) + 3\ H_3O^+(aq) \rightleftharpoons H_3PO_4(aq) + 3\ H_2O(l)$,
$K_{eq} = \dfrac{[H_3PO_4]_{eq}}{[H_3O^+]_{eq}^3[PO_4^{3-}]_{eq}}$

**13.13** (a) $p = 1$ bar for $F_2$ and $IF_5$, $X = 1$ for $I_2$;
(b) $p = 1$ bar for $O_2$, $X = 1$ for $P_4$ and $P_4O_{10}$;
(c) $p = 1$ bar for CO, $X = 1$ for others;
(d) $p = 1$ bar for CO and $H_2$, $X = 1$ for $CH_3OH$;
and (e) $c = 1$ M for $H_3PO_4$, $H_3O^+$, and $PO_4^{3-}$, $X = 1$ for $H_2O$

**13.15** Your views should be like those in Figure 13-4, showing two different-sized samples with equal-sized portions of each highlighted to show that equal volumes contain equal numbers of atoms:

**13.17** $9.9 \times 10^6$

**13.19** (a) $3.3 \times 10^3$; (b) $1 \times 10^{90}$; (c) $3 \times 10^{-60}$; and
(d) $2.9 \times 10^{39}$

**13.21** $2 \times 10^{109}$; $3 \times 10^{-75}$

**13.23** $1.8 \times 10^3$; $9.2 \times 10^{18}$

**13.25** (a), (b), and (d), adding CO causes the reaction to go to the right, forming products; (c), adding CO causes the reaction to proceed to the left, forming reactants.

**13.27** (a) no effect; (b) more solid will dissolve; (c) some solid will precipitate; and (d) no effect

**13.29** This reaction can be driven to the left by removing $SO_2$, removing $Cl_2$, adding $SO_2Cl_2$, or increasing the temperature.

**13.31** 0.92

**13.33** $1.3 \times 10^{-5}$

**13.35** $(p_{H_2})_{eq} = (p_{Br_2})_{eq} = 2.5 \times 10^{-2}$ bar; $(p_{HBr})_{eq} = 10.0$ bar

**13.37** $(p_{CO_2})_{eq} = 1.4$ bar and $(p_{CO})_{eq} = 3.6$ bar

**13.39** (a) $H_2O$ and $CH_3CO_2H$; (b) $H_2O$, $NH_4^+$, and $Cl^-$; (c) $H_2O$, $K^+$, and $Cl^-$; (d) $H_2O$, $Na^+$, and $CH_3CO_2^-$; and (e) $H_2O$, $Na^+$, and $OH^-$

**13.41** (a) $CH_3CO_2H\,(aq)\ H_2O\,(l) \rightleftharpoons CH_3COO^-(aq) + H_3O^+(aq)$; (b) $NH_4^+(aq) + H_2O\,(l) \rightleftharpoons NH_3\,(aq) + H_3O^+(aq)$; (c) no equilibria other than the water equilibrium; (d) $CH_3CO_2^-(aq) + H_2O\,(l) \rightleftharpoons CH_3CO_2H(aq) + OH^-(aq)$; and (e) no equilibria other than the water equilibrium

**13.43** (a) $(CH_3)_2CO$ (acetone) and $H_2O$; (b) $H_2O$, $K^+$, and $Br^-$; (c) $H_2O$, $Li^+$, and $OH^-$; and (d) $H_2O$, $H_3O^+$, and $HSO_4^-$

**13.45** (a) $K_{eq} = \dfrac{[ClO_2^-][H_3O^+]}{HClO_2} = K_a$;

(b) $K_{eq} = \dfrac{1}{[Fe^{3+}][OH^-]^3} = \dfrac{1}{K_{sp}}$; and

(c) $K_{eq} = \dfrac{[HCN]}{[CN^-][H_3O^+]} = \dfrac{1}{K_a}$

**13.47** (a) $Na^+$; (b) $Cl^-$ and $K^+$; and (c) $K^+$ and $NO_3^-$

**13.49**; (a) $K_{eq} = (p_{CO_2})_{eq}\,(p_{H_2O})_{eq}$; (b) $K_{eq} = \dfrac{(p_{NH_3})^4_{eq}\,(p_{O_2})^3_{eq}}{(p_{N_2})^2_{eq}}$;

(c) $K_{eq} = \dfrac{(p_{CH_3CHO})^2_{eq}}{(p_{C_2H_4})^2_{eq}\,(p_{O_2})_{eq}}$;

(d) $K_{eq} = [Ag^+]^2_{eq}\,[SO_4^{2-}]_{eq}$; and

(e) $K_{eq} = (p_{H_2S})_{eq}\,(p_{NH_3})_{eq}$

**13.51** $(p_{CO})_{eq} = 7.2 \times 10^{-2}$ bar; $(p_{COCl_2})_{eq} = 6.2$ bar

**13.53** The molecular picture should show 2 each of the reactants and 10 each of product:

**13.55** Adding more solute to a solution that is saturated increases the total surface area, so the overall rate at which molecules enter the solution becomes greater. However, the rate at which molecules are captured also becomes greater. The rate per unit surface area is unchanged, so the concentration remains the same.

**13.57** 7.34 bar

**13.59** 3.0 at 298 K; $2.1 \times 10^{-4}$ at 525 K

**13.61** (a) $(CH_3)_3N + H_2O \rightleftharpoons (CH_3)_3NH^+ + OH^-$,

$$K_b = \frac{[(CH_3)_3NH^+]_{eq}\,[OH^-]_{eq}}{[(CH_3)_3N]_{eq}}$$

(b) $HF + H_2O \rightleftharpoons F^- + H_3O^+$,

$$K_a = \frac{[F^-]_{eq}\,[H_3O^+]_{eq}}{[HF]_{eq}};$$

(c) $CaSO_4\,(s) \rightleftharpoons Ca^{2+}(aq) + SO_4^{2-}(aq)$,

$$K_{sp} = [Ca^{2+}]_{eq}\,[SO_4^{2+}]_{eq}$$

**13.63** (a) goes to right; (b) goes to left; and (c) no effect

**13.65** $4.7 \times 10^5$; less soluble at higher $T$

**13.67** 0.75

**13.69** (a) $1.6 \times 10^{11}$; (b) 792 K; and (c) $2.2 \times 10^{-2}$

**13.71** (a) $K_{eq} = \dfrac{[CO_3^{2-}]_{eq}}{(p_{CO_2})_{eq}\,[OH^-]^2_{eq}}$;

(b) increases; and (c) increases

**13.73** $1.20 \times 10^2$ bar

**13.75** $4.4 \times 10^{-29}$ bar; $9.4 \times 10^5$ L

**13.77** After determining the reaction order, rate constant can be determined from the slope of the linear plot (as described in Example 12-7 in the text).

(a)

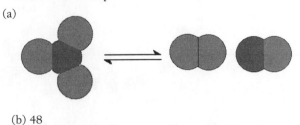

(b) 48

# Chapter 14

**14.1**

**14.3** $[H_3O^+] = 1.25 \times 10^{-3}$ M; $[OH^-] = 8.00 \times 10^{-12}$ M

**14.5** $[H_3O^+] = [Cl^-] = 0.121$ M; $[OH^-] = 8.26 \times 10^{-14}$ M

**14.7** $[H_3O^+] = 4.12 \times 10^{-2}$ M; $[OH^-] = 2.43 \times 10^{-13}$ M

**14.9** (a) $-0.60$; (b) 5.426; (c) 2.32; and (d) 3.593

**14.11** (a) 14.60; (b) 8.574; (c) 11.68; and (d) 10.407

**14.13** (a) 0.22 M; (b) $1.4 \times 10^{-8}$ M; (c) $2.1 \times 10^{-4}$ M; and (d) $4.7 \times 10^{-15}$ M

**14.15** (a) $4.6 \times 10^{-14}$ M; (b) $7.1 \times 10^{-7}$ M; (c) $4.8 \times 10^{-11}$ M; and (d) 2.1 M

**14.17** (a) 9.70; (b) 10.04; and (c) 1.80

**14.19**

**14.21** (a) major species: $HN_3$ and $H_2O$, minor species: $N_3^-$, $H_3O^+$ and $OH^-$; (b) $[H_3O^+] = [N_3^-] = 6.1 \times 10^{-3}$ M and $[HN_3] = 1.50$ M; (c) 2.21; and (d)

**14.23** 0.41 %

**14.25** (a) major species: $N(CH_3)_3$ and $H_2O$, minor species: $HN(CH_3)_3^+$, $OH^-$ and $H_3O^+$; (b) $[OH^-] = [HN(CH_3)_3^+] = 4.8 \times 10^{-3}$ M, $[N(CH_3)_3] = 0.345$ M; (c) 11.68;

and (d)

**14.27** 0.49 %

**14.29** (a) weak base; (b) weak acid; (c) weak acid; and (d) strong base

**14.31**    Propanoic acid    Phosphoric acid    Methylamine
                (acid)            (acid)            (base)

**14.33** $C_5H_5N$ conjugate acid is $C_5H_5NH^+$; $HONH_2$ conjugate acid is $HONH_3^+$; $HCO_2H$ conjugate base is $HCO_2^-$.

**14.35** (a)

Base            Acid
$NH_3 + H_2O \rightleftharpoons NH_4^+ + OH^-$
Acid            Base

(b)

Base            Acid
$H_2O + HCNO \rightleftharpoons H_3O^+ + CNO^-$
Acid            Base

(c)

Base            Acid
$H_2O + HClO \rightleftharpoons H_3O^+ + ClO^-$
Acid            Base

(d)

Base            Acid
$OH^- + H_3O^+ \rightleftharpoons H_2O + H_2O$
Acid            Base

**14.37** (a) major species are $Na^+$, $SO_3^{2-}$, and $H_2O$; (b) $H_2O + SO_3^{2-} \rightleftharpoons HSO_3^- + OH^-$; and (c) 10.43

**14.39** (a) major species are $NH_4^+$, $NO_3^-$, and $H_2O$; (b) $H_2O + NH_4^+ \rightleftharpoons NH_3 + H_3O^+$; and (c) 5.63

**14.41** (a) basic, $H_2O + HS^- \rightleftharpoons OH^- + H_2S$; (b) basic, $H_2O + OI^-(aq) \rightleftharpoons OH^- + HOI(aq)$; (c) neutral, $H_2O(l) + H_2O(l) \rightleftharpoons OH^-(aq) + H_3O^+(aq)$; and (d) acidic, $H_2O(l) + HC_5H_5N^+(aq) \rightleftharpoons H_3O^+(aq) + C_5H_5N(aq)$

**14.43** $NaI < NaF < NaC_6H_5CO_2 < Na_3PO_4 < NaOH$

**14.45** (a) $H_2SO_4$ (anions are poorer proton donors than neutrals); (b) $HClO$ (Cl is more electronegative); and (c) $HClO_2$ (more O atoms)

**14.47** (b)

**14.49**

Charge in the phenolate anion is distributed around the ring, stabilizing the anion.

**14.51** $[Na^+] = [C_2H_3O_2^-] = 0.250$ M; $[OH^-] = 1.2 \times 10^{-5}$ M; $[H_3O^+] = 8.3 \times 10^{-10}$ M

**14.53** $[H_3O^+] = [HCO_3^-] = 8.4 \times 10^{-5}$ M; $[CO_3^{2-}] = 4.7 \times 10^{-11}$ M; $[OH^-] = 1.2 \times 10^{-10}$ M

**14.55** $[Na^+] = 0.11$ M; $[OH^-] = [HCO_3^-] = 3.2 \times 10^{-3}$ M; $[CO_3^{2-}] = 0.052$ M; $[H_2CO_3] = 2.2 \times 10^{-8}$ M; $[H_3O^+] = 3.1 \times 10^{-12}$ M

**14.57** (a)

(b) 5.28

**14.59** 10.81

**14.61** $9.1 \times 10^{-11}$

**14.63** 0.010 %

**14.65**

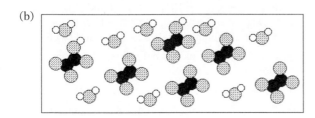

**14.67**  8.30

**14.69**  $[H_3O^+] = 2.01$ M; $[SO_4^{2-}] = 1.0 \times 10^{-2}$ M; $[HSO_4^-] = 1.99$ M

**14.71**  (a) major species: $H_2O$, $HSO_4^-$, and $H_3O^+$;
$H_2O + HSO_4^- \rightleftharpoons SO_4^{2-} + H_3O^+$

(b) major species: $H_2O$, $SO_4^{2-}$, and $Na^+$;
$SO_4^{2-} + H_2O \rightleftharpoons HSO_4^- + OH^-$

(c) major species: $H_2O$, $HSO_4^-$, and $Na^+$;
$H_2O + HSO_4^- \rightleftharpoons SO_4^{2-} + H_3O^+$

(d) major species: $H_2O$, $Cl^-$, and $NH_4^+$;
$H_2O + NH_4^+ \rightleftharpoons NH_3 + H_3O^+$

**14.73**  $K_{eq} = 48$

**14.75**

(a)

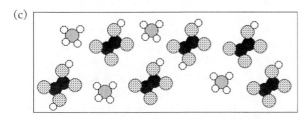

(b)

(c)

**14.77**  (a) $H_2O$ and $H_3PO_4$; (b) $H_3O^+$, $H_2PO_4^-$, $HPO_4^{2-}$, $OH^-$, and $PO_4^{3-}$; and (c) pH = 1.92

**14.79**

**14.81**  (a) $OH^- + C_6H_5CO_2H \rightleftharpoons H_2O + C_6H_5CO_2^-$;
(b) $H_3O^+ + (CH_3)_3N \rightleftharpoons H_2O + (CH_3)_3NH^+$;
(c) $SO_4^{2-} + CH_3CO_2H \rightleftharpoons HSO_4^- + CH_3CO_2^-$;
(d) $OH^- + NH_4^+ \rightleftharpoons H_2O + NH_3$;
(e) $HPO_4^{2-} + NH_3 \rightleftharpoons PO_4^{3-} + NH_4^+$

# Chapter 15

**15.1**  (a) no buffer properties; (b) buffer, 9.25; (c) no buffer properties; and (d) buffer, 9.25

**15.3**  8.88

**15.5**  $1.4 \times 10^{-3}$ mol

**15.7**  $H_3O^+(aq) + CO_3^{2-}(aq) \rightleftharpoons H_2O(l) + HCO_3^-(aq)$;
$OH^-(aq) + HCO_3^-(aq) \rightleftharpoons H_2O(l) + CO_3^{2-}(aq)$

**15.9**  (a) buffer containing $HCO_3^-$ and $CO_3^{2-}$; (b), (c) no buffer; and (d) buffer containing $HCO_3^-$ and $CO_3^{2-}$

**15.11**  For pH = 3.50, the $HCO_2H/HCO_2^-$ system should be used. Sodium formate could be used along with HCl solution. For pH = 12.60, the $HPO_4^{2-}/PO_4^{3-}$ system should be used. Potassium phosphate could be used along with HCl solution.

**15.13**  37 g

**15.15**  55 mL

**15.17**  55.7 g sodium carbonate, 0.789 L HCl solution

**15.19**  1.5 g, still a buffer

**15.21**  (a) $NH_4^+(aq) + H_2O(l) \rightleftharpoons NH_3(aq) + H_3O^+(aq)$, pH < 7; (b) $H_2O(l) + H_2O(l) \rightleftharpoons OH^-(aq) + H_3O^+(aq)$; pH = 7; and (c) $CH_3CO_2H(aq) + H_2O(l) \rightleftharpoons CH_3CO_2^-(aq) + H_3O^+(aq)$, pH < 7

**15.23**  (a) 2.8; (b) 7.8; and (c) 3.5

**15.25**  Phenol red

**15.27**  8.80; thymol blue

**15.29**  (a) 4.05; (b) 7.21; (c) 9.76; (d) 12.32; and (e) 12.62

**15.31**  (a) $AgCl(s) \rightleftharpoons Ag^+(aq) + Cl^-(aq)$,
$K_{sp} = [Ag^+]_{eq} [Cl^-]_{eq}$;
(b) $BaSO_4(s) \rightleftharpoons Ba^{2+}(aq) + SO_4^{2-}(aq)$,
$K_{sp} = [Ba^{2+}]_{eq} [SO_4^{2-}]_{eq}$;
(c) $Fe(OH)_2(s) \rightleftharpoons Fe^{2+}(aq) + 2\ OH^-(aq)$,
$K_{sp} = [Fe^{2+}]_{eq} [OH^-]_{eq}^2$; and
(d) $Ca_3(PO_4)_2(s) \rightleftharpoons 3\ Ca^{2+}(aq) + 2\ PO_4^{3-}(aq)$,
$K_{sp} = [Ca^{2+}]_{eq}^3 [PO_4^{3-}]_{eq}^2$

**15.33**  (a) $9.1 \times 10^{-4}$ g; (b) $1.2 \times 10^{-3}$ g; (c) $9.8 \times 10^{-5}$ g; and (d) $1.6 \times 10^{-5}$ g

**15.35**  $2.3 \times 10^{-9}$

**15.37**  0.53 g

**15.39**  $1.6 \times 10^3$

**15.41**  0.125 M

**15.43**  (a) $[Fe(CN)_6]^{3-}$; (b) $[Zn(NH_3)_4]^{2+}$; and (c) $[V(en)_3]^{3+}$

**15.45** $[Zn^{2+}] = 4.9 \times 10^{-9}$ M;
$[Zn(NH_3)_4{}^{2+}] = 5.4 \times 10^{-3}$ M; $[NH_3] = 0.228$ M

**15.47** $[Cl^-] = 0.14$ M; $[Pb^{2+}] = 8.5 \times 10^{-15}$ M;
$[Na^+] = 0.17$ M; $[PbCl_4{}^{2-}] = 0.0075$ M

**15.49** $Mn^{2+} + en \rightleftharpoons [Mn(en)]^{2+}$
$[Mn(en)]^{2+} + en \rightleftharpoons [Mn(en)_2]^{2+}$
$[Mn(en)_2]^{2+} + en \rightleftharpoons [Mn(en)_3]^{2+}$

**15.51** 7.5 g

**15.53** 2.0;

**15.55** All the salt dissolves.

**15.57** (a) 6.81; (b) 0.06; and (c) $7.5 \times 10^{-3}$ mol

**15.59** (a) yellow; (b) purple; (c) purple; and (d) purple

**15.61** $4.4 \times 10^{-3}$

**15.63** Precipitate forms.

**15.65** $3.2 \times 10^{-8}$

**15.67** (a) 9.43; (b) 0.050 mol; and (c) 9.32

**15.69** $2.0 \times 10^{-33}$

**15.71** (a) 8.52; (b) 3.74; (c) 2.23; and (d) thymol blue

**15.73** 2.16 g; $[Mn^{2+}] = 3.5 \times 10^{-10}$ M

**15.75** $2.0 \times 10^{-4}$ g

**15.77** Mix together 15 g sodium acetate, 0.17 L of 1.0 M acetic acid, and enough water to make 1.0 L.

**15.79** $[C_2O_4{}^{2-}] = 0.237$ M; $[Zn^{2+}] = 2.4 \times 10^{-9}$ M;
$[[Zn(C_2O_4)_3]^{4-}] = 4.5 \times 10^{-3}$ M

**15.81** (a) $MgF_2(s) \rightleftharpoons Mg^{2+}(aq) + 2F^-(aq)$,
$K_{sp} = [Mg^{2+}]_{eq} [F^-]_{eq}{}^2$; (b) $5.93 \times 10^{-9}$;
and (c) $2.1 \times 10^{-10}$

**15.83** $2.6 \times 10^2$ L

**15.85** (a) $6.2 \times 10^{-15}$ M; and (b) $1.0 \times 10^3$ g

**15.87** At 25 °C, $K_{sp} = 1.29 \times 10^{-9}$, $\Delta G° = 50.7$ kJ/mol;
at 100 °C, $K_{sp} = 2.5 \times 10^{-5}$, $\Delta G° = 33$ kJ/mol

**15.89** $1.1 \times 10^{-4}$ M

**15.91** More soluble in acidic solution: $Ag_2CO_3$, $Ag_2SO_4$, and $Ag_2S$; independent of pH: AgBr and AgCl

**15.93** (a) 7.79; and (b) 7.65

**15.95** Point A: Major species are $H_2O$ and $HCO_2H$, dominant equilibrium is
$$H_2O + HCO_2H(aq) \rightleftharpoons HCO_2{}^-(aq) + H_3O^+(aq),$$
$$K = \frac{[HCO_2{}^-][H_3O^+]}{[HCO_2H]};$$

Point B: major species are $H_2O$, $HCO_2H$, and $HCO_2{}^-$, dominant equilibrium is
$$H_2O + HCO_2H(aq) \rightleftharpoons HCO_2{}^-(aq) + H_3O^+(aq),$$
$$K = \frac{[HCO_2{}^-][H_3O^+]}{[HCO_2H]};$$

Point C: major species are $H_2O$ and $HCO_2{}^-$, dominant equilibrium is
$$H_2O + HCO_2{}^-(aq) \rightleftharpoons HCO_2H(aq) + H_3O^-(aq),$$
$$K = \frac{[HCO_2H][OH^-]}{[HCO_2{}^-]};$$

Point D: major species are $H_2O$, $OH^-$, and $HCO_2{}^-$, dominant equilibrium is
$$H_2O + HCO_2{}^-(aq) \rightleftharpoons HCO_2H(aq) + OH^-(aq),$$
$$K = \frac{[HCO_2H][OH^-]}{[HCO_2{}^-]}.$$

The cation of the strong base will also be present as a major species at points B, C, and D.

**15.97** (a) 4.27;
(b)

and (c)

**15.99** $K_{eq} = 5.5 \times 10^6$; 2.54 g

# Chapter 16

**16.1** (a) O is $-2$, H is $+1$, and Fe is $+3$; (b) F is $-1$, N is $+3$; (c) O is $-2$, H is $+1$, C is $-2$; (d) $K^+$ is $+1$, O is $-2$, and C is $+4$; (e) in $NH_4{}^+$, H is $+1$ and N is $-3$; in $NO_3{}^-$, O is $-2$ and N is $+5$; (f) Cl is $-1$, Ti is $+4$; (g) Pb is $+2$; in sulfate, O is $-2$, S is $+6$; and (h) P is 0

**16.3** (a) not redox; (b) redox; (c) not redox; (d) redox; and (e) redox

**16.5** (a) $+5$; (b) $-1$; (c) $+7$; (d) 0; (e) $+3$; and (f) $+1$

**16.7** (a) $Na \rightarrow Na^+ + e^-$ and $2 H_2O + 2 e^- \rightarrow H_2 + 2 OH^-$;
(b) $Au + 4 Cl^- \rightarrow [AuCl_4]^- + 3 e^-$ and $NO_3{}^- + 4 H_3O^+ + 3 e^- \rightarrow NO + 6 H_2O$; (c) $MnO_4{}^- + 8 H_3O^+ + 5 e^- \rightarrow Mn^{2+} + 12 H_2O$ and $C_2O_4{}^{2-} \rightarrow 2 CO_2 + 2 e^-$; and (d) $C + 3 H_2O \rightarrow CO + 2 H_3O^+ + 2e^-$ and $2 H_3O^+ + 2e^- \rightarrow H_2 + 2 H_2O$

**16.9** (a) $Cu^+ + 3 H_2O \rightarrow CuO + 2 H_3O^+ + e^-$;
(b) $S + 2 H_3O^+ + 2 e^- \rightarrow H_2S + 2 H_2O$;
(c) $Ag_2O + H_2O + 2 e^- \rightarrow 2 Ag + 2 OH^-$;
(d) $I^- + 6 OH^- \rightarrow IO_3{}^- + 3 H_2O + 6 e^-$;
(e) $IO_3{}^- + 2 H_2O + 4 e^- \rightarrow IO^- + 4 OH^-$; and
(f) $H_2CO + 5 H_2O \rightarrow CO_2 + 4 H_3O^+ + 4 e^-$

**16.11** (a) $2 Cu^+ + 4 H_2O + S \rightarrow 2 CuO + 2 H_3O^+ + H_2S$;
(b) $3 Ag_2O + I^- \rightarrow 6 Ag + IO_3{}^-$;
(c) $2 I^- + IO_3{}^- \rightarrow 3 IO^-$; and
(d) $2 S + H_2CO + H_2O \rightarrow 2 H_2S + CO_2$

**16.13** (a) $Cl^-$; (b) $MnO_4{}^-$; (c) $MnO_4{}^-$; (d) $Cl^-$; (e) $MnO_4{}^-$; and (f) $Cl^-$

**16.15** (a) $3\,CN^- + 2\,MnO_4^- + H_2O \rightarrow 3\,CNO^- + 2\,MnO_2 + 2\,OH^-$; (b) $4\,As + 3\,O_2 + 2\,H_2O \rightarrow 4\,HAsO_2$; (c) $Br^- + 2\,MnO_4^- + H_2O \rightarrow BrO_3^- + 2\,MnO_2 + 2\,OH^-$; (d) $3\,NO_2 + 3\,H_2O \rightarrow 2\,NO_3^- + NO + 2\,H_3O^+$; (e) $ClO_4^- + 6\,Cl^- + 6\,H_3O^+ \rightarrow ClO^- + 9\,H_2O + 3\,Cl_2$; and (f) $AlH_4^- + 4\,H_2CO + 4\,H_2O \rightarrow Al^{3+} + 4\,CH_3OH + 4\,OH^-$

**16.17** (a), (b), (c) and (d) spontaneous

**16.19**

**16.21**

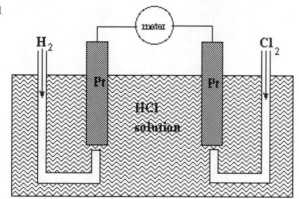

**16.23** The silver–silver chloride electrode (16.19) is active, whereas the platinum electrodes (16.21) are passive.

**16.25;** The shorthand notation is always written as anode to cathode (left to right).
(a) $Ag\,(s)|Ag^+\,(aq)\|Fe^{2+}\,(aq)|Fe\,(s)$
(b) $Pt\,(s)|H_2O_2\,(aq),\,H_3O^+\,(aq)|O_2\,(g)\|NO_3^-\,(aq),\,H_3O^+\,(aq)|NO\,(g)|Pt\,(s)$
(c) $Pt\,(s)|H_2\,(g)|H_3O^+\,(aq)\|Cl_2\,(g)|Cl^-\,(aq)|Pt\,(s)$

**16.27** (a) 1.565 V; (b) 0.981 V; and (e) 0.002 V

**16.29** $8\,NO + 3\,H_2O \rightarrow 3\,N_2O + 2\,NO_3^- + 2\,H_3O^+$; $E°_{cell} = 0.634$ V

**16.31** Set up a standard hydrogen electrode on one side and a Pt electrode immersed in 1 M solution of NaF with $F_2$ bubbling over the electrode. The $H_2/H_3O^+$ electrode will be the anode.

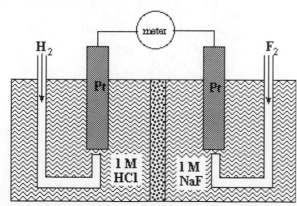

**16.33** Ba, Ca, Cs, Li, Mg, K, and Na. All these metals lie in the s block of the periodic table and easily lose 1 or 2 electrons.

**16.35** (a) −906.0 kJ; (b) $-1.14 \times 10^3$ kJ; and (e) −1 kJ

**16.37** $E = 1.35$ V

**16.39** $m_{Pb} = 9.5 \times 10^{-2}$ g; $m_{PbO_2} = 0.11$ g

**16.41** $1.35 \times 10^3$ s

**16.43** 1.98 g

**16.45** 0.039 V

**16.47** (a) 0.137 V; (b) 0.0186 V; (c) −0.159 V

**16.49** $Q = \dfrac{1}{[H_3O]^2[HSO_4^-]^2}$; as the battery operates, these concentrations decrease, $Q$ and $\ln Q$ increase, and the potential of the battery decreases with use.

**16.51** Zn oxidizes more easily and will preferentially corrode, protecting the steel propeller.

**16.53** Lead storage batteries are heavy. For space flights, weight is a more important consideration than rechargeability, so lead batteries are not suitable.

**16.55** 19.9 g

**16.57** The oxidation reaction is $Cu \rightarrow Cu^{2+} + 2\,e^-$. The reduction of $H_3O^+$ occurs: $2\,H_3O^+ + 2\,e^- \rightarrow 2\,H_2O + H_2$ $(E° = 0$ V)

**16.59** (a) Attach the negative wire to the Cd electrode; $Cd\,(s) + 2\,OH^-\,(aq) \rightleftharpoons Cd(OH)_2\,(s) + 2\,e^-$; (b) 4.56 hr

**16.61** $1.89 \times 10^3$ C; 2.10 A

**16.63**

**16.65** $K_{eq} = 1.0 \times 10^{-14}$

**16.67** $2.5 \times 10^9$

**16.69** (a) $3\,H_2 + 2\,Cr(OH)_3 \rightarrow 6\,H_2O + 2\,Cr$; (b) −0.65 V; and (c) $3.8 \times 10^2$ kJ

**16.71** (a) $Cr_2O_7^{2-} + 14\,H_3O^+ + 6\,e^- \rightarrow 2\,Cr^{3+} + 21\,H_2O$; and (b) 16.2 g

**16.73** (a) on the left, $Ni^{2+} + 2e^- \rightarrow Ni$, $E° = -0.257$ V; on the right, $Fe^{2+} + 2e^- \rightarrow Fe$, $E° = -0.447$ V; (b) $E° = 0.190$ V; (c) Fe electrode is the anode and Ni is the cathode; and (d)

**16.75** $[Ni^{2+}] = 3.8 \times 10^{-7}$ M gives $E = 0.0$ V.

**16.77** (a) $2 MnO_4^- + 5 H_2SO_3 + H_3O^+ \rightarrow 2 Mn^{2+} +$
$5 HSO_4^- + 4 H_2O$; (b) $2 MnO_4^- + 5 SO_2 + H_2O +$
$H_3O^+ \rightarrow 2 Mn^{2+} + 5 HSO_4^-$; (c) $8 MnO_4^- + 5 H_2S +$
$19 H_3O^+ \rightarrow 8 Mn^{2+} + 5 HSO_4^- + 31 H_2O$; and
(d) $8 MnO_4^- + 5 H_2S_2O_3 + 14 H_3O^+ \rightarrow 8 Mn^{2+} +$
$10 HSO_4^- + 21 H_2O$

**16.79** 80.4 min

**16.81**

**16.83** (a) $MnO_4^-$; (b) $O_2$; and (c) $Sn^{2+}$

**16.85** $O_2$ can oxidize Cu, Ag, $Fe^{2+}$, $H_2$, and $I^-$.

**16.87** (a) $Zn(NO_3)_2$ soln (cathode): $Zn^{2+} + 2 e^- \rightarrow Zn$; $ZnCl_2$
soln (anode): $Zn \rightarrow Zn^{2+} + 2 e^-$; (c) 0.0207 V; and

(b)

**16.89** $2.3 \times 10^2$ hr

**16.91** $5.6 \times 10^{-8}$ M

**16.93** (a) $Na^+ + e^- \rightarrow Na$ and $2 Cl^- \rightarrow Cl_2 + 2 e^-$; (b) cathode
is the Pt electrode connected to the negative pole, and
anode is the Pt electrode connected to the positive pole of
the battery;

(c)

**16.95**

**16.97** 0.368 V

## Chapter 17

**17.1**

(a) Amine

(b) Alcohol

(c) Ketone

(d) Alkene — Carboxylic acid

**17.3**

(a) Amine — Phenyl

(b) Carboxylic acid — Carboxylic acid

(c) Alkene — Thiol      (d) OH Alcohol

**17.5**

(a) Amide

(b) Ester

**(c)** Ether

**(d)** Ether

**17.7**

**(a)**

Other isomers are also possible.

**(b)**

Other isomers are also possible.

**(c)**

Others are also possible.

**(d)**

Others are also possible.

**17.9**

**(a)**

**(b)**

**(c)**

**17.11**

**17.13**

**17.15** (a) (b)

**17.17**

Polybutadiene contains double bonds, whereas polyethylene does not.

**17.19**

**17.21**

**17.23** Automobile tire

**17.25** (a) elastomers; (b) fibres; and (c) plastics

**17.27** Dioctylphthalate reduces the amount of cross-linking and adds a fluid component to the polymer, so the polymer becomes more flexible.

**17.29**

$\alpha$-talose

**17.31**

β-talose

**17.33** Polymers of α-glucose (e.g., glycogen) coil on themselves; polymers of β-glucose (e.g., cellulose) form planar sheets.

**17.35** T-T-A-C-G-T-G-A-C

**17.37**

**17.39** The complementary sequence is T-A-G.

**17.41**

(a)

Tyr

(b)

Phe

(c)

Glu

(d)

Met

**17.43** Hydrophilic side chains: Tyr and Glu; hydrophobic side chains: Met and Phe

**17.45** (a) Glycine, hydrophobic side chain; (b) serine, hydrophilic; and (c) cysteine, hydrophilic

**17.47** Six combinations: Met-Ala, Ala-Met, Ala-Glu, Glu-Ala, Met-Glu, and Glu-Met. Only Ala-Glu is shown here:

**17.49** $2.09 \times 10^4$ g/mol

**17.51** 16 777 216

**17.53** (a) random;

(b)

and (c) the backbone of this polymer is hydrophobic, but its side chains contain highly polar $C{=}O$ groups that interact with one another and with polar and H-bonding groups on hair.

**17.55**  C - G

**17.57**

Initiation

Propagation

Termination:

**17.59**

**17.61**  Both contain amide linkage groups. Nylons contain one or at most two different monomers, each of which typically contains several carbon atoms that form part of the backbone of the polymer. Proteins contain backbones that are absolutely regular repetitions of amide—C—amide bonding, but the carbon atoms in the backbone have a variety of substituent groups attached to them.

**17.63**  (a) terephthalic acid and phenylenediamine; (b) PET is made from ethylene glycol and terephthalic acid; and (c) styrene

**17.65**  High-density polyethylene is all straight chains that nest together readily; low-density polyethylene has many side chains that cannot nest easily, so low-density polyethylene has more open space.

**17.67**  Complementary base pairing requires that the molar ratios of A to T and G to C be 1.0. Chargoff's observations indicate that this relationship holds even though DNA from different sources has different sequences, some AT-rich and others GC-rich.

**17.69**  1 = cytosine; 2 = uracil; 3 = adenine; 4 = guanine

**17.71**  Hydrophilic amino acids are most stable when in contact with aqueous solvent, while hydrophobic amino acids are most stable when in contact with the protein backbone. The tertiary structure of a protein is the manner in which the individual amino acids orient themselves, so solvent interactions are the primary determinants of tertiary structure.

**17.73**

**17.75**

**17.77**

**17.79** Aspartic acid and phenylalanine.

**17.81**

**17.83** $C_3H_5NO$, $M = 71$ g/mol; 17 repeat units of alanine

# Chapter 18

**18.1**  (a) Mn is +2; (b) Mo is +5; (c) V is +5; (d) Au is +3; and (e) Fe is +3

**18.3**  (a) chromium, Cr; (b) cadmium, Cd; and (c) copper, Cu

**18.5**  (a) $3d^5$; (b) $5d^6$; (c) $3d^8$; and (d) $4d^4$

**18.7**  (a) Pd; (b) Au; and (c) Co

**18.9**  (a) Ru has oxidation state +2, $d^6$; (b) Cr has oxidation state +3, $d^3$; (c) Pd has oxidation state +2, $d^8$; (d) Ir has oxidation state +3, $d^6$; and (e) Ni has oxidation state 0, $s^2d^8$

**18.11**  (a) hexaammineruthenium(II) chloride; (b) *trans*-(ethylenediamine)diiodochromium(III) iodide; (c) *cis*-dichlorobis(trimethylphosphine)palladium(II); (d) *fac*-triamminetrichloroiridium(III); and (e) tetracarbonylnickel(0)

**18.13**

(a) Six $NH_3$ in an octahedron around the central Ru:

(b) Two I at opposite ends of one axis, two en in a square plane around the central Cr:

(c) Square planar arrangement about Pd, with two Cl adjacent each other:

(d) Octahedral arrangement around Ir, with three Cl in a triangular face:

(e) Tetrahedral arrangement of $C{\equiv}O$ about a central Ni:

**18.15** (a) cis-$[Co(NH_3)_4ClNO_2]^+$; (b) $[PtNH_3Cl_3]^-$; (c) trans-$[Cu(en)_2(H_2O)_2]^{2+}$; and (d) $[FeCl_4]^-$

**18.17** (a)

(b)

(c)

(d)

**18.19** (a) and (b)

(c) and (d)

**18.21** (a) diamagnetic; (b) paramagnetic with four unpaired electrons; (c) paramagnetic with two unpaired electrons; and (d) diamagnetic

**18.23** The colours of transition-metal complexes are generally determined by $d-d$ transitions, but $Zr^{4+}$ has no valence electron that can undergo a transition that would absorb visible light.

**18.25** (a) 6; (b) +3, $3d^5$; (c) octahedral; (d) paramagnetic; and (e) one

**18.27**

Small splitting ($H_2O$)      Large splitting ($CN^-$)

**18.29** 257 kJ/mol, orange

**18.31** Haemoglobin has four subunits, whereas myoglobin has just one. As a consequence, haemoglobin has a more complex cooperative chemical behavior than myoglobin.

**18.33** Remove 1 electron

**18.35** (a) $CuFeS_2 + 2 O_2 \rightarrow CuO + FeO + 2 SO_2$; (b) $Si + O_2 + CaO \rightarrow CaSiO_3$; and (c) $TiCl_4 + 4 Na \rightarrow 4 NaCl + Ti$

**18.37** $1.06 \times 10^6$ g; $2.05 \times 10^5$ L

**18.39** 183.3 kJ; 63.3 kJ

**18.41** The coinage metals are copper, silver, and gold. They are used for money, electrical wire, and decorative objects such as jewelry.

**18.43** Titanium is used as an engineering metal because of its relatively low density, high bond strength, resistance to corrosion, and ability to withstand high temperatures, all of which make it a favored structural material.

**18.45**

(a)

(b)

(c)

(d)

**18.47** Bidentate ligands form complexes with two links. Each Fe ion forms an octahedral complex with three oxalate anions:

**18.49** (a) Cr is $[Ar]4s^13d^5$; $Cr^{2+}$ is $[Ar]3d^4$; $Cr^{3+}$ is $[Ar]3d^3$; (b) V is $[Ar]4s^23d^3$; $V^{2+}$ is $[Ar]3d^3$; $V^{3+}$ is $[Ar]3d^2$; $V^{4+}$ is $[Ar]3d^1$; $V^{5+}$ is $[Ar]$; and (c) Ti is $[Ar]4s^2 3d^2$; $Ti^{2+}$ is $[Ar]3d^2$; $Ti^{4+}$ is $[Ar]$

**18.51** (a) *cis*-tetraaquadichlorochromium(III) chloride; (b) bromopentacarbonylmanganese(I); and (c) *cis*-di-amminedichloroplatinum(II)

**18.53** $2 O_2^- + 2 H_3O^+ \xrightarrow{SOD} O_2 + H_2O_2 + 2 H_2O$

**18.55** Cu(II) in water forms an aqua complex. Addition of fluoride produces the insoluble green salt $CuF_2$, while addition of chloride produces the tetrachlorocopper(II) complex $[CuCl_4]^{2-}$:

$Cu^{2+}(aq) + 2 F^-(aq) \longrightarrow CuF_2(s)$

$Cu^{2+}(aq) + 4 Cl^-(aq) \longrightarrow [CuCl_4]^{2-}(aq)$

**18.57** Silver is difficult to oxidize, so it has good resistance to corrosion and is suitable for jewelry. Vanadium is readily oxidized, so it corrodes rapidly and is unsuitable for jewelry.

**18.59**

**18.61** Silver sulphide $4 Ag (s) + 2 H_2S (g) + O_2 (g) \rightarrow 2 Ag_2S (s) + 2 H_2O (l)$

**18.63** $Cr^{3+}$ is $d^3$, each electron in a different orbital regardless of the splitting energy

**18.65**

$d_{z^2}$         $d_{x^2-y^2}$

**18.67** (a) 2+, 3+, 3+; (b) red-violet, orange, orange; and (c) 232 kJ/mol, 249 kJ/mol, 257 kJ/mol

**18.69** $[Ni(CO)_4]$ has tetrahedral geometry, colourless; $[Zn(CN)_4]^{2-}$ has tetrahedral geometry, colourless

**18.71** Brass is an alloy of zinc and copper, and superoxide dismutase contains zinc and copper ions in its reaction center.

**18.73** $[Ni(CN)_4]^{2-}$ is square planar and $[NiCl_4]^{2-}$ is tetrahedral.

**18.75** When ferritin is neither empty nor filled to capacity, it can provide iron as needed for haemoglobin synthesis, or store iron if an excess is absorbed by the body.

**18.77** Because $Zn^{2+}$ has $d^{10}$ configuration, its $d$ orbitals are completely filled and the lowest unoccupied orbital is quite high in energy. $Co^{2+}$ has $d^7$ configuration, giving this cation unfilled $d$ orbitals. Consequently, metalloproteins that contain $Co^{2+}$ absorb visible light, making it possible to study them with visible spectroscopy.

**18.79** Zn displaces any metal with a less negative standard potential from the solution. The following are two examples of metals that can be displaced by Zn:

$Co^{2+} + 2e^- \rightleftharpoons Co$     $E° = -0.28$ V

$Fe^{2+} + 2e^- \rightleftharpoons Fe$     $E° = -0.447$ V

## Chapter 19

**19.1** (a) Lewis acid is Ni, Lewis base is CO; (b) Lewis acid is $SbCl_3$, Lewis base is $Cl^-$; (c) Lewis acid is $AlBr_3$, Lewis base is $(CH_3)_3P$; and (d) Lewis acid is $BF_3$, Lewis base is $ClF_3$.

**19.3** (a) acceptor: $3p$ orbital (b) acceptor: $3d$ orbital

(c) acceptor: delocalized $\pi$ orbital

**19.5** In the first step, an electron pair from the O atom of $H_2O$ displaces a $\pi$ bond in Lewis acid-base adduct formation. Then a proton from $H_2O$ migrates to a C—O oxygen atom:

Adduct formation      Proton transfer

**19.7**   $Fe^{3+}$ (high charge) < $V^{3+}$ (lowest $Z$, high charge) < $Fe^{2+}$ < $Pb^{2+}$ (large $n$)

**19.9**   (a) hardest is $BF_3$, then $BCl_3$, and $AlCl_3$ is softest; (b) the hardest acid is $Al^{3+}$ (row 3), then $Tl^{3+}$ (row 6), and $Tl^+$ (low charge) is softest; and (c) polarizability increases with $n$, so hardest is $AlCl_3$, then $AlBr_3$, and $AlI_3$ is softest.

**19.11**   When an electronegative O atom bonds to a less negative S atom, it withdraws electron density from S, decreasing the polarizability about S and increasing the hardness of the base.

**19.13**   (a) metathesis occurs, giving $AlCl_3$ and $NaI$; (b) no reaction occurs; (c) metathesis occurs, giving $H_2O$ and $CaS$; (d) metathesis occurs, giving $(CH_3)_3P$ and $LiCl$; and (e) no reaction occurs.

**19.15**   $Al_2Cl_6$ contains two "bridging" chlorine atoms. Standard procedures would predict tetrahedral geometry about all inner atoms, but the Al—Cl—Al bond angles of 91° indicate that Cl uses $p$ orbitals. Each Al atom can be described as using $sp^3$ hybrids to form four $\sigma$ bonds to four different Cl atoms. In Lewis acid–base terms, the bridged molecule forms from two $AlCl_3$ units linking together in double-adduct formation between the Al Lewis-acid atoms and two Cl Lewis-base atoms.

**19.17**   Thallium lies below indium and gallium, so its properties should be similar to those metals: valence of 3, soft Lewis acid. Like its neighbour Pb, it is toxic.

**19.19**   $SnCl_4$ can function as a Lewis acid because Sn has empty $d$ orbitals from which it can accept electrons to form more bonds:

**19.21**   Lewis structure:

The B and N atoms have bonding and geometry that can be described using $sp^2$ hybrid orbitals, resulting in three $\sigma$ bonds around each ring atom. In addition, there is a delocalized $\pi$ bonding network encompassing all six ring atoms and containing six electrons.

**19.23**   Si has a larger band gap.

**19.25**   $2\ C_2H_5Cl + Si(Cu) \longrightarrow (C_2H_5)_2SiCl_2 + Cu$
$(C_2H_5)_2SiCl_2 + 2\ H_2O \longrightarrow (C_2H_5)_2Si(OH)_2 + 2\ HCl$

Condensation will eliminate water to give the polymer:

**19.27**   $P_4O_{13}^{6-}$:

**19.29**   Brønsted acid–base reaction:
$$Ca_5(PO_4)_3F\,(s) + 5\ H_2SO_4\,(aq) \longrightarrow$$
$$3\ H_3PO_4\,(aq) + 5\ CaSO_4\,(s) + HF\,(aq)$$

**19.31**

**19.33**   There are two industrial reactions in Section 19.6 in which sulphuric acid acts as a Brønsted acid:
$$CaF_2\,(s) + H_2SO_4\,(l) \longrightarrow 2\ HF\,(g) + CaSO_4\,(s)$$
$$FeTiO_3\,(s) + 2\ H_2SO_4\,(aq) + 5\ H_2O\,(l) \xrightarrow{150-180°C}$$
$$TiOSO_4\,(aq) + FeSO_4 \cdot 7H_2O\,(s)$$

**19.35**

**19.37**   $TiO_2\,(s) + C\,(s) + 2\ Cl_2\,(g) \longrightarrow TiCl_4\,(l) + CO_2\,(g)$

C is oxidized from 0 to +4 and is the reducing agent; Cl is reduced from −1 to 0 and is the oxidizing agent.

**19.39**   $8.7 \times 10^3$ kg bauxite rock

**19.41**   Sulphur has $d$ orbitals available for bond formation, allowing the formation of $SF_6$, in which the bonding can be described using $d^2sp^3$ hybrid orbitals on the S atom. The Br atom is too large for six Br atoms to be accommodated around a central S atom. Oxygen has no valence $d$ orbitals available.

**19.43**   The As atom in $AsCl_3$ has a lone pair that it donates, giving this compound Lewis-base character. The As atoms also can accommodate additional electron pairs by using valence $d$ orbitals, giving this compound Lewis-acid character.
$AsCl_3 + BF_3 \longrightarrow Cl_3As\text{—}BF_3$ (As acts as Lewis base.)
$AsCl_3 + Cl^- \longrightarrow AsCl_4^-$ (As acts as Lewis acid.)

**19.45**   $Al + 4\ OH^- \longrightarrow AlO_2^- + 2\ H_2O + 3\ e^-$; (a) $8\ Al + 5\ OH^- + 3\ NO_3^- + 2\ H_2O \longrightarrow 3\ NH_3 + 8\ AlO_2^-$; (b) $2\ Al + 2\ OH^- + 2\ H_2O \longrightarrow 3\ H_2 + 2\ AlO_2^-$; and (c) $4\ Al + 3\ SnO_3^{2-} + H_2O \longrightarrow 3\ Sn + 2\ OH^- + 4\ AlO_2^-$

**19.47**   Nitrogen is a nonmetal that forms polar bonds most readily with other nonmetals (B, C, O, the halogens). Phosphorus has a similar pattern of reactivity but can also involve $d$ orbitals in its bonding. Arsenic and antimony are metalloids with useful semiconductor properties, and bismuth is metallic.

**19.49** $Al(O)OH(s) + OH^-(aq) + H_2O \longrightarrow [Al(OH)_4]^-(aq)$

$[Al(OH)_4]^-(aq) \longrightarrow Al(OH)_3(s) + OH^-(aq)$

$2\,Al(OH)_3(s) \xrightarrow{1250°C} Al_2O_3 + 3\,H_2O(g)$

$2\,Al_2O_3(s) + 3\,C(s) \longrightarrow 3\,CO_2(g) + 4\,Al(l)$

**19.51** Silicon dioxide is first converted into silicon tetrachloride by reaction with molecular chlorine:

$SiO_2(s) + 2\,C(s) + 2\,Cl_2(g) \xrightarrow{High\ T} SiCl_4(g) + 2\,CO(g)$

The $SiCl_4$ is purified by distillation and then reduced by reaction with Mg metal:

$SiCl_4 + 2\,Mg \longrightarrow Si + 2\,MgCl_2$

**19.53** Mercury is a very soft Lewis acid, while zinc is intermediate in hard/soft character. Thus mercury preferentially associates with sulphur, a soft Lewis base, while zinc forms stable bonds with hard (O-containing) Lewis bases.

**19.55** Gaseous $Al_2Cl_6$ is in equilibrium with gaseous $AlCl_3$:

$Al_2Cl_6 \rightleftharpoons 2\,AlCl_3$

Le Châtelier's principle predicts that because the forward reaction is endothermic (bonds must be broken), an increase in temperature shifts the position of the equilibrium to the right. Thus, as temperature increases, the number of moles of gaseous substance increases, so the pressure increases faster than would be predicted by the ideal gas equation.

**19.57** $Ca_5(PO_4)_3F(s) + 7\,H_3PO_4(aq) \rightarrow 5\,Ca(H_2PO_4)_2(s) + HF(aq)$; 26.47%

**19.59** (a)

$$:\ddot{O}=\overset{\ddot{C}\!l:}{\underset{\ddot{C}\!l:}{S}}$$

(b) $6\,SOCl_2(l) + FeCl_3 \cdot 6H_2O(s) \longrightarrow$
$6\,SO_2(g) + 12\,HCl(g) + FeCl_3(s)$

**19.61** As the size of an atom increases, so does its polarizability, and greater polarizability leads to larger dispersion forces. Thus, the larger the atoms in a molecule, the larger the intermolecular forces and the easier it is to condense the substance. In the sequence $BCl_3$, $BBr_3$, $BI_3$, the halogens are increasing in size, accounting for the different stable phases at room temperature.

**19.63** The phosphate anion is the conjugate base of a weak acid, $HPO_4^{2-}$:

$PO_4^{3-} + H_2O \longrightarrow HPO_4^{2-} + OH^-$;
$pK_b = 14.00 - pK_a = 14.00 - 12.32 = 1.68$

**19.65** $2\,Al_2O_3 + 6\,Cl_2 \longrightarrow 4\,AlCl_3 + 3\,O_2$;
$AlCl_3 + 3\,Na \longrightarrow Al + 3\,NaCl$

**19.67** (a) $AlCl_3 + 3\,LiCH_3 \longrightarrow Al(CH_3)_3 + 3\,LiCl$;
(b) $SO_3 + H_2O \longrightarrow H_2SO_4(aq)$;
(c) $SbF_5 + LiF \longrightarrow Li(SbF_6)$; and (d) no reaction

**19.69** $P_4O_{10} + 6\,H_2O \longrightarrow 4\,H_3PO_4$

# Chapter 20

**20.1** 

| Part: | (a) | (b) | (c) |
|---|---|---|---|
| $Z$ | 10 | 82 | 40 |
| $A$ | 20 | 205 | 90 |
| $N$ | 10 | 123 | 50 |

**20.3** $^{53}$Mn: unstable, too few neutrons; $^{54}$Mn: unstable, odd–odd; $^{55}$Mn: stable; and $^{56}$Mn: unstable, too many neutrons and odd–odd

**20.5** $\Delta E = 8.99 \times 10^{16}$ kJ

**20.7** $\Delta E = -1.079 \times 10^{11}$ kJ/mol;
$\Delta E$ (per nucleon) $= -8.11 \times 10^8$ kJ/mol nucleon

**20.9** The repulsive barrier is greater for $^4$He + $^4$He $\rightarrow$ $^8$Be than for $^1$H + $^6$Li $\rightarrow$ $^7$Be; $^8$Be is more stable.

**20.11**

| Description | Symbol | Name |
|---|---|---|
| (a) high-energy photon | $\gamma$ | gamma ray |
| (b) mass number 4 | $\alpha$ | alpha particle |
| (c) positive particle with $m_e$ | $\beta^+$ | positron |

**20.13** (a) $^{125}_{52}$Te; (b) $^{123}_{51}$Sb; and (c) $^{127}_{53}$I

**22.15** $^{53}$Mn, $\beta^+$ or EC; $^{54}$Mn, $\beta$, $\beta^+$ or EC; $^{55}$Mn is stable; $^{56}$Mn, $\beta$

**20.17** $t_{1/2} = 1.2$ yr

**20.19** $^{232}_{90}$Th $\rightarrow \alpha + {}^{228}_{88}$Ra $\rightarrow \beta + {}^{228}_{89}$Ac $\rightarrow \beta + {}^{228}_{90}$Th $\rightarrow$

$\alpha + {}^{224}_{88}$Ra

$^{224}_{88}$Ra $\rightarrow \alpha + {}^{220}_{86}$Rn $\rightarrow \alpha + {}^{216}_{84}$Po $\rightarrow \beta + {}^{216}_{85}$At $\rightarrow \alpha + {}^{212}_{83}$Bi

$^{212}_{83}$Bi $\rightarrow \beta + {}^{212}_{84}$Po $\rightarrow \alpha + {}^{208}_{82}$Pb

**20.21** (a) $^{12}_{6}$C + $^1_0$n $\rightarrow (^{13m}_{6}$C$) \rightarrow {}^{12}_{5}$B + $^1_1$P;
(b) $^{16}_{8}$O + $^4_2\alpha \rightarrow (^{20m}_{10}$Ne$) \rightarrow {}^{20}_{10}$Ne + $\gamma$;
(c) $^{247}_{96}$Cm + $^{11}_{5}$B $\rightarrow (^{258m}_{101}$Md$) \rightarrow {}^{255}_{101}$Md + 3 $^1_0$n

**20.23** The reaction in this problem is

$^{14}_{7}$N + $\alpha \rightarrow (^{18}_{9}$F$) \rightarrow {}^{18}_{8}$O + $\beta^+$

$\bigcirc$ = proton      $\bigcirc$ = neutron

**20.25** Cs, Ba, La, Ce

**20.27** $\Delta E = -1.648 \times 10^{10}$ kJ/mol, somewhat less than Section Exercise 20.4.1 result

**20.29** The core of a nuclear reactor generates radiation that converts any material in its vicinity into radioactive substances. Thus, the heat exchanger in immediate contact with the core becomes radioactive and must be separated from the turbine that generates electricity. The primary heat exchanger transfers energy to a secondary heat exchanger, which does not become radioactive and can safely drive the turbine.

**20.31** $\Delta E_{\text{total}} = -2.4 \times 10^9$ kJ

**20.33** $v = 5.3 \times 10^6$ m/s

**20.35** Your description should include the extremely high energies required to initiate fusion and the difficulties in containing the fusion components at the temperature required for nuclei to have these high energies.

20.37

| Stage | Temperature | Composition |
|-------|-------------|-------------|
| H-burning | $4 \times 10^7$ K | C, N, H, He, $e^-$ |
| He-burning | $10^8$ K | He, Be, C, and O |
| C-burning | $10^9$ K | All nuclides from $Z = 6$ (c) up to $Z = 26$ (Fe) |

20.39 Your description should include the fact that elements beyond $Z = 26$ are less stable than Fe, so they cannot be generated by fusion of lighter elements.

20.41 $E = 4.4 \times 10^{-4}$ J

20.43 $3.41 \times 10^4$ $\beta$/s

20.45 Exposure to radiation results first in damage to those cells that reproduce most quickly, including the white blood cells that are responsible for fighting infection and the mucous membrane lining of the intestinal tract. Thus the early symptoms of radiation exposure include reduced resistance to infection and nausea due to disruption of the digestive tract.

20.47 $3.5 \times 10^9$ yr

20.49 Add a radioactive iron isotope to the manufacturer's iron waste stream before treatment, and monitor the iron content of the river downstream. If radioactivity appears in the river water, the source is the manufacturer; if not, the iron in the river comes from some other source.

20.51 Run the hydrolysis in water that is enriched with a radioactive isotope of oxygen. Isolate the two products and analyze them for radioactivity. Whichever product shows radioactivity is the one whose added oxygen atom comes from water.

20.53 $\Delta E = -1.583 \times 10^{11}$ kJ/mol; $\Delta E$ (per nucleon) $= -7.574 \times 10^8$ kJ/mol nucleon

20.55 $N = 0.80$ mg; $\dfrac{\Delta N}{\Delta t} = 1.3 \times 10^{11}$/s

20.57 The other product is an electron; $E_{kinetic}$ (electron) $= 1.25 \times 10^{-13}$ J

20.59 1% decays in 24 yr; 1% remains after $1.1 \times 10^4$ yr

20.61 Both isotopes are neutron-deficient and lie below (to the right of) the belt of stability: $^{11}C \rightarrow \beta^+ + {}^{11}B$ and $^{15}O \rightarrow \beta^+ + {}^{15}N$

20.63 70 hr

20.65 (a) $^{238}_{92}U \rightarrow \alpha + {}^{234}_{90}Th$; (b) $^{60}_{28}Ni + n \rightarrow p + {}^{60}_{27}Co$; (c) $^{239}_{93}Np + {}^{12}_{6}C \rightarrow 3\,n + {}^{248}_{99}Es$; (d) $^{35}_{17}Cl + p \rightarrow {}^{32}_{16}S + \alpha$; and; (e) $^{60}_{27}Co \rightarrow \beta + {}^{60}_{28}Ni$

20.67 $^{256}_{104}Rf$

20.69 $\Delta E$ (per event) $= -4.9 \times 10^{-13}$ J; fraction carried off by the $\alpha$ particle $= 0.73$

20.71 (a) too many neutrons; (b) $Z > 83$; (c) odd–odd; and (d) too many neutrons

20.73 In a nuclear reactor, the moderator serves to slow down fast neutrons so they are more efficiently captured by the nuclear fuel. This reduces the amount of fuel required to sustain the reaction. The control rods serve to absorb some of the neutrons, allowing the reactor to operate just below its critical point and generate large quantities of heat without heating up beyond control.

20.75 If $t_{1/2} = 1100$ s, $N = 36$; if $t_{1/2} = 876$ s, $N = 2$

20.77 $^{10}_{5}B + n \rightarrow {}^{11}_{5}B \rightarrow {}^{7}_{3}Li + \alpha$; the reaction does not pose a significant health hazard.

20.79 $1.4 \times 10^4$ yr

20.81 $5.2 \times 10^4$ yr

20.83 (a) $^{90}_{38}Sr \rightarrow {}^{90}_{39}Y + \beta$; $^{90}_{39}Y + {}^{90}_{40}Zr + \beta$; (b) Sr has the larger mass; (c) $\Delta E = -1.7 \times 10^8$ kJ/mol

20.85 $1.5 \times 10^2$ hr

20.87 $^{99}_{43}Tc$

20.89 Prepare a sample of the alcohol that is enriched in radioactive $^{18}O$ and run the reaction using this sample. Separate the products and measure the radioactivity of the ester and the water. If the C—OH bond in the alcohol breaks during the condensation, the $^{18}O$ will appear in the water, while if the C—OH bond in the carboxylic acid breaks during the condensation, the $^{18}O$ will appear in the ester.

20.91 The precipitate will contain radioactive Na.

## Chapter 9

Page 480-481: kristian sekulic/Shutterstock; Page 481 (inset): ilian studio/Getstock.com: Page 483: Jean-Pierre Pieuchot/Getty Images; Page 484: Courtesy of John Olmsted: Page 485: Courtesy of John Olmsted; Page 486: Courtesy of John Olmsted; Page 491: Charles D. Winters/Science Photo Library; Page 492: Courtesy of John Olmsted; Page 494: Ed Reschke/Peter Arnold, Inc.; Page 497 (top right): © Richard Megna/Fundamental Photographs, New York: Page 497 (middle right): Courtesy of Stephen Frisch: Page 498: © DK Limited/Corbis; Page 500: Monkey Business Images/Shutterstock.com: Page 501: Ocean Image Photography/Shutterstock.com: Page 510: Michael Neelon/Getstock.com: Page 512 (top left): Dr. David Phillips/Visuals Unlimited, Inc.: Page 512 (middle left): Dr. David Phillips/Visuals Unlimited, Inc.: Page 512 (bottom left): Dr. David Phillips/Visuals Unlimited, Inc.: Page 514: Courtesy the City of Melbourne, FL; Page 515: Courtesy of John Olmsted: Page 517: © Michael Rutherford/SuperStock; Page 518 (top right): Adam Jones/Photo Researchers, Inc.; Page 518 (middle right): Satellite image courtesy of NASA Earth Observatory and GeoEye Inc.: Page 520: Dr. Patricia Schulz/Peter Arnold, Inc.

## Chapter 10

Page 530-531: Oleg Nekhaev/Shutterstock.com: Page 531 (inset): Scott Camazine/Getstock.com: Page 532: Ustyuzharin Andrey Anatolyevitch/Shutterstock.com: Page 544: Charles D. Winters/Photo Researchers, Inc.; Page 552: Science Museum/Science & Society Picture Library; Page 560 (top left): Tatiana Popova/Shutterstock.com; Page 560 (middle left): Valentyn Volkov/Shutterstock.com: Page 560 (bottom left): Valentyn Volkov/Shutterstock.com: Page 569: SPL/Photo Researchers, Inc.; Page 574: SPL/Photo Researchers, Inc.; Page 584 (top): igor kisselev/Shutterstock.com: Page 584 (middle): Jonathan Vasata/Shutterstock.com.

## Chapter 11

Page 608-609: Dan Jurak/First Light/Getty Images; Page 608 (inset): Tischenko Irina/Shutterstock.com; Page 611: Courtesy of Patrick Watson; Page 612 (top left): Pelana/Shutterstock.com; Page 612 (bottom left): Courtesy of Andy Washnik; Page 626: Courtesy of J.F Allen; Page 630: Nikola Bilic/Shutterstock.com; Page 634 (bottom left): Courtesy of Stephen Frisch; Page 634 (bottom middle): Courtesy of John Olmsted; Page 634 (bottom right): Ustyuzharin Andrey Anatolyevitch/Shutterstock.com; Page 638 (top): Courtesy of Gael M. Reed; Page 638 (bottom): Courtesy of Linwood Mining; Page 643: Courtesy of John Olmsted; Page 647: Rob Bryon/Shutterstock.com; Page 648: © Pat O'Hara/CORBIS; Page 649: © Lester V. Bergman/ CORBIS; Page 653: © W. Cody/CORBIS; Page 655: Ted Kinsman/Photo Researchers, Inc.

## Chapter 12

Page 670-671: NASA; Page 672 (top): © Richard Megna/Fundamental Photographs, New York; Page 672 (middle): © Richard Megna/Fundamental Photographs, New York; Page 676 (middle): © Kevin Fleming/CORBIS; Page 676 (bottom): Philip and Karen Smith/Iconica/Getty Images; Page 693: © Yulia Butyrina/iStockphoto.com; Page 697: James Thew/Shutterstock.com; Page 722: Courtesy of Stephen Frisch; Page 726: Courtesy of Walker Manufacturing; Page 727 (middle): Shoot PTY/Getty Images; Page 727 (bottom): Health Protection Agency/Science Photo Library; Page 728: Courtesy of Englehard Corp.; Page 731: Courtesy of Thomas Streitz.

## Chapter 13

Page 748-749: Rob Bryon/Shutterstock.com; Page 749 (inset): © Craig Aurness/CORBIS; Page 768: Courtesy of Cargill Inc.; Page 785: Courtesy of John Olmsted; Page 795: Courtesy of John Olmsted; Page 797: Lindsey Moore/Shutterstock.com; Page 798: © David Muench/CORBIS; Page 799 (middle right): © Joshua Haviv/iStockphoto.com; Page 799 (bottom right): © Joe McDaniel/iStockphoto.com.

## Chapter 14

Page 810-811: Pichugin Dmitry/Shutterstock.com; Page 811 (inset): Marie C. Fields/Shutterstock.com; Page 820 (top right): © Alina Hart/iStockphoto.com; Page 820 (middle right): Courtesy of John Olmsted; Page 821: Courtesy of John Olmsted; Page 824: © Michael Dalton/Fundamental Photographs, New York; Page 826 (bottom left): © George Resch/Fundamental Photographs, New York; Page 826 (bottom middle): © George Resch/Fundamental Photographs, New York; Page 826 (bottom right): © George Resch/Fundamental Photographs, New York; Page 837: OPC, Inc.; Page 846: Courtesy of Andy Washnik; Page 851: Courtesy of John Olmsted; Page 854: Courtesy of John Olmsted.

## Chapter 15

Page 862-863: © Robert Holmes/CORBIS; Page 863 (inset): Juergen Berger/Science Photo Library; Page 873: Courtesy of John Olmsted; Page 879: Courtesy of Stephen Frisch; Page 895: Courtesy of Patrick Watson; Page 898 (top left): Paul Chesley/Getty Images; Page 898 (middle left): © Ralph White/CORBIS; Page 898 (middle left): © General Electric Company. All rights reserved; Page 898 (bottom left): © Steve Hamblin/Corbis; Page 914 (middle left): Courtesy of John Olmsted; Page 914 (top): Jenna E. Davies; Page 915: Creatas/ Fotosearch, LLC; Page 916: Inga Spence/Visuals Unlimited; Page 917 (top): Roman Farion/Shutterstock.com; Page 917 (bottom): © Ron Sanford/CORBIS; Page 919: Jenna E. Davies.

## Chapter 16

Page 928-929: Wes Wombara/Getstock.com; Page 929 (inset): © Todd Gipstein; RMS Titanic, Inc./CORBIS; Page 930: © Richard Megna/Fundamental Photographs, New York; Page 936: Charles D. Winters/Photo Researchers, Inc.; Page 942: Courtesy of John Olmsted; Page 944 (bottom left): Courtesy of Stephen Frisch; Page 944 (bottom right): Courtesy of Tom Pantages; Page 945: Charles D. Winters/Photo Researchers, Inc.; Page 947: Charles D. Winters/ Photo Researchers, Inc.; Page 954: © Richard Megna/Fundamental Photographs, New York; Page 955: © Thomas Del Brase; Page 957: Courtesy of Andy Washnik; Page 962: Courtesy of Tom Pantages; Page 964: The Royal Institution/Bridgeman Art Library; Page 970: Charles D. Winters/Photo Researchers, Inc.; Page 971: Tony Freeman/ PhotoEdit; Page 973 (top left): Courtesy of Patrick Watson; Page 973 (top right): Courtesy of Patrick Watson; Page 973 (bottom right): Tony Freeman/PhotoEdit; Page 977: Courtesy of Peter G. Gordon; Page 978: Urbanmyth/ Getstock.com; Page 979: Anobis/ Shutterstock.com; Page 981: Courtesy of Stephen Frisch; Page 985: © Patrick Ward/CORBIS; Page 987 (left): Charles D. Winters/Photo Researchers, Inc.; Page 987 (right): Tony Freeman/Photoedit.

## Chapter 17

Page 996-997: Claude Muridsany & Marie Perennou/Science Photo Library; Page 997 (inset): Marco Secchi/Getstock.com; Page 999: Tom Ulrich/Visuals Unlimited; Page 1004: Felicia Martinez Photography/PhotoEdit; Page 1007 (top): © Owen Franken/CORBIS; Page 1007 (bottom): © Laurence Fordyce/ Eye Ubiquitous/CORBIS; Page 1010: Courtesy of Andy Washnik; Page 1011: © BERNARD ANNEBICQUE/CORBIS SYGMA; Page 1012: Courtesy of Tom Pantages; Page 1015 (top left): © Charles Orrico/SuperStock; Page 1015 (top right): Shantell/iStockphoto.com; Page 1016: © Nora Scarlett/SuperStock; Page 1018 (top left): Arthur Tilley/Getty Images; Page 1018 (middle left): Andrew Lambert Photography/Science Photo Library; Page 1019: Brad Wynnyk/Shutterstock.com; Page 1023 (top right): Digital Vision/Getty Images; Page 1023 (bottom right): ahnhuynh/Shutterstock.com;

Page 1025: Scimat/Photo Researchers, Inc.; Page 1026 (bottom right): James M. Bell/Photo Researchers, Inc.; Page 1026 (bottom left): javarman/Shutterstock.com; Page 1030: Courtesy of the University of Cincinnati; Page 1032 (right): Mopic/Shutterstock.com; Page 1040: Courtesy of Patrick Watson.

## Chapter 18

Page 1054-1055: Courtesy of John Olmsted; Page 1057: Courtesy of John Olmsted; Page 1061: Courtesy of Stephen Frisch; Page 1067: Jenna E. Davies; Page 1069: Courtesy of John Olmsted; Page 1077: Courtesy of Stephen Frisch; Page 1079: Courtesy of Patrick Watson; Page 1080: Courtesy of Patrick Watson; Page 1090: © Rob Belknap/iStockphoto.com; Page 1093: Phil Degginger/Getty Images; Page 1095: © James L. Amos/CORBIS; Page 1096 (top): Charles D. Winters/Photo Researchers, Inc.; Page 1096 (bottom): "Active Metal Mines, 2004", Atlas of Canada, http://atlas.nrcan.gc.ca/site/english/maps/economic/ mining/metal_mines/ metal_mines_2004/active_metal_2004. Reproduced with the permission of the Minister of Public Works and Government Services, 2009; "Selected Metal-production Facilities, 2004", Atlas of Canada, http://atlas.nrcan.gc.ca/site/english/maps/ economic/ mining/ metal_mines/metal_mines_2004/smelt_refine_2004. Reproduced with the permission of the Minister of Public Works and Government Services, 2009; Page 1097: By Edward Burtynsky, courtesy of Nicholas Metivier Gallery, Toronto, Ontario; Page 1098 (bottom left): © Will and Deni McIntyre/CORBIS; Page 1098 (bottom middle): Colorstock/ Getty Images; Page 1098 (bottom right): Mark R./ Shutterstock.com; Page 1099 (top): Sasha Radosavljevich/Shutterstock.com; Page 1099 (bottom): GWImages/Shutterstock.com; Page 1100 (top): Courtesy of Stephen Frisch; Page 1100 (bottom): © Adrian Assalve/ iStockphoto.com; Page 1101: David P. Lewis/ Shutterstock.com; Page 1102: Peter Cade/Getty Images; Page 1103: © James L. Amos/CORBIS.

## Chapter 19

Page 1112-1113: Byung-Je Sung, Kwang Yeon Hwang, Young Ho Jeon, Jae Il Lee, Yong-Seok Heo, et al. (2003). "Structure of the catalytic domain of human phosphodiesterase 5 with bound drug molecules", *Nature* 42, 6953: 98-102. Reprinted by permission from Macmillan Publishers Ltd. Image courtesy of Seonggu Ro; Page 1125 (top right): © James L. Amos/CORBIS; Page 1125 (middle right): Tony Freeman/PhotoEdit; Page 1125 (bottom): Courtesy of Alcoa Technical Center; Page 1127 (top left): oleg kuzminov/Shutterstock.com; Page 1127 (top right): Mark Boulton/Photo Researchers, Inc.; Page 1127 (middle left): Charles D. Winters/Photo Researchers, Inc.; Page 1127

(middle right): TOSP Photo/Shutterstock.com; Page 1129: © Gianni Dagli Orti/CORBIS; Page 1130: © David J. & Janice L. Frent Collection/CORBIS; Page 1136 (middle left): © The Gallery Collection/ CORBIS; Page 1135 (bottom left): © Richard Megna/Fundamental Photographs, New York; Page 1135 (bottom middle): Courtesy of Andy Washnik; Page 1136 (top right): Courtesy of Stephen Frisch; Page 1142 (top): Ronald Sumners/Shutterstock.com; Page 1142 (bottom left): © Martin McCarthy/iStockphoto.com; Page 1142 (bottom right): Reprinted with permission from Daniel R. Mitchell et al. (2001). " The synthesis of megatubes: New dimensions in carbon materials", *Inorganic Chemistry* 40 (12): 2751-2755. Copyright 2001 American Chemical Society. Image courtesy of Dwight Romanovicz; Page 1143 (left): Albert G. Nasibulin, Peter V. Pikhitsa, Hua Jiang, David P. Brown, Arkady v. Krasheninnikov, et al. "A novel hybrid carbon material", *Nature Nanotechnology* 2 (2007): 156-161. Reprinted by permission from Macmillan Publishers Ltd: Nature Nanotechnology, 2009. Image courtesy of Arkady Krasheninnikov; Page 1144 (top): Todd D. Krauss, "Biosensors: Nanotubes light up cells", *Nature Nanotechnology* 4, 2 (2009): 85-86. Reprinted by permission from Macmillan Publishers Ltd: Nature Nanotechnology, 2009; Page 1144 (bottom left): Gary Retherford/Science Photo Library; Page 1148: Charles D. Winters/Science Photo Library; Page 1150: Courtesy of Regal Ware, Inc.; Page 1152: Stockbyte/Getty Images; Page 1163 (lower inset): LBNL/Science Source/Photo Researchers, Inc.

## Chapter 20

Page 1162-1163: Courtesy of SOHO/EIT consortium. SOHO is a project of international cooperation between ESA and NASA; Page 1163 (upper inset): Dept of Nuclear Medicine, Charing Cross Hospital/Science Photo Library; Page 1168: NASA; Page 1174: Photo by Hulton Archive/Getty Images; Page 1184: Courtesy of the University of Technology, Eindhoven, The Netherlands; Page 1185: Courtesy of Fermilab National Accelerator Laboratory - Batavia, IL; Page 1191 (top): © EPA/Corbis; Page 1191 (middle): © George D. Lepp/CORBIS; Page 1191 (bottom): Dick Loek/Getstock.com; Page 1193: Ria Novosti/Science Photo Library; Page 1195: © ITER Organization. Used with permission; Page 1196: Courtesy of the Princeton University Plasma Physics Lab and the US Department of Energy; Page 1197: Courtesy David Malin, Anglo-Australian Telescope Board; Page 1200: Photo by Popperfoto/Getty Images; Page 1204: NASA; Page 1206: © P. Deliss/Godong/Corbis; Page 1208: Courtesy of MDS Noridon.

# GLOSSARY

**Absolute entropy.** (11.3) The amount of disorder contained in a chemical substance.

**Absorption spectrum.** (4.3) The distribution of wavelengths of light absorbed by a species.

**Accuracy.** (1.2) How close a measurement is to the true value.

**Achiral.** (10.4) A molecule that can be superimposed on its mirror image.

**Acid.** (Brønsted definition) (14.1) A substance that acts as a proton donor.

**Acid hydrolysis constant ($K_a$).** (13.6) The equilibrium constant for proton transfer between an acid and water $K_a = \dfrac{[H_3O^+]eq\,[A^-]eq}{[HA]eq}$.

**Acid rain.** (2.8) Rain whose pH is lower than 5.6, which is the pH of water saturated with carbon dioxide.

**Actinide element.** (18.1) Any of the elements in the $5f$ block of the periodic table, between $Z = 89$ and $Z = 102$.

**Activated complex.** (12.6) The unstable molecular species through which an elementary reaction proceeds as it evolves from reactant(s) to product(s).

**Activation energy ($E_a$).** (12.6) The minimum amount of energy reactants must possess to react to form products.

**Activation energy diagram.** (12.6) A graph showing how the energy of a reacting set of molecules varies with the course of the reaction.

**Active electrode.** (16.3) An electrode whose chemical constituents take part in the redox reaction that occurs at its surface.

**Actual yield.** (1.9) The amount of chemical product formed in a chemical reaction.

**Adduct.** (19.1) A compound formed by bond formation between a Lewis base and a Lewis acid.

**Adenine.** (17.6) A two-ring organic base found in DNA and RNA.

**Adenosine triphosphate (ATP).** (11.6) A biochemical molecule, instrumental in biochemical energetics, containing adenine and three phosphate groups.

**Adhesive force.** (8.3) A force of attraction between molecules in one phase and different molecules in another phase.

**Adsorption.** (12.7) The physical attachment of molecules to a surface.

**Alcohol.** (10.3) An organic compound that contains a hydroxyl group, $—OH$.

**Aldehyde.** (10.3) An organic compound that contains the carbonyl group ($C=O$) attached to at least one hydrogen atom.

**Alkali metal.** (5.6) Any of the Group 1 elements, all of which are reactive metals with $s^1$ valence configurations.

**Alkaline dry cell.** (16.6) A galvanic cell whose working components are Zn and $MnO_2$.

**Alkaline earth metal.** (5.6) Any of the Group 2 elements, all of which are reactive metals with $s^2$ valence configurations.

**Alkane.** (6.3) A compound containing only carbon and hydrogen, with general formula $C_nH_{2n-2}$, in which all carbon atoms have four single bonds.

**Alkene.** (6.4) A compound containing only carbon and hydrogen in which there is at least one $C—C$ double bond.

**Alkyl group.** (10.1) One hydrogen less than an alkane molecule ($C_nH_{2n+1}$), attached as a side group to a hydrocarbon.

**Alkyne.** (10.1) A compound containing only carbon and hydrogen in which there is at least one $C—C$ triple bond.

**Allotropes.** (19.6) Forms of an element that are structurally different from one another (e.g. carbon and graphite).

**Alloy.** (9.2) A metallic solution or mixture of two or more metals.

**Alpha emission.** (20.2) Nuclear decay in which a helium nucleus is emitted by the decaying nucleus.

**Alpha particle ($\alpha$).** (20.2) An energetic helium nucleus emitted by a radioactive nuclide.

**Amalgam.** (9.2) A solution of a metal in mercury.

**Amide.** (10.3) An organic compound that contains the R—CO—N linkage.

**Amine.** (10.3) A compound whose molecules can be viewed as ammonia with one or more N—H bonds replaced by N—C bonds.

**Amino acid.** (17.7) A compound that contains both an amine ($-NH_2$) and a carboxylic acid ($-CO_2H$) group.

**Amorphous.** (11.5) Without any organized regular repeating pattern.

**Ampere (A).** (1.5) The base unit of electric current in the SI system $1 A = 1 C s^{-1}$.

**Amphiprotic.** (14.1) Able to act as both an acid and a base.

**Amplitude.** (4.2) The height of a wave.

**Analytical balance.** (1.2) An accurate mass-measuring instrument.

**Anion.** (1.5) An ion that possesses negative charge.

**Anode.** (16.3) The electrode at which oxidation occurs.

**Antibonding molecular orbital.** (7.4) A molecular orbital that has electron density concentrated outside the bonding region, making it less stable than the atomic orbitals from which it forms.

**Aqueous.** (1.6) A solution with water as the solvent.

**Arrhenius equation.** (12.6) The equation describing how rate constants depend on temperature and activation energy $k = Ae^{-E_a/RT}$.

**Atmosphere (atm).** (2.1) A unit of pressure based on the normal pressure exerted by the Earth's atmosphere at sea level. 1 atm = 101.325 kPa.

**Atom.** (1.1) The basic unit of chemical matter, consisting of a nucleus and enough electrons to convey electrical neutrality.

**Atomic number (Z).** (20.1) The number of protons in an atomic nucleus.

**Atomic orbital.** (4.5) A description of an atomic electron that provides the distribution of electron density about the nucleus.

**Atomic radius.** (5.4) The distance from the nucleus of an atom at which electron-electron repulsion prevents closer approach of another atom.

**Atomic symbol.** (1.1) The letter designation for an element.

**Atomic theory.** (1.7) The description of matter as composed of atoms that retain their identities during all physical and chemical processes.

**Atomic view.** (1.1) How matter looks when viewed at the level of atoms.

**Aufbau principle.** (5.2) The statement that the most stable arrangement of electrons in an atom results from placing each successive electron in the most stable available atomic orbital.

**Avogadro's number ($N_A$).** (1.4) The number of particles contained in 1 mol, $6.022142 \times 10^{23}$ particles/mol.

**Axial position.** (6.4) A position along the z axis in a trigonal bipyramid.

**Azeotropes.** (9.4) A liquid mixture that is not possible to separate by distillation.

**Azimuthal quantum number (l).** (4.5) The quantum number, restricted to integers from 0 to $n - 1$, that indexes the shape of an atomic orbital.

**Balanced chemical equation.** (1.7) A description of a chemical reaction using chemical formulas in which the coefficients describe the ratios of molecules of each species that react.

**Ball-and-stick model.** (1.1) A representation of a molecule that shows the atoms as small balls and the chemical bonds as sticks.

**Band gap.** (7.7) The difference in energy between the highest filled orbital and the lowest vacant orbital in a solid.

**Band theory.** (7.7) The description of bonding in solids using delocalized orbitals.

**Bar.** (2.1) A pressure unit equal to $10^5$ pascals.

**Barometer.** (2.1) An instrument for measuring atmospheric pressure.

**Base.** (Brønsted definition). (14.1) A substance that acts as a proton acceptor.

**Base hydrolysis constant ($K_b$).** (13.6) The equilibrium constant for the proton transfer from water to a

base, $Kb = \dfrac{[BH^+]_{eq} [OH^-]_{eq}}{[B]_{eq}}$.

**Base pairing.** (17.6) The formation of hydrogen bonds between strands of DNA or RNA.

**Battery.** (16.3) One or more galvanic cells that serve as a source of electric current or voltage.

**Belt of stability.** (20.1) On a graph of $N$ vs. $Z$, the region that contains the stable nuclides.

**Bent shape.** (6.3) The arrangement of a triatomic molecule having a bond angle less than $180°$.

**Benzene.** (10.2) An aromatic compound with the molecular formula $C_6H_6$ that has alternating double and single bonds between the carbon atoms in a resonance structure.

**Beta emission.** (20.2) Nuclear decay in which an energetic electron is emitted by the decaying nucleus.

**Beta particle ($\beta$).** (20.2) An energetic electron emitted by a radioactive nuclide.

**Bidentate.** (15.5) A ligand that can bind to a metal through two sites.

**Bilayer.** (9.5) A double layer of atoms or molecules.

**Bimolecular reaction.** (12.1) An elementary step in which two molecules collide and undergo a chemical reaction.

**Body centered cubic (BCC).** (8.5) The crystal structure whose unit cell is a cube with identical atoms at its corners and at its center.

**Boiling point.** (8.1) The temperature at which the vapour pressure of a liquid matches the external pressure.

**Boiling point elevation constant.** (9.4) A factor relating to the increase in boiling point of a liquid caused by the presence of nonvolatile solutes in solution.

**Bond.** (6.1) A strong attractive force generated by sharing of electrons between atoms or electrical attraction between ions in a substance.

**Bond angle.** (6.5) The angle between two atoms that are bonded to a third atom.

**Bond energy.** (3.3) The energy (usually expressed per mole) required to break one particular chemical bond in a gaseous substance.

**Bond length.** (6.1) The average distance between the nuclei of two bonded atoms in a molecule.

**Bond order.** (7.4) The net number of pairs of bonding electrons in a chemical bond.

**Bonding molecular orbital.** (7.4) An orbital formed from atomic orbitals that conveys bonding by placing electron density between the bonded atoms.

**Bonding orbital.** (7.1) An orbital with high electron density between atoms, formed by constructive interaction between atomic orbitals.

**Boundary.** (3.2) The imaginary line separating a system from its surroundings.

**Brønsted-Lowry theory.** (14.1) The proton transfer definition of acids and bases.

**Buffer capacity.** (15.2) The amount of added hydronium or hydroxide ions that can be added to a buffer solution without exceeding a specified pH range.

**Buffer equation.** (15.1) The equation linking the pH of a buffer solution with the $pK_a$ and concentrations of the conjugate acid-base pair.

**Buffer solution.** (15.1) A solution containing both a weak acid and its conjugate base, whose pH changes slowly on addition of small amounts of an acid or base.

**Buret.** (15.3) A volume-delivering device that is calibrated to allow measurement of the amount of fluid delivered.

**Calorimeter.** (3.4) An apparatus for measuring the amount of heat absorbed or emitted during a chemical process.

**Capillary action.** (8.3) The upward movement of a liquid in a narrow tube against the force of gravity.

**Carbohydrate.** (17.5) A biochemical substance whose empirical formula is $(CH_2O)_n$.

**Carbonyl group.** (10.3) The $C{=}O$ functional group.

**Carboxylic acid.** (10.3) An acid that contains the $CO_2H$ functional group.

**Catalyst.** (12.7) A substance that increases the rate of a chemical reaction without being consumed or produced in the reaction.

**Cathode.** (16.3) The electrode at which reduction occurs.

**Cation.** (2.5) A positively charged ion.

**Cell potential.** (16.4) The difference in electrical potential (voltage) between the two electrodes in a galvanic cell.

**Cellulose.** (17.1) The polymer, composed of sugar monomers, that makes up the woody tissue of plants.

**Celsius scale.** (1.2) The temperature scale defined by $0$ °C as the normal freezing point and $100$ °C as the normal boiling point of water.

**Change of state.** (3.2) Any process in which there is a change in one or more of the variables ($P$, $V$, $T$, $n$) describing a system.

**Chelate effect.** (15.5) The stabilization of a metal complex by a multidentate ligand.

**Chelating ligand.** (15.5) A ligand containing two or more donor atoms.

**Chemical energy.** (3.1) Energy stored in bonds between atoms or ions.

**Chemical equation.** (1.7) An equation using chemical formulas that describes the identities and relative amounts of reactants and products in a chemical reaction.

**Chemical formula.** (1.1) An expression using chemical symbols that describes the identities and relative amounts of elements in a substance.

**Chemical property.** (1.2) A property that can only be observed during a chemical transformation.

**Chemical transformation.** (1.2) A process in which chemical substances react to form other chemical substances.

**Chiral molecule.** (10.4) A molecule that has four different groups or atoms bound to one carbon atom.

**Chirality centre.** (10.4) An atom that has four different groups or atoms bound to it.

**Close packed.** (8.5) The most efficient arrangement for packing atoms, molecules, or ions in a regular crystal.

**Cohesive force.** (8.3) A force of attraction between like molecules in the same phase.

**Colligative property.** (9.4) A property of a solution that is proportional to the concentration of solute species.

**Colloid.** (9.5) A particle whose dimensions are between 1 nm and 1 $\mu$m.

**Colloidal suspension.** (9.5) A colloid–liquid mixture that does not readily separate into its components.

**Combustion.** (3.3) A reaction of a substance with oxygen that releases chemical energy.

**Common ion effect.** (15.4) Reduction in solubility caused by the presence of one of the ions contained in a salt.

**Complementary base pair.** (17.6) Two nitrogen-containing organic bases that hydrogen bonds in DNA or RNA.

**Complementary colours** (18.3) Colours that are related in that if light of one colour is absorbed by a substance, the substance displays the complementary color.

**Complex ion.** (13.6) A species formed by bonding of species containing lone pairs of electrons to a metal ion.

**Compound.** (1.1) A substance formed of atoms of two or more elements chemically bonded in fixed proportions.

**Compound nucleus.** (20.3) The transient nucleus that forms from the collision of a nuclear projectile with another nucleus.

**Concentration.** (1.6) The amount of a solute contained in a standard amount of a solution.

**Concentration table.** (13.5) A table of amounts for a chemical reaction.

**Condensation.** (8.1) Conversion of a gas into its liquid.

**Condensation reaction.** (17.1) A reaction that joins two molecules, accompanied by the elimination of a small molecule such as water.

**Conduction band.** (7.7) The set of delocalized orbitals in a solid that, when partially occupied, contains mobile electrons and conveys high electrical conductivity.

**Conductor.** (7.7) A substance that is a good conductor of electrical current.

**Conjugate acid-base pair.** (14.1) A Brønsted acid and the base that results when the acid loses one proton.

**Conjugate base.** (14.1) The chemical species formed by removal of a proton from an acid.

**Conjugated $\pi$ system.** (7.6) A set of $\pi$ bonds that is spread over more than two atoms.

**Conservation law.** (1.7) A law stating that the number of atoms of each element is conserved in any chemical reaction.

**Constant-pressure calorimetry.** (3.4) Calorimetry where the pressure of the system is fixed, often used for studying reactions in liquid solutions.

**Constant-volume calorimetry.** (3.4) Calorimetry where the volume of the system is fixed, often used to study reactions that involve gases.

**Conversion ratio.** (1.2) A ratio equal to 1 that converts a quantity from one unit to another.

**Coordination complex.** (18.2) A combination of a metal atom or cation with two or more species (ligands) that bind covalently to the metal.

**Coordination number.** (18.2) The number of atoms bonded to an inner atom in a molecule or metal complex.

**Copolymerization.** (17.2) The formation of a polymer from two or more different monomers.

**Core electrons.** (5.2) The inner atomic electrons with a principal quantum number less than that of the valence electrons.

**Corrosion.** (16.6) The oxidation of a metal in contact with its environment.

**Coupled reactions.** (11.6) A pair of reactions, one of which is driven in what is otherwise its nonspontaneous direction by the influence of the second, highly spontaneous reaction.

**Covalent bond.** (6.1) A bond resulting from the sharing of electrons between atoms.

**Critical mass.** (20.4) The minimum mass of fissionable material required to generate a self-sustaining nuclear fission reaction.

**Critical point.** (8.6) The temperature and pressure above which the distinction between the liquid and vapour phases disappears.

**Cross-linking.** (17.2) The formation of additional chemical bonds between the chains of a polymer.

**Crystal field splitting energy.** (18.3) The difference in energy between atomic d-orbitals created by the effects of neighbouring ions in a crystal.

**Crystal field theory.** (18.3) A model describing the electronic structure of transition metal complexes in terms of the interactions between metal d-orbitals and the electric field of the ligands.

**Crystalline.** (8.5) Containing a regular array of atoms, molecules, or ions.

**Crystalline defect.** (8.5) An imperfection in a regular solid.

**Cubic close packing.** (8.5) The close packing arrangement in which hexagonal layers are stacked in an ABCABC. . . pattern.

**Cyclotron.** (20.3) A particle accelerator used to generate high energy beams of nuclear particles.

**Cytosine.** (17.6) A one-ring organic base found in DNA and RNA.

**Dalton's law of partial pressures.** (2.3) The law stating that in a gas mixture, each gas exerts a pressure equal to the pressure that it would exert if present by itself under otherwise identical conditions.

**de Broglie equation.** (4.4) The equation describing the wave nature of particles, $\lambda = h/mv$.

**Decay constant.** (20.2) The rate constant in a nuclear decay process.

**Dehydrohalogenation.** (10.6) The elimination of HX from adjacent atoms of an alkyl halide.

**Delocalized bond.** (7.1) A bond in which electron density is distributed over more than two atoms.

**Delocalized $\pi$-orbital.** (7.5) An orbital, generally formed by side-to-side overlap of atomic p orbitals, that spreads over more than two atoms.

**Density ($\rho$).** (1.3) The ratio of an object's mass to its volume, $m/V$.

**Deoxyribonucleic acid (DNA).** (17.6) A polymer of nucleotide units that stores genetic information in the chromosomes.

**Desorption.** (12.7) Detachment of an adsorbed molecule from a surface.

**Deuterium (D).** (20.1) An atom (an isotope of hydrogen) containing one proton and one neutron in its nucleus.

**Diamagnetic.** (5.3) The quality of being repelled by a magnetic field. Substances with no unpaired electrons are diamagnetic.

**Diastereomers.** (10.3) Stereoisomers that are not minor images of one another.

**Diatomic molecule.** (1.1) A molecule that contains exactly two atoms.

**Diffusion.** (2.6) The gradual mixing of a solute in a solution by molecular motion.

**Dilution.** (1.6) Production of a less concentrated solution by addition of solvent.

**Diol.** (10.3) The IUPAC name for a glycol, which is a molecule containing two hydroxyl groups.

**Dipolar force.** (8.2) The attractive force between polar molecules that results from the negative end of one molecule aligning with the positive end of its neighbour.

**Dipole moment.** (6.5) The net electrical character arising from an asymmetric charge distribution.

**Diprotic acid.** (14.7) A molecule containing two acidic hydrogen atoms.

**Dispersion force.** (8.2) The net attractive force between molecules generated by temporary dipole moments arising from polarization of their electron clouds.

**Disulphide bridge.** (17.7) An S—S bond linking two portions of a protein molecule.

**Dominant equilibrium.** (13.6) The equilibrium that is most important in determining equilibrium concentrations of major species in a solution.

**Donor atom.** (18.2) An atom that possesses one or more lone pairs of electrons that it can contribute to form a covalent bond.

**Doped semiconductor.** (7.7) A metalloid to which a small amount of another element has been added.

**Double bond.** (6.2) A chemical bond containing two pairs of bonding electrons.

**Double helix.** (17.6) The pair of intertwined spirals that is the secondary structure of DNA.

**Ductile.** (18.1) Able to be drawn into tubes or wires.

**Dynamic equilibrium.** (2.8) The condition in which a forward and a reverse process occur at equal rates, so the system undergoes no net change.

**Effective nuclear charge.** (5.1) The net positive charge that an electron in an atomic orbital experiences, equal to the nuclear charge minus the effects of screening.

**Effusion.** (2.6) The escape of a gas through a pinhole from a container into a vacuum.

**Elastomer.** (17.4) A polymer that can be stretched without breaking.

**Electrical Energy.** (3.1) Energy of an object or system created by the action of electrical force.

**Electrical force.** (3.1) The attraction or repulsion between objects caused by their electrical charge.

**Electrode.** (16.3) An electrical conductor that establishes an interface between an external circuit and an electrochemical cell.

**Electrolysis.** (16.7) The use of electrical energy to drive a nonspontaneous chemical reaction.

**Electrolytic cell.** (16.7) An electrochemical cell whose redox reaction is driven by an externally applied electrical potential.

**Electromagnetic radiation.** (4.2) Wave phenomena having both electrical and magnetic components; visible light is one type.

**Electron.** (2.2) The smallest unit of negative electrical charge, whose mass is $9.11 \times 10^{31}$ kg and whose charge is $1.602 \times 10^{19}$ C.

**Electron affinity (EA).** (5.4) The energy change accompanying the attachment of an electron to an atom or anion.

**Electron capture.** (20.2) The mode of nuclear decay in which an unstable nucleus captures a core electron and converts a proton into a neutron.

**Electron configuration.** (5.3) The distribution of electrons among the various orbitals of an atom, molecule, or ion.

**Electron contour drawing.** (4.6) A depiction of an orbital that shows a contour surface within which most of the electron density is located.

**Electron density.** (4.6) The distribution of electron probability in space around an atom or molecule.

**Electron group.** (6.3) A set of from one to six valence electrons that occupies a particular region around an atom.

**Electron group geometry.** (6.3) The shape created by the electron groups around an atom.

**Electronegativity.** (6.1) A measure of the ability of an atom in a molecule to attract the shared electrons in a chemical bond.

**Electroplating.** (16.7) The electrolytic deposition of a thin metal film on the surface of an object.

**Element.** (1.1) A substance composed of atoms all of whose nuclei have the same amount of positive charge.

**Elemental symbol.** (1.1) A letter or letters designating a particular chemical element.

**Elementary reaction.** (12.1) One of the individual steps by which a chemical reaction occurs.

**Elimination reaction.** (10.6) A reaction in which two fragments of a molecule, Y and Z, are removed as YZ and a double bond forms.

**Emission spectrum.** (4.3) The distribution of wavelengths of light given off by a species in an excited state.

**Enantiomers.** (10.4) Molecules that are non-superimposable mirror images of one another.

**Endothermic.** (3.4) Accompanied by the absorption of energy.

**Energy.** (3.1) The ability to do work.

**Energy level diagram.** (4.3) A diagram of the specific energy levels of an atom, showing increasing energy on the y axis.

**Enthalpy (H).** (3.5) The energy-related thermodynamic state function defined by the equation $H = E + pV$.

**Enthalpy of vapourization.** (3.5) The enthalpy change accompanying the conversion of a liquid into a gas.

**Entropy (S).** (11.2) The thermodynamic state function that measures the degree of dispersal of matter and energy.

**Enzyme.** (17.7) A biological catalyst.

**Equatorial position.** (6.4) A position within the trigonal plane of a trigonal bipyramid.

**Equilibrium.** (13.1) The state in which the concentrations of reactants and products do not change because the forward and reverse reaction rates are identical.

**Equilibrium constant ($K_{eq}$).** (13.1) The value of the ratio of equilibrium concentrations of products to equilibrium concentrations of reactants, each raised to the power equal to its stoichiometric coefficient.

**Ester.** (10.3) An organic compound that has the general formula RCOOR′.

**Ether.** (10.3) An organic compound that has the general formula R—O—R′.

**Evapouration.** (2.8) The escape of molecules from the liquid to the gas phase.

**Excited state.** (4.3) A description of an atom in which one or more electrons has more energy than in the most stable (ground) state.

**Exothermic.** (3.4) Accompanied by the release of energy.

**Extensive property.** (3.4) A property that increases in proportion to the amount of material in the sample.

**Face-centered cubic.** (8.5) A crystalline structure whose cubic unit cell has four atoms inside: $\frac{1}{8}$ of an atom inside the unit cell at each corner plus an additional $\frac{1}{2}$ of an atom inside the unit cell on each face of the cube.

**Faraday constant (F).** (16.5) The electrical charge of 1 mol of electrons, 96,485 C.

**Fibre.** (17.4) A long thin strand of polymeric material.

**Fibrous protein.** (17.7) A protein that forms long linear strands or coils that are water insoluble.

**First-generation star.** (20.5). A star formed from a collapsing cloud that contains only hydrogen and electrons.

**First law of thermodynamics.** (3.2) The statement that energy is conserved; often expressed as an equation for change in energy, $\Delta E = q + w$.

**First order.** (12.3) Dependent on concentration to the first power.

**Fission.** (20.4) A nuclear reaction in which one nucleus fragments into two smaller ones.

**Flotation.** (18.5) The metallurgical process in which water is used to separate components of an ore based on their densities.

**Formal charge.** (6.2) The difference between the number of valence electrons in the free atom and the number of electrons assigned to that atom in the Lewis structure.

**Formation constant.** (13.6) The equilibrium constant for the formation of a complex ion.

**Formation reaction.** (3.5) A reaction in which 1 mol of a compound forms from chemical elements in their standard states.

**Fractional distillation.** (9.4) A process of separating a mixture of volatile components by performing repeated evaporation and condensation cycles.

**Free energy (G).** (11.4) The energy related thermodynamic state function defined by the equation, $G = H - TS$.

**Free radical.** (17.2) A highly reactive chemical species containing one or more unpaired electrons.

**Free radical polymerization.** (17.2) Polymer formation that proceeds by a mechanism that involves free radicals.

**Freezing point.** (8.1) The temperature at which a liquid crystallizes to form a solid.

**Freezing point depression.** (9.4) The reduction in freezing point caused by the presence of solutes in a solution.

**Freezing point depression constant.** (9.4) A factor relating the reduction in freezing point to the total molality of solutes present.

**Frequency (v).** (4.2) The number of wave crests of a wave that pass a fixed point per unit time, usually expressed in $s^{-1}$ (or Hz).

**Functional group.** (10.3) An atom or small group of atoms in a molecule that gives the molecule characteristic chemical properties.

**Fusion.** (20.5) A nuclear reaction in which two nuclei combine (fuse) into one more massive nucleus.

**Galvanic cell.** (16.3) An electrochemical cell operating spontaneously to produce an electrical current from a chemical reaction. Batteries are galvanic cells.

**Gamma ray (γ).** (20.2) A high-energy photon emitted in the course of nuclear decay.

**Gas.** (2.1) A fluid in which intermolecular forces are so small that the substance expands, contracts, and flows freely to take on the size and shape of its container.

**Geometric isomer.** (18.2) One of two or more isomers that have identical chemical bonds distributed differently in space.

**Glass.** (8.5) An amorphous solid that has cooled to a solid state without crystallizing.

**Globular protein.** (17.7) A protein with a compact, roughly spherical tertiary structure.

**Glycol.** (10.3) An alcohol containing two hydroxyl groups.

**Gravitational energy.** (3.1) Energy that is created by gravitational force.

**Gravitational force.** (3.1) The attraction between objects caused by their mass.

**Greenhouse effect.** (2.8) The heating of the Earth's surface caused by gases in the atmosphere that absorb outgoing infrared radiation.

**Greenhouse gases.** (2.8) Gases, primarily carbon dioxide and methane, that trap radiation and lead to heating of the Earth's surface.

**Ground state.** (4.3) A description of an atom in which all electrons are in the lowest energy state.

**Ground state configuration.** (5.2) The most stable orbital arrangement of electrons in an atom or molecule.

**Group.** (5.2) The elements in a column of the periodic table.

**Guanine.** (17.6) A two-ring organic base found in DNA and RNA.

**Haber process.** (13.1) The industrial process for preparing ammonia from molecular nitrogen and hydrogen.

**Half-life ($t_{1/2}$).** (12.4) The time required to consume half the initial concentration of a reactant.

**Half-reaction.** (16.2) A reaction that isolates a reduction or oxidation and explicitly shows the electrons involved.

**Halide.** (5.5) An anion of a group 17 atom, or a binary compound formed from one of these anions and a metal cation.

**Hall-Héroult process.** (19.3) The industrial process for producing aluminum electrolytically, which was discovered by two 22-year-old scientists at the same time on opposite sides of the ocean.

**Haloalkane.** (10.3) An alkane with one or more hydrogen atoms substituted by a halogen atom.

**Halogen.** (5.5) Any of the elements in Group 17 of the periodic table.

**Hard Lewis acid.** (19.2) A Lewis acid whose acceptor atom has a low polarizability.

**Hard Lewis base.** (19.2) A Lewis base whose donor electron pairs are tightly bound, resulting in low polarizability.

**Heat ($q$).** (3.2) The quantity of heat transferred between a system and its surroundings.

**Heat capacity.** (3.2) The amount of energy required to raise the temperature of a given quantity of a substance by 1 K (or 1 °C).

**Heat of dilution.** (9.3) The enthalpy change accompanying a dilution process.

**Heat of solution.** (9.3) The enthalpy change accompanying the dissolving of a solute.

**Helix.** (17.7) A regular, spirally twisted shape taken on by nucleic acids and proteins.

**Henry's law.** (9.3) The statement that the concentration of a gas in solution is directly proportional to the partial pressure of that gas above the solution.

**Henry's law constant.** (9.3) The proportionality constant relating the solubility of a gas in a solution to its partial pressure above that solution.

**Hess' law.** (3.5) The statement that the enthalpy change for any process is equal to the sum of the enthalpy changes for any set of steps that leads from the initial to final conditions.

**Heterogeneous catalyst.** (12.7) A catalyst that exists in a different phase than the one where the chemical reaction of interest is taking place.

**Hexagonal close packing.** (8.5) The close packing arrangement in which hexagonal layers are stacked in an ABAB . . . pattern.

**High-spin complex.** (18.3) A complex in which the maximum number of electrons is unpaired.

**Homogeneous catalyst.** (12.7) A catalyst that is contained in the phase where the chemical reaction of interest is taking place.

**Hund's rule.** (5.3) The observation that the most stable arrangement of electrons among orbitals of equal energy is the one that maximizes the number of unpaired electron spins.

**Hybrid orbital.** (7.2) An atomic orbital obtained by combining two or more valence orbitals on the same atom.

**Hybridization.** (7.2) The formation of a set of hybrid orbitals with favourable directional characteristics by mixing together two or more valence orbitals of the same atom.

**Hydrate.** (1.5) A solid compound or complex ion that has water molecules incorporated into its structure.

**Hydration.** (10.7) The addition of water to the double bond of an alkene or alkyne.

**Hydrocarbon.** (10.1) A compound that contains only hydrogen and carbon.

**Hydrogen bond.** (8.2) A moderately strong intermolecular attraction caused by the partial sharing of non-bonding electrons on an electronegative atom such as O, N, S or F with a positively polarized H-atom on another molecule.

**Hydrophilic.** (9.5) Attracted to water.

**Hydrophobic.** (9.5) Repelled by water.

**Hydroxyl group.** (10.3) The —OH group.

**Ideal gas.** (2.5) A gas in which molecular volumes and intermolecular forces both are negligible.

**Ideal gas equation.** (2.2) The equation describing the behaviour of an ideal gas, $pV = nRT$.

**Indicator.** (15.3) A substance that can be used to identify the stoichiometric point of an acid-base titration because its color is sensitive to pH changes.

**Induced dipole.** (8.2) A dipole caused in a molecule by the approach of another polar molecule or an ion.

**Induced nuclear reaction.** (20.3) A reaction that occurs as a result of a nuclear projectile colliding with an atomic nucleus.

**Initial concentration.** (13.5) The concentration of a reagent immediately after a system is prepared but before any chemical reaction takes place.

**Initiation.** (17.2) A chemical reaction that begins the process of polymerization.

**Initiator.** (17.2) A substance that starts a polymerization chain reaction by creating free radicals.

**Inner atom.** (6.3) Any atom in a molecule that is bonded to more than one other atom.

**Inner transition metal.** (18.1) The lanthanide and actinide elements.

**Insoluble.** (9.2) Having a very low solubility in a liquid. Substances having $K_{sp} < 10^{-5}$ are generally classified as being insoluble in water.

**Intensity.** (4.2) The brightness of light; the number of photons per unit time.

**Intensive property.** (3.4) A property whose magnitude is independent of the amount of the substance present.

**Intermediate.** (12.1) A species that is not part of the overall stoichiometry but is produced in an early step of a reaction mechanism and consumed in a later step.

**Intermolecular forces.** (8.1) Forces that exist between molecules, usually attractive.

**Ion.** (1.5) A charged species resulting from the gain or loss of electrons from a neutral atom or molecule.

**Ionic compound.** (6.1) A neutral compound composed of cations and anions.

**Ionization energy.** (5.1) The energy required to remove an electron from an atom or molecule.

**Isoelectronic.** (5.3) Having the same electron configuration.

**Isolation method.** (12.4) An experimental technique in chemical kinetics that sets the initial concentration of one reactant substantially lower than those of all others.

**Isomers.** (10.1) Molecules that have the same chemical formulas but different molecular structures.

**Isotonic.** (9.4) Having identical total molarity of solutes.

**Isotopes.** (1.4) Atoms of the same element whose nuclei contain different numbers of neutrons.

**IUPAC** International Union of Pure and Applied Chemistry.

**Joule (J).** (3.1) The SI energy unit, defined to be $1\ J = 1\ kg\ m^2\ s^2$.

**Kelvin (K).** (1.5) The SI temperature unit, based on absolute zero, with a unit size equal to 1/100 the difference between the normal freezing point and normal boiling point of water.

**Ketone.** (10.3) An organic compound that contains the carbonyl group ($C{=}O$) linked to two carbon atoms.

**Kilogram (kg).** (1.5) The SI unit of mass.

**Kinetic energy.** (3.1) The energy of motion of an object, $E_{kinetic} = \frac{1}{2}mv^2$.

**Kinetics.** (12.2) The study of chemical reaction rates and mechanisms.

**Lanthanide element.** (18.1) Any of the elements in the $4f$ block of the periodic table, between $Z = 57$ and $Z = 70$.

**Lattice.** (8.5) An organization of ions in a regular, alternating array.

**Lattice energy.** (8.5) The energy released when individual gaseous cations and anions condense to form a solid lattice.

**Law of conservation of energy.** (3.2) The statement that energy cannot be created or destroyed.

**Leaching.** (18.5) The metallurgical process in which one component of an ore mixture is separated by dissolving.

**Leaving group.** (10.5) The group that leaves in a substitution reaction.

**Le Châtelier's principle.** (13.4) The observation that when a change of conditions is applied to a system at equilibrium, the system responds in the manner that reduces the amount of change.

**Lead storage battery.** (16.6) A voltaic cell, widely used for automobile storage batteries, in which both half-reactions involve reactions of lead.

**Lewis acid.** (19.1) An electron-pair acceptor.

**Lewis base.** (19.1) An electron-pair donor.

**Lewis structure.** (6.2) A representation of covalent bonding in a molecule that uses symbols for the elements, dots for nonbonding valence electrons, and lines for pairs of bonding valence electrons.

**Ligand.** (6.3) An atom or group of atoms that is bonded to an inner atom; (18.2) a species that has lone pairs of electrons available to donate to a metal atom or cation.

**Light.** (4.2) Electromagnetic radiation in the visible portion of the spectrum, having wavelengths between 400 and 700 nm.

**Limiting reactant.** (1.10) The reactant whose amount falls to zero in a chemical reaction.

**Linear accelerator.** (20.3) An apparatus that accelerates atomic nuclei to high energy along a linear trajectory.

**Linear geometry.** (6.4) A straight line molecular shape.

**Line structure.** (1.5) A compact representation of a carbon-containing molecule that shows the molecular structure in a simplified fashion.

**Linkage group.** (17.1) A chemical group that joins two monomers in a polymer molecule.

**Linkage isomers.** (18.2) Metal complexes with identical chemical formulas that differ in the way in which one or more ligands binds to the metal.

**Liquid.** (8.3) The phase of matter in which intermolecular forces are large enough that molecules cannot escape into space but small enough that molecules move freely past one another. A liquid has a distinct volume but no distinct shape.

**Litre (L).** (1.2) A unit of volume equal to 1000 cm$^3$ ($10^{-3}$ m$^3$).

**Localized bond.** (7.1) A chemical bond that involves only two atoms.

**Lone pair (nonbonding pair).** (6.2) A pair of valence electrons that is localized on an atom rather than involved in bonding.

**Low-spin complex.** (18.3) A complex in which the maximum number of electrons is paired.

**Macromolecule.** (17.1) A molecule that contains a large number of atoms (typically, more than several hundred).

**Macroscopic.** (1.1) Visible to the naked eye.

**Magnetic quantum number ($m_l$).** (4.5) The quantum number, restricted to integers between $-l$ and $l$, that indexes the orientation in space of an atomic orbital.

**Magnitude.** (1.2) A number assigned to a measured quantity, such that it may be compared to others.

**Main group element.** (5.2) Any of the elements in the $s$-block or $p$-block of the periodic table.

**Main group metal.** (19.3) Any element displaying metallic properties and located in the $p$ block of the periodic table.

**Major species.** (13.5) The species (cations, anions, molecules) present in relatively high concentration in aqueous solution.

**Malleable.** (18.1) Able to be formed into various shapes, including thin sheets.

**Manometer.** (2.1) A bent tube, containing a liquid and open at both ends, used to measure differences in pressure.

**Markovnikov's rule.** (10.7) in the addition of HX to an alkene, the hydrogen atom adds to the carbon atom of the double bond that already had the greater number of hydrogen atoms.

**Mass.** (1.2) The quantity of matter in a substance.

**Mass Defect.** (20.1) The change in mass that accompanies a nuclear reaction.

**Mass number (A).** (20.1) The total number of protons and neutrons in a nucleus.

**Mass spectrometer.** (2.6) An instrument that separates the atomic/molecular components of a sample according to their masses.

**Matter.** (1.1) Anything that occupies space and possesses mass.

**Mechanism.** (12.1) The detailed molecular processes by which a chemical reaction proceeds.

**Meniscus.** (8.3) The curved surface of a liquid contained in a narrow tube.

**Mesosphere.** (4.7) The region of the Earth's atmosphere between 50 km and 85 km above its surface.

**Metabolism.** (11.6) The process of breaking down organic molecules in biochemical cells.

**Metal.** (5.2) An element that is lustrous, conducts heat and electricity well, and tends to lose electrons to form cations.

**Metallic solid.** (8.4) An elemental solid whose atoms are held together by valence electrons occupying delocalized orbitals spread across the entire solid.

**Metalloid.** (7.7) An element with properties intermediate between those of metals and nonmetals.

**Metalloprotein.** (18.4) A biological macromolecule made of long chains of amino acids and containing at least one metal atom.

**Metallurgy.** (18.5) The science of extracting, purifying, and forming useful objects from metals.

**Metathesis reaction.** (19.2) A chemical reaction in which there is an exchange of bonding partners between two atoms or ions.

**Method of initial rates.** (12.4) A way to simplify the behaviour of a reaction based on measuring the change in concentration over time immediately after the reactants are mixed.

**Metre (m).** (1.2) The SI unit of length.

**Micelle.** (9.5) A cluster of molecules in aqueous solution organized with their hydrophobic ends pointing inward and their hydrophilic ends pointing outward.

**Microscopic.** (1.1) Requiring magnification to be seen.

**Midpoint.** (15.3) The point in a titration where the amount of added titrant is exactly half the amount required for complete reaction.

**Minor species.** (13.6) The species (ions, molecules) present in relatively low concentration in aqueous solution.

**Miscible.** (9.2) Soluble in all proportions.

**Moderator.** (20.4) A substance used to slow down neutrons in a nuclear reactor.

**Molality.** (9.1) The concentration measure defined as moles of solute divided by mass of solvent in kilograms.

**Molar heat capacity (C).** (3.2) The amount of energy required to raise the temperature of one mole of a substance by 1 °C (1 K).

**Molar heat of solution.** (9.3) The net energy change that occurs as a mole of substance dissolves.

**Molar mass (M).** (1.4) The mass in grams of 1 mol of atoms or molecules.

**Molar volume.** (2.2) The volume of one mole of gas.

**Molarity.** (1.6) A concentration unit equal to the number of moles of solute dissolved in 1 L of solution.

**Mole.** (1.4) The amount of a substance that contains the same number of units as the number of atoms in exactly 12 g of carbon-12, $6.022 \times 10^{23}$.

**Mole fraction.** (2.3) The ratio of the number of moles of one component to the total number of moles of all components.

**Molecular orbital.** (7.4) A three dimensional wave that encompasses an entire molecule and describes a bound electron.

**Molecular orbital diagram.** (7.4) An energy level diagram displaying the molecular orbitals of a particular molecule.

**Molecular shape.** (6.3) The overall shape of a molecule, arising from the relative positions of its atomic nuclei.

**Molecular solid.** (8.4) A solid containing discrete molecules that do not have chemical bonds between them.

**Molecule.** (1.1) A group of atoms linked together by chemical bonds.

**Monodentate.** (15.5) A ligand with one donor atom.

**Monolayer.** (9.5) A single two dimensional layer of atoms or molecules.

**Monomer.** (17.1) A molecule from which a polymer is synthesized.

**Monosaccharide.** (17.5) A carbohydrate (sugar) molecule containing between three and six carbon atoms.

**n-type semiconductor.** (7.7) A metalloid that contains a dopant that gives it excess valence electrons.

**Nernst equation.** (16.5) The equation that relates the potential of an electrochemical cell to temperature and concentrations.

**Net ionic equation.** (13.6) A chemical equation showing the actual participants in a reaction of ions in solution.

**Network solid.** (8.4) A solid made up of atoms held together in a three dimensional crystalline array by covalent bonds.

**Neutral.** (14.1) A solution is neutral if it contains equal concentrations of $H_3O^+(aq)$ and $OH^-(aq)$

**Neutral atom.** (4.1) An atom containing the same number of electrons as it has protons in its nucleus.

**Neutralization reaction.** (15.3) Proton transfer between an acid and a base to generate a pair of neutral molecules.

**Neutron.** A fundamental nuclear particle that possesses mass but zero electric charge.

**Newton.** (2.1) The SI unit of force, equal to $1 \text{ kg m s}^{-2}$.

**Nickel–cadmium battery.** (16.6) A galvanic cell whose working components are NiO(OH) and Cd.

**Nitrogen fixation.** (11.5) The conversion of molecular nitrogen into nitrogen-containing species such as ammonia or nitrate ion.

**Noble gas.** (5.3) Any of the elements of Group 18 of the periodic table.

**Noble gas configuration.** (5.3) The portion of the distribution of electrons among the orbitals of an atom that matches that of one of the noble gases.

**Node.** (4.6) A point, line, or surface where the electron density of an orbital is zero.

**Nomenclature.** (10.1) A systematic procedure for naming chemical compounds.

**Nonbonding electron.** (6.2) A valence electron that does not participate in bond formation.

**Nonbonding molecular orbital.** (7.5) A delocalized orbital whose occupancy by electrons neither stabilizes nor destabilizes the molecule.

**Normal boiling point.** (8.1) The boiling point of a substance under one atm of pressure.

**Normal freezing point.** (8.1) The freezing point of a substance under one atm of pressure.

**Nuclear binding energy.** (20.1) The amount of energy per nucleon that binds an atomic nucleus together.

**Nuclear decay.** (20.2) The spontaneous decomposition of an unstable nucleus.

**Nucleic acid.** (17.6) A biochemical macromolecule containing nucleotide units.

**Nucleon.** (20.1) One of the protons or neutrons in a nucleus.

**Nucleotide.** (17.6) The repeating unit in DNA and RNA, containing a base, a five-carbon sugar, and a phosphate group.

**Nucleus.** The central core of an atom, where nearly all the mass is concentrated.

**Nuclide.** (20.1) A particular nucleus, characterized by its atomic number ($Z$) and mass number ($A$).

**Octahedral.** (6.4) Having the shape of a regular octahedron, with eight triangular faces and six vertices.

**Octet.** (6.2) A set of eight valence electrons associated with one atom.

**Orbital.** (4.5) A three-dimensional wave describing a bound electron.

**Orbital density picture.** (4.6) A two dimensional dot drawing representing the distribution of electron density in an orbital.

**Orbital mixing.** (7.4) Interactions among $s$ and $p$ atomic orbitals that change the relative energies of the molecular orbitals resulting from these atomic orbitals.

**Orbital overlap.** (7.1) The extent to which two orbitals on different atoms interact.

**Order of reaction.** (12.3) The exponent to which a concentrations is raised in a rate law.

**Organic chemistry.** The chemistry of carbon and its compounds.

**Osmosis.** (9.4) The net movement of solvent molecules through a semipermeable membrane.

**Osmotic pressure ($\Pi$).** (9.4) The pressure difference that must be applied to a solution to prevent osmosis from pure solvent, $\Pi = MRT$.

**Outer atom.** (6.2) Any atom in a molecule that bonds to only one other atom.

**Overall order.** (12.3) The sum of the reaction orders of all species in a rate law.

**Oxidation.** (16.1) Loss of electrons by a chemical species.

**Oxidation number.** (16.1) The charge that an atom would have if each of its bonding electrons were assigned to the more electronegative atom involved in the bond.

**Oxidation-reduction reactions.** (16.1) A class of chemical reactions that involves transfers of electrons between chemical species.

**Oxidizing agent.** (16.1) A chemical species gains electrons from another substance.

**Oxoacid.** (14.4) An acid containing an inner atom bonded to OH groups and O atoms.

**Ozone layer.** (4.7) The region of the Earth's atmosphere that contains ozone ($O_3$) and absorbs potentially lethal ultraviolet radiation.

***p*-type semiconductor.** (7.7) A metalloid containing a dopant that gives it a deficiency of valence electrons.

**Pairing energy.** (18.3) The electron-electron repulsion energy arising from placement of two electrons in the same orbital.

**Paramagnetic.** (7.4) Attracted by a magnetic field, as a consequence of having unpaired electron spins.

**Partial pressure.** (2.3) The pressure exerted by one component of a gaseous mixture.

**Parts per billion.** (2.3) How many of one particular component are present in one billion objects.

**Parts per million.** (2.3) How many of one particular component are present in one million objects.

**Pascal (Pa).** (2.1) The SI unit of pressure, one newton per square meter.

**Passive electrode.** (16.3) An electrode that serves only to conduct electrons between a wire and a solution; its chemical constituents do not take part in the redox reaction that occurs at its surface.

**Path function.** (3.2) A quantity whose magnitude depends on the path along which the change takes place.

**Pauli exclusion principle.** (5.2) The requirement that no two electrons in a chemical species can be described by the same wave function.

**Peptide.** (17.7) A small polymer of amino acids.

**Peptide linkage.** (17.7) The amide group that links two amino acids in a polypeptide.

**Percent yield.** (1.9) The actual yield of a reaction divided by its theoretical yield and multiplied by 100%.

**Periodic table.** The table of the chemical elements arranged in rows of increasing atomic number and columns of similar chemical behaviour.

**pH.** (14.2) The negative logarithm of the hydronium ion concentration in mol $L^{-1}$.

**pH meter.** (16.5) A device for measuring solution pH.

**Phase.** (8.1) Any of the three states of matter: gas, liquid, or solid.

**Phase change.** (8.6) Transformation from one state of matter to another.

**Phase diagram.** (8.6) A pressure-temperature graph showing the conditions under which a substance exists as solid, liquid, or gas.

**Phenyl group.** (10.2) a benzene ring attached as a side group to an organic molecule.

**Phospholipid.** (9.5) A biochemical surfactant molecule that is one component of cell membranes.

**Photochemical smog.** (2.8) The mixture of pollutants that results from the action of sunlight on the emissions from internal combustion engines.

**Photoconductor.** (7.7) A substance which becomes electrically conductive upon absorption of sufficiently energetic photons.

**Photoelectric effect.** (4.2) The ejection of electrons from a metal surface by light.

**Photoelectron spectroscopy.** (5.1) The technique that analyzes the kinetic energies of electrons that are ejected from gaseous atoms or molecules by photons of a specific frequency.

**Photon.** (4.2) A particle of light, characterized by energy $E = h\nu$.

**Pi ($\pi$) bond.** (7.3) A chemical bond formed by side-by-side orbital overlap so that electron density is concentrated above and below the bond axis.

**Physical transformation.** A change in properties that is not accompanied by chemical rearrangements.

**Physical property.** (1.2) A property that can be observed without causing any chemical rearrangements.

**Planck's constant (h).** (4.2) The physical constant, 6.63 $\times 10^{-34}$ J s, that relates the energy of a photon to its frequency.

**Plastic.** (17.4) A polymer that exists as blocks or sheets.

**Plasticizer.** (17.4) A substance added to a plastic to make it more flexible.

**Pleated sheet.** (17.7) The protein secondary structure formed like a corrugated plane.

**Polar covalent bond.** (6.1) A bond that possesses an asymmetric distribution of electrons.

**Polarizability.** (8.2) The ease with which the electron density about an atom or molecule can be distorted.

**Polyamide.** (17.3) A polymer containing the amide linkage group.

**Polyatomic ion.** (1.5) A charged chemical species containing more than one atom.

**Polyester.** (17.3) A polymer containing the ester linkage group.

**Polyethylene.** (17.2) The macromolecule formed by the combination of many ethylene molecules.

**Polymer.** (17.1) A molecule that contains a large number of identical individual units (monomers) linked together.

**Polypeptide.** (17.7) A protein molecule.

**Polyprotic acid.** (14.4) A molecule that contains more than one acidic hydrogen atom.

**Polysaccharide.** (17.5) A macromolecule formed by the condensation of many monosaccharide units.

**Positron.** (20.2) A subatomic particle with the mass of an electron but a positive unit charge.

**Positron emission.** (20.2) Nuclear decay by emission of a positron, which decreases the atomic number by one unit without changing the mass number.

**Potential energy.** (3.1) The energy an object has by virtue of some force (for example, gravitational or electrical) acting on it.

**Precipitate.** (15.4) An insoluble solid that separates from a solution.

**Precipitation.** (15.4) Formation of an insoluble solid that separates from a solution.

**Precision.** (1.2) How reproducible an experimental quantity is.

**Pressure.** (2.1) Force per unit area.

**Primary alkyl halide.** (10.3) An organic molecule where the carbon atom that is bonded to a halogen atom is bonded to only one other carbon atom.

**Primary carbon atom.** (10.3) A carbon atom that is connected to only one other carbon atom in an organic molecule.

**Primary structure.** (17.6) The sequence of monomer units making up DNA or a protein.

**Principal quantum number ($n$).** (4.5) The quantum number, restricted to positive integers, that indexes the energy and size of an atomic orbital.

**Product.** (1.7) A substance that is formed in a chemical reaction.

**Propagation.** (17.2) The continuation of a free-radical polymerization chain reaction.

**Protein.** (17.7) A biochemical polymer composed of amino acids.

**Proton.** The subatomic constituent of nuclei that possesses unit positive charge, $1.602 \times 10^{-19}$ C, and a mass of $1.673 \times 10^{-27}$ kg.

**Proton transfer.** (14.1) The transfer of $H^+$ from one chemical species to another.

**Quantized.** (4.3) Having only discrete allowed values.

**Quantum number.** (4.5) The allowed values of some quantized property.

**Radiant energy.** (3.1) The energy possessed by electromagnetic radiation (photons).

**Radioactive.** (20.2) Unstable nuclides that will undergo nuclear decay.

**Radiation therapy.** (20.7 The use of radioactive nuclides to treat cancers.

**Radioactivity.** The spontaneous breakdown of a nucleus by giving off energetic particles.

**Raoult's law.** (9.4) The equation that relates vapour pressure of a solution to the mole fractions of the solution components.

**Rate.** (12.1) Change per unit time.

**Rate constant ($k$).** (12.3) The constant of proportionality linking a reaction rate with concentrations of reagents.

**Rate law.** (12.3) An expression relating the rate of a reaction to the concentrations of reagents.

**Rate-determining step.** (12.1) The slowest step in a reaction mechanism.

**Reactant.** (1.7) A species that is consumed in a chemical reaction.

**Reaction coordinate.** (12.6) The course of a reaction as reactants are converted to products.

**Reaction mechanism.** (12.1) The detailed molecular processes by which a chemical reaction proceeds.

**Reaction quotient ($Q$).** (11.4) The ratio of concentrations of products to concentrations of reactants, each raised to its stoichiometric coefficient.

**Reagent.** (1.7) Any reactant or product of a chemical reaction.

**Redox reaction.** (16.1) A reaction in which electrons are transferred between species.

**Reducing agent.** (16.1) A substance that loses electrons during a reaction.

**Reduction.** (16.1) Gain of electrons.

**Refining.** (18.5) The metallurgical process by which an impure metal is purified.

**Resonance.** (6.2) The use of two or more equivalent Lewis structures to describe a substance that contains delocalized electrons.

**Resonance structure.** (6.2) One of two or more Lewis structures that are equivalent to one another.

**Reverse osmosis.** (9.4) The forced passage of solvent out of a solution through a semi-permeable membrane.

**Reversible reaction.** (12.5) A molecular process which can proceed readily in both directions.

**Reversibility.** (13.1) The ability of any elementary chemical reaction to proceed in either direction.

**Ribonucleic acid (RNA).** (17.6) A polymer of nucleotide units that transmits genetic information in the cell.

**Roasting.** (18.5) The metallurgical process by which a metal ore is oxidized by heating in air.

**Root-mean-square speed.** (2.6) Average speed obtained by taking the square root of the mean value of the squares of the individual speeds.

**Rounding.** The procedure for expressing a numerical result with the correct precision (number of significant figures).

**Saturated.** (9.2) A solution containing the maximum possible concentration of a solute. (10.1) A hydrocarbon containing the maximum number of hydrogen atoms.

**Schrödinger equation.** (4.5) The equation describing the behaviour of electrons in atoms and molecules.

**Scientific notation.** (1.2) Expression of a number as a value multiplied by the appropriate power of ten.

**Screening.** (5.1) The reduction in effective nuclear charge caused by electrons in orbitals.

**Second (s).** (1.2) The SI base unit of time.

**Second law of thermodynamics.** (11.2) The assertion that entropy (dispersal of matter and energy) always increases in the Universe.

**Second-generation star.** (20.5) A star formed from the collapse of interstellar matter, including (besides hydrogen and electrons) elements with $Z > 26$ that are the debris from supernovae explosions of first-generation stars.

**Second-order reaction.** (12.3) A reaction whose rate law has an overall order equal to two.

**Secondary alkyl halide.** (10.3) an organic molecule that has a carbon that is attached to a halogen atom and two other carbon atoms.

**Secondary carbon atom.** (10.3) a carbon atom that is connected to two other carbon atoms in an organic molecule.

**Secondary structure.** (17.6) The structural arrangement of DNA or a string of amino acids.

**Seesaw shape.** (6.4) The molecular shape that resembles a seesaw.

**Semiconductor.** (7.7) A substance that is intermediate in electrical conductivity between metals (good conductors) and nonmetals (poor conductors).

**Semipermeable membrane.** (9.4) A thin sheet that allows the passage of some types of molecules (typically, solvent) but prevents the passage of others (typically, solutes).

**Sievert (Sv).** (20.6) a unit that measures the effect of radiation on the human body.

**SI (Système International).** (1.5) The system of units that has been adopted by scientists for general use.

**Sigma ($\sigma$) bond.** (7.3) A bond formed by end-on overlap of atomic orbitals, giving electron density that is concentrated along the bond axis.

**Significant figure.** (1.3) A digit in a numerical value that is known with certainty or has an uncertainty of one unit.

**Silicate.** (6.3) A compound containing one or more metal cations and a network of Si-O bonds.

**Simple cubic.** (8.5) The crystal form whose unit cell contains 1/8 of an atom inside the unit cell at each corner.

**Single bond.** (6.2) A chemical bond formed by one pair of electrons shared between two atoms.

**Slightly soluble.** (15.4) Able to dissolve to a modest extent. Salts whose $K_{sp}$ lies between $10^{-2}$ and $10^{-5}$ are classified as slightly soluble in water.

**$S_N1$ mechanism.** (10.5) Organic reaction mechanism where the bond between a carbon atom and the leaving group breaks first.

**$S_N2$ mechanism.** (10.5) Organic reaction mechanism where the bond between a carbon atom and the leaving group breaks at the same time that the new bond forms.

**Soft Lewis acid.** (19.2) A Lewis acid whose acceptor atom has a high polarizability.

**Soft Lewis base.** (19.2) A Lewis base whose donor electron pairs are loosely bound, resulting in high polarizability.

**Solid.** (8.4) The state of matter characterized by a defined volume and shape.

**Solubility.** (9.2) The amount of a solute that will dissolve in a given amount of solution.

**Solubility product ($K_{sp}$).** (13.6) The equilibrium constant for the solubility equilibrium of an ionic compound in water.

**Soluble.** (9.2) Able to dissolve in a liquid. Salts with $K_{sp} > 10^{-2}$ are classified as soluble in water.

**Solute.** (9.1) A substance that dissolves in a solvent to form a solution.

**Solution.** (9.1) A homogeneous mixture of two or more substances.

**Solvent.** (9.1) The component of a solution that defines its phase. Generally, the solvent is the component present in largest amount.

***sp* hybrid orbitals.** (7.2) Two atomic orbitals constructed by the interaction between an *s*-orbital and a *p*-orbital on the same atom.

***sp²* hybrid orbitals.** (7.2) Three atomic orbitals constructed by the interactions among an *s*-orbital and two *p*-orbitals on the same atom.

***sp³* hybrid orbitals.** (7.2) Four atomic orbitals constructed by the interactions among an *s*-orbital and three *p*-orbitals on the same atom.

***sp³d* hybrid orbitals.** (7.2) Five atomic orbitals constructed by the interactions among an *s*-orbital, three *p*-orbitals, and one *d*-orbital on the same atom.

***sp³d²* hybrid orbitals.** (7.2) Six atomic orbitals constructed by the interactions among an *s*-orbital, three *p*-orbitals, and two *d*-orbitals on the same atom.

**Space-filling model.** (6.3) A representation of a molecule that shows the space occupied by its electron cloud.

**Spectator ions.** (13.6) Ions that are present in a solution but do not participate in a chemical reaction.

**Spectrochemical series.** (18.3) The listing of ligands in order of increasing energy-level splitting.

**Spectrum.** (4.2) A graph of the intensity of light as a function of either frequency or wavelength.

**Spin.** (4.4) The intrinsic angular momentum of electrons and protons that gives them magnetism.

**Spin orientation quantum number ($m_s$).** (4.5) The quantum number, restricted to either $+\frac{1}{2}$ or $-\frac{1}{2}$, that indexes the orientation of electron spin.

**Spontaneous.** (11.1) Able to occur without outside intervention.

**Square planar.** (6.4) A flat, square molecular shape.

**Square pyramidal.** (6.4) A pyramid with a square base molecular shape.

**Standard atmosphere (atm).** (2.1) An atmospheric pressure of 760 Torr (1.01325 bar).

**Standard conditions.** (3.5) Unit concentrations (1 M for solutes, 1 bar for gases). Unless otherwise specified, standard conditions also means 298 K.

**Standard enthalpy of formation ($\Delta H_f^0$).** (3.5) The enthalpy change accompanying a formation reaction.

**Standard hydrogen electrode (SHE).** (16.3) The reference standard for standard reduction potentials, with a defined value of exactly 0 V. The electrode is a platinum wire immersed in an acid solution that is 1.00 M in hydronium ion, over which hydrogen gas passes at a pressure of 1.00 bar.

**Standard molar free energy of formation ($\Delta G_f^0$).** (11.4) The change in thermodynamic free energy accompanying the reaction in which one mole of a substance forms from elements in their standard states.

**Standard reduction potential ($E°$).** (16.4) The electrode potential for reduction under standard conditions.

**Standard solution.** (15.3) A solution whose concentration is accurately known.

**Standard state.** (3.5) The most stable phase of a substance under standard conditions.

**Standardization.** (15.3) Accurate determination of the concentration of a solution.

**Starch.** (17.5) The carbohydrate that plants use to store chemical energy.

**State function.** (3.2) A property that depends only on the present state but not on the previous history of the system.

**Steric number.** (6.3) The sum of the coordination number and number of lone pairs for an inner atom.

**Stereochemistry.** (10.4) The study of relative spatial arrangements of atoms.

**Stereoisomers.** (10.4) Isomers that have their atoms connected in the same way as one another, but differ in the arrangement of their atoms in space.

**Stoichiometric coefficient.** (1.7) An integer giving the relative number of molecules of a species that react in a chemical reaction.

**Stoichiometric point.** (15.3) The point in a titration at which the amount of added titrant is exactly enough to react completely with the species being titrated.

**Stoichiometric ratio.** (1.8) The ratio of coefficients of two reagents in a balanced chemical reaction.

**Stoichiometry.** (1.8) The amount relationships among chemical substances undergoing reactions.

**Stratosphere.** (4.7) The region of the Earth's atmosphere between 10 km and 50 km above its surface.

**Strong acid.** (14.1) An acid that generates virtually stoichiometric amounts of hydronium ions in water.

**Strong base.** (14.1) A base that generates virtually stoichiometric amounts of hydroxide ions in water.

**Strong nuclear force.** (20.1) The force that bind nucleons together in the nucleus.

**Structural formula.** (1.5) A molecular formula that shows how atoms are bonded together.

**Structural isomers.** (6.3) Compounds that have identical chemical formulas but different molecular structures.

**Sublimation.** (8.6) The phase change between solid and vapour.

**Substrate.** (10.5) The reactant that contains the leaving group.

**Superconductor.** (8.4) A substance which conducts electricity without resistance.

**Supernova.** (20.5) An exploding star that produces unusually bright emission.

**Surface tension.** (8.3) The resistance of a liquid to an increase in its surface area.

**Surfactant.** (9.5) A molecule containing hydrophilic and hydrophobic parts that is used to modify the behaviour of aqueous solutions.

**Surroundings.** (3.2) All of the universe outside of a system.

**Système Internationale (SI).** (1.2) The set of metric units that have been chosen by international agreement for expressing scientific quantities.

**System.** (3.2) Any specific, well-defined part of the universe that is of interest.

**Temperature.** (1.2) The property of an object that measures the amount of random energy of motion of its molecules and determines the direction of spontaneous heat flow.

**Terminal.** (10.1) At the end of a hydrocarbon chain.

**Termination.** (17.2) The completion of a polymerization chain reaction by the reaction of two radicals to create a non-radical.

**Termolecular reaction.** (12.1) An elementary reaction in which three molecular species collide and react.

**Tertiary alkyl halide.** (10.3) An organic molecule in which the halogen atom is bonded to a carbon which is itself bonded to three other carbon atoms.

**Tertiary carbon.** (10.3) A carbon atom that is bonded to three other carbon atoms.

**Tertiary structure.** (17.7) The overall shape of a protein molecule.

**Tetrahedron.** (6.3) A pyramid with four identical faces that are equilateral triangles.

**Theoretical yield.** (1.9) The amount of a product that would be formed if a reaction proceeded to completion without any losses.

**Thermal energy.** (3.1) Energy associated with the random motion of atoms and molecules.

**Thermal pollution.** (11.5) Heating of the environment as a byproduct of industrial operations.

**Thermodynamics.** (3.2) The scientific study of the relationships among heat and other forms of energy.

**Thermoplastic.** (17.4) A plastic that melts or deforms upon heating.

**Thermosetting.** (17.4) A plastic that retains its structural integrity upon heating.

**Thermosphere.** (4.7) The region of the Earth's atmosphere more than 90 km above its surface.

**Thiol.** (17.1) An organic compound that contains the -SH group.

**Third law of thermodynamics.** (11.3) The statement that the entropy of any pure, perfect crystalline substance is zero at 0 K.

**Third-generation star.** (20.5) A star formed from the collapse of interstellar matter, including (besides hydrogen and electrons) elements with all $Z$ values that are the debris from supernovae explosions of second-generation stars.

**Thymine.** (17.6) A one-ring organic base found in DNA.

**Titrant.** (15.3) The liquid solution added during a titration.

**Titration.** (15.3) The gradual addition of measured amounts of one solution to another until a chemical reaction between them is complete.

**Torr.** (2.1) The pressure unit that is 1/760 of a standard atmosphere.

**Transition state.** (10.5, 12.6) The fleeting, instantaneous configuration of atoms at the top of the activation energy barrier, during the conversion from reactants with one set of bonds into products with a different set of bonds

**Transition metal.** (18.1) Any of the elements in the *d* block of the periodic table.

**Trigonal bipyramidal.** (6.4) The molecular shape in which a central atom is bonded to three atoms in the same plane, as well as two atoms in axial positions, one above and one below the plane.

**Trigonal planar.** (6.4) The molecular shape in which a central atom is bonded to three other atoms lying in the same plane at 120° angles to one another.

**Trigonal pyramid.** (6.3) The molecular shape in which a central atom is bonded to three outer atoms. The central atom does not lie in the plane of the outer atoms.

**Triple bond.** (6.2) A bond between two atoms consisting of three pairs of bonding electrons.

**Triple point.** (8.6) The temperature and pressure at which solid, liquid, and gas can coexist at equilibrium.

**Troposphere.** (2.8) The region of the Earth's atmosphere between its surface and an altitude of 10 km.

**T-shaped.** (6.4) The molecular shape that resembles the letter T.

**Uncertainty principle.** (4.4) The assertion that position and momentum cannot both be exactly known.

**Unimolecular reaction.** (12.1) An elementary reaction in which there is only one reactant molecule.

**Unit.** (1.2) An agreed-upon standard with which other values are compared.

**Unit cell.** (8.5) The simplest repeating unit of a regular pattern, such as an atomic or molecular crystal.

**Universal gas constant (R).** (2.2) $8.314 \ kPa \ L \ mol^{-1} \ K^{-1}$ ($0.08314 \ bar \ L \ K^{-1} \ mol^{-1}$), the constant of proportionality in the ideal gas equation.

**Unsaturated.** (10.1) an organic compound with one or more multiple carbon-carbon bonds.

**Uracil.** (17.6) A one-ring organic base found in RNA.

**Valence electrons.** (5.2) The electrons of an atom that occupy orbitals of highest principal quantum number and incompletely filled orbitals.

**Valence shell electron pair repulsion (VSEPR).** (6.3) A system for predicting molecular geometry which minimizes electron-electron repulsion by placing electron pairs as far apart as possible.

**van der Waals equation.** (2.7) An equation that corrects the ideal gas equation for the effects of molecular size and intermolecular forces.

**Vapour.** (2.8) The gas phase of a substance.

**Vapour pressure.** (2.8) The partial pressure of a gas in equilibrium with a liquid phase.

**Vapourization.** (8.1) The conversion of a liquid into a gas.

**Vesicle.** (9.5) An enclosed bilayer made up of surfactant molecules.

**Vicinal.** (10.7) A molecule with two substituent groups (e.g. halogens) bonded to adjacent carbon atoms.

**Viscosity.** (8.3) The resistance to flow of a fluid.

**Visible light.** (4.2) Photons in the wavelength range between 400 and 700 nm, to which the human eye is sensitive.

**Volt (V).** (19.4) The SI unit for electrical potential $1 \ V = 1 \ J/C$.

**Volume.** (1.2) The amount of three-dimensional space occupied by something.

**Volumetric flask.** (1.6) A vessel calibrated to hold a specified volume of liquid.

**Water autohydrolysis equilibrium constant ($K_w$).** (14.1) The equilibrium constant for proton transfer between two water molecules. $K_w = 1.0 \times 10^{-14}$ at 25 °C.

**Wave.** (4.2) A periodic variation that can be described by amplitude, wavelength, and frequency.

**Wave properties.** (4.2) Properties characteristic of a wave, such as amplitude, frequency, and wavelength.

**Wavelength.** (4.2) The distance between successive crests of a wave.

**Weak acid.** (14.3) An acid that undergoes incomplete proton-transfer in water.

**Weak base.** (14.3) A base that undergoes incomplete proton-transfer in water.

**Work (w).** (3.2) Energy transfer that is described by the product of a force times a displacement, $w = Fd$. In a chemical reaction, work can be defined in terms of the pressure and the change in volume: $w = -p\Delta V$.

**Yield.** (1.9) The amount of a product obtained from a chemical reaction.

**Zinc–air battery.** (16.6) A galvanic cell whose working components are Zn and $O_{2(g)}$ from the air.

# INDEX

# INDEX OF CHEMICAL SUBSTANCES

Chemical substances in this book have been organized into four categories in this part of the index: organic, inorganic, organo-metallic and biochemically-relevant. This organization allows the reader to quickly find information on a particular substance or group of substances. All entries also appear in the main index.

## I. ORGANIC SUBSTANCES

# II. INORGANIC SUBSTANCES

## III. ORGANOMETALLIC SUBSTANCES

## IV. BIOCHEMICALLY SIGNIFICANT SUBSTANCES